Handbook of
Photosynthesis

Second Edition

Handbook of Photosynthesis

Second Edition

Edited by

Mohammad Pessarakli

University of Arizona
Tucson, Arizona, U.S.A.

Taylor & Francis
Taylor & Francis Group

Boca Raton London New York Singapore

A CRC title, part of the Taylor & Francis imprint, a member of the
Taylor & Francis Group, the academic division of T&F Informa plc.

Published in 2005 by
CRC Press
Taylor & Francis Group
6000 Broken Sound Parkway NW, Suite 300
Boca Raton, FL 33487-2742

No claim to original U.S. Government works
Printed in the United States of America on acid-free paper
10 9 8 7 6 5 4 3 2 1

International Standard Book Number-10: 0-8247-5839-0 (Hardcover)
International Standard Book Number-13: 978-0-8247-5839-4 (Hardcover)
Library of Congress Card Number 2004059310

Library of Congress Cataloging-in-Publication Data

Handbook of photosynthesis / edited by Mohammad Pessarakli.--2nd ed.
   The rise of the superconductors / P.J. Ford and G.A. Saunders.
      p. cm.--(Books in soils, plants, and the environment)
   Includes bibliographical references and index.
   ISBN 0-8247-5839-0 (alk. paper)
   1. Photosynthesis. I. Pessarakli, Mohammad, 1948- II. Series.

QK882.H23 2004
572'.46--dc22   2004059310

Taylor & Francis Group is the Academic Division of T&F Informa plc.

**Visit the Taylor & Francis Web site at
http://www.taylorandfrancis.com**

**and the CRC Press Web site at
http://www.crcpress.com**

Dedication

To my brother Haj Ghorban and my sisters Hajiyeh Layla, Maassumeh, and Zahra who have always supported and encouraged me to take risks and challenges for success. This successful work has certainly resulted from their continuous support and encouragement.

Preface

Since photosynthesis has probably been given more attention than any other physiological processes in plant physiology, there have been hundreds of articles published on this topic since the first edition of this book was published in 1997. Therefore, I felt it is necessary that this book be revised and some of these recent and relevant findings be included in the new volume. For revising the book, I have eliminated some of the old chapters and included several new ones in the revised volume. Some of the previous chapters which are included in the revised volume have been extensively revised. Therefore, the new volume looks like a new book.

Photosynthesis is by far the most spectacular physiological process in plant growth and productivity. Due to this fact, the study of photosynthesis has captivated plant physiologists, botanists, plant biologists, horticulturalists, agronomists, agriculturalists, crop growers, and most recently, plant molecular and cellular biologists around the world.

From an aesthetic perspective, I thought that it would be wonderful to include many of the remarkable findings on photosynthesis in a single inclusive volume. In such an album, selected sources could be surveyed on this most magnificent subject. With the abundance of research on photosynthesis available at present, an elegantly prepared exhibition of the knowledge on photosynthesis is indeed in order. Accordingly, one mission of this collection is to provide an array of information on photosynthesis in a single and unique volume. Ultimately, this unique and comprehensive source of intelligence will both attract the beginning students and stimulate further exploration by their educators. Furthermore, since more books, papers, and articles are currently available on photosynthesis than on any other plant physiological processes, preparation of a single volume by inclusion of the most recent and relevant issues and information on this subject can be appreciably useful and substantially helpful to those seeking specific information.

I see from a scientific perspective that the novelty of photosynthesis and its attraction for researchers from various disciplines has resulted in a voluminous, but somewhat scattered, database. However, none of the available sources comprehensively discusses the topic. The sources are either too specific or too general in scope. Therefore, a balanced presentation of the information on this subject is necessary. Accordingly, another main objective of this collection is to provide a balanced source of information on photosynthesis.

Now, more than ever, the excessive levels and exceedingly high accumulation rates of CO_2 due to the industrialization of the nations have drawn the attention of scientists around the globe. If the current accumulation rates of carbon dioxide along with the consequence of imbalance between the atmospheric O_2 and CO_2, continues, all of the living organisms including human beings and animals would be endangered. The only natural mechanism known to utilize atmospheric CO_2 is photosynthesis by the green plants. Therefore, another purpose of preparing this volume is to gather the most useful and relevant issues on photosynthesis on selected plant species. In this regard, we must consider plant species with the most efficient photosynthetic pathways to reduce the excess atmospheric CO_2 concentrations. The use of such plants will result in balanced O_2 and CO_2 concentrations and will reduce toxic levels of atmospheric CO_2. This higher consumption of atmospheric CO_2 by plants through the photosynthetic process not only reduces the toxic levels of CO_2, but will also result in more biomass production and higher crop yields.

To adequately cover many of the issues related to photosynthesis and for the advantage of easy accessibility to the desired information, the volume has been divided into several sections. Each section includes one or more chapters that are closely related to each other.

Like other physiological processes, photosynthesis differs greatly among various plant species, particularly between C_3 and C_4 plants, whether growing under normal or under stressful conditions. Therefore, examples of plants with various photosynthetic rates and different responses are presented in different chapters and included in this collection.

Now, it is well established that any plant species during its life cycle, at least once, is subjected to environmental stress. Since any stress alters the normal course of plant growth and development, metabolism, and other physiological processes, photosynthesis is also subject to this alteration and severely affected under stressful conditions. Therefore, a portion of this volume discusses plant photosynthesis under stressful conditions.

Hundreds of tables and figures are included in the volume to facilitate understanding and comprehension of the information presented throughout the text. Thousands of references have been used to prepare this unique collection. Several hundreds of index words are provided to promote accessibility to the desired information throughout the book.

Mohammad Pessarakli
University of Arizona
Tucson, Arizona

Editor

Dr. Mohammad Pessarakli, the editor, is a research faculty and lecturer in the Department of Plant Sciences, College of Agriculture and Life Sciences at the University of Arizona, Tucson. Dr. Pessarakli is the editor of the *Handbook of Plant and Crop Stress* and the *Handbook of Plant and Crop Physiology* (both titles published by Marcel Dekker, currently part of the Taylor & Francis Group), and is a member of the editorial board of the *Journal of Plant Nutrition*, *Communications in Soil Science and Plant Analysis*, and the *Journal of Agricultural Technology*. He is the author or co-author of over 50 journal articles. Dr. Pessarakli is an active member of the Agronomy Society of America, Crop Science Society of America, and Soil Science Society of America, among others. He is a member of the executive board of the American Association of the University Professors, Arizona Chapter. Dr. Pessarakli is an esteemed member (in-vited) of Sterling *Who's Who*, Marques *Who's Who*, Strathmore *Who's Who*, and Madison *Who's Who*, as well as numerous honor societies. He is a certified professional agronomist, certified professional soil specialist, and certified professional soil scientist (CPAg/SS), designated by the American Registry of the Certified Professionals in Agronomy, Crop Science, and Soil Science. He is a United Nations consultant in agriculture for underdeveloped countries. He received a B.S. degree (1977) in environmental resources in agriculture and an M.S. degree (1978) in soil management and crop science from Arizona State University, Tempe, and Ph.D. degree (1981) in soil and water science from the University of Arizona, Tucson.

For more information about the editor, please visit http://ag.arizona.edu/pls/faculty/pessarakli.htm

Acknowledgments

I would like to express my appreciation for the secretarial and the administrative assistance that I received from the secretarial and administrative staff of the Department of Plant Sciences, College of Agriculture and Life Sciences, the University of Arizona. The continuous encouragement and support of the department head, Dr. Robert T. Leonard, for my editorial work, especially the books, is always greatly appreciated.

In addition, my sincere gratitude is extended to Russell Dekker of Marcel Dekker who supported this project from its initiation to its completion. Certainly, this job would not have been completed as smoothly and rapidly without Dekker's most valuable support and sincere efforts.

I am indebted to the production editors, Dana Bigelow, and Balaji Krishnasamy (Kolam [SPI Publisher Services]) for the professional and careful handling of the volume. Bigelow, many thanks to you for your extra ordinary patience and carefulness in handling this huge volume.

The collective sincere efforts and invaluable contributions of several (83) competent scientists, specialists, and experts from 18 scientifically and technologically most advanced countries in the field of photosynthesis made it possible to produce this unique source that is presented to those seeking information on this subject. Each and every one of these contributors and their contributions are greatly appreciated.

Last, but not least, I thank my wife, Vinca, and my son, Mahdi, who supported me during the course of the completion of this work.

Mohammad Pessarakli
University of Arizona
Tucson, Arizona

Contributors

Carlos Santiago Andreo
Centro de Estudios Fotosintéticos y Bioquímicos
Facultad de Ciencias Bioquímicas y Farmacéuticas
Universidad Nacional de Rosario
Rosario, Argentina

Muhammad Ashraf
Department of Botany
University of Agriculture
Faisalabad, Pakistan

Habib-ur-Rehman Athar
Institute of Pure and Applied Biology
Bahauddin Zakariya University
Multan, Pakistan

Rita Barr
Department of Biological Sciences
Purdue University
West Lafayette, Indiana

W. Berry
Department of Organismic Biology
Ecology and Evolution
University of California
Los Angeles, California

Martine Bertrand
Institut National des Sciences et Techniques
   de la Mer
Conservatoire National des Arts et Métiers
Cherbourg, France

Anil S. Bhagwat
Molecular Biology Division
Bhabha Atomic Research Centre
Mumbai, India

Swapan K. Bhattacharjee
Devi Ahilya University
Indore, India

Basanti Biswal
Laboratory of Biochemistry and Molecular Biology
School of Life Sciences
Sambalpur University
Jyotivihar, Orissa, India

Dennis E. Buetow
Department of Molecular and Integrative
   Physiology
University of Illinois
Urbana, Illinois

Robert Carpentier
Groupe de recherche en biologie végétale
Université du Québec à Trois-Rivières
Québec, Canada

Frederick L. Crane
Department of Biological Sciences
Purdue University
West Lafayette, Indiana

J.J. Crouch
International Crops Research Institute for the
   Semi-Arid Tropics
Patancheru, Andhra Pradesh, India

Iliya Dimitrov Denev
Plant Physiology and Molecular Biology Department
University of Plovdiv "Paisii Hilendarski"
Plovdiv, Bulgaria

Ian C. Dodd
Department of Biological Sciences
Lancaster Environment Centre
University of Lancaster
Lancaster, United Kingdom

Rama Shanker Dubey
Department of Biochemistry
Faculty of Science
Banaras Hindu University
Varanasi, India

Stefan Dukiandjiev
Department of Plant Physiology and Molecular
   Biology
University of Plovdiv
Plovdiv, Bulgaria

Maria J. Estrella
Instituto Tecnológico de Chascomús
Chascomús, Argentina

Ilya Gadjev
Department Molecular Biology of Plants
Researchschool GBB
University of Groningen
Haren, The Netherlands

Eliška Gálová
Department of Genetics
Comenius University
Bratislava, Slovak Republic

José L. Garrido
Instituto de Investigaciónes Mariñas
Vigo, Spain

Tsanko Gechev
Department Molecular Biology of Plants
Researchschool GBB
University of Groningen
Haren, The Netherlands

Johannes Geiselmann
Unité Adaptation et pathogénie des Microorganismes
Université Joseph Fourier
CERMO, Grenoble, France

Bernard Grodzinski
Department of Plant Agriculture
Division of Horticultural Science
University of Guelph
Ontario Agricultural College
Guelph, Ontario, Canada

C.T. Hash
International Crops Research Institute for the
    Semi-Arid Tropics
Patancheru, Andhra Pradesh, India

Bruria Heuer
Institute of Soils, Water and Environmental
    Sciences
Volcani Center
Agricultural Research Organization
Bet Dagan, Israel

Tetsuo Hiyama
Department of Biochemistry and Molecular Biology
Saitama University
Saitama, Japan

Jean Houmard
Ecole Normale Supérieure
Organismes Photosynthétiques et
    Environnement
Paris, France

Bernhard Huchzermeyer
Botany Institute
Hannover College of Veterinary Medicine
Hannover, Germany

Ján Hudák
Department of Plant Physiology
Comenius University
Bratislava, Slovak Republic

Alberto A. Iglesias
Laboratorio de Enzimología Molecular, Bioquímica
    Básica de Macromoléculas
Facultad de Bioquímica y Ciencias Biológicas
Universidad Nacional del Litoral
Santa Fe, Argentina
    and
Grupo de Enzimología Molecular
    Bioquímica Básica de Macromoléculas
Facultad de Bioquímica y Ciencias Biológicas
Universidad Nacional del Litoral
Paraje, Argentina

Osamu Ito
Japan International Research Center for Agricultural
    Sciences
Ohwashi, Tsukuba, Ibaraki, Japan

Emily A. Keller
Department of Plant and Animal Science
Brigham Young University
Provo, Utah

Vladimir L. Kolossov
University of Illinois
Urbana, Illinois

Karen J. Kopetz
University of Illinois
Urbana, Illinois

H.W. Koyro
Botany Institute
Hannover College of Veterinary Medicine
Hannover, Germany

Katarína Král'ová
Institute of Chemistry
Faculty of Natural Sciences
Comenius University,
Bratislava, Slovak Republic

Ho Kwok Ki
Purdue University
Department of Biochemistry
West Lafayette, Indiana

María Valeria Lara
Centro de Estudios Fotosintéticos y Bioquímicos
Facultad de Ciencias Bioquímicas y Farmacéuticas
Universidad Nacional de Rosario
Rosario, Argentina

David W. Lawlor
Crop Performance and Improvement
Rothamsted Research
Harpenden, United Kingdom

Evangelos Demosthenes Leonardos
Department of Plant Agriculture
Division of Horticultural Science
University of Guelph
Guelph, Ontario, Canada

Elena Masarovičová
Department of Plant Physiology
Faculty of Natural Sciences
Comenius University
Bratislava, Slovak Republic

Michael Melzer
Department of Molecular Cell Biology
Institute of Plant Genetics and Crop Plant
   Research
Gatersleben, Germany

Ivan Nikiforov Minkov
Plant Physiology and Molecular Biology
   Department
University of Plovdiv "Paisii Hilendarski"
Plovdiv, Bulgaria

Shruti Mishra
Department of Biochemistry
Faculty of Science
Banaras Hindu University
Varanasi, India

Agnieszka Mostowska
Department of Plant Anatomy and Cytology
Institute of Experimental Biology of Plants
Warsaw University
Warsaw, Poland

Lubomír Nátr
Department of Plant Physiology
Faculty of Science
Charles University
Praha, Czech Republic

Peter Nyitrai
Department of Plant Physiology
Eötvös University
Budapest, Hungary

K. Okada
Crop Production and Environment Division
Japan International Research Center for
   Agricultural Sciences
Ohwashi, Tsukuba, Ibaraki, Japan

Derrick M. Oosterhuis
Department of Crops, Soils, and Environmental
   Science
University of Arkansas
Fayetteville, Arkansas

R. Ortiz
International Institute of Tropical Agriculture
   L.W. Lambourn & Co
Croydon, United Kingdom

Fernando Pieckenstain
Instituto Tecnológico de Chascomús
Chascomús, Argentina

Florencio E. Podestá
Facultad de Ciencias Bioquímicas y
   Farmacéuticas
Universidad Nacional de Rosario
Rosario, Argentina

Jana Pospíšilová
Institute of Experimental Botany
Academy of Sciences of the Czech Republic
Prague, Czech Republic

I. M. Rao
International Center for Tropical Agriculture
Cali, Colombia, South America
   and
Miami, Florida

Ejaz Rasul
Department of Botany
University of Agriculture
Faisalabad, Pakistan

Constantin A. Rebeiz
Department of Natural Resources and
  Environmental Sciences
University of Illinois
Urbana, Illinois

Steven Rodermel
Department of Genetics, Development, and Cell
  Biology
Iowa State University
Ames, Iowa

Anna M. Rychter
Institute of Experimental Plant Biology
Warsaw University
Warsaw, Poland

Jayashree Sainis
Molecular Biology Division
Bhabha Atomic Research Center
Mumbai, India

Éva Sárvári
Department of Plant Physiology
Eötvös Loránd University
Budapest, Hungary

Benoît Schoefs
Dynamique Vacuolaire et Réponses aux Stress de
  l'Environnement
UMR CNRS (5184)/INRA (1088)/
  Université de Bourgogne-Plante-
  Microbe-Environnement
Université de Bourgogne à Dijon
Dijon, France

H. Don Scott
Agribusiness Center
Mount Olive College
Mount Olive, North Carolina

R. Serraj
International Crops Research Institute for the
  Semi-Arid Tropics
Patancheru, Andhra Pradesh, India

Yun-Kang Shen
Shanghai Institute of Plant Physiology
Chinese Academy of Sciences
Shanghai, People's Republic of China

Cosmin Sicora
Institute of Plant Biology
Biological Research Center
Szeged, Hungary

Bruce N. Smith
Department of Plant and Animal Science
Brigham Young University
Provo, Utah

Robert E. Sojka
USDA-ARS Northwest Irrigation and Soils
  Research Laboratory
Kimberly, Idaho

Martin Spalding
Department of Genetics, Development, and Cell
  Biology
Iowa State University
Ames, Iowa

Dan Stessman
University of Illinois at Urbana
Champaign, Illinois

G.V. Subbarao
Crop Production and Environment Division
Japan International Research Center for
  Agricultural Sciences
Ohwashi, Tsukuba, Ibaraki, Japan

Heidi A. Summers
Department of Plant and Animal Science
Brigham Young University
Provo, Utah

András Szilárd
Institute of Plant Biology
Biological Research Center
Szeged, Hungary

Tonya Thygerson
Department of Plant and Animal Science
Brigham Young University
Provo, Utah

S. Tobita
Crop Production and Environment Division
Japan International Research Center for
  Agricultural Sciences
Ohwashi, Tsukuba, Ibaraki, Japan

Imre Vass
Institute of Plant Biology
Biological Research Center
Szeged, Hungary

Joseph C. V. Vu
Crop Physiology and Genetics
Agronomy Department
University of Florida
Gainesville, Florida

Abdul Wahid
Department of Botany
University of Agriculture
Faisalabad, Pakistan

Julian P. Whitelegge
Departments of Psychiatry and Biobehavioral
 Sciences, Chemistry and Biochemistry
David Geffen School of Medicine and the
 College of Letters and Sciences
The Neuropsychiatric Institute, The Brain Research
 Institute and The Molecular Biology Institute
University of California
Los Angeles, California

Da-Quan Xu
Shanghai Institute of Plant Physiology
Chinese Academy of Sciences
Shanghai, People's Republic of China

Galina Teneva Yahubian
Plant Physiology and Molecular Biology Department
University of Plovdiv "Paisii Hilendarski"
Plovdiv, Bulgaria

Yuzeir Zeinalov
Institute of Biophysics
Bulgarian Academy of Sciences
Sofia, Bulgaria

Lenka Zemanová
Department of Plant Physiology
Comenius University
Bratislava, Slovak Republic

Table of Contents

Section I

Principles of Photosynthesis

1 Mechanisms of Photosynthetic Oxygen Evolution and Fundamental Hypotheses of Photosynthesis

Yuzeir Zeinalov
Institute of Biophysics, Bulgarian Academy of Sciences

CONTENTS

I. INTRODUCTION

Intensive investigations on the nature of photosynthetic light reactions during the first half of the 20th century led to several important discoveries and observations that were extremely complicated to explain and resulted in the postulation of two fundamental concepts: the concept of photosynthetic unit (PSU) [1] and the concept of two photosystems [2]. According to the first concept, in all photosynthesizing systems (photosynthesizing bacteria, green unicellular algae, and higher plants), the light-absorbing pigment molecules are divided into two groups. Only one highly specialized pair of chlorophyll molecules (reaction center dimer) present among dozens of bacteria and among hundreds of green photosynthesizing systems could carry out the photochemical (charge separation) reaction, while the essential part of these molecules only absorbs light quanta and transfers the light energy to the reaction centers [1]. According to the second concept, the light-induced linear electron transfer reaction of H_2O to NADP is realized by the serial operation of two different photosynthesizing systems [2].

It is generally believed that these two principal concepts are completely proven and verified and the unsolved problems are connected with the elucidation of the nature of participating components and their mutual relationship. This chapter deals with the basic experiments and results that have led to the concept of the PSU and to the postulation of the concept of photosystems in light-driven photosynthetic reactions and shows that, at the time of their postulation, the existing results and observations were not sufficient.

II. THE CONCEPT OF PHOTOSYNTHETIC UNIT

A. FUNDAMENTAL RESULTS

There is a limited number of experimental data that scientists consider as crucial for the postulation of a given concept. For the concept of the PSU, the following results and observations are significant:

(1) The very high (maximum) quantum efficiency of photosynthesis under limited light intensity conditions, that is, when the probability for light quanta

absorption of a chlorophyll molecule is about one quantum per hour.

This statement has been confirmed by investigations of the dependence of photosynthesis on light (irradiance). It was shown in many experiments that the photosynthetic response to very low light intensities was linear (Figure 1.1, curve A). In a significant number of experiments, the shape of light–response curves had a logarithmic part (Figure 1.1, curve B) with maximum slope (maximum quantum efficiency) at the beginning of curves, that is, when the irradiation was approaching zero. "S"-shaped curves (Figure 1.1, curve C), which indicate that the quantum efficiency under low light intensities tends toward zero, were observed in a limited number of investigations (for review of the early investigations, see [3]). These "S"-shaped curves obtained in green plants were interpreted in favor of the assumption of the existence of a "photic threshold" of photosynthesis. However, this suggestion was not accepted and the results obtained by most researchers were in favor of the linear shape of the light curves of photosynthesis. Under anaerobic conditions, Diner and Mauzerall [4] also observed nonlinear dependence. After the postulation of the concept of the PSU, it was discovered that the initial slope of the light curves below the light compensation point was significantly higher, and nearer to this point on the light curves an abrupt change in the value of quantum efficiency of photosynthesis could be observed [5]. This observation is called "Kok's effect" and was explained by the changes in the rate of dark respiration after irradiation.

(2) The absence of induction period in the process of oxygen evolution or carbon dioxide reduction under very low light intensity conditions was one of the most serious arguments of the PSU concept (Figure 1.2). Five oxygen induction curves were recorded

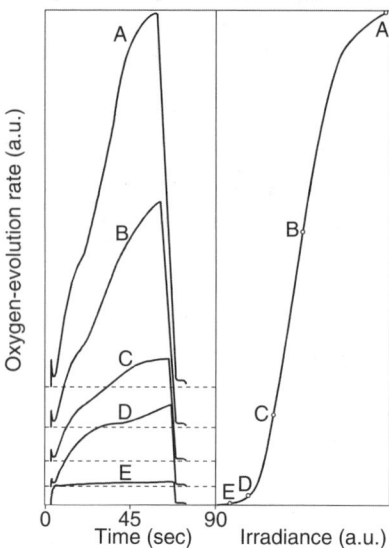

FIGURE 1.2 Oxygen induction curves recorded at different irradiances after 3 min of dark adaptation (left) and the respective "working points" on the "light curve" (right) in *Chlorella pyrenoidosa* suspension with absorbance 0.05. Induction curves A, B, and C are recorded at 8×10^{-8} A/mm and curves D and E at 1.2×10^{-9} A/mm sensitivity of the polarograph (for details see text).

at different irradiances after 3 min of dark adaptation of *Scenedesmus acutus* cell suspension. Curve A was recorded at the maximum irradiance, $I_0 = 135$ W/m^2, corresponding to the oxygen-evolution rate close to saturation (Figure 1.2, right panel point A). Other curves were recorded at $0.76I_0$ (B), $0.46I_0$ (C), $0.19I_0$ (D), and $0.056I_0$ (E). The induction curves indicate that the duration of the induction period decreased simultaneously with decrease in irradiance. Under the lowest irradiance, $0.056I_0$ (E), the rate of oxygen evolution reached its steady state immediately after the light was switched on. This observation is in agreement with the postulate that at low irradiances photosynthesis starts before the absorption of the four quanta needed for the evolution of one oxygen molecule.

(3) Oxygen flash yields depend on the dark intervals between the flashes. The dependence of the oxygen flash yields on the spacing between the saturating flashes was investigated for the first time by Emerson and Arnold [1] with Warburg's manometric apparatus. It was found that the average yields were maximal when intervals between the flashes were about 20 msec.

The dependence of oxygen yields produced by separated flash groups (four saturating short flashes) on the spacing between the flashes in groups and recorded after reaching steady-state yields is pre-

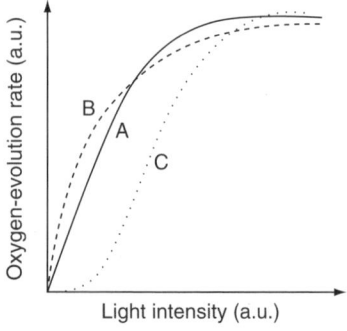

FIGURE 1.1 Different shapes of photosynthetic "light curves": A, linear; B, logarithmic; C, "S"-shaped irradiance dependence of photosynthesis.

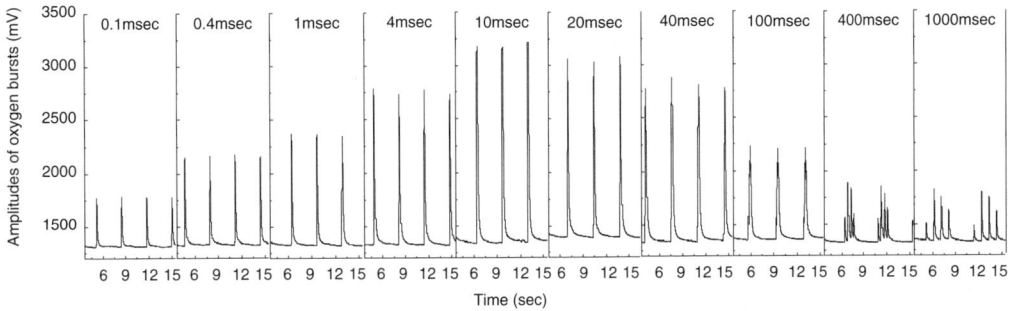

FIGURE 1.3 The steady-state oxygen yields (relative) of groups of four saturating flashes depending on the time between the flashes in the groups in *Scenedesmus obliquus* suspension with absorbance 0.05 (100 mm³ sample volume). The groups of four saturating flashes ($4J$, $t_{1/2} = 8\,\mu sec$) are spaced 3 sec and after reaching the five steady-state several oxygen group yields obtained at different spacing between the flashes are presented.

sented in Figure 1.3. It is clearly seen that amplitudes of oxygen flash yields increase with increase in spacing between flashes up to 10 to 20 msec, after which the yields decrease. Results presented in Figure 1.3 confirm the turnover time of oxygen-evolving centers (2×10^{-2} sec) estimated by Emerson and Arnold [1].

(4) When oxygen flash yields are maximal, the ratio between oxygen molecules evolved per flash and the number of chlorophyll molecules in the investigated suspensions is approximately constant and equal to $1O_2/2500Chl$. For the first time Emerson and Arnold obtained this value in 1932. It was found that in *Chlorella pyrenoidosa* suspensions with different chlorophyll concentrations $4 \times 10^{-4}\,M$ oxygen was evolved from $1\,M$ chlorophyll after every flash.

(5) Earlier studies [6–8] demonstrated that after approximately 5 min of dark adaptation of unicellular algae (e.g., *Chlorella*, *Scenedesmus*) or isolated chloroplast suspensions, the oxygen yield of the first saturating short (10 μsec) flash is zero (Figure 1.4).

(6) Oscillations in the oxygen flash yields (Figure 1.4) with a period of four observed after 5 to 6 min of dark incubation in algae or chloroplast suspensions [9].

At the time of the postulation of the PSU concept only the first four experimental observations were known. Observations 5 and 6 were obtained significantly later and are considered as additional confirmations of the concept of the PSU.

B. PROBLEMS AND HYPOTHESES

Considering the general equation of photosynthesis, it is apparent that for the evolution of one oxygen molecule or for the reduction of one carbon dioxide molecule to the level of carbohydrate, four electrons should be transferred on account of the absorbed

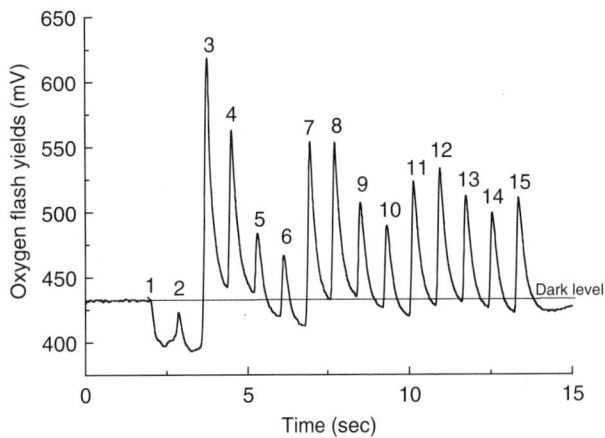

FIGURE 1.4 Oxygen flash yields of isolated pea chloroplast induced by a series of 15 flashes ($4J$, $t_{1/2} = 10\,\mu sec$, spaced 800 msec).

light quanta energy and consequently at least four photons are needed. For understanding and explaining the observed experimental results the following principal questions arise:

1. Whether energy or photoproducts of the four photons absorbed are summarized?
2. Whether oxygen-evolving centers act independently of each other or can exchange energy, or whether the oxygen precursors (positive charges) could migrate and cooperate in the surrounding medium?

It is well known that the average effective cross section for light quanta absorption of a chlorophyll molecule in solution is approximately $0.2 \times 10^{-16}\,cm^{-2}$. This means that under low irradiances, that is, 10^{13} to $10^{14}h\gamma\,cm^{-2}$, the time needed for absorption of four quanta by separated chlorophyll molecules should be approximately 1 h. Under such conditions, if oxygen-evolving centers act independently of each other, the evolution of photosynthetic oxygen should start after a prolonged induction time. This is in contradiction with observation 2 and the results presented in Figure 1.2, which show that in reality photosynthesis starts immediately without any induction period. The flash experiments of Emerson and Arnold (observation 3) show that photosynthesis decreases if the spacing between the flashes is higher than 0.02 sec (Figure 1.3). Therefore, at low irradiances, when the dark intervals between the quanta absorption are of the order of minutes, the effectiveness of photosynthesis should be much lower or tending toward zero. This fact is in contradiction with observation 1, which reflects that quantum efficiency of photosynthesis is very high under low irradiance conditions. Observation 3 as well as additional observations 5 and 6 lead to the conclusion that oxygen-evolving centers operate independently of each other (noncooperative mechanism). This means that every oxygen-evolving center should accept four light quanta (photons) before evolving one oxygen molecule. The results observed could be explained if we assume that oxygen-evolving reaction centers are in a state to conserve some of the oxygen precursors (e.g., positive charges) for several minutes or even hours, and upon subsequent illumination after absorption of the first photons they could immediately start evolving oxygen. This assumption, however, is in contradiction with observations 3, 5, and 6. Hence, we should conclude that the oxygen precursors are unstable in the dark and deactivate for about 100 sec.

If oxygen precursors are unstable in the dark, the observed results, that is, the absence of prolonged induction time and high quantum efficiency of photosynthesis under low light intensities, could be explained by the assumption that even under limited light conditions the oxygen-evolving centers received photons for time intervals of about seconds or even shorter. This assumption could be explained by an additional speculation that hundreds of chlorophyll molecules are functionally or even structurally assembled around a given specialized chlorophyll molecule (named reaction center), which carries out the photochemical reaction and this center is supplied with the photons absorbed by the assembled light-harvesting (antenna) molecules. In this way, the effective cross section of light quanta absorption of the reaction center molecule is increased 100-fold and even under very low light intensity conditions reaction centers received the needed four quanta for intervals 100 times shorter than the intervals of the separated chlorophyll molecules. This assumption explains

both the absence of the prolonged induction period and the high quantum efficiency under low light intensities. In good agreement with this assumption is observation 4, where after every saturating flash only one oxygen molecule is produced from approximately 2500 chlorophyll molecules. This attractive hypothesis was postulated by Emerson and Arnold in 1932 and was accepted immediately. Since then this postulate has been supported in many investigations and especially with the findings of observations 5 and 6. However, a significant number of investigations have shown discrepancies concerning the size and structure of the postulated PSUs [10–12]. This leads to a more ticklish question: Are the above-considered basic arguments sufficient for the postulation of the PSU concept? Careful analysis of these arguments shows that difficulties for a logical explanation of experimental results arise from observations 3, 5, and 6, that is, from the absence of oxygen burst or oxygen yield at the first flash given after several minutes of dark incubation. These observations have led us to reject the presence of any cooperative mechanism in the action of oxygen-evolving centers. The existence of a noncooperative mechanism of oxygen evolution in photosynthesis has been confirmed by observations 5 and 6 as well as by numerous flash experiments. Especially, the model of Kok et al. [13] or the S_i-states model shows that the noncooperative mechanism could explain both the absence of oxygen flash yield at the first saturating flash after prolonged dark incubation and the oscillations in oxygen flash yields with a period of four.

In spite of this, a number of kinetic models have been proposed for the explanation of various complicating phenomena of the oxygen flash yield oscillations [14–16]. According to Lavorel [17,18], a special kind of cooperative action exists in the functioning of S_i-states. In addition, there are some experimental results that cannot be explained by Kok's model (e.g., the linearity of the light curves under very low irradiance conditions). Obviously, the existence of the noncooperative mechanism of oxygen evolution does not exclude the participation and the existence of the cooperative mechanism. On the other hand, the absence of oxygen flash yield after the first flash cannot be considered as a proof for the absence of the cooperative mechanism in oxygen evolution because of the following reasons:

1. The first flash is applied after prolonged dark incubation of algae or chloroplast suspensions that leads to anaerobic conditions in cell and chloroplast volumes.
2. Since the functioning of the cooperative mechanism should be realized by diffusion of oxygen

precursors produced in different oxygen-evolving centers, the rate constant of the reactions leading to oxygen evolution through the cooperative mechanism should be significantly lower than the rate constant of the noncooperative mechanism.

Consequently, it could be concluded that the observation of oxygen burst or oxygen production by the first flash will be difficult and even impossible. Moreover, if we consider observation 1, that is, the linearity of light curves under low light intensity conditions, and observations 3 and 5, that is, the dependence of yields on dark intervals between the flashes and the absence of oxygen yield at the first flash, it is reasonable to conclude that these observations are mutually contradicting. If observations 3 and 5 reflect strictly the photosynthetic oxygen production upon flash irradiation, then even with structures like the postulated PSUs the light curves of photosynthesis (oxygen evolution) should have a nonlinear part under very low light intensity conditions. This means that independently of the existence of photosynthetic units the light curves of photosynthesis should be S-shaped if one assumes that oxygen production is realized only through the noncooperative mechanism and that the defined deactivation reactions exist. Thus, two possibilities could be considered:

1. The cooperative mechanism is functioning simultaneously with the noncooperative mechanism.
2. The light curves of photosynthesis exhibit a nonlinear part at very low light intensity conditions.

The first assumption gives the explanation of the basic arguments that have led to the postulation of the concept of the PSU, that is, observations 1 to 3, while observations 5 and 6 could be explained by the functioning of the noncooperative mechanism. Observation 4 will be reconsidered in Section II.C. If we accept the second possibility, we can explain the "red drop" and "enhancement" effects of Emerson, which are considered as basic observations of the concept of two photosystems, without using this concept. Our investigations during the last 35 years have shown that these two possibilities exist. This means that despite the participation of cooperative and noncooperative mechanisms of photosynthetic oxygen evolution, the irradiance dependence of photosynthesis is a nonlinear function, that is, the "light curves" are "S"-shaped. Probably under low irradiance conditions a significant part of the photosynthetically evolved oxygen is consumed by dark respiration and

under these conditions the registered light curves have low slopes and the quantum efficiency is low.

The following observations could be considered in favor of the cooperative mechanism:

1. In unicellular algal suspensions, prolonged (5 to 20 min) oxygen evolution is registered after switching off the continuous irradiation.
2. Decay kinetics in oxygen flash yields are at least biphasic, probably two different processes exist that lead to oxygen production.
3. One cannot explain the absence of the induction period under low irradiances without the participation of the cooperative mechanism.
4. In some photosynthesizing systems (cyanobacteria) one cannot register any oxygen flash yields, despite the fact that they can produce oxygen at a high rate under continuous irradiation.
5. In our previous studies [19,20] we stressed that the noncooperative oxygen evolving mechanism operates mainly in grana regions while the cooperative mechanism is localized predominantly in stroma thylakoids.

C. Variation in the Number of Effectively Functioning Oxygen-Evolving (Reaction) Centers

If one assumes that a suspension of unicellular algae (*Chlorella*, *Scenedesmus*, etc.) contains N_0 reaction centers, then under very general assumptions it could be shown [21] that the following relationship exists between the number of open reaction centers (N) and the rate of oxygen evolution (photosynthesis) (P):

$$N = N_0 - N_0 P / P_{max} \qquad (1.1)$$

where P_{max} is the saturating (maximum) rate of photosynthesis. Obviously, the N vs. P plot is a straight line (Figure 1.6, curve "c"), crossing the ordinate at $N = N_0$ and the abscissa at $P = P_{max}$.

The experimental determination of the ratio between total and open (unoperative) centers is relatively easy. According to the model of Kok et al. [13], the oxygen-evolving centers exist in five different oxidized states: S_0, S_1^+, S_2^{2+}, S_3^{3+}, and S_4^{4+}. Every center that absorbs one photon will pass to the next higher oxidized state. After reaching state S_4^{4+} one oxygen molecule is produced, and the center returns to the initial S_0 state. It is easy to understand that independently of the oxidation state, every center after absorption of four photons separated by dark intervals equal to or longer than the turnover time τ of the reaction centers will evolve one oxygen molecule and attain its former state. Consequently, the amplitudes of oxygen bursts produced by four saturating flashes will reflect the number of centers in the unoperative (open) state. This means that if the flash groups are given in darkness (when all centers are open) the amplitudes of bursts will reflect the total number of centers.

The results obtained with *C. pyrenoidosa* cells using excitation with groups of four saturating flashes ($t_{1/2}$ = 8 μsec) spaced 20 msec from each other and with 7 sec dark intervals between the groups on the background of a gradually increasing continuous irradiation with achromatic (white) light are shown in

Figure 1.5. In contrast to our expectations, data show that the amplitude of the oxygen bursts produced by the group of flashes in darkness (0) are very small and after continuous background irradiation (1 to 4) a significant increase could be seen. On increasing the intensity of background irradiation (5 to 7) the amplitudes of oxygen bursts decrease and after reaching the saturated background irradiation (7) they are almost invisible.

The relationship between the amplitudes and steady-state oxygen evolution is presented in Figure 1.6. Curve "a" is obtained by increasing the background irradiation from zero to saturation level. Curve "b" is drawn for the reverse direction, that is, with gradually decreasing background irradiation. Obviously, the difference between the two curves reflects an "hysteresis" effect and is more probably a consequence of the induction phenomenon in the photosynthetic process. It should be pointed out that the shapes of curves presented in Figure 1.6 are dependent on the experimental duration and the preceding history of investigated alga suspensions. Nevertheless, an inexplicable difference between the straight line "c," theoretically predicted on the basis of the PSU concept, and curves "a" and "b" still remains. The amplitudes of oxygen burst increase under background irradiation. They reach their maximum at the level of the steady-state oxygen-evolution rate, representing approximately one third of the maximum value of the saturating level. Whenever flash groups are given under low irradiance the lower value of amplitudes reflects the existence of the induction phenomenon. It is obvious that we cannot estimate the exact number of reaction centers from amplitudes of oxygen yield under dark conditions, that is, without background irradiation.

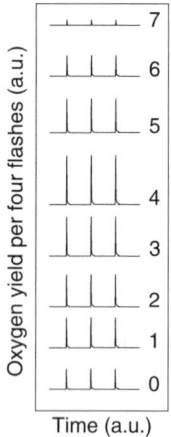

FIGURE 1.5 The amplitudes of oxygen yields (*Chlorella pyrenoidosa*) produced by four saturating flashes (4 J, $t_{1/2}$ = 10 μsec) with 20 msec dark periods between the flashes and 7 sec between the groups depending on the steady-state oxygen evolution rate. The intensities of background light are: 0, 0; 1, 17.0; 2, 25.0; 3, 34.0; 4, 43.0; 5, 52; 6, 82.0; 7, 135 W/m².

FIGURE 1.6 The number of unoperative (open) centers (*Chlorella pyrenoidosa*) depending on the oxygen evolution rate level: (a) experimentally obtained results by increasing the light intensity of background irradiation from 0 to saturation level (O₂ rate from 0 to maximal [saturating-P_{max}] rate); (b) in the opposite direction; and (c) straight line, predicted by the theory of the photosynthetic unit concept.

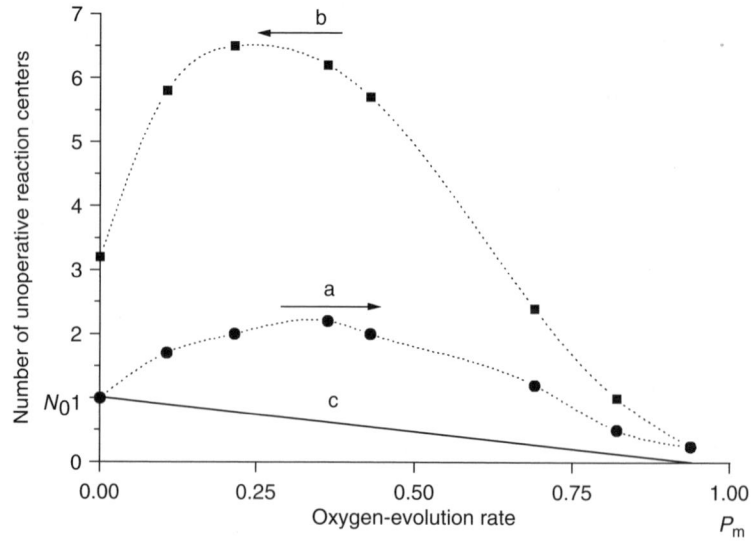

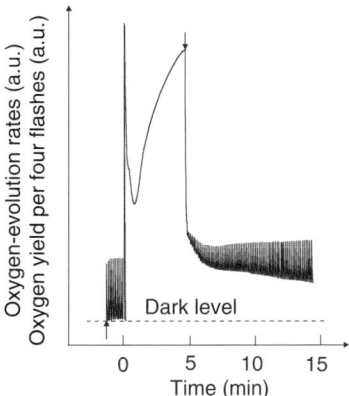

FIGURE 1.7 Variations in the oxygen bursts before, during, and after the induction time of photosynthesis in *Chlorella pyrenoidosa*. The suspension was kept in darkness for 5 min and the groups of four saturating flashes (20 msec spacing between the flashes and 7 sec between the groups) were switched on at the time indicated by "↑." The saturated white light (135 W/m²) was switched on at the time indicated by "0" and switched off at the time indicated by "↓."

The results presented in Figure 1.7 show the changes in amplitudes of oxygen burst in *C. pyrenoidosa* produced by groups of four saturating flashes with 20 msec spacing between the flashes and 7 sec between each flash group before, during, and after the induction time of photosynthesis (irradiation with saturated achromatic [white] light). These results demonstrate well the expressed variation in oxygen yields from flash groups and reflect in fact the number of open reaction centers (oxygen-evolving centers).

The results presented in Figure 1.8, where the oxygen bursts are produced by the same flash groups as in Figure 1.7, show that the effects of flash groups on the background saturating "white light" were negligible. At time 0, the "white light" was switched off and the rate of oxygen evolution decreased sharply to the level indicated by D, after which the process of oxygen evolution in the dark connected with deactivation of S_i states [22] or with the deblocking of inactivated (blocked) states began.

Immediately after switching off the continuous saturating radiation the effect of flash groups was very small and the amplitudes of oxygen yields increased slowly in the dark up to 30 min. Consequently, the increase of amplitude of oxygen group yields in the dark (after switching off the background radiation when all centers are in the open state) showed that the number of effectively working oxygen-evolving centers increased. This number was significantly low immediately after switching off the saturated background radiation and thus one might assume that it had the same low value during the

preceding time of irradiation with saturating "white light." This means that under saturating irradiance conditions the essential parts of the reaction center are in the inactivated (blocked) state. The results in Figure 1.7 show that the initial amplitudes of four flash-induced oxygen bursts are restored approximately 15 min after switching off the continuous saturating irradiation (in the darkness).

It could be shown that the following relationship exists between the number of operating reaction centers (N_c), the amperometric current on the polarograph equipped with oxygen rate electrode (I), the turnover time of reaction centers (τ), and the electric charge of an electron (e):

$$N_c = I\tau/e \qquad (1.2)$$

If one can accept the value of Emerson and Arnold [1] for turnover time of the centers, 2×10^{-2} sec, and for the amperometric current of saturated oxygen-evolution rate in Figure 1.8, 1.32×10^{-5} A, the number of oxygen-evolving centers in the investigated sample can be calculated as

$$
\begin{aligned}
N_c &= I\tau/e \\
&= (1.32 \times 10^{-5}\,\text{A})(2 \times 10^{-2}\,\text{sec})/1.6 \\
&\quad \times 10^{-19}\,\text{C} \\
&= 1.65 \times 10^{12} \qquad (1.3)
\end{aligned}
$$

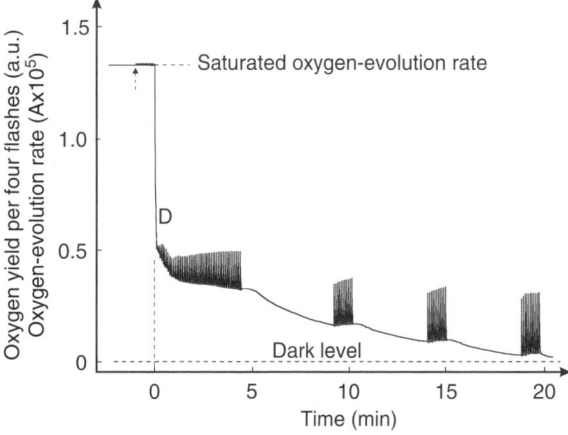

FIGURE 1.8 Oxygen bursts produced by groups of four saturating ($4J$, $t_{1/2} = 8\,\mu\text{sec}$) flashes with 20 msec dark periods between the flashes and 7 sec between the groups. Suspension of *Chlorella pyrenoidosa* (4 mm³, 15 µg Chl. cm⁻³) was irradiated with saturating white light (135 W/m²) and at the time indicated as "0" the saturating light was switched off. The groups of flashes were switched on at the time indicated by "↑" (for details, see text).

A comparison between the number of chlorophyll molecules (N_{Chl}) in the sample ($4\,mm^3$ with $15\,\mu g$ Chl cm^{-3}, that is, 8.8×10^{14} chlorophyll molecules) and the number of oxygen-evolving centers (N_c) leads to

$$P = N_{Chl}/N_c = 8.8 \times 10^{14}/1.65 \times 10^{12}$$
$$= 533\,Chl/1RC \qquad (1.4)$$

If the number of chlorophyll molecules is calculated for one oxygen molecule evolved, the value obtained should be increased four times, that is, about 2130 for one oxygen molecule. Consequently, the value obtained in such a way is in accordance with the value for the PSU of Emerson and Arnold [1].

From the results presented in Figure 1.7 and Figure 1.8, it could be concluded that the number N_c, estimated above, reflects only the number of effectively working reaction centers under saturating irradiance conditions but not their total number. An approximate idea about the total number of oxygen-evolving centers could be obtained if we compare the amplitudes of oxygen yields per four flashes (Figure 1.8) during the irradiation with saturating "white light" with those obtained 20 min after switching off the light: Approximately 200 to 400 times increase was registered after switching the light off. Keeping in mind that the ratio between chlorophyll molecules and the operative reaction center under saturating irradiance conditions is of the order of 500 one can conclude that the total number of reaction centers is practically equal to the number of chlorophyll molecules. This indicates that the usual procedures used for the estimation of the number of PSUs have to be revised. There are mainly two reasons for this:

1. Under high light intensity or frequency of saturating flashes the oxygen flash yields are low

due to inactivation of the essential part of the reaction center.
2. Under low light intensity conditions the oxygen flash yields are low as a consequence of the induction phenomenon.

We found that after switching on the irradiation (during the induction time of photosynthesis), the oxygen absorption reaction occurs connected with the oxidation of oxygen-evolving centers [23]. The amount of oxygen absorbed during the induction time depends on the chlorophyll content and approximately the same amount of oxygen is evolved after switching off the light (in the darkness) (Figure 1.9, Table 1.1).

On the other hand, according to Emerson and Lewis [24] and McAlister [25], the amount of CO_2 burst during the induction period is also of the order of the amount of chlorophyll, which was explained by Franck and Herzfeld [26] as a result of the decomposition of the ACO_2 complex under light (A is the primary acceptor of CO_2 whose quantity is assumed to be equal to the amount of chlorophyll). Thus, it may be assumed that functioning of the oxygen-evolving centers may be presented as follows: in darkness, all oxygen-evolving centers accept CO_2 molecules or HCO_3^- anions. This statement is in agreement with the results of Stemler [27,28]. At low irradiance, every chlorophyll molecule works as a part of the reaction center with low frequency depending on the frequency of the quanta absorbed. If the irradiance is sufficiently high, it leads to the oxidation (blocking) of a significant part of oxygen-evolving centers, a process connected with oxygen consumption and leads to CO_2 evolution from oxygen-evolving centers during the induction time of photosynthesis. At saturating irradiance the number of unoxidized oxygen-evolving (working) centers can

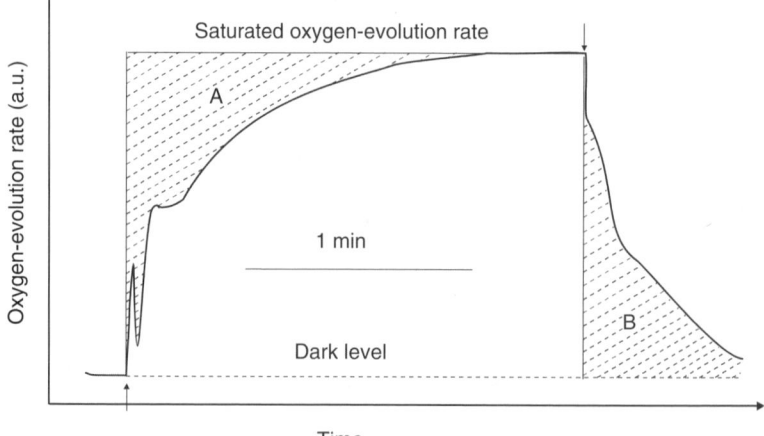

FIGURE 1.9 Induction curve of photosynthesis at *Chlorella pyrenoidosa*, recorded after 5 min dark incubation and after irradiation with $135\,W/m^2$ "white light": "↑" — light on; "↓" — light off. For details see text. The number of oxygen molecules absorbed during the induction time of photosynthesis, calculated from the dashed area "A" and evolved after switching off the irradiation in the dark (dashed area "B") are in order of the number of chlorophyll molecules in suspensions investigated.

TABLE 1.1
The Ratio Between the Number of Oxygen Molecules Absorbed During the Induction Time of Photosynthesis and the Number of Chlorophyll Molecules in the Investigated Suspensions of Scenedesmus Acutus and Chlorella Pyrenoidosa

Samples	O_2/Chl
Scenedesmus	1.1
Scenedesmus	0.9
Scenedesmus	1.0
Chlorella	0.9
Chlorella	0.8

decrease to approximately 1 : 500; thus, the number of oxygen molecules absorbed or CO_2 molecules evolved during the induction time would be practically equal to the number of chlorophyll molecules in the investigated photosynthesizing system. This assertion may explain the observed dependence of induction time on radiation intensity. According to the explanation presented above, if the quanta arrive at oxygen-evolving centers after prolonged intervals (longer than several seconds) the centers cannot reach the higher oxidized states, S_3 or S_4, and oxygen can be evolved by the cooperation of oxygen precursors obtained in different centers, a mechanism considered previously [29,30]. In summary, the following reaction steps could be presumed:

$$Chl \cdot Z + HCO_3^- \rightarrow Chl \cdot Z \cdot HCO_3^- \qquad (a)$$

$$Chl \cdot Z \cdot HCO_3^- + h\nu \rightarrow Chl^* \cdot Z \cdot HCO_3^- \qquad (b)$$

$$Chl^* \cdot Z \cdot HCO_3^- \rightarrow Chl^+ \cdot Z^- \cdot HCO_3^- \qquad (c)$$

$$Chl^+ \cdot Z^- \cdot HCO_3^- + P \rightarrow Chl^+ \cdot Z \cdot HCO_3^- + P^- \qquad (d)$$

$$Chl^+ \cdot Z \cdot HCO_3^- \rightarrow Chl \cdot Z + HCO_3^\bullet \qquad (e)$$

$$4HCO_3^\bullet \rightarrow 2H_2O + 4CO_2 + O_2 \qquad (f)$$

$$H_2O + CO_2 + CA \rightarrow H_2CO_3 \rightarrow H^+ + HCO_3^- \qquad (g)$$

$$4Chl \cdot Z \cdot HCO_3^- + 4O_2 + h\nu \rightarrow 4Chl \cdot$$
$$Z^+ \cdot O_2^{2-} + 4HCO_3^\bullet \qquad (h)$$

$$4HCO_3^\bullet \rightarrow 2H_2O + 4CO_2 + O_2 \qquad (i)$$

During reaction (a), oxygen-evolving centers (i.e., all chlorophyll molecules) capture bicarbonate ions in

the darkness. Reaction (b) reflects the light quanta absorption by the chlorophyll molecule, which forms a complex with the primary electron acceptor (Z). In reaction (c), charge separation is accomplished and one electron is transferred from the chlorophyll molecule to Z. Reaction (d) shows the electron transfer to a component P on the electron transport chain. The electron of the bicarbonate ion fills the missing electron in the chlorophyll molecule and the bicarbonate ion is separated as a radical (reaction [e]). The recombination of four bicarbonate radicals (reaction [f]) accumulated at a given reaction center (in flash experiments or under high irradiation conditions) leads to the evolution of one oxygen molecule, two molecules of water, and four molecules of CO_2 — the so-called noncooperative or Kok's mechanism. Under low irradiances or after switching off the light the cooperation of four bicarbonate radicals, produced in different reaction centers, leads to same reaction — the so-called cooperative mechanism. The restored complex of the chlorophyll molecule and the primary acceptor in reaction (e) and the obtained CO_2 molecules (reaction [f]) after hydration with the participation of carboanhydrase (CA) (reaction [g]) are involved in reaction (a) and the cycle could start again.

Reaction (h) takes place after irradiation and the increased oxygen concentration during the induction time of photosynthesis is connected with the inactivation (blocking) of the oxygen-evolving centers. These processes lead to the liberation of bicarbonate radicals and after their recombination (reaction [i]) the process of CO_2 burst [24] is accomplished. In summary, these two reactions lead to oxygen absorption and CO_2 liberation. Apparently, if the reactions presented above reflect the molecular events in oxygen-evolving centers the isotopic experiments with labeled oxygen will show water as the source of photosynthetic oxygen. Water is included as the ultimate source of electrons in reaction (g) during the hydration of CO_2.

The above interpretation explains the results presented in Figure 1.2. Induction curves showed that the duration of the induction period decreased simultaneously with decrease in irradiation, and under low intensity ($0.056 I_0$) the rate of oxygen evolution reached its steady state very quickly after the light is switched on — reactions (h) and (i) cannot be accomplished as the concentration of oxygen is low (low irradiation). However, under these conditions, the "working point" of the photosynthetic process enters the initial nonlinear part of the curve depicting dependence on irradiance (Figure 1.2, right), which is characterized by a very low quantum efficiency. Analysis of results from flash experiments

[31,32] showed that the linear part of the irradiance curve corresponds to oxygen evolution connected with successive transitions of S_i states from S_0 to S_4^{4+}, while the deactivating back reactions of the oxidized S_i states take place in the region of the initial nonlinear parts of irradiance curves. Thus, at low irradiances when the absorption of four quanta in the individual reaction centers needs a longer time and the centers do not manage to pass over into the S_4^{4+} state, the oxygen evolution is mainly a result of the deactivation of the oxidized S_i states and the cooperation of oxygen precursors (bicarbonate radicals [$HCO_3^\bullet$]) produced from different reaction centers.

The concept of the PSU is now more than 70 years old. During this period, our ideas about the size and the arrangement of these structures have often changed. The most difficult questions still remain: "Are the concepts of Emerson and Arnold [1] or of Gaffron and Wohl [33] sufficiently sound to justify the present day model?" Or "Are there other possibilities for the explanation of the existing observations?" I suppose that if Emerson and Arnold [1] and Gaffron and Wohl [33] have had in their possession the results presented in Figure 1.5–Figure 1.8, which show dramatic changes in the number of oxygen-evolving centers during the induction time, it could hardly be assumed that they would have postulated their hypothesis about the PSU. Unfortunately, all their experiments were performed with Warburg's manometric apparatus. It will be useful to remember the words of Birgit Vennesland [34] concerning the photosynthetic unit concept:

> ...These are (having in view the hypotheses, NB) mainly based on the assumption that a hundred or more chlorophyll molecule operate as a unit to transmit the energy of the absorbed photons to appropriate, hypothetical reaction centers. The flashing light experiments on which this view is based are of dubious significance, and the complexities and detail in which the associated theories have been clothed should not be confused with evidence. Freedom to use a large number of assumptions makes it easy to devise theories and to fit innumerable observations to them. The most valuable experimental facts are those which restrict such flights of the imagination.

The results presented above show the complexity and flexibility of the oxygen-evolving system of photosynthesis. They demonstrate that many of the experimental data obtained cannot be understood within the framework of the postulated PSU. Furthermore, there are many observations whose explanations lead to serious contradictions, which have led

to the proposal of various models. Regarding the basic arguments for the postulation of a PSU one has to admit that the strongest point is the absence of oxygen after the first saturating flash. However, it demands a very careful reconsideration: after prolonged darkness the first flash hits the cells or the chloroplasts in an anaerobic state; the rate constants of reactions leading to oxygen evolution through the cooperative mechanism are significantly lower than those connected with a noncooperative mechanism, since the functioning of a cooperative mechanism requires diffusion of oxygen precursors between different reaction centers. Photosynthetic systems are self-controlled and may attain a modified state after a short saturating flash. This may be connected with oxygen-consuming processes during the induction period and further connected with self-regulating processes that protect the living structure from oxidative damage. This statement is supported by the data of Boitchenko and Efimtcev [35], which prove that under increased oxygen concentrations a significant part of oxygen-evolving (PSII) centers are inactivated (blocked).

Therefore, all three basic arguments about the concept of the PSU could be explained by the existence of two different ways of oxygen evolution in photosynthesis and by the different degrees of inactivation (blocking) of oxygen-evolving centers. In this respect the concept of the PSU should be accepted as a dynamic system rather than as a structural or statistical system.

III. THE CONCEPT OF TWO PHOTOSYSTEMS

A. EXPERIMENTAL GROUNDS

The hypothesis of participation of two photochemical systems in the light-driven reactions of photosynthesis in green plants emerged after the discovery of Emerson's second effect (the "enhancement" effect) and was theoretically substantiated by Hill and Bendall [2] in 1960, who assumed that both photosystems function consecutively. In the course of the following four decades, this hypothesis was supported by a considerable number of experimental facts; that is, the sites of the individual electron carriers were estimated and, along general lines, were accepted by most authors. However, as already pointed out, Emerson's second effect and also the "red drop" of quantum efficiency, which are considered as headstones of this concept, could be explained without resorting to the hypothesis of two photosystems ensuing from the nonlinearity of the light curves of photosynthesis at low light intensities or from the principle of nonadditiveness in the action of light [31]. On the other hand, in the literature there is a great deal of information that cannot be satisfactorily explained by the concept of two photosystems. This is the reason for the existence of several hypotheses about the sequence and the functioning of light reactions in photosynthesis [2,36–39].

The existence of these hypotheses proves the difficulties that different groups of investigators have in interpreting experimental results. Despite the fact that significant differences exist between these hypotheses they all contain at least two different photosystems (PSI and PSII).

The main experimental facts supporting the conception of two photosystems are the following:

1. The quantum efficiency of photosynthesis — 8 to 12 quanta are needed for the reduction of one molecule of CO_2 or for the evolution of one molecule of O_2.
2. The red drop of quantum efficiency of photosynthesis [24].
3. The enhancement effect (Emerson's second effect) [40].
4. The spectral transient effects [41].
5. Myers' and French's effect [42,43].
6. Cytochrome *f* oxidation by light with 700 nm wavelength and its reduction by light with at 680 nm (or shorter wavelength).
7. The existence of alga mutants [44], one of which (mutant no. 8) does not accomplish photolysis of water and does not evolve oxygen (does not show Hill activity) but has the ability to reduce $NADP^+$ and CO_2, while the other (mutant no. 11) evolves O_2 and posseses Hill activity but is not able to reduce $NADP^+$ and CO_2.
8. The existence of chloroplast fragments possessing different activities, that is, some accomplish the Hill activity while the others reduce $NADP^+$.
9. The results of experiments with specific inhibitors of electron transport such as CMU, DCMU, hydroxylamine, and others.
10. Some results obtained by studying photophosphorylation coupled with electron transport in the light reactions of photosynthesis.

Besides the above-cited experimental facts, there are many other results that are interpreted with the aid of the hypothesis of two photosystems, but presumably they could also be explained with the same level of acceptance by leaving out this concept.

The most important experimental result that suggested the idea for two photosystems was Emerson's second effect or the so-called "enhancement effect." As is well known, in 1956 Emerson [40] looked for an explanation of the red drop of quantum efficiency that was observed at wavelengths above 700 nm. During the experiments he observed that if short-wavelength light was added to the less efficient long-wavelength light the efficiency of this light increased. In other words, the effect of simultaneous action of two light beams with different wavelengths is greater than the sum of the effects of their independent action. The principal reason for including the two photosystems in the light induced reactions of photosynthesis is just to explain this nonadditive light action. This raises the question: Is it possible to explain this effect with the operation of a single photosystem? As discussed in Section I, the answer to this question would be positive if one assumes that the light curves of photosynthesis are nonlinear at low light intensities, that is, they are S-shaped.

A suspension of *C. pyrenoidosa* was irradiated with two light beams (Figure 1.10), one of which is 700 nm modulated (1 sec light/1 sec dark) and the second is background light with different wavelengths between 600 and 700 nm. The amplitude of the modulated oxygen rate induced by the 700 nm beam changed after applying background radiation of different wavelengths whose intensities were chosen in such a way as to give an equal oxygen-evolution rate in the linear part of the "light curve." The intensity of the 700 nm modulated beam was kept constant. The amplitude of the modulated oxygen-evolution rate

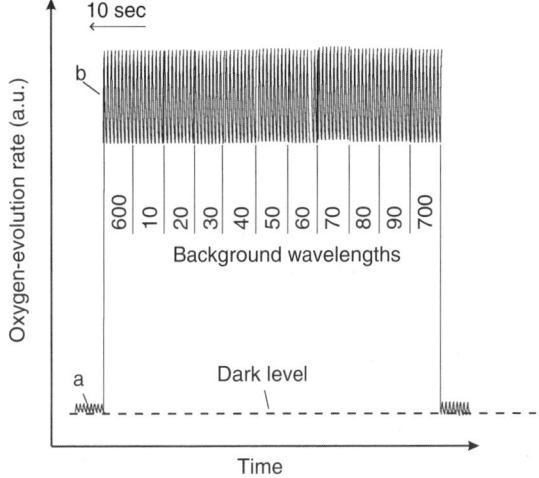

FIGURE 1.10 Amplitudes of the modulated (0.5 Hz) oxygen-evolution rate in *Chlorella pyrenoidosa* induced by a 700 nm beam without background radiation (a) and after compensation of the initial nonlinear part of the "light curve" with background radiation of different wavelengths between 600 and 700 nm (b).

remained constant (in the limit of experimental errors) in all investigated spectral regions (600 to 700 nm). If Emerson's second effect exists as a separate appearance we should not obtain any enhancement in the case of addition of 700 nm background radiation to the 700 nm modulated beam. But this was not observed: the enhancement did not depend on the wavelength of the background radiation but on its intensity and on the obtained oxygen-evolution rate. The equal degree of enhancement with 700 nm and other wavelengths showed that Emerson's second effect is only a particular case of the principle of nonadditive action of radiation in photosynthesis [31] and that it does not exist even as a second-order effect. Obviously, this suggestion is in sharp contradiction with the accepted concepts and literature data. Mann and Myers [45] even obtained a negative enhancement effect in the case of superposition of two beams of the same wavelength. Such a "negative enhancement" (attenuation) exists in different regions of Emerson's second effect action spectra. According to Heath [46], there is no reasonable explanation for this negative effect. Our efforts to find such attenuation after having observed the conditions ensuing from the nonlinearity of the "light curves" were unsuccessful. Probably both absence of enhancement in the case of superposition of two beams of same wavelength and observation of attenuation in different regions of Emerson's second effect action spectra are consequences of reaching saturation with radiant energy. A correct compensation of the initial nonlinear part of the "light curves" is impossible not only in suspensions with high absorbance (>0.5) but also in suspensions with very low absorbance because of the nonhomogeneous distribution of pigments in them (in the cell and the chloroplast volumes). When one tries to compensate the lowest sublayer in suspensions or in chloroplasts of higher absorbance, the oxygen-evolving centers situated in the surface sublayers always reach the region of saturation with radiant energy. Due to the difference in the wavelengths of exciting radiation the distribution of absorbed light quanta in various sublayers of suspension or of chloroplast volumes is also different. This means that the action of light with different absorption coefficients will be different even after equalization of their summary effects.

The graph in Figure 1.11 clearly shows the appearance of the effect of enhancement after excitation of photosynthesis by two continuous monochromatic rays with the same wavelength (650 nm).

Figure 1.12 represents an original protocol from the experiment in which two monochromatic 650 nm light beams are focused on the suspension layer of *C. pyrenoidosa*. One of the beams, I_1, is modulated and the other, I_2, is continuous. In the left part only

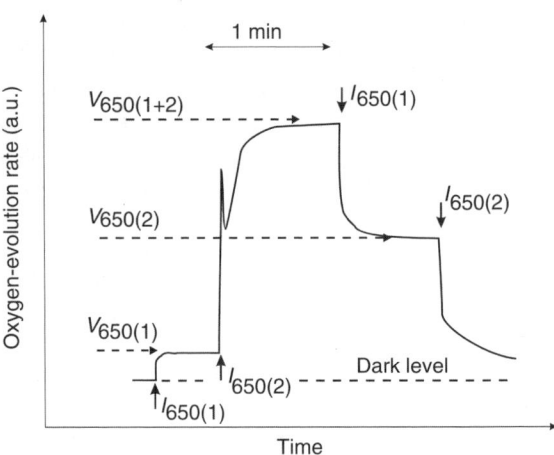

FIGURE 1.11 "Enhancement effect" in *Chlorella pyrenoidosa* obtained by means of two monochromatic light beams of the same wavelength (650 nm): ↑, turning on; ↓, switching off the light beams.

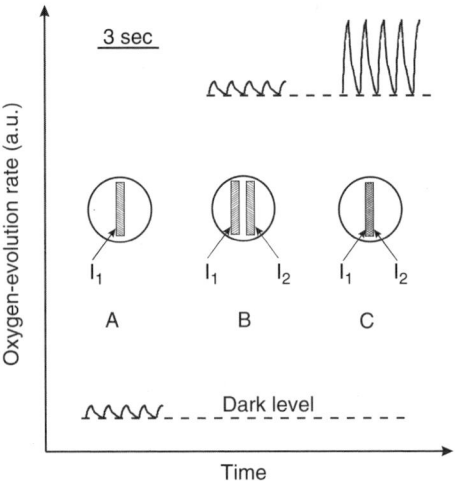

FIGURE 1.12 The effect of two monochromatic 650 nm light beams depending on their positions on the suspension layer of *Chlorella pyrenoidosa* (for details see the text).

the modulated beam is used and the obtained modulated oxygen-evolution rate (designated by "A") is seen on the "zero" dashed line. In the middle part of the figure the continuous light beam I_2 is switched on but is focused on different regions with respect to the modulated beam (I_1). It is seen that the continuous oxygen-evolution rate increases; however, the amplitudes "B" of the modulated oxygen-evolution rate remain unchanged. In the right part of the figure both beams are directed on one and the same surface of the suspension and a significant increase in the amplitudes of the oxygen evolution rate is observed.

The results presented lead to the conclusion that the "enhancement effect" depends on the "working point" of the oxygen-evolving system on the light curve or on some feature belonging to cell or chloroplast structure, but not on the concentration of oxygen in the surrounding volume. The changes of oxygen-evolving amplitudes obtained after irradiation with modulated light beams before switching on the background irradiation, during the induction time (after switching on the background irradiation [arrow "a"]), and in darkness (after switching off the continuous irradiation [arrow "d"]) are presented in Figure 1.13. The wavelength of the two light beams is 650 nm. Arrow "b" indicates switching off and arrow "c" switching on the modulated irradiation. It is seen that the amplitudes of the modulated oxygen-evolution rate do not reach their maximum immediately after the induction of the photosynthetic process. The amplitudes increase simultaneously with increase in the continuous oxygen-evolution rate. After switching off the continuous irradiation the amplitudes do not reach their initial value and during a certain dark period they decrease continuously.

A comparison of the enhancement values (approximately 5 to 10) obtained in our experiments with those in Emerson's second effect investigations (approximately 1.2 to 2.2) shows that the effect provoked by nonlinearity of the irradiance curves is much stronger that that observed for Emerson's en-

hancement effect. Obviously, the effect of irradiance on photosynthesis is nonadditive not only for the beams with different wavelengths (Emerson's enhancement effect) but also for the beams with the same wavelength. This statement was confirmed by Warner and Berry [47] and Milin and Sivash [48]. As pointed out earlier this effect is considered as a "crucial experiment" for the assumption that the electrons from water to NADP are transferred through two consecutive photoacts.

B. PHOTOSYNTHESIS WITH SOLE PHOTOSYSTEM

Figure 1.14 presents a tentative diagram of electron transport light reactions of photosynthesis in green plants by a single photosystem on the basis of the existing diagrams of Hill and Bendall and Arnon's group (cf. Hall and Evans [49]). The best known electron carriers according to their corresponding redox potentials are arranged in three groups. The group of electron carriers at the reduction side of the photosystem, consisting of the primary acceptor of that photosystem Z (FRS; Fe-S), feredoxine (FD), and flavoprotein (fp), is determined

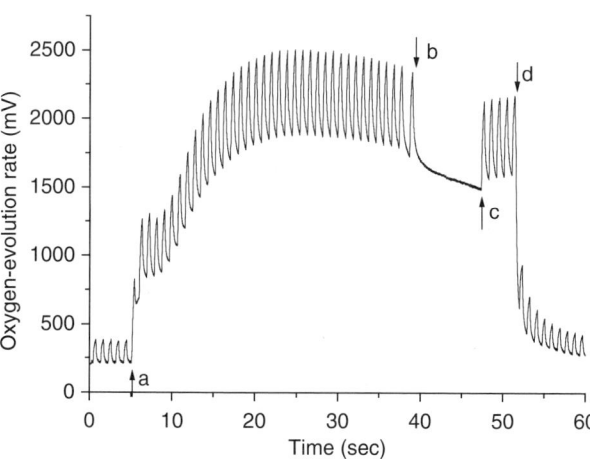

FIGURE 1.13 Dependence of the amplitude of the modulated (0.5 sec light/0.5 sec dark) oxygen evolution in *Scenedesmus obliquus* during the induction time of photosynthesis. The two light beams have the same wavelength (650 nm) and allow 10 and 6 μmol/m²/s irradiances for modulated and continuous beams, respectively. The continuous light is switch on (arrow "a") and switch off (arrow "d.") Arrows "b" and "c" show switching off and switching on of the modulated light beam, respectively.

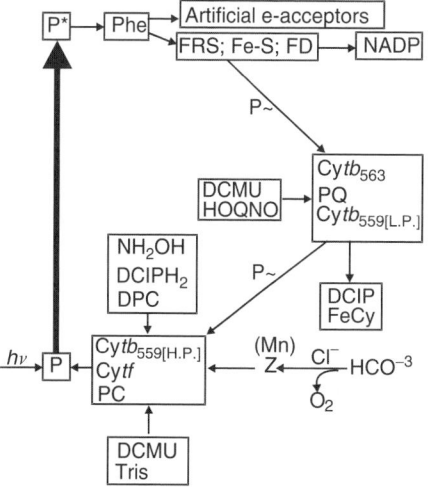

FIGURE 1.14 A tentative model of photosynthetic electron transport with only one photosystem. P, oxygen-evolving (reaction) center; P*, excited state of P; Phe, pheophytin; FRS, ferredoxin reducing substance; Fe-S, bound iron sulfur protein; FD, ferredoxin; NADP, nicotinamide adenine dinucleotide phosphate; DCMU, 3-(3,4-dichlorophenyl)-1, 1-dimethylurea; HOQNO, 2-heptyl-4-hydroxyquinoline-*N*-oxide; PQ, plastoquinone pool; Cyth, cytochromes; DCIP, 2,6-dichlorophenolindophenol; DCIPH₂, reduced form of DCIP; PC, plastocyanin; FeCy, potassium ferricyanide; NH₂OH, hydroxylamine; DPC, 1,5-diphenylcarbazide; DBMIB, 2,5-dibromo-3-methyl-6-isopropyl-*p*-benzoquinone (for detail see the text).

with the highest degree of significance. This group of electron carriers has the highest negative redox potential and is closely connected with the reduction of $NADP^+$. Besides this, it is known that FD participates also in the process of cyclic photophosphorylation, and on this account it is assumed that at this point the electron transport chain branches off toward the cyclic electron transport or the group of electron carriers consisting of cytochrome b_{563}, cytochrome $b_{559(L.P.)}$, and probably plastoquinone (PQ). The group of electron carriers consisting of plastocyanine (PC), cytochrome $b_{559(H.P.)}$, and cytochrome f shows a tendency toward oxidation upon illumination and is probably situated at the donor part of the electron-transport chain. It is possible that some of the carriers of this group take part in the cyclic electron transport.

The figure also shows the possible sites of photophosphorylation at the cyclic electron transport, the expected sites of action of best known inhibitors of the individual reactions, and the artificial electron donors and acceptors. With the exception of the natural electron acceptor in the reducing part of the photosystem the electron carriers and the reactions taking place in this part are relatively well known. All electron carriers shown in the figure are at their respective places in the cyclic transport according to Hill and Bendall [2] and Knaff and Arnon [36]. Of course, many details in both structural and functional aspects should be clarified after a profound analysis of the existing literature data.

Figure 1.14, indicating the functioning of the electron transport reactions of photosynthesis in green plants, could explain the following experimental facts:

1. Emerson's effects, the red drop and the enhancement, are explained by the principle of the nonadditiveness in the action of light during photosynthesis.
2. The existence of mutant forms algae (no. 8 of Bishop) and also of different fragments from chloroplasts (light fragments), which cannot evolve oxygen, could be explained with damages of electron-transport chain in the oxidative part (Z, $Cyth_f$, $Cyth_{559[H.P.]}$, PC, different kinds of polypeptides), and for mutant no. 11 and for heavy chloroplast fragments with destructions in the reduction side (Phe, Fe-S, Fd) [44].
3. The qualitatively different behaviors of the photosynthesizing system toward light of wavelengths over and below 700 nm is probably due to the unequal number of absorbed quanta; hence, depending on the degree of reduction of $NADP^+$, a change occurs in the relative

number of the electrons participating in the cyclic and noncyclic pathways.

4. Depending on the sites of action of the various inhibitors, they will lead to different effects. It is possible that some of these substances may have nonspecific action as well. Certainly, the final effect of the action of individual inhibitors will depend also on the corresponding sequence of the electron carriers in the various groups.
5. As shown in our earlier work [30] the spectral-transient effect of Blinks [41] and Myers and French [42] could be considered as a result of the superposition of the induction-type transient phenomenon observed during oxygen evolution. As a consequence of different permeabilities of the pigmented sections in chloroplasts for light beams with different wavelengths a change occurs in the frequency of turning of the functioning reaction centers and this leads to the difference in oxygen induction curves. The same interpretation will be valid also for the so-called "State 1–State 2" phenomenon. There is no doubt that these effects as well as data obtained upon investigation of photophosphorylation cannot be considered as irrefutable arguments for the serial operation of the two photosystems in the light reactions of photosynthesis.

IV. CONCLUSION

In every field of science the relevant and correct choice of the basic principles or postulates has decisive action on its future progress and development. In photosynthesis, there are still many principal questions concerning the light reactions of photosynthesis that remain unanswered. If the "enhancement effect" is a consequence of the nonlinearity of the irradiance curves under low irradiances, then the idea about the two consecutive photoacts in bringing the electron from the primary electron donor to NADP loses its crucial evidence. However, if the electrons are transferred in only one photoact then a problem from the energetic point of view arises. According to Bolton [50,51], if the photosynthetic process is affected by one photosystem only (using only four photons), then the fraction of photon energy (ε) at λ_{max} (the maximum wavelength at which photosynthesis could be affected) should reach 0.73. This value is approximately equal or even higher than the theoretically calculated thermodynamic limit. As a consequence, it is postulated that the quantum requirement of photosynthesis cannot be less than 8 to 12 quanta per oxygen molecule evolved. However, as pointed out by Brown and Frenkel [52], the experimental determination of the minimum quantum requirement of *Chlorella* photosynthesis has become one of the most strenuously contested problems in all of biology and thus before the acceptance of the idea about the two photosystems there was no real agreement on the value of the quantum efficiency. According to Bell [53], an analysis of the available literature data allowed the drawing of histograms in which from nine studies, four reported quantum requirement less then eight or even seven quanta. I believe that it is possible that this contradiction can be overcome if one accepts the idea of Warburg [54], Metzner [55], and Stemler [28,56] that HCO_3^- is almost certainly the immediate source of photosynthetically evolved oxygen. In this case, the energy of one quantum with wavelength of even 700 to 730 nm will be sufficient. Obviously, if the bicarbonate ions and CO_2 participate only as catalysts (reaction steps [a] to [i] in Section II.C), the experiments with labeled oxygen cannot be considered as evidence in support of the statement that PSII receives its lost electrons directly from water. The only conclusion that could be drawn from these experiments is that the photosynthetic oxygen comes from water, but this does not mean that water is the immediate electron source to the reaction centers of photosynthesis [55]. It seems that we have no decisive experiments to prove the nature of the electron donor of the reaction centers of photosynthesis. It is, therefore, necessary to undertake a thorough study of the arguments considered in favor of the participation of H_2O and against the participation of HCO_3^- ions as an immediate electron source in the process of photosynthesis. Considering this statement the estimated values of the quantum requirement, 5–6–9 quanta per oxygen [57–61], which are lower than the estimated theoretical minimum quantum requirements (maximum efficiency) of photosynthesis (10 quanta per oxygen), predicted by the Z-scheme [62,63] seem entirely correct. Keeping in mind that the entire photosynthetic process contains a significant number of very complicated biochemical steps, it is not possible to understand how every photon is used with almost 100% effectiveness without any losses even while believing that Nature is built absolutely perfectly. The other strange fact is that in many experiments (including Emerson's) on action spectra of photosynthesis it is shown that oxygen evolution could be observed even at wavelengths around 720 to 730 nm where only photosystem I should be active. Obviously, these results are in sharp disagreement with the concept of two photosystems and consequently with the assumption that the oxygen-evolving reaction centers receive their lost electrons immediately from water. Thus, if the energy for electron removal from bicarbonate ions is twice lower [55] than from water molecules and the electrons could be transferred with a single photosystem (with one photon energy) then Nature will use electrons from bicarbonate ions and will not create a second photosystem.

Interpreting the sense of Warburg's statement that "in a perfect nature photosynthesis is perfect too," we can state that Nature is built with maximum simplicity and at minimum expense.

There is no need to point out that the postulate of two photosystems originates from the initial results obtained during the investigation of mechanisms of photosynthetic light reactions and in particular from the results of oxygen evolution. All the other results concerning the structural aspects of the photosynthetic machinery, especially the polypeptide composition of thylakoid membranes and the "water-oxidizing" system, the existence of heavy and light fragments cannot be considered as evidence here. In our previous works [20,64], we hypothesized that a close relationship exists between the grana and stroma localized PSII ($PSII_\alpha$ and $PSII_\beta$ centers) and the participation of two different mechanisms for oxygen evolution. Obviously, during the process of the development of the photosynthetic apparatus the entire electron transport system cannot be constructed simultaneously. Consequently, in every given time we could find different sorts of particles similar to the observed heavy or light particles possessing different functional properties [65]. Moreover, the different kinds of photosystems ($PSII_\alpha$, $PSII_\beta$, PSI_g, and PSI_s centers) should not be on any account considered as artifacts and nonexisting. The main problem is what is the real function of these structures and whether the electron transfer from water (the electrons after all are coming from water) to NADP is accomplished with the participation of two consecutive photoacts or with a single one.

In conclusion, it should be stressed that the rejection of the "generally accepted" hypotheses with more than 40 years of history is a very complicated, difficult, and painful process and needs the cooperation and efforts of many investigators in this field. The aim of this work is only to show that there are serious difficulties concerning the explanation of existing experimental data supporting the concepts of the PSU and the generally accepted "Z" scheme of photosynthesis based on the assumption of two photosystems operating in series [2] but also to emphasize the alternative pathways and mechanisms explaining the basic principles of photosynthetic processes. I hope that the young scientists in the 21th century will reconsider more carefully the basic arguments of these two hypotheses and speed up the understanding of photosynthesis, the unique and important process for life on Earth.

ACKNOWLEDGMENTS

This paper is dedicated to Otto Warburg, Birgit Vennesland, and Helmut Metzner.

This work was supported in part by the National Council for Scientific Investigations (Contract K-808).

REFERENCES

1. Emerson R, Arnold W. A separation of the reactions in photosynthesis by means of intermittent light. *J. Gen. Physiol.* 1932; 15: 391–420.
2. Hill R, Bendall F. Function of the two cytochrome components in chloroplasts: a working hypothesis. *Nature* 1960; 186: 136–137.
3. Rabinowitch E. *Photosynthesis and Related Processes*, Vol. 2. New York: Interscience Publishers, 1951.
4. Diner B, Mauzerall D. Feedback controlling oxygen production in a cross-reaction between two photosystems in photosynthesis. *Biochim. Biophys. Acta* 1973; 305: 329–352.
5. Kok, B. A critical consideration of the quantum yield of *Chlorella* photosynthesis. *Enzymologia* 1948; 13: 1–56.
6. Alen FL, Frank J. Photosynthetic evolution of oxygen by flashes of light. *Arch. Biochem. Biophys.* 1955; 58: 124–143.
7. Whittingham CP, Brown AH. Oxygen evolution from algae illuminated by short and long flashes of light. *J. Exp. Bot.* 1958; 9: 311–319.
8. Joliot P. Cinetique d'induction de la photosynthese chez *Chlorella pyrenoidosa*. II. Cinetique d'emission d'oxygene et fluorescence pendant la phase d'illumination. *J. Chim. Phys.* 1961; 58: 584–595.
9. Joliot, P, Barbieri G, Chabaud R. Un nouveau modele des center photochimique du systeme II. *Photochem. Photobiol.* 1969; 10: 309–329.
10. Tumerman LA, Sorokin EM. Fotosyntheticheskaya edinitsa: "Fizicheskaya" ili "Statisticheskaya" model? (Photosynthetic unit: "Physical" or "Statistical" model?). Molekul. Biol. 1967; 1: 628–638 (in Rusian).
11. Schmid GH, Gaffron H. Fluctuating photosynthetic units in higher plants and fairly constant units in algae. *Photochem. Photobiol.* 1971; 14: 451–464.
12. Lavorel J, Joliot P. A connected model of the photosynthetic unit. *Biophys. J.* 1972; 12: 815–831.
13. Kok B, Forbush B, McGloin M. Co-operation of charges in photosynthetic O_2 evolution. I. A linear four step mechanism. *Photochem. Photobiol.* 1970; 11: 457–475.
14. Delrieu M-J. 3-(3,4-Dichlorphenyl)-1,1-dimethylurea effects on the oxidizing side of photosystem II. *Photobiochem. Photobiophys.* 1981; 3: 137–144.
15. Lavorel J, Lemasson C. Anomalies in the kinetics of photosynthetic oxygen emission in sequences of flashes revealed by matrix analysis. Effect of carbonyl cyanide *m*-chlorphenylhydrazone and variation in time parameters. *Biochim. Biophys. Acta* 1976; 430: 501–516.
16. Lavorel J, Maison-Peteri B. Studies of deactivation of the oxygen-evolving system in higher plant photosynthesis. *Physiol. Veg.* 1983; 21(3): 509–517.
17. Lavorel J. Matrix analysis of the oxygen evolving system of photosynthesis. *J. Theor. Biol.* 1976; 57: 171–185.
18. Lavorel J. On the origin of damping of the oxygen yield in sequences of flashes. In: Metzner H, ed. *Photosynthetic Oxygen Evolution*. New York: Academic Press, 1980: 249–268.
19. Lehoczki E, Zeinalov Yu. Unusual photosynthetic oxygen evolution. I. Cerulenin-induced 3-(3,4-dichlorophenyl)-1,1,-dimethylurea insensitive oxygen evolution in *Chlorella pyrenoidosa. Photobiochem. Photobiophys.* 1984; 7: 135–142.
20. Maslenkova LT, Zanev Yu, Popova LP. Effect of abscisic acid on the photosynthetic oxygen evolution in barley chloroplasts. *Photosynth. Res.* 1989; 21: 45–50.
21. Zeinalov Yu. What does "photosynthetic unit" mean? *Photobiochem. Photobiophys.* 1986; 11: 151–157.
22. Zeinalov Yu, Litvin FF. Oxygen evolution after switching off the light and Si-state deactivation in photosynthesizing systems. *Photosynthetica* 1979; 13(2): 119–123.
23. Zeinalov Yu. On the amount of oxygen taken up during the induction period of photosynthesis in green algae. *Compt. Rend. Acad. Bulg. Sci.* 1979; 32(5): 679–682.
24. Emerson R, Lewis CM. The quantum efficiency of photosynthesis. *Carnegie Inst. Yearbook* 1941; 40: 157–160.
25. McAlister ED. The chlorophyll–carbon dioxide during photosynthesis. *J. Gen. Physiol.* 1939; 22: 613–636.
26. Franck J, Herzfeld KF. Contribution to a theory of photosynthesis. *J. Phys. Chem.* 1941; 45(16): 978–1025.
27. Stemler A. The binding of bicarbonate ions to washed chloroplast grana. *Biochim. Biophys. Acta* 1977; 560: 511–522.
28. Stemler A. Inhibition of photosystem II by formate. Possible evidence for a direct role of bicarbonate in photosynthetic oxygen evolution. *Biochim. Biophys. Acta* 1980; 593: 103–112.
29. Zeinalov Yu. Existence of two different ways for oxygen evolution in photosynthesis and photosynthetic unit concept. *Photosynthetica* 1982; 16: 27–35.
30. Zeinalov Y, Maslenkova L. Mechanisms of photosynthetic oxygen evolution. In: Pessarakli M, ed. *Handbook of Photosynthesis*. New York: Marcel Dekker, 1996: 129–150.
31. Zeinalov Yu. Non-additiveness in the action of light at the photosynthesis of green plants. *Compt. Rend. Acad. Bulg. Sci.* 1977; 30(10): 1479–1482.
32. Zeinalov Yu. The principle of non-additiveness in the action of light and the concept of two photosystems at the photosynthesis in green plants. *Compt. Rend. Acad. Bulg. Sci.* 1977b; 30(11): 1641–1644.
33. Gaffron H, Wohl K. Zur Theorie der Assimilation. *Naturwissenschaften* 1936; 24: 81–103.

34. Vennesland B. The energy conversion reactions of photosynthesis. In: Krogmann DW, Powers WH, eds. *Biochemical Dimensions of Photosynthesis*, Detroit, Wayne State University Publishers, 1965: 48–61.
35. Boitchenko VA, Efimtcev EI. Ingibirovanie aktivnosti fotosistemi II u Chlorelli pri visokih koncentracii kisloroda. (Inhibition of the PSII activity in *Chlorella* under high O_2 concentrations). *Fiziologia Rastenii* 1979; 26(4): 815–823 (in Russian).
36. Knaff DB, Arnon DI. Light-induced oxidation of a chloroplast b-type cytochrome at $-196°C$. *Proc. Natl. Acad. Sci. USA* 1969; 63: 956–962.
37. Park RB, Sane PV. Distribution of function and structure in chloroplast lamelae. *Ann. Rev. Plant Physiol.* 1971; 22: 395–430.
38. Huzisige H, Takimoto N. Analysis of photosystem II using particle II preparation. Role of cytochrome b-559 with different redox potentials and plastocyanin in the photosynthetic electron transport system. *Plant Cell Physiol.* 1974; 15: 1099–1113.
39. Arnon DI, Tsujimoto HY, Tang GM-S. The oxygenic and anoxygenic photosystems of plant photosynthesis: an up-dated concept of light induced electron and proton transport and photophosphorylation. *Proceedings of the V International Photosynthesis Congress*, Vol. II, Halkidiki, Greece, pp. 7–18, 1980.
40. Emerson R. Dependence of yield of photosynthesis in long-wave red on wavelength and intensity of supplementary light. *Science* 1957; 125: 746.
41. Blinks LR. Chromatic transients in photosynthesis of red algae. In: Gaffron H, Brown AH, French CS, Livingston R, Rabinowitch EI, Strehler BL, Tolbert NE, eds. *Research in Photosynthesis*, Papers and Discussions presented at the Gatlinburg Conference, New York, Interscience Publishers, October 25–29, 1955, pp. 444–449 (1957).
42. Myers J, French CS. Evidence from action spectra for a specific participation of chlorophyll *b* in photosynthesis. *J. Gen. Physiol.* 1960; 43: 723–736.
43. Myers J, French CS. Relationships between time course, chromatic transient, and enhancement phenomena of photosynthesis. *Plant Physiol.* 1960; 35: 963–969.
44. Bishop NI. Partial reactions of photosynthesis and photoreduction. *Ann. Rev. Plant Physiol.* 1966; 17: 185–208.
45. Mann JE, Myers J. Photosynthetic enhancement in the diatom *Phaeodactylum tricornutum*. *Plant Physiol.* 1968; 43: 1991–1995.
46. Heath OVS. *The Physiological Aspects of Photosynthesis*. Stanford, CA: Stanford University Press, 1969.
47. Warner JW, Berry RS. Alternative perspective on photosynthetic yield and enhancement. *Proc. Natl Acad. Sci. USA* 1987; 84: 4103–4107.
48. Milin AB, Sivash A. Effect Emersona: Novij podhod. (The effect of Emerson: A new approach). *Fiziologia i biochimiya kulturnih rastenii* 1990; 1: 27–31 (in Russian).
49. Hall DO, Evans MC. Photosynthetic phosphorylation in chloroplasts. *Sub-Cell. Biochem.* 1972; 1: 197–206.
50. Bolton JR. Photochemical conversion and storage of solar energy. *J. Solid State Chem.* 1977; 22: 3–8.
51. Bolton JR. Solar energy conversion efficiency in photosynthesis — Or why two photosystems. In: Hall DO et al., eds. *Proceedings of the IV International Congress on Photosynthesis*. London: The Biochemical Society, 1978: 621–634.
52. Brown AH, Frenkel AW. Photosynthesis. *Ann. Rev. Plant Physiol.* 1953; 4: 23–58.
53. Bell LN. Rastenie kak akumuljator i preobrazovatel solnetcnoj energii (The plants as accumulator and transformer of solar energy). *Vestnik AN SSSR* 1973; 2: 33–41 (in Russian).
54. Warburg O. Prefatory chapter. *Ann. Rev. Biochem.* 1964; 33: 1–14.
55. Metzner H. Oxygen evolution as energetic problem. In: Metzner H, ed. *Photosynthetic Oxygen Evolution*. New York: Academic Press, 1978: 59–76.
56. Stemler A. Forms of dissolved carbon dioxide required for photosystem II activity in chloroplast membranes. *Plant Physiol.* 1980; 65: 1160–1165.
57. Osborne BA, Geider RJ. The minimum photon requirement for photosynthesis. An analysis of data of Warburg and Burk (1950) and Yuan, Evans and Daniels (1955). *New Phytol.* 1987; 106: 631–644.
58. Osborne BA, Raven JA. Light absorption by plants and its implications for photosynthesis. *Biol. Rev.* 1986; 61: 1–61.
59. Osborne BA. Photon requirement for O_2-evolution in red ($\lambda = 680\,nm$) light for some C_3 and C_4 plant and a C_3–C_4 intermediate species. *Plant, Cell Environ.* 1994; 17: 143–152.
60. Pirt SJ, Lee Y-K, Walach MR, Pirt WM, Balyuzi HHM, Bazin MJ. A tubular bioreactor for photosynthetic production of biomass from carbon dioxide: design and performance. *J. Chem. Technol. Biotechnol.* 1983; 33B: 35–58.
61. Pirt SJ. The thermodynamic efficiency (quantum demand) and dynamics of photosynthetic growth. *New Phytol.* 1986; 102: 3–37.
62. Myers, J. On the algae: thoughts about physiology and measurements of efficiency. In: Falkowski PG, ed. *Primary Productivity of Sea*. New York: Plenum Press, 1980: 1–16.
63. Bell LN. *Energetics of Photosynthesizing Plant Cell*. London: Harwood Academic Publishers, 1985.
64. Maslenkova L, Zanev Yu, Popova L. Adaptation to salinity as monitored by PSII oxygen evolving reactions in Barley thylakoids. *J. Plant Physiol.* 1993; 142: 629–634.
65. Ghirardi ML, Melis A. Chlorophyll b deficiency in soybean mutants. Effects on photosystem stoichiometry and chlorophyll antenna size. *Biochim. Biophys. Acta* 1988; 932: 130–137.

2 Thermoluminescence as a Tool in the Study of Photosynthesis

Anil S. Bhagwat and Swapan K. Bhattacharjee
Molecular Biology Division, Bhabha Atomic Research Centre

CONTENTS

I. INTRODUCTION

Thermoluminescence (TL) is defined as a burst of light emission as a function of temperature during the warming of a sample irradiated by light during or before freezing. The energy for emission is supplied by the recombination of positively and negatively charged pairs produced by charge separation in photochemically active centers. The emission originates from heat-activated recombination of electrons and positive holes generated by irradiation that are stabilized in frozen state at low temperature [1]. Arnold and Sherwood were the first to observe TL in green plant materials. This was based on the dis-covery of delayed luminescence by Strehler and Arnold [2]. Their pioneering experiments gave various indications that the emission results from reversal of early reactions in the process of photosynthesis [3]. Besides photosynthetic materials, several minerals show TL in various artificially produced solid states such as semiconductors, organic solids, and complex biological materials. TL can also be used for the detection of irradiated food materials.

In photosynthesis, primary photochemical events of charge separation or the formation of negative and positive charged species occurs by light absorption of the reaction center chlorophyll in the thylakoid membrane. The charges generated in the reaction center and the primary electron acceptor subsequently migrate through the electron transport system. The reducing power of the electrons is stored as NADPH and is used for carbon fixation, whereas the oxidizing power derived from positive holes supplies energy to hydrolyze water molecules by evolving oxygen.

At room temperature, some of the positive and negative charges are metastable and recombine spontaneously with lifetimes several orders of magnitude higher than fluorescence to emit light, which is generally referred to as delayed light emission [4–6]. When chloroplasts were cooled rapidly after or during irradiation or irradiated at certain low temperatures, some of the metastable changes are stabilized. On warming such frozen chloroplasts, the stabilized positive and negative charges can recombine as they are thermally activated over the barrier of activation energy. Thus, light is emitted from the chloroplast molecule that is excited by energy released from charge recombination.

The purpose of this chapter is to provide the readers an overview of the mechanism of TL and its application in the study of primary reactions of photosynthesis. It is a simple and convenient tool to study and delineate early steps of electron transport including the water oxidation complex, as well as primary and secondary electron acceptors. When used in combination with other biophysical techniques like fluorescence and electron spin resonance, the amount of information that can be generated from TL glow curves is really immense. We have described some typical examples of the use of TL in studying the mechanism of the water–oxidase complex and the influence of various biotic and abiotic stresses on the donor and the acceptor sides of photosystem II (PSII), since TL mostly emanates from PSII. The role of various ionic requirements in the water oxidation complex and extrinsic protein were also determined using TL. We shall only briefly touch upon the theory of TL and not go into the details of this process as a number of excellent reviews are available on the subject [7,8]. This chapter mainly focuses on the applications of this powerful technique in study of photosynthesis.

Recently, a new phenomenon termed as "dark TL" was reported by one of us (S.K.B.). The mechanism of light emission in this process seems very different as it does not require any prior illumination [9]. This phenomenon has been described in brief. The theory and other aspects of this phenomenon are still being worked out by one of us (S.K.B.). We may add here that the development and fabrication of a new, highly sensitive, and versatile TL equipment has enabled us to detect this new phenomenon.

II. INSTRUMENTATION AND THEORY

A. THEORY OF THERMOLUMINESCENCE

Since electron traps and luminescence centers are associated with basic and functional membrane structures of chloroplasts, glow curve parameters are useful in determining the electron trap characteristics, such as activation energy, and the mean lifetime of electrons in the trap states. Knowledge of these factors is likely to give insight into the characteristics of the energy storage states as well as the probability of leakage or loss of electrons in nonphotosynthetic events. It is assumed that TL is a reversal of light-induced electron transport similar to the proposals made in several studies to explain the delayed light [6,10–12]. Figure 2.1 shows the basic photochemical reactions of PSII. Electron flow in the reverse direction, through two reaction centers P_{680} and P_{700}, results in the generation of TL. It is assumed that light is emitted from the first excited singlet state or triplet state of the antenna chlorophyll to which the excitation state was transferred after getting generated in the reaction center chlorophyll.

Electron carriers are reaction center components that could trap electrons. Reversal of electron flow, also referred to as the back reaction, causes excitation at the reaction center, which can migrate to the antenna chlorophyll and produce fluorescence. Activation energies (E) for some glow peaks by the application of the Randall–Wilkins theory were subsequently analyzed for all the glow peaks in a comprehensive report in which E values, the preexponential frequency factors, and the lifetime of the electron trap states were calculated by several methods [13–16]. They had used the Arrhenius equation to

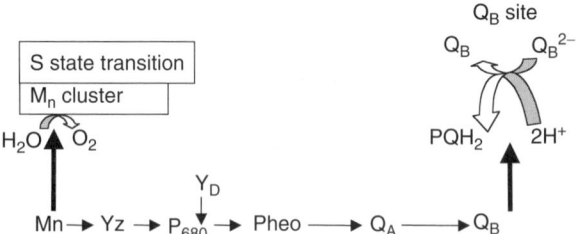

FIGURE 2.1 Outline of the reactions in PSII reaction center leading to oxidation of a water molecule by water–oxidase complex. The S-state transition means S_0 to S_4 redox states of the tetranuclear-Mn. The negative charges are generated by the quinine reduction cycle. The positively charged Yz^+, YD^+, and P_{680}^+ molecules are generated during the electron transfer process reducing the primary electron acceptor quinine Q_A to Q_A^-. Subsequently, Q_A^- donates electrons to Q_B forming semiquinone or quinol molecules.

determine other parameters. These analyses gave rise to some questions regarding the applicability of the Randall–Wilkins theory to explain the phenomenon of thermoluminescence in photodynamic structures like chloroplast membranes.

The intensity *l* of TL is given by the Arrehenius equation

$$I = \phi ns \exp(-E/kT)$$

where ϕ is a constant of proportionality, n is the number of trapped electrons, s is a preexponential frequency factor with dimension S-1, E is the activation energy, k is Boltzmann's constant, and T is the absolute temperature.

B. Setup and Measurement of TL

The main steps of TL measurements are the excitation of the sample and the cooling of the sample at low temperature (liquid nitrogen), which is then followed by heating in the dark and simultaneously recording the luminescence emitted during heating. One simple and very useful cryostat for measuring TL was fabricated in our laboratory in the early 1970s although some other devices are also available [17,18]. For the measurement of steady-state TL, a sample such as a section of leaf, chloroplast, or algal material is placed on the sample holder and is illuminated by white light or light of a particular wavelength through a monochromator. The sample holder is generally made up of copper and is connected to a cold finger that is immersed in liquid nitrogen. A heater coil placed under the sample holder makes it possible to slowly and linearly change the temperature of the sample. A programmable temperature controller ensures the linearity of heating. The thermocouple welded to the sample holder monitors the temperature of the sample and is connected to an X–Y recorder. One of the major problems in TL measurements is that the intensity of the emitted light is very weak and has to be amplified several fold and is measured with a red sensitive photomultiplier tube connected through a differential amplifier to the input of Y-axis of the recorder. The recorded emission intensity against temperature is called the glow curve or a TL band. The shape of the glow curve is strongly influenced by various factors, mainly excitation temperature, time of excitation, heating and cooling rate, and intensity and wavelength of excitation light. The rate of cooling of the sample during freezing and the rate of heating while recording the glow curves are the two most important variables that affect the reproducibility of TL data and mean peak temperatures. This may be largely responsible for the variability reported in the literature by various laboratories.

TL from photosynthetic materials can be easily recorded by a home-made setup like the one described by Tatake et al. [17]. As mentioned earlier, the most essential procedure undertaken in a TL setup is cooling the sample and the photomultiplier tube so as to increase the signal-to-noise ratio followed by controlled slow heating to measure the TL. A small sample holder having minimal heat capacity is recommended. A dark-relaxed leaf disk or a filter paper disk having chloroplast suspension is illuminated mostly with white room light and is quickly cooled to liquid nitrogen temperature. The sample is then placed on the cryostat that was previously cooled to liquid nitrogen temperature. The sample is then heated at a constant rate of 0.5 to 1°C/sec and the TL emission is measured with a red-sensitive photomultiplier tube while recording both the temperature and light emission. The glow curves (TL intensity versus sample temperature) are then plotted. Typical glow curves obtained from spinach chloroplast are shown in Figure 2.2. Single flash illumination or continuous light illumination at low temperature gives only one TL component but continuous illumination during sample cooling gives multiple components.

C. Nomenclature

Basically, two different systems are used in the nomenclature of TL glow curves: alphabetical and numerical. Table 2.1 lists the tentative assignment of glow peaks in these two nomenclatures. The glow curves are characterized by the temperature maximum of the emission band and are assigned to the different charge recombination (Table 2.1). The well-characterized glow curves are peaks II (A), III (B1), IV (B2) and peaks V (C), Z (Z), and I (Zv) according to the two nomenclatures. Thus, about five to six well-resolved peaks or bands are observed in the photosynthetic material. However, peak positions and temperatures of glow curves observed by various authors show a slight variation due to the factors outlined earlier. The B band (peak IV) is the most well-characterized band among all the TL bands. In addition, peaks II (A) and V (C band) have been studied to some extent. It may be noted that peak V (C band) is not related to the photosynthetic electron chain. TL emission from photosystem I has been reported by some workers; however, these peaks have not yet been classified under any nomenclature due to the lack of consensus about their origin [19].

D. Characteristics of Glow Curves

The glow curves of TL obtained from green plants exhibit several peaks (Figure 2.2). In general, the glow

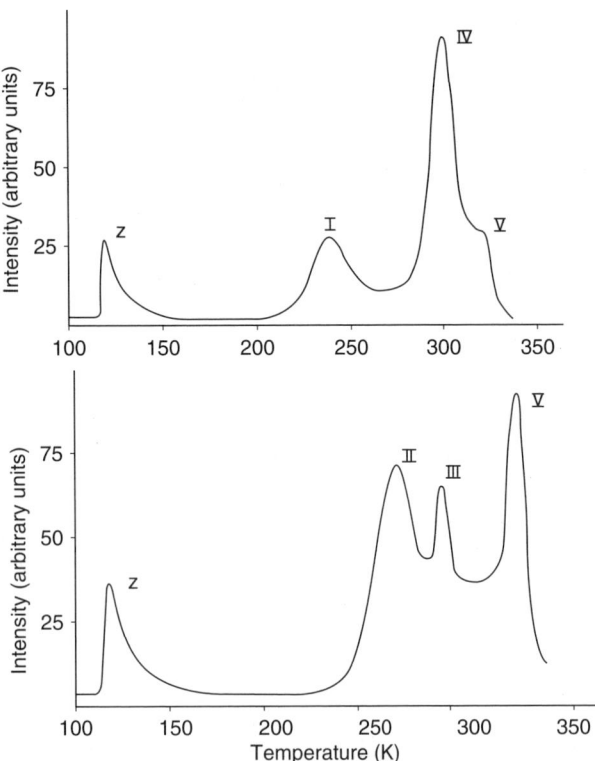

TABLE 2.1
Nomenclature of TL Glow Peaks in Plants

Peak	Approximate Temperature (°C)	Origin	Mean Lifetime (τ, sec)
Z	−160	Chl^+Chl^-	0.2
Z1	−70	$P_{680}^+Q_A^-$	1.3
II (A)	−10	$S_3Q_A^-$	—
III (B1)	+20	$S_3Q_B^-$	—
IV (B2)	+30	$S_2Q_B^-$	29
V (C)	+50	$YD^+Q_A^-$	1062
AG	+45	S_2/S_3Q_B	1.3

Notes: Emission maxima of peak II is at 740 nm and the excitation maxima in the blue region. Peaks Z1, II (A), IV (B2), and V (C) oscillate with flash number and the maxima differs between S3 for peak II, S2/S3 for peak IV (B), and S1 for peak V (C). Some peaks oscillate when diuron was added after excitation, for example, peaks II and V (C).

FIGURE 2.2 (A) A typical glow curve of spinach chloroplast frozen in the presence of intense white light at 77 K. (B) Glow curve of spinach chloroplast — same as above except for pretreatment of the chloroplasts with 10 μ*M* DCMU. (From Tatake VG, Desai TS, Govindjee, Sane PV. *Photochem. Photobiol.* 1981; 33:243–250. With permission.)

The relationship between the functioning of the photosynthetic apparatus and TL glow curves still awaits an explanation about its origin and correlation with various steps of the photosynthetic electron transfer. It is now well accepted that the band obtained at −160°C, the Z band, is not related to the photosynthetic electron transport as it was detected in chloroplasts inactivated by heat treatment at 100°C for 3 to 5 min [24]. It was concluded that this band was due to phosphorescence from the decay of the chlorophyll triplet molecule.

On heating the leaves up to 90°C, peaks II to V were not visible, which was considered as an evidence that these peaks originate from the recombination of charges stabilized on various electron acceptors and donors of the electron transport chain [4]. Illumination of isolated chloroplast with continuous light at −20°C gives a high peak (II) and two lower peaks (III and IV), whereas illumination below −40°C yields two high peaks (III and IV) and a low peak (II). In the presence of 2,5-dibromo-3-methyl-6-isopropyl-*p*-benzoquinone (DBMIB) and at low pH, the main band at +25°C (peak IV) was not visible [25]. Based on several observations, it was concluded that plasto-quinone was involved in the generation of peaks III and IV. Through elegant studies of Demeter and coworkers, the oscillation of the B band was demonstrated and it was concluded that the negative charge of the B band is located on Q_B, the secondary acceptor of PSII [26,27]. In general, peak V is strongly resistant to inhibitors such as DCMU, which block electron transfer from Q_A to Q_B. These observations have led to the conclusion that the negative charge responsible for peak V (C band) is located on Q_A [28,29].

curves are simply characterized by the number of peaks, the position, and relative intensity of the individual bands. The first reports of TL from photosynthetic materials were made by Arnold and Sherwood [1] and Tollen and Calvin [20]. In both the reports dried chloroplasts were used and it seems likely that the observed TL reflected severely damaged systems. These authors also noted that the TL glow curves were also detected in fresh leaves, algae, and many other photosynthetic organisms [21].

The works of Arnold and Azzi [3], Rubin and Vanediktov [22], and DeVault et al. [23] represent a significant step forward in this area of research. In these reports, the first well-resolved TL peaks from photosynthetic materials were presented. Rubin and Vanediktov [22] resolved four peaks between −50°C and +50°C in samples illuminated during cooling and only one band in samples illuminated at −50°C. It was subsequently reported by these researchers that 3-(3,4-dichlorophenyl)-1-1-dimethylurea (DCMU) (which blocks electron flow from Q_A^- to Q_B^-) caused the shift in the +25°C band to 10°C.

Most of the TL bands are closely related to the oxygen evolving system as shown by Inoue and Shibata [30]. However, some genetic studies seem to contradict this generalization. It is now well accepted that Mn^{2+}-containing enzymes participate in the process of oxygen evolution [31]. It is shown that the intensity of peaks I and IV (A and B bands) are extremely low in Mn^{2+}-deficient algae but the addition of Mn^{2+} ions followed by short repetitive flashes restore oxygen evolution and the appearance of glow curves, especially peaks I and IV.

It is well known that oxygen evolution in chloroplast illuminated with very short repetitive flashes shows a period four oscillation [32]. Oscillations were also seen in the case of TL bands, especially peaks III to V. The oscillation of peak IV (B band) of the TL band is the best-characterized band showing maxima at 2, 6, 10, etc., flashes with a periodicity of four [33,34]. Manganese oxidation states S_2/S_3 were found to be the most luminescent states. The different TL peaks attributed to $S_2Q_A^-$ and $S_2Q_B^-$ reflect different activation energies for the recombination reaction to take place in each of these states. This energy difference may, in part, reflect a different midpoint potential between the Q_A/Q_A^- and Q_B/Q_B^- redox couple.

In dark-adapted chloroplasts, the distribution of S_0 and S_1 are 25% and 75%, respectively [35]. Thus, the maxima obtained after the second flash indicate the participation of the S_3 state in the generation of peak IV (B band). Based on several studies it has been concluded that peak III (B_1 band) originates from $S_3Q_B^-$ and peak IV (B_2 band) originates from the recombination of $S_2Q_B^-$.

Peak V (C band) was first observed in DCMU-treated chloroplasts and in etiolated leaves [36]. Since in etiolated leaves the oxygen evolving system is inactive, it has been suggested that peak V is not related to the water splitting enzyme. However, several studies have shown that this peak also undergoes a period four oscillation and it has been proposed that this band may be originating from the charge recombination of the $S_0Q_A^-$ and $S_1Q_A^-$ redox couple [37,38]. Several other observations on peak V using inhibitors like tetranitromethane [39] and o-phthalaldehyde [40] indicate that any block in the transfer of electrons from Q_A to Q_B results in an intensified peak V, thus confirming that this peak originates from the recombination of $S_0Q_A^-$ and $S_1Q_A^-$. In addition, peak II (Q band) was also considerably enhanced under similar conditions [41]. The intensification of peak II depends on several factors such as temperature at which the sample was excited, intensity of excitation, and source and duration of excitation [42,43]. In addition, cooling rate is also an important factor. When the leaf disks are cooled slowly (time taken to cool to 77 K in our setup is about 50 sec), this band reaches its maximal intensity and decays in seconds.

E. RELATIONSHIP BETWEEN TL AND PHOTOSYNTHESIS

The involvement of PSII in TL from green plants was first proposed by Arnold and Azzi [3]. They found that peaks II to V were absent in the *Scenedesmus* mutant deficient in PSII but were present in mutants lacking PSI. Hence, most of the glow peaks seems to originate from PSII; although a few early reports on the origin of one or two glow peaks from PSI are available in the literature, it is now well accepted that these glow peaks could have been due to some artifacts of measurements or incorrect interpretation of data [44]. All the subsequent studies have unequivocally confirmed that all peaks resulting from the charge recombination in the region of $-40°C$ to $+50°C$ have their origin in PSII activity.

This finding has been further corroborated by subsequent experimentation by several workers using bundle sheath chloroplasts of C-4 plants that apparently lack PSII and, therefore, show very weak TL. The inhibition of PSI activity by $HgCl_2$ does not affect the glow peak yield of isolated chloroplast [44,45]. Several other studies using herbicides and inhibitors that interact with PSI electron flow supports this conclusion.

III. A NEW PHENOMENON: QUANTUM CONFINEMENT AS A SOURCE OF TL

Recently, a new phenomenon of TL has been reported both in photosynthetic and nonphotosynthetic biological materials without excitation by any irradiation or external stimulus. It is called dark-TL or "quantum confinement TL" and presumably does not require any charge recombination and therefore rules out the application of the Randall–Wilkins theory to interpret TL [15]. This paper argues that the sources of glow seen by the TL technique may be largely from *in vivo* biological nanoparticles having the property of quantum confinement that entails trapping of energy and delayed emission typical of semiconductor nanoparticles and not solely due to charge recombination. This phenomenon could be observed not only in photosynthetic materials but also in several nonphotosynthetic organisms like bacterial cells and several other biological samples. Arnold and Sherwood [1] also observed this phenomenon in air-dried chloroplast wherein preparations when exposed to light and then allowed to stand in dark for several hours gave some glow. This aspect was not presented in the paper as it was not considered important and hence neglected. However, nonreproducibility of the TL curves from different laboratories still raises some questions. The new phenomenon raises doubt about the interpretation of the results and addresses some of the questions about the origin of TL [9]. A new microprocessor-based instrument was developed to eliminate the uncontrolled variations in the process of light exposure during cooling and relaxation time before light exposure, which seem to influence the details of the glow curve [46]. While testing the instrument, during its development with spinach leaf and culture of cyanobacteria, the appropriate negative controls were difficult to design. This was because a second cycle of cooling and heating of a sample of photosynthetic material glowed at varying temperatures, though at the end of the first cycle the sample reached nearly +110°C. This glow suggested that TL from the reused sample was presumably not due to charge recombination, since all the charges should have been eliminated in the first heating phase after which the electron transfer system should have been destroyed. And since no light was applied before the subsequent heating phase, fresh charge separation could not have occurred and kept stabilized in the cooling phase of the second cycle. Moreover, it was possible to generate glow peaks and bands during repeated cooling and heating cycles at approximately the same temperature range as in the first cycle after light exposure and all subsequent cycles were recorded without any light exposure (for details see legend to Figure 2.3). This clearly raises a difficult question: Are the glow peaks of the earlier reports entirely a consequence of light excitation or are they mixed up with signals also resulting from heat entrapment independent of the energy of the captured light that was delivered with a view to trap charge pairs at low temperature?

IV. APPLICATIONS OF TL IN PSII PHOTOCHEMISTRY

TL as described earlier is a very useful tool for the study of early reactions of photosynthetic electron transfer both at the acceptor and the donor sides of the chain. There are a large number of researches that have used this technique to study the electron transport chain from the water–oxidase complex to PSI (secondary quinone acceptor). The effects of various abiotic and biotic stress factors that influence PSII activity, such as UV, high light, high temperature, drought, viral infection, hormonal effect, have also been studied. The role of various cofactors and ionic requirements were also confirmed by using TL. The role of amino acid residues essential for binding of herbicide was also explained by site-directed mutagenesis studies of *Synechocystis*.

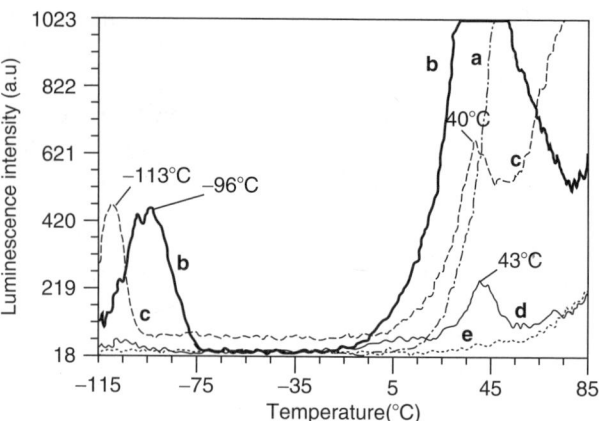

FIGURE 2.3 Superimposed dark-TL signals in the heating phase of first to fifth excursions of the same sample as function of sample temperature. The sample was a circular plug of 15 mm diameter incised from a fresh spinach leaf kept at room temperature exposed to full room light. No idling was done before the first excursion. **a**: first excursion; **b**: second excursion after 50 μl of water was added at the end of the first run; **c**: third excursion. **d**: fourth excursion; **e**: fifth excursion. After adding 50 μl of water at the end of first run, the sample was not disturbed and remained in dark all through till the end of the experiment of five excursions. The peak position is at about 43°C that is within the range of the dark-TL signal from the first run when no water was added. The low temperature bands peak at about −96°C and −113°C in second and third excursions respectively, but the high temperature band is close to 42 °C in all the reruns except the last one when no significant signal was seen. (Adapted after corrections from Bhattacharjee SK, In: *Proceedings of BIOTALK-1*, February 6–7, 2003, pp. 37–43, Hislop School of Biotechnology, Nagpur, India).

Heterogeneity of PSII was also confirmed by TL in addition to other biophysical techniques like fluorescence, electron spin resonance, and pulse amplitude modulated fluorescence (PAM). The effect of glycine–betaine and other solutes on Mn^{2+} depletion of the water oxidation complex was also studied by TL. In the next section, we describe some of the typical applications of TL in studying early reactions of photosynthesis and the effects of various factors affecting photosynthetic electron transfer reactions. The potential of this technique in conjunction with oxygen evolution and fluorescence (both steady state and variable) could provide a wealth of information on the functioning of photosynthetic electron transport.

A. EFFECT OF ELEVATED LIGHT ON PSII

TL has been extensively used to investigate the high light induced fluorescence quenching phenomenon in plants. It is generally accepted that the target of photoinhibition is the D_1 protein or the Q_B binding protein whose turnover is light-dependent. Changes in the properties of the reaction center during photoinhibition in *Chlamydomonas* have been described using TL [47,48]. Photoinhibition shifts peak IV (B band) emission by causing the destabilization of the $S_2Q_B^-$ state and recombination taking place at lower temperature (15°C to 17°C). This correlates with the increase in the value of intrinsic fluorescence F_0 and the decrease in the $S_2Q_A^-$ signal. While at extensive photo inhibitory levels of light the B-type signal was completely lost, $S_2Q_A^-$ band emission remained at about 20%. These events seem to be connected to light-dependent turnover of D_1 protein. The mechanism of photoinhibition was studied using TL as a probe. Light-induced changes were seen in isolated thylakoids such as destabilization of Q_B bound to D_1 protein, which was demonstrated by the reduction in S_2/S_3 charge recombination by TL data [49]. The irreversible light-dependent modification of D_1 protein may serve as the signal for its degradation and may be replaced by newly synthesized molecules. In another interesting study on site-specific mutants of D_1 polypeptide in *Synechocystis* PCC 6803, having deletions on three glutamate residues (242 to 244 from the N terminal), it was shown that the mutations modified the stability of D_1 protein, the manganese transition states, and the charge recombinant S_2Q_A/S_2Q_B states of PSII as demonstrated by TL measurements [50].

Protection of plants against photooxidative damage by violaxanthine has been shown by using TL. The results show that the violaxanthine cycle specifically protects thylakoid membranes against photooxidation by a mechanism involving the partial quenching of a single excited chlorophyll [51]. Lipophilic antioxidants like vitamin E could be involved in high phototolerance [52].

Chlorophyll fluorescence technique is of limited use in distinguishing between different mechanistic models of photodamage; hence, it is necessary to use alternative complementary techniques like TL to unravel processes involved in regulation and damage of PSII by extraneous factors. To investigate the mechanism that potentially protects PSII against high light damage by dissipating part of excess energy as heat, TL has been used as a barometer of chlorophyll fluorescence quenching [48]. The nature and relative intensity of the TL signal provide information about state of PSII.

B. ELUCIDATING THE EFFECT AND ACTION OF ADRY AGENTS

Carbonylcyanide m-chlorophenyl hydrazone (CCCP) is an agents accelerating the deactivation of water splitting enzyme (ADRY) agent whose presence accelerates the deactivation of the water splitting enzyme system. Thus, the higher oxidized "S" states that are created in light quickly revert to the lower "S" state in the presence of CCCP. The data show that the appearance of peaks I and V does not require the formation of the "S" state; however, the formation of the "S" state is absolutely essential for the appearance of peaks II to IV. The molecular mechanism of ADRY agents was also elucidated by excellent TL studies. The nature of oxidizing and reducing (redox) equivalents stored in PSII has been shown by TL studies. These studies showed that the most powerful ADRY agent, ANT-2p, is an inhibitor of R causing detrapping of electrons from B^- and holes from S_2/S_3. It also reduces TL yield due to recombination of the reduced primary plastoquinone acceptor, X-320, and S_2 at room temperature as well as subzero temperatures. The data further confirm that ANT-2p acts as a mobile species that effectively enhances decay of S_2 and S_3. With respect to the mechanism of action of ADRY, it can be concluded that ANT-2p does not affect the quantum yield of exciton formation via recombination of S_2 and S_3 with either X-320 or B^-. ANT-2p specifically accelerates decay of S_2 and S_3 species [53,54].

C. TEMPERATURE STRESS

Temperature is one of the most important factors limiting crop yield. Both low- and high-temperature stress could affect electron transport and carbon fixation reactions of photosynthesis. The effect of chilling stress on TL bands appearing at positive temperatures of 40°C to 50°C was investigated [55]. Far-red light irradiation of leaves induced a positive temperature band (AG band) peaking at 40°C to 45°C together with the B band (20°C to 30°C). Severe stress affects both AG and B bands. The appearance of a low-temperature band indicates lipid peroxidation in membranes. Thus, TL is also useful in studying membrane fluidity and the effect of low temperature on membrane integrity. Chilling-tolerant plants did not show AG band changes, making it a useful indicator for the selection of chilling-tolerant plants. Alteration of PSII activity due to mild and severe heat stress was also investigated [56]. While leaves exposed to mild heat stress retained the ability to withstand transitions, severe heat stress affected the acceptor side of PSII and the donor side remained unaffected. The effect of temperature (low, room, and high temperatures) on photoinhibition was studied in *pothos* leaves using TL as a probe [55]. TL bands III and IV associated with $S_{2/3}Q_A^-$ were more sensitive to photoinhibition at chilling and high temperature, indicating a synergistic effect of these two different types of stresses. Peak V, however, was resistant to photoinhibition; such a behavior can be expected as this peak is not known to be involved in the main chain of the electron transport pathway.

PS II under temperature stress is more susceptible to photoinhibition and osmolytes such as glycine–betaine have been shown to stabilize the oxygen evolving function of the PSII core complex. The stabilization effect is due to the minimization of protein–water interaction as proposed by Akazawa and Timasheff [56]. Decreased PSII activity after thermal stress has been primarily linked to the destruction of the oxygen evolving complex by virtue of the release of Mn^{2+} from the PSII core complex together with the loss of three extrinsic polypeptides. It has been proposed that $His^+Q_A^-$ may be responsible for TL at 30°C and the TL at 55°C may originate from the recombination of Z^+ and the acceptor side of PSII. Osmolyte seems to stabilize the Mn^{2+} cluster and increase the binding of the three extrinsic polypeptides. A similar mechanism is proposed for reduced heat stress sensitivity of PSII in the presence of cosolute [57]. Decreases in the rate of photosynthesis constitute one of the primary symptoms of plant cell damage by high temperature and other abiotic stresses. The integrity of thylakoid membranes is perturbed resulting in damage to the PSII reaction center, which can be easily quantified by using TL. The electron transport is the most heat-sensitive reaction that can be completely studied by TL glow curves using various inhibitors and protective agents like glycine–betaine and other osmolytes that improve osmotic potential and improve heat tolerance of thylakoid membranes *in vivo* [58].

Low-temperature stress (5°C) to *Arabidopsis* plants is associated with changes in the acceptor side of PSII involving redox potentials of Q_A and Q_B, which was indicated by TL studies [59]. It is proposed from the TL data obtained that the population of Q_A^- facilitates back reaction with P_{680}^+ and thus enhancing dissipation of excess energy in PSII. The reasons for the increased resistance of cold-hardened plants to low-temperature photoinhibition were explained using this simple technique [59]. In another study using TL as a tool, Sane et al. [60] have suggested that lowering the redox potential of Q_B by exchanging D_1:1 for D_1:2 imparts the increased resistance to high excitation pressure and temperature stress by specific functional changes in electron transport [60].

Oxidative stress during drought or methyl viologen treatment in plants lacking CDSP32 showed higher lipid peroxidation as compared to the control. Measurements of chlorophyll TL showed the critical component in the defense system [61].

D. Effect of UV Radiation

The effect of UV-A radiation on isolated thylakoid was studied using TL as a probe. The results using flash experiments indicated that UV causes an in-creased amount of the S_0 state in dark, showing the direct effect of UV-A on the water oxidation complex. TL measurements also showed that UV-A induced loss of PSII centers and decreased the amount of Q_B^- relative to Q_B^+, indicating that the reduction of Q_B and oxidation of Q_A^- was affected. Hence, UV-A affects both the water oxidizing complex and the binding site of Q_B quinone [62].

E. Salt and Hormonal Stress

TL parameters of intact leaves of NaCl-stressed seedlings show significant changes in glow curve pattern. Salt stress causes destabilization of Q_A and Q_B, leading to a decrease in the Q and B bands. There were subtle differences in the intact leaf and the isolated thylakoid with respect to the intensity of the two glow curves at +10°C and +32°C. This was explained in terms of aging effect and chlorophyll concentration [63]. In a similar study, it was observed that in aging leaf of Mung bean the TL patterns in leaf and thylakoid were quite different. The aging of leaf brings about a decrease in the B band and an increase in the Q band, indicating a block in Q_A to Q_B transfer [63]. An endogenous electron transport inhibitor was postulated during aging based on the TL data [64]. The effect of jasmonic acid (JA) on the PSII reaction was assessed by TL measurements and oxygen evolution. JA is known to affect plant photosynthesis in general and photosynthetic electron transport in particular; however, the mechanism was elucidated using TL measurements after hormone treatment. JA-treated samples showed reduced efficiency in utilization of oxidizing equivalents and retardation of "S" state transition, especially S_2 and S_3 transition was significantly destabilized [65]. JA has an effect on the PSII donor side, which may be related to specific changes in the polypeptide pattern [66].

F. Indicator of Biotic and Abiotic Stresses in Plants

It has been proposed that the AG band of a TL profile obtained from various green tissues was sensitive to various abiotic stresses and can be a useful indicator of stress effects and response in plants. However, the behavior of the AG band depends on several factors such as leaf age, position, which must be controlled using various means for obtaining meaningful data [67]. In addition, a downshift in the band was also observed during stress such as freezing temperature. Changes in TL characteristics as well as oxygen evolving capacity were used to characterize plants infected with pepper and paprika mottle virus. Electron transfer activity was inhibited by virus infection as shown by the shift in the temperature at which the B band appears from 20°C to 35°C, corresponding to S_3 (S_2) Q_B^- to $S_2Q_B^-$ charge recombination, which showed that the inhibition exists in the formation of the higher "S" state in the water splitting system [68]. Simultaneously, a new band appeared at 70°C due to chemiluminescence of lipid peroxides [69].

Heavy metal exerts multiple inhibitory effects on photosynthesis at different structural and metabolic levels. A strong influence of Cd^{2+} on D_1 protein turnover has been observed. Monitoring the effect of Cu^{2+}, Zn^{2+}, and AS^{2+} on an algal system using advanced and sensitive biophysical techniques such as electron spin resonance, fluorescence, and thermoluminescence have been attempted. PSII can be used as biosensors based on its response to heavy metals in isolated thylakoids and PSII particles as determined by TL glow curve characteristics. This can help in monitoring environmental pollution in aquatic and terrestrial ecosystems [69].

Inhibition of PSII by heavy metals (HMS) is accompanied by several effects on photosynthetic membranes such as disappearance of grana stack and release of some extrinsic polypeptides of the reaction center. Membrane fluidity can be easily studied using TL and the mechanism of heavy metal stress can be delineated. Since there is a close synchronization between the effect of HMS and level of irradiance, these can be studied together by TL using both steady-state and flash-induced glow curves.

Though the photosynthetic reaction centers are known to have good efficiency in forward flow of electrons minimizing the loss of photochemical energy, it is important to know the factors that facilitate back flow of electrons and instability of photosynthetic systems related to charge recombination. Lack of stable charge separation has been one of the major bottlenecks in developing artificial systems that harvest solar energy by mimicking photosynthesis. Since TL occurs by the back reactions of separated charges during electron transfer, it would be a useful tool in understanding and improving the efficiency of artificial photosynthetic systems. It is usually observed that artificial systems are temperature-sensitive and also the transfer times are different. In one such study, it was observed that protein chemical agents can be used to alter the temperature range (making it more optimal) and time period of stable operation of biodevices [69].

G.   REGULATION OF PHOTOSYNTHESIS AND NITROGEN FIXATION

The effect of nitrogen limitation on PSII activity in cyanobacterium was studied using fluorescence and TL measurements. Nitrogen deprivation decreased Fv/Fm, the amplitude of the B band, and the rate of Q_A^- reoxidation. These indicated loss of PSII and the formation of nonfunctional PSI centers and continuous reduction of D_1 protein content [70]. The strong decrease in D_1 protein levels under N-deprivation in *Prochlorococcus marinus* is consistent with results from eukaryotic algae. D_1 protein is the most rapidly turned over component of the thylakoid membrane and its continuous recycling is critical for PSII function. In the case of *P. marinus*, N-limitation blocked de novo synthesis and inhibited PSII repair, leading to progressive inactivation of PSII. In contrast, in *Synecococcus* no significant changes were reported in D_1 content under comparable N-limitation.

Unlike heterocystous cyanobacteria, most of the filamentous nonheterocystous cyanobacteria have the ability to fix nitrogen and carry out photosynthesis by the same undifferentiated cells. The regulation of these two processes was studied using TL as a probe [71]. Since oxygen is inhibitory to nitrogenase and during the photosynthetic phase oxygen is evolved that could be inhibitory to nitrogen fixation, the two phases have to be separated either temporally or spatially as in the case of heterocystous cyanobacteria where a specialized cell type does the nitrogen fixation. On the basis of a detailed TL study on both the acceptor and donor sides of PSII, it was concluded that the redox level of Q_B, the secondary quinine acceptor, regulates the two phases. A shift in peak temperature from 25°C in the P-phase to 10°C in the N-phase is likely to be due to changes in the redox potential of the oxidizing and reducing equivalents involved in generating these band or glow peaks.

Atrazine-resistant species showed a similar shift in the B band from 25 to 15°C, indicating a block in Q_A to Q_B transfer. The decrease in redox potential was from 70 mV in susceptible species to 30 mV in atrazine-resistant species. Peak temperature and stability of the B band is shown to depend on quinine moieties involved in it. The modification of D_1 protein led to a shift in the B band temperature to the lower side. The degree of downshift was related to the stability of the Q_B protein complex.

The donor side was not affected during the nitrogen fixing phase but the downregulation of photosynthesis was brought about by the enhanced degradation of the Q_B protein, as evidenced by the appearance of a strong TL band at +10°C in the N-phase due to recombination of S_2/S_3 Q_A^- instead of S_2/S_3 Q_B^- in the P-phase [71].

H.   HERBICIDE EFFECTS

The resistance to inhibition of electron transport by triazine and other herbicides is due to an alteration in the herbicide binding site, which is clearly shown as the Q_B binding site on D_1 protein. It has been well documented that redox states of primary and secondary quinone acceptors of PSII can be investigated by TL. Using TL, it has been shown that the midpoint oxidation–reduction potential of a secondary quinone acceptor was lowered in herbicide-resistant plants as compared to the susceptible plant types. The midpoint potential can be calculated mathematically from the TL data [72].

Since the oscillation pattern characterizing the "S" states does not change upon addition of DCMU, atrazine, and 4,6 dintro-o-cresol (DNOC), the acceptor side of PSII should be responsible for the differences in peak positions of the bands appearing after herbicide treatment. The results of displacement experiments suggest that DCMU, atrazine, and DNOC have a common binding site in chloroplast membranes and TL bands appearing at $+6°C$, $0°C$, and $-13°C$ can be related to an electron transport component that is located between the site of action of these herbicides and P_{680}.

The difference in peak positions of these bands can be explained in two ways:

1. The structural modification of the protein-aceous component of Q and B, due to binding of DCMU, atrazine, and DNOC, changes the mutual orientation of separation of Q and P_{680} so that the probability of reverse flow of electrons from Q to P_{680} changes. Thus, a change in the position of the TL band is caused.
2. From the theory of TL it follows that the peak position of TL bands is determined by the redox span between the donor and the acceptor molecules, particularly the recombination.

Since the S_3 state is responsible for major glow curves and the addition of herbicide can shift the midpoint redox potential of Q to a different value, the redox state of Q is reflected in the shift of the peak position.

Herbicide-resistant mutants of *Synechocystis* were generated, which showed significant conformational changes in the Q_B binding region of PSII [73]. TL and fluorescence measurements were used to confirm lack of functional PSII activity. TL data showed that Q_A to Q_B transfer was significantly impaired. The mutants also showed increased resistance to trazine. The results further showed that structural changes in the Q_B binding region affected the herbicide and plastoquinone binding and also perturbed the normal regulatory factors that control degradation of D_1 protein.

I. ROLE OF SMALL COMPONENTS OF PSII IN ELECTRON TRANSPORT — A TL STUDY

The PSII complex of photosynthetic oxygen evolving membranes comprises a number of small proteins whose function is still unknown. The TL technique has been effectively used to delineate the function of these small proteins in photosynthetic electron transfer reactions. The role of Cytb559 in PSII was also proposed from TL data [74]. Cytb559 plays an important role in maintaining the plastoquinone pool and thereby the acceptor side of PSII is oxidized in dark. A single alteration in terms of a point mutation (Phe–Ser) inhibits this function. A low molecular weight protein coded by *psb*J gene is an intrinsic component of the PSII complex [75]. TL, fluorescence, and oxygen flash yield studies indicate that inactivation of the gene reduces PSII-mediated oxygen evolution, although PSII can be assembled in the absence of *psb*J. Both the forward electron flow from Q_A to P_Q and the back flow of electrons to Mn(ox) are deregulated in the absence of *psb*J and affects the efficiency of PSII and charge separation.

Analyses of steady-state and flash-induced oxygen evolution and TL profiles demonstrated that *psb*Y mutant cells have normal photosynthetic activities. Thus, psbY protein is not essential for oxygenic photosynthesis and is also not a ligand for Mn^{2+} coordination in the oxygen evolving complex [76]. Chlorophyll florescence, electron paramagnetic resonance spectroscopy, and TL technique have been used to demonstrate that only the dimeric form of CP_{47}–RC complex showed electron transfer activity and Q_A reduction [77].

The gene product of *Psb*U, a 12 kDa extrinsic protein of PSII, seems to be essential for optimizing Ca^{2+} and Cl^- requirements and for maintaining the functional structure of the oxygen evolving complex [78]. A shift in the B and Q bands of TL with a concomitant increase in Q band intensity indicate that the above TL and fluorescence measurements of WT and the mutant of *Synechocystis* sp. PCC 7942 showed that the subunit II of NADH dehydrogenase is essential for functional operation of PSII electron transport at low CO_2 concentrations. The inability to accumulate Ci under air is due to disruption of electron transport in this mutant [79].

The modification of the Q_B binding site by site-directed mutagenesis of essential amino acid residues of D_1 protein seems to influence the binding of Q_B and herbicides, which also induces changes in TL quantum yield and lifetime of S_2 and S_3 of the water oxidation complex [80]. TL data show that Ser_{264} is essential for atrazine and DCMU binding, whereas Phe_{255}, although involved in atrazine binding, does not affect DCMU binding [81]. Arylaminobenzoate derivatives were found to be efficient inhibitors of photosynthetic electron transport at the acceptor side of PSII. This conclusion was supported by TL and other techniques [81]. The molecular mechanism of arylaminobenzoate, which is Cl^- channel inhibitor, blocks PSII activity at low concentration. Its effect is like an herbicide since it also blocks the transfer of electron from Q_A to Q_B at the acceptor side of PSII [81].

J. HETEROGENEITY IN PHOTOSYSTEM II

The measurement of recombination kinetics of $S_2Q_B^-$ using TL revealed that PSII exists in at least two substates with distinct kinetic and thermodynamic behaviors. It is further suggested that heterogeneity probably exists because of two conformational substates of PSII proteins [81]. In principle, a TL band can provide information about the enthalpies of activation, the intrinsic rate constants, and entropic factors for charge recombination. However, previous attempts were only partially successful. The measurements presented by Townsend et al. [78] provided the method for deriving quantitative data from TL curves. It allows the resolution of the TL band into components representing different substates. TL signals were recorded from grana stacks, margins, and stroma lamellae from fractionated and dark-adapted thylakoid membranes of spinach to demonstrate heterogeneity of PSII and the mechanism of photoinhibition. Stroma lamellae mainly gave rise to a C band having emission at 42°C and 52°C in the absence and in the presence of DCMU. This resulted in inactive PSII centers [82].

K. REDOX STATES OF ELECTRON TRANSFER IN CRASSULACEAN ACID METABOLISM (CAM) AND C-3 PLANTS

TL signals were measured in leaves of facultative CAM plants *Mesembryanthemum crystallinum* L. following induction of CAM by salt treatment. The TL measurements were made during and after CAM induction. The results show that the 46°C TL band was an indicator of the metabolic state of leaf originating from PSII centers in the S_2/S_3 Q_B oxidation state. The intensity of the 46°C band shows diurnal rhythm and maximum intensity were observed in the morning and in the evening. TL can be a very useful tool in studying rhythmicity in plant systems. The redox state of the electron transport chain is different in CAM condition as compared to C-3 and changes induced by CAM can be monitored by measuring the amplitude of the TL band at 46°C by flash excitation [83].

L. IONIC REQUIREMENT OF WATER–OXIDASE SYSTEM

TL measurements clearly showed that the normal course of charge accumulation is impaired by the removal of Cl^- from the PSII reaction center. The sensitive step is the formation of the S4 state that is capable of producing oxygen. In addition, S_2 and S_3 states formed in Cl^--deficient enzyme have profound altered properties. Ca^{2+} is required for maintaining the conformation of all polypeptides, and TL patterns of Ca^{2+}-depleted thylakoids may show changes in TL glow curves as removal of 18 and 23 kDa polypeptides [84].

Superoxide formation during photosynthesis seems to contribute to rapid inactivation of the secondary donor of PSII. The donor side becomes selectively inactivated by photodamage, which may have been initiated by overreduction of Q_A, and results in superoxide formation. This was demonstrated by TL measurements of inactivation at the donor site and also over reduction of Q_A [82,85].

V. CONCLUDING REMARKS

The phenomenon of thermoluminescence in photosynthetic materials, discovered some 46 years ago, has immensely helped in furthering our knowledge on many redox reactions of PSII. The role of several small molecular weight proteins, which are intrinsically part of the PSII complex and whose functional identities were not known, could be assigned a function based on the data obtained using TL. The instrumentation is relatively simple and can be easily fabricated even in laboratories having minimal infrastructural support. The method can be applied to study almost all redox components of PSII in both intact leaves and isolated system. A shift in the peak position of the TL band indicates change in the redox distance between the positively charged donor and the negatively charged acceptor. The oscillation in the amount of oxidized donor or reduced acceptor molecule undergoing charge recombination can be followed by flash-dependent amplitude change in TL. On the basis of the oscillation pattern of TL, a block in the "S" state transition can be demonstrated along with the threshold temperature of the "S" state transition.

The disappearance of the TL band with a concomitant intensification of another one indicates the block in the electron transport chain and accumulation of charges on new components located before the site of the block. TL characteristics may help in identifying new site(s) of action of herbicides and other agents.

However, the method has some limitations as this cannot be applied to study PSI reactions and also bacterial reaction centers. The other drawback is the shift in the peak temperature for a particular peak. This largely depends on the instrumentation, illumination temperature, and several other parameters that are usually not indicated clearly. TL is a very useful technique in delineating the effects of various herbicides and other biotic and abiotic stresses on early reaction of photosynthesis both at the donor and the acceptor sides. The phenomenon of "dark-TL" reported here may also be an useful tool in understanding the mechanism of TL and also photosynthetic systems. The new approach may provide better comprehension of the energetics involving light energy, storage systems, and regulation of energy conversion. This may open up the possibility of designing more efficient light-harvesting systems using biomolecules.

REFERENCES

1. Arnold W, Sherwood HK. Are chloroplasts semiconductors? *Proc. Natl Acad. Sci. USA* 1957; 43:105–114.
2. Strehler BL, Arnold W. Light production by green plants. *J. Gen. Physiol.* 1951; 34:809–815.
3. Arnold W, Azzi JR. Chlorophyll energy levels and electron flow in photosynthesis. *Proc. Natl. Acad. Sci. USA* 1968; 61:29–38.
4. Jursinic P, Govindjee. Thermoluminescence and temperature effects on delayed light emission in DCMU treated algae. *Photochem. Photobiol.* 1972; 29:47–63.
5. Amez J, Van Gorkom HJ. Delayed fluorescence in photosynthesis. *Annu. Rev. Plant Physiol.* 1978; 29:47–63.
6. Malkins S. Delayed luminescence. In: Barbar J, ed. *Primary Process in Photosynthesis.* Elsevier, Amsterdam, 1977:349–368.
7. DeVault D, Govindjee. Photosynthetic glow peaks and their relationship with free energy changes. *Photosynth. Res.* 1990; 24:175–181.
8. Chen R, Krisch Y. *International Series on the Science of the Solid State*, Vol. 15. Pergamon Press, Oxford, 1981.
9. Bhattacharjee SK. Thermoluminescence from spinach leaf without excitation by any radiation or external stimuli: stimulatory role of thermal fluctuations. *Curr. Sci.* 2003; 84:1419–1427.
10. Lavorel J, Govindjee. *Bioenergetics and Photosynthesis.* Academic Press, New York, 1975.
11. Malkins S. Delayed luminescence. In: Avron M, Trebst A, eds. *Encyclopaedia of Plant Physiology and Photosynthesis*, Vol. 5. Springer, Berlin, 1977:493–523.
12. Rutherford AN, Inoue Y. Oscillation of delayed luminescence from PSII: recombination of $S_2Q_B^-$ and $S_3Q_B^-$. *FEBS Lett.* 1884; 165:163–170.
13. Lurie S, Beerstsch W. Thermoluminescence studies on photosynthetic energy conversion. I. Evidence for three types of energy storage by photoreaction II of higher plants. *Biochim. Biophys. Acta* 1974; 357:420–429.
14. Fleischman D. Delayed fluorescence and the reversal of primary photochemistry in *Rhodopseudomonas viridis*. *Photochem. Photobiol.* 1974; 19:59–68.
15. Randall JT, Wilkins HF. Phosphorescence and electron traps. I. The study of trap distribution. *Proc. R. Soc. A* 1945; 184:366–389.
16. Tatake VG, Desai TS, Govindjee, Sane PV. Energy storage states of photosynthetic membranes: activation energy and life time of electron trap states by thermoluminescence. *Photochem. Photobiol.* 1981; 33:243–250.
17. Tatake VG, Desai TS, Bhattacharjee SK. A variable temperature cryostat for thermoluminescence studies. *J. Phys. E: Sci. Instrum.* 1971; 4:755–762.
18. Manche EP. Thermoluminescence. In: Ewing GW, ed. *Topics in Chemical Instrumentation. J. Chem. Educ.* 1979; 56A:273.
19. Chen R, McKeever SWS. *Theory of Thermoluminescence and Related Phenomena.* World Scientific, Singapore, 1997.
20. Tollen G, Calvin M. The luminescence of chlorophyll containing plant material. *Proc. Natl. Acad. Sci. USA* 1957; 43:895–908.
21. Arnold W. Light reactions in green plant photosynthesis: a method of study. *Science* 1966; 154:1046–1049.
22. Rubin AB, Vanediktov PS. Storage of light energy by photosynthesizing organisms at low temperature. *Biofizika* 1969, 14:105–109.
23. DeVault D, Govindjee, Arnold W. Energetics of photosynthetic glow peaks. *Proc. Natl. Acad. Sci. USA* 1983; 80:953–957.
24. Demeter S, Herczeg T, Droppa M, Horvath G. Thermoluminescence characteristics of agranal and granal chloroplasts of maize. *FEBS Lett.* 1979; 100:321–329.
25. Sane PV, Tatake VG, Desai TS. Detection of triplet state of chlorophyll *in vivo*. *FEBS Lett.* 1974; 45:290–299.
26. Demeter S, Vass I. Charge accumulation and recombination in photosystem II studies by thermoluminescence. I Participation of the primary acceptor Q and secondary acceptor B in the generation of thermoluminescence of chloroplasts. *Biochim. Biophys. Acta* 1984; 764:24–32.
27. Demeter S. Binary oscillation of thermoluminescence of chloroplasts preilluminated by flashes prior to inhibitor addition. *FEBS Lett.* 1982; 144:97–106.
28. Rutherford AW, Crofts AR, Inoue Y. Thermoluminescence as a probe of PSII chemistry. The origin of flash induced glow peaks. *Biochim. Biophys. Acta* 1982; 682:457–469.
29. Demeter S, Vass I, Horvath G, Laufer A. Charge accumulation and recombination in PSII. Studies by thermoluminescence II Oscillation of C band influences by flash excitation. *Biochim. Biophys. Acta* 1984; 764:33–43.
30. Inoue Y, Shibata K. Thermoluminescence from photosynthetic apparatus. In: Govindjee, ed. *Photosynthesis: Energy Conversion by Plants and Bacteria*, Vol. 1. Academic Press, New York, 1982:508–539.

31. Inoue Y. Manganese catalyst as a possible cation carrier in thermoluminescence from green plants. *FEBS Lett.* 1976; 72:279–286.

32. Joliot P, Barbara G, Chabaud R. UN noveau modele des centres photochimiques des system II. *Photochem. Photobiol.* 1969; 10:309–318.

33. Inoue Y, Shibata K. Oscillation of thermoluminescence at medium–low temperature. *FEBS Lett.* 1978; 85: 193–197.

34. Demeter S, Droppa M, Vass I, Horvath G. The thermoluminescence of chloroplasts in the presence of Photosystem II herbicides. *Photochem. Photobiol.* 1982; 4:163–168.

35. Kok B, Forbrush B, McGloin M. Cooperation of charges in photosynthetic O_2 evolution. 1. A linear 4 step mechanism. *Photochem. Photobiol.* 1970; 11:457–469.

36. Vass I, Horvath G, Herczeg T, Demeter S. Photosynthetic energy conservation investigated by thermoluminescence. Activation energy and half life of thermoluminescence bands of chloroplasts determined by mathematical resolution of glow curves. *Biochim. Biophys. Acta* 1981; 634:140–152.

37. Demeter S, Govindjee. Thermoluminescence in plants. *Physiol. Plantarum* 1989; 75:121–130.

38. Sane PV, Desai TS, Tatake VG. Characterization of glow peaks of chloroplast membranes. I. Relationship with "S" states. *Indian J. Exp. Biol.* 1983; 21:396–405.

39. Sane PV, Johanningmeir U. Inhibition by tetranitromethane of photosynthetic electron transport from water to PSII in chloroplasts. *Z. Naturforsch.* 1980; 35C:293–298.

40. Desai TS, Bhagwat AS, Mohanty P. Thermoluminescence investigation on the site of action of o-phthalaldehyde in photosynthetic electron transport. *Photosynth. Res.* 1996; 48:213–220.

41. Ichikawa T, Inoue Y, Shibata K. Characterization of thermoluminescence band of intact leaves and isolated chloroplasts in relation to water splitting activity in photosynthesis. *Biochim. Biophys. Acta* 1975; 408: 228–239.

42. Desai TS, Sane PV, Tatake VG. Thermoluminescence studies on spinach leaves and Euglena. *Photochem. Photobiol.* 1975; 21:345–350.

43. Ohad I, Koike H, Shochat S, Inoue Y. Changes in the properties of reaction centre II during the initial stages of photoinhibition as revealed by thermoluminescence measurements. *Biochim. Biophys. Acta* 1988; 933: 288–298.

44. Sane PV, Desai TS, Tatake VG. On the origin of glow peaks in Euglena cells, spinach chloroplasts and subchloroplast fragments enriched in system I and II. *Photochem. Photobiol.* 1977; 26:33–39.

45. Horvath G, Droppa M, Mustardy LA, Faludi-Daniel A. Functional characteristics of intact chloroplasts isolated from mesophyll and bundle sheath cells of maize. *Plant* 1978; 141:239–251.

46. Bhatnagar R, Saxena P, Vora HS, Dubey VK, Sarangapani KK, Shirke ND, Bhattacharjee SK. Design and performance of versatile computer controlled instrument for studying low temperature thermoluminescence

from biological samples. *Meas. Sci. Tech.* 2002; 13:2017–2026.

47. Walters RG, Johnson GN. The effect of elevated light on photosystem II function: a thermoluminescence study. *Photosynth. Res.* 1997; 54:169–183.

48. Janda T, Szalai G, Paladi E. Thermoluminescence investigation of low temperature stress in maize. *Photosynthetica* 2000; 38:635–639.

49. Ohad I, Adir N, Koike H, Kyle DJ, Inoue Y. Mechanism of photoinhibition *in vivo*. A reversible light induced conformational change of reaction centre II is related to an irreversible modification of D1 protein. *J. Biol. Chem.* 1990; 265:1972–1979.

50. Maenpaa P, Miranda T, Tyystjarvi E, Tyystjarvi T, Govindjee, Ducrvet JM, Etienne M, Kirilovsky D. A mutation in D-de loop of D1 modifies the stability of $S_2Q_A^-/S_2Q_B^-$ states in photosystem II. *Plant Physiol.* 1995: 107:187–197.

51. Havaux M, Nyogi KK. The violaxanthine cycle protects plants from photooxidative damage by more than one mechanism. *Proc. Natl. Acad. Sci. USA* 1999; 96:8762–8767.

52. Havaux M, Bonfils J, Lutz C, Niyogi KK. Photodamage of the photosynthetic apparatus and its dependence on the leaf developmental stage in the npq1 Arabidopsis mutant deficient in xanthophylls cycle enzymes violaxanthine de-epoxidase. *Plant Physiol.* 2000; 124:273–284.

53. Renger G, Inoue Y. Studies on the mechanism of ADRY agents (agents accelerating the deactivation of reaction of water splitting enzyme system) on thermoluminescence emission. *Biochim. Biophys. Acta* 1983; 725:146–154.

54. Joshi MK, Desai TS, Mohanth P. Temperature dependent alterations in the pattern of photochemical and non-photochemical quenching and associated changes in the photosystem II condition of leaves. *Plant Cell Physiol.* 1995; 36:1221–1227.

55. Misra AN, Ramaswamy NK, Desai TS. Thermoluminescence studies on the photoinhibition of pothos leaf discs at chilling and room temperature. *Photochem. Photobiol.* 1997; 38:164–168.

56. Akazawa T, Timasheff SN. The stabilization of proteins by osmolytes. *Biophys. J.* 1985; 47:411–414.

57. Williams WP, Gounaris K. Stabilization of PSII mediated transport in oxygen-evolving PSII core preparation by the addition of compatible cosolutes. *Biochim. Biophys. Acta* 1992; 1106:92–97.

58. Allakhverdiev SI, Feyziev YM, Ahmed A, Hayashi H, Aliev JA, Klomov VV, Murata N, Carpentier R. Stabilization of oxygen evolution and primary electron transport reactions in PSII against heat stress with glycine–betaine and sucrose. *J. Photochem. Photobiol.* 1996; 34:149–157.

59. Sane PV, Ivanov AG, Hurry V, Huner NPA, Oquist G. Changes in redox potential of primary and secondary electron accepting quinines in PSII confer increased resistance to photoinhibition in low temperature acclimated *Arabidopsis*. *Plant Physiol.* 2003; 126:2144–2151.

60. Sane PV, Ivanov AG, Sveshnikov D, Huner PA, Oquist G. A transient exchange of photosystem II re-

action centre protein D1:1 with D1:2 during low temperature stress of *Synechococcus* sp. PCC 7942 in the light lower the redox potential of QB. *J. Biol. Chem.* 2002; 277:32739–32745.

61. Broin M, Rey P. Potato plants lacking the CDSP32 plastidic thioredoxin exhibit over oxidation of BASI-2-cysteine peroxidation and increased lipid peroxidation in thylakoid under photooxidative stress. *Plant Physiol.* 2003; 132:1335–1343.

62. Turcsanyi E, Vass I. Effect of UV A radiation on photosynthetic electron transport. *Acta Biologica Szegediansis* 2002; 46:171–173.

63. Biswal AK, Dilnawaz F, Ramaswamy NK, David KAV, Misra AN. Thermoluminescence characteristics of sodium chloride salt-stressed Indian mustard seedlings. *Luminescence* 2002; 17:135–140.

64. Biswal AK, Dilnawaz F, David KAV, Ramaswamy NK, Misra AN. Increase in the intensity of thermoluminescence Q band during leaf aging is due to block in the electron transfer from Q_A to Q_B. *Luminescence* 2001; 16:309–313.

65. Maslenkova L, Zeinalov Y. Thermoluminescence and oxygen evolution in JA-treated barley (*Hordeum vulgare* L). *Bulg. J. Plant Physiol.* 1999; 25:58–64.

66. Maselnkova L, Toncheva S, Zeinalov YU. Effect of absssisic acid and jasmonic acid on photosynthetic electron transport and oxygen evolving reactions in pea plants. *Bulg. J. Plant Physiol.* 1995; 21:48–55.

67. Tanda T, Szalai G, Giauffret C, Paldi E, Ducruet J. The thermoluminescence "*after glow*" band as a sensitive indicator of abiotic stress in plants. *Z. Naturforsch.* 1999; 54c:629–633.

68. Rahoutei J, Baron M, Garcia-Luque I, Droppa M, Nenmenyi A, Horvath G. Effect of tobamovirus infection on thermoluminescence characteristics of chloroplasts from infected plants. *Z. Naturforsch.* 1999; 54c:634–639

69. Thomas S, Banerjee M, Vidyasagar PB, Shaligram AD. Instability in photosynthetic systems and its relevance to biodevices. *Mater. Sci. Eng.* 1995; C3:223–226.

70. Steglich C, Behrenfeld M, Koblizek M, Claustre H, Penno S, Prasil O, Partensky F, Hess WR. Nitrogen deprivation strongly affects PSII but not phycoerythrin level in divinyl-chlorophyll containing cyanobacterium *Prochlorococcus marinus*. *Biochim. Biophys. Acta* 2001; 1503:341–349.

71. Misra HS, Desai TS. Involvement of acceptor side component of PSII in the regulatory mechanism of *Plactonema boryanum* grown photoautotrophically under diazotrophic conditions. *Biochem. Biophys. Res. Commun.* 1993; 194:1001–1007.

72. Demeter S, Hideg E, Sallai A. Comparative thermoluminescence study of triazine-resistant and susceptible biotypes of *Erigeron Canadensis* L. *Biochim. Biophys. Acta* 1985; 806:16–24.

73. Andree S, Weis E, Krieger A. Heterogeneity and photoinhibition of photosystem II: studies with thermoluminescence. *Plant Physiol.* 1998; 116:1053–1061.

74. Regel RE, Ivleva NB, Zer J, Meurer J, Shestaakov SV, Hermann RG, Pakrasi HB, Ohad I. Deregulation of electron flow within photosystem II in the absence of PsbJ protein. *J. Biol. Chem.* 2001; 276: 41473–41478.

75. Meetam M, Keren N, Ohad I, Pakrasi HB. The PsbY protein is not essential for oxygenic photosynthesis in the cyanobacterium *Synechocystis* sp. PCC 67803. *Plant Physiol.* 1999; 121:1267–1272.

76. Bianchetti M, Zheleva D, Deak Z, Zharmuhamedov S, Klimov V, Nugent J, Vass I, Barber J. Comparison of functional properties of the monomeric and dimeric forms of the isolated CP47-reaction centre complex. *J. Biol. Chem.* 1998; 273:16128–16133.

77. Shen J, Ikeuchi M, Inoue Y. Analysis of the *Psb*U gene encoding the 12 kDa extrinsic protein of PSII and studies of its role in deletion mutagenesis in *Synechocystis* sp. PCC 6803. J. Biol. Chem. 1997; 272:17821–17826.

78. Townsend JS, Kanazawa A, Kramer DM. Measurement of S2QB⁻ recombination by delayed thermoluminescence reveal heterogeneity in PSII energetics. *Phytochemistry* 1998; 47:641–649.

79. Geliter HM, Ohad N, Koike H, Hirschberg J, Renger G, Inoue Y. Thermoluminescence and flash induced oxygen yield in herbicide resistant mutant s of D1 protein in *Synechococcus* pcc 7942. *Biochim. Biophys. Acta* 1992; 1140:135–143.

80. Bock A, Krieger-Liszakay A, Zarara IB, Schonknecht G. Cl⁻ channel inhibitors of arylamino benzoate type act as PSII herbicides: a functional and structural study. *Biochemistry* 2001; 40:3273–3281.

81. Krieger A, Bolte S, Dietz K, Ducruet J. Thermoluminescence studies on the facultative crassulacean acid metabolism plant *Mesembryanthemum crystallinum* L. *Planta* 1998; 205:587–594.

82. Bondarava N, Pascalis L, Al-Babili, Goussias C, Golecki J, Beyer P, Bock R, Krieger-Liszkey A. Evidence that cytochrome b559 mediate the oxidation of reduced plastoquinone in the dark. *J. Biol. Chem.* 2003; 278:13554–13560.

83. Homann PH, Inoue Y. The anion and cation requirement of the photosynthetic water splitting complex. In: Papageorgiou GC, Barbar J, Papa S, eds., *Ion Interactions in Energy Transfer in Biomembranes*. New York: Plenum Press, 1986:229–290.

84. Chen GX, Blubaugh DJ, Homann PH, Golbeck JH, Cheniae GM. Superoxide contributes to the rapid inactivation of specific secondary donors of the PSII reaction centre during photodamage of manganese-depleted PSII membranes. *Biochemistry* 1995; 34:2317–1232.

85. Litvin S. Study of glow curves in photosynthesis. *Biochim. Biophys. Acta* 1979; 421:321–333.

Section II

Biochemistry of Photosynthesis

3 Chlorophyll Biosynthesis — A Review

Benoît Schoefs
Dynamique Vacuolaire et Résponses aux Stress de l'Environnement,
UMR CNRS 5184/INRA 1088/Université de Bourgogne Plante-Microbe-Environnement,
Université de Bourgogne à Dijon

Martine Bertrand
Institut National des Sciences et Techniques de la Mer, Conservatoire National des Arts
et Métiers

CONTENTS

I. INTRODUCTION

In photosynthetic organisms, at least three distinct classes of tetrapyrroles coexist. They are closed tetrapyrroles chelated with either Mg^{2+} (chlorophyll [Chl] family) or Fe^{2+}/Fe^{3+} (heme family), and open tetrapyrroles (phytochromobilins). There is increasing evidence suggesting that most of them are synthesized inside plastids, with some of them eventually exported to other cell compartments.

As a main component of the photosynthetic apparatus Chl (and bacteriochlorophyll [Bchls]) molecules play major roles in the development and maintenance of life. Despite the importance of Chl molecules for our world, the intimate mechanism of the reactions leading to their formation has not yet been fully elucidated. The regulation of Chl biosynthesis is only beginning to be investigated.

The initial substrate for tetrapyrrole synthesis inside plastids is the activated form of glutamate (Glu), namely, GLU-tRNA^{Glu}, which is also used for protein synthesis. The Glu moiety is reduced by glutamyl-tRNA reductase (Glu-R) to form glutamic acid 1-semialdehyde (GSA), which is rearranged resulting in δ-aminolevulinic acid (ALA) (Figure 3.1). This pathway is known as the Beale pathway. In α-proteobacteria, yeast, and animal cells, ALA is formed through the Shemin pathway, by condensation of glycine and succinyl-CoA.

Two molecules of ALA are condensed to porphobilinogen (PBG). Four molecules of PBG are condensed to form a linear tetrapyrrole, namely hydroxymethylbilane, which, in turn, is cyclized into uroporphyrinogen (Uro) III. The acetic acid side chains of Uro-III are reduced to methyl groups, yielding coproporphyrinogen (Copro) III. Then, the

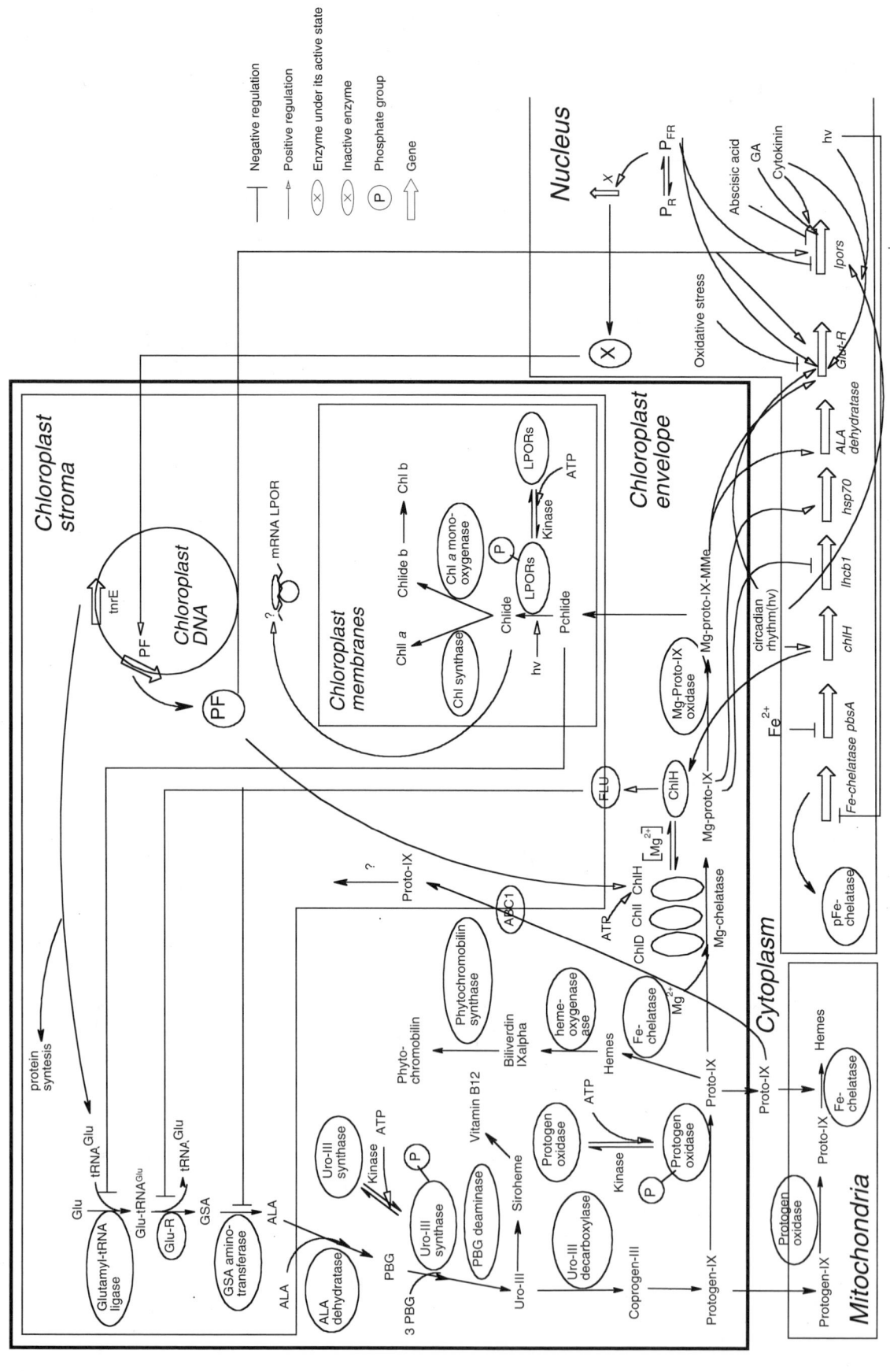

FIGURE 3.1 Scheme of the metabolic and regulation networks of Chl biosynthesis.

propionic acid side chains are reduced to vinyl ones, resulting in the formation of protoporphyrinogen (Protogen) IX, which is subsequently oxidized to protoporphyrin (Proto) IX. After insertion of a Mg^{2+} ion in the center of Proto-IX, Mg–Proto-IX is methylated to yield Mg–Proto-IX monomethyl ester (Mg–Proto-IX-MMe). In the subsequent steps, the isocyclic ring is formed, resulting in protochlorophyllide (Pchlide) synthesis. Pchlide is reduced to chlorophyllide (Chlide), which is either esterified to chlorophyll (Chl) *a* or oxidized to Chlide *b*. Chlide *b* is then esterified to Chl *b*. Since the publication of the first edition of the *Handbook of Photosynthesis* [1], much progress in biochemistry, biophysics, physiology, and molecular biology of Chl biosynthesis has been realized (reviewed in Refs. [2–11]). A large number of papers on the topic covered by this review have been published since 1997. Each of them cannot be cited, and we apologize for this. This chapter summarizes the main findings in the field and is the continuation of the 1997 chapter. Therefore, it has been organized similarly.

II. THE FORMATION OF ALA

Chl biosynthesis is heavily compartmentalized: (i) each gene encoding the enzymes involved in the pathway is encoded in the nucleus (except the light-independent NADPH:Pchlide oxidoreductase [DPOR], as reviewed in Refs. [2,5]); (ii) synthesis of ALA takes place in the plastid stroma; (iii) Protogen oxidase is bound to the envelope and plastid membranes; and (iv) Pchlide reduction occurs in the plastid membranes.

The initial substrate for tetrapyrrole synthesis is the activated form of Glu, namely, $tRNA^{Glu}$. The Glu residue is reduced by Glu-R to form GSA, which, in turn, is transformed to ALA through the catalytic action of GSA amino transferase (Figure 3.1). The three-dimensional structure of the Glu-R has been predicted and a putative heme-binding site suggested [12]. Modeling suggests that binding of the heme molecule to the His99[1] residue would inhibit the formation of a thioester between this residue and the Cys550 residue of the active site. According to this model, a heme-insensitive truncated Glu-R has been described [13]. As Glu-R does not contain the only typical heme regulatory motif identified so far (10 amino acids length) [14], the mechanism of action of heme remains to be determined.

GSA aminotransferase gene expression has been shown to be induced by blue light in *Chlamydomonas reinhardtii*. Light induction of the gene in a *C. reinhardtii* strain deficient in carotenoid allowed Hermann et al. [15] to exclude this family of compounds as a part of the putative photoreceptor. The complete inhibition of the light induction by the flavin antagonist, diphenyleneiodonium, indicates that the light-harvesting pigment in the photoreceptor is a flavin.

Specific inhibitors of Glu-R and GSA aminotransferase have been recently described [16].

III. FROM ALA TO PROTO-IX

A. Uro-III Formation

Two molecules of ALA are condensed to yield one PBG molecule. The reaction is catalyzed by ALA dehydratase. Uro-III is derived by the enzymatic cyclization and rearrangement of the D ring of hydroxymethylbilane through the catalytic action of PBG deaminase and Uro-III synthase (reviewed in Ref. [17]) (Figure 3.1). The nonenzymatic cyclization without rearrangement results in the toxic isomer Uro-I. CsCl inhibits the reaction through an unknown mechanism [18].

PBG deaminase is a unique enzyme in that it contains a covalently dipyrromethane cofactor, which acts as a primer during the enzymatic reaction [17]. So far, the structure of the active form of PBG deaminase from *Escherichia coli* has been reported [19,20]. These studies fully confirmed the position of the cofactor deduced from labeling and degradative studies (reviewed in Ref. [21]). The protein is formed by three flexible domains, which together with the other structural details allow speculations about the action mechanism of the enzyme.

Uro-III synthase is an unstable enzyme, the structure of which has been solved from animal cells. In these cells, a decrease in the activity of Uro-III synthase leads to the autosomal recessive disorder congenetic erythropoietic porphyria. One understands easily the considerable interest of medical research in this enzyme [22,23]. The protein folds into a two-domain structure connected by a two-strand antiparallel β-ladder, which probably contains the catalytic site. Each domain consists of parallel β-sheets surrounded by α-helixes. Domain 1 (residues 1 to 35 and 173 to 260), which belongs to a flavodoxin-like fold family, comprises a five-strand parallel β-sheet surrounded by five α-helixes. Domain 2, which belongs to a DNA glycosylase-like fold family, comprises a four-strand parallel β-sheet surrounded by seven α-helixes. A structural similarity search using domain 1 identified the vitamin B12

---

[1]Numbering according to the enzyme from barley.

binding domain of methionine synthase as the most structurally similar protein. The polypeptide most similar to domain 2 is that of the NAD-binding domain of flavohemoglobin. This structural similarity, however, does not seem to reflect a functional similarity, as Uro-III synthase does not utilize NAD as cofactor.

As already mentioned, the catalytic site of Uro-III synthase is thought to be localized between the domains, where many surface-exposed conserved amino acid residues are localized. So far, several mutations (Thr103Ala, Tyr168Ala, Thr228Ala) significantly alter the catalytic activity of the enzyme [23], whereas mutations Ser68Ala and Ser194Ala cause missfolding [24].

Methylation of Uro-III is the first step of the siroheme pathway. Sirohemes are precursors of vitamin B12 (Figure 3.1, chloroplast envelope).

B. Copro-III Formation

Copro-III is catalyzed by Uro-III decarboxylase, which catalyzes the decarboxylation of the four acetate side chains of the substrate molecule. The gene from human cells has been cloned in *E. coli*, and the enzyme has been purified to homogeneity. The purified protein was crystallized in space group P3(1)21 or P3(2)21 with unit cell dimensions $a = b = 103.6 \text{ Å}$, $c = 75.2 \text{ Å}$ [25]. To the best of our knowledge, there are no data about the enzyme from a photosynthetic organism.

C. Proto-IX Formation

All the steps, including Proto-IX formation, are confined to plastids [26]. Formation of Proto-IX is catalyzed by Protogen-IX oxidase. To date, twelve Protogen-IX oxidases have been determined from various sources, each of them sharing low amino acid identities among different organisms, but a high homology between closely related families exists [27,28]. Two genes encoding this enzyme have been identified in the nuclear genome of tobacco. One product is imported in mitochondria, while the second product is imported in chloroplasts [27].

Protogen-IX oxidase is generally sensitive to herbicides but not that isolated from *Bacillus subtilis* [29]. This pecularity was used to confer herbicide resistance to tobacco plants [30]. The relative increase (twofold at $100 \mu M$ oxyfluorfen) in herbicide resistance suggested that the tobacco cells expressed the *B. subtilis* Proto-IX oxidase gene. However, it remains unclear whether the resistance was localized in the plastids. To solve this question, Lee et al. [31] prepared rice transgenic lines transformed with either the

B. subtilis Protogen-IX oxidase or the *B. subtilis* Protogen-IX oxidase fused to transit peptide allowing the protein to be imported into the chloroplast. The resistance of the different "cytoplasmic" and "plastid" transgenic lines is higher than in the nontransgenic lines, with the highest and homogenous resistances found in the plastid lines [31]. These authors have proposed a model to explain the differences between the resistance of the "cytoplasmic" and "plastid" transgenic lines.

IV. FROM PROTO-IX TO PCHLIDE

A. Magnesium Insertion

The first unique step in (B)Chl biosynthesis is the insertion of Mg^{2+} into Proto-IX. The reaction is catalyzed by the Mg chelatase, a heteromultimeric enzyme composed of three subunits, whose molecular masses are somewhat different in photosynthetic bacteria (*Rhodobacter capsulatus*) and higher plants (*Nicotiana tabacum*): BChlD/ChlD (60/83 kDa), BChlH/ChlH (140/154 kDa), and BChlI/ChlI (40/42 kDa) [32–35] (reviewed in Ref. [36]). The three subunits are encoded by the genes *chlD*, *chlH*, and *chlI*, respectively. Molecular studies revealed that mutations in these individual genes were previously described in barley as mutants *xantha-f*, *xantha-g*, and *xantha-h* [37,38].

In vitro assays established that the stoichiometric amount of each subunit is 4-ChlH/2-ChlI/1-ChlD [39]. More recently, the structure of the ChlI has been published [40]. The diffraction data reveal that ChlI presents a structural homology to AAA-type ATPases (Mg^{2+}-dependent ATPases) and that 6-ChlI subunits assemble to form a ring. The N-terminal domain, which contains the Walker A and B motifs, is connected with a C-terminal four-helix bundle by a long helical region. Three mutations (xantha-h[clo-125], xantha-h[clo-157], xantha-h[clo-161], see Table 3.1) are located in the interface between two neighboring subunits of AAA+ hexamer and close to the parts forming the ATP-binding pocket [41]. These mutations, which are semidominant, confer to the plant tissue a pale green phenotype due to an inhibition of the enzymatic activity of Mg chelatase [42].

By comparison with the already elucidated enzymatic mechanism of AAA+-ATPases, a mechanism for the reaction catalyzed by Mg chelatase was proposed. First, the ChlI hexamer is formed and the subunit D binds the hexamer. Mg–ATP binding is required for this step [41]. Binding of ChlD to the ChlI–hexamer–ATP complex occurs in the presence of ATP and prevents ATP hydrolysis. Hansson et al.

TABLE 3.1

Identification of the Subunit of Mg-Chelatase Affected in Mutants of Barley

Mutant Name	Subunits of Mg-Chelatase Affected			Reference
	ChlD	ChlH	ChlI	
Xantha-f	−	+	−	Jensen et al. 1996
Xantha-g	+	−	−	Jensen et al. 1996
Xantha-h	−	−	+	Jensen et al. 1996
Chlorina-125	−	−	+	Hansson et al. 1999
Chlorina-157	−	−	+	Hansson et al. 1999
Chlorina-161	−	−	+	Hansson et al. 1999

[43] recognized an ATPase function for the ChlD–ChlI complex. ChlD then binds Mg^{2+} atoms, while ChlH binds Proto-IX through an ATP-dependent reaction [39,44], and the complex Proto-IX–ChlH joins to the ChlD–ChlI complex in the presence of a local elevated Mg^{2+} concentration. The binding triggers a conformational modification allowing the ATP of the ChlI–ChlD complex to be hydrolyzed while ChlH protein inserts the Mg^{2+} atom into Proto-IX. After Mg^{2+} insertion the ternary complex is thought to dissociate into two complexes that are ChlH–Mg–Proto-IX on the one hand and ChlI–ChlD–ADP on the other [36]. The postulated conformational change would involve conserved arginine (Arg) residues, the so-called "Arg finger" and "sensor Arg" [45,46]. These Arg residues are placed at the interface between two subunits of the hexamer as they interact with ATP and thereby trigger the conformational change [46]. Directed mutagenesis does not influence ATP binding and the formation of the hexamer but inhibits ATP hydrolysis [41]. During ATP hydrolyzation a nitrogenous base–Mg^{2+}–porphyrin complex is formed. The most likely candidate for the nitrogenous base is one of the conserved histidine (His) residues His679, His683, or His829[2] [47]. Modification of the cysteine (Cys) residues of ChlI leads to inactivation of the Mg chelatase activity with respect to the association of ChlI–Mg–ATP, ATP hydrolysis, and interaction of ChlH with Mg–ATP and Proto-IX [48]. ChlH subunit contains more Cys residues than ChlI. Among them, only three (Cys722[3], Cys896, and Cys1037) are conserved in all organisms [48]. Directed mutagenesis should help in the identification of the Cys residues implied in the binding of nucleotide and in the subunit association as well.

As mitochondria need hemes to synthesize their cytochromes, part of the synthesized Proto-IX should be transported out of the plastid. So far, only a putative ABC-like protein (atABC1 protein) located in the stromal side has been involved in the transport of Proto-IX. On the basis of sequence homology with other ABC-like proteins no membrane-spanning domains but several homologies with ABC proteins from lower eukaryotes have been found. Because the atABC1 protein lacks membrane-spanning domains, it is likely involved in an import mechanism of Proto-IX into the chloroplast. At present, it is not known whether the at ABC1 protein is implied in a reimport process of Proto-IX or in a mechanism correcting the Proto-IX amount in the plastid envelope. In the *laf6* mutant of *Arabidopsis thaliana*, in which the *atABC1* gene has been disrupted, Proto-IX accumulates and a preferential insensitivity to far-red light has been observed. These findings demonstrate that the atABC1 protein is involved in the signaling of PHYA but not PHYB phytochrome protein [49]. In this respect, the latter hypothesis — corrections of the amount of Proto-IX in the envelope — seems to be sufficient to explain the modification in PHYA signaling (Figure 3.1).

V. FROM PCHLIDE TO CHLIDE *a*

It has been shown that the synthesis of Pchlide from Mg–Proto-IX is heterogeneous, as a photosynthetic tissue may synthesize monovinyl or divinyl compounds. However, the accumulation of DV-Chl is lethal except in some marine prochlorophytes (reviewed in Ref. [50]). Therefore, in other organisms the DV intermediates should be converted to MV ones. The links between the DV and MV routes are ensured through the enzymatic activity of four enzymes, namely [4-vinyl] Mg–Proto-IX reductase [51], [4-vinyl] Pchlide *a* reductase [52], [4-vinyl] Chlide *a* reductase [53], and [4-vinyl] Chl *a* reductase [54]. The [4-vinyl] Chlide *a* reductase is the most potent of the 4-vinyl reductase activities. It is a membrane-bound NADPH-dependent enzyme that rapidly converts nascent DV-Chlide *a* to MV-Chlide *a* but is inactive toward DV-Pchlide *a* [53]. Its activity appears to be regulated by a complex interaction of stromal and plasmid membrane components as well as the availability of NADPH [55]. Partial purification of [4-vinyl] Chlide *a* reductase from etiolated barley leaves has been reported [56].

Pchlide reduction can be performed by two families of enzymes. The reaction consists of the hydrogenation of the $C_{17}=C_{18}$ double bond of Pchlide molecule yielding Chlide. One type of enzyme requires light to function, whereas the second does not. Both enzymes are usually present in photosynthetic

---

[2]Numbering based in the *C. reinhardtii* sequence as published by Chekounova et al. [47].
[3]Numbering according the *Synechocystis* ChlI sequence.

cells except angiosperms, which only contain the light-dependent form. As Pchlide reduction is the topic of Chapter 5 by Bertrand and Schoefs, this step will not be discussed here.

VI. CHL *b* FORMATION

Until very recently, Chl *b* formation has remained obscure (reviewed in Refs. [6,57]). Chlide *a* monooxygenase (CAO), the enzyme catalyzing the oxidation of Chlide *a* to Chlide *b*, has been identified in higher plants, green algae [58–61], and in two Prochlorophytes (*Prochlorothrix hollandica* and *Prochloron didemni*) but not in *Prochlorococus* MED4 and *Prochlorococus* MIT 9313, although the last two organisms are able to synthesize Chl *b*. The CAO enzyme is composed of 463 amino acids and has a MW of approximately 51 kDa. The comparison of the amino acid sequences indicates a putative Rieske [2Fe–2S] center and a mononuclear iron [58]. The meaning of this result is discussed below. The CAO enzyme mechanism consists of a particular two-step oxygenase reaction [61]. These studies established that the true substrate of the enzyme is Chlide *a* and confirmed an earlier observation made during the greening of bean leaves with etioplasts [62]. CAP, which is probably localized in chloroplast membranes, catalyzes the transformation of Chlide *a* to [7-CH$_2$OH]-Chlide *a*. Then, the *gem* diol, [7-CH(OH)$_2$]-Chlide *a* spontaneously dehydrates to form Chlide *b*. Then, this compound is phytylated to Chl *b*.

VII. REGULATION

The knowledge of Chl regulation is very important, not only for its basic aspect but also in applied science and agriculture. Deregulation of this pathway or that of hemes can have tremendous effects on the physiology of plants. For instance, an increase in the amount of free tetrapyrrole molecules triggers deleterious photodynamic damages due to the accumulation of porphyrin intermediates (e.g., [63]; reviewed in Ref. [64]). In fact, the photosentization may be so high [65] that the level of the enzymes, catalase, superoxide dismutase, and ascorbate peroxidase, which remove the reactive oxygen species from the chloroplast, decreases [66]. It has been firmly established that a number of components required for plastid structure and development are encoded in the nucleus genome. Most of these components belong to the metabolic network, that is, the set of biochemical reactions ensuring the metabolic activity.

There is a considerable body of evidence that suggests that the proper and timely expression of these genes requires a tight and efficient signaling between chloroplast, mitochondria, and nucleus. The components that participate in this activity are members of the regulatory networks that control the metabolic activity of the cell. The major points where the regulation takes place are (i) the expression of genes, (ii) the posttranslational modification(s) of the enzymes, (iii) the beginning of a metabolic pathway for channeling substrate into the pathway and for defining the overall synthesis rate, (iv) the branching points for controlling the distribution of common intermediates, and (v) the formation of the final products, which may limit the metabolic flow through a feedback mechanism. In the following paragraphs the regulation of Chl formation is reviewed. For easier comprehension we have treated separately the regulation of the Chl biosynthesis itself and the interactions of intermediates of the Chl pathway in the regulation of other biosynthetic routes.

A. REGULATION OF THE CHLOROPHYLL BIOSYNTHETIC PATHWAY

The reaction catalyzed by Glu-R (*hemA* gene) (Figure 3.1, chloroplast stroma) is known to be the limiting step of the tetrapyrrole pathway. The mRNA and protein levels for the reductase oscillate in a phase similar to that of overall ALA synthesis, reaching a maximum in the early hours of illumination [34,67–69]. Plant genomes contain two *hemA* genes. Expression of the *hemA1* gene is regulated at the transcriptional level by light, including high-fluence far-red light and a plastid signal [68,70–72]. The expression of *hemA* gene is repressed under photooxidative conditions [71]. Expression of the *hemA2* gene was so far only observed in roots of seedlings and it is not light regulated. Dissection of the promoter of *hemA1* shows that the −199/+252 fragment, which contains a GT-1/I-box and a CCA-1 binding site, is sufficient to confer the full light responsiveness to the GUS reporter gene expression [72]. McCormac and Terry [73] found that a continuous far-red light illumination blocks subsequent greening through two different responses. The first response is detected after 1 day of continuous far-red illumination. It consists of a white light intensity-dependent incomplete loss of greening capacity with retention of *hemA1* and *lhcb* gene expressions but not that of *lpor* (transcriptionally uncoupled response). This response is prevented in a *phyA* mutant of *Arabidopsis*, by cytokinin treatment [73] and by *lpor* overexpression [74]. The second response is observed later, that is, after 3 days of continuous far-red illumination. It consists of a white light intensity-independent complete loss of the ability to green. Expression of *hemA1* and *lhcb* after transfer to white light were totally lost. This type of response is inhibited by sucrose and *lpor* overexpression [74], and it is also absent in a *phyA* mutant (transcriptionally coupled response). These results have established the involvement of phytochrome in the regulation of *hemA1* and *lhcb* genes through a high-fluence far-red signaling pathway, which includes a plastid signal (denoted PF—for plastid factor—in Figure 3.1) [73]. It follows from the light regulation of these gene expressions that the production of Chl precursors is higher in the first hours of the light period. Reports on the induction of Glu-R by light, temperature, cytokinin, and circadian rhythms [68,75–78] suggest a very complex control at this level. Expression of the GSA aminotransferase gene for *C. reinhardtii* is induced by blue light [15].

Mitochondria contains Protogen oxidase [27,79] and ferrochelatase [80] but not the enzymes catalyzing the earlier steps, which, therefore, appeared to be only localized in the chloroplasts. Consequently in addition to the general supply of precursors, the distribution of tetrapyrrole intermediates should be directed towards Chl and heme synthesis. Thus the substrates of both enzymes, namely Protogen-IX and Proto-IX should be exported from chloroplasts to mitochondria.

In plastids Proto-IX is the substrate of Mg chelatase and Fe chelatase. The activities of these enzymes have antagonistic rhythmicity—Mg chelatase activity is the major one at the transition from dark to light, while the Fe-chelatase displays its highest activity at the transition from light to dark [81]. In addition, ATP, which is a cofactor of Mg chelatase, reduces the activity of pea Fe-chelatase [82] (Figure 3.1). Altogether, these findings suggest that Mg chelatase plays a crucial role in determining how much Proto-IX is directed into heme and Chl biosynthetic pathways (Figure 3.1) [81,83]. The diurnal activity profile of Mg chelatase does not entirely correspond to the expression pattern of the three genes that encode the subunits of Mg chelatase: minor diurnal variations are observed at the levels of ChlD and ChlI mRNAs, whereas the amount of ChlH mRNA oscillates drastically in higher plants. In fact, the level of the ChlH transcript is very low during the dark phase and increases just prior to the start of the next light period, reaching its maximum in the first half of the light period [81,83,84]. As CHLH is the subunit that brings Proto-IX for catalysis, one can expect that CHLH plays a major role in diverting the pool of Proto-IX between the Chl and heme pathways. On the basis of the *in vitro* heme inhibition of ALA formation, it was proposed that hemes regulate ALA synthesis *in vivo* through a feedback mechanism. However, in a *chlH* antisense mutant of tobacco, the Mg chelatase activity was reduced and the levels of Mg tetrapyrroles were low, but no accumulation of Mg–Proto-IX or heme occured. The latter observation resulted from a reduction of the expression of the nuclear genes encoding Glu-R and ALA dehydratase [85]. Therefore, implication of heme in the control of Chl synthesis through a feedback analysis seems unlikely under basic metabolic activity. This conclusion is supported by the fact that the Glu-R and ALA dehydratase do not contain the heme-binding regulatory element found in heme-regulated proteins [14]. Rather Meskauskiene et al. [69] proposed that the activity of Glu-R is regulated by the nuclear-encoded chloroplast-imported protein FLU (Figure 3.1, chloroplast envelope/chloroplast stroma). A mechanism of activation of FLU would involve the release of the CHLH subunit of Mg chelatase from the envelope, which occurs at low Mg^{2+} concentration. Changes in Mg^{2+} concentration that affect the reversible attachment of CHLH to the membrane surface are within the physiological concentration range stroma in the dark and in the light. Then, the activated FLU could bind Glu-R [86]. FLU would be necessary to bridge the gap between the membrane and the stroma. This model is supported by the fact that FLU, which is firmly attached to the membranes [69], contains two different regions in its hydrophylic part that are predicted to contain coiled-coil and tetratricopeptide repeat domains. Both domains are implicated in protein–protein interactions [87,88]. A truncated form of FLU was expressed in yeast, and a strong interaction was found between the truncated protein and Glu-R. This interaction is no longer observed when mutations are introduced in either region [86].

Impairment of the synthesis of phytochromobilins from hemes may affect the heme pool and therefore their regulatory activity. For instance, mutants for heme oxygenase or phytochromobilin synthase accumulate reduced amounts of Chl or Pchlide [89–95]. This observation can be easily explained if an accumulation of heme molecules affects the enzymatic activity of these enzymes. The excess of heme may then repress ALA synthesis through a feedback mechanism [96].

In organisms that contain two or more *lpor* genes, the LPOR proteins seem structurally very similar, judging from the high-sequence homology of the mature proteins (reviewed in Ref. [10]). However, their amount and the corresponding mRNA are differentially regulated by light: LPORA transcription is strongly inhibited by light, while LPORB is constitutively expressed [97,98]. In addition, the amount of LPORA drops very quickly below the limit of detection under illumination due to regulation at the transcriptional and proteolytical levels [99]. Similar behaviors of LPORA and LPORB were recently found in *Pinus mungo* (Swiss mountain pine [100]) and *Pinus taeda* (loblolly pine [101]). In contrast, the transcript level of *Arabidopsis* LPORC, which is not detected in the dark, increases under illumination [102]. Different responses have been found in organisms that have only one *lpor* gene. LPOR mRNA accumulation was unaffected (pea [103,104]), enhanced (cucumber [105,106]; squash [107]), or depressed (cucumber [108]) by light. In cucumber, the unique *lpor* gene expression was controlled by diurnal and circadian rhythms. In this organism, the level of LPOR protein is regulated transcriptionally and posttranscriptionally [107]. As LPOR enzymes are encoded in the nucleus, they have to be imported in the chloroplast. The import is an energy-dependent mechanism [109,110]. As the majority of cytoplasmically synthesized proteins have to be imported into the chloroplasts, the N-terminal part of the LPOR sequence is extended by a transit peptide, which is necessary for the binding of the protein precursor to a receptor located at the external envelope and which mediates the import [111]. The precursor is then imported over the two envelopes. In fact, the receptor is part of a protein complex formed by several subunits, the so-called TIC–TOC complex (translocons at the inner or outer envelope membranes of the chloroplasts) [109,111] (reviewed in [112]), which actually constitutes the general gate for protein import into the plastids [113]. In contrast to the translocation of the small subunit of RuBisCO [114] the import of pLPOR would not require the Protein Import Related Anion Channel (PIRAC) [115]. This suggests that the import of LPOR may occur through an original pathway. It has been also suggested that the import of LPORA, but not LPORB, from barley requires the presence of Pchlide in the envelope [116]. In this respect, the LPORA import pathway would differ from all other known nuclear-encoded plastid-imported proteins. Trials to obtain similar results with pea chloroplasts failed [117]. This difference in the mechanism of LPOR import could have been related to the absence of several *lpor* genes in pea. Reexamination of this discrepancy with barley plastids, which contain both LPORA and LPORB, indicated that there is no strict correlation between Pchlide concentration and the import capacity of the plastids [113].

One of the most striking feature of the Chl biosynthesis pathway is the so-called Pchlide–Chlide cycle. The different reactions composing the cycle have been described in detail in Chapter 5 by Bertrand and Schoefs. One of the major aspects of the cycle resides in the fact that Chlide can be released from the LPOR catalytic site along two metabolic routes. Conse-quently, two pathways can be followed to regenerate the large aggregates of photoactive Pchlide. One of the authors proposed that the "choice" between the different routes is controlled by the actual and local ratio of newly formed Chlide to nonphotoactive Pchlide. This ratio was denoted as R [8]. When R is high the large aggregates are dislocated into dimers, whereas when R is weak they are not.

ATP has no effect on ALA dehydratase, PBG deaminase, or Copro-III oxidase (Figure 3.1, chloroplast envelope) activities but stimulated Uro-III decarboxylase and Protogen oxidase, probably through a kinase-mediated phosphorylation of the enzymes [118]. The phosphorylation state, however, seems important in the case of LPOR as only the phosphorylated enzyme can form large aggregates and insert into the plastid membranes [119,120] (Figure 3.1, chloroplast membrane).

Hormone status influences greening. For instance, cytokinins stimulate Chl synthesis (e.g., Ref. [121]). This augmentation is due to an increase in the activity and mRNA level of Glu-R. The expression of the *lpor* gene is also strongly increased by cytokinins (cucumber [122], moss [123], *Lupinus* [124], tobacco [125]). Cytokinin regulation involves a *cis* element [126] (see above). As the increase in the amount of LPOR mRNA is about four times greater than that of LPOR protein level, it has been suggested that some regulation at the translational or/and posttranslational levels occured.

In the *slender* mutant of barley (a gibberellin [GA]-insensitive overgrowth mutant), the level of LPOR is severely depressed [127]. The decrease affects both LPORA and LPORB mRNAs but not the distribution of the transcripts throughout the leaf. However, the amount of LPOR was not affected and the dark-grown leaves contained plastids with apparently normal prolamellar bodies [128]. As the slender mutant has low levels of biologically active GAs (compared to the wild type), one can hypothesize that in this species the expression of *lpor* is due to the altered hormonal status of the mutant plants. This is confirmed by the increase of *lpor* gene expression observed in cucumber treated with GA.

Except in angiosperms, photosynthetic organisms have at their disposal two enzymatic systems to reduce Pchlide to Chlide: LPOR and DPOR enzymes. Obviously, in the dark only the DPOR can reduce Pchlide, whereas light acts as an on/off switch of the LPOR. Thus, a priori light per se does not impact DPOR activity. So, it is interesting to examine whether LPOR and DPOR can cooperate to supply Chl under illumination.

A study comparing the effects of light intensity on Pchlide reduction in the LPOR-less mutant YF12 and the DPOR-less mutant YFC2 of the cyanobacterium *Plectonema boryanum* demonstrated that DPOR is active when the light intensity is low (approximately 25 $\mu mol\,m^{-2}\,s^{-1}$). Below this value and up to 130 $\mu mol\,m^{-2}\,s^{-1}$, both DPOR and LPOR participate in Chl synthesis, but the activity of DPOR decreases when the light intensity is further increased. Above 130 $\mu mol\,m^{-2}\,s^{-1}$, only LPOR is involved in Pchlide photoreduction [129]. The decrease of the DPOR activity with the increase of the light intensity is not surprising as it will increase the photosynthetic oxygen production to which DPOR is sensitive (see above) [130].

The influence of light intensity on the synthesis of DPOR was investigated in a "yellow-in-the-dark" mutant of the green algae *Chlamydomonas*. In this organism, the synthesis of the subunit ChlL of DPOR is also controlled by the light intensity at the translation level, while the synthesis of the other two polypeptides (ChlB and ChlN) composing DPOR is not modified. The light control would be exerted through the energy state or the redox potential within the chloroplasts [131].

The Shibata shift is inhibited only by low temperature [132], whereas Chlide esterification is inhibited by both low temperature [133] and water deficit [134] (Figure 3.2). A detailed spectroscopic study on the effects of a water deficit on the course of the Shibata shift allowed Le Lay et al. [135] to find an intermediate during the transformation of the large aggregates of Chlide–LPOR–NADPH ternary complexes into dimers (Figure 3.2). This intermediate emits fluorescence at 692 nm.

B. Interactions of Tetrapyrroles with Other Biosynthetic Pathways

As Chl in its free form can cause extensive photooxidative damage under illumination (reviewed in Ref. [64]), Chl formation should be closely coordinated to the synthesis of carotenoids and that of pigment-binding proteins as well (reviewed in Refs. [10,11]).

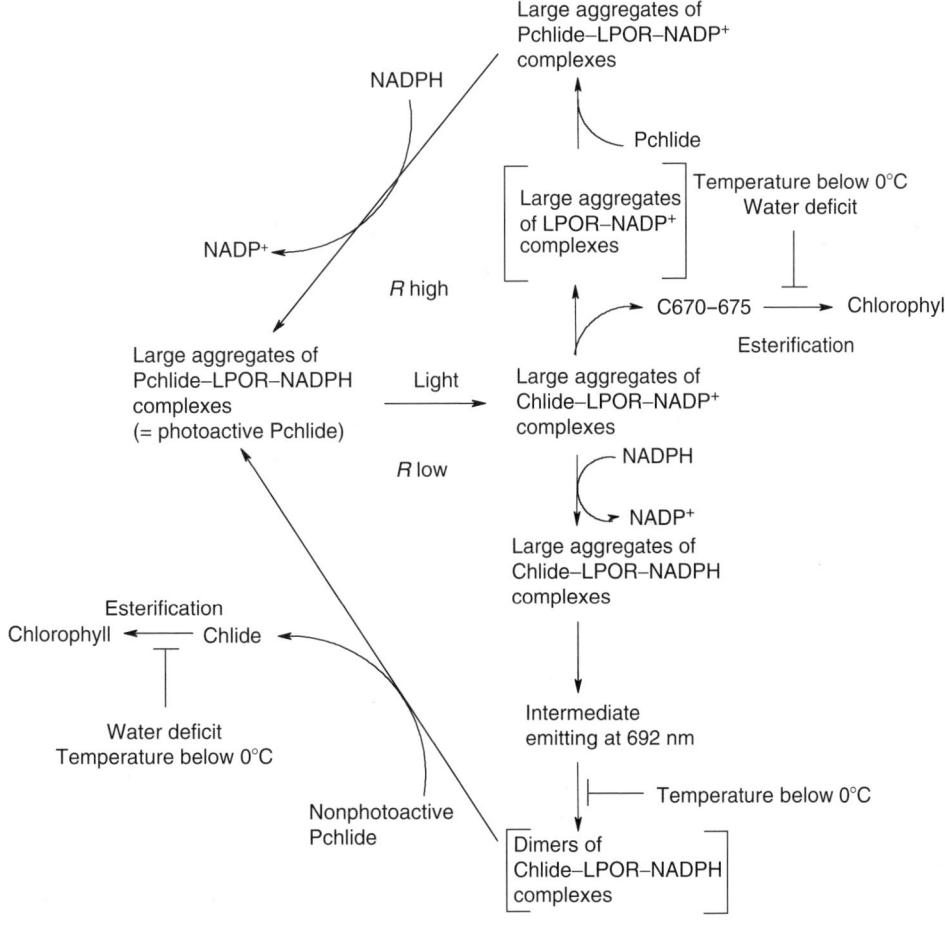

FIGURE 3.2 The Pchlide–Chlide cycles. The brackets indicate a transient state of the pigments. For the others symbols, see Figure 3.1.

As early as 1973, it was shown that accumulation of tetrapyrrole intermediates represses the synthesis of LHCB1 proteins, encoded by a nuclear gene, in dark-grown *C. reinhardtii* [136]. Repression was suppressed in the presence of chloramphenicol and, therefore, one can predict the involvement of a chloroplast-encoded protein in the regulation pathway of *lhcb1* gene expression [136,137]. Later, Mg–Proto-IX-MMe was shown to specifically inhibit expression of *lhcb1* and *rbcS* genes [138–140]. The instability of the mRNA could explain the loss of protein [141]. The inhibition by Mg–Proto-IX-MMe was alleviated under incubation with compounds inhibiting ALA formation [142]. The decrease in ALA would also decrease the formation of Mg–Proto-IX-MMe. Similar results were obtained with cress seedlings [143,144]. Altogether, these experiments have established that Chl precursors are implied in the regulation of the expression of genes involved in other pathways. More recently, it was found in a *chlH* antisense mutant of tobacco that the level of Proto-IX is low, but the *lhcb1* gene expression is also depressed [85]. This experiment establishes that the subunit H of Mg chelatase is also involved in the regulation of *cab* gene expression. This was confirmed by the fact that when Chl synthesis is strongly depressed as in a GSA aminotransferase antisense mutant, the *lhcb1* gene is not affected [145]. Finally, in the *laf6* mutant, in which the level of Proto-IX is high, a high-fluence far-red light reduces the expresion of *lhcb1* gene but remains unaffected under blue, red, low-fluence far-red or white light [49]. Regulation of *lhcb* genes is also mediated by the redox status of plastoquinone [146]. Therefore, *lhcb1* expression does not solely depend on photosynthesis [147]. Altogether, these results suggest that ChlH and Proto-IX-MMe are involved in phytochrome A signaling (Figure 3.1). In contrast to higher plants, a *Chlamydomonas* mutant was defective in the H subunit of Mg chelatase, did not accumulate Proto IX and no reduction in the capacity of ALA synthesis was observed [47]. In the mutant *brs-1* of *Chlamydomonas*, which codes for CHLH but with a +1 frameshift in exon 10 of CHLH, light induction of the chaperone genes *hsp70A* and *hsp70B* is not observed [148,149]. Feeding the mutant with Mg–Proto-IX and Mg–Protogen-IX DME, but not the other tetrapyrroles, mimics the light-activation of the *hsp70* genes and therefore both molecules substitute for the light signal [148]. On the basis of these results, Kropat et al. [150] suggested that Mg–Proto-IX and the ChlH subunit of Mg chelatase take part in the signaling pathway between cytoplasm and nucleus. Regulation of *hsp70* gene expression by Chl precursors was also investigated in *Arabidopsis* lines of mutants presenting defects in ChlH and ChlI [137,151]. The results confirmed the involvement of the H subunit, but not the I subunit, in the nucleus-to-chloroplast signaling. The involvement of Mg–Proto-IX and Mg–Proto-IX-MMe in the regulation of *hsp70* is further indicated by the fact that when dark-grown green algae are transfered to the light, the levels in these precursors increase before that of the corresponding mRNA [150]. This increase in tetrapyrrole precursors does not occur in the presence of cycloheximide. In organisms able to synthesize Chl only in the light, the regulation network may be more complex than in other photosynthetic organisms; in the former organisms the absence of light results in Pchlide accumulation (see Chapter 5). Interestingly, Pchlide may inhibit glutamyl-tRNA ligase [152], an enzyme involved in the synthesis of ALA (Figure 3.1, chloroplast stroma). This way of regulation may not exist in organisms that synthesize Chl in the dark and the eventual accumulation of Pchlide is prevented.

The results, briefly summarized above, have established that Mg–Proto-IX and Mg–Proto-IX-MMe have a role in regulating the expression of some nuclear genes. At present, it is not certain whether the same regulation pathway is used for the regulation of the *cab* and *hsp70* genes. If so, one can propose a federative model explaining the positive and negative effects of these precursors on the expression of nuclear genes. This model is presented in Figure 3.1.

Under a high-light fluence far-red light, phytochrome P_{fr}-form somehow activates the transcription of a putative nuclear-encoded gene (*x* in Figure 3.1, nucleus), which, in turn, activates the transcription of a putative chloroplast-encoded gene (*pf* in Figure 3.1). The product of the *pf* gene would allow the accumulation of Mg–Proto-IX and Mg–Proto-IX-MMe outside the chloroplast (CPL). There the tetrapyrrole precursors may activate positive and negative regulators of the *hsp70* and *cab* gene translations, respectively. The *x* and *pf* genes are postulated since chloramphenicol and cycloheximide block the synthesis of X and PF proteins induced by Chl precursors (see above). In the absence of these proteins the activation of *hsp70* and repression of *lhcb1* genes is not observed. Alternatively, the precursors may be involved in separate ways of regulation.

Under photooxidative conditions, the synthesis of Mg–Proto-IX and Mg–Proto-IX-MMe is reduced and the level of *HEMA* mRNA as well [71]. Therefore, and according to our model, the inhibition of *lhcb1* expression should not be repressed. Quantification of *lhcb1* mRNA shows that under photooxidative stress the cells only contain a low amount of mRNA [153]. Therefore, either the photooxidative *lhcb1* mRNA is highly unstable or there is another regulation pathway for the expression of the *lhcb1* gene. This may involve the subunit H of Mg chelatase [137]. Mochizuki et al. have proposed that ChlH measures the flux at the beginning of the Chl biosynthetic pathway and sends the information about the rate of Chl synthesis to the nucleus. How this occurs remains unclear. It may involve the different "states" of the ChlH subunit, which can exist as a free polypeptide or bound to Proto-IX or Mg–Proto-IX [137].

It has been demonstrated that Chlide (+phytol) is the factor that releases the block in the mRNA translation of plastid-encoded proteins (D1, D2, PSAA, PSAB, etc.) of the photosynthetic apparatus (angiosperms [154–156]) (Figure 3.1). The interdependence of the synthesis of Chl and Chl-binding proteins provides a pool to keep the Chl stable and nontoxic for the cells (see above). As this mechanism of regulation also exists in cyanobacteria [157,158], we can consider it as a "universal" mechanism that photosynthetic cells have evolved to preserve themselves from the production of activated oxygen species produced by free Chl pigments (see also below). In gymnosperms, which synthesize Chl molecules in the dark, this block does not exist in practice, and, therefore, the complete set of pigment–protein complexes composing the photosynthetic apparatus are synthesized in the dark [159].

VIII. EVOLUTION

The reconstruction of the evolution steps of photosynthesis is a difficult task, as it has been evolving since approximately 3.5 billion years. On the basis of the biochemical pathway of Bchl and Chl it was proposed that the actual photosynthetic apparatus derived from green or green-sulfur bacteria. This proposal is known as the Granick hypothesis. Recently, Xiong et al. [160] built phylogenic trees from the comparison of the sequences of genes coding for enzymes involved in Bchl and Chl pathways. They found that the first branching gave purple bacteria, a result that challenges the Granick hypothesis. Reservations about the conclusions of Xiong and collaborators work have been published by Green and Gantt [161].

The study of biological evolution and the understanding of some mechanisms involved in the appearance of new structures with new functions indicates that LPOR might have had another major role in plants than the one observed today (for a review, see Ref. [10]). The identification of two or more expressed forms in pine species suggests that gene duplication and divergence of LPORA and LPORB function may have taken place prior to the divergence of gymnosperms and angiosperms. Evidence of gene duplication and divergence in function prior to the angiosperm–gymnosperm split has been previously reported for several other gene families encoding photosynthesis-related proteins (e.g., LHCb [162]).

Chl *a* monooxygenase, the enzyme catalyzing the oxidation of Chl *a* to Chl *b*, has been identified in higher plants [58,61] and in two Prochlorophytes, namely, *P. hollandica* and *P. didemni*, but not in *Prochlorococus* MED4 and MIT 9313, although the last two organisms are able to synthesize Chl *b*. This finding is in direct conflict with the endosymbiotic theory, which teaches that ancestral genes entered eukaryotes via the cyanobacterial-like endosymbionic progenitor to plastids [60]. As Chl *a* monooxygenase has a particular enzymatic mechanism [61,163], a search for all putative oxygenases genes in the *Prochlorococus* genomes that could show some — even weak — homologies with the Chl monooxygenase gene was performed. One candidate with putative binding sites for [2Fe–2S] Rieske center and mononuclear iron was found. Both domains are essential for Chl *a* monooxygenase activity. The sequence of this gene can only be used for phylogenetic analysis if the most variable regions are taken out. Under this condition, a stable position for the *Prochlorococus* Chl *a* monooxygenases was found. The tree branches at the base, but the *Prochlorococus* Chl *a* monooxygenases are part of the same sequence cluster [163]. Such a level of similarity could have been driven by the constraints of this particular biochemical reaction alone, starting with a gene coding for some kind of monooxygenase. That such a hypothetical convergent evolution did not result in an enzyme more related to the other Chl *a* monooxygenases may be explained by the fact that *Prochlorococus* Chl *a* monooxygenase uses DV-Chl *a*, whereas the other enzymes utilize MV-Chl *a* as substrate [163].

IX. PERSPECTIVES

Much progress has been made in the understanding of the mechanisms of enzymatic conversion of intermediates of the Chl biosynthetic pathway and of its regulation. It has become evident that some intermediates, like Proto-IX, are involved in the signaling pathway between the chloroplast and the nucleus. Additional work is now needed to determine whether other components of the pathways — like the subunit H of Mg chelatase — are involved in the regulation network of tetrapyrrole synthesis.

Some progress in the understanding of the formation of the large aggregates of photoactive Pchlide have been obtained using mathematical analysis of spectroscopic data. Although it seems obvious that the spectral characteristics of the pigment must reflect its immediate environment, the relationship between absorption and emission maxima on the one hand and the molecular composition and organization of the pigment–protein complexes on the other can be difficult to establish. Additional work will be necessary to isolate and characterize the different spectral forms of pigment–LPOR complexes to correlate them with their spectroscopic properties. The fact that the same spectral forms of Pchlide are found in angiosperm and in gymnosperm tissues suggests that the large aggregates of Pchlide–LPOR complexes are formed along a conserved process transmitted from gymnosperms. It would be interesting to determine if this pathway has been inherited from lower organisms like ferns, algae, cyanobacterium, etc., which also have LPOR but usually do not accumulate Pchlide.

REFERENCES

1. Schoefs B, Bertrand M. Chlorophyll biosynthesis. In: Pessarakli M, ed. *Handbook of Photosynthesis*. New York: Marcel Dekker, 1997:49–69.
2. Armstrong GA. Greening in the dark: light-independent chlorophyll biosynthesis from anoxygenic bacteria to gymnosperms. *J. Photochem. Photobiol. B* 1998; 43:87–100.
3. Lebedev N, Timko MP. Protochlorophyllide reduction. *Photosynth. Res.* 1998; 58:5–23.
4. Beale SI. Enzymes of the chlorophyll biosynthesis. *Photosynth. Res.* 1999; 60:43–73.
5. Schoefs B. The light-dependent and the light-independent reduction of protochlorophyllide *a* to chlorophyllide *a*. *Photosynthetica* 1999; 36:481–496.
6. Porra RJ, Scheer H. ^{18}O and mass spectrometry in chlorophyll research: derivation and loss of oxygen atom at the periphery of the chlorophyll macrocycle during biosynthesis, degradation and adaptation. *Photosynth. Res.* 2000; 66:159–175.
7. Papenbrock J, Grimm B. Regulatory network of tetrapyrrole biosynthesis — studies of intracellular signalling involved in metabolic and developmental control of plastids. *Planta* 2001; 213:667–681.
8. Schoefs B. The light-dependent protochlorophyllide reduction: from a photoprotecting mechanism to a metabolic reaction. In: Panadalai, ed. *Recent Research in Photosynthesis*. Vol. 2. Trivandrum: Research Signpost, 2001:241–258.
9. Schoefs B. The protochlorophyllide-chlorophyllide cycle. *Photosynth. Res.* 2001; 70:257–271.
10. Schoefs B, Franck F. Chlorophyll biosynthesis: light-dependent and light-independent protochlorophyllide reduction. *Bull. Cl. Sci. Acad. R. Belgique* 2002; 13:113–157.
11. Brusslan JA, Peterson MP. Tetrapyrrole regulation of nuclear gene expression. *Photosynth. Res.* 2002; 71:185–194.
12. Brody SS, Gough SP, Kannangara CG. Predicted structure and fold recognition for the glutamyl-TRNA reductase family of protein. *Proteins* 1999; 37:483–493.
13. Vothknecht UC, Kannangara CG, von Wettstein D. Barley glutamyl tRNA^{Glu} reductase: mutations affecting haem inhibition and enzyme activity. *Phytochemistry* 1998; 47:513–519.
14. Zhang L, Guarente L. Heme binds to a short sequence that serve as regulatory function in diverse proteins. *EMBO J.* 1995; 14:313–320.
15. Hermann CA, Im CS, Beale SI. Light-regulated expression of the *GSA* gene encoding the chlorophyll biosynthetic enzyme glutamate 1-semialdehyde aminotransferase in carotenoid-deficient *Chlamydomonas reinhardtii* cells. *Plant Mol. Biol.* 1999; 30:289–297.
16. Loida PJ, Thompson RL, Walker DM, Jacob CA. Novel inhibitors of glutamyl-tRNA(Glu) reductase activity by benzyladenine in greening cucumber cotyledons. *Plant Cell Physiol.* 1999; 36:1237–1243.
17. Shoolingin-Jordan PM, Cheung K-M. Biosyntheis of heme. In: Barton D, Nakanishi K, Meth-Cohn O, eds. *Comprehensive Natural Products Chemistry*. Oxford: Elsevier Science, 1999:61–107.
18. Shalygo NV, Mock HP, Averina NG, Grimm B. Photodynamic action of uroporphyrin and protochlorophyllide in greening barley leaves treated with cesium chloride. *J. Photochem. Photobiol. B* 1998; 42:151–158.
19. Louie GV, Brownlie PD, Lambert R, Cooper JB, Blundell TL, Wood SP, Malashkevich VN, Hädener A, Warren MJ, Shoolingin-Jordan PM. The three-dimensional structure of *Escherichia coli* porphobilinogen deaminase at 1.76-Å resolution. *Proteins Struct Funct Genet* 1996; 25:48–78.
20. Hädener A, Matzinger PR, Battersby AR, McSweeney S, Thompson AW, Hammersley, Harrop SJ, Cassetta A, Deacon A, Hunter WN, Nieh YP, Raffery J, Hunter N, Halliwell JR. Determination of the structure of selenomethionine-labelled hydroxymethylbilane synthase in its active form by multi-wavelength anomalous dispersion. *Acta Crystallogr.* 1999; 55D:631–643.
21. Battersby AR, Leeper FJ. Biosynthesis of the pigments of life — Mechanistic studies on the conversion of porphobilinogen to uroporphyrinogen III. *Chem. Rev.* 1990; 90:1261–1267.
22. Phillips JD, Parker TL, Schubert HL, Whitby FG, Kushner JP. Functional consequences of naturally occurring mutations in human uroporphyrinogen decarboxylase. *Blood* 2001; 98:3179–3185.
23. Mathews MAA, Schubert HL, Whitby FG, Alexander KJ, Schadick K, Bergonia HA, Phillips JD, Hill CP. Crystal structure of human uroporphyrinogen III synthase. *EMBO J.* 2001; 20:5832–5839.
24. Roessner CA, Ponnamperuma K, Scott AI. Mutagenesis identifies a conserved tyrosine residue important for the activity of uroporphyrinogen III synthase from *Anacystis nidulans*. *FEBS Lett.* 2002; 525:25–28.

25. Phillips JD, Whitby FG, Kushner JP, Hill CP. Characterization and crystallization of human uroporphyrinogen decarboxylase; X-ray diffraction. *Protein Sci.* 1997; 6:1343–1346.

26. Santana MA, Tan F-C, Smith AG. Molecular characterisation of coproporphyrinogen oxidase from soybean (*Glycine max*) and *Arabidopsis thaliana*. *Plant Physiol. Biochem.* 2002; 40:289–298.

27. Lermontova I, Kruse E, Mock HP, Grimm B. Cloning and characterization of plastidial and mitochondrial isoform of tobacco protoporphyrinogen IX oxidase. *Proc. Natl. Acad. Sci. USA* 1997; 94:8895–8900.

28. Adomat C, Böger P. Cloning, sequence and characterization of protoporphyrinogen IX oxidase from chicory. *Pestic. Biochem. Physiol.* 1999; 66:49–62.

29. Dailey HA, Meisner P, Dailey HA. Expression of a clones protoporphyrinogen oxidase. *J. Biol. Chem.* 1994; 271:8714–8718.

30. Choi KW, Han O, Lee HJ, Yun YC, Moon YM, Kim M, Kuk YI, Han SU, Guh JO. Generation of resistance to the diphenyl herbicide, oxyfluorfen, via expression of *Bacillus subtilis* protoporphyrinogen oxidase gene in transgenic tobacco plants. *Biosci. Biotechnol. Biochem.* 1998; 62:558–560.

31. Lee HJ, Lee SB, Chung JS, Han SU, Han O, Guh JO, Jeon J, An G, Back K. Transgenic rice plants expressing a *Bacillus subtilis* protoporphyrinogen oxidase gene are resistant to diphenyl ether herbicide oxyfluorfen. *Plant Cell Physiol.* 2000; 41:743–749.

32. Zsebo KM, Hearst, JE. Genetic-physical mapping of a photosynthetic gene cluster in *R. capsulatus*. *Cell* 1984; 37:937–947.

33. Willows RD, Gibson LC, Kannangara CG, Hunter CN, von Wettstein D. Three separate proteins constitute the magnesium chelatase of *Rhodobacter sphaeroides*. *Eur. J. Biochem.* 1996; 235:438–443.

34. Kruse E, Mock HP, Grimm B. Isolation and characterisation of tobacco (*Nicotiana tabacum*) cDNA clones encoding proteins involved in magnesium chelation into protoporphyrin IX. *Plant Mol. Biol.* 1997; 35:1053–1056.

35. Papenbrock J, Gräfe S, Kruse E, Hänel F, Grimm B. Mg-chelatase of tobacco: identification of a *ChlD* cDNA sequence encoding a third subunit, analysis of the interaction of the three subunits with the east two-hybrid system and reconstitution of the enzyme activity by co-expression of recombinant ChlD, ChlH and ChlI. *Plant J.* 1997; 12:981–990.

36. Walker CJ, Willows RD. Mechanism and regulation of Mg-chelatase. *Biochem. J.* 1997; 327:321–333.

37. Jensen PE, Willows RD, Petersen BL, Vothknecht UC, Stummann BM, Kannangara CG, von Wettstein D, Henningsen KW. Structural genes for Mg-chelatase subunits in barley: Xantha-f, -g and -h. *Mol. Gen. Genet.* 1996; 250:383–394.

38. Kannangara CG, Vothknecht UC, Hansson M, von Wettstein D. Magnesium chelatase association with ribosomes and mutant complementation studies identify barley subunit Xantha-G as a functional counter-

part of *Rhodobacter* subunit BchD. *Mol. Gen. Genet.* 1997; 254:85–92.

39. Jensen PE, Gibson LCD, Hunter CN. Determinants of catalytic activity with the use of purified I, D and H subunits of the magnesium protoporphyrin IX chelatase from *Synechocystis* PCC6803. *Biochem. J.* 1998; 334:335–344.

40. Fodje MN, Hansson M, Hansson N, Olsen JG, Gough S, Willows RD, Al-Karadaghi S. Interplay between an AAA module and an integrin domain may regulate the function of magnesium chelatase. *J. Mol. Biol.* 2001; 311:111–122.

41. Hansson A, Willows RD, Roberts TH, Hansson M. Three semidominant barley mutants with single amino acid substitutions in the smallest magnesium chelatase subunit form defective AAA$^+$ hexamers. *Proc. Natl. Acad. Sci. USA* 2002; 99:13944–13949.

42. Hansson M, Kannangara CG, von Wettstein D, Hansson M. Molecular basis for semidominance of missense mutations in the XANTHA-H (42 kDa) subunit of magnesium chelatase. *Proc. Natl. Acad. Sci. USA* 1999; 96:1744–1749.

43. Hansson M, Kannangara CG. ATPases and phosphate exchange activities in magnesium chelatase subunits of *Rhodobacter sphaeroides*. *Proc. Natl. Acad. Sci. USA* 1997; 94:13351–13356.

44. Karger GA, Reid JD, Hunter CN. Characterization of the binding of deuteroporphyrin IX to the magnesium chelatase H subunit and spectroscopic properties of the complex. *Biochemistry* 2001; 40:9291–9299.

45. Ogura T, Wilkinson AT. AAA$^+$ superfamily ATPases: common structures — diverse function. *Genes Cells* 2001; 6:575–587.

46. Song HK, Hartmann C, Ramachandran R, Bochtler M, Behrendt R, Morodes L, Huber R. Mutational studies of HslU and its docking mode with HslV. *Proc. Natl. Acad. Sci. USA* 2000; 97:14103–14108.

47. Chekounova E, Voronetskaya V, Papenbrock J, Grimm B, Beck CF. Characterization of *Chlamydomonas* mutants defective in the H subunit of Mg-chelatase. *Mol. Gen. Genet.* 2001; 266:363–373.

48. Jensen PE, Reid JD, Hunter CN. Modification of cysteine residues in the CHLI and CHLL subunits of magnesium chelatase results in enzyme inactivation. *Biochem. J.* 2000; 352:435–441.

49. Moller SG, Kunkel T, Chua N-H. A plastidic ABC protein involved in intercompartmental communication of light signaling. *Genes Dev.* 2001; 15: 90–103.

50. Jeffrey SW, Vesk M. Introduction to marine phytoplankton and their pigment signatures. In: Jeffrey SW, Mantoura RFC, Vesk M, eds. *Phytoplankton Pigments in Oceanography*. Paris: UNESCO, 1997:37–84.

51. Kim JS, Rebeiz CA. Origin of the chlorophyll a biosynthetic heterogeneity in higher plants. *J. Biochem. Mol. Biol.* 1996; 29:327–334.

52. Tripathy BC, Rebeiz CA. Chloroplast biogenesis 60. Conversion of divinyl protochlorophyllide to monovinyl protochlorophylide in green(ing) barley, a dark

monovinyl/light divinyl plant species. *Plant Physiol.* 1988; 87:89–94.

53. Parham R, Rebeiz CA. Chloroplast biogenesis 72: a [4-vinyl]chlorophyllide a reductase assay using divinyl chlorophyllide a as an exogenous substrate. *Anal. Biochem.* 1995; 231:164–169.

54. Andra AN, Rebeiz CA. Chloroplast biogenesis 81: transient formation of divinyl chlorophyll *a* following a 2.5 ms light flash treatment of etiolated cucumber cotyledons. *Photochem. Photobiol.* 1998; 68:852–856.

55. Kim JS, Klossov V, Rebeiz CA. Chloroplast biogenesis 76: regulation of 4-vinyl reduction during conversion of divinyl-Mg-protophorphyrin IX to monovinyl protochlorophyllide a is controlled by plastid membranes and stromal factors. *Photosynthetica* 1997; 34:569–581.

56. Kolossov VL, Rebeiz CA. Chloroplast biogenesis 84: solubilization and partial purification of membrane-bound [4-vinyl] chlorophyllide *a* reductase from etiolated barley leaves. *Anal. Biochem.* 2001; 295: 214–219.

57. Shlyk AAL. Biosynthesis of chlorophyll b. *Annu. Rev. Plant Physiol.* 1971; 22:169–184.

58. Tanaka A, Ito H, Tanaka R, Tanaka NK, Yoshida K, Okada K. Chlorophyll a oxygenase (CAO) is involved in chlorophyll b formation from chlorophyll *a. Proc. Natl. Acad. Sci. USA* 1998; 95:12719–12723.

59. Espineda CE, Lindford AS, Devine D, Brusslan JA. The *AtCAO* gene, encoding chlorophyll *a* oxygenase, is required for chlorophyll *b* synthesis in *Arabidopsis thaliana. Proc. Natl. Acad. Sci. USA* 1999; 76: 10507–10511.

60. Tomitani A, Okada K, Miyashita H, Matthijs HC, Ohno T, Tanaka A. Chlorophyll b and phycobilins in the common ancestor of cyanobacteria and chloroplasts. *Nature* 1999; 400:159–162.

61. Oster U, Tanaka R, Tanaka A, Rüdiger W. Cloning and functional expression of the gene encoding the key enzyme for chlorophyll b biosynthesis (CAO) from *Arabidopsis thaliana. Plant J.* 2000; 21:301–305.

62. Schoefs B, Bertrand M, Lemoine Y. Changes in the photosynthetic pigments in bean leaves during the first photoperiod of greening and the subsequent dark-phase. Comparison between old (10-d-old) leaves and young (2-d-old) leaves. *Photosynth. Res.* 1998; 57:203–213.

63. Kruse E, Mock HP, Grimm B. Reduction of coproporphyrinogen oxidase level by antisense RNA synthesis leads to deregulated gene expression of plastid proteins and affects the oxidation defense systems. *EMBO J.* 1995; 14:3710–3720.

64. Bertrand M, Schoefs B. Photosynthetic pigment metabolism in plants during stress. In: Pessarakli M, ed. *Handbook of Plant and Crop Stress.* New York: Marcel Dekker, 1999:527–543.

65. Boo YC, Lee KP, Jung J. Rice plants with a high protochlorophyllide accumulation show oxidative stress in low light that mimics water stress. *J. Plant Physiol.* 2000; 157:405–411.

66. Shalygo NV, Mock HP, Averina NG, Grimm B. Comparative analysis of the low molecular weight and enzymatic antioxidants in response to the phototoxicity of accumulating uroporphyrin and protochlorophyllide in barley leaves treated with cesium chloride. *Photosynth. Res.* 2000; 64:267–276.

67. Kruse E, Grimm B, Beator J, Kloppstech K. Developmental and circadian control of the capacity for δ-aminolevulinic acid synthesis in green barley. *Planta* 1997; 202:235–241.

68. Ilag LL, Kumar AM, Söll D. Light regulation of chlorophyll biosynthesis at the level of 5-aminolevulinic formation in *Arabidopsis. Plant Cell* 1994; 6:265–275.

69. Meskauskiene R, Nater M, Goslings D, Kessler F, Op den Camp R, Apel K. FLU: a negative regulator of chlorophyll biosynthesis in *Arabidopsis thaliana. Proc. Natl. Acad. Sci. USA* 2001; 98:12826–12831.

70. Kumar AM, Chaturvedi S, Söll D. Selective inhibition of *HEMA* gene expression by photooxidation in *Arabidopsis thaliana. Phytochemistry* 1999; 51: 847–850.

71. McCormac AC, Fischer A, Kumar AM, Söll D, Terry MJ. Regulation of *HEMA1* expression by phytochrome and a plastid signal during de-etiolation in *Arabidopsis thaliana. Plant J.* 2001; 25:549–561.

72. McCormac AC, Terry MJ. Light-signalling pathways leading to the co-ordinated expression of *HEMA1* and Lhcb during chloroplast development in *Arabidopsis thaliana. Plant J.* 2002; 32:549–559.

73. McCormac AC, Terry MJ. Loss of nuclear gene expression during the phytochrome A-mediated far-red block of greening response. *Plant Physiol.* 2002; 130:402–414.

74. Sperling U, Franck F, van Cleve B, Frick G, Apel K, Armstrong G. Etioplast differentiation in *Arabidopsis*: both PORA and PORB restore the prolamellar body and photoactive Pchlide-F655 to the *cop1* photomorphogenic mutant. *Plant Cell* 1998; 10:283–296.

75. Bougi O, Grimm B. Members of low-copy number of gene family glutamyl-tRNA reductase are differentially expressed in barley. *Plant J.* 1996; 9:867–878.

76. Kumar AM, Schaub U, Söll D, Ujwal ML. Glutamyl-transfer RNA: at the crossroad between chlorophyll and protein synthesis. *Trends Plant Sci.* 1996; 1:371–376.4

77. Tanaka R, Yoshida K, Nakayashiki T, Masuda T, Tsuji H, Inokuchi H, Tanaka A. Differential expression of two *hemA* mRNAs encoding glutamyl-tRNA reductase proteins in greening cucumber seedlings. *Plant Physiol.* 1996; 110:1223–1230.

78. Tanaka R, Yoshida K, Nakayashiki T, Tsuji H, Inokuchi H, Okada K, Tanaka A. The third member of the *hemA* gene family encoding glutamyl-tRNA reductase is primarily expressed in roots of *Hordeum vulgare. Photosynth. Res.* 1997; 53:161–171.

79. Jacobs JM, Jacobs NJ, DeMaggio AE. Protoporphyrinogen oxidation in chloroplast and plant mitochondria, a step in heme and chlorophyll biosynthesis. *Arch. Biochem. Biophys.* 1982; 218:233–239.

80. Chow K-S, Singh DP, Roper JM, Smith AG. A single precursor protein for ferrochelatase-I form from *Arabidopsis* is imported *in vitro* into both chloroplast and mitochondria. *J. Biol. Chem.* 1997; 272:27565–27571.

81. Papenbrock J, Mock HP, Kruse E, Grimm B. Expression studies in tetrapyrrole biosynthetic: inverse maxima of magnesium chelatase and ferrochelatase activity during cyclic photoperiods. *Planta* 1999; 208:264–273.

82. Cornah JE, Roper JM, Pal Singh D, Smith AG. Measurement of ferrochelatase activity using a novel assay suggests that plastids are the major site of haem biosynthesis in both photosynthetic and non-photosynthetic cells of pea (*Pisum sativum L.*). *Biochem. J.* 2002; 362:423–432.

83. Gibson LCD, Marrisson JL, Leech RM, Jensen PE, Bassham DC, Gibson M, Hunter CN. A putative Mg-chelatase subunit from *Arabidopsis thaliana* cv. C24. *Plant Physiol.* 1996; 111:61–71.

84. Nakayama M, Masuda T, Bando T, Yamagata H, Ohta H, Takamiya K-I. Cloning and expression of the soybean *chlH* gene encoding a subunit of Mg-chelatase and localization of the Mg^{2+} concentration-dependent ChlH protein within the chloroplast. *Plant Cell Physiol.* 1998; 39:275–284.

85. Papenbrock J, Mock HP, Tanaka R, Kruse E, Grimm B. Role of magnesium chelatase activity in the early steps of the tetrapyrrole biosynthetic pathway. *Plant Physiol.* 2000; 122:1161–1169.

86. Meskauskiene R, Apel K. Interaction of FLU, a negative regulator of tetrapyrrole synthesis, with the glutamyl-tRNA reductase requires the tetratricopeptide repeat domain of FLU. *FEBS Lett.* 2002; 532:27–30.

87. Lupas A. Coils coils: new structures and new functions. *Trends Biochem. Sci.* 1996; 21:375–382.

88. Blatch GL, Lässle M. The tetratricopeptide repeat: a structural motif mediating protein-protein interactions. *BioEssays* 1999; 21:932–939.

89. Korneef M, Roff E, Spruit CJP. Genetic control of light inhibited hypocotyl elongation in *Arabidopsis thaliana* (L.) Heynh. *Z. Pflanzenphyiol.* 1980; 100:147–160.

90. Korneef M, Cone JW, Dekens RG, O'Herne-Roberse H, Spruit CJP, Kendrick REC. Photomorphogenic responses of long-hypocotyl mutants of tomato. *J Plant Physiol.* 1985; 120:153–165.

91. Chory J, Peto C, Feinbaum R, Pratt L, Ausubel F. Arabidopsis thaliana mutant that develops as a light-grown plant in the absence of light. *Cell* 1989; 58:991–999.

92. Davis SJ, Kurepa J, Vierstra RDC. The *Arabidopsis thaliana* HY1 locus required for phytochrome-chromophore biosynthesis, encodes a protein related to heme oxygenases. *Proc. Natl. Acad. Sci. USA* 1999; 96:6541–6546.

93. Muramoto T, Kohchi T, Yokota A, Hwang I, Goodman HM. The *Arabidopsis* photomorphogenic mutant *hy1* is deficient in phytochrome chromophore biosynthesis as a result of a mutation in a plastid heme oxygenase. *Plant Cell* 1999; 11:335–348.

94. Brücker G, Zeidler M, Kohdi T, Hartmann E, Lamparter T. Microinjection of heme oxygenase genes rescues phytochrome-deficient mutants of the moss *Ceratodon purpureus*. *Planta* 2000; 210:520–535.

95. Willows RD, Mayer SM, Falk MS, DeLong A, Hansson K, Chory J, Beale SI. Phytobilin synthesis: the *Synechocystis* sp. PCC 6803 heme oxygenase-encoding *ho1* gene complements a phytochrome-deficient *Arabidopsis thaliana hy1* mutant. *Plant Mol. Biol.* 2000; 43:113–120.

96. Terry MJ, Kendrick RE. Feedback inhibition of chlorophyll synthesis in the phytochrome chromophore deficient *aurea* and *yellow-green-2* mutants of tomato. *Plant Physiol.* 1999; 119:143–152.

97. Armstrong GA, Runge S, Frick G, Sperling U, Apel K. Identification of NADPH:protochlorophyllide oxidoreductases A and B branched pathway for light-dependent chlorophyll synthesis in *Arabidopsis thaliana*. *Plant Physiol.* 1995; 108:1505–1517.

98. Holtorf H, Reinbothe S, Reinbothe R, Bereza B, Apel K. Two routes of chlorophyllide synthesis that are differentially regulated by light in barley (*Hordeum vulgare* L.). *Proc. Natl. Acad. Sci. USA* 1995; 92:3254–3258.

99. Reinbothe S, Reinbothe C, Holtorf H, Apel K. Two NADPH:protochlorophyllide oxidoreductase in barley: evidence for the selective disappearance of PORA during the light-induced greening of etiolated seedlings. *Plant Cell* 1995; 7:1933–1940.

100. Forreiter C, Apel K. Light-independent and light-dependent protochlorophyllide-reducing activities and two distinct NADPH-protochlorophyllide oxidoreductase polypeptides in mountain pine (*Pinus mungo*). *Planta* 1993; 190:536–545.

101. Skinner JS, Timko MP. Loblolly pine (*Pinus taeda* L.) contains multiple expressed encoding light-dependent NADPH:protochlorophyllide oxidoreductase (POR). *Plant Cell Physiol.* 1998; 39:795–806.

102. Oosawa N, Masuda T, Awai K, Fusada N, Shimada H, Ohta H, Takamiya K. Identification and light-induced expression of a novel gene of NADPH-protochlorophyllide oxidoreductase isoform in *Arabidopsis thaliana*. *FEBS Lett.* 2000; 474:133–136.

103. Spano AJ, He Z, Michel H, Hunt DF, Timko MP. Molecular cloning, nuclear gene structure and developmental expression of NADPH:protochlorophyllide oxidoreductase in pea (*Pisum sativum* L.). *Plant Mol. Biol.* 1992; 18:967–972.

104. He Z-H, Li J, Sundqvist C, Timko MP. Leaf developmental age controls expression of genes encoding enzymes of chlorophyll and haem biosynthesis in pea (*Pisum sativum* L.). *Plant Physiol.* 1994; 106:537–543.

105. Kuroda H, Masuda T, Ohta H, Shioi Y, Takamiya K. Light-enhanced gene expression of NADPH-protochlorophyllide oxidoreductase in cucumber. *Biochem. Biophys. Res. Commun.* 1995; 210:310–316.

106. Kuroda H, Masuda T, Fusada N, Ohta H, Takamiya K. Expression of NADPH-protochlorophyllide oxidoreductase gene in fully green leaves of cucumber. *Plant Cell Physiol.* 2000; 41:226–229.

107. Fusada N, Masuda T, Kuroda H, Shiraishi T, Shimada H, Ohta H, Takamiya K. NADPH-protochlorophyllide oxidoreductase in cucumber is encoded by a single gene and its expression is transcriptionally enhanced by illumination. *Phytosynth. Res.* 2000; 64:147–154.

108. Yoshida K, Chen R-M, Tanaka A, Teramoto H, Tanaka R, Timko MP, Tsuji H. Correlated changes in the activity, amount of protein, and abundance of transcript of NADPH:protochlorophyllide oxidoreductase and chlorophyll accumulation during greening of cucumber cotyledons. *Plant Physiol.* 1995; 109:231–238.

109. Keegstra K, Cline K. Protein import and routing systems of chloroplasts. *Plant Cell* 1999; 11:557–570.

110. Dalbey RE, Robinson C. Protein translocation into and across the bacterial plasma membrane and the plant thylakoid membrane. *Trends Biochem. Sci.* 1999; 24:17–22.

111. May T, Soll J. Chloroplast precursor protein translocon. *FEBS Lett.* 1999; 452:52–56.

112. Jarvis P, Söll J. Toc, tic, and chloroplast protein import. *Biochim. Biophys. Acta* 2001; 1542:64–79.

113. Aronsson H, Sohrt K, Soll J. NADPH: protochlorophyllide oxidoreductase uses the general import route into chloroplasts. *J. Biol. Chem.* 2000; 381: 1263–1267.

114. Dabney-Smith C, van Den Wijngaard PW, Treece Y, Vredenberg WJ, Bruce BD. The C terminus of a chloroplast precursor modulates its interaction with the translocation apparatus and PIRAC. *J. Biol. Chem.* 1999; 274:32351–32359.

115. Aronsson H, Sundqvist C, Timko MP, Dahlin C. The importance of the C-terminal region and Cys residues for the membrane association of NADPH:protochlorophyllide oxidoreductase in pea. *FEBS Lett.* 2000; 502:11–15.

116. Reinbothe S, Runge S, Reinbothe C, Van Cleve B, Apel K. Substrate-dependent transport of the NADPH-protochlorophyllide oxidoreductase into isolated plastids. *Plant Cell* 1995; 7:161–172.

117. Aronsson H, Almkvist J, Sundqvist C, Timko MP, Dahlin C. Characterization of the plastid import reaction of the pea NADPH:protochlorophyllide oxidoreductase (POR). In: Argyroudi-Akoyunopglou JH, Senger H, eds. *The Chloroplast: from Molecular Biology to Biotechnology.* Dordrecht: Kluwer Academic Publishers, 1999:167–170.

118. Manohara MS, Tripathy BC. Regulation of protoporphyrin IX biosynthesis by intraplastidic compartmentalization and adenosine triphosphate. *Planta* 2000; 212:52–59.

119. Dahlin C, Sundqvist C, Timko MP. The *in vitro* assembly of the NADPH-protochlorophyllide oxidoreductase in pea chloroplasts. *Plant Mol. Biol.* 1995; 29:317–330.

120. Kovacheva S, Ryberg M, Sundqvist C. ADP/ATP and protein phosphorylation dependence of phototransformable protochlorophyllide in isolated etioplast membranes. *Photosynth. Res.* 2000; 64:127–136.

121. Fletcher RAA, McCullogh D. Benzyladenine as a regulator of the chlorophyll synthesis in cucumber cotyledons. *Can. J. Bot.* 1971; 49:2197–2201.

122. Kuroda H, Masuda T, Ohta H, Shioi Y, Takamiya K-I. Effects of light, development age and phytohormones on the expression of NADPH-protochlorophyllide oxidoreductase gene in *Cucumis sativus*. *Plant Physiol. Biochem.* 1996; 34:17–22.

123. Macuka J, Bashiardes S, Ruben E, Spooner K, Cuming A, Knight C, Cove D. Sequence analysis of expressed sequence tags from an ABA-treated cDNA library identifies stress responses genes in the moss *Phycomitrella patens*. *Plant Cell Physiol.* 1999; 40:378–387.

124. Kusnetsov V, Herrmann RG, Kulaeva ON, Oelmüller R. Cytokinin stimulates and abscisic acid inhibits greening in etiolated *Lupinus luteus* cotyledons by affecting the expression of the light-sensitive protochlorophyllide oxidoreductase. *Mol. Gen. Genet.* 1998; 259:21–28.

125. Zavaleta-Mancera HA, Franklin KA, Ougham HJ, Thomas H, Scott IM. Regreening of scenescent *Nicotiana* leaves. I. Reappearance of NADPH-protochlorophyllide oxidoreductase and light-harvesting chlorophyll a/b-binding protein. *J. Exp. Bot.* 1999; 50:1677–1682.

126. Kuroda H, Masuda T, Fusada N, Ohta H, Takamiya K. Cytokinin-induced transcriptional activation of NADPH-protochlorophyllide oxidoreductase gene in cucumber. *J. Plant Res.* 2001; 114:1–7.

127. Schünmann PHD, Ougham HJ. Identification of three cDNA clones expressed in the leaf extension zone and with altered patterns in the *slender* mutant of barley: a tonoplast intrinsic protein, a putative structural protein and protochlorophyllide oxidoreductase. *Plant Mol. Biol.* 1996; 31:529–537.

128. Ougham HJ, Thomas AM, Thomas BJ, Frick GA, Armstrong GA. Both light-dependent protochlorophyllide oxidoreductase A and protochlorophyllide oxidoreductase B are down-regulated in the *slender* mutant of barley. *J. Exp. Bot.* 2001; 52:1447–1454.

129. Fujita Y, Takagi H, Hase T. Cloning of the gene encoding a protochlorophyllide reductase: the physiological significance of co-existence of light-dependent and -independent protochlorophyllide reduction systems in the cyanobacterium *Plectonema boryanum*. *Plant Cell Physiol.* 1998; 39:177–185.

130. Fujita Y, Bauer CE. Reconstitution of light-independent protochlorophyllide reductase from purified BchL and BchN-BchB subunits. *In vitro* confirmation of nitrogenase-like features of a bacteriochlorophyll biosynthesis enzymes. *J. Biol. Chem.* 2000; 275:23583–23588.

131. Cahoon AB, Timko MP. Yellow-in-the-dark mutants of *Chlamydomonas* lacks the ChlL subunit of light-independent protochlorophyllide reductase. *Plant Cell* 2000; 12:559–568.

132. Eullaffroy P, Salvetat R, Franck F, Popovic R. Temperature dependence of chlorophyll(ide) spectral shifts and photoactive protochlorophyllide regeneration

after flash in etiolated barley leaves. *J. Photochem. Photobiol. B* 1995; 62:751–756.

133. Schoefs B, Bertrand M. The formation of chlorophyll from chlorophyllide in leaves containing proplastids is a four-step process. *FEBS Lett.* 2000; 486:243–246 (erratum appears in *FEBS Lett.* 2001; 494:261).

134. Le Lay P, Eullaffroy P, Juneau P, Popovic R. Evidence of chlorophyll synthesis pathway alteration in desiccated barley leaves. *Plant Cell Physiol.* 2000; 41:565–570.

135. Le Lay P, Böddi B, Kovacevic D, Juneau P, Dewez D, Popovic R. Spectroscopic analysis of desiccation-induced alterations of the chlorophyllide transformation pathway in etiolated barley leaves. *Plant Physiol.* 2001; 127:202–211.

136. Hoober JK, Stegman WJ. Control of the synthesis of major polypeptide of chloroplast membranes in *Chlamydomonas reinhardtii*. *J. Cell Biol.* 1973; 56: 1–12.

137. Mochizuki N, Brusslan JA, Larkin R, Nagatani A, Chory J. *Arabidopsis genomes uncoupled 5 (GUN5)* mutant reveals the involvement of Mg-chelatase H subunit in plastid-to-nucleus signal transduction. *Proc. Natl. Acad. Sci. USA* 2001; 98:2053–2058.

138. Johanningmeier U, Howell SH. Regulation of light-harvesting chlorophyll-binding protein mRNA accumulation in *Chlamydomonas reinhardtii*. *J. Biol. Chem.* 1984; 259:13541–13549.

139. Johanningmeier U. Possible control of transcript levels by chlorophyll precursors in *Chlamydomonas*. *Eur. J. Biochem.* 1988; 177:417–424.

140. La Rocca N, Rascio N, Oster U, Rüdiger W. Amitrole treatment of etiolated barley seedlings leafs to deregulation of tetrapyrrole synthesis and to reduced expression of *lhc* and *rbcS* genes. *Planta* 2001; 213:101–108.

141. Herrin DL, Battey JF, Greer K, Schmidt GW. Regulation of chlorophyll apoprotein expression and accumulation. *J. Biol. Chem.* 1992; 267:8260–8269.

142. Jasper F, Quednau B, Kortenjann M, Johanningmeier U. Control of cab gene expression in synchronized *Chlamydomonas reinhardtii* cells. *J. Photochem. Photobiol. B* 1991; 11:139–150.

143. Kittsteiner U, Brunner H, Rüdiger W. The greening process in cress seedlings. II. Complexing agents and 5-aminolevulinate inhibition of *cab*-mRNA coding for the light-harvesting chlorophyll *a/b* protein. *Physiol. Plant.* 1991; 81:190–196.

144. Oster U, Brünner H, Rüdiger W. The greening process in cress seedlings. II. Possible interference of chlorophyll precursors, accumulated after thujaplicin treatment, with light-regulated expression in LHC genes. *J. Photochem. Photobiol.* 1996; 36:255–261.

145. Höfgen R, Axelsen KB, Kannangara G, Schüttke I, Pohlenz H-D, Willmitzer L, Grimm B, von Wettstein D. A visible marker for antisense mRNA expression in plants: inhibition of chlorophyll synthesis with a glutamate-1-semialdehyde aminotransferase antisense gene. *Proc. Natl. Acad. Sci. USA* 1994; 91:1726–1730.

146. Escoubas JM, Lomas M, La Roche J, Falkowski PG. Light intensity regulation of cab gene transcription is signaled by the redox state of the plastoquinone pool. *Proc. Natl. Acad. Sci. USA* 1995; 92:10237–10241.

147. Sullivan JA, Gray JC. Plastid translation is required for the expression of nuclear photosynthesis genes in the dark and in roots in pea lip1 mutant. *Plant Cell* 1999; 11:901–910.

148. Kropat J, Oster U, Rüdiger W, Beck CF. Chlorophyll precursors are signals of chloroplast origin involved in light induction of nuclear heat-shock genes. *Proc. Natl. Acad. Sci. USA* 1997; 94:14168–14172.

149. Kropat J, Oster U, Pöpperl G, Rüdiger W, Beck CF. Identification of Mg-protoporphyrin IX as a chloroplast signal that mediates the expression of nuclear genes. In: Wagner E, Normann J, Greppin H, Hackstein JHP, Herrmann RG, Kowallik KV, Shenk HEA, Seckbach J, eds. *From Symbiosis to Eukaryotism — Endocytobiology VII*. Geneva: Geneva University Press, 1999:341–348.

150. Kropat J, Oster U, Rüdiger W, Beck CF. Chloroplast signalling in the light induction of nuclear HSP70 genes requires the accumulation of chlorophyll precursors and their accessibility to cytoplasm nucleus. *Plant J.* 2000; 24:523–531.

151. Koncz C, Mayerhofer R, Koncz-Kalman Z, Newrath C, Reiss B, Rédei GP, Schell J. Isolation of a gene encoding a novel chloroplast protein by T-DNA tagging in *Arabidopsis thaliana*. *EMBO J.* 1990; 9:1337–1346.

152. Dörnemann D, Kotzabasis K, Richter P, Breu V, Senger H. The regulation of chlorophyll biosynthesis by the action of protochlorophyllide on ${}^{\text{glu}}$t-RNA-ligase. *Bot. Acta* 1989; 102:102–115.

153. Susek RE, Ausubel FM, Chory J. Signal transduction mutant of *Arabidopsis* uncouple nuclear *CAB* and *RBCS* gene expression from chloroplast development. *Cell* 1993; 74:787–799.

154. Klein RR, Mason HS, Mullet JE. Light-regulated translation of chloroplast proteins. I. Transcripts of *psaA-psaB, psbA,* and *rbcL* are associated with polysomes in dark-grown and illuminated barley seedlings. *J. Cell Biol.* 1988; 106:289–301.

155. Eichacker LA, Soll J, Lauterbach P, Rüdiger W, Klein RR, Mullet JE. *In vitro* synthesis of chlorophyll *a* in the dark triggers accumulation of chlorophyll *a* apoproteins in barley etioplasts. *J. Biol. Chem.* 1990; 265:13566–13571.

156. Eichacker LA, Paulsen H, Rüdiger W. Synthesis of chlorophyll *a* regulates translation of chlorophyll *a* apoproteins P700, CP47, CP43 and D2 in barley etioplasts. *Eur. J. Biochem.* 1992; 205:17–24.

157. He Q, Brune D, Nieman R, Vermaas W. Chlorophyll *a* synthesis upon interruption and deletion of *por* coding for the light-dependent NADPH:protochlorophyllide oxidoreductase in a photosystem-I-less/*chlL*-strain of *Synechocystis* sp. PCC 6803. *Eur. J. Biochem.* 1998; 253:161–172.

158. Kada S, Koike H, Satoh K, Hase T, Fujita Y. Arrest of chlorophyll synthesis and differential decrease of photosystem I and II in a cyanobacterial mutant lacking light-independent protochlorophyllide reductase. *Plant Mol. Biol.* 2003; 51:225–235.

159. Schoefs, B, Franck F. Chlorophyll synthesis in dark-grown pine primary needles. *Plant Physiol.* 1998; 118:1159–1168.

160. Xiong J, Fischer WM, Inoue K, Nakahara M, Bauer CE. Molecular evidence for the early evolution of photosynthesis. *Science* 2000; 289:1724–1730.

161. Green BR, Gantt E. Is photosynthesis really derived from purple bacteria? *J Phycol.* 2000; 36:983–985.

162. Peer W, Silverthorne J, Peters JL. Developmental and light-regulated expression of individual members of the light-harvesting complex b gene family in *Pinus palustris. Plant Physiol.* 1996; 111:627–634.

163. Hess WR, Rocap G, Ting CS, Larimer F, Stilwagen S, Lamerdin J, Chisholm SW. The photosynthetic apparatus of *Prochlorococcus*: insights through comparative genomics. *Photosynth. Res.* 2001; 70:53–71.

4 Probing the Relationship between Chlorophyll Biosynthetic Routes and the Topography of Chloroplast Biogenesis by Resonance Excitation Energy Transfer Determinations

Constantin A. Rebeiz, Karen J. Kopetz, and Vladimir L. Kolossov
Laboratory of Plant Biochemistry and Photobiology, Department of Natural Resources and Environmental Sciences, University of Illinois

CONTENTS

I. INTRODUCTION

In an effort to study the relationship of chlorophyll (Chl) biosynthesis to thylakoid membrane biogenesis, we have recently proposed three Chl–protein thylakoid biosynthesis models [1], which are reproduced in Figure 4.1. The models take into account the dimensions of the photosynthetic unit (PSU) [2–5], the biochemical heterogeneity of the Chl biosynthetic pathway [1,6], and the biosynthetic and structural complexity of thylakoid membranes [7]. Within a PSU, the three Chl–protein thylakoid biosynthesis models were referred to as: (a) the single-branched Chl biosynthetic pathway (SBP)-single location model (Figure 4.1A), (b) the SBP-multilocation model (Figure 4.1B), and (c) the multibranched Chl biosynthetic pathway (MBP)-sublocation model (Figure 4.1C).

Within the PSU, the SBP-single location model (Figure 4.1A) was considered to accommodate only one Chl–apoprotein thylakoid biosynthesis center and no Chl–apoprotein thylakoid biosynthesis subcenters. Within the Chl–apoprotein thylakoid biosynthesis center, Chl a and b were formed via a single-branched Chl biosynthetic pathway at a location accessible to all Chl-binding apoproteins. The latter had to access that location in the unfolded state, pick up a complement of monovinyl (MV) Chl a and/or MV Chl b, and undergo appropriate folding. Then the folded Chl–apoprotein complex had to move from the central location to a specific photosystem I (PSI), PSII, or Chl a/b light harvesting Chl (LHC)-protein location within the Chl-apoprotein biosynthesis center over distances of up to about 225 Å or larger (Figure 4.1A). In this model, observation of resonance excitation energy transfer between intermediate metabolic tetrapyrroles (unless proceeded by MV or DV, tetrapyrroles are used generically to designate metabolic pools, that may consist of MV and DV components) and some of the Chl–apoprotein complexes located at distances larger than 100 Å is unlikely. This is because resonance excitation energy transfer can take place only over distances shorter than 100 Å [8].

In the SBP-multilocation model (Figure 4.1B), every location within the PSU is considered to be a Chl–apoprotein thylakoid biosynthesis center [1,9]. In every Chl–apoprotein biosynthesis location, a complete single-branched Chl a/b biosynthetic pathway (Figure 4.1B) is active. Association of Chl a and/or Chl b with specific PSI, PSII, or LHC apoproteins at any location is random. In every Chl–apoprotein biosynthesis center, distances separating metabolic tetrapyrroles from the Chl–protein complexes are shorter than in the single-branched single-location model.

Because of the shorter distances separating the accumulated tetrapyrroles from Chl–protein complexes, resonance excitation energy transfer between various tetrapyrroles and Chl–apoprotein complexes within each center may be observed. However, accumulation of MV Mg-Proto and its monomethyl ester [Mp(e)] is not observed in any pigment–protein complex, since the single-branched Chl biosynthetic pathway does not account for the biosynthesis of MV Mp(e).

In the MBP-sublocation model (Figure 4.1C), the unified multibranched Chl a/b biosynthetic pathway [1] was visualized as the template of a Chl–protein biosynthesis center where the assembly of PSI, PSII, and LHC takes place [1,9]. The multiple Chl biosynthetic routes were visualized, individually or in groups of one or several adjacent routes, as Chl–apoprotein thylakoid biosynthesis subcenters earmarked for the coordinated assembly of individual Chl–apoprotein complexes. Apoproteins destined to some of the biosynthesis subcenters may possess specific signals for specific Chl biosynthetic enzymes peculiar to that subcenter, such as 4-vinyl reductases, Chl a oxygenase, or Chl a and Chl b synthetases. Once an apoprotein formed in the cytoplasm or in the plastid reached its biosynthesis subcenter destination and its signal was split off, it bound nascent carotenoids and Chl formed via one or more biosynthetic routes. During pigment binding, the apoprotein folded properly and acted at that location, while folding or after folding, as a template for the assembly of other apoproteins. Because of the shorter distances separating the accumulated tetrapyrroles from Chl–protein complexes, resonance excitation energy transfer between various metabolic tetrapyrroles and Chl is observed within each subcenter. In this model, both MV and DV Mp(e) were considered to be present in some of the pigment–protein complexes, in particular if more than one Chl biosynthetic route was involved in the formation of the Chl of a particular Chl–protein complex.

In an effort to determine which of the three aforementioned models was likely to be functional during greening, it was conjectured that if resonance excitation energy transfer could be demonstrated from anabolic tetrapyrroles such as protoporphyrin IX (Proto), Mp(e), and protochlorophyllide a (Pchlide a), to Chl a–protein complex constituents of PSI, PSII, and light (LHCs), it may become possible to distinguish between the various Chl-thylakoid protein biosynthesis models by resonance excitation energy transfer manipulations.

Indeed, at 77 K, emission spectra of isolated chloroplasts exhibit emission maxima at 683 to 686, 693 to 696, and 735 to 740 nm. It is believed that the

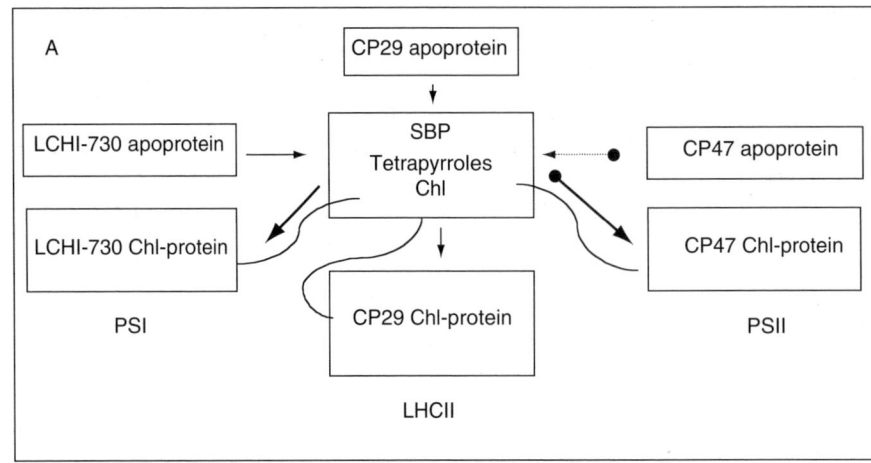

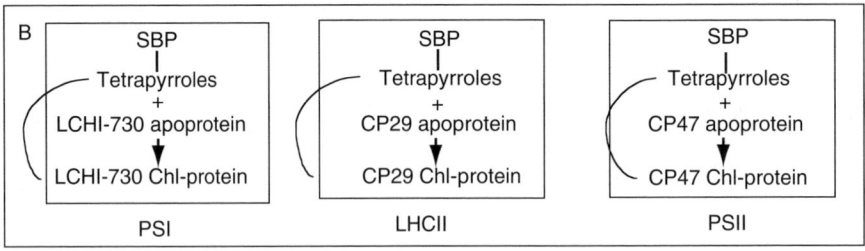

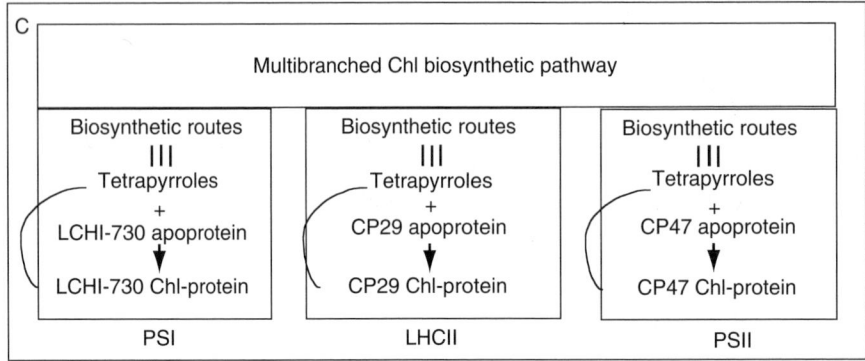

FIGURE 4.1 Schematics of the single- and multibranched Chl–apoprotein thylakoid biosynthesis models in a Chl–protein biosynthesis center, i.e., in a photosynthetic unit: (A) single-branched single-location model; (B) single-branched multilocation model; (C) multibranched sublocation model. As an example, the functionality of the three models was illustrated with the use of three apoproteins, namely, CP29, LCHI-730, and CP47. Abbreviations: SBP = single-branched Chl biosynthetic pathway; PSI = photosystem I; PSII = photosystem II; LHCII = major Chl *a/b* outer light-harvesting Chl–protein antenna. Curved lines indicate putative energy transfer between tetrapyrroles and a Chl–protein complex. (Reproduced from Kolossov VL, Kopetz KJ, Rebeiz CA. *Photochem. Photobiol.* 2003; 78:184–196. With permission.)

fluorescence emitted at 683 to 686 nm arises from the Chl *a* of LHCII, the major light-harvesting Chl–protein complex of PSII, and LHCI-680, one of the LHC antennae of PSI [2]. The fluorescence emitted at 693 to 696 nm is believed to originate mainly from the Chl *a* of CP47 and CP29, two PSII antennae [2]. That emitted at 735 to 740 nm is believed to originate pri-

marily from the Chl *a* of LHCI-730, the LHC antenna of PSI [2]. Since these emission maxima are readily observed in the fluorescence emission spectra of green tissues and are associated with definite thylakoid Chl *a*–protein complexes, it was conjectured that they would constitute a meaningful resource for monitoring the possible occurrence of resonance excitation energy transfer from anabolic tetrapyrroles to representative Chl *a*–protein complexes [9].

DV Proto, Mp(e), and Pchlide *a* were selected as anabolic tetrapyrrole donors of resonance excitation energy. DV Proto is a common precursor of heme and Chl [1]. It is formed from coproporphyrinogen III via protoporphyrinogen IX. As such, it is an early intermediate along the Chl biosynthetic chain and biosynthetically, is several steps removed from the end product, Chl.

Mg-Proto is a mixed MV–DV, dicarboxylic tetrapyrrole pool, consisting of DV and MV Mg-Proto (1). DV Mg-Proto is the first committed Mg-tetrapyrrole intermediate of the Chl biosynthetic pathway. It is formed by insertion of Mg into DV-Proto [1], and is converted to MV Mg-Proto by reduction of the vinyl group at position 4 to ethyl [10]. The formation of DV and MV Mg-Proto are tightly coupled to the formation of DV and MV Mpe by methylation of the carboxylic function at position 6 of the macrocycle [1].

The protochlorophyll (Pchl) of higher plants consists of about 95% Pchlide *a* and 5% Pchlide *a* ester. The latter is esterified with long chain fatty alcohols at position 7 of the macrocycle [1]. While Pchlide *a* ester consists mainly of MV Pchlide *a* ester, Pchlide *a* consists of DV and MV Pchlide *a*. DV Pchlide *a* is formed from DV Mpe via a complex set of reactions that results in the formation of the cyclopentanone ring. On the other hand, MV Pchlide *a* is either formed from MV Mpe via a similar set of reactions, or is formed from DV Pchlide *a* by conversion of the vinyl group at position 4 to ethyl [1,11]. In this work, Pchlide *a* and its minor esterified analog will be referred to collectively as Pchl(ide) *a*. Protochlorophyllide *a* is the immediate precursor of Chlide *a* [1].

In plastid membranes, Pchl(ide) *a* is coordinated to proteins to form pigment–protein complexes referred to as Pchl(ide) *a* holochromes [Pchl(ide) *a* Hs]. The family of Pchl(ide) *a* Hs is extremely heterogeneous with long wavelength (LW) and short wavelength (SW) Pchl(ide) *a* H species. For example, in etiolated cucumber cotyledons, five Pchl(ide) *a* H species were reported with emission maxima at 630, 633, 636, 640, and 657 nm, and excitation maxima at 440, 443, 444, 445, and 450 nm [12,13]. On the other hand, Schoefs et al. [14] reported the occurrence of seven Pchl(ide) *a* H species in bean leaves. The heterogeneous spectroscopic properties of the various Pchl(ide) *a* Hs reflect their different membrane environments. For example, the Pchlide *a* H that exhibits, respectively, *in situ* 77 K excitation and emission maxima at 450 and 657 nm, is a ternary complex that consists of Pchlide *a*, NADPH, and Pchlide *a* oxidoreductase [15,16]. The structure of the other Pchlide *a* Hs is not well understood. An extensive discussion of

this topic can be found on our website at: http://w3.aces. uiuc.edu/NRES/LPPBP/GreeningProcess/Pchl(ide) a holochromes, and in Ref. [1].

Very recently, resonance excitation energy transfers from Proto, Mp(e) and MV, and DV Pchlide *a* donors to various Chl *a*–protein complex acceptors belonging to PSI, PSII, and various LHCs have been described [9]. This, in turn, paved the way for determining the functionality of the three proposed Chl–thylakoid protein biogenesis models. In this undertaking, the SBP-single location model was tested by calculation of resonance excitation energy transfer rates over a range of distances that are likely to separate anabolic tetrapyrroles from the Chl *a* of several Chl–protein complexes within a tightly packed linear array PSU. The investigations were further refined by calculation of the distances separating Proto, Mp(e), and Pchl(ide) *a* donors from Chl *a* acceptors *in situ* [17]. The calculated rates of resonance excitation energy transfer and the distances separating anabolic tetrapyrroles from Chl *a*–protein acceptors were incompatible with the operation of the SBP-single location Chl–protein biosynthesis model, but were compatible with the operation of the MBP-sublocation model. In this chapter, an account of the above work and of the development of analytical techniques that made possible the aforementioned determinations are described.

II. MATERIALS AND METHODS

A. PLANT MATERIAL

Cucumber (*Cucumis sativus* var. Beit alpha) seeds were purchased from Hollar Seeds, Rocky Ford, CO. Barley (*Hordeum vulgare*) seeds were purchased from Murphy Sales Co., Golden Valley, CO. Germination was carried out at 26°C in plastic trays containing wet vermiculite either in darkness or in a growth chamber illuminated by 1000-W metal halide lamps (211 W/m^2) under a 14-h light/10-h dark photoperiod. The incident total spectral irradiance (light intensity) between 400 and 750 nm was determined by numerical integration with an Isco Model SR spectroradiometer and an Isco Model SRC spectroradiometer calibrator. The latter was calibrated against a quartz iodine lamp of known spectral irradiance purchased from the US Bureau of Standards [9].

B. CHEMICALS

δ-Aminolevulinic acid (ALA), was purchased from Biosynth International, Naperville, IL, and 2,2′-dipyridyl (Dpy) was purchased from Sigma Chemical Co., St. Louis, MO. Proto and Mg-Proto were purchased from Porphyrins Products, Logan, UT. DV Pchlide *a* was prepared as described in Refs. [18,19].

C.   INDUCTION OF TETRAPYRROLE ACCUMULATION

Various levels of tetrapyrrole accumulation were achieved by incubation of excised tissues with various concentrations of ALA in the absence and presence of various concentrations of Dpy, for various lengths of time [9,19]. Cucumber cotyledons were used for the induction of Proto, Mp(e), and DV Pchlide *a*, while barley leaves were used for the induction of Proto, Mp(e), and MV Pchlide *a* accumulation. The ALA + Dpy treatment in darkness had no measurable effect on the Chl *a/b* ratio, which remained around a value of three.

One to two grams of 5-day-old cucumber cotyledons excised without hypocotyl hooks, and 1 to 2 g of the top half of 6-day-old barley leaves were incubated in deep Petri dishes (10 cm in diameter), either in 10 ml of water (control) or in 10 ml of 4.5 to 20 m*M* ALA in the absence and presence of various concentrations of Dpy dissolved in 100 μmol of methanol (treated). The Petri dishes were wrapped in aluminum foil, and placed in a dark cabinet for various periods of time. Controls were incubated in distilled water for the same periods of time, under identical conditions. Both green and etiolated tissues were used in these experiments. Under these conditions, per milliliter of undiluted chloroplast suspension, tetrapyrrole accumulation was linear for up to 6600, 1500, and 1200 pmol for Pchlide *a*, Proto, and Mpe, respectively, in cucumber and up to 3000, 1100, and 550 for barley. As a consequence, the mapping of resonance excitation energy transfer sites spanned nonsaturating and saturating tetrapyrrole accumulation conditions [9].

D.   PIGMENT EXTRACTION

At the end of incubation, the dishes were unwrapped in the darkroom under a low irradiance green light that did not photoconvert Pchlide *a* to Chlide *a*. The low irradiance light source had an output maximum at 503 nm, a bandwidth of 40 nm, and a photon density of about 0.01 μmol/m^2/sec. The tissue was blotted dry, and placed in a 40-ml plastic centrifuge tube containing 10 ml of acetone:0.1 *N* NH$_4$OH (9:1 v/v). It was homogenized with a Brinkman Polytron Homogenizer, equipped with a PT 10/35 probe, at 5/10 full intensity for 40 sec. After homogenization, the tubes were centrifuged at 0°C for 12 min at 18,000 rpm in a Beckman J2-21 M/E Centrifuge using a JA-20 angle rotor. After centrifugation, the ammoniacal acetone

supernatant was transferred to a glass tube and stored at –80°C until use [9].

E.   SPECTROFLUOROMETRY

Fluorescence spectra were recorded on a fully corrected photon counting, high-resolution SLM spectrofluorometer Model 8000C, interfaced with an IBM desktop computer [9]. Room temperature spectra were recorded in cylindrical microcells 5 mm in diameter, at emission and excitation bandwidths of 4 nm. Low-temperature fluorescence spectra (77 K) were recorded at emission and excitation bandwidths that varied from 0.5 to 4 nm depending on signal intensity. The photon count was integrated for 0.5 sec at each 1 nm increment. Both fluorescence emission and excitation spectra were recorded at an angle of 90° to the excitation beam.

F.   PARTITIONING OF TETRAPYRROLES BETWEEN HEXANE AND HEXANE-EXTRACTED ACETONE

The acetonic pigment extract was transferred to a graduated conical glass tube and the volume was adjusted to 10 ml with acetone:0.1 *N* NH$_4$OH (9:1 v/v). Six milliliters of supernatant were transferred to a 30-ml separatory funnel, and extracted with an equal volume of hexane. When the phases separated, the hexane-extracted acetone residue (HEAR) hypophase was decanted into a conical glass tube. Fully esterified tetrapyrroles such as Chl and Pchlide *a* ester, partitioned with the hexane epiphase while carboxylic tetrapyrroles such as Mg-Proto and its methyl ester [Mp(e)], Pchlide *a* and Chlide *a* remained in the HEAR hypophase [19]. The HEAR was extracted again with 1/3 volume (2 ml) of hexane. The phases were separated by brief centrifugation at room temperature. The HEAR hypophase was sucked off with a Pasteur pipette and was used for further manipulations and determination of carboxylic tetrapyrroles.

G.   SPECTROFLUOROMETRIC DETERMINATIONS OF TETRAPYRROLES AT ROOM TEMPERATURE

An aliquot of the HEAR was used for determination of the amounts of Proto, Mp(e), Pchlide *a*, and Chlide *a*, by room temperature spectrofluorometry [19]. The amounts of tetrapyrroles were determined using a computer program that converts the fluorescence spectral data into concentrations [20]. The computer program and the various equations used for calculations are described in the Laboratory of Plant Biochemistry and Photobiology (LPBP) website at http://w3.aces.uiuc.edu/NRES/LPPBP/Newsoftware.

H. Acquisition of *In Situ* Emission and Excitation Spectra at 77 K for the Determination of Resonance Excitation Energy Transfer

In situ emission and excitation spectra were recorded on tissue homogenates or isolated plastids as described in Ref. [9]. At the end of dark incubation, the tissue was blotted dry, and homogenized with mortar and pestle in 5 ml of 0.2 *M* Tris–HCl:0.5 *M* sucrose (v/v), pH 8.0, under low irradiance green light. The homogenate was squeezed through two layers of cheesecloth, and 0.3 ml of the filtrate was mixed with 0.6 ml of glycerol. The filtrate–glycerol solutions were diluted with Tris–HCl–sucrose buffer: glycerol (1:2 v/v) to similar Chl concentrations, and subjected to spectrofluorometric analysis at 77 K [13]. Essentially, aliquots were introduced into 2.5-mm diameter glass tubes at room temperature in the darkroom with a Pasteur pipette. This was followed by repeated shaking of the tubes to drive the aliquot to the bottom of the narrow tubes. The tubes were frozen in liquid N_2, and subjected to spectrofluorometric emission and excitation analysis at 77 K. Emission spectra between 580 and 800 nm were elicited by excitation at 400, 420, and 440 nm. Excitation spectra were recorded at all elicited emission peaks. In most cases, excitation spectra were averages of two spectra recorded on two samples of the same aliquot. Spectral averaging was performed with the SLM software [9].

For isolated plastids, 1 g of tissue was homogenized by hand in a chilled mortar in 5 ml of homogenization buffer consisting of 0.5 *M* sucrose, 15 m*M* Hepes, 10 m*M* Tes, 1 m*M* $MgCl_2$, and 1 m*M* EDTA adjusted to pH 7.7 at room temperature. The homogenate was filtered through one layer of Miracloth into cooled 40 ml centrifuge tubes. The homogenate was centrifuged at 200*g* for 3 min in a Beckman JA-20 fixed angle rotor at 1°C. The supernatant was decanted and centrifuged at 1500*g* for 10 min at 1°C. The pelleted plastids were gently resuspended in 2 ml of cold homogenization buffer:glycerol (1:2 v/v). Excitation spectra were recorded as described above for crude homogenates [9].

I. Determination of Resonance Excitation Energy Transfer between Anabolic Tetrapyrroles and Chl A: Experimental Strategy

Before determining whether resonance excitation energy transfer did occur between accumulated anabolic tetrapyrrole donors and various Chl *a* acceptors, it was necessary to: (a) select appropriate and convenient *in situ* Chl *a* acceptors, (b) enhance the detection of putative resonance energy transfer between donors and acceptors by correction for the occurrence of endogenous resonance excitation energy transfers, and (c) generate *in situ* excitation spectra of Proto, Mp(e), and Pchlide *a* to help in locating the tetrapyrrole–Chl *a* resonance excitation energy transfer bands [9].

J. Selection of Appropriate Chl A Acceptors

As mentioned in Section I, the task of selecting appropriate Chl *a* acceptors was facilitated by the fluorescence properties of green plastids which at 77 K, exhibit maxima at 683 to 686 nm (Chl *a* ~F685), 693 to 696 nm (Chl *a* ~F695), and 735 to 740 nm (Chl *a* ~F735). Since these emission maxima are readily observable in the fluorescence emission spectra of green tissues and are associated with definite thylakoid Chl *a*–protein complexes, it was conjectured that they would constitute a meaningful resource for monitoring excitation resonance energy transfer between anabolic tetrapyrroles and representative Chl *a*–protein complexes [9].

To monitor the possible occurrence of resonance excitation energy transfer from accumulated anabolic tetrapyrroles to Chl *a*–protein complexes, excitation spectra were recorded at the respective emission maxima of the selected Chl *a* acceptors, in most cases at 686, 694, and 738 nm. Occurrence of resonance excitation energy transfer between tetrapyrrole donors and Chl *a* acceptors was evidenced by definite excitation maxima that corresponded to absorbance maxima of the various tetrapyrrole donors [9].

K. Correction for Endogenous Resonance Excitation Energy Transfer

Since the detection of resonance excitation energy transfer from anabolic tetrapyrroles to various Chl *a*–protein complexes may be blurred by the occurrence of endogenous resonance excitation energy transfers that occur in all healthy thylakoids, it was necessary to correct for this caveat. For example, in green tissues and isolated chloroplasts, fluorescence excitation spectra, recorded at emission wavelengths of 686 nm (LHCII and LHCI-680), 694 nm (CP47 and CP29), or 738 nm (LHCI-730) exhibit four endogenous resonance excitation energy transfer bands with maxima at 415 to 417, 440, 475, and 485 nm, respectively [21]. The excitation band with a maximum at 415 to 417 nm is attributed to the eta_1 transition of Chl *a*, while the 440-nm band corresponds to the bulk of light absorption by Chl *a* in the Soret region. The excitations with maxima at 475 and 485 nm are resonance excitation energy transfer bands from carotenoids and Chl *b* to Chl *a* [21]. As a consequence, it was realized that the detection of tetrapyrrole donor–Chl *a* acceptor resonance excitation energy transfer bands can be better visualized by eliminating the contribution of the endogenous resonance bands [9]. This was achieved by subtracting a control excitation spectrum from a tetrapyrrole-enriched green thylakoid excitation spectrum. The operation generated an enhanced difference excitation spectrum with optimized detection of accumulated tetrapyrrole donors–Chl *a* acceptors resonance excitation energy transfer bands. The control excitation spectra were recorded on green tissue homogenates or on isolated chloroplasts prepared from tissues that were preincubated in darkness in distilled water, under identical conditions as treated plants, but in the absence of ALA and Dpy. Such tissues contained a normal complement of Chl *a* and carotenoids, but lacked the accumulation of anabolic tetrapyrroles. Both control and treated spectra were recorded on aliquots diluted to the same Chl concentration.

L. GENERATION OF *IN SITU* TETRAPYRROLE EXCITATION SPECTRA

To better locate the wavelength regions where resonance excitation energy transfer bands may be observed, excitation spectra of *in situ* accumulated Proto, Mp(e), and Pchlide *a* were generated [9]. These spectra were recorded at the *in situ* emission maxima of Proto, Mp(e), and Pchlide *a* in dark-prepared homogenates of etiolated cucumber cotyledons or barley leaves preincubated with ALA and Dpy in darkness. The etiolated tissues lacked Chl and Chl-dependent endogenous excitation resonance energy transfer bands, but exhibited pronounced excitation bands corresponding to accumulated Proto, Mp(e), and Pchlide *a*. Since the *in situ* excitation spectrum of a given tetrapyrrole was recorded at the emission maximum of that tetrapyrrole, the most pronounced excitation maximum in the excitation profile corresponded to that particular tetrapyrrole. Other apparent excitation maxima and shoulders of lesser magnitude originated in the other accumulated tetrapyrroles.

M. PROCESSING OF ACQUIRED EXCITATION SPECTRA

To compensate for very small differences in the scatter and Chl concentration of the frozen control and treated samples, excitation spectra of every control and treated pair were normalized to a value of one fluorescence unit at a wavelength of 499 nm [9]. Since the 499-nm wavelength fell outside the Soret

excitation bands of various tetrapyrroles and carotenoids, as a consequence, by normalization to the same value at this wavelength, the difference spectra became more representative of the real differences between control and treated samples. This was because normalization at 499 nm was equivalent to multiplying the fluorescence amplitudes at every wavelength by a constant value. Therefore, this operation did not change the proportion of intraspectral characteristics or amplitudes. Thus, by adjusting two tetrapyrrole excitation spectra to the same amplitude at 499 nm, by normalization, small differences in light scattering and Chl concentrations were eliminated. The resulting difference spectra became authentic reflections of the intraspectral differences between two normalized spectra. The normalized spectra were smoothed five times. For detection of resonance excitation energy transfer bands, control spectra (water incubation) were subtracted from treated spectra.

N. DETERMINATION OF EXCITATION SPECTRA OF RECONSTITUTED TETRAPYRROLE–CHLOROPLAST LIPOPROTEIN COMPLEXES

For comparison purposes, excitation spectra of reconstituted tetrapyrrole–chloroplast lipoproteins were recorded as follows [9]. Plastids were isolated from 10 g of green tissue, as described above. The pelleted plastids were suspended in 2 ml of homogenization buffer. The plastid suspensions were freed of pigments by extraction with 20 ml of acetone:0.1 N NH_4OH (9:1 v/v). The pigment-free plastid lipoproteins were pelleted by centrifugation at 39,000g for 12 min at 1°C. The ammoniacal acetone supernatants containing extracted pigments were discarded and the lipoprotein pellet were suspended in 2 ml of homogenization buffer. Tetrapyrroles were dissolved in 80% acetone. Aliquots of the plastid lipoproteins suspensions (0.95 ml) were placed in 1.5-ml Eppendorf tubes, and 0.025 ml of Proto or Mg-Proto, or 0.5 ml of MV or DV Pchlide *a* acetonic solutions were added, and the total volume was adjusted to 1.0 ml with homogenization buffer. Controls received 0.025 ml of 80% acetone. The tubes were kept on ice for 5 min, after which they were centrifuged at 4°C for 5 min. The pigmented lipoprotein membranes were resuspended in 1 ml of homogenization buffer:glycerol (1:2 v/v). Excitation spectra were recorded at 77 K at emission wavelengths of 686, 694, and 738 nm as described above. Difference spectra of tetrapyrrole-spiked plastid lipoproteins minus plastid lipoproteins devoid of tetrapyrroles were generated as described above.

O. Determination of the Molar Extinction Coefficients of Proto, and Mp(e) in Chloroplast Lipoproteins at 77 K

For the purpose of resonance excitation energy transfer rates and distance calculations, it became necessary to determine the molar extinction coefficients of Proto, and Mp(e) in chloroplast lipoproteins at 77 K. This was achieved as described below.

DV Proto and DV Mg-Proto solutions were dissolved in 80% acetone and absorbance spectra were recorded at room temperature. The concentration of the Proto and Mg-Proto solutions were determined from absorbance values at 402 (Proto) and 417 nm (Mg-Proto) using molar extinction coefficients of 108,244 and 165,900, respectively [19]. Fifty to 100 μl of the acetone solutions containing known amounts of Proto or Mg-Proto were added to 0.75 ml of chloroplast lipoproteins suspended in the homogenization buffer. Total volumes were adjusted to 1 ml with the homogenization buffer. After mixing, the mixtures were centrifuged at 4°C for 10 min, the supernatants were discarded, and the pellets with adsorbed Proto or Mg-Proto were resuspended in 1.5 ml of Tris–HCl buffer:glycerol (1:2 v/v), pH 7.7. Aliquots were introduced into an SLM cold finger absorbance adaptor, with a 2-mm path length, and frozen in liquid N_2. Absorbance spectra were recorded at 77 K from 580 to 700 nm on an SLM-Aminco spectrophotometer Model DW-2000. Blanks consisted of chloroplast lipoprotein suspensions devoid of tetrapyrroles. Molar extinction coefficients at every wavelength were generated by dividing the absorbance values at every wavelength by the molar concentration of the tetrapyrrole in the frozen suspension, and multiplying by a factor of 5 to normalize the data to a 10-mm path length. These operations were carried out with the SLM-Aminco computational modules.

P. Preparation of Monovinyl Pchlide a and Determination of its Molar Extinction Coefficient in Chloroplast Lipoproteins at 77 K

Monovinyl (MV) Pchlide a was prepared from etiolated barley leaves, and was extracted in ammoniacal acetone and transferred to diethyl ether as described elsewhere [19]. The ether extract was dried under N_2 gas and MV Pchlide a was dissolved in a small volume of 80% acetone prior to use. One hundred and fifty microliters of the acetone solutions containing known amounts of MV Pchlide a were added to 0.75 ml of chloroplast lipoproteins suspended in the homogen-

ization buffer. The total volume was adjusted to 1 ml with the homogenization buffer. After mixing, the mixture was centrifuged at 4°C for 10 min, the supernatant was discarded, and the pellet with adsorbed MV Pchlide a was resuspended in 1.5 ml of Tris–HCl buffer:glycerol (1:2 v/v), pH 7.7. Molar extinction coefficients were determined at 77 K at various wavelengths as described for Proto.

Q. Preparation of DV Pchlide a and Determination of its Molar Extinction Coefficient in Chloroplast Lipoproteins at 77 K

DV Pchlide a was prepared from etiolated cucumber cotyledons that were induced to accumulate exclusively DV Pchlide a [19]. This was achieved by excising etiolated cotyledons with hypocotyl hooks, spreading the excised cotyledons on a wet glass plate, and exposure to a 2.5 ms actinic white light flash followed by 60 min of dark incubation. The light–dark treatment was repeated two more times. The light flashes photoconverted Pchlide a to Chlide a and activated the DV Chl a biosynthetic route, which predominates in dark (D) DV, light–dark (LD) DV plant species such as cucumber [22]. The intervening dark periods allowed the regeneration of DV Pchlide a, and conversion of the newly formed Chlide a to Chl a. As a consequence, after three such LD treatments, regenerated Pchlide a consisted exclusively of DV Pchlide a. DV Pchlide a was extracted in ammoniacal acetone and transferred to diethyl ether as described elsewhere [19]. The ether extract was dried under N_2 gas and DV Pchlide a was dissolved in a small volume of acetone prior to use. One hundred and fifty microliters of the acetone solutions containing known amounts of DV Pchlide a were added to 0.75 ml of chloroplast lipoproteins suspended in homogenization buffer. The total volume was adjusted to 1 ml with the homogenization buffer. After mixing, the mixture was centrifuged at 4°C for 10 min, the supernatant was discarded, and the pellet with adsorbed DV Pchlide a was resuspended in 1.5 ml of Tris–HCl buffer:glycerol (1:2 v/v), pH 7.7. Molar extinction coefficients were determined at 77 K at various wavelengths as described for Proto.

R. Determination of the Molar Extinction Coefficients of Total Chl a In Situ at 77 K

In order to calculate the resonance excitation energy transfer distances separating Proto, Mp(e), and Pchl(ide) a donors from Chl a acceptors in situ, molar extinction coefficients of total Chl a and various Chl a acceptors needed to be determined in situ.

The molar extinction coefficient of total Chl a at 77 K was determined *in situ* on green tissue filtrates as follows. Barley and cucumber seedlings were gown in a growth chamber illuminated by 1000-W metal halide lamps (211 W/m^2) under a 14-h light/10-h dark photoperiod. Green barley leaves and cucumber cotyledons were homogenized with mortar and pestle in 5 ml of 0.2 M Tris–HCl, 0.5 M sucrose, pH 7.7. The green homogenates were squeezed through two layers of cheesecloth. Chl a content of the filtrate was determined after extraction in acetone:NH$_4$OH (9:1 v/v) as described in Ref. [19]. The concentration of Chl a in the green filtrates was determined after extraction in ammoniacal acetone, and an absorbance spectrum of the green filtrate was recorded between 580 and 700 nm at room temperature. One volume of the green filtrate was mixed with two volumes of glycerol, and an absorbance spectrum was recorded from 580 to 700 nm at 77 K. Molar extinction coefficients at various wavelengths were determined at 77 K as described for Proto. At 676 nm, the mean of two different determinations of the molar extinction of total Chl a in green barley filtrates amounted to 121,952 ± 5,836. In green cucumber cotyledons filtrates, the mean amounted to 113,694 ± 897.

S. Estimation of the Molar Extinction Coefficients of Chl a ~F685, ~F695, and ~F735 at 77 K

The Chl a species used in the calculation of resonance excitation energy transfer from Proto, Mp(e), and Pchl(ide) a donors to Chl a acceptors *in situ* were as follows: Chl a (E670F685) (i.e., Chl a ~F685), which amounts to about 26% of the total Chl a absorbance area under the Chl a absorbance envelope; Chl a (E677F695) (i.e., Chl a ~F695), which amounts to about 32% of the total Chl a absorbance area; and Chl a (E704F735) (i.e., Chl a ~F735), which amounts to about 2% of the total Chl a absorbance area [23]. In this context, E refers to the absorbance and F to the emission maxima of the Chl a species *in situ* at 77 K. The assignment of emission F values to the absorbance (i.e., excitation) E values was based on the mirror image symmetry of the red excitation and fluorescence emission maxima of Chl a. The molar extinction coefficients of the various Chl a species were estimated from the molar extinction coefficients of total Chl a at 77 K *in situ* and the relative areas and half bandwidths *in situ* of the various Chl a species under the total Chl a envelope as described below.

As an approximation, the area of a Gaussian absorbance band can be characterized in terms of its molar extinction coefficient and its half bandwidth [8], that is,

$$\int \varepsilon_\omega d\omega \cong \varepsilon_{max} \Delta\omega_{1/2}$$

and

$$\varepsilon_{max} \cong \frac{\int \varepsilon_\omega d\omega}{\Delta\omega_{1/2}} \qquad (4.1)$$

where ε_{max} is the molar extinction coefficient at the absorbance maximum, $\int\varepsilon_\omega d\omega$ is the area of the absorbance band, and, $\Delta\omega_{1/2}$ is the half width of the absorbance band.

The total molar extinction coefficients of barley and cucumber total Chl a *in situ* at 77 K and the published *in situ* low-temperature relative areas and half bandwidths of Chl a ~F685, Chl a ~F695, and Chl a ~F735 of a green, higher plant leaf extract [23] were used together with Equation (1) to estimate the low-temperature molar extinction coefficients of Chl a ~F685, Chl a ~F695, and Chl a ~F735, as described below. For example, the *in situ* molar extinction coefficient of Chl a ~F685, in green barley at its absorbance maximum, i.e., at 670 nm, was estimated from the ε_{max} of the total Chl a of green barley at 77 K, which was determined experimentally as described above, and from the *in situ* half bandwidth of total Chl a between 650 and 720 nm reported by French et al. [23], as follows. From Equation (4.1), the integrated total area for total Chl a in green barley amounted to

$$\int \varepsilon_\omega d\omega \cong \varepsilon_{max}\Delta\omega_{1/2total\,Chla} = (121,952)(27.7)$$
$$= 3,378,070$$

where 121,952 is the determined *in situ* ε_{max} value of total Chl a of green barley at 676 nm and 77 K, and 27.7 is the value of $\Delta\omega_{1/2total\,Chla}$, the half bandwidth of total Chl a under the Chl a envelope, as determined by Frech et al. [23].

The area of Chl a ~F685, $\int\varepsilon_\omega d\omega_{Chl\,a-F685}$, is estimated from the total *in situ* Chl a area (i.e., 3,378,070) and the relative *in situ* area of Chl a ~F685, which amounts to 26% of the total Chl a under the Chl a envelope, as reported by French et al. [23], by:

$$(\varepsilon_{max\,Chla-F685})(\Delta\omega_{1/2Chla-F685}) = (3,378,070)(0.26)$$
$$= 878,298$$

From the above equation,

$$\varepsilon_{max\,Chla-F685} = \frac{878,298}{\Delta\omega_{1/2\,Chla-F685}}$$

By substituting $\Delta\omega_{1/2\text{Chl}a\sim\text{F685}}$ by its *in situ* value, which amounts to 9.8 nm as reported by French et al. [23], the above equation yields,

$$\varepsilon_{\text{max Chl}a\sim F685} = \frac{878{,}298}{9.8} = 89{,}622$$

ε_{max} values, calculated by the above procedure, for Chl a ~F685, F~695, and F~735 at 670, 677, and 704 nm, respectively, are reported in Table 4.1.

T. Determination of the Molar Extinction Coefficient of Rhodamine B in Ethanol at Room Temperature

The molar extinction coefficient of rhodamine B in ethanol at room temperature was determined from solutions of rhodamine B of known concentrations and from the absorbance spectra of the rhodamine B solutions as described in Ref. [24]. The mean of three determinations amounted to $81{,}864 \pm 3{,}757$.

U. Calculation of Energy Transfer Rates at Fixed Distances *R*

The rate of resonance excitation energy transfer from a donor D to an acceptor A [25] is given by

$$K_{\text{T}} = \frac{1}{\tau_{\text{D}}}(R_0/R)^6 \qquad (4.2)$$

where K_{T} is the rate constant of resonance excitation energy transfer from an excited donor D* to an unexcited acceptor A, which in the process becomes excited to A*; τ_{D} is the actual mean fluorescence lifetime of the excited donor D*; and R_0 is the critical separation of donor and acceptor for which energy transfer from D* to A and emission from D* to the

ground state amount to 50%, that is, are equally probable. As a consequence, at $R_0 = R$, the energy transfer rate constant is equal to $1/\tau_{\text{D}}$.

R is the separation between the centers of D*, the excited donor, and A the unexcited acceptor.

To calculate the rate constant K_{T} for a given value of R, it is essential therefore to determine the values of R_0 and τ_{D}. Since the occurrence of resonance excitation energy transfer is better observed at low temperatures due to band narrowing, K_{T} was calculated from spectral data recorded at 77 K.

V. Calculation of R_0^6

As described by Equation (4.2), calculation of R_0^6 is needed for the calculation of R, the distance separating the donors from the acceptors. According to Forster [26], for practical applications, R_0^6 can be calculated from an approximate equation, where the emission spectra of donors are expressed in terms of the absorption spectra of the donors by using the approximate mirror-image symmetry of these spectra, namely,

$$R_0^6 \cong \frac{(9)(10^6)(\ln 10)^2\kappa^2 c\tau_{\text{D}}}{16\pi^4\eta^2 N^2\nu_0^2}\int_0^\infty \varepsilon_{\text{A}}(\lambda)\varepsilon_{\text{D}}(2\nu_0 - \nu)\mathrm{d}\nu$$

$$(4.3)$$

where κ is the orientation dipole, c is the velocity of light in vacuum (3.0×10^{10} cm/sec), τ_{D} is the actual mean fluorescence lifetime of the excited donor, i.e., of the excited sensitizer, η is the refractive index which amounts to 1.45 for a membrane environment [27], N is the Avogadro's number (6.02×10^{23} molecules/mole, ν_0 is the wavenumber of the 0–0′ transition of the donor, which is approximated by the arithmetic mean, in wavenumbers, of the donor ab-

---

TABLE 4.1
Estimation of the Molar Extinction Coefficients of Chl *a* ~F685, ~F695, and ~F735 in Green Barley and Cucumber at 77 K

Plant	Chl *a* Species	Absorbance (nm)	Chl *a* (%)	Chl *a* Area $\int\varepsilon_\omega\,d\omega$	$\Delta\omega_{1/2}$ (nm)	ε_{max}
Barley	Total Chl *a*	676	100	3,378,070	27.7	121,952
	Chl *a* F685	670	26	878,298	9.8	89,622
	Chl *a* F695	677	32	1,080,982	9.2	117,498
	Chl *a* F735	704	2	67,561	20.8	3,248
Cucumber	Total Chl *a*	676	100	3,149,324	27.7	113,694
	Chl *a* F685	670	26	818,824	9.8	83,553
	Chl *a* F695	677	32	1,007,784	9.2	109,542
	Chl *a* F735	704	2	62,986	20.8	3,028

Note: Chl *a* area and $\Delta\omega_{1/2}$ values are those reported by French et al. [5] in situ for an unfractionated higher plant leaf extract at 77 K.

Source: Reproduced from Kolossov VL, Kopetz KJ, Rebeiz CA. *Photochem. Photobiol.* 2003; 78:184–196. With permission.

sorption and fluorescence maxima [26], and $\int_0^\infty \varepsilon_A(\lambda) \varepsilon_D(2\nu_0 - \nu)d\nu$ is the overlap integral, J_ν (see below for calculation of J_ν).

By substituting values for the constants, Equation (4.3) can be rewritten as

$$R_0^6 \cong \left\{ \frac{(9)(10^6)(5.3019)(3.0)(e^{10})\text{cm/sec}}{(1.5585e^3)(1.45)^2(6.02e^{23})^2} \kappa^2 \right\} \left\{ \frac{\tau_D J_\nu}{\nu_0^2} \right\}$$

Further calculations reduce the above equation to

$$R_0^6 \cong (1.2055)(10^{-33})\kappa^2 \left[\frac{\tau_D J_\nu}{\nu_0^2} \right] \qquad (4.4)$$

Therefore, to calculate R_0, the following parameters need to be determined: κ, the orientation dipole, J_ν, the overlap integral, ν_0, the arithmetic mean in wavenumbers, of the donor absorption and fluorescence maxima, and τ_D, the actual mean excitation lifetime of the excited donor. The determinations of κ, J_ν, ν_0, and τ_D at 77 K in a chloroplast lipoprotein environment are described below.

W. CALCULATION OF κ, THE ORIENTATION DIPOLE

Determination of the orientation dipole κ is needed for the calculation of R_0 (see Equation (4.4)). The rate of resonance excitation energy transfer from donors D to acceptors A depend upon the orientation of donor and acceptor dipoles and is independent of the polarity of the medium. The orientation dipole is calculated by the following formula [26]:

$$\kappa = \cos \phi_{AD} - 3(\cos \phi_A)(\cos \phi_B) \qquad (4.5)$$

where ϕ_{AD} is the angle between dipoles, ϕ_A is the angle between A and a straight line connecting A to D, and ϕ_D is the angle between D and a straight line connecting D to A.

For two dipoles that are lined up:

$$\phi_{AD} = 180°$$
$$\phi_A = 180°$$
$$\phi_D = 0°$$

By substituting the above values into Equation (4.5), we get $\kappa = \cos 180° - 3(\cos 180°)(\cos 0°)$ which yields, $\kappa = (-1) - 3(-1)(1) = 2$, and $\kappa^2 = 4$.

For adjacent dipoles

$$\phi_{AD} = 0°$$
$$\phi_A = 90°$$
$$\phi_D = 90°$$

By substituting the above values into Equation (4.5), we get $\kappa = \cos 0° - 3(\cos 90°)(\cos 90°)$, which yields, $\kappa = 1 - (3)(0)(0) = 1$ and $\kappa^2 = 1$.

For systems with random dipoles, κ^2 assumes a value of about 0.67 [8,26]. In this work, as in other reported work [27], a random dipole orientation 0.67 will be assumed for κ^2 (see discussion for validation).

X. CALCULATION OF J_ν, THE OVERLAP INTEGRAL AT 77 K

Calculation of the overlap integral J_ν is needed for the calculation of R_0 (see Equation (4.4)). The efficiency of resonance excitation energy transfer from accumulated tetrapyrroles donors to Chl a acceptors depends a great deal on the overlap between the far-red fluorescence vibrational bands of the tetrapyrrole donors, and the red absorbance bands of the Chl acceptors. The overlap between the far-red vibrational bands of the Proto, Mp(e), and Pchlide a donors and the absorbance bands of the Chl a acceptors was complete. For Chl a (E670F685), the tetrapyrrole (donor) emission–Chl a (acceptor) absorbance overlap spanned the wavelength region from 652 to 688 nm. For Chl a (E677F695), the overlap spanned the wavelength region from 660 to 695 nm, and for Chl a (E704F735) it spanned the wavelength region from 692 to 720 nm. The overlaps between the far-red vibrational bands of Proto adsorbed on barley chloroplast lipoproteins and Chl a 670, 677, and 704 are depicted in Figure 4.2.

The overlap integral J_ν (referred to as $J_{(\lambda)}$ by Lakowicz) [28] normalized by the area of the corrected emission spectrum, can be calculated from the following formula:

$$J_{(\lambda)} = \frac{\int_0^\infty F_D(\lambda)\varepsilon_A(\lambda)\lambda^4 d\lambda}{\int_0^\infty F_D(\lambda)d\lambda} \qquad (4.6)$$

where $F_D(\lambda)$ is the corrected fluorescence emission intensity at every wavelength, $\varepsilon_A(\lambda)$ is the molar extinction coefficient of the acceptors as a function of wavelength λ, λ^4 is the wavelength in nanometers in the emission–absorbance overlap region raised to the power 4, and $\int_0^\infty F_D(\lambda)d\lambda$ is the area of the corrected emission spectra.

Two assumptions are made [8] in deriving Equation (4.6). First, it is assumed that the energy available for transfer by donors is that which would otherwise be emitted as fluorescence. As a consequence the transfer probability is stated in terms of the strength of the individual absorbance and emission transitions, and the energy overlap of the emis-

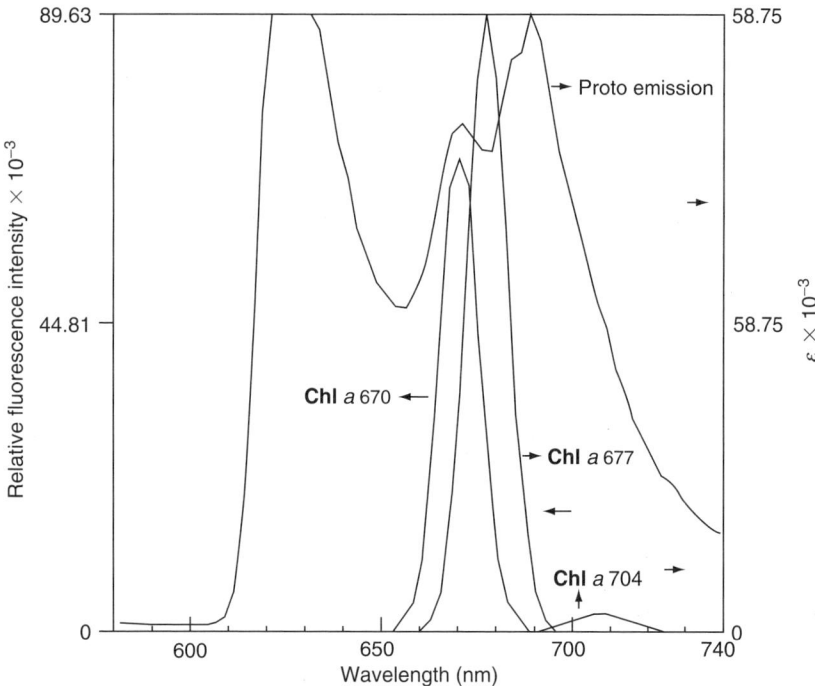

FIGURE 4.2 The overlap between Proto adsorbed on barley chloroplast lipoproteins and Chl *a* 670, 677, and 704 in barley. The Proto emission spectrum was recorded at 77 K on barley chloroplast lipoproteins prepared as described in Section II. It was elicited by excitation at 400 nm and for the purpose of display was arbitrarily normalized to a value 89,622, the molar extinction coefficient of Chl *a* 670. The normalization value of 89,622 was for display purposes only, and had no influence on the calculation of the overlap integral J_v, since the calculation of the latter involved normalization by the area of the corrected emission spectrum. (Lakowicz JR. *Principles of Fluorescence Spectroscopy*. New York: Kluwer Academic/Plenum Press, 1999: pp. 367–394.) The in situ low-temperature absorption spectra of Chl *a* 670, 677, and 704 were taken from Schoch S, Brown JS. Comparative spectroscopy of chlorophyll *a* in daylight and intermittent-light-grown plants. Carnegie Institution of Washington Year Book 1980; pp. 16–20, using SLM software. The Chl *a* peaks correspond to absorbance at the molar extinction maxima for the various Chl *a*. The left ordinate scale refers to relative fluorescence emission units. The right ordinate scale refers to molar extinction coefficients of the various Chl *a* acceptors (From Kopetz KJ, Kolossov VL, Rebeiz CA. *Anal. Biochem.* 2004; 329:207–219. With permission.)

sion band of donors, and the absorption band of acceptors. Second, it is assumed that the transfer time is long relative to vibrational internal conversion processes (i.e., heat dissipation by molecular collision), so that transfer is from the lower vibrational levels (0') of the first excited singlet state of the donor. Calculated J_v values for Proto–Chl *a*, Mp(e)–Chl *a*, and Pchlide *a*–Chl *a* donor–acceptor pairs for barley and cucumber are reported in Table 4.2.

Y. CALCULATION OF ν_0, THE MEAN WAVENUMBER OF ABSORPTION AND FLUORESCENCE PEAKS OF DONORS AT 77 K

Calculation of ν_0, the mean wavenumber of absorption and fluorescence maxima of the donors, is needed for the calculation of R_0 (see Equation (4.4)). It can be determined as follows. The donors are adsorbed to chloroplast lipoproteins prepared from green barley leaves or cucumber cotyledons as described in Section II.C. Their absorbance and fluorescence emission spectra are recorded at 77 K. The absorbance and fluorescence emission maxima are converted to wavenumbers and ν_0, the arithmetic mean of the two wavenumbers is calculated.

For example, for donor Proto adsorbed to chloroplast lipoproteins prepared from green barley leaves at 77 K, ν_0 is calculated as follows. The absorption maximum of Proto in barley chloroplast lipoproteins at 77 K and at 641 nm is 15,601 cm^{-1}. The far-red emission maximum of Proto in the same environment at 77 K and at 687 nm is 14,556 cm^{-1}, and

$$\nu_0 = (15{,}601 + 14{,}556)/2 = 15{,}078 \text{ cm}^{-1}$$

The calculated ν_0 values for Proto, Mp(e), and Pchlide *a* are reported in Table 4.3.

TABLE 4.2
Overlap Integral J_v, for the Proto, Mp(e), MV and DV Pchl(ide) a–Chl a, Donor–Acceptor Pairs at 77 K in Barley and Cucumber

Plant	Tetrapyrrole	Chl a Species	Overlap Integral (J_v) (cm^3/mol)
Barley	Proto	Chl a F685	3.32×10^{12}
		Chl a F695	4.75×10^{12}
		Chl a F735	1.33×10^{11}
	Mp(e)	Chl a F685	1.60×10^{12}
		Chl a F695	1.48×10^{12}
		Chl a F735	4.14×10^{10}
	MV Pchl(ide) a	Chl a F685	2.23×10^{12}
		Chl a F695	2.90×10^{12}
		Chl a F735	1.25×10^{11}
Cucumber	Proto	Chl a F685	1.31×10^{12}
		Chl a F695	1.28×10^{12}
		Chl a F735	3.60×10^{10}
	Mp(e)	Chl a F685	2.79×10^{12}
		Chl a F695	4.24×10^{12}
		Chl a F735	1.49×10^{11}
	DV Pchl(ide) a	Chl a F685	2.22×10^{12}
		Chl a F695	3.84×10^{12}
		Chl a F735	1.36×10^{11}

Note: In the presence of ALA and Dpy, DMV-LDMV plant species such as barley accumulate DV Proto, and about equal amounts of DV and MV Mp(e), and MV Pchlide a, while DDV-LDDV plant species such as cucumber accumulate DV Proto, smaller amounts of MV Mpe, and DV Pchlide a, in darkness.

Source: Reproduced from Kolossov VL, Kopetz KJ, Rebeiz CA. *Photochem. Photobiol.* 2003; 78:184–196. With permission.

TABLE 4.3
Mean Wavenumber ν_0, of Absorbance and Fluorescence Emission Maxima of the Proto, Mp(e), and Pchl(ide) a Donors in Barley and Cucumber Chloroplast Lipoproteins at 77 K

Plant	Donor	Red Absorbance Maximum (cm^{-1})	Far-Red Emission Maximum (cm^{-1})	ν_0 (cm^{-1})
Barley	Proto	15,601	14,556	15,078
	Mp(e)	16,938	15,385	16,161
	MV Pchl(ide) a	15,741	14,706	15,217
Cucumber	Proto	15,564	14,535	15,050
	Mp(e)	16,918	15,408	16,163
	DV Pchl(ide) a	15,728	14,706	15,217

The donors adsorbed to chloroplast-lipoproteins were suspended in 0.2 *M* Tris–HCl, 0.5 *M* sucrose, pH 7.7, diluted 1:2 (v/v) with glycerol. Abbreviations are as in Table 4.1.

Source: Reproduced from Kolossov VL, Kopetz KJ, Rebeiz CA. *Photochem. Photobiol.* 2003; 78:184–196. With permission.

Z. CALCULATION OF τ_0, THE INHERENT RADIATIVE LIFETIME OF DONORS AT 77 K

Determinations of the inherent radiative lifetime of the donors, τ_0, and the relative fluorescence yield of the donors in the presence of acceptors, Fy_{DA}, are needed for the calculation of the actual mean fluorescence lifetimes of excited donors, τ_D (see below). The latter are needed for the calculation of R_0^6 (see Equation (4.4)).

The inherent radiative lifetime of a donor, τ_0, is the inherent radiative lifetime of its excited state. It is the mean time it would take to deactivate the excited state in the absence of radiationless processes such as internal conversion (i.e., heat dissipation) and intersystem crossing (i.e., conversion from a singlet to a triplet excited state) [25]. The measured fluorescence lifetime of an excited donor, τ_D, is determined by the sum of the rates of all processes depopulating the donor excited state. Therefore, in cases where other unimolecular processes (such as intersystem crossing) or bimolecular processes (such as resonance excitation energy transfer), compete with fluorescence, the observed radiative lifetimes τ_D, will be proportionally shorter than the natural fluorescence lifetimes, τ_0 [8]. The inherent radiative lifetime of donors, τ_0, can be calculated as follows [25]:

$$\tau_0 = \frac{3.5 \times 10^8}{(v_m^2)(\varepsilon_m)(\Delta v_{1/2})} \quad (4.7)$$

where v_m is the Soret absorbance maximum of the donors in wavenumbers, ε_m is the molar extinction coefficient at the Soret absorbance maximum of the donors, and $\Delta v_{1/2}$ is the half bandwidth of the Soret absorbance bands of the donors in wavenumbers.

For example, for Proto adsorbed to barley chloroplast lipoproteins at 77 K,

$$v_m = 395.9 \, \text{nm} = 25{,}259 \, \text{cm}^{-1}$$

$$\varepsilon_m = 116{,}751$$

$$\Delta v_{1/2} = 4008 \, \text{cm}^{-1}$$

and

$$\tau_0 = \frac{3.5 \times 10^8}{(25{,}259)^2(116{,}751)(4008)}$$

$$= 1.17248 \times 10^{-9} \, \text{sec or } 1.17 \, \text{nsec}$$

Calculated τ_0 values for Proto, Mp(e), and Pchlide a are reported in Table 4.4.

AA. CALCULATION OF Fy_{DA} THE RELATIVE FLUORESCENCE YIELD OF TETRAPYRROLE DONORS IN THE PRESENCE OF CHL ACCEPTORS *IN SITU* AT 77 K

The relative fluorescence quantum yield Fy_{DA} of donors D in the presence of acceptors A, is needed for the calculation of τ_D, the actual mean fluorescence lifetime of excited donors (see below). The latter are needed for the calculation of R_0^6 (Equation (4.4)).

The absolute fluorescence quantum yields of many compounds have been determined with considerable precision. For example, rhodamine B in ethanol at low concentrations exhibits an absolute fluorescence quantum yield [8] of 0.69. Compounds like rhodamine B are used as actinometers for the determination of the relative fluorescence quantum yield of other compounds as described below.

The relative fluorescence quantum yield, Fy_D, of fluorescent donors D, are related to the absolute fluorescence quantum yield of an actinometer Qy_{act} such as rhodamine B, by the following equation [8]:

$$Fy_D = \frac{(CFI_D)(Qy_{act})}{(CFI_{act})} \quad (4.8)$$

where Fy_D is the relative fluorescence quantum yield of donors D in the absence of acceptors, in a particular solvent, and at a particular temperature; CFI_D is the corrected fluorescence intensity of the red fluorescence emission band of donors D, which is Gaussian (i.e., symmetrical) for all tetrapyrrole donors in chloroplast lipoproteins at 77 K, in the same solvent, and at the same temperature; Qy_{act} is the absolute fluorescence quantum yield of the actinometer, which for rhodamine B in ethanol, at room temperature, has a value of 0.69 [8]; and CFI_{act} is the cor-

---

TABLE 4.4
Inherent Radiative Lifetimes τ_0 of the Proto, Mp(e), and Pchl(ide) a Donors in Barley and Cucumber Chloroplast Lipoproteins at 77 K

Plant	Donor	Soret Absorbance Maximum, v_m (cm^{-1})	ε_m (cm^{-1})	SA at HBW, $\Delta v_{1/2}$ (cm^{-1})	τ_0 (ns)
Barley	Proto	25,259	116,751	4008	1.17
	Mp(e)	23,593	119,000	1597	2.07
	MV Pchl(ide) a	22,512	177,780	939	4.14
Cucumber	Proto	25,707	118,222	4160	1.08
	Mp(e)	23,630	192,827	1592	2.04
	DV Pchl(ide) a	22,212	227,888	862	3.61

Note: The suspension medium consisted of 0.2 M Tris–HCl, 0.5 M sucrose, pH 7.7, diluted 1:2 v/v with glycerol. SA at HBW = Soret absorbance at half bandwidth. Other abbreviations are as in Table 4.1.

Source: Reproduced from Kolossov VL, Kopetz KJ, Rebeiz CA. *Photochem. Photobiol.* 2003; 78:184–196. With permission.

rected intensity of the actinometer at its fluorescence emission maximum at a particular temperature, such as room temperature, recorded in a cell having the same path length as the cell used for recording the fluorescence spectrum of donor D. In this case the actinometer is rhodamine B dissolved in ethanol, at room temperature, which also exhibits a Gaussian emission band. The concentration of the samples should be such that

$$(\varepsilon_D)(C_D) = (\varepsilon_{act})(C_{act}) \qquad (4.9)$$

In Equation (4.9) ε_D is the molar extinction coefficient of donors D in the chosen solvent or matrix, at a particular temperature, say 77 K (see Section II); C_D is the concentration of donors D in the same solvent or matrix, and at the same temperature; ε_{act} is the molar extinction coefficient of the actinometer, in this case rhodamine B, dissolved in ethanol at room temperature (see Section II); and C_{act} is the concentration of the actinometer, i.e., rhodamine B, dissolved in ethanol, at room temperature.

Equation (4.9) is valid when the $\varepsilon c l$ values (l is the optical path length) are = or < 0.02 [8], and when experimental ε values determined in the specifed solvents or matrices are used. In this work, the $\varepsilon c l$ values ranged from a low of 0.0069 to a high of 0.0081.

By substituting 0.69 for Qy_{act}, for the rhodamine B actinometer in Equation (4.8), it transforms to

$$Fy_D = (CFI_D)(0.69)/(CFI_{act}) \qquad (4.10)$$

Equation (4.10), and the values for rhodamine B actinometer dissolved in ethanol at room temperature, can be used for the determination of the relative fluorescence quantum yield of any fluorescent compound or donor, in the presence of an acceptor, in any solvent or matrix at any temperature. For example, the relative fluorescence yield of a tetrapyrrole donor in the presence of a Chl acceptor, Fy_{DA}, at 77 K can be determined via a procedure similar to that described above. The terms in Equation (4.10) are slightly modified, however, to reflect the fact that in this example: (a) the donor is Proto, which was induced to accumulate in barley chloroplast membranes in the presence of the Chl a acceptors, and (b) the actinometer with a calculated quantum yield of 0.69, is rhodamine B dissolved in ethanol at room temperature. The relative fluorescence quantum yield of Proto at 77 K, in the presence of a Chl a acceptor, $Fy_{DAProto77K}$, can be calculated from Equation (4.11) as follows:

$$Fy_{DAProto77K} = (CFI_{Proto77K})(0.69)/CFI_{rdbEt\,RT} \qquad (4.11)$$

where $CFI_{Proto77K}$ is the maximum red fluorescence amplitude in arbitrary number of photons, of the green barley filtrate in Tris–sucrose buffer diluted with glycerol 1:2 (v/v), at 77 K. The filtrate was prepared from green barley leaves induced to accumulate Proto by pretreatment with ALA and, 2,2′-dipyridyl (Dpy) as described in Ref. [9]. $CFI_{rdbEt\,RT}$ denotes the maximum fluorescence amplitude in arbitrary number of photons of rhodamine B dissolved in ethanol at room temperature.

First, 7-day-old, green, photoperiodically grown barley leaves were incubated with ALA and Dpy in darkness for 4 h to induce the accumulation of anabolic tetrapyrroles including Proto [9]. The Proto-enriched tissue was homogenized in Tris–sucrose buffer, pH 7.7, and filtered through two layers of Miracloth as described in Section IIH. The Proto content of the filtrate was determined from an ammoniacal acetone extract as described in Section II. An aliquot of the filtrate was diluted in Tris–sucrose buffer, pH 7.7, and adjusted to 67% glycerol so that its $(\varepsilon_{Proto77K})(C_{Proto77K}) = (\varepsilon_{rdbEt\,RT})(C_{rdbEt\,RT})$. The room temperature corrected emission spectrum of the rhodamine B solution between 400 and 600 nm is elicited by excitation at 400 nm. Fluorescence was monitored at an angle of 90° with respect to the excitation beam. The maximum fluorescence amplitude in arbitrary number of photons, $CFI_{rdbEt\,RT}$, amounted to 0.3516 (Table 4.5). The green barley filtrate in Tris–sucrose–glycerol (1:2 v/v) buffer is cooled down to 77 K. Its 77 K emission spectrum between 500 and 700 nm is elicited by excitation at 400 nm. Likewise, the fluorescence was monitored at an angle of 90° with respect to the excitation beam. The maximum fluorescence amplitude in arbitrary numbers of photons, $CFI_{Proto77K}$, as determined via the SLM software, amounted to 0.1332 (Table 4.5). $Fy_{DaProto77K}$ is calculated from Equation (11) by substituting experimental values for $CFI_{Proto77K}$ and $CFI_{rdbEt\,RT}$, which yields:

$$Fy_{DaProto77K} = (0.1332)(0.69)/(0.3516) = 0.2614$$

The calculated Fy_{DA} values for Proto, Mp(e), and Pchlide a are reported in Table 4.5.

AB. CALCULATION OF τ_D, THE ACTUAL MEAN FLUORESCENCE LIFETIME OF EXCITED DONORS IN THE PRESENCE OF ACCEPTORS AT 77 K

The actual mean fluorescence lifetime of excited donors in the presence of Chl acceptors, τ_D, is needed for the calculation of R_0^6 (see Equation (4.4)). The actual mean fluorescence lifetime of excited donors τ_D, are related to the relative fluorescence yield of donors in the presence of acceptors, Fy_{DA}, by the following equation [25]:

TABLE 4.5
Relative Fluorescence Yields for Proto, Mp(e), and Pchl(ide) *a* Donors In Situ at 77 K

Plant	Donor	Excitation wavelength (nm)	$CFI_{rdbEt\,RT}$	$CFI_{CFIDA\,77K}$	Fy_{DA}
Barley	Proto	400	0.3516	0.1332	0.2614
	Mp(e)	420	0.2081	0.0761	0.2523
	MV Pchl(ide) *a*	440	0.2890	0.0297	0.0709
Cucumber	Proto	400	0.3516	0.0657	0.1289
	Mp(e)	420	0.2081	0.0440	0.1459
	DV Pchl(ide) *a*	440	0.2890	0.0253	0.0604

Note: Barley and cucumber green filtrates in 0.2 *M* Tris–HCl–0.5 *M* sucrose, pH 7.7, were diluted 1:2 (v/v) with glycerol. Rhodamine B was dissolved in ethanol at room trmperature. CFI = corrected fluorescence intensity in arbitrary number of photons; Fy_{DA} = relative fluorescence yield of tetrapyrrole donors in the presence of Chl acceptors. Other abbreviations are as in Table 4.1.

Source: Reproduced from Kolossov VL, Kopetz KJ, Rebeiz CA. *Photochem. Photobiol.* 2003; 78:184–196. With permission.

$$\tau_D = (Fy_{DA})(\tau_0) \qquad (4.12)$$

where Fy_{DA} is the relative fluorescence yield of donors in the presence of acceptors, and τ_0 is the inherent radiative lifetime of donors in the absence of acceptors.

For example, the actual mean fluorescence lifetime of the excited Proto donor in barley chloroplast membranes, τ_D, in the presence of Chl *a* acceptors is calculated as follows. For Proto in barley chloroplast membranes containing Chl *a* acceptors and suspended in 0.2 *M* Tris–HCl, 0.5 *M* sucrose, pH 7.7 diluted with glycerol (1:2 v/v), Fy_{DA} amounted to 0.2614 (Table 4.5).

The inherent radiative lifetime at 77 K of the donor in the absence of acceptor, τ_0, for Proto adsorbed to barley chloroplast lipoproteins and suspended in the same above buffer, amounts to 1.17248×10^{-10} sec (Table 4.4), and

$$\tau_D = (0.2614)(1.17248)(10^{-9})\,s$$
$$= 3.0649 \times 10^{-10}\,s \text{ or } 0.31 \text{ ns}$$

The calculated τ_D values for Proto, Mp(e), and Pchlide *a* are reported in Table 4.6.

AC. Calculation of R_0^6 for Proto, Mp(e), and Pchlide *a* Donor–Chl *a* Acceptor Pairs at 77 K

The critical separation of donors from acceptors, R_0, for which energy transfer from excited donors D* to acceptors A and emission from excited acceptors A* to the ground state, amounts to 50%, i.e., are equally probable, is needed for the calculation of K_T, the rate of resonance excitation energy transfer described by Equation (2), and for *R*, the distance separating donors D from acceptors A (see below). As described by Equation (4), R_0^6 is given by

$$R_0^6 \cong (1.2055)(10^{-33})\kappa^2\left[\frac{\tau_D J_v}{v_0^2}\right]$$

For the Proto–Chl *a* ~F685 pair in barley chloroplast membranes, the following values for the various expressions in Equation (4) are

$$\kappa^2 = 0.67$$
$$\tau_D = 3.0649^{-10}\text{ sec (Table 4.6)}$$
$$J_v = 3.32 \times 10^{12}\text{ cm}^3/\text{mol (Table 4.2)}$$
$$v_0 = 15,078\text{ cm}^{-1}\text{(Table 4.3)}$$

By substituting the above values into Equation (4.4), it reduces to

$$R_0^6 \cong 1.2055 \times 10^{-33} \times 0.67$$
$$\times \frac{(3.0649 \times 10^{-10}\text{ sec})(3.32 \times 10^{12}\text{ cm}^3/\text{mol})}{(15,078\text{ cm}^{-1})^2}$$

or

$$R_0^6 \cong 3.917 \times 10^{-39}$$

and,

$$R_0 \cong 39.17 \times 10^{-8}\text{ cm, i.e., } 39.17\text{Å}$$

TABLE 4.6
Actual Mean Fluorescence Lifetimes τ_D of the Excited Proto, Mp(e), Pchl(ide) a Donors in the Presence of Chl a Acceptors at 77 K

Plant	Donor	Fy_{DA}	τ_0 (ns)	τ_D (ns)
Barley	Proto	0.2614	1.17	0.31
	Mp(e)	0.2523	2.07	0.52
	MV Pchl(ide) a	0.0709	4.14	0.22
Cucumber	Proto	0.1289	1.08	0.14
	Mp(e)	0.1459	2.04	0.30
	DV Pchl(ide) a	0.0604	3.61	0.29

Note: Other abbreviations are as in Table 4.1 to Table 4.4.

Source: Reproduced from Kolossov VL, Kopetz KJ, Rebeiz CA. *Photochem. Photobiol.* 2003; 78:184–196. With permission.

TABLE 4.7
Calculated $R_0{}^6$ and R_0 Values for Anabolic Tetrapyrrole Donor–Chl a Acceptor Pairs at 77 K

Plant	Chl a Species	Chl a Absorbance (nm)	Donor	τ_0 (ns)	J_v (cm^3/mol)	ν_0 (cm^{-1})	$R_0{}^6 \times 10^{-39}$ (cm)	R_0 (Å)
Barley	Chl a F685	670	Proto	0.31	3.32×10^{12}	14,556	3.61	39.17
	Chl a F695	677		0.31	4.75×10^{12}	14,556	5.17	41.58
	Chl a F735	704		0.31	1.33×10^{11}	14,556	0.145	22.92
	Chl a F685	670	Mp(e)	0.52	1.60×10^{12}	16,161	2.59	37.05
	Chl a F695	677		0.52	1.48×10^{12}	16,161	2.40	36.59
	Chl a F735	704		0.52	4.14×10^{10}	16,161	0.067	20.15
	Chl a F685	670	MV Pchl(ide) a	0.22	2.23×10^{12}	15,217	2.28	36.29
	Chl a F695	677		0.22	2.90×10^{12}	15,217	2.97	37.90
	Chl a F735	704		0.22	1.25×10^{11}	15,217	0.127	22.43
Cucumber	Chl a F685	670	Proto	0.14	1.31×10^{12}	15,050	0.677	29.41
	Chl a F695	677		0.14	1.28×10^{12}	15,050	0.632	29.29
	Chl a F735	704		0.14	3.60×10^{10}	15,050	0.018	16.16
	Chl a F685	670	Mpe	0.30	2.79×10^{12}	16,163	2.57	37.02
	Chl a F695	677	Mpe	0.30	4.24×10^{12}	16,163	3.91	39.69
	Chl a F735	704		0.30	1.49×10^{11}	16,163	0.137	22.71
	Chl a F685	670	DV Pchl(ide) a	0.29	2.22×10^{12}	15,217	1.69	34.50
	Chl a F695	677		0.29	3.84×10^{12}	15,217	2.93	37.82
	Chl a F735	704		0.29	1.36×10^{11}	15,217	0.103	21.66

Note: J_v = overlap integral; ν_0 = mean wavenumber. Other abbreviations are as in Table 4.1.

Source: Reproduced from Kolossov VL, Kopetz KJ, Rebeiz CA. *Photochem. Photobiol.* 2003; 78:184–196. With permission.

The calculated R_0^6 and R_0 values for the Proto, Mp(e), and Pchlide a donors–Chl a acceptor pairs are reported in Table 4.7.

AD. SELECTION OF FIXED DISTANCES R SEPARATING ANABOLIC TETRAPYRROLE DONORS FROM CHL A ACCEPTORS

The linear continuous array PSU model, depicts a central cyt b_6 complex flanked on one side by PSI and coupling factor CF1, and on the other side by PSII and LFCII [2]. With this configuration, the shortest distance between the single-branched pathway and PSI, PSII, and LHCII, in the SBP-single location model would be achieved if the single-branched Chl biosynthetic pathway occupied a central location within the PSU. In that case it can be calculated that the core of PSII including CP47 and CP29, would be located about 126 Å away from the SBP. On the other hand, LHCI-730 would be located about 159 Å on the other side of the SBP. The centers of the inner and outer halves of LHCII surrounding the PSII core would be located about 156 Å (outer half) and 82 Å (inner half) from the SBP.

Since the fluorescence emission maxima of Chl a ~F685, ~F695, and ~F735 are readily observed in green tissues and are associated with definite thylakoid Chl a–protein complexes, it was decided to monitor excitation resonance energy transfer rates, K_T from anabolic tetrapyrroles donors to the aforementioned Chl a–protein complexes over distances of 159, 126, and 82 Å, as well as over distances $R = R_0$.

AE. CALCULATION OF K_T AT FIXED DISTANCES R, SEPARATING PROTO, MP(E), AND PCHLIDE A DONORS FROM CHL A ACCEPTORS AT 77 K

As described by Equation (4.2) [25], the rate of resonance excitation energy transfer K_T, is given by

$$K_T = \frac{1}{\tau_D}(R_0/R)^6$$

where τ_D is the actual mean lifetime of excitation of donors D^* in the presence of acceptors A; R_0 is the critical separation of donors from acceptors for which energy transfer from excited donors D^* to unexcited acceptors A and emission from D^* to the ground state D, amount to 50%, i.e., are equally probable; and R is the separation of the centers of D^*, the excited donors, from A, the unexcited centers of acceptors.

In the SBP-single location model, for the Proto–Chl a ~F685 pair in barley chloroplast membranes for example, at 77 K, the following values for Equation (2) have been determined: $\tau_D = 3.0649 \times 10^{-10}$ sec for Proto adsorbed to barley Chloroplast lipoproteins (Table 4.6) and $R_0 = 44.4629 \times 10^{-8}$ cm (Table 4.7).

By substituting the above values in Equation (4.2) for a distance R of 159 Å (159 $\times 10^{-8}$ cm),

$$K_T = (1/3.0649 \times 10^{-10}\,\text{s})(44.4629 \times 10^{-8}\,\text{cm}/159$$
$$\times 10^{-8}\,\text{cm})^6$$
$$= 1.5603 \times 10^6\,\text{s}^{-1}.$$

AF. EXPRESSION OF THE RATES OF RESONANCE EXCITATION ENERGY TRANSFER, K_T, FROM DONORS TO ACCEPTORS AS A PERCENTAGE OF DE-EXCITATION VIA 100% RESONANCE EXCITATION ENERGY TRANSFER

The rates of resonance excitation energy transfer, K_T, from tetrapyrrole donors to the various Chl a acceptors were expressed as a percentage of de-excitation via 100% resonance excitation energy transfer as fol-

lows. For example, for resonance excitation energy transfer from Proto to Chl a (E670F685) at a κ^2 value of 0.67, and at a fixed distance R of 159 Å,

$$R = 1.59 \times 10^{-6}\,\text{cm}$$
$$R^6 = 1.6158 \times 10^{-35}\,\text{cm}$$
$$K_T = 6.64 \times 10^5\,\text{sec}^{-1}$$

At $R_0 = R = 38.56$ Å, $K_T = 3.43 \times 10^9\,\text{sec}^{-1}$ and,

$$K_T\% = \left[\frac{6.64 \times 10^5}{3.43 \times 10^9/50} \times 100\right] \times 100$$
$$= 9.68 \times 10^{-3}\%$$

AG. CALCULATION OF DISTANCES, R, SEPARATING ANABOLIC TETRAPYRROLES FROM VARIOUS CHL A–PROTEIN COMPLEXES

The efficiency of resonance excitation energy transfer, E, from donors D to acceptors A, is directly related to the distance, R, separating donors from acceptors [28], by the following equation:

$$E = R_0^6/(R_0^6 + R^6) \tag{4.13}$$

Equation (4.13) can be rewritten as

$$R^6 = (R_0^6 - ER_0^6)/E$$

or as

$$R^6 = (R_0^6/E) - R_0^6 \tag{4.14}$$

where R is the distance separating donors D from acceptors A, R_0 is the critical separation of donors from acceptors for which energy transfer from excited donors D^* to unexcited acceptors A and emission from D^* to the ground state D amount to 50%, i.e., are equally probable, and E is the efficiency of resonance excitation energy transfer from donors to acceptors.

AH. CALCULATION OF E, THE EFFICIENCY OF ENERGY TRANSFER IN SITU AT 77 K

The efficiency of energy transfer, E, is needed for the calculation of R, the distances separating donors from acceptors. It is calculated from the following equation [28]:

$$E = 1 - \text{Fy}_{DA}/\text{Fy}_D \tag{4.15}$$

where Fy_{DA} is the relative fluorescence yield of donors D in the presence of acceptors A, and Fy_D is the relative fluorescence yield of donors D in the absence of acceptors A. According to Calvert and Pitts [8], Fy_{DA} is given by

$$Fy_{DA} = \frac{(CFI_{DA})(\varepsilon_{act})(C_{act})}{(CFI_{act})(\varepsilon_{DA})(C_{DA})} Qy_{act} \qquad (4.16)$$

where CFI_{DA} is the corrected fluorescence intensity of the fluorescence emission bands of donors D in the presence of acceptors A in a particular solvent or matrix, at a particular temperature; CFI_{act} is the corrected fluorescence intensities of the fluorescence emission band of the actinometer, in a particular solvent or matrix, at a particular temperature; ε_{act} is the molar extinction coefficient of the actinometer; ε_{DA} is the molar extinction coefficient of donors D in the particular solvent or matrix at the same particular donor temperature; C_{act} is the concentration of the actinometer; C_{DA} is the concentration of donors D in the particular solvent or matrix; and Qy_{act} is the absolute fluorescence quantum yield of the actinometer.

Likewise, for donors D in the absence of an acceptor, Fy_D, the latter are given by

$$Fy_D = \frac{(CFI_{DA})(\varepsilon_{act})(C_{act})}{(CFI_{act})(\varepsilon_{DA})(C_{DA})} Qy_{act} \qquad (4.17)$$

If the concentration of the donors are adjusted so that $(\varepsilon_{DA})(C_{DA}) = (\varepsilon_D)(C_D)$, and the emission bands are reasonably Gaussian, then, Fy_{DA}/Fy_D reduces to CFI_{DA}/CFI_D, and $E = 1 - Fy_{DA}/Fy_D$, transforms into

$$E = 1 - CFI_{DA}/CFI_D \qquad (4.18)$$

The calculated efficiencies of resonance excitation energy transfer E for Proto, Mp(e), and MV and DV Pchlide a were calculated from Equation (4.18) as follows. First, green filtrates were prepared from green barley leaves or green cucumber cotyledons incubated for 4 h with ALA and Dpy in darkness to induce the accumulation of Proto, Mp(e), and MV or DV Pchlide a, exactly as described above for Fy_{DA}. Likewise, etiolated filtrates were prepared from etiolated barley leaves or etiolated cucumber cotyledons incubated with ALA and Dpy in darkness for 4 h. The filtrates were diluted with Tris–sucrose buffer, pH 7.7, and adjusted to 67% glycerol so that for every accumulated tetrapyrrole $(\varepsilon_{DA})(C_{DA}) = (\varepsilon_D)(C_D)$. Corrected fluorescence emission spectra were elicited by excitation of the diluted filtrates at 400 nm for Proto, 420 nm for Mp(e), and 440 nm for MV or DV Pchlide a. The CFI_{DA} and CFI_D values for every accumulated tetrapyrrole were determined from the recorded Gaussian emission bands. The calculated efficiencies of energy transfer, E, thus calculated are reported for Proto, Mp(e), and MV and DV Pchlide a in Table 4.8.

AI. SAMPLE CALCULATION OF THE DISTANCE R SEPARATING ANABOLIC TETRAPYRROLE DONORS FROM VARIOUS CHL A ACCEPTORS

As described by Equation (4.14), the distance R separating donors from Chl a acceptors is calculated from

$$R^6 = (R_0^6/E) - R_0^6$$

TABLE 4.8
Relative Fluorescence Intensities and Efficiencies of Energy Transfer E for Proto, Mp(e), and Pchl(ide) a Donors In Situ at 77 K

Plant	Donor	CFI_{DA}	CFI_D	CFI_{DA}/CFI_D	E
Barley	Proto	14.80	30.37	0.49	0.51
	Mp(e)	6.65	15.11	0.44	0.56
	MV Pchl(ide) a	20.50	36.61	0.56	0.44
Cucumber	Proto	20.43	35.26	0.53	0.47
	Mp(e)	18.79	19.38	0.57	0.43
	DV Pchl(ide) a	11.00	28.56	0.53	0.47

Note: Green and etiolated filtrates of barley and cucumber cotyledons in $0.2 M$ Tris–HCl–$0.5 M$ sucrose, pH 7.7, were adjusted to equal donor concentrations and diluted 1:2/ (v/v) with glycerol. CFI_{DA} = corrected fluorescence intensity in arbitrary number of photons of green filtrates; CFI_D = corrected fluorescence intensity in arbitrary number of photons of etiolated filtrates; E = efficiency of energy transfer = $1 - CFI_{da}/CFI_d$. Other abbreviations are as in Table 4.1.

Source: Reproduced from Kolossov VL, Kopetz KJ, Rebeiz CA. *Photochem. Photobiol.* 2003; 78:184–196. With permission.

TABLE 4.9
Calculated R^6 Values for Anabolic Tetrapyrroles–Chl *a* Pairs at 77 K

Plant	Chl *a* Species	Chl *a* Absorbance (nm)	Donor	$R_0{}^6 \times 10^{-39}$ (cm)	E	$R_6 \times 10^{-39}$ (cm)
Barley	Chl *a* F685	670	Proto	3.61	0.51	3.43
	Chl *a* F695	677		5.17	0.51	4.92
	Chl *a* F735	704		0.145	0.51	0.138
	Chl *a* F685	670	Mp(e)	2.59	0.56	2.03
	Chl *a* F695	677		2.40	0.56	1.89
	Chl *a* F735	704		0.067	0.56	0.053
	Chl *a* F685	670	MV Pchl(ide) *a*	2.28	0.44	2.88
	Chl *a* F695	677		2.97	0.44	3.74
	Chl *a* F735	704		0.127	0.44	0.161
Cucumber	Chl *a* F685	670	Proto	0.647	0.47	0.739
	Chl *a* F695	677		0.632	0.47	0.721
	Chl *a* F735	704		0.018	0.47	0.020
	Chl *a* F685	670	Mp(e)	2.57	0.43	3.38
	Chl *a* F695	677		3.91	0.43	5.13
	Chl *a* F735	704		0.137	0.43	0.180
	Chl *a* F685	670	DV Pchl(ide) *a*	1.69	0.47	1.91
	Chl *a* F695	677		2.93	0.47	3.31
	Chl *a* F735	704		0.103	0.47	0.117

Note: R_0 = critical separations of donors from acceptors, taken from Table 4.7; E = the efficiencies E of resonance excitation energy transfer from donors to Chl *a* acceptors, taken from Table 4.8.

Source: Reproduced from Kolossov VL, Kopetz KJ, Rebeiz CA. *Photochem. Photobiol.* 2003; 78:184–196. With permission.

where R is the distance separating the donors from the acceptors; R_0 is the critical separation of the donors from the acceptors; and E is the efficiency of resonance energy transfer from the donors to the acceptors.

For example, the distance R separating Proto from Chl *a* (E670F685) was calculated as follows:

$$R_0^6 = 3.61 \times 10^{-39} \text{ cm}$$

$$E = 0.51$$

Substitution of the appropriate values for R_0^6 and E into Equation (4.14) results in

$$R^6 = (3.61 \times 10^{-39}/0.51) - (3.61 \times 10^{-39})$$

$$= 3.43 \times 10^{-39} \text{cm}$$

and

$$R = [(3.43 \times 10^{-39})^{1/6} \text{ cm}]10^8 \text{Å cm}^{-1} = 38.83 \text{Å}$$

The calculated R^6 values for the Proto, Mp(e), and Pchlide *a* donors in green barley and cucumber cotyledons are reported in Table 4.9.

III. RESULTS

A. DEMONSTRATION OF RESONANCE EXCITATION ENERGY TRANSFER FROM ANABOLIC TETRAPYRROLES TO CHLOROPHYLL *A*–PROTEIN COMPLEXES

Prior to probing the topography of the relationship between various Chl biosynthetic routes and the assembly of Chl–protein complexes, it was mandatory to determine whether resonance excitation energy transfer from anabolic tetrapyrroles to various Chl *a* species did take place *in situ*. This was recently achieved by Kolossov et al. [9] as described below.

1. Excitation Spectra of Accumulated Tetrapyrroles in Isolated Etioplasts

To help locate putative tetrapyrrole resonance excitation energy transfer maxima in green tissue homogenates or isolated chloroplasts, reference was made to excitation spectra of homogenates that were prepared in darkness from etiolated tissues that were induced to accumulate Proto, Mp(e), and Pchlide *a* by incubation with ALA and Dpy in darkness [19].

Proto excitation spectra were recorded at the Proto *in situ* emission maximum, at 630 nm for cucumber and at 627 nm for barley. In etiolated cucumber, Proto excitation appeared as a broad band between 380 and 420 nm with a LW excitation maximum at around 414 nm, and a shorter excitation shoulder at 407 nm (Figure 4.3Aa). In etiolated barley, it appeared as shorter excitation maxima at around 406 and 411 nm (Figure 4.3Ba). It was con-

jectured that the SW and LW Proto excitation maxima emanate from Proto in different *in situ* environments [9]. The other observable excitation maxima and shoulders of lower magnitude between 420 and 465 nm corresponded to excitations of accumulated Mp(e) and Pchlide *a* (Figures 4.3Aa and 4.3Ba). In diethyl ether at 77 K, Proto exhibited a

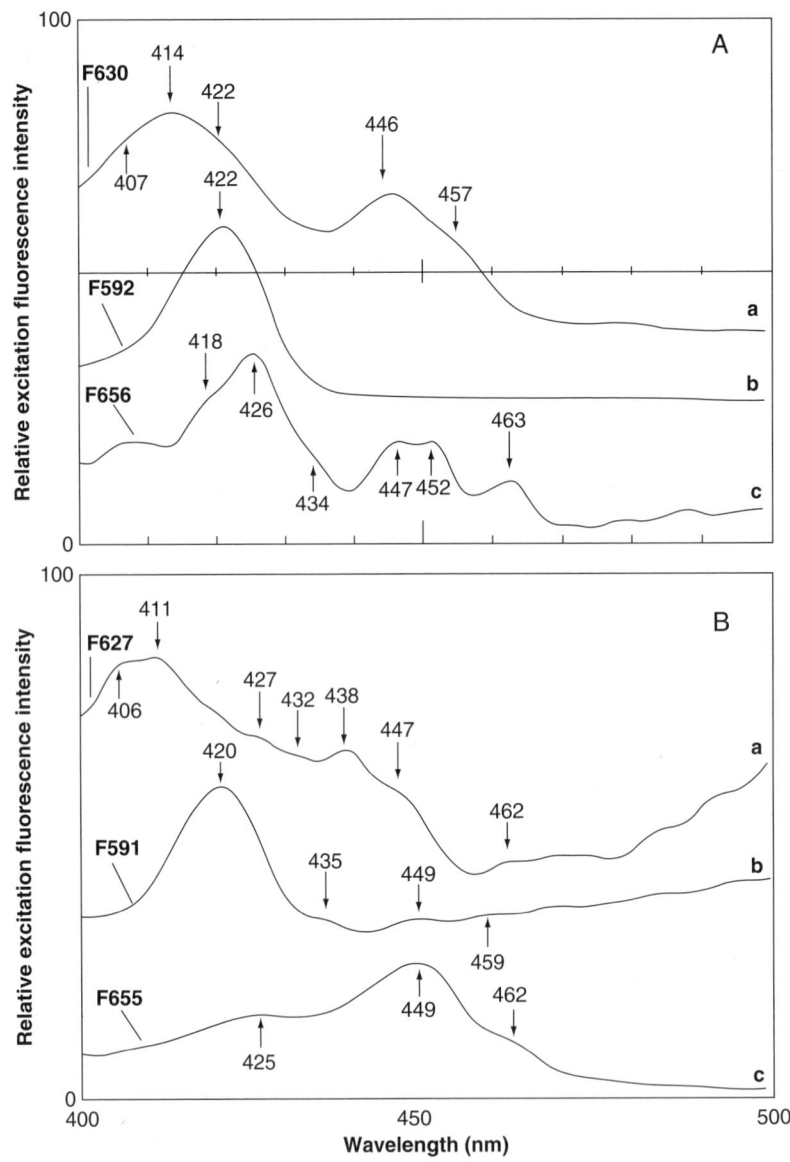

FIGURE 4.3 Excitation spectra recorded at 77 K, on homogenates prepared from (A) etiolated cucumber cotyledons and (B) etiolated barley leaves induced to accumulate Proto, Mp(e), and Pchlide *a* by incubation with 4.5 m*M* ALA + 3.7 m*M* Dpy for 6 h in darkness; (a) Proto–protein complex and Pchlide *a*–protein complex excitation spectra recorded at the emission maximum of the Proto–protein complex and at the SW emission tail of the Pchlide *a*–protein complex at 630 nm (Aa) and 627 nm (Ba); (b) Mp(e)–protein complex excitation spectra recorded at the emission maximum of the Mp(e)–protein complex at 592 nm (Ab) and 591 nm (Bb); (c) Pchlide *a*–protein complex, and Mp(e)–protein complex excitation spectra recorded at the emission maximum of the Pchlide *a*–protein complex and near the LW vibrational emission maximum of the Mp(e)–protein complex at 656 nm (Ac) and at 655 nm (Bc). Arrows point to various wavelengths of interest. (Reproduced from Kolossov VL, Kopetz KJ, Rebeiz CA. *Photochem. Photobiol.* 2003; 78:184–196. With permission.)

red emission, maximum at 629 nm and an excitation maximum at 409 nm [29].

The standard Mp(e) excitation band of etiolated tissue homogenates was elicited by recording excitation spectra at the *in situ* emission maximum of Mp(e) at 592 nm for cucumber, and at 591 nm for barley, or at the LW vibrational Mp(e) maximum at 655 to 656 nm [30]. In the spectra recorded at fluorescence emissions of 592 or 591 nm, Mp(e) exhibited an excitation band between 410 and 440 nm with an excitation maximum at 420 to 422 nm (Figures 4.3Ab and 4.3Bb). In the excitation spectra recorded at emissions of 656 to 655 nm, Mp(e) exhibited a LW excitation maximum at 425 to 426 nm (Figures 3Ac and 3Bc). As was proposed for Proto, it was conjectured that the SW and LW Mp(e) excitation maxima emanated from two Mp(e)s in two different *in situ* environments [9]. Because of emission band broadening it was not possible to distinguish between the two different Mpe environments by their vibrational emission maxima, but they were distinguished by their Soret excitation maxima. This was made possible by the high sensitivity of Soret excitation wavelengths to structural and environmental factors [9,19]. The other observable excitation maxima and shoulders of lower magnitude, between 435 and 465 nm were assigned to excitations of accumulated Pchlide *a*. In diethyl ether, MV Mp(e) exhibited emission and excitation maxima at 589 and 417 nm, respectively, whereas DV Mp(e) exhibited emission and excitation maxima at 591 and 424 nm, respectively [31].

To distinguish between resonance excitation energy transfer from DV and MV Pchl(ide) *a* to various Chl *a*–protein complexes, two different plant species belonging to two different greening groups of plants were used. Cucumber, a DDV–LDDV plant species that formed mainly DV Pchlide *a* in darkness and in light [22,32], allowed the monitoring of resonance excitation energy transfer mainly from DV Pchlide *a*, while barley, a DMV–LDMV plant species that formed mainly MV Pchlide *a* in darkness and in the light [22,32], allowed the monitoring of resonance excitation energy transfer mainly from MV Pchlide *a* to various Chl *a*–protein complexes.

In homogenates prepared from etiolated cucumber cotyledons and barley leaves, preincubated with ALA and Dpy in darkness, Pchl(ide) *a* excitation bands between 434 and 468 nm were elicited by recording an excitation spectrum at a fluorescence emission of 655 to 656 nm, i.e., at the *in situ* emission maximum of the LW emission of Pchl(ide) *a* [12,13]. In etiolated cucumber cotyledon homogenates, three Pchl(ide) *a* excitation maxima were observed, at 447, 452, and 463 nm (Figure 4.3Ac) [9]. In etiolated barley homogenates, excitation maxima were observed at 449 and 462 nm (Figure 4.3Bc). In diethyl ether, MV Pchlide *a* exhibited an emission maximum at 625 nm and a split Soret excitation band with maxima at 437 and 443 nm. DV Pchlide *a* exhibited a similar emission maximum at 625 nm and split Soret excitation maxima at 443 and 451 nm [31].

2. Evidence of Resonance Excitation Energy Transfer from Proto to Chl *a* ~F685

In green cucumber, resonance excitation energy transfer from Proto to Chl *a* ~F685, at low, medium, and high Proto accumulation was manifested by a pronounced resonance excitation energy transfer band between 380 and 420 nm with multiple SW, medium wavelengths (MW), and LW excitation peaks or shoulders between 390 and 417 nm, namely at 390 to 399, 402 to 412, and 415 to 416 nm (Table 4.10, Figure 4.4c) [9]. These resonance excitation energy transfer maxima fell within the Proto excitation band observed in etiolated cucumber cotyledon homogenates (Figure 4.3Aa). The best resolution of resonance excitation energy transfer peaks was achieved at low to medium Proto concentration (54 to 376 pmol/ml suspension) (Table 4.10). At higher Proto concentrations (1046 pmol/ml suspension), the resonance excitation energy transfer band was dominated by a 411 nm peak [9].

In green barley, the most pronounced resonance excitation energy transfer donation appeared to originate from SW Proto sites with excitation maxima at 389 to 391 nm and from MW sites with excitation maxima between 410 and 413 nm (Table 4.10, Figure 4.5c) [9]. Other resonance excitation energy transfer shoulders were observed at 396 to 398 and at 404 nm (Table 4.10, Figure 4.5c).

It was proposed that the observed multiple resonance excitation energy transfer maxima and shoulders indicated different *in situ* environments from which Proto donated its excitation energy to Chl *a* ~F685, namely SW sites with excitation maxima between 389 and 400 nm, MW sites between 402 and 412 nm, and LW sites with excitation maxima between 415 and 416 nm [9].

3. Evidence of Resonance Excitation Energy Transfer from Proto to Chl *a* ~F695

The Chl *a* emission at 694 to 695 nm is believed to originate from CP47 and/or CP29, two PSII antennae [2]. In green cucumber, resonance excitation energy transfer from Proto to Chl *a* ~F695 at low, medium, and high Proto accumulation was manifested by a Proto resonance excitation energy transfer band between 380 and 420 nm which exhibited multiple

TABLE 4.10
Mapping of Resonance Excitation Energy Transfer Maxima to Chl *a* ~F685, Chl *a* ~F695, and Chl *a*~F7335 In Situ

Plant Species	Major Donor	Undil. Donor Conc. (pmol/ml suspension)	Dil. Donor Conc. (pmol/ml suspension)	Excitation Resonance Energy Maxima to: (nm)			Conc. (nM)		Incub. (h)
				Chl *a* F686	Chl *a* F694	Chl *a* F738	ALA	Dpy	
Cucumber	Proto	1620	54	397p, 402p, 410p, 415p	390s, 400p, 409p	390s, 395s, 408p, 417p	4.5	3.7	6
Cucumber	Proto	1242	83	387p, 402p, 412p	392p, 406p	388p, 399p, 403p, 410p, 415p	20	4	6
Cucumber	Proto	1374	92	390p, 399p, 405p, 412p	399p, 409p, 412s	399p, 400p, 416p	20	0	6
Cucumber	Proto	5640	376	395p, 404s, 411p, 416p	395p, 406p, 414p	393p, 400s, 407p	20	16	6
Cucumber	Proto	3138	1046	402s, 411p	404p, 410s, 416p,	399s, 405s, 411p	20	0	12
Barley	Proto	390	13	391, 398s, 404s, 411p	389s, 395p, 406p, 414p	390s, 393p, 400s, 406p, 412p, 416s,	4.5	3.7	6
Barley	Proto	1492	61	389p, 396s, 404s, 410p, 412p	396p, 406p, 412p	389s, 395p, 406s, 410p,	20	16	6
Barley	Proto	966	64	395s, 400p, 405s, 413p	389p, 397s, 403p, 412p	388s, 393p, 400s, 406p, 412p	20	0	6
Barley	Proto	1015	68	389p, 396p, 412p, 413s	389p, 398p, 409p,	396s, 400p, 412p, 414s	20	4	6
Cucumber	Mp(e)	390	26	419p, 431p	422p, 429p, 434p	—	20	0	6
Cucumber	Mp(e)	2490	83	422p, 432p	420p, 425p	417p, 424p, 427s, 429p	4.5	3.7	6
Cucumber	Mp(e)	1374	91	418s, 424p, 433p	419p, 426p	414p, 423p	20	4	6
Cucumber	Mp(e)	1854	185	421p, 427s, 430s	421p, 428s	421p, 430p	20	0	12
Cucumber	Mp(e)	1854	618	421p, 427s, 430s	421p, 427s, 430s	421p, 430p	20	0	12
Barley	Mp(e)	300	10	420p, 428s	424p, 430s	426s, 432s	4.5	3.7	6
Barley	Mp(e)	162	11	423p	418p, 422s, 427p	422s, 426p, 431s	20	0	6
Barley	Mp(e)	378	25	423p, 428s	418p, 430p	426s, 432p	20	4	6
Cucumber	DV Pchlide *a*	1998	133	438p, 446p, 453s, 460s, 467p	440s, 448p, 454s, 460p	448p, 453p, 461p	20	4	6
Cucumber	DV Pchlide *a*	4590	153	443p, 449p, 457p	436s, 442p, 453p, 463p	439s, 453p, 457p, 460p	4.5	3.7	6
Cucumber	DV Pchlide *a*	6180	412	437p, 444s, 452p, 458p	435p, 441p, 451p, 462p	437p, 447s, 454s, 457p, 463s	20	0	6
Cucumber	DV	6180	1030	438s, 447p, 452p, 456s, 462s	441s, 447p, 452p, 459p	436p, 448s, 454s, 458s	20	0	6
Cucumber	DV Pchlide *a*	6180	1435	435p, 447s, 453p, 460s	438s, 445s, 452p, 456s, 460s, 462s	436s, 444s, 452s, 458p, 462s	20	0	12
Cucumber	DV Pchlide *a*	14352	4784	440s, 449p, 455s, 460s	434p, 440s, 447s, 452p, 459s	434s, 440p, 447s, 462p	20	0	12
Barley	MV Pchlide *a*	780	26	434s, 441p, 452p, 460p	438s, 445p, 449p, 463p	440p, 449p, 458s, 468p,	4.5	3.7	6
Barley	MV Pchlide *a*	1554	104	439p, 445s, 450p, 458p, 463s	436s, 447p, 455p, 463s	440p, 450p, 458p	20	4	6
Barley	MV Pchlide *a*	2900	193	439s, 444p, 451p, 462p, 467p	435p, 440s, 446p, 453p, 460p	438s, 453p, 457p, 464s	20	0	6

Note: A dash represents missing data. Undil. = donor concentration before dilution, Dil. = donor concentration after dilution, s = shoulder; p = peak. Only the barley spectra depicted in Figure 4.9 were recorded at the observed peak of Chl *a* emission at F742 nm.

Source: Reproduced from Kolossov VL, Kopetz KJ, Rebeiz CA. *Photochem. Photobiol.* 2003; 78:184–196. With permission.

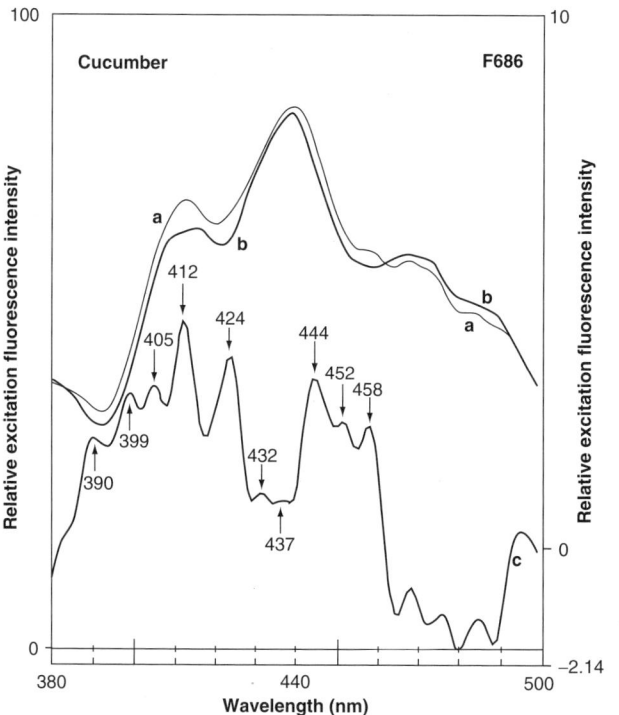

FIGURE 4.4 Excitation energy transfer from anabolic tetrapyrroles to Chl *a* F686 in isolated chloroplasts prepared from green cucumber cotyledons. (a) 77 K excitation spectrum of isolated chloroplasts prepared from green cucumber cotyledons incubated with 20 m*M* ALA for 6 h in darkness. Tetrapyrrole accumulation amounted to 92 (Proto), 26 (Mp(e)), and 412 (Pchlide *a*) pmol/ml of diluted plastid suspension. (b) 77 K excitation spectrum of isolated chloroplasts prepared from green cucumber cotyledons incubated with water for 6 h in darkness. (c) Calculated ALA-treated – water-incubated difference spectrum. Spectra were recorded at an emission wavelength of 686 nm on chloroplast suspensions diluted with glycerol (1:2 v/v), at 77 K. Treated and control chloroplasts were diluted to the same Chl concentration. After smoothing, very small differences in Chl concentrations were adjusted for by normalization to the same value at 499 nm. The left ordinate scale is for the excitation spectra. The right ordinate scale is for the difference spectrum. The upper abscissa scale at an ordinate value of 0 is for the excitation spectra. The lower abscissa scale at an ordinate value of −2.14 is for the difference spectrum. At 499 nm, the difference spectrum intercepts its ordinate at 0.0. Arrows point to wavelengths of interest. Negative peaks in the difference spectra were observed only for cucumber in the carotenoids region. It may be due to specific energy transfer from carotenoids to accumulated tetrapyrroles, which is dissipated as heat by internal conversion. (Reproduced from Kolossov VL, Kopetz KJ, Rebeiz CA. *Photochem. Photobiol.* 2003; 78:184–196. With permission.)

excitation peaks or shoulders between 389 and 416 nm, namely at SW sites between 389 and 400 nm, at MW sites between 406 and 412 nm, and at LW sites be-

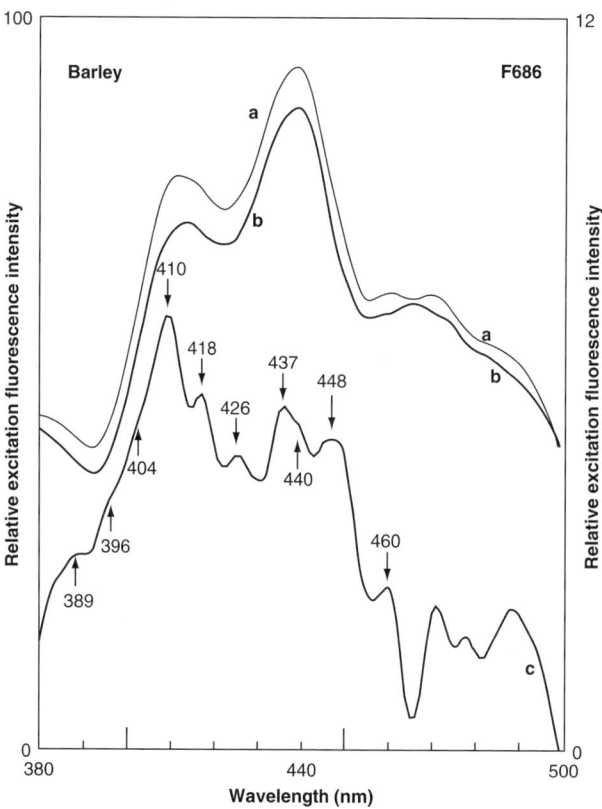

FIGURE 4.5 Excitation energy transfer from anabolic tetrapyrroles to Chl *a* F686 in isolated chloroplasts prepared from green barley leaves. (a) 77 K excitation spectrum of isolated chloroplasts prepared from green barley leaves incubated with 20 m*M* ALA and 16 m*M* Dpy for 6 h in darkness. Tetrapyrrole accumulation amounted to 61 (Proto), 23 (Mp(e)), and 58 (Pchlide *a*) pmol/ml of diluted plastid suspension. (b) 77 K excitation spectrum of isolated chloroplasts prepared from green barley leaves incubated with water for 6 h in darkness. (c) Calculated ALA-treated – water-incubated difference spectrum. Other conditions and conventions are as in Figure 4.4. The abscissa scale at an ordinate value of 0 is for the excitation spectrum and the difference spectrum. Arrows point to wavelengths of interest. (Reproduced from Kolossov VL, Kopetz KJ, Rebeiz CA. *Photochem. Photobiol.* 2003; 78:184–196. With permission.)

tween 414 and 416 nm (Table 4.10, Figure 4.6c) [9]. Resolution of resonance excitation energy transfer peaks was equally good at low medium and high Proto accumulation. In green barley, the most pronounced resonance excitation energy transfer donation appeared to emanate from SW Proto sites at 389 to 396 nm, from MW sites at 403 to 412 nm, and from LW sites with excitation maxima at 414 to 416 nm (Table 4.10, Figure 4.7c).

It was proposed that the observed multiple resonance excitation energy transfer maxima and shoulders

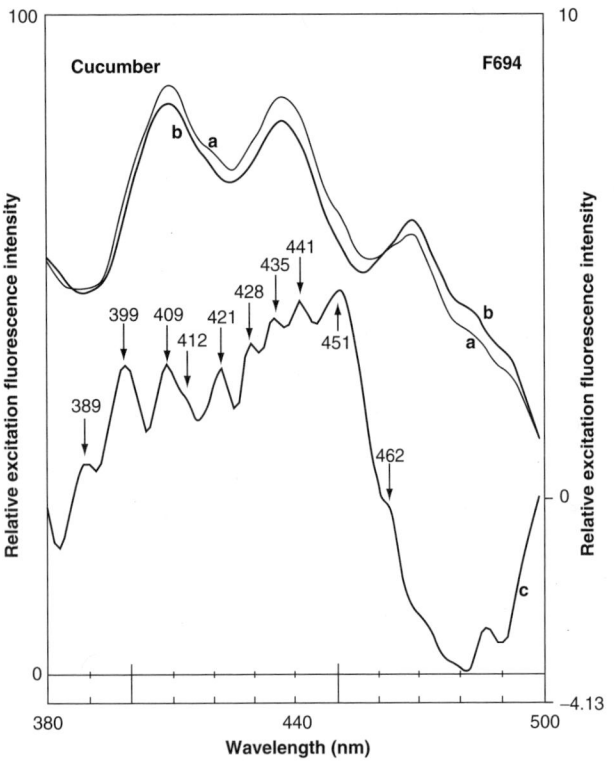

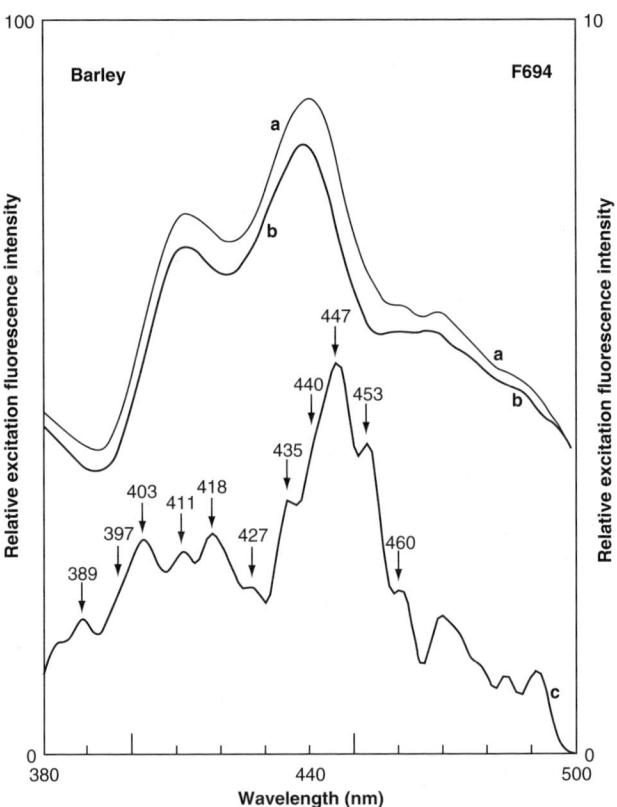

FIGURE 4.6 Excitation energy transfer from anabolic tetra-pyrroles to Chl *a* F694 in isolated chloroplasts prepared from green cucumber cotyledons. (a) 77 K excitation spectrum of isolated chloroplasts prepared from green cucumber cotyledons incubated with 20 m*M* ALA for 6 h in darkness. Tetrapyrrole accumulation amounted to 92 (Proto), 26 (Mp(e)), and 412 (Pchlide *a*) pmol/ml of undiluted plastid suspension. (b) 77 K excitation spectrum of isolated chloroplasts prepared from green cucumber cotyledons incubated with water for 6 h in darkness. (c) Calculated ALA-treated – water-incubated difference spectrum. Other conditions and conventions are as in Figure 4.4. The upper abscissa scale at an ordinate value of 0 is for the excitation spectra. The lower abscissa scale at an ordinate value of −4.13 is for the difference spectrum. At 499 nm, the difference spectrum intercepts its ordinate at 0.0. Arrows point to wavelengths of interest. Negative peaks in the difference spectra were observed only for cucumber in the carotenoids region. It may be due to specific energy transfer from carotenoids to accumulated tetrapyrroles, which is dissipated as heat by internal conversion. (Reproduced from Kolossov VL, Kopetz KJ, Rebeiz CA. *Photochem. Photobiol.* 2003; 78:184–196. With permission.)

FIGURE 4.7 Excitation energy transfer from anabolic tetra-pyrroles to Chl *a* F694 in isolated chloroplasts prepared from green barley leaves. (a) 77 K excitation spectrum of isolated chloroplasts prepared from green barley leaves incubated with 20 m*M* ALA for 6 h in darkness. Tetrapyrrole accumulation amounted to 64 (Proto), 11 (Mp(e)), and 193 (Pchlide *a*) pmol/ml of diluted plastid suspension. (b) 77 K excitation spectrum of isolated chloroplasts prepared from green barley leaves incubated with water for 6 h in darkness. (c) Calculated ALA-treated – water-incubated difference spectrum. Other conditions and conventions are as in Figure 4.4. The abscissa scale at an ordinate value of 0 is for the excitation spectrum and the difference spectrum. Arrows point to wavelengths of interest. (Reproduced from Kolossov VL, Kopetz KJ, Rebeiz CA. *Photochem. Photobiol.* 2003; 78:184–196. With permission.)

4. Evidence of Resonance Excitation Energy Transfer from Proto to Chl *a* ~F735

The Chl *a* emission at F735 to 742 nm is believed to originate from LHCI-730, a PSI antenna [2]. In green cucumber, resonance excitation energy transfer from Proto to Chl *a* ~F735 at low, medium, and high Proto accumulation was manifested by a Proto resonance excitation energy transfer band between 380 and 420 nm which exhibited multiple excitation peaks or shoulders between 388 and 416 nm [9]. SW resonance excitation energy sites were observed at 388 to

indicated different *in situ* environments from which Proto donated its excitation energy to Chl *a* ~F695, namely from SW sites with excitation maxima between 389 and 400 nm, from MW sites between 406 and 412 nm, and from LW sites between 414 and 416 nm [9].

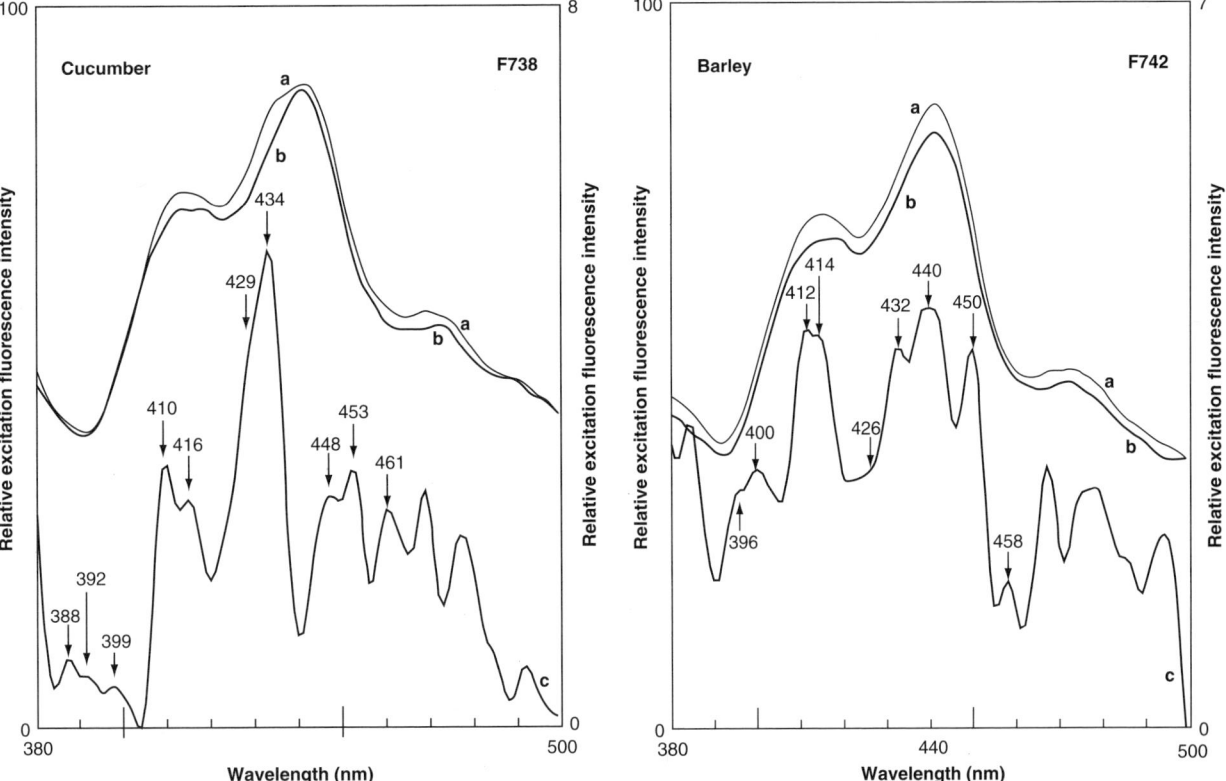

FIGURE 4.8 Excitation energy transfer from anabolic tetrapyrroles to Chl *a* F738 in isolated chloroplasts prepared from green cucumber cotyledons. (a) 77 K excitation spectrum of isolated chloroplasts prepared from green cucumber cotyledons incubated with 20 m*M* ALA and 4 m*M* Dpy for 6 h in darkness. Tetrapyrrole accumulation amounted to 83 (Proto), 91 (Mp(e)), and 133 (Pchlide *a*) pmol/ml of diluted plastid suspension. (b) 77 K excitation spectrum of isolated chloroplasts prepared from green cucumber cotyledons incubated with water for 6 h in darkness. (c) Calculated ALA-treated – water-incubated difference spectrum. Other conditions and conventions are as in Figure 4.4. The abscissa scale at an ordinate value of 0 is for the excitation spectra. The lower abscissa scale at an ordinate value of −1.70 is fused with the upper abscissa scale and is for the difference spectrum. Arrows point to wavelengths of interest. Negative peaks in the difference spectra were observed only for cucumber in the carotenoids region. It may be due to specific energy transfer from carotenoids to accumulated tetrapyrroles in cucumber, which is dissipated as heat by internal conversion. (Reproduced from Kolossov VL, Kopetz KJ, Rebeiz CA. *Photochem. Photobiol.* 2003; 78:184–196. With permission.)

FIGURE 4.9 Excitation energy transfer from anabolic tetrapyrroles to Chl *a* F738–742 in isolated chloroplasts prepared from green barley leaves. This is the only instance where the emission maximum of Chl *a* was observed at 442 instead of 738 nm. (a) 77 K excitation spectrum of isolated chloroplasts prepared from green barley leaves incubated with 20 m*M* ALA and 4 m*M* Dpy for 6 h in darkness. Tetrapyrrole accumulation amounted to 68 (Proto), 25 (Mp(e)), and 104 (Pchlide *a*) pmol/ml of undiluted plastid suspension. (b) 77 K excitation spectrum of isolated chloroplasts prepared from green barley leaves incubated with water for 6 h in darkness. (c) Calculated ALA-treated – water-incubated difference spectrum. Other conditions and conventions are as in Figure 4.4. The abscissa scale at an ordinate value of 0 is for the excitation spectrum and the difference spectrum. Arrows point to wavelengths of interest. (Reproduced from Kolossov VL, Kopetz KJ, Rebeiz CA. *Photochem. Photobiol.* 2003; 78:184–196. With permission.)

400 nm, MW sites were observed at 405 and 411 nm, and LW sites were observed at 415 to 416 nm (Table 4.10, Figure 4.8c). In green barley, the most pronounced resonance excitation energy transfer donation appeared to emanate from SW Proto sites at 389 to 400 nm, from MW sites at 406 to 412 nm, and from

LW sites with excitation maxima at 414 to 416 nm (Table 4.10, Figure 4.9c) [9]. In this case too, it was proposed that the observed multiple resonance excitation energy transfer maxima and shoulders indicated different *in situ* environments from which Proto donated its excitation energy to Chl *a* F738 to 742 [9].

5. Evidence of Resonance Excitation Energy Transfer from Mp(e) to Chl *a* ~F685

Since Mg-Proto and Mp(e) exhibited identical electronic spectroscopic properties and could not be distinguished from one another *in situ*, Mg-Proto and Mp(e) were monitored *in situ*, as a single entity, namely Mp(e). The Mp(e) pool in cucumber and barley consisted mainly of DV Mp(e). In green cucumber, resonance excitation energy transfer from Mp(e) to Chl *a* ~F685, at low, medium, and high Mp(e) accumulation was manifested by a pronounced resonance excitation energy transfer band between 410 and 440 nm with multiple short medium, and LW resonance excitation energy transfer peaks or shoulders at 418 to 422, 421 to 427, and 430 to 433 nm, respectively (Table 4.10, Figure 4.4c) [9]. These resonance excitation energy transfer maxima and shoulders fell within the Mp(e) excitation band observed in etiolated cucumber cotyledon homogenates (Figure 4.3Ab). The best resolution of resonance excitation energy transfer peaks was achieved at low to medium Mp(e) concentrations (26 to 185 pmol/ml diluted suspension) (Table 4.10). At higher Mp(e) concentrations (618 pmol/ml diluted suspension), the resonance excitation energy transfer band was dominated by a SW 421 nm peak (Table 4.10). In green barley, the most pronounced resonance excitation energy transfer donation appeared to originate from SW Mp(e) sites with excitation maxima at 418 to 420 nm, from MW sites with excitation maxima at 423 to 426 nm, and from a LW site at 428 nm (Table 4.10, Figure 4.5c) [9].

It was proposed that the observed multiple resonance excitation energy transfer maxima and shoulders indicated different *in situ* environments from which Mp(e) donated its excitation energy to Chl *a* F686, namely from SW sites with excitation maxima at 418 to 420 nm, MW sites at 423 to 426 nm, and LW sites at 426 to 428 nm [9].

6. Evidence of Resonance Excitation Energy Transfer from Mp(e) to Chl *a* ~F695

In green cucumber, resonance excitation energy transfer from Mp(e) to Chl *a* ~F694, at low, medium, and high Mp(e) accumulation was manifested by a pronounced resonance excitation energy transfer band between 410 and 440 nm with multiple SW, MW, and LW excitation peaks or shoulders at 419 to 421, 425 to 426, and 428 to 430 nm, respectively (Table 4.10, Figure 4.6c) [9]. These resonance excitation transfer maxima fell within the Mp(e) excitation band observed in etiolated cucumber cotyledon homogenates (Figure 4.3Ab). The best resolution of res-

onance excitation energy transfer peaks was achieved at low to medium Mp(e) concentration (26 to 185 pmol/ml diluted suspension) (Table 4.10). At higher Mp(e) concentration (618 pmol/ml diluted suspension), the resonance excitation energy transfer band was dominated by a 421 nm peak (Table 4.12). In green barley, the most pronounced resonance excitation energy transfer donation appeared to emanate from a SW Mp(e) site with an excitation maximum at 418 nm, and from LW sites with excitation maxima at 427 and 430 nm, (Table 4.10, Figure 4.7c) (9). It was proposed that the observed multiple resonance excitation energy transfer maxima and shoulders indicated different *in situ* environments from which Mp(e) donated its excitation energy to Chl *a* F694 [9].

7. Evidence of Excitation Resonance Energy Transfer from Mp(e) to Chl *a* ~F735

In green cucumber, resonance excitation energy transfer from Mp(e) to Chl *a* ~F735, at low, medium, and high Mp(e) accumulation was manifested by a pronounced resonance excitation energy transfer band between 417 and 440 nm with multiple SW, MW, and LW excitation peaks or shoulders at 417 to 421, 424 to 427, and 429 to 430 nm, respectively (Table 4.10) [9]. These resonance excitation transfer maxima and shoulders fell within the Mp(e) excitation band observed in etiolated cucumber cotyledon homogenates (Figure 4.3Ab). At high Mp(e) concentrations (618 pmol/ml diluted suspension), the resonance excitation energy transfer band was dominated by 421 and 430 nm peaks (Table 4.12) [9]. In green barley, the most pronounced resonance excitation energy transfer donation appeared to originate from a MW Mp(e) site with an excitation maximum at 426 nm (Table 4.10), and from a LW site with an excitation maximum at 432 nm (Figure 4.9c) (9). It was proposed that the observed multiple resonance excitation energy transfer maxima and shoulders indicated different *in situ* environments from which Mp(e) donated its excitation energy to ~Chl *a* F735, namely: SW sites at 417 to 422 nm, MW sites at 423 to 427 nm, and LW sites at 429 to 432 nm [9].

8. Evidence of Resonance Energy Transfer from Pchlide *a* to Chl *a* ~F685

In green cucumber, resonance excitation energy transfer from DV Pchlide *a* to Chl *a* F686 at low, medium, and high DV Pchlide *a* accumulation was manifested by a pronounced resonance excitation energy transfer band between 434 and 468 nm with multiple excitation SW, MW, and LW peaks or shoulders at 435 to

438, 440 to 453, and 458 to 462 nm, respectively (Table 4.10, Figure 4.4c) [9]. These resonance excitation transfer maxima and shoulders fell within the DV Pchlide *a* excitation band observed in etiolated cucumber cotyledon homogenates (Figure 4.3Ac). In green barley, the most pronounced resonance excitation energy transfer donation appeared to emanate from SW MV Pchlide *a* sites at 437 to 439 nm, MW MV Pchlide *a* sites with excitation maxima at 441, 448, 451 to 452 nm and from LW sites with excitation maxima at 460 to 462 and 467 nm (Table 4.10, Figure 4.5c) [9]. It was proposed that the observed multiple resonance excitation energy transfer maxima and shoulders indicated different *in situ* environments from which DV and MV Pchlide *a* donated excitation energy to Chl *a* F686, namely: SW sites at 434 to 439, MW sites at 440 to 453 nm, and LW sites at 458 to 467 nm [9].

9. Evidence of Resonance Excitation Energy Transfer from Pchl(ide) *a* to Chl *a* ~F695

In green cucumber, resonance excitation energy transfers from DV Pchlide *a* to Chl *a* ~F695, at low, medium, and high DV Pchlide *a* accumulation were manifested by a pronounced resonance excitation energy transfer band between 434 and 468 nm with multiple SW, MW, and LW resonance excitation energy transfer peaks or shoulders at 435 to 438, 440 to 454, and 458 to 463 nm, respectively (Table 410, Figure 4.6c) [9]. These resonance excitation energy transfer maxima fell within the DV Pchlide *a* excitation band observed in etiolated cucumber cotyledon homogenates (Figure 4.3Ac). In green barley, the most pronounced resonance excitation energy transfer donations appeared to originate from a SW MV Pchlide *a* site with an excitation maximum at 435 nm, from MW MV Pchlide *a* sites with excitation maxima at 445 to 447 and 453 to 455 nm, and from LW sites with excitation maxima at 460 to 463 nm (Table 4.10, Figure 4.7c) [9]. It was proposed that the observed multiple resonance excitation energy transfer maxima and shoulders indicated different *in situ* environments from which DV and MV Pchlide *a* donated excitation energy to Chl *a* ~F695, namely from SW sites at 435 to 438 nm, MW sites at 440 to 453 nm, and LW sites at 458 to 467 nm [9].

10. Evidence of Resonance Excitation Energy Transfer from Pchl(ide) *a* to Chl *a* ~F735

In green cucumber, resonance excitation energy transfer from DV Pchlide *a* to Chl *a* ~F735 at low, medium, and high DV Pchlide *a* accumulation was manifested by a pronounced resonance excitation en-

ergy transfer band between 434 and 468 nm with multiple SW, MW, and LW resonance excitation energy transfer peaks or shoulders at 436 to 439, 440 to 454, and 457 to 463 nm, respectively (Table 4.10, Figure 4.8c) [9]. These resonance excitation energy transfer maxima or shoulders fell within the DV Pchlide *a* excitation band observed in etiolated cucumber cotyledon homogenates (Figure 4.3Ac). In green barley, the most pronounced resonance excitation energy transfer donations appeared to originate from MW MV Pchlide *a* sites with excitation maxima at 440, 449 to 450, and 453 nm and from LW sites with excitation maxima at 458 to 464 nm (Table 4.10, Figure 4.8c) [9]. It was proposed that the observed multiple resonance excitation energy transfer maxima and shoulders indicated different *in situ* environments from which DV and MV Pchlide *a* donated excitation energy to Chl *a* F738–742, namely from SW sites at 436 to 438 nm, MW sites at 440 to 454 nm, and LW sites at 457 to 468 nm [9].

11. Comparison of Excitation Spectra of Reconstituted Tetrapyrroles-Cucumber Plastid Lipoproteins to the Resonance Excitation Energy Transfer Profiles Observed *In Situ*

In an effort to gain a better understanding of the possible relationship between the Soret excitation profiles of Proto, Mg-proto, and DV Pchlide *a*, randomly bound to chloroplast lipoproteins, and the resonance excitation energy transfer profiles observed *in situ* in isolated chloroplasts this issue was investigated as described below.

Isolated cucumber chloroplasts were stripped of their pigments and were complexed to exogenous Proto, Mg-Proto, and DV Pchlide *a* as described in Section II. Excitation spectra of tetrapyrrole-complexed and tetrapyrrole-free lipoproteins were recorded at vibrational emission maxima of 686, 694, and 738 nm, and difference spectra of tetrapyrrole-spiked plastid lipoproteins minus plastid lipoproteins devoid of tetrapyrroles were generated. It was conjectured that if nonspecific tetrapyrrole–lipoproteins binding took place at a highly unspecific binding site, then one would observe a main Soret excitation peak that would overtake and dwarf all others.

As reported in Table 4.11, the putative nonspecific tetrapyrrole–chloroplast lipoprotein binding resulted in very simple Soret excitation profiles, far less complex than the resonance excitation energy transfer profiles reported in Table 4.10 [9]. No corresponding Soret excitation peaks were observed at vibrational emissions of 738 nm.

TABLE 4.11
Mapping of the Soret Excitation Profiles of Exogenous DV Proto,
DV Mg-Proto, and DV Pchlide *a* Complexed to Cucumber Chloroplast
Lipoproteins

Tetrapyrrole	Emission Maximum (nm)	Soret Excitation Maxima Observed at Emissions of		
		686 nm	694 nm	738 nm
DV Proto	624	406p	406p	None
DV Mg-Proto	592	422p, 430p	420s, 425p	None
DV Pchlide *a*	635	446p, 450s	450p	None

Note: p = peak, s = shoulder.

12. Could the Anabolic Tetrapyrroles Have Diffused from Their Enzyme Binding Sites to Bind Nonspecifically to Various Chloroplast Proteins *In Situ*?

It was pointed out by Kolossov et al. [9] that under the present experimental conditions, it was very unlikely for accumulated tetrapyrroles to leave their enzyme binding sites in order to associate at random with membrane lipoproteins. It is noteworthy that for a particular tetrapyrrole, under various incubation conditions, and at various levels of tetrapyrrole accumulation, the heterogeneous resonance excitation energy transfer profiles from SW, MW, and LW sites to a particular Chl *a* species were remarkably preserved (Table 4.10). This was in contrast to the simple Soret excitation profiles which were observed in Table 4.11, for reconstituted cucumber chloroplast lipoproteins–exogenous-tetrapyrrole complexes. This, and the constancy of the resonance excitation donation profiles reported in Table 4.10 over a wide range of tetrapyrrole concentrations, argued against significant tetrapyrrole diffusion and nonspecific binding to proteins (also see Section IV) [9].

Furthermore, the only documented case of a tetrapyrrole leaving its natural enzyme binding site is that of DV Proto which accumulates when protoporphyrinogen IX oxidase activity is inhibited [33–35]. Under these circumstances, protoporphyrinogen IX leaves its enzyme binding site and tunnels its way out of the chloroplast. It was suggested that the tunneling may be caused by the highly flexible structure of protoporphyrinogen IX, which is a reduced tetrapyrrole that lacks the rigid planer structure and alternating double bond system of oxidized tetrapyrroles [9].

Altogether, the above results demonstrated unambiguous resonance excitation energy transfer from anabolic tetrapyrroles to Chl *a*–protein complexes and made it possible to investigate the relationships between Chl biosynthetic routes and the topography of thylakoid membrane biogenesis by resonance excitation energy transfer manipulations, as described below.

B. Calculation of Resonance Excitation Energy Transfer Rates from Anabolic Tetrapyrroles to Chlorophyll a–Protein Complexes at Fixed Distances That May Prevail in a Tightly Packed Linear, Continuous Array PSU

In a first attempt, efforts were made to investigate whether the observed resonance excitation energy transfers described in Section IIIA were compatible with the SBP-single location model described in Figure 4.1A. To this end, resonance excitation energy transfer rates from anabolic tetrapyrroles to various Chl–protein complexes that populated a tightly packed continuous array PSU were calculated over the shortest fixed distances that would prevail in such a model. The results of these calculations are described below.

1. Energy Transfer Rates from Proto to Various Chl a–Protein Species at Fixed Distances *R* That May Prevail in the SBP-Single Location Chl–Thylakoid Apoprotein Biosynthesis Model

As mentioned earlier, DV Proto is an early intermediate along the Chl biosynthetic chains and is several steps removed from the end product, Chl *a*. The detection of excitation resonance energy transfers from Proto to (a) Chl *a* ~F685 (the Chl *a* of LHCII), and to LHCI-680 (the inner LHC antennae of PSI), (b) to Chl *a* ~F695, the Chl *a* of CP47 and CP29 (two PSII antennae), and (c) to Chl *a* ~F735, the Chl *a* of LHCI-730 (the inner PSI antenna) [2], made it possible to investigate whether resonance excitation energy transfer from Proto to the above-mentioned Chl–protein complexes can take place over distances that separate them from a single-branched Chl *a* biosynthetic pathway located in the center of a tightly packed, continuous array PSU model [2]. Indeed, in this model it can be calculated that the core of PSII including CP47 and CP29, would be located about 126 Å away from the SBP. On the other hand, LHCI-730 would be located about 159 Å on the other side of the SBP. The centers of the inner and outer halves of LHCII surrounding the PSII core would be located about 156 Å (outer half) and 82 Å (inner half)

from the SBP. Therefore, energy transfer rates from Proto to the various Chl *a* species over 159, 126, and 82 Å as well as over critical distances $R = R_0$ were calculated.

In Table 4.12 and Table 4.13, the rates of resonance excitation energy transfer, K_T, from Proto to various Chl *a* species are expressed as a percentage of de-excitation via 100% resonance excitation energy transfer. In all cases, rates of excitation resonance energy transfer from Proto to Chl *a* ~F685, ~F695, and ~F730 (i.e., at distances of 159 to 82 Å) were negligible. In other words, resonance excitation energy transfer rates for the SBP-location model from Proto to Chl *a*-protein complexes belonging to PSI, PSII, and LHCII at distances that were likely to prevail in a 130 × 450 Å continuous array PSU were

TABLE 4.12
Rates of Resonance Excitation Energy Transfer, K_T, at Fixed Distances R Separating Proto from Various Chl *a* Species In Situ at 77 K in Green Barley Leaves

Chl *a* Species	Chl *a* Absorbance (nm)	τ_D (ns)	R_0 (Å)	R (Å)	K_T (s⁻¹)	K_T as Percent of 100% Transfer Efficiency
Chl *a* F685 (LHCI-680 + outer half of LHCII)	670	0.31	39.17	159.00	7.29×10^6	1.11×10^{-2}
Chl *a* F685 (inner half of LHCII)	670	0.31	39.17	82.00	3.88×10^7	0.60
Chl *a* F685 at $R_0 = R$	670	0.31	39.17	39.17	3.26×10^9	50
Chl *a* F695 (CP47) + CP29	677	0.31	41.58	126.00	4.22×10^6	0.60×10^{-1}
Chl *a* F695 at $R_0 = R$	677	0.31	41.58	41.58	3.26×10^9	50
Chl *a* F735 (LHCI-730)	704	0.31	22.92	159.00	2.93×10^4	4.49×10^{-4}
Chl *a* F738 at $R_0 = R$	704	0.31	22.92	22.92	3.26×10^9	50

Note: τ_D = actual mean lifetime of excitation of the Proto donor in the presence of the acceptor (Chl *a* species); R_0 = critical separation of Proto donor from Chl *a* acceptors for which energy transfer from the excited Proto donor to the Chl *a* acceptor and emission from the excited donor to the ground state amount to 50% (i.e., are equally probable); R = separation between the centers of the excited Proto donor and the unexcited Chl *a* acceptors.

TABLE 4.13
Rates of Resonance Excitation Energy Transfer, K_T, at Fixed Distances R Separating Proto from Various Chl *a* Species In Situ at 77 K in Green Cucumber Cotyledons

Chl *a* Species	Chl *a* Absorbance (nm)	τ_D (ns)	R_0 (Å)	R (Å)	K_T (s⁻¹)	K_T as Percent of 100% Transfer Efficiency
Chl *a* F685 (LHCI-680 + outer half of LHCII)	670	0.14	29.41	159.00	2.89×10^5	2.00×10^{-3}
Chl *a* F685 (inner half of LHCII)	670	0.14	29.41	82.00	1.53×10^7	0.11
Chl *a* F685 at $R_0 = R$	670	0.14	29.41	29.41	7.20×10^9	50
Chl *a* F695 (CP47) + CP29	677	0.14	29.29	126.00	1.14×10^6	0.79×10^{-2}
Chl *a* F695 at $R_0 = R$	677	0.14	29.29	29.21	7.20×10^9	50
Chl *a* F735 (LHCI-730)	704	0.14	16.16	159.00	7.94×10^3	5.5×10^{-5}
Chl *a* F738 at $R_0 = R$	704	0.14	16.16	16.16	7.20×10^9	50

Note: Abbreviations are as in Table 4.12.

not observable. Yet, as reported elsewhere [9], resonance excitation energy transfers from Proto to Chl a ~F685, ~F695, and ~F730 were very pronounced. These results suggested that in actuality, resonance excitation energy transfers from Proto to Chl a ~F685, ~F695, and ~F738 probably took place over smaller distances, which were more compatible with either the SBP-multilocation, or MBP-sublocation models.

2. Resonance Excitation Energy Transfer Rates from Mg-Proto (Ester) to Chl a ~F685, ~F695, and ~F735 at Fixed Distances R That May Prevail in the SBP-Single Location Chl–Thylakoid Apoprotein Biosynthesis Model

In this instance, two different plant species belonging to two different greening groups of plants were

used: cucumber, a DDV-LDDV plant species [22] that accumulates mainly DV Mp(e), and barley, a DMV-LDMV plant species [22], which usually accumulates larger amounts of MV Mp(e), than cucumber.

As with Proto, rates of resonance excitation energy transfer from Mp(e) to Chl a ~F685, ~F695, and ~F738 were calculated over distances R of 159, 126, and 82 Å, as well as at critical distances $R = R_0$. As reported in Table 4.14 and Table 4.15, the rates of excitation resonance energy transfer, K_T, from Mp(e) to the various Chl a species at 159 to 82 Å were negligible. In this case too the data suggested that actual excitation resonance energy transfer from Mp(e) to various Chl a species probably took place over smaller distances, which are more compatible with the SBP-multilocation, or MBP-sublocation models.

TABLE 4.14
Rates of Resonance Excitation Energy Transfer, K_T, at Fixed Distances R Separating Mp(e) from Various Chl a Species In Situ at 77 K in Green Barley Leaves

Chl a Species	Chl a Absorbance (nm)	τ_D (ns)	R_0 (Å)	R (Å)	K_T (s^{-1})	K_T as Percent of 100% Transfer Efficiency
Chl a F685 (LHCI-680 + outer half of LHCII)	670	0.52	37.05	159.00	3.06×10^5	0.59×10^{-2}
Chl a F685 (inner half of LHCII)	670	0.52	37.05	82.00	1.63×10^7	0.31
Chl a F685 at $R_0 = R$	670	0.52	37.05	37.05	1.91×10^9	50
Chl a F695 (CP47) + CP29	677	0.52	36.59	126.00	1.14×10^6	0.30×10^{-1}
Chl a F695 at $R_0 = R$	677	0.52	36.59	36.59	1.91×10^9	50
Chl a F735 (LHCI-730)	704	0.52	20.15	159.00	3.20×10^4	8.40×10^{-4}
Chl a F738 at $R_0 = R$	704	0.52	20.15	20.15	1.91×10^9	50

Note: Abbreviations are as in Table 4.12.

TABLE 4.15
Rates of Resonance Excitation Energy Transfer, K_T, at Fixed Distances R Separating Mp(e) from Various Chl a Species In Situ at 77 K in Green Cucumber Cotyledons

Chl a Species	Chl a Absorbance (nm)	τ_D (ns)	R_0 (Å)	R (Å)	K_T (s^{-1})	K_T as Percent of 100% Transfer Efficiency
Chl a F685 (LHCI-680 + outer half of LHCII)	670	0.30	37.02	159	5.35×10^5	7.96×10^{-3}
Chl a F685 (inner half of LHCII)	670	0.30	37.02	82	2.84×10^7	0.38
Chl a F685 at $R_0 = R$	670	0.30	37.02	37.02	3.36×10^9	50
Chl a F695 (CP47) + CP29	677	0.30	39.69	126	3.28×10^6	4.88×10^{-2}
Chl a F695 at $R_0 = R$	677	0.30	39.69	39.69	3.36×10^9	50
Chl a F735 (LHCI-730)	704	0.30	22.71	159	2.85×10^4	4.24×10^{-4}
Chl a F738 at $R_0 = R$	704	0.30	22.71	22.71	3.36×10^9	50

Note: Abbreviations are as in Table 4.12.

3. Energy Transfer Rates from Pchlide *a* to Chl *a* ~F685, ~F695, and F~735 at Fixed Distances *R* That May Prevail in the Single-Branched Single-Location Chl–Thylakoid Apoprotein Biosynthesis Model

To distinguish between resonance excitation energy transfer from DV and MV Pchl(ide) *a* to the various Chl *a* species, two different plant species belonging to two different greening groups of plants were used [22]. Cucumber, a DDV-LDDV plant species, which accumulated mainly DV Pchlide *a*, allowed the monitoring of resonance excitation energy transfer mainly from DV Pchl(ide) *a* to the various Chl *a* species. On the other hand, barley, a DMV-LDMV plant species, which accumulated MV Pchlide *a*, allowed the monitoring of excitation resonance energy transfer from MV Pchl(ide) *a* to the various Chl *a* species.

As with Proto and Mp(e), rates of resonance excitation energy transfer from DV and MV Pchl(ide) *a* to Chl *a* ~F686, ~F694, and ~F738 were calculated over distances *R* of 159, 126, and 82 Å, as well as at critical distances $R = R_0$. As shown in Table 4.16 and Table 4.17, the rates of excitation resonance energy transfer, K_T, from DV and MV Pchl(ide) *a* to the various Chl *a* species at 159 to 82 Å were negligible. In this case too, the data suggested that actual resonance excitation energy transfer from Pchl(ide) *a* to various Chl *a* species probably took place over smaller distances which were more compatible with the SBP-multilocation, or MBP-sublocation models.

TABLE 4.16
Rates of Resonance Excitation Energy Transfer, K_T, at Fixed Distances *R* Separating MV Pchlide *a* from Various Chl *a* Species in Situ at 77 K in Green Barley Leaves

Chl *a* Species	Chl *a* Absorbance (nm)	τ_D (ns)	R_0 (Å)	R (Å)	K_T (s^{-1})	K_T as Percent of 100% Transfer Efficiency
Chl *a* F685 (LHCI-680 + outer half of LHCII)	670	0.22	36.29	159	4.82×10^5	0.71×10^{-3}
Chl *a* F685 (inner half of LHCII)	670	0.22	36.29	82	2.56×10^7	0.38
Chl *a* F685 at $R_0 = R$	670	0.22	36.29	36.29	3.41×10^9	50
Chl *a* F695 (CP47) + CP29	677	0.22	37.90	126	2.56×10^6	3.75×10^{-2}
Chl *a* F695 at $R_0 = R$	677	0.22	37.90	37.90	3.41×10^9	50
Chl *a* F735 (LHCI-730)	704	0.22	22.43	159	26.87×10^4	3.90×10^{-3}
Chl *a* F738 at $R_0 = R$	704	0.22	22.43	22.43	3.41×10^9	50

Note: Abbreviations are as in Table 4.12.

TABLE 4.17
Rates of Resonance Excitation Energy Transfer, K_T, at Fixed Distances *R* Separating DV Pchlide *a* from Various Chl *a* Species In Situ at 77 K in Green Cucumber Cotyledons

Chl *a* Species	Chl *a* Absorbance (nm)	τ_D (ns)	R_0 (Å)	R (Å)	K_T (s^{-1})	K_T as Percent of 100% Transfer Efficiency
Chl *a* F685 (LHCI-680 + outer half of LHCII)	670	0.29	34.50	159	47.83×10^5	5.22×10^{-2}
Chl *a* F685 (inner half of LHCII)	670	0.29	34.50	82	2.54×10^7	2.77×10^{-1}
Chl *a* F685 at $R_0 = R$	670	0.29	34.50	34.50	4.58×10^9	50
Chl *a* F695 (CP47) + CP29	677	0.29	37.82	126	3.35×10^6	3.66×10^{-2}
Chl *a* F695 at $R_0 = R$	677	0.29	37.82	37.82	4.58×10^9	50
Chl *a* F735 (LHCI-730)	704	0.29	21.66	159	2.93×10^4	0.3×10^{-3}
Chl *a* F738 at $R_0 = R$	704	0.29	21.66	21.66	4.58×10^9	50

Note: Abbreviations are as in Table 4.12.

C. CALCULATION OF THE DISTANCES THAT SEPARATE PROTO, MP(E), DV PCHLIDE A, AND MV PCHLIDE A FROM VARIOUS CHL A ACCEPTORS IN LATERALLY HETEROGENEOUS PSU

Since resonance excitation energy transfer rates at distances that prevailed in a continuous array PSU were insignificant (see above), an effort was made to calculate the probable distances that separated anabolic tetrapyrroles from Chl *a* receptors in more plausible PSU models. Distances separating Proto, Mp(e), and DV and MV Pchlide *a* from Chl *a* acceptors were therefore determined and were compared to current concepts of PSU structure [3–5] and to the Chl–thylakoid biogenesis models proposed in Refs. [1,9] (see Section IV). The calculated distances separating Proto, Mp(e), and DV and MV Pchlide *a* from various Chl *a* acceptors *in situ* are reported in Table 4.18.

Distances separating anabolic tetrapyrroles from various Chl–protein complexes ranged from a low of 16.52 Å for Proto–Chl *a* ~F735 separation in cucumber, to a high of 41.23 Å for Proto–Chl *a* ~F695 separation in barley (Table 4.18). The magnitude of these distances was compatible with the observation of intense resonance excitation energy transfer reported in Ref. [9].

In cucumber, a DDV-LDDV plant species [22], the distances that separated Proto from Chl *a* acceptors were shorter than those that separated Mp(e) and DV Pchlide *a* from the Chl *a* acceptors (Table 4.18). Since Proto is an earlier intermediate of Chl *a* biosynthesis than Mp(e) and Pchlide *a*, it indicated that in cucumber, the Chl *a*–protein biosynthesis subcenter is a highly folded entity, where linear distances separating intermediates from end products bear little meaning (see Section IV). On the other hands, in barley, a DMV-LDMV plant species [22], distances separating Proto from various Chl *a* acceptors were generally

longer than those separating Mp(e) and MV Pchlide *a* from the Chl *a* acceptors (Table 4.18). This in turn suggested that tetrapyrrole–protein complex folding in cucumber (DV subcenters) is different than in barley (MV subcenters).

In all cases, it was observed that while distances separating anabolic tetrapyrroles from Chl *a* (E670F685) (i.e., Chl *a* ~F685) and Chl *a* (E677F695) (i.e., Chl *a* ~F695), were in the same range, those separating Chl *a* (E704F735) (i.e., Chl *a* ~F735) from anabolic tetrapyrroles were much shorter (Table 4.18). As may be recalled, it is believed that the fluorescence emitted at ~F685 nm arises from the Chl *a* of the light-harvesting Chl–protein complexes (LHCII and LHCI-680), that emitted at ~F695 nm arises mainly from the PSII antenna Chl *a* (CP47 and/or CP29), while that emitted at ~F735 nm arises primarily from the PSI antenna Chl *a* (LHCI-730) [2]. This in turn suggested that in the Chl *a*–protein biosynthesis subcenters, protein folding is such that the PSI antenna Chl *a* (LHCI-730) is much closer to the terminal steps of anabolic tetrapyrrole biosynthesis than the LHCII and LHCI-680 Chl–protein complexes or the CP47 and/or CP29 PSII antenna Chl *a* complexes.

IV. DISCUSSION

Evidence of heterogeneous resonance excitation energy transfer from anabolic tetrapyrroles to Chl–protein complexes was reviewed by describing resonance excitation energy transfer donation from multiple Soret excitation sites to Chl–protein complexes. The accumulation of anabolic tetrapyrroles was induced by treating plant tissues with ALA in the absence and presence of Dpy. Treatment of plant tissues with ALA and/or Dpy resulted in the accumulation of tetrapyrroles [19]. In the light the accumulated tetrapyrroles cause the formation of singlet oxygen that

---

TABLE 4.18
Calculated Distances *R* (Å) that Separate Proto, Mp(e), and Pchlide *a* Donors from Chl *a*–Protein Complexes Acceptors in Barley and Cucumber Chloroplasts at 77 K In Situ

Chl *a* Species	Proto		Mp(e)		MV Pchlide *a*	DV Pchlide *a*
	Barley	Cucumber	Barley	Cucumber	Barley	Cucumber
Chl *a* F685 (LHCI-680 + outer half of LHCII	38.83	30.07	35.60	38.74	37.73	35.22
Chl *a* F695 (CP47) + CP29)	41.23	29.94	35.15	41.53	39.41	30.60
Chl *a* F735 (LHCI-730)	22.72	16.52	19.36	23.76	23.32	22.11

Note: the distances *R* were determined from $[(R^6)^{1/6} cm]10^8$ Å cm^{-1}. The R^6 values were taken from Table 4.9. A κ^2 value of 0.67 was used in the calculations.

destroys all biomolecules including chloroplast pigments [21,36]. However, in darkness, as is the case in this work, induction of tetrapyrrole accumulation left the total Chl profile intact with no obvious alteration in the Chl a/b ratio as reported elsewhere [37].

Since the emission spectrum of isolated chloroplast is flat between 580 and 660 nm, accumulated tetrapyrroles exhibited definite emission maxima in this wavelength region, at 77 K, namely at 591 and 650 nm (Mp(e)), 623 nm (Proto), and 633 and 652 nm (Pchlides). However, the emission peaks were broad. Furthermore, since emission wavelengths are less sensitive to structural and environmental factors than Soret excitation maxima, Soret excitation peaks and Soret resonance excitation energy transfer maxima were more sensitive markers of chemical and site heterogeneity [1] than emission spectra. For example, although MV and DV Pchlide a exhibit identical emission maxima at 625 nm, in ether at 77 K, they exhibit different Soret excitation maxima at 417 and 424 nm, respectively [19].

Demonstration of resonance excitation energy transfers for the purpose of calculating distances separating anabolic tetrapyrroles in their native locations, from Chl–protein complexes, is only meaningful if the accumulated tetrapyrroles occupy their natural positions in the thylakoid membranes [9]. It was argued that the natural positions were most probably binding sites of the enzymes that process various reactions of the Chl biosynthetic pathway [9]. It was also argued that this does not mean that every enzyme-binding site should accumulate stochiometric amounts of tetrapyrroles [9]. It is well known that tetrapyrrole–tetrapyrrole associations via van der Waal forces and/or keto–Mg axial coordination are very ubiquitous in photosynthetic organisms [38–41]. For example, it is very conceivable that Pchlide a accumulation occurs as a shell around a Pchlide a–enzyme binding site, via Pchlide a–Pchlide a–keto–Mg axial coordination. As a consequence excess amounts of Pchlide a per Pchlide a binding site may accumulate. Leaked fluorescence would be emitted by the Pchlide a directly attached to the protein binding site, while the Pchlide a shell would be nonfluorescent or very weakly fluorescent. As a consequence, resonance excitation energy transfer profiles would be relatively independent of the size of the aggregated Pchlide a shell, and as shown in Table 4.10, would be relatively constant over a wide range of tetrapyrrole accumulation. The same reasoning can be extended to other tetrapyrrole side-chains–Mg coordination and/or aggregation via van der Waal forces.

It is also important to point out that, under the present experimental conditions, it was very unlikely for the accumulated tetrapyrroles to leave their enzyme binding sites to associate randomly with membrane lipoproteins. The only documented case of a tetrapyrrole leaving its natural enzyme-binding site is that of protoporphyrinogen IX, which accumulates when protoporphyrinogen IX oxidase activity is inhibited [33–35]. Under these circumstances, protoporphyrinogen IX leaves its enzyme-binding site and tunnels its way out of the chloroplast. The tunneling may be caused by the highly flexible structure of protoporphyrinogen IX, which is a reduced tetrapyrrole that lacks the rigid planar structure and alternating double bond system of oxidized tetrapyrroles. We are unaware of Mg-porphyrins or -phorbins with a rigid planar structure, leaving their enzyme binding sites to be excreted in the incubation medium as is often observed with flexible porphyrinogens such as uroporphyrinogens, coproporphyrinogens, and Protoporphyrinogen IX. It is also noteworthy that for a particular tetrapyrrole, under various incubation conditions, and at various levels of tetrapyrrole accumulation, the heterogeneous resonance excitation energy transfer profiles from SW, MW, and LW sites to a particular Chl a species were remarkably well preserved (Table 4.10). This is in contrast to the simple Soret excitation profiles which were observed during reconstituted binding of cucumber chloroplast lipoproteins to exogenous Proto, Mp(e), and DV Pchlide a (Table 4.11).

Another issue that was addressed in Ref. [9], was the impact of prolamellar body formation on the observed resonance excitation energy transfer profiles. It was reasserted that by the end of the fifth dark cycle of the photoperiod, prolamellar body formation was no longer observed in chloroplasts [42]. However, a very small number of thylakoid plexuses were formed. If as expected, the thylakoid plexuses were devoid of Chl, contribution of plexus-bound tetrapyrroles to resonance excitation energy donation to Chl a ~F685, ~F695, and ~F735 would not be observed.

In Ref. [9], resonance excitation energy transfers between Proto, Mp(e) and Pchlide a, and the Chl a of several Chl–protein complexes of PSI, PSII, and LHCII were clearly demonstrated in the presence of contributions from the vibrational bands of the accumulated tetrapyrroles. That contribution should be considered in the context of the (a) fluorescence intensities of the accumulated tetrapyrrole vibrational bands at ~685, ~695, and ~735 nm, and (b) overlap between the vibrational bands of the accumulated tetrapyrroles and the absorbance bands of Chl a (E671F686), (E677F694), and (E705F738), as discussed in Ref. [9].

First, it was pointed out that the fluorescence intensities at 685, 694, and 738 to 742 nm of the

Mp(e) vibrational band were minimal, and for all practical purposes their contribution to the Soret excitation profile of Mp(e) can be largely ignored. The same held true for the fluorescence intensities at 738 to 742 nm of the Proto and Pchlide *a* vibrational bands. That left the contribution of the fluorescence intensities at 685 and 694 nm of the Proto and Pchlide *a* vibrational bands, to the resonance excitation energy transfer profiles of Proto and Pchlide *a* at ~F685 and ~F695 nm. At these wavelengths the ratio of the vibrational bands emission maxima to the Chl emissions at ~F685 and ~F695 nm is about unity. However, since all excitation spectra were recorded at narrow 0.5 to 4 nm excitation and emission slit widths, one would expect the excitation contribution of the Proto and Pchlide *a* vibrational bands at F686 and F694 to generate single Proto and Pchlide *a* excitation maxima at each wavelength. Such excitation maxima would not be due to resonance excitation energy transfer. Therefore, it was argued that one of the peaks or shoulders reported in Table 4.10 at ~F685 and ~F695, for both Proto and Pchlide *a*, may be true excitation peaks instead of being resonance excitation energy transfer peaks. Nevertheless, that left the majority of the peaks reported in Table 4.10, as authentic resonance excitation energy transfer peaks [9].

Second, it was argued that efficiencies of resonance excitation energy transfer from accumulated tetrapyrrole donors to Chl *a* acceptors depended largely upon the overlap between the fluorescence vibrational bands of the tetrapyrrole donors, and the red absorbance bands of the Chl *a* acceptors. The overlap between the vibrational bands of the Proto and Pchlide *a* donors, and the absorbance bands of the Chl acceptors was complete, as depicted in Figure 4.2 for Proto. The overlap was not as complete for the Mp(e) emission vibrational band. For Chl *a* (E670F686), the tetrapyrrole emission–Chl absorbance overlap spanned the wavelength region from 652 to 688 nm. For Chl *a* (E677F694) the overlap spanned the wavelength region from 660 to 695 nm, and for Chl *a* (E705F738) it spanned the wavelength region from 692 to 720 nm. It was argued that if there were multiple tetrapyrrole fluorescence donor sites with subtle emission wavelength differences in the wavelength regions of the overlap, the resonance excitation energy transfer profiles will exhibit multiple Soret resonance excitation energy transfer peaks that corresponded to the Soret absorbance maxima of the tetrapyrroles emitting from the various sites. This in turn would be compatible with the data reported in Table 4.10. On the other hand, if the observed resonance excitation energy transfer profiles reported in Table 4.10, were only Soret exci-

tation peaks contributed by the Proto and Pchlide *a* vibrational bands at ~685 and ~695 nm, then contrary to the heterogeneous resonance excitation energy transfer profiles reported in Table 4.10, only one Soret excitation maximum per accumulated tetrapyrrole would be observed [9].

It was also pointed out that in view of extensive energy transfer in green systems, input may occur at many different positions and not just at the complexes whose fluorescence was monitored in Ref. [9]. For example, because of the fluorescence–absorbance overlap between Mp(e) fluorescence (as donor) and the red absorbance bands of Proto and Pchlide *a* as acceptors, as well as between Proto fluorescence (as donor) and the red absorbance band of Pchlide *a* as acceptor, resonance excitation energy transfer from Mp(e) to Proto and Pchlide *a*, and from Proto to Pchlide *a* as well as from Proto and Pchlide *a* to the Chl acceptors may be observed. It was argued that this phenomenon was likely to contribute very minimally to the intensities of the resonance excitation energy transfer profiles reported in Table 4.10 for several reasons. First, because of the very low molar extinction coefficients of the red absorbance bands of Proto and Pchlide *a*, the value of the overlap integral would be very small, which will result in turn in poor resonance excitation energy transfer rates between donor Mp(e) and the Proto and Pchlide *a* acceptors. Second, since resonance excitation energy transfer is seldom 100% efficient due to competing nonradiative photochemical processes such as internal conversion and intersystem crossing, the multiple resonance excitation energy transfer steps will result in further losses in resonance excitation energy transfer intensities.

The assignment of *in situ* excitation maxima to the various metabolic tetrapyrroles reported in Ref. [9] was unambiguous except for a few cases at the SW and LW extremes of excitation bands. For example, the 428–433 nm resonance excitation energy transfer maxima were assigned to LW Mp(e) sites, although one can argue that it may belong to SW Pchlide *a* sites. Likewise, the 415–417 nm resonance excitation energy transfer maxima were assigned to LW Proto sites, although one can argues that it may belong to SW Mp(e) sites. In this context, it should be recognized that excitation maxima may be slightly skewed to shorter or longer wavelengths in difference excitation spectra like the ones depicted in Figure 4.4 to Figure 4.9. In spite of these uncertainties, in most cases well-pronounced resonance excitation energy transfer bands with well-defined excitation maxima were observed.

It was most surprising to observe diversity in the various intramembrane environments of Proto,

Mp(e), and Pchl(ide) *a*. This diversity was manifested by a differential donation of resonance excitation energy transfer to different Chl *a*–apoprotein complexes. This observation is highly compatible with the notion of Chl biosynthetic heterogeneity. Consequently, the multibranched Chl *a* biosynthetic pathway depicted in Ref. [19] had to be modified in order to accommodate the existence of multiple donor sites for Proto, Mp(e), and Pchl(ide) *a* [9]. A proposed modification that extends biosynthetic routes 1, 8,

10, 11, and 12 all the way to ALA is reproduced in Figure 4.10 from Ref. [1].

The detection of pronounced excitation resonance energy transfer from Proto, Mp(e), and Pchl(ide) *a* to Chl *a* ~F685, ~F695, and ~F735 indicated that these anabolic tetrapyrrole donors were within distances of 100 Å or less of the immediate Chl *a* acceptors. Indeed, resonance excitation energy transfer is insignificant at distances larger than 100 Å, since dipole–dipole energy transfer may only occur up to a separ-

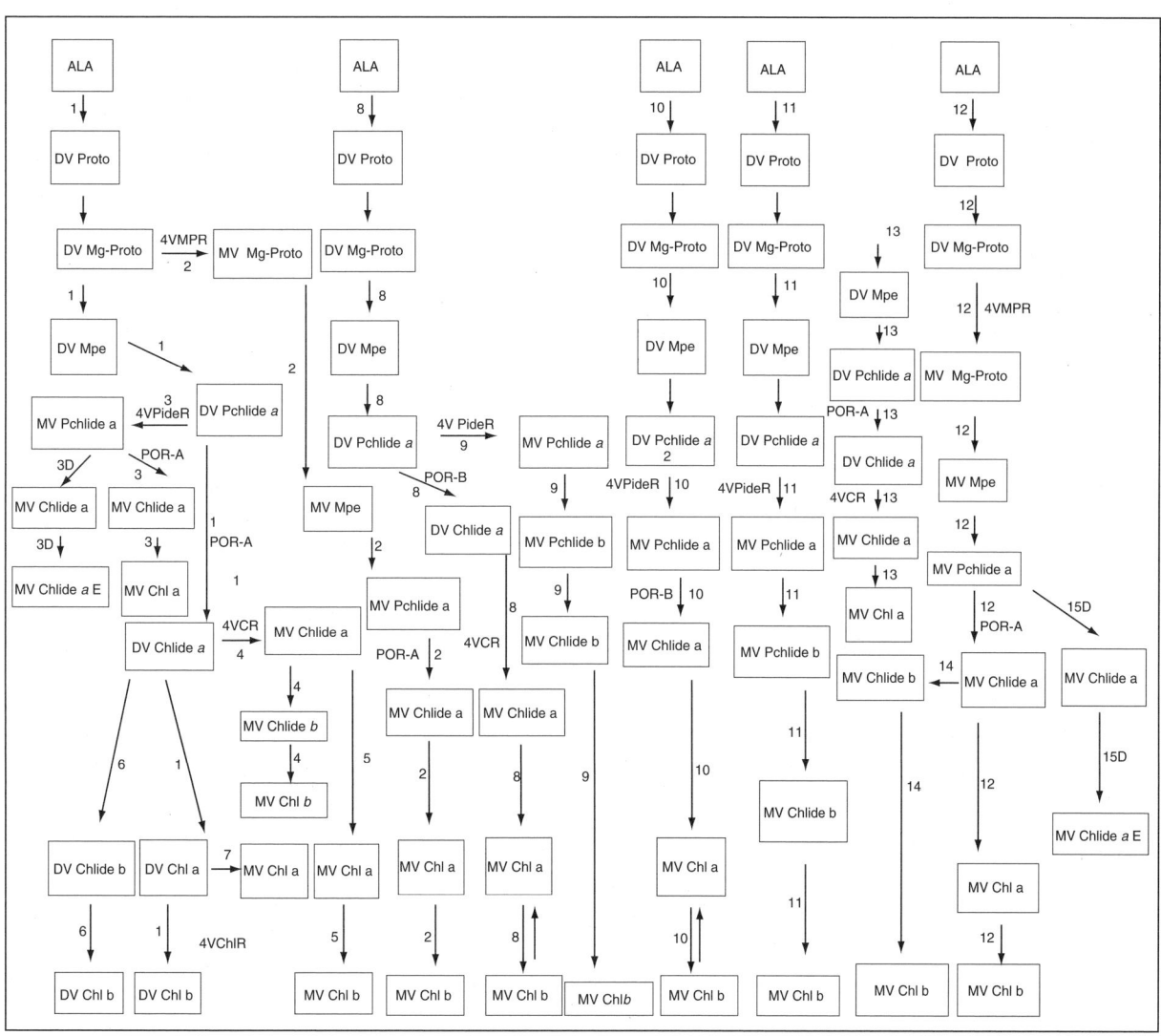

FIGURE 4.10 Modified integrated Chl *a/b* biosynthetic pathway. Routes 8, 10, 11, and 12 were extended all the way to ALA to accommodate the occurrence of multiple Proto, Mp(e), and Pchlide *a* excitation resonance energy transfer donor sites. DV = divinyl, MV = monovinyl, ALA = delta-aminolevulinic acid, Proto = protoporphyrin IX, Mpe = Mg-Proto monomethyl ester, Pchlide = protochlorophyllide, Chlide = chlorophyllide, Chl = chlorophyll, 4VMPR: [4-vinyl] Mg-Proto reductase, 4VPideR = [4-vinyl] protochlorophyllide *a* reductase, 4VCR = [4-vinyl] chlorophyllide *a* reductase, 4VChlR = [4-vinyl] Chl reductase, POR = Pchlide *a* oxidoreductase, D = reaction occurring in darkness. Arrows joining DV and MV routes refer to reactions catalyzed by [4-vinyl] reductases. Various biosynthetic routes are designated by Arabic numerals.

ation distance of 50 to 100 Å [8]. This observation was documented quantitatively as discussed below.

The dimensions of a tightly packed, continuous array PSU that consisted of PSI, PSII, and LHC Chl–apoprotein antenna complexes is approximately 130 × 450 Å [2]. Theoretically, resonance excitation energy transfer is inversely proportional to the power 6 of the distance separating donors from acceptors [8,25]. It was conjectured that calculation of resonance excitation energy transfer rates from anabolic tetrapyrroles to various Chl–protein complexes within a continuous array PSU may determine their possible compatibility with the operation of the single-branched single location model within a Chl–apoprotein biosynthesis center. For these calculations, the choice of a tightly packed continuous array PSU model over the laterally heterogeneous models [3–5] was motivated by the fact that in the latter longer distances would separate a SBP from Chl a acceptors.

In Table 4.12 to Table 4.17, calculated excitation resonance energy transfer rates from anabolic tetrapyrroles to Chl a acceptors were converted into percentages of the 100% energy transfer rates that would be observed if no de-excitation other than resonance energy transfer took place. The calculation of these values was made possible by determination of the resonance excitation energy transfer rates at 50% efficiency that were observed at critical distances $R = R_0$.

As shown in Tables 4.12 to Table 4.17, the rates of resonance excitation energy transfer from Proto, Mp(e), and Pchl(ide) a to Chl a acceptors at distances of 159 to 82 Å, which would include excitation resonance energy transfer to PSI, PSII, and LHCII Chl–protein complexes, were far below those found at critical distances, $R = R_0$, and for all practical purposes were insignificant. Since pronounced resonance excitation energy transfers from Proto, Mp(e), and Pchl(ide) a to Chl a ~F685, ~F695, and ~F735 have been observed [9], the results reported in Table 4.12 to Table 4.17 imply that for a tightly packed 130 × 450 Å continuous array PSU, there must exist more than one location where anabolic tetrapyrrole biosynthesis and resonance excitation energy transfer to nearby Chl a–protein complexes took place over distances shorter than 82 Å. Such a scenario is more compatible with the SBP-multilocation or MBP-sublocation Chl–protein biosynthesis models, and with the observation of multiple resonance excitation energy transfer cites as reported in Ref. [9].

Since the resonance excitation energy transfer rate calculations argued against the operation of the SBP-single location Chl–protein biosynthesis model, the question arises as to which of the other two models, namely the SBP-multilocation and MBP-sublocation model is functional in nature. This issue was addressed by drawing (a) on the wealth of experimental evidence supporting the operation of a multibranched Chl biosynthetic pathway in green plants, and (b) by calculation of the probable distances that separate anabolic tetrapyrroles from Chl a acceptors in recently proposed PSU models [3–5].

The early concept of a PSU consisting of about 500 antenna Chl per reaction center has evolved into two pigment systems each with its own reaction center and antenna Chl [4]. The early visualization of the two photosystems consisted of various pigment–protein complexes arrayed into a tightly packed linear PSU (the continuous array model), about 450 Å in length and 130 Å in width [2]. In the PSU, the LHCII was depicted as being shared between the two photosystems [2]. More recent models, however, favor the concept of a laterally heterogeneous PSU [3–5]. In these models, LHCII shuttles between PSI and PSII upon phosphorylation and dephosphorylation [3]. In all these models, PSI, PSII, and LHCII are depicted as spatially discrete globular entities. While PSII is considered to be located mainly (but not exclusively) in appressed thylakoid domains, PSI is considered to be located in nonappressed stroma thylakoids, grana margins, and end membranes [4,5].

The shorter distances separating anabolic tetrapyrroles from Chl–protein complexes reported in Table 4.18 are compatible with the SBP-multilocation and MBP-sublocation models. However, the SBP-multilocation model implies a random association of pigments including Chl with thylakoid apoproteins which is a very unlikely possibility. Furthermore, since overwhelming experimental evidence argues against the operation of a single-branched Chl biosynthetic pathway in plants [1], that leaves the MBP-sublocation model as a valid working hypothesis.

The MBP-sublocation model is very compatible with the lateral heterogeneity of the PSU [1,6]. In this model, the unified multibranched Chl a/b biosynthetic pathway is visualized as the template of a Chl–protein biosynthesis center where the assembly of discrete PSI, PSII, and LHC entities takes place. In each of these entities, multiple Chl biosynthetic routes may be visualized, in groups of two or several adjacent routes, as Chl–apoprotein biosynthesis subcenters earmarked for the coordinated assembly of the particular Chl–apoprotein complexes that make up PSI, PSII, or LHCII. Apoproteins destined to some of the subcenters may possess specific polypeptide signals for specific Chl biosynthetic enzymes peculiar to that subcenter, such as 4-vinyl reductases, formyl synthetases or Chl a and Chl b synthetases. Once an apoprotein formed in the cytoplasm or in the plastid reaches its subcenter destination and its signal is split off, it binds nascent Chl formed via one or

TABLE 4.19

Calculated Distances R that Separate Proto from Chl a–Protein Complex Acceptors in Barley and Cucumber Chloroplasts at 77 K In Situ for Various Values of the Orientation Dipole κ^2

Chl a Species	Chl a Absorbance	κ^2	Proto–Chl a Acceptor Distances R (?) In Situ in	
			Barley	Cucumber
Chl a F685 (LHCI-680 + outer half of LHCII)	670	0.67	38.83	30.07
Chl a F685 (LHCI-680 + outer half of LHCII)	670	1	41.52	32.14
Chl a F685 (LHCI-680 + outer half of LHCII)	670	4	52.31	40.49
Chl a F695 (CP47) + CP29)	677	0.67	41.23	29.94
Chl a F695 (CP47) + CP29)	677	1	44.08	29.24
Chl a F695 (CP47) + CP29)	677	4	55.54	32.01
Chl a F735 (LHCI-730)	704	0.67	22.72	16.50
Chl a F735 (LHCI-730)	704	1	24.29	17.66
Chl a F735 (LHCI-730)	704	4	30.61	21.25

Note: κ^2 values of 0.67, 1, and 4 are for random, lined up, and adjacent dipole orientations, respectively.

more biosynthetic routes, as well as carotenoids. During pigment binding, the apoprotein folds properly and acts at that location, while folding or after folding, as a template for the assembly of other pigment-proteins. Such a model can readily account for: (a) the observed resonance excitation energy transfer from distinct and separate multiple sites [9], such as PSI, PSII, and LHCII, and (b) the short distances separating anabolic tetrapyrroles from Chl–protein complexes in the distinct PSI, PSII, and shuttling LHCII entities that compose the PSU (Table 4.18).

In calculating the excitation resonance energy transfer rates reported in Table 4.12 to Table 4.17 and the actual distances separating anabolic tetrapyrrole donors from Chl a acceptors (Table 4.18), two type of parameters were used: (a) parameters determined *in situ*, i.e.,on thylakoid membranes suspended in Tris–HCl:glycerol (1:2 v/v), pH 7.7, and cooled to 77 K, such as fluorescence yields, and corrected fluorescence intensities, and (b) parameters determined in chloroplast lipoprotein membranes such as molar extinction coefficients of donors and acceptors (ε_m), mean wavenumber of absorbance and fluorescence emission maxima of donors (ν_0), Soret absorbance maxima of donors (ν_m), Soret absorbance half bandwidth of donors ($\Delta\nu_{1/2}$), and red absorbance maxima of donors. Under ideal conditions, these parameters should be determined *in situ*, i.e.,in the native environment of the thylakoid membranes at 77 K. Techniques are presently being developed for the generation of such data. At this stage, however, an approximation was made by deriving the above parameters from spectra recorded in chloroplast lipoproteins suspended in Tris–HCl:glycerol (1:2 v/v) buffer, pH 7.7. It was conjectured that the polarity of this

environment is an acceptable approximation of the thylakoid *in situ* environment.

Finally, in calculating the orientation dipole κ^2 of donor and acceptor pairs, a random dipole orientation value of 0.67 was used, as proposed by others [27]. In order to determine whether the use of other κ^2 orientation dipole values were likely to drastically change the calculated distances reported in Table 4.18, calculations with extreme κ^2 values of 1 (lined up dipoles) and 4 (adjacent dipoles) were performed on the Proto–Chl a pairs, which exhibited the largest tetrapyrrole–Chl a–protein separation distances. As shown in Table 4.19, the calculated distances separating anabolic tetrapyrroles from Chl a acceptors increased slightly with increasing values of κ^2. However, even at the highest κ^2 value of 4, the calculated distances remained far below those that would have prevailed in the SBP-single location model for a packed continuous array model, where distances separating tetrapyrrole donors from Chl a acceptors would have ranged from 156 to 82 Å, or the longer distances that would have prevailed in the laterally heterogeneous models.

REFERENCES

1. Rebeiz CA, Kolossov VL, Briskin D, Gawienowski M. Chloroplast Biogenesis: Chlorophyll biosynthetic heterogeneity, multiple biosynthetic routes and biological spin-offs. In: (Nalwa, HS, ed.) *Handbook of Photochemistry and Photobiology*, Vol. 4: Los Angeles, CA.: American Scientific Publishers, 2003: pp. 183–248.
2. Bassi R, Rigoni. F, Giacometti GM. Chlorophyll binding proteins with antenna function in higher plants and green algae. *Photochem. Photobiol.* 1990; 52:1187–1206.

3. Allen JF, Forsberg J. Molecular recognition in thylakoid structure and function. *Trends Plant Sci.* 2001; 6:317–326.

4. Anderson JM. Changing concepts about the distribution of photosystem I and II between grana-appressed and stroma-exposed thylakoid membranes. *Photosynth. Res.* 2002; 73:157–164.

5. Staehelin LA. Chloroplast structure: from chlorophyll granules to supra-molecular architecture of thylakoid membranes. *Photosynth. Res.* 2003; 76:185–196.

6. Rebeiz CA, Ioannides IM, Kolossov V, Kopetz KJ. Chloroplast Biogenesis 80: Proposal of a unified multi-branched chlorophyll *a/b* biosynthetic pathway. *Photosynthetica* 1999; 36:117–128.

7. Sundqvist C, Ryberg M (eds.) *Pigment–Protein Complexes in Plastids: Synthesis and Assembly*, New York: Academic Press, 1993.

8. Calvert JG, Pitts JN. *Photochemistry*. New York: John Wiley & Sons, 1967.

9. Kolossov VL, Kopetz KJ, Rebeiz CA. Chloroplast Biogenesis 87: Evidence of resonance excitation energy transfer between tetrapyrrole intermediates of the chlorophyll biosynthetic pathway and chlorophyll *a*. *Photochem. Photobiol.* 2003; 78:184–196.

10. Kim JS, Rebeiz CA. Origin of the chlorophyll *a* biosynthetic heterogeneity in higher plants. *J. Biochem. Mol. Biol.* 1996; 29:327–334.

11. Tripathy BC, Rebeiz CA. Chloroplast Biogenesis 60: Conversion of divinyl protochlorophyllide to monovinyl protochlorophyllide in green(ing) barley, a dark monovinyl/light divinyl plant species. *Plant Physiol.* 1988; 87:89–94.

12. Cohen CE, Rebeiz CA. Chloroplast Biogenesis 22: Contribution of short wavelength and long wavelength protochlorophyll species to the greening of higher plants. *Plant Physiol.* 1978; 61:824–829.

13. Cohen CE, Rebeiz CA. Chloroplast Biogenesis 34: Spectrofluorometric characterization *in situ* of the protochlorophyll species in etiolated tissues of higher plants. *Plant Physiol.* 1981; 67:98–103.

14. Schoefs B, Bertrand M, Franck F. Spectroscopic properties of protochlorophyllide analyzed *in situ* in the course of etiolation and in illuminated leaves. *Photochem. Photobiol.* 2000; 72:85–93.

15. Sironval C, Kuyper Y, Michel JM, Brouers M. The primary photoact in the conversion of protochlorophyllide into chlorophyllide. *Stud. Biophys.* 1967; 5:43–50.

16. Griffiths WT. Source of reducing equivalents for the *in vitro* synthesis of chlorophyll from protochlorophyll. *FEBS Lett.* 1974; 46:301–304.

17. Kopetz KJ, Kolossov VL, Rebeiz CA. Chloroplast Biogenesis 89: Development of analytical tools for probing the biosynthetic topography of photosynthetic membranes by determination of resonance excitation energy transfer distances separating metabolic tetrapyrrole donors from chlorophyll *a* acceptors. *Anal. Biochem.* 2004; 329:207–219.

18. Tripathy BC, Rebeiz CA. Chloroplast Biogenesis: Demonstration of the monovinyl and divinyl monocarboxylic routes of chlorophyll biosynthesis in higher plants. *J. Biol. Chem.* 1986; 261:13556–13564.

19. Rebeiz CA Analysis of intermediates and end products of the chlorophyll biosynthetic pathway. In: (Smith, A, G., Witty, M., eds.) *Heme Chlorophyll and Bilins, Methods and Protocols*, Totowa, NJ: Humana Press, 2002: pp. 111–155.

20. Rebeiz CA, Saab DG. Porphyrin Analytical Tools, Copyrighted software, 1995.

21. Amindari SM, E. SW, Rebeiz CA Photodynamic effects of several metabolic tetrapyrroles on isolated chloroplasts. In: (Heitz, JR, Downum, KR, eds.) *Light-Activated Pest Control*, Vol. 616, Washington, DC: American Chemical Society, 1995: pp. 217–246.

22. Abd-El-Mageed HA, El Sahhar KF, Robertson KR, Parham R, Rebeiz CA. Chloroplast Biogenesis 77: Two novel monovinyl and divinyl light-dark greening groups of plants and their relationship to the chlorophyll *a* biosynthetic heterogeneity of green plants. *Photochem. Photobiol.* 1997; 66:89–96.

23. French CS, Brown JS, Lawrence MC. Four universal forms of chlorophyll *a*. *Plant Physiol.* 1972; 49:421–429.

24. Shedbalkar VP, Rebeiz CA. Chloroplast Biogenesis: Determination of the molar extinction coefficients of divinyl chlorophyll *a* and *b* and their pheophytins. *Anal. Biochem.* 1992; 207:261–266.

25. Turro NJ. *Molecular Photochemistry*. London: W. A. Benjamin, 1965.

26. Forster TH. Transfer mechanisms of electronic excitation energy. *Radiat. Res.* 1960; Suppl. 2:326–339.

27. Wu C-W, Stryer L. Proximity relationship in rhodopsin. *Proc. Natl. Acad. Sci. USA* 1972; 69:1104–1108.

28. Lakowicz JR. *Principles of Fluorescence Spectroscopy*. New York: Kluwer Academic/Plenum Press, 1999: pp. 367–394.

29. Rebeiz CA, Juvik JA, Rebeiz CC. Porphyric insecticides 1. Concept and phenomenology. *Pestic. Biochem. Physiol.* 1988; 30:11–27.

30. Rebeiz CA, Mattheis JR, Smith BB, Rebeiz CC, Dayton DF. Chloroplast Biogenesis. Biosynthesis and accumulation of Mg-protoporphyrin IX monoester and longer wavelength metalloporphyrins by greening cotyledons. *Arch. Biochem. Biophys.* 1975; 166:446–465.

31. Tripathy BC, Rebeiz CA. Chloroplast Biogenesis: Quantitative determination of monovinyl and divinyl Mg-protoporphyrins and protochlorophyll(ides) by spectrofluorometry. *Anal. Biochem.* 1985; 149:43–36.

32. Ioannides IM, Fasoula DM, R. RK, Rebeiz CA. An evolutionary study of chlorophyll biosynthetic heterogeneity in green plants. *Biochem. Syst. Ecol.* 1994; 22:211–220.

33. Matringe M, Scalla R. Studies on the mode of action of acifluorfen-methyl in nonchlorophyllous soybean cells. Effects of acifluorfen-methyl on cucumber cotyledons: porphyrin accumulation. *Plant Physiol.* 1988; 86:619–622.

34. Matringe M, Camadro JM, Labbe P, Scalla R. Protoporphyrinogen oxidase inhibition by three peroxidizing herbicides: oxadiozon, LS 82-556 and M&b 39279. *FEBS Lett.* 1989; 245:35–38.

35. Matringe M, Camadro JM, Labbe P, Scalla R. Protoporphyrinogen oxidase as a molecular target for diphenyl ether herbicides. *Biochem. J.* 1989; 260:231–235.

36. Rebeiz CA, Reddy KN, Nandihalli UB, Velu J. Tetrapyrrole-dependent photodynamic herbicides. *Photochem. Photobiol.* 1990; 52:1099–1117.

37. Rebeiz CA Tetrapyrrole-dependent photodynamic herbicides and the chlorophyll biosynthetic pathway. In: (Pell, E, Steffen, K, eds.) *Active Oxygen/Oxidative Stress and Plant Metabolism*, Rockville, MD: American Society of Plant Physiologists, 1991: pp. 193–203.

38. Boucher LJ, Katz JJ. Aggregation of metalloporphyrins. *J. Am. Chem. Soc.* 1967; 89:4703–4708.

39. Deisenhofer J, Michel H Crystallography of chlorophyll proteins. In: (Scheer, H, ed.) *Chlorophylls*, Boca Raton, FL: CRC Press, 1991: pp. 613–625.

40. Fong FK, Koester VJ. Bonding interactions in anhydrous and hydrated chlorophyll *a. J. Am. Chem. Soc.* 1975; 97:6888–6890.

41. Lutz M, Breton J. Chlorophyll associations in the chloroplast: resonance Raman spectroscopy. *Biochem. Biophys. Res. Commun.* 1973; 53:413–418.

42. Rebeiz CC, Rebeiz CA Chloroplast Biogenesis 53: Ultrastructural study of chloroplast development during photoperiodic greening. In: (Akoyunoglou, G, Senger, H, eds.) *Regulation of Chloroplast Differentiation*, New York: Alan Liss, 1986: pp. 389–396.

5 Protochlorophyllide Photoreduction — A Review

Martine Bertrand

Institut National des Sciences et Techniques de la Mer, Conservatoire National des Arts et Métiers

Benoît Schoefs

Dynamique Vacuolaire et Réponses aux Stress de l'Environnement,
UMR CNRS 5184/INRA 1088/Université de Bourgogne Plante-Microbe-Environnement,
Université de Bourgogne à Dijon

CONTENTS

I. INTRODUCTION

As the main component of the photosynthetic apparatus, Chl (and bacteriochlorophylls) molecules a play major role in the development and maintenance of life since its appearance. Even though the importance of Chl molecules for our world is known, it is obvious that the intimate mechanism of the reactions leading to their formation has not been fully elucidated yet. The regulation of Chl biosynthesis has only begun to be investigated. One of the most attractive reactions of the pathway (see Chapter 3 by Schoefs and Bertrand for a review) is the reduction of protochlorophyllide (Pchlide) to chlorophyllide (Chlide). Pchlide reduction can be performed by two families of enzymes. The enzymatic reaction consists of the reduction of the $C_{17}=C_{18}$ double bond of Pchlide molecule yielding Chlide. One enzyme requires light to function, whereas the second does not. Both enzymes are usually present in every photosynthetic cell except in angiosperms, which only contain the light-dependent enzyme.

In this chapter we have reviewed the recent data concerning the transformation of Pchlide to Chlide reaction.

II. LIGHT-DEPENDENT CHL *a* FORMATION

The light-dependent Chlide formation is catalyzed by the light-dependent Pchlide oxidoreductase (LPOR), which reduces Pchlide and oxidizes NADPH. In the dark the LPOR enzymes are inactive and form stable ternary complexes with both Pchlide and NADPH (or $NADP^+$; see Figure 5.1). In the following the new data dealing with LPOR, LPOR–Pchlide complexes, and their fate under illumination are summarized.

A. THE NADPH:PCHLIDE REDUCTASES

LPOR is accumulated in the dark. Therefore, in etioplasts the protein is in excess of the minimum requirement for normal plastid development [1]. However, this large excess of enzymes is of great help in experiments designed to isolate and purify the enzyme.

Oliver and Griffiths [2] and Apel et al. [3] independently identified LPOR as a polypeptide of 36 kDa by SDS-PAGE electrophoresis. Comparison of the LPOR sequences with the already known sequences revealed that LPOR belongs to the alcohol dehydrogenase family and is not a flavoprotein [4] (reviewed in

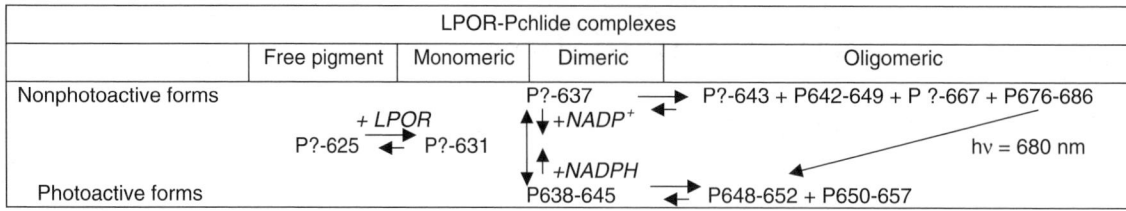

FIGURE 5.1 Scheme of the formation of the LPOR–Pchlide–nucleotide ternary complexes and their aggregation forms.

Ref. [5]). Exploring the possibility that several LPOR proteins could simultaneously occur in plastids, Apel and colleagues identified two genes coding for LPOR in *Arabidopsis thaliana* (mouse-ear cress [6]), *Hordeum vulgare* [7], and *Pinus mungo* (mountain pine [8]). The two corresponding LPOR proteins were denoted LPORA and LPORB. A recent search for the presence of genes encoding LPORA and LPORB in loblolly pine indicates that there are actually many more genes (more than 10) encoding LPOR enzyme in this plant [9]. At present, it is not known whether this situation is common or exceptional in gymnosperms. In the light of this last result, it is not completely surprising that an additional gene, encoding a third *A. thaliana* LPOR protein (LPORC), has been recently found [10]. The occurrence of more than one LPOR gene is, however, not a general rule, and organisms containing one single *lpor* gene have been detected within several taxonomic groups: cyanobacteria (*Synechocystis* sp. strain PCC 6803 [11,12], *Plectonema boryanum* [13]); Chlorophyta (*Chlamydomonas reinhardtii* [14]); and angiosperms (*Cucumis sativus* [10,15], *Pisum sativum* [16]). In organisms that contain two or more LPOR genes, the LPOR proteins seem structurally very similar, judging from the high-sequence homology of the mature proteins. However, their amount and the corresponding mRNA are differentially regulated by light: LPORA transcription is strongly inhibited by light, while LPORB is constitutively expressed [6,7]. In contrast, the transcript level of *A. thaliana* LPORC, which is undetectable in the dark, increases under illumination [10]. Different responses have been found in organisms that have only one *lpor* gene. LPOR mRNA accumulation was unaffected (pea [16,17]), enhanced (cucumber [15,18]), or depressed (cucumber [19]) by light. The regulation of *lpor* gene expression in photosynthetic organisms seems therefore highly variable. This is confirmed by a report on the *lpor* content of tobacco leaves. In this organism, two distinct LPOR cDNAs (LPOR1 and LPOR2) have been isolated. From their expression profile, LPOR1 is similar to *A. thaliana* LPORB, while LPOR2 is similar to *A. thaliana* LPORC [20].

The expression of the *lpor* gene is also regulated by cytokinins. Transient expression assays indicated that the 5′ region upstream of the *lpor* gene is responsible for the transcriptional activation. This suggests that this region contains a *cis*-acting element for cytokinin. A sequence 5′-TGACG-3′, similar to the cytokinin sensitivity motif (5′-AAGATTGATGAG-3′) of hydroxypyruvate reductase gene [21], has been found upstream of the *lpor* sequence [22]. Gibberellin also increases *lpor* gene expression, whereas abscisic acid downregulates its expression [23]. The action of these hormones may involve additional *cis*-acting elements that remain to be identified.

As already pointed out, all the LPOR proteins characterized up to now are very similar (reviewed in Refs. [5,24,25]). Each LPOR polypeptide sequence displays a Gly-X-X-X-Gly-X-Gly motif associated with the β1-αB-β2 binding domain (Rossman fold), which mostly constitutes the NADPH binding pocket. Mutations within the Rossman fold, within the helixes αE or αF (likely constituting the Pchlide binding pocket), or within the helix αH impair LPOR assembly with plastid membranes or Chlide formation [26]. LPOR mutants with Ser instead of Cys residues fail to associate to thylakoids [27]. In the cyanobacterium *Synechocystis* the amino acids beyond residue 111 are necessary for Pchlide binding and LPOR activity [28]. A strong functional similarity between the different LPOR proteins was demonstrated by cloning LPORA or LPORB genes in the *cop1* mutant of *A. thaliana*. *cop1* mutant is affected by pleiotropic phenotypes (reviewed in Ref. [29]), for instance, its inability to accumulate LPOR in the form of photoactive ternary complexes (see below) and to form prolamellar bodies (PLBs) in the dark. The insertion of either LPORA or LPORB gene in the nuclear genome of *cop1* mutant fully restores these capacities. The spectral forms of photoactive Pchlide (see below) are identical in both cloned and wild-type plants [30]. In addition, the accumulation of photoactive Pchlide–LPOR complexes and the development of PLBs in the dark were found to be independent of the relative expression of LPORA or

LPORB in *A. thaliana* [31]. *In vitro* assays of photoreduction of exogenous Pchlide by overexpressed LPOR proteins showed very similar, if not identical, characteristics [30,32,33]. Therefore, there is a great deal of evidence indicating that the different LPOR enzymes present structural and functional similarities. Consequently, in the following, we will refer collectively to the different enzymes as LPOR, except when a distinction between the different enzymes is necessary.

1. Formation of Photoactive and Nonphotoactive Pchlide Aggregates

It was believed that five different spectral forms of Pchlide coexisted in nonilluminated leaves [34–36]. Using a combination of Gaussian deconvolutions and calculations of the fourth derivative spectrum to analyze 77 K fluorescence spectra of leaves at different developmental stages, Schoefs et al. [37] established that not less than ten spectral forms of Pchlide are simultaneously present in nonilluminated leaves. Recently, Ignatov and Litvin [38] refined the analysis in the region $\lambda > 660$ nm and obtained evidence for a new spectral form of Pchlide, absorbing at 676 nm and emitting fluorescence at 686 nm. Thus, there are three forms of photoactive Pchlide[1] and eight forms of nonphotoactive Pchlide. Some progress has been made in the biochemical characterization of the native LPOR–Pchlide–NADPH ternary complexes as Ouazzani Chahdi et al. [39] purified P638–645 and P650–657. These authors established that the former spectral form of photoactive Pchlide is a dimer of the Pchlide–LPOR–NADPH ternary complex, whereas the later is a much larger aggregate. Both P638–645 and P650–657 contain the same set of carotenoids. The most abundant were violaxanthin, antheraxanthin, and zeaxanthin [39]. As all the spectral forms of Pchlide are not yet characterized at the biochemical level, it is very convenient to use their spectral properties to refer to each of them (Table 5.1). Obviously, the spectral characteristics of Pchlide in the different LPOR–Pchlide complexes reflect the immediate environment of the pigments. As the relationship between the spectral characteristics, and the molecular composition and organization of the pigment–protein complexes is not straightforward, no definitive assignment of the different *in situ* Pchlide spectral forms to precise states of the pigment–protein complexes can be done at present. Nevertheless, reasonable hypotheses on the routes leading to the formation of the large aggregates of photoactive and nonphotoactive LPOR–Pchlide complexes can be proposed on the basis of a few assumptions, which relate the spectroscopic shifts of Pchlide with its binding to LPOR, change of redox state of the cofactor, phosphorylation of the enzyme, and formation of aggregates (Figure 5.1). Some of the assumptions used to build the model shown in Figure 5.1 have received experimental support from *in vitro* studies: (i) the redshift due to pigment–pigment interactions (caused by the formation of LPOR dimers or oligomers) is amply demonstrated by studies on Pchlide aggregation in nonpolar solvents [40,41], (ii) *in vitro* reconstitution of long-wavelength photoactive forms from the short-wavelength one [42], (iii) the nonphotoactive Pchlide P?–P625 is mostly not bound to LPOR [43], and (iv) P638–645 and P650–657 have been isolated, partially purified, and their molecular weight determined [39]. As *in vitro* experiments showed that NADPH is not necessary for the firm binding of Pchlide to LPOR [12,44], we hypothesized that the pigment binds first the enzyme (Figure 5.1). Spectroscopic data suggest that these Pchlide–LPOR complexes are monomeric and not well ordered [43]. Klement et al. [45] deduced from their study of LPOR substrate specificity that the side groups around the D ring and the isocyclic ring, and the metal chelate together with the orientation of the $C13^2$ side groups are essential for the correct positioning of Pchlide in the catalytic site. The photoactive LPOR catalytic site contains amino acids with specific charges [46]. The smallest photoactive Pchlide form, P638–645[2], is a dimer [39,47]. As Pchlide is not required for membrane association of LPOR [48], no clear description is found at present on the location at which Pchlide and NADPH binding occurs. It can be deduced from *in vitro* results [32] that the dimers assemble spontaneously, probably through interactions between dimerization domains, localized between the α-helix F and the β-sheet 5. The dimerization domain is composed of 35 hydrophobic residues. Alternatively, this loop could also serve to anchor the protein in the membrane. Correct positioning of the Pchlide molecule in the catalytic center may await LPOR maturation or nucleotide binding. Aggregation of LPOR–Pchlide complex dimers may require ATP [49,50] and LPOR phosphorylation [51].

Each of the spectral forms of photoactive Pchlide has its nonphotoactive Pchlide counterpart. The slight redshift to the positions of the absorption and emission maxima of Pchlide is due to the fact that the complexes

---

[1]A photoactive protochlorophyllide is a protochlorophyllide that is transformed to chlorophyllide during a short illumination (e.g., 5 ms).

[2]The suffix numbers relate to wavelengths of absorption and emission maxima, respectively.

TABLE 5.1
Spectral Heterogeneity of the Nonphotoactive and Photoactive Pchlide Emission Bands. The Question Marks Indicate that the Absorbance Maxima are not yet Determined Symbols: ⊣ Negative Regulation → Positive Regulation

Notation	Maxima		Photoactivity
	Absorbance	Fluorescence	
P?–625[2]	?	625	−
P?–631	?	631	−
P?–637	?	638	−
P?–643	?	644	−
P642–649	642	650	−
P638-645	638	645	+
P648–652	648	652	+
P?–656	?	656	−
P650-657	650	657	+
P?–667	?	667	−
P676–686	676	686	−

Source: Prepared using data from Schoefs B, Bertrand M, Franck F. *Photochem. Photobiol.* 2000; 2:85–93 and Ignatov NV, Litvin FF. *Photosynth. Res.* 2002; 71:195–207.

contain $NADP^+$ instead of NADPH, as demonstrated by the transition from P642–650 to P650–657 upon the reversible replacement of $NADP^+$ by NADPH in isolated etioplast membranes [35,52]. Ignatov and Litvin [38] described a new nonphotoactive Pchlide–LPOR complex, P676–686, which is highly aggregated. Upon a monochromatic illumination at 680 nm, P676–686 partially disaggregates and yields P648–652, the main photoactive Pchlide spectral form in nonilluminated leaves with proplastids [37].

From spectroscopic analysis on isolated PLBs or prothylakoids (PTs) from etioplasts [53,54], it was concluded that aggregated, photoactive Pchlide forms accumulate in PLBs, whereas nonphotoactive and less aggregated photoactive forms predominate in PTs. There is a strong correlation between the development of PLBs and the accumulation of P650–657, as indicated also by studies on mutants unable to accumulate photoactive Pchlide [55] and with plants in which the expression level of LPOR has been manipulated [30,31]. The coexistence of the different Pchlide forms in well-differentiated etioplasts of dark-grown leaves probably indicates the occurrence of dynamic equilibria between these forms [37]. Local conditions may displace these equilibria toward free Pchlide or aggregated, photoactive ternary complexes.

It is noteworthy that the same spectral forms of nonphotoactive and photoactive Pchlide were found in leaves from dark-grown or naturally greening dicotyledons or monocotyledons [37] as well as in dark-grown primary needles of gymnosperms [56] and in the seed coat of the honey locust (*Gleditsia triacanthos* [25]). Therefore, we conclude that the large aggregates of Pchlide–LPOR complexes are formed along a conserved process transmitted from gymnosperms. It is not clear whether a similar process exists in the other groups of organisms like ferns, algae, cyanobacterium, which also have LPOR. In these organisms, Pchlide accumulation is usually not observed.

It has been known for a long time that at 77 K, the excitation energy can be transferred between the Pchlide–LPOR–NADPH ternary complexes composing the aggregates, the so-called energy transfer unit (reviewed in Ref. [5]). However, it was not clear if nonphotoactive Pchlide was also able to participate in this transfer, and if so, whether the ratio between nonphotoactive Pchlide and photoactive Pchlide is fixed. The answer to these two questions came from a study of the relationship between the molecular ratio of nonphotoactive and photoactive Pchlide and their respective fluorescence intensities as measured *in situ* during the course of etiolation. It is important to emphasize that during this process the molecular ratio of nonphotoactive and photoactive Pchlide changes dramatically [37]. A linear relationship between the nonphotoactive Pchlide to photoactive Pchlide ratio and the amount of photoactive Pchlide was found and it was calculated that statistically, there is one nonphotoactive Pchlide for eight photoactive Pchlide molecules in the aggregate. This result was confirmed using transgenic *Arabidopsis* cotyledons with under- or overexpressed *lpora* or *lporb* genes [31]. Altogether, these data show that the organization of photoactive Pchlide does not depend on the amounts of pigment and of enzyme molecules present in the plastid. Another consequence of these studies is that nonphotoactive Pchlide (regardless of its molecular structure) has probably a very minor role in the excitation of photoactive Pchlide, as already deduced from the photoreduction kinetic studies [57]. It is important to note that the searches for Pchlide *b* in etiolated plants have failed so far [58,59].

2. The First Products of Photoreduction, the Spectral Shifts, and the Regeneration of Photoactive Pchlide

Some progress has been made in the understanding of the intimate mechanism of the reaction catalyzed by LPOR: An electron paramagnetic resonance (EPR) study shows that the formation of short-lived paramagnetic intermediates, formed quickly after light absorption by Pchlide, requires the direct transfer of the hydride from the NADPH bound to LPOR [60]. The transfer of the second hydrogen ion would not require light and spontaneously occurs at temperatures higher than 193 K [32,61]. Unfortunately, the attribution of the EPR to a specific spectral form of the pigment is ambiguous as the reconstituted Pchlide–LPOR–NADPH ternary complexes used by different teams do not have the same spectral properties. The use of a more standardized procedure for reconstitution of the complexes or, alternatively, the use of isolated "native" Pchlide–LPOR–NADPH ternary complexes, like those prepared according to Ouazzani Chahdi et al. [39], would help in the clarification of this particular point (see Ref. [62]).

Site directed mutagenesis of the highly conserved Tyr275 (Y275F) and Lys279 (K279I, K279R) residues in the catalytic center demonstrates that the presence of these two amino acids dramatically increases the probability of the formation of the photoactive state. At the same time, they destabilize the enzyme and increase its denaturation. The two amino acids (Tyr and Lys) are not involved in binding the LPOR substrates (Pchlide and NADPH). However, the presence of Tyr275 is absolutely necessary for the second step of photoreduction, that is, the conversion of the intermediate into the first Chlide product [46].

As discussed above, nonilluminated leaves contain three spectral forms of photoactive Pchlide, which are transformed under illumination to three distinct Chlide spectral forms [37,38]. The study of the modifications of the spectral properties of Chlide arising from the photoreduction of photoactive Pchlides in leaves at different stages of development allowed

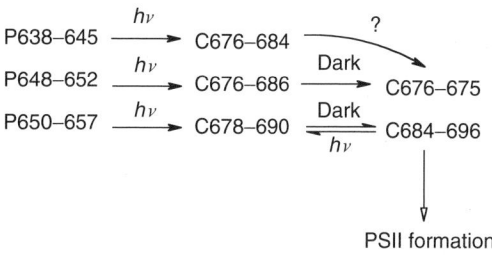

FIGURE 5.2 Scheme of the formation of the three first products resulting from the photoreduction of photoactive Pchlide. The open arrow indicates a positive regulation of the process.

Schoefs [63] to partially clarify the fate of the first products (Figure 5.2). As explained by him, the two first pathways form short wavelengths absorbing Chlide, whereas the third pathway ends with the formation of long wavelengths absorbing Chlide. Although the way of regulation of the formation of either Chlide spectral form remains unclear, it seems that the actual and local ratio between the amount of first Chlide products and the amount of nonphotoactive Pchlide plays a major role in this process. This ratio was denoted R by Schoefs [24,64]. The impact of modifications of R on the regulation of the Pchlide–Chlide cycle is discussed elsewhere (see Chapter 3 by Schoefs and Bertrand). During the spectral shifts, Chlide molecules are released from the LPOR catalytic site. On the basis of spectroscopic data recorded *in situ* and *in vitro*, it was concluded that two different mechanisms are available for this purpose [65] (Figure 5.3). The first pathway, denoted A in Figure 5.3, consists of the direct and fast release of Chlide molecules from the LPOR catalytic center without disaggregation of the large aggregates, while in the second pathway, denoted B, disaggregation of the large aggregates to dimers precedes the release of the Chlide molecules from the enzyme catalytic site [39]. Depending on the value of R, either pathway is used *in vivo*. Once Chlide has left the LPOR catalytic site, it is esterified through a four-step process, identical in leaves with proplastids and in leaves with etioplasts [66,67]. Binding of geranylgeraniol to the carboxyl group of ring D of Chlide is catalyzed by Chl synthase (*chlG* gene) [68]. After the release of Chlide molecules, the catalytic site can be reoccupied by new Pchlide molecules. This leads to the regeneration of LPOR–Pchlide complexes. As two pathways for the release of Chlide are possible, there are also two possible ways to regenerate the photoactive Pchlide complexes:

1. The direct release of Chlide from the catalytic site results in the transient formation of large

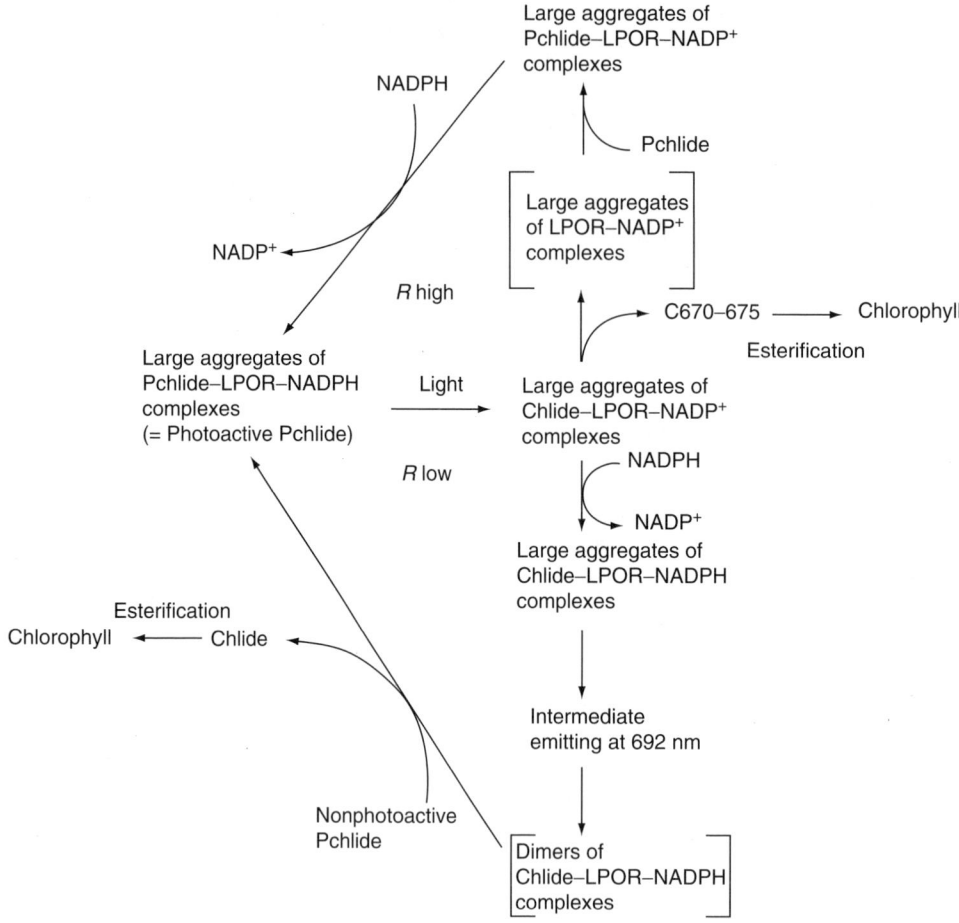

FIGURE 5.3 The Pchlide–Chlide cycles. The brackets indicate a transient state of the pigments.

aggregates of LPOR–NADP⁺ complexes, which after binding new molecules of Pchlide, give large aggregates of Pchlide–LPOR–NADP⁺ ternary complexes (P642–649 in Table 5.1) [52,65]. The NADP⁺ is progressively replaced by NADPH and the large aggregates of photoactive Pchlide are regenerated. As the large aggregates are not dislocated through this pathway, regeneration of photoactive Pchlide through this pathway should not require ATP or phosphorylation. Altogether, this Pchlide cycle is a fast process (second timescale).

2. The dissociation of the large aggregates results in the formation of LPOR–NADPH dimers, which upon binding with new molecules of Pchlide regenerate P638–645 [39,65]. The aggregation of P638–645 together or with ternary complexes of nonphotoactive Pchlide present before the illumination regenerates the large aggregates of photoactive Pchlide [65]. According to Kovacheva et al. [69], ATP has a positive effect on the re-formation of the large aggre-

gates of photoactive Pchlide. The process is inhibited by the kinase inhibitor K252a. However, *in vitro*, re-formation of the large aggregates of photoactive Pchlide can occur in the absence of added ATP [70–72]. Therefore, the involvement of an LPOR kinase in the regeneration of the large aggregates of photoactive Pchlide remains questionable. This Pchlide cycle is slow (minute timescale).

So far the Pchlide–Chlide cycle could only be studied *in situ* or in isolated membranes. Detailed kinetic and structural studies are now necessary for further understanding of the LPOR catalytic mechanism. This requires an abundant source of pure enzyme. Overexpression of LPOR from several sources (pea, barley, *Synechocystis*) has been successful as maltose-binding protein [4,32,47]. However, this procedure presents a major drawback, which is that the maltose moiety cannot be cleaved. Therefore, a procedure using cleavable His-tag for purification of LPOR should be preferred [12]. Using this method,

Heyes et al. [12] determined the apparent K_m and specific activity of the LPOR. The values found differ significantly from those obtained previously, with LPOR–membrane assemblies, suggesting that the membranous environment modifies tremendously the enzymatic properties of the enzyme.

Despite the good progress made in the elucidation of the components implied in the formation of Chlide, one central question has remained unanswered for quite a long time: What is the role of the individual spectral forms of Chlide? The first answer came from the elegant experiments performed by Franck et al. [73], who demonstrated that the formation of a definite amount of C684–696 — a Chlide spectral intermediate during the dislocation of the large aggregates of Chlide–NADPH–LPOR complexes (reviewed in Refs. [24,64,74]) (Figure 5.3) — is a sine qua non condition for the formation of photoactive photosystem II (PSII). In juvenile plants, which only produce a low amount of C684–696 [73,75], PSII is formed with a very low efficiency [76,77].

B. Light-Independent Chl a Formation

This reaction is found in every photosynthetic organism except those belonging to the group of angiosperms (reviewed in Refs. [24,64,78,79]; however, see Ref. [80]). The enzyme is composed of three subunits, ChlL, ChlB, and ChlN. *In vitro* reconstitution of the enzyme has confirmed its nitrogenase-like features, for example, oxygen sensitivity, deduced from the sequence homologies [81]. There is evidence for the fact that ChlL polypeptide is not absolutely required for Chl synthesis in the dark; its presence, however, strongly increases Chl production [82].

III. CHLOROPHYLL BIOSYNTHESIS IN GREENING AND IN GREEN LEAVES

It is known that the requirement of Chl during leaf greening is high. Green leaves also require Chl as the Chl-binding proteins of the photosynthetic apparatus turn over. The same spectral forms of photoactive Pchlide as those found in nonilluminated leaves (see above and Table 5.1) are responsible for Pchlide photoreduction in greening and green leaves [37,38,74]. Only low amounts of the spectral forms of nonphotoactive Pchlide P?-631 and P?-643 were accumulated in green leaves replaced in the dark [38]. As *lpor* gene expression is downregulated and the proteolytic degradation of LPORA occurs during the first hours of illumination (see above), the enzymes that ensure Chl synthesis during greening are LPORB and LPORC. At the earliest stage of greening, LPOR is localized preferentially in the appressed

thylakoids even though a significant amount of the enzyme is present in the nonappressed thylakoids [83]. In mature leaves, the enzyme is exclusively localized in grana. It has been shown that inhibition of Chl synthesis rapidly causes an inhibition of PSII activities and a loss of PSII components [84] as PSII repair would require new Chl molecules. This suggests a major role of LPOR in these regions of the photosynthetic membranes, where a fast PSII reaction center turnover takes place. An alternative, but not exclusive, explanation would involve the presence of a chloroplast stroma light-induced nucleus-encoded protease, which degrades LPOR–Chlide complexes [85]. This action would deplete the nonappressed thylakoids in LPOR.

During senescence, Chl *a/b*-binding proteins and LPOR levels decline considerably leading to a progressive degreening of the photosynthetic tissues. The seed coat of the Cesalpinacea *G. triacanthos* is green and contains Chl *a* and Chl *b*, and several spectral forms of nonphotoactive Pchlide have been observed [25].

During regreening the levels of Chl *a/b*-binding proteins and LPOR increase. The increase in LPOR is accelerated by cytokinin [86].

In the natural environment, plants continuously undergo changes in light intensity. These changes trigger adaptation mechanisms such as the modifications in the Chl *a*/Chl *b* ratio. It is tempting to speculate that the monooxygenase catalyzing the conversion of Chl *a* to Chl *b* is implied in this process. As monooxygenases catalyze strongly exothermic reactions, Chlide *a* to Chlide *b* reaction is a irreversible process. Chl *a* to Chl *b* interconversion may occur through the Chl *a*/Chl *b* cycle, first proposed by Oster et al. [87], as a link between the biosynthetic and degradation pathways for Chl molecules (see also Ref. [88]). The Chl degradation pathway has been reviewed by Bertrand and Schoefs [89]. In fact, it has been shown that overexpression of CAO broke the limit of the Chl *a*/Chl *b* ratio. This suggests that CAO is the primary factor determining antenna size in green tissues [90].

IV. CONCLUSION AND PERSPECTIVES

Progress in the understanding of the formation of the large aggregates of photoactive Pchlide has been made using mathematical analysis of spectroscopic data. Although it seems obvious that the spectral characteristics of the pigment must reflect its immediate environment, the relationship between absorption and emission maxima on the one hand and the molecular composition and organization of the pigment–protein complexes on the other can be difficult to establish. Additional work will be necessary to isolate and characterize the different spectral forms of pigment–LPOR complexes to correlate them with their spectroscopic properties. The fact that the same spectral forms of Pchlide are found in angiosperm and gymnosperm tissues suggests that the large aggregates of Pchlide–LPOR complexes are formed along a conserved process transmitted from gymnosperms. It would be interesting to determine if this pathway has been inherited from lower organisms like ferns, algae, cyanobacterium, which also have LPOR but usually do not accumulate Pchlide.

REFERENCES

1. Ougham HJ, Thomas AM, Thomas BJ, Frick GA, Armstrong GA. Both light-dependent protochlorophyllide oxidoreductase A and protochlorophyllide oxidoreductase B are down-regulated in the *slender* mutant of barley. *J. Exp. Bot.* 2001; 52:1447–1454.

2. Oliver RP, Griffiths WT. Identification of the polypeptides of NADPH:protochlorophyllide oxidoreductase. *Biochem. J.* 1980; 191:277–280.

3. Apel K, Santel H-J, Redlinger TE, Falk H. The protochlorophyllide holochrome of barley (*Hordeum vulgare* L.). Isolation and characterization of the NADPH:protochlorophyllide oxidoreductase. *Eur. J. Biochem.* 1980; 111:251–258.

4. Townley HE, Griffiths WT, Nugent JP. A reappraisal of the mechanism of the photoenzyme protochlorophyllide reductase based on studies with the heterologously expressed protein. *FEBS Lett.* 1998; 422:19–22.

5. Schoefs B, Franck F. Chlorophyll biosynthesis: light-dependent and light-independent protochlorophyllide reduction. *Bull. Cl. Sci. Acad. R. Belgique* 2002; 13:113–157.

6. Armstrong GA, Runge S, Frick G, Sperling U, Apel K. Identification of NADPH:protochlorophyllide oxidoreductase A and B branched pathway for light-dependent chlorophyll synthesis in *Arabidopsis thaliana*. *Plant Physiol.* 1995; 108:1505–1517.

7. Holtorf H, Reinbothe S, Reinbothe C, Bereza B, Apel K. Two routes of chlorophyllide synthesis that are differentially regulated by light in barley (*Hordeum vulgare* L.). *Proc. Natl. Acad. Sci. USA* 1995; 92:3254–3258.

8. Forreiter C, Apel K. Light-independent and light-dependent protochlorophyllide-reducing activities and two distinct NADPH-protochlorophyllide oxidoreductase polypeptides in mountain pine (*Pinus mungo*). *Planta* 1993; 190:536–545.

9. Skinner JS, Timko MP. Loblolly pine (*Pinus taeda* L.) contains multiple expressed gene encoding light-dependent NADPH:protochlorophyllide oxidoreductase (POR). *Plant Cell Physiol.* 1998; 39:795–806.

10. Oosawa N, Masuda T, Awai K, Fusada N, Shimada H, Ohta H, Takamiya K. Identification and light-induced expression of a novel gene of NADPH-protochloro-

11. Kaneko T, Sato S, Kotani H, Tanaka A, Asamizu E, Nakamura Y, Miyajima N, Hirosawa M, Sugiura M, Sasamoto S, Kimura T, Hosouchi T, Matsuno A, Muraki A, Nakazaki N, Naruo K, Okamura S, Shimpo S, Takeuchi C, Wada T, Watanabe A, Yamada M, Ysaduda M, Tabata S. Sequence analysis of the genome of the unicellular cyanobacterium *Synechocystis* sp. strain PCC6903. II. Sequence determination of the entire genome and assignment of potential protein-coding regions. *DNA Res.* 1996; 3:109–136.

phyllide oxidoreductase isoform in *Arabidopsis thaliana*. *FEBS Lett.* 2000; 474:133–136.

12. Heyes DJ, Martin GEM, Reid RJ, Hunter CN, Wilks HM. NADPH:protochlorophyllide oxidoreductase from *Synechocystis*: overexpression, purification and preliminary characterization. *FEBS Lett.* 2000; 483:47–51.

13. Fujita Y, Takagi H, Hase T. Cloning of the gene encoding a protochlorophyllide reductase: the physiological significance of the coexistence of light-dependent and -independent protochlorophyllide reduction systems in the cyanobacterium *Plectonema botyanum*. *Plant Cell Physiol.* 1998; 39:177–185.

14. Li J, Timko MP. The *pc-1* phenotype of *Chlamydomonas reinhardtii* results from a deletion mutation in the nuclear region for NADPH:protochlorophyllide oxidoreductase. *Plant Mol. Biol.* 1996; 30:15–37.

15. Kuroda H, Masuda T, Fusada N, Ohta H, Takamiya K. Expression of NADPH-protochlorophyllide oxidoreductase gene in fully green leaves of cucumber. *Plant Cell Physiol.* 2000; 41:226–229.

16. Spano AJ, He Z, Michel H, Hunt DF, Timko MP. Molecular cloning, nuclear gene structure, and developmental expression of NADPH:protochlorophyllide oxidoreductase in pea (*Pisum sativum* L.). *Plant Mol. Biol.* 1992; 18:967–972.

17. He Z-H, Li J, Sundqvist C, Timko MP. Leaf developmental age controls expression of genes encoding enzymes of chlorophyll and haem biosynthesis in pea (*Pisum sativum* L.). *Plant Physiol.* 1994; 106:537–543.

18. Kuroda H, Masuda T, Ohta H, Shioi Y, Takamiya K. Light-enhanced gene expression of NADPH-protochlorophyllide oxidoreductase in cucumber. *Biochem. Biophys. Res. Commun.* 1995; 210:310–316.

19. Yoshida K, Chen R-M, Tanaka A, Teramoto H, Tanaka R, Timko MP, Tsuji H. Correlated changes in the activity, amount of protein, and abundance of transcript of NADPH:protochlorophyllide oxidoreductase and chlorophyll accumulation during greening of cucumber cotyledons. *Plant Physiol.* 1995; 109:231–238.

20. Masuda T, Fusada N, Shiraishi T, Kuroda H, Awai K, Shimada H, Ohta H, Takamiya K. Identification of two differentially regulated isoforms of protochlorophyllide oxidoreductase (POR) from tobacco revealed a wide variety of light- and development-dependent regulations of POR gene expression among angiosperms. *Photosynth. Res.* 2002; 74:165–172.

21. Jin G, Davey MC, Ertl JR, Chen R, Yu ZT, Daniel SG, Becker WM, Chen CM. Interaction of DNA-binding proteins with the 5′-flanking region of a cytokinin-

responsive cucumber hydroxypyruvate reductase gene. *Plant Mol. Biol.* 1998; 38:713–723.

22. Kuroda H, Masuda T, Fusada N, Ohta H, Takamiya K. Cytokinin-induced transcriptional activation of NADPH-protochlorophyllide oxidoreductase gene in cucumber. *J. Plant Res.* 2001; 114:1–7.

23. Kuroda H, Masuda T, Ohta H, Shioi Y, Takamiya K-I. Effects of light, development age and phytohormones on the expression of NADPH-protochlorophyllide oxidoreductase gene in *Cucumis sativus. Plant Physiol. Biochem.* 1996; 34:17–22.

24. Schoefs B. The protochlorophyllide–chlorophyllide cycle. *Photosynth. Res.* 2001; 70:257–271.

25. Schoefs B. Pigment composition and location in honey locust (*Gleditsia triacanthos*) seeds before and after desiccation. *Tree Physiol.* 2002; 22:285–290.

26. Dahlin C, Aronsson H, Wilks HM, Lebedev N, Sundqvist C, Timko MP. The role of protein surface charge in catalytic activity and chloroplast membrane association of the pea NADPH:protochlorophyllide oxidoreductase (POR) as revealed by alanine mutagenesis. *Plant Mol. Biol.* 1999; 39:309–323.

27. Aronsson H, Sundqvist C, Timko MP, Dahlin C. The importance of the C-terminal region and Cys residues for the membrane association of NADPH:protochlorophyllide oxidoreductase in pea. *FEBS Lett.* 2000; 502:11–15.

28. He Q, Brune D, Nieman R, Vermaas W. Chlorophyll *a* synthesis upon interruption and deletion of *por* coding for the light-dependent NADPH:protochlorophyllide oxidoreductase in a photosystem-I-less/*chlL⁻* strain of *Synechocystis* sp. PCC 6803. *Eur. J. Biochem.* 1998; 253:161–172.

29. von Armin A, Deng X-W. Light control of seedlings development. *Annu. Rev. Plant Physiol. Plant Mol. Biol.* 1996; 47:215–243.

30. Sperling U, Franck F, van Cleve B, Frick G, Apel K, Armstrong G. Etioplast differentiation in *Arabidopsis*: both PORA and PORB restore the prolamellar body and photoactive Pchlide-F655 to the *cop1* photomorphogenic mutant. *Plant Cell* 1998; 10:283–296.

31. Franck F, Sperling U, Frick G, Pochert B, van Cleve B, Apel K, Armstrong GA. Regulation of etioplast pigment–protein complexes, inner membrane architecture, and protochlorophyllide *a* chemical heterogeneity by light-dependent NADPH:protochlorophyllide oxidoreductases A and B. *Plant Physiol.* 2000; 124:1678–1696.

32. Lebedev N, Timko MP. Protochlorophyllide oxidoreductase B-catalyzed protochlorophyllide photoreduction *in vitro*: insight into the mechanism of chlorophyll formation in light-adapted plants. *Proc. Natl. Acad. Sci. USA* 1999; 96:9954–9959.

33. Su Q, Frick G, Armstrong G, Apel K. PORC of *Arabidopsis thaliana*: a third light- and NADPH-dependent protochlorophyllide oxidoreductase that is differentially regulated by light. *Plant Mol. Biol.* 2001; 47:805–813.

34. Böddi B, Ryberg M, Sundqvist C. Identification of four universal protochlorophyllide forms in dark-grown leaves by analyses of the 77 K fluorescence emission spectra. *J. Photochem. Photobiol. B* 1992; 12:389–401.

35. El Hamouri B, Sironval C. NADP⁺/NADPH control of the protochlorophyllide-chlorophyllide proteins in cucumber etioplasts. *Photobiochem. Photobiophys.* 1980; 1:219–223.

36. Böddi B, Franck F. Room temperature fluorescence spectra of protochlorophyllide and chlorophyllide forms in etiolated bean leaves. *J. Photochem. Photobiol. B* 1997; 41:73–83 (erratum appears in *J. Photochem. Photobiol. B* 1998; 42:168).

37. Schoefs B, Bertrand M, Franck F. Spectroscopic properties of protochlorophyllide analyzed *in situ* in the course of etiolation and in illuminated leaves. *Photochem. Photobiol.* 2000; 72:85–93.

38. Ignatov NV, Litvin FF. A new pathway of chlorophyll biosynthesis from long-wavelength protochlorophyllide Pchlide 686/676 in juvenile etiolated plants. *Photosynth. Res.* 2002; 71:195–207.

39. Ouazzani Chahdi MA, Schoefs B, Franck F. Isolation and characterization of photoactive complexes of NADPH:protochlorophyllide oxidoreductase from wheat. *Planta* 1998; 206:673–680.

40. Seliskar CJ, Ke B. Protochlorophyllide aggregation in solution and associated spectral changes. *Biochim. Biophys. Acta* 1968; 153:685–691.

41. Brouers M. Optical properties of *in vitro* aggregates of protochlorophyllide in non-polar solvents. I. Visible and fluorescence spectra. *Photosynthetica* 1972; 6:415–423.

42. Brouers M, Sironval C. Resaturation of a P657-647 form from P643-638 in extracts of etiolated primary bean leaves. *Plant Sci. Lett.* 1975; 4:175–181.

43. Böddi B, Kipetik K, Kapori AD, Fidy J, Sundqvist C. The two spectroscopically different short wavelength protochlorophyllide forms in pea epicotyls are both monomeric. *Biochim. Biophys. Acta* 1998; 1365:531–540.

44. Richards WR, Walker CJ, Griffiths WT. The affinity chromatographic purification of NADPH:protochlorophyllide oxidoreductase from etiolated wheat. *Photosynthetica* 1987; 21:462–471.

45. Klement H, Helfrich M, Oster U, Schoch S, Rüdiger W. Pigment-free NADPH:protochlorophyllide oxidoreductase from *Avena sativa* L. Purification and substrate specificity. *Eur. J. Biochem.* 1999; 265:862–874.

46. Lebedev N, Karginova O, McIvor W, Timko MP. Tyr275 and Lys279 stabilize NADPH within the catalytic site of NADPH:protochlorophyllide oxidoreductase and are involved in the formation of the enzyme photoactive state. *Biochemistry* 2001; 40:12562–12574.

47. Martin GEM, Timko MP, Wilks HM. Purification and kinetic analysis of pea (*Pisum sativum* L.) NADPH:protochlorophyllide oxidoreductase expressed as fusion with maltose-binding protein in *Escherichia coli. Biochem. J.* 1997; 325:139–145.

48. Aronsson H, Sundqvist C, Timko MP, Dahlin C. Characterization of the assembly pathway of the pea NADPH:protochlorophyllide (Pchlide) oxidoreductase (POR), with emphasis on the role of its substrate, Pchlide. *Physiol. Plant.* 2000; 111:239–244.

49. Horton P, Leech RM. The effect of ATP on photoconversion of protochlorophyll to chlorophyll in isolated etioplasts. *FEBS Lett.* 1972; 26:277–280.

50. Horton P, Leech RM. The effect of adenosine 5′-triphosphate on the Shibata shift and the associated structural changes in the conformation of the prolamellar body in isolated maize etioplasts. *Plant Physiol.* 1975; 55:393–400.

51. Ryberg M, Sundqvist C. The regular ultrastructure of isolated prolamellar bodies depends on the presence of membrane-bound NADPH-protochlorophyllide oxidoreductase. *Physiol. Plant.* 1988; 73:218–226.

52. Franck F, Bereza B, Böddi B. Protochlorophyllide-NADP$^+$ and protochlorophyllide-NADPH complexes and their regeneration after flash illumination in leaves and etioplast membranes of dark-grown wheat. *Photosynth. Res.* 1999; 59:53–61.

53. Ryberg M, Sundqvist C. Characterization of prolamellar bodies and prothylakoids fractionated from wheat. *Physiol. Plant.* 1982; 56:125–132.

54. Ryberg M, Sundqvist C. Spectral forms of protochlorophyllide in prolamellar bodies and prothylakoids fractionated from wheat etioplasts. *Physiol. Plant.* 1982; 56:133–138.

55. Henningsen KW, Boynton JE, von Wettstein D. Mutants at *xantha* and *albina* loci in relation to chloroplast biogenesis in barley (*Hordeum vulgare* L.). *Biol. Skrift. Kgl. Dan. Vid. Selsk.* 1993; 42:1–293.

56. Schoefs B, Franck F. Chlorophyll synthesis in dark-grown pine primary needles. *Plant Physiol.* 1998; 118:1159–1168.

57. Schoefs B, Garnir H-P, Bertrand M. Comparison of the photoreduction of protochlorophyllide to chlorophyllide in leaves and cotyledons from dark grown bean as a function of age. *Photosynth. Res.* 1994; 41:405–417.

58. Schoefs B, Bertrand M, Lemoine Y. Changes in the photosynthetic pigments in bean leaves during the first photoperiod of greening and the subsequent dark-phase. Comparison between old (10-d-old) leaves and young (2-d-old) leaves. *Photosynth. Res.* 1998; 57: 203–213.

59. Scheumann V, Klement H, Helfrich M, Oster U, Schoch S, Rüdiger W. Protochlorophyllide *b* does not occur in barley leaves. *FEBS Lett.* 1999; 445:445–448.

60. Griffiths WT, McHugh T, Blankenship E. The light intensity dependence of protochlorophyllide photoconversion and its significance to the catalytic mechanism of protochlorophyllide reductase. *FEBS Lett.* 1996; 398:235–238.

61. Belyaeva OB, Griffiths WT, Kovalev JV, Timofeev KN, Litvin FF. Participation of free radicals in photoreduction of protochlorophyllide to chlorophyllide in an artificial pigment–protein complex. *Biochemistry (Mosc.)* 2001; 66:217–222.

62. Raskin VI, Schwartz A. The charge transfer complex between protochlorophyllide and NADPH: an intermediate in protochlorophyllide photoreduction. *Photosynth. Res.* 2002; 74:181–186.

63. Schoefs B. Photoreduction of protochlorophyllide *a* to chlorophyllide *a* during the biogenesis of the photosynthetic apparatus in higher plants. Dissertation.com. ISBN: 1-58112-097-4, 2000.

64. Schoefs B. The light-dependent protochlorophyllide reduction: from a photoprotecting mechanism to a metabolic reaction. In: Panadalai, ed. *Recent Research in Photosynthesis.* Vol. 2. Trivandrum: Research Signpost, 2001:241–258.

65. Schoefs B, Bertrand M, Funk C. Photoactive protochlorophyllide regeneration in cotyledons and leaves from higher plants. *Photochem. Photobiol.* 2000; 72:660–668.

66. Schoefs B, Bertrand M. The transformation of chlorophyllide to chlorophyllide in leaves with proplastids is a four step reaction. *FEBS Lett.* 2000; 486:243–246 (erratum appears in *FEBS Lett.* 2001; 494:261).

67. Domanskii VP, Rüdiger W. On the nature of the two pathways in chlorophyll formation from protochlorophyllide. *Photosynth. Res.* 2001; 68:131–139.

68. Oster U, Bauer CE, Rüdiger W. Characterization of chlorophyll *a* and bacteriochlorophyll *a* synthase by heterologous expression of *Escherichia coli.* *J. Biol. Chem.* 1997; 272:9671–9676.

69. Kovacheva S, Ryberg M, Sundqvist C. ADP/ATP and protein phosphorylation dependence of phototransformable protochlorophyllide in isolated etioplast membranes. *Photosynth. Res.* 2000; 64:127–136.

70. Griffiths WT. Source of reducing equivalents for the *in vitro* synthesis of chlorophyll from protochlorophyll. *FEBS Lett.* 1974; 26:301–304.

71. Brodersen P. Factors affecting the photoconversion of protochlorophyllide to chlorophyllide in etiolated membranes. *Photosynthetica* 1976; 10:33–39.

72. Selstam E, Schelin J, Brain T, Williams WP. The effects of low pH on the properties of protochlorophyllide oxidoreductase and the organization of prolamellar bodies of maize (*Zea mays*). *Eur. J. Biochem.* 2002; 269:2336–2346.

73. Franck F, Eullaffroy P, Popovic R. Formation of long-wavelength chlorophyllide (Chlide695) is required for the assembly of photosystem II in etiolated barley leaves. *Photosynth. Res.* 1997; 51:107–118.

74. Schoefs B, Bertrand M. Chlorophyll biosynthesis. In: Pessarakli M, ed. *Handbook of Photosynthesis.* New York: Marcel Dekker, 1997:49–69.

75. Schoefs B, Franck F. Photoreduction of protochlorophyllide to chlorophyllide in 2-d-old dark-grown bean (*Phaseolus vulgaris* cv. Commodore) leaves. Comparison with 10-d-old dark-grown (etiolated) leaves. *J. Exp. Bot.* 1993; 44:1053–1056.

76. Schoefs B, Franck F. Photosystem II assembly in 2-day-old bean leaves during the first 16 h. of greening. *C. R. Acad. Sci. Paris Ser. III* 1991; 314:441–445.

77. Schoefs B, Bertrand M, Franck F. Plant greening: the case of bean leaves illuminated shortly after the germination. *Photosynthetica* 1992; 27:497–504.

78. Armstrong GA. Greening in the dark: light-independent chlorophyll biosynthesis from anoxygenic bacteria to gymnosperms. *J. Photochem. Photobiol. B* 1998; 43:87–100.

79. Schoefs B. The light-dependent and the light-independent reduction of protochlorophyllide *a* to chlorophyllide *a.* *Photosynthetica* 1999; 36:481–496.

80. Adamson HY, Hiller RG, Walmsley J. Protochloro-phyllide reduction and greening in angiosperms: an evolutionary perspective. *J. Photochem. Photobiol.* 1997; 41:201–221.
81. Fujita Y, Bauer CE. Reconstitution of light-independent protochlorophyllide reductase from purified BchL and BchN-BchB subunits. *In vitro* confirmation of nitrogenase-like features of bacteriochlorophyll biosynthesis enzymes. *J. Biol. Chem.* 2000; 275:23583–23588.
82. Wu QY, Yu JJ, Zhao NM. Partial recovery of light-independent chlorophyll biosynthesis in the ChlL-deletion mutant of *Synechocystis* sp. PCC 6803. *IUBMB Life* 2001; 51:289–293.
83. Barthélemy X, Bouvier G, Radunz A, Docquier S, Schmid GH, Franck F. Localization of NADPH-protochlorophyllide reductase in plastids of barley at different greening stages. *Photosynth. Res.* 2000; 64:63–76.
84. Feirabend J, Dehne S. Fate of the porphyrin cofactors during the light-dependent turnover of catalase and of the photosystem II reaction-center protein D1 in mature rye leaves. *Planta* 1996; 198:413–422.
85. Reinbothe S, Reinbothe C, Runge S, Apel K. Enzymatic product formation impairs both the chloroplast receptor-binding function as well as translocation competence of the NADPH:protochlorophyllide oxidore-ductase, a nuclear-encoded plastid precursor protein. *J. Cell. Biol.* 1995; 129:299–308.
86. Zavaleta-Mancera HA, Franklin KA, Ougham HJ, Thomas H, Scott IM. Regreening of scenescent *Nicotiana* leaves. I. Reappearance of NADPH-protochlorophyllide oxidoreductase and light-harvesting chlorophyll a/b-binding protein. *J. Exp. Bot.* 1999; 50:1677–1682.
87. Oster U, Tanaka R, Tanaka A, Rüdiger W. Cloning and functional expression of the gene encoding the key enzyme for chlorophyll b biosynthesis (CAO) from *Arabidopsis thaliana*. *Plant J.* 2000; 21:301–305.
88. Porra RJ, Scheer H. ^{18}O and mass spectrometry in chlorophyll research: derivation and loss of oxygen atom at the periphery of the chlorophyll macrocycle during biosynthesis, degradation and adaptation. *Photosynth. Res.* 2000; 66:159–175.
89. Bertrand M, Schoefs B. Photosynthetic pigment metabolism in plants during stress. In: Pessarakli M, ed. *Handbook of Plant and Crop Stress.* New York: Marcel Dekker, 1999:527–543.
90. Tanaka R, Koshino Y, Sawa S, Ishiguro S, Okada K, Tanaka A. Overexpression of the chlorophyllide a oxygenase (CAO) enlarges the antenna size of photosystem II in *Arabidopsis thaliana*. *Plant J.* 2001; 26:365–373.

6 Formation and Demolition of Chloroplast during Leaf Ontogeny

Basanti Biswal

Laboratory of Biochemistry and Molecular Biology, School of Life Sciences,
Sambalpur University

CONTENTS

I. INTRODUCTION

The development of chloroplast from proplastid during leaf formation and the subsequent transformation of the chloroplast to gerontoplast (senescing chloroplast) during leaf yellowing have been extensively examined (for a review, see Ref. [1]). Various studies indicate that the biogenesis of chloroplast, both formation and demolition, is tightly coupled to leaf ontogeny.

The development of the photosynthetic organelle from proplastid is accompanied by the accumulation of pigments, proteins, lipids, and other cofactors required for the facilitation of photosynthesis. During rapid chloroplast development, the rates of transcription and translation, the level of mRNAs, the total content of organelle ribosomes, as well as the level of polysomes remain high, which, however, maintain a steady level in fully mature leaves. On the other hand, the levels of different inclusions like pigments, proteins, and other constituents of the organelle start declining, which results in the formation of gerontoplast during leaf senescence. The events associated with both development and senescence are perfectly coordinated and regulated by genes.

Recent data on the synthesis and assembly of different thylakoid complexes, demolition of these complexes leading finally to their degradation, and the coordinated action of nuclear and plastid genes regulating the biogenesis events of the organelle are critically discussed in this review. Few questions related to the nature and transduction of signals that regulate these events are also addressed.

II. ORGANIZATION AND FORMATION OF CHLOROPLAST DURING LEAF DEVELOPMENT

The transformation of proplastid to chloroplast involves the formation of mature stacked thylakoids from structurally simple membrane precursors. There structural changes are linked to the accumulation of photosynthetic pigments.

A. ACCUMULATION OF GREEN PIGMENTS: BIOLOGY OF NADPH–PROTOCHLOROPHYLLIDE OXIDOREDUCTASE AND CHLOROPHYLL BIOSYNTHESIS

Leaf greening is the visible symptom of chlorophyll accumulation in developing chloroplasts. The biosynthesis of the pigment involves several steps including the formation of 5-ALA and a pyrrole ring with a conjugate bond system, insertion of magnesium, synthesis of protochlorophyllide, and its subsequent reduction to chlorophyllide followed by phytylation. Most of the enzymes involved in the biosynthetic pathway have been characterized and their molecular biology is known (for a review, see Ref. [2]). Among all the enzymes, NADPH–protochlorophyllide oxidoreductase (POR) has been extensively examined [3]. Its molecular biology and photoregulation are considered to be very exciting and fascinating areas of research in plant science. In addition to its role in chlorophyll biosynthesis, POR is reported to play a role in processing and transformation of precursors of thylakoids to their mature form during the development of the photosynthetic organelle. In the biosynthesis of the pigment, the enzyme mediates the light-dependent photoreduction of protochlorophyllide to chlorophyllide. The protochlorophyllide complexed with POR acts as the photoreceptor. The photoreduction step brings about the structural modulation of the membranes, resulting in the formation of lamellar systems

of the chloroplasts. Three types of the enzyme, POR A, B, and C, were isolated and characterized. These enzymes exhibit differential modes of processing and targeting [1,3]. The genes coding for the three POR species are differentially regulated by light and developmental factors. The coordinated action of POR and chlorophyll synthase during the final stages of chlorophyll biosynthesis has been critically discussed [4]. The *in vivo* stabilities of both chlorophyll and carotenoid primarily depend on their insertion to apoprotien, forming pigment–protein complexes of thylakoid membranes. The apoproteins, after synthesis, are targeted, pigmented, and inserted at the proper location of the thylakoid membranes.

B. CHLOROPLAST DNA, PROTEIN SYNTHESIS, AND TARGETING OF THE NUCLEAR ENCODED CHLOROPLAST PROTEINS

Plastid DNA has a circular structure and ranges in size from 100 to 180 kb. The DNA is cloned and sequenced in many plant systems [5]. As mentioned earlier, the biogenesis of the photosynthetic organelle requires the participation of both the plastid genes and the nuclear genes. The plastid genes are normally classified into two major classes — those coding for photosynthetic components and those required for different components of the protein synthesis process of the organelle itself. The plastid and nuclear genes encoding the proteins involved in the biogenesis of chloroplast are shown in Table 6.1 and Table 6.2.

The chloroplast proteins encoded by the nuclear genes are synthesized in the cytoplasm as high molecular weight precursors, processed, and targeted to the organelle through importing mechanisms associated with the organelle envelope. The entire process involves several steps including recognition and binding of the precursor proteins to the import machine, transport through the envelope utilizing energy and various modulators, proteolytic cleavage of the transit sequence, and, finally, insertion of mature proteins at the proper location [6,7]. Because of different locations of the nuclear encoded chloroplast proteins, the targeting follows different paths of transport including ΔpH pathway, Sec-like pathway, and signal recognition particle (SRP)-like pathway. The transport involves energy in different forms for different pathways. The proteins synthesized in chloroplasts also follow regulated transport pathways and are targeted to the correct locations [6].

C. ASSEMBLY OF THYLAKOID COMPLEXES

There are more than 60 thylakoid proteins that constitute four major complexes: PSII, cytochrome

TABLE 6.1
Proteins of Thylakoid Complexes and Rubisco Encoded by Chloroplast Genes

Thylakoid Complexes and Rubisco	Gene	Protein	Function
PSII	psbA	D1	RC II core
	psbB	CP47	Antenna
	psbC	CP43	Antenna
	psbD	D2	RC II core
	psbE	Cyt $b_{559\alpha}$	RC II core heme protein
	psbF	Cyt $b_{559\beta}$	Photoprotection by cyclic electron flow in PSII (?)
	psbH	PSII-H	Photoprotection
	psbI	PSII-I	RCII core?
	psbJ	PSII-J	PSII assembly
	psbK	PSII-K	PSII assembly and stability
	psbL	PSII-L	Involved in Q_A function.
	psbM	PSII-M	?
	psbN	PSII-N	?
PSI	psaA	PSI-A	RC I core
	psaB	PSI-B	RC I core
	psaI	PSI-I	?
	psaJ	PSI-J	Interacts with PSI-E and F
	psaC	PSI-C	[4Fe–4s] electron acceptor, FeS-A and FeS-B.
Cyt b_6/f	petA	Cyt f	c-Type heme protein
	petB	Cyt b_6	b-Type heme protein
	petD	Subunit IV	Quinone binding protein
	petE	Subunit V	Involved in Q_A function.
ATP synthase	atpA	CF_1-α	Regulation
	atpB	CF_1-β	Catalytic site
	atpE	CF_1-ε	Inhibitor of ATPase
	atpF	CF_0-I(b)	Binding CF_0 and CF_1
	atpH	CF_0-III(c)	Rotor complex (9–12 subunits)
	atpI	CF_0-IV(a)	Proton translocation
Rubisco	rbcL	LSU_8	Large subunit of Rubisco enzyme

b/f complex, PSI, and ATPase. In addition to these complexes, plastocyanin, ferredoxin, and ferredoxin–NADP–oxidoreductase (FNR) are the major redox components of the electron transport chain of thylakoids [8]. Among the stroma proteins, Rubisco, a multimeric protein complex, has been well studied [1].

There are several factors that regulate transcriptional, posttranscriptional, translational, and posttranslational processes for the formation and processing of chloroplast proteins during organelle biogenesis [1,9,10].

Existing literature suggests the temporal appearance of the activities of thylakoid complexes during leaf greening [1,11,12]. Ohashi et al. [13] have examined in detail the sequence of assembly of PSI, PSII, electron transport complexes connecting these photosystems, and the partial electron transport systems associated with the individual photosystem during the greening of etiolated barley leaves. However, the sequence of appearance of PSI, PSII, and other com-

plexes varies with plant species and the environmental conditions the plants experience [11,12].

1. Organization and Assembly of PSII

Among the individual complexes, the assembly of PSII has been widely studied in recent years [1,11]. The major intrinsic protein subunits of the PSII complex such as D1, D2, cytochrome b_{559}, CP_{43}, and CP_{47} are encoded by chloroplast genes (Table 6.1), synthesized in the organelle, processed on membranes, and transported within the thylakoids from stroma lamellae to stacked grana regions where they are inserted with other proteins and nonprotein components to form the final stable assembly. On the other hand, the extrinsic proteins of molecular weights 33, 23, and 16 kDa are encoded by nuclear genes (Table 6.2), synthesized in cytoplasm as high molecular weight precursors, processed, and transported through the chloroplast envelope and the thylakoid membrane. Finally, the proteins reach the lumen and are attached

TABLE 6.2
Proteins of Thylakoid Complexes and Rubisco Encoded by Nuclear Genes

Thylakoid Complexes and Rubisco	Gene	Protein	Function
PSII	psbR	PSII-R	Docking extrinsic subunits
	lhcb 1	LHCII b	
	lhcb2	LHCII b	
	lhcb3	LHCII b	
	lhcb4	LHCII a	Light harvesting
	lhcb5	LHCII c	
	lhcb6	LHCII d	
	psbO	33 kD	
	psbP	23 kD	Extrinsic proteins
	psbQ	16 kD	
PSI	psaF	PSI-F	Plastocyanin docking
	psaG	PSI-G	?(in green plants only)
	psaK	PSI-K	Interacts with PSI-A and -B
	psaL	PSI-L	Trimer formation
	psaO	PSI-O	?(in green plants only)
	lhca1	LHCI-I	
	lhca2	LHCI-II	
	lhca3	LHCI-III	Light harvesting
	lhca4	LHCI-IV	
	psaD	PSI-D	Ferredoxin docking
	psaE	PSI-E	Cyclic electron transport Binding of ferredoxin
	psaH	PSI-H	?(in green plants only)
	petG	Ferredoxin	FeS protein
	petH	FNR	Ferredoxin $-NADP^+$ reductase
	petI	FNR binding	Binding FNR
	petF	Plastocyanin	Electron donating to RC I
Cyt b_6/f	petC	Rieske	[2Fe–2S] protein
ATP synthase	atpC	CF_1-y	
	atpD	CF_1-δ	Regulation
	atpG	CF_0-II(b′)	Binding CF_0 and CF_1
Rubisco	rbcS	SSU_8	Small subunit of Rubisco enzyme

to the intrinsic core complex. It is proposed that some of the protein subunits may remain stable in the absence of other subunits of the complex but cannot have a proper orientation on lamellar bilayer membranes [14]. The synthesis, regulation, and assembly of both intrinsic and extrinsic proteins and their final insertion to the PSII core complex were examined in detail in both *in vitro* and *in vivo* conditions (for a review, see Ref. [1]).

2. Assembly of PSI, Cytochrome b/f Complex, and ATPase

The assembly of PSI involves the synthesis of several proteins encoded both by plastid and nuclear genes (see Table 6.1 and Table 6.2; see Refs. [1,8]). It is a heteromultimeric protein complex with different pigments and several redox centers. The assembly process is known to be regulated by the nuclear gene products.

Similarly, the assembly of the cytochrome b/f complex and ATPase requires the proteins that are encoded by the chloroplast and nuclear genes (Table 6.1 and Table 6.2; see Refs. [1,8]). Steps like heme attachment, synthesis and binding of the iron–sulfur centers, and other cofactors modulate the assembly of the cytochrome b/f complex [1]. On the other hand, both the nuclear and plastid factors are shown to regulate the synthesis of protein subunits and assembly of ATPase [1].

3. LHC Assembly

PSI and PSII light-harvesting systems of the thylakoid membrane consists of several distinct pigment–protein complexes. These are predominantly integral protein complexes of lamellar systems both in green algae and higher plants. The complexes associated with PSI and PSII are referred to as light-harvesting chlorophyll protein complex I (LHC I) and light-harvesting chlorophyll protein complex II (LHC II), respectively. Literature on the expression of the nuclear genes coding for LHC apoproteins is extensive. Most of these genes, as shown in Table 6.1 and Table 6.2, are isolated, sequenced, and characterized from different plant species. LHCs are synthesized as high molecular weight precursor proteins, which are processed and transported to the thylakoids of the organelle [6]. Usually, the LHCs degrade when the proteins are not complexed with chlorophylls and carotenoids. Although the precise nature of sequential events leading to the assembly of LHCs is not clear, Dreyfuss and Thornber [15,16] have examined in detail the formation, organization, and sequential assembly of light-harvesting complexes of both the photosystems during the biogenesis of plastids of barley leaves. Their work provides relevant information in understanding the manner in which various components of the complex assemble, particularly the manner in which the sequential assembly of supra-intrinsic LHC IIb occurs in the organelle. The synthesis of protein subunits and their binding with chlorophylls and carotenoids were shown to lead to the formation of LHC IIb monomers. The monomers along with other minor light-harvesting complexes were demonstrated to appear during the early hours followed by the formation of LHC IIb trimers and their subsequent assembly to form a supra-complex with the PSII core during the late hours of greening. The assembly is suggested to be stabilized by different photosynthetic pigments, particularly by chlorophyll *b* and carotenoids. Specific fatty acids in the organelle also appear to play a significant role in the stability of the final assembly of the supra-complex. Similarly, during the early phase of greening, the newly synthesized LHCs I exist as monomers, which subsequently aggregate to form trimers that are finally inserted to the core complex to form a complete PSI assembly of thylakoids [15,16].

The LHC genes are known to be regulated by tissue specificity and light through the action of different photoreceptors [17]. The differential response of individual members of the gene family to different light regimes has been worked out in detail [18]. The expression of genes is also known to be controlled by plastid factors [19].

D. RUBISCO: SYNTHESIS AND REGULATION

Rubisco, an important enzyme of the Calvin cycle, has been extensively studied form various angles including its study as a model for coordinated interaction of nuclear and plastid genes. Its structure–function relationship and regulation were recently reviewed [20]. The enzyme has a hexadecamer structure and is composed of equal numbers of large subunits (LSUs) and small subunits (SSUs). The LSU is encoded by a chloroplast genome and the SSU by a multigene family in the nucleus. The SSUs are synthesized as precursors in the cytoplasm, processed, and transported to the organelle, where they bind with LSUs and take up a hexadecameric form of the holoenzyme. The assembly of Rubisco is suggested to be modulated by chaperonins, which may bind with the LSU of the enzyme immediately after its synthesis in chloroplasts and process it for final assembly in the holoenzyme [21]. Although the synthesis and processing of the chaperonins have been well characterized in the recent years, their precise role in the assembly process still remains unclear.

The regulation of biogenesis of Rubisco is very complex. The assembly of the enzyme was demonstrated to be regulated by different factors. Extensive reports are available on the photoregulation of the synthesis of SSUs and LSUs of the enzyme. The light effect is mainly mediated through the participation of phytochrome and blue light receptors [17]. The expression of the plastid gene coding for the LSU of the enzyme is known to be regulated by nuclear gene products. Similarly, the nuclear gene, coding for the SSU of the enzyme, is regulated by the so-called plastid factor [22]. The other factors that regulate the accumulation of SSUs and LSUs have been well reviewed [1,17].

III. DEMOLITION OF CHLOROPLAST DURING LEAF SENESCENCE

The events associated with the demolition of the chloroplast are reported to be sequential and well coordinated (for a review, see Ref. [23]). The precise mechanism of the induction of leaf senescence leading to the disorganization of the organelle and consequently the loss of photosynthetic activity largely remains unclear.

A. LEAF SENESCENCE IS GENETICALLY PROGRAMMED

The process of leaf senescence involves downregulation of photosynthetic genes and upregulation of senescence associated genes (SAGs) [1,24–27]. Chloroplast is the major source of protein and other nutrients in green plants. Therefore, its demolition during leaf senescence is physiologically significant, particularly in nutrient salvation processes.

TABLE 6.3

Classification of Senescence Associated Genes

Senescence-Related Metabolism	Senescence Associated Genes (SAGs)	References
	Homologs of genes for serine protease	See Roberts et al. [28] and cross-references therein
Protein degradation	Homologs of gene for cysteine proteases and aspartic proteases	
	Homologs of gene for ubiquitin	
Nitrogen mobilization	Glutamine synthatase and aspargine synthatase	
		See the specific references from the review by Buchanan-Wollaston [24] and the book by Biswal et al. [1]
Carbolydrate metabolism	Homologs of genes for β-glucosidase, pyruvate-O phosphate dikinase, and β-galactosidase	
Lipid metabolism and mobilization	Homologs of genes for phospholipase-D, phosphoenol pyruvate carboxykinase, NAD-malate dehydrogenease, isocitrate lyase, and malate synthase	
Defense metabolism	Homologs of genes for PR like proteins, various metallotheonines	

The organelle is dismantled along with the degradation of other cellular components [23]. The degradation of macromolecules, their subsequent conversion to useable forms of nutrients, and transport to growing parts of the plant for reuse are well regulated. The genes that are upregulated to facilitate these processes include those that code for proteases, lipases, and regulatory proteins relating to transport (Table 6.3; see Refs. [24,25,28]). The senescing leaves can carry out this process only when they remain viable and healthy with an effective defense mechanism against pathogen attack and environmental stresses. The genes that are upregulated to provide protection to the senescing cells against these unfavorable conditions are shown in Table 6.3, which also shows other upregulated genes responsible for the conversion of lipids and other metabolites to respiratory substrates for providing energy to facilitate the senescence process. This is necessary because of senescence-induced loss in photosynthesis, the primary source of energy in green leaves.

B. COORDINATED REGULATION OF PIGMENT BREAKDOWN AND ULTRASTRUCTURAL CHANGES OF CHLOROPLAST DURING LEAF SENESCENCE

In addition to the loss of proteins and green pigments, the level of carotenoids also decreases during leaf senescence [29,30]. The carotenoids, however, are shown to degrade slowly compared to chlorophylls [30]. But the general kinetic pattern of loss in pigments and membrane proteins remains more or less the same, suggesting a common point in their degradation mechanisms [30]. Since these pigments exist in the form of complexes with proteins, dislocation or breakdown of any individual component may lead to the collapse of the complex. The dismantling of the complex is the prerequisite for enzymatic degradation of individual components. It appears that the structural status of different pigment–protein complexes may play a key role in coordinating the loss of photosynthetic pigments and proteins during senescence. The possibility of senescence-induced modification in the structure of the light-harvesting protein complex and a change in the topology of the pigments on the protein with consequent loss of pigments has been proposed in the chloroplasts of wheat leaves [31]. But a question still remains unanswered: What really triggers disassembly of the complex and which component of the complex degrades first?

1. Enzymatic Degradation of Photosynthetic Pigments

Reports published thus far on the enzymatic degradation of individual pigments were recently reviewed [1,32].

a. Degradation of Chlorophyll

The degradation of chlorophyll has been considered as a major symptom of thylakoid disorganization during leaf senescence. The enzymes that participate in stepwise degradation of the pigment [32] are described as per the following scheme:

$$\text{Chlorophyll} \xrightarrow{\text{Chlorophyllase}} \text{Chlorophyllide}$$

$$\text{Chlorophyllide} \xrightarrow{\text{Mg-dechelatase}} \text{Pheophorbide}$$

$$\text{Pheophorbide} \xrightarrow[\text{and stroma protein}]{\text{Pheophorbide a oxygenase}} \text{Fluorescent chlorophyll catabolites}$$

$$\text{Fluorescent chlorophyll catabolites} \xrightarrow{\text{Modifications and conjugations}} \text{Nonfluorescent chlorophyll catabolites}$$

The enzyme chlorophyllase, basically a hydrophobic protein, is suggested to be attached to the chloroplast envelope. It is responsible for the hydrolysis of chlorophyll into chlorophyllide and phytol, the first step in the breakdown of the pigment. In the next step, Mg-dechelatase acts on chlorophyllide and removes Mg^{2+} from it, which results in the formation of pheophorbide. The enzyme Mg-dechelatase is also bound to the organelle membrane. The next step in the chlorophyll degradation pathway involves the participation of pheophorbide a oxygenase, which in combination with another enzyme, red chlorophyll catabolite reductase (RCC reductase), is responsible for the opening of the ring structure of the pigment and gives the product RCC. The cleavage of the ring results in the loss of green color of the pigment. The enzyme is specific to the senescence process. The product RCC, in a series of subsequent reactions, is converted to fluorescent chlorophyll catabolites (FCCs), which are subsequently modified and converted to nonfluorescent chlorophyll catabolites (NCCs). The final disposal of chlorophyll catabolites in NCCs may occur in the cytoplasm (for a review, see Ref. [32]).

b. Carotenoid Degradation

Not much is known about the enzymes that participate in the degradation of carotenoids although reports are available on qualitative changes of the pigment-like formation of carotenoid esters and epoxides. The possibility of enzymatic participation, identification of the enzymes, and their regulation for quantitative loss of these pigments were recently described by Biswal et al. [1].

2. Ultrastructural Changes of Thylakoid Membranes

The ultrastructural modifications and changes in molecular composition of thylakoids during leaf senescence have been extensively examined by electron microscopy, x-ray diffraction, immunological techniques, and absorption and fluorescence techniques in different plant systems [1,23]. Membrane disorganization of the organelle as probed by electron microscopy

appears to be sequential starting with the unstacking of grana thylakoids as the first event that is followed by the formation of loose and elongated lamellae. These loose lamellae subsequently undergo massive degradation with the concomitant formation of plastoglobuli, the degradation products of thylakoids [23,33]. The details of the sequential changes in the ultrastructures of thylakoids are shown in Figure 6.1.

C. DISASSEMBLY OF THYLAKOID COMPLEXES AND LOSS IN PRIMARY PHOTOCHEMICAL REACTIONS

Thylakoid complexes were reported to be destabilized during leaf senescence, most likely in an ordered sequence [23,33]. In most of the plant systems, leaf senescence is demonstrated to cause earlier and rapid loss of photochemical activities associated with PSII compared to PSI activities [1]. There could be several factors contributing to the rapid degradation of the PSII of chloroplasts. A significant decline in oxygen evolution and restoration in the loss of PSII mediated 2,6-dichlorophenol indophenol photoreduction in chloroplasts with an exogenous electron donor like diphenyl carbazide during leaf senescence may indicate severe damage of the oxygen evolving system [23]. The restoration of dye reduction is suggestive of the relative stability of the PSII reaction center. The exact nature of senescence-induced loss in the oxygen evolving capacity of chloroplasts is not known. The release of Mn during leaf senescence as observed by Margulies [34] may be a factor directly affecting oxygen evolution. The loss of Mn may be the consequence of the senescence-induced loss of a 33 kDa extrinsic protein that is known to stabilize Mn binding on thylakoids. The loss of this extrinsic protein, as immunologically probed by western blots, has been clearly demonstrated during leaf senescence of *Festuca pratensis* [35]. Experiments conducted during leaf senescence of barley also suggest a parallel loss of extrinsic proteins and a decline in oxygen evolution [36]. The decline in the content of protein is attributed to senescence-induced loss in the quantity of its transcripts [37]. It is assumed that a loss of the proteins may lead to destabilization of Mn clusters, resulting in the inactivation of the oxygen evolv-

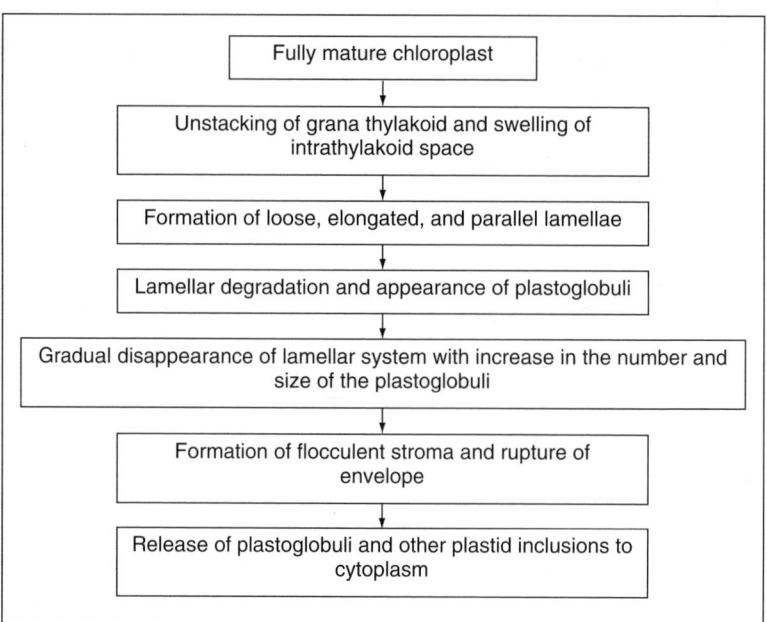

FIGURE 6.1 Ultrastructural changes of chloroplast during leaf senescence.

ing system. With the advancement of senescence, the reaction center core complex may start showing signs of deterioration contributing to the total loss of PSII photochemistry. The core complex may be damaged either by quantitative loss of reaction center proteins [38,39] or their structural modification [40]. Senescence-induced loss and disorganization of the light-harvesting system may be another factor contributing to the loss in the primary photochemistry of the photosystem [41].

It is assumed that the disassembly of PSII occurs in a sequence with disorganization of its oxygen evolving system as the first event followed by damage of the reaction center core complex and finally loss in the light-harvesting systems.

Although relatively stable, the photochemical reactions associated with PSI decline in senescing chloroplasts and the decline is attributed to the inactivation and loss of plastocyanin and NADP reductase [23]. Senescence-induced impairment of electron transport that links two photosystems could be attributed to the quantitative loss or inactivation of plastoquinones and plastocyanines, the shuttling molecules that mediate transfer of electrons between PSII and PSI via the cytochrome b/f complex [23,42,43]. The precise nature of dismantling of the coupling factor complex is not known, in spite of the availability of reports suggesting senescence-induced loss in photophosphorylation and loss of some of the protein subunits of the complex [1].

The existing data on dismantling of thylakoid bound complexes during leaf senescence, although extensive, do not provide any definite clue for understanding the nature of initial events that ultimately lead to the disorganization of complexes. In our earl-

ier review, we have proposed several models of triggering mechanisms that might be operating during senescence [23].

D. DECLINE IN RUBISCO ACTIVITY AND LOSS IN THE ENZYME PROTEIN

The changes in activities of many enzymes located in the stroma were examined in different plant systems during leaf senescence and Rubisco was proposed to be the most susceptible one to senescence [23,42,44]. Extensive literature is available on the loss of activity of the enzyme during the process [42,44,45]. The loss in enzyme activity may be attributed to the quantitative loss of the enzyme protein [42]. The loss in the level of the protein reflects both proteolytic degradation of the enzyme and impairment of its synthesis [1,42]. The proposition that the enzyme protein significantly degrades without much of its synthesis during senescence was reported extensively by many authors (for a review, see Refs. [1,23,46]). The mechanism of impairment of the synthesis of the enzyme during senescence is not clearly understood. Senescence is shown to cause a decline in the LSU and SSU levels of the enzyme [1,42,47]. Further analysis of their corresponding transcripts by Dot and Northern blots clearly suggests the regulation at the level of transcription or posttranscriptional modifications resulting in a loss of mRNAs, one of the limiting factors for the synthesis of enzyme proteins [1,37,42,45]. It seems logical to suggest a senescence-induced alteration in the turnover rate of the enzyme. Once the photosynthetic organelle is mature and shows signs of senescence, the turnover should preferentially shift more toward degradation than synthesis, thereby causing a loss in the level of the enzyme protein. The degradation of the protein could be attributed to senescence-induced activity of specific proteases [1,48,49].

E. DIFFERENTIAL LOSS IN PRIMARY PHOTOCHEMICAL REACTIONS AND THE ACTIVITY OF RUBISCO: PHYSIOLOGICAL SIGNIFICANCE

The data on temporal loss in the efficiency of photo-electron transport of thylakoid membranes and the activity of Rubisco for carbon dioxide fixation during leaf senescence suggest an early and rapid loss of the latter. Since Rubisco is the major source of nitrogen in green leaves, rapid degradation of the enzyme protein is essential so that senescing leaves can act as the source of nitrogen. At the same time, the transport of nutrients from senescing leaves to other growing parts of the plant needs energy, which is likely to be supplied by the relatively stable photoelectron transport system of thylakoid membranes. Reports are available on the relative stability of light-harvesting pigment complexes and reaction centers of the photosystems. The PSI, which is involved in cyclic electron flow for the production of ATP, exhibits remarkable stability during leaf senescence. The relative stability of the so-called light reactions (primary photochemistry) compared to the dark reaction relating to carbon dioxide fixation thus can be considered as a physiological strategy of green plants to provide the requisite energy for nutrient mobilization.

IV. SIGNALS FOR CHLOROPLAST BIOGENESIS

The chloroplast genome has limited genetic information, which can code for about 100 polypeptides and possesses only a few regulatory genes. Nuclear genes, in addition to coding for several protein components of chloroplasts, also code for the proteins that control the location, time of gene expression, processing, and targeting of the organelle proteins. The possible signal transduction systems for coordinated assembly and disassembly of chloroplast complexes as mediated by the gene products of both nuclear and plastid genomes are briefly described. The biogenesis of chloroplast as regulated by photosignals and signals from the developmental program of the organelle are also critically discussed in this section.

A. SIGNALS CONTROLLING PLASTID GENE EXPRESSION

Extensive reports are available on the regulation of plastid gene expression, RNA processing, translation,

and posttranslational modifications by nuclear gene products [10,22]. Many nuclear mutants were isolated, identified, and demonstrated to block synthesis of proteins encoded by the organelle genome [10,50]. For example, a nuclear mutant of *Chlamydomonas*, a green alga, has been shown to lack the ability to synthesize the LSU of Rubisco encoded by the plastid gene in spite of the synthesis of the SSU encoded by the nuclear gene and other plastid proteins [51]. The specific effect of the nuclear gene product on the synthesis of the LSU may suggest that the signal from the nuclear genome has a target site on the plastid for the expression of specific gene(s). Analysis of the nuclear mutants also reveals the control of nuclear gene products on the accumulation of other proteins including core proteins of the PSII reaction center [10].

In addition to nuclear signal, plastid gene expression is also known to be regulated by its own developmental process [52]. The accumulation of transcripts for the synthesis of several intrinsic proteins associated with the core complex of the reaction centers of PSI and PSII is greatly influenced by the aging and functional status of developing chloroplasts [52]. The tissue and organ specificity is another factor assumed to control plastid gene expression [1,22,53]. The levels of transcripts of several plastid genes remain low in plastids of roots compared to their levels in the leaves. The nature of tissue-specific signals and signals originating from the sequences of organelle development are yet to be explained.

B. SIGNALS THAT REGULATE NUCLEAR GENE EXPRESSION FOR THE SYNTHESIS OF CHLOROPLAST PROTEINS

A plastid signal otherwise known as plastid factor, extensively studied during last few years, is shown to regulate nuclear gene expression; that is, the expression of the genes coding for LHCs and SSUs and some of the genes for proteins of the oxygen evolving complex [1,19]. This proposition is supported by the observation that photooxidative damage of chloroplast with possible loss of the signal results in a block in transcription of these genes. The nature of the signal remains unclear. The signal's behavior varies in different phases of plastid development. During the early stages of development, the signal exhibits strong effects on the nucleus in accumulating a high level of transcripts for LHCs and SSUs. It was shown that a quantitative loss or a structural modification during senescence may lead to the switching off of the gene expression.

Nuclear gene expression for chloroplast proteins also appears to be modulated by tissue characteristics.

Differential expression of photosynthetic genes in bundle sheath and mesophyll cells in the leaves of higher plants supports this proposition [10]. However, the nature of the tissue-specific signal remains obscure.

C. LIGHT AS A COMMON SIGNAL FOR COORDINATED EXPRESSION OF NUCLEAR AND PLASTID GENES

Among all the environmental factors, light is considered to be the most important and well studied factor. It acts as a common signal for activating gene expression in the nucleus and in chloroplasts [54,55]. Light is believed to modulate posttranscriptional events in the chloroplasts. On the other hand, it directly controls the transcription during nuclear gene expression [17,56]. Light reportedly acts through two major photoreceptors: phytochrome and blue light receptors [56]. It has been proposed that the light signal in a signal transduction cascade is received by the photoreceptors and is transmitted in the cascade finally to control the transcription or posttranscription modifications. However, the nature of signal transduction that couples light perception by photoreceptors and the final expression of genes still remains a mystery except for the some recent findings that there are some light regulatory elements in the promoter regions that possibly receive the photoreceptor processed signal(s) for gene activity [55]. The possibility of G-proteins (GTP binding proteins) in phytochrome-mediated response cannot, however, be ruled out [57,58].

D. SIGNALING SYSTEMS ASSOCIATED WITH LEAF SENESCENCE

In spite of the presence of large amount of data in the area of molecular biology of senescence, the precise nature of the signaling systems associated with its induction and progress in green leaves remains unclear [1]. As discussed earlier, many genes responsible for macromolecular degradation and nutrient salvation were identified, cloned, and characterized [24,48,49,59]. But the genes that initiate and regulate the process are still unidentified.

Developmental factors, phytohormones, and stresses (both biotic and abiotic) are suggested to bring changes in the metabolic threshold, initiating the signal cascade for senescence induction. The metabolic changes are likely to result in the downregulation of photosynthetic genes and upregulation of senescence associated genes, which subsequently carry out the process of nutrient salvation leading to the death of the organ (Figure 6.2; see Refs. [1,25]).

The loss of photosynthesis as a signal for the induction of senescence in green leaves has been suggested by many authors (for a review, see Refs. [1,25,48]). During progressive senescence of many plants, the lower leaves receive light that is different in quality and quantity when compared to the light received by the upper leaves in the canopy of the plant body. The light transmitted through and reflected from the upper leaves is enriched by the far red component with a loss in photosynthetically active radiation. This may result in the downregulation of photosynthesis and causes induction of senescence.

E. NUCLEAR FACTOR FOR CHLOROPLAST DEGRADATION

Literature is available on the communication system, between nuclear and plastid genomes, for the highly ordered breakdown of the photosynthetic organelle during leaf senescence. The nucleus may have a control of the organelle degradation and the nuclear factor has been proposed to constitute a part of the signal cascade for chloroplast break down. The following experimental findings support the proposition:

1. The senescence-induced degradation is remarkably delayed in cell-free chloroplasts or chloroplasts in the cells devoid of nucleus [23,48].
2. Eukaryotic transcription and translation inhibitors have been demonstrated to arrest chloroplast senescence. Prokaryotic inhibitors fail to exhibit a similar response [23,60].
3. Mutation of the nuclear gene is known to prevent chloroplast degradation [61,62]. A nuclear mutant known as *sid* (senescence-induced degradation), a gene mutant of *Festuca pratensis*, does not show symptoms of degreening and remains green for quite a long time compared to its wild-type counterpart [62]. We have shown a block in the disappearance of PSII reaction center proteins of thylakoids in this mutant during senescence [38].

It was shown that the signal for chloroplast degradation is a protein and is encoded by the nuclear DNA. This proposition is further supported by the findings of Kawakami and Watanabe [37], who have demonstrated the efficient import of a senescence-related protein encoded by the nuclear gene to chloroplasts. The question of what really triggers the expression of the nuclear gene for chloroplast degradation remains unanswered. In the background of the findings on the role of the plastid signal regulating nuclear gene expression for the proteins necessary for its own development, it is quite logical to argue in favor of a

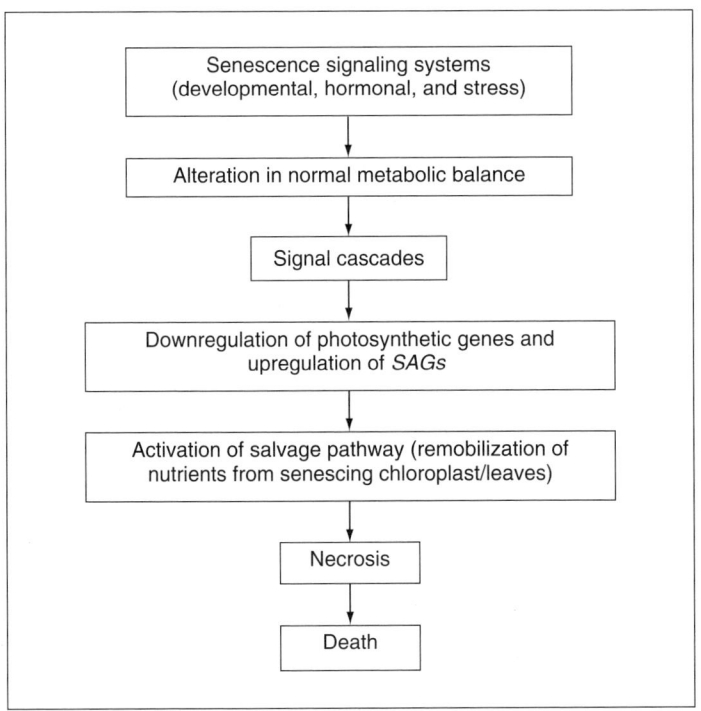

FIGURE 6.2 Signal transduction during leaf senescence.

signal of chloroplast origin that could send a message to the nucleus and initiates its own degradation.

V. THE FUTURE

In spite of significant accumulation of data in the areas of chloroplast development and senescence, there are many questions that need to be addressed for future studies. Some of the new and challenging areas in the field that require further study are as follows:

1. The multimeric thylakoid and stroma complexes are well characterized. Both the nuclear and plastid genomes are known to be involved in the biogenesis of these complexes but the nature of coordination between these two remains unclear. Targeting of the nuclear encoded proteins, the role of transport modulating proteins, and the factor(s) that determine the specific location of the assembly of the organelle complex are poorly understood and therefore need more experimentation.

2. Light is thought to be the major factor in regulating the synthesis of organelle proteins. However, the precise molecular mechanism of photoregulation at the gene levels largely remains unclear. Whether light regulates at transcription, posttranscription, or at both levels has to be resolved. The differential rates of gene expression by light at different stages of plastid development have to be explained.

3. Data are available on the nature and location of the enzymes involved in the synthesis of proteins and pigments during chloroplast development, but the enzymes responsible for the degradation of individual components of multimeric proteins, both in thylakoid and stroma, are poorly identified. There was a study of the participation of enzymes in chlorophyll degradation during leaf senescence [32], but almost nothing is known about the catabolism of carotenoids, a problem that requires serious attention [1]. We also need a better understanding of the mechanism of protein degradation in the organelle. Nevertheless, the preliminary data available on the proteolytic degradation of Rubisco are quite encouraging and provide a base for further research in this area [28,63,64].

4. Leaf senescence is known to be controlled by genes but the question that has to be addressed is whether the senescence program could be genetically altered in a regulated way. The success in the control of fruit ripening, comparable to leaf senescence in many ways, by genetic manipulation may be the beginning of this highly fascinating and applied area of senescence research. Currently, successful attempts have been made in producing "stay green" mutants that exhibit a significant delay in leaf yellowing, but a link between the "stay green" character and ultimate plant productivity in the field is yet to be established.

5. The communication systems operating between the chloroplast and nucleus for the coordinated synthesis of chloroplast complexes are known and the control of nuclear gene products in chloroplast gene expression was extensively examined. On the other hand, the role of the plastid factor in nuclear gene expression for organelle proteins during greening has also been recorded. The triggering mechanisms, in both the cases, however, remain obscure. The nature of the plastid factor still needs clarification.

6. The signaling system associated with chloroplast development and senescence has not been properly identified. Although hormones, developmental factors, other cellular factors, and light are considered to be the major signals, the concept of the coupling between these signals and chloroplast biogenesis remains unclear.

REFERENCES

1. Biswal UC, Biswal B, Raval MK. *Chloroplast Biogenesis: From Proplastid to Gerontoplast.* Dordrecht, The Netherlands: Kluwer Academic Publishers, 2003.
2. Suzuki JY, Bollivar DW, Bauer CE. Genetic analysis of chlorophyll biosynthesis. *Annu. Rev. Genet.* 1997; 31: 61–89.
3. Aronsson H, Sundqvist C, Dahlin C. POR-import and membrane association of a key element in chloroplast development. *Physiol. Plant.* 2003; 118: 1–9.
4. Sundqvist C, Dahlin C. With chlorophyll pigments form prolamellar bodies to light harvesting complexes. *Physiol. Plant.* 1997; 100: 748–759.
5. Sugiura M, Hirose T, Sugita M. Evolution and mechanism of translation in chloroplast. *Annu. Rev. Genet.* 1998; 32: 437–459.
6. Keegstra K, Cline K. Protein import and routing systems of chloroplasts. *Plant Cell* 1999; 11: 557–570.
7. Schleiff E, Soll J. Travelling of proteins through membranes: translocation into chloroplasts. *Planta* 2000; 211: 449–456.
8. Ke B. *Photobiochemistry and Photobiophysics.* Advances in Photosynthesis, Vol. 10. Dordrecht, The Netherlands: Kluwer Academic Publishers, 2001.
9. Tyagi AK, Grover M, Choudhury A, Kapoor S, Kelkar NY, Maheshwari SC. Influence of light and development on expression of genes encoding photosynthesis-related proteins. In: Tewari KK, Singhal GS, eds. *Plant Molecular Biology and Biotechnology.* New Delhi, India: Narosa Publishing House, 1997: 101–114.
10. Goldschmidt-Clermont M. Coordination of nuclear and chloroplast gene expression in plant cells. *Int. Rev. Cytol.* 1998; 177: 115–180.
11. Nyitrai P. Development of functional thylakoid membranes: regulation by light and hormones. In: Pessarakli M, ed. *Handbook of Photosynthesis.* New York, USA: Marcel Dekker, Inc., 1999: 391–406.
12. Biswal B. Greening of leaves and its modulation by various factors. *Indian Rev. Life Sci.* 1985; 5: 35–57.
13. Ohashi K, Tanaka A, Tsuji H. Formation of the photosynthetic electron transport system during the early phase of greening in barley leaves. *Plant Physiol.* 1989; 91: 409–414.
14. Sutton A, Sieburth LE, Bennett J. Light dependent accumulation and localization of photosystem II proteins in maize. *Eur. J. Biochem.* 1987; 164: 571–578.
15. Dreyfuss BW, Thornber JP. Assembly of light harvesting complexes (LHCs) of photosystem II. Monomeric LHC II b complexes are intermediates in the formation of oligomeric LHC II b complexes. *Plant Physiol.* 1994; 106: 829–839.
16. Dreyfuss BW, Thornber JP. Organization of the light-harvesting complex of photosystem I and its assembly during plastid development. *Plant Physiol.* 1994; 106: 841–848.
17. Batschauer A, Gilmartin PM, Nagy F, Schafer E. The molecular biology of photoregulated genes. In: Kendrick RE, Kronenberg GHM, eds. *Photomorphogenesis in Plants*, 2nd Edition. Dordrecht, The Netherlands: Kluwer Academic Publishers, 1994: 559–599.
18. White MJ, Kaufman LS, Horwitz BA, Briggs WR, Thompson WF. Individual members of Cab gene family differ widely in fluence response. *Plant Physiol.* 1995; 107: 161–165.
19. Oelmüller R. Photooxidative destruction of chloroplast and its effects on nuclear gene expression and extraplastidic enzyme levels. *Photochem. Photobiol.* 1989; 49: 229–239.
20. Spreitzer RJ, Salvucci ME. RUBISCO: structure, regulatory interactions, and possibilities for a better enzyme. *Annu. Rev. Plant Physiol. Plant Mol. Biol.* 2002; 53: 449–475.
21. Roy H, Gilson M. Rubisco and the chaperonins. In: Pessarakli M, ed. *Handbook of Photosynthesis.* New York, USA: Marcel Dekker, Inc., 1997: 295–304.
22. Taylor WC. Regulatory interactions between nuclear and plastid genomes. *Annu. Rev. Plant Physiol. Plant Mol. Biol.* 1989; 40: 211–233.
23. Biswal UC, Biswal B. Ultrastructural modifications and biochemical changes during senescence of chloroplasts. *Int. Rev. Cytol.* 1988; 113: 271–321.
24. Buchanan-Wollaston V. The molecular biology of leaf senescence. *J. Exp. Bot.* 1997; 48: 181–199.
25. Biswal B. Senescence-associated genes of leaves. *J. Plant Biol.* 1999; 26: 43–50.
26. Chandlee JM. Current molecular understandings of the genetically programmed process of leaf senescence. *Physiol. Plant.* 2001; 113: 1–8.
27. Yoshida S. Molecular regulation of leaf senescence. *Curr. Opin. Plant Biol.* 2003; 6: 79–84.
28. Roberts IN, Murray PF, Caputo CP, Passeron S, Barneix AJ. Purification and characterization of a subtilisin like serine protease induced during the senescence of wheat leaves. *Physiol. Plant.* 2003; 118: 483–490.
29. Biswal UC, Biswal B. Photocontrol of leaf senescence. *Photochem. Photobiol.* 1984; 39: 875–879.

30. Biswal B. Carotenoid catabolism during leaf senescence and its control by light. *J. Photochem. Photobiol.: B. Biol.* 1995; 30: 3–13.

31. Joshi PN, Biswal B, Kulandaivelu G, Biswal UC. Response of senescing wheat leaves to ultraviolet A light: changes in energy transfer efficiency and PS II photochemistry. *Radiat. Environ. Biophys.* 1994; 33: 167–176.

32. Matile P, Hortensteiner S, Thomas H. Chlorophyll degradation. *Annu. Rev. Plant Physiol. Plant Mol. Biol.* 1999; 50: 67–95.

33. Biswal UC, Biswal B. Leaf senescence induced changes in primary photochemistry of chloroplasts. In: Jaiswal VS, Rai AK, Jaiswal U, Singh JS, eds. *The Changing Scenario in Plant Sciences.* New Delhi, India: Allied Publishers Limited, 2000: 159–174.

34. Margulies MM. Electron transport properties of chloroplasts from aged bean leaves and their relationships to the manganese content of the chloroplasts. In: Forti G, Avron M, Melandri A, eds. *Proceedings of the Second International Congress on Photosynthesis Research.* The Hague: W. Junk Publishers, 1971: 539–545.

35. Nock LP, Rogers LJ, Thomas H. Metabolism of protein and chlorophyll in leaf tissue of *Festuca pratensis* during chloroplast assembly and senescence. *Phytochemistry* 1992; 31: 1465–1470.

36. Choudhury NK, Imaseki H. Loss of photochemical functions of thylakoid membranes and PS 2 complex during senescence of barley leaves. *Photosynthetica* 1990; 24: 436–445.

37. Kawakami N, Watanabe A. Translatable mRNAs for chloroplast targeted proteins in detached radish cotyledons during senescence in darkness. *Plant Cell Physiol.* 1993; 34: 697–704.

38. Biswal B, Rogers LJ, Smith AJ, Thomas H. Carotenoid composition and its relationship to chlorophyll and D1 protein during leaf development in a normally senescing cultivar and a stay green mutant of *Festuca pratensis.* *Phytochemistry* 1994; 37: 1257–1262.

39. Prakash JSS, Baig MA, Mohanty P. Differential changes in the steady state levels of thylakoid membrane proteins during senescence in *Cucumis sativus* cotyledons. *Z. Naturforsch.* 2001; 56c: 585–592.

40. Joshi PN, Ramaswamy NK, Raval MK, Desai TS, Nair PM, Biswal UC. Alteration in photosystem II photochemistry of thylakoids isolated from senescing leaves of wheat seedlings. *J. Photochem. Photobiol.: B. Biol.* 1993; 20: 197–202.

41. Prakash JSS, Baig MA, Mohanty P. Senescence induced structural reorganization of thylakoid membranes in *Cucumis sativus* cotyledons. LHC II involvement in reorganization of thylakoid membranes. *Photosynth. Res.* 2001; 68: 153–161.

42. Grover A. How do senescing leaves lose photosynthetic activity? *Curr. Sci.* 1993; 64: 226–233.

43. Mae T, Thomas H, Gay AP, Makino A, Hidema J. Leaf development in *Lolium temulentum:* photosynthesis and photosynthetic proteins in leaves senescing under different irradiances. *Plant Cell Physiol.* 1993; 34: 391–399.

44. Lauriere C. Enzymes and leaf senescence. *Physiol. Veg.* 1983; 21: 1159–1177.

45. Miller A, Schlagnhaufer C, Spalding M, Rodermel S. Carbohydrate regulation of leaf development: prolongation of leaf senescence in Rubisco antisense mutants of tobacco. *Photosynth. Res.* 2000; 63: 1–8.

46. Biswal UC, Biswal B. Plant senescence and changes in photosynthesis. *Biol. Edn.* 1990; 7: 56–72.

47. Kasemir H. Plant senescence as a developmental strategy. In: Biswal UC, Britton G, eds. *Trends in Photosynthesis Research.* Bikaner, India: Agro Botanical Publishers, 1989: 231–244.

48. Smart CM. Gene expression during leaf senescence. *New Phytol.* 1994; 126: 419–448.

49. Dangl JL, Dietrich RA, Thomas H. Senescence and programmed cell death. In: Buchanan B, Gruissem W, Jones R, eds. *Biochemistry and Molecular Biology of Plants.* Rockville, MD, USA: American Society of Plant Physiologists, 2000: 1044–1099.

50. Barkan A, Voelker R, Mendel-Hartvig J, Johnson D, Walker M. Genetic analysis of chloroplast biogenesis in higher plants. *Physiol. Plant.* 1995; 93: 163–170.

51. Hong S, Spreitzer RJ. Nuclear mutation inhibits expression of the chloroplast gene that encodes the large subunit of ribulose-1, 5-bisphosphate carboxylase/oxygenase. *Plant Physiol.* 1994; 106: 673–678.

52. Kapoor S, Maheswari SC, Tyagi AK. Developmental and light dependent cues interact to establish steady state levels of transcripts for photosynthesis related genes (*psbA, psbD, psaA* and *rbcL*) in rice (*Oryza sativa* L.). *Curr. Genet.* 1994; 25: 362–366.

53. Kapoor S, Maheshwari SC, Tyagi AK. Organ specific expression of plastid-encoded genes in rice involves both quantitative and qualitative changes in m-RNAs. *Plant Cell Physiol.* 1993; 34: 943–947.

54. Gray JC. Regulation of expression of nuclear genes encoding polypeptides required for the light reactions of photosynthesis. In: Ort. DR, Yocum CF, eds. *Oxygenic Photosynthesis: The Light Reactions.* Dordrecht, The Netherlands: Kluwer Academic Publishers, 1996: 621–641.

55. Tyagi AK, Dhingra A, Raghuvanshi S. Light regulated expression of photosynthesis-related genes. In: Yunus M, Pathre U, Mohanty P, eds. *Probing Photosynthesis, Mechanisms, Regulation and Adaptation.* London, UK: Taylor & Francis, 2000: 324–341.

56. Khurana JP, Kochhar A, Tyagi AK. Photosensory perception and signal transduction in higher plants — molecular genetic analysis. *Crit. Rev. Plant Sci.* 1998; 17: 465–539.

57. Romero LC, Biswal B, Song PS. Protein phosphorylation in isolated nuclei from etiolated *Avena* seedlings. Effects of red/far red light and cholera toxin. *FEBS Lett.* 1991; 282: 347–350.

58. Kevei E, Nagy F. Phytochrome controlled signaling cascades in higher plants. *Physiol. Plant.* 2003; 117: 305–313.

59. Scharrenberg C, Falk J, Quast S, Haussühl K, Humbeck K, Krupinska K. Isolation of senescence-related

cDNAs from flag leaves of field grown barley plants. *Physiol. Plant.* 2003; 118: 278–288.

60. Behera YN, Biswal B. Differential response of fern leaves to senescence modulating agents of angiospermic plants. *J. Plant Physiol.* 1990; 136: 480–483.

61. Guiamet JJ, Schwartz E, Pichersky E, Nooden LD. Characterization of cytoplasmic and nuclear mutations affecting chlorophyll and chlorophyll binding proteins during senescence in soyabean. *Plant Physiol.* 1991; 96: 227–231.

62. Thomas H, Smart CM. Crops that stay green. *Ann. Appl. Biol.* 1993; 123: 193–219.

63. Minamikawa T, Toyooka K, Okamoto T, Hara-Nishimura I, Nishimura M. Degradation of ribulose-bisphosphate carboxylase by vacuolar enzymes of senescing French bean leaves: immunocytochemical and ultrastructural observations. *Protoplasma* 2001; 218: 144–153.

64. Grbic V. SAG2 and SAG12 protein expression in senescing *Arabidopsis* plants. *Physiol. Plant.* 2003; 119: 263–269.

7 Role of Phosphorus in Photosynthetic Carbon Metabolism

Anna M. Rychter
Institute of Experimental Plant Biology, Warsaw University

I. M. Rao
Centro Internacional de Agricultura Tropical (CIAT)

CONTENTS

I. INTRODUCTION

Phosphorus (P) is a major mineral nutrient for plants and is required in many compounds in cells and organelles [1]. These compounds are associated with numerous components of metabolism (sugar phosphates, nucleic acids, nucleotides, coenzymes, phospholipids) and are closely associated with energy transfer (triphosphonucleotides) and genetic material (nucleic acids). The covalent ester bond between two P atoms is at a higher "energy level" than the covalent bonds between many other kinds of atoms. That is, it takes more energy for these compounds to be synthesized, and conversely they release more energy when they are either hydrolyzed or participate in alternative reactions such as P addition to other molecules. Plants must have P for plant growth and development. Limited inorganic phosphate (Pi) supply results in numerous perturbations in plant growth and development and strongly affects plant yields [2].

Photosynthesis is the primary physiological process whereby CO_2 diffuses down a concentration gradient from the atmosphere, through the epidermis, and into chloroplasts, where energy derived photochemically is used to assimilate CO_2 in the formation of organic compounds (Figure 7.1). In algae and higher plants there is only one primary carboxylating mechanism, which results in the net synthesis of carbon compounds. The photosynthetic carbon reduction (PCR) cycle is common to all plants (C_3, C_4, and crassulacean acid metabolism [CAM]) although C_4 and CAM plants have auxiliary mechanisms of carbon fixation [3].

During photosynthesis, carbon is fixed through the PCR cycle in the chloroplast, and is then exported to the cytosol as triose phosphate (triose-P). The triose-P is then converted to sucrose in the cytosol, releasing Pi, which is then available to allow further export of triose-P from the chloroplast. If there is any restriction of sucrose synthesis in the cytosol, it will

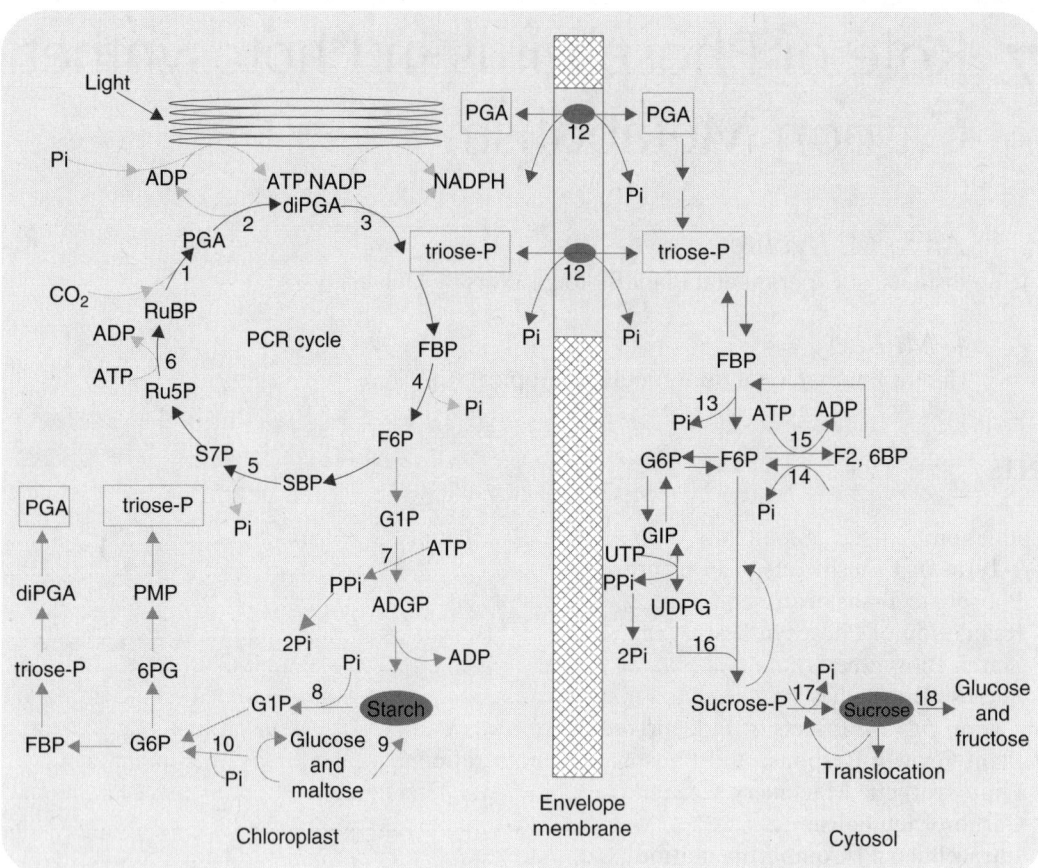

FIGURE 7.1 A simplified model depicting the reactions of photosynthetic carbon metabolism in which Pi has a regulating function or in which energy-rich phosphates and the corresponding phosphate esters are involved. Because of these functions, strict compartmentation and regulation of the Pi level in the metabolic pool are essential for photosynthesis in leaf cells. Fixed carbon inputs and reducing equivalents converge in the PCR cycle. Two major branch points of the PCR cycle lead to the production of starch in the chloroplast and the export of triose-P to the cytosol through the Pi translocator, located on the inner envelope of the chloroplast membrane. Synthesis of sucrose in the cytosol is linked to the release of Pi that is returned to the stroma through the Pi translocator in exchange for triose-P. The dashed arrows indicate possible feedback mechanisms. The reactions are catalyzed by enzymes numbered as follows: 1, Rubisco; 2, PGA kinase; 3, NADP-G3P dehydrogenase; 4, FBPase; 5, SBPase; 6, Ru5P kinase; 7, ADPG PPase; 8, phosphorylase; 9, β-amylase; 10, hexokinase; 11, NADP-G6P dehydrogenase; 12, Pi translocator; 13, FBPase; 14, F2,6BPase; 15, F6P-2-kinase; 16, SPS; 17, SPPase; and 18, invertase.

lead to a decreased export of triose-P from the chloroplast, so more photosynthate is retained in the stroma for conversion to starch (Figure 7.1). Chloroplastic starch degradation may be closely related to internal factors in the cell such as the supply and demand of carbon substrates. Orthophosphate (Pi), together with CO_2 and H_2O, is a primary substrate of photosynthesis [4] according to the overall equation:

$$3CO_2 + 6H_2O + Pi \xrightarrow{hv} triose\text{-}P + H_2O + 3O_2$$

Within the chloroplast, Pi is involved in organic combination during photophosphorylation, as a proton gradient is discharged through an ATPase into the

chloroplast stroma. In the stroma, ATP is consumed by the PCR cycle. Nine molecules of Pi are consumed for every three molecules of CO_2 fixed and three molecules of O_2 evolved. Eight molecules of Pi are released in the PCR cycle and the remaining molecule of Pi is incorporated into triose-P, which is transported to the cytosol in exchange for imported Pi. Sucrose synthesis in the cytosol releases Pi and thereby recycles Pi. Four molecules of Pi must enter the chloroplast for every molecule of sucrose synthesized in the cytosol. Adequate supply of Pi is essential for the assimilation of photosynthetic carbon in plants [4] and there has been a great deal of interest for the past two decades related to the idea that the level of Pi in plant tissues may regulate various aspects of photosynthesis and the flow of carbon between starch and sucrose biosynthesis [5–17]. In addition, it has been proposed that Pi may be involved in the partitioning of photosynthates between plant parts [18–22].

The rate of photosynthesis is dependent on the ATP/reductant (NADPH, NADH, and ferredoxin) balance, which can be stabilized by extrachloroplastic compartments such as mitochondria [23]. At the whole plant level, photosynthesis is regulated by sink demand [24]. In P-deficient plants, low sink strength imposes the primary limitation on photosynthesis [16]. Therefore, the response of photosynthesis to phosphate limitation is a "whole plant" one and depends on the dynamic interactions between sink and source tissues [16,24]. The decrease in phosphate concentration due to limited Pi supply from the growth medium involves several changes not only in the photosynthetic process but also in glycolysis, respiration, and nitrogen metabolism, which affect the rate of net photosynthesis. Metabolic aspects of the phosphate-starvation response were reviewed recently by Plaxton and Carswell [14].

Inadequate supply of Pi limits photosynthesis because of its large demand for adenylate energy and the role of phosphorylated intermediates in the PCR cycle [15]. The inhibition of photosynthesis due to Pi deprivation results from both short- and long-term effects of Pi on photosynthetic carbon assimilation and carbon partitioning processes [13]. In this chapter, we review the research progress that contributed to our present understanding of the role of Pi in photosynthetic carbon metabolism. To illustrate the effects of Pi deprivation on photosynthesis and partitioning of photosynthates, a simplified outline is presented of the short-term *in vitro* effects of Pi deprivation, followed by long-term *in vivo* effects of Pi deprivation, the recovery of plants from P deficiency, and the acclimation and adaptive responses of plants to P deficiency.

II. SHORT-TERM *IN VITRO* EFFECTS OF Pi DEPRIVATION

The evidence for a crucial role of Pi in the regulation of photosynthesis arose from the studies of photosynthetic induction. It was demonstrated that in isolated chloroplasts the induction period is due to a need to build up the pool sizes of the intermediates of the PCR cycle [25,26]. The interrelationships between Pi and induction, together with the demonstration that isolated chloroplasts require Pi for the continuation of photosynthesis, led to the concept that C_3 chloro-

plast is not a fully self-sufficient photosynthetic organelle [27].

Experimental observations on the photosynthetic induction period have led to the view that the chloroplast produces triose-P, glyceraldehyde-3-phosphate (G-3-P), and dihydroxyacetone phosphate (DHAP), which it exchanges for Pi from the cytoplasm of the cell [28,29]. Subsequent research work indicated that light activation of key enzymes [30–32] may be involved along with the autocatalytic build-up of metabolites [33] to overcome the lag period in photosynthetic CO_2 fixation [34]. However, experimental verification of these hypotheses with intact wheat leaves suggested that light activation of enzymes may not be a limiting factor during photosynthetic induction [35].

Studies of the short-term effects of Pi on photosynthesis, based on *in vitro* experiments, have shown the inhibition of triose-P export from the chloroplast to the cytosol through the Pi translocator leading to the build-up of starch and a decrease in the rate of photosynthesis [34,36–38]. It was demonstrated that in isolated chloroplasts the increase in Pi concentration in incubation medium up to 1 mM stimulated net photosynthesis and lowered starch production whereas low Pi concentration in external medium increased starch synthesis despite a low photosynthetic rate [39–41]. Low supply of Pi might restrict photophosphorylation, which should lead to increased energization of the thylakoid membrane, decreased electron flow, and associated inhibition of photosynthesis. At high Pi supply, triose-P export competes with ribulose 1,5-bisphosphate (RuBP) regeneration and the rate of photosynthesis can be diminished.

Optimal photosynthesis of isolated chloroplasts requires a finely balanced concentration of Pi in the cytosol [42]. This optimal concentration may be maintained by transport to and from the vacuole and by metabolic processes causing changes in the rate of sucrose synthesis [18,42]. Over the short term, low Pi in the cytosol decreases the export of triose-P from the chloroplast, which leads to the inhibition of sucrose synthesis in the cytosol [9,34,43,44].

A. PHOSPHATE TRANSLOCATORS

In higher plants, photosynthesis is compartmentalized in the chloroplast, which is bounded by the envelope membranes that serve both as a barrier separating the chloroplast stroma from the cytoplasm and a bridge enabling rapid exchange of specific metabolites between the two (Figure 7.1) [45–47]. The outer envelope membrane is nonspecifically permeable to all molecules, both charged and uncharged. The impermeability of the inner envelope membrane to hydrophilic solutes such as Pi, phosphate esters, dicarboxylates, and glucose is overcome by translocators that catalyze specific transfer of metabolites across the envelope [46,47]. The energy-transducing thylakoid membranes, located within the chloroplasts, are distinct from the envelope membranes.

The mechanism by which external Pi influences photosynthesis has been attributed to the operation of the Pi translocator, an antiport located in the inner membrane of the chloroplast envelope that facilitates a rapid counterexchange of Pi, triose-P, and 3-phosphoglyceric acid (PGA) [39,46,47]. The major flow of metabolites across the chloroplast envelope is mediated by the Pi translocator, which enables the specific transport of Pi and phosphorylated compounds such that photosynthetically fixed carbon in the form of triose-P can be exported from the stroma to the cytosol in a one-to-one stoichiometric and obligatory exchange for Pi [48]. The Pi released during biosynthetic processes is shuttled back through the Pi translocator into the chloroplasts for the formation of ATP catalyzed by the thylakoid ATPase [49].

If triose-P is regarded as the end product of the PCR cycle (Figure 7.1), then one molecule of Pi must be made available for incorporation into triose-P for every three molecules of CO_2 fixed. Some Pi will be released within the stroma as triose-P is utilized for starch synthesis, but starch synthesis is usually slower (by a factor of 3 to 4) than maximal CO_2 fixation. Virtually all the remaining Pi must enter the chloroplast in exchange for exported triose-P [46–48]. In the short term, a sudden decrease in the Pi concentration in the cytosol of photosynthetic mesophyll cells will have a direct effect on the triose-P and Pi exchange between the chloroplast and the cytosol, decreasing the availability of Pi in the chloroplast and thus decreasing the production of ATP needed in the turnover of the PCR cycle.

Triose phosphate/phosphate translocator (TPT) was the first phosphate transporter to be cloned from plants [50]. The activity of TPT is closely associated with photosynthetic carbon metabolism and the expression of the TPT gene is observed only in photosynthetic tissues [41]. Its importance in *in vivo* communication between chloroplast and cytosol was demonstrated in transgenic potato plants with reduced expression of the TPT at both RNA and protein levels due to antisense inhibition [51]. Four different groups of Pi transporters have been described so far in plastids and one among them is phosphoenolpyruvate/phosphate transporter, which transports Pi out of the chloroplast into cytosol under most physiological conditions [52].

Recently, Versaw and Harrison [53] described a low-affinity Pi transporter PHT2;1, H^+/Pi symporter, located in the inner envelope of the chloroplast. The identification of the null mutant of *Arabidopsis thaliana*, *pht2;1-1*, revealed that the PHT2;1 transporter affects Pi allocation and modulates Pi-starvation responses including the expression of genes and the translocation of Pi within leaves [53]. The presence of several transporters indicates highly controlled transport of phosphate into and out of the chloroplast.

The synthesis of sucrose from triose-P is believed to make the major contribution to the recycling of Pi (Figure 7.1). Sucrose synthesis releases Pi due to the action of a phosphatase and rapid export of sucrose from the cytoplasm will make Pi available as fast as the plant can synthesize triose-P; little or none will be available for storage within the stroma as starch. If the demand for sucrose by growing sinks is less however, excess triose-P would be stored as starch and the rate of photosynthesis possibly diminished.

Another important function of the Pi translocator is to link intra- and extrachloroplast pyridine nucleotide and adenylate systems through shuttles involving the exchange of DHAP and PGA. Photosynthetically produced ATP and NADPH are not directly available to the extrachloroplastic compartments due to the low permeability of the inner envelope membrane to these compounds in mature tissue. The Pi translocator provides an indirect shuttle system for transferring ATP and NADPH to the cytoplasm involving exchange of triose-P and PGA. This shuttle can operate in either direction depending on the redox potential of the pyridine nucleotides in the cytoplasm and stroma [46].

Gerhardt et al. [54] observed asymmetric distribution of DHAP and 3-PGA across the chloroplast envelope in spinach leaves and suggested that the Pi translocator may be kinetically limiting *in vivo*. The reduction of TPT activity *in vivo* by antisense repression of chloroplast TPT resembles the situation of chloroplasts performing photosynthesis under Pi limitation [39]. To examine more specifically the role of the Pi translocator in assimilate partitioning in photosynthetic tissues, Barnes et al. [55] transformed tobacco plants with sense and antisense constructs of a cDNA encoding the tobacco Pi translocator. Although the transformed plants showed a 15-fold variation in Pi translocator activity, the growth and development and the rate of photosynthesis showed no consistent differences between antisense and sense transformants. In contrast, the distribution of assimilate between starch and sugar had been altered with no change in the amount of sucrose in leaves, suggesting a homeostatic mechanism for maintaining sucrose concentrations in the leaves at the expense of glucose and fructose. However, in potato plants antisense repression of the triose-P translocator affected carbon partitioning as chloroplasts isolated from such plants showed reduced import of Pi, reduced rate of photosynthesis, and change in carbon partitioning into starch at the expense of sucrose and amino acids [56]. Published evidence indicates that TPT exerts a considerable control on the rate of both CO_2 assimilation and sucrose biosynthesis [41].

B. REGULATION OF PHOTOSYNTHESIS

Since Pi, triose-P, and PGA are exchanged through the Pi translocator, changes in the Pi concentration outside the chloroplast could affect the PCR cycle indirectly by altering the amount of intermediates within the chloroplast. Pi might also have direct effects on PCR cycle enzymes through the level of activation. Heldt et al. [57] indicated that Pi is required for light activation of ribulose 1,5-bisphosphate carboxylase/oxygenase (Rubisco). Later, Bhagwat [58] showed that Pi is an activator of Rubisco. However, Machler and Nösberger [59] showed that although the activity of Rubisco decreased with decreased stromal Pi concentration, they believed this to be an indirect effect mediated through the changes in stromal pH.

The activation of fructose-1,6-bisphosphatase (FBPase) [60] and of sedoheptulose 1,7-bisphosphatase (SBPase) [61] is strongly inhibited by Pi concentrations in the range of 5 to 10 mM. Pi inhibited the PCR cycle turnover in thiol-activated stromal extracts; this inhibition was due primarily to effects on the SBPase [62]. Another PCR cycle enzyme, the light-activated form of ribulose-5-phosphate kinase (Ru5Pkinase), is inhibited by the monovalent ionic species of Pi [63]. The decrease in the concentration of stromal Pi, which occurs upon illumination, is therefore likely to enhance the activity of the PCR cycle.

The reduction in photosynthetic rate that occurs when cytoplasmic Pi is decreased, for example, when Pi is sequestered in the cytoplasm by mannose [64] or glycerol [65], might be explained in terms of an end-product inhibition [66]. This end-product inhibition could be due to high concentrations of triose-P. Because the properties of the Pi translocator dictate that the total Pi (inorganic plus organic) within the chloroplast is relatively constant [48], high triose-P is automatically coupled with low Pi, which in turn could limit photosynthesis [6,10,43].

The consumption of Pi as a substrate of photosynthesis [27] could decrease photosynthesis by a direct effect of low stromal Pi concentration on Rubisco

[57]. Low stromal Pi concentration, together with the accumulation of triose-P, might influence the activation state of Rubisco by various mechanisms [6]. Rubisco could be inactivated by the build-up of various intermediates, for example, ribose-5-phosphate [67,68] and other chloroplast metabolites [69]; or, it may be inactivated by the build-up of PGA [67]. Another possibility is that the pH of the stroma could be changed [70,71].

Alternatively, inhibition of photosynthesis might occur due to a drop in the ATP/ADP ratio [72]. A decrease in stromal Pi concentration could diminish the rate of photophosphorylation and thereby reduce the rate of carbon fixation because of the sensitivity of the PCR cycle to the ATP/ADP quotient. Such a reduction is readily demonstrated with isolated chloroplasts photosynthesizing in a medium containing suboptimal Pi concentrations. The reduced concentration of Pi leads to a reduction in ATP/ADP, which could restrict the activity of Rubisco activase and therefore Rubisco carbamylation [73].

Robinson and Giersch [74] determined the concentration of Pi in the stroma of isolated chloroplasts during photosynthesis under Pi-limited and Pi-saturated conditions. They used colorimetric and ^{32}P labeling techniques in their study and found that when chloroplasts are illuminated in the absence of added Pi, photosynthesis declines rapidly due to Pi depletion in the stroma, which was estimated to be 1.4 mM by the colorimetric method and 0.2 mM by ^{32}P high-performance liquid chromatography. With optimal concentrations of Pi added to the medium, the stromal Pi concentration was estimated to be 2.6 and 1.6 mM with the colorometric and ^{32}P methods, respectively. This study demonstrated that any decrease in the supply of Pi from the medium leads to a rapid decrease in stromal Pi to the point where photophosphorylation may become Pi-limited, decreasing the rate of photosynthesis.

C. STARCH BIOSYNTHESIS

The important role of Pi in starch synthesis stems from the elegant work of Preiss and colleagues [5,75] that ADP-glucose pyrophosphorylase (ADPG PPase), the key regulatory enzyme for starch synthesis, is stimulated by high triose-P/Pi levels. In the chloroplast, the concentration of these effector molecules was postulated to vary due to the physiological conditions to which the plant was exposed [5]. It has been shown that starch synthesis is greatly increased in those plant species where mannose-phosphate accumulates as a result of mannose feeding, which serves to lower the cytoplasmic Pi concentration [76].

A specific effect of Pi ions is exerted through the control of the distribution of newly fixed carbon between starch synthesis in the chloroplasts and the transfer of triose-P to the cytoplasm followed by synthesis of sucrose [48]. In isolated chloroplasts, low Pi slows photosynthesis and shifts the flow of carbon toward starch [48]. In some leaves mannose feeding produces the same effect by sequestering Pi as an abnormal hexokinase reaction becomes linked to oxidative phosphorylation [64,76]. Low levels of phosphate and high levels of sugars in phosphate-limited plants will lead to increased levels of ADP-glucose pyrophosphorylase transcript, which could contribute to increase in starch accumulation [77].

The starch deposited in the chloroplasts is usually degraded during the subsequent night period (Figure 7.1). An increased stromal Pi level favors starch breakdown [78]. Glucose-1-phosphate, the product of phosphorylytic starch degradation, is transformed through the oxidative pentose phosphate pathway [79,80] and also through phosphofructokinase [81] to triose-P or further to PGA [48,82,83]. The Pi translocator catalyzes the export of these phosphate esters into the cytosol.

The influence of Pi concentrations outside the chloroplast on the steady-state concentrations of various stromal metabolites and the corresponding rates of CO_2 fixation and starch production was determined using a kinetic model [84] based on control theory [85]. This kinetic analysis indicated that PGA and Pi play an important role in regulating starch synthesis and that ATP, glucose-1-P, and fructose-6-P make significant contributions. Since these metabolites are either substrates or effectors of the ADPG PPase, the analysis is consistent with the view that Pi is a negative effector and PGA is a positive effector of ADPG synthesis and that the PGA/Pi ratio therefore regulates starch synthesis [75].

D. SUCROSE BIOSYNTHESIS

Sucrose is a major product of photosynthesis. In many plants it is the main form in which carbon is translocated through the phloem of the vascular system from the leaf to other parts of the plant, but sucrose and other sugars may also be isolated and stored in vacuoles in the mesophyll cells. Sucrose is not merely a crucial sugar of vascular plants but is preeminently *the* sugar of vascular plants [86]. The rate of sucrose synthesis is a function of the carbon fixation rate, chemical partitioning of carbon between starch and sucrose, and the rate of sucrose export from the leaf [87]. Several processes may be involved in regulating the movement of carbon from the chloroplast to the vascular tissue [88]. It is

not possible in this review to present a complete analysis.

Sucrose formation occurs exclusively in the cytoplasm [89]. Substantial progress has been made in elucidating the biochemical mechanisms that control sucrose formation in leaves [9,10,86,90,91]. The cytosolic sucrose formation pathway starts with triose-P exported from the chloroplast, which are converted to hexose phosphate (hexose-P) and ultimately to sucrose (Figure 7.1). The key enzymes involved in the synthesis of sucrose from triose-P are cytoplasmic FBPase and sucrose-phosphate synthase (SPS) [90,92–97]. It is now recognized that there are at least two key aspects of the regulation of the pathway of sucrose biosynthesis: (i) control of cytosolic FBPase by the regulatory metabolite fructose-2,6-bisphosphate (F2,6BP) [96] and (ii) control of SPS activity by allosteric effectors and protein phosphorylation [87,90,97]. Although control of the sucrose biosynthesis pathway is shared between cytosolic FBPase and SPS, it appears that SPS probably exerts more of a limitation to the maximal rate of sucrose synthesis than does FBPase [95]. However, recently it was found that decreased expression of these two enzymes in antisense *Arabidopsis* lines has different consequences for photosynthetic carbon metabolism [98]. In transformants with decreased expression of SPS there was a slight inhibition of sucrose synthesis, no accumulation of phosphorylated intermediates, and carbon partitioning was not redirected to starch. This indicates that decreased expression of SPS triggers compensatory responses that favor sucrose synthesis, which included an increase of UDP-glucose/hexose-P ratio and decrease of pyrophosphate concentration. Strand et al. [98] conclude that these responses are presumably triggered when sucrose synthesis is decreased both in light and dark conditions. Decreased expression of cytosolic FBPase represented a passive response to the lower rate of sucrose synthesis and lead to accumulation of phosphorylated intermediates, Pi limitation of photosynthesis, and high rates of starch synthesis [98].

Regulation of FBPase received increased attention with the discovery of F2,6BP in plants [96]. The extensive studies of Stitt and coworkers showed that F2,6BP plays a key regulatory role in sucrose biosynthesis [9,93–96,99–101]. In plants the level of F2,6BP responds to changes in light, specific metabolites, sugars, and CO_2. F2,6BP is a potent inhibitor of cytoplasmic FBPase and sensitizes FBPase to the effects of FBP and Pi. F2,6BP decreases when triose-P becomes available for sucrose synthesis and it increases when hexose-P accumulates in the cytosol. The response of the cytosolic FBPase to a rising supply of triose-P has been described in a semiempirical model [9,102]. This model predicts how cytosolic FBPase activity responds to a rising rate of photosynthesis and relates closely with the actual response of sucrose synthesis *in vivo*.

UDP-glucose pyrophosphorylase is an important enzyme producing UDP-glucose for sucrose synthesis in leaves. The UDP-ase encoding gene of *A. thaliana* was suggested as a possible regulatory entity that is closely involved in the readjustment of plant response to environmental signaling [103]. In *Arabidopsis* mutants (*pho 1–2*) impaired in Pi status *Ugp* was found to be upregulated by conditions of phosphate deficiency [104]. Ciereszko et al. [104] concluded that under Pi deficiency, UGP-ase represents a transcriptionally regulated step in sucrose synthesis/metabolism, and that it is involved in homeostatic mechanisms for adjusting to the nutritional status of the plant.

Huber and coworkers have documented the role of SPS in the regulation of photosynthetic sucrose synthesis and partitioning in leaves [90,97,105–111]. SPS is minimally regulated at three levels. The steady-state level of the SPS enzyme protein is regulated developmentally during leaf expansion [108]. There are two distinct mechanisms to control the enzyme activity of the SPS protein: (i) allosteric control by G6P (activator) and Pi (inhibitor) and (ii) protein phosphorylation (covalent modification). These two mechanisms are often referred to as "fine" and "coarse" controls, respectively. There are apparent differences among species in the properties of SPS that may reflect different strategies for the control of carbon partitioning [109]. The importance of SPS in the regulation of carbon partitioning in leaves has been confirmed using recombinant DNA technology [112]. Although SPS is not the only determinant of the rate of sucrose synthesis; in some cases, the growth rate of the whole plate is correlated closely with SPS activity in leaves [113].

Sucrose synthesis is a Pi-liberating process (net reaction, $4\,triose\text{-}P + 3H_2O = 1\,sucrose + 4Pi$). The liberation of Pi in the cytoplasm during sucrose synthesis favors continued triose-P export from the chloroplast by counterexchange through the Pi translocator. Thus, under conditions that favor sucrose synthesis, triose-P molecules are partitioned away from the starch biosynthetic pathway that resides in the chloroplast. If sucrose synthesis in the cytoplasm is reduced, triose-P remains within the chloroplast for starch synthesis. The resulting increase in PGA within the chloroplast stroma (high PGA/Pi ratio) also favors starch synthesis by allosterically activating the starch-synthesizing enzyme ADPG PPase [75,114].

Pi may be involved in determining the proportion of the flux of photosynthetically fixed carbon between starch synthesis and export from the chloroplast [115]. As an inhibitor of SPS and cytosolic FBPase [116] and an activator of fructose-6-phosphate-2-kinase [117], Pi plays a critical role in regulating the rate of sucrose synthesis. When sucrose synthesis in the cytosol is restricted, there can indeed be substantial changes of the stromal Pi in leaves [43]. The rate of sucrose synthesis may also have an indirect control over the synthesis and accumulation of starch in leaves. Cytosolic FBPase and SPS, when acting in coordination with the Pi translocator, may represent an important link between sink demand and rates of carbon partitioning into starch and sucrose [118,119]. A change of partitioning does not necessarily imply that the rate of photosynthesis has been inhibited [10]. However, Pieters et al. [16] found that low sink strength lowers sucrose synthesis and restricts the recycling of Pi back to the chloroplast thus limiting the rate of net photosynthesis.

Four lines of evidence suggest that short-term availability of Pi in the cytosol may restrict sucrose synthesis and can limit the maximal rate of photosynthesis in saturating light and CO_2 [9]. The first approach is based on the manipulation of leaf material through Pi or mannose feeding [4,8]. A second approach is based on observations that the net rate of CO_2 assimilation does not always increase in C_3 plants when the O_2 concentration is decreased from 21% to 2% to suppress photorespiration, which is generally known as "O_2 insensitivity" [6,120,121]. A third approach involves using a brief interruption of photosynthesis to transiently increase the Pi level in the cytoplasm of the leaf [122]. A fourth line of evidence comes from the study of photosynthetic oscillations that can be triggered by increasing the CO_2 or lowering O_2 [4], or by a short period in the dark [122]. These oscillations are decreased when Pi is supplied to leaves and increase when mannose is supplied to sequester Pi. As sucrose is the major end product of photosynthesis, it is likely that a restriction in sucrose synthesis can limit photosynthesis through short-time limitation of Pi in the cytosol.

III. LONG-TERM *IN VIVO* EFFECTS OF Pi DEPRIVATION

The view that Pi is an important regulator of the rate of photosynthesis and of the partitioning of triose phosphates between starch biosynthesis and sucrose biosynthesis is to a large extent based on research carried out with in *in vitro* systems involving the use of isolated chloroplasts, enzyme systems, protoplasts, and with detached leaves or leaf disks fed with mannose to induce Pi deprivation. All of these studies point to the fact that the concentration of Pi in the cytosol versus that in the chloroplast is what potentially controls the intracellular flow and distribution of triose-P, and possibly, of the rate of photosynthesis itself.

Studies of long-term limitations of Pi on photosynthesis and carbon partitioning based on *in vivo* experiments using low phosphate (low P) plants have shown that the inhibition of photosynthesis was to a large extent due to limitations imposed on the PCR cycle in terms of RuBP regeneration [7,19,123–134] while the changes in carbon partitioning could be influenced in part by the relative capacities of the enzymes involved in starch and sucrose metabolism [134]. Recently, it was shown by Pieters et al. [16] that during Pi deficiency low rates of sucrose synthesis due to low demand from sinks limits Pi recycling to chloroplast and restricts photosynthesis.

A. PLANT GROWTH RESPONSE AND PHOSPHATE CONCENTRATION

Long-term P deficiency greatly affects the plant growth processes at subcellular, cellular, and whole organ levels of organization [1]. The growth of several plant species tested was greatly reduced by P deficiency. Leaf area, leaf number, and shoot dry matter per plant were found to be more sensitive to P deficiency than photosynthetic rate per unit leaf area [19,20,130,132,135,136]. Effects of P deficiency were similar in C_3 (sunflower and wheat) and C_4 (maize) species [130]. In Pi withdrawal experiments of the range of C_3, C_{3-4} intermediate, C_4 annual, and peren-

nial monocotyledons and dicotyledons species, it was shown that C_3 and C_4 species had similar photosynthetic P use efficiency but the growth of C_3 species was more affected by Pi supply than C_4 species; moreover, leaf photosynthetic rates were not correlated with growth response [137]. These results indicated that the relative growth rate decreased before any significant effect on photosynthesis [137]. Growth analysis of maize field crops under P deficiency supported the idea that P deficiency affects plant growth, especially leaf growth, earlier and to a greater extent than photosynthesis per unit leaf area [138]. Jacob and Lawlor [130] showed that the extreme P deficiency reduced plant height by 52%, leaf area per plant by 95%, and shoot dry weight per plant by 93% in sunflower (Table 7.1). The respective reductions were 57%, 89%, and 90% in maize and 53%, 91%, and 93% in wheat. P-deficient leaves contained more and smaller cells per unit leaf area. The mean cell volume and specific leaf weight were reduced to a smaller extent by P deficiency.

A typical response to phosphate deficiency is the increase of root mass/shoot mass ratio resulting from the decrease of shoot growth and the increase of root growth. The increase in root elongation and growth is probably a plant adaptive response to low Pi in surrounding medium and some kind of P searching strategy [139–143]. From the studies on bean plants it was found that the relative growth rate (RGR) of phosphate-deficient roots was higher only at the beginning of phosphate starvation and after 2 weeks (with severe

TABLE 7.1
Effect of Extreme P Deficiency on Plant Characteristics of Sunflower, Maize, and Wheat

Plant Growth Characteristics	Treatment	Sunflower	Maize	Wheat
Plant height (cm)	Control	46	103	68
	P-deficient	22	44	32
Leaf area per plant (cm²)	Control	895	708	232
	P-deficient	41	79	21
Shoot dry matter per plant (g)	Control	6.0	4.0	2.8
	P-deficient	0.4	0.4	0.2
Number of cells per m² leaf area ($\times 10^7$)	Control	723	152	98
	P-deficient	967	172	105
Mean cell volume (pl)	Control	27	43	144
	P-deficient	19	38	131
A_{mes}/A_{leaf} [a]	Control	31	9	14
	P-deficient	34	10	13
Specific leaf weight (g fresh wt. per m²)	Control	235	145	204
	P-deficient	222	121	201

[a] A_{mes}, mesophyll surface area; A_{leaf}, leaf surface area.

Source: From Jacob J, Lawlor DW. *J. Exp. Bot.* 1991; 42:1003–1011. With permission.

P deficiency) RGR was significantly lower as a result of decreasing ATP concentration in the roots [144].

To assess the importance of increased carbon allocation to roots for the adaptation of plants to low P availability, Nielsen et al. [145] constructed carbon budgets for four common bean genotypes with contrasting adaptation to low P availability in the field ("P efficiency"). They found that P-efficient genotypes allocated a larger fraction of their biomass to root growth, especially under low-P conditions. They also found that efficient genotypes had lower rates of root respiration than inefficient genotypes, which enabled them to maintain greater root biomass allocation than inefficient genotypes without increasing overall root carbon costs. Hogh-Jensen et al. [146] tested the influence of P deficiency on growth and nitrogen fixation of white clover plants. Their results indicated that nitrogen fixation did not limit the growth of clover plants experiencing P deficiency. A low-P status induced changes in the relative growth of roots, nodules, and shoots rather than changes in nitrogen and carbon uptake rates per unit mass or area of these organs.

The extent to which plant growth might be affected by P supply may depend on the sink–source status of the examined plant and how this is regulated [147]. The reduction in shoot biomass production in low-P plants may be attributed to a lower rate of leaf expansion, which may be induced by lower hydraulic conductance of the root system and a lower leaf water potential [19,20,148,149]. Using experimental and simulation techniques Rodriguez et al. [150] identified the existence of direct effects of P deficiency on individual leaf area expansion. Recently, Chiera et al. [151] found that expansion of soybean leaves under P stress was limited by the number of cell divisions,

which would imply control of cell division by a common regulatory factor within the leaf canopy. The reduction in leaf expansion in low-P sugar beet plants was associated with a 30% increase in leaf dry weight per unit area. Only 9% (or less) of the increase in dry weight in low-P leaves was due to starch [129]. Most of the remainder of the increase in dry weight may be attributed to other structural carbohydrates (e.g., cellulose and hemicelluloses). But our knowledge is limited regarding the effects of P on cell wall properties, especially those affecting cell division and cell wall expansion.

Jacob and Lawlor [130] reported that the extreme P deficiency in nutrient solution not only diminishes plant growth but also drastically reduces the total and inorganic P contents of leaves of sunflower, maize, and wheat (Table 7.2). The concentration of Pi in the leaf water decreased as the Pi content per unit leaf area decreased. Soluble protein content was lower in P-deficient leaves of all the three species while chlorophyll content was reduced in sunflower and maize.

Under Pi deficiency, the concentration of Pi in leaves depends mainly on the transport from the roots and mobilization of stored phosphate from older leaves [152]. Short-term phosphate starvation tends to maintain constant cytoplasmic Pi concentration at the expense of the vacuolar pool [153]. To regulate Pi homeostasis, plants develop signaling mechanisms [154]. It was recognized since many years that restriction of P, nitrate, or sulfur influences transpiration, stomatal conductance, and root hydraulic conductivity. The experiments with *Lotus japonicus* indicated that roots are capable, by a completely unknown mechanism, of monitoring the nutrient content of the solution in the root apoplasm and of initiating responses that anticipate by hours or

TABLE 7.2
Effect of Extreme P Deficiency on the Leaf Composition in Sunflower, Maize, and Wheat Plants

Leaf Composition	Treatment	Sunflower	Maize	Wheat
Total P content (mmol/m^2)	Control	7.20	4.80	5.90
	P-deficient	1.81	0.51	0.97
Pi content (mmol/m^2)	Control	1.65	0.58	0.65
	P-deficient	0.21	0.11	0.21
Concentration of Pi in leaf tissue water (mol/m^3)	Control	7.81	4.51	4.00
	P-deficient	1.20	1.01	1.21
Total chlorophyll (g/m^2)	Control	0.56	0.42	0.50
	P-deficient	0.48	0.26	0.54
Total soluble protein (g/m^2)	Control	12.2	5.8	6.28
	P-deficient	6.7	1.1	5.83

Source: From Jacob J, Lawlor DW. *J. Exp. Bot.* 1991; 42:1003–1011. With permission.

days any metabolic changes resulting from nutrient deficiency [155].

Reduced hydraulic conductance resulting from phosphate deficiency may affect the distribution of phosphate and nitrate ions between shoot and root [156]. In phosphate-sufficient bean plants the equal Pi distribution between shoot and root was noted, whereas in plants grown on a Pi-deficient solution almost 70% was partitioned to the shoot [157]. It seems that during moderate phosphate deficiency the leaf Pi pool remains relatively more stable mainly due to the possible effect of Pi recycling processes [158].

B. PHOTOSYNTHETIC MACHINERY

Several studies conducted with isolated chloroplasts, thylakoid membranes, and pigment systems have shown that the primary processes of light reactions of photosynthesis and photosynthetic electron transport were relatively little affected by long-term Pi deprivation [7,37,124]. However, it was shown that phosphate availability may change thylakoid membrane lipid composition by replacing some phospholipids for galactolipid digalactosyldiacylglycerol [159,160]. Significant changes in plasma membrane phospholipid composition were also observed in bean roots during prolonged phosphate deficiency [157].

Investigations on changes in photochemical apparatus organization and function in relation to leaf P status in sugar beet revealed the following: low-P leaves exhibited increased levels of chlorophyll/area, PSI/area, LHCP/area, Cyt-b_{563}/area, and Cyt-f/area while PSII, Cyt-b_{559}, and Q per area were not much affected [124]. PSII electron transport was slightly decreased per area while PSI electron transport was slightly increased so that the ratio of PSII/PSI is decreased.

It is generally believed that the results from *in vitro* studies with external supplies of artificial electron donors and acceptors and possibly damaged or atypical membranes may not always represent the *in vivo* situation. Light scattering and modulated chlorophyll *a* fluorescence have been successfully employed by several research workers as experimental probes for analyzing the state of the photosynthetic apparatus *in vivo* [161]. Rao et al. [123] measured the changes in light scattering *in vivo* during photosynthetic induction with variation in the leaf Pi status. Light scattering was markedly increased during photosynthetic induction in low-P leaves. This effect was reversible, disappearing within 24 h after P resupply. Measurements of *in vivo* fluorescence at room temperature and fluorescence at 77 K suggested that the low-P

leaves had less mobility of the antenna, which may be due to (i) the enhanced phosphatase activity leading to dephosphorylation of the antenna and (ii) the large proton gradient may promote dephosphorylation [162]. However, low-P leaves, to overcome this difficulty, developed a larger permanent antenna [124].

Using modulated chlorophyll *a* fluorescence techniques, the effects of extreme P deficiency during growth in the *in vivo* photochemical activity of PS II were determined in leaves of sunflower and maize [163]. In both species, long-term P deficiency decreased the efficiency of excitation energy capture by open PSII reaction centers, the photochemical quenching coefficient of PSII fluorescence and the *in vivo* quantum yield of PSII photochemistry, and increased the nonphotochemical dissipation of excitation energy. Observations from PSII fluorescence from intact leaves suggested that P deficiency causes photoinhibition of PSII. Furthermore, their calculations showed that there was a relatively higher rate of electron transport across PSII per net CO_2 assimilated in extreme P-deficient leaves. Most of these photosynthetic electrons that are not used for CO_2 reduction are diverted to photorespiration leading to proportionately more photorespiration and less CO_2 fixation in P-deficient leaves [164]. The important role of photorespiration for supporting photosynthesis when isolated chloroplasts were incubated at a low Pi level was shown by Usuda and Edwards [42]. Heber et al. [165] proposed that photorespiration substantially increases Pi availability for photosynthesis in the leaves of spinach. Unicellular green algae, *Chlorella vulgaris*, was used to study the effect of low-phosphate supply on glycolate metabolism [166]. P deficiency did not change chlorophyll concentration but with subsequent medium alkalization, dissolved inorganic carbon increased the photosynthetic O_2 evolution and intrachloroplast oxygen concentration resulting in enhancement of glycolate production [167]. The study of postillumination burst (PIB) of CO_2, which is interpreted as short-lived continuation of photorespiration in dark, indicated that the photorespiratory potential activity of P-deficient bean leaves is enhanced [168]. The importance of photorespiratory metabolism in Pi balance in bean plants under moderate phosphate deficiency was also suggested by Kondracka and Rychter [158] but the elucidation of its role needs further studies.

Plesnicar et al. [133] evaluated the efficiency of PSII photochemistry and electron transport, and light utilization capacity of sunflower leaves grown under sub- to supraoptimal Pi supply conditions. The apparent quantum yield (based on the initial slope of the relationship between photon flux density and rate of O_2 evolution) and the maximum (light and CO_2-saturated) rates of photosynthesis were the highest with the plants that were grown in optimal ($0.5\,mol\,m^{-3}$ Pi and $1.0\,mol\,m^{-3}$) Pi concentrations in nutrient solution. The photosynthetic efficiency was decreased by sub- or supraoptimal supply of Pi in nutrient solution. They suggested that the processes associated with nonphotochemical energy dissipation could modify the efficiency with which the reaction centers can capture and utilize excitation energy during Pi limitation of photosynthesis. This downregulation of the efficiency of PSII photochemistry by nonphotochemical energy was attributed to the adjustment of the rate of photochemistry to match that of photosynthetic carbon metabolism in order to avoid overexcitation of the PSII reaction centers.

C. CARBON METABOLISM

Several studies have shown that P deficiency in leaves decreases the rate of net CO_2 assimilation by intact leaves of C_3 and C_4 plants. This decline in net photosynthesis with long-term inadequate supply of Pi may result from a decrease in the conductance of CO_2 from the atmosphere to the chloroplasts; from a detrimental effect on the photosynthetic mechanism (mesophyll activity) itself; or from a combination of the two. It is often associated with decreases in Rubisco activity, RuBP concentration, rate of RuBP regeneration, stomatal conductance, and an increase in mesophyll resistance [7,126,130,131,169,170].

Phosphorus deficiency reduced the rate of photosynthesis in leaves by reducing the carboxylation efficiency and apparent quantum yield [7,127,133] by its influence on leaf metabolism, and also by decreasing leaf conductance [20,126]. Jacob and Lawlor [130] analyzed the effects of P deficiency on stomatal and mesophyll limitations of photosynthesis in sunflower, maize, and wheat plants. They found that stomatal conductance did not restrict the CO_2 diffusion rate; rather the metabolism of the mesophyll was the limiting factor. This was shown by poor carboxylation efficiency and decreased apparent quantum yield for CO_2 assimilation, both of which contributed to the increase in relative mesophyll limitation of photosynthesis in P-deficient leaves.

Brooks [7] attempted to determine which aspects of photosynthetic metabolism are affected when spinach plants are grown with inadequate P supply. P deficiency caused reductions in Rubisco activity, RuBP regenerating capacity, and quantum yield. The reduction in quantum yield was accompanied by changes in chlorophyll fluorescence of PSI and PSII measured at $77\,K$. The levels of RuBP and PGA were significantly reduced than the control

leaves while the response of photosynthesis to low $[O_2]$ was similar to control leaves, indicating that the photosynthesis is not limited by triose-P utilization. Dietz and Foyer [8] also observed decreased levels of phosphorylated metabolites in leaves as a result of P deficiency. The decrease in phosphorylated sugar levels was also observed in roots despite the increased sugar concentrations, which indicates that sugar phosphorylation may be limited by lower activity of fructokinase and hexokinase [171].

Rao and Terry [20] explored the changes in the activity of PCR cycle enzymes in relation to leaf Pi status. Low-P leaves exhibited increased levels in total activity of Rubisco, FBPase, and Ru5PKinase while the activity of PGA kinase, G-3-P-dehydrogenase, trannsketolase, and FBP aldolase decreased. The percentage light activation of Rubisco, PGA kinase, G-3-P-dehydrogenase, FBPase, SBPase, and R5PKinase was lower in low-P leaves (Table 7.3). Jacob and Lawlor [131] have also shown that P deficiency decreased the RuBP content of the leaf more than it decreased Rubisco. They suggested that the decreased specific activity of Rubisco found in Pi-deficient sunflower leaves is a consequence of the decreased ratio of RuBP to RuBP binding sites observed in such leaves allowing inhibitors to bind to the active sites of the enzyme.

It has been shown that long-term inadequate supply of Pi decreases the rate of photosynthesis by limiting the capacity for regeneration of RuBP, although decreased activation of Rubisco may play a part [7,20,130,131]. Rao et al. [126] measured a number of metabolites in low-P leaves, including RuBP, PGA, triose-P, FBP, F6P, G6P, adenylates, nicotinamide nucleotides, and Pi (Table 7.3). They suggested that RuBP regeneration in moderately P-deficient leaves is limited by decreased supply of carbon due to increased diversion of assimilated carbon for starch synthesis rather than by the decreased supply of ATP.

What are the precise metabolic control points that diminish regeneration of RuBP in P-deficient leaves? Several factors, including the initial activity of PCR cycle enzymes, the supply of ATP and NADPH, and the availability of fixed carbon, all affect the RuBP regeneration capacity of leaves. At moderate P-deficient conditions, RuBP regeneration of sugar beet leaves may be limited by the supply of Ru5P and the initial activity of the Ru5P kinase [126,134]. The conditions necessary to alter the RuBP pool size by this mechanism are yet to be clearly understood. According to Jacob and Lawlor [163], it is more probable that a deficiency of ATP in severely Pi-deficient leaves slows down the PCR cycle activity and thus decreases the regeneration of ATP. They found marked reductions in the amounts of ATP, ADP, and oxidized

TABLE 7.3
Effect of Low-P Treatment on the Percent Light Activation of Certain PCR Cycle Enzymes and Pool Sizes of Sugar Phosphates in Leaves of 5-week-old Sugar Beet Plants

PCR Cycle Enzymes and Metabolites	Control	Low-P
Light activation of PCR cycle enzymes (%)		
Rubisco	82	73
PGA kinase	78	65
NADP-G3PD	34	10
FBPase	33	39
SBPase	82	82
Ru5P kinase	34	23
Pool size of sugar phosphates (μmol/m^2)		
RuBP	66	32 (48)[a]
PGA	125	38 (30)
Triose-P	21	10 (48)
FBP	27	18 (67)
F6P	18	2 (11)
G6P	4	7 (16)

[a]Figures in parenthesis represent percentage of control values.

Source: From Rao IM, Terry N. *Plant Physiol.* 1989; 90:814–819 and Rao IM, Arulanantham AR, Terry N. *Plant Physiol.* 1989; 90:820–826. With permission.

pyridine nucleotides per unit leaf area in extremely Pi-deficient sunflower and maize leaves (Table 7.4).

As pointed out by Noctor and Foyer [23], a small change in the ratio of ATP and NADPH production during photosynthesis relative to the ratio of their consumption has an impact on cell adenylate and redox status. In bean leaves, during moderate phosphate deficiency, the net photosynthesis rate was lower and the concentration of NADPH increased; the ratio of NAD(P)H/NAD(P) also increased [172]. At the same time, leaf ATP concentration was reduced by 50% [173]. The reduction in leaf ATP concentration was comparable in light and dark periods. The determinations of ATP in leaf extracts during the light period reflect chloroplastic, mitochondrial, and cytosolic pools of ATP, whereas the leaf extracts from the dark period reflect mainly cytosolic and mitochondrial ATP pools. The ATP produced during photophosphorylation may be immediately utilized in the chloroplasts to support CO_2 fixation and chloroplast synthetic processes [174]. ATP synthesized in mitochondria can be transported to cytosol to support cytosolic reactions connected with sucrose synthesis [174]. Therefore, small differences between light and dark concentrations of ATP in phosphate-deficient leaves may reflect the determination of only the cytosolic pool being strongly dependent on the efficiency of mitochondrial ATP production [173]. It was found by Rychter's group that the efficiency of mitochondrial ATP production in bean plants during

phosphate deficiency is lower due to increased participation of a cyanide resistant, alternative pathway (AOX) [173–177], which bypasses two respiratory chain phosphorylation sites. The determinations of actual participation of AOX and ATP efficiency of respiratory chain phosphorylations in bean, tobacco, and *Gliricidia sepium* leaves revealed that during prolonged phosphate deficiency AOX expression is species dependent and is not observed in tobacco or *G. sepium* [178].

The rates of photosynthesis in C_3 plants have been modeled on Rubisco kinetics and the supply of CO_2, RuBP, and Pi [6,11,73,179–182]. It seems clear that at all levels regulation is serving to maximize efficiency while striving to avoid damage to the photochemical apparatus [179]. In general, nonlimiting processes of photosynthesis are regulated to balance the capacity of limiting processes [180]. When photosynthesis is limited by the capacity of Rubisco, the activities of electron transport and Pi regeneration are downregulated so that the rate of RuBP regeneration matches the rate of RuBP consumption by Rubisco. Similarly, when photosynthesis is limited by electron transport or Pi regeneration, the activity of Rubisco is downregulated to balance the limitation in the rate of RuBP regeneration.

It is important to understand that several parameters interact and a change in any one will result in a change in the activity of the others [85,183]. When the activity or level of any one of the components is

TABLE 7.4
Effect of P Deficiency on Adenylates and Nicotinamide Nucleotides of the Third Fully Expanded Leaves of Sunflower and Maize Grown at P Sufficient (10 mM Pi) or P Deficient (0 mM Pi) Conditions (values indicate pool sizes in μmol/m²)

Leaf Metabolites	Sunflower		Maize	
	P Sufficient	P Deficient	P Sufficient	P Deficient
Adenylates				
ATP	19.8	8.9	22.4	5.7
ADP	13.5	6.5	14.5	8.6
AMP	7.4	6.5 NS	9.1	8.8 NS
Total	40.7	21.9	46.0	23.1
ATP/ADP	1.5	1.4	1.5	0.7
Nicotinamide nucleotides				
NAD+	13.9	5.9	19.7	11.5
NADP+	12.6	7.1	16.8	9.8
NADH	3.2	4.4 NS	7.3	5.2
NADPH	4.5	4.1	8.9	9.4 NS
Total	34.2	21.4	52.7	35.9
NADPH/NADP+	0.36	0.58	0.53	0.96

NS = not significant at $p = .05$.

Source: From Jacob J, Lawlor DW. *Plant Cell Environ.* 1993; 16:785–795. With permission.

reduced (Ru5P kinase or RuBP), that component temporarily assumes an increased importance until equilibrium is restored. The enzymes of the PCR cycle, the pool sizes of sugar phosphates, along with the flux of ATP and NADPH, interacting as a system, share control over the rate of photosynthesis. None of these system elements controls the rate but all regulate jointly. It is the self-regulated lowering of the RuBP pool and not the inability to regenerate it faster that is a major factor in restoring and maintaining metabolic balance [182].

D. INTRACELLULAR PI COMPARTMENTATION

In order to prove that Pi regulation of photosynthesis occurs *in vivo*, it will be essential to demonstrate that cytosolic and chloroplastic Pi concentrations vary sufficiently to bring about changes in the flow and distribution of triose-P within the cell. There are practical problems to overcome in determining Pi compartmentation between chloroplast, cytoplasm, and vacuole. An additional problem is that there may be an internal Pi buffering mechanism. For example, if a mechanism for regulated transport of Pi across the tonoplast membrane were present, then the vacuole could act as a Pi reserve for the cytosol. More general evidence for a cytosolic Pi buffering mechanism arises from studies on P-deficient plants, which

appear to maintain the cytosolic Pi level at the expense of vacuolar Pi [154,184].

Methods have been developed for the assay of subcellular metabolite levels using leaf protoplasts. The protoplasts were ruptured by passage through a nylon net or a capillary tube. This was followed by immediate filtration of the particles (formed after rupture of the protoplasts) through a layer of silicone oil [72,185,186] or a combination of membrane filters [187]. Unfortunately, it has proved experimentally difficult to accurately determine chloroplastic and cytosolic Pi concentrations. Part of the problem relates to the presence, in the leaf cell vacuole, of a comparatively large amount of Pi [188], which masks the much smaller amount present in the cytosol. Furthermore, protoplasts are of limited value since their carbohydrate metabolism is almost certainly affected by the lack of sucrose export to the phloem.

^{31}P-nuclear magnetic resonance (^{31}P-NMR) spectroscopy can provide information on the relative concentration of Pi in the different cellular compartments [189]. A characteristic feature of the ^{31}P-NMR spectra of most plants tissues is the detection of two clearly resolved Pi signals, assigned to the cytoplasmic and vacuolar pools. *In vivo* ^{31}P-NMR provides an important method for studying the interaction between the two pools under different physiological

TABLE 7.5

^{31}P-NMR Determination of P Compartmentation in Leaves of Reproductive Soybeans as Affected by P Nutrition

Growth Stage	Phosphate Pools[a]	P Supply to Plants (mM)			
		0.05	0.10	0.20	0.45
		Pool size (mM)			
Full flower	HMP	0.54	2.11	5.75	7.65
	P_c	0.23	0.87	3.56	8.32
	P_v	< 0.05	< 0.10	3.50	8.01
Full pod	HMP	0.42	0.81	1.24	8.98
	P_c	0.23	0.69	0.93	7.59
	P_v	< 0.025	< 0.005	0.51	13.56
Full seed	HMP	< 0.01	0.39	0.78	4.10
	P_c	< 0.01	0.21	0.72	5.63
	P_v	< 0.01	< 0.05	< 0.10	7.65

[a]HMP, hexose monophosphate; P_c, cytoplasmic inorganic phosphate; P_v, vacuolar inorganic phosphate.

Source: From Lauer MJ, Blevins DG, Sierzputowska-Gracz H. *Plant Physiol.* 1989; 89:1331–1336. With permission.

conditions. With ^{31}P-NMR spectra it is possible to determine the absolute concentrations of Pi in the cytosol and the vacuole and thus to assess the extent to which the Pi distribution across the tonoplast reaches electrochemical equilibrium under different nutritional conditions [189–191].

The ^{31}P-NMR technique has been applied extensively in studies of chloroplasts, protoplasts, cell suspensions, leaves, and roots [18,19,127,153,192–203]. Foyer and Spencer [18] determined the intracellular distribution of Pi in barley leaves grown under different Pi regimes. They showed large differences in the vacuolar Pi content between the plants grown at different levels of P supply. In contrast, the cytosolic Pi level was similar in the leaves of plants grown at 1 and 25 mM Pi. Based on these data, they suggested that in leaves as in isolated cells the cytoplasmic Pi level is maintained constant as far as is possible, while the vacuolar Pi pool is allowed to fluctuate in order to buffer the Pi in the cytoplasm [192,193]. Several studies suggest the role of the vacuole in homeostasis of the cytoplasmic Pi concentration. Under different external phosphate levels, the cytoplasmic phosphate concentration remains relatively stable at the expense of the vacuolar pool, which decreases under Pi deficiency [153,154]. The mechanisms that control Pi transport from and to the vacuole are not clear, but changes in cytosolic and vacuolar Pi concentrations are considered as a signal for triggering different starvation response systems [154].

Using ^{31}P-NMR, Lauer et al. [127] determined P compartmentation in leaves of reproductive soybeans

as affected by P supply in nutrient solution (Table 7.5). As the concentration of P in nutrient solution increased from 0.05 to 0.45 mM, the vacuolar P pool size increased relative to cytoplasmic and hexose monophosphate P pools. Under low-P supply (0.05 mM), cytoplasmic P pool size was greatly reduced at full flower and full seed growth stages. This study indicated that the cytoplasmic P pool and leaf carbon metabolism dependent on it are buffered by the vacuolar P pool until the late stages of reproductive growth of soybeans.

Kerr et al. [106] found that the rates of net fixation of carbon, assimilate export, and net starch accumulation are not constant in continuous light. Since cytoplasmic concentrations of key regulatory metabolites such as F2,6BP and Pi could fluctuate as photosynthetic rates change [100], it may be possible that changes in intracellular Pi compartmentation could alter endogenous rhythms of photosynthesis and SPS activity.

E. CARBON PARTITIONING AND EXPORT

The partitioning of photosynthate between starch and sucrose appears to be strictly regulated at both genetic and biochemical levels [5]. There is a distinct interspecific variation in the ratio of starch: sucrose synthesized in leaves of different species [92,118]. This genetically determined predisposition allows classification of plants as high (e.g., soybean), intermediate (e.g., spinach), or low (e.g., barley) starch formers. P deficiency increased the starch synthesis relative to sucrose in soybean, spinach, and barley leaves although the accompanying limitation on photosynthetic capacity varied considerably between the species [18]. Usuda and Shimogawara [204] measured carbon fixation, carbon export, and carbon partitioning in maize seedlings in the early morning and at noon in P-adequate and P-deficient leaves (Table 7.6). P deficiency caused marked reductions in carbon fixation and carbon export and changed the partitioning of fixed carbon between starch and sucrose.

Long-term P deficiency causes increased starch concentrations in organs of several plant species [19,129,136]. These elevated starch concentrations in P-deficient plants may result from increased partitioning of photosynthetically fixed carbon into starch at the expense of sucrose synthesis in leaves [19,129] and decreased starch utilization in plant organs during the dark phase of the diurnal cycle [136]. Accumulation of high starch concentration in leaves and stems and decreased starch utilization in the dark in P-deficient soybean plants indicated that growth was restricted to a greater degree than photosynthetic capacity [136]. However, in barley plants omission of Pi from the growth medium resulted in increase in fructan concentration whilst little or no effect on starch, sucrose, glucose, and fructose was observed, which indicates that in some plants the mechanism for carbon partitioning into fructans is more sensitive toward low-P conditions than the mechanism for carbon partitioning into starch [205].

The work of Qiu and Israel [21] addressed the issue of whether increased starch accumulation is the cause or the result of decreased growth in P-deficient soybean plants. During onset of P deficiency, significant decreases in relative growth rate and in day and night leaf elongation rate occurred before or at the same time as significant increases in stem, leaf, and root starch concentrations. Based on these data, they concluded that disruption of metabolic functions associated with growth impairs utilization of available nonstructural carbohydrate in plants adjusting to P-deficiency stress.

Pieters et al. [16] studied the importance of sink demand on photosynthesis limitation during low-Pi conditions. The source–sink ratio was altered by darkening of all but two source leaves and compared to fully illuminated leaves of tobacco plants grown in Pi-sufficient and Pi-deficient conditions. They concluded that in tobacco plants grown in phosphate-deficient conditions low demand for assimilate (low sink strength) is the primary reason for photosynthesis limitation. Pi deficiency drastically decreased RuBP content in the Pi-deficient leaves and hence the rate of photosynthesis. This decrease was the result of end-product limitation since decreased sucrose synthesis restricted Pi recycling to chloroplast, thereby limiting ATP synthesis and RuBP regeneration.

In P-deficient sugar beet leaves, large accumulations of not only starch, but also sucrose and glucose were observed. This accumulation was associated with a marked reduction in carbon export from the leaves [129]. P deficiency also increased the levels of starch, sucrose, and glucose of petioles, storage root, and fibrous roots of sugar beet [134]. In contrast to sugar beet, P deficiency in soybean leaves caused a significant decrease in sucrose concentration together

TABLE 7.6
Carbon Fixation, Carbon Export, and Carbon Partitioning Between Starch and Sucrose in the Middle Part of Third Leaves of 18- or 19-Day Old Maize Plants

Measurement Period	Measurement[a]	Treatment	
		Control	Low-P
Early morning	Carbon fixed[b]	293	110
	Carbon exported[c]	221	81.9
	Carbon partitioning[d]	0.191	0.051
Around noon	Carbon fixed	214	112
	Carbon exported	158	103
	Carbon partitioning	0.253	0.103

[a]Measurement conditions were 1400 μmol/m²/s PAR and 33 Pa ambient CO_2 concentration.
[b]Matom/m²/2 h.
[c]Matom/m²/2 h.
[d]Carbon partitioning was expressed as a ratio of carbon atom accumulated in starch to carbon atom accumulated in sucrose (including transported sucrose).

Source: From Usuda H, Shimogawara K. *Plant Cell Physiol*. 1991; 32:497–504. With permission.

with a decrease in the activity of SPS [19,21]. The apparent carbon export rate from leaves was also restricted in soybean but the assimilate transport to stems and roots exceeded assimilate utilization in these organs, which implies that carbohydrate availability was not the primary factor limiting the growth of nonphotosynthetic organs of P-deficient plants [21]. Recently, De Groot et al. [206] investigated growth and dry mass partitioning in tomato as affected by P nutrition and light. They found that at mild P limitation, transport and utilization of assimilates in growth, not the production of assimilates, results in an increase in starch accumulation, and at severe P limitation, the production of assimilates is limited.

In bean leaves, sucrose concentration increased but light-promoted accumulation of sucrose was lower than in control leaves [158]. It is consistent with the observation of enhanced sucrose translocation from shoots to roots during phosphate deficiency [22,207–209]. The increase in soluble sugars in bean roots is believed to be not only the result of greater assimilate transport from leaves to roots but also higher hydrolysis of sucrose [210] and decrease in hexose phosphorylation [171]. Typical responses to phosphate-deficiency stress in root meristematic tissue include increase in sugar concentration, increase in the size of the vacuolar compartment, and changes in factors that control the rate of respiration [211].

IV. RECOVERY OF PLANTS FROM PHOSPHATE DEFICIENCY

Several researchers tested the reversibility of the long-term Pi-deprivation effects on plant processes such as Pi transport, photosynthesis, carbon partitioning, and growth. Leaf Pi levels of P-deficient plants raised markedly when the Pi supply was increased to spinach [212], potato [213], barley [214], maize, and soybean [215] due to an enhanced P uptake system. Obviously, a transport system with a large capacity for Pi uptake was induced in the root system when the plants were deprived of Pi. This system may catalyze a rapid accumulation of Pi in the leaves once the Pi availability is improved [201]. Based on the comparison of the results of the long-term experiment with those of the short-term uptake experiments, Jungk et al. [215] concluded that plants markedly adapt P uptake kinetics to their P status.

Increased Pi supply to low-P plants should increase leaf RuBP and should eliminate the inhibition of photosynthesis. It should also lower the pool sizes of storage carbohydrates (mainly starch) due to the recovery in leaf expansion and plant growth. Under-

standing the changes involved in the reversibility of low-P effects is important in predicting long-term plant growth and yield because of the varying sink strengths during plant development. The ability to reduce accumulations of starch and to relieve photosynthetic inhibition can significantly restore photosynthetic rates and increase the amount of photosynthate available for the actively growing sinks.

The changes in photosynthesis and carbon partitioning induced by low-P treatment could be due both to structural modifications induced by long-term phosphate stress and to metabolic changes accommodating the shortage of Pi as a reactant in biochemical pathways. These effects may be distinguished since the latter should be readily reversible when the supply of Pi is restored. The effects of P deficiency on photosynthesis were shown to be rapidly reversible with the resupply of P to the P-deficient plants or Pi feeding [7,8,21,123,134,212].

Brooks [7] reported that when low-P spinach plants were returned to nutrient solutions with adequate Pi, the percentage activation of Rubisco, amounts of RuBP and PGA, quantum yield, and maximal RuBP regeneration rate were increased within 24 h. The rapid increase of leaf RuBP and other sugar phosphates, which occurred as a consequence of increased Pi supply to low-P plants, substantiates the claim that the photosynthesis in low-P leaves was limited by RuBP regeneration [126].

Rao and Terry [134] monitored changes in photosynthesis, carbon partitioning, and plant growth in sugar beet by increasing the Pi supply to low-P plants. Within 72 h of increased Pi supply, low-P plants developed very high leaf blade Pi concentrations (up to sixfold of control levels). This dramatic increase in leaf blade Pi concentration was associated with a rapid increase in leaf sugar phosphates (especially RuBP), ATP, and total adenylates, which led to the rapid recovery (within 4 h) of the rate of photosynthesis. Increased Pi supply to low-P plants also decreased the amount of carbon accumulation in leaf blades in the form of starch, sucrose, and glucose, but this decrease was found to be slower than the recovery of photosynthesis. These results suggest that the effects of low P on photosynthetic machinery and the partitioning of fixed carbon are reversible. The rapid recovery of photosynthesis may be attributed to the lack of marked effects of low P on the structure and function of the photosynthetic membrane system [124].

Compared to the recovery of photosynthesis, the recovery in leaf expansion and other plant growth parameters were found to be slower in sugar beet [134]. When P-deficient soybean plants were supplied with adequate P, starch concentrations in leaves and stems decreased to the levels of P-sufficient plants within 3 days [21]. Thus, starch stored in leaves and stems is ready to be utilized in the synthesis of structural biomass during the time required for activation and development of additional photosynthetic capacity.

In the context of whole plant growth, plants may have developed an ability to buffer photosynthetic metabolism against decreases in P supply using Pi stored in the vacuoles. The poor correlations between short-term measurements of photosynthetic rate and long-term plant growth [216] may be due to the buffering power of the vacuoles [149]. Therefore, the primary influence of P deficiency on plant growth may be through a reduction in leaf expansion rather than through a marked reduction in photosynthetic capacity.

V. ACCLIMATION AND ADAPTATION OF PLANTS TO PHOSPHATE DEFICIENCY

Deficiency of phosphate in the growth medium creates a stress condition for growing plants. Recent investigations indicated that phosphate deficiency stress, as most if not all stresses, involves also a mild oxidative stress [217,218]. Plants can achieve tolerance to stress either by adaptation or by acclimation. Adaptation refers to heritable modifications, whereas acclimation refers to nonheritable modifications in metabolism and morphology of plant that is subjected to stressful conditions [219]. Both terms are often confusing in literature. It is important to note that many researchers describe acclimation process and refer it as adaptation to phosphate deficiency.

The effect of phosphate deficiency on photosynthesis depends on capability of plant metabolism to acclimate to low internal Pi supply. The current picture of the acclimation of plants to P deficiency is complex and involves integrated cellular, tissue, and whole plant responses [14,52,152,154,220]. Plants acclimate to P stress by changes in the pattern of growth, changes in the activity of Pi transport system, and changes in the physiological and metabolic activities. Changes in the pattern of growth and root architecture can be achieved by the increase in extension rates of roots, root hairs, and lateral root formation [152]. Some plant species develop the cluster or proteoid roots, releasing organic acids and phosphatases to growth media or form the symbiotic associations of roots with mycorrhizae. All those responses are presumed to enhance Pi acquisition from the soil and involve altered gene expression. In acclimation of plants to low-Pi environment over 100 genes may be involved, the expression of some of those genes was described recently by Abel et al. [220] and Raghothama [154]. Sensing a low-Pi environment involves not only the changes in Pi uptake and transport system [221] but also remobilization of phosphate from roots and older leaves to growing leaves to maintain the rate of net photosynthesis [154,222].

The metabolic changes that occur in response to P stress may be part of an acclimation of plants to low-P environments. This physiological and metabolic adjustment increases the amount of Pi available for photosynthesis and other essential physiological functions [14]. In photosynthetic carbon metabolism, Pi is liberated during the synthesis of carbohydrates, organic acids, and amino acids, and during photorespiration. In different plant species, under Pi deficiency, some of the above-mentioned reactions may be enhanced and thereby temporarily serve as an additional Pi source [132,158,164,223]. Also, an enhanced activity of phosphoenolpyruvate carboxylase and changes in PEP metabolism were observed [224,225]. Kondracka and Rychter [158] indicated the crucial role of PEP carboxylase, PEP metabolism, and enhanced amino acid synthesis for Pi recirculation during photosynthesis under moderate phosphate deficiency in bean leaves. It seems that the extent of acclimation of plants to low-phosphate conditions depends on individual plant species and serves primarily to maintain the rate of net photosynthesis through internal Pi recycling processes.

VI. CONCLUSIONS

The pioneering work of Walker and colleagues demonstrated that the isolated chloroplast requires a continuous supply of Pi in order to sustain photosynthesis. The Pi imported into the chloroplasts from the cytosol in exchange for triose-P and the Pi released from metabolic intermediates in the chloroplast stroma is available for photophosphorylation, which generates ATP for utilization in the PCR cycle. Thus, an adequate supply and internal cycling of Pi in the cell are essential for the regeneration of RuBP in the PCR cycle, which is a major limitation to maintain the rate of photosynthesis under Pi deprivation. The view that Pi supply is maintained *in vivo* by sucrose synthesis within the cytosol has been strengthened by substantial experimental evidence. The subcellular compartmentation of reactions and the resulting conservation of stromal and cytosolic Pi play an important role in the regulation of photosynthesis and carbon partitioning in leaves. Further, the rapid recovery of photosynthesis after P resupply to low-P leaves provides the direct evidence for the Pi regulation of photosynthesis *in vivo*.

Inadequate supply of Pi to plants limits the rate of photosynthesis due to both short- and long-term influences of Pi on the development of the photosynthetic machinery and metabolism. In the short term, low Pi might restrict photophosphorylation, which should lead to increased energization of the thylakoid membrane, decreased electron flow, and associated inhibition of photosynthesis. Inadequate supply of Pi over the long term decreases the rate of photosynthesis by limiting the capacity for regeneration of RuBP in the PCR cycle. However, the precise mechanisms that control RuBP regeneration under Pi deprivation are yet to be elucidated.

The research reviewed here suggests the following: (i) Pi deprivation does not affect photosynthetic electron transport; (ii) Pi deprivation reduces photosynthesis through the limitation of RuBP regeneration and not through Rubisco; (iii) RuBP regeneration may be limited by the supply of ATP and by increased partitioning of sugar phosphates to starch and sucrose synthesis; (iv) Pi deprivation affects leaf area most and photosynthesis to a lesser extent; (v) Pi deprivation diminishes carbon export more than the rate of photosynthesis; (vi) carbon accumulates in leaves of Pi-deprived plants; (vii) Pi-deprivation effects on photosynthesis and carbon partitioning are reversible; and (viii) sink strength imposes the most important regulatory role on photosynthesis *in vivo* during phosphate deficiency.

During the last decade, the use of *Arabidopsis* mutants with increased or decreased Pi level in the shoots and transgenic plants with altered gene expression served as powerful tools for studying the *in vivo* effect of Pi on photosynthetic carbon metabolism. Phosphate concentration in the leaves depends strongly on long- and short-distance transport processes and the efficiency of the uptake process [52]. The expression of genes encoding high-affinity root phosphate transporters is regulated by the phosphate status of the plant [221]. Overexpressing genes encoding high-affinity phosphate transporters may be one of the strategies for increasing Pi uptake and in consequence leaf Pi concentration. The recently described novel chloroplast phosphate transporter (PHT2;1) may be a key component in coordinating Pi acquisition and also Pi allocation toward the demands of photosynthetic carbon metabolism [53].

The precise mechanisms of the control of photosynthesis *in vivo* by Pi under a variety of environmental conditions are yet to be defined. It would of great interest to learn more details about the influence of various environmental factors such as light intensity, temperature, ambient CO_2 concentration, water, and

nutrient stress on Pi compartmentation in mesophyll cells to determine whether cytosolic Pi is important in mediating plant photosynthetic response to these environmental factors. Increased P requirement of pine species at elevated CO_2 has been clearly demonstrated [226]. *Arabidopsis* mutants with decreased and increased shoot Pi concentrations were used to demonstrate that low Pi triggers cold acclimatization of photosynthetic carbon metabolism leading to an increase of Rubisco expression, changes in Calvin cycle enzymes, and increased expression of enzymes of sucrose biosynthesis [227]. These results suggest that low-Pi levels resulting from low rates of sucrose synthesis can induce long-term changes in photosynthesis at the level of gene expression.

Phosphite (Phi), the analog of phosphate, is known to interfere with many Pi-starvation responses and could serve as an interesting tool to study plant responses to phosphate starvation. Varadarajan et al. [228] recently provided molecular evidence that Phi suppresses expression of several Pi-starvation-induced genes. They suggest that suppression of multiple Pi-starvation responses by Phi may be due to inhibition of primary Pi-starvation response mechanisms and therefore could serve as a tool in dissecting the Pi-starvation-induced molecular changes [228].

Our intention was not to make an exhaustive review of all the work carried out so far on Pi regulation of photosynthesis, but rather to evaluate the role of Pi in the regulation of photosynthetic carbon metabolism and to point out where our understanding is limited. It is clear that the Pi concentration in the cytosol is what potentially controls the rate of photosynthesis *in vivo* and partitioning of photoassimilates between starch and sucrose. Even though our knowledge from isolated chloroplasts provides substantial basis for the role for Pi in the control of photosynthesis, and undoubted importance of Pi to the life of "higher" plants, advanced theories concerning the mechanisms of Pi control of photosynthesis *in vivo* remain to be fully tested experimentally.

ACKNOWLEDGMENTS

It is a pleasure to thank Professor Norman Terry for discussions about some of the ideas presented in this chapter.

REFERENCES

1. Bieleski R, Ferguson IB. Physiology and metabolism of phosphate and its compounds. In: Bieleski R, Ferguson IB, eds. *Encyclopedia of Plant Physiology*. New York: Springer-Verlag, 1983:422–449.

2. Moorby J, Besford RT. Mineral nutrition and growth. In: Läuchli A, Bieleski RL, eds. *Encyclopedia of Plant Physiology*. New York: Springer-Verlag, 1983:481–527.

3. Malkin RNK. Photosynthesis. In: Buchanan BB, Gruissem W, Jones RL, eds. *Biochemistry and Molecular Biology of Plants*. Rockville, MD: American Society of Plant Physiologists, 2000:586–628.

4. Sivak MN, Walker DA. Photosynthesis *in vivo* can be limited by phosphate supply. *New Phytol*. 1986; 102:499–512.

5. Preiss J. Regulation of the biosynthesis and degradation of starch. *Ann. Rev. Plant. Physiol*. 1982; 33:431–454.

6. Sharkey TD. Photosynthesis in intact leaves of C_3 plants: physics, physiology and rate limitations. *Bot. Rev*. 1985; 51:53–105.

7. Brooks A. Effects of phosphorus nutrition on ribulose-1,5-biophosphate carboxylase activation, photosynthetic quantum yield and amounts of some Calvin-cycle metabolites in spinach leaves. *Aust. J. Plant Physiol*. 1986; 13:221–237.

8. Dietz KJ, Foyer C. The relationship between phosphate status and photosynthesis in leaves — reversibility of the effects of phosphate deficiency on photosynthesis. *Planta* 1986; 167:376–381.

9. Stitt M, Huber S, Kerr P. Control of photosynthetic sucrose synthesis. In: Hatch MD, Boardman NK, eds. *The Biochemistry of Plants. A Comprehensive Treatise*. New York: Academic Press, 1987:327–409.

10. Stitt M, Quick W. Photosynthetic carbon partitioning: its regulation and possibilities for manipulation. *Physiol. Plant*. 1989; 77:633–641.

11. Woodrow IE, Berry JA. Enzymatic regulation of photosynthetic CO_2, fixation in C_3 plants. *Ann. Rev. Plant Physiol. Plant Mol. Biol*. 1988; 39:533–594.

12. Terry N, Rao IM. Nutrients and photosynthesis: iron and phosphorus as case studies. In: Porter JR, Lawor DW, eds. *Plant Growth: Interactions with Nutrition and Environment*. Cambridge: Cambridge University Press, 1991:55–79.

13. Rao IM. Role of phosphorus in photosynthesis. In: Pessarakli M, ed. *Handbook of Photosynthesis*. New York: Marcel Dekker, 1996:173–193.

14. Plaxton WC, Carswell MC. Metabolic aspects of the phosphate starvation response in plants. In: Lerner R, ed. *Plant Responses to Environmental Stresses. From Phytohormones to Genome Reorganization*. New York, Basel: Marcel Dekker, 1999:349–372.

15. Rao IM, Terry N. Photosynthetic adaptation to nutrient stress. In: Yunus M, Pathre U, Mohanty P, eds. *Probing Photosynthesis: Mechanism, Regulation and Adaptation*. London: Taylor & Francis, 2000:379–397.

16. Pieters AJ, Paul MJ, Lawlor DW. Low sink demand limits photosynthesis under Pi deficiency. *J. Exp. Bot*. 2001; 52:1083–1091.

17. De Groot CC, Marcelis LFM, Van den Boogaard R, Lambers H. Growth and dry-mass partitioning in tomato as affected by phosphorus nutrition and light. *Plant Cell Environ*. 2001; 24:1309–1317.

18. Foyer C, Spencer C. The relationship between phosphate status and photosynthesis in leaves. Effects on intracellular ortho-phosphate distribution, photosynthesis and assimilate partitioning. *Planta* 1986; 167:369–375.

19. Fredeen AL, Rao IM, Terry N. Influence of phosphorus nutrition on growth and carbon partitioning in glycine max. *Plant Physiol*. 1989; 89:225–230.

20. Rao IM, Terry N. Leaf phosphate status, photosynthesis, and carbon partitioning in sugar-beet I. Changes in growth, gas-exchange, and Calvin cycle enzymes. *Plant Physiol*. 1989; 90:814–819.

21. Qiu J, Israel DW. Carbohydrate accumulation and utilization in soybean plants in response to altered phosphorus nutrition. *Physiol. Plant*. 1994; 90:722–728.

22. Ciereszko I, Gniazdowska A, Mikulska M, Rychter AM. Assimilate translocation in bean plants (*Phaseolus vulgaris* L.) during phosphate deficiency. *J. Plant Physiol*. 1996; 149:343–348.

23. Noctor G, Foyer CH. Homeostasis of adenylate status during photosynthesis in a fluctuating environment. *J. Exp. Bot*. 2000; 51:347–356.

24. Paul MJ, Foyer CH. Sink regulation of photosynthesis. *J. Exp. Bot*. 2001; 52:1383–1400.

25. Walker DA. Photosynthetic induction. In: Akoyonoglou G, ed. *Photosynthesis IV. Regulation of Carbon Metabolism*. Philadelphia: Balaban, 1981:189–202.

26. Lilley R, Chon C, Mosbach A, Heldt H. The distribution of metabolites between spinach chloroplasts and medium during photosynthesis *in vitro*. *Biochim. Biophys. Acta A* 1977; 460:259–272.

27. Walker DA, Herold, A. Can chloroplasts support photosynthesis unaided? In: Fugita Y, Fatoh S, Shibita K, Miyachu S, eds. *Plant Cell Physiol*. Special Issue. *Photosynthetic Organells: Structure and Function*. Tokyo: Japanese Society of Plant Physiologists and Centre for Academic Publication, 1977:295–310.

28. Giersch C, Heber U, Kaiser G, Walker DA, Robinson SP. Intracellular metabolite gradients and flow of carbon during photosynthesis of leaf protoplasts. *Arch. Biochem. Biophys*. 1980; 205:246–259.

29. Robinson SP, Walker DA. Photosynthetic carbon reduction cycle. In: Stumpf PK, Conn EE, eds. *The Biochemistry of Plants: A Comprehensive Treatise*. New York: Academic Press, 1981:193–236.

30. Stitt M, Wirtz W, Heldt H. Metabolite levels during induction in the chloroplast and extrachloroplast compartments of spinach protoplasts. *Biochim. Biophys. Acta* 1980; 593:85–102.

31. Marques IA, Anderson LE. Changing kinetic properties of fructose-1,6-bisphosphatase from pea chloroplasts during photosynthetic induction. *Plant Physiol*. 1985; 77:807–810.

32. Marques I, Ford D, Muschinek G, Anderson LE. Photosynthetic carbon metabolism in isolated pea chloroplasts: metabolite levels and enzyme activities. *Arch. Biochem. Biophys*. 1987; 252:458–466.

33. Leegood RC, Walker DA. Regulation of fructose-1, 6-biphosphatase activity in intact chloroplasts. Studies

of the mechanism of inactivation. *Biochim. Biophys. Acta* 1980; 593:362–370.

34. Leegood RC, Walker DA. Regulation of the Benson–Calvin cycle. In: Barber J, Barber NR, eds. *Photosynthetic Mechanisms and the Environment.* New York: Elsevier, 1985:188–258.

35. Kobza J, Edwards GE. The photosynthetic induction response in wheat leaves: net CO_2 uptake, enzyme activation, and leaf metabolites. *Planta* 1987; 171:549–559.

36. Giersch C, Robinson SP. Regulation of photosynthetic carbon metabolism during phosphate limitation of photosynthesis in isolated spinach-chloroplasts. *Photosynth. Res.* 1987; 14:211–227.

37. Furbank RT, Foyer CH, Walker DA. Regulation of photosynthesis in isolated spinach-chloroplasts during ortho-phosphate limitation. *Biochim. Biophys. Acta* 1987; 894:552–561.

38. Heineke D, Stitt M, Heldt HW. Effects of inorganic phosphate on the light dependent thylakoid energization of intact spinach chloroplasts. *Plant Physiol.* 1989; 91:221–226.

39. Flugge UI. Phosphate translocation in the regulation of photosynthesis. *J. Exp. Bot.* 1995; 46:1317–1323.

40. Flugge UI. Metabolite transporters in plastids. *Curr. Opin. Plant Biol.* 1998; 1:201–205.

41. Flugge UI. Phosphate translocators in plastids. *Ann. Rev. Plant Physiol. Plant Mol. Biol.* 1999; 50:27–45.

42. Usuda H, Edwards GE. Influence of varying CO_2 and ortho-phosphate concentrations on rates of photosynthesis, and synthesis of glycolate and dihydroxyacetone phosphate by wheat chloroplasts. *Plant Physiol.* 1982; 69:469–473.

43. Sharkey TD, Vanderveer PJ. Stromal phosphate concentration is low during feedback limited photosynthesis. *Plant Physiol.* 1989; 91:679–684.

44. Stitt M. Fructose-2,6-bisphosphate as a regulatory molecule in plants. *Ann. Rev. Plant Physiol. Plant Mol. Biol.* 1990; 41:153–185.

45. Heber U, Walker D. The chloroplast envelope, barrier or bridge? *Trends Biochem. Sci.* 1979; 4:252–256.

46. Heber U, Heldt HW. The chloroplast envelope: structure, function, and role in leaf metabolism. *Ann. Rev. Plant Physiol.* 1981; 32:139–168.

47. Flugge UI, Heldt HW. The phosphate-triose phosphate-phosphoglycerate translocator of the chloroplast. *Trends Biochem. Sci.* 1984; 9:530–533.

48. Heldt HW, Chon CH, Maronde D, Herold A, Stankovic AZ, Walker DA, Kraminer A, Kirk MR, Heber U. Role of orthophosphate and other factors in the regulation of starch formation in leaves and isolated chloroplasts. *Plant Physiol.* 1977; 59:1146–1155.

49. Flugge UI, Heldt HW. Metabolite translocators of the chloroplast envelope. *Ann. Rev. Plant Physiol. Plant Mol. Biol.* 1991; 42:129–144.

50. Flugge UI, Fischer K, Gross A. Molecular-cloning and *in vitro* expression of the chloroplast phosphate translocator protein. *Biol. Chem.* 1989; 370:643–644.

51. Riesmeier JW, Flugge UI, Schulz B, Heineke D, Heldt HW, Willmitzer L, Frommer WB. Antisense repression of the chloroplast triose phosphate translocator affects carbon partitioning in transgenic potato plants. *Proc. Natl Acad. Sci. USA* 1993; 90:6160–6164.

52. Rausch C, Bucher M. Molecular mechanisms of phosphate transport in plants. *Planta* 2002; 216:23–37.

53. Versaw WK, Harrison MJ. A chloroplast phosphate transporter, PHT2;1, influences allocation of phosphate within the plant and phosphate-starvation responses. *Plant Cell* 2002; 14:1751–1766.

54. Gerhardt R, Stitt M, Heldt HW. Subcellular metabolite levels in spinach leaves. Regulation of sucrose synthesis during diurnal alternation in photosynthetic partitioning. *Plant Physiol.* 1987; 83:399–407.

55. Barnes SA, Knight JS, Gray JC. Alteration of the amount of the chloroplast phosphate translocator in transgenic tobacco affects the distribution of assimilate between starch and sugar. *Plant Physiol.* 1994; 106:1123–1129.

56. Riesmeier JW, Flugge UI, Schulz B, Heineke D, Heldt HW, Willmitzer L, Frommer WB. Antisense repression of the chloroplast triose phosphate translocator affects carbon partitioning in transgenic potato plants. *Proc. Natl Acad. Sci. USA* 1993; 90:6160–6164.

57. Heldt HW, Chon CJ, Lorimer H. Phosphate requirement for the light activation of ribulose-1,5-biphosphate carboxylase in intact spinach chloroplasts. *FEBS Lett.* 1978; 92:234–240.

58. Bhagwat AS. Activation of spinach ribulose 1,5-bisphosphate carboxylase by inorganic phosphate. *Plant Sci. Lett.* 1981; 23:197–206.

59. Machler F, Nösberger J. Influence of inorganic-phosphate on photosynthesis of wheat chloroplasts. 2. Ribulose bisphosphate carboxylase activity. *J. Exp. Bot.* 1984; 35:488–494.

60. Charles SA, Halliwell B. Properties of freshly purified and thiol-treated spinach chloroplast fructose bisphosphatase. *Biochem. J.* 1980; 185:689–693.

61. Laing WA, Stitt M, Heldt, HW. Control of CO_2 fixation. Changes in the activity of ribulosephosphate kinase and fructose and sedoheptulose-bisphosphatase in chloroplasts. *Biochim. Biophys. Acta* 1981; 637:348–359.

62. Furbank R, Lilley R. Effects of inorganic phosphate on the photosynthetic carbon reduction cycle in extracts from the stroma of pea chloroplasts. *Biochim. Biophys. Acta* 1980; 592:65–75.

63. Gardemann A, Stitt M, Heldt HW. Control of CO_2 fixation. Regulation of spinach ribulose-5-phosphate kinase by stromal metabolite levels. *Biochim. Biophys. Acta* 1983; 722:51–60.

64. Harris GC, Cheesbrough JK, Walker DA. Effects of mannose on photosynthetic gas exchange in spinach leaf discs. *Plant Physiol.* 1983; 71:108–111.

65. Leegood RC, Labate C A, Huber SC, Neuhaus HE, Stitt M. Phosphate sequestration by glycerol and its effects on photosynthetic carbon assimilation by leaves. *Planta* 1988; 176:117–126.

66. Azcon-Bieto J. Inhibition of photosynthesis by carbohydrates in wheat leaves. *Plant Physiol.* 1983; 73:681–686.

67. Hatch AL, Jensen RG. Regulation of ribulose-1, 5-bisphosphate carboxylase from tobacco: changes in pH response and affinity for CO_2 and Mg^{2+} induced by chloroplast intermediates. *Arch. Biochem. Biophys.* 1980; 205:587–594.

68. Jordan DBC, Ogren WL. Binding of phosphorylated effectors by active and inactive forms of ribulose-1, 5-biphosphate carboxylase. *Biochemistry* 1983; 22:3410–3418.

69. Badger MR, Lorimer GH. Interaction of sugar phosphates with the catalytic site of ribulose-1,5-bisphosphate carboxylase. *Biochemistry* 1981; 20:2219–2225.

70. Enser U, Heber U. Metabolic regulation by pH gradients. *Biochim. Biophys. Acta* 1980; 592:577–591.

71. Flugge UI, Freisl M, Heldt HW. The mechanism of the control of carbon fixation by the pH in the chloroplast stroma. *Planta* 1980; 149:48–51.

72. Robinson SP, Walker DA. The control of 3-phosphoglycerate reduction in isolated chloroplasts by the concentrations of ATP, ADP and 3-phosphoglycerate. *Biochim. Biophys. Acta* 1979; 545:528–536.

73. Portis AR. Regulation of ribulose 1,5-bisphosphate carboxylase oxygenase activity. *Ann. Rev. Plant Physiol. Plant Mol. Biol.* 1992; 43:415–437.

74. Robinson SP, Giersch C. Inorganic-phosphate concentration in the stroma of isolated-chloroplasts and its influence on photosynthesis. *Aust. J. Plant. Physiol.* 1987; 14:451–462.

75. Preiss J. Regulation of the C_3 reductive cycle and carbohydrate synthesis. In: Tolbert NE, Preiss J, eds. *Regulation of Atmospheric CO_2 and O_2 by Photosynthetic Carbon Metabolism.* New York: Oxford University Press, 1994:93–102.

76. Chen-she S-H, Lewis DH, Walker DA. Stimulation of photosynthetic starch formation by sequestration of cytoplasmic orthophosphate. *New Phytol.* 1975; 74:383–392.

77. Nielsen TH, Krapp A, Roper-Schwarz U, Stitt M. The sugar-mediated regulation of genes encoding the small subunit of Rubisco and the regulatory subunit of ADP glucose pyrophosphorylase is modified by phosphate and nitrogen. *Plant Cell Environ.* 1998; 21:443–454.

78. Steup M. Starch degradation. In: Davies DD, ed. *The Biochemistry of Plants.* New York: Academic Press, 1988:255–296.

79. Stitt M, ap Rees T. Estimation of the activity of the oxidative pentosephosphate pathway in pea chloroplasts. *Phytochemistry* 1980; 19:1583–1585.

80. Stitt M, Rees AA. Carbohydrate breakdown by chloroplasts of *Pisum sativum. Biochim. Biophys. Acta* 1980; 627:131–143.

81. Kelly GJ, Latzko E. Chloroplast phosphofructokinase. *Plant Physiol.* 1977; 60:290–294.

82. Stitt M, Heldt HW. Physiological rates of starch breakdown in isolated intact spinach chloroplasts. *Plant Physiol.* 1981; 68:755–761.

83. Dennis DT, Miernyk JA. Compartmentation of nonphotosynthetic carbohydrate metabolism. *Ann. Rev. Plant Physiol.* 1982; 33:27–50.

84. Pettersson G, Ryde-Pettersson U. Metabolites controlling the rate of starch synthesis in chloroplast of C_3 plants. *Eur. J. Biochem.* 1989; 179:169–172.

85. Kacser H. Control of metabolism. In: Davied DD, ed. *The Biochemistry of Plants.* New York: Academic Press, 1987:39–67.

86. ap Rees T. Sucrose metabolism. In: Lewis DH, ed. *Storage Carbohydrates in Vascular Plants.* Cambridge: Cambridge University Press, 1984:53–73.

87. Winter H, Huber SC. Regulation of sucrose metabolism in higher plants: localization and regulation of activity of key enzymes. *Crit. Rev. Biochem. Mol. Biol.* 2000; 35:253–289.

88. Wardlaw IF. The control of carbon partitioning in plants. *New Phytol.* 1990; 116:341–381.

89. Stitt M, Wirtz W, Heldt HW. Metabolite levels during induction in the chloroplast and extrachloroplast compartments of spinach protoplasts. *Biochim. Biophys. Acta* 1980; 593:85–102.

90. Huber SC, Huber JL. Role of sucrose-phosphate synthase in sucrose metabolism in leaves. *Plant Physiol.* 1992; 99:1275–1278.

91. Scott P, Lange AJ, Pilkis SJ, Kruger NJ. Carbon metabolism in leaves of transgenic tobacco (*Nicotiana tabacum* L.) containing elevated fructose 2,6-bisphosphate levels. *Plant J.* 1995; 7:461–469.

92. Huber SC. Role of sucrose-phosphate in partitioning of carbon in leaves. *Plant Physiol.* 1983; 71:818–821.

93. Stitt M, Wirtz W, Heldt HW. Regulation of sucrose synthesis by cytoplasmic fructosebisphosphatase and sucrose phosphate synthase during photosynthesis in varying light and carbon-dioxide. *Plant Physiol.* 1983; 72:767–774.

94. Stitt M, Gerhardt R, Kurzel B, Heldt HW. A role for fructose 2,6-bisphosphate in the regulation of spinach leaves. *Plant Physiol.* 1983; 72:1139–1141.

95. Stitt M. Control analysis of photosynthetic sucrose synthesis — assignment of elasticity coefficients and flux-control coefficients to the cytosolic fructose 1, 6-bisphosphatase and sucrose phosphate synthase. *Philos. Trans. R. Soc. (Lond.) Ser. B — Biol. Sci.* 1989; 323:327–338.

96. Stitt M. Fructose-2,6-bisphosphate as a regulatory molecule in plants. *Ann. Rev. Plant Physiol. Plant Mol. Biol.* 1990; 41:153–185.

97. Huber SC, Huber JL. Role and regulation of sucrose-phosphate synthase in higher plants. *Ann. Rev. Plant Physiol. Plant Mol. Biol.* 1996; 47:431–444.

98. Strand A, Zrenner R, Trevanion S, Stitt M, Gustafsson P, Gardestrom P. Decreased expression of two key enzymes in the sucrose biosynthesis pathway, cytosolic fructose-1,6-bisphosphatase and sucrose phosphate synthase, has remarkably different consequences for photosynthetic carbon metabolism in transgenic *Arabidopsis thaliana. Plant J.* 2000; 23:759–770.

99. Stitt M, Cseke C, Buchanan BB. Regulation of fructose 2,6-bisphosphate concentration in spinach leaves. *Eur. J. Biochem.* 1984; 143:89–93.

100. Stitt M, Herzog B, Heldt HW. Control of photosynthetic sucrose synthesis by fructose-2,6-bisphosphate.

I. Coordination of CO_2 fixation and sucrose synthesis. *Plant Physiol.* 1984; 75:548–553.

101. Stitt M, Herzog B, Heldt HW. Control of photosynthetic sucrose synthesis by fructose 2,6-bisphosphate. V. Modulation of the spinach leaf cytosolic fructose 1,6-bisphosphatase activity *in vitro* by substrate, products, pH, magnesium, fructose 2,6-bisphosphate, adenosine-monophosphate, and dihydroxyacetone phosphate. *Plant Physiol.* 1985; 79:590–598.

102. Stitt M, Gerhardt, R, Wilke I, Heldt HW. The contribution of fructose-2,6-bisphosphate to the regulation of sucrose synthesis during photosynthesis. *Physiol. Plant.* 1987; 69:377–386.

103. Ciereszko I, Johansson H, Kleczkowski L. Sucrose and light regulation of a cold-inducible UDP-glucose pyrophosphorylase gene via a hexokinase-independent and abscisic acid-insensitive pathway in *Arabidopsis*. *Biochem. J.* 2001; 354:67–72.

104. Ciereszko I, Johansson H, Hurry V, Kleczkowski LA. Phosphate status affects the gene expression, protein content and enzymatic activity of UDP-glucose pyrophosphorylase in wild-type and pho mutants of *Arabidopsis*. *Planta* 2001; 212:598–605.

105. Doehlhert DC, Huber SC. Phosphate inhibition of spinach leaf sucrose phosphate synthase as affected by glucose-6-phosphate and phosphoglucoisomerase. *Plant Physiol.* 1984; 76:250–253.

106. Kerr PS, Rufty TW, Huber SC. Endogenous rhythms in photosynthesis, sucrose phosphate synthase activity, and stomatal-resistance in leaves of soybean (*Glycine max* L. Merr.). *Plant Physiol.* 1985; 77:275–280.

107. Kerr PS, Huber SC. Coordinate control of sucrose formation in soybean leaves by sucrose-phosphate synthase and fructose-2,6-bisphosphate. *Planta* 1987; 170:197–204.

108. Walker JL, Huber SC. Regulation of sucrose-phosphate synthase activity in spinach leaves by protein level and covalent modifications. *Planta* 1989; 177:116–120.

109. Huber SC, Huber JL, Pharr DM. Variation among species in light activation of sucrose-phosphate synthase. *Plant Cell Physiol.* 1989; 30:277–285.

110. Huber SC. Biochemical mechanism for regulation of sucrose accumulation in leaves during photosynthesis. *Plant Physiol.* 1989; 91:656–662.

111. Winter H, Huber SC. Regulation of sucrose metabolism in higher plants: localization and regulation of activity of key enzymes. *Crit. Rev. Biochem. Mol. Biol.* 2000; 35:253–289.

112. Worrell AC, Bruneau JM, Summerfelt K, Boersig M, Voelker TA. Expression of a maize sucrose phosphate synthase in tomato alters leaf carbohydrate partitioning. *Plant Cell* 1991; 3:1121–1130.

113. Rocher JP, Prioul JL, Lecharny A, Reyss A, Joussaume M. Genetic variability in carbon fixation, sucrose-P-synthase and ADP glucose pyrophosphorylase in maize plants of differing growth rate. *Plant Physiol.* 1989; 89:416–420.

114. Preiss J, Levi C. Starch biosynthesis and degradation. In: Preiss J, ed. *The Biochemistry of Plants: A Comprehensive Treatise. Carbohydrates: Structure and Function.* New York: Academic Press, 1980:371–423.

115. Woodrow IEM, Walker DA. Regulation of photosynthetic carbon metabolism. The effect of inorganic phosphate on stromal sedoheptulose-1,7-bisphosphatase. *Eur. J. Biochem.* 1983; 132:121–126.

116. Harbron S, Foyer CH, Walker DA. The purification and properties of sucrose phosphate synthetase from spinach leaves: the involvement of this enzyme and fructose bisphosphatase in the regulation of sucrose biosynthesis. *Arch. Biochem. Biophys.* 1981; 212:237–246.

117. Cseke C, Buchanan BB. An enzyme synthesizing fructose-2,6-bis-phosphate occurs in leaves and is regulated by metabolic effectors. *FEBS Lett.* 1983; 155:139–142.

118. Huber SC. Interspecific variation in activity and regulation of leaf sucrose-phosphate synthase. *Z. Pflanzenphysiol.* 1981; 102:443–450.

119. Rufty TW, Huber SC. Changes in starch formation and activities of sucrose phosphate synthase and cytoplasmic fructose-1,6-bisphosphatase in response to source-sink alternations. *Plant Physiol.* 1983; 72:474–480.

120. Sharkey TD, Stitt M, Heineke D, Gerhardt R, Raschke K, Heldt HW. Limitation of photosynthesis by carbon metabolism. II. O_2 insensitive CO_2 uptake results from limitation of triose phosphate utilization. *Plant Physiol.* 1986; 81:1123–1129.

121. Leegood RC, Furbank RT. Stimulation of photosynthesis by 2-percent oxygen at low-temperatures is restored by phosphate. *Planta* 1986; 168:84–93.

122. Stitt M. Limitation of photosynthesis by carbon metabolism. I. Evidence for excess electron transport capacity in leaves carrying out photosynthesis in saturating light and CO_2. *Plant Physiol.* 1986; 81:1115–1122.

123. Rao IM, Abadia J, Terry N. Leaf phosphate status and photosynthesis *in vivo*: changes in light-scattering and chlorophyll fluorescence during photosynthetic induction in sugar-beet leaves. *Plant Sci.* 1986; 44:133–137.

124. Abadia J, Rao IM, Terry N. Changes in leaf phosphate status have only small effects on the photochemical apparatus of sugar beet leaves. *Plant Sci.* 1987; 50:49–55.

125. Brooks A, Woo KC, Wong SC. Effects of phosphorus nutrition on the response of photosynthesis to CO_2 and O_2, activation of ribulose bisphosphate carboxylase and amounts of ribulose bisphosphate and 3-phosphoglycerate in spinach leaves. *Photosynth. Res.* 1988; 15:133–141.

126. Rao IM, Arulanantham AR, Terry N. Leaf phosphate status, photosynthesis and carbon partitioning in sugar beet II. Diurnal changes in sugar phosphates, adenylates, and nicotinamide nucleotides. *Plant Physiol.* 1989; 90:820–826.

127. Lauer MJ, Blevins DG, Sierzputowska-Gracz H. ^{31}P-Nuclear magnetic resonance determination of phosphate compartmentation in leaves of reproductive

soybeans (*Glycine max* L.) as affected by phosphate nutrition. *Plant Physiol.* 1989; 89:1331–1336.

128. Fredeen AL, Raab TK, Rao IM, Terry N. Effects of phosphorus nutrition on photosynthesis in *Glycine max*. *Planta* 1990; 181:399–405.

129. Rao IM, Fredeen AL, Terry N. Leaf phosphate status, photosynthesis, and carbon partitioning in sugar beet III. Diurnal changes in carbon partitioning and carbon export. *Plant Physiol.* 1990; 92:29–36.

130. Jacob J, Lawlor DW. Stomatal and mesophyll limitations of photosynthesis in phosphate deficient sunflower, maize and wheat plants. *J. Exp. Bot.* 1991; 42:1003–1011.

131. Jacob J, Lawlor DW. Dependence of photosynthesis of sunflower and maize leaves on phosphate supply, ribulose-1,5-bisphosphate carboxylase oxygenase activity, and ribulose-1,5-bisphosphate pool size. *Plant Physiol.* 1992; 98:801–807.

132. Rao IM, Fredeen AL, Terry N. Influence of phosphorus limitation on photosynthesis, carbon allocation and partitioning in sugar beet and soybean grown with a short photoperiod. *Plant Physiol. Biochem.* 1993; 31:223–231.

133. Plesnicar M, Kastori R, Petrovic N, Pankovic D. Photosynthesis and chlorophyll fluorescence in sunflower (*Helianthus annuus* L) leaves as affected by phosphorus-nutrition. *J. Exp. Bot.* 1994; 45:919–924.

134. Rao IM, Terry N. Leaf phosphate status, photosynthesis, and carbon partitioning in sugar-beet IV. Changes with time following increased supply of phosphate to low-phosphate plants. *Plant Physiol.* 1995; 107:1313–1321.

135. Lynch J, Lauchli A, Epstein E. Vegetative growth of the common bean in response to phosphorus nutrition. *Crop Sci.* 1991; 31:380–387.

136. Qiu J, Israel DW. Diurnal starch accumulation and utilization in phosphorus-deficient soybean plants. *Plant Physiol.* 1992; 98:316–323.

137. Halsted M, Lynch J. Phosphorus responses of C-3 and C-4 species. *J. Exp. Bot.* 1996; 47:497–505.

138. Plenet D, Etchebest S, Mollier A, Pellerin S. Growth analysis of maize field crops under phosphorus deficiency I. Leaf growth. *Plant Soil* 2000; 223:117–130.

139. Rao IM, Borrero V, Ricaurte J, Garcia R, Ayarza MA. Adaptive attributes of tropical foliage species to acid soils III. Differences in phosphorus acquisition and utilization as influenced by varying phosphorus supply and soil type. *J. Plant Nutr.* 1997; 20:155–180.

140. Rao IM, Friesen DK, Osaki M. Plant adaptation to phosphorus-limited tropical soils. In: Pessarakli M, ed. *Handbook of Plant and Crop Stress*. New York: Marcel Dekker, 1999:61–96.

141. Mollier A, Pellerin S. Maize root system growth and development as influenced by phosphorus deficiency. *J. Exp. Bot.* 1999; 50:487–497.

142. Ciereszko I, Janonis A, Kociakowska M. Growth and metabolism of cucumber in phosphate-deficient conditions. *J. Plant Nutr.* 2002; 25:1115–1127.

143. Vance C, Uhde-Stone C, Allen DL. Phosphorus acquisition and use: critical adaptations by plants for securing a nonrenewable resource. *New Phytol.* 2003; 157:423–447.

144. Gniazdowska A, Mikulska M, Rychter AM. Growth, nitrate uptake and respiration rate in bean roots under phosphate deficiency. *Biol. Plant* 1998; 41:217–226.

145. Nielsen K, Eshel A, Lynch JP. The effect of phosphorus availability on the carbon economy of contrasting common bean (*Phaseolus vulgaris* L.) genotypes. *J. Exp. Bot.* 2001; 52:329–339.

146. Hogh-Jensen H, Schjoerring JK, Soussana J-F. The influence of phosphorus deficiency on growth and nitrogen fixation of white clover plants. *Ann. Bot.* 2002; 90:745–753.

147. Stitt M. Rising CO_2 levels and their potential significance for carbon flow in photosynthetic cells. *Plant Cell Environ.* 1991; 14:741–762.

148. Radin JW, Eidenbock MP. Carbon accumulation during photosynthesis in leaves of nitrogen- and phosphorus-stressed cotton. *Plant Physiol.* 1986; 82:869–871.

149. Hart AL, Greer DH. Photosynthesis and carbon export in white clover plants grown at various levels of phosphorus supply. *Physiol. Plant.* 1988; 73:46–51.

150. Rodriguez D, Zubillaga MM, Ploschuk EL, Keltjens WG, Goudriaan A, Lavado RS. Leaf area expansion and assimilate production in sunflower (*Helianthus annuus* L.) growing under low phosphorus conditions. *Plant Soil* 1998; 202:133–147.

151. Chiera J, Rufty T. Leaf initiation and development in soybean under phosphorus stress. *J. Exp. Bot.* 2002; 53:473–481.

152. Schachtman DP, Reid RJ, Ayling SM. Phosphorus uptake by plants. From soil to cell. *Plant Physiol.* 1998; 116:447–453.

153. Mimura T, Sakano K, Shimmen T. Studies on distribution, retranslocation and homeostasis of inorganic phosphate in barley leaves. *Plant Cell Environ.* 1996; 19:311–320.

154. Raghothama KG. Phosphate acquisition. *Ann. Rev. Plant Physiol. Plant Mol. Biol.* 1999; 50:665–693.

155. Clarkson DT, Carvajal M, Henzler T, Waterhouse RN, Smyth AJ, Cooke, DT, Steudle E. Root hydraulic conductance; diurnal aquaporin expression and the effects of nutrient stress. *J. Exp. Bot.* 2000; 51:61–70.

156. Radin JW, Mathews MA. Water transport properties of cortical cells in roots of nitrogen- and phosphorus-deficient cotton seedlings. *Plant Physiol.* 1989; 89:264–268.

157. Gniazdowska A, Szal B, Rychter AM. The effect of phosphate deficiency on membrane phospholipid composition of bean (*Phaseolus vulgaris* L.) roots. *Acta Physiol. Plant.* 1999; 21:263–269.

158. Kondracka A, Rychter AM. The role of Pi recycling processes during photosynthesis in phosphate-deficient bean plants. *J. Exp. Bot.* 1997; 48:1461–1468.

159. Essigmann B, Guler S, Narang RA, Linke D, Benning C. Phosphate availability affects the thylakoid lipid composition and the expression of SQD1, a gene required for sulfolipid biosynthesis in *Arabidopsis thaliana*. *Proc. Natl Acad. Sci. USA* 1998; 95:1950–1955.

160. Andersson X, Stridh MH, Larsson KE, Liljenberg C, Sandelius AS. Phosphate-deficient oat replaces a major portion of the plasma membrane phospholipids with the galactolipid digalactosyldiacylglycerol. *FEBS Lett.* 2003; 537:128–132.

161. Krause GH, Weis E. Chlorophyll fluorescence and photosynthesis: the basics. *Ann. Rev. Plant Physiol. Plant Mol. Biol.* 1991; 42:313–349.

162. Horton P. Interactions between electron transport and carbon assimilation: regulation of light harvesting. In: Briggs WR, ed. *Photosynthesis*. New York: Alan R. Liss, Inc., 1989:393–406.

163. Jacob J, Lawlor DW. *In vivo* photosynthetic electron-transport does not limit photosynthetic capacity in phosphate-deficient sunflower and maize leaves. *Plant Cell Environ.* 1993; 16:785–795.

164. Jacob J, Lawlor DW. Extreme phosphate deficiency decreases the *in vivo* CO_2/O_2 specificity factor of ribulose 1,5-bisphosphate carboxylase-oxygenase in intact leaves of sunflower. *J. Exp. Bot.* 1993; 44:1635–1641.

165. Heber U, Viil J, Neimanis S, Mimura T, Dietz KJ. Photoinhibitory damage to chloroplasts under phosphate deficiency and alleviation of deficiency and damage by photorespiratory reactions. *Z. Naturforsc., J. Biosci.* 1989; 44:524–536.

166. Kozlowska B, Maleszewski S. Low-level of inorganic orthophosphate in growth-medium increases metabolism and excretion of glycolate by chlorella-vulgaris cells cultivated under air conditions. *Plant Physiol. Biochem.* 1994; 32:717–721.

167. Kozlowska-Szerenos B, Zielinski P, Maleszewski S. Involvement of glycolate metabolism in acclimation of Chlorella vulgaris cultures to low phosphate supply. *Plant Physiol. Biochem.* 2000; 38:727–734.

168. Hauschild T, Ciereszko I, Maleszewski S. Influence of phosphorus deficiency on post-irradiation burst of CO_2 from bean (*Phaseolus vulgaris* L) leaves. *Photosynthetica* 1996; 32:1–9.

169. Sicher RC, Kremer DF. Effects of phosphate deficiency on assimilate partitioning in barley seedlings. *Plant Sci.* 1988; 57:9–17.

170. Usuda H, Shimogawara K. Phosphate deficiency in maize. 2. Enzyme activities. *Plant Cell Physiol.* 1991; 32:1313–1317.

171. Rychter AM, Randall DD. The effect of phosphate deficiency on carbohydrate metabolism in bean roots. *Physiol. Plant.* 1994; 91:383–388.

172. Juszczuk IM, Rychter AM. Changes in pyridine nucleotide levels in leaves and roots of bean plants (*Phaseolus vulgaris* L.) during phosphate deficiency. *J. Plant Physiol.* 1997; 151:399–404.

173. Mikulska M, Bomsel JL, Rychter AM. The influence of phosphate deficiency on photosynthesis, respiration and adenine nucleotide pool in bean leaves. *Photosynthetica* 1998; 35:79–88.

174. Krömer S. Respiration during photosynthesis. *Ann. Rev. Plant Physiol. Plant Mol. Biol.* 1995; 46:45–70.

175. Rychter AM, Mikulska M. The relationship between phosphate status and cyanide-resistant respiration in bean roots. *Physiol. Plant.* 1990; 79:663–667.

176. Rychter AM, Chauveau M, Bomsel JL, Lance C. The effect of phosphate deficiency on mitochondrial activity and adenylate levels in bean roots. *Physiol. Plant.* 1992; 84:80–86.

177. Juszczuk IM, Wagner AM, Rychter AM. Regulation of alternative oxidase activity during phosphate deficiency in bean roots (*Phaseolus vulgaris*). *Physiol. Plant.* 2001; 113:185–192.

178. Gonzalez-Meler MA, Giles L, Thomas RB, Siedow JN. Metabolic regulation of leaf respiration and alternative pathway activity in response to phosphate supply. *Plant, Cell Environ.* 2001; 24:205–215.

179. Foyer CH, Furbank R, Harbinson J, Horton P. The mechanism contributing to photosynthetic control of electron transport by carbon assimilation in leaves. *Photosynth. Res.* 1990; 25:83–100.

180. Sage RF. A model describing the regulation of ribulose-1,5-bisphosphate carboxylase, electron-transport, and triose phosphate use in response to light-intensity and CO_2 in C-3 plants. *Plant Physiol.* 1990; 94:1728–1734.

181. Bowes G. Facing the inevitable: plants and increasing atmospheric CO_2. *Ann. Rev. Plant Physiol. Plant Mol. Biol.* 1993; 44:309–332.

182. Geiger DR, Servaites JC. Diurnal regulation of photosynthetic carbon metabolism in C3 plants. *Ann. Rev. Plant Physiol. Plant Mol. Biol.* 1994; 45:235–256.

183. Servaites JC, Shieh WJ, Geiger DR. Regulation of photosynthetic carbon reduction cycle by ribulose bisphosphate and phosphoglyceric acid. *Plant Physiol.* 1991; 97:1115–1121.

184. Bieleski RL. Phosphate pools, phosphate transport, and phosphate availability. *Ann. Rev. Plant Physiol.* 1973; 24:225-252.

185. Hampp R. Rapid separation of the plastid, mitochondrial, and cytoplasmic fractions from intact leaf protoplasts of Avena. *Planta* 1980; 150:291–298.

186. Wirtz W, Stitt M, Heldt HW. Enzymic determination of metabolites in the subcellular compartments of spinach protoplasts. *Plant Physiol.* 1980; 66:187–193.

187. Lilley RMC, Stitt M, Mader G, Heldt HW. Rapid fractionation of wheat leaf protoplasts using membrane filtration. *Plant Physiol.* 1982; 70:965–970.

188. Hamp R, Goller M, Zeigler H. Adenylate levels, energy charge and phosphorylation potential during dark-light and light-dark transition in chloroplasts, mitochondria, and cytosol of mesophyll protoplasts from *Avena sativa* L. *Plant Physiol.* 1982; 69:448–455.

189. Ratcliffe RG. *In vivo* NMR studies of higher plants and algae. *Adv. Bot. Res.* 1994; 20:43–123.

190. Lee RB, Ratcliffe RG, Southon TE. ^{31}P-NMR measurement of the cytosolic and vacuolar Pi content of mature roots: relationship with phosphorus status and phosphate fluxes. *J. Exp. Bot.* 1990; 41:1063–1078.

191. Lee RB, Ratcliffe RG. Subcellular distribution of inorganic phosphate and levels of nucleoside triphosphate in mature maize roots at low external phosphate concentrations: measurements with ^{31}P-NMR. *J. Exp. Bot.* 1993; 44:587–598.

192. Foyer CH, Walker D, Spencer C, Mann B. Observations on the phosphate status and intracellular pH of intact cells, protoplasts and chloroplasts from photosynthetic tissue using phosphorus-31 nuclear magnetic resonance. *Biochem. J.* 1982; 202:429–434.

193. Rebeille F, Blingy R, Martin J-B, Douce R. Relationship between the cytoplasm and vacuole phosphate pool in *Acer pseudoplatanus* cells. *Arch. Biochem. Biophys.* 1983; 225:143–148.

194. Waterton JC, Bridges IA, Irwing MP. Intracellular compartmentation detected by ^{31}P-NMR in intact photosynthetic wheat-leaf tissue. *Biochim. Biophys. Acta* 1983; 763:315–320.

195. Lee RB, Ratcliffe RG. Phosphorus nutrition and the intracellular if inorganic phosphate in pea root tips: a quantitative study using ^{31}P-NMR. *J. Exp. Bot.* 1983; 34:1222–1224.

196. Mitsumori F, Ito I. Phosphorus-31 nuclear magnetic resonance studies of photosynthesizing Chlorella. *FEBS Lett.* 1984; 174:248–252.

197. Bligny R, Foray M, Roby C, Douce R. Transport and phosphorylation of choline in higher plant cells. Phosphorus-31 nuclear magnetic resonance studies. *J. Biol. Chem.* 1989; 264:4888–4895.

198. Loughman BC, Ratcliffe RG, Southon TE. Observations on the cytoplasmic and vacuolar orthophosphate pools in leaf tissues using *in vivo* ^{31}P-NMR spectroscopy. *FEBS Lett.* 1989; 242:279–284.

199. Lundberg P, Weich RG, Jensen P, Vogel HG. Phosphorus-31 and nitrogen-14 NMR studies of the uptake of phosphorus and nitrogen compounds in the marine microalgae *Ulva lactuca. Plant Physiol.* 1989; 89:1380–1387.

200. Bligny R, Gardestrom P, Roby C, Douce R. ^{31}P NMR studies of spinach leaves and their chloroplasts. *J. Biol. Chem.* 1990; 256:1319–1326.

201. Mimura T, Dietz K-J, Kaiser W, Schramm M.J, Kaiser G, Heber U. Phosphate transport across biomembranes and cytosolic phosphate homeostasis in barley leaves. *Planta* 1990; 180:139–146.

202. Hentrich S, Hebeler M, Grimme LH, Leibfritz D, Mayer A. ^{31}P-NMR Saturation-transfer experiments in *Chlamydomonas reinhardtii* — evidence for the NMR visibility of chloroplastidic-Pi. *Eur. Biophys. J., Biophys. Lett.* 1993; 22:31–39.

203. Lee RB, Ratcliffe RG. Nuclear magnetic resonance studies of the location and function of plant nutrients *in vivo. Plant Soil* 1993; 155/156:45–55.

204. Usuda H, Shimogawara K. Phosphate deficiency in maize.1. Leaf phosphate status, growth, photosynthesis and carbon partitioning. *Plant Cell Physiol.* 1991; 32:497–504.

205. Wang C, Tillberg J-E. Effects of short-term phosphorus deficiency on carbohydrate storage in sink and source leaves of barley (*Hordeum vulgare*). *New Phytol.* 1997; 136:131–135.

206. De Groot CC, Marcelis LFM, Van Den Boogaard R, Lambers H. Growth and dry matter partitioning in tomato as affected by phosphorus nutrition and light. *Plant Cell Environ.* 2001; 24:1309–1317.

207. Cakmak I, Hengeler C, Marschner H. Partitioning of shoot and root dry matter and carbohydrates in bean plants suffering from phosphorus, potassium and magnesium deficiency. *J. Exp. Bot.* 994; 45:1245–1250.

208. Ciereszko I, Milosek I, Rychter AM. Assimilate distribution in bean plants (*Phaseolus vulgaris* L.) during phosphate limitation. *Acta Soc. Bot. Poloniae* 1999; 68:269–273.

209. Ciereszko I, Farrar JF, Rychter AM. Compartmentation and fluxes of sugars in roots of *Phaseolus vulgaris* under phosphate deficiency. *Biol. Plant* 1999; 42:223–231.

210. Ciereszko I, Zambrzycka A, Rychter A. Sucrose hydrolysis in bean roots (*Phaseolus vulgaris* L.) under phosphate deficiency. *Plant Sci.* 1998; 133:139–144.

211. Wanke M, Ciereszko I, Podbielkowska M, Rychter AM. Response to phosphate deficiency in bean (*Phaseolus vulgaris* L.) roots. Respiratory metabolism, sugar localization and changes in ultrastructure of bean root cells. *Ann. Bot.* 1998; 82:809–819.

212. Dietz KJ. Recovery of spinach leaves from sulfate and phosphate deficiency. *J. Plant Physiol.* 1989; 134:551–557.

213. Cogliatti DH, Clarkson DT. Physiological-changes and phosphate-uptake by potato plants during development of, and recovery from phosphate efficiency. *Physiol. Plant.* 1983; 58:287–294.

214. Drew MC, Saker LR, Barber SA, Jenkins W. Changes in the kinetics of phosphate and potassium absorption in nutrient-deficient barley roots measured by solution-depletion technique. *Planta* 1984; 1984:490–499.

215. Jungk A, Asher CJ, Edwards DG, Mayer D. Influence of phosphate status on phosphate uptake kinetics of maize (*Zea mays*) and soybean (*Glycine max*). *Plant Soil* 1990; 124:175–182.

216. McGraw JB, Wulf RD. The study of plant growth: a link between the physiological ecology and population biology of plants. *J. Theor. Biol.* 1983; 103:21–28.

217. Juszczuk I, Malusa E, Rychter AM. Oxidative stress during phosphate deficiency in roots of bean plants (*Phaseolus vulgaris* L.). *J. Plant Physiol.* 2001; 158:1299–1305.

218. Malusa E, Laurenti E, Juszczuk I, Ferrari RP, Rychter AM. Free radical production in roots of *Phaseolus vulgaris* subjected to phosphate deficiency stress. *Plant Physiol. Biochem.* 2002; 40:963–967.

219. Bray EA, Bailey-Serres J, Weretilnyk E. Responses to abiotic stresses. In: Buchanan BB, Gruissem W, Jones RL, eds. *Biochemistry and Molecular Biology of Plants.* Rockville, MD: American Society of Plant Physiology, 2000:1158–1203.

220. Abel S, Ticconi C, Delatorre CA. Phosphate sensing in higher plants. *Physiol. Plant.* 2002; 115:1–8.

221. Smith FW. The phosphate uptake mechanism. *Plant Soil* 2002; 245:105–114.

222. Mimura T. Homeostasis and transport of inorganic phosphate in plants. *Plant Cell Physiol.* 1995; 36:1–7.

223. Dietz K-J, Heilos L. Carbon metabolism in spinach leaves as affected by leaf age and phosphorus and sulfur metabolism. *Plant Physiol.* 1990; 93:1219–1225.

224. Duff SMG, Moorhead GBG, Lefebvre DD, Plaxton WC. Phosphate starvation inducible "bypasses" of adenylate and phosphate-dependent glycolytic enzymes in *Brassica nigra* suspension cells. *Plant Physiol.* 1989; 90:1275–1278.

225. Juszczuk IM, Rychter AM. Pyruvate accumulation during phosphate deficiency stress of bean roots. *Plant Physiol. Biochem.* 2002; 40:783–788.

226. Conroy JP, Milham PJ, Reed ML, Barlow EW. Increases in phosphorus requirements for CO_2-enriched pine species. *Plant Physiol.* 1990; 92:977–982.

227. Hurry V, Strand A, Furbank R, Stitt M. The role of inorganic phosphate in the development of freezing tolerance and the acclimatization of photosynthesis to low temperature is revealed by the pho mutants of *Arabidopsis thaliana. Plant J.* 2000; 24:383–396.

228. Varadarajan DK, Karthikeyan AS, Matilda PD, Raghothama KG. Phosphite, an analog of phosphate, suppresses the coordinated expression of genes under phosphate starvation. *Plant Physiol.* 2002; 129:1232–1240.

8 Inhibition or Inactivation of Higher-Plant Chloroplast Electron Transport

Rita Barr and Frederick L. Crane
Department of Biological Sciences, Purdue University

CONTENTS

I. INTRODUCTION

Photosynthesis takes place in a unique, highly organized organelle, the cholorplast of higher plants. Two photosystems, PS I and PS II, participate in light energy absorption, charge separation, water oxidation, and electron transport reactions, according to a basic "Z-scheme" [1] proposed half a century ago. When electron transport runs its course, reducing equivalents are produced in the form of NAD(P)H in PS I to be used by the Calvin cycle enzymes. On the other side of the equation, protons from PS II are released in the chloroplast lumen to be utilized by the chloroplast ATP synthase, CF_0F_1, to make ATP, an indispensable high-energy source for many different cell functions.

It is not the aim of this review to describe the details of the structural components of PS I or PS II and their function, for which numerous excellent reviews are available in the literature [2–13]. Likewise, we omit all mutant studies, which would require more space than a single chapter. However, we would like to provide a basic overview of various inhibitors and treatments used in chloroplast research in the last 10 to 15 years. Many of these inhibitors were discovered as long as 40 to 50 years ago, but their mode of action has been clarified only recently, as chloroplast structure has been refined on a molecular basis. For a review of earlier chloroplast electron transport inhibitors see Barr and Crane in the *Handbook of Photosynthesis*, first edition [14].

II. THE DONOR SIDE OF PS II

A. PS II REACTION CENTER AND WATER OXIDATION

The PS II core complex where charge separation and water oxidation take place [15–18] consists of up to 25 different integral membrane proteins and light-harvesting chlorophyll–protein complexes (CP43 [Psb C] and CP47 [Psb B]). It includes the reaction center polypeptides D1 (Psb A) and D2 (Psb D) and redox cofactors cytochrome c_{559} (Psb E and Psb F plus heme) and Psb I. The Psb O protein stretches across the surface of the reaction center with its N-terminal and C-terminal domains located toward CP47 and CP43, respectively [19].

The manganese cluster, in which water oxidation takes place, is ligated to the D1 protein and is stabilized by the extrinsic 33-kDa protein (Psb O). Two other extrinsic proteins, the 23-kDa (Psb P) and the 17-kDa proteins (Psb Q), are also involved. They also aid in retaining Cl^- and Ca^{2+} ions necessary for water oxidation.

When light strikes the chloroplast antenna light-harvesting chlorophylls, the light energy is passed to the special reaction center chlorophyll P680$^+$, where charge separation takes place, when P680$^+$ is aided by Y_2, a tyrosine residue on D1 polypeptide, which is the component that switches on the proton currents necessary for water oxidation [20]. After each of four successive charge separations, P680$^+$ abstracts one electron from the four manganese clusters by means of the redox-active tyrosine residue T_Z. Finally, the four positive charges accumulated in the manganese cluster oxidize two water molecules and release one oxygen molecule and four protons [21]. It is also possible for charges to recombine, but this is not the normal case because P680$^+$ oxidizes Y_Z in the 10^{-8} to 10^{-4} time range [22]. P680$^+$ transfers its electron to pheophytin, which, in turn, reduces a bound plastoquinone (PQ) Q_A, located on the D2 polypeptide in PS II. The electron from Q_A is transferred to another PQ molecule, Q_B, forming the plastosemiquinone $Q_{\tilde{B}}$. After another successive electron transfer from Q_A, Q_B is reduced to plastoquinol with the uptake of $2H^+$. The plastoquinol is then exchanged for another PQ from the PQ pool.

Treatments (Table 8.1) that inactivate water oxidation are popular subjects of study. The manganese cluster, located on the lumenal side of the PS II reaction center complex, is inactivated by treatments that remove Mn^{2+}, Ca^{2+}, or Cl^-. The most commonly used inhibitors of water oxidation are hydroxylamine or azide (Table 8.2).

The water oxidizing complex comprises five oxidation states, designated as S_0 to S_4. S_0 is the resting state in the dark. Each turnover in the reaction center of PS II advances the oxidation state from S_0 to S_4. Oxygen release occurs at the end of the cycle with the conversion of S_4 back to S_0 in the dark [38]. Recent studies have also focused on the size of the water cluster around the water-splitting enzyme [39–43].

B.   Cytochrome b_{559}

Cytochrome b_{559} is closely associated with the PS II reaction center [44]. It is not directly involved in the linear electron transport from PS II to PS I, but it may provide a cyclic electron pathway around PS II [45,46].

Cytochrome b_{559} consists of two small subunits: α (9 kDa) and β (4 kDa). The α subunit is the product of Psb E gene; the β subunit of Psb F gene. The heme of this cytochrome is located between the two subunits [43]. Cytochrome b_{599} has two different redox potentials: a low form (0 to 80 mV) or the high potential form (370 to 485 mV). In oxygen-evolving PS II membranes, an intermediate redox form of the

enzyme can also be detected [47,48]. The high-potential form can be converted to the low-potential form in presence of carbonyl cyanide-p-trifluoromethoxyphenyl hydrazone (FCCP), but it can be stabilized by ligation with calcium [49].

Two different functions for cytochrome b_{599} have been shown: it may provide a cyclic electron pathway around PS II [45,50] or it relieves photoinhibition under high-light conditions [51–57].

Alternatively, it may be bicarbonate that protects PS II against photoinhibition [57,58].

III.   THE ACCEPTOR SIDE OF PS II

A.   Bicarbonate

Bicarbonate is an essential component of PS II reaction centers. It facilitates electron transport through PS II [59,60–62]. Bicarbonate has two separate effects on PS II [61]: (1) on water oxidation where it binds to the manganese cluster [63,64] on the donor side of PS II and (2) on the iron–quinone site on the acceptor side of PS II between Q_A and Q_B [60,61,64]. The bicarbonate effect on water oxidation can be shown by replacement of bicarbonate with other ions, such as formate treatment of thylakoids or isolated PS II reaction centers [66,67].

Bicarbonate, which is required for PS II activity [66] on the acceptor side of PS II, binds to the nonheme iron located between Q_A and Q_B [68]. Other anions, such as formate, oxalate, glycolate, or glyoxylate, compete with bicarbonate for its binding site to the nonheme iron [69]. Bicarbonate may have a dual role at this site as a ligand for the nonheme iron and assisting in protonation of Q_B [18,70,71]. Mutants of *Chlamydomonas* are known, which have lost inhibition by formate [72]. Bicarbonate may protect the donor side of PS II against photoinhibition [73]. It may be required by the water-oxidizing complex for its assembly and stabilization through binding other components [74].

B.   Nonheme Iron

Nonheme iron is located between Q_A and Q_B sites in PS II. It may also serve as a ligand for bicarbonate [61,62], but it is not thought to be directly involved in the linear electron transport from PS II to PS I. However, when studied by electron paramagnetic resonance (EPR) spectroscopy, it gives a $g = 6$ signal, which correlates with Fe^{2+} oxidation by PQ or oxygen [18]. In absence of oxygen, the $g = 6$ EPR signal is inhibited. Yet, a high-spin EPR signal ($g = 1.6$) given by a nonheme iron in PS II of chloroplasts involves an interaction between semiquinones $Q_{\tilde{A}}$ and Q_B. The nonheme iron in PS II can be affected by inhibitors of the Q_B site,

TABLE 8.1
PS II Treatments Affecting Water Oxidation

Treatment	Site of Action	Plant Material	Conditions	Ref.
Mn^{2+} depletion	O_2 evolution	PS II particles from spinach	10 mM NaCl wash	23
Mn depletion	Water oxidation complex	PS II membranes from wheat seedlings	PS II particles incubated with 5 mM NH_2OH for 60 min in the dark	24
Removal of Ca^{2+}	$S_2 \to S_3$ transition in water oxidation	PS II membranes from spinach	Membranes suspended in 40 mM sucrose, 20 mM NaCl, and 20 mM citrate–NaOH at pH 3	25
Ca^{2+} depletion	Water oxidation complex	Spinach PS II particles	Particles incubated with 30 mM NaCl, 25 mM Mes, pH 6.5, and 50 μM EGTA	26
Sodium acetate	Inhibits O_2 evolution	Spinach PS II particles	Spinach BBY particles	27
Trichloroacetate	Releases extrinsic polypeptides 33, 23, and 17 kDA from PS II	Spinach chloroplasts	Chloroplasts incubated in dim light at 0°C for 30 min	28
Mn^{2+} depletion	Water oxidation complex	Spinach chloroplasts	Chloroplasts incubated with 5 mM hydroxylamine in darkness for 1 hr at 273 K	29
Mn^{2+} depletion	Water oxidation	Spinach BBY particles	0.8 M Tris, pH 8.5, under room light, 20°C	30
Mn^{2+} depletion	Inactivation of water oxidation	Thylakoid membranes from spinach	0.8 M Tris–HCl, pH 8, for 30 min under dim light	31
Ca^{2+} depletion	Extraction of Ca^{2+} from all S states without extracting Mn^{2+}	PS II core particles from pea seedlings	Low-pH citrate treatment	32
Formate treatment	Inhibits the S_2 state multiline signal	Spinach BBY particles	Incubation with 25 μM to 500 mM formate	33
Mn^{2+} depletion	Water oxidation complex	Spinach chloroplasts	Incubation with 800 mM Tris, pH8, at room light at 0°C for 30 min	34
Mn^{2+} depletion	O_2 evolving complex	PS II core complex from peas	0.8 M Tris buffer, pH 8.8	35
Mn^{2+} depletion	Water oxidation	Spinach PS II core complex	10 mM hydroxylamine	36

TABLE 8.2
Inhibition of Water Oxidation

Inhibitor	Site of Action	Plant Material	Conditions	Ref.
Hydroxylamine	Inactivates the S_2 state of water oxidation complex	Spinach chloroplasts	4–10 μM	29
Acetone hydrozone	Binds to water oxidation complex, followed by photoreversible reduction of Mn^{2+}; loss of $S_1 \rightarrow S_2$ transition due to extraction of Mn^{2+}	PS II membrane from spinach	1–2 mM	30
Antimycin A	Inhibits q_E	Thylakoid membranes	200 nM	31
Tetracyane ethylene	Inhibits O_2 evolution	PS II membrane fragments of chloroplasts	$C_{1/2} \approx 3\ \mu M$	32
Hydroxyurea (photooxidized)	HO aminoxy radical modifies Y_2^*	Incubation at 4°C for 2 hr in the dark	20 mM	33
Azidyl radical	Inhibits tyrosine Z photooxidation	Spinach chloroplasts		34
Hydroxylamine	Inactivation of O_2 evolution	Spinach chloroplast O_2-evolving membranes with hydroxylamine for 45 min in the dark at 4°C	3 mM	35
Trinitrophenol, promoxynil, dinoseb	Effects on $S_1 \rightarrow S_2$ state transition	Pea chloroplasts	Various concentrations	36
Azide	In the presence of chloride, azide suppressed the formation of the multiline and $g = 4.1$ EPR signals normally shown by the S_2 state	Spinach thylakoids	25 mM chloride plus 10 mM NaN_3	37
Azide	O_2 evolution	Spinach thylakoids	20 mM	37

such as 3-(3'4'-dichlorophenyl)-1, 1-dimethylurea (DCMU), 2-chloro-4-ethylamino-6-isopropylamino-5-triazine (atrazine), 2-(*tert*-butylamino-4-ethylamine)-6-methyl-thio-5-triazine, or 2-*sec*-butyl-4,6-dinitrophenol (dinoseb) [75,76].

The midpoint redox potential of the nonheme iron couple (Fe^{2+}/Fe^{3+}) at pH 7 is $+400\,mV$ with a pH difference of $60\,mV$ per pH unit from pH 6 to pH 8.5 [18]. This indicates that the reduction of the nonheme iron is associated with proton binding. Since electron transport may function normally in the absence of the nonheme iron, its function may be different from straight electron transfer. Carboxylate anions, such as glycolate, glyoxylate, or oxalate, can bind to the iron in the state $Q_A^-Fe^{2+}$, replacing bicarbonate [69,77].

The nonheme iron of PS II can also reversibly bind small molecules, such as nitric oxide (NO) [78,79] or sodium cyanide (CN) [80]. Addition of NO to spinach chloroplasts induces an EPR signal at $g = 4$. This signal is small in states $Q_A^-Q_B$, $Q_A^-Q_B^-$, and $Q_AQ_B^-$ but large in states Q_AQ_B and $Q_AQ_BH_2$ on the acceptor side of the $Fe^{2+}-NO$ adduct [78,79]. Competition experiments with CN^- and NO show that $50\,mM$ CN^- at pH 6.5 eliminates the EPR signal at $g = 4$, which arises from the $Fe^{2+}-NO$ adduct [81].

Several functions have been suggested for the nonheme iron in PS II [82]:

1. It maintains a favorable position or a favorable midpoint potential for the acceptor side of PS II.
2. It could also be involved in an oxidase function with access to the PQ pool via Q_B.
3. It could also act as a catalase, since it can react with hydrogen peroxide.

NO and CN can bind to the nonheme iron with various consequences to the PS II reaction sites between Q_A and Q_B (Table 8.3).

C. PLASTOQUINONE

PQ is closely associated with the PS II reaction center. It participates in the linear electron transport chain from water to NADP. It can act as an electron donor (Q_A) and acceptor (Q_B). There is also a mobile PQ pool in thylakoid membranes. After charge separation and water oxidation by the PS II reaction center after illumination, the primary electron donor, chlorophyll $P680^+$, transfers one electron to pheophytin, which reduces Q_A to its semiquinone form. After four successive accumulations of oxidizing equivalents from the water-oxidizing complex, one oxygen

molecule is created. Q_A^- can reduce the secondary PQ accepter Q_B, first to its semiquinone form and then to a quinol after a second charge accumulation. The quinol takes up two protons at the same time to generate the neutral form of the quinone, QH_2, which dissociates from the reaction center and is replaced by a quinone from the membrane quinone pool [75]. A nonheme iron facilitates the transfer from Q_A to Q_B. The electron transfer from Q_A to Q_B can be inhibited by urea-type inhibitors, such as DCMU [82–84].

Table 8.4 cites only a few recent publications where Q_B site inhibitors have been used, since there are too many references over the last 50 years to be cited individually.

The nonheme iron located between Q_A and Q_B has been studied by EPR spectroscopy in regard to photoinhibition [103], which leads to the degradation of the D1 protein in the PS II reaction center. According to chlorophyll fluorescence kinetics, the initial event during photinhibition is an overreduction of the quinone pool, which leaves the Q_B site inoperational. When the Q_B site is nonfunctional, Q_A shows a longer lifetime, thereby forming a semistable F_{oi} form, which leads to light-induced chlorophyll triplet formation. In the presence of oxygen, singlet oxygen species arise that are toxic to the chloroplast. In *Chlamydomonas* cells [104] step 1 leading to D1 degradation under photoinhibition is PQ overreduction, followed by irreversible modification of the D1 protein. The regular cleavage process of D1 is interrupted when the Q_B site is occupied by PQ, PQH_2, or diuron leading to D1, CP43, and CP47 protein degradation. The phenol-type inhibitor of the Q_B site, *N*-octyl-3-nitro-2,4,6-trihydroxybenzamide, prevents D1 degradation into 23- and 9-kDa fragments [95]. DCMU in the Q_B site also prevents D1 from degradation.

The Q_B site in PS II is also known as the herbicide binding site [105,106]. The amino acid sequence 211 to 275 on the D1 protein, encoded by the Psb A gene, provides the dimensions of the herbicide binding site. Only one herbicide molecule binds to the D1 protein, competing with the reversibly bound Q_B. This prevents the oxidation of the firmly bound Q_A on the D2 protein, which means that electron transport through PS II is interrupted. The various classes of herbicides that compete with PQ for the Q_B binding site include ^{14}C-azido atrazine [108] as a representative of the urea/triazine family of herbicides. DCMU or diuron is the most frequently used inhibitor of this group. Another herbicide group includes nitrophenols, azaphenanthrines, hydroxypyridines, and others. This is known as the phenol family of inhibitors [109]. The Q_B site is occupied by DCMU/triazine-type inhibitors.

TABLE 8.3
Inhibition of Nonheme Iron in PS II

Inhibitor	Site of Action	Plant Material	Conditions	Refs.
Nitric oxide	Inhibits electron transport between Q_A and Q_B (reversed by bicarbonate but not by formate)	Spinach chloroplast BBY particles	$30\,\mu M$	78
CN^-	Eliminates EPR signal at $g = 4$ from Fe^{2+}–NO adduct	Spinach chloroplast BBY particles	$50\,mM$ at pH 6.5	79
NaCN	Conversion of $g = 1.98$ to $g = 140$ EPR signal	BBY spinach preparation	30–$300\,mM$	80
Carboxylate ions (oxalate, glycolate, glyoxylate)	Various effects on the EPR signal from Fe^{2+}	BBY spinach particles	$40\,mM$	69,80

PQ also participates in the regulation of electron transport through the state transitions [62] to adjust electron flow between the two photosystems according to the available light. If electron carriers in PS II become more reduced, more excitation energy is transferred to PS I (state 2). In the case of the opposite situation, where electron carriers in PS II become more oxidized, excitation energy is transferred to PS II (state 1). Thus, the redox state of electron carriers in the electron transport chain determines the rate of electron transfer in the system.

A quinol binding site in the cytochrome b_6f complex has been implicated as a trigger for the state transitions [96]. The actual mechanism whereby the regulation of light energy distribution between the two photosystems is carried out by phosphorylation of the light-harvesting protein complexes is clear now. An overreduced PQ bound to the PS II reaction center can activate a thylakoid protein kinase, which catalyzes the phospharylation of light-harvesting complex II (LHC II). This increases the LHC II affinity for PS I. The phospho-LHC II can diffuse in the membrane to PS I, thus equalizing the energy distribution between the two photosystems [110]. Actually, at least two protein kinases with molecular masses of 53 and 66 kDa with different modes of action are known [111]. Other PS II peptides can also be phosphorylated (D2, CP43, and Psb H) in a redox-controlled manner [112]. Phosphorylation of the LHC II and PS II core complex proteins can be inactivated by exposure to high light intensities [111] *in vivo* in pumpkin and spinach leaf disks. This may be due to reduction of thiol groups in the LHC II kinase. All these proteins become phosphorylated at an N-terminal threonine residue exposed to the thylakoid surface [113].

PQ may be distributed differently between appressed or grana thylakoid membranes and nonappressed or stroma lamellae [114–116] according to different reduction rates in the light. The fast PQ pool in PS II reaction centers is reduced in 25 to 60 msec, while the slow pool reacts in 0.8 to 1 sec.

Recent studies also implicate the PQ pool in PS II as a nonphotochemical quencher of fluorescence [114]. 2,5-Dibromo-3-methyl-6-isopropyl benzoquinone (DBMIB), a well-known inhibitor of electron transport in chloroplasts, can suppress F_o fluorescence and retard the light-induced rise of F_v. It was also found to be an efficient energy quencher in PS II in the dark [116]. 5-Hydroxy-1,4-naphthoquinone can serve as a model for nonphotochemical fluorescence quenching in spinach thylakoids [117]. The PQ pool can also control chloroplast gene expression [118].

D. CYTOCHROME b_6f COMPLEX

The cytochrome b_6f complex transfers electrons from reduced PQ to a soluble electron carrier, plastocyanin or a c-type cytochrome, which then carries electrons to PS I. Electron transfer through the b_6f complex is accompanied by translocation of protons across the thylakoid membrane into the lumen to be used by the chloroplast ATP synthase.

The cytochrome b_6f complex is made of seven subunits: the Rieske iron–sulfur protein containing a 2F–2S cluster ($E_{\widetilde{m}}$ +300 to 370 mV), encoded by the pet C gene; a c-type cytochrome; cytochrome f, encoded by the pet A gene ($E_{\widetilde{m}}$ +300 to 370 mV); a b-type cytochrome; cytochrome b_6, encoded by the pet B gene, which comprises two b-hemes, defined by their midpoint potential, b_h ($E_{\widetilde{m}}$ −50 to −80 mV) and b ($E_{\widetilde{m}}$ −160 to −170 mV); and subunit IV (su IV), encoded by the pet D gene. The cytochrome b_6f complex binds PQ at the Q_0 site. Several small subunits have recently been identified for this complex: pet G, pet M, pet L gene products (for details see Refs. [9,119–121]).

TABLE 8.4
Inhibition of Chloroplast Electron Transport

Inhibitor	Site of Action	Plant Material	Conditions	Ref.
Azidoatrazine	Inhibits at the Q_B site	Spinach chloroplasts	0.59 μM	85
Stigmatellin	Inhibits at the reducing side of PS II as DCMU	Spinach chloroplasts	52.5 nM	86
2-(3-Chloro-4-trifluoromethyl)anilin-o-3,5-dinitrothiophene (ANT 2p)	Inhibits water oxidation	PS II particles from peas	0.5 μM	87
Hydroxylamine	Binding to the primary electron Q_A^-Fe	Spinach chloroplasts	10 μM	88
2,3,4-Trichloro-1-hydroxyxantha-quinone	Inhibits in D1 protein	Spinach chloroplasts	pI_{50} value 7.75	89
Aurachin C	Inhibits at the Q_B site	Spinach chloroplasts	pI_{50} value 7.2 μM	90
Various phloroglucinol derivatives	Inhibit PS II electron transport like DCMU and atrazine, but some derivatives could act as phenol-type inhibitors	Chloroplasts from *Brassica napus*	Various concentrations	91
Derivatives of 5-propionyl-3-[1-(3,4-dichlorobenzyl)amino-propylidene]-4-hydroxy-2H-pyron-2,6(3H)-dione (PT 72)	Inhibit PS II electron transport like phenylureas	Spinach chloroplasts	Various concentrations	92
Acridones, xanthones, quinones	Inhibit at the Q_B site	Spinach and Chlamydomonas chloroplasts	Various concentrations	93
4-Hydroxypyridines	Upon halogination 4-hydroxypyridines changed their mode of action from PQ pool inhibitors to phenol-type inhibitors	Spinach thylakoids	Various concentrations	94
DCMU	Inhibits electron transport between Q_A and Q_B	Spinach thylakoids	10^{-5} M	82
Azide or azidyl radical	Inhibits between Y_z and Q_A	Spinach chloroplast membranes	3 mM	34
PNO 8	Inactivates O_2 evolution when bound to Q_B site and degrades D1 into 23- and 9-kDa fragments	Spinach thylakoid membranes	10 μM	95
DCMU	Inhibits electron transport in PS II at Q_B site	Spinach thylakoid	20 μM	83
DBMIB	Acts at the Q_0 site	Spinach thylakoids	3-18 μM	96
O-Phenanthrolene, atrazine	Prevents light-induced oxidation of PS II Fe when bound at the Q_B site	Spinach BBY particles	2.5 mM 30 μM	97
Tricolorin A	Inhibits electron transport between Q_A and Q_B	Spinach chloroplasts	10 μM	98
Trinitrophenol (TNP), 4-hydroxy-3,5-dibromobenzonitrile, dinoseb	Inhibit at the Q_B site (also influence S_1 and S_2 state transitions)	Chloroplasts and maize leaves	I_{50} 175–225 nM for TNP	39
Heterocyclic orthoquinones	Inhibit Q_B site on D1 protein	Spinach chloroplasts	pI_{50} values from 5.19 to 7.51	99
Various quinolones	Inhibit electron transport at the herbicide binding site, Q_B, shown by displacement of [^{14}C]atrazine	Spinach chloroplasts	Various concentrations	100
2-(4-Promobenzyl-amino)-4-methyl-6-trifluromethyl-pyrimidine	Inhibits electron transport at Q_B site	Spinach chloroplasts	Various concentrations	101
Tetraphenylboron (TPB) plus DCMU	Minimize the presence of Q_A^-	Spinach chloroplast incubated in the dark with TPB and DCMU for 15 min	25 μM TPB, 5 μM DCMU	22
Phenolic inhibitors (TNP, ioxymil, dinoseb)	$S_2Q_A^-$ state is tenfold less stable when phenolic inhibitors bind to Q_B site	Spinach chloroplasts	Various concentrations	102

The cytochrome $b_6 f$ complex in highly active state has been purified from spinach [122]. The best-known quinone-type inhibitor of the Q_0 site is DBMIB, which inhibits quinone oxidation. This site is also affected by scores of other inhibitors, including 2-iodo-2',4,4'-trinitro-3-methyl-6-isopropyldiphenyl ether (DNP-INT), 4-hydroxyquinoline N-oxide (HQNO), stigmatellin, aurachins C and D, quinolones, 5n-undecyl-6-hydroxy-4,7-dioxobenzothiazole (UHDBT), E-β-methoxyacrilate-stilbene (MOA-stilbene), and heterocyclic and tertiary amines (Table 8.5).

The cytochrome $b_6 f$ complex is also involved in cyclic electron transfer around PS I. The same electron transport inhibitors as mentioned in Table 8.5 also inhibit cyclic electron transport around PS I.

E. PLASTOCYANIN

Plastocyanin, a 10-kDa copper-containing mobile protein, couples electron transfer from PS II to PS I [136–138]. It is located in the thylakoid lumen and transfers electrons between the reduced cytochrome of the $b_6 f$ complex and the photooxidized chlorophyll special pair P700$^+$ of PS I [119,139–141].

The atomic structure of plastocyanin is described as a β-barrel with hydrophilic residues in the interim of the protein [136,137]. Plastocyanin shows two conserved surface regions, the so-called "eastern" and "northern" protein patches. The eastern patch is a negatively charged region, which participates in electrostatic interactions with its electron transfer partners. The northern patch is hydrophobic and is involved in electron transfer through the copper-bound His^{86}. Both electrostatic and hydrophobic interactions are involved in electron transfer between plastocyanin and PS I [138]. After being reduced by cytochrome c of the $b_6 f$ complex, plastocyanin docks in PS I and reduces P700. The two highly conserved negative patches are essential for electron transfer from cytochrome f to plastocyanin and from plastocyanin to PS I. The hydrophobic flat "north" surface of plastochanin close to His^{87} is essential [137,138].

Plastocyanin binds to a small cavity on the lumenal side of PS I with a slight bias toward the Psa L subunit complex [140,141].

Plastocyanin can be replaced by cytochrome c_6 found in *Arabidopsis* [142]. Higher plants also contain a modified cytochrome c_6 [143].

The plastocyanin molecule can be modified by treatment with ethylenediamine plus carbodiimide with replacement of the negatively charged carboxylate group with the positively charged amino group [145], with the result of inhibiting cytochrome f oxidation.

The plastocyanin pool size in several soybean cultivars varied considerably between 0.1 and 1.3 mol plastocyanin (mol/PS II) [146]. Such variations could influence the photosynthetic capacity of these plants.

IV. PHOTOSYSTEM I

PS I of higher plants is found at the edges of the grama stacks and the stroma lamellae of thylakoid membranes [146]. It consists of 11 to 17 polypeptide subunits with cofactors including about 90 chlorophyll a and b, 10 to 15 β-carotene, 2 phylloquinone molecules, and 3 iron–sulfur centers [147–152]. The molecular structure of PS I has been described in detail [153–156].

The major subunits of PS I are two ≈80-kDa proteins (PS I-A and PS I-B). They bind most of the pigments and members of the electron transport pathway, but the 9-kDa (PS I-C) subunit carries the iron–sulfur centers (4Fe–4S) and some members of the electron transport chain. Polypeptides PS I-D, - E, and −H help maintain the functional integrity of PS I on the lumenal side. PS I also carries four light-harvesting chlorophyll a/b binding proteins. The PS I pigment–protein complex functions as a plastocyanin:ferredoxin oxidoreductase [157].

The electron transfer components of PS I are P700, the reaction center chlorophyll as a dimer, the primary electron acceptor A_0, which is also a chlorophyll molecule, the secondary acceptor A, a phylloquinone molecule [158,159], and the iron–sulfur centers F_x, F_B, and F_A. There are six chlorophyll a molecules, two phylloquinones, and three Fe_4S_4 clusters associated with the PS I reaction center [159].

Light harvesting in PS I is accomplished by four LHC I polypeptides. The genes for encoding the different PS I polypeptides are summarized (14/76). The light-harvesting proteins of PS I from different plant species are described [160] and also energy transfer in PS I [161].

Under illumination with wavelengths shorter than 700 nm, PS I performs a transmembrane electron transfer from the primary electron donor, P700$^+$, through a chain of intermediate electron acceptors to the 4Fe–4S clusters named F_A and F_B [162]. $(F^A F^B)^-$ is a strong reductant (midpoint redox potential ≈−540 mV), which donates its electron to NADP$^+$ via ferredoxin located on the stromal side of the membrane [148]. In the meantime, P700$^+$ (E_m ≈ 490 mV) receives an electron from PS II by way of the cytochrome $b_6 f$ complex and mobile plastocyanin [136].

P700 is a dimer of chlorophyll a, which acts as an electron donor to another chlorophyll a molecule, A_0 [148]. The secondary electron acceptor A_1, is a phylloquinone or vitamin K_1, which has been extracted

TABLE 8.5
Inhibition of the Cytochrome b_6f Complex

Inhibitor	Site of Action	Plant Material	Conditions	Ref.
DBMIB	Inhibits plastoquinol–cytochrome c_{552} oxidoreductase	Spinach chloroplasts used for isolation of the cytochrome b_6f complex	$pI_{50} = 7.6$	123
DNP-INT	Inhibits plant quinone–plastocyanin oxidoreductase	Spinach chloroplasts used for isolation of the cytochrome b_6f complex	$10\,\mu M$	124
Stigmatellin	Inhibits cytochrome b_6f complex (same as DBMIB)	Spinach chloroplasts	$I_{50}\ 59.0\,nM$	86
Stigmatellin	Inhibits plastocyanin oxidoreductase	Isolated b_6f complex	Between 10^{-8} and $10^{-7}\,M$	86
Stigmatellin	Inhibits at the same site as DBMIB	Spinach chloroplasts	$I_{50}\ 59.0\,nM$	124
DNP-INT	Inhibits Rieske iron–sulfur centers in b_6f complex	Isolated b_6f complex from spinach chloroplasts	$5–10\,nM$	125
Stigmatellin	Inhibits Rieske iron–sulfur centers in b_6f complex	Isolated b_6f complex from spinach chloroplasts	$5–10\,nM$	125
Halogenated 1,4-benzoquinones	Bind to Rieske iron–sulfur proteins and to cytochrome f in b_6f complex	Spinach chloroplasts	Various concentrations	126
HQNO	Inhibits quinone reductase site on stroma side	Spinach thylakoids	$1\,\mu M$	127
DBMIB	Inhibits reduction of cytochrome b_6	Spinach thylakoids	$I_{50} = 80\,nM$	127
Stigmatellin	Inhibits cytochrome b_6f complex (Rieske Fe–S centers affected; reduction potential changed from 326 to 460 mv in cytochrome f by quinone)	Cytochrome b_6f complex	$20\,\mu M$	128
Aurachin C	b_6f Complex	Isolated cytochrome b_6f complex	$pI_{50} = 7\,\mu M$	129
Aurachin D	b_6f Complex	Isolated cytochrome b_6f complex	$pI_{50} = 7.49\,\mu M$	129
DBMIB	Inhibits electron transport through b_6f complex	Pea chloroplasts	$0.5\,\mu M$	130
MOA-stilbene	Inhibits cytochrome b_6f complex	Pisum sativum chloroplasts	$40\,\mu M$	131
Cu^{2+}	Inhibits cytochrome f in the b_6f complex	Thylakoids or isolated b_6f complex	$0.3–5\,\mu M$	132
DBMIB (reduced)	Binds to Q_0 site	Purified b_6f complex	$15\,\mu M$	133
DBMIB	Modified cytochrome b_6 at positions D148, A154, and S159	Less sensitivity to DBMIB in mutants A154G and S159A		134
Quinolones or acridon	Cytochrome b_6f complex	Spinach thylakoids	I_{50} values given for 12 different derivatives	135

from spinach chloroplasts with diethyl ether [163]. Reconstitution with phylloquinone and other substituted naphthoquinones has also been shown [164]. The existence of two quinone molecules Q_K and Q_{K1} has been verified on an electron density map [159]. The next members of the PS I electron transport chain are three 4Fe–4S clusters, F_A, F_B, and F_X [164]. Treatment of spinach chloroplasts [166], Synechoeoecus [167], Synechocystis [168], Chlamydomonas, and other mutant cells [169] destroys the F_B cluster and inactivates electron transfer to ferredoxin and, hence, photoreduction of $NADP^+$. These studies and others propose that the sequence of the iron–sulfur clusters is as follows: $F_X \rightarrow F_A \rightarrow F_B \rightarrow$ Fd. Other investigators [149,170] advocate a split pathway of electron transport through PS I. Electrons can be diverted from $NADP^+$ by spraying *Erigeron canadensis* biotypes *in vivo* with paraquat, with production of toxic oxygen species [171].

It has recently been shown that PS I can be destroyed by photoinhibition. In *Cucumis sativus* L. leaves, for example, exposed to low-light intensity and 4°C temperature for 5 hr, the quantum yield of PS I decreased to 20–30% of untreated control leaves due to destruction of P700 [172].

Isolated chloroplast PS I can also be photoinhibited, as shown [150,173] with inactivation of the iron–sulfur clusters first on the acceptor side, leading to later destruction of the reaction center and degradation of the Psa B gene product. After 4 hr of exposure to photoinhibitory light, spinach PS I formed oligomers containing CP1, LHC I-680, and LHC-730 [174]. Photoinhibition in PS I has also been observed in the common bean [175] or pumpkin [176].

PS I polypeptides, carotenoids, and lipids have been characterized by their antisera [177].

A. Cyclic Electron Transport in PS I

Cyclic electron transport in higher-plant chloroplasts utilizes the same electron carriers of PS I and the cytochrome b_6f complex as the linear electron transport system from PS II to PS I. In contrast to the linear electron transport, which produces both ATP and $NADP^+$ when both photosystems are involved, cyclic electron transport by PS I provides only ATP. The stoichiometry between the two photosystems has to be poised for less efficiency by PS II, so that the PS I cyclic system can predominate [178,179]. The cycle starts with reduced ferredoxin and ferredoxin–PQ reducatase. This enzyme can be inhibited by tetrabromo-4-hydroxypyridine, DBMIB, dimaleimide, and heparin [180]. Alternatively, ferredoxin–$NADP^+$ reductase (FNR) may be involved in the PS I cyclic electron transfer [178,180–182]. These two pathways

may be parallel [183]. FNR has been shown to be a 35-kDa subunit of the cytochrome b_6f complex, located on the stromal side of the thylakoid membrane [184].

The cyclic PS I pathway is sensitive to antimycin A inhibition [181,182,185]. It is also impaired in tobacco chloroplasts by disruption of the *ndhB* gene [186,187]. In barley leaves FNR was found to be associated with the chloroplast pyridine nucleotide dehydrogenase complex as shown by antibodies against barley FNR [188].

In studies with extremely high CO_2-tolerant green microalgae, growth under 40% CO_2 in the presence of DCMU showed a higher relative quantum yield of PS I, suggesting an increase in cyclic electron transport around PS I [189].

Cyclic PS I electron transport supports a ΔpH gradient across the thylakoid membranes used for the synthesis of ATP [190]. The calculated rate of PS I-dependent proton transport was found to be 220 μmol protons/mg chlorophyll/h in intact spinach chloroplasts [191], but an active Mehler peroxidase can prevent cyclic electron transport in the presence of oxygen [192]. Cyclic electron transport is known to regulate the quantum yield of PS II by decreasing the intrathylakoid pH, when availability of electron carriers in PS I is limited, as under stress conditions [190]. Downregulation of PS II as a result of the pH gradient generated by cyclic electron transport around PS I also protects PS II against photoinhibition [193].

B. Ferredoxin and FNR

Ferredoxin is the terminal electron acceptor in the linear electron transfer chain from PS II to PS I. It reduces $NADP^+$ to NADPH in a one-electron transfer reaction.

Ferredoxin is a water-soluble protein (11 kDa) found on the stroma side of thylakoid membranes [194,195]. Psa L, Psa D, and Psa E subunits of PS I are mainly required for ferredoxin docking [196–201]. Arginine 39 of the Psa E subunit provides a positive charge for interaction with ferredoxin [202]. From Fourier difference analysis it is seen that ferredoxin is bound on top of the stromal ridge principally interacting with the extrinsic PS I subunits Psa C and Psa E [201].

Ferredoxin reduces $NADP^+$ via the flavo enzyme FNR, which is a 37-kDa protein in spinach chloroplasts. The structural aspects of FNR are found in Refs. [198,199]. Spectral and kinetic studies reveal the existence of several PS I–ferredoxin complexes [200]. Mung bean seedlings also show two isoforms of FNR [203].

Electron flow from NADPH to ferredoxin can also support NO_2 reduction [204].

Ferredoxion–NADP$^+$ oxidoreductase has at least three different locations in chloroplasts: (1) it is associated with PS I on the stromal side where it reduces NADP$^+$ [205], (2) it is associated with the cytochrome $b_6 f$ complex as a 35-kDa protein complex [184], and (3) in barley leaves it is associated with chloroplastic pyridine nucleotide dehydrogenase complex [188]. FNR is a flavoprotein with multiple functions, including a reverse reaction as follows: 2 FdFe^{2+} + NADP$^+$ + H$^+$ ⇆ 2 FdFe^{3+} + NADPH. The plant-type FNR has a multiplicity of functions [206].

Spinach FNR shows three binding sites for substrates: NADP(H), Fd-cytochrome e, quinone/2,6-dichlorophenol indophenol (DCIP) [207]. A specific inhibitor for FNR is disulfodisalicylidenepropane-1,2-diamine as well as maleimides [208].

REFERENCES

1. Hill R, Bendall F. Function of two cytochrome components in chloroplasts: a working hypothesis. *Nature* 1960; 1186:136–137.
2. Debus RG. The manganese and calcium ions of photosynthetic oxygen evolution. *Biochim. Biophys. Acta* 1992; 1102:269–352.
3. Diner BA, Babcock GT. Structure, dynamics, and energy conversion efficiency in photosytem II. In: Ort D, Yocum C, eds. *Oxygenic Photosynthesis: The Light Reactions.* Dordrecht: Kluwer Academic Publishers, 1996:213–247.
4. Hankamer B, Barber G, Boekema AJ. Structure and membrane organization of photosytem II in green plants. *Annu. Rev. Plant Physiol. Plant Mol. Biol.* 1997; 48:641–671.
5. Barber J, Nield J, Morris EP, Zheleva D, Hankamer B. The structure, function and dynamics of photosytem II. *Physiol. Plant.* 1997; 100:817–827.
6. Barber J. Photosystem II. *Biochim. Biophys. Acta* 1998; 1365:269–277.
7. He WZ, Malkin R. Photosystems I and II. In: Raghavendra AS, ed. *Photosynthesis: A Comprehensive Treatise.* Cambridge: Cambridge University Press, 1998:29–43.
8. Whitmarsh J. Electron transport and energy transduction. In: Raghavendra AS, ed. *Photosynthesis: A Comprehensive Treatise.* Cambridge: Cambridge University Press, 1998:87–107.
9. Wollman FA, Minai L, Nechushtai R. The biogenesis and assembly of photosynthetic proteins in thylakoid membranes. *Biochim. Biophys. Acta* 1999; 1411:21–85.
10. Debus RJ. The polypeptides of photosystem II and their influence on manganotyrosyl-based oxygen evolution. In: Sigel A, Sigel H, eds. *Metal Ions in Biological Systems.* New York: Marcel Dekker, 2000:657–711.
11. Dekker JP, van Grondelle R. Primary charge separation in photosystem II. *Photosynth. Res.* 2000; 63:195–208.
12. Britt RD, Peloquin JM, Campbell KA. Pulsed and parallel-polarization EPR characterization of the photosystem II oxygen-evolving complex. *Annu. Rev. Biophys. Biomol. Struct.* 2000; 29:463–495.
13. Rhee K-H. Photosystem II: the solid structural era. *Annu. Rev. Biophys. Biomol. Struct.* 2001; 30:307–328.
14. Barr r, Crane FL. Chloroplast electron transport inhibitors. In: Pessarakli M, ed. *Handbook of Photosynthesis.* New York: Marcel Dekker, 1997:95–112.
15. Debus RJ. Amino acid residues that modulate the properties of tyrosine Y_z and the manganese cluster in the water oxidizing complex of photosystem II. *Biochim. Biophys. Acta* 2001; 1503:164–186.
16. Diner BA. Amino acid residues involved in the coordination and assembly of the manganese cluster of photosystem II. Proton-coupled electron transport of the redox-active tyrosine and its relationship to water oxidation. *Biochim. Biophys. Acta* 2001; 1503:147–163.
17. Diner BA, Rappaport F. Structure, dynamics and energetics of the primary photochemistry of photosystem II of oxygenic photosynthesis. *Annu. Rev. Plant Biol.* 2002; 53:551–580.
18. Nugent JHA, Rich AM, Evans MCW. Photosynthetic water oxidation: towards a mechanism. *Biochim. Biophys. Acta* 2001; 1503:138–146.
19. Nield J, Balsera M, Las Rivas JD, Barber J. Three-dimensional electron cryo-microscopy study of the extrinsic domain of the oxygen-evolving complex of spinach. *J. Biol. Chem.* 2002; 277:15006–15012.
20. Tommos C, Babcock GT. Proton and hydrogen currents in photosynthetic water oxidation. *Biochim. Biophys. Acta* 2000; 1458:199–219.
21. Zouni A, Witt KT, Kern J, Fromme P, Krauss N, Saenger W, Orth P. Crystal structure of photosystem II from Synechoeoccus elongates at 3.8 A resolution. *Nature* 2001; 409:739–743.
22. de Wijn R, van Gorkom HG. The role of charge recombination in photosystem II. *Biochim. Biophys. Acta* 2002; 1553:302–308.
23. Miyao M, Murata N. Role of the 33-kDa polypeptide in preserving Mn in the photosynthetic oxygen-evolution system and its replacement by chloride ions. *FEBS Lett.* 1984; 170:350–354.
24. Tamura N, Cheniae G. Photoactivation of the water-oxidizing complex in photosystem II membranes depleted of Mn and extrinsic proteins. I. Biochemical and kinetic characterization. *Biochim. Biophys. Acta* 1987; 890:179–194.
25. Ono T, Inoue Y. Removal of Ca by pH 3 treatment inhibits S_2 to S_1 transition in photosynthetic oxygen evolving photosystem II. *Biochim. Biophys. Acta* 1989; 973:443–449.
26. Boussac A, Zimmermann J-L, Rutherford AW. Factors influencing the formation of modified S_2 EPR signal and the S_3 EPR signal in Ca^{2+}-depleted photosystem II. *FEBS Lett.* 1990; 277:69–74.
27. Maclachlan DJ, Nugent JHA. Investigation of the S_3 electron paramagnetic resonance signal from the oxygen-evolving complex of photosystem II: effect of in-

hibition of oxygen evolution by acetate. *Biochemistry* 1993; 32:9772–9780.

28. Xu C, Li R, Shen Y, Govindjee. The sequential release of three extrinsic polypeptides in PS II particles by high concentrations of trichloroacetate. *Naturwissenschaften* 1995; 82:477–478.

29. Sivaraja M, Dismukes GC. Binding of hydroxylamine to the water-oxidizing complex and the ferroquinone electron acceptor of spinach photosystem II. *Biochemistry* 1988; 27:3467–3475.

30. Tso J, Petrouleas V, Dismukes GC. A new mechanism-based inhibitor of photosynthetic water oxidation: acetone hydrazone. 1. Equilibrium reactions. *Biochemistry* 1990; 29:7759–7767.

31. Mano J. Ushimaru T, Asada K. Ascorbate in thylakoid lumen as an endogeneous electron donor to photosystem II: protection of thylakoids from photoinhibition and regeneration of ascorbate in stroma by dehydroascorbate reductase. *Photosynth. Res.* 1997; 53:197–204.

32. Haumann M, Junge W. Evidence for impaired hydrogen-bonding of tyrosine Y_z in calcium depleted photosystem II. *Biochim. Biophys. Acta* 1999; 1411:121–133.

33. Kawamoto K, Chen G-X, Mano J, Asada K. Photoinactivation of photosystem II by *in situ*-photoproduced hydroxyurea radicals. *Biochemistry* 1994; 334:10487–10493.

34. Kawamoto K, Mano J, Asada K. Photoproduction of the azidyl radical from the azide anion on the oxidizing side of photosystem II and suppression of photooxidation of tyrosine Z by the azidyl radical. *Plant Cell Physiol.* 1995; 36:1121–1129.

35. Mino H, Kawamori A, Ono T-A. pH dependent characteristics of Y_2 radical in Ca^{2+}-depleted photosystem II studied by CW-EPR and pulsed ENDOR. *Biochim. Biophys. Acta* 2000; 1457:157–165.

36. Ahlbrink R, Semin BK, Mulkidjanian AY, Junge W. Photosystem II of peas: effects of added divalent cations Mn, Fe, Mg and Ca on two kinetic components of P_{680}^{+4} reduction in Mn-depleted core particles. *Biochim. Biophys. Acta* 2001; 1506:117–126.

37. Haddy A, Kimel RA, Thomas R. Effects of azide on the S_2 state EPR signals from photosystem II. *Photosynth. Res.* 2000; 63:35–45.

38. Evans MCW, Gourovskaya K, Nugent JHA. Investigation of the interaction of the water oxidizing manganese complex of photosystem II with the aqueous solvent environment. *FEBS Lett.* 1999; 450:285–288.

39. Roberts A, Townsend JS, Kramer DM. Evidence that phenolic inhibitors of Q_B have long-range effects on the S-state transitions. In: Garab G, ed. *Photosynthesis: Mechanisms and Effects.* Vol. V. Dordrecht: Kluwer Academic Publishers, 1998:3889–3892.

40. Burda K, Bader KP, Schmidt CH. An estimation of the size of the water cluster present at the cleavage site of the water splitting enzyme. *FEBS Lett.* 2001; 491:81–84.

41. Burda K, Schmid GH. Heterogeneity of the mechanism of water splitting in photosystem II. *Biochim. Biophys. Acta* 2001; 1506:47–54.

42. Burda K, Bader KP, Schmid GH. ^{18}O isotope effect in the photosynthetic water splitting process. *Biochim. Biophys. Acta* 2003; 1557:77–82.

43. Foyer CH, Noctor G. Oxygen processing in photosynthesis: regulation and signaling. *New Phytol.* 2000; 146:359–388.

44. Stewart DH, Brudvig GW. Cytochrome B_{559} of photosystem II. *Biochim. Biophys. Acta* 1998; 1367:63–87.

45. Miyake C, Schreiber V, Asada K. Ferredoxin-dependent and antimycin A-sensitive reduction of cytochrome b-559 by far red light in maize thylakoids; participation of a menadiol-reducible cytochrome b-559 in cyclic electron flow. *Plant Cell Physiol.* 1995; 36:743–748.

46. Heimann S, Schreiber V. Cyt b-559 (Fd) participating in cyclic electron transport in spinach chloroplasts: evidence for kinetic connection with the cyt b_6f complex. *Plant Cell Physiol.* 1999; 40:818–824.

47. Thompson LK, Miller A-F, Buser CA, de Paula JC, Brudnig GW. Characterization of the multiple forms of cytochrome b_{559} in photosystem II. *Biochemistry* 1989; 28:8048–8056.

48. Iwasaki I, Tamura N, Okayama S. Effects of light stress on redox potential forms of cyt b-559 in photosystem II membranes depleted of water-oxidizing complex. *Plant Cell Physiol.* 1995; 36:583–589.

49. McNamara VP, Gounaris K. Granal photosystem II complexes contain only the high redox potential form of cytochrome b-559 which is stabilized by ligation of calcium. *Biochim. Biophys. Acta* 1995; 1231:289–296.

50. Falkowski, PG, Fujita Y, Ley A, Mauzerall D. Evidence for cyclic electron flow around photosystem II in Chlorella pyrenoidosa. *Plant Cell Physiol.* 1986; 81:310–312.

51. Thompson LK, Brudvig GW. Cytochrome b-559 may function to protect photosystem II from photoinhibition. *Biochemistry* 1988; 27:6653–6658.

52. Barber J, De Las Rivas J. A functional model for the role of cytochrome b-559 in the photoprotection against donor and acceptor side photoinhibition. *Proc. Natl. Acad. Sci. USA* 1993; 90:10942–10946.

53. Whitmarsh J, Samson G, Poulson M. Photoprotection in photosystem II-the role of cytochrome b559. In: Baker NR, Boyer JR, eds. *Photoinhibition of Photosynthesis: From Molecular Mechanisms to the Field.* Oxford: BIOS Sci Publ, 1994:75–93.

54. Poulson M, Samson G, Whitmarsh J. Evidence that cytochrome b-559 protects photosystem II against photoinhibition. *Biochemistry* 1995; 34:10932–10938.

55. Magnuson A, Rova M, Mamedov F, Fredriksson PO, Styring S. The role of cytochrome b559 and tyrosine D in protection against photoinhibition during *in vivo* photoactivation of photosystem II. *Biochim. Biophys. Acta* 1999; 1411:180–191.

56. Whitmarsh J, Pakrasi HB. Form and function of cytochrome b_{559}. In: Ort DR, Yocum CF, eds. *Oxygenic Photosynthesis: The Light Reactions.* Amsterdam: Kluwer Academic Publishers, 1996:249–264.

57. Klimov VV, Baranov SV, Allakhverdiev SI. Bicarbonate protects the donor side of photosystem II against photoinhibition and thermoinactivation. *FEBS Lett.* 1997; 418:243–246.
58. Sundby C, Mattson M, Schidt T. Effects of bicarbonate and oxygen concentration on photoinhibition of thylakoid membranes. *Photosynth. Res.* 1992; 34:263–270.
59. van Rensen JJS. Role of bicarbonate at the acceptor side of photosystem II. *Photosynth. Res.* 2002; 73:185–192.
60. van Rensen JJS, Xu C, Govindjee. Role of bicarbonate in photosystem II, the water-plastoquinone oxidoreductase of plant photosynthesis. *Physiol. Plant.* 1999; 105:585–592.
61. Nixon PJ, Mullineaux CW. Regulation of photosynthetic electron transport. In: Aro E-M, Andersson BA, eds. *Regulation of Photosynthesis. Advances in Photosynthesis and Respiration* Vol. 11. Dordrecht: Kluwer Academic Publishers, 2001:533–555.
62. Blubaugh DJ, Govindjee. The molecular mechanism of the bicarbonate effect at the plastoquinone reductase site of photosynthesis. *Photosynth. Res.* 1988; 19:85–128.
63. Yruela I, Allakhverdiev SI, Ibara JV, Klimov VV. Bicarbonate binding to the water-oxidizing complex in the photosystem II. A Fourier transform infrared spectroscopy study. *FEBS Lett.* 1998; 425:396–400.
64. Klimov VV, Baranov SV. Bicarbonate requirement for the water-oxidizing complex of photosystem II. *Biochim. Biophys. Acta* 2001; 1503:187–196.
65. Easton-Rye JJ, Govindjee. Electron transfer through the quinone acceptor complex of photosystem II after one or two actinic flashes in bicarbonate-depleted spinach thylakoid membranes. *Biochim. Biophys. Acta* 1988; 935:237–257.
66. Feyziev YM, Yoneda D, Yosin T, Katsuta N, Kawamori A, Wanatable Y. Formate-induced inhibition of the water-oxidizing complex of photosystem II studied by EPR. *Biochemistry* 2000; 39:3848–3855.
67. Govindjee, Xu C, van Rensen JJS. On the requirement of bound bicarbonate for photosystem II activity. *Z. Naturforsch.* 1997; 52c:24–32.
68. Diner BA, Petrouleas V. Formation of NO of nitrosyl adducts of redox components of the photosystem II reaction center. II. Evidence that HCO_3^-/CO_2 binds to the acceptor-side non-heme iron. *Biochim. Biophys. Acta* 1990; 1015:141–149.
69. Petrouleas V, Deligiannakis Y, Diner BA. Binding of carboxylate ions at the non-heme Fe(II) of PS II. *Biochim. Biophys. Acta* 1994; 1188:271–277.
70. Whitmarsh J, Govindjee. The photosynthetic process. In: Singhal GS, Renger G, Sopory SK, Irrgang K-D, Govindjee, eds. *Concepts in Photobiology. Photosynthesis and Photomorphogenesis.* New Delhi: Narosa Publishing House, 1999:11–51.
71. Hienerwadel R, Berthomiew C. Bicarbonate binding to the non-heme iron of photosystem II investigated by Fourier transform infrared difference spectroscopy

and ^{13}C-labeled bicarbonate. *Biochemistry* 1995; 34:16288–16297.
72. Xiong J, Minagawa J, Crofts A, Govindjee. Loss of inhibition by formate in newly constructed photosystem II D1 mutants, D1-R257E and D1-R257M, of *Chlamydomonas reinhardii. Biochim. Biophys. Acta* 1998; 1365:473–491.
73. Klimov VV, Baranov SV, Allakhverdiev SI. Bicarbonate protects the donor side of photosystem II against photoinhibition and thermoinactivation. *FEBS Lett.* 1997; 418:243–246.
74. Klimov VV, Baranov SV. Bicarbonate requirement for the water-oxidizing complex of photosystem II. *Biochim. Biophys. Acta* 2001; 1503:187–196.
75. Diner BA, Petrouleas V. Q_{400}, the non-heme iron of the photosystem II iron-quinone complex. A spectroscopic probe of quinone and inhibitor binding to the reaction center. *Biochim. Biophys. Acta* 1987; 895:107–125.
76. Diner BA, Petrouleas V, Wendoloski JJ. The iron-quinone electron-acceptor complex of photosystem II. *Physiol. Plant.* 1991; 81:423–436.
77. Petrouleas V, Deligiannakis Y, Diner BA. Binding of carboxylate ions at the non-heme Fe (II) of PSII. *Biochim. Biophys. Acta* 1994; 1188:271–277.
78. Petrouleas V, Diner BA. Formation by NO of nitrosyl adducts of redox components of the photosystem II reaction center. I. NO binds to the acceptor-side non-heme iron. *Biochim. Biophys. Acta* 1990; 1015:131–140.
79. Diner BA, Petrouleas V. Formation by NO of nitrosyl adducts of redox components of the photosystem II reaction center II. Evidence that HCO_3^-/CO_2 binds to the acceptor-side non-heme iron. *Biochim. Biophys. Acta* 1990; 1015:141–149.
80. Koulougliotis D, Kostopoulos T, Petrouleas V, Diner BA. Evidence for CN-binding at the PS II non-heme Fe^{2+}. Effects on the EPR signal for Q_AFe^{2+} and on Q_A/Q_B electron transfer. *Biochim. Biophys. Acta* 1993; 1141:275–282.
81. Sanakis Y, Petrouleas V, Diner B. Cyanide binding at the non-heme Fe^{2+} of the iron-quinone complex of photosystem II: at high concentrations, cyanide converts the Fe^{2+} from high (S = 2) to low (S = O) spin. *Biochemistry* 1994; 33:9922–9928.
82. Diner BA, Petrouleas V. Light-induced oxidation of the acceptor-side Fe (II) of photosystem II by exogenous quinones acting through the Q_B binding site. II. Blockage by inhibitors and their effects on the Fe (III) EPR spectra. *Biochim. Biophys. Acta* 1987; 893:138–148.
83. Kirilovsky D, Rutherford Aw, Etienne A-L. Influence of diuron and ferricyanide on photo-damage in photosystem II. *Biochemistry* 1994; 33:3087–3093.
84. Trebst A. Inhibitors of electron flow: tools for the functional and structural localization of carriers and energy conservation sites. In: San Pietro A, ed. *Photosynthesis and Nitrogen Fixation.* San Diego: Academic Press, 1980:675–715.
85. Wolber PK, Steinback KE. Identification of the herbicide binding region of the Q_B-protein by photo affinity

labeling with azidoatrazine. *Z. Naturforsch.* 1984; 39c:425–429.

86. Oettmeier W, Godde D, Kunze B, Hofle G. Stigmatellin, a dual type inhibitor of photosynthetic electron transport. *Biochim. Biophys. Acta* 1985; 807:216–219.

87. Packham NK, Ford RC. Deactivation of the photosystem II oxidation (S) states by 2-(3-chloro-4-4trifluoromethyl)anilino-3,5-dinitnothiophene (ANT2p) and the putative role of a carotenoid. *Biochim. Biophys. Acta* 1986; 852:183–190.

88. Sivaraja M, Dismukes GC. Inhibition of electron transport in photosystem II by NH_2OH: further evidence for two binding sites. *Biochemistry* 1988; 27:6297–6306.

89. Oettmeier W, Masson K, Donner A. Anthraquionone inhibitors of photosystem II electron transport. *FEBS Lett.* 1988; 231:259–262.

90. Oettmeier W, Dostatui R, Majewski C, Höfle G, Fecker T, Kunze B, Reichenbach H. The aurachins, naturally occurring inhibitors of photosynthetic electron flow through photosystem II and the cytochrome b_6f-complex. *Z. Naturforsch.* 1990; 45c:322–328.

91. Yonegama K, Konnai M, Honda I, Yoshida S, Takahashi N, Koike H, Inone Y. Phloroglucinol derivatives as potent photosystem II inhibitors. *Z. Naturforsch.* 1990; 45c:317–321.

92. Yonegama K, Nakajima Y, Konnai M, Iwamura H, Asami T, Takahashi N, Yoshida S. Structure-activity relationships in photosystem II inhibition by 5-acyl-3-amino alkylidene)-4-hydroxy-2H-pyran-2,6(3H)-dione derivatives. *Pestic. Biochem. Physiol.* 1991; 41:288–295.

93. Oettmeier W, Masson K, Kloos R, Reil E. On the orientation of photosystem II inhibitors in the Q_B-binding niche: acridines, xanthones and quinones. *Z. Naturforsch.* 1993; 48c:146–151.

94. Asami T, Baba M, Koike H, Inoue Y, Yoshida S. Halogenation enhances the photosystem II inhibitory activity of 4-hydroxy pyridines: structure-activity relationships and their mode of action. *Z. Naturforsch.* 1993; 48c:152–158.

95. Nakajima Y, Yoshida S, Inoue Y, Yoneyama K, Ono T. Selective and specific degradation of the D1 protein induced by binding of a novel photosystem II inhibitor to the Q_B site. *Biochim. Biophys. Acta* 1995; 1230:38–44.

96. Vener A, van Kam PJ, Rich R, Ohad I, Andersson B. Plastoquinol at the Q_0-site of reduced cytochrome b/f mediates signal transduction between light and thylakoid phosphorylation: thylakoid protein kinase deactivation by a single turnover flash. *Proc. Natl. Acad. Sci. USA*, 1997; 94:1585–1590.

97. Diner BA, Petrouleas V. Light-induced acceptor-side Fe II of photosystem II by exogenous quinones acting through the Q_B binding site II. Blockage by inhibitors and their effects on the Fe (III) EPR spectra. *Biochim. Biophys. Acta* 1987; 893:138–148.

98. Achnine L, Pereda-Miranda R, Iglesias-Prieto R, Lotina-Hennsen B. Impairment of photosystem II acceptor side of spinach chloroplasts induced by tricolorin A. In: Garab G, ed. *Photosynthesis: Mechanisms and Effects*. Vol. V. Dordrecht: Kluwer Academic Publishers, 1998:3877–3880.

99. Oettmeier W, Masson K, Hedit H-J. Heterocyclic orthoquinones, a novel type of photosystem II inhibitors. *Biochim. Biophys. Acta* 2001; 1504:346–351.

100. Reil E, Höfle G, Draber W, Oettmeier W. Quinolones and their N-oxides as inhibitors of photosystem II and the cytochrome b_6/f complex. *Biochim. Biophys. Acta* 2001; 1506:127–132.

101. Ohki S, Takahashi H, Kuboyama N, Koizumi K, Kohno H, van Rensen JJS, Wakabayashi K, Böger P. Photosynthetic electron transport inhibition by pyrimidines and pyrines substituted with benzamino, methyl and trifluoromethyl groups. *Z. Naturforsch.* 2001; 56c:203–210.

102. Roberts AG, Gregor W, Britt RD, Kramer DM. Acceptor and donor-side interactions of phenolic inhibitors in photosystem II. *Biochim. Biophys. Acta* 2003; 1604:23–32.

103. Zer H, Ohad I. Photoinactivation of photosystem II induces changes in the photochemical reaction center II abolishing the regulatory role of the Q-b site in the D1 protein degradation. *Eur. J. Biochem.* 1995; 231:448–451.

104. Zer H, Prasil O, Ohad I. Role of plastoquinol oxidation in regulation of photochemical reaction center II D1 protein turnover Italic NOT ALLOWEDin vivo. J. Biol. Chem. *1994; 269:17670–17676.*

105. Böger P, Sandmann G. Modern herbicides affecting typical plant processes. In: Bowers WS, Ebing W, Martin D, Wegler R, eds. *Chemistry of Plant Protection*. Berlin: Springer-Verlag, 1990:173–216.

106. Böger P, Sandmann G. Action of modern herbicides. In: Raghavendra AS, ed. *Photosynthesis: A Comprehensive Treatise*. Cambridge: Cambridge University Press, 1998:337–351.

107. Satoh K, Katoh S, Dostatni R, Oettmeier W. Herbicide and plastoquinone-binding proteins of photosystem II reaction center complexes from the thermophilic cyanobacterium, *Synechoeoceus* sp. *Biochim. Biophys. Acta* 1986; 851:202–208.

108. Trebst A. The three-dimensional structure of the herbicide binding niche on the reaction center polypeptides of photosystem II. *Z. Naturforsch.* 1987; 42c:742–750.

109. Trebst A, Depka B, Kraft B, Johanningmeier V. The Q_B site modulates the conformation of the photosystem II reaction center polypeptides. *Photosynth. Res.* 1988; 18:163–177.

110. Allen JF, Nilsson A. Redox signaling and the structural basis of regulation of photosynthesis by protein phosphorylation. *Physiol. Plant.* 1997; 100:863–868.

111. Rintamäki E, Salonen M, Souranta V-M, Carlberg I, Andersson B, Aro E-M. Phosphorylation of light-harvesting complex II and photosystem II core proteins shows different irradiance-dependent regulation *in vivo. J. Biol. Chem.* 1997; 272:30476–30482.

112. Gal A, Zer H, Ohad I. Redox controlled thylakoid protein kinase(s). News and views. *Physiol. Plant.* 1997; 100:869–885.

113. Michel HP, Hunt DF, Shabarkowitz J, Bennett J. Tandem mass spectroscopy reveals that three photosystem II proteins of spinach chloroplasts contain N-acetyl-O-phosphothreonine at their NH_2 termini. *J. Biol. Chem.* 1988; 263:1123–1130.

114. Kurreck J, Schödel R, Renger G. Investigation of the plastoquinone pool size and fluorescence quenching in thylakoid membranes and photosystem II (PS II) membrane fragments. *Photosynth. Res.* 2000; 63:171–183.

115. Joliot P, Lavergne J, Beal D. Plastoquinone compartmentation in chloroplasts. I. Evidence for domains with different rates of photoreduction. *Biochim. Biophys. Acta* 1992; 1101:1–12.

116. Bukhov NG, Sridharan G, Egorova EA, Carpenter R. Interaction of exogenous quinones with membranes of higher plant chloroplasts: modulation of quinone capacities as photochemical and non-photochemical quenchers of energy in photosystem II during light-dark transitions. *Biochim. Biophys. Acta* 2003; 1604:115–123.

117. Vasil'ev S, Wiebe S, Bruce D. Non-photochemical quenching of chlorophyll fluorescence in photosynthesis 5-hydroxy-1,4-naphthoquinone in spinach thylakoids as a model for antenna based quenching mechanisms. *Biochim. Biophys. Acta* 1998; 1363:147–156.

118. Pfannschmidt T, Nilsson A, Allen JF. Photosynthetic control of chloroplast gene expression. *Nature* 1999; 397:625–628.

119. Hope AB. The chloroplast cytochrome bf complex: a critical focus on function. *Biochim. Biophys. Acta* 1993; 1143:1–22.

120. Cramer WH, Soriano GM, Ponomarev M, Huang D, Zhang H, Martinez SE, Smith JL. Some new structural aspects and old controversies concerning the cytochrome b_6f complex of oxygenic photosynthesis. *Annu. Rev. Plant Physiol. Plant Mol. Biol.* 1996; 47:477–508.

121. Cramer WH, Scriano GM, Zhang H, Ponomarev MV, Smith JL. The cytochrome b_6f complex. Novel aspects. *Physiol. Plant.* 1997; 100:852–862.

122. Dietrich J, Kühlbrandt W. Purification and two-dimensional crystallization of highly active cytochrome b_6f complex from spinach. *FEBS Lett.* 1999; 463:97–102.

123. Hurt E, Hauska G. A cytochrome f/b_6 complex of five polypeptides with plastoquinol-plastocyanin-oxidoreductase activity from spinach chloroplasts. *Eur. J. Biochem.* 1981; 117:591–599.

124. Oettmeier W, Kude C, Soll H-J. Phenolic herbicides and their methylethers: binding characteristics and inhibition of photosynthetic electron transport and photophosphorylation. *Pestic. Biochem. Physiol.* 1987; 27:50–60.

125. Malkin R. Interaction of stigmatellin and DNP-INT with the Rieske iron-sulfur center of the chloroplast cytochrome b_6f complex. *FEBS Lett.* 1986; 208:317–320.

126. Oettmeier W, Masson K, Dostatni R. Halogenated 1,4-benzoquinones as irreversibly binding inhibitors of photosynthetic electron transport. *Biochim. Biophys. Acta* 1987; 890:260–269.

127. Jones RW, Whitmarsh J. Inhibition of electron transfer and the electrogenic reaction in the cytochrome b/f complex by 2-n-noxyl-4-hydroxyquinolene N-oxide (NQNO) and 2,5-dibromo-3-methyl-6-isopropyl-p-benzoquinone (DBMIB). *Biochim. Biophys. Acta* 1988; 933:258–268.

128. Nitschke W, Hauska G, Rutherford AW. The inhibition of quinol oxidation by stigmatellin is similar in cytochrome bc, and b_6f complexes. *Biochim. Biophys. Acta* 1989; 974:223–226.

129. Oettmeier W, Dostatni R, Majewski C, Höfle G, Fecker T, Kunze B, Reichenbach H. The aurachins, naturally occurring inhibitors of photosynthetic electron flow through photosystem II and the cytochrome b_6/f-complex. *Z. Naturforsch.* 1990; 45c:322–328.

130. Rich PR, Madgwick SA, Moss DA. The interactions of duroquinol, DBMIB and NQNO, with chloroplast cytochrome bf complex. *Biochim. Biophys.* Acta 1991; 1058:312–328.

131. Manasse RS, Bendall DS. Characteristics of cyclic electron transport in the cyanobacterium *Phormidium laminosum*. *Biochim. Biophys. Acta* 1993; 1183:361–368.

132. Sudha Rao BK, Tyryshkin AM, Bowman MK, Kramer DM. Bound Cu^{2+} as a structural and functional probe of the cytochrome b_6f complex. In: Garab G, ed. *Photosynthesis: Mechanisms and Effects.* Vol. III. Dordrecht: Kluwer Academic Publishers, 1998:1569–1572.

133. Schaepp B, Brugna M, Riedel A, Nitschke W, Kramer DM. The Q_O-site inhibitor DBMIB favours the proximal position of the chloroplast Rieske protein and induces a pK-shift of the redox-linked proton. *FEBS Lett.* 1999; 450:245–250.

134. Lee H-Y, Hong Y-N, Chow WS. Putative effects of pH in intrachloroplast compartments on photoprotection of functional photosystem II complexes by photoinactivated neighbours and on recovery from photoactivation in *Capsicum annuum* leaves. *Funct. Plant Biol.* 2002; 29:607–619.

135. Reil E, Höfle G, Draber W, Oettmeier W. Quinolones and their N-oxides as inhibitors of photosystem II and the cytochrome b_6/f complex. *Biochim. Biophys. Acta* 2001; 1506:127–132.

136. Sigfridsson K. Plastocyanin, an electron-transfer protein. *Photosynth. Res.* 1998; 57:1–28.

137. Fromme P, Schubert W-D, Krauss N. Structure of photosystem I: suggestion on the docking sites for plastocyanin, ferredoxin and coordination of P700. *Biochim. Biophys. Acta* 1994; 1187:99–105.

138. Haehnel W, Jansen T, Gause K, Klösgen RB, Stahl B, Michl D, Huvermann B, Karas M, Herrmann RG. Electron transfer from plastocyanin to photosystem I. *EMBO J.* 1994; 13:1028–1038.

139. Ubbink M, Egdebäck M, Karlsson BG, Bendall DS. The structure of the complex of plastocyanin and cytochrome f determined by paramagnetic NMR and restrained rigid-body molecular dynamics. *Structure* 1998; 6:323–335.

140. Hippler M, Reichert J, Sutter M, Zak E, Altschmied I, Schreiber V, Hermann RG, Haehnel W. The plastocyanin binding domain in photosystem I. *EMBO J.* 1996; 15:6374–6384.

141. Rufflet SV, Mustafa AO, Kitmitto A, Holzenburg A. The location of plastocyanin in vascular plant photosystem I. *J. Biol. Chem.* 2002; 277:25692–25696.

142. Gupta R, He Z, Luan S. Functional relationship of cytochrome c_6 and plastocyanin in *Arabidopsis*. *Nature* 2002; 417:567–571.

143. Wastl J, Bendall DS, Howe CJ. Higher plants contain a modified cytochrome c_6. *Trends Plant Sci.* 2002; 7:244–245.

144. Lee CH, Durell S, Anderson LB, Gross EL. The effect of ethylenediamine chemical modification of plastocyanin on the rate of cytochrome of oxidation and P-700^+ reduction. *Biochim. Biophys. Acta* 1987; 894:386–398.

145. Burkey KO, Gizlice Z, Carter TE, Jr. Genetic variation in soybean photosynthetic electron transport capacity is related to plastocyanin concentration in chloroplast. *Photosynth. Res.* 1996; 49:141–149.

146. Anderson JM. Changing concepts about the distribution of photosystems I and II between grana-appressed and stroma-exposed thylakoid membranes. *Photosynth. Res.* 2002; 73:157–164.

147. Scheller HV, Naves H, Møller BL. Molecular aspects of photosystem I. *Physiol. Plant.* 1997; 100:842–851.

148. Brettel K. Electron transfer and arrangement of the redox cofactors in photosystem I. *Biochim. Biophys. Acta* 1997; 1318:322–373.

149. Brettel K, Leibl W. Electron transfer in photosystem I. *Biochim. Biophys. Acta* 2001; 1507:100–114.

150. Hihara Y, Sonoike K. Regulation, inhibition and protection of photosystem I. In: Andersson B, Aro E-M, eds. Advances in Photosynthesis. Vol. XI. *Regulation of Photosynthesis*. Dordrecht: Kluwer Academic Publishers, 2001:507–531.

151. Chitnis PR. Photosystem I: function and physiology. *Annu. Rev. Plant Physiol. Plant Mol. Biol.* 2001; 52:593–626.

152. Nelson N, Ben-Shem A. Photosystem I reaction center: past and future. *Photosynth. Res.* 2002; 73:193–206.

153. Fromme P, Witt HT. Improved isolation and crystallization of photosystem I for structural analysis. *Biochim. Biophys. Acta* 1998; 1365:175–184.

154. Fromme P. Biology of photosystem I: structural aspects. In: Singhal GS, Renger G, Sopory SK, Irrgang K-D, Govindjee, eds. *Concepts in Photobiology: Photosynthesis and Photomorphogenesis*. New Delhi: Narosa Publishing House, 1999:181–220.

155. Fromme P, Jordan P, Krauss N. Structure of photosystem I. *Biochim. Biophys. Acta* 2001; 1507:5–31.

156. Fromme P, Bottin H, Krauss N, Setif P. Crystallization and electron paramagnetic resonance characterization of the complex of photosystem 1 with its natural electron acceptor ferredoxin. *Biophys. J.* 2002; 83:1760–1773.

157. Hope AB. Electron transfers amongst cytochrome f, plastocyanin and photosystem I: kinetics and mechanisms. *Biochim. Biophys. Acta* 2000; 1456:5–26.

158. Klughammer C, Pace RJ. Photoreduction of the secondary photosystem I electron acceptor vitamin K_1 in intact spinach chloroplasts and cyano-bacteria *in vivo*. *Biochim. Biophys. Acta* 1997; 1318:133–144.

159. Klukas O, Schubert WD, Jordan P, Krauss N, Fromme P, Witt HT, Saenger W. Localization of two phylloquinones Q_K and Q_K^1, in an improved electron density map of photosystem I at 4-Å resolution. *J. Biol. Chem.* 1999; 274:7361–7367.

160. Zolla I, Rinalducci S, Timperic AM, Huber CG. PSI proteomics of light-harvesting proteins in different plant species: analysis and comparison by liquid chromatography-electrospraying ionization mass spectrometry. Photosystem I. *Plant Physiol.* 2002; 130:1938–1950.

161. Melkozernov AN. Excitation energy transfer in photosystem I from oxygenic organisms. *Photosynth. Res.* 2001; 70:129–153.

162. Sētif P, Fischer N, Lagoutte B, Bottin H, Rochaix J-D. The ferredoxin docking site in photosystem I. *Biochim. Biophys. Acta* 2002; 1555:204–209.

163. Ikegami I, Itoh S, Iwaki M. Selective extraction of antenna chlorophylls, carotenoids and quinones from photosystem I reaction center. *Plant Cell Physiol.* 2000; 41:1085–1095.

164. Itoh S, Iwaki M, Ikegami I. Modification of photosystem I reaction center by the extraction and exchange of chlorophylls and quinones. *Biochim. Biophys. Acta* 2001; 1507:115–138.

165. Sakurai H, Inoue K, Fujii T, Mathis P. Effects of selective destruction of iron-sulfur center B on electron transfer and charge recombination in photosystem I. *Photosynth. Res.* 1991; 27:65–71.

166. He W-Z, Malkin R. Reconstitution of iron-sulfur center B of photosystem I damaged by mercuric chloride. *Photosynth. Res.* 1994; 41:381–388.

167. Shinkarev VP, Vassiliev IR, Golbeck JH. A kinetic assessment of the sequence of electron transfer from F_x to F_A and further to F_B in photosystem I: the value of the equilibrium constant between F_x and F_A. *Biophys. J.* 2000; 78:363–372.

168. Diaz-Quintana A, Leibl W, Bottin H, Sētif P. Electron transfer in photosystem I reaction centers follows a linear pathway in which iron-sulfur duster F_B is the immediate electron donor to soluble ferredoxin. *Biochemistry* 1998; 37:3429–3439.

169. Golbeck JH. A comparative analysis of the spin state distribution of *in vitro* and *in vivo* mutants of PsaC. A biochemical argument for the sequence of electron transfer as $F_x \rightarrow F_A \rightarrow F_B \rightarrow$ ferredoxin. *Photosynth. Res.* 1999; 61:107–144.

170. Guergova-Kuras M, Boudreaux B, Joliot A, Joliot P, Redding K. Evidence for two active branches for electron transfer in photosystem I. *Proc. Natl. Acad. Sci. USA* 2001; 98:4437–4442.

171. Cseh R, Almási L, Lehoezki E. Effect of paraquat measured via *in vivo* P-700 oxidation at 820 nm on

paraquat-susceptible and resistant Eriger on Canadensis (CRONQ) biotypes. In: Garab G, ed. *Photosynthesis: Mechanisms and Effects*. Vol. V. Dordrecht: Kluwer Academic Publishers, 1998:3905–3908.

172. Terashima I, Funayama S, Sonoike K. The site of photoinhibition in leaves of *Cucumis sativus* L. at low temperatures is in photosystem I, not photosystem II. *Planta* 1994; 193:300–306.

173. Sonoike K. Photoinhibition of photosystem I: its physiological significance in the chilling sensitivity of plants. *Plant Cell Physiol.* 1996; 37:239–247.

174. Rajagopal S, Bukhov NG, Carpentier R. Photoinhibitory light-induced changes in the composition of chlorophyll-protein complexes and photochemical activity in photosystem-1 submembrane fractions. *Photochem. Photobiol.* 2003; 77:284–291.

175. Sonoike K, Wanatable IM. Chilling sensitive steps in leaves of *Phaseolus vulgaris* L. Examination of the effects of growth irradiances on PSI photoinhibition. In: Mathis P, ed. *Photosynthesis: From Light to Biosphere*. Vol. IV. Dordrecht: Kluwer Academic Publishers, 1998:2533–2536.

176. Barth C, Krause GH. Effects of light stress on photosystem I in chilling-sensitive plants. In: Garab G, ed. *Photosynthesis: Mechanisms and Effects*. Vol. IV. Dordrecht: Kluwer Academic Publishers, 1998:2533–2536.

177. Makenicz A, Radunz A, Schmidt GH. Comparative immunological detection of lipids and carotenoids on polypeptides of photosystem I from higher plants and cyanobacteria. *Z. Naturforsch.* 1996; 51c:319–328.

178. Bendall DS, Manasse RS. Cyclic photophosphorylation and electron-transport. *Biochim. Biophys. Acta* 1995; 1229:23–38.

179. Joliot P, Joliot A. Cyclic electron transfer in plant leaf. *Proc. Natl. Acad. Sci. USA* 2002; 99:10209–10214.

180. Cleland RE, Bendall DS. Photosystem I cyclic electron transport: measurement of ferredoxin-plastoquinone reductase activity. *Photosynth. Res.* 1992; 34:409–418.

181. Endo TM, Mi H, Shikanai T, Asada K. Donation of electrons to plastoquinone by NAD(P)H dehydrogenase and ferredoxin-quinone reductase in spinach chloroplasts. *Plant Cell Physiol.* 1997; 38:1272–1277.

182. Ivanov B, Kobayashi Y, Bukhov NG, Heber U. Photosystem I-dependent cyclic electron flow in intact spinach chloroplasts: occurrence, dependence on redox conditions and electron acceptors and inhibition by antimycin A. *Photosynth. Res.* 1998; 57:61–70.

183. Endo T, Shikanai T, Sato F, Asada K. NAD(P)H dehydrogenase-dependent, antimycin A-sensitive electron donation to plastoquinone in tobacco chloroplasts. *Plant Cell Physiol.* 1998; 39:1226–1231.

184. Zhang H, Whitelegge JP, Cramer WA. Ferredoxin:NADP$^+$ oxidoreductase is a subunit of the chloroplast cytochrome b$_6$f complex. *J. Biol. Chem.* 2001; 276:38159–38165.

185. Scheller HV. *In vitro* cyclic electron transport in barley thylakoids follows two independent pathways. *Plant Physiol.* 1996; 110:187–194.

186. Joet T, Cournac L, Horvath EM, Medgyesy P, Peltier G. Increased sensitivity of photosynthesis to antimycin A induced by inactivation of the chloroplast ndhB gene. Evidence for a participation of the NADH-dehydrogenase complex to cyclic electron flow around photosystem I. *Plant Physiol.* 2001; 125:1919–1929.

187. Shikanai T, Endo T, Hashimoto T, Yamada Y, Asada K, Yokota A. Directed disruption of the tobacco ndh B gene impairs cyclic electron flow around photosystem I. *Proc. Natl. Acad. Sci. USA* 1998; 95:9705–9709.

188. Quiles MI, Cuello J. Association of ferredoxin-NADP oxidoreductase with chloroplastic pyridine nucleotide dehydrogenase complex in barley leaves. *Plant Cell Physiol.* 1998; 117:235–244.

189. Satoh A, Jurano N, Senger H, Miyashi S. Regulation of energy balance in photosystems in response to changes in CO$_2$ concentrations and light intensities during growth in extremely-high CO$_x$-tolerant microalgae. *Plant Cell Physiol.* 2002; 43:440–451.

190. Cornic G, Bukhov NG, Wiese C, Bligry R, Heber U. Flexible coupling between light-dependent electron and vectorial proton transport in illuminated leaves of C-3 plants. Role of photosystem I-dependent proton pumping. *Planta* 2000; 210:468–477.

191. Kobayashi Y, Heber U. Rates of vectorial proton transport supported by cyclic electron flow during oxygen reduction by illuminated intact chloroplasts. *Photosynth. Res.* 1994; 41:419–428.

192. Hormann H, Neubauer C, Schreiber U. An active Mehler peroxidase reaction sequence can prevent cyclic PS I electron transport in the presence of dioxygen in intact spinach chloroplasts. *Photosynth. Res.* 1994; 41:429–437.

193. Hihara Y, Sonoike K. Regulation, inhibition and protection of photosystem I. In: Aro E-M, Andersson B, eds. *Regulation of Photosynthesis*. Dordrecht: Kluwer Academic Publishers, 2001:507–531.

194. Ruffle SV, Mustafa AO, Kitmitto A, Holzenburg A, Ford RC. The location of the mobile electron carrier ferredoxin in vascular plant photosystem I. *J. Biol. Chem.* 2000; 275:36250–36255.

195. Setif P. Ferredoxin and flavodoxin reduction by photosystem I. *Biochim. Biophys. Acta* 2001; 1507:161–179.

196. Rousseau F, Sētif P, Lagoutte B. Evidence for the involvement of PSI-E subunit in the reduction of ferredoxin by photosystem I. *EMBO J.* 1992; 12:1755–1765.

197. Andersson B, Scheller HV, Møller BL. The PSI-E subunit of photosystem I binds ferredoxin-NADP$^+$-oxidoreductase. *FEBS Lett.* 1992; 311:169–173.

198. Karplus PA, Bruns CM. Structure-function relations for ferredoxin reductase. *J. Bioenerg. Bioemembr.* 1994; 26:89–99.

199. Deng Z, Aliverti A, Zanetti G, Arakaki AK, Ottado J, Orellano EG, Calcaterra NB, Ceccarelli EA, Carillo N, Karplus PA. A productive NADP$^+$ binding mode of ferredoxin-NADP$^+$ reductase revealed by protein engineering and crystallographic studies. *Nat. Struct. Biol.* 1999; 6:847–853.

200. Setif PQY, Bottin H. Laser flash absorption spectroscopy study of ferredoxin reduction by photosystem I: spectral and kinetic evidence for the existence of several photosystem I-ferredoxin complexes. *Biochemistry* 1995; 34:9059–9070.

201. Setif P, Fisher N, Lagoutte B, Bottin H, Rochaix J-D. The ferredoxin docking site in photosystem I. *Biochim. Biophys. Acta* 2002; 1555:204–209.

202. Barth P, Guillouard I, Sētif P, Lagoutte B. Essential role of a single arginine of photosystem I in stabilizing the electron transfer complex with ferredoxin. *J. Biol. Chem.* 2000; 275:7030–7036.

203. Jin T, Morigasaki S, Wada K. Purification and characterization of two ferredoxin-NADP$^+$ oxidoreductase isoforms from the first foliage leaves of mung beans (*Vigna radiata*) seedlings. *Plant Physiol.* 1994; 106:697–702.

204. Jin T, Huppe HC, Turpin DH. Electron flow from NADPH to ferredoxin in support of NO$_2$ reduction.

In: Garab G, ed. *Photosynthesis: Mechanisms and Effects.* Vol. 5. Dordrecht: Kluwer Academic Publishers, 1998:3625–3628.

205. Pschorn R, Ruhle W, Wild A. Structure and function of ferredoxin-NADP$^+$-oxidoreductase. *Photosynth. Res.* 1988; 17:217–229.

206. Arakaki AK, Ceccarelli EA, Carillo N. Plant-type ferredoxin-NADP$^+$ reductases: a basal structural framework and a multiplicity of functions. *FASEB J.* 1997; 11:133–140.

207. Bojko M, Wieckowski S. Three substrate binding sites on spinach ferredoxin: NADP oxidoreductase. Studies with selective inhibitors. *Photosynthetica* 2001; 39:553–556.

208. Shahak Y, Crowser D, Hind G. The involvement of ferredoxin-NADP$^+$ reductase in cyclic electron transport in chloroplasts. *Biochim. Biophys. Acta* 1981; 636:234–243.

Section III

*Molecular Aspects of Photosynthesis:
Photosystems, Photosynthetic
Enzymes and Genes*

9 Photosystem I: Structures and Functions

Tetsuo Hiyama

Department of Biochemistry and Molecular Biology, Saitama University

CONTENTS

I. HISTORICAL BACKGROUND AND OVERVIEW

In the photosynthetic electron transport of plant-type oxygenic photosynthesis, the concept of a photochemical reaction center pigment is central to the two-photosystem (PS) theory, that is, the "Z-scheme." Historically, the discovery of P700 [1] preceded not only the Z-scheme but also bacterial and photosystem II (PS II) reaction centers.

In contrast to PS II, whose reaction center had been for a long time only a vague hypothetical one, the reaction center of photosystem I (PS I) has been P700 from the beginning. The definition of P700 was well defined by Kok [1,2]: a photosynthetic pigment that is reversibly oxidized by excitation with photons. Upon oxidation, P700 decreases its absorbance characteristically around 700 nm (after which it was named). Another peak is around 430 nm. Moreover, its photochemical oxidation/reduction was proved experimentally by demonstrating that identical spectral changes could be induced by chemical oxidation/reduction. Kok's original reports described all these. In the following decade, Witt's group, using flash spectrophotometry, confirmed these findings and established more solid pictures of P700 (chlorophyll a_{I} by their definition) and the electron transport mechanism around it [3].

A photochemical reaction center is not complete without its primary electron acceptor, a chemical entity that must be photoreduced concomitantly with the photooxidation of the reaction center pigment. Numerous candidates for the primary electron acceptor of PS I had been proposed — pteridines, cytochrome-reducing substance (CRS), and ferredoxin-reducing substance (FRS), among others — before the so called membrane-bound ferredoxin of Malkin and Bearden [4] and P430 of Hiyama and Ke [5] were proposed in 1971. Their pieces of evidence were more solid than those of their predecessors, though neither is considered to be the true primary acceptor any longer; other components found later are more primarily photoreduced as will be shown later.

The main function of PS I is the generation of $NADPH_2$. The enzymatic mechanism of the final stage was well characterized in the early 1960s by Arnon's group [6], who established participation of an iron–sulfur protein (ferredoxin) and a flavin enzyme (ferredoxin:NADP oxidoreductase). The donor side of PS I had been speculated to be either cytochrome f or plastocyanin for a long time. Only recently [7], plastocyanin, a copper protein [8], has been established as the direct donor to P700 of the electron from PS II via the cytochrome b_6/cytochrome f complex (b_6/f complex).

Efforts to isolate the PS I activity in the form of a complex from thylakoid membranes started in 1960s. An earlier work on detergents of Shibata's group [9] was followed by one of the first successful PS I particle preparations of Anderson and Boardman [10]. In the following years, many types of PS I particle were prepared, mainly for optical measurement of kinetics. As their goal at that time was to lower the chlorophyll-to-P700 ratio to facilitate optical monitoring of electron carriers, little attention was paid to their protein constituents.

At the end of the 1970s, in an effort to obtain PS I complexes of simple and minimal subunit compositions, Nelson's group showed for the first time that the PS I complex was composed of several protein subunits [11,12]. They proposed rightly that the large subunit of more than 60 kDa would be the host of the reaction center of PS I, and presented some speculations on the roles of other small subunits smaller than 20 kDa. Since then, a great number of different preparations have been reported from numerous photosynthetic organisms. The trouble was that their subunit compositions varied tremendously even within the same species, not only because preparation methods were different but also because the resulting patterns of sodium dodecyl sulfate polyacrylamide gel electrophoresis (SDS-PAGE), a technique used exclusively for separation and detection of the subunits, could be notoriously variable among workers and laboratories. As a result, one could hardly compare each other's work. Later, N terminal amino acid sequencing of SDS-PAGE bands has opened up possibilities of defining each subunit in terms of its primary structure. Above all, techniques of cloning and sequencing of genes by means of molecular genetics have revealed the entire amino acid sequences as well as the gene structures of those subunits. The most notable were perhaps the sequencing of the genes for the large subunits, now designated as PsaA and PsaB, by Fish et al. [13], and the determinations of the whole nucleotide sequences of tobacco and liverwort chloroplast DNA, by Sugiura's group [14] and Ohyama and Ozeki's group [15], respectively. Numerous reports have appeared since, and we now have an almost complete set of the primary structures of the PS I subunits, as summarized in Table 9.1.

Functionally, PS I can be defined as "a pigment–protein complex embedded in thylakoid membranes that can photoreduce ferredoxin by electrons from PS II fed through plastocyanin." In short, it may also be called a "light-driven plastocyanin:ferredoxin oxidoreductase" [16], although its inherently irreversible nature might not fit well the word "oxidoreductase" in its enzymological sense. The core of the complex is a heterodimer of the two 80 kDa polypeptides (the large

TABLE 9.1
PS I Subunits and Peripheral Proteins

Protein	Synonym		Gene	Location
PsaA	PSI-A	Subunit I$_a$	*psaA*	(Chl)
PsaB	PSI-B	Subunit I$_b$	*psaB*	(Chl)
PsaC	PSI-C	Subunit VII	*psaC*	(Chl)
PsaD	PSI-D	Subunit II	*psaD*	(Nuc)
PsaE	PSI-E	Subunit IV	*psaE*	(Nuc)
PsaF	PSI-F	Subunit III	*psaF*	(Nuc)
PsaG	PSI-G	Subunit V	*psaG*	(Nuc*)
PsaH	PSI-H	Subunit VI	*psaH*	(Nuc*)
PsaI	PSI-I	Subunit X	*psaI*	(Chl)
PsaJ	PSI-J	Subunit IX	*psaJ*	(Chl)
PsaK	PSI-K	Subunit VIII	*psaK*	(Nuc)
PsaL	PSI-L	Subunit V'	*psaL*	(Nuc)
PsaM	PSI-M		*psaM*	(Chl)
PsaN			*psaN*	(Nuc)
PsaX			*psaX*	(*)
PsaY(PsbW)			*psbW*	(Chl**)
Ferredoxin (Fd)			*petF*	(Nuc)
Plastocyanin (PC)			*petE*	(Nuc)
Ferredoxin:NADP$^+$ oxidoreductase (FNR)			*petH*	(Nuc)

Chl: chloroplast-DNA encoded.
Nuc: nuclear genome encoded.
*Cyanobacteria only, so far.
**Higher plants only, so far.

subunits: PsaA and PsaB). This core binds P700, two molecules of phylloquinone (vitamin K$_1$), an iron–sulfur cluster and a number of light-harvesting chlorophyll molecules (mostly chlorophyll *a*). So far, as many as 15 other subunits smaller than 20 kDa have been claimed to be members of the PS I complex (Table 9.1). Recently, a much more elaborate and detailed picture has emerged as a result of high-resolution crystallography, as will be described later.

As stated above, PS I activities, usually represented by photooxidation of P700, can be isolated as pigment–protein complexes from thylakoid membranes by means of detergent solubilization. The most common type of PS I complex consists of, besides the large subunits, PsaC, PsaD, PsaE and a group of other polypeptides smaller than 20 kDa. This type will be categorized later as Type II. Some of the simplest compositions are seen in Triton X-100 treated spinach preparations [17]. More complex compositions are common. Among those 15 polypeptides proposed as the small subunits of PS I, PsaC is most certainly an essential component, which hosts two iron–sulfur clusters. Complexes that contain this component can photoreduce ferredoxin. Thus, a

hypothetical minimal PS I complex would consist of PsaA, PsaB and PsaC. However, no PsaC-containing complex without PsaD and PsaE has been isolated so far, which suggests that PsaD and PsaE help binding those subunits to the complex and stabilizing the complex. Those complexes can be categorized roughly into the following three types:

Type I: complex with "complete" set of PS I subunits including light-harvesting chlorophyll proteins (LHCPs) and pigments
Type II: Type I minus LHCPs and sometimes some of the small subunits
Type III: core complexes that consist only of the large subunits (PsaA and PsaB)

Type I complexes contain typically as many as 200 chlorophyll *a*/*b* per P700 and are sometimes designated as PSI-200 [18]. This type of preparation has been prepared by using mild detergents like digitonin [10], or low concentrations of Triton X-100 [18].

Type II is the most common preparation and can be prepared readily by using Triton X-100, the almost exclusively used solubilizing detergent, followed by ion exchange column chromatography, density gradient centrifugation, and other protein purification techniques. There have been numerous reports on this type of preparation.

It should be noted, however, that there always remains a question of what is the real PS I complex *in vivo* or *in situ* on the thylakoid membranes. In those complexes solubilized from any membranous structures, there are always some possibilities of missing or contamination of certain components. One has to be very careful in deciding a certain subunit to be assigned to a certain system. For that matter, complexes obtained by a number of different methods should be reexamined and compared with each other carefully before the final conclusion. Recently, some cyanobacterial preparations have been crystallized. One of the most successful ones has allowed us to obtain a 3-D structure [19]. This particular crystal was reported to contain PsaA, B, C, D, E, F, I, J, K, L, M, and X [20]. According to this, most of those subunits reported so far seem to belong to PS I after all. The roles of these subunits are not well known except for PsaA, PsaB, and PsaC. Molecular genetics that allows creation of deletion mutants and site-specific mutagenesis have been contributing tremendously in this field. The primary structures and possible roles of the individual subunits will be discussed later.

Both ferredoxin and plastocyanin are peripheral to the thylakoid membrane. These loosely bound proteins as well as ferredoxin:NADP oxidoreductase, another peripheral component, can be included as one of those components that the PS I complex is composed of. PS I preparations, however, usually do not contain these proteins because they are easily

released from thylakoid membrane when cells are broken for preparations.

A Type III preparation from spinach was first reported in 1987 [21], and a cyanobacterial preparation followed [22]. Either strong detergents like sodium/lithium dodecyl sulfate or chaotropic agents have been used to remove the smaller subunits (for a spinach preparation, see Ref. [17]). This type of complex, however, cannot photoreduce ferredoxin, though electrons from plastocyanin can be accepted.

II. FUNCTIONS AND KINETICS

A. Oxidizing Side

1. Reaction Center/Primary Electron Donor

The reaction center pigment of PS I is P700 as stated above (Figure 9.1). More about P700 will appear in the following sections.

2. Physiological (Secondary) Electron Donors

Plastocyanin is most likely the electron carrier that directly donates an electron to P700 [7]. Cytochrome *f* provides electrons to plastocyanin. Recent advances in this field are summarized in Ref. [23]. It is known that in cyanobacteria and red algae under special conditions, such as a copper-deficient growth medium, certain *c*-type cytochromes may replace plastocyanin.

3. Artificial Electron Donors

Ascorbate, although a potentially strong reductant of P700, is a rather poor electron donor by itself,

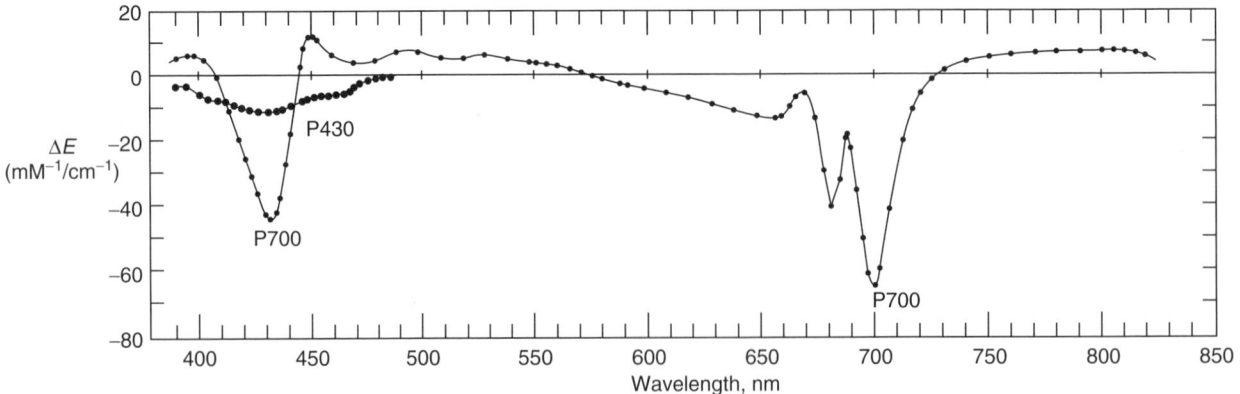

FIGURE 9.1 Light-minus-dark difference spectra of P700 (small circles) and P430 (large circles). A short xenon flash (100 μsec) was applied to a reaction mixture containing digitonin-treated PS I particles from spinach, TMPD, ascorbate, and methylviologen as in Ref. [24]. The P430 spectrum was obtained by subtracting absorbance changes in a sample without methylviologen from those of P700 as in Ref. [33]. See the text for ΔE (difference extinction coefficient) and details of kinetical analysis. Refer to Figure 9.3 as well.

perhaps due to its poor accessibility to the hydrophobic environment of thylakoid membranes. By adding some redox dyes such as 2,6-dichlorophenolindophenol (DCIP), the reduction of P700 by ascorbate becomes extremely rapid. Phenazine methosulfate (PMS) is even more efficient for this purpose. N,N,N',N'-tetramethylphenylenediamine (TMPD) is another convenient artificial electron donor. The combination of TMPD and ascorbate is a recommended reductant for the chemical reduction of P700 for recording a difference spectrum and for flash spectrophotometry [24–26]. So far, plastocyanin, the physiological electron donor, is the most efficient reductant *in vitro* in the presence of excess amounts of ascorbate.

B. REDUCING SIDE

1. Primary and Other Electron Acceptors and Carriers

As stated above, good evidence on this matter emerged in the early 1970s when a thylakoid-bound ferredoxin-type electron paramagnetic resonance (EPR) signal (later designated as Center A) and P430 were reported. Since then, several other entities have been proposed: Center B [27], Component X [28], A_1 [29], A_2 [29], A_0 [30], and vitamin K_1 (phylloquinone) [21]. Those can be reclassified according to evidence accumulated so far as follows.

A_0: the "real" primary acceptor, a chlorophyll *a* bound to the PsaA/PsaB heterodimer protein pigment complex (see the discussion in Refs. [20,31])

A_1: vitamin K_1 (phylloquinone) bound to the PsaA/PsaB heterodimer protein pigment complex [20,22].

A_2: originally called Component X, a 4Fe-4S iron sulfur cluster bound to the PsaA/PsaB heterodimer protein pigment complex; often abbreviated as FeS_x (or F_X), also called P430 [28,29,32]. (A difference spectrum of P430 is shown in Figure 9.1, together with that of P700. An EPR spectrum of component X, represented by a $g = 1.78$ signal, is shown in Figure 9.2. More to come later.)

Centers A/B: 4Fe–4S iron–sulfur clusters on the PsaC subunit, often abbreviated as $FeS_A(F_A)$ and FeS_B (F_B). g values of 2.03, 1.94, and 1.86 are assigned for FeS_A and 2.03, 1.92, and 1.89 for FeS_B (Figure 9.2).

At present, it is thought that electron flows on the reducing side of PS I as follows:

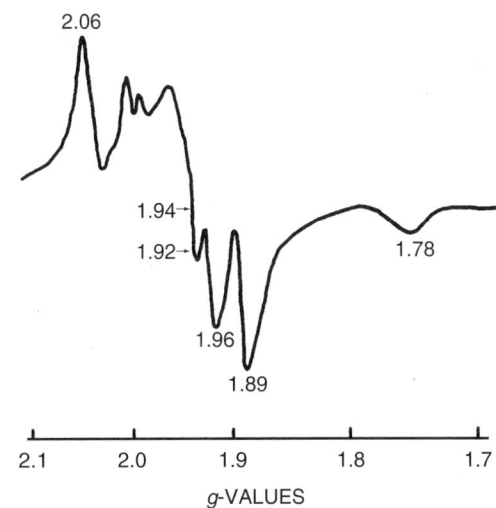

FIGURE 9.2 Low-temperature EPR spectrum of PS I particles. The preparation was a crude membrane fraction from *Nostoc* [32]. The reaction mixture was supplemented with sodium dithionite and illuminated during the entire freezing procedure in liquid nitrogen. Temperature, 15 K; power of X-band microwave, 20 mW. g values for the signals are listed conventionally: 2.05, 1.89, 1.86, and 1.78, measured at the peaks (troughs) of the derivative absorption spectra, and 1.94 and 1.92 as the points of inflexion.

$$(PS\ II \rightarrow b_6f \rightarrow plastocyanin \rightarrow P700) \rightarrow$$
$$A_0 \rightarrow A_1 \rightarrow A_2 \rightarrow FeS_A/FeS_B \rightarrow$$
$$(Ferredoxin \rightarrow NADP)$$

2. Artificial Electron Acceptors

A number of redox dyes have been used as artificial electron acceptors of PS I [33]. Among them, perhaps, methylviologen (1,1'-dimethyl-4,4'-bipyridinium dichloride) is one of the most frequently used acceptors. Readily available as the main ingredient of a widely used but highly toxic herbicide (Paraquat), methylviologen is convenient and quite specific for PS I because of its extremely low redox potential (−446 mV) — so low that PS II cannot photoreduce methylviologen. Benzylviologen, though less electronegative (−360 mV), and Safranin T (−290 mV) are also specific for PS I. The site of the photoreduction of methylviologen on the reducing side of PS I has been shown to be A_2 (FeS_x or P430) rather than FeS_A/FeS_B [17].

As the reducing power of PS I is extremely high, almost any oxidant can potentially be photoreduced by PS I. Methylene blue (+11 mV), DCIP (+217 mV), TMPD (+260 mV), PMS (+80 mV), and ferricyanide

(+360 mV) are among them [33]. They are indeed the so called Hill reagents (oxidants).

3. Physiological Electron Acceptors

Ferredoxin, a 2Fe–2S iron–sulfur protein, accepts electrons from PS I. Ferredoxin is known to form a complex with ferredoxin:NADP oxidoreductase (FNR) to reduce NADP eventually. In cyanobacteria, flavodoxin replaces ferredoxin under iron-deficient growth conditions.

C. MEASUREMENT OF LIGHT-INDUCED REACTIONS AND KINETICS

The reactions (oxidations and reductions) of these electron carriers have been monitored most readily using absorbance spectroscopy in the visible wavelength region. It should be noted that, in photosynthetic systems, background absorbances due to antenna pigments are usually very high, which makes it very difficult in certain wavelength regions, notably around 400 to 450 nm and 650 to 700 nm. It is also noteworthy that fluorescence emission excited by actinic light would become quite a nuisance in the red region (650 to 700 nm). For these reasons, most measurements have used preparations partially depleted of chlorophylls.

1. Optical Properties of P700

As stated above, P700 was first discovered as a component that changes (decreases) the absorbance around 700 nm upon photooxidation. The oxidized form can be rereduced readily by electrons provided by appropriate electron donors in the medium, either physiological or artificial. A typical difference spectrum (photooxidized-minus-reduced or light-minus-dark) is shown in Figure 9.1. Typical difference extinction coefficients at several representative wavelengths [24] are summarized in Table 9.2. Three distinct peaks (troughs) are noteworthy, namely

those at 700, 682, and 430 nm. It should also be noted that there are several isosbestic points, notably one at 444 nm, which is quite useful for monitoring P430 (Figure 9.1, larger dots) independent of P700, and another at 575 nm, which is convenient for monitoring blue colored electron carriers such as plastocyanin, TMPD, and DCIP. It should be added that the quantum efficiencies of the P700 photooxidation in the far red regions have been measured to be close to unity in a PS I complex [25].

2. Quantitative Determination of P700

By using the extinction coefficients shown in Table 9.2, the concentration of P700 can be determined from difference spectra (oxidized-minus-reduced). A commercially available recording spectrophotometer with a computerized data processing system, a rather common feature of a modestly priced spectrophotometer for a biochemistry laboratory nowadays, can be readily used for this purpose [26]. The chemical oxidation is achieved by using ferricyanide and the reduction by using TMPD-ascorbate.

A more sensitive and, once set up, quick method is flash spectrophotometry. Unfortunately, there has been almost no instrument for this purpose commercially available so far. Apparata for flash spectroscopy on the market are all designed for nonbiological photochemistry, where quantum yields are much lower and much less sensitivities are required. Thus, they usually cannot be used for measuring flash-induced absorbance changes in biological photosynthetic systems without extensive modifications. Construction of an instrument set-up for P700 measurement may not be as painstaking as it used to be, since low-cost, high-performance digital oscilloscopes with signal-averaging capability are readily available. Computer interfacing is no longer a state-of-the-art technique; a number of plug-in boards and software packages are presently available for personal computers for this purpose. With a xenon flash, a time resolution of a millisecond would be enough for quantitative determination of P700 and P430. One of the most important points leading to successful monitoring of light-induced absorbance changes is the combination of optical filters for actinic light (flash) and those for protecting a measuring device like a photomultiplier and a photodiode. The best combinations of these complementary filters (e.g. red and blue) are not many; one can refer to Refs. [17,24,25,32,33] for these matters.

Continuous illumination is much easier to obtain and could be useful for determination of P700. Use of fiber optics, a tungsten–halogen lamp, an appropriate filter combination, and a mechanical shutter would

TABLE 9.2
Difference Extinction Coefficients of P700

Wavelength (nm)	$\Delta\varepsilon$ (mM^{-1} cm^{-1})
430	44
444	0
575	0
682	40
700	64
810	8

permit an actinic illuminator to cross-illuminate a sample cuvette. High-intensity light emitting diodes (LEDs), now widely available, may be good choices for light sources. Modification of a common spectrophotometer for this purpose would not be too complicated. Again, the filters are very important, though not as stringent as in the case of flash spectroscopy. As the timescales are in seconds rather than milliseconds, a chart recorder connected to the output of the spectrophotometer would suffice. Another important point is the intensity of actinic light, which has to be checked carefully so that the intensity is saturated. The magnitudes of the light-induced steady state changes are reflections of the balance of photooxidation, which depends on the light intensity, and the reduction by the reductant present in the system. Thus, the intensity required for saturation depends on the concentration and reducing power of the reducing system in the reaction mixture.

Three wavelength regions have been used in most cases. The largest changes, around 700 nm, have several advantages: a high extinction coefficient, low background absorbances, and a high specificity. No light-induced absorbance changes due to components other than P700 can be anticipated around 700 nm. A disadvantage is fluorescence interference in this region, particularly in the case of relatively crude preparations. Fluorescence interference in some cases can be minimized by using a sharp cut-off filter

system or a monochromator between the cuvette and the photodetector. The only advantage of using wavelengths around 430 nm is escaping fluorescence interference. The disadvantages are high background absorbance and coincidental changes due to other components, notably P430. A near-infrared region (800 to 830 nm) [24] has been used in some cases. The advantages here are an almost null fluorescence and very low background absorbance, which might well compensate for otherwise disadvantageous low extinction coefficients in this region. Other merits would be that any actinic wavelengths, either red or blue, can be used for excitation.

3. Kinetics of Flash-Induced Absorbance Changes

In Type II preparations with an electron donor system just enough to keep P700 reduced under a weak measuring beam, a pulse of a saturating actinic flash (pulse width several microseconds to several hundred microseconds) induces typical absorbance changes. At 700 and 430 nm, these are absorbance decreases, and at 820 nm, it is an increase. A typical case is shown in Figure 9.3. These changes are almost instantaneous in a millisecond timescale and are followed by a much slower relaxation (recovery) phase with a half time ranging from 30 to 100 msec (Figure 9.3, left). The half decay time varies from preparation to preparation. This half time does not depend on the

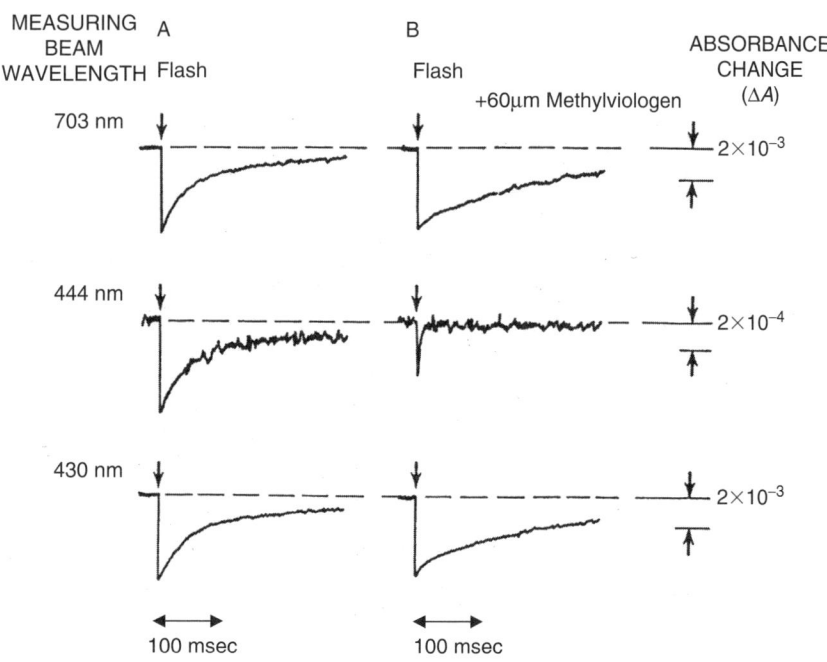

FIGURE 9.3 Flash-induced absorbance changes in a Type II PS I preparation. A, without methylviologen; B, with 60 mM methylviologen. Measuring beam wavelengths: 703 nm, top; 44 nm, middle; 430 nm, bottom. Flashes are applied as indicated by arrows. For experimental details and interpretations, see the text and [17,24,32,33].

concentration of the donor system, typically TMPD with an excess amount of ascorbate. This recovery phase is not exponential but hyperbolic, reminiscent of a typical bimolecular second order reaction; reciprocal plots would give a straight line [33]. A very similar decay is observed at 444 nm, an isosbestic point of P700 where no change due to P700 is expected. When an artificial electron acceptor, typically methylviologen, is added to this reaction mixture (Figure 9.3, right), a remarkable difference is observed in the recovery kinetics, with no appreciable difference in the extent of the initial fast changes. At 700 and 820 nm, the recovery becomes usually slower and now dependent on the concentration of the donor system. At 430 nm, the recovery phase becomes biphasic: a faster and smaller phase is followed by a slower phase. This latter slower and exponential phase is dependent on the concentration of the donor system and kinetically identical with those at 700 and 820 nm, where only one phase is observed [5,33].

The above observations have been interpreted as follows [5,33]: the absorbance changes at 700 and 810 nm are solely due to P700 and those recoveries are dependent on donor concentrations, and represent the rereduction of the flash-oxidized P700 in the dark after the flash. Without any externally supplemented artificial electron acceptor, the electron from a photoreduced molecule, which has accepted the electron from P700, goes directly back to P700, which otherwise would have gone to an artificial acceptor. Although this has been called a ``back reaction'' or a charge recombination, this reaction must be an *inter*photosystem reaction, that is a diffusion-dependent collision of two different PS I particles suspended in an aquatic medium, rather than a charge recombination within a PS I complex. At 444 nm, an isosbestic point of P700, an identical kinetics is observed in the absence of the acceptor, while in the presence of the acceptor, the kinetics becomes more like that of the faster phase at 430 nm. This monophasic recovery at 444 nm, which becomes exponential in the presence of the added acceptor, is dependent on the concentration of the acceptor: the higher, the faster. The absorbance changes at 444 nm and the faster recovery phase at 430 nm thus represent a molecule that has been photoreduced concomitantly with P700, and was originally designated as P430 [5], and later assigned to FeS$_x$ [17,32].

In Type III preparations, the half times of the back reaction are much faster. In a carefully prepared photochemically active preparation, the half recovery time was 8 msec [17], but usually much faster. Otherwise, the kinetics are basically similar to those in the case of Type II preparations [17].

4. Other Electron Carriers

Cytochromes can be measured fairly specifically in their alpha band, where the background is minimal. In intact or nearly intact systems, this region (500 to 550 nm), however, is often dominated and interfered by huge changes, the so called P520, a membrane potential indicator due perhaps to carotenoids, so huge that cytochrome changes often cannot be measured at all. P520 is absent in cyanobacteria.

Although the extinction coefficient of plastocyanin (oxidized form) is quite low ($9.8 \, mM^{-1} \, cm^{-1}$ at 597 nm) due to its broad nature, 575 nm, an isosbestic point of P700, can be used as a measuring beam wavelength. Upon reduction, the absorbance decreases.

Absorbance changes (decrease upon reduction) due to iron–sulfur clusters (FeS$_x$, FeS$_A$, and FeS$_B$) are somewhat confusing and controversial. When P430 was first proposed [5], it was not assigned to any chemical entity except for Center A, which had been reported as a low temperature EPR signal [4]. In the following year, Center B, another EPR signal, was discovered [27], and then P430 was somehow automatically assigned thereafter to ``FeS$_A$/FeS$_B$'' without much substantial evidence. Later, Component X (FeS$_x$), another EPR signal with a presumably much lower redox potential, was proposed [28]. Hiyama and Fork examined both optical absorbance changes and low-temperature EPR signals in a cyanobacterial thylakoid preparation, and concluded that P430 can be equated with Component X (FeS$_x$, A$_2$) rather than with FeS$_A$/FeS$_B$ [32]. Results with preparations devoid of PsaC, the host of FeS$_A$/FeS$_B$, clearly showed a P430-like difference spectrum [17,22] and support an earlier contention that P430 is FeS$_x$. Unfortunately, most reviews still refer to P430 as FeS$_A$/FeS$_B$. The difference spectrum of FeS$_A$/FeS$_B$ is not clear at the moment except for a crude one, which looks quite different from that of P430 [32]. At present, there is another (and perhaps good) possibility that P430 is A$_1$ (phylloquinone, vitamin K$_1$). Evidence in Refs. [17,22] is not inconsistent with this possibility. The difference extinction coefficients of P430 are approximately $12 \, mM^{-1} \, cm^{-1}$ at 430 nm and $6 \, mM^{-1} \, cm^{-1}$ at 444 nm [33].

5. EPR Signals

Iron–sulfur clusters like Center A (FeS$_A$), Center B (FeS$_B$), and Component X (FeS$_x$) can be detected by using low temperature EPR. Figure 9.2 shows a typical X-band EPR spectrum of a Type II preparation reduced by a strong reductant, sodium dithionite, under anaerobic conditions. Optimal temperatures for measurements of FeS$_A$ and FeS$_B$, which are represented by characteristic g values of 2.03, 1.94, and 1.86 for FeS_A and 2.03, 1.92, and 1.89 for FeS_B, are around 20 K, while that for FeS_x, represented by a $g = 1.78$ signal, is lower (near 10 K). Microwave power saturation is achieved at a very high energy, beyond the high end (20 mW) of most commercial instruments [32].

III. STRUCTURAL ASPECTS

A. PROTEIN SUBUNITS AND PROSTHETIC GROUPS

So far, more than 17 polypeptides have been reported as subunits of the PS I reaction center complex. Table 9.1 summarizes those subunits whose amino acid sequences have been reported, together with peripheral proteins. As stated above, it should be noted that most of these subunits have not been well established as actual members of PS I complex *in vivo*. Some of them appear only in certain preparations and cannot be found in others [45]. Notable exceptions are PsaA, PsaB, PsaC, PsaD, PsaE, and possibly PsaL, which are omnipresent in Type II preparations. A recent crystallographical study [20] revealed that nine polypeptides with transmembrane α-helices (PsaA, PsaB, PsaF, PsaI, PsaJ, PsaK, PsaL, PsaM, and PsaX) and three stromal subunits (PsaC, PsaD, and PsaE) in a Type II preparation from a thermophillic cyanobacterium (*Synechococcus* (*Thermosynechococcus*) *elongatus*), which was originally isolated by Sakae Katoh's group from Beppu Hot Spa, Japan. Although the molecular ratios (stoichiometries) of these subunits in a complex were the subject of a few studies long ago [43,100], one each of these subunits seems to be present for one reaction center according to the crystallography. The description of each subunit follows.

1. PsaA (Subunit I_a) and PsaB (Subunit I_b)

The amino acid sequences deduced from the corresponding genes for these proteins (*psaA* and *psaB*) were first reported in maize [13]. Since then, numerous sequences have been reported and registered in data banks. Figure 9.4 shows representative sequences of PsaA and PsaB (spinach). Amino acid residues conserved within 13 species listed are indicated by bold letters. These two genes, located on the chloroplast DNA in higher plants, form an operon with the exception of *Chlamydomonas* [34]. The N terminals of both PsaA and PsaB, as isolated by SDS-PAGE using urea, are usually blocked and cannot be cleaved by Edman degradation chemistry for N terminal sequencing. Fish and Bogorad isolated a peptide fragment by using high performance liquid chromatography (HPLC) from a cyanogen bromide digest of a maize PsaB preparation,

which showed that the N terminal sequence of PsaB is just as predicted from the gene except for the N terminal methionine [35]. A similar fragment with the predicted N-terminal sequence of PsaA without the N-terminal methionine has been isolated by using HPLC from a *Staphylococcus* V8 protease digest of a spinach PsaA/PsaB preparation (A. Ohinata, H. Hirata, H. Hiraiwa, and T. Hiyama, unpublished results). These results suggest that the N terminal residues of the mature PsaA and PsaB are possibly unprocessed formylmethionine. From these sequences, the molecular weights of these two polypeptides would be calculated as 82,000 to 83,000 with 750 to 800 amino acid residues. These two have some 40% homologies to each other.

An earlier computer analysis predicted that each polypeptide had 11 membrane-spanning α-helix domains [36]. The results of x-ray crystallography mostly support this presumption [20,37]. Three and two cysteine residues are conserved in PsaA and PsaB, respectively. Of these, Cys^{604} and Cys^{613} of PsaA and Cys^{568}, and Cys^{577} of PsaB have been implicated as ligands for FeS_x (F_X: component X), a 4Fe–4S iron–sulfur cluster [38,39].

There are 36 and 32 conserved histidine residues in PsaA and PsaB, respectively. These are implicated as ligands to chlorophylls (mostly chlorophyll a). Some of them could be ligands to P700, a possible chlorophyll a and chlorophyll a' heterodimer, as will be discussed later.

According to the recent crystallography, PsaA and PsaB, which share similarities in protein sequence and structure, contain 11 transmembrane helices each that are divided into an N terminal domain and a C terminal domain [20]. The C terminal domain forms two interlocked semicircles enclosing the electron transport cofactors (phylloquinone, etc.). This core structure is separated from the N terminal-helices and the transmembrane-helices of the smaller PSI subunits by an elliptically distorted cylindrical region bridged by-helices and harboring a large number of the antenna Chl a molecules and carotenoids [20].

Chemical analyses and the amino acid composition of a reaction center preparation consisting of PsaA and PsaB alone (Type III) showed previously that there are four iron, four sulfur, and one phylloquinone molecules as well as one each of PsaA and PsaB per P700 [44]. The number of phylloquinone molecules per P700 is usually two in most PS I preparations that contain other low molecular weight subunits (Type II). crystallographical analyses revealed two quinone planes are π-stacked with indole rings of well-conserved tryptophan residues (Trp^{697} of PsaA and Trp^{677} of PsaB) [20].

PsaA

```
  1 MIIRSPEPE- -------VKI LVDRDPVKTS FEAWAKPGHF SRTIAKG-PE
 51 TTTWIWNLHA DAHDFDSHTS DLEEISRKIF SAHFGQLSII FLWLSGMYFH
101 GARFSNYEAW LSDPTHIGPS AQVVWPIVGQ EILNGDVGGG FRGIQITSGF
151 FQIWRASGIT SELQLYCTAI GALVFAALML FAGWFHYHKA APKLAWFQDV
201 ESMLNHHLAG LLGLGSLSWA GHQIHVSLPI NQFLNAGVDP KEIPLPHELI
251 LNRDLLAQLY P----SFAEG ATPFFTLNWS KYADFLTFRG GLDPVTGGLW
301 LTDTAHHHLA IAILFLIAGH MYRTNWGIGH GLKDILEAHK GPFTGQGHKG
351 ---------- -------LY EILTTSWHAQ LALNLAMLGS LTIVVAHHMY
401 AMPPYPYLAT DYGTQLSLFT HHMWIGGFLI VGAAAHAAIF MVRDYDPTTR
451 YNDLLDRVLR HRDAIISHLN WACIFLGFHS FGLYIHNDTM SALGRPQDMF
501 SDTAIQLQPV FAQWIQNTHA LAPSATAPGA TASTSLTWGG SDLVAVGGKV
551 ALLPIPLGTA DFLVHHIHAF TIHVTVLILL KGVLFARSSR LIPDKANLGF
601 RFPCDGPGRG GTCQVSAWDH VFLGLFWMYN SISVVIFHFS WKMQSDVWGS
651 ISDQGVVTHI TGGNFAQSSI TINGWLRDFL WAQASQVIQS YGSSLSAYGL
701 FFLGAHFVWA FSLMFLFSGR GYWQELIESI VWAHNKLKVA PATQPRALSI
751 VQGRAVGVTH YLLGGIATTW AFFLARIIAV G/..................
```

PsaB

```
  1 MALR-FPRFS QGLAQDPTTR RIWFGIATAH DFESHDDITE ERLYQNIFAS
 51 HFGQLAIIFL WTSGNLFHVA WQGNFESWVQ DPLHVRPIAH AIWDPHFGQP
101 AVEAFTRGGA LGPVNIAYSG VYQWWYTIGL RTNEDLYTGA LFLLFLSVIS
151 LLGGWLHLQP KWKPSVSWFK NAESRLNHHL SGLFGVSSLA WTGHLVHVAI
201 P-GSRGEYVR WNNFLDVLPH PQGLGPLFTG QWNLYAQNPD SSSHLFGTSQ
251 GAGTAILTLL GGFHPQTQSL WLTDMAHHHL AIAFVFLVAG HMYRTNFGIG
301 HSMKDLL--- -EAHIPPGGR LGRGHKGLYD TINNSLHFQL GLALASLGVI
351 TSLVAQHMYS LPAYAFIAQD FTTQAALYTH HQYIAGFIMT GAFAHGAIFF
401 IRDYNPEQNE DNVLARMLDH KEAIISHLSW ASLFLGFHTL GLYVHNDVML
451 AFGTPEKQIL IEPIFAQWIQ SAHGKTSYGF DVLLS---ST SGPAFNAGRS
501 IWLPGWLNAV NENSNSLFLT IGPGDFLVHH AIALGLHTTT LILVKGALDA
551 RGSKLMPDKK DFGYSFPCDG PGRGGTCDIS AWDAFYLAVF WMLNTIGWVT
601 FYWHWKHITL WQGNVSQFNE SSTYLMGWLR DYLWLNSSQL INGYNPFGMN
651 SLSVWAWMFL FGHLVWATGF MFLISWRGYW QELIETLAWA HERTPLANLI
701 RWRDKPVALS IVQARLVGLA HFSVGYIFTY AAFLIASTSG KFG/......
```

FIGURE 9.4 Amino acid sequences of the PS I large subunits, PsaA and PsaB, of spinach [43]. Residues conserved throughout 14 species are written in bold letters. Those species are *Marchantia polymorpha* (liverwort, Ref. [15]); *Oryza sativa* (rice, Ref. [40]); *Pisum sativum* (pea, Ref. [41]); *Spinacia oleracea* (spinach, Ref. [42,43]); *Nicotiana tobacum* (tobacco, Ref. [14]); *Chlamydomonas reinhardtii* [34]; *Euglena gracilis* [46]; *Zea mays* (maize, Ref. [13]); *S. elongatus* [47]; *Synechococcus vulcanus* [48]; *Synechocystis* sp. PCC 6803 [49]; *Synechococcus* sp. PCC 7002 (*Agmenellum quadruplicatum*, Ref. [50]); *Anabaena variabilis* [51]. Sequence information has been updated using the BLAST (NCBI) and FASTA (DDBJ) databases.

The separation of PsaA and PsaB has been achieved only by using SDS-PAGE with urea containing gel [35]. It should be noted that an apparent separation achieved with SDS-PAGE without urea in an earlier pioneering report [71] was wrong. N terminal sequencing and immunoblotting of the two separated bands revealed that the lower band obtained by that method was a mixture of degraded PsaA and PsaB, while the upper band was the unresolved mixture of PsaA and PsaB (H. Hiraiwa, H. Hirata, and T. Hiyama, unpublished results).

2. PsaC (Subunit VII)

This 9 kDa protein is now widely believed to be the host of two 4Fe–4S iron–sulfur clusters, FeS$_A$ (F$_A$:

Center A) and FeS$_B$ (F$_B$: Center B). The apoproteins were first isolated and sequenced independently at three different laboratories [53,55,56]. The genes were found in chloroplast genomes of tobacco and liverwort. Since then, a number of sequences from various organisms have been reported. Figure 9.5 shows a representative spinach sequence and indicates (by bold letters) conserved amino acid residues in 22 species in data banks. From the results of a series of studies using site-specific mutagenesis, Golbeck's group recently suggested that those cysteines at positions 11, 14, 17, 58 are ligands for FeS$_B$ and 21, 48, 51, 54 for FeS$_A$ [31]. The overall primary structure resembles those of bacterial ferredoxins with two 4Fe–4S iron–sulfur clusters. Among them, a three-dimensional structure of a crystallized ferredoxin from *Peptococcus*

PsaC

```
01 SHSVKIYDTC IGCTQCVRAC PTDVLEMIPW DGCKAKQIAS APRTEDCVGC
51 KRCESACPTD FLSVRVYLWH ETTRSMGLGY/
```

FIGURE 9.5 A representative amino acid sequence of PsaC from spinach. Residues conserved throughout 22 species are written in bold letters. Species covered are: *Z. mays* (maize) [49]; *N. tabacum* (tobacco) [52]; *Triticum aestivum* (wheat) [54]; *Hordeum vulgare* (barley) [55]; *Oryza sativa* (rice) [57]; *P. sativum* (garden pea) [54]; *S. oleracea* (spinach) [58]; *M. polymorpha* (liverwort) [56]; *Antithamnion* sp. [59]; *C. reinhardtii* [60]; *E. gracilis* [61]; *Fremyella diplosiphon* (*calothrix* PCC 7601) [62]; *Nostoc* sp. PCC 8009 [63]; *Cyanophora paradoxa* [64]; *Calothrix* sp. PCC 7601 [65]; *Anabaena* sp. PCC 7120 [66]; *S. elongatus* [67]; *S. vulcanus* [68]; *Synechococcus* sp. PCC 7002 (*Agmenellum quadruplicatum*) [64]; *Synechocystis* sp. PCC 6803 [69]; *Synechococcus* sp. PCC6301 [70]. Sequence information has been updated using the BLAST (NCBI) and FASTA (DDBJ) databases.

aerogenes has been proposed on the basis of x-ray crystallography [72]. Based on this structure, a number of workers came up with possible three-dimensional structures of the PsaC holoprotein [58, 73,74]. The crystallography of a cyanobacterial Type II preparation mentioned before has also supported these earlier contentions and revealed more solid structural features [20]: Though PsaC harboring two Fe_4S_4 clusters exhibits pseudo-twofold symmetry similar to that of bacterial $2Fe_4S_4$ ferredoxins, it contains an insertion of ten amino acids in the loop connecting the iron–sulfur cluster binding motifs and extensions of the N and C termini by two and 14 amino acids, respectively. As the insertion extrudes as a large loop, it may be engaged in docking of ferredoxin or flavodoxin. The long C terminus of PsaC interacts with PsaA/B/D and appears to be important for the proper assembly of PsaC into the PSI complex [20].

3. PsaD (Subunit II)

Lately, the role of this subunit, once thought to be essential, may not seem as important as those chloroplast genome encoded subunits described above. As shown in Figure 9.6, the sizes and amino acid sequences of this subunit, like other smaller subunits, are quite diverse among species, in contrast to those core subunits described above (PsaA, PsaB, and PsaC). The degrees of homology are fairly low among higher plants and also among cyanobacteria. It was first reported that a mutant of a cyanobacterium that lacked *psaD*, the corresponding gene, could not grow autotrophically [88]. But under more controlled conditions, the same strain seemed to survive well in the light without an organic carbon source (H. Nakamoto et al., unpublished results). Golbeck's group first reported that PsaD was essential for reconstitution of a PS I complex using PsaA, PsaB, and PsaC [89], but later said that it was needed only for a "stable" binding of PsaC [90]. Nevertheless, the ubiquitous presence of this subunit as well as the other

two (PsaE and PsaL) in purified preparations of the PS I complex [17] indicates that these polypeptides are essential constituents of PS I and are required at least in higher plants for the integrity and stability of the complex. The crystallography has again revealed that PsaD forms an antiparallel, four-stranded β-sheet, in which the loop connecting the third and fourth strands contains an α-helix, followed by a two-stranded β-sheet [20]. The loop segment extending from His^{95} to the C terminus is attached by numerous hydrogen bonds to the sides of PsaC and PsaE exposed to stroma [20].

4. PsaE (Subunit IV)

The sequences are shown in Figure 9.7. The corresponding gene, *psaE,* is nucleus encoded in higher plants. The overall homology among species is no better than that in PsaD and other nucleus encoded subunits. A cyanobacterial mutant that lacks this protein grows well under autotrophical conditions [86]. The fact that this subunit remains to be bound even in the simplest Type II preparation [18], nevertheless, suggests an essential role of this subunit. The structure of PsaE consists of a five-stranded antiparallel β-barrel [20].

5. PsaF (Subunit III)

The sequences are shown in Figure 9.8. The corresponding gene, *psaE,* is nucleus encoded in higher plants. This subunit is usually removed in the first step of Triton treatment of higher plant chloroplasts and does not remain in final preparations [12]. In cyanobacteria, however, the protein seems to be bound tightly to thylakoids [102]. The role of this subunit remains unclear despite an earlier claim of it being a plastocyanin-docking protein [12]. The protein was even implicated as a part of other complexes: a ferredoxin:plastoquinone oxidoreductase complex [104] and a light harvesting complex [105]. In the thermophilic cyanobacterial Type II preparation

PsaD

```
 1 Cucumis sativus (cucumber):        . . . . . . . . . .  . .EAETSV-- -------EAP AGFSPPELDP STPSPIFAGS TGGLLRKAQV
 2 Hordeum vulgare (barley):          AAAPDTPAPA APPAEP---- -------AP AGFVPPQLDP STPSPIFGGS TGGLLRKAQV
 3 Lycopersicon esculentum (tomato):  . . . . . . . . . .  . . .AEEAP-A AT-EEKP-AP AGFTPPQLDP NTPSPIFGGS TGGLLRKAQV
 4 Nicotiana sylvestris (wood tobacco): . . . . . . . . .  . . .AEE---A AT-KEA-EAP VGFTPPQLDP NTPSPIFGGS TGGLLRKAQV
 5 Spinacia oleracea (spinach):       . . . . . . . . . .  AAAAEGK--A ATPTETKEAP KGFTPPELDP NTPSPIFGGS TGGLLRKAQV

 6 Fremyella diplosiphon:             . . . . . . . . . .  . . . . . . .A ET-L-S---- -G-Q------ -T--PLFAGS TGGLLKKAEV
 7 Synechococcus elongatus:           . . . . . . . . . .  . . . . . . .  .TTL-T---- -G-QP----- ----PLYGGS TGGLLSAADT
 8 Synechococcus PCC6301:             . . . . . . . . . .  . . . . . . .A ET-L-T---- -G-K------ -T--PVFGGS TGGLLKSAET
 9 Anabaena variabilis:               . . . . . . . . . .  . . . . . . .  .TTL-T---- -G-QP----- ----PLYGGS TGGLLSAADT
10 Synechocystis PCC6803:             . . . . . . . . . .  . . . . . . .  .TEL-S---- -G-QP----- ----PKFGGS TGGLLSKANR
11 Synechococcus PCC7002:             . . . . . . . . . .  . . . . . . .S NE-L-T---- -G-K------ -T--PKFGGS TGGLLAAAET
12 Synechococcus vulcanus :           . . . . . . . . . .  . . . . . . .  .TTL-T---- -G-QP----- ----PLYGGS TGGLLSAADT
13 Nostoc sp PCC8009 :                . . . . . . . . . .  . . . . . . .MA E-QL-S---- -G-K-------T--PLFAGS TGGLLTKANV

 1 Cucumis sativus (cucumber):        EEFYVITWES PKEQIFEMPT GGAAIMREGP NLLKLARKEQ CLALGTR-LR SKYK--I-KY
 2 Hordeum vulgare (barley):          EEFYVITWTS PKEQVFEMPT GGAAIMREGP NLLKLARKEQ CLALGNR-LR SKYK--I-AY
 3 Lycopersicon esculentum (tomato):  EEFYVITWES PKEQIFEMPT GGAAIMRQGP NLLKLARKEQ CLALGTR-LR SKYK--I-NY
 4 Nicotiana sylvestris (wood tobacco): EEFYVITWES PKEQIFEMPT GGAAIMREGA NLLKLARKEQ CLALGTR-LR SKYK--I-NY
 5 Spinacia oleracea (spinach):       EEFYVITWES PKEQIFEMPT GGAAIMREGP NLLKLARKEQ CLALGTR-LR SKYK--I-KY

 6 Fremyella diplosiphon:             EEKYAITWTS PKAQVFELPT GGAATMQQGQ NLLYLARKEY GIALGG-QLR -KFK--ITDY
 7 Synechococcus elongatus:           EEKYAITWTS PKEQVFEMPT AGAAVMREGE NLVYFARKEQ CLALAAQQLR PR-K--INDY
 8 Synechococcus PCC6301:             EEKYAITWTS TKEQVFELPT GGAAVMHEGD NLLYFARKEQ ALALGT-QLR TKFKPKIESY
 9 Synechococcus vulcanus:            EEKYAITWTS PKEQVFEMPT AGAAVMREGE NLVYFARKEQ CLALAAQQLR PR-K--INDY
10 Synechocystis PCC6803:             EEKYAITWTS ASEQVFEMPT GGAAIMNEGE NLLYLARKEQ CLALGT-QLR TKFKPKIQDY
11 Synechococcus PCC7002:             EEKYAITWSS AKEQVFELPT GGAAVMNEGD NLMYFARKEQ CLALGT-QLK TQFKPKITDY
12 Synechococcus vulcanus :           EEKYAITWTS PKEQVFEMPT AGAAVMREGE NLVYFARKEQ CLALAAQQLR PR-K--INDY
13 Nostoc sp.PCC8009 :                EEKYAITWTS PKAQVFELPT GGAATMNQGE NLLYLARKEQ GIALGGQ-LR -KFK--ITDY

 1 Cucumis sativus (cucumber):        QFYRVFPNGE VQYL-HPKDG VYPEKVNPGR EGVGQNFRSI GKNVSPIEVK FTGKQVYDL/
 2 Hordeum vulgare (barley):          QFYRVFPNGE VQYL-HPKDG VYPEKVNAGR QGVGQNFRSI GKNVSPIEVK FTGKNSFDI/
 3 Lycopersicon esculentum (tomato):  QFYRVFPNGE VQYL-HPKDG VYPEKVNPGR EGVGQNFRSI GKNKSAIEVK FTGKQVYDI/
 4 Nicotiana sylvestris (wood tobacco): RFYRVFPNGE VQYL-HPKDG VYPEKVNAGR QGVGQNFRSI GKNKSPIEVK FTGKQVYDL/
 5 Spinacia oleracea (spinach):       QFYRVFPSGE VQYL-HPKDG VYPEKVNPGR QGVGLNMRSI GKNVSPIEVK FTGKQPYDL/

 6 Fremyella diplosiphon:             KIYRILPGGE -TTLIHPADG VFPEKVNAGR EKVRFVPRRI GENPNPSAIK FSGKYTYDA/
 7 Synechococcus elongatus:           KIYRIFPDGE -TVLIHPKDG VFPEKVNKGR EAVNSVPRSI GQNPNPSQLK FTGKKPYDP/
 8 Synechococcus PCC6301:             KIYRIFPGGD VQYL-HPKDG VFPEKVNEGR SFAGKVDRRI GQNPNPATIK FTANSPTRL/
 9 Synechococcus vulcanus:            KIYRIFPDGE -TVLIHPKDG VFPEKVNKGR EAVNSVPRSI GQNPNPSQLK FTGKKPYDP/
10 Synechocystis PCC6803:             KIYRVYPSGE VQYL-HPADG VFPEKVNEGR EAQGTKTRRI GQNPEPVTIK FSGKAPYEV/
11 Synechococcus PCC7002:             KIYRVFPPSE TQFLY-PLDG VPSEKVNEGR EYKGKVDRNI GSNPEPATLK FSGVAPYEA/
12 Synechococcus vulcanus :           KIYRIFPDGE -TVLIHPKDG VFPEKVNKGR EAVNSVPRSI GQNPNPSQLK FTGKKPYDP/
13 Nostoc sp. PCC8009 :               KIYRIFPNGE -TTFIHPADG VFPEKVNEGR EKVRFVPR-I GQNPSPAQLK FSGKYTYDA/
```

FIGURE 9.6 Amino acid sequences of PsaD subunits: *Cucumis sativus* (cucumber) [75]; *H. vulgare* (barley) [76]; *Lycopersicon esculentum* (tomato) [77]; *Nicotiana sylvestris* (wood tobacco) [78]; *S. oleracea* (spinach) [79]; *P. sativum* (garden pea) [80]; *Fremyella diplosiphon* (*Calothrix*) PCC 7601 [81]; *A. variabilis* [82]; *S. elongatus* [83]; *Synechococcus* sp. PCC 6301 [84]; *S. vulcanus* [85]; *Synechocystis* sp. PCC 6803 [86]; *Synechococcus* sp. PCC 7002 [87]; *Nostoc* sp. PCC8009 [88]. Bold letters represent homologous residues among either higher plants or cyanobacteria; underlined ones are common to all species. Sequence information has been updated using the BLAST (NCBI) and FASTA (DDBJ) databases.

[20], PsaF is tightly bound and contributes prominent structural features to this surface of PSI with two hydrophilic α-helices at the N terminus of a transmembrane helix. As the shortest distance between their helix axes and the pseudo-C_2 axis is 27 Å, direct interaction with cytochrome c_6 or plastocyanin is unlikely [20].

6. PsaG (Subunit V) and PsaH (Subunit VI)

Homologs of these two nucleus coded subunits have not been reported in cyanobacteria. As seen in Figure 9.9 and Figure 9.10, the sequences of PsaG and PsaH

of *Chlamydomonas*, a green algae, are remarkably different from those of their higher plant homologs. They are so different that there is even some possibility that the *Chlamydomonas* PsaG and PsaH may not be the homologs of the corresponding proteins of higher plants. On the other hand, the homologies among higher plants are very good. The roles of these subunits have yet to be elucidated.

7. PsaI (Subunit X) and PsaJ (Subunit IX)

These two subunits are usually blocked at the N terminal and, as a consequence, were not recognized

PsaE

```
 1 Hordeum vulgare (barley):        AEEPTAAAPA EPAPAADEKP EEAAVATKEPA KAKPPPRGPK RGTKVKILRR ESYWYNGTGS
 2 Spinacia oleracea (spinach):     AAEEAAAAPA AASPEGEAPK --------A AAKPPPIGPK RGSKVRIMRK ESYWYKGVGS
 3 Chlamydomonas reinhardtii:       .......... .......... ......EEVK AAPKKEVGPK RGSLVKILRP ESYWFNQVGK

 4 Synechococcus PCC7002:           .......... .......... .......AIE RGSKVKILRK ESYWYGDVGT
 5 Synechococcus PCC6301:           .......... .......... .......AIA RGDKVRILRP ESYWFNEVGT
 6 Synechocystis PCC6803:           .......... .......... .......ALN RGDKVSIKRT ESYWYGDVGT
 7 Fremyella diplosiphon:           .......... .......... ........VQ RGSKVRILRP ESYWFQDIGT
 8 Porphyra umbilicalis:            .......... .......... ........ME RGSKVKILRK ESYWYQEVGT
 9 Anabaena variabilis:             .......... .......... ........VQ RGSKVRILRP ESYWFQDVGT
10 Synechococcus elongatus:         .......... .......... ........VQ RGSKVKILRP ESYWYNEVGT
11 Nostoc sp. PCC8009:              .......... .......... ........VQ RGSKVRILRP ESYWFQDVGT


 1 Hordeum vulgare (barley):        VVTVDQDPNT RYPVVVRFAK VNYAGVSTNN YALDEIKEVA A/
 2 Spinacia oleracea (spinach):     VVAVDQDPKT RYPVVVRFNK VNYANVSTNN YALDEIQEVA /
 3 Chlamydomonas reinhardtii:       VVSVDQ-SGV RYPVVVRFEN QNYAGVTTNN YALDEVVAAK /

 4 Synechococcus PCC7002:           VASVD-QSGI KYPVVVRFEK VNYNGFSGSA GGLNTNNFAE HELEVVG/
 5 Synechococcus PCC6301:           VASVE-KSGI LYPVIVRFDR VNYNGFSGSD GGVNTNNFAE AELQVVAAAA KK/
 6 Synechocystis PCC6803:           VASID-QSGI KYSVIVRFDK VNYNGFSGSA SGVNTNNFAE DELLFAE NELELVQAAA K/
 7 Fremyella diplosiphon:           VAAMD-KSGI KYPVLVRFEK VNY------ NNVNTNSFAD NELIEVA PPAAK/
 8 Porphyra umbilicalis:            VASVD-QSGI KYPVIVRFDK VNY------ SGINTXNFAV XELIDLG K/
 9 Anabaena variabilis:             VASVD-QSGI KYPVIVRFDK VNY------ SGINTXNFAV XELIN/
10 Synechococcus elongatus:         VASIDK-SGI IYPVIVRFNK VNYTGYSGSA SGVNTNNFAL HEVQEVAPPK KGK/
11 Nostoc sp. PCC8009:              VASVD-QSGI KYPVIVRFEK VNY------ SGINTNNFAE DELVEVE APKAKPKK/
```

FIGURE 9.7 Amino acid sequences of PsaE subunits: *H. vulgare* (barley) [92]; *S. oleracea* (spinach) [79]; *C. reinhardtii* [93]; *Synechococcus* PCC 7002 [95]; *Synechococcus* PCC 6301 [97]; *Synechocystis* PCC 6803 [91]; *F. diplosiphon* [81]; *Porphyra umbilicalis* [98]; *A. variabilis* [82]; *S. elongatus* [99]; *Nostoc* sp. PCC 8009 [63]. Bold letters represent homologous residues among either higher plants or cyanobacteria; underlined ones are common to all species. Sequence information has been updated using the BLAST (NCBI) and FASTA (DDBJ) databases.

PsaF

```
 1 Hordeum vulgare (barley):        ADIAGLTPCK ESKAFAKREK QSVKKLNSSL KKYAPDSAPA LAIQATIDKT KRRFENYGKF
 2 Spinacia oleracea (spinach):     ADIAGLTPCK ESKQFAKREK QALKKLQASL KLYADDSAPA LAIKATMEKT KKRFDNYGKY
 3 Chlamydomonas reinhardtii.:      ADIAGLTPCS ESKAYAKLEK KELKTLEKRL KQYEADSAPA VALKATMERT KARFANYAKA

 4 Synechococcus elongatus:         .DVAGLVPCK DSPAFQKRAA AAVNTT---- ----AD--PA SG-------Q K-RFERYSQA
 5 Synechocystis sp. PCC 6803:      DDFANLTPCS ENPAYLAKSK NFLNTT---- ----ND--PN SG-------- KIRAERYASA
 6 Synechococcus sp. PCC 7002:      DSLSHLTPCS ESAAYKQRAK NFRNTT---- ----AD--PN SG-------Q -NRAAAYSEA
 7 Synechococcus sp. PCC 6301:      .DVAGLTPTS ESPRFIQRAE --AAAT---- --------PQ --AKAR...
 8 Anabaena variabilis              .LGADLTPXA ENPAFQALAK NARNTT---- ----AD--PQ SGQK-RFEXY S...


 1 Hordeum vulgare (barley):        GLLCGSDGLP HLI----VSG DQRHWGE-F- ITPGVLFLYI AGWIGWVGRS YLIAVSGEKK
 2 Spinacia oleracea (spinach):     GLLCGSDGLP HLI----VSG DQRHWGE-F- ITPGILFLYI AGWIGWVGRS YLIAIRDEKK
 3 Chlamydomonas reinhardtii.:      GLLCGNDGLP HLIADPGLAL KYGHAGEVF- I-PTFGFLYV AGYIGYVGRQ YLIAVKGEAK

 4 Synechococcus elongatus:         --LCGEDGLP HLVVD-GRLS R---AGD-FL I-PSVLFLYI AGWIGWVGRA YLIAVRNSGE
 5 Synechocystis sp. PCC 6803       --LCGPEGYP HLIVD-GRFT ---HAGD-FL I-PSILFLYI AGWIGWVGRS YLIEIRES-K
 6 Synechococcus sp. PCC 7002:      --LCGPEGLP HLIVD GRLD ---HAGE-FL I-PSLLFLYI AGWIGWAGRA YLIAVRDE-K


 1 Hordeum vulgare (barley):        -------PAM REIIIDVELA ARIIPRGFIW PVAAYRELIN GDLVVDDADI GY/
 2 Spinacia oleracea (spinach):     -------PTQ KEIIIDVPLA SSLLFRGFSW PVAAYRELLN GELV--DNNF /
 3 Chlamydomonas reinhardtii.:      FAWPLAAPTD KEIIIDVPLA TKL-----AW QG--LKELAS GELTAKDNEI -TVSPR/

 4 Synechococcus elongatus:         -ANE------ KEIIIDVPLA IKCMLTGVQE -LQR------ GTLLEKEENI -TVSPR/
 5 Synechocystis sp. PCC 6803       -NAAM----- QEVVINVPLA IKKMLGGFLW PLAAVGEYTS GKLVMKDSEI PT-SPR/
 6 Synechococcus sp. PCC 7002:      -DPEM----- QEVVVNVPRA FSKMLAGFAW PLAALKEFTS GELVVKDADV PT-SPR/
```

FIGURE 9.8 Amino acid sequences of PsaF subunits: *C. reinhardtii* [93]; *H. vulgare* (barley [94]); *Synechocystis* sp. PCC 6803 [100]; *S. oleracea* (spinach, Ref. [101]); *S. elongatus* [96]; *Synechococcus* PCC7002 [102]; *A. variabilis* [82]; *Synechococcus* sp. PCC 6301 [103]. Bold letters represent homologous residues among either higher plants or cyanobacteria; underlined ones are common to all species. Sequence information has been updated using the BLAST (NCBI) and FASTA (DDBJ) databases.

PsaG

```
1 Spinacia oleracea (spinach):   ELSPSLVISL STGLSLFLGR FVFFNFQREN MAKQ-VPEQN GMSHFEAGDT RAKEYVSLLK
2 Hordeum vulgare (barley):      ALEPSVVISL STGLSLVMGR FVFFNFQREN VAKQ-VPEQN GKTHFEAGDE RAKEFAGILK
3 Pisum sativum (garden pea):    ELNPSLVISL STGLSLFLGR FVFFNFQREN VAKQGLPEQ...
4 Chlamydomonas reinhardtii:     ALDPQIVISG STAAFLAIGR FVFLGYQRRE ANFDSTVGPK TTGATYFDDL QKNSTIFATN
```

```
1 Spinacia oleracea (spinach):   SNDPVGFNIV DVLAWGSIGH IVAYYILATA S-NGYDPSFF/
2 Hordeum vulgare (barley):      SNDPVGFNLV DVLAWGSIGH IVAYYILATT S-NGYDPPFFG/
4 Chlamydomonas reinhardti:      DPA--GFNII DVAGWGALGH AVGFAVLAIN SLQGANLS/
```

FIGURE 9.9 Amino acid sequences of PsaG subunits: *C. reinhardtii* [106]; *H. vulgare* (barley) [107]; *P. sativum* (garden pea) [80]; *S. oleracea* (spinach) [101]. Sequence information has been updated using the BLAST (NCBI) and FASTA (DDBJ) databases.

PsaH

```
1 Spinacia oleracea (spinach):   KYGDKSVYFD LEDIANTTGQ WDVYGSDAPS PYNSLQSKFF ETFAAPFTKR GLLLKFLILG
2 Oryza sativa indica (rice) :   KYGEKSVYFD LEDIGNTTGQ WDLYASDAPS PYNPLQSKFF ETFAGPFTKR GLLLKFLLLG
3 Hordeum vulgare (barley) :     KYGEKSVYFD LDDIANTTGQ WDLYGSDAPS PYNGLQSKFF NTFAAPFTKR GLLLKFLLIG
4 Pisum sativum (garden pea):    KYGDKSVYFD LEDIGNTTGQ WDLYGSDAPS PYSXLQ...

5 Chlamydomonas reinhardtii:     KYGENSRYFD LQDMENTTGS WDMYGVDEKK RYPDNQAKFF TQATDIISRR ESLRALVALS
```

```
1 Spinacia oleracea (spinach):   G-GSLLTYVS ANAPQDVLPI TRGPQQPPKL GPRGKI/
2 Oryza sativa indica (rice) :   G-GSLVAYVS ASASPDLLPI KKGPHVPPTP GPRGKI/
3 Hordeum vulgare (barley) :     G-GSLVAYVS ASASPDLLPI KKGPQLPPTP GPRGKI/

5 Chlamydomonas reinhardtii:     GIAAIVTYGL KGAKDADLPI TKGPQTT--- GENGKGGSVR SRL/
```

FIGURE 9.10 Amino acid sequences of PsaH subunits: *C. reinhardtii* [106]; *H. vulgare* (barley) [108]; *O. sativa indica* (rice) [109]; *P. sativum* (garden pea) [80]; *S. oleracea* (spinach) [110]. Sequence information has been updated using the BLAST (NCBI) and FASTA (DDBJ) databases.

as PS I subunits until recently. The sequences of cyanobacterial "homologs" only slightly resemble those of higher plants as seen in Figure 9.11 and Figure 9.12. The corresponding genes of the higher plant polypeptides are encoded in chloroplast DNA. Although the roles of these two subunits are not known yet, PsaI seems to be a part of cyanobacterial complexes [20].

8. PsaK (Subunit VIII)

This nucleus encoded subunit seems to be bound to thylakoid membranes, sometimes tightly [128,129] and sometimes loosely [121]. Again, the role is not clear yet. Cyanobacterial homologs are not exactly homologous to those of higher plants as seen in Figure 9.13. The only exceptions are remarkably homologous N terminal sequences (more than 30 residues).

9. PsaL (Subunit V')

This nucleus encoded subunit had long been neglected until recently despite its distinct presence, because the N termini are blocked in most cases. Although the role is not clear yet, this subunit is almost as ubiqui-

tous as PsaD and PsaE. In a spinach preparation, PsaL can be removed exclusively by heat treatment [17]. Possible homologs in cyanobacteria have been reported as seen in Figure 9.14, although the degrees of homology are low. It has been suggested that PsaL is necessary for forming a trimeric complex in cyanobacteria [20].

10. PsaM

In the EMBL data bank, a group of short polypeptides are listed as PsaM (Figure 9.15). The corresponding gene, *psaM*, was found in chloroplast DNA of *Marchantia polymorpha* [138] and of *Euglena gracilis* [139]. No homologous genes (ORFs) have been found in the chloroplast DNA of either tobacco or rice, yet. Nor has any similar polypeptide been reported to be expressed in any higher plants yet. Despite all these, PsaM may be an essential part of cyanobacterial complexes as revealed by the crystallographical study [20].

11. PsaN

One set of amino acid sequences is listed under the name PsaN in the PIR protein sequence database

PsaI

```
 1 Hordeum vulgare (barley):        M-------TD -LNLP----- -----SIFVP LVGLVFPAIA MTSLFLYVQK ------KKIV/
 2 Zea mays (maize):                M-------TD F-NLP----- -----SIFVP LVGLVFPAIA MTSLFLYVQK ------NKIV/
 3 Marchantia polymorpha (liverwort): M----TASY- ---LP----- -----SIFVP LVGLIFPAIT MASLFIYIEQ ------DEIL/
 4 Oryza sativa nipponbare (rice):  MM----- D-NLP----- -----SIFVP LVGLVFPAIA MASLFLYVQK ------NKIV/
 5 Angiopteris lygodiifolia (turnip fern): M---TASY-- ---LP----- -----SIFVP LVGLVFPAIT MASLSIYIEQ ------DEIV/
 6 Nicotiana tabacum (tobacco):     M---TN---- -LNLP----- -----SIFVP LVGLVFPAIA MASLFLHVQK ------NKIV/
 7 Triticum aestivum (wheat):       M-------TD -LNLP----- -----SIFVP LVGLVFPAIA MTSLFLYVQK ------NKIV/
 8 Pisum sativum (garden pea):      M--I------ --NLP----- ---------- ---------- ----FLHVEK RLLFSTKKIN/
 9 Spinacia oleracea (spinach):     M--------- --NFP----- -----SIFVP LVGLVFPAI...

10 Synechococcus elongatus:         MMGSYAAS-- -F-LP----- -----WIFIP VVCWLMPTVV MGLLFLYIEG EA/
11 Synechocystis PCC 6803:          M---NGAYAA SF-LP----- -----VILVP LAGVVFPALA MGLLFNYIES DA/
12 Synechococcus PCC7002:           M---DGSYAA SY-LP----- -----VILIP MVGWLFPAVT MGLLFIHIES EGEG/
13 Anabaena variabilis:             M--ATA---- -F-LPSILAD ASFLSSIFVP VIGWVVPIAT FSFLFLYIEG EDVA/
```

FIGURE 9.11 Amino acid sequences of PsaI subunits: *A. variabilis* ATCC 29413 [111]; *Angiopteris lygodiifolia* (turnip fern) [112]; *H. vulgare* (barley) [113]; *Z. mays* (maize) [114]; *M. polymorpha* (liverwort) [115]; *O. sativa* Nipponbare (rice) [116]; *P. sativum* (garden pea) [117,118]; *S. elongatus* [120]; *N. tabacum* (tobacco) [121]; *T. aestivum* (wheat) [125]; *Synechocystis* PCC 6803 (H. Nakamoto, unpublished data); *Synechococcus* PCC7002 [95]; *A. variabilis* [111]. Bold letters represent homologous residues among either higher plants or cyanobacteria; underlined ones are common to all species. Sequence information has been updated using the BLAST (NCBI) and FASTA (DDBJ) databases.

PsaJ

```
 1 Zea mays (maize):                M--------- RDIKTYLSVA PVLSTLWFGA LAGLLIEINR LFPDALSFPF F/
 2 Oryza sativa Nipponbare (rice):  M--------- RDIKTYLSVA PVVSTLWFGA LRGLLIEINR LFPDALSFPF FSF/
 3 Spinacia oleracea (spinach):     M--------- RDFKTYLSVA PVLXT...
 4 Nicotiana tabacum (tobacco):     M--------- RDLKTYLSVA PVLSTLWFGA LAGLLIEINR FFPDALTFPF FSF/
 5 Pisum sativum (garden pea):      M--------- RDLKTYLXVA PV...
 6 Marchantia polymorpha (liverwort): M--------- QDVKTYLSTA PVLATLWFGF LAGLLIEINR FFPDALVLPF F/
 7 Euglena gracilis:                M--------- KYFTTYLSTA PVVAVLWFTL TASLLIEINR FFPDIL/

 8 Synechococcus elongatus Naegeli: M--------- KHFLTYLSTA PVLAAIWMTI TAGILIEFNR FYPDLLFHPL /
 9 Synechococcus vulcanus:          M--------- KHFLTYLSTA PVLA------ LAGLLIEINR FFPDALTFPF FSF/
10 Synechococcus sp. PCC7002:       D--------- K-F---LSSA PVLLTAMMVF TAGLLIEFNR FFPDLLFHP/
11 Synechocystis sp. PCC 6803:      D-------GL KSF---LSTA PVMIMALLTF TAGILIEFNR FYPDLLFHP/
12 Anabaena variabilis:             ADKADQSSYL IKF---ISTA PVAATIXLII TAGILIEFNX FFPXLL...
```

FIGURE 9.12 Amino acid sequences of PsaJ subunits: *E. gracilis* [126]; *Z. mays* (maize) [115]; *M. polymorpha* (liverwort) [116]; *O. sativa* Nipponbare (rice) [117]; *P. sativum* (garden pea) [121]; *S. elongatus* [119,123]; *S. vulcanus* [127]; *N. tabacum* (tobacco) [124]; *S. oleracea* (spinach, partial) [122]; *Synechococcus* sp.PCC7002 and *Synechocystis* sp. PCC 6803 [102]; *A. variabilis* [111]. Bold letters represent homologous residues among either higher plants or cyanobacteria; underlined ones are common to all species. Sequence information has been updated using the BLAST (NCBI) and FASTA (DDBJ) databases.

PsaK

```
1: D-YIGSSTNLIMVTTTTLMLFAGRFGLAPSANRKATAGLKLEARESGLQ----TGDPAGFTLADTLACGAVGHIMGVGIVLGLKNTGVLDQIIG/
2: .........IMVTTTTLMLFAGRFGLAPSANRKATAGLKLEARDSGLQ----TGDPAGFTLADTLACGAVGHILGVGIVLGLKNTGALDQIIG/
3: D-FIGSSTNLIMVTSTTLMLFAGRFGLAPSANRKATAGLRLEARDSGLQ----TGDPAGFTLADTLACGTVGHIIGVGVVLGLKNIGAI/
4: .........IMVTTTTLMLFAGRFGLAPSANRKSTAGLKLEARDSGLQ----TGDPAGFTLADTLACGAVGHIMGVGVVLGLKNIGVLDQIIG/
5: D-FIGSPTNLIMVTSTSLMLFAGRFGLAPSANRKATAGLKLEVRDSGLQ----TGDPAGFTLADTLACGVVGHIIGVGVVLGLKNIGAL/
6: DGFIGSSTNLIMVASTTATLAAARFGLAPTVKKNTTAGLKL-V-DSKNSAGVISNDPAGFTIVDVLAMGAAGHGLGVGIVLGLKGIGAL/

7: -TLPDTTWTP-S--VGLVVILCNLFAIALGRYAIQSRGKGPGLPIALPALFE------GFGLPELLATTSFGHLLAAGVVSGLQYAGAL/
8: AAVP-TTMA-WSPKVAVVMVICNVLAIAIGKATIKHPSEGPELP-M-PDMFG------GMGLPALLATTSFGHILGVGVILGLGSMGAI/
9: ----------------GMIACNILAIAFGKLTIKQQNVG--TPMPSSNFFG------GFGLGAVLGTASFGHILGAGVILGLANMGVL/
```

Higher plants: *1, Hordeum vulgare; 2, Zea mays; 3, Arabidopsis thaliana; 4, Oryza sativa; 5, Nicotiana tabacum; 6, Chlamydomonas.*

Cyanobacteria: *7, Synechococcus elongatus; 8, Synechococcus PCC7002; 9, Synechocystis PCC 6803.*

FIGURE 9.13 Amino acid sequences of PsaK subunits: *C. reinhardtii* [100]; *H. vulgare* (barley) [130]; *P. sativum* (garden pea, partial) [111]; *S. oleracea* (spinach, partial) [115]; *S. elongatus* [118]; *S. vulcanus* (partial) [111]; *A. variabilis* (partial) [119]; *Synechococcus* PCC7002 [95]. Bold letters represent homologous residues among either higher plants or cyanobacteria; underlined ones are common to all species. Sequence information has been updated using the BLAST (NCBI) and FASTA (DDBJ) databases.

PsaL

```
 1 : Hordeum vulgare (barley)      KAVKSDKPTY  QVVQPINGDP  FIGSLETPVT  SSPLVAWYLS  NLPAYRTAVS  PLLRGIEVGL
 2 : Spinacia oleracea (spinach)   RAIKTEKPTY  QVIQPLNGDP  FIGGLETPVT  SSPLIAWYLS  NLPAYRTAVN  PLLRGVEVGL
 3 : Oryza sativa                  KAIQSEKPTY  QVVQPINGDP  FIGSLETPVT  SSPLVAWYLS  NLPAYRTAVS  PLLRGIEVGL
 4 : Zea mays                      KAIQPEKATY  QVVQPINGDP  FIGSLETPVT  SSPLVAWYLS  NLPAYRTAVS  PLLRGVEVGL

 5 : Cyanidium caldarium           ..........  ..IKPYGSNP  FVGNLSTPVN  SSKVTIWYLK  NLPIYRRGLS  PLLRGLEIGM
 6 : Porphyra purpurea             ..........  EFIKPYNDDP  FVGNLSTPVS  TSSFSKGLLG  NLPAYRRGLS  PLLRGLEIGM
 7 : Prochlorococcus marinus       ..........  ..VTPVS-DP  CVGNLSTPVN  SGYFTKAFLN  NLPFYRGGLS  PNFRGLEVGA

 8 : Synechococcus elongatus       ......AEE.  .LVKPYNGDP  FVGHLSTPIS  DSGLVKTFIG  NLPAYRQGLS  PILRGLEVGM
 9 : Synechocystis PCC 6803        .....MAESN  QVVQAYNGDP  FVGHLSTPIS  DSAFTRTFIG  NLPAYRKGLS  PILRGLEVGM
10 : Nostoc sp. PCC 7120           KNLPSDPRNR  EVVFPAGRDP  QWGNLETPVN  ASPLVKWFIN  NLPAYRPGLT  PFRRGLEVGM
11 : Synechocystis sp.             ..........  QVVQAYNGDP  FVGHLSTPIS  DSAFTRTFIG  NLPAYRKGLS  PILRGLEVGM
12 : Synechococcus PCC7942         ..........  .......GTP  EIGNLATPIN  SSPFTRTFIN  ALPIYRRGLS  SNRRGLEIGM


 1 : Hordeum vulgare (barley)      AHGYLLVGPF  ALTGPLRNTP  VHGQAGTLGA  IGLVSILSVC  LTMYGVASFN  EGEPSTAPVL
 2 : Spinacia oleracea (spinach)   AHGFLLVGPF  VKAGPLRNTE  YAGAAGSLAA  AGLVVILSMC  LTMYGIASFK  EGEPSIAPAL
 3 : Oryza sativa                  AHGYLLVGPF  ALTGPLRNTP  VHGQAGALGA  AGLVAILSVC  LTMYGVASFG  EGEPSTAPTL
 4 : Zea mays                      AHGYLLVGPF  ALTGPLRSTP  VHGQAGALGA  AGLVAILSVC  LTMYGVASFN  EGDPSTAPTL

 5 : Cyanidium caldarium           AHGYFIIGPF  YKLGPLRNTD  LSLLSGLIAA  IGLIIISSIA  MIIYGIVTFD  NSENN-----
 6 : Porphyra purpurea             AHGYFLIGPF  DKLGPLRGTD  VALLAGFLSS  VGLIIILTTC  LSMYGNVSFT  ----------
 7 : Prochlorococcus marinus       AFGYLLYGPY  AMTGPLRNTD  YALTAGLLGT  IGAVHILTAL  LVLYNA----  PGKAPNIQPS

 8 : Synechococcus elongatus       AHGYFLIGPW  VKLGPLRDSD  VA----NLG-  -GLISGIALI  LV--ATACL-  -AAYGLV
 9 : Synechocystis PCC 6803        AHGYFLIGPW  TLLGPLRDSE  YQ----YIG-  -GLIGALALI  LV--ATAAL-  -SSYGLV
10 : Nostoc sp. PCC 7120           AHGYFLFGPF  AKLGPLRDAA  NANLAGLLGA  IGLVVLFTLA  LSLYA----N  SNPPTALASV
11 : Synechocystis sp.             AHGYFLIGPW  TLLGPLRDSE  YQYIGGLIGA  LALILVATAA  L---------  -SSYGLVTFQ
12 : Synechococcus PCC7942         AHGFLLYGPF  SILGPLRNTE  TAGSAGLLAT  VGLVVILTVC  LSLYG----N  AGSGPSAAES


 1 : Hordeum vulgare (barley)      TLTGRKKEAD  KLQTAEGWSQ  FTGGFFFGGVS  GAVWAYFLL  YVLDLPYFFK/
 2 : Spinacia oleracea (spinach)   TLTGRKKQPD  QLQSADGWAK  FTGGFFFGGVS  GVTWACFLM  YVLDLPYYFK/
 3 : Oryza sativa                  TLTGRKKEAD  KLQTADGWAK  FTGGFFFGGIS  GVLWAYFLL  YVLDLPYFFK/
 4 : Zea mays                      TLTGRKKEAD  KLQTADGWAK  FTGGFFFGGIS  GVLWAYFLL  YVLDLPYFFK/

 5 : Cyanidium caldarium           ---------D  KLQTANGWRQ  LASGFLL----  GAVGGAGFA  YILL/
 6 : Porphyra purpurea             ----RSDSKD  PLQTAEGWGQ  FTAGFLV----  GAVGGSGFA  YLLL/
 7 : Prochlorococcus marinus       DCTINNPPAD  -LFTRSGWAD  FTSGFWLGGCG  GAV----FA  WLLCGTLHL/

 8 : Synechococcus elongatus       SFQKGGSSSD  PLKTSEGWSQ  FTAGFFVGAMG  SAFVAFFLL  ENFLLSMAS/
 9 : Synechocystis PCC 6803        TFQGEQGSGD  TLQTADGWSQ  FAAGFFVGGMG  GAFVAYFLL  EN-L-SVVDG  IFRGLFN/
10 : Nostoc sp. PCC 7120           TV---PNPPD  AFQSKEGWNN  FASAFLIGGIG  GAVVAYFL/
11 : Synechocystis sp.             ---GEQGSGD  TLQTADGWSQ  FAAGFFVGGMG  GAFVAYFLL/
12 : Synechococcus PCC7942         TVT-TPNPPQ  ELFTKEGWSE  FTSGFILGGLG  GAFFAFYL/
```

FIGURE 9.14 Amino acid sequences of PsaL subunits: *H. vulgare* (barley) [133]; *S. oleracea* (spinach) [134,137]; *C. caldarium* [131]; *A. variabilis* [82]; *Synechococcus* sp. PCC 6301 (partial) [103]; *S. elongatus* [135]; *S. vulcanus* (partial) [127]; *Synechocystis* sp. PCC 6803 [136]; underlined ones are common to all species. Sequence information has been updated using the BLAST (NCBI) and FASTA (DDBJ) databases.

(National Biomedical Research Foundation). These are sequences of "9 kDa polypeptides" [143], which had been tentatively designated as "PsaO" by Bryant [140]. Since then, homologous genes, *psaN*, have been cloned and sequenced in several higher plants as shown in Figure 9.16. No cyanobacterial homolog has been reported so far. This is another subunit whose function is unknown.

12. PsaX and PsaY

Two partial amino acid sequences were originally listed in the PIR data bank under the name of PsaX.

Now, two complete sequences are available at data banks. These are all from cyanobacteria (Figure 9.17). Recently, it was found that a substantial amount of another small (5 kDa) subunit was tightly bound to Type III PS I preparations from spinach and radish [178] as shown in Figure 9.18. The N terminal sequence of a similar, and most likely identical, polypeptide was reported some time ago in a crude PS II preparation from spinach [152], and has been designated as PsbW. Homologs of this polypeptide have been found in other species, and corresponding genes have been cloned and sequenced from many species, though details have not been

PsaM

```
1 Euglena gracilis:                      -EITTNQVY IALLASLIPA FFAFKLGKSL NQ/
2 Marchantia polymorpha (liverwort):     TSISDSQIV I-LVF--ITS ILALRLGKEL YQ/
3 Synechococcus elongatus naegeli:       -ALTDTQVY VALVIALLPA VLAFRLSTEL YK/
4 Synechococcus PCC7002:                 -GISDTQVL VALAIALIPG VLAFRLSTEL YK/
5 Synechococcus PCC6803:                 -ALSDTQIY AALVVALLPA FLAFRLSTEL YK/
6 Cyanophora paradoxa:                   --LADGQIF TALAVALVPG ILALRLALEL YKF/
7 Anthoceros formosae                    TSISDGQIV VALISAFIIV ILASRLGKEL YQ/
```

FIGURE 9.15 Amino acid sequences of PsaM subunits: *E. gracilis* [139]; *M. polymorpha* (liverwort) [138]; *S. elongatus* [177]; *Cyanophra paradoxa* [140]; *Synechococcus* PCC7002 [102]; *Synechococcus* PCC6803 [141]. Bold letters represent homologous residues among either higher plants or cyanobacteria; underlined ones are common to all species. Sequence information has been updated by using BLAST (NCBI) and FASTA (DDBJ) databases.

PsaN

```
Hordeum vulgare:      SVFDEYLEKS KLNKELNDKK RAATSGANFA RAYTVQFGSC KFPYNFTGCQ DLAKQKKVPF ITDDLEIECE GKEKFKCGSN VFWKW/
Zea mays:             TIFDEYLEKS KANKELNDKK RLATSGANFA RAYTVEFGSC QFPYNFTGCQ DLAKQKKVPF ISDDLEIECE GKEKFKCGSN VFWKW/
Arabidopsis thaliana: GVIDEYLERS KTNKELNDKK RLATSGANFA RAFTVQFGSC KFPENFTGCQ DLAKQKKVPF ISEDIALECE GKDKYKCGSN VFWKW/
Phaseolus vulgaris:   GVIEEYLEKS KTNKELNDKK RLATTGANFA RAYTVEFGSC KFPENFTGCQ DLAKQKKVPF LSDDLDLECE GKDKYKCGSN VFWKW/
Chlamydomonas:        GVVEDLQAKS AANKALNDKK RLATSYANLA RSRTVYDGTC TFPENFFGCE ELAFNKGVKF IAEDIKIECE GKTAKECGSK FTLRSN/
```

PsaN′ (PsaO)

```
Marchantia polymorphosa: MTIAFQLAVF ALIAISFLLV IGVPVVLASP EGWSSNKNVVF SGASLWIGL VFLVGILNSF IS/
Nicotana tobacum:        MTLAFQLAVF ALIATSLILL ISVPVFASP DGWSSNKNVVF SGTSLWIGL VFLVGILNSL IS/
Triticum aestivum:       MTIAFQLAVF ALIATSSILL ISVPVFASP DGWSSNKNVVF SGTSLWLGL VFLVAILNSL IS/
Spinacia oleracea:       MTIAFQLAVF ALIATSSILL ISVPVVFASP DGWSSNKNIVF SGTSLWLGL VFLVGILNSL IS/
Zea mays:                MNIAFQLAVF ALIATSSILL ISVPVVFASP DGWSSNKNIVF SGTSLWLGL VFLVAILNSL IS/
Pisum sativum:           MTIAFQLAVF ALIVTSSILL ISVPVVFASP DGWSSNKNVVF SGTSLWIGL VFLVGILNSL IS/
Hordeum vulgare:         MTIAFQLAVF ALIVTSSILL ISVPVVFASP DGWSSNKNVVF SGTSLWIGL VFLVAILNSL IS/
Oryza sativa:            MTIAFQLAVF ALIVTSSILL ISVPLVFASP DGWSNNKNVVF SGTSLWIGL VFLVAILNSL IS/

Cyanophora paradoxa:     MLIAFQGAVF ALVLLSFVLI VAVPVALASP GEWERSQRLI YAGAALWTSL IIVIGVLDSV VANQA/
```

FIGURE 9.16 Amino acid sequences of PsaN and PsaN′ (O): *H. vulgare* (barley) [142,145,149]; *S. oleracea* (spinach) [15,151] and *P. sativum* (garden pea) [143,148]; *C. sativus* (cucumber, partial) [144]; *O. sativa* (rice) [40,57]; *M. polymorpha* (liverwort) [15]; *Z. mays* (maize) [146]; *T. aestivum* (wheat) [147]; *C. paradoxa* [150]; *N. tobacum* [14]. Sequence information has been updated by using BLAST (NCBI) and FASTA (DDBJ) databases.

PsaX

```
Anabaena variabilis:                -AKAKTPAVAN TGAKPPYTFR TAXALLLLGV NFLVAAYYF...
Synechococcus vulcanus:             -A-TK------ -SAKPTYAFR TFXAVLLLAI NFLVAAY...
Nostoc sp. PCC 7120:                MAKAKISPVAN TGAKPPYTFR TGWALLLLAV NFLVAAYYFH IIQ/
Thermosynechococcus elongatus BP-1: MS-TM----AT KSAKPTYAFR TFWAVLLLAI NFLVAAYYFG ILK/
```

FIGURE 9.17 Amino acid sequences of PsaX subunits. References: *A. variabilis* (partial) [132]; *S. vulcanus* [127]. Sequence information has been updated using the BLAST (NCBI) and FASTA (DDBJ) databases.

published yet. All these have been listed as PS II subunits. It should be noted, however, that they have not been found in "purified" PS II preparations so far.

13. Plastocyanin

A representative sequence from a higher plant (spinach) is given in Figure 9.19. There are three groups: plant, algal, and cyanobacterial types. Although their sequences differ considerably among these groups, the homologies are high within a group. Well-conserved His[42], Cys[92], His[95], and Met[100] (shown by asterisks) are implicated as ligands for coordinating a copper atom. In mechanically broken chloroplast preparations, plastocyanin is usually still bound to thylakoids membranes. High concentrations of salt, sonication, or mild detergents release plastocyanin,

PsaY (PsbW)

Spinach	LVDERMSTEG	TGLPFGLSNN	LLGWILFGVF	GLIWALYFVY	ASGLEEDEES	G--LSL/
Arabidopsis	LVDERMSTEG	TGLPFGLSNN	LLGWILFGVF	GLIWTFFFVY	TSSLEEDEES	GSGLSL/
Retama raetum	LVDERLSTEG	TGLPFGLSNN	LLGWILFVVF	GLIWTFYFIY	ASSLEEDEES	G--LSL/
Zea mays	LVDERMSTEG	TGLSLGLSNN	LLGWILLGVF	GLIWSLYTVY	TSTLDKDDDS	G--LSL/
Oryza sativa japonica	LVDERMSTEG	TGLSLGLSNN	LLGWILLGVF	GLIWSLYTIY	TSDLEEDEES	G-GLSL/
Haynaldia vilosa	LVDERMSTEG	TGLSLGLSNN	LLGWILLGVF	GLIWSLYTVY	TSGLDEDEES	G-GLSL/
Physcomitrella patens	LVDERLSTEG	TGLGLGISNT	KLTWILVGVT	ALIWALYFSY	SSTLPEGDDD	S-GLDL/
Bigelowiella natans	LVDSRLNGDG	AGIPLGLNDS	RLFFILAGVF	TTVWGVYATS	VKGISDNDDE	DSGMGL/
Radish	LVDDRMSTEG	TGLPFGLSNN	LLGXI...			
Chlamydomonas	LVDERMSTEG	TGLPFGLSNN...				
Wheat	LVDERMSTEG	TGLSLGLSNN...				
Chingensai	LVDDRMSTEG	TGLPFGLSNN	LL...			
Parsley	LVDERLSTEG	TGLPFGLSNN...				
Phytolacca	LVDERM...					

FIGURE 9.18 Amino acid sequences of PsaY subunits. References: spinach [178]; spinach (PS II, Ref. [152]); *Arabidopsis* (gene, Ref. [153]); radish (PS I, partial) [178]; wheat (PS II, partial) [152]; *Chlamydomonas* (PS II, partial) [154]. Sequence information has been updated using the BLAST (NCBI) and FASTA (DDBJ) databases.

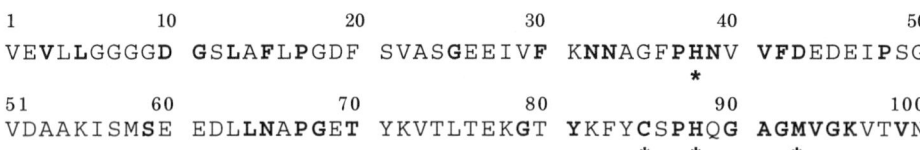

```
      1         10            20            30            40            50
     VEVLLGGGGD GSLAFLPGDF SVASGEEIVF KNNAGFPHNV VFDEDEIPSG
                                                    *
     51        60            70            80            90           100
     VDAAKISMSE EDLLNAPGET YKVTLTEKGT YKFYCSPHQG AGMVGKVTVN
                                          *    *          *
```

FIGURE 9.19 Amino acid sequences of plastocyanins. References: *C. reinhardtii* [157–159]; *S. oleracea* (spinach) [160]; *A. variabilis* [161]. Cu-coordinating residues are marked by asterisks.

which thereafter becomes a "soluble" protein [155]. For more details including three-dimensional structures, refer to a review [156]. The gene, *petE*, is nucleus encoded in eukaryotes [160].

14. Ferredoxin

This is one of the earliest proteins to be sequenced; numerous ferredoxins have been registered in data banks (for a review, see Ref. [162]). The genes are nucleus encoded in higher plants, and in most cases two forms are present. Figure 9.20 shows two isoforms of spinach ferredoxin. The prosthetic group is a 2Fe–2S iron–sulfur cluster coordinated by four cysteine residues (shown in Figure 9.20 by asterisks). The protein is small (a little over 10,000 kDa). A three-dimensional structure of *Anabaena* ferredoxin has been proposed [162]. Ferredoxin can be prepared readily from soluble fractions of plant and algal materials [165]. The gene, *petF*, is nucleus encoded in eukaryotes [166].

15. Ferredoxin:NADP Oxidoreductase

This flavoprotein, often called FNR (ferredoxin-NADP reductase), is fairly tightly bound to thylakoid membranes in higher plants, but readily solubilized by acetone treatment [167]. Once solubilized, this enzyme is soluble in water without any detergent and readily purified [167]. The amino acid sequence of the spinach enzyme is shown in Figure 9.21. The prosthetic group is flavin adenine dinucleotide (FAD). A three-dimensional structure has been proposed from x-ray crystallography with 2.6 Å resolution [168]. For more about structures and functions, refer to a review [169]. The gene, *petH*, is nucleus encoded in eukaryotes [170]. Although this protein is believed to be peripheral and is located on the stromal side, it has been reported to be complexed with some other thylakoid constituents: with the b_6/f complex [179,182], with a 17.5 kDa protein [180,181], and with a 33 kDa protein (H. Yamazaki, T. Hiyama, unpublished result). More work has to be done on these matters, since FNR has often been implicated as a part of the cyclic electron transport [183].

B. What is P700?

Although little substantial evidence had been available, it had long been speculated that P700 was a

```
                                                        *       *   *
Fd1: AAYKVTLVTP  TGNVEFQCPD  DVYILDAAEE  EGIDLPYSCR  AGSCSSCAGK
Fd2: ATYKVTLVTP  SGSQVIECGD  DEYILDAAEE  KGMDLPYSCR  AGACSSCAGK
       1         10          20          30          40         50
                                            *
Fd1: LKTGSLNQDD  QSFLDDDQID  EGWVLTCAAY  PVSDVTIETH  KEEELTA/
Fd2: VTSGSVDQSD  QSFLEDGQME  EGWVLTCIAY  PTGDVTIETH  KEEELTA/
      51         60          70          80          90
```

FIGURE 9.20 Amino acid sequences of two ferredoxins from spinach. References: ferredoxin I [162]; ferredoxin II [164].

```
..1:  QIASDVEAPP  PAPAKVEKHS  KKMEEGITVN  KFKPKTPYVG  RCLLNTKITG  DDAPGETWHM
.61:  VFSHEGEIPY  REGQSVGVIP  DGEDKNGKPH  KLRLYSIASS  ALGDFGDAKS  VSLCVKRLIY
121:  TNDAGETIKG  VCSNFLCDLK  PGAEVKLTGP  VGKEMLMPKD  PNATIIMLGT  GTGIAPFRSF
181:  LWKMFFEKHD  DYKFNGLAWL  FLGVPTSSSL  LYKEEFEKMK  EKAPDNFRLD  FAVSREQTNE
241:  KGEKMYIQTR  MAQYAVELWE  MLKKDNTYFY  MCGLKGMEKG  IDDIMVSLAA  AEGIDWIEYK
301:  RQLKKAEQWN  VEVY/
```

FIGURE 9.21 Amino acid sequences of ferredoxin:NADP oxidoreductase (FNR) from spinach [173]. The corresponding gene has been reported [174].

chlorophyll *a* dimer in a specialized environment created by some special proteins. Watanabe's group proposed that P700 was a heterodimer of chlorophyll *a'* (a chlorophyll *a* epimer present in a variety of PS I preparations; see Figure 9.22) and chlorophyll *a* [171]. Hiyama et al. further showed that, by adding chlorophyll *a'* to a Type III preparation that had been exhaustively treated by strong detergent to remove most of chlorophylls as well as P700 activity, a P700-like pigment was formed [172]. This pigment underwent photooxidation as well as chemical oxidation, yielding difference spectra strongly reminiscent of those of P700. X-ray crystallography now shows clearly that the reaction center special pair consists of one chlorophyll *a'* and one chlorophyll *a* [20], supporting the above hypothesis.

It is of particular interest that the recently found *Acariochloris marina*, a type of cyanobacterium that has chlorophyll *d* in place of chlorophyll *a*, has a small number of a chlorophyll *d* epimer (chlorophyll *d'*, see Figure 9.22). Their photosystem resembles that of PS I, particularly in terms of its strong reductant-generating capacity to reduce NADP. With its absorbance maxima shifted to longer wavelengths in both the blue and red bands, the P700-like absorbance changes also shifted to a longer wavelength [184]. Preliminary analysis suggested that this P700-like reaction center is a heterodimer composed of chlorophyll *d* and chlorophyll *d'* [184]. This is in contrast to the reaction centers of heliobacteria and green sulfur bacteria, which are considered to be homodimers of bacteriochlorophyll *g'* and bacteriochlorophyll *a'*, respectively [185]. These photosynthetic bacteria are presently regarded as precursors of PS I since they also directly reduce NAD(P) [185].

IV. CONCLUDING REMARKS

Due to the space limitation, several subjects have not been covered in this review, LHCPs are one of them, but perhaps somewhat deliberately. The author feels that, as far as PS I is concerned, most of the light energy is harvested by the large subunits and the so called LHCPIs may not have a significant role except for some regulatory ones. Again, this hypothesis is supported by recent crystallographical results that show as many as 100 chlorophyll molecules are bound mostly on the large subunits [20].

Another topic that should have been covered in this review is cyclic electron transport/photophosphorylation. For this increasingly important aspect, the readers should refer to an excellent review by Bendall and Manasse [176].

The present review is admittedly biased and not well balanced, reflecting the author's long indulgence in this field since the 1960s. The emphasis is sometimes on the historical side rather than on hot news items, which appeared often too hot to handle for the present author. An old Chinese proverb says, "Digging into classic literature provides useful hints and often leads to a new discovery." It may not be a waste of time to look back at the past once in a while. It may also be true that "there's many a good tune played on an old fiddle." Some unpublished results in the author's hand have also been included here to back up the author's views. The readers might as well refer to excellent reviews for more details, for subjects not covered here, and for sometimes different and perhaps more "balanced" views in this field [16,31,38, 74,87,140,175,176].

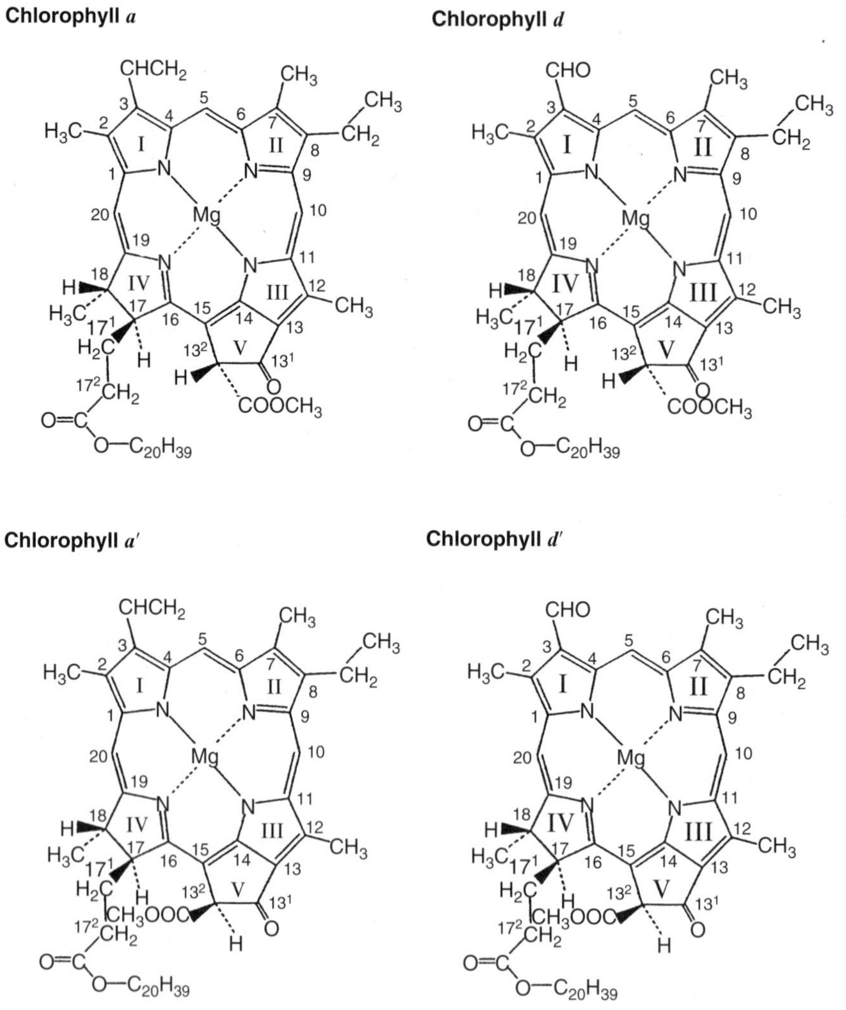

Chlorophyll *a*

Chlorophyll *d*

Chlorophyll *a'*

Chlorophyll *d'*

FIGURE 9.22 Structures of chlorophylls *a, a', d,* and *d'*.

For the present revision, the author has deliberately left out many parts unchanged; some are historical accounts and others are what have been valid throughout these years and most likely will not change in the future as well. Certainly, the recent presentation of three-dimensional structures more elaborate [20] than the previous one [19] is revolutionary and seems to have solved most of the problems. It should be noted, however, that this cyanobacterial PS I has certain differences, though seemingly subtle, such as subunit composition, trimer formation, and donor specificity (*c*-type cytochrome in place of plastocyanin). Primary structures of many subunits as shown in this chapter, notably those nuclear-encoded, are so different from higher plant counterparts that, in some cases, the present designation of some polypeptides may not be valid after all. The advent of complete genome sequences of higher plants (*Arabidopsis,* rice, and more) and cyanobac-

teria (*Synechocystis* 6803, *T.* (*S.*) *elongatus* among others) have also opened up a new era. In addition to x-ray crystallography and NMR, postgenome state-of-the-art technologies such as DNA arrays and numerous proteome techniques will contribute tremendously to our understanding of the structures and functions of PS I. Looking forward to seeing another great leap forward in the coming years, I would like to say once again, "Bring an old chest to new light and find treasures glimmering in the dark."

ACKNOWLEDGMENTS

I dedicate the present article to my teachers Britton Chance, C. Stacy French, Daniel I. Arnon, and Mitsuo Nishimura, without whom I could not have started and continued the studies of photosynthesis. Thanks to my colleague Dr. Hitoshi Nakamoto and numerous graduate and undergraduate students who worked with me for the past 25 years at Saitama University. The work was partly supported by a Grant from T.H. Foundation.

REFERENCES

1. Kok B. *Biochim. Biophys. Acta* 1956; 22: 399–401.
2. Kok B. *Acta Bot. Neer.* 1957; 6: 316–336.
3. Rumberg B, Witt HT. *Z. Naturforsch.* 1964; 19b: 693–699.
4. Malkin R, Bearden AJ. *Proc. Natl. Acad. Sci. USA* 1971; 68: 16–19.
5. Hiyama T, Ke B. *Proc. Natl. Acad. Sci. USA* 1971; 68: 1010–1013.
6. Shin M, Tagawa K, Arnon DI. *Biochem. Z.* 1963; 338:84–89.
7. Haehnel W, Pröpper A, Krause H. *Biochim. Biophys. Acta* 1980; 593: 384–399.
8. Katoh S. *Nature* 1960; 186: 533.
9. Ogawa T, Obata F, Shibata K. *Biochim. Biophys. Acta* 1966; 112: 223–234.
10. Anderson JM, Boardman NK. *Biochim. Biophys. Acta* 1966; 112: 403–421.
11. Bengis C, Nelson N. *J. Biol. Chem.* 1975; 250: 2783–2788.
12. Bengis C, Nelson N. *J. Biol. Chem.* 1977; 252: 4564–4569.
13. Fish LE, Kück U, Bogorad L. *J. Biol. Chem.* 1985; 260:1413–1421.
14. Shinozaki K, Ohme M, Tanaka M, Wakasugi T, Hayashida N, Matsubayashi T, Zaita N, Chunwongse J, Obokata J, Yamaguchi-Shinozaki K, Ohta C, Torazawa K, Meng BY, Sugita M, Deno H, Kamogashira T, Yamada K, Kusuda J, Takaiwa F, Kato A, Tohdoh N, Shimada H, Sugiura M. *EMBO J.* 1986; 5: 2043–2049.
15. Ohyama K, Fukuzawa H, Kohchi T, Shirai H, Sano T, Sano S, Umesono K, Shiki Y, Takeuchi M, Chang Z, Aota S, Inokuchi H, Ozeki H. *Nature* 1986; 322: 572–574.
16. Setif P. In: Barber J, ed. *The Photosystems: Structure and Function and Molecular Biology*. Amsterdam: Elsevier, 1992: 471–499.
17. Hiyama T, Ohinata A, Kobayashi S. *Z. Naturforsch.* 1993; 48c: 374–378.
18. Mullet J, Burke JJ, Arntzen C. *Plant Physiol.* 1980; 65: 814–822.
19. Krauss N, Hinrichs W, Witt I, Fromme P, Pritzkow W, Dauter Z, Betzel C, Wilson KS, Witt HT, SaengerW. *Nature* 1993; 361: 326–331.
20. Jordan P, Fromme P, Witt HT, Klukas O, Saenger W, Krauss N. *Nature* 2001; 411: 909–917.
21. Hiyama T, Katoh A, Shimizu T, Inoue K, Kubo A. In: Biggins J, ed. *Progress in Photosynthesis Research* Vol 2. Dordrecht: Martinus Nijhoff Publishers, 1987: 45–48.
22. Golbeck JH, Parrett KG, Mehari T, Jones KL, Brand J. *FEBS Lett.* 1988; 228: 268–272.
23. Redinbo MR, Yeates TO, Marchant S. *J. Bioenerg. Biomembr.* 1994; 26: 49–66.
24. Hiyama T, Ke B. *Biochim. Biophys. Acta* 1972; 267: 160–171.
25. Hiyama T. *Physiol. Vég.* 1985; 23: 605–612.
26. Markwell JP, Thornber JP, Skrdla MP. *Biochim. Biophys. Acta* 1980; 591: 391–399.
27. Evans MCW, Reeves SG, Cammack R. *FEBS Lett.* 1974; 49: 111–114.
28. MaCintosh AR, Chu M, Bolton JR. *Biochim. Biophys. Acta* 1975; 376: 308–314.
29. Sauer K, Mathis P, Acker S, van Best JA. *Biochim. Biophys. Acta* 1978; 503: 120–134.
30. Bonnerjea J, Evans MCW. *FEBS Lett.* 1982; 148: 313–316.
31. Golbeck JH. *Annu. Rev. Plant Physiol. Plant Mol. Biol.* 1992; 43: 293–324.
32. Hiyama T, Fork DC. *Arch. Biochem. Biophys.* 1980; 199: 488–496.
33. Hiyama T, Ke B. *Arch. Biochem. Biophys.* 1971; 147: 99–108.
34. Kück U, Choquet Y,. Scheider M, Dron M, Bennoun P. *EMBO J.* 1987; 6: 2185–2195.
35. Fish EL, Bogorad L. *J. Biol. Chem.* 1986; 261: 8134–8139.
36. Kirsch W, Seyer P, Herrmann RG. *Curr. Genet.* 1986; 10: 843–855.
37. Krauss N, Hinrichs W, Witt I, Fromme P, Pritzkow W, Duter Z, Betzel C, Wilson KS, Witt HT, Saenger W. *Nature* 1993; 361: 326–331.
38. Golbeck JH, Bryant DA. *Curr. Topics Bioenerg.* 1991; 16: 83–177.
39. Smart LB, Warren PV, Golbeck, JH, McIntosh L. *Proc. Natl. Acad. Sci. USA* 1993; 90: 1132–1136.
40. Hiratsuka J, Shimada H, Whittier R, Ishibashi T, Sakamoto M, Mori M, Kondo C, Honjo Y, Sun C-R, Meng B-Y, Li Y-Q, Kanno A, Nishizawa Y, Hirata A, Shinozaki K, Sugiura M. *Mol. Gen. Genet.* 1989; 217: 185–194.
41. Lehmbeck L, Rasmussen OF, Bookjans GB, Jepsen BR, Stummann BM, Henningsen KW. *Plant Mol. Biol.* 1986; 7: 3–10.
42. Mühlenhoff U, Haehnel W, Witt HT, Herrmann RG. *EMBL* 1992: X63768.
43. Kirsch W, Seyer P, Herrmann RG. *Curr. Genet.* 1986; 10: 843–855.
44. Hiyama T, Yanai N, Takano Y, Ogiso H, Suzuki K, Terakado K. In: Baltscheffsky M, ed. *Current Research in Photosynthesis* Vol 2. Dordrecht: Kluwer Academic Publishers, 1990: 587–590.
45. Bruce BD, Malkin R. *J. Biol. Chem.* 1988; 263: 7302–7308.
46. Cushman JC, Hallick RB, Price CA. *Curr. Genet.* 1988; 13: 159–171.
47. Mühlenhoff U, Haehnel W, Witt HT, Herrmann RG. *Gene* 1993; 127: 71–78.
48. Shimizu T, Hiyama T, Ikeuchi M, Inoue Y. *Plant Mol. Biol.* 1992; 18: 785–791.
49. Smart LB, McIntosh L. *Plant Mol. Biol.* 1991; 17: 959–971.

50. Cantrell A, Bryant DA. *Plant Mol. Biol.* 1987; 9: 453–468.
51. Nuyhus KJ, Sonoike K, Pakrasi HB. In: Bryant DA, ed. *The Molecular Biology of Cyanobacteria.* Dordrecht: Kluwer Academic Publishers, 1994: 331–332.
52. Schantz R, Bogorad L. *Plant Mol. Biol.* 1988; 11: 239–247.
53. Hayashida N, Matsubayashi T, Shinozaki K, Sugiura M, Inoue K, Hiyama T. *Curr. Genet.* 1987; 12: 247–250.
54. Dunn PPJ, Gray JC. *Plant Mol. Biol.* 1988; 11: 311–319.
55. Høj PB, Svendsen I, Scheller HV, Møller BL. *J. Biol. Chem.* 1987; 262: 12676–12684.
56. Oh-oka H, Takahashi Y, Wada K, Matsubara H, Ohyama K, Ozeki H. *FEBS Lett.* 1987; 218: 52–54.
57. Hiratsuka J, Shimada H, Whittier R, Ishibashi T, Sakamoto M, Mori M, Kondo C, Honjo Y, Sun CR, Meng BY, Li Y, Kanno K, Nishizawa Y, Hirai A, Shinozaki K, Sugiura M. *Mol. Gen. Genet.* 1989; 217: 185–194.
58. Oh-oka H, Takahashi Y, Kuriyama K, Saeki K, Matsubara H. *J. Biochem.* 1988; 103: 964–968.
59. Valentin KU, Kostrzewa M, Zetsche K. *Plant Mol. Biol.* 1993; 23: 77–85.
60. Takahashi Y, Goldschmidt-Clermont M, Soen S-Y, Franzen LG, Rochaix J-D. *EMBO J.* 1991; 10: 2033–2040.
61. Hallick RB, Hong L, Drager RG, Favreau M, Monfort A, Orsat B, Spielmann A, Stutz E. *EMBL* 1993: X70810.
62. Mann K, Schlenkrich T, Bauer M, Huber R. *Biol. Chem. Hoppe-Seyler.* 1991; 372: 519–524.
63. Bryant DA, Rhiel E, de Lorimier R, Zhou J, Stirewalt VL, Gasparich GE, Dubbs JM, Snyder W. In: Baltscheffsky M, ed. *Current Research in Photosynthesis* Vol 2. Dordrecht: Kluwer Academic Publishers, 1990: 1–5.
64. Rhiel E, Stirewalt VL, Gasparich GE, Bryant DA. *Gene* 1992; 112: 123–128.
65. Mannan RM, Pakrasi HB. *EMBL* 1991: X57153.
66. Mulligan ME, Jackman DM. *Plant Mol. Biol.* 1992; 18: 803–808.
67. Mühlenhoff U, Haehnel W, Witt HT, Herrmann RG. *EMBL* 1992: X63763.
68. Shimizu T, Hiyama T, Ikeuchi M, Koike H, Inoue Y. *Nucleic Acids Res.* 1990; 18: 3644.
69. Ousseau F, Lagoutte B. *FEBS Lett.* 1990; 260: 241–244.
70. Herman P, Adiwilaga K, Golbeck JH, Weeks DP. In: Bryant DA, ed. *The Molecular Biology of Cyanobacteria.* Dordrecht: Kluwer Academic Publishers, 1994; 344–350.
71. Vierling E, Alberte R. *Plant Physiol.* 1983; 72:625–633.
72. Adman ET, Sieker LC, Jensen LH. *J. Biol. Chem.* 1973; 248: 3987–3996.
73. Hiyama T. *CACS Forum* 1988; 8: 2–8.
74. Almog O, Shoham G, Nechushtai R. In: Barber J, ed. *The Photosystems: Structure, Function and Molecular Biology.* Amsterdam, London, New York, Tokyo: Elsevier, 1992: 443–445.

75. Iwasaki Y, Sasaki T, Takabe T. *Plant Cell Physiol.* 1990; 31: 871–879.
76. Kjarulff S, Okkels JS. *Plant Physiol.* 1993; 101: 335–336.
77. Hoffman NE, Pichersky E, Malik VS, Ko K, Cashmore AR. *Plant Mol. Biol.* 1988; 10: 435–445.
78. Yamamoto Y, Tsuji H, Hayashida N, Inoue K, Obokata J. *Plant Mol. Biol.* 1991; 17: 1251–1254.
79. Münch S, Ljungberg U, Steppuhn J, Schneiderbauer A, Nechushtai R, Beyreuther K, Herrmann RG. *Curr. Genet.* 1988; 14: 511–518.
80. Dunn PPJ, Packman LC, Pappin D, Gray JC. *FEBS Lett.* 1988; 228: 157–161.
81. Mann K, Schlenkrich T, Bauer M, Huber R. *Biol. Chem. Hoppe-Seyler* 1991; 372: 519–524.
82. Nyhus KJ, Ikeuchi M, Inoue Y, Whitmarsh J, Pakrasi HB. *J. Biol. Chem.* 1992; 267: 12489–12495.
83. Kotani N, Enami I, Aso K, Tsugita A. *Protein Seq. Data Anal.* 1991; 4: 81–86.
84. Alhadeff M, Lundell DJ, Glazer AN. *Arch. Microbiol.* 1988; 150: 482–488.
85. Sue S, Sugiya K, Furuki M, Shimizu T, Inoue Y, Nakamoto H, Hiyama T. *Photosynth. Res.* 1995; 46: 265–268.
86. Reilly P, Hulmes JD, Pan Y-CE, Nelson N. *J. Biol. Chem.* 1988; 263: 17658–17662.
87. Bryant DA. In: Bryant DA, ed. *The Molecular Biology of Cyanobacteria.* Dordrecht: Kluwer Academic Publishers, 1994: 348.
88. Chitnis PR, Reily PA, Nelson N. *J. Biol. Chem.* 1989; 264: 18381–18385.
89. Zhao JD, Warren PV, Li N, Bryant D, Golbeck JH. *FEBS Lett.* 1990; 276: 175–180.
90. Li N, Zhao JD, Warren PV, Warden JT, Bryant D, Golbeck JH. *Biochemistry* 1991; 30: 7853–7672.
91. Chitnis PR, Reilly PA, Miedel MC, Nelson N. *J. Biol. Chem.* 1989; 264: 18374–18380.
92. Anandan S, Vainstein A, Thornber JP. *FEBS Lett.* 1989; 256: 150–154.
93. Franzen LG, Frank G, Zuber H, Rochaix JD. *Plant Mol. Biol.* 1989; 12: 463–474.
94. Scott MP, Nielsen VS, Knoetzel J, Ersen R, Moller BL. *EMBL* 1994: U08135.
95. Zhao J, Snyder W, Mühlenhoff U, Rhiel E, Bryant DA. *Mol. Microbiol.* 1993; 9: 183–194.
96. Mühlenhoff U, Haehnel W, Witt HT, Herrmann RG. *EMBL* 1992: X63765.
97. Rhiel E, Bryant DA. *Plant Physiol.* 1993; 101: 701–702.
98. Reith M. *Plant Mol. Biol.* 1992; 18: 773–775.
99. Hatanaka H, Sonoike K, Hirano M, Katoh S. *Biochim. Biophys. Acta* 1993; 1141: 45–51.
100. Chitnis PR, Purvis D, Nelson N. *J. Biol. Chem.* 1991; 266: 20146–20151.
101. Steppuhn J, Hermans J, Nechushtai R, Ljungberg U, Thümmler F, Lottspeich F, Herrmann RG. *FEBS Lett.* 1988; 237: 218–224.
102. Bryant DA. In: Bryant DA, ed. *The Molecular Biology of Cyanobacteria.* Dordrecht: Kluwer Academic Publishers, 1994: 319–360.

103. Li N, Warren PV, Golbeck JH, Frank G, Zuber H, Bryant DA. *Biochim. Biophys. Acta* 1991; 1059: 215–225.
104. Bendall DS, Davies EC. *Physiol. Plant* 1989; 76: A87.
105. Scheller HV, Svendsen I, Møller BL. *J. Biol. Chem.* 1989; 264: 6929–6934.
106. Franzen L-G, Frank G, Zuber H, Rochaix J-D, *Mol. Gen. Genet.* 1989; 219: 137–144.
107. Okkels J, Nielsen V, Scheller H, Møller B. *Plant Mol. Biol.* 1992; 18: 989–994.
108. Okkels JS, Scheller HV, Jepsen LB, Møller BL. *FEBS Lett.* 1989; 250: 575–579.
109. de Pater S, Hensgens LAM, Schilperoort RA. *Plant Mol. Biol.* 1990; 15: 399–406.
110. Steppuhn J, Hermans J, Nechushtai R, Herrmann GS, Herrmann RG. *Curr. Genet.* 1989; 16: 99–108.
111. Sonoike K, Ikeuchi M, Pakrasi HB. *Plant Mol. Biol.* 1992; 20: 987–990.
112. Yoshinaga K, Kubota Y, Ishii T, Wada K. *Plant Mol. Biol.* 1992; 18: 79–82.
113. Scheller HV, Okkels JS, Høj PB, Svendsen I, Røpstorff P, Møller BL. *J. Biol. Chem.* 1989; 264: 18402–18406.
114. Rodermel SR. *EMBL* 1992: X61188.
115. Haley J, Bogorad L. *EMBL* 1989: J04502.
116. Fukuzawa H, Kohchi T, Sano T, Shirai H, Umesono K, Inokuchi H, Ozeki H, Ohyama K. *J. Mol. Biol.* 1988; 203: 333–351.
117. Kjaerulff S, Andersen B, Nielsen VS, Moller BL, Okkels JS. *J. Biol. Chem.* 1993; 268: 18912–18916.
118. Nagano Y, Matsuno R, Sasaki Y. *Curr. Genet.* 1991; 20: 431–436.
119. Mühlenhoff U, Haehnel W, Witt HT, Herrmann RG. *EMBL* 1992: X63765.
120. Smith AG, Wilson RJ, Kaethner TM, Willey DL, Gray JC. *Curr. Genet.* 1991; 19: 403–410.
121. Ikeuchi M, Hirano A, Hiyama T, Inoue Y. *FEBS Lett.* 1990; 263: 274–278.
122. Ayliffe MA, Timmis JN. *Mol. Gen. Genet.* 1992; 236: 105–112.
123. Mühlenhoff U, Haehnel W, Witt HT, Herrmann RG. *EMBL* 1992: X63763.
124. Shimada H, Sugiura M. *Nucleic Acids Res.* 1991; 19: 983–995.
125. Ogihara Y, Terachi T, Sasakuma T. *Genetics* 1991; 129: 873–884.
126. Manzara T, Hallick RB. *Nucleic Acids Res.* 1988; 16: 9866.
127. Koike H, Ikeuchi M, Hiyama T, Inoue Y. *FEBS Lett.* 1989; 253: 257–263.
128. Hoshina S, Sue S, Kunishima N, Kamide K, Wada K, Itoh S. *FEBS Lett.* 1989; 258: 305–308.
129. Wynn RM, Malkin R. *FEBS Lett.* 1990; 262: 45–48.
130. Kjaerulff S, Andersen B, Nielsen VS, Møller BL, Okkels JS. *J. Biol. Chem.* 1993; 268: 18912–18916.
131. Jones CS, Kotani N, Aso K, Yang L, Enami I, Kondo K, Tsugita A. *Protein Seq. Data Anal.* 1991; 4: 327–331.
132. Ikeuchi M, Nyhus KJ, Inoue Y, Pakrasi HB. *FEBS Lett.* 1991; 287: 5–9.
133. Okkels JS, Scheller HV, Svendsen I, Møller BL. *J. Biol. Chem.* 1991; 266: 6767–6773.
134. Hiyama T, Oya T, Kobayashi S, Furuki M, Shimizu T, Senda M, Nakamoto H. In: Murata N, ed. *Research in Photosynthesis* Vol 1. Dordrecht: Kluwer Academic Publishers, 1992: 621–624.
135. Mühlenhoff U, Haehnel W, Witt HT, Herrmann RG. *EMBL* 1992: X63763.
136. Chitnis VP, Xu Q, Yu L, Golbeck JH, Nakamoto H, Xie DL, Chitnis PR. *J. Biol. Chem.* 1993; 268: 11678–11684.
137. Mühlenhoff U, Haehnel W, Witt HT, Herrmann RG. *EMBL* 1992: X59760.
138. Umesono K, Inokuchi H, Shiki Y, Takeuchi M, Chang Z, Fukuzawa H, Kohchi T, Shirai H, Ohyama K, Ozeki H. *J. Mol. Biol.* 1988; 203: 299–331.
139. Hallick RB, Hong L, Drager RG, Favreau MR, Monfort A, Orsat B, Spielmann A, Stutz E. *Nucleic Acids Res.* 1993; 21: 3537–3544.
140. Bryant DA. In: Barber J, ed. *The Photosynthesis: Structure, Function and Molecular Biology.* Amsterdam: Elsevier Science Publishers, 1992: 501–549.
141. Ikeuchi M, Sonoike K, Koike H, Pakrasi HB, Inoue Y. *Plant Cell Physiol.* 1992; 33: 1057–1063.
142. Knoetzel J, Simpson DJ. *Plant Mol. Biol.* 1993; 22: 337–345.
143. Ikeuchi M, Inoue. *FEBS Lett.* 1990; 280: 332–334.
144. Iwasaki I, Ishikawa H, Hibino T, Takabe T. *Biochim. Biophys. Acta* 1991; 1059: 141–148.
145. Knoetzel J, Simpson DJ. *Plant Mol. Biol.* 1993; 22: 337–345.
146. Krebbers ET, PhD Thesis, Harvard University, Cambridge, MA. 1983.
147. Quigley F, Weil JH. *Curr. Genet.* 1985; 9: 495–503.
148. Bookjans G, Stummann BM, Rasmussen OF, Henningsen KW. *Plant Mol. Biol.* 1986; 6: 359–366.
149. Oliver RP, Poulsen C. *Carlsberg Res. Commun.* 1984; 49: 647–673.
150. Evrard JL, Weil JH, Kuntz M. *Plant Mol. Biol.* 1990; 15: 779781.
151. Holscuh K, Bottomley W, Whitfield PR. *Nucleic Acids Res.* 1984; 12: 8819–8834.
152. Ikeuchi M, Takio K, Inoue Y. *FEBS Lett.* 1989; 242: 263–269.
153. Rayal M, Grellet F, Laudie M, Meyer Y, Cooke R, Delsney M. *EMBL* 1992: S29418.
154. Vitry C, Diner BA, Popot J. *J. Biol. Chem.* 1991; 266: 16614–16621.
155. Katoh S. In: San Pietro A, ed. *Methods in Enzymology* Vol 23. New York, London: Academic Press, 1971: 408–412.
156. Redinbo MR, Yeates TO, Merchant S. *J. Bioenerg. Biomembr.* 1994; 26: 49–66.
157. Merchant S, Hill K, Kim JH, Thompson J, Zaitlin D, Bogorad L. *J. Biol. Chem.* 1990; 265: 12372–12379.
158. Quinn J, Li HH, Singer J, Morimoto B, Mets L, Kindle K, Merchant S. *J. Biol. Chem.* 1993; 268: 7832–7841.
159. Redinbo MR, Cascio D, Choukair MK, Rice D, Merchant S, Yeates TO. *Biochemistry* 1993; 32: 10560–10567.

160. Rother C, Jansen T, Tyagi A, Tittgen J, Herrmann RG. *Curr. Genet.* 1986; 11: 171–176.

161. Aitken A. *Biochem. J.* 1975; 149: 675–683.

162. Holden HM, Jacobson BL, Hurley JK, Tollin G, Oh B-H, Skjeldal L, Chae YK, Cheng T, Xia BH, Markley JL. *J. Bioenerg. Biomembr.* 1994; 26: 67–88.

163. Matsubara H, Sasaki RM. *J. Biol. Chem.* 1968; 243: 1732–1757.

164. Takahashi Y, Hase T, Wada K, Matsubara H. *Plant Cell Physiol.* 1983; 24: 189–198.

165. Buchanan BB, Arnon DI. In: San Pietro A, ed. *Methods in Enzymology* Vol 23. New York, London: Academic Press, 1971: 413–439.

166. Wedel N, Bartling D, Herrmann RG. *Bot. Acta* 1988; 101: 295–300.

167. Shin M. In: San Pietro A, ed. *Methods in Enzymology* Vol 23, New York, London: Academic Press, 1971: 413–439.

168. Karplus PA, Daniels MJ, Herriott JR. *Science* 1991; 251: 60–66.

169. Karplus PA, Bruns CM. *J. Bioenerg. Biomembr.* 1994; 26: 89–99.

170. Jansen T, Reilaender H, Steppuhn J, Herrmann RG. *Curr. Genet.* 1988; 13: 517–522.

171. Kobayashi M. *Photosynth. Res.* 1996; 109: 223–230.

172. Hiyama T, Watanabe T, Kobayashi M, Nakazato M. *FEBS Lett.* 1987; 214: 97–100.

173. Karplus PA, Walsh KA, Herriott JR. *Biochemistry* 1984; 23: 6576–6583.

174. Jansen T, Reilaender H, Steppuhn J, Herrmann RG. *Curr. Genet.* 1988; 13: 517–522.

175. Ikeuchi M. *Plant Cell Physiol.* 1992; 33: 669–676.

176. Bendall DS, Manasse RS. *Biochim. Biophys. Acta* 1995; 1229: 23–38.

177. Haehnel WH, Nelson N, Witt I. *EMBL* 1992: X59760.

178. Hiyama T, Yumoto K, Satoh A, Takahashi M, Nishikido T, Nakamoto H, Suzuki K, Hiraide T. *Biochim. Biophys. Acta* 2000; 1459: 117–124.

179. Hurt E, Hauska G. *Eur. J. Biochem.* 1981; 117: 591–599.

180. Vallejos RH, Ceccarelli E, Chan R. *J. Biol. Chem.* 1984; 259: 8048–8051.

181. Matthijs HCP, Coughlan SJ, Hind G. *J. Biol. Chem.* 1986; 261: 12154–12158.

182. Zhang H, Whitelegge JP, Cramer WA. *J. Biol. Chem.* 2001; 276: 38159–38165.

183. Hiyama T, Nishimura M, Chance B. *Plant Physiol.* 1970; 46: 163–168.

184. Hu Q, Miyashita H, Iwasaki I, Kurano N, Miyachi, S, Iwaki M, Itoh, S. *Proc. Natl. Acad. Sci. USA* 1998; 13319–13323.

185. Akiyama M, Miyashita H, Kise H, Watanabe T, Mimuro M, Miyachi S, Kobayashi M. *Photosynth. Res.* 2002; 74: 97–107.

10 Covalent Modification of Photosystem II Reaction Center Polypeptides

Julian P. Whitelegge

Departments of Psychiatry and Biobehavioral Sciences, Chemistry and Biochemistry,
David Geffen School of Medicine and the College of Letters and Sciences,
University of California

CONTENTS

I. INTRODUCTION

A. PHOTOSYSTEM II REACTION CENTER POLYPEPTIDES AND THEIR COFACTORS

Photosystem II (PS II) drives the photooxidation of water generating molecular oxygen, releasing protons to the lumenal side of the thylakoid vesicle, and providing the electrons for the linear photosynthetic electron transport chain. PS II is a largely intrinsic membrane pigment–protein complex consisting of a number of different polypeptides with chlorophyll, pheophytin, β-carotene, heme, plastoquinone, and a number of metal and other ions as cofactors. The activities of PS II can be divided into three functional domains. A light harvesting function is accomplished by a number of peripheral chlorophyll α-binding intrinsic polypeptides (notably CP43 and CP47), which also serve to funnel excitation energy from antenna complexes into the photosynthetic reaction center. The reaction center containing the primary donor P_{680} performs the energy conversion function enabling electrons to be transported to the two-electron gate Q_B via bound pheophytin and plastoquinone molecules. The reaction center also contains the polypeptide tyrosine residue (Y_Z), which is the secondary donor and which in turn accepts electrons from the third functional domain, the oxygen-evolving complex (OEC), which is a four-electron gate. The heart of the OEC is a tetranuclear manganese cluster that is closely associated with the reaction center and stabilized by a number of extrinsic polypeptides as well as calcium and chloride ions [1]. The OEC binds a pair of water molecules and accumulates the four oxidizing equivalents required for their oxidation through five so-called S-states (S_0 to S_4) [2,3]. Both the antenna complexes and the extrinsic polypeptides associated with the OEC vary considerably between the oxygenic prokaryotes and eukaryotes. The reaction center itself, however, is highly conserved.

The PS II reaction center has been isolated [4] and consists of five polypeptides. The D1 and D2 polypeptides bind P_{680}, pheophytin, and the quinone acceptors Q_A and Q_B of linear electron transport in a structure that bears considerable homology to the known structure of the purple bacterial reaction center [5,6]. Polypeptides PS II-E and PS II-F bind the heme and constitute cytochrome b_{559}, which is placed closely to the D1 and D2 polypeptides so that it can both directly donate and accept electrons to the reaction center [7]. The fifth polypeptide PS II-I, though intimately associated with the reaction center [8], has an unknown function and is evidently dispensable *in vivo* [9]. All of the polypeptides of the reaction center are intrinsic; D1 and D2 (~39 kDa each) have

five transmembrane α-helices each, whereas the smaller PS II-E, -F, and -I (~4 to 10 kDa) polypeptides have a single α-helix, each crossing the membrane just once. It is most probable that all five N termini are exposed to the stroma [5,10,11], whereas all C termini are exposed to the lumen. Along with two pheophytin molecules, it is thought that the reaction center contains four to six chlorophyll α and two β-carotene molecules, giving a total molecular weight of a little over 100 kDa.

B. POSTTRANSLATIONAL MODIFICATIONS AND THE ASSEMBLY/REASSEMBLY OF PS II

The PS II reaction center is regularly damaged, presumably as a consequence of the highly oxidizing potential generated by P_{680} (+1.17 V) [12] in order to split water. A complex repair cycle has evolved such that damaged units are replaced via turnover of D1, which is removed from the reaction center and replaced with a newly translated polypeptide [13,14]. If photodamage to PS II exceeds the capacity for its repair, then activity declines in a process called photoinhibition [15–17]. Despite protective mechanisms at every level of plant organization, it is likely that photoinhibition does lead to losses of productivity in the field [18]. Posttranslational modifications to the PS II reaction center polypeptides accompany all stages of the repair cycle; these are discussed in more detail in Section II. Artificially introduced covalent polypeptide modifications and their use in the study of PS II reaction center structure and function are reviewed in Section III.

II. NATURAL COVALENT MODIFICATIONS OF PS II REACTION CENTER POLYPEPTIDES

A. N-TERMINAL PROCESSING

In spinach and other higher plants, the N termini of both D1 and D2 polypeptides are processed. The initiating methionine is removed leaving a threonyl residue at the N terminus that may be both N-acetylated and O-phosphorylated. The wide conservation of threonine 2 of D1 and D2 in all species examined (except *Euglena* D1 [19]) suggests that these modifications may be universal. However, in lower plants, algae, and cyanobacteria the processing of the N termini of both D1 and D2 remains less clearly characterized. The PS II-E, -F, and -I subunits are processed at their N termini but are not widely considered to be phosphorylated. The function of phosphorylation of the reaction center polypeptides is controversial but is probably linked to regulation of PS II activity or the PS II repair cycle.

1. Phosphorylation of PS II Reaction Center Polypeptides

a. Structural determination of phosphorylation sites

Spinach thylakoids were phosphorylated *in vitro*, the N-terminal peptides originating from D1 and D2 were isolated, and their covalent structures were determined by tandem mass spectrometry. The residue corresponding to T2 was demonstrated to be *N*-acetylated and *O*-phosphorylated in both cases [20]. Because the ferric ion affinity chromatography technique was specific for phosphopeptides, it was not possible to determine whether the entire population of the D1 and D2 polypeptides was phosphorylated or whether a significant population remained nonphosphorylated (or nonacetylated/processed).

b. The D1* conformer of D1 is most probably the phosphorylated form of D1

An extended SDS-PAGE run allowed separation of D1 and a slightly more slowly migrating conformer designated D1* to be observed after labeling studies of thylakoids from the aquatic angiosperm *Spirodela* [21]. Further studies provide convincing evidence that D1* is indeed the phosphorylated form of D1 in *Spirodela* [22,23]. The observation that D1* can be converted back to D1 under certain conditions implies that the phosphorylation of D1 is reversible [23]. The appearance of D1* has been observed in other higher-plant species under conditions known to promote phosphorylation [24–26], suggesting that D1 phosphorylation is a widespread phenomenon. However, D1* did not appear in the lower-plant species examined [26], and the authors concluded that D1 phosphorylation was limited to higher-plant species. Since the unicellular green alga *Chlamydomonas reinhardtii* is considered a good model system for the study of PS II structure–function, assembly, and degradation, it is pertinent to consider whether the characteristics of reaction center polypeptide phosphorylation in this and other green algae are similar to those in higher plants.

c. Is D1 phosphorylated in the green algae?

Phosphorylation of *C. reinhardtii* thylakoid polypeptides has been extensively investigated since the early 1980s with no convincing demonstrations of D1 phosphorylation despite both *in vitro* and *in vivo* labeling studies under a variety of conditions including those that led to D1* accumulation in higher plants. A recent detailed analysis of PS II particles isolated from *C. reinhardtii* cells ^{32}P-labeled for 14 hr demonstrated phosphorylation of D2, P6 (PS II-C polypeptide), and three low-molecular-weight polypeptides, but not D1 [27]. It seems unlikely the lack of phosphorylation of D1 is artifactual unless the hypothetical phospho-D1 of *Chlamydomonas* is unusually sensitive to endogenous cellular phosphatases that were not completely inhibited by the 20-mM fluoride present in the isolation buffers. Dephosphorylation of D1 during isolation of thylakoids has been observed, and it is noted that 125 mM NaF was used to prevent dephosphorylation of *Spirodela* D1 [23].

A polypeptide tentatively identified as D1 was observed to be phosphorylated after *in vivo* ^{32}P-labeling of *Dunaliella salina* cells in the light [28]. Phosphorylation of this polypeptide was stimulated under photoinhibitory conditions consistent with the conditions required for D1* formation in higher plants. To conclude, D1 is not phosphorylated across the whole range of green algal species and thus "lower" plants in general.

d. The D2 polypeptide is consistently observed to be phosphorylated

The D2 polypeptide of spinach was shown to be phosphorylated at its N terminus by mass spectrometry [20]. It is also phosphorylated in pea (see Figure 10.1) [31]. In *C. reinhardtii* the phosphorylated form of D2 (D2.1) can be distinguished from the nonphosphorylated form (D2.2) by its slightly lower migration in SDS-PAGE [27,32]. Treatment of phosphorylated PS II particles with alkaline phosphatase removed all signs of phosphopeptides as assessed by autoradiography and led to loss of the D2.1 band observed by staining the polypeptides with Coomassie brilliant blue and a concomitant increase in stain on the D2.2 band [27]. Study of D2 phosphorylation *in vivo* revealed that the polypeptide tended to become phosphorylated under oxidizing conditions rather than the reducing conditions that favor phosphorylation of most other thylakoid polypeptides [33]. *In vitro* redox titrations contradicted this finding, however [34]. Neither D1 nor D2 has been observed to be phosphorylated in the cyanobacteria.

e. Are the low-molecular-weight polypeptides of PS II phosphorylated?

The only low-molecular-weight polypeptides of the reaction center are PS II-E, -F, and -I, none of which are generally considered to be phosphoproteins. Could at least one of them become phosphorylated? de Vitry et al. [27] identified a 5-kDa phosphopeptide of *Chlamydomonas* PS II core particles, which they suggested could be PS II-F or PS II-I. Analysis of the *Chlamydomonas psbI* gene sequence has revealed that the PS II-I protein has a

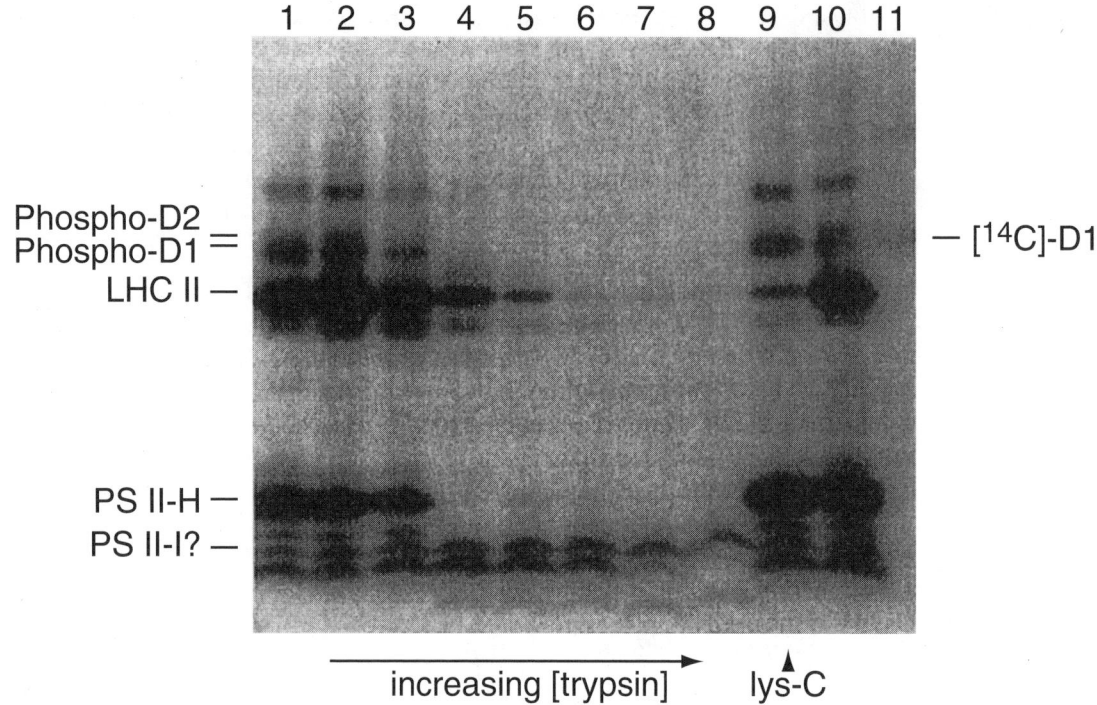

FIGURE 10.1 Pea PS II reaction center polypeptides phosphorylated in vitro. Autoradiograph of pea thylakoid membrane polypeptides subjected to protease treatments after phosphorylation in vitro with [γ^{32}P]-ATP and separation of phospho-peptides using discontinuous tricine SDS-PAGE followed by blotting to nitrocellulose. The phosphorylated D1 (phospho-D1) polypeptide is not degraded by the endoproteinase lys-C because its sequence is devoid of lysyl residues. The phosphorylated D2 polypeptide (phospho-D2), which is observed to migrate more slowly than D1 in this gel system, is degraded by both lys-C and trypsin. Phosphopeptides of LHC II (LHC II) and the 10-kDa *psbH* gene product (PS II-H) are degraded due to the abundant presence of arginyl and lysyl residues. Five low-molecular-weight polypeptides are observed to be phosphorylated, though only one remained resistant to both lys-C and trypsin treatments (PS II-I?). The mobility and protease sensitivity of D1 were confirmed by immunodecoration of the blot using anti-D1 antibodies (not shown) as well as comigration of the [^{14}C]-azidoatrazine labeled D1 polypeptide of *Scenedesmus obliquus* ([^{14}C]-D1; lane 11). Thylakoid membranes were isolated from peas [29] and phosphorylated for 30 min in the presence of 0.5 mM ATP (80 Ci/mol [γ^{32}P]-ATP), 0.5 mg/ml dithionite, and 10 mM NaF. Samples containing 12.5 μg chlorophyll were treated with trypsin or lys-C endopeptidase in 20-μl final volume (lanes 2 to 8: 0.5, 1.0, 5.0, 10, 50, 100, 500 μg/ml trypsin, respectively; lane 9: 500 μg/ml lys-C; lanes 1 and 10, no protease) for 30 min at 37°C prior to solubilization at 80°C for 5 min and tricine–SDS-PAGE 16.5% T, 3% C [30]. These gels are efficient at separating low-molecular-weight peptides. Transfer of the polypeptides to nitrocellulose prior to direct autoradiography proved highly effective for observing the low-molecular-weight phosphopeptides, although it is possible that some larger polypeptides might fail to transfer to nitrocellulose efficiently.

threonine in position 2 that hypothetically could be phosphorylated [9]. However, the sequences of *Chlamydomonas* PS II-E and -F, as translated from their gene sequences, both reveal possible phosphorylation sites at the N termini also [35,36]. It should be noted that the core PS II particles also contain other low-molecular-weight polypeptides, which might be an unidentified small phosphopolypeptide [27] such as the *psbL* gene product that was suggested to be phosphorylated in wheat [37].

Most thylakoid phosphoproteins contain arginine or lysine residues close to their N termini so that the N-terminal phosphate label is removed during trypsin or lys-C endopeptidase treatments. However, there is

a low-molecular-weight phosphoprotein of pea thylakoids that resists both trypsin and lys-C treatments (see Figure 10.1). The sequence of pea PS II-I revealed no arginyl or lysyl residues at the N terminus and threonine at position 2 [8]. Perhaps the PS II-I polypeptide of the reaction center can be phosphorylated with D1 and D2. The identity of the five low-molecular-weight phosphopeptides seen in Figure 10.1 warrants further study.

f. What is the function of PS II reaction center polypeptide phosphorylation?

Current hypotheses involve control of D1 degradation by its phosphorylation [38]. Some predict that

D1 phosphorylation targets the polypeptide for degradation [22], while others suggest that its phosphorylation postpones degradation once damage has occurred [24–26]. The damaged phospho-D1 was proposed to stabilize a dissipative form of PS II involved in protection of the remaining PS II activity against high-light damage [39]. Site-directed mutagenesis of psbA in order to alter the D1 phosphorylation site may provide a handle on this problem.

Phosphorylation of the reaction center polypeptides probably cannot be consistent independently of the observed phosphorylation of other PS II polypeptides such as CP43 and the 10-kDa psbH gene product or the polypeptides of the light harvesting complex (LHC II), all of which tend to be phosphorylated under reducing conditions [38]. It has been suggested that thylakoid polypeptide phosphorylation protects against photoinhibition, and studies have provided some evidence that phosphorylated reaction centers are less likely to be damaged [40].

B. C-TERMINAL PROCESSING

In higher plants and most other species examined, the D1 polypeptide is synthesized with a short C-terminal extension. Structural models place the C terminus of D1 on the lumenal side of the thylakoid such that the newly synthesized C terminus of D1 must transverse the membrane following translation and release from the ribosome sitting on the stromal side of the thylakoid. The C-terminal extension must be removed to allow assembly of the OEC since the mature C terminus is apparently required as a ligand [41]. However, a photochemically competent reaction center is assembled in the LF-1 nuclear mutant of *Scenedesmus obliquus,* which is unable to process the D1 C terminus due to its lack of the appropriate specific protease [42,43]. The PS II membranes isolated from LF-1 can be engineered back to competency in water-splitting by treatment with the protease necessary to process the D1 C terminus followed by assembly of an OEC *in vitro* [44]. A gene encoding a protease apparently specific for D1 C-terminal processing has been sequenced in *Synechocytis* 6803 and designated *ctpA* [45]. A *Synechocystis* mutant in which the *ctpA* gene was inactivated has a phenotype very similar to LG-1 [46]. It is not clear why plants go to the extent of synthesizing the C-terminal extension of D1 and a specific protease for its removal — the sequence of psbA in the green alga *Euglena gracilis* reveals no C-terminal extension and cells that are competent in oxygen evolution [47,48]; removal of the C-terminal extension of *C. reinhardtii* by genetic engineering and chloroplast transformation produced a phenotype indistinguishable from the wild type at least under the

conditions tested [49,50]. The processing does, however, provide a useful means by which the plant nucleus might control the activation of previously assembled reaction centers [44,51]. Other functions might include the possibility that the C-terminal amino acid(s) of D1 are sensitive to nonspecific carboxypeptidase activity or some other modification during the assembly process, which would otherwise waste the entire polypeptide.

The mature C terminus of D1 was confirmed by sequencing studies [52]. Reaction centers isolated from spinach thylakoids were denatured with SDS and the D1 and D2 polypeptides separated by size exclusion chromatography in the presence of 0.2% SDS. Analysis of amino acids released by carboxypeptidase treatment of purified D1 and D2 enabled determination of their C termini revealing the processing site of D1 and the unprocessed D2 C terminus. It is unlikely, though unconfirmed, that PS II-E, -F, and -I are processed at their C termini.

C. METHYLATION

The light-regulated methylation of chloroplast has been documented, but none appeared to be thylakoid membrane proteins [53]. It is possible that D1 is synthesized with a short C-terminal extension because occasional α-carboxymethylation can occur immediately after the polypeptide is synthesized and before the C-terminal domain has been translocated across the thylakoid. The C-terminal processing in the lumen then proceeds once the C terminus is isolated from stromal carboxymethyl transferase activity, allowing 100% of the D1 C termini to bear the free α-carboxy group required for assembly of the OEC. Thus, a single methylation would not waste an entire D1 and tie up other PS II subunits in a complex that could never become active in linear electron transport (see Section II.B).

D. FATTY ACYLATION

When the aquatic angiosperm *Spirodela oligorhiza* was pulse-labeled with [^{3}H]-palmitic acid, a number of chloroplast polypeptides were observed to become labeled. The only thylakoid polypeptide that was observed to be labeled after the 3-min pulse was D1, which was also rapidly synthesized under the conditions. It was confirmed that the acyl group remained as palmitoyl and that a thioester bond linked it to the D1 N-terminal tryptic peptide T22/T20 [54] limiting the modification site to one of only a few methionine or cysteine residues found in this portion of the polypeptide. Since palmitoylation in animals is confined to cysteine [55], the only cysteines of D1, residues 19 and 126, which are highly conserved in the all species examined [19], are strong candidates for the modification site. The palmitoylation event apparently occurred after C-terminal processing of D1 and translocation to the granal lamellae [56], though it is also possible that palmitoylation immediately preceded translocation as the authors concluded [54]. The function of the transient palmitoylation remains obscure.

The palmitoylation studies above also revealed that the large subunit of Rubisco and the chloroplast acyl carrier protein were similarly modified [54]. A more general investigation of plant protein acylation has revealed that many plant proteins from several different organelles, particularly the mitochondria and the nucleus, can be modified with farnesyl, geranylgeraniol, phytol, and other isoprenoids [57]. It seems that the study of plant protein lipidation is in its infancy, and further investigations of thylakoid membrane proteins might be productive.

E. DAMAGE, OXIDATION, AND DEGRADATION

It has been known for some years that PS II is sensitive to electromagnetic radiation of both visible and ultraviolet wavelengths, particularly UV_B [58]. The molecular basis of this sensitivity is under investigation and has revealed several different mechanisms for the deleterious effects of illumination. Loss of activity is often accompanied by polypeptide cleavage, but it is not clear whether the reaction center is designed to promote controlled peptide cleavage or whether such cleavage is simply the gross observable result of extensive polypeptide damage. Until the covalent modifications accompanying activity loss are carefully characterized, it will not be possible to fully understand the mechanisms underlying inhibition.

1. Photosynthetically Active Radiation — Imbalance of Electron Transport

Photodamage of the PS II reaction center is a regular consequence of its function, requiring a sophisticated mechanism for the removal and replacement of D1 polypeptide from damaged PS II units such that the number of active PS II units remains constant. If light-induced damage exceeds the repair capacity, then overall activity drops in a phenomenon called photoinhibition [14–17]. Photodamage to the reaction center appears to involve two separate mechanisms, the first of which is observed when the donor side of the reaction center is unable to supply enough electrons for the rapid reduction of P_{680}^+ (donor-side photoinhibition); the second type results when the acceptor side cannot transfer electrons away from the reaction center fast enough, leading to what is thought to be the double reduction of the primary quinone acceptor Q_A and elevated charge recombination (acceptor-side photoinhibition). Both donor- and acceptor-side photoinhibition can lead to chlorophyll oxidation and cleavage of the D1 polypeptide [59]. However, such polypeptide cleavage, which has been observed *in vivo*, does not lead to immediate destruction of the reaction center [59]. It can be speculated that structural alterations resulting from polypeptide cleavage result in targeting of the reaction center either for disassembly and replacement or for conversion to an energy-dissipating form depending on the prevailing conditions. It is postulated that phosphorylation of the D1 polypeptide may be important in determining the immediate fact of the reaction center [39].

The D1 polypeptide cleavage is not random but results in distinct fragments depending on whether it results from donor- or acceptor-side photoinhibition [59]. These fragments have been identified based on their size and antigenicity: acceptor-side photoinhibition leads to primary cleavage in the region between the fourth and fifth membrane-spanning α-helices giving 23-kDa N-terminal and 10-kDa C-terminal fragments, whereas donor-side photoinhibition leads to primary cleavage in the region of the second transmembrane α-helix giving 9-kDa N-terminal and 24-kDa C-terminal fragments. Since the 10-kDa C-terminal fragment is most often observed *in vivo*, it is inferred that the prevalent mode of damage *in vivo* is via the acceptor-side mechanism. The precise cleavage sites, if indeed they are precise, have not been determined, and the mechanisms of polypeptide cleavage are unclear. In the case of acceptor-side photodamage, the mechanism apparently involves singlet oxygen ($^1O_2$) formation [60], but donor-side damage may occur even in the absence of oxygen [59]. Furthermore, it seems likely that other kinds of damaging oxidation that do not result in cleavage may occur. Some evidence for the formation of a bityrosine cross-link between neighboring segments of the D1 polypeptide has been discussed [14]. Evidence is accumulating that D1 may form cross-links to other PS II polypeptides under conditions of photodamage also [61].

The D2 polypeptide can probably suffer photodamage also since its rate of turnover may also be somewhat accelerated under photoinhibitory conditions [14]. The PS II-E, -F, and -I polypeptides are probably not photodamaged but are recycled through the turnover cycle, unlike D1, which is replaced along with D2 if required.

2. Ultraviolet Radiation

The PS II reaction center is especially sensitive to UV$_B$ irradiation, resulting in inactivation of electron transport activity [61,62]. The D1 polypeptide cleavage can accompany damage both *in vivo* and *in vitro* [64]. A-20 kDa C-terminal fragment is observed after UV$_B$ treatments, suggesting a cleavage site within the second transmembrane helix of the reaction center [64]. How polypeptide cleavage occurs is not known, but the requirement for manganese associated with the OEC [64] hints at a novel mechanism worthy of further investigation. Degradation requiring the presence of plastoquinone bound at the Q$_B$ site has also been discussed in terms of cleavage between the fourth and fifth transmembrane helices of D1 [65], but it is argued that this is not the prominent mode of UV$_B$ damage *in vivo* [64].

Plastoquinone is highly sensitive to UV$_B$, and a significant proportion of PS II inactivation results due to a general loss of plastoquinone [66] as well as the bound Q$_A$ [63]. Recently, degradation of the D2 polypeptide under UV$_B$ has been observed in a process that apparently involves the bound plastoquinone Q$_A$ [67]. A specific D2 cleavage site in the hydrophilic loop connecting transmembrane helices 4 and 5 was inferred from the observed 22-kDa N-terminal fragment and the pair of 10- and 12-kDa C-terminal fragments (seen only in the presence of the artificial quinone acceptor 2,5-dibromo-3-methyl-6-isopropyl benzoquinone [DBMIB]). It was implied that *in vivo* the bound semiquinone Q$_A^-$ is the vulnerable species, with polypeptide cleavage resulting from a novel mechanism independent of oxygen or proteolytic activity [67].

3. Degradation

Degradation of D1 polypeptide is thought to limit the rate at which active PS II units are recovered via translation of a new polypeptide [68]. The initial steps in degradation are probably polypeptide cleavage events as discussed above, but these do not necessarily lead to immediate destabilization and disassembly of the reaction center. The steps leading to degradation of the D1 polypeptide as assessed by its turnover have been summarized [65]. It was demonstrated that occupancy of the Q$_B$ site with quinone or inhibitors modulates primary D1 degradation in this region of the polypeptide. It would be surprising if no proteases were involved in the degradation process, and evidence has been presented that the CP43 polypeptide of the PS II core possesses protease activity [69]. Evidence for the involvement of a nuclear-encoded degradation system also remains compelling [70]. Control over degradation of D2 remains unclear.

Once targeted polypeptides or peptide fragments are removed from the reaction center, they are rapidly broken down, presumably by protease activity.

4. Localization

Several recent studies have indicated that PS II is in fact dimeric [71–76]. Current hypotheses suggest that active PS II units are found in dimers in the appressed granal thylakoid regions, whereas inactive units are found in their monomeric form in the nonappressed stromal membrane regions where degradation and translation of new polypeptides take place [59]. The relationship between membrane localization/aggregation state and posttranslational modifications should help clarify degradation pathways and associated control mechanisms.

III. STRUCTURE–FUNCTION STUDIES USING DIRECTED/ENGINEERED COVALENT MODIFICATIONS

A. INTRODUCTION

With the goal of relating the structure of PS II to its function, a common experimental approach introduces specific alterations at known sites within the reaction center and examines functional consequences. Earlier studies relied on directed chemical modification techniques, which always suffered from the criticism that observed functional alterations may have resulted from an unpredicted modification. Dissection of spontaneous or induced genetic alterations in photosynthesis mutants provided important advances but lacked the goal of the ability to choose the alteration. The development of genetic engineering and transformation techniques allowing site-directed modification of the genes encoding reaction center polypeptides in some model photosynthetic species has effectively provided a potentially more rigorous approach to directed modification, that is, the *in vivo* biosynthesis of reaction centers altered only by a single specific amino acid chosen by manipulation of the genetic code. Both chemical and genetic methods have provided important and often complementary information on PS II structure and function.

B. CHEMICAL MODIFICATIONS TO PS II REACTION CENTER POLYPEPTIDES

1. Controlled Protease Treatments Can Be Used to Modify PS II Activity

Controlled protease treatments of PS II do not lead to destabilization of the complex provided they are not too severe and can be used to gain structure–function information. It was the discovery of a specific protease treatment of thylakoid membranes that modulated electron transport through PS II and herbicide binding that first led to the hypothesis that a "proteinaceous shield" was associated with PS II [77]. Many studies have examined the effect of controlled proteolysis with specific effects on both donor and acceptor sides having been documented (e.g., Refs. [78,79]). Cleavage of D1 and D2 in the regions between their fourth and fifth membrane-spanning α-helices is implicated in modification of the acceptor side [80], whereas perturbation of the donor side probably arises from cuts to polypeptides associated with the OEC.

2. Covalent Modification of PS II Reaction Center Polypeptides with Organic Agents

Phenylglyoxal has been used to modify the arginine residues of PS II with demonstrated effects on both donor and acceptor sites [81,82]. Diethylpyrocarbonate (DEPC) has been used to modify histidine residues with effects on both donor and acceptor sites of PS II [83–85]. Tetranitromethane, which can modify both sulfhydryl and tyrosine residues, appears to affect the donor side of PS II, but it is not clear whether this effect is specifically due to tyrosine or –SH modification [86,87]. Modification of carboxyl groups by 1-ethyl-3-[3-(dimethylamino)propyl]carbodiimide (EDC) has been used to study the high-affinity manganese-binding site of the PS II donor side incorporating suitable controls to diminish the possibility that the observed effects were due to cross-linking or –SH modifications [88]. The results suggested that the site modified was the other half of the high-affinity manganese site that was insensitive to DEPC treatment [89], and protection of the modification site by Mn^{2+} implied that lumenal carboxyl groups provide ligands to manganese bound at this site [88]. Identification of the polypeptide amino acid residue(s) protected from EDC modification by Mn^{2+} would provide an elegant conclusion to this work. Controlled proteolysis experiments indicated that H337 of D1 was one of the DEPC-sensitive ligands, though residues on other polypeptides cannot be ruled out [84].

3. Covalent Modification of PS II Reaction Center Polypeptides with Inorganic Agents

Iodide (I^-) is able to donate electrons to PS II that lack a functional OEC in a light-dependent reaction that iodinates a tyrosine residue on D1. A tyrosine residue on D2 is iodinated in the dark [90,91]. It was concluded from peptide-mapping studies that Y161

of D1 (Y_Z) and Y160 of D2 (Y_D) were probably the modified residues [92,93].

4. Photoaffinity Labeling of PS II Reaction Center Polypeptides with Herbicide Analog

Since photoaffinity labeling of thylakoid membranes with 2-azido-4-ethylamino-6-isopropylamino-s-triazine (azidoatrazine) was used to identify the 32-kDa herbicide receptor protein of PS II [94], this technique has enjoyed considerable focus. The identification of photoaffinity labeling sites combined with genetic analysis of herbicide-resistant mutants provided chemical and genetic proof that the herbicide receptor was indeed the D1 polypeptide that along with D2 formed a heterodimeric reaction center homologous in structure to the solved crystal structure of the purple bacterial reaction center. Peptide-mapping studies [95] and peptide-sequencing studies [29] support modification of M214 of D1 by azidoatrazine. Sequencing studies showed that Y237 and Y254 of D1 were modified by azidomonuron, an analog of the herbicide diuron-[3-(3,4-dichlorophenyl)-1,1-dimethylurea] (DCMU) [96]. 2-Azido-3,5-diiodo-4-hydroxybenzonitrile (azidoioxynil) labeled V249 of D1 [97]. Several other compounds have also been observed to photoaffinity label D1 and other reaction center polypeptides [98,99].

5. Chemical Cross-Linking of PS II Reaction Center Polypeptides

In the absence of a solved crystal structure for the PS II reaction center, chemical cross-linking studies can be used to probe nearest-neighbor relationships of the polypeptides in isolated PS II. This is particularly meaningful with regard to the interface between the PS II-E, -F, and -I polypeptides and the D1/D2 heterodimer, which is predicted to form a structure similar to that of the purple bacterial reaction center. The bifunctional reagents 3,3'-(dithiobis)succinimidyl propionate (DSP) and 1,6-hexamethylene diisocyanate (HMDI) have been used to cross-link PS II reaction centers, suggesting that K4 of PS II-I is close to a stromal loop lysine of D2 as well as the N terminus of PS II-E [100] and that the C-terminal domains of D1 and D2 are in close proximity [101]. PS II particles can be cross-linked using a procedure involving adducts of the photoaffinity reagents succinimidyl [(4-azidophenyl)dithio]propionate (SADP) [102] and sulfosuccinimidyl[(4-azidophenyl)dithio]propionate (SSADP) [103], although the cross-linking sites have not been characterized. Interestingly, D1 is completely resistant to chemical cross-linking using agents such as glutaraldehyde in intact thylakoids unless pretreated with octyl β-D-glycoside [104]. Cross-linking studies have also been used to probe changes in spatial relationships of polypeptides in PS II membranes in response to protein phosphorylation [105].

C. IDENTIFICATION OF SPECIFIC MODIFICATION SITES OF PS II REACTION CENTER POLYPEPTIDES

1. Detection of Specific Modifications

Most of the covalent modifications to PS II reaction center polypeptides have been analyzed by gel electrophoresis (SDS-PAGE) and labeling studies. Antibodies to known epitopes have been useful in identifying specific proteolytic fragments, and sequencing studies have enabled the identification of some photoaffinity labeling sites. As the demand for accurate characterization of modification sites increases, more precise methods of analysis will be required. Structural determinations by x-ray or electron diffraction studies of crystals are one means of characterizing modifications, but the PS II reaction center has not yet yielded to such methods at the levels of resolution required. The reaction center is too big for structural analysis with current nuclear magnetic resonance (NMR) methodologies. The most promising method for accurate analysis of all PS II reaction center polypeptide modifications is mass spectrometry, which can yield primary structure information. Along with primary structures predicted from gene sequences, accurate mass determination can reveal the presence of modifications, and detailed structural determination can then be used to characterize the modification site. The solving of the nature of N-terminal processing of D1 and D2 [20] provides an example of such methodology and highlights some of the technical difficulties that must be overcome to make mass spectrometry more broadly applicable.

2. Characterization of Modification Sites

Mass spectrometric analysis requires moderate quantities of material, highly purified using high-performance liquid chromatography (HPLC) or capillary electrophoresis. Though masses in the range of individual PS II polypeptides can now be accurately measured, much smaller peptides are required for structural information to be obtained. The extreme hydrophobicity of most of the peptides derived from the PS II reaction center makes them difficult to handle without resorting to SDS. The N-terminal phosphopeptides of D1 and D2 are quite hydrophilic, enabling their purification by ferric ion affinity chromatography and standard HPLC techniques [20], though the use of a method of isolation specific for the phosphate group

eliminates the chance to observe the nonphosphorylated form if indeed it exists.

An important breakthrough was made by Whitelegge et al. [29], who used one of the new generation of macroporous poly(styrene/divinylbenzene) chromatography supports combined with a formic acid/isopropanol solvent system to isolate hydrophobic peptides originating from intrinsic α-helical regions of the D1 polypeptide. These peptides were suitable for both sequencing studies and mass-spectrometric analysis. Use of the poly(styrene/divinylbenzene) support has been extended to intact thylakoid membrane proteins [106]. Some cyanogen bromide fragments derived from D1 and D2 were separated on a C8 silica column [101] using a trifluoroacetic acid (TFA)/acetonitrile solvent system. D1 was first isolated by HPLC using a C18 silica column [107].

D. SITE-DIRECTED MUTAGENESIS AND THE COVALENT MODIFICATION OF PS II REACTION CENTER POLYPEPTIDES

1. Introduction

The most elegant method of introducing specific covalent modifications to PS II reaction center polypeptides is surely site-directed mutagenesis. In principle, by altering the appropriate gene it is possible to alter single or multiple amino acid residues or introduce or remove sections of polypeptide of varying lengths. Unfortunately, such goals can only be accomplished in the few species currently amenable to transformation. Furthermore, even single amino acid alterations are frequently sufficient to destabilize the reaction center so that very little or no modified complexes accumulate precluding functional analysis. Despite these drawbacks, it is most likely that site-directed mutagenesis will remain the most important means of modifying reaction center polypeptides for many years to come.

Of the wide range of organisms capable of oxygenic photosynthesis, both prokaryotic cyanobacteria, such as *Synechocystis* PPC 6803, and the eukaryotic green algal species *C. reinhardtii* are transformable to the extent that any of the five PS II reaction center polypeptides can be potentially altered at will. This objective is facilitated in *Chlamydomonas* by the fact that these polypeptides are encoded within the chloroplast genome, which can be conveniently engineered in contrast to its nuclear genome. Importantly, both of the above-mentioned species will grow using heterotrophic metabolism such that mutations that cripple photosynthetic production do not kill the transformed organism, thus overcoming a significant barrier to site-directed mutagenesis of nearly all higher-plant species. Nevertheless, development of a workable chloroplast transformation system for manipulation of PS II reaction center polypeptides in a higher-plant species remains an important priority.

The choice of host species for transformation depends on the type of analysis to be performed upon mutants. Biophysical analysis of the primary reactions of electron transport by PS II can be conveniently accomplished in either *Synechocystis* or *Chlamydomonas* since reaction centers [108,109], oxygen-evolving core particles [27,110,111], or PS II–enriched membranes (BBYs) [112,113] can be isolated from either species in broadly comparable yields. Comparison of the sequences of D1 and D2 reveals a very high homology between the prokaryote and the eukaryote [19], and similarly PS II-E, -F, and -I [9,35,36,114] are also quite highly conserved, suggesting a similar function of the reaction center in both. The OECs of both host types function comparably, yet it is known that extrinsic polypeptides of the OEC, which are thought to stabilize the tetranuclear manganese cluster, do vary considerably between the species, with *Synechocystis* displaying a rather different arrangement from that observed in eukaryotes [1]. The extrinsic phycobilisome light harvesting antenna of the cyanobacteria is also very different from the intrinsic LHC II found associated with PS II in algae and higher plants. Whether such differences between OEC or antenna are significant with regard to the primary function of the reaction center is doubtful. What is clear is that the physiologies of the two host types are quite different and the choice of host for studies with a more physiological bias should be carefully considered. Even *Chlamydomonas,* whose chloroplasts are similar to higher plants in many ways, cannot be regarded a perfect model species.

Undoubtedly, the most engineered species with regard to PS II reaction center polypeptides, *Synechocystis* PCC 6803, offers several features that make it highly attractive to the genetic engineer. Probably the most significant of these is its ability to take up small pieces of homologous DNA and recombine them into its genome [115]. With the appropriate use of heterologous selectable markers, engineering of PS II reaction center polypeptides is accomplished with ease [116,117]. Furthermore, *in situ* complementation [118] achieved by spotting appropriate DNA solutions onto a lawn of mutant cells provides a powerful means of visualizing growth phenotype as well as confirming mutant genotype [119,120]. *C. reinhardtii* PS II reaction center polypeptides have been somewhat less engineered, and I shall here review the subject in more detail to supplement the indispensable "Chloroplast Transformations in *Chlamydomonas*" [121] and *The Chlamydomonas Sourcebook* [122].

2. Manipulation of Chloroplast PS II Electron Transport in *C. reinhardtii* Using Site-Directed Mutagenesis

While *C. reinhardtii* PS II reaction center polypeptides are encoded in the chloroplast genome, the assembly of PS II complexes *in vivo* requires the coordinated expression of many nuclear genes as well [123]. The discovery that DNA could be introduced to the chloroplast via the particle gun and that homologous recombination of transforming DNA with the chloroplast genome occurred [124] paved the way for efficient engineering of chloroplast-encoded PS II polypeptides. The nuclear-encoded polypeptides cannot yet be engineered with precision, although nuclear DNA may be transformed [125], and progress has been made in directing transformation to specific loci as well as accomplishing homologous recombination of transforming DNA with target nuclear genes [126–128].

a.   Choice of hosts

One of the most significant advances of *C. reinhardtii* as a model organism is its ability to synthesize chlorophyll in the dark, unlike nearly all higher-plant species. Thylakoid membranes and associated chlorophyll–protein complexes are thus nearly fully assembled in the dark. PS II is fully assembled, except for the photoactivation (assembly) of the OEC. Consequently, *Chlamydomonas* can assemble its PS II reaction center in complete darkness allowing an otherwise impossible study of superphotosensitive mutants, as well as the study of the photoactivation process *in vivo*. The ability of *C. reinhardtii* to synthesize chlorophyll in the dark is lost quite easily if cells are stored in the light, so it is wise to obtain a green-in-the-dark (GID) line and keep it in the dark. The author's favorite wild-type strain is 2137, which forms compact, very dark green colonies on agar and deep green liquid cultures even when grown in darkness. Other wild-type host varieties have also been used successfully [49,129–132].

An alternative host variety for transformation is deleted in all or part of the gene to be engineered. When using such a host, transformation can be used to replace a missing gene or gene segment with a piece of engineered DNA resulting in restoration of an otherwise wild-type gene bearing the desired alteration. Whitelegge et al. [132] used such a technique to successfully engineer *psbA* site-directed mutants. Alternatively, the piece of DNA used for gene replacement can be more highly engineered. For example, a recent study has produced a single plasmid suitable for all manipulations of *Chlamydomonas psbA* by splicing out the four *psbA* introns, introducing unique restriction sites for more convenient engineering and adding a heterologous selectable marker [133]. These strategies are summarized in Figure 10.2.

b. DNA constructs for transformation

The major DNA constructs used for transformation of *psbA* in *Chlamydomonas* are summarized in Figure 10.2. The chloroplast restriction fragments R16 9pRR, which contain *psbA* exons 1 to 4, and R24, which contain exon 5 (in the pRX subclone), were first isolated and sequenced in the Rochaix laboratory [135,136]. As shown in Figure 10.2, smaller subclones are usually used for genetic manipulation followed by further subcloning into larger constructs. Removal of *psbA* introns by splicing and engineering of unique restriction sites along with the insertion of the aadA cassette has generated a single plasmid (pBA157) that can be used for any *psbA* alteration without the need to subclone or use a second plasmid containing a selectable marker [133].

The *psbD* gene, which does not contain introns, is contained within restriction fragments R3 and R06 [135,137]. The *psbE* and *psbF* genes are found on chloroplast restriction fragment PstI-4 (p074) [35,36]. The *psbI* gene is found on chloroplast restriction fragment R7 [9,135].

c. Transformation method

The method of choice for chloroplast transformation in *C. reinhardtii* is the particle gun. Transforming DNAs are coated on tungsten or gold microprojectiles, which are fired at high velocity into target cells using gunpowder charge or compressed gas [121,124,138,139]. Transformation efficiency is rather low (10^{-4} is around the highest reported) but nevertheless results in up to several thousand successful transformations per individual target of approximately 2 million cells. This success rate is often lessened, depending on the transforming DNA. The high velocity of the microprojectiles ensures that the transforming DNA enters the cell regardless of the presence of the cell wall. It is assumed that the particle leading to successful transformation also penetrate the membranes surrounding the single chloroplast of the *Chlamydomonas* cell, allowing interaction between the transforming DNA and the 50 to 100 copies of the chloroplast genome. Homologous recombination between transforming DNA and the chloroplast genome results in incorporation of foreign DNA into one or more chloroplast genome copies. Cell division and replication eventually allow the segregation of some homoplasmic cell lines where all copies of the chloroplast genome bear the modified DNA sequence.

Unfortunately, the particle gun is a rather specialized piece of equipment not widely available to all researchers, and its price presents a barrier to most individual laboratories. Other techniques for chloroplast transformation have consequently been developed. Vortexing of cells with transforming DNA and glass beads has proved successful provided that the host strain is cell wall minus (e.g., CW15) or the cell walls are removed [140]. To overcome the problem of the cell wall minus requirement, it has recently been reported that the glass beads can be replaced with silicon carbide "whiskers" allowing successful transformation of wild-type strains [141]. Thus, there are other methods for successful chloroplast transformation that can be used instead of the particle gun provided they are not overly efficient at transforming the nucleus. Transformation of the nucleus with heterologous DNA leads to random insertions often accompanied by neighboring deletions [142], therefore it is important to ascertain that the mutant phenotype obtained is truly the result of the designed chloroplast alteration and not the result of an altered nuclear genotype. Of course, such a consideration is also required for mutants obtained using the particle gun.

d. Segregation

Due to the polyploid nature of the chloroplast genome of *Chlamydomonas,* a single transformed cell is likely to contain a mixture of wild-type and mutant genome copies. This transient heteroplasmic state is apparently rapidly replaced by the segregation of homoplasmic siblings after several rounds of cell division. If the mutant genome is providing resistance to some kind of selection pressure (e.g., a drug resistance marker), then it is likely that all surviving siblings will be mutant. If, however, there is no selection pressure for the mutation, then both wild-type and mutant siblings would be expected. Such a state of affairs is observed after a cotransformation experiment like that shown in Figure 10.2. Only transformants bearing the selectable marker mutation in the 168 rRNA gene survive during segregation, but not all of these contain the second mutation, the desired *psbA* alteration. The double mutants with the desired *psbA* alteration as well as the selectable marker must then be identified among the different siblings of the initial transformant, if indeed any contain the second mutation. Fortunately, cotransformation frequencies are often quite high (up to 25%) [132]. If the deletion mutant host is used, then only mutant copies of *psbA* will be found in the segregating population. If a wild-type host is used, then it is possible that both wild-type and mutant copies of *psbA* are found during segregation. Any phenotype observed might arguably result from a mixed genotype

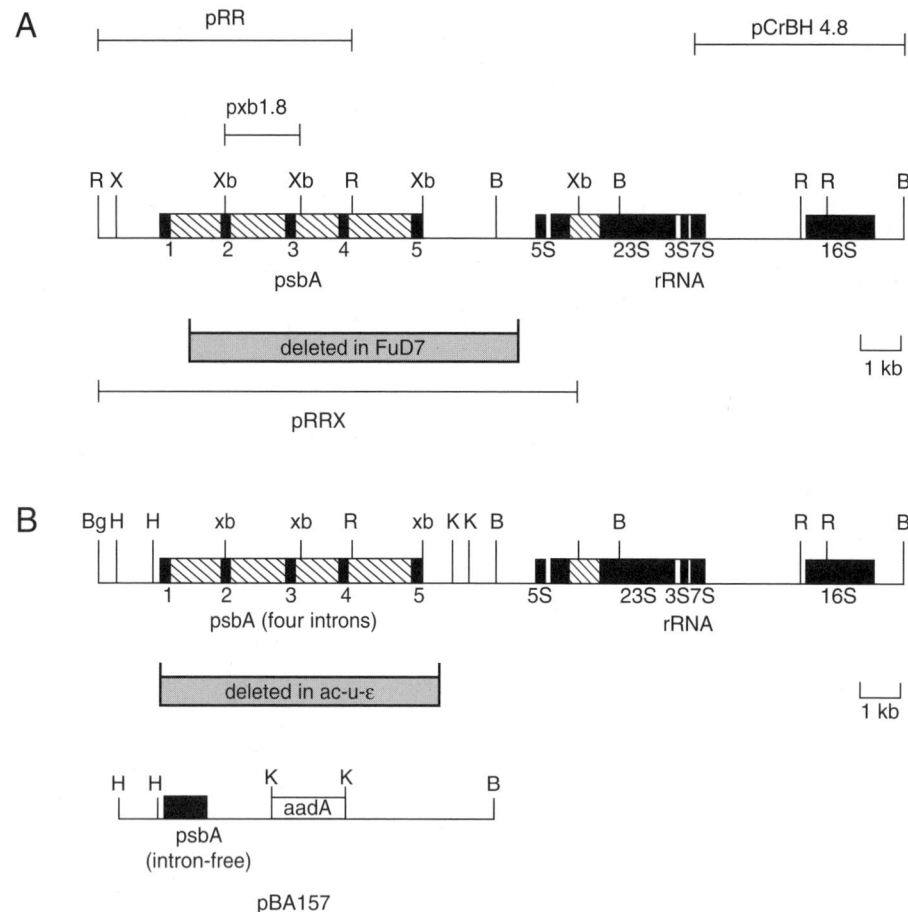

FIGURE 10.2 Strategies for the transformation of *psbA* in *Chlamydomonas reinhardtii*. (A) Transformation using a homologous selectable marker. Plasmid insert pCrBH4.8 is used to introduce a single-point mutation in the 16S rRNA gene that confers spectinomycin resistance upon successful transformants. In cotransformation strategies a second plasmid is introduced to cells along with the spectinomycin resistance marker (pCrBH4.8). The plasmid insert pRR can be used to transform a wild-type host (above the map), but the larger pRRX plasmid insert is required to replace *psbA* in the FuD7 deletion mutant (below the map; note the deletion in FuD7-speckled box). Since both pRR and pRRX are too large for convenient engineering techniques such as site-directed mutagenesis, a smaller plasmid insert must be used for manipulations. For example, to engineer specific alterations to codon asp170 of *psbA* exon 3, Whitelegge et al. [132] used the pXb1.8 insert (shown above the map), which required subcloning into the larger constructs pRR and pRRX for transformation of wild-type and FuD7 deletion hosts, respectively. Transformants containing the desired *psbA* alteration must be identified among those bearing the spectinomycin resistance marker, with observed cotransformation frequencies in the 1% to 25% range. Final transformants contain solely alterations to a maximum of three base pairs per altered codon of *psbA* and a single base-pair alteration to the 16S rNA (adapted from Ref. [132]). (B) Transformation using a heterologous selectable marker. The heterologous aadA cassette (open box) confers resistance to spectinomycin upon expression of the aadA gene in transformants [134]. Alteration of *psbA* is achieved in the intron-free *psbA* gene in plasmid insert pBA 157, which also contains the spectinomycin resistance marker aadA. Linkage of the two genes in this way results in efficient transformation of a deletion host such as ac-u-;ys (below the map; speckled box) with approximately 100% of spectinomycin-resistant transformants also carrying the desired alteration to *psbA* [133]. Final transformants contain the spliced intronless *psbA* gene with chosen codon alterations as well as silent changes used to introduce restriction sites and express the heterologous aadA gene in their chloroplasts. (Adapted from J. Minagawa and A. R. Crofts, *Photosynth. Res., 42*:121 (1994)).

leading to interpretation problems. It is thus necessary to demonstrate that segregation is complete, particularly if using a wild-type host. It is believed that a fully segregated mutant contains identical copies of *psbA* in each half of the inverted repeat resulting from a copy correction mechanism such that all chloroplast gen-

ome copies are identical [121]. Thus, it should be experimentally verified at a sensitivity of around 1:200 that all *psbA* copies are mutant if conclusions regarding phenotype are to be considered valid. Obviously, the same mutation can be constructed in a deletion host to confirm a particular phenotype [132].

e. Controls

The ideal controls to use when examining the phenotype of transformants include the host strain when a wild-type host is used or a transformant bearing a wild-type replacement gene if a deletion host is used [132]. It should also be confirmed that the selectable marker mutation does not perturb whatever aspect of phenotype is examined. Since the particle gun can induce both chloroplast and nuclear mutations, it is preferable to examine phenotype in two or three independent transformants for each alteration studied to absolutely confirm that the observed phenotype results from the desired alteration and not from another unsuspected mutation.

f. Reduction of chloroplast copy number

Many chloroplast transformation protocols suggest growing host cells in 5-fluoro-2′-deoxyuridine to decrease the chloroplast copy number and increase chloroplast transformation efficiency [121,140,143]. Since the treatment is mutagenic toward chloroplast DNA [144] as well as personnel, it is desirable to avoid the use of this chemical. Transformation and cotransformation efficiencies apparently remain satisfactorily high even when FdUrd treatments are not used [132].

g. Maintenance of mutant lines and storage

Transformant lines are kept in darkness to avoid any selection pressure for revertants. Mutants are usually kept growing on agar plates since cold storage of *Chlamydomonas* cells usually kills them. Fortunately, protocols for the successful long-term freezing of cells are under development [145] that will hopefully eliminate the tedious task of keeping all cell lines on agar.

h. C. reinhardtii site-directed mutants with modified PS II reaction center polypeptides

Only a limited number of studies have so far examined the effect of site-directed mutations on PS II structure and function. Whitelegge et al. [131,132] have examined the role of D170 of D1 in the assembly of the OEC (see Figure 10.3). Roffey et al. have made alterations at D1 codons 195 [130] and 190 [146,147] to probe electron donation within the reaction center. Przibilla et al. [129] have examined the effect of twin alterations to D1 codons 266 and 264 and a triple alteration (D1 codons 266, 264, and 259) on herbicide sensitivity of PS II. Lers et al. [49] engineered a mutant that lacked the D1 C-terminal extension.

IV. CONCLUSIONS

The many possible combinations of posttranslational modifications to PS II reaction center polypeptides may underlie the difficulty in obtaining high-resolution three-dimensional structural information from crystallographic studies. When the structure is solved, it will aid our understanding of how the dynamic nature of covalent modification relates to all aspects

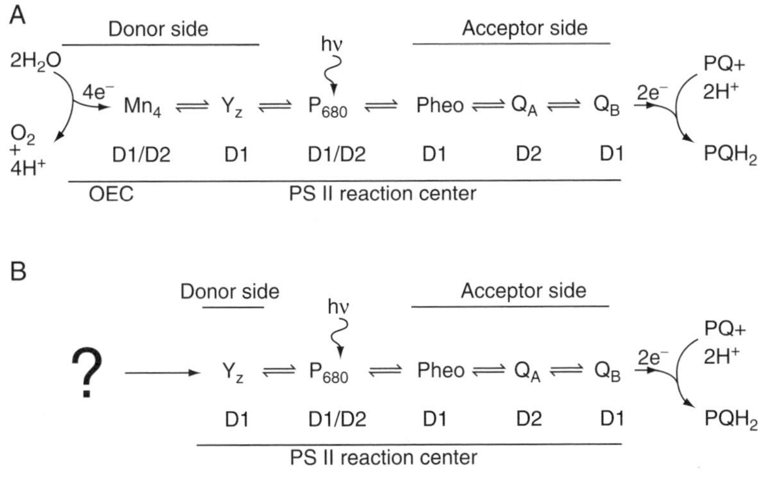

FIGURE 10.3 Manipulation of electron transport through PS II in vivo. The linear electron transport pathway through PS II is shown for wild-type reaction center (A) and those where D1 codon 170 has been covalently modified via site-directed mutagenesis (B). In *Chlamydomonas reinhardtii* Whitelegge et al. [132] have demonstrated that such modifications lead to either partial or complete loss of the ability to assemble the OEC, thus generating a shortage of electrons for reduction of the primary and secondary donors, P_{680}^+ and Y_Z^+. Alternative donors such as cytochrome b_{559} or Y_D may provide some electrons, but it is also likely that the lifetime of P_{680}^+ will be increased leading to oxidation of chlorophyll, carotenoids, and amino acid residues. Manipulation of the PS II electron transport pathway in vivo provides an exciting tool for the dissection of damage and protection mechanisms. (Adapted from J. P. Whitelegge, D. Koo, B. A. Diner, I. Domain, and J. M. Erickson, *J. Biol. Chem., 270*:225 (1995)).

of PS II physiology and ultimately plant productivity. A deeper comprehension of processes such as the PS II repair cycle and functional heterogeneity as well as their intimate relationship to the supermolecular organization of the thylakoid will require further careful analysis of covalent modifications. Controlled modification via site-directed mutagenesis will prove invaluable for the testing of hypotheses not only concerning PS II physiology but also with regard to the biophysics of energy conversion by the photosynthetic reaction center. Recent advances have been reviewed [148].

REFERENCES

1. R. J. Debus, *Biochim. Biophys. Acta, 1102*:269 (1992).
2. J. Barber, *Biochem. Soc. Trans., 22*:313 (1994).
3. J. H. A. Nugent, P. J. Bratt, M. C. W. Evans, D. J. MacLachlan, S. E. J. Rigby, S. V. Ruffle, and S. Turconi, *Biochem. Soc. Trans., 22*:327 (1994).
4. O. Nanba and K. Satoh, *Proc. Natl. Acad. Sci. USA, 84*:109 (1987).
5. A. Trebst, *Z. Naturforsch., 42c*:742 (1987).
6. H. Michel and J. Deisenhofer, *Biochemistry, 27*:1 (1988).
7. J. Barber and J. De Las Rivas, *Proc. Natl. Acad. Sci. USA, 90*:10942 (1993).
8. A. N. Webber, L. Packman, D. J. Chapman, J. Barber, and J. C. Gray, *FEBS Lett., 242*:259 (1989).
9. P. Künstner, A. Guardiola, Y. Takahashi, and J.-D. Rochaix, *J. Biol. Chem., 270*:9651 (1995).
10. A. Trebst, *Z. Naturforsch, 41c*:240 (1986).
11. G.-S. Tae and W. A. Cramer, *Biochemistry, 33*:10060 (1994).
12. V. V. Klimov, S. I. Allakhverdiev, S. Demeter, and A. A. Krasnovskii, *Dokl. Akad. Nauk SSSR, 249*:227 (1979).
13. A. K. Mattoo, J. B. Marder, and M. Edelman, *Cell, 56*:241 (1989).
14. O. Prášil, N. Adir, and I. Ohad, in *The Photosystems: Structure, Function and Molecular Biology* (J. Barber, ed.), Elsevier Science Publishers B.V., New York, 1992, p. 295.
15. S. B. Powles, *Annu. Rev. Plant Physiol., 35*:15 (1984).
16. D. J. Kyle, in *Photoinhibition* (D. J. Kyle, C. B. Osmond, and C. J. Arntzen, eds.), Elsevier Science Publishers B.V., New York, 1987, p. 197.
17. J. Barber and B. Andersson, *TIBS, 17*:61 (1992).
18. N. R. Baker and D. R. Ort, in *Crop Photosynthesis: Spatial and Temporal Determinants* (N. R. Baker and H. Thomas, eds.), Elsevier Science Publishers, B.V., New York, 1992, p. 289.
19. B. Svensson, I. Vass, and S. Styring, *Z. Naturforsch., 46c*:765 (1991).
20. H. Michel, D. F. Hunt, J. Shabanowitz, and J. Bennett, *J. Biol. Chem., 263*:1123 (1988).
21. F. E. Callahan, M. L. Ghirardi, S. K. Sopory, A. M. Mehta, M. Edelman, and A. K. Mattoo, *J. Biol. Chem., 265*:15357 (1990).
22. T. D. Elich, M. Edelman, and A. K. Mattoo, *J. Biol. Chem., 267*:3523 (1992).
23. T. D. Elich, M. Edelman, and A. K. Mattoo, *EMBO J., 12*:4857 (1993).
24. E.-M. Aro, I. Virgin, and B. Andersson, *Biochim. Biophys. Acta, 1143*:113 (1993).
25. A. J. Syme, H. R. Bolhar-Nordenkampf, and C. Critchley, *Z. Naturforsch., 48c*:246 (1993).
26. E. Rintamäki, R. Salo, E. Lehtonen, and E.-M. Aro, *Planta, 195*:379 (1995).
27. C. de Vitry, B. A. Diner, and J.-L. Popot, *J. Biol. Chem., 266*:16614 (1991).
28. J. P. Whitelegge, The Role of Protein Phosphorylation in Photosynthetic Light Acclimation, Ph.D. thesis, University of London, 1989.
29. J. P. Whitelegge, P. Jewess, M. G. Pickering, C. Gerrish, P. Camilleri, and J. R. Bowyer, *Eur. J. Biochem., 207*:1077 (1992).
30. H. Schagger and G. von Jagow, *Anal. Biochem., 166*:368 (1987).
31. P. A. Millner, J. B. Marder, K. Gounaris, and J. Barber, *Biochim. Biophys. Acta, 852*:30 (1986).
32. P. Delèpelaire, *EMBO J., 3*:701 (1984).
33. P. Delèpelaire and F.-A. Wollman, *Biochim. Biophys. Acta, 809*:277 (1985).
34. T. Silverstein, L. Cheng, and J. F. Allen, *Biochim. Biophys. Acta, 1183*:215 (1993).
35. S. Alizadeh, R. Nechushtai, J. Barber, and P. Nixon, *Biochim. Biophys. Acta, 1188*:439 (1994).
36. T. S. Mor, I. Ohad, J. Hirschberg, and H. Pakrasi, *Mol. Gen. Genet., 246*:600 (1995).
37. A. N. Webber, S. M. Hird, L. C. Packman, T. A. Dyer, and J. C. Gray, *Plant Mol. Biol., 12*:141 (1989b).
38. J. F. Allen, *Physiol. Plant., 93*:196 (1995).
39. C. Critchley and A. W. Russel, *Physiol. Plant., 92*:188 (1994).
40. M. A. Harrison and J. F. Allen, *Biochim. Biophys. Acta, 1058*:289 (1991).
41. P. J. Nixon, J. T. Trost, and B. A. Diner, *Biochemistry, 31*:10859 (1992).
42. M. A. Taylor, P. J. Nixon, C. M. Todd, J. Barber, and J. R. Bowyer, *FEBS Lett., 235*:109 (1988).
43. B. A. Diner, D. F. Ries, B. N. Cohen, and J. G. Metz, *J. Biol. Chem., 263*:8972 (1988).
44. J. R. Bowyer, J. C. L. Packer, B. A. McCormack, J. P. Whitelegge, C. Robinson, and M. A. Taylor, *J. Biol. Chem., 267*:5424 (1992).
45. S. V. Shestakov, P. R. Andudurai, G. E. Stanbekova, A. Gadzhiev, and H. Pakrasi, *J. Biol. Chem., 269*:19354 (1994).
46. P. R. Anbudurai, T. S. Mor, I. Ohad, S. V. Shestakov, and H. Pakrasi, *Proc. Natl. Acad. Sci. USA, 91*:8082 (1994).
47. M. Keller and E. Stutz, *FEBS Lett., 175*:173 (1984).
48. G. D. Karabin, M. Farley, and R. B. Hallick, *Nucleic Acids Res., 12*:5801 (1984).
49. A. Lers, P. B. Heifitz, J. E. Boynton, N. W. Gillham, and C. B. Osmond, *J. Biol. Chem., 267*:17494 (1992).
50. S. Schrader and U. Johanningmeier, *Plant Cell Physiol., 31*:273 (1990).

51. P. J. Nixon and B. A. Diner, *Biochem. Soc. Trans., 22*: 338 (1994).

52. Y. Takahashi, N. Nakane, H. Kojima, and K. Satoh, *Plant Cell Physiol., 31*:273 (1990).

53. M. T. Black, D. Meyer, W. R. Widger, and W. A. Cramer, *J. Biol. Chem., 262*:9803 (1987).

54. A. K. Mattoo and M. Edelman, *Proc. Natl. Acad. Sci. USA, 84*:1497 (1987).

55. P. J. Casey, *Science, 268*:221 (1995).

56. F. E. Callahan, M. Edelman, and A. K. Mattoo, in *Progress in Photosynthesis Research* (J. Biggins, ed.), Nijhoff, The Hague, The Netherlands, 1987, p. 799.

57. C. A. Shipton, I. Palmryd, E. Swiezewska, B. Andersson, and G. Dallner, *J. Biol. Chem., 270*:566 (1995).

58. L. W. Jones and B. Kok, *Plant Physiol., 41*:1037 (1966).

59. J. Barber, *Aust. J. Plant Physiol., 22*:201 (1994b).

60. A. Telfer, S. M. Bishop, D. Phillips, and J. Barber, *J. Biol. Chem., 269*:13244 (1994).

61. H. Mori, Y. Yamashita, T. Akasaka, and Y. Yamamoto, *Biochim. Biophys. Acta, 1228*:37 (1995).

62. B. M. Greenburg, V. Gaba, O. Canaani, S. Malkin, A. K. Mattoo, and M. Edelman, *Proc. Natl. Acad. Sci. USA, 86*:6617 (1989).

63. A. Melis, J. A. Nemson, and M. A. Harrison, *Biochim. Biophys. Acta, 1100*:312 (1992).

64. R. Barbato, A. Frizzo, G. Friso, F. Figoni, and G. M. Giacometti, *Eur. J. Biochem., 227*:723 (1995).

65. M. A. K. Jansen, B. Depka, A. Trebst, and M. Edelman, *J. Biol. Chem., 268*:21246 (1993).

66. A. Trebst and B. Depka, *Z. Naturforsch., 45c*:765 (1990).

67. G. Friso, R. Barbato, G. M. Giacometti, and J. Barber, *FEBS Lett., 339*:217 (1994).

68. N. Adir, S. Schochat, and I. Ohad, *J. Biol. Chem., 265*:12563 (1990).

69. A. H. Salter, J. Virgin, A. Hagman, and B. Andersson, *Biochemistry, 31*:3990 (1992).

70. E. Bracht and A. Trebst, *Z. Naturforsch., 49c*:439 (1994).

71. J. P. Dekker, E. J. Boekema, H. T. Witt, and M. Rögner, *Biochim. Biophys. Acta, 936*:307 (1988).

72. G. F. Peter and J. P. Thornber, *J. Biol. Chem., 266*:16745 (1991).

73. R. Bassi and P. Dainese, *Eur. J. Biochem., 204*:317 (1992).

74. C. Santini, V. Tidu, G. Tognon, A. Ghirette Magaldi, and R. Bassi, *Eur. J. Biochem., 221*:307 (1994).

75. M. Seibert, *Aust. J. Plant Physiol., 22*:161 (1994).

76. E. J. Boekema, B. Hankamer, D. Bald, J. Kruip, J. Nield, A. F. Boonstra, J. Barber, M. Rögner et al., *Proc. Natl. Acad. Sci. USA, 92*:175 (1995).

77. G. Renger, *Biochim. Biophys. Acta, 440*:287 (1976).

78. M. Völker, G. Renger, and A. W. Rutherford, *Biochim. Biophys. Acta, 851*:424 (1986).

79. R. Fromme and G. Renger, *Z. Naturforsch., 45c*:373 (1990).

80. A. Trebst, B. Depka, B. Kraft, and U. Johanningmeier, *Photosynth. Res., 18*:163 (1988).

81. M. Kuhn, A. Thiel, and P. Boger, *Physiol. Veg., 24*:485 (1986).

82. K. Csatorday, S. Kumar, and J. T. Warden, *Biochim. Biophys. Acta, 890*:224 (1987).

83. T. Ono and Y. Inoue, *FEBS Lett., 278*:183 (1991).

84. C. Preston and M. Seibert, *Biochemistry, 30*:9625 (1991).

85. U. Hegde, S. Padhye, L. Kovacs, A. Zozar, and S. Demeter, *Z. Naturforsch., 48c*:896 (1993).

86. P. V. Sane and U. Johanningmeier, *Z. Naturforsch., 35c*:293 (1979).

87. Y. Gingras, J. Harnois, G. Ross, and R. Carpentier, *Photochem. Photobiol., 61*:183 (1995).

88. C. Preston and M. Seibert, *Biochemistry, 30*:9615 (1991).

89. M. Seibert, N. Tamura, and Y. Inoue, *Biochim. Biophys. Acta, 974*:185 (1989).

90. Y. Takahashi, M. Takahashi, and K. Satoh, *FEBS Lett., 208*:347 (1986).

91. M. Ikeuchi and Y. Inoue, *FEBS Lett., 210*:71 (1987).

92. M. Ikeuchi and Y. Inoue, *Plant Cell Physiol., 29*:695 (1988).

93. Y. Takahashi and K. Satoh, *Biochim. Biophys. Acta, 973*:138 (1989).

94. K. Pfister, K. Steinback, G. Gardner, and C. J. Arntzen, *Proc. Natl. Acad. Sci. USA, 78*:981 (1981).

95. P. K. Wolber, M. Eilmann, and K. Steinback, *Arch. Biochem. Biophys., 248*:224 (1986).

96. R. Dostatni, H. E. Meyer, and W. Oettmeier, *FEBS Lett., 239*:207 (1988).

97. W. Oettmeier, K. Masson, J. Hohfeld, H. E. Meyer, K. Pfister, and H. P. Fischer, *Z. Naturforsch., 44c*:444 (1989).

98. J. R. Bowyer, P. Camilleri, and W. F. J. Vermass, in *Herbicides* (N. R. Baker and M. Percival, eds.), Elsevier Science Publishers B.V., New York, 1991, p. 27.

99. W. Oettmeier, in *The Photosystems: Structure, Function and Molecular Biology* (J. Barber, ed.), Elsevier Science Publishers B.V., New York, 1992, p. 349.

100. T. Tomo, I. Enami, and K. Satoh, *FEBS Lett., 323*:15 (1993).

101. T. Tomo and K. Satoh, *FEBS Lett., 351*:27 (1994).

102. N. R. Bowlby and W. D. Frasch, *Biochemistry, 25*:1402 (1986).

103. R. Mei, J. P. Green, R. T. Sayre, and W. D. Frasch, *Biochemistry, 28*:5560 (1989).

104. N. Adir and I. Ohad, *J. Biol. Chem., 263*:283 (1988).

105. P. A. Millner and J. Barber, *Physiol. Veg., 23*:767 (1986).

106. I. Damm and B. R. Green, *J. Chromatogr. A, 664*:33 (1994).

107. G. F. Wildner, C. Fiebig, N. Dedner, and H. E. Meyer, *Z. Naturforsch., 42c*:739 (1987).

108. M. Orenshamir, P. S. M. Sai, M. Edelman, and A. Scherz, *Biochemistry 34*:5523 (1995).

109. S. Alizadeh, P. J. Nixon, A. Telfer, and J. Barber, *Photosynth. Res., 26*:223 (1990).

110. J. P. Dekker, E. J. Boekema, H. T. Witt, and M. Rögner, *Biochim. Biophys. Acta, 936*:307 (1988).

111. X.-S. Tang and B. A. Diner, *Biochemistry, 33*:4594 (1994).

112. H. Shim, J. Cao, Govindjee, and P. G. Debrunner, *Photosynth. Res., 26*:223 (1990).

113. F. Nilsson, K. Gounaris, S. Styring, and B. Andersson, *Biochim. Biophys. Acta, 1100*:251 (1992).

114. H. Pakrasi, J. G. K. Williams, and C. J. Arntzen, *EMBO J., 7*:325 (1988).

115. J. G. K. Williams, *Methods Enzymol., 167*:766 (1988).

116. P. J. Nixon, D. A. Chisholm, and B. A. Diner, in *Plant Protein Eng.* (P. Shrewry and S. Gutteridge, eds.), Cambridge University Press, Cambridge, 1992, p. 93.

117. W. Vermaas, *Annu. Rev. Plant Physiol. Plant Mol. Biol., 44*:457, (1993).

118. V. A. Dzelzkalns and L. Bogorad, *EMBO J., 7*:333 (1988).

119. S. R. Mayes, K. M. Cook, S. J. Self, Z. H. Zhang, and J. Barber, *Biochim. Biophys. Acta, 1060*:1 (1991).

120. S. R. Mayes, J. M. Dubbs, I. Vass, E. Hidge, L. Nagy, and J. Barber, *Biochemistry, 32*:1454 (1993).

121. J. E. Boynton and N. W. Gillham, *Methods Enzymol., 217*:510 (1993).

122. E. Harris, *The Chlamydomonas Sourcebook,* Academic Press, San Diego, 1989.

123. J. M. Erickson and J.-D. Rochaix, in *The Photosystems: Structure, Function and Molecular Biology* (J. Barber, ed.), Elsevier Science Publishers B.V., New York, 1992, p. 101.

124. J. E. Boynton, N. W. Gillham, E. H. Harris, J. P. Hosler, A. M. Johnson, A. R. Jones, B. L. Randolph-Anderson, D. Robertson, T. M. Klein, K. B. Shark, and J. C. Sanford, *Science, 240*:1534 (1988).

125. K. L. Kindle, R. A. Schnell, E. Fernandez, and P. A. Lefebvre, *J. Cell Biol., 109*:2589 (1989).

126. O. A. Sodeinde and K. L. Kindle, *Proc. Natl. Acad. Sci. USA, 90*:9199 (1993).

127. N. J. Gumpel, J.-D. Rochaix, and S. Purton, *Curr. Genet., 26*:438 (1994).

128. K. P. VanWinkle-Swift, *Nature, 358*:106 (1992).

129. E. Przibilla, S. Heiss, U. Johanningmeier, and A. Trebst, *Plant Cell, 3*:169 (1991).

130. R. A. Roffey, J. H. Golbeck, C. R. Hille, and R. T. Sayre, *Proc. Natl. Acad. Sci. USA, 88*:9122 (1991).

131. J. P. Whitelegge, D. Koo, and J. M. Erickson, in *Current Research in Photosynthesis,* Vol. II, (M. N. Murata, ed.), Kluwer Academic Publishers, Dordrecht, 1992, p. 151.

132. J. P. Whitelegge, D. Koo, B. A. Diner, I. Domain, and J. M. Erickson, *J. Biol. Chem., 270*:225 (1995).

133. J. Minagawa and A. R. Crofts, *Photosynth. Res., 42*:121 (1994).

134. M. Goldschmidt-Clermont, *Nucleic Acids Res., 19*:4083 (1991).

135. J.-D. Rochaix, *J. Mol. Biol., 126*:597 (1978).

136. J. M. Erickson, M. Rahire, and J.-D. Rochaix, *EMBO J., 3*:2753 (1984).

137. J. M. Erickson, M. Rahire, P. Malnoe, J. Girard-Bascou, Y. Pierre, P. Bennoun, and J.-D. Rochaix, *EMBO J., 5*:1745 (1986).

138. G. Zumbrunn, M. Schneider, and J.-D. Rochaix, *Technique, 1*:204 (1989).

139. J. J. Finer, P. Vain, M. W. Jones, and M. D. McMullen, *Plant Cell Rep., 11*:323 (1992).

140. K. L. Kindle, K. L. Richards, and D. B. Stern, *Proc. Natl. Acad. Sci. USA, 88*:1721 (1991).

141. T. Dunahay, *BioTechniques, 15*:452 (1993).

142. L.-W. Tam and P. A. Lefebvre, *Genetics, 135*:375 (1993).

143. S. M. Newman, J. E. Boynton, N. W. Gillham, B. L. Randolf-Anderson, A. M. Johnson, and E. H. Harris, *Genetics, 135*:875 (1990).

144. E. A. Wurtz, B. B. Sears, D. K. Rabert, H. S. Shepherd, N. W. Gillham, and J. E. Boynton, *Mol. Gen. Genet., 170*:235 (1979).

145. D. E. Johnson and S. K. Dutcher, *Trends Genet., 9*:194 (1993).

146. R. A. Roffey, K. J. van Wijk, R. T. Sayre, and S. Styring, *J. Biol. Chem., 269*:5115 (1994).

147. R. A. Roffey, D. M. Kramer, Govindjee, and R. T. Sayre, *Biochim. Biophys. Acta, 1185*:257 (1994).

148. J. P. Whitelegge, *Photosynth. Res., 78*:265 (2003).

11 Reactive Oxygen Species as Signaling Molecules Controlling Stress Adaptation in Plants

Tsanko Gechev and Ilya Gadjev
Department of Molecular Biology of Plants, University of Groningen

Stefan Dukiandjiev and Ivan Minkov
Department of Plant Physiology and Molecular Biology, University of Plovdiv

CONTENTS

I. INTRODUCTION

Reactive oxygen species (ROS) are constantly produced during normal cellular metabolism. Originally regarded mainly as toxic by-products of metabolism, nowadays their diverse and indispensable role in numerous aspects of plant growth and development is fully appreciated. Alterations in ROS levels can act as the signals that switch on developmental programs or regulate physiological processes such as adaptation to abiotic stress, resistance to pathogens, cross-tolerance, and programmed cell death (PCD) (Figure 11.1).

Because of their role in such profound processes and their toxicity at high concentrations, the levels of ROS are kept under stringent control [1]. Dramatic increases in ROS lead to a phenomenon referred to as oxidative stress. Severe or persistent oxidative stress eventually results in PCD. Many adverse environmental factors, including extreme temperatures, salt, and drought, can cause oxidative stress and PCD [2–4]. On the other hand, deliberate production of ROS, known as oxidative burst, is essential for triggering the hypersensitive response (HR), a defense reaction against pathogens [5,6]. Likewise, moderate transient elevations of ROS levels are necessary for

switching on protective mechanisms leading to stress adaptation [7]. The transient kinetics of the ROS changes is indeed very important, ensuring that the protective mechanisms are switched on and are operational only when needed. Constant elevation of ROS even at a moderate rate under nonstressful conditions would have a negative effect, as illustrated by the growth suppression in ascorbate peroxidase (APx)-deficient plants [8].

The essential role of ROS in plant growth and development is further substantiated by the interplay of ROS with a number of plant hormones. H_2O_2 mediates the effect of MeJa during wounding [9], ABA and stomatal closure [10], and the auxin-mediated root gravitropism [11]. On the other side, H_2O_2 can repress the auxin signaling via an MAP kinase cascade [12]. Other important compounds like salicylic acid, NO, and ozone also act through formation or interaction with H_2O_2 [2,13–15].

Chloroplasts are the main sources of ROS in photosynthetically active organisms. ROS produced in chloroplasts can damage the photosynthetic apparatus but they can also diffuse out, causing damage to other cellular compartments and eventually cell death [16]. At the same time, ROS generated in the

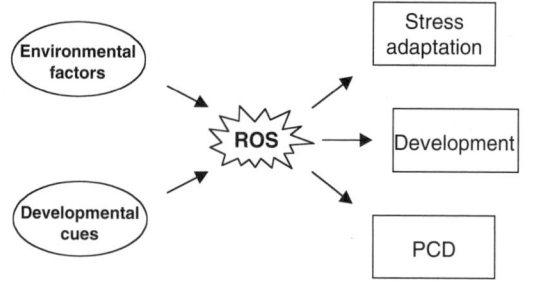

FIGURE 11.1 Biological effects mediated by oxidative stress (H_2O_2). H_2O_2 resulting from various developmental cues, including plant hormones, or generated in response to environmental factors (abiotic and biotic stress), mediates a number of important biological processes related to plant stress adaptation, development, or PCD. The stress adaptation may include antioxidant enzyme activation, inhibition of photosynthesis, accumulation of HSPs, PR, and other host defense genes, cell wall cross-linking, phytoalexin biosynthesis, and stomatal closure. Examples of developmental programs related with ROS signaling include root gravitropism, peroxisome biogenesis, as well as the PCD in barley aleurone cells and during aging/senescence. HR, occurring in some incompatible plant–pathogen interactions, is also a type of PCD.

chloroplasts are important signals for the communication of the plastids with the nucleus [17].

Not surprisingly, plants have evolved elaborate mechanisms to regulate their ROS homeostasis. These include a sophisticated antioxidant system that can scavenge the excess ROS levels produced under stress and a number of ROS generating systems that can raise the ROS levels when necessary. Apparently, plants can sense the changes in ROS levels very efficiently and respond to those changes accordingly. The signals originating from the changes in ROS levels are transduced via an extensive stress signaling network. Essential components of this network are the oscillations in Ca^{2+} fluxes that can trigger various cellular responses through diverse Ca^{2+} binding proteins, alterations in the redox status of the cell, and various protein kinase cascades. Recent studies revealed that the eventual activation of stress-regulated transcription factors results in massive transcriptional reprogramming and dramatic biological effects as described above. In the past few years it has become increasingly clear that selective degradation of key regulatory proteins is equally as important and acts in concert with the upregulation of stress-related genes to fine tune the biological response.

II. PRODUCTION AND DETOXIFICATION OF ROS

The most important biochemical property of ROS is their reactivity with other biomolecules, which determines their half-life and the ability to diffuse away from the site of their production. The first and the only endothermic step in the reduction of molecular dioxygen leads to the formation of superoxide ($O_2^{·-}$) or hydroperoxyl ($HO_2^{·-}$) radicals. During its relatively short life (half-life 2 to 4 µs), $O_2^{·-}$ can oxidize amino acids like histidine, metionine, and tryptophan or reduce quinones and transition metal complexes of Fe^{3+} and Cu^{2+}, thus affecting the activity of metal-containing enzymes [1]. Its protonated form, the hydroperoxyl radical, is predominant in acidic environment. It can cross biological membranes and subtract hydrogen atoms from polyunsaturated fatty acids, thus initiating lipid auto-oxidation. The second step leads to the formation of hydrogen peroxide (H_2O_2), a moderately active, relatively stable and therefore long-lived molecule with a half-life of 1 ms. Because of these properties, H_2O_2 can migrate quite some distance from the site of its production and is therefore the best candidate for a signaling molecule. In addition to its well-known ability to inactivate enzymes by oxidizing their thiol groups (e.g., enzymes from the Calvin cycle, Cu/Zn-superoxide dismutase [SOD], phosphotyrosine phosphatases), it can also form hydroxyl radicals in the presence of Fe^{2+} or Cu^+. The hydroxyl radical is the most reactive of all ROS with a half-life of less than 1 µs. It can react with and damage all biological molecules and ultimately cause cell death. Due to its extreme reactivity, cells do not have enzymatic mechanisms to detoxify it, so care should be taken to avoid its production. $O_2^{·-}$ and H_2O_2 can also initiate cascade reactions leading to the formation of lipid peroxides [18]. Singlet oxygen ($^1O_2$), a ROS arising from quenching of P680 triplet, is also very dangerous. It can either transfer its excitation energy to other biological molecules or react with them, thus forming endoperoxides or hydroperoxides, and can trigger, for instance, degradation of the D1 protein and subsequent destruction of PSII [19].

Chloroplasts are the major sources of ROS in plants, especially under conditions limiting CO_2 fixation [1]. Superoxide radicals are formed during electron leakage to oxygen from the Fe–S centers, the reduced ferredoxin, and thioredoxin. The produced $O_2^{·-}$ is then rapidly converted to H_2O_2 by SOD. Although production of ROS is generally considered detrimental, in this case the ability of oxygen to accept excess electrons prevents overreduction of the electron transport chain, thus minimizing the risk of formation of activated singlet oxygen [1]. Other major sources of H_2O_2 are glycolate oxidase in peroxisomes and fatty acid β-oxidase in glyoxysomes. Mitochondria, the main ROS producing organelles in animals, also generate ROS in the plant cell. NAD(P)H–oxidase complex is the primary ROS generating system during the oxidative burst in plant–pathogen interactions. In addition, a number of cell wall peroxidases and germin-like oxalate oxidases also contribute to the oxidative burst. ROS are also produced by xanthine oxidase during the catabolism of purines ($O_2^{.-}$, H_2O_2), ribonucleotid reductase during deoxyribonucleotide synthesis ($O_2^{.-}$), and various amine and flavine oxidases.

To keep ROS under control, plants have evolved a very efficient antioxidant system comprising antioxidants and antioxidant enzymes. Antioxidants are components capable of quenching ROS without themselves being destroyed or converted to destructive radicals. Antioxidants are water-soluble (ascorbate, glutathione) or lipid-soluble (α-tocopherol, carotenoids). The antioxidant enzymes catalyze the quenching of ROS directly or with the help of the antioxidants. The most important antioxidant enzymes include catalase, SODs, the enzymes of the ascorbate–glutathione cycle, glutathione peroxidase

(GPx), glutathione-S-transferases (GSTs), and guaiacol peroxidases. Catalases decompose H_2O_2 to water and oxygen without any reducing substrates. They are mainly found in peroxisomes and glyoxysomes (mitochondria in some plants) and function as a cellular sink for H_2O_2 [20]. SODs catalyze the immediate dismutation of $O_2^{.-}$ to H_2O_2 and oxygen at the site of its production. As these are the only plant enzymes that convert $O_2^{.-}$, they are distributed in all cellular compartments. Based on their metal cofactor, three groups can be distinguished in plants: FeSOD in chloroplasts, MnSOD in mitochondria and peroxisomes, and Cu/ZnSOD in cytosol and chloroplasts. APx, monodehydroascorbate reductase (MDHAR), dehydroascorbate reductase, and glutathione reductase (GR) form the so-called ascorbate–glutathione cycle [18], which is found in the chloroplasts, cytosol, mitochondria, and peroxisomes [18,21]. This cycle converts H_2O_2 to water using the reducing power of ascorbate, glutathione, and ultimately NADPH (Figure 11.2).

Other antioxidant enzymes that have attracted more attention recently are thioredoxins and peroxiredoxins. Thioredoxins belong to an ancient group that also includes glutaredoxins and protein disulfide isomerases [22]. Together with thioredoxin reductases they are electron donors to peroxiredoxins, low-molecular-weight peroxidases present in all kingdoms

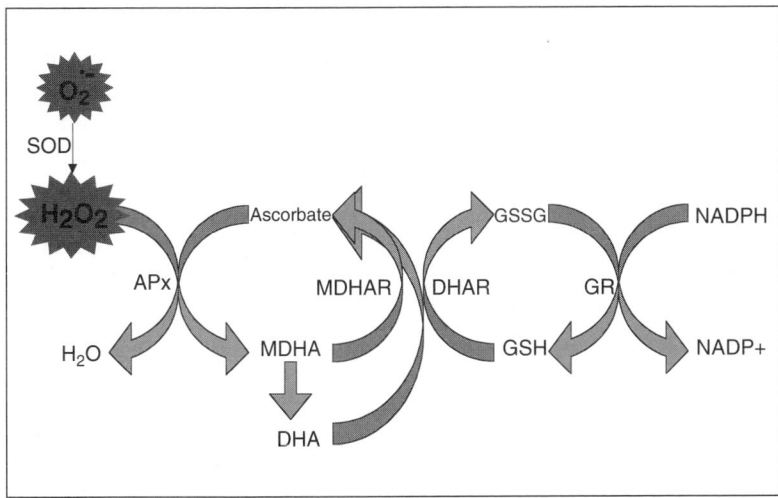

FIGURE 11.2 Ascorbate–glutathione cycle. Hydrogen peroxide, produced nonenzymatically or by various enzymes (SOD, oxidases), is reduced to water by APx acting with ascorbate as electron donor. During that process, the monodehydroascorbate radical (MDHA) is formed. MDHA can be reduced back to ascorbate by monodehydroascorbate reductase (MDHAR) or reduced ferredoxin (not shown here). Alternatively, MDHAR can spontaneously disproportionate to ascorbate and dehydroascorbate (DHA). DHAR is reduced to ascorbate by dehydroascorbate reductase (DHAR) utilizing reduced glutathione as electron donor. The reduced glutathione is recovered by GR and the ultimate electron donor NADPH. Such a cycle operates in cytosol, in chloroplasts, in mitochondria, and in a slightly modified version in peroxisomes. While SOD is the only plant enzyme capable of detoxifying superoxide radicals, hydrogen peroxide can be also scavenged by catalase, GPx, and various other nonspecific peroxidases (please see the text for more explanations).

[23–25]. Peroxiredoxins are important for antioxidant defense, at least in the chloroplasts [26]. Their substrate specificity can be rather broad and includes alkyl hydroperoxides as well as H_2O_2 [27].

III. ROS MEDIATED SIGNAL TRANSDUCTION IN PLANTS

Our knowledge of ROS signal transduction is much more advanced in microorganisms and animals than in plants. In bacteria ROS are sensed directly by transcription factors or repressors. For example, in *Escherichia coli* OxyR is activated by H_2O_2 through formation of intramolecular disulfide bonds [28], while O_2^{-} activates SoxS by oxidizing the Fe–S cluster of its repressor, SoxR [29]. In *Bacillus subtilis*, OhrR repressor senses organic hydroperoxides by reversible formation of cys-sulphenic (–SOH) acid derivatives [30], and in *Streptomyces coelicolor* σ^R is activated by H_2O_2-dependent oxidation of its anti-sigma repressor, RsrA [31]. In Eukaryota oxidative stress is sensed by redox-sensitive components and then signal transduced to the nucleus, though direct activation of transcription factors may also occur. In yeast, genes induced by redox signals consist of a complex network of different regulons [32]. In animals, more than half of the oxidative stress events are mediated by MAP kinase or NF-kB signaling pathways [33].

Although less studied, the available data suggest that the pathways in plants are as complex as those in animals. The plant cell can also sense different ROS like O_2^{-} and H_2O_2 [34,35] and even respond differently to increasing concentrations of H_2O_2 [7]. Generally, very little increases in H_2O_2 have no effect while moderate doses lead to regulatory effects, for example, acclimation to certain stress factors. High doses of H_2O_2 can trigger PCD or cause necrotic damage. How can such a simple molecule cause so many different biological effects? The recent work of Quinn et al. [36] partly answered that question, unraveling how distinct regulatory proteins control the graded transcriptional response to increasing H_2O_2 levels in the fission yeast *Schizosaccharomyces pombe*. In this study, two histidine kinases sense low doses of H_2O_2 and activate a MAPK cascade. The MAPK cascade eventually phosphorilates and activates a transcription factor called Pap1, which in turn regulates the antioxidant genes thioredoxin peroxidase and catalase. At high doses of H_2O_2, other yet unidentified factors progressively activate the MAPK cascade, but at the same time the nuclear translocation of Pap1 is somehow prevented. As a result, another transcription factor called Atf is activated, and

a different set of genes are transcribed. Similar mechanisms may be present in plants. There are 60 MAPKKKs, 10 MAPKKs, and 20 MAPKs in *Arabidopsis thaliana*, which means that these may be convergence and divergence points of the stress signaling [37]. In *A. thaliana* H_2O_2 activates an MAPKKK called ANP1, which in turn activates two downstream MAPKs — AtMPK3 and AtMPK6. These two MAPKs eventually lead to upregulation of the stress-related genes GST6 and Hsp18.2 [12]. This is in accordance with the observation that H_2O_2 can induce the GST6 promoter [38]. AtMPK3 and AtMPK6 can also be activated in response to flagelin, although in this case a different set of genes are transcribed [39]. More recently, *Arabidopsis* NDPK2 kinase has been found to be strongly induced by H_2O_2, and yeast two-hybrid assays suggested that AtNDPK2 kinase interacts with AtMPK3 and AtMPK6 [40]. Mutants lacking AtNDPK2 accumulated ROS, while AtNDPK2 overexpressors had lower levels of ROS and were more tolerant to cold, salt stress and methyl viologen [40]. H_2O_2 and O_2^{-} can also activate the tobacco ortholog of AtMPK6, SIPK [41]. Interestingly, both overexpression and suppression of SIPK result in ozone sensitivity [42]. It is also possible that MAPKs themselves can increase H_2O_2 levels, as suggested by Ren et al. [43]. In this work, overexpression of two MAPKKs, AtMEK4 and AtMEK6, activates a downstream MAPK, the prolonged activation of which leads to generation of H_2O_2 and subsequent triggering of HR-like PCD.

Ca^{2+} is an important second messenger in plants. Increased H_2O_2 levels lead to Ca^{2+} mobilization [44]. Ca^{2+} signals are generated through opening of Ca^{2+}-permeable ion channels in plasmalema, endoplasmatic reticulum, and vacuole, while Ca^{2+} pumps and H^+/Ca^{2+} antiporters maintain the Ca^{2+} homeostasis [45]. Ca^{2+} is a point where a cross talk between different stress factors occurs or specificity by a particular Ca^{2+} signature can be exerted [46]. Elevation in cytosolic Ca^{2+} can lead to activation of the NADPH oxidase complex directly or indirectly through activation of NAD kinase, thus amplifying the H_2O_2 signal (Figure 11.3). NADPH oxidase is essential for the oxidative burst during HR [47]. NADPH oxidase and Ca^{2+} are also involved in the regulation of H_2O_2 production at low oxygen concentrations, when Rop signaling plays a key role [48]. At the same time, the rise in cytosolic Ca^{2+} may activate plant catalases through interaction with calmodulin (CAM), which would have the opposite effect on H_2O_2 levels [49]. In addition to catalases, plants possess a unique set of Ca^{2+} and $Ca^{2+}/$ CAM binding proteins that influence numerous aspects of plant stress physiology. Examples of such

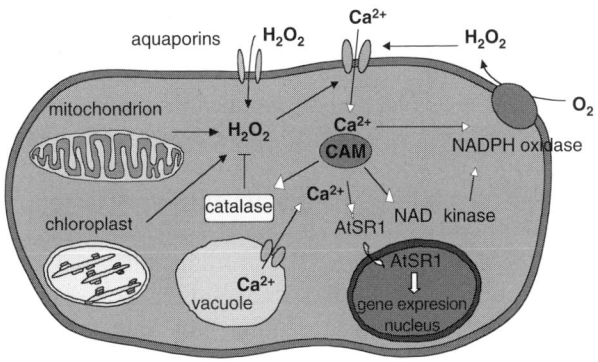

FIGURE 11.3 Interplay between H_2O_2 and Ca^{2+}. H_2O_2 produced at different locations (oxidative stress) increases cytosolic Ca^{2+} through Ca^{2+} mobilization from external or internal sources (vacuole, endoplasmatic reticulum). Ca^{2+} then can activate NADPH oxidase complex directly by binding to the EF-hand of one of its subunits or indirectly through binding to CAM. The Ca^{2+}/CAM complex then activates NAD kinase, generating more substrate molecules for NADPH oxidase. Increased activity of NADPH oxidase leads to more H_2O_2 formed in the apoplast. H_2O_2 can migrate via peroxiporins inside the plant cell, thus over-amplifying the oxidative stress signal. On the other hand, the Ca^{2+}/CAM complex can activate catalases, thus reducing H_2O_2 levels. In addition, Ca^{2+} or Ca^{2+}/CAM can bind to and activate Ca^{2+} or Ca^{2+}/CAM dependent protein kinases (not shown in the figure for clarity) or the stress-inducible Ca^{2+}/CAM binding transcription factor AtSR1, and in this way influence a wide spectrum of other genes.

stress-related proteins include SOS3 (salt overly sensitive), GAD, and SCaBP5 [50–52]. Ca^{2+}/CAM can also bind to all six *Arabidopsis* signal-responsive genes (*AtSR1-6*), named so because they are rapidly and differentially induced by a variety of stresses, including H_2O_2 [53]. AtSR1 is in fact a DNA binding protein that recognizes a novel CGCG box. Such boxes are found in many genes, including *ein3*, *TCH4*, as well as genes encoding transcription factors and heat shock proteins (HSPs) [53]. Plants also possess calcium/CAM dependent protein kinases and an impressive number of calcium dependent protein kinases (CDPK), the latter binding Ca^{2+} ions directly [54]. They mediate a wide variety of growth and developmental processes and are deeply involved in abiotic stress and pathogen defenses. Overexpression of a rice CDPK was able to confer tolerance to both cold and salt/drought [55]. Interestingly, in this experiment the overexpression of CDPK induced a distinct set of genes in response to the salt/drought treatment, but the same genes were not induced in response to the cold, suggesting that different signaling pathways function downstream from the CDPK. In another experiment, a tomato CDPK was systemically induced

upon wounding [56]. As wounding generates the second messenger H_2O_2 [9], it was demonstrated that indeed H_2O_2 alone can also upregulate the mRNA levels of the tomato CDPK, and this correlated with increased CDPK activity [56]. CDPKs are indispensable for mounting HR in *Nicotiana benthamiana* in response to Avr–cf. interactions [57]. The two CDPKs studied, NtCDPK2 and NtCDPK3, show rapid activation in response to Avr9 race-specific response, and silencing of the two genes compromised the HR reaction. It is now clear that both MAPKs and CDPKs are mediators of the ROS signals and both are essential for such processes as pathogen defense; however, the exact interrelation between MAPKs and CDPKs is still not well understood.

H_2O_2 signal can be mediated through alterations in the glutathione homeostasis of the plant cell. In addition to the role of a substrate of various enzymes, glutathione itself can be a signaling molecule and can regulate, for example, the synthesis of a number of enzymes [58,59]. Under normal conditions, glutathione redox state is constant with almost all of the glutathione in a reduced state (GSH). However, oxidative stress and many extreme environmental factors like high light and cold can cause increases in the glutathione pool as well as alterations in the reduced/oxidized GSH/oxidized glutathione (GSSG) ratio [59,60]. Elevation of H_2O_2 levels by the catalase inhibitor aminotriazole can also cause rapid stimulation of glutathione synthesis and accumulation of GSSG [60]. It seems that both the size of the GSH pool as well as the GSH/GSSG ratio are very important in conveying the oxidative stress signal [61]. In *Arabidopsis*, excess light can induce *APx1* and *APx2* via signals originating from the photosynthetic electron transport in chloroplasts [62]. Treatment with GSH can completely abolish that induction, suggesting a primary role of the redox poise in the regulation of these genes. It has been speculated that H_2O_2 and GSH have opposite effects on a chloroplast sensor that controls nuclear and chloroplast expression [63]. On the other hand, GSH treatment has been shown to upregulate the transcription from the parsley chalcone synthase promoter through the GST1-dependent mechanism [64]. Alterations in the GSH/GSSG ratio can modulate the activity of the enzymes as well. In Scots pine, lowering the GSH/GSSG ratio by exogenous application of GSSG results in increase in GR activity without any apparent increase in GR mRNA or protein levels [65]. In the same experiments, exogenous application of GSH resulted in increased GSH/GSSG ratio and decreased Cu/ZnSOD levels without any alterations in the enzyme activity.

Transgenic tobacco seedlings overexpressing an enzyme with both GST and GPx activity demonstrated higher GST- and GPx-specific activities and grew significantly faster than control seedlings under chilling or salt stress [66]. Interestingly, the levels of GSSG were significantly higher in transgenic seedlings than in wild types. In agreement with that observation, growth of wild-type seedlings was accelerated by treatment with GSSG, while treatment with GSH or other sulfhydryl-reducing agents inhibited growth. In this case the oxidation of the glutathione pool observed in the GST/GPx transgenic plants can stimulate seedling growth under stress.

Plants can respond to stress conditions by slowing down growth and saving energy for mounting defense responses. Both abiotic and biotic stresses can repress cell cycle genes and arrest cell division at specific checkpoints [61]. Such cell growth arrest and blocked cell division is associated with low GSH/GSSG ratio and GSH depletion. *Arabidopsis* plants deficient in GSH due to a mutation in a gene of the GSH biosynthetic pathway (γ-glutamylcysteine synthetase) are sensitive to CAM and are unable to develop normal meristems in the roots [67]. A similar phenotype can be obtained with the inhibitor of γ-glutamylcysteine synthetase buthionine sulfoximine, while the mutant phenotype can be rescued by exogenous application of GSH. GSH, as well as other redox agents, can also promote cell proliferation and hair tip growth in *Arabidopsis* [68].

IV. ROS ARE INVOLVED IN PLANT ADAPTATION TO STRESS

In the last few years a number of publications have demonstrated that relatively low sublethal doses of either $O_2^{.-}$ or H_2O_2 can protect against subsequent oxidative stress or play an essential role in plant adaptation to abiotic and biotic stress. Pretreatment with H_2O_2 can induce tolerance to high temperatures in potato and to chilling stress in maize and mungbean [69–72], as well as to high light intensities in *Arabidopsis* [73]. Pretreatment with the superoxide generating compound menadione also induced chilling tolerance in maize [74]. More recently, methyl viologen, another superoxide generating agent, applied at low doses was able to render tobacco leaf disks resistant to subsequent oxidative stress generated by high doses of the same compound [34]. In addition, a number of other compounds or acclimation treatments can also induce stress tolerance through transient accumulation of ROS. Acclimation of mustard plants at elevated temperatures for a short time results in acquiring thermotolerance, and salicylic acid has been found to transiently accumulate during the acclimation period [75]. Indeed, exogenous

application of salicylic acid can also induce thermotolerance, and the induced thermotolerance was associated with short, transient elevation in the endogenous H_2O_2 levels [72]. Similar thermoprotective results were obtained with salicylic acid and potato [69,72].

The adaptation to the different stress factors is concomitant with global and specific switches in gene expression [34,76,77], including alterations in the expression of specific transcription factors [78]. The changes in transcriptome can lead to both short-term and long-term protective effects through induction of stress-related genes encoding antioxidant enzymes, dehydrins, cold-responsive, heat shock, and pathogenesis related proteins, downregulation of elements of the photosynthetic apparatus, and others. In tobacco, H_2O_2 can induce a set of antioxidant enzymes, including catalase, APx, GPx, and guaiacol peroxidases, and protect against subsequent exposure to oxidative stress generated by high light or the catalase inhibitor aminotriazole [7]. Similarly, the tolerance to low temperatures in H_2O_2 treated or acclimated maize plants is associated with higher activities of the antioxidant enzymes catalase and guaiacol peroxidases [74]. In agreement with the role of antioxidant enzymes in stress tolerance, a number of stress-tolerant species or cultivars have increased antioxidant capacities compared with the stress-sensitive ones. Manipulation of the various components involved in the ROS signaling is an indispensable tool for studying the enormous complexity of that network. It is also an attractive approach to enhance the tolerance to a number of stress factors and thus to generate plants with better agricultural properties. All components of a stress signaling cascade can be manipulated to achieve stress tolerance: upstream events like the levels of ROS that trigger the cascade, the various kinases or phosphatases that are involved in the transduction of the signal, the specific transcription factors that switch the expression pattern of the cell, and the downstream genes that are ultimately responsible for acquiring the stress tolerance. Generally, manipulating the early steps can have multiple effects on different stresses, because parallel signal transduction pathways may be affected. These pathways often converge and diverge in a complex network, as is the case with the MAP kinase network or Ca^{2+} fluxes. Transgenic tobacco plants with reduced catalase activity accumulate H_2O_2 under high-light conditions and express antioxidant and defense-related proteins, including APx, GPx, and PR-1 [79]. Induction of PR-1 is independent of leaf damage and is associated with increased resistance against the bacterial pathogen *Pseudomonas syringae pv. syringae*. In similar experiments, transgenic tobacco plants with severely reduced catalase activity expressed very high levels of PR-1 proteins and showed enhanced resistance to tobacco mosaic virus [80]. In another experiment, antisense suppression of *Arabidopsis* ankyrin repeat-containing protein AKR2 resulted in small necrotic areas in leaves accompanied by higher production of H_2O_2, similar to the HR to pathogen infection in plant disease resistance [81]. The elevation of H_2O_2 levels was concomitant with increased transcripts of PR-1 and GST6, as well as with a ten-fold resistance to a bacterial pathogen. Transgenic plants that express glucose oxidase also accumulate H_2O_2 and are more tolerant to pathogens [82]. At the same time, plants with a compromised ROS scavenging system are more susceptible to abiotic stresses like high light intensities [20]. Interestingly, double-antisense tobacco plants lacking the two major H_2O_2 detoxifying enzymes APX and CAT were shown to have reduced susceptibility to oxidative stress [83]. A possible explanation of this phenomenon is the fact that the double-antisense plants were able to switch on alternative metabolic pathways, including induction of pentose phosphate pathway genes, MDHAR, IMMUTANS — a chloroplastic homolog of mitochondrial alternative oxidase (AOX), and to suppress photosynthetic activity. Suppression of photosynthesis seems to be a general response under stress, allowing plants to minimize chloroplastic ROS production and to activate various defense mechanisms [84]. An integral part of the defense mechanisms is mitochondrial AOX. H_2O_2 as well as salicylic acid and actinomycin A, a mitochondrial electron transport inhibitor, can induce AOX, thioredoxin peroxidase, and a number of PCD-related genes [85]. Although not a typical antioxidant enzyme, AOX can minimize mitochondrial ROS production by diverting electrons from the electron transfer chains directly to oxygen [86,87]. AOX seems to be crucial in preventing cell death as transgenic plants lacking this enzyme are much more sensitive to PCD induced by H_2O_2 or salicylic acid [88]. A distantly related chloroplastic homolog of this enzyme — IMMUTANS — diverts electrons from the flow between photosystem II and photosystem I, acting as a terminal oxidase by reducing O_2 into water at the plastoquinone step and thus decreasing the overall ROS production in chloroplasts [89,90]. The important role of IMMUTANS makes it essential also for chloroplast biogenesis [89].

HSPs can be induced by heat shock as well as by other stress factors [91]. Their biological functions are diverse, but the common feature is their ability to act as molecular chaperones and protectors against stress. In tomato cell suspension culture, mild H_2O_2 pretreatment and heat shock can induce tolerance against oxidative stress [92]. Both treatments induced a number of HSPs, among which the main protein identified was HSP22. It is believed that the induction of the HSPs and HSP22 in particular plays a major role in the tolerance against oxidative stress. In agreement with that, oxidative stress (H_2O_2 or methyl viologen) can upregulate the mRNA levels of a rice HSP, Oshsp26 [93]. In *Arabidopsis* the developmentally and environmentally regulated HSP101 is a crucial regulator of thermotolerance. Antisense inhibition or cosuppression of HSP101 results in higher sensitivity to elevated temperatures, while overexpression of the same gene leads to increased thermotolerance without any detrimental effects on normal growth or development [94]. Upregulation of HSPs can be exerted by the HSP transcription factors, HSFs, while selective protein degradation may account for reduction in HSP levels. As in the case of HSPs, plants possess a much larger number of HSFs than any other kingdoms. Humans and animals have four different HSFs, while in *Arabidopsis* they are 21 [95]. Interestingly, HSFs regulate the expression not only of HSPs but also of other stress protective proteins like APx. *Arabidopsis APx1* gene contains a functional heat shock element in its promoter region [96], and the mRNA level of APx is upregulated by H_2O_2 as well as by excess excitation energy [73].

H_2O_2 is second messenger for the induction of proteinase inhibitors and polyphenol oxidase in response to wounding, systemin, and MeJa in tomato [9]. The induction probably depends on H_2O_2 generation arising at least partially from the NADPH oxidase complex, as the NADPH oxidase inhibitor diphenylene iodonium can completely prevent it. The authors also showed that the same genes can be induced by the H_2O_2 generating system glucose + glucose oxidase [9].

Another example of acquiring multiple stress resistance is the overexpression of the upstream MAPKK kinase ANP1, which leads to increased tolerance to salt and heat stress [12]. In this case, no negative side effects have been reported. The multiple effects can be explained by the activation of a number of downstream genes, in particular Hsp18.1 and GST6. A similar effect can also be achieved by overexpression of transcription factors that control expression of important stress protective genes. Heterologous expression of *Arabidopsis* C-repeat/dehydration response element binding factor 1 (CBF1) in tomato conferred enhanced tolerance against chilling and methyl viologen [97]. This was accompanied by induction of catalase, linking the oxidative stress signaling and abiotic tolerance. CBF1 binds to DRE promoter element found in the complex promoter region of a number of stress-responsive genes [98], and its overexpression induces an array of cold-regulated (COR) genes [99]. Another two transcription factors from the same family, DREB1A and DREB2A, can also bind to DRE and mediate drought, cold, and salt tolerance [100]. Overexpression of these two genes under a constitutive promoter results in growth retardation. However, when overexpressed under control of the stress-inducible gene *rd29A*, DREB1A can protect against drought, salt, and freezing with no obvious negative side effects [101]. These as well as other unfavorable abiotic conditions can cause oxidative stress, as pointed out earlier. In addition to *rd29A*, a number of other stress-inducible genes possess promoter elements that can be activated by different stress-inducible transcription factors. The transcription factors themselves can be regulated by multiple stress factors, as is the case with the drought- and salt-inducible DREB2 [100] or hormone and stress-inducible AtSR1 [53]. Like MAP kinases and Ca^{2+} fluxes, these promoter elements and transcription factors can be convergence points and provide additional insights into the phenomenon called cross-tolerance [46,50,102]. The regulation of the transcription factors can be positive as well as negative. Interestingly, DREB1 may be negatively regulated by selective ubiquitin-dependent protein degradation of upstream signaling components, as revealed by the cloning of HOS1 locus [103]. HOS1 contains a RING finger motif similar to that found in IAPs and probably acts as E3 ubiquitin ligase to target regulatory proteins for proteasome degradation. The ubiquitin–proteasome pathway is a highly complex system involved in many housekeeping functions as well as in a number of developmental processes and responses to stress [104]. Plants also possess a group of small ubiquitin-like proteins with a role not only in protein degradation but also mostly in protein modification and regulation. Members of that family include Nedd8 and small *u*biquitin-like modifier (SUMO) [105,106]. Recently, H_2O_2 and other stress factors were reported to induce rapid SUMOylation of proteins in *Arabidopsis*, suggesting that this type of regulation can also mediate the H_2O_2 signal [107].

V.   CONCLUSION

The immense research on ROS in recent years revealed the multilateral effects these compounds have on virtually all aspects of plant physiology. Their interaction with many plant hormones further adds to the complexity of the ROS signaling. O_2^- and H_2O_2 play essential roles in plant development, stress adaptation, and PCD. Low levels of these ROS serve as signals that induce stress protective mechanisms. If the protective mechanisms fail, further accumulation of ROS trig-

gers PCD. We can also distinguish this "accidental" or "unwanted" PCD from the cases where we have deliberate production of ROS and PCD, as in barley aleurone cells during embryo development or in HR. Chloroplasts have key roles in regulating these processes as they are the most significant source of ROS in plants. Moreover, often it is the ROS from chloroplasts that communicate with the nucleus and other cell compartments to trigger adaptive responses. The responses to the ROS derived signals are carried out by an array of proteins and genes that interact to form a complex signaling network. It is amazing how such simple molecules can be so pleyotropic and at the same time so specific in their biological effects. Such different outcomes of ROS signaling are often determined by the whole cellular context. To understand this complexity, we need to know more about the primary sensing mechanisms for ROS, as well as more about the intermediate and downstream network components of the signaling network leading to gene regulation.

Combined genetic, molecular biological, and physiological approaches are already revealing the picture. Microarray studies showed us the large number of genes responsive to elevated ROS levels, with some of these genes never associated with stress responses before. Extensive proteome research will not only identify new proteins involved in plant stress adaptation but also add to our knowledge of how selective protein degradation contributes to the regulation and execution of these processes. Then, the real challenge will be to integrate this vast information into a model that can unravel the multifunctionality of ROS signaling.

REFERENCES

1. Dat J, Vandenabeele S, Vranová E, Van Montagu M, Inzé D, Van Breusegem F. Dual action of the active oxygen species during plant stress responses. *Cell. Mol. Life Sci.* 2000; 57(5):779–795.
2. Huh GH, Damsz B, Matsumoto TK, Reddy MP, Rus AM, Ibeas JI, et al. Salt causes ion disequilibrium-induced programmed cell death in yeast and plants. *Plant J.* 2002; 29(5):649–659.
3. Koukalová B, Kovarik A, Fajkus J, Široký J. Chromatin fragmentation associated with apoptotic changes in tobacco cells exposed to cold stress. *FEBS Lett.* 1997; 414(2):289–292.
4. Long SP, Humphries S, Falkowski P.G. Photoinhibition of photosynthesis in nature. *Annu. Rev. Plant Physiol. Plant Mol. Biol.* 1994; 45:633–662.
5. Levine A, Tenhaken R, Dixon R, Lamb C. H_2O_2 from the oxidative burst orchestrates the plant hypersensitive disease resistance response. *Cell* 1994; 79(4): 583–593.

6. Levine A, Belenghi B, Damari-Weisler H, Granot D. Vesicle-associated membrane protein of *Arabidopsis* suppresses Bax-induced apoptosis in yeast downstream of oxidative burst. *J. Biol. Chem.* 2001; 276(49):46284–46289.

7. Gechev T, Gadjev I, Van Breusegem F, Inzé D, Dukiandjiev S, Toneva V, et al. Hydrogen peroxide protects tobacco from oxidative stress by inducing a set of antioxidant enzymes. *Cell. Mol. Life Sci.* 2002; 59(4):708–714.

8. Pnueli L, Liang H, Rozenberg M, Mittler R. Growth suppression, altered stomatal responses, and augmented induction of heat shock proteins in cytosolic ascorbate peroxidase (Apx1)-deficient *Arabidopsis* plants. *Plant J.* 2003; 34(2):187–203.

9. Orozco-Cárdenas M, Narváez-Vásquez J, Ryan CA. Hydrogen peroxide acts as a second messenger for the induction of defense genes in tomato plants in response to wounding, systemin, and methyl jasmonate. *Plant Cell* 2001; 13(1):179–191.

10. Pei ZM, Murata Y, Benning G, Thomine S, Klusener B, Allen GJ, et al. Calcium channels activated by hydrogen peroxide mediate abscisic acid signalling in guard cells. *Nature* 2000; 406(6797):731–734.

11. Joo JH, Bae YS, Lee JS. Role of auxin-induced reactive oxygen species in root gravitropism. *Plant Physiol.* 2001; 126(3):1055–1060.

12. Kovtun Y, Chiu WL, Tena G, Sheen J. Functional analysis of oxidative stress-activated mitogen-activated protein kinase cascade in plants. *Proc. Natl. Acad. Sci. USA.* 2000; 97(6):2940–2945.

13. Van Camp W, Van Montagu M, Inzé D. H_2O_2 and NO: redox signals in disease resistance. *Trends Plant Sci.* 1998; 3:330–334.

14. Pellinen RI, Korhonen MS, Tauriainen AA, Palva ET, Kangasjarvi J. Hydrogen peroxide activates cell death and defense gene expression in birch. *Plant Physiol.* 2002; 130(2):549–560.

15. Neill SJ, Desikan R, Clarke A, Hurst RD, Hancock JT. Hydrogen peroxide and nitric oxide as signalling molecules in plants. *J. Exp. Bot.* 2002; 53(372):1237–1247.

16. Zolla L, Rinalducci S. Involvement of active oxygen species in degradation of light-harvesting proteins under light stresses. *Biochemistry* 2002; 41(48):14391–14402.

17. Rodermel S. Pathways of plastid-to-nucleus signaling. *Trends Plant Sci.* 2001; 6(10):471–478.

18. Noctor G, Foyer CH. Ascorbate and glutathione: keeping active oxygen under control. *Annu. Rev. Plant Physiol. Plant Mol. Biol.* 2003; 49:249–279.

19. Trebst A, Depka B, Hollander-Czytko H. A specific role for tocopherol and of chemical singlet oxygen quenchers in the maintenance of photosystem II structure and function in *Chlamydomonas reinhardtii*. *FEBS Lett.* 2002; 516(1–3):156–160.

20. Willekens H, Chamnongpol S, Davey M, Schraudner M, Langebartels C, Van Montagu M, et al. Catalase is a sink for H_2O_2 and is indispensable for stress defence in C3 plants. *EMBO J.* 1997; 16(16):4806–4816.

21. Jimenez A, Hernandez JA, Del Rio LA, Sevilla F. Evidence for the presence of the ascorbate-glutathione cycle in mitochondria and peroxisomes of pea leaves. *Plant Physiol.* 1997; 114(1):275–284.

22. Jacquot JP, Gelhaye E, Rouhier N, Corbier C, Didierjean C, Aubry A. Thioredoxins and related proteins in photosynthetic organisms: molecular basis for thiol dependent regulation. *Biochem. Pharmacol.* 2002; 64(5–6):1065–1069.

23. Dietz KJ. Plant peroxiredoxins. *Annu. Rev. Plant Biol.* 2003; 54:93–107.

24. Rhee SG, Kang SW, Chang TS, Jeong W, Kim K. Peroxiredoxin, a novel family of peroxidases. *IUBMB Life* 2001; 52(1–2):35–41.

25. Broin M, Cuine S, Eymery F, Rey P. The plastidic 2-cysteine peroxiredoxin is a target for a thioredoxin involved in the protection of the photosynthetic apparatus against oxidative damage. *Plant Cell* 2002; 14(6):1417–1432.

26. Konig J, Baier M, Horling F, Kahmann U, Harris G, Schurmann P, et al. The plant-specific function of 2-Cys peroxiredoxin-mediated detoxification of peroxides in the redox-hierarchy of photosynthetic electron flux. *Proc. Natl. Acad. Sci. USA* 2002; 99(8):5738–5743.

27. Bryk R, Griffin P, Nathan C. Peroxynitrite reductase activity of bacterial peroxiredoxins. *Nature* 2000; 407(6801):211–215.

28. Zheng M, Aslund F, Storz G. Activation of the OxyR transcription factor by reversible disulfide bond formation. *Science* 1998; 279(5357):1718–1721.

29. Demple B, Hidalgo E, Ding H. Transcriptional regulation via redox-sensitive iron-sulfur centres in an oxidative stress response. *Biochem. Soc. Symp.* 1999; 64:119–128.

30. Fuangthong M, Helmann JD. The OhrR repressor senses organic hydroperoxides by reversible formation of a cysteine-sulfenic acid derivative. *Proc. Natl. Acad. Sci. USA* 2002; 99(10):6690–6695.

31. Kang JG, Paget MS, Seok YJ, Hahn MY, Bae JB, Hahn JS, et al. RsrA, an anti-sigma factor regulated by redox change. *EMBO J.* 1999; 18(15):4292–4298.

32. Jamieson DJ. Oxidative stress responses of the yeast *Saccharomyces cerevisiae*. *Yeast* 1998; 14(16):1511–1527.

33. Allen RG, Tresini M. Oxidative stress and gene regulation. *Free Radic. Biol. Med.* 2000; 28(3):463–499.

34. Vranová E, Atichartpongkul S, Villarroel R, Van Montagu M, Inzé D, Van Camp W. Comprehensive analysis of gene expression in *Nicotiana tabacum* leaves acclimated to oxidative stress. *Proc. Natl. Acad. Sci. USA* 2002; 99(16):10870–10875.

35. Desikan R, Mackerness S, Hancock JT, Neill SJ. Regulation of the *Arabidopsis* transcriptome by oxidative stress. *Plant Physiol.* 2001; 127(1):159–172.

36. Quinn J, Findlay VJ, Dawson K, Millar JB, Jones N, Morgan BA, et al. Distinct regulatory proteins control the graded transcriptional response to increasing H_2O_2 levels in fission yeast *Schizosaccharomyces pombe*. *Mol. Biol. Cell* 2002; 13(3):805–816.

37. Jonak C, Okresz L, Bogre L, Hirt H. Complexity, cross talk and integration of plant MAP kinase signalling. *Curr. Opin. Plant Biol.* 2002; 5(5):415–424.

38. Chen W, Singh KB. The auxin, hydrogen peroxide and salicylic acid induced expression of the *Arabidopsis* GST6 promoter is mediated in part by an ocs element. *Plant J.* 1999; 19(6):667–677.

39. Asai T, Tena G, Plotnikova J, Willmann MR, Chiu WL, Gomez-Gomez L, et al. MAP kinase signalling cascade in *Arabidopsis* innate immunity. *Nature* 2002; 415(6875):977–983.

40. Moon H, Lee B, Choi G, Shin D, Prasad DT, Lee O, et al. NDP kinase 2 interacts with two oxidative stress-activated MAPKs to regulate cellular redox state and enhances multiple stress tolerance in transgenic plants. *Proc. Natl. Acad. Sci. USA* 2003; 100(1):358–363.

41. Samuel MA, Miles GP, Ellis BE. Ozone treatment rapidly activates MAP kinase signalling in plants. *Plant J.* 2000; 22(4):367–376.

42. Samuel MA, Ellis BE. Double jeopardy: both over-expression and suppression of a redox-activated plant mitogen-activated protein kinase render tobacco plants ozone sensitive. *Plant Cell* 2002; 14(9):2059–2069.

43. Ren D, Yang H, Zhang S. Cell death mediated by MAPK is associated with hydrogen peroxide production in *Arabidopsis*. *J. Biol. Chem.* 2002; 277(1):559–565.

44. Price AH, Taylor A, Ripley SJ, Griffiths A, Trewavas AJ, Knight MR. Oxidative signals in tobacco increase cytosolic calcium. *Plant Cell* 1994; 6(9):1301–1310.

45. Sanders D, Pelloux J, Brownlee C, Harper JF. Calcium at the crossroads of signaling. *Plant Cell* 2002; 14(Suppl):S401–S417.

46. Knight H, Knight MR. Abiotic stress signalling pathways: specificity and cross-talk. *Trends Plant Sci.* 2001; 6(6):262–267.

47. Yoshioka H, Numata N, Nakajima K, Katou S, Kawakita K, Rowland O, et al. Nicotiana benthamiana gp91phox homologs NbrbohA and NbrbohB participate in H_2O_2 accumulation and resistance to Phytophthora infestans. *Plant Cell* 2003; 15(3):706–718.

48. Baxter-Burrell A, Yang Z, Springer PS, Bailey-Serres J. RopGAP4-dependent Rop GTPase rheostat control of *Arabidopsis* oxygen deprivation tolerance. *Science* 2002; 296(5575):2026–2028.

49. Yang T, Poovaiah BW. Hydrogen peroxide homeostasis: activation of plant catalase by calcium/calmodulin. *Proc. Natl. Acad. Sci. USA* 2002; 99(6):4097–4102.

50. Zhu JK. Salt and drought stress signal transduction in plants. *Annu. Rev. Plant Biol.* 2002; 53:247–273.

51. Baum G, Lev-Yadun S, Fridmann Y, Arazi T, Katsnelson H, Zik M, et al. Calmodulin binding to glutamate decarboxylase is required for regulation of glutamate and GABA metabolism and normal development in plants. *EMBO J.* 1996; 15(12):2988–2996.

52. Guo Y, Xiong L, Song CP, Gong D, Halfter U, Zhu JK. A calcium sensor and its interacting protein kinase are global regulators of abscisic acid signaling in *Arabidopsis*. *Dev. Cell* 2002; 3(2):233–244.

53. Yang T, Poovaiah BW. A calmodulin-binding/6CGCG box DNA-binding protein family involved in multiple signaling pathways in plants. *J. Biol. Chem.* 2002; 277(47):45049–45058.

54. Cheng SH, Willmann MR, Chen HC, Sheen J. Calcium signaling through protein kinases. The *Arabidopsis* calcium-dependent protein kinase gene family. *Plant Physiol.* 2002; 129(2):469–485.

55. Saijo Y, Hata S, Kyozuka J, Shimamoto K, Izui K. Over-expression of a single Ca^{2+}-dependent protein kinase confers both cold and salt/drought tolerance on rice plants. *Plant J.* 2000; 23(3):319–327.

56. Chico JM, Raices M, Tellez-Inon MT, Ulloa RM. A calcium-dependent protein kinase is systemically induced upon wounding in tomato plants. *Plant Physiol.* 2002; 128(1):256–270.

57. Romeis T, Ludwig AA, Martin R, Jones JD. Calcium-dependent protein kinases play an essential role in a plant defence response. *EMBO J.* 2001; 20(20):5556–5567.

58. Noctor G, Gomez L, Vanacker H, Foyer CH. Interactions between biosynthesis, compartmentation and transport in the control of glutathione homeostasis and signalling. *J. Exp. Bot.* 2002; 53(372):1283–1304.

59. Kocsy G, Galiba G, Brunold C. Role of glutathione in adaptation and signalling during chilling and cold acclimation in plants. *Physiol. Plant.* 2001; 113(2):158–164.

60. Smith I. Stimulation of GSH synthesis in photorespiring plants by catalase inhibitors. *Plant Physiol.* 1985; 79:1044–1047.

61. May M, Vernoux T, Leaver C, Van Montagu M, Inzé D. Glutathione homeostasis in plants: implications for environmental sensing and plant development. *J. Exp. Bot.* 1998; 49:649–667.

62. Karpinski S, Escobar C, Karpinska B, Creissen G, Mullineaux PM. Photosynthetic electron transport regulates the expression of cytosolic ascorbate peroxidase genes in *Arabidopsis* during excess light stress. *Plant Cell* 1997; 9(4):627–640.

63. Karpinska B, Wingsle G, Karpinski S. Antagonistic effects of hydrogen peroxide and glutathione on acclimation to excess excitation energy in *Arabidopsis*. *IUBMB Life* 2000; 50(1):21–26.

64. Loyall L, Uchida K, Braun S, Furuya M, Frohnmeyer H. Glutathione a6nd a UV light-induced glutathione S-transferase are involved in signaling to chalcone synthase in cell cultures. *Plant Cell* 2000; 12(10):1939–1950.

65. Wingsle G, Karpinski S. Differential redox regulation by glutathione of glutathione reductase and CuZn-superoxide dismutase gene expression in *Pinus sylvestris* L. needles. *Planta* 1996; 198(1):151–157.

66. Roxas VP, Smith RK Jr, Allen ER, Allen RD. Over-expression of glutathione S-transferase/glutathione peroxidase enhances the growth of transgenic tobacco

seedlings during stress. *Nat. Biotechnol.* 1997; 15(10):988–991.

67. Vernoux T, Wilson RC, Seeley KA, Reichheld JP, Muroy S, Brown S, et al. The ROOT MERISTEMLESS1/CADMIUM SENSITIVE2 gene defines a glutathione-dependent pathway involved in initiation and maintenance of cell division during postembryonic root development. *Plant Cell* 2000; 12(1):97–110.

68. Sanchez-Fernandez R, Fricker M, Corben LB, White NS, Sheard N, Leaver CJ, et al. Cell proliferation and hair tip growth in the *Arabidopsis* root are under mechanistically different forms of redox control. *Proc. Natl. Acad. Sci. USA* 1997; 94(6):2745–2750.

69. Lopez-Delgado H, Dat JF, Foyer CH, Scott IM. Induction of thermotolerance in potato microplants by acetylsalicylic acid and H_2O_2. *J. Exp. Bot.* 1998; 49(321):713–720.

70. Prasad DT, Anderson M, Steward C. Acclimation, hydrogen peroxide, and abscisic acid protect mitochondria against irreversible chilling injury in maize seedlings. *Plant Physiol.* 1994; 105(619):627.

71. Yu C, Murphy T, Sung W, Lin C. H_2O_2 treatment induces glutathione accumulation and chilling tolerance in mung bean. *Funct. Plant Biol.* 2002; 29(9):1081–1087.

72. Dat JF, Lopez-Delgado H, Foyer CH, Scott IM. Parallel changes in H_2O_2 and catalase during thermotolerance induced by salicylic acid or heat acclimation in mustard seedlings. *Plant Physiol.* 1998; 116(4):1351–1357.

73. Karpinski S, Reynolds H, Karpinska B, Wingsle G, Creissen G, Mullineaux P. Systemic signaling and acclimation in response to excess excitation energy in *Arabidopsis*. *Science* 1999; 284(5414):654–657.

74. Prasad DT, Anderson M, Martin B, Steward C. Evidence for chilling-induced oxidative stress in maize seedlings and a regulatory role of hydrogen peroxide. *Plant Cell* 1994; 6:65–74.

75. Dat JF, Foyer CH, Scott IM. Changes in salicylic acid and antioxidants during induced thermotolerance in mustard seedlings. *Plant Physiol.* 1998; 118(4):1455–1461.

76. Seki M, Narusaka M, Ishida J, Nanjo T, Fujita M, Oono Y, et al. Monitoring the expression profiles of 7000 *Arabidopsis* genes under drought, cold and high-salinity stresses using a full-length cDNA microarray. *Plant J.* 2002; 31(3):279–292.

77. Schenk PM, Kazan K, Wilson I, Anderson JP, Richmond T, Somerville SC, et al. Coordinated plant defense responses in *Arabidopsis* revealed by microarray analysis. *Proc. Natl. Acad. Sci. USA* 2000; 97(21):11655–11660.

78. Chen W, Provart NJ, Glazebrook J, Katagiri F, Chang HS, Eulgem T, et al. Expression profile matrix of *Arabidopsis* transcription factor genes suggests their putative functions in response to environmental stresses. *Plant Cell* 2002; 14(3):559–574.

79. Chamnongpol S, Willekens H, Moeder W, Langebartels C, Sandermann H Jr, Van Montagu M, et al.

Defense activation and enhanced pathogen tolerance induced by H_2O_2 in transgenic tobacco. *Proc. Natl. Acad. Sci. USA* 1998; 95(10):5818–5823.

80. Takahashi H, Chen Z, Du H, Liu Y, Klessig DF. Development of necrosis and activation of disease resistance in transgenic tobacco plants with severely reduced catalase levels. *Plant J.* 1997; 11(5):993–1005.

81. Yan J, Wang J, Zhang H. An ankyrin repeat-containing protein plays a role in both disease resistance and antioxidation metabolism. *Plant J.* 2002; 29(2):193–202.

82. Wu G, Shortt BJ, Lawrence EB, Leon J, Fitzsimmons KC, Levine EB, et al. Activation of host defense mechanisms by elevated production of H_2O_2 in transgenic plants. *Plant Physiol.* 1997; 115(2):427–435.

83. Rizhsky L, Hallak-Herr E, Van Breusegem F, Rachmilevitch S, Barr JE, Rodermel S, et al. Double antisense plants lacking ascorbate peroxidase and catalase are less sensitive to oxidative stress than single antisense plants lacking ascorbate peroxidase or catalase. *Plant J.* 2002; 32(3):329–342.

84. Mysore KS, Crasta OR, Tuori RP, Folkerts O, Swirsky PB, Martin GB. Comprehensive transcript profiling of Pto- and Prf-mediated host defense responses to infection by *Pseudomonas syringae pv. tomato*. *Plant J.* 2003; 32(3):299–315.

85. Maxwell DP, Nickels R, McIntosh L. Evidence of mitochondrial involvement in the transduction of signals required for the induction of genes associated with pathogen attack and senescence. *Plant J.* 2002; 29(3):269–279.

86. Maxwell DP, Wang Y, McIntosh L. The alternative oxidase lowers mitochondrial reactive oxygen production in plant cells. *Proc. Natl. Acad. Sci. USA* 1999; 96(14):8271–8276.

87. Mittler R. Oxidative stress, antioxidants and stress tolerance. *Trends Plant Sci.* 2002; 7(9):405–410.

88. Robson CA, Vanlerberghe GC. Transgenic plant cells lacking mitochondrial alternative oxidase have increased susceptibility to mitochondria-dependent and -independent pathways of programmed cell death. *Plant Physiol.* 2002; 129(4):1908–1920.

89. Wu D, Wright DA, Wetzel C, Voytas DF, Rodermel S. The IMMUTANS variegation locus of *Arabidopsis* defines a mitochondrial alternative oxidase homolog that functions during early chloroplast biogenesis. *Plant Cell* 1999; 11(1):43–55.

90. Josse E, Simkin A, Gaffe J, Laboure A, Kuntz M, Carol P. 8A plastid terminal oxidase associated with carotenoid desaturation during chromoplast differentiation. *Plant Physiol.* 2000; 123:1427–1436.

91. Sun W, Van Montagu M, Verbruggen N. Small heat shock proteins and stress tolerance in plants. *Biochim. Biophys. Acta* 2002; 1577(1):1–9.

92. Banzet N, Richaud C, Deveaux Y, Kazmaier M, Gagnon J, Triantaphylides C. Accumulation of small heat shock proteins, including mitochondrial HSP22, induced by oxidative stress and adaptive response in tomato cells. *Plant J.* 1998; 13(4):519–527.

93. Lee BH, Won SH, Lee HS, Miyao M, Chung WI, Kim IJ, et al. Expression of the chloroplast-localized small heat shock protein by oxidative stress in rice. *Gene* 2000; 245(2):283–290.

94. Queitsch C, Hong SW, Vierling E, Lindquist S. Heat shock protein 101 plays a crucial role in thermotolerance in *Arabidopsis*. *Plant Cell* 2000; 12(4):479–492.

95. Nover L, Bharti K, Doring P, Mishra SK, Ganguli A, Scharf KD. *Arabidopsis* and the heat stress transcription factor world: how many heat stress transcription factors do we need? *Cell Stress Chaperones* 2001; 6(3):177–189.

96. Storozhenko S, De Pauw P, Van Montagu M, Inzé D, Kushnir S. The heat-shock element is a functional component of the *Arabidopsis* APX1 gene promoter. *Plant Physiol.* 1998; 118(3):1005–1014.

97. Hsieh TH, Lee JT, Charng YY, Chan MT. Tomato plants ectopically expressing *Arabidopsis* CBF1 show enhanced resistance to water deficit stress. *Plant Physiol.* 2002; 130(2):618–626.

98. Stockinger EJ, Gilmour SJ, Thomashow MF. *Arabidopsis thaliana* CBF1 encodes an AP2 domain-containing transcriptional activator that binds to the C-repeat/DRE, a cis-acting DNA regulatory element that stimulates transcription in response to low temperature and water deficit. *Proc. Natl. Acad. Sci. USA* 1997; 94(3):1035–1040.

99. Jaglo-Ottosen KR, Gilmour SJ, Zarka DG, Schabenberger O, Thomashow MF. *Arabidopsis* CBF1 overexpression induces COR genes and enhances freezing tolerance. *Science* 1998; 280(5360):104–106.

100. Liu Q, Kasuga M, Sakuma Y, Abe H, Miu6ra S, Yamaguchi-Shinozaki K, et al. Two transcription factors, DREB1 and DREB2, with an EREBP/AP2 DNA binding domain separate two cellular signal transduction pathways in drought- and low-temperature-responsive gene expression, respectively, in *Arabidopsis*. *Plant Cell* 1998; 10(8):1391–1406.

101. Kasuga M, Liu Q, Miura S, Yamaguchi-Shinozaki K, Shinozaki K. Improving plant drought, salt, and freezing tolerance by gene transfer of a single stress-inducible transcription factor. *Nat. Biotechnol.* 1999; 17(3):287–291.

102. Iba K. Acclimative response to temperature stress in higher plants: approaches of gene engineering for temperature tolerance. *Annu. Rev. Plant Biol.* 2002; 53:225–245.

103. Lee H, Xiong L, Gong Z, Ishitani M, Stevenson B, Zhu JK. The *Arabidopsis* HOS1 gene negatively regulates cold signal transduction and encodes a RING finger protein that displays cold-regulated nucleocytoplasmic partitioning. *Genes Dev.* 2001; 15: 912–924.

104. Ingvardsen C, Veierskov B. Ubiquitin- and proteasome-dependent proteolysis in plants. *Physiol. Plant.* 2001; 112(4):451–459.

105. Hellmann H, Estelle M. Plant development: regulation by protein degradation. *Science* 2002; 297(5582):793–797.

106. Yeh ET, Gong L, Kamitani T. Ubiquitin-like proteins: new wines in new bottles. *Gene* 2000; 248(1-2):1–14.

107. Kurepa J, Walker JM, Smalle J, Gosink MM, Davis SJ, Durham TL, et al. The small ubiquitin-like modifier (SUMO) protein modification system in *Arabidopsis*. Accumulation of SUMO1 and -2 conjugates is increased by stress. *J. Biol. Chem.* 2003; 278(9):6862–6872.

12 Plastid Morphogenesis

Ján Hudák, Eliška Gálová, and Lenka Zemanová
Faculty of Natural Sciences, Comenius University

CONTENTS

I. INTRODUCTION

Plastids are typical cell organelles of the plant body. Their presence or absence divides living organisms into two categories: autotrophs and heterotrophs. Different types of plastids occur in plant cells. The most distinctive plastid types are the chloroplasts, which are discrete cell organelles in which photosynthesis is carried out. Plastid morphogenesis is the result of mutual cooperation of the nuclear and plastid genomes, carried out under the influence of internal and external factors. The series of steps involved in plastid development can be interrupted at a certain stage of differentiation, resulting in the creation of a specialized type of plastids. Plastid morphogenesis has been studied extensively for many years, and there are several reviews describing the structure, morphology, and function of plastids [1–3].

Much of the material in other sections of this book concerns the physical, biochemical, and physiological processes involved in photosynthesis. In the present chapter, we will describe plastid ultrastructure, variability, and ontogenesis.

II. PLASTIDS

A. CLASSIFICATION AND DISTRIBUTION OF PLASTIDS

There are several types of plastids that are more or less related to one another developmentally. Criteria used to classify them vary. The best-known plastid classification is based on color, including the colorless plastids named leucoplasts, green chloroplasts, and yellow and red chromoplasts. Leucoplasts occur mostly in roots and in meristematic tissues, whereas chloroplasts are found in leaves, superficial tissues of stems, undifferentiated flowers, and unripe fruits. Chromoplasts occur in flowers, fruits, and occasionally in roots of carrot. The inner membrane system is best developed in chloroplasts, whereas leucoplasts and chromoplasts are scarce in the membranes. On the basis of photosynthetic ability, plastids can be divided into two groups: photosynthetic (chloroplasts) and nonphotosynthetic (leucoplasts and chromoplasts).

The unpigmented plastids, a special category, contain different storage products such as the amyloplasts, proteinoplasts, and elaioplasts. Amyloplasts contain starch in the form of starch grains (Figure 12.1). The starch in amyloplasts can occur either as a single large grain or as a number of granules of variable size. Due to the presence of numerous and large starch grains, the amyloplast shape is irregular. Starch grains often almost completely fill the whole volume of amyloplasts, and therefore it is very difficult to recognize other structural components in the plastid stroma.

Amyloplasts occur in storage tissues, meristems, and specialized cells. In the central part of root caps, the columella, there are specialized cells called statocytes, which possess gravity-sensitive bodies, statoliths, which are actually starch grains located in the amyloplasts. The first person to observe the active role of amyloplasts in root gravitropism was the Czech botanist B. Němec in 1900. Amyloplasts are located in the distal (lower) part of statocytes, where they sediment and press on the cisternae of the endoplasmic reticulum and plasma membrane. It has been suggested that the interaction of these three compartments (amyloplasts, endoplasmic reticulum, and plasma membrane) is responsible for the positive gravitropism of the roots [4].

Chemically, starch is made from two substances: amylose and amylopectin. Amylose may be absent in starch grains. A high content of amylopectin is noted in the amyloplasts of the sieve elements. Reaction of such starch grains with iodine does not give a typical blue-violet coloration but rather a red one.

Generally, amyloplasts are achlorophyllous, but it is well known that peripheral cell layers of potato tubers turn green when they are kept for some time in the light. The greening is accompanied by the transformation of the amyloplasts into chloroamyloplasts. Detectable traces of chlorophylls and thylakoids arranged in small grana occur in the amylochloroplasts after only 2 days of illumination [5]. The process of amyloplast transformation and chlorophyll synthesis in potato tubers is not as intense as it is during the formation of the photosynthetic apparatus in etiolated leaves after illumination. This slow rate of plastid transformation is also typical for plastids in greening roots. Plastid transformation in potato tubers and roots is probably governed differently from that in leaves.

Under certain circumstances, chloroplasts can also accumulate a great deal of starch and then originate transitional types of plastids, chloroamyloplasts. Chloroamyloplasts appear, for example, during spring in mesophyll cells of evergreen plants, and bundle sheath chloroplasts of C_4 plants are in fact also chloroamyloplasts.

Protein inclusions can occur in plant cells freely in the cytosol or they can be present in plastids. Plastids containing protein inclusions are called proteino-

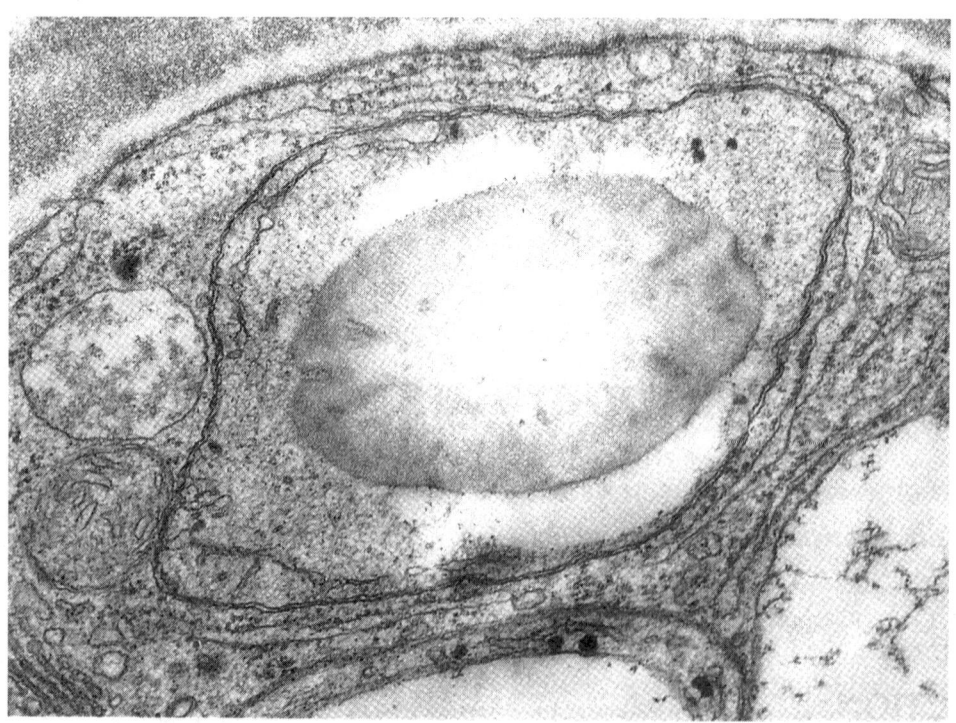

FIGURE 12.1 Amyloplast from the stylar tissue of *Brugmansia suaveolens* (28,000×).

plasts. Proteinoplasts have been observed in different types of cells, for example, in plastids of meristematic cells, epidermal cells, and root tip cells, in plastids of heterotrophic plants, and in chloroplasts at different stages of development [6].

In the stroma, protein inclusions are defined by a membrane. It is a generally accepted view that storage material present in the membrane-bound bodies of nongreen plastids is used in the differentiation of plastid membranes, but proteins present in the intrathylakoidal space of chloroplasts have been identified as the enzyme ribuloso 1,5-bisphosphate carboxylase [7].

The striking accumulation of protein can be also observed in plastids of sieve elements. Sieve element plastids possess either proteins or starch (see above). According to the presence of storage material, sieve elements plastids have been classified into two fundamental types, the P (protein) type and the S (starch) type [8]. The proteins present in sieve element plastids look like crystalloids (Figure 12.2), which are not limited by a membrane. P plastids have been observed only in the sieve elements of monocotyledons. It has been claimed that the protein inclusions together with callose play an active role in plugging the sieve plate pores of injured sieve tubes [9,10].

Leucoplasts can serve also as a reservoir of lipids, and such plastids have been called elaioplasts. Lipids

are present in plastid stroma in the form of globules. Numerous plastoglobuli are present in un-differentiated chloroplasts and in chromoplasts with degenerated membranes. The striking occurrence of plastoglobuli is typical for superficial tissues of cacti stems. It has been found that these plastoglobuli store photosynthetically bound carbon. It is commonly known that lipids present in the plastoglobuli are used in plastid membrane differentiation and released lipids from disintegrated membranes are placed back into the plastoglobuli [11].

The plastid stroma may also contain deposits of phytoferritin (Figure 12.3). Phytoferritin occurs mostly in nonphotosynthetic plastids, for example, proplastids, amyloplasts, etioplasts, and senescent plastids. Phytoferritin in plastids has a similar structure to ferritin in animal cells. Fe–protein complex is made of electron-dense nucleoid, which comprises around 4000 to 5000 Fe atoms. The nucleoid is covered by apoferritin envelope made up of 20 to 24 protein subunits [12]. It is accepted that the phytoferritin in plastids represents a reservoir of nontoxic iron, which is later utilized in enzymatic processes.

Different cells contain leucoplasts of variable structure and function. The plastid is probably the best named of cell organelles, for the name indicates the plasticity of both its structure and its function [6]. Leucoplasts are involved in different metabolic

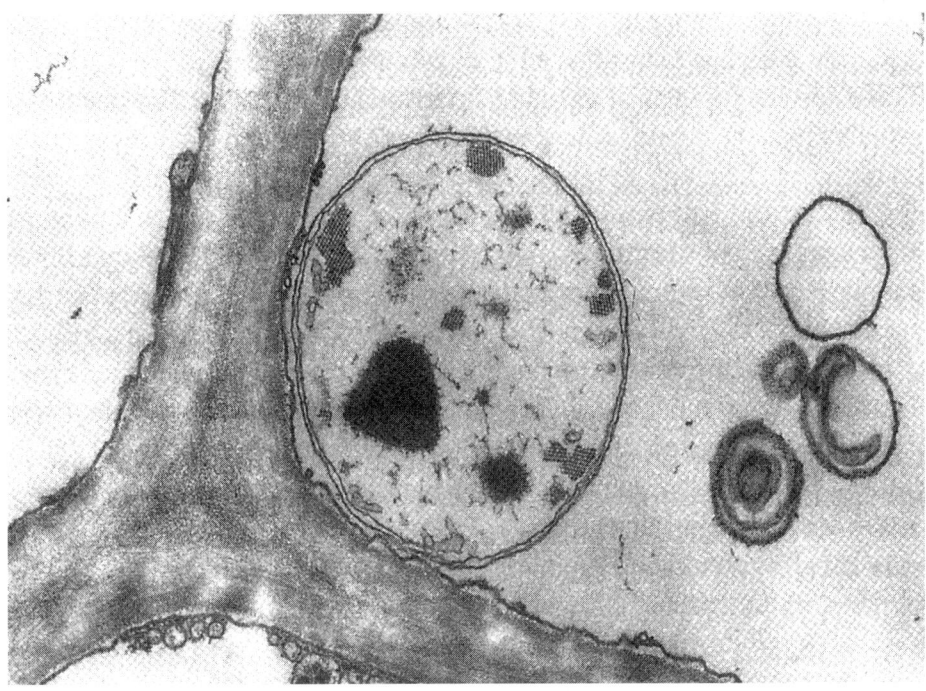

FIGURE 12.2 Plastid in a fully mature sieve element of *Aegilops comosa* with two kinds of crystalloids (40,800×). (From Binns AN. *Annu. Rev. Plant Physiol. Plant Mol. Biol.* 1994; 45: 173–196. With permission.)

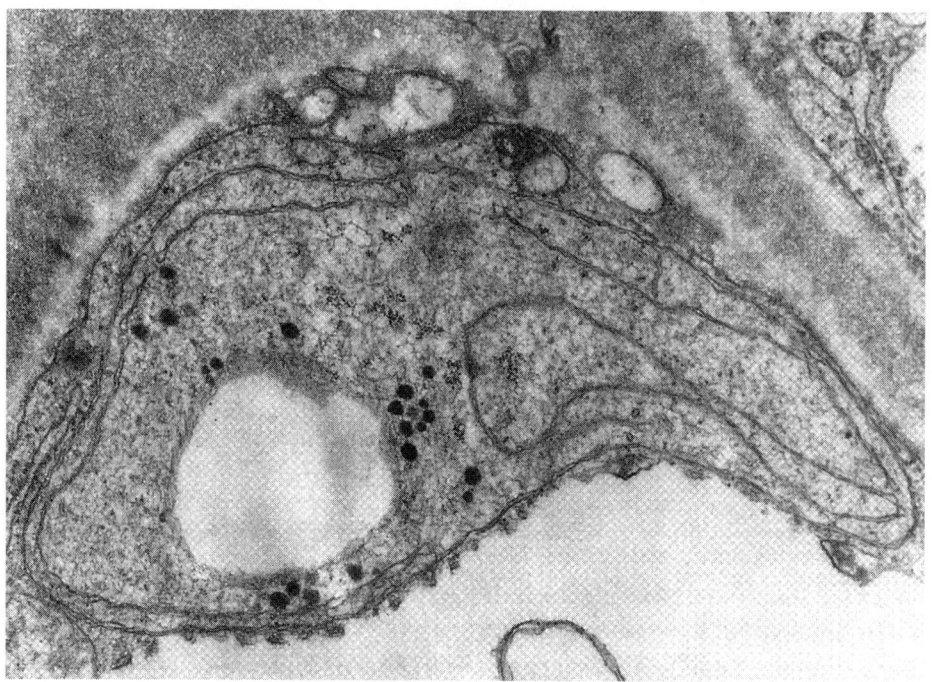

FIGURE 12.3 Leucoplast from the stylar tissue of *Brugmansia suaveolens* with phytoferritin inclusions (25,000×).

processes, for example, synthesis of carbohydrates, amino acids, some proteins, lipids, and isoprenoids. Therefore, we must think of the leucoplast as a specialized type of plastid in a certain stage of plastid development and not only as an enlarged proplastid [13]. Plastid differentiation is carried out under the influence of internal and external factors. The series of steps involved in plastid development can be interrupted at a certain stage of differentiation, resulting in the creation of a specialized type of plastid. One of the factors that fundamentally affect plastid diversity is the degree of cell differentiation. Structural heterogeneity of plastids is the expression of the cell type wherein they occur.

It is quite common for two neighboring cells to have plastids with different inner architecture. A good example of these are the assimilatory leaves. Leaf tissues, in fact, represent a mosaic of cell diversity. Different groups of cells contain heterogenous plastid populations.

It is well known that more than one type of chloroplast exists within the same leaf blade in C_4 plants. In the leaves of *Amaranthus retroflexus* as many as seven distinct types of chloroplasts have been observed [14].

Leaf epidermal cells contain either leucoplasts with protein inclusions, plastoglobuli, and reduced membrane systems or chloroplasts with a different degree of chloroplast membrane differentiation. Chloroplasts are invariably present in stomatal cells.

A great variety of plastid modification can be observed in vascular tissues. In dicotyledons both vascular parenchyma cells and companion cells have chloroplasts, but in monocotyledons chloroplasts are absent, for example, plastids of vascular parenchyma cells in leaves of *Ophrys sphegodes* lack any traces of thylakoids (Figure 12.4) [15]. Plastids in sieve elements, as mentioned above, store either starch or proteins. In tracheal elements plastids occur only in the early phases of their development. These plastids are leucoplasts with prominent starch grains. During subsequent development of xylem cells, up to the stage of secondary wall formation, plastids gradually lose starch. The starch is utilized for secondary wall formation. The first signs of plastid degeneration appear when autolytic processes in the protoplast of xylem cells are activated. Plastid degeneration during xylem formation is a part of programmed senescence of these cells [16].

The striking plastid polymorphism caused by a different stage of cell differentiation is well observed in the ribbon leaves of some monocotyledons (e.g., barley, maize, wheat). It is the case of a linear gradient of cell and plastid differentiation. Cells of the expanding monocot leaves are produced primarily from a meristem located at the leaf base. Therefore, more differentiated cells and better-developed plastids (chloroplasts) are located close to the leaf tip. Juvenile (meristematic) cells on the leaf base contain undeveloped plastids (proplastids). This gradient of

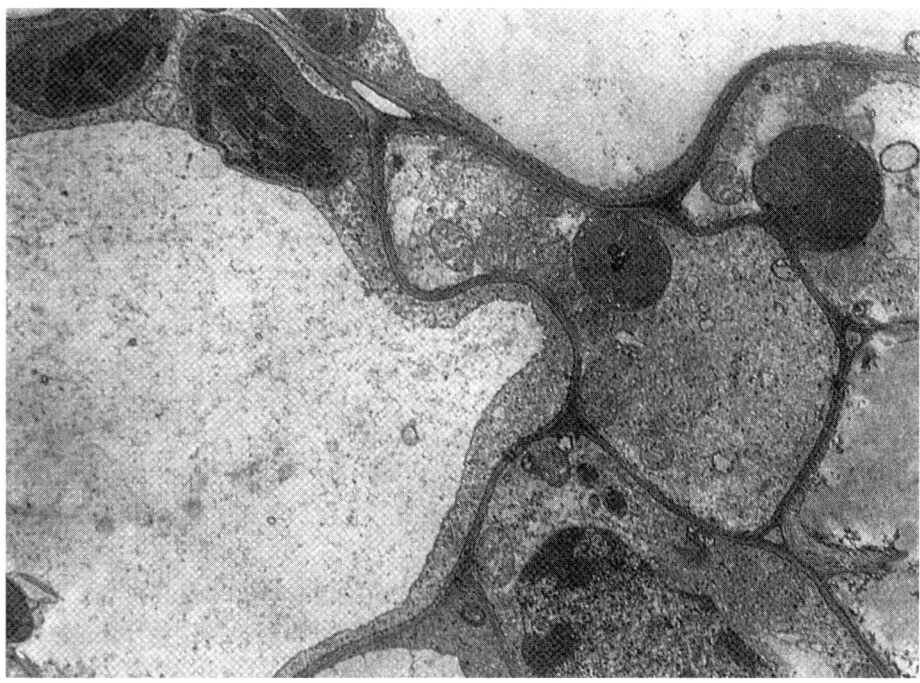

FIGURE 12.4 Different types of plastids in mesophyll cells and vascullar parenchyma cells of *Ophrys sphegodes* (3500×). (From Dahline C, Cline K. *Plant Cell* 1991; 3: 1131–1140. With permission.)

plastid differentiation in the leaves of monocots is well observed not only in the light but also in the dark [17]. The occurrence of ameboid plastids also contributes to the plastid heterogeneity. During plastid development the shape of plastids alters. Originally they are spherical, and subsequently they transform to discoid shape of the mature chloroplasts. In addition to this typical plastid shape, ameboid or pleomorphic plastids often occur. Ameboid plastids are of irregular shape, they make protrusions into cytoplasm, and cytoplasmic inclusions can be seen in their stroma (cup-shaped plastids). The position of the pleomorphic plastids in the pattern of plastid biogenesis is uncertain. Do they represent a real step in plastid differentiation or do they occur in cells as a result of metabolic changes? The shift in plastid form indicates (1) a change in the sol–gel state of the stroma, (2) a change in the character of the envelope, (3) a change in the ratio of volume to surface area of the plastid (as during the loss of starch from a distended amyloplast), or (4) a combination of these three [6]. Ameboid plastids have been occasionally observed in meristemic tissues where they might be considered as an optional stage of plastid development [18,19]. Plastids require components synthesized in the cytoplasm for their development, and pleomorphic forms conspicuously increase the plastid surface area over which such exchange of metabolites can take place [20]. However, ameboid plastids

occur also in cells engaged in the secretion of different substances, in senescent leaves, in the leaves during their regreening, in early phases of plastid development in tissue cultures, in the actinorhizal root nodules, and in degenerated leaves after the effect of herbicides, antibiotics, and heavy metals [11,19, 21–25]. These findings indicate that formation of pleomorphic plastids is a metabolic response induced by environmental factors. The presence of ameboid plastids contributes to the structural heterogeneity in plastid population.

Plastid biogenesis in higher plants is influenced also by external factors (nutrition, light), and their effect on this process is discussed in other parts of this chapter.

B. PLASTID ONTOGENY

Plastid ontogenesis is considered as a chain of structural and functional processes that represent changes in plastid development from structurally simple proplastids via chloroplasts (or other specialized types of plastids) to the last phase of their existence — plastid senescence. Every developmental plastid stage is characterized by a certain level of membrane differentiation.

The pattern of plastid development is similar in different higher plants. It is, therefore, suggested that the changes in plastid structure that take place during maturation may be permanent (proplastids, the change of size and shape, the origin of plastid membranes, and grana formation) or optional (determined by species and tissue specificity and environmental factors).

The basic precursors of all plastid types (leucoplasts, chloroplasts, and chromoplasts) are proplastids. These are present in zygotes and in root, stem, leaf, and flower meristems. Proplastids are usually small (0.4 to 1 μm in diameter), spherical organelles. They are separated from the cytoplasm by a double-membrane envelope. The internal structure of proplastids is very simple (Figure 12.5). In proplastid stroma, a few single thylakoids, vesicles, and small plastoglobuli are present. Proplastids can also contain minute starch grains, which are present in the root meristematic cells. In proplastid stroma, there are low amounts of plastid DNA, RNA, ribosomes, and soluble proteins. This indicates that a basal level of plastid gene expression is active in the dividing cells of the meristem [3]. During ontogenesis of the cells into differentiated forms, proplastids are gradually transformed into specialized types of plastids, for example, leucoplasts, which have already been described.

1. Plastid Differentiation in Light

From the view of photosynthesis, the most important plastid type is chloroplast. As meristem cells divide and develop into leaf cells, proplastids differentiate into chloroplasts. Subsequent development of chloroplasts is connected with gradual differentiation of leaf meristems into mesophyll cells.

The process of gradual transformation of proplastids into chloroplast requires light. If the light conditions are sufficient, chlorophylls are synthesized and the membrane system is differentiated.

The chloroplast membrane system is derived from the envelope. The inner membrane of the plastid envelope at many places makes invaginations into plastid stroma (Figure 12.6). These protrusions often appear long and thin, and from their ends different vesicles are released. While one protrusion is still in contact with the envelope, the others are freely scattered in plastid stroma. The process of invagination continues until there are many thylakoids in the stroma. Vesicles and tiny thylakoids coalesce and form primary membranes. As differentiation proceeds, the number of thylakoids in stroma increases. Many thylakoids occur in stacks of two or three, representing immature grana. With further chloroplast differentiation, the number of thylakoids in each stack increases until the typical grana of mature chloroplasts are produced. Single grana are interconnected by thylakoids, which pass from one to the other [1,17,21].

As already noted, the pathway of chloroplast development can be strikingly influenced by tissue specificity. In many developing leaves chloroplasts can reach different developmental stages in adjacent cells,

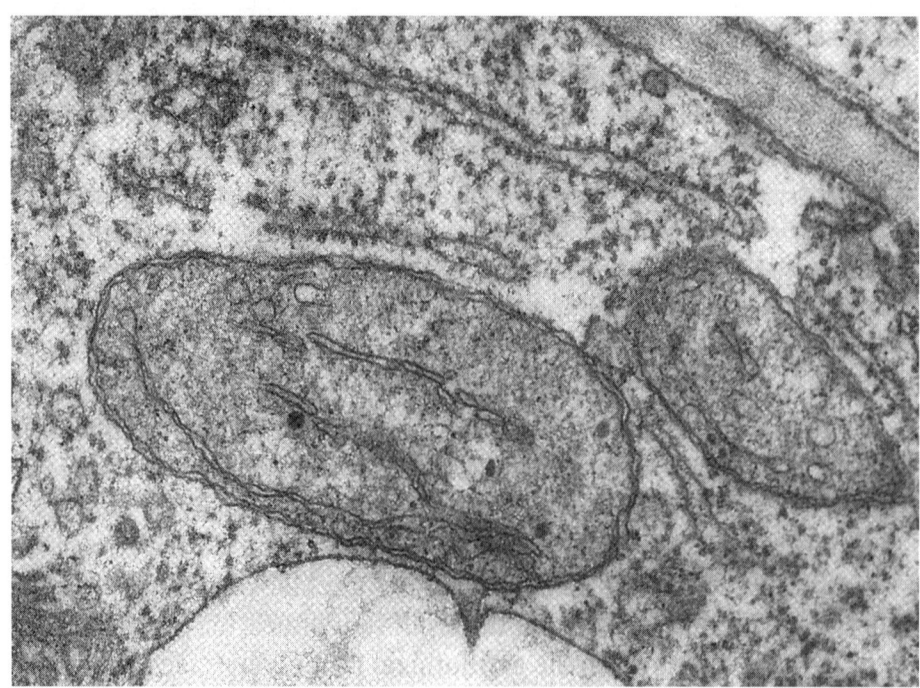

FIGURE 12.5 Proplastid from the transmitting tissue of *Brugmansia suaveolens* (30,000×).

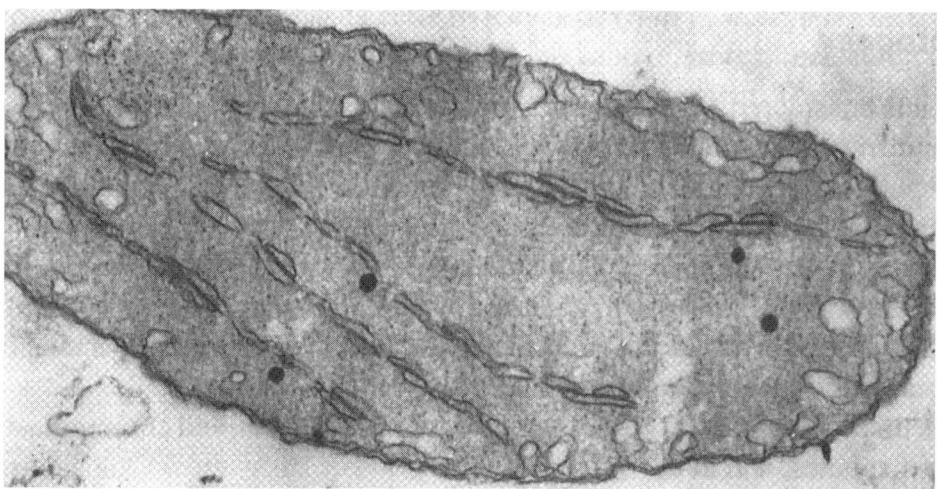

FIGURE 12.6 Chloroplast membrane differentiation in mesophyll cells of *Zea mays* (27,000×). (From Oross JW, Possingham JV. *Protoplasma* 1989; 150: 131–138. With permission.)

namely, in dicotyledons, whose leaf tissue is a mosaic of cells in different phases of development and in which islands of young, still dividing cells are surrounded by regions of cells that have completed their expansion. The strap-shaped leaves of many monocotyledons are convenient for the study of the sequential changes during chloroplast development. A linear gradient of cell and plastid differentiation occurs in these leaves. Young cells on the leaf base have proplastids, but older cells close to the tip leaves contain chloroplasts [17].

Proplastids during leaf development in the light are transformed into structurally and functionally mature chloroplasts. However, not all building material of newly arisen membranes comes from the plastid envelope. A substantial part of structural and functional proteins is synthesized in the process of chloroplast differentiation. Accumulation of the light-harvesting chlorophyll *a/b* protein complexes can be first detected when formation of grana starts, and the increases continue until chloroplast development is complete. The accumulation of the membrane lipid components also becomes maximal as granal stacking progresses [26]. During this time, the plastids accumulate thylakoid membrane proteins involved in the light reactions of photosynthesis and soluble proteins that participate in CO_2 fixation as well as other metabolic pathways.

Plastids are semiautonomous organelles having their own DNA but strongly dependent on the nuclear DNA and the cytosolic translation system. Approximately 80% to 85% of chloroplast proteins are encoded on nuclear genes, and the remaining 15% to 20% are encoded by plastid genes [27]. The majority of plastid proteins are synthesized in cytosol in the form of precursors, and these are transported into the plastids. Protein transport comprises the following steps [28]:

1. Association of the precursor with the outer envelope membrane
2. Translocation of the polypeptide across the inner and outer envelope membranes, perhaps at contact sites
3. Proteolytic removal of the transit peptide by the stromal processing protease
4. Further sorting of modified precursor to other chloroplastic compartments, followed by further proteolytic processing (if necessary)
5. Association with other polypeptides to form multimeric protein complexes (if necessary).

Simultaneously, with chloroplast membrane differentiation, the plastid population per cell increases. It is generally accepted that all plastids arise from the division of preexisting plastids. Plastids can divide at any stage of their development from the proplastid to the recently mature chloroplast, and all plastids appear to be capable of division. Two types of plastid division have been observed: binary fission and partition. In binary fission, a constriction of the entire plastid gradually divides the organelle into two daughter plastids. In plastid partition, only the inner membrane of the plastid envelope forms an invagination that progressively divides stroma into nearly equal parts. The number of plastids per cell increases proportionally with increase in cell size. Plastids divide and expand as long as they do not occupy a constant proportion of the mesophyll cell surface [29–32].

2. Etioplasts

When the plants have been cultivated several days in the absence of light or in weak light, their leaves are achlorophyllous and mesophyll cells contain plastids named etioplasts. Etioplasts are typical for the leaves of etiolated plants. During dark growth, the leaf proplastids are not transformed into chloroplasts, but they take an alternative route of plastid development via the temporal stage, which results in the formation of plastids with peculiar architecture. Thus, light is one of the external factors that directly affects plastid biogenesis. In the dark, plastid volume increases and stroma exhibit the prominent structure of etioplasts, called the prolamellar body (Figure 12.7). Each etioplast contains one or more prolamellar bodies of paracrystalline appearance consisting of interconnected membranous tubules [1,33]. Single thylakoids can occur on the periphery of prolamellar bodies, and where there is more than one prolamellar body in the etioplast, these thylakoids may extend from one to the other.

Besides the prolamellar bodies in the etioplast stroma, there are also plastoglobuli gathered into groups, ribosomes, and small starch grains (Figure 12.8). The membranes of etioplasts have no chlorophyll but contain protochlorophyllide. When etiolated plants are illuminated, the protochlorophyllide is immediately converted into chlorophyll and prolamellar bodies are gradually transformed into the membrane system of mature chloroplasts. If the plants are cultivated permanently at low light intensity, etioplast transformation into chloroplasts is incomplete. The chloroplasts of these plants have developed grana, but instead of stroma lamellae, small prolamellar bodies are present called chloroetioplasts [21].

3. Ability of Gymnosperms to Form Chloroplasts in the Dark

Angiosperms synthesize chlorophylls and form chloroplasts only in the light. In complete darkness chlorophyll synthesis is blocked at the level of protochlorophyllide, and plastids are differentiated as etioplasts. Among the angiosperms, the ability to form chlorophyll in the dark is very rare and is confined to the embryos, and on the basis of this plants are divided into Chloroembryphyta and Leucoembryphyta [34]. This striking phenomenon is observable in onion bulbs, where quite frequently green leaf primordia occur. These green tissues inside the bulbs contain chlorophyll a and b and plastids, besides prominent prolamellar bodies possess grana composed of up to ten thylakoids. This example confirms that under certain circumstances angiosperms can synthesize chlorophyll and also form chloroplast in the dark, but the meaning of this ability is unclear [35]. In

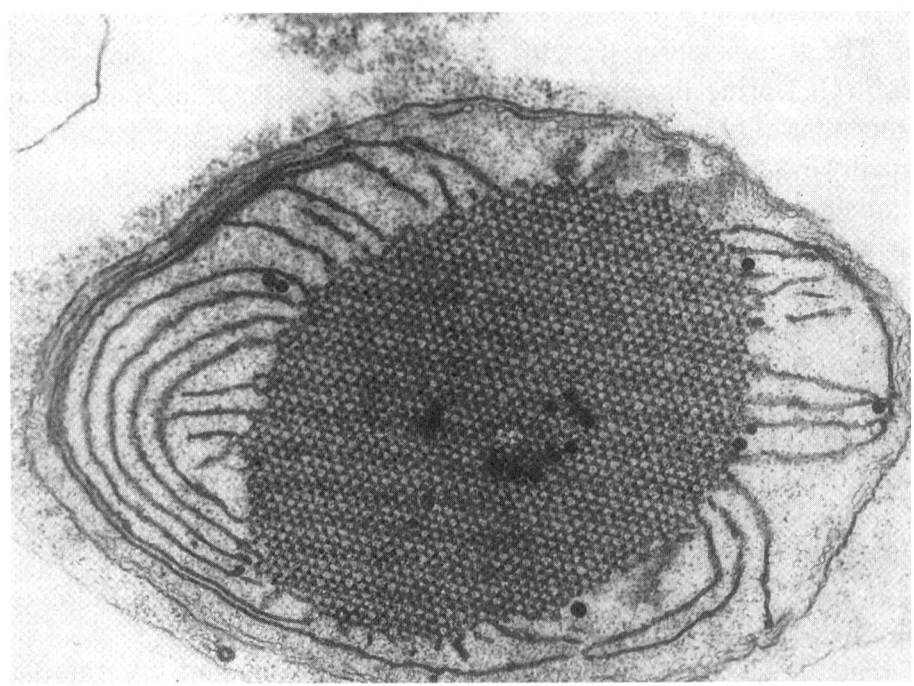

FIGURE 12.7 Etioplast in a palisade parenchyma cell of *Zea mays* (30,000×). (From Oross JW, Possingham JV. *Protoplasma* 1989; 150: 131–138. With permission.)

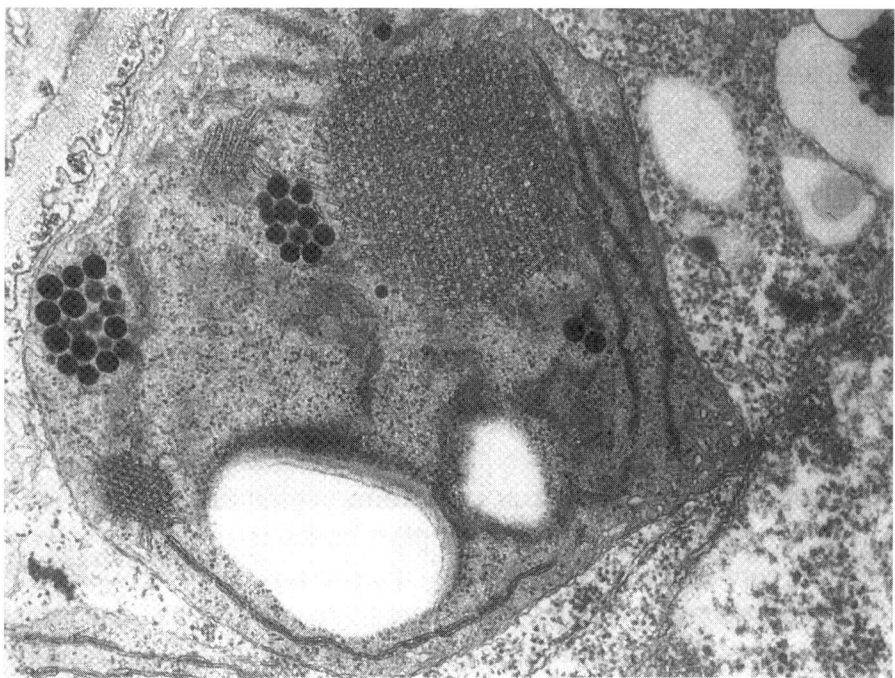

FIGURE 12.8 Etioplast of dark-grown seedling of *Larix decidua* with prolamellar bodies, starch grains, plastoglobuli, and plastid ribosomes (35,000×). (From Tevini M, Steinmüller D. *Planta* 1985; 163: 91–96. With permission.)

contrast, gymnosperms form chloroplasts and chlorophylls in the dark as well as in the light. The ability to synthesize chlorophylls in the absence of light appears to be confined to the cotyledons of gymnosperms. When mature branches of conifers are allowed to form new needles in the dark, these contain almost no chlorophyll [1,36].

There is a considerable variation among different species in the structural organization of the dark formed chloroplasts and in the ability to form chlorophylls. Of the different species of gymnosperms investigated after germination and growth in darkness *Ephedra twediana*, *Picea excelsa*, *Abies alba*, *Pinus nigra*, and *Pinus mugo* form chloroplasts, while *Gnetum montana*, *Welwitschia mirabilis*, *Larix deciduas*, and *Pinus sylvestris* form only etioplasts under the same conditions [37–39]. The structural differences in the chloroplast architecture are significant. *L. decidua* plastids have immature lamellar systems with minute grana, each of which contains only two or three thylakoids; prominent plastoglobuli are assembled into groups and large prolamellar bodies (Figure 12.8). *P. excelsa*, *A. alba*, and *P. mugo* have chloroplasts, where the large grana may each contain up to ten thylakoids (Figure 12.9). If prolamellar bodies are present, they occur in the place of future stroma lamellae, they are smaller, and their number is higher than in the case of larch etioplasts [38].

Differences exist not only in the chloroplast ultrastructure, but also in the ability to synthesize chlorophylls. Seedlings of *L. decidua* appear to be much less effective than seedlings of *P. excelsa* and *P. mugo* in synthesizing chlorophylls in the dark. They contain far less of both chlorophylls than the other two species. When etiolated seedlings are exposed to light, *P. excelsa* and *P. mugo* immediately show a net oxygen release, while *L. decidua* exhibits a net oxygen uptake until 6 hr of light [39].

These results indicate that chloroplasts in dark grown seedlings of *P. excelsa* and *P. mugo* are structurally and functionally well developed. Both the reaction centers and the light-harvesting complexes are formed and regularly assembled in the membranes.

As already noted, many gymnosperm species (their seedlings) and lower plants can synthesize chlorophyll in the dark. There is evidence that these plants possess two reductive pathways, one protochlorophyll(ide) oxidoreductase, which does not require the presence of light, and the light-dependent protochlorophyll(ide) reductase [37,40,41]. Classical and molecular-genetic studies of anoxygenic photosynthetic bacteria, cyanobacteria, and green algae, combined with plastid genome analyses of algae and higher plants, proved crucial to identifying the *chlB*, *chlL*, and *chlN* genes required for light-independent Pchlide reduction. These genes have been revealed to

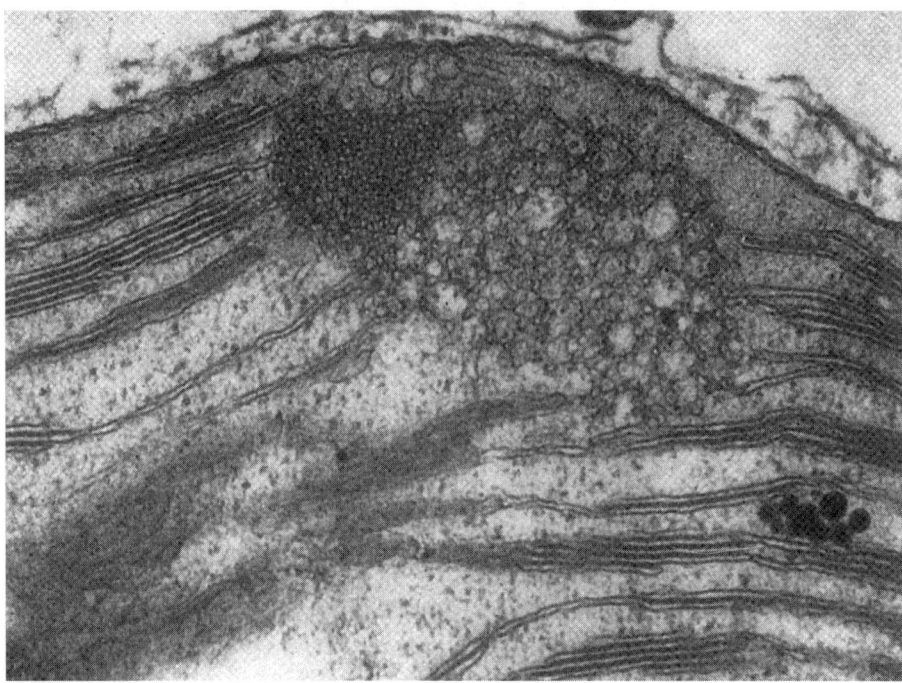

FIGURE 12.9 Detail of the *Picea abies* chloroplast with grana and prolamellar body. This figure demonstrates the remarkable ability of spruce to differentiate thylakoids in the absence of light (35,000×).

encode a multibisubunit light-independent Pchlide oxidoreductase [42].

In spite of these findings, why these plants are equipped with this ability remains unclear. One explanation is that the signal for chlorophyll and chloroplast formation in the dark originates in the endosperm via the effect of cytokinins [43]. The available data from DNA–DNA hybridization studies, plastid genome analyses, and characterization of PCR-amplified gene fragments support the hypothesis that angiosperms and the few other eukaryotic organisms that do not green in the dark have, in most cases, lost *chlB*, *chlL*, and *chlN* during evolution [42]. Another assumption takes into consideration light requirements for normal growth of the species. It is a well-known fact that *P. excelsa* and *A. alba* belong to the group of shadow-tolerating trees. Moreover, these two species and *P. mugo* compose very dense stands. When seeds from the three species germinate, the seedlings grow under conditions of very low light intensity. Therefore, it is believed that the ability of these seedlings to form chlorophyll and chloroplast is due to their developmental adaptation of very low-light conditions.

C. CHLOROPLASTS

The process of photosynthesis is carried out within a specific cytoplasmic compartment, the chloroplast.

Chloroplasts are the best studied of all plastid types, and there are numerous reviews describing their structure and functions [1–3,21].

Most chloroplasts are present in the mesophyll cells of the leaves. They also occur in the outer stem cells, in guard cells, in immature flowers, and fruits; however, the internal organization of their thylakoid system in these tissues is variable [1,44].

The occurrence of chloroplasts in root tissues is rare. Plant roots grow underground as heterotrophic organs and have little ability to turn green and form chloroplasts after illumination. However, if the roots are grown in root cultures, they maintain their typical root anatomy, but in the cortical cells well-developed chloroplasts are present [45].

The number of chloroplasts per cell fluctuates from one plant species to another but generally increases with the cell size. A striking variation can also be seen in plastid shape and size. Algal chloroplasts can have very bizarre shapes, but higher plant leaves show a characteristic lens shape for chloroplasts, which are usually 5 to 10 μm in diameter.

Mesophyll cells are highly vacuolized, and chloroplasts are found within the cytoplasm, usually around the cell periphery close to the plasma membrane. Distribution of chloroplasts inside the plant cell varies according to the light conditions. Under low light intensity, chloroplasts are lined up along anticlinal walls of palisade cells where there is more light, but under high light intensity they are placed along the inner walls where the light is weaker [46].

Chloroplasts are plant cell organelles with highly organized internal architecture. The structural modification of the inner membrane system of chloroplasts is influenced by different factors and one of these is the mode of photosynthesis.

1. Chloroplast of C$_3$ Plants

The first initial products after carboxylation of the CO_2 acceptor, ribulose 1,5-bisphosphate are two molecules of 3-phosphoglycerate. The presence three-carbon compounds leads to the name of this group, which contains monocotyledonous and dicotyledonous plants.

The distinct photosynthetic tissue in the leaves of C$_3$ plants is mesophyll. The mesophyll cells are spread out between the upper and lower epidermis (Figure 12.10) [47] and are made up of palisade and spongy cells. C$_3$ plants usually have uniform-appearing chloroplasts throughout the leaf. Chloroplast from the leaf mesophyll cells of both C$_3$ and C$_4$ plants exhibit similar internal membrane organization (Figure 12.12).

On the basis of numerous electron microscopic investigations, we can distinguish three major structural regions of the chloroplasts: double-outer-membrane envelope, chloroplast stroma, and highly organized lamellar system.

The chloroplast envelope consists of two membranes separated by a translucent gap of about 10 nm. This gap regulates the movement of carbon intermediate products in and out of the chloroplasts, it is the site of biosynthesis of galactolipids, and necessary proteins synthesized in cytoplasm are transported across the envelope. The envelope does not contain chlorophyll but possesses carotenoids that probably protect chloroplasts against photooxidation.

Inside the chloroplasts, there is a proteinaceous stroma. The stroma surrounds the thylakoids and is the site of biochemical (dark) reactions of photosynthesis. The prominent chemical substance of the chloroplast stroma is the enzyme ribulose bisphosphate carboxylase, which catalyzes carboxylation of ribulose bisphosphate. Ribulose bisphosphate carboxylase is composed of large and small subunits. The large subunit is encoded by the nuclear genome and synthesized by cytoplasmic ribosomes. The small subunit is encoded by the nuclear genome and synthesized by the cytoplasmic ribosomes. The proteins of the small subunit are transported across the envelope and assembled in the stroma into functional molecules of the enzyme [48].

The chloroplast stroma also contains a number of discrete particles. Chloroplast DNA appears as a mesh of 2.5-nm fibrils, and the area in which the fibrils are present is called nucleoid. The molecule of chloroplast DNA is of circular configuration,

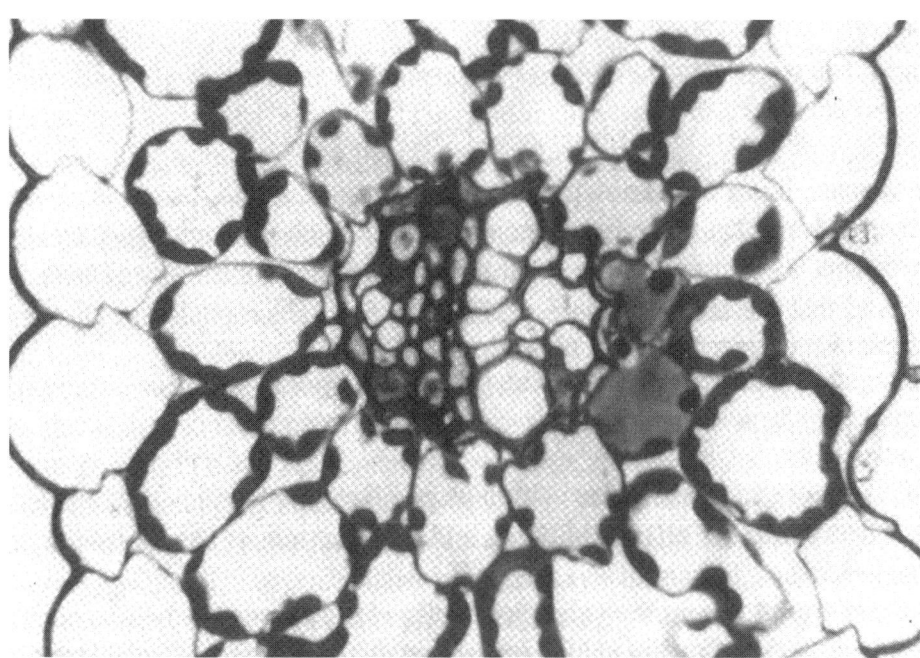

FIGURE 12.10 Leaf anatomy of C$_3$ plant *Hordeum vulgare* (580×). (From Benková E, Van Dongen W, Kolář J, Motyka V, Brzobohaty B, Van Onckelen HA, Macháčková I. *Plant Physiol.* 1999; 121: 245–251. With permission.)

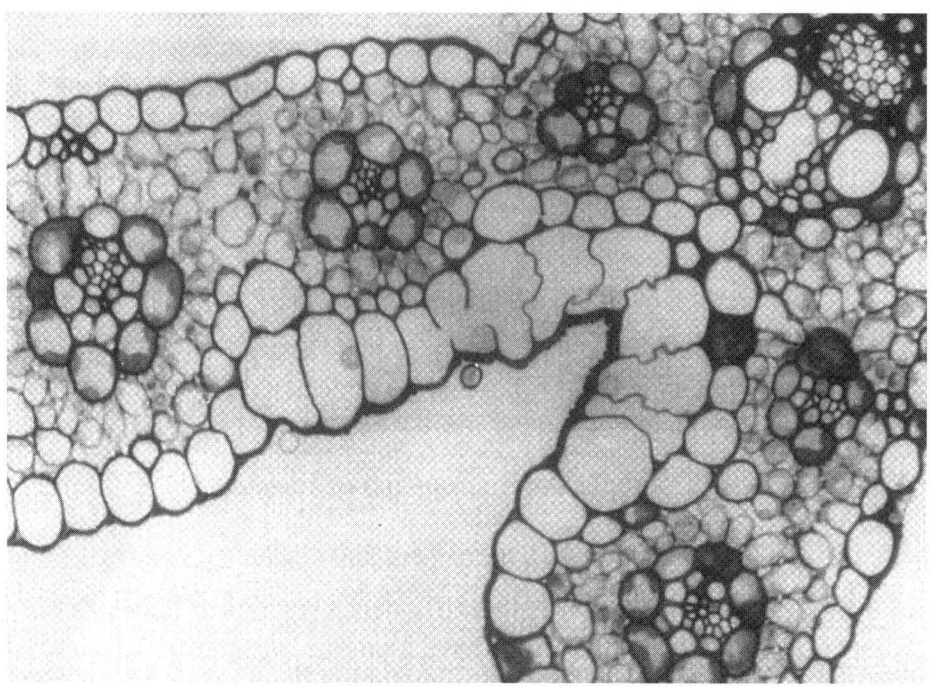

FIGURE 12.11 Leaf anatomy of C₄ plant *Chrysopogon gryllus*. Chloroplasts are located centrifugally in the bundle sheath cells (360×). (From Armstrong GA. *J. Photochem. Photobiol. B* 1998; 43: 87–100. With permission.)

and it occurs in all plastid types. Ribosomes are present in varying abundance in the stroma of higher-plant chloroplasts. They are either free in the stroma or bound to the chloroplast membranes. Plastoglobuli in chloroplast with a highly developed membrane system are regularly spread over the stroma. Starch grains are also often present in the stroma, which in general represent transitionally stored photosynthate. The number of starch grains greatly varies in chloroplasts; however, spongy mesophyll cells contain invariably more starch than palisade cells.

The light reactions of photosynthesis are localized in the chloroplast membranes. The internal membrane system of the chloroplasts includes grana and stroma thylakoids (Figure 12.12). The internal membranes are shaped like disks and are often stacked together, forming a granum. Each disk is vesiculated or saclike and is termed a thylakoid. A granum is made of at least two or three thylakoids. The number of thylakoids per granum varies considerably within the same chloroplast. The thylakoids that traverse the stroma and interconnect the grana are called stroma thylakoids or stroma lamellae. The number as well as the size of the grana are variable, depending on cell type and light conditions where the plants are cultivated. For example, spongy mesophyll cells have bigger grana stacks than palisade cells, and shade plant chloroplasts have larger grana with more thylakoids,

while chloroplasts from plants grown in the sun contain poorly stacked grana [49].

Granal thylakoids have their own substructure. The areas of paired membranes brought about by the close contact or adhesion of the surfaces of the adjacent thylakoid layers within the granum are termed partitions. The membranes exposed to the stroma at the edge of the granal thylakoids are termed margins. The partitions plus the margins enclose the electron-translucent space or lumen [50].

This substructure of granal thylakoids is useful in locating different proteins and photosystems in the thylakoid. A wide variety of proteins essential to photosynthesis are embedded in the thylakoid membrane. In many cases portions of these proteins extend into the aqueous regions of both sides of the thylakoid.

Integral membrane proteins of the chloroplasts often have a unique orientation within the membrane. Thylakoid membrane proteins have one region pointing toward the stromal side of the membrane and the other oriented toward the interior portion of the thylakoid, the lumen.

In recent years it has been established that the photosystem II reaction center, along with its antenna chlorophylls and associated electron transport proteins, is located predominantly in the stacked regions of the grana thylakoids. The photosystem I reaction center and its associated antenna pigments and elec-

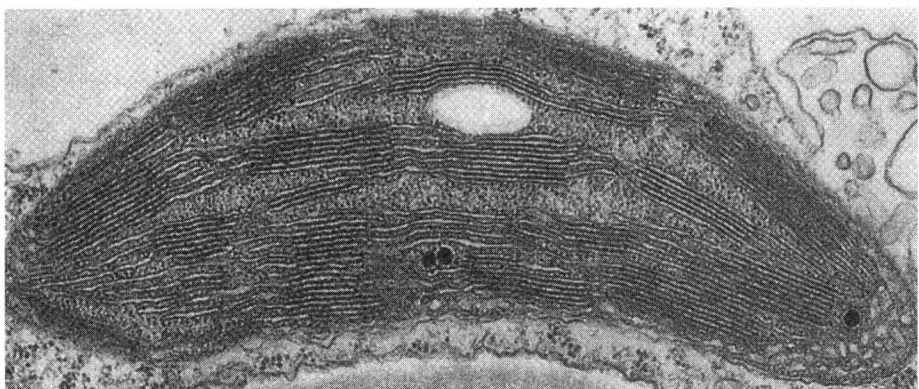

FIGURE 12.12 A mesophyll cell chloroplast of *Andropogon ischaemum* (30,000×). (From Armstrong GA. *J. Photochem. Photobiol. B* 1998; 43: 87–100. With permission.)

tron transfer proteins, as well as the coupling factor enzyme that catalyzes the formation of ATP, are found almost exclusively in the stroma lamellae and at the edges of the grana thylakoids. The cytochrome b_6–f complex that connects the photosystems is evenly distributed. Thus, the two photochemical events that take place in O_2-producing photosynthesis are spatially separated [50].

The various photosynthetic pigments involved in the absorption of light are part of the thylakoids. Higher plants have two groups of photosynthetic pigments: chlorophylls and carotenoids. There are two types of chlorophylls, chlorophyll *a* and chlorophyll *b*, in the higher plants. In algae and photosynthetic bacteria, bacteriochlorophylls are present. Chlorophyll *a* is the major pigment and is found in all photosynthetic organisms that produce oxygen. It has various forms with different absorption maxima. The short-wavelength Chl *a* forms are predominantly present in photosystem II. The long-wavelength forms are mostly present in photosystem I. The major portion of chlorophyll *b* is present in photosystem II. The chlorophylls are noncovalently bound to protein in the thylakoid membrane forming chlorophyll proteins.

Carotenoids are the yellow and orange pigments found in most photosynthetic organisms. There are two classes of carotenoids: carotens, for example, α and β carotene and lycopene, and xanthophylls (containing a hydroxyl group), for example, zeaxanthin, antheraxanthin, and violaxanthin. It is generally accepted that most of the carotenes are present in photosystem I, while the xanthophylls are involved in photosystem II [51].

Carotenoids are usually intimately associated with both the antenna and the reaction center pigment proteins and are integral constituents of the membrane. The energy of the light absorbed by carotenoids is rapidly transferred to chlorophylls, so carotenoids are termed accessory pigments. Carotenoids also play an essential role in photoprotection [50].

2. Chloroplasts of C_4 Plants

The basic characteristic of the C_4 plants is that the primary initial products of CO_2 fixation are the four-carbon dicarboxyl acids — oxaloacetate, malate, and aspartate. Both monocotyledons and dicotyledons from this group have a striking leaf anatomy and chloroplast architecture. The most prominent characteristic of the C_4 plant leaves is the organization of the chlorenchymatous tissue in concentric layers around the vascular tissue — Kranz-type (wreathlike) anatomy. This peculiar leaf anatomy was first described and named by the German botanist Haberlandt in 1904. As we have already noted, a cross section of C_3 plant leaf reveals essentially only one type of photosynthetic tissue containing chloroplasts — mesophyll. In contrast, the C_4 plant leaf has two distinct tissues containing chloroplasts — mesophyll and the bundle sheath (Figure 12.11) [52]. There is a considerable variation in the arrangement of the bundle sheath cells with respect to the mesophyll and vascular tissue [53].

Chloroplasts from the leaf mesophyll cells of C_4 plants exhibit grana similar to other higher-plant chloroplasts. However, the chloroplasts of the neighboring bundle sheath cells of these plants often have different chloroplast organization.

Originally, it was thought that the chloroplasts of C_4 bundle sheath cells were agranal. This assumption was supported by the observations of chloroplast ultrastructure of C_4 plants such as corn and sugarcane. But further evaluation of many C_4 plants showed that bundle sheath chloroplasts often possess grana [51,54].

On the basis of numerous physiological and structural studies of C$_4$ plants, it has been suggested that they can be divided into three distinct subgroups. The sorting of plants into these subgroups is based on the presence of the enzymes that catalyze their decarboxylation reactions, and they are also named after these enzymes. In decarboxylating mechanisms, NADP-malic enzyme (NADP-ME), NAD-malic enzyme (NAD-ME), and phosphoenolpyruvate carboxykinase (PEP-CK) enzymes are involved. The chloroplast organization in the single groups is as follows.

Chloroplasts in NADP-ME subgroup are agranal, and they are located centrifugally in the bundle sheath cells. Grana, if present, are few and are composed of two to four thylakoids. Examples of plants with these chloroplasts are corn, sugarcane, and sorghum (Figure 12.13).

Chloroplasts in NAD-ME subgroup contain numerous and well-developed grana, and they have a centripetal position in the bundle sheath cells. Plants like pigweed, purslane, and millet belong to this subgroup.

Chloroplasts in the PEP-CK subgroup possess grana, and their position in the bundle sheath cells is centrifugal. Plants that belong to this subgroup include guinea grass and Rhodes grass [40].

From this minireview it is quite obvious that chloroplast position in the vascular bundle sheath cells is variable. Disposition of bundle sheath chloroplasts changes during leaf development. Young chloroplasts of finger millet are almost evenly distributed along the cell walls in bundle sheath cells of folded immature leaves. Above the elongation zone,

the bundle sheath chloroplasts tend to lie along radial walls and the walls adjacent to the vascular bundle. They further migrate close to the vascular bundle, finally establishing a centripetal arrangement [55]. Bundle sheath chloroplasts typically have a high accumulation of the starch. However, if translocation of photosynthetic products is inhibited, numerous starch grains are present in the mesophyll chloroplasts after the bundle sheath chloroplasts are first loaded [51].

In the periphery of C$_4$ plant chloroplasts, a complex of vesicles and tubules occurs called the peripheral reticulum. The peripheral reticulum, which initially was thought to be unique to the bundle sheath and mesophyll cell chloroplasts, has also been found in the mesophyll cell chloroplasts of a number of C$_3$ plants. The peripheral reticulum is continuous with the chloroplast envelope and possibly with the thylakoid system. For these reasons it has been suggested that the peripheral reticulum may be involved in the rapid transport of metabolites between thylakoids and the chloroplast envelope [14,56]. In addition to starch grains, in plastoglobuli, ribosomes and regions with DNA fibrils can be observed in the stroma of bundle sheath chloroplasts. Variation in the chloroplast organization of C$_4$ plants is a good example of the influence of cell differentiation and function on plastid biogenesis.

D. CHROMOPLASTS

Chromoplasts represent a group of plastids that lack chlorophyll but accumulate carotenoids. They provide the bright red, yellow, and orange colors of

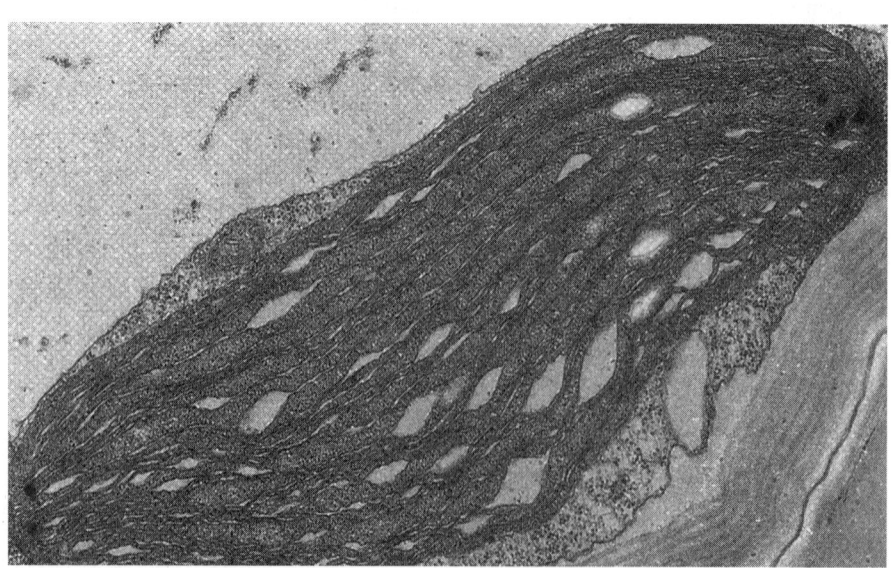

FIGURE 12.13 Chloroplast from a bundle sheath cell of *Chrysopogon gryllus* (21,300×). (From Armstrong GA. *J. Photochem. Photobiol. B* 1998; 43: 87–100. With permission.)

many flowers, old leaves, fruits, and some roots [6]. Morphologically, chromoplasts are very heterogenous. The original lens shape changes into elongated, spindle-shaped, and irregular ameboidal shape.

Chromoplasts can develop from chloroplasts or leucoplasts. When chloroplasts are transformed into chromoplasts, the membranes are broken down, and simultaneously the number of plastoglobuli increases. The course of chromoplast development in fruits and in flowers is similar to chromoplast differentiation in senescent leaves. Membrane breakdown takes place in the granal and stroma thylakoids but not in the plastid envelope.

During chloroplast transformation into chromoplasts in the tissues of fruits and flowers, there are transitional plastid chlorochromoplasts. Chlorochromoplasts contain both chlorophyll and carotenoids.

Fully differentiated chromoplasts lack chlorophylls and have a poor membrane system, but they have the ability to produce new types of carotenoids. The carotenoid present in green, unripe fruits and undeveloped flowers are those characteristic of the chloroplast. However, in the course of ripening, different carotenoids are formed, for example, lycopene in tomato and capsanthin in red pepper [1].

There is great variation not only in chromoplast shape and size but also in their ultrastructure, which varies in different fruits and flowers. This morphological variability has led to classifying chromoplasts as follows: globulous, membranous, tubulous, reticulotubulous, and crystallous [57]. The sorting of chromoplasts into different classes is done on the basis of morphological differences in the carotenoid containing structures.

The most frequent chromoplast type is globulous. Carotenoids of these chromoplasts are bound to globules of variable size. Globulous chromoplasts are present, for example, in fruits of *Solanum luteum*, bananas, oranges, cucumbers, and in flowers of *Ranunculus repens*, in tulips, *Chrysosplenium alternifolium*, and in senescent leaves (Figure 12.14).

Membranous chromoplasts are characterized by having multiple layers of membranes, which contain the carotenoid pigments. Such chromoplasts have been observed, for example, in the flowers of narcissus and in tomato fruits.

Tubulous chromoplasts typically exhibit tubulous and fibrillar structures, whereas carotenoids are bound. Tubules are often organized into the bundles, which are separated by single thylakoids. There is close contact between tubules and plastoglobuli. Tubulous chromoplasts occur, for example, in the fruits of red pepper and in cucumber flowers.

Crystallous chromoplasts contain their carotenoids (β-carotene and lycopene) in crystals. They are formed within or in association with the thylakoidal membrane. They occur in carrot roots, tomatoes, and in leaves of the lycopenic maize mutant (Figure 12.15) [58].

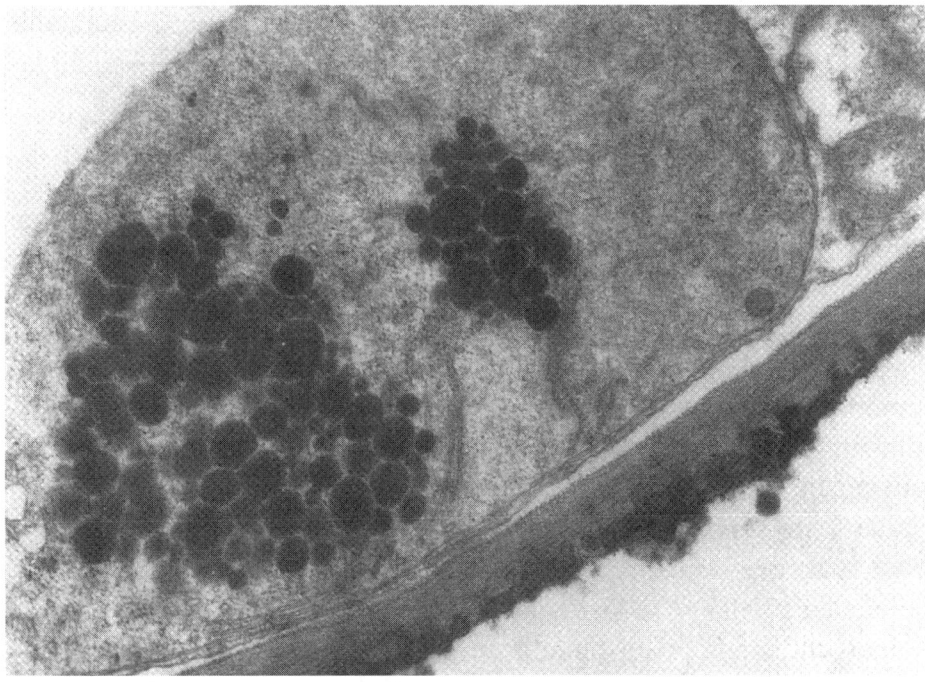

FIGURE 12.14 Globulous chromoplast of *Aucuba japonica* (27,000×).

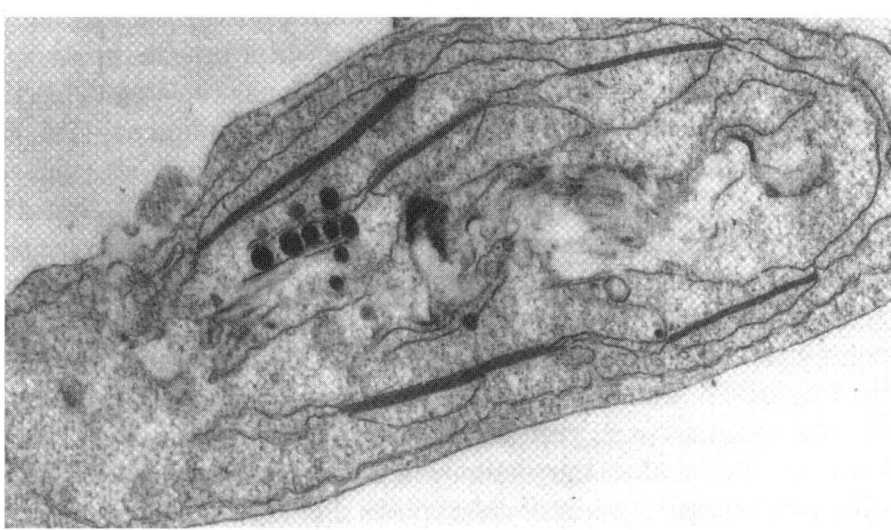

FIGURE 12.15 Crystallous chromoplast with lycopenic crystals of *Zea mays* lycopenic mutant (40,000×). (From Tevini M, Steinmüller D. *Planta* 1985; 163: 91–96. With permission.)

Reticulotubulous chromoplasts contain mutually connected tubules branched in different ways composing a network of tubules of variable size. Such chromoplasts have been observed in *Typhonium divaricatum*.

The ability of chromoplasts to synthesize new carotenoids indicates that metabolically they are not inactive organelles. The total content of proteins decreases due to the thylakoid breakdown, but they still contain DNA [59].

It is generally accepted that chromoplasts in fruits and flowers serve as attractants for pollinators and seed distributors.

E. PLASTID SENESCENCE

Senescence is the last phase in the ontogeny of a whole organism, organ, cell, or organelle. It is basically a degenerative process that leads to the death of a living system. Senescence of the leaf is controlled by nuclear genes and is accompanied by decreased expression of genes related to photosynthesis and protein synthesis and increased expression of senescence-associated genes (SAGs) [60–65].

Different tissues and cells of leaves have their own pattern and timing of senescence. The first leaves formed by a plant generally begin to senesce first, for example, cotyledons in dicotyledonous plants. Leaf senescence is commonly caused by shading as the canopy thickens above the early leaves, but it can also be caused by developmental changes taking place elsewhere in the plant, such as in the formation of seeds; by competition between the mature leaves and the growing shoot; or by environmental factors,

which can bring about the synchronous senescence of the leaves of deciduous trees in autumn [11].

Leaf senescence is accompanied by loss of proteins and chlorophyll and by extensive degradation of chloroplast membranes. A change in leaf color is the first symptom of leaf senescence. Disappearance of chlorophyll in senescent leaves is attributed to the action of chlorophyllase. This enzyme is an intrinsic thylakoid protein, therefore its activity is modulated by the membrane environment. Chlorophylase under normal conditions is in an inactive form in the membrane. Senescence-induced changes in the thylakoid organization may lead to the activation of chlorophyllase, which subsequently breaks down chlorophyll [66].

The fate of carotenoids is questionable. They are part of chlorophyll–protein complexes, and therefore their degradation is possible only with the destruction of these complexes. Compared to chlorophyll, carotenoids are quite stable. It is suggested that during the breakdown of membranes, the fatty acids released interact with liberated carotenoids to form carotenoid esters in plastoglobuli, thus keeping the pigments stable [4,67]. In connection with carotenoids, it is necessary to take into consideration the ability of chromoplasts to synthesize new forms of carotenoids not present in the chloroplasts [1].

During leaf senescence the original green color changes to yellow. A gradual deepening of this yellow color is accompanied by alterations in the chloroplast architecture. Yellow-green leaves possess a transitional type of plastid chlorochromoplasts. These plastids have the features of both chloroplasts (with degenerating membranes) and chromoplasts (with numerous plastoglobuli).

Ultrastructure degradation of senescent chloroplasts consists of three major events: thylakoid breakdown, formation of plastoglobuli, and rupture of the envelope [4].

At the beginning of chloroplast senescence the stroma thylakoids are destroyed first and the number of plastoglobuli increases with advanced membrane destruction (Figure 12.16). Gathering of plastoglobuli during chloroplast senescence and their closeness to degenerated membranes have led to the suggestion that plastoglobuli contain released lipids from destroyed thylakoids [68,69]. After stroma thylakoids break down, grana disorganization begins. The degradation of grana is induced by the loss of chlorophyll *b* and light-harvesting complex, which are known to be responsible for grana stacking [60].

Senescent yellow leaves may have either regular a green stripes or green spots on their margins. The yellow regions contain chromoplasts, but the green regions have chloroplasts with grana that are remarkable for their size and the number of thylakoids (they are also called giant grana) [11]. Chloroplasts with a similar lamellar organization occur in green islands in barley leaves (e.g., after infection with powdery mildew) [70].

The late phase of chloroplast senescence is characterized by both a change in shape and extensive vacuolation. At the beginning of plastid vacuolation many electron-transparent vacuoles appear in plastid stroma, which gradually fuse and finally occupy almost the entire plastids. Unlike during induced senescence, when vacuoles are formed by the hypertrophy of intrathylakoidal space [21,71], during natural senescence vacuoles originate from the local lysis of plastid stroma. The origin of vacuoles is the result of the activity of hydrolytic enzymes (e.g., proteases, Chl-degrading enzymes, galactolipase, and other enzymes), which are able to carry out the degradation within the chloroplasts.

The pattern of ultrastructural changes of chloroplasts differs with plant species and depends on the conditions under which senescence is carried out. However, generally we can conclude that complete plastid destruction during plastid senescence precedes the extensive vacuolation that results in the rupture of the plastid envelope and the decay of plastids.

F. Plastids of Heterotrophic Plants

The mode of plant nutrition significantly influences plastid morphogenesis. Among the plants with nonautotrophic nutrition are saprophytic and parasitic plant species from different families. Observations of different nongreen species indicate that the pathway of plastid differentiation differs from that in autotrophic plants. Plastid ultrastructure may be species specific in saprophytic and parasitic plants.

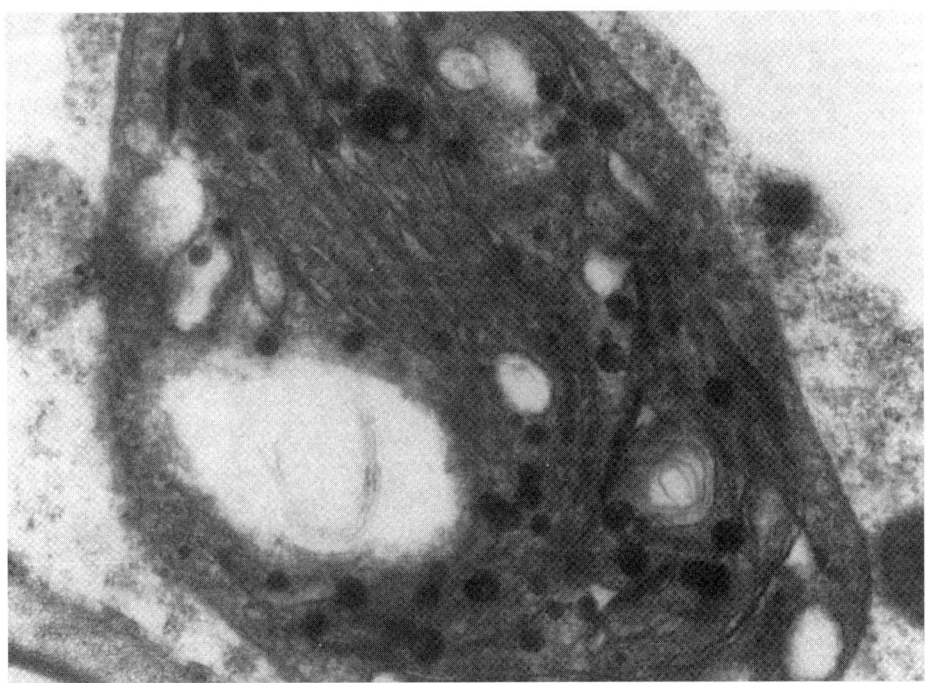

FIGURE 12.16 Senescent plastid of *Limodorum abortivum* with numerous plastoglobuli, dilated thylakoids, and stroma vacuolation (23,000×).

Semiparasitic plants constitute a special category. Chloroplasts of semiparasitic *Viscum album* are developed as in other autotrophic plants; their structure does not depend on the seasonal activity of the host. Green leaves of semiparasitic plants, both in winter (even at −7°C) and in summer, possess chloroplasts with a well-developed thylakoid system [72,73].

Saprophytic *Neottia nidus-avis* contains plastids with coiled and branched thylakoids, plastoglobuli, as well as starch grains in flowers, stalks, and scales (Figure 12.17) [74,75].

Parasitic plants (e.g., *Epifagus virginiana, Orobanche fuliginosa, Orobanche hederae, Lathrea squamaria, Cuscuta epithymum, Aeginatia indica,* and *Cuscuta europaea*) possess plastids with strongly reduced membrane systems. Besides single thylakoids, plastoglobuli and prominent protein inclusions and starch grains can occur in these plastids (Figure 12.18) [72,76,77]. It is of considerable interest that plastids of both parasitic and saprophytic plants are often developed as amyloplasts. The presence of large starch grains indicates that plastids of both saprophytes and parasites may serve as compartments of starch synthesis and storage. The precursors for starch synthesis in the case of saprophytes are taken from the substrate where they grow or from the host in the case of parasites.

The effect of heterotrophy on the sieve element plastids is interesting. While typical proteinoplasts (P-plastids) are present in the phloem cells of *N. nidus-avis,* the sieve elements of *E. virginiana* plastids are present only in the early stages of sieve element ontogeny, but they are absent in virtually all mature sieve elements [77,78].

Heterotrophic plants usually appear yellow, brown, or purple. Pigment analysis of the holoparasite *O. fuliginosa* has shown the presence of different carotenoids. The presence of carotenoids together with the simple inner organization of plastids led to their classification as chromoplasts [72,76].

The presence of chlorophyll in parasitic plants is questionable. Detectable levels of chlorophyll have not been observed in the parasites *E. virginiana, C. europaea,* and *O. fuliginosa* [76,79,80].

But in other parasites like *Cuscuta reflexa, Cuscuta campestris, L. squamaria,* and *Orobanche lutea* both chlorophylls are detectable [79]. Explaining these discrepancies in the chlorophyll content of parasitic plants may be difficult. For instance, chlorophyll content can be influenced by the sensitivity of apparatuses used for chlorophyll detection. Originally, in *N. nidus-avis* only chlorophyll *a* was observed, but from recent results it is obvious that chlorophyll *b* is also present in detectable amounts [75].

Small differences among the various species may be caused by special growth conditions. For example, the chlorophyll content in the stems of *Cuscuta australis* grown in the darkness is two times higher than in illuminated stems [81]. The ontogenetic stage of plant development is probably also import-

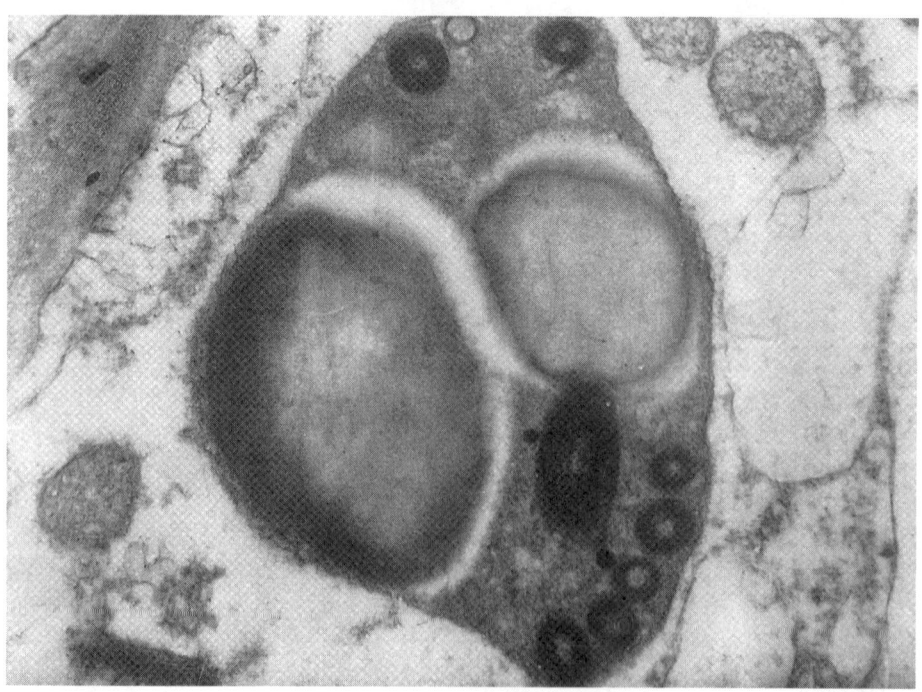

FIGURE 12.17 Plastid of saprophytic *Neottia nidus-avis* with coiled thylakoids and starch grains (21,000×).

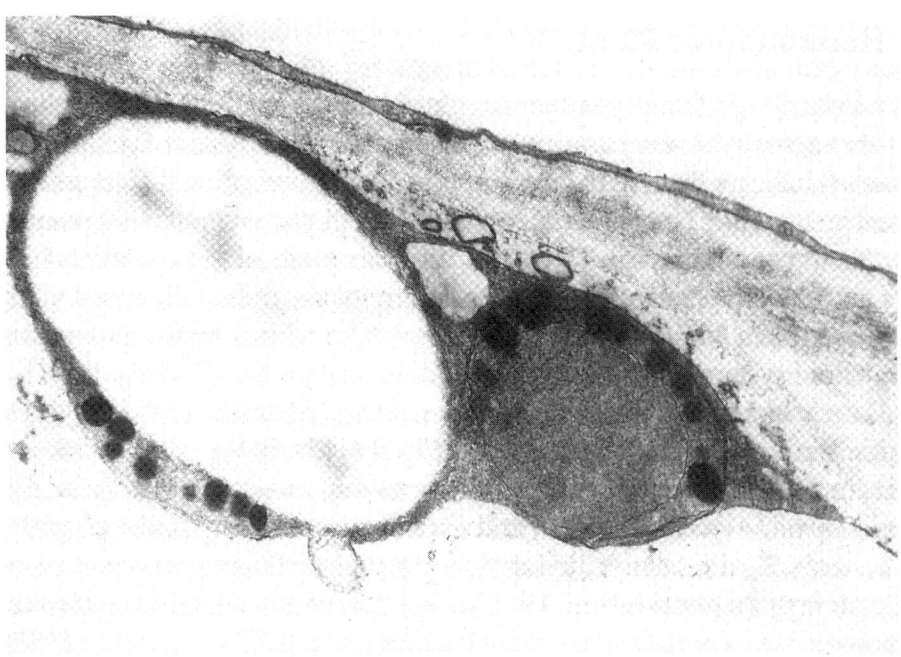

FIGURE 12.18 Plastids of holoparasite *Aeginatia indica* (27,000×).

ant. Parasitic plants, during flowering, contain more chlorophyll than during their vegetative phase. Suprisingly, a high chlorophyll content has been found in the pistils of *C. campestris*. Slight photosynthetic activity together with the presence of chlorophyll has been also identified in some parasites [79].

It is generally accepted that heterotrophic plants have evolved from autotrophs. Therefore, the question arises: Why do parasitic plants possess chlorophylls and exhibit low photosynthetic activities? Perhaps it is only a relic of their evolution from an autotrophic to a parasitic way of life, or it may be connected with the higher organic requirements of parasites during seed production (increased content of chlorophyll during flowering). The extent to which a parasite extracts organic compounds from its hosts presumably depends on the extent to which the parasite can satisfy its organic needs by its own ability to carry out photosynthesis.

G. Plastids of Evergreen Plants

Juvenile plastids of annual and deciduous plants in spring are gradually transformed into fully differentiated chloroplasts, and at the end of the vegetative period they mature and change into chromoplasts. Seasonal variation occurs not only in the chloroplast structure but also in their photosynthetic capacity [82]. In species in which the leaves persist for several years, the chloroplasts may undergo seasonal changes

in the structure of the thylakoidal system and its function [83–85].

In summer mesophyll cells possess well-developed chloroplasts with a rich membrane system. Grana are made of numerous thylakoids and are interconnected with the stroma lamellae. Chloroplasts are lined up along the cell walls. Summer chloroplasts contain a variable number of starch grains.

In autumn the membrane system of chloroplasts is not changed substantially. A peculiar shift occurs in the reduction of both the number and the size of the starch grains.

Conspicuous changes in chloroplasts take place in winter. The starch completely disappears. Chloroplasts usually are not regularly distributed along the cell walls but create irregular formations in different parts of cells. The membrane system of chloroplasts is often located in one part of the plastid, and in the other there is only membrane-free stroma. The numbers of grana and stroma lamellae are reduced, and they are not as compact as during summer (Figure 12.19). Both grana and stroma thylakoids swell slightly. To what extent chloroplast membranes are altered during winter is probably determined by the sensitivity of plant species to low temperatures.

In spring, the plastids are again distributed along the cell walls. The striking feature of these plastids is the presence of a great number of starch grains. Numerous and large starch inclusions make the plastid shape irregular. The membrane system of the spring plastids in evergreen plants does not show a regular

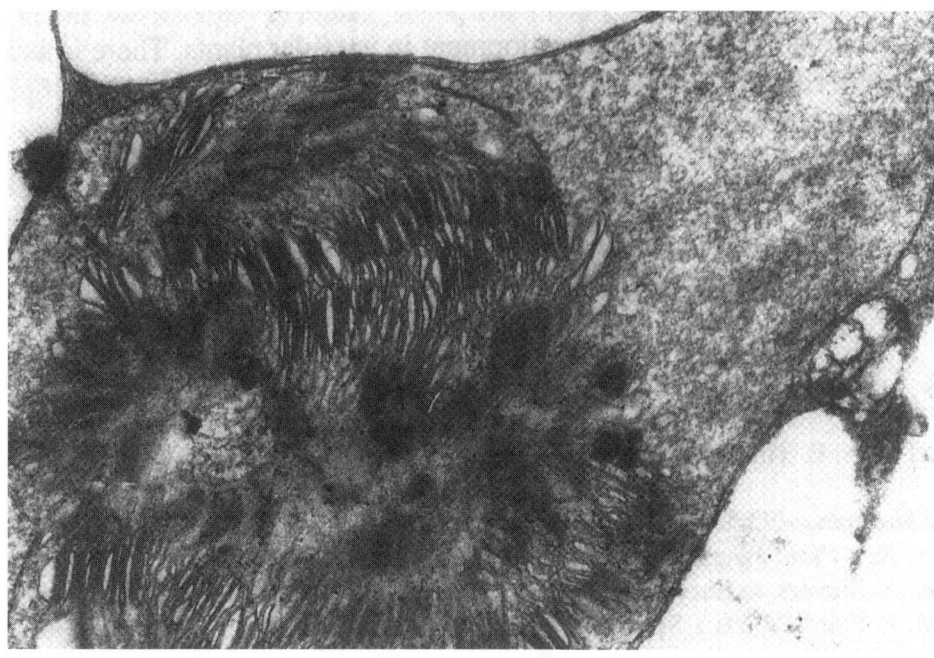

FIGURE 12.19 Chloroplast of *Aucuba japonica* in winter with slightly dilated membranes and membrane-free stroma (22,000×).

organization (typical for chloroplasts in other seasons and strongly limited by starch grains). These plastids are typical chloroamyloplasts. Such a high content of starch in the plastids has been observed only during spring. It is therefore possible to conclude that in spring these plastids function as amyloplasts and provide reserve material for cell division and differentiation in forming new shoots.

Studies of chloroplast alterations during the annual cycle of evergreen plants confirm the plasticity of the plastid membrane system, which is able to suitably respond to changing environmental conditions during the annual cycle.

H. PLASTID REGENERATION

It is obvious that various plastid types may in principle be reversibly transformed into one another. In different tissues, under certain conditions, transitional types of plastids (e.g., chloroamyloplast, chloroetioplasts, and chlorochromoplasts) occur.

In this regard it is necessary to ask if chromoplasts from both mature flowers and fruits or from the senescent leaves can be transformed into fully functional chloroplasts. Regeneration of senescent plastids into chloroplasts is very interesting and important for a better understanding of plastid biogenesis. The study of plastid transformation can indicate the approaches of the possible manipulation of leaf senescence [86]. Photosynthetically active chloro-

plasts lose their membrane system in the process of senescence, and gradually they are developed into senescent plastids — gerontoplasts [11,60]. During the reversion process chromoplasts are transformed into chloroplasts. Leaf regreening is accompanied by structural changes of plastids (new membranes are differentiated) and synthesis of chlorophyll. The results of this process are full transformation of chromoplasts into chloroplasts.

Several attempts have been made to induce plastid regeneration. Reversion of chromoplasts into chloroplasts has been described in Valencia orange fruits, in greening carrot roots [87,88], in pumpkin fruits [89], in the spathe of *Zantedeschia elliotiana*, and in the sepals of *Nuphar luteum* [90], in soybean cotyledons [91], in the leaves of tobacco and blackberries [92], in Buxus leaves [93], in lemon fruits [94], and in *in vitro* cultures of pericarp segments from fruits of *Citrus sinensis*.

The cotyledons have proved to be very useful plant material for the study of the regreening process, especially for their short life span. During the early phase of plant development cotyledons are photosynthetically active. Later, when the stems with leaves are differentiated, the cotyledons turn yellow and fall down. The senescence of cotyledons can be delayed when the differentiating stems are permanently removed from the seedlings, the cotyledons are green (evergreen), their size is bigger and even their life span is longer than those cotyledons with stems. In the time

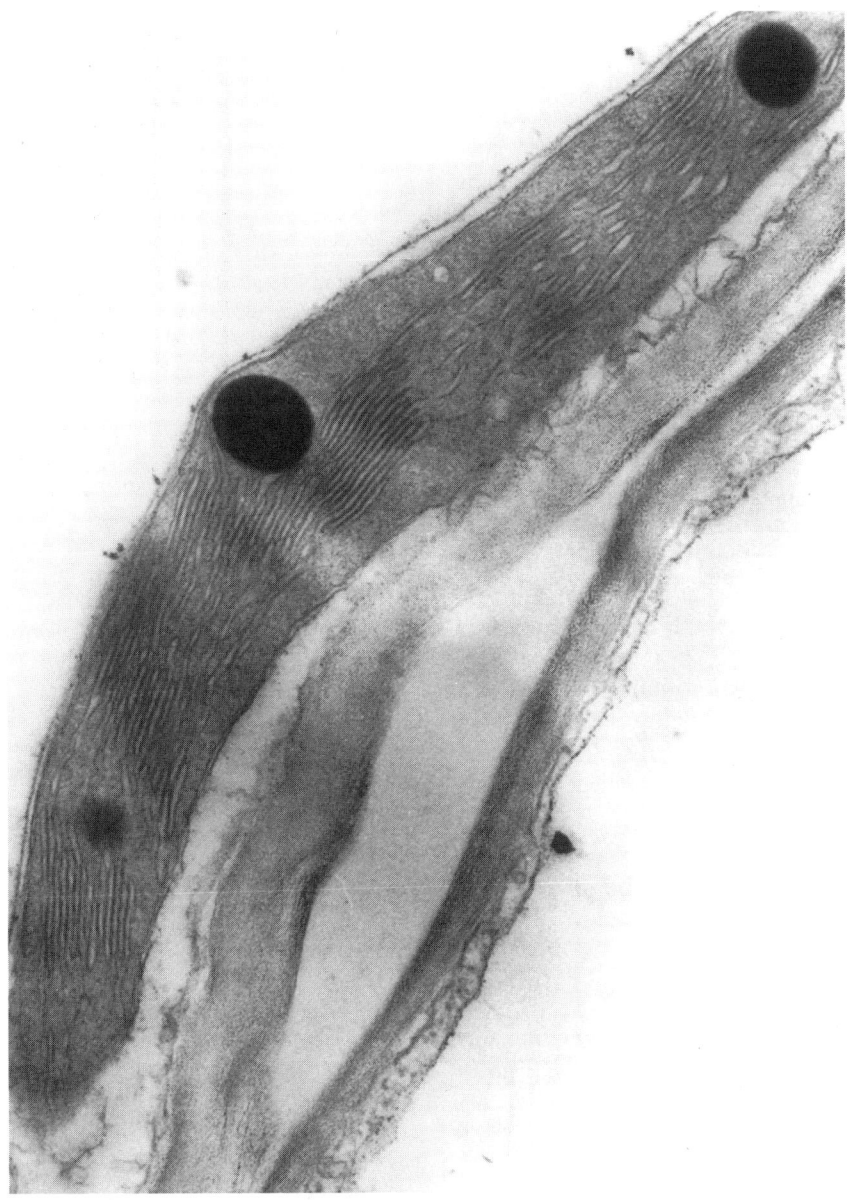

FIGURE 12.20 Regenerated chloroplast from mustard cotyledons (40,000×).

when the seedlings possess only green cotyledons, they are the only sink of the endogenous cytokinins. During the phase of stem and leaf differentiation all cytokinins are utilized in this process. When the stems from the seedlings are removed, the transport and new sink of cytokinins is blocked and they begin to accumulate in the yellow cotyledons. A positive influence after external application of cytokinins in delaying of plastid senescence and in plastid differentiation has been reported in numerous observations [61,95–100]. Therefore, it is supposed that the accumulation of cytokinins in senescent cotyledons might have induced their regreening.

During the transformation of chromoplasts into chloroplasts, chlorophyll accumulates and new thylakoids are formed within the original chromoplasts [91]. The protein spectrum is also changed during senescence and after regreening. Significant changes occur especially in the case of both subunits of RUBISCO. In senescent chloroplasts there are only traces of RUBISCO, but the regreened cotyledons are interesting for strong expression of both large and small RUBISCO subunits. Ultrastructural observations have revealed that during regreening, chloroplasts appeared to be formed by reversion of chromoplasts. No proplastids have been observed in the regreening tissues. New thylakoids are formed either by invagination of the inner membrane of the plastid envelope [89,90,92] or by multiplication of pre-existing thylakoids as in the case of mustard cotyledons [101]. If new thylakoids are differentiated from present membranes at the beginning, the membranes are swollen, but later they assume the shape of normal thylakoids and form typical grana (Figure 12.20). Both residual plastid structures (thylakoid membranes) and preservation of the plastid envelope integrity seem to be a prerequisite for the regreening phenomenon [91,102,103]. Plastoglobuli formed in chromopast during the disintegration of their membranes are reduced in number and size after extensive regreening. Factors influencing the reversal transformation of chromoplasts into chloroplasts are variable.

Suitable light and temperature conditions can cause regreening of tissues [87,89,93]. In the mustard cotyledons, after excision of the epicotyls, a direct correlation between the extent of regreening and the cytokinin content of the cotyledons has been observed. A similar course of plastid reversion can be also seen after external cytokinin treatment of senescent mustard cotyledons [25]. Plastid regeneration can be obtained not only in the case of natural senescence but even after the harmful effect of cadmium on cotyledons. Yellow Cd-treated cotyledons, after transferring 10^{-5} benzylaminopurine, solution are recovered after 72 hr of cultivation, and regreened cotyledons possess plastids with a well-developed membrane system [104].

Although plastid reversion occurs only under special experimental circumstances, the significance of this phenomenon lies in the fact that plastids are organelles that can progressively change their architecture according to the growth conditions.

III. SUMMARY

In this chapter we have focused on plastid development in vascular plants. There is a close interrelationship between plastid differentiation and cell differentiation, the tissue specificity, the mode of photosynthesis and nutrition, and different factors of the environment.

Plastids are flexible plant cell organelles that are able to dynamically respond to the changing conditions of plant growth. Plastid structural heterogeneity reflects their different functions in both plant species and plant tissues.

This contribution describes the fundamental principles of plastid differentiation in higher plants and brings new data regarding plastid regeneration.

REFERENCES

1. Kirk JTO, Tilney-Bassett RAE. *The plastids*. Elsevier/North Holland Biomedical Press, Amsterdam, 1978: 650.
2. Schnepf E. Types of plastids: their development and interconversions. In: Reinert J, ed. *Chloroplasts*. Springer Verlag, Berlin, 1980: 1–27.
3. Mullet JE. Chloroplast development and gene expression. *Annu. Rev. Plant Physiol. Plant Mol. Biol.* 1988; 39: 475–502.
4. Whatley JM. The ultrastructure of plastids in roots. *Int. Rev. Cytol.* 1983; 85: 175–220.
5. Muraja-Fras J, Krsnik-Rasol M, Wrischer M. Plastid transformation in greening potato tuber tissue. *J. Plant Physiol.* 1994; 144: 58–63.
6. Thomson WW, Whatley JM. Development of non-green plastids. *Annu. Rev. Plant Physiol.* 1980; 36: 569–593.
7. Sprey B. Intrathylakoidal occurrence of ribulose 1, 5-diphosphate carboxylase in spinach chloroplasts. *Z. Pflanzenphysiol.* 1976; 78: 85–89.
8. Behnke HD. Sieve-tube plastids in relation to angiosperm systematics. An attempt towards a classification by ultrastructural analysis. *Bot. Rev.* 1972; 38: 155.
9. Eleftheriou EP. Monocotyledons. In: Behnke HD, Sjölund RD, eds. *Sieve Elements*. Springer-Verlag, Berlin, 1990: 140–159.
10. Eleftheriou EP. A Light and Electron Microscopy Study on Phloem Differentiation of the Grass *Aegilops comosa* var. *thessalica*. Thesis. Univ. Thessaloniki, 1992: 95.
11. Hudák J. Plastid senescence. I. Changes of chloroplast structure during natural senescence in cotyledons of *Sinapis alba* L. *Photosynthetica* 1981; 15: 174–178.
12. Sprey B, Gliem G, Jánossy AGS. Iron containing inclusions in chloroplasts of *Nicotiana clevelandii* and *Nicotiana glutinosa* I. X-ray microanalysis und ultrastructure. *Z. Pflanzenphysiol.* 1976; 79: 165–176.
13. Cheniclet C, Carde JP. Differentiation of leucoplasts: comparative transition of proplastids to chloroplasts or leucoplasts in trichomes of *Stachys lanata* leaves. *Protoplasma* 1988; 143: 74–83.
14. Fisher DG, Evert RF. Studies on the leaf of *Amaranthus retroflexus* (*Amaranthaceae*): Chloroplast polymorphism. *Bot. Gaz.* 1982; 143: 146–155.
15. Lux A, Hudák J. Plastid dimorphism in leaves of the terrestrial orchid *Ophrys sphegodes* Miller. *New Phytol.* 1987; 107: 47–51.
16. Lux A. Changes of plastids during xylem differentiation in barley root. *Ann. Bot.* 1986; 58: 547–550.
17. Robertson D, Laetsch WM. Structure and function of developing barley plastids. *Plant Physiol.* 1974; 54: 148–159.
18. Whatley JM. Variations in the basic pathway of chloroplast development. *New Phytol.* 1977; 78: 407–420.
19. Whatley JM. Plastids in a changing environment. In: Bock JH, Linhart YB, eds. The evolutionary ecology of plants. Westview Press Co., Boulder, CO, 1989: 7–35.

20. Newcomb EH. Fine structure of protein-storing plastids in bean root tips. *J. Cell Biol.* 1967; 33: 143–163.

21. Hudák J, Herich R, Bobák M. *The Plastids.* Veda Sav, Bratislava, 1983; 101.

22. Jásik J, Hudák J. Plastid polymorphism in grape-vine (*Vitis labrusca* x *V. riparia*) tissue culture. *Photosynthetica* 1987; 21: 179–181.

23. Gardner IC, Abbas H, Scott AS. The occurrence of amoeboid plastids in the actinorhizal root nodules of *Alnus glutinosa* (L.) Gaertn. *Plant Cell Environ.* 1989; 12: 205–211.

24. Jásik J, Lux A, Hudák J, Mikuš M. Plastid degeneration induced by vanadium. *Biol. Plant.* 1987; 29: 73–75.

25. Ballová J, Hudák J, Cholvadová B. Effect of cytokinins on chlorophyll content and chloroplast ultrastructure in excised cotyledons of *Sinapis alba* L. In: *Book of Abstracts.* Symposium IXth Days of Plant Physiology, September 17–21, 2001, České Budějovice, 45.

26. Leech R.M. Chloroplast development in angiosperms: Current knowledge and future prospects. In: Baker NR, Barber J, eds. *Chloroplast Biogenesis.* Elsevier Science Publishers, Amsterdam, 1984: 2–21.

27. Dahline C, Cline K. Developmental regulation of the plastid protein import apparatus. *Plant Cell* 1991; 3: 1131–1140.

28. Keegstra K, Olsen LJ, Theg SM. Chloroplastic precursors and their transport across the envelope membranes. *Annu. Rev. Plant Physiol. Plant Mol. Biol.* 1989; 40: 471.

29. Whatley JM. Mechanisms and morphology of plastid division. In: Boffey SA, Lloyd D, eds. *Division and Segregation of Organelles.* Cambridge University Press, Cambridge, 1988; 63–83.

30. Oross JW, Possingham JV. Ultrastructural features of the constricted region of dividing plastids. *Protoplasma* 1989; 150: 131–138.

31. Modrušan Z, Wrischer M. Studies on chloroplast division in young leaf tissues of some higher plants. *Protoplasma* 1990; 154: 1–7.

32. Hudák J. Contribution to proplastid division. *Acta Biol. Acad. Sci. Hung.* 1974; 25: 135–137.

33. Gunning BES, Jagoe MP. The prolamellar body. In: Goodwin TW, ed. *Biochemistry of Chloroplasts.* Academic Press, London, 1967; 2: 655–676.

34. Čiamporová M, Pretová A. Ultrastructural changes of plastids in flax embryos cultivated *in vitro. New Phytol.* 1981; 87: 473–479.

35. Hudák J, Lojanová Z. Plastid ultrastructure and chlorophyll content in leaf primordia in bulbs of *Allium cepa* L. In: *Book of Abstracts.* Symposium IXth Days of Plant Physiology, September 17–21, 2001, České Budějovice, 165.

36. Stabel P, Sundas A, Engström P. Cytokinin treatment of embryos inhibits the synthesis of chloroplast proteins in Norway spruce. *Planta* 1991; 183: 520–527.

37. Laudi G, Medeghini-Bonatti P. Ultrastructure of chloroplast of some Chlamidospermae (*Ephedra twediana, Gnetum montana, Welwitschia mirabilis*). *Caryologia* 1973; 26: 107–114.

38. Walles B, Hudák J. A comparative study of chloroplast morphogenesis in seedlings of some conifers (*Larix decidua, Pinus sylvestris* and *Picea abies*). *Stud. For. Suec.* 1975; 127: 1–22.

39. Mariani P, DeCarli ME, Rascio N, Baldan B, Casadoro G, Gennari G, Bodner M, Laveher W. Synthesis of chlorophyll and photosynthetic competence in etiolated and greening seedlings of *Larix decidua* as compared with *Picea abies. J. Plant Physiol.* 1990; 137: 5–14.

40. Castelfranco PA, Beale SI. Chlorophyll biosynthesis. In: Hatch MD, Boardman NK, eds. *The Biochemistry of Plants.* Vol. 8. *Photosynthesis.* Academic Press, New York, 1981: 375–421.

41. Li J, Goldshmidt-Clermont M, Timko MP. Chloroplast-encoded chlB is required for light-independent protochlorophyllide reductase activity in *Chlamydomonas reinhardtii. Plant Cell* 1993; 5: 1817–1829.

42. Armstrong GA. Greening in the dark: light-independent chlorophyll biosynthesis from anoxygenic photosynthetic bacteria to gymnosperms. *J. Photochem. Photobiol. B* 1998; 43: 87–100.

43. Jelić G, Bogdanović M. The relationship between chlorophyll accumulation and endogenous cytokinins in the greening cotyledons of *Pinus nigra* Arn. *Plant Sci.* 1990; 71: 153–157.

44. Larcher W, Lütz C, Nagele M, Bodner M. Photosynthetic functioning und ultrastructure of chloroplasts in stem tissues of *Figus sylvatica. J. Plant Physiol.* 1988; 132: 731–737.

45. Flores HE, Dai Y, Cuello JL, Maldonado-Mendoza IE, Loyola-Vargas VM. Green roots: Photosynthesis and photoautotrophy in an underground plant organ. *Plant Physiol.* 1993; 101: 363–371.

46. Psaras GK. Chloroplast arrangement along intercellular spaces in the leaves of a mediterranean subshrub. *J. Plant Physiol.* 1986; 126: 189–193.

47. Hudák J, Lux A, Masarovičová E, Böhmová B. Plastid ultrastructure and pigment content in barley mutant induced by ethylnitrosourea. *Acta Physiol. Plant.* 1993; 15: 155–161.

48. Schmidt GW, Mishkind ML. The transport of proteins into chloroplasts. *Annu. Rev. Biochem.* 1986; 55: 879–912.

49. Boardman NK. Comparative photosynthesis of sun and shade plants. *Annu. Rev. Plant Physiol.* 1977; 28: 355–377.

50. Taiz L, Zeiger E. Cytokinins. In: Taiz L, Zeiger E, eds. *Plant Physiology.* The Benjamin/Cummings Publishing Company, Inc., Menlo Park, CA, 1991: 559.

51. Jensen RG. Biochemistry of the chloroplast. In: Tolbert NE, ed. *The Biochemistry of Plants.* Vol. 1. *The Plant Cell.* Academic Press, New York, 1980: 274–313.

52. Eleftheriou E., Noitsakis B. A comparative study of the leaf anatomy of the grasses *Andrapogon ischaenum* and *Chrysopogon gryllus. Phyton* 1978; 19: 27–36.

53. Laetsch WM. The C_4 syndrome a structural analysis. *Annu. Rev. Plant Physiol.* 1974; 25: 27–52.

54. Edwards GE, Huber SC. The C_4 pathway. In: Hatch MD, Boardman NK, eds. *The Biochemistry of Plants.*

Vol. 8. *Photosynthesis*. Academic Press, New York, 1981: 238–281.

55. Miyake H, Yamamoto Y. Centripetal disposition of bundle sheath chloroplasts during the leaf development of *Eleusine coracana*. *Ann. Bot.* 1987; 60: 641–647.

56. Sprey B, Laetsch WM. Structural studies of peripheral reticulum: in C₄ plant chloroplasts of *Portulacca oleracea* L. *Z. Pflanzenphysiol.* 1978; 87: 37–53.

57. Sitte P. Plastiden-metamophose und chromoplasten by *Chrysosplenium*. *Z. Pflanzenphysiol.* 1974; 73: 243–265.

58. Walles B, Hudák J. Etioplast and chromoplast development in the lycopenic mutant of maize. *J. Submicrosc. Cytol.* 1975; 75: 325–334.

59. Wuttke HG. Circular DNA in chromoplasts of *Tulipa gsneriana*. *Planta* 1976; 132: 317–319.

60. Biswal VC, Biswal B. Ultrastructural modification and biochemical changes during senescence of chloroplasts. *Int. Rev. Cytol.* 1988; 113: 271–321.

61. Gan S, Amasino RM. Making sense of senescence. Molecular genetic regulation and manipulation of leaf senescence. *Plant Physiol.* 1997; 113: 313–319.

62. Smart CM, Hosken SE, Thomas H, Greaves JA, Blair BG, Schuch W. The timing of maize leaf senescence and characterisation of senescence-related cDNAs. *Physiol. Plant.* 1995; 93: 673–682.

63. Martins L.M, Earnshaw WC. Apoptosis: alive and kicking in 1997. *Trends Cell Biol.* 1997; 7: 111–114.

64. Nam HG. Molecular genetic analysis of leaf senescence. *Curr. Opin. Biotechnol.* 1997; 8: 200–207.

65. Noodén LD, Guiamét JJ, John I. Senescence mechanisms. *Physiol. Plant.* 1997; 101: 746–753.

66. Takamiya K-I, Tsuchiya T, Ohta H. Degradation pathway(s) of chlorophyll: what has gene cloning revealed? *Trends Plant Sci.* 2000; 5: 426–431.

67. Guiamét JJ, Pichersky E, Noodén LD. Mass exodus from senescing soybean leaves. *Plant Cell Physiol.* 1999; 40: 986–992.

68. Tevini M, Steinmüller D. Composition and function of plastoglobuli. II. Lipid composition of leaves and plastoglobuli during leaf development. *Planta* 1985; 163: 91–96.

69. Lichtenthaler HK. Die Plastoglobuli von Spinat, ihre Grosse. Isolierung und Lipochinon zusammensetzung. *Protoplasma* 1969; 68: 65–77.

70. Camp RR, Whittingham WF. Fine structure of chloroplasts in "green islands" and in surrounding chlorotic areas of barley leaves infected by powdery mildew. *Am. J. Bot.* 1975; 62: 403–409.

71. Hudák J, Herich R. Effect of boron on the ultrastructure of sunflower chloroplasts. 1976; *Photosynthetica* 10: 463–465.

72. Dodge JD, Lawes GB. Plastid ultrastructure in some parasitic and semi-parasitic plants. *Cytobiologie* 1974; 9: 1–9.

73. Hudák J, Lux A. Chloroplast ultrastructure of semi-parasitic *Viscum album* L. *Photosynthetica* 1986; 20: 223–224.

74. Reznik H, Lichtenthaler HK, Paveling E. Untersuchungen uber den Lipochinon-Pigment-Gehalt und die Struktur der Plastiden von *N. nidus-avis* (L.) L. C. Rich. *Planta* 1969; 86: 353–359.

75. Masarovičová E, Lux A, Hudák J.Chlorophyll content, CO₂– exchange and plastid structure of the saprophyte *Neottia nidus-avis* (L.) L. C. Rich. *Biol Plant.* 1992; 34: 505.

76. Kollmann R, Kleinig H, Dörr I. Fine structure and pigments of plastids in *Orobanche*. *Cytobiologie* 1969; 1: 152–158.

77. Walsh MA, Rechel EA, Popovich TM. Observations on plastid fine-structure in the holoparasitic angiosperm *Epifagus virginiana*. *Am. J. Bot.* 1980; 67: 833–837.

78. Hudák J, Lux A, Masarovičová E. Plastid ultrastructure and carbon metabolism of the saprophytic species *Neottia nidus-avis*. *Photosynthetica* 1997; 33: 587–594.

79. Machado MA, Zetsche K. A structural, functional and molecular analysis of plastids of the holoparasites *Cuscuta reflexa* and *Cuscuta europaea*. *Planta* 1990; 181: 91–96.

80. de Pamphilis CW, Palmer JD. Loss of photosynthetic and chlororespiratory genes from the plastid genome of a parasitic flowering plant. *Nature* 1990; 348: 337–339.

81. Laudi G, Bonati BM, Fricano G. Ultrastructure of plastids of parasitis igher plants. V. Influence of light on *Cuscuta* plastids. *Isr. J. Bot.* 1974; 23: 145–150.

82. Vapaavuori E, Nurmi A, Vuorinen H, Kangas T. Comparison between the structure and function of chloroplasts at different levels of willow canopy during a growing season. In: Dreyre E, et al., eds. *Forest Tree Physiology*. Elsevier, Amsterdam. *Annu. Sci. Forum* 1989; 46: 815–818.

83. Senser F, Schotz F, Beck M. Seasonal changes in structure and function of spruce chloroplasts. *Planta* 1975; 126: 1–10.

84. Hudák J, Salaj J. Seasonal changes in chloroplast structure in mesophyll cells of *Acuba japonica*. *Photobiochem. Photobiophys.* 1986; 12: 173–176.

85. Hudák J, Salaj J. Seasonal changes in chloroplast structure in mesophyll cells of *Prunus laurocerasus* L. *Photosynthetica* 1990; 24: 105–109.

86. Thomas H, Howarth CJ. Five ways to stay green. *J. Exp. Bot.* 2000; 51: 329–337.

87. Grönegress P. The regreening of chromoplasts in *Daucus carot. Planta* 1971; 98: 274–278.

88. Wrischer M. Plastid transformation in carrot roots induced by different lights. *Acta Bot. Croat.* 1974; 33: 53–61.

89. Devidé Z, Ljubešić N. The reversion of chromoplasts to chloroplasts in pumpkin fruits. *Z. Pflanzenphysiol.* 1974; 73: 296–306.

90. Grönegress PL. The structure of chloroplasts and their conversion to chloroplasts. *J. Microsc.* 1974; 19: 183–192.

91. Huber DJ, Newman DW. Relationships between lipid changes and plastid ultrastructural changes in senescing and regreening soybean cotyledons. *J. Exp. Bot.* 1976; 27: 490–511.

92. Wrischer M, Ljubešić N, Marčenko E, Kunst J, Hlovšek-Radojčić A. Fine structural studies of plastids

during their differentiation and dedifferentiation. *Acta Bot. Croat.*1986; 45: 43–54.

93. Koiwa H, Ikeda T, Yoshida Y. Reversal of chromoplasts to chloroplasts in Buxus leaves. *Bot. Mag. Tokyo* 1986; 99: 233–240.

94. Ljubešić N. Structural and functional changes of plastids during yellowing and regreening of lemon fruits. *Acta Bot. Croat.* 1984; 43: 25–30.

95. Richmond AE, Lang A. Effect of kinetin on protein content and survival of detached Xanthium leaves. *Science* 1957; 125: 650–651.

96. Badenoch-Jones J, Parker CW, Letham DS, Singh S. Effect of cytokinins supplied via the xylem at multiples of endogenous concentrations on transpiration and senescence in derooted seedlings of oat and wheat. *Plant Cell Environ.* 1996; 19: 504–516.

97. Smart CM. Gene expression during leaf senescence. *New Phytol.* 1994; 126: 419–448.

98. Thomas H, Stoddart JL. Leaf senescence. *Annu. Rev. Plant Physiol.* 1980; 31: 83–111.

99. Binns AN. Cytokinin accumulation and action: biochemical, genetic and molecular approaches. *Annu. Rev. Plant Physiol. Plant Mol. Biol.* 1994; 45: 173–196.

100. Benková E, Van Dongen W, Kolář J, Motyka V, Brzobohaty B, Van Onckelen HA, Macháčková I. Cytokinins in tobacco and wheat chloroplasts. Occurrence and changes due to light/dark treatment. *Plant Physiol.* 1999; 121: 245–251.

101. Saganová, L, Hudák, J, Luxová, M, Malbeck, J. Reversion of senescent plastids into chloroplasts in *Sinapis alba* L. cotyledons. In: *Book of Abstracts.* Symposium IXth Days of Plant Physiology, September 17–21, 2001, České Budějovice, 54.

102. Hudák J, Vizárová G, Šikulová J. Ovečková O. Effect of cytokinins produced by strains of *Agrobacterium tumefaciens* with binary vectors on plastids in senescent barley leaves. *Acta Physiol. Plant.* 1996; 18: 205–210.

103. Devidé Z, Ljubešić N. Plastid transformation in greening scales of the onion bulb (*Allium cepa, Alliaceae*) *Plant Syst. Evol.* 1989; 165: 85–89.

104. Hudák J, Eleftheriou EP, Moustakas M, Bobák M. Plastid recovery from Cd-treated mustard cotyledons. In: *Book of Abstracts.* Symposium IXth Days of Plant Physiology, September 17–21, 2001, České Budějovice, 50.

13 Plastid Proteases

Dennis E. Buetow
Department of Molecular and Integrative Physiology, University of Illinois

CONTENTS

I. INTRODUCTION

Plastids form a major proteinaceous compartment in plant and algal cells. For example, chloroplasts contain 2000 to 2500 proteins [1] and account for 75% to 80% of the total nitrogen in a leaf [2]. Also, the dry weight of a chloroplast is 50% to 60% protein [3]. It is well known that the concentrations of apparently all proteins in plant cells result from both synthesis and degradation of the individual proteins [4]. Thus, plastid proteins, like those in other compartments of eukaryotic cells, are continually synthesized and subsequently degraded by a variety of proteases. Also, many plastid proteins are first synthesized as precursor molecules that are then processed to mature forms by proteases in the organelle (e.g., Ref. [5]).

Plastid proteins originally were proposed to be degraded by proteases located in a plant cell's vacuole, the presumed plant counterpart [6] of the animal cell lysosome. This proposal, however, became untenable when studies on senescing cells showed a large loss of protein occurred within chloroplasts before any loss in number of the organelles took place [7]. The latter indicated that, though final destruction of

247

senesced chloroplasts may be accomplished within the vacuole, proteolysis indeed does occur in the plastids themselves. The idea also arose that plastid proteins were degraded by a ubiquitin-type system within the organelles (e.g., Refs. [8,9]), but this idea was discounted when it was clearly shown that there are no ubiquitinated proteins in chloroplasts [10].

It is now well known that chloroplasts contain multiple proteases. Proplastids, etioplasts, and chromoplasts also contain proteases but the proteolytic complement of these plastids has been little studied. The topic of plastid proteases was reviewed for the first time in 1996 [5] and covered the published literature to about mid-1995. The present chapter is an updated version of the earlier one [5] and emphasizes the relevant literature published from mid-1995 to mid-2003. A number of relevant publications also have appeared in recent years [11–19] and these should be consulted for additional information on plastid proteases. In the present chapter, the terms "protease" and "peptidase" are used interchangeably.

II. PROTEASE FAMILIES

A. CLP PROTEASES

The ATP-dependent Clp proteases (caseinolytic protease) are soluble multisubunit protein complexes in both prokaryotes and chloroplasts. A Clp protease was first discovered in *Escherichia coli* and subsequently in chloroplasts (e.g., Ref. [5]). A typical Clp protease in chloroplasts consists of two types of subunit: ClpP (now known as ClpP1, see Table 13.1), which is a protease, and ClpC, which is an ATPase that regulates proteolysis by activating the ClpP protease. ClpC is encoded in the nuclear genome and ClpP in the plastid genome [5,20]. In plastids, ClpP1 is the homolog of the bacterial ClpP and ClpC is the homolog of the bacterial ClpA [13,21].

Clp protease genes and subunits are found in a wide variety of higher plants and green algae [5,18–28]. ClpC and ClpP genes and polypeptides are found in all tissues, including roots, of *Arabidopsis thaliana* [21,28] and pea seedlings [24], with ClpC mRNA and protein being the most abundant in green leaves compared to etiolated ones [24].

The Clp enzyme has been implicated in the degradation of the cytochrome b_6f complex [29] and suggested to perform a vital housekeeping function [30] including the degradation of proteins misdirected to the wrong compartment of the plastid [26]. The level of ClpC decreases during senescence in *Arabidopsis* leaves [31] while the level of the plastid-

TABLE 13.1
Clp Subunits from *Arabidopsis* Chloroplasts

Encoding Genome	Proposed Name[a]	Other Names and References
Plastid	ClpP1	pClpP [30,31] ClpP (generic for ClpP isomers)
Nucleus	ClpP3	nClpP3 [31] nClpP4 [30]
	ClpP4	nClpP3 [30] nClpP4 [31]
	ClpP5	nClpP1 [31] nClpP5 [30]
	ClpP6	nClpP1 [30] nClpP6 [31]
	ClpC1, C2	C1 and C2 not usually distinguished in the literature; most often either one is called ClpC
	ClpD	ERD1 [22,32,43]
	ClpR1	nClpP5 [31]
	ClpR2	nClpP2 [31]
	ClpR3	
	ClpR4	

[a]Adapted from Refs. [17,18].

encoded ClpP1 is reported both to remain unchanged in one study [31] and to decrease in another [32]. Any role for the Clp protease during senescence remains to be defined.

The Clp protease is reportedly essential in several cases for chloroplast function and for the growth and viability of algae and plant cells. Inactivation of the plastid-encoded *ClpP1* prevents growth in *Chlamydomonas* [33] and normal chloroplast development in tobacco [25]. Inactivation of the gene by insertional mutagenesis results in failure to produce homoplasmic transformants in tobacco [25] and *Chlamydomonas* [29]. Reduced levels of ClpP1 result in arrested chloroplast development in tobacco [27]. However, a requirement of *ClpP1* for plant cell viability has been questioned recently: two nonphotosynthetic lines of maize suspension cells lack the gene but still grow [34]. Therefore, a functional *ClpP1* does not seem to be required for general plant survival but has been suggested to be essential for the development and function of plastids [34]. However, a plastid-encoded ClpP is not essential in all plastids because the plastid genomes of the green algae, *Euglena gracilis* [35] and *Odontella sinensis* [36], and of the red alga, *Porphyra purpurea* [37], do not contain a gene for ClpP, yet the plastids of these algae are functional.

In bacteria [38], there is one gene for each of the ATPase subunits present, ClpA and ClpX, and one gene for the proteolytic subunit, ClpP. In contrast in *Arabidopsis,* multiple Clp isomers are encoded. *Arabidopsis* has become the standard organism for studies on molecular biology in higher plants, and especially since its genome has been completely sequenced [39]. Useful nomenclature for the previously otherwise-named Clp isomers in *Arabidopsis* has been proposed [17,18]. This nomenclature includes ClpP1, which is encoded in the plastid genome, as well as multiple subunits encoded in the nuclear genome, i.e., four more ClpP proteolytic subunits (ClpP3, 4, 5, and 6), three ATPase subunits (two near identical ClpC isomers, i.e., ClpC1 and C2, and ClpD), plus ClpR1, R2, R3, and R4. All of these Clp subunits locate in the chloroplasts (Table 13.1) and an additional ClpP subunit locates in the mitochondria [17,18]. ClpC1, C2, and D belong to the heat shock protein (Hsp) family of chaperones [18,28,40,41]. The high number of *Clp* genes in *Arabidopsis* is the result of gene duplication during evolution [42].

Aside from the ClpC and ClpP1 subunits that form the "Clp protease" in chloroplasts, little is known about the other Clp subunits. ClpD was first identified in *Arabidopsis* as a desiccation-induced protein, ERD1 [22]. ClpD also is induced by a high salt concentration, by dark-induced etiolation and by senescence [43]. The function of the ClpR isomers is not known [17]; however, some of them apparently interact with ClpP proteins in a large 350-kDa complex on thylakoid membranes in *Arabidopsis* [44]. This complex contains the chloroplast-encoded ClpP1, the nuclear-encoded ClpP proteins, two additional and unassigned ClpP homologs (ClpP7 and 8), and another Clp protein (ClpS1), which does not belong to any of the known Clp gene families. A ClpC subunit also is reported to interact with the import apparatus of the chloroplast [40,41,45].

Clp proteins generally are considered to be located in the chloroplast stroma [17,19,28]. However, other locations also have been noted. The thylakoid-membrane-associated 350-kDa Clp complex and the import-apparatus-associated ClpC mentioned above are two cases in point. Also, in both *E. gracilis* [46] and wheat [47], ClpP is more abundant in the chloroplast membrane fraction than in the stromal fraction.

B. DEG PROTEASES

The Deg proteases are nuclear-encoded serine-proteases, comprising three families [19]. The *Arabidopsis* genome contains 14 genes for Deg isomers of which four (DegP1, P2, P5, P8) encode proteases that are known to locate in the chloroplasts and two (DegP6

TABLE 13.2

Known and possible chloroplast Deg proteases in *Arabidopsis*[a]

Protease	Location
Known	
DegP1	Luminal side, thylakoid membrane
DegP2	Stromal side, thylakoid membrane
DegP5	Thylakoid lumen
DegP8	Thylakoid lumen
Possible	
DegP6	
DegP14	

[a]Adapted from Refs. [17,19].

and 14) encode proteases that are thought to locate in chloroplasts ([17,19]; see also Table 13.2). DegP1 is a thylakoid lumen protease [15,48,49] that is tightly associated with thylakoid membranes [48]. DegP1 is expressed constitutively and its level increases in plants exposed to high temperatures [48,50]. DegP1 can degrade both plastocyanin and OEC33 suggesting that it may be a general protease [50].

Deg P2 is active on the stromal side of thylakoid membranes and has been reported to initiate the GTP-dependent [51,51] degradation of photodamaged D1 protein, at least *in vitro* ([52]; but see also Section II.C). DegP5 and DegP8 are thylakoid lumen proteases [49]. Little is known about Deg P6 and Deg P14.

C. FTSH PROTEASES

In chloroplasts, the FtsH proteases are AAA proteases [53,54]. AAA enzymes (ATPase associated with a variety of cellular activities) form a novel class of conserved ATP-dependent proteases that are embedded in the membranes of chloroplasts, mitochondria, and bacteria and recognize membrane proteins as substrates [53].

Arabidopsis has 16 homologous nuclear-encoded FtsH genes. The protein products of 13 of these genes are known or suspected to locate in chloroplast membranes while the products of two locate in mitochondria and the product of one is unknown in location [19]. Of the 13 FtsH protein isomers in chloroplasts (Table 13.3), nine contain a catalytic zinc-binding site and are proteolytically active. The remaining four, designated FtsHi, lack the zinc-binding motif and are proteolytically inactive [19].

An FtsH protease was first identified in spinach and pea thylakoid membranes, and the expression of

TABLE 13.3
FtsH in *Arabidopsis* Chloroplasts[a]

FtsH

Proteolytic	FtsH 1
	FtsH 2
	FtsH 5
	FtsH 6
	FtsH 7
	FtsH 8
	FtsH 9
	FtsH 11
	FtsH 12
Non-proteolytic	FtsHi 1
	FtsHi 2
	FtsHi 3
	FtsHi 4

[a]Adapted from Ref. [19].

its gene was shown to be light inducible [55]. This protease was proposed to degrade unassembled Rieske FeS proteins in pea seedling thylakoids, but this is questionable because the protease activity measured was ATP independent [56]. However, if the unassembled FeS protein is already unfolded, then the ATPase function of FtsH may not be needed because ATP is thought necessary for unfolding polypeptides but not for cleaving peptide bonds [56]. Later, Spetea et al. [51] showed that a plastid ATP-dependent protease degrades the 23-kDa fragment resulting from the primary degradation of photodamaged D1 protein. In a following study, Lindahl et al. [57] reported that the latter enzyme is an FtsH protease.

In tobacco mosaic virus-infected tobacco leaf cells, a decrease of FtsH in chloroplasts leads to an acceleration of the hypersensitive reaction [58]. The hypersensitive reaction is defined as the rapid death of infected cells accompanying the formation of necrotic lesions [59]. Clearly, further work is needed to elucidate how a decreased level of FtsH is related to an amplification of the hypersensitive reaction.

FtsH1, also called Var1 [55,57,60], is located on the stromal side of thylakoid membranes in red pepper chromoplasts [61] and *Arabidopsis* chloroplasts [55]. FtsH2 in *Arabidopsis*, also called VAR2, is highly diverged from FtsH1 at both its amino- and carboxyl-termini and is required for plastid differentiation [62,63]. With *Arabidopsis* mutants studied *in vivo*, Bailey et al. [54,64] showed that FtsH2 is required for the efficient turnover of the D1 protein during photoinhibition and presented evidence that

FtsH2 does the initial cleavage of photodamaged D1. Further, Silva et al. [65] showed that an FtsH enzyme plays a similar role in the degradation of photodamaged D1 in the cyanobacterium *Synechocystis*. These latter results [54,64,65] contrast with the results of Haussühl et al. [52] who reported that DegP2 does the initial cleavage of photodamaged D1 (see Section II.B).

In sum, other than FtsH1 and FtsH2, very little is known about the FtsH proteases. The FtsHi isomers in particular are little understood but are speculated to have evolved from the FtsH isomers by gene duplication accompanied by changes (unknown) in function [19].

III. PROTEIN PROCESSING

A. BACKGROUND

Some chloroplast proteins are encoded in the organelle's own DNA while most are encoded in nuclear DNA. Those encoded in the organelle and destined for the chloroplast stroma are synthesized as the mature form. Others are destined for the thylakoid lumen and may require some processing. For example, the organelle-encoded D1 protein is synthesized with a C-terminal extension (e.g., Ref. [66]), which is subsequently removed by a C-terminal processing enzyme (e.g., Ref. [19]). Proteins encoded in the nucleus must be directed not only into the chloroplast but also to their proper locations within the organelle (e.g., [67,68]). The large majority of the nuclear-encoded proteins are synthesized as precursors with a cleavable N-terminal transit peptide that targets the proteins to the chloroplast. Nuclear-encoded precursors that lack additional targeting information are deposited into the organelle's stroma where a stromal processing protease removes the transit peptide. For nuclear-encoded precursors that are to be inserted into membranes, additional targeting information often is contained in the mature region of the protein. This appears to be the case for proteins targeted to the thylakoid membranes and the inner membrane of the chloroplast. Some nuclear-encoded proteins may require a stop-transfer signal for localization to the outer membrane of the chloroplast. Other outer membrane proteins lack a cleavable transit peptide and are inserted directly into the outer membrane without being first imported into the organelle. Precursors that are destined for the thylakoid lumen require a bipartite transit peptide. The first part is removed by the stromal processing protease. The second part, located just behind the first in amino acid sequence, is removed by a second protease as the protein enters the thylakoid lumen.

B. ENZYMES PROCESSING NUCLEAR-ENCODED PLASTID PROTEINS

1. Processing in the Chloroplast Stroma

In higher plants, a zinc-binding, stromal-processing enzyme (SPP) removes the transit peptide from a nuclear-encoded chloroplast protein [69–71]. If the enzyme is rendered nonfunctional, chloroplast development is affected with altered plastid division and chlorotic leaves resulting [72].

This general SPP binds to the transit peptide of a nuclear-encoded precursor protein and proteolytically removes it [73,74]. The SPP then fragments the transit peptide but does not further degrade the resulting fragments. Instead, the fragments are degraded [73,74] by a separate ATP-dependent, soluble metalloprotease with broad optimum pH and temperature but which is not FtsH (Section II.C). As previously reported for *Chlamydomonas* [75], more than one SPP with differing specificities for nuclear-encoded precursor proteins may exist in higher-plant chloroplasts [71].

E. gracilis presents an interesting situation. Its light-harvesting chlorophyll *a/b*-binding protein (LHCPII) is synthesized as a polyprotein precursor composed of eight LHCPIIs covalently joined by a decapeptide. This precursor is processed in chloroplasts to mature LHCPII molecules by a stromal thiol protease that differs from SPP [76].

2. Processing in the Chloroplast Thylakoids

Pea thylakoid membranes contain a processing protease that cleaves the thylakoid-transfer domain from the nuclear-encoded precursor to the mature 23-kDa extrinsic protein of photosystem II [77]. A cDNA encoding a similar thylakoid processing protease from *Arabidopsis* has been identified [78]. A possibly related protease was described in photosystem II membranes prepared from spinach thylakoids. This latter enzyme is a metalloprotease and exists as interconvertible 39-kDa monomers and 159-kDa tetramers but its role was not determined [79].

Interestingly, the heterokont alga, *Heterosigma akashiwo* [80], possesses a thylakoid processing protease with substrate specificity similar to the plant enzyme [77]. However, the algal enzyme matures a nuclear-encoded protein destined for the thylakoid lumen by cleaving, in a single step, the entire presequence including both the stromal- and the thylakoid-targeting domains.

C. ENZYMES PROCESSING PLASTID-ENCODED PROTEINS

1. Background

Plastid-encoded proteins, destined to locate in the stroma of the chloroplast, are translated as mature-sized molecules. Others, for example, those destined to locate in the thylakoid lumen, are translated as precursors with C-terminal extensions that are removed by a processing protease.

2. C-Terminal Processing Proteases

The C-terminal processing proteases of *Arabidopsis* consist of CtpA, CtpB, and CtpC [19]. A proteome analysis showed that all three Ctps are located in the thylakoid lumen of *Arabidopsis* chloroplasts [81].

A CtpA-type protease processes the C-terminal extension of the D1 precursor protein in barley [82], pea [83], and spinach [84,85]. cDNAs for spinach and barley CtpAs [86,87] have been isolated and sequenced. Steady-state CtpA mRNA levels are strongly light regulated [87]. The CtpA protease appears to have a unique catalytic center because the enzyme is not a serine-, aspartate-, or cysteine-type endoprotease nor a metalloprotease [87–89]. More studies are needed to define molecular mechanisms of action of CtpA. Further, very little is known about CtpB or CtpC.

IV. INITIAL DEGRADATION BY ACTIVE OXYGEN SPECIES

A. BACKGROUND

Although widely searched for, so far a protease has not been found to be responsible for the initial denaturation–degradation step of certain proteins in plastids incubated in light. In these cases, active oxygen species are said to be responsible for the initial alteration of the protein in question. The transport of electrons through the thylakoids and the oxidative events associated with this transport lead to the formation of active oxygen species and possibly other highly oxidizing species. Then, subsequent to or possibly concomitant with the action of active oxygen, the affected protein appears to become susceptible to chloroplast proteases [90,91].

B. PROTEINS

1. Ribulose-1,5-Bisphosphate Carboxylase/Oxygenase (Rubisco)

Rubisco is normally found as a soluble enzyme in a chloroplast in the light and in the dark. However, oxidative treatment in the light stimulates the association of Rubisco with the insoluble fraction of the organelle and also, at least, leads to the partial fragmentation of the enzyme's large subunit (LS; Ref. [92]). Active oxygen, for example, breaks down the LS into 36 to 37 and 16 to 20 kDa fragments representing the N- and the C-terminal portions, respectively, of the subunit [92–96]. Desimone et al. [93] reported that it is the Rubisco holoenzyme which, upon exposure to active oxygen species, then is degraded by proteases in the chloroplast stroma and that this proteolysis proceeds in an ATP-dependent manner.

The LS apparently is fragmented differently in the dark. Lysates of chloroplasts incubated in the dark degrade the LS to a 44-kDa fragment that lacks the N-terminal portion of the subunit [97]. This degradation is thought to be triggered by an unknown protease.

2. Glutamine Synthase

Under conditions of oxidative stress in the light, wheat chloroplasts and chloroplast lysates apparently use active oxygen species to fragment glutamine synthase into degradation products of 39 and 31 kDa [98,99].

3. D1

Light-induced inactivation of photosystem electron transfer, i.e., photoinhibiton, appears to be a prerequisite for D1 protein degradation (e.g., Ref. [100]). Active oxygen species or other highly oxidizing species generated within photosystem II are thought to be responsible for the initial "photodamage" to the D1 protein (e.g., Refs. [13,101]). The nature of the initial photodamage during photoinhiibition remains unsettled, however. Photoinhibition has been reported to be accompanied by the fragmentation of the D1 protein (e.g., Refs. [102,103]). However, it has been claimed that photochemical reactions arising during photoinhibition do not directly cleave D1 but rather alter it via a conformational change, which then turns the protein into a substrate for proteolysis [101]. Recent experimental evidence is in line with this claim and indicates that the photodamaged/conformationally altered D1 protein is then initially cleaved by the DegP2 (Section II.B) or the FtsH2 protease (Section II.C).

During photoinhibition in spinach chloroplasts, the D1 protein cross-links covalently or aggregates noncovalently with nearby polypeptides in photosystem II complexes [104]. These adducts are degraded by multiple, sodium dodecyl sulfate resistant proteases and most prominantly by a 15-kDa protease. In the case where D1 protein cross-links to cytochrome b_{559}, a 41-kDa product forms [105]. A chloroplast stromal extract, enhanced by ATP or GTP and containing mainly a 15-kDa protease, degrades the 41-kDa product and enhances the degradation of the D1 protein itself.

V. MISCELLANEOUS PLASTID ENZYMES

A. AMINOPEPTIDASES

Aminopeptidases (AP) are a class of enzymes involved in the removal of N- or C-terminal amino acid residues from proteins or peptides [106]. In recent years, several N-terminal APs have been identified in the stroma of chloroplasts in several plants: a leucine aminopeptidase in potato [107], two leucine APs in tomato [108–110], a methionine AP in *Arabidopsis* [111], and an alanine AP in barley [112]. However, the significance of N-terminal processing by aminopeptidases in chloroplasts is not known.

B. GLUTAMYL PROTEASE

A glutamyl protease was partially purified from spinach chloroplasts [113]. This protease is located in the chloroplast stroma, has a high molecular weight (350 to 380 kDa), is optimally active at about pH 8.0, and depends on chloride ions for activity.

C. PHOTOSYSTEM II-PARTICLE PROTEASE

A 43-kDa metalloprotease has been purified from photosystem II particles prepared from spinach [114]. Its function was not determined.

D. NADPH: PROTOCHLOROPHYLLIDE REDUCTASE DEGRADATION

Two different light-dependent NADPH: protochlorophyllide oxidoreductases, i.e., PORA and PORB, control chlorophyll synthesis in barley plastids [115–117]. PORA is present in large amount in etioplasts but selectively disappears shortly after the start of illumination. In the dark, complexes containing PORA, protochlorophyllide, and NADPH, form. In the light, these complexes photoreduce their protochlorophyllide to chlorophyllide and simultaneously become susceptible to degradation by plastid proteases. The PORA-degrading activity is not detected in etioplasts but is induced during illumination. In contrast, PORB remains functional in the light and leads to chlorophyll production. The PORA-degrading activity is composed of multiple constituents comprising both aspartic- and cysteine-type proteases.

E. LIGHT-HARVESTING CHLOROPHYLL *a/b*-BINDING PROTEIN DEGRADATION

Plants adapted to low or high light intensities contain larger or smaller light-harvesting antennas, respectively (e.g., Ref. [13]). Plants acclimated to low light contain more light-harvesting chlorophyll *a/b*-binding protein (LHCPII) per photosystem II reaction center while plants transferred to high light reduce their content of this protein. When spinach leaves are transferred from low to high light, LHCPII is degraded by an ATP-dependent serine- and/or cysteine-type protease associated with thylakoid membranes [118]. The LHCPII targeted for degradation laterally migrates from its functional site with PSII in the appressed regions of grana stacks to the stroma-exposed thylakoid regions where the protease is located [119].

A possibly related protease has been solubilized from thylakoids of etiolated *Phaseolus vulgaris*. This latter enzyme is a serine-type protease that degrades LHCPII and increases in activity when etiolated plants are exposed to light [120,121]. The enzyme cycles between the stroma and the thylakoids depending upon the local magnesium ion concentration [122]. Another possibly related serine-type protease called SppA has been isolated from thylakoid membranes of *Arabidopsis* [123]. This 68-kDa protease is light inducible, is speculated to be involved in the degradation of light-harvesting complexes, and may associate with thylakoid membranes as a tetramer.

A cysteine-type protease closely associated with the light-harvesting complex of photosystem II (LHCII) is reported to "self-digest" the LHCII as well as the D1 and D2 proteins of this photosystem [124]. This 114-kDa protease is membrane bound and light inducible.

F. EARLY LIGHT-INDUCIBLE PROTEIN DEGRADATION

Early light-inducible proteins (ELIP) are expressed transiently in etioplasts during the greening of etiolated seedlings and also are expressed in the chloroplasts of mature leaves exposed to a high light stress [13]. The ELIPs are stably maintained at high light but are rapidly degraded in the dark [125,126]. The ELIP-degrading activity is of the serine type, is ATP independent, and is located at the outer membrane surface of the stroma-exposed regions of thylakoids [126,127].

G. TRUNCATED D1 DEGRADATION

When a plasmid containing a deletion in the reading frame of *psbA* (encodes the D1 protein of photosystem II) is inserted into the chloroplast of *Chlamydomonas*, a truncated protein is synthesized but does not accumulate [128]. Instead, the truncated protein is rapidly degraded in the chloroplast by an ATP- and metal-dependent protease.

VI. CONCLUSIONS AND FUTURE CHALLENGES

A growing number of proteases are now known to be present in plastids but how many proteolytic reactions and pathways exist in these organelles remains an open question. Best understood so far are the proteolytic pathways involved in the processing of precursor proteins to mature and functional molecules.

In several cases, active oxygen species appear to initiate the degradation of a specific protein by altering its structure by such as a change in conformation. The altered molecule then seems to be marked for degradation, but all the proteolytic enzymes involved in the degradation are not well defined.

Several protease families are present in plastids including the Clp, DegP, and FtsH families. Of particular note are the numerous isomers that exist in these families, but it is not yet clear whether these isomers have overlapping activities or, at least in some cases, have distinct properties such as substrate specificity or pattern of expression. Indeed, substrate specificity has not yet been defined for most of the known plastid proteases.

Besides their substrate specificities and patterns of expression, much else remains to be discovered about plastid proteases. For example, the molecular structures of even the known proteases and the mechanisms whereby their activities are regulated remain to be defined. Major challenges remain in elucidating all the proteases that exist in plastids and then determining how their individual functions are related to regulatory events associated with the physiological responses and changes that characterize plants and algae.

REFERENCES

1. Abdallah F, Salamini F, Leister D. A prediction of the size and evolutionary origin of the proteome of chloroplasts of *Arabidopsis. Trends Plant Sci* 2000; 5:141–142.
2. Makino A, Osmond B. Effects of nitrogen nutrition on nitrogen partitioning between chloroplasts and mitochondria in pea and wheat. *Plant Physiol* 1991; 96:355–362.
3. Kirk JTO, Tilney-Bassett RAE. *The Plastids: Their Chemistry, Structure, Growth and Inheritance.* 2nd Ed. Amsterdam: Elsevier/North-Holland Biomed Press, 1978, p. 12.

4. Vierstra RD. Protein degradation in plants. *Annu Rev Plant Physiol Plant Mol Biol* 1993; 44:385–410.

5. Buetow DE. Plastid proteases. In: Pessarakli M, ed. *Handbook of Photosynthesis*. New York: Marcel Dekker, 1996:315–330.

6. Matile P. Biochemistry and functions of vacuoles. *Annu Rev Plant Physiol* 1978; 29:193–213.

7. Dalling MJ, Nettleton AM. Chloroplast senescence and proteolytic enzymes. In: Dalling MJ, ed. *Plant Proteolytic Enzymes*, Vol. II. Boca Raton, FL: CRC Press, 1986:125–153.

8. Veierskov B, Ferguson IB. Conjugation of ubiquitin to proteins form green plant tissues. *Plant Physiol* 1991; 96:4–9.

9. Hoffman NE, Ko K, Milcowski D, Pichersky E. Isolation and characterization of tomato CDNA and genomic clones encoding the ubiquitin gene *ubi3*. *Plant Mol Biol* 1991; 17:1189–1201.

10. Beers E, Moreno TN, Callis JA, Subcellular localization of ubiquitin and ubiquitinated proteins in *Arabidopsis*. *J Biol Chem* 1992; 267:15432–15439.

11. Adam Z. Protein stability and degradation in chloroplasts. *Plant Mol Biol* 1996; 32:773–783.

12. Adam, Z. Chloroplast proteases: Possible regulators of gene expression? *Biochimie* 2000; 82:647–654.

13. Andersson B, Aro E-M. Proteolytic activities and proteases of plant chloroplasts. *Physiol Plant* 1997; 100:780–793.

14. Choquet Y, Vallon O. Synthesis, assembly and degradation of thylakoid membrane proteins. *Biochimie* 2000; 82:615–634.

15. Adam Z and Ostersetzer O. Degradation of unassembled damaged thylakoid proteins. *Biochem Soc Trans* 2001; 29:427–430.

16. Estelle M. Proteases and cellular regulation in plants. *Curr Opinion Plant Biol* 2001; 4:254–260.

17. Adam Z, Adamska I, Nakabayashi K, Ostersetzer O, Haussuhl K, Manuell A, Zheng B, Vallon O, Rodermel SR, Shinozaki K, Clarke AK. Chloroplast and mitochondrial proteases in *Arabidopsis*. A proposed nomenclature. *Plant Physiol* 2001; 125:1912–1918.

18. Adam Z, Clarke AK. Cutting edge of chloroplast proteolysis. *Trends Plant Sci* 2002; 7:451–456.

19. Sokolenko A, Pojidaeva E, Zinchenko V, Panichkin V, Glazer VM, Herrmann RG, Shestakov SV. The gene complement for proteolysis in the cyanobacterium *Synechocystis* sp. PCC6803 and *Arabidopsis thaliana* chloroplasts. *Curr Genet* 2002; 41:291–310.

20. Sokolenko A, Lerbs-Mache S, Altschmied L, Hermann, RG. Clp protease complexes and their diversity in chloroplasts. *Planta* 1998; 207:286–295.

21. Shanklin J, DeWitt ND, Flanagan JM. The stroma of higher plant plastids contain ClpP and ClpC, functional homologs of *Escherichia coli* ClpP and ClpA: An archetypal two-component ATP-dependent protease. *Plant Cell* 1995; 7:1713–1722.

22. Kiyosue T, Yamaguchi-Shinozaki K, Shinozaki K. Characterization of cDNA for a dehydration-inducible gene that encodes a ClpA, B-like protein in *Arabidopsis thaliana* L. *Biochem Biophys Res Commun* 1993; 196:1214–1220.

23. Benesova M, Durcova G. Kuzela S. Kutejova E, Psenak M. Spinach chloroplast ATP-dependent endoprotease: Ti-like protease. *Phytochemistry* 1996; 41:65–69.

24. Ostersetzer O, Adam Z. Effects of light and temperature on expression of ClpC, the regulatory subunit of chloroplastic Clp protease, in pea seedlings. *Plant Mol Biol* 1996; 31:373–376.

25. Shikanai T, Shimizu K, Ueda K, Nishimura Y, Kuroiwa T, Hashimoto T. The chloroplast *clpP* gene, encoding a proteolytic submit of ATP-dependent protease, is indispensable for chloroplast development in tobacco. *Plant Cell Physiol* 2001; 42:264–273.

26. Halperin T Ostersetzer O, Adam Z. ATP-dependent association between subunits of Clp protease in pea chloroplasts. *Planta* 2001; 213:614–619.

27. Kuroda H, Maliga P. Overexpression of the *clpP* 5′-untranslated region in a chimeric context causes a mutant phenotype, suggesting competition for a *clpP*-specific mRNA maturation factor in tobacco chloroplasts. *Plant Physiol* 2002; 129:1600–1606.

28. Zheng B, Halperin T, Hruskova-Heidingsfeldova O, Adam Z, Clarke AK. Characterization of chloroplast Clp protein in *Arabidopsis*: Localization, tissue specificity and stress responses. *Physiol Plant* 2002; 114:92–101.

29. Majeran W, Wollman F-A, Vallon O. Evidence for a role of ClpP in the degradation of the chloroplast cytochrome b_6f complex. *Plant Cell* 2000; 12:137–149.

30. Clark AK. ATP-dependent Clp protease in photosynthetic organisms — a cut above the rest! *Ann Bot* 1999; 83:593–599.

31. Nakabayashi K, Ito M, Kiyosue T, Shinozaki K, Watanabe A. Identification of *clp* gene expressed in senescing *Arabidopsis* leaves. *Plant Cell Physiol* 1999; 40:504–514.

32. Weaver LM, Froechlich JE, Amasino RM. Chloroplast-targeted ERDI protein declines but its mRNA increases during senescence in *Arabidopsis*. *Plant Physiol* 1999; 119:1209–1216.

33. Huang C, Wang S, Chen L, Lemieux C, Otis C, Turmel M, Liu X-Q, The *Chlamydomonas* chloroplast ClpP gene contains translated large insertion sequences and is essential for cell growth. *Mol Gen Genet* 1994; 244:151–159.

34. Cahoon AB, Cunningham KA, Stern DB. The plastid *clpP* gene may not be essential for plant viability. *Plant Cell Physiol* 2003; 44:93–95.

35. Hallick RB, Hong L, Drager GR, Favreau MR, Monfort A, Orsat B, Spielmann A, Stutz E. Complete sequence of *Euglena gracilis* chloroplast DNA. *Nucleic Acids Res* 1993; 21:3537–3544.

36. Kowallik KV, Stoebe B, Schaffran I, Kroth-Pancic P, Frier U. The chloroplast genome of a chlorophyll a+c containing alga, *Odontella sinensis*. *Plant Mol Biol Rep* 1995; 13:336–342.

37. Reith M, Munholland J. Complete nucleotide sequence of the *Porphyra purpurea* chloroplast genome. *Plant Mol Biol Rep* 1995; 13:333–335.

38. Gottesman S. Proteases and their targets in *Escherichia coli*. *Annu Rev Genet* 1996; 30:465–506.

39. *Arabidopsis* Genome Initiative. Analysis of the genome sequence of the flowering plant *Arabidopsis thaliana*. *Nature* 2000; 408:796–815.

40. Akita M, Nielsen E, Keegstra K. Identification of protein transport complexes in the chloroplastic envelope membranes via chemical cross-linking. *J Cell Biol* 1997; 136:983–994.

41. Nielsen E, Akita M, Davila-Aponte J, Keegstra K. Stable association of chloroplastic precursors with protein translocation complexes that contain proteins from both envelope membranes and a stromal Hsp100 molecular chaperone. *EMBO J* 1997; 16:935–946.

42. Vision TJ, Brown DG Tanksley SD. The origins of genomic duplications in *Arabidopsis*. *Science* 2000; 290:2114–2117.

43. Nakashima K, Kiyosue T, Yamaguchi-Shinozaki K, Shinozaki K. A nuclear gene *erd1*, encoding a chloroplast-targeted Clp protease regulatory subunit homolog is not only induced by water stress but also developmentally up-regulated during senescence in *Arabidopsis thaliana*. *Plant J* 1997; 12:851–861.

44. Peltier J-B, Ytterberg J, Liberles DA, Roepstorff P, vanWijk KJ. Identification of a 350-kDa ClpP protease complex with 10 different Clp isoforms in chloroplasts of *Arabidopsis thaliana*. *J Biol Chem* 2001; 276:16318–16327.

45. Jackson-Constan D, Keegstra K. *Arabidopsis* genes encoding components of the chloroplastic protein import apparatus. *Plant Physiol* 2001; 125:1567–1576.

46. Erdös G, Buetow DE. Chloroplasts of *Euglena gracilis* contain a Clp-like protease. In: Murata N, ed. *Research in Photosynthesis*, Vol. III. Dordrecht: Kluwer, 1992: 295–298.

47. Weiss-Wichert C, Altenfeld U, Johanningmeir U. Detection of the P-subunit of the Clp-protease in chloroplasts. In: Mathis P, ed. *Photosynthesis: From Light to Biosphere*, Vol. III. Dordrecht: Kluwer, 1995:787–790.

48. Itzhaki H, Naveh L, Lindahl M, Cook M, Adam Z. Identification and characterization of DegP, a serine protease associated with the luminal side of the thylakoid membrane. *J Biol Chem* 1998; 273:7084–7098.

49. Schubert M, Petersson UA, Haas BJ, Funk C, Schroder WP, Kieselbach T. Proteome map of the chloroplast lumen of *Arabidopsis thaliana*. *J Biol Chem* 2002; 277:8354–8365.

50. Chassin Y, Kapri-Pardes E, Sinvany G, Arad T, Adam Z. Expression and characterization of the thylakoid lumen protease DegP1 from *Arabidopsis*. *Plant Physiol* 2002; 130:857–864.

51. Spetea C, Hundal T, Lohmann F, Andersson B. GTP bound to chloroplast thylakoid membranes is required for light-induced, multienzyme degradation of the photosystem II D1 protein. *Proc Natl Acad Sci USA* 1999; 96:6547–6552.

52. Haüssuhl K, Andersson B, Adamska I. A chloroplast DegP2 protease performs the primary cleaveage of the photodamaged D1 protein in plant photosystem II. *EMBO J* 2001; 20:713–722.

53. Langer T. AAA proteases: cellular machines for degrading membrane proteins. *Trends Biochem Sci* 2000; 25:247–251.

54. Bailey S, Thompson E, Nixon PJ Horton P, Mullineaux CW, Robinson C, Mann NH. A critical role for the Var2 FtsH homolog of *Arabidposis thaliana* in the photosystem II repair cycle *in vivo*. *J Biol Chem* 2002; 277:2006–2011.

55. Lindahl M, Tabak S, Cseke L, Pichersky E, Andersson B, Adam Z. Identification, characterization, and molecular cloning of a homologue of the bacterial FtsH peptidase in chloroplasts of higher plants. *J Biol Chem* 1996; 271:29329–29334.

56. Ostersetzer O, Adam Z, Light-stimulated degradation of an unassembled Rieske FeS protein by a thylakoid-bound protease: the possible role of the FtsH protease. *Plant Cell* 1997; 9:957–965.

57. Lindahl M, Spetea C, Hundal T, Oppenheim AB, Adam Z Andersson B. The thylakoid FtsH protease plays a role in the light-induced turnover of the photosytem II D1 protein. *Plant Cell* 2000; 12:419–431.

58. Seo S, Okamoto M, Iwai T, Iwano M, Fukui K, Isogai A, Nakajima N, Ohashi Y. Reduced levels of chloroplast FtsH protein in tobacco mosaic virus-infected tobacco leaves accelerate the hypersensitive reaction. *Plant Cell* 2000; 12:917–932.

59. Goodman RN, Novacky AJ. *The Hypersensitive Reaction in Plants to Pathogens: A Resistance Phenomenon*. St Paul, MN: American Phytopathological Society Press, 1994.

60. Sakamoto W, Tamura T, Hanba-Tomita Y, Murata M. The *VAR1* locus of *Arabidopsis* encodes a chloroplastic FtsH and is responsible for leaf variegation in the mutant alleles. *Genes Cells* 2002; 7:769–780.

61. Huegueney P, Bouvier F, Badillo A, d'Harlingue A, Kuntz M, Camara B. Identification of a plastid protein involved in vesicle fusion and/or membrane protein translocation. *Proc Natl Acad Sci USA* 1995; 92:5630–5634.

62. Chen M, Choi Y, Voytas D, Rodermel S. Mutations in the *Arabidopsis* VAR2 locus cause leaf variegation due to the loss of chloroplast FtsH peptidase. *Plant J* 2000; 22:303–313.

63. Takeuchi K, Sodmergen, Murata M, Motoyoshi F, Sakamoto W. The *yellow 2 variegated (Var2)* locus encodes a homolgue of FtsH, an ATP-dependent protease in *Arabidopsis*. *Plant Cell Physiol* 2000; 41:1334–1346.

64. Bailey S, Silva P, Nixon P, Mullineaux C, Robinson C, Mann N. Auxiliary functions in photosynthesis: The role of the FtsH protease. *Biochem Soc Trans* 2001; 29:455–459.

65. Silva P, Thompson E, Bailey S, Kruse O, Mullineaux CW, Robinson C, Mann NH, Nixon PJ. FtsH is involved in the early stages of repair of photosystem

II in *Synechocystis* PC 6803. *Plant Cell* 2003; 15: 2152–2164.

66. Svensson B, Vass I, Styring S. Sequence analysis of the D1 and D2 reaction center proteins of photosystem II. *Z Naturforsch Sec C Biosci* 1991; 46:765–776.

67. Keegstra K, Cline K. Protein import and routing systems of choroplasts. *Plant Cell* 1999; 11:557–570.

68. Raikhel N, Chrispeels MJ. Protein sorting and vesicle traffic. In: Buchanan BB, Gruissem W, Jones RL, eds. *Biochemistry and Molecular Biology of Plants*. Rockville, MD: American Society of Plant Physiologists, 2000:160–201.

69. VanderVere PS, Bennett TM, Oblong JE, Lamppa GK. A chloroplast processing enzyme involved in precursor maturation shares a zinc-binding motif with a recently recognized family of metalloendopeptidases. *Proc Natl Acad Sci USA* 1995; 92:7177–7181.

70. Koussevitzky S, Ne'eman E, Sommer A, Steffens JC, Harel E. Purification and properties of a novel chloroplast stromal peptidase. *J Biol Chem* 1998; 273: 27064–27069.

71. Richter S, Lamppa GK. A chloroplast processing enzyme functions as the general stromal processing peptidase. *Proc Natl Acad Sci USA* 1998; 95:7463–7468.

72. Wan J, Bringloe D, Lamppa GK. Disruption of chloroplast biogenesis and plant development upon down-regulation of a chloroplast processing enzyme involved in the import pathway. *Plant J* 1998; 15:459–468.

73. Richter S, Lamppa GK. Stromal processing peptidase binds transit peptides and initiates their ATP-dependent turnover in chloroplasts. *J Cell Biol* 1999; 147: 33–43.

74. Richter S, Lamppa GK Determinants for removal and degradation of transit peptides of chloroplast precursor proteins. *J Biol Chem* 2002; 277:43888–43894.

75. Rüfenacht A, Boschetti A. Specificity of processing enzymes in chloroplasts of *Chlamydomonas reinhardii*. In: Mathis P, ed. *Photosynthesis: From Light to Biosphere*, Vol. III. Dordrecht: Kluwer, 1995:767–770.

76. Enomoto T, Sulli C, Schwartzbach SD. A soluble chloroplast protein processes the *Euglena* polyprotein precursor to the light harvesting chlorophyll *a/b* binding protein of photosystem II. *Plant Cell Physiol* 1997; 38:743–746.

77. Barbrook AC, Packer JCL, Howe CJ. Inhibition by penem of processing peptidases from cyanobacteria and chloroplast thylakoids. *FEBS Lett* 1996; 398: 198–200.

78. Chaal BK, Mould RM, Barbrook AC, Gray JC, Howe CJ. Characterization of a cDNA encoding the thylakoidal peptidase from *Arabidopsis thaliana*. *J Biol Chem* 1998; 273:689–692.

79. Kuwabara T. The 60-kDa precursor to the dithiothreitol-sensitive tetrameric protease of spinach thylakoids: structural similarities between the protease and polyphenol oxidase. *FEBS Lett* 1995; 317:195–198.

80. Chaal BK, Ishida K, Green BK. A Thylakoidal processing peptidase from the heterokont alga *Heterosigma akashiwo*. *Plant Mol Biol* 2003; 52: 463–472.

81. Schubert M, Petersson UA, Haas BJ, Funk C, Schröder WP, Kieselbach T. Proteome map of the chloroplast lumen of *Arabidopsis thaliana*. *J Biol Chem* 2002; 277:8354–8365.

82. Pakrasi HB, Oelmüller R, Herrman RG, Shestakov SV. Molecular analysis of CtpA, the carboxyl-terminal procesing protease for the D1 protein of photosystem II, in higher plants and cyanobacteria. In: Mathis P, ed. *Photosynthesis: From Light to Biosphere*, Vol. III. Dordrecht: Kluwer, 1995: 719–724.

83. Magnin N, Hunt A, Camilleri R, Thomas P, Ridley S, Bowyer J. Purification and characterization of the carboxyl terminal processing protease of the D1 protein of photosystem II from *Pisum sativum*. In: Mathis P, ed. *Photosynthesis: From Light to Biosphere*, Vol. III. Dordrecht: Kluwer, 1995:831–834.

84. Inagaki N, Mori H, Fujita S, Yamamoto Y, Satoh K. Carboxyl-terminal processing protease for D1 precursor protein in spinach. In: Mathis P, ed. *Photosynthesis: From Light to Biosphere*, Vol. III. Dordrecht: Kluwer, 1995:783–786.

85. Fujita S, Inagaki N, Yamamoto Y, Taguchi F, Matsumoto A, Satoh K. Identification of the carboxyl-terminal processing protease for the D1 precursor protein of the photosystem II reaction center of spinach. *Plant Cell Physiol* 1995; 36:1169–1177.

86. Inagaki N, Yamamoto Y, Mori H, Satoh K. Carboxyl terminal processing protease for the D1 precursor protein: Cloning and sequencing of the spinach cDNA. *Plant Mol Biol* 1996; 30:39–50.

87. Oelmüller R, Hermann RG, Pakrasi HB. Molecular studies of CtpA, the carboxyl-terminal processing protease for the D1 protein of the photosystem II reaction center in higher plants. *J Biol Chem* 1996; 271:21848–21852.

88. Yamamoto Y, Taguchi F, Satoh K. Recognition signal for processing protease on D1 precursor protein of PSII reaction center. In: Mathis P, ed. *Photosynthesis: From Light to Biosphere*, Vol. III. Dordrecht: Kluwer, 1995:771–774.

89. Yamamoto Y, Inagaki N, Satoh K. Overexpression and characterization of carboxyl-terminal processing protease for precursor D1protein. *J Biol Chem* 2001; 276: 7518–7520.

90. Stadtman ER. Covalent modification reactions are marking steps in protein turnover. *Biochemistry* 1990; 29:6323–6331.

91. Moreno J, Peñarrubia L, Garcia-Ferris C. The mechanism of redox regulation of ribulose-1,5-bisphosphate carboxylase/oxygenase turnover. A hypothesis. *Plant Physiol Biochem* 1995; 33:121–127.

92. Desimone M, Henke A, Wagner E. Oxidative stress induces partial degradation of the large subunit of ribulose-1,5-bisphosphate carboxylase/oxygenase in isolated chloroplasts of barley. *Plant Physiol* 1996; 111:789–796.

93. Desimone M, Wagner E, Johanningmeier U. Degradation of active-oxygen-modified ribulose-1,5-bispho-

sphate carboxylase/oxygenase by chloroplastic proteases requires ATP-hydrolysis. *Planta* 1998; 205:459–466.

94. Ishida H, Nishimori Y, Sugisawa M, Makino A, Mae T. The large subunit of ribulose-1,5-bisphosphate carboxylase/oxygenase is fragmented into 37-kDa and 16-kDa polypeptides by active oxygen in the lysates of chloroplasts from primary leaves of wheat. *Plant Cell Physiol* 1997; 38: 471–479.

95. Ishida H, Shimizu S, Makimo A, Mae T. Light-dependent fragmentation of the large subunit of ribulose-1,5-bisphosphate carboxylase/oxygenase in chloroplasts isolated from wheat leaves. *Planta* 1998; 204:305–309.

96. Ishida H, Makino A, Mae T. Fragmentation of the large subunit of ribulose-1,5-bisphosphate carboxylase by reactive oxygen species occurs near gly-329. *J Biol Chem* 1999; 274:5222–5226.

97. Kokubun N, Ishida H, Makino A, Mae T. The degradation of the large subunit of ribulose-1,5-bisphosphate carboxylase/oxygenase into the 44-kDa fragment in the lysates of chloroplasts incubated in darkness. *Plant Cell Physiol* 2002; 43:1390–1395.

98. Stieger PA, Feller U. Requirements for the light-stimulated degradation of stromal proteins in isolated pea (*Pisum sativum* L.) chloroplasts. *J Exp Bot* 1997; 48:1639–1654.

99. Palatnik JF, Carrillo N, Valle EM. The role of photosynthetic electron transport in the oxidative degradation of chloroplastic glutamine synthetase. *Plant Physiol* 1999; 121:471–478.

100. Tyystjärvi E, Aro E-M. The rate constant of photoinhibition, measured in lincomycin-treated leaves, is directly proportional to light intensity. *Proc Natl Acad Sci USA* 1996; 93:2213–2218.

101. Anderson B, Barber J. Mechanisms of photodamage and protein degradation during photoinhibition of photosystem II. In: Baker NR, ed. *Photosynthesis and the Environment*. Dordrecht: Kluwer, 1996:101–121.

102. Miyao M, Ikeuchi M, Yamamoto N, Ono T. Specific degradation of the D1 protein of photosystem II by treatment with hydrogen peroxide in darkness: Implications for the mechanism of degradation of the D1 protein under illumination. *Biochemistry* 1995; 34:10019–10026.

103. Kettunen R, Tyystjärvi E, Aro E-M. Degradation pattern of photosystem II reaction center protein D1 in intact leaves. *Plant Physiol* 1996; 111:1183–1190.

104. Ishikawa Y, Nakatami E, Henmi T, Ferjani A, Harada Y, Tamura N, Yamamoto Y. Turnover of aggregates and cross-linked products of the D1 protein generated by acceptor-side photoinhibition of photosystem II. *Biochim Biophys Acta* 1999; 1413:147–158.

105. Ferjani A, Abe S, Ishikawa Y, Henmi T, Tomokawa Y, Nishi Y, Tamara N, Yamamoto Y. Characterization of the stromal protease(s) degrading the cross-linked products of the D1 protein generated by photoinhibition of photosystem II. *Biochim Biophys Acta* 2001; 1503:385–395.

106. Taylor A, ed. *Aminopeptidases*. Austin, TX: RG Landes, 1996.

107. Herbers K, Prat S, Willmitzer L. Functional analysis of a leucine aminopeptidase from *Solanum tuberosum* L. *Planta* 1994; 194:230–240.

108. Gu Y-Q, Chao WS, Walling LL. Localization and post-translational processing of the wound-induced leucine aminopeptidase proteins of tomato. *J Biol Chem* 1996; 271:25880–25887.

109. Gu Y-Q, Walling LL. Specificity of the wound-induced leucine aminopeptidase (LAP-A) of tomato activity on dipeptide and tripeptide substrates. *Eur J Biochem* 2000; 267:1178–1187.

110. Tu C-J, Park S-Y, Walling LL. Isolation and characterization of the neutral leucine aminopeptidase (*LapN*) of tomato. *Plant Physiol* 2003; 132:243–255.

111. Giglione C, Serero A, Pierre M, Boisson B, Meinnel T. Identification of eukaryotic deformylases reveals universality of N-terminal processing mechanisms. *EMBO J* 2000; 19:5916–5929.

112. Desimone M, Kruger M, Wessel T, Wehofsky M, Hoffman R, Wagner E. Purification and characterization of an aminopeptidase from the chloroplast stroma of barley leaves by chromotographic and electrophoretic methods. *J Chromatogr B Biomed Sci Appl* 2000; 737:285–293.

113. Laing WA, Christeller JT. A plant chloroplast glutamyl proteinase. *Plant Physiol* 1997; 114:715–722.

114. Zhang L-X, Wang J, Wen J-Q, Liang H-G, Du L-F. Purification and partial characterization of a protease associated with photosystem II particles. *Physiol Plant* 1995; 95:591–595.

115. Reinbothe C, Apel K, Reinbothe S. A light-produced protease from barley plastids degrades NADPH: protochlorophyllide oxidoreductase complexed with chlorophyllide. *Mol Cell Biol* 1995; 15:6206–6212.

116. Reinbothe S, Reinbothe C, Holtorf H, Apel K. Two NADPH: protochorophyllide oxidoreductases in barley: evidence for the selective disappearance of PORA during the light-induced greening of etiolated seedlings. *Plant Cell* 1995; 7:1933–1940.

117. Reinbothe S, Reinbothe C. Regulation of chlorophyll biosynthesis in angiosperms. *Plant Physiol* 1996; 111:1–7.

118. Lindahl M, Yang D-H, Andersson B. Regulatory proteolysis of the major light-harvesting chlorophyll *a/b* protein of photosystem II by light-induced membrane-associated enyzmic system. *Eur J Biochem* 1995; 231:503–509.

119. Yang D-H, Webster J, Adam Z, Lindahl M, Andersson B. Induction of acclimative proteolysis of the light-harvesting chlorophyll *a/b* protein of photosystem II in response to elevated light intensities. *Plant Physiol* 1998; 118:827–834.

120. Anastassiou R, Argyroudi-Akoyunoglou JH. Thylakoid-bound proteolytic activity against LHCII apoprotein in bean. *Photosynth Res* 1995; 43:241–250.

121. Tziveleka AL, Argyroudi-Akoyunoglou JH. Implication of a developmental-stage-dependent thylakoid-bound protease in the stabilization of the

light-harvesting pigment-protein complex serving photosystem II during thylakoid biogenesis in red kidney bean. *Plant Physiol* 1998; 117:961–970.

122. Tziveleka AL, Argyroudi-Akoyunoglou JH. Cations control the association of a stroma protease to thylakoids. Involvement of the proteolytic activity in "low-salt"-induced grana unstacking and pigment-protein complex organization? In: Mathis P, ed. *Photosynthesis: From Light to Biosphere*, Vol. III. Dordrecht: Kluwer, 1995:835–838.

123. Lensch M, Herrman, RG, Sokolenko A. Identification and characterization of SppA, a novel light-inducible chloroplast protease complex associated with thylakoid membranes. *J Biol Chem* 2001; 276:33645–33651.

124. Georgakopoulos JH, Sokolenko A, Arkas M, Sofou M, Herrmann RG, Argyroudi-Akoyunoglou JH. Proteolytic activity against the light-harvesting complex

and the D1/D2 core proteins of photosytem II in close association to the light-harvesting complex II trimer. *Biochim Biophys Acta* 2002; 1556:53–64.

125. Adamska J, Kloppstech K, Ohad I. Early light-inducible protein in pea is stable during light-stress but is degraded during recovery at low light intensity. *J Biol Chem* 1993; 268:5438–5444.

126. Adamska J, Lindahl M, Roobol-Bóza M, Andersson B. Degradation of the light-stress protein is mediated by an ATP-independent, serine-type protease under low-light conditions. *Eur J Biochem* 1996; 236:591–599.

127. Adamska J. ELIPs — Light-induced stress proteins. *Physiol Plant* 1997; 100:794–805.

128. Preiss S, Schrader S, Johanningmeier U. Rapid, ATP-dependent degradation of a truncated protein in the chloroplast. *Eur J Biochem* 2001; 268:4562–4569.

14 Supramolecular Organization of Water-Soluble Photosynthetic Enzymes along the Thylakoid Membranes in Chloroplasts

Jayashree K. Sainis
Molecular Biology Division, Bhabha Atomic Research Center

Michael Melzer
Department of Molecular Cell Biology, Institute of Plant Genetics and Crop Plant Research

CONTENTS

I. INTRODUCTION

The concept that cells are bags full of freely diffusing macro- and micromolecules, where the precise chemical reactions occur by unintended encounter of these molecules, forms the basis of modern biochemistry and molecular biology. However, this view has been seriously questioned many times. The problems of protein concentration, solvation capacity and evidence for enzyme–enzyme interaction have given rise to the proposal that sequential enzymes of a metabolic pathway should exist *in vivo* as loosely organized complexes which remain associated with subcellular structures as well as with membranes. The existence of such supramolecular organization among sequential enzymes should result in channeling and facilitate delivery of substrates, ensuing the efficiency of the metabolic processes in water-limited and protein-crowded environment *in vivo* [1]. It is also known that intermediates of different metabolic pathways,

ATP, NADPH, GTP, and several other micromolecules in cells exist in separate pools, which explains why enzymes exhibiting differences in affinities for identical substrates can function efficiently in the same cell. However, it has been extremely difficult to prove the concept that the soluble macro- and micromolecules are spatially organized in the cell with the prevailing technologies in biochemistry and molecular biology. In fact, supramolecular organization and its metabolic significance has habitually been a subject of debate [2]. Though, recently scientific world seems to be realizing the importance of such kind of organization [3].

Several efforts were made in the past to demonstrate the organization of sequential enzymes in different metabolic pathways. Noteworthy results for the Krebs cycle [4], glycolysis [5], Calvin cycle [6], and several other pathways [7] have been obtained. These studies were done using conventional methods such as copurification, kinetic analyses, cross-linking for nearest neighbor analysis, fluorescence measurements, and countercurrent distribution etc. [6–8].

The importance of protein–protein interactions is becoming apparent in the postgenomic-proteomic era. Recent advances in biotechnology have opened up new high throughput technologies to study interaction among proteins. The important among these are use of yeast two-hybrid system, mass spectrometry of purified complexes, FRET assays, correlated expression of genes, genetic interactions, and in silico analysis, etc. Most of these studies have been done in the eukaryotic model system of yeast where over 80,000 interactions between proteins have been predicted [9]. The comparative protein–protein interaction maps have been obtained by developing a technology involving tandem affinity purification and mass spectrometry in high throughput approach to characterize supramolecular complexes in *Saccharomyces cerevisiae* in order to study functional organization of the yeast proteome by systematic analysis of protein complexes [10,11]. Since it is now slowly realized that most of the cellular process are carried out by multiprotein complexes either permanent or transient, scientists are getting engaged in evolving newer methods to study such interactions especially in models systems such as yeast or higher eukaryotes including some mammals [3].

In contrast, information on supramolecular organization of enzymes in plant metabolic pathways is relatively limited [12] and is mostly confined to components of the electron transport chains. It is noteworthy that the ultimate aim of research in plant biochemistry revolves around the idea of improving the efficiency of the metabolic processes including photosynthesis in crop plants. It is, therefore, crucial that regulation of essential metabolic pathways *in situ* needs to be understood for any ingenious genetic reconstruction to alter metabolic pathways *in vivo* in plants. Such study will, therefore, require understanding of the supramolecular organization of enzymes of different metabolic pathways. In this article we present evidence for the need for organization among soluble enzymes in stroma and describe the data gathered by different groups working on supramolecular organization among Calvin cycle enzymes and their distribution along the thylakoid membranes.

II. PROTEIN-CROWDED ENVIRONMENT IN THE CHLOROPLAST MATRIX

The chloroplast matrix (stroma) encompasses all the structures and molecules located between inner envelope membrane and thylakoid membranes. Most of these molecules are water soluble. Enzymes of Calvin cycle (reductive pentose phosphate pathway) especially Ribulose bisphosphate carboxylase/oxygenase (Rubisco) mainly contribute to the bulk of protein in the stroma of C-3 chloroplasts. Besides, the enzymes of other metabolic pathways such as oxidative pentose–phosphate cycle, nitrite reduction and ammonia assimilation, amino acid biosynthesis, fatty acid biosynthesis, ribosomes, and the entire protein synthesis machinery, multiple copies of DNA, ribosomes, mRNA, starch grains, and platsoglobuli are also present in stroma. Conventionally, all these proteins and the metabolites present in stroma are considered to be freely mobile and evenly distributed through out the chloroplast matrix in green plants and algae. The protein concentration of the chloroplast matrix is around 400 mg/ml. Out of this around 250 mg/ml alone are due to Rubisco [13]. This concentration is similar to that found in Rubisco crystals [14]. Since such high protein concentrations are difficult to be attained even *in vitro*, a tight packing of stromal enzymes has to be assumed. Employing cryo-scanning electron microscopy at $-190°C$, dense packing of large stroma protein clusters can be actually observed on extended areas of chloroplast matrix, which do not possess thylakoid membranes (Figure 14.1). These protein clusters lead to the formation of solvent channels, which may allow specific translocation of proteins, ions, and metabolites. This suggests that the random mobility of proteins within stroma may be very low, if at all possible. Interestingly, it was observed in *Avena sativa* that when chloroplasts lose water during wilting, Rubisco molecules aggregate in stroma to form whorl-like masses of tightly packed fibrils termed as stroma centers [15]. Sprey [16] also had observed crystalline Rubisco in water-stressed spinach leaves. Although the principles governing the distribution and supramolecular organization of stroma proteins still remain to be elucidated, these ultrastructural observations point to the existence of properly arranged enzyme clusters rather than that of randomly distributed water-soluble chloroplast proteins (K.H. Süss, personal communication, 2000). The prevailing doubts about the supramolecular organization of Calvin cycle enzymes are mainly due to the experience that stromal proteins are easily released into solution by osmotic shock of intact chloroplasts and can be subsequently purified to homogeneity following salt treatments. It should be noted that osmotically shocked chloroplasts do not carry out CO_2 fixation, unless supplied by RuBP [17]. Though significant progress has been made in elucidation of structure of pigment–protein complexes involved in electron transport chain of chloroplast thylakoid membranes [18], the concept of supramolecular organization and

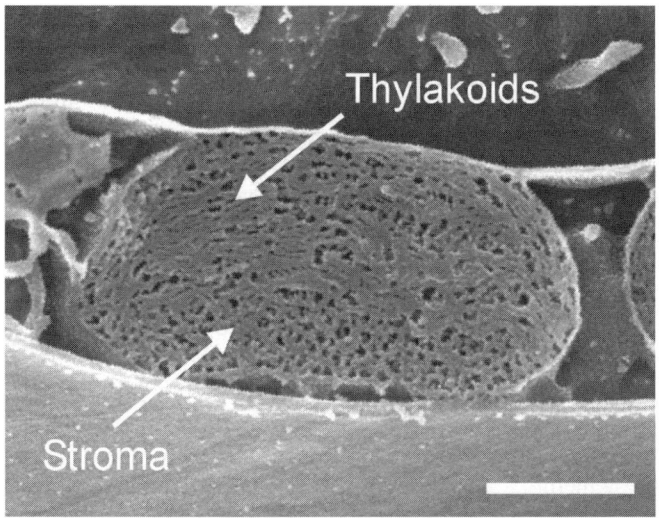

FIGURE 14.1 Cryo-scanning electron micrograph of a tobacco chloroplast. Freeze fracturing after rapid freezing of a piece of leaf to expose the substructure of the chloroplast. Note the highly packed protein clusters. Proteins present in these clusters will be highly immobile and may not diffuse randomly in chloroplasts. The network of water-filled channels may be important for translocation of proteins, water and solutes. Bar = 1 μm.

coupling of electron transport components with soluble enzymes in stroma is never mentioned or considered. The conviction that substrate specificity of sequential enzyme reactions is sufficient to maintain metabolic pathways in the crowded atmosphere in stroma sustains the universal philosophy in biochemistry which does not take into consideration the fact that same substrate is shared by various enzymes of different metabolic pathways concurrently. By not considering the possibility that enzyme pairing at high protein concentrations may be lost upon aqueous dilution of stromal proteins, due to disintegrating action of aqueous dipoles, the research on regulation of chloroplast metabolism remained grounded in chemistry of dilute solutions and assumption of freely diffusing molecules [19]. Thus, as acknowledged by Ellis [20] macromolecular crowding, though perceptible, remains mostly unacknowledged by most of the biologists, biochemists and biotechnologists.

III. VOLUME CHANGES IN CHLOROPLASTS

Chloroplasts not only have high concentration of macromolecules and limited solvation capacity, but they also show light-induced structural changes. Chloroplast volume changes were demonstrated *in vitro* using particle volume counters such as coulter counter and in living cells by scattering of light beams of 546 nm [21]. These and related studies showed that chloroplasts have larger volume in dark while illumination decreases their volume by 20% to 40%. The half time for these changes is 3 min with concomitant increase in rate of photoassimilation [22]. It was observed that in general chloroplasts are more spherical in dark and flatten in light resulting in increase in concentration of metabolites and enzyme active sites in light. Chloroplasts were shown to selectively allow

free movement of water across the membranes but restrict movement of solutes like sugars, amino acids, intermediates of Calvin cycle, etc. Light was shown to cause extrusion of K^+ and Cl^- from chloroplasts along with efflux of 32% of free water. Light also induces changes in ultrastructure of thylakoid membranes [23]. Inhibitors of electron transport and phosphorylation were shown to inhibit light-induced changes, indicating a role for these processes in structural changes of chloroplast volume.

Such volume changes in chloroplasts, resulting in selective extrusion of water will also be important in structural reorganization of enzyme systems in light and dark and are expected to play an important role in the regulation of assimilation of light energy by photosynthetic machinery in chloroplasts. Crowding of macromolecules will be much higher in light and free water may be highly limited in chloroplasts during operation of Calvin cycle. Thus, mere substrate specificity of cognate enzyme may not be sufficient to maintain activities of the various metabolic pathways in chloroplast stroma especially in face of such drastic volume alterations in light. Supramolecular organization will be the key to the precision of metabolism in these circumstances.

IV. ISOLATION AND CHARACTERIZATION OF CALVIN CYCLE MULTIENZYME COMPLEXES

Among the metabolic pathways occurring in chloroplasts, Calvin cycle is the most studied one and several groups have investigated the supramolecular organization among the sequential enzymes of this cycle. The existence of a CO_2 fixing complex was predicted in early 1960s especially when it was observed that some enzymes of photosynthetic carbon reduction cycle remain associated even after isolation [24,25]. Muller [26] had suggested in 1972 that some of the CO_2 fixing enzymes might be associated in the form of a labile complex. However, with the advances in protein purification techniques, several publications on purified enzymes accumulated and these initial observations and hypotheses were pushed into oblivion. Regardless of the ever increasing knowledge about several purified enzymes of Calvin–Benson cycle, the answer to the key question in plant biology "What controls rates of photosynthesis?" remained elusive [27]. Considering the discrepancies in the *in vitro* and *in vivo* conditions, several laboratories sought to revive the investigations in multienzyme organization among stromal enzymes of chloroplasts. Since the interactions among the soluble enzymes can be generally weak and transient, the study of their supramolecular organization is often tricky and at times perplexing. The multimolecular associations among Calvin cycle enzymes have now been discovered by a variety of procedures. Sainis and Harris [28,29] observed that Rubisco fractions isolated on sucrose density gradient showed R-5-P + ATP-dependent carboxylase activity. This indicated that phosphoriboisomerase (RPI) and phosphoribulokinase (RPK) must be copurifying with Rubisco. Later it was observed that almost all the RPK activity was associated with carboxylase on sucrose gradients. However, if the stromal extracts were precipitated with ammonium sulfate prior to density gradient centrifugation, RPK was dissociated from the complex. Rubisco purified using method of PEG precipitation was also associated with RPI and RPK [30].

Gontero et al. [31] have purified a functional five enzyme complex of the consecutive enzymes of Calvin cycle, viz., RPI, RPK, Rubisco, PGK and GAPDH by using DEAE Tris-acryl, Sephadex G-200, and hydroxyapaptite. The homogeneity of the complex was tested by analytical centrifugation. Studies on the structural and functional properties of this multienzyme complex from spinach chloroplasts indicated that the phosphoribulokinase, which was inserted in the complex, showed reduced autooxidation [32]. Analysis of this complex showed that the quaternary structure of the enzymes in the complex was different than that reported for isolated and purified enzymes [33]. Kinetic investigation showed that the enzymes in the complex had higher V_{MAX} and lower K_M [34]. Based on the statistical thermodynamics, interactions among these enzymes in the complex have been shown to exert stabilization, modulation of their reaction rates and result in information transfer of their altered kinetic parameters. Gontero et al. had reported that these effects on conformation stabiliza-

tion in the complexes are unusual as compared to the standard effects on channeling of intermediates in multienzyme system [35,36].

Later, this group used *Chlamydomonas* chloroplasts to isolate and purify a bienzyme complex of RPK and GAPDH, which are the nonsequential enzymes in photosynthetic carbon reduction cycle [37,38]. In 1991, Nicholson et al. [39] had also reported a stable complex between GAPDH and RPK from chloroplasts. These two enzymes remain associated and influence each other's kinetic properties. Stabilization or destabilization of the complex is produced by conformational changes generated by protein–protein interaction and results in creating imprints of their association. Both the enzymes were found to carry memory of these imprints even after dissociation, which was studied using thermodynamics of the conformational changes, resulting in alteration of kinetic properties with respect to cofactors [37,38,40,41]. Mouche et al. in 2002 [42] used a multitechnique approach to study multienzyme complex of GAPDH and RPK. The dimers of RPK are supposed to undergo a remarkable change in the activity during binding and detaching from GAPDH. The authors have reported striking structural changes in the isolated and modeled RPK dimer and the counterpart in the three-dimensional reconstruction volume of whole complex, obtained using cryoelectron microscopy and image processing. This bienzyme complex uses ATP and NADPH produced by the primary reactions in photosynthesis. The authors envisage that this bienzyme complex may allow concerted regulation of two enzymes as "Unit Control" — a starting point for the regulation of Calvin cycle by light, pH, and metabolites. Thus, protein–protein interactions may provide for a fine control of Calvin cycle [40].

Several other groups have also worked on the multienzyme organization among Calvin cycle enzymes. Persson and Johansson [43] had reported partition behavior of six Calvin cycle enzymes using countercurrent distribution in the aqueous two-phase system that suggested a trend to exist as a protein–protein complex among these enzymes. The enzymes involved were Rubisco, PGK, GAPDH, TPI, aldolase, and FBPase. Association between RPI and RPK has also been predicted from the kinetic studies [44]. This was further confirmed by studies with countercurrent distribution and copurification [45,46].

Süss et al. [47] were able to isolate a multienzyme complex containing RPI, RPK, Rubisco, GAPDH, sedoheptulose-1,7-bisphosphatase, and also ferredoxin NADP reductase (FNR) on FPLC using molecular sieve and anion exchange chromatography. The multienzyme complex had a molecular weight of 900 kDa and accommodated 80% of RPK and GAPDH and also catalyzed R-5-P + ATP-dependent CO_2 fixation.

Thus, there is adequate data from these *in vitro* studies to demonstrate that many of the sequential Calvin cycle enzymes can be isolated as supramolecular complexes. The differences in the constituent enzymes among various complexes isolated *in vitro* may be ascribed to the variations in extraction and purification procedures employed and also to the loose-fitting as well as dynamism in multimolecular associations. Therefore, the exact stoichiometric ratios of the components cannot be predicted. The complex of Calvin cycle enzymes showed 530-kDa band on nondenaturing polyacrylamide gels that cross-reacted with antibodies against Rubisco, RPK, and GAPDH. The densitometric analysis of Coomassie blue-stained polypeptides suggested that there are two RPK, two subunits of RPI, two large and four small subunits of Rubisco, along with one subunit of PGK, and two subunits of RPI in enzyme complex [33]. Hosur et al. [48] attempted to crystallize the complex which shows R-5-P + ATP-dependent CO_2 fixation activity. Preliminary X-ray diffraction analysis of such crystals showed that besides normal L_8S_8 form of Rubisco these crystals have extra density, which may be due to the other protein in the complex.

The stoichiometric ratios of the component enzymes are vital for the formation of supramolecular complexes even *in situ*. However, they have been somewhat elusive. The protein complexes comprising stromal and membrane bound enzymes are probably arranged as protein networks rather than as a "chaotic" random distribution of single-enzyme components *in vivo*. Since multienzyme complexes tend to dissociate into their constituents when dissolved in aqueous media, only the most stable enzyme aggregates may sustain the unphysiological aqueous conditions after extraction and can be further isolated and characterized. Additionally, it is not known how the rates of synthesis and degradation of these enzymes are coordinated in the chloroplasts in response to various physical and physiological factors. However, an analysis of the protein composition of chloroplast stroma extracts and thylakoid membranes of spinach did not result in any qualitative and quantitative differences in the protein composition [6] when simple analysis by SDS–urea PAGE was carried out for plants of the same variety grown under different environmental conditions in field.

Isolation of stromal enzymes is no proof, however, that Calvin cycle enzyme complex do not represent isolation artifact caused by uncontrolled aggregation of partially unfolded enzymes in the course of several experimental manipulations. A sensitive test is therefore necessary to reveal complexation of enzymes in fresh aqueous organelle extracts prior to any other manipulation. Süss et al. [47] employed a limited proteolysis combined with immunoblotting to demonstrate enzyme pairing in solution. This method is based on the assumption that complementary protein interfaces, while perhaps accessible and cleaved by specific proteases in the case of isolated enzymes, should not be susceptible to proteolysis if these components are organized into multienzyme system. Trypsin, which specifically cleaves arginine and lysine residues, has proved suitable for this purpose and was successfully used to show that RPK and GAPDH are complexed in freshly prepared stromal extracts. Moreover, dissociation of Calvin cycle enzyme complexes into their components at different ionic strengths could be followed by employing the same technique [47]. It was obvious from these results that any treatment of stromal extracts, including aqueous dilution, may cause dissociation of Calvin cycle multienzyme complexes. This provides the explanation why high and low molecular mass forms of Calvin cycle enzyme complexes have been isolated from same extracts.

Like the Calvin cycle, the sequential enzymes of other metabolic pathways also have to be organized in chloroplasts so that all metabolic reactions occur in a coordinated manner to make efficient use of energy and reducing power generated by light reaction. However, not much information is available about the organization of enzymes in other metabolic pathways.

V. SUBSTRATE CHANNELING AND ADVANTAGES OF ORGANIZED STATE

Supramolecular organization is considered to result in substrate channeling, which will have obvious advantage in crowded atmosphere *in vivo*. Substrate channeling between components of CO_2 fixing complexes of Calvin cycle enzymes has been observed in vitro [49,50]. The multienzyme complex containing RPI, RPK, and Rubisco shows R-5-P + ATP-dependent CO_2 fixation activity. The observed transient time for the above linked reaction is much less than that expected from the K_M and V_{MAX} of individual enzymes. The rates of R-5-P + ATP-dependent reactions are 70% to 80% of RuBP dependent rates even though free RuBP concentration is very low. R-5-P + ATP-dependent activity is stable and linear for much longer time unlike RuBP-dependent activity. The dark-inactivated Rubisco from bean leaves was found to be activated *in vitro* in R-5-P + ATP-dependent assay [51]. Kinetic analysis of this complex by Gontero et al. in 1993 [34] showed that catalytic efficiency of Rubisco is increased when it is present in the complex, which may be due to an increase in V_{MAX} per active site and a decrease in K_M.

Rubisco shows a progressive decrease in activity during catalysis, which is called fall-over due to the production of catalytic inhibitor [52]. In R-5-P + ATP-dependent assay this fall-over was not observed [50]. It is highly unlikely that the rapid inactivation of Rubisco observed during catalysis in the *in vitro* assays could be tolerated in the chloroplast during the active periods of photosynthesis. In fact, it has been shown that Rubisco does not undergo any inactivation *in vivo* [53]. Hence, such fall-over may be an *in vitro* artifact. RPK in the complex can be more rapidly activated by reduced ferredoxin as compared to free RPK indicating fine-tuning of regulation of enzyme activity [34].

Proteins are known to bind water molecules. It is known that the properties of water molecules *in vivo* are different compared to normal water molecules [54]. Pulsed NMR studies have revealed that water molecules *in situ* in leaves and chloroplasts experience severe restriction in mobility [49] as compared to aqueous buffers used for enzyme assays. Such aqueous environment results in almost near-crystalline state for several enzymes *in vivo*. The *in vivo* state of water molecules can be simulated *in vitro* by addition of water-binding macromolecules to the enzyme assay mixtures of the multienzyme complex of Calvin cycle enzymes. Channeling was found to offer advantage to the sequential enzymes when diffusion was restricted *in vitro* due to hydrophilic macromolecules [49] in enzyme assay mixtures.

Several metabolic pathways share the same intermediates and operate simultaneously *in vivo* at a given time in such crowded environment. The microcompartmentation of pathways and channeling of substrates among the sequential enzymes is very essential and obvious. This will minimize competition for common substrates and cofactors and efficient functioning of these metabolic pathways. The organization and networking of Calvin cycle enzymes with thylakoid membranes will aid in accessibility of NADPH and ATP produced by electron transport to the respective enzymes and will avoid the problem of nonspecific binding of intermediates [55].

VI. ARE THERE TWO POPULATIONS OF RUBISCO?

It is now realized that some proteins can have functions, other than those assigned to them traditionally by enzymology. This phenomenon is currently described as moonlighting. Rubisco is the most abundant enzyme present in chloroplast stroma of C-3 plants. Some of the Rubisco molecules may be moonlighting and can get engaged in activities other than their normal role in Calvin cycle. Analogous to the dynamic behavior of hemoglobin molecules, these Rubisco molecules along with Rubisco activase and carbonic anhydrase may function in a CO_2 concentrating mechanism. The function of Rubisco activase may be to keep active conformation of Rubisco to bind CO_2 and Mg^{2+}. This is consistent with the view that Rubisco activase binds close to loop 6, which represents a flexible domain of catalytic large subunit of Rubisco. Perhaps bound CO_2 is further translocated from catalytic site along the interdimer interface of large subunit towards the solvent filled channel, which is 1.5 nm and extends through the center of the Rubisco molecule (K.H. Süss, personal communication, 2001). The abundance of Rubisco molecules over and above that needed for photosynthetic carbon reduction has been suggested by the fact that in tobacco transgenic plants where Rubisco amount was decreased by over 50%, rates of CO_2 fixation were not affected at low light intensities [56]. At higher intensities, CO_2 concentrating mechanisms become rate limiting for realizing higher carboxylation efficiencies. In nature, C-4 plants have managed very high CO_2 fixation rates even at high light intensities with much lesser amounts of carboxylases, albeit with an alternative CO_2 concentrating mechanism. If C-3 plants are grown at higher CO_2 concentrations, the amount of Rubisco in leaves decreases [57]. It may be that CO_2-concentrating mechanism becomes redundant at higher CO_2 concentrations. However, more direct proof for the role of Rubisco as a tool for concentrating CO_2 is awaited. Another interesting observation on the transgenic tobacco lines is that under high nitrogen conditions, in the normal wild type tobacco plants the amount of Rubisco increases, whereas in the tobacco transgenics, the excess of nitrogen is stored in other proteins. Rubisco thus may be acting as nitrogen-store under excess nitrogen conditions [58].

Two forms of Rubisco differing in *in situ* localization [59] have been shown in chloroplasts of green algae. In chloroplasts of *Chlamydomonas reinhardtii* one of the forms is associated with thylakoid membranes and inner surface of pyrenoid tubules, whereas another enzyme form is confined to the crystalline pyrenoid matrix. Analysis of multienzyme complex of Calvin cycle enzymes also has suggested a possibility of two populations of Rubisco differing in their subunit composition [33]. These results have demonstrated that the soluble enzymes can exist in complexed and uncomplexed forms and may serve different functions depending on the need.

VII. ASSOCIATION OF WATER-SOLUBLE ENZYMES WITH THE THYLAKOID MEMBRANES

Protein extraction studies have indicated partial binding of Rubisco and non abundant Calvin cycle enzymes FBPase, RPK, GAPDH, PGK, and RPI to chloroplast membranes [60–62]. It was also observed that H^+ ATP synthase may be a possible membrane attachment site for Rubisco [62]. However, in the absence of information on binding of Calvin cycle enzymes *in situ*, the possibility of nonspecific enzyme adsorption to thylakoid membranes cannot be ruled out. This problem was partially solved by employing immunoelectron microscopy of cryo-fixed and cryo-substituted leaf sections. Earlier immunochemical studies had revealed a random distribution of Rubisco molecules throughout the chloroplast matrix *in situ* [63,64]. Other Calvin cycle enzymes were also thought to be evenly distributed. Unexpectedly, immunoelectron microscopy on cryo-fixed ($-185°C$) and cryo-substituted preparations called this assumption into question and showed that nonabundant enzymes of Calvin cycle like RPK, GAPDH, SBPase, and FNR, the terminal electron transport enzyme, are predominantly associated with nonappressed thylakoid membranes [47]. Moreover, predominant thylakoid binding of RPK in microalgae [65] and of FBPase and thioredoxin has been shown [66]. The same *in situ* localization for Calvin cycle enzymes RPI, FBPase, aldolase, and PGK has been observed in chloroplasts of green alga *Chlamydomonas reinhardtii* [67]. Since spatial distribution of Calvin cycle enzymes does not differ in illuminated and darkened leaves, it has been inferred that light-dependent energization–deenergization of thylakoid membranes is not accompanied by spatial relocalization of stromal enzymes *in situ*. These observations anticipate that Calvin cycle enzymes may be permanently located with nonappressed thylakoid membranes. It should be emphasized that this is also true of those Calvin cycle enzymes, which did not coisolate with CO_2-fixing multienzyme complexes [68]. The ferredoxin–thioredoxin reductase complex is responsible for the light-dependent activation of several Calvin cycle enzymes and H^+-ATP synthase. The complex which is also localized at the surface of thylakoid membranes [47,67], close to the Calvin cycle or even more likely, is itself a constituent of Calvin cycle multienzyme complex *in vivo*.

Melzer, Süss, and Sainis (unpublished) used immunoelectron microscopy to study the *in situ* localization of several soluble enzymes in the dimorphic chloroplasts of maize leaves. Antibodies against Rubisco, Rubisco activase, RPI, RPK, aldolase, pentose-5-phosphate 3-epimerase, NADP-malic enzyme, pyruvate phosphate dikinase (PPDK), PGK, GAPDH, transketolase, H^+-ATP synthase, FNR, chaperonin 60, ribosomes, DNA ligase, and glutamate-1-semialdehyde aminotransferase (heme synthesis) were used along with protein A–gold labeling. Figure 14.2 shows the expected labeling pattern for antibodies against the large subunit of Rubisco with specific signals only in bundle sheath chloroplasts. Interestingly around 80% of the gold particles for Rubisco in bundle sheath chloroplasts could be located along the thylakoid membranes (Figure 14.3). The same or even higher percentage of preferential location adjacent to thylakoid membranes could be shown for all other soluble chloroplast enzymes.

The *in situ* localization of Calvin cycle enzymes suggests that most of the reactions of the Calvin cycle may be occurring close to the membranes. It is unlikely that Calvin cycle intermediates will be diffusing randomly in the stroma to find the sequential enzymes of this cycle. If the existence of stroma protein clusters is taken into consideration, the distance over which free intermediates would have to diffuse will be very large. Moreover, any collision of RuBP and other intermediates of the Calvin cycle with other macromolecules will significantly decrease the diffusion rate and thus slow down the cycle. The supramolecular organization among enzymes of the Calvin cycle will make sense only if the reactions of photosynthetic carbon reduction proceed sequentially in the restricted spaces close to the thylakoid membranes and therefore at least a fraction of enzymes involved in the Calvin cycle will be membrane bound. The isolation of partial Calvin cycle complexes and their association with nonappressed chloroplast thylakoid membranes *in situ* lend support to the idea that enzymes catalyzing photosynthetic dark and light reactions may be organized as supercomplexes *in situ*. The evidence for such enzyme supercomplexes is tentative and the idea of cofactor and metabolite channeling in photosynthesis is still a matter of discussion. However, an association between photosystem I (PSI) and Calvin cycle can be inferred from the observations that FNR is a thylakoid-bound electron transport component *in situ*, but a portion of this enzyme can be coisolated with Calvin cycle multienzyme complexes also comprising of GAPDH [47]. The association of GAPDH, the only enzyme catalyzing NADPH-dependent reduction in Calvin cycle with FNR can be shown by immunoaffinity chromatography of stromal extracts using anti-FNR and anti-GAPDH antibody (C. Arkona and K.H. Süss, unpublished results). The major binding site for FNR at thylakoid membranes was shown to be the E-subunit of PSI [69]. It appears, therefore, that FNR does not

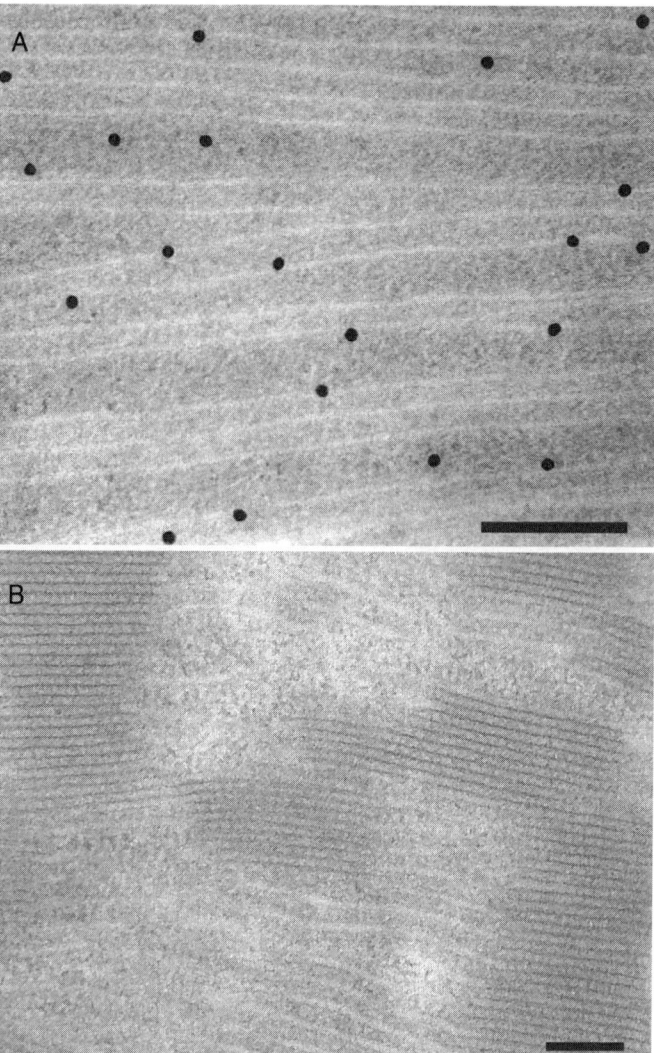

FIGURE 14.2 Transmission electron microscopy. Immunolabeling of a typical leaf thin section from maize with anti-Rubisco IgG and 10 nm protein A–gold. The absence and presence of grana thylakoid membranes in bundle sheath (A) and mesophyll (B) chloroplasts, respectively, is evident. Compared to the high labeling of bundle sheath chloroplasts, mesophyll chloroplasts did not show any significant signals. Bars = 0.1 μm.

simply function as a membrane linker of Calvin cycle complexes to PSI, but in association with GAPDH is thought to enable channeling of the cofactor pair $NADP^+$/NADPH. Hence, the FNR–GAPDH pair along with ATP synthase and kinase may actually represent the linking element between light and dark reactions of photosysnthesis.

A functional connection between PSI and Calvin cycle is also strengthened by the observations on photosynthesizing leaves. *In vivo* measurements have shown that the level of oxidized PSI complexes is probably related to the rate of CO_2 fixation in intact leaves [70].

This strengthens the view that chloroplast metabolism may be performed by a thin enzyme layer on the membranes. Such an enzyme organization may facilitate metabolite and nucleotide channeling between membrane-associated and integral membrane enzymes, and also interconnect metabolic pathways. Membrane-associated enzyme assemblages may also

account for an efficient coupling between photosynthetic electron transport and CO_2 fixation in higher plants. Besides, membrane attachment of interconnected enzymes would limit water diffusion into the enzyme layer and create a more nonpolar environment at catalytic sites to facilitate enzyme catalysis. Indeed, nonaqueous media do significantly increase the stability and turnover number of soluble enzymes. Moreover, the lifetime of enzymes bound to membrane support increases considerably, because it prevents enzyme aggregation and inactivation [71]. The membrane attachment can prevent degradation of the cross-connected enzymes and integral membrane proteins covered by them by housekeeping proteases. It has been shown that surface-exposed lysine and arginine residues of Calvin cycle enzymes are not susceptible to trypsinolysis as long as they are assembled into multienzyme complexes [47,72].

The rate of CO_2 fixation in broken chloroplasts is orders of magnitudes lower, if at all detectable, than in

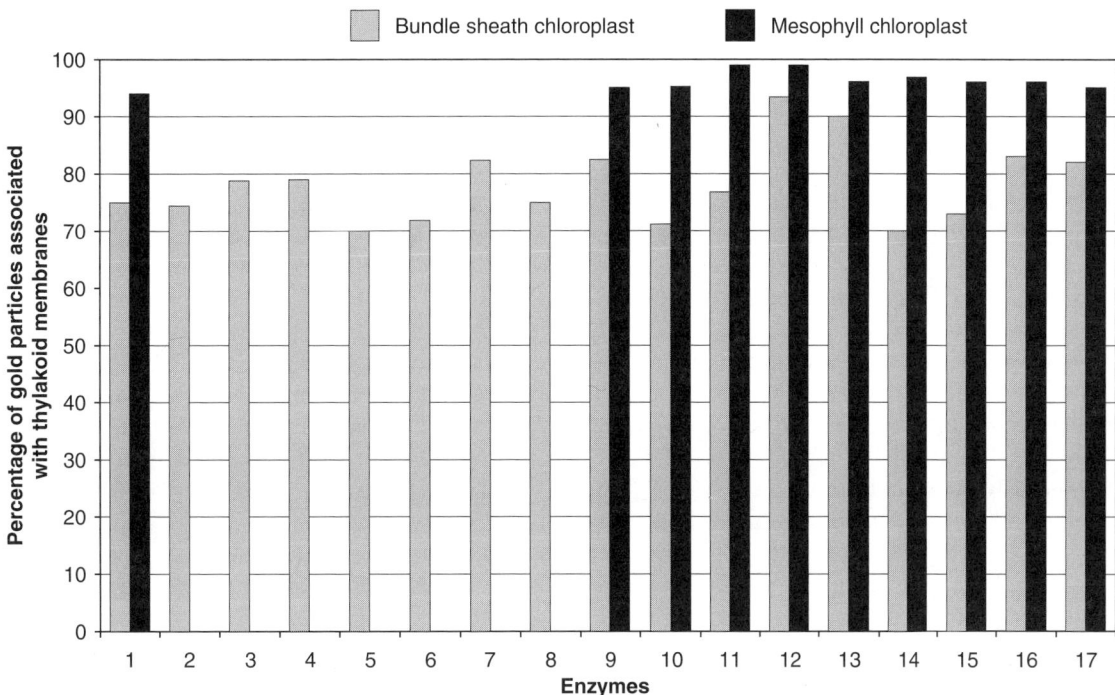

FIGURE 14.3 Immunogold labeling of the chloroplasts of bundle sheath and mesophyll cells in maize leaves. Antibodies against 1: PPDK, 2: NADP ME, 3: RPI, 4: RPK, 5: Rubisco (large subunit), 6: Rubisco activase, 7: aldolase, 8: epimerase, 9: PGK, 10: GAPDH, 11: transketolase, 12: CF1, 13: FNR, 14: ribosomes, 15: DNA ligase, 16: chaperonin 60 and 17: glutamate-1-semialdehyde aminotransferase (heme synthesis) were used for immunolabeling of thin sections of maize leaf samples. The sections were contrasted with uranyl acetate for visualization under TEM. The number of gold particles along the thylakoid membrane were counted and expressed as percent of total.

intact chloroplasts [17]. This indirectly indicates that photosynthetic enzymes are spatially organized *in vivo*. Aqueous media favor dissociation of enzyme assemblages from membranes and disintegration into partial structures and free enzymes [6].

VIII. QUANTASOMES, PHOTOSYNTHESOMES, METABONUCLEONS

In the early years, the search for a morphological and functional photosynthetic unit led to the detection of quantasomes [73]. Electron microscopy was used to visualize the quantasomes in thylakoid membranes and quantasomes were regarded as smallest units, which perform light reactions of photosynthesis. Although the quantasome concept was of fundamental importance, it did not stand the test of time [74]. The supercomplexes of Calvin cycle enzymes along with the components of electron transport system in thylakoid membranes can be termed as photosynthesomes [6]. However, the evidence on the membrane association of several soluble enzymes belonging to

protein biosynthesis, protein folding and DNA metabolism along with Calvin cycle enzymes, solicit for the explanations concerning the organization of whole chloroplast metabolic system. The sequential enzymes of a metabolic pathway may be tightly and orderly packed to form a set of connections on the surface of thylakoid membranes. The peripheral components in these complexes may associate with enzymes of different metabolic pathways, resulting in large supramolecular structures performing all the linked functions in chloroplasts including photosynthesis, DNA and protein synthesis as well as fatty acid synthesis. The term metabolon would be inappropriate to describe these structures because it defines that only enzymes of one pathway are assembled [75,76]. A term Metabonucleon had been proposed to describe the putative superstructures along the thylakoid membranes (K.H. Süss, personal communication, 2002). In the following, we describe the concept of structure and function of Metabonucleon as visualized by Süss in his own words. "A delicate but definite interaction among different enzymes will assemble as Metabonucleon-like structures. The Metabonucleons will perform all the pathways necessary for the synthesis of chloroplast-made proteins and nucleic acids, which concomitantly serve as receptors for chloroplast proteins of nuclear-cytosolic origin. Environmental adaptation and transgenic effects can be coherently explained in terms of multiple, albeit metabolically interdependent chloroplast Metabonucleons. For instance, stress factors such as UV-light causing positive mutations in chloroplast and nuclear genes encoding chloroplast proteins would favor formation of functionally improved offspring Metabonucleons either due to the incorporation of more active enzymes or an advanced spatial arrangement of their components. The later may facilitate metabolite channeling between sequential enzymes and pathways, but also limit degrading processes. In contrast, negative gene mutations will cause offspring Metabonucleons with partially or completely inactive pathways, because enzymes are either not synthesized or assembled as inactive components. In the worst case that DNA nucleoids bound to some, but not those attached to other Metabonucleons are severely damaged, the former entities cannot contribute further to chloroplast biogenesis and will be degraded. Accordingly, Metabonucleon-like structures would enable Darwinian evolution, i.e., evolution that favors the most vital self-reproducing enzyme assemblages and extinguishes the worse. Such superstructures may also account for the maintenance, multiploidy, and maternal inheritance of chloroplast-encoded genes, because the proteins encoded by them serve as receptors for nuclear-encoded chloroplast proteins." The Metaboucleon hypothesis can also provide an explanation for the common observation that the level of a particular soluble enzyme can be drastically lowered by antisense-mRNA expression with only marginal effects on photosynthesis and plant growth [77]. Those entities lacking a particular enzyme are affected in one or more pathways, but may use metabolites set free in the stroma by other Metabolonucleons to perform partial sequences of the affected pathways eventually to reproduce chloroplast components. A thin enzyme layer on the surface of thylakoid membranes can ensure that free metabolites can be made available to the interconnected enzymes. If so, up-regulation of an enzyme by sense-mRNA expression would not improve metabolic pathways, because the principle of interlocking pairing determines the number of potential binding sites for a particular enzyme in Metabonucleons and in turn the quantitative ratios between chloroplast proteins. However, plastid differentiation may require that Metabonucleons are flexible structures that can either loose or adopt enzyme complexes to fulfill their functions. These principles may apply similarly to mitochondria and enzyme assemblages

in other cell compartments. The Metabonucleon hypothesis can be tested by a combination of transgenic, biochemical, and ultrastructural approaches. The formation of Metabonucleons may not only facilitate CO_2 fixation through channeling of cofactor pairs ($NADP^+$/NADPH, ATP/ADP) and enzyme intermediates at least in partial reaction sequences of photosynthesis, but may also cause an enzyme-enclosed microspace where intermediates can accumulate and migrate preferentially among associated enzymes. The FNR system may be localized to a similar microspace to allow for light-mediated activation of several enzymes of Calvin cycle. The hydrophobic environment in the neighborhood of membranes can facilitate the organization and the functioning of enzymes performing sequential reactions.

Thus, the organized system of enzymes will confer several advantages to the living organism such as cofactor recycling, prevention of competition with enzymes of other metabolic pathways for intermediates, synchronization of enzyme turnover rates probably through substrate dependent conformational changes in the enzymes, protection against chemical denaturation as well as uncontrolled proteolytic degradation to increase the biological lifetime of sequential enzymes and precise functioning of sequential reactions in crowded environment *in vivo*.

IX. FUTURE DIRECTIONS IN THE STUDIES ON SUPRAMOLECULAR ORGANIZATION

These experimental findings have revealed a new perspective in research of the mechanism of the regulation of photosynthetic carbon reduction cycle. The reductive pentose phosphate pathway is a unique process in plant anabolism responsible for the assimilation of carbon dioxide and also, in turn, related to the total biomass and productivity. Several attempts have been made to understand the regulation of this pathway, by studying the properties of individual enzymes of this cycle in isolation. Since Rubisco is considered as a "rate liming" enzyme, it has been characterized extensively at protein and genetic level [78]. However, the observations by Quick et al. [58] regarding the redundancy of Rubisco put the theory of "rate-limiting" concept in question. In this context, as aptly discussed by Srere [79], it is now important to reconsider whether the controls for the metabolic processes should still theoretically rest on the activities of singular enzymes of a metabolic pathway. *In vivo*, the supramolecular organizations and microcompartmentalization of metabolites as well as their transport may be playing important part in regulating the metabolic processes. It is, therefore, important to be able to monitor reactions of entire metabolic pathway *in situ* by entering the living cells gently.

Recently, an attempt was made to monitor sequential reactions of the Calvin cycle in cells of *Anacystis nidulans*, which were differentially permeabilized with lysozyme, toluene, toluene–triton, and toluene–triton–lysozyme [80]. Transmission electron microscopy showed that cells permeabilized with only lysozyme or toluene showed the typical concentric arrangement of thylakoid membranes. However, when toluene-treated cells were further treated with triton and lysozyme the thylakoid membranes were disrupted. Sequential reactions of the Calvin cycle were examined in these differentially permeabilized cells *in vivo* by monitoring CO_2 fixation, using various intermediates such as 3-PGA, GA-3-P, FDP, SDP, R-5-P, RuBP and cofactors like ATP, NADPH depending on the requirement. RuBP and R-5-P+ATP-dependent activities could be observed in all types of permeabilized cells. Sequential reactions of the entire Calvin cycle using 3-PGA could be detected only in cells that had retained the internal organization of the thylakoid membranes after permeabilization. The results indicated that integrity of thylakoid membranes might be necessary for the organization as well as functioning of sequential enzymes of the Calvin cycle *in vivo*. The conclusive demonstration of the precise role of thylakoid membranes in the organization of soluble Calvin cycle enzymes may be possible in future when the technology of cryoelectron tomography and use of structural signatures to localize the enzymes *in vivo* will be feasible [81]. The functional significance of organization can be examined in future by generation of precise mutations using site-directed mutagenesis, where membrane location of the Calvin cycle enzymes will be specifically disrupted.

The concept of organization may answer several perplexing questions in biology. The observation that enzymes can exist in complexed and uncomplexed states can explain the irregular molar ratios of sequential enzymes of a metabolic pathway *in vivo*. Supramolecular organization may be helping enzymes to function in the water-restricted, protein-crowded milieu *in vivo* with much too less concentration of intermediates. The metabolic pathways could be controlled by dynamic association–dissociations of these complexes in response to environmental factors. The associations between sequential proteins will need specific structural features for individual proteins, which can explain why there are isozymes and isoforms, which are proteins with simi-

lar catalytic functions, but different structural parameters. The occurrence of multigene families can also be justified by the divergent structural needs to pack the proteins *in vivo* suitably. The enzymes, therefore, will have to be structurally defined by sites, which interface with other proteins, and not merely by their active site residues. The structural compatibility of sequential enzymes may be playing substantial role in the precise regulation of metabolic reactions of living systems.

In short, with minor modifications of our tools and techniques, but with major deviation from the contemporary philosophy of biochemistry and molecular biology, the research in supramolecular complexes will take us nearer to the goal of understanding chemistry of biological systems [3].

ACKNOWLEDGMENTS

This article is dedicated to the memory of Dr. Karl-Heinz Süss who was the co-author of this chapter in the first edition of *Handbook of Photosynthesis* and was deeply involved in the research on supramolecular organization in chloroplasts. We are indebted to him for several ideas and opinions, which he used to share with us freely and frequently, during our never-ending scientific discussions, some of which are included here.

The authors wish to express their sincere thanks to authorities in BARC for the kind permission to write this review and also to Miss Diksha Dani for her help in preparation of manuscript. We gratefully acknowledge Waltraud Panitz and Bernhard Claus for their excellent technical assistance, Dr. Twan Rutten (IPK-Gatersleben) for the SEM work and Prof. Dr. Isabella Prokhorenko (Institute of Basic Biological Problems, Pushtchino, Russia) for a critical reading of the manuscript and helpful discussions.

ABBREVIATIONS

FNR: Ferredoxin NADP reductase
FBPase: Fructose-1,6-diphosphatase
GAPDH: Glyceraldehyde-3-phosphate dehydrogenase
PGK: 3-phosphoglycerate kinase
PSI: Photosystem I
RPI: Phosphoriboisomerase
RPK: Phosphoribulokinase
Rubisco: Ribulose bisphosphate carboxylase/oxygenase
R-5-P: Ribose-5-phosphate
RuBP: Ribulose-5-phosphate
TPI: Triose-phosphate isomerase

REFERENCES

1. Srere PA. Complexes of sequential enzymes. *Annu. Rev. Biochem.* 1987; 56:89–124.
2. Ovadi J. Physiological significance of metabolite channeling. *J. Theor. Biol.* 1991; 152:1–22.
3. Huber LA. Is proteomics heading in the wrong direction? *Nat. Rev. Mol. Cell Biol.* 2003; 4:1–23.
4. Srere PA. The infrastructure of the mitochondrial matrix. *Trends Biochem. Sci.* 1980; 5:120–121.
5. Ovadi J. Old pathway — new concept. Control of glycolysis by metabolite-modulated dynamic enzyme association. *Trends Biochem. Sci.* 1988; 13:486–490.
6. Süss KH, Sainis JK. Supramolecular organization of water-soluble photosynthetic enzymes in chloroplasts. In: Pessarakli M, eds. *Handbook on Photosynthesis.* New York: Marcel Dekker, 1997:305–314.
7. Keleti T, Ovadi J, Batke J. Kinetic and physico-chemical analysis of enzyme complexes and their possible role in control of metabolism. *Prog. Biophys. Mol. Biol.* 1989; 53:105–152.
8. Sainis JK. Supramolecular organization in living cells. *Proc. Indian Natl. Sci. Acad.* 1998; B64:197–212.
9. Mering C, von Krauss R, Snel B, Cornell M, Oliver SG, Fields S, Bork P. Comparative assessment of large-scale data sets of protein-protein interactions. *Nature* 2002; 417:399–403.
10. Gavin AC, Bosche M, Krause R, Grandi P, Marzioch M, Bauer A, Schultz J, Rick JM, Michon AM, Cruciat CM, Remor M, Hofert C, Schelder M, Brajenovic M, Ruffner H, Merino A, Klein K, Hudak M, Dickson D, Rudi T, Gnau V, Bauch A, Bastuck S, Huhse B, Leutwein C, Heurtier MA, Copley RR, Edelmann A, Querfurth E, Rybin V, Drewes G, Raida M, Bouwmeester T, Bork P, Seraphin B, Kuster B, Neubauer G, Superti-Furga G. Functional organization of the yeast proteome by systematic analysis of protein complexes. *Nature* 2002; 415:141–145.
11. Ho Y, Gruhler A, Heilbut A, Bader GD, Moore L, Adams SL, Millar A, Taylor P, Bennett K, Boutilier K, Yang L, Wolting C, Donaldson I, Schandorff S, Shewnarane J, Vo M, Taggart J, Goudreault M, Muskat B, Alfarano C, Dewar D, Lin Z, Michalickova K, Willems AR, Sassi H, Nielsen PA, Rasmussen KJ, Andersen JR, Johansen LE, Hansen LH, Jespersen H, Podtelejnikov A, Nielsen E, Crawford J, Poulsen V, Sorensen BD, Matthiesen J, Hendrickson RC, Gleeson F, Pawson T, Moran MF, Durocher D, Mann M, Hogue CW, Figeys D, Tyers M. Systematic identification of protein complexes in *Saccharomyces cerevisiae* by mass spectrometry. *Nature* 2002; 415:180–183.
12. Hrazdina G, Jensen RA. Spatial organization of enzymes in plant metabolic pathways. *Annu. Rev. Plant Physiol. Plant Mol. Biol.* 1992; 43:241–261.
13. Robinson SP, Walker DA. Photosynthetic carbon reduction cycle. In: Hatch MD, Boardman NK, eds. *The Biochemistry of Plants in a Comprehensive Treatise.* Vol. 8. New York: Academic Press, 1981:193–236.
14. Bowien B, Mayer F, Spiess E. Pahler AE, Englisch E, Saenger W. On the structure of crystalline ribulose bisphosphate carboxylase from *Alcaligenes eutrophus. Eur. J. Biochem.* 1980; 106:405–410.
15. Gunning BES. The fine structure of stroma following aldehyde-osmium tetraoxide fixation. *J. Cell Biol.* 1965; 24:79–93.
16. Sprey B. Lamellae-bound inclusions in isolated spinach chloroplasts. I Ultrastructure and isolation. *Z. Pflanzenphysiol.* 1977; 83:159–179.
17. Edwards G, Walker D. *C3, C4: Mechanism, and Cellular and Environmental Regulation of Photosynthesis.* Blackwell Scientific Publications, Oxford, 1983.
18. Staehelin. LA, Arntzen CJ. *Encyclopedia of Plant Physiology. New Series. Photosynthesis III: Photosynthetic Membranes and Light Harvesting Systems.* Springer-Verlag, Berlin, 1986.
19. Woodrow IE, Berry JA. Enzymatic regulation of photosynthetic CO_2 fixation in C_3 plants. *Annu. Rev. Plant Physiol. Plant Mol. Biol.* 1988; 39:533–594.
20. Ellis RJ. Macromolecular crowding; obvious but under-appreciated. *Trends Biochem. Sci.* 2001; 26:1–16.
21. Dilley RA, Vernon LP. Ion water transport processes related to light dependent shrinkage of spinach chloroplasts. *Arch. Biochem. Biophys.* 1965; 111:365–375.
22. Nobel PS, Chang DT, Wang CT, Smith SS, Barcus DE. Initial ATP formation, CO_2 fixation, NADP reduction and chloroplast flattening upon illumination of pea leaves. *Plant Physiol.* 1969; 44:655–661
23. Miller MM, Nobel PS. Light induced changes in the ultrastructure of pea chloroplasts *in vivo. Plant Physiol.* 1971; 49:535–541.
24. Van Noort G, Wildman SG. Proteins of green leaves IX. Enzymatic properties of fraction-1 protein isolated by specific antibody. *Biochim. Biophys. Acta,* 1964; 90:309–317.
25. Mendiola L, Akazawa T. Partial purification and enzymatic nature of fraction I protein of rice leaves. *Biochemistry* 1964; 3:174–179.
26. Muller B. A labile CO_2 fixing enzyme complex from spinach leaves. *Z. Naturforsch.* 1972; 27B:925–932.
27. Walker DA, Leegood RG, Sivak M. Ribulose bisphosphate carboxylase-oxygenase: its role in photosynthesis. *Philos. Trans. R. Soc. Lond. B* 1986; 313:305–322.
28. Sainis JK, Harris GC. The association of ribulose-1,5-bisphosphate carboxylase with phosphoriboisomerase and phosphoribulokinase. *Biochem. Biophys. Res. Commun.* 1986; 139:947–954.
29. Sainis JK, Harris GC. Studies on multienzyme complex containing RuBP carboxylase phosphoriboisomerase and phosphoribulokinase. In: Biggens J, eds. *Progress in Photosynthesis Research.* Vol. III-6. Netherlands: Martinus Nijhoff Publishers. 1987:491–494.
30. Sainis JK, Merriam K, Harris GC. The association of ribulose-1,5-bisphosphate carboxylase/oxygenase with phosphoribulokinase. *Plant Physiol.* 1989; 89:368–374.
31. Gontero B, Cardenas ML, Ricard J. A functional five-enzyme complex of chloroplasts involved in the Calvin cycle. *Eur. J. Biochem.* 1988; 173:437–443.
32. Rault M, Gontero B, Ricard J. Thioredoxin activation of phosphoribulokinase in a chloroplast multi-enzyme complex. *Eur. J. Biochem.* 1991; 197:791–797.

33. Rault M, Giudici-Orticoni M-T, Gontero B, Ricard J. Structural and functional properties of a multi-enzyme complex from spinach chloroplasts. 1. Stoichiometry of the polypeptide chains. *Eur. J. Biochem.* 1993; 217:1065–1073.

34. Gontero B, Mulliert G, Rault M, Giudici-Orticoni M-T, Ricard J. Structural and functional properties of a multi-enzyme complex from spinach chloroplasts. 2. Modulation of the kinetic properties of enzymes in the aggregated state. *Eur. J. Biochem.* 1993; 217:1075–1082.

35. Ricard J, Giudici-Orticoni M-T, Gontero B. The modulation of enzyme reaction rates within multi-enzyme complexes. 1. Statistical thermodynamics of information transfer through multi-enzyme complexes. *Eur. J. Biochem.* 1994; 226:993–998.

36. Gontero B, Giudici-Orticoni M-T, Ricard J. The modulation of enzyme reaction rates within multi-enzyme complexes. 2. Information transfer within a chloroplast multi-enzyme complex containing ribulose bisphosphate carboxylase-oxygenase. *Eur. J. Biochem.* 1994; 226:999–1006.

37. Lebreton S, Gontero B, Avilan L, Ricard J. Information transfer in multienzyme complexes — 1. Thermodynamics of conformational constraints and memory effects in the bienzyme glyceraldehyde-3-phosphate-dehydrogenase-phosphoribulokinase complex of *Chlamydomonas reinhardtii* chloroplasts. *Eur. J. Biochem.* 1997; 250:286–295.

38. Avilan L, Gontero B, Lebreton S, Ricard J. Information transfer in multienzyme complexes — 2. The role of Arg64 of *Chlamydomonas reinhardtii* phosphoribulokinase in the information transfer between glyceraldehyde-3-phosphate dehydrogenase and phosphoribulokinase. *Eur. J. Biochem.* 1997; 250:296–302.

39. Nicholson S, Easterby JS, Powls R. Properties of multimeric protein complex from chloroplasts possessing potential activities of NADPH dependent glyceraldehyde-3-phoshate dehydrogenase and phosphoribulokinase. *Eur. J. Biochem.* 1991; 162:423–431.

40. Lebreton S, Gontero B. Memory and imprinting in multienzyme complexes. Evidence for information transfer from glyceraldehyde-3-phosphate dehydrogenase to phosphoribulokinase under reduced state in *Chlamydomonas reinhardtii*. *J. Biol. Chem.* 1999; 274:20879–20884.

41. Graciet E, Lebreton S, Camadro JM, Gontero B. Thermodynamic analysis of the emergence of new regulatory properties in a phosphoribulokinase-glyceraldehyde 3-phosphate dehydrogenase complex. *J. Biol. Chem.* 2002; 277:12697–12702.

42. Mouche F, Gontero B, Callebaut I, Mornon JP, Boisset N. Striking conformational change suspected within the phosphoribulokinase dimer induced by interaction with GAPDH. *J. Biol. Chem.* 2002; 277:6743–6749.

43. Persson L-O, Johansson G. Studies of protein-protein interaction using countercurrent distribution in aqueous two-phase systems. Partition behavior of six Calvin-cycle enzymes from a crude spinach (*Spinacia oleracea*) chloroplast extract. *Biochem. J.* 1989; 259:863–870.

44. Anderson LE. Ribose–5-phosphate isomerase and ribulose-5-phosphate kinase show apparent specificity for specific ribulose-5-phosphate species. *FEBS Lett.* 1987; 212:45–48.

45. Skrukrud CL, Gordon IM, Dorwin S, Yuan X-H, Johansson G, Anderson LE. Purification and characterization of pea chloroplastic phosphoriboisomerase. *Plant Physiol.* 1991; 97:730–735.

46. Skrukrud CL, Anderson LE. Chloroplast phosphoribulokinase associates with yeast phosphoriboisomerase in presence of substrate. *FEBS Lett.* 1991; 280:259–261.

47. Süss KH, Arkona C, Manteuffel R, Adler K. Calvin cycle multienzyme complexes are bound to chloroplast thylakoid membranes of higher plants *in situ. Proc. Natl. Acad. Sci. USA* 1993; 90:5514–5518.

48. Hosur MV, Sainis JK, Kannan KK. Crystallization and X-ray analysis of a multienzyme complex containing Rubisco and RuBP. *J. Mol. Biol.* 1993; 234:1274–1278.

49. Sainis JK, Srinivasan VT. Effect of the state of water as studied by pulsed NMR on the function of RUBISCO in a multienzyme complex. *J. Plant Physiol.* 1993; 142:564–568.

50. Sainis JK, Jawali N. Channeling of the intermediates and catalytic facilitation to Rubisco in a multienzyme complex of Calvin cycle enzymes. *Indian J. Biochem. Biophys.* 1994; 31:215–220.

51. Sainis JK, Jawali N. Reactivation of dark-inactivated RuBP carboxylase from *Phaseolus vulgaris in vitro*. In: Baltscheffsky M, eds. *Current Research in Photosynthesis III*. Vol. 11. Netherlands: Kluwer Academic Publishers, 1990:407–410.

52. Edmondson DL, Badger MR, Andrews TJ. A kinetic characterization of slow inactivation of RuBP carboxylase during catalysis. *Plant Physiol.* 1990; 93:1376–1382.

53. Sicher RC, Hatch AL, Stump DK, Jensen RG. Ribulose-1,5-bisphosphate carboxylase & activation of carboxylase in chloroplasts. *Plant Physiol.* 1981; 68:252–255.

54. Mathur-De Vre R. The NMR studies on water in biological systems. *Prog. Biophys. Mol. Biol.* 1979; 35:103–134.

55. Ashton AR. A role for ribulose-1,5-bisphosphate carboxylase as metabolite buffer. *FEBS Lett.* 1982; 145:1–7.

56. Quick WP, Schurr U, Schiebe R, Schulze E-D, Rodermel SR, Bogorad L, Stitt M. Decreased ribulose carboxylase-oxygenase in transgenic tobacco transformed with "antisense." *Planta* 1991; 183:542–554.

57. Bowes G. Growth at elevated CO_2: photosynthesis responses mediated through Rubisco. *Plant Cell Environ.* 1991; 14:795–806.

58. Quick WP, Fichtner K, Schulze E-D, Wendler R, Leegood RC, Mooney H, Rodermel SR, Bogorad L, Stitt M. Decreased ribulose-1,5-bisphosphate carboxylase/oxygenase in transgenic tobacco transformed with "antisense" rbcS IV. Impact on photosynthesis and plant growth at altered nitrogen supply. *Planta* 1992; 188:522–531.

59. Lacoste-Royal G. Gibbs SP. Immunocytochemical localization of ribulose-1,5-bisphosphate carboxylase in

the pyrenoid and thylakoid region of the chloroplasts of *Chlamydomonas reinhardtii*. *Plant Physiol.* 1987; 83:602–606.

60. Mori H, Takabe T, Akazawa T. Loose association of ribulose-1,5-biphosphate carboxylase with thylakoid membranes. *Photosynth. Res.* 1984; 43:17–28.

61. Hermoso R, Felipe MR, Vivo A, Chueca A, Lazaro J, Gorge JL. Immunogold localization of photosynthetic fructose-1,6-bisphosphatase in pea leaf tissue. *Plant Physiol.* 1989; 89:381–385.

62. Süss KH. Ribulose-1,5-bisphosphate carboxylase-binding chloroplast membrane proteins. *In vitro* evidence that H^+ ATP synthase may serve as a membrane receptor. *Z. Naturforsch.* 1990; 45C:633–637.

63. Vaughn KC. Two immunological approaches to the detection of ribulose-1,5-bisphosphate carboxylase in guard cell chloroplasts. *Plant Physiol.* 1987; 84:188–196.

64. Shaw P.J, Henwood JA. Immunogold localization of cyt f, light harvesting complex, ATP synthase and RuBP carboxylase. *Planta* 1985:165:333–339.

65. Mangeney E, Hawthornthwaite AM, Codd C, Gibbs SP. Immunocytochemical localization of phosphoribulokinase in cyanelles of *Cyanophora pardoxa* and *Glaucocystis nostochinearum*. *Plant Physiol.* 1987; 84:1028–1032.

66. Hermoso R, Fonolla J, Rosario de Felipe M, Vivo A, Chueaca A, Lazaro JJ, Lopez George J. Double immunogold localization of thioredoxin f and photosynthetic fructose-1,6-bisphosphate in spinach leaves. *Plant Physiol.* 1992; 30:39–46.

67. Süss KH, Prokhorenko I, Adler K. *In situ* association of Calvin cycle enzymes ribulose-15-bisphosphate carboxylase/oxygenase, activase, ferredoxin-NADP reductase and nitrite reductase with thylakoid and pyrenoid membranes of *Chlymydomonas reinhardtii* chloroplasts as revealed by immunoelectron microscopy. *Plant Physiol.* 1995; 107:1387–1397.

68. Teige M, Melzer M, Süss KH. Purification, properties and *in situ* localization of the amphibolic enzymes D-ribulose-5-phosphate 3-epimerase and transketolase from spinach chloroplasts. *Eur. J. Biochem.* 1998; 252:237–244.

69. Andersen B, Scheller HV, Moller-Lindberg B. The PSI-E subunit of photosystem I binds ferredoxin: $NADP^+$ oxidoreductase. *FEBS Lett.* 1992; 311:169–173.

70. Harbinson J, Hedley CL. Changes in P_{700} oxidation during early stages of induction of photosynthesis. *Plant Physiol.* 1993; 103:649–660.

71. Bell G, Halling PJ, Moore BD, Patridge J, Rees DG. Biocatalysts behaviour in low water systems. *Trends Biotech.* 1995; 13:468–473.

72. Jebanathirajah JA, Coleman JR. Association of carbonic anhydrase with a Calvin cycle enzyme complex in *Nicotiana tabacum*. *Planta* 1998; 204:177–182

73. Park RB, Biggins J. Quantasomes size and composition. *Science* 1964; 144:1009–1011.

74. Ball R, Richter M, Wild A. What are quantasomes? The background of nearly forgotten term. *Photosynthetica* 1994; 30:161–173.

75. Srere PA. Organization of proteins within the mitochondrion. In: Welch GR, eds. *Organized Multienzyme Systems: Catalytic Properties.* New York: Academic Press, 1985:1–61.

76. Srere PA. Wanderings (wonderings) in metabolism. 17th Fritz Lipmann lecture. *Biol. Chem. Hoppe-Seyler* 1993; 374:833–842.

77. Stitt M, Sonnewald U. Regulation of metabolism in transgenic plants. *Annu. Rev. Plant Physiol. Plant Mol. Biol.* 1995; 46:341–368.

78. Spreitzer RJ, Salvucci ME. Rubisco: structure, regulatory interactions and possibility for a better enzyme. *Annu. Rev. Plant Biol.* 2002; 53:449–475.

79. Srere PA. Complexities of metabolic regulation. *Trends Biochem. Sci.* 1994; 19:519–520.

80. Sainis JK, Dani DN, Dey GK. Involvement of thylakoid membranes in supramolecular organization of Calvin cycle enzymes in *Anacystis nidulans*. *J. Plant Physiol.* 2003; 160/1:23–32.

81. Baumeister W, Steven AC. Macromolecular electron microscopy in the era of structural genomics. *Trends Biochem. Sci.* 2000; 25:624–631.

15 Cytochrome c_6 Genes in Cyanobacteria and Higher Plants

Ho Kwok Ki

Department of Biochemistry, Purdue University

CONTENTS

I. INTRODUCTION

This chapter presents new information about cytochrome c_6 in the areas of genomic sequence and location. Complete genomic sequencing has revealed that multiple copies of cytochrome c_6 genes occur in both filamentous and unicellular cyanobacteria (Table 15.1). The location of one of the cytochrome c_6 genes next to a plastocyanin appears to be a feature in three of the genomes studied (Table 15.1). These findings, together with the recent realization of a higher plant gene encoding a cytochrome c_6-like protein [1], offer new opportunities to study the role and regulation of cytochrome c_6 using molecular approaches. For many years this protein has been thought to be involved in photosynthesis [2,3], respiration [4], and anoxygenic photosynthesis [5,6]. It remains unclear whether these processes involve a single or multiple forms of this protein. The available gene sequences of different cytochrome c_6 from the same organism will be useful molecular tools to address this problem. The discovery of a cytochrome c_6-like gene in higher plants dispels a long-held belief that this cytochrome was lost during evolution and provides a unique opportunity to study the role of a very ancient protein in these organisms. The detection of isoforms of cytochrome c_6 in the same organism has so far failed based on conventional biochemical techniques [2,3]. For this purpose, a mass spectrometric method made feasable by organic solvent extraction has been introduced here. The organic solvent extraction provides a simple method for sample preparation and does not require preliminary chromatographic separation. Initial studies aimed at the detection of cytochrome c_6 in *Lyngbya* spp. have been successful. The final part of this chapter describes what is known now about heterologous expression of cytochrome c_6 in bacteria. This method will be useful for producing material for structure and function analysis.

II. CYTOCHROME c_6 GENES

A. GENOME CONTENT AND ORGANIZATION

The genomes of several different cyanobacteria have already been sequenced, and the genome sequences can be obtained from databases in Kazusa DNA Research Institute (http://www.kazusa.gov.jp/cyano), Oak Ridge National Laboratory (http://genome.ornl.gov/microbial), and National Institutes of Health (http://www.ncbi.nlm.nih.gov/Genbank). From sequence information, a total of 17 cytochrome (cyt) c_6 genes have been identified in nine genomes, four of which are from filamentous cyanobacteria and the rest from unicellular ones. The filamentous cyanobacteria include *Anabaena* sp. PCC7120 [7–9], *Gloeobacter violaceus* PCC7421 [10,11], *Nostoc punctiforme* ATCC29133 [12], and *Trichodesmium erythraeum*

273

TABLE 15.1

Genomic Sequences and Locations of Cytochrome c_6

Genome	#	ID	INT	TER	*petE*	INT	TER
Anabaena sp. PCC 7120	3	all0161	166206	165871			
(www.kazusa.or.jp/cyano/Anabaena)		alr4251	5100219	5100554			
		asl0256	276831	276724	all0258	277816	277397
Gloeobacter violaceus PCC 7421	2	gll1980	2129049	2128705	gll2341	2502200	2501811
(www.kazusa.or.jp/cyano/Gloeobacter)		glr1906	2027275	2027613			
Nostoc punctiforme ATCC 29133	3	Contig					
(genome.ornl.gov/microbial/npun)		350Gene11	1939	1664			
		477Gene104	31674	31339			
		497Gene57	75088	75408	497Gene56	74399	74818
Prochlorococcus marinus MIT 9313	2	Gene					
(genome.ornl.gov/microbial/pmar_mit)		PMT0462	516010	516387	PMT0447	505204	505563
		PMT0509	560175	560513			
Prochlorococcus marinus SS 120	1	COG2863	547525	547139	COG3794	1003534	1003893
(NCBI access number; www.ncbi.nlm.nih.gov)		(NP_874970)			(NP_875473)		
Synechococcus sp. WH 8102	2	Syn_wh					
(genome.ornl.gov/microbial/syn_wh)		Gene448	1462484	1462816			
		Gene434	1451769	1452125	Gene433	1451285	1451644
Synechocystis sp. PCC 6803	1	Sll1796	846328	845966	Sll0199	2526207	2525827
(www.kazusa.or.jp/cyano/Synechocystis)							
Thermosynechococcus elongatus BP-1	1	tll1283	1336182	1335844			
(www.kazusa.or.jp/cyano/Thermo)							
Trichodesmium erythraeum IMS 101	2	Contig122					
(genome.ornl.gov/microbial/tery)		Gene6986	1403730	1404065			
		Gene7635	581790	581395	Gene7632	585086	584685

Key: INT, initial nucleotide; TER, terminal nucleotide; *petE*, plastocyanin gene.

IMS101 (for preliminary sequence files, see its website listed in Table 15.1) and the unicellular ones are *Prochlorococcus marinus* MIT9313 [13], *Prochlorococcus marinus* SS120 [14], *Synechococcus* sp. WH8102 [15], *Synechocystis* sp. PCC6803 [16–18], and *Thermosynechococcus elongatus* BP-1 [19,20]. Table 15.1 lists the different cyt c_6 gene sequences identified in each genome and the genome websites. The positions of the initial and terminal nucleotides for the genes of cyt c_6 and plastocyanin in the genome are also listed. It is evident that six genomes have multiple copies of cyt c_6 gene sequences, belonging to four filamentous cyanobacteria and two unicellular ones. Among the filamentous cyanobacteria, *Anabaena* sp. PCC7120 and *N. punctiformec* have three copies each whereas *G. violaceus* and *T. erythraeum* have two each. The two unicellular cyanobacteria, *P. marinus* MIT9313 and *Synechococcus* sp. WH8102, have two copies each.

The genome organization of different cyt c_6 genes is similar in different cyanobacteria. Regardless of the number of cyt c_6 genes in a particular cyanobacterium, they appear to scatter over the genome. For example, the two cyt c_6 genes in *G. violaceus* are separated by as many as 373,106 base pairs while the smaller number of 10,000 base pairs is found to separate the two genes in *Synechococcus* sp. WH8102. An interesting difference emerges between different cyanobacteria, when a plastocyanin (PC) gene is used as a reference gene to locate the different cyt c_6 genes. PC is a copper protein that can replace cyt c_6 in photosynthetic electron transport. In *Anabaena* sp. PCC7120, *N. punctiforme*, and *Synechococcus* sp. WH8102, all have one cyt c_6 gene located just downstream from a PC gene. While the two genes are separated by 566 and 270 base pairs in *Anabaena* sp. PCC7120 and *N. punctiforme*, respectively, they are found even closer together in *Synechococcus* sp. WH8102 and separated by only 125 base pairs. Such close proximity between a cyt c_6 gene and a PC gene is not found in the other three cyanobacteria that have two cyt c_6 genes each. Instead, the cyt c_6 gene that is close to the PC gene is separated from the PC gene by 2,895, 10,447, and 372,762 base pairs in *T. erythraeum*, *P. marinus* MIT 9313, and *G. violaceus*, respectively. This type of large separation between a cyt c_6 gene and a PC gene is also observed in *P. marinus* SS120 and *Synechocystis* sp, PCC6903, both of which have only a single cyt c_6 gene. It is interesting to note that *T. elongatus* has a single cyt c_6 gene but no PC gene. Thus, it is not possible to compare the gene locations in this organism with the other cyanobacteria.

It is worth noting that one of the cyt c_6 gene sequences, identified in the *Anabaena* sp. PCC7120 genome and assigned with an accession code asl0256 in the Kazusa database, is actually incomplete. The sequence has only 108 base pairs with ATG as the initiation codon, encoding a short protein of 35 amino acids. This protein sequence aligns well with a carboxyl-terminal portion of other complete cyt c_6 sequences, with its amino-terminal residue matching a conserved methionine located near the carboxyl-terminus. In the present presentation, the gene sequence corresponding to as10256 has been extended to include an unannotated genome sequence of 258 base pairs located upstream next to the initial nucleotide of its 108 base pairs sequence. The expanded sequence assembles a full-length cyt c_6 gene with three potential initiation translation sites. Two of the potential initiation codons start with TTG and one with ATT. These are rare codons first discovered in the *Synechocystis* spp. genome sequence [21]. Throughout this work, as10254 is referred to the full-length cyt c_6 gene as described here.

B. PROPERTIES OF THE MATURE CODING SEQUENCES

Table 15.2 shows some properties of mature protein sequences derived from different cyt c_6 genes. Despite sequence variability, all the mature protein sequences have a similar mass ranging from 9,000 to 10,000 Da and contain 86 to 96 amino acids. In *Anabaena* sp. PCC7120 as well as in *N. punctiforme*, two of the mature protein sequences are basic and a third one is acidic. Pair wise comparisons among the three

TABLE 15.2
Some Properties of the Mature Protein Sequences Derived from Different cyt c_6 Genes

Cyanobacteria	#AA	MW (Da)	pI	% AA identity	
				1	2
Anabaena sp. PCC 7120					
(1) ana_all0161	86	9669	8.05		
(2) ana_alr4251	86	9130	8.94	52.3	
(3) ana_asl0256	96	10733	5.35	41.0	50
Cloeobacter violaceus PCC 7421					
(1) Glo_1906	88	9330	5.6		
(2) Glo_1980	88	9518	9.55	55.7	
Nostoc punctiforme ATCC 29133					
(1) Nos_Gene57	89	9631.8	5.6		
(2) Nos_Gene11	86	9597.9	8.65	50	
(3) Nos_Gene104	86	8999.3	9.4	55.6	54.65
Trichodesmium erythraeum IMS 101					
(1) Tri_7635	86	9230	9.43		
(2) Tri_6986	86	9848	9.55	40.7	
Prochlorococcus marinus MIT 9313					
(1) Pro9313_A	89	8751.6	4.54		
(2) Pro9313_J	88	9491.6	6.92	35.9	
Prochlorococcus marinus SS 120	88	9734	8.01		
Synechococcus sp. WH 8102					
(1) Syn8102_gene434	92	10142.3	4.6		
(2) Syn8102_gene448	88	9166.3	8.1	31.6	
Synechocystis sp. PCC 6803	85	8688.8	5.5		
Thermosynechococcus elongatus BP-1	87	9178.4	5.5		

Key: AA, amino acid; MW, molecular weight; pI, isoelectric point. The pI and MW values were predicted by a Compute pI/Mw software listed in the ExPASy molecular biology server (us.expasy.org/tools/pi_tool.html).

sequences reveal an identity of 40% to 50% for Anabaena spp. and 50% to 55% for *N. punctiforme*. For
G. violaceus, *T. erythraeum*, *P. marinus* MIT 9313,
and *Synechococcus* sp. WH8102, they are characterized by a combination of one acidic and one basic,
one acidic and one neutral, or two basic mature protein sequences depending on the species. A pair wise
comparison between the two mature protein sequences in each of these cyanobacteria results in an
identity ranging from 32% to 56% depending on the
species. For other cyanobacteria identified with a
single copy of cyt c_6 gene, the charge property of the
mature protein sequences is either basic or acid depending on the species.

Of the nine genomes, three have a cyt c_6 gene
next to a PC gene. The three belong to *Anabaena* sp.
PCC7120, *N. punctiforme*, and *Synechococcus* sp.
WH8102 and their respective genes are listed as
asl0256, Contig 497Gene 57, and Syn_wh Gene434
(Table 15.1). Sequence comparison among the mature protein sequences deduced from different cyt c_6
genes (Figure 15.1) reveals that cyt c_6 genes of this
group share a high degree of similarity and that
among the nine genomes, generally cyt c_6 genes
from the same genome are more similar to some
other cyt c_6 genes than they are to each other. Figure
15.1 shows the alignment of mature protein sequences deduced from 19 cyt c_6 genes. The latter
include two cyt c_6 genes cloned from *Synechococcus*
sp. PCC7002 [22], the genome of which is not fully
sequenced. The alignment appears to encompass
three sequence groups, with each group having a
different pair of *Anabaena* sp. PCC7120 and *N. punctiforme* sequences and one sequence each from some
other cyanobacteria. Group 1 contains sequences
from eight cyanobacteria (sequences 1 to 8 in Figure
15.1) except *P. marinus* SS120 and *T. erythraeum*.
The protein sequences corresponding to asl0256,
Contig 497Gene 57, and Syn_wh Gene434 are close
together in this group and occupy the first three
positions of the alignment pattern. A common feature of this group is that all of its sequences have an
acidic isoelectric point, ranging from pH 4.6 to 5.6.
Group 2 contains four sequences from different filamentous cyanobacteria (sequences 9 to 12) and they
share a basic isoelectric point, ranging from pH 9.0
to 9.5. A member of this group, *G. violaceus*, is
thought to be among the earliest branched cyanobacteria [23]. It has no thylakoid membranes and the
plasma membrane possesses the catalysts for both
photosynthetic and respiratory pathways. Group 3
contains sequences from seven cyanobacteria (se-
quences13 to 19) and excludes *G. violaceus*, *Synechocystis* sp. PCC6803, and *T. elongatus*. While one
sequence from *P. marinus* MIT9313 is neutral, all
the other sequences are basic with an isoelectric
point ranging from pH 8.0 to 9.5.

C. ROLE OF SPECIFIC CYT c_6 GENES

The facts that of nine completely sequenced genomes,
six contain more than one copy of cyt c_6 gene and that
the six genomes include both filamentous and unicellular cyanobacteria suggest isogenes may be a feature
of a large group of cyanobacteria. Isogenes raise the
possibility that each gene may have evolved for a
specific function. Cyt c_6 can contribute to different
metabolic pathways through electron transport. One
is photosynthesis [2,3]. Cyt c_6 has long been implicated in photosynthetic electron transport as it can
catalyze the transport of electrons from the cyt b_6f
complex into the PSI complex. A second pathway is
respiration [4]. Cyt c_6 is thought to act as an electron
donor to the cyt oxidase complex in the thykaloids as
well as the plasma membrane. A third mechanism
[5,6] involving cyt c_6 is anoxygenic photosynthesis in
which hydrogen sulfide acts as an electron donor. Cyt
c_6 may act as an electron carrier between some quinones and iron–sulfur centers during anaerobic sulfide
oxidation. An analysis of the functional role of specific cyt c_6 isogenes in photosynthetic electron transport has been conducted with *Synechococcus* sp. PCC
7002 [22]. Two cyt c_6 genes, designated as petJ1 and
petJ2 (Figure 15.1), were inactivated by intersposon
mutagensis and used for cell transformation. The
transformants were then selected for genetic and
growth characterization. The results suggested that
petJ1 has an essential function in electron transport
under normal photoautotrophic or photoheterotrophic growth conditions and that petJ2 has no effect in photoautotrophic growth. A major difference
between petJ1 and petJ2 is the charge property of the
gene product. For petJ1, the mature protein sequence
of cyt c_6 is acidic, whereas for petJ2, it is basic (Table
15.2). This agrees well with an earlier report that an
acidic cyt c_6 of unicellular cyanobacteria can act as a
competent electron carrier to both photosynthetic
and respiratory pathways [24,25]. For functional
characterization of cyt c_6 isogenes, *N. punctiforme*
provides the advantage of its genome having been
completely sequenced. In addition, much information
is available about its physiology and different modes
of growth [12]. *N. punctiforme* can grow in both
photoautotrophic and heterotrophic conditions. In
prolonged darkness and provided with an appropriate organic substrate, it can maintain a growth rate
less than half of that in the light. *N. punctiforme* is
likely to perform anoxygenic photosynthesis with
hydrogen sulfide as an electron donor in view of the
following considerations. First, it can survive in

```
                            10              30              50        60
 1.  ana_asl0256    ---AELPTGAKIFNNNCASCHIGGGNILISEKTLKKEALLKYLEDYETN---SIQAIIHQ
 2.  Nos_Gene57     ---AETSNGAKIFEANCASCHIGGGNILISQKTLKKEALSKYLENYNSD---SIEAIIHQ
 3.  Syn8102_gene434 LESALIEQGEQIFSSNCAACHMGGGNVIRANRSLKIRDLNAHLEEYQQD---PLEAIEHQ
 4.  Syn7002_1      ---ADAAAGAQVFAANCAACHAGGNNAVMPTKTLKADALKTYLAGYKDGSKSLEEAVAYQ
 5.  Pro9313_A      ---ADSAHGGQVFSSTCAACHAGGGNIVDPAKTLQKAALEATLSNYGSG---HEEAIVAQ
 6.  Synechystis    ---ADLAHGKAIFAGNCAACHNGGLNAINPSKTLKMADLEANGK-------NSVAAIVAQ
 7.  Thermosyn_BP1  ---ADLANGAKVFSGNCAACHMGGGNVVMANKTLKKEALEQFGMY-------SEDAIIYQ
 8.  Glo_1906       --QPDLAAGEKIFKANCAACHAGGNNIVEPEKTLKKEALAHFGMG-------SPAAIIQQ

 9.  ana_alr4251    ---ADSVNGAKIFSANCASCHAGGKNLVQAQKTLKKADLEKYGMY-------SAEAIIAQ
10.  Nos_Gene104    ---GDAVSGAKVFSANCASCHAGGKNLVQAAKNLKKEALEKYGLY-------SAEAIIAQ
11.  Tri_7635       ---ADIASGKGVFQGNCAACHIGGKNNINPAKTLQKSDLEKYGMF-------AAEKIIYQ
12.  Glo_1980       ----DLAQGEKVFKANCAMCHAGGRNTVNPAKTLKIEDLKKYKMD-------TAAAISAQ

13.  ana_all0161    ---ENTINGEQIFSVHCAGCHINGSNIIRRGKNLQKKTLKKYGMD-------SLEAIEAI
14.  Nos_Gene11     ---ADIVNGEQIFSLHCAGCHINGSNIVRRGKNLKKQALKKYGMD-------SIEAVTSI
15.  Syn7002_2      ---ADLDQGAQIFEAHCAGCHLNGGNIVRRGKNLKKRAMAKNGYT------SVEAIANL
16.  Tri_6986       ---TEITQGAEVFQIHCAGCHAKGGNIVKWWKNLKIRTLKRNKLD------SVEAIAYL
17.  Pro1375_c6     ---NNQTNGERLFIENCAGCHINGGNIIRRSKTLRLKDLHRNGLD------NADAIAKI
18.  Pro9313_J      ---LDTDAGGSLFKQHCSGCHVNGGNIIRRNKTLRLKALERNGLD------NPQAIARV
19.  Syn8102_gene448 ---ETSGEGAVLFGQHCAGCHVNGGNIIRRGKNLKLATLKRQGLD------STEAIASI
                     *   :*    *:  **   * *   :    :.*:    :

                            70              90
 1.  ana_asl0256    IQYGKNAMPAFKD--KLSTEEILEVAAYIFQKAEKDWSNLEKEG
 2.  Nos_Gene57     VQNGKNAMPAFKG--KLSAEEILDVAAYVFQNAEQGW-------
 3.  Syn8102_gene434 IEAGKNAMPSYEG--KLTEAEIIAVATYVEQQAELGW-------
 4.  Syn7002_1      VTNGQGAMPAFGG--RLSDADIANVAAYIADQAENNKW------
 5.  Pro9313_A      VTNGKGGMPSFAD--VLSAADIADVAAYVEAQASSGW-------
 6.  Synechystis    ITNGNGAMPGFKG--RISDSDMEDVAAYVLDQAEKGW-------
 7.  Thermosyn_BP1  VQHGKNAMPAFAG--RLTDEQIQDVAAYVLDQAAKGWAG-----
 8.  Glo1906_c553   VTGGKNAMPAFGG--ELSTEEIRQVASYVLEMADKDWQK-----

 9.  ana_alr4251    VTNGKNAMPAFKG--RLKPEQIEDVAAYVLGKADADWK------
10.  Nos_Gene104    VTNGKNAMPAFGK--RLKADQIENVAAYVLSQADKGWK------
11.  Tri_7635       VTNGKNAMPAFGR--RLKPQQIENVAAYVMAQAEGGWK------
12.  Glo1980_c6     LYNGKGAMPAFGKNGKLKQDQIDSVTAYVLDQANKGWKK-----

13.  ana_all0161    VTNGKNNMSAYKD--RLSEQEIQDVAAYVLEQAEKGWR------
14.  Nos_Gene11     VTNGKNNMSAYKD--RLTEQQITDVAAYVLEQAEKDWR------
15.  Syn7002_2      VTQGKGNMSAYGD--KLSSEQIQAVSQYVLQQSQTDWKS-----
16.  Tri_6986T      VKNGKNNMSAYKD--RLTEIEIQTVSAYVLKQAENRWN------
17.  Pro1375_c6     AKEGIGIMSGYKD--VLGENGDNLVANWIWEQSQKAWVQG----
18.  Pro9313_J      AREGIGQMSGYED--VLGDSGDQLVAAWIWAQAQNAWTQG----
19.  Syn8120_gene448 ARKGIGQMSGYGD--KLGEGGDQLVAGWILEQAQNAWTQG----
                     *  . *..:    :         *: ::    :
```

FIGURE 15.1 An alignment of the mature protein sequences deduced from different cyanobacterial cytochrome c_6 genes. The translated sequences as reported in the data bases were analyzed by a SignalP program (www.cps.dtu.dk/services/SignalP) to determine the cleavage sites of the signal peptides. The mature protein sequences were aligned by a ClustalW program (clustalw.genome.ad.jp). Key (each species name is followed by its database accession code or identification): 1. ana_asl0256, *Anabaena* sp. PCC7120 (asl0256); 2. Nos_Gene57, *Nostoc punctiforme* ATCC29133 (Contig597 Gene57); 3. Syn8102_gene434, *Synechococcus* sp. WH8102 (syn_wh Gene434); 4. Syn7002_1, *Synechococcus* sp. PCC7002 (petJ1 in reference #22); 5. Pro9313_A, *Prochlorococcus marinus* MIT9313 (Gene PMT0462); 6. Synechocystis, *Synechocystis* sp. PCC6803 (sll1796); 7. Thermosyn_BP1, *Thermosynechococcus elongatus* BP-1 (tll1283); 8. Glo_1906, *Gloeobacter violaceus* PCC7421 (glr1906); 9. ana_alr4251, *Anabaena* sp. PCC7120 (alr4251); 10. Nos_Gene104, *Nostoc punctiforme* ATCC29133 (Contig477 Gene104); 11. Tri_7635, *Trichodesmium erythraeum* (Contig122 Gene7635); 12. Glo_1980, *Gloeobacter violaceus* PCC7421 (gll1980); 13. ana_all0161, *Anabaena* sp. PCC7120 (all0161); 14. Nos_Gene11, *Nostoc punctiforme* ATCC29133 (Contig350 Gene11); 15. Syn7002_2, *Synechococcus* sp. PCC7002 (petJ2 in reference #22); 16. Tri_6986, *Trichodesmium erythraeum* (Contig122 Gene6986); 17. Pro1375_c6, *Prochlorococcus marinus* SS120 (COG2863); 18. Pro9313_J, *Prochlorococcus marinus* MIT9313 (Gene PMT0509); 19. Syn8102_gene448, *Synechococcus* sp. WH8102 (syn_wh Gene448). (*) identical residues; (:) and (.) similar residues.

anaerobic or acidic conditions. Second, a search through its genome sequence reveals a gene encoding a putative sulfide–quinone reductase that is thought to catalyze an early step in sulfide oxidation [26]. Lastly, a member of its genus *Nostoc muscorum* has been shown to catalyze H$_2$ production in the presence of sodium sulfide [27]. Thus, a comprehensive approach that relates different modes of growth to expression of different cyt c_6 genes is possible in the same organism and may lead to a better understanding of the roles of individual cyt c_6 genes in different metabolic pathways.

III. HIGHER PLANT CYT c_6-LIKE GENE

Studies of cyt c_6 have always been conducted with cyanobacteria and algae until very recently [1]. From information about genome sequences and cDNA sequences, it is now clear that a gene of a cyt c_6-like protein occurs in a variety of higher plants [28]. *Glycine max* (gi26049241) was probably the first plant identified with such a gene in the NIH Genbank database. Others include *Arabidopsis thaliana* (gi19863220), *Populus tremula × Populus tremuloides* (gi24057988), *Medicago truncatula* (gi11610049), *Antirrhinum majus* (gi31662857), *Solanum tuberosum* (gi13615108), *Lactuce sativa* (gi22411485), *Hordeum vulgare* (gi16311393), *Triticum aestivum* (gi20298797), *Aegilops speltoides* (gi11222607), *Ipomoea nil* (gi27239253), *Nicotiana tabacum* (gi32878225), *Oryza sativa* (gi34899810), *Poncirus trifoliate*

(gi34433042), and *Zea mays* (gi9900484). Most of the higher plant sequences are obtained from cDNA clones and appear to be incomplete. Figure 15.2 shows a comparison between mature protein sequences derived from the cyt c_6 genes of *Anabaena* sp. PCC7120 and the cyt c_6-like genes from *G. max*, *A. thaliana*, and *O. sativa*. The higher plant sequences are all acidic and seem to align better with an alkaline form of the *Anabaena* cyt c_6. They have a higher molecular mass than the *Anabaena* sequences and a 12 amino acid insert that is missing in the *Anabaena* sequences. This insert GFGKEC(M/T)PRGQC has so far been detected in most of the higher plant sequences found in the NIH Genbank database. It has been shown that *Arabidopsis* plants lacking both the cyt c_6-like protein and a plastocyanin are not viable [1]. However, the functional role of the cyt c_6-like protein in photosynthetic electron transport remains unclear [29].

IV. PROPERTIES AND ISOLATION OF CYT c_6

Cyt c_6 has been isolated from a variety of cyanobacteria including species similar to those that are now known to contain multiple cyt c_6 genes. While most studies resulted in the isolation of a single form of cyt c_6, some reported the isolation of isoforms differing in size, charge, hydrophobicity or location [30–34]. The isoforms are generally explained by posttranslational modifications or protein aggregation. In view of the present finding that multiple cyt c_6 genes occur among

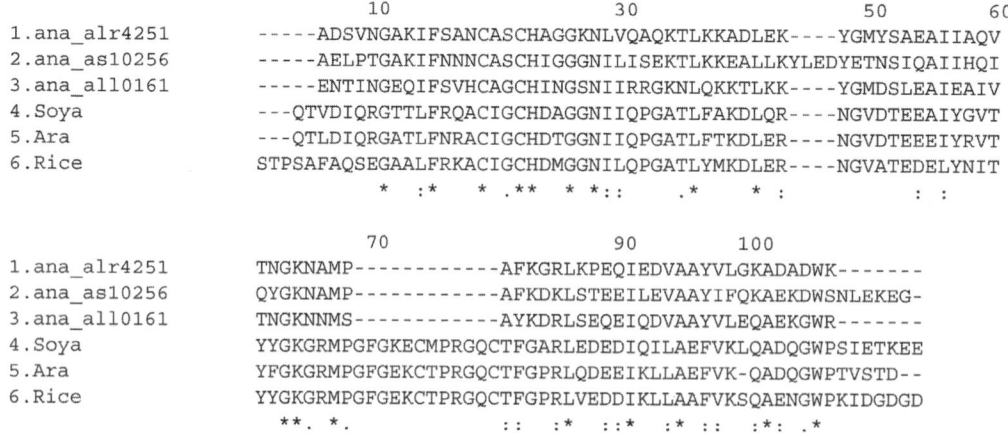

FIGURE 15.2 An alignment of the mature protein sequences deduced from cytochrome c_6 genes of *Anabaena* sp. PCC7120 and higher plants. The translated sequences as reported in the databases were analyzed by a SignalP program (www.cps.dtu.dk/services/SignalP) to determine the cleavage sites of the signal peptides. The mature protein sequences were aligned by a ClustalW program (clustalw.genome.ad.jp). Key (each species name is followed by its database accession code or identification): 1. ana_alr4251, *Anabaena* sp. PCC7120 (alr4251); 2. ana_asl0256, *Anabaena* sp. PCC7120 (asl0256); 3. ana_all0161, *Anabaena* sp. PCC7120 (all0161); 4. Soya, *Glycine max* (gi26049241); Ara, *Aradibopsis thaliana* (gi19863220); Rice, 6. *Oryza sativa* (gi34899810). (*) identical residues; (:) and (.) similar residues.

different species of cyanobacteria, it may be a time to reexamine the isoforms isolated from a few species. *Anacystis nidulans*, also known as *Synechoccocus* PCC6301, was the first to be recognized as having two isoforms of cyt c_6 [30,31]. One acidic form has been sequenced and crystallized for structure analysis. Less is known about the other basic form that is recovered in a small quantity. It may be cyt c_M (personal communication with DW Krogmann). *A. nidulans* is closely related to *Synechococcus* sp. PCC7002. It should be recalled that the latter contains two cyt c_6 genes, one for an acidic form of the protein and the other for a basic form. There is a reasonable chance that *A. nidulans* also contains two such genes and that its two cyt c_6 isoforms represent different gene products. A second study shows an acidic cyt c_6 with a molecular weight of 23,500 from *Oscillatoria* Bo32. Cyt c_6 proteins that have been isolated from other cyanobacteria grown under aerobic conditions generally show a smaller molecular weight ranging from 9,000 to 10,000 Da. It is possible that *Oscillatoria* Bo32 grown under normal aerobic conditions has an additional cyt c_6 similar to those found in other cyanobacteria. A smaller basic cyt c_6 has actually been isolated from a member of the same genus, *Oscillatoria princes* [32], which was collected from the bottom of a shallow pond. There may be an alternate explanation for the molecular weight of cyt c_6 in *Oscillatoria* Bo32 based on the observations made in *A. nidulans* and *Arthrospira maxima*. When purified, the acidic cyt c_6 in *A. nidulans* was reported to have a molecular weight of 23,000 per heme [31]. This value is about twice of that predicted by its protein sequence [35]. In *A. maxima*, a dimeric form of cyt c_6 has been detected on gel filtration chromatography (personal communication). Thus, the acidic cyt c_6 in *Oscillatoria* Bo32 could represent a dimeric form of a smaller protein. Another study reported the location of a basic cyt c_6 in the perisplasmic and intracellular spaces of *Nostoc* MAC [33]. When grown under chemoheterotrophic instead of photoautotrophic conditions, *Nostoc* MAC produced more cyt c_6 in the perisplasmic and intracellular spaces. This resulted in a tenfold increase of the perisplasmic cyt c_6 and a less than twofold increase of the intracellular cyt c_6, suggesting that the former protein might be expressed for dark respiration. Initial characterization of the two proteins did not reveal any difference between their molecular weights and isoelectric points. However, there is a need to further characterize these proteins in view of a previous experience with the cyt c_6 samples of *Oscillatoria princeps* and *Schizothrix calcicola* [32]. These samples, that had been purified by isoelectric focusing gels and judged to be homogenous, were resolved into different bands by re-

versed-phase HPLC. In addition, *N. punctiforme*, a species closely related to *Nostoc* MAC, is known to contain three cyt c_6, two of which encode two basic forms of a mature cyt c_6. From the deduced amino acid sequences, the two mature proteins are predicted to have similar molecular weights and isoelectric points. It will be of interest to see whether the two cyt c_6 proteins from *Nostoc* MAC can be resolved by reversed-phase HPLC prior to any attempt at protein sequencing. All the species discussed above were grown under laboratory conditions. Materials collected in nature have been found to yield isoforms of cyt c_6 as well, including *O. princeps* and *S. calcicola* [32]. These hardy species usually form a surface mat in the diverse environments where they survive. Since cyt c_6 plays an important role in photosynthesis and respiration, different isoforms of cyt c_6 would be expected in these species. The genome information about other filamentous species, *Anabaena* sp. PCC7120 and *N. punctiforme*, tends to support this expectation since the latter two species have more cyt c_6 genes than the less complicated, unicellular species. The work concerning *O. princeps* and *S. calcicola* resulted in the separation of forms differing in hydrophobicity using reversed-phase HPLC. If these isoforms can be verified by protein sequencing to be different, they can be used as tools for studying gene expression in different populations of the two species.

Cyt c_6 is a small water-soluble protein that is characterized by a distinct visible light absorption spectrum. These properties make it simple to isolate. An isolation usually starts with an aqueous extraction of a broken cell mass and follows by the application of a series of steps involving such techniques as ammonium sulfate precipitation, ultrafiltration, and column chromatography in different gel matrices and gel electrophoresis. The number of steps and the techniques being used depend on the scale of preparation. The details about the techniques have already been presented as a chapter in an earlier edition of this book. Despite its ease of isolation, cyt c_6 has been isolated only as a single protein from many different species of cyanobacteria. There are several possible explanations for this observation. First, the organism being studied has a single cyt c_6 gene like *Synechocytis* sp. PCC6803., *T. elongatus*, and *P. marinas* SS120. Second, in *Oscillatoria* Bo35 and *Nostoc* MAC, an increase level of cyt c_6 has been observed in cells grown in acidic sulfide environments or in the dark. This suggests that different cyt c_6 genes in the same organism may have different purposes. Since most of the studies were conducted with cells that were grown under photoautotrophic conditions, other cyt c_6 genes that are not relevant in such conditions might or might not be expressed. If expressed, the level of the gene products might be too low for detection. Third, the isolation protocol may not be applicable to different cyt c_6 proteins, particularly to the ones that are available in small quantities. Lastly, the other cyt c_6 genes may be simply noncoding DNA sequences. Whatever the explanation, it is obvious that there is a need to study the growth conditions for cyt c_6 gene expression as well as to have a more systematic characterization of cyt c_6 isolated from cells grown under different conditions. Here, a simple method has been introduced to isolate highly purified fractions of cyt c_6 from small cell samples in order to facilitate growth studies and the use of mass spectrometric-based protein identification to further characterization. For cyt c_6 purification, the problem is that the initial cell extract contains a large amount of colored pigments, particularly the phycobiliproteins. Methods to remove the colored pigments have been developed for large samples ranging from several liters to hundreds of liters of cells. The present method takes into account the small amounts of starting material and the

problem of phycobiliproteins. It uses a mixture of ethanol and chloroform to remove phycobiliproteins and other colored pigments that normally require repetitive chromatographic separation on gel matrices. The resulting aqueous layer is concentrated and exchanged into an appropriate buffer for mass spectrometry. The protocol described below has been successfully applied to *A. maxima*. Typically, cells harvested from 50 to 100 ml cultures are suspended in a 5 to 10 ml of Tris buffer (50 mM Tris/HCl, 1 mM EDTA, pH 7.5) and subjected to two cycles of freezing and thawing. The broken cell mass is brought to 40% ammonium sulfate saturation and spun at 14,000 rpm for 15 min. The resulting supernatant is brought to 80% ammonium sulfate saturation and spun again to give a final pellet that is resuspended in the Tris buffer. A 1-ml sample corresponding to ~5 to 10 μg chlorophyll is used for organic solvent extraction. A 0.5-ml sample of two parts of chloroform and three parts of ethanol previously chilled at −20°C is mixed with the protein sample kept on ice for

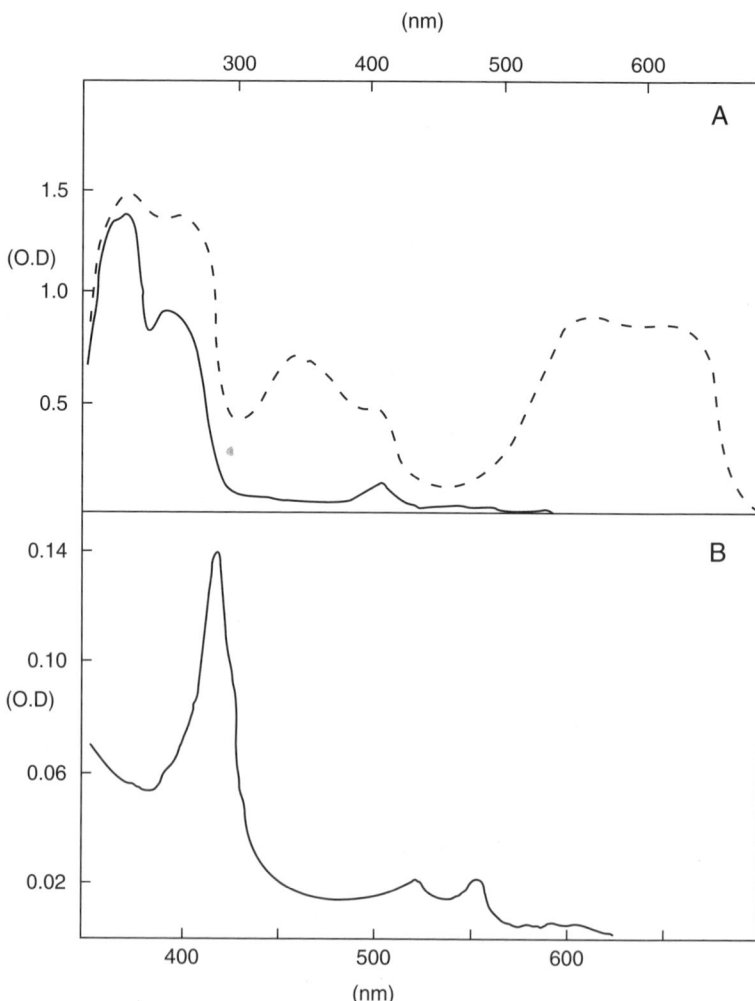

FIGURE 15.3 An absorption spectrum of *Lyngbya* spp. cytochrome c_6 purified by organic extraction as described in the text. (A) –, Prior to organic solvent extraction; after organic solvent extraction. (B) An enlarged view of the absorption spectrum from 700 to 350 nm showing the distinct 553, 523, and 418 nm peaks of the cytochrome c_6 after organic solvent extraction.

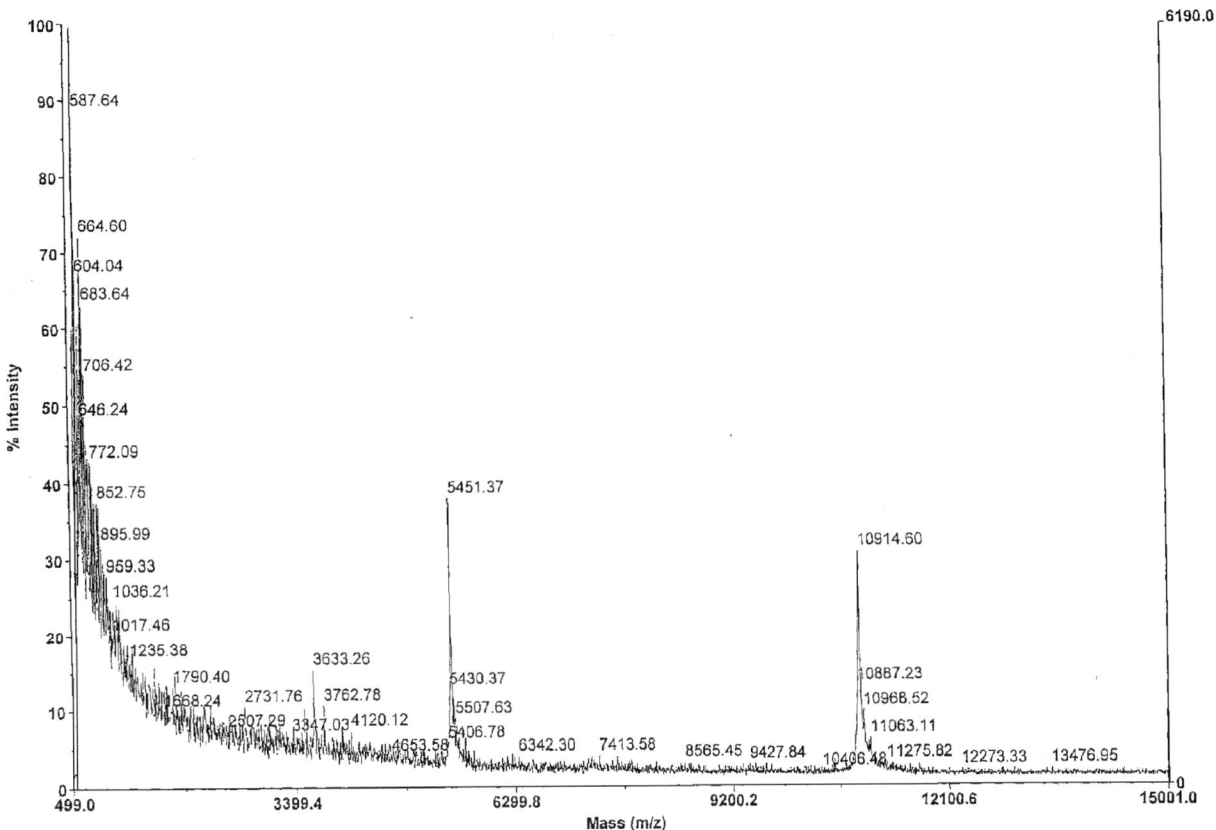

FIGURE 15.4 A matrix-assisted laser desorption/ionization (MALDI) spectrum of *Lyngbya* spp. cytochrome c_6 purified by organic extraction and concentrated as described in the text. The spectrum was produced by a Voyager-DE Pro time-of-flight mass spectrometer. The cytochrome c_6 sample shows a distinct peak with a mass of 10914.60.

~2 min and vortexed for 10 to 15 sec. The resulting mixture is spun at 14,000 rpm in a table top Eppendorf centrifuge for 5 min, and the clear aqueous layer is removed. Figure 15.3 shows the absorption spectra of the sample prior to organic extraction and the clear aqueous layer containing the cyt c_6. The latter is diluted 20 times in distilled water and concentrated to ~1 ml using a centrifugal filter unit (Centriprep YM-3, Amicon). The concentrated sample is diluted ten times in 1 mM Tris/HCl (pH 7.0) and reconcentrated to ~200 µl in the same filter unit. Figure 15.4 shows the result of matrix-assisted laser desorption/ionization (MALDI) mass spectrometry applied to a sample of the concentrated cyt c_6 sample. The peak at 10914.6 corresponds with the calculated mass of the cyt c_6 whose amino acid sequence was done (unpublished result by Krogmann DW). The results indicate that the present method offers a simple way to identify cyt c_6. This method also has the advantage of higher sensitivity and therefore will be useful for detecting isoforms in cyt c_6 samples.

V. HETEROLOGOUS EXPRESSION OF CYT c_6 GENE IN BACTERIA

The electron transport activities of cyt c_6 are achieved through protein–protein interactions between cyt c_6 and its interacting partners in the protein complexes of cyt $b_6 f$, PSI, and cyt oxidase. Comparing the interactions between different isoforms of cyt c_6 and individual complexes can provide information about the functional role of specific isoforms. This approach requires recombinant proteins that are produced by heterologous expression of different cyt c_6 genes in *Escherichia coli*. Cyt c_6 genes have been cloned from *Anabaena* sp. PCC 7119 [36], *T. elongatus* [37], and *Synechocystis* sp. PCC6803 [38] and expressed with some success in the bacteria. In the case of *Anabaena* sp. PCC 7119, the cyt c_6 gene was cloned in a pBluescriptII SK(+) vector and expressed in DH5α strain of *E. coli*. The yield of the purified cyt c_6 was reported to be ~200 µg/l of the cell culture. A similar study of *Synechocystis* sp. PCC6803 produced about the same amount of purified cyt c_6. In a more recent study, the yield of the *Synechocystis* sp. PCC6803. Cyt c_6 has been improved five to ten fold using a different strain of bacteria, *E. coli* MC1061 for protein expression [39]. The yield of cyt c_6 can be improved further by having the bacterial cells cotransformed with a plasmid containing the *E. coli* cyt maturation genes (*ccm*A to H). This system has been used for the production of other *c*-type cytochromes [40].

VI. CONCLUSION

The genome sequences have raised some intriguing questions, not the least being the role of specific cyt c_6 genes. Equally perplexing is why there are more cyt c_6 genes in the filamentous cyanobacteria than in the unicellular ones. One possible explanation is that different gene products are responsible for different functions ascribed to cyt c_6 in filamentous species like *N. punctifirme* and *Anabaena* sp. PCC6803. Biochemical evidence is not available to support this despite many attempts to isolate isoforms of cyt c_6. Clearly, a new approach to detect the isoforms is needed. Mass spectrometry has been used to identify isoforms of proteins associated with the thykaloid membranes [41]. Initial results presented here showed that this method could help identify individual isoforms of cyt c_6. Currently, this method is being extended to *Phormidium* 1058, *Lyngbya* spp. and *A. platensis*. Other questions raised by the genome sequences are why one of the cyt c_6 genes is close to a plastocyanin gene and whether different cyt c_6 genes are regulated differently. Since plastocyanin and cyt c_6 are interchangeable proteins in the photosynthetic pathway it is not unreasonable to assume that the close proximity of the two genes help coordinate their regulation. However, in *Anabaena* sp. PCC7120 the cyt c_6 gene thought to be responsible for photosynthetic electron transport is not the one that is close to the plastocyanin gene. The expression of this cyt c_6 gene is downregulated by copper [42]. Thus, it will be of interest to look at the upstream regions of the coding sequences among the three cyt c_6 genes and to see whether they are all responsive to copper regulation.

ACKNOWLEDGMENTS

The author is indebted to DW Krogmann for his helpful discussions and careful review of this manuscript.

REFERENCES

1. Gupta R, He Z, Luan S. Functional relationship of cytochrome c$_6$ and plastocyanin in *Arabidopsis*. *Nature* 2002; 417:567–571.

2. Kerfeld CA, Krogmann DW. Photosynthetic cytochromes *c* in cyanobacteria, algae, and plants. *Annu Rev Plant Physiol Plant Mol Biol* 1998; 49:397–425.

3. Kerfeld C, Ho KK, Krogmann DW. The cytochrome *c* of cyanobacteria. In: Peschek GA, Löffelhardt W, Schmetterer G, eds. *The Phototrophic Prokaryotes*. New York: Kluwer Academic/Plenum Publishers, 1999:259–268.

4. Peschek GA. Photosynthesis and respiration in cyanobacteria. Bioenergetic significance and molecular interactions. In: Peschek GA, Löffelhardt W, Schmetterer G, eds. *The Phototrophic Prokaryotes*. New York: Kluwer Academic/Plenum Publishers, 1999:201–209.

5. Padan E. Facultative anoxygenic photosynthesis in cyanobacteria. *Annu Rev Plant Physiol* 1979; 30:27–40.

6. Garlick S, Oren A, Padan E. Occurrence of facultative anoxygenic photosynthesis among filamentous and unicellular cyanobacteria. *J Bacteriol* 1977; 129:623–629.

7. Kaneko T, Nakamura Y, Wolk CP, Kuritz T, Sasamoto S, Watanabe A, Iriguchi M, Ishikawa A, Kawashima K, Kimura T, Kishida Y, Kohara M, Matsumoto M, Matsuno A, Muraki A, Nakazaki N, Shimpo S, Sugimoto M, Takazawa M, Yamada M, Yasuda M, Tabata S. Complete genomic sequence of filamentous nitrogen-fixing cyanobacterium *Anabaena* sp. strain PCC 7120. *DNA Res* 2001; 8:205–213.

8. Kaneko T, Nakamura Y, Wolk CP, Kuritz T, Sasamoto S, Watanabe A, Iriguchi M, Ishikawa A, Kawashima K, Kimura T, Kishida Y, Kohara M, Matsumoto M, Matsuno A, Muraki A, Nakazaki N, Shimpo S, Sugimoto M, Takazawa M, Yamada M, Yasuda M, Tabata S. Complete genomic sequence of filamentous nitrogen-fixing cyanobacterium *Anabaena* sp. strain PCC 7120 (Supplement). *DNA Res* 2001; 8:227–253.

9. Ohmori M, Ikeuchi M, Sato N, Wolk P, Kaneko T, Ogawa T, Kanehisa M, Goto S, Kawashima S, Okamoto S, Yoshimura H, Katoh H, Fujisawa T, Ehira S, Kamei A, Yoshihara S, Narikawa R, Tabata S. Filamentous nitrogen-fixing cyanobacterium *Anabaena* sp. strain PCC 7120. *DNA Res* 2001; 8:271–284.

10. Nakamura Y, Kaneko T, Sato S, Mimuro M, Miyashita H, Tsuchiya T, Sasamoto S, Watanabe A, Kawashima K, Kishida Y, Kiyokawa C, Kohara M, Matsumoto M, Matsuno A, Nakazaki N, Shimpo S, Takeuchi C, Yamada M, Tabata S. Complete genome structure of *Gloeobacter violaceus* PCC 7421, a cyanobacterium that lacks thylakoids. *DNA Res* 2003; 10:137–145.

11. Nakamura Y, Kaneko T, Sato S, Mimuro M, Miyashita H, Tsuchiya T, Sasamoto S, Watanabe A, Kawashima K, Kishida Y, Kiyokawa C, Kohara M, Matsumoto M, Matsuno A, Nakazaki N, Shimpo S, Takeuchi C, Yamada M, Tabata S. Complete genome structure of *Gloeobacter violaceus* PCC 7421, a cyanobacterium that lacks thylakoids (Supplement). *DNA Res* 2003; 10:181–201.

12. Meeks JC, Elhai J, Thiel T, Potts M, Larimer F, Lamerdin J, Predki P, Atlas R. An overview of the genome of *Nostoc punctiforme*, a multicellular, symbiotic cyanobacterium. *Photosynth Res* 2001; 70:85–106.

13. Rocap G, Larimer FW, Lamerdin J, Malfatti S, Chain P, Ahlgren NA, Arellano A, Coleman M, Hauser L, Hess WR, Johnson ZI, Land M, Lindell D, Post AF, Regala W, Shah M, Shaw SL, Steglich C, Sullivan MB, Ting CS, Tolonen A, Webb EA, Zinser ER, Chisholm SW. Genome divergence in two *Prochlorococcus* ecotypes reflects oceanic niche differentiation. *Nature* 2003; 424:1042–1047.

14. Dufresne A, Salanoubat M, Partensky F, Artiguenave F, Axmann IM, Barbe V, Duprat S, Galperin MY, Koonin EV, Gall FL, Makarova KS, Ostrowskii M, Oztas S, Robert C, Rogozin IB, Scanlani DJ, Tandeau de Marsac N, Weissenbach J, Wincker P, Wolf YI, Hess WR. Genome sequence of the cyanobacterium *Prochlorococcus marinus* SS120, a nearly minimal oxyphototrophic genome. *Proc Natl Acad Sci USA* 2003; 100:10020–10025.

15. Palenik B, Brahamsha B, Larimer FW, Land M, Hauser L, Chain P, Lamerdin J, Regala W, Allen EE, McCarren J, Paulsen I, Dufresne A, Partensky F, Webb EA, Waterbury J. The genome of a motile marine *Synechococcus* WH8120. *Nature* 2003; 424:1037–1042.

16. Kaneko T, Sato S, Kotani H, Tanaka A, Asamizu E, Nakamura Y, Miyajima N, Hirosawa M, Sugiura M, Sasamoto S, Kimura T, Hosouchi T, Matsuno A, Muraki A, Nakazaki N, Naruo K, Okumura S, Shimpo S, Takeuchi C, Wada T, Watanabe A, Yamada M, Yasuda M, Tabata S. Sequence analysis of the genome of the unicellular cyanobacterium *Synechocystis* sp. strain PCC6803. II. Sequence determination of the entire genome and assignment of potential protein-coding regions. *DNA Res* 1996; 3:109–136.

17. Kaneko T, Sato S, Kotani H, Tanaka A, Asamizu E, Nakamura Y, Miyajima N, Hirosawa M, Sugiura M, Sasamoto S, Kimura T, Hosouchi T, Matsuno A, Muraki A, Nakazaki N, Naruo K, Okumura S, Shimpo S, Takeuchi C, Wada T, Watanabe A, Yamada M, Yasuda M, Tabata S. Sequence analysis of the genome of the unicellular cyanobacterium *Synechocystis* sp. strain PCC 6803. II. Sequence determination of the entire genome and assignment of potential protein-coding regions (Supplement). *DNA Res* 1996; 3:185–209.

18. Nakamura, Y, Kaneko T, Hirosawa M, Miyajima N, Tabata S. CyanoBase, a WWW database containing the complete genome of *Synechocystis* sp. strain PCC 6803. *Nucleic Acids Res* 1998; 26:63–67.

19. Nakamura Y, Kaneko T, Sato S, Ikeuchi M, Katoh H, Sasamoto S, Watanabe A, Iriguchi M, Kawashima K, Kimura T, Kishida Y, Kiyokawa C, Kohara M, Matsumoto M, Matsuno A, Nakazaki N, Shinpo S, Sugimoto M, Takeuchi C, Yamada M, Tabata S. Complete genome structure of the thermophilic cyanobacterium *Thermosynechococcus elongatus* BP-1. *DNA Res* 2002; 9:123–130.

20. Nakamura Y, Kaneko T, Sato S, Ikeuchi M, Katoh H, Sasamoto S, Watanabe A, Iriguchi M, Kawashima K, Kimura T, Kishida Y, Kiyokawa C, Kohara M, Matsumoto M, Matsuno A, Nakazaki N, Shinpo S, Sugimoto M, Takeuchi C, Yamada M, Tabata S. Complete genome structure of the thermophilic cyanobacterium *Thermosynechococcus elongatus* BP-1. (Supplement) *DNA Res* 2002; 9:135–148.

21. Sazuka T, Ohara O. Sequence features surrounding the translation initiation sites assigned on the genome sequence of *Synechocystis* sp. Strain PCC6803 by amino-terminal protein sequencing. *DNA Res* 1996; 3:225–232.

22. Nomura C. Electron Transport Proteins from *Synechococcus* sp. PCC 7002. Ph.D. dissertation, Penn State University, Philadelphia, PA, 2001.

23. Rippka R, Waterbury J, Cohen-Bazire G. A cyanobacterium which lacks thykaloids. *Arch Microbiol* 1974; 100:419–436.

24. Moser D, Nicholls P, Wastyn M, Peschek G. Acidic cytochrome c_6 of unicellular cyanobacteria is an indispensable and kinetically competent electron donor to cytochrome oxidase in plasma and thylakoid membranes. *Biochem Int* 1991; 24:757–768.

25. Nicholls P. Obinger C, Niederhauser H, Peschek GA. Cytochrome c and c-554 oxidation by membranous *Anacystis nidulans* cytochrome oxidase. *Biochem Soc Trans* 1991; 19:252S.

26. Theissen U, Hoffmeister M, Grieshaber M, Martin W. Single eubacterial origin of eukaryotic sulfide:quinone oxidoreductase, a mitochondrial enzyme conserved from the early evolution of eukaryotes during anoxic and sulfidic times. *Mol Biol Evol* 2003; 20:1564–1574.

27. Fry I, Robinson AE, Spath S, Packer L. The role of Na_2S in anoxygenic photosynthesis and H_2 production in the cyanobacterium *Nostoc muscorum*. *Biochem Biophy Res Commun* 1984; 123:1138–1143.

28. Wastl J, Bendall DS, Howe CJ. Higher plants contain a modified cytochrome c_6. *Trends Plant Sci* 2002; 7:244–245.

29. Molina-Heredia FP, Wastl J, Navarro JA, Bendall DS, Hervás M, Howe CJ, De la Rosa MA. Photosynthesis: A new function for an old cytochrome. *Nature* 2003; 424:33–34.

30. Holton RW, Myers. Water-soluble cytochromes from a blue-green algae. I. Extraction, purification, and spectral properties of cytochrome c (549,552 and 554, *Anacystis nidulans*). *Biochim Biophys Acta* 1967; 131:362–374.

31. Holton RW, Myers J. Water-soluble cytochromes from a blue-green algae. II. Physicochemical properties and quantitive relationships of cytochromes c (549,552, and 554, *Anacystis nidulans*). *Biochim Biophys Acta* 1967; 131:375–384.

32. Ho KK, Krogmann DW. Electron donors to P700 in cyanobacteria and algae. An instance of unusual genetic variability. *Biochim Biophys Acta* 1984; 766:310–316.

33. Obinger C, Knepper JC, Zimmermann U, Peschek GA. Identification of a periplasmic C-type cytochrome as electron donor to the plasma membrane-bound cytochrome oxidase of the cyanobacterium *Nostoc* Mac. *Biochem Biophys Res Commun* 1990; 169:492–501.

34. Frier I, Rethmeier J, Fischer U. Molecular properties of soluble cytochrome c-552 and its participation in sulfur metabolism of *Oscillatoria* strain Bo32. In: Peschek GA, Löffelhardt W, Schmetterer G, eds. *The*

Phototrophic Prokaryotes. New York: Kluwer Academic/Plenum Publishers, 1999:275–280.

35. Ludwig ML, Pattridge KA, Powers TB, Dickerson RE, Takano T. Structure analysis of a ferricytochrome *c* from the cyanobacterium, *Anacystis nidulans.* In: Chien Ho, ed. *Interactions between Iron and Proteins in Oxygen and Electron Transport.* New York: Elsevier, 1982:27–32.

36. Molina-Heredia FP, Hervas M, Navarro JA, De la Rosa MA. Cloning and correct expression in *Escherichia coli* of the petE and petJ genes respectively encoding plastocyanin and cytochrome c_6 from the cyanobacterium *Anabaena* sp. PCC 7119. *Biochem Biophys Res Commun* 1998; 243:302–306.

37. Sutter M, Sticht H, Schmid R, Hörth P, Rösch P, Haehnel W. Cytochrome *c*6 from the thermophilic *Synechococcus elongatus.* In: Mathis P. ed. *Photosynthesis: From Light to Biosphere.* Vol. 2. Dordrecht, The Netherlands: Kluwer Academic Publishers, 1995:563–566. 38. Diaz A, Navarro F, Hervas M, Navarro JA, Chavez S, Florencio FJ, De la Rosa MA. Cloning and correct expression in *E. coli* of the petJ gene encoding cytochrome c_6 from *Synechocystis* 6803. *FEBS Lett* 1994; 347:173–177.

39. De la Cerda B, A Díaz-Quintana, JA Navarro, M Hervás, De la Rosa MA. Site-directed mutagenesis of cytochrome c_6 from *Synechocystis* sp. PCC 6803. The heme protein possesses a negatively charged area that may be isofunctional with the acidic patch of plastocyanin. *J Biol Chem* 1999; 274:13292–13297.

40. Cho YS, Pakrasi HB, Whitmarsh J. Cytochrome c_M from *Synechocystis* 6803. Detection in cells, expression in *Escherichia coli*, purification and physical characterization. *Eur J Biochem* 2000; 267:1068–1074.

41. Kieselbach T, Bystedt M, Hynds P, Robinson C, Schröder WP. A peroxidase homologue and novel plastocyanin located by proteomics to the *Arabidopsis* chloroplast thylakoid lumen. *FEBS Lett* 2000; 480:271–276.

42. Ghassemian M, Wong B, Ferreira F, Markley JL, Straus NA. Cloning, sequencing and transcriptional studies of the genes for cytochrome *c*-553 and plastocyanin from *Anabaena* sp. PCC 7120. *Microbiology* 1994; 140:1151–1159.

Section IV

Atmospheric and Environmental Factors Affecting Photosynthesis

16 External and Internal Factors Responsible for Midday Depression of Photosynthesis

Da-Quan Xu and Yun-Kang Shen
Shanghai Institute of Plant Physiology, Chinese Academy of Sciences

CONTENTS

I. INTRODUCTION

Midday depression of photosynthesis occurs in many plants and significantly affects crop yields. Since it was discovered at the beginning of the last century, many studies have been carried out, and several hypotheses, such as feedback inhibition of photosynthesis resulting from assimilate accumula-tion, stomata closure, enzyme deactivation, and reversible decline in photochemical activity, have been proposed to explain the phenomenon [1–4]. In recent years, the midday depression has been scrutinized by modern techniques. However, its causal mechanism is still not established [4]. Based on available data, the ecological, physiological, and biochemical factors related to the midday depression are analyzed and the possible mechanisms and adaptive importance are discussed in this chapter.

II. THE PHENOMENON

A. PATTERN OF DIURNAL VARIATION FOR PHOTOSYNTHESIS

Under natural conditions there are two typical patterns of photosynthetic diurnal course [5]. One is one-peaked, that is, net photosynthetic rate increases gradually with the increase in sunlight intensity in the morning, reaches its maximum around noon, then decreases gradually with the decrease in sunlight intensity in the afternoon. Another is two-peaked, that is, there are two peak values of net photosynthetic rate, one in late morning and the other in late afternoon with a depression around noon, the so-called midday depression of photosynthesis, as shown in Figure 16.1 (curves 1 and 2).

B. MIDDAY DEPRESSION OF PHOTOSYNTHESIS

Midday depression of photosynthesis is a common phenomenon. It may occur in many species of plants including C_3, C_4, and calmodulin (CAM) plants under a particular combination of environmental conditions [6]. In plants that show midday depression, however, it does not necessarily occur in all situ-

ations. For example, in some plants midday depression occurs in summer but not in winter [7,8]. In addition, this phenomenon is remarkable only in the upper-layer leaves of cassava [9].

When the midday depression is serious, no second peak in the diurnal course of photosynthesis appears [10]. The single-peaked curve of the diurnal course of photosynthesis in such cases differs very much from those where the midday depression is absent. For the former the peak value of net photosynthetic rate is often in the morning (Figure 16.1, curve 3), but the peak value is at noon for the latter (Figure 16.1, curve 1).

III. ECOLOGICAL FACTORS RESPONSIBLE FOR MIDDAY DEPRESSION

A. SUNLIGHT

In general, the two-peaked diurnal course of photosynthesis occurs on clear days with intense sunlight, while the one-peaked diurnal course occurs on cloudy days with weak sunlight [2,11]. Naturally, it is assumed that the midday depression is caused by intense light. Nevertheless, it may occur at medium light of about 500 µmol photons/m^2/sec [12,13]. Although intense light is not a necessary condition for midday depression to occur, in fact, intense sunlight is the most important ecological factor for midday depression. In some cases, it may lead indirectly to midday depression through low humidity and high temperature because intense sunlight is the primary driving force of diurnal variation in many environmental conditions. In other cases, it may result in midday depression through downregulation of photosynthetic capacity caused by intense sunlight, as observed in some woody plants [14].

B. AIR TEMPERATURE

Herppich et al. [15] reported that *Protea acaulos*, a prostrate fynbos shrub, often experiences very low air humidity at leaf temperatures over 10°C higher than mean air temperature, and shows a pronounced midday depression of gas exchange at the end of the dry summer season, independent of water supply. However, artificially lowered leaf temperatures in a gas exchange cuvette can prevent this midday depression almost completely under the same light conditions. Therefore, they considered that leaf temperature, directly or via the vapor pressure deficit (VPD) between leaf and air, rather than plant water status, is the determinant of midday depression. Around noon, high temperature can enhance CO_2 efflux from respiration or photorespiration, causing a decline in net

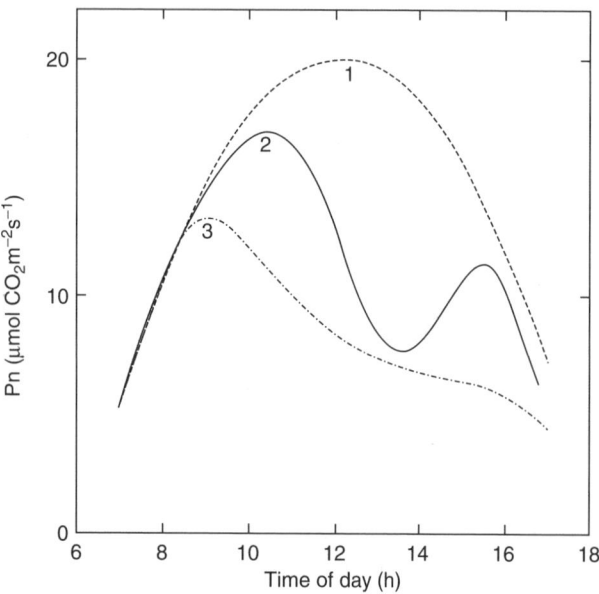

FIGURE 16.1 Schematic diagram of diurnal variation of net photosynthetic rate in plant leaves. Curve 1, one-peaked diurnal course; curve 2, two-peaked diurnal curve; curve 3, one-peaked diurnal course, but with severe midday depression.

photosynthetic rate to some extent. High temperature can also lead to a decrease in activated Rubisco [16]. High VPD can induce stomatal closure, limiting photosynthetic CO_2 uptake due to decreased CO_2 availability and exacerbating photoinhibition due to excessive light energy, thereby leading to a decrease in net photosynthetic rate.

C. AIR HUMIDITY

Photosynthesis in many plants is highly sensitive to changes in air humidity, or, more precisely, VPD. One-peaked diurnal course of photosynthesis could be artificially induced by high air humidity even at the end of the dry season when two-peaked patterns are common in natural weather [17]. Under low air humidity the two-peaked diurnal course of photosynthesis was observed in apricot even when soil water status was good [18]. Net photosynthetic rate in cassava decreased rapidly as VPD increased [19]. In wheat a significant negative correlation between net photosynthetic rate and air saturation deficit was observed. Furthermore, increasing air humidity led to an increase in net photosynthetic rate and to disappearance of midday depression [20]. It was found in maize that both high photon flux density and high air saturation deficit were necessary for afternoon inhibition of photosynthesis to appear [21]. The afternoon declines in canopy CO_2-exchange rates found in a number of species were associated with an increase in VPD [22]. It was observed that enhanced air humidity increased not only net photosynthetic rate but also the optimal temperature of photosynthesis in wheat leaves [23]. The nonstomatal mechanism by which air humidity affects photosynthesis is not clear [24].

Due to its effect on VPD, influence of temperature is often closely linked to air humidity impact on the diurnal course. Raschke and Resemann [13] demonstrated the dominant role of humidity in the induction of midday depression in *Arbutus unedo* leaves. The depression occurred at a constant leaf temperature in their experiment when a threshold in VPD was exceeded, but the depressions were hardly noticeable when VPD was held constant and leaf temperature was allowed to vary within a certain range.

Low air humidity has been considered an important ecological factor responsible for the midday depression [13,25–30].

D. SOIL WATER STATUS

Among environmental factors, soil water status seems to be a decisive factor in midday depression of photosynthesis. For instance, with a decline in soil water potential, a one-peaked diurnal course of photosyn-

thesis in soybean leaves became two-peaked, and midday depression became more severe [31]. After heavy rain, midday depression disappeared almost completely in wheat leaves on the following day [32]. Leaf water potential at dawn is a reflection of soil water status. As the leaf water potential at dawn declined, the pattern of the diurnal course of photosynthesis in soybean leaves changed from one-peaked to two-peaked, and the midday depression gradually became severe [33]. In addition, it was observed that midday depression of photosynthesis occurred in pot-grown, but not in field-grown, wheat under the same aboveground conditions (D.-Q. Xu et al., unpublished data). This difference was also reported between field-grown and pot-grown soybean plants [34]. Of course, the effect of soil water status on leaf photosynthesis is indirect. Many studies have suggested that under drought conditions stomatal closure often plays the main role in the decline in leaf photosynthesis, that is, photosynthetic biochemistry and photochemistry are not impaired by the lack of water [35].

E. CARBON DIOXIDE CONCENTRATION IN THE AIR

Midday depression of photosynthesis is often accompanied by decreased air CO_2 concentration around noon. Some researchers consider the decreased CO_2 concentration as an important ecological factor leading to midday depression [36]. However, according to Xu et al. [20], the extent of the decline in CO_2 concentration did not match the extent of the midday depression. Moreover, the air CO_2 concentration did not increase when the second peak of net photosynthetic rate in the daily course appeared, indicating that the diurnal variation pattern in net photosynthetic rate is not dependent on the air CO_2 concentration. The midday depression in *Quercus suber* persisted even at a CO_2 partial pressure of 250 Pa [37]. It appears that decreased air CO_2 concentration around noon is not an important ecological factor for midday depression.

IV. PHYSIOLOGICAL FACTORS RESPONSIBLE FOR MIDDAY DEPRESSION

A. STOMATAL CLOSURE

In some plants midday closure of stomata occurs [5,38], and it is often coincident with midday depression of photosynthesis [13,18,20]. However, whether the midday closure of stomata is the cause of midday depression of photosynthesis cannot be established only on the basis of a change in stomatal conductance. According to Farquhar and Sharkey [39], stomatal closure can be considered an important cause of decline in photosynthetic rate only when the intercellular space CO_2 partial pressure (C_i) also decreases.

A decreased C_i was observed when midday depressions in net photosynthetic rate and stomatal conductance occurred in bamboo [14,40], wheat [41], soybean [42,43], *Ginkgo biloba* [44], and strawberry [45]. These reports indicate that stomatal partial closure is indeed responsible for midday depression of photosynthesis.

Although among the 37 cases of midday depression investigated, 19 were accompanied by a reduction in C_i of 1 to 3 Pa, Raschke and Resemann [13] concluded that the midday depression of photosynthesis in leaves of *A. unedo* was not caused by stomatal closure. However, it is not clear whether nonuniform stomatal closure occurs in their experiments. Due to the patchy closure of stomata under stress conditions [46–48], overestimated C_i may lead to the misinterpretation that the reduction in photosynthesis caused by stomatal closure results from nonstomatal factors.

In general, C_i is calculated from leaf gas exchange data according to the equation $C_i = C_a - A/G_c$, where C_a and C_i are the partial pressures of CO_2 in the air and inside the leaf, and A and G_c are net photosynthetic rate and stomatal conductance to diffusion of CO_2, respectively [39]. From this equation, it is very clear that C_i decreases rarely in proportion to the decrease in A when A and G_c decrease simultaneously. In fact, during midday depression the magnitude of C_i decrease is often much less than that of the decrease in net photosynthetic rate. For instance, compared with the value of the first peak, net photosynthetic rate in wheat leaves decreased by about 48% during midday depression, while C_i decreased by only 11%, although an analysis showed that stomatal closure was the most important physiological cause of midday depression [41]. It is likely that an increased CO_2 efflux from respiration or photorespiration is responsible for the difference in the extent of decline between A and C_i because the CO_2 efflux leads to a decrease in A and an increase in C_i. Therefore, stomatal limitation of photosynthesis during midday depression cannot be precluded based only on the fact that the extent of C_i decline is less than that of the decline in net photosynthetic rate. Furthermore, when A and G_c evidently decline in a coordinated way, namely, the plot of A against G_c is linear, or patchy closure of stomata occurs, calculated C_i from the equation is unchanged because A/G_c is constant, but actually C_i is changed. Thus, such an apparently constant C_i is likely to mask the fact of stomatal limitation, forming an artifact of nonstomatal limitation of photosyn-

thesis. In other words, only when C_i increases can one confidently say that the decline in net photosynthetic rate results from a nonstomatal factor.

B. ENHANCEMENT OF RESPIRATION AND PHOTORESPIRATION

There is evidence that a rise in respiration or photorespiration near noon is one of the physiological causes of midday depression. Thus, in the leaves of *Q. suber* a substantial increase in the CO_2 compensation point has been observed during midday depression of photosynthesis [37], implying that respiration and photorespiration are enhanced by the higher leaf temperature around noon. In satsuma mandarin (*Citrus unshiu* Marc) midday depression of both net photosynthetic rate and apparent photosynthetic quantum efficiency has been attributed to increased photorespiration around noon [49]. The increased photorespiration may be a response to high light or the decline in C_i due to midday closure of stomata.

C. INCREASE IN MESOPHYLL RESISTANCE

Mesophyll resistance to CO_2 diffusion should be considered when one explores further the physiological causes of midday depression. In soybean leaves both stomatal resistance and mesophyll resistance increased during the midday depression of photosynthesis [33]. Mesophyll resistance seems to play a more important role in some conifers [2].

D. DECREASE IN LEAF WATER POTENTIAL

As a consequence of the larger evaporative demand near noon, there is usually a midday depression of leaf water potential. The diurnal course of change in leaf water potential similar to that of photosynthesis was observed in some conifers [2]. In some experiments with *Helianthus annuus*, however, no unique relationship among stomatal conductance, photosynthetic rate, and leaf water potential was observed, but stomatal conductance and net photosynthetic rate decreased when about two thirds of the extractable water in the soil had been used irrespective of the leaf water potential. Therefore, it was suggested that soil water status, not leaf water status, affected the stomatal behavior and photosynthesis of *H. annuus* [50].

E. DEVELOPMENT STAGE

Gao et al. [51] reported that under high temperature and low humidity midday depression of photosynthesis could occur in spring, summer, and autumn, and it occurred easily at the grain-filling stage in field-grown and pot-grown soybean plants. It is not clear, however, why midday depression occurs easily at this stage. There is a possibility that at this stage a particular microclimate around soybean plants or a combination of light, temperature, and water factors, leads easily to midday depression.

F. CIRCADIAN RHYTHM

Many studies have shown that midday depression is not related to circadian rhythm. Under simulated habitat conditions in a growth chamber, increasing atmospheric stress in the form of higher temperature and lower humidity resulted in midday depression of transpiration rate and net photosynthetic rate of the leaves in *Arbulus unedo* and *Quercus ilex* due to midday stomatal closure, while midday depression did not occur when the atmospheric stress was absent. These experiments were carried out under the same light conditions on four consecutive days [38]. It was demonstrated by experiments in which only one environmental variable changed at a time while all others were held constant that a circadian component was not essential for the development of midday depression in *A. unedo* L. [13]. Obviously, the fluctuation in atmospheric conditions rather than circadian rhythm is responsible for midday depression.

Under constant conditions net photosynthetic rate in peanut (*Arachis hypogaea*) leaf displayed a rhythm change within a period of about 24 hr, but its valley value or depression was at midnight not at midday [52]. Gao et al. [53] reported that under relatively constant conditions of light, temperature, humidity, and CO_2 concentration, net photosynthetic rate and stomatal conductance were lower in the morning and afternoon, and higher around noon, indicating a periodic change, namely circadian rhythm. Nevertheless, the periodic change is not related to midday depression of photosynthesis observed in the field. Their experiments showed that midday depression of photosynthesis was negligible after soybean plants were transferred to relatively constant conditions from field conditions where they often displayed a remarkable midday depression. This fact indicates that under natural conditions the environmental factors rather than circadian rhythm are the determinants for the daily pattern of photosynthesis.

There is another view on the relationship between midday depression and circadian rhythm. On the basis of a remarkable midday depression of photosynthesis in rice plant observed under constant light and temperature conditions, Deng and Chen [54] concluded that midday depression is related to circadian rhythm. However, it is not clear whether air humidity around rice plants was constant during their observation.

V. BIOCHEMICAL FACTORS RESPONSIBLE FOR MIDDAY DEPRESSION

A. PHOTOSYNTHATE ACCUMULATION

In 1868 Boussingault [55] first proposed a hypothesis that the accumulation of assimilates in an illuminated leaf might result in a reduction in net photosynthetic rate. Some investigators are in favor of the hypothesis and consider the photosynthate accumulation to be an important cause of the midday depression of photosynthesis [56]. Nevertheless, some studies have indicated that photosynthate accumulation has no negative effect on photosynthesis under normal conditions without environmental stress or block of assimilate export from leaves [12,57]. Moreover, it has been observed that the photosynthate content in wheat leaves is not higher during midday depression than in the morning when photosynthesis is actively going on. Net photosynthetic rate in wheat leaves decreased by less than 10% even when photosynthate contents were much higher than the control after blocking of photosynthate export from the leaves for 6 hr by heat girdling of the leaf sheath [20]. Undoubtedly, the effect should be even less when photosynthate export is normal. Therefore, photosynthate accumulation is not a likely cause of midday depression.

B. DECREASE IN RUBISCO ACTIVITY

Rubisco is a key enzyme in photosynthetic carbon assimilation. It often limits the maximal net photosynthetic rate [58–60]. However, there is a great deal of evidence indicating that plants may contain excess Rubisco and that photosynthesis may be controlled by several enzymes or processes [61]. Perhaps, the activated amount rather than the total amount of Rubisco often limits the maximal photosynthesis. In consonance with this supposition, a soybean cultivar with a higher net photosynthetic rate had a higher carboxylation efficiency and higher initial activity of RuBP carboxylation of Rubisco [62]. In addition, under unfavorable conditions net photosynthetic rate may be maintained by a greater concentration of Rubisco [63]. A midday decline in carboxylation efficiency, namely, the initial slope of the $A–C_i$ curve, associated with midday depression of photosynthesis has been observed in *Q. suber* leaves [37,64]. Furthermore, Jiang et al. [65] reported that midday depression of net photosynthetic rate was accompanied by a midday decline of Rubisco initial activity in rice flag leaves. It seems that midday depression is related to a decrease in Rubisco activity or content of activated Rubisco. However, one cannot be sure whether the decreased Rubisco activity is the main reason for midday depression of photosynthesis because of the lack of data on diurnal variation in stomatal conductance and intercellular CO_2 concentration measured simultaneously.

C.   ENHANCED ABA BIOSYNTHESIS

There is a possibility that abscisic acid (ABA) is an important biochemical factor responsible for midday depression. In the daily course of ABA content change, a midday peak associated with midday stomatal closure was observed in grape (*Vitis vinifera*) leaves [66]. Unfortunately, the diurnal variation in net photosynthetic rate was not measured simultaneously in this study. Thus, the relationship between ABA and midday depression of photosynthesis is still an open question.

D.   DECLINE IN PHOTOSYSTEM II PHOTOCHEMICAL EFFICIENCY

On clear days the midday decline in photosynthetic efficiency, expressed in apparent quantum yield of CO_2 uptake or chlorophyll fluorescence parameter F_v/F_m, a measure of phostosystem II (PS II) photochemical efficiency, often occurs in plants [67–70]. Naturally, the question arises whether the midday depression of net photosynthetic rate often observed results from the midday decline in photosynthetic efficiency.

Demmig-Adams et al. [3] observed that the midday depressions of net photosynthetic rate and stomatal conductance were accompanied by decreases in F_v/F_m and apparent quantum yield of O_2 evolution in *A. unedo* leaves. However, they were not sure whether this reduction in photochemical efficiency is serious enough to limit CO_2 fixation in high light and thereby to impose a nonstomatal limitation to net CO_2 uptake in *A. unedo* in the field at noon.

It should be pointed out that midday depression of the photosynthetic rate is always observed at saturating light, while the photosynthetic quantum efficiency is often measured at low light intensity. Therefore, decreased efficiency does not necessarily lead to a decrease in light-saturated rate because strong sunlight may compensate for the decline in PS II efficiency to maintain the high rate to some extent. The light-saturated rate of photosynthesis began to decrease when photoinhibition reached a level of 40% to 60%, and at a lower inhibition level the efficiency, but not the light-saturated O_2 production, was affected [71,72]. In wheat flag leaves a midday decline in photosynthetic efficiency was not invariably accompanied by midday depression of net photosynthetic rate. Intercellular CO_2 concentration decreased when midday depression of both the effi-

ciency and the rate occurred simultaneously. Furthermore, photosynthetic rate was correlated with stomatal conductance and intercellular CO_2 concentration to a higher level of significance than with photosynthetic efficiency. These facts indicate that midday decline of photosynthetic efficiency may be, if at all, a less important cause of midday depression of net photosynthetic rate than midday closure of stomata in the case studied [41].

Some woody plants require a lower light intensity (a photon flux density not more than one half of full sunlight) to saturate photosynthesis. Thus, in these plants severe photoinhibition, characterized by a decrease in the quantum efficiency of photosynthetic carbon assimilation and a decline in PS II photochemical efficiency caused by excessive light energy, often occurs around noon on clear days. For these plants the main immediate cause of midday depression may be the decline in PS II photochemical efficiency induced by strong sunlight. In summer, midday depression of both the efficiency and the rate often occurred in the upper leaves of the bamboo canopy, while intercellular CO_2 concentration declined first, and then increased. These facts indicate that midday depression of net photosynthetic rate is related to decline in photochemical efficiency, at least in part [14]. Similarly, midday depression of net photosynthetic rate was accompanied by a pronounced decrease in leaf conductance and a substantial increase in intercellular CO_2 concentration, as well as a considerable decline in PS II photochemical efficiency (F_v/F_m) in *P. acaulos* [15]. Midday depression in tea (*Camellia sinensis*) [11] and grapevine (*Viitis uinifera*) [73] leaves has been attributed to photoinhibition. Results from other studies also show that photoinhibition may be a factor contributing to midday depression of photosynthesis [4,74].

As mentioned above, midday depression of net photosynthetic rate is closely related to many factors such as stomatal partial closure, decreased Rubisco activity, and declined PS II photochemical efficiency. Then, which of them, stomatal or nonstomatal factor, is the main cause of midday depression when these factors exist simultaneously? The data of change in intercellular CO_2 concentration (C_i) during midday depression may help to answer this question. In general, stomatal partial closure or a decrease in stomatal conductance may lead to a decreased C_i, whereas the decline in photosynthetic activity of leaf mesophyll cells such as a decrease in Rubisco carboxylation activity or PS II photochemical efficiency may induce an increase in C_i. The direction, increase or decrease, of change in C_i depends on the predominant one when changes in these factors occur simultaneously. When the decreases in stomatal conductance,

Rubisco activity, and PS II photochemical efficiency occur simultaneously during midday depression, for example, if C_i declines, the main cause of midday depression is the decreased stomatal conductance. On the contrary, if C_i increases, the main cause must be the decreases in Rubisco activity and PS II photochemical efficiency. In this case, the direction rather than the extent of change in C_i is important for making the conclusion [75].

E. POSSIBLE MECHANISMS

From most of the evidence cited above, it is suggested that for midday depression of photosynthesis, strong sunlight, low air humidity or high VPD, and low soil water potential may be the main environmental factors, decreased stomatal conductance may be the most important physiological factor, and increased ABA synthesis and decreased PS II photochemical efficiency may be the most important biochemical factors. Of course, these factors are closely linked to each other. Strong sunlight causes an increase in air temperature and a decrease in air relative humidity and soil water potential because of enhanced plant transpiration. These changes in ecological factors re-

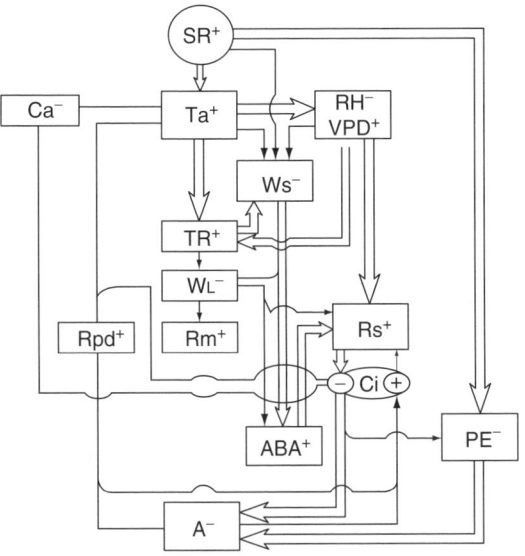

FIGURE 16.2 Possible relationships between ecological, physiological, and biochemical factors and midday depression. SR, solar radiation; Ta, air temperature; Ca, CO_2 concentration in the air; RH, relative humidity; VPD, water vapor pressure deficit from leaf cell to air; Ws, soil water potential; TR, transpiration; W_L, leaf water potential; Rpd, photorespiration and respiration; Rm, mesophyll resistance to CO_2 diffusion; Ci, CO_2 concentration in intercellular space; PE, photochemical efficiency; ABA, abscisic acid; A, net photosynthetic rate. "+" and "−" indicate increase and decrease, respectively. Double-line arrow indicates a strong effect.

sult in variations in the physiological and biochemical factors. Low soil water potential leads to increase in ABA synthesis, and both increased ABA and increased VPD cause a decrease in stomatal conductance, resulting in a decline in net photosynthetic rate due to decreased CO_2 supply. However, midday closure of stomata is not the sole important physiological or biochemical cause of midday depression. In some woody plants such as bamboo and tea, the decline in PS II photochemical efficiency induced by strong sunlight may be the most important biochemical cause of midday depression. Perhaps the main immediate cause and mechanism of midday depression are different for different plant species under various conditions. The factors related to midday depression are shown in Figure 16.2 [76,77].

VI. ADAPTIVE IMPORTANCE

A. ADAPTIVE IMPORTANCE

In many cases midday depression of photosynthesis seems to be a strategy to cope with environmental stresses formed during evolution. Midday stomatal closure and downregulation of photochemical efficiency are effective ways to avoid excess water loss and photodamage of the photosynthetic apparatus under strong sunlight and dry conditions.

Midday stomatal closure may be a response to low air humidity or high VPD. In this case midday closure of stomata is an important physiological cause of midday depression of photosynthesis. Alternately, midday closure of stomata may be a response to increased intercellular space CO_2 concentration due to a decline in mesophyll photosynthetic activity or increase in respiration and photorespiration. In this case midday closure of stomata is the result rather than the cause of decreased photosynthetic rate. In any case, midday stomatal closure always increases the water use efficiency of plants [78–80]. This is because of the predominant occurrence of leaf gas exchange in the morning and in the afternoon when net photosynthetic rate is higher and transpiration rate is lower. Obviously, such stomatal regulation is quick, reversible, and favorable for growth and development of plants under dry conditions of air and soil.

Downregulation of photochemical efficiency around noon is often observed in many plants under field conditions on clear days [69,81]. In some cases it may be responsible for the midday depression, for example, in the leaves of some woody plants such as bamboo and tea. Such downregulation may be due to enhanced thermal energy dissipation related to the xanthophyll cycle or the reversible inactivation of PS II, which is considered to be an important mechanism to protect the photosynthetic apparatus from photo-damage [82–84]. Although there have been many studies, the molecular mechanism of such thermal energy dissipation is not yet clear [85–87].

B. MEASURE OF ALLEVIATION

Midday depression of photosynthesis, as a regulation process, is advantageous for the survival of plants under stress conditions, but it is at the expense of effective use of light energy and plant productivity. Midday depression may decrease crop productivity by 30% to 50% or more. Therefore, it is worthwhile to search for alleviating or eliminating measures. Under strong-light and high-transpiration conditions, midday mist irrigation could increase stomatal conductance and photosynthetic rate in leaves of *Beta vulgaris* despite adequate soil water supply [88]. Mist irrigation for 40 days not only increased the photosynthetic rate in cassava leaves but also increased production of dry roots (91%) and total biomass (27%) [89]. Similar effects of mist irrigation were observed in wheat and soybean plants. Mist irrigation in the grain-filling period increased stomatal conductance and net photosynthetic rate in flag leaves, thus increasing grain yield by about 18% in wheat [32]. Mist irrigation in the seed-filling period increased the seed yield by about 19% in soybean [10].

VII.   CONCLUDING REMARKS

Midday depression of photosynthesis is a common phenomenon in higher plants. It is related to many external and internal factors interacting with each other. Midday stomatal closure or decreased photochemical efficiency may cause the midday depression, depending on plant species and environmental conditions. It may be a strategy of plants to cope with environmental stresses. Further study on the mechanisms of midday depression is required for understanding the regulation of photosynthesis and finding ways to increase plant productivity. Because the present viewpoints and hypotheses about these mechanisms are based on inadequate or incomplete data, in the following studies a better combination of many kinds of experimental methods, such as physiological, biochemical, and biophysical ones, is absolutely necessary for getting more abundant data to reveal exactly these mechanisms.

REFERENCES

1. Kostyschew S, Kudriavzewa M, Messejewa W, Smirnova M. Der tagliche verlauf der photosynthethesc bci landpflanzen. *Planta* 1926; 1: 679–699.

2. Hodges JD. Patterns of photosynthesis under natural environmental conditions. *Ecology* 1967; 43: 234–242.

3. Demmig-Adams B, Adams WWIII, Winter K, Meyer A, Schreiber U, Pereira JS, Kruger A, Czygan F-C, Lange OL. Photochemical efficiency of photosystem II, photon yield of O₂ evolution, photosynthetic capacity, and carotenoid composition during the midday depression of net CO₂ uptake in *Arbutus unedo* growing in Portugal. *Planta* 1989; 177: 377–387.

4. Geiger DR, Servaites JC, Shieh W-J. Balance in the source-sink system: a factor in crop productivity. In: Baker NR, Thomas H, eds. *Crop Photosynthesis: Spatial and Temporal Determinations*. Amsterdam : Elsevier Science Publisher, 1992: 155–176.

5. Schulze E-D, Hall AE. Stomatal responses, water loss and CO₂ assimilation rates of plants in contrasting environments. In: Lange OL, Nobel PS, Osmond CB, Ziegler, eds. *Physiological Plant Ecology II. Encyclopedia of Plant Physiology (NS)*. Vol. 12B. Berlin: Springer-Verlag, 1982: 181–230.

6. Osmond CB, Winter K, Powles SB. Adaptive significance of carbon dioxide cycling during photosynthesis in water-stressed plants. In: Turner NC, Kramer PJ, eds. *Adaptation of Plants to Water and High Temperature Stress*. New York: John Wiley & Sons, 1980: 139–154.

7. Hellmuth EO. Eco-physiological studies on plants in arid and semi-arid regions in western Australia I. Autecology of *Rhagodia baccata* (*Labill*) MOQ. *J. Ecol.* 1968; 56: 319–344.

8. Hellmuth EO. Eco-physiological studies on plants in arid and semi-arid regions in western Australia II. Comparative studies on photosynthesis, respiration and water relations of ten arid zones and two semi-arid zones plants under winter and late summer climatic conditions. *J. Ecol.* 1971; 59: 225–259.

9. San Lose JJ. Diurnal course of CO₂ and water vapor exchange in *Manihot esculenta* Crantz var. *cubana*. *Photosynthetica* 1983; 17: 12–19.

10. Zheng G-S, Zou Q. Effect of spray irrigation on yield of field-grown soybean and diurnal variation of photosynthesis. *Acta Agric. Sin.* 1993; 1: 302–305 (in Chinese).

11. Tao H-Z. Studies on the diurnal variations of photosynthesis of tea plant (*Camellia sinensis*). *Acta Agron. Sin.* 1991; 17: 444–452 (in Chinese).

12. Xu D-Q, Shen Y-G. Preliminary study on the midday depression of photosynthesis of sweet potato (*Ipomoea batatas*) leaves. *Acta Phytophysiol. Sin.* 1985; 11: 423–426 (in Chinese).

13. Raschke K, Resemann A. The midday depression of CO₂ assimilation in leaves of *Arbutus unedo* L.: diurnal changes in photosynthetic capacity related to changes in temperature and humidity. *Planta* 1986; 168: 546–558.

14. Shen Y-G, Qiu G-X, Xu D-Q, Huang Q-M, Yang D-D, Gao A-X, Long SP, Hall DO. Studies on the photosynthesis of bamboo. *Chin. J. Bot.* 1991; 3: 116–121.

15. Herppich M, Herppich WB, von Willert DJ. Influence of drought, rain and artificial irrigation on photosynthesis, gas exchange and water relations of the fynbos

plant *Protea acaulos* (L.) Reich at the end of the dry season. *Bot. Acta* 1994; 107: 440–450.

16. Crafts-Brandner SJ, Salvucci ME. Rubisco activase constrains the photosynthetic potential of leaves at high temperature and CO_2. *Proc. Natl. Acad. Sci.* 2000; 97: 13430–13435.

17. Schulze E-D, Lange OL, Koch W. Eco-physiological investigations on wild and cultivated plants in the Negev Desert. III. Daily courses of net photosynthesis and transpiration at the end of the dry period. *Oecologia* 1972; 9: 317–340.

18. Tenhunen JD, Lange OL, Braun M, Mcyer A, Losch R, Percira JS. Midday stomatal closure in *Arbutus unedo* leaves in a natural macchia and under simulated habitat conditions in an environmental chamber. *Oecologia* 1980; 147; 365–367.

19. El-Sharkawy MA, Cock JH. Water use efficiency of cassava. I. Effects of air humidity and water stress on stomatal conductance and gas exchange. *Crop Sci.* 1984; 24: 497–502.

20. Xu D-Q, Li D-Y, Shen Y-G, Liang G-A. On midday depression of photosynthesis of wheat leaf under field conditions. *Acta Phytophysiol. Sin.* 1984; 10: 269–276 (in Chinese).

21. Bunce JA. Afternoon inhibition of photosynthesis in maize. 2. Environmental causes and physiological symptom. *Field Crop Res.* 1990; 24: 261–271.

22. Pettigrew WT, Hesketh JD, Peters DB, Woolley JT. A vapor pressure deficit effect on crop canopy photosynthesis. *Photosynth. Res.* 1990; 24: 27–34.

23. Xu D-Q. *Photosynthetic Efficiency*. Shanghai: Shanghai Scientific and Technical Publishers, 2002 (in Chinese).

24. Schulze E-D. Carbon dioxide and water vapor exchange in response to drought in the atmosphere and in the soil. *Annu. Rev. Plant Physiol.* 1986; 37: 247–274.

25. Du Z-C, Yang Z-G. Studies on characteristics of photosynthetic ecology in *Leymus chinensis*. *Acta Bot. Sin.* 1983; 25: 370–379 (in Chinese).

26. Resemann A, Raschke K. Midday depressions in stomatal and photosynthetic activity of leaves of *Arbutus unedo* are caused by large water vapor pressure differences between leaf and air. *Plant Physiol.* 1984; 75(Suppl): 66 (abstract).

27. Tang H-S, Liu T-H, Yu Y-B. Studies on ecological factors of the photosynthetic 'nap' in wheat. *Acta Ecol. Sin.* 1986; 6: 128–132 (in Chinese).

28. Beyschlag W, Lange DL, Tenhunen JD. Diurnal patterns of leaf internal CO_2 partial pressure of the sclerophyll shrub *Arbutus unedo* growing in Portugal. In: Tenhunen JD, Catarino FM, Lange OL, Oechel WC, eds. *Plant Response to Stress*. Berlin: Springer-Verlag, 1987: 355–368.

29. Du Z-C, Yang Z-G. The reasons of midday photosynthetic depression in *Aneurolepidiun chinesense* and *Stipa grandis* under sufficient moisture in the soil. *Acta Phytoecol. Geobot. Sin.* 1989; 13: 106–113 (in Chinese).

30. Qi Q-H, Sheng X-W, Jiang S, Jin Q-H, Hong L. A comparative study of the community photosynthesis of *Aneurolepidium chinense* and *Stipa grandis*. *Acta Phytoecol. Geobot. Sin.* 1989; 13: 332–340 (in Chinese).

31. Turner NC, Burch GJ. The role of water in plants. In: Teare ID, Peet MM, eds. *Crop-Water Relations*. New York: Wiley-Interscience, 1983: 73–126.

32. Xu D-Q, Li D-Y, Shen Y-G, Yan J-Y, Zhang Y-G, Zheng Y-S. On the midday depression of photosynthesis of wheat leaf under field conditions II. The effects of spraying water on the photosynthetic rate and the grain yield of wheat. *Acta Agron. Sin.* 1987; 13: 111–115 (in Chinese).

33. Rawson HM, Turner NC, Begg JE. Agronomic and physiological responses of soybean and sorghum crop to water deficits IV. Photosynthesis, transpiration and water use efficiency of leaves. *Aust. J. Plant Physiol.* 1978; 5: 195–209.

34. Gao H-Y, Zhou Q, Cheng B-S. Comparision of diurnal variation of photosynthesis between pot-cultured soybean and field cultured soybean. *J. Aug. 1st Agric. Coll.* 1992; 15: 1–6 (in Chinese).

35. Cornic G. Drought stress inhibits photosynthesis by decreasing stomatal aperture — not by affecting ATP synthesis. *Trends Plant Sci.* 2000; 5: 187–188.

36. Han F-S, Zhao M, Zhao S-S. Study on the causes for photosynthetic decrease of wheat at the middle day (1). *Acta Agron. Sin.* 1984; 10: 137–143 (in Chinese).

37. Tenhunen JD, Lange OL, Gebel J, Beyschlag W, Weber JA. Changes in photosynthetic capacity, carboxylation efficiency, and CO_2 compensation point associated with midday stomatal closure and midday depression of net CO_2 exchange of leaves of *Quercus suber*. *Planta* 1984; 162: 193–203.

38. Tenhunen JD, Lange OL, Braun M. Midday stomatal closure in Mediterranean type sclerophylls under stimulated habitat conditions in an environmental chamber II. Effect of the complex of leaf temperature and air humidity on gas exchange of *Arbutus unedo* and *Quercus ilex*. *Oecologia* 1981; 50: 5–11.

39. Farquhar GD, Sharkey TD. Stomatal conductance and photosynthesis. *Annu. Rev. Plant Physiol.* 1982; 33: 317–345.

40. Xu D-Q, Li D-Y, Qiu G-X, Shen Y-G, Huang Q-M, Yang D-D, Beadle CL. Studies on stomatal limitation of photosynthesis in the bamboo (*Phyllostachys pubescens*) leaves. *Acta Phytophysiol. Sin.* 1987; 13: 154–160 (in Chinese).

41. Xu D-Q, Ding Y, Wu H. Relationship between diurnal variation of photosynthetic efficiency and midday depression of photosynthetic rate in wheat leaves under field conditions. *Acta Phytophysiol. Sin.* 1992; 18: 279–284 (in Chinese).

42. Sun G-Y. Studies on the daily changes of photosynthesis of two soybean cultivars. *Soyb. Sci.* 1989; 8: 33–37 (in Chinese).

43. Zheng G-S, Zou Q. A study on the diurnal variation of photosynthesis of field-grown soybean in different weather conditions. *Sci. Agric. Sin.* 1993; 26: 44–50 (in Chinese).

44. Meng Q-W, Wang C-X, Zhao S-J, Zhao Li-Y. Study on the characteristics of photosynthesis in *Ginkgo biloba* L. *For. Sci.* 1995; 31: 69–71 (in Chinese).

45. Su P-X, Du M-W, Zhang L-X, Bi Y-R, Zhao A-F, Liu X-M. Changes of photosynthetic characteristics and

response to rising CO_2 concentration in strawberry in solar greenhouse. *Acta Hortic. Sin.* 2002; 29: 423–426 (in Chinese).

46. Downton WJS, Loveys BR, Grant WJR. Non-uniform stomatal closure induced by water stress causes putative non-stomatal inhibition of photosynthesis. *New Phytol.* 1988; 110: 503–510.

47. Ward DA, Drake BG. Osmotic stress temporarily reverses the inhibitions of photosynthesis and stomatal conductance by abscisic acid — evidence that abscisic acid induces a localized closure of stomata in intact, detached leaves. *J. Exp. Bot.* 1988; 39: 147–155.

48. Beyschlag W, Pfanz H, Ryel RJ. Stomatal patchiness in Mediterranean evergreen sclerophylls. Phenomenology and consequences for the interpretation of the midday depression in photosynthesis and transpiration. *Planta* 1992; 187: 546–553.

49. Chen Z-H, Zhang L-C. Diurnal variation in photosynthetic efficiency of leaves in satsuma mandarin. *Acta Phytophysiol. Sin.* 1994; 20: 263–271 (in Chinese).

50. Turner NC, Schulze E-D, Gollan T. The responses of stomata and leaf gas exchange to vapour pressure deficits and soil water content II. In the mesophytic herbaceous species *Helianthus annuus. Oecologia* 1985; 65: 348–355.

51. Gao H-Y, Zhou Q, Cheng B-S. The different types of diurnal course of photosynthesis of soybean [*Glycine max* (L). Merr.] leaves and related factors. *Soyb. Sci.* 1992; 11: 219–225 (in Chinese).

52. Pallas JEJR, Samish YB, Willmer CM. Endogenous rhythmic activity of photosynthesis, transpiration, dark respiration, and carbon dioxide compensation point of peanut leaves. *Plant Physiol.* 1974; 53: 907–911.

53. Gao H-Y, Zhou Q, Cheng B-S. Relationship between diurnal variation of photosynthesis and circadian rhythm in soybean leaves. *Plant Physiol. Commun.* 1992; 28: 262–264 (in Chinese).

54. Deng Z, Chen C. Preliminary studies on afternoon-nap in rice photosynthesis. *J. Huazhong Agric. Univ.* 1989; 8: 208–211 (in Chinese).

55. Neales TF, Incoll LD. The control of leaf photosynthesis rate by the level of assimilate concentration in the leaf: a review of the hypothesis. *Bot. Rev.* 1968; 34: 107–125.

56. Yu Y-B, Liu T-H. Study on the ecology of photo-effect on the plants I. Cause of mid-nap in the wheat. *Acta Ecol. Sin.* 1985; 5: 336–342 (in Chinese).

57. Xu D-Q, Shen Y-G. Exploring the relationship between the photosynthate level and the operation of photosynthetic apparatus. *Acta Phytophysiol. Sin.* 1982; 8: 173–186 (in Chinese).

58. Andrews TJ, Lorimer GM. Rubisco: structure, mechanism and prospects for improvement. In: Hatch MD, Boardman NK, eds. *The Biochemistry of Plants.* Vol. 10. New York: Academic Press, 1987: 131–218.

59. Woodrow IE, Berry JA. Enzymatic regulation of photosynthetic CO_2 fixation in C_3 plants. *Annu. Rev. Plant Physiol. Plant Mol. Biol.* 1988; 39: 533–594.

60. Bernacchi CJ, Singsaas EL, Pimentel C, Portis AR, Long SP. Improved temperature response functions for models of Rubisco-limited photosynthesis. *Plant Cell Environ.* 2001; 24: 253–259.

61. Stitt M, Schulze E-D. Does Rubisco control the rate of photosynthesis and plant growth? An exercise in molecular ecophysiology. *Plant Cell Environ.* 1994; 17: 465–487.

62. Jiang H, Xu D-Q. The cause of the difference in leaf net photosynthetic rate between two soybean cultivars. *Photosynthetica* 2001; 39: 453–459.

63. Warren CR, Adams MA, Chen Z. Is photosynthesis related to concentration of nitrogen and Rubisco in leaves of Australian native plants? *Aust. J. Plant Physiol.* 2000; 27: 407–416.

64. Tenhunen JD, Beyschlag W, Lange OL, Harley PC. Changes during summer drought in leaf CO_2 uptake rates of macchia shrubs growing in Portugal: limitations due to photosynthetic capacity, carboxylation efficiency, and stomatal conductance. In: Tenhunen JO, Catarino FM, Lange OL, Oechel WD, eds. *Plant Response to Stress. Functional Analysis in Mediterranean Ecosystems.* Berlin: Springer, 1987: 355–368.

65. Jiang D-A, Lu Q, Weng X-Y, Zheng B-S, Xi H-F. Role of key enzymes for photosynthesis in the diurnal changes of photosynthetic rate in rice. *Acta Agron. Sin.* 2001; 27: 301–307.

66. Loveys BR, During H. Diurnal changes in water relations and abscisic acid in field-grown *Vitis vinifera* cultivars II. Abscisic acid changes under semi-arid conditions. *New Phytol.* 1984; 97: 37–47.

67. Adams WWIII, Diaz M, Winter K. Diurnal changes in photochemical efficiency, the reduction state of Q, radiationless energy dissipation, and non-photochemical fluorescence quenching in cacti exposed to natural sunlight in northern Venezuela. *Oecologia* 1989; 80: 553–561.

68. Xu D-Q, Xu B-J, Shen Y-G. Diurnal variation of photosynthetic efficiency in C_3 plants. *Acta Phytophysiol. Sin.* 1990; 16: 1–5 (in Chinese).

69. Xu D-Q, Chen X-M, Zhang L-X, Wang R-F, Hesketh JD. Leaf photosynthesis and chlorophyll fluorescence in a chlorophyll-deficit soybean mutant. *Photosynthetica* 1993; 29: 103–112.

70. Guo L-W, Xu D-Q, Shen Y-G. The causes for diurnal variation of photosynthetic efficiency in cotton leaves under field conditions. *Acta Phytophysiol. Sin.* 1994; 20: 360–366 (in Chinese).

71. Kok B. On the inhibition of photosynthesis by intense light. *Biochim. Biophys. Acta* 1956; 21: 235–244.

72. Leverenz JW, Falk S, Pilstrom C-M, Samuelsson G. The effect of photoinhibition on the photosynthetic light-response curve of green plant cells (*Chlamydomonas reinhardtii*). *Planta* 1990; 182: 161–168.

73. Zhang D-P, Huang C-L, Wang X-C, Lou C-H. Study of diurnal changes in photosynthetic rate and quantum efficiency of grapevine leaves and their utilization in canopy management. *Acta Bot. Sin.* 1995; 37: 25–33 (in Chinese).

74. Epron D, Dreyer E, Breda N. Photosynthesis of oak trees [*Quarks petrea* (Matt) Liebl] during drought under

field conditions: diurnal course of net CO_2 assimilation and photochemical efficiency of photosystem II. *Plant Cell Environ.* 1992; 15: 809–820.

75. Xu D-Q. Some problems in stomatal limitation analysis of photosynthesis. *Plant Physiol. Commun.* 1997; 34: 241–244 (in Chinese).

76. Xu D-Q. Ecology, physiology and biochemistry of midday depression of photosynthesis. *Plant Physiol. Commun.* 1990; 6: 5–10 (in Chinese).

77. Xu D-Q, Shen Y-K. Midday depression of photosynthesis. In: Pessarakli M, ed. *Handbook of Photosynthesis.* New York: Marcel Dekker, 1997: 451–459.

78. Cowan IR, Farquhar GD. Stomatal function in relation to leaf metabolism and environment. In: Jennings DH, ed. *Integration of Activity in the Higher Plant.* London: Cambridge University Press, 1977: 471–505.

79. Farquhar GD, Schulze E-D, Kuppers M. Responses to humidity by stomata of *Nicotiana glauca* L. and *Corylus avellana* L. Of carbon dioxide uptake with respect to water loss. *Aust. J. Plant Physiol.* 1980; 7: 315–327.

80. Tenhunen JD, Serra AS, Harley PC, Dougherty RL, Reynolds JF. Factors influencing carbon fixation and water use by Mediterranean sclerophyll shrubs during summer drought. *Oecologia* 1990; 82: 381–393.

81. Xu D-Q, Wu S. Three phases of dark recovery course from photoinhibition resolved by the chlorophyll fluorescence analysis in soybean leaves under field conditions. *Photosynthetica* 1996; 32: 417–423.

82. Demmig B, Winter K, Kruger A, Czygan F-C. Photoinhibition and zeaxanthin formation in intact leaves.

A possible role of the xanthophyll cycle in the dissipation of excess light energy. *Plant Physiol.* 1987; 84: 218–224.

83. Demmig-Adams B, Adams WWIII. Photoprotection and other responses of plant to high light stress. *Annu. Rev. Plant Physiol. Plant Mol. Biol.* 1992; 43: 599–626.

84. Hong S-S, Xu D-Q. Light-induced increase in initial chlorophyll fluorescence Fo level and the reversible inactivation of PS II reaction centers in soybean leaves. *Photosynth. Res.* 1999; 61: 269–280.

85. Demmig-Adams B, Adams WWIII, Ebbert V, Logan BA. Ecophysiology of the xanthophyll cycle. In: Frank HA, Yong AJ, Britton G, Cogdell RJ, eds. *The Photochemistry of Carotenoids.* Netherlands: Kluwer Academic Publishers, 1999: 245–269.

86. Xu D-Q. Reversible inactivation of photosystem II reaction centers and its physiological significance. *Plant Physiol. Commun.* 1999; 35: 273–276 (in Chinese).

87. Zhang H-B, Xu D-Q. Different mechanisms for photosystem 2 reversible down-regulation in pumpkin and soybean leaves under saturating irradiance. *Photosynthetica* 2003; 41: 177–184.

88. Miliford GFJ. Effect of mist irrigation on the physiology of sugar beet. *Ann. Appl. Biol.* 1975; 80: 247–250.

89. Cock JH, Porto MCM, El-Sharkawy MA. Water use efficiency of cassava.III. Influence of air humidity and water stress on gas exchange of field grown cassava. *Crop Sci.* 1985; 25: 265–272.

17 Root Oxygen Deprivation and the Reduction of Leaf Stomatal Aperture and Gas Exchange

Robert E. Sojka
Northwest Irrigation and Soils Research Laboratory, Agricultural Research Service, U.S. Department of Agriculture

Derrick M. Oosterhuis
Department of Crops, Soils, and Environmental Sciences, University of Arkansas

H. Dan Scott
Center for Agribusiness and Environmental Policy, Mount Olive College

CONTENTS

I. INTRODUCTION

The most ubiquitous plant abiotic stress in the global environment is generally thought to be water deficit. The opposite of water-deficit stress, flooding, initially involves relief of the abiotic factor of water deficit and only becomes stressful after flooding persists long enough to directly or indirectly interfere with a variety of plant functions via several mechanisms. The relief of stress with short term flooding (typically a day or less) is the principle upon which irrigation hinges. By contrast, the negative impacts of prolonged flooding on ecosystems, and particularly agricultural production systems, are substantial [1] and may be as significant as drought, depending on one's accounting strategy. Much of this impact is the result of the combination of soil and plant chemical, physical, and biological changes that cause stomata to close after prolonged flooding. This contributes significantly to a drastic reduction in photosynthesis and damages many other plant functions by disrupting transpiration and the complex system of hormonal control of plant systems and processes.

Figure 17.1 gives a conceptual diagram of the effects of flooding on the yield potential of a crop and compares the pattern with what is typically seen under drought. With drought stress, onset is very gradual and plant adaptation has ample time to occur at a pace that moderates the impact of the water-deficit stress. Drought would have to persist for weeks in most crops to collapse the yield potential to near-zero levels. Unless water-deficit stress is exceedingly severe and has persisted for weeks, the loss in yield potential is moderate, and relief of the stress can usually bring about substantial recovery in yield potential, even full recovery, although yield components may shift. By contrast, when flooding occurs, plants initially see relief of any water deficit stress they may be experiencing. However, as the oxygen in the root zone is depleted by plant roots and competing soil organisms (usually in the first 24–48 h), the initial boost in yield potential rapidly gives way to a

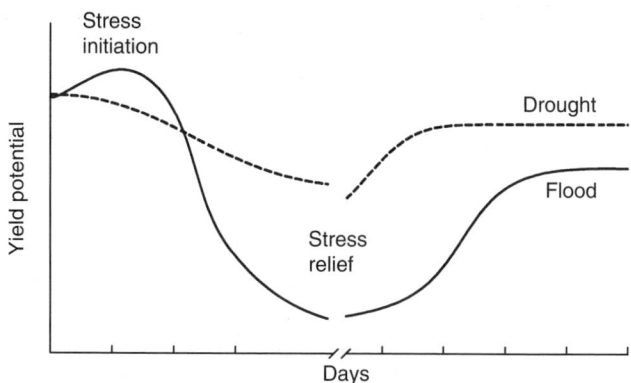

FIGURE 17.1 Conceptual comparison of stress accumulation and stress relief effect on yield potential for flooding stress vs. drought stress.

precipitous drop. Stress relief upon drainage typically produces a far more gradual, and usually less successful recovery than with moderate drought, simply because the plant infrastructure is often far more devastated by the many system impairments that can accumulate with flooding. In our chapter, reference to flooding in the context of this subject matter refers to prolonged flooding, typically 24–48 h or longer, which is about the length of time usually needed for soil organisms to deplete soil water of dissolved oxygen.

It is interesting and curious that common plant reactions to root inundation or prolonged flooding involve several physiological responses much akin to drought stress. This occurs even though plant roots are submerged, i.e., in contact with free water. That wilting and stomatal closure occuring in flooded plants indicate that the physiological responses to flooding are not caused by the energy status of the water, which is the dominant direct mechanism initiating wilting and stomatal closure during drought. The physiological responses to soil hypoxia and flooding have been reviewed by a number of scientists [1–5].

The wilting, stomatal closure, and various other physiological responses to flooding have been explained by several plant response scenarios. These fall into about five categories: obstruction of xylem elements by disease organisms, reduced root system extent or root system/membrane water conductance, altered soil–plant nutritional status, production or imbalancing of plant hormones or biochemical signaling compounds, and the action of soil- or plant-produced toxins [2,6–11].

II. FLOODING AND HYPOXIA EFFECTS ON SOIL PROCESSES

The way in which flooding or waterlogging proceeds along a given scenario or set of scenarios is related to how the physical and chemical properties of water affect soil mineral and biological processes. Ponnamperuma [12] gave an excellent summary of the physicochemical processes that occur in soil upon prolonged flooding, depleting oxygen as an electron acceptor. As reactive oxygen disappears, soil redox potential falls, causing a cascading series of organic and mineral transformations, resulting in the release of numerous soluble chemically reduced minerals, many of which are toxic to plants including methane, sulfides, and reduced forms of iron and manganese.

Water is essential to most soil biological activities. As the amount of water in the soil environment shifts from shortage to plentiful and on to excess, the populations and functional dominance of competing organisms also shift. Under excessively wet or flooded conditions, disease organisms are often favored [8]. Water affects the heat capacity, heat conductivity, and evaporative properties of soil in a way that generally tends to cool soil when wet. Water is a potent solvent, facilitating the mobility of mineral and organic solutes, to the benefit or detriment of a given soil biological process, depending on the intensity and direction of solute movement into or out of an organism's sphere of influence.

Very important to our discussion is the fact that water also changes the net oxygen availability of the soil environment in a temperature-dependent fashion. While soil aeration can be characterized as the volume of gas-filled pore space in a given soil volume, or as the concentration of oxygen (and other gases) within the pores, most edaphologists agree that soil oxygen diffusion rate (ODR) is the best indicator of soil aeration status. This is because ODR gives an indication of the soil's ability to supply oxygen to organisms as a rate function [13]. Rhizosphere ODR is also relatively easy to determine using the platinum microelectrode technique [14,15], and leaves both soil and roots essentially undisturbed. The rate at which soil can supply oxygen must be balanced against the rate at which an organism in soil consumes oxygen. This balance of rates has been the basis of understanding and modeling soil-oxygen-mediated pro-

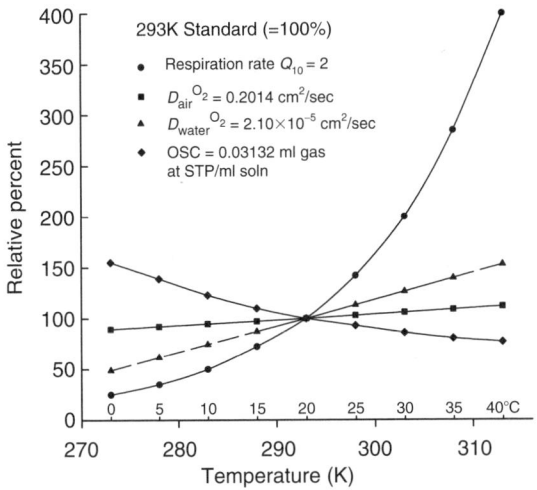

FIGURE 17.2 Relative temperature-related changes in corn root respiration rate (assuming respiration rate doubles for each 10 K increase, i.e., $Q_{10} = 2$), diffusion coefficient (D) of O_2 in air and in water, and O_2 solubility coefficient (OSC) in water. (From Ref. [9], as adapted from Refs. [17,18].)

cesses [16,17]. Luxmoore and Stolzy [18] gave an elegant graphic depiction of this dependency (Figure 17.2).

As soil water content increases, the thickness of water films around soil particles, microorganisms, and surfaces of plant roots also increases. The thickness of these water films greatly influences the transfer of oxygen from the soil environment to respiration sites in roots and microorganisms [8]. Oxygen diffuses 10^4 times more slowly through water than through air [19] and only one-fourth as rapidly through dense protoplasm as through water [20,21]. The physics of this process are described by Fick's first law:

$$J = D_0 \, dC_0/dx$$

where J is the gas flux per unit cross sectional area of soil, C_0 is the concentration of the particular gas in the gas phase of the medium, and D_0 is the apparent diffusion coefficient of the gas in the medium [22,23].

There is a long history and voluminous literature pointing to the direct and indirect roles of rhizosphere oxygen status during flooding as key factors in plant physiological response to flooding. Clements [24] documented that the negative impacts of waterlogging on plants have been recognized for centuries. The specific role of soil oxygen for maintaining plant vigor was noted as early as 1853 [25]. Rhizospere oxygen status appears to affect plant physiological responses both directly (via respiration-mediated metabolic processes in the root) and indirectly (via

cascading chemical, biochemical, and physical processes in the soil, rhizosphere, and the plant).

Our chapter focuses primarily on the role of root zone hypoxia and anoxia in bringing about stomatal closure. While flooding or waterlogging is certainly the most common circumstance limiting root oxygen availability, it is not the sole scenario. Several other examples can be noted. Generous incorporation of fresh organic matter into warm wet soil can stimulate depletion of soil oxygen through the respiration of microorganisms decomposing the fresh substrate. Soil compaction, which reduces average gas-filled soil pore size and total pore space of soil, creates many dead-end soil pores, and favors blockage of the smaller soil pores with water films, restricting diffusion of oxygen through the soil matrix. Oxygen diminishes with soil depth, and if an established plant's roots are buried too deeply under additional soil, the root system can become oxygen limited.

The dominant literature, of course, relates to flooding; however, a number of studies have manipulated soil oxygen independently of flooding, providing important insights to the phenomena [8]. Also, since oxygen unavailability is probably the dominant direct trigger for most of the plant responses that ultimately manifest themselves as familiar visual and otherwise easily monitored physiological responses, it is logical to quantitatively tie measurable physiological responses to rhizosphere ODR values. ODR can be physically predicted with reasonable reliability for a range of soil conditions [26–28]. Thus, the correlation of quantifiable physiological responses to ODR measurements facilitates the normalizing of responses to a reliable soil indicator, allowing species and cultivar response comparisons. Ultimately this approach also enables modeling of physiological responses on a sound physical basis.

In contrasting the effects of flooding and other sources of oxygen exclusion, it is important to remember that flooding causes numerous ancillary changes in the rhizosphere. These include lowered chemical redox potential, resultant specific ion effects, leaching of mobile water-soluble nutrients, metabolic release, and dispersal from microorganisms of organic compounds affecting higher plant function, displacement of soil oxygen with carbon dioxide, ethylene, and other partially water-soluble plant-impacting gases, and promotion of favorable conditions for pathogens. When they occur *en suite*, these multiple rhizosphere changes confound our ability to understand stomatal closure, which so strongly impacts gas exchange and photosynthesis. Direct manipulation of soil atmospheres has been used in many experiments to limit the sources of confounding, and/or reduce their intensity.

III.   SOIL HYPOXIA, THE RHIZOSPHERE, AND PLANT METABOLISM

As the rate of oxygen supply dwindles in a soil system, eventually falling below the demand rate of respiring organisms, a series of consequences often results. Initially, root respiration, lacking sufficient free oxygen, begins to proceed along a fermentative pathway, rapidly consuming the available pool of stored carbohydrates in what is often referred to as the Pasteur effect [29–31]. Under these conditions, oxidative phosphorylation of mitochondria is blocked and the Krebs cycle is bypassed in meeting the demand for adenosine triphosphate (ATP) [32]. Alcohol, rather than carbon dioxide, becomes the dominant metabolic by-product released. The relative amount of energy released in this manner is only about 5% of that liberated by substrates utilized via the aerobic respiration pathway [29]. There can be numerous other alternative pathways, depending on the organism and properties of the soil system [31]. These include reduction of inorganic compounds such as sulfur and production of other by-products, such as methane. The specific biochemical pathways taken under hypoxic conditions probably varies among higher plant species and their complexities are not yet fully understood [33–37]. Several authors have suggested that the alcohol produced under hypoxic conditions does not injure roots because it easily migrates out of and away from the root and perhaps the action of acetaldehyde, rather than alcohol is the injury causing agent in these scenarios [38].

Boamfa et al. [39] showed that oxygen released by photosynthesis in rice (*Oryza sativa*) was completely consumed within the plant and that exposure to light reduced the intensity of the anaerobic metabolic responses. By contrast Luxmoore et al. [16] showed an increase in root porosity and hypoxic symptoms in oxygen-stressed wheat (*Triticum aestivum*) exposed to increasingly higher light intensities. It was their interpretation that under high light intensity there is a large supply of carbohydrate to the root, a high respiration rate, and an "induced oxygen scarcity" to inner root cells resulting in necrosis of some cells and the development of gas spaces.

Generally, as aerobic respiration becomes impaired, energy conversion slows and potentially toxic organic and inorganic wastes begin to accumulate in the rhizosphere and in the plant, impairing various metabolic and membrane functions, particularly in roots. Flooded plants also tend to produce fewer mycorrhizal filaments affecting nutrient and water availability as well as extent of contact surface for diffusion entry of oxygen [40,41]. As a result, in the early stages of root hypoxia, root uptake of nutrients from soil slows and plants begin to experience mobilization and reallocation of existing nutrients from areas of higher concentration (usually from actively growing, more juvenile tissue) to areas of lower concentration [42–44]. Passive transfer of water and nutrients in the xylem stream is also reduced as stomata close and transpiration decreases.

Reviews of physiological response to flooding or hypoxia have usually noted that there is not a consistent co-occurrence of plant water potential shift associated with hypoxia-induced stomatal closure. Even when changes in water potential accompany stomatal response, it is often not clear whether stomata are more directly affecting or affected by the changes in plant water potential. Because of the complicated nature of these environmental alterations and the equally or greater complexity of species-specific plant response to each given hypoxia-dominated scenario, it may well be that different processes dominate under different circumstances.

Eventually with prolonged hypoxia, because energy conversion has become so inefficient, the substrate requirement of roots can only be met by metabolizing less resistant cellular constituents in place. This latter process gradually results in the development of lysigenous zones of intercellular voids, which eventually contribute to improved internal diffusion of oxygen to the roots from the aerial portions of the plant. This constitutes one of the most important adaptive mechanisms of flood resistant plants, allowing survival and eventual return to more normal plant function [16–18,45–58].

If a plant is less capable of shifting metabolic pathways, or if hypoxia persists and the entire soil profile is completely depleted of oxygen, resulting in hypoxia or anoxia that persists for several days, root systems become necrotic. Necrotic tissues lose physical integrity and can provide an easy vector for pathogen and pest invasion. This process, which is sometimes referred to as root pruning, also impairs physiological recovery following improved aeration of the profile — for example, upon drainage following flooding. In this case root extent has been abruptly decreased making plants far more susceptible to subsequent water deficits. The increase in root-to-shoot ratio impairs soil-nutrient and soil water extraction and slows the recovering plant's subsequent growth. In crop plants this usually significantly reduces crop yield [59–67].

IV.   HYPOXIA AND STOMATAL CLOSURE

The effect of flooding on stomatal closure has been recognized directly or indirectly for at least 60 years, however, only a few papers have concentrated on soil oxygen effects per se. Reduced transpiration and photosynthesis was seen by Childers and White [68] within 2 to 7 days of flooding apple trees (*Malus domestica*). They reported slight elevation of transpiration and photosynthesis immediately upon inundation, likely due to initial relief of mild water-deficit stress. But, as in many findings to the present day for many species, after about 48 h leaf expansion ceased and root necrosis became extensive. While their measurements showed no leaf temperature or stomatal aperture differences among treatments, this failure may have been the result of inadequate measurement technology at the time of their work. Reduced stomatal conductance and photosynthesis in soybean (*Glycine max*) 2 days after flooding imposition was reported by Oosterhuis et al. [60,61].

Moldau [69] published the first measurement of increased leaf diffusive resistance (R_L), which is the inverse of leaf conductance (g_s), caused by root waterlogging in common bean (*Phaseolus vulgaris*). Smucker [70] also reported similar findings for navy beans. Regehr et al. [71] reported increased R_L for flooded cottonwood (*Populus deltoides*). Meek et al. [72] reported that R_L was greater for cotton (*Gossypium hirsutum*) with a continuous 30 cm water table depth than with a 90 cm depth, and also noted reduced soil ODR in wetter profiles. These early measurements of increased R_L drew attention to waterlogging's impairment of normal plant control of leaf gas exchange and regulation of water and solute transport. These reports also explained earlier observations of reduced leaf damage by airborne oxidants when exposure occurred during flooding [73,74].

Increased R_L in wheat (*Triticum aestivum*) was measured by Sojka et al. [75] when the wheat was grown at optimal water content but had soil oxygen excluded by continuous flushing of the soil with mixtures of air and nitrogen gas (Figure 17.3). Flushing with ambient air (21% O_2) had the lowest R_L, flushing with pure N_2 produced the highest R_L, and flushing with a 4% oxygen concentration only slightly increased R_L over the air-flushed treatment. In subsequent publications [9,76–78] curvilinear regression demonstrated that R_L could be reliably related to measurements of soil ODR as measured by the platinum microelectrode technique [14] for a number of diverse plant species grown at optimum water contents in controlled soil oxygen chambers. This pattern suggested that stomatal response to soil oxygen availability was abrupt at some threshold value of oxygen availability. The curvilinear regressions of R_L against ODR for numerous species have shown sharp response thresholds occurring at or near ODR values of 20×10^{-8} g/cm^2/min. This same ODR value is a

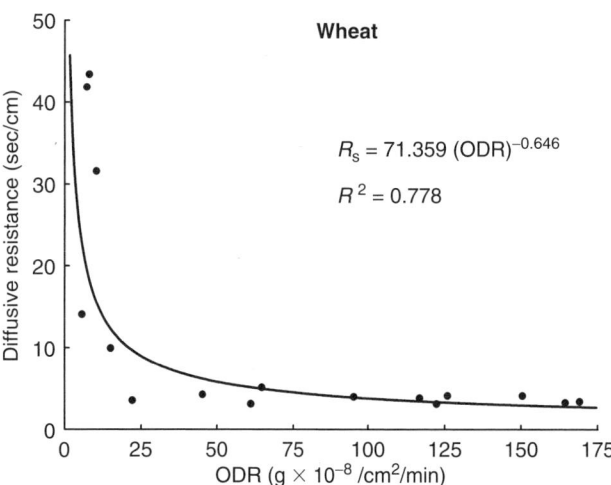

FIGURE 17.3 Diffusive resistance of wheat flag leaves as affected by soil oxygen diffusion rate (ODR). (From Sojka RE, Stolzy LH. *Soil Sci.* 1980; 130:350–358. With permission.)

recognized threshold for a variety of plant growth, physiological and nutritional responses [8,79].

The observations from controlled root atmosphere chambers also suggested that stomatal closure from reduced oxygen in the root zone was largely independent of increases in rhizosphere carbon dioxide or other physiologically active gases such as ethylene. Even though those gases were not measured in the studies, they could not have accumulated significantly in the soil because of the continuous flushing of the root chambers with gas mixtures free of the suspect gases. While various power or exponential equations could provide high correlation of R_L to ODR for a given study, the equation form of the curvilinear relationships observed in these root-gas studies that most often worked well across species and studies was the simple power function:

$$R_L = a(\text{ODR})^b$$

As Figure 17.4 shows, there was also an interaction of stomatal response with root temperature. As root temperature increased, the baseline R_L increased. This would be expected, since as we learned in Figure 17.1 that the respiration requirement increases with temperature. Thus, the adequacy of oxygen availability for roots or root-linked plant functions at any given soil ODR diminishes as temperature in the root environment rises, increasing the demand side of the two rate functions. The expression of this dependency in Figure 17.4 is the increase in R_L with root temperature.

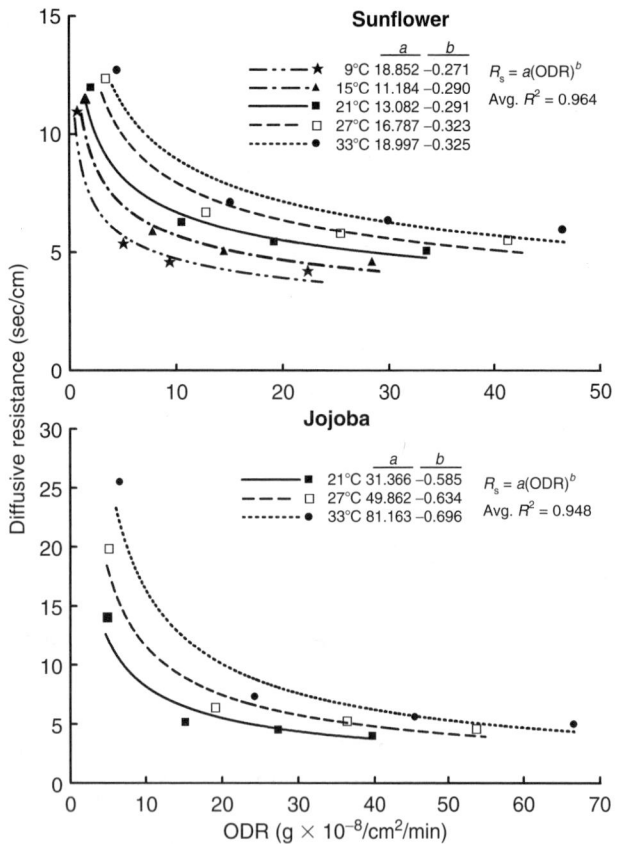

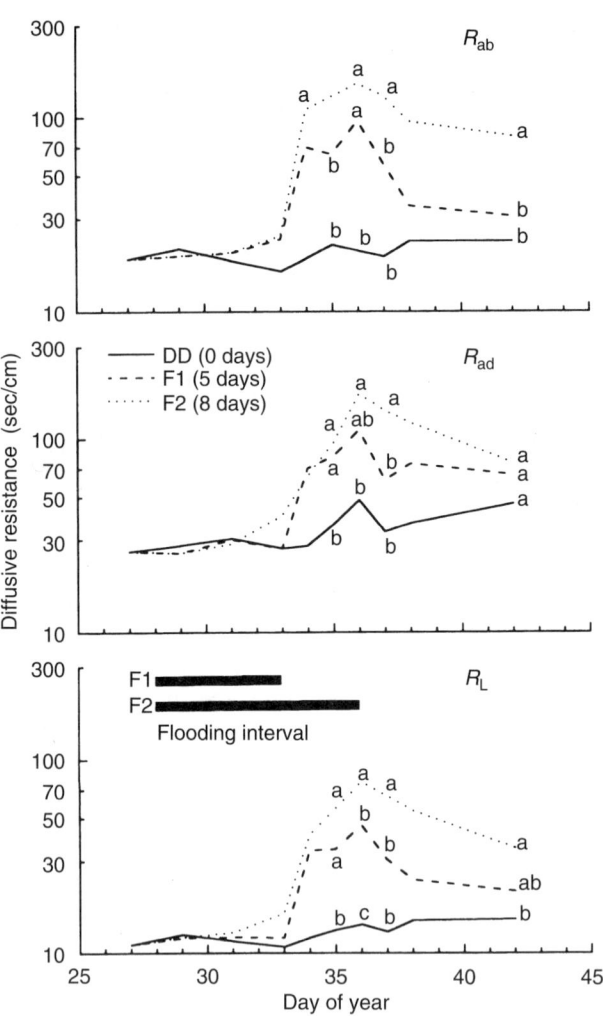

FIGURE 17.4 Leaf diffusive resistance (R_s) as a function of soil oxygen diffusion rate (ODR) at various soil temperatures for sunflower and Jojoba. (From Sojka RE, Stolzy LH. *Soil Sci.* 1980; 130:350–358. With permission.)

In the series of investigations conducted by Sojka and Stolzy, cited above, the value of R_L regressed against ODR was the parallel resistance calculated from the individual adaxial (R_d) and abaxial (R_b) leaf measurements, using the relationship

$$R_S^{-1} = R_{ab}^{-1} + R_{ad}^{-1}$$

In a flooding study of tomato (*Lypersicon esculentum*), Karlen et al. [80] showed that, while adaxial surfaces of control plant leaves had somewhat higher diffusive resistance values than their abaxial surfaces, the diffusive resistance response to flooding regimes of either individual surface or of the calculated parallel resistance were similar in pattern and magnitude (Figure 17.5). One difference was a faster recovery to a normal resistence value for adaxial leaf surfaces.

Figure 17.6 and Figure 17.7 show the stomatal response of soybean (*Glycine max*) to reduction in root zone oxygen availability [78]. Figure 17.6 shows a series of vinyl leaf surface impressions associated

FIGURE 17.5 Time-course of flooding effects on tomato leaf diffusive resistance (R_{ab} = abaxial resistance, R_{ad} = adaxial resistance and R_L = calculated parallel resistance). Flood treatments were well drained (DD), 5-day flooded (F1), or 8-day flooded (F2), where flooding began on day 28. Points with differing letters on a given date in a given figure differ statistically at $P < 0.05$. (From Ref. [9] as adapted from Ref. [80].)

with continuous flushing with varying oxygen mixtures through the sealed cylinders in which the soybean root systems were growing. Figure 17.7 gives the R_L and ODR values generated by the treatment scheme. A key finding of this study was that the R_L increase in the poorly aerated treatments were not due to changes in the stomatal number per unit leaf area. This finding is not entirely consistent among reports of stomatal closure with flooding in the literature. The effect of hypoxia on stomatal distribution and function is likely species dependent and, perhaps more importantly, dependent upon the onset history of flooding treatments. Plants that

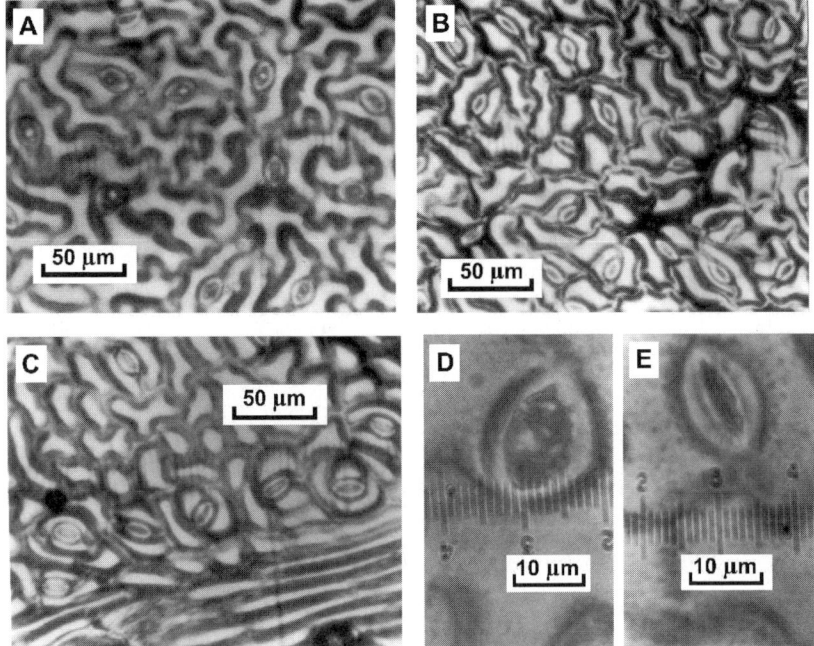

FIGURE 17.6 Photomicrographs of vinyl leaf impressions showing: (A) open, well-aerated (21% O_2) abaxial stomata; (B) closed, more densely distributed abaxial stomata of the poorly aerated (0% O_2) treatment; (C) grouping of adaxial stomata along leaf xylem; (D) enlarged impression of an open (21% O_2) stomate and (E) enlarged impression of a closed (0% O_2) stomate. (From Sojka RE. *Soil Sci.* 1985; 140:333–343. With permission.)

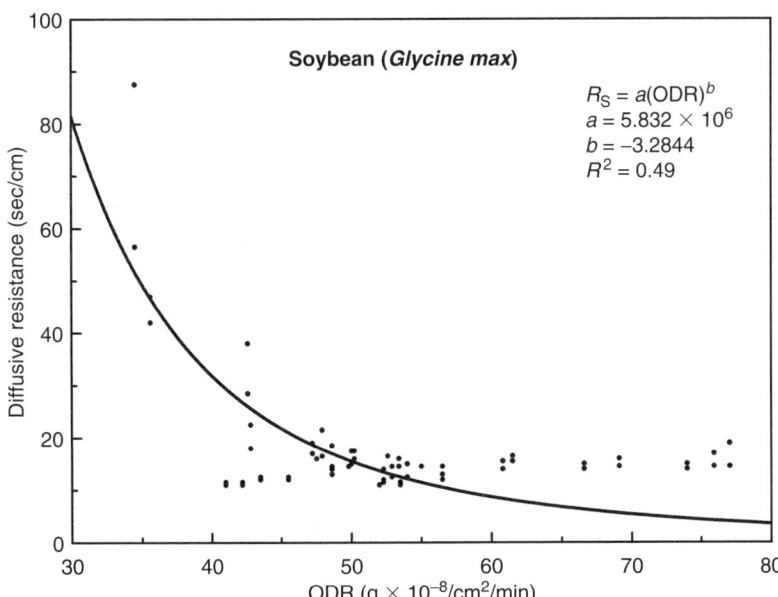

FIGURE 17.7 Parallel leaf diffusive resistance (R_S) as a function of soil oxygen diffusion rate (ODR) measured on several observation dates. Each point is the mean of between 3 and 12 observations. (From Sojka RE. *Soil Sci.* 1985; 140:333–343. With permission.)

are abruptly stressed would have no opportunity to experience changes in leaf expansion or cell differentiation affecting R_L or g_s, and any response in these parameters would have to be physiologically driven rather than morphologically driven. Gradual or repeated onset of stress would provide an opportunity for morphological differentiation. Greater R_L or reduced g_s caused by changes in stomatal distribution or dimensions would have to result from a drop in stomatal density or a reduction in stomatal (i.e., guard cell) size. These morphological changes in re-

sponse to growth-inhibiting stress scenarios have rarely been reported.

There have been extensive observations of increased leaf diffusive resistance, or decreased leaf conductance across scores of plant species (Table 17.1). Not all studies specify whether the resistances reported are abaxial, adaxial, or parallel resistances. Among the studies where abaxial and adaxial responses are observed separately, the most common occurrence is a general similarity of abaxial and adaxial response. However, some cases of surface-

TABLE 17.1
Observed Increase of R_L or Decrease of g_s in Response to Root Flooding or Hypoxia

Species	Stimulus	Refs.	Species	Stimulus	Refs.
Acer rubrum	Soil O_2 + CH_4	[129]	*Phaseolus vulgaris*	Flood	[69,181–183]
Acer rubrum	Flood	[130]		Anoxic soln.	[127,182–185]
Acer saccharum	Soil O_2 + CH_4	[129]	*Picea glauca*	Flood + heat	[186]
Actinidia chinensis	Flood	[132]	*Picea gauca*	Flood	[165]
	Anoxic soln.	[132]	*Picea mariana*	Flood + heat	[186]
Actinidia deliciosa	Flood	[133,134]	*Picea mariana*	Flood	[164]
Apios americana	Flood	[135]	*Pisum sativum*	Flood	[103,187–189]
Avicennia germinans	Flood	[136]	*Poa pratensis*	Compaction +	[190]
Avicennia marina	Flood	[137]		irrigation	
Betula papyrifera	Flood	[138]	*Populus balsamifera*	Flood	[191]
Betula nigra	Flood	[138]	*Populus canadensis*	Flood	[191]
Betula platyphylla	Flood	[139,140]	*Populus deltoides*	Flood	[71,81]
Bruguiera gymnorrhiza	Flood + salt	[128]	*Prioria copaifera*	Flood	[156]
	Flood	[137,141]	*P. trichocarpa × deltoides*	Anoxic soln.	[128,192,193]
Citrus aurantium	Flood	[94,142]	*Prunus armeniaca*	Flood	[194]
Citrus jambhiri	Flood	[94,142]	*Prunus cerasus*	Flood	[115]
Citrus sinensis	Flood	[142]	*Prunus persica*	Anoxic soln.	[147]
Capiscum annuum	Flood	[143]		Flood	[91,175,195]
Carya illinoensis	Soil O_2	[144]	*Pyrus betulaefolia*	Anoxic soln.	[147,196]
	Flood	[82,145]		Flood	[91]
Cucurbita pepo	Flood + salt	[146]	*Pyrus calleryana*	Anoxic soln.	[147,196]
Cydonia oblongs	Anoxic soln.	[147]		Flood	[91]
	Flood	[91]	*Pyrus communis*	Anoxic soln.	[147,196]
Eucalyptus camaldulensis	Flood	[81,148]		Flood	[91,174]
			Pyrus pyrifolia	Flood	[91]
Eucalyptus globulus	Flood	[81,148]	*Pyrus ussuriensis*	Flood	[91]
Eucalyptus obliqua	Flood	[148]	*Quercus alba*	Flood	[149]
Fraxinus pennsylvanica	Flood	[81,149–152]	*Quercus falcata*	Flood	[197,198]
Glycine max	Flood	[60,61]	*Quercus lyrata*	Flood	[198]
	Soil O_2	[78]	*Quercus macrocarpa*	Flood	[152]
Gmelina arborea	Flood	[153,154]	*Quercus nigra*	Flood	[149]
Gossypium barbadense	Soil O_2	[76,155]	*Quercus michauxii*	Flood	[83]
Gossypium hirsutum	Flood	[72]	*Quercus nuttallii*	Flood	[83]
	Soil O_2	[76,155]	*Quercus rubra*	Flood	[81]
Gustavia superba	Flood	[156]	*Rhizophora mangle*	Flood	[136]
Helianthus annuus	Anoxic soln.	[157]	*Rhizophora mucro nata*	Flood	[137]
	Flood	[158,159]	*Salix discolor*	Anoxic soln.	[147]
	Anoxic soln. + salt	[160,161]		Flood	[91]
	Soil O_2 + heat	[76,77,162]	*Salix nigra*	Flood	[81]
Hydrangea macrophylla	Flood	[163]	*Simmondsia chinensis*	Soil O_2 + heat	[76,77,162]
Larix laricina	Flood	[164,165]	*Sorghum bicolor*	Flood	[159]
Liquidambar styraciflua	Flood	[166]	*Taxodium distichum*	Flood + salt	[85]
Lycopersicon esculentum	Flood	[80,102,167–171]	*Tectona grandis*	Flood	[154]
Mangifera indica	Flood	[172,173]	*Theobroma cacao*	Flood	[199]
Malus domestics	Flood	[91,174,175]	*Triticum aestivum*	Soil O_2 + heat	[75–77]
Melaleuca quinquenervia	Flood	[176]		Anoxic soln.	[200]
Momordica charantia	Flood	[177]	*Ulmus americana*	Flood	[81,201]
Nauclea diderrichii	Flood	[154]	*Vaccinium ashei*	Flood	[93,202]
Nyssa aquatica	Flood	[149]	*Vaccinium corymbosum*	Flood	[92,203,204]
Nyssa aquatica	Flood + heat	[178]	*Virola surinamensis*	Flood	[156]
Panicum antidotale	Flood + salt	[179]	*Vitis* sp.	Flood	[175]
Persea americana	Flood	[180]	*Zea mays*	Anoxic soln.	[183,205]

differentiated onset or recovery of stomatal response to hypoxia or flooding have been reported among species with varying degrees of surface differentiation [80–82].

In rare instances, prolonged flooding has been associated with reduced R_L or increased g_s, usually in highly specialized plants, such as bald cypress (*Taxodium distichum*) or rice (*Oryza sativa*), which are specifically adapted to flooded environments [83,84]. We have not attempted to comprehensively catalogue these exceptions, which are not always consistent, even for the particular adapted species [85], but have found a few reports for several species [86–90]. It is not always clear what caused these responses, although factors may include intrinsic species adaptations to hypoxia, gradual exposure allowing adaptation, exposure brevity or an undepleted oxygen supply.

V. STOMATA CLOSURE MECHANISMS

While there is not yet a complete understanding of the physiological and biochemical mechanisms that bring about stomatal closure, several processes are repeatedly implicated in the published literature. A number of studies have shown increased root resistance to water entry to meet transpirational needs [91–94]. This may be the result of loss of root hairs or microrrhiza as hypoxia persists, or changes in membrane properties reducing the hydraulic conductivity of roots. With prolonged flooding disease entry may physically block xylem elements [8].

Potassium ion flux is crucial to regulation of guard cell turgor. Several researchers [9,78,95,96] noted that the single most consistent nutritional shift reported for plant hypoxia and flooding is a drop in leaf or plant potassium concentration. While reviews of nutritional involvement in root hypoxia have noted that several other plant nutrients, particularly nitrogen and phosphorus are often impacted [96], the consistency of response and directness of cause–effect relationship, particularly in the response time frame of stomatal closure is less clear. Because potassium accumulation and retention is an active uptake process requiring outlay of energy [97], it is rapidly disrupted when anaerobic respiration ensues and plants become energy-starved. Loss of potassium ion in the leaves is thought to impair the function of the potassium ion pump responsible for maintaining the turgor of guard cells that opens stomatal pores for gas exchange between the atmosphere and the leaf interior. Peaslee and Moss [98] showed that potassium deficiency alone can impair stomatal opening of corn (*Zea mays*), and Graham and Ulrich [99] showed potassium deficiency reduces sugarbeet root system permeability to water.

Many observations of stomatal closure with root hypoxia or flooding have noted increases in leaf abscisic acid (ABA) concentrations, with the ABA originating in the hypoxic roots and then transferred to leaves [100–109]. Abscisic acid interferes with stomatal control by impairing guard cell accumulation and/or retention of potassium ions [110] and by causing transient potassium and chloride ion efflux [111]. Markart et al. [112] found that ABA affected the root hydraulic conductivity.

Reduction in leaf conductance (g_s), or increase in diffusive resistance (R_L) to water vapor, directly impacts photosynthesis by concomitantly lowering the rate of carbon dioxide exchange (Figure 17.8). However, because the diffusion coefficient of carbon dioxide in air is only about 60% that of water, assuming all other factors equal, there should be a greater incremental effect of stomatal closure on water vapor transfer than on carbon fixation and photosynthesis. The effect of stomatal closure on C3 plant carbon exchange reduction is greater than on C4 plants because of the steeper concentration gradient to sites of carbon fixation in the C4 substomatal mesophyll [113].

However, explaining the effect of root hypoxia on photosynthesis reduction by only considering the effects on gas transfer into and out of the leaf is an oversimplification. Many biochemical processes within flooded plants are affected by root hypoxia, and the intensity and nature of the aberrations vary with stress scenarios and species as the citations in Table 17.1 bear out. Oosterhuis et al. [60,61] essentially demonstrated this point (Figure 17.8) for soybean. Photosynthesis was depressed to a plateau rate by reduction of stomatal conductance in the presence or absence of flooding, however, the flooded plants had a lower plateau value than the nonflooded plants, indicating the involvement of additional factors. Gardiner and Krauss [114] showed that the photosynthetic light response (Figure 17.9) was reduced by nearly half as the result of flooding of cherrybark oak (*Quercus pagoda*). While stomatal closure may be the most significant mechanism restricting photosynthesis in the early hours of root hypoxia, with prolonged oxygen deprivation the rate of photosynthesis declines in response to other inhibitory effects on the photosynthetic process involving changes in carboxylation enzymes and loss of chlorophyll [92,93, 115–117]. Reicosky et al. [118,119] used infrared thermometry to measure increased cotton leaf temperature when plants were flooded. As stomata close, transpirational cooling is reduced. This may also lead to several metabolic stress reactions in addition to de-optimization of photosynthesis if leaf heating

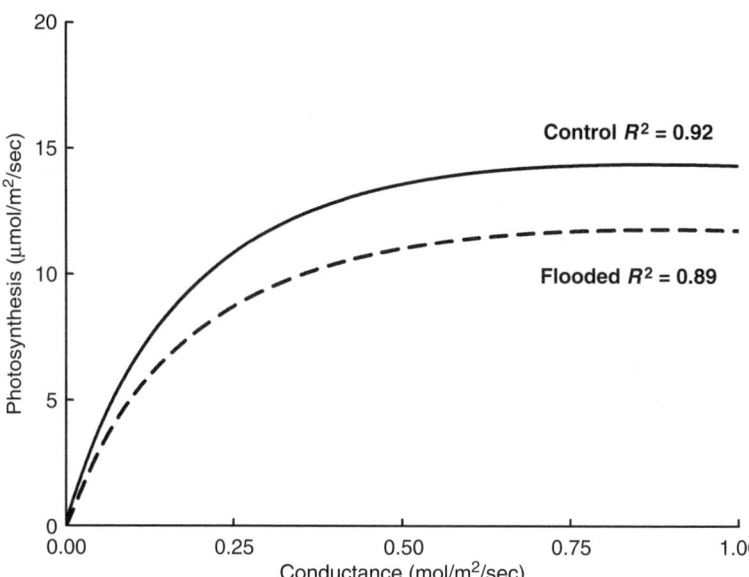

FIGURE 17.8 Difference in the relationship between leaf photosynthetic rate and leaf conductance for flooded vs. nonflooded soybean (From Ref. [9], adapted from Refs. [60,61].)

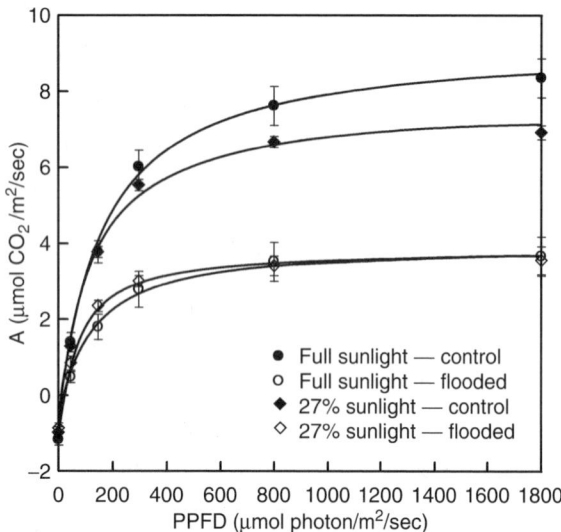

FIGURE 17.9 Net photosynthetic light response of cherrybark oak seedlings grown in full or partial (27%) sunlight and subjected to 30 to 45 days of flooding. Carbon assimilation is reported on a leaf area basis, and each value represents the mean $\forall$ standard error for nine leaves. (From Gardiner ES, Krauss KW. *Tree Physiol.* 2001; 21:1103–1111. With permission.)

causes plants to deviate from their ideal thermal kinetic window [120].

Several other biochemical triggers have been implicated in the closure of stomata of plants exposed to root hypoxia although they have been less intensively researched. These include changes in the nitrogen metabolism of hypoxic plants [121,122], leaf ethylene accumulation [123–127], transport of cytokinin from the roots to the shoot [128], and possibly other as yet unidentified biochemicals acting alone or in concert with other signaling agents [102].

VI. SUMMARY

The negative effects of flooding and root hypoxia on plant performance has been recognized for centuries and the important role of soil oxygen depravation in triggering the metabolic and physiological changes causing damage have been recognized with increasing clarity for nearly a century. Strong quantitative links between the soil oxygen diffusion rate and leaf conductance to water vapor and other gases have been documented. Flooding effects on plant performance are primarily caused by the sharp reduction in oxygen diffusion to roots, with numerous secondary soil physical and chemical and plant biochemical or pathological effects rapidly ensuing as flooding becomes prolonged. Direct manipulation of soil atmospheres at optimal (nonflooded) soil water contents is a powerful tool for studying plant response with minimal interference of ancillary stress-causing factors. Correlation of stomatal hypoxic response to soil ODR is suggested as the most appropriate way to normalize plant response to the primary environmental stimulus that could facilitate discrimination of species and cultivar sensitivity to hypoxia and offer potential for modeling the response.

REFERENCES

1. Kozlowski TT. Extent, causes, and impacts of flooding. In: Kozlowski TT, ed. *Flooding and Plant Growth.* New York: Academic Press, 1984:1–6.

2. Kramer PJ, Jackson WT. Causes of injury to flooded tobacco plants. *Plant Physiol.* 1954; 29:241–245.

3. Kramer PJ. *Plant and Soil Water Relationships: A Modern Synthesis.* New York: McGraw-Hill, 1969.

4. Glinski J, Stepniewski W. *Soil Aeration and Its Role for Plants.* Boca Raton, FL: CRC Press, 1985.

5. Pezeshki SR. Wetland plant responses to soil flooding. *Environ. Exp. Bot.* 2001; 46:299–312.

6. Bradford KJ, Yang SF. Physiological responses of plants to waterlogging. *HortScience* 1981; 16:25–30.

7. Bradford KJ. Regulation of shoot responses to root stress by ethylene, abscisic acid, and cytokinin. In: Warring PF, ed. *Plant Growth Substances.* London: Academic Press, 1982:599–608.

8. Stolzy LH, Sojka RE. Effects of flooding on plant disease. In: Kozlowski TT, ed. *Flooding and Plant Growth.* New York: Academic Press, 1984:222–264.

9. Sojka RE. Stomatal closure in oxygen-stressed plants. *Soil Sci.* 1992; 154:269–280.

10. Kozlowski TT. Responses of woody plants to flooding and salinity. *Tree Physiology Monograph No. 1.* Victoria, Canada: Heron Publishing, 1997. http://www.heronpublishing.com/tp/monograph/kozlowski.pdf

11. Liao CT, Lin CH. Physiological adaptation of crop plants to flooding stress. *Proc. Natl. Sci. Counc. Repub. China Pt B* 2001; 25(3):148–157.

12. Ponnamperuma FN. Effects of flooding on soils. In: Kozlowski TT, ed. *Flooding and Plant Growth.* Orlando, FL: Academic Press, 1984:9–45.

13. Sojka RE, Scott HD. Aeration measurement. In: Lal R, ed. *Encyclopedia of Soil Science.* 1st ed. New York: Marcel Dekker Inc, 2002:27–29.

14. Letey J, Stolzy LH. Measurement of oxygen diffusion rates with the platinum micro-electrode. 1. Theory and equipment. *Hilgardia* 1964; 35:545–554.

15. Birkle DE, Letey J, Stolzy LH, Szuszkiewicz TE. Measurement of oxygen diffusion rates with the platinum microelectrode. II. Factors influencing the measurement. *Hilgardia* 1962; 35:555–566.

16. Luxmoore RJ, Sojka RE, Stolzy LH. Root porosity and growth responses of wheat to aeration and light intensity. *Soil Sci.* 1972; 113:354–357.

17. Luxmoore RJ, Stolzy LH. Oxygen diffusion in the soil-plant system. VI. A synopsis with commentary. *Agron. J.* 1972; 64:725–729.

18. Luxmoore RJ, Stolzy LH. Oxygen diffusion in the soil-plant system. V. Oxygen concentration and temperature effects on oxygen relations predicted for maize roots. *Agron. J.* 1972; 64:720–725.

19. Greenwood DJ. The effect of oxygen concentrations on decomposition of organic materials in soil. Plant Soil 1961; 14:360–376.

20. Krogh A. The rate of diffusion of gases through animal tissues with some remarks on the coefficient of invasion. *J. Physiol. (Lond.)* 1919; 52:391–408.

21. Warburg O, Kubowitz F. Atmung bei sehr kleinen Sauerstoffdrucken. *Biochem. Z.* 1929; 214:5–18.

22. Stolzy LH, Focht DD, Flühler H. Indicators of soil aeration status. *Flora (Jena)* 1981; 171:236–265.

23. Scott HD. *Soil Physics: Agricultural and Environmental Applications.* Ames, IA: Iowa State University Press, 2000.

24. Clements FE. Aeration and Air content. The Role of Oxygen in Root Activity. Publication 315. Washington, DC: Carnegie Institute, 1921.

25. Boussignault J, Lewy A. The composition of air in the cultivated soil. *Ann. Chim. Phys. Serie.* 1853; 3, 37, 5.

26. Stepniewski W. Oxygen diffusion and strength as related to soil compaction. I. ODR. *Pol. J. Soil Sci.* 1980; 13:3–13.

27. Asady GH, Smucker AJM. Compaction and root modifications of soil aeration. *Soil Sci. Soc. Am. J.* 1989; 53:251–254.

28. Wilson GV, Thiesse BR, Scott HD. Relationships among oxygen flux, soil water tension, and aeration porosity in a drying soil profile. *Soil Sci.* 1985; 139:30–36.

29. Wiedenroth EM. Relations between photosynthesis and root metabolism of cereal seedlings influenced by root anaerobiosis. *Photosynthetica* 1981; 15:575–591.

30. Bertani A, Brambilla I, Menegus F. Effect of anaerobiosis on carbohydrate content in rice roots. *Biochem. Physiol. Planz.* 1981; 176:835–840.

31. Wiebe WJ, Christian RR, Hansen JA, King G, Sherr B, Skyring G. Anaerobic respiration and fermentation. In: Pomeroy LR, Wiegert RG, eds. *Ecological Studies: Analysis and Synthesis, Salt Marsh Soils and Sediments: Methane Production.* Vol. 38. New York: Springer-Verlag, 1981:137–159.

32. Davies DD. Anaerobic metabolism and the production of organic acids. In: Davies DD, ed. *The Biochemistry of Plants.* Vol. 2. New York: Academic Press, 1980:581–611.

33. Roberts JKM, Callis J, Wemmer D, Walbot V, Jardetzky O. Mechanism of cytoplasmic pH regulation in hypoxic maize root tips and its role in survival under hypoxia. *Proc. Natl. Acad. Sci. USA* 1984; 81:3379–3383.

34. Menegus F, Cattaruzza L, Chersi A, Fronza G. Differences in the anaerobic lactate-succinate production and in the changes of cell sap pH for plants with high and low resistance to anoxia. *Plant Physiol.* 1989; 90:29–32.

35. Menegus F, Cattaruzza L, Mattana M, Beffagna N, Ragg E. Response to anoxia in rice and wheat seedling. Changes in the pH of intracellular compartments, glucose-6-phosphate level, and metabolic rate. *Plant Physiol.* 1991; 95:760–767.

36. Vanlerberghe CC, Feil R, Turpin DH. Anaerobic metabolism in the N-limited green alga *Selenastrum minutum.* I. Regulation of carbon metabolism and succinate as a fermentation product. *Plant Physiol.* 1990; 94:1116–1123.

37. Su PH, Lin CH. Metabolic responses of luffa roots to long-term flooding. *J. Plant Physiol.* 1996; 148:735–740.

38. Perata P, Alpi A. Ethanol-induced injuries to carrot cells. The role of acetaldehyde. *Plant Physiol.* 1991; 95:748–752.

39. Boamfa EI, Ram PC, Jackson MB, Reuss J, Harren FJM. Dynamic aspects of alcohol fermentation of rice seedlings in response to anaerobiosis and to complete submergence: Relationship to submergence tolerance. *Ann. Bot.* 2003; 91:279–290.

40. Entry JA, Rygiewicz PT, Watrud LS, Donnelly PK. Influence of adverse soil conditions on the formation and function of Arbuscular mycorrhizas. *Adv. Environ. Res.* 2002; 7:123–138.

41. Miller, SP. Arbuscular mycorrhizal colonization of semi-aquatic grasses along a wide hydrologic gradient. *New Phytol.* 2000; 145:145–155.

42. Drew MC, Sisworo EJ. The development of waterlogging damage in young barley plants in relation to plant nutrient status and changes in soil properties. *New Phytol.* 1979; 82:301–314.

43. Trought MCT, Drew MC. The development of waterlogging damage in wheat seedlings (*Triticum aestivum* L.) I. Shoot and root growth in relation to changes in the concentrations of dissolved gases and solutes in the soil solution. *Plant Soil* 1980; 54:77–94.

44. Trought MCT, Drew MC. The development of waterlogging damage in wheat seedlings (*Triticum aestivum* L.). II. Accumulation and redistribution of nutrients by the shoot. *Plant Soil* 1980; 56:187–199.

45. Jensen CR, Luxmoore RJ, Van Gundy SD, Stolzy LH. Root air space measurements by a pycnometer method. *Agron. J.* 1969; 61:474–475.

46. Luxmoore RJ, Stolzy LH. Root porosity and growth responses of rice and maize to oxygen supply. *Agron. J.* 1969; 61:202–204.

47. Yu P, Stolzy LH, Letey J. Survival of plants under prolonged flooded conditions. *Agron. J.* 1969; 61: 844–847.

48. Varade SB, Stolzy LH, Letey J. Influence of temperature, light intensity, and aeration on growth and root porosity of wheat, *Triticum aestivum. Agron. J.* 1970; 62:505–507.

49. Varade SB, Letey J, Stolzy LH. Growth response and root porosity of rice in relation to temperature, light intensity, and aeration. *Plant Soil* 1971; 34:415–420.

50. Luxmoore RJ, Stolzy LH, Joseph H, DeWolfe TA. Gas space porosity of citrus roots. *HortScience* 1971; 6:447–448.

51. Papenhuijzen C, Roos MH. Some changes in the subcellular structure of root cells of *Phaseolus vulgaris* as a result of cessation of aeration in the root medium. *Acta Bot. Neerl.* 1979; 28:491–495.

52. Benjamin LR, Greenway H. Effects of a range of O_2 concentrations on porosity of barely roots and on their sugar and protein concentrations. *Ann. Bot (Lond.)* 1979; 43:383–391.

53. Konings H, Verschuren G. Formation of aerenchyma in roots of *Zea mays* in aerated solutions and its relation to nutrient supply. *Physiol. Plant.* 1980; 49:265–270.

54. Konings H, de Wolf A. Promotion and inhibition by plant growth regulators of aerenchyma formation in seedling roots of *Zea mays. Physiol. Plant.* 1984; 60:309–314.

55. Stelzer R, Laüchli A. Salt and flooding tolerance of *Puccinellia peisonis*. IV. Root respiration and the role of aerenchyma in providing atmospheric oxygen to the roots. *Z. Pflanzenphysiol.* 1980; 97:171–178.

56. Drew MC, Chamel A, Garrec JP, Foucy A. Cortical air spaces (aerenchyma) in roots of corn subjected to oxygen stress. *Plant Physiol.* 1980; 65:506–511.

57. Kawase M, Whitmoyer RE. Aerenchyma development in waterlogged plants. *Am. J. Bot.* 1980; 67:18–22.

58. Konings H. Ethylene promoted formation of aerenchyma in seedling roots of *Zea mays* L. under aerated and non-aerated conditions. *Physiol. Plant.* 1982; 54:119–124.

59. Box JL. Winter wheat grain yield responses to soil oxygen diffusion rates. *Crop Sci.* 1986; 26:355–361.

60. Oosterhuis DM, Scott HD, Hampton RE, Wullschleger SD. Physiological responses of two soybean (*Glycine max* (L.) Merr) cultivars to short-term flooding. *Environ. Exp. Bot.* 1990; 30:85–92.

61. Oosterhuis DM, Scott HD, Wullscleger SD, Hampton RE. Photosynthetic and yield responses of two soybean cultivars to flooding. *Ark. Farm Res.* 1990; 39:11.

62. Sallam A, Scott HD. Effects of prolonged flooding on soybeans during early vegetative growth. *Soil Sci.* 1987; 144:61–66.

63. Sallam A, Scott HD. Effects of prolonged flooding on soybean at the R_2 growth stage. I. Dry matter and N and P accumulation. *J. Plant Nutr.* 1987; 10:657–592.

64. Scott HD, Sallam A. Effects of prolonged flooding on soybean at the R_2 growth stage. II. N and P uptake and translocation. *J. Plant Nutr.* 1987; 10:593–608.

65. Scott HD, DeAngulo J, Daniels MB, Wood LS. Flood duration effects on soybean growth and yield. *Agron. J.* 1989; 81:631–636.

66. Stepniewski W, Przywara G. The influence of oxygen availability on the content and uptake of B, Cu, Fe, Mn and Zn by soybean. *Folia Societatis Scientarum Lublinensis* 1990; 30:79–89.

67. Atwell BJ, Steer BT. The effect of oxygen deficiency on uptake and distribution of nutrients in maize plants. *Plant Soil* 1990; 122:1–8

68. Childers NE, White DG. Influence of submersion of the roots on transpiration, apparent photosynthesis, and respiration of young apple trees. *Plant Physiol.* 1942; 17:603–618.

69. Moldau H. Effects of various water regimes on stomatal and mesophyll conductances of bean leaves. *Photosynthetica* 1973; 7:1–7.

70. Smucker AJM. Interactions of soil oxygen and water stresses upon growth, disease, and production of navy beans. *Rep. Bean Improve. Coop.* 1975:72–75.

71. Regehr DL, Bazzaz FA, Boggess WR. Photosynthesis, transpiration and leaf conductance of *Populus deltoides* in relation to flooding and drought. *Photosynthetica* 1975; 9:52–61.

72. Meek BD, Owen-Bartlett EC, Stolzy LH, Labanauskas CK. Cotton yield and nutrient uptake in relation to water table depth. *Soil Sci. Soc. Am. J.* 1980; 44:301–305.

73. Stolzy LH, Taylor OC, Letey J, Szuszkiewicz TE. Influence of soil-oxygen diffusion rates on susceptibility of tomato plants to airborne oxidants. *Soil Sci.* 1961; 91:151–155.

74. Dugger WM, Ting IP. Air pollution oxidants — their effects on metabolic processes in plants. *Annu. Rev. Plant Physiol.* 1970; 21:215–234.

75. Sojka RE, Stolzy LH, Kaufmann MR. Wheat growth related to rhizosphere temperature and oxygen levels. *Agron. J.* 1975; 67:591–596.

76. Sojka RE, Stolzy LH. Soil-oxygen effects on stomatal response. *Soil Sci.* 1980; 130:350–358.

77. Sojka RE, Stolzy LH. Stomatal response to soil oxygen. *Calif. Agric.* 1981; 35:18–19.

78. Sojka RE. Soil oxygen effects on two determinate soybean isolines. *Soil Sci.* 1985; 140:333–343.

79. Stolzy LH, Letey J. Measurements of oxygen diffusion rates with the platinum microelectrode. III. Correlation of plant response to soil oxygen diffusion rate. *Hilgardia* 1964; 35:567–576.

80. Karlen DL, Sojka RE, Robbins ML. Influence of excess soil-water and N rates on leaf diffusive resistance and storage quality of tomato fruit. *Commun. Soil. Sci. Plant Anal.* 1983; 14:699–708.

81. Pereira JS, Kozlowski TT. Variations among woody angiosperms in response to flooding. *Physiol. Plant.* 1977; 41:184–192.

82. Smith MW, Ager PL. Effects of soil flooding on leaf gas exchange of seedling pecan trees. *HortScience* 1988; 23:370–372.

83. Anderson PH, Pezeshki SR. Effects of flood preconditioning on responses of three bottomland tree species to soil waterlogging. *J. Plant Physiol.* 2001; 158:227–333.

84. Lu J, Ookawa T, Hirasawa T. The effects of irrigation regimes on the water use, dry matter production and physiological responses of paddy rice. *Plant Soil* 2000; 223:207–216.

85. Pezeshki SR, De Laune RD, Choi HS. Gas exchange and growth of bald cypress seedlings from selected U.S. Gulf Coast populations: Responses to elevated salinities. *Can. J. For. Res.* 1995; 25:1409–1415.

86. Harrington CA. Responses of red alder and black cottonwood seedlings to flooding. *Physiol. Plant.* 1987; 69:35–38.

87. Osundina MA, Osonubi O. Adventitious roots, leaf abscission and nutrient status of flooded Gmelina and Tectona seedlings. *Tree Physiol.* 1989; 5:473–483.

88. Pezeshki SR, DeLaune RD, Patrick WH Jr. Differential response of selected mangroves to soil flooding and salinity: Gas exchange and biomass partitioning. *Can. J. For. Res.* 1990; 20:869–874.

89. Javier TT. Effects of adventitious root removal on the growth of flooded tropical pasture legumes *Macroptilium lathyroides* and *Vigna luteola*. *Ann. Trop. Tes.* 1985; 7:12–20.

90. Thorton RK, Wample RL. Changes in sunflower in response to water stress conditions. *Plant Physiol.* 1980; 65(Suppl):7.

91. Andersen PC, Lombard PB, Westwood MN. Leaf conductance, growth, and survival of willow and deciduous fruit tree species under flooded soil conditions. *J. Am. Soc. Hortic. Sci.* 1984; 109:132–138.

92. Davies FS, Flore JA. Flooding, gas exchange and hydraulic root conductivity of highbush blueberry. *Physiol. Plant.* 1986; 67:545–551.

93. Davies FS, Flore JA. Short-term flooding effects on gas exchange and quantum yield of rabbiteye blueberry (*Vaccinium ashei* Reade). *Plant Physiol.* 1986; 81:289–292.

94. Syvertsen JP, Zablotowicz RM, Smith ML Jr. Soil temperature and flooding effects on two species of citrus. I. Plant Growth and hydraulic conductivity. *Plant Soil* 1983; 72:3–12.

95. Drew MC. Effects of flooding and oxygen deficiency on plant mineral nutrition. In: Tinker B, Lauchli A, eds. *Advances in Plant Nutrition*. Vol. III. New York: Praeger Publishers, 1988:115–159.

96. Sojka RE, Stolzy LH. Mineral nutrition of oxygen stressed crops and its relation to some physiological responses. In: Hook DD, ed. *Ecology and Management of Wetlands. Vol. I. Ecology of Wetlands.* Kent, UK: Croom Helm Ltd, 1988:429–440.

97. Fisher HM, Stone EL. Active potassium uptake by slash pine roots from O_2 depleted solutions. *For. Sci.* 1990; 36:582–598.

98. Peaslee DE, Moss DN. Stomatal conductivities in K-deficient leaves of maize (*Zea mays* L.). *Crop Sci.* 1966; 8:427–430.

99. Graham RD, Ulrich A. Potassium deficiency-induced changes in stomatal behavior, leaf water potentials, and root system permeability in *Beta vulgaris* L. *Plant Physiol.* 1972; 49:105–109.

100. Davies WJ, Kozlowski TT. Effects of applied abscisic acid and plant water stress on transpiration of woody angiosperms. *For. Sci.* 1975; 22:191–195.

101. Davies WJ, Kozlowski TT. Effects of applied abscisic acid and silicone on water relations and photosynthesis of woody plants. *Can. J. For. Res.* 1975; 5:90–96.

102. Else MA, Tiekstra AE, Croker SJ, Davies WJ, Jackson MB. Stomatal closure in flooded tomato plants involves abscisic acid and a chemically unidentified anti-transpirant in xylem sap. *Plant Physiol.* 1996; 112:239–247.

103. Jackson MB, Hall KC. Early stomatal closure in waterlogged pea plants is mediated by abscisic acid in the absence of foliar water deficits. Plant Cell Environ. 1987; 10:121–130.

104. Reid DM, Bradford KJ. Effects of flooding on hormone relations. In: Kozlowski TT, ed. *Flooding and Plant Growth*. Orlando, Fl: Academic Press, 1984:195–219.

105. Shaybany B, Martin GC. Abscisic acid identification and its quantitation in leaves of Juglans seedlings during waterlogging. *J. Am. Soc. Hortic. Sci.* 1977; 102:300–302.

106. Wadman-van-Schravendijk H, van Andel OM. Interdependence of growth, water relations and abscisic

acid levels in *Phaseolus vulgaris* during waterlogging. *Physiol. Plant.* 1985; 63:215–220.

107. Zhang J, Davies WJ. Chemical and hydraulic influences on the stomata of flooded plants. *J. Exp. Bot.* 1986; 37:1479–1491.

108. Zhang J, Davies WJ. Changes in concentration of ABA in xylem sap as a function of changing soil water status can account for changes in leaf conductance and growth. *Plant Cell Environ.* 1990; 13:277–286.

109. Zhang J, Schurr U, Davies WJ. Control of stomatal behaviour by abscisic acid which apparently originates in the roots. *J. Exp. Bot.* 1987; 38:1174–1181.

110. Mansfield TA, Jones RJ. Effects of abscisic acid on potassium uptake and starch content of stomatal guard cells. Planta 1971; 101:147–158.

111. MacRobbie EAC. Effects of ABA in isolated guard cells of *Commelina communis* L. *J. Exp. Bot.* 1981; 32:563–572.

112. Markhart AH III, Fiscus EL, Naylor AW, Kramer PJ. Effect of abscisic acid on root hydraulic conductivity. *Plant Physiol.* 1979; 64:611–614.

113. Heatherington AM, Woodward FI. The role of stomata in sensing and driving environmental change. *Nature* 2003; 24:901–908.

114. Gardiner ES, Krauss KW. Photosynthetic light response of flooded cherrybark (*Quercus pagoda*) seedlings grown in two light regimes. Tree Physiol. 2001; 21:1103–1111.

115. Beckman TG, Perry RL, Flore JA. Short-term flooding affects gas exchange characteristics of containerized sour cherry trees. *HortScience* 1992; 27:1297–1301.

116. Crane JH, Davies FS. Flooding responses of Vaccinium species. *HortScience* 1989; 24:203–210.

117. Pezeshki SR. Responses of baldcypress (*Taxodium distichum*) seedlings to hypoxia: leaf protein content, ribulose-1,5-bisphosphate carboxylase/oxygenase activity and photosynthesis. *Photosynthetica* 1994; 30:59–68.

118. Reicosky DC, Meyer WS, Schaefer NL, Sides RD. Cotton responses to short-term waterlogging imposed with a watertable gradient facility. *Agric. Water Manage.* 1985; 10:127–143.

119. Reicosky DC, Smith RCG, Meyer WS. Foliage temperature as a means of detecting stress of cotton subjected to a short-term water-table gradient. *Agric. Meteorol.* 1985; 35:193–203.

120. Mahan JR, Burke JJ, Orzech KA. The thermal dependence of the apparent KM of glutathione reductases from three plant species. *Plant Physiol.* 1990; 93:822–824.

121. Sakihama Y, Murakami S, Yamasaki H. Involvement of nitric oxide in the mechanism for stomatal opening in *Vicia faba* leaves. *Biol. Plant.* 2003; 46:117–119.

122. Hocking P, Reicosky DC, Meyer WS. Nitrogen status of cotton subjected to two short term periods of waterlogging of varying severity using a sloping plot watertable facility. *Plant Soil* 1985; 87:375–391.

123. Hunt PG, Campbell RB, Sojka RE, Parsons JE. Flooding-induced soil and plant ethylene accumula-

tion and water status response of field-grown tobacco. *Plant Soil* 1981; 59:427–439.

124. Pallas JE Jr, Kays SJ. Inhibition of photosynthesis by ethylene-A stomatal effect. *Plant Physiol.* 1982; 70:598–601.

125. Tang ZC, Kozlowski TT. Water relations, ethylene production, and morphological adaptation of *Fraxinus pennsylvanica* seedlings to flooding. *Plant Soil* 1984; 77:183–192.

126. Gunderson CA, Taylor GE Jr. Ethylene directly inhibits foliar gas exchange in *Glycine max. Plant Physiol.* 1991; 95:337–339.

127. Wang TW, Arteca RN. Effects of low O_2 root stress on ethylene biosynthesis in tomato plants (*Lycopersicon esculentum* Mill cv Heinz 1350). *Plant Physiol.* 1992; 98:97–100.

128. Neuman DS, Rood SB, Smit BA. Does cytokinin transport from root-to-shoot in the xylem sap regulate leaf responses to root hypoxia? *J. Exp. Bot.* 1990; 41:1325–1333.

129. Arthur JJ, Leone IA, Flower FB. Flooding and landfill gas effects on red and sugar maples. *J. Environ. Qual.* 1981; 10:431–433.

130. Anella LB, Whitlow TH. Photosynthetic response to flooding of Acer rubrum seedlings from wet and dry sites. *Am. Midl. Nat.* 2000; 14:330–341.

131. Save' R, Serrano L. Some physiological and growth responses of kiwi fruit (*Actinidia chinensis*) to flooding. *Physiol. Plant.* 1986; 66:75–78.

132. Smith GS, Buwalda JG, Green TGA, Clark CJ. Effect of oxygen supply and temperature at the root on the physiology of kiwifruit vines. *New Phytol.* 1989; 113:431–437.

133. Smith GS, Judd MJ, Miller SA, Buwalda JG. Recovery of kiwifruit vines from transient waterlogging of the root system. *New Phytol.* 1990; 115:325–333.

134. Smith GS, Miller SA. Effects of root anoxia on the physiology of kiwifruit vines. *Acta Hort. (ISHS)* 1992; 297:401–408.

135. Musgrave ME, Hopkins AG Jr, Daugherty CJ. Oxygen insensitivity of photosynthesis by waterlogged *Apios americana. Environ. Exp. Bot.* 1991; 31:117–124.

136. Pezeshki SR, De Laune RD, Meeder DF. Carbon assimilation and biomass partitioning in *Avicennia germinans* and *Rhizophora mangle* seedlings in response to soil redox conditions. *Environ. Exp. Bot.* 1997; 37:161–171.

137. Naidoo G. Effects of waterlogging and salinity on plant–water relations and on the accumulation of solutes in three mangrove species. *Aquat. Bot.* 1985; 22:133–143.

138. Norby RJ, Kozlowski TT. Flooding and SO_2 stress interaction in *Betula papyrifera* and *B. nigra* seedlings. *For. Sci.* 1983; 29:739–750.

139. Ranney TG, Bir RE. Comparative flood tolerance of birch rootstocks. *J. Am. Soc. Hortic. Sci.* 1994; 119:43–48.

140. Tsukahara H, Kozlowski TT. Effect of flooding and temperature regime on growth and stomatal resistance

of *Betula platyphylla* var. japonica seedlings. *Plant Soil* 1986; 92:103–112.

141. Naidoo G. Effects of flooding on leaf water potential and stomatal resistance in *Bruguiera gymnorrhiza* (L.) Lam. *New Phytol.* 1983; 93:369–376.

142. Vu JCV, Yelenosky G. Photosynthetic response of citrus trees to soil flooding. *Physiol. Plant.* 1991; 81:7–14.

143. Pezeshki SR, Sundstrom FJ. Effect of soil anaerobiosis on photosynthesis of *Capsicum annuum* L. *Sci. Hortic.* 1988; 35:27–35.

144. Smith MW, Wazir FK, Akers SW. The influence of soil aeration on growth and elemental absorption of greenhouse-grown seedling pecan trees. *Commun. Soil Sci. Plant Anal.* 1989; 20:335–344.

145. Wazir FK, Smith MW, Akers SW. Effects of flooding and soil phosphorous levels on pecan seedlings. *HortScience* 1988; 23:595–597.

146. Huang B, NeSmith DS, Bridges DC, Johnson JW. Responses of squash to salinity, waterlogging, and subsequent drainage. I. Gas exchange, water relations, and nitrogen status. *J. Plant Nutr.* 1995; 18:127–140.

147. Andersen PC, Montano JM, Lombard PB. Root anaerobiosis, root respiration, and leaf conductance of peach, willow, quince, and several pear species. *HortScience* 1985; 20:248–250.

148. Blake TJ, Reid DM. Ethylene, water relations and tolerance to waterlogging of three Eucalyptus species. *Aust. J. Plant Physiol.* 1981; 8:497–505.

149. Gravatt DA, Kirby CJ. Patterns of photosynthesis and starch allocation in seedlings of four bottomland hardwood tree species subjected to flooding. *Tree Physiol.* 1998; 18:411–417.

150. Kozkowski TT, Pallardy SG. Stomatal responses of *Fraxinus pennsylvanica* seedlings during and after flooding. *Physiol. Plant.* 1979; 46:155–158.

151. Sena Gomes AR, Kozlowski TT. Growth responses and adaptations of *Fraxinus pennsylvanica* seedlings to flooding. *Plant Physiol.* 1980; 66:267–271.

152. Tang ZC, Kozlowski TT. Some physiological and morphological responses of *Quercus macrocarpa* seedlings to flooding. *Can. J. For. Res.* 1982; 12:196–202.

153. Osonubi O, Fasehun FE, Fasidi IO. The influence of soil drought and partial waterlogging on water relations on *Gmelina arborea* seedlings. *Oecologia* 1985; 66:126–131.

154. Osonubi O, Osundina MA. Stomatal responses of woody seedlings to flooding in relation to nutrient status in leaves. *J. Exp. Bot.* 1987; 38:1166–1173.

155. Owen-Bartlett EJ. The Effect of Different Oxygen and Salinity Levels in the Rooting Media on the Growth of Cotton (*Gossypium barbadense* and *G. hirsutum* L.). Ph.D. dissertation, University Microfilms #77-27134, University of California at Riverside, Ann Arbor, MI, 1977.

156. Lopez OR, Kursar TA. Flood tolerance of four tropical tree species. *Tree Physiol.* 1999; 19:925–932.

157. Everard JD, Drew MC. Water relations of sunflower (*Helianthus annuus*) shoots during exposure of the root

system to oxygen deficiency. *J. Exp. Bot.* 1989; 40:1255–1264.

158. Guy RD, Wample RL. Stable carbon isotope ratios of flooded and nonflooded sunflowers (*Helianthus annuus*). *Can. J. Bot.* 1984; 62:1770–1774.

159. Orchard PW, Jessop RS, So HB. The response of sorghum and sunflower to short-term waterlogging. *Plant Soil* 1986; 91:87–100.

160. Kriedmann PE, Sands R, Foster R. Salt tolerance and root-zone aeration in *Helianthus annuus* (L.) — Growth and stomatal response to solute uptake. In: Marcelle R, Clijsters H and van Poucke M, eds. Effects of Stress on Photosynthesis. The Hague, Netherlands: Nijhoff M, Junk W, Publishers, 1983:313–324.

161. Kriedmann PE, Sands R. Salt resistance and adaptation to root-zone hypoxia in sunflower. *Aust. J. Plant Physiol.* 1984; 11:287–301.

162. Reyes-Manzanares D. Effects of Soil Aeration and Soil Temperature on Physiology and Nutrition of Tomato, Sunflower and Jojoba. Ph.D. dissertation, University Microfilms #77-21433, University of California at Riverside, Ann Arbor, MI, 1975.

163. Serrano I, Save R, Marfa O. Effects of waterlogging on rooting-capacity of cuttings of *Hydrangea macrophylla* L. *Sci. Hortic.* 1988; 36:119–124.

164. Islam MA, McDonald SE, Zwiazek JJ. Response of black spruce (*Picea mariana*) and tamarack (*Larix laricina*) to flooding and ethylene. *Tree Physiol.* 2003; 23:545–552.

165. Reece cf., Riha SJ. Role of root systems of eastern larch and white spruce in response to flooding. *Plant Cell Environ.* 1991; 14:229–234.

166. Pezeskhi SR, Chambers JL. Stomatal and photosynthetic response of sweet gum (*Liquidambar styraciflua*) to flooding. *Can. J. For. Res.* 1985; 15:371–375.

167. Bradford KJ. Effects of soil flooding on leaf gas exchange of tomato plants. *Plant Physiol.* 1983; 73:475–479.

168. Bradford KJ. Involvement of plant growth substances in the alteration of leaf gas exchange of flooded tomato plants. *Plant Physiol.* 1983; 73:480–483.

169. Bradford KJ, Hsiao TC. Stomatal behavior and water relations of waterlogged tomato plants. *Plant Physiol.* 1982; 70:1508–1513.

170. Lopez MV, del Rosario DA. Performance of tomatoes (*Lycopersicon lycopersicum* (L.) Karsten) under waterlogged condition. *Philipp J. Crop Sci.* 1983; 8:75–80.

171. Poysa VW, Tan CS, Stone JA. Flooding stress and the root development of several tomato genotypes. *HortScience* 1987; 22:24–26.

172. Larson KD, Schaffer B, Davies FS. Flooding, leaf gas exchange, and growth of mango in containers. *J. Am. Soc. Hortic. Sci.* 1991; 116:156–160.

173. Larson KD, Schaffer B, Davies FS, Sanchez CA. Flooding, mineral nutrition, and gas exchange of mango trees. *Sci. Hortic.* 1992; 52:113–124.

174. Lee DK, Lee JC. Studies on the flooding tolerance and its physiological aspects in fruit plants. II. Physiological changes associated with flooding. *J. Kor. Soc. Hortic. Sci.* 1991; 32:97–101

175. Olien WC. Seasonal soil waterlogging influences water relations and leaf nutrient content of bearing apple trees. *J. Am. Soc. Hortic. Sci.* 1989; 114:537–542.

176. Sena Gomes AR, Kozlowski TT. Response of *Melaleuca quinquenervia* seedlings to flooding. *Physiol. Plant.* 1980; 49:373–377.

177. Liao CT, Lin CH. Effect of flooding stress on photosynthetic activities of *Momordica charantia. Plant Physiol. Biochem.* 1994; 32:1–5.

178. Donovan LA, Stumpff NJ, McLeod KW. Thermal flooding injury of woody swamp seedlings. *J. Therm. Biol.* 1989; 14:147–154.

179. Ashraf M. Relationships between leaf gas exchange characteristics and growth of differently adapted populations of Blue panicgrass (*Panicum antidotale* Retz.) under salinity or waterlogging. *Plant Sci.* 2003; 165:69–75.

180. Ploetz RR, Schaffer B. Effects of flooding and phytophthora root rot on net gas exchange and growth of avocado. *Phytopathology* 1989; 79:204–208.

181. Lakitan R, Wolfe DW, Zobel RW. Flooding affects snap bean yield and genotypic variation in leaf gas exchange and root growth response. *J. Am. Soc. Hortic. Sci.* 1992; 117:711–716.

182. Schildwacht PM. Is a decreased water potential after withholding oxygen to roots the cause of the decline of leaf-elongation rates in *Zea mays* L. and *Phaseolus vulgaris* L.? *Planta* 1989; 177:178–184.

183. Singh BP, Tucker KA, Sutton JD, Bhardwaj HL. Flooding reduces gas exchange and growth in snap bean. *HortScience* 1991; 26:372–373.

184. Schumacher TE, Smucker AJM. Ion uptake and respiration of dry bean roots subjected to localized anoxia. *Plant Soil* 1987; 99:411–422.

185. Neuman DS, Smit BA. The influence of leaf water status and ABA on leaf growth and stomata of *Phaseolus* seedlings with hypoxic roots. *J. Exp. Bot.* 1991; 42:1499–1506.

186. Grossnickle SC. Influence of flooding and soil temperature on the water relations and morphological development of cold stored black spruce and white spruce seedlings. *Can. J. For. Res.* 1987; 17:821–828.

187. Jackson MB, Kowalewska AKB. Positive and negative messages from roots induce foliar desiccation and stomatal closure in flooded pea plants. *J. Exp. Bot.* 1983; 34:493–506.

188. Jackson MB, Young SF, Hall KC. Are roots a source of abscisic acid for the shoots of flooded pea plants? *J. Exp. Bot.* 1988; 39:1631–1637.

189. Zhang J, Davies WJ. ABA in roots and leaves of flooded pea plants. *J. Exp. Bot.* 1987; 38:649–659.

190. Agnew ML, Carrow RN. Soil compaction and moisture stress preconditioning in Kentucky Bluegrass. II. Stomatal resistance, leaf water potential, and canopy temperature. *Agron. J.* 1985; 77:878–884.

191. Liu Z, Dickmann DI. Abscisic acid accumulation in leaves of two contrasting hybrid poplar clones affected by nitrogen fertilization plus cyclic flooding and soil drying. *Tree Physiol.* 1992; 11:109–122.

192. Smit B, Stachowiak M. Effects of hypoxia and elevated carbon dioxide concentration on water flux through Populus roots. *Tree Physiol.* 1988; 4:153–165.

193. Smit B, Stachowiak M, van Volkenburgh E. Cellular processes limiting leaf growth in plants under hypoxic root stress. *J. Exp. Bot.* 1989; 40:89–94.

194. Domingo R, Perez-Pastor A, Ruiz-Sanchez MC. Physiological responses of apricot plants grafted on two different rootstocks to flooding conditions. *J. Plant Physiol.* 2002; 159:725–732.

195. Basiouny FM. Response of peach seedlings to water stress and saturation conditions. *Proc. Fl. State Hortic. Soc.* 1977; 90:261–263.

196. Andersen PC, Lombard PB, Westwood MN. Effect of root anaerobiosis on the water relations of several *Pyrus* species. *Physiol. Plant.* 1984; 62:245–252.

197. Pezeshki SR, Chambers JL. Response of cherrybark oak seedlings to short term flooding. *For. Sci.* 1985; 31:760–771.

198. Pezeshki SR, Pardue JH, De Laune RD. Leaf gas exchange and growth of flood-tolerant and flood-sensitive tree species under low soil redox conditions. *Tree Physiol.* 1996; 16:453–458.

199. Sena Gomes AR, Kozlowski TT. The effects of flooding on water relations and growth of *Theobroma cacao* var. Catongo seedlings. *J. Hortic. Sci.* 1986; 61:265–276.

200. Huang B, Johnson JW, NeSmith DS. Response to root zone CO_2 enrichment and hypoxia of wheat genotypes differing in waterlogging tolerance. *Crop Sci.* 1997; 37:464–468.

201. Newsome RD, Kozlowski TT, Tang ZC. Responses of *Ulmus americana* seedlings to flooding of soil. *Can. J. Bot.* 1982; 60:1688–1695.

202. Crane JH, Davies FS. Flooding, hydraulic conductivity, and root electrolyte leakage of rabbiteye blueberry plants. *HortScience* 1987; 22:1249–1252.

203. Crane JH, Davies FS. Periodic and seasonal flooding effect on survival, growth, and stomatal conductance of young rabbiteye blueberry plants. *J. Am. Soc. Hortic. Sci.* 1988; 113:488–489.

204. Abbott JD, Gough RE. Growth and survival of the highbush blueberry in response to root zone flooding. *J. Am. Soc. Hortic. Sci.* 1987; 112:603–608.

205. Wenkert W, Fausey NR, Watters HD. Flooding responses in *Zea mays* L. *Plant Soil* 1981; 62:351–366.

18 Rising Atmospheric CO_2 and C_4 Photosynthesis

Joseph C.V. Vu

Crop Genetics and Environmental Research, U.S. Department of Agriculture —
Agricultural Research Service, and Agronomy Department, University of Florida

CONTENTS

I. INTRODUCTION

With the rapid increase in human population, industrial development, fossil fuel dependence, and changing land-use practices, a doubling of atmospheric carbon dioxide concentration ([CO_2]), currently at about 370 parts per million (ppm), is expected within this century [1–4]. Because CO_2 is responsible for about 61% of global warming [5], a rise in atmospheric [CO_2] and other "greenhouse" gases will increase the mean global temperature, possibly as much as 4°C to 6°C [3,6,7], as well as alter the precipitation patterns in many areas of the world [8,9]. Producing crops under climate change conditions is, therefore, an emerging problem in world agriculture, and new strategies are required to improve and maintain world food supplies and nutrition. As a consequence, the need to enhance the production efficiency of economically important crop plant species and their tolerance of warmer, more arid environment conditions will escalate, as competition for arable land and freshwater increases.

As the present atmospheric [CO_2] limits photosynthesis and growth of many plants [10–12], rising atmospheric [CO_2] could potentially benefit many important agricultural crops. Current knowledge of photosynthetic CO_2 assimilation processes classifies terrestrial plants into three major photosynthetic categories, namely C_3, C_4, and Crassulacean acid metabolism (CAM). Although C_4 plants represent only 1%

of the total plant species, as compared to 95% for the C_3 and 4% for the CAM species, their ecological and economic significance is substantial [13]. On a global basis, about 21% of gross primary productivity is provided by C_4 plants [14,15]. In many tropical regions, the food source is primarily based on C_4 species, which supply grains for human consumption and forage for livestock [16]. Maize, millet, sorghum, and sugarcane are the most important C_4 food crops globally in terms of production. On a land area basis, maize, millet, and sorghum account for 46%, 55%, and 70% of the cereals grown in North America, South America, and Africa, respectively [16].

In C_3 plants, the binding of atmospheric CO_2 to its primary acceptor, ribulose bisphosphate (RuBP), is catalyzed in the chloroplasts of mesophyll cells by the enzyme RuBP carboxylase–oxygenase (Rubisco), and the product of this carboxylation reaction, 3-phosphoglycerate (PGA), is converted to other carbohydrates, including starch, sucrose, and reducing sugars. In addition, Rubisco also catalyzes an oxygenase reaction, widely known as photorespiration, in which O_2 reacts with RuBP to form PGA and phosphoglycolate. This oxygenation process results in the loss of CO_2 and energy and therefore has an adverse effect on the photosynthetic efficiency of C_3 plants. As the balance between carboxylation and oxygenation of RuBP depends on the relative concentration of CO_2 and O_2 at the Rubisco site, a higher atmospheric [CO_2] will reduce photorespiration and enhance leaf photosynthetic CO_2 exchange rate (CER) and growth and yield of C_3 plants.

CAM plants are widely distributed in arid and semiarid regions, where their contribution to community biomass production is significant. CAM plants normally close their stomata during the day to prevent water loss. At night, the stomata are open and atmospheric CO_2 is combined with phosphoenolpyruvate (PEP) in the chloroplast-containing cells of leaf or stem tissues, via PEP carboxylase (PEPC), to form oxaloacetic acid. This C_4 acid is subsequently reduced to malate, which then accumulates in large vacuoles. During the daylight hours, stomata are closed, and malate is transported back into the cytoplasm where it is decarboxylated. The CO_2 just released enters the chloroplasts where it is fixed by Rubisco of the conventional C_3 cycle. Presumably, minimal response to rising atmospheric $[CO_2]$ may be expected for CAM plants, which are capable of raising their daytime internal CO_2 levels as high as 10,000 ppm through decarboxylation of the C_4 malic acid accumulated during the previous evening. Such a presumption, however, is only partially corroborated [11,17].

C_4 plants have developed a CO_2-concentrating mechanism to overcome the limitations of low atmospheric $[CO_2]$ and photorespiration [18–21]. Leaves of C_4 plants feature a Kranz architecture, having both mesophyll cells where atmospheric CO_2 is fixed by PEPC into C_4 acids and bundle sheath cells in which Rubisco refixes the CO_2 released from the C_4 acids. The release of CO_2 is catalyzed by one of the three C_4 acid-decarboxylating enzymes: NADP-malic enzyme (NADP-ME), NAD-malic enzyme (NAD-ME), or PEP carboxykinase (PEPCK) [22]. Most, if not all, C_4 species fit into one of these three groups, namely NADP-ME type, NAD-ME type, and PEPCK type, based on differing C_4 acid decarboxylating systems and leaf ultrastructural features [22,23]. Thus, the reactions that are unique to C_4 photosynthesis can be considered as an additional step to the conventional C_3 pathway. They operate to transfer CO_2 from mesophyll cells to bundle sheath cells through the intermediary of dicarboxylic acids, and consequently increase the concentrations of CO_2 in the bundle sheath cells specifically for refixation via Rubisco in the C_3 photosynthetic pathway [18]. Through this additional step, C_4 plants are able to concentrate $[CO_2]$ at the Rubisco site to levels up to 3 to 20 times higher than atmospheric $[CO_2]$ [18,20,21,24]. Photosynthesis by C_4 plants is therefore near saturation at current atmospheric $[CO_2]$, and a rise in atmospheric $[CO_2]$ presumably may have little or no enhancement on C_4 photosynthesis and growth.

As a result, research on rising atmospheric $[CO_2]$ and climate changes has focused mainly on C_3 species. The existing information on acclimation in leaf photosynthetic capacity under long-term exposure to elevated $[CO_2]$ and the nature of interactive effects of elevated $[CO_2]$ on the fundamental aspects of leaf photosynthesis in plants subjected to global climate changes (elevated air temperature or soil water deficit) are well documented for the C_3 species [12,25–29]. Nevertheless, the literature does reveal a positive growth response of many C_4 plants to elevated atmospheric $[CO_2]$, although to a smaller extent than that of C_3 plants [10,30–38]. Such increases in biomass are not as easily explained, because these C_4 plants often show little or no enhancement in short-term CER measurements of mature leaves at the elevated $[CO_2]$ used for growth, which is in contrast to the C_3 species [10,34,37–41].

This review focuses primarily on our current knowledge of C_4 leaf photosynthesis response to elevated atmospheric $[CO_2]$, with emphasis on economically important annual crops. Comparisons will be made in several instances to similar studies conducted on C_3 crop plants. In addition, interactive effects of elevated $[CO_2]$ with anticipated simultaneous changes in climate, including air temperature and soil water availability, will be discussed.

II. LEAF PHOTOSYNTHESIS ACCLIMATION TO ELEVATED $[CO_2]$

In C_3 plants, current atmospheric concentrations of CO_2 and O_2 and Rubisco specificity factors translate into photorespiratory losses by as much as 40% [42]. Existing research data show that a doubling of the atmospheric $[CO_2]$ would increase CERs of C_3 crops up to 63%, and their growth and yield up to 58% [10,12,30,34,43,44]. However, long-term exposure of C_3 plants to elevated $[CO_2]$ leads to a variety of acclimation effects, including changes in leaf photosynthetic physiology and biochemistry and alterations in plant growth and development [11,12,45,46]. Under long-term growth $[CO_2]$, many C_3 species show decreased leaf photosynthesis [47–49], and carbohydrate source–sink imbalance is believed to have a major role in the regulation of photosynthesis through feedback inhibition [50–53]. With respect to the acclimation in photosynthesis biochemistry, long-term exposure to elevated $[CO_2]$ results in a downregulation of the Rubisco capacity in many C_3 plants [12,25,26,48,54–57]. Both "coarse" control, through lowering of the Rubisco protein concentration, and "fine" control, through decreasing the activation state of the enzyme, play a role in this downregulation. In addition to Rubisco, long-term growth at elevated $[CO_2]$ also affects the regulation of sucrose metabolism enzymes, including sucrose phosphate synthase and acid invertase [58,59].

Reduced expression of the Rubisco genes and differential response of other photosynthetic genes, including chlorophyll binding protein *Cab* and Rubisco activase *Rca*, have been also reported for a variety of C$_3$ crops grown at elevated [CO$_2$] [28,58,60–66]. The expression of several genes coding for key C$_3$ photosynthetic enzymes has been shown to be influenced by the levels of soluble sugars [67–69]. Particularly for Rubisco, transcription of the small subunit (*rbcS*), and to a lesser extent the large subunit (*rbcL*), has been shown to be strongly repressed by sucrose and glucose [60,70]. The buildup in carbohydrates at elevated growth [CO$_2$], however, may signal the repression, but does not directly inhibit the expression, of Rubisco and other proteins that are required for photosynthesis [52,53,71–73]. Although the signal transduction pathway for regulation of the sugar-sensing genes may involve phosphorylation of hexoses derived from sucrose hydrolysis [65,68,74–81], unknown gaps still exist between hexose metabolism and repression of gene expression at elevated [CO$_2$] [53,68,80].

For C$_4$ species, limited research attention has been paid to the photosynthesis mechanisms in response to rising atmospheric [CO$_2$]. Studies on a number of C$_4$ plants grown at elevated [CO$_2$] show either no enhancement or only a minor increase in leaf CER, even though a stimulating effect on biomass does exist [10,30,34,36,37,43,82]. In a study conducted in naturally sunlit greenhouses, long-term double-ambient growth [CO$_2$] (720 ppm) increased sugarcane leaf area 31%, total aboveground dry weight 21%, and main stem juice volume 83%, when compared with plants grown at ambient [CO$_2$] [36,37]. Such increases occurred without enhancement of midday leaf CERs, measured at the growth [CO$_2$] for the most expanded sections of the uppermost leaves. Similarly, a study by Ziska and Bunce [35] showed no differences in leaf CERs of ambient [CO$_2$]- and double-ambient [CO$_2$]-grown plants of maize and sugarcane, although leaf area and total plant biomass of the CO$_2$-enriched plants increased 14%. However, there are reports for some C$_4$ grasses showing that leaf CERs are also responsive, although to a small extent, to elevated growth [CO$_2$] [35,41,83,84]. Light-saturated CERs of mature leaves of maize plants grown and measured at triple-ambient [CO$_2$] (i.e., 1100 ppm) were 10% higher than those of plants grown and measured at ambient [CO$_2$] [38]. The triple-ambient [CO$_2$]-grown maize plants, however, were 20% higher in biomass, 23% higher in leaf area, 85% to 100% higher in dark respiration, 65% to 71% lower in stomatal conductance, and 2- to 3.5-fold higher in water-use efficiency (WUE) [38].

It has been suggested that elevated [CO$_2$] could affect growth of C$_4$ plants via several mechanisms. First, a reduction in stomatal aperture and conductance, which is a common response of plants to elevated growth [CO$_2$], occurs across a variety of both C$_3$ and C$_4$ species, although there are cases of insensitive stomatal responses [11,12,25]. The reduction in stomatal aperture and conductance eventually leads to a reduction in transpiration rates, resulting in improvement of WUE for plants grown under elevated [CO$_2$] [12,85]. For C$_4$ plants, the reduction in stomatal aperture and hence transpiration rate, in response to elevated [CO$_2$], would also increase leaf temperature and enhance leaf CER and plant growth through conserving soil water and improving shoot water relations [82,84]. Second, elevated [CO$_2$] could affect growth of C$_4$ plants by raising the intercellular [CO$_2$] (C_i) and consequently enhancing leaf CER [82]. There is an indication that leaf CERs of some C$_4$ species are likely not saturated at current atmospheric [CO$_2$], thus allowing for some response to rising [CO$_2$] [34]. Even a small, but consistent, percent stimulation in leaf CER throughout the growing season could account for the plant biomass enhancement at final harvest observed in a number of C$_4$ species [34,40]. Third, elevated [CO$_2$] could enhance tillering and leaf area, so that photosynthesis of the whole plant is greater, even without an increase in CER per unit leaf area [11,25,86]. Besides, elevated [CO$_2$] may reduce dark respiration [12] and improve the efficiency in photosynthate partitioning [87], and such factors could contribute to growth enhancement in C$_4$ plants. Fourth, any consideration of elevated [CO$_2$] effect on growth and physiology of C$_4$ species must also address time-dependent changes in the plant growth rate [88]. Leaf photosynthetic rates, often determined through short-term midday measurements on fully expanded leaves of developmentally advanced C$_4$ plants, generally show little or no response to elevated growth [CO$_2$]. However, recent studies show that leaf CERs of sugarcane, sorghum, and maize are responsive to elevated [CO$_2$] at certain growth stages of the leaf or plant [89–91]. Evaluation of the impacts of elevated [CO$_2$] on C$_4$ leaf photosynthesis, therefore, should be carried out at various stages of leaf/plant growth and development, and diurnal variations of leaf CERs for elevated CO$_2$-grown C$_4$ crop plants also should be characterized.

III. PHOTOSYNTHESIS DURING LEAF ONTOGENY AT ELEVATED [CO$_2$]

In C$_3$ plants, interactions exist between leaf ontogeny and the degree of the acclimation response to elevated [CO$_2$] exposure [49,64,88,92–95]. Long-term exposure of a number of annual crops to elevated [CO$_2$] leads to an enhancement of the growth rate in young plants, but not in older plants [96–98]. Similarly, for tree crops, increases in biomass are mostly due to increased growth rates during the first year of elevated [CO$_2$] exposure, and growth is enhanced less or not at all in the subsequent years [99–101]. Leaves of dicots during ontogeny undergo two distinct photosynthetic phases: a phase of increasing CER correlated with leaf expansion and a prolonged senescence phase of declining CER, with a transient peak of maximal CER in between [102]. In tobacco, both ambient (at 350 ppm) and high (at 950 ppm) CO$_2$-grown plants exhibit this photosynthetic pattern during leaf ontogeny [95]. However, the high-CO$_2$ plants, which show a temporal shift to an earlier transition from the increasing-CER first phase to the declining-CER senescence phase, enter the declining-CER phase several days before their ambient-CO$_2$ counterparts. Such changes in leaf CER are controlled largely by Rubisco activity [95]. Similar observations are also reported for tomato during leaf ontogeny [49].

For C$_4$ plants, the causes of the observed growth stimulation by elevated [CO$_2$] remain uncertain. As mentioned earlier, there are studies showing a positive growth response of C$_4$ plants to elevated [CO$_2$] without a concomitant enhancement in leaf CER [36,37,40,82]. Such photosynthetic rates, however, were determined by short-term measurements on fully expanded mature leaves of developmentally advanced plants, and there were no studies on the variations of CER during the day or at various growth stages of the leaf or plant. It is possible that increases in CER at elevated growth [CO$_2$] are only apparent at certain daylight periods, and during early, but not late, development of the leaf or plant [82].

Expression of the C$_4$ photosynthetic characteristics has been shown to be controlled by leaf age. Tremmel and Patterson [103] reported that the young, developing leaves of some C$_4$ species show the normal C$_3$ type of photosynthesis, and this may cause such species to be responsive to high [CO$_2$], at least in the short term. In *Portulaca oleracea*, an NADP-ME C$_4$ dicot, there is a shift in the route of CO$_2$ assimilation toward a limited, direct entry of CO$_2$ into the PCR cycle in senescent leaves [104]. In *Flaveria trinervia*, also a C$_4$ dicot of the NADP-ME type, an estimated 10% to 12% of the CO$_2$ entered the PCR pathway directly in young expanding leaves. However, CO$_2$ is apparently fixed entirely through the C$_4$ pathway in mature expanded leaves, and this partitioning pattern is attributed to the bundle sheath compartment in young leaves, which have a relatively high conductance to CO$_2$ [105].

In maize, an NADP-ME-type monocot, pulse-chase experiments with mature and senescent leaf tissues show that the predominant C$_4$ acids malate and aspartate differ between the two leaf ages [106], and a high CO$_2$ compensation point (~25 ppm) is found in senescent leaves of maize, in contrast to lower values (<10 ppm) for most C$_4$ plants [107]. In addition, the activity of Rubisco during leaf ontogeny in maize parallels the development in activity of this enzyme in C$_3$ plants [106]. The activities of Rubisco and PEPC in maize leaves are found to vary according to leaf position (also related to leaf age), with activity of PEPC less than that of Rubisco in the lower leaves, while the upper leaves exhibit high levels of PEPC [108]. The ^{14}C-labeling patterns of photosynthetic products in different sections of a developing maize leaf suggest that there may be some direct entry of CO$_2$ into the PCR pathway in the young tissues of the basal section, while the C$_4$ pathway functions in the more differentiated tissues of the center and top sections [109]. Besides, bundle sheath cell walls of young and senescent maize leaves have a relatively high conductance, leading to a low capacity for CO$_2$ concentration in these bundle sheath cells during photosynthesis [110]. Diurnal courses of gas exchange measurements for upper-canopy leaves show that CER under elevated [CO$_2$] is significantly greater early, but not late, in the growth season [91]. Such increases in maize CER at elevated [CO$_2$] are associated with reduced stomatal conductance and transpiration and increased C_i and WUE of the leaf.

In sorghum, elevated growth [CO$_2$] is shown to enhance CER of young expanding leaves [89]. At later stages of leaf growth, however, elevated [CO$_2$] has little effect on CER. In sorghum, although activities of Rubisco and PEPC increase rapidly as leaf differentiates and emerges from the surrounding whorl, Rubisco accumulates well before significant amounts of PEPC are detectable in the very early stages of leaf development, suggesting that the youngest leaf tissues are more C$_3$-like and are thus likely more sensitive to elevated [CO$_2$] [111]. The CER enhancement in young leaves of sorghum at elevated growth [CO$_2$] is partially due to suppression of the photorespiration rate. Furthermore, elevated growth [CO$_2$] enhances energy-use efficiency in the young leaves, possibly by reducing both the overcycling of the C$_4$ pump and the amount of CO$_2$ leaking from the bundle sheath cells [89].

In an experiment with sugarcane, grown for a season under field-like conditions at ambient and double-ambient [CO$_2$], leaf CERs increase with leaf development and are highest by 14 days after leaf emergence (DAE) for both [CO$_2$] treatments [90]. Leaf CERs of the CO$_2$-enriched plants, however, are 16% higher than the ambient-CO$_2$ counterpart plants at 7 DAE. At other DAE, the impacts of elevated growth [CO$_2$] on leaf CER are not significant. In contrast, leaf WUE of the CO$_2$-enriched plants is higher for most DAE. Activities of Rubisco and those of the C$_4$ photosynthetic enzymes generally follow leaf CER patterns during leaf ontogeny. At 14 DAE, Rubisco activity of the high-CO$_2$ plants is 12% greater than that of the ambient-CO$_2$ plants. PEPC activity of the high-CO$_2$ plants does not differ from that of the ambient-CO$_2$ plants at 7 or 14 DAE, but is lower at other DAE. Thus, for sugarcane, there are certain growth stages of leaf ontogeny during which leaf photosynthesis could be enhanced by elevated [CO$_2$], even though this enhancement occurs to a lesser degree compared with the C$_3$ rice and soybean [26].

For sugarcane, CER, stomatal conductance, and activities of both PEPC and Rubisco increase from the base to the tip of the leaf [112]. Analyses of a range of leaf developmental stages in maize also indicate that when leaf chlorophyll and Rubisco protein contents are below a critical level, that is, 50% or less compared to those found in mature leaves, the degree of photorespiration could approach that of C$_3$ plants [110]. For the NADP-ME-type maize and sugarcane, chloroplasts of the bundle sheath and mesophyll cells are morphologically similar early in development, i.e., both contain granal stacks [113,114]. Subsequent dedifferentiation of bundle sheath cell chloroplasts results in the agranal bundle sheath chloroplasts seen in the mature leaves [113,114]. The expression of C$_4$ genes does not occur until Kranz anatomy has been established, and exclusive use of the C$_3$ photosynthetic pathway may occur prior to the full differentiation of Kranz anatomy [115]. Therefore, a proposed explanation for the growth enhancement observed in C$_4$ plants at elevated [CO$_2$] is that the "immature" C$_4$ pathway in young C$_4$ leaves has the C$_3$-like characteristics, and thus photosynthesis of these young leaves is responsive to elevated atmospheric [CO$_2$] [40,43,89,103,111,116]. This, however, may be species specific, as gas exchange parameters in young leaves of *Panicum antidotale* (C$_4$, NADP-ME type) and *P. coloratum* (C$_4$, NAD-ME type) do not show the C$_3$-like characteristics of *P. laxum* (C$_3$) [117]. These young C$_4$ *Panicum* leaves have CO$_2$ and light response curves typical of C$_4$ photosynthesis. In addition, over the range of O$_2$ concentrations between 2% and 40%, CERs of both mature and young C$_4$ *Panicum* leaves are hardly affected while those of the C$_3$ *Panicum* leaves decline [117].

Elevated growth [CO$_2$] can induce alterations in C$_4$ leaf ultrastructure, in addition to changes in leaf biochemistry. In sorghum, ultrastructural examination of leaf sections indicates a twofold decrease in the thickness of the bundle sheath cell walls of plants grown at elevated relative to ambient [CO$_2$] [118]. Gas exchange analysis of leaf CER response to growth [CO$_2$] also indicates that there is a significant [CO$_2$] effect on the CER/C_i response curve, and that both carboxylation efficiency and CO$_2$-saturated photosynthetic rate are lower in plants grown at elevated relative to ambient [CO$_2$]. This is accompanied by an almost 50% reduction in PEPC content of leaves in the elevated CO$_2$-grown sorghum plants, but no change in Rubisco content is observed. Despite a lower PEPC content, rates of assimilation are similar when leaves are measured at growth [CO$_2$], and there is no difference in the CO$_2$ compensation point for plants grown at either ambient or elevated [CO$_2$]. In addition, bundle sheath leakiness is higher, and there is a threefold increase in carbon isotope discrimination for leaves of sorghum plants grown at elevated [CO$_2$] [118]. This suggests that acclimation to elevated growth [CO$_2$] could be also encountered in C$_4$ plants. Such acclimation to elevated [CO$_2$] would be species specific and occur at a certain stage of leaf growth and development. In maize, both Rubisco and PEPC activities are generally lower in plants grown at elevated [CO$_2$] than in plants grown at ambient [CO$_2$] [54]. In addition, activities of a number of C$_3$ and C$_4$ cycle enzymes decreased while those of the enzymes required for triose-phosphate utilization increased in maize plants grown at 1100 vs. 350 ppm CO$_2$ [38]. Analyses of leaf cross-sections from maize plants grown under triple-ambient CO$_2$ do not reveal apparent differences in size of mesophyll cells and bundle sheath cells. In the C$_4$ plant *Amaranthus retroflexus*, both activity and protein concentration of Rubisco are lower in plants grown at near-double-ambient [CO$_2$] [119], whereas in the C$_4$ species *F. trinervia*, *Panicum miliaceum*, and *Panicum maximum*, this increase is not associated with any significant change in Rubisco or PEPC activity, despite an enhancement in leaf CER of plants grown at double-ambient [CO$_2$] [83].

The differences in growth response of C$_4$ plants to elevated [CO$_2$] could be related also to the biochemical subtype [41,83], or more specifically to the C$_4$ photosynthetic pathway leakiness, defined as ratio of the rate of CO$_2$ leakage out of the bundle sheath to that of PEP carboxylation [120]. In other words, the growth response of C$_4$ to elevated [CO$_2$] has been thought to increase with leakiness. A recent analysis, however, indicates that there is no defined relationship between leakiness and growth response of C$_4$ plants to elevated [CO$_2$] [82]. Gas exchange determinations on seven C$_4$ grass species containing the three biochemical subtypes NADP-ME, NAD-ME, and PEPCK showed that net CER is only increased in one, but stomatal conductance decreases in five, and WUE increases in all species under elevated growth

[CO_2] [121]. In addition, gas exchange responses for the C_4 grass species are generally poorly related to CO_2 responsiveness, and parameters derived from leaf CER–light response curves are also not differentially influenced by CO_2 treatment. Under elevated [CO_2], significant increases in leaf growth and canopy leaf area are found in two NADP-ME-subtype species, whereas increases in nonleaf, aboveground growth are significant in three species representing all three C_4 subtypes [121]. Therefore, as for C_3, the responses of C_4 plants to elevated [CO_2] are species specific.

IV. RISING ATMOSPHERIC [CO_2] AND ANTICIPATED CLIMATE CHANGES

A. HIGH TEMPERATURE

Photosynthesis of C_3 plants is, in addition to elevated [CO_2], influenced by high growth temperature, and Rubisco plays a central role in these responses [122]. Temperature and CO_2 have interactive effects, because a rise in temperature reduces the activation state of Rubisco [26,123,124] and decreases both the specificity for CO_2 and the solubility of CO_2, relative to O_2 [122,125,126]. The latter two effects result in greater losses of CO_2 to photorespiration as the temperature rises. Consequently, a doubling of atmospheric [CO_2] and the concomitant inhibition of the Rubisco oxygenase reaction should moderate the adverse effects of high temperature on C_3 photosynthesis and result in even greater enhancement of net photosynthesis by elevated [CO_2] as growth temperatures increase [122]. However, the data in this regard are equivocal [51], and species-specific effects may be partly responsible for the differing results. In soybean, the enhancement effect on leaf photosynthetic rate due to doubling the growth [CO_2] increases linearly from 32% to 95% with increase in day temperatures from 28°C to 40°C, whereas with rice it is relatively constant at 60% from 32°C to 38°C [26]. In addition, although both elevated [CO_2] and temperature reduced Rubisco protein concentration and activity, the reduction by either factor is greater for rice than for soybean [26]. Even within the same species, however, plant biomass and grain yield respond differently to increasing growth temperature. In the case of rice, plants grown at 34°C accumulate biomass and leaf area at a faster rate than plants at 28°C, but grain yield declines by about 10% for each 1°C rise above 26°C [127,128]. Similar scenarios have been reported for soybean [129] and wheat [130].

For C_4 species, the interactive effects of elevated [CO_2] and temperature are not well understood. For C_3 and C_4 plants adapted to similar climates, C_4 plants have long been recognized as having a higher temperature optimum for photosynthesis, as well as a higher overall photosynthetic rate at the temperature optimum, than C_3 plants [131–135]. C_4 plants, in addition to the C_3 pathway, use the C_4 photosynthetic cycle to concentrate the [CO_2] at the site of Rubisco and thus suppress its oxygenase activity. This mechanism enables C_4 plants to achieve, at the current atmospheric [CO_2], a much greater photosynthetic capacity than C_3 plants, particularly at elevated growth temperatures, in addition to higher water- and nitrogen-use efficiencies [42]. However, because of their concentrating mechanism capability, it has been generally considered that C_4 plants would show little CO_2 stimulation irrespective of temperature [25]. In spite of that, the degrees of enhancement in growth parameters for C_4 plants may be greater under long-term exposure to both elevated CO_2 and temperature than to elevated [CO_2] alone [36,37]. Besides, factors such as light regime, soil moisture, nutrient status, and plant developmental stage, all modify the interactive responses to elevated [CO_2] and temperature [2,25,40,88,136–138].

There is an indication that the growth response of C_4 plants to elevated [CO_2] is partly due to the reduced transpiration effect on leaf temperature [82]. Under elevated growth [CO_2], leaf temperature may rise as a result of less transpiration, and this could lead to increases in leaf CER and expansion rates. In C_4 grasses grown between 20°C and 36°C, a rise of 1°C in leaf temperature under both ambient and elevated growth [CO_2] conditions leads to an average increase of 2 μmol CO_2/m²/sec in leaf CER [82]. In maize, leaf expansion rates of both field- and chamber-grown plants at temperatures between 13°C and 34°C are closely related to the meristem temperature, and such growth enhancement responses to temperature are the results of simultaneous increases in cell division and cell-wall expansion rates [139]. Such increases in leaf CER and expansion rates may occur due to leaf warming, and this could contribute to an enhancement in biomass accumulation for elevated CO_2-grown C_4 plants [82]. Growth of sugarcane plants at double-ambient [CO_2] and 6°C above ambient temperature increases leaf area 56%, total aboveground dry weight 74%, and juice volume 164%, when compared to plants grown at ambient [CO_2] and 1.5°C above ambient temperature. Such enhancements, however, are only 31%, 21%, and 83%, respectively, for double-ambient [CO_2] plants at 1.5°C above ambient temperature [36,37].

B. LIMITED SOIL WATER AVAILABILITY

As atmospheric [CO_2] rises, potential shifts in regional scale precipitation patterns could result in increased drought conditions in many areas of the world. As discussed earlier, a reduction in stomatal conductance is a common response of many plants to elevated growth [CO_2]. Observations made from a variety of C_3 and C_4 species indicate that a doubling of atmospheric [CO_2] can also double the instantaneous WUE [140]. As the [CO_2] is increased, the improvements in WUE are due to increased assimilation rate and decreased water loss, with the latter being more important under water-deficit situations [25]. Under soil water-deficit conditions, an improvement in WUE as induced by elevated [CO_2] could enhance growth, and this has been suggested as a factor responsible for the improved photosynthesis and increased biomass of some C_4 species [141–144]. As soil water becomes less available, the relative enhancement of photosynthesis and growth by elevated [CO_2] tends to be greater, which can alleviate drought stress and delay its onset [45,46,141,145]. A delay in the adverse effects of soil water deficit on leaf and canopy photosynthesis by elevated [CO_2] has been reported for a variety of crop plant species, including soybean [146,147], sweet potato [148], groundnut [149], rice [27,28,150,151], and sugarcane [87]. In rice, elevated [CO_2] delays the adverse effects of severe drought on *rbcS* transcript abundance and activities of Rubisco, and permits photosynthesis to continue for an extra day during the drought–stress cycle [28]. Similarly in sugarcane, a reduction in transpiration and an enhancement in WUE for plants grown at elevated [CO_2] also help to delay the adverse effects of severe drought and allow the stressed plants to continue photosynthesis during drought [87]. Furthermore, there is evidence that C_4 growth can respond as strongly to elevated CO_2 as do C_3 species under soil water-deficit conditions. In the tallgrass prairie ecosystems, C_4 species show increased productivity under elevated [CO_2] in dry years, but not in wet years [143]. In drying soil, growth of maize and sorghum also responds strongly to CO_2 enrichment [152,153].

V. CONCLUDING REMARKS

There is speculation that competitiveness and productivity of C_4 plants would decrease with future rising atmospheric [CO_2], since the responses of photosynthesis and growth of C_3 species to elevated [CO_2] are much greater than those of C_4 species [16]. However, as increases in atmospheric [CO_2] are likely to be accompanied by increases in air temperature, C_4 plants may retain their competitive advantage, and the areas of their adaptation may even increase [154,155]. As the impact of C_4 agricultural crops on world food supply is enormous, there is a need to understand how they respond to a future rising at-

mospheric [CO_2] and changing global climate. The near saturation of C_4 photosynthesis at current ambient [CO_2] and the positive growth response of several C_4 crop species to elevated [CO_2] offer a challenging opportunity for research to unravel the causes and mechanisms leading to the enhancement of CO_2 in C_4 biomass.

REFERENCES

1. RT Watson, H Rodhe, H Oeschger, U Siegenthaler. Greenhouse gases and aerosols. In: Houghton JT, Jenkins GJ, Ephraums JJ, eds. *Climatic Change. The IPCC Scientific Assessment.* Cambridge University Press, Cambridge, 1990, pp. 1–40.
2. JIL Morison, DW Lawlor. Interactions between increasing CO_2 concentration and temperature on plant growth. *Plant Cell Environ.* 22:659–682, 1999.
3. SH Schneider. What is 'dangerous' climate change. *Nature* 411:17–19, 2001.
4. MI Hoffert, K Caldeira, G Benford, DR Criswell, C Green, H Herzog, AK Jain, HS Kheshgi, KS Lackner, JS Lewis, HD Lightfoot, W Manheimer, JC Mankins, ME Mauel, LJ Perkins, ME Schlesinger, T Volk, TML Wigley. Advanced technology paths to global climate stability: energy for a greenhouse planet. *Science* 298:981–987, 2002.
5. KP Shine, RG Derwent, DJ Wuebbles, JJ Morcrette. Radiative forcing of climate. In: Houghton JT, Jenkins GJ, Ephraums JJ, eds. *Climatic Change. The IPCC Scientific Assessment.* Cambridge University Press, Cambridge, 1990, pp. 41–68.
6. JFB Mitchell, S Manabe, V Meleshko, T Tokioka. Equilibrium climate change and its implications for the future. In: JT Houghton, GJ Jenkins, JJ Ephraums, eds. *Climate Change. The IPCC Scientific Assessment.* Cambridge University Press, Cambridge, 1990, pp. 131–172.
7. A Kattenberg, F Giorgi, H Grassl, GA Meehl, JFB Mitchell, RJ Stouffer, T Tokioka, AJ Weaver, TML Wigley. Climate models — projections of future climate. In: JT Houghton, LG Meira Filho, BA Callendar, N Harris, A Kattenberg, K Maskell, eds. *Climate Change 1995, IPCC.* Cambridge University Press, Cambridge, 1996, pp. 285–357.
8. TML Wigley, SCB Raper. Implications for climate and sea level of revised IPCC emissions scenarios. *Nature* 357:293–300, 1992.
9. CD Keeling, TP Whorf, M Wahlen, J van der Plicht. Interannual extremes in the rate of rise of atmospheric carbon dioxide since 1980. *Nature* 375:660–670, 1995.
10. BA Kimball. Effects of elevated CO_2 and climate variables on plants. *J. Soil Water Conserv.* 48:9–14, 1993.
11. G Bowes. Facing the inevitable: plants and increasing atmospheric CO_2. *Annu. Rev. Plant Physiol. Plant Mol. Biol.* 44:309–332, 1993.
12. BG Drake, MA Gonzalez-Meler, SP Long. More efficient plants: a consequence of rising atmospheric CO_2?

Annu. Rev. Plant Physiol. Plant Mol. Biol. 48:609–639, 1997.

13. RF Sage, RK Monson. *C₄ Plant Biology*. San Diego: Academic Press, 1999, pp. XIII–XV.

14. J Lloyd, GD Farquhar. ^{13}C discrimination during CO_2 assimilation by the terrestrial biosphere. *Oecologia* 99:201–215, 1994.

15. TE Cerling, JM Harris, BJ MacFadden, MG Leakey, J Quade, V Eisenmann, JR Ehleringer. Global vegetation change through the Miocene/Pliocene boundary. *Nature* 389:153–158, 1997.

16. RH Brown. Agronomic implications of C₄ photosynthesis. In: RF Sage, RK Monson, eds. *C₄ Plant Biology*. Academic Press, San Diego, 1999, pp. 473–507.

17. CB Osmond, JAM Holtum. Crassulacean acid metabolism. In: MD Hatch, NK Boardman, eds. *The Biochemistry of Plants. A Comprehensive Treatise*. Vol 8. *Photosynthesis*. Academic Press, New York, 1981, pp. 283–328.

18. MD Hatch. Resolving C₄ photosynthesis: trials, tribulations and other unpublished stories. *Aust. J. Plant Physiol.* 24:413–422, 1997.

19. MD Hatch. C₄ photosynthesis: a historical overview. In: RF Sage, RK Monson, eds. *C₄ Plant Biology*. Academic Press, San Diego, 1999, pp. 17–46.

20. R Kanai, G Edwards. Biochemistry of C₄ photosynthesis. In: RF Sage, RK Monson, eds. *The Biology of C₄ Photosynthesis*. Academic Press, New York, 1999, pp. 49–87.

21. S von Caemmerer, RT Furbank. Modeling C₄ photosynthesis. In: RF Sage, RK Monson, eds. *The Biology of C₄ Photosynthesis*. Academic Press, New York, 1999, pp. 173–211.

22. MD Hatch, T Kagawa, S Craig. Subdivision of C₄ pathway species based on differing C₄ acid decarboxylating systems and ultrastructural features. *Aust. J. Plant Physiol.* 2:111–128, 1975.

23. M Gutierrez, VE Gracen, GE Edwards. Biochemical and cytological relationships in C₄ plants. *Planta* 19:279–300, 1974.

24. JP Maroco, MSB Ku, GE Edwards. Utilization of CO_2 in the metabolic optimization of C₄ photosynthesis. *Plant Cell Environ.* 23:115–121, 2000.

25. G Bowes. Photosynthetic responses to changing atmospheric carbon dioxide concentration. In: NR Baker, ed. *Photosynthesis and the Environment*. Kluwer Academic Publishers, The Netherlands, 1996, pp. 387–407.

26. JCV Vu, LH Allen Jr, KJ Boote, G Bowes. Effects of elevated CO_2 and temperature on photosynthesis and Rubisco in rice and soybean. *Plant Cell Environ.* 20:68–76, 1997.

27. JCV Vu, JT Baker, AH Pennanen, LH Allen Jr, G Bowes, KJ Boote. Elevated CO_2 and deficit effects on photosynthesis, ribulose bisphosphate carboxylase-oxygenase, and carbohydrate metabolism in rice. *Physiol. Plant.* 103:327–339, 1998.

28. JCV Vu, RW Gesch, LH Allen Jr, KJ Boote, G Bowes. CO_2 enrichment delays a rapid, drought-in-

duced decrease in Rubisco small subunit transcript abundance. *J. Plant Physiol.* 139–142, 1999.

29. JCV Vu, RW Gesch, AH Pennanen, LH Allen Jr, KJ Boote, G Bowes. Soybean photosynthesis, Rubisco, and carbohydrate enzymes function at supraoptimal temperatures in elevated CO_2. *J. Plant Physiol.* 158:295–307, 2001.

30. BA Kimball. Carbon dioxide and agricultural yield: an assemblage and analysis of 430 observations. *Agron. J.* 75:779–788, 1983.

31. JP Simon, C Potvin, BR Strain. Effects of temperature and CO_2 enrichment on kinetic properties of phospho-enol-pyruvate carboxylase in two ecotypes of *Echinochloa crus-galli* (L.) Beauv., a C₄ weed grass species. *Oecologia* 63:145–152, 1984.

32. C Potvin, BR Strain. Photosynthetic response to growth temperature and CO_2 enrichment in two species of C₄ grasses. *Can. J. Bot.* 63:483–487, 1985.

33. C Potvin, BR Strain. Effects of CO_2 enrichment and temperature on growth in two C₄ weeds, *Echinochloa crus-galli* and *Eleusine indica*. *Can. J. Bot.* 63:1495–1499, 1985.

34. H Poorter. Interspecific variation in the growth response of plants to an elevated ambient concentration. *Vegetatio* 104/105:77–97, 1993.

35. LH Ziska, JA Bunce. Influence of increasing carbon dioxide concentration on the photosynthetic and growth stimulation of selected C₄ crops and weeds. *Photosynth. Res.* 54:199–208, 1997.

36. LH Allen Jr, JCV Vu, JD Ray. Sugarcane responses to carbon dioxide, temperature, and water table. In: American Society of Agronomy Annual Meeting, Division A-3, Baltimore, MD, October 18–22, 1998, Abstr., p. 20.

37. JCV Vu, LH Allen Jr, G Bowes. Growth and photosynthesis responses of sugarcane to high CO_2 and temperature. *Plant Biology 98*. American Society of Plant Physiologists, Madison, WI, June 27 to July 1, 1998, Abstr. No. 448.

38. JP Maroco, GE Edwards, MSB Ku. Photosynthetic acclimation of maize to growth under elevated levels of carbon dioxide. *Planta* 210:115–125, 1999.

39. DH Greer, WA Laing, BD Campbell. Photosynthetic responses of thirteen pasture species to elevated CO_2 and temperature. *Aust. J. Plant Physiol.* 22:713–722, 1995.

40. O Ghannoum, S von Caemmerer, EWR Barlow, JP Conroy. The effect of CO_2 enrichment and irradiance on the growth, morphology and gas exchange of a C₃ (*Panicum laxum*) and a C₄ (*Panicum antidotale*) grass. *Aust. J. Plant Physiol.* 24:227–237, 1997.

41. DR LeCain, JA Morgan. Growth, gas exchange, leaf nitrogen and carbohydrate concentrations in NAD-ME and NADP-ME C₄ grasses grown in elevated CO_2. *Physiol. Plant.* 102:297–306, 1998.

42. M Matsuoka, RT Furbank, H Fukayama, M Miyao. Molecular engineering of C₄ photosynthesis. *Annu. Rev. Plant Physiol. Plant Mol. Biol.* 52:297–314, 2001.

43. H Poorter, C Roumet, BD Campbell. Interspecific variation in the growth response of plants to elevated

CO_2: a search for functional types. In: C Korner, FA Bazzaz, eds. *Carbon Dioxide, Populations, and Communities.* Academic Press, New York, 1996, pp. 375–412.

44. RJ Norby, SD Wullschleger, CA Gunderson, DW Johnson, R Ceulemans. Tree responses rising CO_2 in the field experiments: implications for the future forest. *Plant Cell Environ.* 22:683–714, 1999.

45. LH Allen Jr. Plant responses to rising carbon dioxide and potential interactions with air pollutants. *J. Environ. Qual.* 19:15–34, 1990.

46. FA Bazzaz. The response of natural ecosystems to the rising global CO_2 levels. *Annu. Rev. Ecol. Syst.* 21:167–196, 1990.

47. EH Delucia, TW Sasek, BR Strain. Photosynthetic inhibition after long-term exposure to elevated levels of atmospheric carbon dioxide. *Photosynth. Res.* 7:175–184, 1985.

48. RF Sage, TD Sharkey, JR Seemann. Acclimation of photosynthesis to elevated CO_2 in five C_3 species. *Plant Physiol.* 89:590–596, 1989.

49. RT Besford, LJ Ludwig, AC Withers. The greenhouse effect: acclimation of tomato plants growing in high CO_2, photosynthesis and ribulose-1,5-bisphosphate carboxylase protein. *J. Exp. Bot.* 41:925–931, 1990.

50. WJ Arp. Effects of source-sink relations on photosynthetic acclimation to elevated CO_2. *Plant Cell Environ.* 14:869–875, 1991.

51. JF Farrar, ML Williams. The effects of increased atmospheric carbon dioxide and temperature on carbon partitioning, source-sink relations and respiration. *Plant Cell Environ.* 14:819–830, 1991.

52. M Stitt. Rising CO_2 levels and their potential significance for carbon flow in photosynthetic cells. *Plant Cell Environ.* 14:741–762, 1991.

53. A Makino, T Mae. Photosynthesis and plant growth at elevated levels of CO_2. *Plant Cell Physiol.* 40:999–1006, 1999.

54. SC Wong. Elevated atmospheric partial pressure of CO_2 and plant growth. I. Interactions of nitrogen nutrition and photosynthetic capacity in C_3 and C_4 plants. *Oecologia* 44:68–74, 1979.

55. CV Vu, LH Allen Jr, G Bowes. Effects of light and elevated atmospheric CO_2 on the ribulose bisphosphate carboxylase activity and ribulose bisphosphate level of soybean leaves. *Plant Physiol.* 73:729–734, 1983.

56. MA Porter, B Grodzinski. Acclimation to high CO_2 in bean. Carbonic anhydrase and ribulose bisphosphate carboxylase. *Plant Physiol.* 74:413–416, 1984.

57. DT Tissue, RB Thomas, BR Strain. Long-term effects of elevated CO_2 and nutrients on photosynthesis and Rubisco in loblolly pine seedlings. *Plant Cell Environ.* 16:859–865, 1993.

58. BD Moore, S-H Cheng, J Rice, J Seemann. Sucrose cycling, Rubisco expression and prediction of photosynthetic acclimation to elevated carbon dioxide. *Plant Cell Environ.* 21:905–915, 1998.

59. MW Hussain, LH Allen Jr, G Bowes. Up-regulation of phosphate synthase in rice grown under elevated CO_2 and temperature. *Photosynth. Res.* 60:199–208, 1999.

60. JJ Van Oosten, RT Besford. Sugar feeding mimics effect of acclimation to high CO_2-rapid down regulation of Rubisco small subunit transcripts but not of the large subunit transcripts. *J. Plant Physiol.* 143:306–312, 1994.

61. JJ Van Oosten, S Wilkins, RT Besford. Regulation of the expression of photosynthetic nuclear genes by high CO_2 is mimicked by carbohydrates: A mechanism for the acclimation of photosynthesis to high CO_2. *Plant Cell Environ.* 17:913–923, 1994.

62. AN Webber, GY Nie, SP Long. Acclimation of photosynthetic proteins to rising atmospheric CO_2. *Photosynth. Res.* 39:413–425, 1994.

63. GY Nie, DL Hendrix, AN Weber, BA Kimball, SP Long. Increased accumulation of carbohydrates and decreased photosynthetic gene transcript levels in wheat grown at an elevated CO_2 concentration in the field. *Plant Physiol.* 108:975–983, 1995.

64. JJ Van Oosten, RT Besford. Some relationships between the gas exchange, biochemistry and molecular biology of photosynthesis during leaf development of tomato plants after transfer to different carbon dioxide concentrations. *Plant Cell Environ.* 18:1253–1266, 1995.

65. S-H Cheng, BD Moore, JR Seemann. Effects of short- and long-term elevated CO_2 on the expression of ribulose-1,5-bisphosphate carboxylase/oxygenase genes and carbohydrate accumulation in leaves of *Arabidopsis thaliana* (L.) Heynh. *Plant Physiol.* 116:715–723, 1998.

66. RW Gesch, KJ Boote, JCV Vu, LH Allen Jr, G Bowes. Changes in growth CO_2 result in rapid adjustments of ribulose-1,5-bisphosphate carboxylase/ oxygenase small subunit gene expression in expanding and mature leaves of rice. *Plant Physiol.* 118:521–529, 1998.

67. J Sheen. C_4 gene expression. *Annu. Rev. Plant Physiol. Plant Mol. Biol.* 50:187–217, 1999.

68. JV Pego, AJ Kortstee, C Huijser, SCM Smeekens. Photosynthesis, sugars and the regulation of gene expression. *J. Exp. Bot.* 51:407–416, 2000.

69. S Smeekens. Sugar-induced signal transduction in plants. *Annu. Rev. Plant Physiol. Plant Mol. Biol.* 51:49–81, 2000.

70. J Sheen. Metabolic repression of transcription in higher plants. *Plant Cell* 2:1027–1038, 1990.

71. J Sheen. Feedback control of gene expression. *Photosynth. Res.* 39:427–438, 1994.

72. JC Jang, J Sheen. Sugar sensing in higher plants. *Plant Cell* 6:1665–1679, 1994.

73. JJ Van Oosten, RT Besford. Acclimation of photosynthesis to elevated carbon dioxide through feedback regulation of gene expression: climate of opinion. *Photosynth. Res.* 48:353–365, 1996.

74. EE Goldschmidt, SC Huber. Regulation of photosynthesis by end-product accumulation in leaves of plants storing starch, sucrose and hexose sugars. *Plant Physiol.* 99:1443–1448, 1992.

75. A Krapp, B Hofmann, C Schafer, M Stitt. Regulation of the expression of rbcS and other photosynthetic genes by carbohydrates: a mechanism for the 'sink' regulation of photosynthesis? *Plant J.* 3:817–828, 1993.

76. JC Jang, J Sheen. Sugar sensing in higher plants. *Trends Plant Sci.* 2:208–214, 1997.

77. JC Jang, P Leon, L Zhou, J Sheen. Hexokinase as a sugar sensor in higher plants. *Plant Cell* 9:5–19, 1997.

78. S Smeekens, F Rook. Sugar sensing and sugar-mediated signal transduction in plants. *Plant Physiol.* 115:7–13, 1997.

79. NG Halford, PC Purcell, DG Hardie. Is hexokinase really a sugar sensor in plants? *Trends Plant Sci.* 4:117–124, 1999.

80. BD Moore, S-H Cheng, D Sims, JR Seemann. The biochemical and molecular basis for photosynthetic acclimation to elevated atmospheric CO_2. *Plant Cell Environ.* 22:567–582, 1999.

81. KE Koch, Z Ying, Y Wu, WT Avigne. Multiple paths of sugar-sensing and a sugar/oxygen overlap for genes of sucrose and ethanol metabolism. *J. Exp. Bot.* 51:417–427, 2000.

82. O Ghannoum, S von Caemmerer, LH Ziska, JP Conroy. The growth response of C_4 plants to rising atmospheric CO_2 partial pressure: a reassessment. *Plant Cell Environ.* 23:931–942, 2000.

83. LH Ziska, RC Sicher, JA Bunce. The impact of elevated carbon dioxide on the growth and gas exchange of three C_4 species differing in CO_2 leak rates. *Physiol. Plant.* 105:74–80, 1999.

84. K Siebke, O Ghannoum, JP Conroy, S von Caemmerer. Elevated CO_2 increases the leaf temperature of two glasshouse-grown C_4 grasses. *Funct. Plant Biol.* 29:1377–1385, 2002.

85. AJ Jarvis, TA Mansfield, WJ Davies. Stomatal behavior, photosynthesis and transpiration under rising CO_2. *Plant Cell Environ.* 22:639–648, 1999.

86. SD Smith, BR Strain, TD Sharkey. Effects of CO_2 enrichment on four Great Basin grasses. *Funct. Ecol.* 1:139–143, 1987.

87. JCV Vu, LH Allen Jr. CO_2 enrichment delays the adverse effects of severe drought on C_4 sugarcane photosynthesis. In: *Plant Biology 2001*, American Society of Plant Biologists, Providence, RI, July 21–25, 2001, Abstr. No. 401.

88. M Stitt, A Krapp. The interaction between elevated carbon dioxide and nitrogen nutrition: the physiological and molecular background. *Plant Cell Environ.* 22:583–621, 1999.

89. AB Cousins, NR Adam, GW Wall, BA Kimball, PJ Pinter Jr, SW Leavitt, RL LaMorte, AD Matthias, MJ Ottman, TL Thompson, AN Webber. Reduced photorespiration and increased energy-use efficiency in young CO_2-enriched sorghum leaves. *New Phytol.* 150:275–284, 2001.

90. JCV Vu, LH Allen Jr. Photosynthesis during leaf ontogeny of sugarcane at elevated CO_2. In: *Plant Biology 2003*, American Society of Plant Biologists, Honolulu, HI, July 25–30, 2003, Abstr. No. 72.

91. A Leakey, C Bernacchi, S Long, D Ort. Will photosynthesis of maize (*Zea mays*) in the US Corn Belt increase in future $[CO_2]$ rich atmospheres? An analysis of diurnal courses of gas exchange under Free-Air Concentration Enrichment (FACE). *Plant Biology 2003*, American Society of Plant Biologists, Honolulu, HI, July 25–30, 2003, Abstr. No. 67.

92. DQ Xu, RM Gifford, WS Chow. Photosynthetic acclimation in pea and soybean to high atmospheric CO_2 partial pressure. *Plant Physiol.* 106:661–671, 1994.

93. GY Nie, SP Long, RL Garcia, BA Kimball, RL LaMorte, PJ Pinter Jr, GW Wall, AN Webber. Effects of free air CO_2 enrichment on the development of the photosynthetic apparatus in wheat, as indicated by changes in leaf proteins. *Plant Cell Environ.* 18:855–864, 1995.

94. M Pearson, G Brooks. The influence of elevated CO_2 on growth and age-related changes in leaf gas exchange. *J. Exp. Bot.* 46:1651–1659, 1995.

95. A Miller, CH Tsai, D Hemphill, M Endres, S Rodermel, M Spalding. Elevated CO_2 effects during leaf ontogeny. A new perspective on acclimation. *Plant Physiol.* 115:1195–1200, 1997.

96. K Garbutt, WE Williams, FA Bazzaz. Analysis of the differential response of five annuals to elevated carbon dioxide during growth. *Ecology* 7:1185–1194, 1990.

97. R Baxter, M Grantley, TW Ashenden, J Farrar. Effects of elevated carbon dioxide on three grass species from montane pasture. II. Nutrient allocation and efficiency of nutrient use. *J. Exp. Bot.* 45:1267–1278, 1994.

98. M Geiger, L Walch-Piu, J Harnecker, ED Schulze, F Ludewig, U Sonnewald, WR Scheible, M Stitt. Enhanced carbon dioxide leads to a modified diurnal rhythm of nitrate reductase activity and higher levels of amino acids in higher plants. *Plant Cell Environ.* 21:253–268, 1998.

99. R Cuelemans, XN Jiang, BY Shao. Growth and physiology of one year old poplar under elevated atmospheric carbon dioxide levels. *Ann. Bot.* 75:609–617, 1994.

100. HSJ Lee, PG Jarvis. Trees differ from crops and from each other in their response to increases in CO_2 concentration. *J. Biogeogr.* 22:323–330, 1995.

101. DT Tissue, RB Thomas, BR Strain. Atmospheric CO_2 enrichment increases growth and photosynthesis of *Pinus taeda*, a 4 year experiment in the field. *Plant Cell Environ.* 20:1123–1134, 1997.

102. S Gepstein. Photosynthesis. In: LD Nooden, AC Leopold, eds. *Senescence and Aging in Plants*. Academic Press, San Diego, 1988, pp. 85–109.

103. DC Tremmel, DT Patterson. Responses of soybean and 5 weeds to CO_2 enrichment under 2 temperature regimes. *Can. J. Plant Sci.* 73:1249–1260, 1993.

104. RA Kennedy, WM Laetsch. Relationship between leaf development, carboxylase enzyme activities and photorespiration in the C_4-plant *Portulaca oleracea* L. *Planta* 115:113–124, 1973.

105. BD Moore, S-H Cheng, GE Edwards. The influence of leaf development on the expression of C_4 metabolism

in *Flaveria trinervia*, a C$_4$ dicot. *Plant Cell Physiol.* 27:1159–1167, 1986.

106. LE Williams, RA Kennedy. Photosynthetic carbon metabolism during leaf ontogeny in *Zea mays* L. Enzyme studies. *Planta* 142:269–274, 1978.

107. LE Williams, RA Kennedy. Relationship between early photosynthetic products, photorespiration and stage of leaf development in *Zea mays*. *Z. Pflanzenphysiol.* 81:314–322, 1977.

108. HM Crespo, M Frean, CF Cresswell, J Tew. The occurrence of both C$_3$ and C$_4$ photosynthetic characteristics in a single *Zea mays* plant. *Planta* 147:257–263, 1979.

109. JT Perchorowicz, M Gibbs. Carbon dioxide fixation and related properties in sections of the developing green maize leaf. *Plant Physiol.* 65:802–809, 1980.

110. Z Dai, MSB Ku, GE Edwards. C$_4$ photosynthesis. The effects of leaf development on the CO$_2$-concentrating mechanism and photorespiration in maize. *Plant Physiol.* 107:815–825, 1995.

111. AB Cousins, NR Adam, GW Wall, BA Kimball, PJ Pinter Jr, MJ Ottman, SW Leavitt, AN Webber. Development of C$_4$ photosynthesis in sorghum leaves grown under free-air CO$_2$ enrichment (FACE). *J. Exp. Bot.* 54:1969–1975, 2003.

112. FC Meinzer, NZ Saliendra. Spatial patterns of carbon isotope discrimination and allocation of photosynthetic activity in sugarcane leaves. *Aust. J. Plant Physiol.* 24:769–775, 1997.

113. WM Laetsch, I Price. Development of the dimorphic chloroplasts of sugar cane. *Am. J. Bot.* 56:77–87, 1969.

114. SJ Kirchanski. The ultrastructural development of the dimorphic plastids of *Zea mays* L. *Am. J. Bot.* 62:695–705, 1975.

115. T Nelson, JA Langdale. Developmental genetics of C$_4$ photosynthesis. *Annu. Rev. Plant Physiol. Plant Mol. Biol.* 43:25–47, 1992.

116. N Sionit, DT Patterson. Responses of C$_4$ grasses to atmospheric CO$_2$ enrichment. I. Effects of irradiance. *Oecologia* 65:30–34, 1984.

117. O Ghannoum, K Siebke, S von Caemmerer, JP Conroy. The photosynthesis of young *Panicum* C$_4$ leaves is not C$_3$-like. *Plant Cell Environ.* 21:1123–1131, 1998.

118. JR Watling, MC Press, WP Quick. Elevated CO$_2$ induces biochemical and ultrastructural changes in leaves of the C$_4$ cereal sorghum. *Plant Physiol.* 123:1143–1152, 2000.

119. D Tissue, K Griffin, R Thomas, B Strain. Effects of low and elevated CO$_2$ on C$_3$ and C$_4$ annuals. II. Photosynthesis and leaf biochemistry. *Oecologia* 101:21–28, 1995.

120. GD Farquhar. On the nature of carbon isotope discrimination in C$_4$ species. *Aust. J. Plant Physiol.* 10:205–226, 1983.

121. SJE Wand, GF Midgley, WD Stock. Growth responses to elevated CO$_2$ in NADP-ME, NAD-ME and PCK C$_4$ grasses and a C$_3$ grass from South Africa. *Aust. J. Plant Physiol.* 28:13–25, 2001.

122. SP Long. Modification of the response of photosynthetic productivity to rising temperature by atmospheric CO$_2$ concentrations: has its importance been underestimated? *Plant Cell Environ.* 14:729–739, 1991.

123. J Kobza, GE Edwards. Influences of leaf temperature on photosynthetic carbon metabolism in wheat. *Plant Physiol.* 83:69–74, 1987.

124. AS Holaday, W Martindale, R Alred, AL Brooks, RC Leegood. Changes in activities of enzymes of carbon metabolism in leaves during exposure of plants to low temperature. *Plant Physiol.* 98:1105–1114, 1992.

125. DB Jordan, WL Ogren. The CO$_2$/O$_2$ specificity of ribulose 1,5-bisphosphate carboxylase/oxygenase. *Planta* 161:308–313, 1984.

126. A Brooks, GD Farquhar. Effect of temperature on the CO$_2$/O$_2$ specificity of ribulose-1,5-bisphosphate carboxylase/oxygenase and the rate of respiration in the light. *Planta* 165:397–406, 1985.

127. JT Baker, LH Allen Jr, KJ Boote. Temperature effects on rice at elevated CO$_2$ concentration. *J. Exp. Bot.* 43:959–964, 1992.

128. JT Baker, LH Allen Jr. Contrasting crop species responses to CO$_2$ and temperature: rice, soybean and citrus. *Vegetatio* 104/105:239–260, 1993.

129. JT Baker, LH Allen Jr, KJ Boote, P Jones, JW Jones. Response of soybean to air temperature and carbon dioxide concentration. *Crop Sci.* 29:98–105, 1989.

130. RAC Mitchell, VJ Mitchell, SP Driscoll, J Franklin, DW Lawlor. Effects of increased CO$_2$ concentration and temperature on growth and yield of winter wheat at two levels of nitrogen application. *Plant Cell Environ.* 16:521–529, 1993.

131. JA Berry, O Bj_rkman. Photosynthetic response and adaptation to temperature in higher plants. *Annu. Rev. Plant Physiol.* 31:491–543, 1980.

132. LH Allen Jr. Carbon dioxide increase: direct impacts on crops and indirect effects mediated through anticipated climatic changes. In: KJ Boote, JM Bennett, TR Sinclair, GM Paulsen, eds. *Physiology and Determination of Crop Yield.* American Society of Agronomy, Madison, WI, 1994, pp. 425–459.

133. MKH Ebrahim, G Vogg, MNEH Osman, E Komor. Photosynthetic performance and adaptation of sugarcane at suboptimal temperatures. *J. Plant Physiol.* 153:587–592, 1998.

134. MKH Ebrahim, O Zingsheim, MN El-Shourbagy, PH Moore, E Komor. Growth and sugar storage in sugarcane grown at temperatures below and above optimum. *J. Plant Physiol.* 153:593–602, 1998.

135. SP Long. Environmental responses. In: RF Sage, RK Monson, eds. *C$_4$ Plant Biology.* Academic Press, San Diego, 1999, pp. 215–249.

136. HM Rawson. Plant responses to temperature under conditions of elevated CO$_2$. *Aust. J. Bot.* 40:473–490, 1992.

137. JS Coleman, FA Bazzaz. Effects of CO$_2$ and temperature on growth and resource use of co-occurring C$_3$ and C$_4$ annuals. *Ecology* 73:1244–1259, 1992.

138. MJ Robertson, GD Bonnett, RM Hughes, RC Muchow, JA Campbell. Temperature and leaf area expansion of sugarcane: integration of controlled-

environment, field and model studies. *Aust. J. Plant Physiol.* 25:819–828, 1998.

139. H Ben-Haj-Salah, F Tardieu. Temperature affects expansion rate of maize leaves without change in spatial distribution of cell length. *Plant Physiol.* 109:861–870, 1995.

140. JIL Morison. Response of plants to CO_2 under water limited conditions. *Vegetatio* 104/105:193–209, 1993.

141. BG Drake, PW Leadley. Canopy photosynthesis of crops and native plant communities exposed to long-term elevated CO_2. *Plant Cell Environ.* 14:853–860, 1991.

142. CE Owensby, PI Coyne, JM Ham, LM Auen, AK Knapp. Biomass production in a tallgrass prairie ecosystem exposed to ambient and elevated CO_2. *Ecol. Appl.* 3:644–653, 1993.

143. AB Samarakoon, RM Gifford. Soil water content under plants at high CO_2 concentration and interactions with the direct CO_2 effects: A species comparison. *J. Biogeogr.* 22:193–202, 1995.

144. HW Polley, HB Johnson, HS Mayeux, CR Tischler. Are some of the recent changes in grassland communities a response to rising CO_2 concentrations? In: GW Koch, HA Mooney, eds. *Carbon Dioxide and Terrestrial Ecosystems.* Academic Press, New York, 1996, pp. 177–195.

145. WJ Arp, JEM Mierlo, F Berendse, W Snijders. Interactions between elevated carbon dioxide concentration, nitrogen and water: effects on growth and water use of six perennial plant species. *Plant Cell Environ.* 21:1–11, 1998.

146. HH Rogers, N Sionit, JD Cure, HM Smith, GE Bingham. Influence of elevated CO_2 on water relations of soybeans. *Plant Physiol.* 74:233–238, 1984.

147. P Jones, JW Jones, LH Allen Jr. Seasonal carbon and water balances of soybeans grown under stress treat-

ments in sunlit chambers. *Trans. Am. Soc. Agric. Eng.* 28:2021–2028, 1985.

148. NC Bhattacharya, DR Hileman, PP Ghosh, RL Musser, S Bhattacharya, PK Biswas. Interaction of enriched CO_2 and water stress on the physiology and biomass production in sweet potato grown in open-top chambers. *Plant Cell Environ.* 13:933–940, 1990.

149. SC Clifford, IM Stronach, AD Mohamed, SN Azam-Ali, NMJ Crout. The effects of elevated atmospheric carbon dioxide and water stress on light interception, dry matter production and yield in stands of groundnut (*Arachis hypogaea* L.). *J. Exp. Bot.* 44:1763–1770, 1993.

150. JT Baker, LH Allen Jr, KJ Boote, NB Pickering. Rice responses to drought under carbon dioxide enrichment. 1. Growth and yield. *Global Change Biol.* 3:119–128, 1997.

151. JT Baker, LH Allen Jr, KJ Boote, NB Pickering. Rice responses to drought under carbon dioxide enrichment. 2. Photosynthesis and evaporation. *Global Change Biol.* 3:129–138, 1997.

152. AB Samarakoon, RM Gifford. Elevated CO_2 effects on water use and growth of maize in wet and drying soil. *Aust. J. Plant Physiol.* 23:53–62, 1996.

153. MJ Ottman, BA Kimball, PJ Pinter, GW Wall, RL Vanderlip, SW Leavitt, RL LaMorte, AD Matthias, TJ Brooks. Elevated CO_2 increases sorghum biomass under drought conditions. *New Phytol.* 150:261–273, 2001.

154. S Henderson, P Hattersley, S von Caemmerer, CB Osmond. Are C_4 plants threatened by global climatic change? In: ED Schulze, MM Caldwell, eds. *Ecophysiology of Photosynthesis.* Springer-Verlag, Berlin, 1995, pp. 529–549.

155. DT Patterson. Weeds in a changing climate. *Weed Sci.* 43:685–701, 1995.

19 Influence of High Light Intensity on Photosynthesis: Photoinhibition and Energy Dissipation

Robert Carpentier
Groupe de recherche en biologie végétale, Université du Québec à Trois-Rivières

CONTENTS

I. INTRODUCTION

Exposure of photosynthesis apparatus to illumination that exceeds the capacity of the carbon cycle or the ability of the light-harvesting system to dissipate light energy not used for photosynthetic functions leads to reversible and irreversible damage. The first stage of this process is related to the destruction of the photosystem II (PSII) reaction centers, and the resulting inhibition of electron transport is referred to as photoinhibition [1,2]. The second stage is due to the lack of dissipation of absorbed energy through photosynthetic energy conversion and concerns the bleaching of the antenna pigments. The latter process requires light and oxygen and is therefore referred to as pigment photooxidation [3,4]. This oxidative process significantly occurs only when the photosynthetic activity is strongly impeded and under saturating illumination.

There exist, however, several mechanisms that prevent or retard the above damaging processes [5].

The most important of these is probably the energy-dependent nonphotochemical fluorescence quenching connected with the so-called violaxanthin cycle [6] and cyclic electron transport pathway around PSII [7]. The dissipation of excessive absorbed energy can also be implemented by special chlorophyll forms absorbing at relatively high wavelengths [4].

II. PHOTOINHIBITION OF PSII

A. MOLECULAR MECHANISM

PSII is the photosystem most sensitive to excessive illumination and is primarily affected by photoinhibition [1]. Photosystem I (PSI) is also affected, and there is some indication that under certain specific conditions PSI may be more affected than PSII [8]. PSI photoinhibition will be discussed in a separate section. In PSII, two mechanisms of photoinactivation are involved, which affect the acceptor and donor sides [9].

Acceptor-side photoinhibition is considered as the predominant photoinactivation mechanism. It is initiated by the fully reduced state of the plastoquinone pool and of the primary quinone acceptors of the photosystem generated under strong light or when the illumination exceeds the capacity of the carbon cycle to reoxidize the reduced intermediates of the electron transport chain. In that case, the Q_B site remains nonfunctional due to the absence of oxidized plastoquinone, and the primary quinone acceptor remains in its reduced state (Q_A^-) [10]. There is a probability that the latter becomes doubly reduced and protonated under conditions where the reoxidation is prevented [11]. The fully reduced and protonated Q_A is supposed to leave its binding site [11], but even if it does not, the protonated state of Q_A should prevent electrostatic repulsion between reduced pheophytin (Pheo$^-$) and reduced Q_A [12]. Under these conditions, charge separation between P680 and Pheo can still occur with a significant yield. Recombination of the primary charge separation products [P680$^+$Pheo$^-$] leads to the formation of P680 in triplet state [13]. Under aerobic conditions, these triplet states are quenched by oxygen and highly reactive singlet oxygen molecules thus appear [14]. These species are thought to irreversibly inactivate the reaction center P680 with the consequent full inhibition of photochemical energy conversion [10].

It is clear that the above process requires the presence of oxygen. The donor-side photoinhibition, however, can proceed under both aerobic and anaerobic conditions because it does not seem to depend on oxygen. Such photoinhibitory conditions are produced when the water-splitting complex is either partially or fully inactivated [15–18]. This situation is created when Mn is depleted from the water-splitting enzyme or if the Cl$^-$ or Ca^{2+} requirement for oxygen evolution is not fully met. The lack of reducing equivalents usually arriving from the water-splitting apparatus leads to formation of strongly oxidizing radicals such as P680$^+$ and Y_Z^+, which is the intermediate electron carrier between the water-splitting site and P680. These radicals are thought to be responsible for photoinhibitory damage. Donor-side photoinhibition also results in the irreversible photobleaching of β-carotene and Chl670 (Chl$_Z$), an accessory chlorophyll of the reaction center complex [19,20]. The above intermediates are oxidized by P680 when Y_Z is kept in its oxidized form.

B. Turnover of the Polypeptide D1

1. D1 Synthesis/Degradation

Photoinhibition in PSII is known to be accompanied by the degradation of the polypeptide D1 (psbA gene product), one of the two polypeptides (D1 and D2) that bear the primary acceptors and donors of the photosystem together with the reaction center P680. This polypeptide is characterized by its fast light-induced turnover rate in comparison with other polypeptides of the thylakoid membrane [21]. At high light intensity, the processing of the polypeptide, its degradation, and its replacement by newly synthesized ones is saturated, a condition that leads to strong photoinhibition [22]. However, it was also demonstrated that the quantum yield for photoinhibition was independent of light intensity [23]. Degradation of D1 thus occurs even at very low light intensity and increases with light intensity [23,24], which is also reflected in the ability to adjust psbA gene expression with light condition to ensure the reconstruction of altered PSII centers [25]. It was clearly shown that both translation and elongation of D1 and its subsequent incorporation in PSII core complexes require light [26]. Incorporation of D1 seems to depend on the primary quinone acceptors redox state, with the accumulation of reduced Q_A specifically inactivating psbA transcription [27]. It has also been reported that D1 degradation is triggered by the inactivation of Q_A reduction [28]. Thus, with oxidized Q_A, psbA transcription is activated and D1 degradation is promoted, which correlates well with the finding that the rate of D1 degradation coincides with the extent of accumulation of radiolabeled methinonine into the D1 polypeptide [29].

It should be mentioned that the processing of D1 is required to protect PSII against abusive photoinhibition. Therefore, mutants deficient in biosynthesis or degradation of D1 were more sensitive to photoinhibition [30]. One factor that delays D1 synthesis is the low content of unsaturated fatty acids in the thylakoid membranes [31–33]. Therefore, the rate of photoinactivation of PSII is greatly enhanced in mutant organisms with deficient saturation levels of membrane fatty acids [34,35]. Various stresses such as oxidative stress and slat stress inhibit the repair mechanism by affecting the transcription and translation of the psbA gene and thus also increase the susceptibility of PSII to photoinhibition [36,37]. Conversely, phosphorylation of D1 was reported to be part of the regulation mechanism of D1 degradation [38]. It was shown that D1 degradation was retarded in its phosphorylated state whether acceptor- or donor-side photoinhibition was studied, and the implication of kinase and phosphatase activity was proposed to be part of the control mechanism for D1 degradation [39]. It was further indicated that a pool of PSII centers with phosphorylated D1 not involved in protein turnover was formed in the central region of the grana [40]. There are some contradictory reports indicating that dephosphorylation is either essential [41] or not [42] for D1 cleavage. However, it seems likely that phosphorylated D1 migrates from the granal section of the thylakoid membrane to the stroma lamellae section where it would be dephosphorylated and submitted to more extensive proteolysis [40,43]. Another means of regulating the degradation of the protein D1 that seems specific for cyanobacteria concerns the exchange of constitutive D1 (D1:1) with another form of D1 (D1:2, a product of *psb*AIII) [44]. The content in D1:2 is increased with closure of the PSII reaction centers [44,45]. Insertion of D1:2 in the PSII centers significantly reduces the sensitivity to photoinhibition [44–46]. This was associated with a greater turnover rate of D1:2 compared to D1:1 [47].

It is suggested that during assembly, the polypeptide D2 is first introduced as a stabilizing component followed by D1 and cytochrome $b559$ (cyt $b559$) [48]. A heat shock protein, HSP70B, has also been implicated in the repair of photodamaged reaction centers. Increasing growth light intensity in green algae cultures corresponded with an increased level of HSP70B [49]. Correspondingly, mutants with underexpression of HSP70 present an increased light sensitivity and those with overexpression are more resistant to photoinhibition, and their reactivation after photoinhibition is more effective [50]. The repair mechanism of PSII core may thus be controlled by chloroplast-specific chaperones.

2. D1 Cleavage Mechanism

The photoinduced breakdown of the D1 polypeptide was shown to occur in various photosynthetic systems such as whole cells, isolated thylakoids, PSII particles, or reaction center preparations. There is some indication that the degradation of D1 may be preceded by the disassembly of the internal antenna complexes CP47 and CP43 from the photosystem [51]. The initial degradation of D1 is characterized by specific breakdown products that can be detected after separation of the polypeptides by gel electrophoresis and subsequent immunoblotting using D1 antiserum. A fragment with an apparent molecular weight of 23.5 kDa was first analyzed by Greenberg et al. [52] as originating from the N terminus of D1. It is believed that the fragments that are generated during photoinhibition are different according to whether the donor- or the acceptor-side mechanisms are involved. Donor-side photoinhibition was shown to produce a 24-kDa fragment originating from the C terminus, with a cleavage site located between transmembrane segments I and II, and another fragment of about 10 kDa derived from the N terminus [53,54].

On the other hand, acceptor-side photoinhibition produced an N-terminal 23-kDa fragment and a C-terminal 10-kDa fragment [55,56]. A 16-kDa fragment was also detected during donor-side photoinhibition and attributed to the C-terminal portion of D1 [57], but another report from the same group identifies the 16-kDa fragment as originating from midway in the protein [58].

The cleavage of D1 has been inferred to be initiated by a conformational change of the polypeptide due to primary damage [10,59]. This change may expose specific sites to proteolysis. Dissociation of the PSII cores during photoinhibition also implicates the proteolysis of the polypeptide D2 and the degradation of cyt $b559$ [60]. Photodegradation also results in the oxidative cross-linking of the associated polypeptides of the core complex D1/D2–cyt $b559$ both *in vitro* and *in vivo* [61,62]. During donor-side photoinhibition, a cross-linked product of D1 and CP43, an inner light-harvesting complex of PSII, also appears, which is absent if the oxygen evolving complex is preserved by the extrinsic polypeptide OEC33 [63].

It is still unclear if proteolysis is affected by active oxygen species or by the enzymatic activity of endogenous proteases. The current idea is that an oxidative mechanism initiates the degradation process, which is then completed by proteases. Although the proteinase that catalyzes D1 degradation has not been identified, it has been suggested that a serine-type proteinase or a thiol-endoproteinase could be responsible [64,65]. A proteinase in charge of catalyzing the degradation of damaged proteins, such as the protein complex Clp A/P, which is a chloroplast serine-type proteinase, could be involved. However, because D1 degradation can lead to multiple fragments and proceed differently depending on whether the photoinhibitory action is located on the acceptor or the donor side of the photosystem, more than one protease should be involved. Reports of the retardation of D1 light-dependent degradation by inhibitors of chloroplast transcription and translation and inhibition of D1 degradation by inhibitors of serine-type proteases seem to constitute a good argument in favor of the enzymatic cleavage of the polypeptide [64,66,67].

However, D1 degradation has been demonstrated in submembrane fractions, and even in the isolated reaction center complex of PSII [54]. This indicates the involvement of a proteinase in charge of D1 degradation that would be a membrane-bound protein in close relation to PSII [68–70]. The above correlates with the inhibition of D1 fragmentation by inhibitors of serine-type proteinase in the isolated D1/D2–cyt b-559 reaction center complex [71]. The 43-kDa protein of the internal antenna complex was also proposed to have proteolytic activity [67]. More recently, a GTP-dependent DegP2 protease homologous to prokaryotic trypsin-type Deg/Htr serine proteases associated with the outer surface of the thylakoid membrane was shown to cleave damaged D1 on its stromal D–E loop [72]. An ATP-dependent zinc metalloprotease, FtsH, was also shown to participate in D1 degradation targeting the 23-kDa primary light-induced cleavage product [73]. The involvement of FtsH in the early step of D1 degradation was strongly suggested as it is present in purified PSII preparations [74]. Further to the cleavage of D1, other unidentified stromal proteases, presumably of serine type, are proposed to be involved in the degradation of the cross-linked products generated during photoinhibition [61,75,76].

Another approach to explain the mechanism of D1 degradation is to consider the involvement of the active oxygen species generated during photoinhibition. These species may be involved in an initial conformational change that exposes the protein to the proteinase activity. Alternatively, the initial cleavage of D1 may be due to a reaction between susceptible protein segments and active oxygen species. In support of the above mechanism, Miyao [77] has suggested that the inhibitory action of protease inhibitors on D1 degradation may be due to their capacity for scavenging active oxygen species owing to their nucleophilic character.

Singlet oxygen was clearly detected by electron paramagnetic resonance spectroscopy in thylakoid membranes and PSII core complexes affected by acceptor-side photoinhibition, and hydroxyl radicals were present in the preparations submitted to donor-side photoinhibition [78,79]. Hydroxyl radicals are able to react directly with protein bonds, and singlet oxygen can generate alkoxyl radicals that can also react with the polypeptides [80,81]. Scavengers of oxygen radicals such as mannitol, propyl gallate, and uric acid as well as the scavengers of singlet oxygen, histidine and rutin, were shown to reduce the light-induced degradation of D1 [82–84]. Exposure of PSII core complexes to artificially generated singlet oxygen increases the rate of D1 fragmentation [79,85]. It was verified in isolated PSII reaction center preparations that singlet oxygen generation correlated with the production of P680 triplet states during recombination between oxidized P680 and reduced pheophytin [86]. Singlet oxygen was also detected in intact leaf segments exposed to photoinhibitory treatment [87,88], though superoxide ions were not detected in the intact leaves [88]. The latter observation coincides with the lack of protective effect of an overexpression of the enzyme superoxide dismuatase in poplar plants [89]. However, superoxide ions were detected in photoinhibited PSII membrane fragments where

they were proposed to have originated from the Q_B binding site [90]. All of the above point to the involvement of active oxygen species during D1 degradation under aerobic photoinhibition. On the other hand, active oxygen species cannot be created during anaerobic illumination [16], and in this case, the highly oxidizing radicals $P680^+$ and Z^+ generated during donor-side photoinhibition may trigger D1 degradation [15,16].

III. PHOTOINHIBITION IN PSI

Several early *in vitro* studies suggested that PSI photochemical activity could also be inhibited by strong light. Satoh [91,92] reported that the illumination of isolated chloroplasts damaged PSI as well as PSII. In these initial experiments, the photoinactivation of PSI was not observed in the absence of oxygen. Later, Satoh and Fork [93] demonstrated the inactivation of PSI in intact *Bryopsis* chloroplasts under strongly reducing conditions. The authors suggested the photoinactivation site to be either P700 itself [93] or close to the PSI reaction center [94] under both aerobic and anaerobic conditions [93]. Inoue et al. [95] proposed that the main cause of PSI photoinhibition under aerobic conditions was the inactivation of iron–sulfur centers, whereas the anaerobic photoinhibition resulted from the blocking of electron transfer between A_0 and F_X [96].

PSI was also found to be inactivated by light in intact leaves. In *Cucumis sativus* leaves, weak light illumination readily photoinhibited PSI at chilling temperatures ($4°C$), while little damage to PSII was observed [97,98]. Sonoike et al. [98] suggested that the destruction of F_X, F_A, F_B, and A_1 electron carriers was responsible for the photoinhibition of PSI. In barley, it was shown that the degradation started with the centers F_A, F_B, followed by F_X, and then the phylloquinone A_1 and the chlorophyll A_0 [99]. Sonoike and Terashima [100] assumed that the photoinactivation of PSI occurs essentially in three successive steps: (i) inactivation of the acceptor side, (ii) destruction of reaction center chlorophylls, and (iii) degradation of the reaction center subunits.

The general mechanism of preferential photodegradation of PSI under relatively low light and low temperature is proposed to be due to the generation of active oxygen species and inactivation of oxygen-scavenging enzymes at low temperature [101,102]. When the acceptor side of PSI is in a reduced state, the recombination between A_1^- or A_0^- and $P700^+$ generates P700 in triplet state [103]. The triplet states are quenched by oxygen to form singlet oxygen. Further, oxygen can accept electrons at the acceptor side of the photosystem to produce superoxide ions that dismute into hydrogen peroxide. The latter can also react with the iron–sulfur centers of PSI to produce hydroxyl radicals. These active forms of oxygen can be scavenged by active enzyme systems. However, at low temperatures the activities of these enzymes can be lower. It was indicated that in both isolated chloroplasts and PSI submembrane fractions kept under stressful conditions leading to the production of active oxygen species, membrane-bound superoxide dismutase was either denatured or released from the membranes [104,105], which can further enhance the photooxidative processes. Thus, in barley leaves, inhibition of ascorbate peroxidase and superoxide dismutase increases photoinhibitory damage to PSI [99]. Conversely, the addition of oxygen radical scavengers retarded the photoinhibitory effects measured in chloroplasts or PSI particles [106–108].

The explanation above shows the protective effect of DCMU against PSI photoinhibition [97,109] because the inhibition of electron transport from PSII prevents the reduction of the acceptor side of PSI, which is required for the deleterious effects to operate [109]. A similar protecting action of methylviologen, an artificial electron acceptor that keeps the FeS centers in the oxidized state, was also observed [110,111]. Thus, inhibition of carbon fixation at chilling temperature should also enhance PSI photoinhibition [102]. Illumination under strong light mainly inhibits PS II, and the reduced electron transport activity keeps P700 in the oxidized state P700$^+$. It is proposed that this cation radical can convert efficiently the excess absorbed light energy into heat and protect against PSI photoinhibition at high light intensities [112]. The above indicates that photoinhibition of PSII may protect PSI from photodamage [102,113], to the advantage of the whole plant because, in contrast with PSI core proteins, which are processed very slowly [114], the reaction center of PSII can be replaced with a fast turnover rate.

The loss of PSI activity in thylakoids isolated from spinach leaves exposed to weak illumination at room temperature is associated with the degradation of PsaB protein, one of the PSI reaction center subunits [110]. In barley and cucumber leaves exposed to weak light at chilling temperatures, Tjus et al. [115] have shown that the photoinhibition of PSI was due to the damage caused to both reaction center proteins, PsaA and PsaB, and also to the smaller proteins located on the acceptor side of the photosystem, such as PsaD and PsaE. It was reported that active oxygen species could trigger the initial degradation of the polypeptides as in PSII, followed by proteolysis involving a serine-type protease [108,110,116]. It is proposed that the initial degradation of FeS centers by reactive oxygen species may expose PsaB for degrad-

ation [113]. In spinach, the photodegradation of PsaB produced 18-, 45-, and 51-kDa fragments [110]. In barley, the initial degradation of PsaA/B polypeptides under low-temperature photoinhibition resulted in 33- and 35-kDa fragments. These fragments were further degraded only when the plants returned to higher temperatures [116]. In isolated PSI submembrane fractions, the initial effect of photoinhibitory illumination was to induce a photooxidative cross-linking of chlorophyll–protein complexes that has not yet been reported *in vivo* [105,117]. However, such cross-linking of polypeptides was reported for PSII as discussed above [61,62].

IV. CHLOROPHYLL PHOTOBLEACHING AND ENERGY DISSIPATION

A. MECHANISM OF PHOTOBLEACHING

Exposing photosynthetic organisms or isolated photosynthetic materials to excess light is known to induce the photobleaching of pigment molecules. This process has been demonstrated to involve oxygen radicals, and antiradical species — mainly α-tocopherol [118], superoxide dismutase [119], ascorbate [119,120], flavonols [121,122], and carotenoids [3,118,119,123] — exert a protective action against this photooxidation. The mechanism is thought to involve the formation of singlet oxygen through energy transfer from triplet chlorophyll, and chlorophyll is in turn oxidized in its triplet state by singlet oxygen [3,123,124].

Although pigment bleaching can occur in photosynthetically active materials, inhibition of electron transport does accelerate the process. It has been shown that herbicides affecting the reduction of plastoquinone on the acceptor side of PSII can greatly increase the photobleaching rate [125–127], which indicates that when absorbed light energy is not used for photosynthesis, the greater pool of excited chlorophylls allows the formation of a larger population of triplet chlorophylls, and thus leads to the formation of singlet oxygen. In isolated PSI submembrane fractions and complexes, photoinhibition and pigment photobleaching occur almost simultaneously during strong illumination because of the absence of electron donors and acceptors to support photochemical dissipation of absorbed light energy [105,117,128,129].

Even in the presence of active electron transport, an effective dissipation of absorbed energy must prevent pigment bleaching. Absorbed energy migrates through the pigment bed toward pigment absorbing at higher wavelengths. The holochromes located at the end of the migration pathway undergo faster bleaching because they are unable to dissipate the transferred energy. The preferential bleaching of holochromes absorbing at relatively long wavelengths has been shown previously [130–132]. Based on this same idea, chlorophyll *b*, which transfers energy to chlorophyll *a*, was shown to be photobleached at a much slower rate than chlorophyll *a* due to the protective effect of this energy transfer [131–134]. On the other hand, carotenoids can protect chlorophyll *a* from photobleaching because of the energy transfer between the triplet form of these pigments [123,124,134]. Similarly, experiments using isolated chlorophyll incorporated into chloroplast lipid vesicles have shown that at high chlorophyll:lipid ratio, an aggregate absorbing at 700 nm can dissipate the absorbed energy and protect the pigment against photooxidation [135]. Chlorophylls absorbing at wavelengths above 700 nm were found in the outer antenna complexes of PSI and in the core of PSI [136,137]. In vivo, the occurrence of energy-dissipating aggregates between long-wavelength-absorbing chlorophyll molecules and carotenoids was suggested to be involved in photoprotection [138]. The long-wavelength-absorbing chlorophylls have been proposed to act as energy traps to dissipate excess energy [139].

B. ENERGY DISSIPATION PATHWAYS

Migration of absorbed energy usually occurs toward holochromes with higher absorption wavelengths, in which case, as stated above, chlorophyll forms having absorption maxima in the red at a relatively high wavelength are beached first. This phenomenon was clearly shown by a blue shift in the absorption maximum in the red observed during strong illumination of various photosynthetic materials [4,117,134]. Such absorbance shifts were clearly detected in whole thylakoid membranes, PSI submembrane fractions, the pigment–protein complex of PSI, and the antenna complex of CP29 [4,117,134]. Hence, the difference spectrum obtained by subtracting the absorption spectrum of an illuminated (photobleached) sample from that of an untreated sample, representing in fact the absorption spectrum of the bleached pigments, presented an absorbance maximum in the red at a higher wavelength than the spectrum of the untreated sample. In other words, the bleached holochromes absorbed at higher wavelength compare to the bulk of the pigments. It was therefore assumed that the pigment aggregates absorbing at these high wavelengths could be involved in photoprotection.

These shifts, however, were not present in other complexes such as the major light-harvesting complex of PSII, the complexes CP47 and CP43, and the PSII submembrane fractions. The above indicates that the aggregates absorbing in the far red were either absent or present in smaller proportions in these materials. It was shown by deconvolution of the absorption spectra that holochromes absorbing at relatively high wavelengths were also preferentially bleached in PSII submembrane fractions [140].

The bleaching kinetics followed at the absorption maximum in the red often presented an initial lag phase where illumination failed to produce significant bleaching [4,117,134]. This delay was associated with energy-dissipating aggregates that could effectively prevent the bleaching of bulk pigments. The bleaching was assumed to resume at a greater rate after the photodestruction of these dissipating centers [4,117]. The weak photooxidation rates found at the beginning of illumination were thought to be associated with the bleaching of far-red-absorbing aggregates. Accordingly, in isolated PSI core complexes where the chlorophyll:P700 ratio was below 40, there was no absorbance shift of the absorption maximum in the red and no lag phase was detected at the beginning of the bleaching kinetics [141]. It was deduced that far-red-absorbing holochromes were absent in these core complexes and that the reaction center P700 was bleached at a similar rate with the bulk pigments due to a homogenous distribution of excitons [141].

An alternative interpretation of the increased bleaching rates after an initial slower rate is the increased interpigment distance. The efficiency of energy transfer in PSII preparations was shown to be affected only after 30% of the pigments were bleached, and the rate of bleaching kinetics may increase after this point when the pigment network does not allow efficient energy transfer to the protecting traps [141,142].

Another effect of the energy migration pathways on chlorophyll photobleaching is that pigment–protein complexes located at the end of the migration pathway through the light-harvesting complexes, core antenna complexes, and reaction center complexes, are the most photosensitive ones. The peripheral antenna complexes are protected from photobleaching by transferring the excess absorbed energy to the core complexes [4,117].

The above is also at the root of the sensitivity of PSII to photoinhibition. PSI is less affected by photoinhibition than PSII but is more sensitive to pigment photooxidation [4]. This may be explained by the presence of the core antenna complex on the same polypeptide with the reaction center P700 in PSI. This configuration protects the reaction center and the primary electron acceptors from overreduction because the excess energy is dissipated directly into the pigment bed, which favors chlorophyll photobleaching. In fact, it has been demonstrated that in PSI reaction center complexes the excitons are randomly distributed among all the chlorophyll holochromes and that energy transfer to the reaction center P700 was not favored over transfer between antenna pigments [143]. On the other hand, in PSII, the core antenna complex is located apart from the reaction center complex on other polypeptides (CP47 and CP43), and the funneling of absorbed energy toward the reaction center P680 leads to photoinhibition.

V. PHOTOPROTECTION

A. Cyclic Electron Transport around PSII

Because the specific energy migration pathway leads to overreduction of the primary acceptors in PSII, there exist some mechanisms to help in preventing photoinhibition. The first mechanism that will be discussed is cyclic electron transport around the photosystem, which is thought to retard both acceptor- and donor-side photoinhibition [19,144,145]. This cycle would allow the reduction of the donor side of the photosystem when the water-splitting enzyme is not functioning properly, and would increase the oxidation state of the acceptor side in the presence of excessive illumination.

Several different schemes of cyclic electron transport have been suggested [146–149]. The most accepted view of the electron transport pathway involves the reduction of cyt $b559$ by plastoquinone (Q_A or Q_B), followed by subsequent reoxidation by β-carotene and P680. It was also suggested that the low-potential form of the cytochrome can be reduced by the primary acceptor pheophytin [150], and cyclic electron transport was demonstrated in the D1/D2– cyt $b559$ reaction center complex devoid of quinone acceptors [146]. However, it is believed that the reduction of cyt $b559$ by plastoquinone would proceed at a faster and more functional rate [147,148].

The high-potential form of the cytochrome could be converted to the low-potential form following the reduction of the intermediates on the acceptor side of the photosystem by strong illumination as was shown to occur during photoinhibition [11]. The reverse conversion of the low-potential form of cyt $b559$ to its high-potential form was proposed to depend on the ambient redox system [146] and would lead to the reduction of P680 [144].

The electron transfer pathway between an accessory chlorophyll absorbing at 670 nm located on the polypeptide D1, Chl_Z, and P680 has recently been suggested to involve a chain of accessory chlorophylls located on the pigment–protein complexes CP43 and CP47 [149]. The electron then would go to the β-carotene molecule located on D2 [20]. Then it would be transferred to an accessory chlorophyll on the polypeptide D2 (Chl_{D2}) and P680 [149]. The location of cyt $b559$ indicates that it may also transfer electrons to the carotenoid [20,149]. Implication of tyrosine$_Z$ on D1 and tyrosine$_D$ on D2 in this type of electron transfer involved in photoprotection is also suggested [151].

It has been indicated by photoacoustic experiments that the cyclic pathway around PSII would remain active even after complete inhibition of linear electron transport [152]. A direct and efficient protective function of cyclic electron transport against photoinhibition has been demonstrated in PSII submembrane fractions and reaction center complexes [145].

B. Nonphotochemical Fluorescence Quenching

Nonphotochemical fluorescence quenching (qN) is, in contrast to photochemical quenching (qP), not directly related to the photochemical reactions [153]. Several reactions contribute to nonphotochemical quenching. Photoinhibition contributes to qN as the component qI [1], but the most important contribution to qN originates from the energy-dependent fluorescence quenching (qE), which results in thermal dissipation of absorbed energy. Energy-dependent quenching is related to the formation of the pH gradient across the thylakoid membrane during electron transport [154], and in fact, a strictly linear relationship was demonstrated between qE and the intrathylakoid H^+ concentration [155].

The exact mechanism of qE is still debated in the literature. It has been suggested that nonphotochemical quenching could be due to thermal dissipation of absorbed energy by inactive reaction centers [156]. Alternatively, it has been proposed that cyclic electron transport around PSII, possibly involving cyt $b559$, or charge recombination between acceptor- and donor-side intermediates could mediate thermal dissipation [157]. These processes should be promoted by donor-side limitation following impairment of oxygen evolution, a situation shown to occur at pH < 5.5 with the concurrent release of the Ca ions required for an active water-splitting complex [158–160]. A study using time-resolved fluorescence measurements in isolated thylakoid membranes indicated that only a nanosecond decay component is affected by qE, which was first interpreted by the recombination between Q_A^- and P680 during ΔpH-dependent quenching [161]. However, a more recent study has shown that the rate of Q_A^- reoxidation is independent of low pH, and the nonphotochemical quenching was suggested to originate from direct quenching by P680$^+$ [162].

Another interpretation of energy-dependent non-photochemical fluorescence quenching is that the quenching occurs in the light-harvesting system where energy is dissipated as heat following aggregation of the pigment–protein complexes [163,164]. Energy-dependent quenching coincides with the accumulation of the carotenoids zeaxanthin and antheraxanthin, which are formed during the so-called xanthophyll cycle [165–168]. Both structural rearrangements of the light-harvesting complex and zeaxanthin formation are needed in conjunction with an increased ΔpH for optimal formation of qE [169–172].

During the xanthophyll cycle, a zeaxanthin-epoxidase converts zeaxanthin to violaxanthin, and the latter is converted back to zeaxanthin by a de-epoxidase. These reactions include antheraxanthin as an intermediate, and it was shown in a prasinophycean alga where the xanthophyll cycle is incomplete and does not generate zeaxanthin that antheraxanthin can enhance nonphotochemical quenching [173]. Both enzymes were isolated and their genes cloned from various plant species [174–178]. Violaxanthin de-epoxidase is localized in the thylakoid lumen. It has its maximal activity at pH 5.2, requires ascorbate, and is inhibited by dithiothreitol [179]. It was suggested that violaxanthin is present in the light-harvesting complexes in its 15,15′-cis isomer and a cis–trans-isomerization would coincide with its dissociation from the pigment-binding protein forming a rod-like structure when it is located in the lipid phase of the thylakoid membrane [180]. This conformation would make violaxanthin available for introduction into the tubular cavity of the de-epoxidase where the catalytic center for de-epoxidation is located [181]. The major light-harvesting complex of PSII has been reported to exhibit epoxidase activity [182]. However, more recent data indicate that this complex is unlikely to have epoxidase activity and is rather limited to substrate binding [183]. Zeaxanthin-epoxidase activity was indicated to require oxygen, NADPH, ferredoxin, and ferredoxin-like reductase activity [177] and was suppressed by chloroplast phophatase inhibitors showing that dephosphorylation of an as yet unidentified component is also required [184]. Regulation of the conversion of violaxanthin into zeaxanthin in relation to ΔpH is postulated to be associated with the presence of localized and delocalized proton domains. It has been proposed that the protons in the localized domains (at the surface or in membrane proteins) are diverted to the lumen under excessive light intensity. The decrease in pH favors violaxanthin availability with the consequent rise in

zeaxanthin concentration [185]. On the other hand, de-epoxidation activity seems to be controlled by the pH in the localized domains [185].

It has been shown that the removal of zeaxanthin during epoxidase activity reverses qE [186], and, further, the formation of zeaxanthin due to the de-epoxidase activity is strictly correlated with qE and with increased ΔpH in the thylakoid lumen [165,187]. A low intrathylakiod pH was shown to increase the affinity for zeaxanthin and antheraxanthin at their binding site in the chlorophyll–protein complexes, and the binding of these carotenoids decreased fluorescence emission [188]. Thus, it is clear that zeaxanthin is involved in qE formation. However, the exact mechanism of quenching by this xanthophyll is unclear. It has been proposed that the lowest singlet state of zeaxanthin may quench the first singlet state of chlorophyll [189]. This idea is sustained by a lower energy level of the first singlet state of zeaxanthin in comparison with chlorophyll due to the length of the π conjugation of the polyene, whereas the lower energy level of violaxanthin would favor energy transfer to chlorophyll [189]. It should be noted that besides xanthophyll cycle pigments, the carotenoid lutein was also reported to have participated in energy dissipation through nonphotochemical quenching [190–192].

Both mechanisms of qE formation, that is, fluorescence quenching at the reaction center and thermal dissipation in the antenna due to aggregation of the light-harvesting complexes and to zeaxanthin formation are believed to occur [193]. It is probable that both mechanisms are either supplemental to each other or that quenching in the antenna amplifies the quenching occurring at the reaction center level [194–196]. It was shown using barley mutants lacking the light-harvesting complex that qE can be formed without these complexes, which were also shown to contain most of the xanthophyll pigments [197,198]. However, the above mutants may still contain the PsbS protein of 22 kDa that has been demonstrated to be essential for qE [199]. It was proposed that protonation of this and other polypeptides is involved in nonphotochemical quenching by the xanthophylls located either on the PsbS protein or on other light-harvesting subunits such as CP29 and CP26 [200,201]. It has been demonstrated that photoinhibition is retarded by qE formation and zeaxanthin formation [200,202], and it has even been suggested that the xanthophyll cycle would participate in the repair of PSII after photoinhibition [203,204]. It is believed that the xanthophyll cycle directly participates in the removal of excess excitation energy through qE [195,205]. Hence, it was shown in isolated thylakoids that added ascorbate protects against photoinhibition due to its cofactor requirement for de-epoxidation of violaxanthin [206]. The δpH was also correlated with decreased photoinhibition [207,208]. Thus, the release of energy-dependent quenching by uncouplers increases photoinhibition [209].

VI. CONCLUSION

When absorbed light energy exceeds what can be used by the photosynthetic apparatus, specialized processes engage in photoprotection. Processes such as thermal dissipation by quenching of chlorophyll aggregates absorbing at relatively high wavelengths or by non-photochemical quenching of absorbed energy due to proton gradient formation and to the associated xanthophyll cycle seem to effectively retard photodegradation of the electron transport intermediates and photobleaching of photosynthetic pigments.

However, photoprotection mechanisms are not sufficient, and photoinhibition does occur when the replacement of cleaved D1 polypeptides cannot keep up with the degradation process. It appears that PSI is less sensitive to photoinhibition due to homogenization of the excitons in the core antenna complex, which favors pigment photooxidation. In PSII the excess light energy can probably not be returned to the antenna pigment and, depending on the activity of the water-splitting complex, leads to overreduction or overoxidation of electron transport intermediates.

REFERENCES

1. Powles SB. Photoinhibition of photosynthesis induced by visible light. *Annu. Rev. Plant Physiol.* 1984; 35: 15–44.
2. Melis A. Photosystem-II damage and repair cycle in chloroplasts: what modulates the rate of photodamage *in vivo. Trends Plant Sci.* 1999; 4:130–135.
3. Siefermann-Harms D. The light-harvesting and protective functions of carotenoids in photosynthetic membranes. *Physiol. Plant.* 1987; 69:561–568.
4. Miller N, Carpentier R. Energy dissipation and photoprotection mechanisms during chlorophyll photobleaching in thylakoid membranes. *Photochem. Photobiol.* 1991; 90:465–472.
5. Niyogi KK. Photoprotection revisited: genetic and molecular approach. *Annu. Rev. Plant Physiol. Plant Mol. Biol.* 1999; 50:333–359.
6. Demming-Adams B. Carotenoids and photoprotection: a role for xanthophyll zeaxanthin. *Biochim. Biophys. Acta* 1990; 1020:1–24.
7. Barber J, De Las Rivas J. A functional model for the role of cytochrome b-559 in the protection against donor and acceptor side photoinhibition. *Proc. Natl. Acad. Sci. USA.* 1993; 90:10942–10946.
8. Ohad I, Sonoike K, Andersson B. Photoinactivation of the two photosystems in oxygenic photosynthesis: mechanisms and regulation. In: Yunus M, Pathre U,

Mohanty P, eds. *Probing Photosynthesis. Mechanisms, Regulation and Adaptation.* London: Taylor and Francis, 2000:293–309.
9. Aro EM, Virgin I, Andersson B. Photoinhibition of photosystem II: inactivation, protein damage and turnover. *Biochim. Biophys. Acta* 1993; 1143:113–134.
10. Vass I, Styring S, Hundal T, Koivuniemi A, Aro EM, Andersson B. Reversible and irreversible intermediates during photoinhibition of photosystem II: stable reduced Q_A species promote chlorophyll triplet formation. *Proc. Natl. Acad. Sci. USA.* 1992; 89: 1408–1412.
11. Styring S, Virgin I, Ehrenberg A, Andersson B.Strong light photoinhibition of electron transport in photosystem II: impairment of the function of the first quinine acceptor Q_A. *Biochim. Biophys. Acta* 1992; 1015:269–278.
12. Vass I, Styring S. Spectroscopic characterization of triplet forming states in photosystem II. *Biochemistry* 1992; 31:5957–5963.
13. Takahashi Y, Hansson Ö, Mathis P, Katoh K. Primary radical pair in the photosystem II reaction centre. *Biochim. Biophys. Acta* 1987; 893:49–59.
14. Durrant JR, Giorgi LB, Barber J, Klug DR, Porter G. Characterization of triplet states in isolated photosystem II reaction centres — oxygen quenching as a mechanism for photodamage. *Biochim. Biophys. Acta* 1990; 1071:167–175.
15. Jegershöld C, Virgin I, Stryring S. Light-dependent degradation of the D1 protein in photosystem II is accelerated after inhibition of the water splitting reaction. *Biochemistry* 1990; 29:6179–6186.
16. Jegershöld C, Stryring S. Fast oxygen-independent degradation of the D1 reaction center protein in photosystem II. *FEBS Lett.* 1991; 280:87–90.
17. Wang WQ, Chapman DJ, Barber J. Inhibition of water splitting increases the susceptibility of photosystem II to photoinhibition. *Plant Physiol.* 1992; 99: 16–20.
18. Blubaugh DJ, Cheniae GM. Kinetics of photoinhibition in hydroxylamine-extracted photosystem II membranes: relevance to photoactivation and sites of electron donation. *Biochemistry* 1990; 29:5109–5118.
19. Telfer A, De Las Rivas J, Barber J. β-Carotene within the isolated photosystem II reaction centre: photooxidation and irreversible bleaching of this chromophore by oxidised P680. *Biochim. Biophys. Acta* 1991; 1060:106–114.
20. Telfer A. What is β-carotene doing in the photosystem II reaction centre. *Philos. Trans. R. Soc. Lond. B* 2002; 357:1431–1440.
21. Mattoo AK, Hoffman-Falk H, Marder JB, Edelman M. Regulation of protein metabolism: coupling of photosynthetic electron transport to *in vivo* degradation of the rapidly metabolized 32-kilodalton protein of the chloroplast membrane. *Proc. Natl. Acad. Sci. USA* 1984; 81:1380–1384.
22. Flanigan YS, Critchley C. Light response of D1 turnover and photosystem II efficiency in the seagrass *Zostera capricorni. Planta* 1996; 198:319–323.

23. Tyystjärvi E, Aro E-M. The rate constant of photo-inhibition, measured in lincomycin-treated leaves, is directly proportional to light intensity. *Proc. Natl. Acad. Sci. USA* 1996; 93:2213–2218.

24. Jansen MAK, Mattoo AK, Edelman M. D1-D2 protein degradation in the chloroplast. Complex light saturation kinetics. *Eur. J. Biochem.* 1999; 260: 527–532.

25. Kettunen R, Pursiheimo S, Rintamäki E, Van Wijk KJ, Aro E-M. Transcriptional and translational adjustments of *psb*A gene expression in mature chloroplasts during photoinhibition and subsequent repair of photosystem II. *Eur. J. Biochem.* 1997; 247:441–448.

26. Van Wijk KJ, Eichacker L. Light is required for efficient translation elongation and subsequent integration of the D1-protein into photosystem II. *FEBS Lett.* 1996; 388:89–93.

27. Alfonso M, Perewoska I, Constant S, Kirilovsky D. Redox control of *psb*A expression in cyanobacteria *Synechocystis* strains. *J. Photochem. Photobiol.* 1999; 48:104–113.

28. Mulo P, Laakso S, Mäenpää P, Aro E-M. Stepwise photoinhibition of photosystem II. *Plant Physiol.* 1998; 117:483–490.

29. Tyystjärvi T, Mulo P, Mäenpää P, Aro E-M. D1 polypeptide degradation may regulate *psb*A gene expression at transcriptional and translational levels in *Synechocystis* sp. PCC 6803. *Photosynth. Res.* 1996; 47:111–120.

30. Zhang L, Niyogi KK, Nemson JA, Grossman AR, Melis A. DNA insertional mutagenesis for the elucidation of a photosystem II repair process in the green alga *Chalmydomonas reinhardtii*. *Photosynth. Res.* 1997; 53:173–184.

31. Gombos Z, Wada H, Murata N. The recovery of photosynthesis from low-temperature photoinhibition is accelerated by the unsaturation of membrane lipids: a mechanism of chilling tolerance. *Proc. Natl. Acad. Sci. USA* 1994; 91:8787–8791.

32. Kanervo E, Tasaka Y, Murata N, Aro E-M. Membrane lipid unsaturation processing of the photosystem II reaction-center protein D1 at low temperature. *Plant Physiol.* 1997; 114:841–849.

33. Gombos Z, Kanervo E, Tsvetkova N, Sakamoto T, Aro E-M, Murata N. Genetic enhancement of the ability to tolerate photoinhibition by introduction of unsaturated bonds into membrane glycerolipids. *Plant Physiol.* 1997; 115:551–559.

34. Kanervo E, Aro E-M, Murata N. Low unsaturation level of thylakoid membrane lipids limits turnover of the D1 protein of photosystem II at high irradiance. *FEBS Lett.* 1995; 364:239–242.

35. Vijayan P, Browse J. Photoinhibition in mutants of Arabidopsis deficient in thylakoid unsaturation. *Plant Physiol.* 2002; 129:876–885.

36. Nishiyama Y, Yamamoto H, Allakhverdiev SI, Inaba M, Yokota A, Murata N. Oxidative stress inhibits the repair of photodamage to the photosynthetic machinery. *EMBO J.* 2001; 20:5587–5594.

37. Allakhverdiev SI, Nishiyama Y, Miyairi S, Yamamoto H, Inagaki N, Kanesaki Y, Murata N. Salt stress inhibits the repair of photodamaged photosystem II by suppressing the transcription and translation of *psb*A genes in *Synechocystis*. *Plant Physiol.* 2002; 130:1443–1453.

38. Rintamäki E, Salo R, Lehtonen E, Aro E-M. Regulation of D1-protein degradation during photoinhibition of photosystem II *in vivo*: phosphorylation of the D1 protein in various plant groups. *Planta* 1995; 195:379–386.

39. Koivuniemi A, Aro E-M, Andersson B. Degradation of the D1- and D2-proteins of photosystem II in higher plant is regulated by reversible phosphorylation. *Biochemistry* 1995; 34:16022–16029.

40. Ebbert V, Godde D. Phosphorylation of PSII polypeptides inhibits D1 protein-degradation and increases PSII stability. *Photosynth. Res.* 1996; 50:257–269.

41. Rintamäki E, Kettunen R, Aro E-M. Differential D1 dephosphorylation in functional and photodamage photosystem II centers. *J. Biol. Chem.* 1996; 271:14870–14875.

42. Mizusawa N, Yamamoto N, Miyao M. Characterization of damage to the D1 protein of photosystem II under photoinhibitory illumination in non-phosphorylated and phosphorylated thylakoid membranes. *J. Photochem. Photobiol.* 1999; 48:97–103.

43. Baena-González E, Barbato R, Aro E-M. Role of phosphorylation in the repair cycle and oligomeric structure of photosystem II. *Planta* 1999; 208:196–204.

44. Campbell D, Zhou G, Gustafsson P, Öquist G, Clarke AK. Electron transport regulates exchanges of two forms of photosystem II D1 protein in the cyanobacterium *Synechococcus*. *EMBO J.* 1995; 14:5457–5466.

45. Soitamo AJ, Zhou G, Clarke AK, Öquist G, Gustafsson P, Aro E-M. Over-production of the D1:2 protein makes *Synechococcus* cells more tolerant to photoinhibition of photosystem II. *Plant Mol. Biol.* 1996; 30:467–478.

46. Öquist G, Campbell D, Clarke AK, Gustafsson P. The cyanobacterium *Synechococcus* modulates photosystem II function in response to excitation stress through D1 exchange. *Photosynth. Res.* 1995; 46:151–158.

47. Komenda J, Koblízek M, Masojídek J. The regulatory role of photosystem II photoinactivation and de novo protein synthesis in the degradation and exchange of two forms of the D1 protein in the cyanobacterium *Synechococcus* PCC 7942. *J. Photochem. Photobiol.* 1999; 48:114–119.

48. Van Wijk KJ, Roobol-Boza M, Kettunen R, Andersson B, Aro E-M. Synthesis and assembly of the D1 protein into photosystem II: processing of the C-terminus and identification of the initial assembly partners and complexes during photosystem II repair. *Biochemistry* 1997; 36:6178–6186.

49. Yokthongwattana K, Chrost B, Behrman S, Casper-Lindley C, Melis A. Photosystem II damage and repair cycle in the green alga *Dunaliella salina*: involvement of a chloroplast-localized HSP70. *Plant Cell Physiol.* 2001; 42:1389–1397.

50. Schroda M, Vallon O, Wollman F-A, Beck CF. A chloroplast-targeted heat shock protein 70 (HSP70) contributes to the protection and repair of photosystem II during and after photoinhibition. *Plant Cell* 1999; 11:1165–1178.

51. Giardi MT. Phosphorylation and disassembly of the photosystem II core as an early stage of photoinhibition. *Planta* 1993; 190:107–113.

52. Greenberg BM, Gaba V, Mattoo AK, Edelman M. Identification of a primary *in vivo* degradation product of the rapidly-turning-over 32 kd protein of photosystem II. *EMBO J.* 1987; 6:2865–2869.

53. Barbato R, Shipton CA, Giorgio M, Giacometti M, Barber J. New evidence suggests that the initial photo-induced cleavage of the D1-protein may not occur near the PEST sequence. *FEBS Lett.* 1991; 290: 162–166.

54. Shipton CA, Barber J. Photoinduced degradation of the D1 polypeptide in isolated reaction centers of photosystem II: evidence for an autoproteolytic process triggered by the oxidizing side of the photosystem. *Proc. Natl. Acad. Sci. USA* 1991; 88:6691–6695.

55. De Las Rivas J, Andersson B, Barber J. Two sites of primary degradation of the D1-protein induced by acceptor or donor side photo-inhibition in photosystem II core complexes. *FEBS Lett.* 1992; 301: 246–252.

56. Canovas PM, Barber J. Detection of a 10 kDa breakdown product containing the C-terminus of the D1-protein in photoinhibited wheat leaves suggests an acceptor side mechanism. *FEBS Lett.* 1993; 324: 341–344.

57. Barbato R, Frizzo A, Friso G, Rigoni F, Giacometti M. Photoinduced degradation of the D1 protein in isolated thylakoids and various photosystem II particles after donor-side inactivations. *FEBS Lett.* 1992; 304:136–140.

58. Friso G, Giacometti GM, Barber J, Barbato R. Evidence for concurrent donor and acceptor side photo-induced degradation of the D1-protein in isolated reaction centers of photosystem II. *Biochim. Biophys. Acta* 1993; 1144:265–270.

59. Ohad I, Koike H, Shoat S, Inoue Y. Changes in the properties of reaction center II during the initial stages of photoinhibition as revealed by thermoluminescence measurements. *Biochim. Biophys. Acta* 1988; 933: 288–298.

60. Ortega JM, Roncel M, Losada M. Light-induced degradation of cytochrome b559 during photoinhibition of the photosystem II reaction center. *FEBS Lett.* 1999; 458:87–92.

61. Mizusawa N, Tomo T, Satoh K, Miyao M. Degradation of the D1 protein of photosystem II under illumination *in vivo*: two different pathways involving cleavage or intermolecular cross-linking. *Biochemistry* 2003; 42:10034–10044.

62. Vasilikiotis C, Melis A. The role of chloroplast-encoded protein biosynthesis on the rate of D1 protein degradation in *Dunaliella salina. Photosynth. Res.* 1995; 147–155.

63. Henmi T, Yamasaki H, Sakuma S, Tomokawa Y, Tamura N, Shen J-R, Yamamoto Y. Dynamic interaction between the D1 protein, CP43 and OEC33 at the luminal side of photosystem II in spinach chloroplasts: evidence from light-induced cross-linking of the proteins in the donor-side photoinhibition. *Plant Cell Physiol.* 2003; 44:451–456.

64. Virgin I, Salter AH, Ghanotakis DF, Andersson B. Light-induced D1 protein degradation is catalyzed by a serine-type protease. *FEBS Lett.* 1991; 287: 125–128.

65. Wettern M, Galling G. Degradation of the 32-kilodalton thylakoid-membrane polypeptide of *Chlamydomonas reinhardi* Y-1. *Planta* 1985; 166:474–482.

66. Gong H. Light-dependent degradation of the photosystem II D1 protein is retarded by inhibitors of chloroplast transcription and translation: possible involvement of a chloroplast-encoded proteinase. *Biochim. Biophys. Acta* 1994; 1188:422–426.

67. Salter AH, Virgin I, Hagman A, Andersson B. On the molecular mechanism of light-induced D1 protein degradation in photosystem II core particles. *Biochemistry* 1992; 31:3990–3998.

68. Aro E-M, Hundal T, Carlsberg I, Andersson B. In vitro studies on light-induced inhibition of photosystem II and D1-protein degradation at low temperature. *Biochim. Biophys. Acta* 1990; 1019:269–275.

69. Ohad I, Kyle DJ, Hirshberg J. Light-dependent degradation of the Q_B-protein in isolated pea thylakoids. *EMBO J.* 1985; 4:1655–1659.

70. Virgin I, Ghanotakis DF, Andersson B. Light-induced D1-protein degradation in isolated photosystem II core complexes. *FEBS Lett.* 1990; 269:45–48.

71. Misra AN, Hall SG, Barber J. The isolated D1/D2/cyt *b*-559 reaction center complex of photosystem II possesses a serine-type endopeptidase activity. *Biochim. Biophys. Acta* 1991; 1059:239–242.

72. Haußühl K, Andersson B, Adamska I. A chloroplast DegP2 protease performs the primary cleavage of the photodamage D1 protein in plant photosystem II. *EMBO J.* 2001; 20:713–722.

73. Lindahl M, Spetea C, Hundal T, Oppenheim AB, Adam Z, Andersson B. The thylakoid FtsH protease plays a role in the light-induced turnover of the photosystem II D1 protein. *Plant Cell* 2000; 12:419–431.

74. Silva P, Thompson E, Bailey S, Kruse O, Mullinaux CW, Robinson C, Mann NH, Nixon PJ. FtsH is involved in the early stages of repair of photosystem II in *Synechocystis* sp PCC 6803. *Plant Cell* 2003; 15:2152–2164.

75. Fernaji A, Abe S, Ishikawa Y, Henmi T, Tomokawa Y, Nishi Y, Tamura N, Yamamoto Y. Characterization of the stromal protease(s) degrading the cross-linked products of the D1 protein generated by photoinhibition of photosystem II. *Biochim. Biophys. Acta* 2001; 1503:385–395.

76. Yamamoto Y. Quality control of photosystem II. *Plant Cell Physiol.* 2001; 42:121–128.

77. Miyao M. Involvement of active oxygen species in degradation of the D1 protein under strong

illumination in isolated subcomplexes of photosystem II. *Biochemistry* 1994; 33:9722–9730.

78. Hideg E, Spetea C, Vass I. Singlet oxygen end free radical production during acceptor- and donor-side induced photoinhibition. Studies with spin trapping EPR spectroscopy. *Biochim. Biophys. Acta* 1994; 1186:143–152.

79. Mishra RK, Mishra NP, Kambourakis S, Orfanopoulos M, Ghanotakis DF. Generation and trapping of singlet oxygen during strong illumination of a photosystem II core complex. *Plant Sci.* 1996; 115:151–155.

80. Stadtmen ER. Oxidation of free amino acids and amino acid residues in proteins by radiolysis and by metal-catalyzed reactions. *Annu. Rev. Biochem.* 1993; 62:797–821.

81. Eltsner EF. Oxygen activation and oxygen toxicity. *Annu. Rev. Plant Physiol.* 1982; 33:73–96.

82. Sopory SK, Greenberg BM, Mehta RA, Edelman M, Mattoo AK. Free radical scavengers inhibit light-dependent degradation of the 32 kDa photosystem II reaction center protein. *Z. Naturforsch.* 1989; 45c: 412–417.

83. Casano LM, Trippi VS. The effect of oxygen radicals on proteolysis in isolated oat chloroplasts. *Plant Cell Physiol.* 1992; 33:329–332.

84. Mishra NP, Francke C, van Gorkom HJ, Ghanotakis DF. Destructive role of singlet oxygen during aerobic illumination of the photosystem II core complex. *Biochim. Biophys. Acta* 1994; 1186:81–90.

85. Mishra NP, Ghanotakis DF. Exposure of a photosystem II complex to chemically generated singlet oxygen result in D1 fragments similar to the ones observed during aerobic photoinhibition. *Biochim. Biophys. Acta* 1994; 1187:296–300.

86. Telfer A, Oldham TC, Phillips D, Barber J. Singlet oxygen formation detected by near-infrared emission from isolated photosystem II reaction centres: direct correlation between P680 triplet decay and luminescence rise kinetics and its consequences for photoinhibition. *J. Photochem. Photobiol.* 1999; 48:89–96.

87. Hideg E, Kálai T, Hideg K, Vass I. Do oxidative stress conditions impairing photosynthesis in the light manifest as photoinhibition. *Philos. Trans. R. Soc. Lond. B* 2000; 355:1511–1516.

88. Hideg E, Barta C, Kálai T, Vass I, Hideg K, Asada K. Detection of singlet oxygen and superoxide with fluorescent sensors in leaves under stress by photoinhibition and UV radiation. *Plant Cell Physiol.* 2002; 43:1154–1164.

89. Tyystjärvi E, Riikonen M, Arisi A-CM, Kettunen R, Jouanin L, Foyer CH. Photoinhibition of photosystem II in tobacco plants overexpressing glutathione reductase and poplars overexpressing superoxide dismutase. *Physiol. Plant.* 1999; 105:409–416.

90. Zhang S, Weng J, Pan J, Tu T, Yao S, Xu C. Study on the photo-generation of superoxide radicals in photosystem II with EPR spin trapping techniques. *Photosynth. Res.* 2003; 75:41–48.

91. Satoh K. Mechanism of photoinactivation in photosynthetic systems. I. The dark reaction in photoinactivation. *Plant Cell Physiol.* 1970; 11:15–27.

92. Satoh K. Mechanism of photoinactivation in photosynthetic systems. II. The occurrence of and properties of two different types of photoinactivation. *Plant Cell Physiol.* 1970; 11:29–38.

93. Satoh K, Fork DC. Photoinhibition of reaction centers of photosystem I and II in intact *Bryopsis* chloroplasts under anaerobic conditions. *Plant Physiol.* 1982; 70:1004–1008.

94. Satoh K. Mechanism of photoinactivation in photosynthetic systems. III. Site and mode of photoactivation in photosystem I. *Plant Cell Physiol.* 1970; 11:187–197.

95. Inoue K, Sakurai H, Hiyama T. Photoinactivation sites of photosystem I in isolated chloroplasts. *Plant Cell Physiol.* 1986; 27:961–968.

96. Inoue K, Fujii T, Yokoyama E, Matsuura K, Hiyama T, Sakurai H. The photoinhibition site of photosystem I in isolated chloroplasts under extremely reducing conditions. *Plant Cell Physiol.* 1989; 30: 65–71.

97. Terashima I, Funayama S, Sonoike K. The site of photoinhibition in leaves of *Cucumis sativus* L. at low temperatures is photosystem I, not photosystem II. *Planta* 1994; 193:300–306.

98. Sonoike K, Terashima I, Iwaki M, and Itoh S. Destruction of photosystem I iron-sulfur centers in leaves of *Cucumis sativus* L. by weak illumination at chilling temperatures. *FEBS Lett.* 1995; 362:235–238.

99. Tjus SE, Moller BL, Scheller HS. Photosystem I is an early target of photoinhibition in barley illuminated at chilling temperatures. *Plant Physiol.* 1998; 116: 755–764.

100. Sonoike K, Terashima I. Mechanism of photosystem-I photoinhibition in leaves of *Cucumis sativus* L. *Planta* 1994; 194:287–293.

101. Tjus SE, Scheller HV, Andersson B, Moller BL. Active oxygen produced during selective excitation of photosystem I is damaging not only to photosystem I but also to photosystem II. *Plant Physiol.* 2001; 125: 2007–2015.

102. Sonoike K. Various aspects of inhibition of photosynthesis under light/chilling stress: "photoinhibition at chilling temperatures" versus "chilling damage in the light". *J. Plant Res.* 1998; 111:121–129.

103. Golbeck JH, Bryant DA. Photosystem I. *Curr. Top. Bioenerg.* 1991; 16:3–177.

104. Casano LM, Gómez LD, Lascano HR, González CA, Trippi VS. Inactivation and degradation of CuZn-SOD by active oxygen species in wheat chloroplasts exposed to photooxidative stress. *Plant Cell Physiol.* 1997; 38:433–440.

105. Rajagopal S, Bukhov NG, Carpentier R. Photoinhibitory light-induced changes in the composition of chlorophyll-protein complexes and photochemical activity in photosystem-I submembrane fractions. *Photochem. Photobiol.* 2003; 77:284–291.

106. Jacob B, Heber U. Photoproduction and detoxification of hydroxyl radicals in chloroplasts and leaves and relation to photoinactivation of photosystem I and photosystem II. *Plant Cell Physiol.* 1996; 37:629–635.

107. Baba K, Itoh S, Hoshina S. Degradation of photosystem I reaction center proteins during photoinhibition *in vitro*. In: Mathis P, ed. *Photosynthesis: From Light to Biosphere*, Vol II. Dordrecht: Kluwer Academic Publishers, 1995:179–182.

108. Sonoike K. Degradation of *psaB* gene product, the reaction center subunit of photosystem I, is caused during photoinhibition of photosystem I: possible involvement of active oxygen species. *Plant Sci.* 1996; 115:157–164.

109. Velitchkova M, Yruela I, Alfonso M, Alonso PJ, Picorel R. Different kinetics of photoinactivation of photosystem-I mediated electron transport and P700 in isolated thylakoid membranes. *J. Photochem. Photobiol.* 2003; 69:41–48.

110. Sonoike K, Kamo M, Hihara Y, Hiyama T, Enami T. The mechanism of the degradation of psaB gene product, one of the photosynthetic reaction center subunits of photosystem I, upon photoinhibition. *Photosynth. Res.* 1997; 53:55–63.

111. Barth C, Krause GH. Inhibition of photosystem I and II in chilling-sensitive and chilling-tolerant plants under light and low-temperature stress. *Z. Naturforsch.* 1999; 54c:645–657.

112. Barth C, Krause GH, Winter K. Response of photosystem I compared with photosystem II to high-light stress in tropical shade and sun leaves. *Plant Cell Environ.* 2001; 24:163–176.

113. Sonoike K. Photoinhibition of photosystem I: its physiological significance in the chilling sensitivity of plants. *Plant Cell Physiol.* 1996; 37:239–247.

114. Kudoh H, Sonoike K. Irreversible damage to photosystem I by chilling in the light: cause of the degradation of chlorophyll after returning to normal growth temperature. *Planta* 2002; 215:541–548.

115. Tjus SE, Moller BL, Scheller HV. Photoinhibition of photosystem I damaged both reaction centre proteins PSI-A and PSI-B and acceptor-side located small photosystem I polypeptides. *Photosynth. Res.* 1999; 60:75–86.

116. Teicher HB, Moller BL, Scheller HV. Photoinhibition of photosystem I in field-grown barley (*Hordeum vulgare* L.): induction, recovery and acclimation. *Photosynth. Res.* 2000; 64:53–61.

117. Rajagopal S, Bukhov NG, Carpentier R. Changes in the structure of chlorophyll-protein complexes and excitation energy during photoinhibitory treatments of isolated photosystem I submembrane particles. *J. Photochem. Photobiol.* 2002; 62:194–200.

118. Merzlyak MH, Kovrighnikh VA, Reshetnikova IV, Gusev MV. Interaction of dephenylpicrylhydrazyl with chloroplast lipids and the antioxidant function of tocopherol and carotenoids in photosynthetic membranes. *Photobiochem. Photobiophys.* 1986; 11:49–55.

119. Gillam DJ, Dodge AD. Chloroplast protection in greening leaves. *Physiol. Plant.* 1985; 65:393–396.

120. Szigeti Z, Vagujfalvi D. Studies on chlorophyll photooxidation enhanced by benzonitriles *in vivo* and *in vitro*. Photobiochem. Photobiophys. *1984; 7:103–109.*

121. Takahama U. Suppression of carotenoid photobleaching by kaempferol in isolated chloroplasts. *Plant Cell Physiol.* 1982; 23:859–864.

122. Wagner GR, Youngman RJ, Elstner EF. Inhibition of chloroplast photo-oxidation by flavonoids and mechanisms of the antioxidative action. *J. Photochem. Photobiol. B* 1988; 1:451–460.

123. Krinsky NI. Non-photosynthetic functions of carotenoids. *Philos. Trans. R. Soc. Lond. B* 1978; 284:581–590.

124. Koka P, Song P-S. Protection of chlorophyll-*a* by carotenoids from photodynamic decomposition. *Photochem. Photobiol.* 1978; 28:509–515.

125. Yamashita K, Konishi K, Itoh M, Shibata K. Photobleaching of carotenoids related to the electron transport in chloroplasts. *Biochim. Biophys. Acta* 1969; 172:511–524.

126. Elstner EF, Osswald W. Chlorophyll photobleaching and ethane production in dichlorophenyldimethylurea-(DCMU) or paraquat-treated *Euglena gracilis* cell. *Z. Naturforsch.* 1980; 35c:129–135.

127. Ridley SM, Horton P. DCMU-induced fluorescence changes and photodestruction of pigments associated with an inhibition of photosystem I cyclic electron flow. *Z. Naturforch.* 1984; 39c:351–353.

128. Baba K, Itoh S, Hastings G, Hoshina S. Photoinhibition of photosystem I electron transfer activity in isolated photosystem I preparations with different chlorophyll contents. *Photosynth. Res.* 1996; 47:121–130.

129. Hui Y, Jie W, Carpentier R. Degradation of the photosystem I complex during photoinhibition. *Photochem. Photobiol.* 2000; 72:508–512.

130. Brown JS, French CS. Absorption spectra and relative photostability of the different forms of chlorophyll in *Chlorella. Plant Physiol.* 1959; 34:305–309.

131. Thomas JB, Nijhuis HH. Relative stability of chlorophyll complexes *in vivo. Biochim. Biophys. Acta* 1968; 153:868–877.

132. Thomas JB, Bollen MHM, Klijn WJ. Photobleaching and dark-bleaching in *Euglena gracilis* chloroplast fragments. *Acta Bot. Neerl.* 1976; 25:361–369.

133. Sauer K, Calvin M. Absorption spectra of spinach quantasomes and bleaching of pigments. *Biochim. Biophys. Acta* 1962; 64:324–336.

134. Carpentier R, Leblanc RM, Bellemare G. Chlorophyll photobleaching in pigment-protein complexes. *Z. Naturforsch.* 1986; 41c:284–290.

135. Carpentier R, Dijkmans H, Leblanc RM, Aghion J. Chlorophyll *a* in unilamellar vesicles made with chloroplast lipids. Absorbance and photobleaching. *Photobiochem. Photobiophys.* 1983; 5:245–252.

136. Croce R, Zucchelli G, Galaschi FM, Jennings RC. A thermal broadening study of the antenna chlorophylls in PSI-200. *Biochemistry* 1998; 37:17355–17360.

137. Knoetzel J, Bossmann B, Grimme H. *Chlorina* and *viridis* mutants of barley (*Hordeum vulgare* L.) allow assignment of long-wavelength chlorophyll forms to individual Lhca proteins of photosystem I *in vivo*. *FEBS Lett*. 1998; 436:339–342.

138. Oquist G, Samuelson G. Sequential extraction of chlorophyll *a* antenna of photosystem I. *Physiol. Plant*. 1980; 50:63–70.

139. Trissl HW. Long-wavelength absorbing antenna pigments and heterogeneous bands concentrate excitons and increases absorption cross-section. *Photosynth. Res*. 1993; 35:247–263.

140. Zucchelli G, Galarschi FM, Jennings RC. Spectroscopic analysis of chlorophyll photobleaching in spinach thylakoids, grana and light-harvesting chlorophyll *a/b* protein complex. *J. Photochem. Photobiol. B* 1988; 2:483–490.

141. Purcell M, Carpentier R. Homogeneous photobleaching of chlorophyll holochromes in a photosystem I reaction center complex. *Photochem. Photobiol*. 1994; 59:215–218.

142. Jennings RC, Zucchelli G, Garlaschi FM. The influence of reducing the chlorophyll concentration by photobleaching on energy transfer to artificial traps within photosystem II antenna system. *Biochim. Biophys. Acta* 1989; 975:29–33.

143. Croce R, Zucchelli G, Garlaschi FM, Bassi R, Jennings RC. Excited state equilibration in the photosystem I–light–harvesting I complex: P700 is almost isoenergetic with its antenna. *Biochemistry* 1996; 35: 8572–8579.

144. Thompson LK, Brudvig GW. Cytochrome *b*-559 may function to protect photosystem II from photoinhibition. *Biochemistry* 1988; 27:6653–6658.

145. Allkhverdiev SI, Klimov VV, Carpentier R. Evidence for the involvement of cyclic electron transport in the protection of photosystem II against photoinhibition: influence of a new phenolic compound. *Biochemistry* 1997; 36:4149–4154.

146. Barber J, De Las Rivas J. A functional model for the role of cytochrome *b*-559 in the protection against donor and acceptor side photoinhibition. *Proc. Natl. Acad. Sci. USA* 1993; 90:10942–10956.

147. Arnon DI, Tang GM-S. Cytochrome *b*-559 and proton conductance in oxygenic photosynthesis. *Proc. Natl. Acad. Sci. USA* 1988; 85:9524–9528.

148. Buser CA, Diner BA, Brudvig GW. Photooxidation of cytochrome b559 in oxygen-evolving photosystem II. *Biochemistry* 1992; 31:11449–11459.

149. Vassil'ev S, Brudvig GW, Bruce D. The X-ray structure of photosystem II reveals a novel electron transport pathway between P680, cytochrome b_{559} and the energy-quenching cation, Chl^+_Z. *FEBS Lett*. 2003; 543:159–163.

150. Nedbal L, Samson G, Whitmarsh J. Redox state of a one-electron component controls the rate of photoinhibition of photosystem II. *Proc. Natl. Acad. Sci. USA* 1992; 89:7929–7933.

151. Magnuson A, Rova M, Mamedov F, Fredriksson P-O, Styring S. The role of cytochrome b_{559} and

152. Lapointe L, Huner NPA, Leblanc RM, Carpentier R. Possible photoacoustic detection of cyclic electron transport around photosystem II in photoinhibited thylakoid preparations. *Biochim. Biophys. Acta* 1993; 1142:43–48.

153. Krause GH, Weis E. Chlorophyll fluorescence and photosynthesis: the basics. *Annu. Rev. Plant Physiol. Plant Mol. Biol*. 1991; 42:313–349.

154. Krause GH, Vernotte C, Briantais J-M. Photoinduced quenching of chlorophyll fluorescence in intact chloroplasts and algae. *Biochim. Biophys. Acta* 1982; 679:116–124.

155. Briantais J-M, Vernotte C, Picaud M, Krause GH. A quantitative study of the slow decline of chlorophyll *a* fluorescence in isolated chloroplast. *Biochim. Biophys. Acta* 1979; 548:128–138.

156. Weiss E, Berry JA. Quantum efficiency of photosystem II in relation to energy-dependent quenching of chlorophyll fluorescence. *Biochim. Biophys. Acta* 1987; 894:198–208.

157. Schreiber U, Neubauer C. O_2 dependent electron flow, membrane energization and the mechanism of non-photochemical quenching of chlorophyll fluorescence. *Photosynth. Res*. 1990; 25:279–293.

158. Schlodder E, Meyer B. pH-dependence of oxygen evolution and reduction kinetics of photooxidized chlorophyll *a* II (P680) in photosystem II particles from *Synechococcus* sp. *Biochim. Biophys. Acta* 1987; 890:23–31.

159. Ono T, Inoue I. Discrete extraction of Ca atom functional for O_2-evolution in higher plant photosystem II by a simple low pH treatment. *FEBS Lett*. 1988; 227:147–152.

160. Kreiger A, Weiss E. The role of calcium in the pH-dependent control of photosystem II. *Photosynth. Res*. 1993; 37:117–130.

161. Kreiger A, Moya I, Weiss E. Energy-dependent quenching of chlorophyll-*a*-fluorescence: effect of pH on stationary fluorescence and picosecond-relaxation kinetics in thylakoid membranes and photosystem II preparations. *Biochim. Biophys. Acta* 1992; 1102:167–176.

162. Bruce D, Samson G, Carpenter C. The origins of nonphotochemical quenching of chlorophyll fluorescence in photosynthesis. Direct quenching by $P680^+$ in photosystem II enriched membranes at low pH. *Biochemistry* 1997; 36:749–755.

163. Genty B, Briantais J-M, Baker NR. The relationship between quantum yield of photosynthetic electron transport and quenching of chlorophyll fluorescence. *Biochim. Biophys. Acta* 1989; 990:87–92.

164. Horton P, Ruban AV, Rees D, Pascal AA, Noctor G, Young AJ. Control of light harvesting function of chloroplast membranes by aggregation of the LHCII chlorophyll-protein complex. *FEBS Lett*. 1991; 292:1–4.

165. Demming B, Winter K, Kruger A, Czygan FC. Photoinhibition and zeaxanthin formation in intact leaves.

tyrosine$_D$ in protection against photoinhibition during *in vivo* photoactivation of photosystem II. *Biochim. Biophys. Acta* 1999; 1411:180–191.

A possible role of the xanthophyll cycle in the dissipation of excess light energy. *Plant Physiol.* 1987; 84:218–224.

166. Demming-Adams B, Gilmore AM, Adams III WW. *In vivo* functions of carotenoids in higher plants. *FASEB J.* 1996; 10:403–412.

167. Gilmore AM. Mechanistic aspects of xanthophyll cycle-dependent photoprotection in higher plant chloroplasts and leaves. *Physiol. Plant.* 1997; 99: 197–209.

168. Eskling M, Arvidsson P-O, Akerlund H-E. The xanthophyll cycle, its regulation and components. *Physiol. Plant.* 1997; 100:806–816.

169. Lokstein H, Härtel H, Hoffmann P, Woitke P, Renger G. The role of light-harvesting complex II in excess excitation energy dissipation: an *in vivo* fluorescence study on the origin of high-energy quenching. *J. Photochem. Photobiol. B* 1994; 26:175–184.

170. Bilger W, Björkman O. Relationship among violaxanthin deepoxidation, thylakoid membrane conformation, and nonphotochemical chlorophyll fluorescence quenching in leaves of cotton. *Planta* 1994; 193:238–246.

171. Gilmore A, Hazlett TL, Debrunner PG, Govindjee. Photosystem II chlorophyll *a* fluorescence lifetimes and intensity are independent of the antenna size differences between barley wild-type and *chlorina* mutants: photochemical quenching and xanthophyll cycle-dependent nonphotochemical quenching of fluorescence. *Photosynth. Res.* 1996; 48:171–187.

172. Härtel H, Lokstein H, Grimm B, Rank B. Kinetics studies on the xanthophyll cycle in barley leaves. Influence of antenna size and relations to nonphotochemical chlorophyll fluorescence quenching. *Plant Physiol.* 1996; 110:471–482.

173. Goss R, Böhme K, Wilhelm C. The xanthophyll cycle of *Mantoniella squamata* converts violaxanthin into anteraxanthin but not to zeaxanthin: consequences for the mechanism of enhanced non-photochemical energy dissipation. *Planta* 1998; 205:613–621.

174. Rockholm DC, Yamamoto HY. Violaxanthin de-epoxidase. Purification of a 43-kilodalton luminal protein from lettuce by lipid-affinity precipitation with monogalactodyldiacylglyceride. *Plant Physiol.* 1996; 110:697–703.

175. Bugos RC, Yamamoto HY. Molecular cloning of violaxanthin de-epoxidase from romaine lettuce and expression in *Escherichia coli*. *Proc. Natl. Acad. Sci. USA* 1996; 93:6320–6325.

176. Havir EA, Tausta SL, Peterson RB. Purification and properties of violaxanthin de-epoxidase from spinach. *Plant Sci.* 1997; 123:57–66.

177. Bugos RC, Hieber AD, Yamamoto HY. Xanthophyll cycle enzymes are members of the lipocalin family, the first identified from plants. *J. Biol. Chem.* 1998; 273:15321–15324.

178. Bouvier F, d'Harlingue A, Hugueney P, Marin E, Marion-Poll A, Camara B. Xanthophyll biosynhthesis: cloning, expression, functional reconstitution, and regulation of b-cyclohexenyl carotenoids epoxidase from pepper (*Capsicum annuum*). *J. Biol. Chem.* 1996; 271:28861–28867.

179. Hager A, Holocher K. Localization of the xanthophyll-cycle enzyme violaxanthin de-epoxidase within the thylakoid lumen and abolition of its mobility by a (light-dependent) pH increase. *Planta* 1994; 192:581–589.

180. Gruszecki WI, Matula M, Ko-chi N, Koyama Y, Krupa Z. *Cis-trans*-isomerisation of violaxanthin in LHCII: violaxanthin isomerization cycle within the violaxanthin cycle. *Biochim. Biophys. Acta* 1997; 1319:267–274.

181. Grotz B, Molnár P, Stransky H, Hager A. Substrate specificity and functional aspects of violaxanthin-de-epoxidase, an enzyme of the xanthophyll cycle. *J. Plant Physiol.* 1999; 154:437–446.

182. Gruszecki WI, Krupa Z. LHCII, the major light-harvesting pigment-protein complex is a zeaxanthin epoxidase. *Biochim. Biophys. Acta* 1993; 1144:97–101.

183. Färber A, Jahns P. The xanthophyll cycle of higher plants: influence of antenna size and membrane organization. *Biochim. Biophys. Acta* 1998; 1363:47–58.

184. Xu CC, Jeon YA, Hwang HJ, Lee C-H. Suppression of zeaxanthin epoxidation by chloroplast phosphatase inhibitors in rice leaves. *Plant Sci.* 1999; 146:27–34.

185. Pfündel EE, Renganathan M, Gilmore AM, Yamamoto HY, Dilley RA. Intrathylakoid pH in isolated pea chloroplasts as probed by violaxanthin deepoxidation. *Plant Physiol.* 1994; 106:1647–1658.

186. Gilmore AM, Mohanty N, Yamamoto HY. Epoxidation of zeaxanthin and antheraxanthin reverses nonphotochemical quenching of photosystem II chlorophyll *a* fluorescence in the presence of trans-thylakoid ΔpH. *FEBS Lett.* 1994; 350:271–274.

187. Büch K, Stransky H, Bigus H-J, Hager A. Enhancement by artificial electron acceptors of thylakoid lumen acidification and zeaxanthin formation. *J. Plant Physiol.* 1994; 144:641–648.

188. Gilmore AM, Shinkarev VP, Hazlett TL, Govindjee. Quantitative analysis of the effects of intrathylakoid pH and xanthophyll cycle pigments on chlorophyll *a* fluorescence lifetime distributions and intensity in thylakoids. *Biochemistry* 1998; 37:13582–13593.

189. Franck HA, Cua A, Chynwat V, Young A, Gosztola D, Wasielewski MR. Photophysics of the carotenoids associated with the xanthophyll cycle in photosynthesis. *Photosynth. Res.* 1994; 41:389–395.

190. Niyogi KK, Björkman O, Grossman AR. The roles of specific xanthophylls in photoprotection. *Proc. Natl. Acad. Sci. USA* 1997; 94:14162–14167.

191. Pogson BJ, Niyogi KK, Björkman O, DellaPenna D. Altered xanthophyll compositions adversely affect chlorophyll accumulation and nonphotochemical quenching in *Arabidopsis* mutants. *Proc. Natl. Acad. Sci. USA* 1998; 95:13324–13329.

192. Wentworth M, Ruban AV, Horton P. Thermodynamic investigation into the mechanism of the chlorophyll fluorescence quenching in isolated photosystem II light-harvesting complexes. *J. Biol. Chem.* 2003; 278:21845–21850.

193. Wagner B, Goss R, Richter M, Wild A, Holzwarth AR. Picosecond time-resolved study on the nature of high-energy-state quenching in isolated pea thylakoids. Different localization of zeaxanthin dependent and independent quenching mechanisms. *J. Photochem. Photobiol.* 1996; 36:339–350.

194. Demming-Adams B, Adams III WW, Heber U, Neimanis S, Winter K, Kruger A, Czygan F-C, Bilger W, Björkman O. Inhibition of zeaxanthin formation and of rapid changes in radiationless energy dissipation by dithiothreitol in spinach leaves and chloroplasts. *Plant Physiol.* 1990; 92:293–301.

195. Bilger W, Björkman O. Role of xanthophyll cycle in photoprotection elucidated by measurements of light-induced absorbance, fluorescence and photosynthesis in leaves of *Hedera canariensis*. *Photosynth. Res.* 1990; 25:173–185.

196. Gilmore AM, Yamamoto HY. Zeaxanthin formation and energy-dependent fluorescence quenching in pea chloroplasts under artificially mediated linear cyclic electron transport. *Plant Physiol.* 1991; 96:635–643.

197. Briantais J-M. Light-harvesting chlorophyll *a-b* complex requirement for regulation of photosystem II photochemistry by non-photochemical quenching. *Photosynth. Res.* 1994; 40:287–294.

198. Lee AL, Thornber JP. Analysis of the pigment stoichiometry of pigment-protein complexes from barley (*Hordeum vulgare*). The xanthophyll cycle intermediates occurs mainly in the light-harvesting complexes of photosystem I and photosystem II. *Plant Physiol.* 1995; 107:565–574.

199. Li X-P, Björkman O, Shih C, Grossman AR, Rosenquist M, Jansson S, Niyogi KK. A pigment-binding protein essential for regulation of photosynthetic light harvesting. *Nature* 2000; 403:391–395.

200. Li X-P, Müller-Moulé P, Gilmore AM, Niyogi KK. PsbS-dependent enhancement of feedback de-excitation protects photosystem II from photoinhibition.

Proc. Natl. Acad. Sci. 2002; 99:15222–15227.

201. Gastaldelli M, Canino G, Croce R, Bassi R. Xanthophyll binding sites of the CP29 (Lhcb4) subunit of higher plant photosystem II investigated by domain swapping and mutation analysis. *J. Biol. Chem.* 2003; 278:19190–19198.

202. Jahns P, Miehe B. Kinetic correlation of recovery from photoinhibition and zeaxanthin epoxidation. *Planta* 1996; 198:202–210.

203. Jin E, Polle JEW, Melis A. Involvement of zeaxanthin and of Cbr protein in the repair of photosystem II from photoinhibition in the green alga *Dunaliella salina*. *Biochim. Biophys. Acta* 2001; 1506:244–259.

204. Jin E, Yokthongwattana K, Polle JEW, Melis A. Role of the reversible xanthophyll cycle in the photosystem II damage and repair cycle in *Dunaliella salina*. *Plant Physiol.* 2003; 132:352–364.

205. Thiele A, Krause GH. Xanthophyll cycle and thermal energy dissipation in photosystem II: relationship between zeaxanthin formation, energy-dependent fluorescence quenching and photoinhibition. *J. Plant Physiol.* 1994; 144:324–332.

206. Forti G, Barbagallo RP, Inversini B. The role of ascorbate in the protection of thylakoids against photoinactivation. *Photosynth. Res.* 1999; 59:215–222.

207. Krause GH, Laasch H, Weis E. Regulation of thermal dissipation of absorbed light energy in chloroplasts indicated by energy-dependent fluorescence quenching. *Plant Physiol. Biochem.* 1988; 26:445–452.

208. Oxborough K, Horton P. A study of the regulation and function of energy-dependent quenching in pea chloroplasts. *Biochim. Biophys. Acta* 1988; 934:135–143.

209. Krause GH, Behrend U. pH dependent chlorophyll fluorescence quenching indicating a mechanism of protection against photoinhibition of chloroplasts. *FEBS Lett.* 1986; 200:298–302.

20 Development of Functional Thylakoid Membranes: Regulation by Light and Hormones

Peter Nyitrai
Department of Plant Physiology, Eötvös University

CONTENTS

I. INTRODUCTION

Many environmental factors influence plant development, including light, temperature, water availability, abiotic and biotic stress, and nutrient conditions that serve as signals for the activation of endogenous developmental programs. Of these, not only is light used by plants as the source of energy for photosynthesis, but it is also the most important factor in the regulation of the biogenesis of the photosynthetic apparatus in chloroplasts and the adaptation and acclimation of that apparatus to the everchanging environment, as well as actual incident light acting as a signal for various photomorphogenetic responses.

It is well known that photosynthetic machinery can acclimate to its particular light environment fluctuating in intensity, duration and spectral quality by modulating the composition of its photosynthetic membranes and biochemical circuits to enhance the efficiency of photosynthesis and protect itself from photoinhibitory damage caused by excessive irradiance. Light signals are mediated through photoreceptors (protochlorophyllide, phytochromes, UV-blue light receptors) that are already present at the earliest stages of leaf development. Activation of these photoreceptors initiates and continuously regulates the structural and functional responses of chloroplasts.

The development of photosynthetically active chloroplast from progenitor organelles — proplastids and etioplasts — is accomplished by the cooperation of two genetic systems: the nucleus and the plastid genomes. Light interacts with the endogenous developmental program modulating these genes responses and appears to operate at various levels, transcriptionally and posttranscriptionally.

In addition to light-induced and -controlled processes affecting plant genetic program, there are other endogenous factors, mainly phytohormones, which are responsible for the maintenance of the effect of light interacting synergistically or antagonistically with it.

This chapter focuses on the effects of light on the biogenesis of photosynthetic apparatus interacting with hormonal factors, especially cytokinins.

II. DEVELOPMENTAL STAGES OF PLASTIDS

Meristematic cells in the shoot and root apices and leaf primordia contain small, undifferentiated proplastids, where their number per cell is low [1]. The ultrastructure of these nongreen proplastids is simple and consists of the envelope, often with some invaginations from the inner membrane, and a stroma with some ribosomes, nucleoids, plastoglobuli, sometimes some inner membranes, starch grains, and a limited amount of procaryotic type DNA.

Proplastids develop into various types of plastids with different functions. The stage of plastids development closely reflects the developmental program of the entire cell in which the plastids are located. A given cell type has a plastid population that is relatively uniform with regard to its differentiation state [2], depending on the tissues and on the environmental conditions, such as light, temperature, water availability, and nutrient supply [3]. In developing plants, most of proplastids in a leaf meristem develop into chloroplasts. Proplastids can also differentiate into specialized plastid types that assume other functions in nonphotosynthetic plant organs of higher plants, such as amyloplasts in roots and tubers or chromoplasts in many flower petals and fruits.

In those tissues and organs destined to become green and photosynthesizing in light, the proplastids develop into etioplasts in darkness. Etioplasts are the end product of a differentiation route in dark-grown angiosperm seedlings that begins with the proplastid stage. A characteristic feature of the etioplast is the prolamellar body (PLB), which consists of a cubic lattice of interconnected membranous tubule network derived from the inner membrane of plastid envelope forming a paracrystalline structure composed of lipids and proteins. Only a few thylakoid membranes, prothylakoids (PT) extending from the PLB, are usually found. The lipid to protein ratio is higher in PLBs than in PTs, and so is the ratio of monogalactosyldiacylglycerol (MGDG) to digalactosyldiacylglycerol (DGDG). PLBs and PTs have similar polypeptide composition even if NADP-protochlorophyllide oxidoreductase (POR) quantitatively dominates the

PLBs. The basic arrangement of PLBs and PTs can be varied in many different ways resulting in differences between plant species or groups.

Seedlings grown in the dark, or plant tissues that are developing for a prolonged period in darkness, accumulate a larger amount of phytochrome (Phy) and NADP-protochlorophyllide oxidoreductase (POR A) relative to the green chloroplasts. POR A is a protein that has been shown to form complexes with chlorophyll (Chl) precursors, protochlorophyll(ide)s (Pchlide). This protein is stabilized as a ternary complex with Pchlide and NADPH [4]. The stored Pchlide functions not only as a precursor for Chl synthesis but also as a photoreceptor in the etioplast/chloroplast transformation [5]. When illuminated, POR A photoreduces its Pchlide to Chlide, and simultaneously, the PBs begin to disintegrate. During catalysis, POR A is inactivated and subsequently rapidly degraded [6].

Etioplast inner membranes contain 30 to 35 different polypeptides. The dominating polypeptides belong to coupling factor (CF1) γ and δ subunits of ATP synthetase and POR. Some components of the thylakoid electron transport chain, such as cytochrome $b559_{LP}$, cytochrome $b563$, cytochrome f, the precursor form of D1 protein, the Rieske Fe–S protein, plastocyanin, ferredoxin, ferredoxin-NADP oxidoreductase, and the extrinsic polypeptides of the oxygen-evolving system, are synthesized in dark [7–9]. Light-induced components are assembled together with those of preexisting ones after illumination to form a complete electron transport system during greening.

The transition from etioplast to chloroplast by light is to a large extent reversible, thus, chloroplast development via this route can be experimentally influenced in both directions. The transition from etioplast to chloroplast by light is also a favorite system to study the processes of photoconversion of Chl precursors, the light-induced synthesis of some groups of compounds (pigments, nucleic acids, polypeptides, and lipids), and the organization of photosynthetic apparatus during greening.

III. ACTION OF LIGHT ON THE STRUCTURAL AND FUNCTIONAL ORGANIZATION OF ETIOPLASTS

When dark-grown etiolated seedlings are exposed to light, dramatic structural and functional changes take place. The differentiation of etioplast into chloroplast is a multistep process, involving the biosynthesis of photosynthetic pigments, nucleic acids, proteins, and lipids stimulated mainly by light. It requires the close cooperation of both chloroplast and nucleic genetic machineries. Their products are processed and assembled into supramolecular complexes of developing thylakoid membranes.

A. PIGMENT CONVERSION

Photosynthetic organisms have evolved two different strategies to synthesize Chl or bacteriochlorophyll (BChl) [10,11]. In a light-independent reaction, PChlide can be reduced to Chlide through a Pchlide reducing enzyme that requires three different polypeptides encoded by chloroplast DNA [12]. The presence of these genes has been established in cyanobacteria, green algae, bryophytes, pteridophytes, and gymnosperms [10]. In angiosperms, the absence of a light-independent Pchlide-reductase complex is connected with the inability of etiolated angiosperm seedlings to reduce Pchlide to Chlide in the dark [13]. In another way, the first detectable light-dependent step is the phototransformation of Pchlide to Chlide catalyzed by the plastid-localized POR enzyme, which is encoded by nuclear genes [14,15]. It is present in all oxygenic photosynthetic organisms including cyanobacteria, algae, liverworts, mosses, gymnosperms, and angiosperms. Among all enzymes POR is unique because it is a photoenzyme requiring light for its catalytic activity and using Pchlide itself as a photoreceptor [16]. Two distinct POR enzymes can be distinguished on the basis of their *in vivo* activity. The function of POR A is confined to the very early stages of transition from etiolated to light growth. The amounts of both POR A protein and its mRNA decrease drastically soon after the beginning of illumination, due in part to the rapid proteolytic turn over of the enzyme protein stimulated by a light-induced protease, and a negative light-regulation of its synthesis, a process which is accompanied by the degradation of PBs. To perform the reduction of PChlide to Chlide during the final stages of the light-induced greening and to sustain Chl synthesis in fully green, mature leaves, another Pchlide-reducing enzyme has to operate in angiosperms, termed POR B. POR B and its mRNA are present throughout the angiosperm life cycle, and its expression is more or less independent of light [17].

After the light triggered, there is a short lag period in Chl formation followed by a steady increase in the accumulation of Chl until a plateau is reached. Under continuous light, the formation of Chl *a* is followed by the appearance of Chl *b*, however, the mechanism and localization of Chl *b* synthesis are not well understood.

The preexisting components of photosystem I (PSI) and photosystem II (PSII) in etioplasts are complemented with newly synthesized polypeptides coming at first from the plastidic protein-synthesizing system, where the initial signal may be the appearance of Chlide *a* or Chl *a*. During further greening, core complexes of PSI and PSII are supplied with light-harvesting Chl *a/b* complexes (LHCs) arriving from the cytoplasmic protein-synthesizing system switching on through phytochrome action.

B. MEMBRANE DEVELOPMENT

Changes in pigment content are paralleled by changes in protein composition and by structural changes in membrane composition. The PLBs disrupt and are replaced by vesicles, and PTs are converted to primary thylakoids. These membranes serve as a lipid matrix for the assembly of large pigmented (PSI and PSII core complexes, light-harvesting complexes of PSI and PSII), and nonpigmented (cytochrome b_6/f complex, ATP synthetase) complexes, which are made of proteins, pigments, and other cofactors attached and embedded in the lipid bilayer. Thylakoids are organized into appressed and nonappressed regions, which are a consequence of a segregation of complexes showing a lateral heterogeneity between the stroma and grana membranes in rearrangement [18–20].

C. PIGMENT–PROTEIN COMPLEXES

Etioplast do not contain chlorophyll–protein complexes (CPCs). The synthesis of CPCs occurs during the light-induced greening of etiolated seedlings. A large number of CPCs exist in chloroplasts, which differ from one other in molecular weight, function, and amount and quality of associated pigments (chlorophylls, carotenes, and xanthophylls). These CPCs are the components of photosystems, functioning as reaction centers of PSI and PSII and some of the associated light-harvesting chlorophyll *a/b* complexes (LHCI and LHCII inner and outer antenna).

However, some photochemical activities have often been reported to appear at the early stages of greening, even before the end of the lag period of massive Chl synthesis, and CPC formation is required for the appearance of photochemical activities, but none have yet been isolated at such early stages of greening. The accumulation of CPCs in thylakoid membranes during chloroplast development was observed early [21]. The earliest reported appearances of CP1 (P700–Chl *a*–protein complex) [22], "PSI complex" [23], and Cpa (Chl *a*–protein complex of PSII core) [24] are 4, 6, and 13 h after the onset of illumination, respectively. By an improved method of sodium dodecyl sulfate-polyacrylamide gel electrophoresis (SDS-PAGE), 10 min after the onset of illumination, a labile CP complex (CPX) was detected. The CP1 appeared after 45 to 60 min of illumination together with P700 activity, and LHCs began to accumulate at 2.5 h with the beginning of Chl b synthesis [25]. It seems that at early stages of greening, CP1 is the only detectable CPC. Cpa and LHCPs could be detected only after 3 to 4 h of continuous illumination, although PSII activity appeared much earlier [25,26]. These discrepancies between time of detection of photosynthetic activity and of its CPCs are likely caused by the difficulties of detecting a very small amount of these complexes and by the lability of Chl binding at the early stages of greening, but it is generally due to the problems of solubilization processes.

The rapid development of PSI and PSII requires biosynthesis of Chl a-binding polypeptides shortly after Pchlide has been reduced to Chlide a and converted to Chl a. Results demonstrated that POR and Chl-synthetase are key enzymes in the light-induced expression of plastid-encoded polypeptides such as CP47 and CP43 polypeptides of PSII and the P700-apoproteins that are accumulated in plastids. However, when the rate of Chl formation is low (lag period) relative to the other thylakoid components, small PSI and PSII units are formed, containing only the core complexes of these units. When the rate of Chl formation enhanced, the units that are increased in size by incorporation of LHCPs are nuclear encoded. This stepwise formation and growth of the photosynthetic units has been observed in etiolated leaves exposed to intermittent light (IML), millisecond flashes, far-red light, or very low light intensity and then transferred to continuous light of normal intensity [27,28]. Similar information can be obtained by studying the redistribution of Chls among newly synthesized core complexes in dark [29,30].

D. APPEARANCE OF PHOTOSYNTHETIC ACTIVITIES.

Formation of the photosynthetic apparatus during chloroplast development consists of three main processes: the expression of genetic information, conveyance of constitutive subunits to the thylakoid membranes, and the assembly of subunits to form the functional, oligomeric structure of photochemical systems.

When dark-grown plants are illuminated, synthesis of apoproteins of subunits of photosystems is initiated and photochemical activities appear. It has been generally established that PSI activity can be detected at an earlier stage of greening than that of PSII [31,32], but the time of the appearance of these two activities varies greatly with the plant material,

growth conditions, duration of etiolated period, light regimes, and the method used to detect them. Hardly any work has been done to compare the beginning of photochemical activities with the appearance of PSI and PSII obtained using a method suitable for isolating them from tissues at these earliest stages of greening.

The development of photochemical activities in isolated plastids during the early phase of greening has been studied by traditional chemical methods. Photochemical activities of PSI [dichlorophenol-indophenol ($DCPIPH_2$) to methyl viologen (MV)] and PSII [H_2O to DCPIP, diphenylcarbazide (DPC) to DCPIP] appeared 1 and 1.5 h after the onset of illumination, respectively. However, PSI + PSII activity (H_2O to MV, H_2O to $NADP^+$) appeared at 4 h [32,33]. Time-appearance of photochemical activities can also be verified by cytochemical methods by the photooxidation of 3,3'-diaminobenzidine (DAB) for PSI activity and photoreduction of thiocarbamyl nitrotetrazolium blue (TCNB) for PSII activity [34].

Appearance of Chlide a and Chl a within several seconds of exposure to light is sufficient to trigger the formation of very small amounts of functional photosynthetic units. Induction of low level of O_2 evolution and CO_2 fixation could be detected as early as 20 min after a 40-sec light pulse [35]. Higher amounts of O_2 evolution during the second illumination suggest that photoactivation of the water-splitting system in PSII is formed as a result of the first light pulse. However, PSII is very unstable at the early stage of greening. PSII-mediated electron transport could be detected a very short time after illumination using delayed luminescence and fluorescence variation measurements [36]. Similar results were shown in the case of PSI activity [37].

As for the development of PSI after 1 h, it continued as follows: some components may be associated with the reaction center (RC), and they began to reduce ferredoxin after 2 h. At 4 h, ferredoxin:$NADP^+$ oxidoreductase (FNR) became attached to the PSI core allowing reduction of NADP. The whole electron transport from water to NADP was operational after 4 h of development. Once PSI and PSII have appeared, the number of both photosystems increased at the same rates during the greening process. After 6 h, the increase of both photosystems was accelerated, probably because the electron transport activity driven by the two photosystems was sufficient to further support the development of plastids. As for the antenna of PSI, it was proposed that LHCI apoproteins are first assembled into monomeric pigmented complexes after 2 h of illumination and then assemble into a trimer before attaching to the preexisting core complexes to form a complete PSI holocomplex, which in total required more than 4 h of greening [38].

IV. LIGHT REGULATION OF CHLOROPLAST DEVELOPMENT

A. RECEPTOR SIDE OF LIGHT-INDUCED PROCESSES

Developmental processes in plants are regulated not only by a genetic program, but also by a huge amount of environmental information, particularly light signals. The light-dependent development is induced and controlled by the combined action of several distinct photoreceptors including POR, blue and UV light receptors, and the red/far-red light-absorbing receptors, termed phytochromes [39]. These receptors can absorb photons over a wide range of wavelengths, ranging from the far-red to the UV. In addition, plants also sense the duration, intensity, and direction of light using these photoreceptors. Photoreceptors exhibit dual molecular functions: a sensory function responsible for detecting relevant light signals, and a regulatory function in which perceived information is transferred to downstream signal transduction pathway.

Phytochrome is the most intensively studied and characterized red/far-red absorbing chromoproteins and is responsible for initiating a wide range of photomorhogenic events including the expression of nuclear and chloroplastic genes encoding photosynthetic components [40–42]. It exists in two spectrophotometrically different, photointerconvertible forms between its inactive (Pr) and active (Pfr) forms by sequential absorption of red and far-red photons of light, which leads to the isomerization of the phytochrome chromophore [43]. Two types of phytochromes have been defined on the basis of their lability in light. Type I phytochromes (PhyA) are abundant in etiolated seedlings, Type II phytochromes (PhyB-E) are present in much lower amounts, but their stability in the Pfr form ensures that they are predominant in light-grown plants during their life cycle. In higher plants, phytochromes are encoded by a small gene family [40], and are synthesized in its red-absorbing form (Pr), which has a major absorption peak at 660 nm. Illumination with red light converts Pr to far-red absorbing form (Pfr). Irradiation of Pfr with far-red light converts it back to Pr. Pfr is considered the active form of phytochrome, because red light irradiation is correlated with the induction of growth and development.

In contrast to PhyA, which is supposed to be localized mainly in the cytosol, the nucleo/cytoplasmic partitioning of PhyB is a light-regulated process, and the import of PhyB into the nucleus is part of this regulatory circuit [44]. It could be suggested that the cytosolic pool of PhyB Pfr either may have a role in regulating rapid phytochrome effects such as influence of ion fluxes at the plasma membrane [45]. Based on recent observation it is plausible to assume that light quality- and quantity-dependent import of PhyA and PhyB into the nucleus is an essential step in signal transduction controlled by these photoreceptors. Data indicate that light-induced conversion of Pr to Pfr is required for translocation, and only the Pfr forms of PhyA and PhyB are transportable [46].

Genetic studies indicate that light responses are not simply endpoints of linear signal transduction pathways but result from the integration of information from a network of interacting signaling components. The signaling components include the photoreceptors themselves, as well as positive and negative regulatory elements that act downstream from these photoreceptors [41,47]. Many genes have been reported to be activated through more than one photoreceptor [48]. It is conceivable that different light-activated transduction pathways target different transcription factors and/or cis-acting light-responsive elements within the promoter of a given gene, but it is also possible that they target common light-responsive elements.

Individual phytochromes may have at least two separate mechanisms of action: one that results in selective expression of target genes (slow response) and another that rapidly and reversibly operates to modulate cellular ionic balance [49], strongly suggesting that it does not require de novo protein synthesis.

Recent data indicate that PhyA and PhyB have at least two different modes of action. On the one hand, they directly interact with transcription factors and act as transcriptional modulators in the nucleus [50], and on the other hand, they function as light-regulated kinases in cytoplasm [51]. It was shown that the C-terminal domain of PhyA and PhyB play an essential role in mediating phytochrome signal transduction. Several interacting proteins with C-terminal domain of PhyB were isolated like phytochrome interacting factor-3 (PIF3) a member of a small gene family. It has the characteristics of a transcription factor localized in the nucleus. Besides binding to the Pfr form of PhyB, PIF3 also binds to G-box-containing promoter elements of light-regulated genes. These findings point to a straightforward mechanism for a PhyB-dependent signaling cascade [50]. Phytochrome has now been demonstrated to have Ser/Thr protein kinase activity. Phosphorylation is a light- and chromophore-regulated process, and it is believed that after photoconversion of Pr into Pfr, the phytochrome kinase activity initiates light signaling either by phosphorylating downstream elements of the signaling pathway or by phosphospecific interactions with them [51,52].

It has been reported that some molecules as second messengers take part in phytochrome signal transduction. In this way, results have shown that membrane-bound heterotrimeric G-proteins are involved in amplification of PhyA signal transduction. First, cGMP can activate chalcone synthase (CHS) and stimulate anthocyanin biosynthesis [53]. Second, calcium and calmodulin can activate synthesis of light-harvesting complexes, PSII, ribulose bisphosphate carboxylase, and ATPase and, thus, partial chloroplast development. Third, a pathway requiring both calcium and cGMP can activate synthesis of PSI and cytochrome $b_6 f$ complex. The combination of these three pathways therefore leads to produce fully mature chloroplasts [54].

In addition to the components mediating phytochrome-dependent signaling cascade, a type-A response regulator, ARR4, was identified as an interactive component [55], which might be a candidate molecule that is required for regulation of Pfr $\rightarrow$ Pr dark reversion. Expression of ARR4 is regulated not only by PhyB but also by cytokinin [56]. These results provide the first molecular evidence for an interaction between light- and hormone-dependent signaling cascade with ARR4 acting as a novel molecular switch to mediate this cross-talk [55].

Other groups of photoreceptors are responsible for the perception of UV and blue region of incident light called cryptochromes. Large number of blue light responses have been documented in plants, including inhibition of the rate of hypocotyl growth, phototropism, stomatal opening, anthocyanin production, and the induction of gene expression (CHS) [41,57]. Though blue light receptors play a considerably smaller role in the development of photosynthetic apparatus, interactions between phytochrome- and cryptochrome-activated signal transduction pathways can be demonstrated.

B. REGULATION OF LIGHT-INDUCED EXPRESSION OF CHLOROPLAST GENES

The biosynthesis of the photosynthetic apparatus in higher plants and algae depends on the concerted action of two genetic systems located in the nucleus and in the plastid. Approximately 50% of thylakoid membrane proteins are encoded by the plastid genome in plants and green algae, and appear to be synthesized on thylakoid membrane-bound ribosomes. There are some differences between the plastid and nuclear genomes. The informational content of the chloroplast genome is modest, the chloroplast genome is polyploid, it is present in 100 to 10,000 copies per cell, and the nuclear genome contains numerous genes for the construction of chloroplast [58]. Studies have revealed the existence of approximately 100 plastid genes that fall into three major functional categories: (1) half of them encode RNAs and proteins involved in transcription and translation of plastid genome (RNA polymerase subunits, t-RNAs, r-RNAs, ribosomal proteins, initiation and elongation factors, etc.), (2) about 30 genes encode subunits of the five large multicomponent, photosynthetic complexes (PSI and PSII core complexes, cytochrome $b6/f$ complex, ATP synthetase, and the large subunit of Rubisco), and (3) genes encoding proteins of the NADH-dehydrogenase complex, are known to be regulated by light [59]. The other subunits of the photosynthetic complexes are encoded by nuclear genes, translated as precursors on cytosolic ribosomes, processed and imported into the plastid, where they are associated with the chloroplast-encoded subunits, various photosynthetic pigments, and cofactors to form functional multicomplexes. The dual genetic origin and a vast difference of two to three orders of magnitude in gene copy number between nucleus and plastids imply that the synthesis and assembly of thylakoid polypeptides must be highly coordinated in time and space, with regard to stoichiometry and in response to external signals. Genetic and biochemical studies of these processes have revealed several general features:

1. The stable assembly of thylakoid complexes requires that all core components be synthesized. Failure to produce any of the core components usually leads to the rapid degradation of the other subunits of the complex, neither free chlorophylls nor apoproteins of CPCs accumulate.

2. Expression of chloroplast genes depends on many nuclear-encoded factors, and not surprisingly these factors are involved in the expression of specific chloroplast genes. They may act either at the level of RNA processing and splicing, RNA maturation, stability and translation (initiation and elongation), or at the level of the assembly of the photosynthetic complexes [60,61]. The nuclear control is mediated by organellar-targeted proteins that specifically affect the expression of either on gene or of a subset of organellar genes [62,63]. For example, some nuclear mutations interrupt certain processes in chloroplast biogenesis, assembly of multicomponent complexes [64], plastid transcription [65], and plastid RNA processing and translation [66].

3. Expression of certain nuclear genes depends on so-called "plastid factor" that is thought to be reproduced by plastids [60,67, 68]. The chloroplast signal may play an important role in controlling nuclear gene expression during the greening process, and when chloroplast is senescent or develops into another type of plastid [69].

4. Regulation of protein synthesis in plastids by light is realized at various levels including transcriptional and posttranscriptional (translational, posttranslational) mode of actions, but some characteristic differences can be observed between light-regulated synthesis of plastid and nuclear-encoded polypeptides.

5. In most cases, signal transduction chains are still largely unknown, especially in the case of plastids, however, vast amount of knowledge has been accumulated about the perception of light signal as the starting point and the final changes in gene expression as a target point in the synthesis and assembly of polypeptides.

Plastid transcriptional activity and levels of RNAs increase during the early stages of greening, especially for rpoB–rpoC1–rpoC2, rps1b, rRNAs, and some tRNAs relative to genes encoding proteins of the photosynthetic apparatus [70]. This step is followed shortly by the activation of plastid and nuclear genes encoding proteins. The induction of plastid transcription could be resulted from increased DNA template level, and changes in DNA conformation as well as modification of the level of RNA polymerases. After completion of greening of chloroplast a decline in overall plastid transcription is observed [71], but the extent to which illumination controls plastid gene transcription varies depending on the plant species, leaf type, influence of nuclear genome, and the developmental state of the tissue investigated.

The expression of many of the genes encoding chloroplast components is regulated by phytochromes and cryptochromes via direct or indirect way at the transcriptional level; posttranscriptional, translational, and posttranslational regulatory processes have also been detected. While the role of transcription has been documented for several plastid genes [72,73], the major regulatory step in plastid gene expression appears to occur posttranscriptionally [74–77], at the levels of pre-mRNA processing [75,78,79], mRNA stability [80], translation or protein modification or turnover [65,81,82] influenced as a nuclear control. The fact that translational and posttranslational regulatory mechanisms are important in plastid gene expression is evident in many cases from the lack of correspondence between plastid mRNA and protein levels in different growth conditions.

Most of chloroplast-encoded proteins for thylakoid membrane complexes are not detectable in dark-grown plants despite the relatively high levels of their mRNAs, but accumulate rapidly after illumination [76,83]. For example, it is noted that mRNA for the psaA–psaB genes, which encode the 65 to 70-kDa PSI (CP1) reaction center, Chl a-apoproteins were present in dark-grown seedlings even though the PSI Chl a-apoproteins were undetectable. After illumination the rapid synthesis of these apoproteins in the absence of changes in their mRNA level is consistent with the activation of translation. It is evidenced that the synthesis of CP1 polypeptides is arrested at the level of translational elongation in dark-grown barley seedlings [74,84]. This translational block is removed by the synthesis of Chl a in vivo and in isolated plastids, allowing CP1 accumulation. The distribution of mRNAs — localized almost exclusively in large membrane-bound polysomes — indicates that synthesis of Chl a-apoproteins in dark-grown plants is inhibited late in apoprotein translation, elongation or at the point of apoprotein–mRNA–ribosome dissociation [58]. Similar results were obtained where synthesis of Chl a controls the translation of Chl a-apoproteins of P700, CP47, CP43, and D2 [85] and for the translation of the large subunit of Rubisco (rbcL) [86].

Posttranslational control mechanisms in the chloroplast appear to have evolved as a regulatory mechanism that allows the plant to uncouple the synthesis and assembly of photosynthetic proteins from the obligate transcription of their genes, if it is not necessary to be running on.

C. REGULATION OF LIGHT-INDUCED EXPRESSION OF NUCLEAR-ENCODED GENES

It is well accepted that the chloroplast genome originated from endosymbiotic procaryotes. There has been a tendency for the plastid genome to lose gene sequences to the nuclear genome integrating into the eucaryotic cell. So the plastid genome contains relatively little genetic information and its expression in now under the control of the nuclear genome. The transition from an etiolated seedling to a fully green plant is accompanied by a dramatic change of expression of nuclear-encoded genes, many of which are regulated by light through one or more different photoreceptors [87]. Nuclear-encoded thylakoid proteins are synthsized on cytoplasmic ribosomes as precursors containing a cleavable N-terminal extension designated the transit sequence [88]. The photocontrolled appearance and accumulation of nuclear-encoded polypeptides also seem to be regulated both transcriptionally and posttranscriptionally [58,87,89] influencing the activity of plastid genome, and the formation of chloroplast architecture.

The most studied nuclear-encoded chloroplast proteins are the light-harvesting chlorophyll *a/b*-binding (LHCI and LHCII) proteins and the small subunit (SSU) of Rubisco [60,89–92]. The LHCs are encoded in the nucleus by a multigene family. The LHC-apoproteins are translated on cytoplasmic ribosomes as soluble precursor polypeptides and translocated into the chloroplast by an energy-dependent posttranslational process, where they are integrated into the thylakoids with chlorophylls and carotenoids. Their synthesis regulated by light and controlled by the developmental state of the plastid [93]. However, light may also affect the redox state of intersystem electron carriers of the photosynthetic electron transport chains. In addition, light stimulate chlorophyll biosynthesis, a pathway that is coupled by some presently unknown mechanisms to the rate of transcription of the Lhcb genes (i.e., plastid development and the functional state of the chloroplast strongly influence transcription of the Lhcb genes).

The synthesis, accumulation, and assembly of photosynthetic pigments and LHCs require mutual regulatory interactions to coordinate the formation of complete membrane complexes. A posttranslational regulation also operates whenever the rate of synthesis of one or more thylakoid components is relatively low. Such a posttranslational control can be observed in cases when (1) the rate of Chl accumulation is low (IML, far-red, and low-light illumination) or (2) Chl *a* synthesis is abolished in contrast to the other thylakoid components, which continue to be synthesized in darkened conditions. Under IML conditions translatable LHC-mRNA is synthesized, although no LHC is accumulated [94]. Under such condition, LHCII is degraded because pigments are not available in sufficient amounts to bind and stabilize it [95,96]. This posttranslational regulation is also evidenced under greening of low-light irradiance, millisecond flashes, and far-red illumination [97–99]. An alternative explanation for posttranscriptional light regulation of LHCII synthesis, in particular at the translational level, has been suggested [100]. However, IML plants and Chl *b*-less mutants do differ in some aspects of posttranslational cab (Lhcb) polypeptide accumulation. For example, some LHCI polypeptides will accumulate in Chl *b*-less mutants [101], but not in plants grown under IML light regime, which strongly limits Chl *b* synthesis [102,103].

Other cofactors necessary for LHC assembly also require the presence of light. An important light requirement for Chl synthesis is the Pchlide photoconversion and the accumulation of Chl *b*, which is also strongly light dependent. It seems that the appearance of Chl *b* is strictly correlated with the synthesis of LHC polypeptides, and the nature of regulation depends on the actual light regime [104,105].

Another example of regulated translation in plants is that of the SSU of Rubisco. It was found that synthesis of SSU protein declined 10- to 20-fold within 4 h, when light-grown amaranth seedlings were placed in the darkness [86]. In contrast, levels of mRNA encoding this protein declined two- to fourfold at most. Isolation of polysomes indicated that the SSU message remained bound to polysomes for several hours following the shift to darkness, suggesting that translational elongation or initiation was blocked. When dark-grown amaranth seedlings were placed in light, the rate of SSU synthesis increased at least 20-fold, while only a two- to threefold increase in the level of mRNA was observed.

Various types of regulatory mechanisms take part in the light-induced gene expression of chloroplast development during greening. All possible levels of regulation have an important role in contributing to the formation of the photosynthetic apparatus, but they show more or less importance and activity in time and space (i.e., during the development of chloroplast and between the two genetic systems). Transcription activity has an important role being switched on by light through an appropriate photoreceptor at the onset of illumination or reillumination. Beside the primary importance of transcriptional regulation, posttranscriptional mechanisms also play a very great role in the regulation of normal plant gene expression [106–108]. Mechanisms that regulate either translation or transcript stability are inherently more rapid than transcriptional mechanisms, and as such, they may be preferentially employed for those responses that need to be flexible and rapid. For example, upon illumination of the etiolated plant, Pr is converted to Pfr, Pchlide is converted to Chlide, signal transduction is initiated, and light-regulated genes switch on. As time passes the abundance of phytochrome and POR A mRNA is greatly decreased, in part due to the rapid decline in the transcription of both Phy and POR genes [109,110]. In addition, the transcription activity of the plastid and nucleus usually decrease when chloroplast development is completed [71], and other regulatory mechanisms come to the forefront (different levels of posttranscriptional control), increasing the flexibility and rapidity of responses.

D. REGULATION OF THE OPERATION OF PHOTOSYNTHETIC APPARATUS BY LIGHT

To survive and work efficiently under ever-changing light conditions, which vary over two orders of magnitude, photosynthetic organisms have some control mechanisms to optimize and maintain their photosynthetic functions. These control mechanisms are in existence during the greening process, but also operate during the life of mature plants. This strongly suggests that the composition, function, and structure of thylakoid membranes, and related biochemical network are highly dynamic rather than static. The dynamic response of thylakoid membranes and the adjustment of photosystem stoichiometry to different light quality/quantity regimes suggest the existence of a mechanism capable of recognizing imbalance in the rate of light utilization by the two photoreactions and directing cellular metabolic activity for photosystem stoichiometry adjustment [19,82,111–113]. The dynamic adjustment of the photosystem stoichiometry in oxygenic photosynthesis has been documented in cyanophytes [114], red and green algae [115], and higher plant chloroplasts [116]. Since coordination of the action of the two photosystems is the primary determinants of photosynthetic, the adjustment of the photosystem ratio must come about as a result of a feedback control mechanism whose function is to regulate the biosynthesis/assembly and concentration of the two photosystems in the chloroplast thylakoids, as well as to avoid photosynthetic machinery the potentially damaging effects of excess irradiance.

A variety of mechanisms have been proposed for sensing the fluence rate of light. Despite the involvement of light sensors in the expression of genes encoding components of the photosynthetic apparatus, and although changes in chloroplast composition occur in response to the light environment, there is a growing evidence that chloroplast level acclimation is not direct regulated by light. The phytochrome photoreceptors appear to play only very indirect roles in acclimation [117]. However, there is evidence for a specific role of blue light photoreceptors that cause changes in nuclear gene expression.

1. Short-Term Adaptation

Two kinds of timescale adaptations can be distinguished. On short timescales of minutes to hours, plants undergo changes in response to light frequency variations in the light intensity. "Leaf level" responses include changes in leaf thickness, change in orientation of leaves, and in number per cell and per unit leaf area, as well as the ability of plants (shade-avoidance response) to increase their capacity for light capture by detecting and responding to one another [118,119]. "Chloroplast level" responses include the dissipation of excess excitation energy via xantophyll-cycle carotenoids [120] and "state 1–state 2" transitions where reversible phosphorylation of LHCII leads to comple-

mentary alterations in the effective absorption cross section of PSII and PSI [121–123], and as a consequence modulates the extent of grana stacking [20]. It is suggested that the redox state of electron transport chain components (plastoquinone and cytochrome b_6/f complex) between PSII and PSI is the primary signal, which alters the distribution of absorbed light energy between photosystems [124].

2. Long-Term Acclimation

Long-term adaptation, termed acclimation, involve the selective synthesis and degradation of chloroplast components as LHCs, PSI and PSII reaction centers, electron carriers, ATP-synthetase, and other proteins to modulate the composition and function of the photosynthetic apparatus [125]. It occurs on a longer timescale, usually many hours or more. The long-term mechanisms compensate for the unequal utilization of light by the two photosystems with an increase in the relative quantity of PSII in PSI light and the relative quantity of PSI in PSII light [126], sometimes mimicking "shade or low-light" or "sun or high-light" plants developed and genetically fixed during the evolution [20,127]. The explanation for the long-term acclimation phenomenon based on a two-component regulatory system reflecting changes in the state of redox potential of redox components has been discovered at first in bacteria [128]. On the longer scales, photoacclimatory responses can lead to changes in gene expression. This process requires a signal transduction pathway that couples an irradiance sensor, but not definitely photosensor, and its substrate, a response regulator to both nuclear and plastid gene expression. It was demonstrated that the rate of transcription of chloroplast genes encoding the reaction center apoproteins of PSII and PSI also depends on the redox state of plastoquinone. This chloroplast gene expression is direct and rapid, either permitting transcriptional responses within minutes of perturbation of the redox state of electron carriers between the photosystems [129], or it is even thought to be primarily regulated posttranscriptionally via RNA stability and translation [63,78]. Moreover, thioredoxin, which undergoes a reversible, photosynthesis-dependent reduction, has also been implicated in the translational regulation of the chloroplast-encoded psbA transcript (D1) [130].

The rate of transcription of nuclear genes encoding LHCIIb is also controlled by the redox state of plastoquinone in *Dunaliella* species [131]. It was also shown that changes in photosynthetic electron transport can cause dramatic effects on posttranscriptional steps in the expression of nuclear genes for photosynthetic proteins [132,133].

Other molecular pathways, including carbon fixation (activation state of Rubisco), carbohydrate and nitrogen metabolism exert feedback influences on photosynthesis, which are also sensed by the redox state of the intermediate carriers [19,134]. A feedback mechanism is postulated from a light intensity regulation of chlorophyll synthesis, which is confined to the plastid, to either gene expression or protein stability [135]. Therefore, signal transduction pathway could supposedly involve a chloroplast phosphoprotein intermediate (a redox sensor) that is released from the membrane upon phosphorylation. Its substrate, the response regulator, is a sequence-specific DNA-binding protein, whose phosphorylation is required for the binding of RNA polymerase and initiation of transcription at promoters of genes. The mode of signal transduction between the chloroplast and nucleus is not clear, although the phosphorylase inhibitor experiments suggest the involvement of a phosphorylation cascade. This possible feedback control could act at the level of chloroplast transcription or translation by protein phosphorylation targeting chloroplast-encoded (psaA, psaB genes) or nuclear-encoded (cab, Lhc genes) polypeptides [117,123,136].

Regulation of both short-term adaptation and long-term acclimation potentially involves multiple signal tranduction chains, which cross-talk between redox control and other pathways that control photosynthetic gene expression [117].

V. ROLE OF PHYTOHORMONES IN THE DEVELOPMENT OF PHOTOSYNTHETIC APPARATUS

A. CYTOKININS

Cytokinins as members of the phytohormonal network of plants have been implicated in many developmental processes and environmental responses of plants, including regulation of cell division, initialization of shoot meristem, differentiation of leaf and root, bud opening, apical dominance, anthocian production, protection against abiotic oxidative stress, plant–pathogen interaction, some aspects of nutrient metabolism, and control of leaf senescence [137,138]. Furthermore, cytokinins are involved in the induction and regulation of some biochemical and structural steps associated with photomorphogenesis including chloroplast development [139]. These processes are also influenced by various other stimuli (e.g., light and other phytohormones), and the physiological and developmental outcomes reflect a highly integrated response to these multiple stimuli.

Cytokinins are known to affect photosynthetic structures and functions, and to stimulate chloroplast

biogenesis, promoting chlorophyll synthesis, the expression of genes encoding chloroplast polypeptides and influence the architecture of thylakoid membranes [140].

Numerous experimental data are available to demonstrate numerous effects of cytokinins on plastid development at physiological level. Cytokinins can promote chlorophyll synthesis. Hormone treatment abolishes the lag period in chlorophyll synthesis and accelerates its rates [141,142], thus mimicking the effect of a red-light pulse on the greening process [143]. It has been suggested that cytokinins initiate production of the enzymes involved in 5-aminolevulinic acid (ALA) synthesis [141,144] and Pchlide regeneration [141,145].

Cytokinins enhance the steady-state levels of certain mRNAs and proteins in etiolated or green leaves, cotyledons, and cell cultures [146,147] but cause a decline in the levels of others [148]. Cytokinin inducible RNAs include those encoding the light-harvesting chlorophyll a/b-binding protein [149–151], the SSU of Rubisco [152], and nitrate reductase [153]. These hormones induce the synthesis of nuclear and plastid DNA synthesis [154,155], transcriptional activity of nucleus [156], polyribosome formation [157,158], polyadenylation of RNAs [159], translational activity of poly(A)RNAs [160], photosynthetic electron transport [161,162], photosynthetic activities, and protein synthesis [163–165]. All of these effects of cytokinins promote the formation of the photosynthetic apparatus and are more or less synergistic with light and directed to maintain an enhanced developmental status of plastids.

There are much experimental data indicating that cytokinins affect the abundance of transcripts and proteins encoded both by nuclear and plastid genomes [106,166]. The most notable genes are for the SSU of Rubisco and LHCP. In dark-adapted *Lemna gibba* plants, cytokinin treatment increases fivefold the level of transcripts encoding the major chlorophyll a/b-binding protein of LHCII, as measured by RNA blot hybridization [149]. Similarly, in dark-adapted plants subjected to a pulse of red-light, a threefold increase in LHC mRNA was observed in response to cytokinin treatment. The increase in mRNA accumulation levels may be due to a stabilization of LHC transcript in the presence of cytokinin. Control at the level of mRNA stability may contribute to the regulation of the light-responsive genes encoding the SSU of Rubisco. However, cytokinins did not modulate the synthesis of every light-controlled polypeptide, but did modulate the synthesis of some light-independent polypeptides [167]. The mechanism by which these effects occur has not been completely resolved, but there is evidence for different sites of regulatory points [166,168]. It was concluded from the analysis of runoff transcription assays that the cytokinin response of nitrate reductase and hydroxypyruvate reductase genes was at least in part due to a hormonal influence on transcriptional control [169,170]. The transcription rate for the phosphoenol pyruvate-carboxylase (C4Ppc) gene in maize leaves was stimulated by cytokinin more than sixfold as determined by nuclear runoff analysis [171]. Cytokinin-responsive promoters have provided further evidence of cytikinin-regulated transcription of plant genes [172]. However, it has become evident that posttranscriptional mechanisms are also involved in cytokinin regulation of plant gene expression [173]. It was shown that the cytokinin-induced accumulation of mRNAs encoding LHCII apoprotein and the SSU of Rubisco occurs primarily by a posttranscriptional mechanism in Lemna gibba [149,152149]. Differences in the steady-state mRNA levels that accumulated in the dark or light, with or without cytokinin treatment were compared with transcriptional activity as determined by nuclear runoff experiments, which point to a posttranscriptional regulation. This can affect, for example, the ripening process of mRNA, mRNA transport, increased transcript stability, increased level of polyribosome formation increasing the rate of initiation through an enhancement of the affinity of ribosomes for the mRNAs in the cytoplasm [157,166,174], while phytochrome regulates the abundance of these mRNAs at the level of transcription [175,176]. Hormonal influence on the posttranscriptional or posttranslational control of protein synthesis may be mediated by polyribosome formation [177], modification of the secondary structure of poly(A^+)RNA [178], phosphorylation of the ribosomal proteins [179], or regulation of the activity of tRNAs [180].

Chloroplast genes are also affected by cytokinins. Two- to threefold greater steady-state mRNA levels were found for the plastid-encoded LSU of Rubisco gene (rbcl) after cytokinin treatment of etiolated cotyledons of Cucurbita pepo [146,147]. In contrast, in Lemna gibba [149] and cucumber cotyledons [181] there was no noticeable cytokinin influence on the rbcl mRNA level. Kusnetsov et al. [182] conducted a large study on the abundance of 15 chloroplast-encoded genes and their corresponding proteins, which were related to all major thylakoid membrane complexes, in etiolated lupin cotyledons. Transcript levels changed only slightly in response to cytokinins while the corresponding polypeptides accumulated to high levels. This indicates a primary action of cytokinins at the level of mRNA translation in chloroplasts. These conclusions are based on the comparative analysis of rates of mRNA synthesis (nuclear run-on

transcription), and mRNA accumulation (northern analysis) or direct measurements of mRNA half-lives in the presence of potent inhibitors of transcription (cordycepin or actinomycin D) [183,184].

Some data indicate that the mode of action of cytokinins is not always quite the same. Conflicting results arise due to different experimental conditions and plant material. For example, it is well known that light is required for the transcription of the cab (Lhcb) genes, but low levels of LHC-mRNAs have often been detected in dark-grown plants, while the corresponding proteins are absent. Detached watermelon cotyledons seem to be a peculiar system in this regard, since both the protein and its mRNA can be detected in the dark, if exogenous cytokinin is present [150].

Hormone-induced changes in gene expression lead to the increase of some components that are constituents of thylakoid membranes and metabolic machinery of chloroplast, thus it is not surprising to observe changes in the composition and function of the photosynthetic apparatus and related biochemical processes. At the early stages of greening, cytokinins stimulate measurable photosynthetic activities (i.e., oxygen evolution, CO_2 fixation) [161,165], and partial reactions of photosynthetic electron transport [162,185,186].

Stimulating effects of cytokinins on the synthesis of transcripts, corresponding polypeptides and chlorophylls manifest at the level of the composition and function of thylakoid membranes during chloroplast development. It was shown that cytokinin treatment increase the Chl content and decrease the Chl a/b ratio at the early stages of greening and promoted the formation of complete, LHCI-containing PSI particles and the amount of LHCII in developing thylakoids. A larger antenna size for PSI and PSII and a facilitated synthesis of enzymes (Rubisco subunits) taking part in the primary action of the Calvin cycle can explain the increase in photosynthetic activity. Cytokinin treatment promotes the synthesis and assembly of some important components of thylakoid membranes (Chls, apoproteins of LHCs), allowing a more developed structure (i.e., more complete LHC system) and higher efficiency of photosynthetic function at different light regimes [165].

The abovementioned effects of cytokinins show great similarity to the effect of light on the development of chloroplasts indicating a more or less synergistic cooperation between them.

It is well known that light acts via different kinds of photoreceptors, most dominantly phytochromes, so the mode of coaction of cytokinins and phytochromes is controlling plant morphogenesis including chloroplast development and maintenance. It is generally presumed that there are multiple inter-actions or regulatory networks between phyto-chromes and cytokinin action, but the nature of these interactions is not well established. It can be assumed that signal transduction chains are working independently, but common use is also strongly sug-gested (additive and multiplicative responses), or cytokinins may be the elements of the signal trans-duction pathway mediated by light.

A model for the interaction of light signal and cytokinins has been interpreted by Chory and cow-orkers [139,187] based mainly on the study of *Arabi-dopsis thaliana* de-etiolated (det) mutants. Whether light and hormones act independently to influence developmental responses or whether plant hormones are involved in the sequence of events initiated by physiologically active photoreceptors is still un-known.

The activity of any signal molecule must be medi-ated by the interaction of that signal with some per-ceiving factors, which then affect response elements leading to molecular, biochemical, or physiological events. In such a system, molecule signals as cytoki-nins bind to a receptor that may, for example, affect systems controlling ion channels, phosphorylation of regulatory proteins modifying expression of genes then manifest at physiological level.

Despite the wealth of information concerning cytokinin chemistry and physiology, the transition from descriptive studies to molecular biology, includ-ing perception and signal transduction pathways, has been relatively slow compared with other phytohor-mones, particularly ethylene [188,189], and gibberel-lins [190].

In the last decade, genetic and molecular analysis of mutant plants, especially huge pieces of new infor-mation arising from studies of *Arabidopsis* and the completion of its genome sequence have provided valuable insights into the molecular mechanisms underlying the action of cytokinins. Now several re-cent reports have implicated two well-characterized types of signaling pathways in cytokinin action: a G protein-coupled receptor pathway, and a two-component histidine kinase pathway.

Recently, evidence has emerged that cytokinin action may involve a G protein-coupled receptor (GPCR). A gene was identified in *Arabidopsis* that encodes a protein with similarity to GPCRs [191,192]. It might be a component in cytokinin sig-naling, or a receptor for a different ligand, whose signaling pathway interacts with cytokinin signaling [193]. Furthermore, it is not excluded that G protein-coupled signaling pathway takes part in the transduc-tion of other plant hormones like auxin [194] and gibberellins [195] (Figure 20.1).

Rapid progress has recently been made in the characterization of a cytokinin signaling pathway that is similar to two-component systems, which are known as a prevalent procaryotic signaling pathway [199]. Typically, signaling involves two partners, the sensor kinase and the response regulator, and pro-ceeds through and alternating His-Asp phosphoryl-ation. The sensor histidine (His) kinase consists of an N-terminal input and a C-terminal transmitter do-main. Detection of the signal by the input domain controls the activity of the transmitter domain, which is an invariant histidine kinase. Sensor kinases, which associate into dimers, transphosphorylate onto a con-served His residue located in the transmitter domain. The phosphoryl group is then transferred onto the Asp residue within the receiver domain of a cognate response regulator. The phosphorylation state of the receiver domain regulates the more complex version of the two-component system includes multistep phosphorelay circuits, an additional signaling domain known as the histidine-containing phosphotransfer (HPt) domain [196].

At first CKI1, a hybrid histidine (His) protein kinase with a conserved receiver domain, has been implicated in cytokinin responses [200]. Another *Ara-bidopsis* hybrid His protein kinase, CRE1, has been shown to be a cytokinin receptor [201,202]. It binds cytokinin; cytokinin binding induces the His kinase activity of CRE1, and CRE1 can transfer the phos-phoryl group to the intrinsic His phosphotransfer protein (HPt) domain protein initiating multistep, two-component signaling cascade [197]. CRE1 is probably not responsible for mediating all of the plant cytokinin responses [203,204]. Two additional CRE1-like His kinases and further 14 His and hybrid kinase-like proteins are identified in *Arabidopsis* (AHKs, *Arabidopsis* histidine kinase). AHK2 and AHK3, presumably also function as cytokinin recep-tor [197,202,205] (Figure 20.1).

According to a phylogenetic analysis of their amino acid sequence, domain composition and tran-scriptional regulation two types of *Arabidopsis* re-sponse regulators (ARRs) can be classified: the cytokinin inducible A-type and the DNA-binding B-type [206,207]. Recent evidence suggests that B-type response regulators ARR1, ARR2, and ARR10-14 are involved in cytokinin signaling, and they have a long C-terminal extension that mediate sequence-specific DNA binding and transcriptional activity. Type-B ARRs act as positive regulators of cytokinin responsiveness, including the induction of type-A ARR gene expression. Because AHKs are present on an extranuclear membrane, the *Arabidop-sis* HPts (AHPs) domain proteins could perform the information transfer to B-type ARRs.

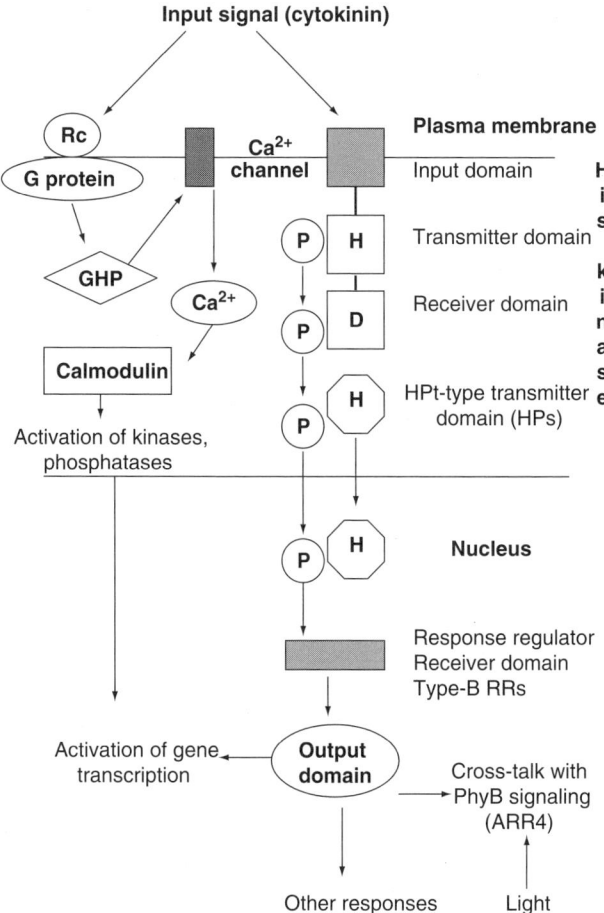

FIGURE 20.1 Schematic model of the multistep two-component and G protein-coupled signaling systems in cytokinin signal transduction. On perception of the cytokinin signal, cytokinin binds to the input domain of a hybrid hystidine kinase (CRE1, AHK1, AHK2). The binding of cytokinin activates the transmitter domain, which autophosphorylates on a His (H) residue. The phosphate is then transferred to an Asp (D) residue on a fused receiver domain and then to a His residue on a HPt-type transmitter domain containing AHP protein, which translocates to the nucleus, where it activates type-B ARRs (phosphorelay cascade). The activated type-B ARR binds to their promoter sequence of target genes inducing the activation of their transcriptional processes [196–198,201,209]. This model is mainly based on findings gained in *Arabidopsis thaliana*. (H, conserved His residue; D, conserved Asp residue; HPt, His-containing phosphotransfer domain). Other signal transduction pathway includes a G protein-coupled receptor (Rc) perceiving cytokinin and/or other signals. G proteins could induce dihydropyridin binding to calcium channels causing calcium penetration into the cytosol. Calcium could result in the enhancement of kinase/phosphatase activities mediated by calmodulin proteins leading to different responses (DHP, dihydropyridins) [193].

Type-A ARR proteins may in turn modulate the output of cytokinin signaling via interaction with other members of the signaling chain, or alternatively, may by themselves regulate other downstream targets. One of the type-A ARR, ARR4 can specifically interact with the plant photoreceptor phytochrome-B [55]. A novel type of cross-talk is suggested between an ancestral two-component system and the phytochrome-B-dependent light-signaling pathway. This cross-talk may enable the integration of signals provided by hormone (cytokinin) and red light (phytochrome-B) signal transduction cascade.

Five genes encoding His-containing phosphotransfer proteins (AHP) have been identified in the *Arabidopsis* genome [208]. AHPs are predicted to mediate the transfer of the phosphoryl group from the receiver domain of an activated hybrid sensor His kinase to the receiver domain of a response regulator in a multistep phosphorelay signal transduction pathway.

In conclusion, the multiple elements of the plant two-component signaling system may offer the diversity required to integrate a simple hormone signal into the diverse developmental processes that are regulated by cytokinins. Differences in signal interpretation and output can be reached by combining different parts of the signaling chains in a cell-specific fashion. This system may also cross-talk with other signaling pathway such as those of known coactors of cytokinin, other phytohormones like ethylene and light.

B.   OTHER PHYTOHORMONES

All of other phytohormones including auxins, gibberellins, abscisic acid, ethylene, jasmonates, brassinosteroids influence and control many aspects of plant development, and show synergistic or antagonistic effect on light, but their role in the formation and maintenance of photosynthetic processes is more or less indirect. However, some direct effects of phytohormones (abscisic acid, jasmonates) are known.

Several lines of evidence have implicated the action of brassinosteroid hormones in light-regulated gene expression [209,210], and data suggest that light may affect the levels of gibberellins as a possible interaction between phytochromes and hormones [211], but the relationship between light and hormonal signal transduction pathways is not understood.

Apart from some other effects of jasmonates, they were shown to have a direct effect on the photosynthetic apparatus. They repress genes encoding protein involved in photosynthetic processes, and reduce the synthesis of LSU and SSU subunits of Rubisco, and suppress translation of LSU mRNA level [212], and also cause a loss of Chl from leaves [213].

There have been a number of reports that abscisic acid (ABA) participates in a wide range of physiological processes including opening and closing of stomata, response to environmental stresses, wound responses, regulation of the expression of ABA-responsive genes, and control of plastid biogenesis [214]. So ABA has some direct effects on photosynthetic processes. This phytohormone inhibits the expression of genes that are known to be positively regulated by phytochromes. It appears to modulate POR gene expression and decreases the steady-state level of POR mRNA [215]. It was shown that ABA inhibits the appearance of rbcS mRNA in germinating wheat seedlings [216], and ABA could repress transcription of both rbcS and Lhcb genes in tomato and soybean [217]. It is suggested that light and ABA effects are antagonistic and it was shown that DNA sequence elements for the two signals are separable in the Lhcb gene [218], while ABA and light have a synergistic effect on phosphoenol pyruvate carboxylase expression [219], but antagonistic effect on chlorophyll *a/b*-binding protein gene transcription [218]. These findings have shown that ABA acts in the physiological integration of light signals in the control of plant development.

VI.   CONCLUDING REMARKS

During recent years, a molecular genetic approach has been used to gain more insight into the relationships between light and phytohormones. Most of the mutants, especially *Arabidopsis* mutants considered are modified in their responses to light and hormones in perception or transduction, but it is not quite clear how light and hormonal regulatory pathways are organized into common network or act independently, and how they exert influence on each other. However, it has been proven that light interferes with hormone signaling by regulating the expression and activity of the elements involved in hormone biosynthesis, and there is growing evidence that other signaling cascades such as those induced by phytohormones modulate phytochrome signaling at various levels.

Further search for new mutants and novel interaction components in combination with molecular genetic and other highly specific techniques would be useful in unraveling the secrets of the light and hormonal regulatory mechanisms and their relationship.

REFERENCES

1. Kirk JTO, Tilney-Basset RAE. *The Plastids: Their Chemistry, Structure, Growth and Inheritance.* Amsterdam/New York: Elsevier/North Holland Biomedical Press, 1978.

2. Rascio N, Mariani P, Casadoro G. Etioplast–chloroplast transformation in maize leaves: effects of tissue age and light intensity. *Protoplasma* 1984; 119:110–120.

3. Sundqvist C, Björn LO, Virgin HI. Factors in chloroplast differentiation. In: Reinert J, ed. *Results and Problems in Cell Differentiation.* Vol. 10. Berlin: Springer-Verlag, 1980:201–224.

4. Griffith WT. Reconstitution of chlorophyllide formation by isolated etioplast membranes. *Biochem. J.* 1978; 174:681–692.

5. Schultz R, Senger H. Protochlorophyllide reductase: A key enzyme in the greening process. In: Sundqvist C, Ryberg M, eds. *Pigment–Protein Complexes in Plastids. Synthesis and Assembly.* San Diego, CA: Academic Press, 1993: 179–218.

6. Furuya M. Phytochromes: their molecular species, gene families, and functions. *Annu. Rev. Plant. Physiol. Plant Mol. Biol.* 1993; 44:617–645.

7. Boardman NK. Thylakoid membrane formation in the higher plant chloroplast, In: Akoyunoglou G, ed. *Chloroplast Development.* Vol. V. Philadelphia, PA: Balaban International Science Services, 1981: 325–339.

8. Herrmann RG, Westhoff P, Alt J, Tittgen J, Nelson N. Thylakoid membrane proteins and their genes. In: van Vloten-Doting L, Groot GSP, Hall TC, eds. *Molecular Form and Function of the Plant Genome.* Amsterdam: Plenum, 1985:233–256.

9. Takabe T, Takabe T, Akazawa T. Biosynthesis of P700-chlorophyll *a* protein complex, plastocyanin, and cytochrome b_6/f complex. *Plant Physiol.* 1986; 81:60–66.

10. Armstrong GA. Greening in the dark: light-independent chlorophyll biosynthesis from anoxygenic photosynthetic bacteria to gymnosperms. *J. Photochem. Photobiol. B: Biol.* 1998; 43:87–100.

11. Fujita Y, Takagi H, Hase T. Cloning of the gene encoding a protochlorophyllide reductase: the physiological significance of the co-existence of light-dependent and -independent protochlorophyllide reduction systems in the cyanobacterium *Plectonema boryanum. Plant Cell Physiol.* 1998; 39:177–185.

12. Fujita Y. Protochlorophyllide reduction: a key step in the greening of plants. *Plant Cell Physiol.* 1996; 37:411–421.

13. Apel K, Motzkus M, Dehesh K. The biosynthesis of chlorophyll in greening barley (*Hordeum vulgare*). Is there a light-independent protochlorophyllide reductase? *Planta* 1984; 161:550–554.

14. Griffith WT, Kay AS, Oliver RP. The presence and photoregulation of protochlorophyllide reductase in green tissues. *Plant Mol. Biol.* 1985; 4:13–22.

15. Apel K, Santel HJ, Falk H. The protochlorophyllide holochrome of barley (*Hordeum vulgare L.*). Isolation and characterization of the NADPH-protochlorophyllide oxidoreductase. *Eur. J. Biochem.* 1980; 111:251–258.

16. Lebedev N, Timko MP. Protochlorophyllide photoreduction. *Photosynth. Res.* 1998; 58:5–23.

17. Reinbothe S, Reinbothe C, Lebedev N, Apel K. POR A and POR B, two light-dependent protochlorophyllide-reducing enzymes of angiosperm chlorophyll biosynthesis. *Plant Cell* 1996; 8:763–769.

18. Barber J. Composition, organization, and dynamics of the thylakoid membrane in relation to its function. In: Hatch MD, Boardman NK, eds. *The Biochemistry of Plants*. Vol. 10. New York: Academic Press, 1987:75–130.

19. Anderson JM, Chow WS, Park Y-L. The grand design of photosynthesis: acclimation of the photosynthetic apparatus to environmental cues. *Photosynth. Res.* 1995; 46:129–139.

20. Anderson JM. Insights into the consequences of grana stacking of thylakoid membranes in vascular plants: a personal perspective. *Aust. J. Plant Physiol.* 1999; 26:625–639.

21. Hiller RG, Pilger TBG, Genge S. Formation of chlorophyll protein complexes during greening of etiolated barley leaves. In: Akoyunoglou G, Argyroudi-Akoyunoglou JH, eds. *Chloroplast Development*. Vol. II. New York: Elsevier/North Holland Biomedical Press, 1978:215–220.

22. Tanaka A, Tsuji H. Changes in chlorophyll *a* and *b* content in dark-incubated cotyledons excised from illuminated seedlings. *Biochim. Biophys. Acta* 1982; 680:265–270.

23. Alberte RS, Thornber JS, Naylor AW. Time of appearance PSI and PSII in chloroplast of greening jack bean leaves. *J. Exp. Bot.* 1972; 23:1060–1069.

24. Argyroudi-Akoyunoglou JH, Akoyunoglou A, Kalosakas K, Akoyunoglou G. Reorganization of photosystem II unit in developing thylakoids of higher plants after transfer to darkness. Changes in chlorophyll *b*, light-harvesting chlorophyll protein content, and grana stacking. *Plant Physiol.* 1982; 70:1242–1248.

25. Tanaka A, Tsuji H. Appearances of chlorophyll-protein complexes in greening barley seedlings. *Plant Cell Physiol.* 1985; 26:893–902.

26. Burkey KO. Chlorophyll-protein composition and photochemical activity in developing chloroplasts from greening barley seedlings. *Photosynth. Res.* 1986; 10:37–49.

27. Mathis JN, Burkey KO. Regulation of light-harvesting chlorophyll protein biosynthesis in greening seedlings, a species comparison. *Plant Physiol.* 1987; 81:971–977.

28. Wellburn AR, Hampp R. Appearance of photochemical function in prothylakoids during plastid development. *Biochim. Biophys. Acta* 1979; 547:380–397.

29. Akoyunoglou G. Biosynthesis of the pigment–protein complexes. In: Sironval C, Brouers M, eds. *Protochlorophyllide Reduction and Greening*. The Hague, Boston, Lancaster: Martinus Nijhoff/Dr. W. Junk Publishers 1984:243–254.

30. Akoyunoglou G, Argyroudi-Akoyunoglou JH. Organization of the photosynthetic units, and onset of electron transport and excitation energy distribution in greening leaves. *Photosynth. Res.* 1986; 10:171–180.

31. Wollman F-A, Minai L, Nechustai R. The biogenesis and assembly of photosynthetic proteins in thylakoid membranes. *Biochim. Biophys. Acta* 1999; 1411:21–85.

32. Ohashi K, Tanaka A, Tsuji H. Formation of the photosynthetic electron transport system during the early phase of greening in barley leaves. *Plant Physiol.* 1989; 91:409–414.

33. Tanaka A, Yamamoto Z, Tsuji H. Formation of chlorophyll-protein complex during greening. 2. Redistribution of chlorophyll among apoproteins. *Plant Cell Physiol.* 1991; 32:195–204.

34. Wrischer M. Ultrastructural localization of photosynthetic activity in thylakoids during chloroplast development in maize. *Planta* 1989; 177:18–23.

35. Franck F. Photosynthetic activities during early assembly of thylakoid membranes. In: Sundqvist C, Ryberg M, eds. *Pigment–Protein Complexes in Plastids: Synthesis and Assembly*. New York: Academic Press, 1993:365–381.

36. Franck F. Development of PSII photochemistry after a single white flash in etiolated barley leaves. In: Baltscheffsky M, ed. *Current Research in Photosynthesis*. Vol. III. Dordrecht, The Netherlands: Kluwer Academic 1990: 751–754.

37. Bertrand M, Bereza P, Dujardin E. Evidence for photoreduction of NADP$^+$ in a suspension of lysed plastids from etiolated bean leaves. *Z. Naturforsch.* 1988; 43c:443–448.

38. Dreyfuss BW, Thornber JP. Organization of the light-harvesting complex of photosystem I and its assembly during plastid development. *Plant Physiol.* 1994; 106:841–848.

39. Kendrick RE, Kronenberg GMH, eds. *Photomorphogenesis in Plants*. 2nd ed. Dordrecht: Kluwer Academic, 1994.

40. Quail PH, Boylan MT, Parks BM, Short TM, Xu Y, Wagner D. Phytochromes: photosensory perception and signal transduction. *Science* 1995; 268:675–680.

41. Frankhauser C, Chory J. Light control of plant development. *Annu. Rev. Cell Dev. Biol.* 1997; 13:203–229.

42. Malakhov M, Bowler C. Phytochrome and regulation of photosynthetic gene expression. In: Aro E-M, Andersson B, eds. *Regulation of Photosynthesis*. Dordrecht, The Netherlands: Kluwer Academic, 2001:51–56.

43. Rüdiger W, Schoch S. Chlorophylls. In: Goodwin TW, ed. *Plant Pigments*. London: Academic Press, 1988:1–25.

44. Yamaguchi R, Nakamura M, Mochizuki N, Kay SA, Nagatani A. Light-dependent translocation of a phytochrome B-GFP fusion protein to the nucleus in transgenic *Arabidopsis*. *J Cell Biol.* 1999; 145:437–445.

45. Kraml M. Light direction and polarization. In: Kendrick RE, Kronenberg GHM. eds. *Photomorphogenesis in Plants*. Dordrecht: Kluwer Academic, 1994:417–443.

46. Nagy F, Schäfer E. Phytochromes control photomorphogenesis by differentially regulated, interacting signaling pathways in higher plants. *Annu. Rev. Plant Biol.* 2002; 53:329–355.

47. Deng XW, Quail PH. Signalling in light-controlled development. *Semin. Cell Dev. Biol.* 1999; 10:121–129.

48. Kuno N, Muramatsu T, Hamazato F, Furuya M. Identification by large-scale screening of phytochrome-regulated genes in etiolated seedlings of *Arabidopsis thaliana* using a fluorescent differential display technique. *Plant Physiol.* 2000; 122:15–22.

49. Smith H. Phytochromes and light signal perception by plants — an emerging synthesis. *Nature* 2000; 407:585–591.

50. Martinez-Garcia JF, Huq E, Quail PH. Direct targeting of light signals to a promoter element-bound transcription factor. *Science* 2000; 288:859–863.

51. Yeh KC, Lagarias JC. Eucaryotic phytochromes: light-regulated serine/threonine protein kinases with histidine kinase ancestry. *Proc. Natl. Acad. Sci. USA* 1998; 95:13976–13981.

52. Frankhauser C, Yeh KC, Lagarias JC, Zhang H, Elich TD, Chory J. PKS1, a substrate phosphorylated by phytochrome that modulates light signaling in *Arabidopsis. Science* 1999; 284:1539–1541.

53. Neuhaus G, Bowler C, Kern R, Chua N-H. Calcium/calmodulin-dependent and -independent phytochrome signal transduction pathway. *Cell* 1993; 73:937–952.

54. Neuhaus G, Bowler C, Hiratsuka K, Yamagata H, Chua N-H. Phytochrome-regulated repression of gene expression requires calcium and cGMP. *EMBO J.* 1997; 16:2554–2564.

55. Sweere U, Eichenberg K, Lohrmann J, Mira-Rodado V, Bäurle I, Kudla J, Nagy F, Schäfer E, Harter K. Interaction of the response regulator ARR4 with phytochrome B in modulating red-light signaling. *Science* 2001; 294:1108–1111.

56. Brandstatter I, Kieber JJ. Two genes with similarity to bacterial response regulators are rapidly and specifically induced by cytokinin in *Arabidopsis. Plant Cell* 1998; 10:1009–1019.

57. Ahmad M, Cashmore AR. Seeing blue: the discovery of cryptochrome. *Plant Mol. Biol.* 1996; 30:851–861.

58. Mullet JE. Chloroplast development and gene expression. *Annu. Rev. Plant Physiol. Plant Mol. Biol.* 1988; 39:475–502.

59. Sugiura M. The chloroplast genome. *Plant Mol. Biol.* 1992;19:149–168.

60. Taylor WC. Regulatory interaction between nuclear and plastid genomes. *Annu. Rev. Plant Physiol. Plant Mol. Biol.* 1989; 40:211–233.

61. Oelmüller R, Levitan I, Bergfeld R, Rajasekhar VK, Mohr M. Expression of nuclear genes as affected by treatments acting on the plastids. *Planta* 1986; 168:482–492.

62. Choquet Z, Goldschmidt-Clermont M, Girard-Bascon J, Kuck U, Bennoun P, Rochaix J-D. Mutant phenotypes support a trans-splicing mechanism for the expression of the tripartite psaA gene in *C. reinhardtii* chloroplast. *Cell* 1988; 52:903–913.

63. Rochaix J-D. Post-transcriptional regulation of chloroplast gene expression in *Chlamydomonas reinhardtii. Plant Mol. Biol.* 1996; 327–341.

64. Barkan A, Miles D, Taylor WC. Chloroplast gene expression in nuclear photosynthetic mutants of maize. *EMBO J.* 1986; 5:1421–1427.

65. Jensen KH, Herrin DL, Plumley G, Schmidt GW. Biogenesis of photosystem II complexes: transcriptional, translational, and posttranslational regulation. *J. Cell Biol.* 1986; 103:1315–1325.

66. Kuchka MR, Mayfield SP, Rochaix J-D. Nuclear mutations specifically affect the synthesis and/or degradation of the chloroplast-encoded D2 polypeptide of photosystem II in *Chlamydomonas reinhardtii. EMBO J.* 1988; 7:319–324.

67. Oelmüller R. Photooxidative destruction of chloroplast and its effect on nuclear gene expression and extraplastidic enzyme levels. *Photochem. Photobiol.* 1989; 49:229–239.

68. Goldschmidt-Clermont M. Coordination of nuclear and chloroplast gene expression in plant cells. *Int. Rev. Cytol.* 1998; 117:115–180.

69. Piechulla B, Pichersky E, Cashmore AR, Gruissem W. Expression of nuclear and plastid genes for photosynthesis-specific proteins during fruit development and ripening. *Plant Mol. Biol.*, 1986; 7:367–376.

70. Baumgartner BJ, Rapp JC, Mullet JE. Plastid genes encoding the transcription/translation apparatus are differentially transcribed early in barley (*Hordeum vulgare*) chloroplast development. *Plant Physiol.* 1993; 101:781–791.

71. Mullet JE, Klein RR. Transcription and RNA stability are important determinants of higher plant chloroplast RNA levels. *EMBO J.* 1987; 6:1571–1579.

72. Kobayashi H, Ngernprasirtsiri J, Akazawa. Transcriptional regulation and DNA methylation in plastids during transitional conversion of chloroplasts to chromoplasts. *EMBO J.* 1990; 9:307–313.

73. Schrubar H, Wanner G, Westhoff P. Transcriptional control of plastid gene expression in greening *Sorghum* seedlings. *Planta* 1990; 183:101–111.

74. Klein RR, Gamble PE, Mullet JE. Light-dependent accumulation of radiolabeled plastid-encoded chlorophyll *a*-apoproteins requires chlorophyll *a. Plant Physiol.* 1988; 88:1246–1256.

75. Gamble PE, Klein RR, Mullet JE. Illumination of eight-day-old dark-grown barley seedlings activates chloroplast protein synthesis; evidence for regulation of translation initiation. In: Briggs WR, ed. *Photosynthesis.* New York: Alan R. Liss, 1989: 285–298.

76. Klein RR, Mullet JE. Regulation of chloroplast-encoded chlorophyll-binding protein translation during higher plant chloroplast biogenesis. *J. Biol. Chem.* 1986; 261:11138–11145.

77. Mullet JE, Klein PG, Klein RR. Chlorophyll regulates accumulation of the plastid-encoded chlorophyll apoproteins CP43 and D1 by increasing apoprotein stability. *Proc. Natl. Acad. Sci. USA* 1990; 87:4038–4042.

78. Stern DB, Higgs DC, Yang J. Transcription and translation in chloroplasts. *Trends Plant Sci.* 1997; 2:308–315.

79. Gillham NW, Boynton JE, Hauser CR. Translational regulation of gene expression in chloroplast and mitochondria. *Annu. Rev. Genet.* 1994; 28:71–93.

80. Stern DB, Jones H, Gruissem W. Function of plastid mRNA 3′ inverted repeats: RNA stabilization and gene-specific protein binding. *J. Biol. Chem.* 1989; 264:18742–18750.

81. Deng X-W, Gruissem W. Control of plastid gene expression during development: the limited role of transcriptional regulation. *Cell* 1987; 49:379–387.

82. Deng X-W, Tonkyn JC, Peter GF, Thornber JP, Gruissem W. Post-transcriptional control of plastid mRNA accumulation during adaptation of chloroplast to different light quality environments. *Plant Cell* 1989; 1:645–654.

83. Herrmann RG, Westhoff P, Link G. Biogenesis of plastids in higher plants. In: Herrmann RG, ed. *Plant Gene Research: Cell Organelles.* Vol. 6. Wien, New York: Springer-Verlag, 1992: 275–349.

84. Kreuz K, Dehesh K, Apel K. The light-dependent accumulation of the P700 chlorophyll a protein of the photosystem I reaction center in barley. *Eur. J. Biochem.* 1986; 159:459–467.

85. Eichacker L, Paulsen H, Rüdiger W. Synthesis of chlorophyll *a* regulates translation of chlorophyll *a* apoproteins P700, CP47, CP43, and D2 in barley etioplasts. *Eur. J. Biochem.* 1992; 205:17–24.

86. Berry JD, Carr JP, Klessig DF. mRNAs encoding ribulose-1,5-bisphosphate carboxylase remain bound to polysome but are not translated in amaranth seedlings transferred to darkness. *Proc. Natl. Acad. Sci. USA* 1988; 85:4190–4194.

87. Thomson WF, White MJ. Physiological and molecular studies of light-regulated nuclear genes in higher plants. *Annu. Rev. Plant Physiol. Plant Mol. Biol.* 1991; 42:423–466.

88. Schnell DJ. Protein targeting to the thylakoid membrane. *Annu. Rev. Plant Physiol. Plant Mol. Biol.* 1998; 49:97–126.

89. Tobin EM, Silverthorne J. Light regulation of gene expression in higher plants. *Annu. Rev. Plant Physiol.* 1985; 36:569–593.

90. Keegstra K, Olsen LJ. Chloroplastic precursors and their transport across the envelope membranes. *Annu. Rev. Plant Physiol. Plant Mol. Biol.* 1989; 40:471–501.

91. Dean C, Pichersky E, Dunsmuir P. Structure, evolution, and regulation of RbcS genes in higher plants. *Annu. Rev. Plant Physiol. Plant Mol. Biol.* 1989; 40:415–439.

92. Manzara T, Gruissem W. Organization and expression of the genes encoding ribulose-1,5-bisphosphate carboxylase in higher plants. *Photosynth. Res.* 1988; 16:117–139.

93. Batschauer A, Mösinger E, Kreuz K, Dorr I, Apel K. The implication of a plastid-derived factor in the transcriptional control of nuclear genes encoding the light-harvesting chlorophyll *a/b* protein. *Eur. J. Biochem.* 1986; 154:625–634.

94. Viro M, Kloppstech K. Expression of genes for plastid membrane proteins in barley under intermittent light condition. *Planta* 1982; 154:18–23.

95. Akoyunoglou G, Argyroudi-Akoyunoglou JH. Post-translational regulation of chloroplast differentiation. In: Akoyunoglou G, Senger H, eds. *Regulation of Chloroplast Differentiation.* New York: Alan R. Liss, 1986:571–582.

96. Mathis JN, Burkey KO. Light intensity regulates the accumulation of the major light-harvesting chlorophyll-protein in greening seedlings. *Plant Physiol.* 1989; 90:560–566.

97. Bennett J, Schwender JR, Shaw EK, Tempel N, Ledbetter M, Williams KS. Failure of corn leaves to acclimate to low irradiance. Role of protochlorophyllide reductase in regulating levels of five chlorophyll-binding proteins. *Biochim. Biophys. Acta* 1987; 892:118–129.

98. Nyitrai P, Sárvári É, Keresztes Á, Láng F. Organization of thylakoid membranes in low-light grown maize seedlings. Effect of lincomycin treatment. *J. Plant Physiol.* 1994; 144:370–375.

99. Marquardt J, Bassi R. Chlorophyll-proteins from maize seedlings grown under intermittent light conditions. *Planta* 1993; 191:265–273.

100. Slovin JP, Tobin EM. Synthesis and turnover of the light-harvesting chlorophyll *a/b*-protein in *Lemna gibba* grown with intermittent red light: possible translational control. *Planta* 1982; 154:465–472.

101. White MJ, Green BR. Polypeptides belonging to each of the three major chlorophyll *a+b* protein complexes are present in a chlorophyll-*b*-less barley mutant. *Eur. J. Biochem.* 1987; 165:531–535.

102. White MJ, Green BR. Intermittent-light chloroplasts are not developmentally equivalent to chlorina f2 chloroplasts in barley. *Photosynth. Res.* 1988; 15: 195–203.

103. Sárvári É, Nyitrai P, Keresztes Á. Relative accumulation of LHCP II in mesophyll plastids of intermittently illuminated maize seedlings under lincomycin treatment. *Biochem. Physiol. Pflanzen* 1989; 184: 37–47.

104. Tzinas G, Argyroudi-Akoyunoglou JH, Akoyunoglou G. The effect of dark interval in intermittent light on thylakoid development: photosynthetic unit formation and light-harvesting protein accumulation. *Photosynth. Res.* 1987; 14:241–258.

105. Mogen K, Eide J, Duysens M, Eskins K. Chloramphenicol stimulates the accumulation of light-harvesting chlorophyll *a/b* protein II by affecting posttranscriptional events in the chlorina CD3 mutant of wheat. *Plant Physiol.* 1990; 92:1233–1240.

106. Gallie DR. Posttranscriptional regulation of gene expression in plants. *Annu. Rev. Plant Physiol. Plant Mol. Biol.* 1993; 44:77–105.

107. Sachs AB. Messenger RNA degradation in eucaryotes. *Cell* 1993; 74:413–421.

108. Vierstra RD. Protein degradation in plants. *Annu. Rev. Plant Physiol. Plant Mol. Biol.* 1993; 44:385–410.

109. Lissemore JL, Quial PH. Rapid transcriptional regulation by phytochrome of the genes for phytochrome and chlorophyll *a/b*-binding protein in *Avena sativa*. *Mol. Cell. Biol.* 1988; 8:4840–4850.

110. Mösinger E, Batschauer A, Vierstra R, Apel K, Schäfer E. Comparison of the effects of exogenous native phytochrome and *in vivo* irradiation on *in vitro* transcription in isolated nuclei from barley (*Hordeum vulgare*). *Planta* 1987; 170:505–514.

111. Glick RE, McCauley SW, Melis A. Effect of light quality on chloroplast membrane organization and function in pea. *Planta* 1985; 164:487–494.

112. Murakami A, Fujita Y. Steady state of photosynthesis in cyanobacterial photosynthetic systems before and after regulation of electron transport composition: overall rate of photosynthesis and PSI/PSII composition. *Plant Cell Physiol.* 1988; 29:305–311.

113. Melis A. Dynamics of photosynthetic membrane composition and function. *Biochim. Biophys. Acta* 1991; 1058:87–106.

114. Manodori A, Melis A. Cyanobacterial acclimation to photosystem I or II light. *Plant Physiol.* 1986; 82:185–189.

115. Melis A, Murakami A, Nemson JA, Aizawa K, Ohki K, Fujita Y. Chromatic regulation in *Chlamydomonas reinhardtii* alters photosystem stoichiometry and improves the quantum efficiency of photosynthesis. *Photosynth. Res.* 1996; 47:253–265.

116. Kim MY, Christopher DA, Mullet JE. Direct evidence for selective modulation of psbA, rpoA, rbcL and 16S RNA stability during barley chloroplast development. *Plant Mol. Biol.* 1993; 22:447–463.

117. Walters RG, Rogers JJM, Shepard F, Horton P. Acclimation of *Arabidopsis thaliana* to the light environment: the role of photoreceptors. *Planta* 1999; 209:517–527.

118. Aphalo PJ, Ballaré CL, Scopel AL. Plant–plant signaling, the shade-avoidance response and competition. *J. Exp. Bot.* 1999; 50:1629–1634.

119. Ballaré CL. Keeping up with the neighbours: phytochrome sensing and other signaling mechanisms. *Trends Plant Sci.* 1999; 4:97–102.

120. Demmig-Adams B, Adams WW. The role of xantophyll cycle carotenoids in the protection of photosynthesis. *Trends Plant Sci.* 1996; 1:21–26.

121. Allen JF. Protein phosphorylation in regulation of photosynthesis. *Biochim. Biophys. Acta* 1992; 1098:275–335.

122. Allen JF. Thylakoid protein phosphorylation, state1–state2 transitions, and photosystem stoichiometry adjustment: redox control at multiple levels of gene expression. *Physiol. Plant.* 1995; 93:196–205.

123. Fujita Y, Murakami A, Aizawa K, Ohki K. Short-term and long-term adaptation of the photosynthetic apparatus: homeostatic properties of thylakoids. In: Bryant DA, ed. *The Molecular Biology of Cyanobacteria*. Dordrecht, The Nertherlands: Kluwer Academic, 1994:677–692.

124. Huner NPA, Öquist G, Sarhan F. Energy balance and acclimation to light and cold. *Trends Plant Sci.* 1998; 3:224–230.

125. Bailey S, Walters RG, Jansson S, Horton P. Acclimation of *Arabidopsis thaliana* to the light environment: the existence of separate low light and high light responses. *Planta* 2001; 213:794–801.

126. Chow WS, Melis A, Anderson JM. Adjustments of photosystem stoichiometry in chloroplast improve the quantum efficiency of photosynthesis. *Proc. Natl. Acad. Sci. USA* 1990; 87:7502–7506.

127. Anderson JM, Andersson B. The dynamic photosynthetic membrane and regulation of solar energy conversion. *Trends Biochem. Sci.* 1988; 13:351–355.

128. Stock JB, Ninfa AM, Stock AM. Protein phosphorylation and regulation of adaptive response in bacteria. *Microbiol. Rev.* 1989; 53:450–490.

129. Pfannschmidt T, Nilsson A, Allen JF. Photosynthetic control of chloroplast gene expression. *Nature* 1999; 397:625–628.

130. Danon A, Mayfield SP. Light-regulated translation of chloroplast messenger RNAs through redox potential. *Science* 1994; 266:1717–1719.

131. Maxwell DP, Laudenbach DE, Huner PA. Redox regulation of light-harvesting complex II and cab mRNA abundance in *Dunaliella salina*. *Plant Physiol.* 1995; 109:787–795.

132. Link G. Redox regulation of photosynthetic genes. In: Aro E-M, Andersson B, eds. *Regulation of Photosynthesis*. Dordrecht, The Netherlands: Kluwer Academic, 2001:85–107.

133. Petracek ME, Dickey LF, Nguyen TT, Gatz C, Sowinski DA, Allen GC, Thompson WF. Ferredoxin-1 mRNA is destabilized by changes in photosynthetic electron transport. *Proc. Natl. Acad. Sci. USA* 1998; 95:9009–9013.

134. Durnford DG, Falkowski PG. Chloroplast redox regulation of nuclear gene transcription during photoacclimation. *Photosynth. Res.* 1997; 53:229–241.

135. Durnfold DG, Price JA, McKim SM, Sarchfield ML. Light-harvesting complex gene expression is controlled by both transcriptional and post-transcriptional mechanisms during photoacclimation in *Chlamydomonas reinhardtii*. *Physiol. Plant.* 2003; 118:193–205.

136. Allen JF. Redox control of gene expression and the function of chloroplast genomes — an hypothesis. *Photosynth. Res.* 1993; 36:95–102.

137. Binns AN. Cytokinin accumulation and action: biochemical, genetic and molecular approaches. *Annu. Rev. Plant Physiol. Plant Mol. Biol.* 1994; 45:173–196.

138. Mok DWS, Mok MC. Cytokinin metabolism and action. *Annu. Rev. Plant Physiol. Plant Mol. Biol.* 2001; 52:89–118.

139. Chory J, Reinecke D, Sim S, Washburn T, Brenner M. A role for cytokinins in de-etiolation in *Arabidopsis*: det mutants have an altered response to cytokinins. *Plant Physiol.* 1994; 104:339–347.

140. Parthier B. Hormone-induced alterations in plant gene expression. *Biochem. Physiol. Pflanzen* 1989; 185:289–314.

141. Lew R, Tsuji H. Effect of benzyladenine treatment duration on δ-aminolevulinic acid accumulation in

the dark, chlorophyll lag phase abolition, and long-term chlorophyll production in excised cotyledons of dark-grown cucumber seedlings. *Plant Physiol.* 1982; 69:663–667.

142. Dei M. A two-fold action of benzyladenine on chlorophyll formation in etiolated cucumber cotyledons. *Physiol. Plant.* 1982; 56:407–413.

143. Cohen L, Arzee T, Zilberstein A. Mimicry by cytokinin of phytochrome-regulated inhibition of chloroplast development in etiolated cucumber cotyledons. *Physiol. Plant.* 1988; 72:57–64.

144. Dei M. Benzyladenine-induced stimulation of 5-aminolevulinic acid accumulation under various light intensities in levulinic acid-treated cotyledons of etiolated cucumber. *Physiol. Plant.* 1985; 64:153–160.

145. Daniell H, Rebeiz CA. Chloroplast culture VIII. A new effect of kinetin in enhancing the synthesis and accumulation of protochlorophyllide *in vitro*. *Biochem. Biophys. Res. Commun.* 1982; 104:837–843.

146. Lerbs S, Lerbs W, Klyachko NL, Romankoe G, Kulaeva ON, Wollgienh R, Parthier B. Gene expression in cytokinin- and light-mediated plastogenesis of *Cucurbita* cotyledons: ribulose-1,5-bisphosphate carboxylase/oxygenase. *Planta* 1984; 162:289–298.

147. Yusibov VM, Chun P, Andrianov VM, Piruzian ES. Phenotypically normal transgenic T-cyt tobacco plants as a model for the investigation of plant gene expression in response to phytohormonal stress. *Plant Mol. Biol.* 1991; 17:825–836.

148. Crowell DN, Amasino RM. Cytokinins and plant gene regulation. In: Mok DWS, Mok MC, eds. *Cytokinins: Chemistry, Activity, and Function*. Boca Raton, FL: CRC Press, 1994:243–254.

149. Flores S, Tobin E. Cytokinin modulation of LHCP mRNA levels: the involvement of post-transcriptional regulation. *Plant Mol. Biol.* 1988; 11:409–415.

150. Longo GP, Bracale M, Rossi G, Longo CP. Benzyladenine induces the appearance of LHCP-mRNA and of the relevant protein in dark-grown excised watermelon cotyledons. *Plant Mol. Biol.* 1990; 14:569–573.

151. Teyssendier de la Serve B, Axelos M, Peaud-Lenoel C. Cytokinins modulate the expression of genes encoding the protein of the light-harvesting chlorophyll *a/b* complex. *Plant Mol. Biol.* 1985; 5:155–163.

152. Flores S, Tobin E. Benzyladenine modulation of the expression of two genes for nuclear-encoded chloroplasts proteins in *Lemna gibba*: apparent post-transcriptional regulation. *Planta* 1986; 168:340–349.

153. Dilworth MF, Kende H. Comparative studies on nitrate reductase in *Agrostemma githago* induced by nitrate and benzyladenine. *Plant Physiol.* 1974; 54:821–825.

154. Kinoshita I, Yokomura E, Tsuji H. Benzyladenine induces doubling of nuclear DNA in intact bean leaves. *Biochem. Physiol. Pflanzen* 1991; 187:167–172.

155. Momotani E, Kinoshita I, Yokomura E, Tsuji H. Rapid induction of synthesis and doubling of nuclear DNA by benzyladenine in intact bean leaves. *Plant Cell Physiol.* 1990; 31:621–625.

156. Schneider J, Gruszka M, Legočka J, Szweykowska A. Role of cytokinins in development and metabolism of barley leaves. VII. Effect of light and kinetin on the transcriptional activity of nuclei from etiolated leaves. *Biochem. Physiol. Pflanzen* 1983; 178:381–390.

157. Gwoždž EA, Wožny A. Cytokinin-controlled polyribosome formation and protein synthesis in cucumber cotyledons. *Physiol. Plant.* 1983; 59:103–110.

158. Ohya T, Suzuki H. Cytokinin-promoted polyribosomes formation in excised cucumber cotyledons. *J. Plant Physiol.* 1988; 133:295–298.

159. Jackowski G, Legočka J, Szweykowska A. Effect of kinetin on the polyadenylated RNA in cucumber cotyledons. *Acta Physiol. Plant.* 1984; 6:181–188.

160. Legočka J. Kinetin-induced changes in the population of translable messenger RNA in cucumber cotyledons. *Biochem. Physiol. Pflanzen* 1987; 182:299–307.

161. Caers M, Vendrig J. Benzyladenine effects on the development of the photosynthetic apparatus in *Zea mays*: studies on photosynthetic activity, enzymes and (etio)chloroplast ultrastructure. *Physiol. Plant.* 1986; 66:685–691.

162. Le Pabic C, Hoffelt M, Roussaux J. Photosynthetic electron transport in plastids in detached cucumber cotyledons treated with 6-benzylaminopurine and potassium. *Plant Cell Physiol.* 1990; 31:333–339.

163. Longo GP, Bracale M, Rossi G, Longo CP. Effect of benzyladenine on the polypeptide pattern of plastids in excised watermelon cotyledons. *Physiol. Plant.* 1987; 68:678–684.

164. Lerbs S, Lerbs W, Wollgienh R, Parthier B. Cytokinin, light, and "developmental control" of protein synthesis in *Cucurbita* cotyledons. *Bot. Acta* 1988; 101:338–343.

165. Nyitrai P. Development of functional thylakoid membranes: regulation by light and hormones. In: Pessarakli M, ed. *Handbook of Photosynthesis*. New York: Marcel Dekker, 1996:391–406.

166. Schmülling T, Schäfer S, Romanov G. Cytokinins as regulators of gene expression. *Physiol. Plant.* 1997; 100:505–519.

167. Abdelghani OM, Suty L, Chen JN, Renaudin JP, Teyssendier de la Serve B. Cytokinins modulate the steady-state levels of light-dependent and light-independent proteins and messenger RNAs in tobacco cell suspensions. *Plant Sci.* 1991; 77:29–40.

168. Link G. Photocontrol of plastid gene expression. *Plant Cell Environ.* 1988; 11:329–338.

169. Lu J-L, Ertl JR, Chen C. Cytokinin enhancement of the light induction of nitrate reductase transcript levels in etiolated barley leaves. *Plant Mol. Biol.* 1990; 14:585–594.

170. Andersen BR, Jin G, Chen R, Ertl JR, Chen C-M. Transcriptional regulation of hydroxipyruvate reductase gene expression by cytokinin in etiolated pumpkin cotyledons. *Planta* 1996; 198:1–5.

171. Suzuki I, Crétin C, Omata T, Sugiyama T. Transcriptional and posttranscriptional regulation of nitrogen-responding expression of phosphoenolpyruvate

carboxylase gene in maize. *Plant Physiol.* 1994; 105:1223–1229.

172. Claes B, Smalle J, Dekeyser R, Van Montagu M, Caplan A. Organ-dependent regulation of a plant promoter isolated from rice by 'Promoter trapping' in tobacco. *Plant J.* 1991; 1:15–26.

173. Abler ML, Green PJ. Control of mRNA stability in higher plants. *Plant Mol. Biol.* 1996; 32:63–78.

174. Flores S, Tobin EM. Cytokinin modulation of LHCP mRNA levels: the involvement of post-transcriptional regulation. *Plant Mol. Biol.* 1989; 11:409–415.

175. Silverthorne J. Tobin E. Demonstration of transcriptional regulation of specific genes by phytochrome action. *Proc. Natl. Acad. Sci. USA* 1984; 81:1112–1116.

176. Schneider J, Legočka J. The role of cytokinins in the development and metabolism of barley leaves. V. Effect of light and kinetin on the transcriptional activity of chromatin in etiolated leaves. *Biochem. Physiol. Pflanzen* 1983; 178:381–390.

177. Ohya T, Suzuki H. Cytokinin-promoted polyribosome formation in excised cucumber cotyledons. *J. Plant Physiol.* 1988; 133:295–301.

178. Jackowski G, Jarmdowski A, Szweykowska A. Kinetin modifies the secondary structure of poly (A)RNA in cucumber cotyledons. *Plant Sci.* 1987; 52:67–70.

179. Yakovleva LA, Klueva NY, Kulaeva NO. Phosphorylation of ribosome proteins as a mechanism of phytohormone regulation of protein synthesis in plants. In: Kaminek M, Mok DWS, Zazimalova E, eds. *Physiology and Biochemistry of Cytokinins in Plants.* The Hague: SPB Academic Publishing, 1992: 169–172.

180. Romanov GA. Cytokinins and tRNAs: a hypothesis on their competitive interaction via specific receptor proteins. *Plant Cell Environ.* 1990; 13:751–754.

181. Ohya T, Suzuki H. The effects of benzyladenine on the accumulation of messenger RNAs that encode the large and small subunits of ribulose-1,5-bisphosphate carboxylase/oxygenase and light-harvesting chlorophyll *a/b* protein in excised cucumber cotyledons. *Plant Cell Physiol.* 1991; 32:577–580.

182. Kusnetsov VV, Oelmüller R, Sarwatt MI, Porfirova SA, Cherepneva GN, Herrmann RG, Kulaeva ON. Cytokinins, abscisic acid and light affect accumulation of chloroplast proteins in *Lupinus luteus* cotyledons without notable effect on steady-state mRNA levels. *Planta* 1994; 194:318–327.

183. Fritz CC, Herget T, Wolter FP, Schell J, Schreier PH. Reduced steady-state levels of rbcs mRNA in plants kept in the dark are due to differential degradation. *Proc. Natl. Acad. Sci. USA* 1991; 88:4458–4462.

184. Taylor CB, Green PJ. Identification and characterization of genes with unstable transcripts (GUTS) in tobacco. *Plant Mol. Biol.* 1995; 28:27–38.

185. Pedhadiya MD, Vaishnav PP, Singh YD. Development of photosynthetic electron transport reactions under the influence of phytohormones and nitrate nutrition in greening cucumber cotyledons. *Photosynth. Res.* 1987; 13:159–165.

186. Behera LM, Choudhury NK. Effect of organ excision and kinetin treatment on chlorophyll content and DCPIP photoreduction activity of chloroplasts of pumpkin cotyledons. *Plant Physiol.* 1990; 137:53–57.

187. Chory J. Out of darkness: mutants reveal pathways controlling light-regulated development of plants. *Trends Genet.* 1993; 9:167–172.

188. Johnson PR, Ecker JR. The ethylene gas signal transduction pathway: a molecular perspective. *Annu. Rev. Genet.* 1998; 32:227–254.

189. Kieber JJ. The ethylene response pathway in *Arabidopsis. Annu. Rev. Plant Physiol. Plant Mol. Biol.* 1997; 48:277–296.

190. Hooley R. Gibberellins: perception, transduction and responses. *Plant Mol. Biol.* 1994; 26:1529–1556.

191. Josefsson LG, Rask L. Cloning of a putative G-protein-coupled receptor from *Arabidopsis thaliana. Eur. J. Biochem.* 1997; 249:415–420.

192. Plakidou-Dymock S, Dymock D, Hooley R. A higher plant seven-transmembrane receptor that influence sensitivity to cytokinins. *Curr. Biol.* 1998; 8:315–324.

193. Hooley R. A role for G protein in plant hormone signalling? *Plant Physiol. Biochem.* 1999; 37:393–402.

194. Napier RM, Venis M. Auxin action and auxin-binding proteins. *New Phytol.* 1995; 129:167–201.

195. Jones HD, Smith SJ, Desikan R, Plakidou-Dymock S, Lovegrove A, Hooley R. Heterotrimeric G proteins are implicated in gibberellin induction of α-amylase gene expression in wild oat aleurone. *Plant Cell* 1998; 10:245–254.

196. Lohrmann J, Harter K. Plant two-component signaling systems and the role of response regulators. *Plant Physiol.* 2002; 128:363–369.

197. Hwang I, Sheen J. Two-component circuitry in *Arabidopsis* cytokinin signal transduction. *Nature* 2001; 413:383–389.

198. Deruère J, Kieber JJ. Molecular mechanism of cytokinin signaling. *J. Plant Growth Regul.* 2002; 21:32–39.

199. Stock AM, Robinson VL, Goudreau PN. Two-component signal transduction. *Annu. Rev. Biochem.* 2000; 69:183–215.

200. Kakimoto T. CKI1, a histidine kinase homolog implicated in cytokinin signal transduction. *Science* 1996; 274:982–985.

201. Inoue T, Higuchi M, Hashimoto Y, Seki M, Kobayashi M, Kato T, Tabata S, Shinozaki K, Kakimoto T. Identification of CRE1 as a cytokinin receptor from *Arabidopsis. Nature* 2000; 409:1060–1063.

202. Suzuki T, Miwa K, Ishikawa K, Yamada H, Aiba H, Mizuno T. The Arabidopsis sensor His-kinase, AHK4, can respond to cytokinins. *Plant Cell Physiol.* 2001; 42:107–113.

203. Schmülling T. CREam of cytokinin signalling: receptor identified. *Trends Plant Sci.* 2001; 6:281–284.

204. Haberer G, Kieber JJ. Cytokinins: New insights into a classic phytohormone. *Plant Physiol.* 2001; 128:354–362.

205. Ueguchi C, Koizumi H, Suzuki T, Mizuno T. Novel family of sensor histidine kinase genes in *Arabidopsis thaliana. Plant Cell Physiol.* 2001; 42:231–235.

206. Urao T, Yakubov B, Yamaguchi-Shinozaki K, Shinozaki K. Stress responsive expression of genes for two-component response regulator-like proteins in *Arabidopsis thaliana*. *FEBS Lett.* 1998; 427:175–178.

207. Sakai H, Aoyama T, Oka A. *Arabidopsis* ARR1 and ARR2 response regulators operate as transcriptional factors. *Plant J.* 2000; 24:703–711.

208. Suzuki T, Ishikawa K, Mizuno T. An Arabidopsis histidine-containing phophotransfer (HPt) factor implicated in phosphorelay signal transduction: overexpression of AHP2 in plants results in hypersensitiveness to cytokinin. *Plant Cell Physiol.* 2002; 43:123–129.

209. Chory J, Li J. Gibberellins, brassinosteroids and light-regulated development. *Plant Cell Environ.* 1997; 20:801–805.

210. Clouse SD, Sasse JM. Brassinosteroids: Essential regulators of plant growth and development. *Annu. Rev. Plant Physiol. Plant Mol. Biol.* 1998; 49:427–451.

211. Mandava NB. Plant growth-promoting brassinosteroids. *Annu. Rev. Plant Physiol. Plant Mol. Biol.* 1988; 39:23–52.

212. Reinbothe S, Reinbothe S, Heintzen C, Seidenbecher C, Parthier B. A methyl jasmonate-induced shift in the length of the 5′ untranslated region impairs translation of the plastid rbcL transcript in barley. *EMBO J.* 1993; 12:1505–1512.

213. Weidhase RAE, Kramell HM, Lehmann J, Liebisch HW, Lerbs W, Parthier B. Methyl jasmonate-induced changes in the polypeptide pattern of senescing barley leaf segments. *Plant Sci.* 1987; 51:177–186.

214. Kusnetsov V, Bolle C, Lubberstedt T, Sopory S, Herrmann RG, Oelmüller R. Evidence that the plastid signal and light operate via the same *cis*-acting elements in the promoters of nuclear genes for plastid proteins. *Mol. Gen. Genet.* 1996; 252:631–639.

215. Kusnetsov VV, Oelmüller R, Makeev AV, Cherepneva GN, Romankoe.g.,Selivankina SY, Mokronosov AT, Herrmann RG, Kulaeva ON. Cytokinin and abscisic acid in regulation of chloroplast protein gene expression and photosynthetic activity. In: Smith AR, Berry AW, Harpham NVJ, Moshkov IE, Novikova GV, Kulaeva ON, Hall MA, eds. Plant hormone signal perception and transduction. Dordrecht: Kluwer Academic, 1996:109–118.

216. Medford JI, Horgan R, El-Sawi Z, Klee HJ. Alterations of endogenous cytokinins in transgenic plants using a chimeric isopentenyl transferase gene. *Plant Cell* 1989; 1:403–413.

217. Bartholomew DM, Bartley GE, Scolnik PA. Abscisic acid control of rbcs and cab transcription in tomato leaves. *Plant Physiol.* 1991; 96:291–296.

218. Weatherwax SC, Ong MS, Degenhardt J, Bray EA, Tobin EM. The interaction of light and abscisic acid in the regulation of plant gene expression. *Plant Physiol.* 1996; 111:363–370.

219. McElwain EF, Bohnert HJ, Thomas JC. Light moderates the induction phosphoenolpyruvate carboxylase by NaCl and abscisic acid in *Mesembryanthenum crystallinum*. *Plant Physiol.* 1992; 99:1261–1264.

Section V

Photosynthetic Pathways in Various Crop Plants

21 Photosynthetic Carbon Assimilation of C_3, C_4, and CAM Pathways

Anil S. Bhagwat
Molecular Biology Division, Bhabha Atomic Research Center

CONTENTS

I. INTRODUCTION

Photosynthetic efficiency sets in the ultimate limit for the productivity of crops for food, fuel, and fiber. Atmospheric carbon dioxide provides through the process of photosynthesis all raw materials for bio-logical activity on our planet. The global uptake of CO_2 in photosynthesis is about 120 gigatons (Gt) of carbon per year and virtually all passes through one enzyme, ribulose 1,5-bisphosphate carboxylase. Increasing photosynthetic efficiency is an important aspect of any program targeted toward increasing plant productivity. Thus, a thorough understanding of the process of carbon fixation is a prime requisite for achieving this goal. In late 1950s, Melvin Calvin and his colleagues had delineated the details of the photosynthetic carbon assimilation process that operates in chlorella by following the fate of labeled carbon dioxide. The essential steps of the pathway have withstood the test of time for over 50 years without any appreciable changes. This process provides the final source of total free energy in all living organisms. Even though the Calvin cycle represents the cardinal pathway whereby net carbon gain is affected, some adaptations of this process have been seen in various plant types. In the late 1960s, Hatch and Slack showed an adaptation in several grasses that had evolved an additional pathway of carbon fixation in conjunction with the special leaf anatomy having two distinct types of photosynthetic cells. This evolutionary pressure was due to the fast depleting concentrations of atmospheric CO_2 and increasing global temperatures. This pathway of photosynthesis was called the dicarboxylic acid (C_4) pathway because the first detectable stable product of carbon fixation was a dicarboxylic acid (C_4) (such as malate and aspartate), in contrast with a C_3 acid, 3-phosphoglycerate, in C_3 plants. One of the neatest elements of the C_4 pathway is the well known division of labor between chloroplast and the spatial separation of the process that permits the working of the CO_2 concentrating mechanism. Another adaptation of the carbon fixation pathway is found in succulent plants that are known as CAM (crassulacean acid metabolism) plants and are popularly called "dark fixers" as they mostly assimilate atmospheric CO_2 during dark periods and store it as malate in vacuoles. This adaptation is purely for conserving water during the day by keeping the stomata closed under desert conditions. About 90% of all terrestrial plant species, which include major crops such as rice, wheat, soybean, and potato, are classified as C_3 plants, and they assimilate CO_2 directly through the Calvin cycle. Even though most photosynthetic systems follow one of the three established pathways for carbon fixation, some variations have been noticed, and I shall briefly touch upon this aspect later.

The scope of this chapter is to provide the reader with an overview of the mechanisms of carbon fixation operating in these three biochemical classes of plant. In addition, I would also highlight the importance of C_3–C_4 intermediate species as they provide an excellent experimental material for a study of the regulation of the process of photorespiration in C_3 as well as C_4 plants and evolution of the C_4 pathway. Molecular evolution and genetic engineering of C_4 photosynthetic enzymes, their cell specific expression in C_3 plants and mimicking of C_4 photosynthesis as operational in some aquatic organisms and in single-cell systems of C_3 plants will be some of the recent developments included in this revised chapter. I shall, however, not go into the details of the description, structure function, and regulatory properties of the enzymes of these pathways, which will be adequately covered in other chapters of this book.

II. C_3 PATHWAY OF PHOTOSYNTHESIS

The C_3 cycle is apparently found in all photoautotrophic green plants [1,2] and also occurs in a variety of photosynthetic bacteria [3,4]. In C_3 plants, the whole process of carbon assimilation takes place in the chloroplasts [5]. The enzymes catalyzing the steps of reductive pentose phosphate (RPP) cycle are generally thought to be water soluble and are located in the stromal region of the chloroplast [6]. However, recent studies indicate that at least some of the enzymes of the Calvin cycle may be membrane bound [7–10]. These studies also provide evidence indicating that in chloroplast stromal extracts, Calvin cycle enzymes such as Rib-5-PI, Ru-5-PK, Rubisco, GAPDH, Sed-1, and 7-BPase are organized into heteromeric CO_2-fixing multienzyme complexes of approximately 900 kDa.

A.   Calvin Cycle

The delineated path of carbon in the Calvin cycle has remained virtually unchanged from that originally outlined by M. Calvin in the late 1950s. The subsequent studies have, no doubt, removed many of the lingering problems such as the insufficient catalytic rates of the enzymes like Rubisco and diphosphatases under *in vitro* conditions. The mechanism of activation and regulation of many of these enzymes became known during this period, which helped us in understanding the process of activation under *in vitro* and *in vivo* conditions. Activation of these enzymes requires either a metal ion and CO_2 or sulfhydryl group reduction by some special systems like the ferredoxin/thioredoxin system in the activation of FDPase and SDPase, which has been clearly established [11–13]. The role of a soluble protein, Rubisco activase in the case of Rubisco, has been well documented [14–16]. Recent studies have indicated that Rubisco activase is actually a molecular chaperone rather than a conventional enzyme. Several of the known biochemical properties of molecular chaperones are found in Rubisco activase [17]. The Calvin cycle consists of three phases and 13 steps. The three phases are (1) the carboxylation phase, (2) the reductive phase, and (3) the regenerative phase. As many as ten out of the total 13 reactions of the cycle are devoted to regenerating the CO_2 acceptor molecule ribulose 1,5 bisphosphate.

1. Autocatalytic Nature of the Calvin Cycle

The appearance of the first stable labeled product other than 3-phosphoglyceric acid detected in plants having the C_4 or CAM pathway of photosynthesis does not necessarily indicate an alternative to the Calvin cycle. Any CO_2 fixation mechanism permitting growth must be autocatalytic in nature and should be able to generate more CO_2 acceptor molecules than present initially [18]. In other words, it must function as a breeder reaction. Several studies have clearly differentiated between the Calvin cycle and the C_4 pathway on this basis and concluded that while the Calvin cycle is autocatalytic in nature, the C_4 pathway is not. This uniqueness of the Calvin cycle as the only presently known pathway for net incorporation of CO_2 for growth is helpful in interpreting much of the data available so far.

2. Possible Alternatives to the Calvin Cycle

A direct reduction of CO_2 to formate has been demonstrated by an aerobic bacterium [19]. This observation was also used in explaining the distribution of labeled CO_2 in organic acids in *Vicea faba* leaves in light. However, reduction of CO_2 to formate in these leaves has not been demonstrated directly. During a study of solanidine biosynthesis in greening potatoes, it was observed that in isolated chloroplasts from greening potato tubers the formation of 3-PGA could not be detected [20,21]. Ribulose 1,5-bisphosphate carboxylase activity was also not detected. Formate was identified as the primary product of CO_2 fixation by potato chloroplast. The ultimate product in the presence of DCMU was shown to be mevalonate, and in its absence the major product was identified as solanidine [22]. The enzyme involved in the CO_2 fixation has been identified as CO_2 reductase that has been partially purified. The CO_2 reductase was found to be activated by light and the activation is mediated through a reversible sulfhydryl group reduction [23].

3. Stoichiometry and Energetics of the Calvin Cycle

There are two steps in the Calvin cycle that require ATP and one step requiring the reducing power of NADPH. Thus, a complete turn of the reductive pentose pathway requires 9 mol of ATP and 6 mol of NADPH to make 1 mol of triose phosphate:

$$6(NADPH) + \tfrac{1}{2}O + H^+ \rightarrow NADP + H_2O \quad \Delta G'$$
$$-325.5\,kcal$$
$$9(ATP) + H_2O \rightarrow ADP + Pi + H^+ \quad \Delta G' - 68.8\,kcal$$
$$\text{Total } \Delta G' - 384.3\,kcal$$
$$3CO_2 + 3H_2O + Pi \rightarrow triose\text{-}P_2 + 3O_2 \quad \Delta G' + 350.4\,kcal$$

Therefore, the net difference in free energy is -33.9 kcal, which is the driving energy for one turn of the cycle. The energy efficiency is $350/384 = 91\%$. However, the actual measurements in photosynthesizing chlorella cells gave a value of 83% [14]. This high efficiency of the basic RPP reactions is offset by the relatively lower efficiency of photosynthesis and plant growth. The production of nine ATP and six NADPH molecules along with oxidation of water to O_2 stores about 384.3 kcal but requires a minimum of 24 moles of photons of PAR. Thus, the efficiency is only 32% for PAR and since PAR is only 43% of the solar spectrum, the overall efficiency of the light reaction is 13.8%. If this is multiplied by a carbon efficiency of 83%, the overall photosynthetic efficiency becomes 11.4% even before any respiratory or photorespiratory losses are taken into account. The highest efficiency for conversion of solar energy to total biomass during the periods of most rapid growth are in the range of 2% to 3% for C_4 plants and 1% to 2% for C_3 plants.

4. Enzymes of the Calvin Cycle

The ATP and NADPH generated by the photosynthetic light reactions are used in green plants to drive CO_2 fixation into sugars by a chain of reactions known as the photosynthetic carbon reduction cycle (Calvin cycle). The ribulose 1,5-bisphosphate carboxylase/oxygenase performs the role of initial acceptor of atmospheric carbon dioxide to give two molecules of three carbon compounds that are metabolized by a subsequent ten reactions resulting in both regeneration of the acceptor molecule and translocation of three molecules of trioses to the cytosol for synthesis of sucrose and starch. For detailed information on the various enzymes of the Calvin cycle including Rubisco, readers are advised to refer to some excellent reviews by Raines et al. [24] on all enzymes in general and by Hartman and coworkers [25,26] for Rubisco in particular.

B. REGULATION OF C_3 PHOTOSYNTHESIS

Many of the regulatory properties of the PCR cycle have been discussed in detail in several earlier reviews [27–30]. Understanding how the Calvin cycle mediates regulatory interactions among the cytosol, electron transport, and Rubisco is extremely important. Conservation of phosphate is of great importance because it requires that a change in the level of any phosphorylated intermediate be compensated by an equal and opposite charge (in terms of phosphate elsewhere in the cycle). Therefore, changes in the activity of any of the PCR enzymes can affect both the substrate concentration and activities of other enzymes in the chloroplast regardless of whether they are adjacent on a metabolic scheme or are connected by the classical mechanism. Studies on the effect of various chloroplastic metabolites on Rubisco activity under suboptimal CO_2 concentrations have shown that both phosphoglycolate and inorganic phosphate activate the enzyme [31,32]. The effectors elicit their response by stabilizing reversibly the binding of a CO_2 and Me^{2+} in much the same manner that CABP does, but unlike CABP these effectors readily dissociate on dilution, yielding an activated ternary complex. The activation by inorganic phosphate appears very interesting because the enzyme seems to have two binding sites for Pi, which appears to be an allosteric activator, and play an important role in regulating Rubisco activity [32]. This conclusion has been subsequently confirmed by Sawada et al. [33,34]. They have emphasized the role of Pi in regulating Rubisco activity, especially under conditions of sink limitations. The second property of PCR cycle involves regulatory responses at the two major branch points. At the first of these, triose-P is withdrawn from the chloroplast for ultimate synthesis of sucrose and the second hexose phosphate is used for synthesis of starch. Thus, in the stroma, the FBPase, SBPase, Ru5P-kinase, and ADP-glucose phosphorylase are important in determining the metabolite partition at both branch points, and in the cytosol FBPase and phosphate translocator can influence the competition.

1. Photochemical Events

Illumination of chloroplast causes large changes in the chemical environment of the stromal compartment, and such changes are likely to influence the operation of the PCR cycle. For plants in full sunlight the rate of photosynthesis will be limited by the rate of reaction of the PCR cycle rather than by the rate of electron transport and phosphorylation. Up to 90% of the chloroplast NADP is in the reduced state in full light intensity, in contrast to 5% to 20% in the dark [35]. The levels of ADP and ATP are variable [36]. Thus, the rise in NADPH and ATP levels in light will exert a major control on photosynthesis as the cycle cannot function without the continued supply of these two metabolites. Second, the electron transport is coupled to the proton uptake into the thylakoid

space and increases the stromal pH from 7.0 in the dark to 8 to 8.5 in light [37]. The stromal enzymes can thus function best in this environment. This is also accompanied by an uptake of Cl^- and efflux of Mg^{2+} in the stroma, and both these ions can influence the activity of the enzymes. The most important regulation mediated through the electron transport chain is through redox control of many enzymes of the Calvin cycle like SDPase and FDPase, which are regulated by the ferredoxin–thioredoxin system. Several other photosynthetic enzymes are regulated through activation of specific protein kinases.

2. Activase as Potential Target for Increasing Photosynthesis

The activation state of Rubisco is modulated through Rubisco activase, which also has ATPase activity, by the supply of ATP generated by photophosphorylation. The active site of Rubisco assumes a closed conformation with certain phosphorylated ligands irrespective of its carbamylation state. The binding of RuBP to the uncarbamylated Rubisco indicates that once closed these sites are very slow to open.

This closed conformation represents a potential dead end for carbamylated sites because these are unable to trigger C—C bond cleavage. In the absence of catalysis, conversion of Rubisco from the closed to the open conformation is extremely slow. To facilitate the process, plants have Rubisco activase, an ATP-dependent enzyme that releases tight-binding sugar phosphates from the Rubisco active site [38].

The activase interacts with Rubisco somehow to release the bound sugar phosphate from the active site by causing a conformational change that promotes opening of the closed configuration. Once the activase opens a closed site, the sugar phosphate can dissociate and activation of the enzyme is possible. Activase protein can be detected in all plant species examined so far including C_3 and C_4 and many algal systems. Because activase is an ATP-dependent enzyme, stromal ATP seems to control the closed and open conformations of Rubisco and regulate its activity by light. The mechanism of activase action is not known; however, the process is coupled to ATP hydrolysis either for priming the activase or its interaction.

The C-terminus of activase is important in recognizing that Rubisco and 50 N terminal amino acids are required for activation but not for ATP hydrolysis. Alternative splicing is generally believed to be involved in generation of activase isoforms; however, in some cases two different genes may be present. Redox control of the longer form of activase is mediated through thioredoxin-f mediated reduction of a pair of cysteins located at the C terminal extension.

Reduction decreases the sensitivity of activase to inhibition by ADP. Redox control/regulation of activase serves a regulatory role. Several questions regarding the regulation of activase remain unanswered as tobacco, tomato, maize, and chlamydomonas have only single activases that do not have cystein at the extension but are still regulated by irradiance levels. The strategies employed so far for increasing photosynthesis generally revolve around Rubisco and controlling the oxygenase activity of the enzyme in particular. However, no significant success is yet in sight in achieving this goal. The Rubisco would require a change in the structure of the enzyme, and it is also necessary to consider how each change will affect the interaction of Rubisco with activase. Thus each strategy for improving Rubisco should include the possible need to co-design activase. Redesigning the activase will require a more complete understanding of the mechanism of action and site of interaction with Rubisco.

Because the activation state of Rubisco limits photosynthesis under conditions of high CO_2 and temperature, improvements in the activase like making it heat stable, especially in C_4 plants, may stimulate photosynthesis. The decrease in Rubisco activation level that occurs at high CO_2 levels appears to involve activase. Heat stress inhibits photosynthesis by reducing the activation of Rubisco using Rubisco activase. The thermal denaturation of activase *in vivo* occurs at a temperature close to those that denature isolated activase and far below those required to denature Rubisco or phosphoribulokinase. Thus, loss of activase during heat stress is caused by an exceptional sensitivity of the protein to thermal denaturation [39].

Using specific amino acid replacement by site-directed mutagenesis of Rubisco activase from *Arabidopsis* showed two- to threefold higher activation by recombinant 43 kDa Rubisco activase. However, the rate of ATP hydrolysis was only marginally greater. The sensitivity to ADP inhibition was also less. The ADP/ATP ratio plays a role in regulation of the Rubisco activation and the photosynthetic rate [40]. In addition to ADP/ATP ratio, regulation of activase by a transthylakoid pH gradient and the reduced state of the acceptor side of PSI and/or the degree of reduction of the thioredoxin is also possible [41]. Thus a balance between electron transport and the consumption of its products may also be important. RuBP modulates Rubisco activity *in vivo* by binding tightly to an uncarbamylated active site blocking the carbamylation process, and removing the RuBP from these sites is the primary function of activase. The rate of RuBP generation is matched by its consumption and decarbamylation by the activase [42].

C. PHOTORESPIRATION

Under most conditions of photosynthesis in chloroplast some glycolate is formed. Under the conditions that favor photorespiration, a large part of the carbon fixed by the PCR cycle can be converted to glycolate [43]. Such conditions include high light, low CO_2, high O_2, and high temperature. The most accepted pathway of glycolate formation, which has been clearly worked out, is through the oxygenase reaction of Rubisco [44,45]. It is generally believed that photorespiration is a necessary evil which is a consequence of the chemistry of this bifunctional enzyme. A detailed consideration of the process of photorespiration can be found in several excellent reviews by Lorimer and others [44–47]. Photorespiration consumes light-generated ATP and NADPH. The CO_2 remains inside the leaf, and it is effectively captured by the rapid photosynthesizing machinery of the chloroplast. However, photorespiration is indeed a wasteful process because the released CO_2 must be refixed in the leaf at an expense of extra energy to the plants, and the net assimilation of CO_2 from the atmosphere is reduced proportionately, affecting the entire yield and productivity. Photorespiration has been assigned a major role in preventing photooxidative damage at low CO_2 and high light intensities [47]. The oxygenase activity of Rubisco may have a major role in maintaining electron transport during periods of water stress, a feature of even well irrigated crops on hot summer days. Since some of the above conclusions are based on extrapolation of laboratory experiments done with isolated chloroplasts, it is argued that these may not hold true under real-life field conditions. However, several studies have shown that prevention of photorespiration could increase the net photosynthesis [48,49]. A recent study by Zelitch has shown that genetically decreasing photorespiration in C_3 species could increase yield without any substantial increase in the inputs. During the selection of oxygen resistant mutants in tobacco it was observed that these mutants had higher levels of catalase activity, up to 40% to 50% higher than the wild type [49]. The mutants showed higher rates of net photosynthesis even under conditions that normally show high rates of photorespiration such as high temperature, high oxygen, and low CO_2, indicating that high catalase activity in the mutant decreases photorespiratory CO_2 losses by utilizing H_2O_2 produced by glycolate oxidase that otherwise would rapidly decarboxylate keto acids such as glyoxylate, producing additional CO_2 and facilitating photorespiratory losses. The hypothesis must now be put to a severe test by cloning the catalase gene and by production of transgenic plants with different levels

TABLE 21.1
Effect of CO$_2$ Concentration on Photorespiration and Quantum Yield in C$_3$ and C$_4$ Plants

Gas Phase	CO$_2$/O$_2$	V_c/V_o[a]	Photorespiration	Quantum Yield[b] CO$_2$ (%)
Air level of CO$_2$ + O$_2$	0.025	2.5	20%	0.05
CO$_2$ 10× in air	0.25	25.0	<2%	~0.08

[a]V_c, carboxylase; V_o, oxygenase.
[b]Mole CO$_2$/mole quanta, 30°C.

of catalase activity to establish the biochemical linkage of these traits. When photorespiration is reduced in C$_3$ plants either by increasing ambient levels of CO$_2$ (see Table 21.1) or reducing levels of O$_2$, both the yield (vegetative dry matter) and nitrogen use efficiency are enhanced in wheat but not in maize [50]. Recent studies have shown that the high levels of NO$_3$ seen in barley and wheat relative to sorghum and maize are related to a carbon deficiency caused by inhibition of the mitochondrial pyruvate dehydrogenase complex by monovalent cations, in particular by ammonium ions produced by photorespiration in C$_3$ plants [51–53]. Ammonium production is lower in C$_4$ plants than in C$_3$ cereals. In addition, NH$_4$ production is localized in bundle sheath cells in C$_4$ plants, whereas NO$_3$ assimilation is found in mesophyll cells: its impact on the carbon flow, which is required for NO$_3$ assimilation should be negligible in C$_4$ plants. The interplay of mitochondrial respiration, nitrogen metabolism, and photosynthesis involves recycling of carbon and nitrogen and reducing equivalents. At least four different metabolic pathways are required to be coordinated for efficient recycling of carbon and nitrogen. Apart from CO$_2$ fixation, the second largest sink for photosynthates is nitrogen (nitrate) metabolism. These processes are coupled. The mitochondrial respiratory chain can play a role in cellular ATP production in light.

III. C$_4$ PATHWAY OF CARBON ASSIMILATION

A. BASIC PHENOMENA

The C$_4$ pathway of photosynthesis has evolved subsequent to the C$_3$ cycle and provides several advantages to the plants in which this cycle operates. The C$_4$ pathway may have evolved about 600 million years ago as a consequence of a dramatic drop in atmospheric CO$_2$ concentrations at the end of the Cretaceous period and the beginning of the Paleozoic era. It was not until the second drop, 6 to 8 million years ago that C$_4$ plants became dominant in grassland

in tropical and subtropical areas all over the world. The CO$_2$ concentrating mechanism, several of the enzymes, and the altered leaf anatomy has evolved several times over during the evolution of C$_4$ plants during angiosperm evolution [54]. The fast evolution of the C$_4$ pathway was due to duplication of genes already present in C$_3$ plants. There have been well defined advances and contractions in the distribution of C$_4$ plants during last full Glacial, 20,000 to 30,000 years ago. Another contraction of C$_4$ plants may begin in the next 50 to 100 years. It may perhaps take the transfer of C$_4$ traits effectively to C$_3$ crops. It is obvious that the low atmospheric CO$_2$ concentration was the major selective pressure favoring C$_4$ photosynthesis, which is, however, vanishing very fast. The industrial revolution is returning several billion years of fossil photosynthesis to the atmospheric CO$_2$ concentration. The so called C$_4$ species have evolved a complex biochemical process, and with the help of some unique modifications in leaf anatomy and associated ultrastructure they concentrate CO$_2$ in the bundle sheath cells at the site of the PCR cycle that is initiated by Rubisco. C$_3$ photosynthesis is the only mode of carbon assimilation in algae, bryophytes, pteridophytes, gymnosperms, and the majority of angiosperm families. Only about ten monocot and dicot families possess the C$_4$ pathway of photosynthesis. These families are not closely related, and it seems that the C$_4$ pathway arose independently in each of them. In some instances more than one separate origin can be documented in a single family. C$_4$ photosynthesis has evolved at least 20 separate times [54]. The existence of this modified cycle was recognized in the mid-1960s, and the basic pathway of carbon assimilation was determined from kinetic analysis of CO$_2$ fixation into intermediates [55–58]. By the early 1970s it was realized that the partitioning of photosynthates between two cell types in C$_4$ plants was a crucial process of the pathway [58]. In the mid-1970s it became evident that the C$_4$ pathway basically functions as a pumping mechanism in increasing large pools of CO$_2$ + HCO$_3$ in bundle sheath cells and this is accomplished by three different decarboxylating enzymes, which formed the biochemical basis of classification of the C_4 plants [59–62]. The developments in the state of our knowledge on the C_4 pathway have been assessed in several excellent reviews that have appeared during last 10 years [63–66]. A common feature of all the enzymes implicated in the C_4 pathway is their high activity in C_4 leaves as compared with that recorded in C_3 plants (15- to 100-fold higher). In addition, C_4 leaves also contain high activities of adenylate kinase and pyrophosphatase (20 to 50 times higher than C_3 plants), largely located in mesophyll chloroplast together with pyruvate Pi dikinase [58]. Figure 21.1 shows the detailed pathways of the carbon metabolism operating in three subgroups of C_4 plants. In general terms C_4 photosynthesis involves the initial assimilation of CO_2 into C_4 acids via PEP carboxylase in mesophyll cells. The decarboxylation of these C_4 acids after transfer to bundle sheath cells is followed by assimilation of the released CO_2 via Rubisco and the PCR cycle. The regeneration of the primary CO_2 acceptor PEP takes place in the mesophyll cells through the unique enzyme of the C_4 pathway, pyruvate Pi dikinase. The subcellular location of the enzymes is listed in Table 21.2. The finding that C_4 plants evolved three distinct options as described below for decarboxylating C_4 acids in bundle sheath cells was one of the many unexpected and surprising features of this process. Under the metabolic conditions which apply in the cytosol, the activity of NAD malate dehydrogenase would limit the rate of conversion of oxaloacetate to malate in the C_4 pathway in all the three plant species. A brief account of the decarboxylating enzymes is given below.

B. BIOCHEMICAL MECHANISM AND ENZYMES

1. NADP — Malic Enzyme Type

In these plant types, e.g. sugarcane, sorghum, maize, the CO_2 is initially fixed in mesophyll cells into oxaloacetate, which is then reduced to malate by NADP–malate dehydrogenase, which moves to bundle sheath cells. In the chloroplasts of these cells the malate is decarboxylated by NAD–malic enzyme giving rise to pyruvate, CO_2 and NADPH, which can be cycled back to NADP by coupling to PGA reduction in the PCR cycle. The pyruvate formed moves back to mesophyll cells, where it is converted to PEP by pyruvate Pi dikinase, thereby regenerating the primary acceptor of CO_2.

2. PEP Carboxykinase Type

A great deal of diversity has been observed among C_4 plants, and in many cases the NADP–malic

enzyme as well as NADP–malate dehydrogenase activity was much lower than that required for efficient functioning of the pathway; however, these plant species exhibited much higher activities of aspartate and alanine aminotransferases. Such species were found to be aspartate-formers rather than malate formers. The enigma was further resolved by the observation that in aspartate formers the activity of PEP carboxykinase was much higher and had low NADP–malic enzyme activity [59]. Later studies have clearly established that PEP carboxykinase is specifically located in the cytosol of bundle sheath cells.

3. NAD–Malic Enzyme Type

The discovery of the above two decarboxylating pathways did not completely resolve the problem because there remained a number of C_4 plants that showed neither high NADP-malic enzyme activity nor PEP carboxykinase. This problem was resolved by Hatch and Kagawa when they showed that such species showed high activities of the NAD–malic enzyme located in mitochondria [63]. In NAD–malic enzyme species, the activity of this enzyme in mitochondria is 10- to 50-fold higher than in other species and the number of mitochondria is also increased by a factor of 3 to 4.

a. Regulation of C_4 pathway

Leaf ontogeny is a useful tool for studying the regulation of the C_4 pathway. Young maize leaves are more C_3 than older leaves, which when fully developed are C_4. A similar tendency was seen in several other C_4 plants. Both C_3 and C_4 photosynthetic characteristics in the same plant have also been reported by several researchers [67–69]. Expression of C_4 characteristics is controlled by leaf position, leaf age, and nutritional status. Such variable C_4 function in leaves is due to the gradient of development of the "Kranz" anatomy as well as the levels of mRNA of the key enzymes of the pathway. The major emphasis during the subsequent years has been in understanding the regulation of the enzymes involved in various mechanism of C_4 photosynthesis. For a detailed account of this aspect please refer to the chapters by Iglesias and Podesta as well as by Andreo and Podesta (this book). The posttranslational regulation of at least two enzymes of NADP-ME type has been understood to a large extent [70,71]. Both PEPC and pyruvate Pi dikinase seem to be regulated by endogenous protein kinase through reversible phosphorylation/dephosphorylation [32,72–74]. Light regulation at the level of transcription is supposedly mediated through phyto-

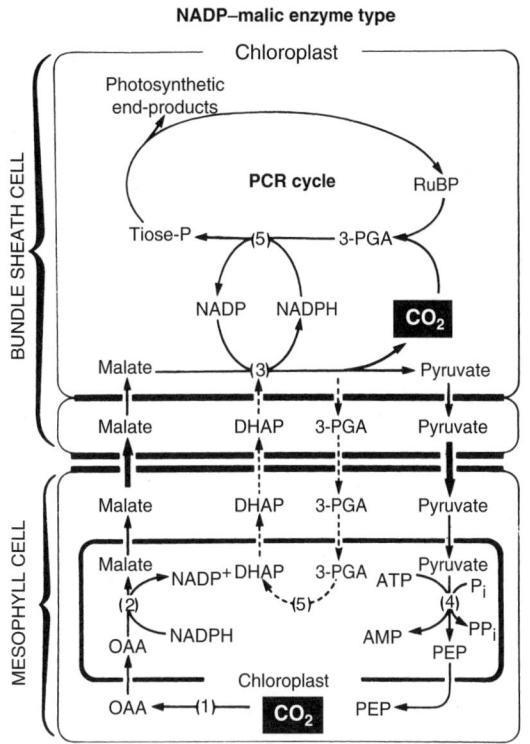

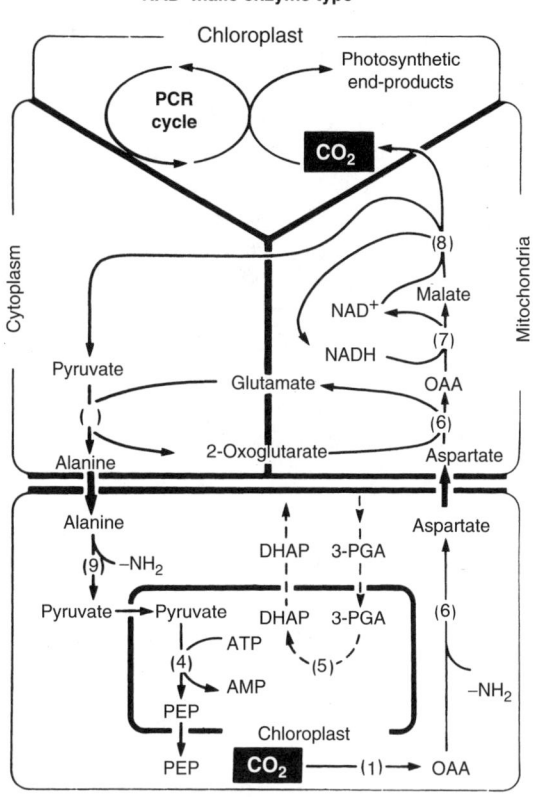

FIGURE 21.1 Pathway of carbon assimilation and localization of reaction intermediates for the three biochemically distinct subgroups of C_4 species: (A) NADP–malic enzyme; (B) NAD–malic enzyme; (C) PEP carboxykinase. The enzymes involved are shown by numbers in the parentheses: 1, PEP carboxylase; 2, NADP-MDH; 3, NADP–malic enzyme; 4, pyruvate Pi dikinase; 5, 3-PGA kinase and GAD dehydrogenase; 6, aspartate amino transferase; 7, NAD malate dehydrogenase; 8, NAD–malic enzyme; 9, alanine aminotransferase; 10, PEP carboxykinase; 11, mitochondrial NADH oxidation system. [Reproduced with permission of the publishers and the author from Hatch MD. C_4 photosynthesis: a unique blend of modified biochemistry anatomy and ultrastructure. *Biochim. Biophys. Acta* 1987; 895:81–106.].

TABLE 21.2
Activity and Localization of Photosynthetic Enzymes in the Three Subgroups of C₄ Plants

ENZ	Activity[a]	Location	Activity	Location	Activity	Location
NADP-ME	25–40	BS, chl	Trace	—	Trace	—
PEP-CK	Trace	—	40–110	BS, cyt	Trace	—
NAD-ME	1	BS, mit	3–10	BS, mit	18–60	BS, mit
Asp trans	1–3	cyt	15–35	cyt	12–27	cyt
Ala trans	1–3	cyt	15–35	cyt	12–27	cyt
NADP-MDH	12–21	mit	2–6	mit chl	1–2	mit chl
Rubisco	4	chl	4–6	chl	5–8	chl
NAD-tri-P	8–11	chl	8–10	chl	8–10	chl
PEPC	16–30	mit cyt	21–34	mit cyt	15–31	mit cyt
PPDK	40–80	mit chl	40–90	mit chl	15–40	mit chl

Note: chl, chloroplast; cyt, cytoplasm; mit, mitochondria; Asp trans, aspartate amino transferase; Ala trans, alanine amino transferase; NAD-MDH, NAD dependent malate dehydrogenase.

[a]Activities are expressed as μmole/mg chl/min.

chrome. The other two enzymes of the pathway, NADP–malic enzyme and NADP–malate dehydrogenase, have been shown to be light regulated through the involvement of sulfhydryl groups by reversible oxidation–reduction that is mediated by the ferredoxin–thioredoxin system [104]. Regulation of PEPC and NADP–malic enzyme by several chloroplastic metabolites has also been demonstrated [75–78].

b. Function of C₄ pathway
of photosynthesis

As mentioned earlier, the C₄ pathway transfers CO_2 from mesophyll cells to bundle sheath cells through a dicarboxylic acid intermediate. The advantage of this additional pathway will be clearly realized when we consider the operation of the Calvin cycle in C₃ plants under normal environmental conditions, whereby these plants lose up to 30% to 50% of the carbon fixed by photorespiration. The high levels of CO_2 maintained through the C₄ pathway helps in inhibiting the oxygenase activity of Rubisco and eliminating the photorespiratory loss. The oxygenase reaction and the associated metabolism have an adverse effect on the photosynthetic efficiency of plants. It is clearly seen from the data compiled in Table 21.2 that increasing the CO_2 concentration to ten times the ambient level virtually eliminates the oxygenase activity and reduces the photorespiratory loss to a mere 2%, thus increasing the quantum yield from 0.05 to 0.08. The question is, does this really resemble the *in vivo* situation? The answer is yes. The CO_2 concentration in C₃ leaves is ten times less than that found in C₄ bundle sheath cells. In addition, Jenkins et al. [79] have developed a model of the

inorganic carbon status in bundle sheath cells that predicted that the total Ci in bundle sheath cytosol is about $150 \mu M$, of which the steady state concentration is at least $30 \mu M$, which is about 20 times the steady state concentration of CO_2 in the adjacent mesophyll cells during active photosynthesis, thereby virtually eliminating the oxygenase function of Rubisco. The photosynthetic efficiency of C₄ plants is virtually unaffected by temperature between 20°C to 45°C, whereas C₃ plants perform much better at the lower temperatures, but at higher temperatures their yields are significantly affected. The biochemical reasons for the higher efficiency of C₄ plants at higher temperatures and light intensities were investigated, and the higher affinity of NADP–malic enzyme for malate at higher temperatures seems to contribute towards a higher fixation of CO_2 [80,81]. Unlike Calvin cycle enzymes, some of the key enzymes of the C₄ pathway are sensitive to low temperatures. It has been shown that at temperatures below the optimum range, these enzymes tend to dissociate into monomeric forms. The temperature inactivation seems to be the primary cause for decreased photosynthetic capacity below 10°C and 15°C and susceptibility to photoinhibition at chilling temperatures as compared with the C₃ plants [82,83]. The water use efficiency and stomatal conductance of C₄ plants are some of the other features that are superior to those of C₃ plants. The C₄ plants are generally twice as efficient in conserving water at 25°C as C₃ plants, although this difference decreases at lower temperatures [84]. The nitrogen use efficiency of C₄ plants is twice that of C₃ plants and has both agronomical and evolutionary implications. Apart from these characteristics, the general permeability of C_4 plants is also at least ten times higher than the maximum metabolite permeability measured for other plant cells. There is strong evidence to show that the increased flux of metabolites in C_4 plants is due to proliferation of plasmodesmeta in the mesophyll/bundle sheath cell walls [85,86]. However, the back flux of CO_2 from the bundle sheath to mesophyll is insignificant because of reduced permeability, which is attributed to suberin lamellae specifically located in the wall separating these cells. Despite these permeability barriers, whatever back flux occurs can be readily reassimilated by PEP carboxylase [87]. Many of the important agricultural crops are C_3 plants, and only a few crops and some fodder grasses are of the C_4 type. A serious problem that is often encountered during crop management has been the draining of soil nutrients by weeds, which are mostly C_4 type. In the U.S. alone, among the top ten weeds that affect plants, six are of C_4 type. Thus, an efficient control of these weeds by making a specific inhibitor of C_4 enzymes like pyruvate Pi dikinase would be extremely beneficial in weed management and overall yield of the crop plants. Recently a specific inhibitor of PEP carboxylase that drastically reduces carbon assimilation in C_4 plants without significantly affecting carbon fixation in C_3 plants has been discovered. The application of such compounds under field conditions will now have to be tested rigorously to determine their efficacy.

c. Expression of C_4 genes

There has been a great deal of information available on the molecular mechanism and signaling pathway that control C_4 photosynthetic gene expression. Current evidence suggests that preexisting genes were recruited for the C_4 pathway after acquiring potent and surprisingly diverse regulatory elements. I will deal briefly with the coordinated nuclear chloroplast action hormonal metabolite, stress and light responses in gene regulation and control of enzyme activity by phosphorylation and the reductive process. In the meantime the notions that the crop yield can be improved through greater photosynthetic capacity and that the C_4 metabolism alone may boost the yield of C_3 crops continue to stimulate creative research. Such projects are exposing the consequences of introduction of C_4 photosynthetic traits into C_3 plants, but evidence of functional C_4 metabolism is yet to be published. Achievement of a high level of expression of C_4 enzymes in rice suggests that *trans* acting factors present in rice recognize C_4 genomic clones and that the mechanism for upregulation of housekeeping genes such as PPC and PDK still exists in C_3 plants. The discovery that overexpres-

sion of maize NADP–malic enzyme in rice chloroplast is accompanied by reduction in PSII activity and reduced granal stacking. This opens up newer possibilities for research into coregulation of unrelated genes.

Mesophyll-specific gene expression is mainly regulated at the transcriptional level, whereas bundle sheath specific expression is likely to be controlled at both the transcriptional and posttranscriptional levels in maize. Most C_4 genes are closely related homologs displaying low ubiquitous expression in C_4 and C_3 plants. C_4 genes are products of recent gene duplication. Drastic changes in regulatory sequences could have caused the changes.

Several of the C_4 genes have been cloned, and these are members of a small gene family whose individual members have independent roles. It has been clearly shown that the cell-specific localization of many of the C_4 enzymes is due to differential expression of the respective genes, which are regulated in the two cell types at the level of transcription. In C_4 plants, there appears to be a spatial reregulation of genes for metabolic enzymes that are also present in C_3 plants i.e. plants not using the C_4 pathway. However, the nature of this reregulation of certain C_4 genes involved in photosynthesis is unknown because the regulation of these C_4 genes in C_3 plants has also not been studied. On the contrary, a large amount of information is available on regulation of C_4 gene expression, especially with regard to the cell specificity and light inducibility [88]. With the advent of *in situ* transient expression assay it has now become possible to study the cell-specific and photoregulated expression of C_4 genes that are also present in C_3 plants, and C_4 plants. The assay would help in distinguishing the C_4 promoters from their C_3 counterparts and in determining whether the upstream sequences have been altered to give rise to cell-specific expression.

4. Leaf Development and Differential Gene Expression

Most C_4 genes examined show strict expression in leaves but not in roots or stem. The expressions of C_4 genes show temporal and spatial regulation that relates developmental stages of leaves. A current model points to unknown regulatory signals generated from veins for the control of bundle sheath and mesophyll differentiation and C_4 pattern gene expression [89].

Several studies have illustrated links between nitrogen and cytokinine signaling in controlling C_4 expression [90,91]. Metabolite repression has been extensively demonstrated in maize [92]. Stress and

ABA repress the expression of genes in C_4 photosynthesis.

a. Metabolites in C_4 regulation

Metabolite repression of C_4 gene transcription has been demonstrated [92,93]. Some promoters are repressed by sucrose, glucose, fructose, acetate, and glycerol. Inhibition of positive elements and not repression of negative elements cause transcriptional repression. Six other promoters have shown global effects of sugars and acetate on gene expression. Metabolite repression overrides light and developmental control and may serve as a mechanism for feedback regulation of photosynthesis.

b. Stress and abscisic acid

Eleocharis vivapara shows Kranz anatomy and C_4 characteristics under terrestrial conditions but develops C_3 traits when submerged in water [94]. Transfer from water to land can cause water stress. The anatomical development and gene expression could represent an adaptation. Abscisic acid is a stress hormone and controls the development of the C_4 pathway. ABA was able to induce the Kranz anatomy and C_4 gene expression, which can be uncoupled [94]. A nitrogen signal that controls C_4 gene expression was discovered in maize [90]. The signaling process does not involve *de novo* protein synthesis and protein phosphorylation [95]. A recent study shows that nitrogen signals are sensed by roots and stimulate accumulation of cytokinines that activate C_4 gene expression [95]. Recent studies have identified a cytokinine-inducible gene in maize and *Arabidopsis* [96,97].

c. Fixation of atmospheric CO_2 by bundle sheath cells

Does the Calvin cycle present in the bundle sheath cells of C_4 plants assimilate the carbon dioxide from the air directly in addition to the CO_2 released from C_4 acids? The answer to this is partially in the affirmative. Labeling studies using several inhibitors of photosynthesis like malonate for PEPC, DL-glyceraldehyde for RuBP generation, and 3-mercaptopicolinic acid and oxalate for C_4 acid decarboxylation showed that in the absence of functional PEP carboxylase and C_4 acid decarboxylation, no label was found in C_4 acids, whereas, the label was detected in a variety of Calvin cycle intermediates (14% to 17% of the control with a fully operational C_4 cycle). This indicates that a small amount of atmospheric CO_2 diffuses into the bundle sheath cells, where it is fixed by the Calvin cycle. The permeability coefficients for CO_2 diffusion in to bundle sheath cells show that these cells are at least 100 times less permeable to CO_2 than are C_3 mesophyll cells. The O_2 permeability barrier allows the development of high CO_2 concentrations in the bundle sheath cells during C_4 photosynthesis.

d. Photorespiration in C_4 plants

Several lines of evidence suggest an apparent lack of photorespiration in C_4 plants. The photorespiration in C_4 may be too low to be detected. The levels of photorespiratory enzymes and pools of metabolites involved in photorespiration are small. In addition to this, whatever CO_2 may be liberated through the photorespiratory cycle is refixed efficiently by PEPC sitting outside in the mesophyll cells. In C_4 plants, both the photoproduction of O_2 and photorespiration are low even at limiting CO_2 concentrations.

e. Energetics of C_4 pathway

The photochemical reactions of C_4 plants are of course the same as those operating in C_3 plants; however, the ratios of PSI and PSII may be different in the two plant types and also among the subgroups of the C_4 plants. In C_4 plants, the relationship of the quantum yield of PSII electron transport to quantum yield of CO_2 fixation is linear; suggesting that photochemical use of energy absorbed by PSII is tightly linked to CO_2 fixation in C_4 plants. This is nearly identical in all the three subgroups and may allow estimates of the photosynthetic rate based on the PSII efficiency. The energy requirement for the classical PCR cycle is 3ATP and 2NADPH/CO_2 fixed. The real cost of oxygenase and photorespiration is much higher. The estimate of energy requirement of C_4 plants is based on the specific demand for carbon metabolism. Since NADP–malic enzyme species exhibit agranal chloroplast, the entire NADPH requirement has to be met by NADP–malic enzyme. However, the latter could provide only half of the total NADPH required, the other half of 3-PGA must be exported to the mesophyll cells for fixation to DHAP, which is then transported back to bundle sheath cells. The relation between carbon metabolism, energy requirement, and quantum yield is given in Table 21.3. The calculated total energy demand for NADP-ME and NAD-ME-type species are 5ATP and 2NADPH/CO_2 fixed, but photoreactions of PCK-type species generate less ATP and more NADPH. Irrespective of this difference, the greater energy efficiency of C_4 photosynthesis is apparent. In terms of ATP, the C_3 photosynthesis requires 14.5ATP/CO_2 fixed as compared with about 11ATP/CO_2 in C_4 photosynthesis (assuming 1NADPH = 3ATP in terms of energy equivalents). The NADPH requirement can be met by eight quanta, but the amount of energy needed to produce ATP depends on the mechanism of proton partitioning across the thylakoid membrane. The

TABLE 21.3
Photosynthetic Traits of C₃, C₄, and CAM

Characteristics	C₃	C₄	CAM
Leaf anatomy	No bundle sheath cell	Well organized bundle sheath, organelle rich	Usually no palisade cells, large vacuoles in mesophyll cells
Carboxylating enzymes	Rubisco	PEP carboxylase	Dark, PEP carboxylase; light, Rubisco
Theoretical energy requirement	1:3:2	1:5:2	1:6.5:2
Transpiration ratio	450–950	250–350	18–25
Leaf chlorophyll *a/b* ratio	2.8	3.9	2.5
CO_2 compensation point	30–70	0–10	0–5 in dark
Photosynthesis inhibited by 21% O_2?	Yes	No	Yes
Photorespiration detectable?	Yes	Only in BS cells	In late afternoon
Optimum temperature (°C)	15–25	30–47	Approximately 35
Dry matter production (tons hectare^{-1} year^{-1})	22	39	Low and highly variable

involvement of the Q cycle in ATP synthesis has been well documented. The existence of the Q cycle mechanism is considered to be largely responsible for the evolution of the C₄ pathway in terms of its energy requirement. The CO_2 leakage from the bundle sheath cells and its recycling would of course put an additional burden on the total energy requirement.

f. Role of carbonic anhydrase in CO_2 fixation

The initial carboxylation reaction of the C₄ pathway utilizes bicarbonate rather than CO_2 as the inorganic carbon substrate [98]. Thus, the atmospheric carbon dioxide must be converted to bicarbonate rapidly. The enzyme carbonic anhydrase, which catalyzes the conversion, is indeed localized in mesophyll cells of C₄ plants, where PEP carboxylase is located [99,100]. The maximum activity of this enzyme when measured at the saturating concentrations of CO_2 at 25°C was found to be 3,000 to 10,000 times the maximum photosynthetic rates in leaves; however, under *in vivo* conditions the rates are just sufficient to maintain the optimal rate of carbon fixation and are not limiting. The role of CA in C₃ plants is still unclear [101]. It may convert HCO_3 to CO_2. CA is indeed found to be associated with Rubisco in carboxysomes in cyanobacteria. A recent study on expression of carbonic anhydrase and Rubisco in a C₃ plant, pea, shows that CA and Rubisco expression are correlated during development and CA protein levels in mature tissue are modulated with respect to Rubisco abundance [102]. The enzyme may simply facilitate diffusion of CO_2 through cytosol and chloroplast stroma. The role of CA in photosyn-

thesis has been reviewed recently by Badger and Price [103].

g. Evolution of C₄ pathway

C₄ plants are known to be of polyphyletic origin and to have evolved independently several times during the evolution of angiosperms. This indicates that the evolution of the C₄ metabolism was quite easy to accomplish. One reason for this may be that all the enzymes of the C₄ pathway of photosynthesis including PEP carboxylase were already present in C₃ plants. The evolution of C₄ photosynthesis therefore took advantage of a set of genes already existing in ancestral C₃ species and used them as a starting point to create genes in this specialized pathway of photosynthesis. The new expression pattern and regulatory elements of these genes were suitably modified during evolution to make them more efficient and spatially regulated. It is believed that C₄ pathway has probably existed at low abundance for past 12–13 million years. Since the time of the fossil grass *Tomlinsomia*, which has a known Kranz anatomy and $\delta^{13}C$ value of −13.7%, much of the evidence from direct and indirect sources dates the explosion of C₄ plant biomass at some 6 to 8 million years ago, when atmospheric CO_2 levels fell to about 200 ppm in air with 20% oxygen. Under these conditions the catalytic shortcomings of Rubisco favor the oxygenation of RuBP. This increases the energy cost of C₃ photosynthesis. The CO_2 concentrating mechanism evolved in C₄ plants gave a competitive edge during atmospheric and warmer periods. C₄ photosynthetic genes had previously been considered to be specific for C₄ plants, since the activities of the corresponding enzymes were low in C_3 plants and their kinetic properties are usually different from those in C_4 plants [104–106]. However, recent comparative studies have revealed that C_3 plants have at least two different types of gene, one encoding enzymes of "housekeeping" function and other very similar to C_4 genes of C_4 plants, although the expression of the latter is very low or even undetectable in C_3 plants. Based on these it is believed that C_4 genes evolved from a set of preexisting counterpart genes in ancestral C_3 plants with modifications in the expression level in leaves and kinetic properties of the enzyme [107]. Maize has three different isoforms of PPDK, namely chloroplastic (involved in C_4 pathway) and two cytosolic. Rice also has three different isoforms. One of the genes of rice is highly homologous to maize pdkI [108]. The reasons for low level of expression of C_4 genes in C_3 plants are because of regulatory elements. Recent studies have clearly shown that a *cis* acting element for light responsive expression is present in the rice promoter but those of cell-specific and high level expression were missing and had to be acquired during evolution. Some of these *cis* acting elements in the promoter have been identified as *trans* acting elements required for expression of C_4 specific genes present in C_3 plants. Thus, modification of pdkI or other genes required for evolution from C_4-like to C_4-specific genes is relatively simple, namely the gain of *cis* acting elements for cell-specific and high level expression in the promoter region.

So far only a limited number of transgenic plants containing a maximum of two C_4 cycle enzymes have been investigated physiologically. Most of the observed effects were based on pleiotropic changes in metabolism. Modification of enzyme activity by covalent modification and the availability of cofactors and substrates ought to be considered as well. It appears to be useful and perhaps even necessary to engineer a larger number of C_4 enzymes or the respective promoters to be better adapted to specific requirements of the C_4-like cycle in a C_3 environment.

h. Possibility of using single-cell C_4-like system in air

Since the discovery of the mechanisms related to the C_4 cycle some 37 years ago, the spatial separation into mesophyll and bundle sheath cells was thought to be a prerequisite for an efficient CO_2 concentrating mechanism. It is therefore surprising that a submerged aquatic plant, *Hydrilla verticillate*, was identified as being capable of inducing a C_4-like metabolism but lacks the Kranz anatomy [113]. The switch from C_3 to C_4 is triggered by low CO_2 concentrations, i.e. at high water temperatures, which

results in increased PEPC, malic enzyme, and pyruvate Pi dikinase activities and causes a substantial drop in the CO_2 compensation point. Further evidence of a C_4-like metabolism without a Kranz anatomy has emerged from *Egeria densa* [114]. Apart from inducible photosynthesis, these species also exhibit a pH polarity of their leaf surface allowing a higher CO_2 concentration at equilibrium [114] CAM plants also exhibit single-cell concentrating mechanisms. Attempts to introduce a single-cell CO_2 concentrating mechanism in terrestrial crops using a transgenic approach is in progress [115]. It is believed that introduction of an intracellular CO_2 pump might improve the efficiency of C_3 plants by suppression of photorespiration. The progress in this direction has been heartening and it is believed that engineering C_4 cycle enzymes driven by more specific promoters may be useful. The terrestrial chenopod *Borszczowia aralocaspica* appears to contain a single-cell concentrating mechanism distributed between the cytosol, mitochondria, and two types of chloroplast. Although this is a very exciting discovery, the anatomy of this plant is adapted to a semi-dry environment with succulent leaves, central vascular bundle water storing cells, and intracellular air spaces. These features are different from typical C_3 plants. It might be necessary to decrease the free air space in the leaves of C_3 crops to minimize stomatal aperture.

A number of attempts are presently made to introduce the enzyme components of C_4 photosynthesis into C_3 plants such as rice, potato, and tobacco. These attempts are based on the *Hydrilla* system, which is single–celled, having PEP carboxylase in the mesophyll cytosol, and the decarboxylation of C_4 acid takes place in the chloroplast with the premise that a CO_2 concentrating system would operate in C_3 mesophyll cells. A single-cell CO_2 concentrating mechanism is effective in algae and cyanobacteria; however, both have an internal compartment within the chloroplast that prevents CO_2 leakage from the site of CO_2 release. How does this system operate in *Hydrilla* in the vicinity of Rubisco? It seems that many of these single-cell CO_2 concentrating mechanisms are able to function because of a long diffusion path from one end of the cell to the other. So the answer to this question is apparently yes.

The major question that needs to be addressed is whether or not it is desirable to engineer C_3 plants to behave biochemically as C_4 plants with higher photosynthetic efficiency [109]. It seems unlikely that attempts to introduce single-cell CO_2 concentrating mechanisms will be successful without introducing some of the structural characteristics of C_4 plants, i.e. a compartment in which CO_2 can be concentrated, and coordinated regulation of all enzymes of the C_4 pathway will be required including posttranslational modification and other regulation. Expressing a single gene or even two genes may be deleterious as in the case of egression of high levels of NDP–malic enzyme, which causes marked changes in the NADP/NADPH ratio in rice plants, causing seriously stunted growth and bleaching of leaf color due to enhanced photoinhibition under natural light conditions. The increase in the NADP/NADPH ratio suppresses photorespiration, rendering photosynthesis more susceptible to photoinhibition.

Apart from photosynthesis, overproduction of a single C_4-specific enzyme seems to have some positive effects on the physiology of C_3 plants. It has been reported that overproduction of the chloroplastic but not the cytosolic PPDK increased the number of seed capsules and the weight of each seed capsule in transgenic tobacco [110]. Overproduction of C_4-specific PEP carboxylase improved the resistance of root elongation to aluminium [111]. It is important to elucidate the mechanism of these effects and to confirm whether or not similar phenomena can be generally observed in various other plant species.

i. C_4 Pathway in nonleaf green tissues of C_3 plants

The role of nonleaf green tissues like pods, pod walls, and ears in carbon assimilation in several C_3 plants has been shown to be extremely important for the overall yield because of their ability to reassimilate respiratory CO_2 and thus help in reducing dry matter losses. The contribution of nonleaf tissues and the mechanism of CO_2 fixation have been studied in pigeon pea and wheat [112]. It has been clearly established that these C_3 species fix CO_2 in nonleaf tissues through the C_4 pathway and utilize PEPC for recapturing respired CO_2; however, the pod wall is impermeable to atmospheric CO_2. Immature pericarps of wheat and barley, both C_3 plants, have been reported to exhibit very high activity of PEPC [112].

IV. PHOTOSYNTHESIS IN C_3–C_4 INTERMEDIATES

C_3–C_4 intermediate species offer an excellent model system for studying the mechanism of photorespiration and the evolutionary aspects of the C_4 pathway [116–119]. Till now 23 species belonging to seven genera from five families have been reported to be C_3–C_4 intermediates. C_3–C_4 intermediate species seem to have arisen during the process of evolution of C_4 plants from the C_3 species, and C_3 photosynthesis is believed to be the phylogenetic precursor of the C_4 pathway [120]. That intermediary characteristics are evolutionary rather than hybrid products is evident from the fact that in the Cruciferae, *Moricandia arvensis* and *Panicum milliodes* are intermediates but C_4 plants have not been identified in this family. A partial "Kranz" anatomy occurs in all the present day C_3–C_4 intermediates [121,122], and efficient recycling of photorespiratory CO_2 results in a low level of photorespiration. A partial reduction of activities of the photorespiratory enzymes in the intermediate species is evident from limitations in both the extent of glycine production and decarboxylation. In addition to this, the enzyme compartmentation may be an important factor in reducing photorespiration. It has been shown recently that bundle sheath mitochondria are enriched with glycine decarboxylase, which would facilitate efficient refixing of CO_2 not only from photorespiration but also dark respiration [123]. The efficiency of water use of C_3–C_4 intermediates is also greater than that of C_3 plants, apparently due to a low CO_2 compensation point [124]. Artificial production of C_3–C_4 intermediates have been attempted by generating hybrids between C_3 and C_4 species of *Atriplex* [125]. A few hybrid individuals resembled the C_4 parents both anatomically and biochemically but still were unable to carry out fully integrated C_4 photosynthesis. Thus, it appears that a proper compartmentation of photosynthetic reactions seems to be a stringent requirement for C_4 photosynthesis, although some studies conducted earlier on *Hydrilla* do no corroborate this conclusion [114]. The inheritance of a complete C_4 pathway is rather complex, even though the number of genes involved for each component may not be large [126,127]. A subsequent study on hybrids between C_3 and C_4 *Atriplex* species showed that the inheritance of the biochemistry of C_4 photosynthesis and of a Kranz anatomy are not closely linked, and segregation of a hybrid possessing complete and fully coordinated C_4 photosynthesis is a rare event. Thus, conventional breeding methods may not be useful in incorporating C_4 traits into C_3 plants. The newer techniques of plant transformation and genetic engineering are opening up new vistas in this endeavor and are under extensive investigation [128]. Of the C_3–C_4 intermediates identified so far, those in the genus *Flaveria*, which appear to be true intermediates, seem to provide the maximum potential for future evolutionary studies of C_4 pathway. Interspecific crosses between C_3 and C_4 and C_3–C_4 intermediate species are quite successful. Future taxonomic, cytogenetic, and genetic studies may be helpful in interpreting the evolutionary data.

V. CRASSULACEAN ACID METABOLISM

The crassulacean acid metabolism was discovered in 1815 by Heyne [129,130]. Intensive investigation of the CAM phenomena during the last three decades has led to the concept that CAM represents a modification of the photosynthetic pathway by which certain terrestrial plants harvest carbon dioxide from the atmosphere. Furthermore, as in the case of C_4 plants, CAM also provides a dramatic example of strategies that enable plants to consolidate their water and carbon balance in an arid environment but CAM is not the ancestor of C_4 photosynthesis. CAM is found in terrestrial angiosperms that have diversified polyphyletically from C_3 ancestors some time during the Miocene as a consequence of reduced CO_2 and water stress. CAM has attracted worldwide attention because of its ecological importance and interesting aspects of comparative biochemistry and physiology. It is therefore not surprising that several reviews and monographs have been published on this aspect during the last two decades [129–133].

A. Basic Phenomena

In contrast to other photosynthetic plants, CAM plants fix CO_2 mainly at night; hence they are popularly known as "dark fixers." They open their stomata at night and keep them closed during the major part of the day. This we know now is true for a very limited number of CAM plants, and there is a wide spectrum of response ranging from no net CO_2 uptake to 24 h continuous CO_2 uptake [134]. The large amount of CO_2 fixed at night is stored as malate via the C_4 pathway of photosynthesis. The accumulation of malate causes the tissue to become acidic, thus the name "acid metabolism." This phenomenon was first observed in the family Crassulaceae. Therefore, the term "crassulacean" was added to the name of the pathway, but it is by no means limited to this family. Some plants are obligate CAM, while in others it is facultative. In the latter, this phenomenon can be induced by the photoperiod, water stress, hormonal treatment, etc. [129–131]. Although CAM is definitely an adaptation to water stress in terrestrial species, it is manifested in a diverse array of species and life forms that make generalization about the pathway difficult. One of the most striking themes to emerge in the recent years is the extent to which the phylogenetic and ecological diversity of CAM plants is also reflected in a remarkable plasticity of the basic metabolic scheme. Genotypic, ontogenetic, and environmental factors such as light intensity, relative humidity, and water availability combine to govern the extent to which the biochemical and physiological attributes of CAM are expressed [135]. Most notable among these CAM plants are pineapple, agave, cacti, and orchids. CAM is correlated with various anatomical or morphological features that minimize water loss including thick cuticles, a low surface to volume ratio, large cells, and vacuoles with enhanced water storage capacity and reduced stomatal size and/or frequency. CAM is found in approximately 7% of vascular plant species yet suffers from being considered as a minor photosynthetic pathway restricted to a small number of desert plants. CAM has evolved independently on numerous occasions in different families and even with individual families. It is curious that CAM is also found in aquatic vascular plants where it presumably enhances inorganic carbon under conditions where CO_2 availability can become rate limiting.

CAM plants exhibit a distinct diurnal oscillation in their malic acid content. The malic acid rhythm is characterized by a nocturnal increase in concentration up to 20 $\mu Eq\ g^{-1}$ fresh wt. and the disappearance of that acid during the day. The diurnal rhythm is coupled to an inverse rhythm of the starch level in the cells.

1. Comparative Aspects of C_3, C_4, and CAM Plants

The major differences between the C_4 and CAM pathways is that the CO_2 concentrating mechanism in C_4 plants is based on spatial separation of C_4 acid decarboxylation, and the enzymes involved in carboxylation–decarboxylation are compartmentalized in adjacent cells, whereas in CAM plants these enzymes are found in the same cells and the two processes are temporally separated, requiring complex control of alternative metabolic pathways in response to light and water stress. A comparative statement on various physiological traits of C_3, C_4, and CAM plants is given in Table 21.3. The carbon dioxide concentration in CAM plants results from a high rate of decarboxylation in light, and internal production of CO_2 (up to 1%) results in stomatal closure. The unique physiological feature of crassulacean plants is their capacity to store and mobilize free malic acid in the cell vacuoles. The uptake of malate into the vacuoles is an energy-dependent process that occurs against a transmembrane concentration gradient and the transport is coupled to the action of an H^+ translocating ATPase [136]. The linkage between net CO_2 fixation, malic acid synthesis, and carbohydrate utilization in dark has been clearly established.

$$Reserve\ Carbohydrate \xrightarrow{\quad CO_2 \quad} C3\ Precursors \longrightarrow Malic\ acid.$$

2. Induction of CAM

Water economy is the most significant functional advantage the pathway provides to the plants. A recent demonstration of induction of CAM in normal succulent C_3 plants under water stress has provided excellent material for the study of induction, regulation, and significance of the CAM pathway. For example, 7 to 14 days after application of water stress, mature leaves of *Mesembryanthemum crystallinum* show substantial dark fixation with malic enzyme synthesis and stomatal closure during acidification [137–140]. The capacity for CAM induction in these plants is lost when the stress is removed. CAM-specific PEPC is elicited by osmotic stress and shows an about 50-fold increases in the enzyme activity. Exposure of roots to low temperature or anoxia also elicits CAM in *M. crystallinum*. Associated with the transition from C_3 to CAM in facultative CAM plants are increases in the activities of the key enzymes like PEP carboxylase, nocturnal opening of stomata, and other CAM associated properties. The increase in PEP carboxylase activity is paralleled by an increase in PEPC mRNA. However, very little is known about the nature of the stimulus perceived by the plants during stress. It is postulated that a common mechanism or signal such as an endogenous growth regulator may play a role in induction [141]. Exogenously applied ABA has been shown to substitute for salinity or drought in inducing CAM [142].

3. Regulation and Adaptive Mechanisms in CAM

Recent studies have illustrated how the behavior of plants using CAM provides adaptations to salinity. Plants having constitutive CAM show adaptations at the whole plant level, involving regulation of stomata, internal CO_2 recycling, and a root physiology associated with salt exclusion. Thus, they are stress avoiders. Annual plants such as *M. crystallinum*, which is an inducible CAM, are salt includers. They are stress tolerant and show regulation at several levels: (i) regulation of turgor and gas exchange at the whole plant level; (ii) metabolic adjustments at the cellular level; (iii) adaptive transport proteins at the membrane level and (iv) at the macromolecular level; and (v) inductive changes at the gene expression level of the enzyme complement for metabolism (in particular involving glycolysis and malic acid synthesis with PEPC as the key enzyme, and gluconeogenesis (with PPDK as the key enzyme) and membrane transport (in particular involving tonoplast ATPase [143]. The development of CAM in inducible CAM plants is under strict developmental control and appears only when a certain stage of development of the whole plant is reached. Higher concentrations of NH_4^+ in hydroponically grown *Kalanchoe blossfeldiana* depressed CAM photosynthesis through suppression of K^+ absorption, and K^+ is also known to play an important role in CAM photosynthesis, particularly in relation to PEPC synthesis [144–146].

4. Control of Malic Acid Synthesis

Malic acid synthesis is influenced by environmental factors like temperature, carbon dioxide, oxygen, and light intensity. An increase in temperature also increases the efflux of malate from vacuoles by changing the properties of the tonoplast. It is assumed that the increased export of malic acid lowers the *in vivo* activity of phosphoenolpyruvate carboxylase by feedback inhibition. Malic acid synthesis is also influenced by internal factors including tissue water potential. Some of these responses are controlled by CO_2 availability through stomatal control. It has also been postulated that reduced sensitivity of guard cells PEPC activity to malate inhibition is an important regulatory feature of stomatal opening, which is associated with malate accumulation [147].

5. Regulation of Carboxylation–Decarboxylation

The rate-limiting enzyme in CAM is phosphofructokinase [148,149]. The activity of this enzyme *in vivo* regulates the flow through glycolysis [150]. One of the known regulators of phosphofructokinase is PEP [151]. The increase in the enzyme activity is accompanied by a decrease in PEP pool size.

6. Regulation of Decarboxylation

There is often a lag of several hours before any significant decarboxylation commences on illumination. This is because of the activation of light-dependent reactions and the metabolite activation of rate-limiting enzymes. Once malic acid is available at the saturating concentrations, it may be decarboxylated by NADP–malic enzyme in cytoplasm or NAD–malic enzyme in mitochondria. The exact distribution of the malic acid through these two decarboxylating pathways is not known [152]. In most of the PEPCK-CAM the activity of NADP–malic enzyme was not negligible, but the pyruvate cannot be converted efficiently to PEP in these systems. An essential component of all hypotheses regarding the regulation of CAM is the inhibition of PEPC during deacidification, which is essential for preventing a futile cycle of carboxylation–decarboxylation. The inhibition of the enzyme by malic acid is generally the accepted mode of regulation. However, several recent studies show that phosphorylation–dephosphorylation and dimer/tetramer conversion may be the primary regulating mechanisms influencing the malate sensitivity [153]. The oligomerization state vis-à-vis phosphorylation is still not clear. The day enzyme was mainly a malate-sensitive homodimer, and during night it is malate insensitive homotetramer [154].The dark form of CAM-PEPC is phosphorylated at the serine residue and is more active and less sensitive to malate as compared with the unphosphorylated day form, which is highly sensitive to malate and less active. Malate reduces binding of PEP, while glucose-6-phosphate increases the binding of the substrate. Glucose-6-phosphate requires magnesium for binding, while malate does not. In view of the light-induced changes in the aggregation state of the holoenzyme, the general mechanism for regulation of PEPC in CAM plants is currently under dispute [154]. Recent studies show that PEP carboxylase is not only controlled by the light dark transition but circadian oscillations may also be involved in flux through PEP carboxylase [155]. The primary effect of circadian oscillation in this system may be at the level of tonoplast, and changes in kinase expression may be secondary to circadian changes in the concentration of the metabolite, perhaps cytosolic malate.

It is possible that there is a direct connection between circadian oscillation and expression of PEP carboxylase kinase in CAM plants that is mediated by a transcription factor similar to CCA1 (circadian clock-associated) protein of *Arabidopsis* [156]. Such a connection could involve cytosolic pH, but this hypothesis could not be substantiated experimentally. The other possibility is that this could be a secondary control. Concentration of malate reduces both kinase mRNA and accumulation of kinase [157].

In well irrigated CAM plants the stomata open in the afternoon following malic acid consumption. The CO_2 fixation involves conventional photosynthesis and assimilation of atmospheric carbon. The initial product of CO_2 fixation is 3-PGA and is subsequently converted to sucrose as in C_3 and C-4 plants. Similarly the CO_2 fixation is also stimulated by reducing the oxygen concentration in CAM plants.

7. Productivity of CAM Plants

CAM species (agave and cactus) are taxonomically at least five times more numerous than C_4 species, but often grow slowly. However, the slow growth is not a necessary corollary of the CAM photosynthetic pathway as can be seen from the energetics of CO_2 fixation. For every CO_2 fixed photosynthetically, C_3 plants require 3ATP and 2NADPH, whereas the add-

itional enzymatic reactions and compartmentation complexity for C_4 plants require 4 or 5ATP and 2NADPH and the CAM plants require 5.5 to 6.5ATP and 2NADPH. Photorespiration in C_3 plants can release part of the CO_2 fixed and also has an energetic cost, whereas photorespiration is much less in C_4 and CAM plants. Therefore, CAM plants can perform net CO_2 fixation 15% more efficiently than C_3 plants, although 10% less efficiently than the C_4 plants. Using a simple model that assumes eight photons per CO_2 fixed and a processing time of 5 msec, a maximum instantaneous rate of 55 msec mol m^{-2} is predicted. Using this model to study the potential productivity of all three classes of plant has clearly shown a high potential productivity of certain CAM species under optimal environmental conditions exceeding that of most C_3 species. This may increase the cultivation of such CAM species in various areas in the future.

8. Circadian Rhythm in CAM Plants

Circadian rhythms are biological oscillations displaying a 24 h period under natural environmental conditions. These rhythms reflect the behavior of endogenous systems or a property called the biological clock. The circadian rhythm of organic acids in the Crassulaceae was classically defined as a night/day two-step mechanism. The malate accumulates during the night and the major part of it is depleted during the day with production of CO_2. Malate and aspartate are allosteric inhibitors of PEPC and the activators of the malic enzyme [158]. The causes of rhythmicity in CAM appear to be very complex, and allosteric inhibition of PEPC by malate seems to be a good candidate for initiating oscillations [159]. The carboxylation rhythm is controlled by the stomatal rhythm. The available data from the experiments done in continuous light and dark suggest that these rhythms are primarily limited by physiological processes such as tonoplast flux and stomatal conductance [159]. One great challenge in understanding circadian regulation of CAM will be dissecting the mechanism responsible for controlling the circadian oscillations in malate uptake and release across the tonoplast membrane. It will be important in particular to understand how tonoplast malate transport is controlled by an underlying nuclear-controlled circadian clock. Rapid molecular identification of malate transport components in the tonoplast and circadian clock components from CAM species will be essential for these efforts. Diel oscillations in the activity of PEPC are controlled in part by circadian changes in its phosphorylation state and play a key role in directing the carbon flux through the CAM pathway by changing the enzyme sensitivity to malate, which in turn is regulated by PEPC kinase.

9. Molecular Genetics of CAM

A large number of enzymes, transporters, and regulatory proteins required for CAM have been identified and characterized. Most studies have been restricted to inducible CAM because of the differential gene expression induced in response to water. CAM induction is in response to salinity, water deficit, osmotic stress, or ABA treatments controlled by transcriptional activation initiated through a signaling cascade with an apparent requirement of calcium and calcium-dependent protein kinase. Changes in mRNA stability and utilization or translational efficiency are also likely to govern gene expression changes during the C_3 to CAM transition. Molecular characterization of vacuolar malate transporters, carriers, and channels for malate influx and efflux has remained a challenge. Recent studies indicate about threefold increase in malate transport on CAM induction.

10. Genetic Model for CAM

C_3 and C_4 plants have well-established genetic models like *Arabidopsis* and *Zea mays,* respectively. However, no genetic model has yet been developed for CAM plants. This has indeed hindered our understanding of many molecular processes that regulate CAM. The common ice plant seems to fulfill many of the conditions required to be called a model plant for molecular and genetic studies including an efficient transformation system and does not involve tissue culture-based methodology.

VI. CONCLUDING REMARKS

Throughout this review I have tried to present an overall perspective of the photosynthetic pathways operating in higher plants. Even though the basic mechanism of the process is known, the finer details on the regulation of the key enzymes involved in carbon fixation, transport, and partitioning of the photosynthates are still not very clear. The present day emphasis has been on understanding the regulation both at the protein level and also at the level of gene expression. Understanding the intricacies of light-regulated gene expression would be useful in deciding the strategies for plant transformation with an ultimate goal of improving potential crop yield. We know from CO_2 and light enrichment studies that crop yield is frequently, perhaps usually, limited by photosynthetic efficiency, yet genetic improvement in the growth rate or photosynthetic rate has not oc-

curred so far. Attempts to improve either the catalytic efficiency of Rubisco or its specificity for CO_2 or O_2 using various methods including site-directed mutagenesis should receive greater attention. Control of plant productivity by regulation of photorespiration is another area whose potential has been well recognized, but the solution still eludes us. Of course, modulating the Rubisco specificity would surely alleviate this problem to a large extent. Evolutionary and regulatory aspects of C_4 and CAM photosynthesis are also topics of present day interest, including the phylogenetic relationship of various PEPC isoforms for gaining insight into the evolution of C_4 and CAM plants.

ACKNOWLEDGMENTS

I am very grateful to all those who supplied me reprints of their papers. I wish to thank Dr. P.V. Sane for introducing me to this area of research.

REFERENCES

1. Bassham JA, Benson AA, Kay LD, Harris AZ, Wilson AT, Calvin M. The path of carbon in photosynthesis. XXI. The Cyclic regeneration of carbon dioxide acceptor. *J. Am. Chem. Soc.* 1954; 76:1760–1770.
2. Bassham JA, Calvin M. *The Path of Carbon in Photosynthesis*, Englewood Cliffs, New Jersey: Prentice Hall, 1957.
3. Calvin M. Forty years of photosynthesis and related activities. *Photosynth. Res.* 1989; 21:3–16.
4. Arnon DI, Allen MB, Whatley FR. Photosynthesis by isolated chloroplasts. *Nature (London)* 1954; 174: 394–396.
5. Allen MB, Whatley FR, Rosenberg JR, Capindale JB, Arnon DI. In: *Research in Photosynthesis* eds Gaffron H, Brown AH, French CC, Livingston R, Rabinowitch EI, Strehler BC, Tolbert NE, New York: Wiley Interscience, 1957:288–310.
6. Gontero B, Cardienas ML, Ricard J. Localization of enzymes of photosynthetic carbon reduction cycle. *Eur. J. Biochem.* 1988; 173:437.
7. Sainis JK, Merriam K, Harris GC. The association of ribulose-1,5-bisphosphate carboxylase/oxygenase with phosphoribulokinase. *Plant Physiol.* 1989; 89: 368–374.
8. Guidici-Orticoni MT, Gontero B, Rault M, Ricard J. Multienzyme complexes in photosynthetic systems. *C. R. Acad. Sci.* 1992; 3:314–319.
9. Suss KH, Arkona C, Manteuffel R, Adler K. Calvin cycle multienzyme complexes are bound to chloroplast thylakoid membrane of higher plants *in situ*. *Proc. Natl. Acad. Sci. USA* 1993; 90:5514–5518.
10. Buchanan BB. Thioredoxin and enzyme regulation in photosynthesis. In: *Modulation of Protein Function* eds

Atkins DE, Fox DB, New York: Academic Press, 1979:93–111.

11. Buchanan BB. Role of light in the regulation of chloroplast enzymes. *Annu. Rev. Plant Physiol.* 1980; 31:341–374.

12. Buchanan BB, Schurmann P, Kalberer PP. Ferridoxin activated fructose diphosphatase of spinach chloroplast. *J. Biol. Chem.* 1971; 246:5952–5959.

13. Bassham JA, Krause GK. Free energy changes and metabolic regulation in steady state photosynthetic caron reduction. *Biochem. Biophys. Acta.* 1969; 189:207–221.

14. Salvucci ME, Portis, AR Jr, Ogren WL. A soluble chloroplast protein catalyses ribulose bisphosphate carboxylase/oxygenase activation *in vivo*. *Photosynth. Res.* 1985; 7:193–201.

15. Somerville CR, Portis AR Jr, Ogren WL. A mutant of *Arabidopsis thaliana* which lacks activation of RuBP carboxylase *in vivo*. *Plant Physiol.* 1982; 70:381–387.

16. Portis AR Jr. Regulation of Rubisco by Rubisco activase. *J. Exp. Bot.* 1995; 46:1285–1291.

17. Jimenez ES, Medrano L, Martinez-Barajas E. Rubisco activase, a possible new member of the molecular chaperone family. *Biochemistry* 1995; 34:2826–2831.

18. Walker DA. In: *Plant Carbohydrate Chemistry Proceedings of the Tenth Symposium of the Photochemical Society*, ed Prindham JB, Edinburgh: Academic Press, 1974:7–12.

19. Anderson LE, Heinrikson RL, Noyes C. Chloroplastic and cytoplasmic enzyme subunit structure of pea leaf aldolase. *Arch. Biochem. Biophys.* 1975; 169:262–271.

20. Ramaswamy NK, Nair PM. Evidence for the operation of a C_1 pathway for the fixation of CO_2 in isolated intact chloroplasts from green potato tubers. *Plant Sci. Lett.* 1984; 34:261–267.

21. Ramaswamy NK, Behre AG, Nair PM. A novel pathway for the synthesis of solanidine in the isolated chloroplasts from greening potatoes. *Eur. J. Biochem.* 1976; 67:275–282.

22. Ramaswamy NK, Sangeeta GJ, Behre AG, Nair PM. Glyoxylate synthetase isolated from green potato tuber chloroplasts catalyzing the conversion of formate to glyoxylate. *Plant Sci. Lett.* 1983; 32:213–220.

23. Arora S, Ramaswamy NK, Nair PM. Partial purification and some properties of a latent CO_2 reductase from green potato tuber chloroplasts. *Eur. J. Biochem.* 1985; 153:509–514.

24. Raines CA, Lloyd JL, Dyer TA. Molecular biology of C-3 photosynthetic carbon reduction cycle. *Photosynth. Res.* 1991; 27:1–14.

25. Hartman. FC, Harpel MR. Structure, function, and assembly of D-ribulose 1,5-bisphosphate carboxylase/oxygenase. *Annu. Rev. Biochem.* 1994; 63:197–234.

26. Hartman FC, Harpel MR. Chemical and genetic probes of the active sites of D-ribulose-1,5-bisphosphate carboxylase: a retrospective based on the three-dimensional structure. *Adv. Enzymol.* 1993; 67:1–75.

27. Walker DA. Regulatory mechanism in photosynthetic carbon metabolism. *Curr. Top. Cell Regul.* 1976; 2:204–241.

28. Edwards GE, Walker DA. C_3, C_4 *Mechanisms, and Cellular and Environmental Regulation of Photosynthesis*, Oxford: Blackwell Scientific Publications, 1983.

29. Leegood RC. Regulation of photosynthetic CO_2-pathway enzymes by light and other factors. *Photosynth. Res.* 1985; 6:247–259.

30. Woodrow IE. Control of the rate of photosynthetic carbon dioxide fixation. *Biochim. Biophys. Acta* 1986; 851:181–192.

31. Bhagwat AS. Activation and inhibition of spinach ribulose-1,5-bisphosphate carboxylase by 2-phosphoglycolate. *Phytochemistry* 1982; 21:285–289.

32. Bhagwat AS. Activation of spinach ribulose-1,5-bisphosphate carboxylase by inorganic phosphate. *Plant Sci. Lett.* 1981; 23:197–206.

33. Sawada S, Usuda H, Hasegawa Y, Tsukui T. Regulation of ribulose-1,5-bisphosphate carboxylase activity in response to changes in the source/sink balance in single-rooted soyabean leaves: The role of inorganic orthophosphate in activation of the enzyme. *Plant Cell Physiol.* 1992; 31:697–704.

34. Sawada S, Usuda H, Tsukui T. Participation of inorganic orthophosphate in regulation of ribulose 1,5-bisphosphate in response to changes in photosynthetic source sink balances. *Plant Cell Physiol.* 1992; 33:943–949.

35. Krause GH, Heber U. In: *Intact Chloroplast*, ed Barber J., Amsterdam: Elsevier, 1976:171–189.

36. Hall DO. In: *Intact Chloroplast*, ed Barber J. Amsterdam: Elsevier, 1976:135.

37. Heldt HW. In: *Intact Chloroplast*, ed Barber J., Amsterdam: Elsevier, 1976:215–235.

38. Salvucci ME, Ogren WL. The mechanism of Rubisco activase: insights from studies of the properties and structure of the enzyme. *Photosynth. Res* 1996; 47:1–11.

39. Salvucci ME, Osteryoung KW, Crafts-Brandner SJ, Vierling E. Exceptional sensitivity of Rubisco activase to thermal denaturation *in vitro* and *in vivo*. *Plant Physiol.* 2001; 127:1053–1064.

40. Kallis RP, Ewy RG, Portis, AR Jr. Alteration of the adenine nucleotide response and increased Rubisco activation activity of *Arabidopsis* Rubisco activase by site directed mutagenesis. *Plant Physiol.* 2000; 123:1077–1086.

41. Ruuska S, Andrews TJ, Badger MR, Price GD, Caemmerer S. The role of chloroplast electron transport and metabolites in modulating Rubisco activity in tobacco. Insight from transgenic plants with reduced amount of cytochrome b/f complex or glyceraldehyde 3-phosphate dehydrogenase. *Plant Physiol.* 2000; 122:491–504.

42. Mate CJ, Van Caemmerer S, Evans RJ, Hudson GS, Andrews TJ. The relationship between CO_2-assimilation rate, Rubisco carbamylation and Rubisco activase in activase deficient transgenic tobacco suggest a simple mode of activase action. *Planta* 1996; 198:604–613.

43. Zelitch I. Photorespiration: studies with whole tissue. In: *Photosynthesis II, Photosynthetic Carbon Metabol-*

ism and Related Processes eds Gibbs M, Latzko E, New York: Springer-Verlag, 1979; 6:353–373.

44. Lorimer GH. The carboxylation and oxygenation of Ribulose1,5-bisphosphate: the primary events in photosynthesis and photorespiration. *Annu. Rev. Plant Physiol.* 1981; 32:349–383.

45. Andrews TJ, Lorimer GH. In: *The Biochemistry of Plants*, eds Hatch MD, Boardman NK, New York: Academic Press, 1988:131–218.

46. Heber U, Schreiber U, Siebke K, Dietz KJ. In: *Perspectives in Biochemical and Genetic Regulation of Photosynthesis*, ed Zelitch I, New York: Wiley Life Sciences 1990:239–254.

47. Ogren WL Chollet R. In: *Photosynthesis, Development of Carbon Metabolism and Plant Productivity*, Vol. II, ed Govindjee, New York: Academic Press 1982: 341–375.

48. Zelitch I. Selection and characterization of tobacco plants with novel O_2 resistant photosynthesis. *Plant Physiol.* 1989; 90:1457–1464.

49. Zelitch I. *Perspectives in Biochemical and Genetic Regulation of Photosynthesis*, New York: Wiley-Liss, 1990.

50. Schuller KA, Randell DD. Regulation of pea mitochondrial pyruvate dehydrogenase complex–Does photo respiratory ammonium influence mitochondrial carbon metabolism? *Plant Physiol.* 1989; 89:1207–1212.

51. Gemel J., Randell DD. Light regulation of leaf mitochondrial pyruvate dehydrogenase complex–role of photorespiratory caron metabolism. *Plant Physiol.* 1992; 100:908–914.

52. Oaks A. Efficiency of nitrogen utilization in C_3 and C_4 cereals. *Plant Physiol.* 1994; 106:407–416.

53. Ehleringer JR, Cerling TE, Helliker BR. C_4 photosynthesis, atmospheric CO_2 and climate. *Oecologia* 1997; 112:285–299.

54. Kellogg EA. In: Phylogenetic aspects of the evolution of C-4 photosynthesis, eds Sage RF, Monson RK, San Diego: Academic Press 2000:411–444.

55. Hatch MD, Slack CR. Photosynthesis in sugar cane leaves: a new carboxylation reaction and the path of sugar formation. *Biochem. J.* 1966; 101:103–111.

56. Hatch MD, Slack CR, Johnson HS. Further studies on a new pathway of photosynthetic carbon dioxide fixation in sugarcane and its occurrence in other plant species. *Biochem. J.* 1967; 102:417–422.

57. Kortschak HP, Hart CE, Burr GO. Carbon dioxide fixation in sugarcane leaves. *Plant Physiol.* 1967; 40:209–213.

58. Slack CR, Hatch MD, Goodchild DJ. Distribution of enzymes in mesophyll and parenchyma sheath chloroplasts of maize leaves in relation to C_4 dicarboxylic acid pathway of photosynthesis. *Biochem. J.* 1969; 114:489–498.

59. Edwards GE, Black CC. In: *Photosynthesis and Photorespiration* eds Hatch MD, Osmond CB, Slayter RO New York: Wiley Interscience 1971:153–175.

60. Hatch MD, Slack CR. Photosynthetic CO_2 fixation pathways. *Annu. Rev. Plant Physiol.* 1970; 21:141–162.

61. Hatch MD. The C-4 pathway of photosynthesis: evidence for an intermediate pool of carbon dioxide and the identity of the donor C-4 dicarboxylic acid. *Biochem. J.* 1971; 125:425–432.

62. Edwards GE, Kanai R, Black CC. Phosphoenol pyruvate carboxykinase in leaves of certain plants which fix CO_2 by C-4 decarboxylic acid cycle of photosynthesis. *Biochem. Biophys. Res. Commun.* 1971; 45: 278–285

63. Hatch MD, Kagawa T. Activity location and role of NAD malic enzyme in leaves with C-4 pathway photosynthesis. *Aust. J. Plant Physiol.* 1974; 1:357–369.

64. Edwards GE, Walker DA. *Mechanism and Cellular and Environmental Regulation*, Oxford: Blackwell Scientific Publications, 1983.

65. Hatch MD. C-4 photosynthesis: a unique blend of modified biochemistry anatomy and ultrastructure. *Biochim. Biophys. Acta* 1987; 895:81–106.

66. Ashton AR, Burnwell JN, Furbank RT, Jenkins CLD, Hatch MD. In: *Methods in Plant Biochemistry* 1990; 3:39–61.

67. Kennedy AR, Laetsch WM. Relationship between leaf development and primary photosynthetic products in the C_4 plant *Portulaca oleracea* L. *Planta* 1973; 115:113–119.

68. Raghavendra AS, Rajendrudu G, Das VSR. Simultaneous occurrence of C-3 and C-4 photosynthesis in relation to leaf position in *Molluga nudicaulis*. *Nature (London)* 1978, 273:143–144.

69. Wolf J. In: *Hand Buch der Pflanzen Physiologie*, ed Ruhland W, Berlin and New York: Springer-Verlag, 1960; 12:809–854.

70. Rajagopalan AV, Tirumala M, Raghavendra AS. Molecular biology of C-4 phosphoenol pyruvate carboxylase: structure, regulation and genetic engineering. *Photosynth. Res.* 1994; 39:115–135.

71. Edwards GE, Nakamoto H, Burnell JN, Hatch MD. Pyruvate, Pi dikinase and NADP-malate dehydrogenase in C_4 photosynthesis: properties and mechanism of light/dark regulation. *Annu. Rev. Plant Physiol.* 1985; 36:255–286.

72. Budde RJA, Chollet R. *In vitro* phosphorylation of maize leaf phosphoenol pyruvate carboxylase. *Plant Physiol.* 1986; 82:1107–1114.

73. Jio AJ, Chollet R. Light/dark regulation of maize leaf phosphoenol pyruvate carboxylase by *in vivo* phosphorylation. *Arch. Biochem. Biophys.* 1988; 261:409–417.

74. Drincovich MF, Andreo CS. Redox regulation of maize NADP malic enzyme by thiol-disulfide interchange: effect of reduced thioredoxin on activity. *Biochim. Biophys. Acta* 1994; 1206:10–16.

75. Bhagwat AS, Sane PV. Studies on the enzymes of C4 pathway: partial purification and kinetic properties of maize phosphoenol pyruvate carboxylase. *Ind. J. Exp. Biol.* 1976; 14:155–158.

76. Jawali N, Bhagwat AS. Inhibition of phosphoenolpyruvate carboxylase from maize by 2-phosphoglycollate. *Photosynth. Res.* 1987; 11:153–159.

77. Bhagwat AS, Mitra J, Sane PV. Studies on enzymes of C-4 pathway. Part III–Regulation of malic enzyme of

Zea mays by fructose-1,6-diphosphate & other metabolites. *Ind. J. Exp. Biol.* 1977; 15:1008–1012.

78. Asami S, Inoue K, Akazawa T. NADP-malic enzyme from maize leaf: regulatory properties. *Arch. Biochem. Biophys.* 1979; 196:581–587.

79. Jenkins CLD, Furbank RT, Hatch MD. Mechanism of C_4 photosynthesis: a model, describing the inorganic carbon pool in bundle sheath cells. *Plant Physiol.* 1989; 91:1372–1381.

80. Bhagwat AS, Sane PV. Studies on enzymes of C-4 pathway. Part II–Regulation of CO_2 fixation by malic enzyme in C-4 plants. *Ind. J. Exp. Biol.* 1976; 14:691–693.

81. Sane PV, Bhagwat AS. An analysis of the C-4 pathway–a possible mechanism for the regulation of CO_2 fixation. *Plant Biochem. J.* 1976; 3:1–10.

82. Hull MR, Long SP, Raines CP. In: *Current Research in Photosynthesis*, ed Baltscheffsky M, Dordrecht, the Netherlands: Kluwer Academic, 1990:675–681.

83. Baker NR, Nie GY, Ortiz-Lopez A, Ort DR, Long SP. In: *Current Research in Photosynthesis* ed Baltscheffsky M, Dordrecht, the Netherlands: Kluwer Academic, 1990:565–570.

84. Downes RW. Differences in transpiration rates between tropical and temperate grasses under controlled conditions. *Planta* 1969; 88:261–275.

85. Gunning BES, Robards AW. *Intercellular Communications in Plants*, Heidelberg: Springer-Verlag, 1976.

86. Hatch MD, Osmond CB. In: *Encyclopedia of Plant Physiology* (new series), Volume 3, eds Stocking CR, Heber U, New York: Springer Verlag, 1976: 144–163.

87. Rathnam CKM. Evidence for reassimilation of CO_2 released from C4 acids by PEP carboxylase in leaves of C_4 plants. *FEBS Letts.* 1977; 82:288–291.

88. Nelson T, Langdale JA. Developmental genetics of C-4 photosynthesis. *Annu. Rev. Plant Physiol.* 1992; 43:25–47.

89. Dengler NG, Nelson T. In: *C-4 Plant Biology* eds Sage RF, Monson RK, San Diego: Academic Press, 1999.

90. Sugiyama T. In: *Stress Responses of Photosynthetic Organisms* eds Satoh S, Murata N, Tokyo: Elsevier 1998:167–180.

91. Taniguchi T, Kiba T, Sakakibara H, Ueguchi C, Mizuno T, Sugiyama T. Expression of *Arabidopsis* response regulator homology is induced by cytokines and nitrate. *FEBS Lett.* 1998; 429:259–262.

92. Sheen J. Metabolic repression of transcription in higher plants. *Plant Cell* 1990; 2:1027–1038.

93. Jang JC, Leon P, Zhou L, Sheen J. Hexokinase as sugar sensor in higher plants. *Plant Cell* 1997; 9:5–19.

94. Agarie S, Kai M, Takatsuji H, Uneo O. Environmental and hormonal regulation of gene expression of C-4 photosynthetic enzymes in the amphibious sedge *Eleocharis vivipara*. *Plant Sci.* 2002; 163:571–580.

95. Sakakibara H, Suzuki M, Takei K, Deji A, Taniguchi M, Sugiyama T. A response-regulator homologue possibly involved in nitrogen signal transduction mediated by cytokinine in maize. *Plant J.* 1998; 14:337–344.

96. Taniguchi M, Kiba T, Sakakibara H, Ueguelis C, Mizuno T, Sugiyama T. Expression of *Arabidopsis* response regulator homologue is induced by cytokinines and nitrate. *FEBS Lett.* 1998; 429:259–262.

97. Takeuchi Y, Akagi H, Kamasawa N, Osumi M, Honda H.Aberrant chloroplasts in transgenic rice plants expressing a high level of maize NADP-dependent malic enzyme. *Planta* 2000; 211:265–274.

98. O'Leary M. Phosphoenolpyruvate carboxylase: an enzymologist's view. *Annu. Rev Plant Physiol.* 1982; 33:297–315.

99. Burnell JN, Hatch MD. Low bundle sheath carbonic anhydrase is apparently essential for effective C-4 pathway operation. *Plant Physiol.* 1988; 86: 1252–1256.

100. Gurierrez M, Huber SC, Ku SB, Edwards GE. Intracellular localization of carbon metabolism in mesophyll cells. In: *Proceedings of Third International Congress on Photosynthesis, Amsterdam, The Netherlands*, ed Avron M, Amsterdam: Elsevier, 1974:1219–1230.

101. Reed ML, Graham D. In: *Progress in Phytochemistry*, Volume 7, eds Reinhold L, Harborne JB, Swain J, Oxford: Pergamon Press, 1980:48–75.

102. Mageau N, Coleman JR. Correlation of carbonic anhydrase and ribulose-1,5-bisphosphate carboxylase/oxygenase expression in pea. *Plant Physiol.* 1994; 104:1393–1399.

103. Badger MR, Price GD. Role of carbonic anhydrase in photosynthesis. *Annu. Rev. Plant Physiol.* 1994; 45: 369–392.

104. Brown RH, Bouton JH. Physiology and genetics of interspecific hybrids between photosynthetic types. *Annu. Rev. Plant Physiol. Plant Mol. Biol.* 1993; 44:435–456.

105. Svensson P, Blasing OE, Westhoff P. Evolution of the enzymatic characteristics of C_4 phosphoenolpyruvate carboxylase–a component of the orthologous *PPCA* phosphoenolpyruvate carboxylase from *Flaveria trinervia* (C_4) and *Flaveriapringlei* (C_3). *Eur. J. Biochem.* 1997; 246:452–460.

106. Ku MSB, Kano-Murakami Y, Matsuoka M. Evolution and expression of C-4 photosynthesis genes. *Plant Physiol.* 1996; 111:949–957.

107. Imaizumi N, Ku MSB, Ishihara K, Samejima M, Kaneko S, Matsuoka M. Characterization of the gene for pyruvate orthophosphate dikinase from rice, a C-3 plant, and a comparison of structure and expression between C-3 and C-4 genes for this protein. *Plant Mol. Biol.* 1997; 34:701–716.

108. Sheny JE, Hardy M. In: *Rice Photosynthesis to Increase Yield*, Philippines: IRR Institute, Amsterdam: Elsevier Publications, 2000.

109. Sheriff A, Meyer H, Riedel E, Schmitt JM, Lapke C. The influence of plant pyruvate, orthophosphate dikinase on a C_3 plant with respect to the intracellular location of the enzyme. *Plant Sci.* 1998; 136:43–47.

110. Miyao K, Fukuyama H, Tamai H, Matsuoka M. High level expression of C-4 photosynthesis enzymes in transgenic rice. In: *Proceedings of the 12th International Congress on Photosynthesis, Canberra*,

Australia CSIRO, eds Osmons CB and Critchley C. CSIRO Publishing, Collingwood, Australia.

111. Luthra YP, Sheron IS, Randhir Singh. Photosynthetic rates and enzyme activities of leaves, developing seeds and pod walls of pigeon pea (*Cajanus cajan* L.) *Photosynthetica* 1983; 17:210–215.

112. Singhal HR, Sheron IS, Singh R. *In vitro* enzyme activities and products of $^{14}CO_2$ assimilation in flag leaf and ear parts of wheat (*Triticum aestivum* L). *Photosynth. Res.* 1986; 8:113–122.

113. Van Ginkel LC, Bowes G, Reiskend J, Prins HBA. A CO_2-flux mechanism operating via pH-polarity in *Hydrilla verticillata* leaves with C_3 and C_4 photosynthesis. *Photosynth. Res.* 2001; 68:81–88.

114. Casati P, Lara MV, Andreo CS. Induction of a C_4-like mechanism of CO_2 fixation in *Egeria densa,* a submerged aquatic species. Plant Physiol. 2000; 123: 1611–1621.

115. Matsuoka M, Furbank RT, Fukayama H, Miyao M. Molecular engineering of C-4 photosynthesis. *Plant Physiol. Plant Mol. Biol.* 2001; 52:297–314.

116. Raghavendra AS, Das VSR. Plant species intermediate between C3 and C4 pathways of photosynthesis: their focus on mechanism and evolution of C4 syndrome. *Photosynthetica* 1980; 14:271–283.

117. Edwards GE, Ku MSB. In: *The Biochemistry of Plants — A Comprehensive Treatise,* Volume 10, eds Hatch MD, Boardman NK, New York: Academic Press, 1987:275–311.

118. Monason RK, Moore BD. C-3, C-4 intermediate as a model for study in evolution. *Plant Cell Environ.* 1989; 12:689–693.

119. Raghavendra AS, Das VSR. In: *Photosynthesis, Photoreactions to Plant Productivity,* eds Abrol YP, Mohanty P, Govindjee, Oxford & IBH Publishing, New Delhi, 1993:317–337.

120. Moore BD, Franceschi VR, Cheng SH, Wu J, Ku MSB. Photosynthetic characteristics of C_3-C_4 intermediates *Parthenium hysterophorus.* Plant Physiol. 1987; 85:978–983.

121. Rajendrudu G, Prasad JSR, Das VSR. C_3-C_4 intermediate species in *Alternenthera* (*Amaranthaceae*). Leaf anatomy, CO_2 compensation point, net CO2 exchange and activities of photosynthetic enzymes. *Plant Physiol.* 1986; 80:409–414.

122. Devi MT, Rajagopalan AV, Raghavendra AS. Predominant localization of mitochondria enriched with glycine-decarboxylating enzymes in bundle sheath cells of *Alternanthera tenella,* a C_3–C_4 intermediate species. *Plant Cell Environ.* 1995; 18:1–6.

123. Devi MT, Raghavendra AS. Partial reduction in activities of photorespiratory enzymes in C_3-C_4 intermediates of *Alternenthera* and *Parthenium. J. Exp. Bot.* 1993; 44:779–784.

124. Kisaki T, Hirabayashi S, Yano N. Effect of age of tobacco leaves on photosynthesis and photorespiration. *Plant Cell Physiol.* 1973; 14:505–514.

125. Bjorkman O. In: *Metabolism and Plant Productivity,* eds Burris RH, Black CC, Baltimore, London, Tokyo: University Park Press, 1976; 287–312.

126. Osmond CB, Bjorkman O, Anderson DJ. *Physiological Processes in Plant Ecology,* New York: Springer-Verlag, 1980; 66–87.

127. Hudspeth RL, Grula JW, Dai Z, Edwards GE, Ku MS. Expression of maize phosphoenolpyruvate carboxylase in transgenic tobacco. *Plant Physiol.* 1992; 98:458–464.

128. Kluge M, Ting I. *Crassulacean Acid Metabolism: Analysis of Ecological Adaptation.* Berlin: Springer-Verlag, 1978:30–61.

129. Ranson SL, Thomas M. Crassulacean acid metabolism. *Annu. Rev. Plant Physiol.* 1960; 11:81–110.

130. Osmond CB. Crassulacean acid metabolism: a curiosity in context. *Annu. Rev. Plant Physiol.* 1978; 29: 379–414.

131. Osmond CB, Holtum JAM. In: *The Biochemistry of Plants,* Volume 8, eds Stumpf PK, Conn EE, New York: Academic Press, 1981:283–321.

132. Nobel PS. Achievable productivities of certain CAM plants: basis for high values compared with C_3 and C_4 plants. *New Phytol.* 1991; 119:183–205.

133. Dodd AN, Borland AM, Haslam RP, Griffins H, Maxwell K. Crassulacean acid metabolism: plastic, fantastic. *J. Exp. Bot.* 2002; 53:569–580.

134. Cushman JC, Borland AM. Induction of Crassulacean acid metabolism by water limitation. *Plant Cell Environ.* 2001; 24:31–40.

135. Cockburn W, Ting IP, Sternberg LO. Relationship between stomatal behaviour and internal carbon dioxide concentration in Crassulacean acid metabolism plant. *Plant Physiol.* 1979; 63:1029–1032.

136. Winter K. Evidence for the significance of crassulacean acid metabolism as an adaptive mechanism to water stress. *Plant Sci. Lett.* 1974; 3:279–281.

137. Winter K. CO_2 fixieringsreaktionen bei der salzpflanzen *Mesembryanthem crystallinum* unter variierten Aussenbedingungen. *Planta* 1973; 114:75–85.

138. Ting IP, Hanscom Z. Induction of acid metabolism in *Portulca riaafra. Plant Physiol.* 1977; 59:511–514.

139. Von Willert DJ, Treichel S, Kirst GO, Curdts E. Environmentally controlled changes of phosphoenolpyruvate carboxylase in *Mesembryanthemum. Phytochemistry* 1976; 15:1435–1436.

140. Winter K. NaCl-induzierter Crassulacean-Saurestoffwecyhselbei der Salzpflanzen *Mesembryanthem crystallinum* Abhangigkeit des CO_2-Gas wechsels von der Tag/Nacht Temeratur und von der Wasserversorgung der pflanzen. *Oecologia* 1974; 15:383–392.

141. Chu C, Dai ZY, Ku MSB, Edwards GE. Induction of Crassulacean acid metabolism in the facultative halophyte *Mesembranthemum crystallinum* by abscisic acid. *Plant Physiol.* 1990; 93:1253–1260.

142. Luettge U. The role of crassulacean acid metabolism (CAM) in the adaptation of plants to salinity. *New Phytol.* 1993; 125:59–71.

143. Herppich W, Herppich M, Von-Willert DJ. The irreversible C_3 to CAM shift in well-watered and salt stressed plants of *Mesembryanthemum crystallinum* is under strict ontogenetic control. *Bot. Acta* 1992; 105:34–40.

144. Ota J, Yamamato Y. Effect of nitrogen sources and concentrations on CAM photosynthesis in *Kalanchoe blossfeldiana. J. Exp. Bot.* 1991; 42:1271–1277.

145. Ota J, Kono Y. Depression of CAM photosynthesis through K+-deficiency in *Kalanchoe Blossfeldiana. Research in Photosynthesis, Proceedings of the IXth International Congress on Photosynthesis, Nagoya, Japan* 1992:III.18.895–898.

146. Friemert V, Heininger D, Kluge M, Ziegler H. Temperature effects on malic acid efflux from the vacuoles and on carboxylation pathway in Crassulacean acid meatabolism plants. *Planta* 1988; 174:453–461.

147. Stutton BG. The path of carbon in CAM plants at night. *Aust. J. Plant Physiol.* 1975; 2:377–387.

148. Pierre JN, Quiroz O. Regulation of glycolysis and level of Crassulacean acid metabolism. *Plant* 1979; 144:143–151.

149. Cockburn W, McAulay A. Changes in metabolite levels in *Kalanchoe daigremontiana* and the regulation of malic acid accumulation in crassulacean metabolism. *Plant Physiol.* 1977; 59:455–458.

150. Kelly GJ, Turner JF. The regulation of pea seed phosphofructokinase by 6-phosphogluconate, 3-phosphoglycerate, 2-phosphoglycerate and phosphoenolpyruvate. *Biochim. Biophys. Acta* 1970; 208:360–371.

151. Spalding MH, Aaron GP, Edwards GE. Malate decarboxylation in isolated mitochondria from crassulacean acid metabolism plant *Sedum praealtum. Arch. Biochem. Biophys.* 1980; 199:448–456.

152. Baur B, Dietz B, Winter K. Regulatory protein phosphorylation of PEP carboxylase in the facultative CAM plants *Mesembrynthemum crystallinum* L. *Eur. J. Biochem.* 1992; 209:95–101.

153. Wu MX, Wedding RT. Diurnal regulation of phosphoenolpyruvate carboxylase from *Crassula. Plant Physiol.* 1985; 77:667–675.

154. Nimmo HG. The regulation of phosphoenolpyruvate carboxylase in CAM plants. *Trends Plant Sci.* 2000; 5:121–124.

155. Wang ZY, Tobin EM. Constitutive expression of the *CIRCARDIAN CLOCK ASSOCIATED 1 (CCA1)* gene disrupts circadian rhythms and suppresses its own expression. *Cell* 1998; 93:1207–1217.

156. Borland AM, Hartwell J, Jenkins GI, Williams MB, Nimmo, HG. Metabolic control overrides circadian regulation of phosphoenolpyruvate carboxylase kinase and CO_2 fixation in crassulacean acid metabolism. *Plant Physiol.* 1999; 121:889–896.

157. Nobel PS. Achievable productivities of certain CAM plants: basis of high values compared with C_3 and C_4 plants. *New Phytol.* 1991; 119:183–205.

158. Kluge M, Osmond CB. Studies on the phosphoenolpyruvate carboxylase and other enzymes of crassulacean acid metabolism of *Bryophyllum tubiflorum* and *Sedum prealtum. Z. Pflanzenphysiol.* 1972; 66:97–105.

159. Kusumi K, Arata H, Iwasaki I, Nishimura M. Regulation of PEP-carboxylase by biological clock in a CAM plant. *Plant Cell Physiol.* 1994; 35:233–242.

22 Photosynthesis in Nontypical C$_4$ Species

María Valeria Lara and Carlos Santiago Andreo
Centro de Estudios Fotosintéticos y Bioquímicos, Facultad de Ciencias Bioquímicas y
Farmacéuticas, Universidad Nacional de Rosario

CONTENTS

I. INTRODUCTION

All plants use the same basic pathway for photosynthetic CO_2 fixation: the C_3 cycle (alternatively called photosynthetic carbon reduction cycle or Calvin and Benson cycle). In this pathway, ribulose bisphosphate carboxylase–oxygenase (RuBisCO) catalyzes the entry of CO_2 into the cycle. At ambient CO_2 and O_2 conditions, the enzyme also acts as an oxygenase

incorporating O_2 into the photorespiratory carbon oxidation cycle with the resultant loss of the fixed carbon [1]. To overcome the effect of O_2 on RuBisCO, some plants have developed ways to increase the level of CO_2 at the location of RuBisCO in the plant, decreasing in this way the oxygenation reaction and thus the carbon flux through the photorespiratory carbon oxidation cycle. Among the different photosynthetic modes are the C_4 cycle and the crassulacean acid metabolism (CAM), which are evolutionarily derived from C$_3$ photosynthesis [2]. The C$_4$ photosynthesis requires the coordination of biochemical functions between two types of cells and the cell type-specific expression of the enzymes involved [1,3]. In these plants, atmospheric CO$_2$ is first incorporated into C$_4$ acids in the mesophyll cells by phosphoenolpyruvate carboxylase. These C$_4$ acids are then transported to bundle sheath cells where they are decarboxylated and the released CO$_2$ is incorporated into the C$_3$ cycle. The C$_4$ system is more efficient under some environmental conditions as it increases the concentration of CO$_2$ in bundle sheath cells, suppressing the oxygenase activity of RuBisCO and thus, photorespiration. On the other hand, CAM is a metabolic adaptation to arid environments: stomata are closed during most of the day and opened at night. Malic acid is accumulated in the vacuoles of mesophyll cells at night as a result of fixation of CO$_2$ by the phosphoenolpyruvate carboxylase. During the day, malic acid is decarboxylated and the released CO$_2$ is refixed in the C$_3$ cycle [4]. Compared with C$_4$ plants, leaves of CAM plants have a simple inner structure [5].

Another mechanism found among photosynthetic organisms that eliminates the O$_2$ inhibition of photosynthesis is the one present in unicellular and multicellular algae [6] and cyanobacteria. In this case, the concentration of inorganic carbon in the site of RuBisCO is the result of different transporters located at the plasma membrane or at the chloroplast envelope and carbonic anhydrase (reviews: Refs. [7–9] and references therein).

Although most C$_4$ plants present Kranz anatomy and C$_4$ biochemical features in a constitutive manner, many variations as well as transitions to and from other photosynthetic modes have been described. These nontraditional C$_4$ plants can be grouped as follows: (a) submersed aquatic species like *Egeria densa*, *Elodea canadensis*, and *Hydrilla verticillata* that show induction of a C$_4$-like metabolism without Kranz anatomy under conditions of high temperatures and light intensities [10,11]; (b) amphibious species like the sedge *Eleocharis vivipara*, which has traits of C$_4$ plant in the terrestrial form and those of a C$_3$ plant in the submerged form [12]; (c) C$_4$ photosynthetic plants belonging to the Chenopodiaceae where C$_4$ photosynthesis functions within a single photosynthetic cell though lacking the Kranz anatomy [13,14]; and (d) C$_4$ succulent species of *Portulaca* that exhibit, under water stress, transition to a crassulacean acid-like metabolism (as in *Portulaca oleracea* [15]) or induction of a CAM-cycling metabolism compartmentalized in a different cell type while the C$_4$ pathway is also operating (*Portulaca grandiflora* [16]).

II. INDUCTION OF A C$_4$-LIKE MECHANISM IN SUBMERSED AQUATIC PLANTS OF THE HYDROCHARITACEAE FAMILY

Submersed aquatic macrophytes are a large group of phototrophs, which include nonvascular plants, primitive vascular plants, and angiosperms. The supply of dissolved inorganic carbon species in water can be limiting because of the high diffusive resistance in water [17]. In this respect, submersed aquatic autotrophs exhibit plasticity in relation to photosynthesis in aspects such as biochemistry, physiology, and anatomy [18] and have developed mechanisms to cope with limiting CO$_2$ and high O$_2$ concentrations, such as CO$_2$-concentrating mechanisms and the ability to use HCO$_3^-$ in photosynthesis, and the presence of carbonic anhydrase in the apoplast [18,19]. In general, submersed aquatic macrophytes exhibit unique characteristics related to their environment, such as low photosynthetic rates [20], low light requirements, very high K_m (CO$_2$/HCO$_3^-$) values [21], and the requirement of high CO$_2$ levels to saturate photosynthesis [17,19,20].

In a variety of submersed aquatic macrophytes low CO$_2$ compensation points are induced by submergence and growth under stress conditions of low CO$_2$ levels, high temperatures, and long photoperiods [18,22–25]. At least three members of the monocot family Hydrocharitaceae, *H. verticillata*, *E. canadensis*, and *E. densa*, have an appreciable Kranz-less C$_4$ acid metabolism in the light.

A. THE CASE OF *E. DENSA*

Among the higher aquatic plants, *E. densa* has been preferred material for a number of different studies in plant physiology. Its leaves contain a single longitudinal vascular bundle, and the blade consists of two layers of cell only, allowing studies of the whole undamaged organ in a natural environment. In this species, heterogeneity is reduced to a minimum; all leaf cells are in direct contact with the external medium and at the same developmental stage and thus in similar physiological condition. These properties, together with the leaf polarity displayed by *E. densa*, represent an advantage for different kinds of research and make this species one of the model organisms of the plant kingdom for experiments such as electrophysiology [26–29].

1. Anatomical Description

E. densa is a submersed rooted aquatic dioecious herb [30]. It is a common waterweed that occurs in streams, ponds, and lakes. The slender stems of

Egeria are usually a foot or two long but can be much longer. The small leaves are strap-shaped, about 1 in. long and 0.25 in. wide. The leaf margins have very fine saw teeth that require a magnifying lens to be seen. Leaves occur in whorls of three to six around the stem. The flowers are on short stalks about 1 in. above the water. Flowers have three white petals and are about 0.75 in. across.

2.   Photosynthetic Modes in *E. densa*

The existence of two photosynthetic metabolic conditions in this plant has been reported. The high photorespiration state with C_3 photosynthesis is characterized by a high CO_2 compensation point, which occurs under conditions of high CO_2 availability; the second condition is a low photorespiration state, which is induced after conditions of high light, temperature, O_2, probably UV-B radiation, and low CO_2 [11,31,32].

3. $^{14}CO_2$ Fixation and CO_2 Compensation Point in *E. densa* under Different Environmental Conditions

The first studies in *E. densa* as well as in *Lagarosiphon major* Moss. [33] showed radiolabeled inorganic carbon fixation into malate and aspartate, especially at low pH and in the short term. Two further reports also indicated that malate is the major product of short-term ^{14}C labeling in *Egeria* [34] and *E. canadensis* Michx. [35]. Later, it was found that the Calvin cycle is the primary carboxylation mechanism, responsible for over 90% of the ^{14}C initially incorporated [34]. Although these primary studies in *E. densa* have produced conflicting results, it was then established that low CO_2 levels influence the products formed, with malate increasing at the expense of the Calvin cycle intermediates [34,36–39].

Then, a CO_2 compensation point of 43 µl CO_2/l (typical of terrestrial C_3 plants) was described when plants were incubated under conditions of low temperature (12°C) and light [22]. But when incubated under high light and temperature (30°C) (conditions that cause a decrease of gas solubility in water), a value of 17 µl CO_2/l was observed. Maximal RuBisCO activity (76.0 and 70.6 µmol/mg Chl/h) under these conditions shows no correlation between the decrease in compensation point and RuBisCO activity.

4.   Study of the Expression and Regulation of the Main Photosynthetic Enzymes

After 23 days under conditions of high light intensity (300 µmol/m²/sec) and high temperature, phosphoe-

nolpyruvate carboxylase and NADP–malic enzyme (ME) increased their activity 3.7 and 3 times, respectively, above the constitutive levels expressed in plants under low light intensity (30 µmol/m²/sec) and low temperature (12°C), which corresponds to conditions of high compensation point [11]. Western blot analysis showed the increased protein levels involved in this process. In contrast, RuBisCO content remained constant during the induction period. After a 23-day induction period, the phosphoenolpyruvate carboxylase/RuBisCO ratio increased but to a lesser extent than that observed in *H. verticillata* [11,22].

A 72-kDa isoform of NADP–ME expressed in *E. densa* leaves was purified. It possesses physical and kinetic properties similar to those of the enzyme from terrestrial C_3 plants. The increase in the amount of NADP–ME after temperature and light induction may facilitate the maintenance of high rates of decarboxylation of malate and delivery of CO_2 to RuBisCO [11].

In the case of phosphoenolpyruvate carboxylase, two immunoreactive bands of 108 and 115 kDa are expressed in plants kept under low light and temperature; after 23 days of induction, the lower-molecular mass form is clearly induced, while the level of the other isoform seems not to be affected by the treatment. The purified inducible 108-kDa isoform has a low K_m for phosphoenolpyruvate and exhibits a hyperbolic response as a function of this substrate [11]. Moreover, the estimated K_m value for HCO_3^- (7.7 µM) is lower than all the reported values for the enzyme from different C_4 species [40]. Thus, this *E. densa* phosphoenolpyruvate carboxylase isoform not only has a high affinity for its substrates but also is induced under conditions of low CO_2 availability [11]. In addition, it shows differential phosphorylation during the day, with the process of phosphorylation–dephosphorylation highly regulated during the induction of this Kranz-less C_4 acid metabolism in *E. densa* [41].

5.   Localization of Key Enzymes Participating in C_4 Photosynthesis

Cellular fractionation using Percoll gradient accompanied by Western blot analysis indicated that phosphoenolpyruvate carboxylase, as in C_4 plants, is located in the cytosol of the photosynthetic cells of *E. densa* leaves, whereas NADP–ME and RuBisCO are located in the chloroplasts [11]. The specific localization of these enzymes is very important for delivering inorganic carbon from the cytosol to the chloroplasts via C_4 acids. The chloroplast is the site of CO_2 generation from C_4 acids and, consequently, of the concentrating mechanism.

6. Induction of Phosphoenolpyruvate Carboxylase and NADP–ME from *E. densa* by Abscisic Acid

Although *E. densa* does not show leaf dimorphism (as in the case of *E. vivipara*, see below), abscisic acid supplied exogenously to plants kept under low light and temperature caused increased activities of phosphoenolpyruvate carboxylase and NADP–ME similar to those occurring after temperature induction. However, the increase in phosphoenolpyruvate carboxylase activity is related to an increase in the 108-kDa as well the 115-kDa isoform. Therefore, different signaling systems may exist in this species in response to high temperature or abscisic acid, both leading to changes in photosynthetic metabolism, and may have evolved in plants that can change the mode of photosynthesis according to environmental fluctuations, like *E. vivipara*, *Mesembryanthemum crystrallinum*, and *E. densa*.

7. Electrophysiological Studies in *E. densa*

E. densa shows leaf pH polarity and values of electric potential difference at the plasmalemma of up to ca. $-300\,mV$ in the light [29]. Under strong illumination, leaf cells acidify the medium on the abaxial side of the leaf and alkalize the adaxial side of the leaf. This acidification is mediated by a H^+ ATPase located in the plasmalemma [42,43] and controlled by the photosynthetic process, apparently by redox regulation [26]. The leaf polarity present in the *E. densa* leaf is proposed to be used for bicarbonate utilization, and it has been described not only in species of the Hydrocharitaceae but also in *Potamogeton* species [44].

At high light intensities and low dissolved carbon concentrations, *Egeria* generates a low pH at the abaxial leaf side for CO_2 uptake. To balance the loss of the H^+ from the symplasm, there is an OH^- efflux at the upper side of the leaf together with the K^+ (Ca^{+2}) flux from the abaxial to the adaxial solution. In this way, the photosynthetic reduction of HCO_3^- by these so-called polar plants produces one OH^- for each CO_2 assimilated. The acidification in the lower side of the leaf results in a shift in the equilibrium from HCO_3^- to CO_2, with CO_2 then entering the abaxial cells by passive diffusion [45]. By this means, *E. densa* cells have extra CO_2 available under conditions where the concentration of this gas is limiting for photosynthesis.

8. CO₂-Concentrating Mechanism in *E. densa* under Conditions of High Light and Temperature

E. densa responds to a decrease in CO_2 concentration by induction of an ancient-like isoform of NADP–ME similar to the one present in C_3 terrestrial species [46]. The regulation of phosphoenolpyruvate carboxylase in *E. densa*-induced plants appears to be similar to that of phosphoenolpyruvate carboxylase from C_4 species, so the 108-kDa isoform is probably participating in a C_4-like mechanism, with a regulation similar to that of phosphoenolpyruvate carboxylase from C_4 plants.

High activity and kinetic and regulatory properties of C_4 enzymes such as phosphoenolpyruvate carboxylase and NADP–ME, and labeling of malate, together with a decrease in the compensation point, are evidence that a C_4-like photosynthetic system operates in *E. densa*, when transferred to conditions of low CO_2 availability (Figure 22.1).

Finally, in *E. densa*, both the C_4-like mechanism and the pH-polarity mechanism contribute to its photosynthetic performance under CO_2-limiting circumstances.

B. THE CASE OF *H. VERTICILLATA*

1. Botanical Description

H. verticillata is a submersed, usually rooted, aquatic perennial herb with slender ascending stems to 30-ft length, heavily branched. The stems from slender rhizomes are often tipped with a small tuber. Leaves are whorled, three to eight per whorl, bearing coarse teeth along the margins. Fleshy axilliary buds are often formed at the leaf axils, to 2-in. length with three sepals and three petals, each about 0.3 in. long, whitish or translucent, floating at the water surface. Male flowers detached and free floating at maturity are white to reddish brown, about 2 mm long. Flowers release floating pollen from stamens when they pop open at the water surface.

2. Physiological and Biochemical Studies

In 1924 it was shown that *H. verticillata* was more acidic in summer than in winter and had a higher malate content and that malate could replace CO_2 in driving photosynthetic O_2 evolution [47]. Later, it was indicated that its CO_2 compensation point was modified depending on growing conditions [48] and correlated with changes in phosphoenolpyruvate carboxylase activity. Plants grown in winter or under short photoperiod accompanied by low temperature $(25°C)$ exhibited higher CO_2 compensation points than those grown in summer or under longer photoperiod and higher temperatures $(27°C)$. The low-compensation point plants incorporated 60% $^{14}CO_2$ into malate and aspartate, with only 16% in sugar phosphates [39]. Moreover, in *H. verticillata* plants, low CO_2 levels influenced the products formed, with

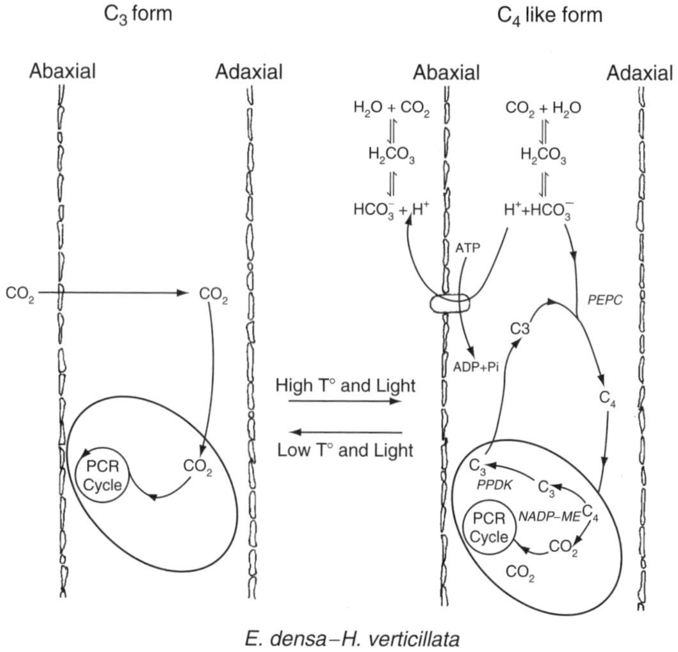

FIGURE 22.1 Proposed photosynthetic mechanisms operating in *Egeria densa* and *Hydrilla verticillata* under conditions of low (induced by high light and temperature) and high (induced by low light and temperature) CO_2 availability. The intracellular localization of the main C_4 photosynthetic enzymes is shown. NADP–ME: NADP–malic enzyme; PCR cycle: photosynthetic carbon reduction cycle; PEPC: phosphoenolpyruvate carboxylase; PPDK: pyruvate orthophosphate dikinase.

malate increasing at the expense of the Calvin cycle intermediates. However, pulse–chase labeling experiments indicated that malate was turned over at a low rate [39].

In high compensation point plants photorespiration, as a percentage of net photosynthesis, was equivalent to that in terrestrial C_3 plants, while plants with decreasing CO_2 compensation points were associated with reduced photorespiration and increase in net photosynthesis rates [22]. In this last group of plants, the phosphoenolpyruvate carboxylase/RuBisCO activity ratio was higher with respect to the high-compensation group of plants, and plants exhibited increased activity of pyruvate orthophosphate dikinase, pyrophosphatase, adenylate kinase, NAD–and NADP–malate dehydrogenase, NAD–and NADP–MEs, and aspartate and alanine aminotransferases [22]. In contrast, the activities of the photorespiratory enzymes phosphoglycolate phosphatase and glycolate oxidase and of phosphoglycerate phosphatase showed no change. The decrease in the CO_2 compensation point reflects at least increased (or refixed) fixation via phosphoenolpyruvate carboxylase. In addition, dark ^{14}C fixation, fixation of CO_2 in the dark, and diurnal fluctuations in the level of titratable acidity have been observed in *H. verticillata* [24,39]. It was then established that malate formation occurred in both photorespiratory states, but reduced photorespiratory states resulted when malate was utilized in the light [10].

3. Anatomical Features and Localization of Key Enzymes Participating in C_4 Photosynthesis

Leaf sections show no evidence of Kranz anatomy and reveal the existence of two layers. The adaxial section is composed of larger cells than the abaxial, with prominent vacuoles. RuBisCO was found in the chloroplasts of both photosynthetic-type cells with an equivalent extent of labeling. As expected, the localization of phosphoenolpyruvate carboxylase was mainly cytosolic, and it was found in all leaf cells [49]. For both enzymes, the localization did not vary when plants under both photorespiratory states were analyzed. Subcellular fractionation of leaves under low CO_2 compensation point confirmed the localization of phosphoenopyruvate carboxylase and showed that NADP–ME as well as pyruvate orthophosphate dikinase were located in the chloroplasts. Most of the NADP–malate dehydrogenase was found in the chloroplasts, while NAD–ME was mitochondrial [50].

Intercellular differentiation of fixation events, found in plants with Kranz anatomy, does not occur in low-CO_2 compensation point *H. verticillata* plants, and it is suggested that intercellular separation of C_3 and C_4 fixation events may account for the low photorespiration state [49]. A C_4-like cycle concentrating CO_2 was proposed to take place in the chloroplast, with phosphoenopyruvate carboxylase fixing inorganic carbon in the cytosol and C_4 acids moving to the chloroplasts and being decarboxylated

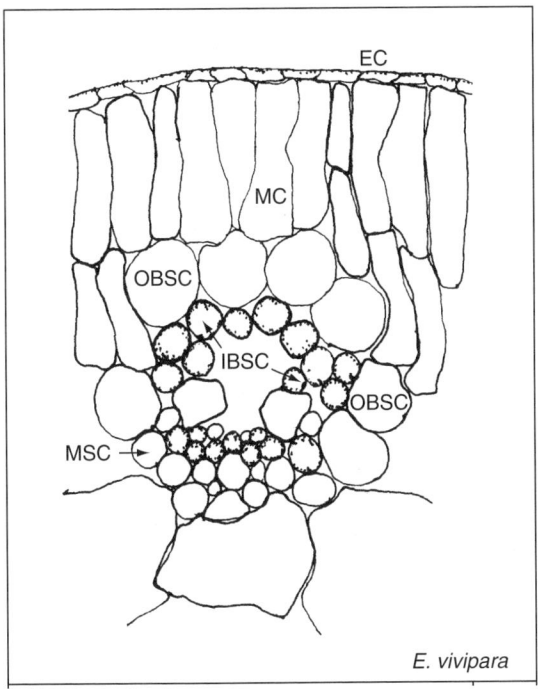

E. vivipara

FIGURE 22.2 Diagram of various cell types in a cross section of *Eleocharis vivipara* grown in terrestrial conditions (Kranz anatomy). EC: epidermal cell; IBSC: innermost bundle sheath cell (Kranz cell); MC: mesophyll cell; MSC: mestome sheath cell; OBSC: outermost bundle sheath cell.

by NADP–ME so as to supply CO$_2$ to RuBisCO [50].

In contrast to NADP–ME terrestrial C$_4$-type plants, *Hydrilla* chloroplasts have grana stacks, and they are presumably not deficient in photosystem II or NADP, and their leaves are not prone to photoinhibition. Thus, oxaloacetate or aspartate, rather than malate, is probably the major imported acid to the chloroplasts [47].

4. Study of the Expression and Regulation of the Main Photosynthetic Enzymes

Later, the participation of phosphoenolpyruvate carboxylase in the expression of low-photorespiratory gas exchange characteristics in *H. verticillata* was demonstrated by directly inhibiting this enzyme in this state, resulting in increases in the CO$_2$ compensation point, O$_2$ inhibition of photosynthesis, and CO$_2$ evolution in the dark by 51% [51]. This also indicates that the use of HCO$_3^-$ is not responsible for concentrating CO$_2$. As the relation between phosphoenolpyruvate carboxylase activity and the CO$_2$ compensation point was nonlinear, the expression of low-photorespiratory gas exchange characteristics

might require the subsequent induction of other C$_4$ photosynthetic enzymes [51].

After a 12-day induction period, increases were observed in phosphoenolpyruvate carboxylase and aspartate and alanine aminotransferase activities (with the major increase occurring in 3 days), with a slower but considerable increase of NADP–ME and pyruvate phosphate dikinase activities, accompanied by constant levels of RuBisCo [52]. These results were also observed by Western blotting, showing that the induction of some C$_4$ enzymes is consistent with a C$_4$-like cycle concentrating CO$_2$ in leaves of *H. verticillata* [50].

The increase of phosphoenolpyruvate carboxylase activity was correlated with de novo synthesis of the protein. The 110-kDa enzyme from C$_4$-type leaf extracts was 14 times more active than that in C$_3$-type leaves, with daytime values 53% higher than at night, resembling the enzyme from C$_4$ plants [47]. During the day the enzyme was less sensitive to malate inhibition as accounted by higher I_{50} values and glucose-6-phosphate acting as a positive effector of the enzyme. In contrast, while this form exhibited upregulation in the light, phosphoenolpyruvate carboxylase from C$_3$ leaves showed no light activation, lower rates, and was virtually insensitive to malate inhibition [47].

Two isoforms of NADP–ME have been identified: the 72-kDa isoform was more similar to the housekeeping isoform of the enzyme in C$_3$ and C$_4$ plants [53].

5. Molecular Studies of Phosphoenolpyruvate Carboxylase and NADP–ME

Three full-length cDNAs encoding phosphoenolpyruvate carboxylases were isolated from *H. verticillata*. *Hvpepc*4 was exclusively expressed in leaves during C$_4$ induction, with kinetic data from the expressed protein consistent with a C$_4$ form of the enzyme, but interestingly the C$_4$ signature serine of terrestrial plant C$_4$ isoforms was absent in all sequences, and an alanine was found in its place [54]. Nevertheless, it contains the putative C$_4$ determinant Lys349. *Hvpepc*3 and *5* have Arg, a common feature of C$_3$ sequences [55]. In phylogenetic analyses, the three sequences grouped with C$_3$, nongraminaceous C$_4$, and CAM phosphoenolpyruvate carboxylases, but not with the graminaceous C$_4$, and formed a clade with a gymnosperm, which is consistent with *H. verticillata* phosphoenolpyruvate carboxylase predating that of other C$_4$ angiosperms [54]. It seems that *Hvpepc*3 and *4* isoforms function in the C$_4$ leaf, with the latter being the photosynthetic form and the other probably participating in the dark fixation. In contrast, *Hvpepc*5 is a root enzyme [54].

A partially deduced amino acid sequence from a cDNA clone of *Hydrilla* C$_4$ leaf NADP–ME shows that the 540 amino acid-long *Hydrilla* sequence shares 85% identity with a 585 amino acid-long *Aloe arborescens* NAD–ME, apparently being more similar to a CAM isoform than to the C$_4$-maize photosynthetic isoform [47].

6. CO$_2$-Concentrating Mechanism Operating in *H. verticillata*

In *H. verticillata*, a C$_4$-like cycle has been described when this plant is grown under summer conditions (high temperature and light), which result in limiting levels of inorganic carbon in water [24]. These effects produce gas exchange and biochemical modifications that demonstrate a shift from C$_3$- to C$_4$-like photosynthesis in the leaves [18], but without the typical Kranz anatomy [25]. This lack of Kranz anatomy is perhaps the major difference with terrestrial C$_4$ species and challenges the notion that Kranz compartmentation is essential for C$_4$ photosynthesis [47]. Direct measurements of internal inorganic carbon pools in C$_3$ and C$_4$ *H. verticillata* leaves [25] demonstrate that the C$_4$ cycle produces a CO$_2$ concentration in this species, even at low pH medium where HCO$_3^-$ is negligible. A CO$_2$ flux mechanism facilitates the entry of inorganic carbon into the cells in both C$_3$ and C$_4$ modes by providing access to the HCO$_3^-$ pool in the medium. This plant is a HCO$_3^-$ user and becomes pH polarized in the light [56]. However, when inorganic carbon is limiting, the effectiveness of this carbon flux mechanism is diminished and the CO$_2$-concentrating mechanism in the chloroplasts based on a C$_4$-like cycle minimizes CO$_2$ losses from respiration in *H. verticillata* [47] (Figure 22.1).

In addition, net photosynthetic rates at limiting CO$_2$ concentrations are substantially greater when shoots are in the C$_4$ mode [22,57]. However, the C$_4$ quantum yield is half of that in the C$_3$ shoots [57], demonstrating that the C$_4$ cycle in *Hydrilla* has a substantial energy cost to it, perhaps more than in a terrestrial C$_4$ plant [47].

C. THE CASE OF *E. CANADENSIS*

First studies suggested that *E. canadensis* was not a C$_4$ plant as 50% of the ^{14}C initially incorporated was in C$_4$ acids but at a low turnover rate [35]. C$_4$ acids were proposed to have a role in a pH-stat mechanism, to act as an anaplerotic carbon source or as a counter ion.

Later, mature shoots of *E. canadensis* grown in artificial pond water exposed to ^{14}CO$_2$ or submersed in solution containing H^{14}CO$_3^-$ for 30 min under photosynthetic active radiation showed label in malic acid, glucose, asparagine, proline, and hexose phosphates, while sucrose appeared only slightly labeled, indicating that bicarbonate was used for photosynthesis, while gaseous ^{14}CO$_2$ was not incorporated [58]. ^{14}C was taken by penetrating the epidermis of the stem and the lower side of the leaves. In the symplast, label was accumulated in the chloroplasts of both epidermal layers and in the plastid envelope of the chloroplasts of cortex cells.

A model has been advanced by Prins et al. [56], which operates with proton pumps, with the energy derived from ATP hydrolysis. The upper leaf surface is negatively charged, while the lower epidermis wall is provided with protons that neutralize OH$^-$ groups derived by the generation of CO$_2$ from HCO$_3^-$. There is an alkalinization at the adaxial side to pH values of 9 to 11 and an acidification at the abaxial side of the leaf to pH values of 4 to 5, which results in a shift of the equilibrium between HCO$_3^-$ and CO$_2$ in the direction of CO$_2$, which then diffuses into the leaf cells [27].

In *E. canadensis*, acidification is induced by conditions of high light and low dissolved inorganic carbon in plants previously grown under conditions favorable for photorespiration, but it is absent in plants grown under high CO$_2$ conditions [26]. Under low CO$_2$ conditions the plants exhibited low CO$_2$ compensation point, indicative of plants in a C$_4$-like state, and an increase in the activity ratio phosphoenolpyruvate carboxylase/RuBisCO, indicating a central role of the former in this metabolism [59]. The very same conditions induced HCO$_3^-$ utilization in this species. In addition, differences in HCO$_3^-$ affinity have been described for *E. canadensis* that correlate with the inorganic carbon concentration in its habitat [60]. Both mechanisms operate in the same plant simultaneously [59]. It is thought that in *E. canadensis*, a carbon-concentrating mechanism activity results in the suppression of photorespiration but has only a marginal effect on the net rate of photosynthesis when the diffusion of CO$_2$ in the leaf is limiting.

D. EVOLUTIONARY CONTEXT

Submersed aquatic species include cyanobacteria, several algal divisions, bryophytes, lower vascular plants, and angiosperms. In marine and freshwater environments, the approximately 50,000 submersed photosynthetic species are taxonomically far more diverse than the estimated 300,000 species of vascular plants, which constitute the major photosynthetic organisms of terrestrial habitats [61]. Viewed in this way, it is not surprising that the variability in carbon acquisition mechanisms would be greater than that among terrestrial species. The macroscopic green alga *Udotea fla-bellum* is the most primitive plant demonstrated to have a C$_4$-based form of photosynthesis [62]. Recently, Reinfelder et al. [63] presented evidence that C$_4$ photosynthesis supports carbon assimilation in the marine diatom *Thalassiosira weissflogii*. Although it is premature to designate marine diatoms as C$_4$ photosynthesizers in a traditional sense [64], there is evidence that marine diatoms can concentrate inorganic carbon for photosynthesis [65]. Diatoms underwent their main evolutionary diversification during the Mesozoic, when the concentration of CO$_2$ was lower than in the earlier eras (Precambrian and Paleozoic) when most photosynthetic microorganisms evolved. Unicellular C$_4$ carbon assimilation may have predated the appearance of multicellular C$_4$ plants. Moreover, Voznesenskaya et al. [13,14] (see below) suggested that Kranz anatomy is not essential for terrestrial C$_4$ plant photosynthesis, providing evidence that C$_4$ photosynthesis can function within a single photosynthetic cell in *Borszczowia aralocaspica* and *Bienertia cycloptera*. Among angiosperms, there are at least three members of the Hydrocharitaceae that have been reported to possess a CO$_2$-concentrating mechanism under certain conditions [22,50]. Like the use of HCO$_3^-$, CO$_2$ fixation into C$_4$ acids could be part of a concentrating mechanism to improve photosynthesis under carbon-limiting conditions in these species. Both mechanisms may be ancient within submersed species, considering that the Hydrocharitaceae is a submersed monocot family that may have its origin 100 million years ago in the Cretaceous period [66,67] and that this family could be more ancient than the C$_4$ monocots, which became more abundant in the Miocene (7 million years ago [2,66]). Fossil evidence has been found for the *Hydrilla* genus from the upper Eocene (40 million years ago [47]). The phylogenetic analysis of *H. verticillata* phosphoenolpyruvate carboxylases [54] as well as the kinetic properties of NADP–ME in *E. densa* [11] also show the ancientness of these plants. The C$_4$-like mechanism in *E. densa* and *H. verticillata*, which takes place in a single cell, may represent an ancient form of C$_4$ photosynthesis compared to that occurring in terrestrial plants. This could have arisen in response to declines in CO$_2$ concentration in water [68], in a similar way as C$_4$ photosynthesis may have evolved in response to decreases in atmospheric CO$_2$ concentration [67].

III.   C$_3$ AND C$_4$ DIFFERENTIATION IN AMPHIBIOUS SEDGES

The monocotyledonous family Cyperaceae includes several genera with C$_3$, C$_4$, and C$_3$–C$_4$ intermediate species [12,69]. A great number of species are hygrophytes growing in wet habitats, and some possess the characteristics of amphibious plants [12]. In this respect, the C$_4$ group of this family represents ecologically unusual C$_4$ plants [70]. *E. vivipara* displays dramatic changes in its photosynthetic and anatomical traits depending on the environmental conditions: it exhibits C$_4$-like traits as well as Kranz anatomy under terrestrial conditions but C$_3$-like characteristics without Kranz anatomy in the submersed form. This plant exhibits heterophylly, the name given to this dimorphism [12,66,71], between aerial and aquatic leaves.

A.   THE CASE OF *E. VIVIPARA*

1.   Gross Morphology of Plants and Anatomy of Mature Tissues

Photosynthesis in *E. vivipara* occurs primarily in the culm. Leaves are not apparent and only short, tubular, bladeless sheaths enclose the base of culms. A culm of this sedge is composed of a single long internode with a crown at the top, which contains axilliary buds. Developmentally, new culms originate as axilliary buds in the crown of the preceding culm and grow by elongation of the single internode. The elongation actually occurs only at the base of the internode where the intercalary meristem produces cells upward [72].

2.   The Beginning: C$_4$–C$_3$ Characterization

The firsts biochemical studies were carried out in the 1980s with *E. vivipara* plants collected from a creek in the vicinity of Tampa, FL, and maintained in the aquarium or transplanted to sandy clay soil for a couple of months [12]. In the terrestrial form the gross morphology is that of the genus, and erect and firm culms possess a Kranz-type anatomy typical of C$_4$ plants, characterized by tightly arranged mesophyll cells and well-developed bundle sheath cells, which consist of three kinds of cell, the innermost Kranz cells containing numerous and large chloroplasts, the middle mestome sheath cells lacking chloroplasts, and the outermost parenchyma sheath cells with chloroplasts [12,73] (Figure 22.2). In contrast, the submersed form shows a hairlike morphology with soft culms with a completely different structure; one layer of spherical mesophyll cells are surrounded by the epidermis, and the vascular bundles are reduced and possess large air cavities. Small bundle sheath cells are present, but with few and small chloroplasts (Figure 22.3). Culms died as a result of rapid desiccation when the submersed form

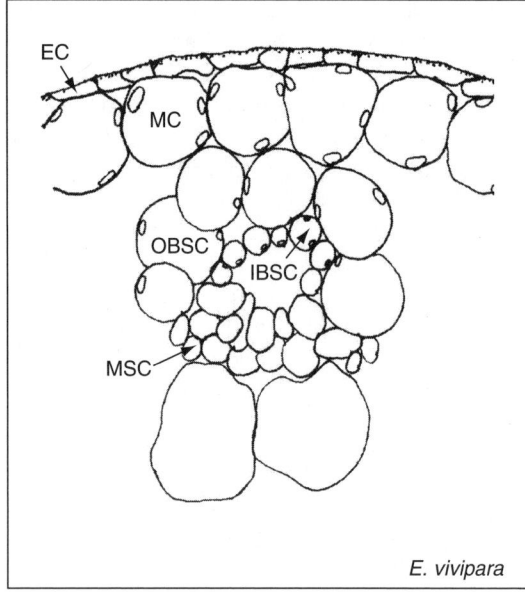

FIGURE 22.3 Diagram of various cell types in a cross section of *Eleocharis vivipara* under submergence (non-Kranz anatomy). EC: epidermal cell; IBSC: innermost bundle sheath cell (Kranz cell); MC: mesophyll cell; MSC: mestome sheath cell; OBSC: outermost bundle sheath cell.

was exposed to air, but the plants would develop new photosynthetic organs. In contrast, when the terrestrial form was submersed, new hairlike culms with intermediate traits were produced and after several months; they changed gradually from the C_4 to the C_3 mode [12].

Under terrestrial conditions, the initial photosynthetic products formed are aspartate (40%) and malate (35%), and the label in C_4 acids "chases" into phosphate esters typical of C_4 plants. This form of *E. vivipara* exhibits a $\delta^{13}C$ value of −13.5%, characteristic of C_4 plants but slightly more negative than the values of other *Eleocharis* spp. [12,69], and presents high activities of the key C_4 photosynthetic enzymes phosphoenolpyruvate carboxylase, pyruvate orthophosphate dikinase, and NAD–ME, with a low ratio of RuBisCO/phosphoenolpyruvate carboxylase activity [12]. Altogether, the data suggest that the terrestrial form behaved as a C_4–NAD–ME-type species [12].

Under submergence, when the stomata density is low and they seem to be nonfunctional, the Kranz anatomy is absent, with a higher volume ratio of mesophyll/bundle sheath cells than in the terrestrial form. The pattern of ^{14}C incorporation into photosynthetic products in ^{14}C pusle–^{12}C chase experiments was typical of the C_3 pathway, with 3-phosphoglyceric acid (53%) and sugar phosphates (14%) as the main primary products. A C_3-like $\delta^{13}C$ value

(−25.9%) was also obtained under submergence. In this condition, high activity of RuBisCO and a decrease in the activities of key C_4 enzymes corresponds to the functioning of the C_3 metabolism [12]. Under water conditions of low carbon, the proportion of ^{14}C incorporated into C_4 compounds is higher than under conditions of high carbon. However, the turnover of ^{14}C in C_4 compounds is very slow [74]. In consequence, even though the NAD–ME-dependent C_4 cycle also operates in the submerged form, the contribution to total carbon flux is not large [75].

3. Going Deeper: The Ultrastructural Characterization of Photosynthetic Cells

By the middle of the 1990s there was a great body of information on C_4 anatomical variation in the Cyperaceae [76–78], but up to that time the ultrastructure of the family had received relatively little attention. Before the discovery of NAD–ME *Eleocharis* species with the photosynthetic carbon reduction in the border parenchyma position [79], it had seemed that the biochemical type could be predicted from anatomy alone [80]. Bruhl and Perry [81] carried out ultrastructural characterization of *E. vivipara* but only in the terrestrial form. Sedges, regardless of anatomical and biochemical type, possess a suberized lamella in photosynthetic organs, which is invariably present and confined to the mestome sheath cell walls, occupying the inner and the outer tangential walls of the mestome sheath [69,82]. As the mestome sheath externally envelops the Kranz sheath, it may play a role in suppressing leakage of CO_2 from the Kranz cells [73].

The correlation between biochemical type and chloroplast position in sedges does not always correspond to the situation seen in grasses [83]. In contrast with other C_4 sedges, NAD–ME *Eleocharis* species, in the terrestrial form, possess abundant mitochondria and chloroplasts with well-stacked grana in the photosynthetic carbon reduction at the bundle sheath cells [81]. Peripheral reticulum is well developed in both types of chloroplasts but differs from that seen in other sedges (fimbristyloid and chlorocyperoid) [81].

Latter, Ueno [73] found clear differences between the submerged and terrestrial forms, with the former showing a higher ratio of fresh weight to dry weight than the terrestrial form, probably due to vacuolated mesophyll cells and to the reduced vascular tissues and sclerenchyma in the submerged form [73]. In the latter state, the ratio of surface area to dry weight of the culms was about five times that in the terrestrial form. A few stomata were found at the tips of the culms of the submerged form, while large numbers of stomata were uniformly distributed on the epidermis of the terrestrial form, suggesting that this form probably fixes dissolved inorganic carbon in water via the epidermis like the submersed aquatic species [66,73]. The mesophyll cells of the terrestrial form were not organized in the radial arrangement typical of leaves of many C_4 plants; these cells were just inside the epidermis and tightly packed, and the inner mesophyll cells exhibited a tendency toward a radial arrangement. In contrast, in the culms of the submerged form, mesophyll cells were inside the epidermis forming a sheath-like ring, appearing circular in transverse sections, differing from those in the terrestrial form [73]. The Kranz cells of the terrestrial form had basically the structural characteristics of plants of the NAD–ME type, with the exception of the intracellular location of the organelles. The Kranz cells of the terrestrial form contained many organelles scattered randomly, the chloroplasts had well-developed grana and a large number of mitochondria with well-organized cristae. Both parenchyma sheath and mesophyll cells contained chloroplasts with well-developed grana, but smaller in parenchyma sheath that in mesophyll cells, and possessed a lower density of mitochondria than the Kranz cells. The Kranz cells of the submersed form were few in number and included only a few organelles, although chloroplasts also showed well-developed grana. The size (length of the long axis) of chloroplasts of the Kranz cells of the submerged form was 64% to 70% that of the terrestrial form.

4. *In Situ* Immunolocalization Studies

Since the photosynthetic tissues of *E. vivipara* have a complex anatomical structure [12,73,81], the photosynthetic cells being difficult to separate, *in situ* immunolocalization has been proved to be very useful in determining the specific location of an enzyme [84].

In the terrestrial form both pyruvate orthophosphate dikinase (chloroplastic) and phosphoenolpyruvate carboxylase (cytosol) were expressed, in a gradient manner, in the mesophyll as well as in the parenchyma sheath cells (stronger) [85] but were absent in the Kranz cells. By contrast, RuBisCO, which was localized in all photosynthetic cells, showed the lowest label in the parenchyma sheath cells. This pattern differs from that in C_4 plants [3], and parenchyma sheath cells but not mesophyll cells have stronger C_4 features. In C_4 species of the Cyperaceae, which also show unusual Kranz anatomy, RuBisCO is restricted to only the innermost bundle sheath cells [86]. In conclusion, the terrestrial form of *E. vivipara* possesses C_4-like characteristics rather than complete C_4 traits, because of the accumulation of RuBisCO in the mesophyll cells as well as in the Kranz cells [85] (Figure 22.4).

In the submerged form RuBisCO was densely distributed in the chloroplasts of all the photosynthetic cells, but pyruvate orthophosphate dikinase (chloroplastic) and phosphoenolpyruvate carboxylase

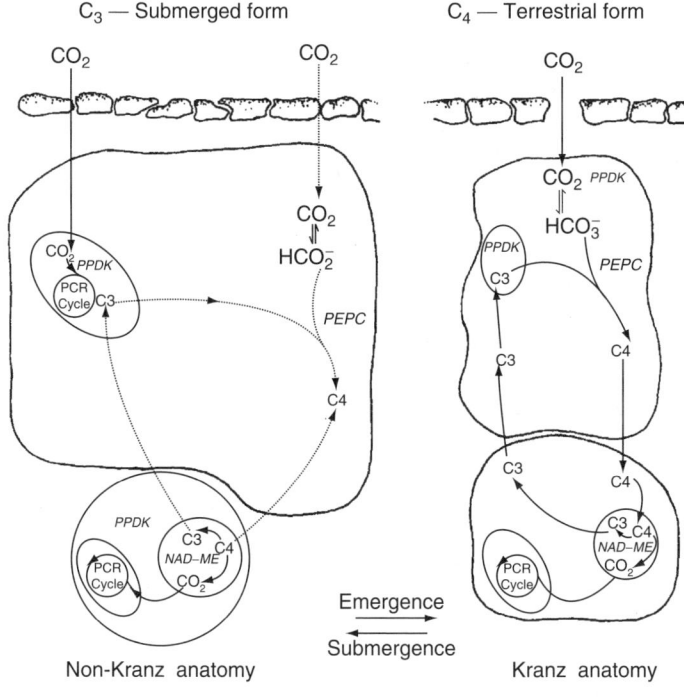

FIGURE 22.4 Schematic diagram of the photosynthetic metabolisms operating in *Eleocharis vivipara* under submerged and terrestrial conditions. Mesophyll (upper) and innermost (lower) bundle sheath cells are shown. The minor carbon flux is shown in dashed lines. The intracellular localization of the main C_4 photosynthetic enzymes is shown. NAD-ME: NAD-malic enzyme; PCR cycle: photosynthetic carbon reduction cycle; PEPC: phosphoenolpyruvate carboxylase; PPDK: pyruvate orthophosphate dikinase.

(cytosol) were found in both the mesophyll and parenchyma sheath cells and at lower levels than in the terrestrial form [85], in agreement with the lower activities previously determined [12].

In the terrestrial form, an NAD–ME-type C_4-like metabolism operates in *E. vivipara* although RuBisCO is also present in mesophyll and parenchyma sheath cells. In the submersed form, the levels and activities of NAD–ME, pyruvate orthophosphate dikinase, and phosphoenolpyruvate carboxylase declined with modification in the structure (Figure 22.4). Up to now, although there is no direct evidence that RuBisCO in the mesophyll cells and the parenchyma sheath cells of the submerged form is active, the enzyme in these cells would be responsible for the operation of the C_3 pathway in this state [85].

5. Regulation of Differentiation by Abscisic Acid in *E. vivipara*

It is known that abscisic acid is a stress hormone in plants [87] and is involved in the determination of leaf identity in aquatic plants showing dimorphism between the terrestrial and the submerged form [88,89], as well as in the induction of CAM in succulent plants [90,91]. When the submerged form of *E. vivipara* form was grown in water containing $5\,\mu M$ abscisic acid, it developed new photosynthetic tissues with Kranz anatomy, forming well-developed Kranz cells that contained many organelles and expressing C_4-like biochemical traits [74]. The abscisic acid-induced tissues accumulated large amounts of phosphoenolpyruvate carboxylase, pyruvate orthophosphate dikinase, and NAD–ME. The pattern of cellular accumulation of photosynthetic enzymes was basically similar to that found in the terrestrial form. However, the accumulation pattern of pyruvate orthophosphate dikinase appeared to be a combination of the patterns in the terrestrial and submerged forms. The tissues had 3.4 to 3.8 times more C_4 enzyme (phosphoenolpyruvate carboxylase, pyruvate orthophosphate dikinase, and NAD–ME) activity than did the tissues of the untreated submerged plants. The activity of RuBisCO in the abscisic acid-induced culms was higher than in both the terrestrial and the submerged forms (1.6 times). Consequently, in the abscisic acid-induced culms the ratio of RuBisCO/phosphoenolpyruvate carboxylase activity was intermediate between the ratios in the submerged and terrestrial forms. The abscisic acid-induced tissues fixed higher amounts of ^{14}C into C_4 compounds and lower amounts of ^{14}C into C_3 compounds as initial products than did the submerged plants, and they exhibited a C_4-like pattern of carbon fixation under aqueous conditions of low carbon, suggesting higher C_4 capacity in the tissues. In this way, in the presence of abscisic acid, photosynthetic tissues with Kranz anatomy and C_4-like biochemical traits are developed in the submerged form of *E. vivipara*.

6. Structure and Analysis of the Expression of Photosynthetic Enzymes

Pyruvate orthophosphate dikinase catalyzes the ATP- and P_i-dependent synthesis of phosphoenolpyruvate in C_4-ME plants and also participates in the recycling of the pyruvate generated by malate decarboxylation in the light in ME-CAM plants [92]. Two isoforms of the pyruvate orthophosphate dikinase have been described in mesophyll cells of NAD(P)–ME-CAM plants [93]. In species like *Mesembryanthemun crystallinum* this enzyme is exclusively localized in chloroplasts, while in species like *Kalanchoe* it is found in chloroplasts as well as in cytosol. On the other hand, a cytosolic isoform has been found in nonphotosynthetic organs of plants C_3 and C_4 [94], and is also induced by drought stress in rice roots [95]. In C_3 leaves the enzyme has also been found in chloroplasts [85,96,97].

Using a cDNA library prepared from culms of the terrestrial form, two highly homologous (95%) full-length cDNAs, *ppdk*1 and *ppdk*2, have been characterized. The deduced sequence for the PPDK1 protein contained an extra domain at the amino terminus of 69 aminoacids that would correspond to a chloroplast transit peptide; in contrast, *ppdk*2 encodes for a cytosolic protein. In this way, the two transcripts found in *E. vivipara* are derived from different genes [98]. In maize pyruvate two genes encode orthophosphate dikinase, one encoding a cytosolic isoform and the other a chloroplastic and cytosolic enzyme as a consequence of differential splicing [99–101]. Southern blot analysis indicated that a small family of genes, probably two, would encode pyruvate orthophosphate dikinase [98]. Northern blot analysis indicated that both isoforms are expressed simultaneously in the culms and the levels are more abundant in the terrestrial form than in the submerged form, in agreement with higher activities and *in situ* immunolocalization studies [12,85,98]. In addition, in the terrestrial form *ppdk*1 was expressed at somewhat higher levels than *ppdk*2, and *ppdk*2 was expressed at higher levels than *ppdk*1 in the submersed form [98]. It was also found that *ppdk*2 transcript is expressed in roots, in agreement with the cytosolic location of the enzyme. Thus, the expression pattern of *ppdk*1 and *ppdk*2 would be related to its specific localization.

A full-length cDNA (*pep*1) encoding a 698-aminoacid phosphoenolpyruvate carboxylase (PEP1) with a predicted molecular mass of $110\,kDa$ was characterized from the terrestrial form of *E. vivipara* [102]. A phylogenetic analysis indicated that the protein is located between a cluster of C_4-form phosphoenolpyruvate carboxylase of C_4 grasses (maize and sorghum) and another large cluster including C_3- and CAM-form phosphoenolpyruvate carboxylase of dicots and monocots and C_4-form phosphoenolpyruvate carboxylase from *Flaveria* [102]. The gene *pep*1 has similar behavior to the isogene for C_4 phosphoenolpyruvate carboxylase, although it holds high homology to the C_3 form. Apart from other conserved sequence motifs, PEP1 contains the putative phosphorylation motif (serine-11) for serine kinase [103]. The terrestrial form of PEP1 has kinetic properties that differ from those of the typical C_4-phosphoenolpyruvate carboxylase and are intermediate between the C_3 and C_4 forms of the enzyme [75].

7. Expression of Isogenes Encoding for Phosphoenolpyruvate Carboxylase, Pyruvate Orthophosphate Dikinase, and RuBisCO

Northern blot studies indicated that higher levels of phosphoenolpyruvate carboxylase and pyruvate orthophosphate dikinase were present in the terrestrial form than in the submerged form [102]. The expression of *pep*1 is regulated by light as its transcript is reduced when plants grown under light conditions are transferred to darkness for up to 3 days.

The transcript level of *ppdk*1, *ppdk*2, and *pep*1 is controlled by the water environment even in a single plant growing both underwater and above the water surface and having aerial secondary culms with Kranz anatomy, underwater culms without this anatomy, and secondary floating culms showing various intermediate anatomies [102]. Northern blot studies indicated that *pep*1 was more abundant in the above-water secondary culms and less abundant in the underwater culms. In the secondary culms floating at the water surface the level was intermediate. In addition, *ppdk* transcripts exhibited a similar behavior; however, the difference in *ppdk*2 transcript level in the three types of culms was smaller than that for *ppdk*1 [102].

Abscisic acid induces the expression of *ppdk* isogenes and *pep*1, even in mature photosynthetic tissues without Kranz anatomy [102], indicating that the expression of the enzymes under these conditions is regulated largely at the transcriptional level [72,98, 102–104]. Thus, anatomical and biochemical components of the C_4 traits are controlled independently, although both components seems to be well coordinated in naturally growing environments.

Studies of mRNA *in situ* hybridization indicated that the expression of RuBisCO small subunit (*RuBisCOss*) and *pepc* occurs mainly in bundle sheath cells and mesophyll cells, respectively [72]. In the submerged form, *RuBisCOss* was expressed in both bundle sheaths and mesophyll cells, and no expression of *pepc* was observed. In the immature internodal region with undeveloped bundle sheath cells, both life forms showed the same expression pattern as in C_3 plants. *RuBisCOss* expression was localized in mesophyll cells, and no phosphoenolpyruvate carboxylase was observed [72]. The C_3-type expression pattern of *RuBisCOss* and *pepc* in the intercalar meristem, where the photosynthetic tissue is located, represents a ground state in both terrestrial and submerged forms. In the terrestrial form, the transition from C_3- to C_4-type expression pattern occurs during tissue maturation, concomitantly with the development of bundle sheath cells [72]. The transition of the expression pattern of *RuBisCOss* occurs earlier than that of *pepc*. In addition, in response to submergence, the transition of the gene expression of the two carboxylases from C_4- to C_3-type pattern is not concomitant with the change in the anatomy of the internode [72].

In summary, the C_4-type pattern of expression does not directly depend on the anatomy but probably depends on an independent mechanism that does not correlate to the formation of Kranz anatomy. The good correlation between the C_4 pattern of gene expression and the anatomical development during the formation of Kranz anatomy suggests that the two events proceed with a casual relationship. However, the correlation during the reversal of Kranz anatomy is not rigid. The C_4-type gene expression pattern does not depend on Kranz anatomy, a fact that has never been observed in other C_4 plants [72].

8. Regulation of Differentiation

There is no evidence of a single gene capable of setting in motion the entire C_4 photosynthesis machinery. C_4 photosynthesis appears to be a combination of independently inherited characteristics [105]. It appears that abscisic acid may stimulate one single signaling system that leads to the entire differentiation or separate signaling systems that are responsible for the individual differentiation of the anatomical and biochemical traits [74]. At present, it is unclear why the structural features and the photosynthetic traits of *E. vivipara* are modified under submerged conditions [85]. What is certain is that in *E. vivipara* the structural and biochemical characteristics of C_4 photosynthesis are not always differentiated in a coordinated manner, implying that separate signaling systems are responsible for both traits.

B. STUDIES IN OTHER *ELEOCHARIS* SPECIES

Two other *Eleocharis* species, *E. baldwinii* and *E. retroflexa* subsp. *Chaetaria,* show NAD–ME C$_4$ traits in their terrestrial forms [69]. The culms also exhibit unusual Kranz anatomy: the mestome sheath is interposed between the mesophyll cell and the bundle sheath cells. The δ^{13}C values of the terrestrial forms of these two species are less negative than those found in *E. vivipara*, but still within the range. However, they differ in their responses to aquatic environments [106].

Under submergence conditions *E. baldwinii* express C$_3$–C$_4$ intermediate characteristics with the reduction of the bundle sheath cells and the development of the mesophyll cells [106]. The amounts of phosphoenolpyruvate carboxylase, pyruvate orthophosphate dikinase, and NAD–ME are lower in the submersed form, but NADP–ME exhibits an opposite trend. The amount of RuBisCO is higher in the submerged form than in the terrestrial form. RuBisCO is accumulated in the mesophyll cell chloroplasts and in the bundle sheath chloroplasts, increasing the amount of enzyme in these last cells under submergence conditions [107]. NAD–ME was located in the mitochondria of bundle sheath cells, and pyruvate orthophosphate dikinase and phosphoenolpyruvate carboxylase were detected in the mesophyll cells. In contrast, NADP–ME was found in the chloroplasts of both mesophyll and bundle sheath cells [107]. It is suggested that under submergence conditions *E. baldwinii* could fix some CO$_2$ through a C$_4$-like metabolism as in the case of the aquatic plants *E. densa* and *H. verticillata*, which show a C$_4$-like metabolism without Kranz anatomy under conditions of low carbon availability [11,18,25,32,107].

On the other hand, *E. retroflexa* maintains the C$_4$ characteristics when it is submersed, the Kranz-like anatomy with the C$_4$ biochemical traits [74], with slightly higher levels of RuBisCO under this condition.

Although it is interesting that there is a gradient in the degree of expression of C$_4$ characteristics among the three terrestrial forms of these *Eleocharis* species, it should not be surprising considering that the C$_4$ photosynthesis in *Eleocharis* spp. evolved recently in the genus, generating various intermediate stages [69].

C. CONCLUDING REMARKS

The high developmental plasticity shown by *E. vivipara* makes this species a suitable system for studying the genetics and development of C$_4$ photosynthesis in response to environmental factors, as well as the ecological and adaptive aspects of C$_4$ species [98]. Ecological considerations suggest that the extreme and changeable microenvironments inhabited by its species may have been decisive in the appearance of these variations in the photosynthetic pathway [81]. The dimorphism of this species is of considerable interest because it provides a system in which to study the relationship between the development of Kranz anatomy and C$_4$-type cell-specific expression of genes involved in photosynthetic carbon metabolism [72].

IV. C$_4$ PHOTOSYNTHESIS IN THE CHENOPODIACEAE FAMILY

The Chenopodiaceae family has C$_3$, C$_4$, and C$_3$–C$_4$ intermediate metabolism species [108–112], with the most C$_4$ species among the dicot families [113]. There are interesting anatomical variations in the photosynthetic apparatus of leaves/stems and cotyledons of representative species of the tribe Salsoleae [114]. Apart from the Kranz-like organs with mesophyll cells and bundle sheath cells, in many species a circular anatomy that includes water storage and subepidermal cells has been described [114]. In addition, the species *B. cycloptera* and *B. aralocaspica* show appreciable C$_4$ photosynthesis in a single chlorenchyma cell [13,14]. In spite of C$_4$ photosynthesis being accomplished by more than a dozen biochemical and anatomical combinations that arose independently [116,117], all C$_4$ plants share the common initial step of phosphoenolpyruvate carboxylation, and, until recently, it was thought that Kranz anatomy was also essential for C$_4$ photosynthesis in terrestrial plants [116]. The new findings in *B. cycloptera* and *B. aralocaspica* challenge this view [118].

A. THE CASE OF *B. CYCLOPTERA*

1. Occurrence and Description

Bienertia is a monotypic genus of the tribe Suaedeae within the Chenopodiaceae family. It is an Irano-Turanian floristic element distributed in the central Iranian deserts and subdeserts within northern and southern radiation in the Persian Gulf countries and Central Asia [118]. *Bienertia* grows as a halophytic annual on clay, silt, or sandy alluvial soils in depressions and along lagoons. Climatic conditions are marked by hot, sunny, and dry summers and by winters with no or little to moderate frost and mean precipitation from ca. 100 to 200 mm [119].

The plants are 10 to 60 cm in height and richly branched from the base. Leaves (2–15 mm × 2.5 mm × 1.5–2.5 mm) are usually glaucous, narrow to broad oblong, slightly narrow at the base, and obtuse at the apex. Young leaves are covered by small, shortly stalked vesicular hairs [119].

2.   Structure and Ultrastructure of the Leaf

Five variants of Kranz anatomy, Atriplicoid, Koichioid, Salsoloid, Kranz-Suaedoid [109], and Conospermoid [120], were identified in the family. In the case of *Bienertia* it represents a new leaf type named Bienertiod, and in the overall layout it resembles most of the C_3 species of *Suaeda*, except for the cytological structure and arrangement of the chlorenchyma and its strict separation from the water storage tissue [119] (Figure 22.5).

The epidermis is one-layered with large individual cells in which chloroplasts are missing, except for the small and slightly sunken guard cells.

The chlorenchyma consists of two to three layers of radially elongated cells that are arranged in short rows between the epidermis and the aqueous tissue. Cells are unique in containing two cytoplasmic compartments, a peripheral cytoplasmic layer with scattered chloroplasts, and a large globular cytoplasmic body located in the center of the cell, completely surrounded by the vacuole and densely packed with starch producing chloroplasts and joined by the nucleus [119]. Both compartments are connected by cytoplasmic channels that contain a few chloroplasts

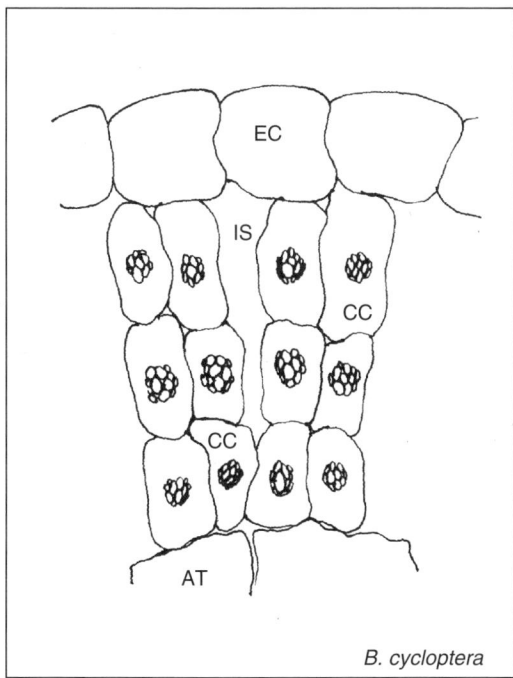

FIGURE 22.5 Diagram of various cell types in a cross section of *Bienertia cyclopetera*. EC: epidermal cell; IS: intercellular space; CC: chlorenchyma cell; AT: aqueous tissue.

with a development of grana intermediate between those of both cytoplasmic compartments [14]. The central compartment is filled with mitochondria, with an extensive system of tubules and lamellae, which encircle the granal chloroplasts. In contrast, the peripheral compartment lacks mitochondria and has agranal chloroplasts that exhibit lower granal index, density of appressed thylakoids, and ratio of appressed to nonappressed thylakoids than the chloroplasts in the central compartment [14]. Even after long light exposure periods starch was absent in the chloroplasts of the peripheral cytoplasmic compartment [119].

The aqueous tissue together with the embedded vascular bundles accounts for about two thirds to three fourths of the leaf volume (Figure 22.5). It consists of one to two layers of large cells with thin walls and thin peripheral cytoplasmic layers without chloroplasts. The vascular system consists of one large central bundle and a varying number of smaller secondary bundles [119].

3.   Biochemical Studies: Carbon Isotope Composition and $^{14}CO_2$ Fixation

Studies involving leaf anatomy observation and the finding that after 10 sec of $^{14}CO_2$ exposure 45% of the total radioactivity was present in sugar phosphate and only 19/11% in malate/aspartate indicated that the plant was exclusively C_3 [110]. However, later, $\delta^{13}C$ isotope values of -15.4% [112] and 14.3% [108] were reported, which are similar to those of C_4 plants with Kranz anatomy. Recently, Freitag and Stichler [119] measured $\delta^{13}C$ isotope discrimination values in *Bienertia* leaves under a wide range of conditions. They found values of -13.4 to -15.5% in leaves from natural habitats (typical of C_4 species), but values ranging from -15.5 to -21.1% were measured in newly formed leaves in the greenhouse. In addition, Voznesenskaya et al. [14] using mature leaves found values (-13.5%) similar to that reported for plants collected in the natural habitat, and also reported more negative values for younger leaves. These results suggest that *Bienertia* is a C_4/C_3 facultative species and explain the differences in the results reported by the first authors. With respect to the photosynthetic carbon labeling pattern, considering than the diffusion pathway is shorter within a single cell rather than between mesophyll and bundle sheath cells, Voznesenskaya et al. [14] used a shorter pulse (3 sec) to identify the C_4 acids. Under this experimental condition, 50% of the initial products were in malate and aspartate and 13% in 3-phosphoglycerate.

4. Biochemical Studies: Analysis of the Main Photosynthetic Enzymes

RuBisCO was mainly found in the chloroplasts of the central compartment, and only a weak signal was found in the chloroplasts of the peripheral compartment. Although phosphoenolpyruvate carboxylase was found through all the cytosol, the most intensive labeling was found in the peripheral cytoplasm. The levels of these enzymes are similar to those in C_4 chenopods and also exhibited a phosphoenolpyruvate carboxylase/RuBisCO ratio greater than 1. In contrast to RuBisCO, pyruvate orthophosphate dikinase was located in the peripheral chloroplasts. Again, activity and immunoreactive protein levels were similar to those found in the C_4 plant *Salsola laricina*. Higher activity levels of NAD–ME relative to NADP–ME were measured in *Bienertia*. While the former was immunodetected, the NADP-dependent enzyme was not. Immunoelectron microscopy indicated that NAD–ME and glycine decarboxylase were localized in the mitochondria of the central compartment [14].

5. Gas Exchange Analysis

Its photosynthetic response to varying CO_2, with 21% versus 3% O_2, is typical of C_4 plants with Kranz anatomy. In addition, O_2 did not inhibit photosynthesis in *Bienertia*. The lack of night-time CO_2 fixation indicates that this species does not undergo CAM [14].

6. Biochemical Pathway in *B. cycloptera*

On the basis of carbon isotope values, anatomical studies although excluding Kranz anatomy [13,120], products of $^{14}CO_2$ fixation, immunolocalization studies, and facultative C_3/C_4 expression [120], it can be concluded that *B. cycloptera* exhibits a novel solution to C_4 photosynthesis without Kranz anatomy. This metabolism is carried out in a single chlorenchyma cell through spatial compartmentation of dimorphic chloroplasts, other organelles, and photosynthetic enzymes in distinct cell positions that mimics the spatial separation of Kranz anatomy [14] (Figure 22.6). Thus, the peripheral and central cytoplasmic compartments may be functionally equivalent in photosynthesis to palisade and Kranz cells in the Kranz-type C_4 plants [119]. The vacuole appears to be the resistant barrier minimizing CO_2 efflux, and the cytoplasmic strands are the channels for metabolite flux [117]. CO_2 would be initially fixed by phosphoenolpyruvate carboxylase localized in the peripheral compartment. Then, C_4 acids would be transported through the cytoplasmic channel to the mitochondria in the central cytoplasmic compartment where NAD–

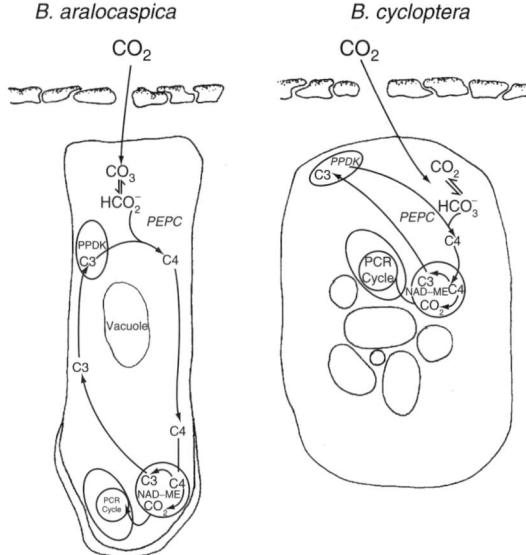

FIGURE 22.6 Proposed scheme of C_4 photosynthesis within a single cell in *Borszczowia aralocaspica* and *Bienertia cycloptera*. The intracellular localization of the main C_4 photosynthetic enzymes is shown within the chlorenchyma cell. NAD-ME: NAD-malic enzyme; PCR cycle: photosynthetic carbon reduction cycle; PEPC: phosphoenolpyruvate carboxylase; PPDK: pyruvate orthophosphate dikinase.

ME acting as decarboxylase would provide RuBisCO in the neighboring chloroplasts with CO_2. Three carbon compounds would then return to the peripheral cytoplasmic compartment where pyruvate orthophosphate dikinase located in the chloroplasts would regenerate phosphoenolpyruvate. As this last enzyme is not found in the central cytoplasmic compartment, and thus the presence of phosphoenolpyruvate is unlikely to occur there, phosphoenolpyruvate carboxylase in the central compartment probably would not be involved in this metabolism [14].

B. cycloptera is also remarkable for its ability to adapt its photosynthetic machinery to environmental conditions as a facultative C_4/C_3 species. Leaves developed under conditions of high light and temperature carry out C_4 photosynthesis, but under conditions of low light and temperature C_3 photosynthesis is performed [119].

B. The Case of *B. aralocaspica*

1. Occurrence and Description

B. aralocaspica Bunge is a Central Asian floristic element with scattered distribution through the semideserts from the northeastern Caspian lowlands to western China and southern Mongolia. It belongs to the strongest halophytes and is found in monospecific stands in the foremost zone of higher vegetation around salt lakes and deep depressions. The habitat is usually flooded in spring and covered with a thick salt crust during summer and autumn [120].

B. aralocaspica is a succulent species with unusual chlorenchyma. Plants are 4 to 50 cm in height and richly branched from the base. Succulent glabrous leaves (10–17 mm × 1.5–2.2 mm × 0.5–1.1 mm) are isolateral with a pronounced tendency toward a centric arrangement of tissues [120].

2. Structure and Ultrastructure of the Leaf

Similar to the C$_4$ Salsoloid type *B. aralocaspica* has particular cytological features with only a single layer of unusual palisade-shaped chlorenchyma cells located between the central water storage tissue and the hypodermal cells [13] (Figure 22.7). It is unique because all known C$_4$ species have two, more rarely three, layers of chlorenchyma, which are highly differentiated in cytological [121] and physiological respects [122,123]. The pattern of vascularization with numerous small peripheral bundles attached to the inner surface of the chlorenchyma makes *B. aralocaspica* similar to the Salsolid type of C$_4$ chenopods [120].

Hypodermal cells are large with thin walls and no chloroplasts. The central aqueous tissue comprises one third of the volume of the cell, and it is consti-

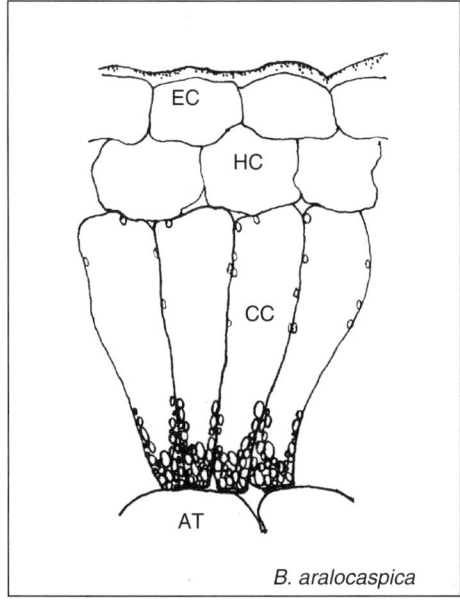

FIGURE 22.7 Diagram of various cell types in a cross section of *Borszczowia aralocaspica*. CC: chlorenchyma cell; EC: epidermal cell; HC: hypodermal cell; AT: aqueous tissue.

tuted by large polyhedral cells with thin walls, a thin protoplast, and few or no chloroplasts [120].

Chlorenchyma cells have a cylindrical shape in the upper two thirds to three fourths, but the basal parts of the cells are trapezoid. The layer is compact to the center of the leaf, with intercellular spaces in the outermost part of the chlorenchyma (Figure 22.7). The cell wall is thicker in the basal parts and has perfect straight orientation. The protoplast has also differential features in the upper and basal parts of the chlorenchyma cells [120]. The basal parts possess a much smaller central vacuole, a higher number of chloroplasts per unit volume, chloroplasts with abundant starch grains, higher density of cytoplasm, and firmer attachment of plasmalemma to the cell walls. The nucleus is invariably located at or just above the border of the two macrocompartments of the chlorenchyma cells [120]. Large mitochondria are in the proximal position (from the vascular bundle) [13], and chloroplasts in the distal part of the cell lack grana and are without starch, while those in the proximal part have grana and contain starch. Thus, *B. aralocaspica* has leaf anatomy similar to C$_4$ *Salsola*, but it has dimorphic chloroplasts in a single photosynthetic cell instead of in two cell types with Kranz anatomy [13].

3. Biochemical Studies: Carbon Isotope Composition

B. aralocaspica carbon isotope ratios of −13.78 [120] and −13.37‰ δ^{13}C [13] not only matches with the those described for C$_4$ plants (−10 to −14‰, 124) but also with those reported from 198 C$_4$ chenopods, −9.27 to −15.06‰, [108]. In contrast, the C$_3$ species *Suaeda heterophylla* exhibited values of −25.34 and −27.28‰ [13] as in this plant RuBisCO fixes atmospheric CO$_2$. Thus, although lacking Kranz anatomy *B. aralocaspica* probably fixes atmospheric CO$_2$ by phosphoenolpyruvate carboxylase that does not discriminate against ^{13}CO$_2$.

4. Biochemical Studies: Analysis of Main Photosynthetic Enzymes

Voznesenskaya et al. [13] found biochemical compartmentation in the chlorenchyma cells for the carbon assimilation enzymes. RuBisCO was found in the chloroplasts in the proximal part of the cell, while pyruvate orthophosphate dikinase was located in the chloroplasts in the distal cytosol. RuBisCO was also detected in the few and small chloroplasts of water storage and hypodermal cells. Phosphoenolpyruvate carboxylase was visualized in all the cytosol. NAD–ME is located in the mitochondria in the proximal part of the cells.

The activities of RuBisCO, phosphoenolpyruvate carboxylase, NAD–ME, and pyruvate orthophosphate dikinase are high enough to support the measured rates of CO_2 assimilation and except for NAD–ME, were similar to the values measured in the C_4 plant *S. laricina*.

5. Gas Exchange Analysis

As the response of photosynthesis to varying CO_2 and O_2 is a diagnostic tool for discriminating C_3 versus C_4 photosynthesis, the response of *B. aralocaspica* against different O_2 concentrations was evaluated, and it was found that photorespiration is restricted in this species [13]. While ambient levels of O_2 (21%) inhibited photosynthesis in C_3 *S. heterophylla*, it did not have an effect on photosynthesis in the C_4 species *S. laricina* or *B. aralocaspica*.

6. Biochemical Pathway in *B. aralocaspica*

B. aralocaspica presents a unique pattern of enzyme distribution, with some of the C_4 enzymes present to the same extent as in C_4 plants. Gas exchange analysis and carbon isotope composition indicate that *B. aralocaspica* behaves like a C_4 plant. Overall, *Borszczowia* is a C_4 plant, although it lacks Kranz anatomy, in which the distal and proximal cytosolic compartments would probably act as the mesophyll and bundle sheath in C_4 plants with Kranz anatomy (Figure 22.6). In this novel photosynthetic metabolism atmospheric CO_2 enters the cell at the distal part thorough the thin cell wall and is fixed by phosphoenolpyruvate carboxylase. Then, C_4 acids would be transported to the proximal part of the cell where the decarboxylation in the mitochondria by the action of NAD–ME provides RuBisCO with CO_2. In this proximal part of the cell CO_2 leakage is restricted by the thicker cell wall. Thus, the radial arrangement of the elongated chlorenchyma cell is of great importance to prevent the CO_2 efflux [117]. As pyruvate orthophosphate dikinase is located in the chloroplasts in the distal cytosol, phosphoenolpyruvate would be regenerated in this part of the cell, providing the substrate for phosphoenolpyruvate carboxylase. Distal generation of phosphoenolpyruvate and the thick cell wall at the proximal part of the cell make improbable that phosphoenolpyruvate carboxylase in the proximal cytosol fixes atmospheric CO_2.

C. Evolution of C_4 Photosynthesis in *B. cycloptera* and *B. aralocaspica*

C_4 photosynthesis in terrestrial plants was thought to require Kranz anatomy. However, it is now known that in *B. cycloptera* and *B. aralocaspica* the compartmentation of the photosynthetic enzymes, and thus the avoidance of futile cycles, is achieved through the separation of two types of chloroplasts and diverse organelles within the photosynthetic cell. In these species the evolution of cytoplasmic organization and C_4 photosynthesis has taken distinctly different paths. These species are more distantly related, and their C_4 leaf types have evolved along independent evolutionary lines from the basal Austrobassioid type, which is still present in C_3 species of *Suaeda* [14]. The tribe Suaedeae comprises four genera: *Suaeda* (100 species, 60% C_4), *Alexandra* (one C_3 species), *Bienertia* (one C_4 species), and *Borszczowia* (one C_4 species) [113,120,125]. Within the Suaedeae, five distinct C_4 origins are suspected, three in *Suaeda*, and one in *Bienertia* and one in *Borszczowia* [119]. It seems that the present tribe is very prolific in evolving C_4 photosynthesis [117]. As the Suaedeae species grow in extreme saline soils where the interspecific competition is restricted and which favor characteristics enhancing water use efficiency, *Bienertia* and *Borszczowia* might have had a competitor-free space in which to evolve [117].

D. Perspectives for Biotechnological Approaches to Improve C_3 Photosynthesis

The efficiency by which aquatic plants are capable of switching from a C_3-type to a C_4-like metabolism in single cells when the availability of CO_2 declines is encouraging because it might be possible to establish a similar system in terrestrial C_3 crops by the transgenic approach [126]. The discovery of chenopods containing a single CO_2-concentrating mechanism [13,14,119,120] opens a new possibility in engineering C_4 plants from C_3 species, which is easier than was previously believed [127]. However, it should be considered that the anatomy of these halophytic chenopods is adapted to semidry environments [126], even though engineering a single-celled C_4 system would require a reworking of C_3 leaf structure [128]. Nevertheless, it would not require the development of two photosynthetic cells [13].

V. STUDY OF THE TRANSITION FROM C_4 PHOTOSYNTHESIS TO CRASSULACEAN ACID-LIKE METABOLISM IN THE *PORTULACA* GENUS

In general, leaves of a plant fix CO_2 through one type of photosynthetic pathway. However, some plants can shift their photosynthetic mode depending on the age or environment, like the shift from C_3 photosynthesis to CAM [129,130]. Current mechanistic understanding of the induction of CAM is largely based on studies with facultative species that present a transition from C$_3$ photosynthesis. In contrast, the shift from C$_4$ photosynthesis to CAM has been described only in some succulent C$_4$ plants belonging to the genus *Portulaca* [15,16,131–135]. *P. oleracea*, *P. grandiflora*, and *P. mundula* were reported to express CAM characteristics when subject to drought stress conditions or short photoperiods [15,16,131–136].

Although most CAM and C$_4$ species occur in a few families where they are mutually exclusive, five plant families share genera containing species that are CAM or C$_4$. The photosynthetic diversity of the Portulacaceae is considerable despite the small size of the family, with species C$_3$, C$_4$, and facultative CAM plants. Moreover, the occurrence of the decarboxylating enzymes reinforces this diversity. *P. grandiflora* is an NADP–ME species [134], *P. oleracea* is an NAD–ME subtype [137], and *Portulacalaria afra* is a phosphoenolpyruvate carboxykinase plant [138].

CAM induction is a complex reaction needing the coordination of many changes in metabolism [139]. The study of the transition from one type of photosynthesis to another is of great interest since it can provide insight into the different expression–regulation of the enzymes involved in each type of photosynthesis. For example, the transition from C$_4$ to CAM photosynthesis may involve the expression of different isoenzymes or the occurrence of an opposite regulation of the same enzyme, like phosphoenolpyruvate carboxylase, which is opposite regulated in both metabolisms [140,141]. Investigations were performed in *Portulaca* species under different stress conditions, mainly drought stress [15,16,131–135].

A. THE CASE OF *P. OLERACEA*

1. Distribution and General Features

P. oleracea is found in places as diverse as gardens and cultivated fields, or in agriculturally poor habitats such as roadsides and out-rocks, but it is usually found in hot, dry conditions and in high-light intensity environments [142].

The firsts works done with respect to this species remarked the weediness exhibited by it. By the 1970s, *P. oleracea* was the eighth most common plant on earth and was classified as a weed, in part because of its process and pattern of growth, which gave the plant quick response capability [142]. Nowadays, this plant is listed as a noxious weed by the U.S. Federal Government. Zimmerman [142] pointed out that *P. oleracea* uses a wide variety of photoperiods, and capsule numbers are correlated with the amounts of

light received. This weed is widely tolerant of light intensities, temperature regimes, and soil types, and the plants produce adequate levels of capsules over a wide range of these factors. The plasticity shown by *P. oleracea* with respect to its habitats would be just a feature of its flexibility with respect to its photosynthetic metabolism.

2. First Studies on Photosynthetic Metabolism

Early studies have classified *P. oleracea* as a C$_4$ plant, although the stage of leaf development was one of the most important factors determining the operation of C$_4$ photosynthesis. Young and mature leaves fixed $^{14}CO_2$ primarily into organic and aminoacids, and less than 2% of the $^{14}CO_2$ fixed appeared in phosphorylated compounds. Young leaves produced more malate than aspartate, whereas mature leaves produced more aspartate than malate [143]. In addition, young and mature leaves exhibited typical C$_4$-plant light/dark $^{14}CO_2$ evolution ratios [144]. In contrast, senescent leaves had a relatively large amount of C$_3$ photosynthesis, as accounted by a quantitative shift of primary products toward phosphorylated compounds (18% in phosphoglyceric acid) with a concomitant reduction of the label residing in malate and aspartate [143]. Senescent leaves had photorespiration ratios similar to C$_3$ plants. In this state, *P. oleracea* leaves had a less absolute amount of ribulose-1,5-bisphosphate, phosphoenolpyruvate carboxylase (10% of mature leaves), and RuBisCO (27% of mature leaves) than mature leaves [144].

P. oleracea has small succulent leaves with Kranz anatomy, characteristic of other C$_4$ species, and it also possesses large water storage. Total chlorophyll content is low, but as leaves of *P. oleracea* are relatively succulent, this value is still within the reported range for several other C$_4$ plants [143].

3. Conditions for CAM Induction in *P. oleracea*

Studies of a wide range of environments in which *P. oleracea* grows, the succulence of its leaves, the particularly high water use efficiency, and its location within a family that contains species exhibiting C$_3$, C$_4$, or CAM metabolism led to the investigation of the possible occurrence of CAM or facultative CAM in this species by analyzing diurnal acid fluctuation, CO_2 gas exchange, and leaf resistance under various photoperiods and water regimes [131]. The data presented strongly suggested that CAM occurred in *P. oleracea* leaves and stems. Under short photoperiods or water stress, *P. oleracea* presented a CAM-like pattern of acid fluctuation accompanied by low nocturnal leaf resistance, in the first case, and this feature plus net dark CO_2 uptake and daytime CO_2 release, in the second [131].

Later, Koch and Kennedy studied the occurrence of CAM under natural environmental conditions but protected from the rain. This study confirmed the results obtained previously in drought stressed plants and established that malate was the predominant compound labeled during the night, with some citrate and aspartate [132]. Under natural environment conditions, CAM contributed to the carbon balance and water retention in the C_4 dicot *P. oleracea*. Stomatal closure in the light reduced water loss from plants, with insoluble compounds synthesized from the CO_2 assimilated during the night. They also determined that not only drought stress conditions but also photoperiod, developmental state, and diurnal temperature changes appear to be important in CAM expression in *P. oleracea* [132]. Similar results have been reported by Kraybill and Martin [133]. In either case, the diurnal fluctuation of titratable acidity, accounted for by malic acid, suggests that a crassulacean acid-like metabolism is effectively induced in *P. oleracea* upon water stress treatment.

4. Induction of a Crassulacean Acid-Like Metabolism in *P. oleracea* under Drought Stress Conditions

Recently, Lara et al. [15] studied the induction of a crassulacean acid-like metabolism in *P. oleracea* under drought stress conditions and a 12-h photoperiod by stopping watering of young plants. After 23 days of drought stress a diurnal change in titratable acidity was observed, evidenced by a seven- to eightfold increase of the fluctuation in titratable (day/night) acidity. The stressed leaves presented an increase in malate levels of almost two orders of magnitude between the end of the day and night periods relative to well-watered plants, suggesting the induction of a crassulacean acid-like metabolism. This was also confirmed by limited or nil CO_2 assimilation in the light and slow CO_2 rates of uptake in the dark by *P. oleracea* stressed plants.

5. Leaf Anatomy Studies in Control and Water Stressed Plants

The C_4 dicot *P. oleracea* possesses succulent leaves with branched veination and composed of various types of cells (Figure 22.8). Bundle sheath cells are around the vascular bundles, and their chloroplasts are located in a centripetal position (toward the vascular tissue). Mesophyll cells completely surround the bundle sheath cells. A third type of cell constitutes two or three layers between the epidermis and the parenchyma. These cells, called water storage cells, are particularly large, vacuolated, and contain a few small chloroplasts [15].

Drought had an important effect on the leaf structure of *P. oleracea* with the distortion of the mesophyll cells and the displacement of its chloroplasts toward the bundle sheath cells, probably due to water loss, accompanied by a general loss of the chloroplast number through the tissue. Partial collapse of the water storage cells and the loss of organization of this tissue was also observed.

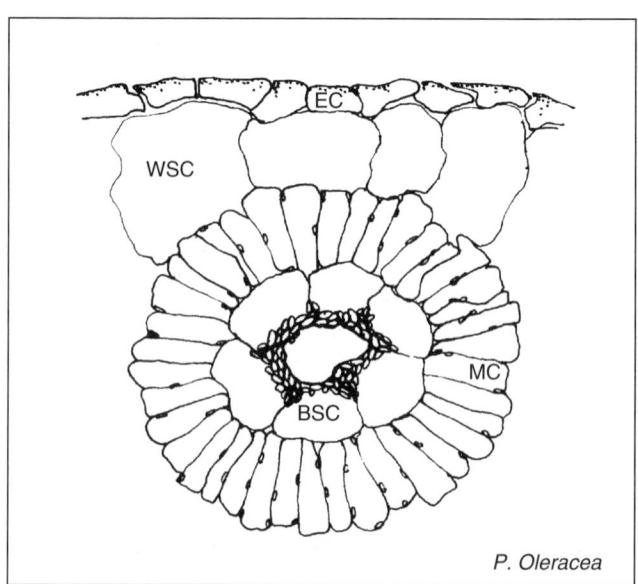

FIGURE 22.8 Diagram of various cell types in a cross section of *Portulaca oleracea*. EC: epidermal cell; BSC: bundle sheath cell; MC: mesophyll cell; WSC: water storage cell.

6. Phosphoenolpyruvate Carboxylase in Control and Water Stressed Plants: Characterization of the Different Isoforms

The regulatory properties of phosphoenolpyruvate carboxylase were also studied under both conditions. The enzyme presented a subunit mass of 110 kDa and exhibited changes in the isoelectric point and electrophoretic mobility of the native enzyme [15].

In C$_4$ and CAM plants the enzyme is regulated by a mechanism of phosphorylation/dephosphorylation [145]. In the phosphorylated state, feedback inhibition by L-malate is severely diminished [145–147]. *In vivo* phosphorylation and native isoelectrofocusing studies indicated that phosphoenolpyruvate carboxylase activity and regulation are modified upon drought stress treatment in a way that allows *P. oleracea* to undergo a crassulacean acid-like metabolism. Stressed leaves contained less (lower specific activity and lower enzyme content on a protein basis) of a noncooperative form of phosphoenolpyruvate carboxylase with different kinetic properties, such as a higher affinity for phosphoenolpyruvate and more sensitivity to malate inhibition. In summary, the regulation of the enzyme from *P. oleracea* control plants appears to be similar to that occurring in C$_4$ species, while the regulation of the enzyme from drought stressed plants resembles that of the CAM type [15].

Other work has shown that the induction of CAM in the C$_4$ *P. oleracea* was accompanied by an increase in the levels of activity and quantity of phosphoenolpyruvate carboxylase [135]. This finding was rather surprising since, although this enzyme should be active at different periods of the day in C$_4$ and CAM plants, one biochemical similarity between these types of photosynthesis is the high phosphoenolpyruvate carboxylase activity in mesophyll cells.

7. Study of Decarboxylating Systems, RuBisCO, and Pyruvate Orthophosphate Dikinase in *P. oleracea* Control and Water Stressed Plants

As the function of a C$_4$ and CAM metabolism implies a different spatial and temporal participation of the enzymes involved in the CO$_2$ fixation, the study of these enzymes in both tissue conditions was of great interest. Lara et al. analyzed the activity and levels of the decarboxylation enzymes. They confirmed that this species is a NAD–ME-subtype C$_4$ plant and that the enzyme is upregulated by light in this photosynthetic mode. However, as in the case of maize, phosphoenolpyruvate carboxykinase may contribute, along with NAD–ME, to the concentrating mechanism [148].

After 23 days of drought stress, a general decrease in the photosynthetic metabolism was found, as accounted for by the decrease in the net CO$_2$ fixation [15] and in the activity of enzymes related to that metabolism, such as RuBisCO, phosphoenolpyruvate carboxylase, pyruvate orthophosphate dikinase, phosphoenolpyruvate carboxykinase, and NAD–ME. In contrast, NADP–ME shows no variation in the activity or immunoreactive levels. Western blot analysis indicated that changes in the activities were correlated with the levels of immunoreactive proteins, except for phosphoenolpyruvate carboxykinase activity in which putative phosphorylation of the enzyme would be responsible for the changes in the activity. Pyruvate orthophosphate dikinase showed two immunoreactive bands the levels of which varied not only between day and night but also with the water status; a higher level of immunoreactive protein was found in control samples than in stressed plants, with the lower-molecular mass protein being the more abundant and possibly being involved in phosphoenolpyruvate regeneration in control plants performing C$_4$ photosynthesis. In contrast, as the levels of the higher-molecular mass isoform were higher in stressed leaves, this isoform could be related to the operation of CAM [149].

8. *In Situ* Immunolocalization of Phosphoenolpyruvate Carboxylase, RuBisCO, and NAD–ME

RuBisCO is located in the bundle sheath chloroplasts of *P. oleracea* leaves, while NAD–ME is found in the mitochondria of the same cells. In contrast, phosphoenolpyruvate carboxylase was found in the cytosol of mesophyll and water storage cells. Although the localization of the enzymes was the same for control and stressed plants, the amount of immunogold particles varied in the same way as the amount of protein previously determined. The main change was observed for phosphoenolpyruvate carboxylase, which exhibited higher levels of immunocomplexes in the cytosol of the water storage cell in stressed samples in comparison with control leaves [149].

9. Carbohydrate Metabolism-Related Enzymes

Different patterns of assimilate partitioning are described among CAM species. As in *M. crystallinum* [4], in *P. oleracea* changes have been reported in the day/night activities and in the level of immunoreactive protein of some of the enzymes involved in glycolysis, in gluconeogenesis, and in the generation of reduction power when CAM is induced. While some enzymes (ATP-dependent phosphofructokinase,

PP$_i$-dependent phosphofructokinase, NAD– and NADP–malate dehydrogenase, and nonphosphorylating glyceraldehyde-3-phosphate dehydrogenase) increased under stress conditions; others (aldolase and glucose-6-phosphate dehydrogenase) decreased. These changes in the activities could be of great importance in the establishment of CAM in *P. oleracea* [149].

10. Biochemical Pathways Operating in Control and Stressed *P. oleracea* Leaves

In control *P. oleracea* plants under well-watered conditions a C$_4$ photosynthetic metabolism operates with NAD–ME as the major decarboxylating enzyme, accompanied by phosphoenolpyruvate carboxykinase (Figure 22.9). Additionally, as phosphoenolpyruvate carboxylase was also found in the water storage cells, these cells could also contribute to the primary CO$_2$ fixation by producing oxaloacetate, which could be then transformed in malate by malate dehydrogenase in these cells and then transported to the bundle sheath cells through the mesophyll cell, thus increasing the malate pool given by these last cells.

An inducible CAM may be a physiological feature in leaves of water stressed *P. oleracea*, which could account for the relative long duration of this species in habitats characterized by extended dry periods [131]. In these periods a general decrease of CO$_2$ would occur. A spatial separation of the primary and secondary carbon fixation would be accompanied by a temporal separation of these processes. According to this, during the night CO$_2$ would be fixed through phosphoenolpyruvate carboxylase in the cytosol of mesophyll cells yielding oxaloacetic acid, which would be then kept in the vacuoles in the form of malate. In the following light period, the released malate would be transported to the bundle sheath cells where NAD–ME, together with other decarboxylating enzymes, would provide RuBisCO with CO$_2$. Probably, a cytosolic CAM-specific pyruvate orthophosphate dikinase could participate in the phosphoenolpyruvate regeneration. Although the role of water storage cells is still uncertain, an enhanced activity of these cells could take place under this condition by generating and storing malate. Alternatively, these cells act as a water reservoir for mesophyll cells and bundle sheath cells for

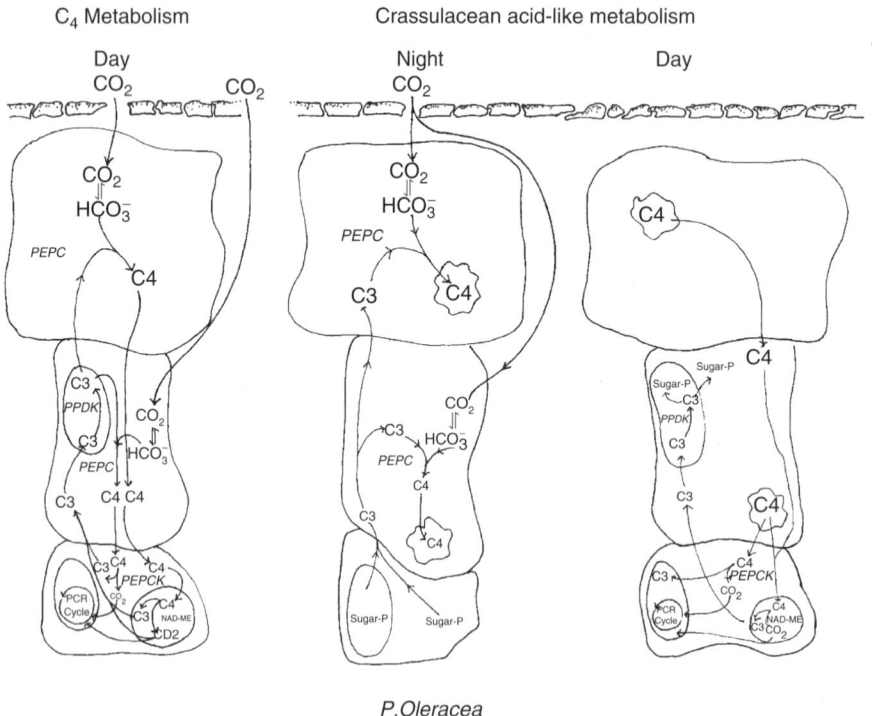

P.Oleracea

FIGURE 22.9 Proposed photosynthetic metabolisms operating in *Portulaca oleracea* in well-irrigated (C$_4$ metabolism) and water stressed plants (crassulacean acid-like metabolism). From top to bottom water storage, mesophyll and bundle sheath cells are shown. The intracellular localization of the main C$_4$ photosynthetic enzymes is shown. NAD–ME: NAD-malic enzyme; PCR cycle: photosynthetic carbon reduction cycle; PEPC: phosphoenolpyruvate carboxylase; PEPCK: phosphoenolpyruvate carboxykinase; PPDK: pyruvate orthophosphate dikinase.

conditions of limited water supply; the presence of phosphoenolpyruvate carboxylase in water storage cells reveals a possible remaining ancestral CAM metabolism.

B. THE CASE OF *P. GRANDIFLORA*

P. grandiflora Hook is an ornamental dicot that grows best in sandy soil originally from northern South America [150]. It has been reported as a C$_4$ plant based on the measurement of $\delta^{13}C$ [151] and biochemical studies [137]. The occurrence of *P. grandiflora* in an open stand community with xeric and high light conditions, its succulent cylindrical leaves having a large volume of water tissue and a low surface area to volume ratio, which aids water retention, and its phylogenetic position (it belongs to Portulacaceae) motivated the investigation of the possible occurrence of CAM in this species, especially under drought stress conditions.

1. Induction of a Crassulacean Acid-Like Metabolism in *P. grandiflora* under Drought Stress Conditions

Ku et al. [134] measured greater diurnal acid fluctuations and malic acid concentration in stems and leaves of well-watered plants of *P. grandiflora* than in drought stressed individuals. These parameters were much reduced under severe drought conditions. In contrast, Kraybill and Martin [133] indicated that no significant diurnal malic acid fluctuations occurred under well-watered conditions but under drought stress significant diurnal malic acid fluctuations were measured. They also found that no net CO$_2$ uptake took place under drought stressed conditions and suggested that this species underwent CAM cycling. Subsequent research by Guralnick and Jackson [152] confirmed that *P. grandiflora* maintained high organic acid levels and large diurnal acid fluctuation when water stressed.

Recently, diurnal fluctuation in acidity in leaves that was not accompanied by a net gain or loss of CO$_2$ at night and a decrease in net CO$_2$ fixation during the day showed the upregulation of the CAM pathway after water stress [16]. After 8 days of water stress the leaves presented CAM-cycling activity and suggested that this feature may occur completely in water storage tissue. On the basis of diurnal fluctuations in acidity in stems and no net carbon gain during the day or night, as the stems lack stomata, they also proposed that CAM-idling photosynthesis could take place in stems, which may have an important role in recycling carbon and conserving water to support photosynthetic activity in leaves during water stress [16].

2. Anatomical and Ultrastructural Studies

P. grandiflora has a slightly different Kranz leaf anatomy (Figure 22.10). The succulent, cylindrical leaves of this dicot species possess three distinct green cell types: bundles sheath cells in radial arrangement around the vascular bundles, mesophyll cells in an outer layer adjacent to the bundle sheath cells, and water storage cells in the leaf center. This last type of cell constitutes the water tissue and contains scattered chloroplasts and a large vacuole. Unlike typical Kranz leaf anatomy, the mesophyll cells do not surround the bundle sheath tissue but occur only in the area between the bundle sheath and the epidermis [134]. While bundle sheath cells around the peripheral vascular bundles possess agranal chloroplasts that are centripetally oriented, as is typical of NADP–ME plants [153,154], mesophyll cells have granal chloroplasts [154]. Nevertheless, granal development is suppressed in the chloroplasts of the innermost water storage cells adjacent to the central vascular bundle. Chloroplasts in the water storage cells just adjacent to the central vascular bundle show a tendency to be located near the vascular bundles [154].

When leaf anatomy is examined in plants under drought stress the collapse of the water storage cells is observed, with a C$_4$ tissue still hydrated and functioning. Thus, under limited water supply, the water storage tissue transfers water to the mesophyll and bundle sheath cells [16]. Similar to *P. oleracea* mesophyll cells, the chloroplasts of cortical cells of *P. grandiflora*

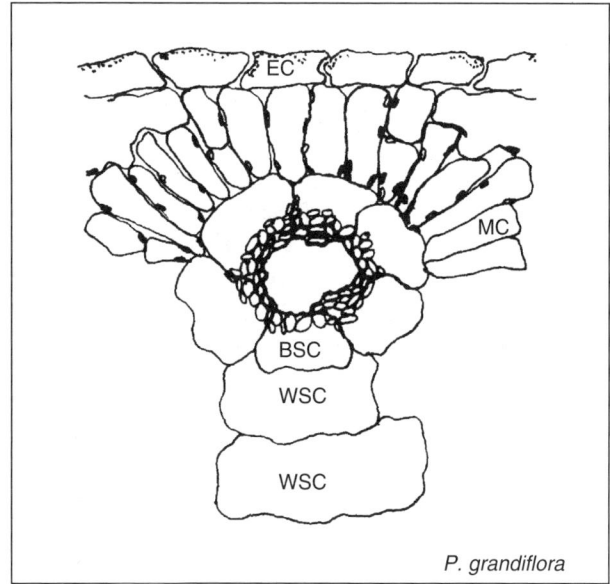

P. grandiflora

FIGURE 22.10 Diagram of various cell types in a cross section of *Portulaca grandiflora*. EC: epidermal cell; BSC: bundle sheath cell; MC: mesophyll cell; WSC: water storage cell.

became clustered in comparison to well-watered leaves [16].

3. Intercellular and Intracellular Location of Photosynthetic Enzymes

Using protoplasts it was found that RuBisCO and NADP–ME were located in the bundle sheath cell chloroplasts, but pyruvate orthophosphate dikinase is mainly found in mesophyll cell chloroplasts [134]. Although pyruvate orthophosphate dikinase, phosphoenolpyruvate carboxylase, and NADP–malate dehydrogenase were present in all three green cell types [134], it can be proposed that this ubiquitous localization of the enzyme is the consequence of cross-contamination of the protoplast fractions. In contrast, and unlike other C_4 plants, other enzymes such as NADP–malate dehydrogenase, aspartate aminotransferase, 3-phophoglicerate kinase, and NADP–triose-phosphate-dehydrogenase were equally distributed between chloroplasts from mesophyll and bundle sheath cells. Alanine aminotransferase and NAD–malate dehydrogenase were mainly present in the cytosol of both cell types [134].

Nishioka et al. [154] investigated the location of RuBisCO using immunogold labeling and found that the enzyme was accumulated in bundle sheath chloroplasts and in the chloroplasts of water storage cells adjacent to the central vascular bundle. Labeling of RuBisCO was markedly reduced in mesophyll chloroplasts and in the chloroplasts of the remaining water storage cells. Gradient in labeling of RuBisCO and granal development is not observed in the remaining water storage cells. Thus, accumulation of RuBisCO and the suppression of granal development is restricted to the chloroplasts in the cells adjacent to the vascular bundles [154].

In situ immunolocalization in well-watered leaves showed that phosphoenolpyruvate carboxylase was located in mesophyll and water storage cells, with lower density in the latter type of cell. As expected, RuBisCO was found in chloroplasts of bundle sheath cells. In stems, RuBisCO and phosphoenolpyruvate carboxylase were located in cortical cells. Finally, Guralnick et al. [16] performed tissue printing and found an increase in phosphoenolpyruvate carboxylase in the water storage tissue of leaves and the cortex of stems.

4. Study of Phosphoenolpyruvate Carboxylase, NADP–ME, Pyruvate Orthophosphate Dikinase, and RuBisCO in *P. grandiflora* Control and Water Stressed Plants

Western blot analysis showed that stressed leaves had a slight decrease in the C_4-CAM pathway proteins

NADP ME and pyruvate orthophosphate dikinase, while a new isoform of NADP–ME appeared and an increase of phosphoenolpyruvate carboxylase was observed after 10 days of water stress treatment. RuBisCO content remained constant during the water stress period in leaves as well as in stems. Interestingly, Fathi and Schnanenberg [155] found that NADP–ME of *P. grandiflora* displayed an immunochemical cross-reaction more similar to that of C_3 and CAM plants than to that of C_4 species.

Stems exhibited increases in the levels of phosphoenolpyruvate carboxylase, NADP–ME, and pyruvate orthophosphate dikinase when subjected to drought stress [16].

5. Biochemical Pathways Operating in Control and Stressed *P. grandiflora* Leaves

P. grandiflora leaves have shown CAM-cycling activity, while the stems exhibit CAM-idling metabolism, both upregulating the CAM pathways under water stress. Guralnick et al. [16] proposed that both C_4 and CAM metabolisms operate simultaneously in stressed plants as the diurnal course of acid fluctuation commences in leaves as the light period begins, as does CO_2 uptake. Nevertheless, the photosynthetic role of the innermost water storage cells is questionable because of their distance from mesophyll cells and the sparse occurrence of chloroplasts in these cells [154]. However, some kind of activity may take place in the outermost water storage cells as accounted for by the presence of RuBisCO in these cells and the increase of phosphoenolpyruvate carboxylase protein in the same cells after drought stress treatment [16].

C. The Case of *Portulaca mundula*

P. mundula is a small succulent plant native to the central and southwestern United States that grows in shallow gravelly soil over limestone outcrops [156]. On the basis of anatomical [157] and physiological characterization [158] and $\delta^{13}C$ of -13.8% [133], *P. mundula* is clearly a C_4 plant.

In contrast to *P. oleracea* and *P. grandiflora*, there are only a few scattered reports with respect to *P. mundula* metabolism plasticity. Kraybill and Martin [133] studied the occurrence of a CAM-type metabolism under drought stress conditions.

Net CO_2 uptake in well-watered and drought stress plants occurred only during the day, although small amounts of night-time CO_2 uptake occurred in a few stressed individuals. Generally, CO_2 assimilation rates were lower in stressed than in control plants during the day. Stomatal conductances in well-watered and drought stress *P. mundula* were higher during the night than during the day, though these higher conductances were generally not accompanied by net CO_2 assimilation. In addition, no significant diurnal acid fluctuations occurred under well-watered conditions; however, under drought stress conditions significant diurnal malic acid fluctuations occurred in *P. mundula* [133]. In consequence, the primary source of carbon for nocturnal malic production acid was presumed to be respiratory CO_2, as occurs in C_3 plants that undergo CAM cycling.

D. EVOLUTIONARY MATTERS

It has been proposed that the species belonging to *Portulaca* constitute an advanced line in the Portulacaceace family, which have derived from CAM ancestors [152]. It is suggested that during *Portulaca* evolution C_4 metabolism could have probably been developed from a preexistent CAM tissue and that CAM has been weakened and released to a minor function, for example, under adverse environmental conditions [117]. The occurrence of a crassulacean acid-like metabolism, operating throughout mesophyll and bundle sheath cells in *P. oleracea* leaves, could be the consequence of the expression of the ancestral CAM with the more evolved Kranz anatomy. The existence of CAM and C_4 photosynthesis in the same cell within a leaf is likely incompatible due to futile cycles and loss of metabolic control [117]. Thus, this combination is not possible when considering the simultaneous functioning of both photosynthetic metabolisms. However, it can be accepted that a transition from one mode to another can occur under different environmental conditions, as in the case of *P. oleracea* [15], and in different cells within a tissue, as in *P. grandiflora*, where CAM cycling may take place in water storage cells [16].

E. CONCLUDING REMARKS

Different metabolic modifications can be induced in *Portulaca* due to water stress: transition to a crassulacean acid-like metabolism (as in *P. oleracea* [15]) or induction of a CAM-cycling metabolism compartmentalized in a different cell type while the C_4 pathway is also operating (*P. grandiflora* [16]). Inducible CAM may be a physiological feature in leaves of water stressed *Portulaca* spp. that could account for their relative long duration in habitats characterized by extended dry periods. In addition, comparative studies indicated that *P. mundula* exhibited a higher degree of acid fluctuations and a higher water use efficiency than *P. oleracea* and *P. grandiflora*, and it took more time to effect an appreciable decrease in the shoot water potential and net CO_2 exchange rates

were much less affected [133], showing, as this species tends to grow in the most arid environments, a correlation between photosynthetic mode and habitat in which the species is found and thus indicating that the transition from C_4 metabolism to a type of CAM is an adaptation to drought stress. Moreover, due to the biochemical similarities between CAM and C_4 plants, both exploiting a common set of enzymes and facilitating CO_2 capture and concentration around the active site of RuBisCO, the transition between both pathways should not be surprising.

Thus, *Portulaca* spp. constitute attractive biological systems to study the molecular and biochemical modifications underlying the shift from C_4 photosynthesis to CAM and offer an alternative field of investigation on the different CO_2 fixation mechanisms [15]. The capacity for acid metabolism may be dependent on endogenous species and developmental stage as well as on environmental conditions such as photoperiod, temperature, and light intensity [134].

VI. SUMMARY AND CONCLUDING REMARKS

Photosynthesis forms the basis on which all ecosystems on earth function. Most plants are C_3 plants, in which the first product of photosynthesis is a three-carbon compound. C_4 plants, in which the first product is a four-carbon compound, have evolved independently many times and are found in at least 18 families [113,159]. These distinctive properties enable the capture of CO_2 and its concentration in the vicinity of RuBisCO. In addition, photosynthetic organs of C_4 plants show alteration in their anatomy and ultrastructure, [3] and thus C_4 photosynthesis in terrestrial plants was thought to require Kranz anatomy because the cell wall between mesophyll and bundle sheath cells restricts leakage of CO_2.

Recent work with ancestral Asian chenopods shows that C_4 photosynthesis functions efficiently in individual cells containing both C_4 and the C_3 cycles [13]. In addition, in the aquatic plants *E. densa* and *H. verticillata*, a single-celled C_4-type CO_2-concentrating mechanism has been described [10,11]. Diatoms like *Thalassiosina weissflogii* are also reported to operate a complete C_4 photosynthesis cycle in individual photosynthetic cells [63]. Thus, even though the vast majority of C_4 plants exhibit Kranz anatomy, it would be not a sine qua non for the C_4 photosynthesis. These discoveries provide new inspiration for efforts to convert C_3 crops into C_4 plants because the anatomical changes required for C_4 photosynthesis might be less stringent than was previously thought

[126]. To engineer even the simplest theoretical single-celled system into a C_3 leaf requires a deep knowledge of the operation of the CO_2-concentrating mechanisms within a single cell.

C_4 photosynthesis is more plastic than was previously thought. Amphibious species like *E. vivipara* exhibit changes in their photosynthetic and anatomical traits depending on the environmental conditions, shifting between forms with C_4-like traits with Kranz anatomy and those with C_3-like characteristics without Kranz anatomy, in the terrestrial or submersed habitats. In contrast, C_4 plants with Kranz anatomy such as *P. oleracea* and *P. grandiflora* can shift to a kind of CAM when subjected to drought stress conditions.

In summary, since Kortschak and coworkers performed $^{14}CO_2$ pulse–chase experiments with sugarcane and found a different pattern from the one they were expecting, giving the basis for the discovery of the C_4 metabolism, much progress has been made in the study of CO_2-concentration mechanisms. Nevertheless, as science keeps in motion new variations in the photosynthetic modes are found and not everything is as was previously stated or thought. Fortunately for those who are studying the photosynthetic mechanisms, there is still a long way to the complete characterization of photosynthesis and a lot of work is waiting to be done.

REFERENCES

1. Edwards GE, Walker DA. C_3, C_4: Mechanisms, and Cellular and Environmental Regulation of Photosynthesis. Oxford: Blackwell Scientific, 1983.
2. Ehleringer JR, Monson RK. Evolutionary and ecological aspects of photosynthetic pathway variation. *Annu. Rev. Ecol. Syst.* 1993; 24: 411–439.
3. Hatch MD. C_4 photosynthesis: a unique blend of modified biochemistry, anatomy and ultrastructure. *Biochim. Biophys. Acta* 1987; 895: 81–106.
4. Holtum JAM, Winter K. Activity of enzymes of carbon metabolism in *Mesembryanthemun crystallinum* by salt stress. *Plant Cell Physiol.* 1982; 37: 257–262.
5. Kluge M, Ting IP. *Crassulacean Acid Metabolism. Analysis of an Ecological Adaptation.* New York: Springer, 1978.
6. Raven JA. CO_2-concentrating mechanisms: a direct role for thylakoid lumen acidification? *Plant Cell Environ.* 1997; 2: 147–154.
7. Moroney JV, Somanchi A. How do algae concentrate CO_2 to increase the efficiency of photosynthetic carbon fixation? *Plant Physiol.* 1999; 119: 9–16.
8. Badger MR, Spalding MH. CO_2 acquisition, concentration and fixation in cyanobacteria and algae. In: Leegood RC, Sharkey TD, von Caemmerer S, eds. *Photosynthesis: Physiology and Metabolism.* Dordrecht: Kluwer Academic Publishers, 2000: 369–397.
9. Badger MR, Price GD. CO_2 concentrating mechanisms in cyanobacteria: molecular components, their diversity and evolution. *J. Exp. Bot.* 2003; 383: 609–622.
10. Salvucci ME, Bowes G. Two photosynthetic mechanisms mediating the low photorespiratory state in submersed aquatic angiosperms. *Plant Physiol.* 1983; 73: 488–496.
11. Casati P, Lara MV, Andreo CS. Induction of a C_4-like mechanism of CO_2 fixation in *Egeria densa*, a submersed aquatic species. *Plant Physiol.* 2000; 123: 1611–1622.
12. Ueno O, Samejima M, Muto S, Miyachi S. Photosynthetic characteristics of an amphibious plant, *Eleocharis vivipara*: expression of C_4 and C_3 modes in contrasting environments. *Proc. Natl. Acad. Sci. USA* 1988; 85: 6733–6737.
13. Voznesenskaya EV, Franceschi VR, Kiirats O, Freitag H, Edwards GE. Kranz anatomy is not essential for terrestrial C_4 plant photosynthesis. *Nature* 2001; 414: 543–546.
14. Voznesenskaya EV, Franceschi VR, Kiirats O, Artyushevae G, Freitag H, Edwards GE. Proof of C_4 photosynthesis without Kranz anatomy in *Bienertia cycloptera* (Chenopodiaceae). *Plant J.* 2002; 31: 649–662.
15. Lara MV, Disante KB, Podestá FE, Andreo CS, Drincovich MF. Induction of a crassulacean acid like metabolism in the C_4 succulent plant, *Portulaca oleracea* L.: physiological and morphological changes are accompanied by specific modifications in phosphoenolpyruvate carboxylase. *Photosynth. Res.* 2003; 77: 241–254.
16. Guralnick LJ, Edwards GE, Ku MSB, Hockema B, Franceschi VR. Photosynthetic and anatomical characteristics in the C_4-crassulacean acid metabolism-cycling plant, *Portulaca grandiflora. Funct. Plant Biol.* 2002; 29: 763–773.
17. Madsen JS, Sand-Jensen K. Photosynthetic carbon assimilation in aquatic macrophytes. *Aquat. Bot.* 1991; 41: 5–40.
18. Bowes G, Salvucci ME. Plasticity in the photosynthetic carbon metabolism of submersed aquatic macrophytes. *Aquat. Bot.* 1989; 34: 233–266.
19. Raven JA. Exogenous inorganic carbon sources in plant photosynthesis. *Biol. Rev.* 1970; 45: 167–221.
20. Van TK, Haller WT, Bowes G. Comparison of the photosynthetic characteristics of three submersed aquatic plant. *Plant Physiol.* 1976; 58: 761–768.
21. Maberly SC. Photosynthesis by *Fontinalis antipyretica*: II. Assessment of environmental factors limiting photosynthesis and production. *New Phytol.* 1985; 100: 141–155.
22. Salvucci ME, Bowes G. Induction of reduced photorespiratory activity in submerse and amphibious aquatic macrophytes. *Plant Physiol.* 1981; 67: 335–340.
23. Salvucci ME, Bowes G. Ethoxyzolamide repression of the low photorespiration state in two submersed angiosperms. *Planta* 1983; 158: 27–34.
24. Holaday AS, Salvucci ME, Bowes G. Variable photosynthesis/photorespiration ratios in *Hydrilla* and other

submersed aquatic macrophyte species. *Can. J. Bot.* 1983; 61: 229–236.

25. Reiskind JB, Madsen TV, van Ginkel LC, Bowes G. Evidence that inducible C$_4$-type photosynthesis is a chloroplastic CO$_2$-concentrating mechanism in *Hydrilla*, a submersed monocot. *Plant Cell Environ.* 1997; 20: 211–220.

26. Elzenga JTM, Prins HBA. Light-induced polar pH changes in leaves of *Elodea canadensis. Plant Physiol.* 1989; 91: 62–67.

27. Prins HBA, Elzenga JTM. Bicarbonate utilization: function and mechanism. *Aquat. Bot.* 1989; 34: 59–83.

28. Rascio N, Mariani P, Tommasini E, Bodner M, Larcher W. Photosynthetic strategies in leaves and stems of *Egeria densa. Planta* 1991; 185: 297–303.

29. Buschmann P, Sack H, Köhler AE, Dahse I. Modeling plasmalemma ion transport of the aquatic plant *Egeria densa. J. Membr. Biol.* 1996; 154: 109–118.

30. Watson L, Dallwitz MJ. The Families of Flowering Plants: Descriptions, Illustrations, Identification, and Information Retrieval. Version: 19th. 1999. http.//biodiversity.uno.edu/delta/.

31. Casati P, Lara MV, Andreo CS. Regulation of enzymes involved in C$_4$ photosynthesis and the antioxidant metabolism by UV-B radiation in *Egeria densa*, a submersed aquatic species. *Photosynth. Res.* 2000; 71: 251–264.

32. Lara MV, Casati P, Andreo CS. CO$_2$ concentration mechanisms in *Egeria densa*, a submersed aquatic species. *Physiol. Plant.* 2002; 115:487–495.

33. Brown JMA, Dromgoole FI, Towsey MW, Browse J. Photosynthesis and photorespiration in aquatic macrophytes. In: Bieleski RL, Ferguson AR, Cresswell MM, eds. *Mechanism of Regulation of Plant Growth*. Wellington: The Royal Society of New Zealand, 1974: 243–249.

34. Browse JA, Dromgoole FI, Brown JMA. Photosynthesis in the aquatic macrophyte *Egeria densa*. I. ^{14}CO$_2$ fixation at natural CO$_2$ concentrations. *Aust. J. Plant Physiol.* 1977; 4: 169–176.

35. DeGroote D, Kennedy RA. Photosynthesis in *Elodea canadensis* Michx. Four carbon acid synthesis. *Plant Physiol.* 1977; 59: 1133–1135.

36. Browse JA, Dromgoole FI, Brown JMA. Photosynthesis in the aquatic macrophyte *Egeria densa*. II. The effects of inorganic carbon conditions on ^{14}C-fixation. *Aust. J. Plant Physiol.* 1979; 6: 1–9.

37. Browse JA, Dromgoole FI, Brown JMA. Photosynthesis in the aquatic macrophyte *Egeria densa*. III. Gas exchange studies. *Aust. J. Plant Physiol.* 1979; 6: 1133–1135.

38. Browse JA, Brown JMA, Dromgoole FI. Malate synthesis and metabolism during photosynthesis in *Egeria densa* Planch. *Aquat. Bot.* 1980; 8: 295–305.

39. Holaday AS, Bowes G. C$_4$ metabolism and dark CO$_2$ fixation in a submersed aquatic macrophyte (*Hydrilla verticillata*). *Plant Physiol.* 1980; 89: 1231–1237.

40. Bauwe H. An efficient method for the determination of K$_m$ values for HCO$_3^-$ of phosphoenolpyruvate carboxylase. *Planta* 1986; 169: 356–360.

41. Lara MV, Casati P, Andreo CS. *In vivo* phosphorylation of phosphoenolpyruvate carboxylase in *Egeria densa*, a submersed aquatic species. *Plant Cell Physiol.* 2001; 42: 141–145.

42. Miedema H, Prins HBA. pH-dependent proton permeability of the plasma membrane is a regulating mechanism of polar transport through the submerged leaves of *Potamogeton lucens. Can. J. Bot.* 1991; 69: 1116–1122.

43. Miedema H, Staal M, Prins HBA. pH-induced proton permeability changes of plasma vesicles. *J. Membr. Biol.* 1996; 152: 159–167.

44. van Ginkel LC, Prins HBA. Bicarbonate utilization and pH polarity. The response of photosynthetic electron transport to carbon limitation in *Potamogeton lucens* leaves. *Can. J. Bot.* 1998; 76: 1018–1024.

45. Staal M, Elzenga JTM, Prins HBA. ^{14}C fixation by leaves and leaf cell protoplasts of the submerged aquatic angiosperm *Potamogeton lucens* L.: carbon dioxide or bicarbonate? *Plant Physiol.* 1989; 90: 1035–1040.

46. Drincovich MF, Casati P, Andreo CS. NADP-malic enzyme from plants: a ubiquitous enzyme involved in different metabolic pathways. *FEBS Lett.* 2001; 460: 1–6.

47. Bowes G, Rao SK, Estavillo GM, Reiskind JB. C$_4$ mechanisms in aquatic angiosperms: comparisons with terrestrial C$_4$ systems. *Funct. Plant Biol.* 2002; 29: 379–392.

48. Bowes G, Holaday AS, Van TK, Haller WT. Photosynthetic and photorespiratory carbon metabolism in aquatic plants. In: Hall DO, Coombs J, Goodwin, eds. *Photosynthesis 77*. Proceedings of the Fourth International Congress on Photosynthesis. London: The Biochemical Society, 1978: 289–298.

49. Reiskind JB, Berg RH, Salvucci ME, Bowes G. Immunogold localization of primary carboxylases in leaves of aquatic and a C$_3$-C$_4$ intermediate species. *Plant Sci.* 1989; 61: 43–52.

50. Magnin NC, Cooley BA, Reiskind JB, Bowes G. Regulation and localization of key enzymes during the induction of Kranz-less, C$_4$ type photosynthesis in *Hydrilla verticillata. Plant Physiol.* 1997; 115: 1681–1689.

51. Spencer WE, Wetzel RG, Teeri J. Photosynthetic phenotype plasticity and the role of phosphoenolpyruvate carboxylase in *Hydrilla verticillata. Plant Sci.* 1996; 118: 1–9.

52. Ascencio J, Bowes G. Phosphoenolpyruvate carboxylase in *Hydrilla* plants with varying CO$_2$ compensation points. *Photsynth. Res.* 1983; 4: 151–170.

53. Estavillo GE, Rao SK, Reiskind JB, Bowes G. Molecular studies of an inducible C$_4$-type photosynthetic system: NADP-ME isoforms. *Plant Mol. Biol. Rep.* 2000; 18: S21–S23.

54. Rao SK, Magnin NC, Reiskind JB, Bowes J. Photosynthetic and other phosphoenolpyruvate carboxylase isoforms in the single-cell, facultative C$_4$ system of *Hydrilla verticillata. Plant Physiol.* 2002; 130: 876–886.

55. Blässing OE, Weshoff P, Svensson P. Evolution of C_4 phosphoenolpyruvate carboxylase in *Flaveria*, a conserved serine residue in the carboxyl-terminal part of the enzyme is a major determinant of C_4-specific characteristics. *J. Biol. Chem.* 2000; 275: 27917–27923.

56. Prins HBA, Snel JFH, Zanastra PE, Helder RJ. The mechanism of bicarbonate assimilation by the polar leaves of *Potamogeton* and *Elodea*. CO_2 concentrations at the leaf surface. *Plant Cell Environ.* 1982; 5: 207–214.

57. Spenser W, Teeri J, Wetzel RG. Acclimation of photosynthetic phenotype to environmental heterogeneity. *Ecology* 1994; 75: 301–314.

58. Krabel D, Eschrich W, Gamelei YV, Fromm, Ziegler H. Acquisition of carbon in *Elodea candensis* Michx. *J. Plant Physiol.* 1995; 145: 50–56.

59. van Ginkel LC, Schütz I, Prins HBA *Elodea canadensis* under N and CO_2 limitation: adaptive changes in RuBisCO and PEPCase activity in a bicarbonate user. *Phyton* 2000; 40: 133–143.

60. Sand-Jensen K, Gordon DM. Variable HCO_3^- affinity of *Elodea canadensis* Michaux in response to different HCO_3^- and CO_2 concentrations during growth. *Oecologia* 1986; 71: 426–432.

61. Raven JA. Carbon fixation and carbon availability in marine phytoplankton. *Photsynth. Res.* 1994; 39: 259–273.

62. Reiskind JB, Bowes G. The role of phosphoenolpyruvate carboxykinase in a marine macroalga with C_4-like photosynthetic characteristics. *Proc. Natl. Acad. Sci. USA* 1991; 88: 2882–2887.

63. Reinfelder JR, Kraepiel AM, Morel FMM. Unicellular C_4 photosynthesis in a marine diatom. *Nature* 2000; 407: 996–999.

64. Johnston AM, Raven JA, Beardall, Leegood RC. Photosynthesis in a marine diatom. *Nature* 2001; 412: 40–41.

65. Raven JA. Inorganic carbon acquisition by marine autotrophs. *Adv. Bot. Res.* 1997; 27: 85–209.

66. Sculthorpe CD. *The Biology of Aquatic Vascular Plants*. London: Arnold E, 1967.

67. Sage RF. Environmental and evolutionary preconditions for the origin and diversification of the C_4 photosynthetic syndrome. *Plant Biol.* 2001; 3: 202–213.

68. Bowes G. Growth in elevated CO_2: photosynthetic responses mediated through RuBisCO. *Plant Cell Environ.* 1991; 14: 795–806.

69. Ueno O, Samejima M, Koyama T. Distribution and evolution of C_4 syndrome in *Eleocharis*, a sedge group inhabiting wet and aquatic environments, based on culm anatomy and carbon isotope ratios. *Ann. Bot.* 1989; 64: 425–438.

70. Ueno O, Takeda T. Photosynthetic pathways, ecological characteristics and the geographical distribution of the Cyperaceae in Japan. *Oecologia* 1992; 89: 195–203.

71. Smith LG, Hake S. Initiation and determination of leaves. *Plant Cell* 1992; 4: 1017–1027.

72. Uchino A, Sentoku N, Nemoto K, Ishii R, Samejima M, Matsuoka M. C_4-type gene expression is not directly dependent on Kranz anatomy in an amphibious sedge *Eleocharis vivipara* Link. *Plant J.* 1998; 14: 565–572.

73. Ueno O. Structural characterization of photosynthetic cell in an amphibious sedge, *Eleocharis vivipara*, in relation to C_3 and C_4 metabolism. *Planta* 1996; 199: 382–393.

74. Ueno O. Induction of Kranz anatomy and C_4-like biochemical characteristics in a submersed amphibious plant by abscisic acid. *Plant Cell* 1998; 10: 571–583.

75. Ueno O. Environmental regulation of C_3 and C_4 differentiation in the amphibious sedge *Eleocharis vivipara*. *Plant Physiol.* 2001; 127: 1524–1532.

76. Bruhl JJ, Watson L, Dallawitz MJ. Genera of Cyperaceae: interactive identification and information retrieval. *Taxon* 1992; 41: 225–234.

77. Sharma OP, Mehra PN. Systematic anatomy of *Fimbristylis* Vahl (Cyperaceae). *Bot. Gaz.* 1972; 133: 87–95.

78. Takeda T, Ueno O, Samejima M, Ohtani T. An investigation for the occurrence of C_4 photosynthesis in the Cyperaceae from Australia. *Bot. Gaz. Tokyo* 1985; 98: 393–411.

79. Bruhl JJ, Stone NE, Hattersley PW. C_4 acid decarboxylation enzymes and anatomy in sedges (Cyperaceae): first record of NAD-malic enzyme species. *Aust. J. Plant Physiol.* 1978; 14: 719–728.

80. Ueno O, Takeda T, Murata T. C_4 acid decarboxylating enzyme activities of C_4 species possessing different Kranz anatomical types in the Cyperaceae. *Photosynthetica* 1986; 20: 111–116.

81. Bruhl JJ, Perry S. Photosynthetic pathway-related ultrastructures of C_3, C_4 and C_3-like C_3-C_4 intermediate sedges (Cyperaceae), with special reference to *Eleocharis*. *Aust. J. Plant Physiol.* 1995; 22: 521–530.

82. Carolin RC, Jacobs SWL, Vesk M. The ultrastructure of Kranz cells in the family Cyperaceae. *Bot. Gaz.* 1977; 138: 413–419.

83. Hattersley PW. Variations in photosynthetic pathway. In: Soderstrom TR, Hilu KW, Campbell SC, Barkworth ME, eds. *Grass Systematics and Evolution*. Washington, DC: Smithsonian Institution Press, 1987: 49–64.

84. Edwards GE, Franceschi VR, Ku MSB, Voznesenskaya EV, Pyankov VI, CS Andreo. Compartmentation of photosynthesis in cells and tissues of C_4 plants. *J. Exp. Bot.* 2001; 52: 577–590.

85. Ueno O. Immunocytochemical localization of enzymes involved in the C_3 and C_4 pathways in the photosynthetic cells of an amphibious sedge, *Eleocharis vivipara*. *Planta* 1996; 199: 394–403.

86. Ueno O. Immunogold localization of photosynthetic enzymes in leaves of various C_4 planta, with particular reference to pyruvate orthophosphate dikinase. *J. Exp. Bot.* 1998; 49: 1637–1646.

87. Hartung W, Davis WJ. Drought-induced changes in physiology and ABA. In: Davis WJ, Jones HG, eds. *Abscisic Acid: Physiology and Biochemistry*. Oxford: BIOS Scientific Publishers, 1991: 63–79.

88. Anderson LWJ. Abscisic acid induces formation of floating leaves in the heterophyllous aquatic angiosperm *Potamogeton nodosus. Science* 1978; 201: 1135–1138.

89. Goliber TE, Feldman LJ. Osmotic stress, endogenous abscisic acid and the control of leaf morphology in *Hippuris vulgaris* L. *Plant Cell Environ.* 1989; 12: 163–171.

90. Dai Z, Ku MSB, Zhang D, Edwards GE. Effects of growth regulators on the induction of Crassulacean acid metabolism in the facultative halophyte *Mesembryantemum crystallinum* L. *Planta* 1994; 192: 287–294.

91. Taybi T, Sotta B, Gehring H, Güclü S, Kluge M, Brulfert J. Differential effects of abscisic acid on phosphoenolpyruvate and CAM operation in *Kalanchoë blossfeldiana. Bot. Acta* 1995; 108: 240–246.

92. Edwards GE, Nakamoto H, Burnell JN, Hatch MD. Pyruvate orthophosphate dikinase and NADP-malate dehydrogenasse in C$_4$ photosynthesis: properties and mechanism of light/dark regulation. *Annu. Rev. Plant Physiol.* 1985: 36; 255–286.

93. Kondo A, Nose A, Ueno O. Leaf inner structure and immunolgold localization of some key enzymes involved in carbon metabolism in plants. *J. Exp. Bot.* 1998; 49: 1953–1961.

94. Aoyagi K, Chua N-H. Cell specific expression of pyruvate, Pi dikinase: *in situ* mRNA hybridization and immunolocalization labeling of protein in wheat seed. *Plant Physiol.* 1988; 86: 364–368.

95. Moons A, Valcke R, Van Montagu M. Low-oxygen stress and water deficit induce cytosolic pyruvate orthophosphate dikinase (PPDK) expression in roots of rice, a C$_3$ plant. *Plant J.* 1998; 15: 89–98.

96. Aoyagi K, Bassham JA. Pyruvate orthophosphate dikinase of C$_3$ seeds and leaves as compared to the enzyme form maize. *Plant Physiol.* 1984; 75: 387–392.

97. Aoyagi K, Bassham JA. Synthesis and uptake of cytoplasmically synthesized pyruvate orthophosphate dikinase polypeptide by chloroplasts. *Plant Physiol.* 1985; 78: 807–811.

98. Agarie S, Kai M, Takatsuji H, Ueno O. Expression of C$_3$ and C$_4$ photosynthetic characteristics in the amphibious plant *Eleocharis vivipara*: structure and analysis of the expression of isogenes for pyruvate, orthophosphate dikinase. *Plant Mol. Biol.* 1997; 34: 363–369.

99. Glackin C, Grula JW. Organ-specific transcripts of different size and abundance derive from same pyruvate, orthophsophate dikinase gene in maize. *Proc. Natl. Acad. Sci. USA* 1990; 87: 30004–30008.

100. Hudspeth RL, Glackin CA, Bonner J, Grula JW. Genomic and cDNA clones for maize phosphoenolpyruvate carboxylase and pyruvate, orthophosphate dikinase: expression of different gene-family members in leaves and roots. *Proc. Natl. Acad. Sci. USA* 1986; 83: 2884–2888.

101. Sheen J. Molecular mechanisms underlying the differential expression of maize pyruvate, orthophosphate dikinase genes. *Plant Cell* 1991; 3: 225–245.

102. Agarie S, Kai M, Takatsuji H, Ueno O. Environmental and hormonal regulation of gene expression of C$_4$ photosynthetic enzymes in the amphibious sedge *Eleocharis vivipara. Plant Sci.* 2002; 163: 571–580.

103. Jiao JA, Chollet R. Regulatory phosphorylation of serine-15 in maize phosphoenolpyruvate carboxylase by a C$_4$-leaf protein serine kinase. *Arch. Biochem. Biophys.* 1990; 283: 300–305.

104. Baba A, Agarie S, Takatsuji H, Ueno O. Expression of C$_3$ and C$_4$ characteristics in the amphibious sedge *Eleocharis vivipara*: cellular pattern of expression of photosynthetic genes. *Plant Physiol.* 1997; 114: S-212 (abstr. No. 1065-9).

105. Brown RH, Bouton JH. Physiology and genetics of interspecific hybrids between photosynthetic types. *Annu. Rev. Plant Physiol. Plant Mol. Biol.* 1993; 44: 435–456.

106. Uchino A, Samejima M, Ishii R, Ueno O. Photosynthetic carbon metabolism in an amphibious sedge, *Eleocharis baldwinii* (Torr.) Chapman: modified expression of C$_4$ characteristics under submerged aquatic conditions. *Plant Cell Physiol.* 1995; 36: 229–238.

107. Ueno O. Cellular expression patterns of C$_3$ and C$_4$ photosynthesis enzymes in the amphibious sedge *Eleocharis baldwinii. Plant Cell Physiol.* 2000; supplement 41: s113.

108. Akhani H, Trimborn P, Zeigler H. Photosynthetic pathways in Chenopodeaceae from Africa, Asia and Europe with their ecological, phytogeographical and taxonomical importance. *Plant Syst. Evol.* 1997; 206: 187–221.

109. Carolin RC, Jacobs SWL, Vesk M. Leaf structure in Chenopodiaceae. *Bot. Jahrb. Syst. Pflanzegesh. Planzengeogr.* 1975; 95: 226–255.

110. Glagoleva TA, Chulanovskaya MV, Pakhomova MV, Voznesenskaya EV, Gamaley YV. Effect of salinity on the structure of the assimilatory organs and C labelling patterns in C$_3$ and C$_4$ plants of Ararat plain. *Photosynthetica* 1992; 26: 363–369.

111. Pyankov VI, Voznesenskaya EV, Kuz'min AN, Ku MSB, Ganko E, Franceschi VR, Black CC Jr, Edwards GE. Occurrence of C$_3$ and C$_4$ photosynthesis in cotyledons and leaves of *Salsola* especies (Chenopodiaceae). *Photosynth. Res.* 2000; 63: 69–84.

112. Winter K. C$_4$ plants of high biomass in arid regions of Asia: occurrence of C$_4$ photosynthesis in Chenopodiaceae and Polygonaceae from the Middle East and USSR. *Oecologia* 1981; 48: 100–106.

113. Sage RF, Li MR, Monson RK. The taxonomic distribution of C$_4$ photosynthesis. In: Sage RF, Monson RK, eds. *C$_4$ Plant Biology*. New York: Academic Press, 1999: 551–584.

114. Voznesenskaya EV, Franceschi VR, Pyankov VI, Edwards GE. Anatomy, chloroplast structure and compartmentation of enzymes relative to photosynthetic mechanisms in leaves and cotyledons of species in the tribe Salsolaeae (Chenopodiaceae). *J. Exp. Bot.* 1999; 50: 1779–1795.

115. Kellogg EA. Phylogenetic aspects of the evolution of C$_4$ photosynthesis. In: Sage RF, Monson RK, eds. *C$_4$*

Plant Biology. San Diego: Academic Press. 1999: 411–444.

116. Sage RF. Environmental and evolutionary preconditions for the origin and diversification of the C$_4$ photosynthesis syndrome. *Plant Biol*. 2001; 3: 202–213.

117. Sage RF. C$_4$ photosynthesis in terrestrial plants does not require Kranz anatomy. *Trends Plant Sci*. 2002; 7: 283–285.

118. Akhani H, Ghobadnejhad M, Hashemi SMH. Ecology, biogeography and pollen morphology of *Bienertia cycloptera* Bung ex Boiss. (Chenopodiaceae), an enigmatic C$_4$ plant without Kranz anatomy. *Plant Biol*. 2003; 5: 167–178.

119. Freitag H, Stichler W. *Bienertia cycloptera* Bunge ex Boiss., Chenopodiaceae, another C$_4$ plant without Kranz tissues. *Plant Biol*. 2002; 4: 121–132.

120. Freitag H, Stichler W. A remarkable new leaf type with unusual photosynthetic tissue in a central Asiatic genus of Chenopodiaceae. *Plant Biol*. 2000; 2: 154–160.

121. Dengler NC, Nelson T. Leaf structure and development in C$_4$ plants. In: Sage RF, Monson RK, eds. *C$_4$ Plant Biology*. San Diego: Academic Press, 1999: 133–172.

122. Kanai R, Edwards GE. The biochemistry of C$_4$ photosynthesis. In: Sage RF, Monson RK, eds. *C$_4$ Plant Biology*. San Diego: Academic Press, 1999: 449–487.

123. Leegood RC, Walker RP. Regulation of the C$_4$ pathway. In: Sage RF, Monson RK, eds. *C$_4$ Plant Biology*. San Diego: Academic Press, 1999: 89–131.

124. Cerling ThE. Peleorecords of C$_4$ plants and ecosystems. In: Sage RF, Monson RK, eds. *C$_4$ Plant Biology*. San Diego: Academic Press, 1999: 448–469.

125. Kühn U. Chenopodiaceae. In: Kubitzki K, ed. *The Families and Genera of Glowering Plants*. New York: Springer-Verlag, 1993: 253–281.

126. Häusler RE, Hirsch H-J, Kreuzaler F, Peterhánsenl C. Overexpression of C$_4$-cycle enzymes in transgenic C$_3$ plants: a biotechnological approach to improve C$_3$-photosynthesis. *J. Exp. Bot*. 2002; 53: 591–607.

127. Surridge C. The rice squad. *Nature* 2002; 416: 576–578.

128. Leegood RC. C$_4$ photosynthesis: principles of CO$_2$ concentration and prospects for its introduction into C$_3$ plants. *J. Exp. Bot*. 2002; 53: 581–590.

129. Osmond CB. Crassulacean acid metabolism. A curiosity in context. *Annu. Rev. Plant Physiol*. 1978; 29: 379–414.

130. Winter K, Smith JAC. An introduction to crassulacean acid metabolism. Biochemical principles and ecological diversity. In: Winter K, Smith JAC, eds. *Crassulacean Acid Metabolism: Biochemistry, Ecology and Evolution. Ecological Studies*. Vol. 114. New York: Springer-Verlag,1996: 1–10.

131. Koch K, Kennedy RA. Characteristics of crassulacean acid metabolism in the succulent C$_4$ dicot, *Portulaca oleracea* L. *Plant Physiol*. 1980; 65: 193–197.

132. Koch KE, Kennedy RA. Crassulacean acid metabolism in the succulent C$_4$ dicot, *Portulaca oleracea* L under natural environmental conditions. *Plant Physiol*. 1982; 69: 757–761.

133. Kraybill AA, Martin CE. Crassulacean acid metabolism in three species of the C$_4$ genus *Portulaca*. *Int. J. Plant Sci*. 1996; 157: 103–109.

134. Ku S, Shie Y, Reger B, Black CC. Photosynthetic characteristics of *Portulaca grandiflora*, a succulent C$_4$ dicot. *Plant Physiol*. 1981; 68: 1073–1080.

135. Mazen AMA. Changes in levels of phosphoenolpyruvate carboxylase with induction of crassulacean acid metabolism (CAM)-like behavior in the C$_4$ plant *Portulaca oleracea*. *Physiol. Plant*. 1996; 98: 111–116.

136. Gurlanick LJ, Jackson MD. Crassulacean acid metabolism activity in the family Portulacaceae. *Plant Physiol*. 1993; 102 (suppl): 139.

137. Gutierrez M, Gracen VE, Edwards GE. Biochemical and cytological relationships in C$_4$ plants. *Planta* 1974; 119: 279–300.

138. Guralnick LJ, Ting IP. Seasonal patterns of water relations and enzyme activity in the facultative CAM plant *Portulacalaria afra* (L.) Jacq. *Plant Cell Environ*. 1988; 811–818.

139. Cushman J, Bohnert H. Molecular genetics of crassulacean acid metabolism. *Plant Physiol*. 1997; 113: 667–676.

140. Chollet R, Vidal J, O'Leary MH. Phosphoenolpyruvate carboxylase: a ubiquitous, highly regulated enzyme in plants. *Annu. Rev. Plant Physiol. Plant Mol. Biol*. 1996; 47: 273–298.

141. Nimmo HG. The regulation of phosphoenolpyruvate carboxylase in CAM plants. *Trends Plant Sci*. 2000; 2: 75–80.

142. Zimmerman CA. Growth characteristics of weediness in *Portulaca oleracea* L. *Ecology* 1976; 57: 964–974.

143. Kennedy RA, Laetsch WM. Relationship between leaf development and primary photosynthetic products in the C$_4$ plant *Portulaca oleracea*. *Planta* 1973; 115: 113–124.

144. Kennedy RA. Relationship between leaf development, carboxylase enzyme activities and photorespiration in the C$_4$ plant *Portulaca oleracea* L. *Planta* 1976; 128: 149–154.

145. Vidal J, Chollet R. Regulatory phosphorylation of phosphoenolpyruvate carboxylase. *Trends Plant Sci*. 1997; 2: 230–237.

146. Andreo CS, González DH, Iglesias A. Higher plant phosphoenolpyruvate carboxylase: structure and regulation. *FEBS Lett*. 1987; 213: 1–8.

147. Duff SMG, Andreo CS, Pacquit V, Lepiniec L, Sarath G, Codon SA, Vidal J, Gadal P, Chollet R. Kinetic analysis of the non-phosphorylated, *in vitro* phosphorylated, and phosphorylation-site-mutant (Asp8) forms of the intact recombinant C$_4$ phosphoenolpyruvate carboxylase from sorghum. *Eur. J. Biochem*. 1995; 228: 92–95.

148. Walker RP, Acheson RM, Técsi LI, Leegood RC. Phosphoenolpyruvate carboxykinase in C$_4$ plants: its

role and regulation. *Aust. J. Plant Physiol.* 1997; 24: 459–468.

149. Lara MV, Drincovich MF, Andreo CS. Induction of a crassulacean acid like metabolism in the C$_4$ succulent plant, *Portulaca oleracea* L.: study of enzymes involved in carbon fixation and carbohydrate metabolism. *Plant Cell Physiol.* 2004; 45: 618–626.

150. Bailey LH, Bailey EZ. Hortus third: a concise dictionary of plants cultivated in the United States and Canada. New York: MacMillan, 1976.

151. Troughthon JH, Card KA, Hendy CH. Photosynthetic pathways and carbon isotope discrimination by plants. Carnegie Inst Wash Yearbook 1974; 73: 768–780.

152. Guralnick LJ, Jackson MD. The occurrence and phylogenetics of crassulacean acid metabolism in the Potulacaceae. *Int. J. Plant Sci.* 2001; 162: 257–262.

153. Nishioka D, Brisibe EA, Miyake H, Taniguchi T. Ultrastructural observations in the suppression of granal development in bundle sheath chloroplasts of NAD-ME type C$_4$ monocot and dicot species. *Jpn. J. Crop Sci.* 1993; 62: 621–627.

154. Nishioka D, Miyake H, Taniguchi T. Suppression of granal development and accumulation of Rubisco in different bundle sheath chloroplasts of the C$_4$ succulent plant *Portulaca grandiflora. Ann. Bot.* 1996; 77: 629–637.

155. Fathi M, Scharrenberger. Purification by immunoadsorption and immunochemical properties of NAD-dependent malic enzymes from leaves of C$_3$, C$_4$ and crassulacean acid metabolism plants. *Plant Physiol.* 1990; 92: 710–717.

156. Matthews JF, Levins PA. The genus *Portulaca* in the south-eastern United States. *Costanea* 1985; 50: 96–104.

157. Welkie GM, Caldwell M. Leaf anatomy of species in some dicotyledon families as related to the C$_3$ and C$_4$ pathways of carbon fixation. *Can. J. Bot.* 1970; 48: 2135–2146.

158. Tregunna EB, Downton J. Carbon dioxide compensation in members of the Amaranthaceae and some related families. *Can. J. Bot.* 1967; 45: 2385–2387.

159. Moore PD. Evolution of photosynthetic pathways in flowering plants. *Nature* 1982; 310: 696.

Section VI

Photosynthesis in Lower and
Monocellular Plants

23 Regulation of Phycobilisome Biosynthesis and Degradation in Cyanobacteria

Johannes Geiselmann
Adaptation et Pathogénie des Microorganismes, Université Joseph Fourier

Jean Houmard
Organismes Photosynthétiques et Environnement, Ecole Normale Supérieure

Benoît Schoefs
Dynamique Vacuolaire et Réponses aux Stress de l'Environnement,
UMR CNRS 5184/INRA 1088/Université de Bourgogne Plante-Microbe-Environnement,
Université de Bourgogne à Dijon

CONTENTS

I. INTRODUCTION

Extant cyanobacteria derive from the oldest oxygen-evolving photosynthetic organisms that appeared on Earth some billion years ago. Their photosynthetic apparatus is very similar to the one of green plants and was at its origin. Indeed the chloroplasts of green plants and algae all derive from a unique endosymbiotic event that occured between a cyanobacterial ancestor and a primitive eukaryotic cell (Ref. [1] and references therein). Although oxygen-evolving photosynthesis is achieved by the same mechanism, differences exist in the light-harvesting apparatus between phylogenetic groups. Instead of the integral membrane light-harvesting complexes that surround photosystems I and II (PS

I and PS II), most cyanobacteria harvest light through an extrinsic membrane-anchored, water-soluble multiprotein assembly, the phycobilisome (PBS). This structure transfers the light energy to the underlying PSs [2]. PS II uses the harvested energy to create a charge separation at the level of special chlorophyll (Chl) molecules. The electron is then transferred to PS I through the chain of electron transporters while the electrical neutrality of the special Chl molecules is restored by the transfer of one electron coming from the oxidation of a water molecule.

Survival of free-living organisms in a changing environment relies upon their capacity to modulate their cellular metabolism according to the external conditions. As energy production relies on photosynthesis, it clearly appears that the effective absorption of light by PBSs is a critical step in the photosynthetic process and thus for the physiology of the photosynthetic organisms. Light flux to the PSs must therefore be strongly regulated. In cyanobacteria, most of this regulation occurs by controlling the transcription of the genes coding for the PBS components. The purpose of this contribution is to: (i) review the literature dealing with the regulation of PBS biosynthesis and degradation; and (ii) present models for the biochemical and regulatory networks involved in this process.

The organization of the cyanobacterial cells, their photosynthetic apparatus, as well as their physiology may differ between taxons (even at the species level) [3–5]. Responses specific and adapted to changes in the environmental parameters have been described for strains able to differentiate, fix atmospheric dinitrogen and perform complementary chromatic adaptation (e.g.,[6,7]). Extensive structural and functional analyses of PBS have been performed, and different shapes have been observed. We will focus this chapter on the most common hemidiscoidal PBS that consists of a central tricylindrical core from which six rods radiate [2].

II. PIGMENT, POLYPEPTIDE ORGANIZATION, AND FUNCTIONING OF THE PHYCOBILISOMES

PBSs form an ordered spatial arrangement, which covers the outer surface of the photosynthetic membranes. In electron micrographs, PBSs generally appear as being made up of two discrete subdomains: the "core" and the "peripheral rods". In a front view, the core presents itself as three stacked cylinders that form a triangle (or, for a few strains, only as two stacked cylinders arranged side-by-side; reviewed in Ref. [4]). The composition and length of the rods depend on cell growth conditions. For the hemidiscoidal PBSs, both the core cylinders and the rods are composed of stacked disks, connected by the so-called linker polypeptides (reviewed in Ref. [4]). The polypeptides required for the building up of functional PBSs or associated with them may be grouped into three classes (Table 23.1):

Phycobiliproteins (PB): These proteins carry open tetrapyrrolic pigments and represent about 85% of the total PBS proteins. Typically, cyanobacteria contain two main PBs, namely phycocyanin (PC), which forms the rods and allophycocyanin (AP), which forms the core [2]. Some strains may additionally possess, as part of the rods, either a second type of PC, phycoerythrin (PE) or phycoerythrocyanin

(PEC). Most PBSs are heteromonomers composed of equimolar amounts of α and β subunits. The α and β subunits of AP and the α subunit of PC and PEC bind one chromophore molecule whereas the β subunit of PC and PEC, as well as the α subunit of PE bind two chromophore molecules, three chromophores being bound to the β subunit of PE (Table 23.1) [2,8]. Some marine strains contain yet another type of PE that carries PUB chromophores [2]. The α and β subunits assemble to form either trimers in the core or hexamers $(\alpha\beta)_6$ in the rods. The most recent refined structure reported for a PB is the one of c-PC from *Thermosynechococcus vulcanus*, solved to 1.6 Å resolution [9].

Linker polypeptides: They account for approximately 10 to 20% of the total PBS proteins. The generally accepted nomenclature uses the abbreviation L_X^Y, with X referring to the location of the linker within the structure (C for core, R for rod, and M for membrane), and Y to its apparent molecular mass (if it is a number) or the PB with which it is associated (if letters, PC, PE, or PEC). Most of the linker polypeptides are nonpigmented, exceptions being: (i) the largest one (L_{CM}), which always carries a phycocyanobilin and serves as terminal energy acceptor for the PBS; and (ii) the recently discovered PE-associated rod linkers of the marine *Synechococcus* sp. WH8102 (F. Partensky, personal communication). The linker polypeptides have different functions: they are involved in the association of the PB trimers and they modify the absorption properties of the PBSs and thus have to perform a unidirectional transfer of the excitation energy within the PBS structure [2]. The above mentioned L_{CM}, besides its role as terminal energy acceptor, plays a key role in the assembly of the PBS core. It has been shown that it is the scaffold onto which AP subunits assemble, its size determining the shape of the core [10,11]. Most of the linkers induce a face-to-face aggregation of trimers and tail-to-tail joining of hexamers in the peripheral rods. It has been suggested that the linker proteins occupy positions running through the internal cavities of the disks. The small L_C has been crystallized with AP trimers and the structure showed it lying within the cavity, in contact with two AP αβ monomers [12]. On the basis of crystallographic data and calculation, it has been proposed that a special PC isoform, together with specific linkers (L_{RC}), make the contact between the PC rods and the AP core in *T. vulcanus* [13]. Models have been

TABLE 23.1
Phycobilisome Components and Associated Proteins

Protein	Pigment	Function	Gene Names
Phycobiliproteins			
α^{AP-B}	1 PCB	Core terminal energy acceptor	*apcD*
α^{AP}	1 PCB	Allophycocyanin α subunit	*apcA*
β^{AP}	1 PCB	Allophycocyanin β subunit	*apcB*
$\beta^{18.3}$	1 PCB	Allophycocyanin β-type subunit	*apcF*
α^{PC}	1 PCB	Phycocyanin α subunit	*cpcA*
β^{PC}	2 PCB	Phycocyanin β subunit	*cpcB*
α^{PE}	2 PEB	Phycoerythrin α subunit	*cpeA*
β^{PE}	3 PEB	Phycoerythrin β subunit	*cpeB*
α^{PEC}	1 PCB	Phycoerythrocyanin α subunit	*pecA*
β^{PEC}	1 PCB + 1 PXB	Phycoerythrocyanin β subunit	*pecB*
Linker polypeptides			
L_{CM}	1 PCB	Large core linker with a PB domain acting as terminal energy acceptor	*apcE*
L_C	None	Small core linker	*apcC*
L_{RC}	None	Core–rod linker	*cpcG*
L_R^{PC}	None	Rod linker for PC	*cpcC* or *cpcH* or *cpcI*
L_R^{10}	None	Small rod linker	*cpcD*
L_R^{PE}	None	Rod linker for PE	*cpeC* or *cpeD* or *cpeE*
L_R^{PEC}	None	Rod linker for PEC	*pecC*
Associated proteins			
CpcE	None	α Subunit of the phycocyanobilin lyase	*cpcE*
CpcF	None	β Subunit of the phycocyanobilin lyase	*cpcF*
CpeY	None	Putative α subunit of the phycoerythrobilin lyase	*cpeY*
CpeZ	None	Putative β subunit of the phycocyanobilin lyase	*cpcZ*
PecE	None	α Subunit of the phycoerythrocyanobilin lyase	*pecE*
PecE	None	β Subunit of the phycoerythrocyanobilin lyase	*pecF*
PcyA	None	Phycocyanobilin:ferredoxin oxidoreductase	*pcyA*
NblA	None	Polypeptide involved in PBS degradation	*nblA*
FNR or PetH	None	Ferredoxin-NADP oxidoreductase	*petH*

proposed to explain the close to 100% transfer of excitation energy all along the supramolecular structure: first from the distal to the proximal hexamer within the rods [2,14], from the rods to the core, and then to the reaction centers that are embedded into the thylakoid membranes (for a model of the interaction between AP and PS II [15].

PBS-associated proteins: Although tetrapyrroles can spontaneously form adducts with PB apoproteins, it has been shown that specific lyases are necessary for the attachment of phycocyanobilin to the α subunits of PC [16]. Similarly, another lyase is required to produce holo-PEC from the apo-PEC α subunit [17]. Genes sharing similarities with the ones that code for these lyases have been reported, and their products would catalyze the covalent attachment of phycoerythrobilin to the apo-PE α subunit [18]. All

of these lyases are heterodimeric enzymes. On the other hand, the copurification of the ferredoxin:$NADP^+$ oxidoreductase with the PBSs strongly suggests that a certain amount of this enzyme is bound to the PBSs [19]. Another protein, NblA, first identified as required for PBS degradation [20], also copurifies with PBSs [21].

III. PIGMENT BIOSYNTHESIS

The pigments attached to the PBs are linear tetrapyrroles, called phycobilins. They are bound to the protein moiety at conserved positions by cysteinyl thioether linkages through the vinyl substituent of the pyrrole ring A. Occasionally, a second linkage is established through the vinyl substituent of the pyrrole ring D [2]. Phycobilin synthesis follows the same pathway as Chl until the metal chelation step. As these steps are described in detail in another chapter (see Chapter 3), they will be only summarized here. Like for the production of Chl molecules, the phyco-bilin biosynthetic pathway starts with the synthesis of δ-aminolevulinic acid molecules, which are formed along the Beale pathway [22]. This pathway requires the activity of three enzymes, namely tRNA$^{Glu(UUC)}$ ligase, also termed glutamyl-tRNA synthetase [23], a NADPH:glutamate tRNA dehydrogenase [24], and a glutamate 1-semialdehyde aminotransferase (reviewed in Chapter 3). The dehydrogenase catalyzes the conversion of Glu-tRNA^{Glu} to glutamate 1-semi-aldehyde and the glutamate 1-semialdehyde amino-transferase converts glutamate 1-semialdehyde into δ-aminolevulinic acid through transaminations. The aminotransferase from *Synechococcus* sp. PCC 6301 has been purified, and its N-terminal amino acid sequence show significant similarities with that from barley [24]. Subsequent intermediates are uropor-phyrinogen III, coproporphyrinogen III, and proto-porphyrin (Proto) IX (*Cyanidium caldarium*, [25]). The route by which Proto-IX is converted into linear bilins was resolved in 1981 after administration of [^{14}C] heme to *Cyanidium caldarium*. The resultant phycocyanobilin was radiolabeled [26]. The direct proof that heme is a biosynthetic intermediate in phycobilin synthesis implied the existence of the en-zymes ferrochelatase, catalyzing iron insertion into Proto-IX, and the NADPH:heme oxygenase, per-forming heme oxidative degradation to biliverdin IXα. Such enzymes have indeed been found in the genomes of completely sequenced cyanobacteria (http://www.kazusa.or.jp/cyano/). Using $^{18,18}O_2$, it was shown that a heme oxygenase opens the heme molecules by incorporating two different oxygen mol-ecules to yield biliverdin IXα [27,28]. The enzyme is soluble [27]. Full activity requires two reductants: ferredoxin and soluble vitamin E or ascorbate, with the vitamin as the most efficient cofactor [29]. Heme oxygenase is encoded by the gene *ho1* [30]. Light regulates the production and activity of ferrochela-tase [31]. Biliverdin IXα is then transformed to 3Z-isomers of phycocyanobilin [29], which, in turn, bind to the apoprotein [27] via a thioether linkage to cystei-nyl residues [32–34]. Phycoerythrobilin was shown to be an intermediate in this pathway [35], but an alter-native route has recently been described [36]. *Syne-chocystis* PCC 6803 cells can accumulate only pigmented PBS proteins [29]. Chromophore attach-ment seems to stabilize the structure and enhance subunit binding of the αβ monomers [37]. How the synthesis of pigments and apoproteins are coupled is still under debate. Whether the pigmented proteins and the nonpigmented ones assemble spontaneously or require chaperones remains to be determined. One

model for the assembly of PBSs, based on numerous experimental observations obtained with mutants, proposed that the main control would operate at the level of the formation of the αβ monomers [38]. In agreement with this, it was found that for the synthe-sis of PEC the release of the holo-α subunit from its complex with the chromophore lyase would be medi-ated by the formation of the holo-α–holo-β hetero-dimer [17].

IV. GENOME ORGANIZATION

The number of components, as well as the number of transcriptional units encoding the PBS components varies between species. Table 23.1 lists the compon-ents that can be found in PBSs together with the corresponding gene names. Gene clustering also widely differs: the 22 PBS-related genes of *Anabaena* PCC 7120, as well as the 15 of *Themosynechococcus elongatus* BP-1, are grouped into five clusters, whereas nine clusters exist for the 15 genes of *Syne-chocystis* PCC 6803 (Table 23.2). The genes that code for the α and β subunits of a given PB are usually adjacent and cotranscribed. For the core AP, *apcA* (α) is located upstream of *apcB* (β), while for the rod PBs (PC, PE, and PEC) the gene coding for β pre-cedes the one coding for α. The genes for the linker polypeptides are often adjacent and cotranscribed with those encoding the PB with which they are spe-cifically associated [4,6]. The genes coding for the two subunits of the PB α-subunit phycobilin lyases, which attach the chromophore to the α apoproteins, are often adjacent to and part of the corresponding phy-cobiliprotein operon.

Although the clustering and location of the genes on the chromosome is highly variable, all of the genes required for the building up of functional PBSs are regulated tightly in a coordinated manner. No free PBs or phycobilins are found in cyanobacter-ial cells under standard conditions. Differences have been reported in the level of stable transcripts corre-sponding to the genes that constitute an operon. As an example, *apcEABC*, *apcABC*, *apcAB*, and *apcC* mRNAs have been found in *Tolypothrix* PCC 7601, with the same 5′-end for the *apcABC* and *apcAB* transcripts [39]. Whether they correspond to start sites originating from internal promoters or to stable processed products remains to be established. It is worth noting that the relative ratio between the dif-ferent mRNAs reflects the relative abundance of the gene products within the PBS, the *apcAB* transcripts being by far the most abundant. Gene families have been found for a few genes but the specific roles of each of the gene products is largely unknown:

TABLE 23.2
Physical Organization of the Genes Related to Phycobilisome Biosynthesis in Three Fully Sequenced Cyanobacterial Geneomes. Gene Assignments were taken from Cyanobase (http://www.kazusa.or.jp/cyano/)

Cluster	Gene	Gene Product	Gene Assignment
Anabaena/Nostoc sp. PCC 7120			
7120 cluster 1	*apcF*	$\beta^{18.3}$ Allophycocyanin β-type subunit	*all2327*
7120 cluster 2	*apcE*	L_{CM} terminal energy acceptor	*alr0020*
	apcA	α^{AP} Allophycocyanin α subunit	*alr0021*
	apcB	β^{AP} Allophycocyanin β subunit	*alr0022*
	apcC	L_C small core linker	*asr0023*
7120 cluster 3	*apcD*	α^{APB} Allophycocyanin B	*all3653*
7120 cluster 4	*apcA2*	Allophycocyanin α-type subunit	*all0450*
7120 cluster 5	*pecB*	β^{PEC} Phycoerythrocyanin beta chain	*alr0523*
	pecA	α^{PEC} Phycoerythrocyanin alpha chain	*alr0524*
	pecC	L_R^{PEC} PEC-associated rod linker protein	*alr0525*
	pecE	Phycobiliviolin lyase α subunit	*alr0526*
	pecF	Phycobiliviolin lyase β subunit	*alr0527*
	cpcB	β^{PC} Phycocyanin beta chain	*alr0528*
	cpcA	α^{PC} Phycocyanin alpha chain	*alr0529*
	cpcC	L_R^{PC} PC-associated rod linker protein	*alr0530*
	cpcD	L_R^{PC} small rod linker polypeptide	*asr0531*
	cpcE	Phycocyanobilin lyase α subunit	*alr0532*
	cpcF	Phycocyanobilin lyase β subunit	*alr0533*
	cpcG1	L_{RC} rod–core linker polypeptide	*alr0534*
	cpcG2	L_{RC} rod–core linker polypeptide	*alr0535*
	cpcG3	L_{RC} rod–core linker polypeptide	*alr0536*
	cpcG4	L_{RC} rod–core linker polypeptide	*alr0537*
Thermosynechococcus elongatus BP-1			
BP-1 cluster 1	*apcA*	α^{AP} Allophycocyanin α subunit	tll0957
	apcB	β^{AP} Allophycocyanin β subunit	tll0956
	apcC	L_C small core linker	tsl0955
BP-1 cluster 2	*apcD*	α^{APB} Allophycocyanin B	tll1551
BP-1 cluster 3	*apcE*	L_{CM} terminal energy acceptor	tll2365
BP-1 cluster 4	*apcF*	$\beta^{18.3}$ Allophycocyanin β-type subunit	tlr2034
BP-1 cluster 5	*cpcB*	β^{PC} phycocyanin beta chain	tlr1957
	cpcA	α^{PC} Phycocyanin alpha chain	tlr1958
	cpcC	L_R^{PC} PC-associated rod linker protein	tlr1959
	cpcD	L_R^{PC} small rod linker polypeptide	tlr1960
	cpcE	Phycocyanobilin lyase α subunit	tlr1961
	cpcF	Phycocyanobilin lyase β subunit	tlr1962
	cpcG1	L_{RC} rod–core linker polypeptide	tlr1963
	cpcG2	L_{RC} rod–core linker polypeptide	tlr1964
	cpcG4	L_{RC} rod–core linker polypeptide	tlr1965
Synechocystis sp. PCC 6803			
6803 cluster 1	*cpcB*	β^{PC} Phycocyanin beta chain	sll1577
	cpcA	α^{PC} Phycocyanin alpha chain	sll1578
	cpcC2	L_R^{PC} PC-associated rod linker protein	sll1579
	cpcC1	L_R^{PC} PC-associated rod linker protein	sll1580
	cpcD	L_R^{PC} small rod linker polypeptide	ssl3093
6803 cluster 2	*cpcG2*	L_{RC} rod–core linker polypeptide	sll1471
6803 cluster 3	*cpcG1*	L_{RC} rod–core linker polypeptide	slr2051
6803 cluster 4	*cpcE*	Phycocyanobilin lyase α subunit	slr1878
6803 cluster 5	*cpcF*	Phycocyanobilin lyase β subunit	sll1051
6803 cluster 6	*apcD*	α^{APB} Allophycocyanin B	sll0928
6803 cluster 7	*apcE*	L_{CM} terminal energy acceptor	slr0335
6803 cluster 8	*apcF*	$\beta^{18.3}$ Allophycocyanin β-type subunit	slr1459
6803 cluster 9	*apcA*	α^{AP} Allophycocyanin α subunit	slr2067
	apcB	β^{AP} Allophycocyanin β subunit	slr1986
	apcC	L_C small core linker	ssr3383

cpcG1-2 in *Synechocystis* PCC 6803 code for slightly different core — rod linkers (up to four have been found in *Anabaena* PCC 7120; [40]); *apcA2* code for a second α-type AP subunit in *Tolypothrix* PCC 7601 and *Anabaena* PCC 7120, but no function has yet been attributed to ApcA2 ([41]; http://www.kazusa. or.jp/cyano/Anabaena/) and, *nblA1-2* specify two NblA polypeptides [42].

V. TRANSCRIPTIONAL REGULATIONS INDUCED BY ENVIRONMENTAL CHANGES: A COMPLEX INTRICATE NETWORK

It is well established that the composition and the number of PBSs per surface unit of thylakoidal membrane vary in response to environmental changes such as intensity and spectral quality of light and nutrient availability (e.g., [6]). The effects of these two kinds of stress will be discussed.

A. NUTRIENT AVAILABILITY

Nitrogen is a key element that accounts for approximately 10% of the dry weight of a cyanobacterial cell [43]. Under standard conditions, cyanobacteria use ammonium, nitrate, urea, or amino acids (in decreasing order of preference) to satisfy their nitrogen requirements, a number of strains also being able to grow on molecular dinitrogen [44]. Whatever the initial source, ammonium is produced and assimilated through the central glutamine synthetase/glutamate synthase cycle, so-called GS-GOGAT pathway (Figure 23.1). Nitrogen starvation has pleiotropic effects on cell metabolism: it triggers the expression of genes involved in nitrate and nitrite uptake and assimilation, and the degradation of the PBSs. Indeed, because PBSs may constitute nearly half of the soluble proteins of a cyanobacterial cell, they represent an important source of nitrogen (reviewed in Ref. [45]). Degradation of PBSs leads to the depigmentation of the cells, which turn from blue-green to yellow-green, a phenomenon known as chlorosis [46,47]. Electrophoretic studies have revealed a sequential degradation from the distal end of the rods towards the central core of the PBS [47], i.e., essentially the reverse of the assembly process [4,48].

Mutants of *Synechococcus* sp. PCC 7942 that do not bleach when grown under nitrogen starvation conditions have been selected and have led to the identification of the so-called *nbl* (nonbleaching) genes: *nblA* encodes an ~7-kDa polypeptide, *nblB* the product of which shares similarities with phycobilin lyases, *nblR* and *nblS* code, respectively, for a response regulator

and a histidine kinase, typical of bacterial two-component systems [20,49–51]. In *Synechococcus* PCC 7942, the *nblA* gene is transcribed under nitrogen- and sulfur-limiting growth conditions [20], whereas in *Synechocystis* PCC 6803 the transcription of the two tandem copies of *nblA* is only activated by nitrogen deficiency [42]. *nblA* genes are present in all PBS-containing strains with the exception of the marine strains (http://www.kazusa. or.jp/cyano and http://www.jgi.doe.gov/JGI_microbial/html), may be because the latter live in a more stable environment (Figure 23.1). Although an *nblA* gene is necessary for PBS degradation, it may not be the triggering factor since it was found to be expressed in *Tolypothrix* sp. PCC 7601 cells grown under nitrogen-replete conditions [21]. The precise mechanism by which NblA proteins act in PBS degradation remains to be elucidated. NblA may tag or provoke a conformational change of the PBSs, which would then be degraded by a protease induced by starvation conditions. On the other hand, it has been shown that *nblA* mutants of both *Synechocystis* PCC 6803 and *Synechococcus* PCC 7942 enter, under N-limiting conditions, a nondividing dormant state which suggests pleiotropic effects of NblA on cell physiology [52,53]. For *Synechococcus* PCC 7942, the transcription of *nblA* is controlled by NblR and NtcA, a global nitrogen regulator (see below), under nitrogen starvation, but only NtcA under conditions of sulfur deficiency [54].

The function of NblB still remains to be determined. Its similarities with the phycobilin lyases suggest that it could take the chromophore off the PBSs, thereby rendering these PBSs susceptible to proteolysis. *nblR* codes for a response regulator typical of the two-component systems, and NblS appears to be the cognate histidine kinase that will form with NblR a signal transduction pathway controlling general acclimation responses [51]. The global regulator NtcA [55] is present in all cyanobacteria examined so far ([44], http://www.kazusa.or.jp/cyano and JGI). It belongs to the cAMP receptor protein (CRP) family of transcriptional regulators [56]. It bears close to its C-terminal end a helix–turn–helix motif for interaction with DNA, and acts as a dimer, each subunit making contact with one half of a palindromic recognition sequence $GTA-N_8-TAC$ [57]. NtcA binding sites have been found in the promoter regions of *nblA* genes, but the role of NtcA in controlling *nblA* transcription has only been demonstrated to date in *Synechococcus* PCC 7942 [42,54].

B. STRESS BY LIGHT

While sunlight provides the energy for photosynthesis, high visible light intensity and UV light injure

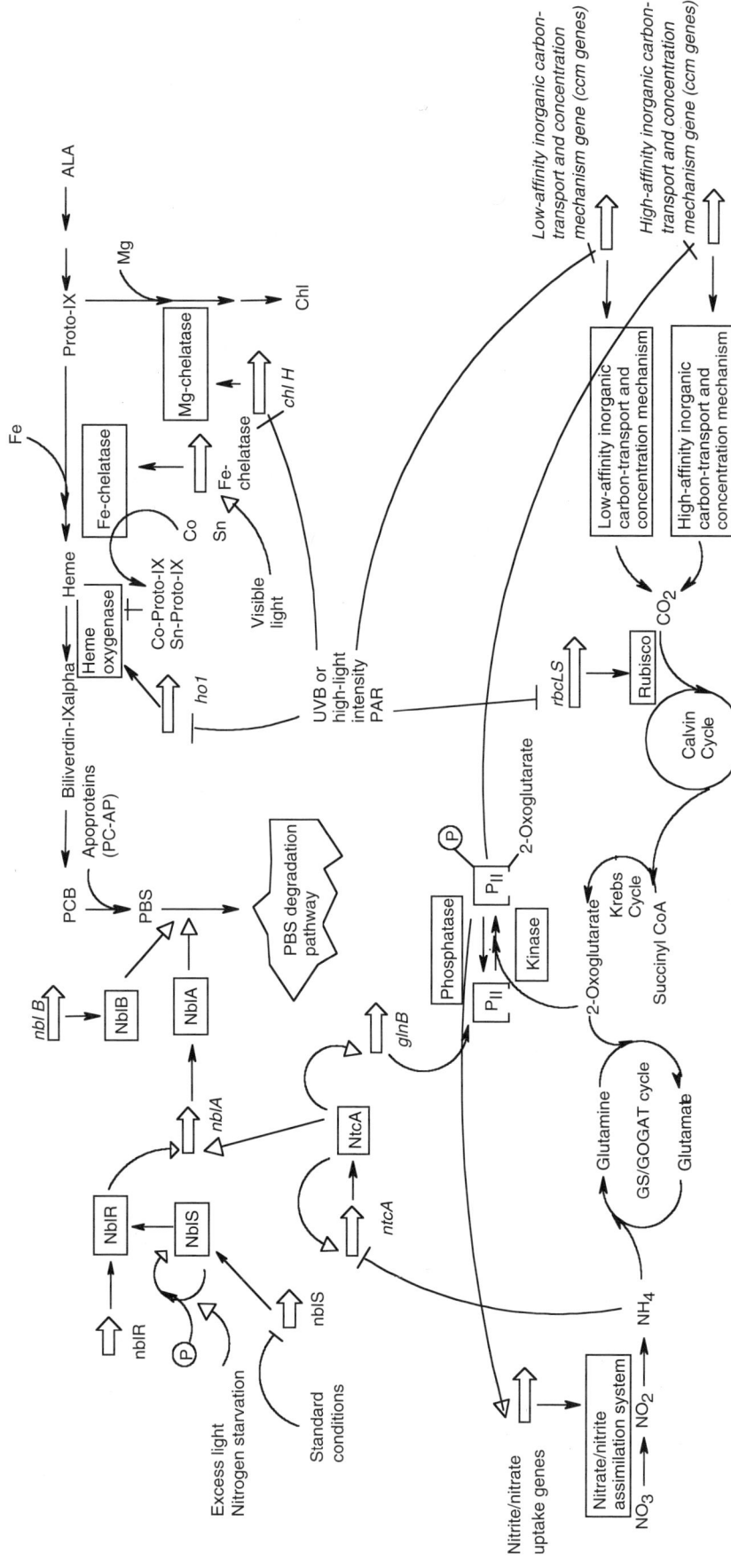

FIGURE 23.1 Schematic representation of the links between metabolic and regulatory pathways for the control of PBS biosynthesis and degradation following nitrogen or light stresses. Arrows represent controls either at a transcriptional or at a post-transcriptional level. It should be noted that the scheme has been built using data from several species of cyanobacteria. All links have not yet been directly ascertained and it is known that there exist differences between cyanobacterial species that likely are related to the ecological niches in which they are living.

many organisms, and their responses to such stresses vary with strains (for a review, see Ref. [6]). Grossman et al. [58] have recently reviewed the effect of the light environment on PBS composition, in particular during complementary chromatic adaptation, and we will therefore not discuss this phenomenon here. Changes in the spectral quality or intensity of visible light trigger another cellular response known as the "state transition". This cellular defense mechanism is rapidly initiated in order to regulate the transfer of light energy between the two PSs. The state transition model predicts that the "excess" energy absorbed by PBSs associated with PSII is directed preferentially to PSI; cells are then said to be in state 2. Three models have been proposed: the "mobile PBS model" (movement of the PBS between PSII and PSI), the "spillover" model (change in the rate of energy transfer from PSII to PSI chlorophyll molecules), and the "detachment model" (detachment of PBSs from PSII) (reviewed in Ref. [59]). Fluorescence recovery experiments after photobleaching treatment have indicated that PBSs could be more mobile than PSII [60]. However, experimental evidences such as (i) state transitions that occur in mutants devoid in PBS [61], (ii) the absence of reversible phosphorylation–dephosphorylation process, which, in chloroplasts, is involved in state transitions [62], and (iii) *in vitro* experiments performed with isolated PSI and PSII [63] make the "spillover" model the most likely. Mutants of the *apcD* gene (Table 23.1) have been shown to be impaired in state transition and appear to be blocked in state 1 (*Synechococcus* sp. PCC 7002: [64]; *Synechocystis* PCC 6803: [65]). Insertional inactivation of the *Synechocystis* PCC 6803 ORF*sll1926* (*rpaC* = regulator of PBS association) also prevents state transitions [66]. Rigidification of membranes, occurring naturally under low-temperature stress or provoked by engineering strains, was shown to influence state transitions [67]. State transitions are an example of the physiological changes that occur in photosynthetic organisms to cope with changes in the light quality and intensity of UV-B and white light, but little is known about the regulatory networks controlling the response. It is worth mentioning that recent data showed that the transfer of excitation energy harvested by the PBSs directly to PSI has been highly underestimated [68].

Global analyses of the transcriptional modifications triggered by high-intensity visible light or UV-B irradiation, using the cDNA microarray technology, have been reported recently [69,70]. In addition to the downregulation of PS I gene expression, there was a coordinated decrease of most of the genes encoding structural subunits of PBSs, including *apcA*, *apcB*, *apcC*, *apcD*, *apcE*, *apcF*, *cpcB*, *cpcC*,

cpcD, and *cpcG*. The genes encoding key enzymes of the biosynthetic pathway for tetrapyrroles, i.e., *hemA* (NADPH:glutamate-tRNA^{Glu} reductase), *hemF* (coproporphyrinogen oxidase), and *ho* (heme oxygenase) are also downregulated. In contrast, the expression of the genes encoding PC-phycocyanobilin lyase (*cpcE*, *cpcF*) is not affected. Interestingly, the expression of *nblA* is upregulated by a factor of 3 to 8 but whether this activation is directly triggered by the UV-B irradiation remains to be elucidated. UV-B downregulates genes involved in CO_2 fixation, i.e., Rubisco (*rbcL* and *rbcS* genes) [70] (Figure 23.1). It is interesting to note that in *Anabaena* PCC 7120 the *rbcL* promoter presents a site at which NtcA can bind [71,72]. UV-B also represses the constitutive low-affinity inorganic carbon-transport and concentrating mechanism proteins (*ccm* genes) (Figure 23.1). Therefore, under UV-B irradiation, the cellular CO_2 income would decrease. Consequently, the cellular concentration in 2-oxoglutarate would also be progressively reduced since the Krebs cycle will be less productive as a result of the reduction in the Calvin cycle activity. Because 2-oxoglutarate is at the heart of the GS/GOGAT cycle, through which all cellular nitrogen is incorporated (Figure 23.1), UV-B irradiated cells would be rapidly depleted in nitrogen. As described above (Section V.A), nitrogen deficiency triggers the activation of the *ntcA* gene and PBS degradation, as well as the activation of the synthesis of the nitrate assimilation system. The latter would however not occur if the cellular 2-oxoglutarate concentration is low because it requires the phosphorylated form of the P_{II} (GlnB) protein, the formation of which requires 2-oxoglutarate [73] (Figure 23.1).

The increase in the intensity of visible light from 25 to 200 $\mu mol/m^2/s$ produces responses similar to those induced by an UV-B irradiation at 60 $\mu E/m^2/s^1$ [70]. This suggests that the regulatory networks controlling PBS biosynthesis and functioning under these two conditions involve common intermediates. Time-course studies using quantitative RT-PCR, as well as analyses of the proteome by two-dimensional gels will be necessary to elucidate the order of the events that allow cells to appropriately modify their metabolism and cope with the changes that occur in their environment.

VI. REGULATORY LOGIC

As pointed out in the preceding sections, the synthesis and degradation of PBSs is highly regulated and must respond to multiple environmental influences. The regulatory network controlling PBS synthesis and degradation dynamically integrates these environmental parameters, most important among which are light intensity and quality, as well as availability of carbon, nitrogen, and other essential nutriments. In order to fully understand the functioning of this regulatory network we would need to know all (or at least most) of the regulatory components involved, their interactions, as well as the kinetic constants describing these interactions. Unfortunately, we are still far from this goal and at present we have to resort to a more intuitive understanding of the regulatory logic.

The analysis is complicated by the fact that the PBS system combines genetic regulatory controls (gene expression, protein stability, etc.) and metabolic control (intracellular concentration of metabolites that influence enzymatic reaction rates). This latter type of control acts on a much faster timescale than the former and we consider metabolic adaptation as essentially instantaneous in our discussion, which therefore focuses mainly on genetic control.

A. CONTROL OF PBS SYNTHESIS AND DEGRADATION

The rate of synthesis of the PBSs is determined by the metabolic state of the cell, as described in the previous paragraph. The genetic control of PBS synthesis is primarily exerted at the level of expression of the *apc* and *cpc* genes. Although some of the transcription start sites have been mapped, no in depth studies of the promoter regions have been performed, and little is known about transcriptional control of these genes. Their expression is greatly diminished under conditions of nitrogen starvation [74]. In *Synechocystis* PCC 6803, the transcription of the photosynthetic genes (*psa*, *psb*, *apc*, and *cpc*), as well as of *nblA*, is at least in part controlled by the two-component response regulator RppA [75]. At present it is not clear what physiological or environmental signal is exactly detected by the putative sensor kinase RppB.

Induction of NblA expression promotes, directly or indirectly, the degradation of PBS; NblA being a protein that is necessary, but not sufficient, for the degradation of PBS. The less well characterized protein NblB also participates in the degradation of PBS together with at least one other protein, a protease, which remains to be discovered. Analysis of the transcriptional control mechanisms of the *nblA* gene shows a direct influence of NtcA and NblR [54]. These multiple, often redundant or parallel sensory inputs into the control of PBS synthesis and degradation, connect the PBS to all major physiological and environmental parameters sensed by the cells. Figure 23.2 shows a summary of some of the known regulatory connections.

B. THE KEY REGULATORY PROTEINS

At least three global regulators are intimately involved in the regulatory network controlling PBS expression: NtcA, NblR, and SigE. Even though each one of these regulators acts on the PBS system, they each have multiple other regulatory connections to different control circuits in the cell. NtcA has been described as the central regulator of nitrogen metabolism, and its homologs are found in many cyanobacteria [44]. The *ntcA* gene is transcribed from a single or multiple promoters, depending on the cyanobacterial species, and its expression is repressed by NH$_4$. For the marine *Synechococcus* WH7803 this occurs at the posttranscriptional level through a reduction in the mRNA half-life [76]. Further control of NtcA activity may be exerted through 2-oxoglutarate, which modulates the affinity of NtcA to its DNA-binding sites. The control circuits are rather complex; the transcription of the *ntcB* gene, for example, is activated by NtcA and both proteins activate simultaneously the expression of the *nir* operon, i.e., the genes responsible for nitrate uptake and assimilation.

Furthermore, many other genes possessing NtcA binding sites have been identified in different cyanobacteria. In *Anabaena* PCC 7120, NtcA is involved in the control of heterocyst formation, may control the global regulator HU [77], and binds to the promoter of the *rbcL* gene (ribulose bisphosphate carboxylase), thereby linking carbon and nitrogen metabolism. Furthermore, NtcA activates the transcription of *sigE* in *Synechocystis* PCC 6803. This sigma factor, also called rpoD2-V, is important for survival under conditions of nitrogen starvation but deletion of this gene affects the transcription of many other genes unrelated to nitrogen metabolism (unpublished results). As all class II sigma factors, SigE directs the transcription of a well-defined class of genes and thereby changes profoundly the transcriptional profile of the cell. Identification of the members of this regulon is underway and preliminary results show a connection between SigE and the transcription of genes involved in photosynthesis (unpublished results).

Another central building block of the regulatory system is the response regulator NblR. The importance of this regulator lies in the fact that the activity of this protein, its phosphorylation state, is affected by a number of different physiological conditions. An *nblR* mutant rapidly dies during sulfur or nitrogen starvation or after exposure to high light. NblR is thus the target of kinases that detect various stresses to the cell. The sensor kinase NblS may be the cognate histidine kinase transferring the phosphate to NblR, even though this hypothesis has not yet been demonstrated experimentally [51]. This sensor kinase

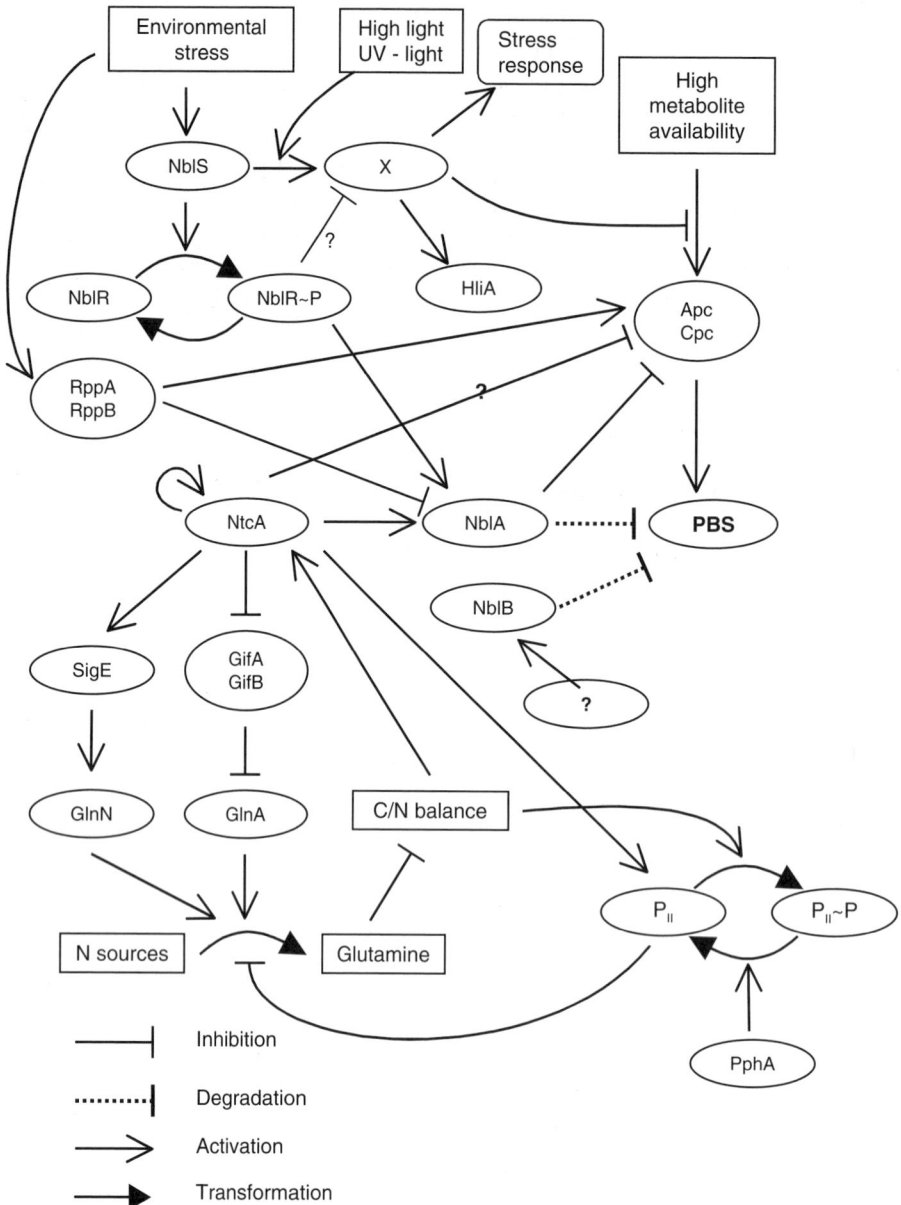

FIGURE 23.2 Regulatory connections controlling the synthesis and degradation of PBSs. The same symbols are used for activating of repressing interactions irrespective of the molecular mechanism (transcription, translation, activity, etc.). Transformations of one molecule into another, e.g., phosphorylation, are denoted by a full arrowhead. Most of the known regulatory interactions are indicated and some of the putative or missing interactions are denoted by a question mark. Future studies will further complicate this already highly connected network.

appears to detect multiple stresses such as UV light and nutrient deprivation, as well as the redox state of the cell. The signals emanating from other sensor kinases may also converge to NblR. Even though the only well-established target of NblR is NblA, indirect evidence shows that NblR controls many other genes. For example, an NblA mutant grows in high light, whereas an NblR mutant does not. This shows that signal transduction through NblR does

not necessarily pass on to NblA. Most of the targets of the global transcription regulator NblR remain to be identified.

C. COMMON THEMES OF REGULATION

The regulatory networks controlling the expression of the PBSs are highly interconnected and controls are exerted at all levels: enzyme activity, transcription, mRNA, and protein stability. Much of the genetic control converges to NblA. However, at present we do not yet know the molecular basis of many of the interactions that constitute this intricate regulatory circuit. Some of the uncertainties are indicated in Figure 23.2 by question marks. This lack of information precludes a detailed analysis of the dynamical properties of the regulatory network. However, certain themes of construction can already be seen in the still sketchy and incomplete scheme of Figure 23.2.

A large number of the signal transduction pathways utilize two-component systems (histidine kinase–response regulator). Since a sensor kinase never has an absolute specificity for only one response regulator there is necessarily crosstalk between the different systems. The path of signal transduction is therefore never strictly linear; the response regulators rather integrate environmental and physiological signals originating from very different sources. Further integration is achieved by multiple controls of key regulators, such as NblA or the PBS genes and proteins themselves.

The response to a particular stress or physiological state is not channeled to the key regulator only via a single signal transduction pathway. On the contrary, the network architecture ensures that one particular signal is distributed to numerous subsystems of the organism. For example, the carbon–nitrogen balance of the cell is probably detected via the intracellular concentration of 2-oxoglutarate. This "signal" is perceived not only by NtcA, which by itself controls a large number of diverse target genes [44], but also by PII (GlnB). One of the NtcA targets is the sigma factor SigE (*sll1689*). The specific promoter elements recognized by this sigma factor have not yet been characterized, but being the subunit of the RNA polymerase that confers specificity for gene transcription, induction of this gene will certainly profoundly affect the global transcriptional program of the cell.

A third striking feature of the phycobilisome regulatory network is redundancy: the same signal is often perceived by and/or acting on more than one transduction pathway. For example, exposure to high light results in the degradation of the PBS via the NblR–NblA pathway as well as by an NblR-dependent but NblA-independent route [49]. The upregulation of nitrogen assimilation genes passes not only through the activation of SigE and GlnN, but also through the parallel pathway that relieves the inhibitory influences of NtcA on GifAB, and in turn that of GifAB on GlnA.

The high connectivity within this regulation network, and our limited knowledge of the interactions precludes precise model building of the dynamics for this system. Much of what we know is based on the investigation of individual interactions that are mostly organized in a linear manner. However, the global behavior of the network, the stable states, and the transitions between these states, must emerge from intertwined positive and negative feedback loops. At present we can apprehend the response of the system in very specific conditions, but a true understanding of the dynamics has to await the discovery of at least some of the missing connections. The recent development of DNA microarrays and quantitative RT-PCR allow us to perform the global analyses required for a more comprehensive understanding of the regulatory networks. In addition, we expect that further levels of control, e.g., at translation, mRNA stability, etc., will emerge as this system is further investigated.

REFERENCES

1. Martin W, Rujan T, Richly T, Hansen T, Cornelsen S, Lins T, Leister D, Stoebe B, Hasegawa M, Penny D. Evolutionary analysis of *Arabidopsis*, cyanobacterial, and chloroplast genomes reveals plastid phylogeny and thousands of cyanobacterial genes in the nucleus. *Proc. Natl. Acad. Sci. USA* 2002; 99:12246–12251.
2. Glazer AN. Light guides. Directional energy transfer in a photosynthetic antenna. *J. Biol. Chem.* 1989; 264:1–4.
3. Cohen-Bazire G, Bryant DA. Phycobilisomes: composition and structure. In: Carr NG, Whiton BA, eds. *The Biology of Cyanobacteria Botanical Monographs*. Vol. 19. Oxford: Blackwell Scientific, 1982:143–190.
4. Sidler WA. Phycobilisome and phycobiliprotein structure. In: Bryant DA, ed. *The Molecular Biology of Cyanobacteria*. Dordrecht: Kluwer Academic, 1994:139–216.
5. Ting CS, Rocap G, King J, Chisholm SW. Cyanobacterial photosynthesis in the oceans: the origins and significance of divergent light-harvesting strategies. *Trends Microbiol.* 2002; 10:134–142.
6. Tandeau de Marsac N, Houmard J. Adaptation of cyanobacteria to environmental stimuli: new steps towards molecular mechanisms. *FEMS Microbiol. Rev.* 1993; 104:119–190.
7. Grossman AR, Bhaya D, He Q. Tracking the light environment by cyanobacteria and the dynamic nature of light harvesting. *J. Biol. Chem.* 2001; 276:11449–11452.
8. MacColl R. Cyanobacterial phycobilisomes. *J. Struct. Biol.* 1998; 124:311–334.
9. Adir N, Vainer T, Lerner N. Refined structure of c-phycocyanin from the cyanobacterium *Synechococcus vulcanus* at 1.6 angstrom: insights into the role of solvent molecules in thermal stability and cofactor structure. *Biochim. Biophys. Acta* 2002; 1556:168–174.
10. Capuano V, Thomas JC, Tandeau de Marsac N, Houmard J. An *in vivo* approach to define the role of the LCM, the key polypeptide of cyanobacterial phycobilisomes. *J. Biol. Chem.* 1993; 268:118277–118283.

11. Gindt YM, Zhou J, Bryant DA, Sauer KJ. Core muta-tions of *Synechococcus* sp. PCC 7002 phycobilisomes: a spectroscopic study. *Photochem. Photobiol. B* 1992; 15:75–89.

12. Reuter W, Wiegand G, Huber R, Than ME. Structural analysis at 2.2 Å of orthorhombic crystals presents the asymmetry of the allophycocyanin-linker complex, AP·L$_C^{7,8}$, from phycobilisomes of *Mastigocladus laminosus*. *Proc. Natl. Acad. Sci. USA* 1999; 96:1363–1368.

13. Adir N, Lerner N. The crystal structure of a novel unmethylated form of C-phycocyanin, a possible connector between cores and rods in phycobilisomes. *J. Biol. Chem.* 2003; 278:25926–25932.

14. Xie, Zhao JQ, Peng CH. Analysis of the disk-to-disk energy transfer process in C-phycocyanin complexes by computer simulation technique. *Photosynhetica* 2002; 40:251–257.

15. Barber J, Morris EP, da Fonseca PCA. Interaction of allophycocyanin core complex with photosystem II. *Photochem. Photobiol. Sci.* 2003; 2:536–541.

16. Tooley AJ, Cai YA, Glazer AN. Biosynthesis of a fluorescent cyanobacterial C-phycocyanin holo-α sub-unit in a heterologous host. *Proc. Natl. Acad. Sci. USA* 2001; 98:10560–10565.

17. Tooley AJ, Glazer AN. Biosynthesis of the cyanobac-terial light-harvesting polypeptide phycoerythrocyanin holo-alpha subunit in a heterologous host. *J. Bacteriol.* 2002; 184:4666–4671.

18. Kahn K, Mazel D, Houmard J, Tandeau de Marsac N, Schaefer MR. A role for cpeYZ in cyanobacterial phycoer-ythrin biosynthesis. *J. Bacteriol.* 1997; 179:998–1006.

19. van Thor JJ, Gruters OW, Matthijs HC, Hellingwerf KJ. Localization and function of ferredoxin:NADP$^+$ reductase bound to the phycobilisomes of *Synechocys-tis*. *EMBO J.* 1999; 18:4128–4136.

20. Collier JL, Grossman AR. A small polypeptide triggers complete degradation of light-harvesting phycobilipro-teins in nutrient-deprived cyanobacteria. *EMBO J.* 1994; 13:1039–1047.

21. Luque I, Ochoa de Alda JAG, Richaud C, Zabulon G, Thomas J-C, Houmard J. The NblAI protein from the filamentous cyanobacterium *Tolypothrix* PCC 7601: regulation of its expression and interactions with phycobilisome components. *Mol. Microbiol.* 2003; 50:1043–1054.

22. Rieble S, Beale SI. Enzymatic transformation of glu-tamate to δ-aminolevulinic acid by soluble extracts of *Synechocystis sp.* 6803 and other oxygenic prokaryotes. *J. Biol. Chem.* 1988; 263:8864–8871.

23. Schneegurt MA, Rieble S, Beale SI. The tRNA required for *in vitro* δ-aminoleculinic acid formation from glu-tamate in *Synechocystis* extracts. *Plant Physiol.* 1988; 88:1358–1366.

24. Grimm B, Bull A, Welinder KG, Gough SP, Kannan-gara CG. Purification and partial amino acid sequence of the glutamate-1-semialdehyde aminotransferase of barley and *Synechococcus*. *Carlsberg Res. Commun.* 1989; 59:67–79.

25. Troxler RF, Bogorad L. Studies on the formation of phycocyanin, porphyrins and a blue phycobilin by wild type and mutant straions of *C. caldarium*. *Plant Physiol.* 1966; 41:491–499.

26. Brown SB, Holroyd JA, Troxler RF, Offner GD. Bile pigment synthesis in plants: incorporation of heme into phycocyanobilin and phycobiliproteins in *Cyanidium caldarium*. *Biochem. J.* 1981; 194:137–147.

27. Beale SI, Cornejo J. Enzymatic haem oxygenase activ-ity in soluble extracts of the unicellular red alga, *Cyanidium caldarium*. *Arch. Biochem. Biophys.* 1984; 235:371–384.

28. Brown SB, Holroyd JA. Biosynthesis of the chromo-phore of phycobiliproteins. A study of mesohaem and mesobiliverdin as possible intermediates and further evidence for an algal haem oxygenase. *Biochem. J.* 1984; 217:265–272.

29. Cornejo J, Beale SI. Phycobilin biosynthesis reactions in extracts of cyanobacteria. *Photosynth. Res.* 1997; 51:223–230.

30. Cornejo J, Willows RD, Beale SI. Phytobilin biosyn-thesis: cloning and expression of a gene encoding sol-uble ferredoxin-dependent heme oxygenase from *Synechocystis* sp. PCC 6803. *Plant J.* 1998; 15:99–104.

31. Brown SB, Holroyd JA, Vernon DL, Jones OTG. Fer-rochelatase activity in the photosynthetic alga *Cyani-dium caldarium*. *Biochem. J.* 1984; 220:861–863.

32. Cole WJ, Chapman DJ, Siegelman HW. The struc-ture of phycocyanobilin. *J. Am. Chem. Soc.* 1967; 89:3643–3645.

33. Crespi WJ, Chapman DJ, Siegelman HW. The chromo-phore–protein bonds in phycocyanin. *Phytochemistry* 1970; 9:205–212.

34. Lagarias JC, Klotz AV, Dallas JL, Glazer AN, Bishop JE, O'Connell JF, Rapaport H. Exclusive A ring linkage for singly attached phycocyanobilins and phycoerythro-bilins in phycobiliproteins: absence of singly D-ring linked bilins. *J. Biol. Chem.* 1988; 263:12977–12985.

35. Beale SI, Cornejo J. Enzymic transformation of biliver-din to phycocyanobilin by extracts of the unicellular red alga, *Cyanidium caldarium*. *Plant Physiol.* 1984; 16:7–15.

36. Frankenberg N, Lagarias JC. Phycocyanobilin:ferre-doxin oxidoreductase of *Anabaena* sp. PCC 7120. Bio-chemical and spectroscopic. *J. Biol. Chem.* 2003; 278:9219–9226.

37. Toole CM, Plank TL, Grossman AR, Anderson LK. Bilin deletions and subunit stability in cyanobacterial light-harvesting proteins. *Mol. Microbiol.* 1998; 30:475–486.

38. Anderson LK, Toole CM. A model for early events in the assembly pathway of cyanobacterial phycobili-somes. *Mol. Microbiol.* 1998; 30: 467–474.

39. Houmard J, Capuano V, Colombano MV, Coursin T, Tandeau de Marsac N. Molecular characterization of the terminal energy acceptor of cyanobacterial phycobilisomes. *Proc. Natl. Acad. Sci. USA* 1990; 87:2152–2156.

40. Glauser M, Stirewalt VL, Bryant DA, Sidler W, Zuber H. Structure of the genes encoding the rod–core linker polypeptides of *Mastigocladus laminosus* phycobili-somes and functional aspects of the phycobiliprotein/

linker–polypeptide interactions. *Eur. J. Biochem.* 1992; 205:927–937.

41. Houmard J, Capuano V, Coursin T, Tandeau de Marsac N. Genes encoding core components of the phycobilisome in the cyanobacterium *Calothrix* sp. strain PCC 7601: occurrence of a multigene family. *J. Bacteriol.* 1988; 170: 5512–5521.

42. Richaud C, Zabulon G, Joder A, Thomas JC. Nitrogen or sulfur starvation differentially affects phycobilisome degradation and expression of the *nblA* gene in *Synechocystis* strain PCC 6803. *J. Bacteriol.* 2001; 183: 2989–2994.

43. Wolk CP. Physiology and cytological chemistry of blue-green algae. *Bacteriol. Rev.* 1973; 37:32–101.

44. Herrero A, Muro-Pastor AM, Flores EJ. Nitrogen control in cyanobacteria. *J. Bacteriol.* 2001; 183:411–425.

45. Bhaya D, Schwartz R, Grossman A.R. Molecular responses to environmental stresses. In: Whitton BA, Potts M, eds. *The Ecology of Cyanobacteria.* Dordrecht: Kluwer Academic, 2000:397–442.

46. Allen MM, Smith AJ. Nitrogen chlorosis in blue-green algae. *Arch. Microbiol.* 1969; 69:114–120.

47. Yamanaka Y, Glazer AN. Dynamic aspects of phycobilisome structure. *Arch. Microbiol.* 1980; 124:39–47.

48. Anderson LK, Toole CM. A model for early events in the assembly pathway of cyanobacterial phycobilisomes. *Mol. Microbiol.* 1992; 30:467–474.

49. Schwarz R, Grossman AR. A response regulator of cyanobacteria integrates diverse environmental signals and is critical for survival under extreme conditions. *Proc. Natl. Acad. Sci. USA* 1998, 95:1108–1113.

50. Dolganov N, Grossman AR. A polypeptide with similarity to phycocyanin alpha-subunit phycocyanobilin lyase involved in degradation of phycobilisomes. *J. Bacteriol.* 1999; 181:2610–2617.

51. van Waasbergen LG, Dolganov N, Grossman AR. nblS, a gene involved in controlling photosynthesis-related gene expression during high light and nutrient stress in *Synechococcus elongatus* PCC 7942. *J. Bacteriol.* 2002; 184:2481–2490.

52. Li H, Sherman LA. Characterization of *Synechocystis* sp. strain PCC 6803 and deltanbl mutants under nitrogen-deficient conditions. *Arch. Microbiol.* 2002; 178:256–266.

53. Gorl M, Sauer J, Baier T, Forchhammer K. Nitrogen-starvation-induced chlorosis in *Synechococcus* PCC 7942: adaptation to long-term survival. *Microbiology* 1998; 144:2449–2458.

54. Luque I, Zabulon G, Contreras A, Houmard J. Convergence of two global transcriptional regulators on nitrogen induction of the stress-acclimation gene *nblA* in the cyanobacterium *Synechococcus* sp. PCC 7942. *Mol. Microbiol.* 2001; 41:937–947.

55. Vega-Palas MA, Madueno F, Herrero A, Flores E. Indentification and cloning of a regulatory gene for nitrogen assimilation in the cyanobacterium *Synechococcus* sp. Strain PCC 7942. *J. Bacteriol.* 1990; 172:643–647.

56. Vega-Palas MA, Flores E, Herrero A,. NtcA, a global nitrogen regulator from the cyanobacterium *Synechococcus* that belongs to the Crp family of bacterial regulators. *Mol. Microbiol.* 1992; 6:1853–1859.

57. Luque I, Flores F, Herrero A. Molecular mechanism for the operation of nitrogen control in cyanobacteria. *EMBO J.* 1994; 13:2862–2869.

58. Grossman AR, Bhaya D, He Q. Tracking the light environment by cyanobacteria and the dynamic nature of light harvesting. *J. Biol. Chem.* 2001; 276: 11449–11452.

59. Mullineaux CW. How do cyanobacteria sense and respond to light? *Mol. Microbiol.* 2001 41: 965–971.

60. Mullineaux CW, Tobin MJ, Jones MR. Mobility of photosynthetic complexes in thylakoids membranes. *Nature* 1997; 390:421–424.

61. Olive J, Ajlani G, Astier C, Recouvreur M, Vernotte C. Ultrastructure and light adaptation of phycobilisome mutants of *Synechocystis* PCC 6803. *Biochim. Biophys. Acta* 1997; 1319:275–282.

62. Mullineaux CW. The thylakoid membranes of cyanobacteria: structure, dynamics and function. *Aust. J. Plant Physiol.* 1999; 26:671–677.

63. Federman S, Malkin S, Scherz A. Excitation energy transfer in aggregates in photosystem I and photosystem II of the cyanobacterium *Synechocystis* sp. PCC 6803: can assembly of the pigment–protein complexes control the extent of spillover? *Photosynth. Res.* 2000; 64:199–207.

64. Zhao J, Zhao J, Bryant DA. Energy transfer processes in phycobilisomes as deduced from analyses of mutants of *Synechococcus* sp. PCC 7002. In: Murata N, ed. *Research in Photosynthesis.* Vol. 1. Dordrecht: Kluwer Academic, 1992: 25–36.

65. McConnell MD, Koop R, Vasil'ev S, Bruce D. Regulation of the distribution for chlorophyll and phycobilin-absorbed excitation energy in cyanobacteria. A structure-based model for the light state transition. *Plant Physiol.* 2002; 130:1201–1212.

66. Emlyn-Jones D, Ashby MK, Mullineaux CW. A gene required for the regulation of photosynthetic light harvesting in the cyanobacterium *Synechocystis* sp. PCC 6803. *Mol. Microbiol.* 1999; 33:1050–1058.

67. Sakamoto T, Murata N. Regulation of the desaturation of fatty acids and its role in tolerance to cold and salt stress. *Curr. Opin. Microbiol.* 2002; 5:208–210.

68. Rakhimberdieva MG, Boichenko VA, Karapetyan NV, Stadnichuk IN. Interaction of phycobilisomes with photosystem II dimers and photosystem I monomers and trimers in the cyanobacterium *Spirulina platensis.* *Biochemistry* 2001; 40:15780–15788.

69. Hihara Y, Kamei A, Kanehisa M, Kaplan A, Ikeuchi M. DNA microarray analysis of cyanobacterial gene expression during acclimation to high light. *Plant Cell* 2001; 13:793–806.

70. Huang L, McCluskey MP, Ni H, LaRossa RA. Global gene expression profiles of the cyanobacterium *Synechocystis* sp. Strain PCC 6803 in response to irradiation with UV-B and white light. *J. Bacteriol.* 2002; 184:6845–6858.

71. Chastain CJ, Brusca JS, Ramasubramanian TS, Wei T-F, Golden JW. A sequence-specific DNA-binding

factor (VF1) from *Anabaena* sp strain PCC7120 vegetative cells binds to three adjacent sites in the *xisA* upstream region. *J. Bacteriol.* 1990; 172:5044–5051.

72. Ramasubramanian TS, Wei T-F, Golden JW. Two *Anabaena* sp. strain PCC 7120 DNA-binding factors interact with vegetative cell- and heterocyst-specific genes. *J. Bacteriol.* 1994; 176:1214–1223.

73. Lee HM, Flores E, Forchhammer K, Herrero A, Tandeau de Marsac N. Phosphorylation of the signal transducer PII protein and an additional effector are required for the PII-mediated regulation of nitrate and nitrite uptake in the cyanobacterium *Synechococcus* sp. PCC 7942. *Eur. J. Biochem.* 2000; 267:591–600.

74. Johnson TR, Haynes JL, Wealand JL, Yarbrough LR, Hirschberg R. Structure and regulation of genes encoding phycocyanin and allophycocyanin from *Anabaena variabilis* ATCC 29413. *J. Bacteriol.* 1988; 170:1858–1865.

75. Li H, Sherman LA. A redox-responsive regulator of photosynthesis gene expression in the cyanobacterium *Synechocystis* sp. Strain PCC 6803. *J. Bacteriol.* 2000; 182:4268–4277.

76. Lindell D, Padan E, Post AF. Regulation of *ntcA* expression and nitrite uptake in the marine *Synechococcus* sp. strain WH 7803. *J. Bacteriol.* 1998; 180:1878–1886.

77. Khudyakov I, Wolk CP. Evidence that the *hanA* gene coding for HU protein is essential for heterocyst differentiation in, and cyanophage A-4(L) sensitivity of *Anabaena* sp. strain PCC 7120. *J. Bacteriol.* 1996; 178:3572–3577.

Section VII

Photosynthesis in Higher Plants

24 Short-Term and Long-Term Regulation of Photosynthesis during Leaf Development

Dan Stessman
University of Illinois

Martin Spalding and Steve Rodermel
Department of Genetics, Development, and Cell Biology, Iowa State University

CONTENTS

I. INTRODUCTION

Normal growth of plants depends on the coordinated regulation of sink and source metabolism. The growth of new sinks, such as fruits or new leaves, and the demand of nonphotosynthetic tissues, such as roots, must be balanced with the source acquisition of nutrients, such as carbon assimilation during photosynthesis in fully expanded leaves. This balance ensures efficient use of all nutrients by all parts of the plant. Short-term changes in this balance can result in regulation of enzyme activities by metabolic intermediates, while perturbing long-term growth conditions can result in regulation of gene expression and a reallocation of nutrients. For example, C3 photosynthesis is thought to be limited by CO_2 at atmospheric concentrations [1]. Therefore, C3 plants usually have a large excess of ribulose-1,5-bisphosphate carboxylase/oxygenase (Rubisco), presumably to enable the plant to respond to rapidly fluctuating irradiance [2] and as a storage form of nitrogen [3]. Typically, when C3 plants are grown at elevated CO_2 levels there is an initial increase in photosynthetic rate as a response to the increased substrate availability. However, after a short period the photosynthetic rate decreases to lower levels than plants grown at normal CO_2 levels. This acclimation response is controlled at the level of gene expression, with decreased expression of photosynthetic genes, such as *RbcS* transcripts and decreases in Rubisco protein [4–7]. This response also ensures reallocation of nitrogen from Rubisco to other parts of the plant. Changes in growth conditions that lower sink demand usually result in downregulation of photosynthesis [8], while increases in sink demand result in upregulation of photosynthesis [9,10]. These changes are controlled at the level of enzyme activities and gene transcription.

II. SHORT-TERM CONTROL OF PHOTOSYNTHESIS

Leaf development involves a balance between sink demand and source strength that can be affected by a variety of factors [11]. Source photosynthesis and sink utilization occur in a coordinated fashion with fine control of enzyme activity by metabolic intermediates and coarse control by changes in photosynthetic gene expression [12]. During photosynthesis in source leaves, Rubisco incorporates CO_2 into ribulose-1,5-bisphosphate (RuBP) to form two molecules of 3-phosphoglycerate (3-PGA) and ultimately triose phosphate (TP) (Figure 24.1). While some of this TP is used to regenerate RuBP, the remaining TP is

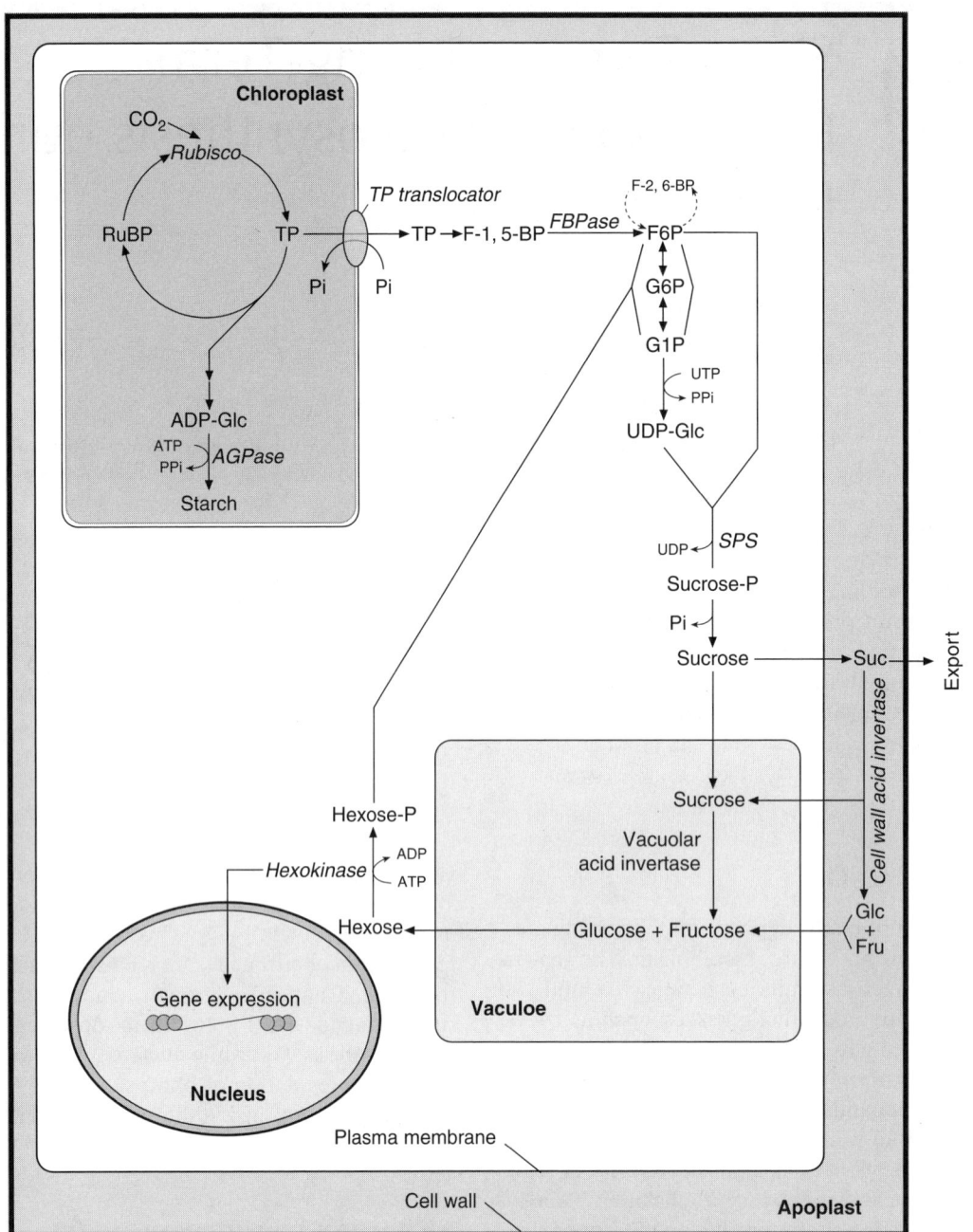

FIGURE 24.1 Schematic of fine and coarse control points of carbon partitioning in a source leaf cell. CO_2 is fixed by Rubisco to form triose phosphate (TP), which can either be converted to starch in the chloroplast, or exit the chloroplast via the TP translocator. In the cytosol, TP is converted to sucrose involving several steps, with controlling reactions catalyzed by FBPase and SPS. The sucrose is then exported to sink tissues. If sucrose accumulates, it is hydrolyzed by the cell wall or vacuole form of acid invertase and re-enters the cytosol where it is phosphorylated by hexokinase. The phosphorylation of hexoses by hexokinase sends a signal which results in downregulation of genes involved in photosynthesis [5,12,14,15]. Abbreviations: RuBP, ribulose-1,5-bisphosphate; TP, triose-phosphate; ADP-Glc, adenine-5-diphosphate glucose; AGPase, adenine-5-diphosphate glucose pyrophosphorylase; Pi, inorganic phosphate; PPi, Pyrophosphate; F-1,5-BP, fructose-1,5-bisphosphate; FBPase, fructose-1,5-bisphosphatase; F6P, fructose-6-phosphate; F-2,6-BP, fructose-2,6-bisphosphate; G6P, glucose-6-phosphate; G1P, glucose-1-phosphate; SPS, sucrose phosphate synthase; Suc, sucrose; Glc, glucose; Fru, fructose.

exported from the chloroplast to the cytosol via the triose phosphate translocator with a mandatory exchange of inorganic phosphate (Pi) [13]. TP is then synthesized into sucrose in the cytosol and ultimately exported to developing and nonphotosynthesizing parts of the plant, e.g., roots, fruits, and young, emerging leaves.

When sink demand declines, the need for source photosynthetic capacity decreases in a coordinated fashion. This sink regulation of source strength is thought to be triggered by a buildup of carbohydrates, namely sucrose, in the source tissues. In the short term, the accumulation of sucrose causes a downregulation of sucrose phosphate synthase (SPS) [16] and an increase in glucose and fructose as the accumulated sucrose is hydrolyzed by apoplastic or vacuolar invertases. Hexose phosphates begin to accumulate due to the decreased SPS activity and possibly the phosphorylation of free hexoses by hexokinase [17]. An increase in fructose-6-phosphate stimulates fructose-6-phosphate, 2-kinase (Fru-6-P,2K) to form fructose-2,6-bisphosphate (F-2,6-BP), which in turn inhibits the cytosolic fructose-1, 5-bisphosphatase (cytFBPase) [18]. The overall result is an accumulation of TP and a decrease in free Pi in the cytosol, which limits the export of TP from the chloroplast via the TP translocator. In the chloroplast, the increased ratio of 3-PGA/Pi stimulates ADP glucose pyrophosphorylase activity, the controlling step in starch synthesis [19,20], resulting in increased starch in the source leaf. Hence, the overall short-term result of limiting sink utilization of sucrose is an increase in starch accumulation in the source leaf. Starch synthesis thus provides a buffer to a short-term decrease in demand for photosynthate, and provides a temporary "sink" for TP to maintain a high rate of photosynthetic activity in the leaf [12].

III. LONG-TERM CONTROL OF PHOTOSYNTHESIS

The long-term adjustment to decreased sink demand involves a coarse control of photosynthesis via a downregulation of gene expression for photosynthetic genes such as *RbcS* and *LhcB* [5,14,15,21]. This downregulation appears to be linked to the accumulation of carbohydrates, but the mechanism is not well understood. Studies of plants grown in enriched CO_2 environments demonstrate this kind of regulation [22–24]. For example, in similar studies where plants have been transferred from ambient to elevated CO_2 environments [25], there is an initial short-term burst of enhanced photosynthetic activity followed by a downregulation of photosynthesis, or acclimation response, in which carbon exchange rate (CER) val-

ues are lower than the ambient control. This decrease in photosynthesis is a result of decreased Rubisco activity, which is a result of decreased amounts of Rubisco protein as well as transcript levels for the nuclear-encoded gene for the Rubisco small subunit (*rbcS*) and the plastid-encoded gene for the Rubisco large subunit (*rbcL*). This downregulation appears to be in response to an accumulation of all carbohydrates in the source leaf [24,25].

To study the acclimation response we examined several photosynthetic parameters in a single leaf of tobacco grown at either ambient or elevated CO_2 concentrations [22]. We selected leaf 10 (in order of emergence) because of its large size at full expansion. When tobacco plants are grown at ambient CO_2 (350 μl/l), individual leaves show a phase of increasing photosynthetic activity, followed by a brief period of maximal photosynthetic activity at day 12 (coincident with full leaf expansion), and ending with a long senescent phase of photosynthetic decline, during which time leaf yellowing occurs and nutrients are reallocated to the rest of the plant (Figure 24.2) [26–28]. This pattern is similar for chlorophyll concentrations, total soluble proteins, and Rubisco protein contents and activities. To examine photosynthesis at elevated levels of CO_2, ambient-grown tobacco plants were transferred to controlled-CO_2 growth chambers (350 μl/l) when leaf 10 had reached 1 cm in length. This leaf initially shows an accelerated increase in photosynthesis when compared to ambient conditions (Figure 24.2). However, the photosynthetic rate reaches an earlier photosynthetic maximum and begins to decline even before the leaf is fully expanded. The senescent decline appears to proceed at the same rate as in ambient-grown tobacco, but with the CER reaching zero at approximately day 25 as opposed to day 35 in the tobacco grown at ambient CO_2. This pattern is similar for chlorophyll concentrations and for Rubisco contents and activities.

Based on these results, we proposed a "temporal shift model" to explain the acclimation response in plants grown at high CO_2 [22] (Figure 24.3). In this model, there is a shift in timing of leaf development in high-CO_2 grown plants with an earlier peak photosynthetic rate and a shift to an earlier onset of senescence. The lower CER observed in high-CO_2 grown plants is due to a further progression of the senescent phase when compared to ambient-grown plants examined at the same time point.

If increased source strength results in an earlier onset of senescence, then decreased source strength might be expected to delay the onset of senescence. To test this hypothesis, we examined tobacco plants transformed with the *RbcS* gene in the antisense orientation [29]. These plants have reduced amounts

FIGURE 24.2 Photosynthetic rates (CERs) of tobacco grown at ambient and elevated CO_2 levels, and Rubisco antisense tobacco grown at ambient CO_2. For comparisons of wild type and antisense, plants were maintained under identical grown conditions. For tests at high CO_2, plants were transferred to elevated CO_2 when leaf 10 had reached 1 cm in length. CER measurements were taken from developmentally similar leaves of the different treatments throughout leaf ontogeny. Leaf 10 was used for the ambient and high-CO_2-grown wild type plants, and leaf 13 for the antisense plants. Each point represents the average ($\pm$ SD) of multiple measurements from at least four different plants. (Adapted from Miller A, Tsai CH, Hemphill D, Endres M, Rodermel S, Spalding M. *Plant Physiol.* 1997; 115:1195–1200 and Miller A, Schlagnhaufer C, Spalding M, Rodermel S. *Photosynth. Res.* 2000; 63:1–8. With permission.).

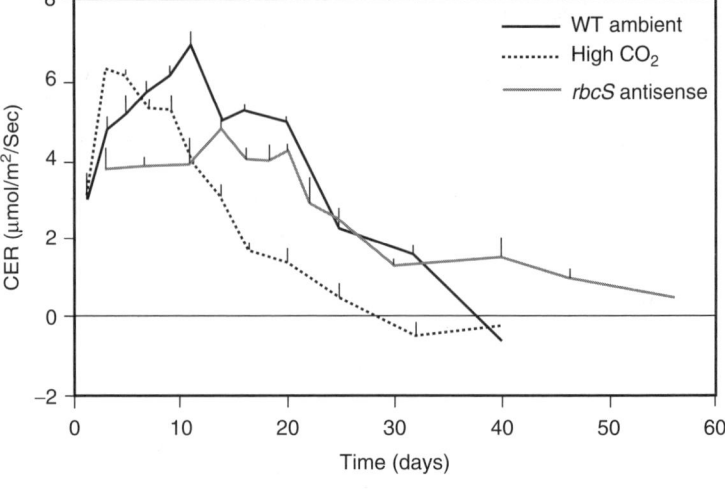

of Rubisco protein and decreased leaf carbohydrates. These plants also have an early slow-growth phenotype with a retarded leaf emergence and a greater number of smaller leaves. This is followed by a normal phase of growth similar at flowering to wild type plants. To study "developmentally similar" leaves, we examined leaf 13 of antisense plants, as opposed to leaf 10 of wild type, due to similar expansion and size of the leaves. Photosynthetic rates of the antisense plants were lower than wild type and reached a peak at day 20, as opposed to day 12 in wild type (Figure 24.2). A prolonged senescent phase was also observed in antisense plants during which CER did not fall below zero before day 55, compared to wild type around day 35. A similar pattern was observed with chlorophyll concentrations, Rubisco contents, and Rubisco activities. Taken together with the studies of high-CO_2-grown plants, we hypothesize that there is a threshold level of carbohydrate production which initiates the senescent phase of leaf development. This threshold level perhaps represents the point at which leaf source strength exceeds sink demand for photosynthate. When this occurs, an accumulation of carbohydrates may signal a downregulation of photosynthesis and a progression into the senescent phase of leaf development.

A. CARBOHYDRATE CONTROL OF GENE EXPRESSION

Carbohydrates have been shown to modulate the expression of genes in plants, including photosynthetic genes [20,30–33]. Gene expression appears to be controlled by a "feast versus famine" condition that may exist in the plant. Feast gene expression favors those

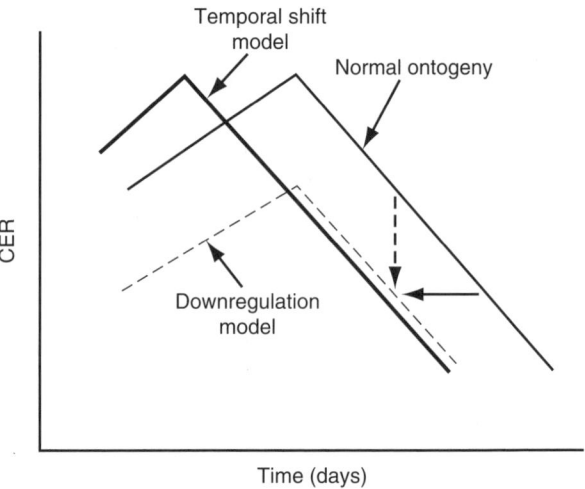

FIGURE 24.3 Comparison of temporal shift model as compared to the downregulation model of acclimation to high CO_2. The inner arrows represent the perceived shift in photosynthetic rate observed under elevated CO_2 concentrations. The downregulation model explains acclimation as an overall decrease in photosynthesis at all time points. The temporal shift model explains acclimation as a shift to an accelerated leaf development, with the decrease in photosynthesis due to an earlier peak in photosynthetic rate. (From Miller A, Tsai CH, Hemphill D, Endres M, Rodermel S, Spalding M. *Plant Physiol.* 1997; 115:1195–1200. With permission.)

genes that involve storage and utilization of carbohydrates, while famine favors expression of genes for photosynthesis, reserve mobilization, and export processes [30]. The coordination between the two ensures a balance of utilization and acquisition of resources.

Much of the attention about gene expression responses to carbohydrates has focused on hexokinase-mediated signaling. It is known that yeast hexokinase has a dual function of catalysis and signaling. Jang et al. [21] and Jang and Sheen [34] have proposed that hexokinase plays a similar role in plants, with a catalytic activity and a hexose-sensing or signaling activity. The hexose-sensing ability appears to be dependent on metabolism of hexoses rather than on the concentrations of the substrates or products. This was demonstrated in a maize protoplast expression system, in which sugars that are substrates for HXK were able to repress expression of photosynthetic genes [20,34]. Glucose analogs that are transported across the plasma membrane but are not phosphorylated by HXK did not repress photosynthetic gene expression. The same is true of hexose phosphates that were delivered by electroporation into the protoplasts. 2-deoxyglucose and mannose, which are phosphorylated by HXK but not further metabolized, were able to repress gene expression. Mannoheptulose, an inhibitor of HXK, is able to block the repression caused by mannose [35].

Experiments with transgenic *Arabidopsis* also demonstrate the signaling effects of hexokinase. *Arabidopsis* contains two genes encoding hexokinase, *AtHXK1* and *AtHXK2*. Plants that overexpress *AtHXK1* or *AtHXK2* displayed a hypersensitive response when grown on glucose media, with enhanced repression of *RBCS* gene expression [21]. Antisense hexokinase plants had a hyposensitive response with reduced repression of *RBCS* expression. Mutants overexpressing the sense yeast *HXK2* gene also showed a hyposensitive response even though they had an elevated phosphorylating activity. This occurred presumably because the yeast signaling activity is not recognized by the plant system, and the increased catalytic activity reduced the available substrates for the endogenous enzyme. Overexpression of the *Arabidopsis AtHXK1* in tomato shows a decrease in photosynthesis and an accelerated senescence [36].

B. INVERTASE, SUCROSE CYCLING, AND A POSSIBLE CONTROL POINT OF PHOTOSYNTHESIS

Invertase is an enzyme that catalyzes the irreversible reaction converting sucrose to glucose and fructose. In plants there are two forms of the enzyme, an acid and a neutral/alkaline form, each reflecting the pH optimum of the enzyme [37,38]. The neutral form exists in the cytosol while the acid form is located in both the apoplast and the vacuole. The acid invertases are encoded by a small gene family, with separate and multiple genes for both the apoplastic and vacuolar

forms. Both forms of invertase, along with sucrose synthase, are associated with sink tissues that require a constant influx of sucrose for metabolism [39]. Invertase maintains this flux by hydrolyzing sucrose and thus maintaining a gradient of sucrose from source to sink.

A perplexing observation is that source leaves have acid invertase activity [40]. This would seem to confound the ability of the source leaf to export sucrose if it is hydrolyzed. However, it has been proposed that acid invertase is part of a sensing system that initiates sink regulation of photosynthesis [23,41,42]. According to this proposal, limited utilization of sucrose by sinks would cause an accumulation of sucrose in source leaves. This sucrose would then be hydrolyzed into hexoses by either the apoplastic or vacuolar acid invertases, and the free hexoses would enter the cytosol and be phosphorylated by hexokinase. The hexose-phosphates would be resynthesized into sucrose. As stated previously, this would ultimately shift partitioning toward starch synthesis. Indeed, species that have high activities of vacuolar acid invertase tend to store higher amounts of starch [43]. Also, the net effect of sucrose hydrolysis by invertase and synthesis through hexokinase would generate a signal to downregulate photosynthesis (Figure 24.1).

Experiments with tobacco that overexpress a yeast invertase provide an extreme example of the possible effects of source invertase activity. These plants develop pale sectors in the leaves that contain low levels of chlorophyll and accumulate large amounts of all carbohydrates due to an inhibition of sucrose export [17]. These leaves also have lower rates of photosynthesis with decreased levels of Calvin cycle enzymes. Interestingly, SPS activity is increased in these plants, this may be indicative of an increased cycling of sucrose synthesis and hydrolysis. Also interesting is that the pale sectors do not appear until the leaf has made the sink to source transition. This enhances the role of carbohydrates in regulating photosynthesis, and possibly a role for acid invertase in generating the signal by sucrose hydrolysis.

Antisense experiments with acid invertase, however, have not yielded the opposite effect, that being an increase in photosynthesis by decreased generation of hexoses or an increased sucrose/hexose ratio. Rather, these experiments have demonstrated that vacuolar acid invertase controls the ratio of sucrose to hexose stored in tomato fruits and tomato leaves [44,45], and that there appears to be a threshold level of activity above which hexoses rather than sucrose accumulate [41,44]. This also occurs over a range of species, with those that have low vacuolar acid invertase storing sucrose [23].

Moore et al. [23] have shown that the acclimation response to high CO_2 is more pronounced in species with high vacuolar acid invertase. In a wide range of species examined, there was no correlation between the photosynthetic decline and the amount of carbohydrate that accumulated in leaves, nor was it correlated with any particular sugar. Rather, high acid vacuolar acid invertase activity was always associated with a decrease in photosynthetic capacity, and those species that lacked an acclimation response had low invertase activity. These data also showed a threshold response.

The evidence for a role of invertase and sucrose cycling in the regulation of photosynthesis is indirect at best. To address this question, we pulse-fed $^{14}CO_2$ to *Arabidopsis* during different time points of development [46]. We chose *Arabidopsis* because of its wide use as a model organism, with a wealth of genetic information available. The feeding experiments had two goals: (1) to characterize photosynthesis during development of a single leaf, and (2) to examine the short-term partitioning or flux of newly synthesized carbohydrates which may correlate with regulation of photosynthesis. For this experiment, we exposed a whole *Arabidopsis* plant to $^{14}CO_2$ in an enclosed chamber for a 10 min followed by a chase in unlabeled air for 10 min to allow time for partitioning of label into the different carbohydrate fractions. After the chase, leaf 8 was harvested and evaluated for partitioning of ^{14}C into ethanol-soluble and ethanol-insoluble fractions. The soluble fraction was further fractionated into neutral, anionic, and cationic fractions, with the neutral fraction containing sucrose, glucose, and fructose. Partitioning into sucrose, glucose, and fructose was analyzed using thin layer chromatography (TLC) plates.

Our experiments demonstrated that in *Arabidopsis*, the photosynthetic rate declines from the first time points measured (Figure 24.4B), as do chlorophyll concentrations and total protein levels. This decline occurs even as the leaf is still expanding and increasing in fresh weight. This is similar to results obtained from studies of photosynthetic rates using the entire rosette of *Arabidopsis* [28], but is in contrast to the previously mentioned tobacco experiments, which initially have a period of increasing photosynthetic rate followed by a senescent decline [22]. The expression of *LhcB* declines slightly during development, with a dramatic decline after full leaf expansion is attained (Figure 24.5). The expression of the senescence-associated gene, *SAG12*, dramatically increased at the last time point measured (Figure 24.5).

The majority of ^{14}C incorporation was into the neutral soluble sugar fraction at each time point measured. Further analysis of this fraction revealed that the majority of the label was present in the

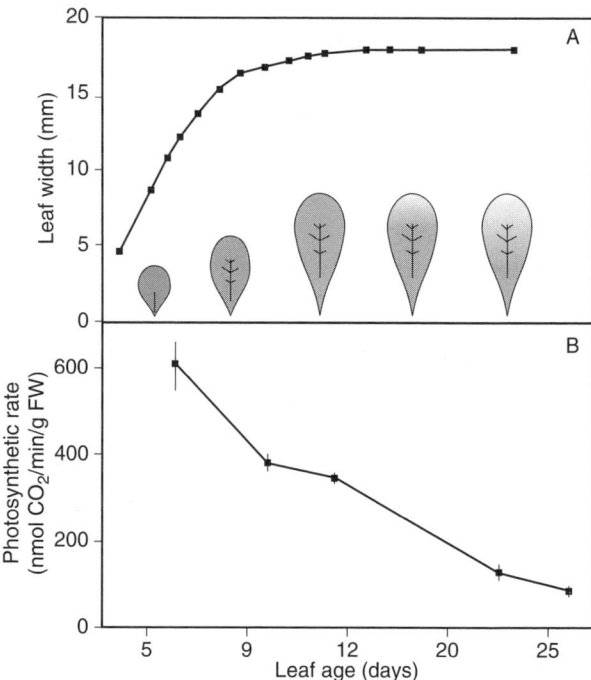

FIGURE 24.4 Changes in leaf expansion and photosynthetic rates during the development of leaf 8 of *Arabidopsis* grown under constant light conditions. Leaf expansion measurements were initiated when the leaves were 5 mm wide (day 1). Each point is the average of 31 plants. Photosynthetic rates were determined using five or six leaves per time point. (From Stessman D, Miller A, Spalding M, Rodermel S. *Photosynth. Res.* 2002; 72:27–37. With permission.)

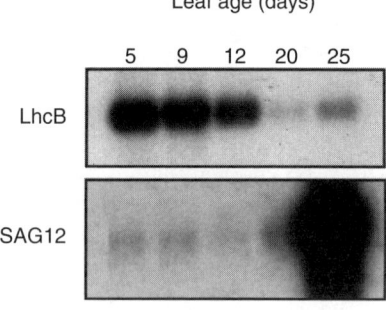

FIGURE 24.5 Northern blot of total RNA from leaf 8 of Arabidopsis grown under constant light. Blots were probed with the photosynthetic gene, *LhcB*, and the senescence-associated gene, *SAG12*. Each lane contains 5 μg of total RNA. (From Stessman D, Miller A, Spalding M, Rodermel S. *Photosynth. Res.* 2002; 72:27–37. With permission.)

sucrose fraction (Figure 24.6B). However, there was an increasing amount of label incorporated into the hexose fraction until full leaf expansion was attained, after which the partitioning of label into hexoses

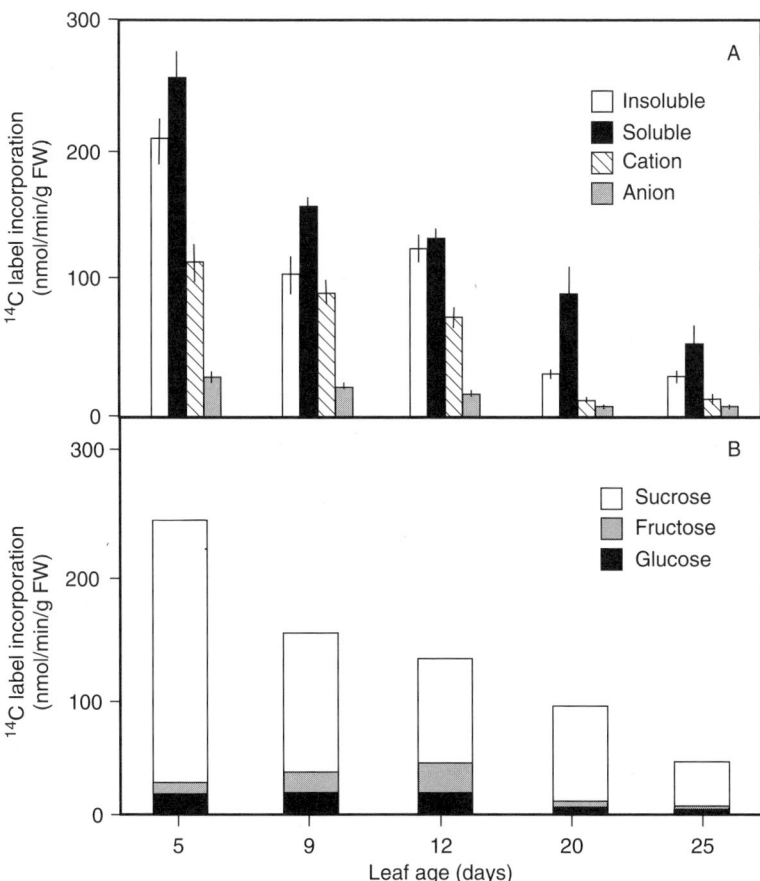

FIGURE 24.6 Carbon partitioning during the ontogeny of Arabidopsis leaf 8 grown under constant light conditions. (A) Incorporation of $^{14}CO_2$ into ethanol-insoluble and ethanol-soluble fractions. The soluble fraction was further fractionated into neutral, cationic, and anionic fractions. Each time point represents five leaf samples. Error bars are the standard error for each point. (B) The neutral fraction was further analyzed for sucrose, glucose, and fructose. (From Stessman D, Miller A, Spalding M, Rodermel S. *Photosynth. Res.* 2002; 72:27–37. With permission.)

declined (Figure 24.6B). This increase was not reflected in the total pool sizes of hexoses in the leaf. Rather, the maximum flux of label into hexoses coincided with the strong decrease in *LhcB* expression. At this time the leaf may have reached a threshold level of hexose metabolism or sucrose cycling, which would support the hypothesis that flux of carbohydrate through invertase and hexokinase signals a decrease in photosynthetic gene expression, since it is the short-term partitioning into hexoses and not a change in total pool size that correlates with gene expression changes. However, this evidence is indirect and cannot account for the decline in photosynthetic rate that occurs from the beginning of *Arabidopsis* leaf development.

Interestingly, these experiments showed no change in acid invertase or hexokinase activities early in leaf development when partitioning of labeled ^{14}C into hexoses increases, indicating there is probably some other factor controlling partitioning between sucrose and hexose at this time. However, others have shown that the source leaf acid invertase activity may serve as a mechanism to detect accumulation of carbohydrates under conditions of stress rather than the normal developmental process. It is known that acid invertase expression is induced by pathogen infection or wounding [47–51]. This may serve as a signal to downregulate photosynthesis to shift metabolic needs towards defense mechanisms.

IV. CONCLUSIONS

The balance between sink utilization of photosynthate and source strength is carefully controlled by both a short-term control of sucrose–starch partitioning by intermediate metabolites and by a long-term regulation by gene expression. Perturbing the sink–source balance can interrupt the partitioning of carbohydrates and lead to an accumulation of carbohydrates in source leaves. The accumulation of carbohydrates, specifically sucrose, may generate a signal to downregulate photosynthesis through metabolism by acid invertase and hexokinase [23]. There are also nonhexokinase mediated signaling pathways and sucrose signaling pathways that may also contribute to the regulation of carbohydrate production and its mobilization to sinks [30,33,52]. To date these pathways are not well characterized. Several mutants have

been identified that display a sugar-responsive phenotype, but very few have linked the signal with changes
in gene expression. Rather, most have identified
points of cross-talk with hormone signaling pathways. The existence of multiple gene families and
redundancy of metabolic pathways probably makes
direct identification of mutants involved in signaling
difficult to obtain or characterize. The combination of
activation-tagging, selection of sugar-responsive mutants, T-DNA knockout lines, and information from
genome sequencing projects will continue to aid in
the dissection of the pathways relating carbohydrate
signaling and gene expression responses.

REFERENCES

1. Bowes G. Growth at elevated CO_2: photosynthetic responses mediated through Rubisco. *Plant Cell Environ.*
1991; 14:795–806.
2. Stitt M, Quick WP, Schuur U, Schulze ED, Rodermel
SR, Bogorad L. Decreased ribulose-1,5-bisphosphate
carboxylase/oxygenase in transgenic tobacco transformed with 'antisense' rbcS. II. Flux control coefficients for photosynthesis in varying light, CO_2 and air
humidity. *Planta* 1991; 183:555–566.
3. Quick WP, Schuur U, Fichner K, Schulze ED, Rodermel SR, Bogorad L, Stitt M. The impact of decreased
Rubisco on photosynthesis, growth, allocation and
storage in tobacco plants which have been transformed
with antisense rbcS. *Plant J.* 1991; 1:51–58.
4. Van Oosten JJ, Besford RT. Acclimation of photosynthesis to elevated CO_2 through feedback regulation of
gene expression: climate of opinion. *Photosynth. Res.*
1996; 48:353–365.
5. Moore BD, Cheng SH, Sims D, Seemann JR. The
biochemical and molecular basis for photosynthetic
acclimation to elevated atmospheric CO_2. *Plant Cell
Environ.* 1999; 22:567–582.
6. Stitt M. Rising CO_2 levels and their potential significance for carbon flow in photosynthetic cells. *Plant Cell
Environ.* 1991; 14:741–762.
7. Bowes G. Facing the inevitable: plants and increasing
atmospheric CO_2. *Annu. Rev. Plant Physiol. Plant Mol.
Biol.* 1993; 44:309–332.
8. Plaut Z, Mayoral ML, Reinhold L. Effect of altered
sink:source ratio on photosynthetic metabolism of
source leaves. *Plant Physiol.* 1987; 85:786–791.
9. Lauer M, Schibles R. Soybean leaf photosynthetic
response to changing sink demand. *Crop Sci.* 1987;
27:1197–1201.
10. Diethelm R, Schibles R. Relationship of enhanced sink
demand with photosynthesis and amount and activity
of ribulose-1,5-bisphosphate carboxylase in soybean
leaves. *J. Plant Physiol.* 1989; 134:70–74.
11. Van Lijsebettens M, Clarke J. Leaf development in
Arabidopsis. Plant Physiol. Biochem. 1998; 36:47–60.
12. Sonnewald U. Sugar sensing and regulation of photosynthetic carbon metabolism. In: Aro EM, Andersson

B, eds. *Regulation of Photosynthesis.* Dordrecht:
Kluwer Academic, 2001:109–120.
13. Flügge UI, Heldt HW. Metabolite translocators of the
chloroplast envelope. *Annu. Rev. Plant Physiol. Plant
Mol. Biol.* 1991; 42:129–144.
14. Paul MJ, Foyer CH. Sink regulation of photosynthesis.
J. Exp. Bot. 2001; 52:1383–1400.
15. Jang J-C, León P, Zhou L, Sheen J. Hexokinase as a
sugar sensor in higher plants. *Plant Cell* 1997; 9:5–19.
16. Winter H, Huber SC. Regulation of sucrose metabolism in higher plants: localization and regulation of
activity of key enzymes. *Crit. Rev. Biochem. Mol. Biol.*
2000; 35:253–289.
17. Stitt M, von Schaewan A, Willmitzer L. "Sink" regulation of photosynthetic metabolism in transgenic tobacco plants expressing yeast invertase in their cell
wall involves a decrease of the Calvin-cycles enzymes
and an increase of glycolytic enzymes. *Planta* 1990;
183:40–50.
18. Stitt M. Fructose-2,6-bisphosphate as a regulatory molecule in plants. *Annu. Rev. Plant Physiol. Plant Mol.
Biol.* 1990; 41:153–185.
19. Heldt HW, Chon J, Maronde D, Stan Kovic ZS,
Walker DA, Kraminer A, Kirk MR, Heber V. Role of
orthophosphate and other factors in the regulation of
starch formation in leaves and isolated chloroplasts.
Plant Physiol. 1977; 59:1146–1155.
20. Preiss J, Levi C. Starch biosynthesis and degradation.
In: Stumpf PK, Conn EE, eds. *The Biochemistry of
Plants.* New York: Academic Press, 1980:371–423.
21. Sheen J. Feedback control of gene expression. *Photosynth. Res.* 1994; 39:427–438.
22. Miller A, Tsai CH, Hemphill D, Endres M, Rodermel
S, Spalding M. Elevated CO_2 effects during leaf ontogeny — a new perspective on acclimation. *Plant Physiol.* 1997; 115:1195–1200.
23. Moore BD, Cheng SH, Seemann JR. Sucrose cycling,
Rubisco expression, and prediction of photosynthetic
acclimation to elevated atmospheric CO_2. *Plant Cell
Environ.* 1998; 21:905–915.
24. Van Oosten JJ, Besford RT. Some relationships between the gas exchange, biochemistry and molecular
biology of photosynthesis during leaf development
of tomato plants after transfer to different carbon
dioxide concentrations. *Plant Cell Environ.* 1995;
18:1253–1266.
25. Cheng SH, Moore BD, Seemann JR. Effects of short-
and long-term elevated CO_2 on the expression of ribu-
lose-1,5-bisphosphate carboxylase/oxygenase genes and
carbohydrate accumulation in leaves of *Arabidopsis
thaliana* (L.) Heynh. *Plant Physiol.* 1998; 116:715–723.
26. Gepstein S. Photosynthesis. In: Noodén LD, Leopold
AC, eds. *Senescence and Aging in Plants.* San Diego,
CA: Academic Press Publishers, 1988:85–109.
27. Matile P. Chloroplast senescence. In: Baker NR, Thomas
H, eds. *Crop Photosynthesis: Spatial and Temporal Determinants.* Amsterdam: Elsevier Science 1992: 413–441.
28. Lohman KN, Gan S, Manorama CJ, Amasino RM.
Molecular analysis of natural leaf senescence in *Arabidopsis thaliana. Plant Physiol.* 1994; 92:322–328.

29. Miller A, Schlagnhaufer C, Spalding M, Rodermel S. Carbohydrate regulation of leaf development: Prolongation of leaf senescence in Rubisco antisense mutants of tobacco. *Photosynth. Res.* 2000; 63:1–8.

30. Koch KE. Carbohydrate-modulated gene expression in plants. *Annu. Rev. Plant Physiol. Plant Mol. Biol.* 1996; 47:509–540.

31. Smeekens S. Sugar-induced signal transduction in plants. *Annu. Rev. Plant Physiol. Plant Mol. Biol.* 2000; 51:49–81.

32. Gibson SI. Plant sugar-sensing pathways. Part of a complex regulatory web. *Plant Physiol.* 2000; 124:1532–1539.

33. Rolland F, Moore B, Sheen J. Sugar sensing and signaling in plants. *Plant Cell* 2002; (Suppl):S185–S205.

34. Jang J-C, Sheen J. Sugar sensing in higher plants. *Plant Cell* 1994; 6:1665–1679.

35. Pego JV, Weisbeek PJ, Smeekens SCM. Mannose inhibits *Arabidopsis* germination via a hexokinase-mediated step. *Plant Physiol.* 1999; 119:1017–1023.

36. Dai N, Schaffer A, Petreikov M, Shahak Y, Giller Y, Ratner K, Levine A, Granot D. Overexpression of *Arabidopsis* hexokinase in tomato plants inhibits growth, reduces photosynthesis, and induces rapid senescence. *Plant Cell* 1999; 11:1253–1266.

37. Tymowska-Lalanne Z, Kreis M. The plant invertases: physiology, biochemistry and molecular biology. *Adv. Bot. Res.* 1998; 28:72–117.

38. Sturm A. Invertases. Primary structures, functions, and roles in plant development and sucrose partitioning. *Plant Physiol.* 1999; 121:1–7.

39. Sung SS, Sheih WJ, Geiger DR, Black CC. Growth, sucrose synthase, and invertase activities of developing *Phaseolus vulgaris* L. fruits. *Plant Cell Environ.* 1994; 17:419–426.

40. Kingston-Smith AH, Walker RP, Pollock CJ. Invertase in leaves: conundrum or control point? *J. Exp. Bot.* 1999; 50:735–743.

41. Huber SC. Biochemical mechanism for regulation of sucrose accumulation in leaves during photosynthesis. *Plant Physiol.* 1989; 91:656–662.

42. Nguyen-Quoc B, Foyer CH. A role for 'futile cycles' involving invertase and sucrose synthase in sucrose metabolism of tomato fruit. *J. Exp. Bot.* 2001; 52:881–889.

43. Goldschmidt EE, Huber SC. Regulation of photosynthesis by end-product accumulation in leaves of plants storing starch, sucrose, and hexose sugars. *Plant Physiol.* 1992; 99:1443–1448.

44. Scholes J, Bundock N, Wilde R, Rolfe S. The impact of reduced vacuolar invertase activity on the photosynthetic and carbohydrate metabolism of tomato. *Planta* 1996; 200:265–272.

45. Ohyama A, Ito H, Sato T, Nishimura S, Imai T, Hirai M. Suppression of acid invertase activity by antisense RNA modifies the sugar composition of tomato fruit. *Plant Cell Environ.* 1995; 36:369–376.

46. Stessman D, Miller A, Spalding M, Rodermel S. Regulation of photosynthesis during *Arabidopsis* leaf development in continuous light. *Photosynth. Res.* 2002; 72:27–37.

47. Sturm A, Chrispeels MJ. cDNA cloning of carrot extracellular β-fructosidase and its expression in response to wounding and bacterial infection. *Plant Cell* 1990; 2:1107–1119.

48. Roitsch T. Source-sink regulation by sugar and stress. *Curr. Opin. Plant Biol.* 1999; 2:198–206.

49. Herbers K, Meuwly P, Frommer WB, Metraux JP, Sonnewald U. Systemic acquired resistance mediated by the ectopic expression of invertase: possible hexose sensing in the secretory pathway. *Plant Cell* 1996; 8:793–803.

50. Roitsch T, Balibrea ME, Hofmann M, Proels R, Sinha AK. Extracellular invertase: key metabolic enzyme and PR protein. *J. Exp. Bot.* 2003; 54:513–524.

51. Rosenkranz H, Vogel R, Greiner S, Rausch T. In wounded sugar beet (*Beta vulgaris* L.) tap-root, hexose accumulation correlates with the induction of a vacuolar invertase isoform. *J. Exp. Bot.* 2001; 52:2381–2385.

52. Chiou TJ, Bush DR. Sucrose is a signal molecule in assimilate partitioning. *Proc. Natl. Acad. Sci. USA* 1998: 95:4784–4788.

25 Recent Advances in Chloroplast Development in Higher Plants

Iliya D. Denev, Galina T. Yahubian, and Ivan N. Minkov
Department of Plant Physiology and Molecular Biology,
University of Plovdiv

CONTENTS

I. INTRODUCTION

In angiosperms, chloroplast development is a light-dependent process [1]. It begins from small (0.4 to 1 μm in diameter) spherical organelles called proplastids (Figure 25.1), which can be observed in zygotes and cells of meristem tissues [2,3]. Like all types of plastids they possess a double-membrane envelope. The internal space is filled with stroma, which contains a few single thylakoids, vesicles, plastoglobuli, and often starch grains [4].

Under normal light conditions the proplastids of leaf tissues differentiate into structurally and functionally mature chloroplasts. The differentiation processes include a gradual increase of plastid size (5 to 10 times) and number of plastids per cell (from 10–15 up to 150) [5]; intensive synthesis and accumulation of photosynthetic pigments (chlorophylls [Chls], carotenoids) and pigment–protein complexes; building of photosynthetically active thylakoid membranes; accumulation of the enzymes of the biochemical machinery responsible for carbon dioxide fixation [6], etc.

Under optimal conditions the transformation of proplastids to photosynthetically active chloroplasts takes about 6 h [1]. Electron micrographs of fully developed chloroplasts reveal that they consist of a double-membrane build envelope enclosing a complex of inner membranes known as thylakoids (Figure 25.1). The thylakoid membrane system of one chloroplast is believed to be formed from one continuous membrane, which divides the inner chloroplast volume into two separated spaces: extrathylakoid — stroma — and intrathylakoid — thylakoid lumen [7–9]. The biochemical part of photosynthesis takes place in the stroma, which contains all the enzymes of the CO_2 fixation pathway. The thylakoids of higher-plant chloroplasts are probably the most complexly organized of all biological membranes. Their main function is to capture light quanta and to drive a

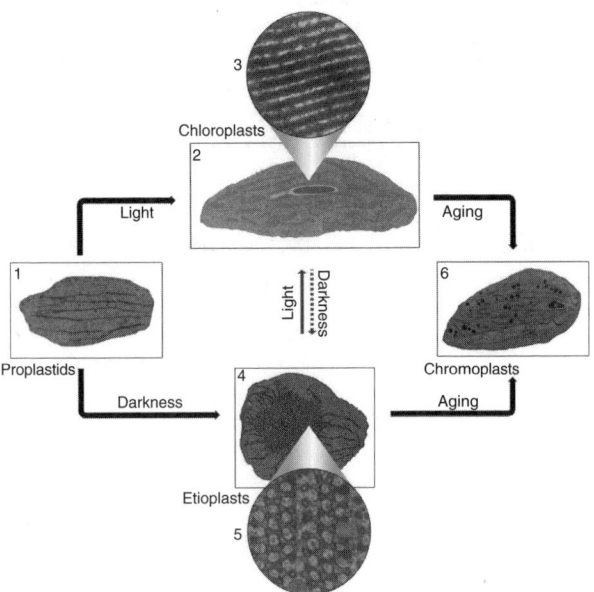

FIGURE 25.1 Simplified scheme of the development of plastids based on the presence of light and the stage of development. The numbers in the figure represent: 1 — proplastid; 2 — mature chloroplast; 3 — granal thylakoids; 4 — etioplast; 5 — crystalline-like structure of PLBs; 6 — chromoplast.

series of redox reactions, which produce ATP and oxygen and reduce ferredoxin [10].

In the absence of light or in weak light the chloroplast development is retarded and the proplastids develop into achlorophyllous plastids named etioplasts [3,11]. Etioplasts accumulate precursors of Chl in a concentration of about 1% of the normal Chl content of the green plant. These precursors are represented mainly by protochlorophyllide (Pchlide) and esterified protochlorophyll (Pchl) in amounts 5% to 15% of the precursors' pool [12–14], as well as traces of other precursors, such as Mg–protoporphyrine ester [15].

Instead of the arrangement of grana and stroma lamellae seen in light-grown chloroplasts, the etioplast inner membranes form one or more semicrystalline prolamellar bodies (PLBs) connected with simple unbranched lamellae named prothylakoids (PTs) (Figure 25.1) [3,11,16]. The PLBs are unique structures with a number of typical characteristics. Nevertheless, the recognition of PLBs and PTs as separate membrane systems is artificial, and no distinct border between them can be seen *in situ*. Together they constitute a dynamic interconnected membrane system where the relative amount of the two structures depends on plant species and age and growth conditions of the plant [17–20]. The PLBs contain the predominant part of the Pchlide, highly organized in pig-

ment–protein complexes [20]. Upon illumination they ensure rapid and very efficient conversion of Pchlide to chlorophyllide (Chlide) [21–26]. Soon after the Pchlide reduction the PLBs disappear, and Chlide is esterified to Chl and channeled to Chl-binding proteins [20,27].

The regular structure of PLBs is a prerequisite for their specialized functions [28]. Electron micrographs revealed that they are composed of a highly regular network of tubular membranes, which resemble a bicontinuous cubic Q224 lipid phase organization [20,29]. The high percentage (up to 60 mol%) of monogalactosyldiacyl glycerol (MGDG) in plastid lipids is regarded to be important for PLB formation since the cone shape of the MGDG molecules favors rounded or tubular lipid assembly [30,31]. However, the typical PLB structure cannot be achieved only on the base of lipid mixtures [32,33]. The PLBs consist of about 50% proteins, mainly (up to 90% of proteins) NADPH–PChlide oxidoreductase (POR, EC 1.3.1.33) [34,35]. POR is a nuclear-encoded, 36-kDa protein that catalyzes the light-dependent reduction of Pchlide to Chlide [20,34]. Two forms of POR — PORA and PORB, were identified in *Arabidopsis* [36] and in barley [37]. Recently, a third putative isoform of POR (PORC) was reported in *Arabidopsis* [38]. Both PORA and PORB are capable of binding stoichiometrically their substrate (Pchlide) and cofactor (NADPH) in photoactive ternary complexes. Several authors [20,34,35,39,40] have suggested that POR may bind substoichiomentric amounts of nonphotoactive Pchlide to a secondary site. In that way also PLBs can have defined pigment stoichiometry within minimal structural units where a small amount of nonphotoactive Pchlide transmits excitation energy to a large excess of photoactive Pchlide [40]. The POR–Pchlide complexes accumulate in PLBs as large aggregates [41–44]. The aggregation of POR complexes decreases the lifetime of excited triplet Pchlide forms with about an order of magnitude [28]. POR is a peripheral membrane protein, tightly associated with lipids [29,31]. Many authors [28,29,31,43,45–47] consider that the aggregated POR complexes tightly associated with lipids are sufficient to provoke the formation of PLBs. However, some newly obtained results indicate that the other big group of plastid pigments — carotenoids — also plays a role in PLB formation and disassembly; particularly, zeaxanthin and violaxanthin molecules might be associated with the photoactive POR–Pchlide–NADPH complexes [48]. Recently, Park et al. [49] identified a novel class of mutations in *Arabidopsis* gene encoding carotenoid isomerase (crtISO) and reported that the etioplasts of these mutants accumulate acyclic poly-*cis*-carotenoids and lack PLBs.

PLBs can be found also in young chloroplasts [50,51] during the chloroplast–chromoplast conversion [52], in plastids of epidermal cells [53] and in some meristem tissues like vascular cambia [54].

Initially, PLBs have been considered only as a laboratory artifact and a temporal storage of lipids and pigments [55]. Recently, it became clear that PLBs act as a defense system that protects plants against photooxidative damage at early stages of their life [20–22,28]. PLBs are probably more deeply, than thought before, involved in the regulation of the chemical heterogeneity of Chl pigments, for example, in the ratio between mono and divinyl forms [40]. Therefore, knowledge of the mechanisms of PLB formation and disassembly is important for better understanding the regulation of the formation of photosynthetic machinery. Applying membrane fluorescent probes for studying plastid membrane architecture and creating artificial membrane systems led to obtaining of more straightforward data about the membrane structure, the distribution of pigment–protein complexes within the lipid phase, and their role in the formation of membrane structures.

II. BIOCHEMICAL AND SPECTRAL CHARACTERISTICS OF INNER ETIOPLAST MEMBRANES

A. Pigments

1. Pchlide as the Main Chl Precursor of Dark-Grown Leaves

Pchlide is the immediate precursor of Chl in the inner membranes of a dark-grown leaf, and it is also coupled to the normal processes of greening. Pchlide has been proved to have several other functions such as that of a photoreceptor in etioplast–chloroplast transformation [56,57], a protector against photoinduced damage [21,22], or a regulator of plant adaptation to different light intensities [58].

The combination of different spectroscopic methods (absorption spectroscopy, low-temperature fluorescence emission and excitation spectroscopy, circular dichroism) and computational programs (SPSERV for Gaussian deconvolution of spectra) gives information about the variety of Pchlide spectral forms that reflect the native arrangement of the pigment and its molecular environment.

In etiolated leaves Pchlide excitation at 440 nm results at low temperature (77 K) in three major spectral bands with emission maxima at 633, 657 (of highest intensity), and 670 nm [59–63].

Using different excitation wavelengths Cohen and Rebeiz [59] showed that the spectral band at 633 nm consisted of four forms with emission maxima at 630, 633, 636, and 640 nm. Later, Böddi and coworkers [62–64] did a more detailed spectroscopic analysis and interpreted the band at 633 as an envelope of three emission bands with maxima at 628, 632, and 642 nm. Pchlide 628 is presented by "free" Pchlide molecules that are not connected to the protein, and in wheat leaves their small amount contributes to a very weak peak [65]. Pchlide 633 was attributed to monomeric Pchlide or Pchl molecules possibly bound to a protein of PT [66] and reported as photochemically inactive at flash irradiation [67]. The presence of Pchlide 645 in the spectra of holochrome preparations and after detergent treatment [68], the strong energy transfer to Pchlide 657 [62,65], and its situation on the edges of the PLBs prove the association of this form with POR in small and loose aggregates [62,69].

The dominating Pchlide form, which has absorption maximum at 650 nm and fluorescence emission maximum at 657 nm (Pchlide 650 to 657), is referred to as phototransformable for its ability to transform light-dependently to Chlide. Pchlide 657 forms ternary complexes with POR and NADPH [35] that are specifically arranged in the regular structure of the PLBs.

Site selective excitation revealed the composite nature of the band at 657 nm, namely, that it is organized in donor and acceptor structures [64]. Kis-Petik et al. [64] observed that when the excitation was in the blue side of the band, the emission spectrum had its position at a well-observable redshifted position relative to the energy of excitation. It was considered that the emitting electronic state is not the one that is directly excited, but the energy is transferred to always the same specific type of molecules with an emitting state of lower energy. When the excitation was in the red side of the band, the emission spectrum shifted with the tuning of the exciting laser frequency. This shows that in each case, the excitation selects certain groups of molecules from an inhomogenous population of chromophore dimers.

When the temperature of the spectroscopic assays is raised from 10 to 100 K the energy absorbed in the small-size aggregates and in the donor–acceptor complexes of the main structural form is transferred to nonfluorescent intermediates in a thermally activated excited state reaction [64,70]. In experiments at room temperature by flash light excitation, short-lived intermediates of Pchlide phototransformation with absorption bands at wavelengths longer than those of the excitation were detected in bean leaves [71].

Fluorescence line narrowing revealed the Pchlide 670 as a vibronic satellite of the band at 657 nm, but containing also a small band with an absorption maximum at 674 nm. Such a small contribution may arise from the presence of a very small amount of random Pchlide aggregates [64].

Many plants accumulate a mixture of monovinyl (MV)- and divinyl (DV)-Pchlide in darkness but wheat, oats, barley, and maize accumulate almost exclusively MV-Pchlide [72]. Kotzabasis et al. [73] suggested that the two main fluorescent maxima at 633 and 657 nm in etiolated leaves represent aggregates of MV- and DV-Pchlide, respectively. The accumulation pattern of both forms in different species, though, implies that the DV precursors could be reduced at the Pchlide or Chlide level to form the functional MV form of Chl [20].

2. Light-Dependent Transformation of Pchlide to Chlide

Irradiation of dark-grown angiosperms starts Chl biosynthesis primarily by photoreduction of accumulated Pchlide 657 and secondarily by further transformation of continuously newly synthesized Pchlide 657 [74]. The initial photoreaction is followed by the successive formation of light-independent short-lived intermediates [28,75,76]. The number and kind of the intermediates strongly depend on the temperature or the light intensity used for the Pchlide to Chlide transformation.

The formation of a nonfluorescent intermediate with absorption band at 690 nm was shown to be a result of primary photoreaction [77,78] that was proved to be dark reversible [79,80]. Measurement of electron spin resonance spectra showed that this intermediate is a free radical. Comparison of the fluorescence and absorption spectra of leaves irradiated at low temperature showed that this intermediate has the same absorption bands as the active Pchlide forms [81,82]. In these studies it was revealed that the first nonfluorescent intermediate is the sum of two components with absorption maxima at 697 and 688 nm. Thus, both forms can be transformed to Chlide only together and are interpreted as a mix dismutation of the free radicals of the carbon atoms 17 and 18 of the Pchlide molecule [83].

The next step is a light-independent transformation of the nonfluorescent intermediate to several short-lived fluorescent Chlide forms [75,76,79,84]. The primary fluorescent Chlide form was shown to have an emission maximum at 688 nm and to represent the complex POR–Chlide–NADP$^+$ [85]. Given that exogenous NADPH is present in etioplasts and PLBs, the unstable Chlide 688 undergoes a long-wavelength shift in 30 sec to give rise to Chlide 696 consisting of POR–Chlide–NADPH complexes [86,87]. The Chlide 688 transformation leads also to the formation of a minor short-wavelength form,

Chlide 675. This form is likely to be dissociated Chlide or a mixture of Chl *a* and Pheo *a* derived from part of the Chlide formed [88]. This heterogeneity in Chlide forms shortly after the photoreduction step is most probably associated with the heterogeneity of the etioplast inner membrane structure. Short-wavelength Chlide is preferentially formed in the PTs, while in the PLBs mainly long-wavelength Chlide is observed [89]. A similar formation of pheophytin can also occur during irradiation with low light intensities [88]. It is speculated that this may be a specific route for the formation of Pheo used for pigment ligation to the D1/D2 proteins in the reaction centers of photosystem II (PS II).

During prolonged incubation in darkness after a brief saturating irradiation, Chlide 696 is shifted within some minutes to Chlide 682, a process well known as a Shibata shift [90]. This shift is connected to the disaggregation of the complex between Chlide and POR [86], dispersal of the PLBs [91], and esterification of Chlide via Chl-synthetase [92,93]. A final slow shift with a fluorescence maximum at 685 nm has been shown to parallel the appearance of PS II activity and to be strongly dependent on plastid protein synthesis [67].

Recently, Belyaeva and Sundqvist [94] showed that the formation of the first fluorescent Chlide forms from the nonfluorescent intermediates might include several dark reactions with different temperature dependencies. When the temperature of samples that had been illuminated at 77 K is increased to 190 K, the dark reactions are slowed down and four primary Chlide forms are found with fluorescence emission maxima at 690, 696, 684, and 706 nm. During the prolonged dark exposure at 253 K, Chlide 684 remains stable while Chlide 696 and Chlide 706 are transformed to Chlide 672 and Chlide 684, respectively. The fate of Chlide 690 is unclear.

The Shibata shift rate depends on the Pchlide content, and a high Pchlide/Chlide ratio results in very fast formation of Chlide 682 [86,95]. The shift cannot be found if a small part of Pchlide (5% to 10%) is transformed with a short light impulse of 15 msec [96] or with low-intensity light [97], and in 10 sec after insufficient irradiation the newly formed Chlide already has absorption at 672 nm [96]. The red light can trigger a shift from 684 to 678 nm [98,99].

The Shibata shift appears to be regulated by the plastid NADPH/NADP$^+$ ratio [100]. According to the scheme of Oliver and Griffiths [86] the ternary complex POR–Pchlide–NADPH has to be formed before the Pchlide phototransformation. After enzyme oxidation or in the presence of surplus of NADP$^+$, the absorption peak is shifted to around 642 nm but the complex is not phototransformable

[87,101]. So, together with the "free" Pchlide, a "non-phototransformable" Pchlide might be formed and found in etiolated leaves. The first stable product is Chlide, absorbing at 678 nm, which is bound to the oxidized enzyme. This complex is stable *in vitro* especially in the presence of NADP$^+$, while in the presence of NADPH a form with an absorption maximum at 684 nm is formed instead. The next step is a slow shift to 672 nm, which can be stimulated *in vitro* by the Chlide being released from the complex [86]. The authors consider the form of 672 nm as a free Chl(ide), a mix of Chlide and Chl, that has been displaced *in vivo* from the pigment–protein complexes by Pchlide molecules of the inactive pool. The last shift from 672 to 678 nm, although it does not take place *in vitro*, is shown *in vivo* in isolated re-formed PLBs [102]. Ryberg and Sundqvist [103] have shown that the presence of NADPH stabilizes the PLBs and prevents their further transformation in intact leaves.

3. Spectral Changes after Irradiation of Re-Etiolated Leaves

Upon irradiation of etiolated leaves the Pchlide-to-Chlide transformation triggers PLB disintegration and thylakoid formation. If the plants are returned to darkness their leaves become re-etiolated and ac-

cumulate new PLBs, called "re-formed" PLBs, along with the thylakoids already formed during the irradiation. The main Pchlide form of the re-formed PLBs, isolated and purified from etiolated leaves illuminated for 4 h and grown in darkness for 16 h, has a fluorescence emission maximum at 657 nm and is bound to POR and NADPH [102,104,105]. Almost no Pchlide with fluorescence emission maximum at 633 nm could be found there [105]. In the thylakoids, however, most of the Pchlide is nontransformable. Both membrane systems contain also Chl(ide) synthesized during irradiation and having *in vitro* a fluorescence emission maximum at 680 nm [105].

We have found multiple Pchlide forms existing in isolated re-formed PLBs. High-pressure liquid chromatography (HPLC) analysis of pigments extracted from re-formed PLBs in 50% methanol revealed two Pchlide peaks instead of the one peak that could be seen in 80% methanol. After irradiation with three flashes in the presence of NADPH, only the Pchlide peak with lower mobility decreased while the peak of the newly formed Chlide increased. The irradiation did not affect the other Pchlide peak significantly, and it remained almost unchanged (Figure 25.2).

At the beginning of redarkening no large amount of Pchlide was found to resynthesize in bean leaves [90]. Using the *in vivo* spectrophotometry Wolf and Price [106] and Madsen [107] have clearly shown the

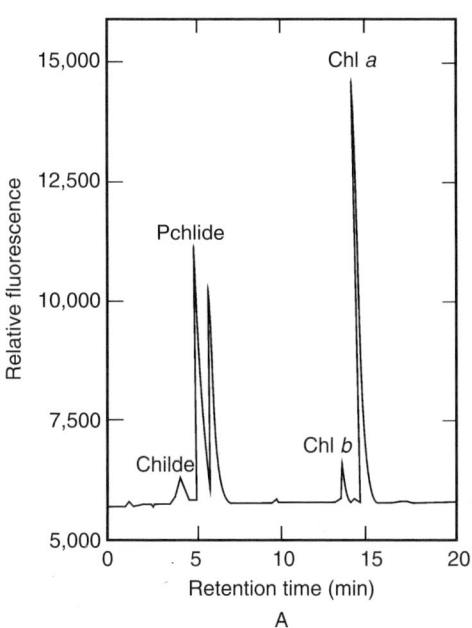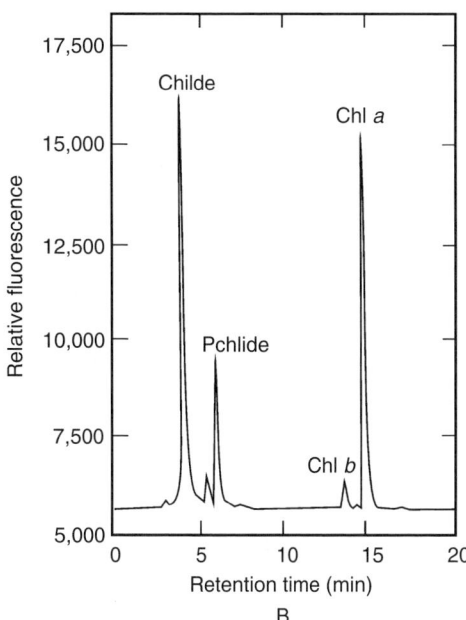

FIGURE 25.2 HPLC of 50% pigment methanol extracts from isolated re-formed PLBs (A) and irradiated with three flashes of saturating white light (B). A linear gradient 20% to 80% ethyl acetate in methanol:water (80:20, v/v) with a flow rate of 1 ml/min for 20 min was used for the HPLC. Two distinct peaks of Pchlide are present in nonirradiated PLBs, which are transformed to Chlide after irradiation.

S-shaped character of the Pchlide regeneration curve. It seems that the lack or the presence of a lag-phase depends on plant age. In bean leaves, regeneration could not be registered before the fifth day of growth, after which it appeared and was enhanced with age [108].

Under continuous irradiation the Pchlide reaccumulation is also accomplished as an S-shaped curve with an initial lag-phase [74,109], and the Pchlide resynthesis in the dark closely coincides with the curve of Chl accumulation in the first hours of greening [74,110]. In that sense, the lag-phase of Chl accumulation might be due to the lag-phase of Pchlide synthesis or actually of its resynthesis [111]. According to Smith and Young [112] the accelerated synthesis of Chl after the lag-phase is an autocatalytic process, connected with photosynthesis, but this does not explain the S-shaped curve of Pchlide regeneration in darkness. The lag-phase might be overcome by the addition of exogenous 5-aminolevulinic acid (ALA), which implies that the rate of Chl formation likely depends on the rate of ALA synthesis [113,114].

The Pchlide resynthesis is closely related to the Shibata shift, and it is probably not possible in plants that have not undergone this shift. In bean leaves the Pchlide resynthesis does not start if the absorption of Chl(ide) is at 682 nm. The Pchlide regeneration was not observed in barley mutants that are unable to undergo the Shibata shift [93,115]. In PLBs isolated from barley plastids the shift of Chl(ide) from 682 to 672 nm blocked the secondary accumulation of Pchlide at 650 nm [116].

After flash irradiation of isolated re-formed PLBs the phototransformable Pchlide 657 disappears and the fluorescence maximum of Chl(ide) shifts to from 680 to 690 nm [102]. Unlike the original PLBs, the Chlide of the re-formed PLBs has spectral characteristics that are independent of NADPH. This might suggest that there should be an energy transfer from the Chl present already before the second irradiation to the Chlide formed after it; and hence that the Chl should be located close to POR. Ryberg and Sundqvist [103] assumed that in PLBs the newly synthesized Chl(ide) remains bound to their membranes after irradiation since there is hardly any Pchlide available to replace the Chl(ide).

4. Carotenoid Contribution to Plastid Development

Carotenoids play a key role in light harvesting, photoprotection, singlet oxygen scavenging, excess energy dissipation, and membrane stabilization [117]. In the past decade the participation of carotenoids in photosynthesis was well identified, while

their importance for processes of Chl accumulation still unclear.

Lütke-Brinkhaus and Kleinig [118], performing plastid subfractionation experiments, showed that the phytoen–synthase complex in etioplasts and etiochloroplasts is present in a soluble form in the stroma, whereas the subsequent enzymes, that is, the dehydrogenase, cis–trans isomerase, and cyclase, are integral membrane proteins and occur in both membrane fractions, the PLBs/PTs and envelope membranes, with an eightfold higher specific activity for the PLBs/PTs. The presence of carotenoids in PLBs [119] and the association of antheraxantin and zeaxanthin with POR [48] have also been demonstrated.

Recent studies demonstrate that PLB formation requires carotenoid biosynthesis and reveal the key role of carotenoids in plastid photomorphogenesis. Recently, a novel class of mutations, carotenoid, and, chloroplast regulation (ccr) was identified in *Arabidopsis* [49]. These mutations disrupt carotenoid synthesis resulting in the accumulation of acyclic carotene isomers in the etioplasts and the reduction of lutein in the chloroplasts. The molecular basis of the three alleles of ccr2 unequivocally confirms it as the gene encoding for crtISO. Etioplasts of dark-grown crtISO mutants accumulate acyclic poly-cis-carotenes instead of cyclic all-trans-carotenes and lack PLBs. It is proposed that the role of all-trans-carotenoids in PLB assembly may be to stabilize or facilitate the curved membranes that form as a result of interaction between membranes, POR:Pchlide, and carotenoids. The stepped shape of the poly-cis-carotenoids may destabilize membrane curvature by altering membrane fluidity [20,120].

The Pchlide–Chlide interactions during the phototransformation process might be affected by carotenoids [121]. The partial Pchlide photoreduction and the successive formation of long-wavelength Chlide forms in wheat leaves with norflurazon-induced carotenoid deficiency were studied by low-temperature florescence spectroscopy (77 K). There were significant differences between the fluorescence emission spectra (the position and height of the peaks) of dark-grown normal and carotenoid-deficient leaves irradiated with nonsaturating white light of increasing intensity. The successive appearances of the newly formed Chlide species varied — the long-wavelength Chlide forms appeared first in the leaves nearly devoid of carotenoids, then in the leaves with carotenoid deficiency, and finally in the normal leaves. These findings are in agreement with the findings of Koski et al. [122] that the low efficiency of Pchlide to Chlide phototransformation in normal etiolated leaves in the blue region can be attributed to the competitive absorption of light by carotenoid pigments. In the leaves devoid of carotenoids or with carotenoid deficiency, the Pchlide molecules were the main light absorbing molecules. Thus, many more Pchlide molecules absorb light quanta, pass into the excited state, and trigger the photochemical reaction. As a result, a greater number of Pchlide molecules are transformed to Chlide per flash, which can be the cause of the higher effectiveness of partial Pchlide phototransformation in carotenoid-deficient leaves than in normal ones.

However, the question is whether the competitive absorption is the only way of carotenoid interference in Pchlide phototransformation. Taking into account that carotenoids absorb light in the blue spectral region, etiolated normal and carotenoid-deficient leaves were irradiated with red light to avoid carotenoid absorption. Under these conditions a light dose causing partial Pchlide photoreduction also caused a more effective formation and accumulation of Chlide species in leaves with a low carotenoid content. This finding demonstrates the role of carotenoids in different light screenings. The faster appearance of newly formed Chlide at partial Pchlide reduction in carotenoid-deficient leaves may be due to a more efficient energy transfer from Pchlide to Chlide in the pigment–protein complexes in PLBs. The absence of energy transfer from carotenoids to Pchlide [83] does not exclude the ability of carotenoids to influence the Pchlide–Chlide energy migration. Both the carotenoid dependence of partial Pchlide phototransformation demonstrated here and the carotenoid association with POR complexes found by Chahdi et al. [48] suggest the location of some carotenoids close to Pchlide or Chlide in the pigment–protein complexes of PLBs. These carotenoids might also somehow delay the transformation process.

B. Proteins of the Etioplast Membranes

Isolated inner etioplast membranes contain 30 to 35 different proteins [123]. During greening about 15 of them disappear and another 20 appear [124]. The prevailing proteins of the inner etioplasts membranes are the α- and β-subunits of the chloroplast-binding factor 1 (CF1) and POR [123,125]. Polypeptides with a molecular mass of 10 to 15 kDa have often been observed and are thought to play a role in the early stages of light-harvesting complex formation [126]. Traces of a number of polypeptides typical for the fully developed chloroplast membranes, such as cytochromes *f*, *b*-6, *b*-559, the apoprotein of Fe–S Riske center, have also been found [127,128]. There are several studies revealing a kind of proteolitic activity bound to the inner etioplast membranes as well [129–133].

The protein composition of PTs is dominated by the α- and β-subunits of CF1 [134–136]. Silver staining of electrophoretically separated samples indicates the presence of other proteins having molecular masses of 28, 34, and 86 kDa [136]. The 33- and 24-kDa proteins of the water-splitting system were found in the inner PT lumen [137,138]. Other findings showed that the enzymes involved in the Chl biosynthetic pathway after coproporphyrinogen III formation are associated with the thylakoids [139–141], though it remains unclear whether they are organized in supramolecular complexes identical to the centers for Chl biosynthesis proposed by the group of Shlyk [142].

Separation of PLB proteins by polyacrylamide gel electrophoresis revealed that their main component has a molecular mass of 36 kDa, comprises up to 98% of the total protein, and is identified as NADPH-dependent POR (EC 1.3.1.33 [143,144]). Some other proteins such as Chl-synthetase, chlorophylase, and carotenoid have also been detected as bound to the PLB membrane [20,49,93].

1. POR Structure and Enzymatic Activity

POR is one of only two enzymes known to require light for their catalysis; the other one is the DNA photolyase [145]. On the basis of sequence comparison, POR has been shown to be a member of the family of short-chain dehydrogenase or RED (reductases/epimerases/dehydrogenases) enzymes [146,147]. The enzymes of this family are all single-domain dinucleotid-binding oxidoreductases. They are generally dimers or tetramers, with the tetrameric form being essentially dimers of dimers. The cross-linking studies revealed that POR from wheat is present in PLBs as aggregates and the fundamental aggregated unit is a dimer [43]. From x-ray structural analysis of several RED proteins, the dimer interface has been identified as two long conserved parallel helices (α-4 and α-5) that correspond to residues 190 to 208 and 271 to 292 in pea POR [148]. The N terminus of POR is likely to be exposed on the surface of the protein, away from the dimerization interface, so that proteolytic processing of the enzyme on import into plastids may occur.

The reaction catalyzed by POR is a light-dependent *trans*-reduction of the double bond in ring D of the tetrapyrrole ring system of Pchlide [149] and requires NADPH as cosubstrate [34,123,150]. Using barley etioplast membrane preparations, Griffiths [150] determined the $K_m^{Pchlide}$ to be 0.46 μM and K_m^{NADPH} to be 35 μM. The enzyme seems to prefer DV-Pchlide [151] but both forms, MV- and DV-Pchlide, can be photoconverted to Chlide [152].

Blue-light excitation of Pchlide triggers the photo-chemical act [71,153]. During reduction, a hydride is suggested to be transferred from NADPH to the C-17 of Pchlide [149,154,155] to form a Pchilde-H⁻ anion [156]. A proton is then transferred from Tyr-275 (pea POR) to the C-18 position of Pchlide.

The catalytic site is located in a pocket where the Pchlide is inserted with the C and D rings at the bottom and the A and B rings protruding [157]. Within the catalytic site the Tyr-275 and Lys-279 stabilize NADPH [158], and the close proximity of Lys-279 is considered to be necessary to lower the pK_a of Tyr-275 to facilitate deprotonation of the phenolic group [146]. Although the Cys-308 is not located within the active site, it is considered close enough to exert an effect on Pchlide binding and reduction [47].

According to Oliver and Griffiths [86] NADP⁺ is rapidly displaced by fresh NADPH. In isolated etio-plasts, there is no measurable decrease in the level of NADPH during the phototransformation of Pchlide to Chlide, indicating a rapid regeneration of reduced NADPH [159]. During the Shibata shift Chlide re-leases the POR–Chlide–NADPH complex. After that a new molecule of Pchlide binds to the POR–NADPH complex to re-form the photoactive ternary complex, and a new reaction of photoreduction takes place. On the contrary, Reinbothe et al. [160] reported that Chlide is not easily dissociated from the enzyme and considered that each PORA molecule may be used for catalysis only once. Nevertheless, later find-ings have supported the idea that POR is able to carry out multiple turnovers of substrate binding, product formation, and product release [161].

2. POR is Encoded in the Nucleus, Translated in the Cytosol, and Transported into Plastids

The enzyme POR is encoded by the nucleic DNA. Full-length cDNAs encoding POR were first isolated from barley and oat by differential screening of etiol-ated and light-grown plants [162,163]. Later, Spano et al. [164,165] isolated POR cDNAs from pea and pine by immunoscreening of expression libraries. Based on the sequence similarities more POR cDNAs were isolated from *Arabidopsis thaliana*, wheat, and barley [36,37,166,167]. The deduced amino acid sequences revealed that the products of isolated cDNA clones result in a protein with a mo-lecular mass between 41 and 43 kDa, and consist of 36-kDa mature products and a transit peptide of varying size (74 amino acids for barley [165], 64 amino acids for pea [164]).

The comparison of sequences revealed two differ-ent cDNA clones in barley encoding for PORA [162]

and PORB [37] and three different cDNAs in *Arabi-dopsis* for PORA [36], PORB [166], and PORC [38]. The homology between sequences of mature POR proteins is 75% to 88% [36,38,166]. Within the cereals it is even greater — 98% between wheat and barley [167], which suggests high conservation in the gene organization within the angiosperms.

The first identification of the nuclear genes encod-ing POR was done by Spano et al. [164,165] in pea and loblolly pine. The genes have four introns, two of which are situated in the area encoding the transit peptide [28,164,165]. The genes encoding PORA and PORB in *Arabidopsis* are situated in chromosomes 5 and 4, respectively, while the PORC was mapped in chromosome 1 [38,168].

Despite the high degree of sequence similarities and identical functions, the different POR isoforms have different expression patterns. The expression of PORA is negatively regulated by light. PORA is strongly expressed in etiolated seedlings and its mRNA abundance is much higher than that of PORB in barley [28]. In etiolated *Arabidopsis* the amounts of both PORA and PORB mRNAs are similar [40]. The exposure to continuous light causes a dramatic decrease in PORA protein and mRNA amounts [123]. The experiments with the effects of red and far-red light on the PORA mRNA level demonstrate that PORA expression is under the con-trol of phytochrome A [169–173]. Unlike PORA, the mRNA and protein of PORB remain at approxi-mately constant levels during the plant's transition from dark growth to light. Study of the effects of plant age on POR gene expression show that PORA is expressed nearly exclusively in young seedlings, while PORB is expressed both in seedlings and in adult plants [28,36,174]. PORA mRNA is much less stable than PORB mRNA, which is explained by the presence of the plant-specific DownStream Element (DST) in 3′-UTR of PORA mRNA [175].

PORC, like PORA and PORB, catalyzes light-dependent reduction of Pchlide to Chlide with cofac-tor NADPH. The expression of its mRNA is light inducible. There are no indications whether any en-zyme accumulates in etiolated seedlings. The tran-script and the protein appear after transition of the etiolated plants to light [38]. When light-grown seed-lings of *Arabidopsis* are transferred to the dark, PORB mRNA levels do not change, while the con-centration of PORC mRNA rapidly declines. In more mature, light-adapted plants these differences in the expression of the PORB and PORC genes are less apparent, and transcripts of both POR genes were reported to accumulate in all photosynthetically ac-tive organs of *Arabidopsis* [38]. However, under dif-ferent light regimes, the transcript levels of PORB gene remain unaffected, while those of PORC increase with the increase in light intensities [168].

On the protein level, a few minutes of exposure to continuous light causes a decrease of both the enzymatic activity (with 80% to 90%) and the amounts of immuno-detectable PORA protein (with 60%) [176]. This is due to the rapid degradation of PORA caused by the activity of nuclear-encoded proteases whose ability to degrade PORA is considered to depend on conformational changes in the enzyme following Pchlide photoreduction [28]. At the same time the level of PORB protein remains unchanged. A similar effect of light on PORA protein and mRNA levels has been observed in different dicotyledonous and monocotyledonous species [177].

The different light responses suggest that the functions of the three POR enzymes of *Arabidopsis* are not completely redundant, but may allow the plant to adapt its needs for Chl biosynthesis more selectively by using preferentially one of the three enzymes under a given light regime. It was hypothesized that PORA plays an important role as a photoprotectant in the initial phase of the light-induced chloroplast formation, while PORB seems to maintain a certain level of Chl synthesis at all stages of chloroplast development [25,178]. On the other hand, PORC starts to accumulate only at the end of rapid Chl accumulation, so one can expect that the two enzymes, PORB and PORC, are required to jointly meet the changing demands for Chl synthesis in mature chloroplasts or act at different sites within the photosynthetic membranes [168].

3. POR Import and Assembly into Plastids

Plastid proteins are products of two genetic systems. The plastid genome encodes only about 100 proteins [179], while the vast majority of plastid proteins are nuclear encoded. The latter are translated by cytosolic 80S ribosomes as precursors. These precursors have on their N-terminal ends targeting sequences with highly variable lengths (from 20 to 120 amino acids), enriched with basic amino acids and a high content of serine and threonine [180,181]. The targeting sequences do not fold in aqueous surroundings, but in a hydrophobic environment form amphipathic β-strands or α-helices [182,183]. The transport across the double-envelope membranes mediated by stromal-targeting sequences is a common step for all proteins, regardless of their ultimate destination in organelles. Most of the proteins studied to date are imported via a single transport system, referred to as the general import pathway (GIP) [184,185]. Although most of the results obtained so far indicate that the vast majority of the plastid proteins are

imported via GIP, it is likely that other routes exist for entry into chloroplasts [186].

Protein translocation across the plastid envelope is a highly energy-dependent process and occurs probably at the regions where the two membranes are in close contact [180,182]. The translocon at the outer membrane of chloroplasts (Toc) is composed of at least four membrane proteins — Toc 159 (earlier known by its degradation product as Toc 86), Toc 75, Toc 34, and Toc 36. One of the functions of Toc 159 and Toc 75 is to bind the precursor protein — the receptor function [180]. They also form *trans*-membrane pores of diameter 0.8 to 0.9 nm, through which the precursor protein translocates [182]. Toc 159 and Toc 34 bind also GTP and have intrinsic GTPase activities. The functions of Toc 36 are unclear, but it is likely required for optimal protein translocation. It was shown that there is also a membrane-bound Com70 protein (from the Hsp70 family) that has ATP-dependent protein unfoldase activity [187,188]. Another Hsp70-like protein is localized on the inner face of the outer membrane and facilitates the translocation of the proteins from the Toc complex to the translocon at the inner membrane of the chloroplasts (Tic) complex [180,182]. The Tic complex is composed of Tic 110, Tic 22, and Tic 20. Tic 110 binds the precursor protein during the translocation and translocates it in close association with Toc 22 and Toc 20 [189]. The three proteins do not form stable associations except when interacting with the Toc complex. Insertion of precursor proteins requires ATP hydrolysis within the stroma [180]. Two other proteins were reported as components of the Tic complex: Tic 55 [190], which possesses an iron–sulfur center, and Tic 44 [191]. However, the questions about the roles of these proteins and whether they indeed are part of the Tic complex remain to be elucidated [186]. The molecular chaperones Clpc and Cpn60 are associated with Tic 110. Clpc might be required for protein transport across the membrane, while Cpn60 assists in protein refolding in the stroma [186,188].

When a precursor protein reaches the stroma, the signal peptide removes the precursor sequence completely or partly [180]. The partial removal occurs in proteins confined to the inner space of thylakoids, and such targeting sequences are known as bipartite transit peptides [186,192,193].

Like all other nuclear-encoded proteins, POR is synthesized in cytosol as a precursor protein (pPOR). Some authors report that the POR mRNAs gather close to the plastid surface and are translated there by cytosolic 80S ribosomes [194,195]. The pPOR possesses at its N-terminal end a targeting sequence whose length and primary structure vary among the plant species and between POR isoforms. There is no common view about the mechanisms of POR import into the plastids. Using cross-linking during the early stages of the import, Aronsson et al. [196] found pea pPOR cross-linked to Tic 75. The addition of a large excess of pSS (precursor of the small subunit of Rubisco which is imported via GIP) out-competed the import of pea POR and barley pPORA and pPORB [196]. Thus, there are a number of indications that pPOR proteins are imported in plastids via GIP [167,196–199].

Reinbothe and coworkers [160,200–202] have suggested that the import of barley pPORA depends on the Pchlide content in envelope membranes. pPORA was readily imported into plastids enriched with Pchlide (by etiolation, re-etiolation or δ-aminolevolinic acid treatment), while chloroplasts lacking Pchlide were unable to import pPORA. *In vitro* processing experiments showed that when pPORA is bound to Chlide the targeting sequence is masked and is unable to interact physically with the outer plastid envelope membrane. In contrast, the pPORA alone as well as pPORA–Pchlide and pPORA–Pchlide–NADPH have their transit peptide exposed and able to interact with the plastid envelope [24,160,200]. The role of the transit peptide was clarified when the transit peptides of barley pPORB and dihydrofilate reductase (DHFR) were replaced by that of pPORA. Under these conditions both pPORB and DHFR showed a Pchlide-dependent pattern of import [201]. Later, other groups performed import studies with barley pPORA and pPORB in the presence and absence of Pchlide and could not confirm the Pchlide dependence of pPORA import [196,198]. However, a kind of Pchlide-stimulated membrane association was observed for barley pPORA in barley and pea chloroplast lysates [203]. Obviously, more investigations are needed to determine whether the pPORA import is Pchlide dependent, Pchlide stimulated, or completely independent of Pchlide.

The other contradictory problem is the localization of POR toward membrane lipids. Little is known about the interaction of POR with the membrane lipids, and there are almost no data about the POR of re-formed PLBs. Oliver and Griffiths [204] showed that POR is only loosely attached to the PLB membrane and can be easily washed off the membrane. On the basis of the high hydrophobic amino acid ratio in the POR molecule, Röpper et al. [205] suggested that it is an integral protein. Based on the hydrophobicity of POR, Selstam and Widell-Wigge [206,207] concluded that POR is an amphiphilic membrane protein tightly associated with the membrane and thus most probably is an integral membrane protein [208]. However, the secondary structure prediction based on the amino acid composition revealed no obvious membrane-spanning region [163]. Hydropathy profiles also do not indicate any clear membrane-spanning or thylakoid transfer domain [164,165], which might indicate that the enzyme is peripherally associated with the membrane on the stromal side. The comparison of the deduced POR protein sequences allowed identification of a group of similar enzymes — short-chain alcohol dehydrogenases [146]. The secondary structure prediction based on similarities between enzymes of this group revealed about seven β-sheets connected with eight to nine α-helices. The circular dichroism study of purified POR showed that 33% of amino acids are organized in α-helices and 19% in β-sheets, 20% form turns and 28% random coils [209]. Later studies suggested that the POR secondary structure consists of a central β-sheet built of seven β-strands, surrounded by nine α-helices [47,210].

The attachment of POR to the plastid inner membranes seems to involve first interactions of charged amino acids of POR with membranes [47]. Mutagenesis performed on POR to substitute uncharged alanine for charged amino acids revealed that charges in the central region of POR (between amino acid residues 86 and 342) or close to amino acids involved in NADPH or Pchlide binding were essential for the membrane association of POR in a thermolysin resistant way [47]. The charged amino acids in the N- and C-terminal regions of the mature protein did not significantly affect membrane association [47]. However, studies with POR deletion mutants revealed that amino acids in the range of 362 to 395 are vital for membrane association [148]. Both hydrolysable ATP and a minor δpH are required to stimulate association of POR [197]. Both PORA and PORB are capable of binding stoichiometrically their substrate (Pchlide) and cofactor (NADPH) in photoactive ternary complexes [20,34,35,39]. The cofactor NADPH is vital for the membrane association of POR [47,148,197,199]. The POR–Pchlide–NADPH complexes accumulate in PLBs as large aggregates [41–44]. The phosphorylation of POR protein seems to favor its membrane aggregation [45]. It is possible that the reversible phosphorylation can regulate the POR aggregation and disassembly during the formation of PLBs and their light-induced disappearance, but the exact role of the phosphorylation is still unclear [211]. Many authors consider that the aggregated POR complexes, tightly associated with lipids in membranes reached of the cone shape MGDG molecules, are sufficient to provoke the formation of PLBs [28,29,31,43,45,47].

C. LIPIDS OF INNER PLASTID MEMBRANES

The lipids form about 50% of the mass of inner plastid membranes and act as a fluid matrix for the functional supramolecular complexes. The fatty acid tails shape the hydrophobic, central core of the membrane, and the hydrophilic heads of lipids are situated at the surface. The lipids are not equally distributed between the two monolayers as well as in the lateral direction [212]. Since the lipids are highly unsaturated, the membranes are very fluid at physiological temperatures. Fluidity allows for the high lateral mobility of pigment–protein complexes through the membranes. However, due to the high content of proteins (50% of the mass) the diffusion coefficient of individual molecules is limited to 10^{-10} to 10^{-9} m^2/sec [10,213].

Thylakoid lipids are a complex mixture containing about 80% galactolipids — such as MGDG (50 mol% of total lipids) and digalactosyldiacil glycerol (DGDG, 25 mol%) — that are electrically neutral. The remainder are mainly phosphatidyl glycerol (PG, 10 to 15 mol%) and sulphoquinovosyl diacilglycerol (SQDG, 5 to 10 mol%), charged under physiological pH [212,214]. The predominant fatty acid in plastid inner membrane lipids is linolenic (C18:3). However, C16:3 fatty acids are also present in some groups of plants [31]. Specific to the thylakoid membrane fatty acid is *trans*-3-hexadecanoil acid (C16:1). It is a component of PG [31,212,214].

D. Structural Properties of Plastid Lipids

Studies on structural properties of plastid lipids revealed that when mixed with water, MGDG forms reverse hexagonal phase (type HII) while DGDG, SQDG, and PG form lamellar (Lα) phase [215–218]. The ability to form lipid phases with different organization is due to the shape of the molecule. MGDG with its small polar head group and large wider hydrophobic part formed by the hydrocarbon chains of the fatty acids has a cone shape, whereas DGDG with a larger polar head group is a more cylindrically shaped molecule [32,218].

Mixtures of isolated chloroplast lipids (MGDG, DGDG, SQDG, and PG) are able to form different cubic phases. When the water content is low (5% to 13%), they form cubic phase identified as bicontinuous reversed Ia3d phase [33], which is composed of three armed units. Higher water content results also in cubic bicontinuous reversed phases, but they vary between Pn3m and Im3m types — having four-armed and six-armed structural units, respectively [33]. The description of different bicontinuous cubic phases was made using an infinite periodic minimal surface model [219,220]. Structures made of polyhedra, in comparison with the lipid–water phase made of aggregated 12- or 14-armed lipid micelles (the cubic phase Q223 [221]), although about eight times smaller, resemble much more closely the PLBs than other MGDG–DGDG-composed phases [29,217,222]. Nevertheless, the typical geometry for open-type PLBs cannot be deduced based only on lipid–water mixtures [32,200,223].

III. ULTRASTRUCTURE AND BIOGENESIS OF DEVELOPING PLASTIDS

A. Structure of the PLBs

Etioplasts or etiochloroplasts arise when normal chloroplast development is curtailed by lack of light [16]. The etioplasts contain two interconnected inner-membrane systems: PLBs and PTs. The PLBs are relatively small, regular "paracrystalline" structures (2 to 3 μm). The membrane architecture of PLBs has been analyzed in detail by means of ultrathin sections [16,224–230], freeze fracture [231,232], x-ray diffraction [44], and mathematical analysis [233,234]. Granick first proposed a simple cubic lattice model, according to the suggestion of von Vettstein, in which tubular units with six arms at right angles are connected to each other [235]. Based on ultrastructural studies, the structure of PLBs is considered to consist of four- or six-armed units [224–226,229,230]. Gunning and Jagoe [236] on the basis of ultrathin sections from *Avena* etioplasts also considered the six-armed model, but later they explained the structure of *Avena* PLBs [230], using the model of Wermayer [224–226]. According to this model the PLBs from *Phaseolus* consist of four-armed tubular units.

Use of high-resolution scanning electron microscopy has shown that the basic unit of the squash PLB is a tetrapodal structure, which has four short tubular arms meeting at one point with equal angles. This structure displays three lattice forms, hexagonal, square, and zigzag (distorted hexagonal), and is referred to as zincblende type [232].

The PLBs from different plants differ. The oat etioplasts contain highly organized paracrystaline PLBs, while other plants (barley and wheat) exhibit loosely packed and irregular structures. The types are referred to as "narrow" for the paracrystalline PLBs and "wide" for the type in wheat [237,238]. According to Gunning and Jagoe [236] 99% of the PLBs show paracrystaline structure.

There are data that some tissues grown in darkness lack PLBs. Wellburn [239] reports the absence of photoconversion of Pchlide to Chlide in etiolated tobacco tissue culture, which is connected with the absence of PLBs. Some conifers (*Pinus*) develop their photosynthetic system in darkness but, nevertheless, contain PLBs [17]. After irradiation of wheat plants with intermittent light they develop thylakoids and grana and also keep their PLBs [240].

On the basis of the investigation of the lipids in the PLBs one might consider their structure as built through connection of the lipid phase and the POR. The cubic phase, which is typical for the membrane structure of the PLBs [220], is the thermodynamically preferred organization of the POR complex and the plastid lipids [31]. It is known that on the places where POR is found bound to the glycolipids, the membrane builds a cubic phase of the lipids [135,241,242]. The formation of such regular crystalline PLBs is also proven by model investigations with artificial lipids and lipid–protein phases, which build spontaneously a regular structure with two independent canal systems, resembling the PLB structure [33,219,220,243]. This shows that the way of building the PLBs is a spontaneous process of organizing lipids and proteins by the principle of Schwarts for infinite periodic minimal surface [244]. The POR in the planar membrane of the envelope [245] is not organized in a cubic phase because of the different lipids in this membrane.

B.   RE-FORMED PLBS

When dark-grown plants are irradiated and Pchlide is converted to Chlide, the "original" PLBs disintegrate and thylakoids are synthesized. When the plants are returned to darkness, new PLBs appear, called "re-formed" PLBs [236,237,246]. Their structure is mainly of "narrow" type [102,104,247]. This recrystallization is accompanied by new accumulation of Pchlide depending on the leaf age [108,236,237,248]. The rate of Pchlide resynthesis found in young leaves allows very fast recrystallization of PLBs and even inhibits the initial destruction of PLBs after irradiation [91]. In plants grown in darkness, have maximum amount of Pchlide, the lag-phase of Pchlide resynthesis follows the photoconversion of the pigment [108,236,237] with simultaneous destruction of the PLBs. After the lag-phase a re-formation of new PLBs is initialized together with Pchlide resynthesis [237]. Such PLBs have a higher degree of crystallinity, and the process is known as re-etiolation. Berry and Smith have shown that the re-etiolated etioplasts from barley have highly organized tubular structures that differ from the "wide" PLBs in the etioplasts [238]. Later, those investigations were proved for wheat [104,105,249] and for rye and oat [249]. The PLBs that are re-formed in wheat and rye after re-etiolation for 30 to 60 min cannot be distinguished from the original PLBs in oat.

The re-formed PLBs might also appear in young etioplasts in a weak white light or under a red light independent of the Pchlide reduction [4,238,250–255].

These structures were referred to in the beginning as "PLBs of the weak light" [252]. On studying these PLBs Ikeda found that they have a structure that is different from the original PLBs formed in darkness [228]. The investigations have shown that the PLBs that appear under these conditions depend on the light intensity. They also react faster than the original PLBs to external changes, and they might be useful for studying the nature of PLBs [102,256]. The sensitivity of these PLBs might depend on the higher amount of pigments in their membranes compared with the original PLBs formed in the dark [91]. In some cases the re-formation might not be connected with Pchlide regeneration [250].

At the same time there are re-formed PLBs that appear at night and disappear after sunrise [256]. In young maize seedlings grown in a dark–light cycle, crystalline PLBs are formed during the dark period and are destructed during the light period [257]. Plastids from the primary leaves in barley that have formed grana after 6 h of light following 6 h of darkness form a number of small crystalline PLBs [237]. The crystalline PLBs have been found in detached leaves or in intact bean seedlings, which are greened or kept in weak light [251,252]. Plastids that contain crystalline PLBs and grana have also been found in leaves of maize chloroplast mutants [258,259] and barley, and the presence of Pchlide was shown [237].

In fully green and redarkened plants the crystalline structures found are re-formed PLBs [236,237,246,260,261]. In our investigations we showed that all the original PLBs in etiolated wheat leaves disappeared after 4 h of irradiation and new, re-formed PLBs appeared after redarkening of the plants for 16 h [104,105]. We have applied a method for isolating PLBs from re-etiolated wheat leaves by differential centrifugation, and PLBs were separated on a 10% to 50% continuous sucrose density gradient. The isolated re-formed PLBs were found at the same density (1.17 g/ml) as the PLBs from dark-grown material [60]. Electron microscopy showed a large similarity between re-formed PLBs and PLBs from dark-grown material; even the re-formed PLBs were mainly of the narrow type (Figure 25.3) and offered the possibility of controlling the cross-contamination of PLBs and thylakoids during isolation.

IV.   USE OF FLUORESCENCE PROBES FOR INVESTIGATING PIGMENT–PROTEIN COMPLEXES IN PLASTID MEMBRANES

The fluorescent probes are low-molecular weight fluorescent dyes, which during the interactions with membrane structures change their fluorescence by a

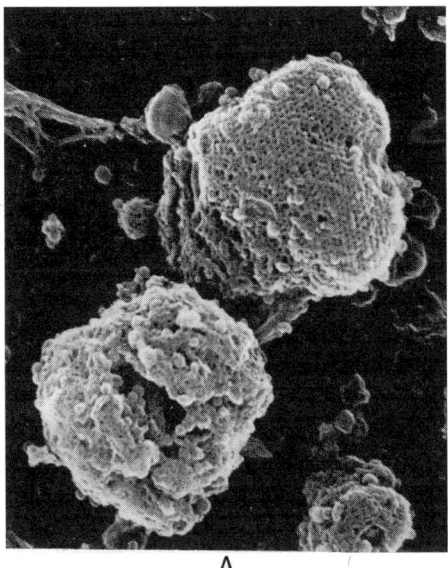

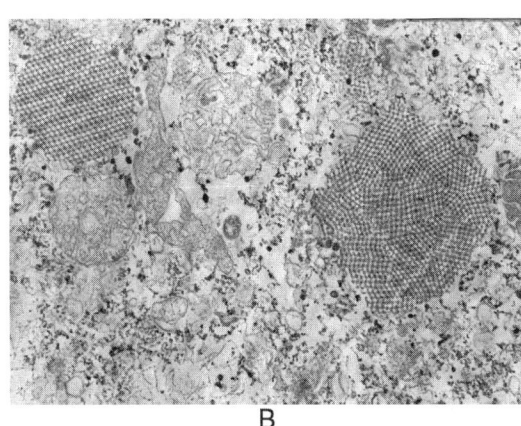

FIGURE 25.3 Electron micrographs of re-formed PLBs from 7-day-old wheat seedlings irradiated for 4 h with white light with intensity of 4 W/m^2 and redarkened by keeping in darkness for 16 h. (A) Scanning micrograph of isolated re-formed PLBs. A bottom-loaded 1% to 50% sucrose gradient was used for isolation. (B) Transmission micrographs of a cross-section of isolated re-formed PLBs.

certain pattern. The changes in the fluorescence spectra can give us information about the structures and the processes in the membranes [262]. Here, we applied double-membrane fluorescence probes to examine POR organization and its changes after brief irradiation in PLBs isolated from etiolated plants (oPLBs) and from plants treated with 10 μM norflurazone and therefore lacking carotenoids (cdPLBs). The localization of the fluorescent probe 1,8-ANS (Mg salt of 1-aniline-8-naphthalene sulfonate) in membranes is known — the probe partly binds to the membrane proteins and partly integrates with the membrane lipids at the level of their polar heads [263]. The nonpolar hydrophobic probe pyrene is localized in the fatty acid region of the membranes [264]. The energy transfer from Trp residues of the membrane proteins to ANS and pyrene can be used to obtain information about the protein localization in membranes [265]. This approach is very promising in the study of PLBs, where the only main protein is POR (90% of the total proteins) [144,204]. According to Birve et al. [209] POR is anchored to the lipid phase with β-sheets or α helices, which contain the Trp residues of the molecule. Therefore, by determination of Trp localization one can estimate the localization of the whole POR molecule. The energy transfer was measured by quenching of tryptophan fluorescence at 328 nm after excitation at 294 nm. The energy transfer occurs if the distance between Trp and the probe is less than 1.2 times the Förster radius (R_0), equal to

2.6 nm for the pair Trp–ANS and 2.8 nm for the pair Trp–pyrene [265]. Based on the distance between probes and Trp the latter could be divided into two parts — accessible for quenched β and nonquenched $(1 - β)$ by the probe. Dobretsov et al. [265] described a method for calculating β:

$$\beta = a/[F_0/(F_0 - F)]_{\min}$$

where a is a constant equal either to 0.75, if the acceptors are in the volume of the lipid phase (pyrene), or to 0.65, if they are on the surface (ANS) [266]. The $[F_0/(F_0-F)]_{\min}$ values were obtained by using the modified Stern–Folmer plot [267]. They represent the maximum possible quenching of tryptophan fluorescence in the largest concentration (infinite) of the acceptor (the probe).

The distances between the Trp and probes were calculated using the equations given by Dobretsov and coworkers [265,266]. First, the percentage of Trp accessible for quenching by the surface localized 1,8-ANS ($β$) was calculated by the method described above [266]. The relative localization and average distance (X) between Trp and the probe was determined according to the following equation [265]:

$$F/F_0 = (1 - \beta) + \beta \exp(-\alpha Ca B(X) Na\pi R_0^2 / S)$$

where F is the fluorescence intensity of the pyrene monomers in the presence of ANS in concentration

Ca; β has been previously determined; α represents the orientation of donor and acceptor molecules and is equal to 1.35 [265]; S is the estimated total membrane surface; and $B(X)$ is a function describing the relation between the quenching rate and the ratio X/R_0. After determination of $B(X)$ using the descriptions given by Dobretsov et al. [265] the X values were calculated. They represent the average distance between Trp and ANS molecules. The complete description of that function was given by Dobretsov and coworkers [265,266]:

$$X = \int\limits_{+1.2R_0}^{0} XN(X)\mathrm{d}X \Big/ \int\limits_{+1.2R_0}^{0} N(X)\mathrm{d}X$$

The same calculations were used to determine the average distance between Trp and pyrene, but the function in this case is:

$$X = \int\limits_{+1.2R_0}^{-1.2R_0} XN(X)\mathrm{d}X \Big/ \int\limits_{+1.2R_0}^{-1.2R_0} N(X)\mathrm{d}X$$

where $N(X)$ is the amount of donor molecules (Trp) separated by distance X from the acceptor molecule (probe).

On the other hand, the quenching of pyrene monomers by ANS was used to determine the average distance between the two probes.

We found that in nonirradiated cdPLBs and oPLBs the average distances between Trp and ANS were similar: -0.5 and 0.3 nm, respectively. The average distances between Trp and pyrene were -1.2 nm in both cdPLBs and oPLBs. The negative value means that the Trp is situated beyond a border, which cannot be crossed by pyrene molecules due to thermodynamical limitations. In the case of biomembranes hydrophobic pyrene molecules are situated in the fatty acid region of lipid phase and have very limited ability to cross beyond the glycerol backbone of the membrane lipids. Therefore, the distance -1.2 nm means that the Trp residues are situated above the glycerol backbone, within the area of lipid polar heads.

The average distance between pyrene monomers and ANS was 2.1 nm.

The flash irradiation did not change significantly the distances between pyrene monomers and ANS — it increased to 2.4 nm in oPLB and 2.3 nm in cdPLB (Figure 25.4). In oPLBs the distance between Trp and pyrene increased from -1.2 to -1.9 nm and between Trp and ANS from 0.3 to 1.0 nm. This probably means that Trp residues (resp. POR) were relocalized closer to the surface but still remain within the polar

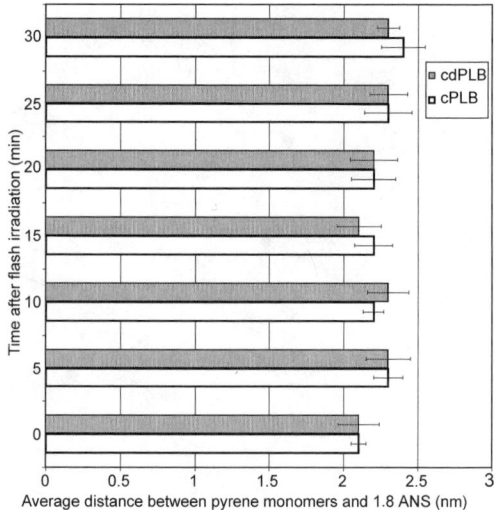

FIGURE 25.4 Light-induced changes in average distances between pyrene and 1,8-ANS probes in isolated cdPLB and oPLB. The distances were calculated by fluorescence of pyrene monomers at 390 nm (excitation 330 nm) in the presence of different concentrations of 1,8-ANS.

head of the lipids (Figure 25.5A). Irradiation caused drastic changes in the localization of Trp in cdPLBs: the distance between Trp and pyrene increased from -1.2 to -4.2 nm and between Trp and ANS from 0.5 to 2.1 nm (Figure 25.5B). Such distances are equal to the thickness of the lipid monolayer and could be an indication that a substantial part of Trp residues move closer to the membrane surface or even leave the lipid phase.

It is known that POR, together with NADPH and Pchlide, is organized in large aggregates and that the irradiation triggers dissociation of the aggregates into smaller units [20,34,35,39]. Our results suggest that the dissociation of POR complexes in cdPLB is accompanied by relocalization of POR closer to the lipid surface or separation from the lipids, a process different from that registered in oPLBs [268]. Taking into account all these findings, we consider carotenoids as an important factor for the stable association of POR to the lipid phase of PLBs.

We also used fluorescent probes in combination with large unilamellar vesicles (LUVETs) to study the interactions of *in vitro* synthesized POR with lipids and pigment [269].

We have developed an *in vitro* system based on LUVETs containing fluorescent probes to study early interactions between POR and membrane lipids. The great advantage of LUVETs in comparison with isolated plastids and plastid membranes is that by using the miniextruder of Avestin Ltd. (Canada) we can produce LUVETs of the desired size (100 to 400 nm)

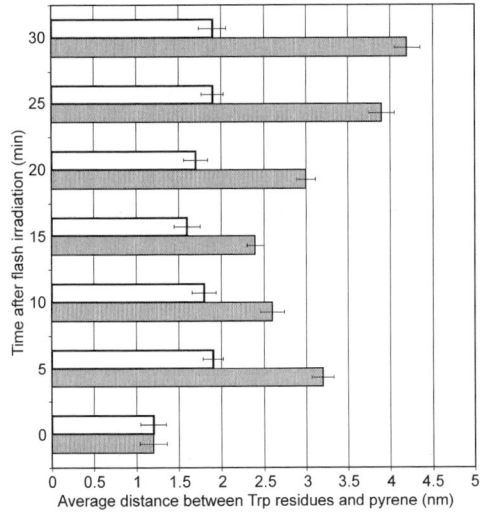
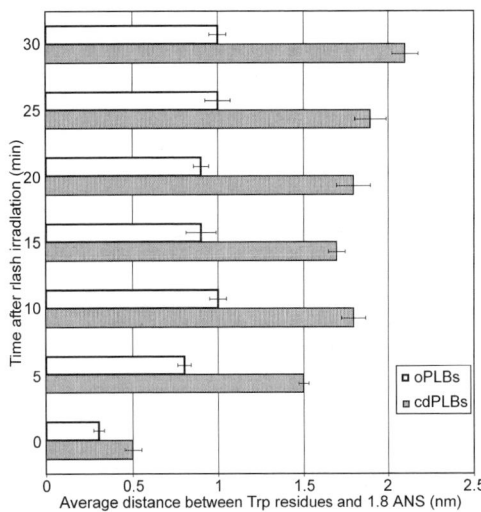

FIGURE 25.5 Light-induced changes in average distances between tryptophan residues in membrane proteins of cdPLB and oPLB and fluorescent probes pyrene (A) and 1,8-ANS (B). The quenching of tryptophan fluorescence at 330 nm (excitation 290 nm) in the presence of different concentrations of fluorescent probes was used to calculate the distances.

and composition of lipid, pigments, and proteins. The presence of membrane fluorescent probes allows us to study localization of proteins, organization and dynamic properties of lipid phase, etc. We used these artificial membranes to study the interaction of *in vitro* synthesized pea pPOR and mPOR with LUTETs. The LUVETs were prepared by the method of Mayer et al. [270] and MacDonald et al. [271], and the proteins were synthesized and added to LUVETs following the assembly protocol of Dahlin et al. [197].

First, we studied whether the LUVETs remain intact during all the treatments included in the integration protocol. With this aim we loaded 1 μM 6-carboxy fluorecein within the LUVETs. This probe has a high level of self-quenching. The loaded concentration is not high enough to get complete self-quenching, but it is enough for registration of any probe leakage through the bilayer. Samples were taken at any one of the critical steps of the assembly protocol. The results indicate no leakage and no destruction of LUVETs during the experimental steps (Figure 25.6). Next, the specificity of the LUVET–POR interactions was studied. The assembly reaction was run only with translation mixture, followed by thermolysin treatment. The results clearly indicate that water-soluble proteins cannot remain attached to LUVETs after the washing procedures (Figure 25.7).

The localization of *in vitro* synthesized protein after assembly reaction and thermolysin treatment was examined by two independent methods:

1. Using electrophoresis followed by fluorography — the fluorograms showed that after the

assembly reaction all systems contained radiolabeled proteins. Thermolysin treatment removed all proteins, which indicated that they have been attached to the LUVET surface without a penetrating bilayer.
2. Using the energy transfer from Trp to probes, we calculated the average distance between them.

The two-membrane fluorescence probes, 1,8-ANS and pyrene, were used to study the membrane localization of POR. The energy transfer from Trp to probes was used to determine the localization of proteins.

The results obtained allowed us to draw the following conclusions:

1. LUVETs are a suitable artificial system for studying the interactions between POR and galatolipids.
2. The association of the proteins to LUVETs shifts the maximum of tryptophan fluorescence from 345 to 330 nm. This shift is an indication for more hydrophobic surroundings of tryptophan residues.
3. The quenching accessibility of tryptophan residues for probes shows that they are situated on the level of the lipids' polar heads in all variants.
4. The amount of proteins associated to galactolipid LUVETs is higher than that associated to the phospholipid LUVETs and is proportional to the presence of nonbilayer monogalactosyl diacylglycerol.

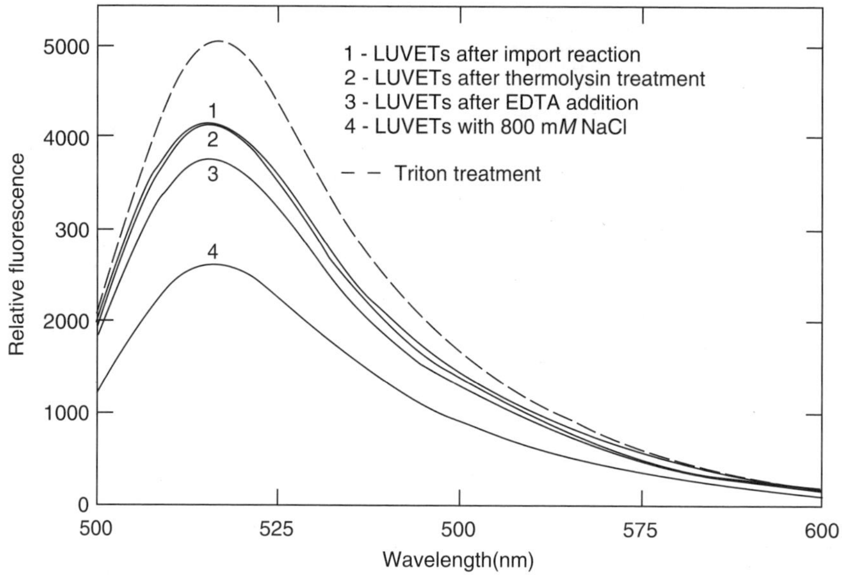

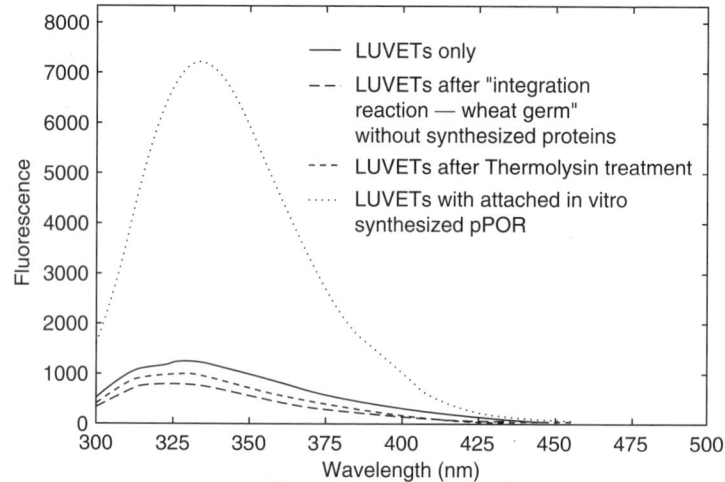

FIGURE 25.6 Changes in fluorescence of 6-carboxyfluorescein (6-cf.) trapped in LUVETs during different steps of protein assembly reactions and postassembly treatments according to the assembly protocol given by Dahlin et al. [197]. The release of 6-cf. with Triton X-100 treatment proves that the LUVETs remain intact through all the steps of the assembly protocol.

FIGURE 25.7 Tryptophan fluorescence of LUVETs measured at 330 nm (excitation 290 nm). The assembly reaction with LUVETs was run with translation mixture only, followed by thermolysin treatment. The lack of Trp fluorescence in all other variants compared with LUVETs to which pPOR was associated indicates that water-soluble proteins cannot remain attached to LUVETs after the washing procedures.

5. The precursor proteins are found deeper in the zone of the lipids' polar heads than mature PORA, probably due to their transit sequences.

6. Thermolysin treatment destroys all proteins but Trp-containing fragments remain anchored to the lipids.

7. By incorporation of Pchlide into the LUVETs we stimulated the association of POR to the lipid phase. We assumed that the effect is specific for POR and is based on conformational changes in POR protein after binding of its substrate — 2 Pchlide. The presence of Pchlide did not influence the association of pSS to the LUVETs.

8. Carotenoids have a much higher effect on protein–lipid interactions than was considered before. We assume that this effect is not specific

and is limited only to POR because the proteins with transit peptides (pPOR and pSS) were associated deeper than the mature form of POR. The presence of more MGDG increases this effect. Probably, the presence of carotenoids and nonbilayer MGDG influences the membrane geometry of the lipid bilayer and facilitate the lipid–protein interactions. Our data are in agreement with the recent results of Park et al. [49]. They studied the crtISO mutants of *Arabidopsis*, which accumulate acyclic poly-*cis*-carotenes in place of cyclic all-*trans*-carotenes and lack PLBs. They [49] proposed that the role of all-*trans*-carotenoids in PLB assembly may be to stabilize or facilitate the curved membranes that form as a result of interaction between membranes, POR:Pchlide, and carotenoids.

REFERENCES

1. Leech RM. Chloroplast development in angiosperm: Current knowledge and future prospects. In: *Chloroplast Biogenesis* (Baker NR, Barber J eds, Elsevier, Amsterdam), 1984: 1–21.

2. Whatley JM. Variation in the basic pathway of chloroplast development. *New Phytol.* 1977; 78: 407–420.

3. Khandakar K, Bradbeer JW. Primary leaf growth in bean (*Phaseolus vulgaris*) II. Cell and plastid development during growth in darkness and after transfer to illumination at various stages of dark growth. *Bangladesh J. Bot.* 1988; 17: 173–188.

4. Ryberg H, Axelsson L, Widell KO, Virgin HI. Chlorophyll b accumulation and grana formation in low intensity of red light. *Physiol. Plant.* 1980; 49: 431.

5. Whatley JM. Plastid growth and division in *Phaseolus vulgaris*. *New Phytol.* 1980; 86: 1–16.

6. Douce R, Joyard J. The regulatory role of the plastid envelope during development. In: *Topics in Photosynthesis, vol V: Chloroplast Biogenesis* (Baker NR, Barber J eds, Elsevier Science Publishers, Amsterdam), 1984: 72–132.

7. Heslop-Harrison J. Structure and morphogenesis of lamellar system in grana-containing chloroplasts. Membrane structure and lamellar architecture. *Planta* 1963; 60: 243–260.

8. Thorne SW, Duniec JT. The physical principles of energy transduction in chloroplast thylakoid membranes. *Q. Rev. Biophys.* 1983; 16: 197–278.

9. Arvidsson PO, Sundby C. A model for the topology of the chloroplast thylakoid membrane. *Aust. J. Plant Physiol.* 1999; 26: 687–694.

10. Albertsson PA. The structure and function of the chloroplast photosynthetic membrane — a mode for the domain organization. *Photosynth. Res.* 1995; 4: 141–149.

11. Bradbeer JW, Gyldenholm AO, Irelend HMM, Smith JW, Rest J, Edge HJW. Plastid development in primary leaves of *Phaseolus vulgaris*. VIII. The effect of the transfer of dark-brown plants to continuous illumination. *New Physiol.* 1974; 73: 271–279.

12. Schöch S, Lempert U, Rudiger,W. Uber die letzen Stufen der Chlorophyll-Biosynthese Zwischenprodukte und zwischen Chlorophyllid und phytolhaltigen Chlorophyll. *Z. Pflanzenphysiol.* 1977; 83: 427–436.

13. Shioi Y, Sasa T. Separation of protochlorophylls esterified with different alcohols from inner seed coats of three Cucurbitaceae. *Plant Cell Physiol.* 1982; 23: 1315.

14. Shioi Y, Sasa T. Formation and degradation of protochlorophylls in etiolated and greening cotyledons of cucumber. *Plant Cell Physiol.* 1983; 24: 835.

15. Rebeiz CA, Wu SM, Kuhadja M, Daniel H, Perkins EJ. Chlorophyll b biosynthetic routes and chlorophyll a chemical heterogeneity in plants. *Mol. Cell. Biochem.* 1983; 57: 97–125.

16. Gunning BES. Membrane geometry of "open" prolamellar bodies. *Protoplasma* 2001; 215(1–4): 4–15.

17. Lütz C. On the significance of prolamellar bodies in membrane development of etioplasts. *Protoplasma* 1981; 108: 99–115.

18. Bergweiler P, Röper U, Lütz C. The development and ageing of membranes from etioplasts of Avena sativa. *Physiol. Plant.* 1984; 60: 395–400.

19. Frances S, White MJ, Edgerton MD, Jones AM, Elliott RC, Thompson WF. Initial characterization of a pea mutant with light-independent photomorphogenesis. *Plant Cell* 1992; 4: 12: 1519–1530.

20. Sundqvist C, Dahlin C. With chlorophyll pigments from prolamellar bodies to light-harvesting complexes. *Physiol. Plant.* 1997; 100(4): 748–759.

21. Franck F, Schoefs B, Barthelemy X, Mysliwa KB, Strzalka K, Popovic R. Protection of native chlorophyll(ide) forms and of photosystem II against photodamage during early stages of chloroplast differentiation. *Acta Physiol. Plant.* 1995; 17: 123–132.

22. Lebedev N, Van Cleve B, Armstrong GA, Apel K. Chlorophyll synthesis in deetiolated (det 340) mutant of *Arabidopsis* without NADPH-protochlorophyllide (Pchlide) oxidoreductase (POR) A and photoactive Pchlide-P655. *Plant Cell* 1995; 7(12): 2081–2090.

23. Reinbothe S, Reinbothe C, Apel K, Lebedev N. Evolution of chlorophyll biosynthesis — the challenge to survive photooxidation. *Cell* 1996; 86: 703–705.

24. Reinbothe S, Reinbothe C, Neumann D, Apel K. A plastid enzyme arrested in the step of precursor translocation *in vivo*. *Proc. Natl. Acad. Sci.* 1996; 93: 12026–12030.

25. Sperling U, Van Cleve B, Frick G, Apel K, Armstrong GA. Overexpression of light-dependent PORA or PORB in plants depleted of endogenous POR by far-red light enhances seedling survival in white light and protects against photooxidative damage. *Plant J.* 1997; 12(3): 649–658.

26. Sperling U, Frick G, van Cleve B, Apel K, Armstrong GA. Pigment-protein complexes, plastid development and photooxidative protection: the effects of PORA and PORB overexpression of *Arabidopsis* seedlings shifted from far-red to white light. In: *The Chloroplast: From Molecular Biology to Biotechnology* (Argyroudi-Akoyonoglou JH, Senger H eds, Kluwer Academic Publishers, Dordrecht), 1999: 97–102.

27. von Wettstein D, Gough S, Kannangara CG. Chlorophyll biosynthesis. *Plant Cell* 1995; 7: 1039–1057.

28. Lebedev N, Timko MP. Protochlorophyllide photoreduction. *Photosynth. Res.* 1998; 58(1): 5–23.

29. Bruce BD. The role of lipids in plastid protein transport. *Plant Mol. Biol.* 1998; 38(1–2): 223–246.

30. Ryberg M, Sandelius AS, Selstam E. Lipid composition of prolamellar bodies and prothylakoids of wheat etioplasts. *Physiol. Plant.* 1983; 57: 555–560.

31. Selstam E, Widell-Wigge A. Chloroplast lipids and the assembly of membranes. In: *Pigment-Protein Complexes in Plastid: Synthesis and Assembly* (Sundqvist C, Ryberg M eds, Academic Press, New York), 1993: 241–277.

32. Rilfors L, Lindblom G, Wieslander A, Christiansson A. Lipid bilayer stability in biological membranes. In: *Membrane Fluidity* (Kates M, Manson LA eds, Plenum Press, New York), 1984: 205–245.

33. Lindblom G, Rilfors L. Cubic phases and isotropic structures formed by membrane lipids — possible biological relevance. *Biochim. Biophys. Acta* 1989; 988: 221–256.

34. Apel K, Santel HJ, Redlinger TE, Falk H. The Pchlide holochrome of barley (*Hordeum vulgare* L.). Isolation and characterization of the NADPH:protochloriphyllide oxidoreductase. *Eur. J. Biochem.* 1980; 111: 251–258.

35. Griffiths WT. Protochlorophyllide photoreduction. In: *Chlorophylls* (Scheer H ed, CRC Press, Boca Raton, FL), 1991: 443–449.

36. Armstrong GA, Runge S, Frick G., Sperling U, Apel K. Identification of NADPH:protochlorophyllide oxidoreductases A and B: a branched pathway for light-dependent chlorophyll biosynthesis in *Arabidopsis thaliana*. *Plant Physiol.* 1995; 108: 1505–1517.

37. Holtorf H, Reinbothe S, Reinbothe C, Bereza B, Apel K. Two routes of chlorophyllide synthesis that are differentially regulated by light in barley (*Hordeum vulgare* L.). *Proc. Natl. Acad. Sci.* 1995; 92: 3254–3258.

38. Oosawa N, Masuda T, Awai K, Fusada N, Shimada H, Ohta H, Takamiya KH. Identification and light-induced expression of a novel gene of NADPH-protochlorophyllide oxidoreductase isoform in *Arabidopsis thaliana*. *FEBS Lett.* 2000; 474: 133–136.

39. Armstrong GA, Apel K, Rüdiger W. Does a light-harvesting protochlorophyllide a/b-binding protein complex exist? *Trends Plant Sci.* 2000; 5: 40–44.

40. Franck F, Sperling U, Frick G, Pochert B, van Cleve B, Apel K, Armstrong GA. Regulation of etioplast pigment-protein complexes, inner membrane architecture, and protochlorophyllide a chemical heterogeneity by light-dependent NADPH-protochlorophyllide oxidoreductase. *Plant Physiol.* 2000; 124: 1678–1696.

41. Ryberg M, Dehesh K. Localization of NADPH-protochlorophyllide oxidoreductase in dark-grown wheat (*Triticum aestivum*) by immuno-electron microscopy before and after transformation of the prolamellar bodies. *Physiol. Plant.* 1986; 66: 616–624

42. Artus NN, Ryberg M, Lindsten A, Ryberg H, Sundqvist C. The shibata shift and the transformation of etioplasts to chloroplasts in wheat with clomazone fmc 57020 and amiprophos-methyl Tokunol M. *Plant Physiol.* 1992; 98: 253–263.

43. Wiktorsson B, Engdahl S, Zhong LB, Boddi B, Ryberg M, Sundquvist C. The effect of cross-linking of the subunits on NADPH-protochlorophyllide oxidoreductase on the aggregational state of protochlorophyllide. *Photosynthetica* 1993; 29(2): 205–218.

44. Williams W P, Selstam E, Brain T. X-ray diffraction studies of the structural organisation of prolamellar bodies isolated from *Zea mays*. *FEBS Lett.* 1998; 422(2): 252–254.

45. Wiktorsson B, Ryberg M, Sundqvist C Aggregation of NADPH-protochlorophyllide oxidoreductase-pigment complexes is favoured by protein phosphorylation. *Plant Physiol. Biochem. Paris* 1996; 34(1): 23–34.

46. Younis S, Ryberg M, Sundqvist C. Plastid development in germinating wheat (*Triticum aestivum*) is enhanced by gibberellic acid and delayed by gabaculine. *Physiol. Plant.* 1995; 3: 336–346.

47. Dahlin C, Aronsson H, Wilks HM, Lebedev N, Sundqvist C, Timko MP. The role of protein surface charge in catalytic activity and chloroplast membrane association of the pea NADPH:protochlorophyllide oxidoreductase (POR) as revealed by alanine scanning mutagenesis. *Plant Mol. Biol.* 1999; 39(2): 309–323.

48. Chahdi MA, Schoefs B, Franck F. Isolation and characterization of photoactive complexes of NADPH:-protochlorophyllide oxidoreductase from wheat. *Planta* 1998; 206: 673–680.

49. Park H, Kreunen SS; Cuttriss AJ, DellaPenna D, Pogson BJ. Identification of the carotenoid isomerase provides insight into carotenoid biosynthesis, prolamellar body formation, and photomorphogenesis. *Plant Cell* 2002; 14(2): 321–332.

50. Smith BB, Rebeiz CA. Chloroplast biogenesis XXIV. Intrachloroplastic localization of the biosynthesis and accumulation of protoporphyrin IX, magnesium-protoporphyrin monoester, longer wavelength metalloporphyrins during greening. *Plant Physiol.* 1979; 63: 227.

51. Rascio N. Chloroplast differentiation. In: *Cell Interaction and Differentiation* (Ghiara G ed, University of Naples, Naples), 1988: 159.

52. Casadoro G, Rascio N, Pagiusco M, Ravagnan N. Flowers of *Orontium aquaticum* L.: membrane rearrangement in chloroplast-chromoplast interconversions. *J. Ultrastruct. Res.* 1982; 81: 202.

53. Finer JJ, Smith RH. Structure and development of plastids in epidermal cells of African violet Saintpaulia-ionantha cultivar marge-winters in culture. *Ann. Bot.* 1983; 51(6): 691–696.

54. Rao KS and Dave YS. Ultrastructure of active and dormant cambial cells in teak tectona-grandis. *New Phytol.* 1983; 93(3): 447–456.

55. Liljenberg C. The structural organization of chloroplast lipids *in vivo* and in model systems: some aspects. In: *Biogenesis and Function of Plant Lipids* (Mazliak P, Benveniste P, Costes C, Douce R eds, Elsevier, Amsterdam), 1980: 29–38.

56. Virgin HI, Kahn A, Wettstein D. The physiology of Chl formation in relation to structural changes in chloroplasts. *Photochem. Photobiol.* 1963; 2: 83–91.

57. Schulz R, Senger H. Protochlorophyllide reductase: a key enzyme in the greening process. In: *Pigment-Protein Complexes in Plastid: Synthesis and Assembly* (Sundqvist C, Ryberg M eds, Academic Press, New York), 1993: 179–218.

58. Franck F, Strzalka K. Detection of the protochlorophyllide–protein complex in the light during the greening of barley. *FEBS Lett.* 1992; 309: 73–77.

59. Cohen CE, Rebeiz CA. Chloroplast biogenesis 34. Spectrofluorometric characterization *in situ* of the protochlorophyll species in etiolated tissues of higher plants. *Plant Physiol.* 1981; 67: 98.

60. Ryberg M, Sundqvist C. Characterization of prolamellar bodies and prothylakoids fractionated from wheat etioplasts. *Physiol. Plant.* 1982; 56: 125–132.

61. Ryberg M, Sundqvist C. Spectral forms of protochlorophyllide in prolamellar bodies and prothylakoids fractionated from wheat etioplasts. *Physiol. Plant.* 1982; 56: 133–138.

62. Böddi B, Ryberg M, Sundqvist C. Identification of four universal protochlorophyllide forms in dark-grown leaves by analyses of the 77 K fluorescence emission spectra. *J. Photochem. Photobiol. B* 1992; 12: 389–401.

63. Böddi B; Kis-Petik K; Kaposi AD; Fidy J, Sundqvist C. The two spectroscopically different short wavelength protochlorophyllide forms in pea epicotyls are both monomeric. *Biochim. Biophys. Acta Bioenerg.* 1998; 1365(3): 531–540.

64. Kis-Petik K, Boddi B, Kaposi AD, Fidy J. Protochlorophyllide forms and energy transfer in dark-grown wheat leaves. Studies by conventional and laser excited fluorescence spectroscopy between 10K-100K. *Photosynth. Res.* 1999; 60: 87–98.

65. Kahn A, Boardman NK, Thorne SW. Energy transfer between protochlorophyllide molecules: evidence for multiple chromophores in the photoactive protochlorophyllide-protein complex *in vivo* and *in vitro*. *J. Mol. Biol.* 1970; 48: 85–101.

66. Böddi B, Lindsten A, Ryberg M, Sundqvist C. The aggregational state of protochlorophyllide in isolated prolamellar bodies. *Physiol. Plant.* 1989; 76: 135–143.

67. Franck F, Barthelemy X, Strzalka K. Spectroscopic characterization of protochlorophyllide photoreduction in the greening leaf. *Photosynthetica* 1993; 29: 185–194.

68. Wiktorsson B, Ryberg M, Gough S, Sundqvist C. Isoelectric focusing of pigment-protein complexes solubilized from non-irradiated and irradiated prolamellar bodies. *Physiol. Plant.* 1992; 85(4): 659–669.

69. Böddi B, McEwen B, Ryberg M, Sundqvist C. Protochlorophyllide forms in non-greening epicotyls of dark-grown pea (*Pisum sativum*). *Physiol. Plant.* 1994; 92(1): 160–170.

70. Ignatov NV, Litvin FF. Light-regulated pigment interconversion in pheophytin/chlorophyll-containing complexes formed during plant leaves greening. *Photosynth. Res.* 1995; 46: 445–453.

71. Franck F, Mathis P. A short-lived intermediate in the photoenzymatic reduction of protochlorophyll(ide) into chlorophyll(ide) at a physiological temperature. *Photochem. Photobiol.* 1980; 32: 799–803.

72. Carey EE, Rebeiz CA. Chloroplast biogenesis. 49. Differences among angiosperms in the biosynthesis and accumulation of monovinyl and divinyl protochlorophyllide during photoperiodic greening. *Plant Physiol.* 1985; 79: 1–9.

73. Kotzabasis K, Miyachi S, Sendger H. Influence of calcium on formation and reduction of protochlorophyllide in the pigment mutant C-2A′ of *Scenedesmus obliquus*. *Plant Cell Physiol.* 1990; 31: 419–422.

74. Virgin HI. Protochlorophyll formation and greening in etiolated barley leaves. *Physiol. Plant.* 1955; 8: 630–643.

75. Inoue Y, Kobayashi T, Ogawa T, Shibata K. A short intermediate in the photoconversion of protochlorophyllide to chlorophyllide a. *Plant Cell Physiol.* 1981; 22(2): 197.

76. Dujardin E, Franck F, Gysemberg R, Sironval C. The protochlorophyllide chlorophyllide cycle and photosynthesis. *Photochem. Photophys.* 1986; 12: 97–105.

77. Sironval C, Kuyper Y. The reduction of protochlorophyllide and chlorophyllide IV. The nature of the intermediate P688-676. *Photosynthetica* 1972; 6: 254–275.

78. Dujardin E, Correia M. Long-wavelength absorbing pigment-protein complexes as fluorescence quenchers in etiolated leaves illuminated in liquid nitrogen. *Photobiochem. Photobiophys.* 1979; 1: 25–32.

79. Belyaeva OB, Litvin FF. Primary reactions of protochlorophyllide into chlorophyllide phototransformation at 77 K. *Photosynthetica* 1981; 15: 210–215.

80. Litvin FF, Ignatov NV, Belyaeva OB. Photoreversibility of transformation of protochlorophyllide into chlorophyllide. *Photobiochem. Photobiophys.* 1981; 2: 233.

81. Belyaeva OB, Timofeev KN, Litvin FF. The primary reactions in the protochlorophyll(ide) photoreduction as investigated by optical and ESR spectroscopy. *Photosynth. Res.* 1988; 15: 247–256.

82. Belyaeva OB, Ignatov NV, Litvin FF. Investigation of the primary intermediates of chlorophyllide photobiosynthesis at low temperature. In: *Research in Photosynthesis* (Murata N ed, Kluwer Academic Publishers, Dordrecht), Proceedings of IXICP, 1992: 75–78.

83. Ignatov NV, Belyaeva OB, Litvin FF. Possible role of a flavin component of protochlorophyllide-protein complexes in primary processes of protochlorophyllide photoreduction in etiolated plants. *Photosynthetica* 1993; 29: 235–241.

84. Belyaeva OB, Litvin FF. New intermediate reactions in the process of protochlorophyllide photoreduction. *Biofizika* 1980; 25: 617–623.

85. Schoefs B, Franck F. Photoreduction of protochlorophyllide to chlorophyllide in 2-d-old dark-grown bean (*Phaseolus vulgaris* cv. Commodore) leaves: comparison with 10-d-old dark-grown (etiolated) leaves. *J. Exp. Bot.* 1993; 44: 1053–1057.

86. Oliver RP, Griffiths WT. Pigment-protein complexes in illuminated etiolated leaves. *Plant Physiol.* 1982; 70: 1019–1025.

87. El Hamouri B, Sironval C. NADP+/NADPH control of the protochlorophyllide-chlorophyllide proteins in cucumber etioplasts. *Photobiochem. Photobiophys.* 1980; 1: 219–223.

88. Ignatov NV, Litvin FF. Photoinduced formation of pheophytin/chlorophyll-containing complexes during the greening of plant leaves. *Photosynth. Res.* 1994; 42: 27–35.

89. Böddi B, Ryberg M, Sundqvist C. The formation of a short-wavelength chlorophyllide form at partial

phototransformation of protochlorophyllide in etioplast inner membranes. *Photochem. Photobiol.* 1991; 53: 667–673.

90. Shibata K. Spectroscopic studies on chlorophyll formation in intact leaves. *J. Biochem. (Tokyo)* 1957; 44: 147–173.

91. Henningsen KW, Boynton JE. Macromolecular physiology of plastids. VII. Pigment and membrane formation in plastids of barley greening under low light intensity. *J Cell Biol.* 1970; 44: 290–304.

92. Sironval C, Michel-Wolwertz MR, Madsen A. On the nature and possible functions of the 673- and 684-nm forms *in vivo* of chlorophyll. *Biochem. Biophys. Acta* 1065; 94: 344.

93. Henningsen KW, Thorne SW. Esterification and spectral shifts of chlorophyll(ide) in wild type and mutant seedlings developed in darkness. *Plant Physiol.* 1974; 30: 82.

94. Belyaeva OB, Sundqvist C. Comparative investigation of the appearance of primary chlorophyllide forms in etiolated leaves, prolamellar bodies and prothylakoids. *Photosynth. Res.* 1998; 55(1): 41–48.

95. Griffiths WT, Oliver RP, Kay SA. A critical appraisal of the role and regulation of NADPH-protochlorophyllide oxidoreductase in greening plants. In: *Protochlorophyllide Reduction and Greening* (Sironval C, Brouers M eds, Martinus Nijhoff, The Hague), 1984: 19–29.

96. Klockare B, Virgin HI. Chlorophyll(ide) forms after partial phototransformation of protochlorophyll(ide) in etiolated wheat leaves. *Physiol. Plant.* 1983; 57: 28.

97. Ogawa M, Konishi M. Effects of illumination on absorption peak shifts in spectra of intact etiolated cotyledons of *Pharbitis nil*. I. Existence of two kinds of shift patterns. *Plant Cell Physiol.* 1977; 18: 303.

98. Franck F, Inoe Y. Light-driven reversible transformation of chlorophyllide P$_{696,682}$ into chlorophyllide P$_{688,678}$ in illuminated etiolated bean leaves. *Photochem. Photobiophys.* 1984; 8: 85.

99. Ikeuchi M, Inoue Y, Murakami S. Characterization of two forms of Pchlid (Pchlid 650 and 640) in Pchlid holochrome solubilized with detergents from squash etioplasts membranes. In: *Advances in Photosynthesis Research* (Sybesma C ed, Martinus Nijhoff, The Hague), 1984.

100. Mapleston RE, Griffiths TW. Effects of illumination of etiolated leaves on the redox state of NADP in the plastids. *FEBS Lett.* 1978; 92: 168–172.

101. El Hamouri B, Sironval C. Pigment-protein complexes and NADPH binding peptides of two etioplast membrane fractions. *Photobiochem. Photobiophys.* 1983; 2: 263–272.

102. Minkov IN, Sundqvist C, Ryberg M. Spectral changes of chlorophyll pigments in irradiated reformed prolamellar bodies isolated from wheat etiochloroplasts. *Photosynthetica* 1989; 23(3): 306–313.

103. Ryberg M, Sundqvist C The regular ultrastructure of isolated prolamellar bodies depends on the presence of membrane-bound NADPH-protochlorophyllide oxidoreductase. *Physiol. Plant.* 1988; 73: 218–226.

104. Ryberg M and Minkov I. Characteristics of in-vivo reformed prolamellar bodies after isolation. In: *Advances in Photosynthesis Research*, Vol. IV (Sybesma C ed, Martinus Nijhoff, The Hague), 1984: 633.

105. Minkov IN, Ryberg M, Sundqvist C. Properties of reformed prolamellar bodies from illuminated and redarkened etiolated wheat plants. *Physiol. Plant.* 1988; 72: 725–732.

106. Wolf JB, Price L. Terminal steps of chlorophyll a biosynthesis in higher plants. *Arch. Biochim. Biophys.* 1957; 72: 293.

107. Madsen A. On the formation of chlorophyll and initiation of photosynthesis in etiolated plants. *Photochem. Photobiol.* 1963; 2: 93–100.

108. Akoyunoglou G, Siegelman HW. Protochlorophyll(ide) resynthesis in dark-grown bean leaves. *Plant Physiol.* 1968; 43(1): 66–68.

109. Smith JHC. Processes accompanying chlorophyll formation. In: *Photosynthesis in Plants* (Franck J, Lomis WE eds, Iowa State College Press, Ames, IA), 1949: 209.

110. Rudoi AB. Final stages of chlorophyll formation. In: *Biogenesis of the Pigment Apparatus of Photosynthesis* (Litvin FF ed, Nauka i Tekhnika, Minsk), 1988: 78–103 (in Russian).

111. Virgin HI. Studies on the formation of protochlorophyll and chlorophyll under varying light treatments. *Physiol. Plant.* 1958; 11: 347.

112. Smith JHC, Young VMK. Chlorophyll formation and accumulation in plants. In: *Radiation Biology* (Hollaender A ed, McGraw-Hill, New York), 1965; 3: 393.

113. Nadler K, Granick S. Controls on chlorophyll synthesis in barley. *Plant Physiol.* 1970; 46(2): 240–246.

114. Sisler EC, Klein WH. The effect of age and various chemicals on the lag phase of chlorophyll synthesis in dark grown bean seedlings. *Physiol. Plant.* 1963; 16(2): 315.

115. Goedheer JC. Effect of changes in chlorophyll concentration on photosynthetic properties. I. Fluorescence and absorption of greening bean leaves. *Biochim. Biophys. Acta* 1961; 51(3): 494–504.

116. Brodersen P. Factors affecting the photo conversion of proto chlorophyllide to chlorophyllide in etioplast membranes isolated from barley. *Photosynthetica* 1976; 10(1): 33–39.

117. Frank H and Cogdell R. Carotenoids in photosynthesis. *Photochem. Photobiol.* 1996; 63(3): 257–264.

118. Lütke-Brinkhaus F, Kleinig H. Carotenoid and chlorophyll biosynthesis in isolated plastids from mustard seedling cotyledons (*Sinapis alba* L.) during etioplasts-chloroplast conversion. *Planta* 1987; 171(3): 121–129.

119. Bahl J. Chlorophyll, carotenoid, and lipid content in *Triticum sativum* L. Plastid envelopes, prolamellar bodies, stroma lamellae, and grana. *Planta* 1977; 136: 21–24.

120. Hyde S, Andersson S, Larsson K, Blum Z, Landh T, Lidin S, Ninham BW. *The Language of Shape: The Role of Curvature in Condensed Matter: Physics, Chemistry and Biology*. Elsevier Science Publishers, Amsterdam, 1997.

121. Yahubyan G, Minkov I, Sundqvist C. Carotenoid dependence of protochlorophyllide to chlorophyllide phototransformation in dark-grown wheat seedlings. *J. Photochem. Photobiol. B* 2002; 65: 171–176.

122. Koski VM, French CS, Smith JHC. The action spectrum for the transformation of protochlorophyll to chlorophyll *a* in normal and albino corn seedlings. *Arch. Biochem. Biophys.* 1951; 31(1): 1–17.

123. Santel HJ, Apel K. The protochlorophyllide holochrome of barley (*Hordeum vulgare* L.). The effect of light on the NADPH:protochlorophyllide oxidoreductase. *Eur. J. Biochem.* 1981; 120: 95–103.

124. Hoyer-Hansen G, Simpson DJ, Changes in the polypeptide composition of internal membranes of barley plastids during greening. *Carlsberg Res. Commun.* 1977; 42: 379.

125. Lütz C, Roper U, Beer NS, Griffiths T. Sub-etioplast localization of the enzyme NADPH protochlorophyllide oxidoreductase. *Eur. J. Biochem.* 1981; 118: 347–353.

126. Droppa M, Masojidek J, Horvath G. Changes of the polypeptide composition in thylakoid membranes during differentiation. *Z. Naturforsch. (C)* 1990; 45(3–4): 253–257.

127. Savchenko GE, Chajka MT. Biogenesis of photosynthetic membrane. In: *Biogenesis of the Pigment Apparatus of Photosynthesis* (Litvin FF ed, Nauka i Tekhnika, Minsk), 1988: 232–250 (in Russian).

128. Ohashi K, Murakami A, Tanaka A, Tsuji H, Fujita Y. Developmental changes in amounts of thylakoid components in plastids of barley leaves. *Plant Cell Physiol.* 1992; 33(4): 371–377.

129. Hampp R, De Filippis LF. Plastid protease activity and prolamellar body transformation during greening. *Plant Physiol.* 1980; 65: 663.

130. Dehesh K, Apel K. The function of proteases during the light-dependent transformation of etioplasts to chloroplasts in barley *Hordeum vulgare* L. *Planta* 1983; 157: 381.

131. Hauser I, Dehesh K, Apel K. The proteolytic degradation *in vitro* of the NADPH-protochlorophyllide oxidoreductase of barley (*Hordeum vulgare* L). *Arch. Biochem. Biophys.* 1984; 228: 577–586.

132. Walker CJ, Griffiths WT. Light independent proteolysis of protochlorophyllide reductase. In: *Regulation of Chloroplast Differentiation* (Akoyunoglou G, Senger H eds, Alan R. Liss Inc., New York, 1986: 99.

133. Honda T, Tanaka A, Tsuji H. Proteolytic activity in intact barley etioplasts: endoproteolysis of NADPH protochlorophyllide oxidoreductase protein. *Plant Sci.* 1994; 97: 129–135.

134. Wellburn AR. Distribution of chloroplast coupling factor cf-1 particles on plastid membranes during development. *Planta* 1977; 135(2): 191–198.

135. Shaw P, Henwood J, Oliver R, Griffits T. Immunogold localization of protochlorophyllide oxidoreductase in barley etioplasts. *Eur. J. Cell Biol.* 1985; 39: 50–55.

136. Lindsten A, Ryberg M, Sundqvist C. The polypeptide composition of highly purified prolamellar bodies and prothylakoids from wheat (*Triticum aestivum*) as revealed by silver staining. *Physiol. Plant.* 1988; 72: 167–176.

137. Franck F, Schmid GH. Interaction of hydroxylamine with the water-oxidizing complex at oxidation states S-1, S-2 and S-3 in etiochloroplasts of oat. *Biochim. Biophys. Acta* 1989; 977(2): 215–218.

138. Hashimoto A, Akasaka T, Yamamoto Y. Characteristics of the assembly of the 33 kDa oxygen-evolving complex protein in the etioplasts and the developing chloroplasts of barley seedlings. *Biochim. Biophys. Acta* 1993; 1183(2): 397–407.

139. Jacobs JM, Jacobs NJ. Oxidation of protoporphyrinogen to protoporphyrin, a step in chlorophyll and haem biosynthesis. *Biochem. J.* 1987; 244: 219.

140. Lee H.J, Ball MD, Rebeiz CA. Intraplastid localization of the enzymes that convert delta-aminolevulinic acid to protoporphyrin IX in etiolated cucumber cotyledons. *Plant Physiol.* 1991; 96(3): 910–915.

141. Walker CJ, Weinstein JD. Further characterization of the magnesium chelatase in isolated developing cucumber chloroplasts: substrate specificity, regulation, intactness and ATP requirements. *Plant Physiol.* 1991; 95(4): 1189–1196.

142. Fradkin LI. Structural localization of chlorophyll biosynthesis. In: *Biogenesis of the Pigment Apparatus of Photosynthesis* (Litvin FF ed, Nauka i Tekhnika, Minsk), 1988: 164–195 (in Russian).

143. Ikeuchi M, Murakami S. Separation and characterization of prolamellar bodies and prothylakoids from squash etioplasts. *Plant Cell Pysiol.* 1983; 24: 71–80.

144. Dehesh K, Ryberg M. The NADPH-protochlorophyllide oxidoreductase is the major protein constituent of prolamellar bodies in wheat *Triticum aestivum* L. *Planta* 1985; 164: 396.

145. Sankar GB, Sankar A. Structure and function of DNA photolyases. *Trends Biochem. Sci.* 1987; 12: 259–261.

146. Wilks HM, Timko MP. A light-dependent complementation system for analyses of NADPH:protochlorophyllide oxidoreductase: identification and mutagenesis of two conserved residues that are essential for enzyme activity. *Proc. Natl. Acad. Sci.* 1995; 92: 724–728.

147. Labesse G, Vidal-Cros A, Chomilier J, Gaudry M, Mornon JP. Structural comparisons lead to the definition of a new superfamily of NAD(P)(H)-accepting oxidoreductases: The single-domain reductases/epimerases/dehydrogenases (the 'RED' family). *Biochem. J.* 1994; 304: 95–99.

148. Aronsson H, Sundqvist C, Timko MP, Dahlin C. The importance of the C-terminal region and Cys residues for the membrane association of the NADPH:protochlorophyllide oxidoreductase (POR) in pea (*Pisum sativum*). *FEBS Lett.* 2001; 502: 11–15.

149. Begley TP, Young H. Protochlorophyllide reductase.1. Determination of the regiochemistry and the stereochemistry of the reduction of protochlorophyllide to chlorophyllide. *J. Am. Chem. Soc.* 1989; 111: 3095–3096.

150. Griffiths WT. Reconstitution of chlorophyllide formation by isolated etioplast membranes. *Biochem. J.* 1978; 681–692.
151. Whyte BJ, Griffiths WT. 8-Vinyl reduction and chlorophyll a biosynthesis in higher plants. *Biochem. J.* 1993; 291: 939–944.
152. Knaust R, Urbig T, Senger H. Photoreduction of monovinyl- and divinyl-protochlorophyllide *in vitro* and *in vivo* in the light-dependent greening mutant C-2A' of *Scenedesmus obliquus*. *Physiol. Plant.* 1995; 95: 134–140.
153. Sironval C, Michel JM. On a "photoenzyme" or, the mechanism of the protochlorophyllide-chlorophyllide photoconversion. In: *Photochemistry and Photobiology in Plant Physiology*. Europe Photobiology Symposium, Hvar, Yugoslavia, September 19–22, 1967: 105–108.
154. Valera V, Fung M, Wessler AN, Richards WR. Synthesis of 4*R*- and 4*S*-tritium labeled NADPH for the determination of the coenzyme stereospecificity of NADPH:protochlorophyllide oxidoreductase. *Biochem. Biophys. Res. Commun.* 1987; 148: 515–520.
155. Nayar P, Begley TP. Protochlorophyllide reductase. III: synthesis of a protochlorophyllide-dihydriflavin complex. *Photochem. Photobiol.* 1996; 63: 100–105.
156. Griffiths WT, McHugh T, Blankenship E. The light intensity dependence of protochlorophyllide photoconversion and its significance to the catalytic mechanism of protochlorophyllide reductase. *FEBS Lett.* 1996; 398: 235–238.
157. Klement H, Helfrich M, Oster U, Schoch S, Rüdiger W. Pigment-free NADPH:protochlorophyllide oxidoreductase from Avena sativa L. *Eur. J. Biochem.* 1999; 265: 862–874.
158. Lebedev N, Karginova O, McIvor W, Timko MP. Tyr275 and Lys279 stabilize NADPH within the catalytic site of NADPH:protochlorophyllide oxidoreductase and are involved in the formation of the enzyme photoactive state. *Biochemistry* 2001; 40: 12562–12574.
159. Peine G, Pilz K, Walter G Dujardin EC, Hoffman P. Pyridine nucleotide contents and activity of protochlorophyllide oxidoreductase *in vivo*. 2. Comparative investigations in protoplasts and etioplasts of *Avena sativa* L. during greening. *Photosynthetica* 1993; 29: 13–24.
160. Reinbothe S, Reinbothe C, Runge S, Apel K. Enzymatic product formation impairs both the chloroplast receptor binding function as well as translocation competence of the NADPH:protochlorophyllide oxidoreductase, a nuclear-encoded plastid precursor protein. *J. Cell Biol.* 1995; 129: 299–308.
161. Martin GE, Timko MP and Wilks HM. Purification and kinetic analysis of pea (Pisum sativum L.) NADPH:protochlorophyllide oxidoreductase expressed as a fusion with maltose-binding protein in *Escherichia coli*. *Biochem. J.* 1997; 325: 139–145.
162. Schulz R, Steinmuller K, Klaas M, Forreither C, Rasmussen S, Hiller C, Apel K. Nucleotide sequence of a cDNA coding for the NADPH-protochlorophyllide oxidoreductase (P?R) of barley *Hordeum vulgare* L. and its expression in *Escherichia coli*. *Mol. Gen. Genet.* 1989; 217: 355–361.
163. Darrah P, Kay SA, Teakle GR, Griffiths WT. Cloning and sequencing of protochlorophyllide reductase. *Biochem. J.* 1990; 265: 789–798.
164. Spano AJ, He Z, Michel H, Hunt DF, Timko MP. Molecular cloning, nuclear gene structure, and developmental expression of NADPH-protochlorophyllide oxidoreductase in pea (*Pisum sativum* L.). *Plant Mol. Biol.* 1992; 18: 967–972.
165. Spano AJ, He Z, Timko MP. NADPH:protochlorophyllide oxidoreducatses in white pine (*Pinus strobes*) and loblolly pine (*P. taeda*). Evidence for light and developmental regulation of expression and conservation in gene organization and protein structure between angiosperms and gymnosperms. *Mol. Gen. Genet.* 1992; 236: 86–95.
166. Benly M, Schulz R, Apel K. Effect of light on the NADPH protochlorophyllide oxidoreductase of *Arabidopsis thaliana*. *Plant Mol. Biol.* 1991; 16: 615–626.
167. Teakle GR, Griffiths WT. Cloning, characterization and import studies on protochlorophyllide reductase from wheat (*Triticum aestivum*). *Biochem. J.* 1993; 296: 225–230.
168. Su Q, Frick G, Armstrong G, Apel K. POR C of *Arabidopsis thaliana*: a third light- and NADPH-dependent protochlorophyllide oxidoreductase that is differentially regulated by light. *Plant Mol. Biol.* 2001; 47: 805–813.
169. Apel K. The protochlorophyllide holochrome of barley (*Hordeum vulgare* L.) Phytochrome-induced decrease of translatable mRNA coding for the NADPH-protochlorophyllide oxidoreductase. *Eur. J. Biochem.* 1981; 120: 89–96.
170. Galanger TF, Ellis JR. Light-stimulated transcription of genes for two chloroplast polypeptides in isolated pea leaf nuclei. *EMBO J.* 1982; 1: 1943–1952.
171. Batschauer A, Apel K. An inverse control by phytochrome of the expression of two nuclear genes in barley (*Hordeum vulgare* L.). *Eur. J. Biochem.* 1984; 143: 593.
172. Silverthorne J, Tobin EM. Demonstration of transcriptional regulation of specific genes by phytochrome action. *Proc. Natl. Acad. Sci.* 1984; 81: 1112.
173. Mosinger E, Batschauer A, Schafer E, Apel K. Phytochrome controlof *in vitro* transcription of specific genes in isolated nuclei from barley (*Hordeum vulgare*). *Eur. J. Biochem.* 1985; 147: 137.
174. Schunmann PH, Ougham HJ. Identification of three cDNA clones expressed in the leaf extension zone and with altered patterns of expression in the slender mutants of barley: a tonoplast intrinsic protein, a putative structural protein and protochlorophyllide oxidoreductase. *Plant Mol. Biol.* 1996; 31: 529–537.
175. Holtorf H, Apel K. Transcripts of the two NADPH protochlorophyllide oxidoreductase genes porA and porB are differentially degraded in etiolated barley seedlings. *Plant Mol. Biol.* 1996; 31: 387–392.

176. Mapleston RE, Griffiths T. Light modulation of proto-chlorophyllide reductase. *Biochem. J.* 1980; 189: 125.

177. Forreiter C, van Cleve B, Schmidt A, Apel K. Evidence for a general light-dependent negative control of NADPH-protochlorophyllide oxidoreductase in angiosperms. *Planta* 1991; 183(1): 126–132.

178. Runge S, Sperling U, Frick G, Apel K, Armstrong GA. Distinct roles for light-dependent NADPH protochlorophyllide oxidoreductase (POR) A and B during greening in higher plants. *Plant J.* 1996; 9: 513–523.

179. Sugiura M. The chloroplast chromosomes in land plants. *Annu. Rev. Cell Biol.* 1989; 5: 51–70.

180. Chen DD, Schnell DJ. Insertion of the 34-kDa chloroplast protein import component, IAP34, into the chloroplast outer membrane is dependent on its intrinsic GTP-binding capacity. *J. Biol. Chem.* 1997; 272: 6614–6620.

181. Schnell DJ. Shedding light on the chloroplast protein import machinery. *Cell* 1995; 83: 521–524.

182. May T, Soll J. Chloroplast precursor protein translocation. *FEBS Lett.* 1999; 452: 52–56.

183. Wienk HLJ, Czisch M, de Kruijff B. The structural flexibility of the preferredoxin transit peptide. *FEBS Lett.* 1999; 453(3): 318–326.

184. Cline K, Henry R. Import and routing of nucleus-encoded chloroplast proteins. *Annu. Rev. Cell Dev. Biol.* 1996; 12: 1–26.

185. Heins L, Collinson I, Soll J. The protein translocation apparatus of chloroplast envelopes. *Trends Plant Sci.* 1998; 3: 56–61.

186. Keegstra K, Cline C. Protein import and routing systems of chloroplasts. *Plant Cell* 1999; 11(4): 557–586.

187. Koutrtz L, Ko K. The early stage of chloroplast protein import involves Com70. *J. Biol. Chem.* 1997; 272: 2808–2813.

188. Agarraberes FA, Dice JF. Protein translocation across membranes. *Biochim. Biophys. Acta* 2001; 1513: 1–24.

189. Wu C, Seibert FS, Ko K. Identification of chloroplast envelope proteins in close physical proximity to a partially translocated chimeric precursor protein. *J. Biol. Chem.* 1994; 269: 32264–32271.

190. Caliebe A, Grimm R, Kaiser G, Lubeck J, Soll J, Heins L. The chloroplastic protein import machinery contains a Rieske-type iron-sulfur cluster and a mononuclear iron-binding protein. *EMBO J.* 1997; 16: 7342–7350.

191. Ko K, Budd D, Wu CB, Seibert F, Kourtz L, Ko ZW. Isolation and characterization of a cDNA clone encoding a member of the Com44/Cim44 envelope components of the chloroplast protein import apparatus. *J. Biol. Chem.* 1995; 270: 28601–28608.

192. VanderVere PS, Bennett TM, Oblong JE, Lamppa GK. A chloroplast processing enzyme involved in precursor maturation shares a zinc-binding motif with a recently recognized family of metalloendopeptidases. *Proc. Natl. Acad. Sci.* 1995; 92: 7177–7181.

193. Chaal BK, Mould RM, Barbrook AC, Gray JC, Howe CJ. Characterization of a cDNA encoding the thylakoidal processing peptidase from *Arabidopsis thaliana* — implications for the origin and catalytic mechanism of the enzyme. *J. Biol. Chem.* 1998; 273: 2715–2722.

194. Batschauer A, Santel HJ, Apel K. The presence and synthesis of the NADPH-protochlorophyllide oxidoreductase in barley leaves with a high-temperature-induced deficiency of plastid ribosomes. *Planta* 1982; 154: 459–464.

195. Marrison JL, Schünmann PHD, Ougham HJ, Leech RM. Subcellular visualization of gene transcripts encoding key proteins of the chlorophyll accumulation process in developing chloroplasts. *Plant Physiol.* 1996; 110: 1089–1096.

196. Aronsson H, Sohrt K, Soll J. NADPH: protochlorophyllide oxidoreductase uses the general import pathway. *Biol. Chem.* 2000; 381: 1263–1267.

197. Dahlin C, Sundqvist C, Timko MP. The *in vitro* assembly of the NADPH-protochlorophyllide oxidoreductase in pea chloroplasts. *Plant Mol. Biol.* 1995; 29(2): 317–330.

198. Dahlin C, Aronsson H, Almkvist J, Sundqvist C. Protochlorophyllide-independent import of two NADPH: Pchlide oxidoreductase proteins (PORA and PORB) from barley into isolated plastids. *Physiol. Plant.* 2000; 109: 298–303.

199. Aronsson H, Sundqvist C, Timko MP, Dahlin C. Characterisation of the assembly pathway of the pea NADPH:protochlorophyllide (Pchlide) oxidoreductase (POR), with emphasis on the role of its substrate, Pchlide. *Physiol. Plant.* 2001; 111: 239–245.

200. Reinbothe S, Runge S, Reinbothe C, van Cleve B, Apel K. Substrate-dependent transport of the NADPH:protochlorophyllide oxydoreductase into isolated plastids. *Plant Cell* 1995; 7: 161–172.

201. Reinbothe C, Lebedev N, Apel K, Reinbothe S. Regulation of chloroplast protein import through a protochlorophyllide-responsive transit peptide. *Proc. Natl. Acad. Sci.* 1997; 94: 8890–8894.

202. Reinbothe S, Mache R, Reinbothe C. A second, substrate-dependent site of protein import into chloroplasts. *Proc. Natl. Acad. Sci.* 2000; 97: 9795–9800.

203. Aronsson H. Plastid import and membrane association of NADPH:protochlorophyllide oxidoreductase (POR) — a key enzyme in chloroplast development. Ph D dissertation. Göteborg University, Sweden. ISBN 91-88896-30-7, 2001.

204. Oliver RP, Griffiths WT. Identification of the polypeptides of NADPH-protochlorophyllide oxidoreductase. *Biochem. J.* 1980; 191: 227.

205. Röper U, Prinz H, Lütz C. Amino acid composition of the enzyme NADPH:protochlorophyllide oxidoreductase. *Plant Sci.* 1987; 52: 15.

206. Selstam E, Widell-Wigge A. Hydrophobicity of protochlorophyllide oxidoreductase characterized by means of Triton X-114 partitioning of isolated etioplasts membrane fractions. *Physiol. Plant.* 1989; 77(3): 401–406.

207. Widell-Wigge A, Selstam E. Effect of salt wash on the structure of the prolamellar body membrane and the membrane binding of NADPH-protochlorophyllide oxidoreductase. *Physiol. Plant.* 1990; 78(3): 315–323.

208. Grevby C, Engdahl S, Ryberg M, Sundqvist C. Binding properties of NADPH-protochlorophyllide oxidoreductase as revealed by detergent and ion treatments of isolated and immobilized prollamelar bodies. *Physiol. Plant.* 1989; 77: 493–503.

209. Birve SJ, Selstam E, Johansson LBA. Secondary structure of NADPH:protochlorophyllide oxidoreductase examined by circular dichroism and prediction methods. *Biochem. J.* 1996; 317(2): 549–555.

210. Townley HE, Sessions RB, Clarke AR, Dafforn TR, Griffiths WT. Protochlorophyllide oxidoreductase: a homology model examined by site-directed mutagenesis. *Proteins* 2001; 44: 329–335.

211. Kovacheva S, Ryberg M, Sundqvist C. ADP/ATP and protein phosphorylation dependence of phototransformable protochlorophyllide in isolated etioplast membranes. *Photosynth. Res.* 2000; 64(2–3): 127–136.

212. Anderson JM. Photoregulation of the composition, function and structure of thylakoid membranes. *Annu. Rev. Plant Physiol.* 1986; 37: 93–136.

213. Blackwell M, Gibas C, Gyqax S, Roman D, Wagner B. The plastoquinone diffusion coefficient in chloroplasts and its mechanistic implications. *Biochim. Biophys. Acta* 1994; 1183: 533–543.

214. Benson A. Plant membrane lipids. *Annu. Rev. Plant Physiol.* 1974; 15: 1–16.

215. Shipley GG, Green JP, Nichols BW. The phase behavior of monogalactosyl, digalactosyl and sulphoquinovosyl diglycerides. *Biochim. Biophys. Acta* 1973; 311: 531–544.

216. Quinn PJ and Williams WP. The structural role of lipids in photosynthetic membranes. *Biochim. Biophys. Acta* 1983; 737: 223–266.

217. Sprague SG, Staehelin LA. Effect of reconstitution method on the structural organization of isolated chloroplast membrane lipids. *Biochim. Biophys. Acta* 1984; 777: 306–322.

218. Brentel I, Selstam E, Lindblom G. Phase equilibria of mixtures of plant galactolipids. The formation of a bicontinuous cubic phase. *Biochim. Biophys. Acta* 1985; 818: 816.

219. Andersson S, Hyde ST, Larsson K, Lidin S. Minimal surfaces and structures: from inorganic and metal crystals to cell membranes and biopolymers. *Chem. Rev.* 1988; 88: 221–242.

220. Lindstedt I, Liljenberg C. On the periodic minimal surface structure of the plant prolamellar body. *Physiol. Plant.* 1990; 80: 1–4.

221. Lazzati V, Vargas R, Mariani P, Gulik A, Delacroix H. Cubic phases of lipid-containing systems: elements of a theory and biological connotations. *J. Mol. Biol.* 1993; 229: 540–551.

222. Murphy DJ. The molecular organization of the photosynthetic membranes of higher plants. *Biochim. Biophys. Acta* 1986; 864: 33–94.

223. Ericsson B, Larsson K, Fontell K. A cubic protein-monoolein-water phase. *Biochem. Biophys. Acta* 1983; 729: 23–27.

224. Wehrmeyer W. Zur Kristallgitterstruktur der sogenannten Prolamellarkörper in proplastiden etiolierter

Bohnen I: Pentagondodekaeder als Mittelpunkt konzentrischer Prolamellarkörper. *Z. Naturforsch.* 1965; 20b: 1270–1278.

225. Wehrmeyer W. Zur Kristallgitterstruktur der sogenannten Prolamellarkörper in proplastiden etiolierter Bohnen II: Zinkblendegitter als Muster tubularer Anordnungen in Prolamellarkörper. *Z. Naturforsch.* 1965; 20b: 1278–1288.

226. Wehrmeyer W. Zur Kristallgitterstruktur der sogenannten Prolamellarkörper in proplastiden etiolierter Bohnen III: Wurtzitgitter als Muster tubularer Anordnungen in Prolamellarkörper. *Z. Naturforsch.* 1965; 20b: 1288–1296.

227. Wehrmeyer W. Prolamellar bodies: structure and development. In: *Croissance et Vieillissement des Chloroplasts* (Masson et Cie, Paris), 1967: 62–67.

228. Ikeda T. Analytical studies on the structure of prolamellar body in *Phaseoulus vulgaris* by electron microscopy of chloroplasts. *Bot. Mag. Tokyo* 1968; 81: 517–527.

229. Gunning BES. The greening process in plastids 1: the structure of prolamellar body. *Protoplasma* 1965; 60: 111–130.

230. Gunning BES and Steer MP. *Ultrastructure and the Biology of Plant Cells.* Arnold Publishers, London, 1975: 111–115, 254–257.

231. Osumi M, Yamada N, Nagano M, Murakami S, Baba N, Oho E, Kanaya K. 3 Dimensional observation of the prolamellar bodies in etioplasts of squash Cucurbita moschata var melonaeformis cultivar Tokyo. *Scan. Electron Microsc.* 1984; 1: 111–120.

232. Murakami S, Yamada N, Nagano M, Osumi M. Three dimensional structure of the prolamellar body in squash etioplasts. *Protoplasma* 1985; 128(2–3): 147–156.

233. Israelachvili JN, Wolfe J. Membrane geometry of the pro lamellar body. *Protoplasma* 1980; 102 (3–4): 315–322.

234. Charvolin J, Sadoc JF. Ordered bicontinuous films of amphiphiles and biological membranes. *Philos. Trans. R. Soc. Lond. A* 1996; 345: 2173–2192.

235. Granick S. Magnesium protoporphyrin monoester and protoporphyrin monomethylester in chlorophyll biosynthesis. *J. Biol. Chem.* 1961; 236: 1168.

236. Gunning BES, Jagoe MP. The prolamelar body. In: *Biochemistry of Chloroplasts*, Vol 2 (Goodwin TW ed, Academic Press, London), 1967: 655–676.

237. Henningsen KW, Boynton JE. Macromolecular physiology of plastids. VII. The effect of brief illumination on plastids of dark-grown barley leaves. *J. Cell Sci.* 1969; 5: 757–793.

238. Berry DR, Smith H. Red light stimulation of prolamellar body recrystallization and thylakoid formation in barley etioplasts. *J. Cell Sci.* 1971; 8: 185–200.

239. Wellburn AR. Ultrastructural, respiratory and metabolic changes associated with chloroplast development. In: *Chloroplast Biogenesis* (Baker NR, Barber J eds., Elsevier Science Publishers, Amsterdam), 1984: 253.

240. Bahl J, Lechevallier D, Moneger R. Etude compare de l'evolution et l'obscurite et la lumiere des lipides plas-

tidiaux des feuilles etiolees de BIB. *Physiol. Veg.* 1974; 12: 229.

241. Böddi B, Soos J, Lang F. Protochlorophyll forms with different molecular arrangements. *Biochim. Biophys. Acta* 1980; 593: 158.

242. Selstam E, Widell A. Characterization of prolamellar bodies from dark-grown seedlings of Scots pine *Pinus sylvestris* containing light-dependent and NADPH-dependent protochlorophyllide oxidoreductase. *Physiol. Plant.* 1986; 67(3): 345–352.

243. Larsson K. Cubic lipid-water phases: structures and biomembrane aspects. *J. Phys. Chem.* 1989; 93: 7304.

244. Larsson K, Fontell K, Krog N. Structural relationships between lamellar cubic and hexagonal phases in mono glyceride water systems possibility of cubic structures in biological systems. *Chem. Phys. Lipids* 1980; 27(4): 321–328.

245. Joyard J, Block M, Pineau B, Albrieux C, Douce R. Envelope membranes from mature spinach chloroplasts contain an NADPH:protochlorophyllide reductase on the cytosolic side of the outer membrane. *J. Biol. Chem.* 1990; 265(35): 21820–21827.

246. von Wettstein DV, Kahn A. Macromolecular physiology of plastids. In: *Proceedings of the European Regional Conference on Electron Microscopy*, Delft, Vol II (Houvink AL, Spit BJ eds), 1960: 1051.

247. Treffry T. Phytylation of chlorophyllide and prolamellar body transformation in etiolated peas. *Planta* 1970; 91: 279.

248. Henningsen KW, Stummann BM. Use of mutants in the study of chloroplast biogenesis. In: *Encyclopedia of Plant Physiology*, Vol 14B (Pathier B, Boutler D eds, Springer-Verlag, Berlin), 1982: 597.

249. Kesselmeier J, Laudenbach U. Prolamellar body structure composition of molecular species and amount of galactolipids in etiolated greening and reetiolated primary leaves of oat Avena sativa cultivar Regent wheat Triticum aestivum cultivar Kolibri and Rye secale cereale cultivar Halo. *Planta* 1986; 168(4): 453–460.

250. Treffry T. Chloroplast development in etiolated peas: reformation of prolamellar bodies in red light without accumulation of protochlorophyllide. *J. Exp. Bot.* 1973; 24(78): 185–195.

251. Elam Y, Klein S. The effect of light intensity and sucrose feeding on the fine structure in chloroplasts and on the chlorophyll content of etiolated leaves. *J. Cell Sci.* 1962; 14: 169.

252. Dujardin E, Sironval C. The reduction of protochlorophyllide to chlorophyllide III. The phototransformability of the protochlorophyllide-lipoprotein complex found in dark. *Photosynthetica* 1970; 4: 129–138.

253. Wrischer M. Neubildung von prolamellarkörpern in chloroplasten. *Z. Pflanzenphysiol.* 1966; 55: 296.

254. De Greef J, Butler WL, Roth TF. Greening of etiolated bean leaves in far red light. *Plant Physiol.* 1971; 47: 457.

255. Kasemir H, Bergfeld R, Mohr H. Phytochrome mediated control of prolamellar body reorganization and plastid size in mustard cotyledons. *Photochem. Photobiol.* 1975; 21: 111.

256. Ikeda TSO. Prolamellar body formation under different light and temperature conditions. *Bot. Mag. (Tokyo)* 1971; 84: 363–375.

257. Signol M. Action du rythme nycthemeral sur la formation du granum primaireau cours de la differenciation des chloroplastes de *Zea mays*. *C. R. Hebd. Seances Acad. Sci. Paris* 1961; 252: 4177.

258. Faludi-Daniel A, Fridvalszki L, Gyurjan I. Pigment composition and plastid structure in leaves in carotenoid mutant of maize. *Planta* 1968; 78: 184.

259. Robertson DS, Bachmann MD, Anderson IC. The effect of modifier genes on the plastid fine structure of albino mutants of maize. *Genetics* 1968; 60: 216.

260. Kahn A. Developmental physiology of bean leaf plastids. III. Tube transformation and protochlorophyll(ide) photoconversion by flash irradiation. *Plant Physiol.* 1968; 43: 1781–1785.

261. Klein S, Schiff JA. The correlated appearance of prolamellar bodies, protochlorophyll(ide) species, and the Shibata shift during development of bean etioplasts in the dark. *Plant Physiol.* 1972; 49: 619.

262. Weber G, Laurence DJR. Fluorescent indicators of absorption in aqueous solution and on the solid phase. *Biochem. J.* 1954; 56: XXXI.

263. Slavik J. Anilinonaphtalene sulphonate as a probe of membrane composition and function. *Biochim. Biophys. Acta* 1982; 694: 1–25.

264. Podo F, Blasie JK. Nuclear magnetic resonance studies of lecethin bimolecular leaflets with incorporated fluorescent probes. *Proc. Natl. Acad. Sci. USA* 1977; 74(3): 1032–1036.

265. Dobretsov GE, Spirin MM, Chekrygin OV. A fluorescent study of a protein localization in relation to lipids in serum low density lipoproteins. *Biochim. Biophys. Acta* 1982; 710: 172–180.

266. Dobretsov GE. *Fluorescence Probes in Investigation of Cells, Membranes and Lipoproteins* (Vladimirov JY ed, Nauka Publ., Moscow), 1989.

267. Eftink MR, Ghiron CA. Fluorescence quenching studies with proteins. *Anal. Biochem.* 1981; 114: 199–210.

268. Denev I, Minkov I. Membrane protein localization in non-irradiated and flash irradiated prolamellar bodies and prothylakoids, isolated from wheat, measured by fluorescence probes 1-aniline-8-naphthalene sulfonate and pyrene. *Bulg. J. Plant Physiol.* 1996; 22: 40–52.

269. Denev I, Dahlin C, Minkov I, Sunqvist C, Timko M. Interactions of *in vitro* synthesized precursor and mature protochlorophylide oxidoreductase with artificial membranes revealed by two membrane fluorescence probes. — plenary lecture. *11th Congress of the Federation of European Societies of Plant Physiology*, September, 1998.

270. Mayer LD, Hope MJ, Cullis PR. Vesicles of variable sizes produced by a rapid extrusion procedure. *Biochem. Biophys. Acta* 1986; 858: 161–168.

271. MacDonald RC, MacDonald RI, Menco BPhM, Takeshita K, Subbarao NK, Hu L. Small-volume extrusion apparatus for preparation of large, unilamellar vesicles. *Biochem. Biophys. Acta* 1991; 1061: 297–303.

Section VIII

Photosynthesis in Different Plant Parts

26 Photosynthesis in Leaf, Stem, Flower, and Fruit

Abdul Wahid and Ejaz Rasul
Department of Botany, University of Agriculture

CONTENTS

I. INTRODUCTION

The site of photosynthesis in plants is predominantly the green leaf, and its productivity directly depends upon the chlorophyll bearing surface area, irradiance, and the potential to utilize CO$_2$ [1,2]. Although leaf is the main organ contributing to carbon budget of plant throughout life cycle, other vegetative and reproductive parts also fix carbon and contribute to plant growth. In some xerophytic deciduous plants, photosynthesis takes place in chlorophyll bearing phylloclades [3,4]. Herbaceous and woody stems in the temperate and tropical plants [5], twigs and branches of trees may also be photosynthetic [6,7]. Plant parts other than leaf that can retain or develop chlorophyll such as stem, branches, floral parts, fruits, and some aerial roots, also photosynthesize [8]. Some tropical and temperate fruits stay green even after ripening (e.g., varieties of tomato, mango, guava, and apple). Besides these, the calyx, glumes, tiny stamen parts, and carpels are also photosynthetic and contribute to carbon economy, although in low to negligible amounts [9]. Green streaks in veins and splashed occurrence of chlorophyll in petals of petunia not only beautify it, but also contribute to photosynthesis — a clever utilization of surface area. Photosynthesis in the stem is similar to that in the leaf when it contains stomata, but photosynthesis in cortex and wood tissues is entirely different from normal photosynthesis [5,8]. Similarly, photosynthesis reported in a great variety of fruits, although physiologically different from both leaf and stem, also has a great participation in their growth and size attainment [10,11]. A comparative account of photosynthetic processes operative in various plant parts and their contribution to the carbon economy is given below.

II. PHOTOSYNTHESIS IN LEAVES

Leaf lamina is the main photosynthetic organ of most plants to trap light energy and convert into chemical energy. The gaseous exchange and water loss take place through stomatal pores, varying widely in number on the abaxial and adxial surfaces (20 to 300 mm^{-2}) in xeric and nonxeric leaves. A mesophyll cell generally contains about 20 to 100 chloroplasts, which can shift their position according to the intensity and direction of illumination for optimum interception of radiations. This expanded surface area increases the area in contact with CO_2, while the intracellular spaces facilitate CO_2 diffusion. Photosynthesis can differ in various plants or their parts because of following diverse

carboxylation reactions, and as summarized in Table 26.1.

A. C₃ PLANTS

Majority of the dicots and monocots of temperate regions belong to this category. They contain a three-carbon compound, phosphoglycerate, as the initial stable product of CO_2 fixation. The chloroplasts of the mesophyll cell have well-developed grana and stromal lamellae where the light is harnessed, and stroma where CO_2 is fixed by ribulose-1,5-bisphosphate carboxylase/oxygenase (Rubisco). Plants with this pathway have low water use efficiency and a high photorespiratory rate, wasting 30% to 50% of assimilates, and thus termed less efficient.

B. C₄ PLANTS

Most of the grasses and cereals generally endemic to tropics are C₄ plants. They are more efficient than C₃ plants because they apparently lack photorespiration and have a mechanism for maintaining high CO_2 concentration at the site of Rubisco action. This process can work at very low levels of CO_2 even if the stomata are partially or fully closed. They show the formation of C₄ dicarboxylate as a result of initial CO_2 fixation, i.e., oxaloacetate and malate. Because of distinct Kranz-type anatomy [12], the photosynthetic process is spatially located in the mesophyll and bundle sheath cells. Mesophyll chloroplasts possessing well-developed grana are the site of initial carboxylation where phosphoenolpyruvate (PEP)

TABLE 26.1
Comparison of Photosynthetic Pathways Operative in Leaves of Plants Exhibiting Various Categories of Carboxylation Reactions

Characteristics	C₃	C₃–C₄	C₄	CAM
Origin	Temperate	Subtropical and tropical	Tropical	Arid zone/desert
Anatomy	Non-Kranz	Kranz	Kranz	Non-Kranz
Interveinal distance (μm)	Large (190–300)	Small (72–120)	Small (72–120)	—
Carboxylation pathway	C₃	C₃	C₃–C₄	C₃–C₄
Carboxylating enzymes	Rubisco	Rubisco	PEPCase, Rubisco	PEPCase, Rubisco
Carboxylation products	Phosphoglycerate	Phosphoglycerate	Oxaloacetate, malate	Oxaloacetate, malate
CO_2 compensation point (ppm)	30–50	10–20	5–10	5–10
Water use efficiency (g dry matter/kg H_2O transpired)	3–5	5–6	6–10	2–3
Rate of photosynthesis (μmol CO_2/m²/sec)	24-'43	24–50	21–59	4–26
Site of photosynthesis	Mesophyll cells	Mesophyll and bundle sheath cells	Mesophyll and bundle sheath cells	Mesophyll cells
Separation of photosynthetic processes	—	Spatial	Spatial	Temporal

carboxylase (PEPCase) converts PEP to C_4 dicarboxylate. Subsequently, the carboxylation of released CO_2 occurs in chloroplasts of bundle sheath cells, which generally lack grana and contain Rubisco in the stroma. PEPCase has a high affinity for CO_2 and is less affected by high O_2 concentration. The C_4 plants are very productive as they show higher water use efficiency compared with other photosynthetic category plants.

C. C₃–C₄ Intermediate Plants

A form of higher plants, intermediate to C_3 and C_4, exists (e.g. *Flaveria*, *Panicum* spp.), which depicts Kranz-type anatomy like that in C_4 plants but lacks a functional C_4 cycle [13]. The photosynthetic and photorespiratory pathways are identical to those of C_3 leaves but Rubisco is found in both mesophyll and bundle sheath, cells and carboxylation and oxygenation properties are similar [2]. Oxygen inhibition of net photosynthesis is less than those of C_3 plants primarily due to the lack of leakage of CO_2 from leaf to the ambient atmosphere [14]. CO_2 evolved during light or dark respiration is refixed efficiently with the help of glycine carboxylase located in the bundle sheath mitochondria. This results in low CO_2 compensation point. They are more productive than C_3 plants due to greater water use efficiency [15].

D. Crassulacean Acid Metabolism Plants

Crassulacean acid metabolism (CAM), occurring in succulent plants in arid environments and some aquatics, shows temporally separated CO_2 fixing processes. The leaves possess well-packed parenchyma with large vacuoles [2]. The mesophyll cells containing Rubisco are not generally differentiated into palisade and spongy layers. The stomata are fewer in number and remain open during the night to replenish the CO_2 supply of the leaf. PEP is converted into oxalate and malate with the help of PEPCase and malate dehydrogenase respectively. Malate is ultimately stored in the vacuole. During the day, the stomata close, malate is converted into PEP, and CO_2 released is used in light-dependent carboxylation by Rubisco. Two modifications in their gas-exchange parameters include (i) CAM-cycling — diurnal acid fluctuation due to recycling of respired CO_2 and (ii) CAM-idling — drought-induced stomatal closure. During drought periods, glucan and mucilage are stored, mostly in young cortex [16], and are used as substrates for malate synthesis during CAM-cycling. These plants are inefficient because they show photorespiration and much reduced water use efficiency.

E. Leaf Photosynthesis — Plant Factors

1. Leaf Position

Presence of leaves at a certain position is very important to perform optimum photosynthesis. The leaves at the top of the canopy usually have higher values of electron transport due to optimal absorption of photosynthetically active radiations, and Rubisco activity for assimilation of CO_2 and acquisition of nutrients [17,18]. The lower leaves are at a disadvantage due to nonavailability of light as a sole source [19]. For example, second, third, and lower leaves of wheat exhibit much reduced photosynthesis compared with the flag leaf lamina, sheath, and stem internode [20].

2. Leaf Age

The rate of photosynthesis declines steadily when the leaves become aged or senesced [21,22]. A comparison of leaves of different ages manifests a significant decrease in the $^{14}CO_2$ photosynthates [20]. Rawson and Constable [23] found that younger leaves could not become light-saturated at $1800\,\mu mol/cm^2/sec$, while older ones were about 90% light-saturated at $600\,\mu mol/cm^2/sec$ indicating that leaf age, and not light intensity, determines the photosynthetic rate.

Aged maize leaves show higher photorespiration than younger ones because lower senescing leaves have a reduced capacity to concentrate CO_2 in the bundle sheath. When the chlorophyll and Rubisco contents reduce to 50% or less than that of mature maize leaves, the degree of photorespiration could reach that of C_3 plants due to photosynthetic inhibition under atmospheric levels of CO_2 [24]. Xu et al. [6] attributed a decrease in gross photosynthesis of tomato leaves to a decrease in the content rather than activity of Rubisco. Photosynthetically active life of aged leaves may be prolonged if they (i) are shaded and (ii) do not accumulate sugars during the early phase of senescence [25].

3. Plant Gender

Recently, attention has been given to the differences in leaf photosynthesis of male and female plants at different phonological phases [26]. Gynodioecious *Sidalcea hirtipes* (Malvaceae) shows a well-defined sexual dimorphism in photosynthetic gas exchange [27]. Wang and Griffin [28] did not find any difference in the male and female leaf photosynthesis during the vegetative phase. However, the leaf photosynthesis increased by 82% in males and 97% in females under elevated CO_2 during reproductive phase in *Silence latifolia*. Likewise, in *Sipuruna grandifloa* (a dioecious shrub) males had a higher photosynthetic capacity at the leaf level, while females had a higher capacity at the crown level [22]. This indicates a shift of photosynthesis in favor of reproductive growth.

F. LEAF PHOTOSYNTHESIS — ENVIRONMENTAL FACTORS

1. CO_2 Concentration

Carbon dioxide is of fundamental importance as a substrate in dark reactions of photosynthesis. Low CO_2 concentration lowers the photosynthetic rate, while its elevation enhances it. As ribulose-1, 5-bisphosphate (RuBP) is the acceptor of CO_2 in the Calvin cycle, changes in the level and activity of Rubisco are of great significance. Elevated levels of CO_2 do not change the levels of soluble proteins, Rubisco protein and chlorophyll $a + b$ content [29]. However, conflicting reports are available on the activity of Rubisco, which decreases, remains unchanged [29] or increases under high levels of CO_2 [30]. Application of external sucrose, but not the glucose, and high CO_2 reduces Rubisco content and photosynthesis of leaves [31]. The supplementation of sucrose presumably reduces the photosynthesis as a feedback inhibition of the activity of sucrose phosphate synthase (SPS). Higher assimilatory carbon flux under elevated CO_2 is accompanied by a higher activation state rather than level of SPS [32].

In addition to biochemical changes, CO_2 causes certain allometric, anatomical, and physiological changes at the whole leaf level. Leaf area and leaf fresh weight are greater in plants grown in CO_2-enriched air [31]. Exposure of leaves to an abrupt increase in CO_2 greatly enhances the photosynthesis than a gradual increase [33]. Loblolly pine under the application of elevated CO_2 for 4 years produced 90% more biomass and that the photosynthetic rate was always higher in summer [34]. This may be due to the acclimation response of leave to high CO_2, as reported for transgenic lines of tobacco [35]. Campbell et al. [36] reported that exposure of soybean to higher levels of CO_2 produces thicker leaves with a higher number of palisade cells, showing increased net photosynthesis and enhanced yield up to 44%.

2. Temperature

Both the photochemical and biochemical reactions of photosynthesis are affected by temperature fluctuations. Photosystem-II (PSII) is highly labile and its activity is greatly reduced or even partially lost under high temperature [37,38]. However, PSI-driven cyclic electron pathway capable of contributing to thylakoid proton gradient are activated [38]. Increased temperature strongly influences the photosynthetic capacity of leaves with the exception of C_4 plants. Changes in temperature affect the activities of the Calvin cycle enzymes including Rubisco [2,39] and the rate of RuBP regeneration [40]. Moreover, starch or sucrose synthesis is also greatly influenced as seen from the reduced activity of SPS [41], ADPG pyrophosphorylase, and invertase [42]. Increased temperature curtails photosynthesis and increases the CO_2 transfer conductance between intercellular spaces and carboxylation sites [43].

Low temperature affects the chlorophyll content, chlorophyll a/b ratio, and larger total carotenoid pool size as well as their composition [44]. Low temperature alters leaf gas exchange that is related to reduced chlorophyll florescence [45]. In terms of leaf anatomy, growth at low temperature results in increased leaf thickness, which may hamper CO_2 diffusion [46].

3. Irradiance

Any change in light intensity changes leaf photosynthesis. With a gradual increase in irradiance, there is a relatively large increase in photosynthesis up to the light compensation point. Beyond this point, there is saturation in photosynthesis of C_3 plants but not of C_4 plants. Any abrupt decrease in irradiance instantly decreases the photosynthesis of the individual leaves [47].

Long-term changes in photon flux density (PFD) alter the anatomical, physiological, and biochemical properties of leaf. Plants, when grown under reduced radiation fluxes, show a reduced carbon exchange rate, low stomatal conductance, and reduced mesophyll area, resulting in slower photosynthetic rate [48]. Consequently, this reduces the growth rate and dry matter production in C_4 compared to C_3 plants, although the morphological responses are almost similar [49]. A comparison of tall fescue leaves grown under low (30%) and high (full) sunlight showed a 25% reduction in CO_2 exchange rate (CER) per unit leaf area under low light intensity [50]. High PFD grown vines of kiwifruit (*Actinida deliciosa*) show greater leaf area, leaf, and stem biomass than low-PFD vines [51].

Biochemically, small modulations in PFD disturb the balance of RuBP and other metabolites of the Calvin cycle, as reflected by a change in carboxylation rate. Reduced PFD instantly decreases RuBP concentrations, but with optimum PFD, its concentration rises again to steady level [52]. Changes in PFD have a direct relationship with the soluble protein content of the leaf, but not with a change in CO_2 levels [53]. Carboxylation efficiency of Rubisco increases up to PFD of 550 μmol/m^2/sec and to promote it further, higher Rubisco is a prerequisite. The efficiency to reduce the CO_2 so attained could be explained by a mesophyll CO_2 barrier associated with a high chlorophyll and protein content of leaves [54].

4. Shade

Shade is an ecological factor that is important for interception of light in accruing optimal photosynthesis [55]. Lower shaded leaves are largely affected as far as photosynthesis is related [56]. Severe shading at a PFD of 85 to 95 μmol/m^2/sec induces greatest reduction in the maximum photosynthesis [57], while partial shading reduces the percentage of photosynthetic area [58]. A comparison of differentially shade tolerant *Acer* species indicated that light-demanding species have much higher water use efficiency compared with intermediate and shade tolerant species [59]. Soybean plants receiving 63% less light due to shading display a 9% to 23% less seed growth. This indicates that plants become source limited due to shading [60].

Shading not only causes a reduction in Rubisco activity, maximum electron transport, and triosephosphate activity in grapevine [56], but also induces changes in the allometric and anatomical characteristics, as is evident from the increased leaf size and branching, but reduced specific leaf weight [61]. Shading up to 90% leads to increased senescence of leaves [62]. Some plants may show survival under shaded conditions. For example deep shade was highly detrimental to growth of six green-stemmed leguminous species while moderate shade promoted their survival [63].

III. PHOTOSYNTHESIS IN THE STEM

Although the observation on stem photosynthesis of various plants was made at the beginning of the twentieth century [64,65], the efforts were continued subsequently to quantitatively prove its occurrence and contribution to carbon economy in many plants from diverse origins, e.g., aspen bark [66,67], trees [68,69], cladodes [70,71], etc. These research efforts increased manifold particularly during the past decade [4,5,8,72–74]. Green stems of herbaceous (monocots and dicots), shrubs and woody plants, and cladodes of the CAM plants photosynthesize and contribute to carbon economy, while wood photosynthesis is of greater ecophysiological significance [8,75]. So far, stem photosynthesis has been reported in more or less 36 plant families [5]. Various stems and their tissues display varied patterns and mechanisms of photosynthesis. For photosynthesis to take place, the presence of pertinent photosynthetic machinery is imperative, which is present in all stems. Aschan and Pfanz [8] have conveniently categorized the stems into different types based on the pathways of carbon fixation and the specificity of these pathways to various tissues.

A. STEM PHOTOSYNTHESIS — CO_2 ASSIMILATION

1. Nonsucculent Green Stems

Herbaceous and shrubby stems of temperate and tropical species show the presence of stomata in epidermal layer for gas exchange [76–79] and Rubisco for the CO_2 fixation [80]. The chlorophyll and carotenoids are present throughout the thickness of the stem and maximum chlorophyll a/b ratio (i.e., 3.8) matches the minimum value for leaves [81]. This type of stem photosynthesis is identical to leaf photosynthesis in many respects, for instance, the use of C_3 pathway, presence of abundant stomata in the epidermis and similar responses to environmental changes. These stems therefore contribute to net or positive carbon gain [8,82].

Determination of net photosynthesis using carbon isotope discrimination has a more negative value for stem $^{13}C/^{12}C$ than for leaves on the same plant [83]. Maximum net photosynthesis in the green-stemmed nonsucculent leguminous plants species ranges from 1.7 to 11.6 μmol CO_2/m^2/sec [5,8], and it varies substantially in other species (Table 26.2). Although the stem photosynthesis is more sensitive to changes in low water potential and temperature [82], it remains positive annually. This is because the leaves of woody legumes are small and ephemeral and plant relies on the carbon gain by the stem photosynthesis round the year [89].

2. Cladodes

The plants confined to desert environment show photosynthesis in green flattened stems without leaves (cladodes) or with highly reduced leaves. This type of photosynthesis has been reported in many members of the families Asclepiadaceae, Asteraceae, Cactaceae, Crassulaceae, and Euphorbiaceae [3,90]. Net CO_2 is assimilated during night when the stomata are open and malate is produced from PEP by PEP-Case. Net CO_2 assimilation ranges from 10 to 26 μmol/m^2/sec (Table 26.2) [4,85]. As a result, stem of a highly productive CAM, *Opuntia ficus-indica* yields biomass up to 47 metric tons/ha/year, which is greater than the productivity of most of the C_3 and some C_4 species [75].

During the leafless state of plant growth, stem photosynthesis becomes increasingly important in the carbon economy [84]. During the initial phases of development, although the newly emerging (daughter) cladodes are photosynthetically active but act as sink, and import up to 60% of the photosynthates from the basal cladodes. After 6 months the daughter cladodes also become source [91]. Maximum net photosynthesis is comparable to C_3 leaves (Table 26.2). Coincidence of high glucose content of the daughter cladodes with the SPS activity also suggests that these cladodes act as strong sinks for sucrose during initial stages of growth [4]. The cladodes [85,92] or stems [93] usually persist for long time and act as reservoir of carbohydrates to support the growth of new cladodes in subsequent years.

B. STEM PHOTOSYNTHESIS — CO_2 REFIXATION

But in contrast to stomatal-dependent carbon gain in the stems of tropical and temperate species, woody stems do not have stomata [5,94]; rather they contain chlorophyll in the bark and cortical layers, referred to as chlorenchymal tissues, and perform photosynthesis. This is different from normal leaf and CAM photosynthetic pathway [95] in that these tissues rely on the refixation of respired CO_2 in the cortical or other living cells within the wood [74]. Woody stem chlorenchyma lies some 80 to 150 μm underneath the periderm or rhytidome, which offer high resistance to the diffusion of gases [76,95,96]. Chloroplasts in the chlorenchyma of the outer bark region are high in number and resemble those of shaded leaves [97,98], but they show a greater number of plastglobulii, unlike shaded leaves [99]. The number and density of grana stacks are greater in phelloderm and chlorophyllous outer cortex and decrease towards the middle. In the perimedullary region of the stem, the chloroplasts show slightly greater number of grana stacks than wood rays [100].

The distribution pattern of starch grains in the chloroplasts of cortical cells differs from the shade type leaves. The chloroplasts from the stem of plants of all regions contain starch grains, but their size increases from the outer cortex to the pith. Both the density and size of grana stacks and size of starch grains indicate a higher photosynthetic capacity in the phellodermal and outer cortical region than in the chloroplasts of more inwardly situated cells [95].

The chlorophyll content of stem resembles those of shaded leaves when expressed on fresh mass basis. However, total chlorophyll content of stem bark is much less than those of leaves on the same plant. It accounts for up to 70% of the leaf when expressed on surface area basis [101,102]. The chlorophyll *a/b* ratio ranges between 1.8 and 2.7 in different species [95]. The chlorophyll is distributed throughout the deeper tissues in the wood, but its concentration may be

TABLE 26.2
Photosynthesis in the Green Stems of Nonsucculent, Succulent, and Woody Plants Species

Species	Maximum Net Rate of CO_2 Uptake (μmol/m²/sec)	Ref.
Nonsucculent green stem		
Caesalpinia virgata	3.5–7.8	[82]
Cytisus scoparius	1.7–11.6	[84]
Spartinum junceum	6.5	[78]
Syringa vulgaris	6.0–8.0	[79]
Succulent stems (cladodes)		
Opuntia ficus-indica (28 days old)	Approx. 12.0	[4]
Retama sphaerocarpa	10.0–26.0	[85]
Woody stems (bark and twigs)		
Populus tremula (bark)	7–10[a]	72
Pinus monticola (twigs)	0.64[b]	86
Populus tremuloides (bark)	2.8[b]	87
Ilex equifolium (bark)	Approximately 2[a]	88

[a]In terms of photosynthetic O_2 evolution.
[b]Gross photosynthesis in terms of CO_2 fixation.

much lower [103]. Chlorophyll content of various photosynthetic tissues depends on the age of the bark and exposure to light. Full sunlight exposed side of the *Populus tremula* twigs show greater chlorophyll content than on the side exposed to 20% of full sunlight. Furthermore, 1-year-old twigs contain lower chlorophyll content than 2-year-old twigs [7].

The chlorenchymal tissue of the stem bark is shaded by epidermis, periderm and rhytidome. For photosynthesis to occur, enough light has to penetrate all these tissues to reach light-harvesting complexes in the thylakoid membranes, but these structures tend to reduce the transmission of light to reach the photosynthetic light harvesting systems. Determination of light transmission to various depths reveals that after passing through shading layers, 0.2% to 1% of the light reaches the chlorenchyma, while 0.01% to 0.2% reach the center of the stem. This, however, depends upon the age of the twig and intensity of light [73,95]. The light can penetrate through the natural openings in the outer bark, lenticels, and bark valleys or cracks in the absence of stomata [104,105].

Carbon isotope discrimination data shows a more negative value of $^{13}C/^{12}C$ for outer than the inner bark region [87]. Of the two types of photosynthetically active radiations, blue light is mainly absorbed in the outer bark region while red light penetrates the deeper tissues [73,102]. The light funneling system at the PSII in the thylakoid membrane [2] might be one reason for a greater photosynthetic rate of the outer layer of the chlorenchymal tissue in the stem bark.

Since the stomata are typically absent in the stem bark and there is no stomatal gas exchange, the refixation of CO_2, can take place in the dark [106], which accounts for up to 55% of the leaves at 1000 μmol PAR/m²/sec [86]. This is achieved by the PEPCase that can fix CO_2 (actually HCO_3^-) in quite high amounts converting PEP to malate [15]. The refixation rate is higher in young twigs (~6 μmol CO_2/m²/sec) than in the older ones (~1.3 μmol CO_2/m²/sec) [8] as measured from photosynthetic oxygen evolution (Table 26.2). Like other plant parts, the mitochondria perform dark respiration in the wood parenchyma and pith and produce CO_2 in quite high amounts (i.e., 500 to 800 times higher than the ambient air). This amount of CO_2 is an ideal prerequisite for PEPCase activity within the stem [95]. Any depletion in oxygen levels due to respiration is compensated by the photosynthesis operative in the chlorenchymatous tissues [6,8].

From the above it is evident that bark photosynthesis is different from C_3, C_4, and CAM pathways of photosynthesis in that it does not involve the exchange of gases from and to the ambient air, rather

relies upon the internally produced CO_2 due to respiration and consumption of O_2 produced due to photosynthesis (Table 26.2). Although there is no net or positive photosynthesis in the woody stems, it has ecophysiological implications in having (i) a very little associated water loss [86], (ii) an increased carbon use efficiency to offset the respiratory losses [96], (iii) a greater importance for C-budget in the leafless state of plants [8], and (iv) a diminished danger of anaerobiosis that leads to the fermentation and production of ethanol or lactate to toxic levels [95].

IV. PHOTOSYNTHESIS IN FLOWERS

A. FLORAL PARTS

Photosynthesis in floral organs is important in reducing the cost of reproduction [8]. Early reports on the flower photosynthesis date back to the findings of Arditti [107], emphazing that flowers have green pigment and the capability to fix $^{14}CO_2$ in the dark [108]. It is now established that flowers and their floral parts have necessary apparatus for photosynthesis. Stomata are present in the epidermis of floral parts for the gaseous exchange, but their density varies widely depending on the part [9]. For instance, green carpels of Alpine butter cup have one to two stomata [110], sepals of apple have 200 stomata/mm² [111] and stomata on the bracts of *Spiranthes cernua* are comparable to those on the leaf [112].

The chloroplasts of floral parts in *Petunia* are similar to leaves showing PSI activity at the young stage, but remain no longer photosynthetic and are converted to chromoplasts on maturity [113]. Chloroplasts of *Dendrobium* flower are smaller with fewer thylakoid membrane and grana compared to leaf. These chloroplasts degenerate towards maturity and become less distinct with dilated thylakoid membranes [114]. PSII photochemical efficiency of *Dendrobium* flower is lower than leaf with much reduced chlorophyll content [115]. Rubisco activity of the flower declines towards maturity with an increase in PEPCase activity and nocturnal malate accumulation, representing a CAM-like pathway [114].

All nonleaf green parts of tomato photosynthesize and contribute to the fruit growth [116]. Floral parts like calyx, green shoulder, and pericarp fix CO_2 and contribute up to 14% to the fruit growth [9]. In addition to CO_2 fixation, sepals show some adaptations to contribute to the fruit development of different plants. (i) With the expansion of tomato fruit, the sepals align vertically, so that at any one time the upper surface of some and lower surface of the other parts are maximally exposed to light [9]. (ii) During the course of flower senescence, the photosynthetic activity of sepals is important for the final yield of the Meadowfoam seed [117]. Among the various parts, the calyx has the highest net photosynthetic capacity followed by green shoulder (Table 26.3).

B. INFLORESCENCE

Inflorescence and ear are important sites of photosynthesis, because they form the canopy for maximum exposure to radiation flux. Large ear cultivars show greater rate of net photosynthesis and grain yield than small ear type due to optimum interception of photosynthetically active radiations [123]. Various parts in the inflorescence of grasses show net photosynthesis, albeit in lower amounts (~10% of the leaf) [124], and most of the carbohydrate requirement for grain filling is fulfilled from the flag leaf [118].

Different parts of the ear including rachis, lemma, palea, awn, glume, and even the panicle, photosynthesize and contribute to grain filling [123,125,126]. Net photosynthesis by these parts, though lower, has a significant contribution to the flower and fruit growth (Table 26.3). Carbon isotope discrimination value is more negative for the glume than for the leaves. Glume has PEPCase to refix the respired CO_2 and accumulates sucrose for partitioning to the developing grain [127]. In addition, palea, lemma, pericarp, and panical also contain PEPCase and refix the respired CO_2 during reproductive development [126]. The developing wheat grains also show gas exchange as they have up to 10 to 30 stomata in their greenish epidermis. However, the chloroplasts of these stomata disintegrate at maturity 128]. The

grains contain PEPCase in the scutellum and aleurone layer for the refixation of respired CO_2, as seen from the continued accumulation of transcript throughout the grain development [129]. Bort et al. [130] have shown that $^{14}CO_2$ evolved by respiration of ^{14}C-labeled sucrose fed to the grain while in the ears of wheat and barley is trapped (about 55% to 75%) by the closely located structures of the ear.

The awn plays a key role in ear photosynthesis. Olugbemi et al. [121] established that awnless wheat lines contribute only 10% to the grain filling, while the awned ones contribute up to 18%; thus a strong correlation exists between the final weight of the grain and the length and area of the awn. This correlation is due to the fact that presence of the awn increases the photosynthetic area and at the same time economizes on total ear transpiration during water stress [131,132].

V. PHOTOSYNTHESIS IN FRUIT

After the fertilization has taken place, the next stage is the development of fruit and seed set for continuity in generations. During early stages of development, the fruits are usually green and photosynthetic [10]. Nevertheless, the fruits remain heterotrophic and rely upon the supply of photoassimilates from the adjacent leaves [9,122,133]. At maturity, the fruits are no more photosynthetic and show variegated pattern of the flesh in different species.

While green in the early ontogeny, the fruit are endowed with stomata in the epidermis. The density

TABLE 26.3
Photosynthesis in Flowers and Floral Parts of Various Plant Species

Species	Part	Maximum Net CO₂ Assimilation (μmol/m²/sec)	Ref.
Acephylla glaucescens	Inflorescence	11.0	[118]
Helleborus niger	Sepals	295	[119]
Helleborus viridus	Sepals	2.3	[8]
Lilium hybr.enchantment	Anther	2.3	[120]
Spiranthes cernua	Flower	2.5	[112]
	Bud	3.7	
	Inflorescence	0.2	
Limnanthes alba	Rosette leaves	7.9	[117]
Lycopersicon esculentum	Calyx	88–154[a]	[9]
	Green shoulder	97[a]	
Triticum aestivum	Awn	5.2	[121]
Rubus ursinus	Fruiting canes	3–7	[122]
Rubus discolor	Fruiting canes	10–14	[122]

[a]Measured as electron transport activity.

of stomata is 10 to 100 times lower than the abaxial surface of corresponding leaves [8,10]. Size, shape, and function of these stomata, as well as their sensitivity to environmental stresses resemble C_3 leaves [134]. The stomatal density varies greatly in different species. For instance, pea pod contains 25% lesser stomata than lower surface of corresponding leaf [135]. Different varieties of currant (*Rabies* sp.) show 4 to 18 stomata per berry [134]. Developing wheat and barley have 10 to 30 stomata per single fruit [128]. In apple fruit, after a week of petal fall, the stomatal frequency is maximum and uniform over the entire surface, but decreases as the fruit expands. In midseason, these stomata are transformed into lenticels for gas exchange [134]. In tomato fruit, the stomata are absent but the CO_2 fixation is achieved by trapping respired CO_2 from the fruit inside [136].

Microscopic observations in different fruits and pods reveal the presence of chloroplasts [137,138]. In apple, chloroplasts are found in green hypodermal and inner perivascular tissue but they differ in structure. The chloroplasts of perivascular tissue are larger than leaf, contain no starch grain and show C_4 photosynthesis [10,139]. On the contrary, the chloroplasts of hypodermal layer are relatively small, contain grana and starch grains comparable to mesophyll cells, and perform photosynthesis-like C_3 pathway. These chloroplasts become vacuolated 60 days after full bloom, show markedly increased number and size of starch grains after 120 to 145 days, and deteriorate on full maturity [137]. Similar to apple, the grana containing chloroplasts are highest in the exocarp, low in mesocarp, and lowest in the endocarp of pea [138]. The epidermal chloroplasts are comparable in size to leaf, but those in the central region resemble amyloplasts and are quite large due to increased quantity of starch [135].

The chlorophyll content of the fruit peel varies greatly in different species, being lowest (7 μg/g fresh weight) in apple and highest (850 μg/g fresh weight) in cucurbits [8]. Apple peel contains chlorophyll *a*, *b*, PSI, PSII, and other apparatus required for energy utilization, but the rate of photosynthesis is proportionate to the amount of pigments. The chlorophyll content gradually declines towards maturity but never approaches zero [10]. However, the chlorophyll *a/b* ratio of 2.4 at unripe stage approaches to 0.5 at overripe stage, indicating a more pronounced loss of chlorophyll "*a*" towards maturity [11]. At this stage the carotenoid and anthocyanins contents increase at the cost of chlorophyll [140]. The increase in carotenoid content is due to the activity of phytoene synthase located in the plastid stroma [141].

Gas exchange at the fruit surface takes place through stomata and is positively correlated with its size, chlorophyll content, CO_2 concentration, light intensity, and low diffusive conductance of epidermis [133,134,140,142]. Like the leaf, the fruit has surface cuticle but it is up to 10 times more permeable to inflow of CO_2 than the leaf [142]. An exception is the tomato fruit, which has no net CO_2 uptake but shows photochemical activity and an effective electron transport system [11]. Stomatal conductance is high during initial fruit formation and declines as the fruit grows further [143–145]. The ambient CO_2 is not reflective of the fruit's internal CO_2 concentration. The outflow of CO_2 from fruit is much less, which builds up the internal CO_2 concentration as high as 0.3% to 2%, that is, 7- to 60-fold greater than ambient CO_2 [146]. This clearly indicates the CO_2 recycling within the fruit is similar as seen for woody stems [8].

The ambient CO_2 taken up by the fruit is assimilated by Rubisco, which has been detected in various fruits and pods, but with reduced activity [11,147]. The activity of Rubisco is 10–100 times less than in the subtending leaf [10]. In contrast, PEPCase has been found in all fruit tissues in high amounts with increased capacity to recapture respired CO_2 [138,148]. A high ratio of PEPCase to Rubisco (4:5) suggests a different CO_2 assimilation mechanism compared to leaf, where this ratio is 1:10 [10]. In avocado fruit, the general pattern of PEPCase inhibition and partial relief from inhibition by glucose-6-phosphate is similar to C_3, C_4, and CAM leaves. However, its pH sensitivity (below 7.00) supports a non-C_4 and non-CAM behavior [149].

Bravdo et al. [150] reported a decreased RUBP carboxylase and increased RUBP oxygenase activity with the progressive ripening of tomato fruit. High activities of malate dehydrogenase and glycolate oxidase indicated a higher respiratory potential as compared to leaf. The activity of carboxylating enzymes coupled with those of CO_2-releasing enzymes would be considerably higher in the pod wall than in the leaf, thus stimulating the PEPCase when there is net loss of CO_2 [151]. Substantial evidence indicates that high PEPCase activity is associated with the C_4 and CAM photosynthetic systems and is also involved in the development of plants with a C_3 type of photosynthesis. The photosynthetic response of peach fruit to temperature, light, and CO_2 levels does not show C_4 photosynthesis, as the CO_2 response curve are more similar to C_3 photosynthesis.

Based on differences in the photosynthetic patterns, Blanke and Lenz [10] suggested a novel name for CO_2 assimilation in fruits — fruit photosynthesis — which is dissimilar to C_3, C_4, or CAM pathways in many respects (Table 26.4). It differs from C_3 pathway in having β-carboxylation, high internal CO_2 concentration, and high diffusive resistance of fruit

TABLE 26.4
Difference of Fruit Photosynthesis from C$_3$, C$_4$, and CAM Pathways

Characteristics	Fruit	C$_3$	C$_4$	CAM
β-Carboxylation	+	−	+	+
Chloroplast dimorphism	−	−	+	−
Kranz anatomy	−	−	+	−
Stomatal frequency	Low	High	High	Low
Chl a/b ratio	1	2.6	3.9	−
Direction of CO$_2$ gradient	Inside to outside	Outside to inside	Outside to inside	Outside to inside
Nocturnal malate accumulation	+	−	−	+
Diurnal malate fluctuation	−	−	−	+
Concentration of PEPCase, PEPCK, malic enzyme, NAD-MDH	High	Low	High	High
K_m of PEPCase for PEP	Low	Low	High	High
Decarboxylation types	Central perivascular tissue (apple): i. Early stage PEPCK ii. Later stage NADP-ME	−	PEPCK, NADP-ME, NAD-ME	NADP-ME
Tissues involved	Pericarp, locular chlorenchyma	Mesophyll	Mesophyll and bundle sheath	Mesophyll
Source of CO$_2$	Respiratory	Ambient	Ambient	Ambient
Net rate of photosynthesis	Negligible	Low–high	High	Low

Source: Partly adapted from Blanke MM, Lenz F. *Plant Cell Environ.* 1989; 12:31–46.

epidermis to CO_2 evolution. Variation from the C_4 pathway is due to non-Kranz anatomy, pH sensitivity, and CO_2 affinity being constant for PEP carboxylase and from the CAM pathway because of the absence of pH fluctuation specific to this pathway.

VI. PHOTOSYNTHETIC RELATIONSHIPS BETWEEN LEAF, STEM, FLOWER, AND FRUIT

Although various chlorophyllous tissues contribute to the carbon economy of the plants, the fact remains that leaves are an exclusive source of net carbon gain and supporting the growth of other parts. Photoassimilates partitioning is under the control of two component system; source — the site of photoassimilate production, and sink — the site of utilization or storage of assimilates [152]. The partitioning of assimilates from a source to a sink continues to occur throughout the life of plant and is important for both vegetative and reproductive growth.

Roots, stems, and young emerging leaves are competitive sinks for assimilates at the vegetative stage. After passing through a number of structural and physiological changes, the leaf attains the status of an active source and becomes the exporter of carbohydrates to the organs where they are required [2,153]. Green stems most of the time and roots all the time are importers of assimilates for respiration and storage. Defoliation (source removal) has a negative effect on the growth of *Vitis vinifera* vines [154]. The carbohydrates stored in the stem and roots of tea plant are utilized for the production of new shoots following pruning [155]. Sometimes feedback inhibition of photosynthesis occurs when the sink is removed or its capacity is not enough to accept the photoassimilates [156].

Partitioning of photosynthetic products is crucially important to the reproductive growth. Sustained supply of photoassimilates is needed for flowering, fruit, and seed set [124,157,158]. It is generally accepted that the altered status of both source and sink limits the yield [159]. However, most of the recent reports indicate that sink strength is more important than source [157,158,160, 161]. Low sink activity results in poor translocation and partitioning of assimilates to grain, leading to allocation of more resources for vegetative growth [161]. Demand of sink triggers the changes in the physiological processes of different organs. The presence of fruit increases the stomatal opening and net photosynthesis of apple leaves [162]. Similarly, the rate of leaf photosynthesis enhances significantly during blooming and rapid fruit expansion without any change in the light

and dark respiration, leaf conductance, and transpiration [163].

Translocation of ^{14}C-assimilates in leaves of the extension shoot or of spur without fruit as compared to the fruit on the spur was promoted by decreasing the leaf to fruit ratio. Faster transport took place on the side where the leaf to fruit ratio was the lowest. Attraction of photosynthates depends upon the spur size and its location on the branch [164]. Photoassimilate contribution by proximal leaves or leaves subtending the fruit is much greater than distal ones [157,158,165]. It is plausible that the sink has direct vascular connections with the source (leaf) tissue [166,167], which helps in the catchment of photosynthates for the growth of the sink (fruit) tissue [168].

Manipulation of the source–sink ratio determines the dry matter partitioning, as the removal of one hampers the other's activity. Partial defoliation results in increased photosynthesis of the remaining leaves. Sink removal, on the other hand, greatly decreases the net photosynthesis and final grain yield of wheat [160]. The removal of fruit at anthesis in garden pea altered the photoassimilate distribution pattern of associated leaflets, revealing that fruit growth substantially controls the pattern of photosynthesis [169]. Removal of spikelets form the ear reduces the flag leaf photosynthesis due to feedback inhibition [170] as a result of excess of photosynthates accumulation [156]. Likewise, removal of ear from monotillered plant brings about a 50% reduction in the net photosynthesis of flag leaf but this is not the case for ear removal from a multitillered plant. This suggests that the remaining tillers of the same plant are connected by phloem via roots. This is verified from the presence of radiolabeled compounds in the tillers other than that exposed to $^{14}CO_2$ in the same plant [171].

VII. CONTRIBUTION OF LEAF, STEM, FLOWER, AND FRUIT PHOTOSYNTHESIS TO PRODUCTIVITY

Leaves initially contribute to vegetative growth and then to reproductive growth during whole of the life cycle. Leaf contribution to the vegetative growth varies substantially and is related to position on the plant, availability of light, sink strength, etc. Nevertheless, the leaves are important in fulfilling the carbohydrate needs and contributing to the net carbon budget from as low as 10% to as high as 98% in various plant species (Table 26.5).

The leaf photosynthetic activity becomes crucially important for the reproductive growth [8,10,164].

TABLE 26.5
Photosynthetic Contribution of Various Parts to Plant Biomass Production

Plant Species	% Contribution				
	Leaves	Stem	Floral Parts	Fruits	Ref.
Triticum aestivum	66–84[a]	—	9–18 (ear)	—	[121]
	55–75	—	9–11 (awn)	—	[121]
	80–92	—	—	—	[171]
	10–20	—	17–30 (ear)	—	[172]
Hordeum vulgare	40–50	—	50–60 (ear)	—	[172]
	94	—	5 (ear)	—	[173]
	Up to 70	3–40	—	—	[174]
Oryza sativa	60	17	23 (ear)	—	[175]
Glycine max	70–85	—	—	4	[1]
	60–70	—	—	16–20	[138]
Pisum sativum	69	—	—	16	[169]
Vigna unguiculata	85–95	—	—	4–13	[176]
Phaseolus vulgaris	90–95	—	—	5–9	[135]
Malus sylvestris	30–40	3–6	—	55–60	[164]
	65–75	—	—	27–30	[133]
	70	—	15–33	—	[177]
Prunus persica	70–78	—	—	3–15	[142]
Cymbidium spp.	90–93	—	7–10	—	[108]
Rubus ursinus	50	<50	—	—	[122]
Olea sp.	30–40	—	—	40–80	[145]
Mangifera indica	>97	—	—	1	[178]
Lycopersicon esculantum	56	—	29	15	[116]
Cucumis sativus	70–80	—	—	20–30	[179]
Opuntia sp.	—	90	—	10	[180]
Spiranthes cernua	91.6	—	8.4	—	[112]

[a] Includes contribution by other vegetative parts if not mentioned otherwise.

Photosynthetic activity of flag leaf in monocots is always crucial that contributes from 50% [172] to 94% [173] to grain filling. However, towards senescence, the tendency decreases considerably, and any extra nitrogen supplied to grain is by the degradation of proteins mainly Rubisco [181]. Like lamina, flag leaf sheath also participates in the grain filling. It photosynthesizes when green but the CO_2 uptake rate is one third that of the leaf [182,183]. It stores assimilates (usually nonstructural carbohydrates) during active periods and transfers them to grain during senescence [184–186].

Sustained assimilate supply from proximal leaves or those subtend the floral parts or fruit is crucial in the provision of carbohydrates and nutrients for the flower formation and fruit growth [158,187,188]. Tomato leaves contribute about 56% to fruit growth [116], while this value is higher, i.e., 70% to 80% in cucumber [179] and even higher, i.e., 91% in *Spiranthes cernua* [112] (Table 26.5). Nearly 90% of the ^{14}C-assimilates by an apple leaf is translocated to the nearby fruit during midseason [164].

Herbaceous stems, nonsucculent green stems, and succulent stems contain stomata and show net photosynthesis, but the contribution of these stems is too low to support their own carbohydrate requirement (Table 26.5). Although rice stem contributes up to 17% to the plant carbon budget [175], it is just 3% for the apple stem [164]. *Opuntia ficus-indica,* by virtue of CAM photosynthesis, annually yields up to 47 tons/ha above ground dry matter [75,180]. Stem of *Trillium* sp. acts as temporary reservoir of carbohydrates and supports the fruit growth when completely defoliated [93]. Woody stems, on the other hand, are devoid of stomata but perform photosynthesis by capturing most part (55% to 90%) of the internally produced CO_2 by pathway similar to CAM [8]. Gross rate of photosynthesis in cortex is always positive, but no net photosynthesis, as determined from O_2 evolution [72,88]. These stems therefore partly rely on the photosynthates imported from leaves.

Reproductive parts including awns, glumes, lemma, palea, ear, and developing grains in cereals, and buds, sepals, petals, anthers, green shoulder, and rosette leaves in other species possess stomata and show net CO_2 uptake [9,10,110,130]. Out of net photosynthesis (18%) of wheat ear, 11% was contributed by awn [121]. Nonfoliar green parts in tomato contribute up to 29% [116], and that of *Spiranthes cernua* contribute approximately 9% to the net photosynthesis, and partly support fruit growth and seed set [113]. Flower photosynthesis is advantageous to apple fruit growth when rosette leaves are no longer exporters of carbohydrates [177].

Tree fruits, legume pods, and developing cereal grains while at immature stage are photosynthetic and show net carbon gain. However, as their growth advances, the photosynthesis by fruits is insufficient to meet the carbohydrate demand and they bank upon assimilates supplied by leaves [10]. With the advancing age, the pod/fruit photosynthesis contributes to the carbon economy by fixing internally respired CO_2 both in the light and the dark with the help of PEPCase. This contributes up to 20% to pea pod [138], 40% to 80% to olive fruit [145], and 10% to 60% to orchids [189]. Ears of cereals refix CO_2 and contribute up to 75% (barley) and 63% (wheat) to gross photosynthesis [130]. Tomato fruit, although devoid of stomata, shows up to 15% of the gross photosynthesis by the PEPCase activity in the pericarp and locular tissues [9].

VIII. CONCLUSIONS

Photosynthesis in leaf, stem, flower, and fruit differs considerably with regard to mechanisms and contribution to carbon budget. In leaves, C_3 pathway operates in temperate plants, while C_4 pathway is characteristic of tropical ones. Another group of tropical and subtropical plants intermediate to C_3 and C_4 pathway has Kranz anatomy but lacks a functional C_4 cycle, whereas the CAM pathway is characteristic of succulent desert plants. Dry matter production is relatively lower in C_3 plants, greater in C_3–C_4 intermediate plants and highest in C_4 plants. CAM plants, on the other hand, show very low productivity due to reduced water use efficiency.

Leaf photosynthesis is modulated by plant as well as environmental factors. Plants factors include position of leaf on the plant for the interception of radiations. Ageing affects net photosynthesis and assimilate partitioning. Leaf photosynthesis might also be different at various phonological stages. A greater and net CO_2 assimilation in the leaves of female plant during flowering indicates a shift of photosynthesis to reproductive growth. Among the important environmental factors, increased CO_2 levels enhance the photosynthetic rate resulting in more dry matter. Temperature adversely affects the activities of photosystems and enzymes of CO_2 assimilation. Sub- or supraoptimal light intensity reduces the rate of leaf photosynthesis by affecting the gas exchange properties. Similarly, lower leaves shaded by canopy or immediately upper leaves in a stand show a diminished rate of photosynthesis and hence senesce readily.

Stem photosynthesis, which is important to the plant carbon economy, is different in plants of different origins. Green stems of nonsucculent plants and cladodes of desert plants possess stomata and show net photosynthesis, although at much lower rate than leaves. Green bark of woody plants, on the contrary, does not have stomata but photosynthesis takes place in the chlorophyllous cortical layer by virtue of PEP-Case that can trap the internally produced CO_2. Stem photosynthesis is important in increased carbon use efficiency to offset the respiratory losses and avoid the anaerobiosis, in addition to curtailing surface water loss.

Reproductive parts, including floral parts, inflorescence of grasses, and fruits, all photosynthesize. The photosynthetic tendency of these structures is greater during early phenology when they are green and decreases towards ripening. They contribute to the net CO_2 fixation, but in very low amounts compared to respective leaves. Fruits are heterotrophic in nature and possess stomata in the outer wall and chloroplasts in the green parts. Their photosynthetic pathway is different from the leaf photosynthesis in many physiological and biochemical respects.

The leaf acts as a major source tissue during vegetative and reproductive growth, whereas other parts are sink most of the time. However, at reproductive growth all green parts contribute to the grain development, seed filling, or fruit growth depending upon the species. It is estimated that flag leaf in grasses and leaf proximal to the fruit in other species contribute up to 60% to 75%, floral parts 8% to 26%, and pods or fruit 1% to 50% to the carbon budget during reproductive development.

REFERENCES

1. Edwards GE, Walker DA. *C3, C4: Mechanisms and Cellular and Environmental Regulation of Photosynthesis*. Oxford: Blackwell Scientific, 1983.
2. Taiz L, Zeiger E. *Plant Physiology*. 3rd ed.. Sunderland, MA: Sinauer Associates, 2002.
3. Ting IP. Crassulacean acid metabolism. *Annu. Rev. Plant Physiol*. 1985; 36:595–622.
4. Wang N, Zhang H, Nobel PS. Carbon flow and carbohydrate metabolism during sink to source transition for developing cladodes of *Opuntia ficus-indica. J. Exp. Bot*. 1998; 49:1835–1843.

5. Nilsen ET. Stem photosynthesis extent, patterns and role in plant carbon economy. In: Gartner B, ed. *Plant Stems: Physiological and Functional Morphology*. San Diego, CA: Academic Press, 1995:223–240.

6. Xu H-L, Gauthier L. Desjardins Y, Gosselin A. Photosynthesis in leaves, fruits, stem and petioles of greenhouse grown tomato plants. *Photosynthetica* 1997; 33:113–123.

7. Wittmann C, Aschan G, Pfanz H. Leaf and twig photosynthesis of young beech (*Fagus sylvatica*) and aspen (*Populus tremula*) trees grown under different light intensity regimes. *Basic Appl. Ecol.* 2001; 2: 145–154.

8. Aschan H, Pfanz H. Non-foliar photosynthesis — a strategy of additional carbon acquisition. *Flora* 2003; 198:81–97.

9. Smillie RM, Hetherington SE, Davies WJ. Photosynthetic activity of the calyx, green shoulder, pericarp and locular parenchyma of tomato fruit. *J. Exp. Bot.* 1999; 50:707–718.

10. Blanke MM, Lenz F. Fruit photosynthesis — a review. *Plant Cell Environ.* 1989; 12:31–46.

11. Carrara S, Pardossi A, Soldatini GF, Tognoni F, Giudi L. Photosynthetic activity of ripening tomato fruit. *Photosynthetica* 2001; 39:75–78.

12. Ali I, Rasul E, Wahid A, Zafar Y, Malik KA. General survey of C-4 grasses and sedges in Faisalabad. *Pak. J. Agric. Sci.* 1986; 23:11–18.

13. Brown RH, Brown WV. Photosynthetic characteristics of *Panicum milioides*, a species with reduced photorespiration (compared to *Panicum maximum*, *Panicum laevifolium*, *Festuca arundinacea*). *Crop Sci.* 1975; 15:681–685.

14. Ku MSB, Wu J, Dai J, Scott RA, Chu C, Edwards EG. Photosynthetic and photorespiratory characteristics of *Flaveria* species. *Plant Physiol.* 1990; 96: 518–528.

15. Bhagwat AS. Photosynthetic carbon assimilation of C-3, C-4, and CAM pathways. In: Passarakli M, ed. *Handbook of Photosynthesis*. New York: Marcel Dekker, 1997:461–480.

16. Sutton BG, Ting IP, Sutton R. Carbohydrate metabolism of cactus in a desert environment. *Plant Physiol.* 1981; 68:784–787.

17. Gonzalez-Real MM, Baille A. Changes in leaf photosynthetic parameters with leaf position and nitrogen content within a rose plant canopy (*Rosa hybrida*). *Plant Cell Environ.* 2000; 23:351–363.

18. Adam NR, Wall GW, Kimball BA, Pinter Jr PJ, LaMorte RL, Hunsaker DJ, Adamsen FJ, Thompson T, Matthias AD, Leavitt SW, Webber AN. Acclimation response of spring wheat in free air CO_2 enrichment (FACE) atmosphere with variable soil nitrogen regimes. 1. Leaf position and phenology determine acclimation response. *Photosynth. Res.* 2000; 66: 65–77.

19. Proietti P, Palliotti A, Famiani F, Antognozzi E, Ferranti F, Andreutti R, Frenguelli G. Influence of leaf position, fruit and light availability on photosynthesis of two chestnut genotypes. *Sci. Hort.* 2000; 85:63–73.

20. Wang Z, Fu J, He M, Yin Y, Cao H. Planting density effects on assimilation and partitioning of photosynthates during filling in the late-sown wheat. *Photosynthetica* 1997; 33:199–204.

21. Ryle GJA, Powell CE, Gordon AJ. Pattern of ^{14}C-labelled assimilate partitioning in red and white clover during vegetative growth. *Ann. Bot.* 1981; 47:505–514.

22. Nicotra AB, Chazdon RL, Montgomery RA. Sexes show contrasting patterns of leaf and crown carbon in a dioecious rainforest shrub. *Am. J. Bot.* 2003; 90:347–355.

23. Rawson HM, Constable GA. Carbon production of sunflower cultivars in field and controlled environments. I. Photosynthesis and transpiration of leaves, stems and heads. *Aust. J. Plant Physiol.* 1980; 7:555–573.

24. Dai Z, Ku MSB, Edwards GE. C_4 photosynthesis. the effects of leaf development on the CO_2-concentrating mechanism and photorespiration in maize. *Plant Physiol.* 1995; 107:815–825.

25. Paul MJ, Pellny TK. Carbon metabolite feedback regulation of leaf photosynthesis and development. *J. Exp. Bot.* 2003; 54:539–547.

26. Rowland DL. Diversity in physiological and morphological characteristics of four cottonwood (*Populus deltoides* var. Wislizenii) populations in New Mexico: evidence for a genetic component of variation. *Can. J For. Res.* 2001; 31:845–853.

27. Schultz ST. Sexual dimorphism in gynodioecious *Sidalcea hirtipes* (Malvaceae). I. Seed, fruit and ecophysiology. *Int. J. Plant Sci.* 2003; 164:165–173.

28. Wang X, Griffin KL. Sex-specific physiological and growth responses to elevated atmospheric CO_2 in *Silene latifolia* Poiret. *Glob. Change Biol.* 2003; 9:612–618.

29. Sicher RC, Bunce JA. Photosynthetic enhancement and conductance to water vapor of field-grown *Solanum tuberosum* (L.) in response to CO_2 enrichment. *Photosynth. Res.* 1999; 62:155–163.

30. Reddy AR, Reddy KR, Hodges HF. Interactive effects of elevated carbon dioxide and growth temperature on photosynthesis in cotton leaves. *Plant Growth Regul.* 1998; 26:33–40.

31. de la Viña G, Pliego-Alfaro F, Driscoll SP, Mitchell VJ, Parry MA, Lawlor DW. Effects of CO_2 and sugars on photosynthesis and composition of avocado leaves grown *in vitro*. *Plant Physiol. Biochem.* 1999; 37:587–595.

32. Isopp H, Frehner M, Long SP, Nosberger J. Sucrose-phosphate synthase responds differently to source-sink relations and to photosynthetic rates: *Lolium perenne* L. growing at elevated P_{CO2} in the field. *Plant Cell Environ.* 2000; 23:597–607.

33. Hui D, Sims DA, Johnson DW, Cheng W, Luo Y. Effect of gradual versus step increase in carbon dioxide on *Plantago* photosynthesis and growth in a microcosm study. *Environ. Expt. Bot.* 2000; 47:51–66.

34. Tissue DT, Thomas RB, Strain BR. Atmospheric CO_2 enrichment increases growth and photosynthesis of *Pinus taeda*: a 4 year experiment in the field. *Plant Cell Environ.* 1997; 20:1123–1134.

35. Sicher RC, Kramer DF, Rodermel SR. Photosynthetic acclimation to elevated CO_2 occurs in transformed tobacco with decreased ribulose-1,5-bisphosphate carboxylase/oxygenase content. *Plant Physiol.* 1994; 104:409–415.

36. Campbell WJ, Allen LH, Bowes G. Effect of CO_2 concentration on Rubisco activity, amount and photosynthesis on soybean leaves. *Plant Physiol.* 1988; 88:1310–1316.

37. McDonald GK, Paulsen GM. High temperature effects on photosynthesis and water relations of grain legumes. *Plant Soil* 1997; 196:47–58.

38. Bukhov NG, Wiese C, Neimanis S, Heber U. Heat sensitivity of chloroplasts and leaves: leakage of protons from thylakoids and reversible activation of cyclic electron transport. *Photosynth. Res.* 1999; 59: 81–93.

39. Holaday AS, Martindale W, Alred R, Brooks AL, Leegood RC. Changes in activities of enzymes of carbon metabolism in leaves during exposure of plants to low temperature. *Plant Physiol.* 1992; 98:1105–1114.

40. Ferrar PJ, Slattyer RO, Vranjic NA. Photosynthetic temperature acclimation in *Eucalyptus* species from diverse habitats and a comparison with *Nerium oleander. Aust. J. Plant Physiol.* 1989; 16:199–217.

41. Chaitanya KV, Sundar D, Reddy AR. Mulberry leaf metabolism under high temperature stress. *Biol. Plant.* 2001; 44:379–384.

42. Vu JCV, Gesch RW, Pennanen AH, Allen LHJ, Boote KJ, Bowes G. Soybean photosynthesis, Rubisco and carbohydrate enzymes function at supra-optimal temperatures in elevated CO_2. *J. Plant Physiol.* 2001; 158:295–307.

43. Makino A, Nakano H, Mae T. Effects of growth temperature on the responses of ribulose-1,5-biphosphate carboxylase, electron transport components, and sucrose synthesis enzymes to leaf nitrogen in rice, and their relationships to photosynthesis. *Plant Physiol.* 1994; 105:1231–1238.

44. Haldimann P. How do changes in temperature during growth affect leaf pigment composition and photosynthesis in *Zea mays* genotypes differing in sensitivity to low temperature? *J. Exp. Bot.* 1999; 50:543–550.

45. Ying J, Lee EA, Tollenaar M. Response of maize leaf photosynthesis to low temperature during the grain-filling period. *Field Crop Res.* 2000; 68:87–96.

46. Boes SR, Huner NPR. Effect of growth temperature and temperature shifts on spinach leaf morphology and photosynthesis. *Plant Physiol.* 1990; 94:1830–1836.

47. Kikuzawa K. Phenological and morphological adaptations to the light environment in two woody and two herbaceous plant species. *Funct. Ecol.* 2003; 17:29–38.

48. Gutierrez MV, Meinzer FC. Carbon isotope discrimination and photosynthetic gas exchange in coffee hedgerows during canopy development. *Aust. J. Plant Physiol.* 1994; 21:207–219.

49. Kephert KD, Buxton RD, Taylor SE. Growth of C_3 and C_4 perennial grasses under reduced irradiance. *Crop Sci.* 1992; 32:1033–1038.

50. Allard G, Nelson CJ, Pallardy SG. Shade effects on growth of tall fescue: I. Leaf anatomy and dry matter partitioning. *Crop Sci.* 1991; 31:163–167.

51. Greer DH. Photon flux density dependence of carbon acquisition and demand in relation to shoot growth of kiwifruit (*Actinida deliciosa*) vines grown in controlled environments. *Funct. Plant Biol.* 2001; 28:111–120.

52. Pearcy RW. Sunflecks and photosynthesis in plant canopies. *Annu. Rev. Plant. Physiol. Plant Mol. Biol.* 1990; 41:421–453.

53. Bunce JA. Contrasting effects of carbon dioxide and irradiance on the acclimation of photosynthesis in developing soybean leaves. *Photosynthetica* 2000; 38:83–89.

54. Vadell J, Socias FX, Medrano H. Light dependency of carboxylation efficiency and ribulose-1,5-bisphosphate carboxylase activation in *Trifolium subterraneum* leaves. *J. Exp. Bot.* 1993; 44:1757–1762.

55. Rosati A, De Jong TM. Estimating photosynthetic radiation use efficiency using incident light and photosynthesis of individual leaves. *Ann. Bot.* 2003; 91: 869–877.

56. Schultz HR. Extension of a Farquhar model for limitations of leaf photosynthesis induced by light environment, phenology and leaf age in grapevine (*Vitis vinifera* L. cvv. White Riesling and Zinfandel). *Funct. Plant Biol.* 2003; 30:673–687.

57. Peri PL, McNeil DL, Moot DJ, Varella AC, Lucas RJ. Net photosynthetic rate of cocksfoot leaves under continuous and fluctuating shade conditions in the field. *Grass Forage Sci.* 2002; 57:157–170.

58. Walcroft AS, Whitehead D, Kelliher FM, Arneth A, Silvester WB. The effects of long-term partial shading on growth and photosynthesis in *Pinus radiate* D. Don trees. *For. Ecol. Manage.* 2002; 28:151–163.

59. Hanba YT, Kogami H, Tarashima I. The effect of growth irradiance on leaf anatomy and photosynthesis in *Acer* species differing in light demand. *Plant Cell Environ.* 2002; 25:1021–1030.

60. Egli DB. Variation in leaf starch and sink limitations during seed filling in soybean. *Crop Sci.* 1999; 39:1361–1368.

61. Oguchi R, Hikosaka K, Hirose T. Does the photosynthetic light-acclimation need change in leaf anatomy? *Plant Cell Environ.* 2003; 26:505–512.

62. Vos J, van der Putten PEL. Effect of partial shading of the potato plant on photosynthesis of treated leaves, leaf area expansion and allocation of nitrogen and dry matter in component plant parts. *Eur. J. Agron.* 2001; 14:209–220.

63. Valladares F, Hernandez LG, Dobarro H, Garcia-Perez C, Sanz R, Pugnaire FI. The ratio of leaf to total photosynthetic area influences shade survival and plastic response to light of green-stemmed leguminous shrub seedlings. *Ann. Bot.* 2003; 91:577–584.

64. Cannon W. The topography of the chlorophyll apparatus in desert plants. *Carnegie Inst. Wash. Pub.* 98 1908.

65. Scott DG. On the distribution of chlorophyll in the young shoots of woody plants. *Ann. Bot.* 1907; 21:437–439.

66. Pearson LC, Lawrence DB. Photosynthesis in aspen bark. *Am. J. Bot.* 1958; 45:383–387.

67. Foote KC, Schaedle M. The contribution of aspen bark photosynthesis to the energy balance of the stem. *For. Sci.* 1978; 24:569–573.

68. Strain BR, Johnson PL. Corticular photosynthesis and growth on *Populus tremuloides*. *Ecology* 1963; 44:581–584.

69. Kriedmann PE, Buttrose MS. Chlorophyll content and photosynthetic activity within woody shoots of *Vitis vinifera* (L.). *Photosynthetica* 1971; 5:22–27.

70. Hartsock Tl, Nobel PS. Watering converts a CAM plant to daytime CO_2 uptake. *Nature* 1976; 262:574–576.

71. Klug M, Ting IP. *Crassulacean Acid Metabolism. Analysis of an Ecological Adaptation.* New York: Springer-Verlag, 1978.

72. Solhaug KA, Haugen J. Seasonal variation of photoinhibition of photosynthesis in bark from *Populus tremula* L. *Photosynthetica* 1998; 35:411–417.

73. Pfanz H, Aschan G. The existence of bark and stem photosynthesis and its significance for the overall carbon gain. An ecophysiological and ecological approach. *Prog. Bot.* 2001; 62:477–510.

74. Hibberd JM, Quick WP. Characteristics of C_4 photosynthesis in stems and petioles of flowering plants. *Nature* 2002; 415:451–454.

75. Nobel PS, Garcia-Moya E, Quero E. High annual productivities of certain agaves and cacti under cultivation. *Plant Cell Environ.* 1992; 15:329–335.

76. Gibson A. Anatomy of photosynthetic old stems of nonsucculent dicotyledons from North America desert. *Bot. Gaz.* 1983; 144:347–362.

77. Osmond CB, Smith SD, Gai-Ying B, Sharkey TD. Stem photosynthesis in a desert ephemeral *Erigonum inflatum*. Characterization of leaf and stem CO_2 fixation and H_2O vapor exchange under controlled conditions. *Oecologia* 1987; 72:542–549.

78. Nilsen ET, Bao Y. The influence of water stress on stem and leaf photosynthesis in *Glycine max* and *Spartium junceum* (Leguminoseae). *Am. J. Bot.* 1990; 77:1007–1015.

79. Pilarski J. Diurnal and seasonal changes in the intensity of photosynthesis in stems of lilac (*Syringa vulgaris* L.). *Acta Physiol. Plant.* 2002; 24:29–36.

80. Brayman AA, Schaedle M. Photosynthesis and respiration of developing *Populus tremuloides* internodes. *Plant Physiol.* 1982; 69: 911–915.

81. Pilarski J. Gradient of photosynthetic pigments in the bark and leaves of lilac (*Syringa vulgaris* L.). *Acta Physiol. Plant.* 1999; 21:365–373.

82. Nilsen ET, Sharifi MR. Seasonal acclimation of stem photosynthesis in woody legume species from the Mojave and Sonoran deserts of California. *Plant Physiol.* 1994; 105:1385–1391.

83. Nilsen ET, Sharifi MR. Carbon isotopic composition of legumes with photosynthetic stems from Mediterranean and desert habitats. *Am. J. Bot.* 1997; 84:1707–1713.

84. Bossard CC, Rejmanek M. Why have green stems? *Funct. Ecol.* 1992; 6:197–205.

85. Haase P, Pugnaire FI, Clark SC, Incoll LD. Diurnal and seasonal changes in cladode photosynthetic rate in relation to canopy age structure in the leguminous shrub *Retama sphearocarpa*. *Funct. Ecol.* 1999; 13:640–649.

86. Cernusak LA, Marshall JD. Photosynthetic refixation in branches of Western White Pine. *Funct. Ecol.* 2000; 14:300–311.

87. Cernusak LA, Marshall JD, Comstock JP, Blaster NJ. Carbon isotope discrimination in photosynthetic bark. *Oecologia* 2001; 128:24–35.

88. Schmidt J, Batic F, Pfanz H. 2000. Photosynthetic performance of leaves and twigs of evergreen holly (*Ilex equifolium* L.). *Phyton* 2000; 40:179–190.

89. Nilsen ET, Rundel PW, Sharifi MR. Diurnal gas exchange characteristics of two stem photosynthesizing legumes in relation to the climate at two contrasting sites in the California desert. *Flora* 1996; 191:105–116.

90. Nobel PS, Hartsock T. Leaf and stem CO_2 uptake in the three subfamilies of the Cactaceae. *Plant Physiol.* 1986; 80:913–917.

91. Lou Y, Nobel PS. Carbohydrate partitioning and compartmental analysis of a highly productive CAM plant *Opuntia ficus-indica*. *Ann. Bot.* 1992; 70:551–559.

92. Nilsen ET, Sharifi MR, Rundel PW, Virginia RA. Phenology of warm desert phreatophytes: seasonal growth and herbivory in *Prosopis glandulosa* var. *Torreyana* (honey mesquite). *J. Arid Environ.* 1987; 13:217–229.

93. Lapointe L. Fruit development in *Trillium*. Dependence on stem carbohydrate reserves. *Plant Physiol.* 1998; 117:183–188.

94. Foote KC, Schaedle M. Diurnal and seasonal patterns of photosynthesis and respiration by stems of *Populus tremuloides Michx*. *Plant Physiol.* 1976; 58:651–655.

95. Pfanz H, Aschan G, Langenfeld-Heyser R, Wittmann C, Loose M. Ecology and ecophysiology of tree stem: corticular and wood photosynthesis. *Naturwissen* 2002; 89:147–162.

96. Aschan G, Wittmann C, Pfanz H. Age dependent bark photosynthesis of aspen twigs. *Trees* 2001; 15:431–437.

97. Szujko-Lacza J, Rakovan JN, Horvath G, Fekete G, Faludi-Daniel A. Anatomical, ultrastructural and physiological studies on one year old *Euonymus europeus* bark displaying photosynthetic activity. *Acta Agron. Acad. Sci. Hung.* 1971; 20:247–260.

98. Langenfeld-Heyser R, Ebrahim-Nesbat F. CO_2-fixation and assimilate transport in the stem of *Picea abies* (L.) Karst. In: Edelin C, ed. *L'arbre: Biologie et Developpment*. Montpellier: Naturalia Monspeliensia, 1991.

99. Larcher W, Lutz C, Nagele M, Bodner M. Photosynthetic functioning and ultrastructure of chloroplasts in stem tissue of *Fagus sylvatica*. *J. Plant Physiol.* 1988; 132:731–737.

100. Buns R, Acker G, Beck E. The plastids of the yew tree (*Taxus haccata* L.): Ultrastructure and immunocyto-chemical examination of chloroplastic enzymes. *Bot. Acta* 1993; 106:32–41.

101. Kharouk VI, Middleton EM, Spencer SL, Rock BN, Williams DL. Aspen bark photosynthesis and its significance to remote sensing and carbon budget estimate in the boreal ecosystem. *Water Air Soil Pollut.* 1995; 82:483–497.

102. Solhaug KA, Gauslaa Y, Haugen J. Adverse effects of epiphytic crustose lichens upon stem photosynthesis and chlorophyll of *Populus tremula* L. *Bot. Acta* 1995; 108:233–239.

103. Van Cleve B, Forretter C, Sauter JJ, Apel K. Pith cells of poplar contain photosynthetically active chloroplasts. *Planta* 1993; 189:70–73.

104. Langenfeld-Heyser R, Schella B, Buschmann K, Speck F. Microautoradiographic detection of CO_2 fixation in lenticels chlorenchyma of young *Fraxinus excelsior* L. stem in early spring. *Tree Struct. Funct.* 1996; 10:255–260.

105. Nedroff JA, Ting IP, Lord EM. Structure and function of the green stem tissue in ocotillo (*Fouquieria splendens*). *Am. J. Bot.* 1985; 72:143–151.

106. Höll W. Dark CO_2 fixation by cell free preparation of wood of *Robinia pseudacacia*. *Can. J. Bot.* 1974; 52:727–734.

107. Arditti J. The green pigment in *Cymbedium* flowers. What is it? *Cymb. Soc. News* 1965; 20:10–11.

108. Dueker J, Arditti J. Photosynthetic $^{14}CO_2$ fixation by green *Cymbidium* (Orchidaceae) flowers. *Plant Physiol.* 1968; 43:130–132.

109. He J, Khoo GH, Hew CS. Susceptibility of CAM *Dendrobium* leaves and flowers to high light and high temperature under normal tropical conditions. *Environ. Exp. Bot.* 1998; 40:255–264.

110. Galen C, Dawson TE, Stanton ML. Carpels as leaves: meeting the carbon cost of reproduction in an alpine buttercup. *Oecologia* 1993; 95:187–193.

111. Vemmos SN, Goldwin GK. Stomatal and chlorophyll distribution of Cox's Orange Pippin apple flowers relative to other cluster parts. *Ann. Bot.* 1993; 71:245–250.

112. Antifinger AE, Wendel LF. Reproductive effort and floral photosynthesis in *Spiranthes cernua* (Orchidaceae). *Am. J. Bot.* 1997; 84:769–780.

113. Weiss D, Schönfeld M, Halevy AH. Photosynthetic activity in the *Petunia* corolla. *Plant Physiol.* 1988; 87:666 670.

114. Khoo GH, Hew CS. Developmental changes in chloroplast ultrastructure and carbon fixation metabolism of *Dendrobium* flowers (Orchidaceae). *Int. J. Plant Sci.* 1999; 160:699–705.

115. Khoo GH, He J, Hew CS. Photosynthetic utilization of radiant energy by CAM *Dendrobium* flowers. *Photosynthetica* 1997; 34:367–376.

116. Hetherington SE, Smillie RM, Davies WJ. Photosynthetic activities of vegetative and fruiting tissues of tomato. *J. Exp. Bot.* 1998; 49:1173–1181.

117. Seddish M, Jolliff GD, Breen PJ. Characterization of Meadofoam CO_2 exchange rates. *Crop Sci.* 1993; 33:515–519.

118. Hogan KP, Garcia MB, Cheeseman JM, Loveless MD. Inflorescence photosynthesis and investment in reproduction in the dioecious species *Aciphylla glaucescens* (Apiaceae). *N.Z. J. Bot.* 1998; 36:653–660.

119. Salopek-Sondi B, Kovac M, Ljubesic N, Magnus V. Fruit initiation in *Helleborus niger* L. triggers chloroplast formation and photosynthesis in the perianth. *J. Plant Physiol.* 2000; 157:357–364.

120. Clement C, Mischler P, Burrus M, Audran JC. Characteristics of the photosynthetic apparatus and CO_2-fixation in the flower bud of *Lilium* I. Corolla. *Int. J. Plant Sci.* 1997; 58:794–800.

121. Olugbemi LB, Bingham J, Austin RB. Ear and flag leaf photosynthesis of awned and awnless *Triticum* species *Ann. Appl. Biol.* 1976; 84:231–240.

122. McDowell SCL, Turner DP. Reproductive effort in invasive and non-invasive *Rubus*. *Oecologia* 2002; 133:102–111.

123. Wang Z-M, Wel A-L, Zheng D-M. Photosynthetic characteristics of non-leaf organs of winter wheat cultivars differing in ear type and their relationship with grain mass per ear. *Photosynthetica* 2001; 39:239–244.

124. Yin Y, Wang Z, He M, Lu S. Postanthesis allocation of photosynthates and grain growth in wheat cultivars as affected by source/sink change. *Biol. Plant.* 1998; 41:203–209.

125. Wirth E, Kelly GK, Fischbeck G, Latzko E. Enzyme activities and products of CO_2 fixation in various photosynthetic organs of wheat and oat. *Z. Pflanzenphysiol.* 1977; 82:78–87.

126. Imaizumi N, Samejima M, Ishihara K. Characteristics of photosynthetic carbon metabolism of spikelets in rice. *Photosynth. Res.* 1997; 52:75–82.

127. Gebbing T, Schnyder H. ^{13}C labeling kinetics of sucrose in glumes indicates significant refixation of respiratory CO_2 in wheat ear. *Aust. J. Plant Physiol.* 2001; 28:1047–1053.

128. Duffus CM, Nutbeam AR, Scragg SA. Photosynthesis in the immature cereal pericarp in relation to grain growth. In: Jeffcoat B, Hawkins AF, Stead Ad, eds. *British Plant Growth Regulator Group Monograph.* Vol. 12. Publication of Long Ashton Research Station. Bristol: University of Bristol, 1985:243–256.

129. Gonzalez M-C, Osuna L, Echevarria C, Vidal J, Cejudo FJ. Expression and localization of phosphoenolpyruvate carboxylase in developing and germinating wheat grains. *Plant Physiol.* 1998; 116:1249–1258.

130. Bort J, Brown RH, Araus JL. Refixation of respiratory CO_2 in the ears of C_3 cereals. *J. Exp. Bot.* 1996; 47:1567–1575.

131. Blum A. Photosynthesis and transpiration in leaves and ear of wheat and barley varieties. *J. Exp. Bot.* 1985; 36:432–440.

132. Vehrich RA, Carner BF, Martin BC. Photosynthesis and water use efficiency of awned and awnletted lines of hard red winter wheat. *Crop Sci.* 1995; 35:172–176.

133. Pavel EW, De Jong TM. Estimating the photosynthetic contribution of developing peach (*Prunus persica*) fruits to their growth and maintenance carbohydrate requirements. *Physiol. Plant.* 1993; 88:331–338.

134. Blanke MM. Stomata of currant fruits. *Angew. Bot.* 1993; 67:1–2.

135. Crookston RK, O'Toole J, Ozbun JL. Characterization of bean pod as a photosynthetic organ. *Crop Sci.* 1974; 14:708–712.

136. Laval-Martin D, Farineau J, Diamond J. Light versus dark carbon metabolism and photosynthetic activity in cherry tomato fruits I. Occurrence of photosynthesis. Study of intermediates. *Plant Physiol.* 1977; 60:872–876.

137. Clijster H. On the photosynthetic activity of developing apple fruit. *Qual. Plant Mater. Veget.* 1999; 19:129–140.

138. Atkins CA, Kuo J, Pate JS, Flinn AM, Steele TW. Photosynthetic pod wall of pea (*Pisum sativum* L.). *Plant Physiol.* 1977; 60:779–786.

139. Phan CT. Photosynthetic activity of fruit tissues. *Plant Cell Physiol.* 1970; 11:823–825.

140. Jones HG. Carbon dioxide exchange of developing apple (*Malus pumila* Mill.) fruits. *J. Exp. Bot.* 1981; 32:1203–1210.

141. Fraser DP, Truesdale MR, Bird CR, Scuch W, Bramley PM. Carotenoid biosynthesis during tomato fruit development. Evidence for tissue-specific gene expression. *Plant Physiol.* 1994; 105:405–413.

142. Moreshet S, Green GC. Photosynthesis and the diffusion conductance of the Valencia orange fruit under field conditions. *J. Exp. Bot.* 1980; 31:15–27.

143. Banks NH, Nicholson SE. Internal atmosphere composition and skin permeance to gases of pepper fruit. *Postharvest Biol. Technol.* 2000; 18:33–41.

144. Blanke MM, Whiley AW. Bioenergetics, respiration cost and water relations of developing avocado fruit. *J. Plant Physiol.* 1995; 145:87–92.

145. Proietti P, Famiani F, Tombesi A. Gas exchange in olive fruit. *Photosynthetica* 1999; 36:423–432.

146. Blanke MM, Holthe PA. Bioenergetics, maintenance respiration and transpiration of pepper fruit. *J. Plant Physiol.* 1997; 150:247–250.

147. Vu JCV, Yelnosky G, Bausher MG. Photosynthetic activity in the flower bud of "Valencia" orange (*Citrus sinensis* L. Osbeck). *Plant Physiol.* 1985; 78:420–423.

148. Chopra J, Kaur N, Gupta AK. Changes in the activities of carbon metabolizing enzymes with pod development in lentil (*Lens culinaris* L.). *Acta Physiol. Plant.* 2003; 25:234–240.

149. Notton BA, Blanke MM. Phosphoenolpyruvate carboxylase in avocado fruit. Purification and properties. *Phytochemistry* 1993; 33:1333–1337.

150. Bravdo B-A, Palgi A, Lurie S. Changing ribulose diphosphate carboxylase/oxygenase activity in ripening tomato fruit. *Plant Physiol.* 1977; 60:309–312.

151. Hedley CL, Harvey DM, Keely RJ. Role of PEP carboxylase during seed development in *Pisum sativum. Nature* 1975; 258:352–354.

152. Jeuffory M-H, Warembourg FR. Carbon transfer and partitioning between vegetative and reproductive organs in *Pisum sativum* L. *Plant Physiol.* 1991; 97:440–448.

153. Gardner FP, Pearcy RB, Mitchell RL. *Physiology of Crop Plants.* Ames, IA: Iowa State University Press, 1985.

154. Petrie PR, Trought MCT, Howell GS, Buchan GD. The effect of leaf removal and canopy height on whole-vine gas exchange and fruit development of *Vitis vinifera* L. Sauvignon Blanc. *Funct. Plant Biol.* 2003; 30:711–717.

155. Sivaplan K. Photosynthetic assimilation of $^{14}CO_2$ by mature brown stem of tea plant (*Camellia sinensis* L.). *Ann. Bot.* 1975; 39:137–140.

156. Mondal MH, Burn WA, Brenner ML. Effect of sink removal on photosynthesis and senescence in leaves of soybean (*Glycine max* L.) plants. *Plant Physiol.* 1978; 61:394–397.

157. Borras L, Otegui ME. Maize kernel weight response to postflowering source–sink ratio. *Crop Sci.* 2001; 41:1816–1822.

158. Wahid A, Bukhari S, Rasul E. Inter-specific differences in cotton for nutrient partitioning from subtending leaves to reproductive parts at various developmental stages: consequences for fruit growth and yield. *Biol. Plant.* 2003/04; 47:379–385.

159. Evans LT. *Crop Evolution, Adaptation and Yield.* Cambridge, UK: Cambridge University Press, 1993.

160. Wang Z, Fu J, He M, Tian Q, Cao H. 1997. Effects of source sink manipulation on net photosynthetic rate and photosynthate partitioning during grain filling in winter wheat. *Biol. Plant.* 1997; 39:379–385.

161. Yang J, Peng S, Zheng Z, Wang Z, Visperas RM, Zhu Q. Grain and dry matter yields and partitioning of assimilates in Japonica/Indica hybrid rice. *Crop Sci.* 2002; 42:766–772.

162. Vemmos SN. Net photosynthesis, stomatal conductance, chlorophyll content and specific leaf weight of pistachio trees (cv. Aegenes). *J. Hort. Sci.* 1994; 69:775–782.

163. Fukii JA, Kennedy RA. Seasonal changes in the photosynthetic rate in apple tree. A comparison between fruiting and non-fruiting trees. *Plant Physiol.* 1985; 78:519–524.

164. Hansen P. ^{14}C-studies on apple trees. IV. Photosynthetic consumption in fruits in relation to the leaf-fruit ratio and to the leaf-fruit position. *Physiol. Plant.* 1969; 22:186–198.

165. Ruan Y-L, Llewellyn DJ, Furbank RT. Pathway and control of sucrose import into initiating cotton fiber cells. *Aust. J. Plant Physiol.* 2000; 27:795–800.

166. Turgeon R. Plasmodesmata and solute exchange in the phloem. *Aust. J. Plant Physiol.* 2000; 27:521–529.

167. Offler CE, Thorp MR, Patrick JW. Assimilate transport and partitioning: Integration of structure, physiology and molecular biology. *Aust. J. Plant Physiol.* 2000; 27:473–476.

168. Bertin N. Competition for assimilates and fruit position affect fruit set in indeterminate greenhouse tomato. *Ann. Bot.* 1995; 75:55–65.

169. Flinn AM. Regulation of leaflet photosynthesis by developing fruit in the pea. *Physiol. Plant.* 1974; 31:275–278.

170. Cruz-Aguado JA, Reyes F, Rodes R, Perez, I, Dorado M. Effect of source-to-sink ratio on partitioning of dry matter and ^{14}C-assimilates in wheat during grain filling. *Ann. Bot.* 1999; 83:655–665.

171. Austin RB, Edrich J. Effect of ear removal on photosynthesis, carbohydrate accumulation and on the distribution of assimilated ^{14}C in wheat. *Ann. Bot.* 1975; 39:141–152.

172. Thorne GN. Varietal differences in photosynthesis of ears and leaves of barley. *Ann. Bot.* 1963; 27:155–174.

173. Walpole RP, Morgan DG. Physiology of grain filling in barley. *Nature* 1972; 240:416–417.

174. Bonnett GD, Incoll LD. The potential pre-anthesis and post-anthesis contribution of stem internodes to grain yield in crops of winter barley. *Ann. Bot.* 1992; 69:219–225.

175. Enyi BAC. The contribution of different organs to grain weight in upland and swamp rice. *Ann. Bot.* 1962; 26:525–531.

176. Peoples MB, Pate JS, Atkins CA, Murray DR. Economy of water carbon and nitrogen in the developing cowpea fruit. *Plant Physiol.* 1985; 77:142–147.

177. Vemmos SN, Goldwin GK. The photosynthetic activity of Cox's orange pippin apple flowers in relation to fruit setting. *Ann. Bot.* 1994; 73:385–391.

178. Chauhan PS, Pandey RM. Relative ^{14}CO$_2$ fixation by leaves and fruits, and translocation of ^{14}C-sucrose in mango. *Sci. Hortic.* 1984; 22:121–128.

179. Marcelis LFM, Hofman-Eijer LRB. The contribution of fruit photosynthesis to the carbon requirement of cucumber fruits as affected by irradiance, temperature and ontogeny. *Physiol. Plant.* 1995; 93:476–483.

180. Nobel PS, De la Barrera E. Carbon and water balances for young fruits of platyopuntia. *Physiol. Plant.* 2000; 109:160–166.

181. Moutot F, Heut JC, Morot-Gaudry JF, Pernollet JC. Relationship between photosynthesis and protein synthesis in maize I. Kinetics of the translocation of photoassimilates from ear leaf to the seed. *Plant Physiol.* 1986; 80:211–215.

182. Evans LT, Rawson MH. Photosynthesis and respiration by the flag leaf and component of the ear during grain development in wheat. *Aust. J. Biol. Sci.* 1970; 23:245–254.

183. Petterson TG, Moss DN, Burn WA. Enzymatic changes during the senescence of field grown wheat. *Crop Sci.* 1980; 20:15–18.

184. Judal GK, Mengel K. Effect of shading on non-structural carbohydrates and their turnover in culm and leaves during the grain filling period of spring wheat. *Crop Sci.* 1980; 22:958–962.

185. Araus JL, Tapia L. Photosynthetic gas exchange characteristics of wheat flag leaf blades and sheaths during grain filling: the case of a spring crop grown under Mediterranean climate conditions. *Plant Physiol.* 1985; 85:667–673.

186. Prioul J-L, Schwebel-Dugue N. Source–sink manipulations and carbohydrate metabolism in maize. *Crop Sci.* 1992; 32:751–756.

187. Wullschlenger SD, Oosterhuis DM. Photosynthetic carbon production and use by developing cotton leaves and bolls. *Crop Sci.* 1990; 30:1259–1264.

188. Unruh BL, Silvertooth JC. Comparison between upland and a pima cotton cultivar: II. Nutrient uptake and partitioning. *Agron. J.* 1996; 88:589–595.

189. Zotz G, Vollrath B, Schmidt G. Carbon relations of fruits of epiphytic orchids. *Flora* 2003; 198:98–105.

Section IX

Photosynthesis and Plant/Crop Productivity and Photosynthetic Products

27 Photosynthetic Plant Productivity

Lubomír Nátr
Department of Plant Physiology, Faculty of Science, Charles University

David W. Lawlor
Rothamsted Research, Harpenden

CONTENTS

Feeding the world's population is considered to be one of the major challenges of the 21st century. The response to this challenge consists of several factors.

1. Policy makers and governments should make a major effort to alleviate hunger all over the world, particularly in developing countries where it is worst, by encouraging local production.

2. Economic and technological means should be used to maximize plant production in developing countries and to optimize it to resources in developed ones, without the distortions produced by subsidies and market protection.

3. Farmers should be provided with current knowledge about yield formation and how to increase efficiency of utilization of all the resources and technology, including use of

improved genotypes, fertilization, crop protection, and irrigation methods to preserve soil fertility and biodiversity, etc.

4. Scientists should enhance understanding of crop growth and development with special emphasis on the quantification of productive ideotypes for particular environments, with optimal efficiency in mineral nutrient use and dry matter allocation, and with resistance to various biotic and abiotic stresses.

This chapter does not deal with the first three aspects. Nevertheless, it is important to stress that science itself cannot solve the deplorable problem of food shortage. Scientists have to fulfill their duties to develop the understanding and techniques, enabling an environmentally friendly, and biologically effective, increase in crop production at all locations, not only those with optimal climate and soil conditions but also those with less favorable conditions.

This chapter deals with principles of photosynthetic productivity. It is evident that the scientific base of crop production involves many branches of science including pedology, chemistry, genetics, and plant physiology. Of all the aspects of plant processes contributing to production, photosynthesis has to be considered the most relevant. This is demonstrated by the following.

Crop biomass produced in the field consists of water and dry matter. Considering the energy and nutritional needs of people and other animal consumers, it is the dry mass which is important, particularly in the case of cereal and legume grains, the staple

food of humanity. Farmers harvest this biomass with various amounts of water simply because in many cases, it is not convenient or possible to leave the water on the field. Dry mass consists of organic substances (carbohydrates, proteins, lipids) and mineral elements (phosphorus, potassium, calcium, magnesium, etc.). These mineral constituents contribute only about 5% to the total dry mass. Hence, the majority of dry matter consists of assimilates synthesized in photosynthesis. This is not to say that both water and minerals in biomass are useless. But it is essential to emphasize that the amount (mass) of assimilate produced plays a decisive role in the formation of economic yield. Because assimilates are synthesized in photosynthesis, using energy derived from the sun and consuming carbon dioxide from the atmosphere, the amount of absorbed and fixed solar energy, as well as the amount of incorporated carbon dioxide, are the biological basis of crop production.

In this chapter, several features of photosynthetic productivity will be illustrated by considering cereals. This is because of their importance in nutrition (Figure 27.1) and because many principles of yield formation have been deduced from studies with cereal species, predominantly wheat, rice, and maize.

I. DRY MATTER PRODUCTION IN A CROP CANOPY

In most crops, economic yield is represented by only one structural part (grain, tubers, roots). The ratio of

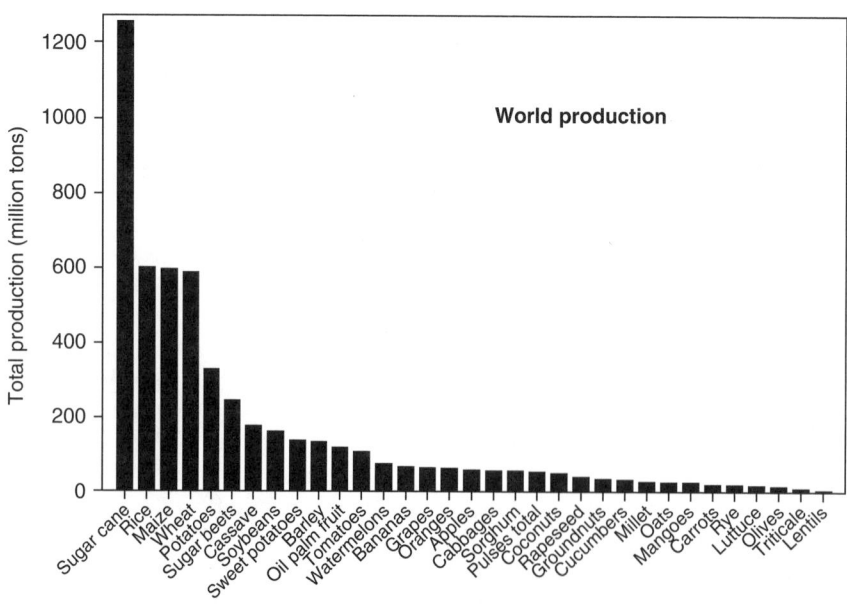

FIGURE 27.1 World production (10^6 tons) of the main crops. FAO data.

dry mass of the main economic yield (Y) to the total plant or canopy dry mass (W) is called the harvest index (HI). Hence,

$$Y = W \cdot \text{HI}$$

Instead of total plant dry mass (W), only shoot (above ground) dry mass is used because measuring root dry mass is often difficult.

It follows that yield could be increased either by increasing harvest index or total plant dry matter. Intuitively, HI cannot reach unity. This would be possible only for crops in which the main economic yield is represented by the total plant. In most crops, HI increases with selection of improved crop varieties and can be optimized for particular environmental conditions, so it is not a fixed value. However, it is assumed that HI has biological limits, the exact values of which are not known. For example, in cereals, HI ranges from about 0.4 to 0.6, indicating, that up to 60% of the total or above ground dry matter accumulates in grain.

The other possibility of increasing yield is by an enhancing total plant dry mass, which is determined by the length of the growing period (D), leaf area of the canopy (L), and rate of net photosynthesis per unit of leaf area (P). On the other hand, total plant dry mass (W) is reduced by losses due to respiration (R). Hence (Figure 27.2),

$$W = (D \cdot L \cdot P) - R$$

This expression is simplified, neglecting the variability of each of the components, but it clearly indicates the main factors determining both the absorption and utilization of solar radiation in a canopy. Next, the individual components are treated in more detail.

A. Length of the Growing Period

There is no doubt that the longer the duration of active photosynthesis by a crop, the more the produc-

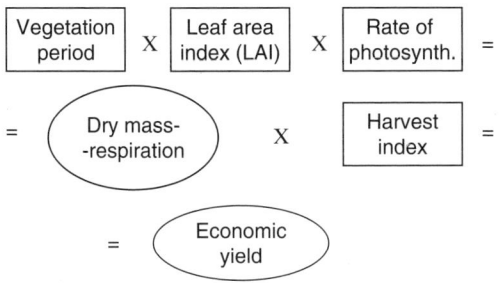

FIGURE 27.2 Schematic illustration of the participation of the main plant and canopy parameters in yield formation.

tion of assimilates. However, the period suitable for crop growth is mainly correlated with the genetically determined life-span of the crop and by the geographical location; the latter can hardly be altered although selective adaptation of wheat, for example, to grow at high latitudes with extended daylight in summer, is associated with increased yields. There are at least two other possibilities of how to extend the growing period:

1. The selection of genotypes (plant species or varieties) that have the capacity for growth and photosynthesis over longer periods, and are, for example, less sensitive to low temperature, could enhance the growing period in cold climates. Similarly, in climates where summer heat restricts the growing season, more drought and heat resistant genotypes could use more days of a year to capture solar radiation and use it for photosynthesis.
2. The use of all the days of a growing season by minimizing the time lag between the cultivation of successive crops. Bare soil, without a full green surface cover of growing plants, should be avoided as much as possible by suitable management practices.

Often, the final dry mass of crops depends on the length of their vegetation period (Figure 27.3) According to Monteith and Elston [1], the mean values of daily dry matter production of a canopy for the two plant groups are:

$$C_3 \text{ plants}: 13.0 \pm 1.6 \, \text{g m}^{-2} \, \text{day}^{-1}$$

$$C_4 \text{ plants}: 22.0 \pm 3.6 \, \text{g m}^{-2} \, \text{day}^{-1}$$

The values are realistic, obtained under conditions suitable for growth of the particular types and with no nutrient or water limitation. Similar values of the maximum rate of daily plant dry matter production have been given by de Wit in the 1960s (Table 27.1).

Obviously, in the term "vegetation period," solar radiation plays a decisive role as the amount of dry matter produced by a canopy, well supplied with water and nutrients, depends on the amount of solar radiation absorbed. Photosynthetic efficiency of a canopy is the amount of dry matter produced per unit of absorbed energy (Table 27.2). This efficiency is relatively constant during the period when the canopy is closed, and is mostly maintained by gradual replacement of old senescing leaves of lower insertions with newly developed leaves of higher insertion in the well-lit upper canopy.

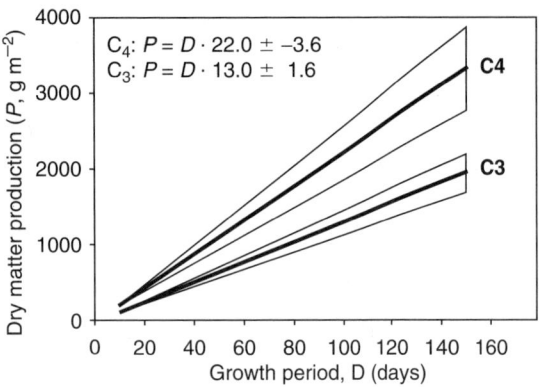

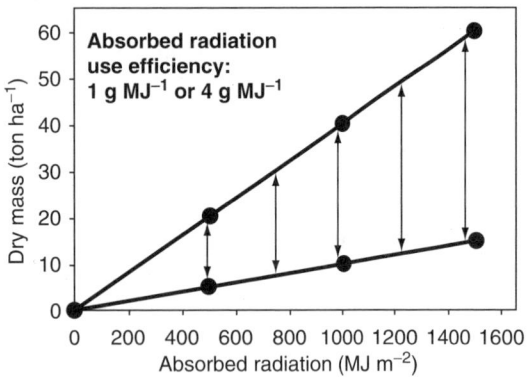

FIGURE 27.3 (Top) Dependence of canopy dry mass on the duration of growth of C_3 and C_4 plants. Equations (included in the graph) taken from Ref. [1]. (Bottom) The dependence of canopy dry mass on absorbed radiation, which is used with the efficiency of 1 or 4 g/MJ.

TABLE 27.1
Potential (Theoretical Maximum) Production of Assimilates ($t\,ha^{-1}\,month^{-1}$). The Values in the Numerator and Denominator are for a Clear and a Cloudy Day, Respectively. Calculated with Data from [56]

Latitude	January	April	July
0°	12.8/6.8	12.8/6.8	12.8/6.8
40°	6.8/3.1	12.8/6.7	15.4/8.2
85°	0/0	10.0/4.0	19.6/9.2

B. LEAF AREA INDEX

Quanta of solar energy incident on a canopy must be absorbed as effectively as possible and then utilized with the maximum efficiency in photosynthetic processes for maximizing production (Figure 27.4). Photosynthesizing organs, mostly leaf blades but including leaf sheaths, stem, internodes, ears, etc., con-

TABLE 27.2
Efficiency with which Photosynthetically Active Radiation is Used to Form Dry Matter of Several Plant Species [100]. One Gram of Dry Matter Corresponds to About 17 kJ and so a 100% Use Efficiency would give Values of About 60 g Dry Matter per 1 MJ

Plant Species	g (dry matter) MJ^{-1}
Soybean (*Glycine max*)	1.3
Clover (*Trifolim subterraneum*)	1.6
Cotton (*Gossypium hirsutum*)	2.5
Sunflower (*Helianthus annuus*)	2.6
Maize (*Zea mays*)	3.4
Rice (*Oryza sativa*)	4.2

tribute to light absorption and photosynthesize to different extents. We should speak about the area of photosynthesizing or assimilating plant parts, but for simplicity and with respect to tradition, we shall call it leaf area. It has been shown that the size of individual leaves, or total leaf area per plant, is not so important as the total leaf area of a canopy per unit ground area, which is called the leaf area index (LAI), and is most often expressed as m^2 (leaf area) per m^2 (ground area). The concept of LAI proved to be extremely useful in assessing the most important parameter of the photosynthetic activity of a canopy [4,5].

It is interesting to recall that LAI is one of several parameters used in growth analysis, which was developed in England in the 1920s, and further elaborated and used, especially in the 1950s, by Watson and his coworkers [6,7]. At the same time, Ničiporovič represented a Soviet school [8,9], which also contributed substantially to our understanding of the importance of canopy leaf area.

Radiant energy incident on a canopy (Q) is either reflected (Q_R), absorbed (Q_A), or transmitted (Q_T). The coefficients of reflection (r), absorption (a), and transmission (t) are then given by Q_R/Q, Q_A/Q, and Q_T/Q, respectively. The mean values for these are 0.15, 0.75, and 0.1, respectively, for photosynthetically active radiation (PAR). However, it should be taken into account that the values of the coefficients also depend on the spectrum (wavelength) of the incoming radiation. Because leaves absorb blue and red wavelengths predominantly, in comparison with absorption of the green part of the spectrum, vegetation appears green to the human eye. In the ultraviolet part of the spectrum (wavelengths shorter than 380 nm), the absorption coefficient reaches nearly 1. On the other hand, in the near infrared part (750 to 1200 nm), absorption is rather low with values of the

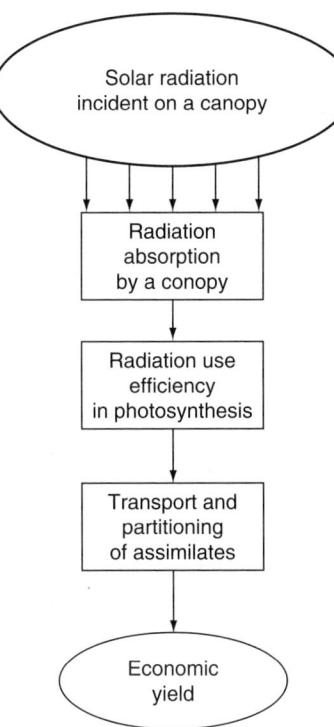

FIGURE 27.4 Consecutive steps in the use of solar radiation for production of crop yield.

absorption coefficient only 5% to 25%. And finally, long-wave radiation (above 4000 nm) is nearly completely absorbed by leaves.

C. PENETRATION OF SOLAR RADIATION INTO THE CANOPY

Leaf absorption characteristics are important photosynthetic features, but the penetration of radiation into the canopy is of more importance for photosynthesis of the whole stand. Japanese researchers Monsi and Saeki [10] quantitatively described penetration of radiation into the canopy. Irradiance at a particular positioning the canopy (I) is related to the cumulative LAI from the top of the canopy (L) by:

$$I = I_0 e^{-kL}$$

where I_0 is the solar radiation incident on the top of the canopy (Wm^{-2}), k is the extinction coefficient of the canopy, and L is the cumulative LAI from the top of the canopy down to the appropriate height.

According to a review [11], the extinction coefficients vary in canopies of different species (Table 27.3) from about 0.4 to 0.9. If the extinction coefficient equals 0.5, then the irradiance at the bottom of a canopy with LAI 3 equals $1 \times e^{-0.5.3}$, i.e., 22%. Similarly for $L = 5$ irradiance at the bottom of the

TABLE 27.3
Extinction Coefficients of Radiation Penetration into the Canopy of Several Plant Species [11]

Plant Species	Coefficient
Reed (*Phragmites communis*)	0.51
Fescue (*Festuca arundinacea*)	0.57
Sugar cane (*Saccharum officinarum*)	0.59
Sorghum (*Sorghum vulgare*)	0.61
Clover (*Trifolium repens*)	0.76
Rape (*Brassica napus*)	0.84
Alfalfa (*Medicago sativa*)	0.88
Cowpea (*Vigna unguicaulata*)	0.93

canopy corresponds to only 8.5% of the irradiance at the top (see Figure 27.5).

The extinction coefficient for canopies with vertically oriented leaves (monocots — cereals, grasses) is generally less than 0.6. On the other hand, canopies of dicots, with large leaves in a more horizontal orientation, have extinction coefficients larger than 0.7.

Knowing the proportion of reflected radiation, it is possible to calculate the amount of absorbed radiation by individual canopy layers on the basis of their cumulative LAI and extinction coefficients. The mean proportion of radiation reflected from a canopy is in the range of 15% to 25%. A canopy with predominantly vertical orientation of narrow leaf blades reflects less radiation, especially if the sun is high above the horizon. In this case, radiation penetrates deeper into the canopy. On the other hand, if the sun is low above the horizon, and particularly in canopies with large leaf blades, the reflection increases.

In order to determine the suitability of a canopy to maximize capture of the incident solar radiation, the static value of the LAI is not sufficient. The dynamics of the development of LAI during the whole growth period determines the total energy captured by the canopy.

D. DYNAMICS OF LEAF AREA INDEX

At sowing, the bare soil absorbs all of the incident radiation, thus increasing evaporation. After emergence, the leaves appear: If they are distributed horizontally, this enables a more rapid cover of the soil compared with the more vertical orientation. Hence, at this stage of canopy development, rapid growth of horizontally oriented leaves is required to achieve a highly productive canopy. When LAI reaches values of about 3 and higher, most of the incident radiation is absorbed by the leaves. At this stage, optimum LAI must be maintained as long as possible. It is evident

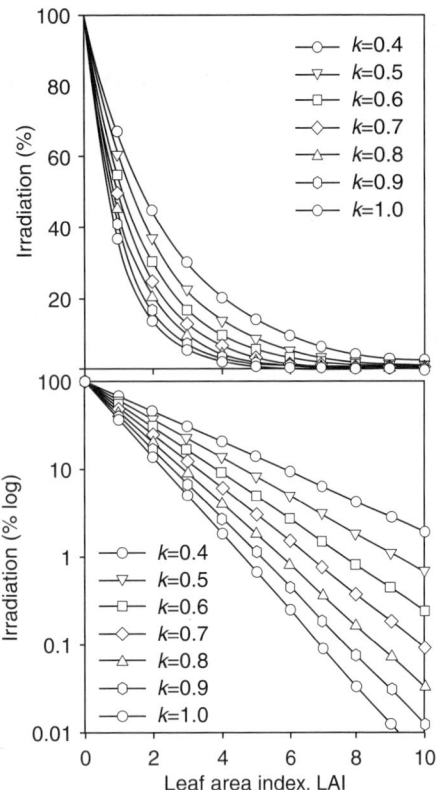

FIGURE 27.5 Radiation (I, % of the incident radiation I_0) within the canopy at a height expressed by cumulative leaf area index (L, from the canopy top) and its dependence on the canopy extinction coefficient (k) calculated as $I = I_0 \cdot e^{(-Lk)}$. For details, see text.

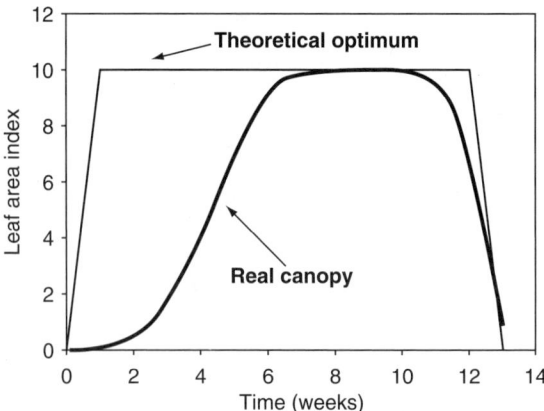

FIGURE 27.6 Schematic of the time course of development of the leaf area index in the real canopy (thick line) and theoretical optimum (thin line) for maximum absorption of incident solar radiation during the growing period.

that in most crops, an optimum LAI is not maintained by the same leaves. The older leaves become senescent and die and must be regularly replaced by leaves of higher insertions. In cereals and other crops cultivated not for the whole shoot biomass but only for certain organs, the leaves should be active as long as possible. Theoretically, they should senescence just before harvest (Figure 27.6).

As seen from the description of the dynamics of LAI, there are three growth stages suitable for genetic or physiological manipulation. (1) The period from emergence to the time when optimum LAI is reached should be as short as possible. Otherwise, part of incident solar radiation is absorbed by the soil surface and lost for production. (2) The period of optimum LAI should last as long as possible. The question about the optimum value itself is dealt with in the next chapter. (3) Leaf senescence should be prevented, for as long as possible, in order to produce the maximum of carbohydrates. However, this statement should be taken with caution. In many crops, leaves

contain a considerable amount of protein, which is broken down during senescence and the nitrogen compounds translocated into organs of economic importance (grain). Hence, this third stage represents a compromise between the need to prolong assimilate production in the leaves whilst allowing enough time for the break down of proteins and their transport out of senescent leaves.

E. RADIATION USE EFFICIENCY

According to many authors (see Ref. [12]), the radiation use efficiency of crops has not changed during the last century, a period over which yields increased considerably. The major contribution to this yield increase was modification of dry matter allocation, shown by an increase in HI of cereals from about 0.3 at the beginning of the last century up to the current values approaching about 0.6. However, the possibility of a further enhancement of HI is obviously limited because a plant cannot consist only of grains. Therefore, any further yield increase will depend on the increase in the efficiency with which the absorbed radiation is used in photosynthesis, to increase biomass production. However, this also requires that the capacity of the grain to accumulate dry matter must also increase.

Efficiency can be considered in the following way. For fixation of 1 mol CO_2, some 8 to 10 mol of PAR quanta are needed, which approaches 20% of the utilization of the absorbed radiation energy at maximum efficiency. The global annual utilization of incident solar radiation corresponds to about 0.1% of total radiation, or 0.2% of the PAR. In canopies of various crops, the utilization efficiency varies from about 0.5% to some 3%.

Depending on the CO_2 concentration and temperature, the amount of dry matter produced per unit of radiation ranges from 1 to 3 μg/ J, corresponding to 3 g/MJ (Table 27.2; [13]). In legumes, the values are somewhat lower because of the higher content of proteins, the synthesis of which requires more energy. For example, Jeuffroy and Ney [14] found values in the range from 0.96 to 1.42 g dry mass per megajoule incident PAR.

Theoretical analysis of radiation use efficiency in a closed canopy and its dependence on quantum requirement has been published [15]. The calculations show (Table 27.4) that a closed canopy could achieve radiation use efficiency from 4.2 to 5.8 g dry mass per megajoulePAR absorbed. There is no doubt that an increase in the radiation use efficiency achieved either by breeding or agronomic management is a major challenge to the plant sciences, and represents one of the most promising tools to obtain yield increase in the future.

F. EFFICIENCY OF SOLAR ENERGY UTILIZATION IN CANOPIES

Leaves absorb about 75% of the incident solar radiation. If we suppose that in a canopy, leaves with horizontal spatial arrangement are uniformly distributed, then with an LAI value of 3, nearly all incident radiation penetrating into the upper layers of a canopy would be absorbed. Hence, an LAI of 3 should be sufficient. Therefore, it is surprising that highly productive canopies have been characterized by LAIs of 8 and more. The explanation becomes apparent, if we take into consideration not only maximization of solar energy capture, but also maximization of the use of absorbed energy in photosynthesis.

The rate of net photosynthesis per unit area of individual leaves increases with increasing irradiation, reaching saturation at about PAR 600 and 900 $\mu mol\ m^{-2}\ sec^{-1}$ for C_3 and C_4 plants, respectively. In regions with moderate climate, the rate of canopy net photosynthesis is not saturated even at the highest summer irradiances. Leaves at the top of the canopy receive more light than can be efficiently used in photosynthesis. The amount of assimilate produced per unit leaf area increases progressively less with the increase in irradiance, so the radiation use efficiency per joule of absorbed radiation is highest at lowest irradiances and decreases with an increase in irradiation (Figure 27.7). This point explains the need to increase the LAI of highly productive canopies well above values that would assure the maximum physically possible absorption of the incident radiation.

It follows that in a productive canopy, not only the time course of the LAI but also optimization of the spatial distribution of leaves — mainly blades is of prime importance [16]. Varying spatial arrangement enables the absorption of energy by the individual leaves to be varied and optimized for photosynthesis.

Let us suppose that several types of canopies differ in the vertical inclination of their leaves. With an angle of 90° and uniform horizontal distribution, an LAI of

---

TABLE 27.4
Calculation of the Radiation use Efficiency (RUE, g(dry matter) MJ^{-1}) of a Closed Maize Canopy (C_4 plant) for Three Different Quantum Requirements (Quantum 15 is almost the Minimum for the Photosynthetic Fixation of 1 CO_2). Assumed Rate of Maintenance Respiration is 0.5 mmol (CH_2O) per g (Dry Mass) per Day, which is Equivalent to 0.015 g g^{-1} Day^{-1} of the 1400 g Canopy Dry Mass per m^2, and Incident Solar Radiation of 28 MJ m^{-2} Day^{-1}. Growth Efficiency, i.e. Produced Plant Dry Mass per Unit Consumed Glucose, is 0.74, which Corresponds to Dry Matter Containing 43% Carbon and 1.1% Nitrogen. Assumptions: Intercepted Solar Radiation: 2.20 mol (1.00 MJ), Radiation Reflection from the Canopy: –0.13 mol, PhAR Quanta Absorbed by the Canopy: 2.07 mol [13]

Calculation of the RUE

Quantum requirement	14	16	18
The amount of produced CH_2O (mmol)	148	129	115
CH_2O consumed for maintenance (mmol)	25	25	25
CH_2O available for growth (mmol/g)	123/3.69	104/3.12	90/2.70
RUE for intercepted solar radiation (g/MJ^{-1})	2.7	2.3	2.0
RUE for PhAR			
g (dry matter) MJ^{-1} (incident/absorbed PhAR)	5.5/5.8	4.6/4.9	4.0/4.2

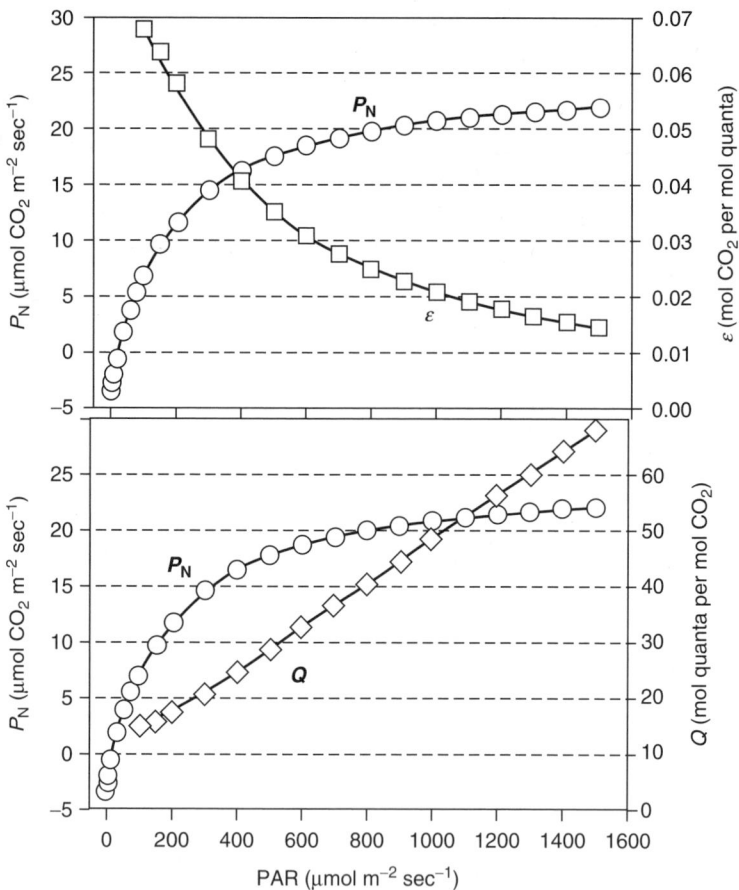

FIGURE 27.7 The effect of PAR flux on the rate of photosynthesis (P_N, µmol m^{-2} sec^{-1}) and quantum yield (ε, mol CO_2 per mol quanta) or (below) quantum requirement (mol quanta per mol CO_2).

unity is sufficient to cover the ground. All the leaves will be exposed to direct radiation of, for example, 1000 W m^{-2} from above (Figure 27.8). In this case, its leaves will be irradiated with 1000 W m^{-2} intensity (for simplicity neglecting reflectance), and their rate of photosynthesis will be high but their radiation use efficiency rather low. If the canopy is composed of plants with a uniform leaf angle of 20°, i.e., nearly vertical position, then its rate of photosynthesis will be only about 15 µmol m^{-2} sec^{-1}, but its radiation use efficiency will be 0.05 mol CO_2/mol quanta (Figure 27.9). Furthermore, in such a canopy, the LAI will be about 3 and the total photosynthetic absorption of CO_2 will reach 45 µmol m^{-2} sec^{-1}. Figure 27.10 illustrates the effect of leaf angle on irradiation of individual leaves and on the rate of photosynthesis per unit leaf area and also per unit ground area. The relatively low rates of photosynthesis per leaf area of nearly vertical leaves is more than compensated by their potential maximum LAI (Figure 27.8) resulting in a large rate of photosynthesis per unit ground area (Figure 27.10). It is also important to stress that erect leaf blades are irradiated from both adaxial and abaxial surfaces, which further improves their radiation use efficiency [17].

As shown by Kuroiwa [18], using a theoretical model of canopy structure consisting of leaves with various inclinations, leaves of the highest insertion should be in a near vertical position, absorbing only a minor part of the incident radiation. The lower the insertion, the more horizontal the leaf position should be, with leaves of the lowest insertion being horizontal in order to absorb the remaining radiation penetrating deeply into the canopy.

Let us now recall the time course of the LAI during the vegetation. After sowing, rapid growth of horizontally oriented leaves is desirable in order to fully cover the whole ground surface. When this is achieved, leaves should take more vertical position in order to maximize not only energy absorption but also photosynthesis (see above).

Of course, leaves will not change their spatial orientation according to our theoretical consideration. But plant breeders have been successful in selecting genotypes, the first leaves of which are oriented horizontal, while the subsequent ones are much more vertical [19]. This is well demonstrated by the canopy structures of old and current cereal varieties. The old varieties were not only tall (and lodging prone), but also had large leaves, which

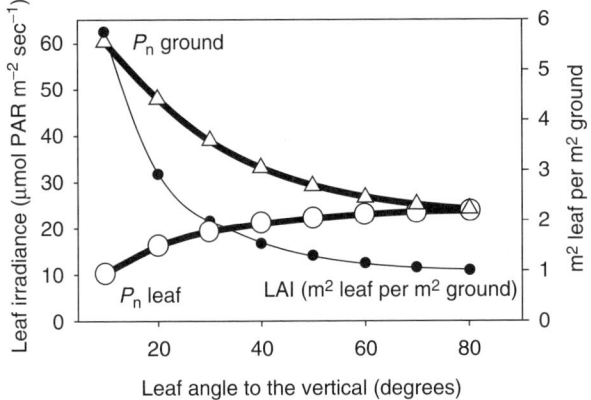

FIGURE 27.8 The effect on the rate of photosynthesis per unit leaf area (P_N leaf; μmol m^{-2} sec^{-1}) and per unit ground area (P_N ground; μmol m^{-2} sec^{-1}) of the angle between the leaf and the vertical, and the maximum leaf area index (m^2 leaf per m^2 ground) that would uniformly cover the ground by one leaf layer of the appropriate inclination,. For details, see text.

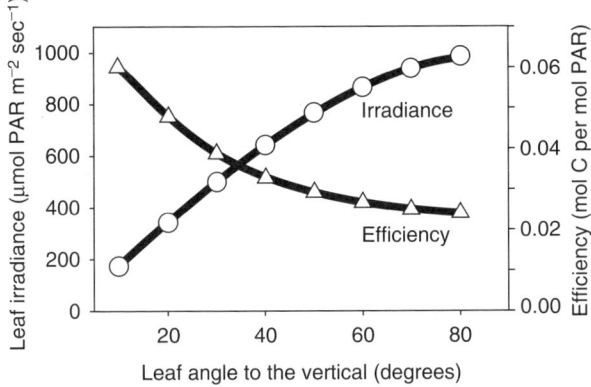

FIGURE 27.9 The effect of leaf angle, on the radiation incident on the leaf and on the leaf's radiation use efficiency.

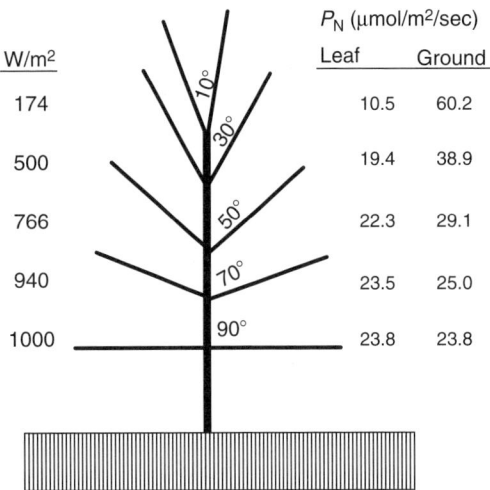

FIGURE 27.10 Schematic of the effect of leaf inclinations, from 10° to 90°, on PAR fluxes received (W m^{-2}) and on P_N (the rate of photosynthesis per unit leaf area, μmol m^{-2} sec^{-1}). The rate per unit ground area was calculated on the assumption that the ground is uniformly covered by one leaf layer of the appropriate inclination.

were unable to maintain a vertical orientation. In contrast, modern wheat or barley varieties are not only much shorter (in order to prevent lodging) but also their leaf blades are small and practically vertical. The optimum spatial arrangement of leaves of cereal varieties has been supported by fact that their leaf sheaths and internodes are vertical and also contribute substantially to total assimilate production. In this way, the LAI, including the total assimilating area, of these crops reaches values well above 6.

A remarkable confirmation of the advantage of vertical orientation of the leaves in the canopy was provided by Blackmann [20]. Under the climatic conditions of Oxford, U.K., dry matter production of several plant species cultivated for 101 days was com-

pared. Surprisingly, the maximum dry matter production corresponding to 30 ton ha^{-1} was reached by Gladiolus, the canopy of which consisted of vertical leaves and reached an LAI above 20.

The great increases in production and efficiency that may result from the growing conditions is shown by Polonskij and Lisovskij [21]. Wheat plants were cultivated in environmentally controlled chambers at 20°C to 21°C, relative air humidity 50 to 60%, 0.4 to 0.9% of CO_2 concentration, and irradiance up to 1300 W m^{-2} PAR. There were 2000 plants m^{-2} and the vegetation period was 63 to 65 days. At highest irradiance, the LAI reached 33.2, the total plant biomass 8 kg m^{-2}, and grain yield 3.3 kg m^{-2}. These correspond to 80 and 33 ton ha^{-1}. Their experiments clearly demonstrate the importance of high values of LAI. Furthermore, the authors also document that even the current varieties are able to produce unexpected high yields, if cultivated at optimum conditions. In a similar way, Angus et al. [22] confirmed that modern varieties of barley, with more or less vertical spatial arrangement of their leaves, have to be cultivated at very high densities, either by high sowing rate or by producing a large number of tillers.

In fact, the LAI of canopies with prevailing vertical orientation of leaves cannot be too high. From this point of view, the optimum LAI represents an equilibrium between two contrasting processes:

1. In a theoretical canopy with vertical leaves, the LAI could be so high that the irradiation of

leaves would approach irradiance compensation point. In this case, the energy utilization efficiency and canopy production would be the highest although the rate of photosynthesis per unit leaf area would be very low.

2. Increase in LAI increases the construction costs (e.g., for carbon, nitrogen, and energy) of leaf formation. Such an increase is desirable only if the subsequent production of the constructed leaves is high enough to ensure high economic yield [23].

The fact that the daily course of irradiance varies must also be taken into consideration. The LAI cannot be too large, because at lower irradiance (cloudy days, during the early morning and late afternoon) the leaves of lower insertion could not be sufficiently irradiated and their loss of assimilate due to respiration could be higher than their gain in photosynthesis. Thus, the optimization must take into account the dynamics of plant growth and environmental conditions.

Finally, leaves of different insertion differ in their physiological properties [24]. The leaves adapt to their immediate environment. The adaptation irradiance, which is defined as an irradiance at which the radiation use efficiency of the leaf reaches its maximum, was calculated [25]. The adaptation irradiance of the upper leaves should be higher than that of the lower ones because of the light environment. In this case, an optimum canopy structure should ensure that irradiance of the leaves at various depths in the canopy would correspond to their appropriate adaptation irradiance. This concept extends the features of an optimum structure from one of light absorption and includes the physiological considerations.

G. The Role of Respiration

Mitochondrial respiration decreases the amount of accumulated dry matter. However, this physiological

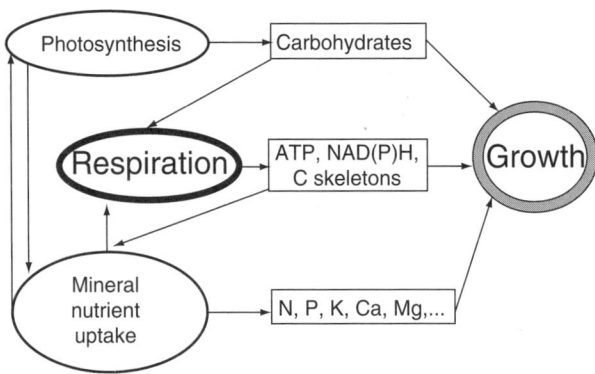

FIGURE 27.11 Schematic of the role of respiration in the provision of chemical energy (ATP), reducing equivalents (NAD(P)H), and carbon skeletons for both growth and nutrient uptake. The mutual interdependence of photosynthesis and nutrient uptake is also shown.

process cannot be considered negative for yield formation. Respiration is needed for the production of both various metabolites in the whole plant and of energy for growth and maintenance (Figure 27.11). There is no longer the tendency to suggest minimization of respiration in order to maximize production of plant organic matter. Growth efficiency is used to assess the efficiency with which the primary photosynthetic products are transformed into plant structures. Its value (ε) is calculated as the ratio between plant dry mass (W) and the total assimilates (P), i.e., $\varepsilon = W/P$.

Growth efficiency is often expressed in percent. The value of growth efficiency varies from about 50% to 70% (Figure 27.12), meaning that about 30% to 50% of assimilates are respired in the processes of plant growth. Because of the higher energy demand for protein and lipid synthesis, it is not surprising that growth efficiency is negatively related to the content of these substances (Figure 27.13).

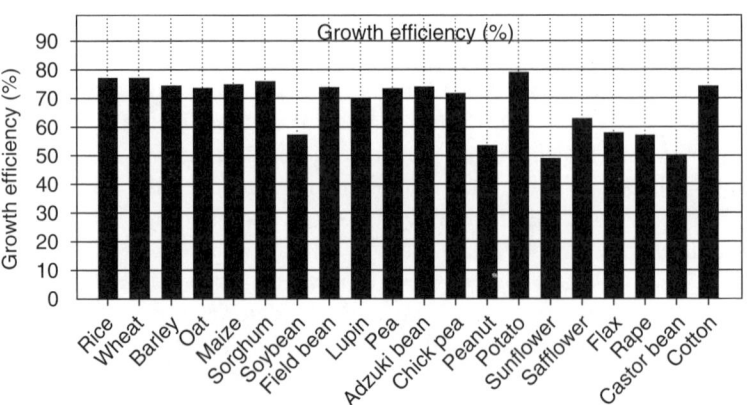

FIGURE 27.12 Growth efficiency expressed as the ratio of plant dry mass to plant dry mass + plant dry mass respired, in 20 crop species. Data from Ref. [26].

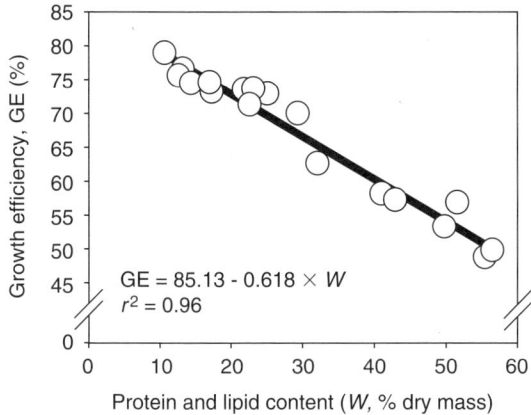

FIGURE 27.13 The effect of the protein and lipid content in plant dry matter (%) on the growth efficiency expressed as described in Figure 27.12. Data from Ref. [26].

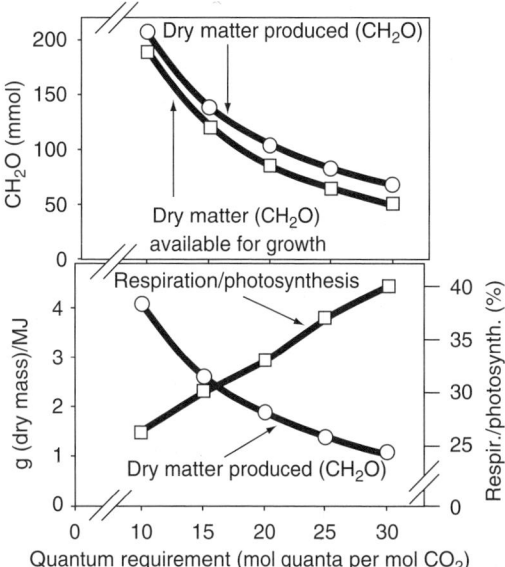

FIGURE 27.14 The effect of quantum requirement (top) on the amount (mmol) of assimilate produced and on the amount of assimilate (mmol) available for growth assuming that maintenance respiration consumes 18 mmol assimilate and (bottom) on the amount of plant dry mass produced per megajoule solar radiation and on the ratio of the rate of total respiration to the rate of gross respiration. Data from Ref. [12].

In a more detailed analysis, it is useful to distinguish several components of respiration. Often the most important components are [27]:

1. Respiration for synthesis of new structures of a growing plant.
2. Maintenance respiration, which produces necessary energy and turnover of metabolites for a nongrowing plant, i.e., when no substance for growth is available or required.
3. Respiration involved in the active processes of mineral nutrition, especially nitrogen metabolism, and active uptake of ions by plant roots.
4. Respiration providing carbon skeletons, ATP and NAD(P)H for the assimilation of mineral nutrients, predominantly nitrate, into organic components.
5. Respiration needed for the active loading and unloading of carbohydrates and amino acids into the phloem.

It is possible to quantify the individual components of respiration in model situations, but more research is needed in order to understand energy costs of the individual metabolic processes enabling a plant to grow.

The relationships between photosynthesis and respiration in the formation of plant dry matter are illustrated in Figure 27.14. Increasing the quantum requirement for CO_2 fixation decreases dry matter production and increases the ratio between the rates of respiration and photosynthesis. Because of the relatively constant maintenance respiration, a decrease in radiation use efficiency decreases final plant dry matter production.

II. TRANSPORT AND DISTRIBUTION OF ASSIMILATES

Not only the economic yield but also the total dry matter production could be modified substantially by the allocation of assimilate. Simply, if a plant invests most of its new assimilates into leaf growth, then it enhances the growth of the whole much more than if the majority of assimilates were allocated to roots. Optimization of this distribution depends on the environment; where water or nutrients are scarce, the investment in roots will be of greater benefit. The importance of assimilate distribution was identified in the 1970s and the 1980s. Previously, it was believed that yields could be enhanced if photosynthetic production was increased by any means (e.g., Refs. [28,29]). However, in the 1970s the concept of the sink emerged [30,31]. Finally, it was recognized that, depending on the genotype and external conditions, the yield is determined by the interplay between the source and the sink. Most recently, the importance of translocation capacity has been identified.

A. PHOTOSYNTHESIS AND SINKS

Sinks could be defined as places within a plant (tissue, organ) where assimilates are consumed (growing young leaves, tillers, buds) or accumulated (grains, bulbs, tubers). While the principles of the capacity to produce assimilates in the source (mostly leaves) have been fully understood (e.g., Ref. [32]), the mechanisms underlying sink attraction capacity or sink strength are still unknown. But it is well documented that sufficient sink size, expressed as number of growing or accumulating organs, is a necessary prerequisite for high yields. It has also been recognized that sinks are not only able to attract assimilates, but also alter the rate of photosynthesis [30,31] by mechanisms that are still poorly understood. For example, King et al. [30] found a decrease in the rate of photosynthesis of the main source (flag leaf blade), when the main sink (ear) was removed: experiments [33] confirmed this. On the other hand, a close relationship between sink size and rate of photosynthesis is not always observed (e.g., Ref. [34]). Mokronosov [31,35] explained the dependence of the rate of photosynthesis on the "need" of the plants. Relatively early, the effect of sink on the regulation of photosynthesis was ascribed to phytohormones [36,37].

At present, photosynthesis remains the center of attempts to enhance crop production. However, it is not just a problem of optimization of the external conditions or photosynthetic properties of the genotype. It is evident that understanding of the mechanisms of growth, morphogenesis, and organogenesis are decisive, not only because these processes are strongly dependent on production of assimilates but also because they regulate photosynthetic properties of the plant. Hence, photosynthesis, allocation, and accumulation of assimilates must be considered as integrated activities of the whole plant which determine the growth and productivity. From this point of view, the rate of photosynthesis under certain conditions could be better understood and subsequently regulated by manipulating seemingly independent processes of leaf growth, reproductive organ initiation, plant hormone regulation, etc.

B. ASSIMILATE DISTRIBUTION AND DRY MATTER PRODUCTION

In most crops, economic yield results from assimilate allocation into particular, mostly reproductive or storage, organs. It is obvious that the translocation capacity could alter the total amount of assimilate moving into the organs providing the economic yield (see next chapter). However, the model of assimilate allocation within a plant is also decisive for the total plant production. This is illustrated with a theoretical example as presented by Good and Bell [38].

Plant dry mass (W) may be separated into dry matter of assimilating organs (W_P) and the remaining

organs (W_S). The rate of production of the new biomass is expressed by the relative growth rate (RGR), which calculates the dry mass increase per unit plant dry mass and unit of time (t), i.e.

$$\mathrm{RGR} = \frac{\mathrm{d}W}{\mathrm{d}t}\frac{1}{W}$$

Let us also define α as that ratio of the newly synthesized assimilate, which is used for the construction of photosynthesizing organs (W_P). Then the rate of dry mass increase in photosynthesizing organs is:

$$\frac{\mathrm{d}W}{\mathrm{d}t} = \alpha \cdot \mathrm{RGR} \cdot W_P$$

By integrating this equation, an expression describing dry mass of the photosynthesizing organs (W_P) at any time (t) is obtained, i.e.,

$$W_P(t) = W_{P0} \cdot \mathrm{e}^{\alpha t \cdot \mathrm{RGR}}$$

where W_{P0} is the dry mass of the photosynthesizing organs at time $t = 0$.

Let us suppose that the total plant dry matter increase is distributed both into the production of new assimilating structures (W_P) and into the other (supporting, accumulating, etc.) organs and tissues (W_S). Then the ratio W_P/W_S is proportional to the ratio $\alpha/(1 - \alpha)$. It can be shown that:

$$W_P + W_S = \frac{1}{\alpha}W_{P0}\mathrm{e}^{\alpha T \cdot \mathrm{RGR}} - \frac{1 - \alpha}{\alpha}W_{P0}$$

Figure 27.15 illustrates this relationship. It is evident that the total plant dry mass depends not only on the rate of photosynthesis, in this example represented by RGR, but also on the dry matter allocation (α). Any factor decreasing the investment of newly synthesized assimilates into photosynthetic organs also decreases the potential increase in the total plant dry matter although the rate of photosynthesis remains constant. The importance of dry matter partitioning increases with the increase in the assimilate production, as illustrated by the curves in the right-hand part of Figure 27.15.

Under nonlimiting conditions, the source production of assimilate corresponds to the sink capacity of the plant. This equilibrium is determined by unknown mechanisms, with which the sink endogenously regulates the longevity, size or activity of the source organs. Alterations of this equilibrium by both abiotic and biotic stresses could result in changes in importance of either source (nitrogen or water limitation, premature senescence, attack of pathogens) or sink

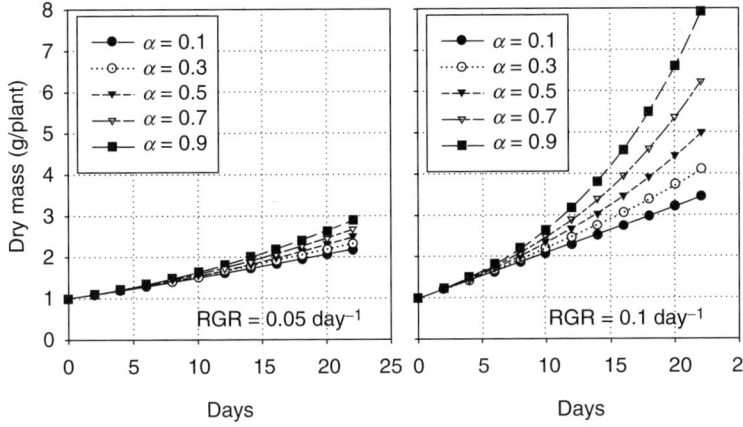

FIGURE 27.15 The effect on total plant dry mass of the way in which assimilates are partitioned between photosynthetic and non-photosynthetic organs. Relative growth rate (RGR) equals 0.05 day^{-1} (left) or 0.1 day^{-1} (right). See text for details.

(disturbance of flowering, embryo or tuber development). From the agronomic point of view, it is important to identify whether yield has been limited by source or sink limitation In the former case, management should improve leaf development and activity, while in the latter case, stimulation of branching, tillering, and flower initiation could help.

C. ASSIMILATE TRANSLOCATION

In most crops, economic yield is represented by particular organs (grains, tubers), the dry matter of which has been accumulated via translocation from other organs. With increases in productivity, both source production and sink accumulation are enhanced to such a degree that translocation capacity between the source and sink could become limiting. This translocation capacity is determined by both size (cross-sectional area) of the phloem tissue and the speed of the translocation.

There is some evidence that the amount of translocated assimilates could be correlated with the cross-sectional area of the phloem. Using three wheat genotypes with potentially very high grain dry mass per ear has shown [39] (Figure 27.16) that grain dry mass per ear closely correlates with the number of sheath bundles, their cross-sectional area and cross-sectional area of phloem. Subsequently, using 26 winter wheat varieties, it was shown [40] (Figure 27.17) that the cross-sectional area of phloem of the last internode (the one below the ear) is a very useful indicator of the maximum translocation capacity of the genotypes. As illustrated in Figure 27.17, the values for some genotypes strongly deviate from the general regression, for all the data, of grain dry mass per ear on phloem cross-sectional area. If the limiting lines of all but one value have been calculated, a maximum translocation capacity was obtained which was not surpassed by any variety (with one exception). It is concluded that in varieties approaching the limiting

regression value, further yield enhancement is limited by the translocation capacity.

Obviously, cross-sectional area of translocation tissue need not be the best expression of the translocation capacity. However, it is a first attempt to quantify this plant characteristic. More sophisticated measurements are needed, which would characterize not only the size but also the actual ability to translocate assimilates. The main difficulty arises from the fact that the basic mechanism of phloem movement, as described long ago by Münch [41], is still poorly understood [42]. An understanding of the mechanism of phloem translocation is of prime importance, as is the establishment of quantitative parameters suitable for the assessment of the translocation ability [43].

D. HARVEST INDEX

The final result of dry matter production, translocation, distribution, and allocation is manifested in the HI. It may be determined at the end of the growing season and, in a certain way, it integrates the whole of the preceding period. This is its major advantage — it sums up the whole growing period — but also its disadvantage — it cannot identify changes during plant growth.

It is generally accepted that the increase in HI has been the main reason for the enormous increase in economic yields in developed agriculture in the last century. This increase has been mainly due to the reduction in crop height, which simultaneously increased lodging resistance and allowed an increase in the use of fertilizers [44]. For example [45], an old variety (Nürnberg from the year 1832) was compared with newly released spring barley varieties. Total dry matter production was comparable, but the HI of the modern varieties increased considerably. As shown by Wacker et al. [46], the dry matter distribution among leaves, stems, and roots did not change during the domestication of wheat and barley varieties.

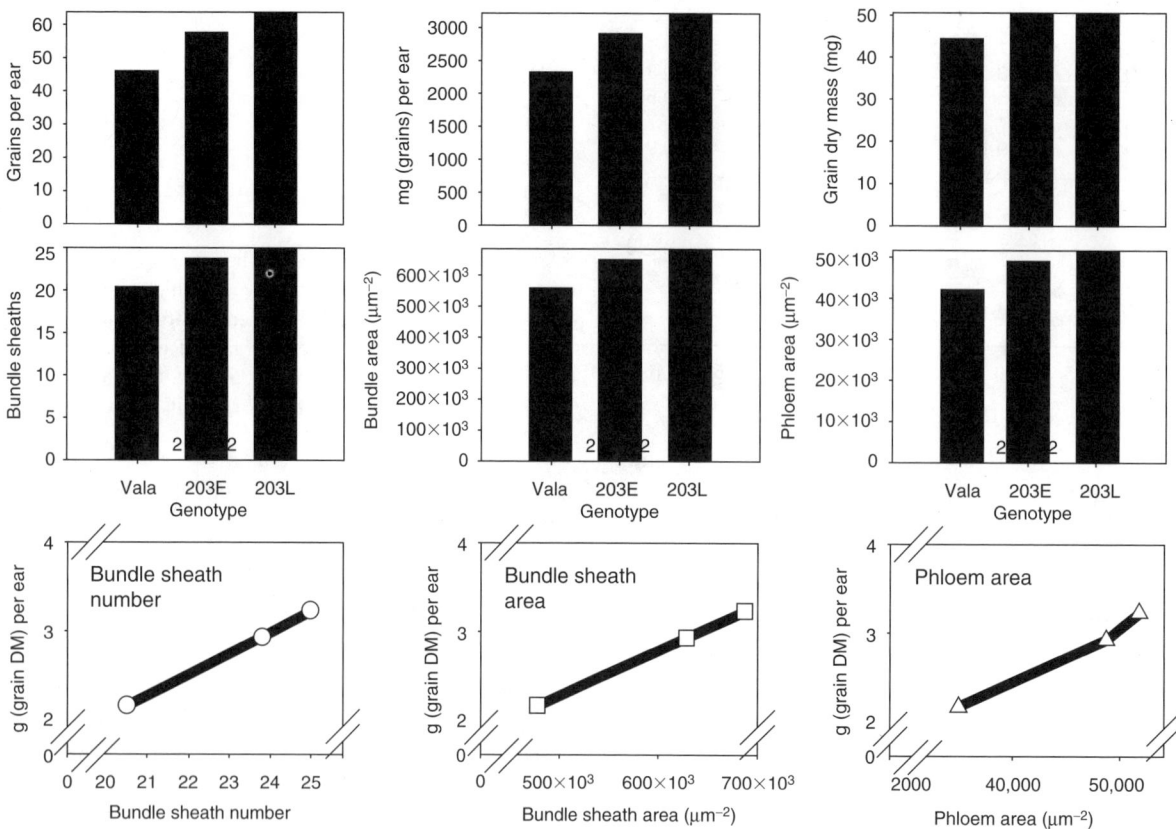

FIGURE 27.16 (Top) Productivity characteristics of three winter wheat genotypes: grain number per ear, grain dry mass per ear, and dry mass per grain. (Middle) Characteristics of the conducting tissue in the last internode (below the ear): number of bundle sheaths, cross-sectional area of the bundle sheaths and of the phloem. (Bottom) The dependence of the grain dry mass per ear on characteristics of the conducting tissue as given above. Data from Ref. [39].

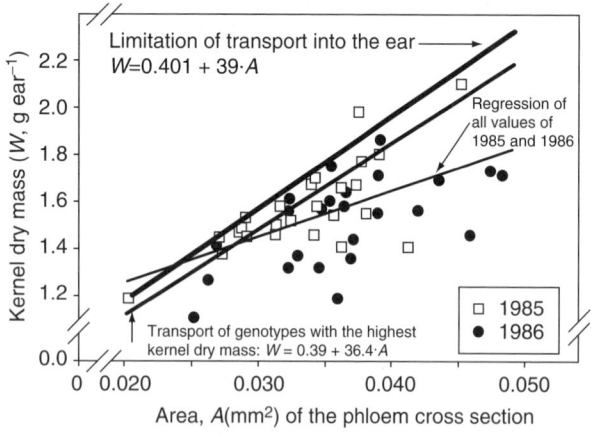

FIGURE 27.17 Dependence of the grain dry mass per ear on the cross-sectional area of the phloem in all bundle sheaths in the last internode of 26 winter wheat varieties cultivated in 1985 and 1986. For details see text. Data from Ref. [40].

Characteristic values of HI of various crops are given in Table 27.5. In some crops, more than 50% of the total plant dry matter is accumulated in the eco-

nomic yield. There is an upper limit for HI. In most cases, there are no useless organs present on a plant even if we take economic yield as the main criterion of usefulness. We could think of the possible reduction of all the unfertile tillers or florets in cereals. However, a plant has to produce roots, stem and leaves in order to be able to produce sufficient assimilates to be accumulated in the organs with most important economic yield. In cereals, the upper limit for HI could be in the range of about 0.6 to 0.65. In the individual varieties, its value will also depend on the longevity and photosynthetic activity of the assimilating tissue.

III. PLANT PHOTOSYNTHETIC CHARACTERISTICS AND YIELD IMPROVEMENT

Before agriculture was developed, natural food resources were able to support only several million people globally. At present, modern agriculture can support, reasonably well, the current 6 billion people, if economic and political conditions are suitable. And

TABLE 27.5
Characteristic Values of Harvests Index of Several Crop Species Cultivated in the 1980s (simplified from a review [44])

Plant Species	Harvest Index
Wheat (*Triticum aestivum*)	0.31–0.54
Barley (*Hordeum vulgare*)	0.33–0.63
Triticale	0.45–0.47
Rice (*Oriza sativa*)	0.35–0.62
Maize (*Zea mays*)	0.36–0.57
Soybean (*Glycine max*)	0.35–0.53
Cickpea (*Cicer arietinum*)	0.28–0.36
Cowpea (*Vigna unguicelata*)	0.15–0.64
Rape (*Brassica napus*)	0.22–0.38
Cassava (*Manihot esculenta*)	0.30–0.65
Potato (*Solanum tuberosum*)	0.47–0.62

in the future, agriculture must produce food for 8 to 12 billion humans.

Photosynthesis is no doubt the very basis of any economic yield. In the past, yields have been mainly increased by the change of assimilate allocation favoring the accumulation in the organs of prime economic importance. This increase in HI has been achieved predominantly by: (1) breeding and (2) fertilizer use. By emphasizing these two factors it is not the intention to negate the importance of other factors, such as plant protection, irrigation, soil management, etc. But it seems that the former two played a decisive role. Furthermore, they could be of prime importance in future yield improvement, which will be achieved by the enhancement of photosynthetic production and not only by the assimilate allocation.

A. HIGHLY PRODUCTIVE VARIETIES

The discovery of the basic genetic laws by Mendel in the middle of the 19th century [47] enabled their application in crop selection in the first half of the last century. In the 1950s and the 1960s, the role of photosynthesis in yield formation was elucidated and parameters of the optimum leaf area size and spatial distribution formulated. This enabled the unexpected rise of yields in the second half of the 20th century.

One of the most important achievements was the Green Revolution, with the use of new, short stem, and photoperiodically neutral wheat varieties selected by Borlaug (awarded the Nobel Prize in 1970), which subsequently increased yields in Mexico, India, Bangladesh, Pakistan, and other developing countries [48].

An important milestone reached by the tremendous yield increase in the 20th century was the selec-

tion of new rice varieties in the International Rice Research Institute in the Philippines. Their success is attested by the fact that in Asia, since 1962, the population increased from 1.6 to 3.7 billion and rice production increased by 170% [17,49].

Another achievement, which however, has not yet been fully exploited in breeding was the concept of ideotype described by Donald [50]. According to him, selection aimed only at removing defects in existing varieties or simply at yield improvement by any means. Donald emphasizes that it is time to adopt a new strategy, namely selection on the basis of an ideotype, i.e., "a form denoting an idea." Now, some 35 years after the Donald's definition of an ideotype, the time is here to use it both in research in the photosynthetic basis of yield formation, as well as in practical breeding. The concept required adaptation to include the interaction of plant with environment, and indeed the development of an ideotype has been achieved in several complex dynamic mathematical models, which could well be used as a starting point for the research into the quantitative parameters for breeding.

Donald [50] also described his idea of an ideotype of spring wheat, which was characterized as a single shoot (uniculm) plant. His idea has been criticized as unrealistic, because tillering is considered a necessity to fill in gaps in a canopy, caused, for example, by plant death. However, the time to appreciate Donald's uniculm plant is approaching and modern wheats have very limited tillering, with production of yield mainly from the main stem ear. Combining such a variety with methods of precision agriculture could be very promising [51].

As regards the future possibilities for improvement of photosynthetic characteristics of more productive varieties, there are many, some that could be considered as ready for practical breeding or at least highly desirable for applied research are mentioned here.

1. Plant Morphology

Plants are characterized by their morphology, which includes number and size of individual organs, their longevity, activity, and spatial arrangement. Compared with varieties from some 50 years ago, current varieties are shorter, with smaller leaf blades, which are more vertically oriented. This trend will continue. More information is needed about optimum stature of the plant and the whole canopy best adapted to current local climate and soil conditions. Knowledge about canopy LAI and its time course should be better exploited with the aim of maximizing solar energy absorption by the leaves and increasing efficiency in photosynthesis.

Using the example of rice, the yield potential of the modern high-yielding varieties, cultivated under optimum conditions, is 10 ton ha^{-1}, which corresponds to about 20 ton dry matter ha^{-1} with an HI of 0.5. The need to further increase food availability has stimulated researchers at the International Rice Research Institute (IRRI) to attempt to develop a new plant type, which would further increase total plant dry matter production by enhancing the rate of photosynthesis and which would have an HI of 0.6. Such varieties would have a yield potential of about 13 ton ha^{-1}. It is also expected that the use of these new plant type varieties could be used to produce hybrid rice, exceeding the best of its parents by about 25% and thus reaching a yield potential of 15 ton/ha. It is important to mention that among the most important attributes of the new plant type, the following have been included: lower tillering capacity with no unproductive tillers, short growth duration which is needed in order to get high values of HI, and dark green, thick, erect leaves [17,49,52].

2.   Rate of Photosynthesis

Current techniques enable measurement of the rate of photosynthesis, its individual processes or spectral characteristics even under field conditions, and on hundreds of plants [53,54]. Varietal differences in the rate of photosynthesis have been known for a long time [28,55]. Although this parameter depends on external conditions (irradiation, temperature, nutrient and water availability), selection could stimulate progress in total plant dry matter production [56].

3.   Dependence of the Rate of Photosynthesis on CO_2 Concentration

The current CO_2 concentration (370 μmol mol^{-1}) exceeds that of the 19th century by about one third and the rise will continue. Remarkable variety differences have been found in the effect of enhanced CO_2 concentration on the rate of photosynthesis and especially on the plant dry matter production. As far as we are aware, no attempt has been described in the literature about the practical use of these differences.

4.   Mathematical Modeling

Mathematical modeling has not been restricted to photosynthetic aspects of crop production, but has proved to be extremely useful in explaining photosynthesis at the organ, plant, and canopy level by incorporating biophysical and biochemical principles of energy and CO_2 absorption and assimilation.

Since the 1970s, mathematical modeling has been used both as a tool for integrating knowledge across several hierarchical levels of plant functioning and as a heuristic means to stimulate formulation, as well as testing, of new scientific hypothesis [2,57,58]. Mathematical modeling is the only means for the objective integration of the steadily growing knowledge of plant growth. Furthermore, only by combining scientific knowledge with management practices, will further progress in crop productivity be achieved. In a stimulating review [59], two major goals for extending modeling activity in the future were identified:

1. Heuristic role supporting further scientific activity aimed at facilitating decision making by farmers as well as education of students. The authors also emphasize that modeling will be needed in order to integrate crop and landscape management.
2. Simulation to enhance understanding of the genetic regulation of plant growth and crop improvement.

5.   Introduction of C_4 Photosynthesis into C_3 Plants

The most productive and efficient crops — maize and sugar cane — have C_4 photosynthesis. With the advances in molecular genetics [60], attempts have been made to introduce C_4 photosynthesis into the less efficient C_3 plants (rice, wheat, etc.). The C_4 carboxylating enzyme phosphoenopyruvate carboxylase has been successfully expressed in the C_3 plants [61–63], and in some plants not only higher rate of photosynthesis but also increased yield has been found [64,65]. Until recently, it has been argued that such an introduction of the C_4 photosynthesis into C_3 plants will be extremely difficult because of the special leaf anatomy ("Kranz" type) of C_4 plants. However, in CAM plants, practically the C_4 type of photosynthesis occurs in the same cells in which the two carboxylating reactions take place at different times — C_4 fixation at night and C_3 during the day. Furthermore, it has recently been found that some other species — not of the CAM type — exist, in which the typical C_4 photosynthesis takes place in the same cells [66,67]. Understanding of this type of one-cell-C_4 photosynthesis will further facilitate the possibility of genetically introducing the C_4 type of photosynthesis into C_3 plants of economic value [68].

This procedure has been hailed as the most promising breakthrough in enhancing the rate of dry matter production in the most important crops — rice and wheat. However, confirmation of such work is required. Also, even if achievable, an increase in the potential rate of photosynthesis need not be automatically coupled with an increase in grain yield. C_4 photosynthesis is more energy demanding and its temperature optimum is higher than that of C_3 photosynthesis. Furthermore, with future increase in the atmospheric CO_2 concentration the photosynthetic predominance of the C_4 photosynthesis over C_3 type will decrease [69]. However, introduction of C_4 photosynthesis into C_3 plants would greatly increase the potential for selection of important crops with improved production.

6. Plant Biotechnology Potential

Modern plant biotechnology has provided new transgenic varieties grown on millions of hectares. They are mostly improved in their resistance to pathogens or herbicides. Some of the biotechnology achievements represent varieties with improved nutritional properties, as the well-known "golden rice" with enhanced content of provitamin A and several other [70–72]. However, the value of this is contested because of the large amounts of the grain that would be needed to provide for daily need. It is believed that the use of plants for pharmaceutical purposes will rise and could be a considerable part of agriculture activities in the next decades [73].

Considerable enhancement of yield from biotechnology alone is not to be expected in the near future [74]. Yield represents a very complex phenomenon, involving many metabolic processes in the source and sink organs as well as translocation of large amounts of sugars and amino acids over long distance. Plant metabolic engineering is far less understood than the genetic manipulation of only one or few genes, e.g., of herbicide resistance. It is extremely difficult to predict the effects of alteration of appropriate metabolic route in cells with thousands of metabolites, if only the concentration of one enzyme is modified by changing gene expression [75]. In the case of yield, not only thousands of metabolites in one cell, but also billions of cells within the organism have to be taken into account. This is not to negate the present and future role of biotechnologies [76,77]. But we fully agree with the statement of Morandini and Salamini [75] that classical "plant breeding will not be substituted in a few years by plant biotechnology, rather the two different approaches are — and will be — cooperating for years to come." And in this connection, the following quotation concerning the role of mathematical modeling in connecting plant breeding and functional genomics from Ref. [59] is of special importance: "This frontier provides a unique opportunity for crop modeling to play a significant role in enhancing the integration of molecular genetics with crop improvement whilst offering new intellectual

challenges to those who assemble logically constructed frameworks of how plant systems work." Present knowledge represents a very good starting point for such constructions [78,79].

7. Other Possibilities for Future Plant Use

There is a shortage of food in large parts of the globe (particularly with extreme environmental conditions) with many millions of people lacking adequate nutrition for growth and energy (and this becomes particularly acute in times of civil unrest). However, crop production will continue to be oriented not only toward producing food, but also toward production of animal feed, raw materials, and biomass for energy [80]. In any orientation of crop production, high rate of photosynthesis will always be a desirable trait [81]. The main differences will consist of dry matter allocation between individual organs. Maximizing solar energy utilization in photosynthetic CO_2 fixation will remain the ultimate goal.

This chapter was devoted to crop photosynthetic production, but it is worth mentioning the role of photosynthesis of algae, cultivation of which is very efficient and could play an important role in the production of food and feed as well as a non-CO_2 producing source of energy [82–84].

B. Mineral Nutrients

Mineral nutrients, and especially nitrogen, play an important role both in plant growth and crop productivity (Figure 27.18).

1. About half of the soluble nitrogen in the leaf is present in the main carboxylating enzyme — Rubisco, the content of which may reach 8 g m^{-2} [32,79]. There is much evidence that the rate of net photosynthesis is directly proportional to the content of N in the leaf [85] over a broad range of N content, with an upper limit. C_4 plants, which are more efficient in photosynthetic CO_2 assimilation than C_3 plants, also use N with higher efficiency, expressed as the amount of produced dry matter per unit nitrogen in the photosynthetic tissue [32].

2. In cereals, grains contain some 2% to 3% N per unit dry mass. Hence, removing a yield of about 5 tons grain per hectare each year means removal of some 100 to 150 kg nitrogen. There is no evidence that in the long term, high yields can be achieved without the use of the appropriate applications of fertilizers [86]. Although the Green Revolution was achieved predominantly by the use of new short-stem

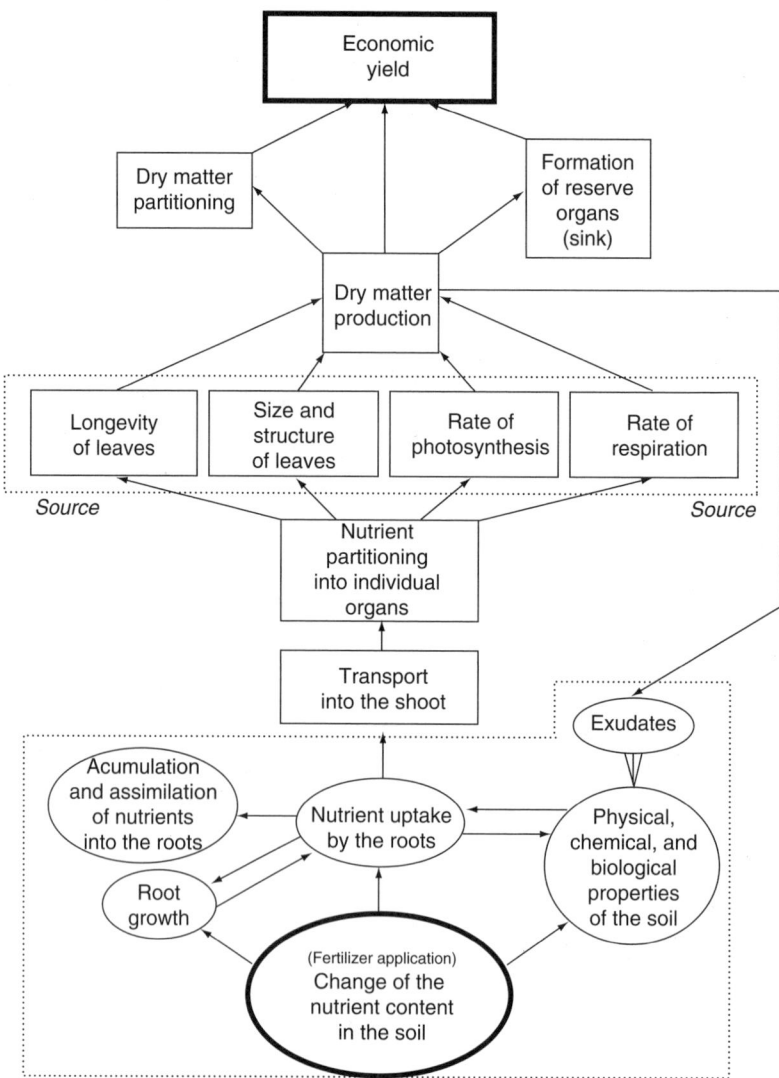

FIGURE 27.18 Schematic of the changes induced by the alteration of mineral nutrients in the soil (addition of fertilizers), leading to a sequence of structural and metabolic changes in the roots and shoots, resulting finally in changes to economic yield.

varieties, their successful cultivation was only possible by use of higher fertilization applications [87,88]. The content of N in the plant dry matter steadily decreases during growth period and this general course has been also successfully modeled (Figure 27.19). Such a decrease is well explained by the continuous accumulation of both structural (cellulose in cell walls) and nonstructural carbohydrates (starch).

1. Mineral Nutrients and Yield

Mineral nutrients do not affect yield directly (Figure 27.18). They may alter, directly (Figure 27.20):

1. duration of the vegetation period by delaying or accelerating senescence,
2. LAI by modifying the rate of appearance of individual leaves and by modifying their size,

3. rate of photosynthesis predominantly by an adequate supply of nitrogen and phosphorus,
4. allocation of assimilates by encouraging translocation into shoot and reproductive organs, if mineral nutrient and water supply is adequate.

An interplay of the four factors leads to the final plant dry matter, a proportion of which is the economic yield.

Uptake of mineral nutrients depends not only on nutrient availability in the soil, but also on their rate of absorption by the roots, which may be limited by assimilate availability (see Figure 27.11). Assimilates are needed as a source of energy (ATP) and reducing equivalents (NAD(P)H) for active ion uptake. Furthermore, carbon skeletons, produced mainly in the processes of plant respiration, are needed for the incorporation — assimilation — of the mineral nutrients into organic substances. Again,

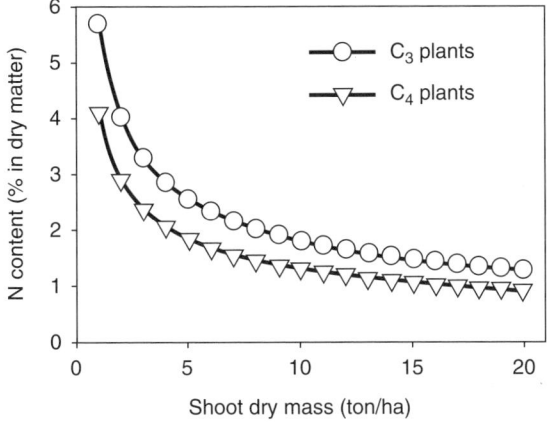
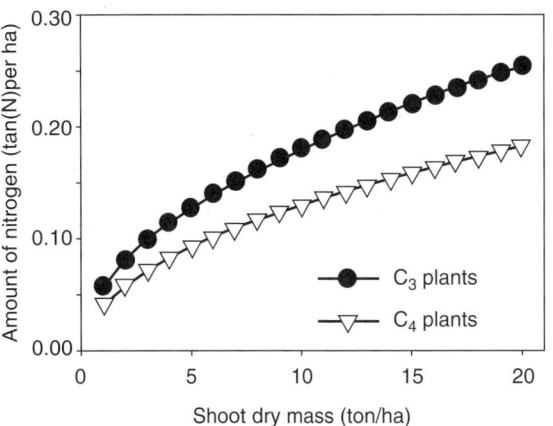

FIGURE 27.19 Dependence of the nitrogen content (left) and nitrogen amount (right) in dry matter on shoot dry mass during growth in C_3 and C_4 plants. Figures constructed with the use of equations from Ref. [89].

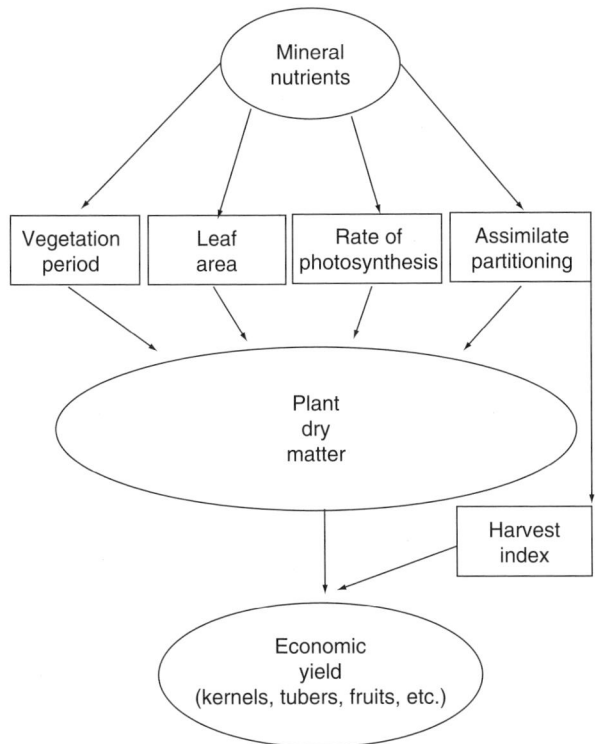

FIGURE 27.20 Schematic of the effect of mineral nutrients on parameters determining plant dry matter production and subsequently the economic yield.

nitrogen, which is taken up predominantly in the form of nitrate ions, requires large amount of energy for its reduction and subsequent incorporation into amino acids.

Unfortunately, fertilizer production requires huge amount of energy. In this respect, nitrogen fertilizers are of special importance. Generally, fertilizers repre-

sent the major part of the input energy, which is needed to support crop growth. This energy is provided by fossil fuel burning. The highest efficiency of input energy use is obtained at very low levels of this input energy, i.e., in primitive, extensive agriculture (Figure 27.21). However, high yields need considerable support in various forms (fertilizers, mechanization, soil management, pesticides) all of which require energy, although their efficiency of use decreases steadily with increasing input. This is another important reason why agriculture should shift from "oil assisted" to "solar powered" forms. In another words, future increase in crop yields should rely mostly on an enhanced rate of photosynthesis. To increase the photosynthetic rate without increasing capacity (more biochemical component of the photosynthetic system per unit leaf area) is essential if nutrient requirements are not to increase. This means that the components and the system as a whole must work more efficiently. This is a major challenge to the plant sciences, breeders, and genetic engineers.

2. Nitrogen and Photosynthesis

There is a general relationship between the rate of photosynthesis (yield, dry matter accumulation) and the content of mineral nutrients in the plant [90]. An increase in nutrient content in the region where it is deficient increases the rate of photosynthesis considerably [91]. Photosynthesis reaches a plateau with final decline when the nutrients reach such concentrations at which they become toxic. Because of the difficulty in determining the beginning of the saturating content of the nutrient, the so-called critical content or critical concentration is often taken as the value at which 90% of the maximum rate is reached [92].

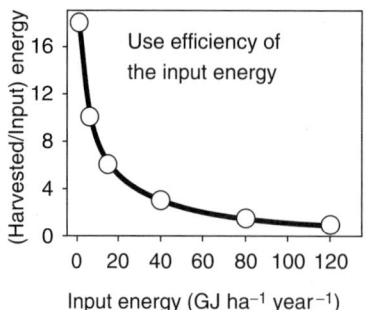

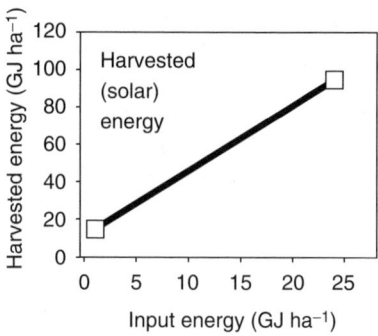

FIGURE 27.21 The effect of input energy on the ratio of solar to input energy (left) and on the amount of harvested energy (right). Data for graph construction taken from Ref. [4].

Although the value of the optimum nutrient content in the crops is of prime importance for crop management [93], physiologically it is not a constant value, as it depends on radiation, CO_2 concentration, temperature, etc. Furthermore, it has been shown that the mineral nutrient content varies considerably, even in different tissue of a leaf [94]. Sufficient data and understanding of the biochemical system are available for making an attempt to construct a generally valid model, although many aspects, including the most important external and internal (age, leaf insertion) factors modifying the optimum content of nitrogen and other biogenic elements in leaves, plants, or canopies are poorly understood.

3. Nitrogen Allocation

Shoot and root growth are colimited by the availability of assimilate provided by the shoot and availability of nutrients and water supplied by the roots. Primarily, local carbon and nitrogen availability control dry matter distribution between the shoot and the leaf. A relative surplus of nitrogen within the plant will enhance dry matter allocation into the shoot because this organ is able to restore the balance between C and N. In contrast, with N limitation, roots assume priority as assimilate sinks because they can decrease the N/C imbalance. However, knowledge of how these imbalances are sensed and of the mechanisms regulating assimilate partitioning is required [95] within a framework allowing for interpretation.

Leaves represent the basic unit of photosynthesis of the whole plant or canopy. Leaf structure and heterogeneity of chemical components have been often studied in order to establish the optimum for maximizing the rate of CO_2 assimilation. Although the rate of photosynthesis is most often expressed per unit leaf area surface, it is also related to dry matter or leaf volume. Each of these relationships has some advantages and shortcomings. Leaf surface area is the most natural because it determines the amount of absorbed solar radiation as well as the diffusion of CO_2. On the other hand, leaf volume may better

express the internal (intercellular) leaf surface, which reflects the final diffusion step for the CO_2. Dry matter may reflect the amount of enzymes available for photosynthetic CO_2 fixation. For example, Garnier et al. [96] demonstrated that rate of photosynthesis per unit leaf area was positively correlated with leaf thickness and with the amount of mesophyll tissue. On the other hand, rate of photosynthesis per unit leaf volume negatively correlated with leaf thickness, but positively with leaf organic nitrogen. A positive correlation occurred between the rate of photosynthesis per unit leaf mass and the relative growth rate in several *Poa* species [97]. Such aspects should not be an impediment to developing effective models of crop photosynthesis.

Rubisco and other enzymes of CO_2 assimilation and carbohydrate metabolism, each containing nitrogen, have to be distributed within the leaf to make most effective use of the captured light energy, Considerable differences in irradiance occur over a leaf and the chloroplasts are spread along the mesophyll cell surface exposed to intercellular airspaces [98]. A high ratio of the mesophyll cell surface to volume thus facilitates diffusion of CO_2 to the carboxylation sites in the chloroplast stroma. Figure 27.22 demonstrates that high diffusion conductance for CO_2 does not necessarily correlate with large amounts of Rubisco (potentially large rates of CO_2 assimilation), and with large nitrogen allocation that is required. There is an optimum "cooperation" between nitrogen and carbon metabolism and light energy capture and use. At the top of the canopy, bright light and availability of CO_2 require photosynthetic systems of large capacity, whereas at the bottom, light and capacity are much smaller. This is the reason, as Sheehy et al. [99] concluded, that radiation absorption within a leaf declines exponentially with cumulative chlorophyll content measured from the irradiated surface. Hence, the light absorption profile is dominated by the distribution of pigments. Similarly, the distribution of nitrogen within the canopy is of comparable importance. More knowledge on the carbon and nitrogen metabolism and their relationships is needed [32]. In the near

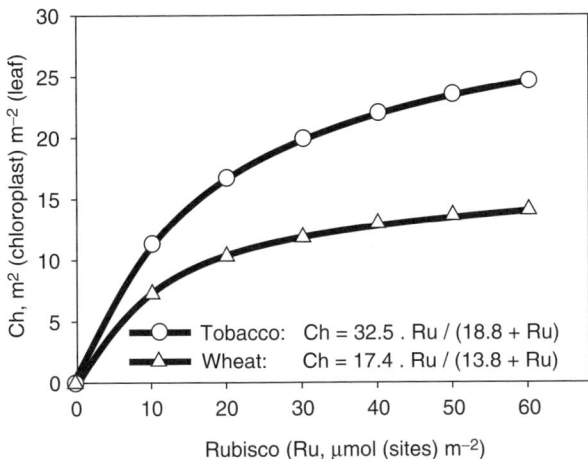

FIGURE 27.22 The relationships between chloroplast surface area exposed to intercellular air spaces per unit leaf area (Ch) and the amount of rubisco per unit leaf area (Ru) in tobacco and wheat. Graphs constructed with the use of equations taken from Ref. [98].

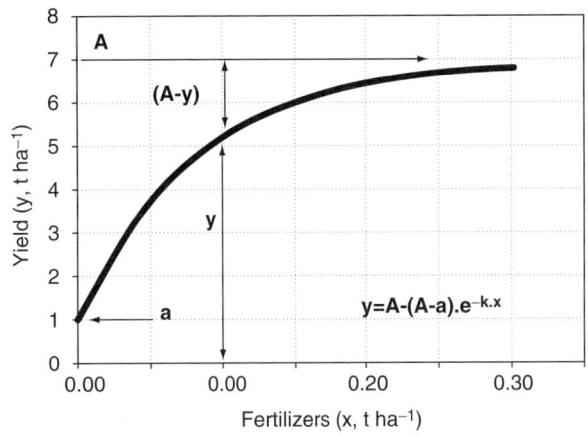

FIGURE 27.23 Illustration of the Mitscherlich equation (in the figure) expressing the dependence of yield on applied fertilizers. Explanation of symbols is given in the figure.

future, current use of empirical correlations between yield and mineral nutrient content in the plant needs to be replaced by the quantitative description of events starting with ion uptake by root cells, leading up to the appropriate modification of photosynthesis with subsequent assimilate partitioning.

IV. CONCLUSION

Innumerable experiments have been carried out for analyzing the effect of increasing amount of fertilizer on yield. Figure 27.23 illustrates the well-known concept of Mitscherlich. These types of experiment were needed in the early stages of fertilizer use between 1850 and 1950 and offered valuable advice to farmers, and stimulation to research. It was soon realized, however, that considerable differences could exist between the amount of nutrients in the soil and in the plant itself. Hence, by the middle of the 20th century, much attention was paid toward establishing suitable criteria for the assessment of nutritional status to achieve an appropriate crop canopy [100]. At the same time, considerable progress has been made in the elucidation of both photosynthetic carbon fixation and nutrient assimilation. Biophysical, biochemical, and physiological principles of photosynthesis have already enabled agricultural scientists to construct mechanistic models describing the effects of the main climatic factors on canopy photosynthesis. However, similar progress has not been made in understanding the mechanisms of absorption and assimilation of mineral nutrients. This field of plant biology is just beginning to be studied in detail.

However, taking the formation of economic yield into consideration, even good knowledge of both carbon and nitrogen metabolism will not be sufficient to replace — within a realistic time frame — Mitscherlich's equation with a mechanistic model of crop growth and yield. Individual yield components (e.g., ear number, grains per year, grain dry mass) are established at various times during crop development (tillering and tiller death, flowering, grain set, and filling). Each of these components depends on both genetic determination and on environmental factors, including weather and soil conditions, determining nutrient and water supply. The interplay between carbon and nitrogen metabolism represents the fundamental process of dry matter production. Its understanding will enhance not only our ability to describe plant growth, but also offer valuable hints for crop management. Nevertheless, mechanisms controlling dry matter partitioning within plants, as well as all the processes of morphogenesis and developmental regulation, also await improved understanding. In summary, we repeat from Ref. [32]: "The processes involved in crop production are very complex and multilayered, ranging from the molecular to the whole organism, and environmental factors affect all levels of organization."

REFERENCES

1. Monteith JL, Elston J. Performance and productivity of foliage in the field. In: Dale JE, Milthorpe FL, eds. *The Growth and Functioning of Leaves.* Cambridge, UK: Cambridge University Press, 1983:499–518.

2. De Wit CT. Photosynthesis of leaf canopies. *Agric. Res. Reports.* Wageningen, 1965.

3. Charles-Edwards DA. *Physiological Determinants of the Crop Growth.* Sydney; Academic Press, 1982.

4. Evans LT. *Crop Evolution, Adaptation and Yield.* Cambridge, UK: Cambridge University Press, 1996.

5. Buermann W, Wang YJ, Dong JR, Zhou LM, Zeng XB, Dickinson RE, Potter CS, Myneni RB. Analysis of a multiyear global vegetation leaf area index data set. *J. Geophys. Res. D Atmos.* 2002; 107:4641–4646.

6. Watson DJ. A prospect of crop physiology. *Ann. Appl. Biol.* 1968; 62:1–9.

7. Watson DJ, Thorne GN, French SAW. Physiological causes of differences in grain yield between varieties of barley. *Ann. Bot. NS* 1958; 22:321–352.

8. Ničiporovič AA. *KPD zelenogo lista.* Moskva: Izd Znanije, 1964.

9. Ničiporovič AA, Strogonova LE, Tschmora SN, Vlasova MP. *Fotosintětičeskaja dějatělnost rastěnij v posevach.* Moskva: Izd Akad Nauk SSSR, 1961.

10. Monsi M, Saeki T. Uber den Lichtfaktor in den Pflanzengesellschaften und seine Bedeutung für die Stoffproduktion. *Jpn. J. Bot.* 1953; 14:22–52.

11. Varlet-Grancher C, Gosse G, Chartier M, Sinoquet H, Bonhomme R, Allirand JM. Mise au point: rayonnement solaire absorbé ou intercepté par un couvert végétal. *Agronomie* 1989; 9:419–439.

12. Reynolds MP, Ginkel van M, Ribaut JM. Avenues for genetic modification of radiation use efficiency in wheat. *J. Exp. Bot.* 2000; 51:459–473.

13. Rizzalli RH, Villalobos FJ, Orgaz F. Radiation interception, radiation-use efficiency and dry matter martitioning in garlic (*Allium sativum* L.). *Eur. J. Agron.* 2002; 18:33–43.

14. Jeuffroy MH, Ney B. Crop physiology and productivity. *Field Crops Res.* 1997; 53:3–16.

15. Loomis RS, Amthor JS. Yield potential, plant assimilatory capacity, and metabolic efficiencies. *Crop Sci.* 1999; 39:1584–1596.

16. Begna SH, Hamilton RI, Dwyer LM, Stewart DW, Smith DL. Variability among maize hybrids differing in canopy architecture for above-ground dry matter and grain yield. MAYDICA 2000; 45:135–141.

17. Peng S, Khush GS, Cassman KG. Evolution of the new plant type for increased yield potential. In: Cassman KG, ed. *Breaking the Yield Barrier.* Manila: International Rice Research Institute, 1994:5–20.

18. Kuroiwa S. Total photosynthesis of a foliage in relation to inclination of leaves. In: Šetlík I, ed. *Prediction and Measurement of Photosynthetic Productivity.* Wageningen: Centre Agric. Publ. Document., 1970:79–89.

19. Nátr L. Physiological principles of the yield formation of cultivated plants. In: Petr J, Černy V, Hruška L, et al. eds. *Yield Formation in the Main Field Crops.* Amsterdam: Elsevier. 1988:43–55.

20. Blackmann GE. The application of the concepts of growth analysis to the assessment of productivity. In: Eckardt FE, ed. *Functioning of Terrestrial Ecosystems at the Primary Production Level. Proceedings of the Copenhagen Symposium.* Paris: UNESCO, 1968:243–259.

21. Polonskij VI, Lisovskij GM. Net production of wheat crop under high PhAR irradiance with artificial light. *Photosynthetica* 1980; 14:177–181.

22. Angus JF, Jones R, Wilson JH. A comparison of barley cultivars with different leaf inclinations. *Aust. J. Agric. Res.* 1972; 23:945–957.

23. Givnish, TJ. *On the Economy of Plant Form and Function.* London: Cambridge University Press, 1986.

24. Pazourek J, Nátr L. Changes in the anatomical structure of the 1st two leaves of barley caused by the absence of nitrogen or phosphorus in the nutrient medium. *Biol. Plant.* 1981:296–301.

25. Tooming H. *Solněčnaja Radiacija I Formirovanie Urožaja.* Leningrad: Gidrometeoizdat, 1977.

26. Shinano T, Osaki M, Tadano T. Comparison of production efficiency among field crops related to nitrogen nutrition and application. *Plant Soil* 1993; 155/156:207–210.

27. Amthor JS. The McCree–de Wit–Penning de Vries–Thornley respiration paradigms: 30 years later. *Ann. Bot.* 2000; 86:1–20.

28. Stoy V. Photosynthesis, respiration, and carbohydrate accumulation in spring wheat in relation to yield. *Physiol. Plant.* 1965; 4(Suppl):1–125.

29. Apel P. Über Wechselbeziehungen zwischen Assimilatbildung und Speicherung. *Tag Ber. Akad. Landw. Wiss. DDR* 1972; 119:41–46.

30. King RW, Wardlaw IF, Evans LT. Effects of assimilate utilization on photosynthetic rate in wheat. *Planta* 1967; 77:261–276.

31. Mokronosov AT: Endogennaja determinacija fotosintěza v sistěmě rastěnija. *Uč Zap Ural Gos. Univ. Sverdlovsk* 1970; 113:3–19.

32. Lawlor DW. Carbon and nitrogen assimilation in relation to yield: mechanisms are the key to understanding production systems. *J. Exp. Bot.* 2002; 53:773–787.

33. Romer W. Untersuchungen über die Auslastung des Photosyntheseapparates bei Gerste (*Hordeum distichon* L. und weissem Senf (*Sinapis alba* L.) in Abhängigkeit von den Umweltbedingungen. *Arch. Bodenfrucht Pflanzenprodukt.* 1971; 15:415–423.

34. Criswell JG, Shibles RM: Influence of sink-source on flag-leaf net photosynthesis in oats. *Iowa State J. Sci.* 1972; 46:405–415.

35. Mokronosov AT. Transport assimiljatov kak factor endogennoj reguljacii fotosintěza. *Trudy Biol-počv Inst. Sverdlovsk* 1973; 20:76–83.

36. Michael G. Seiler-Kelbisch H. Cytokinin content and kernel size of barley grains as affected by environment and genetic factors. *Crop Sci.* 1972; 12:162–165.

37. Borzenkova RA. Gormonalnaja reguljacija fotosinteza. In: Anon., ed. *Voprosy reguljacii fotosinteza.* Vol. 3. Sverdlovsk: Ural. Univ., 1973: 45–57.

38. Good NE, Bell DH. Photosynthesis, plant productivity, and crop yield. In: Carlson PS, ed. *The Biology of Crop Productivity.* New York: Academic Press, 1980:3–15.

39. Nátrová Z. Wheat stem vascular size in relation to kernel number and weight. *Acta Univ. Agric. Brno Fac. Agron.* 1985; 33:643–646.

40. Nátrová Z, Nátr L. Limitation of kernel yield by the size of conducting tissue in winter wheat varieties. *Field Crop Res.* 1993; 31:121–130.

41. Münch E. *Die Stoffbewegungen in der Pflanze.* Jena: Verlag von Gustav Fischer, 1930.

42. Bancal P, Soltani F. Source–sink partitioning. Do we need Munch? *J. Exp. Bot.* 2002; 53:1919–1928.

43. Lalonde S, Tegeder M, Throne-Holst M, Frommer WB, Patrick JW. Phloem loading and unloading of sugars and amino acids. *Plant Cell Environ.* 2003; 26:37–56.

44. Hay RKM. Harvest index: a review of its use in plant breeding and crop physiology. *Ann. Appl. Biol.* 1995; 126:197–216.

45. Petr J, Lipavsky J, Hradecká D. Production process in old and modern spring barley varieties. *Die Bodenkultur* 2002; 53:19–27.

46. Wacker L, Jacomet S, Körner C. Trends in biomass fractionation in wheat and barley from wild ancestors to modern cultivars. *Plant Biol.* 2002; 4:258–265.

47. Mendel G.: Versuche über Pflanzenhybriden. *Verhandl Naturforsch. Vereines Brno* 1865; 4:3–46. Reprinted by Tschermak E. Leipzig: Verlag von W. Engelmann, 1901.

48. Tilman D. The greening of the green revolution. *Nature* 1998; 396:211–212.

49. Leung H, Hettel GP, Cantrell RP. International Rice Research Institute: roles and challenges as we enter the genomics era. *Trends Plant Sci.* 2002; 7:139–142.

50. Donald CM. The breeding of crop ideotypes. *Euphytica* 1968; 17:385–403.

51. Stafford JV. Implementing precision agriculture in the 21st century. *J. Agric. Eng. Res.* 2000; 76:267–275.

52. Khush GS. Breaking the yield frontier of rice. *GeoJournal* 1995; 35(3):329–332.

53. Earl HJ, Tollenaar M. Maize leaf absorptance of photosynthetically active radiation and its estimation using a chlorophyll meter. *Crop Sci.* 1997; 37:436–440.

54. Chapman SC, Barreto HJ. Using a chlorophyll meter to estimate specific leaf nitrogen of tropical maize during vegetative growth. *Agron. J.* 1997; 89:557–562.

55. Nátr L. Varietal differences in the intensity of photosynthesis. *Rost. Vyr. (Praha)* 1966; 12:163–178 (in Czech).

56. Evans JR. Photosynthetic characteristics of fast- and slow-growing species. In: Lambers H, Poorter H, Van Vuuren MMI, eds. *Inherent Variation in Plant Growth. Physiological Mechanisms and Ecological Consequences.* Leiden: Backhuys Publ., 1998:101–117.

57. Duncan WG, Loomis RS, Williams WAQ, Hanau R. A model for simulating photosynthesis in plant communities. *Hilgardia* 1967; 38:181–205.

58. Duncan WG. Leaf angles, leaf area, and canopy photosynthesis. *Crop Sci.* 1971; 11:482–485.

59. Hammer GL, Kropff MJ, Sinclair TR, Porter JR. Future contributions of crop modeling – from heuristics and supporting decision making to understand genetic regulation and aiding crop improvement. *Eur. J. Agron.* 2002; 18:15–31.

60. Matsuoka M, Furbank RT, Fukayama H, Miyao M. Molecular engineering of C$_4$ photosynthesis. *Annu. Rev. Plant. Physiol. Plant Mol. Biol.* 2001; 52:297–314.

61. Matsuoka M, Nomura M, Agarie S, Miyao-Tokutomi M, Ku MSB. Evolution of C$_4$ photosynthetic genes and overexpression of maize C$_4$ genes in rice. *J. Plant Res.* 1998; 111:333–337.

62. Matsuoka M, Sanada Y. Expression of photosynthetic genes from the C$_4$ plant, maize, in tobacco. *Mol. Gen. Genet.* 1991; 225:411–419.

63. Suzuki S, Murai N, Burnell JN, Arai M. Changes in photosynthetic carbon flow in transgenic rice plants that express C$_4$-type phosphoenolpyruvate cyrboxykinase from *Urochloa panicoides*. *Plant Physiol.* 2000; 124:163–172.

64. Ku MSB, Cho DH, Li X, Jiao DM, Pinto M, Miyao M, Matsuoka M. Introduction of genes encoding C$_4$ photosynthesis enzymes into rice plants: physiological consequences. *Novartis Found. Symp.* 2001; 236:100–116.

65. Agarie S, Miura A, Sumikura R, Tsukamoto S, Nose A, Arima S, Matsuoka M, Miyao-Tokutomi M. Overexpression of C$_4$ PEPC caused O$_2$-insensitive photosynthesis in transgenic rice plants. *Plant Sci.* 2002; 162:257–265.

66. Voznesenskaya EV, Franceschi VR, Kiirats O, Freitag H, Edwards GE. Kranz anatomy is not essential for terrestrial C$_4$ plant photosynthesis. *Nature* 2001: 414:543–546.

67. Sage RF. C$_4$ photosynthesis in terrestrial plants does not require Kranz anatomy. *Trends Plant Sci.* 2002; 7:283–285.

68. Häusler RE, Hirsch HJ, Kreuzaler F, Peterhanzel C. Overexpression of C$_4$ cycle enzymes in transgenic C$_3$ plants: a biotechnoloogical approach to improve C$_3$ photosynthesis. *J. Exp. Bot.* 2002; 53:591–607.

69. Edwards GE, Furbank RT, Hatch MD, Osmond CB. What does it take to be C$_4$? Lessons from the evolution of C$_4$ photosynthesis. *Plant Physiol.* 2001; 125:46–49.

70. Goto F, Yoshihara T, Shigemoto N, Toki S, Takaiwa F. Iron fortification of rice seed by the soybean ferritin gene. *Nat. Biotechnol.* 1999; 17:282–286.

71. Grusak MA, DellaPenna D. Improving the nutrient composition of plants to enhance human nutrition and health. *Annu. Rev. Plant Physiol. Plant Mol. Biol.* 1999; 50:133–161.

72. Vasconcelos M, Datta K, Oliva N, Khalekuzzaman M, Torrizo L, Krishnan S, Oliveira M, Goto F, Datta SK. Enhanced irone and zink accumulation in transgenic rice with the ferritin gene. *Plant Sci.* 2003; 164:371–378.

73. Raskin I, Ribnicki DM, Komarnytsky S, Ilic N, Poulev A, Borisjuk N, Brinker A, Moreno DA, Ripoll C, Yakoby N, O'Neal JM, Cornwell T, Pastor I, Fridlender B. Plants and human health in the twenty-first century. *Trends Biotechnol.* 2002; 20:522–531.

74. Tilman D, Cassman KG, Matson PA, Naylor R, Polasky S. Agricultural sustainability and intensive production practices. *Nature* 2002; 418:671–677.

75. Morandini P, Salamini F. Plant biotechnology and breeding: allied for years to come. *Trends Plant Sci.* 2003; 8:70–75.

76. Daniell H, Khan MS, Allison L. Milestones in chloroplast genetic engineering: an environmentally friendly era in biotechnology. *Trends Plant Sci.* 2002; 7:84–91.

77. Koebner RMD, Summers RW. 21st century wheat breeding: plot selection or plate detection? *Trends Biotechnol.* 2003; 21:59–63.

78. Tooming H. Mathematical description of net photosynthesis, growth and adaptation processes in the photosynthetic apparatus of plant communities. In: Šetlík I, ed. *Prediction and Measurement of Photosynthetic Productivity*. Wageningen: Centre Agric. Publ. Document., 1970:103–113.

79. Lawlor DW. *Photosynthesis*. 3rd ed. Oxford: BIOS Sci Publ Ltd, 2001.

80. Kern J. Food, feed, fibre, fuel and industrial products of the future: challenges and opportunities. Understanding the strategic potential of plant genetic engineering. *J. Agron. Crop Sci.* 2002; 188:291–305.

81. Nátr L, Nonhebel S. Bioresources — present state and outlook. *Zemed Technika (Praha)* 1995; 41:83–84.

82. Melis A, Zhang L, Forestier M, Ghirardi ML, Seibert M. Sustained photobiological hydrogen gas production upon reversible inactivation of oxygen evolution in the green alga *Chlamydomonas reinhardtii*. *Plant Physiol.* 2000; 122:127–135.

83. Melis A, Happe T. Hydrogen production. Green algae as a source of energy. *Plant Physiol.* 2001; 127:740–748.

84. Happe T, Hemschemeier A, Winkler M, Kaminski A. Hydrogenases in grean algae: do they save the algae's life and solve our energy problems? *Trends Plant Sci.* 2002; 7:246–250.

85. Nátr L. Influence of mineral nutrition on photosynthesis and the use of assimlates. In: Cooper JP, ed. *Photosynthesis and Productivity in Different Environments*. Cambridge, UK: Cambridge University Press, 1975:537–556.

86. Greenwood DJ. Nitrogen supply and crop yield: the global scene. *Plant Soil* 1982; 67:45–59.

87. Mann C. Reseeding the Green Revolution. *Science* 1997; 277:1038–1043.

88. Anonymous. Green revolution. Curse or blessing? International Food Policy Research Institute. http://www.ifpri.org (March 19, 2003).

89. Sheehy JE, Dionora MJA, Mitchel PL, Peng S, Cassman KG, Lemaire G, Williams RL. Critical nitrogen concentrations for high-yielding rice (*Oryza sativa* L.) cultivars in the tropics. *Field Crops Res.* 1998; 59:31–41.

90. Nátr L. The effects of mineral nutrients on photosynthesis. *Acta Univ. Carol. Biol.* 1997; 41:145–156.

91. Nátr L. Mineral nutrients — a ubiquitous stress factor for photosynthesis. *Photosynthetica* 1992; 27:271–294.

92. Nátr L. Influence du déficit en éléments minéraux sur la production de la matière sèche, l'intensité de la photosynthèse et la quantité de N, P et K dans les plantes. *Physiol. Végét.* 1970; 8:573–583.

93. Sinclair AG, Morrison JD, Smith LC, Dodds KG. Determination of optimum nutrient element ratios in plant tissue. *J. Plant Nutr.* 1997; 20:1069–1083.

94. Karley AJ, Leigh RA, Sanders D. Differential ion accumulation and ion fluxes in the mesophyll and epidermis of barley. *Plant Physiol.* 2000; 122:835–844.

95. Andrews M, Raven JA, Sprent JI. Environmental effects on dry matter partitioning between shoot and root of crop plants: relations with growth and shoot protein concentration. *Ann. Appl. Biol.* 2001; 138:57–68.

96. Garnier E, Salager JL, G, Sonié L. Relationships between photosynthesis, nitrogen and leaf structure in 14 grass species and their dependence on he basis of expression. *New Phytol.* 1999; 143:119–129.

97. Atkin OK, Westbeek MHM, Cambridge ML, Lambers H, Pons TL. Leaf respiration in light and darkness. *Plant Physiol.* 1997; 113:961–965.

98. Evans JR. Leaf anatomy enables more equal access to light and CO_2 between chloroplasts. *New Phytol.* 1999; 143:93–104.

99. Evans JR. Carbon fixation profiles do reflect light absorption profiles in leaves. *Aust. J. Plant Physiol.* 1995; 22:865–873.

100. Nátr L. *Photosynthetic Production and Mankind Nutrition*. Praha: Nakl ISV, 2002 (in Czech).

28 Photosynthate Formation and Partitioning in Crop Plants

Alberto A. Iglesias
Laboratorio de Enzimologia Molecular Grupo de Enzimología Molecular, Bioquímica Básica de Macromoléculas, Facultad de Bioquímica y Ciencias Biológicas, Universidad Nacional del Litoral

Florencio E. Podestá
Facultad de Ciencias Bioquímicas y Farmacéuticas, Universidad Nacional de Rosario

CONTENTS

I. INTRODUCTION

The photosynthetic process sustains life on our planet since it is the mechanism by which energy from sunlight is used to synthesize biomolecules. Broadly speaking, photosynthesis may be defined as the metabolism by which atmospheric CO_2 is fixed into carbohydrates (photosynthates), with sucrose (Suc) and starch being the quantitative major end-products of the process. The production of these two metabolites provides organic carbon source for the synthesis of other cellular constituents such as proteins and lipids. In higher plants, photosynthesis takes place in green tissues, mainly in leaves, with the photosynthetic machinery localized within the chloroplast where the newly fixed carbon (namely triose-phosphate or triose-P) can be initially found [1–5]. Triose-P is the immediate source for starch synthesis within the chloroplast; alternatively, this photosynthate can be exported to the cytoplasm and utilized for Suc production [1,4,6]. With some exceptions, starch is the principal storage carbohydrate that accumulates in photosynthetic and nonphotosynthetic tissues of higher plants [7–9]. Leaf starch constitutes a transient pool of carbohydrate accumulation, the level of which varies within the day [8]. Long-term storage of starch is confined to plastids (amyloplasts) of cells in specialized nonphotosynthetic organs (roots, seeds, tubers) [7,8]. Suc, meanwhile, is the major form by which carbon is moved from photosynthetic to reserve tissues, although some plants may also use it for long-term storage [1,8].

From the above discussion, it follows that photosynthates are partitioned between a rather stationary (starch) and a mobile (Suc) form. The partitioning occurs not only at the intracellular level, between plastid and cytosol; but also affects the plant as a whole, as determined by its relative abundance in source and sink tissues [6]. The importance of the knowledge of the processes involved in photosynthates formation, partitioning, and storage by plants is clear from the fact that most crops are dedicated to the production of Suc or starch [8]. Be it for animals or humans, starch is the main resource obtained from plants as a food [8,10]. The efficiency in the control of assimilated carbon partitioning is crucial with regard to the productivity of a crop plant. Such a control is the result of many different factors acting at a local, cellular level or systematically, affecting the carbon and energy demand in different organs. A great deal of effort on plant research has been dedicated to individualize and characterize the factors governing carbon partitioning. Successful results allow the manipulation of the partition of assimilated carbon in certain crop plants in order to improve productivity [6,8–13].

II. PHOTOSYNTHATE FORMATION AND PARTITIONING IN THE LEAF

A. STARCH SYNTHESIS AND BREAKDOWN

Leaf starch metabolism is a dynamic process occurring within the chloroplast organelle [3,7,8,14] (Figure 28.1). Starch is predominantly synthesized during the day from sugar-P intermediates produced by photosynthetic carbon assimilation. In the light, up to 30% of the CO_2 fixed by leaves is incorporated into starch. The polysaccharide accumulates in the chloroplast during the light period forming relatively small granules, whose shape is not species specific. During darkness, leaf starch is actively degraded and so utilized to provide photoassimilates to the whole plant [1,3, 6–8,14].

A key intermediate metabolite of the autotrophic Benson–Calvin cycle, fructose-6-phosphate (Fru6P), serves to initiate synthesis of starch in photosynthetic cells [1,3]. Chloroplastic glucose-6-phosphate (Glc6P) isomerase (EC 5.3.1.9; Equation (28.1)) and P-glucomutase (EC 5.4.2.2; Equation (28.2)) interconvert Fru6P, Glc6P, and Glc1P. It is regarded that hexose-P pool is close to equilibrium in plant cells.

$$Fru6P \Rightarrow Glc6P \qquad (28.1)$$

$$Glc6P \Rightarrow Glc1P \qquad (28.2)$$

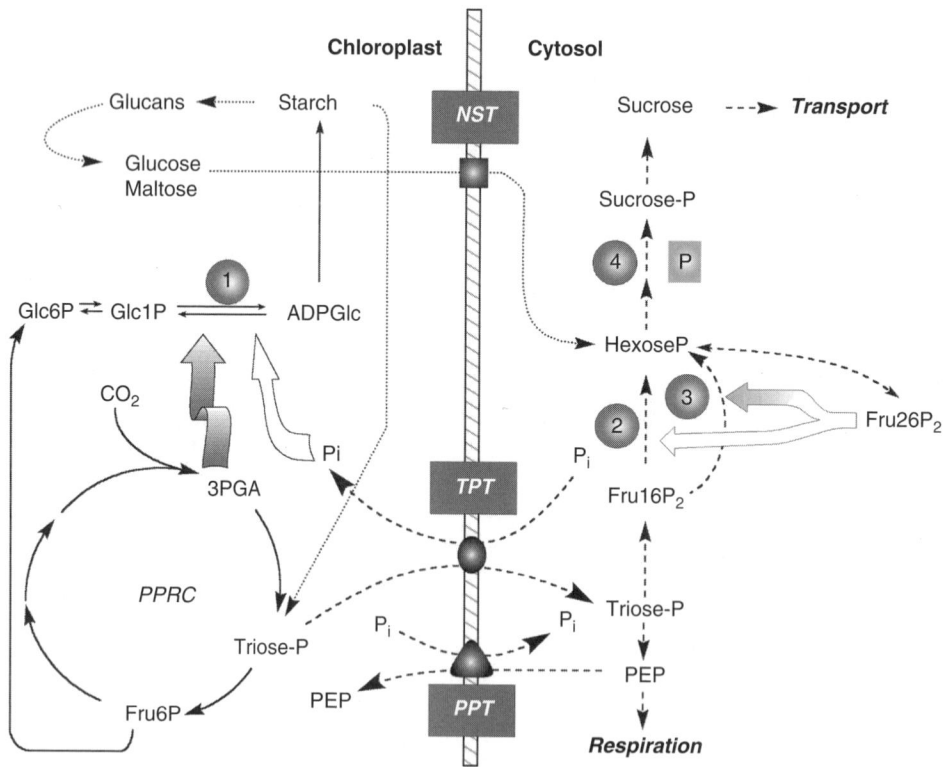

FIGURE 28.1 Diagram of carbon flow in photosynthetic tissues. Pathways leading to starch or Suc formation or degradation are depicted as solid lines for processes taking place in the light, as dotted lines for those occurring predominantly in the dark, and as dashed lines for processes occurring throughout the day. The modes of regulation of some of the key control points are indicated. Dotted arrows indicate inhibition, whereas empty arrows indicate activation. The shadowed **P**s indicates control by phosphorylation. Numbers indicate reactions catalyzed by: 1, ADPGlc pyrophosphorylase; 2, PFP; 3, cFru1,6bisPase; 4, SPS. NST, neutral sugar transporter; TPT, triose-phosphate transporter; PPT, phosphoenolpyruvate transporter.

The predominant metabolic route for starch synthesis from Glc1P in photosynthetic cells is the ADPGlc pathway [1,3–5,7,8,13–15]. The biosynthetic metabolism involves three enzymatic steps (Equations (28.3)–(28.5)), respectively catalyzed by ADPGlc synthase or ADPGlc PPase (EC 2.7.7.27), starch synthase (EC 2.4.1.21), and branching enzyme (EC 2.4.1.18).

$$\text{Glc1P} + \text{ATP} \Leftrightarrow \text{ADPGlc} + \text{PP}_i \tag{28.3}$$

$$\begin{aligned} \text{ADPGlc} + (\alpha\text{-1,4-glucan})_n \Rightarrow \\ \text{ADP} + (\alpha\text{-1,4-glucan})_{n+1} \end{aligned} \tag{28.4}$$

$$\begin{aligned} \text{Linear } \alpha\text{-1,4-glucanchain} \Rightarrow \alpha\text{-1,4-glucan chain} \\ \text{with } \alpha\text{-1,6 linkage branch point} \end{aligned} \tag{28.5}$$

Since the first report on the occurrence of ADPGlc PPase activity made by Espada in 1962 [16], numerous experimental evidences have been accumulated strongly supporting that the metabolic route for starch buildup is confined to the chloroplast, and that mainly (if not solely) occurs through the ADPGlc pathway. The autonomy of the organelle to synthesize the polysaccharide was established after evidences showing that isolated chloroplasts accumulate starch when incubated with CO_2 in the light, with no requirement of other subcellular components [17]. The kinetic, regulatory, and structural characterization of the enzymes involved further sustain the ADPGlc pathway as the predominant route for starch synthesis in chloroplasts [8,14,15].

The key role of Fru6P as a branch point between the Benson–Calvin cycle and transitory starch synthesis has been demonstrated by analysis of antisense RNA transformed potato plants that express reduced levels of chloroplastic fructose-1,6-bisphosphate 1-phosphatase (Fru1,6bisPase) [18]. The restriction to produce Fru6P by these transgenic plants was correlated with a reduction of starch accumulation in the light and a consequent increase in carbohydrate partitioning toward soluble sugars. In a similar way, mutants with reduced activity of Glc6P isomerase and P-glucomutase demonstrate the importance of these enzymes in photosynthate partitioning as well as their involvement in the route for starch synthesis [19,20].

In addition to this common core biochemical pathway, other enzymes of malto-oligosaccharide metabolism are required for normal starch metabolism. As recently reviewed in great detail [21,22], the functioning of these additional enzymes is particularly important to understand how the semicrystalline structure of the polysaccharide and the starch granule are formed.

Conclusive experimental evidence also demonstrated that ADPGlc PPase is a key regulatory enzyme [5,8,12–15,23]. As critically analyzed by Preiss et al. (see Ref. [4,8] and references therein), genetic data from different starch deficient mutants, molecular biology studies as well as analysis on flux control and metabolism clearly establish that, in vivo, chloroplastic starch is synthesized via ADPGlc PPase, with this enzyme catalyzing the first committed step of the metabolic pathway.

The ADPGlc PPases from different plant leaves, green algae, and cyanobacteria have been characterized, and recent extensive reviews on the enzyme from bacteria and plants are available [14,23]. ADPGlc PPase catalyzes the synthesis of ADPGlc and PP_i from Glc1P and ATP (see Equation (3)), in the presence of a divalent metal ion (physiological Mg^{2+}). Although in vitro the reaction is freely reversible, the hydrolysis of PP_i by a high activity of plastidic alkaline pyrophosphatase certainly pulls the equilibrium far toward synthesis of ADPGlc in vivo. An almost irreversible operation of ADPGlc PPase under physiological conditions agrees with the rationale of it being the regulatory step of starch synthesis.

All ADPGlc PPases from photosynthetic tissues of higher plants characterized so far are allosterically regulated by 3-phosphoglycerate (3PGA) and inorganic phosphate (P_i), as the more important activator and inhibitor, respectively [2–4,8,12,14,15,23]. The enzyme is also subject to transcriptional regulation [24]. Studies on the structure–function relationships show that the enzyme from higher plants is a heterotetrameric protein of molecular mass around 210 kDa, composed of two subunits (α and β) that give rise to an $\alpha_2\beta_2$ quaternary structure [14,23]. Subunits α and β are different in size, immunogenicity, and amino acid sequence [8,23,25]. Similar structural properties have been found for the ADPGlc PPase from the green alga Chlamydomonas reinhardtii [26]. The enzyme from cyanobacteria seems to be the only ADPGlc PPase regulated by 3PGA and P_i that is a homotetramer (as corresponds to the structure of heterotrophic and anoxygenic photosynthetic bacteria) [14,15]. However, the cyanobacterial protein is immunologically more related and shares higher sequence homology with the plant than with the enzyme from heterotrophic bacteria [27]. Independently of similitudes or differences, it is clear that regulation by 3PGA and P_i is a characteristic of ADPGlc PPase from cells performing oxygenic photosynthesis.

Activation of plant ADPGlc PPase by 3-phosphoglycerate (3PGA) not only increases V_{max} and the affinity of the enzyme by its substrates, ATP and Glc1P, but also decreases sensitivity towards inhibition by P_i [14]. For the enzyme from different plants leaves, maximal activation by 3PGA varies between 5- and 100-fold, depending on pH conditions; whereas the concentration of 3PGA causing half this stimulatory effect ($A_{0.5}$) ranges from 0.01 to 0.5 mM [8,14]. On the other hand, values of $I_{0.5}$ (concentration required for 50% inhibition) between 0.02 and 0.2 mM are reported for the inhibition of the enzyme by P_i. These values are effectively increased by the presence of 3PGA [2,3,8,12].

Recent studies have underscored the relevance of the cross-talk between allosteric regulators of ADPGlc PPase from photosynthetic cells in the fine modulation of the enzyme activity [28–31]. Studies performed with the ADPGlc PPase from the cyanobacterium *Anabaena* PCC 7120 have shown that P_i and molecular crowding conditions (those resembling high concentration of macromolecules, mainly proteins, found in the chloroplast stroma) elicit an ultrasensitive response of the enzyme toward the activator 3PGA [29]. As defined by Koshland et al. [32,33], ultrasensitivity is a type of amplification by which a biological system exhibits a sharp response (manifold increase in the signal) after a narrow variation range of the stimulus. Ultrasensitivity adds a level of complexity in the interplay between allosteric regulators of ADPGlc PPase, enhancing their respective effects. Thus, activation of the cyanobacterial enzyme by 3PGA exhibits a cooperative behavior in the presence of P_i (and in molecularly crowded media), as relatively small changes of the allosteric activator produce manifold increases in the enzyme activity [28]. The cooperativity in the saturation curve for 3PGA increases as the concentration of P_i and crowded conditions is increased. Exemplifying this, for a ninefold (from 10% to 90% of V_{max}) activation of ADPGlc PPase, it is necessary to increase 3PGA concentration by 200-fold in a medium without P_i and molecular crowding; whereas in crowded media containing 5 mM P_i the same activation of the enzyme is exerted by only sevenfold increase in the activator level [29,30]. More recently [31], it has been shown that such an ultrasensitive behavior of ADPGlc PPase is operative under intracellular conditions occurring in cyanobacteria and probably in plastids of higher plant cells. In addition, a mathematical model was developed and experimental data agree with the fact that the whole process of starch synthesis occurs with ultrasensitivity.

A rational picture for the physiological regulation of starch synthesis in leaves through modulation of ADPGlc PPase activity by the 3PGA:P_i ratio can be established [1–3,8,14,17,20]. Active photosynthesis produces an increase of 3PGA, the primary CO_2

fixation product, and a decrease in the P_i level because of photophosphorylation [1,17,20]. In this scenario, the high 3PGA:P_i ratio in the chloroplast activates ADPGlc PPase and, consequently, the increased levels of sugar-P and ATP are channeled towards synthesis of starch. Conversely, in the dark P_i concentration increases, ATP decreases and the Benson–Calvin cycle is inactive with the consequent decrease in 3PGA levels. Under these dark conditions, starch synthesis is effectively inhibited at the level of ADPGlc PPase. The physiological significance of this regulatory mechanism has been demonstrated by different studies, including determination of metabolite levels as well as analysis of mutants from plant leaves and green algae with altered capacity to accumulate starch [11,20,34–36]. The numerous experimental data available clearly establish that the 3PGA:P_i ratio within the chloroplast regulates starch synthesis via modulation of ADPGlc PPase activity. The latter regulation is thus critical for the accurate partitioning of photosynthates in leaf cells [8,11,35].

Studies performed with ADPGlc PPase from potato tuber [37,38] have demonstrated that reduction by dithiothreitol or thioredoxin of an intermolecular disulfide bridge (linking the enzyme small subunits) produces an activation (operating in addition to the effect of 3PGA) of the enzyme (for more details see Section IV.B.). Since the cysteine residues involved in the disulfide bridge are conserved in leaf ADPGlc PPases, it has been proposed that the same regulatory mechanism could operate via the ferredoxin–thioredoxin system in chloroplasts [14,38]. According to this model, starch metabolism is regulated by covalent modification of the enzyme on the basis of a light–dark cycle. Upon illumination, reduced thioredoxin modifies ADPGlc PPase to a more active state thus stimulating starch synthesis; with the mechanism being reversed during the dark period [14,38]. This regulatory mechanism is complementary to that involving ultrasensitive allosteric modulation of ADPGlc PPase activity by the cross-talk between 3PGA and P_i, and it could ensure a more fine control the synthesis of the polysaccharide.

Concerning starch breakdown, there are two possible routes of operation in leaves: the hydrolytic and the phosphorolytic [3,4,7,8,39]. It is not completely clear which of these pathways for starch degradation is more important *in vivo*. It is thought that in leaves starch is catabolized through the combination of both the hydrolytic and the phosphorolytic route. The preeminence of one or the other catabolism seems to be dependent on the level of P_i in the chloroplast stroma [2,3,8]. The products of starch phosphorolysis are triose-P and 3PGA, which can be exported from the chloroplast via the P_i-translocator system. The hydrolytic route renders Glc and maltose, which are released through the hexose transporter of the chloroplast envelope [39,40].

Chloroplastic starch phosphorylase (EC 2.4.1.1) phosphorolytically cleaves α-1,4-glucosyl bonds from the nonreducing end of an α-glucan chain according to Equation (28.6)

$$(\alpha\text{-}1,4\text{-glucan})_n + P_i \Rightarrow (\alpha\text{-}1,4\text{-glucan})_{n-1} + \text{Glc1P} \quad (28.6)$$

Since the substrates for starch phosphorylase are α-1,4-glucan larger than maltotetraose, complete phosphorolysis of amylose/amylopectine requires the additional action of a glucosyltranferase (EC 2.4.1.2; namely D-enzyme) and the debranching enzyme (EC 3.2.1.4). D-enzyme catalyzes condensation of donor and acceptor α-1,4-glucan with the release of free Glc

$$(\alpha\text{-}1,4\text{-glucan})_d = (\alpha\text{-}1,4\text{-glucan})_a \Rightarrow (\alpha\text{-}1,4\text{-glucan})_{d+a-1} + \text{Glc} \quad (28.7)$$

whereas the debranching enzyme hydrolyzes α-1,6-glucosyl bonds at the branch points of the amylopectin molecule [2,3,7,8,22].

On the basis of their affinities for glucans, molecular mass, and subcellular localization, starch phosphorylases has been classified into two types [41]. One type localizes in plastids (chloroplast, amyloplast) and utilizes long linear α-glucan chains as the preferred substrates. The second type comprises cytosolic phosphorylases exhibiting high affinity for branched glucans. Four different forms of starch phosphorylase, differing in optimal pH, primer dependence, and developmental expression have been found in maize. A plastidic starch phosphorylase, constituting a major protein in the amyloplast stromal fraction of maize has been identified [42,43]. The enzyme was purified and characterized as a 112-kDa protein, composed of two identical subunits, and exhibiting an eightfold higher affinity toward amylopectin than for glycogen. When malto-oligosaccharides are used as substrates, the purified enzyme catalyzes the phosphorolytic reaction with preference over the starch synthesis reaction [43]. However, the exclusive involvement of phosphorylase in starch degradation has been challenged by studies showing that expression of the plastidial enzyme correlates with starch biosynthesis [44,45].

The hydrolytic pathway for starch breakdown in chloroplasts mainly involves α-amylase (EC 3.2.1.1) [2,3,8]. The enzyme possesses an endoamylolytic activity hydrolyzing α-1,4-glucosyl bonds of starch and rendering a mixture of linear and branched oligosaccharides, branched dextrins, maltotriose, maltose, and Glc [2,3]. Complete degradation of starch through the hydrolytic route can be achieved with the complementary action of debranching enzyme and α-glucosidase (EC 3.2.1.20) [3]. The latter hydrolyzes α-1,4-glucan linkages of dextrins, maltose, and short maltosaccharides, attacking from the nonreducing end and liberating Glc [2,3]. Of the three types of α-glucosidases classified, enzymes of type III mainly hydrolyze maltose, isomaltose, and starch [46].

Starch can also be hydrolyzed by β-amylase (EC 3.2.1.2), an enzyme with exoamylolytic activity cleaving maltosyl residues from amylose starting from the nonreducing end [3,7,8]. Although β-amylase has been reported to be present in leaves of different plants [7], its involvement in starch metabolism is unlikely after studies demonstrating the extrachloroplastic localization of the enzyme [2,7]. A different class of starch/glycogen degrading enzyme, namely α-1,4-glucan lyase, has been characterized in red algae [47]. The enzyme catalyzes the degradation of maltose, malto-oligosaccharides, starch, and glycogen to produce 1,5-anhydrofructose. The importance of this enzyme in red algae metabolism as well as its occurrence in other organisms remains to be elucidated. Although red algae are a very distinct group of organisms and phylogenetically distant from green plants, the absence or presence of α-1,4-glucan lyase in higher plants needs to be determined to have a clear picture of all the possible starch catabolic pathways.

Regulation of the routes for starch degradation in the chloroplast is not as well understood as its synthesis is. It is unlikely that the diurnal dynamic process of starch buildup/breakdown is regulated only at the synthetic level. Simultaneous operation of starch synthesis and degradation would constitute a futile cycle and thus both pathways must be finely and coordinately regulated [3,7,8]. Moreover, the absence of detectable starch turnover when leaves are exposed to light indicates a direct regulation of starch breakdown operating *in vivo* [3]. The function of the phosphorolytic pathway may be restricted during the light period due to the low P_i levels available as substrate for the phosphorylase [1,3,8]. However, this control point cannot by itself account for an adequate regulation of starch catabolism [8].

A possible regulatory mechanism of starch degradation could be associated with photosynthetically driven pH change in the chloroplast stroma. Activity of most α-glucosidases, α-amylases, glucosyltransferases, and debranching enzymes markedly decrease at pH values higher than the optimum value, which is at or below 6.0 [7,8]. Accordingly, hydrolytic starch degradation would be significantly more active at the stromal pH of 7 found in the dark than at pH 8 occurring in the light [2,3,8]. To a lesser extent, this effect could also apply to starch phosphorylase, since the pH–activity curve of the enzyme shows a sharp optimum around pH 6.5 [7].

Recently [39], a protein that is bound to potato starch granule, namely R1, has been characterized and proposed to be involved in the metabolism of the polysaccharide. R1 is a 120-kDa protein that exhibits glucan water dikinase activity and thus was identified as the starch phosphorylating enzyme. Phosphorylation is the only covalent modification found in natural starch. Depending on the botanical origin, the polysaccharide (specifically the amylopectin fraction) contains different quantities (from minute amounts in cereals to 0.2% to 0.4% w/w in potato) of phosphate groups monoesterifying C-6 and C-3 positions of the Glc units [39]. The phosphorylation degree is determinant of the starch physical properties and thus is a relevant issue for the industrial use of the polymer. Characterization of the mechanism of starch phosphorylation has opened up new avenues in plant carbohydrate research [39]. A dependence of phosphorylation on the starch structure has been shown. It is now considered that phosphorylation constitutes an integral part of the biosynthesis of starch. Additionally, and despite the apparent lack of amylolytic activity of R1, it is thought that phosphorylation is required for normal starch degradation [39]. This is suggested by results obtained with potato antisense R1 mutants that present excess levels of starch in leaves. Thus, new insights are emerging that seems to be relevant for the complete elucidation of the reactions involved in normal biosynthesis, and (mainly) temporary mobilization or degradation of starch.

The possible occurrence of other molecular components involved in starch metabolism and regulation should not been ruled out. In this way, a recent report pointed out the presence of regulatory 14-3-3 proteins within starch granules of Arabidopsis chloroplasts [48]. Consequently, the understanding of the mechanisms operating and regulating leaf starch metabolism remains incomplete and somehow speculative, waiting for further experimental work.

B. TRANSPORT OF METABOLITES ACROSS THE CHLOROPLAST ENVELOPE

The occurrence of separate but coordinate carbon metabolisms in the chloroplast and the cytosol makes obvious the importance of metabolite transport between both cellular compartments in a photosynthetic cell. Chloroplasts are enclosed by two galactolipid-rich membranes, which are different in terms of both their physical and functional properties [49–53]. The outer membrane, believed to be derived from the endoplasmic reticulum, has a low protein:lipid ratio [53]. Pore-forming proteins (porins) are responsible for the permeability of the envelope outer membrane. Porins are similar to proteins found in mitochondria and Gram-negative bacteria allowing the free passage of molecules smaller than 10 kDa in mass [52,53]. The former idea that porin-like proteins of the outer envelope membrane are unspecific has been reassessed [40,51,54]. After characterization of outer envelope proteins it became clear that they exhibit substrate specificity. In addition, the functioning of these membrane proteins seems to be under regulation (probably including voltage-independent gating control) [40,54]. The new picture proposes that the chloroplast outer envelope is a molecular sieve and, consequently, the interenvelope space is not freely accessible [40,51,54]. In this way, the chloroplast permeability is controlled by the outer and inner envelope membranes [54].

The inner envelope membrane is strictly selective with respect to metabolite flux because of the presence of specific translocators [40,49–54]. The P_i-translocator is, quantitatively, the major protein of the chloroplast envelope inner membrane [49,52,53]. It amounts for about 15% of the total envelope protein and constitutes the main transport system between chloroplast and cytosol, playing a key role for the functioning and regulation of carbon metabolism in leaf cells of C_3, C_4, and CAM plants [49,50,52,55] (Figure 28.1). Despite its relative abundance in the inner chloroplast envelope, the P_i-translocator may be kinetically limiting in vivo, at least under certain circumstances [50,56]. Transformation of plants with sense and antisense constructs of cDNA encoding for the tobacco P_i-translocator underscored the importance of the transport protein in photosynthate partitioning. Studies with transgenic plants showed that the higher the amount of functional translocator expressed, the lower the starch to soluble sugars ratio found, and vice versa [56].

The P_i-translocator from spinach chloroplasts has been characterized at the structural and functional level [49,50,52,54]. The isolated protein reveals as a single band in SDS-PAGE with an apparent molecular mass of 29 kDa. Hydrodynamic studies, carried out with the purified protein reconstituted into liposomes, demonstrated that the functional P_i-translocator is a homodimer possessing a prolate ellipsoidal shape with semiaxes 6.6 and 1.6 nm length and protruding from both sides of the membrane.

The P_i-translocator from C_3 plants, also known as triose-phosphate translocator or TPT, is highly specific for P_i or three-carbon chains possessing a P-ester group at the end of the molecule; namely P_i, triose-P, and 3PGA, the main metabolites serving as substrates *in vivo* [49,50,52,54]. These compounds compete with each other for the binding to the transport protein. The P_i-translocator facilitates the passive transport of its substrates by a mechanism implying a strict and stoichiometric counterexchange, via a ping-pong type of reaction mechanism. Thus, for each molecule of P_i, triose-P, or 3PGA transported inwards another one is transported outwards. The translocator from C_3 plants is also selective for the transport of double-negatively charged anions, which is of physiological relevance [50,52–54]. On the other hand, electrophysiological studies have shown that the Pi-translocator can also behave as a voltage-dependent ion channel, with preference to permeate anions [50,54].

The chloroplast envelope of mesophyll cells from C_4 and CAM plants contains a different form of the P_i-translocator, the PPT, which displays a high affinity towards phosphoenolpyruvate (PEP) [52,55]. Effective export of PEP from mesophyll C_4 chloroplasts, occurring in the light period, indicates that the PPT possesses different specificity, counterexchanging a three-carbon molecule with a P-group attached at position 2, that at the pH of the stroma of illuminated chloroplasts is an anion with three negative charges [55]. In 1997, the PPT was also found to be operative in the chloroplast envelope of C_3 plants [57]. This translocator mainly interchanges PEP and P_i, exhibiting a poor efficiency for the transport of triose-P and 3PGA. The PPT was found in different subtype of plastids, including those from nonphotosynthetic plant cells. The functioning of this transport system fulfills carbon demands of different metabolisms, as fatty acids synthesis and the shikimate pathway, conducting to the production of aromatic amino acids and secondary metabolites [50,54,57].

Isolation and sequencing of cDNA clones coding for the P_i-translocator from different plant tissues gave further information on the structure, synthesis, and assembly of this ubiquitous protein [50,58,59]. The translocator protein is coded for by the nuclear DNA. The functional, mature protein is derived after synthesis of a higher molecular mass precursor on soluble cytosolic ribosomes followed by posttranslational import into chloroplast and cleavage of the transit peptide by a specific protease [50,58,59]. The N-terminal extension of P_i-translocator precursors from spinach and pea involving 80 and 72 amino acid residues, respectively, contain a positively charged amphiphilic α-helix, which directs the polypeptides to the envelope inner membrane [50]. There is a high sequence similarity between TPT

transporters in different plants or tissues, but not with the PPT or hexose-P_i translocator or GPT (see Section IV.B), with homologies not higher than 35% and restricted to a few regions of the protein [50].

The molecular mass of the mature protein, obtained from amino acid sequence, is around 36 kDa, a value higher than that calculated from SDS-PAGE [58,59]. Differences probably reflect an enhanced capacity of the protein to bind detergent due to the high number of nonpolar amino acid residues it contains. Hydrophobicity distribution analysis showed that the mature P_i-translocator contains at least seven membrane-spanning segments anchoring the protein to the membrane [49,50.54]. These segments form α-helical structures and the amphiphilic character found in some of them would form a hydrophilic channel functional for the translocation process.

Specific translocators of the inner membrane mediate the transport through the chloroplast envelope of other compounds involved in different metabolic pathways. Transfer of neutral sugars, mono and dicarboxylates and adenylates is thus catalyzed by the following transport systems:

1. A selective carrier (neutral sugar transporter or NST) can transfer several neutral sugars such as hexoses and pentoses (D-Glc, D-Fru, D-Rib, D-Xyl, D-Man) as well as maltose. The NST seems to be important for the export of Glc from the chloroplast in the dark [52,53]. Nuclear magnetic resonance studies of intracellular carbon distribution performed in deuterium-labeled water have demonstrated physiologic export of Glc, maltose, and higher maltodextrin from chloroplasts at night [60].
2. Transport of dicarboxylates and amino acids (L-malate, 2-oxoglutarate, L-aspartate, L-glutamate, and L-glutamine) occurs by a counterexchange mechanism catalyzed by two different translocators, one transporting preferentially glutamine while the other is specific for dicarboxylates [50,52–54]. These translocators are mainly related to nitrogen assimilation and photorespiration metabolisms [54].
3. Oxaloacetate transport is mediated by a highly specific translocator, which plays an important role in C_4 plants, by importing the ketoacid into the chloroplast of mesophyll cells [52].
4. The pyruvate translocator found in bundle-sheath and mesophyll chloroplasts of C_4 plants also plays a key role for the functioning of carbon metabolism. Pyruvate transport is uncoupler-sensitive, since it is an active process

driven by a light-dependent cation gradient across the envelope [52,53].

5. The glycolate–glycerate translocator operates through a proton symport or hydroxyl antiport mechanism. This system is important for the photorespiratory metabolism where two glycolate molecules are exported from chloroplast in exchange for one D-glycerate molecule. The translocator also binds glyoxylate and D-lactate [52].

6. An ATP/ADP translocator of the chloroplast envelope, different from the mitochondrial carrier, has been characterized. The role of this transporter in protein translocation into chloroplasts has been demonstrated; whereas its involvement in the energy import into the organelle in the dark was proposed from the preference to catalyze uptake of ATP instead of ADP as well as the high transport activity found in young immature leaves [52,53].

C. SUCROSE SYNTHESIS AND BREAKDOWN

Sucrose (Suc) is one of the main end-products of photosynthesis and the prevalent sugar mobilized in plants, by which carbohydrate is distributed from source to sink tissues [1]. Exceptionally, Suc can also be used as long-term carbohydrate reserve by some plants (i.e. sugarcane). Suc synthesis is a cytosolic process [1] that must be maintained at levels that ensure a ready carbon source for export throughout the day.

In photosynthetic tissue, Suc is derived mainly from the triose-P (DHAP) synthesized in the chloroplast by the Benson–Calvin cycle turnover and exported to the cytosol through the P_i translocator (Figure 28.1). After isomerization of DHAP to Ga3P, mediated by the triose-P isomerase (EC 5.3.1.1) (Equation (28.8)), these compounds are converted to Fru1,6P_2 by Fru1,6P_2 aldolase (EC 4.1.2.13) (Equation (28.9)), followed by hydrolysis mediated by the cytosolic Fru1,6bisPase (cFru16bisPase, EC 3.1.3.11) (Equation (28.10)) or by the reversible conversion catalyzed by PFP (EC 2.7.1.90, pyrophosphate:fructose-6-phosphate 1-phosphotransferase) (Equation (28.11)). The Fru6P thus produced is converted to Glc6P and Glc1P by the sequential action of hexose phosphate isomerase and phosphoglucomutase (Equations (28.1) and (28.2), respectively). Glc1P is uridylated in a reaction catalyzed by UDP-glucose pyrophosphorylase (EC 2.7.7.9) (Equation (28.12)). Suc-P synthesis proceeds then via sucrose phosphate synthase (SPS; EC 2.4.1.14) (Equation (28.13)) and Suc is finally formed by hydrolytic cleavage of Suc-P by Suc-P phosphatase (EC 3.1.3.24) (Equation (28.14)).

$$DHAP \Rightarrow Ga3P \qquad (28.8)$$

$$DHAP + Ga3P \Rightarrow Fru1,6P_2 \qquad (28.9)$$

$$Fru1,6P_2 \Rightarrow Fru6P \qquad (28.10)$$

$$Fru1,6P_2 + P_i \Leftrightarrow Fru6P + PP_i \qquad (28.11)$$

$$Fru6P \Rightarrow Glc6P \qquad (28.1)$$

$$Glc6P \Rightarrow Glc1P \qquad (28.2)$$

$$Glc1P + UTP \Rightarrow UDPGlc + PP_i \qquad (28.12)$$

$$UDPGlc + Fru6P \Rightarrow Suc-P + UTP \qquad (28.13)$$

$$Suc-P + H_2O \Rightarrow Suc + P_i \qquad (28.14)$$

Several regulatory mechanisms are responsible for sustaining cytoplasmic Suc in photosynthetic tissues at levels allowing an uninterrupted flow of carbon to sink tissues, and these act mainly upon cFru1,6bisPase and SPS [61,62].

A metabolite-based type of fine control is responsible for the regulation of both the cFru1,6bisPase (by the Fru2,6P_2 system, AMP, and P_i) [63] and SPS (by P_i and Glc6P) [64]. Additionally, cFru1,6bisPase [65], and SPS [66] protein levels are developmentally regulated and, most notably, the latter protein is subject to reversible phosphorylation, which accounts for its coarse regulation, depending on factors such as illumination, nitrogen metabolism, and carbohydrate accumulation [67]. The participation of PFP in this metabolism is dubious [68,69], since severe reduction of PFP levels in tobacco photosynthetic tissue did not alter the rate of sucrose synthesis. Rather than an alternative to FBPase, PFP seems to act in the glycolytic direction in this tissue [69].

The signal metabolite Fru2,6P_2 strongly inhibits cFru1,6bisPase [70], which catalyzes the first nonreversible reaction of the Suc formation pathway. However, it does not affect the activity of cytosolic PFK. The effect of Fru2,6P_2 on cFru1,6bisPase is to induce sigmoidal kinetics for Fru1,6P_2, increasing the $S_{0.5}$ for this substrate and the sensitivity of the enzyme towards the inhibitors P_i and AMP [71,72]. The presence of a high level of Fru2,6P_2 would thus prevent Suc formation. In turn, levels of Fru2,6P_2 are dictated by the relative activities of the Fru6P 2-kinase and Fru2,6P_2 bisphosphatase, located in a bifunctional single polypeptide [73]. The kinase activity is inhibited by three-carbon phosphorylated metabolites and PP_i and activated by P_i [73]. The Fru2,6P_2 bisphosphatase is subject to product inhibition by both Pi and Fru6P [73]. A high cytosolic concentration of DHAP (as expected to emerge from a "surplus" in carbon fixation by the Benson–Calvin cycle, which would concomitantly raise the 3PGA concentration), would have a dual effect in promoting its own use: first, through the release of Fru2,6P$_2$ inhibition of cFru1,6bisPase, since at high DHAP (and mainly 3PGA) the Fru6P 2-kinase would be inhibited; and second, obviously, due to an increase of the substrate for cFru1,6bisPase, since cytosolic conversion of triose-P to Fru1,6P$_2$ is a near-equilibrium reaction *in vivo*. A rise in Suc synthesis would also liberate P$_i$, shifting the Fru2,6P$_2$ metabolism towards degradation. Conversely, accumulation of hexose phosphates in the cytosol would lower the rate of Fru1,6P$_2$ utilization, since Fru6P activates Fru6P 2-kinase [73], restricting the metabolic flow in the direction of Suc formation. This balancing mechanism is thought to be responsible for the shift observed in metabolism to net starch accumulation during the photoperiod [74]. Work on plant mutants with altered contents of phosphoglucoisomerase show that increased levels of Fru6P are in correspondence with a higher Fru2,6P2 content and a change in partitioning in favor of starch [75].

This correlative evidence of Fru2,6P$_2$ regulating carbon partitioning has been confirmed by experiments showing that direct manipulation of Fru2,6P$_2$ levels in tobacco [76] or *Arabidopsis* [77] leaves affects carbon partitioning. Transgenic tobacco plants regenerated from leaf disks that had been transformed with heterologous Fru6P 2-kinase/Fru2,6P$_2$ phosphatase constructs showing only one of the two activities showed that increased Fru2,6P$_2$ content shifts the balance of photosynthetically assimilated carbon allocation toward starch, with a concomitant decrease in Suc accumulation. The high Fru2,6P2 transgenic lines also show a gradual accumulation of starch during growth, which has been attributed to a decreased capacity for starch breakdown [76]. More recently, it has been shown that in plants with reduced content of Fru2,6P$_2$ the production of Suc, Glc, or Fru is clearly favored with respect to starch, albeit with a concomitant reduction in CO$_2$ assimilation [77,78]. Changes effected by Fru2,6P$_2$ levels are thought to be produced by activation of PFP [79].These results also show that Fru2,6P$_2$ definitely affects the coordination of cytosolic and plastid carbon metabolisms.

Evidence exists that cFru1,6bisPase from sugar beet leaves is subject to indirect light regulation. During a diurnal cycle, cFru1,6bisPase activity and transcripts were higher at the end of light and lower at the end of the dark period. No corresponding changes were observed in cFru1,6bisPase protein levels [65]. Etiolated leaves showed little or no cFru1,6bisPase protein and transcript levels, but both increased

within 24 h of exposure to light [65]. Current thinking on the regulation of the enzyme strongly acknowledges the possibility of posttranslational modification of cFru1,6bisPase as a regulatory mechanism [65,80].

Interspecies variations in the properties of the cFru1,6bisPase allow the regulation of the pathway for Suc synthesis to operate in systems in which concentration of metabolites assume extreme values. This is the case in maize leaves, for instance. In the mesophyll cells, where Suc synthesis takes place, DHAP levels are very high compared to C$_3$ plants [81] due to the existence of a DHAP/3PGA shuttle that reduces 3PGA in the mesophyll and sends the DHAP back to the photosystem II-deficient bundle sheath cells [1], where definitive carbon fixation by the Benson–Calvin cycle takes place. The shuttle is driven by metabolite concentration gradients [81,82], and hence the elevated levels of DHAP in the mesophyll. However, as noted above, a high DHAP level would trigger a response of low Fru2,6P$_2$/increased cFru1,6bisPase activity/high Suc formation rate. This situation could enable enough DHAP to be removed from the shuttle as to thwart the C$_4$ cycle operation. However, maize cFru1,6bisPase has a higher threshold for Fru1,6P$_2$ utilization compared to its C$_3$ counterparts (i.e., S$_{0.5}$ 7- to 15-fold higher), and therefore Suc synthesis proceeds only when DHAP levels are correspondingly in excess [82].

Suc-P synthesis in leaves is catalyzed by SPS, an allosteric, rate limiting enzyme that shares the control of Suc formation with cFru1,6bisPase. SPS is thought to be a determinant factor in controlling the partitioning of carbon in the leaf and the whole plant [83–85]. Work with tomato plants expressing the maize SPS show that high SPS activities are not necessarily accompanied by a parallel compensating restriction in cFru1,6bisPase activity, which would imply that regulation of SPS is more important than that of cFru1,6bisPase in determining the rate of Suc synthesis [85]. Recent research in transgenic tobacco overexpressing maize SPS demonstrated that a higher sucrose synthesis capacity accelerates plant development without altering total biomass production, accelerates photosynthesis, and increases the sucrose:starch ratio, suggesting a major shift in carbohydrate metabolism [83]. As mentioned before, SPS is subject to fine and coarse types of regulation, afforded respectively by allosteric control by Glc6P (activator) and P$_i$ (inhibitor) and by reversible protein phosphorylation (which deactivates SPS). Work by Neuhaus et al. [86] on the rate of carbon flux in spinach leaves has shed light on the interaction between the feedforward (i.e., control of photosynthetic rate through irradiance) and feedback (i.e., dependent on Suc content of the leaf) regulation of photosynthesis. The authors found that absolute rates of Suc formation were always lower in leaves with higher Suc content. Differences in rates were correlated with large changes in Fru2,6P$_2$ concentration and the activation state of SPS and smaller changes in metabolites concentrations. Forcing high photosynthetic rates provoked stimulated responses of Suc synthesis to changing metabolites concentrations (i.e. rising Glc6P and decreasing P$_i$) [86]. This effect could arise as a consequence of the dual effect of the SPS allosteric effectors, which also act in the regulation of the enzymes that regulate the SPS activation state, as described below.

These enzymes are a light-inducible, okadaic acid-sensitive type 2A protein phosphatase and SNF1-related protein kinases that copurifies with SPS and phosphorylates it in Ser residues [64,87]. Phosphorylation in two sites (Ser-424 and Ser-158 in spinach) is responsible for the light/dark modulation and osmotic stress activation [88,89]. A third site promotes inhibitory interaction with 14-3-3 proteins [90,91]. Changes in activity are evident at suboptimal substrate levels in the spinach SPS, and consist of an increase in the affinities for the substrates (about twofold) and effectors (fourfold) [92]. Maize SPS is regulated in a manner analogous to that of the spinach enzyme, but dephosphorylation also has an effect on V_{max} (two- to threefold increase) [93]. Phosphorylation of the enzyme occurs at multiple sites, but only some of these sites are involved in SPS regulation, while the others appear to be constitutively phosphorylated [87]. The soybean SPS, instead, does not appear to be regulated by phosphorylation and is only weakly affected by metabolites [87]. Glc6P inhibits the SPS kinase, while P$_i$ has an opposite effect by decreasing the rate of dephosphorylation [87], so that the allosteric effectors that affect directly SPS activity also act upon the processes regulating its activation state with the same overall effect.

In addition to its pivotal role in regulating carbon partitioning, SPS may be involved in the cross-talk between C and N metabolism, as suggested by the apparent coordinate operation of Suc biosynthesis, the reduction of nitrate and the rate of photosynthesis [94], processes that are driven or stimulated by light. The link between these related processes could be a common regulation of the reversible phosphorylation processes affecting each particular metabolic route (both proteins are probably dephosphorylated by the same enzyme [87]). Van Quy and Champigny [95] have suggested that NO$_3^-$, or a product of its metabolism, favors the deactivation of SPS by promoting the light activation of the SPS protein kinase without affecting the phosphatase activity.

Early reports on the presence of Suc in chloroplasts have been reinforced by introducing enzymes that metabolize the disaccharide in plastids [96]. The latter studies do not provide any clues as to the metabolic function of Suc within these organelles.

D. BALANCE OF CARBON DISTRIBUTION BETWEEN SUCROSE, STARCH, AND RESPIRATORY METABOLISM

So far, the central metabolic pathways in leaves, leading to carbohydrates formation and their biochemical regulation, have been examined. The quantitative allocation of assimilated carbon is an interactive process determined by the coordinated function of the above regulatory mechanisms. Factors influencing carbon partitioning in the leaf cell include:

1. Rates of transport to sink tissues
2. Variable energetic needs during the day cycle (respiratory metabolism)
3. Drainage of carbon for intermediate metabolisms and other biosynthetic pathways
4. Variations in photosynthetic rates
5. Availability of nutrients by the plant.

In general, regulation of carbon partitioning in source tissues must accommodate to sustain a permanent (although variable) translocation of assimilates [97,98]. Several examples illustrate the relevance of a sustained export rate. For instance, actively growing parts of the plant show different diurnal growth rates and thus impose fluctuating demands of carbon and energy [97]. The developmental state of a plant also affects the timing and supply of assimilates from leaves [70,98]. As reported by Huber et al. [97], the rate of assimilate export from leaves of vegetative soybean leaves tends to decrease during the day but, contrarily, it increases in leaves of reproductive plants.

Starch metabolism is governed primarily by the concentrations of chloroplastic metabolites (levels of substrates, 3PGA and P$_i$) and is then highly dependent on the photosynthetic rate. During the light period, starch is actively synthesized due to the high levels of assimilated carbon and energy as well as the high 3PGA:P$_i$ ratio within the chloroplast; while it is degraded in the dark. Conversely, Suc synthesis is controlled somewhat independently of photosynthetic rates. Different studies show that starch synthesis can be affected by alterations in Suc formation, whereas the reverse is not necessarily true [74].

Starch and Suc synthesis are also subject to endogenous rhythms as revealed by studies on plants maintained under a continuous light regime showing:

1. Starch accumulation slowed down at a time near the usual day's end and ceased within 4 h

2. A parallel increase in Suc formation and export rate;
3. A constant photosynthetic activity during the usual light period, which decreased gradually at the beginning of the extended illumination [98,99].

During the course of a normal day period, starch levels show sinusoidal variation, although the general trend is a net accumulation in the light and degradation in the night [98]. With regards to Suc, as previously mentioned, production levels are maintained fairly constant throughout the diurnal cycle. However, the carbon source for Suc synthesis switches from newly fixed carbon in the light to carbon derived from starch during the night.

Part of the photoassimilated carbon remains in the photosynthetic tissue and is utilized to satisfy the many biosynthetic reactions occurring within the leaf cell. In particular, much of the carbon flows into the respiratory metabolism [100]. Respiration plays an important function in green cells, even under light conditions. For instance, routing carbon to the tricarboxylic acid cycle provides carbon skeletons for amino acid biosynthesis. The relevance of this metabolism is such that the rate of Suc synthesis may be sharply affected by nitrate availability [94]. The diversion of photosynthetic carbon from carbohydrate toward amino acid synthesis results after N-signaled phosphorylation of SPS and PEP carboxylase [95]. As a consequence, SPS activity is impaired and PEP carboxylase is activated. In addition to its anapleurotic role, respiration can also furnish ATP for biosynthetic reactions taking place outside the chloroplast through the mitochondrial electron transport [101]. Actually, mitochondrial activity may be necessary to sustain the photosynthetic rate [101].

In synthesis, carbon assimilation and partitioning is regulated by multiple factors arising from the interaction of all metabolic pathways occurring inside the leaf cell.

III. TRANSPORT OF CARBON FROM SOURCES TO SINKS

The yield of crop plants has a direct correspondence with the ability of the plant to allocate carbon from source leaves to the economically important sink tissue. In addition, growth and energy needs from other sink tissues must also be conveniently satisfied by the plant. The most widely accepted form of movement of photoassimilated carbon (of which Suc is frequently the major component) from sources to sinks is the pressure flow mechanism through the phloem [102].

According to this mechanism, active loading of Suc into the phloem at the source site increases the hydrostatic pressure and drives the subsequent flow [102]. A turgor-dependent Suc movement has been shown in developing bean seeds [103].

While this model is widely recognized, the pathway through which Suc is finally loaded in the phloem has been subject to some controversy. Suc is thought to move symplastically from one mesophyll cell to the next until it reaches the sieve tubes. Two modes of loading of the phloem are possible at this point. The entirely symplastic model calls for symplastic (through plasmodesmata) connections between the mesophyll cells and the phloem cells. These plasmodesmatal connections have been observed in adult leaves [104], but there is not certainty as to whether they are actually functional or not [104,105]. The alternative model of phloem loading proposes an apoplastic transport step prior to entry of Suc into the minor veins [104], and requires a carrier-mediated, active transport of this metabolite [104]. Both mechanisms have their advocates, and yet other researchers think of the problem as being species-dependent and proposing that both ways could be operative [106,107].

Those who argue in favor of an apoplastic step being involved in phloem loading cite a low symplastic connectivity and the inherent difficulty of moving Suc from the relatively dilute mesophyll cytosol (1 to 20 mM) to the highly concentrated phloem sap (500 to 1000 mM) as arguments against a totally symplastic phloem-loading model [104,108].

In recent years, work with transgenic plants has yielded compelling evidence favoring the apoplastic loading hypothesis. For instance, expression of yeast invertase in tobacco [109,110], tomato [105], and potato [111] results in an increased carbohydrate contents in leaves and remarkable variations in the plant's phenotype. Expression of apoplastic invertase in tomato plants reduces growth in a fashion which is correlated with the level of invertase activity [105]. Tobacco plants also show reduced growth and bleached leaf areas in which a high accumulation of Glc, Fru, Suc, and starch can be detected [109]. Green areas do not degrade starch at night. Accumulation of proline suggest that transformed plants are subject to water stress, as expected from the higher concentration of soluble carbohydrates [111]. From the above described results it follows that Suc transport probably needs from an apoplastic step [105,109] in which a transporter, specific for this metabolite, plays a pivotal role [112]. This translocator shows a high specificity for Suc and has been described as a Suc-H$^+$ cotransporter with a 1:1 stoichiometry [113]. The H$^+$ gradient is linked to a plasmalemma-bound H$^+$-ATPase extruding protons into the apoplast [104].

Suc transporter-like proteins are under hormonal control and Suc itself can act as a hormone-like signal [114]. Transporter levels are enhanced by auxin-type hormones and light [114]. Roblin et al. [115] have shown that phosphorylation inhibits sucrose transport, adding a further level to the complex Suc movement regulation.

Suc unloading in sink cells is also apoplastic in some tissues like the developing embryo or the endosperm, which lack of symplastic continuity with the parent plant [107]. However, in roots and expanding leaves this process may be completely symplastic [107]. The unloading of Suc at the sink and its subsequent transformation or storage in the vacuole provides the means of maintaining the steep concentration gradient needed to support the carbon flow. Vacuolar Suc transport is a well studied, pH-dependent process, which, unlike the plasma membrane transport, responds to a proton antiport mechanism [114]. More recent results indicate a facilitated diffusion mechanism in some species [116].

Clearly the transport processes occurring at the whole-plant level need further work before they are thoroughly understood. These unsolved aspects of plant physiology could hold important clues for the enhancement of plant breeding.

IV. CARBON PARTITIONING IN SINK TISSUES

A. Carbon Metabolism in the Cytosol

As discussed in the preceding section, photoassimilated carbon, mainly in the form of Suc, is distributed through the veinal network to sink tissues, i.e., developing leaves, the apex, shoots, fruits, etc. Upon uptake of the mobilized Suc, it undergoes different changes: conversion to storage forms, like starch, fructans or stored as such in the vacuole, or it may be degraded to provide energy through respiration or to supply carbon skeletons for different biosynthetic reactions. The degree of participation of each route of Suc utilization depends on the tissue involved; in some tissues most of the carbohydrate will be dedicated to support growth or elongation, while in others, such as in tubers, both growth and starch synthesis will occur simultaneously [117]. In any case, sink tissues are heterotrophic, and cytosolic carbohydrate metabolism proceeds mainly in the direction of Suc breakdown, or storage, as opposed to Suc formation.

In this section, the two main metabolic routes of Suc utilization in the cytosol of sink cells will be considered, namely its conversion to starch precursors and its use as a glycolytic start-point. It should be borne in mind that these two processes may take place simultaneously and that the separation is solely for the purpose of a better understanding of the issue, and does not necessarily reflect a real separation in a sink cell. The glycolytic route has been previously reviewed [118] and will not be considered in detail here, except for those features relevant to the control of carbon partitioning.

The study of animal glycolysis has always considered Glc as a starting point of the metabolic pathway ending in pyruvate. The carbon currency in plants, instead, is not Glc but Suc or one of its derivatives [70], and then plant glycolysis should therefore begin with this metabolite. The ending point of glycolysis may also differ in plant when compared to animals, since, in addition to pyruvate kinase, another cytosolic enzyme, namely PEP carboxylase, may use PEP and transfer the product of its transformation (oxaloacetate in this case) to the Krebs cycle, behaving as an anapleurotic enzyme.

The first step in Suc utilization in the cell is its cleavage in its monosaccharide components (or its derivatives). This is accomplished by two different enzymes activities, invertases (β-D-fructofuranosidases, EC 3.2.1.26 [Equation (28.15)] or Suc synthase [SuSy, EC 2.4.1.13] [Equation (28.16)]:

$$Suc + H_2O \Rightarrow Glc + Fru \qquad (28.15)$$

$$Suc + UDP \Rightarrow UDP\text{-}Glc + Fru \qquad (28.16)$$

Invertases cleave Suc into Glc and Fru and the energy of the glycosydic bond is lost. Invertases can be classified into several groups [119]. First, a distinction can be made according to the pH optimum, and so acid (pH 4.5 to 5.0) and alkaline or neutral (pH 7.0 to 7.8) forms are found. Second, these enzymes may be found in the intra- or extracellular milieu. Finally, invertases may be soluble or be bound to the cell wall. Neutral invertases are mainly cytosolic. Their acid counterparts share similar kinetic properties (low K_m values, product inhibition) and may be found in the vacuole or in the extracellular space [119]. The precise role of invertases in Suc breakdown is still a matter of debate [106,119,120], partly due to insufficient characterization of the properties of these enzymes and the metabolic implications of the widely varying proportions of acid or neutral invertases in different tissues [119,121–123]. There are a few studies that indicate that neutral invertases are associated with mature tissues in which Suc is being actively stored [122,124], while others have inversely correlated acid invertase activity with Suc contents in fruits [123,125,126]. Reduction of invertase activity by genetic manipulation results in the conversion of the fruit from sugar-storing to Suc-storing [119].

The other way of Suc degradation is through SuSy. This enzyme catalyzes a reversible reaction, but is generally thought to be involved in Suc breakdown [127,128], in part because it is predominantly present in tissues in which Suc breakdown occurs and that its activity is higher than that of the invertases in certain tissues also concerned with Suc utilization [119,120]. While the products of the invertases may be converted to Fru1,6P$_2$ by the classical glycolytic pathway, that is, through hexokinase or fructokinase, sucrolysis by SuSy, instead, follows a pathway based on uridine-diphosphate and inorganic PP$_i$ [128]. The pathway's main characteristics are the preferential use of uridylates over adenylates, and the cycling of Fru6P and Fru1,6P$_2$ by the ATP-dependent PFK and PFP to provide PP$_i$. As sketched in Figure 28.2, the pathway calls for a leading role of UDPGlc pyrophosphorylase, which utilizes PP$_i$ to generate UTP and Glc1P; and PFP, which regenerates the PP$_i$ utilized in the former reaction. Alternatively, uridylates may cycle via fructokinase using UTP (Figure 28.2). Recent reports indicate that indeed uridine nucleotides play an important role in controlling sucrose degradation and starch synthesis in sliced potato tubers [129]. Uridilates may be a limiting factor of this process since the K_m for UDP of SuSy (100 to 700 μM) and reported levels for this metabolite do not exceed 60 μM. Adenilate levels are more important in the control of respiration and ADPGlc PPase activity [129,130].

The hypothesis of a SuSy-linked degradation of Suc is backed by the presence of appreciable amounts of PP$_i$ in the cytosol [68,104,105], by the fact that PFK may use UTP instead of ATP and finally because transgenic potato lines with low SuSy levels are characterized by decreased starch and tuber yield [68]. Since PFP seems to play a central role in this route, the system would be expected to be subject to control by Fru2,6P$_2$. Recent work has provided some evidence on this. First, transgenic potato tubers containing decreased PFP activity, obtained by antisense inhibition of the enzyme, show an increased level of hexose-P and decreased level of triose-P but no modification of the flux of glycolysis [131]. Less starch was synthesized only when PFP was extremely reduced, probably as a result of a decrease of 3-PGA, which acts as an activator of ADPGlc PPase [68]. Notably, these plants also show an increased content of Fru2,6P$_2$, suggesting that the plant's response is to activate the remaining PFP. This constitutes a further proof of the flexibility of plant metabolism, in the sense that plants possess many alternatives to circumvent steps otherwise considered as essential.

Concurrently, another study using transgenic potato plants, in which high levels of Fru2,6P$_2$ were induced in the cytosol, demonstrate the opposite effect, that is decreased hexose-P and increased triose-P pools [76], suggesting an increment in glycolytic flux brought about by Fru2,6P$_2$.

Depending on the tissue, carbon from degraded Suc may be used for starch synthesis or continue the glycolytic pathway. Plastid starch synthesis is

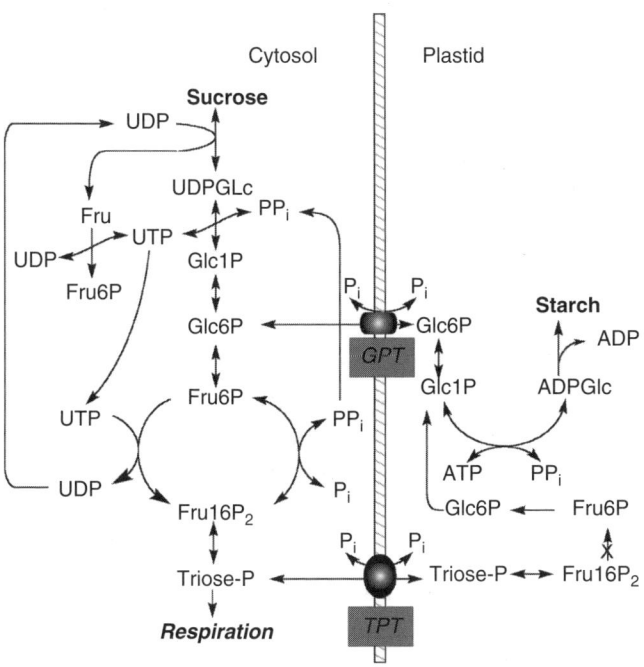

FIGURE 28.2 Simplified diagram of carbon metabolism in sink tissues. Question marks indicate transport of triose-P by the P$_i$-translocator (TPT) and the reaction of the plastidic Fru1,6bisPase; both processes are not fully established to take place in plastids.

supported by import of hexose-P, as will be dealt with in the next section. The same hexose-P pool constitutes the major source of cytosolic glycolysis in plants.

The first regulatory step in plant glycolysis is, as in animals, the conversion of Fru6P to Fru1,6P$_2$. However, it cannot be said to be the committed step, since this reaction can be catalyzed by either PFK (EC 2.7.1.11) (Equation (28.17)), which makes it irreversible, or by PFP, which catalyzes a reaction that is close to equilibrium *in vivo* (Equation (28.11)) [118].

$$Fru6P + ATP \Rightarrow Fru1,6P_2 + ADP \qquad (28.17)$$

The fact is that in many tissues PFK and PFP coexist at comparable activity levels. As discussed by Kruger [132] , based on the known properties of PFK and the correlation between glycolytic flux and PFK activity in several plants, there is little doubt that PFK contributes to the entry of hexose-P to glycolysis. Although there are many reasons to assume that PFP plays an important role in plant metabolism, such as a the strong activation by nanomolar amounts of Fru2,6P$_2$ [70,133], an activity often equal or exceeding that of PFK [134,135] and levels that depend on the developmental stage or nutrition status [134,136], the role of this enzyme has been debated for some time. The main reason is the fact that physiological concentrations of Fru2,6P$_2$ are on average 2000-fold greater than its K_a to activate PFP. Therefore, PFP would seem as a permanently activated (and thus nonregulated) enzyme. However, new evidence suggests that *in vivo* concentration of substrates, especially Pi, and phosphorylated intermediates raise the K_a to values that allow variations in Fru2,6P$_2$ concentration to effectively regulate PFP [137]. A good example of a definite function for this enzyme in plant metabolism is that of pineapple leaf PFP. Pineapple, a crassulacean acid metabolism (CAM) plant, degrades mainly soluble sugars at night to provide the carbon skeletons necessary for CO$_2$ fixation by PEP carboxylase [138]. PFP activity in this tissue is about 50- to 70-fold higher than PFK [135] and, according to the known properties of PFP and PFK and the environmental situation favoring CAM in pineapple, only PFP is expected to contribute to the glycolytic flux [135]. However, as noted by Hajirezai et al. [131], these "correlative" approaches to the study of plant metabolism may often lead to unclear and contradictory results. Therefore, clarification of these unsolved aspects must wait for studies combining molecular genetics and biochemistry.

With respect to the factors governing the amount of carbon dedicated to respiration or starch synthesis, a major factor to be taken into account is the availability of N. Increased N supply has been reported to accelerate the respiratory rate in a number of plants, cell suspension cultures, and green algae [131,139,140]. This response has been linked to an increased demand for carbon skeletons for amino acid synthesis and the associated higher ATP demand [134,139]. Both terminal enzymes of glycolysis in plants, namely PEP carboxylase (EC 4.1.1.31) (Equation (28.18)) and pyruvate kinase (EC 2.7.1.40) (Equation (28.19)) show regulatory properties, which makes them well suited to their apparent major role in controlling glycolytic flux in response to varying N assimilation rates [134].

$$PEP + HCO_3^- \Rightarrow oxaloacetate + P_i \qquad (28.18)$$

$$PEP + ADP \Rightarrow pyruvate + ATP \qquad (28.19)$$

A coordinate control of both enzymes has been proposed based on detailed kinetic studies of the enzymes extracted from the same tissues. In general, PEP carboxylase is sensitive to feedback inhibition by the amino acids glutamate and aspartate and malate, and feedforward activated by Glc6P [118,141,142]. Pyruvate kinase, on the other hand, is inhibited by glutamate but activated by aspartate [141]. This mode of regulation is relevant not only for the control of these enzymes (and the glycolytic flux) but also provides a link between the use of Suc and N assimilation, which is dependent on an adequate provision of carbon skeletons.

B.   CARBON METABOLISM IN AMYLOPLASTS

Plastids are double-membrane-limited organelles characteristically occurring in plant cells. Except for autotrophic chloroplasts, all the other (nongreen) plastids depend on the cytoplasm for providing carbon and energy [143]. The biochemical machinery, distinctive of each type of plastid, determines the specialization of the organelle to synthesize and accumulate specific biomolecules. Next, we will discuss carbon metabolism in amyloplasts, which are plastids largely present in sink organs and specialized in starch accumulation [8,21,143,144].

In reserve tissues, starch accumulates in amyloplasts as a long-term form of carbon storage. In certain organs (i.e. potato tubers, cereal endosperm), the polysaccharide piles up to constitute between 50% and 80% of the dry weight [144]. Amyloplastic starch granules vary in size, shape, composition (relative amounts of amylose and amylopectin), and properties, although they are generally higher than chloroplastic granules. Moreover, shape, size, and other fine features of amyloplastic starch granules are specific of the plant; and thus, from their microscopic examination, it is possible to identify the botanical source of the polysaccharide [8,22,144].

In amyloplasts, the metabolic route leading to starch synthesis, as well as its regulation, has been less characterized than in chloroplasts. As a result, the flow of carbon in the nongreen plastid has been a matter of some controversy [145]. Although it is clear that ADPGlc serves as the glucosyl donor for starch synthesis in amyloplasts, two pathways have been proposed for the biosynthetic process. The route characterized earlier is supported by major experiments (consensus route, see Ref. [145]) and is similar to that described in chloroplasts.

According to the consensus route, amyloplastic starch is synthesized from Glc1P by the series of reactions, shown in Equations (28.3)–(28.5) (see Section II.A), all occurring within the plastid and catalyzed, respectively, by ADPGlc PPase, starch synthase, and branching enzyme [3–5,8,13,22,143, 145]. On the other hand, the alternative pathway assumes that ADPGlc is synthesized in the cytoplasm, either by sucrose synthase [146] or by a cytosolic isoform of ADPGlc PPase [147]. Then, the sugar-nucleotide is imported into the plastid, via an adenylate translocator, where it serves as a substrate for starch synthase [143,145,146]. It has been proposed that the alternative route for starch synthesis is operative in cereal endosperm [148], where it could function in addition to the consensus pathway and its occurrence may be dependent on the development stage of the storage tissue [143].

Figure 28.2 shows the flow of carbon in cytoplasm and amyloplast in cells of plant reserve tissues, as supported by the bulk of experimental data and conducting to starch synthesis through the consensus route. Different studies have shown that transport of triose-P through amyloplast envelope is not a significant source for starch synthesis inside the plastid. Experiments using Glc or Fru labeled with ^{13}C in carbons 1 or 6, showed that the radioactive label found in amyloplastic starch redistributed only partially (12% to 20%) [149]. Based on this, it was suggested that hexose-P, rather than triose-P, is the metabolite mainly transported across the amyloplast envelope for starch synthesis. In agreement with the above, it was observed that only Glc1P is actively used to synthesize starch by amyloplasts from wheat endosperm, and nuclear magnetic resonance studies demonstrated that hexose-P is effectively transported into nongreen plastids [143,150]. In addition, amyloplasts lack Fru1,6P$_2$ phosphatase, an enzyme necessary to convert triose-P into hexose-P [151].

The above picture is strongly supported by the molecular characterization of the GPT present in the envelope of nongreen plastids [152]. This transport system imports hexose-P (preferentially Glc6P) into the amyloplast in exchange with Pi or triose-P [50,54,152]. Transport of Glc6P was also found in chloroplasts of guard-cells (which also lack Fru1,6P$_2$ phosphatase as well as chloroplasts of detached leaves supplied with Glc [50]. Thus, the entrance of hexose-P (mainly Glc6P) is the main pathway conducting to starch synthesis in amyloplasts, with triose-P transport (if it occurs) being quantitatively less important for the process (Figure 28.2).

Regulatory and structural properties of amyloplastic ADPGlc PPase can be assumed to be similar to those previously described for the enzyme from different sources (see Section II.A). A well-characterized amyloplastic ADPGlc PPase is the potato tuber enzyme, which is a heterotetramer of molecular mass 210 kDa, being allosterically regulated by 3PGA (activator) and P$_i$ (inhibitor) [153–155].

It has been reported that the potato tuber ADPGlc PPase is also subject to regulation by chemical modification [37,38]. Cysteine-12 forms an intermolecular disulfide bridge that binds two small subunits in the heterotetrameric enzyme. Reduction of this disulfide by dithiothreitol [37], or more efficiently by thioredoxins f and m [38], activates the enzyme at low concentrations of 3PGA. Oxidized thioredoxin reverses activation of the enzyme. This suggests that potato tuber ADPGlc PPase could be regulated by a system sensing the reductive status in the amyloplast [38]. Concurrently, studies performed on potato tubers detached from the growing plant have shown inhibition of starch synthesis caused by changes in the structure of ADPGlc PPase [156]. In the detached tubers the small subunit was found in a dimeric state and the enzyme exhibited decreased activity and affinity for substrates and activator. Interestingly, incubation of detached tubers with either dithiothreitol or sucrose reduced dimerization of the enzyme and increased the synthesis of starch [156]. These *in vivo* studies strongly support the assumption that of starch synthesis in potato tuber amyloplasts via the reductive activation of ADPGlc PPase is of physiological significance.

Reports on the regulatory properties of ADPGlc PPase from some reserve tissues have been subject of some discrepancies. Many studies showing a relative insensitivity to 3PGA and P$_i$ of the enzyme from endosperm of different plant species have been reported [157–161]. These results can, in part, be attributed to proteolytic degradation undergoes by the enzyme during purification [158,162]. Characterization of ADPGlc PPase purified from wheat endosperm has shown that the enzyme exhibits distinctive regulatory properties, being coordinated regulation exerted by a series of metabolites [161]. Thus, the wheat endosperm enzyme is sensitive to allosteric inhibition by Pi, ADP, and $Fru1,6P_2$, with 3PGA and Fru6P being effective toward reverse inhibition. Although the enzyme is not affected by 3PGA alone, it is still sensitive to the 3PGA:Pi ratio within the endosperm amyloplast. In fact, this behavior constitutes one of the variations in the interaction between 3PGA and Pi found to regulate ADPGlc PPases from plants (reviewed in Ref. [14]).

The presence of multiple isoforms of starch synthase and branching enzyme has been established in different reserve tissues. The properties of the different isoenzymes and their involvement in the formation of amylose and amylopectin, with the consequent determination of the specific structure of starch granules, have been comprehensively reviewed [5,21,22].

V. CONCLUDING REMARKS AND PERSPECTIVES

As we have seen in this chapter, the success of a crop plant is dictated by its aptitude to efficiently assimilate carbon followed by its mobilization and storage in the agronomically important sink organ. Biochemical and genetic work is gradually leading to a better understanding of the events participating in plant metabolism. This is potentially relevant for the targeted manipulation of selective parts of the biosynthetic machinery determining plant efficiency and productivity. Although plant science is still midway in elucidating many of these aspects, some remarkable results have already been achieved in plant transformation. It is predictable that in the not so distant future many of the most important crops will be genetically engineered so as to introduce beneficial characteristics like improved productivity and quality of natural products.

ACKNOWLEDGMENTS

The authors are research members from CONICET (Argentina). Research carried out by the authors has been supported by grants from Fundación Antorchas and from ANPCyT, PICT'99 1–6074 and 1–03397.

REFERENCES

1. Edwards GE, Walker DA. *C_3, C_4: Mechanisms of Cellular and Environmental Regulation of Photosynthesis.* Berkeley, CA: University of California Press, 1983.
2. Preiss J. Regulation of the biosynthesis and degradation of starch. *Annu. Rev. Plant Physiol.* 1982; 54:431–454.
3. Preiss J. Biosynthesis of starch and its regulation. In: Preiss J, ed. *The Biochemistry of Plants: Carbohydrates, Structure and Function.* New York: Academic Press, 1988:184–254.
4. Preiss J, Sivak MN. Starch and glycogen biosynthesis. In: Pinto BM, ed. *Comprehensive Natural Products Chemistry.* Oxford: Pergamon Press, 1998:441–495.
5. Ball S. Regulation of starch biosynthesis. In: Rochaix JD, Goldschmidt-Clermont M, Merchant S, eds. *The Molecular Biology of Chloroplasts and Mitochondria in Chlamydomonas.* Dordrecht: Kluwer Academic, 1998:549–567.
6. Fernie AR, Willmitzer L, Trethewey N. Sucrose to starch: a transition in molecular plant physiology. *Trends Plant Sci.* 2002; 7:35–41.
7. Beck E, Ziegler P. Biosynthesis and degradation of starch in higher plants. *Annu. Rev. Plant Physiol. Plant Mol. Biol.* 1989; 40:95–117.
8. Sivak MN, Preiss J. Starch: basic science to biotechnology. In: Taylor SL, ed. *Advances In Food and Nutrition Research.* San Diego, CA: Academic Press, 1998:1–199.
9. Cortassa S, Aon MA, Iglesias AA, Lloyd D. *An Introduction to Metabolic and Cellular Engineering.* Singapore: World Scientific, 2002:248.
10. Morrison WR, Karkalas J. Starch. In: Dey PM, ed. *Methods in Plant Biochemistry.* London: Academic Press, 1990:323–352.
11. Stark DM, Timmerman KP, Barry GF, Preiss J, Kishore GM. Role of ADPglucose pyrophosphorylase in regulating starch levels in plant tissues. *Science* 1992; 258:287–292.
12. Preiss J. ADPglucose pyrophosphorylase: basic science and applications in biotechnology. In: El-Gewely, ed. *Biotechnology Annual Review.* Amsterdam: Elsevier Science, 1996:259–279.
13. Kossmann J, Lloyd J. Understanding and influencing starch biochemistry. *Crit. Rev. Plant Sci.* 2000; 19:171–226.
14. Ballicora MA, Iglesias AA, Preiss J. ADPglucose pyrophosphorylase, a regulatory enzyme for plant starch synthesis. *Photosynth. Res.* 2004; 79:1–24.
15. Iglesias AA, Preiss J. Bacterial glycogen and plant starch biosynthesis. *Biochem. Educ.* 1992; 20:196–203.
16. Espada J. Enzymic synthesis of adenosine diphosphate glucose from glucose-1-phosphate and adenosine triphosphate. *J. Biol. Chem.* 1962; 237:3577–3581.
17. Heldt HW, Chon CJ, Maronde D, Herold A, Stankovic ZS, Walker DA, Kraminer A, Kirk MR, Heber U. Role of orthophosphate and other factors in the regulation of starch formation in leaves and isolated chloroplasts. *Plant Physiol.* 1977; 59:1146–1155.
18. Kossmann J, Sonnewald U, Willmitzer L. Reduction of chloroplastic fructose-1,6-bisphosphatase in transgenic potato plants impairs photosynthesis and plant growth. *Plant J.* 1994; 6:637–650.
19. Kruckeberg AL, Neuhaus HE, Feil R, Gottlieb LD, Stitt M. Decreased activity mutants of phosphoglucose isomerase in the cytosol and chloroplast of *Clarkia xantiana.* Impact on mass-action ratios and fluxes

to sucrose and starch, and estimation of flux control coefficients and elasticity coefficients. *Biochem. J.* 1989; 261:457–467.

20. Neuhaus HE, Stitt M. Control analysis of photosynthate partitioning: Impact of reduced activity of ADPglucose pyrophosphorylase or plastid phosphoglucomutase on the fluxes to starch and sucrose in *Arabidopsis*. *Planta* 1990; 182:445–454.

21. Smith AM, Denyer K, Martin C. The synthesis of the starch granule. *Annu. Rev. Plant Physiol. Plant Mol. Biol.* 1997; 48:67–87.

22. Ball SG, Morell MK. From bacterial glycogen to starch: understanding the biogenesis of the plant starch granule. *Annu. Rev. Plant Biol.* 2003; 54:207–233.

23. Ballicora MA, Iglesias AA, Preiss J. ADPglucose pyrophosphorylase; a regulatory enzyme for bacterial glycogen synthesis. *Microbiol. Mol. Biol. Rev.* 2003; 67:213–225.

24. Koch KE. Carbohydrate-modulated gene expression in plants. *Annu. Rev. Plant Physiol. Plant Mol. Biol.* 1996; 47:509–540.

25. Smith-White BJ, Preiss J. Comparison of proteins of ADP-glucose pyrophosphorylase from diverse sources. *J. Mol. Evol.* 1992; 34:449–464.

26. Iglesias AA, Charng YY, Ball S, Preiss J. Characterization of the kinetic, regulatory, and structural properties of ADP-glucose pyrophosphorylase from *Chlamydomonas reinhardtii*. *Plant Physiol.* 1994; 104:1287–1294.

27. Iglesias AA, Kakefuda G, Preiss J. Regulatory and structural properties of the cyanobacterial ADPglucose pyrophosphorylases. *Plant Physiol.* 1991; 97:1187–1195.

28. Aon MA, Gómez Casati DF, Iglesias AA, Cortassa S. Ultrasensitivity in (supra)molecularly organized and crowded environments. *Cell Biol. Int.* 2001; 25:1091–1099.

29. Gomez-Casati DF, Aon MA, Iglesias AA. Ultrasensitive glycogen synthesis in Cyanobacteria. *FEBS Lett.* 1999; 446:117–121.

30. Gomez-Casati DF, Aon MA, Iglesias AA. Kinetic and structural analysis of the ultrasensitive behavior of cyanobacterial ADP-glucose pyrophosphorylase. *Biochem. J.* 2000; 350:139–147.

31. Gomez-Casati DF, Cortassa S, Aon MA, Iglesias AA. Ultrasensitive behavior in the synthesis of storage polysaccharides in cyanobacteria. *Planta* 2003; 261:969–975.

32. Goldbeter A, Koshland DE, Jr. Sensitivity amplification in biochemical systems. *Q. Rev. Biophys.* 1982; 15:555–591.

33. Koshland DE. Switches, thresholds and ultrasensitivity. *Trends Biochem. Sci.* 1987; 12:225–229.

34. Ball S, Marianne T, Dirick L, Fresnoy M, Delrue B, Decq A. A *Chlamydomonas reinhardtii* low-starch mutant is defective for 3-phosphoglycerate activation and orthophosphate inhibition of ADPglucose pyrophosphorylase. *Planta* 1991; 185:17–26.

35. Stitt M. The first will be last and the last will be first: non-regulated enzymes call the tune. In: Bryant JA,

Burrell MM, Kruger NJ, eds. *Plant Carbohydrate Biochemistry*. Oxford: BIOS Scientific Publishers Ltd, 1999:1–16.

36. Van den Koornhuyse N, Libessart N, Delrue B, Zabawinski C, Decq A, Iglesias AA, Carton J, Preiss J, Ball S. Control of starch composition and structure through substrate supply in the monocellular alga *Chlamydomonas reinhardtii*. *J. Biol. Chem.* 1996; 271:16281–16287.

37. Fu Y, Ballicora MA, Leykam JF, Preiss J. Mechanism of reductive activation of potato tuber ADP-glucose pyrophosphorylase. *J. Biol. Chem.* 1998; 273:25045–25052.

38. Ballicora MA, Frueauf JB, Fu Y, Schurmann P, Preiss J. Activation of the potato tuber ADP-glucose pyrophosphorylase by thioredoxin. *J. Biol. Chem.* 2000; 275:1315–1320.

39. Blennow A, Engelsen SB, Nielsen TH, Baunsgaard L, Mikkelsen R. Starch phosphorylation: a new front line in starch research. *Trends Plant Sci.* 2002; 7:445–450.

40. Flügge UI. Transport in and out of plastids: does the outer envelope membrane control the flow? *Trends Plant Sci.* 2000; 5:135–137.

41. Steup M. Starch degradation. In: Preiss J, ed. *The Biochemistry of Plants*. San Diego, CA: Academic Press, 1988:255–296.

42. Yu Y, Mu HH, Wasserman BP, Carman GM. Identification of the maize amyloplast stromal 112-kDa protein as a plastidic starch phosphorylase. *Plant Physiol.* 2001; 125:351–359.

43. Mu HH, Yu Y, Wasserman BP, Carman GM. Purification and characterization of the maize amyloplast stromal 112-kDa starch phosphorylase. *Arch. Biochem. Biophys.* 2001; 388:155–164.

44. van Berkel J, Conrads-Strauch J, Steup M. Glucanphosphorylase forms in cotyledons of *Pisum sativum* L.: localization, developmental change, *in vitro* translation, and processing. *Planta* 1991; 185:432–439.

45. Duwenig E, Steup M, Kossmann J. Induction of genes encoding plastidic phosphorylase from spinach (*Spinacia oleracea* L.) and potato (*Solanum tuberosum* L.) by exogenously supplied carbohydrates in excised leaf discs. *Planta* 1997; 203:111–120.

46. Frandsen TP, Svensson B. Plant alpha-glucosidases of the glucoside hydrolase family 31: molecular properties, substrate specificity, reaction mechanism, and comparison with family members of different origin. *Plant Mol. Biol.* 1998; 37:1–13.

47. Yu S, Kenne L, Pedersen M. Alpha-1,4-glucan lyase, a new class of starch/glycogen degrading enzyme. I. Efficient purification and characterization from red seaweeds. *Biochim. Biophys. Acta* 1993; 1156:313–320.

48. Sehnke PC, Chung H-J, Wu K, Ferl RJ. Regulation of starch accumulation by granule-associated plant 14-3-3 proteins. *Proc. Natl. Acad. Sci. USA* 2001; 98:765–770.

49. Flugge UI. Phosphate translocation in the regulation of photosynthesis. *J. Exp. Bot.* 1995; 46:1317–1323.

50. Flugge UI. Phosphate translocators in plastids. *Annu. Rev. Plant Physiol. Plant Mol. Biol.* 1999; 50:27–45.

51. Neuhaus HE, Wagner R. Solute, pores, ion channels, and metabolite transporters in the outer and inner envelope membranes of higher plant plastids. *Biochim. Biophys. Acta* 2000; 1465:307–323.

52. Flugge UI, Heldt HW. Metabolite translocators of the chloroplast envelope. *Annu. Rev. Plant Physiol. Plant Mol. Biol.* 1991; 42:129–144.

53. Heber U, Heldt HW. The chloroplast envelope: structure, function, and role in leaf metabolism. *Annu. Rev. Plant Physiol.* 1981; 32:139–168.

54. Soll J, Bölter B, Wagner R, Hinnah SC. The chloroplast outer envelope: a molecular sieve? *Trends Plant Sci.* 2000; 5:137–138.

55. Iglesias AA, Plaxton WC, Podestá FE. The role of inorganic phosphate in the regulation of C_4 photosynthesis. *Photosynth. Res.* 1993; 35:205–211.

56. Barnes SA, Knight JS, Gray JC. Alteration of the amount of the chloroplast phosphate translocator in transgenic tobacco affects the distribution of assimilate between starch and sugar. *Plant Physiol.* 1994; 106:1123–1129.

57. Fischer K, Kammerer B, Gutensohn M, Arbinger B, Weber A, Hausler RE, Flugge UI. A new class of plastidic phosphate translocators: a putative link between primary and secondary metabolism by the phosphoenolpyruvate/phosphate antiporter. *Plant Cell* 1997; 9:453–462.

58. Willey DL, Fischer K, Wahcter E, Link TA, Flugge UI. Molecular cloning and structural analysis of the phosphate translocator from pea chloroplasts and its comparison to the spinach phosphate translocator. *Planta* 1991; 183:451–461.

59. Wagner R, Apley EC, Gross A, Flugge UI. The rotational diffusion of chloroplast phosphate translocator and of lipid molecules in bilayer membranes. *Eur. J. Biochem.* 1989; 182:165–173.

60. Schleucher J, Vanderveer PJ, Sharkey TD. Export of carbon from chloroplasts at night. *Plant Physiol.* 1998; 118:1439–1445.

61. Stitt M, Heldt HW. Control of photosynthetic sucrose synthesis by fructose-2,6-bisphosphate: intercellular metabolite distribution and properties of the cytosolic fructosebisphosphatase in leaves of *Zea mays* L. *Planta* 1985; 164:179–188.

62. Strand A, Zrenner R, Trevanion S, Stitt M, Gustafsson P, Gardestrom P. Decreased expression of two key enzymes in the sucrose biosynthesis pathway, cytosolic fructose-1,6-bisphosphatase and sucrose phosphate synthase, has remarkably different consequences for photosynthetic carbon metabolism in transgenic *Arabidopsis thaliana*. *Plant J.* 2000; 23:759–770.

63. Trevanion SJ. Regulation of sucrose and starch ynthesis in wheat (*Triticum aestivum* L.) leaves: role of fructose 2,6-bisphosphate. *Planta* 2002; 215: 653–665.

64. Huber JL, Huber SC. Site-specific serine phosphorylation of spinach leaf sucrose-phosphate synthase. *Biochem. J.* 1992; 283 (Pt 3):877–882.

65. Khayat E, Harn C, Daie J. Purification and light-dependent molecular modulation of the cytosolic fruc-

tose-1,6-bisphosphatase in sugarbeet leaves. *Plant Physiol.* 1993; 101:57–64.

66. Walker JL, Huber SC. Regulation of sucrose-phosphate synthase activity in spinach leaves by protein level and covalent modificaton. *Planta* 1989; 177:116–120.

67. Campbell WH. Nitrate reductase structure, function and regulation: bridging the gap between biochemistry and physiology. *Annu. Rev. Plant Physiol. Plant Mol. Biol.* 1999; 50:277–303.

68. Stitt M. Pyrophosphate as an alternative energy donor in the cytosol of plant cells: an enigmatic alternative to ATP. *Bot. Acta* 1998; 111:167–175.

69. Paul M, Sonnewald U, Hajirezaei M, Dennis DT, Stitt M. Transgenic tobacco plants with strongly decreased expression of pyrophosphate: fructose-6-phosphate 1 phosphotransferase do not differ significantly from the wild type in photosynthate partitioning, plant growth or ability to cope with limiting phosphate, limiting nitrogen and suboptimal temperatures. *Planta* 1985; 196:277–283.

70. Dennis DT, Blakeley SD. Carbohydrate metabolism. In: Buchanan BB, Gruissem W, Jones RL, eds. *Biochemistry and Molecular Biology of Plants.* Rockville, MD: American Society of Plant Physiologists, 2000:630–675.

71. Herzog B, Stitt M, Heldt HW. Control of photosynthetic sucrose synthesis by fructose 2,6-bisphosphate. III. Properties of the cytosolic fructose-1,6-bisphosphatase. *Plant Physiol.* 1984; 75:561–565.

72. Stitt M, Herzog B, Heldt HW. Control of photosynthetic sucrose synthesis by fructose 2,6-bisphosphate. V. Modulation of the spinach leaf cytosolic fructose 1,6 bisphosphatase *in vitro* by substrate, products, pH, magnesium, fructose 2,6 bisphosphate, adenosine monophosphate and dihydroxyacetone phosphate. *Plant Physiol.* 1985; 79:590–598.

73. Markham JE, Kruger NJ. Kinetic properties of bifunctional 6-phosphofructo-2-kinase/fructose-2,6-bisphosphatase from spinach leaves. *Eur. J. Biochem.* 2002; 269:1267–1277.

74. Stitt M, Quick WP. Photosynthetic carbon partitioning: its regulation and possibilities for manipulation. *Physiol. Plant.* 1989; 77:633–641.

75. Neuhaus HE, Kruckeberg AL, Feil R, Stitt M. Reduced activity mutants of phosphoglucose isomerase in the cytosol and chloroplast of *Clarkia xantiana*. *Planta* 1989; 178:110–122.

76. Kruger NJ, Scott P. Manipulation of fructose 2,6-bisphosphate levels in transgenic plants. *Biochem. Soc. Trans.* 1994; 22:904–909.

77. Draborg H, Villadsen D, Nielsen TH. Transgenic *Arabidopsis* plants with decreased activity of fructose-6-phosphate,2-kinase/fructose-2,6-bisphosphatase have altered carbon partitioning. *Plant Physiol.* 2001; 126: 750–758.

78. Scott P, Lange AJ, Kruger NJ. Photosynthetic carbon metabolism in leaves of transgenic tobacco (*Nicotiana tabacum* L.) containing decreased amounts of fructose 2,6-bisphosphate. *Planta* 2000; 211:864–873.

79. Fernie AR, Roscher A, Ratcliffe RG, Kruger NJ. Fructose 2,6-bisphosphate activates pyrophosphate: fructose-6-phosphate 1-phosphotransferase and increases triose phosphate to hexose phosphate cycling in heterotrophic cells. *Planta* 2001; 212:250–263.

80. Harn C, Daie J. Regulation of the cytosolic fructose-1,6-bisphosphatase by post-translational modification and protein level in drought-stressed leaves of sugarbeet. *Plant Cell Physiol.* 1992; 33:763–770.

81. Leegood R. The intercellular compartmentation of metabolites in leaves of *Zea mays* L. *Planta* 1985; 164:163–171.

82. Stitt M, Heldt HW. Generation and maintenance of concentrating gradients between the mesophyll and bundle sheath in maize leaves. *Biochim. Biophys. Acta* 1985; 808:400–414.

83. Baxter CJ, Foyer CH, Turner J, Rolfe SA, Quick WP. Elevated sucrose-phosphate synthase activity in transgenic tobacco sustains photosynthesis in older leaves and alters development. *J. Exp. Bot.* 2003; 54:1813–1820.

84. Worrell AC, Bruneau J-M, Summerfelt K, Boersig M, Voelker TA. Expression of a maize sucrose phosphate synthase in tomato alters leaf carbohydrate partitioning. *Plant Cell* 1991; 3:1121–1130.

85. Galtier N, Foyer CH, Huber JL, Voelker TA, Huber SC. Effects of elevated sucrose-phosphate synthase activity on photosynthesis, assimilate partitioning, and growth in tomato (*Lycopersicon esculentum* var UC82B). *Plant Physiol.* 1993; 101:535–543.

86. Neuhaus HE, Quick WP, Siegl G, Stitt M. Control of photosynthate partitioning in spinach leaves: Analysis of the interaction between feedforward and feedback regulation of sucrose synthesis. *Planta* 1990; 181:583–592.

87. Kaiser WM, Huber SC. Post-translational regulation of nitrate reductase: mechanism, physiological relevance and environmental triggers. *J. Exp. Bot.* 2001; 52:1981–1989.

88. McMichael RW, Klein RR, Salvucci ME, Huber SC. Identification of the major regulatory phosphorylation site in sucrose-phosphate synthase. *Arch. Biochem. Biophys.* 1993; 307:248–252.

89. Toroser D, Huber SC. Protein phosphorylation as a mechanism for osmotic-stress activation of sucrose-phosphate synthase in spinach leaves. *Plant Physiol.* 1997; 114:947–955.

90. Toroser D, Athwal GS, Huber SC. Site-specific regulatory interaction between spinach leaf sucrose-phosphate synthase and 14-3-3 proteins. *FEBS Lett.* 1998; 435:110–114.

91. Moorhead G, Douglas P, Cotelle V, Harthill J, Morrice N, Meek S, Deiting U, Stitt M, Scarabel M, Aitken A, MacKintosh C. Phosphorylation-dependent interactions between enzymes of plant metabolism and 14-3-3 proteins. *Plant J.* 1999; 18:1–12.

92. Siegl G, Stitt M. Partial purification of two forms of spinach leaf sucrose-phosphate synthase which differ in their kinetic properties. *Plant Sci.* 1990; 66:205–210.

93. Huber SC, Nielsen TH, Huber JL, Pharr DM. Variation among species in light activation of sucrose-phosphate synthase. *Plant Cell Physiol.* 1989; 30:277–285.

94. Van Quy L, Lamaze T, Champigny M-L. Short-term effects of nitrate on sucrose synthesis in wheat leaves. *Planta* 1991; 185:53–57.

95. Van Quy L, Champigny M-L. NO_3 enhances the kinase activity for phosphorylation of phosphoenolpyruvate carboxylase and sucrose phosphate synthase proteins in wheat leaves. *Plant Physiol.* 1992; 99:344–347.

96. Gerrits N, Turk SCHJ, van Dun KPM, Hulleman SHD, Visser RGF, Weisbeck PJ, Smeekens CM. Sucrose metabolism in plastids. *Plant Physiol.* 2001; 125:926–934.

97. Huber SC, Phillip SK, Kalt-Torres K. Regulation of carbon partitioning in photosynthetic tissue. In: Heath RL, Preiss J, eds. *Proceedings of the Eighth Annual Symposium on Plant Physiology*. Baltimore, MD: Waverly Press, 1985:199.

98. Geiger DR, Servaites JC. Diurnal regulation of photosynthetic carbon metabolism in C_3 plants. *Annu. Rev. Plant Physiol. Plant. Mol. Biol.* 1994; 45:235–256.

99. Li B, Geiger DR, Shieh WR. Evidence for circadian regulation of starch and sucrose synthesis in sugar beet leaves. *Plant Physiol.* 1992; 99:1393–1399.

100. Huppe HC, Turpin DH. Integration of carbon and nitrogen metabolism in plant and algal cells. *Annu. Rev. Plant Physiol. Plant Mol. Biol.* 1994; 45:577–607.

101. Krömer S, Stitt M, Heldt HW. Mitochondrial oxidative phosphorylation participating in photosynthetic metabolism of a leaf cell. *FEBS Lett.* 1988; 226:352–356.

102. Wyse R, Briskin D, Aloni B. Regulation of carbon partitioning in photosynthetic tissue. In: Heath RL, Preiss J, eds. *Proceedings of the Eighth Annual Symposium on Plant Physiology*. Baltimore, MD: Waverly Press, 1985:231.

103. Walker N, Zhang W-H, Harrington G, Holdaway N, Patrick J. Effluxes of solutes from developing seed coats of *Phaseolus vulgaris* L. and *Vicia faba* L.: locating the effect of turgor in a coupled chemiosmotic system. *J. Exp. Bot.* 2000; 51:1047–1055.

104. Giaquinta RT. Phloem loading of sucrose. *Annu. Rev. Plant Physiol.* 1983; 34:347–387.

105. Dickinson CD, Altabella T, Chrispeels MJ. Slow-growth phenotype of transgenic tomato expressing apoplastic invertase. *Plant Physiol.* 1991; 95:420–425.

106. Eschrich W, Fromm J. Evidence for two pathways of phloem loading. *Physiol. Plant.* 1994; 90:699–707.

107. Hawker JS, Jenner CF, Niemietz CM. Sugar metabolism and compartmentation. *Aust. J. Plant. Physiol.* 1991; 18:227–237.

108. Lerchl J, Geigenberger P, Stitt M, Sonnewald U. Inhibition of long distance sucrose transport by inorganic pyrophosphatase can be complemented by phloem specific expression of cytosolic yeast-derived invertase in transgenic plants. *Plant Cell* 1995; 7:259–270.

109. von Schaewen A, Stitt M, Scmidt R, Sonnewald U, Willmitzer L. Expression of a yeast-derived invertase in the cell wall of tobacco and *Arabidopsis* plants leads to accumulation of carbohydrate, inhibition of photosynthesis and strongly influences growth and phenotype of transgenic tobacco plants. *EMBO J.* 1990; 9:3033–3044.

110. Sonnewald U, Brauer M, von Schaewen A, Stitt M, Willmitzer L. Transgenic tobacco plants expressing yeast-derived invertase in either the cytosol, vacuole or apoplast: a powerful tool for studying sucrose metabolism and sink/source interactions. *Plant J.* 1990; 1:95–106.

111. Heineke D, Sonnewald U, Büssis D, Günter G, Leidreiter K, Wilke I, Raschke K, Willmitzer L, Heldt HW. Apoplastic expression of yeast-derived invertase in potato. *Plant Physiol.* 1992; 100:301–308.

112. Riesmeier JW, Willmitzer L, Frommer WB. Evidence for an essential role of the sucrose transporter in phloem loading and assimilate partitioning. *EMBO J.* 1994; 13:1–7

113. Bush DR. Proton-coupled sugar and amino acid transporters in plants. *Annu. Rev. Plant Physiol. Plant Mol. Biol.* 1993; 44:513–542.

114. Kühn C. A comparison of the sucrose transporter systems of different plant species. *Plant Biol.* 2003; 5:215–232.

115. Roblin G, Sakr S, Bonmort J, Delrot S. Regulation of a plant membrane sucrose transporter by phosphorylation. *FEBS Lett.* 1998; 424:165–168.

116. Echeverria E, Gonzalez PC, Brune A. Characterization of proton and sugar transport at the tonoplast of sweet lime (*Citrus limettioides*) juice cells. *Physiol. Plant.* 1997; 101: 291–300.

117. Burton WG. *The Potato.* Harlow, UK: Longman Scientific and Technical, 1991.

118. Plaxton WC. The organization and regulation of plant glycolysis. *Annu. Rev. Plant Physiol. Plant Mol. Biol.* 1996; 47:185–214.

119. Sturm A. Invertases. Primary structures, functions, and roles in plant development and sucrose partitioning. *Plant Physiol.* 1999; 121:1–7.

120. Sung S-J, Xu D-P, Black CC. Identification of actively filling sucrose sinks. *Plant Physiol.* 1989; 89: 1117–1121.

121. Ranwala AP, Iwanami SS, Masuda H. Acid and neutral invertases in the mesocarp of developing muskmelon (*Cucumis melo* L. cv Prince) fruit. *Plant Physiol.* 1991; 96:881–886.

122. Kato T, Kubota S. Properties of invertase in sugar storage tissues of citrus fruit and changes in their activities during maturation. *Physiol. Plant.* 1978; 42:67–72.

123. Miron D, Schaffer AA. Sucrose phosphate synthase, sucrose synthase, and invertase activities in developing fruit of *Lycopersicon esculentum* Mill. and the sucrose accumulating *Lycopersicon hirsutum* Humb. and Bonpland. *Plant Physiol.* 1991; 95:623–627.

124. Ricardo CPP. Alkaline β-fructosidases of tuberous roots; possible physiological function. *Planta* 1974; 118:333–343.

125. Yelle S, Cretelat RT, Doris M, DeVerna JW, Bennet AB. Sink metabolism in tomato fruit. IV. Genetic and biochemical analysis of sucrose accumulation. *Plant Physiol.* 1991; 95:1026–1035.

126. Klann EM, Chetelat RT, Bennet AB. Expression of acid invertase gene controls sugar composition in tomato (*Lycopersicon*) fruit. *Plant Physiol.* 1993; 103:863–870.

127. Morrell S, ap Rees T. Sugar metabolism in developing tubers of *Solanum tuberosum*. *Phytochemistry* 1986; 25:1579–1585.

128. Huber SC, Akazawa T. A novel sucrose synthase pathway for sucrose degradation in cultured sycamore cells. *Plant Physiol.* 1986; 81:1008–1013.

129. Geigenberger P. Regulation of sucrose to starch conversion in growing potato tubers. *J. Exp. Bot.* 2003; 54:457–465.

130. Loef L, Stitt M, Geigenberger P. Orotate leads to a specific increase in uridine nucleotide levels and stimulation of sucrose degradation and starch synthesis in discs from growing potato tubers. *Planta* 1999; 209:314–323.

131. Hajirezaei M, Sonnewald U, Viola R, Carlisle S, Dennis DD, Stitt M. Transgenic potato plants with strongly decreased expression of pyrophosphate: fructose-6 phosphate phosphotransferase show no visible phenotype and only minor changes in metabolic fluxes in their tubers. *Planta* 1994; 192:16–30.

132. Kruger NJ. Carbohydrate synthesis and degradation. In: Dennis DT, Turpin DH, eds. *Plant Physiology, Biochemistry and Molecular Biology.* London: Longman, 1990:59–76.

133. Podestá FE, Plaxton WC. A fluorescence study of ligand binding to potato tuber pyrophosphate-dependent phosphofructokinase. Evidence for competitive binding between fructose-1,6-bisphosphate and fructose-2,6-bisphosphate. *Arch. Biochem. Biophys.* 2003; 414:101–107.

134. Podestá FE, Plaxton WC. Regulation of carbon metabolism in germinating Ricinus communis cotyledons. I. Developmental profiles for the activity, concentration, and molecular structure of pyrophosphate- and ATP-dependent phopshofructokinases, phosphoenolpyruvate carboxylase, and pyruvate kinase. *Planta* 1994; 194:374–380.

135. Trípodi KEJ, Podestá FE. Purification and structural and kinetic characterization of the pyrophosphate: fructose-6-phosphate 1-phosphotransferase from the crassulacean acid metabolism plant, pineapple. *Plant Physiol.* 1997; 113:779–786.

136. Theodorou ME, Cornel FA, Duff SM, Plaxton WC. Phosphate starvation-inducible synthesis of the alpha-subunit of the pyrophosphate-dependent phosphofructokinase in black mustard suspension cells. *J. Biol. Chem.* 1992; 267:21901–21905.

137. Theodorou ME, Kruger NJ. Physiological relevance of fructose 2,6-bisphosphate in the regulation of spinach leaf pyrophosphate:fructose 6-phosphate 1-phosphotransferase. *Planta* 2001; 213:147–157.

138. Carnal NW, Black CC. Soluble carbohydrates as the carbohydrate reserve for CAM in pineapple leaves. Implications for the role of pyrophosphate:6-phosphofructokinase in glycolysis. *Plant Physiol.* 1989; 90:91–100.

139. Turpin DH, Botha FC, Smith RG, Feil R, Horsey AK, Vanlerberghe GC. Regulation of carbon partitioning to respiration during dark ammonium assimilation by the green alga *Selenastrum minutum*. *Plant Physiol.* 1990; 93:166–175.

140. Geigenberger P, Stitt M. Regulation of carbon partitioning between sucrose and nitrogen assimilation in cotyledons of germinating *Ricinus communis*. *Planta* 1991; 185:563–568.

141. Smith CR, Knowles VL, Plaxton WC. Purification and characterization of cytosolic pyruvate kinase from *Brassica napus* (rapeseed) suspension cell cultures: implications for the integration of glycolysis with nitrogen assimilation. *Eur. J. Biochem.* 2000; 267:4477–4485.

142. Podestá FE, Plaxton WC. Regulation of carbon metabolism in germinating *Ricinus communis* cotyledons. II. Properties of phosphoenolpyruvate carboxylase and cytosolic pyruvate kinase associated with the regulation of glycolysis and nitrogen assimilation. *Planta* 1994; 194:406–417.

143. Neuhaus HE, Emes MJ. Nonphotosynthetic metabolism in plastids. *Annu. Rev. Plant Physiol. Plant Mol. Biol.* 2000; 51:111–140.

144. Smith AM, Denyer K, Martin C. What controls the amount and structure of starch in storage organs? *Plant Physiol.* 1995; 107:673–677.

145. Okita TW. Is there and alternative pathway for starch synthesis? *Plant Physiol.* 1992; 100:560–564.

146. Pozueta-Romero J, Frehner M, Viale A, Akazawa T. Direct Transport of ADPglucose by an adenylate translocator is linked to starch biosynthesis in amyloplasts. *Proc. Natl. Acad. Sci. USA* 1991; 88:5769–5773.

147. Denyer K, Dunlap F, Thorbjornsen T, Keeling P, Smith AM. The major form of ADP-glucose pyrophosphorylase in maize endosperm is extra-plastidial. *Plant Physiol.* 1996; 112:779–785.

148. Kleczkowski LA. Back to the drawing board: Redefining starch synthesis in cereals. *Trends Plant Sci.* 1996; 1:363–364.

149. Keeling P, Wood JR, Tyson HW, Bridges IG. Starch biosynthesis in developing wheat grain. Evidence against the direct involvement of triose phosphates in the metabolic pathway. *Plant Physiol.* 1988; 87:311–319.

150. Viola R, Davies HV, Chudeck AR. Pathways of starch and sucrose biosynthesis in developing tubers of potato (*Solanum tuberosum* L.) and seeds of faba bean (*Vicia faba* L.). Elucidation by ^{13}C-nuclear-magnetic-resonance spectroscopy. *Planta* 1991; 183:202–208.

151. Entwistle G, apRees T. Lack of fructose-1,6-bisphosphatase in a range of higher plants that store starch. *Biochem. J.* 1990; 271:467–472.

152. Kammerer B, Fischer K, Hilpert B, Schubert S, Gutensohn M, Weber A, Flugge UI. Molecular characterization of a carbon transporter in plastids from heterotrophic tissues: the glucose 6-phosphate antiporter. *Plant Cell* 1998; 10:105–118.

153. Nakata PA, Greene TW, Anderson JM, Smith-White BJ, Okita TW, Preiss J. Comparison of the primary sequences of two potato tuber ADPglucose pyrophosphorylase subunits. *Plant Mol. Biol.* 1991; 17:1089–1093.

154. Iglesias AA, Barry GF, Meyer C, Bloksberg L, Nakata PA, Greene T, Laughlin MJ, Okita TW, Kishore GM, Preiss J. Expression of the potato tuber ADP-glucose pyrophosphorylase in *Escherichia coli*. *J. Biol. Chem.* 1993; 268:1081–1086.

155. Ballicora MA, Laughlin MJ, Fu Y, Okita TW, Barry GF, Preiss J. Adenosine 5′-diphosphate-glucose pyrophosphorylase from potato tuber. Significance of the N terminus of the small subunit for catalytic properties and heat stability. *Plant Physiol.* 1995; 109:245–251.

156. Tiessen A, Hendriks JH, Stitt M, Branscheid A, Gibon Y, Farre EM, Geigenberger P. Starch synthesis in potato tubers is regulated by post-translational redox modification of ADP-glucose pyrophosphorylase: a novel regulatory mechanism linking starch synthesis to the sucrose supply. *Plant Cell* 2002; 14:2191–2213.

157. Hylton C, Smith AM. The *rb* mutation of peas causes structural and regulatory changes in ADP-Glc pyrophosphorylase from developing embryos. *Plant Physiol.* 1992; 99:1626–1634.

158. Kleczkowski LA, Villand P, Lüthi E, Olsen OA, Preiss J. Insensitivity of barley endosperm ADP-glucose pyrophosphorylase to 3-phosphoglycerate and orthophosphate regulation. *Plant Physiol.* 1993; 101:179–186.

159. Rudi H, Doan DNP, Olsen OA. A (His)₆-tagged recombinant barley (*Hordeum vulgare* L.) endosperm ADPglucose pyrophosphorylase expressed in the baculovirus-insect cell system is insensitive to regulation by 3-phosphoglycerate and inorganic phosphate. *FEBS Lett.* 1997; 419:124–130.

160. Weber H, Heim U, Borisjuk L, Wobus U. Cell-type specific, coordinate expression of two ADPglucose pyrophosphorylase genes in relation to starch biosynthesis during seed development in *Vicia faba* L. *Planta* 1995; 195:352–361.

161. Gomez-Casati DF, Iglesias AA. ADP-glucose pyrophosphorylase from wheat endosperm. Purification and characterization of an enzyme with novel regulatory properties. *Planta* 2002; 214:428–434.

162. Plaxton WC, Preiss J. Purification and properties of non-proteolytically degraded ADPglucose pyrophosphorylase from maize endosperm. *Plant Physiol.* 1987; 83:105–112.

Section X

Photosynthesis and Plant Genetics

29 Crop Radiation Use Efficiency and Photosynthate Formation — Avenues for Genetic Improvement

G.V. Subbarao and O. Ito
Crop Production and Environment Division, Japan International Research Center for Agricultural Sciences (JIRCAS)

W. Berry
Department of Organismic Biology, Ecology and Evolution, University of California

CONTENTS

I. INTRODUCTION

From an ecological perspective, plants can be considered to be "biological machines" that capture and convert solar energy into chemically available forms (i.e., biomass and grain) using a process unique to plants — photosynthesis. Earth receives solar energy at an average mean rate of $1.36\,kJ/m^2/sec$, known as the "solar constant," which is the energy equivalent of about 26,000 t biomass/ha/year [1]. Worldwide the production of agricultural crops ranges from a maximum of 30 to 60 t/ha/year under the best of conditions to less than 1 t/ha/year under the marginal conditions of subsistence farming. Thus, only a very small fraction of the total sunlight received by the earth is converted into edible biomass for human use by our present agricultural systems. This represents an efficiency of only between 0.2% and 0.004% [1]. If this light use efficiency were to be increased by only a small fraction, the effort required for food production would be reduced drastically, thereby providing many more options for society. As a biological machine, the efficiency of crop production can be looked at as the ratio of energy output (carbohydrate or biomass produced) to energy input (solar radiation) [1]. Thus, radiation use efficiency (RUE) is basically a measure or index of the growth efficiency of field-grown crops. Monteith [2] is one of the pioneers in the use of the concept of RUE as a factor to help explain the variability in biomass production of various crop species grown in different environments.

Crop biomass production can be portrayed as the product of two major components; the amount of accumulated intercepted radiation and the efficiency with which the intercepted radiation is converted into biomass, i.e., RUE [2,3]. The amount of radiation that is intercepted by a canopy depends on the level of incident radiation, the proportion of that radiation intercepted by the photosynthetically active surfaces of the crop (i.e., leaf area), and the length of the growing season. Light interception by crop canopies depends on its architecture, which is to some extent determined genetically, but can also be manipulated to some degree by crop management (such as by the manipulation of plant density). Breeders have improved cumulative intercepted radiation in crops mostly through modifications in the leaf area index (LAI) of crop species. Another major part of the breeding efforts in crop improvement has been directed toward improving the harvest index (HI), which has been increased to the degree that it is now approaching its theoretical maximum. It appears that future efforts toward crop improvements should be directed toward improving RUE as this seems to have the most potential to increase the total biomass production and also to stabilize crop production over a range of production environments where nutrient and water stress are the major constraints.

Despite four decades of sporadic efforts to study the RUE phenomenon in crops and other plants, our understanding of this important critical attribute is very limited. Most crop models by default consider RUE as a plant-specific coefficient, mainly because of our very limited understanding of how RUE is genetically influenced by crop variety (or genotype) and environmental or agronomic factors. Our objective in writing this chapter is to examine some of the factors contributing to RUE along with the options and strategies available to crop scientists to possibly improve genetical crop response to these factors.

II. COMPONENTS OF RUE

The efficiency (ε) with which crops produce biomass as defined by Monteith [1] is the net amount of solar energy stored by photosynthesis divided by the solar constant integrated over the same period. RUE can also be expressed as the product of seven factors [1]:

$$\varepsilon = \varepsilon(g,r) = \varepsilon_g \varepsilon_a \varepsilon_s \varepsilon_q \varepsilon_i \varepsilon_d \varepsilon_r$$

where ε_g is the geometrical factor, ε_a the atmospheric transmission factor, ε_s the spectral factor, ε_q the photochemical efficiency, ε_d the diffusion efficiency, ε_i the interception efficiency and ε_r the respiration factor

Geometrical factor ε_g: The geometrical factor (ε_g) is defined as the ratio of the solar energy received (outside the atmosphere on a plane parallel to the Earth's surface) to the solar constant integrated over the same period, and this solar constant depends only on latitude and season. Monteith [1] estimated that the average value of ε_g decreases from about 0.3 in the tropics to 0.2 in temperate latitudes.

Atmospheric transmission factor ε_a: Solar radiation is absorbed and scattered by gases, clouds, and aerosols (such as soil, smoke, and salt particles) while passing through the earth's atmosphere. By taking into account the water vapor and ozone in the atmosphere, it is possible to calculate the amount of solar energy received at the surface of the earth in the absence of clouds, and express this solar energy as a fraction of the extraterrestrial radiation on a horizontal surface (i.e., ε_g times the solar constant). The mean transmissivity of a cloudless atmosphere is relatively constant during the year in the tropics compared to temperate latitudes [1], and ranges from 0.48 (cloudy locations) to about 0.58 (cloud-free locations).

Spectral factor ε_s: Only a small fraction of the solar radiation (wave band of 0.4 to 0.7 μm) is utilized for photosynthesis, which is referred to as photosynthetically active radiation (PAR). At the earth's surface, the intensity of PAR depends on the extent to which the solar spectrum is modified by absorption and scattering and can be calculated for a given solar spectrum as a function of water vapor and dust content of the atmosphere. At solar elevations of 30° and above, the fraction of PAR to the total solar radiation is estimated as 45% [4]. When solar elevation exceeds 40°, the estimated ratio of PAR to total radiation in the diffuse component is about 60%. Combining the direct and diffuse components in appropriate proportions, Monteith [5] suggested that the ratio of PAR to total radiation should be close to 0.5. Integrating over the spectrum from 0.4 to 0.7 μm, the fraction of PAR absorbed by leaves is usually between 80% and 90%. Thus, the fraction of whole-spectrum radiation absorbed by green leaves is close to $0.5 \times 0.85 = 0.425$, which is denoted by ε_s.

Photochemical efficiency ε_q: The efficiency of photosynthesis can be defined as the ratio of energy stored in the formation of carbohydrates to the absorbed radiant energy. Most of the absorbed energy is used in biochemical cycles involving many intermediate compounds that act as carriers of energy. In very weak light (i.e., when the rate of photosynthesis is limited only by the supply of light quanta and not by other factors such as the supply of carbon dioxide molecules to the chloroplasts), nearly 20% of the absorbed energy is stored in the final products of the photochemical system; the remaining 80% is used to form intermediates that are used in the formation of proteins and fats; the rest is either rejected in the form of heat or used for plant structure. The synthesis of one molecule of carbohydrate requires one molecule of CO_2 and the energy of ten light quanta [6]. The average energy content of one quantum of PAR is 3.6×10^{-19}. Monteith [1] has shown that the maximum photochemical efficiency (ε_q) of a ten-quantum process in sunlight is

$$\varepsilon_q = 7.7 \times 10^{-19}/10(3.6 \times 10^{-19}) = 0.215$$

The energy content of 1 g of biomass (for plants that do not store oil) is approximately 18 to 20 kJ [7–9]. Based on these above calculations [$(0.425 \times 0.215)/16.7$], 1 kJ of solar radiation is assumed to be equivalent to 54 mg of dry matter [1].

Diffusion efficiency ε_d: Intercellular CO_2 decreases as the photosynthetic rate (which is driven by intensity of irradiance) increases from near zero to high values as a result of the limited rate at which CO_2 can be transported to the chloroplasts by diffusion from the external air or from respiring mitochondria. This decrease in the availability of intercellular CO_2 is responsible for the characteristic shape of the photosynthesis–light curve (Figure 29.1). "Light saturation" is achieved when an increase in irradiance no longer increases the rate of photosynthesis.

Interception efficiency ε_i: The interception efficiency of a crop canopy is defined as the ratio of actual photosynthesis by a crop canopy to the maximum rate achieved at full light interception [1]. ε_i is a function of LAI of the canopy, canopy coverage, and light extinction coefficient (k), and can be estimated from the light intercepted by the crop canopy.

Respiration factor ε_r: The respiration factor is defined as the fraction of assimilates used in respir-

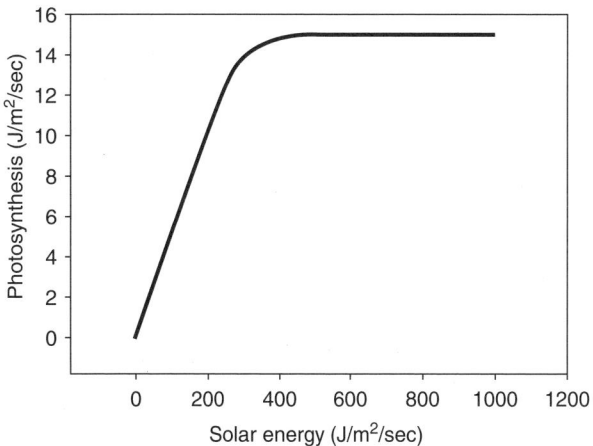

FIGURE 29.1 Photosyntheis (in terms of energy trapped in stored carbohydrate, i.e., $J/m^2/sec$) is plotted against solar energy ($J/m^2/sec$). (Adapted from Chrispeels MJ, Sadava DE. *Plants, Genes and Crop Biotechnology*. Boston, MA: Jones and Bartlett, 2003.)

ation, which ranges from 0.25 to 0.50 as estimated in several field crops [11–13]:

$$\varepsilon_r = 1 - R/P$$

where R is the weight of the carbohydrate used for respiration, and P is the weight of the carbohydrate produced by the photosynthesis.

For temperate crops, ε_r will be lower than for tropical crops on an average; however, for short-term crops, the temperature of the actual growing season may be the determining factor. Monteith [1] suggested that for the tropics, ε_r should be about 0.50.

III. EFFICIENCY OF PHOTOSYNTHESIS IN CONVERTING ENERGY INTO BIOMASS

The efficiency with which plants convert intercepted solar energy into biomass rarely exceeds 5% because of a number of inherent limitations associated with the photosynthetic process and the biological systems in general (Figure 29.2). Some of them are

1. Only about 50% of the solar energy (i.e., wavelengths between 400 and 700 nm) can be used for photosynthesis by crops, and much of the remaining (i.e., >700-nm wavelengths) is in the near-infrared part of the spectrum, which does not have sufficient energy per quantum to drive photosynthesis.

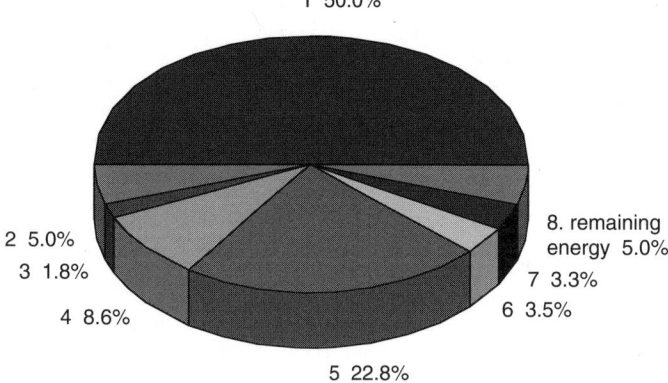

% solar energy lost due to
1. -unavailability for photosynthesis
2. -Reflection from crop canopy
3. -Inactive absorption by the crop canopy
4. -Photochemical inefficiency
5. -Carbohydrate synthesis
6. -Photorespiration
7. -Dark respiration
8. % Remaining energy stored by the plants

1 50.0%

8. remaining
energy 5.0%

2 5.0%
3 1.8%
7 3.3%
6 3.5%
4 8.6%
5 22.8%

FIGURE 29.2 Energy losses during conversion of intercepted solar energy into plant biomass. (Adapted from Chrispeels MJ, Sadava DE. _Plants, Genes and Crop Biotechnology_. Boston, MA: Jones and Bartlett, 2003.)

2. Of the PAR (about 50% of the total radiation intercepted by the crops) absorbed by plants, about 10% is not used in photochemistry. Although violet photons have 70% more energy than red photons, the photosynthetic outputs in terms of ATP for red photon and violet photon are exactly the same. The additional energy of the violet photon is lost as heat, representing an intrinsic photochemical inefficiency of photosynthesis [14].

3. About 6% of the absorbed radiation is reflected back to the sky from a closed canopy [15].

4. Leaf pigments such as anthocyanins absorb light, but they are not part of the photosynthetic chain, thus this energy is not directly productive in terms of biomass production.

5. Transfer of energy from the short-term energy stores of the chloroplasts, ATP and NADPH, to the synthesis of carbohydrates (i.e., the long-term energy stores of most plants) proceeds only at the cost of energy, with an efficiency of about 35% in C_3 plants.

Theoretically, only 5% of the radiation intercepted by the crop canopies can be converted into biomass under even the best agronomic and climatic conditions as has been demonstrated with a number of field and pasture crops [16,17].

A. DETERMINATION OF RUE

RUE as used by many crop physiologists is defined as the ratio of chemical energy stored in assimilates (i.e., biomass produced by the crop) to radiant energy intercepted and absorbed by the crop canopy during the growing season. RUE is derived from measurements of accumulated crop mass and intercepted radiation. However, the measurements needed for calculating RUE are not standardized. Thus, the estimates of RUE reported in the literature differ for a number of reasons:

1. Some are expressed on a PAR basis, while others are expressed on a short-wave radiation basis. Transmission of PAR through canopies differs from that of short-wave radiation [18].

2. Some are expressed on an intercepted basis, while others are expressed on an absorbed basis.

3. Some are based on net aboveground dry matter production with variable leaf losses, while others are expressed on a total dry matter production basis including roots.

4. Some are based on differences between two discrete samplings at different stages of crop growth, which is subject to large sampling errors.

5. RUE estimations are based on the fitted slope from many crop growth samplings and cumu-

lative intercepted radiation, perhaps the most reliable estimation among all, but labor and resource intensive.

6. Some estimates of RUE are based on short-term gas exchange measurements.

7. There are also differences in the time frame used for the calculations of RUE ranging from minutes to hours to full growing season.

These different methods of measuring RUE are not necessarily wrong, for they are often the result of studies set up to look at different aspects of RUE. Short-term measurements of RUE are appropriate for biochemical studies, but long-term measurements are better suited for the evaluation of the overall efficiency of the crop.

1. Measurement of Crop Biomass

Usually the net aboveground biomass is used for the calculation of RUE as recovering root biomass under field conditions can be extremely difficult. RUE estimates based on consecutive field samplings can be highly variable when based on the difference between two discrete samplings [3]. Use of small plots and small sampling areas can lead to edge effects, which will bias estimates when there is a significant amount of lateral radiation [19]. In some crop species, RUE declines during the reproductive growth stage because of the remobilization and loss of nitrogen from the leaves to the reproductive tissue even under high nitrogen input conditions, with the subsequent loss of photosynthetic capacity [20]. This should be taken into account while obtaining RUE estimates based on regression of dry matter accumulation based on several samplings during the growing season and cumulative intercepted radiation. In these cases, RUE estimations should be made only on dry matter estimations up to flowering [21]. Also, it is necessary to consider the energy content of plant mass. For high energy content of oily seed crops such as soybean, peanut, and sunflower, the energy content of the grain needs to be adjusted upward (about 1.3 times that of the vegetative material) when calculating RUE [22–27]. For other legumes such as cowpea or mungbean, the energy content of seed and vegetative material can be considered the same as that of the rest of the plant biomass [28].

2. Measurement of Incident and Intercepted Solar Energy

Solar radiation can be expressed as either total solar radiation (0.4 to 3 µm) or that in the wavebands of PAR (0.4 to 0.7 µm) [2]. The ratio of PAR to total

radiation in the direct solar beam is between 0.44 and 0.45 when the sun is more than $30°$ above the horizon [4], and a figure of 0.45 has often been used by biologists to calculate PAR from the flux of total radiation recorded with a solarimeter [29]. When the solar elevation exceeds $40°$, the estimated ratio of PAR to total radiation and the diffuse component is about 0.60. Combining the direct and diffuse components in appropriate proportions, the ratio of PAR to total solar radiation is close to 0.50, which is considered to be appropriate for the tropics and temperate latitudes [1]. Thus, to calculate RUE values on a total solar radiation basis, it is appropriate to multiply those estimates based on intercepted PAR by 0.5 [1,30]. The amount of radiation intercepted by the crop canopy can be determined based on the received and transmitted radiation, which can be measured by placing the tube solarimeters beneath and above the canopy [31]. Tube solarimeters measure irradiance received, and the information can be stored on a data logger to provide continuous diurnal measurement of the fraction of incoming radiation that is intercepted by the crop canopy [32].

B. RUE in Various Crop Species

Crop species differ substantially in RUE (Table 29.1). The plant's photosynthetic pathway has a major impact on RUE. In general, C_4 crop species have the highest RUE, followed by nonleguminous species, with leguminous species having the lowest [33,36]. Among C_4 crop species, sugarcane has been reported to have the highest RUE, with values approaching close to 2.0 g/MJ. Among C_3 crop species, potato has RUE values ranging from 1.6 to 1.75 g/MJ. Among cereals (wheat, barley, and rice), RUE ranges from 1.3 to 1.5 g/MJ. Sunflower is reported to have the highest RUE among oilseed crops. Most of the grain legumes have the lowest RUE among crop species with values below 1.0 g/MJ [33]. Some of the reported values for various crop species are presented in Table 29.1. Early in the season many crops grow rapidly, thus higher efficiencies can be achieved for a few days. For the comparative evaluation of crop species, RUE averaged over the life of a crop would be appropriate. For C_3 crops, the highest efficiencies are about 3.5% and for C_4 about 4.3%.

There are only a few systematic evaluations of genotypic variation in crop species for RUE, a prerequisite for future genetic improvement efforts. Nearly 30% variation in RUE among peanut genotypes was reported [67]. For pigeonpea, about 29% variation in RUE was reported among six genotypes tested under irrigated and drought conditions (Figure 29.3) [69]. Similarly, genotypic variation in RUE has

TABLE 29.1
RUE in Various Crop Species

Crop/Plant Species	RUE Based on Total Radiation (g/MJ)	RUE based on PAR (g/MJ PAR)	Ref.
Sorghum	1.25–1.30	2.8	[32–34]
Wheat	−1.25	1.68–3.82	[3,32,35–38]
Barley		1.79–2.90	39,40
Maize	1.26–1.75	3.0–4.14	32,33,36,41–46
Sugarcane	1.72–1.96		47–49
Rice	0.93	2.2–3.28	33,50,51
Pearlmillet	2.4	4.1	18
Potato	1.6–1.76		52,53
Sunflower		1.77–3.13	27,33,54
Soybean	0.60–1.20	2.04–2.4	28,55–61
Cowpea	1.05		28
Fababean		2.06–4.80	62,63
Mungbean	0.94	2.17	28,64
Peanut	0.98–2.24	2.5–3.04	18,25,65–67
Pigeonpea	0.83	1.70–2.19	68–70
Lentil		2.14	71
Chickpea	0.67–0.78		72
Garlic		2.9	73
Cassava	0.90		74

been reported in sorghum [75], potato [76], wheat [38,77], barley [40], soybean [60], peanut [64], fababean [63], chickpea [78], and pigeonpea [69,70].

IV. FACTORS THAT INFUENCE RUE

RUE is highly dependent on the photosynthetic performance of crop canopies and can be influenced by several factors, namely, extremes of temperature [79,80], water, and nutrient status [59,69,81,82]. This is indicated by the variation reported in RUE among and within crop species and across locations and growing environments. Of the factors that influence RUE in crop canopies, water availability, temperatures, vapor pressure deficits, radiation levels, and crop nutritional status are some of the environmental factors that affect the RUE of crop species. Crop-specific factors such as photosynthetic metabolism types (i.e, C_3 vs. C_4), growth stage, source–sink status, and canopy- and biochemical-attributes (that determine photochemical efficiency) are some of the biological factors that are important in influencing the RUE of crops. Some of these factors are closely linked to the growing environment, and thus can be manipulated through agronomic management to optimize RUE and thus crop production at any given location. Other factors are more crop/genotype specific, and thus amenable to genetic manipulations using traditional breeding or molecular biological

tools. RUE often is calculated from the aboveground biomass production, assuming that the root biomass is a constant proportion of the total biomass (usually 10%). However, it has been shown in several studies that the root/shoot ratio will increase substantially under marginal environments where drought and nutrient stress are integral features of the production systems.

Crop species differ in their adaptation to water deficits and nutrient utilization efficiencies, which are some of the reasons for the expected genetic differences within and among crop species with regard to RUE. Earlier researchers have assumed that RUE in crop species is a conservative feature and thought to be little influenced by either stage of growth, or environmental or agronomical factors [2]. This has been subsequently shown not to be the case. Indeed, RUE was shown in a number of crop species to change with crop growth stage and environmental and agronomic growing conditions. Some of the important crop, agronomic, and environmental factors that influence RUE are as follows.

A. GROWTH STAGE

For a given crop species, RUE has been widely regarded as a stable entity during various stages of growth in the absence of water deficits, inadequate nutrition, pests, and diseases [3,83]. However, subsequent studies have shown that biomass production of

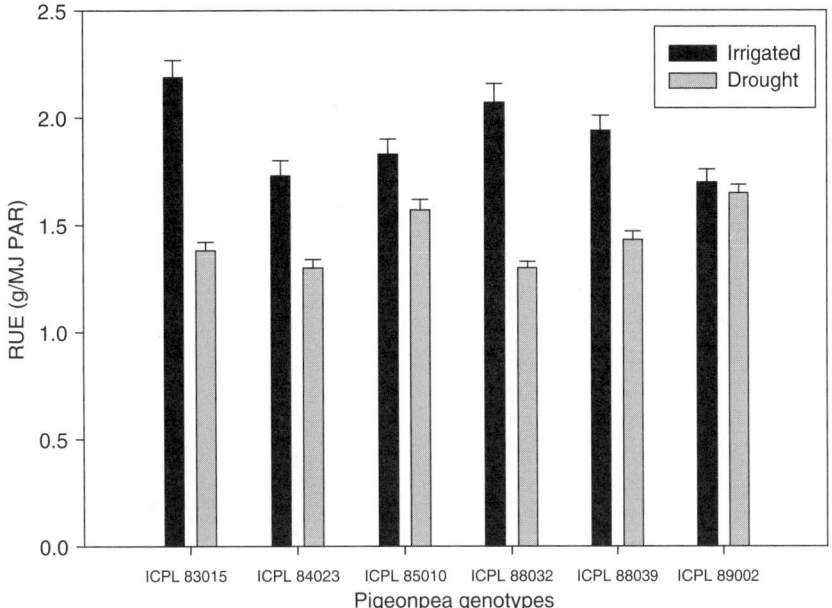

FIGURE 29.3 RUE (g dry matter per MJ PAR) of six genotypes grown under irrigated and drought conditions. (Drought was imposed by withholding water to the field plots from 50 days after planting and continued until physiological maturity; the experiment was conducted in automated rainout shelter, which excludes the rainfall during the growing season). (Adapted from Nam NH, Subbarao GV, Chauhan YS, Johansen C. *Crop Sci.* 1998; 38:955–961.)

crops commonly increases linearly with the amount of solar radiation and that the RUE of many crops reaches its maximum just before the onset of reproductive growth, thus RUE is not as stable as it was once assumed to be [30,69]. RUE has been reported to decline during the reproductive stage in sunflower [26,54], maize [32,38,84,85], and sorghum [32,38,84]. In many leguminous crops including soybean, field pea, cowpea, mungbean, and pigeonpea, RUE has declined from flowering until physiological maturity [21,58,86,87] (G.V. Subbarao and N.H. Nam, unpublished data). In pigeonpea (based on the evaluation of six short-duration varieties), RUE was shown to increase up to flowering, then declined and reached its lowest values during grain filling, under both nonlimiting and limiting water conditions (Figure 29.4).

The decline in RUE of many leguminous crops is related to the mobilization of nitrogen (which includes nitrogen from Rubisco) from leaves to the reproductive tissue, which resulted in loss of the photosynthetic capacity and hence a decline in RUE [81,88–90]. The close link between leaf nitrogen and photosynthetic capacity in general [81] suggests that such a decrease in RUE during the reproductive stage could be a general phenomenon in many crops [21]. A number of studies with soybean have shown that photosynthetic rates of individual leaves decline during seed filling [91,92]. This decline in photosynthetic rate correlates well with the decline in leaf nitrogen

[87,91], leaf protein loss, and RuBPcase protein activity. It is suggested that this loss of nitrogenous compounds could be the trigger that induces leaf

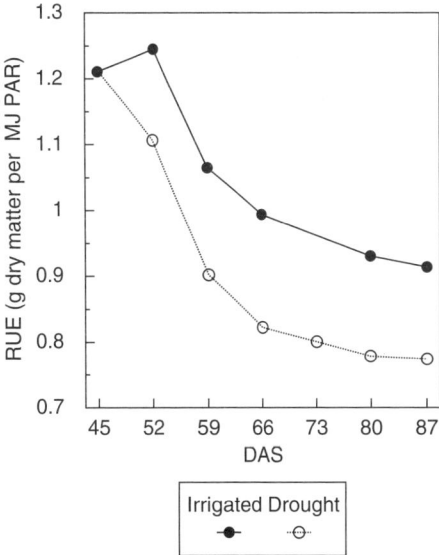

FIGURE 29.4 Changes in RUE (g dry matter per MJ PAR) of pigeonpea (mean of six genotypes) at various growth stages under irrigated and drought (drought stress imposed from 50 days after sowing and continued until physiological maturity) growing conditions. (From G.V. Subbarao and N.H. Nam, unpublished data.)

senescence [22], thus limiting yield by cutting short the time available for seed filling [93]. During the early stages of growth when water and nutrients are most abundant, the relationship between intercepted solar radiation and biomass production is about linear, and has been recognized in various crops [30]. Nevertheless, RUE is influenced by the crop growth stage; growth stages control the progression of growth and development and therefore photosynthesis, respiration, and partitioning biomass to roots. As a result of this realization, changes in RUE during different stages of crop growth are now being considered in some of the crop models [59].

B. NITROGEN STATUS

High crop RUE is directly dependent on obtaining the maximum leaf photosynthetic rate [81,94]. Nearly 70% of the soluble protein in the leaf is concentrated in the carboxylation enzymes (i.e., Rubisco), thus the relation between nitrogen availability, canopy photosynthetic efficiency, and RUE is not unexpected [95–101]. A positive relationship between leaf nitrogen content per unit area (specific leaf nitrogen) and photosynthetic rates has been reported for a number of crops including wheat, maize, sorghum, rice, soybean, potato, sunflower, peanut, and sugarcane [20,60,81,94,97,102–108]. The quantum yield of CO_2 assimilation, which is one of the major determinants of the photosynthetic efficiency of crop canopies, reportedly decreases under N deficiency [108]. Levels of photoinhibition also increase under N deficiency [109]. Thus, a favorable crop nitrogen status appears to be necessary for the realization/expression of maximum RUE in a given crop species.

Several studies have reported a positive response of RUE to N fertilization in a number of crops [20,26,110]. Nitrogen deficiency should decrease the range where there is a linear response between PAR and increased light and thus the range of maximum RUE [43,111]. A substantial decrease in RUE under nitrogen stress has been reported for maize [32,85], sorghum [43], kenaf [112], wheat [110], sunflower [26,113,114], and peanut [25].

There are major differences between C_4 and C_3 metabolism in relation to their response to CO_2. This is reflected in the amount of N required for photosynthesis and thus in RUE for a given amount of N. Under low-N environments, the C_4 species maintain a higher level of RUE compared to the C_3 species [115]. A leaf N content of about 0.5 g N per square meter of leaf area will sustain a RUE of about 0.5 g/MJ in crops such as soybean and rice, whereas for maize, a similar nitrogen level will sustain a RUE of

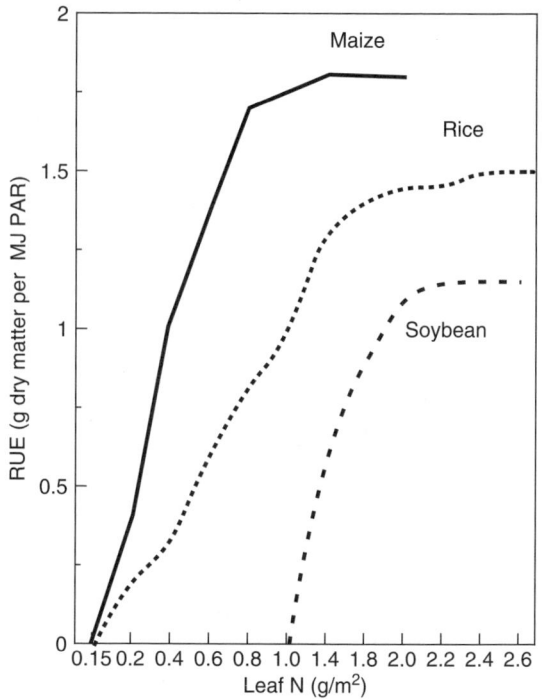

FIGURE 29.5 Theoretical relationship between canopy leaf N content and the RUE in maize, rice, and soybean. (Adapted from Sinclair TR, Horie T. *Crop Sci.* 1989; 29:90–98. With permission.)

about 1.50 g/MJ [116] (Figure 29.5). The differences between C_3 and C_4 species tend to diminish under high N production environments. Under conditions that provide high leaf N for each species (i.e., maize, rice, and soybean), RUE in maize reaches a maximum of 1.70 g/MJ, in rice about 1.4 g/MJ, and in soybean about 1.2 g/MJ [81,116] (Figure 29.5).

Nitrogen distribution among leaves in a crop canopy changes with LAI [117] or leaf age when N availability is limited. Below the top zone of the canopy, leaf nitrogen content declines with increased cumulative LAI [21]. The distribution of N in the crop canopy (i.e., the vertical distribution of N in the canopy) is one of the factors responsible for the genotypic differences in RUE under optimal N and growing conditions. Leaf N content largely determines the maximum photosynthetic rate at high irradiances [98,118,119]. Genotypes/varieties that can preferentially allocate more N to the most illuminated part of the canopy will have higher levels of RUE than genotypes that lack such a strategy [118,120]. The interactions between RUE and photosynthetic rates is possibly one of the principal underlying mechanisms for the differences in RUE among genotypic/cultivar variations or among crop species [81].

C. WATER STATUS

Soil water and the resulting plant water status play a key role in determining stomatal conductance and canopy photosynthesis. Soil water deficit results in plant water deficits that lead to stomatal closure and reduced photosynthesis, and results in loss of photosynthetic efficiency of the canopy and thus to a decrease in RUE [2]. Plants have developed a number of adaptive mechanisms to cope with water deficits to minimize the impact on their productivity [121,122]. Nearly a 70% decline in RUE due to drought stress was observed in a number of grain legumes [58]. Though RUE of C_4 crop species is generally higher than that of C_3 crop species, this photosynthetic advantage disappears as the water stress increases. In the very dry environments of West Africa, the RUE of C_4 crops spcies such as pearl millet and sorghum was often similar to that of C_3 crop species such as cowpea [21].

When drought stress is imposed from flowering until physiological maturity, a 25% decline in RUE has resulted in pigeonpea [69] (Figure 29.6). Significant genotypic variation in RUE under both water-nonlimiting and water-limiting environments has been reported (Figure 29.3) [69]. The growth of many field crops can be slowed down or even stopped by a relatively moderate water stress such as 0.2 to 0.5 MPa (2 to 5 bar) [123]. Stress of this magnitude develops following only a few days without rain, resulting in stomata closure, thus limiting photosynthesis [124]. For rice, wheat, maize, sorghum, and pearl millet, drought stress has been reported to decrease RUE [3,21,51,84,125–127]. A variety of mechanisms that include leaf movements (that can reduce the radiation load on the canopy when exposed to water deficits) and osmotic adjustment, and root attributes (that can maintain water supply during drought spells) play a major role in maintaining high levels of RUE during water stress [70]. For example, RUE of six pigeonpea genotypes under water deficit conditions was correlated with the relative leaf water contents (Figure 29.7).

Also, drought stress alters the partitioning of biomass between root and shoot, with a tendency to increase the root/shoot ratio, thus more carbon is partitioned into root dry matter under water stress [74]. Since RUE is usually calculated based on aboveground dry matter production, the calculated effect of drought stress on RUE is further exacerbated. This apparent reduction in RUE is in addition to the actual decline in carbon fixation from the decrease in photosynthetic rates and canopy size. The extent of reduction in RUE because of drought is dependent on the severity and time of drought stress during the growth cycle of the crop. Many crop plants respond to water deficits by limiting the leaf area expansion as

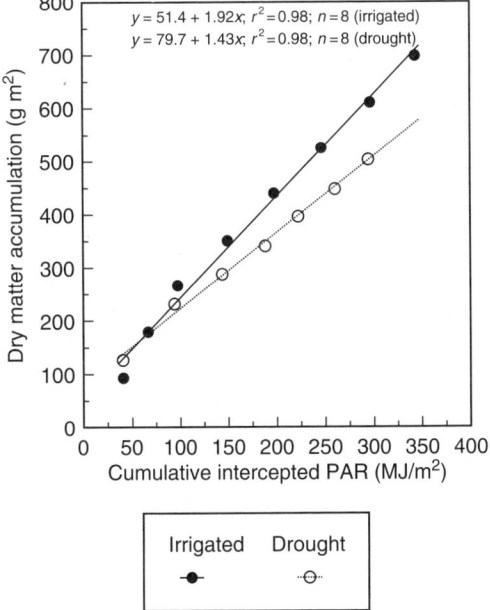

FIGURE 29.6 Relationship between cumulative intercepted radiation and dry matter production in pigeonpea (mean of six genotypes) under irrigated and drought growing conditions (the slope value b is considered as RUE). (Adapted from Nam NH, Subbarao GV, Chauhan YS, Johansen C. *Crop Sci.* 1998; 38:955–961. With permission.)

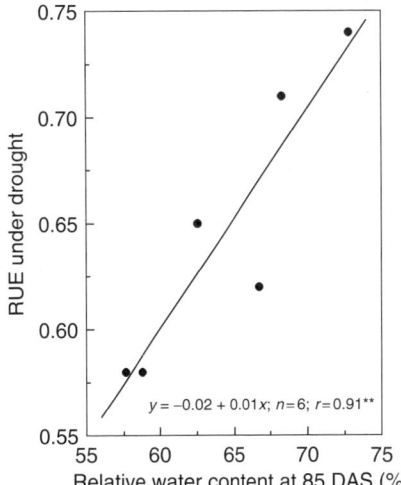

FIGURE 29.7 Relationship between relative leaf water content (i.e., plant water status) and RUE in six pigeonpea genotypes under drought conditions. (Adapted from Subbarao GV, Nam NH, Chauhan YS, Johansen C. *J. Plant Physiol.* 2000; 157:651–659. With permission.)

the first stage of adaptation, thus canopy size, and and the resulting decrease in light interception, is the principal factor that is adjusted by many plants to compensate for limited water availability. However, if the water stress develops after the canopy is fully formed (i.e., during the time of reproductive development), then the effect of drought stress directly affects RUE and depends on the degree of internal water stress the crop experiences. The timing of drought stress thus is very important in determining the degree of its impact on RUE in crops [74].

When crops are grown on stored water, like in many Mediterranian environments, the leaf area expansion and canopy size are determined by the amount of stored water in the soil. The effects of slowly developing water deficits on RUE may not be detectable until some threshold is reached. For example, RUE of chickpea is independent of the fraction of extractable water until it reaches a threshold value of 30% in the soil; then there is a marked decline in RUE [72]. This study demonstrated that the response of RUE to water deficits is quantitatively dependent on the severity of soil water deficit.

D. Vapor Pressure Deficit

Vapor pressure deficits in the growing environment could affect the water status of the plant and thus canopy photosynthesis. As would be expected, a reduction in RUE occurs when the vapor pressure deficits are sufficient to induce water stress [30]. The influence of vapor pressure deficit on leaf photosynthesis tends to be most evident at high vapor pressure deficits (>2 kPa), but modest decreases in leaf photosynthesis can affect RUE [128]. For sorghum and maize, an increase in environmental vapor deficits has been shown to decrease RUE [128]. Similarly, it was also shown that vapor pressure deficits reduce the RUE in barley. In one season there was a vapor pressure deficit of 1.06 kPa, with a RUE of 0.67 g/MJ, whereas in another growing season where the vapor pressure deficit was 0.68 kPa, the RUE was 1.30 g/MJ [40].

E. Shading

Decreasing solar radiation through shading can result in improvements in RUE as the photosynthesis response to radiation at higher levels is curvilinear. In many plants half the intensity of sunlight is more than adequate to reach light saturation and any subsequent increases in radiation over saturation do not improve the canopy photosynthesis [129]. However, increasing RUE may not be beneficial to the overall growth of the crop if the increase in RUE was because

the overall growing system had become light limiting. Hammer and Wright [94] have provided a detailed analysis of the importance of the radiation environment for RUE. By altering the atmospheric transmission ratio, changes in radiation levels and its effects on RUE are simulated. It was shown that RUE increased by about 0.4 g/MJ from a clear day to a cloudy day [94]. It was shown that the increasing fraction of diffuse component on the cloudy day accounted for a RUE increase of 0.15 g/MJ. Thus, higher RUE values can be expected for low-radiation production environments and will be of significance when comparing RUE of a target crop species across a range of production environments.

Several studies have reported that reducing the intensity of radiation of the exposed canopy through shading has increased RUE up to 72% in comparison to the unshaded control [50] (Figure 29.8). Shading has been reported to increase RUE in sunflower [114]. This phenomenon is exploited in cereal legume intercropping (such as sorghum/pigeonpea, pearlmillet/pigeonpea, pearlmillet/cowpea, and several other combinations of legume cereal intercropping). This intercropping results in a higher combined system

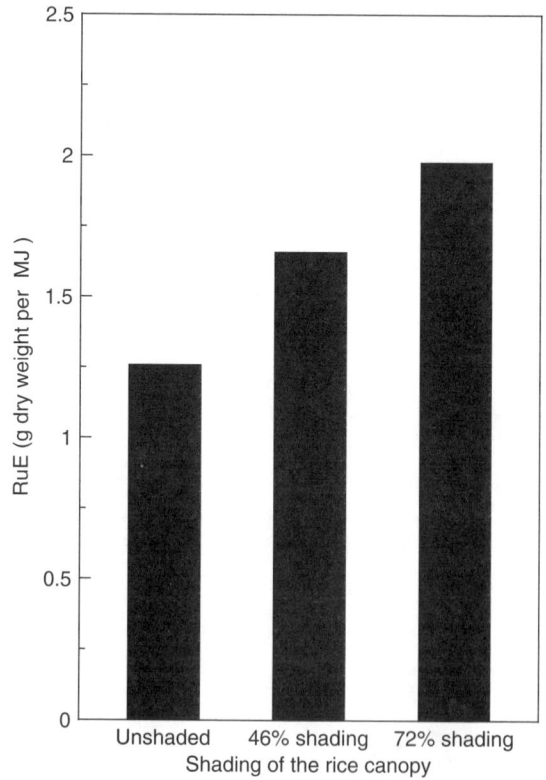

FIGURE 29.8 Shading effects on RUE of rice canopies. (Adapted from Horie T, Sakuratani T. *Jpn. Agric. Meteorol.* 1985; 40:331–342.)

RUE than that of straight cereal or legume production systems, even including the period near the end of the season when the legume of the intercrop is growing alone. Consequently, mixed crops produced more dry matter than either crop grown singly. They intercepted more solar radiation than either of the crops alone and used it more efficiently. In three cereal/legume mixed crops, dry matter production was 1.1 to 1.9 times that of the cereal and 2 to 2.5 times that of the legume. In effect, the mixed crops grew for the duration of the legume at a rate more nearly that of a C_4 than that of a C_3 [74].

Generally, the RUE of the cereal component of the intercropping system is little influenced in the intercropping system as the cereal is always fast growing and taller than the legume component. The legume in the intercropping component usually grows more slowly in the beginning and receives only the diffusive light (the light that is passed through the cereal canopy) and thus has higher RUE, which improves the overall RUE of the intercropping system. The intercropped plant system uses solar radiation with efficiency comparable to that of a C_4 species. In groundnut, RUE during vegetative growth was 1.3 g/MJ when groundnut was grown alone but 2.0 g/MJ when it was grown beneath pearl millet — almost as efficient as the millet itself [74]. The higher RUE of groundnut in the pearl millet/groundnut intercropping was solely due to the shading effect from pearl millet and not due to any other interaction effect from millet [74]. This was demonstrated by using other types of shading. A similar effect of shade on RUE has been shown on cassava [74].

F. SOURCE–SINK ISSUES

The ability of the plant to use or translocate photosynthates can limit the rate of photosynthesis [130,131]. Thus, sink strength of a crop can be hypothesized to influence RUE of a canopy [132,133]. This is based on the hypothesis that the photosynthetic system has excess capacity that is presently not utilized mostly because of the local buildup of excess photosynthates (i.e., negative feedback) in the leaves [134]. Such excess capacity of the photosynthetic system could be exploited through genetic manipulations by creating additional sink capacity [135–138]. The unloading of leaf sucrose is necessary to maximize photosynthesis, and sucrose loading into the phloem is stimulated by an increased sink demand [139]. Different growth rates of sinks and thus photosynthate demand were shown to influence photosynthetic rate [131]. The rapid use or unloading of photosynthates stimulates photosynthesis by avoiding negative feedback limitation of chloroplast activity due to the local

accumulation of sucrose [130,140]. It has been shown that photosynthesis will respond to altered sink demand in several crop species including soybean [135,141,142]. Increased yield potential can also be achieved through simultaneously increasing the capacity for both photoassimilation and sink strength [143–145]. To optimize the balance between source and sink throughout the life cycle of a crop is a challenging task for breeders and physiologists.

There is sufficient evidence to indicate that sink strength is a major regulator of photosynthetic activity in crop canopies [146,147]. Positive associations between sink size, canopy photosynthesis, and RUE have been shown for wheat [148] and sunflower [149]. The dwarfing genes in wheat have altered the source–sink balance (i.e., improvements in the sink strength), and this has resulted in improvements in RUE during the postanthesis phase of growth in many modern semidwarf wheat varieties [150–153]. This was further evident from a study contrasting old (tall) and modern (semidwarf) wheat in relation to RUE at various growth phases. This study showed that there were no major differences in RUE between modern cultivars and old cultivars during the preanthesis period, but during the postanthesis period, RUE of the modern wheats was substantially higher than RUE of the old cultivars [38,154]. The higher RUE during postanthesis in the modern cultivars appears to be entirely a sink-driven stimulation of canopy photosynthesis [38,144,155–157]. Positive relationships between grain yield and postanthesis RUE have been shown in wheat, which is a reflection of the influence of sink strength driving canopy photosynthesis in the modern semidwarf varieties. Thus, genetic increase in the competitiveness of the desired sink will allow the genetically increased photosynthesis to be used more effectively and also unlock the reserve photosynthetic potential that seems to exist in many crop species [133,158].

V. OPPORTUNITIES TO IMPROVE RUE IN CROPS

The theoretical assessments of the capacity of plants to convert intercepted radiation into biomass is close to 5% and could possibly be larger if the biological system is operating at maximum efficiency. The realized efficiencies of radiation use in the present agricultural production systems vary from 0.5 to 2.0 g/MJ of the intercepted radiation. Agricultural systems produce biomass ranging from 30 to 60 tons/ha/year under optimal conditions to less than 1 tons/ha/year under subsistence farming. As a fraction of the integrated solar constant, the efficiencies of agricultural production systems lie between 0.2 and 0.004, which is a factor of 50 between excellent production systems and subsistence level farming systems. There are many factors that influence the rate of carbon fixation and RUE in plants including photosynthetic metabolism (C_3 vs. C_4), canopy architecture and light distribution in the canopy, photosynthetic rates, photorespiration, photoinhibition, Rubisco specificity factor, and maintenance respiration. The interaction of climatic, edaphic, and environmental factors with a range of physiological, morphological, and phenological mechanisms will also modulate the realizable RUE in a given production environment. Agronomic management of nutrient and water can alleviate some of the environmental limitations found in nature and allow expression of the genetic potential of RUE. This is the case for many of the well-developed agricultural production systems, but often this increase is at the cost of relatively high inputs. Some researchers considered RUE as a very conservative feature in crop species that varied little with cultivars, species, or within the same photosynthesis group [2,83,155]. However, this view has recently been challenged as there are now a number of studies indicating that RUE is not a conservative feature and that a range of genetic and environmental factors influence RUE resulting in genetic/genotypic variation in many crops.

Because of the many factors involved, there are many options available to improve plant performance. Many of these options directly or indirectly influence the RUE of the crop. This effort can be directed either to improve the genetic yield potential of a target crop species or bridge the gap between realizable and potential levels of RUE. This section evaluates various components where genetic interventions are reasonably possible or theoretically plausible to improve the productivity of agricultural systems.

A. Improving the Light Distribution in the Canopy

A mature healthy crop may have three or more layers of leaves; that is, above each square meter of soil there will be the equivalent of $3\,m^2$ of leaves; this ratio of leaves to surface area is described as an LAI of 3. If the leaves are horizontal, the uppermost layer will intercept most of the direct light, about 10% may penetrate to the next layer, and 1% will penetrate to the layer below that. In most dicots, a major portion of the intercepted radiation is absorbed by the upper portion of the canopy, with only 10% to 20% of the radiation penetrating beyond about 2 LAI units [159]. The response of photosynthesis to solar radiation is often hyperbolic, where the photosynthesis is saturated often at one third of the natural light levels (Figure 29.1). When the sun is directly overhead, the PAR intercepted per unit leaf area by a horizontal leaf at the top of a plant canopy would be $900\,J/m^2/sec$, or about three times that required to saturate leaves for photosynthesis (Figure 29.1). Thus, about two thirds of the energy intercepted by the uppermost leaves is not useable by them for photosynthesis.

Once the crop canopy closes, nearly 90% of the intercepted radiation is intercepted by the top of the canopy, leaving the leaves that are located at the lower levels with insufficient radiation to be productive or even to sustain themselves. In addition, the upper portion of the canopy becomes light saturated (because of the light–radiation response relationship — see Figure 29.1), and a significant portion of the radiation that is intercepted does not contribute to photosynthesis; thus, it becomes a wasted resource for the crop [160]. Also, in many tropical environments, the high-intensity radiation that is intercepted by the upper surface of the canopy can contribute to photoinhibition [161–163]. This further reduces the photosynthetic capability of the crop canopy to fix carbon efficiently [164]. High levels of irradiance along with high temperatures are shown to cause metabolic imbalances [165], deleterious effects on thylakoid function [166], enhanced photoinhibition [167], and increased photorespiration [168,169].

One of the strategies to overcome the above constraints would be for the upper leaf layer to intercept a smaller fraction of light, allowing more light to reach the lower leaves, facilitating a more uniform distribution of the intercepted radiation across the canopy (i.e., optimizing the light distribution in the canopy) [170–175]. This could be achieved by a more vertical leaf angle, which would result in a reduction in the number of leaves that are light saturated, while allowing more radiation to penetrate into the deeper layers of the canopy. Given the appropriate morphology, this would lead to a more uniform contribution of the various layers of leaves to the overall photosynthesis. The amount of sunlit area at the bottom of the canopy would be increased, thus increasing the number of leaves receiving radiation at levels most efficient for photosynthesis [170,176,177]. Leaf orientation also influences the amount of light absorbed by altering both the level of reflectance and the available cross-sectional area [178,179]. Vertical orientation of the leaves will also facilitate better air movement within the canopy and create a microclimate that is not as favorable for many diseases and insect pests as horizontal and droopy leaf canopy types [180,181]. Several theoretical and computer simulation models have shown that an erect leaf angle is an essential characteristic of any model of canopy architecture producing high RUE values [182].

Genetic variability for canopy photosynthesis or RUE in wheat and winter cereals was associated with different patterns of radiation distribution within the canopy [155,173,176,183–185]. In summer crops such as rice, a positive effect of leaf erectness on RUE and yield has been consistently shown [172,186,187]. Field studies using vertical leaf orientation types have shown consistently that canopy photosynthesis and RUE were higher in canopy types with leaves having an upright angle than in those with horizontal or droopy leaves for a range of crop species that include sugar beet, wheat, maize, and rice [176,183,188–193]. The light was more evenly distributed in the canopy types with erect leaves and thus used more efficiently than the horizontal canopy types [191,194,195]. Using ^{14}C, it was demonstrated in barley that the canopy photosynthesis of the erect leaf type was higher than that of the horizontal droopy leaf variety (4.3 vs. 3.8 g $CO_2/m^2/h$). The traditional rice cultivar "Peta" has a high concentration of leaves near the top of the canopy, which results in rapid decay of light intensity on the leaves below the top of the canopy. This is in contrast to the modern cultivar IR-8, which has a high concentration of erect leaves near the center of the canopy and a more uniform distribution of light throughout the entire canopy than Peta; IR-8 also has higher values of RUE than the Peta variety. The advantages of erect leaf posture for photosynthetic efficiency and RUE were also demonstrated using

semidwarf wheat genotypes with erect leaf type vs. lax leaf habit [173].

The beneficial effects of erect leaf posture on photosynthetic efficiency and crop growth are only evident above a certain threshold level of LAI. On the basis of crop/canopy modeling studies, it was shown that erect leaf posture would be beneficial to improvements in RUE and productivity when LAI of a crop canopy reaches above 4 to 5 [196–198]. Nearly 80% of the light can be intercepted with an LAI of about 3.0 (Figure 29.9) in pigeonpea; further increases in LAI of up to 5 could only improve the light interception to 95%, thus as the LAI increases above 3.0, leaf orientation would play more of a role in the distribution of light across the LAI of the canopy (Figure 29.9). Thus, leaf angle and LAI should be considered together in determining the importance of leaf angle on the photosynthetic efficiency of crop canopies [170]. Also, when the ambient light exceeds that required for light saturation of the top leaves, canopy architecture and leaf angles can play a significant role in the utilization of that excess light.

The canopy light extinction coefficient (k) is determined by the LAI and leaf angle. The more acute the leaf angle, the lower is the k value of the canopy. Generally, a k value of 0.3 corresponds to a canopy with predominantly erect leaves, and a k value of 0.9 represents predominantly horizontal leaves [196,197]. In perennial rye grass with different growth habits, genetic stocks with erect leaves have a k value of 0.3, while horizontal leaf types have a k value of 0.7 [199]. Growth rates of six forage grasses in simulated

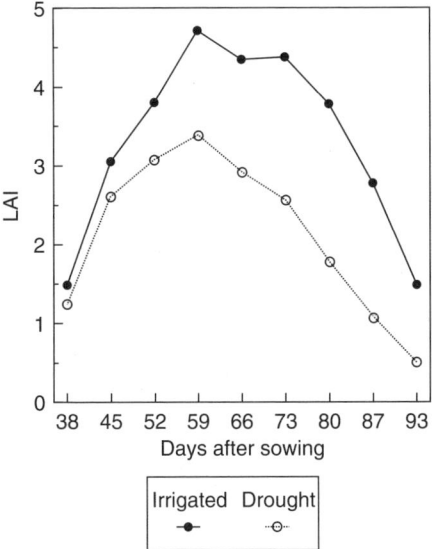 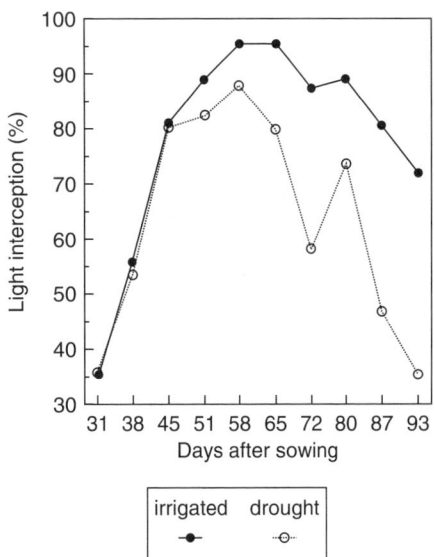

FIGURE 29.9 Changes in LAI and light interception of pigeonpea (mean of six genotypes) grown under irrigated and drought conditions using rainout shelters (From G.V. Subbarao and N.H. Nam, unpublished data.)

swards were shown to be strongly correlated with crop growth rate (CGR) and RUE with k values ranging from 0.3 to 0.9; the CGR of swards was nearly doubled when k was improved from 0.9 (horizontal orientation) to 0.3 (vertical orientation) [200]. Similar results were reported in winter wheat, where cultivars with low k values had a higher level of RUE than cultivars with high k values [155] (Figure 29.10).

Developing crops with erect leaf angles also have an interaction with N storage in leaves. Many of the high-yielding crops also require large amounts of N to be stored in their leaves before the reproductive (i.e., seed growth) growth phase. In order to retain sufficient leaf tissue to store adequate N, it is essential that even the lower leaves in the canopy receive sufficient light to remain functional [201]. For the leaves at the bottom of the canopy to receive the required minimum amount of light, it is necessary for the leaves higher in the canopy to be displayed at an erect angle. In addition to the advantage of erect leaves in increasing RUE, there is also the need for a larger functioning canopy to provide stored N for any increase in seed development [201].

Another advantage of vertical leaf orientation is that this type of canopy structure permits narrower planting rows, facilitating a higher plant density, which improves both LAI and RUE and results in a higher yield potential for the target crop. This concept has been demonstrated in maize genotypes where vertical leaf orientation permitted higher planting densities of 75,000 to 90,000 plants/ha compared to a normal planting density of 60,000 plants/ha of the

normal leaf types (i.e., horizontal leaf types). This higher planting density has given significantly higher biomass, grain yields, and RUE [191,195]. Genetic manipulation of leaf angle is not complex and is controlled by only two or three genes [202]. In wheat it was shown that improving the leaf angle could permit further gains in RUE over the current high-yielding agronomic types [202]. Using simulation studies, it was shown that by improving leaf angle further in rice genetic stocks where LAI exceeds 8, increases in RUE and yield potential are possible for high-radiation environments [203].

B. IMPROVEMENTS IN PHOTOSYNTHETIC PERFORMANCE IN CROPS

Leaf photosynthetic rate is important in determining the photosynthetic efficiency of the canopy, thus it is one of the major crop genetic factors that influence RUE [30,118,204]. Theoretical analyses have consistently indicated a dependence of RUE on leaf photosynthetic activity [30,205]. Between C_3 and C_4 photosynthetic pathways, photosynthetic efficiency has been known to differ and has been well established [95]. Also, within a photosynthetic group (i.e., C_3 or C_4), genetic variability in leaf photosynthetic rates has been reported by several researchers in many field crops that include rice, wheat, barley, maize, and soybean [157,206]. The relationship between photosynthesis and irradiance is such that there are two opportunities for the genetic modifications of this relationship to improve photosynthetic efficiency and RUE in a crop. There is room for potential improvement in efficiency at low levels of irradiance (apparent quantum yield) and also in the rate of photosynthesis at saturating irradiance (A_{max}).

1. Photosynthetic Efficiency at Low Levels of Irradiance

Most of the canopy photosynthesizes at nonsaturating light levels [207] (as a result of poor light penetration past the very top of the canopy). Thus, there appears to be a larger potential for improving RUE and production in leaves having low light intensities rather than focusing only on A_{max} at high light intensities [133,160]. The quantum yield of C_3 (*Triticum aestivum*) and C_4 (*Zea mays*) plants ranges from 0.054 to 0.059 mol CO_2/E. Also, very limited genetic variation exists in quantum yield of photosynthesis among the 22 crop species tested [133,208]. It appears that there is only a limited scope for improving quantum yield through genetic inventions [12,209]. The theoretical upper limit for the quantum yield is about 0.067 mol CO_2/E, and some C_4 plants of

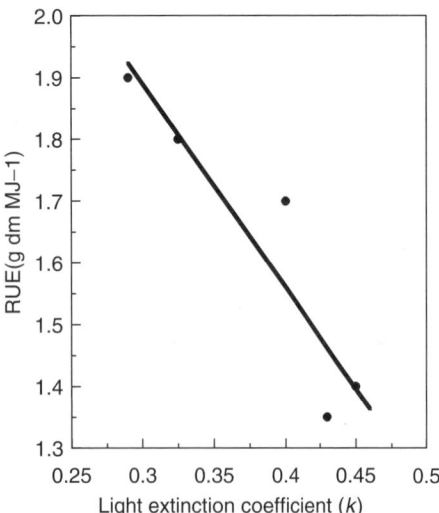

FIGURE 29.10 Relationship between canopy light extinction coefficient (k) and RUE (g dry matter per MJ PAR) in five cultivars of wheat. (From Green CF. *Field Crops Res.* 1989; 19:285–295. With permission.)

NADP–malic enzyme types were reported to have quantum yields close to the theoretical upper limit [210,211]. An alternative approach is to optimize the photosynthetic apparatus, such as adjusting N distribution in the canopy to make leaf photosynthesis efficient at different light intensities and thus increasing overall canopy photosynthetic efficiency [160]. In Lucerne, it was shown that the leaf N levels and chlorophyll $a{:}b$ ratios declined with depth in the canopy, which facilitates better capture of the limited light (found in the deeper layers of the canopy). This is accomplished by altering the investment of N in the chlorophyll associated with the light antennae, relative to the reaction centers. This is consistent with a lower total N to chlorophyll N ratio, reflecting a smaller investment in soluble protein associated with CO_2 fixation. Consequently, lower leaves have a reduced overall photosynthetic capacity under normal light, but are more efficient in light capture per unit N at the low light intensities experienced toward the bottom of the canopy [138]. Consequently, several crop models support the advantage of optimizing the vertical distribution of canopy nitrogen levels [177].

2. Selection for Photosynthesis at Saturating Irradiance (A_{max})

Direct measurement of canopy photosynthesis on multiple genotypes is costly, therefore crop physiologists have measured maximum leaf photosynthetic rate as a surrogate [160]. Also, very little genetic variation for photosynthetic rate at subsaturating light intensities has been reported [208,212]. Because of the above reasons, a major portion of the genetic improvement efforts in photosynthetic rates of crop species has been directed toward improving A_{max}. Genetic variation in A_{max} has been reported for many crop species including wheat [133,138,160,213–218], soybean [142,219], peas [220], and tall fescue [221,222]. In high-radiation environments such as in Israel, up to 50% of the genetic differences in leaf photosynthesis of C_3 crops like wheat can be realized in the measured improvements in canopy photosynthesis [133]. For soybean, it is estimated that about 40% of the improvements in the single-leaf photosynthesis (i.e., A_{max}) can be realized in the canopy photosynthesis [93].

C. Improvements in Photosynthetic Duration of the Canopy

Leaf area duration is one of the most critical factors in the light harvesting process of the canopy to determine the canopy photosynthetic efficiency integrated over the entire growth period of the crop. The stay-green trait is a plant attribute that has attracted attention from physiologists and breeders and has the potential to improve RUE during the reproductive phase of growth. The decline in RUE during the reproductive phase is partly triggered by the remobilization of carbon and nitrogen from leaves to the reproductive tissue. This triggers the start of canopy senescence, inducing a significant decrease in RUE. It has been noticed that genetic stocks that keep a photosynthetically productive and active canopy throughout the reproductive phase of development (i.e., the stay-green trait) maintain higher RUE. By genetically transferring the stay-green trait from *Avena sterilis* into the breeding lines of oat (*Avena sativa*), biomass production, RUE, and yielding ability in oat have been improved to 20% over the normal oat, and this has been largely attributed to improvements in the leaf area duration rather than other photosynthetic rates [223–225].

D. Can Photorespiration be Suppressed?

One of the ways to improve RUE is to reduce or suppress photorespiration; this increases quantum yield and stimulates net assimilation rates under both light limitation and light saturation environments [160]. Nearly 30% of the carbohydrate formed in C_3 photosynthesis can be lost via photorespiration, and this amount increases with an increase in temperature and could reach up to 50% in warm tropical environments or during hot summer weather in temperate climates [226]. The kinetic properties of Rubisco determine the partitioning of ribulose 1,2-biphosphate between carboxylation and oxygenation, and thus the amount of fixed carbon lost through photorespiration. The carboxylase and oxygenase reactions involve the competition of molecular O_2 and CO_2 for an activated enediol form of ribulose 1,2-biphosphate, which is generated at the active site [227,228]. There was no active site detected in the Rubisco for the substrates CO_2 or O_2. In the absence of a formal binding site on the enzyme for CO_2 and O_2, partitioning between the two reactions (i.e., carboxylation and oxygenation) (i.e., the Rubisco specificity factor) should be the same irrespective of the source from which this enzyme is isolated [228]. However, substantial variation in the Rubisco specificity factor was discovered for Rubisco isolated from a wide range of photosynthetic organisms [229–234] (Table 29.2).

The most primitive form of Rubisco (Rubisco form-I), isolated from prokaryotic photosynthetic organisms, contains two large subunits and has a higher specificity factor for oxygenation (i.e., most of the

TABLE 29.2
Specificity Factor of Rubisco from Various Species

Species	Substrate Specificity Factor
Rhodospirillum rubrum	9.0 ± 0.9
Anacystis nidulans	52 ± 4.2
Pastinaca lucida	84.7 ± 1.0
Hippocrepis balearica	94.4 ± 1.7
Medicago arborea L. ssp. *citrine*	96.4 ± 2.2
Triflium subterraneum L.	96.5 ± 4.9
Ceratonia siligua	98.1 ± 2.9
Triticum aestivum	100.0 ± 0.7
Chrysantehmum coranarium L.	106.6 ± 1.9
Lotus creticus L. ssp. *Cytisoides*	106.9 ± 3.5
Helleborus lividus Aiton ssp. *Corsicus*	107.5 ± 3.8
Maize	92 ± 6.6
Tobacco	89 ± 4.1
Sunflower	104 ± 4.1

Note: Based on Refs. [232,235].

substrate is directed toward the oxygenation process rather than the carboxylation process). This is in contrast to Rubico form-II, which is isolated from higher plants, which have both large and small subunits that differ in specificity toward CO_2 and O_2. These differences in specificity are thought to be the basis for the variation in Rubisco specificity among species and possibly genotypes of some species. This could open up possibilities for manipulating enzyme structure and engineering a superior carboxylase that would have a higher specificity for CO_2 and a lower specificity for O_2, thus improving the photosynthetic efficiency of the biological systems [227,231,234,235–238]. The cyanobacterial Rubisco is closely related to that of higher plants, but it has a lower specificity for CO_2, thus it will partition more substrate through oxygenation than carboxylation. Also, Rubisco from *Rhodospirillum rubrum* is structurally the simplest form of Rubisco, and the specificity factor is similar to that of Rubisco from cyanobacteria.

Among the C_3 plants studied, nearly 20% of the variation was in Rubisco specificity among species [232,239]. Rubisco from tobacco has the lowest specificity. Rubisco specific factor in wheat, sunflower, *Chrysanthemum coronarium, Lotus creticus*, and *Helleborus lividus* has higher levels of specificity than that of tobacco [232,239]. Rubisco specificity of some marine red algae is substantially higher (about 195 compared to 95 in wheat) than that reported for C_3 plants [233]. Thus, a search for more efficient Rubisco in crop species may be worthwhile, particularly in crop species that have adapted to or evolved in high-temperature environments. Rubisco specificity factor decreases as

temperature increases because of the inherent properties of this enzyme. Also, the relative solubility of CO_2 and O_2 favors the oxygenase reaction at high temperatures [240–242]. Because of the above reasons, it could be expected that more efficient forms of Rubisco may have evolved in plant/crop species that were evolved in high-temperature environments as a normal part of their adaptation [243]. Improvements in the Rubisco specificity factor (and the associated reductions in photorespiration) are widely believed to have significant impacts on yield under high production, as well as in more marginal environments. Also, molecular techniques may offer the possibility of genetically transforming wheat Rubisco from its current specificity (i.e., V_c/V_o) of 95 to a value of 195, which corresponds to that of the thermophilic alga (*Galderia partite*) [244]. If this were to be achieved in field crops, improvements in photosynthetic rates up to 20% from their current levels are predicted using biochemical models for CO_2 assimilation [245].

Another way of improving canopy photosynthesis is to optimize the composition of the photosynthetic apparatus, as well as N distribution, throughout the canopy, so that leaf photosynthesis is equally efficient throughout the canopy and at different light intensities. This phenomenon was investigated in Lucerne [138], where leaves showed a clear tendency for reduced total leaf N at greater depth in the canopy. In addition, chlorophyll *a:b* ratios declined with depth, indicating an increased ability to capture scarce light by an increased investment in chlorophyll associated with the light antennae, relative to the reaction centers. This was consistent with a lower total N to chlorophyll N ratio, reflecting a smaller investment in soluble protein associated with CO_2 fixation [138].

VI. IMPORTANCE OF RUE IN CROP PRODUCTIVITY AND YIELD POTENTIAL

RUE is a critical crop genetic component determining yield potential and stability in performance over a range of production environments [21]. Since there are a number of genetic and environmental factors that can potentially influence the crop's ability to utilize radiation efficiently for production, there are several options or strategies that can be deployed to improve RUE in crops. Crop yield can be considered as a function of

$$Y = RI \times RUE \times HI$$

where Y is the grain yield, RI is the intercepted radiation, HI is the harvest index, and RUE is the radiation use efficiency.

Because of its central role in evaluating the yield and productivity of crops, RUE is an integral feature of all crop models. A wide range of morphological (canopy attributes that determine the canopy architecture for optimum distribution of intercepted light) and biochemical traits and a number of environmental factors influence in varying degrees the photosynthetic performance of crop canopies, which in turn largely determines the RUE. Our research on pigeonpea has shown that RUE is a key index for evaluating the potential of dry matter production and grain yield under optimum to deficit water environments (Figure 29.11 and Figure 29.12). Osmotic adjustment, a key metabolic strategy for adapting to water deficits, was also found to be an important mechanism contributing to RUE under water deficits in pigeonpea as RUE was found to be linearly correlated with relative leaf water content in pigeonpea (Figure 29.7).

Several attempts have been made by earlier researchers to target genetic improvement in specific components that affect RUE, such as photosynthetic rates (A_{max}), Rubisco levels, and canopy attributes such as leaf angle. Because of the limitations in understanding the function of specific traits/mechanisms in improving grain yield under field conditions, it has been difficult to identify the specific genetic interventions that are responsible for the improved RUE in crops [246,247]. Nevertheless, there have been no systematic research efforts to directly improve the RUE in any of the crops to our knowledge. As explained earlier, RUE is an integrated crop attribute where a

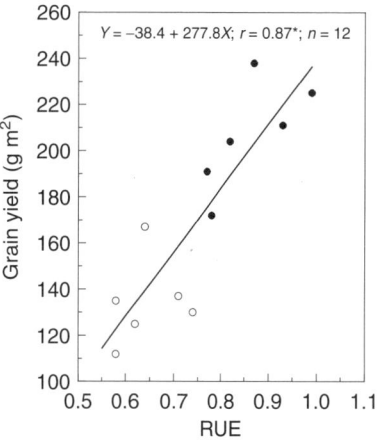

FIGURE 29.12 Relationship between RUE (g dry matter per MJ PAR) and grain yield (g/m^2) in six pigeonpea genotypes under irrigated (closed circles) and drought conditions (open circles). (From Nam NH, Subbarao GV, Chauhan YS, Johansen C. *Crop Sci.* 1998; 38:955–961.)

number of morphological and biochemical traits contribute to determine the observed phenotype. Large-scale systematic evaluation of genetic stocks for RUE has been difficult to undertake from a breeder's perspective. Presently, the only practical way to deal with this issue is to target genetic improvements on specific components with the assumption that they would have a measurable effect on the total phenotype under the right environmental conditions. However, the new generation of growth chambers presently available may make it feasible to evaluate the RUE of the total canopy for at least the final plant selections.

Despite several decades of research on RUE, this important crop attribute is perhaps one of the least understood phenomena in crop physiology. This is underscored by the fact that in many crop models, RUE is considered as a constant for a given crop species (i.e., generally a crop-specific coefficient) with little or no consideration as to variation among genotypes, or changes in RUE during the different growing phases of the crop. The various nutritional and environmental factors that affect RUE are seldom taken into consideration in many of the crop models [21]. Thus, a better understanding and appreciation of RUE are needed before genetic and management strategies to improve RUE in crops as a means of improving their yield potential and stability become routine.

It is rather surprising to see that there has been very little genetic improvement in RUE of the major food crops in the last four decades of breeding. During this same time period, breeding has resulted

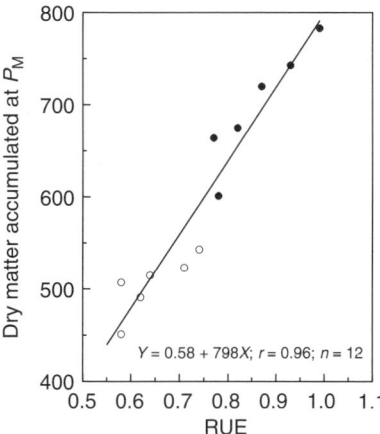

FIGURE 29.11 Relationship between RUE (g dry matter per MJ PAR) and dry matter accumulated at physiological maturity (g dry matter per m^2) in six pigeonpea genotypes under irrigated (closed circles) and drought conditions (open circles). (From Nam NH, Subbarao GV, Chauhan YS, Johansen C. *Crop Sci.* 1998; 38:955–961. With permission.)

in nearly doubling the yield of crops, but with relatively no change in crop RUE [154,157,206]. This has been indicated by comparative studies involving wheat varieties that were released between 1960 and 1990, which showed very little change in RUE [38,77,154,155,206,248]. However, the flag leaf photosynthetic rates, stomatal conductance, and canopy temperature depression have been improved substantially in some of the new varieties of wheat when compared to the older varieties (i.e., those released in the 1960s) [217]. The canopy photosynthetic rates and RUE of the modern wheat varieties appear to have been higher in the semidwarf varieties of wheat postanthesis, but not during the preanthesis period of growth; this change is largely driven by the improved sink size of the new varieties [218]. Nevertheless, some of the canopy attributes in the modern high-yielding cultivars of wheat have been improved, where the light penetration into the interior of the canopies is much higher in some of the semidwarf varieties than in the older taller varieties [249].

Recent yield improvements have been achieved largely through improved partitioning of crop biomass into grain [206,248,250–252]. The HI of many field crops is approaching very high levels (about 60%) [250]; further yield improvements most likely will have to come from improving the total biomass production, and RUE most likely will have to be improved and maintained over the entire season to accomplish this task [250,253]. It has been shown that canopy attributes such as leaf angle and leaf arrangement are some of the key plant attributes that need to be improved for better light distribution. This type of modification is an essential feature for the new ideotypes that are proposed by the breeders and physiologists to develop new plant types for rice, wheat, and maize to break the current yield barriers. It has been argued that to improve the present levels of grain yield in rice of about 10 to 15 tons/ha, the canopy architecture needs to be modified substantially to utilize the light more efficiently for the production of biomass [187,254–256]. In order to realize the full genetic potential of an increased RUE (including the reserve photosynthetic capacity of the canopy) [158], sink size needs to be further improved in many crops. This is in addition to increasing the photosynthetic duration of the canopy by introducing novel traits such as stay-green characteristics of the canopy [254,257,258]. Increased utilization of stem nutrient reserves is another important attribute of yield determination and stability that needs to be improved in order to exploit any improved RUE for higher grain yields [258].

VII. CHALLENGES AND CONSTRAINTS TO IMPROVE RUE

Improvements in a crop RUE largely depend on the genetic manipulation of the overall photosynthetic output of the canopy [177]. Improving leaf photosynthesis and other biochemical attributes can improve the photosynthetic performance of crops, and thus their biomass production and possibly RUE, often considered for improvement in a number of crop species [147,259]. However, improving photosynthetic rates of specific leaves has not resulted in improvements in biomass in a number of crops, as individual leaf photosynthetic rate is only one of the attributes that determine canopy photosynthetic efficiency and RUE [214,260,261]. Often leaf photosynthetic rates are negatively associated with leaf size, with no benefits to the overall canopy photosynthetic efficiency [147,262]. Similarly, Rubisco levels are often negatively correlated with leaf expansion rate and leaf size [262]. Nevertheless, a number of reports indicate that genetic variation in leaf photosynthesis is independent of specific leaf weight; thus, improvements in specific leaf weight (often a function of increased Rubisco levels) could independently be genetically altered [221,222,263].

Evaluating a large number of genetic stocks for canopy photosynthetic rates under field conditions is not presently feasible because of the lack of techniques that are suitable for large-scale evaluation of genetic stocks [264–266]. Because of these inherent limitations associated with evaluating RUE or canopy photosynthetic efficiency for a large number of genetic stocks, research specifically aimed at improving RUE has not been undertaken so far, to our knowledge. Other biochemical attributes such as improving the Rubisco specificity factor may have potential in future crop improvement efforts. Also, some of the new molecular tools that can potentially modify Rubisco could have a major impact on canopy photosynthetic efficiency in crops. Of the various attributes that can be genetically manipulated to improve RUE, the light extinction coefficient (k), largely determined by the vertical orientation of the leaf angle and leaf arrangement, would perhaps be one of the most practical ways of improving RUE in many of the current crop varieties. Despite the limited amount of definitive evidence to show that improving leaf orientation would improve canopy photosynthetic efficiency and RUE of crops, there are sufficient reasons to consider this trait to be important and potentially a practical way to improve crop RUE. An evaluation of the value of vertical leaf orientation needs to be considered along with higher plant density and in high-radiation environments.

VIII. CONCLUDING REMARKS AND FUTURE OUTLOOK

Net canopy photosynthesis is a function of many interacting physiological attributes that involve both photosynthesis and metabolism (sink strength). There are many individual components involved such as (a) photosynthetic metabolism, (b) canopy structure, (c) sink strength, (d) rooting attributes that supply nutrients and water, (e) respiration costs, (f) buffering of environmental fluxes, and (g) tolerance mechanisms for water and nutrient stress (particularly water and nitrogen), all of which must function together to make up the overall structure necessary for the utilization of light. Conceptually, improvements in a crop's RUE can be through the genetic manipulation of any of the traits/characters that influence the above processes as their interactions determine net assimilation rate and thus RUE. As mentioned earlier, the last 85 years (since the 1920s or so) of wheat breeding have not resulted in any significant improvements in RUE in the modern semidwarf wheat varieties when compared to the traditional tall land races [267]. However, improvements in the RUE of some of the modern high-yielding rice varieties indicate that breeding can improve RUE through changes in some of these physiological characters that contributed toward RUE [175].

Canopy attributes such as vertical leaf orientation would have a large impact on the canopy photosynthesis if they could be introduced into agronomically elite materials. However, demonstrating the unequivocal beneficial effects of leaf orientation on yield potential during the early stages of breeding is still a challenge to breeders and physiologists. This is largely due to the existence of allometric relationships between leaf erectness and smaller leaves, spikes and stems, which are associated with agronomicaly poor phenotypes [192]. Also, genetic stocks with erect canopy structure require different environmental conditions (such as narrowly spaced planting) to take advantage of any such improvements in canopy structure. It is very difficult to take into consideration all possible environmental conditions during the early screening and evaluation of these materials [192,268].

More highly focused efforts are needed to develop the elite genetic stocks containing the desirable components and traits that shape RUE. Such efforts need to be carefully planned and will require a long-term commitment of resources, along with the joint involvement of genetists and crop physiologists in the breeding efforts. As mentioned earlier the best potential for further yield improvements of major food crops (such as wheat, rice, and maize) seems to be in improving RUE, as it appears to be the only viable option for major improvements in biomass production for many crops. This also puts a high priority on the various physiological mechanisms/traits that directly or indirectly influence RUE (such as tolerance to water or nutrient stress) and contribute to adaptation to marginal (stressful) environments. Improvements in these factors would also contribute to yield stability across a range of production environments. Currently, there is only limited understanding of the underlying reasons for the variation of RUE, which is often observed across a range of production environments, or of the functioning of the physiological mechanisms that contribute to the improvement of RUE in these marginal production environments. Although RUE is not a physical parameter, it is a sensitive biological index that basically integrates the overall efficiency of the plant, providing key information about the production potential of the plant.

Breaking the current yield barriers to production for some of the major food crops will require modifications in crop canopy architecture, sink strength, photosynthetic rates, tolerance to nutrient and water stress, and increased photosynthetic duration. Improvements in these traits taken either together or individually will contribute to improvements in RUE of crops [187,254,258]. Genetic interventions are possible at various biological levels to improve RUE. Introduction of the C_4 photosynthetic pathway into major food crops such as rice and wheat, which have the C_3 pathway, is one such avenue that has been shown to be conceptually possible and technically feasible and can have a major upward impact on RUE of these crops. Such interventions will hopefully lead to the development of new plant types with traits that address the underlying phenomenon responsible for RUE. Crop modeling, biotechnology, and physiological breeding will become increasingly important for targeting, evaluating, and incorporating desirable traits into new plant ideotypes of the major food crops, which will be able to use intercepted radiation more efficiently to produce greater amounts of biomass than the current crop varieties.

REFERENCES

1. Monteith JL. Solar radiation and productivity in tropical ecosystems. *J. Appl. Ecol.* 1972; 9:747–766.
2. Monteith JL. Climate and the efficiency of crop production in Britain. *Philos. Trans. R. Soc. Lond. Ser. B* 1977; 281:277–294.
3. Gallagher JN, Biscoe PV. Radiation absorption, growth and yield of cereals. *J. Agric. Sci. Camb.* 1978; 91:47–60.
4. Moon P. Proposed standard radiation curves. *J. Franklin Inst.* 1940; 230:583–617.

5. Monteith JL. Light interception and radiative exchange in crop stands. In: Eastin JD, ed. *Physiological Aspects of Crop Yield*. Madison, WI: American Society of Agronomy, 1970:89–109.

6. Hill R. Bioenergetics of photosynthesis at the chloroplast and cellular level. IBP/UNESCO Meeting on "*Productivity of Tropical Ecosystems*," Makerere University, Uganda, 1970.

7. Westlake DF. Comparison of plant productivity. *Biol. Rev.* 1963; 38:385–425.

8. Hadley EB, Kieckhefer BJ. Productivity of two prairie grasses in relation to fire frequency. *Ecology* 1963; 44:389–395.

9. Wiegert RG, Evans FC. Primary production and the disappearance of dead vegetation on an old field in southeaster Michigan. *Ecology* 1964; 45:49–62.

10. Chrispeels MJ, Sadava DE. *Plants, Genes and Crop Biotechnology*. 2nd ed. Boston, MA: Jones and Bartlett, 2003.

11. Watson DJ, Hayashi K. Photosynthetic respiratory components of the net assimilation of sugar beet and barley. *New Phytol.* 1965; 64:38–47.

12. Ludlow MM, Wilson GL. Studies on the productivity of tropical pasture plants. *Aust. J. Agric. Res.* 1968; 19:35–45.

13. Monteith JL. Analysis of the photosynthesis and respiration of field crops. *UNESCO Symposium on Functioning of Terrestrial Ecosystems*, 1968:349–356.

14. Seybold A. Uber die optischen Eigenschaften der Laubblatter IV. *Planta* 1933; 21:251.

15. Goudriaan J, van Laar HH. *Modelling Potential Crop Growth Processes: A Textbook with Exercises. Current Issues in Production Ecology*. Vol. 2. Dordrecht: Kluwer Academic Publishers, p 238, 1994.

16. Cooper JP. Potential production and energy conversion in temperate and tropical grasses. *Herb. Abstr.* 1970; 40:1–3.

17. Loomis RS, Amthor JS. Limits to yield revisited. In: Reynolds MP, Rajaram S, McNab A, eds. *Increasing Yield Potential in Wheat: Braking the Barriers*. Mexico: CIMMYT, 1996:76–89.

18. Marshall B, Willey RW. Radiation interception and growth in an intercrop of pearl millet/groundnut. *Field Crops Res.* 1983; 7:141–160.

19. Monteith JL. Reassessment of maximum growth rates for C_3 and C_4 crops. *Expl. Agric.* 1978; 14:1–5.

20. Muchow RC, Sinclair TR. Nitrogen response of leaf photosynthesis and canopy radiation use efficiency in field-grown maize and sorghum. *Crop Sci.* 1994; 34:721–727.

21. Lecoeur J, Ney B. Change with time in potential radiation use efficiency in field pea. *Eur. J. Agron.* 2003; 19:91–105.

22. Sinclair TR, de Wit CT. Photosynthate and nitrogen requirements for seed production by various crops. *Science* 1975; 189:565–567.

23. Penning De Vries FWT, van Laar HH, Chardon MCM. Bioenergetics of growth of seeds, fruits and storage organs. In: *Proceedings of the Symposium of Potential Productivity of Field Crops under Different Environments*. Los Banos, The Philippines: The International Rice Research Institute, 1983:37–59.

24. Bell MJ, Wright GC, Hammer GL. Night temperature affects radiation-use efficiency in peanut. *Crop Sci.* 1992; 32:1329–1335.

25. Wright GC, Bell MJ, Hammer GL. Leaf nitrogen content and minimum temperature interactions affect radiation-use efficiency in peanut. *Crop Sci.* 1993; 33:476–481.

26. Hall AJ, Connor DJ, Sadras VO. Radiation use efficiency of sunflower crops: effects of specific leaf nitrogen and ontogeny. *Field Crops Res.* 1995; 41:65–77.

27. Flenet F, Kiniry JR. Efficiency of biomass accumulation by sunflower as affected by glucose requirement of biosynthesis and leaf nitrogen content. *Field Crops Res.* 1995; 44:119–127.

28. Muchow RC, Robertson MJ, Pengelly BC. Radiation-use efficiency of soybean, mungbean and cowpea under different environmental conditions. *Field Crops Res.* 1993; 32:1–16.

29. Meek DW, Hatfield JL, Howell TA, Ido SB, Reginato RJ. A generalized relationship between photosynthetically active radiation and solar radiation. *Agron. J.* 1984; 76:939–945.

30. Sinclair TR, Muchow RC. Radiation use efficiency. *Adv. Agron.* 1999; 65:215–265.

31. Monteith JL, Gregory PJ, Marshall B, Ong CK, Saffell RA, Squire GR. Physical measurements in crop physiology. 1. Growth and gas exchange. *Expl. Agric.* 1981; 17:113–126.

32. Muchow RC, Davis R. Effect of nitrogen supply on the comparative productivity of maize and sorghum in a semi-arid tropical environment. II. Radiation interception and biomass accumulation. *Field Crops Res.* 1988; 18:17–30.

33. Gosse G, Varlet-Grancher C, Bonhomme R, Chartier M, Allirand JM, Lemaire G. Maximum dry matter production and solar radiation intercepted by a canopy. *Agronomie* 1986; 6:47–56.

34. Muchow RC, Coates DB. An analysis of the environmental limitation to yield of irrigated grain sorghum during the dry season in tropical Australia using a radiation interception model. *Aust. J. Agric. Res.* 1986; 37:135–148.

35. Garcia R, Kanemasu ET, Blad BL, Bauer A, Hatfield JL, Major DJ, Reginato RJ, Hubbard KG. Interception and use efficiency of light in winter wheat under different nitrogen regimes. *Agric. For. Meteorol.* 1988; 44:175–186.

36. Kiniry JR, Jones CA, O'Toole JC, Blanchet R, Cabelguenne M, Spanel DA. Radiation-use efficiency in biomass accumulation prior to grain-filling for five grain-crop species. *Field Crops Res.* 1989; 20: 51–64.

37. Gregory PJ, Eastham J. Growth of shoots and roots, and interception of radiation by wheat and lupin crops on a shallow, duplex soil in response to time of sowing. *Aust. J. Agric. Res.* 1996; 47:427–447.

38. Calderini DF, Dreccer MF, Slafer GA. Consequences of breeding on biomass, radiation interception and

radiation-use efficiency in wheat. *Field Crops Res.* 1997; 52:271–281.

39. Gregory PJ, Tennant D, Belford RK. Root and shoot growth, and water and light use efficiency of barley and wheat crops grown on a shallow duplex soil in a Mediterranean-type environment. *Aust. J. Agric. Res.* 1992; 43:555–573.

40. Goyne PJ, Milroy SP, Lilley JM, Hare JM. Radiation interception, radiation use efficiency and growth of barley cultivars. *Aust. J. Agric. Res.* 1993; 44: 1351–1366.

41. Williams WA, Loomis RS, Lepley CR. Vegetative growth of corn as affected by population density. I. Productivity in relation interception of solar radiation. *Crop Sci.* 1965; 5:211–215.

42. Sivakumar MVK, Virmani SM. Crop productivity in relation to interception of photosynthetically active radiation. *Agric. For. Meteorol.* 1984; 31:131–141.

43. Muchow RC. Effect of nitrogen supply on the comparative productivity of maize and sorghum in a semiarid tropical environment. 1. Leaf growth and leaf nitrogen. *Field Crops Res.* 1988; 18:1–16.

44. Tollenaar M, Bruulsema TW. Efficiency of maize dry matter production during periods of complete leaf area expansion. *Agron. J.* 1988; 80:580–585.

45. Otegui ME, Nicolini MG, Ruiz RA, Dodds PA. Sowing date effects on grain yield components for different maize genotypes. *Agron. J.* 1995; 87:29–33.

46. Westgate ME, Forcella F, Reicosky DC, Somsen J. Rapid canopy closure for maize production in the northern US corn belt: radiation-use efficiency and grain yield. *Field Crops Res.* 1997; 47:249–258.

47. Muchow RC, Spillman MF, Wood AW, Thomas MR. Radiation interception and biomass accumulation in a sugarcane crop grown under irrigated tropical conditions. *Aust. J. Agric. Res.* 1994; 45:37–49.

48. Muchow RC, Evensen CI, Osgood RV, Robertson MJ. Yield accumulation in irrigated sugarcane: II. Utilization of intercepted radiation. *Agron. J.* 1997; 89:646–652.

49. Robertson MJ, Wood AW, Muchow RC. Growth of sugarcane under high input conditions in tropical Australia. I. Radiation use, biomass accumulation and partitioning. *Field Crops Res.* 1996; 48:11–25.

50. Horie T, Sakuratani T. Studies on crop-weather relationship model in rice. (1) Relation between absorbed solar radiation by the crop and the dry matter production. *Jpn. Agric. Meteorol.* 1985; 40:331–342.

51. Inthapan P, Fukai S. 1988. Growth and yield of rice cultivars under sprinkler irrigation in South Eastern Queensland. 2. Comparison with maize and grain sorghum under wet and dry conditions. *Aust. J. Exp. Agric.* 1988; 28:243–248.

52. Allen EJ, Scott RK. An analysis of growth of the potato crop. *J. Agric. Sci. Camb.* 1980; 94:583–606.

53. Jefferies RA, Mackerron DKL. Radiation interception and growth of irrigated and droughted potato (*Solanum tuberosum*). *Field Crops Res.* 1989; 22:101–112.

54. Trapani N, Hall AJ, Sadras VO, Vilella F. Ontogenetic changes in radiation use efficiency of sunflower

(*Helianthus annuus* L.) crops. *Field Crops Res.* 1992; 29:301–316.

55. Shibles RM, Weber CR. Interception of solar radiation and dry matter production by various soybean planting patterns. *Crop Sci.* 1966; 6:55–59.

56. Nakaseko K, Gotoh K. Comparative studies on dry matter production, plant type and productivity in soybean, azuki bean and kidney bean. VII. An analysis of the productivity among three crops on the basis of radiation absorption and its efficiency for dry matter accumulation. *Jpn. J. Crop Sci.* 1983; 52:49–58.

57. Unsworth MH, Lesser VM, Heagle AS. Radiation interception and the growth of soybeans exposed to ozone in open-top field chambers. *J. Appl. Ecol.* 1984; 21:1059–1079.

58. Muchow RC. An analysis of the effects of water deficits on grain legumes grown in a semi-arid tropical environment in terms of radiation interception and its efficiency of use. *Field Crops Res.* 1985; 11:309–323.

59. Sinclair TR. Water and nitrogen limitations in soybean grain production. 1. Model development. *Field Crop Res.* 1986; 15:125–141.

60. Sinclair TR, Shiraiwa T. Soybean radiation-use efficiency as influenced by nonuniform specific leaf nitrogen distribution and diffuse radiation. *Crop Sci.* 1993; 33:808–812.

61. Rochette P, Desjardins RL, Pattey E, Lessard R. Crop net carbon dioxide exchange rate and radiation use efficiency in soybean. *Agron. J.* 1995; 87:22–28.

62. Fasheun A, Dennett MD. 1982. Interception of radiation and growth efficiency in field beans (*Vicia faba* L.). *Agric. Meteorol.* 1982; 26:221–229.

63. Silim SN, Saxena MC. Comparative performance of some faba bean (*Vicia faba*) cultivars of contrasting plant types. 2. Growth and development in relation to yield. *J. Agric. Sci. Camb.* 1992; 118:333–342.

64. Muchow RC, Charles-Edwards DA. An analysis of the growth of mung beans at a range of plant densities in tropical Australia. I. Dry matter production. *Aust. J. Agric. Res.* 1982; 33:41–51.

65. Bell MJ, Muchow RC, Wilson GL. The effect of plant population on peanuts (*Arachis hypogaea*) in a monsoonal tropical environment. *Field Crops Res.* 1987; 17:91–107.

66. Stirling CM, Williams JH, Black CR, Ong CK. The effect of timing of shade on development, dry matter production and light-use efficiency in groundnut (*Arachis hypogaea* L.) under field conditions. *Aust. J. Agric. Res.* 1990; 41:633–644.

67. Bell MJ, Roy RC, Tollenaar M, Michaels TE. Importance of variation in chilling tolerance for peanut genotypic adaptation to cool, short-season environments. *Crop Sci.* 1994; 34:1030–1039.

68. Hughes G, Keatinge JDH, Scott SP. Pigeonpea as a dry season crop in Trinidad, West Indies. II. Interception and utilization of solar radiation. *Trop. Agric. Trinidad* 1981; 58:191–199.

69. Nam NH, Subbarao GV, Chauhan YS, Johansen C. Importance of canopy attributes in determining dry

matter accumulation of pigeonpea under contrasting moisture regimes. *Crop Sci.* 1998; 38:955–961.

70. Subbarao GV, Nam NH, Chauhan YS, Johansen C. Osmotic adjustment, water relations and carbohydrate remobilization in pigeonpea under water deficits. *J. Plant Physiol.* 2000; 157:651–659.

71. McKenzie BA, Hill GD. Intercepted radiation and yield of lentils (*Lens culinaris*) in Canterbury, New Zealand. *J. Agric. Sci. Camb.* 1991; 117:339–346.

72. Singh P, Sri Rama YV. Influence of water-deficit on transpiration and radiation use-efficiency of chickpea (*Cicer arietinum L.*). *Agric. For. Meteorol.* 1989; 48: 317–330.

73. Rizzalli RH, Villalobos FJ, Orgaz F. Radiation interception, radiation-use efficiency and dry matter partitioning in garlic (*Allium sativum L.*). *Eur. J. Agron.* 2002; 18:33–43.

74. Squire GR. *The Physiology of Tropical Crop Production.* Wallingford, U.K.: CAB International, 1990.

75. Hammer GL, Vanderlip RL. Genotype-by-environment interaction in grain sorghum. I. Effects of temperature on radiation use efficiency. *Crop Sci.* 1989; 29:370–376.

76. Burstall L, Harris PM. The physiological basis for mixing varieties and seed 'ages' in potato crops. *J. Agric. Sci. Camb.* 1986; 106:411–418.

77. Yunusa IAM, Siddique KHM, Belford RK, Karimi MM. Effect of canopy structure on efficiency of radiation interception and use in spring wheat cultivars during the pre-anthesis period in a Mediterranean-type environment. *Field Crops Res.* 1993; 35:113–122.

78. Leach GJ, Beech DF. Response of chickpea accessions to row spacing and plant density on a Vertisol on the Darling Downs, south-eastern Queensland. 2. Radiation interception and water use. *Aust. J. Exp. Agric.* 1988; 28:377–383.

79. Ritchie JT, Otter S. Description and Performance of CERES-Wheat: A User-Oriented Wheat Yield Model. Vol. 38. Washington, DC: U.S. Department of Agriculture, ARS, 1985:159–175.

80. Sharpley AN, Williams JR. EPIC-Erosion/Productivity Impact Calculator: 1. Model Documentation. U.S. Department of Agriculture Technical Bulletin No. 1768, 1990.

81. Sinclair TR, Horie T. Leaf nitrogen, photosynthesis, and crop radiation use efficiency: a review. *Crop Sci.* 1989; 29:90–98.

82. Kiniry JR, Landivar JA, Witt M, Gerik TJ, Cavero J, Wade LJ. Radiation-use efficiency response to vapor pressure deficit for maize and sorghum. *Field Crop Res.* 1998; 56:265–270.

83. Monteith JL, Elston J. Performance and productivity of foliage in the field. In: Dale JE, Milthorpe FL, eds. *The Growth and Functioning of Leaves.* London: Cambridge University Press, 1983:449–518.

84. Muchow RC. Comparative productivity of maize, sorghum and pearl millet in a semiarid tropical environment. 1. Yield potential. *Field Crops Res.* 1989; 20:191–205.

85. Muchow RC. Effect of nitrogen on yield determination in irrigated maize in tropical and subtropical environments. *Field Crops Res.* 1994; 38:1–13.

86. Littleton EJ, Dennett MD, Monteith JL, Elston J. The growth and development of cowpeas (Vigna unguiculata) under tropical field conditions. 2. Accumulation and parition of dry weight. *J. Agric. Sci. Camb.* 1979; 93:309–320.

87. Wittenbach VA, Ackerson RC, Giaquinta RT, Hebert RR. Changes in photosynthesis, ribulose biphosphate carboxylase, proteolytic activity, and ultrastructure of soybean leaves during senescence. *Crop Sci.* 1980; 20:225–231.

88. Egli DB, Leggett JE, Duncan WG. Influence of N stress on leaf senescence and N redistribution in soybeans. *Agron. J.* 1978; 70:43–47.

89. Zeiher C, Egli DB, Leggett JE, Reicosky DA. Cultivar differences in nitrogen redistribution in soybeans. *Agron. J.* 1982; 74:375–379.

90. Preeda B-L, Egli DB, Leggett JE. Leaf N and photosynthesis during reproductive growth in soybeans. *Crop Sci.* 1983; 23:617–620.

91. Boote KJ, Gallaher RN, Robertson WK, Hinson K, Hammond LC. Effect of foliar fertilization on photosynthesis, leaf nutrition, and yield of soybeans. *Agron. J.* 1978; 70:787–791.

92. Sinclair TR. Leaf CER from post-flowering to senescence of field-grown soybean cultivars. *Crop Sci.* 1980; 20:196–200.

93. Sinclair TR, de Wit CT. Analysis of the carbon and nitrogen limitations to soybean yields. *Agron. J.* 1976; 68:319–324.

94. Hammer GL, Wright GC. A theoretical analysis of nitrogen and radiation effects on radiation use efficiency in peanut. *Aust. J. Agric. Res.* 1994; 45:575–589.

95. Brown RH. A difference in N use efficiency in C_3 and C_4 plants and its implications in adaptation and evolution. *Crop Sci.* 1978; 18:93–98.

96. Schmitt MR, Edwards GE. Photosynthetic capacity and nitrogen use efficiency of maize, wheat, and rice: a comparison between C_3 and C_4 photosynthesis. *J. Exp. Bot.* 1981; 128:459–466.

97. Evans JR. Nitrogen and photosynthesis in the flag leaf of wheat (*Triticum aestivum L.*). *Plant Physiol.* 1983; 72:297–302.

98. Evans JR. Photosynthesis and nitrogen relationships in leaves of C_3 plants. *Oecologia* 1989; 78:9–19.

99. Sage RF, Pearcy RW. The nitrogen use efficiency of C_3 and C_4 plants. I. Leaf nitrogen, growth and biomass partitioning in *Chenopodium album L.* and *Amaranthus reflexus.* *Plant Physiol.* 1987; 84:954–958.

100. Evans JT, Terashima I. Photosynthetic characteristics of spinach leaves grown with different nitrogen treatments. *Plant Cell Physiol.* 1988; 29:157–165.

101. Terashima I, Evans JR. Effects of light and nitrogen nutrition on the organization of the photosynthetic apparatus in spinach. *Plant Cell Physiol.* 1988; 29:143–155.

102. Marshall B, Vos J. The relation between the nitrogen concentration and photosynthetic capacity of potato

(*Solanum tuberosum L.*) leaves. *Ann. Bot.* 1991; 68: 33–39.

103. Conner DJ, Hall AJ, Sadras VO. Effect of nitrogen content on the photosynthetic characteristics of sunflower leaves. *Aust. J. Plant Physiol.* 1993; 20:251–263.

104. Giminez C, Connor DJ, Rueda F. Canopy development, photosynthesis and radiation use efficiency in sunflower in response to nitrogen. *Field Crops Res.* 1994; 38:15–27.

105. Anten NPR, Schieving F, Werger MJA. Patterns of light and nitrogen distribution in relation to whole canopy carbon gain in C_3 and C_4 mono- and dicotyledonous species. *Oecologia* 1995; 101:504–513.

106. Peng S, Cassman KG, Kropff MJ. Relationship between leaf photosynthesis and nitrogen content of field-grown rice in the tropics. *Crop Sci.* 1995; 35:1627–1630.

107. Vos J, Van Der Putten PEL. Effect of nitrogen supply on leaf growth, leaf nitrogen economy and photosynthetic capacity of potato. *Field Crops Res.* 1998; 59: 63–72.

108. Meinzer FC, Zhu J. Nitrogen stress reduces the efficiency of the C_4 CO_2 concentrating system and, therefore, the quantum yield, in *Saccharum* (sugarcane) species. *J. Exp. Bot.* 1998; 49:1227–1234.

109. Henley WJ, Levavasseur G, Franklin LA, Osmond CB, Ramus J. Photoacclimation and photoinhibition in *Ulva rotundata* as influenced by nitrogen availability. *Planta* 1991; 184:235–243.

110. Green CF. Nitrogen nutrition and wheat in growth in relation to absorbed solar radiation. *Agric. For. Meteorol.* 1987; 41:207–248.

111. Sinclair TR. Nitrogen influence on the physiology of crop yield. In: Rabbinge R, Goudriaan J, van Keulen H, Penning de Vries T, van Laar HH, eds. *Theoretical Production Ecology: Reflections and Prospects*. Wageningen: PUDOC, 1990:41–55.

112. Muchow RC. Effects of water and nitrogen supply on radiation interception and biomass accumulation of kenaf (*Hibiscus cannabinus*) in a semi-arid tropical environment. *Field Crops Res.* 1992; 28:281–293.

113. Bange MP, Hammer GL, Rickert KG. Effect of specific leaf nitrogen on radiation use efficiency and growth of sunflower. *Crop Sci.* 1997; 37:1201–1207.

114. Bange MP, Hammer GL, Rickert KG. Effect of radiation environment on radiation use efficiency and growth of sunflower. *Crop Sci.* 1997; 37:1208–1214.

115. Brown RH. Growth of C_3 and C_4 grasses under low N levels. *Crop Sci.* 1985; 25:954–957.

116. Sinclair TR, Vadez V. 2002. Physiological traits for crop yield improvement in low N and P environments. *Plant Soil* 2002; 245:1–15.

117. Shiraiwa T, Sinclair TR. Distribution of nitrogen among leaves in soybean canopies. *Crop Sci.* 1993; 33:804–808.

118. Field C, Mooney HA. Leaf age and seasonal effects on light, water and nitrogen use efficiency in a California shrub. *Oecologia* 1983; 56:348–355.

119. Van Keulen H, Goudriaan J, Seligman NG. Modelling the effects of nitrogen on canopy development and crop growth. In: Russell G, Marshall LB, Jarvis PG, eds. *Plant Canopies: Their Growth, Form and Function.* Cambridge: Cambridge University Press, 1989:83–104.

120. Hirose T, Werger MJA. Maximizing daily canopy photosynthesis with respect to the leaf nitrogen allocation pattern in the canopy. *Oecologia* 1987; 72: 520–526.

121. Subbarao GV, Johansen C, Slinkard AE, Nageswara Rao RC, Saxena NP, Chauhan YS. Strategies for improving drought resistance in grain legumes. *Crit. Rev. Plant Sci.* 1995; 14:469–523.

122. Turner NC. Further progress in crop water relations. *Adv. Agron.* 1997; 58:293–337.

123. Boyer JS. Leaf enlargement and metabolic rates. *Plant Physiol.* 1970; 46:233–235.

124. Sheehy J, Green R, Robson M. The influence of water stress on the photosynthesis of a simulated sward of perennial ryegrass. *Ann. Bot.* 1975; 39:387–401.

125. Whitfield DM, Smith CJ. Effects of irrigation and nitrogen on growth, light interception and efficiency of light conversion in wheat. *Field Crops Res.* 1989; 20:279–295.

126. Robertson MJ, Giunta F. Responses of spring wheat exposed to pre-anthesis water stress. *Aust. J. Agric. Res.* 1994; 45:19–35.

127. Jamieson PD, Martin RJ, Francis GS, Wilson DR. Drought effects on biomass production and radiation use efficiency in barley. *Field Crops Res.* 1995; 43: 77–86.

128. Stockle CO, Kiniry JR. Variability in crop radiation-use efficiency associated with vapor-pressure deficit. *Field Crops Res.* 1990; 25:171–181.

129. Murata Y. Dependence of potential productivity and efficiency for solar energy utilization on leaf photosynthetic capacity in crop species. *Jpn. J. Crop Sci.* 1981; 50:223–232.

130. Dickson CD, Altabella T, Chrispeels, MJ. Slow growth phenotype of transgenic tomato expressing apoplastic invertase. *Plant Physiol.* 1991; 95:420–425.

131. Farquhar GD, Sharkey TD. Photosynthesis and carbon assimiltion. In: Boote KJ, Bennett JM, Sinclair TR, Paulsen GM, eds. *Physiology and Determination of Crop Yield.* Madison, WI: American Society of Agronomy, 1994:187–210.

132. Hein MB, Brenner ML, Brun WA. Accumulation of ^{14}C-radiolabel in leaves and fruits after injection of [^{14}C]tryptophan into seeds of soybean. *Plant Physiol.* 1986; 82:454–456.

133. Nelson CJ. Genetic associations between photosynthetic characteristics and yield: review of the evidence. *Plant Physiol. Biochem.* 1988; 26:543–554.

134. Herold A. Regulation of photosynthesis by sink activity — the missing link. *New Phytol.* 1980; 86:131–144.

135. King RW, Wardlaw IF, Evans LT. Effect of assimilate utilization on photosynthetic rate in wheat. *Planta* 1967; 77:261–276.

136. Dornhoff GM, Shibles RM. Varietal differences in net photosynthesis of soybean leaves. *Crop Sci.* 1970; 10:42–45.

137. Evans LT, Rawson HM. Photosynthesis and respiration by the flag leaf and components of the ear during

grain development in wheat. *Aust. J. Biol. Sci.* 1970; 23:245–254.

138. Evans JR. Photosynthetic acclimation and nitrogen partitioning within a Lucerne canopy. 1. Canopy characteristics. *Aust. J. Plant Physiol.* 1993; 20:55–67.

139. Estruch JJ, Pereto JG, Vercher Y, Beltram JP. Sucrose loading in isolated veins of *Pisum sativum*: regulation by abscisic acid, gibberellic acid, and cell turgor. *Plant Physiol.* 1989; 91:259–265.

140. Jeffroy MH, Ney B. Crop physiology and productivity. *Field Crops Res.* 1997; 53:3–16.

141. Thorne JH, Koller HR. Influence of assimilate demand on photosynthesis, diffusive resistances, translocation, and carbohydrate levels of soybean leaves. *Plant Physiol.* 1974; 54:201–207.

142. Buttery BR, Buzzell RI, Findlay WI. Relationships among photosynthetic rate, bean yield and other characters in field-grown cultivars of soybean. *Can. J. Plant Sci.* 1981; 61:191–198.

143. Richards RA. Defining selection criteria to improve yield under drought. *Plant Growth Regul.* 1996; 20: 57–166.

144. Slafer GA, Calderini DF, Miralles DJ. Yield components and compensation in wheat: opportunities for further increasing yield potential. In: Reynolds MP, Rajaram S, McNab A, eds. *Increasing Yield Potential in Wheat: Breaking the Barriers*. Mexico: CIMMYT, 1996:101–134.

145. Kruck BC, Calderini DF, Slafer GA. Grain weight in wheat cultivars released from 1920 to 1990 as affected by post-anthesis defoliation. *J. Agric. Sci.* 1997; 128:273–281.

146. Neales TF, Incoll LD. The control of leaf photosynthesis rate by the level of assimilate concentration in the leaf: a review of the hypothesis. *Bot. Rev.* 1968; 34:107–125.

147. Austin RB. Genetic variation in photosynthesis. *J. Agric. Sci.* 1989; 112:287–294.

148. Blum A, Mayer J, Golan G. The effect of grain number per ear (sink size) on source activity and its water relations in wheat. *J. Exp. Bot.* 1988; 39:106.

149. Sadras VO. Transpiration efficiency in crops of semi-dwarf and standard-height sunflower. *Irrig. Sci.* 1991; 12:87–91.

150. Gale MD, Youssefian S. Dwarfing genes of wheat. In: Russell GE, ed. *Progress in Plant Breeding*. London: Butterworth, 1985:1–35.

151. Allan RE. Agronomic comparison among wheat lines nearly isogenic for three reduced-height genes. *Crop Sci.* 1986; 26:707–710.

152. Allan RE. Agronomic comparison between Rht1 and Rht2 semidwarf gene in winter wheat. *Crop Sci.* 1989; 29:1103–1108.

153. Miralles DJ, Slafer GA. Yield, biomass and yield components in dwarf, semidwarf and tall isogenic lines of spring wheat under recommended and late sowings. *Plant Breed.* 1995; 114:392–396.

154. Calderini DF, Reynolds MP, Slafer GA. Genetic gains in wheat yield and main physiological changes associated with them during the 20th century. In: Satorre

EH, Slafer GA, eds. *Wheat: Ecology and Physiology of Yield Determination*. New York: Food Products Press, 1999:351–377.

155. Green CF. Genotypic differences in the growth of *Triticum aestivum* in relation to absorbed solar radiation. *Field Crops Res.* 1989; 19:285–295.

156. Slafer GA, Andrade FH. Genetic improvement in bread wheat (*Triticum aestivum, L.*) yield in Argentina. *Field Crops Res.* 1989; 21:289–296.

157. Slafer GA, Savin R. Grain mass change in a semi-dwarf and a standard-height wheat cultivar under different sink-source relationships. *Field Crops Res.* 1994; 37:39–49.

158. Cheeseman JM, Clough BF, Carter, DR, Lovelock CE, Eong OJ, Sim RG. The analysis of photosynthetic performance in leaves under field conditions — a case study using *Bruguiera* mangroves. *Photosynth. Res.* 1991; 29:11–22.

159. Loomis RS, Williams WA, Hall AE. Agricultural productivity. *Annu. Rev. Plant Physiol.* 1971; 22:431–468.

160. Reynolds MP, van Ginkel M, Ribaut JM. Avenues for genetic modification of radiation use efficiency in wheat. *J. Exp. Bot.* 2000; 51:459–473.

161. Weis E, Berry J. Quantum efficiency of PSII in relation to energy dependent quenching of chlorophyll fluorescence. *Biochim. Biophys. Acta* 1989; 894:198–208.

162. Osmond CB. What is photoinhibition? Some insights from comparison of sun and shade plants. In: Baker NR, Boyer JR, eds. *Photoinhibition: Molecular Mechanisms to the Field*. Oxford: Bios Scientific Publications, 1994:1–24.

163. Murchie EH, Chen Y, Hubbart, S, Peng S, Horton P. Interactions between senescence and leaf orientation determine *in situ* patterns of photosynthesis and photoinhibition in field-grown rice. *Plant Physiol.* 1999; 115:553–563.

164. Park YI, Chow WS, Anderson J. Light inactivation of functional photosystem II in leaves of pea grown in moderate light depends on photon exposure. *Planta* 1995; 196:401–411.

165. Pastenes C, Horton P. Effect of high temperature on photosynthesis in beans. II. CO_2 assimilation and metabolite contents. *Plant Physiol.* 1996; 112:1253–1260.

166. Pastenes C, Horton P. Effect of high temperature on photosynthesis in beans. 1. Oxygen evolution and Chl fluorescence. *Plant Physiol.* 1996; 112:1245–1251.

167. Fuse T, Iba K, Satoh H, Nishimura M. Characterisation of a rice mutant having an increased susceptibility to light stress at high temperature. *Physiol. Plant.* 1993; 89:799–804.

168. Bjorkman O, Demmig-Adams B. Regulation of photosynthetic light energy capture, conversion, and dissipation in leaves of higher plants. In: Schulze E-D, Caldwell MM, eds. *Ecophysiology of Photosynthesis. Ecology Studies 100*. Berlin: Springer-Verlag, 1994: 17–70.

169. Leegood RC, Edwards G. Carbon metabolism and photorespiration: temperature dependence in relation to other environmental factors. In: Baker NR, ed.

Photosynthesis and the Environment. Dordrecht: Kluwer Academic Publishers, 1996:191–221.

170. Duncan WG. Leaf angles, leaf area, and canopy photosynthesis. *Crop Sci.* 1971; 11:482–485.

171. Yoshida S. Physiological aspects of grain yield. *Annu. Rev. Plant Physiol.* 1972; 23:437–464.

172. Trenbath BR, Angus JF. Leaf inclination and crop production. *Field Crops Abstr.* 1975; 28:231–244.

173. Austin RB, Ford MA, Edrich JA, Hooper BE. Some effects of leaf posture on photosynthesis and yield in wheat. *Ann. Appl. Biol.* 1976; 83:425–446.

174. Ledent JF. Anatomical aspects of leaf angle changes during growth in wheat. *Phytomorphology* 1976; 26:309–314.

175. Yoshida S. Growth and development of the rice plant. In: *Fundamentals of Rice Crop Science.* Los Banos, The Philippines: The International Rice Research Institute, 1981:1–61.

176. Angus JF, Jones, R, Wilson JH. A comparison of barley cultivars with different leaf inclinations. *Aust. J. Agric. Res.* 1972; 23:945–957.

177. Dreccer MF, Slafer GA, Rabbinge R. Optimization of vertical distribution of canopy nitrogen: an alternative trait to increase yield potential in winter cereals. *J. Crop Prod.* 1998; 1:47–77.

178. He J, Chee CW, Goh CJ. 'Photoinhibition' of Heliconia under natural tropical conditions: the importance of leaf orientation for light interception and leaf temperature. *Plant Cell Environ.* 1996; 19:1238–1248.

179. Valladares F, Pearcy RW. Interactions between water stress, sun-shade acclimation, heat tolerance and photoinhibition in the sclerophyll *Heteromeles arbutifolia. Plant Cell Environ.* 1997; 20:25–36.

180. Akiyama T, Yingchol P. Studies on response to nitrogen of rice plant as affected by difference in plant type between Thai native and improved varieties. *Proc. Crop Sci. Soc. Jpn.* 1972; 41:126–132.

181. Yoshida S, Coronel V. Nitrogen nutrition, leaf resistance, and leaf photosynthetic rate of the rice plant. *Soil Sci. Soc. Plant Nutr.* 1976; 22:207–211.

182. Hawkins AF. Light interception, photosynthesis and crop productivity. *Outlook Agric.* 1982; 11:104–113.

183. Innes P, Blackwell RD. Some effects of leaf posture on the yield and water economy of winter wheat. *J. Agric. Sci.* 1983; 101:367–376.

184. Rasmusson DC. An evaluation of ideotype breeding. *Crop Sci.* 1987; 27:1140–1146.

185. Aikman DP. Potential increase in photosynthetic efficiency from the redistribution of solar radiation in a crop. *J. Exp. Bot.* 1989; 40:855–864.

186. Chang TT, Tagumpay O. Genotypic association between grain yield and six agronomic traits in a cross between rice varieties of contrasting plant types. *Euphytica* 1970; 19:356–363.

187. Peng S, Khush GS, Cassman KG. Evolution of the new plant ideotype for increased yield potential. In: *Breaking the Yield Barrier: Proceedings of a Workshop on Rice Yield Potential in Favorable Environments.* Cassman KG, ed. Los Banos, The Philippines: The International Rice Research Institute, 1994:5–20.

188. Watson DJ, Witts KJ. The net assimilation rates of wild and cultivated beets. *Ann. Bot.* 1959; 23:431–439.

189. Tanner JW, Gardner CJ, Stoskopf NC, Reinbergs E. Some observations on upright-leaf-type small grains. *Can. J. Plant Sci.* 1966; 46:690.

190. Hadfield H. Leaf temperature, leaf pose and productivity of the tea bush. *Nature Lond.,* 1968; 219:282–284.

191. Pendleton JW, Smith GE, Winter SE, Johnston TJ. Field investigations of the relationships of leaf angle in corn (*Zea mays L.*) to grain yield and apparent photosyntheis. *Agron. J.* 1968; 60:422–424.

192. Araus JL, Reynolds MP, Acevedo E. Leaf posture, grain yield, growth, leaf structure and carbon isotope discrimination in wheat. *Crop Sci.* 1993; 33:1273–1279.

193. Fischer RA. Wheat physiology at CIMMYT and raising the yield plateau. In: Reynolds MP, ed. *Increasing Yield Potential in Wheat: Breaking the Barriers.* Mexico: CIMMYT, 1996:150–166.

194. Verhagen AM, Wilson JH, Britten EJ. Plant production in relation to foliage illumination. *Ann. Bot.* 1963; 27:627–640.

195. Lambert RJ, Johnson RR. Leaf angle, tassel morphology and the performance of maize hybrids. *Crop Sci.* 1978; 18:499–502.

196. Monteith JL. Light and crop production. *Field Crops Abstr.* 1965; 18:213–219.

197. Monteith JL. Light distribution and photosynthesis in field crops. *Ann. Bot.* 1965; 29:17–37.

198. Duncan WG, Loomis RS, Williams WA, Hanau R. A model for simulating photosynthesis in plant communities. *Hilgardia* 1967; 38:181–205.

199. Stern WR. Light measurement in pastures. *Herb. Abstr.* 1962; 32:91–96.

200. Skeehy JE, Cooper J. Light interception and photosynthesis in grass crops. *J. Appl. Ecol.* 1973; 10:239–250.

201. Sinclair TR, Sheehy JE. Erect leaves and photosynthesis in rice. *Science* 1999; 283:1456–1457.

202. Carvakgi FIF, Qualset CO. Genetic variation for canopy architecture and its use in wheat breeding. *Crop Sci.* 1978; 18:561–567.

203. Van Keulen H. *A Calculation Method for Potential Rice Production.* Vol. 21. Bogorra: Center for Research, Institute for Agriculture, p 26, 1976.

204. Loomis RS. Optimization theory and crop improvement. In: Buxton DR, Shibles R, Forsberg RA, Blad BL, Asay KH, Paulsen GM, Wilson RF, eds. *International Crop Science I.* Madison, WI: Crop Science Society of America, 1993:583–588.

205. Nasyrov YS. Genetic control of photosynthesis and improving of crop productivity. *Annu. Rev. Plant Physiol.* 1978; 29:215–237.

206. Calderini DF, Dreccer MF, Slafer GA. Genetic improvement in wheat yield and associated traits. A reexamination of previous results and the latest trends. *Plant Breed.* 1995; 114:108–112.

207. Ort DR, Baker NR. Consideration of photosynthetic efficiency at low light as a major determinant of crop photosynthetic performance. *Plant Physiol. Biochem.* 1988; 26:555–565.

208. McCree KJ. The action spectrum, absorbance and quantum yield of photosynthesis in crop plants. *Agric. Meteorol.* 1971; 9:191–216.

209. Robichaux RH, Pearcy RW. Photosynthetic responses of C_3 and C_4 species from cool shaded habitats in Hawaii. *Oecologia* 1980; 47:106–109.

210. Ehleringer J, Bjorkman O. Quantum yields for CO_2 uptake in C_3 and C_4 plants. *Plant Physiol.* 1977; 59: 86–90.

211. Ehleringer J, Pearcy RW. Variation in quantum yield for CO_2 uptake among C3 and C_4 plants. *Plant Physiol.* 1983; 73:555–559.

212. Charles-Edwards DA. An analysis of the photosynthesis and productivity of vegetative crops in the United Kingdom. *Ann. Bot.* 1978; 42:717–731.

213. Evans LT, Dunstone RL. Some physiological aspects of evolution in wheat. *Aust. J. Biol. Sci.* 1970; 23: 725–741.

214. Austin RB, Morgan CL, Ford MA, Bhagwat SG. Flag leaf photosynthesis of *Triticum aestivum* and related diploid and tetraploid species. *Ann. Bot.* 1982; 49: 177–189.

215. Day W, Chalabi ZS. Use of models to investigate the link between the modification of photosynthetic characteristics and improved crop yields. *Plant Physiol. Biochem.* 1988; 26:511–517.

216. Reynolds MP, Balota M, Delgado MIB, Amani I, Fischer RA. Physiological and morphological traits associated with spring wheat yield under hot, irrigated conditions. *Aust. J. Plant Physiol.* 1994; 21:717–730.

217. Fischer RA, Rees D, Sayre KD, Lu ZM, Condon AG, Saavedra AL. Wheat yield progress associated with higher stomatal conductance and photosynthetic rate, and cooler canopies. *Crop Sci.* 1998; 38:1467–1475.

218. Reynolds MP, Rajaram S, Sayre KD. Physiological and genetic changes of irrigated wheat in the post-green revolution period and approaches for meeting projected global demand. *Crop Sci.* 1999; 39:1611–1621.

219. Secor J, McCarty DR, Shibles R, Green DE. Variability and selection for leaf photosynthesis in advanced generations of soybeans. *Crop Sci.* 1982; 22:255–259.

220. Hobbs SLA, Mahon JD. Variation, heritability, and relationship to yield of physiological characters in peas. *Crop Sci.* 1982; 22:773–779.

221. Nelson CJ, Sleeper DA. Using leaf area expansion rate to improve yield of tall fescue. In: Smith JA, Hays VW, eds. *Proceedings of the XIV International Grassland Congress.* Boulder: Westview Press, 1983:413–416.

222. Wilhelm WW, Nelson CJ. Carbon dioxide exchange rate of tall fescue. Leaf area vs. leaf weight basis. *Crop Sci.* 1985; 25:775–778.

223. Frey KJ. Plant breeding in the seventies: useful genes from wild plant species. *Egypt. J. Genet. Cytol.* 1976; 5:460–482

224. Brinkman MA, Frey KJ. Flag leaf physiological analysis of oat isolines that differ in grain yield from their recurrent parents. *Crop Sci.* 1978; 18:69–73.

225. Takeda K, Frey KJ. Contributions of vegetative growth rate and harvest index to grain yield of progenies from *Avena sativa* × *A. sterilis* crosses. *Crop Sci.* 1976; 16:817–821.

226. Zelitch I. The close relationship between net photosynthesis and crop yield. *BioScience* 1982; 32:796–802.

227. Gutteridge S, Parry MAJ, Schmidt CNG, Feeney J. An investigation of ribulose biphosphate carboxylase activity by high resolution ^{1}H NMR. *FEBS Lett.* 1984; 170:355–359.

228. Pierce J, Lorimer GH, Reddy GS. Kinetic mechanism of ribulose biphosphate carboxylase: evidence for an ordered, sequential reaction. *Biochemistry* 1986; 25:1636–1644.

229. Jordan DB, Ogreen WL. Species variation in the specificity factors of ribulose 1,5-biphosphate carboxylase/oxygenase. *Nature* 1981; 291:513–515.

230. Jordan DB, Ogren WL. Species variation in kinetic properties of ribulose 1,5 biphosphate carboxylase/oxygenase. *Arch. Biochem. Biophys.* 1983; 227: 425–433.

231. Gutteridge S, Phillips AL, Kettleborough CA, Parry MAJ, Keys AJ. Expression of bacterial genes in *Escherichia coli. Philos. Trans. R. Soc. Lond., Ser. B* 1986; 313:433–435.

232. Parry MAJ, Keys AJ, Gutteridge S. Variation in the specificity factor of C3 higher plant Rubiscos determined by the total consumption of ribulose-P2. *J. Exp. Bot.* 1989; 40:317–320.

233. Read BA, Tabita FR. High substrate specificity factor ribulose biphosphate carboxylase/oxygenase from eukaryotic marine algae and properties of recombinant cyanobacterial Rubisco containing 'algal' residue modifications. *Arch. Biochem. Biophys.* 1994; 312:210–218.

234. Hartman FC, Harpel MR. Structure, function, regulation, and assembly of D-ribulose biphosphate carboxylase/oxygenase. *Annu. Rev. Biochem.* 1994; 63:197–234.

235. Delgado E, Medrano H, Keys AJ, Parry MAJ. Species variation in rubisco specificity factor. *J. Exp. Bot.* 1995; 46:1775–1777.

236. Terzaghi BE, Laing WA, Christeller JT, Petersen GB, Hill DF. Ribulose 1,5-biphosphate carboxylase: effect on the catalytic properties of changing methionine 330 to leucine in *Rhodospirillum rubrum* enzyme. *Biochem. J.* 1986; 235:839–846.

237. Voordouw G, De Vries PA, Van Den Berg WAM, De Clerck EPJ. Site directed mutagenesis of the small subunit of ribulose 1,5-biphosphate carboxylase/oxygenase. *Eur. J. Biochem.* 1987; 163:591–598.

238. Kettleborough CA, Parry MAJ, Burton S, Gutteridge S, Keys AJ, Phillips AL. The role of the N-terminus of the large subunit of ribulose bisphosphate carboxylase investigated by construction and expression of chimaeric genes. *Eur. J. Biochem.* 1987; 170:335–342.

239. Kane HJ, Viil J, Entsch B, Paul K, Morell MK, Andrews TJ. An improved method for measuring the CO_2/O_2 specificity of ribulose bisphosphate carboxylase/oxygenase. *Aust. J. Plant Physiol.* 1994; 21:449–461.

240. Ku SB, Edwards GE. Oxygen inhibition of photosynthesis. III. Temperature dependence of quantum yield and its relation to CO_2/O_2 solubility ratio. *Planta* 1978; 140:1–6.

241. Jordan DB, Ogren WL. 1984. The carbon dioxide/oxygen specificity of ribulose-1,5-biphosphate carboxylase/oxygenase. *Planta* 1984; 161:308–313.

242. Brooks A, Farquhar GD. Effect of temperature on the CO_2/O_2 specificity of ribulose-1,5-biphosphate carboxylase/oxygenase and the rate of respiration in the light. Estimates from gas exchange measurements on spinach. *Planta* 1985; 165:397–406.

243. Estelle M, Hanks J, McIntosh L, Somerville C. Site specific mutagenesis of ribulose-1,5-bisphosphate carboxylase/oxygenase. *J. Biol. Chem.* 1985; 260:9523–9526.

244. Uemura K, Anwaruzzaman S, Miyachi S, Yokota A. Ribulose-1,5-biphosphate carboxylase/oxygenase from thermophillic red algae with a strong specificity for CO_2 fixation. *Biochem. Biophys. Res. Commun.* 1997; 233:568–571.

245. Farquhar GD, von Caemmerer S, Berry JA. A biochemical model of photosynthetic CO_2 assimilation in leaves of C_3 species. *Planta* 1980; 149:78–90.

246. Evans LT. Assimilation, allocation, explanation, extrapolation. In: Rabbinge R, Goudriaan J, van Keulen H, Penning de Vries FWT, van Laar HH, eds. *Theoretical Production Ecology: Reflections and Prospects.* Wageningen: PUDOC, 1990:77–87.

247. Lawlor DW. Photosynthesis, productivity and environment. *J. Exp. Bot.* 1995; 46:1449–1461.

248. Siddique KHM, Belford RK, Perry MW, Tennant D. Growth, development and light interception of old and modern wheat cultivars in a Mediterranean-type environment. *Aust. J. Agric. Res.* 1989; 40:473–487.

249. Takeda T. Physiological and ecological characteristics of high yielding varieties of lowland rice. Proceedings of the International Crop Science Symposium, Fukuoka, Japan, October 17–20, 1984.

250. Austin RB, Bingham J, Blackwell RD, Evans LT, Ford MA, Morgan CL, Taylor M. Genetic improvement in winter wheat yield since 1900 and associated physiological changes. *J. Agric. Sci. Camb.* 1980; 94:675–689.

251. Cox TS, Shroyer RJ, Ben-Hui L, Sears RG, Martin TJ. Genetic improvement in agronomic traits of hard red winter wheat cultivars from 1919 to 1987. *Crop Sci.* 1988; 28:756–760.

252. Sayre KD, Rajaram S, Fischer RA. Yield potential progress in short bread wheats in northwest Mexico. *Crop Sci.* 1997; 37:36–42.

253. Slafer GA, Andrade FH. Changes in physiological attributes of the dry matter economy of bread wheat (*Triticum aestivum L.*) through genetic improvement of grain yield potential at different regions of the world. A review. *Euphytica* 1991; 58:37–49.

254. Setter TL, Peng S, Kirk GJD, Virmani SS, Kropff MJ, Cassman KG. Physiological considerations and heterosis to increase yield potential. In: Cassman KG, ed. *Breaking the Yield Barrier: Proceedings of a Workshop on Rice Yield Potential in Favorable Environments.* Los Banos, The Philippines: The International Rice Research Institute, 1994:39–62.

255. Setter TL, Conocono EA, Egdane JA, Kropff MJ. Possibility of increasing yield potential of rice by reducing panicle height in the canopy. 1. Effects of panicles on light interception and canopy photosynthesis. *Aust. J. Plant Physiol.* 1995; 22:441–451.

256. Setter TL, Peng S, Khush GS, Kropff MJ, Cassman KG. *Yield potential of rice: past, present and future perspectives. Trop. Agric.* 1997; spl issue:80–95.

257. Bennett J, Brar DS, Khush GS, Huanh N, Setter TL. Molecular approaches to yield potential. In: Cassman KG, ed. *Breaking the Yield Barrier: Proceedings of a Workshop on Rice Yield Potential in Favorable Environments.* Los Banos, The Philippines: The International Rice Research Institute, 1994:63–76.

258. Kropff MJ, Cassman KG, Peng S, Matthews RB, Setter TL. Quantitative understanding of rice yield potential. In: Cassman KG, ed. *Breaking the Yield Barrier: Proceedings of a Workshop on Rice Yield Potential in Favorable Environments.* Los Banos, The Philippines: The International Rice Research Institute, 1994:21–38.

259. Carver BF, Nevo E. Genetic diversity of photosynthetic characters in native populations of *Triticum dicoccoides. Photosynth. Res.* 1990; 25:119–128.

260. Johnson RC, Kebede H, Mornhinweg DW, Carver BF, Rayburn AL, Nguyen HT. Photosynthetic differences among *Triticum* accessions at tillering. *Crop Sci.* 1987; 27:1046–1050.

261. Carver BF, Johnson RC, Rayburn AL. Genetic analysis of photosynthetic diversity in hexaploid and tetraploid wheat and their interspecific hybrids. *Photosynth. Res.* 1989; 20:105–118.

262. Bhagsari AS, Brown RH. Leaf photosynthesis and its correlation with leaf area. *Crop Sci.* 1986; 26:127–131.

263. Nelson, CJ, Sleper DA, Coutts JH. Field performance of tall fescue selected for leaf area expansion rate. In: *Proceedings of XV International Grassland Congress.* Nishinasuno, Tochigi-ken, Japan: Japanese Society of Grassland Science, 1985:320–322.

264. Puckridge DW. Photosynthesis of wheat under field conditions. III. Seasonal trends in carbon dioxide uptake of crop communities. *Aust. J. Agric. Res.* 1971; 22:1–9.

265. Harrison SA, Boerma HR, Ashley DA. Heritability of canopy apparent photosynthesis and its relationship to seed yield in soybeans. *Crop Sci.* 1981; 21:222–226.

266. Gent MPN, Kiyomoto RK. Physiological and agronomic consequences of Rht genes in wheat. *Crop Sci. Recent Adv.* 1998; 27:46.

267. Slafer GA, Andrade FH, Satorre, EH. Genetic-improvement effects on pre-anthesis physiological attributes related to wheat grain yield. *Field Crops Res.* 1990; 23:255–263.

268. Loomis RS, Gerakis RA. Productivity of agricultural ecosystems. In: Cooper JP, ed. *Photosynthesis and Productivity in Different Environments.* Cambridge: Cambridge University Press, 1975:145–173.

30 Physiological Perspectives on Improving Crop Adaptation to Drought — Justification for a Systemic Component-Based Approach

G.V. Subbarao and O. Ito
Crop Production and Environment Division, Japan International Research Center for Agricultural Sciences

R. Serraj and J. J. Crouch
International Crops Research Institute for the Semi-Arid Tropics

S. Tobita and K. Okada
Crop Production and Environment Division, Japan International Research Center for Agricultural Sciences

C. T. Hash
International Crops Research Institute for the Semi-Arid Tropics

R. Ortiz
International Institute of Tropical Agriculture,
L.W. Lambourn & Co.

W. L. Berry
Department of Organismic Biology, Ecology and Evolution,
University of California

CONTENTS

I. INTRODUCTION

Adaptation to drought stress is a very complex process. Drought stress is the result of numerous climatic, edaphic, and agronomic factors that are frequently further confounded by major variations in their timing, duration, and intensity. The combinations of morphological, physiological, and phenological traits/attributes required for optimal adaptation to drought stress varies with each local environment. Overall, the genetic contribution to drought adaptation is based on a combination of constitutive and induced physiological and biochemical traits. Furthermore, new plant varieties must have an array of biotic stress resistances, while still retaining product quality traits which are required for the farmers' adoption of any new cultivar. This may be of paramount importance as it can interact with the expression of drought tolerance. The understanding of the full interaction of a complex collection of drought adaptation traits is much more difficult than understanding the functioning of each individual trait. The accurate assessment of a cultivar's drought adaptation may not be possible from a single season evaluation. Similarly, there generally is no single trait that breeders can deploy to improve the productivity of a given crop under all water-limiting conditions. Development of overall drought adaptation for a plant is most generally the result of the collective expression of many plant characteristics in the appropriate environment. The genetic fragments defining these traits may be distributed across various locations over the entire genome. Even if one is fortunate enough to find the right combination of underlying loci in a single accession, it is very difficult to transfer such a collection of traits intact from a donor parent into the targeted local varieties. For these reasons, the use of standard procedures to directly breed for drought adaptation using classical empirical screening approaches, which rely heavily on yield or yield-derived indices have generally had at best only modest return for most crops [1–7].

A potential alternative is a systemic component-based approach. The underlying concept of this approach is that many of the various morphological, physiological, and biochemical components that improve plants adaptation to drought can be collected in one plant genotype where they can contribute to drought tolerance either by virtue of their direct effects or through their interactions with other loci. It is now possible and practical to pyramid many of these individual traits in a genetic stock/variety through marker-assisted breeding. However, permutations and combinations of pyramided genes will need to be evaluated through replicated multilocational trials in order to select the most appropriate interaction effects. In nature, these traits are not usually all found within a single plant genotype but are dispersed in a number of different genetic stocks derived from different geographical areas where that specific type of drought stress is the limiting factor. These individual drought adaptation traits need to be located and identified, in the varieties where they originated. Once these traits have been identified it is then possible to introduce a number of them into a locally adapted variety. The precise traits or components required to improve a crop's drought adaptation depends on the environment of the target growth area. Thus, identification of the appropriate genetic traits needed for improvement of drought tolerance must be based on an understanding of the eco-physiology of the area where the trait developed as well as the eco-physiology of the target area where the crop is to be grown. Only incremental progress should be expected from each of the independent drought adaptation component traits introduced into the local variety. This is due to the fact that many different environmental factors contribute to drought stress and that many adaptation traits are effective only for certain aspects of drought and often only over a limited range of drought stress. Moreover, the new interaction effects between introgressed genes and their new genetic background may indeed at times result in "negative progress." However, it should be possible to achieve greater and broader improvement of drought adaptation as more and more drought adaptation traits are incorporated into the locally desired plant genotype.

Because the diverse environmental factors inducing drought may change from season to season, it is not reasonable to expect that all traits would be equally functional every season. This is one of the reasons why it may be difficult to quickly reach the full potential of these approaches in achieving a high degree of drought adaptation.

II. RELEVANCE OF DROUGHT STRESS TO THE SEMIARID TROPICS AND POTENTIAL YIELD GAINS FROM CROP IMPROVEMENT

Grain losses of maize in tropics alone because of drought exceed 20 million tons per year and this amounts to 17% of their realizable potential yield (i.e., well-watered conditions) [2]. For Southern Africa, nearly 60% of the potential maize yield can be lost due to drought where it is severe [2]. Drought is considered to be one of the major abiotic constraints to crop production in the Guinea Savanna belt of West and Central Africa. The risk of drought stress is particularly high in the Sudan Savanna zone because rainfall there is unpredictable both in quantity and distribution. Even in lowland locations where there is generally adequate precipitation for growth of maize, periodic droughts can occur during flowering and grain-filling stages [4]. These growth stages are the most sensitive phases of maize to moisture deficits. When drought stress coincides with flowering and grain-filling stages of maize, the resultant yield losses range from 20% to 50% [8]. Similarly, cowpea, which is also widely grown in the semiarid tropics of Africa and Asia, also have drought as one of their major production constraints. Although, cowpeas are considered to be one of the most drought tolerant legumes grown in dry-savanna of Africa, the severe droughts of the Sahel, can still substantially limit cowpea production [9,10].

There are two notable successes from ICRISAT's cereal breeding programs, in terms of the impact of genetic enhancement of crop yield under drought-prone rainfed conditions, the release of pearl millet variety Okashana 1 in Namibia and sorghum variety S 35 in Chad and Cameroon. The pearl millet variety "Okashana 1," bred at ICRISAT (Patancheru, India), is grown on almost 50% of the pearl millet area in Namibia, where the main limitations to crop yield are low rainfall, frequent drought, and low-input agronomic conditions [11]. This variety is early maturing, has good terminal drought tolerance, and is generally adapted to marginal production environments [12,13]. The development and dissemination of this variety has contributed to substantial improve-

ments in pearl millet production and overall food security in Namibia [11]. However, this composite and the base population from which it was bred were selected during the highest rainfall years at ICRISAT and thus were more fortuitous than strictly a product of precisely controlled knowledge-led selection trials. Nevertheless, Okashana 1 and its sister variety ICMV 221, which was bred by progeny-based selection under managed terminal drought stress and has been released in Eritrea, Kenya and India is a clear indication that genetic improvement of drought tolerance is possible within economically productive and market acceptable genetic backgrounds.

The sorghum variety S 35 was originally a photoperiod-insensitive, high-yielding, early-maturing, and drought tolerant pure-line developed from ICRISAT's breeding program in India. Its subsequent introduction into drought-prone areas of Chad has been very successful, resulting in an estimated yield advantage of about 50% over the farmers' local varieties [14].

For upland rice, West African Rice Development Association (WARDA) has released several rice varieties that are suitable for upland conditions (i.e. Guinea Savanna), where drought stress is a major constraint on production. The genetic stocks WAB56-104 and WAB56-50 also showed good adaptation to drought-prone sites [15]. Another successful example of drought tolerance is the rice variety 'NERICA', which was developed through interspecific hybridization using *Oryza sativa* and *Oryza glaberrima* [16,17]. This rice variety has been reported to have stable yields across a range of production environments in West Africa, where drought stress is an integral feature of the environment [16,17].

Until recently, the maize breeding program at the International Institute of Tropical Agriculture (IITA, Ibadan, Nigeria) screened germplasm for tolerance to drought under naturally occurring drought stress at a location in the Sudan savanna. However, because the nonmanaged drought stress in these fields was not always consistent, the screening for drought tolerance was not very effective. To effectively differentiate between tolerant and sensitive genotypes, selection needs to be made under controlled conditions where known levels of drought stress can be reliably induced. Consequently, since 1997, IITA has been screening diverse maize germplasm under drought stress at a location carefully selected where drought stress occurs predictably during flowering and grain-filling stages of the crop. Using such a site, late- and early-maturing broad-based populations were improved for drought tolerance using recurrent selection schemes. The Japan International Research for

Agricultural Sciences (JIRCAS) and IITA have jointly developed efficient screening methods for drought tolerance in cowpea [9,10,18,19].

Based on empirical field evaluations, the chickpea germplasm line ICC 4958 has been considered as drought tolerant in relation to other legume crops [20]. In chickpea, large root systems are considered to be the single most important component of terminal drought tolerance in areas where the crop is grown on residual soil moisture during the postrainy season. However, the drought tolerance of this line has been definitively shown not be due to a significant difference in relative root size (Kashiwagi, personal communication). Yet, plant breeders continue to routinely use this genotype as it has good general combining ability for conferring drought tolerance. Thus there is clearly a great deal more to learn about the underlying basis of drought tolerance in crops of the semiarid tropics (SAT). Nevertheless, there appears to be no shortage of genetic variation in the germplasm collections for this trait. Groundnut varieties ICGS 11, ICGS 37 are considered tolerant to end-of-season drought; ICGS 44, ICGS 76 and ICGS 10 are considered tolerant to mid-season drought patterns [21]. Here, water use efficiency has been implicated as playing a major role. For pigeonpea, ICPL 87 and UPAS 120 are considered to have tolerance to mid-season drought [22].

For the SAT, drought stress is an integral part of the overall agricultural production ecosystems. Thus, any genetic improvement program that targets the SAT region should include adaptation to drought stress as one of the primary selection criteria. The general problem of drought in the SAT is further compounded with erratic, unpredictable rainfall, high temperatures, high levels of solar radiation, and poor soil characteristics. Considering all of these factors, drought stress is still considered to be the most limiting factor for achieving enhanced yield potential in the SAT (Table 30.1). Any increase in yield resulting from improved adaptation to drought would have enormous economic benefits to the resource-poor farmers in the SAT region (Table 30.1). A cost–benefit analysis of various potential ICRISAT research themes in this area indicates that the potential returns on investment in drought research could be very substantial and will clearly have long-term impacts [23].

Unlike disease, insect or parasitic weed resistances, which tend to become nonfunctional with time due to the evolution of new virulent pest biotypes, the progress made in improving drought adaptation is likely to remain effective over relatively long periods of time. This is because these adaptations are responding to climatic patterns, which do not readily change in response to changes in plant response, but rather change as a function of geology in the slow time frame, which is a characteristic of geology. However, with the increasing threat of human-induced global climate change, it appears that the dry regions of Africa and South Asia may become even drier, thus increasing the need for drought tolerant crops. The agro-climatic and production system environments of the SAT regions are very diverse and thus the inherent water constraints that limit crop production are also very variable. The first step necessary to characterize the drought patterns of these environments should be long-term studies of water-balance modeling using existing weather datasets and geographical information systems (GIS) tools. The accurate assessment of moisture availability in these environments is critical for identifying crop genotypes adapted to such drought-prone environments. However, even with these challenges of drought complexity, a systemic component-based approach should provide a valuable tool for achieving crop improvement for drought-prone environments such as the SAT [24–27].

III. CROP YIELD AND WATER USE

The relationship between carbon fixation and transpirational water loss has been well established in plants [28]. The slope of this relationship, termed transpiration efficiency, varies substantially among and within plant species and with phase of growth [26,29–32]. Research over the last 100 years comparing water use to crop growth has shown an intimate

TABLE 30.1
Yield Loss Due to Drought Stress and Potential Gain from Crop Improvement for ICRISAT Mandated Crops

Crop	Yield Loss due to Drought Stress (US$ million)	Potential Yield Gain from Crop Improvement (US$ million)
Pearl-millet	630	142
Sorghum	1744	143
Chickpea	1058	525
Groundnut	520	208
Pigeonpea	570	92
Total	Yield loss (in US$ million): 4.522	Yield gain (in US$ million) : 1.110

Source: ICRISAT. Medium Term Plan 1994–1998. Vol. 1. Main Report. Patancheru, India: ICRISAT, 1992. With permission.

and predictable relationship between plant growth and transpirational water-use after correcting for variations in atmospheric humidity [33]. However, there remains substantial heritable genetic variation reported for transpiration efficiency (see Refs. [32,34] for further discussion) in many C_3 crop plants (and perhaps C_4 crop plants as well [35]), suggesting that there is room for improvement in this phenomenon.

Generally, the response of plants to soil water deficits can be related to a sequence of three successive stages of soil dehydration [27,33] (Figure 30.1). Stage I occurs at high soil moisture when water is still freely available from the soil and both stomatal conductance and water vapor loss are not limited by soil water availability. The transpiration rate during this stage is therefore determined by environmental conditions around the leaves. Stage II starts when the rate of water uptake from the soil cannot match the potential transpiration rate. Stomatal conductance declines, limiting the transpiration rate to a rate similar to that of uptake of soil water, resulting in the maintenance of the water balance of the plant. Finally, stage III begins when the stomatal adjustment is no longer sufficient to maintain a positive water balance and plant's survival depends on other drought adaptation mechanisms being available.

In the absence of drought adaptation mechanisms, virtually all major processes contributing to crop yield including leaf photosynthetic rate, leaf expansion and growth are inhibited late in stage I or in stage II of soil drying [36.37]. At the end of stage II, these growth-supporting processes have reached zero and no further net growth (i.e., increase in biomass) occurs in the plants. The focus of stage III is survival and water conservation mechanisms that will allow

the plant to endure these severe conditions until they are relieved and improved moisture availability permits growth to resume. Plant survival is a critical trait in natural dry-land ecosystems and perennial crops, but for most agricultural situations, stage III has little relevance to questions about increasing crop yield. Consequently, the amount of water available up to the end of stage II for all practical purposes determines the cumulative growth and yield on a particular soil. Recovery from stage III can only have relevance to yield performance if water is added to the system while there is still sufficient time for growth. Therefore, options involving mechanisms to enhance crop survival do not usually mean any increase in crop yield for annual crops under severe drought stress conditions [38]. Increased crop yields and water use efficiency generally require the optimization of the physiological processes involved in the critical early stages (mainly stage II) of plant response to soil dehydration.

Several physiological, morphological, and phenological traits/mechanisms/attributes (see Refs. [6,36,39,40–45] for further discussion) can have a significant role in adaptation to drought stress induced by either stage I or stage II of the soil drying process (i.e., to avoid the onset of internal moisture stress, and thus maintain growth rate). Extracting soil moisture from deeper or different soil layers can be a major strategy for such an adaptation. Many high-yielding short-duration crop varieties have shallow root systems, thus cannot extract water effectively beyond 50 cm. (see Ref. [46] for further discussion). Osmotic adjustment can also have a substantial role in increasing productivity under stage III drought stress. Osmotic adjustment provides additional functional metabolic time so that the stored carbohydrates in stems can be remobilized to help in the grain-filling process. Stored carbohydrates in stems play a substantive role in grain filling under terminal drought stress. The contribution from the pre-anthesis stored carbon to the grain-filling process during stage III can often exceed 80% of the total grain yield in cereals (where the current photosynthate production can be very limited or almost nil during the grain-filling phase of growth because of the severity of drought stress during stage III) (i.e. *rabi* season) [36,39,41,47].

IV. DECENTRALIZATION IN BREEDING FOR DROUGHT TOLERANCE

A. THE OLD PARADIGM

Finding a *magic* phenotype that has complete tolerance to drought in terms of yielding ability under high drought stress conditions has not been successful. The

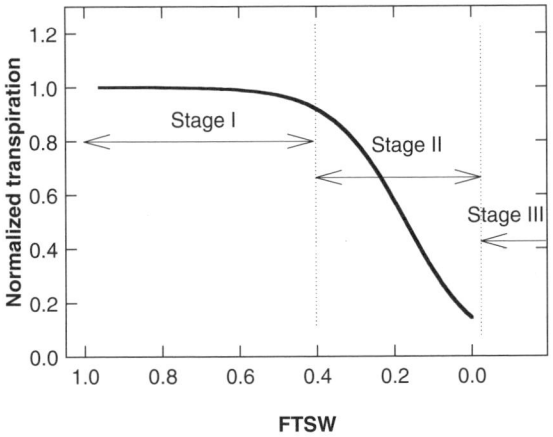

FIGURE 30.1 Normalized leaf transpiration (NTR) against the fraction of transpirable soil water (FTSW). (From Sinclair TR, Ludlow MM. *Aust. J. Plant Physiol.* 1986; 13:329–341. With permission.)

biggest hurdle with this approach is even if one is sufficiently fortunate to find such a *magic* phenotype, it is extremely difficult to transfer all of the desirable attributes intact to high-yielding cultivar, because of the complex nature of these drought characters (as discussed earlier). Many IARCs have spent several decades in the past, using enormous amounts of resources to screen world germplasm, searching for such "elusive phenotypes" with only limited success. Breeding directly for drought tolerance has made relatively little headway because of inherent weaknesses associated with the perceived assumption that drought tolerance is a single or a single composite plant characteristic and can be handled like any other trait (such as disease resistance). In addition, this old paradigm is closely associated with the traditional centralized breeding approach to develop crop varieties that have a wide adaptation and are suitable for a large range of geographical regions. But, the complex nature of drought adaptation makes this simple approach of limited value as there is no one trait able to address the many aspects of drought found in a large area like the SAT.

B. THE NEW PARADIGM

The new paradigm revolves around "traits or components" that have a functional role in improving adaptation to one or more aspects of drought stress rather than an all-encompassing phenotype (i.e., drought tolerance). The underlying implication here is that breeding for drought tolerance under the new paradigm can be a decentralized process where the breeding efforts would be focused in the National Agricultural Research Systems (NARS) that have responsibility to develop crop varieties that have specific adaptation to locally or nationally important agro-ecologies. The inherent hypothesis of the new paradigm is that specific components of each individual drought adaptation trait need to be selectively incorporated into varieties and breeding populations that are locally adapted and are currently under improvement. This should be based on a thorough analysis of what needs to be improved in the locally adapted varieties. It will also be dependent on which physiological/biochemical/morphological/ phenological traits are available. Perhaps, the shift to this new paradigm will provide an alternative means of identifying novel physiological mechanisms or other morphological traits that contribute towards the improvements in adaptation of crops to drought. The role of IARCs or other research institutions with a global mandate to address the issue of drought adaptation would concentrate on providing the components of drought tolerance or genetic stocks with specific traits linked to improving crop performance under drought stress. Thus, the IARCs would develop the ingredients and information (technical knowhow) to pass on to the NARS and then assist them in developing varieties that have specific adaptations required for their target agro-ecologies.

C. STRATEGY OF THE NEW PARADIGM

The complexity of the drought syndrome should be tackled from a holistic perspective. This means that physiological/mechanistic dissection and molecular genetic analysis of tolerance/adaptation at key phenological stages should be integrated with the development of agronomic practices that in the overall lead to better conservation and utilization of soil moisture in time and space. This endeavor would identify the most appropriate combination of crop genotypes and management systems for maximum stable productivity under the target environmental profile. A critical element to achieve this would be the development and improvement of screening tools, along with the appropriate protocols for characterization of these component traits in multilocational environments. This would necessitate the identification of genetic stocks, and the evaluation of the functional relationships of relevant traits to crop adaptation for various types of drought stress. The development of working protocols for evaluating the value of various combinations of morphological/biochemical/physiological/developmental traits for different target drought-prone environments would also be required. It would be highly desirable to develop ideotype concepts for various crops in an appropriate range of agro-ecologies [25,48]. A good starting point for choosing appropriate ideotype/s for modeling studies can come from Farmer-Participatory Assessment of a wide range of plant architectures and phenologies in the crop(s) target range of drought-prone environments.

Recent advances in molecular techniques have contributed significantly to germplasm utilization and enhancement along with innovative approaches to plant breeding under stress conditions. These molecular tools, including high-throughput DNA marker genotyping, can be used for diversity analysis, genetic linkage mapping, and marker-assisted selection of crops for improved tolerance to drought [49–51]. Identification and genetic mapping of quantitative trait loci (QTL) for specific components is currently used to dissect the genetic basis of various traits associated with crop performance, including drought tolerance [52–68]. As the genotyping process is developed and refined, it is essential that similar refinements of the necessary phenotyping tools are also developed concurrently. Many crops of the SAT have not yet been intensively studied by molecular biologists and, therefore, have only a limited number of DNA markers and other genomic tools available. Nevertheless, there are large public domain databases accumulating for a number of model plant and crop species that offer great opportunities for rapid advances in orphan crops through synteny and sequence alignment studies [69,70]. This approach is presently fueling rapid progress in chickpea research through association with *Medicago*, *Phaseolus*, and soybean.

Marker-aided genetic analysis suggests that a large proportion of the variation for response to drought or water-use efficiency may be accounted for by a few QTL of large effect plus many others of relatively small effect [8,67,71]. This may mean that significant genetic gains can be made in marker-assisted selection programs when focused on a relatively few target loci which have the largest favorable effects. Careful phenotyping is critical in the QTL mapping of drought tolerance components. To a certain extent, cross-breeding assisted by selection with DNA markers could be a promising strategy for a fast and objective selection of new cultivars with enhanced adaptation to drought. In addition, advances in molecular biology and crop model systems offer a number of new avenues to address the issue of drought adaptation. For instance, cowpea would be a suitable species to determine the genetic potential of drought adaptation in legume crops using QTL analysis, and germplasm characterization; *Medicago truncatula* would be the suitable species among legumes for assessing whole-genome transcriptional responses to drought. Some of the desirable characteristics of interest in a drought tolerant "concensus legume" species are root architecture, transcriptional pathways (e.g., dehydrin proteins), physiological parameters (e.g., osmotic adjustment), and plant development (earliness). Comparative mapping will be the means to assess gene synteny of drought tolerance loci across crop legume genomes. Forward, and reverse genetics (in these legume species) may identify key regulators of drought tolerant genotypes. The outputs of such legume genomic research are genetically defined loci controlling the trait, candidate genes (as defined by mapping, mutation, and transcriptional investigations) for drought tolerance, and DNA markers for assisted-selection or aided-introgression and germplasm management for improvement of drought adaptation.

Participation of NARS and private sector breeding programs from various target production areas in the mapping of drought tolerance components and the development of component trait-specific genetic

stocks will be highly desirable. They can play an essential role in verifying and refining DNA markers and molecular breeding strategies for rapidly improving crop plants' adaptation to drought stress. Not only is such interaction necessary for this level of scientific endeavor, it is also vital to appropriately orientate the research goals such that there is a high level of adoption of the drought-adapted crops. Such collaborative programs could also lead to the strengthening of NARS and an enhanced linkage between public and private sector institutions.

V. CONCEPTUAL FRAMEWORK FOR TRAIT-BASED IMPROVEMENT

Though the following two approaches are being discussed separately for the relative ease of presentation, in practice the approach that needs to be adopted is a combination of these two to derive the best from both for the rapid trait-based improvement of crop adaptation to drought stress.

A. BLACKBOX APPROACH

Before the development of specific genetic markers, plants with different adaptation to drought should be identified so that independent drought tolerance mechanisms can be studied. The logical route of separation is from phenotypic performance, to the underlying reasons, to the mechanisms behind them and finally to genetic markers (or actual genes) for the individual mechanisms. Identification of QTL for the superior phenotype (i.e., drought tolerance), and the understanding of the functional role of the individual QTL on the phenotype is critical to this approach. Once this is accomplished, one can selectively incorporate individual QTL either together or independently into crop genotypes of several genetic backgrounds to evaluate the functional and adaptive significance for drought adaptation of individual QTL (i.e., development of near-isogenic lines [NILs] in several genetic backgrounds). One of the major advantages of using NILs is that once a favorable QTL has been identified, it is already fixed in an elite recipient line and the initial cycle of breeding work to incorporate the trait into crop varieties of interest to farmers is essentially completed for that specific trait. Also, lines with favorable QTL alleles can be easily maintained and then used for pyramiding favorable alleles at several QTL into a single genetic stock. Nevertheless, the amount of resources (both man power and financial) required for this approach can be substantial and thus should not be underestimated. Initially, only those traits that strongly influence adaptation to drought would have a high priority, with traits having less of an influence following.

It is not surprising that different genetic stocks, with a similar degree of drought tolerance may have achieved their tolerance through entirely different physiological mechanisms. For instance, one genotype may have realized a given degree of drought tolerance by better osmotic adjustment, whereas another one could have achieved a similar degree of drought tolerance by a better rooting depth. For example, rice varieties "Moroberekan" and "Azucena" derive most of their drought adaptation through their deeper rooting attributes, whereas rice variety CT9993 accomplishes a similar degree of adaptation to drought through better leaf osmotic adjustment [61,62]. Similarly, in short-duration phenotypes of pigeonpea where the root systems are often shallow [46], osmotic adjustment is a prevalent form of drought adaptation [46,72,73], whereas for long-duration pigeonea phenotypes where roots go deeper than 2 m, often there is no osmotic adjustment in leaves even during terminal drought (Subbarao and Chauhan, unpublished). The challenge is to bring together appropriate complementary levels of such independent attributes that have the potential to complement each other functionally, or even have a synergistic effect when brought together within a single genetic stock. Similar arguments can be made regarding the interaction between WUE and the deep rooting characteristics in groundnut (see Ref. [32] for further discussion on this).

B.   IDEOTYPE APPROACH

The assumption of the ideotype approach is that the morphological, physiological, or biochemical traits that influence adaptation to drought stress need not influence directly the desired yield formation at least in the donor parent where a particular trait has been sought. This is in contrast to the blackbox approach where yield performance under drought stress conditions is the primary criteria in selecting a genetic stock to unravel the mechanisms contributing to it. The ideotype approach is focused on traits that in theory have a functional role in adaptation to drought stress, which could be combined into a genetic stock (either in adapted breeder's lines or germplasm lines). Adaptation of a genetic stock to the experimental site (where the genetic stock is evaluated) may be independent of this process. For example, transpiration efficiency (TE) (evaluated based on ^{13}C discrimination analysis) is high (about 30%) in many of the land-races in durum wheat in comparison to the high yielding varieties [74; Subbarao, unpublished results].

Though TE is an important component of drought adaptation, these land-races when evaluated for drought tolerance based on grain yield, cannot be ranked as drought tolerant because other characteristics such as low harvest index (HI), and low early above-ground growth rate. Thus, though some of these land-races could be an excellent source for improving TE in modern durum cultivars with high grain yield potential, they are unlikely to be selected as source material for improving drought tolerance of these high-yielding cultivars if evaluations are based solely on yield or yield-derived indices in a single stress environment. Residual transpiration, a physiological trait in durum wheat that has been implicated in adaptation to drought stress would be effectively eliminated in germplasm evaluations because of poor agronomic score [40,75]. Similarly in groundnut, some of the genetic stocks that have high levels of TE have low HI, and thus would not be selected if drought evaluations were based on pod yield or yield-derived indices [32,76–78].

In crops such as wheat, barley and beans, selection for high TE can lead to low dry matter production, thus low potential productivity under water nonlimiting conditions [7,31,79]. Because of this, it was suggested that selection for high TE would improve adaptation to drought [80], whereas selection for low TE should improve yield potential [79]. However, there is no theoretical reason that genotypes have to comply with this general relationship. For example, in barley, although, there is generally a negative relationship between TE and dry matter accumulation among the genotypes tested, but certain genotypes deviate from this relationship (Figure 30.2) [81]. For crops such as groundnut and cool-season grasses, where photosynthetic rates are the main source of variation in TE, selection for high TE should lead to genotypes with high dry matter production capabilities irrespective of the water regime in which they are grown [76,77,82,83]. It is interesting to note that the usefulness of selection for high TE could vary depending on the crop species and the target environment; in one case it could improve productivity, and in other cases it could be detrimental to productivity.

Thus, the ideotype approach has the advantage of focusing on traits that are expected to have a functional role in adaptation irrespective of their direct influence on economic yield. Some of the steps involved in this approach are:

1. Define the target drought-prone environments, and identify the predominant types of drought stress in each environment. Identify the crop species most likely to be able to con-

Transpiration efficiency (g/dm/kg H₂O/kPa VPD)

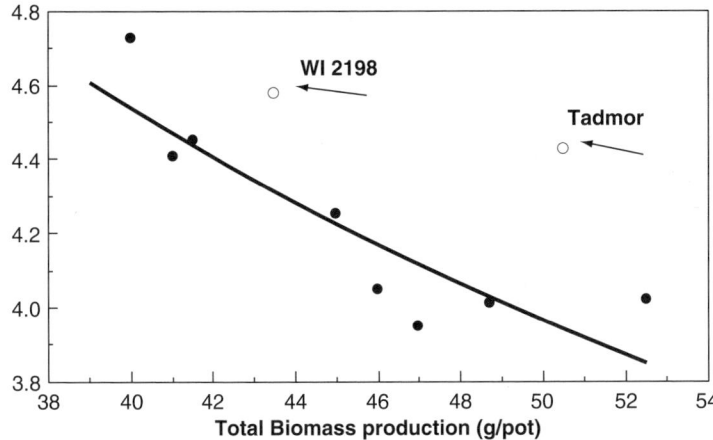

FIGURE 30.2 Transpiration efficiency and total biomass production in barley genotypes grown in a greenhouse. (From Acevedo E. In: Acevedo E, Conesa AP, Monneveux P, Srivastava JP, eds. *Physiology: Breeding of Winter Cereals for Stressed Mediterranean Environments*. Paris: INRA, 1991. With permission.)

tribute to improved productivity and stability in each target environment.

2. Define the possible morphological, physiological, phenological, and developmental traits that could contribute substantially to adaptation to drought stresses in each target environment.

3. Develop working hypotheses regarding the combinations of traits required for a given target environment (farmer participation should be encouraged at this stage).

4. Identify the genetic stocks for various putative constitutive and inducible traits in the germplasm and establish genetic correlations between the traits of interest and the degree of adaptation to the targeted drought stress.

5. Identify appropriate screening methodologies and protocols for characterizing selected genetic stocks that could act as donor parents for traits of interest.

6. Develop genetic markers for traits that are of critical nature for drought tolerance.

7. For a number of putative morphological traits such as leaf size, orientation, waxiness, cuticular transpiration, canopy temperature differentials, and developmental plasticity, markers perhaps need not be developed as the traits can be scored in a conventional breeding program with relative ease. However, if this were enough for them to be successfully used in applied breeding for improved drought tolerance, there would be clearly demonstrable cases of drought tolerant improved cultivars based on selection for these traits and this is seldom, if ever, the case. There may be no need to even include these traits unless opportunities arise to cost-effectively exploit the existing mapping populations to identify markers for these traits.

8. Incorporate some of the components of relevant physiological traits into various elite agronomic genetic backgrounds to provide a range of materials with specific traits of interest (i.e., developing NILs and/or RILs) to the NARS for improving drought adaptation of locally adapted varieties.

VI. HOW DO WE IDENTIFY TRAITS RELEVANT FOR DROUGHT ADAPTATION?

A. FUNCTIONAL RELATIONSHIPS AND THEIR SIGNIFICANCE FOR THE TRAITS OF INTEREST

"Passioura's concept" considers that yield (Y) can be expressed as $Y = W \times \text{WUE} \times \text{HI}$ (Y = yield; W = water transpired; WUE = water use efficiency; HI = harvest index) under water-limiting conditions [84]. Those traits (whether physiological or morphological) that can be shown to contribute to any of these three components that determine yield under drought stress conditions will be a trait of interest for improving drought tolerance. It is even possible that some of these traits will have synergistic effects when brought together e.g., osmotic adjustment (OA) and deep rooting attributes; glaucousness and deep rooting pattern; wax bloom and reduced leaf area; WUE and deep rooting; vertical leaf orientation and small leaf size; reduced cuticular transpiration and deep rooting; carbohydrate remobilization from stems and OA, etc. (see Refs. [24–26,47,48,85] for further discussion on trait based approaches). It is also necessary to characterize the target environment (e.g., soil type, water holding capacity, soil depth, soil moisture index, radiation load, relative humidity, evaporative demand, rainfall pattern, and amount) during the growing season. It is these environmental factors that determine the type and combination of traits required in the local variety undergoing improvement. For example, for shallow Vertisols of *rabi* (postrainy season where a crop is raised mostly on stored soil moisture) production systems in India, deep rooting habit will have no advantage; perhaps OA and morphological traits that reduce water loss or minimize water use during prereproductive stages of crop growth would be more relevant. Water use efficiency would be more relevant for *rabi* production environments than for the *kharif* (rainy season) crops, where water stress is intermittent and unpredictable. For crops such as pearl millet that are predominantly grown in sandy soils of low water-holding capacity, osmotic adjustment will have little advantage after the initiation of flowering.

VII. NECESSITY OF A THOROUGH UNDERSTANDING OF PHYSIOLOGICAL MECHANISMS TO DETERMINE THEIR ADAPTIVE ROLE

Very often, researchers pay inadequate attention while evaluating the adaptive role of a particular physiological stress tolerance mechanism to grain yield or biomass production under drought stress. Also, there is no unanimity in relation to what constitutes drought tolerance [41]. It is particularly challenging to evaluate the claims reported by a number of researchers, particularly where the conclusions were based on pot-grown plants. Also, research on physiological mechanisms needs to be conducted on a number of genotypes from any given crop to understand the relative importance of various mechanisms to the adaptation to drought. This is not often the case, and in many cases where a number of genotypes were evaluated for a given set of physiological mechanisms, the measurements were not carried out at various growth stages in order to assess the full implications of a particular mechanism contribution to drought adaptation. This can be illustrated using our experience with osmotic adjustment in pigeonpea. We have observed a widespread occurrence of osmotic adjustment in extra-short duration pigeonpea genotypes under drought [72]. The degree of de-osmotic adjustment during active grain-filling period is as important as the extent of osmotic adjustment during flowering and pod-setting period in determining grain yield under drought (Table 30.2). Two distinct patterns of osmotic adjustment were noticed (Figure 30.3). One set of genotypes, where the osmotic adjustment continued until physiological maturity, had the

TABLE 30.2
Forward Stepwise Multiple Regression of Osmotic Adjustment (OA) at Various Growth Periods to Total Dry Matter and Grain Yield of Pigeonpea Genotypes Under Drought ($n = 26$)[a]

Variable Added	Model r^2
Total dry matter at harvest	0.362**
OA 72 DAS	
Grain yield at maturity	
OA 72 DAS	0.161*
OA 82 DAS	0.068
OA 92 DAS	−0.210*
OA 72 + OA 82 + OA 92 DAS	0.407**

[a]Contributions of added variable (partial r^2) significant at *$P < 0.05$ or **$P < 0.01$.

Source: Subbarao GV, Chauhan YS, Johansen C. *Eur. J. Agron.* 2000; 12:239–249.

lowest grain yields (ranged from 1.0 to 1.34 Mg/ha) (Figure 30.3). In another set of genotypes, where osmotic adjustment reached its peak at the beginning of active grain-filling period, but dropped rapidly during the active grain-filling period attained the highest yields among the 26 genotypes (ranging from 1.30 to 1.80 Mg/ha) (Figure 30.3). The de-osmotic adjustment during active grain-filling period facilitated the remobilization of carbon and nitrogen re-

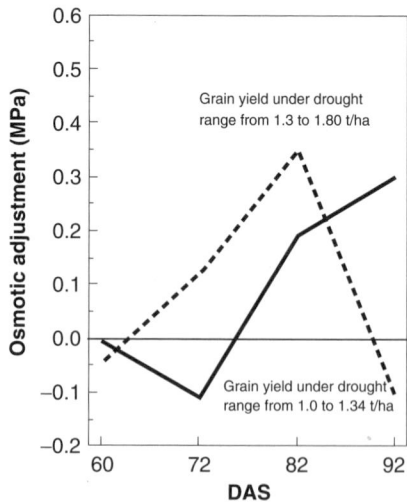

FIGURE 30.3 Developmental patterns of osmotic adjustment between two sets of pigeonpea genotypes and their grain yield under drought. (From Subbarao GV, Chauhan YS, Johansen C. *Eur. J. Agron.* 2000; 12:239–249. With permission.)

serves from the rest of the plant for the grain-filling process (even the solutes contributing to the osmotic adjustment were utilized for the grain production. Thus the early initiation of this mechanism is as important as the early termination of this mechanism during active grain-filling stage if this is going to have a maximum impact on grain yield under drought conditions. The above case illustrates the importance of a thorough understanding of a given mechanism before one can make a full assessment of its relative contribution to the adaptation to drought conditions.

VIII. MOLECULAR MARKERS AND THEIR IMPLICATIONS FOR A TRAIT-BASED APPROACH FOR GENETIC IMPROVEMENT IN DROUGHT ADAPTATION

A. MOLECULAR MARKERS

Molecular marker-assisted selection (MAS) is a powerful tool for plant breeding programs [49–51,54,55,86–88]. Screening at the molecular level is independent of environment and plant developmental stage. Thus, if genetic markers can be found that are associated with components of drought adaptation, this would resolve many of the technical, time, and expense problems associated with field screening. There are a wide range of DNA markers, but not all are suitable for application in plant breeding programs. Restriction fragment length polymorphism (RFLP) markers rely on hybridization of probe DNA with plant DNA. Although, RFLP markers provide high quality data and can often be used for comparative mapping across species, low throughput potential is one of their serious limitations for applied plant breeding. Random amplified polymorphic DNA (RAPD) was the first polymerase chain reaction (PCR)-based assay to receive widespread attention in plant fingerprinting and mapping studies. The problems of reproducibility within and between populations and between laboratories are the major drawbacks associated with this assay. Amplified fragment length polymorphism (AFLP) markers are very useful for simultaneously detecting a large number of nonspecific polymorphisms as in fingerprinting and diversity analysis. The AFLP assay is particularly useful as it requires no prior knowledge of the genome but offers more stringent amplification than RAPD. However, their direct application in MAS programs is not presently cost-effective and they must first be converted to allele-specific simple PCR tests. Sequence tagged microsatellite site (STMS) or simple sequence repeat (SSR) markers based on variable numbers of tandem repeats (VNTR), are much more reliable PCR-based markers as they are based on stringent amplification of known DNA sequences. STMS markers are highly polymorphic, provide codominant data (essential for plant breeders to distinguish between heterozygous and homozygous individuals), and thus remain the assay of choice for marker-assisted selection systems. Microsatellite markers (STMS) are expensive and time consuming to develop for practical breeding programs, but once available, they are very cost-effective to use.

The DNA chip-based technologies such as microarrays can be used to survey the expression of candidate genes from different genotypes under drought stress [89–92], but may have difficulties in producing repeatable results. Techniques such as "Diversity Array Technology, DArT® [93] appear to offer low-cost, high-throughput, and reliable assays, with minimal DNA sample requirement, and are capable of providing good coverage of the genome without prior sequence information for germplasm characterization, gene tagging, and MAS. If prior sequence information is available, then serial analysis of gene expression (SAGE) can be used to analyze patterns of expression. Expression analysis by a semiautomated DNA fragment analyzer is one of the major advantages of SAGE over traditional microarray technology for established molecular breeding programs [94].

The use of DNA markers for indirect selection offers greatest gains for quantitative traits with low heritability, as these are the most difficult characters to work with in the field when using conventional phenotypic selection [95,96]. However, this type of trait is also amongst the most difficult to develop effective marker-assisted selection systems. This is largely due to the effects of genotype-by-environment (G/E) interaction and epistasis. Precise phenotypic evaluation in several locations and seasons is essential to obtain unbiased estimates of the genetic variation controlling traits with low heritabilities and obtain estimates of the mean performance of individual mapping progeny in order to accurately estimate the relative contribution and stability of component QTL [65]. However, there is increasing support for the idea that the use of much larger populations is an even more important factor. The dissection of quantitative traits such as physiological mechanisms using DNA markers can force an increasing dependence on ever more complex biometric tools to facilitate interpretation and manipulation of datasets to identify the underlying genetic factors controlling the traits of interest. This requirement for assessing large progeny numbers in many environments can be a negative factor in this approach as it dramatically increases phenotyping costs. Nevertheless, the techniques that have allowed traditional plant breeders to deal with other complex phenotypes (such as yield in non-stressed environments) can be adopted for this new field of molecular breeding.

B. MAS-Based Approaches for Drought Adaptation Components

Trait-based approaches have been advocated by crop physiologists for quite some time, nevertheless, the approach has not been adequately integrated into many public sector breeding programs that target drought adaptation. Most of these physiological traits are complex and require carefully controlled environmental conditions to meaningfully evaluate expression. The measurement of these traits also requires an additional level of training, and in addition, the number of samples that can be handled in a reasonable amount of time is limited. This makes it very difficult to routinely integrate them into large-scale breeding programs. Given the current advances in the development of technology for molecular markers, it seems that it is now the right time to use molecular markers to integrate some of these physiological mechanisms into applied breeding programs addressing drought adaptation [52,53,56,61–65,68,96,97].

Phenotyping of most physiological traits across sufficiently large progeny populations and with adequate replication to permit effective QTL mapping is still a challenge, but it is an attainable target with new semiautomated technologies, if applied within carefully formulated experimental designs amenable to spatial analyses [95]. Once a QTL is identified for a given physiological trait, the use of its respective DNA markers in breeding programs is practical and desirable as these markers are not influenced by the growing environment. Once molecular markers (i.e., for a QTL) are linked to specific physiological mechanisms, it would be possible to move these various traits/characters into adapted cultivars or other agronomic backgrounds through marker-assisted breeding approaches. As mentioned earlier the effectiveness of a particular combination of drought tolerance component traits, be it biochemical, physiological, morphological, or developmental, will be determined by the locally adapted variety under improvement and the environment where the crop will be grown. Thus, with recent advances in marker-assisted selection systems, it is now possible and indeed an opportune time to adapt/investigate trait-based approaches for improving drought adaptation in crops of dry-land production systems.

IX. VISION AND RESEARCH PRIORITIES IN DROUGHT RESEARCH

The primary research mandate of International agricultural research centers that are located in the SAT region (or mandated to address drought adaptation) is to improve genetic yield potential and yield stability in dry-land crops (such as sorghum, millets, wheat, barley, chickpea, pigeonpea groundnut, cassava, beans, lentil, cowpea, upland rice, etc.). Drought stress is a major limiting factor in the SAT, preventing realization of genetic yield potential. Earlier research efforts aimed at improving levels of drought tolerance in staple food crops around the world have relied mostly on empirical approaches, which will provide the necessary foundation for future research that may include a substantial trait-based component. These empirical approaches resulted in the identification of several genetic stocks with varying degrees of adaptation to drought-prone production environments. These genetic stocks can be utilized for comparative physiological studies to unravel the mechanisms causing these differences in adaptation to drought stress (biotech-based analytical approach). Also, defining iso-drought environments (by using emerging research tools such as satellite image analysis, GIS, and soil water-balance modeling) in order to match traits with the appropriate niche-production environments, should receive a high priority in any new drought research program (see Ref. [98] for further discussion).

Breeders and crop physiologists need to work closely in testing the viability/validity of the trait-based approaches for improving drought tolerance. This has not happened to any great extent previously, thus missing a good opportunity for advancement. Applied breeding programs always want a clear demonstration of the value of a trait (morphological/physiological/biochemical) before considering incorporating it as a selection criterion. However, without breeder's participation, development of genetic stocks with specific traits incorporated into locally adapted genetic backgrounds to test the hypothesis of the value of a trait is very difficult. Thus, despite many decades of research on drought tolerance in several crops, only limited progress has been reported in developing NILs or RILs for specific component traits of drought tolerance [47,48,65,99–101]. Another limitation is that laboratory physiologists have been reluctant to adapt their trait assessment protocols to allow phenotyping of sufficiently large numbers of individual plants that segregating progenies can be assessed in screening experiments to demonstrate that their target traits are indeed heritable. Plant breeders on the other hand are unwilling to invest much of their time and resources pursuing such traits without this basic information. The few exceptions are often cases where the breeder has physiological expertise or vice versa [99,102,103].

Developing genetic stocks with specific traits or combination of traits in the adapted genetic background is critical in testing hypotheses related to the value of these traits in adaptation to drought-prone environments. Also, trait-based approaches can be used to delineate the very nature of those site-specific problems of drought adaptation from a breeding perspective [104]. Information of this type would facilitate transferring traits from germplasm of one variety to another. Thus IARCs should be aiming at providing genetic stocks with specific traits for components of drought tolerance that can be utilized by NARS breeding programs for improving drought adaptation of the local varieties. Since IARCs have a global mandate on several crops requiring improvement for drought stress, adding a larger component of this trait-based approach would provide a strong perspective for these institutes in addressing improvement of drought tolerance in the staple food crops grown under dry-land conditions in their mandated regions of SAT without constraint by site-specific adaptation problems.

A long-term vision is required while addressing the research agenda for improving drought adaptation in these international institutes that have a global mandate to reduce the vulnerability of staple food crops to drought stress. Incorporated into such a long-term vision must be trait identification, characterization, and evaluation, including the development of tools and genetic markers for specific physiological traits. Once the research has generated the necessary information, these international institutes could also act as resource centers for NARS of the SAT in developing and providing genetic stocks with specific traits that are components of drought adaptation in a range of genetic backgrounds, along with consulting services to assist the transfer of technology for marker-assisted and phenotypic selection of key drought tolerance component traits to NARS.

X. CONCLUDING REMARKS

It is hoped that this concept review will stimulate further discussion on determining the future direction of drought research for development in the SAT. Some initial work on trait-based approaches is ongoing at ICRISAT (pearl millet, sorghum, groundnut, and chickpea), JIRCAS, IRRI, WARDA (upland rice), and CIMMYT (maize and wheat). This has been facilitated by advances in DNA marker technologies that now allow relatively low-cost screening

of the large-scale mapping populations ($n = 250$ to 750) that are essential for precise QTL mapping drought tolerance components. In turn it is now time to reassess the instrumentation and methodologies of physiological research, so that these too can be readily applied to large-scale, replicated, multilocational trials. The limiting factor for molecular breeding programs of very many traits is now the quality of the phenotype data on which linkage mapping must be based. Only with equal advances in all components of the process will molecular breeding be able to have a substantial impact on the genetic improvement of drought tolerance. Characterization of the target drought environments where the crops are grown must also be done precisely and systematically to enable appropriate targeting of drought tolerance traits. This can be achieved through using historical climatic data series, GIS tools, water-balance, and crop simulation models [98]. With improved knowledge of probable soil moisture availability over time, it also becomes easier to further exploit the drought escape option, considering the spectrum of crop duration and germplasm availability.

The ideotype approach for incorporating the relevant drought tolerance traits requires a better knowledge of the physiological mechanisms involved in drought tolerance and their genetic control. Simple mechanistic models that can reliably simulate crop growth and yield in different environments can also be used for the assessment of the putative drought tolerance traits in a wide range of target environments. Despite the difficulties associated with genetic enhancement of root systems to make them more effective in water extraction, this would seem a high-priority effort for rainfed chickpea and extra-short-duration pigeonpea. Dissection of root traits and development of a screening system relevant to field conditions are therefore needed, in parallel with extensive genotyping, use of functional genomics and search for molecular markers. Other promising integrated traits for improving drought tolerance and crop water productivity include panicle harvest index in pearl millet, stay-green in sorghum, xylem exudation rate in upland rice, and transpiration use efficiency in groundnut. There seems to be much scope for improving such characters, using QTL mapping and molecular breeding techniques aided by physiological characterization and conventional breeding, to significantly improve the ability of staple food crops to withstand drought stress in defined target environments.

The challenges associated with trait-based analytical approaches, particularly for physiological traits are further complicated by the fact that the degree of expression for these physiological traits needed for a given target environment may vary depending on the type and severity of drought they experience [75]. To some extent, this can be resolved by using simulation modeling to determine the degree of expression needed in any target environment, based on the historical rainfall and water-balance information. Also, germplasm enhancement efforts through the generation of genetic stocks and population improvement for specific traits, with varied genetic backgrounds, are essential resources. These clearly take an enormous amount of effort and should therefore be generated by the international institutes through crop-based consortia. The training required for regional breeding program personnel to undertake such a knowledge-based breeding requires a "new paradigm shift" in our way of thinking. The issue of drought tolerance is of fundamental importance to dry-land agriculture; thus it will have to be resolved as crop adaptation to moisture deficits is the key factor for improving the crop productivity in the SAT, where drought is an integral part of any dry-land agricultural production environment. A second green revolution cannot be achieved without improving crop productivity of these dry-land crops, particularly as water scarcity becomes an increasing threat throughout many of the countries in the SAT.

One of the most critical components of this approach is the assumption that MAS-based methodologies can and will be suitable for handling physiological traits now or in the near future. We understand that there are still practical problems and challenges associated with MAS-based methodologies for their widespread adoption in practical public sector plant breeding (see Ref. [54] for further discussion). However, the technological advances driven by the genomics revolution will have substantial impact on even these underfunded public sector plant breeding programs. A major portion of the easily achievable yield gains have been accomplished by empirical breeding approaches in the past. The next phase of improvements should attempt to break barriers on yield potential or to bridge the gap between this potential and realized yields of crops grown in drought-prone environments. This requires knowledge-based breeding where analytical skills provided by the crop physiologists coupled with the tools of modern genomics will provide the synergistic strengths, which are crucial for crop improvements in the 21st century.

ACKNOWLEDGMENTS

The authors wish to thank Dr. Chris Johansen for many useful suggestions on early drafts and Dr. Hutokshi Buhariwalla (ICRISAT) for helpful discussions on the application of DarT and SAGE.

REFERENCES

1. Rosenow DT, Quisenberry JE, Wendt CW, Clark LE. Drought tolerant sorghum and cotton germplasm. *Agric. Water Manag.* 1983; 7:207–222.

2. Edmeades GO, Bolanos J, Lafitte HR. Progress in breeding for drought tolerance in maize. In: Wilkinson DB, ed. *Proceedings of the 47th Annual Corn & Sorghum Research Conference.* Washington, DC: ASTA, 1992:93–111.

3. Bolanos J, Edmeades GO. 1993. Eight cycles of selection for drought tolerance in lowland tropical maize. II. Responses in reproductive behavior. *Field Crops Res.* 1993; 31:253–268.

4. Heisey PW, Edmeades GO. Part I. Maize production in drought stressed environments: technical options and research resource allocation. In: CIMMYT, ed. *World Maize Facts and Trends 1997/98.* Mexico: CIMMYT, 1999:1–36.

5. Edmeades GO, Cooper M, Lafitte R, Zinselmeier C, Ribaut JM, Habben JE, Loffler C, Banziger M. Abiotic stresses and staple crops. In: Nosberger J, Geiger HH, Struik PC, eds. *Crop Science: Progress and Prospects.* Wallingford, U.K.: CAB International, 2001:137–154.

6. Araus JL, Slafer GA, Reynolds MP, Royo C. Plant breeding and drought in C_3 cereals: what should we breed for? *Ann. Bot.* 2002; 89:925–940.

7. Condon AG, Richards RA, Rebetzke GJ, Farquhar GD. Improving intrinsic water-use efficiency and crop yield. *Crop Sci.* 2002; 42:122–131.

8. Ortiz R, Ekanayake I, Mahalakshmi V, Menkir A, Nigam SN, Saxena NP, Singh BB. Development of drought resistant and water stress tolerant crops through traditional breeding. In: Yajima M, Okada K, Matsumoto N, eds. *Water for Sustainable Agriculture in Developing Regions.* Tsukuba, Japan: Japan International Research Center for Agricultural Sciences (JIRCAS), 2002:11–21.

9. Singh BB, Mai-Kodomi Y, Terao T. A simple screening method for drought tolerance in cowpea. *Indian J. Genet.* 1999; 59:211–220.

10. Singh BB, Mai-Kodomi Y, Terao T. Relative drought tolerance of major rainfed crops of the semi-arid tropics. *Indian J. Genet.* 1999; 59:437–444.

11. Rohrbach DD, Lechner WR, Ipinge SA, Monyo ES. *Impact from Investments in Crop Breeding: The Case of Okashana 1 in Namibia.* ICRISAT Impact Series No. 4. Patancheru, India: ICRISAT, 1999.

12. Witcombe JR, Rao MNVR, Lechner WR. Registration of ICMV 88908 pearl millet. *Crop Sci.* 1995; 35:1216–1217.

13. Witcombe JR, Rao MNVR, Raj AGBR, Hash CT. Registration of ICMV 88904 pearl millet. *Crop Sci.* 1997; 37:1022–1023.

14. Yapi AM, Debrah SK, Dehala G, Njomaha C. *Impact of Germplasm Research Spillovers: The Case of Sorghum Variety S 35 in Cameroon, and Chad.* ICRISAT Impact Series No. 3. Patancheru, India: ICRISAT, 1999.

15. Jones MP, Mande S. Breeding Drought-Resistant Upland Rice Varieties. WARDA Annual Report. Cote d'Ivoire: West Africa Rice Development Association, ISBN 92 9113 0656, 1994.

16. Jones MP. Food security and major technological challenges. The case rice in sub-Saharan Africa. *Jpn. J. Crop Sci.* 1999; 67(extra issue 2):57–64.

17. Tobita S, Ookawa T, Audebert AY, Jones MP. Xylem exudation rate: a proposed screening criterion for drought resistance in rice. *Proceedings of the 6th Symposium of the International Society of Root Research*, Nagoya, Japan, ISBN4-931358-07-1, 2001.

18. Mai-Kodomi Y, Singh BB, Myers O Jr, Yopp JH, Gibson PJ, Terao T. Two mechanisms of drought tolerance in cowpea. *Indian J. Genet.* 1999; 59:309–316.

19. Mai-Kodomi Y, Singh BB, Terao T, Myers Jr O, Yopp JH, Gibson PJ. Inheritance of drought tolerance in cowpea. *Indian J. Genet.* 1999; 59:317–323.

20. Saxena NP. Management of drought in chickpea — a holistic approach. In: Saxena NP, ed. *Management of Agricultural Drought — Agronomic and Genetic Options*. New Delhi: Oxford & IBH Publishing Co. Pvt. Ltd., 2003.

21. Nageswara Rao RC, Nigam SN. Genetic options for drought management in groundnut. In: Saxena, NP, ed. *Management of Agricultural Drought — Agronomic and Genetic Options*. New Delhi: Oxford & IBH Publishing Co. Pvt. Ltd., 2001.

22. Chauhan YS, Wallace DH, Johansen C, Singh L. Genotype-by-environment interaction effect on yield and its physiological bases in short-duration pigeonpea. *Field Crops Res.* 1998; 59:141–150.

23. ICRISAT. Medium Term Plan 1994–1998. Volume 1. Main Report. Patancheru, India: ICRISAT, 1992.

24. Richards RA, Rebetzke GJ, Appels R, Condon AG. Physiological traits to improve yield of rainfed wheat. Can molecular genetics help? CIMMYT Workshop on Molecular Approaches for the Genetic Improvement of Cereals for Stable Production in Water-Limited Environments, El Batan, Mexico, June 21–25, 1999 (website only).

25. Richards RA. Selectable traits to increase crop photosynthesis and yield of grain crops. *J. Exp. Bot.* 2000; 51:447–458.

26. Turner NC, Wright GC, Siddique KHM. Adaptation of grain legumes (pulses) to water-limited environments. *Adv. Agron.* 2001; 71:193–231.

27. Serraj R, Bidinger FR, Chauhan YS, Seetharama N, Nigam SN, Saxena NP. Management of drought in ICRISAT cereal and legume mandate crops. In: Kijne JW, ed. *Water Productivity in Agriculture: Limits and Opportunities for Improvement*. Wallingford, U.K.: CAB International, 2002:127–144.

28. Wong SC, Cowan IR, Farquhar GD. Stomatal conductance correlates with photosynthetic capacity. *Nature* 1979; 282:424–426.

29. Ramos C, Hall AE. Relationships between leaf conductance, intercellular CO_2 partial pressure and CO_2 uptake rate in two C_3 and two C_4 plant species. *Photosynthetica* 1982; 16:343–355.

30. O'Leary MHO, Treichel I, Rooney M. Short-term measurement of carbon isotope fractionation in plants. *Plant Physiol.* 1986; 80:578–582.

31. Rebetzke GJ, Condon AG, Richards RA, Farquhar GD. Selection for reduced carbon isotope discrimination increases aerial biomass and grain yield of rainfed bread wheat. *Crop Sci.* 2002; 42:739–745.

32. Subbarao GV, Johansen C. Transpiration efficiency — avenues for genetic improvement in crops. In: Pessarakli M, ed. *Handbook of Plant and Crop Physiology*. Second edition. New York: Marcel Dekker, 2002:835–856.

33. Sinclair TR, Ludlow MM. Influence of soil water supply on the plant water balance of four tropical grain legumes. *Aust. J. Plant Physiol.* 1986; 13:329–341.

34. Turner NC. Water use efficiency of crop plants: potential for improvement. In: Buxton DR, Shibles R, Forsberg RA, Blad BL, Asay KH, Paulsen GM, Wilson RF, eds. *International Crop Science I*. Madison, WI: Crop Science Society of America, 1993:75–81.

35. Payne WA, Gerard B, Klaij MC. Subsurface drip irrigation to evaluate transpiration ratios of pearl millet. In: Lamm FR, ed. *Microirrigation for a Changing World: Conserving Resources/Preserving the Environment*. Proceedings of Fifth International Microirrigation Congress, Orlando, FL, April 2–6, 1995. St. Joseph, MI: American Society of Agricultural Engineers, 1995.

36. Turner NC. Further progress in crop water relations. *Adv. Agron.* 1997; 58:293–337.

37. Serraj R, Allen HL, Sinclair TR. Soybean leaf growth and gas exchange response to drought under carbon dioxide enrichment. *Global Change Biol.* 1999; 5:283–292.

38. Serraj R, Sinclair TR. Osmolyte accumulation: can it really help increase crop yield under drought conditions? *Plant Cell Environ.* 2002; 25:335–341.

39. Ludlow MM, Muchow RC. A critical evaluation of traits for improving crop yields in water-limited environments. *Adv. Agron.* 1990; 43:107–153.

40. Clarke JM, Romagosa I, DePauw RM. Screening durum wheat germplasm for dry growing conditions. *Crop Sci.* 1991; 31:770–775.

41. Subbarao GV, Johansen C, Slinkard Al, Rao RCN, Saxena NP, Chauhan YS. Strategies for improving drought resistance in grain legumes. *CRC Crit. Rev. Plant Sci.* 1995; 14:469–523.

42. Campbell SA, Close TJ. Dehydrins: genes, proteins, and associations with phenotypic traits. *New Phytol.* 1997; 137:61–74.

43. Nam NH, Subbarao GV, Chauhan YS, Johansen C. Importance of canopy attributes in determining dry matter accumulation of pigeonpea under contrasting moisture regimes. *Crop Sci.* 1998; 38:955–961.

44. Turner NC. Optimizing water use. In: Nosberger J, Geiger HH, Struik PC, eds. Crop Science. London: CAB International, 2001:119–135.

45. Subbarao GV, Levine LH, Stutte GW, Wheeler RM. Glycine betaine accumulation: its role in stress resistance in crop plants. In: Pessarakli M, ed. *Handbook of*

Plant and Crop Physiology. Second edition. New York: Marcel Dekker, 2002:881–907.

46. Subbarao GV, Nam NH, Chauhan YS, Johansen C. Osmotic adjustment, water relations, and carbohydrate remobilization of pigeonpea under water deficits. *J. Plant Physiol.* 2000; 157:651–659.

47. Morgan JM. Osmoregulation as a selection criterion for drought tolerance in wheat. *Aust. J. Agric. Res.* 1983; 34:607–614.

48. Richards R, Passioura JB. A breeding program to reduce the diameter of the major xylem vessel in the seminal roots of wheat and its effect on grain yield in rain-fed environments. *Aust. J. Agric. Res.* 1989; 40:943–950.

49. Lee M. DNA markers and plant breeding programs. *Adv. Agron.* 1995; 55:265–344.

50. Mohan M, Nair S, Bhagwat A, Krishna TG, Yano M, Bhatia CR, Sasaki T. Genome mapping, molecular markers and marker-assisted selection in crop plants. *Mol. Breed.* 1997; 3:87–103.

51. Nguyen HT, Babu RC, Blum A. Breeding for drought resistance in rice: physiology and molecular genetics considerations. *Crop Sci.* 1997; 37:1426–1434.

52. Ribaut JM, Hoisington DA, Deutsch JA, Jiang C, Gonzalez-de-Leon D. Identification of quantitative trait loci under drought conditions in tropical maize. I. Flowering parameters and the anthesis-silking interval. *Theor. Appl. Genet.* 1996; 92:905–914.

53. Ribaut JM, Jiang C, Gonzalez-de-Leon D, Edmeades GO, Hoisington DA. Identification of quantitative trait loci under drought conditions in tropical maize. 2. Yield components and marker-assisted selection strategies. *Theor. Appl. Genet.* 1997; 94:887–896.

54. Ribaut JM, Hoisington DA. Marker-assisted selection: new tools and strategies. *Trends Plant Sci.* 1998; 3:236–239.

55. Ribaut JM, Betran FJ. Single large-scale marker-assisted selection (SLS-MAS). *Mol. Breed.* 1999; 5:531–541.

56. Jones NH, Ougham H, Thomas H. Markers and mapping: we are all geneticists now. *New Phytol.* 1997; 137:156–177.

57. Prioul JL, Quarrie S, Causse M, de Vienne D. Dissecting complex physiological functions through use of molecular quantitative genetics. *J. Exp. Bot.* 1997; 48:1151–1163.

58. Crusta OR, Xu WW, Rosenow DT, Mullet J, Nguyen HT. Mapping of post-flowering drought resistance traits in grain sorghum: association between QTLs influencing premature senescence and maturity. *Mol. Gen. Genet.* 1999; 262:579–588.

59. Ito O, O'Toole J, Hardy B. *Genetic Improvement of Rice for Water-Limited Environments*. Los Banos, The Philippines: The International Rice Research Institute, 1999.

60. Kebede H, Subudhi PK, Rosenow DT, Nguyen HT. Quantitative trait loci influencing drought tolerance in grain sorghum (*Sorghum bicolor* L. Moench). *Theor. Appl. Genet.* 2001; 103:266–276.

61. Zhang J, Chandra Babu R, Pantuwan G, Kamoshita A, Blum A, Wade L, Sarkarung S, O'Toole JC, Nguyen HT. Molecular dissection of drought tolerance in rice: from physio-morphological traits to field performance. In: Ito O, Hardy J, eds. *Genetic Improvement of Rice for Water-Limited Environments*. Los Banos, Philippines: International Rice Research Institute, 1999:331–343.

62. Zhang J, Nguyen HT, Blum A. Genetic analysis of osmotic adjustment in crop plants. *J. Exp. Bot.* 1999; 50:291–302.

63. Zhang J, Zheng HG, Aarti A, Pantuwan G, Nguyen TT, Tripathy JN, Sarial AK, Robin S, Babu RC, Nguyen BD, Sarkarung S, Blum A, Nguyen HT. 2001. Locating genomic regions associated with components of drought resistance in rice: comparative mapping within and across species. *Theor. Appl. Genet.* 2001; 103:19–29.

64. Ribaut JM, Poland D. Molecular approaches for the genetic improvement of cereals for stable production in water-limited environments. A strategic planning workshop held at CIMMYT, El Batan, Mexico, June 21–25 1999, 2000.

65. Tao YZ, Henzell RG, Jordan DP, Butler DG, Kelly AM, McIntyre CL. Identification of genomic regions associated with stay-green in sorghum by testing RILs in multiple environments. *Theor. Appl. Genet.* 2000; 100:1125–1232.

66. Crouch JH, Serraj R. DNA marker technology as a tool for genetic enhancement of drought tolerance at ICRISAT. In: Saxena NP, ed. *International Workshop on Field Screening for Drought Tolerance in Rice*. Patancheru, India: ICRISAT, 2002:155–170.

67. Yadav RS, Hash CT, Cavan GP, Bidinger FR, Howarth CJ. Quantitative trait loci associated with traits determining grain and stover yield in pearl millet under terminal drought stress conditions. *Theor. Appl. Genet.* 2002; 104:67–83.

68. Haussmann B, Mahalakshmi V, Reddy BVS, Seetharama N, Hash CT, Geiger HH. QTL mapping of stay-green in two sorghum recombinant inbred populations. *Theor. Appl. Genet.* 2003; 106:133–142.

69. Mahalakshmi V, Ortiz R. Plant genomics and agriculture: from model crops to other crops, the role of data mining for gene discovery. *Electron. J. Biotechnol.* 2001; 4: http://ejb.ucv.cl.content.vol4/issue3/full/5/index.html.

70. Mahalakshmi V, Aparna P, Ramadevi S, Ortiz R. Genomic sequence derived simple sequence repeat markers — case study with Medicago sp. *Electron. J. Biotechnol.* 2002; 5(3):13–14.

71. Subudhi PK, Rosenow DT, Nguyen HT. Quantitative trait loci for the stay green trait in sorghum (*Sorghum bicolor* L. Moench). *Theor. Appl. Genet.* 2000; 101:733–741.

72. Subbarao GV, Chauhan YS, Johansen C. Patterns of osmotic adjustment in pigeonpea — its importance as a mechanism of drought resistance. *Eur. J. Agron.* 2000; 12:239–249.

73. Morgan JM. Osmoregulation and water stress in higher plants. *Annu. Rev. Plant Physiol.* 1984; 35:299–319.

74. Ehdaie B, Hall AE, Farquhar GD, Nguyen HT, Waines JG. Water-use efficiency and carbon isotope discrimination in wheat. *Crop Sci.* 1991; 31:1282–1288.

75. Yang RC, Jana S, Clarke JM. Phenotypic diversity and associations of some potentially drought-responsive characters in durum wheat. *Crop Sci.* 1991; 31:1484–1491.

76. Hubick KT, Shorter R, Farquhar GD. Heritability and genotype X environment interactions of carbon isotope discrimination and transpiration efficiency in peanut (*Arachis hypogaea*). *Aust. J. Plant Physiol.* 1988; 15:799–813.

77. Wright GC, Hubick KT, Farquhar GD. Discrimination in carbon isotopes of leaves correlates with water-use efficiency of field-grown peanut cultivars. *Aust. J. Plant Physiol.* 1988; 15:815–825.

78. Nageswara Rao RC, Williams JH, Wadia KDR, Hubick KT, Farquhar GD. Crop growth, water-use efficiency and carbon isotope discrimination in groundnut (*Arachis hypogaea L.*) genotypes under end-of season drought conditions. *Ann. Appl. Biol.* 1993; 122:357–367.

79. Condon AG, Richards RA, Farquhar GD. Carbon isotope discrimination is positively correlated with grain yield and dry matter production in field-grown wheat. *Crop Sci.* 1987; 27:996–1001.

80. Farquhar GD, Richards RA. Isotopic composition of plant carbon correlates with water-use efficiency of wheat genotypes. *Aust. J. Plant Physiol.* 1984; 11:539–552.

81. Acevedo E. Improvement of winter cereal crops in Mediterranean environments. Use of yield, morphological and physiological traits. In: Acevedo E, Conesa AP, Monneveux P, Srivastava JP, eds. *Physiology: Breeding of Winter Cereals for Stressed Mediterranean Environments.* Paris: INRA, 1991.

82. Read JJ, Johnson C, Carver BF, Quarrie SA. Carbon isotope discrimination, gas exchange, and yield of spring wheat selected for abscisic acid content. *Crop Sci.* 1991; 31:139–146.

83. Johnson RC, Bassett LM. Carbon isotope discrimination and water use efficiency in four cool-season grasses. *Crop Sci.* 1991; 31:157–162.

84. Passioura JB. Grain yield, harvest index and water use of wheat. *J. Aust. Inst. Agric. Sci.* 1977; 43:117–120.

85. Shan BR, Carlton GP, Siddique KHM, Regan KL, Turner NC, Anderson WK. Integration of breeding and physiology: lessons from a water-limited environment. In: Buxton DR, Shibles R, Forsberg RA, Blad BL, Asay KH, Paulsen GM, Wilson RF, eds. *International Crop Science I.* Madison, WI: Crop Science Society of America, 1993:607–614.

86. Crouch JH. Molecular Marker-Assisted Breeding: A Perspective for Small to Medium-Sized Plant Breeding Companies. Asia and Pacific Seed Association Technical Report No. 30, 2001:1–14.

87. Hash CT, Schaffent RE, Peacock JM. Prospects for using conventional techniques and molecular biological tools to enhance performance of orphan crop plants on soils low in available phosphorus. *Plant Soil* 2002; 245:135–146.

88. Sharma HC, Crouch JH, Sharma KK, Seetharama N, Hash CT. Applications of biotechnology for crop improvement: prospects and constraints. *Plant Sci.* 2002; 163:381–395.

89. Shinozaki K, Yamaguchi-Shinozaki K. Molecular responses to drought and cold stress. *Curr. Opin. Biotechnol.* 1996; 7:161–167.

90. Shinozaki K, Shinozaki-Yamaguchi K. Gene expression and signal transduction in water-stress response. *Plant Physiol.* 1997; 115:327–334.

91. Bohnert H, Fischer R, Kawasaki S, Michalowski C, Wang H, Yale J, Zepeda G. Cataloging stress-inducible genes and pathways leading to stress tolerance. In: Ribaut JM, Poland D, eds. *Molecular Approaches for the Genetic Improvement of Cereals for Stable Production in Water-Limited Environments.* A strategic planning workshop held at CIMMYT, El Batan, Mexico, June 21–25, 1999. Mexico: CIMMYT, 2000:156–161.

92. Seki M, Narusaka M, Abe H, Kasuga M, Yamaguchi-Shinozaki K, Carninci P, Hayashizaki Y, Shinozaki K. Monitoring the expression pattern of 1300 Arabidopsis genes under drought and cold stress by using a full-length cDNA microarray. *Plant Cell* 2001; 13:61–72.

93. Jaccoud D, Peng K, Feinstein D, Kilian A. Diversity arrays: a solid state technology for sequence information independent genotyping. *Nucleic Acids Res.* 2001; 29:1–7.

94. Matsumura H, Nirasawa S, Terauchi R. Transcript profiling in rice (*Oryza sativa L.*) seedlings using serial analysis of gene expression (SAGE). *Plant J.* 1999; 20:716–726.

95. Moreau L, Charcosset A, Hospital F, Gallais A. Marker-assisted selection efficiency in populations of finite size. *Genetics* 1998; 148:1353–1365.

96. Ribaut JM, Banziger M, Hoisington D. 2002. Genetic Dissection and Plant Improvement under Abiotic Stress Conditions: Drought Tolerance in Maize as an Example. JIRCAS Working Report. Tsukuba, Japan: JIRCAS, 2002:85–92.

97. Tuberosa R, Salvi S, Sanguineti MC, Landi P, Maccaferri M, Conti S. Mapping QTLs regulating morpho-physiological traits and yield: case studies, shortcomings, and perspectives in drought stressed maize. *Ann. Bot.* 2002; 89:941–963.

98. Subbarao GV, Kumar Rao JVDK, Johansen C, Deb UK, Ahmad I, Krishnarao MV, Venkatratnam L, Hebbar KR, Sai MVSR, Harris D. *Spatial Distribution and Quantification of Rice-Fallows in South Asia — Potential for Legumes.* Patancheru, India: International Crops Research Institute for the Semi-Arid Tropics, 2001.

99. Morgan JM, Tan MK. Chromosomal location of a wheat osmoregulation gene using RFLP analysis. *Aust. J. Plant Physiol.* 1996; 23:803–806.

100. Ray JD, Yu LX, McCouch SR, Champoux MC, Wang G, Nguyen HT. Mapping quantitative trait loci associ-

ated with root penetration ability in rice (*Oryza sativa L.*). *Theor. Appl. Genet.* 1996; 42:627–636.

101. Mahalakshmi V, Bidinger FR. Evaluation of stay-green sorghum germplasm lines at ICRISAT. *Crop Sci.* 2002; 42:965–974.

102. Morgan JM. Pollen grain expression of a gene controlling differences in osmoregulation in wheat leaves: a simple breeding method. *Aust. J. Agric. Res.* 1999; 50:953–962.

103. Morgan JM. Changes in ecological properties and endosperm peroxidase activity associated with breeding for an osmoregulation gene in bread wheat. *Aust. J. Agric. Res.* 1999; 50:963–968.

104. Wade LJ, McLaren CG, Quintana L, Harnpichitvitaya D, Rajatasereekul S, Singh SAK. Genotype by environment interactions across diverse rainfed lowland rice environments. *Field Crops Res.* 1999; 64:33–50.

Section XI

*Photosynthetic Activity
Measurements and Analysis
of Photosynthetic Pigments*

31 Whole-Plant CO_2 Exchange as a Noninvasive Tool for Measuring Growth

Evangelos D. Leonardos and Bernard Grodzinski
Department of Plant Agriculture, University of Guelph

CONTENTS

I. INTRODUCTION

Plant growth and productivity are frequently quantified on the basis of dry matter accumulation. Dry matter accumulation is dependent on photosynthesis. Carbon, oxygen, and hydrogen represent approximately 95% of the dry mass [1]. Net C uptake is derived from photosynthetic fixation of CO_2 via the C_3 cycle [2–4]. However, it is difficult to correlate quantitatively whole-plant productivity to photosynthesis particularly that of a single leaf [5–9]. Most crops achieve less than 50% of their photosynthetic potential due to mutual shading, limitations of CO_2, nutrient and water supply during plant development [5,10–13]. Respiratory losses occur from all tissues and constitute an important limitation to seasonal C gain [14–16]. The balance of daytime C gain and nighttime C loss from the whole plant determines the rate of daily C accumulation which subsequently controls plant growth and development [17–20]. Plant biomass production can be correlated to whole-plant and canopy net CO_2 exchange rates if the duration of the leaf canopy is known and nighttime respiratory losses are assessed [15,21–25].

In this chapter, we discuss two fundamental approaches that have been used to estimate growth and productivity. The traditional approach involves measurement of plant dry weight and leaf area following destructive harvests [26]. However, the emphasis in this chapter is on the value of noninvasive continual whole-plant CO_2 exchange measurements as a tool for quantifying biomass gain and growth. As outlined below there are advantages and disadvantages of both approaches to understanding the effects of environmental factors, plant pathogen interactions, and gene regulation on whole-plant net C gain and productivity. The primary advantages of measurements of whole-plant gas exchange and C gain are that the analysis: (1) is made nondestructively, (2) requires fewer plants than does traditional destructive analytical procedures, and (3) permits direct and accurate comparisons of photosynthetic C gain and nighttime respiratory C loss in real time that can be correlated with development. A problem with net C gain estimated by CO_2 exchange alone is that partitioning of reduced C, N, and S compounds and ultimately the growth, form, and development of sinks still need to be determined. Quantitative analysis of assimilate allocation and development of sinks still rely on additional experimental approaches often requiring invasive procedures analysis.

II. GROWTH ANALYSIS

A. INVASIVE ANALYSIS OF BIOMASS GAIN

Traditional measurements of plant dry mass require destructive harvest of the plants [26,27]. Although large numbers of samples are generally required for statistical accuracy, the great advantage of conventional plant growth analysis are that it provides accurate measurements of whole-plant biomass and the opportunity to determine dry matter allocation to developing sinks [28,29]. Samples collected during destructive plant growth analysis are available for more detailed analyses of other variables such as leaf area, organ size, chemical composition, metabolites, proteins, and genes. A historic advantage of destructive analyses is that, although sample handling can be very labor intensive, for assessing general growth patterns the procedures do not require sophisticated equipment. Modern equipment such as leaf area meters, flat bed scanners, and electronic balances greatly enhance the speed and accuracy with which measurements can be made and data analyzed. Destructive analytical procedures have been applied extensively in agricultural and ecological studies. Readers are directed to several key references for more detailed discussions of methodologies [26–32].

Two important parameters that are frequently determined following conventional destructive harvests are dry mass and leaf area [26,30,33]. Total dry weight is a measurement of photosynthetic accumulation of biomass corrected for respiratory loss over time [26]. Leaf area measurements provide a means of expressing the photosynthetic potential. For example, the net assimilation rate (NAR) is the rate of dry weight production per unit leaf area per unit time. Normally, the time intervals are several days or weeks. Another important calculation which is derived from the leaf area and dry weight data is the relative growth rate (RGR), which simply defined is the increase in dry weight per unit dry weight, per unit time. RGR expressed on a biomass basis is the product of NAR and leaf area ratio (LAR), which is the ratio between the total leaf area and the total plant dry weight.

Plant productivity is significantly correlated with RGR in many species [34]. In both agricultural and ecological studies, it is valuable to compare the RGR of different species [34–36] during development, as influenced by exposure to different environmental conditions [26,32]. Grime and Hunt [35] conducted extensive examinations of the RGR of 132 species of flowering plants from contrasting habitats. Between 2 and 5 weeks after germination, RGR ranged from 0.22 to 2.20 g/g/week. Herbaceous species tended to have higher RGR than woody species. Sun plants had higher RGR than shade-adapted species. In a study of 24 C_3 species with varying RGR, Poorter et al. [37] reported that short-term rates of shoot net photosynthesis, dark respiration, and root respiration were all positively correlated with RGR on a dry weight basis. Fast growing species fixed more CO_2 per unit total plant dry weight than slow growing species. In addition, fast growing plants allocated a lower percentage of their fixed C to shoot and root respiration and more C to leaves.

Environmental conditions such as light, CO_2 concentration, temperature, and nutrient supply affect both NAR and RGR. Differences in RGR are due to variations in NAR and or LAR. Poorter [36] concluded that LAR is a very important parameter determining the inherent differences in RGR among species and NAR is of secondary importance. The relationship between NAR and canopy photosynthesis may be altered by changes in dark respiration [8]. On a leaf area basis, NAR has been found to correlate with leaf photosynthesis, leaf N content, and dark respiration of shoots and roots [38]. A decrease in canopy NAR is usually compensated by an increase in LAR. It has been reported that in many herbaceous C_3 species, an increase in LAR accounts for 80% to 90% of the increase in RGR while higher NAR only accounts for 10% to 20% higher RGR [36].

Daily growth rate is the balance between daytime dry weight gain (primarily from C fixation) and nighttime dry weight loss as a result of dark respiration. Diurnal patterns of plant dry weight change can only be detected by destructive growth analysis when sampling is frequent (e.g., hourly) and large number of plants are sacrificed [39,40]. Seedlings of the grass *Holcus lanatus* were grown in full or limited nutrient medium and harvested hourly for dry weight measurement over a 3-day period [39]. A diurnal pattern of dry weight change was distinguished using regression analysis, with the maximum growth occurring during the period of illumination. The RGR in the dark was not significantly different than zero, suggesting the plants did not respire during the nighttime. However, random variability in the primary dry weight data was high. Wickens and Cheeseman [40] also applied destructive analysis techniques in a short-term study with seedlings of *Spergular marina* L. and *Lactuca sativa* L. grown in nutrient solutions in controlled environments. It was found that the RGR was higher during the nighttime than during the daytime which the authors reasoned was unrealistic. Alternatively, measurement of whole-plant CO$_2$ exchange has been adopted as a more sensitive method to measure small changes in dry matter accumulation occurring during short-term studies.

B. NONINVASIVE ANALYSIS

Unlike destructive analyses that can be relatively inexpensive, noninvasive analyses based on net C exchange rate (NCER) generally require specialized equipment for the measurement of CO$_2$ fluxes and environmentally controlled plant holding chambers.

1. Chamber and System Design

There have been many gas exchange systems developed for measuring whole-plant biomass accumulation based on analysis of CO$_2$ exchange. It is beyond the scope of this chapter to describe these systems and materials used in their construction in any detail. The reader is directed to specific references [18,19,21,22, 41–54]. Although each system may vary in the degree of automation and complexity of its design, whole-plant CO$_2$ exchange systems have two basic components: (1) the plant holding chamber with its associated environmental control systems, and (2) the gas mixing and analysis systems. In addition, specialized hardware and software are used to integrate environmental control with data collection and analysis.

The size of the assimilation chambers varies greatly depending on the number of plants and the characteristics of the canopy which are investigated. The light source will also vary depending on the style and the objective of the experiment. In chambers designed to enclose a portion of the canopy in field [43,48] or in the greenhouse [18,45,46,49,50,55] natural radiation alone may be used. Artificial irradiation, usually supplied from overhead lights (e.g., high-intensity discharge lamps) is a common feature of many laboratory systems. In addition, novel systems have been design with the capacity to provide inner canopy lighting [54]. Materials commonly used to construct the chambers and the gas analysis system may affect plant growth, development, and gas exchange [47,53,56–58]. For example, plastics can release volatile hydrocarbons such as C$_2$H$_4$ gas, which can alter plant development, canopy architecture, and net CO$_2$ exchange (see below).

The CO$_2$ level inside the chamber will either decrease due to net photosynthesis or increase in the dark as a result of respiration. The rate of change in

CO$_2$ levels depends on chamber volume, canopy size, environmental conditions, and the design of the gas analysis system. The gas mixing and analysis system may be described "closed," "semiclosed," or "open" [22,47,59]. A "closed" system is one which is completely isolated from outside air and chamber air is recycled after analysis. A "closed" system is not suitable for long-term whole-plant gas exchange or growth studies because CO$_2$ is rapidly depleted during photosynthesis. An example of a "semiclosed" system is one in which the plant chamber is closed to the outside atmosphere, but CO$_2$ is added to compensate for the depletion of CO$_2$ which occurs during plant photosynthesis [18,59,60]. In an "open" system, there is continuous flow of air through the chamber and gases are not recycled to the plants [22,59,61,62]. The selection of the most suitable system for measuring whole-plant growth and productivity depends on the research objectives.

2. Net Carbon Exchange Rate and Net Carbon Gain

Unlike field production, modern greenhouses provide an opportunity for control of the aerial and root environments of selected high value crops [63,64]. Our whole-plant gas analysis system was initially designed to develop environmental algorithms which could be used to predict growth of greenhouse crops and to design larger commercial scale test in which productivity and crop yield could be assessed [18,65].

Whole-plant NCER response to irradiance (I), CO$_2$ concentration, and temperature (T) can be expressed by the following polynomial function:

$$NCER = \beta_1 + \beta_2 I + \beta_3 CO_2 + \beta_4 T + \beta_5 I^2 \\ + \beta_6 CO_2^2 + \beta_7 T^2 + \beta_8 IT + \beta_9 ICO_2 \\ + \beta_{10} TCO_2 + \beta_{11} ITCO_2$$

where NCER is in units of μmol CO$_2$/m^2/sec, I is in units of μmol photon/m^2/sec photosynthetically active radiation (400 to 700 nm, PAR), CO$_2$ unit is μl/l, T is in unit of °C, and β_1 to β_{11} are coefficients.

There is a strong interaction between the influence of irradiance, CO$_2$ concentration, and temperature on whole-plant net photosynthesis, daily C gain, and growth. Irradiance is the most important determinant of whole-plant net photosynthesis followed by CO$_2$ and temperature. In a woody ornamental species, roses, irradiance, CO$_2$, and temperature accounted for 70%, 20%, and 5%, respectively, of the variance in whole-plant NCER [66]. In a herbaceous C$_3$ crop, *Alstroemeria*, irradiance accounted for almost 60% of the variation in whole-plant NCER, whereas CO$_2$ and temperature accounted for 23% and 14%, respectively

[65]. This crop is more sensitive to fluctuations in temperature due to its large rhizomes. The polynomial equation helps to describe how environmental variables affect net C gain due to photosynthesis. By knowing the duration of the photoperiod, as well as the respiration rate of the plants at different night temperatures one can predict how daily C gain and growth will be affected in controlled environments [65,66].

Whole-plant dark respiration rate generally increases exponentially with increasing temperature [67]. The Arrhenius equation can be used to predict the rate of whole-plant dark respiration rate at different temperatures:

$$NCER_d = Ae^{-Ea/RT}$$

where $NCER_d$ is dark respiration rate of whole plants in units of $\mu mol\ CO_2/m^2/sec$, A is a constant, e is the base of natural logarithms with a value of 2.718, E_a is the apparent active energy in units of cal/mol, R is the gas constant with a value of 1.987 cal/mol/ K, and T' is the Kelvin temperature (K).

Over a 24-h day/night period, daily C balance of the whole plant includes the daytime net photosynthesis and the nighttime respiration (Figure 31.1). Daily plant growth (i.e., increase in dry mass) can be estimated from NCERs and the C content [18,68].

Daytime net C gain (C_d) is the integrated NCER during the day:

$$C_d = \sum_i^m (NCER_{i \times t_i})$$

where $NCER_i$ is the whole-plant net photosynthetic rate over a period of time (t_i).

Total nighttime respiratory C loss of whole plant is integrated as:

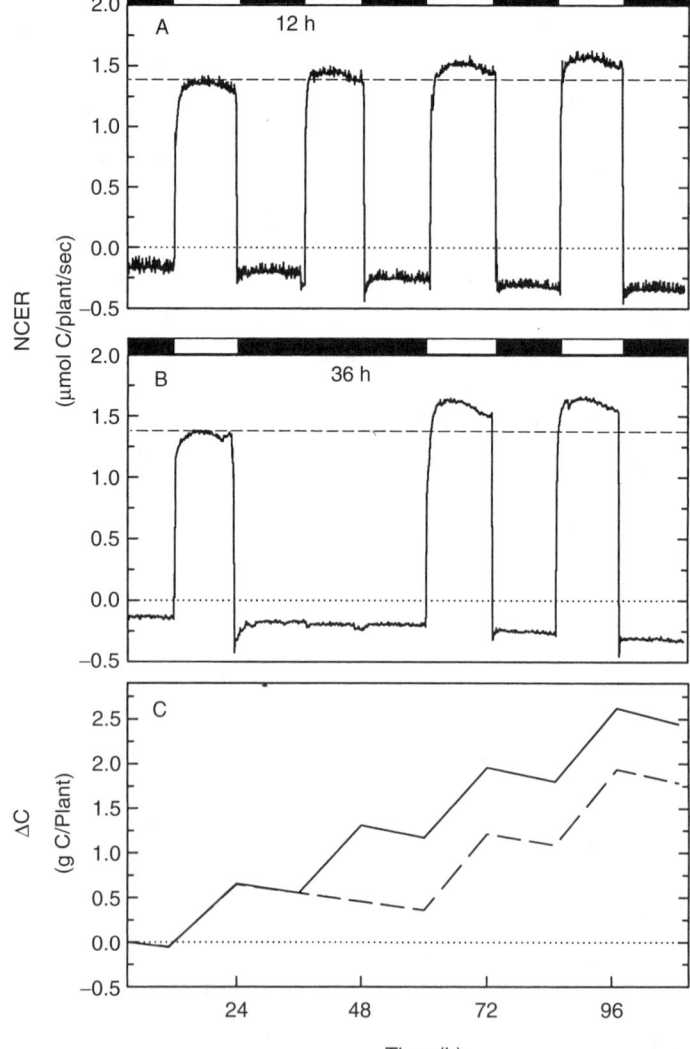

FIGURE 31.1 Diurnal patterns of net C exchange rate (NCER) of greenhouse peppers (*Capsicum annum* 'Cubico') maintained at (A) 12 h/12 h day/ night, (B) an extended dark period of 36 h, and (C) daily C gain (ΔC) calculated from A (solid line) and from B (broken line). In both sets of plants, CO_2 concentration during the experiment was maintained at 350 μl/l, irradiance was 1150 μmol/m²/sec PAR during the 12 h daytime period. Temperature was 22°C during the daytime and 18°C at night. (Adapted from Watts B. *The Effects of Temperature and CO₂ Enrichment on Growth and Photoassimilate Partitioning in Peppers (Capsicum annuum L.).* Guelph, Canada: University of Guelph, 1995. With permission.)

$$C_n = \sum_{j}^{n} (NCER_{d_j \times t_j})$$

where $NCER_{d_j}$ is whole-plant dark respiration rate during a period of time (t_j). Whole-plant daily net C gain (ΔC) is calculated as:

$$\Delta C = C_d - C_n$$

The NCER of a common greenhouse sweet pepper shows a clear diurnal pattern (Figure 31.1A). The significance of nighttime C losses on daily C gain (ΔC) is clearly illustrated in Figure 31.1C. Panels A and B show the NCER of two similar populations (i.e., A and B) of pepper plants. The only difference between the two populations was the A population was maintained in a 12/12 h day/night regimes throughout the experiment, whereas the B population was subjected to a 36 h uninterrupted dark period, which corresponded to 24 to 60 h into the experiment. During the first dark period of the experiment, $NCER_d$ (negative values) of the two populations were similar. The rates of whole-plant net photosynthesis (positive values) were also similar (approximately 1.3 µmol C/plant/sec) during the first 12-h light period. Thus, at the end of the first 24-h period, the ΔC was virtually identical in the two populations (Figure 31.1C). Stated in conventional terms used in destructive growth analysis, these data show that RGRs of the two populations were the same.

During the next 3-day period (24 to 72 h), population A increased its daytime NCER and nighttime $NCER_d$, consistent with increases in net photosynthesis and dark respiration as plants increased in size and new sinks developed. In comparison, population B, which was maintained in total darkness for a 36-h period, lost biomass (Figure 31.1C). During the extended dark period, there was a reduction in specific leaf weight, which corresponded with a reduction in stored reserves of sucrose and starch [69]. Figure 31.1C shows the effect of darkness on productivity. Interestingly, however, as a result of the extended dark period in population B, net photosynthesis on the third day increased more dramatically than that of the control population (Figure 31.1A). By the end of the fourth day, leaf starch reserves were replenished [69]. The data in Figure 31.1 represent a study of a relatively simple environmental perturbation in which only the length of a single dark period was altered. As many as 50 to 100 times more plants would have been required to obtain a similar data set if a destructive growth analysis protocol had been used.

III. AGRICULTURAL AND ECOLOGICAL CASE STUDIES

In agricultural and ecological studies, light intensity, CO$_2$ level in the atmosphere, temperature, and nutrient availability are all important environmental variables affecting source and sink development that determines net CO$_2$ exchange. Plant growth regulators and pathogens affect sink–source relationships in part by altering canopy architecture, development, and allocation of assimilates within the plant. Integrated analyses of development of sources and sinks are required for a full appreciation of the value of whole-plant NCER measurements for studying growth and productivity.

A. Light

Light intensity and quality inside a canopy fluctuates dramatically due to time of day, cloud cover, canopy density, and season [10,63,70,71]. As illustrated in Figure 31.1, the most dramatic changes in C gain occur diurnally. Due to mutual shading within the canopy, differences in leaf position, orientation, age, and dark respiration of different organs, whole-plant NCER is a better measurement of whole-plant growth response to light than that obtained from single leaf studies [65,72–76]. Figure 31.2A shows whole-plant photosynthesis of greenhouse roses. Leaf photosynthesis is saturated at lower light intensities than are required to saturate NCER of whole plants. The maximum rate of leaf photosynthesis is much higher than that of whole plants. Furthermore, the light compensation point (LCP) (i.e., the light intensity at which C gain and C loss are balanced), is lower for the leaves than for whole plants. Increasing the irradiance from 0 to 1200 µmol/m^2/sec PAR resulted in a marked increase in whole-plant NCER (Figure 31.2A) and ΔC (Figure 31.2B) calculated for a 24-h period consisting of a 10 h day and a 14 h night. ΔC was linearly proportional to daytime whole-plant NCER (Figure 31.2C). We define the LCP for ΔC as that light intensity required during daytime hours to sustain photosynthetic C gain which will balance nighttime (dark) respiratory C losses. The LCP for ΔC (Figure 31.2B) was greater than the LCP for the whole-plant NCER (Figure 31.2A). The difference between LCP of the whole-plant NCER (Figure 31.2A) and that of ΔC (Figure 31.2B) was primarily due to the duration of the night period and the magnitude of nighttime respiration.

In a different experiment in which rose plants were either irradiated over a 12-h light period or continuously for 24 h with half the irradiance, but with the same total radiant energy input of 17.6 mol/m^2, net C

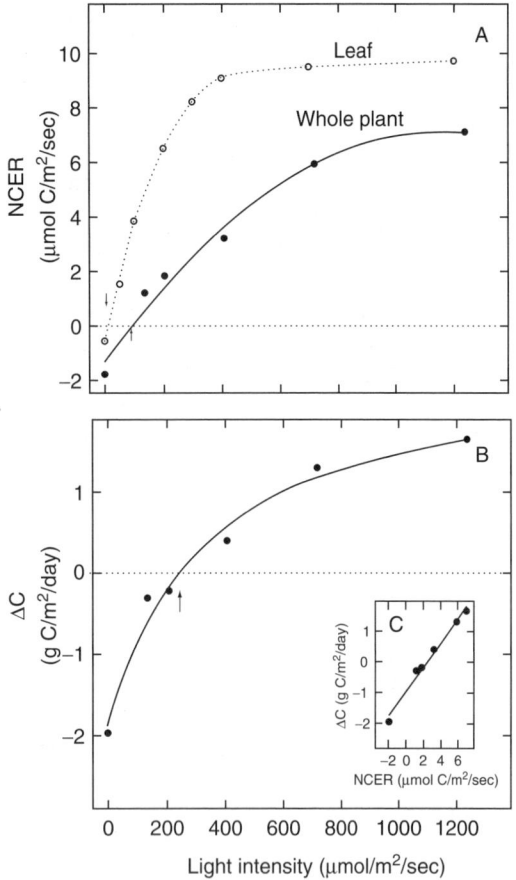

FIGURE 31.2 The effect of light intensity on (A) net C
exchange rate (NCER) of a leaf (broken line, open symbols)
and a whole plant (solid line, solid symbols), (B) whole-
plant net C gain (ΔC), and (C) the relationship between
plant ΔC and plant NCER of a greenhouse rose, *Rosa
hybrida* 'Samantha'. Values of ΔC were calculated from
plant NCER measurements for a 10 h/14 h day/night period.
The arrows in panels A and B indicate light compensation
points (LCPs) of leaf photosynthesis, of whole-plant photosynthesis, and of whole-plant C gain.

gains during the periods of illumination were identical
[77]. In both cases, approximately 1.8 g C/m² was
assimilated in the light period. However, when the
plants exposed to a 12 h daytime period were placed
in the dark for 12 h, ΔC over the 24-h period was
reduced to 1 g C/m². The length of light period not
only affected the total canopy C assimilation during
the day, but also influenced C loss through respiration in the subsequent dark period. Commercially
grown greenhouse roses are frequently provided with
artificial lighting at night to offset nighttime respiratory loss, even though the irradiance levels achieved
at canopy level are well below those achieved when
natural sunlight is available.

Leaf and canopy photosynthetic rates and C gain
change in close relation to changes in incident radi-

ation (Figure 31.2) [24, 78–80]. Productivity of a
plant community is generally proportional to PAR
absorbed as long as the canopy photosynthesis is
not light saturated for long periods during the growing season [8]. There are many factors which affect
the annual radiation absorbed by plant canopies during the season [10,75]. The geographic and demo-
graphic (e.g., urban environments) locations of the
plants will determine the seasonal variation of incident radiation [63]. The location of the plants will also
determine water and nutrient availability. Annual
temperature fluctuations affect leaf emergence, expansion, and leaf area duration (LAD), each of
which influences canopy light absorption and seasonal canopy productivity. Crop physiologists have
understood for many years that the structure of plant
canopy (e.g., leaf orientation), plays an important
role in determining the absorption of long and short
wave radiation [10,63]. An important determinant of
productivity is LAD which describes the total time
that the crop is photosynthetically active. Plant productivity is limited by the efficiency of light utilization
which is also influenced by the genetic and morphological differences among plant species [12].

The rate of CO_2 fixation is determined by the
efficiency of the light reactions, the activity of
Rubisco, the concentration of CO_2, and the ribulose
bisphosphate (RuBP) level in the chloroplast [2–4,81–
83]. It has been suggested that RuBP regeneration is
generally the dominant limitation to leaf photosynthesis under low light intensity [82,84] and in canopies
due to self-shading [85]. The upper leaf canopy, which
intercepts most of the incident light, contributes most
to whole-plant net photosynthesis [72]. Because of
light acclimation and aging, leaves in the lower canopy have lower maximum photosynthetic rates which
are saturated at lower irradiance levels. Sims and
Pearcy [86] reported that *Alocasia macrorrhiza* plants
grown in high irradiance have higher canopy photosynthetic rate than plants grown in low irradiance
when measured at high irradiance. However, when
measured at low irradiance, plants grown at low irradiance had a much larger daily C balance than did
plants adapted to the high irradiance environment.
Plants grown at higher irradiance levels generally displayed higher shoot respiration. Plant productivity
depends on the efficiency of light interception of the
canopy as well as the efficiency of carboxylation processes and subsequent respiratory losses due to
growth and maintenance respiration [12,13,16].

B. CO_2 CONCENTRATION

At present, suboptimal atmospheric CO_2 (approximately 370 μl/l) and inhibitory O_2 (21%) levels, rates

of CO$_2$ assimilation in C$_3$ plants are limited by CO$_2$ availability [86–89]. C$_3$ species grown under present atmospheric conditions, lose as much as 40% of CO$_2$ assimilated as a result of photorespiration which is regulated by CO$_2$ and O$_2$ concentrations and temperature [88,90]. There are two direct effects of increasing CO$_2$ concentration on increasing net photosynthesis [91,92]. One is the direct increase of the primary substrate CO$_2$ for carboxylation. Stomata tend to close in response to high CO$_2$, however, the increase in gradient between atmospheric CO$_2$ and leaf internal CO$_2$ concentration under CO$_2$ enrichment offsets the inhibiting effect of stomatal closure, resulting in higher rates of CO$_2$ fixation. The increase in the CO$_2$ concentration at the site of fixation in the chloroplast has a second direct effect on carboxylation efficiency of the chloroplast. Oxygenase activity of Rubisco is reduced and the flow of C to the glycolate pathway (i.e., photorespiration) is reduced. The benefits of CO$_2$ enrichment can come directly from the enhanced photosynthetic rate per unit leaf area or indirectly as a more long-term consequence of an increased total plant leaf area and altered pattern of carbon partitioning among developing sinks [43,91,92]. One of the consequences of CO$_2$ enrichment of young tomato seedlings, for example, is an increase in the allocation of assimilates to the roots [93]. Healthy root establishment is a fundamental objective during transplant production. Many bedding plants are grown commercially in greenhouses under CO$_2$ enrichment (normally 1000 to 2000 μl/l) to establish vigorous root systems that will improve the degree of hardiness of these transplants when they are exposed to field conditions [64].

In greenhouse production systems, daytime CO$_2$ enrichment is commonly used to stimulate growth and enhance crop yield [63,64,94]. Typical leaf and whole-plant net photosynthetic responses of a herbaceous C$_3$ greenhouse crop, *Alstroemeria* to varying levels of CO$_2$ are shown in Figure 31.3. The major differences between leaf and whole-plant CO$_2$ exchange was the higher rate of leaf gas exchange when comparisons were made at the same CO$_2$ concentration [65]. Leaf NCER was 18 μmol CO$_2$/m^2/sec at 1500 μl/l CO$_2$ under 1000 μmol/m^2/sec PAR whereas whole-plant NCER was 9 μmol CO$_2$/m^2/sec at 1500 μl/l CO$_2$ under 1200 μmol/m^2/sec PAR. The lower rates of CO$_2$ fixation by whole-plant NCER were primarily due to mutual shading and respiratory activity of sinks. Nevertheless, CO$_2$ enrichment marginally reduces the LCP in some crops [50] and substantially increases the optimum irradiance for conversion efficiency as well as the maximum conversion efficiency [76,95]. Quantum yields of C$_3$ leaves are dependent on CO$_2$ concentration, leaf tempera-

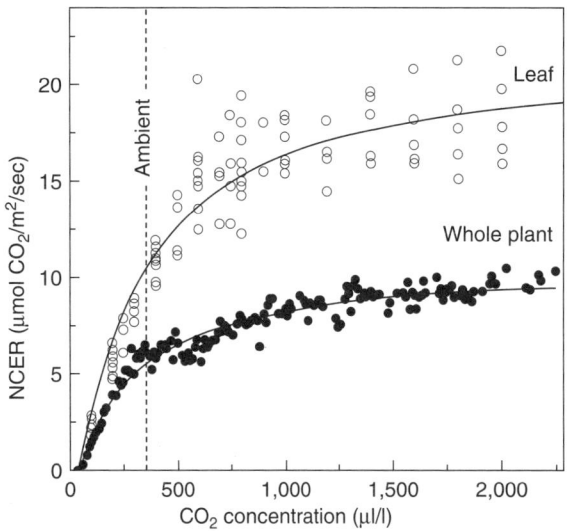

FIGURE 31.3 Effect of CO$_2$ concentration on leaf (open symbols) and plant (closed symbols) net C exchange rate (NCER) of *Alstroemeria sp.* 'Jacqueline'.

ture, and O$_2$ while quantum yields of C$_4$ leaves are independent of these factors [78,96].

Although the extent of photosynthetic and growth responses to CO$_2$ enrichment varies with plant species and depends on other environmental variables including light, water, nutrients, and temperature, increases in photosynthesis, growth, and productivity have been observed in nearly all C$_3$ species tested [64,85,87,94,97–99]. In C$_4$ species, a natural mechanism of CO$_2$ enrichment in the bundle sheath cells exists reducing photorespiration [87,92,100]. Net photosynthesis is usually much higher in C$_4$ than in C$_3$ plants at ambient CO$_2$ [88,101]. When there is a positive growth response to elevated CO$_2$ by C$_4$ plants, that response is usually less than that observed among C$_3$ plants [64,87,94,96,99,102–104]. The CO$_2$ exchange response of CAM plants to CO$_2$ enrichment also depends on species, developmental stage, and environmental factors such as light, temperature, nutrients, and water availability [105–108]. Under conditions of water stress, stomata are a major limitation to C assimilation in CAM, C$_4$, and C$_3$ plants [104]. In C$_3$ plants at ambient CO$_2$, stomatal limitation is about 30% of the total limitation of leaf photosynthesis [109]. Elevated levels of atmospheric CO$_2$ increases the CO$_2$ gradient between the atmosphere and the fixation site of the chloroplasts, high CO$_2$ generally reduces stomatal conductance and increase water use efficiency in C$_3$ and C$_4$ plants [110–112].

During long-term exposures to elevated CO$_2$, photosynthesis on a leaf and on a whole plant basis is altered in a species specific manner [87,92,113]. In roses grown at 1000 μl/l CO$_2$ for several weeks whole-plant net photosynthesis was identical to that of plants grown under ambient CO_2 conditions, indicating no inhibiting effect of long-term CO_2 enrichment. Similarly, lettuce plants grown under CO_2 enrichment showed no decrease in canopy photosynthesis under high CO_2 [95]. Nevertheless, in some herbaceous species even in canopies open to the light and with healthy developing root systems prolonged exposure to high CO_2 results in reduced rates of mature leaf photosynthesis when comparisons were made at ambient CO_2 and O_2 [64]. These observations can, in part, be explained by the fact that under CO_2 enrichment plant growth is enhanced and new sinks (e.g., leaves) [91,92] place a heavy demand on the key nutrients such as N [64,87,92,114–119]. In some species grown at high CO_2, a reduction in leaf photosynthesis can be attributed to a reduction in key enzymes associated with the fixation of CO_2 (e.g., Rubisco) [87,92,115,116]. These enzymes are a major source of N and their levels tend to be reduced during senescence as N is reallocated to growing sinks. As mentioned above, the magnitude of photorespiration relative to that of photosynthesis is reduced at high CO_2 [87,88,92]. However, there is no reduction in glycolate oxidase activity, a key enzyme of the photorespiratory pathway at high CO_2 [88,115]. There is growing evidence that dark respiratory processes are altered in leaves of plants grown at elevated levels of CO_2 [16,17,87,92]. For example, enhanced whole-plant dark respiration following growth under CO_2 enrichment has been observed in several species and can in part be attributed to the increased level of carbohydrates present in leaf tissue of plants grown at high CO_2 levels [17,120]. In wheat, the number of mitochondria in the leaf mesophyll appears to increase during growth at high CO_2 [121]. These observations together serve to illustrate that during CO_2 enrichment there will be profound changes in both photosynthetic and respiratory CO_2 fluxes as developmental processes are generally accelerated compared to growth at ambient CO_2 [92]. However, the effects of short- and long-term high CO_2 on respiration rates need to be carefully examined in view of limitations and errors that can arise using CO_2 gas exchange systems [122–125].

Studies with several greenhouse crops show daytime starch accumulation at high CO_2 [126–129]. Furthermore, the increase in photoassimilate storage supports an enhanced nighttime carbon export rate from the leaves, which, in part, explains the faster growth rate of plants exposed to elevated daytime CO_2 [3,69,92,128,130,131]. Storage of carbohydrates may compete with leaf and root growth and reduce the maximum growth rate [132]. The increase in stored carbohydrates (e.g., starch levels) during

long-term CO_2 enrichment represents a problem in equating net CO_2 exchange rates obtained nondestructively to growth and development. Daytime and nighttime CO_2 exchanges are dependent on the partitioning and the allocation of the stored reserves such as starch. Starch in the source leaves definitely represent biomass accumulation [120,133]. However, growth and development requires further partitioning and allocation of these reserves to developing sinks such as the roots and reproductive structures. In a survey of 42 C_3, C_3–C_4 intermediate, and C_4 photosynthetic types the linear relationship between leaf photosynthesis and C export at ambient CO_2 breaks down at elevated CO_2 [101,134–136]. Of the leaf parameters tested, C export correlated best with whole-plant RGR obtained noninvasively using whole-plant CO_2 exchange analysis [9].

Long-term exposure to elevated CO_2 has also been investigated to understand how different plant communities will grow if atmospheric levels continue to rise [43,137]. In a field study of pine, seedlings were grown in soil and enclosed within open-top-plastic chambers [43]. Plants were subjected to various CO_2 enrichment treatments for 15 months. Canopy CO_2 exchange was measured using an "open" system within the CO_2 enrichment chambers for a period of 4 to 5 days. Although, the contribution of root respiration to canopy gas exchange was not determined, canopy net photosynthesis increased with increasing CO_2 concentrations. CO_2 enrichment also increased total leaf number and leaf area. Reid and Strain [138] studied the effect of CO_2 enrichment on whole-plant C budgets of seedlings of beech and sugar maple grown in low irradiance. At ambient CO_2, photosynthetic rate per unit mass of beech was lower than for sugar maple, whereas elevated CO_2 enhanced the photosynthesis of beech only. Elevated CO_2 preferentially enhanced net C gain of beech by increasing net photosynthesis and reducing respiration. Above ground (i.e., shoot) respiration per unit mass decreased with CO_2 enrichment for both species while root respiration per unit mass decreased for sugar maple only. C losses per plant to nocturnal shoot and root respiration were similar for both species. Under elevated CO_2, C uptake was similar for both species, indicating a significant increase in whole-seedling NCER with CO_2 enrichment for beech but not for sugar maple. Total C loss per plant to shoot respiration was reduced for beech only because the increase in sugar maple leaf mass counterbalanced a reduction in respiration rates. The RGR estimated by destructive analysis indicated that the biomass accumulation was not affected by CO_2 enrichment in either species possibly because of the slow growth rate at low irradiance used to grow these plants. In both species, the greatest C loss occurred from the roots, indicating the importance of below ground biomass (sinks) in estimates of plant net C gain. This study illustrates how whole-plant gas exchange used to estimate total C gain can be affected by different photosynthetic activity of the source and respiratory balances of the sinks.

C. TEMPERATURE

One of the advantages of using whole-plant NCER as a tool to study growth is that the effects of temperature during the dark can be differentiated from those in the light. Light and CO_2 are environmental parameters which primarily affect photosynthesis and photorespiration of the leaves. However, temperature affects all aspects of metabolism, growth and development of all organs.

Leaf net photosynthesis of most C_3 plants has an optimal temperature range of 20°C to 35°C at ambient CO_2 level and saturating light [139]. Leaf photorespiration increases sharply at temperatures above 30°C due to decreases in CO_2 solubility [140] and in CO_2/O_2 specificity of Rubisco [141–143]. In addition, as we have outlined elsewhere temperature can dramatically alter C export rates from leaves [128,129,135,144,145]. Thus, temperature moderates source–sink relationships by affecting fluxes of metabolites as well as photosynthetic and respiratory metabolism more directly.

Because CO_2 enrichment reduces photorespiration in C_3 plants, the optimal temperature for whole-plant photosynthesis is usually shifted a few degrees higher than at ambient CO_2 [85,98,146,147]. Both the LCP [146] and the CO_2 compensation point [148] of whole plants increase with an increase in temperature because respiration from all tissues are increased. The effect of temperature on ΔC is the balance among its effect on photosynthesis, photorespiration, and dark respiration. Both daytime and nighttime temperatures are important in relation to plant daily C gain. In white clover plants, ΔC at 30°C/

10°C (day/night) was higher than plants maintained at 30°C/20°C [149].

The importance of nighttime temperature on daily net C gain during greenhouse rose production is shown in Table 31.1. Roses were maintained at either 27°C/27°C (day/night) or at 27°C/17°C (day/night). The respiration rate of rose plants maintained at 27°C during the night was twice that of plants maintained at 17°C during the night. Daily net C gain maintained at 27°C/17°C was 50% higher than that of plants maintained at the same 27°C/27°C day/night temperature. Dark respiration rate is more sensitive to changes in temperature than photosynthesis. Respiration rate generally increases exponentially with increasing temperature [67] with a Q_{10} of about 2 [14]. However, the rate varies with developmental stage of specific tissues. For example, in a flowering rose shoot, the respiration rate of the flower bud on a dry weight basis is three to four times higher than that of leaves and accounts for half of the total respiratory C loss from the shoot [146].

In greenhouses, root zone temperatures can be controlled by bench heating and cooling systems. Lower root zone temperatures stimulate flowering in *Alstroemeria*, which alters the growth and development pattern of the whole plant. The NCER of *Alstroemeria* is very sensitive to changes in aerial and root zone temperature [65]. The optimal temperature for leaf photosynthesis under ambient CO_2 level and saturating light is about 20°C whereas that of whole-plant NCER is only 10°C to 12°C, which, in part, reflects the metabolism of the rhizomes. Whole-plant gas exchange measurements have also been used to discriminate between growth and maintenance respiration and how these processes relate to C use efficiency [150,151].

D. C/N BALANCE

Carbon is a major nutrient obtained by reduction of atmospheric CO_2 during photosynthesis. The supply of any of the major mineral nutrients, such as N can

---

TABLE 31.1
Effect of Night Temperature on Whole-Plant Daily C Gain (ΔC) of Greenhouse Roses (*Rosa Hybrida* 'Samantha') Maintained at 12/12 h Day/Night

Temperature (day/night) in °C	NCER (μmol C/m^2/sec)	$NCER_d$ (μmol C/m^2/sec)	C_d (g C/m^2)	C_n (g C/m^2)	ΔC	C_n/C_d (%)
27/27	6.3 ± 0.35	2.6 ± 0.16	3.3 ± 0.15	1.5 ± 0.09	1.8 ± 0.16	45 ± 2.4
27/17	6.9 ± 0.35	1.3 ± 0.07	3.5 ± 0.16	0.8 ± 0.05	2.7 ± 0.17	23 ± 1.3

Source: Adapted from Jiao J, Tsujita MJ, Grodzinski B. *Can. J. Plant Sci.* 1991; 71:245–252. With permission.

have profound effects on plant metabolism, C alloca-tion, and canopy development [90,117–119,152,153]. An improved N status generally results in an increase in the above ground parts of the plant including a larger leaf area [154], which in turn, increases the opportunity for greater canopy light interception and CO_2 assimilation. Leaf photosynthetic capacity and leaf N content are correlated in many species [155–160]. Noninvasive approaches such as those based on CO_2 exchange analysis provide a means of describing how respiratory demands and photosyn-thesis are related to overall plant productivity, par-ticularly when attempts are also made to monitor separately the CO_2 exchanges of the canopy and the root zone [37,138,161,162].

Measurement of whole-plant net CO_2 exchange becomes a very useful tool in assessing N and C economies in relation to plant growth. Many plants adjust the amount of N which is allocated to the leaf canopy according to a pattern which tends to opti-mize the absorption of light energy within the canopy and maintain the highest possible rates of canopy photosynthesis [157,159,163,164]. For example, in dense stands of Carex acutiformis, canopy NCER was markedly affected by the pattern of leaf N allo-cation whereas in open stands the net daily canopy photosynthesis was essentially independent of leaf N distribution [159]. Leaf N content is generally higher in upper canopy leaves, especially in dense canopies in which the light gradient is steep [159,165]. Upper canopy leaves also contribute more to whole-plant net photosynthesis than do lower canopy leaves [72], however, the photosynthetic contribution of the inner canopy (shade) leaves can be significant [71]. Increas-ing leaf N content also enhances dark respiration [159]. Increases in dark respiration with higher N content may be the consequence of higher daytime photosynthetic rates, which provide more photoassi-milates as substrates for respiration or a change in enzyme levels [16,17]. The effect of N availability on photosynthetic and respiratory gas exchanges is driven in part by the demand for reduced forms of N (e.g., protein synthesis) and the supply of photo-assimilates (e.g., sucrose), which act both as sources of energy for N reduction and as the C-skeletons for the synthesis of the primary amino acids in the shoot and in the roots. The demand for N will vary with growth rate, the developmental stage of the plant, and the environmental conditions such as light, CO_2, and temperature each of which modifies photosynthesis, respiration, and RGR [162]. Higher values of RGR are generally associated with higher plant N content [37,154,166,167].

Long-term exposure to elevated atmospheric CO_2 reveals the importance of N availability. A significant stimulation of both light saturated and daily integrals of photosynthesis in Lolium perenne sward were maintained over a period of 10 years managed as a herbage crop grown in open field conditions in spe-cially designed exposure chambers (Free Air CO_2 Enrichment) [137]. The acclimation of photosynthesis during long-term exposure to elevated CO_2 depended on development and growth of new sinks which were limited by N availability. This study also provides evidence that stimulation of photosynthesis at high CO_2 in not a transient phenomenon.

E. GROWTH REGULATORS

The use of plant growth regulators and herbicides for controlling vegetative growth of agricultural crops has increased dramatically in recent decades [63,168,169]. Ethylene is a natural plant growth regu-lator which is produced in nonphotosynthetic organs (e.g., flowers and fruits) as well as in photosynthetic leaf tissues [169]. Because of our interest in CO_2 en-richment in closed environments we began to investi-gate the relationship between C_2H_4 and CO_2 gas exchange in photosynthetic tissue [64,92,170]. The stimulatory effect of high CO_2 levels on C_2H_4 release from photosynthetic tissue during short-term expos-ures (1 to 8 h) has been demonstrated in intact plants, in detached leaves, and in excised leaf tissue [171–175]. The CO_2 levels which affect C_2H_4 release from leaf tissue during short-term incubations (i.e., 50 to 5000 μl/l) [176] parallel those encountered by leaf tis-sue in closed greenhouse environments or in tissue cultures [177]. Active photosynthesis under high ir-radiance can deplete the CO_2 to below ambient levels in protected environments [63,64]. Therefore, CO_2 is added to supplement growth. Interestingly, predicted future global CO_2 concentrations also fall within this range [87,92]. We know from earlier studies that C_2H_4 release from C_3 and C_4 leaf tissue during short-term exposures to varying CO_2 are different [172,173]. Long-term exposure to elevated CO_2 concentrations modifies endogenous C_2H_4 metabolism and affects plant growth and development [92,176]. Prolonged growth at high CO_2 results in a persistent increase in the rate of endogenous C_2H_4 release which can, only in part, be attributed to an increase of the endogenous pools of C_2H_4 pathway intermediates [176]. During acclimation to high CO_2 leaves appear to have higher levels of ethylene forming enzyme activity [175,176]. Photosynthetically active young leaves contribute most of the C_2H_4 emanating form the canopy [176,178,179].

All lower and higher plant tissues produce C_2H_4, which can elicit a wide range of biochemical and morphological responses [169]. For example, leaf ontogenesis and maturation are correlated with changing rates of both C$_2$H$_4$ emanation and sensitivity to exogenously supplied C$_2$H$_4$ [176,177,180]. Ethylene can modify leaf and whole-plant photosynthesis [181–184]. Vegetative growth measured nondestructively as net C gain was reduced 50% within 24 h of C$_2$H$_4$ exposure in *Lycopersicon esculentum* L. [183] and 35% in *Xanthium strumarium* L. [184]. Similar results have been obtained with destructive analysis of tomato [185] and corn [186]. The observed decrease was attributed to well-known morphological responses exhibited by these plants when treated with C$_2$H$_4$ [185,187]. The reduction in whole-plant NCER were attributed to (1) epinastic changes in leaf angle (i.e., light interception patterns) [183,184], and (2) alteration of sink–source relations [185]. For example, when the leaves that showed C$_2$H$_4$ induced epinasty were repositioned with respect to the overhead light source in the analysis chamber, an NCER comparable to that of the untreated plants was observed. The reduction in C gain associated with C$_2$H$_4$ is an indirect effect of C$_2$H$_4$ on canopy photosynthesis since it is a consequence of C$_2$H$_4$ induced epinastic responses, which alter the orientation of the leaves and light interception [183–185].

The role of C$_2$H$_4$ in regulating the CO$_2$ exchange of varying plant density was further tested by treating model canopies of tomato seedlings with C$_2$H$_4$. Plants were exposed to a 12 h/12 h, day/night regime, during which NCER was measured after treatment with C$_2$H$_4$ [188]. The critical leaf area index (LAI) at which 95% of the maximum rate of canopy photosynthesis was achieved corresponded to a value of about 5. When a well-developed (i.e., dense) canopy, with a LAI of about 6, was treated with C$_2$H$_4$ there was no change in the photosynthesis of the stand. However, when the LAI was only 4, treatment with C$_2$H$_4$ resulted in a 20% decrease in canopy photosynthesis. These studies with model canopies support our earlier conclusion that the effects of C$_2$H$_4$ on photosynthesis and C gain [183–185] can be ascribed to classical hormonal responses such as epinastic development which result in altered light interception within the canopy.

Light interception is a major determinant of canopy photosynthesis. Endogenously produced C$_2$H$_4$ can accumulate to physiologically active levels in plant canopies [169]. For example, C$_2$H$_4$ concentrations sufficient to stimulate premature cotton boll abscission have been documented in field cotton canopies [189]. In closed greenhouse environments in which crops are growing, C$_2$H$_4$ levels of 10 to 15 ppb have been detected [176], which can be attributed to production by the plant tissue. In ongoing closed environment studies with lettuce, wheat and

soybean canopies at the National Aeronautics and Space Administration (NASA, USA) [42,178] and at the University of Guelph [179], C$_2$H$_4$ levels of up to 100 ppb in air samples have been detected. The productivity of wheat and rice are also affected by low concentration of ethylene that might accumulate in closed environments [190]. Collectively, these observations support the view that C$_2$H$_4$ from the plants accumulates in closed environments and can modify canopy architecture and photosynthesis. Canopy density (i.e., LAI) and light interception patterns are important factors in determining the extent to which exposure to this growth regulator will alter daily C gain [54,179].

F. CANOPY ARCHITECTURE

During the early stages of plant growth, LAI is low and the efficiency of light interception is almost entirely dependent on the total leaf area of the canopy. Therefore, leaf area production is the most important factor determining growth of plants during early stages of development [10]. There is a good correlation between RGR and leaf expansion rate at early stages of plant growth [191]. In cotton, plant growth and canopy photosynthesis per plant has been found proportional to total leaf area during early growth [192]. The relationship between canopy photosynthesis and total leaf area diminished as leaf area approached maximum values because of increased mutual shading of canopy leaves. The results indicate that in a closed canopy, photosynthesis was not limited by total leaf area, but more by canopy architecture and reduced light interception at lower canopy due to mutual shading.

The LAI, distribution, size, and orientation of the leaves (i.e., canopy architecture) will determine the amount of incident radiation intercepted by the plants, the canopy leaf temperature, and gas exchange characteristics [73,193–197]. An accurate description of the three-dimensional distribution of leaves in plant canopies is very difficult. The techniques generally used for determination of canopy structure have been discussed by Campbell and Norman [198]. The orientation of branches and leaves has a significant effect on light penetration through the canopy and allow a large LAI to be sustained by plants [75]. Canopy architecture varies greatly with plant species [75], as well as during plant development and ontogeny [199]. Seasonal changes of LAI depend on planting density, the pattern of leaf initiation, growth, and senescence. Plants with different canopy architecture have significant influence on light use efficiency and productivity. In some species such as evergreen forest and pasture grasses with relatively constant

LAI year around, annual PAR absorption by the canopy is primarily dependent on the annual PAR receipts and the mean absorptance. In addition to the seasonal variation of incident radiation, leaf area affects annual canopy radiation absorption and many acclimation processes such as cold acclimation. Pine and winter wheat have quite different leaf form and canopy architecture. Recently, we have been able to utilize diel patterns of whole-plant NCER to help explain different strategies for cold acclimation and winter survival in these two species [200].

Plant canopies are composed of a population of leaves of different ages and of different exposure to light [71,201]. In a sugar maple forest canopy, the top of the canopy intercepted approximately 60% of the total light received by the whole canopy and contributed 37% of the daily C gain, even though the leaf area at the top canopy represented only 11% of the total leaf area [202]. The lower canopy, accounted for more than 50% of the total leaf area, but received less than 10% of the daily irradiance. Both leaf mass per leaf area and N per leaf area were 50% lower at the bottom canopy than at the top of the tree. It was suggested that the differences in leaf traits along the vertical canopy gradient were mainly structural in nature. Leaf orientation may itself be affected by the light environment in which the plants are growing. Sun leaves tend to have more vertical orientation than do shade leaves, which are more horizontally positioned [70,195]. Compact and isolated tussock grasses with more horizontally oriented leaves received more light during the midday than tussocks with more steeply oriented foliage. The lower midday incident radiation in tussocks with a steep foliage orientation may reduce photoinhibition compared to plants with the more horizontal leaf orientation. However, the advantage of steep foliage orientation in daily C gain over plants with more horizontal leaves depends on plant species and changes with time of the day [193].

Leaf morphology also influences canopy light interception and, consequently, canopy photosynthesis. In near-isogenic cotton lines, variations in leaf morphologies (i.e., size and shape) resulted in different LAI and leaf dry weight [199]. The genotypic variation in LAI of different lines caused differences in light penetration through the canopy, integrated canopy apparent photosynthesis and limit yield. "Afila" mutants of peas, in which a single gene modification results in replacement of the laminar shaped leaflets by cylindrically shaped tendrils [203–205], have high plant NCER in the light because more light penetrates the leaf canopy than is the case with conventional leafy cultivars. In the semileafless "afila" phenotypes the tendrils and laminar shaped stipules accounted for approximately 60% and 40%,

respectively, of the total plant photosynthesis even though on a chlorophyll or area basis the tendrils were predicted to account for only 30% of whole-plant photosynthesis. These values were derived from two different sets of experiments. In one set of experiments $^{14}CO_2$ was supplied to whole plants for 1 min after which the plants were rapidly killed to prevent translocation of ^{14}C-labeled assimilates. The $^{14}CO_2$ fixed in the different photosynthetic organs was measured following destructive analysis. In a parallel experiment similar values for the contributions of leaflets, stipules, and tendrils to plant CO_2 exchange were determined by measuring whole-plant NCER before and after surgical removal of the tendrils or the stipules or the leaflets [206]. These experiments demonstrate the importance of tendril structures in peas and the heterogeneous nature of leaf canopies. They also serve to underscore the need for a greater degree of resolution in monitoring gas exchanges from different parts of the canopy if we are to fully describe how canopy architecture contributes to canopy photosynthesis, growth, and development through the season.

In this chapter, we have focused on the value of direct measurements of whole-plant NCER for estimating plant growth and productivity. This approach is generally limited to plants of small sizes and to small populations. It is very difficult to measure directly whole canopy NCER of a large population of plants such as a forest and relate these values to seasonal estimates of C gain and productivity. Many researchers have developed models to predict canopy gas exchange based on the knowledge and information of changes of NCER of individual leaves and activity of layers of canopy in relation to environmental changes [193,207–212]. Canopy photosynthesis models are helpful in understanding and predicting the C balance of plant canopies and communities under natural environmental conditions. Model predictions need to be tested with independent field studies such as those using eddy correlation micrometeorological methods to estimate mass and energy exchange between atmosphere and plant canopies [213–217]. Seasonal canopy photosynthesis models usually combine microclimate submodels with photosynthesis submodels [208–211]. However, it is becoming evident that progress in canopy level photosynthesis and ecological modelling has increased the demand for advanced description of canopy architecture [218].

IV. CONCLUSIONS

It is naive to suggest that any single technique such as classical destructive growth analysis or a less invasive analysis based on CO$_2$ gas exchange can provide all the data necessary to describe the complex sequence of events that occur during plant development. Future advances in imaging and remote sensing procedures will undoubtedly facilitate a more complete analysis of plant biomass accumulation and canopy development. Many experimental approaches need to be applied simultaneously at each stage of the vegetative and the reproductive development of a crop to fully assess primary gas exchanges, source–sink interactions and their impact on productivity.

Although there are limitations to all procedures used to investigate whole-plant growth and productivity, measurements of CO$_2$ exchange made in real time provides valuable information for many levels of inquiry. Analysis of diel patterns of CO$_2$ gas exchange is by no means limited to studies of photosynthetic and respiratory metabolism. The application of using gas signatures to quantify growth are endless [18–20]. Numerous case studies within our group alone demonstrate the value of gas exchange in obtaining algorithms for modeling and optimizing productivity of crops in controlled environments [54,65,66]; assessing the role of canopy architecture and form on productivity [183,184,206]; comparing the productivity of natural photosynthetic variants [9]; correlating whole-plant productivity with specific leaf functions such as C-fixation or export [9]; evaluating the impact of specific gene alterations on growth [219]; assessing plant pathogen interactions [220]; and investigating acclimation processes such as hardening of overwintering perennials [200].

REFERENCES

1. Epstein E. *Mineral Nutrition of Plants: Principles and Perspectives.* New York: John Wiley & Sons, 1972.
2. Raines C. The Calvin cycle revised. *Photosynth. Res.* 2003; 75:1–10.
3. Geiger DR, Servaites JC. Diurnal regulation of photosynthetic carbon metabolism in C3 plants. *Annu. Rev. Plant Physiol. Plant Mol. Biol.* 1994; 45:235–256.
4. Woodrow IE, Berry JA. Enzymatic regulation of photosynthetic CO$_2$ fixation in C3 plants. *Annu. Rev. Plant Physiol. Plant Mol. Biol.* 1988; 39:533–594.
5. Bunce JA. Measurements and modeling of photosynthesis in field crops. *CRC Crit. Rev. Plant Sci.* 1986; 4:47–77.
6. Elmore CD. The paradox of no correlation between leaf photosynthetic rates and crop yields. In: Hesketh JD, Jones JW, eds. *Predicting Photosynthesis for Ecosystem Models.* Vol. II. Boca Raton, FL: CRC Press, 1980:155–167.
7. Gifford RM, Jenkins CLD. Prospects of applying knowledge of photosynthesis towards improving crop production. In: Govindjee, ed. *Photosynthesis Devel-opment, Carbon Metabolism, and Plant Productivity.* New York: Academic Press, 1982:419–457.
8. Pereira JS. Gas exchange and growth. In: Schulze E-D, Caldwell MM, eds. *Ecophysiology of Photosynthesis.* Berlin: Springer-Verlag, 1994:147–181.
9. Leonardos ED, Grodzinski B. Correlating source leaf photosynthesis and export characteristics of C3, C3–C4 intermediate and C4 Panicum and Flaveria species with whole-plant relative growth rate, *Proceedings for the 12th International Congress on Photosynthesis,* PS2001, Brisbane, Australia, 2001. CSIRO Publishing.
10. Hay PKM, Walker AJ. *An Introduction to the Physiology of Crop Yield.* London, UK: Longman Sci. & Tech., 1989.
11. Ort DR, Baker NR. Consideration of photosynthetic efficiency at low light as a major determinant of crop photosynthetic performance. *Plant Physiol. Biochem.* 1988; 26:555–565.
12. Horton P. Prospects for crop improvement through the genetic manipulation of photosynthesis: morphological and biochemical aspects of light capture. *J. Exp. Bot.* 2000; 51:475–485.
13. Lawlor DW. Photosynthesis, productivity and environment. *J. Exp. Bot.* 1995; 46:1449–1461.
14. Lambers H. Respiration in intact plants and tissues. *Encyclopedia of Plant Physiology.* Vol. 18. Berlin: Springer-Verlag, 1985:418–473.
15. Zelitch I. The close relationship between net photosynthesis and crop yield. *BioScience* 1982; 32:796–802.
16. Amthor JS. The McCree–de Wit–Penning de Vries–Thornley respiration paradigms: 30 years later. *Ann. Bot.* 2000; 86:1–20.
17. Amthor JS. *Respiration and Crop Productivity.* New York: Springer-Verlag, 1989.
18. Dutton R, Jiao J, Tsujita MJ, Grodzinski B. Whole plant CO$_2$ exchange measurements for nondestructive estimation of growth. *Plant Physiol.* 1988; 86:355–358.
19. McCree KJ. Measuring the whole-plant daily carbon balance. *Photosynthetica* 1986; 20:82–93.
20. Penning de Vries FWT, Brunsting AHM, van Larr HH. Products, requirements and efficiency of biosynthetic process: a quantitative approach. *J. Theor. Biol.* 1974; 45:377–399.
21. Bate GC, Canvin DT. A gas-exchange system for measuring the productivity of plant populations in controlled environments. *Can. J. Bot.* 1971; 49:601–608.
22. Bugbee B. Steady-state canopy gas exchange: System design and operation. *HortSci* 1992; 27:770–776.
23. Christy AL, Porter CA. Canopy photosynthesis and yield in soybean. In: Govindjee, ed. *Photosynthesis Development, Carbon Metabolism, and Plant Productivity.* New York: Academic Press, 1982:449–511.
24. Charles-Edwards DA, Acock B. Growth response of a chrysanthemum [morifolium] crop to the environment. II. A mathematical analysis relating photosynthesis and growth. *Ann. Bot.* 1977; 41:49–58.
25. Wheeler RM, Mackowiak CL, Sager JC, Yorio NC, Knott WM, Berry WL. Growth and gas exchange by lettuce stands in a closed, controlled environment. *J. Am. Soc. Hort. Sci.* 1994; 119:610–615.

26. Evans GC. *The Quantitative Analysis of Plant Growth.* Oxford: Blackwell Scientific Publications, 1972.

27. van der Werf A. Growth analysis and photoassimilate partitioning. In: Zamski E, Schaffer AA, eds. *Photoassimilate Distribution in Plants and Crops: Source–Sink Relationships.* New York: Marcel Dekker, 1996:1–20.

28. Hunt R. *Plant Growth Curves.* London: Arnold Publishers, 1982.

29. Poorter H, Garnier E. Plant growth analysis: an evaluation of experimental design and computational. *J. Exp. Bot.* 1996; 47:1343–1351.

30. Hunt R. *Basic Growth Analysis.* London: Unwin Hyman, 1990.

31. Beadle CL. Growth analysis. In: Hall DO, Scurlock JM, Bolhar-Nordenkampf HR, Leegood RC, Long SP, eds. *Photosynthesis and Production in a Changing Environment: A Field and a Laboratory Manual.* London: Chapman and Hall, 1993:36.

32. Causton DR, Venus JC. *The Biometry of Plant Growth.* London: Edward Arnold, 1981.

33. Roberts MJ, Long SP, Tieszen LL, Beadle CL. Measurement of plant biomass and net primary production. In: Coombs J, Hall DO, Long SP, Scurlock JMO, eds. *Techniques in Bioproductivity and Photosynthesis.* Oxford: Pergamon Press, 1987:1–19.

34. van Andel J, Biere A. Ecological significance of variability in growth rate and plant productivity. In: Lambers H, Cambridge ML, Konings H, Pons TL, eds. *Causes and Consequences of Variation in Growth Rate and Productivity of Higher Plants.* The Hague, The Netherlands: SPB Academic Publishing, 1990:257.

35. Grime JP, Hunt R. Relative growth-rate: Its range and adaptive significance in a local flora. *J. Ecol.* 1975; 63:393–422.

36. Poorter H. Interspecific variation in relative growth rate:on ecological causes and physiological consequences. In: Lambers H, Cambridge ML, Konings H, Pons TL, eds. *Causes and Consequences of Variation in Growth Rate and Productivity of Higher Plants.* The Hague, The Netherlands: SPB Academic Publishing, 1990:45–68.

37. Poorter H, Remkes C, Lambers H. Carbon and nitrogen economy of 24 wild species differing in relative growth rate. *Plant Physiol.* 1990; 94:621–627.

38. Konings H. Physiological and morphological differences between plants with a high NAR or a high LAR as related to environmental conditions. In: Lambers H, Cambridge ML, Konings H, Pons TL, eds. *Causes and Consequences of Variation in Growth Rate and Productivity of Higher Plants.* The Hague, The Netherlands: SPB Academic Publishing, 1990:101–123.

39. Hunt R. Diurnal progressions in dry weight and short-term plant growth studies. *Plant Cell Environ.* 1980; 3:475–478.

40. Wickens LK, Cheeseman JM. Application of growth analysis to physiological studies involving environmental discontinuities. *Physiol. Plant.* 1988; 73:271–277.

41. Acock B, Charles-Edwards DA, Hearn AR. Growth response of a chrysanthemum (Morifolium) crop to the environment. I. Experimental techniques. *Ann. Bot.* 1977; 41:41–48.

42. Corey KA, Wheeler RM. Gas exchange capabilities in NASA's plant biomass production chamber. *BioScience* 1992; 42:503–509.

43. Garcia RL, Idso SB, Wall GW, Kimball BA. Changes in net photosynthesis and growth of *Pinus eldarica* seedlings in response to atmospheric CO_2 enrichment. *Plant Cell Environ.* 1994; 17:971–978.

44. Hand DW. A null balance method for measuring crop photosynthesis in an airtight day-lit controlled-environment cabinet. *Agric. Meteor.* 1973; 12:259.

45. Hand DW, Clark G, Hannah MA, Thornley JHM, Warren Wilson J. Measuring the canopy net photosynthesis of glasshouse crops. *J. Exp. Bot.* 1992; 43:375–381.

46. Lawlor DW, Mitchell RAC, Franklin J, Mitchell VJ, Driscoll SP. Facility for studying the effects of elevated carbon dioxide concentration and increased temperature on crops. *Plant Cell Environ.* 1993; 16:603–608.

47. Long SP, Hällgren J-E. Measurements of CO_2 assimilation by plants in the field and the laboratory. In: Photosynthesis and production in a changing environment. In: Hall DO, Scurlock JM, Bolhar-Nordenkampf HR, Leegood RC, Long SP, eds. *Photosynthesis and Production in a Changing Environment: A Field and a Laboratory Manual.* London: Chapman and Hall, 1993:129–167.

48. Louwerse W, Eikhoudt JW. A mobile laboratory for measuring photosynthesis, respiration and transpiration of field crops. *Photosynthetica* 1975; 9:31–34.

49. Mortensen LM. Growth responses of some greenhouse plants to environment. I. Experimental techniques. *Sci. Hort.* 1982; 16:39–46.

50. Nederhoff EM, Vegter JG. Photosynthesis of stands of tomato, cucumber and sweet peper measured in greenhouse under various CO_2-concentrations. *Ann. Bot.* 1994; 73:353–361.

51. Poorter H, Welschen RAM. Variation in RGR underlying carbon economy. In: Hendry GAF, Grime JP, eds. *Methods in Comparative Plant Ecology, A Laboratory Manual.* London: Chapman and Hall, 1993:107.

52. van Iersel MW, Bugbee B. A multiple chamber, semicontinuous, crop carbon dioxide exchange system: Design, calibration, and data interpretation. *J. Am. Soc. Hort. Sci.* 2000; 125:86–92.

53. Wheeler RM, Stutte GW, Subbarao GV, Yorio NC. Plant growth and human life support for space travel. In: Pessarakli M, ed. *Handbook of Plant and Crop Physiology.* 2nd ed. New York: Marcel Dekker, 2002:925–941.

54. Stasiak MA, Cote R, Dixon MAD, Grodzinski B. Increasing plant productivity in closed environments with inner canopy illumination. *Life Support Biosph. Sci.* 1998; 5:175–182.

55. Lake JV. Measurement and control of the rate of carbon dioxide assimilation by glasshouse crops. *Nature* 1966; 209:97.

56. Bloom A, Mooney, Björkman O, Berry J. Materials and methods for carbon dioxide and water exchange analysis. *Plant Cell Environ.* 1980; 3:371–376.

57. Knight SL. Constructing specialized plant growth chambers for gas exchange research: considerations and concerns. *HortSci* 1992; 27:767–769.

58. Tibbitts TW, McFarlane JC, Krizek DT, Berry WL, Hammer PA, Hodgson RH, Langhans RW. Contaminants in plant growth chambers. *HortSci* 1977; 12:310–311.

59. Mitchell CA. Measurement of photosynthetic gas exchange in controlled environments. *HortSci* 1992; 27:764–767.

60. Wheeler RM. Gas-exchange measurements using a large, closed plant growth chamber. *HortSci* 1992; 27:777–780.

61. Donahue RA, Poulson ME, Edwards GE. A method for measuring whole plant photosynthesis in *Arabibopsis thaliana. Photosynth. Res.* 1997; 52:263–269.

62. Miller DP, Howell GS, Flore JA. A whole-plant, open, gas-exchange system for measuring net photosynthesis of potted woody plants. *HortSci* 1996; 31:944–946.

63. Hanan JJ. *Greenhouses. Advanced Technology for Protected Horticulture.* New York: CRC Press, 1998.

64. Porter MA, Grodzinski B. CO₂ enrichment of protected crops. *Hort. Rev.* 1985; 7:345–398.

65. Leonardos ED, Tsujita MJ, Grodzinski B. Net carbon dioxide exchange rates and predicted growth patterns in Alstroemeria "Jacqueline" at varying irradiances, carbon dioxide concentrations and air temperatures. *J. Am. Soc. Hort. Sci.* 1994; 119:1265–1275.

66. Jiao J, Tsujita MJ, Grodzinski B. Optimizing aerial environments for greenhouse rose production utilizing whole-plant net CO₂ exchange data. *Can. J. Plant Sci.* 1991; 71:253–261.

67. Johnson IR, Thornley JHM. Temperature dependence of plant and crop processes. *Ann. Bot.* 1985; 55:1–24.

68. Ho LC. Variation in the carbon/dry matter ratio in plant material. *Ann. Bot.* 1976; 40:163–165.

69. Watts B. *The Effects of Temperature and CO₂ Enrichment on Growth and Photoassimilate Partitioning in Peppers (Capsicum annuum L.).* Guelph, Canada: University of Guelph, 1995.

70. Björkman O, Demming-Adams B. Regulation of photosynthetic light energy capture, conversion, and dissipation in leaves of higher plants. In: Schulze E-D, Caldwell MM, eds. *Ecophysiology of Photosynthesis.* Berlin: Springer-Verlag, 1994:17–47.

71. Pearcy RW, Pfitsch WA. The consequences of sunflecks for photosynthesis and growth of forest understory plants. In: Schulze E-D, Caldwell MM, eds. *Ecophysiology of Photosynthesis.* Berlin: Springer-Verlag, 1994:343–359.

72. Acock B, Charles-Edwards DA, Fitter DJ, Hand DW, Ludwig LJ, Wilson WJ, Withers AC. The contribution of leaves from different levels within a tomato crop to canopy net photosynthesis: an experimental examination of two canopy models. *J. Exp. Bot.* 1978; 29:815–827.

73. Duncan WG. Leaf angles, leaf area and canopy photosynthesis. *Crop Sci.* 1971; 11:482–485.

74. Reddy VR, Baker DN, Hodges HF. Temperature effects on cotton canopy growth, photosynthesis, and respiration. *Agron. J.* 1991; 83:699–704.

75. Russell G, Jarvis PG, Monteith JL. Absorption of radiation by canopies and stand growth. In: Russell G, Marshall B, Jarvis PG, eds. *Plant Canopies: Their Form and Functions.* Cambridge, UK: Cambridge University Press, 1989:21–39.

76. Warren Wilson J, Hand DW, Hannah MA. Light interception and photosynthetic efficiency in some glasshouse crops. *J. Exp. Bot.* 1992; 43:363–373.

77. Jiao J, Tsujita MJ, Grodzinski B. Influence of temperature on net CO₂ exchange in roses. *Can. J. Plant Sci.* 1991; 71:235–243.

78. Ehleringer J, Björkman O. Quantum yields for CO₂ [carbon dioxide] uptake in C3 and C4 plants: dependence on temperature, CO₂, and O₂ [oxygen] concentration. *Plant Physiol.* 1977; 59:85–90.

79. Björkman O. Responses to different quantum flux densities. In: Lange OL, Nobel PS, Osmond CB, Ziegler H, eds. *Physiological Plant Ecology I.* Vol. 12A. Berlin: Springer-Verlag, 1981:57–107.

80. Ögren E, Evans JR. Photosynthetic light-response curves. 1. The influence of CO₂ partial pressure and leaf inversion. *Planta* 1993; 189:182–190.

81. Poolman MG, Fell DA, Thomas S. Modelling photosynthesis and its control. *J. Exp. Bot.* 2000; 51:319–328.

82. Mott KA, Woodrow IE. Modelling the role of Rubisco activase in limiting non-steady-state photosynthesis. *J. Exp. Bot.* 2000; 51:399–406.

83. Mitchell RCA, Theobald JC, Parry MAJ, Lawlor DW. Is there scope for improving balance between RuBP-regeneration and carboxylation capacities in wheat at elevated CO₂? *J. Exp. Bot.* 2000; 51:391–397.

84. Mott KA, Jensen RG, O'Leary JW, Berry JA. Photosynthesis and ribulose 1,5-bisphosphate concentrations in intact leaves of *Xanthium strumarium* L. *Plant Physiol.* 1984; 76:968.

85. Kirschbaum MUF. The sensitivity of C3 photosynthesis to increasing CO₂: a theoretical analysis of its dependence on temperature and background CO₂ concentration. *Plant Cell Environ.* 1994; 17:747–754.

86. Sims DA, Pearcy RW. Scaling sun and shade photosynthetic acclimation of *Alocasia macrorrhiza* to whole-plant performance — I. Carbon balance and allocation at different daily photon flux densities. *Plant Cell Environ.* 1994; 17:881–887.

87. Bowes G. Facing the inevitable: plants and increasing atmospheric CO₂. *Annu. Rev. Plant Physiol. Plant Mol. Biol.* 1993; 44:309–332.

88. Zelitch I. Control of plant productivity by regulation of photorespiration. *BioScience* 1992; 42:510–517.

89. Drake BG, Gonzales-Meler M, Long SP. More efficient plants: A consequence of rising atmospheric CO₂. *Annu. Rev. Plant Physiol. Plant Mol. Biol.* 1997; 48:609–639.

90. Keys AJ, Leegood RC. Photorespiratory carbon and nitrogen cycling: evidence from studies of mutant and trangenic plants. In: Foyer CH, Noctor G, eds. *Photosynthetic Nitrogen Assimilation and Associated Carbon and Respiratory Metabolism.* Kluwer Academic, 2002:115–134.

91. Kramer PJ. Carbon dioxide concentration, photosynthesis, and dry matter production. *BioScience* 1981; 20:1201–1208.

92. Grodzinski B. Carbon dioxide enrichment: plant nutrition and growth regulation. *BioScience* 1992; 42:517–525.

93. Woodrow L, Liptay A, Grodzinski B. The effects of CO_2 enrichment and ethephon application on the production of tomato transplants. *Acta Hort.* 1987; 201:133–140.

94. Kimball BA, Idso SB. Increasing atmospheric CO_2 effects on crop yield, water use and climate. *Agric. Water Manag.* 1983; 7:55–72.

95. Caporn SJM, Hand DW, Mansfield TA, Wellburn AR. Canopy photosynthesis of CO_2-enriched lettuce (*Lactuca sativa* L.). Response to short-term changes in CO_2, temperature an oxides of nitrogen. *New Phytol.* 1994; 126:45–52.

96. Ehleringer JR, Cerling TE, Helliker BR. C4 photosynthesis, atmospheric CO_2, and climate. *Oecologia* 1997; 112:285–299.

97. Wittwer SH. Carbon dioxide levels in the biosphere: effects on plant productivity. *CRC Crit. Rev. Plant Sci.* 1985; 2:171–198.

98. Long SP. Modification of the response of photosynthetic productivity to rising temperature by atmospheric CO_2 concentrations: has its importance been underestimated? *Plant Cell Environ.* 1991; 14:729–739.

99. Dippery JK, Tissue DT, Thomas RB, Strain BR. Effects of low and elevated CO_2 on C3 and C4 annuals. 1. Growth and biomass allocation. *Oecologia* 1995; 101:13–20.

100. Hatch MD. C4 photosynthesis: a unique blend of modified biochemistry, anatomy and ultrastructure. *Biochim. Biophys. Acta* 1987; 895:81–106.

101. Leonardos ED, Grodzinski B. Photosynthesis, immediate export and carbon partitioning in source leaves of C3, C3–C4 intermediate and C4 Panicum and Flaveria species at ambient and elevated CO_2 levels. *Plant Cell Environ.* 2000; 23:839–851.

102. Kellogg EA, Farnsworth EJ, Russo ET, Bazzaz F. Growth responses of C4 grasses of contrasting origin to elevated CO_2. *Ann. Bot.* 1999; 84:279–288.

103. Wand SJE, Midgley GF, Stock WD. Growth responses to elevated CO_2 in NADP-ME, NAD-ME and PCK C4 grasses and a C3 grass from South Africa. *Aust. J. Plant Physiol.* 2001; 28:13–25.

104. Sage RF, Monson RK. *C4 Plant Biology.* New York: Academic Press, Harcourt Brace & Company, Publishers, 1999.

105. Osmond CB, Björkman O. Pathways of CO_2 fixation in the CAM plant *Kalanchoë daigremontiana*. II. Effects of O_2 and CO_2 concentration on light and dark CO_2 fixation. *Aust. J. Plant Physiol.* 1975; 2:155–162.

106. Huerta AJ, Ting IP. Effects of various levels of CO_2 on the induction of Crassulacean acid metabolism in *Portulacaria afra* (L.) Jacq. *Plant Physiol.* 1988; 88:183–188.

107. Hogan KP, Smith AP, Ziska LH. Potential effects of elevated CO_2 and changes in temperature on tropical plants. *Plant Cell Environ.* 1991; 14:763–778.

108. Cui M, Nobel PS. Gas exchange and growth responses to elevated CO_2 and light levels in the CAM species *Opuntia ficus-indica*. *Plant, Cell Environ.* 1994; 17:935–944.

109. Jones HG. *Plants and Microclimate.* Cambridge, UK: Cambridge University Press, 1983.

110. Lawlor DW, Mitchell RAC. The effects of increasing CO_2 on crop photosynthesis and productivity: a review of field studies. *Plant Cell Environ.* 1991; 14:807–818.

111. Wolfe DW. Physiological and growth response to atmospheric carbon dioxide concentration. In: Pessarakli M, ed. *Handbook of Plant and Crop Physiology.* New York: Marcel Dekker, 1995:223.

112. Ghannoum O, von Caemmerer S, Conroy JP. Plant water use efficiency of 17 Australian NAD-ME and NADP-ME C4 grasses at ambient and eleveted CO_2 partial pressure. *Aust. J. Plant Physiol.* 2001; 28:1207–1217.

113. Gifford RM. Whole plant respiration and photosynthesis of wheat under increased CO_2 concentration and temperature: long-term vs short-term distinctions for modelling. *Global Change Biol.* 1995; 1:385–396.

114. Besford RT, Ludwig LJ, Withers AC. The greenhouse effect: acclimation of tomato plants growing in high CO_2, photosynthesis and ribulose-1,5-bisphosphate carboxylase protein. *J. Exp. Bot.* 1990; 41:925–931.

115. Porter MA, Grodzinski B. Acclimation to high CO_2 in bean. *Plant Physiol.* 1984; 74:413–416.

116. Sage RF, Sharkey TD, Seamann JR. Acclimation of photosynthesis to elevated CO_2 in five C3 species. *Plant Physiol.* 1989; 89:590–596.

117. Stitt M, Muller C, Matt P, Gibon Y, Carillo, Morcuende R, Scheible W-R, Knapp A. Steps towards an integrated view of nitrogen metabolism. *J. Exp. Bot.* 2002; 53:959–970.

118. Paul MJ, Pellny TK. Carbon metabolite feedback regulation of leaf photosynthesis and development. *J. Exp. Bot.* 2003; 54: 539–547.

119. Lawlor DW. Carbon and nitrogen assimilation in relation to yield: mechanisms are key to understanding production systems. *J. Exp. Bot.* 2002; 53:773–788.

120. Thomas RB, Griffin KL. Direct and indirect effects of atmospheric carbon dioxide enrichment on leaf respiration of *Glycine max* (L.) Merr. *Plant Physiol.* 1994; 104:355–361.

121. Robertson EJ, Williams M, Hardwood JL, Linsday JG, Leaver CJ, Leech RM. Mitochondria increase threefold and mitochondrial proteins and lipid change dramatically in postmeristematic cells in young wheat leaves grown in elevated CO_2. *Plant Physiol.* 1995; 108:469–474.

122. Jahnke S. Atmospheric CO$_2$ concentration does not directly affect leaf respiration in bean or poplar. *Plant Cell Environ.* 2001; 24:1139–1151.

123. Jahnke S, Krewitt M. Atmospheric CO$_2$ concentration may directly affect leaf respiration measurement in tobacco, but not respiration itself. *Plant Cell Environ.* 2002; 25:641–651.

124. Long SP, Bernacchi CJ. Gas exchange measurements, what can they tell us about the underlying limitations to photosynthesis? Procedures and sources of error. *J. Exp. Bot.* 2003; 54:2393–2401.

125. Hunt S. Measurements of photosynthesis and respiration in plants. *Physiol. Plant.* 2003; 117:314–325.

126. Madore M, Grodzinski B. Photosynthesis and transport of ^{14}C-labelled in a dwarf cucumber cultivar under CO$_2$ enrichment. *J. Plant Physiol.* 1985; 121:59–71.

127. Madsen E. Effect of CO$_2$ concentration on the accumulation of starch and sugar in tomato leaves. *Physiol. Plant.* 1968; 21:168–175.

128. Jiao J, Grodzinski B. Environmental influences on photosynthesis and carbon export in greenhouse roses during development of the flowering shoot. *J. Am. Soc. Hort. Sci.* 1998; 123:1081–1088.

129. Leonardos ED, Tsujita MJ, Grodzinski B. The effect of source or sink temperature on photosynthesis and 14C-partitioning in, and export from a source leaf of *Alstroemeria* sp. cv. Jacqueline. *Physiol. Plant.* 1996; 97:563–575.

130. Grange RI. Carbon partitioning in mature leaves of pepper: effects of daylength. *J. Exp. Bot.* 1985; 36:1749–1759.

131. Grimmer C, Komor E. Assimilate export by leaves of *Ricinus communis* L. growing under normal and elevated carbon dioxide concentrations: the same rate during the day, a different rate at night. *Planta* 1999; 209:275–281.

132. Chapin FSI, Schulze ED, Mooney HA. The ecology and economics of storage in plants. *Annu. Rev. Ecol. Syst.* 1990; 21:423–447.

133. Schulze W, Schulze E-D. The significance of assimilatory starch for growth in Arabidopsis thaliana wildtype and starchless mutants. In: Schulze E-D, Caldwell MM, eds. *Ecophysiology of Photosynthesis.* Berlin: Springer-Verlag, 1994:123–131.

134. Grodzinski B, Jiao J, Leonardos ED. Estimating photosynthesis and concurrent export rates in C3 and C4 species at ambient and elevated CO$_2$. *Plant Physiol.* 1998; 117:207–215.

135. Leonardos ED, Grodzinski B. Quantifying immediate C export from source leaves. In: Pesarakli M, ed. *Hanbook of Plant and Crop Physiology.* 2nd ed. New York: Marcel Dekker, 2002:407–420.

136. Lalonde S, Tegeder M, Throne-Holst M, Frommer WB, Patrick JW. Phloem loading and unloading of sugars and amino acids. *Plant Cell Environ.* 2003; 26:37–56.

137. Ainsworth EA, Davey PA, Hymus GJ, P. Osborne C, Rogers A, Blum H, Nösberger J, Long SP. Is stimulation of leaf photosynthesis by elevated carbon dioxide concentration maintained in the long term? A test with *Lolium perenne* grown for 10 years at two nitrogen fertilization levels under Free Air CO$_2$ Enrichment (FACE). *Plant Cell Environ.* 2003; 26:705–714.

138. Reid CD, Strain BR. Effects of CO$_2$ enrichment on whole-plant carbon budget of seedlings of *Fagus grandifolia* and *Acer saccharum* in low irradiance. *Oecologia* 1994; 98:31–39.

139. Berry JA, Björkman O. Photosynthetic response and adaptation to temperature in higher plants. *Annu. Rev. Plant Physiol. Plant Mol. Biol.* 1980; 31:491–543.

140. Ku SB, Edwards GE. Oxygen inhibition of photosynthesis. I. Temperature dependence and relation to O$_2$/CO$_2$ solubility ratio. *Plant Physiol.* 1979; 59:991.

141. Brooks A, Farquhar GD. Effects of temperature on the CO$_2$/O$_2$ specificity of ribulose-1,5-bisphosphate carboxylase/oxygenase and the rate of respiration in the light. *Planta* 1985; 165:397–406.

142. Jordan DB, Ogren WL. The CO$_2$/O$_2$ specificity of ribulose-1,5-bisphosphate carboxylase/oxygenase. Dependence on ribulose bisphosphate concentration, pH and temperature. *Planta* 1984; 161:308.

143. Bernacchi CJ, Portis AR, Nakano H, von Caemmerer S, Long SP. Temperature response of mesophyll conductance. Implications for the determination of rubisco enzyme kinetics and for limitations to photosynthesis *in vivo*. *Plant Physiol.* 2002; 130:1992–1998.

144. Jiao J, Grodzinski B. The effect of leaf temperature and photorespiratory conditions on export of sugars during steady-state photosynthesis in *Salvia splendens*. *Plant Physiol.* 1996; 111:169–178.

145. Leonardos ED, Savitch LV, Huner NPA, Öquist G, Grodzinski B. Daily photosynthetic and C-export patterns in winter wheat leaves during cold stress and acclimation. *Physiol. Plant.* 2003; 117:521–531.

146. Jiao J, Tsujita MJ, Grodzinski B. Influence of irradiation and CO$_2$ enrichment on whole plant net CO$_2$ exchange in roses. *Can. J. Plant Sci.* 1991; 71:245–252.

147. Woo KC, Wong SC. Inhibition of CO$_2$ assimilation by supraoptimal CO$_2$: effect of light and temperature. *Aust. J. Plant Physiol.* 1983; 10:75–85.

148. Nilwik HJM. Photosynthesis of whole sweet pepper plants 2. Response to CO$_2$ (carbon dioxide) concentration, irradiance and temperature as influenced by cultivation conditions. *Photosynthetica* 1980; 14:382–391.

149. McCree KJ, Amthor ME. Effects of diurnal variation in temperature on the carbon balances of white clover plants. *Crop Sci.* 1982; 22:822.

150. van Iersel MW. Carbon use efficiency depends on growth respiration, maintenance respiration, and relative growth rate. A case study with lettuce. *Plant Cell Environ.* 2003; 26:1441–1449.

151. van Iersel MW, Seymour L. Growth respiration, maintenance respiration, and carbon fixation of vinca (*Catharanthus roseus* L.) G. Don.: a time series analysis. *J. Am. Soc. Hort. Sci.* 2000; 125:702–706.

152. Marshner H. *Mineral Nutrition in Higher Plants.* London: Academic Press, 1986.

153. Kumar PA, Parry MAJ, Mitchell CA, Ahmad A, Abrol YP. Photosynthesis and nitrogen use efficiency. In: Foyer CH, Noctor G, eds. *Photosynthetic Nitrogen Assimilation and Associated Carbon and Respiratory Metebolism.* Kluwer Academic, 2002:23–34.

154. McDonald AJS. Phenotypic variation in growth rate as affected by N-supply: its effects on net assimilation rate (NAR), leaf weight ratio (LWR) and specific leaf area (SLA). In: Lambers H, Cambridge ML, Konings H, Pons TL, eds. *Causes and Consequences of Variation in Growth Rate and Productivity of Higher Plants.* The Hague, The Netherlands: SPB Academic Publishing, 1990:35–44.

155. van Keulen H, Goudriaan J, Seligman NG. Modelling the effects of nitrogen on canopy development and crop growth. In: Russell G, Marshall B, Jarvis PG, eds. *Plant Canopies: Their Growth, Form and Function.* Cambridge, UK: Cambridge University Press, 1989: 83–104.

156. Evans JR. Photosynthesis and nitrogen relationships in leaves of C3 plants. *Oecologia* 1989; 78:9.

157. Pons TL, Schieving F, Hirose T, Werger MJA. Optimization of leaf nitrogen allocation for carbon photosynthesis in *Lysimachia vulgaris.* In: Lambers H, Cambridge ML, Konings H, Pons TL, eds. *Causes and Consequences of Variation in Growth Rate and Productivity of Higher Plants.* The Hague, The Netherlands: SPB Academic Publishing, 1990:175–186.

158. Sage RF, Pearcy RW. The nitrogen use efficiency of C3 and C4 plants. II. Leaf nitrogen effects on the gas exchange characteristics of *Chenopodium album* (L.) and *Amaranthus retroflexus* (L.). *Plant Physiol.* 1987; 84:959–963.

159. Schieving F, Pons TL, Werger MJA, Hirose T. The vertical distribution of nitrogen and photosynthetic activity at different plant densities in *Carex acutiformis. Plant Soil* 1992; 14:9–17.

160. Seemann JR, Sharkey TD, Wang J, Osmond CB. Environmental effects on photosynthesis, nitrogen-use efficiency, and metabolite pools in leaves of sun and shade plants. *Plant Physiol.* 1987; 84:796–802.

161. Hunt S, Layzell DB. Gas exchange of legume nodules and the regulation of nitrogenase activity. *Annu. Rev. Plant Physiol. Plant Mol. Biol.* 1993; 44:483–511.

162. Clarkson DT. Regulation of the absorption and release of nitrogen by plant cells: a review of current ideas and methodology. In: Lambers H, Neetson JJ, Stulen I, eds. *Fundamental, Ecological, and Agricultural Aspects of Nitrogen Metabolism in Higher plants.* Dordrecht: Martinus Nijhoff, 1986:3–27.

163. Chapin FSI, Bloom AJ, Field CB, Waring RH. Plant responses to multiple environmental factors. *BioScience* 1987; 37:49–57.

164. Field C. Allocating leaf nitrogen for the maximization of carbon gain: leaf age as a control on the allocation program. *Oecologia* 1983; 56:341–347.

165. Werger MJA, Hirose T. Leaf nitrogen distribution and whole canopy photosynthetic carbon gain in herbaceous stands. *Vegetatio* 1991; 97:11–20.

166. Lambers H, Freijsen N, Poorter H, Hirose T, van der Werf A. Analysis of growth based on net assimilation rate and nitrogen productivity. Their physiological background. In: Lambers H, Cambridge. ML, Konings H, Pons TL, eds. *Causes and Consequences of Variation in Growth Rate and Productivity of Higher Plants.* The Hague, The Netherlands: SPB Academic Publishing, 1990:1–17.

167. Freijsen AHJ, Veen BW. Phenotypic variation in growth as affected by N-supply: nitrogen productivity. In: Lambers H, Cambridge ML, Konings H, Pons TL, eds. *Causes and Consequences of Variation in Growth Rate and Productivity of Higher Plants.* The Hague, The Netherlands: SPB Academic Publishing, 1990:19–33.

168. Davis TD, Curry EA. Chemical regulation of vegetative growth. *Crit. Rev. Plant Sci.* 1991; 10:151–188.

169. Abeles FB, Morgan PW, Saltvein ME Jr. *Ethylene in Plant Biology.* New York: Academic Press, 1993.

170. Grodzinski B. Enhancement of ethylene release from leaf tissue during glycolate decarboxylation: a possible role for photorespiration. *Plant Physiol.* 1984; 74:871–876.

171. Dhawan KR, Bassi PK, Spencer MS. Effects of carbon dioxide on ethylene production and action in intact sunflower plants. *Plant Physiol.* 1981; 68:831–834.

172. Grodzinski B, Boesel I, Horton RF. Light stimulation of ethylene release from leaves of *Gomphrena globosa* L. *Plant Physiol.* 1983; 71:588–593.

173. Grodzinski B, Boesel I, Horton RF. Ethylene release from leaves of *Xanthium strumarium* L. and *Zea mays* L. *J. Exp. Bot.* 1982; 33:344–354.

174. Kao CH, Yang SF. Light inhibition of the conversion of 1-aminocyclopropane-1-carboxylic acid (ACC) to ethylene in leaves is mediated through carbon dioxide. *Planta* 1982; 155:261–266.

175. Philosoph-Hadas S, Aharoni N, Yang SF. Carbon dioxide enhances the development of the ethylene-forming enzyme in tobacco leaf discs. *Plant Physiol.* 1986; 82:925–929.

176. Woodrow L, Grodzinski B. Ethylene exchange in *Lycopersicon esculentum* Mill. Leaves during short- and long-term exposures to CO_2. *J. Exp. Bot.* 1994; 44:471–480.

177. Woodrow L, Jiao J, Tsujita MJ, Grodzinski B. Photoautotrophic systems: ethylene and carbon dioxide interactions from the callus to the canopy. In: Flores HE, Arteca RN, Shannon JC, eds. *Polyamines and Ethylene: Biochemistry, Physiology, and Interactions.* Rockville, MD: American Society of Plant Physiologists, 1990:91–100.

178. Wheeler RM, Peterson BV, Sager JC, Knott WM. Ethylene production by plants in a closed environment. *Adv. Space Res.* 1996; 18:193–196.

179. Stasiak MA. Effect of Inner Canopy Irradiation on Plant Productivity in a Sealed Environment. PhD Thesis, Department of Plant Agriculture. Guelph, Canada: University of Guelph, 2002.

180. Osborne DJ. Ethylene in leaf ontogeny and abscission. In: Mattoo AK, Suttle JC, eds. *Ethylene.* Boca Raton, FL: CRC Press, 1990:193–214.

181. Kays SJ, Pallas JE Jr. Inhibition of photosynthesis by ethylene. *Nature* 1980; 285:51–52.

182. Taylor GE, Gunderson CA. The response of foliar gas exchange to exogenously applied ethylene. *Plant Physiol.* 1986; 82:653–657.

183. Woodrow L, Grodzinski B. An evaluation of the effects of ethylene on carbon assimilation in *Lycopersicon esculentum* Mill. *J. Exp. Bot.* 1989; 40:361–368.

184. Woodrow L, Jiao J, Tsujita MJ, Grodzinski B. Whole plant and leaf steady state gas exchange during ethylene exposure in *Xanthium strumarium* L. *Plant Physiol.* 1989; 90:85–90.

185. Woodrow L, Thompson RG, Grodzinski B. Effects of ethylene on photosynthesis and partitioning in tomato *Lycopersicon esculentum* Mill. *J. Exp. Bot.* 1988; 39:667–684.

186. Cliquet J-B, Boutin J-P, Deleens E, Morot-Gaudry J-F. Ethephon effects on translocation and partitioning of assimilates in *Zea mays*. *Plant Physiol. Biochem.* 1991; 29:623–630.

187. Crocker W, Zimmerman PW, Hitchcock AE. *Contrib. Boyce Thompson Inst.* 1932; 4:177.

188. Grodzinski B, Woodrow L, Leonardos ED, Dixon M, Tsujita MJ. Plant responses to short- and long-term exposures to high carbon dioxide levels in closed environments. *Adv. Space Res.* 1996; 18:203–211.

189. Heiman MD, Meredith FI, Gonzalez CL. Ethylene production in the cotton plant (*Gossypium hirsutum* L.) canopy and its effect on fruit abscission. *Crop Sci.* 1971; 11:25–27.

190. Klassen SP, Bugbee B. Sensitivity of wheat and rice to low levels of atmospheric ethylene. *Crop Sci.* 2002; 42:746–753.

191. Potter JR, Jones JW. Leaf area partitioning as an important factor in growth. *Plant Physiol.* 1977; 59:10–14.

192. Wells R, Meredith J, W. R., Williford JR. Heterosis in upland cotton. II. Relationship of leaf area to plant photosynthesis. *Crop Sci.* 1988; 28:522–525.

193. Boote KJ, Loomis RS. Modeling crop photosynthesis — from biochemistry to canopy. *Proceedings of a Symposium*, Madison, WI, Crop Science Society of America, 1991.

194. Ryel RJ, Beyschlag W, Caldwell MM. Foliage orientation and carbon gain in two tussock grasses as assessed with a new whole-plant gas-exchange model. *Funct. Ecol.* 1993; 7:115–124.

195. Werk KS, Ehleringer JR. Non-random leaf orientation in *Lactuca serriola* L. *Plant Cell Environ.* 1984; 7:81–87.

196. Oker-Blom P, Kellomäki S. Effect of angular distribution of foliage on light absorption and photosynthesis in the plant canopy: theoretical computations. *Agric. Meteor.* 1982; 26:105–116.

197. Wells R. Soybean growth response to plant density: relationships among canopy photosynthesis, leaf area, and light interception. *Crop Sci.* 1991; 31:755–761.

198. Campbell GS, Norman JM. The description and measurement of plant canopy structure. In: Russell G, Marshall B, Jarvis PG, eds. *Plant Canopies: Their Growth, Form and Functions.* Cambridge, UK: Cambridge University Press, 1989:1–19.

199. Wells R, Meredith Jr. WR, Williford JR. Canopy photosynthesis and its relationship to plant productivity in near-isogenic lines differing in leaf morphology. *Plant Physiol.* 1986; 82:635–640.

200. Savitch LV, Leonardos ED, Krol M, Jansson S, Grodzinski B, Huner NPA, Öquist G. Two different strategies for light utilisation in photosynthesis in relation to growth and cold acclimation. *Plant Cell Environ.* 2002; 25:761–771.

201. Happer JL. Canopies as populations. In: Russell G, Marshall B, Jarvis PG, eds. *Plant Canopies: Their Growth, Form and Functions.* Cambridge, UK: Cambridge University Press, 1989:105–128.

202. Ellsworth DS, Reich PB. Canopy structure and vertical patterns of photosynthesis and related leaf traits in a deciduous forest. *Oecologia* 1993; 96:169–178.

203. Côté R, Jerrath JM, Posluszny U, Grodzinski B. Comparative leaf development of conventional and semileafless peas (*Pisum sativum*). *Can. J. Bot.* 1992; 70:571–580.

204. Gould KS, Cuttere G, Young JPW. Does growth rate determine leaf form in *Pisum sativum*? *Can. J. Bot.* 1989; 67:2590–2595.

205. Marx GA. A suite of mutants that modify pattern formation in pea leaves. *Plant Mol. Biol. Rep.* 1987; 5:311–335.

206. Côté R, Grodzinski B. Improving light interception by selecting morphological leaf phenotypes: a case study using "alifa" pea mutant. AES Technical Papers ES-288, 1999.

207. Amthor JS. Scaling CO$_2$-photosynthesis relationships from the leaf to the canopy. *Photosynth. Res.* 1994; 39:321–350.

208. Beyschlag W, Ryel RJ, Caldwell MM. Photosynthesis of vascular plants: assessing canopy photosynthesis by means of simulation models. In: Shulze ED, Caldwell MM, eds. *Ecophysiology of Photosynthesis.* Berlin: Springer-Verlag, 1994:409–430.

209. Evans JR, Farquhar GD. Modeling canopy photosynthesis from the biochemistry of the C3 chloroplast. In: Boote KJ, Loomis RS, eds. *Modeling Crop Photosynthesis from Biochemistry to Canopy.* Madison, WI: Crop Science Society of America, 1991:1–15.

210. Forseth IN, Norman JM. Modelling of solar irradiance, leaf energy budget and canopy photosynthesis. In: Hall DO, Scurlock JM, Bolhar-Nordenkampf HR, Leegood RC, Long SP, eds. *Photosynthesis and Production in a Changing Environment: A Field and a Laboratory Manual.* London: Chapman and Hall, 1993:207.

211. Thornley JHM, Johnson IR. *Plant and Crop Modelling.* Oxford: Clarendon Press, 1990:243.

212. Werner C, Ryel RJ, Correia O, Beyschlag W. Effects of photoinhibition on whole-plant carbon gain assessed with a photosynthetic model. *Plant Cell Environ.* 2001; 24:27–40.

213. Baldocchi D. A comparative study of mass and energy exchange over a closed C3 (wheat) and an open C4 (corn) canopy: I. The partitioning of available energy into latent and sensible heat exchange. *Agri. For. Meteor.* 1994; 67:191–220.

214. Baldocchi D. A comparative study of mass and energy exchange rates over a closed C3 (wheat) and an open C4 (corn) crop: II. CO_2 exchange and water use efficiency. *Agri. For. Meteor.* 1994; 67:291–321.

215. Baldocchi D. Assessing the eddy covariance technique for evaluating carbon dioxide exchange rates of ecosystems: past, present and future. *Global Change Biol.* 2003; 9:479–492.

216. Wofsy SC, Goulden ML, Munger JW, Fan S-M, Bakwin PS, Daube BC, Bassow SL, Bazzaz FA. Net exchange of CO_2 in a mid-latitude forest. *Science* 1993; 260:1314–1317.

217. Field CB, Gamon JA, Peñuelas J. Remote sensing of terrestrial photosynthesis. In: Schulze E-D, Caldwell MM, eds. *Ecophysiology of Photosynthesis.* Berlin: Springer-Verlag, 1994:511.

218. Chen MJ, Liu J, Leblanc SG, Lacaze R, Roujean JL. Mutli-angular optical remote sensing for assessing vegetation structure and carbon absorption. *Remote Sensing Environ.* 2003; 84:516–525.

219. Grodzinski B, Jiao J, Knowles VL, Plaxton WC. Photosynthesis and carbon partitioning in transgenic tobacco plants deficient in leaf cytosolic pyruvate kinase. *Plant Physiol.* 1999; 120:887–895.

220. Johnstone M, Yu H, Liu W, Leonardos ED, Sutton J, Grodzinski B. Physiological changes associated with pythium root rot in hydroponic lettuce. *Acta Hort.* 2004; 635:67–75.

32 Approaches to Measuring Plant Photosynthetic Activity

Elena Masarovičová
Department of Plant Physiology, Faculty of Natural Sciences,
Comenius University

Katarina Kráľová
Institute of Chemistry, Faculty of Natural Sciences,
Comenius University

CONTENTS

I. INTRODUCTION

This chapter provides some basic theoretical knowledge and techniques (methodical approach in the laboratory and in the field) for the study of plant photosynthetic activity. It is not intended to be a complete manual of techniques. Not only will methodical approaches and published results be described, but problems the scientist should be aware of when planning experiments will be discussed. Measurements of environmental or plant parameters must be recorded in some manner in order to obtain the data that meet the research objective. However, each type of measurement requires a different approach.

Almost 35 years ago, one of the most cited methodological monographs, *Plant Photosynthetic Production: Manual of Methods*, was published [1], which includes a rich source of background material. Handbooks published later [2,3] or recent manuals [4,5] describe general measurement principles, techniques, and devices that form system components. Now, highly sophisticated systems capable of accurate measurements with a variety of transducers are widely in use. These systems (e.g., portable photosynthesis and transpiration systems, steady state porometer, portable chlorophyll fluorometer, plant canopy analyzer) have permitted research that was previously impossible or too difficult to attempt.

II. IN THE LABORATORY

A. MEASUREMENTS OF HILL ACTIVITY USING ARTIFICIAL ELECTRON ACCEPTOR 2,6-DICHLOROPHENOLINDOPHENOL

The Hill reaction is formally defined as the photoreduction of an electron acceptor by the hydrogens of water, connected with the evolution of oxygen. *In vivo*, or in the plant organism, the final electron acceptor is NADP$^+$. The rate of the Hill reaction can be measured in isolated chloroplasts [6]. A number of artificial electron acceptors can replace the natural acceptors and allow electron transport to proceed in the light. The artificial electron acceptor intercepts the electrons before they cascade down to photosystem 1 (PSI), but after they have gone down the electron transport chain. This procedure uses a dye as an artificial electron acceptor that changes color as it is reduced. Various dyes can be used as the artificial electron acceptor (A), so that the general equation, known as the Hill reaction, can be written as follows:

$$H_2O + A \longrightarrow AH_2 + {}^1/_2 O_2$$

As a suitable artificial electron acceptor, which can accept electrons from the electron transport chain of chloroplasts, 2,6-dichlorophenol indophenol (DCPIP) can be used [7]. When it accepts electrons, DCPIP becomes reduced and changes from blue (oxidized form) to colorless (reduced form DCPIP/H$_2$). This color change can be measured spectrophotometrically at it can be used to measure the rate of the Hill reaction. The oxidized form of DCPIP has an absorption maximum at 600 nm, while the reduced form does not absorb at this wavelength. The extent of the color change is proportional to the number of electrons transferred, or more precisely to the rate of photosynthetic electron transport (PET). The change in absorbance will be measured at certain intervals (e.g., 30 sec) of exposure to an intense light source. Since the DCPIP will begin to revert to its oxidized (blue) state as soon as the chloroplasts in the reaction vessel are removed from the light path, it is essential that all absorbance readings be taken as quickly as possible. The rate of electron transport can be determined by calculating the number of molecules of DCPIP reduced per minute. For this purpose the following equation can be used:

$$\text{PET rate} = (\Delta A_{600}/\text{min}/\varepsilon) \times 10^6 \ \mu\text{mol/mol} \times V$$

where ΔA_{600}/min is the rate of absorbance change at 600 nm, ε is the extinction coefficient for DCPIP at 600 nm (21,000 dm^3/mol/cm), and V is the volume of the reaction mixture in dm^{-3}. This parameter is usually expressed per chlorophyll content unit.

Šeršeň et al. [8] isolated chloroplasts from spinach leaves using the following procedure: 80 g of leaf tissue (minus petioles and midrib veins) were rinsed in ice water, then the tissue was blotted and cut into pieces about 1 cm^2. Cold conditions during isolation

were necessary to maintain good activity of the chloroplasts. The leaf pieces were placed in a pre-chilled blender cup containing 200 cm^3 of ice-cold isolation medium (20 mmol/dm^3 Tris, 5 mmol/dm^3 MgCl$_2$ and 15 mmol/dm^3 NaCl, and 0.4 mol/dm^3 sucrose). The leaves were blended for 30 sec at top speed, and after a break of 10 sec they were blended again for 30 sec. The resulting homogenate was squeezed through eight layers of nylon cloth and a 10-mm layer of cotton wool into the prechilled beaker. The green filtrate was centrifuged at 500×g for 5 min. This removes unwanted whole cells and groups of unbroken cells and cell wall debris, but most of the chloroplasts remain suspended in the supernatant solution. The decanted supernatant was then centrifuged at 5000×g for 10 min to sediment the chloroplasts. Afterward, the supernatant solution was discarded and the chloroplast precipitates were immediately resuspended in a small amount of buffer. For determination of the chlorophyll (Chl) content in the chloroplast suspension, 50 μl of chloroplasts were pipetted into 10 ml of 80% acetone, the solution was mixed, and after filtering the absorbance at 652 nm was measured. The Chl content was evaluated according to Arnon [9].

The effects of different inhibitors of PET on the oxygen evolution rate (OER) in spinach chloroplasts prepared according to the above procedure were investigated spectrophotometrically in the presence of DCPIP (30 μmol/dm^3) [10–12]. Before the measurements the chloroplasts were resuspended in phosphate buffer (20 mmol/dm^3; pH 7.2) containing 5 mmol/dm^3 MgCl$_2$ and 15 mmol/dm^3 NaCl. The Chl content in the suspension was adjusted to 30 mg Chl/dm^3. Samples were irradiated from a distance of 1 dm with a halogen lamp (250 W) through a 4-cm water filter to prevent overheating of the samples. This photochemical assay was carried out under saturating irradiance of "white light" (≈900 μmol/m^2/sec photosynthetically active radiation [PAR]) at 25°C. The inhibitory activity of the inhibitors studied was expressed in terms of IC$_{50}$ values (eventually in terms of their negative logarithms) corresponding to molar concentrations of inhibitors causing a 50% decrease of OER with respect to the untreated control sample. Compounds with low aqueous solubility were dissolved in dimethyl sulfoxide. The applied solvent content (up to 4 vol%) did not affect the photochemical activity of spinach chloroplasts. Table 32.1 presents IC$_{50}$ values related to OER inhibition in spinach chloroplasts by substituted benzanilides [10]. The photosynthesis-inhibiting activity of the benzanilides studied showed quasiparabolic dependence on the sum of lipophilicity of R^1 and R^2 substituents expressed as $(\pi_1{}^- + \pi_2{}^-)$. The π^- parameters express-

TABLE 32.1
IC$_{50}$ Values Related to Inhibition of OER in Spinach Chloroplasts by the Benzanilides Studied (Me = CH$_3$; iPr = CH(CH$_3$)$_2$; Bu = CH$_2$CH$_2$CH$_2$CH$_3$)

R^1—〈 〉—CONH—〈 〉—R^2

Compound	R^1	R^2	IC$_{50}$ (μmol/dm^3)	Compound	R^1	R^2	IC$_{50}$ [μmol dm^{-3}]
I	H	3-NO$_2$	374	X	3-F	H	324
II	3-Br	3-F	86	XI	3-F	3-Cl	71
III	4-Cl	3-NO$_2$	73	XII	3-F	4-Me	193
IV	H	4-iPr	50	XIII	3-F	3-NO$_2$	126
V	4-OMe	4-iPr	53	XVI	3-F	4-NO$_2$	109
VI	3-Br	4-iPr	67	XV	3-F	4-OMe	484
VII	3-NO$_2$	4-iPr	41	XVI	3-F	3-OMe	263
VIII	H	4-Bu	48	XVII	4-F	H	497
IX	3-Br	4-Bu	357	XVIII	4-F	3-OMe	365

Source: From Král'ová K, Seršeň F, Kubicová L, Waisser K. *Chem. Pap.* 1999; 53: 328–331. With permission.

ing lipophilicity of the substituents on the aromatic ring were taken from Norrington et al. [13] (Figure 32.1). The results of statistical analysis confirmed that Hansch's parabolic model is suitable for description of the correlation between photosynthesis-inhibiting activity and lipophilicity of the benzanilides studied. These compounds were found to interact with the intermediates D$^+$, i.e., tyrosine radicals Y$_D$$^+$ that are situated in the 161st position in D$_2$ protein on the donor side of PS 2 [10].

Piperidinoethyl esters of 2-, 3- and 4-alkoxy substituted phenylcarbamic acids (PAPC; alkyl = methyl to decyl) inhibited OER in spinach chloroplasts and their inhibitory activity depended on the alkyl chain length as well as on the position of the alkoxy substituent on the benzene ring of the effector (Table 32.2) [11]. The OER-inhibiting activity showed quasi parabolic course on the lipophilicity of PAPC, expressed by lipophilicity characteristics log k' (from high-performance liquid chromatography [HPLC]) and Kovats indices K_{10} (from gas chromatography). The lowest OER-inhibiting activity exhibited 2-alkoxy substituted derivatives. The highest biological activity showed compounds with heptyloxy, octyloxy, and nonyloxy substituents.

An expressive dependence of the OER-inhibiting activity in spinach chloroplasts on the alkyl chain length of the 2-alkylsulfanyl substituent showed 2-alkylsulfanyl-6-R-benzothiazoles with R = formamido (n-alkyl = ethyl to nonyl), acetamido (ethyl, butyl to hexyl, octyl, nonyl), benzoylamino (methyl to butyl, hexyl to nonyl), bicyclo[2.2.1]hept-5-ene-2, 3-dicarboximido (methyl to nonyl), and bicyclo[2.2.1]- hept-5-ene-2,3-dicarboximidomethylamino (methyl to pentyl, heptyl) (Figure 32.2) [12]. A quasiparabolic

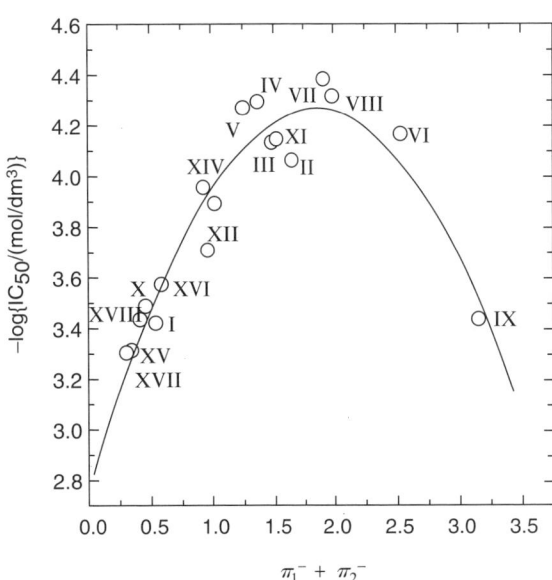

FIGURE 32.1 The dependence of inhibition of OER in spinach chloroplasts on the lipophilicity of the substituents R^1 and R^2 expressed by π^- parameters of substituents on the aromatic ring taken from Norrington et al. [13]. (From Král'ová K, Seršeň F, Kubicová L, Waisser K. *Chem. Pap.* 1999; 53: 328–331. With permission.)

TABLE 32.2
Negative Logarithms of IC$_{50}$ Values (in mol/dm^3) Related to OER Inhibition in Spinach Chloroplasts and Physicochemical Characteristics, log k' (from HPLC) and Kovats Indices K_{10} (from Gas Chromatography), of Piperidinoethyl Esters of Alkoxyphenylcarbamic Acids (m — No. of Carbon Atoms in the Alkoxy Substituent)

Substituted Position	m	log(1/IC$_{50}$)	log k'	K_{10}
	1	1.220	0.0969	2.341
	2	1.782	0.1207	2.468
	3	2.114	0.1761	2.566
	4	2.591	0.2540	2.622
2	5	3.262	0.3319	2.694
	6	3.431	0.4409	2.782
	7	3.485	0.5187	2.924
	8	—	0.6110	3.022
	9	3.252	0.7175	3.088
	10	3.002	0.8342	3.188
	2	2.388	0.0717	2.564
	3	2.581	0.1514	2.653
	4	3.117	0.2398	2.750
3	5	3.563	0.3079	2.857
	6	3.965	0.4239	2.973
	7	4.048	0.5279	3.061
	8	4.351	0.5836	3.104
	9	3.875	0.7342	3.247
	10	4.020	0.8691	3.354
	1	1.477	0.0111	2.552
	2	2.204	0.0717	2.653
	3	2.609	0.1532	2.709
	4	3.262	0.2540	2.847
4	5	3.184	0.3446	2.919
	6	3.720	0.4464	3.029
	8	4.712	0.6612	3.207
	9	4.079	0.7744	3.293
	10	3.876	0.9020	3.424

Source: From Král'ová K, Loos D, Čižmárik J. *Collect. Czech. Chem. Commun.* 1994; 59: 2293–2302. With permission.

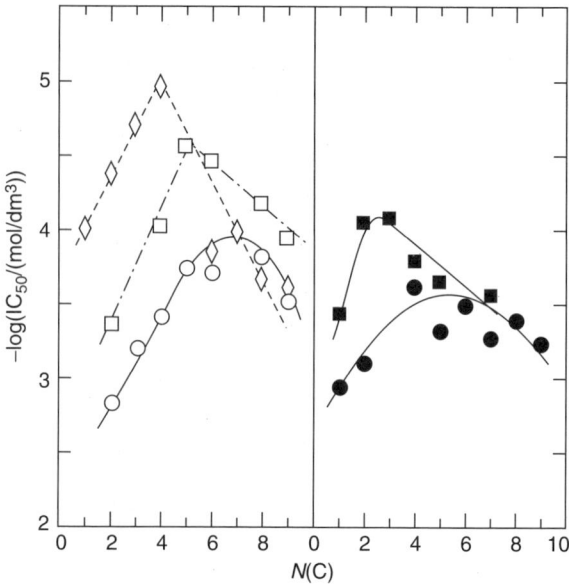

FIGURE 32.2 Dependence of the negative logarithm of IC$_{50}$ values related to OER in spinach chloroplasts by Hill reaction on the number of carbons of the alkyl chain of 2-alkythio-6-R-benzothiazoles (R = formamido [circles], acetamido [squares], benzoylamino [diamonds], bicyclo[2.2.1]hept-5-ene-2,3-dicarboximido [filled circles], and bicyclo[2.2.1]hept-5-ene-2,3-dicarboximidomethylamino [filled squares] groups). (From Král'ová K, Šeršeň F, Sidóová E. *Chem. Pap.* 1992; 46: 348–350. With permission.)

perturbation and subsequent changes in the biological function of the membrane occur.

B. MEASUREMENTS OF OXYGEN EVOLUTION RATE BY CLARK ELECTRODE

The OER could be also measured by Clark's electrode [14,15]. This electrode consists of an anode and a cathode in contact with an electrolyte solution. It is covered at the tip by a semipermeable membrane, usually polypropylene or teflon membrane, which is permeable to gases but not to contaminants and reducible ions of the sample. The cathode is in a glass envelope in the body of the electrode. The anode has a larger surface that provides stability and guards against drift due to concentration of the electrolyte (usually potassium chloride, 0.1 M). This silver/silver chloride (Ag/AgCl) anode provides electrons for the cathode reaction. The Clark electrode measures oxygen tension amperometrically. The pO$_2$ (partial pressure of oxygen) electrode produces current at a constant polarizing voltage (usually 0.6 V vs. Ag/AgCl), which is directly proportional to the pO$_2$ diffusing to the reactive surface of the electrode. Silver at the anode becomes oxidized.

course of these dependences is typical for all the series studied. This "cutoff" effect — a decreased activity for the more lipophilic substances within the homologous series is caused by the interaction of the alkyl substituent with constituents of biological membranes, mainly lipids. Due to this interaction,

$$\text{Ag anode: } 4Ag + 4Cl^- \rightarrow 4AgCl + 4e^-$$

$$\text{Pt cathode: } O_2 + 4H^+ + 4e^- \rightarrow 2H_2O$$

Reduction of oxygen occurs at the surface cathode, which is exposed at the tip of the electrode. Oxygen molecules diffuse through the semipermeable membrane and combine with the KCl electrolyte solution. The current produced is a result of the following reduction of oxygen at the cathode. Production of four electrons accompanies each molecule reduced. A simple scheme of a Clark electrode is shown in Figure 32.3.

Šeršeň et al. [16] and Král'ová et al. [17] measured OER in algal suspension of *Chlorella vulgaris* at 24°C using a Clark-type electrode (SOPS 31 atp. Chemoprojekt, Prague) in a chamber constructed according to the method of Bartoš et al. [18]. The liquid medium of the suspension contained 20 mmol KNO_3, 2.5 mmol KH_2PO_4, 4.0 mmol $MgSO_4 \cdot 7H_2O$, 7.0 µmol $CaCl_2 \cdot 6H_2O$, 34.6 µmol $FeSO_4$, 34.6 µmol Na_2 EDTA, 50.0 µmol H_3BO_3, 5.0 µmol $ZnSO_4 \cdot 5H_2O$, 5.0 µmol $CuSO_4 \cdot 5H_2O$, 5.0 µmol $MnCl_2 \cdot 4H_2O$, 5.0 µmol $CoCl_2 \cdot 6H_2O$, and 1.5 µmol $(NH_4)_6Mo_7$ $O_{24} \cdot 4H_2O$ in 1 dm^{-3} of H_2O, pH 7.2. Irradiation was carried out with a 250-W halogen lamp through a water filter (irradiance 450 µmol/m^2/sec PAR). Before OER measurements the algal suspension was accommodated in the dark for 4 h.

Immediate effect of the local anesthetic trimecaine on *C. vulgaris* was compared with its long-term effect on the chlorophyll content in statically cultivated *C. vulgaris* suspensions (14 days, 16 h light/8 h dark,

24°C) (Figure 32.4) [16]. Depending on the concentration of trimecaine, two different effects on OER were observed — at low trimecaine concentrations the effect was stimulating, at higher concentrations this effect was inhibitory.

The inhibitory effects on OER in spinach chloroplasts exhibited by structurally similar local anesthetics used at sufficiently high concentrations were observed by Král'ová et al. [11,19,20]. It was found that local anesthetics caused inhibition of OER in plant chloroplasts by damaging the manganese containing protein in the PS 2 with subsequent release of Mn^{2+} ions into the interior of the thylakoid membrane. The stimulating effect of trimecaine, i.e., the enhancement of OER in algae, can be connected with the uncoupling of photophosphorylation in algal chloroplasts or with changes in the arrangement of the thylakoid membranes. Based on the results of Gallová et al. [21], who found that carbisocaine at low concentrations decreases the microviscosity of phosphatidylcholine membranes, it could be assumed that due to incorporation of trimecaine into thylakoid membranes changes occur in their arrangement, leading to the enhancement of photosynthetic activity. The migration of the plastoquinone pool between the photosynthetic centers becomes easier, which enables faster photosynthetic electron transport. The concentration of trimecaine causing death of *C. vulgaris*

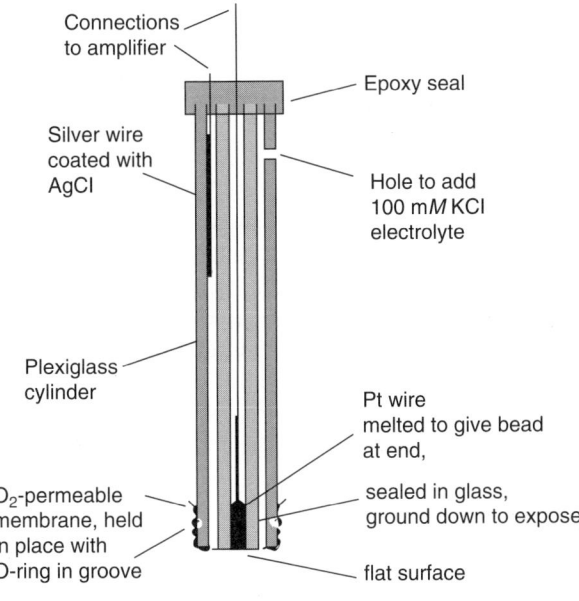

FIGURE 32.3 A simple scheme of a Clark electrode.

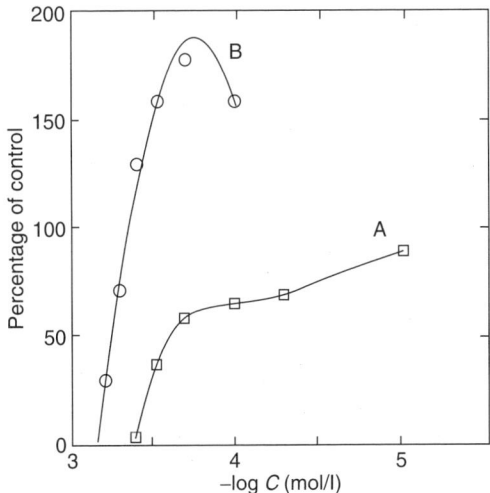

FIGURE 32.4 The dependence of Chl synthesis (curve A) and OER (curve B) in *C. vulgaris* on the concentration of trimecaine (the parameters studied are expressed as a percentage of the control samples). The Chl content in the algal suspension was 6.3 mg/dm^3 for OER and 6.1 mg/dm^3 (starting concentration) for the growth experiments in statically cultivated algae. (From Šeršeň F, Král'ová K. *Gen. Physiol. Biophys.* 1994; 13: 329–335. With permission.)

in long-term action causes enhancement of OER immediately after the treatment.

A similar experiment focused on OER inhibition in *C. vulgaris* was carried out with a series of Cu(II) complexes with biologically active ligands of the type $CuX_2 \cdot H_2O$ and CuX_2L_y, where X = flufenamate (N-(α,α,α-trifluoro-m-tolyl)anthranilate), mefenamate (2-((2,3-dimethylphenyl)amino)benzoate)), niflumate (2-(α,α,α-trifluoro-m-toluidino) nicotinate), naproxenate (6-methoxy-α-methyl-2-naphthaleneacetate); L = nicotinamide, N,N-diethylnicotinamide, ronicol (3-hydroxymethylpyridine), caffeine, methyl-3-pyridylcarbamate; and y = 1 or 2. In this experiment the Chl content in the algal suspensions was 20 mg/dm^3 [17]. The corresponding IC_{50} values monitoring the immediate effect of the Cu(II) complexes on the PET in *C. vulgaris* are summarized in Table 32.3. The anionic X ligands increased the inhibitory effect while the effect of the L ligands was not significant. Taking into account the X ligands, the inhibitory activity decreased in the order flufenamate ~ niflumate > mefenamate > naproxenate, i.e., the most active inhibitors were compounds containing fluoro atoms in their molecules. However, the differences between the corresponding IC_{50} values for the set of compounds studied were relatively small and varied in a relatively narrow concentration range of 0.976 ($Cu(fluf)_2 \cdot H_2O$) to 2.291 mmol/dm^3 $Cu(nap)_2(caf)$. These values are also comparable with the corresponding IC_{50} value determined for $CuSO_4$ (2.49 mmol/dm^3), indicating the predominant role of Cu^{2+} ions in OER inhibition. On the other hand, the IC_{50} values for OER inhibition by the complexes under study in the suspension of spinach chloroplasts at comparable Chl content in the suspension (30 mg/dm^3) varied in the range of 6.3 to 14.5 μmol/dm^3 (Table 32.3) [22], and so they were approximately two to three orders lower than those determined for OER inhibition in *C. vulgaris*. Similar results have been obtained with diaqua(4-chloro-2-methylacetato)copper(II) complex (IC_{50} = 1.46 mmol/dm^3 [OER in *C. vulgaris*] [23] or 15.38 μmol/dm^3 [OER in spinach chloroplasts]) [24]. These differences in IC_{50} values could be explained as follows: whereas in *C. vulgaris* for reaching the site of action the inhibitor must penetrate through the outer and inner algal membranes, in partially broken spinach chloroplasts (used in the above study) this inhibitor could directly interact with the thylakoid membranes.

C. DETERMINATION OF EXTENT OF PS 2 DAMAGE BY ELECTRON SPIN RESONANCE SPECTROSCOPY

Chloroplasts of higher plants exhibit electron spin resonance (ESR) signals belonging to both photosystems

TABLE 32.3
Concentrations of Cu(II) Compounds Causing 50% Decrease of OER in the Suspensions of *Chlorella vulgaris* and Spinach Chloroplasts. The Chlorophyll Concentration of the Algal Suspensions was 20 mg/dm^3, that of Spinach Chloroplasts was 30 mg/dm^3

Compounds	$IC_{50} \pm C.L._{0.05}$ (mmol/dm^3), *Ch. vulgaris*	IC_{50} (μmol/dm^3), Spinach Chloroplasts
$Cu(fluf)_2 \cdot H_2O$	0.976 (0.937–1.076)	9.4
$Cu(fluf)_2(ron)_2$	1.213 (1.146–1.300)	6.6
$Cu(fluf)_2(Et_2nia)_2$	1.068 (1.026–1.128)	6.7
$Cu(mef)_2 \cdot H_2O$	1.401 (1.261–1.554)	8.9
$Cu(mef)_2(ron)_2$	1.632 (1.485–1.779)	6.3
$Cu(mef)_2(Et_2nia)$	1.664 (1.634–1.761)	13.0
$Cu(nif)_2 \cdot H_2O$	1.143 (1.111–1.194)	8.4
$Cu(nif)_2(mpc)_2$	1.109 (1.026–1.173)	n.d.
$Cu(nif)_2(nia)_2$	1.180 (1.102–1.246)	6.7
$Cu(nap)_2 \cdot H_2O$	2.261 (2.085–2.392)	14.2
$Cu(nap)_2(caf)$	2.291 (2.135–2.382)	9.0

Source: From Kráľová K, Šeršeň F, Melník M. *J. Trace Microbe Tech.* 1998; 16: 491–500 and Kráľová K, Šeršeň F, Melník M, Fargašová A. *Progress in Coordination and Organometallic Chemistry*, Slovale Technical University Press, Bratislava, 1997; 233–238. With permission.

(the so-called signal I and signal II) in the region of free radicals ($g \approx 2.00$) [25]. Signal I is situated in the region with g = 2.002 and its half-width is $\Delta B_{PP} \approx$ 0.9-mT. This signal has been identified as P^+_{700}, i.e., the oxidized primary donor in PS 1. Signal II is a broader signal with side "lobes," centered around g = 2.004 ($\Delta B_{PP} \approx$ 2 mT) and it is associated with PS 2. Signal II has two components, identified from the decay as signal II_s (slow) and signal II_{vf} (very fast) [25]. The latter was observed only in preparations with inhibited O_2 evolution, leading to the concept that it represented an intermediate between the oxygen evolving complex and P^+_{680} (oxidized primary donor in PS 2). With the availability of sequences, and the application of molecular engineering to PS 2, it was demonstrated that signal II_s came from a redox active tyrosine-161 in D_2 protein (Y_D), and signal II_{vf} from tyrosine-161 in D_1 protein (Y_Z) [26]. The ESR spectrum of Y_Z^+ is normally measured as the light–dark difference spectrum after a relatively short dark time, so that the spectrum due to Y_D^+, which is relatively stable in the dark, can be subtracted out.

The form of the ESR signals I and II of chloroplasts treated by compounds causing inhibition of PET usually differs from that of untreated ones. From the changes of the intensity and the shape of ESR signals in the presence of the inhibitor, its site and size of action in the photosynthetic apparatus can be determined. Within a homologous series with the same site and mechanism of action the extent of PET-inhibiting activity for individual compounds can be expressed by the P parameter [27,28], which can be evaluated from the intensities of ESR signals measured in the dark and in the light according to the following formula:

$$P\text{parameter} = [(I_{(\text{inhib.})\text{light}} : I_{(\text{inhib.})\text{dark}})/(I_{(\text{control})\text{light}} : I_{(\text{control})\text{dark}})] [C_{\text{Chl}}]^{-1}$$

where I represents the intensities of ESR signals of the control and of inhibitor-treated chloroplasts in the dark and in the light and C_{Chl} is the chlorophyll content in the sample (in mg). The values of the P parameter for untreated plant chloroplasts are usually in the range 1.5 to 2.0.

Král'ová et al. [27] investigated the effects of the amphiphilic compounds 1-alkyl-1-ethylpiperidinium bromides (C_6 to C_{18}) (AEPBr) and 1-alkylpiperidine-N-oxide (C_8 to C_{18}) (APNO) on the photosynthetic apparatus of spinach chloroplasts using ESR spectroscopy. The spinach chloroplasts applied for ESR measurements were prepared using the procedure described above. The ESR spectra of the untreated suspensions of spinach chloroplasts in phosphate buffer ($0.02\,\text{mol/dm}^3$, pH 7.2) containing sucrose ($0.4\,\text{mol/dm}^3$), $MgCl_2$ ($0.005\,\text{mol/dm}^3$), and NaCl ($0.015\,\text{mol/dm}^3$) and in the presence of inhibitors ($0.05\,\text{mol/dm}^3$) were recorded with an ESR 230 instrument (WG AdW, Berlin) operating in X-band at 5 mW of microwave power and 0.5 mT modulation amplitude. ESR spectra of all samples were recorded in the dark and in the light. The samples were irradiated with $\approx 400\,\mu\text{mol/m}^2/\text{sec}$ PAR directly in the resonator cavity using a 250-W halogen lamp from 0.5-m distance through a 5-cm water filter.

Figure 32.5 presents the ESR spectra of untreated spinach chloroplasts (lines A) and chloroplasts treated with 1-octylpiperidine-N-oxide and 1-dodecylpiperidine-N-oxide (lines B and C) in the dark (full

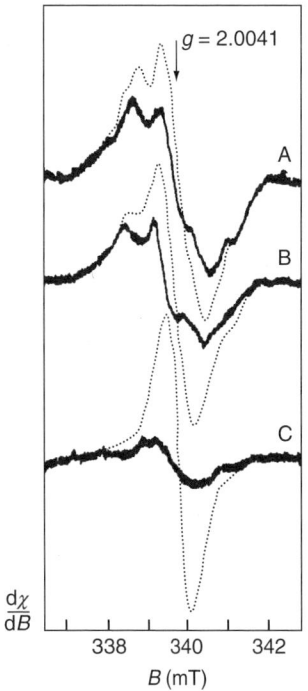

FIGURE 32.5 ESR spectra of spinach chloroplasts recorded in the dark (full line) and under irradiation (dotted line) for the control sample (A) and treated with 0.01 M 1-octadecylpiperidine-N-oxide (B) or 1-dodecylpiperidine-N-oxide (C) (dotted line: magnification 0.5×). B is the magnetic induction (in mT) and $d\chi/dB$ is the first derivative of the imaginary part of magnetic susceptibility χ with respect to B. (From Král'ová K, Seršeň F, Mitterhauszerová L', Krempaská E, Devinsky F. *Photosynthetica* 1992; 26: 181–187. With permission.)

lines) and in the light (dashed lines). In the presence of inhibitor a decrease of ESR signal II intensity and an increase of the corresponding signal I on irradiation could be observed (Figure 32.5, lines B and C). The damaged PS 2 could not supply electrons to PS 1 and thus a great rise of the signal I under irradiation was recorded (Figure 32.5, line C). The changes in ESR signal intensities were used for evaluation of the P parameter. From the dependence of the P parameter on the number of C atoms in the alkyl chain of the surfactants (Figure 32.6) it is evident that the most active inhibitors were surfactants with alkyl = decyl to tetradecyl. A similar quasiparabolic course showed dependence of the Hill reaction rate in spinach chloroplasts expressed by IC_{50} values on the number of carbon atoms in the alkyl chain of AEPBr and APNO (Figure 32.7). A very sharp dependence of the parameter P on the alkyl chain length was also found for N-alkyl-N,N-dimethylamine oxides (alkyl = hexyl to hexadecyl) [8].

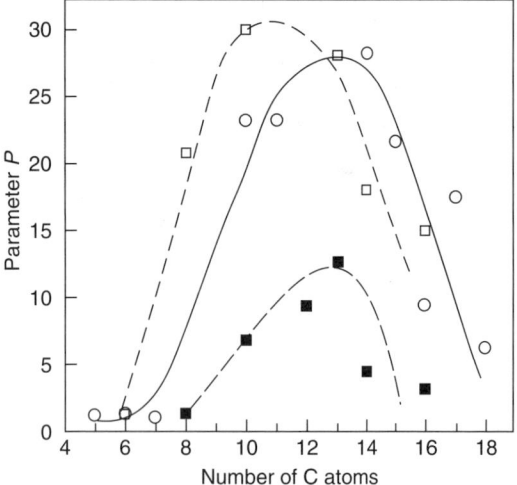

FIGURE 32.6 Inhibition of PET in spinach chloroplasts expressed by the parameter P evaluated from ESR measurements in the presence of AEPBr (circles) and APNO (squares) on the number of carbon atoms in surfactant alkyl chain; the applied constant surfactant concentration was $0.05 \, mol/dm^3$ (empty symbols) or $0.01 \, mol/dm^3$ (filled symbols). (From Kráľová K, Šeršeň F, Mitterhauszerová Lʼ, Krempaská E, Devinsky F. *Photosynthetica* 1992; 26: 181–187. With permission.)

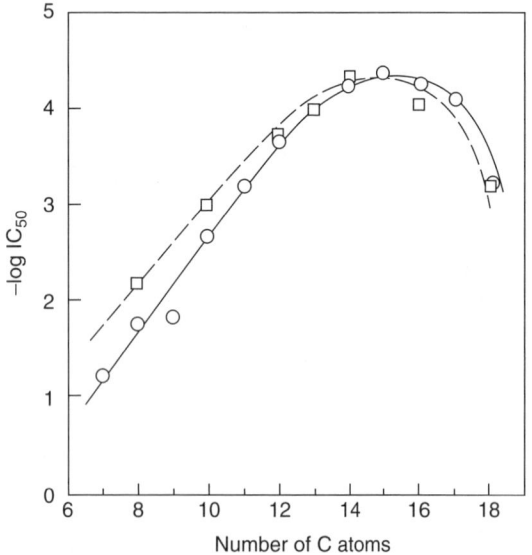

FIGURE 32.7 Dependence of Hill reaction rate in spinach chloroplasts expressed by IC_{50} values on the number of carbon atoms in the alkyl chain of AEPBr (circles) and APNO (squares). (From Kráľóvá K, Šeršeň F, Mitterhauszerová Lʼ, Krempaská E, Devinsky F. *Photosynthetica* 1992; 26: 181–187. With permission.)

P parameters were also evaluated from ESR spectra of horse bean chloroplasts treated with 22 substituted aryloxyaminopropanols (Table 32.4) [28]. All these compounds exhibited inhibitory effects on the PS 2 of the photosynthesizing apparatus. An exponentially increasing inhibitory influence was observed for R^1 substituents ranging from methyl to pentyl. This effect was particularly pronounced in compounds with substituents in the *para* position of the benzene ring. Branching of the alkyl group in the esteric substituent was associated with decreased inhibitory activity (compounds 2-i33, 3-i33, 4-i33 with R^1 = isopropyl) as compared to the corresponding compounds with linear alkyl chains (compounds 2-33, 3-33, 4-33 with R^1 = propyl). The inhibitory activity also decreased when the i-propyl group of amine nitrogen was replaced by i-butyl (R^2 substituent).

D. Ribulose-1,5-bisphosphate Carboxylase/Oxygenase Activity

Net carbon dioxide fixation in photosynthetic organisms is due to the action of ribulose-l,5-bisphosphate carboxylase/oxygenase (Rubisco), the bifunctional enzyme that catalyzes the initial steps in both the Calvin cycle and photorespiration. These two processes are initiated when Rubisco either carboxylates or oxygenates the common substrate, RuBP. Since photorespiration results in a net loss of CO_2, Rubisco catalyzes two fundamentally opposing reactions [29]. Much of our present knowledge includes the structure, mechanisms, and activity of this important enzyme (for details see Ref. [30]).

A rapid method to determine the CO_2/O_2 specificity factor of Rubisco was found. The assay measures the amount of CO_2 and O_2 fixation of varying CO_2/O_2 ratios to determine the relative rates of each reaction. Carbon dioxide fixation is measured by the incorporation of the moles of $^{14}CO_2$ into 3-phosphoglycerate, while O_2 fixation is determined by subtraction of the moles of CO_2 fixed from the moles of RuBP consumed in each reaction. By analyzing the inorganic phosphate specifically hydrolyzed from RuBP under alkaline conditions, the amount of RuBP present before and after catalysis by Rubisco can be determined. Changes in Rubisco activity have been found to be a valuable tool in the field of "stress physiology." Temperature, activating metal ions, and amino acid substitutions are known to influence the CO_2/O_2 specificity of Rubisco [31]. However, an understanding of the physical basis for enzyme specificity has been elusive. It has been estimated [29] that the temperature dependence of CO_2/O_2 specificity can be attributed to a difference between the free energies of activation for the carboxylation and oxygenation

TABLE 32.4
Inhibition of PS 2 of Horse Bean Chloroplasts by Substituted Aryloxyaminopropanols; Changes in ESR Spectra of Chloroplasts are Expressed by the Parameter *P*

OH
|
OCH₂CHCH₂NHR²

.HCl

NHCOOR¹

Compound	Substituent Position	R¹	R²	Parameter P
2–13	2	CH₃	CH(CH₃)₂	3.89
2–23	2	C₂H₅	CH(CH₃)₂	3.72
2–33	2	C₃H₇	CH(CH₃)₂	8.20
2–43	2	C₄H₉	CH(CH₃)₂	9.20
2–53	2	C₅H₁₁	CH(CH₃)₂	14.23
2–i33	2	CH(CH₃)₂	CH(CH₃)₂	2.36
2–55	2	C₅H₁₁	C(CH₃)₃	18.10
3–13	3	CH₃	CH(CH₃)₂	1.27
3–23	3	C₂H₅	CH(CH₃)₂	3.05
3–33	3	C₃H₇	CH(CH₃)₂	3.64
3–43	3	C₄H₉	CH(CH₃)₂	5.27
3–53	3	C₅H₁₁	CH(CH₃)₂	17.99
3–i33	3	CH(CH₃)₂	CH(CH₃)₂	1.69
3–25	3	C₂H₅	C(CH₃)₃	1.60
3–45	3	C₄H₉	C(CH₃)₃	3.17
3–55	3	C₅H₁₁	C(CH₃)₃	5.11
4–13	4	CH₃	CH(CH₃)₂	1.27
4–23	4	C₂H₅	CH(CH₃)₂	1.59
4–33	4	C₃H₇	CH(CH₃)₂	8.71
4–43	4	C₄H₉	CH(CH₃)₂	16.41
4–53	4	C₅H₁₁	CH(CH₃)₂	30.65
4–i33	4	CH(CH₃)₂	CH(CH₃)₂	1.27

Source: From Mitterahuszerová L, Král'ová K, Seršeň F, Blanáriková V, Csöllei J. *Gen. Physiol. Biophys.* 1991; 10: 309–319. With permission.

partial reactions. The reaction between the 2,3-enediolate of RuBP and O_2 has a higher free energy of activation than the corresponding reaction of this substrate with CO_2. Thus, oxygenation is more responsive than carboxylation to temperature. Furthermore, the reduction in CO_2/O_2 specificity that is observed when activator Mg^{2+} is replaced by Mn^{2+} may be due to Mg^{2+} being more effective in neutralizing the negative charge of the carboxylation transition state, whereas Mn^{2+} is a transition metal ion that can overcome the triplet character of O_2 to promote the oxygenation reaction.

Recently, the biochemistry of C_3 photosynthesis in high CO_2 concentration (in relation to the so-called "greenhouse effect") has been intensively studied [32–34]. It was found that during long-term exposure to high ambient CO_2 concentration, the initial stimulation of photosynthesis decreases or disappears. This means that one must distinguish between the short-term effect (with stimulation of net photosynthetic CO_2 fixation rate) and the above-mentioned long-term effect of CO_2 enhancement on photosynthesis. Regulation of photosynthesis with the short-term effect is determined by interactions among the capacities of light harvesting electron transport, Rubisco, and orthophosphate (Pi) regeneration during starch and sucrose synthesis. Photosynthesis under high CO_2 conditions is limited by either electron transport or Pi-regeneration capacities, and Rubisco is deactivated to maintain a balance between each step in the photosynthetic pathway. Long-term CO_2 enhancement leads to carbohydrate accumulation. However, accumulation of carbohydrates is not associated with a Pi-regeneration limitation on photosynthesis, and this limitation is apparently removed during long-term exposure to high CO_2. Enhanced CO_2 does not affect Rubisco content and electron transport capacity for a given leaf nitrogen content. In addition, the deactivated Rubisco immediately after exposure to high CO_2 does not recover during the subsequent prolonged exposure. Such evidence may indicate that plants do not necessarily have an ideal acclimation response to high CO_2 at the biochemical level [35].

There are some difficulties in estimating both the Rubisco activity and the protein content in the broad-leaf forest trees that usually have higher amounts of phenolic compounds. Therefore, the methodological procedures must be modified. Results have been obtained with three Slovakian autochthonous oak species, *Quercus cerris* L., *Q. robur* L., and *Q. dalechampii* Ten., from forest stands with different degrees of pollution damage (Rimavska Sobota Enterprise, Central Slovakia) (see Ref. [36]). In addition to mature trees, seedlings of the oak species examined were also available at the research areas. The experimental materials were

1. Leaves of seedlings (the same age as control seedlings) transferred in spring and transplanted outdoors to permit them to grow under the same conditions as the control seedlings
2. Leaves of seedlings that were processed immediately after sampling (July).

Contents of metallic and nonmetallic elements were also estimated in the leaves of both control and damaged seedlings.

According to several investigators [37–40], the Rubisco assays in leaf extracts were not successful. In oak leaves the bulk of interfering substances, especially of phenolic nature, must be removed. Of the procedures tested, the method using dry plant homogenates (acetone dried powder) was the most successful.

Preparation of acetone dried powder was performed as follows: 1 g of freshly cut oak leaf material was ground in $20 \, cm^3$ of cold ($-20°C$) acetone in a chilled mortar and pestle with acid-washed sand. The homogenate was filtered through a glass sinter No. 2 under vacuum. The insoluble material was handled two or three times in the same way until the Chls were completely extracted. The acetone dried powder was dried in open air for 30 min and stored at $-20°C$ for more than 4 weeks.

For the enzyme assay, acetone dried powder of oak leaves was extracted at 0°C for 40 min using a magnetic stirrer with $6 \, cm^3$ of extraction medium (M): 0.1 Tris–HCl, pH 7.8, 0.01 $MgCl_2$, 0.002 EDTA, 0.02 2-Me, 0.002 DTE, 0.02 $NaHCO_3$, and 1% Tween 80, 1% PVP The tissue suspension was centrifuged at $10,000 \times g$ for 10 min. The supernatant was used as a crude enzyme source.

Assay of Rubisco (E.C.4.1.1.39) was carried out according to Stiborová et al. [37,38] with some modifications. To $0.1 \, cm^3$ of buffer solution (M) (0.1 Tris–HCl, pH 8.0, 0.002 EDTA, 0.001 DTE, 0.03 $MgCl_2$, and $0.07 \, cm^3$ 0.005 M Na_2 $^{14}CO_3$ [total disintegrations/sec = 933.3]), $0.1 \, cm^3$ crude enzyme extract was added, and the mixture was incubated at 37°C for 15 min. The enzymatic reaction was started by the addition of substrate (0.002 M Rubisco). The reaction was stopped by adding $0.5 \, cm^3$ 6 M HCl after 5 min of incubation, and the mixture was left standing for 12 h.

Incorporation of $^{14}CO_2$ was measured after the addition of $10 \, cm^3$ Instagel, a scintillation cocktail of Packard Instruments, into the KLB Wallace 1217 liquid scintillation counter. The incorporation of $^{14}CO_2$ was linear for about 6 min. The specific enzyme activity of the sample was expressed as $^{14}CO_2$ incorporation per second per milligram of protein. Total protein contents of the extract were determined spectrophotometrically according to the method of Bradford [41].

Both the Rubisco activity and the protein content in the leaves of oak seedlings were significantly lowered (Table 32.5). A lower content of nonmetallic and a higher content of metallic elements were also found in the damaged leaves (Table 32.6). The changes in Rubisco activity and protein content may be used as a sensitive diagnostic parameter in ascertaining the negative effects of abiotic and biotic factors in the environment (for details, see Ref. [36]).

E. PIGMENT ANALYSIS USING HIGH-PERFORMANCE LIQUID CHROMATOGRAPHY

It is widely accepted that the primary function of photosynthetic pigments (chlorophylls and accessory pigments, carotenoids, and phycobilins) is the conversion of light energy to chemical energy by forming chemical bonds. This conversion of energy from one form to another is a complex process that depends on cooperation between a large number of pigment molecules and a group of electron transfer proteins. Protection of the photosynthetic apparatus against excess light energy is achieved through the xanthophyll cycle: photoconversion and de-epoxidation of violaxanthin via antheraxanthin to zeaxanthin [42]. Enhanced epoxidation of the xanthophyll cycle (increased conversion of zeaxanthin to antheraxanthin and violaxanthin) correlated with the increase in the endogenous levels of abscisic acid induced by leaf senescence. This, in turn, arises from fluctuations in carotenoid turnover; therefore, abscisic acid production and xanthophyll cycle are not independent reactions in the process of leaf senescence. The protective function of carotenoids in the desiccated leaves of some plants may play an essential role in the reorganization of chloroplasts and of the whole photosynthetic apparatus at leaf rehydration [43]. Some carotenoid compounds are metabolized to retinol, which is a physiologically active form of vitamin A [44]. β-Carotene (the most plentiful carotenoid) has potential vitamin A activity; it can be cleaved to form two molecules of retinol. It was also confirmed that β-carotene possesses a protective function against reactive products or reactive forms of oxygen (mainly superoxides) [45,46]. The carotenoids efficiently quench singlet oxygen and free radicals that could otherwise initiate reactions such as lipid peroxidation [47]. β-Carotene is a potent free radical scavenger. The composition and contents of photosynthetic pigments appear to be important for taxonomic classification as well as for the determination of physiological characteristics of different groups of algae [48,49]. In recent years, plant pigments (especially carotenoids) became interesting from a commercial point of view as substances used in the food or cosmetic industries [50]. In spite of the abovementioned importance of plant pigments, little attention was devoted to their structure and function or the methodological procedures for their detection in the older or recent sources on photosynthesis [2,4,5].

Characteristics of the Chls and carotenoids as well as some methodological procedures for their qualitative and quantitative estimations are briefly described in the following sections.

TABLE 32.5
Values of Specific RuBPC Activity and Protein Content in Different Leaves

	Specific Activity (Bq/μg protein)			Protein Content (μg/mm³)		
		Damaged Samples			Damaged Samples	
Healthy Samples		A	B	Healthy Samples	A	B
Quercus dalechampii						
	58.55	37.42	29.96	7.20	1.92	0.29
	107.42	31.13	30.41	5.56	1.88	0.32
	97.22	27.78	20.01	2.38	1.88	0.45
	104.52	—	19.34	3.01	—	0.49
				2.82		
$\bar{x}$	91.93	32.11	24.93	4.19	1.89	0.39
$s_{\bar{x}}$	11.33	2.83	3.04	0.93	0.01	0.05
Quercus robur						
	41.98	25.41	20.10	6.90	1.16	0.45
	69.95	33.60	31.15	4.86	1.76	0.33
	83.77	17.78	28.87	4.38	1.16	0.39
	—	—	31.76	3.42	—	0.49
		25.39				0.57
$\bar{x}$	65.23	25.60	27.45	4.89	1.36	0.45
$s_{\bar{x}}$	12.29	4.57	2.15	0.73	0.20	0.04
Quercus cerris						
	24.58	27.72	27.81	6.04	1.28	0.45
	34.63	20.78	30.53	4.47	1.36	0.73
	27.74	17.38	31.67	3.12	1.36	0.84
	26.57	—	23.50	2.84	—	0.77
	—	—	19.55	1.92	—	0.77
$\bar{x}$	28.38	21.96	26.61	3.68	1.33	0.71
$s_{\bar{x}}$	2.18	3.04	2.26	0.72	0.03	0.07

Note: Sample A: leaves of the seedlings transferred in the spring of 1987 from damaged forest stand and transplanted in the garden; sample B: leaves of the seedlings processed immediately after sampling (July 1987).

Source: From Konecná B, Frič F, Masarovičová E. *Photosynthetica* 1989; 23: 566–574. With permission.

The photosynthetic pigments, Chls and carotenoids, belong to the group of isoprenoid plant lipids later named prenyl lipids. Chls *a* and *b* are mixed prenyl lipids. They have an isoprenoid phytyl chain, which is bound to a nonisoprenoid porphyrin ring system. This phytyl side chain, which is esterified to the carboxyl group of the ring, gives the Chls their lipid character. Carotenoids as tetraterpenoids are simple or pure prenyl lipids, with carbon skeletons made up solely of isoprenoid units. Because of their biogenetic relationship (isopentenoid pathway), Chls and carotenoids are also called prenyl pigments.

The Chls of higher plants, ferns, mosses, and green algae consist of Chls *a* as the major pigment and Chls *b* as an accessory pigment. Both Chls are genuine components of the photosynthetic membranes and occur in the ratio *a:b* of approximately 3:1. Chl content as well as Chl *a:b* ratio can be modified by both internal factors and the environmental conditions [51].

More than 450 carotenoids occur in nature. The carotenoids can be divided into oxygen-free carotenes and xanthophylls, which contain oxygen in different forms, such as one or several hydroxy or epoxy groups. α-Carotene has one ε-ionone and one β-ionone ring, whereas β-carotene has two β-ionone rings. Introduction of hydroxy and epoxy functions into α-carotene (ionone rings) gives rise to lutein and lutein epoxide. β-Carotene is the precursor of the pigments of the xanthophyll cycle: violaxanthin, antheraxanthin, and zeaxanthin. The carotenoids of functional chloroplasts include β-carotene, lutein, violaxanthin, and neoxanthin as the major and regular components of the photochemically active

TABLE 32.6
Contents of Metallic and Nonmetallic Elements in Healthy and Damaged Oak Trees

Species	Content of Elements (mg/kg)								
	K	Na	Ca	Mg	S	Al	Cu	Zn	Pb
Healthy samples									
Q. cerris	8,750	75	12,820	1,671	800	3,727	55	36	6
Q. robur	8,470	100	24,120	2,023	810	1,090	78	62	3
Q. dalechampii	6,750	50	15,580	2,673	836	1,878	463	63	3
Damaged samples									
Q. cerris	8,000	80	8,086	4,376	632	6,738	212	26	7
Q. robur	6,620	100	8,288	851	824	1,697	68	23	2
Q. dalechampii	3,720	60	12,330	1,823	989	2,109	97	22	3

Source: Ref. 36.

thylakoids of chloroplasts of higher plants and green algae. The composition of carotenoids can vary with both environmental and internal conditions [46,51–53].

The use of HPLC for the analysis of Chls, carotenoids, and other natural plant pigments is rapidly replacing classical gravity-flow column chromatographic methods (see Section III.A). The reason for the increasing use of HPLC in pigment analysis lies in the rapid, nondestructive, and improved analytical nature of these methods [54]. This method of chromatography, using small-diameter columns, fine particle size, and rapid flow rate, is now widely used; its theory is discussed in articles by Snyder and Kirkland [55]. It has the advantage of speed and sensitivity, in addition to protecting pigments from degradation by oxygen, and it can be used for preparative chromatography. Although HPLC can resolve carotenoids and chlorophyllous pigments into sharp peaks, certain separations, as with thin-layer chromatography (TLC), are not always possible with some of the methods used (in particular, for lutein and zeaxanthin; diadinoxanthin, dinoxanthin, and fucoxanthin; and Chls c_1 and c_2). An outstanding resolution has been achieved by Wright and Shearer [56], who separated a mixture of 44 chlorophyllous and carotenoid pigments [57]. On the other hand, analysis of variance showed no significant differences between the results (Chls and carotenoids content) given by the TLC and HPLC methods [58].

HPLC was used for qualitative and quantitative analysis of pigments in six strains of xanthophyceae algae belonging to the genera *Goniochloris, Pleurochloris*, and *Heterothrix* and in one green algae species, *Scenedesmus quadricauda*, which is widely used for laboratory and outdoor experiments [49]. For quantification of antheraxanthin, neoxanthin, and

violaxanthin, the standard of pigment zeaxanthin was used, since the pigments have the same absorbance at a wavelength of 450 nm. For the spectrophotometrical quantification of Chl *a* and Chl *b* in 80% acetone, the equation by Lichtenthaler [53] was used. For all pigment standards, the specific absorption coefficients were measured in the following solvents: 80% acetone, 90% acetone, 100% acetone, chloroform, ethanol, diethyl formamide, dimethyl sulfoxide, and diethyl ether.

Figure 32.8 presents an HPLC chromatogram of *Pleurochloris* sp. Table 32.7 gives the values of pigment concentrations of the examined algae, and Table 32.8 shows pigment contents (in %) of the total dry

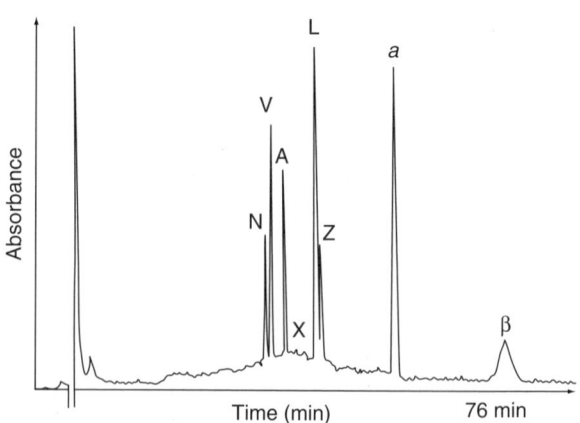

FIGURE 32.8 HPLC chromatogram of *Pleurochloris* sp. (acetone pigment extract). Peak identification: abscissa — retention time (min), ordinate — absorbance. N, neoxanthin; V, violaxanthin; A, antheraxanthin; L, lutein; Z, zeaxanthin; *a*, Chl *a*; β, β-carotene. (From Krasnovská E, Masarovičová E, Hindák F. *Biologia* (*Bratisl.*) 1994; 4: 501–509. With permission.)

TABLE 32.7

Values of Pigment Concentrations of Examined Xanthophycean Algae and *Scenedesmus quadricauda* in Acetone Extracts

Strain	Pigment Concentration (μg/ml)							
	N	V	A	L	Z	*b*	*a*	β
Scenedesmus	0.59	0.47	0.49	0.84	0.13	1.00	3.28	0.15
Goniochloris sculpta	0.56	0.45	0.63	2.77	0.12	—	6.51	0.30
Pleurochloris sp.	0.96	1.48	1.29	3.50	0.90	—	7.15	0.30
Heterothrix musicola	0.19	0.16	0.19	1.65	0.12	—	5.43	0.25
Heterothrix sp. 1	0.97	0.15	0.67	3.16	1.05	—	9.46	0.53
Heterothrix sp. 2	0.32	0.04	0.32	1.82	0.18	—	4.63	0.22
Heterothrix sp. 3	0.84	0.07	0.84	1.35	0.20	—	3.29	0.16

Note: N, neoxanthin; V, violaxanthin; A, anteraxanthin; L, lutein; Z, zeaxanthin; *b* Chl. *b*; *a*, Chl. *a*; β, β-carotene.

Source: From Krasnovská E, Masarovičová E, Hindák F. *Biologia* (*Bratisl.*) 1994; 4: 501–509. With permission.

TABLE 32.8

Pigment Contents of Total DM in the Xanthophycean Algae Studied and *Scenedesmus quadricauda*

Strain	Pigment Contents (%)							
	N	V	A	L	Z	*b*	*a*	β
Scenedesmus	0.195	0.102	0.177	0.295	0.015	0.311	1.016	0.124
Goniochloris sculpta	0.173	0.138	0.195	0.860	0.037	—	2.019	0.093
Pleurochloris sp.	0.313	0.662	0.579	1.577	0.401	—	3.221	0.136
Heterothrix musicola	0.013	0.011	0.013	0.113	0.085	—	0.372	0.016
Heterothrix sp. 1	0.102	0.016	0.071	3.340	0.111	—	1.000	0.039
Heterothrix sp. 2	0.606	0.028	0.225	1.281	0.127	—	3.261	0.162
Heterothrix sp. 3	0.643	0.050	0.596	0.957	0.141	—	2.276	0.117

Note: For abbreviations see Table 32.7.

Source: From Krasnovská E, Masarovičová E, Hindák F. *Biologia* (*Bratisl.*) 1994; 4: 501–509. With permission.

mass. In Table 32.9 the specific absorption coefficients of used pigment standards are presented [49].

F. CO₂ EXCHANGE IN OPEN AND CLOSED SYSTEMS

The methods applied to the estimation of plant photosynthesis may be divided, in principle, into gravimetric and gasometric methods. The gasometric method allows nondestructive measurements within brief time intervals on the same plant, and quick changes of CO_2 exchange due to environmental conditions can also be registered. On the other hand, the disadvantage of this method lies in the well-known problem of the "cuvette effect," which has already been partly solved by the water thermo-stabilized assimilation chamber or by an automatized chamber allowing the simulation of actual environmental conditions. The principle of gasometric measurement of CO_2 exchange may be divided into closed, semiclosed or null-balance systems (see Ref. [5]).

This section will deal mainly with the CO_2 exchange of natural plants, especially forest herbs and trees. In general, to obtain objective data a comprehensive methodological approach is required.

For the correct measurement and registration of the parameters followed (CO_2 and O_2 concentration, registration of micrometeorological factors), choosing a suitable apparatus (with sufficient sensibility and measurement accuracy) is essential. A brief description of the measuring devices, registering equipment, and accessories used for the investigation

TABLE 32.9
Specific Absorption Coefficients of Used Pigment Standards, Calculated in Different Solvents

Solvent	Specific Absorption Coefficients, A (l/g cm)		
	β-Carotene	Zeaxanthin	Lutein
100% acetone	2520	2600	2440
90% acetone	2430	2780	2444
80% acetone	2390	2780	2420
Ethanol	2390	2800	2454
Dimethyl formamide	2280	2640	2350
Dimethyl sulfoxide	2080	2600	2430
Diethyl ether	2520	2640	2424

Source: From Krasnovská E, Masarovičova E, Hindák F. *Biologia* (*Bratisl.*) 1994; 4: 501–509. With permission.

of CO_2 exchange in beech (*Fagus sylvatica* L.) seedlings is presented herewith.

To study the CO_2 exchange of 3-year-old beech seedlings, a special assimilation chamber was constructed (Figure 32.9). It consisted of two independent parts. The lower section contained the lower part of the seedlings and enough water to ensure soil and plant saturation. The upper section of the chamber contained the upper portion of the seedling. The two parts of the chamber were separated by a rubber lining with a surface layer of waterproof Ramsay vaseline and were hermetically separated by a plexiglass partition wall and interconnected with metal clamps that isolated the chamber from its surroundings.

The joints of the partition wall between the upper and lower parts of the chamber were sealed with special plastics. Figure 32.10 presents a schematic draft of the chamber with its basic constructional elements. The assimilation chamber was thermostabilized with a water bath. In the upper part of the chamber brass pipes were installed, providing gas input and output into the CO_2 and O_2 analyzers as well as the sites for the electric wiring of the photodiode, thermocouple, resistance thermometer, and microfans.

The radiation source consisted of 1000-W halogen lamps installed in a reflector with a parabolic mirror. The infrared radiation was absorbed by an 80-mm layer of chilled circulating water. The spectral characteristics of the irradiation source were measured with a monochromator with a prism and nonselective thermocouple detector. Irradiance was measured with a silicon photodiode.

Air temperature in the assimilation chamber was measured with a platinum ceramic resistor, type PtK_m 100 W at 0°C. The same type of resistor (dry and wet configuration) was also used for psychrometric measurement of air humidity. The leaf surface temperature was measured by the copper–constant (Cu–Const.) thermocouple. Gas transflux (air, CO_2, nitrogen) through both the open and the closed systems was measured using a flowmeter [59]. Figure 32.11 shows the light and temperature curves of the net photosynthetic rate (P_N) of leaves of 3-year-old beech seedlings.

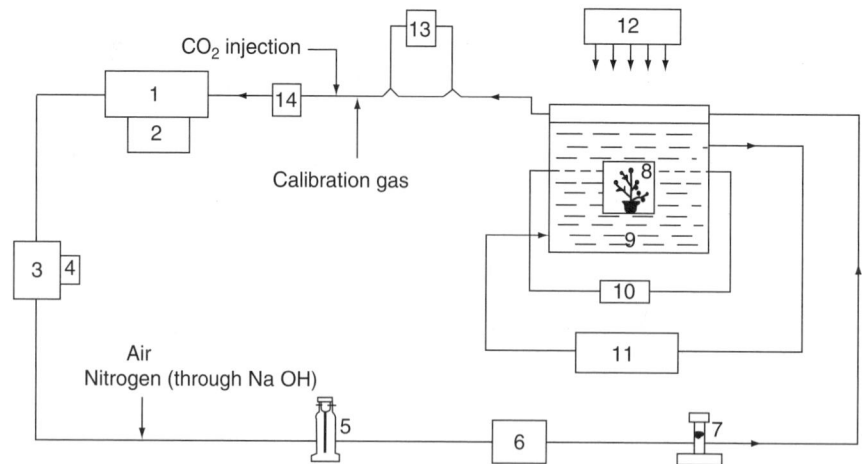

FIGURE 32.9 Diagram of the measuring system. 1 — IRGA, Irex; 2 — recorder (vpm CO_2); 3 — O_2 analyzator, Permolyt 2; 4 — recorder (vol.% O_2); 5 — overflow; 6 — pumps; 7 — flowmeter; 8 — assimilation chamber with accessories; 9 — water bath; 10 — recorders for registration of micrometeorological factors; 11 — adjacent circuit with Ascarit; 14 — drier with $ZnCl_2$. (From Masarovičová E. *Gasometrical Investigation into CO_2 Exchange of the Fagus sylvatica* L. *Species under Controlled Conditions*, Veda, Publishing House of the Slovak Academy of Sciences, Bratislava, 1984. With permission.)

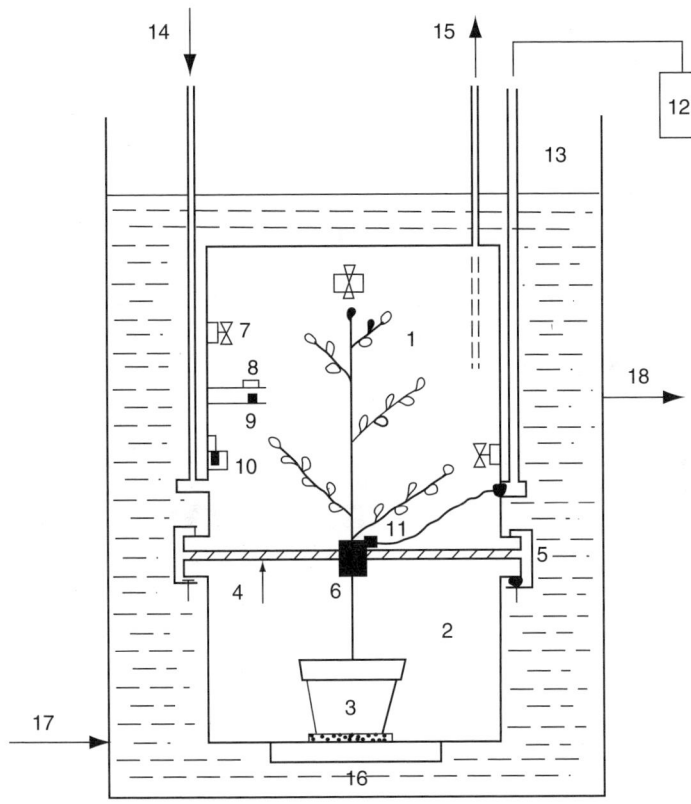

FIGURE 32.10 Schematic draft of the thermostabilized assimilation chamber. 1 — Upper part of the assimilation chamber; 2 — lower part of the assimilation chamber; 3 — potted seedling; 4 — packing with Ramsay vaseline; 5 — metal clamps; 6 — slit for the stem closed with Colorplast; 7 — microfans; 8 — filter; 9 — Si photodiode; 10 — Pt ceramic-resistant thermometer (PtK_m); 11 and 12 — thermocouple; 13 — electrical lead to the recorders. (From Masarovičová E. *Gasometrical Investigation into CO_2 Exchange of the* Fagus sylvatica L. *Species under Controlled Conditions*, Veda, Publishing House of the Slovak Academy of Sciences, Bratislava, 1984. With permission.)

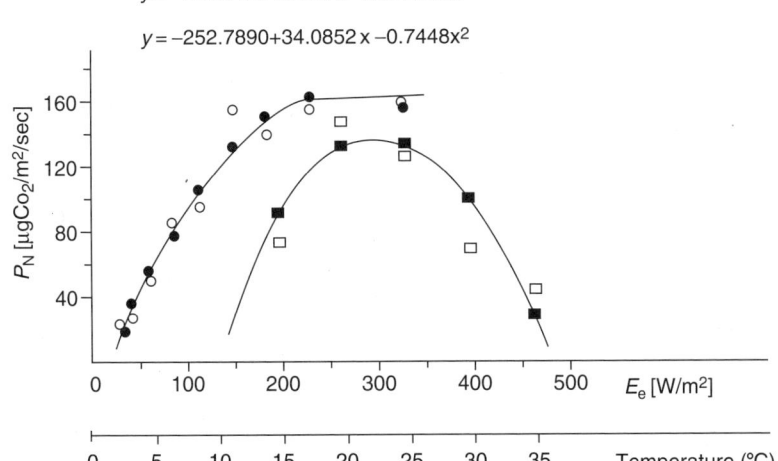

$$y = -21.2548 + 1.3627 x - 0.002498 x^2$$

$$y = -252.7890 + 34.0852 x - 0.7448 x^2$$

FIGURE 32.11 Light curve of net photosynthetic rate (P_N) at optimal temperature $19 \pm 0.5°C$ (filled circles and squares, calculated data; empty circles and squares, measured data) and temperature curve of P_N at saturating irradiance 235 W m^2 (filled circles and squares, calculated data; empty circles and squares, measured data) in the leaves of Young European beech plants. (From Masarovičová E. *Gasometrical Investigation into CO_2 Exchange of the* Fagus Sylvatica L. *Species under Controlled Conditions*, Veda, Publishing House of the Slovak Academy of Sciences, Bratislava, 1984. With permission.)

It follows from the investigated dependence that the saturation of the photochemical reactions of photosynthesis was at an irradiance of 235 W/m^2 and maximal P_N was reached at 270 W/m^2. The highest values of P_N were found at a surface temperature of the abaxial leaf side of 22.8°C. However, the optimal temperature for assimilation processes (90% max. P_N) was 19.5°C. The values of saturation irradiance and optimal temperature were used in photorespiration estimation [59].

To measure CO_2 exchange in the leaves of forest herbs (*Mercurialis perennis, Arum maculatum, Corydalis cava, Symphytum tuberoswn, Aegopodiwn podagraria, Impatiens parviflora*) and forest trees (*F. sylvatica, Quercus cerris, Quercus petraea,* and *Quercus dalechampii*), other methodological approaches and types of assimilation chambers were used (Figure 32.12).

Plants (herb species) were collected from natural forest conditions with the whole rhizosphere and after

FIGURE 32.12 Thermostabilized assimilation chamber for measuring of CO_2 exchange under laboratory conditions.

transfer were planted into pots, supplied with original forest soil, and acclimatized for a few days outside. Measurements made immediately after transfer showed symptoms of shock.

In the case of adult forest trees (43-year-old beech) growing in a natural forest stand, branches (~3 m) were cut from parts of the tree crowns both in the sun (top) and in the shade (bottom). From these, smaller shoots were cut (~1 m) to prevent disturbing the coherent water column in the vascular bundles. The cut areas were wrapped in cotton wool and plunged into PVC sacs with water. The experimental material was then transferred from the forest area into the garden of the research institute and kept outside. After 2 to 5 days of adaptation CO_2 exchange measurements were made [60], together with other quantitative analyses of the leaves (Chl content, specific leaf mass, stomata density, etc.)

The same plants (forest herbs and young forest trees — saplings) were used for ecophysiological measurements. The leaves of the plants were exposed in the simple assimilation chamber for a short period (approximately 5 to 10 min) (Figure 32.13). Carbon dioxide concentration was measured by infrared gas analyser (Infralyt 4, VEB Junkalor, Desseau, Germany). Simultaneously with the ecophysiological measurements, the basic meteorological factors (air temperature, relative air humidity, wind speed, etc.) were recorded. The measurements and equipment used have been described in detail by Masarovičová [61–65], Masarovičová and Eliáš [66], and Masarovičová and Stefančík [67].

Table 32.10 presents values of saturating (I_s), adaptation (I_a), and compensating (I_c) irradiances,

photosynthetic efficiency (α), net photosynthetic rate at I_s ($P_{N,sat}$), and dark respiration rate (R_D) of *M. perennis* [65]. Some of photosynthetic characteristics of the above-mentioned forest herbs are given in Table 32.11 [66].

In beech leaves in the sun significantly higher rates of photosynthesis, photorespiration and dark respiration, photosynthetic CO_2-fixation capacity, and photosynthetic productivity, and higher values of I_s, I_a, and I_c were found than in shaded leaves (Figure 32.14 and Figure 32.15, Table 32.12 and Table 32.13 [67].

Mean and maximal daily net photosynthetic rates, shoot length, leaf area, and stomatal density in the various growth phases (polycyclic growth) of *Quercus robur* were compared (Figure 32.16, Table 32.14 [62].

A number of models for describing the irradiance response curve for CO_2 uptake ("light response curve of photosynthesis") are extant (e.g., Ref. [68]). The rectangular hyperbola, given by the formula:

$$P_N = \frac{(\alpha I P_{N,max})}{(\alpha I + P_{N,max})} - R_D$$

is routinely used by plant physiologists because of its simple formulation and the fact that each parameter has a straightforward interpretation: α is the initial slope of the $P_N(I)$ curve (photochemical efficiency of photosynthesis at low values of irradiance); it provides the number of moles of CO_2 assimilated per mole of absorbed quanta. R_D is the mitochondrial respiration in the dark, and $P_{N,max}$ is the net photosynthetic rate at saturating I. However, Marshall and Biscoe [69] note some problems in estimating values

A   B

FIGURE 32.13 Simple assimilation chamber for measuring CO_2 exchange under field conditions.

for these parameters using standard nonlinear least-square fitting procedures: α is overestimated, $P_{N,max}$ is greatly overestimated, and P_N at the shoulder of the curve, i.e., the area between the initial and saturating rates, is always underestimated. To avoid these problems, Marshall and Biscoe [69] suggested the use of the nonrectangular hyperbola:

$$P_N = (\alpha I + P_{N,max} + R_D)$$
$$\frac{-\sqrt{(\alpha I + P_{N,max} + R_D)^2 - 4\alpha I \Theta (P_{N,max} + R_D)}}{2\Theta} - R_D$$

where Θ describes the degree of curvature at the shoulder of the $P_N (I)$ curve (later called the convexity), i.e., the ratio of physical to total diffusion resistance to CO_2; α, $P_{N,max}$, and R_D are defined in the first formula. The advantage of this formula is its simplicity and the fact that initial estimates of each parameter can be readily obtained directly from photosynthetic I response data. The advantage of the second formula is its increased flexibility in describing the observed photosynthetic data. However, Θ, unlike the other parameters, does not have a straightforward geometric interpretation and cannot

be estimated from the I response data without a computer [70].

According to Leverenz [71], the light response curve of photosynthesis can be considered to consist of four parts:

1. One part exists below the Kok effect, where changes in respiration appear to have a considerable influence [72,73], in addition to the effects of light absorption and factors determining the quantum yield of photosynthesis.
2. A linear part exists immediately above the Kok inflection where the quantum yield of photosynthesis is measured [73] and where the slope is proportional to absorptance.
3. There is a nonlinear part above the initial slope, but below light saturation where both light absorption and distribution of light within the leaf can affect the rates of photosynthesis.
4. There exists a light saturation region.

For a complete understanding, it is important to know how various factors affect photosynthesis in

TABLE 32.10

Saturating (I_s), Adaptation (I_a), and Compensating (I_c) Irradiances, Photosynthetic Efficiency (α), Net Photosynthetic Rate at I_s ($P_{N,sat}$), Dark Respiration Rate (R_D), Specific Leaf Area (SLA), Specific Leaf Mass (SLW), Average Leaf Area, and Dry Matter per Shoot

Date	I_s	I_a (W/m²)	I_c	α (μg CO₂/l)	$P_{N,sat}$ (μg CO₂/m²/sec)	R_D (mg CO₂/kg/sec)	SLA (dm²/g¹)	SLW (g/dm²)	Average A per shoot (dm²)	Average W per Shoot (g)
May 10	403	117	20	0.67	56	6			0.65	0.15
	406	122	21	0.60	2.50	0.26				
May 17	500	176	33	0.70	68	9			0.61	0.14
	503	192	33	0.65	3.05	0.4				
May 27	412	203	60	0.92	59	20			0.68	0.15
	414	208	62	0.78	2.63	0.9				
	$\bar{x}$ = 438	165	38	0.76	61	12	$\bar{x}$ = 4.48		$\bar{x}$ = 0.648	$\bar{x}$ = 0.145
	$\bar{x}$ = 441	174	39	0.68	2.73	0.52		$\bar{x}$ = 0.224		
May 16	283	45	13	2.60	94	12			1.52	0.49
	283	60	12	1.88	2.9	0.42				
August 20	283	88	21	2.75	106	21			1.13	0.39
	283	92	21	1.60	3.11	0.63				
	$\bar{x}$ = 283	66	17	2.67	100	16	$\bar{x}$ = 3.01		$\bar{x}$ = 1.326	$\bar{x}$ = 0.440
	$\bar{x}$ = 283	76	16	1.74	3.01	0.53		$\bar{x}$ = 0.333		
August 13	288	164	60	1.73	73	41				
	285	169	60	1.57	3.42	1.9				
October 14	300	146	41	1.43	69	21				
	295	143	42	1.05	2.55	0.8				
	$\bar{x}$ = 294	155	51	1.58	71	31	$\bar{x}$ = 4.19			
	$\bar{x}$ = 290	156	51	1.31	2.98	1.35		$\bar{x}$ = 0.242		

Notes: Upper values are related to the leaf area (A) and lower to the DM (W). $\bar{x}$, Mean.

Source: From Masarovičová E. *Bot. Közlem.* 1993; 80: 61–72. With permission.

TABLE 32.11
Maximum Daily Values of Net Photosynthetic Rate ($P_{N,max}$) and Stomatal Conductance ($g_{s, max}$) in Forest Herbaceous Plants in Spring and Summer

Species	Date	Time of Day (h)	$P_{N,max}$ (mg CO_2/m²/sec)	Time of Day (h)	$g_{s,max}$ (mm/sec)
Early spring species					
Arum maculatum	April 21	09.45	0.295	11.30	14.490
	April 25	10.10	0.277	09.24	20.34[a]
				(09.21)	(31.45)[a]
Corydalis cava	April 21	—		12.07	14.14[a]
	April 25	09.15	0.082		—
Symphytum tuberosum	April 21	09.30	0.315	11.30	21.540
	April 25	08.45	0.237	13.38	18.32[a]
				(09.11)	(47.62)
Summer species					
Aegopodium podagraria	April 21	10.45	0.245	15.00	19.51[a]
	April 25	09.30	0.217	09.38	25.360
	July 5	08.30	0.059		—
	July 19	09.45	−0.042	09.31	3.29[a]
				(09.28)	(5.01)[a]
	August 8	09.15	0.020	09.25	4.150
Impatiens parviflora	April 25	—		11.36	23.88[b]
	July 5	09.25	0.139		—
	July 8	09.30	0.074		—
	July 19	10.50	0.039	09.34	3.58[a]
	August 16	9.40–11.08	−0.0097 to −0.0088	09.47	0.57[a]
		11.20	0.013	09.44	3.29[a]

[a]Sunflecks and shade.
[b]Cotyledons.
Note: Numbers in parentheses indicate extreme values of series of measurements.

Source: From Masarovičová E, Eliáš P. *Photosynthetica* 1986; 20: 187–195. With permission.

each of these four regions. According to the values of convexity the following were defined: Blackman response curve ($\Phi = 1$), rectangular hyperbola ($\Phi = 0$), and nonrectangular hyperbola (Φ between 0 and 1) [74].

Tooming [75,76] analyzed the light response curve of photosynthesis and determined the following parameters: compensating irradiance (I_c), adaptation irradiance (I_a), saturation irradiance (I_s), and net photosynthetic rate at the saturating irradiance ($P_{N,sat}$). Compensating irradiance (formerly called light compensation point) is defined explicitly by both the dark respiration rate and the photochemical (or quantum) efficiency. Adaptation irradiance is the PAR at which the rate of efficiency of PAR energy conversion for the leaf area is at its maximum [76]. Saturating irradiance is the light energy at which the photosynthetic (or assimilation) processes are saturated. $P_{N,sat}$ is the maximum, or asymptotic, rate of CO_2 assimilation, frequently called the light saturated rate of photosynthesis. This is the most frequently

quoted single parameter, but it is highly nonspecific since it may be limited by an almost infinite number of steps downstream from the light harvesting processes. The light saturated rate of photosynthesis is primarily useful for categorizing plants broadly as shade tolerant or shade intolerant [77]. (The use of a nonrectangular hyperbola for describing the irradiance response curve for CO_2 uptake is presented in Section III.B.)

The relationship between P_N and intercellular CO_2 concentration at an optimal temperature and a particular irradiance is expressed by the CO_2 curve of P_N. In general, the relationship of these two parameters has a linear character up to ambient CO_2 concentration of approximately 300 μl CO_2 per liter. Figure 32.17 shows the CO_2 curve of P_N in the leaves of a beech seedling measured at an abaxial leaf surface temperature of 19.5°C at different irradiances (63 to 330 W/m²) [78]. The slope of these curves represents the carboxylation efficiency, and the curves also give the values of CO_2 compensation

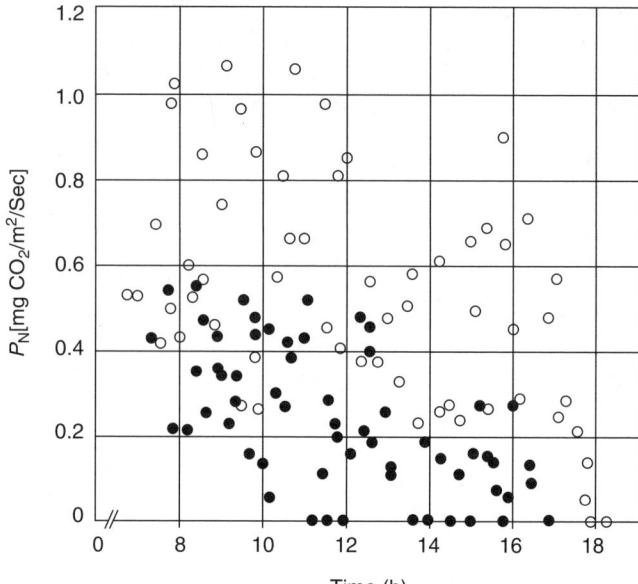

FIGURE 32.14 Daily course of the net photosynthetic rate (P_N) in leaves in the sun (open symbols) and shaded leaves (filled symbols) of tall beech trees. (From Masarovičová E, Štefančík L. *Biol. Plant.* 1990; 32: 374–387. With permission.)

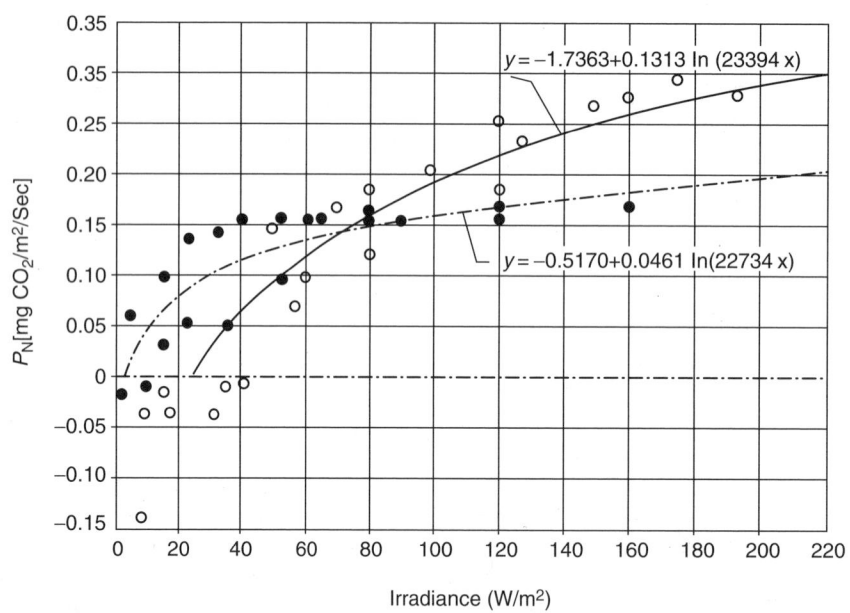

FIGURE 32.15 Irradiance response curves for CO_2 uptake in leaves in the sun (open symbols) and shaded leaves (filled symbols) of tall beech trees. (From Masarovičová E, Štefančík L. *Biol. Plant.* 1990; 32: 374–387. With permission).

concentration, Γ. Since Γ is a function of photosynthesis, photorespiration, and mesophyll resistance, all factors influencing these physiological parameters also affect Γ. CO_2 compensation concentration is one of the physiological characteristics according to which plants are categorized into C_3, C_4, or C_3–C_4 intermediate species [78,59].

For the estimation of photosynthetic rate and its dependence on CO_2 concentration, a closed gas exchange system is often used. Kotvalt and Hák [79] evaluated and analyzed methods for the mathemat-

ical estimation of CO_2 response curve parameters based on a closed-system measurement. A mathematical model (program "FOTOS") was used to calculate the following parameters of the CO_2 curve P_N in spruce needles: maximal net photosynthetic rate, CO_2 compensation concentration, mesophyll conductance, and convexity [80].

Photorespiration rate (R_L) is an important characteristic of the CO_2 exchange between the plant and the environment. The photorespiration rate can be measured (1) by the extrapolation of the P_N to inter-

TABLE 32.12
Physiological Characteristics of Leaves of Tall Beech Trees in the Sun and Shade Measured under Field Conditions

Physiological Characteristics		Leaves in the Sun	Shaded Leaves
Mean daily net photosynthetic rate, P_N (mg CO_2/m²/sec)	$\bar{x}$	0.578	0.287**
	$s_{\bar{x}}$	± 0.032	± 0.019
Maximal daily net photosynthetic rate, $P_{N,max}$ (mg CO_2/m²/sec)	$\bar{x}$	0.964	0.470**
	$s_{\bar{x}}$	± 0.025	± 0.027
Dark respiration rate, R_D (mg CO_2/m²/sec)	$\bar{x}$	0.261	0.292
	$s_{\bar{x}}$	± 0.011	± 0.033
Photosynthetic CO_2-fixation capacity (mg CO_2/g (Chl a+Chl b)/sec)		1.004 (for 1)	0.659 (for 1)**
		1.675 (for 2)	1.078 (for 2)**
Photosynthetic productivity (mg DM/m²/sec)		0.617	0.301**
Saturating irradiance, I_s (W/m²)	range	110–200	55–95
Compensating irradiance, I_c (W/m²)	range	25–50	10–25

* Significant differences at $p = .05$; **, significant differences at $p = .01$.

Note: $\bar{x}$, Mean, s_x, standard error.

Source: From Masarovičová E, Štefančik L. *Biol. Plant.* 1990; 32: 374–387. With permission.

TABLE 32.13
Physiological Characteristics of Leaves of Beech Trees in the Sun and Shade Measured under Controlled Conditions

Physiological Characteristics		Leaves in the Sun	Shaded Leaves
Net photosynthetic rate at saturating irradiance, $P_{N,sat}$ (mg CO_2/m²/sec)	$\bar{x}$	0.270	0.160*
	$s_{\bar{x}}$	± 0.023	± 0.005
Photorespiration rate, R_L (mg CO_2/m²/sec)	$\bar{x}$	0.049	0.025**
	$s_{\bar{x}}$	± 0.005	± 0.001
Dark respiration rate, R_D (mg CO_2/m²/sec)	$\bar{x}$	0.040	0.026*
	$s_{\bar{x}}$	± 0.005	± 0.001
Photosynthetic CO_2-fixation capacity (mg CO_2/g (Chl a + Chl b)/sec)		0.422	0.367
Photosynthetic productivity, α (μg CO_2/J)		1.428	4.011**
CO_2 compensation concentration, Γ (10^{-6} kg CO_2/m³)	$\bar{x}$	108	87
	$s_{\bar{x}}$	± 12.9	± 10.5
Saturating irradiance, I_s (W/m²)		Close to 200	Close to 110**
Adaptation irradiance, I_a (W/m²)		Close to 63	Close to 8**
Compensating irradiance, I_c (W/m²)		Close to 25	Close to 5**

*Significant differences at $p = .05$; **significant differences at $p = .01$.

Note: $\bar{x}$, Mean, $s_{\bar{x}}$, standard error.

Source: From Masarovičová E, Štefančík L. *Biol. Plant.* 1990; 32: 374–387.

cellular CO_2 concentration (c_i) to zero c_i; (2) from the P_N difference in 1% and 21% oxygen (the so-called Warburg effect); (3) from the postillumination burst, or CO_2 effusion after the light fades away. To estimate R_L in beech seedlings, the following two principles were used: the Warburg effect and extrapolation of the P_N curve in 2% to 3% O_2 and 21% O_2 to zero CO_2 concentration (Figure 32.17).

The photorespiration rate was then calculated as follows:

$$R'_L = P_N(2\text{–}3\%O_2) - P_N(21\%O_2)$$
$$R_L = R'_L + 0.1R'_L + R_M$$

where $0.1R'_L$ is the correction to the hypothetical concentration of 0% O_2, $R_M \sim 0.25\ R_D$, and R_L

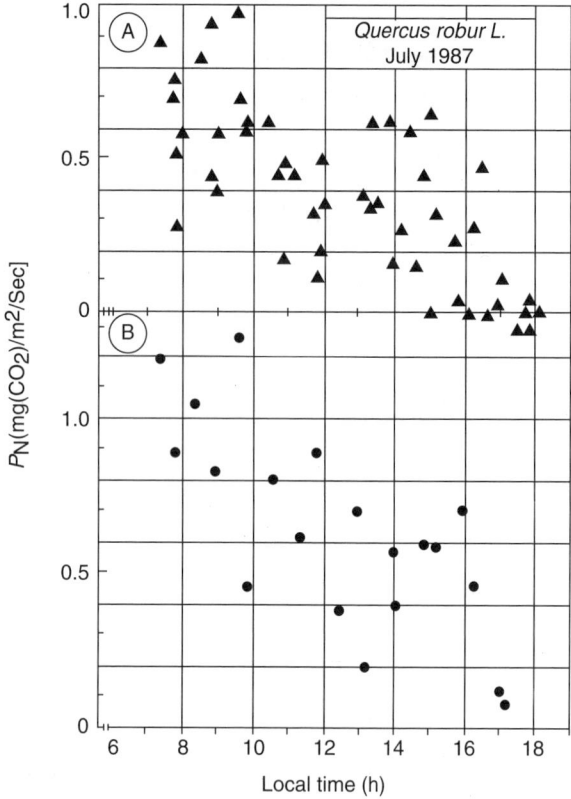

FIGURE 32.16 Daily course of net photosynthetic rate (P_N) in leaves of the first (A) and second (B) growth phase shoots. (From Masarovičová E. *Biol. Plant.* 1991; 33: 495–500. With permission.)

represents the total amount of CO_2 released by photorespiration and mitochondria! respiration in the light [59,81–83]. As mitochondrial respiration in the light is approximately one fourth of that occurring in the dark ($R_m \sim \frac{1}{4} R_D$) [59,82–84], it is also necessary to consider the amount of CO_2 released in the light by this pathway.

G. QUANTITATIVE PHOTOSYNTHETIC PARAMETERS IN MATHEMATICAL MODELS

Empirical models are usually based on the analysis of experimental data and on estimations in the form of an equation or a system of equations that may be used as a mathematical model that can be adapted to data. In certain cases this method is an adequate way of evaluating data in the given problem. If experimental data are well expressed by applying an empirical approach, then it is possible to analyze the mechanism that can give rise to the recorded response. However, in assessing the results of the model, it is necessary to avoid an extreme schematization of the given reality and an overestimation of the

model results, because the model is only a simplified image of the recorded reality [85].

The models of photosynthesis, as of most important physiological processes, have evolved using various levels of organization. The first models of this type were on molecular and cellular levels (e.g., Refs. [86,87]). Subsequently, attention was paid to individual leaves or shoots (e.g., Refs. [88,89]), to whole individuals (e.g., Refs. [90,91]), and to stands [92,93]. Indeed, the most complex models represent CO_2 exchange of complicated natural systems exemplified by forest ecosystems [94].

The presented empirical model of CO_2 exchange in young *F. sylvatica* plants, was evolved from the basic photosynthetic characteristics obtained over several years of experimental studies under controlled conditions — for details, see Ref. [59]).

We defined the following photosynthetic parameters:

P_N — net photosynthetic rate
P_G — gross photosynthetic rate
R_L — photorespiration rate (amount of CO_2 released by photorespiration, i.e., metabolism of glycolic acid in peroxisome, and by mitochondrial respiration in the light)
R_D — mitochondrial respiration rate in the dark
R_M — mitochondrial respiration rate in the light

Carbon dioxide exchange (photosynthesis, respiration) was measured at nine x_1 levels (PAR of 32, 44, 63, 86, 118, 55, 190, 235, and 330 W/m²), in five temperature regimes, x_2 (15°C, 20°C, 25°C, 30°C, and 35°C), in an environment with ambient CO_2 concentration (from 0 to 330 μl CO_2 per liter, 0% O_2, x_3, and 21% O_2, x_4).

In elaborating the model, an attempt was made to use a minimal number of parameters to quantify P_N as the function of irradiance (x_1), temperature (x_2), ambient CO_2 concentration (x_3), and O_2 concentration (x_4), assuming that

1. R_D rises exponentially with temperature increase
2. P_N declines linearly with rising O_2 concentration
3. P_G (1) increases with the rise of irradiance until it reaches a plateau (saturation of photochemical processes), and declines subsequently; (2) rises until it reaches the maximum in dependence on temperature, and declines subsequently; (3) rises nonlinearly to saturation in dependence on ambient CO_2 concentration; and (4) declines linearly in dependence on O_2 concentration.

TABLE 32.14

Net Photosynthetic Rate (P_N) and Some Quantitative Leaf Characteristics of the First (SGP-1) and Second (SGP-2) Growing Phase Shoot in Common Oak Saplings

Parameter		SGP-1	SGP-2
Mean daily P_N (mg CO_2/m^2/sec)	July	0.482 ± 0.045[a]	0.678 ± 0.076[**,a]
	August	—	0.724 ± 0.077
	September	0.437 ± 0.041	0.567 ± 0.051**
$P_{N,max}$ (mg CO_2/m^2/sec)	July	0.737 ± 0.042	1.046 ± 0.083**
	August	—	1.196 ± 0.107
	September	0.670 ± 0.025	0.804 ± 0.071**
Leaf area per tree (dm^2)		69.72 ± 14.49	101.31 ± 16.58**
		Total 180 ± 0.24	
Leaf dry mass per tree (g)		49.93 ± 10.38	60.32 ± 9.87
		Total 116.92	
Length of shoot (m)		0.22	0.5
Stomata density (mm^{-2})		384 ± 5	471 ± 4**
Stomata length (μm)		33.48 ± 0.21	$33.04 \pm 0.1\,7$
Stomata width (μm)		22.07 ± 0.15	22.54 ± 0.16
Chl a content (g/m^2)		0.425 ± 0.004	0.481 ± 0.027
Chl b content (g/m^2)		0.107 ± 0.005	0.118 ± 0.018
Chl $(a + b)$ content (g/m^2)		0.533 ± 0.008	0.600 ± 0.033
Chl a content (g/kg DM)		5.593 ± 0.109	8.071 ± 0.233**
Chl b content (g/kg DM)		1.502 ± 0.087	2.002 ± 0.305*
Chl $(a + b)$ content (g/kg DM)		7.440 ± 0.186	10.073 ± 0.390**
Chl $a{:}b$ ratio		3.97	4.08

[a]Mean $\pm$ standard error.

*Significant differences at $p = .05$; **significant differences at $p = .01$.

Source: From Masarovičová E. *Biol. Plant.* 1991; 33: 495–500. With permission.

The model takes the following form:

$$P_N = P_G - R_L =$$

$$\mu_i \frac{x_i}{x_{1\,opt}} \exp\left(1 - \frac{x_1}{x_{1\,opt}}\right) \frac{x_2}{x_{2\,opt}} \exp\left(1 - \frac{x_2}{x_{2\,opt}}\right) \frac{x_3(\mu_4)}{1 + \mu_3 x_3}(\mu_4 - x_4) - (\mu_2 2^{\alpha x_2 + \beta} - \mu_5)$$

Least-square estimates for constants μ_1, μ_3, α, and β were calculated from measurements of P_N and R_D. P_N was measured in 25 cases at various values of x_1, x_2, x_3, and x_4; R_D was measured at four values of x_2. The correlation coefficients between the measured and predicted values were 0.940 and 0.986 for P_N and R_D, respectively (for details, see Ref. [85]).

In conclusion, it is necessary to stress that the question of model compartmentalization is often a practical issue of compromise between the practicalities of measuring the necessary parameters and the desire to include the highest possible number of basic parameters in the model. The notion of a model implies the idea of simplification. This simplification, however, must not result in the distortion of the investigated system. There is a limit within the scope of simplification to which the similarity of the model to real behavior can be restricted. Each model should lie within this limit [95].

III. IN THE FIELD

A. QUALITATIVE AND QUANTITATIVE ESTIMATION OF PHOTOSYNTHETIC PIGMENTS

Chromatography in its various forms is now the major method of separating and purifying lipid-soluble pigments, often preceded by saponification when only carotenoids are to be examined or measured. Chromatographic separations are based either on differential adsorption of mixtures of compounds between a stationary phase and a moving phase (in columns, paper, or thin layers) or on the differential portion of the mixture between a stationary liquid

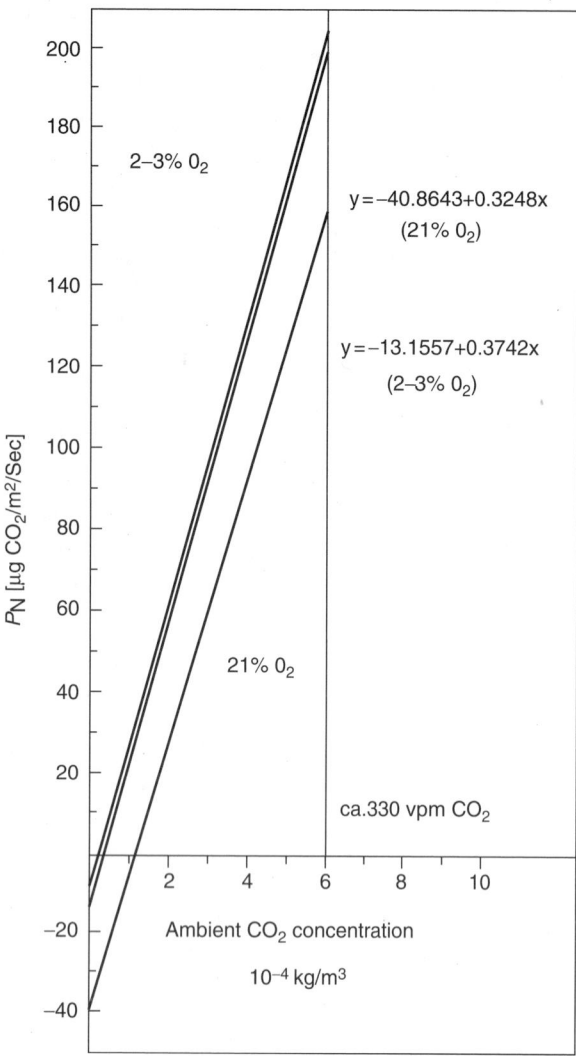

FIGURE 32.17 Net photosynthetic rate (P_N) in physiological adult leaves of Young European beech plants at saturating irradiance (235 W/m^2) and optimal temperature (19 ± 0.5°C) as affected by ambient CO_2 concentration in 21% and 2–3% O_2. The thin line represents the curve in zero oxygen concentration. (From Masarovičová E. *Gasometrical Investigation into CO$_2$ Exchange of the* Fagus sylvatica *L. Species under Controlled Conditions*, Veda, Publishing House of the Slovak Academy of Sciences, Bratislava, 1984. With permission.)

phase (supported on an inert material in a column, fine tube [HPLC], filter paper, or thin layer) and a moving phase. These are known as adsorption and partition chromatography, respectively. Chromatography is known as "reverse phase" when the support (paper, thin layer, or powder) is impregnated with a liquid, giving the reverse type of retention to the untreated supports [57].

TLC is still a frequently used method. Thin-layer chromatograms can run in one or two dimensions,

and soaking the material of the layer in organic liquids allows reverse-phase chromatography by partition of the pigment between a stationary phase and a moving phase. Šesták [96] confirmed that the advantages of TLC outweigh those of paper chromatography for separating plastid pigments; possible degradation of pigments by the adsorbent is the only serious disadvantage. However, choosing an inert adsorbent or neutralizing its acidity can overcome this problem [97,98]. One advantage of TLC over paper chromatography lies in the wide choice of adsorbents. Some types of silica gels are still frequently used.

In order to investigate the dynamics of changes in the amounts of β-carotene and lutein, we selected the plant dominant in the herbaceous undergrowth, *Pulmonaria officinalis* L. Samples were taken at 10-day intervals during the entire growing season. The quantitative estimation of β-carotene and lutein was carried out by TLC on silica gel plates, which consisted of MN-Kieselgel G (23 g), CaSO$_4$·2H$_2$O p.a. (3.3 g), starch powder p.a. (0.4 g), and distilled water (75 ml). The following solution for pigment separation was used: petrol p.a. (120 ml), isopropylalcohol p.a. (10 ml), and distilled water (0.25 ml). β-Carotene was dissolved in *n*-hexane p.a. and lutein in ethyl alcohol p.a. Measurements of absorbance were made on a UNICAM SP 800 recording spectrophotometer. Table 32.15 shows seasonal changes in the contents of β-carotene and lutein with regard to the ontogenesis of *P. officinalis* L. A close relationship was found between pigment content and various phenological phases. The seasonal changes in carotenoid content were also considerably influenced by climatic factors (global radiation and air temperature) [99].

Chl contents (*a*, *b*, *a* + *b*) were determined directly in acetone extracts (80% acetone p.a.) of forest plant leaves (herbs, shrubs, trees). Measurements of absorbance (Chl *a* at 665 nm, Chl *b* at 649 nm) were made on a UNICAM SP 800 recording spectrophotometer. The Chl contents were calculated according to Vernon [100]. The results were expressed per dry mass and leaf area unit.

Chl contents were estimated in the summer period for 19 herbaceous species growing in a temperate hardwood deciduous (oak–hornbeam) forest in southwest Slovakia. Chl *a* content varied between 11.0 and 19.2 g/kg dry mass DM (0.26 to 0.45 g/m^2), Chl *b* content between 3.9 and 8.2 g kg DM (0.09 to 0.19 g/m^2), and total Chl (*a* + *b*) between 15.0 and 27.3 g/kg DM (0.23 to 0.31 g/m^2). Chl *a*:*b* ratio ranged from 2.5 to 2.8. Within the same plant species, the variations in Chl contents in stem and ground leaves and in apical and basal parts of the leaf blade were determined (Table 32.16) [101].

TABLE 32.15
Quantitative Changes in Carotene and Lutein in *Pulmonaria officinalis* Leaves

Day	Carotene (mg/g DW)	Carotene (mg/dm^2)	Lutein (mg/g DW)	Lutein (mg/dm^2)
April 5	0.5551	0.1382	0.8975	0.2235
April 15	0.7418**	0.1246	1.1307*	0.1899
April 25	0.5751**	0.1277	0.9950*	0.2209
May 5	0.42 11**	0.1078	0.7844**	0.2008
May 18	0.5657**	0.1589	0.9077*	0.2551
May 29	0.5801	0.1560	1.0305	0.2772
June 10	0.5659	0.1443	1.1172	0.2849
July 1	0.6326*	0.1689	0.1758	0.3139
July 12	0.5643**	0.1642	1.1001	0.3201
July 27	0.5898	0.1634	1.0210	0.2828
August 9	0.5152*	0.1870	0.9874	—
August 22	0.6408**	0.1871	1.0896	0.3182
September 4	0.6708	0.2000	1.1523	0.3434
September 15	0.6746	0.2489	1.0731	0.3960
September 27	0.6502*	0.1918	1.2678*	0.3740
October 10	0.5320**	0.1713	1.3297	0.4282
October 25	0.4702*	0.1749	1.0419**	0.3876

*Significant differences at $p = .05$; **high significant differences at $p = .01$.

Notes: Spring caulis *of Pulmonaria officinalis* = April 5 to 25. DW, dry weight.

Source: From Masarovičová E, Duda M. *Biologia* (Bratisl.) 1976; 31: 15–23. With permission.

In the same oak–hornbeam forest stand, Chl contents in the leaves of nine shrub species (Table 32.17) and four tree species (Table 32.18) were also estimated. For shrubs, Chl *a* content varied between 7.3 and 12.6 g/kg DM (0.29 to 0.45 g/m^2), Chl *b* between 2.4 and 4.7 g/kg DM (0.09 to 0.17 g/m^2), and Chl *a* + *b* between 10.5 and 17.4 g/kg DM (0.29 to 0.62 g/m^2). The Chl *a:b* ratio ranged from 2.5 to 3.4 [102].

For tree species, Chl contents were determined in the leaves of *Carpinus betulus* L., *Q. cerris* L., *Q. petraea* Liebl., and *Acer campestre* L. growing in various strata in the above-mentioned oak–hornbeam forest. The Chl contents expressed on a DM basis in sun leaves on tall, dominant, and codominant trees (forming an active surface of the canopy) were half of those found in leaves from the interior of the stand, i.e., in the lower crown regions of dominant and partially codominant trees, in intermediate and undertopping trees, as well as in young individuals of tree species forming the undergrowth, e.g., Chl *a* + *b*, 4.3 to 7.8 g/kg DM in the first case and 9.1 to 16.6 g/kg DM in the second.

Relatively small variations in Chl content expressed per leaf area unit were caused by the compensating influence of a specific leaf area. Chl *a:b* ratio was usually above 3.0 in leaves in the sun and often below 3.0 in shaded leaves [103].

The total amounts of Chls per forest area unit were computed for individual species by multiplying data on Chl content and leaf mass or area. The standing crop of Chl *a* + *b* for trees was 3.1 g/m^2, for shrubs 0.04 g/m^2, and for herbs 1.5 g/m^2. The total amounts of Chl per forest area unit varied between 3.5 and 5.5 g/m^2, depending on the calculation methods used [104].

In the vertical profile of the stand, maximum Chl *a* + *b* was concentrated in the canopy layer at 15 to 19 m (about 40% of the total amount), i.e., in the upper canopy of the 22-m-high stand. There was also a relatively high Chl content in the lower tree canopy, at a height between 9 and 15 m. Large differences in vertical distribution of Chl among five tree species forming the stand reflect adaptation of the species to irradiance [105].

B. Measurement of Photosynthesis in the Forest Stand

An understanding of forest stand productivity requires both a qualitative and a quantitative analysis of forest growth and an understanding of many processes that contribute to the growth of trees [106]. However, the relationships among stand structure, energy interception, and stand productivity for deciduous forests are still poorly understood.

TABLE 32.16
Chl Contents per Unit DM and Chl *a:b* Ratio in Leaves of Herbaceous Plants Growing in an Oak–Hornbeam Forest

Species	Chl *a*	Chl *b*	Chl (*a* + *b*)	Chl *a:b*
Herbs				
Ajuga reptans L.	13.45 ± 0.00[a]	5.38 ± 0.00	18.83 ± 0.00	2.50
	14.04 ± 0.30[b]	5.02 ± 0.02	19.05 ± 0.47	2.78
Asperula odoraia L.	11.80 ± 0.06[a]	4.52 ± 0.04	16.32 ± 0.21	2.61
	14.32 ± 0.22[b]	5.23 ± 0.04	19.56 ± 0.19	2.74
Convallaria majalis L.	13.24 ± 0.05[a]	4.28 ± 0.46	17.51 ± 0.78	3.10
	13.82 ± 0.26[b]	5.07 ± 0.06	18.89 ± 0.56	2.73
Fragaria moschata DLJCH.	10.38 ± 0.05[a]	3.93 ± 0.00	14.31 ± 0.09 ,	2.64
	11.33 ± 0.13[b]	4.44 ± 0.15	15.77 ± 0.28	2.55
Galeobdolon luteum Huds	17.90 ± 0.37[a]	7.23 ± 0.04	25.13 ± 0.67	2.48
	19.15 ± 0.53[b]	8.17 ± 0.22	27.32 ± 0.42	2.34
Geum urbanum L.	11.38 ± 0.77[a]	4.10 ± 0.07	25.48 ± 1.28	2.78
	11.06 ± 0.21[b]	3.97 ± 0.02	15.02 ± 0.32	2.79
Clechoma hirsuta W. et K.	14.12 ± 0.38[a]	5.15 ± 0.10	19.27 ± 0.81	2.74
	14.08 ± 0.83[b]	4.69 ± 0.01	18.77 ± 0.26	3.00
Mercurialis perennis L.	11.82 ± 0.00[a]	4.22 ± 0.00	16.05 ± 0.00	2.80
	12.33 ± 0.02[b]	4.62 ± 0.03	16.95 ± 0.11	2.67
Pulmonaria officinalis L.	13.60 ± 0.04[a]	5.23 ± 0.03	18.82 ± 0.14	2.60
	13.19 ± 0.37[b]	4.32 ± 0.07	17.51 ± 0.74	3.05
Viola sylvatica L.	12.76 ± 0.01[a]	4.89 ± 0.02	17.65 ± 0.06	2.60
	15.67 ± 0.00[b]	5.91 ± 0.00	21.58 ± 0.00	2.65
Viola mirabilis L.	12.32 ± 0.04[a]	4.71 ± 0.00	17.03 ± 0.06	2.81
	13.39 ± 0.27[b]	4.88 ± 0.03	18.27 ± 0.52	2.75
Campanula trachelium L.	17.76 ± 0.00[a]	6.50 ± 0.05	24.26 ± 0.04	2.73
Lamium macalatum L.	16.06 ± 0.78[a]	6.45 ± 0.52	22.51 ± 2.57	2.49
Polygonatum odoratum Druce	13.03 ± 0.07[a]	4.64 ± 0.03	17.67 ± 0.1 8	2.81
Sanicula europea L.	12.45 ± 0.20[b]	4.67 ± 0.01	17.12 ± 0.07	2.67
Grasses				
Bromus benekenii (Lange)	13.10 ± 0.09[a]	5.04 ± 0.06	18.14 ± 0.24	2.60
Trimen	15.80 ± 0.27[b]	6.34 ± 0.02	22.15 ± 0.46	2.49
Dactylis polygama.	14.12 ± 0.19[a]	5.37 ± 0.06	19.50 ± 0.47	2.63
Horvatovszky	14.10 ± 0.01[b]	5.10 ± 0.00	19.21 ± 0.00	2.76
Melica uniflora Retz.	13.33 ± 0.07[a]	5.04 ± 0.01	18.37 ± 0.13	2.64
	15.16 ± 0.00[b]	5.96 ± 0.00	21.13 ± 0.00	2.54
Brachypodium sylvaticum L.	14.29 ± 0.70	5.71 ± 0.27	20.11 ± 1.82	2.50

[a] July 16.
[b] August 13.

Source: From Masarovičová E, Eliáš P. *Photosynthetica* 1980; 14: 580–588. With permission.

A series of micrometerological and ecophysiological measurements were made in an unevenly aged, multispecies oak–hornbeam forest in Báb, southwest Slovakia. The aim of this work was to improve our understanding of the physiological processes (photosynthesis, respiration, and transpiration) of adult trees and their microclimate, to collect data for the simulation of canopy (stand) photosynthesis, and to study the ecological synthesis of the functioning of the forest ecosystem [107].

Vertical and diurnal variations in PAR, air temperature (AT) and relative air humidity (RH), wind speed (WS), and CO_2 concentration in and above the forest were characterized for the fully leaved season using diurnal courses, vertical profiles, and isodiagrams (isopleths) [107]. The data obtained were used for simulating the daily course of photosynthetic rate and stomatal conductance of leaves in the sun and in the shade from tall trees (*Q. cerris* L., *C. betulus* L.) using a mathematical model [108].

TABLE 32.17
Chl Contents per Unit DM and Chl *a:b* Ratios in Leaves of Shrub Species Growing in an Oak–Hornbeam Forest

| Species | Chl Content (g/kg) | | | |
	Chl *a*	Chl *b*	Chl (*a* + *b*)	Chl *a:b*
Cerasus avium L.	11.97 ± 0.48[a]	4.37 ± 0.28	16.34 ± 0.77	2.73
Cornus mas L.	10.71 ± 0.55[b]	3.98 ± 0.11	14.97 ± 1.08	2.69
	9.06 ± 2.61[c]	3.31 ± 0.53	12.37 ± 5.54	2.74
	11.54 ± 0.25[a]	4.34 ± 0.01	15.91 ± 0.35	2.65
Crataegus leavigata (Poir) DC.	8.10 ± 0.01[c]	2.43 ± 0.01	10.54 ± 0.03	3.38
	7.99 ± 0.72[a]	2.97 ± 0.08	10.96 ± 1.28	2.69
Euonymus europea L.	9.86 ± 0.09[b]	3.27 ± 0.16	13.13 ± 0.37	3.01
	9.32 ± 0.01[a]	3.41 ± 0.01	12.73 ± 0.04	2.74
Euonymus verrucosa Scop.	12.05 ± 0.63[c]	4.74 ± 0.13	16.80 ± 1.31	2.54
	11.12 ± 0.56[a]	4.33 ± 0.10	15.45 ± 1.13	2.57
Hedera helix L.	8.14 ± 0.35[c]	3.23 ± 0.50	11.38 ± 0.64	2.52
Ligustrum vulgare L.	7.29 ± 0.12[b]	2.59 ± 0.03	9.88 ± 0.25	2.82
	7.77 ± 0.05[c]	2.86 ± 0.04	10.63 ± 0.18	2.72
	9.85 ± 0.00[a]	3.50 ± 0.00	13.34 ± 0.01	2.82
Sorbus torminalis L	11.50 ± 0.01[a]	4.38 ± 0.00	15.86 ± 0.01	2.63
Ulmus campestris L.	12.65 ± 0.00[c]	4.75 ± 0.00	17.40 ± 0.00	2.67

[a]August 13.
[b]June 20.
[c]July 11.

Source: From Masarovičová E, Eliáš P. *Photosynthetica* 1981; 15: 16–20. With permission.

Vertical profiles of the micrometerological factors were determined at two positions above the hornbeam canopy and at nine levels in the forest: 1, 4, 7, 10, 13, 16, 19, 22, and 25 m above the ground. A 30-m-tall steel meteorological tower was used for the installation of instruments and for the measurement of microclimate (Figure 32.18). Global radiation was measured at 22 m above the ground, with a U-200SB pyranometer sensor connected to a Li-Cor 185 B quantum-radiometer-photometer, and with at 15, 12, and 9 m above the ground, a Kipp-Zonen pyranometer. During the day, measurements were taken at 30-min intervals and mean values were calculated from the data sets. The PAR was measured with a Li-Cor 190S-1 quantum sensor connected to the Li-185B quantum-radiometer-photometer. The position of the sensors in relation to the mast, frequency of measurement, and calculation of mean values were similar to those used for global radiation measurements. AT and RH were monitored with thermohygrographs 1, 7, 13, 16, and 22 m above the ground. During field measurements, the thermohygrographs were periodically calibrated by Assman aspiration psychrometers. WS profiles were estimated with sensitive four-cup anemometers with a range of 1–20 = 0.8 m sec, or with three-cup anemometers from Rauchfuss Instruments Division (Australia), which have an accuracy of 1%. The anemometers were placed 1, 13, 16, 19, 22, and 25 m above the ground. Vertical profiles of air CO_2 concentration were measured with the Li-Cor 6000 or Li-Cor 6200 Portable Photosynthesis System 1, 3, 6, 9, 12, 15, 18, 21, and 24 m above the ground. The accuracy of the measurements was ±2 vpm. The measurements were made every 2 h (for details, see Ref. [107]).

The aim of this ecophysiological research was to compare the photosynthetic activity of Turkey oak (*Q. cerris* L.) as the dominant and hornbeam (*C. betulus* L.) as the codominant forest tree species. The crown shape of these trees makes it possible to divide crowns into two main layers: the upper "sun" layer with "sun" foliage (upper canopy layer [UCL]) and the lower "shade" layer with "shade" foliage (lower canopy layer [LCL]). These crown layers form the main sites of tree photosynthetic productivity. Comparison of the photosynthetic features of these layers and estimation of the effects of basic microclimatic factors will make it possible to appreciate the contribution of different types of foliage to whole-tree photosynthesis.

TABLE 32.18
Chlorophyll (Chl) Contents per Unit DM and Chl *a:b* Ratios in Leaves of Tree Species Growing in Various Strata of an Oak–Hornbeam Forest

Species	Chl Content (g/kg)			
	Chl *a*	Chl *b*	Chl (*a* + *b*)	Chl *a:b*
Tall trees				
Upper crown region (leaves in the sun)				
Acer campestre L. (codominant tree)	4.59 ± 0.02[a]	1.42 ± 0.01	6.01 ± 0.07	3.24
	5.13 ± 0.06[b]	1.77 ± 0.02	6.90 ± 0.15	2.91
	5.38 ± 0.09[c]	1.84 ± 0.01	7.22 ± 0.16	2.93
Carpinus betulus L. (codominant tree)	3.76 ± 0.20[a]	1.14 ± 0.04	4.89 ± 0.40	3.31
	4.73 ± 0.01[b]	1.49 ± 0.01	6.23 ± 0.03	3.17
	3.34 ± 0.07[c]	0.10 ± 0.02	4.32 ± 0.17	3.41
Quercus cerris L. (dominant tree)	6.16 ± 0.01[a]	1.67 ± 0.00	7.83 ± 0.01	3.68
	5.72 ± 0.41[b]	1.72 ± 0.54	7.44 ± 0.50	3.32
	5.00 ± 0.01[c]	1.20 ± 0.03	6.20 ± 0.01	4.15
Quercus petraea Liebl. (dominant tree)	5.82 ± 0.01[b]	1.91 ± 0.02	7.73 ± 0.05	3.04
	5.35 ± 0.60[c]	1.36 ± 0.01	6.71 ± 0.80	3.39
Middle crown region				
A. campestre L.	9.03 ± 0.35[b]	3.38 ± 0.04	12.41 ± 0.35	2.67
C. betulus L.	6.88 ± 0.00[b]	3.32 ± 0.00	10.21 ± 0.00	2.07
Lower crown region				
A. campestre L.	7.50 ± 0.54[a]	2.59 ± 0.07	10.09 ± 0.94	2.90
	9.09 ± 0.07[b]	3.36 ± 0.00	12.45 ± 0.11	2.70
	9.34 ± 1.52[c]	3.63 ± 0.56	12.97 ± 3.91	2.57
C. betulus L.	7.00 ± 0.01[b]	2.33 ± 0.00	9.34 ± 0.01	3.00
	8.29 ± 0.19[c]	2.99 ± 0.08	11.29 ± 0.51	2.67
Q. cerris L. (dominant tree)	7.04 ± 0.09[a]	2.09 ± 0.02	9.13 ± 0.18	3.37
	5.70 ± 0.02[b]	1.93 ± 0.01	7.63 ± 0.05	3.00
	5.99 ± 0.89[c]	2.00 ± 0.12	7.99 ± 1.65	3.00
Q. petraea Liebl.	10.65 ± 0.00[b]	3.59 ± 0.00	14.23 ± 0.01	3.00
	7.03 ± 0.02[c]	2.18 ± 0.01	9.21 ± 0.04	3.22
Shrub-size individuals growing in shrub layer (up to 3.0 m)				
A. campestre L.	9.45 ± 0.05[b]	3.64 ± 0.01	13.09 ± 0.08	2.60
	7.71 ± 0.10[c]	2.55 ± 0.02	10.26 ± 0.03	3.02
C. betulus L.	8.31 ± 0.00[b]	3.13 ± 0.00	11.44 ± 0.00	2.83
	8.45 ± 0.07[c]	2.94 ± 0.00	11.38 ± 0.09	2.88

Source: From Eliáš P, Masarovičová E. *Photosynthetica* 1980; 14: 604–610. With permission.

Gas exchange measurements were carried out on physiologically adult leaves from June to August 1987. Field data were collected during 1 week of each month. Tree canopies were divided into the above-mentioned two layers: UCL and LCL leaves. The classification of leaves was based on earlier investigations of leaf characteristics, such as specific leaf area [109], leaf Chl content [103], and light conditions within the crowns [107].

The CO_2-exchange measurements were made with the Li-6200 systems (Li-Cor, U.S.). The gas analyzer was calibrated against dilutions of CO_2 in nitrogen. Gas mixtures were generated by a gas-mixing pump (Wösthoff, Bochum, Germany).

Photosynthetic characteristics of oak and hornbeam leaves were estimated in relation to the main environmental factor — the photon flux rate (I). The CO_2 concentration within the assimilation chamber of the Li-6200 system was set using a flow switch. The flow switch allows the Li-6200 system to be toggled between the open and closed modes of operation. The open mode is useful for reaching equilibrium between the CO_2 concentration and air humidity within the measuring system and the ambient air of the crown space (320 to 360 μl CO_2 per liter) [108].

The results of measurements of the relationships between P_N and I were processed by an empirical mathematical model [69,79,110]. The model was

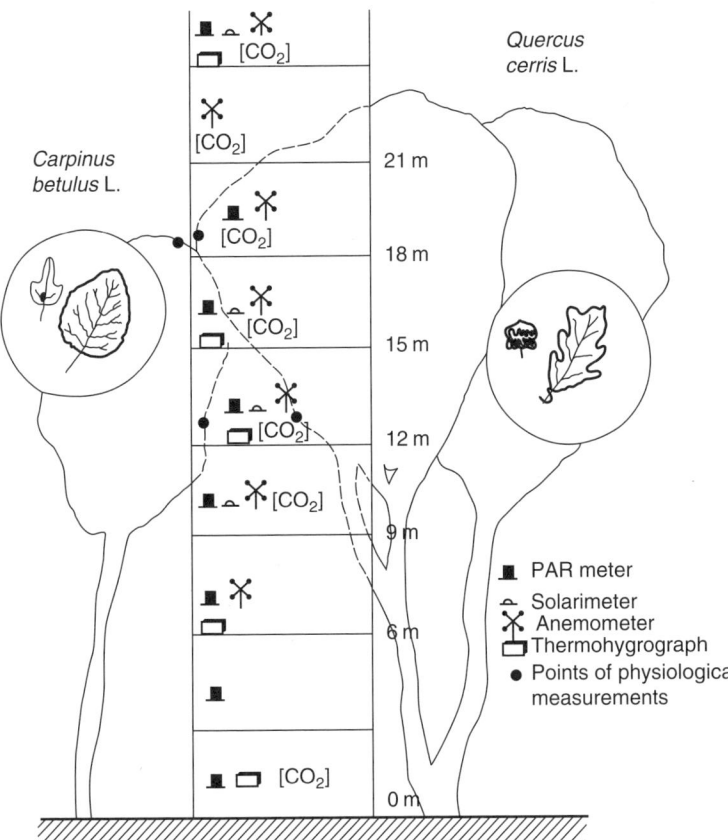

Carpinus betulus L.

Quercus cerris L.

21 m

18 m

15 m

12 m

9 m

6 m

0 m

■ PAR meter
△ Solarimeter
✕ Anemometer
▱ Thermohygrograph
● Points of physiological measurements

FIGURE 32.18 The measuring tower for measurements of basic micrometeorological parameters and leaf physiological characteristics (solid symbols are the points of the physiological measurements) of forest trees. (From Marek M, Masarovičová E, Kratochvílová I, Eliáš P, Janouš D. *Trees* 1989; 4: 234–240. With permission.)

applied as an analytical tool for summarizing information about relations between P_N and I. Kotvalt and Hák [79] published a method for fitting an implicit function directly into the primary experimental data. The implicit function used for fitting was the model of Marshall and Biscoe [69]. The results of field measurements are composed of discrete clusters of data. The possibility of constructing, mathematically, the response curve on the basis of the abovementioned model, using some physiologically interpretable parameters, was considered advantageous. A computer was used for these calculations. The light response curve for CO_2 uptake was calculated from whole data sets for both types of leaf [108].

Stomatal conductance values measured with the Li-6200 system and maintenance respiration rate (R_m) were estimated from measurements of CO_2 efflux after a period of prolonged darkness (48 h [111,112]). The diurnal courses of P_N were found from the combination of calculated P_N, using the above-mentioned mathematical model, and I measurements at the relevant leaf position in the tree crown (see Ref. [107]). For every leaf position and time interval, a sufficient number of measurements (90 on average) were obtained (for details, see Ref. [108]).

The values presented in Figure 32.19 and Figure 32.20 of PAR, AT, RH, WS, and CO_2 concentration in and above the forest are characterized for fully developed leaves during the season, using diurnal courses, vertical profiles, and isodiagrams (isopleths). Approximately 50% of incident PAR was absorbed by the upper 4 to 5 m of leaves, and only approximately 5% or less penetrated to the forest floor. Vertical gradients of AT and RH were generally low, but large differences in diurnal ranges of AT and RH were observed between vertical levels. The ULC greatly reduced WS, and at a height of about 14 m above the ground it was close to 0. The highest diurnal CO_2 concentration and variations occurred at 1 m above the ground, and the lowest above the forest. In favourable light conditions, the entire leaf canopy of the forest (overstory and understory canopy) is a large sink of CO_2. At night the forest stand is a source of CO_2, with the soil and forest floor as the largest internal source [107].

The average photosynthetic rate of oak foliage was higher than that of hornbeam. Net photosynthetic rate of hornbeam at saturating photon flux ($P_{N,max}$) amounted to only 60% that of oak for UCL leaves and 67% for LCL leaves (Figure 32.21). In the summer months, the main photosynthetic activity of this deciduous stand was focused upon the

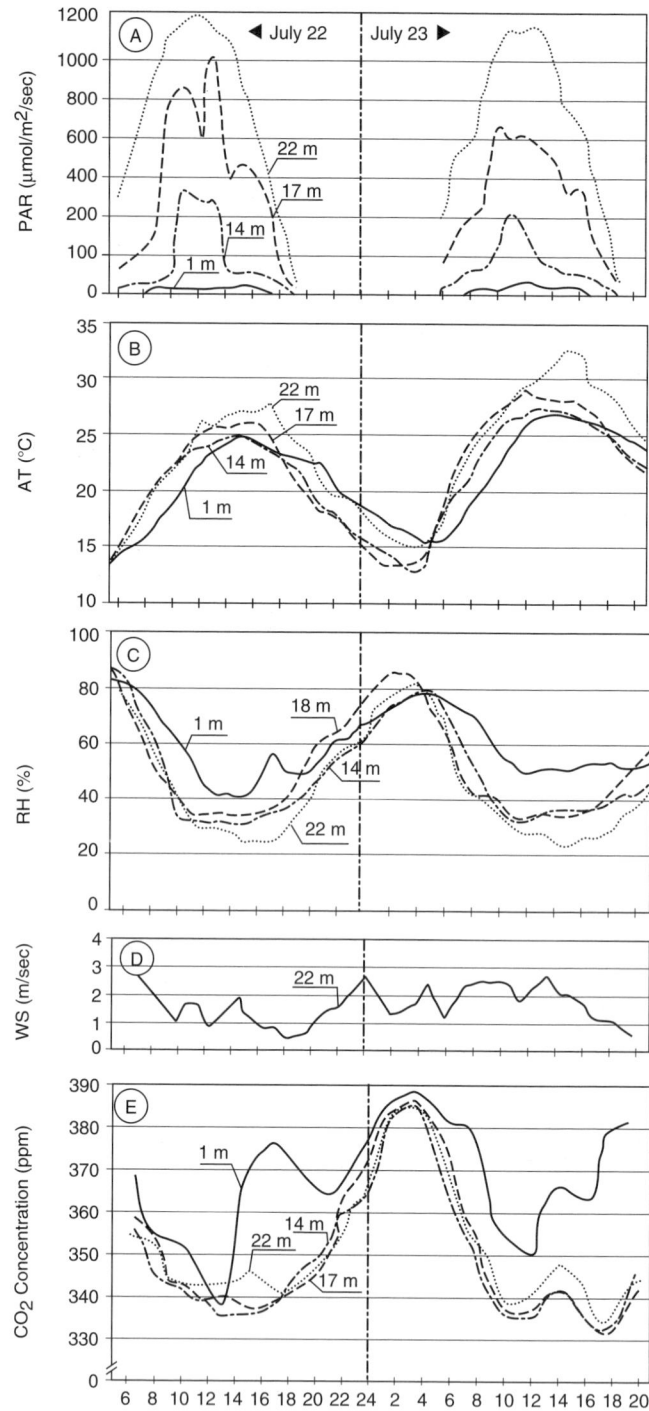

FIGURE 32.19 Diurnal courses of micrometeorological elements in and above the forest at Báb. (A) PAR, (B) AT, (C) RH, (D) WS, and (E) CO_2 concentration. (From Eliáš P, Kratochvílová I, Janouš D, Marek M, Masarovičová E. *Trees* 1989; 4: 227–233. With permission.)

UCL leaves and oak species. The relationship between P_N and photon flux rate, as well as the diurnal course of P_N and stomatal conductance (g_s), was calculated using a mathematical model. The diurnal course for P_N and g_s was similar for both tree species and both types of leaf. Maximal g_s values were observed at noon (Figure 32.22).

The lower values of compensation photon flux rate (Γ_I) and photosynthetic efficiency (α)

but higher values of the maintenance respiration rate (R_M) confirmed the higher shade tolerance of hornbeam. The dark respiration rate (R_D) of the UCL leaves was higher than that of the LCL leaves (Table 32.19). Various photosynthetic features and production capacities of the abovementioned types of leaves indicated adaptation pressures to radiation conditions. In the stand studies, the primary production of the greater part of the

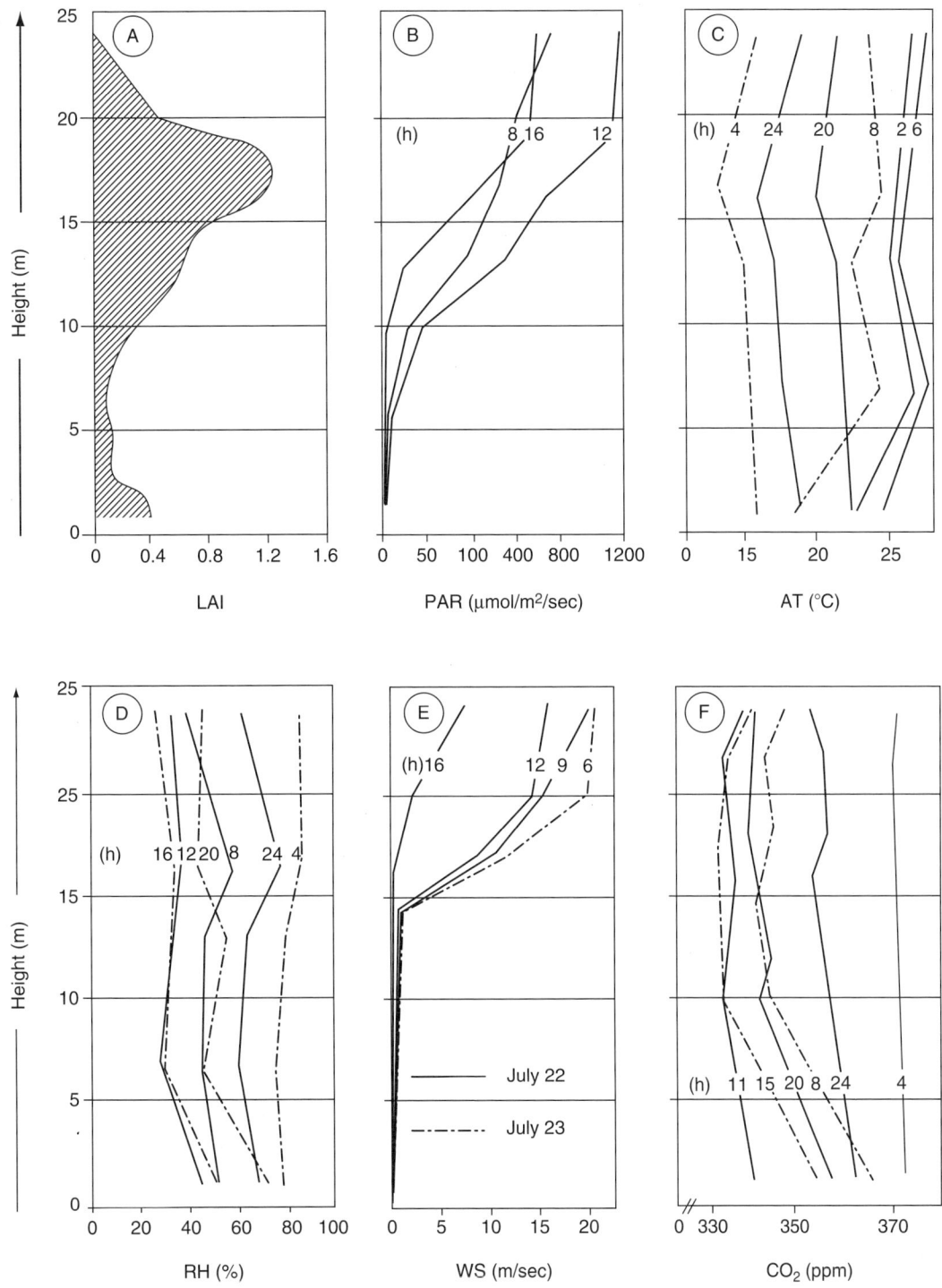

FIGURE 32.20 Vertical profiles of stand and microclimatic elements in and above the forest at Báb. (A) Leaf area index (LAI), (B) PAR, (C) AT, (D) RH, (E) WS, and (F) CO$_2$ concentration. (From Eliáš P, Kratochvílová I, Janouš D, Marek M, Masarovičová E. *Trees* 1989; 227–233. With permission.)

crown depended on the vertical foliage distribution and on light penetration during the midday hours [108].

The balance of CO$_2$ exchange may be calculated on the basis of the dependence of physiological pro-

cesses on ecological factors, on stand structure, and on the diurnal course of ecological factors. Using these, both the seasonal balance of CO$_2$ exchange of the stand and the annual dry mass production can be estimated [108].

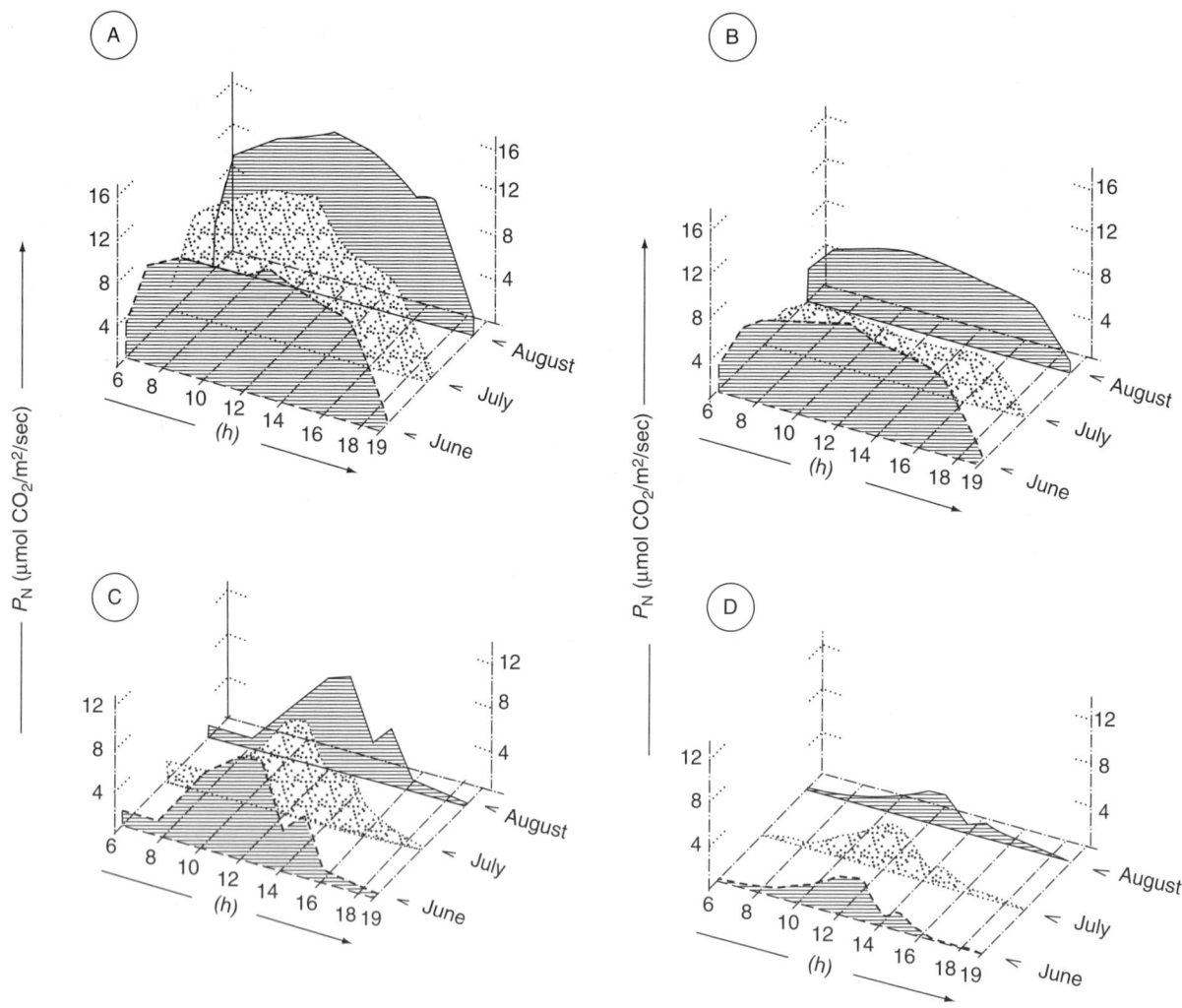

FIGURE 32.21 Net photosynthetic rate of *Quercus cerris* L. (A, UCL leaves; C, LCL leaves) and *Carpinus betulus* L. (B, UCL leaves; D, LCL leaves). (From Marek M, Masarovičová E, Kratochvílová I, Eliáš P, Janouš D. *Trees* 1989; 4: 234–240. With permission.)

C. GROWTH ANALYSIS METHOD

An excellent and wide-ranging review, covering both the classical and the functional approaches to plant growth analysis, was published by Šesták et al. [113]. After 25 years it is still frequently used and cited as a manual for methodology. Later the methods of growth analysis were improved not only for individuals but also for populations and communities [114]. One of the most recent papers [115] deals with a short but clear explanation of the assumptions involved in the use of the classical formulas and a brief introduction to the functional approach.

Growth analysis represents the first step in the study of primary production using the technique of direct harvesting, mathematical procedures, and the application of the growth analysis method to investi-

gations of photosynthetic production. One advantage of growth analysis is that the primary values (DM of whole plant or their parts and dimensions of the assimilatory organs) are relatively easy to obtain without great demands on laboratory equipment. Although the methods of plant growth analysis seemed nearly complete a number of years ago, new aspects have emerged, especially in mathematical and computer techniques. This section will discuss the basic concepts and methical procedures in the study of growth processes of plant individuals using the components of classical growth analysis.

The basic component of growth analysis is the relative growth rate (R, in kg/kg/day) of the plant. This is defined at any instant in time (t) as the increase in the material present and is the only component of growth analysis that does not require

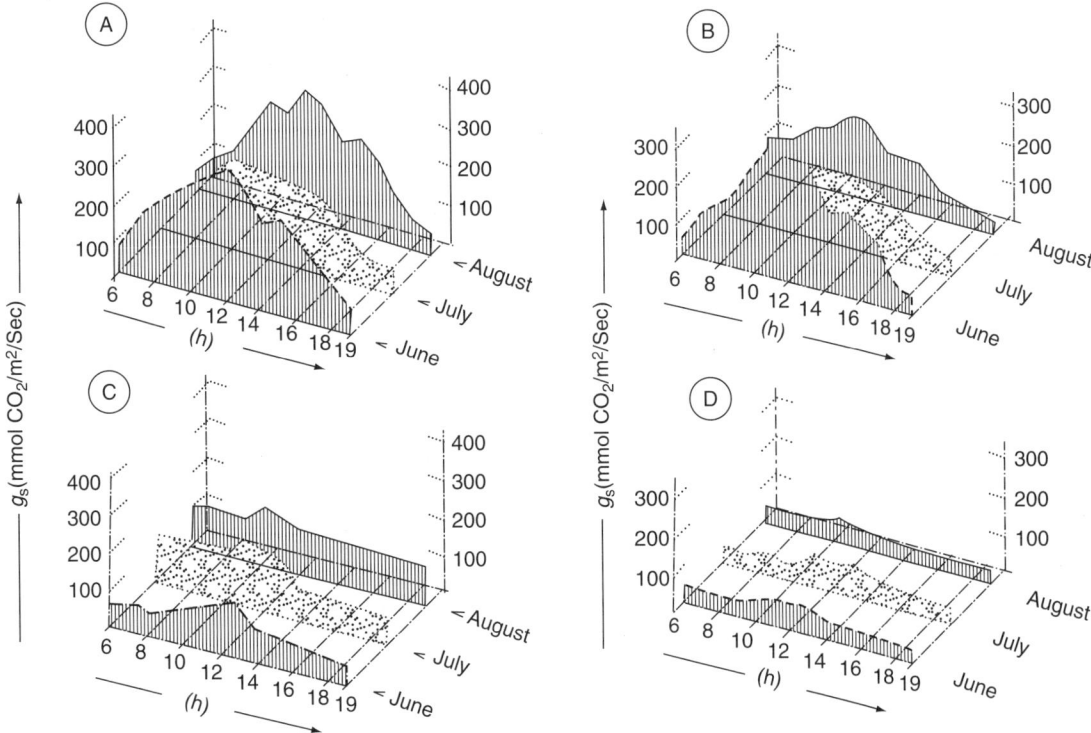

FIGURE 32.22 Stomatal conductance of *Quercus cerris* L. (A, UCL leaves; C, LCL leaves) and *Carpinus betulus* L. (B, UCL leaves; D, LCL leaves). (From Marek M, Masarovičová E, Kratochvílová I, Eliáš P, Janouš D. *Trees* 1989; 4: 234–240. With permission.)

knowledge of the size of the assimilatory system. Thus,

$$R = \frac{1}{W}\frac{dW}{dt} = \frac{d}{dt}(\ln W)$$

where W is the plant dry weight (kg DW). The relative growth rate is therefore the dry weight increase per unit of dry weight present per unit of time. The mean relative growth rate, $\bar{R}$ is measured over a discrete time interval, t_1 to t_2, which is usually no less than 1 day. $\bar{R}$ is defined as

$$\bar{R} = \frac{\ln W_2 - \ln W_1}{t_2 - t_1}$$

The relative growth rate serves as a fundamental measure of DM production and can be used to compare the performance of species or the effect of treatments under defined conditions.

The unit leaf rate, E (kg/m²/day), of a plant at any instant in time (t) is defined as the increase of plant material (kg DW) per unit of assimilatory material, s (m²), per unit of time:

$$E = \frac{1}{s}\frac{dW}{dt}$$

The term unit leaf rate is often used interchangeably with net assimilation rate (NAR), but the former is now preferred. It measures the net gain in dry weight of the plant per unit leaf area per unit of time (kg/m²/day) and differs from the photosynthetic rate, which measures the net carbon gain during the light period. The mean unit leaf rate, $\bar{E}$ between t_1 and t_2 is given by the formula:

$$\bar{E} = \frac{(W_2 - W_1)(\ln s_2 - \ln s_1)}{(s_2 - s_1)(t_2 - t_1)}$$

There are differences in E between plant species with different carbon metabolisms (e.g., C_3 and C_4 species), and E will also vary with age and the growing environment.

The leaf area ratio, F (m²/kg⁻¹), of a plant at any instant in time (t) is the ratio of the assimilatory material (m²) per unit of plant material (kg DW) and is defined as

TABLE 32.19
Photosynthetic Features of Turkey Oak and Common Hornbeam Leaves Measured in the Summer of 1987

	Turkey Oak						Common Hornbeam					
	UCL leaves			LCL leaves			UCL leaves			LCL leaves		
	June	July	August	June	July	August	June	July	August	June	July	August
$P_{N,max}$	13.4	18.5	19.3	10.3	14.1	13.5	12.1	10.8	8.2	10.9	8.1	6.6
Γ_l	20.8	21.6	18.6	16.3	15.1	13.6	17.6	18.2	19.8	12.1	14.2	15.4
α	19.2	23.1	24	14.1	15.6	18.6	19.2	16.8	12.2	13.8	15.3	15.7
g_m	52.2 ± 9.4	79.3 ± 10.2	49.2 ± 7.6	14.2 ± 3.8	17.1 ± 2.6	18.8 ± 3.6	23.7 ± 6.4	41.2 ± 8.2	42.5 ± 9.1	13.9 ± 3.1	14.5 ± 2.5	15.7 ± 6.1
R_D	0.78	0.71	0.75	0.42	0.52	0.78	0.67	0.66	0.65	0.49	0.5	0.36
R_M	1.35 ± 0.05			1.22 ± 0.11			1.73 ± 0.18			1.44 ± 0.09		
R_M as %R_D	45.6			68.0			77.3			69.7		

Notes: $P_{N,max}$ = net photosynthetic rate at saturating photon flux rate (μmol CO$_2$/m^2/sec); Γ_l = compensation photon flux rate (μmol CO$_2$/m^2/sec); α = photosynthetic efficiency (quanta/mol CO$_2$); g_m = mesophyll conductance (mmol CO$_2$/m^2/sec) measured (mean ± standard error, n = 90); R_D = dark respiration rate (μmol CO$_2$/m^2/sec); R_M = maintenance respiration rate (g CO$_2$/m^2/day) measured (mean ± standard error, n = 5); R_M as % R_D = R_M as percentage of R_D. Γ_l, $P_{N,max}$, R_D are calculated using the mathematical model. UCL leaves and LCL leaves are leaves of the upper canopy layer and lower canopy layer, respectively.

Source: From Marek M, Masarovičová E, Kratochvílová I, Eliáš P, Janouš D. *Trees* 1989; 4: 234–240. With permission.

$$F = \frac{s}{W}$$

The mean leaf area ratio, $\bar{F}$, is given by:

$$F = \frac{(s_2 - s_1)(\ln W_2 - \ln W_1)}{(W_2 - W_1)(\ln s_2 - \ln s_1)}$$

Using the above-mentioned growth parameters, R can be defined as

$$R = E \times F = \frac{dW}{dt} = \frac{1}{s}\frac{dw}{dt}\frac{s}{W}$$

The relative growth rate, therefore, consists of two components, which measure the efficiency of the plant as a producer of dry weight (E) and as a producer of leaf area (F). Leaf area ratio can be redefined as

$$F = \frac{W_1}{W} \times \frac{s}{W_1}$$

where W_1 is the dry weight of the leaves. The two components W_1/W and s/W_1 are called the leaf weight ratio (LWR) and the specific leaf area (SLA), respectively. LWR (kg/kg) measures the leafiness of the plant on the basis of its total dry weight. It also defines the partitioning of dry weight to leaves, a parameter that determines the potential capacity of the plant to support the existing dry weight and to further increase its dry weight through photosynthesis. The SLA (m^2/kg) measures the leafiness of a plant on a dry weight basis. For a given light environment, species with leaves having higher values of SLA (i.e., less carbon invested per unit of area) will have a higher relative growth rate (R).

The reciprocal parameter of SLA, the specific leaf weight (in kg/m^2), is a measure of the weight of leaf material per unit of leaf area. It tends to be positively correlated with leaf thickness. All of the above-mentioned components of growth analysis have been characterized and defined in detail by Beadle [115].

This method of growth analysis was successfully used in the study of *Smyrnium perfoliatum* L., a strongly endangered species of the flora of Slovakia, since this is the only locality where it is believed to be an autochthonous species. The other central European localities are considered to be secondary. *S. perfoliatum*, a conspicuous and aromatic species, is used in natural medicine and in homeopathy, mainly in southern Europe. The plant forms a storage tap root and a rosette of compound leaves in the first year. In the second year (after the growth of the rosette with compound leaves is complete), shoots with three morphologically different types of leaves (basal compound leaves, upper simple amplexicaul leaves, and bracts), inflorescences, and fruits appear.

Relative growth rate and DM partitioning into the shoot and root were established in the second year of ontogenesis. The intensive growth of the inflorescence (sink of assimilates) conspicuously affected DM partitioning in the whole plant. Table 32.20, Table 32.21, and Figure 32.23 present the values of NAR, relative growth rate of both leaf (RGR_L) and shoot (RGR_w), leaf weight ratio, SLA, shoot and root dry weights, and shoot:root (S:R) ratio, which confirm that *S. perfoliatum* L. should be characterized as a fast-growing species that rapidly increases in size and occupies a large space in the early phase of the growing season (for details, see Refs. [116–118]).

IV. CONCLUSIONS

In general, a positive correlation between whole-plant photosynthesis and growth or biomass production has been established. However, growth rates based on CO_2 exchange must take into account respiratory losses. On the other hand, CO_2 exchange (photosyn-

TABLE 32.20
Values of Net Assimilation Rate (NAR) and Relative Growth Rate of Shoot (RGR_w), Root (RGR_R), Leaf Area (RGR_A), and Leaf Dry Weight (RGR_L) of *Smyrnium pefoliatrum* L. in the Growing Season of 1993

Time of Sampling	NAR (g/m^2/day)	RGR_w (g/g/day)	RGR_R (g/g/day)	RGR_A (m^2/m^2/day)	RGL_L (g/g/day)
April 18–30	4.8803	0.0716	0.0266	0.0500	0.0581
April 30 to May 5	4.8281	0.0596	−0.0549	0.0375	−0.0514
May 5–12	−0.0132	−0.0016	−0.0203	0.0050	−0.0191
May 12–21	3.4584	0.0355	0.0320	−0.0033	−0.0059
May 21 to June 6	−2.9775	−0.0246	−0.0394	−0.0320	−0.0379

Source: From Masarovičová E, Lux A, Kobelová G. *Biol. Plant.* 1994; 36(suppl): S283. With permission.

TABLE 32.21
Mean Values of Shoot:Root Ratio (S:R), Leaf Area Ratio (LAR), Leaf Weight Ratio (LWR), Root Weight Ratio (RWR), Specific Leaf Weight (SLW), and Specific Leaf Area (SLA) of *Smymium perfoliatum* L. in the Growing Season of 1993

Time of Sampling	S:R	LAR (m^2/g)	LWR (g/g)	RWR (g/g)	SLW (g/dm^2)	SLA (dm^2/g)
April 18	3.079	0.0129	0.564	0.256	0.295	3.903
April 30	5.526	0.0113	0.540	0.158	0.245	4.119
May 5	12.085	0.0107	0.335	0.097	0.196	5.141
May 12	10.682	0.0114	0.293	0.088	0.206	5.044
May 21	11.239	0.0080	0.206	0.085	0.249	4.028
June 4	12.399	0.0073	0.173	0.069	0.236	4.250

Source: From Masarovičová E, Lux A, Kobelová G. *Biol. Plant.* 1994; 36(suppl): S283. With permission.

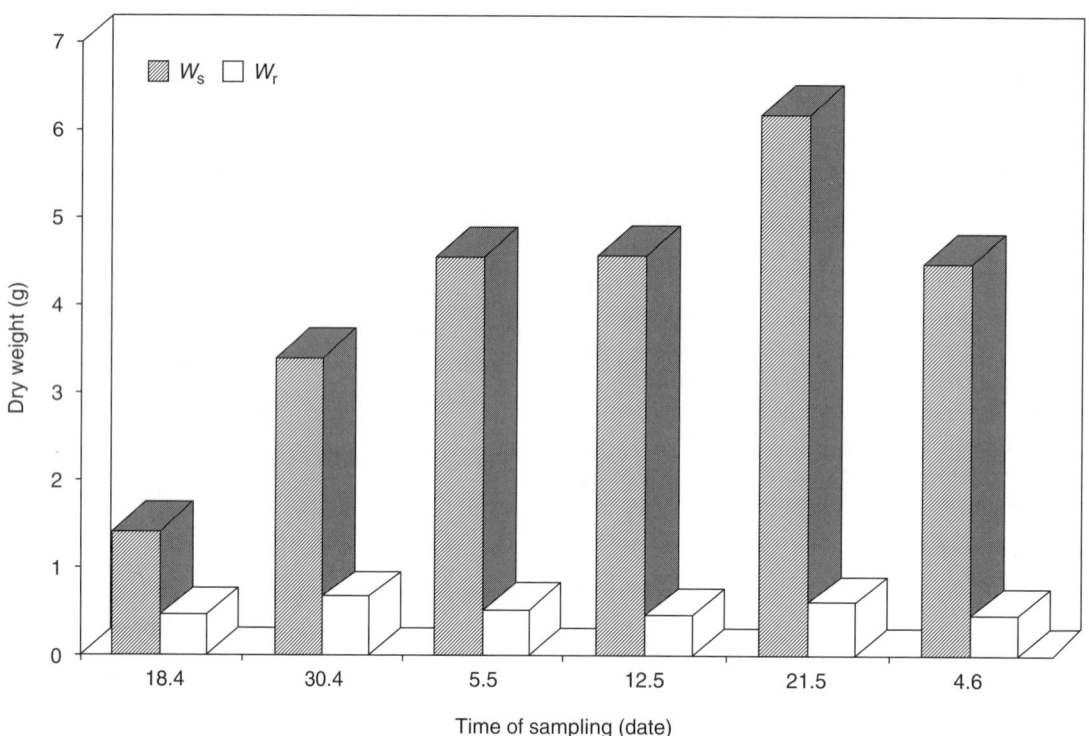

FIGURE 32.23 Values of the shoot dry weight (W_s) and root dry weight (W_r) of *Smymium perfoliatum* L. in the 1993 growing season. (From Masarovičová E, Lux A, Kobelová G. *Biol. Plant.* 1994; 36 (Suppl): S283.)

thesis, respiration) is plant specific and depends on both internal (stage of plant development) and external factors (physical factors of the environment). Therefore, photosynthetic activity can be studied on different levels — from cell, tissue, organ, whole plant, and population to the ecosystem. Depending on the above-mentioned facts, a specific methodical approach to measuring plant photosynthetic activity under controlled conditions or in the field has to be chosen. This chapter provides some basic theoretical knowledge and techniques (methodical approach in the laboratory and in the field) for the study of plant photosynthetic activity.

REFERENCES

1. Šesták Z, Čatský J, Jarvis PG. *Plant Photosynthetic Production, Manual of Methods*, Dr. W Junk NV Publishers, The Hague, 1971.
2. Coombs J, Hall DO, Long SP, Scurlock JMO. *Techniques in Bioproductivity and Photosynthesis*, Pergamon Press, Oxford, 1985.

3. Marshall B, Woodward FI. *Instrumentation for Environmental Physiology*, Cambridge University Press, Cambridge, 1985.

4. Pearcy RWJ, Ehleringer R, Mooney HA, Rundel PW. *Plant Physiological Ecology, Field Methods and Instrumentation*, Chapman and Hall, London, 1989.

5. Hall DO, Scurlock JMO, Bolhár-Nordenkampf HR, Leegood RC, Long SP. *Photosynthesis and Production in a Changing Environment. A Field and Laboratory Manual.* Chapman and Hall, London, 1993.

6. Walker DA. And whose bright presence" — an application of Robert Hill and his reaction. *Photosynth. Res.* 2002; 73: 51–54.

7. Izawa S. Acceptors and donors of chloroplast electron transport. In: San Pietro A, ed. *Methods in Enzymology.* V 69. Photosynthesis and Nitrogen Fixation, Part C. Academic Press, New York, 1980; 413–434.

8. Šeršeň F, Gabunia G, Krejčířová E, Král'ová K. The relationship between lipophilicity of N-alkyl-N, N-dimethylamine oxides and their effects on the thylakoid membranes of chloroplasts. *Photosynthetica* 1992; 26: 205–212.

9. Arnon DI. Copper enzymes in isolated chloroplasts. Polyphenoloxidase in *Beta vulgaris. Plant Physiol.* 1949; 24: 1–15.

10. Král'ová K, Šeršeň F, Kubicová L, Waisser K. Inhibitory effects of substituted benzanilides on photosynthetic electron transport in spinach chloroplasts. *Chem. Pap.* 1999; 53: 328–331.

11. Král'ová K, Loos D, Čižmárik J. Quantitative relationships between inhibition of photosynthesis and lipophilicity of piperidinoethyl alkoxyphenylcarbamates. *Collect. Czech. Chem. Commun.* 1994; 59: 2293–2302.

12. Král'ová K, Šeršeň F, Sidóová E. Photosynthesis inhibition produced by 2-alkylthio-6-R-benzothiazoles. *Chem. Pap.* 1992; 46: 348–350.

13. Norrington FE, Hyde RM, Williams SG, Wooten RJ. Physicochemical activity relationship pratice. 1. A rational and self consistent data bank. *J. Med. Chem.* 1975; 18: 604–607.

14. Clark LC. Monitor and control of blood and tissue oxygen tensions. *Trans. Am. Soc. Artif. Intern. Organs* 1956; 2: 41–48.

15. Buschmann C, Grumbach K. *Physiologie der Photosynthese.* Springer Verlag, Berlin, 1985.

16. Šeršeň F, Král'ová K. Mechanism of inhibitory action of the local anaesthetic trimecaine on the growth of *Chlorella vulgaris* algae. *Gen. Physiol. Biophys.* 1994; 13: 329–335.

17. Král'ová K, Šeršeň F, Melník M. Inhibition of photosynthesis in *Chlorella vulgaris* by Cu(II) complexes with biologically active ligands. *J. Trace Microprobe Tech.* 1998; 16: 491–500.

18. Bartoš J, Berková E, Šetlík I. A versatile chamber for gas exchange measurements in suspensions of algae and chloroplasts. *Photosynthetica* 1975; 9: 395–406.

19. Král'ová K, Šeršeň F, Čižmárik J. Inhibitory effect of piperidinoethylesters of alkoxyphenylcarbamic acids on photosynthesis. *Gen. Physiol. Biophys.* 1992; 11: 261–267.

20. Král'ová K, Šeršeň F, Čižmárik J. Dimethylaminoethyl alkoxyphenylcarbamates as photosynthesis inhibitors. *Chem. Pap.* 1992; 46: 266–268.

21. Gallová J, Uhríková D, Balgavý P. Biphasic effect of local anaesthetic carbisocaine on fluidity of phosphatidylcholine bilayer. *Pharmazie* 1992; 47: 444–448.

22. Král'ová K, Šeršeň F, Melník M, Fargašová A. Inhibition of photosynthetic electron transport in spinach chloroplasts by anti-inflammatory Cu(II) compounds. In: Ondrejovič G, Sirota A, eds. *Progress in Coordination and Organometallic Chemistry.* Slovak Technical University Press, Bratislava, 1997; 233–238.

23. Šeršeň F, Král'ová K, Blahová M. Photosynthesis of *Chlorella vulgaris* as affected by diaqua(4-chloro-2-methyl-phenoxyacetato)copper(II) complex. *Biol. Plant.* 1996, 38: 71–75.

24. Král'ová K, Šeršeň F, Bláhová M. Effects of Cu(II) complexes on photosynthesis in spinach chloroplasts. Aqua(aryloxyacetato)copper(II) complexes. *Gen. Physiol. Biophys.* 1994; 13: 483–491.

25. Hoff AJ. Application of ESR in photosynthesis. *Phys. Rep.* 1979; 54: 75–200.

26. Svenson B, Vass I, Styring S. Sequence analysis of the D_1 and D_2 reaction center proteins of photosystem II. *Z. Naturforsch.* 1991; 46: 765–776.

27. Král'ová K, Šeršeň F, Mitterhauszerová L', Krempaská E, Devínsky F. Effect of surfactants on growth, chlorophyll content and Hill reaction activity. *Photosynthetica* 1992; 26: 181–187.

28. Mitterahuszerová L, Král'ová K, Šeršeň F, Blanáriková V, Csöllei J. Effects of substituted aryloxyaminopropanols on photosynthesis and photosynthesizing organisms. *Gen. Physiol. Biophys.* 1991; 10: 309–319.

29. Chen Z, Spreitzer RJ. How various factors influence the CO_2/O_2 specificity of ribulose-1,5-bisphosphate carboxylase/oxygenase. *Photosynth. Res.* 1992; 31: 157–164.

30. Hatch MD, Boardman NK. *The Biochemistry of Plants*, Academic Press, New York, 1987.

31. Ghashghaie J, Cornic G. 1994. Effects of temperature on partitioning of photosynthetic electron flow between CO_2 assimilation and O_2 reduction and on the CO_2/O_2 specificity of Rubisco. *J. Plant Physiol.* 1994; 143: 643–650.

32. Bowes, G. Growth at elevated CO_2: photosynthetic responses mediated through Rubisco. *Plant Cell Environ.* 1991; 14: 795–806.

33. Stitt M. Rising CO_2 levels and their potential significance for carbon flow in photosynthetic cells. *Plant Cell Environ.* 1991; 4: 741–762.

34. Ceulemans R, Mousseau M. Effects of elevated atmospheric CO_2 on woody plants. *New Phytol.* 1994; 127: 425–446.

35. Makino A. Biochemistry of C_3-photosynthesis in high CO_2. *J. Plant Res.* 1994; 107: 79–84.

36. Konečná B, Frič F, Masarovičová E. Ribulose-1, 5-bisphosphate carboxylase activity and protein content in pollution damaged leaves of three oak species. *Photosynthetica* 1989; 23: 566–574.

37. Stiborová M, Doubravová M, Březinová A, Fridrich A. Effect of heavy metal ions on growth and biochemical characteristics of photosynthesis of barley (*Hordeum vulgaris* L.) *Photosynthetica* 1986; 20: 418–425.

38. Stiborová M, Doubravová M, Leblová S. A comparative study of the effect of heavy metal ions on ribulose-1,5-bisphosphate carboxylase and phosphoenolpyruvate carboxylase. *Biochem. Physiol. Pflanzen.* 1986; 181: 373–379.

39. Gezelius K. Ribulose bisphosphate carboxylase, protein and nitrogen in Scots pine seedlings cultivated at different nutrient levels. *Physiol. Plant.* 1986; 68: 245–251.

40. Gezelius K, Widell A. Isolation of ribulose bisphosphate carboxylase-oxygenase from non-hardened and hardened needles of *Pinus sylvestris*. *Physiol. Plant.* 1986; 67: 199–204.

41. Bradford M. A rapid and sensitive method for the quantitation of microgram quantities of protein utilising the principle of protein-dye binding. *Anal. Biochem.* 1976; 72: 248–254.

42. D' Ambrosio N, Schindler CH, De Santo AV, Lichtenthaler H. Carotenoid composition in green leaf and stem tissue of the CAM-plant *Cissus quinquangularis* Chiov. *J. Plant Physiol.* 1994; 143: 508–513.

43. Tuba Z, Lichtenthaler HK, Csintalan Z, Pocs T. Regreening of desiccated leaves of the poikilochlorophyllous *Xerophyta scabrida* upon rehydration. *J. Plant Physiol.* 1993; 142: 103–108.

44. Willett WC. Vitamin A and lung cancer. *Nutr. Rev.* 1990; 48: 201–211.

45. Cadenas E. Biochemistry of oxygen toxicity. *Annu. Rev. Biochem.* 1989; 58: 79–110.

46. Taiz L, Zeiger E. *Plant Physiology*. The Benjamin/Cummings Publishing Company, Inc., Redwood City, CA, 1991.

47. Peto R, Doll R, Buckley JD, Sporn MB. Can dietary beta-carotene materially reduce human cancer rates? *Nature* 1981; 290: 201–208.

48. Škoda B. Contribution to the biochemical taxonomy of the genus *Chlorella* BEIJERINCK sensu lato — pigment composition. *Algol. Stud.* 1992; 53: 19–35.

49. Krasnovská E, Masarovičová E, Hindák F. Pigment composition of six xanthophycean algae and *Scenedesmus quadricauda*. *Biologia (Bratisl.)* 1994; 4: 501–509.

50. Hall DO, Scurlock JMO, Bolhár-Nordenkampf HR, Leeggod RC, Long SP. *Photosynthesis and Production in a Changing Environment, A Field and Laboratory Manual*, Chapman and Hall, London, 1993.

51. Wellburn AR. The spectral determination of chlorophylls *a* and *b*, as well as total carotenoids, using various solvents with spectrophotometers of different resolution. *J. Plant Physiol.* 1994; 144: 307–313.

52. Haspelová-Horvatovičová A. *Assimilation Pigments in the Healthy and Diseased Plants* (in Slovak), Veda, Publishing House of the Slovak Academy of Sciences, Bratislava, 1981.

53. Lichtenthaler HK, Chlorophylls and carotenoids: pigments of photosynthetic biomembranes. *Methods Enzymol.* 1987; 148: 350–382.

54. Kalasz H, Ettre LS. *Chromatography*, Akademiai Kiadó, Budapest, 1988.

55. Snyder LR, Kirkland JJ. *Introduction to Modern Liquid Chromatography*, J. Wiley and Sons, New York, 1979.

56. Wright SW, Shearer JD. Rapid extraction and high-performance liquid chromatography of chlorophylls and carotenoids from marine phytoplankton. *J. Chromatogr.* 1984; 294: 281–295.

57. Rowan KS. *Photosynthetic Pigments of Algae*, Cambridge University Press, Cambridge, 1989.

58. Minguez-Mosquera MI, Gandul-Rojas B, Montano-Asquerino A, Garrido-Fernandez J. Determination of chlorophylls and carotenoids by high-performance liquid chromatography during the olive lactic fermentation. *J. Chromatogr.* 1991; 585: 259–266.

59. Masarovičová E. *Gasometrical Investigation into CO₂ Exchange of the Fagus sylvatica L. Species under Controlled Conditions*, Veda, Publishing House of the Slovak Academy of Sciences, Bratislava, 1984.

60. Barden JA, Love JM, Porpiglia PJ, Marini RP, Caldwell JD. Net photosynthesis and dark respiration of apple leaves are not affected by shoot detachment. *HortScience* 1980; 15: 595–597.

61. Masarovičová E. Comparative study of growth and carbon uptake in *Fagus sylvatica* L. trees growing under different light condition. *Biol. Plant.* 1988; 30: 2285–293.

62. Masarovičová E. Leaf shape, stomatal density and photosynthetic rate of the common oak leaves. *Biol. Plant.* 1991; 33: 495–500.

63. Masarovičová E. Morphological, physiological, biochemical and productional characteristics of three oak species. *Acta Physiol. Plant.* 1992; 14: 99–106.

64. Masarovičová E, Eliáš P. Seasonal changes in the photosynthetic response of *Mercurialis perennis* plants from different light regime conditions. *Biol. Plant.* 1985; 27: 41–50.

65. Masarovičová E. Interspecific and intraspecific differences of physiological and growth characteristics in Slovakian autochtonous oak species. *Bot. Közlem.* 1993; 80: 61–72.

66. Masarovičová E, Eliáš P. Photosynthetic rate and water relations in some forest herbs in spring and summer. *Photosynthetica* 1986; 20: 187–195.

67. Masarovičová E, Štefančík L. Some ecophysiological features in sun and shade leaves of tall beech trees. *Biol. Plant.* 1990; 32: 374–387.

68. Thornley JHM. *Mathematical Models in Plant Physiology*, Academic Press, London, 1976.

69. Marshall B, Biscoe PV. A model for C₃ leaves describing the dependence of net photosynthesis on irradiance. I. Derivation. *J. Exp. Bot.* 1980; 31: 29.

70. Lieth JH, Reynolds JF. The nonrectangular hyperbola as a photosynthetic light response model: geometrical interpretation and estimation of the parameter. *Photosynthetica* 1987; 21: 363–366.

71. Leverenz JW. Chlorophyll content and the light response curve of shade-adapted conifer needles. *Physiol. Plant.* 1987; 71: 20–29.

72. Cornic G, Jarvis PG. Effect of oxygen on CO_2 exchange and stomatal resistance in Sitka spruce and maize at low irradiances. *Photosynthetica* 1972; 6: 225–239.

73. Sharp RE, Matthews MA, Boyer JS. Kok effect and the quantum yield of photosynthesis: light partially inhibits dark respiration. *Plant Physiol.* 1984; 75: 95–101.

74. Terashima I, Saeki T. A new model for leaf photosynthesis incorporating the gradients of light environment and of photosynthetic properties of chloroplasts within a leaf. *Ann. Bot.* 1985; 6: 489–499.

75. Tooming H. Mathematical description of net photosynthesis and adaptation processes in the photosynthetic apparatus of plant communities. In: Šetlík I, ed. *Prediction and Measurements of Photosynthetic Productivity*, PUDOC, Wageningen, 1970; 103–113.

76. Tooming H. *Ecological Principles of the Maximal Productivity of Plant Stands* (in Russian), Gidrometeoizdat, Leningrad, 1984.

77. Jarvis PG, Sandford AP. Temperate forests. In: Baker NR, Long SP, eds. *Photosynthesis in Contrasting Environments*, Elsevier Science Publ. B. V., Amsterdam, 1986; 199–236.

78. Masarovičová E. Relationship between the CO_2 compensation concentration, the slope of CO_2 curves of net photosynthetic rate and energy of irradiance. *Biol. Plant.* 1979; 27: 434–439.

79. Kotvalt V, Hák R. Method of mathematical estimation of CO_2 response curve parameters based on closed system measurement. *Photosynthetica* 1987; 21: 92–95.

80. Klimo E, Materna J. *Verification of Hypotheses on the Mechanisms of Damage and Possibilities of Recovery of Forest Ecosystem*, University of Agriculture, Brno, 1990.

81. Čatský J, Tichá I. A closed system for measurement of photosynthesis, photorespiration and transpiration rates. *Biol. Plant.* 1975; 17: 405–410.

82. Masarovičová E. Photosynthesis, photorespiration and mitochondrial respiration of *Fagus sylvatica* L. seedlings: effects of temperature and oxygen concentration. *Photosynthetica* 1980; 14: 321–325.

83. Masarovičová E. Seasonal curves of mitochondrial respiration and photorespiration of European beech seedlings. Biologia, Bratislava, 1981; 36: 833–839.

84. Dickmann DI, Gjerstad DH, Gordon JC. Developmental pattern of CO_2 exchange, diffusion resistance and protein synthesis in leaves of *Populus euramericana*. In: Marcelle R, ed. *Environmental and Biological Control of Photosynthesis*. Dr. W. Junk NV Publishers, The Hague, 1975: 171–181.

85. Julinyová M., Masarovičová E. Empirical model of carbon dioxide exchange in young European beech plants under controlled conditions. *Ekológia* 1984; 3: 149–158.

86. Farquhar GD, von Caemmerer S, Berry JA. A biochemical model of photosynthetic CO_2 assimilation in the leaves of C_3 species. *Planta* 1980; 749: 78–90.

87. Lawlor DW, Pearlman JG. Compartmental modelling of photorespiration and carbon metabolism of water stressed leaves. *Plant Cell Environ.* 1981; 4: 37–52.

88. Marshall B, Biscoe PV. A model for C_3 leaves describing the dependence of net photosynthesis on irradiance. I. Derivation. *J. Exp. Bot.* 1980; 31: 29–39.

89. Tenhunen JD, Meyer A, Lange OL, Gates DM. Development of a photosynthesis model with an emphasis on ecological applications. V. Test of the applicability of a steady-state model to description of net photosynthesis of *Prunus armeniaca* under field condition. *Oecologia* 1980; 45: 147–155.

90. Hunt WF, Loomis RS. Respiration modelling and hypothesis testing with a dynamic model of sugar beet growth. *Ann. Bot.* 1979; 44: 5–17.

91. Reynolds JF, Cunningham GL, Syvertsen JP. A net CO_2 exchange for *Larrea tridentata*. *Photosynthetica* 1979; 13: 279–286.

92. Ågren GI, Axelson B. Population respiration: a theoretical approach. *Ecol. Modell.* 1980; 11: 39–54.

93. Parsons AJ, Robson M.J. Seasonal changes in the physiology of 24 perennial ryegrass (*Lolium perenne* L.). 2. Potential leaf and canopy photosynthesis during the transition from vegetative to reproductive growth. *Ann. Bot.* 1981; 47: 249–258.

94. Troeng E. *Some Aspects on the Annual Carbon Balance of Scots Pine*, Sveriges Lantbruksuniversitet, Uppsala, 1981.

95. Ondok JP, Gloser J. The photosynthetic model of the reed stand of *Phragmites communis* Trin (in Czech). *Acta Ecol.* 1978; 8: 43–69.

96. Šesták Z. Thin layer chromatography of chlorophylls. Part 1. *Photosynthetica* 1967; 1: 269–292.

97. Šesták Z. Thin layer chromatography of chlorophylls. Part 2. *Photosynthetica* 1982; 16: 568–617.

98. Heftmann E. *Fundamentals and Applications of Chromatographic and Electrophoretic Methods*, Elsevier, Amsterdam, 1983.

99. Masarovičová E, Duda M. Some ecophysiological aspects of quantitative changes of carotenoids in the leaves of *Pulmonaria officinalis* L. *ssp. maculosa* (Hayne) Gams. *Biologia (Bratisl.)* 1976; 31: 15–23.

100. Vernon LP. Spectrophotometric determination of chlorophylls and pheophytins in plant extracts. *Anal. Chem.* 1960; 32: 1144–1150.

101. Masarovičová E, Eliáš P. Chlorophyll content in leaves of plants in an oak-hornbeam forest. 1. Herbaceous species. *Photosynthetica* 1980; 14: 580–588.

102. Masarovičová E,. Eliáš P. Chlorophyll content in leaves of plants in an oak-hornbeam forest. 2. Shrub species. *Photosynthetica* 1981; 15: 16–20.

103. Eliáš P, Masarovičová E. Chlorophyll content in leaves of plants in an oak-hornbeam forest. 3. Tree species. *Photosynthetica* 1980; 14: 604–610.

104. Eliáš P, Masarovičová E. Chlorophyll content in leaves of plants in an oak-hornbeam forest. 4. Amounts per stand area unit. *Photosynthetica* 1985; 19: 49–55.

105. Eliáš P, Masarovičová E. Vertical distribution of leaf-blade chlorophylls in a deciduous hardwood forest. *Photosynthetica* 1985; 19: 43–48.

106. Landsberg JJ. *Physiological Ecology of Forest Production*, Academic Press, London, 1986.

107. Eliáš P, Kratochvílová I, Janouš D, Marek M, Masarovičová E. Stand microclimate and physiological activity of tree leaves in an oak-hornbeam forest. I. Stand microclimate. *Trees* 1989; 4: 227–233.

108. Marek M, Masarovičová E, Kratochvílová I, Eliáš P, Janouš D. Stand microclimate and physiological activity of tree leaves in an oak-hornbeam forest. II. Leaf photosynthetic activity. *Trees* 1989; 4: 234–240.

109. Huzulák J, Eliáš P. Contribution to the water regime in *Quercus cerris*. In: Biskupský V, ed. *Research Project Báb*. Progr. Rep. II. Veda, Publishing House of the Slovak Academy of Science, Bratislava, 1975; 503–511.

110. Palovský R., Hák R. A model of light-dark transition of CO_2 exchange in the leaf (post-illumination burst of CO_2). Theoretical approach. *Photosynthetica* 1988; 22: 423–430.

111. Šetlík I. *Prediction and Measurements of Photosynthetic Productivity*, PUDOC, Wageningen, 1970.

112. Challa H. *An Analysis of the Diurnal Course of Growth, Carbon Dioxide Exchange and Carbohydrate Reserve Content of Cucumber*, PUDOC, Wageningen, 1976.

113. Šesták Z, Čatský J, Jarvis PG. *Plant Photosynthetic Production, Manual of Methods*, Dr. W Junk NV Publishers, The Hague, 1971.

114. Hunt R. *Plant Growth Analysis*. E. Arnold Publishers, London, 1978.

115. Beadle CL. Growth Analysis. In: Hall DO, Scurlock JMO, Bolhár-Nordenkampf HR, Leegood RC, Long SP, eds. *Photosynthesis and Production in a Changing Environment, A Field and Laboratory Manual*. Chapman and Hall, London, 1993; 36–46.

116. Lux A, Masarovičová E, Hudák J. Physiological and structural characteristics of root and shoot in *Smyrnium perfoliatum* L. (Apiaceae). 4th International Symposium on Structure and Function of Roots, Stará Lesná, Slovak Republic, June 20–26, 1993.

117. Lux A, Masarovičová E, Oláh R. Structural and physiological characteristics of the tap root of *Smyrnium perfoliatum* L. (Apiaceae). In: Baluška F, Čiamporová M, Gašparíková O, Barlow P, eds. *Structure and Function of Roots*, Kluwer Academic Publishers, The Netherlands, 1995; 99–105.

118. Masarovičová E, Lux A, Kobelová G. Growth and dry mass partitioning in *Smyrnium perfoliatum L. Biol. Plant*. 1994; 36 (suppl): S283.

33 Analysis of Photosynthetic Pigments: An Update

Martine Bertrand
Institut National des Sciences et Techniques de la Mer,
Conservatoire National des Arts et Métiers

José L. Garrido
Instituto de Investigacións Mariñas

Benoît Schoefs
Dynamique Vacuolaire et Résponses aux Stress de l'Environnement,
UMR CNRS 5184/INRA 1088/Université de Bourgogne Plante-Microbe-Environnement,
Université de Bourgogne à Dijon

CONTENTS

I. INTRODUCTION

The molecules that appear colored to our eyes are named pigments. Among them are photosynthetic pigments. They include at least 50 chlorophylls (Chls), their precursors and derivatives (green tetrapyrrole rings), 600 carotenoids (yellow to red isoprenoids), and 10 phycobilins (red and blue open tetrapyrroles). Usually, several kinds of pigments coexist *in situ*, and the most abundant ones determine the color. A demonstrative example is that of endive. The top of leaves of dark-grown plants is yellow due to the presence of carotenoids; after some hours of illumination, they turn green; this change in color is due to the fact that Chl has accumulated faster than carotenoids.

Color is a trading argument for fresh food but also for processed products — who would buy yellowish spinach or brown tomato puree? In some processed products, pigments are present as additives to give the color expected by the customer (e.g., β-carotene in margarine, paprika in sweets). The up-to-date functional food contains additives known for their health benefits; among them one finds natural pigments such as β-carotene or lycopene as antioxidants.

In addition to protocols developed in basic research, pigment extraction and analyses are routinely performed by food and feed industries and phamaceutical industries as well as by control services [1,2]. Due to the appearance of new pigments, on the one hand, and to technological progress, on the other, pigment analysis is continuously necessary to increase our knowledge on the topic.

In the first edition of *Handbook of Photosynthesis* we presented a chapter entitled "Working with Photosynthetic Pigments : Problems and Precautions" [3]. The present contribution complements this chapter as it details pigment analysis, relating some examples of the past years.

II. NONINVASIVE MEASUREMENTS

The global color of a plant organ is strongly influenced by its physiological status, like greening, ripening, senescence, etc. Therefore, to characterize these biological processes, it is important to study the changes in pigment composition.

A. COLOR MEASUREMENT

The setup of reflected-light colorimeters has allowed us to measure the external color that the human eye perceives (for a complete description of color measurement principle, see Refs. [4,5]). To characterize the influence of different kinds of drying on the color of basil leaves, Di Cesare et al. [6] used a chromameter. The greener leaves were obtained with microwave dried samples. Quantitative high-performance liquid chromatography (HPLC) Chl measurements indicated that Chl is best preserved during this drying procedure. Carotenoids are interesting flavor precursors. Therefore, they contribute to the organic properties of food product such as wine. By color measurements of grapes, Razungles et al. [7] established a correlation between a change in color and a

decrease of carotenoids quantified by HPLC. As it appears from the above two examples, a chromatographic analysis is usually necessary to correlate color measurement with pigment nature and content. Other noninvasive methods such as *in situ* absorbance or fluorescence spectroscopy can be used to get indications of the pigment composition (see Sections II.B and II.C).

B. ABSORBANCE SPECTROSCOPY

The principle of absorbance spectrophotometry is described elsewhere [8,9]. As this spectroscopic method requires that light travels through matter, the sample must be thin (e.g., a leaf) or not too concentrated (e.g., a suspension of microalgae). The information contained in *in vivo* absorbance spectra is usually rather poor. However, this spectroscopy is still used for samples where transformation of pigments is expected, for instance, transformation of protochlorophyllide to chlorophyllide from etiolated leaves, or to follow the quantitative evolution of protochlorophyllide forms [10]. What is now more often found in the literature is the absorbance difference between two spectra relating to slight changes in a sample. Using this method, Bertrand et al. [11] detected the inhibition of the xanthophyll cycle in diatoms by cadmium. The change in carotenoid composition was confirmed by further pigment identification and quantification.

C. FLUORESCENCE SPECTROSCOPY

Fluorescence consists of radiations emitted during the de-excitation of pigments that have been excited by absorption of visible or UV photons. The particular tetrapyrrole structure of Chls and phycobilins makes these pigments fluoresce [12]. Technological progress has led to set-up devices for recording images, spectra, and kinetics of fluorescent objects. As fluorescence intensity is low, it is necessary to perform measurements in the dark. It has been established that the Chl fluorescence yield varies at room temperature during an actinic illumination (reviewed in Refs. [13,14]; see Section II.C.3). Therefore, when different samples have to be compared, it is always necessary to verify that the results have been obtained under similar conditions.

1. Imaging

With the development of the CCD camera, we have now the opportunity to visualize the global fluorescence emission of an organ such as a leaf or a fruit. Not only pigments fluoresce in plants, but Buschmann et al. [14] mentioned, for instance, ferulic acids. These molecules emit blue-green photons when Chls are present. Imaging allows us to detect the presence of a fluorescing compound and its location within a cell or an organ. This new dimension of measurement differs from the classical one used during the last decades, which consisted of focusing on a single leaf point or cell [3]. Fluorescence imaging has many applications. For instance, Nedbal et al. [15] developed a strategy to measure the decrease of Chl fluorescence from lemons during ripening. On the basis of fluorescence data, they defined robust parameters allowing the prediction of damage at the lemon surface before visible signs appear. This methodology can also be applied to trace Chl in highly colored plant tissues (e.g., red tomatoes). The development of Chl fluorescence imaging in fields other than agriculture is very promising too, as it has been already used to study pollutions and even to visualize the stress induced by insect footsteps on leaves [16].

2. Spectra

Fluorescence spectra have been in use since a long time. The advantage of fluorescence spectroscopy over absorbance spectroscopy is twofold: first, the investigated molecule can be excited selectively and second, fluorimetry is more sensitive than absorbance.

The main applications of *in situ* fluorescence spectroscopy are the characterization of the state of the photosynthetic apparatus [17–19] and the identification of the different spectral forms of Chl precursors [10,20,21]. Authors have defined parameters to characterize physiological states. For instance, the Chl fluorescence ratio F_{690}/F_{730} in green leaves is used to follow the biogenesis of the photosynthetic apparatus. This ratio decreases during greening and development of leaves [22] and increases during the autumnal Chl breakdown [23].

3. Kinetics

Fluorescence kinetics is the most popular method to study the photochemistry of photosynthesis. Because the chlorophyll fluorescence yield is influenced by the photosynthetic activities, fluorescence kinetics reveals the health status of the photosynthetic apparatus (reviewed in Refs. [13,24]). To relate the changes to a particular aspect of photosynthesis, several parameters, such as photochemical quenching and nonphotochemical quenching, have been defined (reviewed in Refs. [13,25]). For instance, a decrease in the combinations between kinetics and imaging is possible so that F_0 relaxation time can be correlated to the inhib-

ition of the xanthophyll cycle by cadmium in diatoms, while other parameters remain unaffected [11].

III. INVASIVE MEASUREMENTS: ANALYTICAL METHODS

Pigment analysis is a biochemical obligatory step when details on the pigment composition are needed during biosynthetic pathways [26,27], secondary metabolism [28–30], and degradation pathways [31–33] or when the effects of pollutants or xenobiotics [34,35] are studied.

A. General Precautions and Considerations

Because of the presence of numerous double bonds, pigments are very sensitive to light, high temperature, oxygen, and acids. Therefore, it is recommended to maintain them in darkness, at low temperature, under nitrogen, and in slightly alkaline conditions [3]. Sometimes compounds are added in order to stabilize pigments: sodium or magnesium carbonate for Chl [6,36,37] or 2,6-di-*tert*-butyl-4-methylphenol for carotenoids[38]. When working with a mixture of pigments, it is advisable to test the possible negative effects of the added compounds on the pigments under study.

B. Extraction

To extract photosynthetic pigments, the operator may consider two facts: the hardness of the sample (e.g., seeds > leaves) and the relative polarity of pigments (e.g., phycobilins > Chls > carotenes). Numerous methods are described in the literature. They can be divided into (1) mechanical modes (e.g., grinding in a mortar [7,39], ball mill or glass beads [36], French press [12,40,41], sonication [42], osmotic shock [43,44]); (2) chaotropic treatments, e.g., repeated freezing–thawing [45,46]; and (3) chemical ways: use of organic solvents [35,47,48]. Sometimes different methods are even combined. Among these methods, it is difficult to know which is the best one, and several tests should be performed (e.g., [28,46]). In fact, there is no reference protocol, except the one used in hydrobiology for Chl *a* using a 90% ethanol/water (v/v) mixture [47]; however, this has been criticized by Papista et al. [48], who argued that ethanol/water mixture ensures a poor extraction yield in the case of numerous alga taxa. According to Skidmore et al. [49] and Lean and Pick [50], methanol/water mixtures eventually containing dimethylsulfoxide are preferred because their extraction power is higher. Kopecky et al. [28] studied the effects of the solvents and the initial sample size on the extraction yield of secondary carotenoids from algae. Dimethylsulfoxide was found to be the best one, just ahead of tetrahydrofuran, and far from 90% acetone. The highest amount of extracted carotenoids was obtained when the initial biomass was below 8 mg dry weight. Ideally, this kind of verification should be systematically performed before quantitative analysis. This is especially true when the pigments are contained in a particular matrix or have undergone treatments that can affect the extractability of pigments. It is, for instance, the case for carotenoids from frozen carrot [51] or high-pressure processed tomato [52]. Differences in Chl extraction have also been reported in heat-treated spinach leaves [31] or dry basil leaves [6]. The extraction protocol should be modified to get the same extraction yield as in the absence of treatment (for a review, see Ref. [2]).

C. Quantification of a Pigment in a Crude Extract

Once extracted, pigments in a mixture can usually be characterized and quantified by absorbance or fluorescence spectroscopy without further analysis. For this purpose equation sets have been designed in order to calculate the concentration of each chromophore [3,53]. When using fluorescence, care should be brought to ensure that the pigment concentration is low enough so that the emission intensity is proportional to the concentration of the investigated molecules.

D. Separation by Chromatography

Photosynthetic pigments differ in their polarity. This property is used to separate them by adsorption chromatography with a polar or nonpolar stationary phase. To optimize the separation, the mobile phase consists of a mixture of two to three solvents of different polarities.

1. Low-Pressure Chromatography

Open column chromatography (OCC) is adapted for a coarse separation from a crude concentrated extract of some milliliters [38,51]. The collected fractions can be submitted to thin layer chromatography (TLC) or HPLC. Low-pressure chromatography is well adapted to preparative purification of pigments. Bermejo et al. [43] used it to isolate B-phycoerythrin. For such a polar molecule (actually a protein–pigment complex), the developer is a buffer. The purity of the phycobiliproteins can be tested using sodium dodecylsulfate polyacrylamide gel electrophoresis [43,54].

2. Thin Layer Chromatography

TLC can separate different kinds of Chls and carotenoids (for a review, see Refs. [55,56]). TLC is especially adapted to mixtures containing only a few pigments, adequately concentrated in a few microliters. The main advantages of this method over OCC is that it is rapid, inexpensive, and allows the separation of numerous samples at the same time in strictly comparable conditions. In that optic, Kopecky et al. [28] compared the secondary carotenoids from 25 stressed microalgae. The plates were developed by a three-stage procedure differing in proportions of hexane/acetone/2-propanol (Figure 33.1). A similar procedure was used to identify the reaction products of putative enzymes of the carotenoid bixin pathway overexpressed in *Escherichia coli*. These experiments allowed Bouvier et al. [27] to establish the sequence of reactions involved in this biosynthetic pathway.

The greatest disadvantage of TLC is its poor resolution. In some cases this can be overcome using chemical derivatization of pigments. An example of this procedure is detailed by Guerra-Vargas et al. [51], who wanted to identify carotenoids differing in the number of hydroxyl groups from canned pickled peppers and carrots. An efficient separation was obtained after methylation of allylic hydroxyl groups and acetylation of primary and secondary hydroxyl groups. The modified molecules had higher R_f and were separated according to the number of acetyl and methyl groups added. Another way to increase the resolution consists in the use of HPTLC. This quantitative method is carried out on layers composed of particles with a smaller diameter (5 μm compared to 12 to

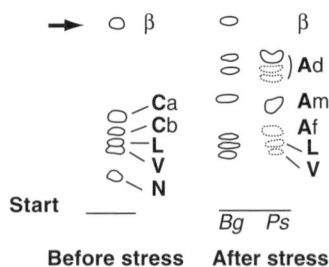

FIGURE 33.1 Thin layer chromatograms of pigments from *Bracteacoccus grandis* (*Bg*) and *Pleurastrum sarcinoideum* (*Ps*) before (left) and after (right) stress (combination of light and nitrogen starvation). The arrow indicates the solvent front. Abbreviations: Ad = astaxanthin diester, Af = free astaxanthin, Am = astaxanthin monoester, β = β-carotene, Ca = chlorophyll *a*, Cb = chlorophyll *b*, L = lutein, N = neoxanthin, V = violaxanthin. (Adapted from Kopecky J, Schoefs B, Loest K, Stys D, Pulz O. *Algolog. Stud.* 2000; 98:153–168. With permission.)

20 μm for conventional TLC), therefore offering greater separation efficiency, faster separation, and improved detection limits [56]. Consequently it is also more expensive.

When using TLC, the analyst must be aware of possible artifacts. For instance, residual chlorine ions on a plate can be transferred during chromatography from the plate to Chl [57]. During carotenoid analysis, rearrangement of epoxide–furanoxide in the presence of NaOH is the major cause of artifact formation [58].

3. High-Performance Liquid Chromatography

The particle size of the stationary phase is as small as the ones of HPTLC, and therefore a better resolution power than that with classical TLC is obtained. This is obvious when the TLC (three spots) and HPLC (12 peaks) chromatograms of the pigments from pumpkin seed oil are compared [59]. Today, HPLC is widely used for pigment separation, and numerous protocols are available in the literature (reviewed in Refs. [1,2,60]). Before injection of a sample, it is advisable to clean and equilibrate the column by running the eluent, which is used at the beginning of the elution programme. Usually pigment analysis requires an absorbance detector, which most frequently is hyphenated to the HPLC column. The fact that detectors present a good sensitivity explains that low concentrations of analytes can be detected. In some cases, it is interesting to use fluorescence detectors.

Reversed phases (RPs) are especially adapted to pigment separation [61] as they allow chlorophylls and carotenoids to develop more interactions with the phase. Among the RPs, octadecyl-bonded stationary phases (C18) are the most used. Stecher et al. [62] compared different conditions of mobile phase, temperatures, and flow rates for the separation of carotenoids by RP C18-HPLC. In order to separate the numerous cis–trans-carotenoid isomers, a C30 RP was created. Using this material, Lee et al. [63] separated 25 carotenoids from a sweet orange within 40 min due to a ternary gradient elution. This material seems particularly suitable to separate the numerous carotenoid isomers [64]. Monolithic stationary phases have attracted considerable attention in the sphere of liquid chromatography in recent years due to their simple preparation procedure, unique properties, and excellent performance (for reviews, see Refs. [65–68]). A monolithic column consists of "one piece of solid that possesses interconnected skeletons and interconnected flow paths through the skeletons." This unique architecture allows shorter retention times compared to particle-packed columns. Svec [67] listed the stationary phases that are now

commercially available. Garrido et al. [69] reported the resolution of eight Chls and derivatives in less than 5 min through a monolithic silica C18 column (Figure 33.2).

The changes in phase manufacturing as well as the discovery of new pigments force researchers to modify older elution programs or develop new ones to improve pigment separation, for example, zeaxanthin/lutein [70], α-/β-carotene/pheophytin [70], cis- and trans-isomers of α-/β-carotenes [71], Chl b allomers [72], mono- and divinyl Chl forms [73].

Numerous problems may arise using HPLC. A troubleshooting guide is summarized in Ref. [61]. It is essential to take time at each step when working with HPLC or testing a new protocol. For example, water or aqueous solutions added to acetone or methanol extract, to avoid distortion of early-eluting pigments due to differences in solvent viscosity, can produce losses of the most nonpolar pigments [74].

E. Molecular Identification and Quantification

Parameters such as R_f for TLC and the selectivity factor ($\log k'$) for HPLC have been defined to characterize the relative polarity of pigments. However, the determination of the nature and the quantification of the separated pigment require additional measurements.

Purified pigments can be readily identified from their absorbance or fluorescence spectrum. Of course, a comparison of spectra with standards and data from the literature is necessary to come to a conclu-

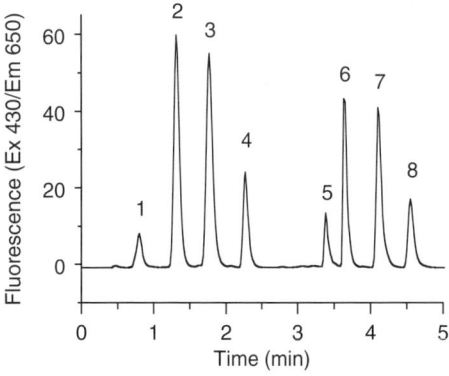

FIGURE 33.2 Fluorescence chromatogram of standards of chlorophylls a and b and their derivatives after their HPLC separation using a monolithic silica C18 column and a pyridine-containing mobile phase. Peaks: 1 = chlorophyllide b; 2 = chlorophyllide a; 3 = pheophorbide b; 4 = pheophorbide a; 5 = chlorophyll b; 6 = chlorophyll a; 7 = pheophytin b; 8 = pheophytin a. (Adapted from Garrido JL, Rodriguez F, Campana E, Zapata M. *J. Chromatogr.* 2003; A994:85–92. With permission.)

sion. When a molecule is not clearly identified by this way or if additional structural information is needed, other methods should be used such as infrared (IR), circular dichroism, Raman, or nuclear magnetic resonance (NMR) spectroscopy, or mass spectrometry (MS). The use of these techniques should then provide specific information on particular bounds, functional groups, or radicals, and therefore help in structural elucidation. For instance, the complete identification of the 13 carotenoids from passion fruit was only possible after analysis by electron impact (EI) MS, UV–visible absorbance spectroscopy, and ^{1}H and ^{13}C NMR spectroscopy of HPLC-separated pigments [75]. When possible, hyphenated devices should be used to allow a complete online analysis that reduces risks of alteration and interpretation errors. All these methods may be also used to detect adulteration in food products [76].

1. Standards

More and more pigments, such as authentic standards, can be easily purchased, but some, such as protochlorophyll a esters [26,77], metal-free pheophytins and some metalloporphyrin analogs [78], neochrome [7], and β-citraurin [75], should be prepared. When standards are stored dried, under nitrogen (or better, argon) at low temperature (below −30°C), they are stable for months.

2. UV–Visible Absorbance Spectroscopy

The identification of pigments is based on the wavelengths of maximal absorbance in the UV–visible region and on the overall shape of the spectrum. Quantification requires knowledge of the coefficient extinctions, each specific to a pigment in a defined solvent and at a precise wavelength [79]. Extinction coefficients of Chls, Chl precursors, most carotenoids, and phycobilins are given in Refs. [80–83], respectively. In the case of phycobilins, which are often isolated as bound to a protein [84], one can estimate the purity of the preparation by measuring the absorbance ratio AX/A_{280}, where X is the absorbance maximum wavelength of the pigment (see Ref. [45]).

3. Fluorescence Spectroscopy

Some photosynthetic pigments emit fluorescence, as mentioned in Section II.C. Pigment quantification by fluorescence is more delicate as fluorescence intensity is proportional only to very low concentrations of pigments. In any case, this concentration range should be determined using a calibration curve established with a standard.

4. IR and Resonance Raman Spectroscopies

IR spectroscopy is used to determine the presence or the absence of functional groups. This method has revealed for the first time the presence of an allelic group in fucoxanthin, alloxanthin, and bastaxanthin c (reviewed in Ref. [85]). IR was also used to establish the details of the light-induced oxygen-dependent bleaching of the food colorant chlorophyllin [86]. The mechanism involves oxidation of a vinyl side group together with aggregation of oxidized chlorophyllin [87]. This last point might be investigated using circular dichroism spectroscopy. Resonance Raman spectroscopy is complementary to IR as it can be also used for in situ studies to confirm the presence of identified functional groups. For instance, absorbance shifts of photosynthetic pigments were used to get important details on the protein structural environment [40]. The Raman spectroscopy of HPLC-purified Chl d from the marine prokaryote Acaryochloris marina has been reported for the first time in Ref. [88]. The formyl group at the C-3 position, typical of Chl d and BChl a, gives a specific Raman peak at 1659 cm^{-1}. As for Chl a, Chl b, and BChl a, there are many strong Raman signals in the range 800–1800 cm^{-1}, which are mainly due to the CH$_3$ bend, CH bending, and CO, CC, CN stretching vibrations. Raman spectroscopy also allows the in situ analysis of carotenoids in complex matrices [89,90].

5. NMR Spectroscopy

NMR spectroscopy is based on the fact that several atomic nuclei (^{1}H, ^{15}N, ^{13}C, etc.) may be oriented by a strong magnetic field and will absorb radiofrequency radiations at characteristic frequencies. The technique carries information about the chemical environment of the nucleus being studied and, by extension, information on the molecular structure or conformation. In the carotenoid field, it is mainly used to localize the cis (Z)-carotenoid isomers.

NMR has been also used to detect adulteration of food products. For instance, it is known that Chl molecules are sometimes added to virgin olive oil in order to improve its color [76]. The presence of the added Chl may be detected by NMR as it gives a different signature from that of pheophytin, the regular tetrapyrrole pigment in olive oils (the central NH groups of the Mg-depleted Chl having a specific resonance signal).

6. Mass Spectrometry

When the identity of a certain pigment has to be established, MS is the first technique selected as it can provide information about both its molecular weight and its structural features. MS is based on sample ionization and subsequent separation of the ions thus formed, depending on their mass-to-charge ratio, using the forces exerted by magnetic and electric fields in a system under high vacuum. The ions must be introduced in the gas phase into the vacuum system of the mass spectrometer. This is easily done for gaseous or heat-volatile samples using classical EI or chemical ionization (CI) techniques, but thermally labile analytes require either desorption methods (like field desorption, fast atom bombardment [FAB] matrix assisted laser desorption ionization [MALDI]) or desolvation methods (like electrospray ionization [ESI] and atmospheric pressure chemical ionization [APCI]). Some MS techniques can be coupled to HPLC to obtain separation of different pigments or to remove interfering contaminants prior to ionization and detection. General information on MS and on the different ionization methods is given in Refs. [91,92]. The general precautions to be taken when working with photosynthetic pigments also apply for MS analysis, and it is especially important to consider that certain additives of common use, for example, those employed to improve the ionization, can lead to the formation of degradation products (e.g., the addition of volatile acids to promote the formation of protonated pseudomolecular ions in ESI–MS easily produces the demetallated derivatives of chlorophylls).

Carotenoid analysis by MS can be performed using classical techniques like EI and CI, which produce abundant, structurally informative fragment ions, but usually weak (or even absent) molecular ions. The exhaustive work by Enzell and Back [93] reviews fragmentation patterns and spectra interpretation. Ionization techniques like FAB, MALDI, ESI, and APCI provide information on the molecular weight as they reduce the fragmentation of the molecular ion, which can be increased if structural information is required by employing collision-induced dissociation and tandem MS (MS/MS) [93–96].

Most of the recent applications of MS for carotenoid analysis employ LC–MS systems. Thus, Goericke et al. [97] applied ESI–MS to identify carotenol chlorin esters formed in marine sediments, whereas APCI–MS has been employed in the determination of lutein and zeaxanthin stereoisomers [98], in the differentiation between lutein monoester regioisomers [99], and to study the specificity of gastric lipases on carotenol fatty acid esters [100]. In a recent study, Hornero-Méndez and Britton [101] employed labeling with stable isotopes and LC–MS to study the cyclization reaction of carotenoid biosynthesis. Classical EI–MS continues to be a powerful tool in carotenoid research, for example, for the identification of gyro-

xanthin, the first allenic acetylenic carotenoid described [102].

Due to their high masses, low volatility, and thermal instability, chlorophylls were an analytical challenge to MS for a long time [1]. The introduction of desorption and desolvation techniques generalized the study of Chl-related pigments by MS and LC–MS (reviewed in Refs. [1,103]). Most of the Chls render good spectra in the positive ion mode, but the acidic pigments of the Chl c group, characteristic of the chromophyte algae, are best analyzed in the negative ion mode, providing good $[M-H]^-$ ions that can be cleaved to give characteristic fragments corresponding to formal losses of CO_2 ($[M-H-44]^-$) and CH_3OH ($[M-H-44-32]^-$) (Figure 33.3).

The application of MS recently allowed the characterization of naturally occurring Chls. Garrido et al. [104] and Zapata et al. [105] employed FAB–MS to describe two new Chl c_2-monogalactosyldiacylglyceride esters isolated from marine micralgae, and Airs et al. [106] identified the bacteriochlorophyll homologs of *Chlorobium phaeobacteroides* by APCI–LC–MS/MS.

MS has also been applied in studies on the formation of Chl degradation products. Chl allomers have been identified by APCI–MS [107,108], and the allomerization pathways traced by ^{18}O labeling and ESI–MS [109]. Gautier-Jacques et al. [110] also employed HPLC–APCI–MS/MS to track Chl degradation during the processing of spinaches and green beans. The kinetics of Chl a demetallation was studied by ESI–MS [111].

A method for the analysis of phycobilins from a cyanobacterium by HPLC–ESI–MS has been recently published [44].

IV. CONCLUSION

Pigments have been studied for a long time, and they will probably still interest basic and applied research for many years for the following reasons: studies on not well-known material can lead to the discovery of new pigments; new technologies allow us more and more to characterize *in situ* molecules or to detect small amounts of compounds. From the economic point of view, the few examples that we have mentioned in this chapter show the impact of color and the importance of detection of adulterated food.

ACKNOWLEDGMENTS

This chapter is dedicated to our friends Dr. Pavel Siffel and Dr. Gulya Siffelova (Czech Academy of

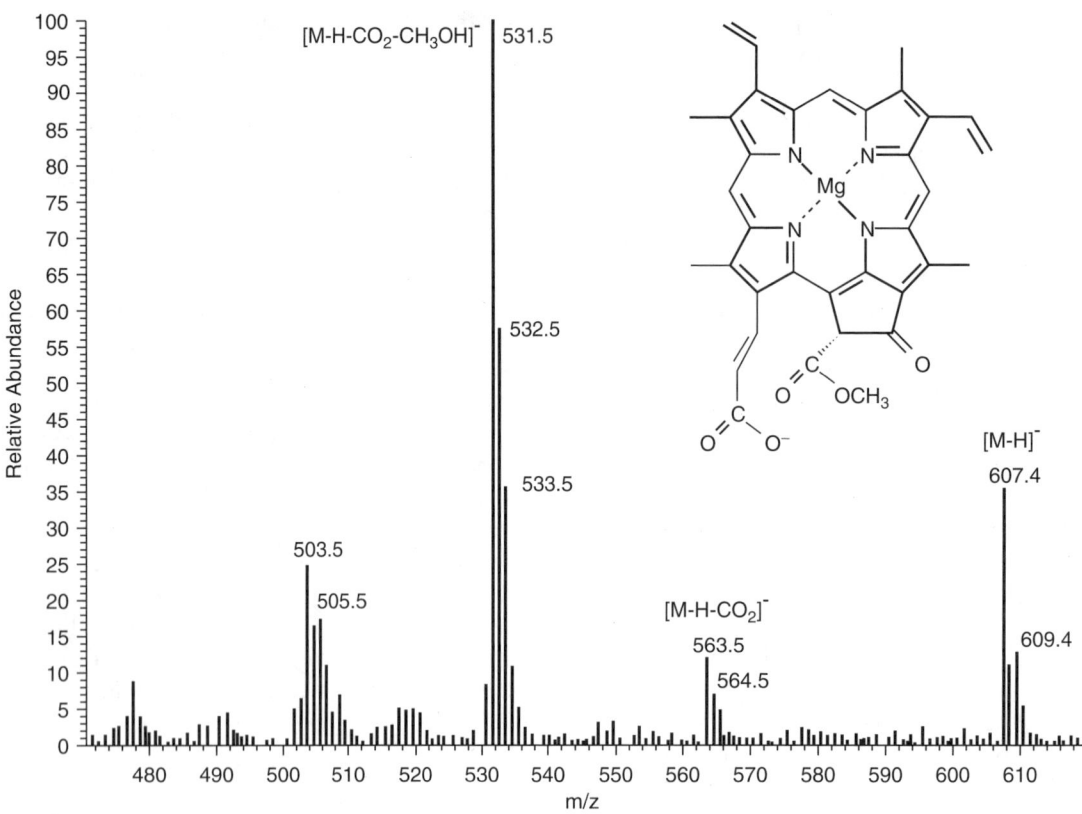

FIGURE 33.3 Negative ion electrospray mass spectrum of chlorophyll c_2.

Sciences), who disappeared suddenly during the summer of 2003.

REFERENCES

1. Schoefs B. Chlorophyll and carotenoid analysis in food products. Properties of the pigments and methods of analysis. *Trends Food Sci. Technol.* 2002; 13:361–371.
2. Schoefs B. Chlorophyll and carotenoid analysis in food products. A case by case. *Trends Anal. Chem.* 2003; 22:335–339.
3. Bertrand M, Schoefs B. Working with photosynthetic pigments: problems and precautions. In: Pessarakli M, ed. *Handbook of Photosynthesis.* New York: Marcel Dekker, 1997:151–172.
4. Hunter RS. Scales for the measurements of color difference. In: *The Measurements of Appearance.* New York: John Wiley & Sons, 1975:133–140.
5. Wyszecki G, Stiles WS. *Colour Science.* New York: John Wiley & Sons, 1987.
6. Di Cesare LF, Forni E, Viscardi D, Nani RC. Changes in the chemical composition of basil caused by different drying procedures. *J. Agric. Food Chem.* 2003; 51:3575–3581.
7. Razungles AJ, Babic I, Sapis JC, Bayonove CL. Particular behavior of epoxy xanthophylls during verai-

son and maturation of grape. *J. Agric. Food Chem.* 1996; 44:3821–3825.
8. Rabinovich EI. *Photosynthesis and Related Processes.* Vol. 2. Part 2. New York: Interscience Publishers, 1956.
9. Rabinovich EI, Govindjee. *Photosynthesis.* New York: John Wiley & Sons, 1969.
10. Schoefs B, Bertrand M, Funk C. Photoactive protochlorophyllide regeneration in cotyledons and leaves from higher plants. *Photochem. Photobiol.* 2000; 72:660–668.
11. Bertrand M, Schoefs B, Siffel P, Rohacek K, Molnar I. Cadmium inhibits epoxidation of diatoxathinn to diadinoxanthin in the xanthophylls cycle of the marine diatom *Phaeodactylum tricornutum. FEBS Lett.* 2001; 508:153–156.
12. Sun L, Wang S. A phycoerythrin-allophycocyanin complex from the intact phycobilisomes of the marine red alga *Polysiphonia urceolata. Photosynthetica* 2000; 38:601–605.
13. Rohacek K. Chlorophyll fluorescence parameters: the definitions, photosynthetic meaning, and mutual relationships. *Photosynthetica* 2002; 40:13–29.
14. Buschmann C, Langsdorf G, Lichtenthaler HK. Imaging of the blue, green, and red fluorescence emission of plants: an overview. *Photosynthetica* 2000; 38:483–491.
15. Nedbal L, Soukupova J, Whitmarsh J, Trtilek M. Postharvest imaging of chlorophyll fluorescence from

lemons can be used to predict fruit quality. *Photosynthetica* 2000; 38:571–579.

16. Bown AW, Hall DE, MacGregor KB. Insect footsteps on leaves stimulate the accumulation of 4-aminobutyrate and can be visualized through increased chlorophyll fluorescence and superoxide production. *Plant Physiol.* 2002; 129:1430–1434.

17. Raskin VI, Marder JB. Chlorophyll organization in dark-grown and light-grown pine (*Pinus brutia*) in barley (*Hordeum vulgare*). *Physiol. Plant.* 1997; 101:620–626.

18. Doan JM, Schoefs B, Ruban AV, Etienne AL. Changes in the LHCI aggregation state during iron repletion in the unicellular red alga *Rhodella violacea*. *FEBS Lett.* 2003; 533:59–62.

19. Lamote M, Darko E, Schoefs B, Lemoine Y. Assembly of the photosynthetic apparatus in embryos from *Fucus serratus* L. *Photosynth. Res.* 2003; 77:45–52.

20. Lebedev NN, Siffel P, Krasnovsky AA. Detection of protochlorophyllide forms in illuminated bean leaves and chloroplasts bv diffference fluorescence spectroscopy at 77 K. *Photosynthetica* 1985; 19:183–197.

21. Schoefs B, Bertrand M, Franck F. Spectroscopic properties of protochlorophyllide analyzed *in situ* in the course of etiolation and in illuminated leaves. *Photochem. Photobiol.* 2000; 72:85–93.

22. Hak R, Lichtenthaler HK, Rinderle U. Decrease of the chlorophyll fluorescence ratio F690/F730 during greening and development of leaves. *Radiat. Environ. Biophys.* 1990; 29:329–336.

23. D'Ambrosio N, Szabo K, Lichtenthaler HK. Increase of the chlorophyll fluorescence ratio F690/F735 during the autumnal chlorophyll breakdown. *Radiat. Environ. Biophys.* 1992; 31:51–62.

24. Sestak Z, Siffel P. Leaf-age related differences in chlorophyll fluorescence. *Photosynthetica* 1997; 33:347–369.

25. Karukstis KK. Chlorophyll fuorescence as a physiological probe of the photosynthetic apparatus. In: Scheer H, ed. *Chlorophylls*. Boca Raton, FL: CRC Press, 1991.

26. Schoefs B, Bertrand M. The formation of chlorophyll from chlorophyllide in leaves containing proplastids is a four-step process. *FEBS Lett.* 2000; 486:243–246.

27. Bouvier F, Dogbo O, Camara B. Biosynthesis of the food and cosmetic plant pigment bixin (annatto). *Science* 2003; 300:2089–2091.

28. Kopecky J, Schoefs B, Loest K, Stys D, Pulz O. Microalgae as a source for secondary carotenoid production: a screening study. *Algolog. Stud.* 2000; 98:153–168.

29. Orosa M, Torres E, Fidalgo P, Zbalde J. Production and analysis of secondary carotenoids in green algae. *J. Appl. Phycol.* 2000; 12:553–556.

30. Hagen X, Grunewald K. Fosmidomycin as an inhibitor of the non-mevalonate terpenoid pathway depresses synthesis of secondary carotenoids in flagellates of the alga *Haematococcus pluvialis*. *J. Appl. Bot. Angew. Bot.* 2000; 74:137–140.

31. Choe E, Lee J, Park K, Lee S. Effects of heat pretreatment on lipid and pigments of freeze-dried spinach. *J. Food Sci.* 2001; 66:1074–1079.

32. Fish WW, Davis AR. The effects of frozen storage conditions on lycopene stability in watermelon tissue. *J. Agric. Food Chem.* 2003; 51:3582–3585.

33. Oberhuber M, Berghold J, Breuker K, Hörtensteiner S, Kräutler B. Breakdown of chlorophyll: a nonenzymatic reaction accounts for the formation of the colorless "nonfluorescent" chlorophyll catabolites. *Proc. Natl. Acad. Sci. USA* 2003; 100:6910–6915.

34. Steiger S, Schäfer, Sandmann G. High-light-dependent upregulation of carotenoids and their antioxidative properties in the cyanobacterium *Synechocystis* PCC 6803. *J. Photochem. Photobiol. B* 1999; 52:14–18.

35. Chamovitz E, Sandmann G, Hirschberg J. Molecular and biochemical characterization of herbicide-resistant mutants of cyanobacteria reveals that phytoene desaturation is a rate-limiting step in carotenoid biosynthesis. *J. Biol. Chem.* 1993; 268:17348–17353.

36. Dahlman L, Persson J, Nasholm T, Palmqvist K. Carbon and nitrogen distribution in the green algal lichen *Hypogymnia physodes* and *Platismatia glauca* in relation to nutrient supply. *Planta* 2003; 217:41–48.

37. Cano MP, Monreal M, de Ancos B, Alique R. Effects of oxygen levels on pigment concentrations in cold-stored green beans (*Phaseolus vulgaris* L. Cv. Perona). *J. Agric. Food Chem.* 1998; 46:4164–4170.

38. Lopez-Hernandez E, Ponce-Alquicira E, Cruz-Sosa F, Guerrero-Legarreta I. Characterization and stability of pigments extracted from *Terminalia catappa* leaves. *J. Food Sci.* 2001; 66:832–836.

39. Schoefs B. Pigment composition and location in honey locust (*Gleditsia triacanthos*) seeds before and after desiccation. *Tree Physiol.* 2002; 22:285–290.

40. Gall A, Robert B, Cogdell RJ, Bellissent-Funel MC, Fraser NJ. Probing the binding sites of exchanged chlorophyll *a* in LH2 by Raman and site-selection fluorescence spectrocopies. *FEBS Lett.* 2001; 491:143–147.

41. Schoefs B, Rmiki, NE, Rachadi J, Lemoine Y. Astaxanthin accumulation in *Haemococcus* requires a cytochrome P450 hydrolyse and an active synthesis of fatty acids. *FEBS Lett.* 2001; 500:125–128.

42. Sauer J, Schreiber U, Schmid R, Völker U, Forchhammer K. Nitrogen starvation-induced chlorosis in *Synechococcus* PCC 7942. Low-level photosynthesis as a mechanism of long-term survival. *Plant Physiol.* 2002; 126:233–243.

43. Bermejo Rr, Acien FG, Ibanez MJ, Fernandez JM, Molina E, Alvarez-Pez JM. Preparative purification of B-phycoerythrin from the microalga *Porphyridium cruentum* by expanded-bed adsorption chromatography. *J. Chromatogr. B Anal. Technol. Biomed. Life Sci.* 2003; 790:317–325.

44. Zolla L, Bianchetti M. High-performance liquid chromatography coupled on-line with electrospray ionisation mass spectrometry for the simultaneous separation and identification of the *Synechocystis*

PCC 6803 phycobilisome proteins. *J. Chromatogr.* 2001; A912:269–279.

45. Jaouen P, Lépine B, Rossignol N, Royer R, Quéméneur F. Clarification and concentration with membrane technology of phycocyanin solution extracted from *Spirulina platensis. Biotechnol Tech.* 1999; 13:877–881.

46. Abalde J, Betancourt L, Torres E, Cid A, Barwell C. Purification and characterization of phycocyanin from the marine cyanobacterium *Synechococcus* sp. IO9201. *Plant Sci.* 1998; 136:109–120.

47. ISO 10260. *Water Quality, Measurement of Biochemical Parameters; Spectrometric Determination of the Chlorophyll-a Concentration.* Berlin: Beuth Verlag GmbH, 1992.

48. Papista E, Acs E, Böddi B. Chlorophyll-*a* determination with ethanol — a critical test. *Hydrobiologia* 2002; 485:191–198.

49. Skidmore R, Maberly SC, Whitton BA. Patterns of spatial and temporal variation in phytoplankton chlorophyll *a* in the River Tent and its tributaries. *Sci. Total Environ.* 1998; 210/211:357–365.

50. Lean DR, Pick FR. Photosynthetic response to nutrient enrichment: a test for nutrient limitation. *Limnol. Oceanogr.* 1981; 26:1001–1019.

51. Guerra-Vargas M, Jaramillo-Flores ME, Dorantes-Alvarez L, Hernandez-Sanchez H. Carotenoid retention in canned pickled Jalapeno peppers and carrots as affected by sodium chloride, acetic acid, and pasteurisation. *J. Food Sci.* 2001; 66:620–626.

52. Fernandez-Garcia A, Butz P, Tauscher B. Effects of high-pressure processing on carotenoid extrability, antioxidant activity, glucose diffusion, and water binding of tomato puree (*Lycopersicon esculentum* Mill.). *J. Food Sci.* 2001; 66:1033–1038.

53. Kouril R, Ilik P, Naus J, Schoefs B. On the limits of applicability of spectrophotometric and spectrofluorimetric methods for the determination of chlorophylls *a*/b ratios. *Photosynthesis Res.* 1999, 62:107–116.

54. Sauer J, Schreiber U, Schmid R, Völker U, Forchhammer K. Nitrogen starvation-induced chlorosis in *Synechococcus* PCC 7942. Low-level photosynthesis as a mechanism of long-term survival. *Plant Physiol.* 2001; 126:233–243.

55. Kirchner JG. *Techniques of Chemistry. Vol. 14: Thin-Layer Chromatography.* St Louis, MO: Sigma, 1990.

56. Sherma J. Thin-layer chromatography in food and agricultural analysis. *J. Chromatogr.* 2000; A880:129–147.

57. Senge M, Dörnemann D, Senger H. The chlorinated chlorophyll RC1, a preparation artefact. *FEBS Lett.* 1988; 234:215–217.

58. Schiedt K, Liaaen-Jensen S. Isolation and analysis. In: Britton G, Liaaen-Jensen S, Pfander H, eds. *Carotenoids. Vol 1A: Analysis.* Basel: Birkhäuser, 1995:81–108.

59. Schoefs B. Analyse des Farbstoffe in Kürbiskernöl. *Lebensm. Biotechnol.* 2001; 1:8–10.

60. Sadek PC. *The HPLC Solvent Guide.* New York: Chichester, 1996.

61. van Breemen RB. Chromatographic separation of chlorophylls. In: *Current Protocols in Food Analytical Chemistry Online.* Chapter F4. Unit 4. New York: John Wiley & Sons, 2003.

62. Stecher G, Huck CW, Stöggl WM, Bonn GK. Phytoanalysis: a challenge in phytomics. *Trends Anal. Chem.* 2003; 22:1–14.

63. Lee HS, Castle WS, Coates GA. High-performance liquid chromatography for the characterization of carotenoids in the new sweet orange (Earlygold) grown in Florida, USA. *J. Chromatogr.* 2001; A913:371–377.

64. Przybyciel M. Novel phases for HPLC separations. *LC–GC Eur.* 2003; 16(6a):29–32.

65. Zou H, Huang X, Ye M, Luo Q. Monolithic stationary phases for liquid chromatography and capillary electrochromatography. *J. Chromatogr.* 2002; A954:5–32.

66. Tanaka N, Kobayashi H, Ishizuka N, Minakuchi H, Nakanishi K, Hosoya K, Ikegami T. Monolithic silica columns for high-efficiency chromatographic separations. *J. Chromatogr.* 2002; A965:35–49.

67. Svec F. Porous monoliths: the newest generation of stationary phases for HPLC and related methods. *LC–GC Eur.* 2003; 16(6a):24–28.

68. Garrido JL, Zapata M. Chlorophyhll analysis by new HPLC methods. In: Grimm B, Porra RJ, Rudiger W, Scheer H, eds. *Chlorophylls and Bacteriochlorophylls: Biochemistry, Biophysics and Biological Function.* Dordrecht: Kluwer Academic Publishers. In press.

69. Garrido JL, Rodriguez F, Campana E, Zapata M. Rapid separation of chlorophylls *a* and *b* and their demetallated and dephytylated derivatives using a monolithic silica C18 column and a pyridine-containing mobile phase. *J. Chromatogr.* 2003; A994:85–92.

70. Darko E, Schoefs B, Lemoine Y. Improved liquid chromatographic method for the analysis of photosynthetic pigments of higher plants. *J. Chromatogr.* 2001; A876:111–116.

71. Bononi M, Commissati I, Lubian E, Fossati A, Tateo F. A simplified method for the HPLC resolution of α-carotene and β-carotene (*trans* and *cis*) isomers. *Anal. Bioanal. Chem.* 2002; 372:401–403.

72. Hyvärinen K, Hynninen PH. Liquid chromatographic separation and mass spectrometric identification of chlorophyll b allomers. *J. Chromatogr.* 1999; A837:107–116.

73. Zapata M, Rodriguez F, Garrido JL. Separation of chlorophylls and carotenoids from marine phytoplankton: a new HPLC method using a reversed phase C8 column and pyridine-containing mobile phases. *Mar. Ecol. Prog. Ser.* 2000; 195:29–45.

74. Latasa M, Van Lenning K, Garrido JL, Scharek R, Estrada M, Rodriguez FF, Zapata M. Losses of chlorophylls and carotenoids in aqueous acetone and methanol extracts prepared for RPHPLC analysis of pigments. *Chromatographia* 2001; 53:385.

75. Mercadante AZ, Britton G, Rodriguez-Amaya DB. Carotenoids from yellow passion fruit (*Passiflora edulis*). *J. Agric. Food Chem.* 1998; 46:4102–4106.

76. Mannina L, Sobolev AP, Segre A. Olive oil as seen by NMR and chemometrics. *Spectrosc. Eur.* 2003; 15/3:6–14.

77. Schoefs B, Bertrand M, Lemoine Y. Changes in the photosynthetic pigments in bean leaves during the first photoperiod of greening and the subsequent dark-phase. Comparison between old (10-d-old) leaves and young (2-d-old) leaves. *Photosynth. Res.* 1998; 57:203–213.

78. Schwartz SJ. High performance liquid chromatography of zinc and copper pheophytins. *J. Liq. Chromatogr.* 1984; 7:1673–1683.

79. Seely GR, Jensen RG. Effect of solvent on the spectroscopy of chlorophyll. *Spectrochim. Acta* 1965; 21:1835–1845.

80. Porra RJ, Thompson WA, Kriedmann PE. Useful data to correct Arnon-derived data. Millimolar and specific coefficients for chlorophylls *a* and b. *Biochim. Biophys. Acta* 1989; 975:384–394.

81. Brouers M, Wolwertz MR. Estimation of protochlorophyll(ide) contents in plant extracts: re-evaluation of the molar absorption coefficient of protochlorophyllide. *Photosynth. Res.* 1983; 4:265–270.

82. Britton G, Liaaen-Jensen S, Pfander H. *Carotenoids.* Vol. 2. Basel: Birkhäuser Verlag, 1996.

83. Stadnichuk IN. Phycobiliproteins: determination of chromophore composition and content. *Phytochem. Anal.* 1995; 6:281–288.

84. Sidler WA. Phycobilisome and phycobiliprotein structure. In: Bryant DA, ed. *The Molecular Biology of Cyanobacteria.* Chapter 7. Dordrecht: Kluwer Academic Publishers, 1994:139–216.

85. Eugster CH. History: 175 years of carotenoid chemistry. In: Britton G, Liaaen-Jensen S, Pfander H, eds. *Carotenoids. Vol 1A: Isolation and Analysis.* Basel: Birkhäuser Verlag, 1995.

86. Salin ML, Avarez ML, Lynn BC, Habulihaz B, Fountain AW. Photooxidative bleaching of chlorophyllin. *Free Radic. Res.* 1999; 31:97–105.

87. Yagai S, Miyatake T, Tamiaki H. Self-assembly of synthetic 81-hydroxy-chorophyll analogues. *J. Photochem. Photobiol. B* 1999; 52:74–85.

88. Cai ZL, Zeng H, Chen M, Larkum AWD. Raman spectroscopy of chlorophyll d from *Acaryochloris marina. Biochim. Biophys. Acta* 2002; 1556:89–91.

89. Withnall R, Chowdhry BZ, Silver J, Edwards HGM, de Oliveira LFC. Raman spectra of carotenoids in natural products. *Spectrochim. Acta* 2003; A59:2207–2212.

90. Edwards HGM, Newton EM, Wynn-Williams DD, Lewis-Smith RI. Non-destructive analysis of pigments and other organic compounds in lichens using Fourier-transform Raman spectroscopy: a study of Antarctic epilithic lichens. *Spectrochim. Acta* 2003; A59:2301–2309.

91. Watson, JT. *Introduction to Mass Spectrometry.* 3rd ed. Philadelphia, PA: Lippincott-Williams and Wilkins, 1997.

92. Niessen, WMA. *Liquid Chromatography-Mass Spectrometry.* New York: Marcel Dekker, 1999.

93. Enzell, C.R., Back, S. *Mass Spectrometry.* In: Britton G., Liaaen-Jensen S, Pfander H, eds. *Carotenoids. Vol. 1B: Spectroscopy.* Basel: Birkhäuser, 1995.

94. van Breemen RB. Innovations in carotenoid analysis using LC/MS. *Anal. Chem.* 1996; 68:299A–304A.

95. van Breemen RB. Liquid chromatography/mass spectrometry of carotenoids. *Pure Appl. Chem.* 1997; 69:2061–2066.

96. van Breemen RB. Mass spectrometry of carotenoids. In: *Current Protocols in Food Analytical Chemistry Online.* Chapter F2. Unit 4. New York: John Wiley & Sons, 2003.

97. Goericke R, Shankle A., Repeta DJ. Novel carotenol chlorin esters in marine sediments and water column particulate matter. *Geochim. Cosmochim. Acta* 1999; 63:2825–2834.

98. Dachtler M., Glaser T., Kohler K., Albert K. Combined HPLC-MS and HPLC-NMR On-line coupling for the separation and determination of lutein and zeaxanthin stereoisomers in spinach and in retina. *Anal. Chem.* 2001; 73:667–674.

99. Breithaupt DE, Wirt U, Bamedi A. Differentiation between lutein monoester regioisomers and detection of lutein diesters from marigold flowers (*Tagetes erecta* L.) and several fruits by liquid chromatography-mass spectrometry. *J. Agric. Food Chem.* 2002; 50:66–70.

100. Breithaupt DE, Bamedi A., Wirt U. Carotenol fatty acid esters: easy substrates for digestive enzymes? *Comp. Biochem. Physiol.* 2002; B132:721–728.

101. Hornero-Méndez D, Britton G. Involvement of NADPH in the cyclization reaction of carotenoid biosynthesis. *FEBS Lett.* 2002; 515:133–136.

102. Bjørnland T, Fiksdahl A, Skjetne T, Krane J., Liaaen-Jensen S. Gyroxanthin-the first allenic acetylenic carotenoid. *Tetrahedron* 2000; 56:9047–9056.

103. van Breemen RB. Mass spectrometry of chlorophylls. In: *Current Protocols in Food Analytical Chemistry Online.* Chapter F4. Unit 5. New York: John Wiley & Sons, 2003.

104. Garrido JL, Otero J, Maestro MA, Zapata M. The main nonpolar chlorophyll *c* from *Emiliania huxleyi* (Prymnesiophyceae) is a chlorophyll c_2-monogalactosyldiacylglyceride ester: a mass spectrometry study. *J. Phycol.* 2000; 36:497–505.

105. Zapata M, Edvardsen B, Rodríguez F, Maestro MA, Garrido JL. Chlorophyll c_2-monogalactosyldiacylglyceride ester (chl c_2-MGDG). A novel pigment marker for *Chrysochromulina* species (Haptophyta). *Mar. Ecol. Prog. Ser.* 2001; 219:85–98.

106. Airs RL, Borrego CM, García-Gil J, Keely BJ. Identification of the bacteriochlorophyll homologues of *Chlorobium phaeobacteroides* strain UdG6053 grown at low light intensity. *Photosynth. Res.* 2001; 70:221–230.

107. Jie C, Walker JS, Keely BJ. Atmospheric pressure chemical ionization normal phase liquid chromatography mass spectrometry and tandem mass spectrometry of chlorophyll *a* allomers. *Rapid Commun. Mass Spectrom.* 2002; 16:473–479.

108. Walker JS, Jie C, Keely BJ. Identification of diastereomeric chlorophyll allomers by atmospheric pressure chemical ionization liquid chromatography/tandem mass spectrometry. *Rapid Commun. Mass Spectrom.* 2003; 17:1125–1131.

109. Hynninen PH, Hyvärinen K. Tracing the allomerization pathways of chlorophylls by ^{18}O labeling and mass spectrometry. *J. Org. Chem.* 2002; 67:4055–4061.

110. Gautier-Jacques A, Bortlik K, Hau J, Fay LB. Improved method to track chlorophyll degradation. *J. Agric. Food Chem.* 2001; 49:1117–1122.

111. Kolakowski BM, Konermann L. From small-molecule reactions to protein folding: studying biochemical kinetics by stopped-flow electrospray mass spectrometry. *Anal. Biochem.* 2001; 292:107–117.

Section XII

Photosynthesis and Its Relationship with Other Plant Physiological Processes

34 Photosynthesis, Respiration, and Growth

Bruce N. Smith
Brigham Young University

CONTENTS

I. PHOTOSYNTHESIS

A. CHLOROPLASTS

Photosynthesis supports all life on earth and in eukaryotes occurs exclusively in chloroplasts. In higher plants, all green tissues contain chloroplasts and perform, to some degree, photosynthesis. Most photosynthesis, by far, occurs in the chloroplasts of leaves. The only cells in the epidermis that contain chloroplasts are the stomatal guard cells. Photosynthesis in guard cells has to do with stomatal opening and not with photosynthate production and export. About 100 to 150 chloroplasts are found in each mesophyll or palisade parenchyma cell of the leaf. Approximately the same number of mitochondria are present also. Chloroplasts are a bit larger than mitochondria and can be separated from them by centrifugation of leaf brei [1].

Chloroplasts are bound by a double membrane but have considerable internal structure as well. Thylakoids are roughly circular but flattened bags made of thin (~90 Å) membranes. Stacks of thylakoids called grana are the site of the light-dependent (or light) reactions of photosynthesis. The region of the chloroplast outside of the grana is called the stroma, where the light-independent (or dark) reactions of photosynthesis take place. About 60% of the volume of the chloroplast is stroma, with the rest being grana [1].

B.   LIGHT-HARVESTING SYSTEMS

With very dense suspensions of cells and weak incident light intensity, as much as 17% of the light can be converted to chemical potential energy. In natural systems, however, only 1 to 2% of solar energy is converted to plant material. Most of the light energy is reflected away or reradiated as heat from soil, plant, and other surfaces. If more light is absorbed than can be transformed into chemical energy, the excess energy may result in the formation of free radicals, which can destroy membranes in the chloroplast and leaf. Light is required for the transformation of proplastids into chloroplasts as well as the constant synthesis of chlorophyll, which has a high turnover rate [2]. Nevertheless, leaves exposed to high-intensity light (sun-leaves) have many fewer chloroplasts and much less chlorophyll than leaves exposed to low-intensity or intermittent light (shade-leaves).

Thylakoid membranes are approximately 50% protein, 40% lipid, and 10% chlorophyll by weight. Much of the membrane protein is associated with chlorophyll. About 8% of the light energy absorbed by chlorophyll is reradiated as red light by fluorescence. If you look through a microscope, at high magnification, at cells with large chloroplasts (e.g., the moss *Mnium affine*) you can see glowing red dots in the chloroplast. This is fluorescence from the grana. While purified chlorophyll *a* absorbs blue and red light, the action spectrum for intact chloroplasts covers most of the visible light range. Other pigments including carotenoids, xanthophylls, other chlorophylls (*b, c, d*), etc., can absorb light and transfer excited electrons to chlorophyll *a*. The chlorophyll *a* and most of the protein and lipid of the thylakoid are organized in light-harvesting complexes containing 250 to 300 chlorophyll molecules. The light-harvesting antennae are intimately associated with electron and proton-transfer systems called photosystems I and II. Photosystem II (PS II) is larger and has a higher proportion of accessory pigments than does photosystem I (PS I). Excited electrons produced by light absorption of pigment molecules can be passed

from molecular to molecule until finally a chlorophyll *a* molecule transfers the electron to an iron–sulfur protein (PS I) or to plastoquinone (PS II). Energy can be transferred from PS II to PS I but not in the other direction. Water is split (in PS II) to release oxygen and protons in the inner thylakoid space. Proteins are transferred from the stromal side of the membrane into the inner thylakoid space by plastoquinone. Electrons are transferred from plastoquinone to cytochrome *f* and on to plastocyanin. From plastocyanin the electrons are passed to PS I. Excited electrons from either chlorophyll *a* or plastocyanin in PS I are passed to an iron–sulfur protein, to ferredoxin, and eventually to a flavoprotein involved in the reduction of nicotinamide adenine dinucleotide phosphate (NADP). The proton gradient can promote formation of ATP via ATP synthetase, which is partially imbedded in the thylakoid membrane. Both ATP and NADPH + H$^+$ are conveniently produced on the stromal side of the thylakoid [3].

The light-dependent processes of photosynthesis are biophysical in nature and involve electron and proton transfer. They are not temperature dependent and are very rapid, being complete in 10^{-3} to 10^{-4} sec. In summary,

$$Light + H_2O \xrightarrow{\text{thylakoid membranes}} ATP + NADPH + H^+ + O_2$$

C.   RIBULOSE BISPHOSPHATE CARBOXYLASE

The first step of the light-independent (dark) reactions of photosynthesis is catalyzed by ribulose bisphosphate (RuBP) carboxylase, also known as Fraction I protein or Rubisco.

$$RuBP + CO_2 \xrightarrow{\text{RuBP carboxylase}} 2PGA$$

This enzyme was called Fraction I protein because when first isolated as much as 50% of the soluble protein in the leaf was in the form of this enzyme. Since it is confined to the stroma of the chloroplast, it is probably more useful to think of it as a solid-state system than as a soluble enzyme. In higher plants, the enzyme has a molecular weight of 550,000 Da and is composed of eight large subunits (55,000 Da each) and eight small subunits (15,000 Da each), all bound together by magnesium. Catalytic sites on the large subunits are coded in the chloroplast DNA. Allosteric control sites are on the small subunits, which are coded in the nuclear DNA, synthesized in the cytoplasm, transported via companion proteins into the chloroplasts, and there assembled into the functional enzyme [4].

Recently, it has been shown that RuBP carboxylase must be enzymatically activated with ATP before catalysis can begin [5]. The activated enzyme has a high affinity for CO_2, but if the CO_2 concentration is very low and the O_2 concentration is high, photorespiration can result. Despite great effort, it has proven difficult to decrease photorespiration while maintaining high rates of photosynthesis. It appears that O_2 may bind to the same amino acid residues in the active site of the large subunit as CO_2. Oxygen not only competes competitively with CO_2 but results in non-energy-conserving CO_2 evolution in the light. Fortunately, RuBP carboxylase has a much higher affinity for CO_2 than it does for O_2.

D. THE CALVIN CYCLE AND PENTOSE PHOSPHATE PATHWAY

While the initial reaction of the light-independent (dark) processes of photosynthesis catalyzed by RuBP carboxylase does not require energy input, subsequent conversion of phosphoglyceric acid (PGA) to sugar requires energy conserved from the light-dependent reactions. Once sugar phosphates are available, transaldolases and transketolases (group-specific, not substrate-specific enzymes) transfer 2- and 3-carbon fragments to and from various sugars, forming a number of 4-, 5-, 6-, and 7-carbon sugars [6]. One of these, ribulose phosphate, receives a phosphate from ATP to produce RuBP ready to combine with CO_2 via RuBP carboxylase. The energy from the light-dependent reactions required for fixing one CO_2 is $3ATP + 2NADPH + 2H^+$.

Chloroplasts kept in the dark but supplied with ATP and $NADPH + H^+$ can fix carbon. The light-independent reactions represent the synthesis part of photosynthesis and are enzyme catalyzed, temperature dependent, and relatively slow (10^{-2} sec and longer). One must remember that chloroplasts are great centers of synthesis — not only of carbohydrates but also of lipids, proteins, nucleotides, and secondary metabolites. Biosynthesis does occur at many places within the plant, but energy and building blocks come from the chloroplasts. Leaves are the source for all of the many sinks (flowers, fruit, seed, storage, and growth) in the plant. Not only are producers the base of the food chain, but photosynthesis is the source of useful energy for all of the processes within the plant itself.

E. C_3, C_4, AND CAM PLANTS

Anciently in all plants and in most plants today, the light-independent reactions are as outlined above with the first product after CO_2 addition being PGA, which has three carbons. This common ances-

tral type is thus called C_3 photosynthesis. Relatively recently, possibly in response to lower CO_2 concentrations, C_4 photosynthesis evolved as a means of channeling carbon to RuBP carboxylase and thus reducing photorespiration (see Table 34.1). Concentric cylinders of tissue surrounding the vascular tissue characterize C_4 photosynthesis with the bundle sheath cells nearest to the leaf traces and the morphologically distinct mesophyll cells external to that. Initial carbon fixation occurs in the cytoplasm of the mesophyll cells:

$$CO_2 + PEP \xrightarrow{\text{PEP carboxylase}} OAA \rightarrow \text{malate or aspartate}$$

Oxaloacetic acid (OAA), malate, and aspartate all have four carbons, hence the designation C_4 [7].

The C_4 plants are sun-plants, some of which are not light-saturated at full sunlight. It has been suggested that the light-harvesting systems in C_4 plants are smaller but more numerous, with about 75 chlorophyll molecules, as compared with 25 to 300 chlorophyll molecules in C_3 plants. Photosynthetic rates of C_4 plants are often double those of C_3 plants. Water use is more efficient in C_4 plants as well, even though more ATP is required per amount of CO_2 fixed. It is a relatively recent adaptation to high light, warm temperatures, and low moisture as well as low CO_2 concentrations.

Crassulacean acid metabolism (CAM) is an even more extreme adaptation to hot, dry conditions. These plants have the capacity to keep their stomates closed during the day and open them during the cooler night period for CO_2 fixation and storage in the vacuole as malate. This is an adaptation for water conservation, not rapid growth. While some plants are obligate, others are facultative and use the CAM

TABLE 34.1

Distinguishing Characteristics of Three Types of Photosynthetic Plants

Characteristics	Plants		
	C_3	C_4	CAM
Dark respiration (μg CO_2/g fr. wt./min)	+5–15	+5–15	+5–15
Photorespiration (μg CO_2/g fr. wt/min)	+50–75	0	0
Photosynthesis (μg CO_2/g fr. wt/min)	−100 to 150	−200 to 300	−10 to 30
Water loss (mg H_2O/g fr. wt/min)	4.5–6.0	2.5–3.5	0.25–1.5

mode only under harsh conditions, with C_3 photosynthesis during moderate times [8].

F. Variation in Photosynthetic Rates

The C_4 plants have a photosynthesis rate that is two to three times faster than that of C_3 plants and 100-fold faster than CAM plants [9]. Many weed species are C_4 plants. Within each of these groups there is a great deal of variability in photosynthetic rate (see Table 34.1). Much effort has been made to correlate these smaller differences with growth rate, but no consistent results have been obtained. Photosynthesis is surely the basis for all subsequent metabolic events in the plant. The process of photosynthesis seems to have been perfected to the point that any genetic change is negative.

G. Ecology

Differences between sun and shade leaves are well known both on a single plant and in different species. Much work has been done on sun-flecks and the fact that considerable photosynthetic activity can be generated from brief exposure to full sunlight. In the same environment, C_3 plants will grow during early spring and summer while C_4 plants appear later in the season. The C_4 plants show a preference for open spaces and the edge of the forest. All of the deep forest species are C_3. Although different survival strategies may be used by different photosynthetic types, all are well adapted to their present circumstances [10].

II. RESPIRATION

A. Cytoplasm

1. Glycolysis and Fermentation

Stepwise oxidation of carbohydrates begins in the cytoplasm. Glucose monomers are released from starch by α-amylase, β-amylase, or phosphorylase, and then step by step taken to pyruvate. The entire sequence is controlled by phosphofructokinase and the incessant demand for energy. An alternate route to pyruvate is the pentose phosphate pathway, which also has interesting controls and great flexibility. In the presence of oxygen, pyruvate moves to the mitochondrion and greater glory. In the absence of air, pyruvate goes to lactate in some species and to ethanol and CO_2 in others [11]. Energy is conserved as 2 ATP per glucose molecule. While many plant tissues can survive short periods of time without oxygen, only about 5% of plants can endure prolonged anoxia.

2. Free Radicals

Stress caused by drought, temperature extremes, air pollution, heavy metals, etc., results in free radical formation. Superoxide, super hydroxide, hydrogen peroxide, etc., can oxidize the fatty acids in membranes, resulting in leakage and eventually death. Plant defenses against free radicals and potentially harmful organisms than do more primitive types of plants.

B. Mitochondria

1. The Krebs Cycle

The pyruvate dehydrogenase complex produces acetyl coenzyme A (acetyl CoA), which condenses with OAA to form citric acid in the inner matrix of the mitochondrion, then on to aconitase and isocitric acid. The allosteric enzyme isocitric dehydrogenase controls the rate of the Krebs cycle in response to demand in the cell for ATP. The cycle continues to eventually regenerate OAA. Energy is conserved as $NADH + H^+$ at several steps: $FADH + H^+$ at one step and ATP at another [11].

2. Electron Transport

In the inner mitochondrial membrane, $NADH_2$ and $FADH_2$ start electron and proton transport, resulting in the formation of ATP — 18-fold more than from fermentation. This is often termed the cytochrome pathway since several cytochromes are involved.

3. Alternative Pathways

Most plant tissues either have or can develop an alternative pathway for electron transport. If the cytochrome pathway is inhibited by cyanide or azide and the alternative pathway is operative, both CO_2 evolution and O_2 uptake usually show an increase. Much less energy is conserved via the alternative path [12]. This pathway is active in thermogenic tissues. Plants that produce cyanogenic compounds may find an alternative path worthwhile. The alternative pathway might represent a sink for energy that could otherwise go towards free radical formation. The physiological role of the alternative pathway in most plants is really unknown.

4. The Pasteur Effect and Respiratory Control

Pasteur observed that yeast cells produced as much or more CO_2 in the absence of oxygen as in air. However, in air more yeast cells were produced. In the presence of substances ("uncouplers") that destroy the proton gradient across the inner mitochondrial membrane, CO_2 production and O_2 consumption increase. Isolated mitochondria exhibit respiratory control, responding with increased O_2 consumption to the addition of ADP. This evidence implies that in most cases rates of catabolism are controlled by anabolic demand. The sites of control are phosphofructokinase for glycolysis and isocitric dehydrogenase for the Krebs cycle [3].

C. Microbodies

1. Lipid Metabolism

Microbodies or glyoxysomes are single membrane-bound organelles that are the site of β-oxidation of fatty acids in plants. In seeds and other tissues that store triglycerides, the glyoxylate shunt develops. This consists of de novo synthesis of two enzymes: isocitritase and malate synthetase. They allow the process of gluconeogenesis to begin. Cell walls can be formed from fatty acids [5].

2. Free Radical Control

Microbodies (glyoxysomes or peroxysomes) are organelles where preformed enzymes useful in control of free radicals are sequestered. All plant tissues have large pools of catalase and peroxidase in microbodies [13]. This is somewhat analogous to the role that microbodies play in detoxification in mammalian liver cells.

III. TRANSPORT AND PARTITIONING OF PHOTOSYNTHATE

A. Transfer Cells

Photosynthesis is often limited by the availability of CO_2. It may also be limited by the rate of movement from source photosynthetic cells to sink cells where growth or storage takes place. Starch grains will be formed in the chloroplasts during periods of rapid photosynthesis when transport cannot keep up with synthesis. Gradually the starch is transformed into sucrose and moved to sink tissues via the phloem [14].

A key role in the transport process is played by transfer cells. These cells have projections and protuberances that greatly increase the surface area of the plasmalemma and facilitate transport of sucrose and amino acids into and out of the phloem. Transfer cells are located in association with both source and sink cells and tissues [15].

B. Phloem

Phloem transports primarily sucrose and amino acids through sieve tubes from source to sink. Partitioning of the photoassimilate depends on vascular connections, proximity, and sink strength. The sink will differ depending on season, stage of development, etc. At various times vegetative growth, reproductive development (including fruit development), or storage might be dominant [15].

C. Storage

Photoassimilate may be stored in the form of starch, lipid, or protein. Storage may occur in root, stem, fruit, or seed endosperm.

IV. SECONDARY METABOLISM

The origin of angiosperms somewhat predates but in many ways parallels the origin of mammals. In addition to the flowering habit, angiosperms developed a wide array of compounds that are not part of internal (primary) metabolism but designed to influence other organisms. These substances have been called secondary metabolites and include a wide array of terpenoids, phenolics, alkaloids, nonprotein amino acids, cyanogenic glycosides, etc. Some of these substances are toxins, others are feeding deterrents, and still others are mammalian and insect hormones. Some substances protect against pathogens, others against plant competitors, and still others against herbivores [16]. Of the approximately 300,000 species of flowering plants on Earth, only 33 species are used by humans as food. The rest are poisonous or noxious to some degree. Even food plants must often be prepared in such a way as to remove harmful substances. Through long experience humankind has learned which part of the plant to eat. For example, the potato tuber is edible but leaves and fruit from the same plant are quite toxic.

The advantage to flowering plants is apparently sufficient to warrant expending considerable energy to synthesize secondary metabolites. For some species, it has been calculated that as much as 20% of the total photoassimilate goes into the shikimic acid pathway to produce phenolics [17].

V. MODELS FOR GROWTH

A. Crop Plants

It has long been recognized that both total growth (biomass) or crop yield (seed, fruit, etc.) is ultimately dependent on photosynthesis [18]. Photosynthetic rates do vary a great deal, but no correlation between photosynthesis and growth (or yield) has been found [19]. Good correlations have been found between dark respiration and growth in several instances [20]. Thornley [21] proposed a simple expression:

$$P = PR + R_m + R_g + R_d + R_s + ?$$

where P is total photosynthate, PR is photorespiration, R_m is maintenance respiration necessary for life processes, R_g is energy for growth, R_d is energy for defense, and R_s is energy for stress reduction. While incident light, gross photosynthesis, biomass, and crop yield can be measured, it is more difficult to measure PR. It has not been agreed upon how to measure R_m or the other R values [22]. In this form, the hypothesis is difficult to test.

B. ECOSYSTEM DYNAMICS

In addition to internal factors (photosynthesis, respiration, etc.), both living and nonliving parts of the ecosystem have an impact on productivity. Pathogens, herbivores, competitors, temperature, drought, air pollution, water pollution, heavy metals, excess light, high or low concentrations of CO_2 (including the greenhouse effect), and O_2 all contribute to the formation of free radicals and resulting plant stress [23]. Any of these can thus reduce plant growth.

VI. GROWTH AND EFFICIENCY

Clearly it would be very useful if some metabolic or physiological measure could reliably predict future growth. Despite great effort, photosynthetic variability does not seem to be a good predictor. Farquhar and coworkers [24] have found carbon isotopic ratios to be reliable indicators of the degree of stomatal opening and thus the ratio of carbon assimilation to transpiration water loss. The technique, while somewhat expensive, is very useful under carefully defined conditions. A variety of physical and biological processes can result in isotopic fractionation. Clearly a technique that is cheap, simple, and reliable is still needed.

Yamaguchi [25] introduced the concept of growth efficiency, which is related to biomass production, gross photosynthesis, and respiration. The value of growth efficiency decreased with tissue maturity. Accurate predictions of growth rate based on respiration measurements (see Table 34.2) require a great deal of knowledge about the biology (growth habit, pattern of growth, etc.) of a particular plant [26]. Recently, a model has been proposed that emphasizes growth

TABLE 34.2
Photosynthetic rate (P_{CO_2}) and respiration rate (R_{CO_2}) as a function of leaf number (age) and leaf area from tip to base of a 6-month old seedling of *Atriplex canescens*

| Leaf # | Leaf area (cm²) | μg CO₂/g ft. wt/min | | |
		P_{CO_2}	R_{CO_2}	P_{CO_2}/R_{CO_2}
3, 4	0.75	−96.9	+42.1	2.3
5, 6	1.52	−100.8	+25.2	4.0
7, 8	1.35	−92.1	+20.3	4.5
9, 10	3.08	−128.0	+9.1	14.1
22, 23	5.61	−134.9	+8.1	16.7

CO_2 uptake (−) and CO_2 production (+) are indicated.

Source: Adapted from B. N. Smith, C. M. Lytle, and L. D. Hansen, in *Proceedings: Wildland Shrub and Arid Lane Restoration Symposium*, Las Vegas, NV, INT-GTR-315, 1995, p. 243. With permission.

efficiency and metabolism [27]. Every part of the model is clearly defined and can be measured in the laboratory. It now needs to be tested for a number of different types of plants (Table 34.2). The model will be most useful in assessing the effects of stress on plant growth.

VII. SUMMARY

Since all life depends on it, photosynthesis is the most important process in all biology. After nearly 4.5 billion years of selection, many genetic changes in photosynthesis have negative consequences. For this reason, variations in photosynthetic rates are not predictive of growth. Respiratory metabolism is absolutely dependent on photosynthetic assimilation, but there is a demand-driven balance between catabolic and anabolic processes. Growth, reproduction, adaptations to stress, defense against pathogens and herbivores, etc., are all part of these processes and are reflected in respiratory rates. Photorespiration rates differ widely among species of plants but do not seem to be part of the respiration dynamic. Metabolic comparisons of similar tissues predict plant growth. Much more work is needed to establish this as a generality.

REFERENCES

1. R. G. Hiller and D. J. Goodchild in *The Biochemistry of Plants*, Vol. 8 (M. D. Hatch and N. K. Boardman, eds.), Academic Press, New York, 1981, p. 1.

2. P. A. Castelfranco and S. I. Beale, in *The Biochemistry of Plants*, Vol. 8 (M. D. Hatch and N. K. Boardman, eds.), Academic Press, New York, 1981, p. 375.

3. P. C. Hinkle and R. E. McCarty, *Sci. Am.* 238(3):104 (1978).

4. R. G. Jensen, in *Plant Physiology, Biochemistry, and Molecular Biology*, (D. T. Dennis and D. H. Turpin, eds.), Longman Scientific and Technical, Essex, England, 1990, p. 224.

5. G. T. Byrd, D. T. Ort, and W. L. Ogren, *Plant Physiol., 107*:585 (1995).

6. M. Calvin, *Science, 138*:879 (1962).

7. J. R. Ehleringer, R. F. Sage, L. B. Flanagan, and R. W. Pearcy, *Trends Ecol. Evol., 6*:95 (1991).

8. M. Kluge and I. P. Ting, *Crassulacean Acid Metabolism*, Springer-Verlag, Berlin, 1978.

9. J. R. Ehleringer, *HortScience, 14*:217 (1979).

10. B. N. Smith, *BioSystems, 8*:24 (1976).

11. W. G. Hopkins, *Introduction to Plant Physiology*, John Wiley & Sons, New York, 1995.

12. J. G. Scandalios, *Plant Physiol., 101*:7 (1993).

13. D. A. Day, A. H. Miller, J. T. Wiskich, and J. Whelan, *Plant Physiol., 106*:1421 (1994).

14. S. Wolf, *Potato Res., 36*:253 (1993).

15. B. E. S. Gunning, *Sci. Prog. Oxf., 64*:539 (1977).

16. J. B. Harborne, *Introduction to Ecological Biochemistry*, 2nd ed., Academic Press, London, 1982.

17. K. M. Hermann, *Plant Physiol., 107*:7 (1995).

18. R. K. M. Hay and A. J. Walker, *An Introduction to the Physiology of Crop Yield*, Longman Scientific and Technical, Essex, England, 1989.

19. C. J. Nelson, *Plant Physiol. Biochem., 26*:543 (1988).

20. K. J. McCree, *Crop Sci., 14*:509 (1974).

21. J. H. M. Thornley, *Nature, 227*:304 (1970).

22. R. B. Thomas, C. D. Reid, R. Ybema, and B. R. Strain, *Plant Cell Environ., 16*:539 (1993).

23. C. H. Foyer, M. Lelandais, and K. J. Kunert, *Physiol. Plant., 92*:696 (1994).

24. G. D. Farquhar, M. H. O'Leary, and J. A. Berry, *Aust. J. Plant Physiol., 9*:121 (1982).

25. J. Yamaguchi, *J. Fac. Agric. Hokkaido Univ., 59*:59 (1978).

26. B. N. Smith, C. M. Lytle, and L. D. Hansen, in *Proceedings: Wildland Shrub and Arid Lane Restoration Symposium*, Las Vegas, NV, INT-GTR-315, 1995, p. 243.

27. L. D. Hansen, M. S. Hopkin, D. R. Rank, T. S. Anekonda, R. W. Breidenbach, and R. S. Criddle, *Planta, 194*:77 (1994).

35 Nitrogen Assimilation and Carbon Metabolism

Alberto A. Iglesias
Laboratorio de Enzimología Molecular Grupo de Enzimología Molecular, Bioquímica
Básica de Macromoléculas, Facultad de Bioquímica y Ciencias Biológicas,
Universidad Nacional del Litoral

Maria J. Estrella and Fernando Pieckenstain
Instituto Tecnológico de Chascomús

CONTENTS

I. INTRODUCTION

Nitrogen is one of the most important nutrients necessary for plant growth (frequently it is the major limiting nutrient) and its incorporation from the environment onto biomolecules determines productivity and yield in crops. Nitrogen assimilation is the incorporation of inorganic forms of nitrogen into carbon skeletons, mainly synthesizing amino acids [1]. Ammonium is the most reduced form of nitrogen ultimately utilized by plants for assimilation. Since in nature nitrogen is largely present in more oxidized forms (principally nitrate, nitrite, and dinitrogen), organisms have to expend energy to reduce these nitrogen sources. All higher plants (nonlegumes and legumes) reduce nitrate to ammonium by sequential reactions catalyzed by cytosolic nitrate reductase and plastidial nitrite reductase. Legumes are able to utilize dinitrogen.

Leguminosae (Fabaceae), the third largest family in the angiosperms, includes more than 19,000 species varying from annual herbs to trees, that grow in a wide range of habitats. This widespread distribution is related, at least in part, to the capacity of legumes to grow in soils with low nitrogen content [2]. This ability is due to the fact that legumes are able to establish a symbiotic relationship with nitrogen fixing bacteria present in soil. All of these bacteria (usually known as rhizobia, see Ref. [3]) belong to the family of *Rhizobiaceae*, which includes three genera: *Rhizobium*, *Bradyrhizobium*, and *Azorhizobium*. Rhizobia interact with legumes by characteristically inducing the development of specialized structures, normally not present in the plant: the nodules. These specialized structures are typically formed on the roots, although some aquatic legumes (such as *Sesbania rostrata*) exhibit stem nodules [4]. Subsequently, rhizobia infect and colonize nodules and a metabolic cooperation is established between both symbionts. Bacteria reduce atmospheric nitrogen to ammonia, which is delivered to the plant and subsequently incorporated into organic molecules. On the other hand, the plant provides bacteria with sugars synthesized by carbon dioxide reduction during

TABLE 35.1
Rhizobia–Legumes Symbiotic Associations

Species	Host Legume	Ref.
Rhizobium meliloti	*Medicago sativa, Medicago truncatula, Melilotus albus*	[7]
Rhizobium leguminosarum bv. *viciae*	*Pisum sativum, Vicia sativa*	[8]
Rhizobium leguminosarum bv *trifolii*	*Trifolium* species	[8]
Rhizobium leguminosarum bv. *phaseoli*	*Phaseolus* species	[8]
Rhizobium tropici	*Phaseolus* and *Leucaena* species	[9]
Rhizobium NGR 234	70 genera	[10]
Rhizobium loti	*Lotus* species	[11]
Bradyrhizobium japonicum	*Glycine* and *Vigna* species	[12]
Azorhizobium caulinodans	*Sesbania rostrata*	[4]
Sinorhizobium fredii	*Glycine max, Glycine soja*	[8,10]

photosynthesis. These carbohydrates are utilized by bacteria for carbon and energy requirements [5].

A given bacterial symbiont is able to nodulate a limited range of legume hosts and similarly, a given legume can be nodulated by only a restricted number of bacterial species [6]. However, the degree of specificity varies for different rhizobia [3], with some species having a broad host range, while others have a more limited one. Table 35.1 illustrates about reported rhizobia–legumes symbiotic associations [7–12]. *R. leguminosarum* bv *viciae*, which only nodulates species of European pea (*Pisum sativum*) and vetch (*Vicia sativa*), is a classical example of a narrow host range bacteria [13]. On the other hand, *Rhizobium* sp strain NGR 234 forms nodules in more than 70 legume genera and also in the nonlegume *Parasponia* [14].

Legume–rhizobia symbiosis has been studied with great detail, perhaps much more than any other symbiotic process, mainly because of two reasons. First, the process of nodule development is interesting in itself, after each symbiont influences in the other important events such as gene expression, metabolism, cell division, and differentiation [2]. The second reason is the agricultural and ecological importance of legumes. Although nitrogen fixation occurs in nature in many different ways, symbiotically fixed nitrogen constitutes the most important part of the overall nitrogen fixed [15]. Thus, legume cultivation constitutes a natural way of improving nitrogen content in the soil, with the obvious advantage of avoiding the use of chemical fertilizers, which are expensive and also contribute to environmental pollution.

In this work, we describe the structure of nodules and the biochemistry of nitrogen fixation in legumes, and also briefly describe reduction of inorganic nitrogen forms in nonlegumes. We analyze the metabolism of ammonium assimilation and its relationships with the carbon flux within the plant cell.

II. NODULE DEVELOPMENT, STRUCTURE, AND FUNCTION

A. INITIATION

Initial steps of nodule formation involve an exchange of chemical signals between both partners. Leguminous plant roots exude quimiotactic substances, such as carbohydrates, amino acids, carboxylic acids and flavonoids, which attract rhizobia towards root hairs [16]. Besides their role as chemical attractants, flavonoids also regulate the expression of a set of nodulation genes (*nod*) of the bacteria [5,16]. In many *Rhizobium* species these genes are organized in operons and are located in episomal plasmids, whereas in *Bradyrhizobium* spp. they are chromosomal [17]. Function and regulation of rhizobial *nod* have been extensively described [1]. As a general rule, NodD protein plays a key role in the recognition of the induced flavonoid. Subsequently, NodD acts as a transcriptional activator of other *nod* genes. In general, this process of chemical signals exchange is highly specific, thus resulting in that a given legume is nodulated by only one or a few rhizobial species [18].

Certain carbohydrate-associated plant proteins known as lectins also play putative functions in nodule development [19]. These functions are associated with a differential distribution of lectins during development of legume root nodule [16,19], as follows: (i) lectins distributed at the surface of root hairs may promote the aggregation of rhizobia at the beginning of the infection thread development; (ii) in the nodule, primordium lectins may reduce the threshold of response to nodulation factors, thus stimulating mitotic activity; and (iii) lectins may constitute a reserve of the nitrogen fixed in the mature nodule tissue [19]. An unusual lectin exhibiting apyrase and Nod factor (see below) binding activities has been identified in *Dolichos biflorus* [20]. Although legume lectins have been studied intensively, the complete understanding of their functional role within plant tissues is far from complete. Further studies are necessary to clearly establish if these proteins are absolutely essential for nodulation and if introduction of a legume lectin into a nonlegume would result in effective rhizobial colonization [16].

Proteins coded by *nod* participate in the synthesis of Nod factors, a family of substituted lipo-oligosaccharides that elicit morphological changes in root hairs being critical for nodulation [21,22]. In fact, the responsiveness to Nod factors is one of the key traits that makes a distinction for the nodulating legumes respect to other plant species [16]. Normally straight root hairs become deformed, branched and curled, with Nod factors also inducing mitotic divisions in cortical root cells. Curled root hairs form a pocket-like structure, within which bacteria are entrapped, to subsequently penetrate plant cell walls by means of an infection thread. This tubular structure is mainly constituted of a matrix of plant-derived glycoproteins within which bacteria are enclosed [23]. The thread grows and reaches the cortical layers of cells, where it ramifies. It follows that part of the infection thread is degraded by a still not well-established mechanism and bacteria are taken into host cells by endocytosis or phagocytosis, a process not very common in plant cells [24]. Once there, they are surrounded by a peribacteroid membrane (PBM) derived from host cells and undergo morphological alterations, leading to the formation of bacteroids or symbiosomes. These modified bacteria divide and, depending on the species, each single bacterium remains enclosed within a sac of PBM or, on the contrary, several bacteria share a common one. PBM acts as a physical interface between both symbionts, its integrity being essential for a stable symbiosis. At this stage, bacteroids begin to fix atmospheric nitrogen. Although other strategies of root penetration (not involving infection threads development) exist in certain legumes such as *Arachis* and *Stylosanthes* [14,23], they are very uncommon routes of invasion.

B. Types of Nodules

During the early events of nodule formation summarized above, the sequence of cell division and invasion varies for different legumes. As a consequence, two types of nodule may result: indeterminate or determinate. The former are cylindrical in shape, containing cells undergoing division that are located in inner layers of the cortex, nearby the pericycle. In this kind of nodules, a group of cells of the cortex and the pericycle divide together and remain uninfected, constituting an apical meristem that grows outwards from the root; while another group of cells stop dividing and then are infected by rhizobia. Meristematic activity is permanent, continuously adding new cells to the nodule tissues [23]. Thus, a differentiation of cell types and functions are found along the longitudinal axis of an indeterminate nodule [25]. Indeterminate nodules are typically found in alfalfa, pea, and clover.

Determinate nodules arise in legumes such as soybean, birdsfoot trefoil, and common bean. These are spherical in shape. During development of this type of nodules, cortical cells division occurs in layers located just beneath the epidermis [26] and after that cell invasion by rhizobia took place [27]. In this case, meristematic tissue consists of a combination of infected and uninfected cortical cells, along with uninfected cells of the pericycle. Unlike indeterminate nodules, meristematic activity is transient in determinate ones.

In a given legume, infection and nodule formation always occur by the same mechanism, whichever bacterial species or strain is involved. The kind of nodule is also host determined and it does not depend on the bacterial partner [10].

C. Rhizobial Polysaccharides

This group of biomolecules plays an important role in the establishment of symbiosis. Exopolysaccharides (EPS) are important for the successful formation and invasion of indeterminate nodules [28] provided *Rhizobium* EPS mutants induce no nodules or empty ones in alfalfa and pea [23]. Similar mutations have no effect on the development of determinate nodules in trefoil, soybean, and bean [29]. On the other hand, lipopolysaccharides (LPS) are necessary for the establishment of determinate nodules, but they exhibit variable effects on the development of indeterminate ones. Thus, *Rhizobium* LPS mutants produce empty nodules in soybean and bean [30,31]; whereas they are able to form normal nodules in alfalfa but not in pea, where bacteroids release into host cells is impaired [23].

During the formation of an indeterminate nodule infection threads are the main way of bacterial dissemination in host cells and, consequently, an abnormal development of this structure should negatively affect nodule colonization. The fact that EPS seems to participate in the constitution of the luminal matrix of the thread [32] could explain the previously described lack of effect of EPS on indeterminate nodules development. Regarding development of a determinate nodule, cell to cell spread of bacteria is quite independent on the infection thread [33]. In this case, rhizobia are released in layers just beneath the epidermis [26] and endocytosis of bacteria by host cells occurs very early in the development of the nodule as compared with indeterminate ones [23]. As a consequence, bacterial spread occurs mainly by their division within host cells that are also dividing. LPS could play a role in the process of membrane fusion during endocytosis and bacteria lacking them should not favourably enter host cells. This could be the reason for the observed phenotype of *Rhizobium* LPS mutants cited above.

Cyclic β-glucans, a third group of cell surface bacterial carbohydrates are involved in the infection process. *R. meliloti* strains that do not synthesize these glucans (*ndv* mutants) form empty nodules in alfalfa [34], with a low number of infection threads which further abort early during the development of the nodule. These observations suggest that cyclic β-glucans participate in the late stages of nodulation. The family Rhizobiaceae includes both fast- and slow-growing species, and a relationship between the type of periplasmic cyclic glucan synthesized and the growth rate has been established [35]. Cyclic β-(1,2) or cyclic β-(1,6)-β-(1,3) are the glucans found in fast- or slow-growing species, respectively. Cyclic β-(1,6)-β-(1,3)glucans synthesized by *B. japonicum* elicit in soybean roots production of daidzein, an isoflavonoid that induces *nod* expression in the bacterium [36]. It was proposed that cyclic glucans may serve as modulators of isoflavonoids synthesis in roots, playing the role of suppressing defence response in the host during rhizobial invasion [37].

D. MATURATION

Uninfected cells differentiate into a variety of specialized types. Nodule parenchyma is separated from the outer cortex by nodule endodermis, a single sheet of cells with suberized cell walls that restricts lateral diffusion of solutes. Cell layers from nodule parenchyma immediately beneath the endodermis are also uninfected [38]. These cells are closely packed and few intercellular spaces are present between them. Therefore, they constitute an important barrier to oxygen diffusion, proven that this gas diffuses much more slower in an aqueous phase than through intercellular spaces [39].

Vascular tissue is also found peripherally located in the nodule. Towards the center of the nodule, small uninfected cells are intermingled with larger ones that contain rhizobia inside. The metabolism of uninfected cells differs markedly from that of infected ones [40]. Uninfected cells probably are part of a network that transports carbon substrates from the vascular tissue to the infected cells and organic nitrogen compounds in the opposite direction [20].

Biological nitrogen fixation takes place in the inner part of the nodule, where an oxygen level below 1% of the atmospheric concentration must be maintained due to the fact that although *Rhizobium* spp. are obligate aerobes, the enzyme nitrogenase is irreversibly inactivated by atmospheric oxygen concentrations [2]. The existence of a variable diffusion barrier has been postulated [41–43], although the exact mechanism of its regulation is still not clear [44,45]. In this respect, modifications in pO_2 lead to alterations in the frequency of intercellular spaces and the differentiation of cortical cells, which could be associated with changes in the permeability of cowpea nodules to gas diffusion [46]. Moreover, James et al. [47] demonstrated that rhizosphere O_2 levels affect the content of a glycoprotein that occludes intercellular spaces in the inner cortex of soybean nodules. A later study conducted on white lupin nodules [45] also suggested that cell wall and cell expansion along with glycoprotein mediated occlusion of intercellular spaces are involved in the operation of a variable diffusion barrier.

Leghemoglobins, a group of nodule-specific proteins that are present in high concentrations in the cytoplasm of infected cells, also participate in the regulation of intracellular O_2 concentration [48,49]. These proteins bind oxygen with high affinity and release it when intracellular concentration falls below a certain critical level. In this way, they provide rhizobia with the amount of oxygen necessary for respiration while they keep a low intracellular concentration of this gas in the free state, which would otherwise inactivate nitrogenase complex [48]. Studies carried out in pea have shown the presence of five leghemoglobin genes showing distinct patterns of spatial expression in nodules [49]. These genes were classified into two groups that express leghemoglobins exhibiting different O_2-binding affinities [49]. Another mechanism of regulation of O_2 incorporation into nodules involving the action of ascorbate peroxidase has been proposed by Dalton et al. [50], after determining the presence of high concentrations of this enzyme in the peripheral cell layers of nodules of several legumes. Ascorbate peroxidase prevents oxidative damage in plants by scavenging H_2O_2, a potentially harmful form of activated O_2 that tends to be produced in high quantities in nodules. This enzyme could be part of a diffusion barrier that controls the entry of oxygen into the nodule interior, thus protecting nitrogenase from inactivation.

Alterations of gene expression occurring in plant cells lead to drastic changes in the metabolism of oxygen, carbon, and nitrogen compounds. Many legume proteins are mainly expressed in cells forming part of the nodule, and so they are termed nodulins [51–54]. Examples are proteins that play specific roles in the nodule, as leghemoglobin, enzymes of carbon (sucrose synthase, [55]) or nitrogen (glutamine synthetase and glutamate synthetase, [56–59]) metabolism, proline-rich protein present in plant cell wall [60], and the protein of the symbiosome membrane nodulin 26. It has been proposed that nodulin 26 could be responsible for the movement of NH_3 as well as dicarboxylates across the peribacteriod membrane [61]. More recently, this membrane protein was identified as an aquaporin that is regulated by phosphorylation, and being involved in the response to osmotic changes [62]. Although a number of nodulin genes have been identified on the basis of their exclusive expression in the nodule, it is now clear that the proteins expressed by many of them are also found under nonsymbiotic conditions or in different plant tissues [16].

III. THE METABOLISM OF NITROGEN FIXATION IN LEGUMES

Nodule formation and functioning make legumes able to assimilate atmospheric N_2 to satisfy demands for this elemental nutrient. Symbiotic association between plants and rhizobia operates at a biochemical level. In this way, the plant provides the bacteria with metabolites for their nutrition and in turn the leguminous receives ammonia produced from nitrogen in air by nitrogenase, a prokaryotic enzyme. Ammonia is then metabolized to produce glutamate, one of the first organic forms of assimilated nitrogen, which is then widely utilized for the biosynthesis of different N-containing compounds [1,63].

Assimilation of atmospheric N_2 is a key reaction occurring in biosphere. Although a number of free-living bacteria are nitrogen fixers, the single greatest contribution to the assimilatory process comes from the symbiotic association between rhizobia and legumes [62]. In addition to the relevance for the agriculture, the biological nitrogen fixation process can play a key role in land remediation [64]. N_2 is one of the most inert molecules to react under normal laboratory conditions [65]. Chemical synthesis of ammonia from N_2 is normally produced by a process requiring high temperatures (400°C to 500°C) and several hundred atmospheres of pressure. Biologically, this reaction is very efficiently catalyzed by nitrogenase, a complex enzyme composed of multiple redox centers and found in a relatively few species of microorganisms, all of them prokaryotes [62,63,66].

The nitrogenase complex consists of two iron–sulfur proteins: dinitrogenase reductase and dinitrogenase [66,67]. The first is a homodimer of molecular mass 60 to 62 kDa, containing a single $[Fe_4–S_4]$ redox center and two binding sites for ATP. Dinitrogenase is an $\alpha_2\beta_2$ heterotetramer of molecular mass 200 to 240 kDa (α and β about 56 and 60 kDa, respectively) containing both iron and molybdenum. Redox centers of dinitrogenase have a total of 2 Mo, and between 24 and 32 Fe and S atoms per tetramer distributed in a called P-cluster (located at the $\alpha\beta$ interface) and a FeMo-cofactor [63,66,67].

In an atmosphere containing nitrogen gas, the reaction catalyzed by the nitrogenase complex is an associated reduction of N_2 and H^+, which can be described as follows:

$$N_2 + 10H^+ + 16ATP + 8e^- \Rightarrow 2NH_4^+ + H_2 + 16ADP + 16Pi$$

The mechanism for this reaction has been proposed by Thorneley and Lowe [68] as involving the sequential action of both proteins of the complex. The role of dinitrogenase reductase is to transfer electrons from a high-potential donor (i.e., ferredoxin) to the dinitrogenase component, a process followed by the binding of ATP which produces a conformational change in the protein. Electrons are transferred to the dinitrogenase, and ATP bound to the reductase is hydrolyzed, being the product (ADP) released from the protein. Electrons flow to the P-cluster and then to the FeMo-cofactor in the dinitrogenase, where finally nitrogen fixation takes place. After the transfer of four electrons to the FeMo-cofactor, the state of the dinitrogenase makes possible the binding of N_2 to this cofactor (which weakens the interaction between both N atoms in the molecule) and this results in a concomitant release of H_2. The following transfer of electrons is used for reduction of nitrogen to render ammonia [68].

Nitrogenase complex can also catalyze other reactions, utilizing the flow of electrons to reduce protons to molecular H_2 (in the absence of N_2) or to produce ethylene (in the presence of saturating concentrations of acetylene). The later reaction is usually used to measure nitrogenase activity [69].

Regulation of nitrogenase *in vivo* is exerted at different levels, including transcription, translation, substrate availability, covalent modification, and allosteric effectors [66]. However, the importance of each regulatory mechanism seems to be dependent of the species determining symbiosis. Consequently, a view of the regulation of the enzyme in the general picture of the nitrogen assimilation process was not established. Genes required for nitrogen fixation are organized in a cluster (*nif*) comprising 17 genes that are transcribed in eight adjacent operons. Nitrogenase is synthesized when bacteria are grown under anaerobic, nitrogen-limiting conditions and, contrarily, it is repressed by the presence of an excess of O_2 or nitrogen [63]. O_2 has also an effect at the level of the enzyme activity, since it is an irreversible inhibitor. It has been established that dinitrogenase reductase is the most O_2 labile component in the complex, being inactivated in air with a half-life less than 1 min (dinitrogenase inactivation occurs at about tenfold lower velocity) [69]. The high sensitivity of nitrogenase to O_2 inhibition is paradoxical with the requirement of the enzyme for ATP, since O_2 is also the substrate required for ATP production by oxidative phosphorylation. In legume symbiosis, additions of nitrate or ammonia produce a decrease in nitrogenase activity; with different mechanisms possibly accounting for this inhibition (i.e., disruption of membrane potential that indirectly affects enzyme activity) [66].

Ammonium is the primary stable product of nitrogen fixation and the major (if not the sole) nitrogen source secreted by the bacteroid [62]. Although the latter statement was challenged by some controversial results [62], it was strongly supported by recent studies using *in vivo* nuclear magnetic resonance spectroscopy and liquid chromatography combined with mass spectrometry [70]. In root nodules, fixed ammonium is exported from the bacteroid cytoplasm to the plant cytoplasm by diffusion across the membranes [62]. It has been proposed that the movement of the reduced form of nitrogen, as NH_4^+, could be facilitated by a proton pumping ATPase. Also, the movement of NH_3 through aquaporins (probably nodulin 26) has been shown. The general picture points out the importance of the pH in the different intracellular compartments that could determine distinct routes for the export of the fixed nitrogen [62]. In addition, the high activity of enzymes metabolizing ammonium in the host assure its rapid assimilation (ammonium is a toxic compound) [62,63,66].

As is clear from the reaction of nitrogenase, dinitrogen fixation has a strict requirement of energy. In the symbiotic process, this energetic demand is supplied by plant photosynthates. In this respect, it is important to consider that photosynthesis, respiration and nitrogen assimilation are interrelated processes [71]. In nodules, the flow of photosynthates is relevant not only to support the energy requirements of bacteroids but also to provide carbon skeletons necessary for nitrogen incorporation into organic compounds [62,71,72]. Carbon provided to the nodule by the host cell is derived from sucrose delivered by the sieve tubes. Sucrose is primarily metabolized

by sucrose synthase, an enzyme playing a key role in nitrogen assimilation, being included between nodulins [55,73,74]. Studies carried out with *Pisum sativum* mutants exhibiting severely reduced sucrose synthase activity clearly established the essential involvement of the enzyme to provide carbon skeletons for nitrogen fixation and to allow development of functional nodules [74]. One of the genes involved in nodule metabolism codifies for sucrose synthase and regulation of the enzyme by heme seems to play a role in controlling the flow of carbon [72,73].

Thus, sucrose transported through the phloem from the leaf is incorporated to degradative routes in the sink tissue to supply carbon intermediates to the bacteroid [62]. Main catabolism occurs via glycolysis to phosphoenolpyruvate, which is carboxylated by phosphoenolpyruvate carboxylase, to render keto acids necessary for synthesis of nitrogenated organic compounds. Excess of photosynthates are stored in the nodule as starch. Active starch accumulation occurs during early stages of nodule development, and a positive correlation was shown between the capacity of mature nodules to fix N_2 and their ability to degrade starch in order to supply demands of metabolic energy of bacteroids [72].

Catabolism of dicarboxylic acids is a main source fueling nitrogen fixation in the bacteroid [62]. In this respect, a dicarboxylate transport (Dct) system operative in rhizobia is relevant. In *Rhizobium*, the Dct system was characterized as involving three genes: *dctA*, *dctB*, and *dctD*; coding for a putative transport protein (DctA) and for a sensor-regulation protein pair (DctB plus DctD) involved in the activation of *dctA* transcription after the presence of dicarboxylates [75]. DctA has a typical structure of membrane transport proteins, with 12-membrane spanning helices and the N- and C-termini located in the cytoplasm [76]. Transport of dicarboxylates by DctA involves a H^+ symport mechanism, with a high affinity toward malate, fumarate, and succinate [62,75].

Rhizobia are obligate aerobes, thus having an active tricarboxylic acid (TCA) cycle. The latter metabolic route is mainly involved in the oxidation of dicarboxylic acids in the bacteroid to fuel nitrogen fixation [62]. Different studies have shown that in rhizobia the TCA cycle may be blocked at the 2-oxoglutarate dehydrogenase step and that a full cycle is not necessary for effective nodule function. Most probably, much of the carbon could be routed linearly and also reversibly diverted to pools of polyhydroxybutyrate and glycogen as well as to amino acids [62]. The function of storage pools in bacteroids is not clear. Different studies have shown that polyhydroxybutyrate and glycogen granules are present in early stages of nodule development, suggesting that these polymers could play a role in the process. However, in mature, active nodules, synthesis of these compounds (specially glycogen) could compete with nitrogenase by energetic substrates. Studies with rhizobia mutants with null activity of enzymes of the glycogen have shown contradictory results respect to their abilities to form functional nodules [77,78]. Probably, these discrepancies are related with the association of glycogen and EPS synthesis, since both require of sugar nucleotides, ADPglucose and UDPglucose, respectively [62].

IV. REDUCTION OF NITRATE TO AMMONIUM

Nitrate assimilation in plants is intiated by the import of the anion by cells, a process mediated by specific transporters. Nitrate uptake and assimilation are processes highly regulated in relation with the whole plant metabolism and nutritional status [79,80]. Several genes that code for nitrate transporters have been identified in Arabidopsis [79]. These genes were grouped into two families (*NRT1* and *NRT2*), each including genes differentially regulated and encoding transporters with distinct kinetic properties. Nitrate itself induces *NRT* genes, and this is upregulated by sugars. Recent studies using gene expression have shown that nitrite is able to repress genes involved in nitrogen uptake, mainly from the *NRT1* family [80].

Once in the cytoplasm of plant cells, nitrate is reduced by nitrate reductase (NR):

$$NO_3^- + NADH \Rightarrow NO_2^- + NAD^+ + OH^-$$

This reaction is the rate-limiting, highly regulated step in nitrogen assimilation. In algae and higher plants two forms of NR are found, one NADH-specific (EC 1.6.6.1) and one NAD(P)H-specific form (EC 1.6.6.2) [81]. The enzyme is a homodimer of molecular mass about 200 kDa, with each monomer containing three redox centers: FAD, heme-iron and molybdenum-molybdopterin.

Eight sequence segments have been identified in NADH-NR [81]. One of them is the hinge-1 region that links the molybdenum cofactor and heme-iron domains and contains a serine residue that is phosphorylated by a calmodulin-domain protein kinase (reviewed in Refs. [82,83]). After phosphorylation of the seryl residue the hinge-1 region becomes a recognition site for 14-3-3 proteins. It has been proposed that NR has a second binding site for 14-3-3, although its specific location was not characterized [82]. The binding of these regulatory proteins pro- duces inactivation of NR by blocking the electron flow between the cytocrome- and molybdenum-cofactor domains. Inactivation of NR occurs in darkened leaves. Binding of 14-3-3 exhibits an additional effect since it influences sensitivity of NR to proteolytic degradation [82,83].

NR also catalyzes reduction of nitrite, generating nitric oxide (NO), and peroxinitrite [84]. These additional reactions are particularly important, and it is thought that they mainly occur under stress conditions. NO have been identified as a versatile signal molecule playing key roles in a broad specrum of pathophysiological and developmental processes in plants [85].

Reduction of nitrite to ammonium is the last step of nitrate assimilation. It takes place in the plastid through a reaction catalyzed by nitrite reductase (NiR), an enzyme that utilizes reduced ferredoxin (Fd) as an electron donor [86]:

$$NO_2^- + 6Fd(reduced) \Rightarrow NH_4^+ + 6Fd(oxidized)$$

NiR is a nuclear-encoded enzyme exhibiting a monomeric structure (molecular mass about 60 kDa). The enzyme has two redox centers formed by a siroheme and a $[Fe_4S_4]$ cluster. Binding of both redox cofactors involves conserved cystein residues in the protein [86].

V. INCORPORATION OF AMMONIUM INTO ORGANIC COMPOUNDS

Ammonium produced by nitrate assimilation as well as that derived from nitrogen fixation and exported from the bacteroids to the host plant is rapidly assimilated via the joint action of glutamine synthetase (GS) and glutamate synthase (GOGAT, for glutamate 2-oxoglutarate aminotransferase) [1,62,63,66,87,88]. These two enzymes constitute the so-called GS/GOGAT system (also known as the glutamate synthase cycle), which is the primary pathway for ammonia assimilation in plants:

GS: $NH_4^+ + glutamate + ATP \Leftrightarrow glutamine + ADP + Pi$

GOGAT: $glutamine + 2\text{-}oxoglutarate + 2e^- \Leftrightarrow 2glutamate$

Sum GS/GOGAT: $NH_4^+ + 2 - oxoglutatarate + 2e^- + ATP \Leftrightarrow glutamate + ADP + Pi$

The reaction catalyzed by GS requires a divalent cation and the enzyme exhibits a high affinity for ammonium ($K_m \cong 10$ to $50 \, \mu M$). Different isoforms of GS (including cytosolic and plastidic forms) can be found throughout the plant and root nodules of all legumes contain multiple isoenzymes; being cytosolic forms those mainly involved in the assimilation of ammonia fixed by rhizobia [1,62,66]. GS enzyme can constitute up to 2% of the total soluble protein in organs actively assimilating NH_4^+ and its activity highly increases during the development of legume root nodule [1]. Octameric structures have been established for native GS from plants, the enzyme being composed of a single subunit of molecular mass 38 to 46 kDa depending on wheter localized in the cytosol (GS_1) or in plastids (GS_2) [1,63]. Several genes encoding GS_1 have been sequenced [89]. On the contrary, the GS_2 isoenzyme is encoded by a single gene.

GS catalyzes a key regulatory step in ammonium assimilation in plants. Both GS_1 and GS_2 isoforms are target for regulation via posttranslational modification (phosphorylation) followed by interaction with 14-3-3 proteins [82,83]. Modification of GS_1 is reversible and catalyzed by a protein kinase and a microcystin-sensitive serine/threonine protein phosphatase [83]. Phosphorylation and 14-3-3 binding increase the activity of GS_1 and also reduce susceptibility of the enzyme to proteolytic degradation. The phosphorylation status of GS_1 varies during light–dark transition. Also, phosphorylation of the enzyme increases during senescence and it is thought that this mechanism is important for nitrogen remobilization [83]. GS_2 was also found to interact with 14-3-3, with the binding of these regulatory proteins being associated with an increase in the degradation of the enzyme [82].

Two different isoforms of GOGAT are found in higher plants: ferredoxin-dependent (Fd-) GOGAT (EC 1.4.7.1) and NADH-GOGAT (EC 1.4.1.14) [87,88]. The former is a monomeric (molecular mass 130 to 180 kDa), iron–sulfur flavoprotein mainly involved in assimilation of NH_4^+ generated by photorespiration or derived from NO_3^- reduction. Fd-GOGAT from spinach contains one FMN and one [3Fe–4S] cluster per molecule [88]. Two genes for Fd-GOGAT (*GLU1* and *GLU2*) have been identified, apparently expressing the enzyme in leaves (*GLU1*) and nonphotosynthetic tissues (*GLU2*) [87]. The enzyme has been found localized solely in chloroplasts in leaves and green algae. Concerning heterotrophic plant cells, it has been shown that Fd-GOGAT is also restricted to plastids [88]. Levels of the enzyme are affected by light conditions and availability of nitrogen sources.

NADH-GOGAT is also an iron–sulfur flavoprotein of molecular mass 190 to 230 kDa primarily found in nongreen tissues of plants [87]. The plant enzyme is monomeric in structure, which differs from the bacterial enzyme characterized as a dimer comprised of two dissimilar subunits [88]. NADH-GOGAT is a plastidial enzyme and its activity is dramatically increased during nodule development [87,88]. It has been hypothesised that NADH-GOGAT is involved in the rate-limiting step of ammonia assimilation in root nodules [1] and the differential increase in its expression *in vivo* was shown to be associated with the formation of effective nodules [87]. From these results, it was proposed that maximal gene expression of the enzyme requires active nitrogen fixation, the process being regulated by NH_4^+ or other derived metabolites [87,88].

From the above, it is clear that the GS/GOGAT system utilizes 2-oxoglutarate as the metabolite supplying the carbon skeleton necessary for ammonium assimilation. The exact enzymatic origin of this keto acid in plant metabolism is unknown. It has been proposed that different isoforms os NADP-dependent isocitrate dehydrogenase (present in mitochondria and cytosol) could be involved in such a function [90]. The enzyme catalyzes the reaction:

$$\text{isocitrate} + NADP^+ \Rightarrow \text{2-oxoglutarate} + \\ NADPH + H^+$$

Another key function thought to be played by isocitrate dehydrogenase is the production of NADPH for redox-regulated plant cell metabolism [90].

In plants, it is also found glutamate dehydrogenase (GDH) [91], a mitochondrial enzyme catalyzing synthesis of glutamate as follows:

$$NH_4^+ + \text{2-oxoglutarate} + NAD(P)H + H^+ \Leftrightarrow \\ \text{glutamate} + NAD(P)^+$$

Prior to 1970 (when GOGAT was discovered), GDH was considered the key enzyme for ammonia assimilation. However, the low affinity of plant GDH for NH_4^+ ($K_m > 1 \, mM$, see above for the value corresponding to GS, for a comparison) suggested a minor involvement of the enzyme in nitrogen assimilation, with possible functions in processes of ammonia detoxification [1,66]. Studies using nitrogen isotopes, enzymes inhibitors, and different plant mutants demonstrated that the main route for nitrogen assimilation is the GS/GOGAT system. Possible secondary roles for GDH have been proposed as the enzyme involved in anaplerotic functions, replacing amino acids or producing 2-oxoglutarate (in the reverse reaction) for the replenishing of the TCA cycle during protein catabolism [1,66]. The latter is reinforced by results showing that GDH activity increases during periods of active amino acids catabolism such as germination and senescence [86]. Studies using transformed tobacco and corn plants overexpressing bacterial GDH suggest that the enzyme could play a role in stress conditions [91]. Thus, plants expressing enhanced GDH activity showed an increased tolerance to water stress accompanied by an increase in biomass and yield. From the flexible biochemical properties and catalytic properties exhibited by the enzyme, it has also been proposed its involvement in sensing the redox status of the plant representing a stress monitoring protein [91].

Glutamine and glutamate, the products of primary ammonia assimilation participate as nitrogen donors in many cellular reactions, mainly those catalyzed by aspartate aminotransferase and asparagine synthetase, which synthesize aspartate and asparagine, respectively [1]. Aspartate aminotransferase exists as distinct isoenzymes, which seem to be related to different roles played by the enzyme according to the plant metabolic status. High levels of asparagine synthetase activity were found in nitrogen-fixing root nodules, thus suggesting a key role for the enzyme and the relevance of asparagine as a compound involved in the transport of nitrogen in plants [1,86].

VI. CONCLUDING REMARKS

Assimilation of nitrogen by plants is a main process, mainly because the essentiality and limiting status of this nutrient. The efficiency by which plants incorporate nitrate, nitrite, and dinitrogen are critical in determining growth and yield in crops. From this, the understanding of the functioning of metabolic routes for nitrogen utilization by photosynthetic organisms is of critical relevance. The functioning of nitrate and nitrite metabolism in nonlegumes is a highly regulated process that is coordinately operative with carbon photoassimilation and partitioning. In legumes the metabolic scenario is even more complex since the existence of a synchronized symbiotic association between plant and rhizobia allowing dinitrogen fixation. The isolation and characterization of different genes and enzymes involved in nitrogen assimilation, together with the construction of several mutants and genetically transformed plants have afforded key new insights for the understanding of nitrogen metabolism and its regulation. It is visualized that with the realization of proteomic, transcriptomic, and metabolomic investigations a quite clear map will be available in the near future. This will be relevant for rationally manipulate crops to improve carbon and nitrogen incorporation into biomass.

ACKNOWLEDGMENTS

This work was supported, in part, by a grant from ANPCyT (PICT'99 1-6074). MJE is a member of the research assistant career from Comisión de Investigaciones Científicas (CIC, Bs. As.). A.A.I. is a research staff member from Consejo Nacional de Investigaciones Científicas y Técnicas (CONICET, Argentina).

REFERENCES

1. Lam HM, Coschigano KT, Oliveira IC, Melo-Oliveira R, Coruzzi GM. The molecular genetics of nitrogen assimilation into amino acids in higher plants. *Annu. Rev. Plant Physiol. Plant Mol. Biol.* 1996; 47:569–593.
2. Long S. *Rhizobium*-legume nodulation: life together in the underground. *Cell* 1989; 56: 203–214.
3. Young JM, Kuykendall CE, Martínez-Romero E, Kerr A, Sawada H. A revision of *Rhizobium* Frank 1889, with an emended description of the genus, and the inclusion of all species of *Agrobacterium* Conn 1942 and *Allorhizobium undicola* de Lajudie et al. 1998 as new combinations: *Rhizobium radiobacter*, *R. rhizogenes*, *R. rubi*, *R. undicola* and *R. vitis*. *Int. J. Syst. Evol. Microbiol.* 2001; 51:89–103.
4. Dreyfus B, García JL, Gillis M. Characterization of *Azorhizobium caulinodans* gen. nov., sp. nov., a stem-nodulating nitrogen-fixing bacterium isolated from *Sesbania rostrata*. *Int. J. Syst. Bacteriol.* 1988; 38:89–98.
5. Spaink HP. The molecular basis of infection and nodulation by rhizobia: the ins and outs of sympathogenesis. *Annu. Rev. Phytopathol.* 1995; 33:345–368.
6. Fusher RF, Long SR. *Rhizobium*-plant signal exchange. *Nature* 1992; 357:655–660.
7. Baldwin IL, Fred EB. Nomenclature of the root-nodule bacteria of the Leguminosae. J. *Bacteriol.* 1929; 17:141–150.
8. Jordan DC. Family III. Rhizobiaceae Conn 1938, 321. In: Krieg NR, Holt JG, eds. *Bergey's Manual of Systematic Bacteriology.* Baltimore, MD: Williams and Wilkins, 1984:234–244.
9. Martínez-Romero E, Segovia L, Mercante FM, Franco AA, Graham P, Pardo MA. *Rhizobium tropici*, a novel species nodulating *Phaseolus vulgaris* L. beans and *Leucaena* sp. trees. *Int. J. Syst. Bacteriol.* 1991; 41: 417–426.
10. Trinick MJ. Relationship amongst the fast growing rhizobia of *Lablab purpureus*, *Leucaena leucocephala*, *Mimosa* spp., *Acacia farnesiana* and *Sesbania grandiflora* and their affinities with other rhizobial groups. J. *Appl. Bacteriol.* 1980; 49:39–53.
11. Jarvis BDW, Pankhurst CE, Patel JJ. *Rhizobium loti*, a new species of legume root nodule bacteria. *Int. J. Syst. Bacteriol.* 1982; 32:378–380.
12. Jordan DC. Transfer of *Rhizobium japonicum* Buchanan 1980 to *Bradyrhizobium* gen. nov., a genus of slow growing root-nodule bacteria from leguminous plants. *Int. J. Syst. Bacteriol.* 1982; 32:136–139.

13. Lerouge P. Symbiotyc host specificity between leguminous plants and rhizobia is determined by substituted and acylated glucosamine oligosaccharide signals. *Glycobiology* 1994; 4: 127–134.

14. Dénarié J, Debellé F. *Rhizobium* lipo-chitooligosaccharide nodulation factors: signalling molecules mediating recognition and morphogenesis. *Annu. Rev. Biochem.* 1996; 65:503–535.

15. Gresshoff PM, Roth LE, Stacey G, Newton WE. *Nitrogen Fixation: Achievements and Objectives.* New York: Chapman and Hall, 1990.

16. Hirsch AM, Lum MR, Downie JA. What makes the rhizobia-legume symbiosis so special? *Plant Physiol.* 2001; 127:1484–1492.

17. Pankhurst CE, MacDonald PE, Reeves JM. Enhanced nitrogen fixation and competitiveness for nodulation of *Lotus pedunculatus* by a plasmid-cured derivative of *Rhizobium loti. J. Gen. Microbiol.* 1986; 132:2321–2328.

18. Scott DB, Young CA, Collins-Emerson JM, Terzaghi EA, Rockman ES, Lewis PE, Pankhurst CE. Novel and complex chromosomal arrangement of *Rhizobium loti* nodulation genes. *Mol. Plant. Microbe Interact.* 1996; 9:187–197.

19. Brewin NJ, Kardailsky IV. Legume lectins and nodulation by *Rhizobium. Trends Plant Sci.* 1997; 2:92–98.

20. Etzler ME, Kalsi G, Ewing NN, Roberts NJ, Ezy RB, Murphy JB. A Nod factor binding lectin with apyrase activity from legume roots. *Proc. Natl. Acad. Sci. USA* 1999; 96; 4704–4709.

21. Downie JA, Walker SA. Plant responses to nodulation factors. *Curr. Opin. Plant Biol.* 1999; 2:483–489.

22. Cullimore JV, Ranjeva R, Bono JJ. Perception of lipochitooligosaccharidic Nod factors in legumes. *Trends Plant Sci.* 2001; 6:24–30.

23. Brewin NJ. Development of the legume root nodule. *Annu. Rev. Cell Biol.* 1991; 7:191–226.

24. Udvardy MK, Day DA. Metabolite transport across symbiotic membranes of legume nodules. *Annu. Rev. Plant Physiol. Plant Mol. Biol.* 1997; 48:493–523.

25. Vasse J, de Billy F, Camut S, Truchet G. Correlation between ultrastructural differentiation of bacteroids and nitrogen fixation in alfalfa nodules. *J. Bacteriol.* 1990; 172:4295–4306.

26. Mathews A, Carroll BJ, Gresshoff PM. Development of *Bradyrrhizobium* infections in supernodulating and non-nodulating mutants of soybean (*Glycine max* "L." Merrill). *Protoplasma* 1989; 150:40–47.

27. Rolphe BG, Gresshoff PM. Genetic analysis of legume nodule initiation. *Annu. Rev. Plant Physiol. Plant Mol. Biol.* 1988; 39:297–319.

28. Gray JX, Rolfe BG. Exopolysaccharide production in *Rhizobium* and its role in invasion. *Mol. Microbiol.* 1990; 4:1425–1431.

29. Hotter GS, Scott DB. Exopolysaccharide mutants of *Rhizobium loti* are fully effective on a determinate nodulating host but are ineffective on an indeterminate nodulating host. *J. Bacteriol.* 1991; 173:851–859.

30. Puvanesarajah V, Schell FM, Gerhold D, Stacey G. Cell surface polysaccharide from *Bradyrrhizobium japo-nicum* and a nonnodulating mutant. *J. Bacteriol.* 1987; 169:137–141.

31. Noel KD, Vandenbosch KA, Kulpaca B. Mutation in *Rhizobium phaseoli* that leads to arrested development of infection threads. *J. Bacteriol.* 1986; 168:1392–1401.

32. Keller M, Muller P, Simon R, Puhler A. *Rhizobium meliloti* genes for exopolysaccharide synthesis and nodule infection located on megaplasmid 2 are actively transcribed during symbiosis. *Mol. Plant Microbe Interact.* 1988; 1:267–274.

33. Sprent JI. Which steps are essential for the formation of functional legume nodules?. *New Phytol.* 1989; 111:129–153.

34. Dylan T, Ielpi L, Stanfield S, Kashyap L, Douglas C, Yanofsky M, Nester E, Helinski DR, Ditta G. *Rhizobium meliloti* genes required for nodule development are related to chromosomal virulence genes in *Agrobacterium tumefaciens. Proc. Natl. Acad. Sci. USA* 1986; 83:4403–4407.

35. Estrella MJ, Pfeffer PE, Brouillette JN, Ugalde RA, Iñón de Iannino N. Biosynthesis and structure of cell associated glucans in the slow growing *Rhizobium loti* strain NZP 2309. *Symbiosis* 2000; 29:173–199.

36. Darvill AG, Albersheim P. Phytoalexins and their elicitors a defence against microbial infection in plants. *Annu. Rev. Plant Physiol.* 1984; 35:243–275.

37. Bhagwat AA, Mithöfer A, Pfeffer PE, Kraus C, Spickers N, Hotchkiss A, Ebel J, Keister DL. Further studies of the role of cyclic β-glucans in symbiosis. An *ndvC* mutant of *Bradyrhizobium japonicum* synthesizes cyclodecakis-(1-3)-β-glucosyl. *Plant Physiol.* 1999; 119:1057–1064.

38. van de Wiel C, Scheres B, Franssen H, van Lierop MJ, van Lammerem A, van Kammen A, Bisseling T. The early nodulin transcript ENOD2 is located in the nodule parenchyma (inner cortex) of pea and soybean root nodules. *EMBO J.* 1990; 9:1–7.

39. Parsons R, Day DA. Mechanism of soybean nodule adaptation to different oxygen pressures. *Plant Cell Environ.* 1990; 13:501–512.

40. Scheres B, van Engelen F, van der Knaap E, van de Viel C, van Kammen A, Bisseling T. Sequential induction of nodulin gene expression in the developing pea nodule. *Plant Cell* 1990; 2:687–700.

41. Sheehy JE, Minchin FR, Witty JF. Biological control of the conductance to oxygen flux in nodules. *Ann. Bot.* 1983; 52:565–562.

42. Hunt S, King BJ, Canvin DT, Layzell DB. Steady and non steady state gas exchange characteristics of soybean nodules in relation to the oxygen diffusion "barrier." *Plant Physiol.* 1987; 84:164–172.

43. Weisz PR, Sinclair TR. Regulation of soybean nitrogen fixation in response to rhizosphere oxygen. I. Role of nodule respiration. *Plant Physiol.* 1987; 84:900–905.

44. Serraj R, Roy G, Drevon JJ. Salt-stress induces a decrease in the oxygen uptake of soybean nodules and in their permeability to oxygen diffusion. *Physiol. Plant.* 1994; 91:161–168.

45. Iannetta PPM, James EK, Sprent JI, Minchin FR. Time course of changes involved in the operation of

the oxygen diffusion barrier in white lupin nodules. *J. Exp. Bot.* 1995; 46:565–575.

46. Dakora F, Atkins CA. Morphological and structural adaptation of nodules of cowpea to functioning under sub- and supra-ambient oxygen pressure. *Planta* 1990; 182:572–582.

47. James EK, Sprent JI, Minchin FR, Brewin NJ. Intercellular location of glycoprotein in soybean nodules: effect of altered rhizosphere oxygen concentration. *Plant Cell Environ.* 1991; 14:467–476.

48. Appleby CA. Leghemoglobin and *Rhizobium* respiration. *Annu. Rev. Plant Physiol.* 1984; 35:443–478.

49. Kawashima K, Suganuma N, Tamaoki M, Kouchi H. Two types of pea leghemoglobin genes showing different O2-binding affinities and distinct patterns of spatial expression in nodules. *Plant Physiol.* 2001; 125: 641–651.

50. Dalton DA, Joyner SL, Becana M, Iturbe-Ormaetxe I, Chatfield M. Antioxidant defences in the peripheral cell layers of legume root nodules. *Plant Physiol.* 1998; 116:37–43.

51. Legocki RP, Verma DPS. Identification of "nodule-specific" host proteins (nodulins) in soybean involved in the development of *Rhizobium*-legume symbiosis. *Cell* 1980; 20:153–163.

52. Fuller F, Kunstner PW, Nguyen T, Verma DPS. Soybean nodulin genes: analysis of cDNA clones reveals several major tissue specific sequences in nitrogen-fixing root nodules. *Proc. Natl. Acad. Sci. USA* 1983; 80:2594–2598.

53. Govers F, Gloudemans T, Moerman M, van Kammen A, Bisseling T. Expression of plant genes during the development of pea root nodules. *EMBO J.* 1985; 4:861–867.

54. Vance CP, Boylan KLM, Stade S, Somers DA. Nodule specific proteins in alfalfa (*Medicago sativa* L.) *Symbiosis* 1985; 1:69–84.

55. Thummler F, Verma DPS.. Nodulin-100 of soybean is the subunit of sucrose synthase regulated by the availability of free heme in nodules. *J. Biol. Chem.* 1987; 262:14730–14736.

56. Sengupta-Gopalan C, Pitas JW, Thompson DV, Hoffman LM. Expression of nodule-specific glutamine synthetase genes during nodule development in soybeans. *Plant Mol. Biol.* 1986; 7:189–199.

57. Tingey SV, Walker EL, Coruzzi GM. Glutamine synthetase genes of pea encode distinct polypeptides which are differentially expressed in leaves, roots and nodules. *EMBO J.* 1987; 6:1–9.

58. Cullimore JV, Bennett MJ. The molecular biology and biochemistry of plant glutamine synthtase from root nodules of *Phaseolus vulgaris* L. and other legumes. *J. Plant Physiol.* 1988; 132:387–393.

59. Cullimore JV, Gebhardt C, Saarelainen R, Miflin BJ, Idler KB, Barker RF. Glutamine synthetase from *Phaseolus vulgaris* L.: organ-specific expression of a multigene family. *J. Mol. Appl. Genet.* 1984; 2:589–599.

60. Wilson RC, Long F, Maruoka M, Cooper JB. A new proline-rich early nodulin form Medicago truncatula is highly expressed in nodule meristematic cells. *Plant Cell* 1994; 6:1265–1275.

61. Guenther JF, Chanmanivone N, Galetovic MP, Wallace IS, Cobb JA, Roberts DM. Phosphorylation of soybean nodulin 26 on serine 262 enhances water permeability and is regulated developmentally and by osmotic signals. *Plant Cell* 2003; 15:981–991.

62. Lodwig E, Poole P. Metabolism of *Rhizobium* bacteroids. *Crit. Rev. Plant Sci.* 2003; 22:37–78.

63. Vance CP, Griffith SM. The molecular biology of N metabolism. In: Dennis DT, Turpin DH, eds. *Plant Physiology, Biochemistry and Molecular Biology.* Essex: Longman Scientific & Technical, 1990:373–388.

64. Zahran HH. Rhizobium–legume symbiosis and nitrogen fixation under severe conditions and in an arid climate. *Microbiol. Mol. Biol. Rev.* 1999; 63:968–989.

65. Nishibayashi Y, Iwai S, Hidai M. Bimetallic system for nitrogen fixation: ruthenium-assisted protonation of coordinated N_2 on tungsten with H_2. *Science* 1998; 279:540–542.

66. Layzell DB. N_2 fixation, NO_3^- reduction and NH_4^+ assimilation. In: Dennis DT, Turpin DH, eds. *Plant Physiology, Biochemistry and Molecular Biology.* Essex: Longman Scientific & Technical, 1990:389–406.

67. Lanzilotta WN, Fisher K, Seefeldt LC. Evidence for electron transfer-dependent formation of a nitrogenase iron protein–molybdenum–iron protein tight complex. The role of aspartate 39. J. *Biol. Chem.* 1997; 272:4157–4165.

68. Thorneley RNF, Lowe DJ. Kinetics and mechanisms of the nitrogenase enzyme system. In: Spiro TG, ed. *Molybdenum Enzymes.* New York: John Wiley & Sons, 1985:220–284.

69. Hunt S, Layzell DB. Gas exchange of legume nodules and the regulation of nitrogenase activity. *Annu. Rev. Plant Physiol. Plant Mol. Biol.* 1993; 44: 483–511.

70. Scharff AM, Egsgaard H, Hansen PE, Rosendahl L. Exploring symbiotic nitrogen fixation and assimilation in pea root nodules by *in vivo* [15]N nuclear magnetic resonance spectroscopy and liquid chromatography–mass spectrometry. *Plant Physiol.* 2003; 131:367–378.

71. Turpin DH, Weger HG. Interactions between photosynthesis, respiration and nitrogen assimilation. In: Dennis DT, Turpin DH, eds. *Plant Physiology, Biochemistry and Molecular Biology.* Essex: Longman Scientific & Technical, 1990:422–433.

72. Forrest SI, Verma DPS, Dhindsa RS. Starch content and activities of starch-metabolizing enzymes in effective and ineffective root nodules of soybean. *Can. J. Bot.* 1991; 69:697–701.

73. González EM, Gordon AJ, James CL, Arrese-Igor C. The role of sucrose synthase in the response of soybean nodules to drought. J. *Exp. Bot.* 1995; 46:1515–1523.

74. Gordon AJ, Minchin FR, James CL, Komina O. Sucrose synthase in legume nodules is essential for nitrogen fixation. *Plant Physiol.* 1999; 120:867–877.

75. Jording D, Sharma PK, Schmidt R, Engelke T, Uhde C, Pühler A. Regulatory aspects of the C_4-dicarboxylate transport in *Rhizobium meliloti*-transcriptional activation and dependence on effective symbiosis. *J. Plant Physiol.* 1992; 141:18–27.

76. Jording D, Pühler A. The membrane topology of the *Rhizobium meliloti* C$_4$-dicarboxylate permease (DctA) as derived from protein fusions with *Escherichia coli* K12 alkaline phosphatase (PhoA) and beta galactosidase (LacZ). *Mol. Gen. Genet.* 1993; 241:106–114.

77. Marroquí S, Zorreguieta A, Santamaría C, Temprano F, Soberón M, Megías M, Downie JA. Enhanced symbiotic performance by *Rhizobium tropici* glycogen synthase mutants. J. *Bacteriol.* 2001; 183:854–864.

78. Lepek VC, D'Antuono AL, TOmatis PE, Ugalde JE, Giambiagi S, Ugalde RA. Analysis of *Mesorhizobium loti* glycogen operon: effect of phosphoglucomutase (pgm) and glycogen synthase (glgA) null mutants on nodulation of *Lotus tenuis*. *Mol. Plant Microbe Interact.* 2002; 15:368–375.

79. Forde BG. Local and long-range signaling pathways regulating plant responses to nitrate. *Annu. Rev. Plant Physiol. Plant Mol. Biol.* 2002; 53:203–224.

80. Loqué D, Tillard P, Gojon A, Lepetit M. Gene expresión of the NO$_3^-$ transporter NRT1.1 and the nitrate reductase NIA1 is repressed in Arabidopsis roots by NO$_2^-$, the product of NO$_3^-$ reduction. *Plant Physiol.* 2003; 132:958–967.

81. Campbell WH. Nitrate reductase structure, function and regulation: bridging the gap between biochemistry and physiology. *Annu. Rev. Plant Physiol. Plant Mol. Biol.* 1999; 50:277–303.

82. Huber SC, MacKintosh C, Kaiser WM. Metabolic enzymes as targets for 14-3-3 proteins. *Plant Mol. Biol.* 2002; 50:1053–1063.

83. Comparot S, Lingiah G, Martin T. Function and specificity of 14-3-3 proteins in the regulation of carbohydrate and nitrogen assimilation. *J. Exp. Bot.* 2003; 54:595–604.

84. Yamasaki H, Sakihama Y. Simultaneous production of nitric oxide and peroxynitrite by plant nitrate reductase: *in vitro* evidence for the NR-dependen formation of active nitrogen species. *FEBS Lett.* 2000; 468:89–92.

85. Lamattina L, García-Mata C, Graziano M, Pagnussat G. Nitric oxide: the versatility of and extensive signal molecule. *Annu. Rev. Plant Biol.* 2003; 54:109–136.

86. Fernández E, Galván A, Quesada A. Nitrogen assimilation and its regulation. In: Rochaix JD, Goldschmidt-Clermont M, Merchant S, eds. *The Molecular Biology of Chloroplast and Mitochondria in Chlamydomonas.* Dordrecht: Kluwer Academic Publishers, 1998:637–659.

87. Temple SJ, Vance CP, Gantt JS. Glutamate synthase and nitrogen assimilation. *Trends Plant Sci.* 1998; 3: 51–56.

88. Lea PJ, Miflin BJ. Glutamate synthase and the synthesis of glutamate in plants. *Plant Physiol. Biochem.* 2003; 41:555–564.

89. Ochs G, Schock G, Trischler M, Kosemund K, Wild A. Complexity and expression of the glutamine synthetase multigene family in the amphidiploid crop *Brassica napus*. *Plant Mol. Biol.* 1999; 39:395–405.

90. Hodges M, Flesch V, Gálvez S, Bismuth E. Higher plant NADP-dependent isocitrate dehydrogenases, ammonium assimilation and NADPH production. *Plant Physiol. Biochem.* 2003; 41:577–585.

91. Dubois F, Tercé-Laforgue T, Gonzalez-Moro MB, Estavillo JM, Sangwan R, Gallais A, Hirel B. Glutamate dehydrogenase in plants: is there a new story for an old enzyme? *Plant Physiol. Biochem.* 2003; 41:565–576.

36 Leaf Senescence and Photosynthesis

Agnieszka Mostowska
Department of Plant Anatomy and Cytology, Institute of Experimental Biology of Plants, Warsaw University

CONTENTS

I. INTRODUCTION

Senescence is a developmental stage of the plant life cycle leading to death of specific cells or whole organisms.

Leaf senescence is the last, genetically controlled, phase of leaf ontogenesis. Various aspects of leaf senescence, physiological, biochemical, molecular, and anatomical, were analyzed in the literature (for reviews see Refs. [1–17]). Leaf senescence, as the last phase of leaf ontogenesis, is a period of subsequent series of events involving cessation of many processes, such as chloroplast ultrastructure disintegration, loss of chlorophyll (Chl), breakdown of leaf proteins, diminished level of ribulose-1,5-bisphosphorane carboxylase (RuBisCO), and loss of photosynthetic capability. All this eventually leads to cell death [1,2,8,18,19].

Many authors consider leaf senescence as a type of programmed cell death (PCD) [5,7–9,11,12,20]. The activation of cell-death-associated hydrolytic enzymes, protein degradation, and breakage of nuclear DNA (nDNA) strands are the main symptoms of PCD [9,12]; however, the regulatory mechanisms that control PCD are still not clear.

Leaf senescence can be initiated by endogenous factors connected with regulation of this process as a natural stage of development (developmentally induced senescence). Natural senescence starts when the leaf reaches a certain age or when the plant reaches the reproductive phase. For example, in *Arabidopsis*, a plant with a very short life cycle,

691

senescence starts shortly after the leaf reaches the final dimensions [21]. Leaf senescence can be also induced by external environmental factors that cause premature leaf senescence (stress-induced senescence) [10].

The environmental or even stress factors inducing leaf senescence include drought, temperature extremes, intense light, UV radiation, herbicides, shading, wounding, or pathogen infection, and others [22,23]. Another form of leaf senescence takes place when crop plants, such as lettuce, asparagus, broccoli, cabbage are harvested before maturation. In that case leaves of these plants also show senescence symptoms (postharvest senescence).

Each type of senescence is a long-lasting process with specifically programmed gradual disintegration of cell components, leading to massive release of proteins, phospholipids, galactolipids, and nucleic acids and transport of their products to growing parts of plants or to seeds or cumulative organs [2]. When redistribution of metabolites is completed induction of mechanisms connected with PCD takes place.

All types of senescence require activity of certain genes [23,24].

It is a matter of debate whether molecular programs of different types of senescence or molecular programs of senescence caused by different treatments are the same.

Many genes whose transcripts are upregulated during leaf senescence have been identified over the past several years. Analysis of expression of these genes in response to different types of senescence has been used to distinguish between molecular aspects of age-dependent leaf senescence and those of senescence induced by other factors [13].

Manipulating leaf senescence through breeding or genetic engineering might help to improve crop yields by keeping leaves photosynthetically active for a longer time.

In this chapter a review of recent results on leaf senescence will be presented. Aspects of PCD during leaf senescence process will be emphasized.

II. GENETIC CONTROL OF LEAF SENESCENCE

A. METHODS OF IDENTIFICATION AND CHARACTERIZATION OF SENESCENCE-ASSOCIATED GENES

Initiation of the leaf senescence and its normal course require expression of many specific genes. Most of the genes, whose expression increases during natural leaf

senescence, are called senescence-associated genes (SAGs) [4,6,10,13,16,23,25].

Different groups of experimental studies have been applied in order to identify and characterize SAGs. They include studies with the use of (a) enucleated cells, (b) selective RNA and protein inhibitors, (c) mutations and transgenic plants with defects of the senescence process, and (d) other methods [4]. Examples of these studies and their results are now briefly discussed.

1. Chloroplasts in protoplasts of *Elodea* leaves did not undergo senescence process when deprived of nuclei, whereas those with nuclei senesced normally. Cell cycle of *Acetabularia* was much extended without nuclei [4].
2. Actinomycin D, a selective inhibitor of DNA-dependent RNA synthesis, inhibits leaf senescence. Cycloheximide and similar inhibitors of protein synthesis, acting on cytoplasmic ribosomes, retard leaf senescence. On the other hand, chloramphenicol, inhibiting chloroplast protein synthesis, usually does not retard leaf senescence. This suggests that senescence is driven mainly by nuclear genes and by mRNAs translated in cytoplasm and that chloroplast genes and protein synthesis in chloroplasts may contribute to the senescence process [2,3]. These results suggest also that senescence is not induced primarily by shutting genes off but by turning them on instead [2].
3. Most mutations interfere with hormone or Chl production and cause degeneration and premature death. Senescence-retarding mutations are mainly *nonyellowing* or *stay-green* and are easy to identify [4,8,26–28]. They are mainly recessive mutations and alter the expression of single genes encoding senescence-related enzymes (Table 36.1). Most SAGs are nuclear genes, except for *cytG* — a chloroplast stay-green gene, isolated from soybean leaves [2,4,6,8,10]. The mutations *cytG* and d_1d_2 prevent yellowing of leaves [4,8]. *CytG* partially blocks yellowing of leaves, selectively preserving the light-harvesting complex II (LHCII). This indicates that chloroplast may control its own disintegration. The genes *cytG* and d_1d_2 preserve the photosynthetic capacity, degradation of Chl, of LHC proteins and RuBisCO. The *gf* tomato mutant retains Chl during ripening and shows inhibition of chloroplast degradation and Chl degradation [4,29]. The mutation *Sid* (senescence-induced degradation) prevents leaf yellowing, blocks degradation of Chl [30], but not the decline of photosynthesis and presum-

TABLE 36.1
Examples of Mutations Altering Leaf Senescence

Mutation	Phenotype Effect	Species
Nd	Plant death delayed, leaves last till fruiting	Cowpea
Nd	Plant death delayed, extended cytokinin synthesis	Sorghum
dt, dt₂, e₁, e₂	Plant death delayed, reduced decrease of photosynthesis and of nitrogen fixation	Soybean
e sn hr	Plant death delayed, delayed apex senescence and plant death in short-day photoperiod	Pea
ih (recessive), *gr* (dominant)	Inhibition of degradation of chloroplast, thylakoid membranes, chlorophyll, LHC, and cytochrome *f*	Bean
sid (recessive)	Inhibition of degradation of chloroplast, thylakoid membranes, chlorophyll, LHC, D1 protein, and cytochrome *f*	*Festuca pratensis*
d₁d₂	Inhibition of degradation of chloroplast, chlorophyll-binding proteins, RuBisCO, and soluble proteins	Soybean
g (dominant)	Inhibition of degradation of chloroplast and chlorophyll content in seed coat	Soybean
cytg (chloroplast gene)	Inhibition of degradation of chloroplast, chlorophyll, LHC II, and cytochrome *f*	Soybean
gf (recessive)	Inhibition of degradation of chloroplast, thylakoid membranes, and chlorophyll	Tomato
ab (recessive)	Leaf abcission delayed, in particular under stress	Soybean
etr (dominant)	Leaf senescence delayed, reduced sensitivity to ethylene	*Arabidopsis*
Ore9	Progression of leaf senescence delayed, disturbed hormone signaling	*Arabidopsis*
Nr	Reduced sensitivity to ethylene, delayed chlorophyll degradation	Tomato
Rin	Slowdown of ethylene synthesis, chlorophyll degradation, and cell wall softening	Tomato
det 2 (recessive)	Light-signaling aberration; Chl degradation delayed	*Arabidopsis*

Source: Adapted from Noodén LD, Guiamét JJ. *Handbook of the Biology of Aging*. New York: Academic Press, 1996:94–118.

ably the death of leaves [3,4,31,32]. The *Ore4-1 Arabidopsis* mutant exhibits a delay in leaf senescence during the natural senescence but not during the hormone-induced or dark-induced senescence [16]. Stay green phenotype can be obtained by disabled Chl catabolism, enhanced endogenous cytokinins, or reduced ethylene production [33].

4. Cloning of the senescence-specific gene allows one to obtain information about the timing of expression of the gene, the site of activity, and the possible function of its products.

The levels of the total RNA decrease and the expression of many genes is switched off during senescence. Identification of many senescence-enhanced genes proved that *de novo* transcription of genes is necessary for the initiation and the normal course of the senescence process. Many cDNA clones representing SAGs have been identified (Table 36.2) using differential screening and subtractive hybridization

techniques. cDNA libraries constructed from mRNA isolated from senescing leaves have been screened differentially using labeled cDNA from green or senescing leaves. Clones showing hybridization to the senescing and not to the green probe have been selected [6,31]. Differential screening method is useful only when a gene, represented by a certain cDNA clone, is expressed at fairly high levels in the tissue. Substractive hybridization technique has been used to identify the genes expressed at lower levels [6,34,35].

Over the last 10 years many genes that show increased levels of transcription during senescence, from various plants such as *Arabidopsis thaliana* [16,21,36] and *Brassica napus* [37], have been identified (Table 36.2). Among them are genes encoding the degradative enzymes such as: proteases, nucleases, enzymes involved in lipid and carbohydrate metabolism, and enzymes involved in nitrogen mobilization. All of them create a family of senescence-enhanced genes; in many papers, also in this chapter, they are called SAGs.

TABLE 36.2
cDNA Clones Representing SAGs from Selected Plants

Name of gene	Possible Function	Plant	Characteristics	Ref.
SAG2	Cysteine protease	*Arabidopsis*	Oryzain γ-like	[21]
Seel	Cysteine protease	Maize	Oryzain γ-like	[38]
See1	Cysteine protease	*Lolium multiflorum*	Oryzain γ-like	[58]
LSC7	Cysteine protease	*Brassica napus*	Oryzain γ-like	[6]
See2	Cysteine protease	Maize	Vacuolar processing	[38]
SAG12	Cysteine protease	*Arabidopsis*	Papain-like	[36]
LSC790	Cysteine protease	*B. napus*		[35]
CysP1, CysP2	Cysteine proteinase	Soybean		[39]
LSC760	Aspartic protease	*B. napus*		[35]
UBC4	Ubiquitin carrier protein	*Nicotiana sylvestris*		[40]
UBI7	Polyubiquitin	Potato		[41]
RNS2	S-like ribonuclease	*Arabidopsis*		[42]
AhSL28	S-like ribonuclease	*Antirrhinum*		[43]
MS	Malate synthase	Cucumber		[50]
ICL	ICL	Cucumber		[48]
gMDH	NAD–malate dehydrogenase	Cucumber		[49]
LSC101	Fructose 1,6-bisphosphate aldolase	*B. napus*		[6]
LSC540	Glyceraldehyde-3 phosphate dehydrogenase	*B. napus*		[6]
See3	Puryvate phosphate dikinase	Maize		[38]
pTIP11	β-galactosidase	*Asparagus*	Postharvest	[47]
PLD	Phospholipase D	Castor bean	Detached leaf	[70]
Atgsr2	Glutamine synthetase	*Arabidopsis*		[45]
	Glutamine synthetase	Radish		[46]
	Glutamine synthetase	Rice		[44]
LSC460	Glutamine synthetase	*B. napus*		[35]
pPTIP12	Asparagine synthetase	*Asparagus*	Postharvest	[47]
LSC54	Metallothionein I	*B. napus*		[37]
LSC210	Metallothionein	*B. napus*		[35]
MT3-2	Metallothionein	Oil palm	Heavy metal induced	[51]
LSC30	Ferritin	*B. napus*		[35]
GSTII-27	Glutathione *S*-transferase	Maize		[38]
LSC650	Catalase	*B. napus*		[35]

Source: Adapted from Buchanan-Wollaston V. *J. Exp. Bot.* 1997; 48:181–199.

To identify the precise time during leaf senescence at which the expression of a certain gene is induced, biochemical and physiological changes, such as Chl content and photosynthetic rate during the senescence process, have to be characterized [6,38]. Sometimes it is difficult to determine when the senescence process starts. It appears that different cDNAs representing genes are expressed during the onset of induced senescence as compared to natural senescence [52]. The level of Chl in a leaf is a reasonable estimate of the stage of this leaf senescence. Patterns of gene expression in *B. napus* leaves during development and senescence were used to divide genes into different classes [6,53], presented in Table 36.3. Different classes of genes are expressed at different times during leaf ontogenesis, some of them are not specific to senescence and are expressed at a constant level through-

out the whole life of the plant, others are active before senescence starts and are switched off before any sign of senescence occurs. Some of these genes are specified, others are not (Table 36.3).

An attempt to clarify the role of SAGs in *Arabidopsis* was recently made by Lim et al. [16]. They conceptually categorized the genes that are either involved in initiation or in progression of senescence. The genes involved in initiation are (a) genes that control the developmental aging process, (b) genes that control other endogenous biological processes in addition to leaf senescence, (c) genes that affect senescence in response to environmental factors, (d) regulatory genes that upregulate the senescence-associated activities or downregulate the cellular maintenance activities, and (e) genes that are suggested to be involved in the degradation processes of

TABLE 36.3
Expression of Genes during Leaf Ontogenesis. Patterns of Expression of SAGs during Leaf Ontogenesis are Used to Divide the Genes into Classes

Class	Time of Expression during Leaf Ontogenesis	Characteristics of Class of *Brassica* SAGs
I	Expressed at a constant level during whole ontogenesis	"Housekeeping genes"
II	Expressed in green leaves; switched off before signs of senescence occur	Encoded proteins activated during senescence
III	Expressed in green leaves; switched off before signs of senescence occur	Encoded proteins may cause the initiation of senescence by their absence
IV	Expressed immediately prior to or at the onset of senescence, but for a relatively short time	Regulatory genes
V	Expressed specifically during senescence till the death of the leaf	Genes involved in the mobilization of storage products, LSC54, LSC22, LSC25
VI	Expressed specifically during senescence till the death of the leaf but also during other ontogenesis stages	Genes involved in the mobilization of storage products
VII	Expressed at low level in young leaves, increasing gradually through the senescence stages	LSC7, LSC10, LSC12, LSC460
VIII	Expressed at low level in the early stages of leaf ontogenesis but increasing dramatically at a particular stage of senescence	LSC94
IX	Expressed specifically during some stages of senescence	LSC550, LSC680
X	Expressed strongly early in leaf ontogenesis and again during senescence	LSC8, LSC101

Source: Adapted from Buchanan-Wollaston V. *J. Exp. Bot.* 1997; 48:181–199.

senescence regulation. Another class of genes is involved in the progression of senescence [16].

B. CHARACTERISTICS OF GENES AND THEIR PRODUCTS ENGAGED IN THE DEGRADATION PROCESSES DURING LEAF SENESCENCE

1. Enzymes Involved in Protein Degradation

Protein degradation is one of the most important processes during leaf senescence. The role of proteolytic degradation in leaf senescence was illustrated by the biochemical identification of cysteine protease and serine protease, which catalyze the degradation of RuBisCO [54,55], and by the immunological identification of alkaline endopeptidases [56].

Most (above 60%) of the proteins in the photosynthetic tissues are located in chloroplasts, therefore the proteolysis starts probably within the chloroplasts [6–9]. For example, endoprotease whose activity increased during leaf senescence of *Medicago sativa* was purified and characterized. It appeared that this purified protease is capable of degrading a large subunit of RuBisCO *in vitro* [55].

In *Arabidopsis* the chloroplast subunits ClpP and ClpC of ATP-dependent protease have been identified. The role of this protease is not clear, because the expression of gene encoding this protease takes place during whole-leaf ontogenesis [6,7,57]. Probably pro-

teins designed to be degraded during leaf senescence are transported to the vacuole [4,6,7].

Some of the cysteine proteases showing an enhanced level in different stages of senescing leaves have been identified (Table 36.2) [6,58–60]. Senescence-enhanced protease genes were isolated from maize: *see1* and *see2*, from *Arabidopsis*: clone SAG2, from *B. napus*: clone LSC7, and from *Lolium multiflorum*: clone See1. They show a sequence similarity to seed-specific proteases from cereals, such as oryzain γ protease from rice. Their function is also quite similar — remobilization of storage proteins (Table 36.2) [6,21,36,38,61]. One of the cysteine proteases showing an enhanced level during leaf senescence, represented by the cDNA clone SAG12, has a similar protein sequence to papain-like proteases (Table 36.2) [6,36]. Another cysteine protease from *B. napus*, represented by the cDNA clone LSC790, is expressed at all stages of leaf ontogenesis [6]. The transcript level of this gene is high in young green leaves, subsequently decreases when leaves are mature, and increases significantly during senescence (Table 36.2) [6,35].

Maize cysteine protease encoded by the *see2* gene has two prodomains, indicating that this enzyme is activated by the proteolysis. This protease is similar to enzyme activating proteases contained in the vacuoles of *Ricinus* seeds [38]. It is conjectured that See2 protease can activate other proteases by proteolysis, and in this way triggers a cascade of cysteine protease actions, similarly as in animal cells. The pattern of expression of this *see2* gene suggests that some genes encoding enzymes taking part in degradative processes during senescence are transcribed during the whole ontogenesis, but their products remain inactive inside the vacuoles; their activation starts during certain stages of senescence with the help of specific proteases. Sequence analysis of two cysteine proteases and aspartic protease isolated from *B. napus* indicates that all three have similar hydrophobic N terminal regions, probably responsible for directing proteins to the endoplasmatic reticulum. It is known that cysteine protease encoded by the gene from the cDNA clone LSC7 in senescing leaves of *B. napus* is located in chloroplasts [6] (Table 36.2). Cysteine proteases can also play a regulatory role apart from their proteolytic function.

Recently, the structure and expression of the SAG, *SPG31*, encoding cysteine proteinase precursors of sweet potato have been characterized. Northern blot analysis revealed that the transcripts of *SPG31* are specifically induced in senescing leaves. It appeared that *SPG31* plays an important role in proteolysis and nitrogen remobilization during the leaf senescence process [60]. Also, genes encoding two cysteine proteinases, *CysP1* and *CysP2*, were isolated from senescing soybean cotyledons from the same stage when the Chl content decreased (Table 36.2) [39]. One of three recently identified endopeptidases from cucumber leaves appeared to be highly active in senescing leaves. It seems that the appearance of this enzyme, CEP4.3, is regulated by the presence of sink tissues and is involved in the degradation of proteins in senescing leaves, facilitating nitrogen transfer to upper developing leaves. The activity of another endopeptidase, CEP4.5, correlates with the degradation of RuBisCO [62]. Also, the activity of the 70-kDa serine protease increased considerably parallel to the advance of senescence and the reduction of the protein content of leaves [63].

There are also suggestions that ubiquitin-dependent degradation of proteins takes place during leaf senescence [6]. One of the ubiquitin-activating enzymes, E2, might be involved in the breakdown of proteins during senescence. Expression of the gene encoding the E2 enzyme increases during leaf senescence in *Nicotiana sylvestris* [6,40]. Increased expression of another senescence-related gene — *UB17* encoding polyubiquitin — was registered in senescing potato leaves (Table 36.2) [41].

2.   Enzymes Involved in RNA Degradation

Decrease of RNA during leaf senescence is related to RNase activity. During senescence of *Arabidopsis* leaves, induced by limitation of phosphate, increased expression of three RNase genes was noticed; one of these genes, *RNS2*, is also expressed during the natural leaf senescence (Table 36.2) [42]. Recently, a gene named *AhSL28* encoding an S-like RNase in *Antirrhinum* was cloned (Table 36.2) [43]. It appeared very similar to RNS2 (also S-like RNase). Its RNA transcripts were induced during leaf senescence and phosphate starvation but not by wounding, indicating that *AhSL28* plays a role in remobilizing phosphate and other nutrients. Also, VRN1, encoding an S-like RNase of *Volvoc carteri*, promotes RNA degradation during senescence of somatic cell of this green alga [64]. Its regulation is similar to that of certain senescence-associated RNases in higher plants. Products of these genes play an important role in RNA degradation and in the metabolism of phosphate groups during leaf senescence [6]. The products of degradation such as purines and pirimidines are probably degraded further. Recently, two nucleases, PcNUC1 and PcNUC2, were observed to increase steadily as senescence progressed. Both nucleases were found to be glycosylated and could degrade both RNA and DNA [65].

3.   Enzymes Involved in Nitrogen Metabolism

Direct products of protein and RNA degradation during senescence are amino acids, mainly glutamines, asparagines, and amides that are transported through phloem to the developing parts of the plant [4,44,66,67]. By deamination of amino acids and catabolism of nuclei acids ammonia is released and converted into glutamine by glutamine synthetase. During leaf senescence the activity of the cytosol form of glutamine synthetase (GS1) increases mainly near the vascular bands where transport of glutamine takes place. The activity of the plastidial form (GS2, synthesized in photosynthetic tissues) decreases (Table 36.2) [44]. In mesophyll cells, enhanced expression of GS1 gene was found in several plants species: *Arabidopsis*, radish, rice, *B. napus* (Table 36.2) [35,44–46].

Enhanced expression of the gene encoding asparagine synthetase responsible for asparagine synthesis was found in senescing mesophyll tissue of *Asparagus* (Table 36.2) [6,47].

4.   Enzymes Involved in Lipid Degradation in Peroxisomes/Glioxysomes

The total amount of lipids decreases during senescence. Lipids are released from photosynthetic membranes, undergo modifications, and are a source of energy for the senescing process. In peroxisomes, called glioxysomes in senescing leaves, an increased level of several enzymes was observed during different phases of senescence [6,48,57,67,68]. Among them are catalases, active during the whole ontogenesis of the leaf. Three genes encoding catalases were isolated from *Nicotiana plumbaginifolia*. The increased level of catalase from *B. napus* (Table 36.2) [35] during the last stages of senescence is correlated with decreasing activity of ascorbate peroxidase in chloroplasts. Both these enzymes, as antioxidants, neutralize hydrogen peroxide during senescence.

Enzymes involved in β-oxidation of fatty acids and glioxylate cycle, among them isocitrate lyase (ICL) and malate synthase (Table 36.2), are located in glioxysomes. According to Vicentini and Matile [67] the best substrates for these enzymes are galactolipids from degraded thylakoid membranes. Specific expression of the gene from the cDNA clone ICL encoding ICL, of the gene *ms* encoding malate synthase, and of the gene from the cDNA clone gMDH encoding NAD–malate dehydrogenase from cucumber leaves was shown (Table 36.2) [48–50]. It was reported that the amount of ICL was increased by starvation and during senescence of barley leaves and might be due to the conversion of lipids into organic acids, which are then utilized in the mobilization of amino acids from leaf proteins [69].

Other enzymes connected with lipid degradation and remobilization are: β-galactosidase, released during degradation of galactolipids in senescing *Asparagus*, and phospholipase D, which hydrolyzes phospholipids from the degraded thylakoid membranes of senescing castor bean leaves (Table 36.2) [6,70].

5. Enzymes Involved in Photosynthetic Apparatus Degradation

Proteolytic enzymes, probably serine proteases, are involved in degradation of protein components of thylakoid complexes, mainly LHCII [71]. Disintegration of photosystem I (PSI) and PSII is due at first to the activity of chlorophyllase. Phytol tail is removed by this enzyme when Chl is still bound to the thylakoid membrane [53]. Subsequently, Mg–dechelatase removes the magnesium atom, then dioxygenase opens the ring, and finally Chl-binding proteins are released [72].

Enzymes involved in photosynthetic apparatus degradation, whose synthesis increases during senescence, include also fructose,1,6-bisphospate aldolase and glyceraldehyde-3-phosphate dehydrogenase (Table 36.2). In *B. napus* the relevant genes are expressed in green young leaves; in mature leaves their expression decreases and increases again during leaf senescence. The role of these enzymes is only indir-

ectly connected with degradation processes; they probably play a role in gluconeogenesis in synthesizing sucrose from the components of degraded lipids and proteins [6]. Puryvate orthophosphate dikinase, the enzyme that in C_4 plants normally synthesizes phosphoenolopyruvate from pyruvate, can be also involved in the gluconeogenesis pathway. As with the genes discussed previously, the pattern of expression of the gene encoding puryvate orthophosphate dikinase was observed during senescence of maize (Table 36.2) [38].

C. Characteristics of Genes and Their Products That Have Protective Function during Leaf Senescence

The function of products of some genes that exhibit enhanced expression during leaf senescence is not fully understood. Some of these products protect or detoxify the cell or cellular components. Some of the genes that are induced during leaf senescence are associated with the hypersensitive response (HR) and with the systemic acquired resistance defense programs [13].

Metallothioneins were found in many plant species as a response to different heavy metal treatments; some metallothionein or metallothionein-like genes have been reported in different species [51,73]. For instance, genes LSC210, LSC54 encoding metallothionein-like protein or products with high homology to metallothionein-like protein, and LSC30 encoding ferritin have been detected in senescing leaves of *B. napus* (Table 36.2) [35,37]. The products of all three genes are metal-binding proteins, and their possible function is to bind metal ions released during protein degradation, accumulate them in the vacuole, and thus detoxify the cell. It is known that metallothioneins in animal cells protect DNA from oxidative damage caused by reactive oxygen species (ROS). Degradation of Chl and peroxidation of protein and lipid compounds of thylakoid membranes cause an increase in ROS production [22,37]. Probably enhanced expression of metallothionein-like genes during leaf senescence establishes an antioxidant system protecting DNA and other cellular compounds.

Antioxidant enzymes such as glutatione *S*-transferase, superoxide dismutase, catalase, and ascorbate peroxidase play a similar role [22,67,74]. The last two enzymes were described already as involved, among others, in lipid degradation in chloroplasts. Both of these enzymes detoxify the cell from hydrogen peroxide that is produced during photosynthesis and photorespiration processes or as a response to excess iron, copper, and zinc [74,75].

Enhanced expression of GSTII-27 and higher activity of its product, glutathione *S*-transferase, was found in maize leaves during senescence (Table 36.2) [38]. The function of this enzyme, belonging to antioxidants, is probably to protect the photosynthetic apparatus and the cell against ROS [6,22].

In pea leaves the total activity of manganese superoxide dismutase (Mn-SOD), mainly localized in mitochondria and peroxisomes, increases with senescence, but the expression of mitochondrial and peroxisomal Mn-SOD is regulated differently. The expression of mitochondrial Mn-SOD is induced during the senescence of pea leaves, whereas peroxisomal Mn-SOD is probably posttranslationally activated [76].

Several identified SAGs include genes executing the senescence program, such as genes involved in disintegration or remobilization of macromolecules and genes involved in protecting cell viability for completion of the senescence process. However, SAGs may also include genes involved in the initiation or triggering of leaf senescence and genes controlling the progress and the rate of senescence [77]. Many SAGs are not uniquely induced during senescence but show an induction pattern during leaf development, suggesting that they have other roles in leaf development as compared to senescence. Some SAGs are activated during leaf senescence induced by environmental factors. Some are induced also during fruit ripening and during postharvest induced senescence [24]; however, the definitive identification of key proteins that are involved in the senescence process is still under investigation.

Isolation and characteristics of promoters and transcription factors related to SAGs during the whole senescence will elucidate the mechanisms regulating the senescence process. Some of them have been identified, for example, the promotor of *sag12* gene expressed specifically during *Arabidopsis* leaf senescence [78]. It appeared also that the WRKY transcription factor gene *At*WRKY6, identified in *Arabidopsis*, is involved in the regulation of SAG genes. Its level substantially increases in nuclei of naturally senescing leaves of *Arabidopsis* and during pathogen infection [79,80]. Hinderhofer and Zentgraf [81] identified the transcription factor WRKY53 in leaves of 6-week-old *Arabidopsis* prior to SAG12 expression. This indicates that WRKY53 is expressed very early in leaf senescence and might therefore play a regulatory role in the early events of leaf senescence.

Although many cDNA clones showing enhanced expression during senescence have been identified, the function of gene products is still not clear in many cases (Table 36.2).

III. DEGRADATION OF PHOTOSYNTHETIC APPARATUS DURING LEAF SENESCENCE

The first macroscopic symptom of leaf senescence is leaf yellowing. However, this change does not indicate the onset of senescence but the advance of this process caused by degradation of thylakoid membranes together with loss of Chl. Leaf senescence involves chloroplast ultrastructure disintegration, loss of Chl, breakdown of leaf proteins, and loss of photosynthetic capability; all these processes finally lead to cell death [1,2,8,12,19,82].

First changes at the ultrastructural level during the leaf senescence process concern swelling of thylakoid in still green photosynthetically active leaves. There is also a change in the shape and dimensions of chloroplasts during senescence [18,83]. Together with leaf yellowing, degradation of thylakoid membranes and massive accumulation of plastoglobules take place. Close contact of these plastoglobules with degraded membranes suggests that the lipids they contain come from degraded thylakoids. The released fatty acids interact with carotenoids forming esters located in plastoglobules [84]. Such chloroplasts filled with plastoglobules are often named gerantoplasts [3,57]. The degradation of thylakoid membranes is connected with the disintegration of protein complexes and electron carriers and also with the release of photosynthetic pigments: Chl and carotenoids [85].

Degradation of protein complexes takes place in a certain sequence: first, the cytochrome *b6/f* complex and then PSI, PSII, and synthase ATP complexes are degraded [19]. Complexes are released from the thylakoid membrane as intact units and are gradually degraded later [71]. After that inactivation or release of plastocyanin, ferredoxin, and NADP reductase takes place. Proteolytic enzymes involved in protein complex degradation have already been mentioned.

It is known that the disintegration of thylakoid membrane complexes during leaf senescence is accompanied by Chl release [19,86]. It was established that loss of Chl in senescing leaves is not directly related to the activity of chlorophyllase and that chlorophyllase activity is not altered in the nonyellowing mutant of *Phaseolus vulgaris* [87]. Enzymes engaged in Chl degradation have already been discussed. Probably the products of Chl degradation are transported to the vacuole by ATP transporters localized in the tonoplast [57]. The total content of carotenoids also decreases during leaf senescence, although carotenoids are more stable than Chl [3]. Neoxanthin and β-carotene content decreases concomitantly with Chl, while lutein and xanthophyll cycle pigments are less affected. Chl *a/b* ratio increases while PSII photochemistry decreases with senescence progression. It is suggested that down-regulation of PSII occurs in senescent leaves and that the xanthophyll cycle plays a role in the protection of PSII from inhibitory damage by dissipating excess excitation energy, particularly when exposed to high light [86]. During the first stages of leaf senescence of *Pistacia lentiscus* grown under Mediterranean field conditions, no damage to photosynthetic apparatus occurred; xanthophyll cycle pigments, lutein, neoxanthin, and ascorbate levels were kept constant while β-carotene and α-tocopherol levels increased [88]. By contrast, Chl, carotenoids (neoxanthin, lutein, β-carotene), and ascorbate were degraded during the later stages of leaf senescence. These results demonstrated that the mechanisms of photo- and antioxidative protection may play a role in maintaining chloroplast function during the first stages of senescence, while antioxidant defenses are lost during the later stages [88].

Because Chl and chloroplast breakdown is so prominent, leaf senescence is generally measured in terms of Chl loss [8]. Decline of both Chl and soluble protein levels has been used as a classical indicator of leaf senescence [1]. According to Noodén et al. [8] chloroplast breakdown is accompanied by the degradation of protein complexes or damage of any of its components, which immediately destabilizes the whole complex.

The decrease of Chl is accompanied by a lowered photosynthetic activity. The activity and content of RuBisCO decrease even before the degradation of photosystems takes place, probably due to the chloroplast protease activity [19].

IV. ULTRASTRUCTURAL CHANGES OF MESOPHYLL CELLS DURING LEAF SENESCENCE

At the ultrastructural level one of the main changes during leaf senescence is chloroplast disintegration, which was described in the previous section. Changes in nuclei structure will be described in the next section.

The number of rybosomes diminishes, both in cytoplasm and stroma, as senescence proceeds; also, degradation of Golgi apparatus and endoplasmic reticulum takes place. Multifunctional peroxisomes, abundant and characteristic of senescing mesophyll, take part in the catabolism of purines and lipids deriving from degraded thylakoid membranes [48,67]. Vacuoles in senescing leaf cells contain cya-

nides and flavonoids, responsible for the purple and yellow colors of leaves. Tonoplasts break down causing the release of the vacuole content into the cytoplasm [48]. Plasmolemma loses its integrity, and finally cell death takes place.

V. ASPECTS OF PCD DURING LEAF SENESCENCE

PCD is an active and highly coordinated process, occurring as a part of normal growth and development and also during the response of the plant to pathogen infection and stress factors (reviewed in Refs. [5,7,9,26,59,89–95]). Leaf senescence and cell death during this process are under control of a coordinated signaling pathway. Both leaf senescence and cell death during this process are often triggered by the same inducing environmental factors, initiated or modified by endogenous signals such as hormone levels and ROS, which subsequently stimulate synthesis of many similar enzymes such as cysteine proteases and nucleases [9,26,96].

There are numerous results that prove that leaf senescence is a genetically defined process involving mechanisms of PCD [11,12,26]. A lower rate of cell death is associated with efficient recycling of nutrients that are released during senescence [16].

Many molecular and structural features such as condensation of nuclei chromatin and the subsequent disorganization of the nuclei were identified and recognized as one of the hallmarks of PCD [5,7,11]. Significant chromatin condensation takes place during leaf senescence [8 and references therein,9,97]. Chromatin changes in mesopyll cell nuclei of senescing leaves of *Ornithogalum virens* and *Nicotiana tabacum* were also reported by Simeonova et al. [12].

Relations between PCD, specific nDNA fragmentation, changes in chromatin condensation, and degradation of chloroplast ultrastructure together with decrease of photosynthetic pigment level during leaf senescence were found by Simeonova et al. [12] in two plant species: *O. virens* and *N. tabacum*. In *O. virens* the gradient of leaf development, characteristic of monocotydelons, proceeds from the apical region of the leaf blade, which develops first, to the base of the leaf, which develops later. Development of *N. tabacum* leaf, characteristic of dicotyledons, proceeds at the same rate within the whole leaf blade. Description of leaf development and senescence stages is given below.

In the first stage, mesophyll cells of basal green parts of *O. virens* leaves as well as fully developed mature green *N. tabacum* leaves contain differentiated chloroplasts with numerous grana stacks and single plastoglobuli (Figure 36.1a and b). Protoplasts used for the comet assay, which will be described later, were isolated from the same leaf regions (Figure 36.1a and b, insets). Nuclei of mespohyll cells from the same regions of both plant species contain dispersed chromatin (Figure 36.1c and d) [12].

In the second stage of development, mesophyll cells of the middle part of *O. virens* leaves and of yellowish *N. tabacum* leaves have chloroplasts with large grana. The number of plastoglobuli increase

slightly in *O. virens* chloroplasts and enormously in *N. tabacum* ones (Figure 36.2a and b) as compared to the previous stage. Nuclei of mesophyll cells from the same leaf blade regions of both analyzed plant species contain dispersed chromatin (Figure 36.2c and d) [12].

In the third analyzed stage, mesophyll cells of the apical yellow parts of *O. virens* leaves and *N. tabacum* yellow leaves contain mostly chloroplasts that have

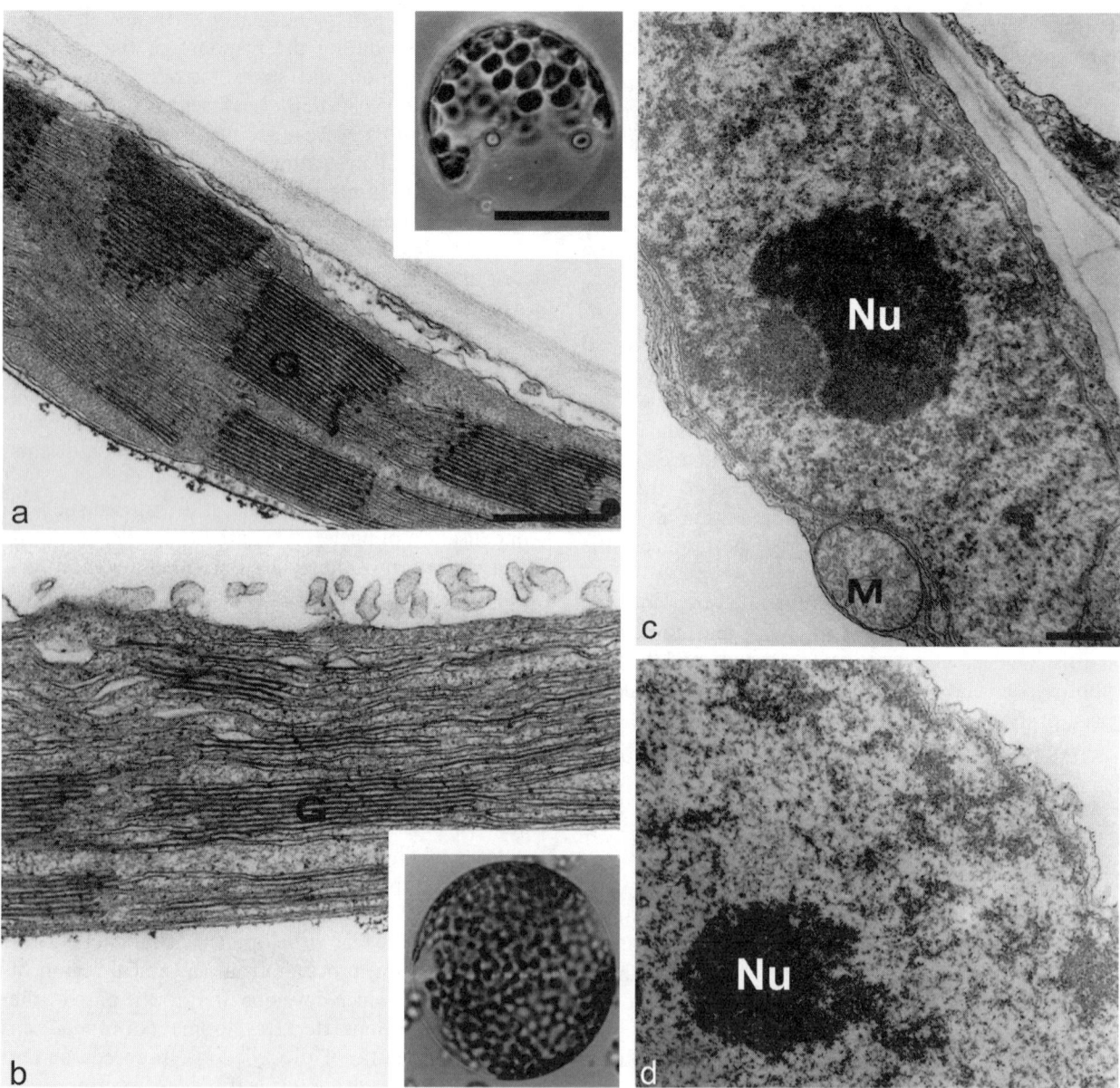

FIGURE 36.1 (a, b) Electron micrographs of the mature chloroplasts with typically organized grana (G), from the green basal part of an *Ornithogalum virens* leaf (a) and the second green leaf of *Nicotiana tabacum* (b); bar: 0.5 μm. Insets: Light microscopic images of protoplasts of both plant species from the respective leaf regions; bar for both insets: 200 μm. (c, d) Electron micrographs of nuclei of mesophyll cells, from the green basal part of the *O. virens* leaf (c) and the second green leaf of *N. tabacum* (d); Nu: nucleolus, M: mitochondrium; bar: 0.5 μm. (From Simeonova E, Sikora A, Charzyñska M, Mostowska A. *Protoplasma* 2000; 214:93–101. With permission.)

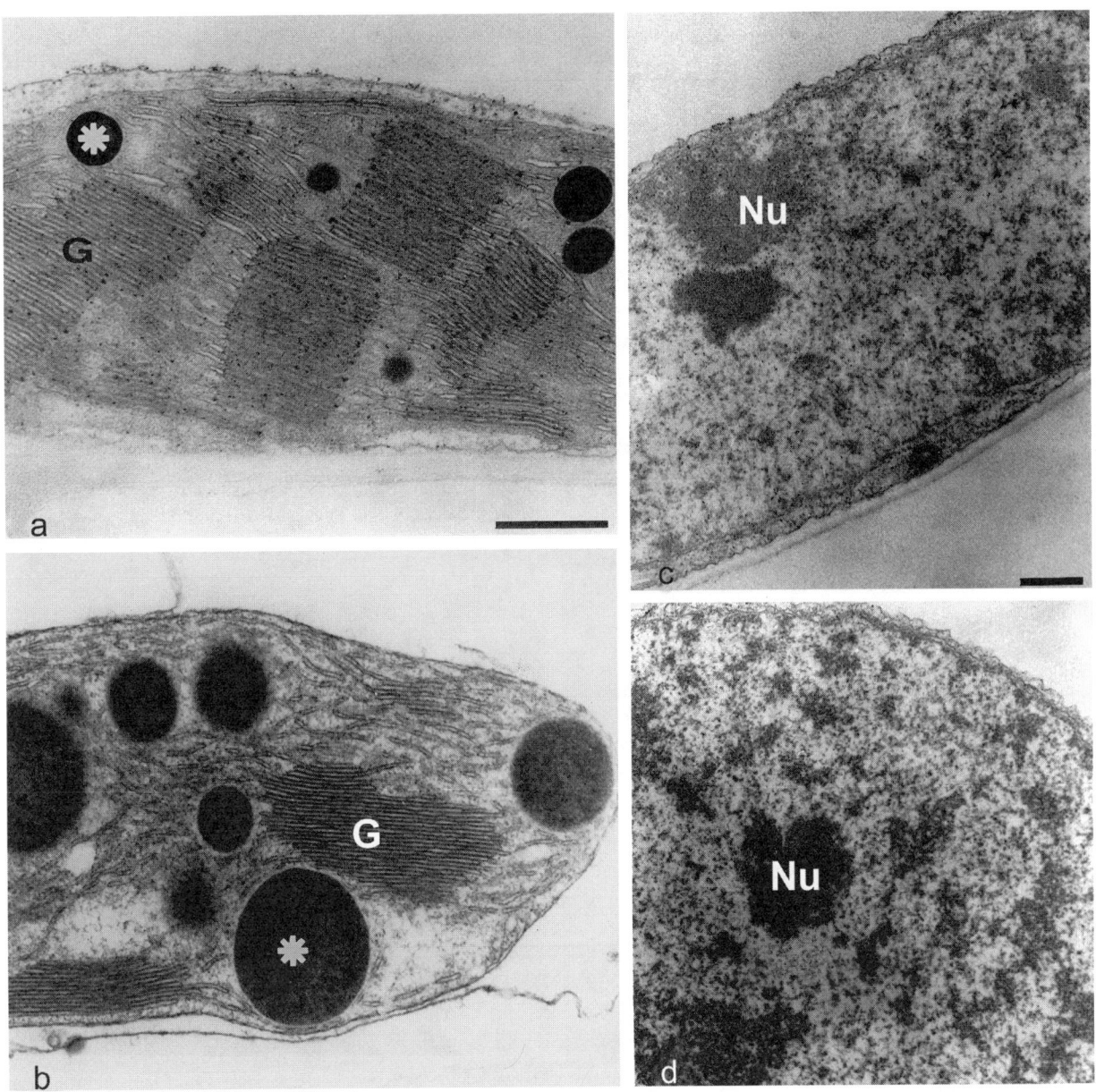

FIGURE 36.2 (a, b) Electron micrographs of chloroplasts containing large grana (G), and plastoglobuli (asterisks) observed in the mesophyll cell of the middle green part of an *Ornithogalum virens* leaf (a) and a yellowish *Nicotiana tabacum* leaf (b); bar: 0.5 μm. (c, d) Electron micrographs of nuclei of the mesophyll cell from the middle green part of the *O. virens* leaf (c) and the yellowish *N. tabacum* leaf (d); Nu: nucleolus; bar: 0.5 μm. (From Simeonova E, Sikora A, Charzyñska M, Mostowska A. *Protoplasma* 2000; 214:93–101. With permission.)

changed their ultrastructure; the stromal and granal thylakoids are characteristically swelled (*O. virens*) or completely degraded (*N. tabacum*) (Figure 36.3a and b). Chloroplasts of *N. tabacum* are filled with large plastoglobuli pushing aside rudimentary thylakoids (Figure 36.3b). Nuclei of mesophyll cells from the same leaf blade regions in both the species studied demonstrate a significant condensation of chromatin (Figure 36.3c and d).

Simultaneous analyses of mesophyll cell ultrastructure and photosynthetic pigment concentration in leaves of *O. virens* and *N. tabacum* have shown that the decrease of Chl *a* and *b* and carotenoid contents of apical yellow parts of *O. virens* leaves and *N. tabacum* yellow leaves is correlated with the gradual disintegration of the thylakoids membranes in chloroplasts, which is characteristic of the progress of senescence [12].

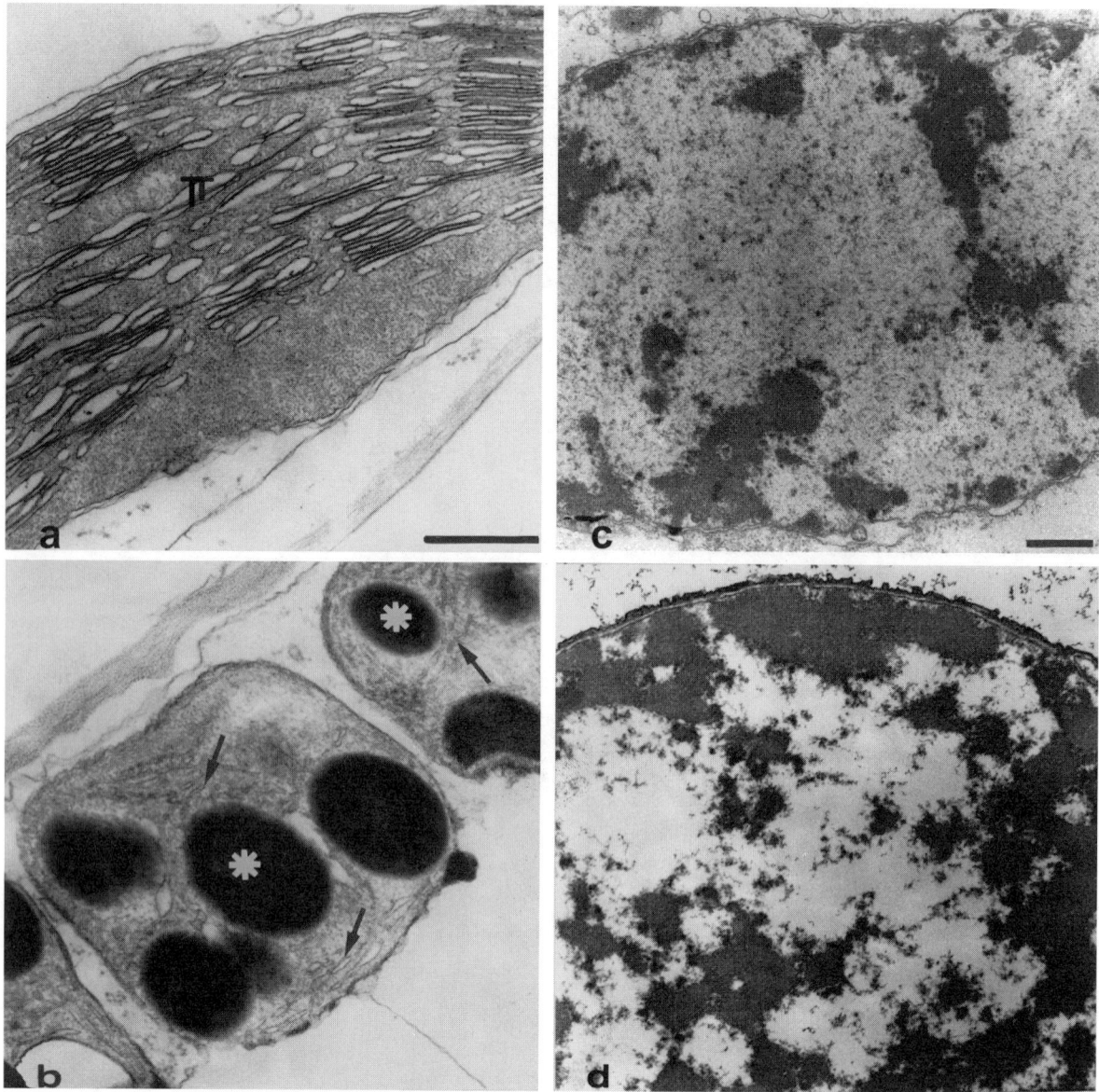

FIGURE 36.3 (a) Electron micrograph of chloroplast with changed ultrastructure observed in the senescing apical part of an *Ornithogalum virens* leaf; visible dilation of thylakoid membranes (T). (b) Electron micrographs of chloroplast with changed shape and containing destroyed thylakoids and large plastoglobuli (asterisks), observed in *Nicotiana tabacum* mesophyll cell of the second yellow leaf; arrows indicate the remains of the thylakoids. Bar for (a) and (b): 0.5 μm. (c, d) Electron micrographs of mesophyll cell nuclei observed in the senescing apical part of the *O. virens* leaf (c) and the *N. tabacum* second yellow leaf (d); distinctly visible condensation of chromatin; bar for c and d: 0.5 μm. (From Simeonova E, Sikora A, Charzyñska M, Mostowska A. *Protoplasma* 2000; 214:93–101. With permission.)

The chloroplast ultrastructure degradation in mesophyll cells shows a similar pattern: dilation and breakage of thylakoids, increase in the number and size of plastoglobuli correlated with the process of Chl degradation [12,98,99].

Kołodziejek et al. [100] found relations between the changes in mesophyll cell ultrastructure and pigment concentration in every region of leaf during senescence in maize and barley. They demonstrated that degrad-

ation of chloroplast structure is not fully correlated with the change in photosynthestic pigment content; Chl and carotenoid content remains still at a rather high level in the final stage of chloroplast destruction. Changes to the mesophyll cell such as chromatin condensation, degradation of thylakoid membranes, increase in the number of plastoglobules, damage to the internal mitochondrial membrane, and chloroplast destruction do not occur at the same time in different parts of the leaf. The senescence damage begins at the base and moves to the top of the leaf. The dynamics of mesophyll cell senescence is different in leaves of both analyzed plant species; in initial stages this process is faster in barley and in later stages, in maize. At the final stage, the oldest barley mesophyll cells are more damaged than maize of the same age [100].

One of the universal hallmarks of PCD is nonrandom, internucleosomal fragmentation of nDNA, occurring as a result of a specific endonuclease activation [101]. It was shown that nonrandom, internucleosomal fragmentation of nDNA also occurs during leaf senescence. The *in situ* detection of DNA fragmentation leading to cell death can be achieved by the terminal deoxynucleotidyl transferase-mediated dUTP nick and labeling of DNA 3′-OH groups (TUNEL method) [102,103].

Comet assay, specific in revealing nonrandom internucleosomal cleavage, was applied for the analysis of nDNA fragmentation in leaf mesophyll [12,104]. Using this method Simeonova et al. [12] proved that nDNA degradation, specific for PCD, occurs during the natural leaf senescence of *O. virens* and *N. tabacum*.

Comet assay performed for isolated single-mesophyll protoplasts did not detect any fragmentation of nDNA either in mesophyll cells of the basal parts (the youngest parts) of *O. virens* leaves or in mesopyll cells of *N. tabacum* green leaves containing already differentiated chloroplasts. Figure 36.4a and b present nuclei of young mesophyll cells after gel electrophoresis, stained with DAPI, observed with a fluorescent microscope. There is no formation of "comets" at this stage of development, either in *O. virens* protoplasts or in *N. tabacum*.

The nuclei from the middle, still green part of *O. virens* leaves do not form comets (Figure 36.4c). However, nuclei from mesopyll cells of yellowish *N. tabacum* leaves give images that resemble comets through the fluorescent microscope (Figure 36.4d), although these nuclei still contain dispersed chromatin. The "head" of the comet visualizes the nDNA, which still remains in the region of the cell nucleus. The "tail" of the comet visualizes negatively charged DNA fragments, liberated from the nucleus, migrating toward the anode. Weak fluorescence of the comet tail indicates that the process of nDNA damage is not advanced yet.

The nuclei of the yellow apical parts of *O. virens* leaves and senescing mesophyll cells of yellow *N. tabacum* leaves are clearly seen as comets in the fluorescent light microscope (Figure 36.4e and f). As opposed to the previous stage there is strong fluorescence of the comet tail indicating the advanced process of nDNA damage.

In mesophyll cells of both plant species the appearance of comets in the apical senescing part of *O. virens* leaves and yellow senescing leaves of *N. tabacum* was followed by changes in chromatin structure, and chloroplast and pigment degradation.

The nDNA fragmentation, typical for PCD, was also detected by TUNEL and "ladder" standard gel electrophoresis in senescent yellow leaves of *Philodendron hastatum*, *Epipremnum aureum*, *Bauhinia purpurea*, *Delonic regia*, and *Butea monosperma* by Yen and Yang [11] and Wang et al. [105]. Gradual nDNA fragmentation, detected by gel electrophoresis, and decrease of the protein level take place before degradation of chloroplast structure, loss of Chl, and decrease of photosynthetic activity during wheat leaf senescence [106].

Comet assay gives a more sensitive and early detection of DNA damage of the viable individual protoplasts isolated from mesophyll tissue during the natural leaf senescence. The nDNA fragmentation, specific for PCD, precedes the condensation of nuclear chromatin. It is possible that by applying the alkaline version of comet assay both single- and double-strand breaks of nDNA were detected. An increase of single strand-preferring nuclease activity that hydrolyzes single-stranded DNA has been observed during dark-induced senescence in barley as well as during the natural senescence of wheat and barley leaves [107].

Simeonova et al. [12] pointed out the sequence of changes during the leaf senescence process:

1. nDNA fragmentation, swelling of thylakoid membranes, slight increase in the number of plastoglobuli, and decrease of pigment contents
2. further nDNA fragmentation, condensation of chromatin, degradation of thylakoid membranes, significant increase in the number and size of plastoglobuli, further decrease of pigment contents.

According to Inada et al. [108,109] each mesophyll cell follows a similar senescence program: chloroplast DNA degradation, condensation of nuclear chromatin, decrease of chloroplast size, degradation of RuBisCO, degeneration of chloroplast inner membranes, and cell disorganization. Degradation of chloroplast nuclei before degeneration of chloroplasts during senescence of rice coleoptiles and leaves was reported also by Sodmergen et al. [110]. There are reports that cleavage of chloroplast DNA occurs before leaf yellowing in peach [99] and rice [111]. It was also reported that proteases involved in the protein degradation in mesophyll tissue are mainly (>60%) located

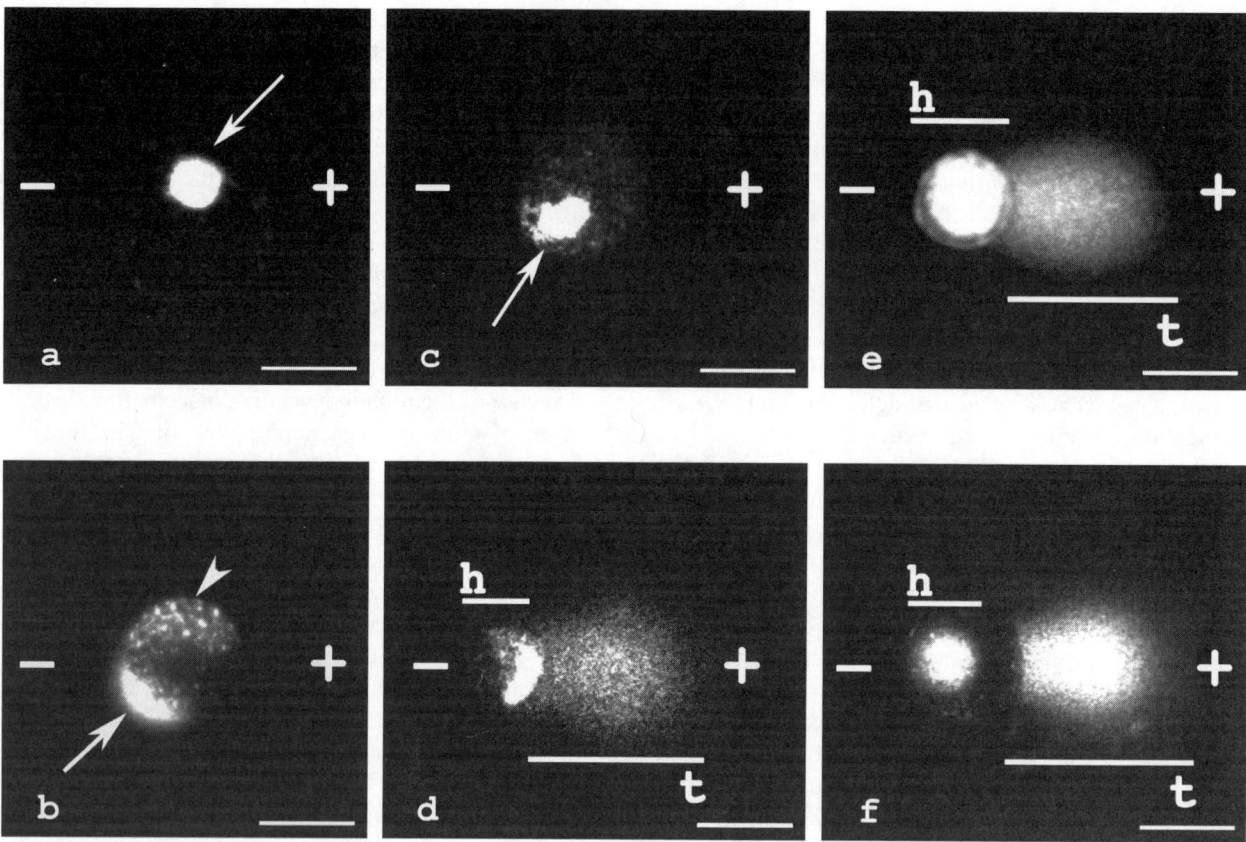

FIGURE 36.4 (a–f) Fluorescence images of nuclei of the individual protoplasts of *Ornithogalum virens* (a, c, e) and *Nicotiana tabacum* (b, d, f) leaves, stained by DAPI after electrophoresis (comet assay); h: comet head, t: comet tail; bar: 10 μm. Nucleus of protoplast isolated from the basal part of the *O. virens* leaf (a) and from a green *N. tabacum* leaf (b). There is no formation of comets indicating that there is no fragmentation of nDNA (a, b). Nucleus of protoplast isolated from the central green part of the *O. virens* leaf (c) and from the yellowish *N. tabacum* leaf (d). The appearance of the comet demonstrates that nDNA fragmentation has already started but much more of the DNA is still tightly associated with the nuclear matrix (d). Nucleus of protoplast isolated from the yellow apical part of the *O. virens* leaf (e) and from a yellow *N. tabacum* leaf (f). Strong fluorescence of the comet tail indicates that the process of DNA degradation has already advanced (e, f). (From Simeonova E, Sikora A, Charzyñska M, Mostowska A. *Protoplasma* 2000; 214:93–101. With permission.)

within the chloroplasts. It is likely that the process of protein degradation starts first in these organelles [6,9].

DNA fragmentation together with other symptoms of PCD, such as nuclear condensation, was reported in *Kalanchoë* leaves exposed to different gravity environments [112]. It was also reported that the formation of nitric oxide (NO), a free radical, is associated with ethylene biosynthesis, drought stress, and cell death and proliferation [112–114]. Exposure to hypergravity caused NO burst, which was histochemically detected *in vivo* using 4,5-diaminofluorescein diacetate (DAF-2 DA), whereas nucleoid and nuclei fragmentation was detected by double-staining TUNEL-DAPI (shown in Figure 36.5A and B as TUNEL-positive cells). Chloroplast DNA fragmentation was detected 10 min after exposure to hypergravity, fragmentation increased intensively more

than 60 min after gravitation treatment (Figure 36.5A). nDNA fragmentation was observed 20 min later (30 min after exposure) and was also increasing, although not so rapidly (Figure 36.5A). Detection of DNA fragmentation 1 day (24 h) after exposure to hypergravity treatment showed that DNA fragmentation increased further, compared to the results taken immediately after exposure, but the nucleoid fragmentation decreased significantly (Figure 36.5B). The highest number of labeled nuclei and nucleoids were visible 60 min after hypergravity treatment and 60 min after exposure the next day (60 min + 24 h) (Figure 36.5A and B). A NO burst preceded a significant increase in nDNA fragmentation. Exposure to hypergravity showed that chloroplast DNA fragmentation occurred prior to nuclear fragmentation, chromatin condensation, and nuclear blebbing.

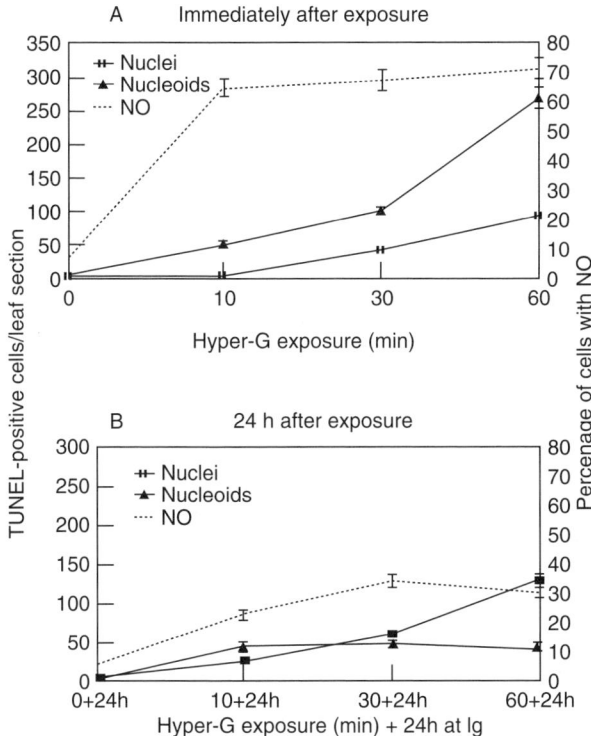

FIGURE 36.5 Quantitation of nitric oxide (NO) formation and of nuclear and chloroplast DNA fragmentation following acute hypergravity treatments. (A) Leaves, collected at the times indicated after hypergravity exposure (150g; hyper-G), were sectioned, stained for 1 h with 10 μm DAF-2 DA for nitric oxide visualization, fixed, processed for TUNEL and counterstained with DAPI. Leaf sections, fresh and fixed, not assayed for TUNEL, treated with DNase-I, and processed without terminal transferase and without DAPI staining, were used as controls of this staining assay. (B) Hyper-G treated leaves, kept for 24 h in 1g under a 16-h photoperiod, were processed as described for (A). Leaves not exposed to hypergravity (0 + 24 h) were also collected and used as controls. Three to five leaf sections (90 μm), performed in at least six clonal leaves per treatment, were used to quantify the percentage of leaf cells with NO (shaded line) and within TUNEL-positive cells, those presenting nucleoid (▲) and nuclear (■) DNA fragmentation. Values are the mean of six independent experiments. Error bars are lower than 7%. (From Peolvoso MC, Durzan D. *Ann. Bot.* 2000; 86:983–994. With permission.)

Chloroplasts were the first visible organelles to show NO production and DNA fragmentation. *Kalanchoë daigremontiana* chloroplasts have small uniformly dispersed nucleoids located in the matrix between thylakoid membranes. Treatments with some chemical agents — NO generator and NO-synthase inhibitor — showed a direct correlation between NO formation and DNA fragmentation in chloroplasts, epidermis, and mesophyll cells. Thylakoid membranes are the first to be under the influence of NO. NO reacts with superoxide (O_2^-) to form peroxynitrite ($OONO^-$) causing degradation of DNA, RNA, proteins, and lipids. Membrane lipids of chloroplasts could be one of the main targets for NO attack. Pedroso and Durzan [112] suggested that NO was involved in signaling pathways leading to PCD in plants.

All these data confirm that the leaf senescence process involves mechanisms of PCD.

There are not many data concerning endonucleases that are responsible for internucleosomal nDNA fragmentation specific for PCD in plants. Two classes of endonucleases have been identified in plants:

1. Zn^{2+}-dependent endonucleases, causing both single- and double-strand nDNA breaks (single-strand-preferring endonucleases [SSNs] and double-strand-preferring endonucleases), isolated from *Aspergillus oryzae*
2. Ca^{2+}-dependent endonucleases, mainly SSNs [115].

Zn^{2+}-dependent endonuclease, named ZEN1, was isolated from a mesophyll cell culture *in vitro*; its activity increased after auxin and cytokinin treatment [116]. ZEN1 mRNA was accumulated during differentiation of tracheid elements. Probably ZEN1 is transported to the vacuole and after the disintegration of tonoplast is responsible for nDNA fragmentation during xylem differentiation [115].

Ca^{2+}-dependent endonucleases, NUC I (100.5 kDa), NUC II (30 kDa), and NUC III (36 kDa), are involved in HR in tobacco leaves. These endonucleases, responsible for HR nDNA fragmentation, are localized in apoplast and transported to the nucleus after plasmalemma disintegration [115].

It is probable that at first the Ca^{2+}-dependent endonucleases and then the Zn^{2+}-dependent endonucleases are involved in nDNA breakage during PCD in plants.

SSNs were also detected in barley leaves during the natural leaf senescence and senescence induced by darkness [107]. The activity of Ca^{2+}- and Zn^{2+}-dependent endonucleases, causing chloroplast DNA degradation, was shown in peach and rice mesophyll cells. Chloroplast DNA degradation precedes ultrastructural and physiological changes during senescence [99,108–111].

Further identification of endonucleases is still needed. In spite of rapid progress many questions concerning endonucleases remain unanswered.

In recent years the role of plant mitochondria in controlling cell death activation was recognized

[117–119]. A specific release of cytochrome *c* from intact mitochondria was described in cucumber cotyledons undergoing PCD. According to Balk et al. [120] and Zhao et al. [121,122], the release of cytochrome *c* into cytosol is an early event in plant PCD.

There is also evidence that caspase-like proteases might participate in PCD in plants [123] and that they might be activated by the release of cytochrome *c* from mitochondria into the cytosol [121,122,124].

The activity of caspase-like proteases during plant PCD was revealed in a cell-free system [121,122,124]. Isolated mouse liver nuclei were incubated in cytosol of carrot cell suspension, enriched with cytochrome *c*. Condensation of chromatin was observed already after 30 min of incubation, and nDNA fragmentation started after 1 h. The treatment with caspase 3 and caspase 1 inhibitor prevented nDNA fragmentation [121]. It appeared that caspase-like proteases were present in plant cytosol and were involved in the onset of apoptosis in the nuclei of mammals [121]. Caspase-like activity was revealed also in a cell-free system induced by a heat shock that resulted in chromatin condensation and nDNA fragmentation [125].

In the case of PCD induced by environmental agents, for example, *in vivo* victorin-induced PCD of young oat leaves, a loss of transmembrane mitochondrial potential ($\Delta\Psi_m$) was reported [126]. Moreover, in isolated oat mitochondria after *in vitro* victorin treatment, a mitochondrial permeability transition occurred, which was accompanied by the release of cytochrome *c* from mitochondria into the cytosol [126].

The ability to regulate plant cell death may have important applications in agriculture and postharvest industries in the foreseeable future. For instance, suppression of PCD induced by pathogens could minimize disease symptoms and may prolong the life of crop plants [93].

VI. LEAF SENESCENCE INDUCED BY ENVIRONMENTAL FACTORS

Various environmental factors like drought, temperature extremes, intense light, UV radiation, herbicides, and pathogens can induce plant response similar to natural senescence [22]. Plants respond rapidly to deteriorating environmental conditions. As opposed to animals they cannot move to escape an unfavorable situation but can eliminate inessential organs or tissues. For example, a diseased leaf will senesce, die, and drop off the plant to prevent the whole plant from being infected by the disease. Drought stress,

darkness or nitrogen deficiency, nutrient deprivation [28], dark-induced senescence [34], low oxygen [83], photodynamic herbicides [127], UV-A, and high temperature [128] can initiate senescence processes that may cause early seed production and shortening of plant life [6].

The mRNA coded for early light-induced protein was detected earlier than Chl loss during tobacco leaf senescence induced by leaf detachment, water stress, and anaerobiosis [129].

The main difference between the natural senescence and senescence induced by environmental factors is that the latter process can be reversible when stress factors are relieved before senescence has progressed beyond a certain phase [1].

It was found that many SAGs are also expressed as a plant response to different internal and external environmental factors, such as heavy metals, darkness, wounding, heat shock, nutrient starvation, and hormones [77]. For example, eight *Arabidopsis* SAGs are induced by ozone [130]. All senescence upregulated mRNAs are expressed in senescing leaves when senescence is induced by drought, increased light, and high temperature [77]. Also, *Arabidopsis sen1* gene is activated during leaf senescence induced by age, darkness, abscisic acid (ABA), or ethylene. The promoter of the *Arabidopsis sen1* gene is activated upon sugar starvation and is repressed by exogenous sugar compounds [131]. It was also discovered that transcripts of the *tbzF* gene of tobacco encoding a basic region leucine zipper protein (bZIP), belonging to the *LIP19* subfamily, accumulate during leaf senescence and also on cold, ABA, or ethylene treatment. It was suggested that the *tbzF* gene possesses a multiple function [132].

According to Lim et al. [16], among the 43 transcription factor genes that were induced during senescence, 28 were also induced by stress treatments, suggesting that there is extensive overlap between the response to natural leaf senescence and the response to stress. Often, several closely related transcription factors have the potential to activate or repress genes through *cis*-acting sequences that respond to specific stresses. These factors may have closely overlapping functions [133].

Many environmental factors cause oxidative stress, that is, stress caused by the excess of ROS overwhelming the system of natural defense. Each cell compartment has its most sensitive target for oxidative stress and its own mechanisms of defense [22].

Chloroplasts are among the first to react when the plant is exposed to environmental stress; they are exposed to oxidative stress more than any other organelle because of the high internal O_2 concentration, inside the thylakoid membranes in particular. Therefore, they are especially prone to generate ROS. Under normal conditions, due to defense mechanisms, chloroplasts minimize the potential damage following from the formation of large amounts of ROS. Oxidative stress occurs when all these protective mechanisms are insufficient. Environmental factors such as high light intensity, UV radiation, air pollutants, herbicides, water, and heavy metal stress, as well as some others, induce oxidative stress and often give similar symptoms of structural damage and dysfunction independent of the primary stressing factor and similar to symptoms of natural senescence [22]. These alterations consist mostly in swelling of thylakoids and membrane damage, intensive plastoglobules and starch accumulation, photodestruction of pigments, and inhibition of photosynthesis [22,83]. Symptoms of photosynthetic apparatus degradation caused by natural senescence are similar to those that occur during natural senescence. ROS are known to be mediators of PCD induced by different endogenous and exogenous factors [134] and seem also to be mediators of a natural program of cell death. It has been shown that different antioxidants can protect cells and tissues under various death-promoting conditions. Production of ROS can be a general alarm signal both to modify cell metabolism and to stimulate antioxidative defense mechanisms. It might indicate that the stressing factors inducing oxidative stress do not act specifically and, therefore, a plant resistant to one stressing factor is quite often resistant to other oxidative stress-inducing factors [22].

VII. REGULATION OF LEAF SENESCENCE PROCESS

All processes taking place during senescence are highly coordinated and involve complex interactions of several factors, such as signal perception and induction of cascade expression of many genes regulated by activator proteins. These proteins are activated and controlled by a variety of internal factors, for example, plant hormones. It is known that plant hormones, especially cytokinins and ethylene, are involved in the senescence process. Their role was determined mainly by using mutants and transgenic plants [53]. However, other signaling pathways can be also involved in regulation of the leaf senescence process.

A. LEVEL OF CYTOKININS

The level of endogenous cytokinins in *Arabidopsis* mutant amp1 is several times higher than in wild

Arabidopsis plants; mutant leaves have significantly reduced senescence compared to wild ones [23]. The treatment with cytokinins delays leaf senescence in many plants; on the other hand, a reduced cytokinin level can induce senescence [2,6,23,26,135]. The isopentenyl-transferase gene, which catalyzes one of the steps in cytokinin biosynthesis, has been cloned under regulation of the promoter for the *Arabidopsis* SAG12 gene encoding protease [6,10,26,136]. Transgenic plants carrying this gene developed normally until the moment when senescence should start. The onset of senescence was significantly retarded, leaves of transgenic plants did not show symptoms of senescence, and the photosynthetic rate was comparable with that of young, green plants. Expression of the SAG12 promoter was induced when senescence started, cytokinins were synthesized, senescence was stopped, and the promoter was switched off [137]. The SAG12 promoter region of *Arabidopsis* was identified, and it was discovered that this gene is expressed only during the natural senescence [78]. A high level of cytokinins or sugars can inhibit expression of this gene.

It seems that the control of senescence by cytokinins is at the transcriptional level [137]. Expression of SAGs is inhibited above a certain level of cytokinins. A critically low level of cytokinins is one of the crucial signal factors inducing senescence.

In conclusion, senescence can be initiated when the level of cytokinins falls below a certain value. A relatively high cytokinin level prevents the onset of senescence.

B. ROLE OF ETHYLENE

It is known that ethylene is involved in plant response to stress and environmental factors and that it plays a regulatory role in different processes such as seed germination, fruit ripening, and flower senescence. Experiments with mutants and transgenic plants were used to elucidate the role of ethylene in leaf senescence. Two enzymes, 1-amino-cyclopropane-1-carboxylate (ACC) synthase and ACC oxidase, are involved in the biosynthesis of ethylene from *S*-adenosyl methionine. At the beginning of tomato leaf senescence an increased expression of the ACC oxidase gene, which preceded Chl degradation, was detectable [138]. Transgenic tomato plants with reduced ethylene production expressing the antisense of the ACC oxidase gene showed a delayed leaf senescence as compared with wild plants. However, once senescence had already started the progress of senescence was similar to that in leaves of wild plants [6,10,26,138]. A similar effect was obtained with etr-1 *Arabidopsis* mutant leaves, which senesce much slower than leaves of the wild type (Table 36.1) [139]. However, when senescence had already started it lasted much longer than in the wild type, but the photosynthesis rate fell with age [6,26]. Recently, a wheat ethylene receptor homolog, W-er1, was isolated using the *Arabidopsis* ETR1 cDNA as a probe [140]. Treatments with jasmonate, ABA, and wounding induced senescence and caused increased accumulation of W-er1 mRNA [140]. Very recently, a new ACC oxidase cDNA clone (CP-ACO2) was isolated from papaya; expression of the gene *cp-aco2* was induced only at a late stage of leaf senescence [141].

Analysis of the ore9 *Arabidopsis* mutant, which exhibits a significant delay in the senescence process, revealed that the ore9 mutant carries a mutation in a gene that encodes an F-box-containing protein (Table 36.1). The ORE9 protein forms an SCF complex and probably works in senescence signaling, which is mediated by ethylene, ABA, and jasmonic acid (JA) derivatives [142].

According to Buchanan-Wollaston [6] treatment with ethylene does not directly induce the transcription of SAGs. However, the presence of ethylene may increase the sensitivity for signaling age-related factors or enhance and accelerate their transmission.

To conclude, reduced ethylene production delays leaf senescence. A relatively high level of ethylene enhances the rate of senescence in leaves where the process of senescence has already started. Probably ethylene does not activate SAGs directly but modifies the activation of genes through other signals. Ethylene may also repress the expression of genes involved in photosynthesis [139].

It is possible that pathogen-stress-induced PCD and natural PCD share the signaling pathway. Ethylene is the best possible candidate for a signaling molecule in multiple PCD programs; it promotes cell death triggered by ozone, some toxins, and pathogens [92].

C. Role of Other Hormones

Some gibberellins, like GA_{4+7}, prevented leaf senescence, while others were not effective [143]. Methyl jasmonate promoted senescence [144]. It is known that ABA, JA, methyl jasmonate, and even brassinosteroids are also involved in the regulation of the natural leaf senescence process and senescence induced by environmental factors and by pathogen attack [6,10,26,145–149].

D. Level of Metabolites

Like plant hormones, the level of metabolites can regulate the expression of genes related to senescence.

Some of the senescence-induced genes involved in the gloxylate pathway are also regulated by the levels of carbon compounds [50]. Stimulation of SAG transcription can be related to changes in the level of metabolites that occur due to the reduced rate of photosynthesis [21]. A decrease of photosynthesis reduces fixed carbon availability and can be a signal for SAG induction [6]. Because sugars are the primary products of photosynthesis, their levels could be a part of the signaling system. Using SAG12, a gene regulated specifically by senescence, it was shown that exogenous sugars can repress gene expression in senescent *Arabidopsis* leaves [78]. This means that the senescence pathway, represented by SAG12 expression, can probably be activated by low sugar levels [13]. Deficiency of sugar may be one of the components regulating leaf senescence. Loss of photosynthesis can influence the integrity of chloroplast photosynthetic membranes and thus produce signals that initiate the senescence program. Results with transgenic tomato plants, which overexpress the gene for hexokinase (HXK), a well-known sugar sensor, suggest that an enhanced sugar signal can induce premature senescence, but an increased sugar level represses the photosynthetic activity by a negative feedback regulation [17,150].

E. Signals from Developing Organs

When there are no signals from developing organs, such as young flowers or fruits, leaf senescence does not occur or is even repressed. Many fruiting plants receive signals from their fruits [6], but *Arabidopsis* leaf senescence does not involve such signals [21].

F. Role of ROS and Nitric Oxide

ROS can serve as direct or indirect mediators of PCD. They can function as a facultative signal, starting the program of cell death caused by different external or internal factors and can also seem to be mediators of the natural program of cell death. ROS can interact directly or indirectly with several other signaling pathways, such as the stress hormones ethylene, JA, and salicylic acid [151].

A possible role of NO as a regulator of leaf senescence was suggested by Leshem et al. [152] and Pedroso and Durzan [112]. Involvement of NO as a trigger of a senescence-like process that exhibits characteristic aspects of PCD [112] has already been discussed in this chapter. NO formation is associated with ethylene biosynthesis [153], drought stress [154,155], cell death, and cell proliferation [113,114] and can be directly or indirectly responsible for irre-

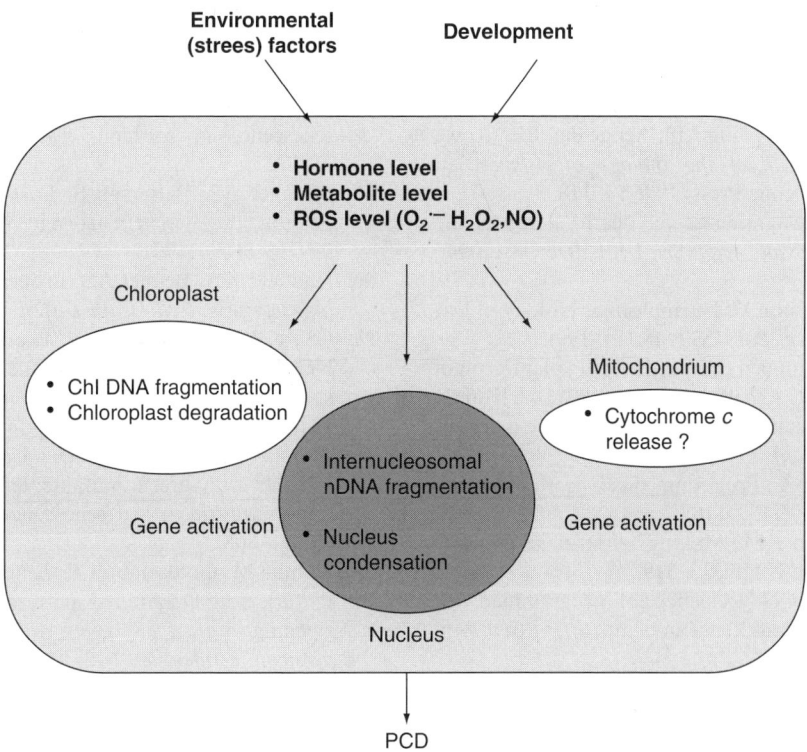

Environmental (strees) factors   **Development**

* **Hormone level**
* **Metabolite level**
* **ROS level ($O_2{}^{.-}$ H_2O_2,NO)**

Chloroplast

* Chl DNA fragmentation
* Chloroplast degradation

Mitochondrium

* Cytochrome *c* release ?

Gene activation

* Internucleosomal nDNA fragmentation
* Nucleus condensation

Gene activation

Nucleus

PCD

FIGURE 36.6 Model of action of possible regulators of leaf senescence either induced by environmental (stress) factors or evoked by natural, developmental processes, leading finally to death of leaf cells.

versible nDNA fragmentation [112]. The role of ROS and NO as signal molecules was also discussed by Corpas et al. [156] and Huang et al. [157].

In conclusion, a complete understanding of regulatory mechanisms underlying senescence can be made possible only by isolation and identification of the promoters controlling SAGs and by analysis of regulatory factors that are associated with these promoters.

Cooperation between ROS, NO, and plant hormones might provoke a life or death decision in plant cells [92]. A model showing the action of possible regulators of leaf senescence induced by environmental (stress) factors or evoked by natural, developmental processes leading finally to the death of leaf cells is proposed in Figure 36.6.

VIII. CONCLUSIONS AND FUTURE CHALLENGES

Although the whole mechanism of senescence regulation is still not understood, numerous molecular data prove that leaf senescence is a genetically defined program of cell death, accompanied by changes in gene expression [10]. The results presented clearly indicate that leaf senescence passes through a certain sequence of changes and that this program involves PCD. An

important unanswered question still remains, what is the mechanism that restricts endogenous signals of cell death to individual cells within the same plant organ?

The application of molecular biology techniques should make a major contribution to understanding the basis for the onset of senescence in plants. To clarify the complex regulatory network of leaf senescence it is necessary to select a diverse range of novel and informative mutants and to identify novel transcription factors. By studying the altered senescence regulation in multiple mutants of SAGs, catabolic processes, and metabolite export in these mutants, a better approach to understanding the nature of senescence will be possible. Experiments with transgenic plants may greatly contribute to the improvement of important agronomic traits, crop yield, and storage of harvested tissues.

REFERENCES

1. Stoddart L, Thomas H. Leaf senescence. In Boulter D, Parthier B, eds. *Encyclopedia of Plant Physiology*. New Series, Vol. 14A. Berlin: Springer Verlag, 1982:592–636.
2. Noodén LD. The phenomena of senescence and aging. In: Noodén LD, Leopold AC, eds. *Senescence and Aging in Plants*. San Diego, CA: Academic Press, 1988:1–50.

3. Matile P. Chloroplast senescence. In: Baker NR, Thomas H, eds. *Crop Photosynthesis: Spatial and Temporal Determinants*. Amsterdam: Elsevier, 1992:413–440.

4. Noodén LD, Guiamét JJ. Genetic control of senescence and aging in plants. In: Schneider EL, Rowe JW, eds. *Handbook of the Biology of Aging*. San Diego, CA: Academic Press, 1996:94–118.

5. Greenberg JT. Programmed cell death: a way of life for plants. *Proc. Natl. Acad. Sci. USA* 1996; 93:12094–12097.

6. Buchanan-Wollaston V. The molecular biology of leaf senescence. *J. Exp. Bot.* 1997; 48:181–199.

7. Beers EP. Programmed cell death during plant growth and development. *Cell Death Differ.* 1997; 4:649–661.

8. Noodén LD, Guiamét JJ, John I. Senescence mechanisms. *Physiol. Plant.* 1997; 101:746–753.

9. Pennel RI, Lamb C. Programmed cell death in plants. *Plant Cell* 1997; 9:1157–1168.

10. Gan S, Amasino RM. Making sense of senescence. *Plant Physiol.* 1997; 113:313–319.

11. Yen CH, Yang CH. Evidence for programmed cell death during leaf senescence in plants. *Plant Cell Physiol.* 1998; 39:922–927.

12. Simeonova E, Sikora A, Charzyñska M, Mostowska A. Aspects of programmed cell death during leaf senescence of mono- and dicotyledonous plants. *Protoplasma* 2000; 214:93–101.

13. Quirino BF, Noh YS, Himelblau E, Amasino RM. Molecular aspects of leaf senescence. *Trends Plant Sci.* 2000; 5:278–282.

14. Jones AM. Programmed cell death in development and defence. *Plant Physiol.* 2001; 125:94–97.

15. Simeonova E, Mostowska A. Biochemical and molecular aspects of leaf senescence. *Post. Biol. Kom.* 2001; 28:17–32.

16. Lim POK, Woo HR, Nam HG. Molecular genetics of leaf senescence in *Arabidopsis. Trends Plant Sci.* 2003; 8:272–278.

17. Yoshida S. Molecular regulation of leaf senescence. *Curr. Opin. Plant Biol.* 2003; 6:79–84.

18. Hudák J. Photosynthetic apparatus. In: Pessarakli M, ed. *Handbook of Photosynthesis*. New York: Marcel Dekker, 1997:27–48.

19. Jackowski G. Rozpad aparatu fotosyntetycznego w trakcie starzenia się siś lisci. *Post. Biol. Kom.* 1998; 25:63–73.

20. Jones AM, Dangl JL. Logjam at the Styx: programmed cell death in plants. *Trends Plant Sci.* 1996; 1:114–119.

21. Hensel LL, Grabić V, Baumgarten DA, Bleecker AB. Developmental and age-related processes that influence the longevity and senescence of photosynthetic tissues in *Arabidopsis. Plant Cell* 1993; 5: 553–564.

22. Mostowska A. Environmental factors affecting chloroplasts.In: Pessarakli M, ed. *Handbook of Photosynthesis*. New York: Marcel Dekker 1997:407–426.

23. Buchanan-Wollaston V, Morris K. Senescence and cell death in *Brassica napus* and *Arabidopsis*. In: Bryant JA, Hughes SG, Garland JM, eds. *Programmed Cell Death in Animals and Plants*. Oxford: BIOS Scientific Publishers 2000:149–162.

24. Page T, Griffiths G, Buchanan-Wollaston V. Molecular and biochemical characterization of postharvest senescence in broccoli. *Plant Physiol.* 2001; 125: 718–727.

25. Bleecker AB, Patterson S. Last exit: senescence, abscission, and meristem arrest in *Arabidopsis. Plant Cell* 1997; 9:1169–1179.

26. Hadfield KA, Bennet AB. Programmed senescence of plant organs. *Cell Death Differ.* 1997; 4:662–670.

27. Cha KW, Lee YJ, Koh HJ, Lee BM, Nam YW, Paek NC. Isolation, characterization, and mapping of the stay green mutant in rice. *Theor. Appl. Genet.* 2002; 104:526–532.

28. Thomas H, Ougham H, Canter P, Donnison I. What stay-green mutants tell us about nitrogen remobilization in leaf senescence. *J. Exp. Bot.* 2002; 53:801–808.

29. Akhtar M, Goldschmidt E, John I, Rodoni S, Matile P, Grierson D. Altered patterns of senescence and ripening in *gf*, a stay-green mutant of tomato (*Lycopersicon esculentum* Mill.). *J. Exp. Bot.* 1999; 50: 1115–1122.

30. Macduff JH, Humphreys MO, Thomas H. Effects of a *stay-green* mutation on plant nitrogen relations in *Lolium perenne* during N starvation and after defoliation. *Ann. Bot.* 2002; 89:11–21.

31. Thomas H, Ougham HJ, Davies TGE. Leaf senescence in non-yellowing mutant of *Festuca pratensis*. Transcripts and translation products. *J. Plant Physiol.* 1992; 139:403–412.

32. Rodoni S, Schellenberg M, Matile P. Chlorophyll breakdown in senescing barley leaves as correlated with phaeophorbide a oxygenase activity. *Plant Physiol.* 1998; 152:139–144.

33. Thomas H, Howarth CJ. Five ways to stay green. *J. Exp. Bot.* 2000; 51(Suppl. 1):329–337.

34. Lee RH, Wang CH, Huang LT, Chen SCG. Leaf senescence in rice plants: cloning and characterization of senescence up-regulated genes. *J. Exp. Bot.* 2001; 52:1117–1121.

35. Buchanan-Wollaston V, Ainsworth C. Leaf senescence in *Brassica napus*: cloning of senescence-related genes by subtractive hybridization. *Plant Mol. Biol.* 1997; 33:821–834.

36. Lohman KN, Gan S, John MC, Amasino RM. Molecular analysis of natural leaf senescence in *Arabidopsis thaliana. Physiol. Plant.* 1994; 92: 322–328.

37. Buchanan-Wollaston V. Isolation of cDNA clones for genes that are expressed during leaf senescence in *Brassica napus*. Identification of a gene encoding a senescence-specific metallothionein-like protein. *Plant Physiol.* 1994; 105:839–846.

38. Smart CM, Hosken SE, Thomas H, Greaves JA, Blair BG, Schuch W. The timing of maize leaf senescence and characterisation of senescence-related cDNAs. *Physiol. Plant.* 1995; 93:673–682.

39. Ling JQ, Kojima T, Shiraiwa M, Takahara H. Cloning of two cysteine proteinase genes, CysP1 and

CysP2, from soybean cotyledons by cDNA representational difference analysis. *Biochim. Biophys. Acta* 2003; 1627:129–139.

40. Genschik P, Durr A, Fleck J. Differential expression of several E2-type ubiquitin carrier protein genes at different developmental stages in *Arabidopsis thaliana* and *Nicotiana sylvestris*. *Mol. Gen. Genet.* 1996; 244:548–556.

41. Garbarino JE, Oosumi T, Belknap WR. Isolation of a polyubiquitin promoter and its expression in transgenic potato plants. *Plant Physiol.* 1995; 109:1371–1378.

42. Taylor CB, Bariola PA, Delcardayre SB, Raines RT, Green PJ. RNS2-a senescence-associated RNAse of *Arabidopsis* that diverged from the sRNAses before speciation. *Proc. Natl. Acad. Sci. USA* 1993; 90:5118–5122.

43. Liang L, Lai Z, Ma W, Zhang Y, Xue Y. AhSL28, a senescence- and phosphate starvation-induced S-like RNase gene in *Antirrhinum*. *Biochim. Biophys. Acta* 2002; 1579:64–71.

44. Kamachi K, Yamaya T, Hayakawa T, Mae T, Ojima K. Changes in cytosolic glutamine synthetase polypeptide and its mRNA in a leaf blade of rice plants during natural senescence. *Plant Physiol.* 1992; 98:1323–1329.

45. Bernhard W, Matile P. Differential expression of glutamine synthetase genes during the senescence of *Arabidopsis thaliana* rosette leaves. *Plant Sci.* 1994; 98: 7–14.

46. Kawakami N, Watanabe A. Senescence-specific increase in cytosolic glutamine synthetase and its mRNA in radish cotyledons. *Plant Physiol.* 1988; 88:1430–1434.

47. King GA, Davies KM, Stewart RJ, Borst WM. Similarities in gene-expression during the post-harvest-induced senescence of spears and natural foliar senescence of asparagus. *Plant Physiol.* 1995; 108:125–128.

48. McLaughlin JC, Smith SM. Glyoxylate cycle enzyme synthesis during the irreversible phase of senescence of cucumber cotyledons. *J. Plant Physiol.* 1995; 146:133–138.

49. Kim DJ, Smith SM. Expression of a single gene encoding microbody NAD-malate dehydrogenase during glyoxysome and peroxisome development in cucumber. *Plant Mol. Biol.* 1994; 26:1833–1841.

50. Graham IA, Denby KJ, Leaver CJ. Carbon catabolite repression regulates glyoxylate cycle gene expression in cucumber. *Plant Cell* 1994; 6:761–772.

51. Abdullah SNA, Cheah SC, Murphy DJ. Isolation and characterisation of two divergent type 3 metallothioneins from oil palm, *Elaeis guineensis. Plant Physiol. Biochem.* 2002; 40:255–263.

52. Becker W, Apel K. Differences in gene-expression between natural and artificially induced leaf senescence. *Planta* 1993; 189:74–79.

53. Smart CM. Gene expression during leaf senescence. *New Phytol.* 1994; 126:419–448.

54. Yoshida T, Minamikawa T. Successive amino-terminal proteolysis of the large subunit of rybulose

55. Nieri B, Canino S, Versace R, Alpi A. Purifaction and characterization of an endoprotease from alfaalfa senescent leaves. *Phytochemistry* 1998; 49:643–649.

56. Morris K, Thomas H, Rogers L. Endopeptidases during the developmental and senescence of *Lolium temulentum* leaves. *Phytochemistry* 1996; 41:377–384.

57. Thomas H, Donnison I. Back from the brink: plant senescence and its reversibility. In: Bryant JA, Hughes SG, Garland JM, eds. *Programmed Cell Death in Animals and Plants*. Oxford: BIOS Scientific Publishers, 2000:149–162.

58. Li Q, Bettany AJE, Donnison I, Griffiths CM, Thomas H, Scott IM. Characterisation of a cysteine protease cDNA from *Lolium multiflorum* leaves and its expression during senescence and cytokinin treatment. *Biochim. Biophys. Acta* 2000; 1492:233–236.

59. Kuriyama H, Fukuda H. Developmental programmed cell death in plants. *Curr. Opin. Plant Biol.* 2002; 5:568–573.

60. Chen GH, Huang LT, Yap MN, Lee RH, Huang YJ, Cheng MC, Chen SCG. Molecular characterization of a senescence-associated gene encoding cysteine proteinase and its gene expression during leaf senescence in sweet potato. *Plant Cell Physiol.* 2002; 43(9): 984–991.

61. Watanabe H, Abe K, Emori Y, Hosoyama H, Arai S. Molecular cloning and gibberellin-induced expression of multiple cysteine proteinases of rice seeds (oryzains). *J. Biol. Chem.* 1991; 266:16897–16902.

62. Yamauchi Y, Sugimoto T, Sueyoshi K, Oji Y, Tanaka K. Appearance of endopeptidases during the senescence of cucumber leaves. *Plant Sci.* 2002; 162:615–619.

63. Jiang WB, Lers A, Lomaniec E, Aharoni N. Senescence-related serine protease in parsley. *Phytochemistry* 1999; 50:377–382.

64. Shimizu T, Inoue T, Shiraishi H. A senescence-associated S-like RNase in the multicellular green alga *Volvox carteri. Gene* 2001; 274:227–235.

65. Canetti L, Lomaniec E, Elkind Y, Lers A. Nuclease activities associated with dark-induced and natural leaf senescence in parsley. *Plant Sci.* 2002; 163:873–880.

66. Feller U, Fischer A. Nitrogen metabolism in senescing leaves. *Crit. Rev. Plant Sci.* 1994; 13:241–273.

67. Vicentini F, Matile P. Gerontosomes, a multifunctional type of peroxisomes in senescent leaves. *J. Plant Physiol.* 1993; 142:50–56.

68. Pastori G, del Rio LA. Natural senescence of pea leaves. An activated oxygen-mediated function for peroxisomes. *Plant Physiol.* 1997; 113:411–418.

69. Chen ZH, Walker RP, Acheson RM, Técsi LI, Wingler A, Lea PJ, Leegood RC. Are isocitrate lyase and phosphoenolpyruvate carboxykinase involved in gluconeogenesis during senescence of barley leaves and cucumber cotyledons? *Plant Cell Physiol.* 2000; 41(8):960–967.

70. Ryu Sb, Wang XM. Expression of phospholipase-D during castor bean leaf senescence. *Plant Physiol.* 1995; 108:713–719.

71. Biswal B. Chloroplast metabolism during leaf greening and degreening. In: Pessarakli M, eds. *Handbook of Photosynthesis*. New York: Marcel Dekker 1997:71–81.

72. Hörtensteiner S. Chlorophyll breakdown in higher plants and algae. *Cell Mol. Life Sci.* 1999; 56:330–347.

73. Ma M, Lau PS, Jia YT, Tsang WK, Lam SKS, Tam NFY, Wong YS. The isolation and characterization of Type 1 metallothionein (MT) cDNA from a heavy-metal-tolerant plant, *Festuca rubra* cv. Merlin. *Plant Sci.* 2003; 164:51–60.

74. Foyer CH, Descourvières P, Kunert KJ. Protection against oxygen radicals: an important defence mechanism studied in transgenic plants. *Plant Cell Environ.* 1994; 17:507–523.

75. Fang WC, Kao CH. Enhanced peroxidase activity in rice leaves in response to excess iron, copper and zinc. *Plant Sci.* 2000; 158:71–76.

76. Del Río LA, Sandalio LM, Altomare DA, Zilinskas BA. Mitochondrial and peroxisomal manganese superoxide dismutase: differential expression during leaf senescence. *J. Exp. Bot.* 2003; 54:923–933.

77. Nam HG. The molecular genetic analysis of leaf senescence. *Curr. Opin. Biotechnol.* 1997; 8:200–207.

78. Noh Y, Amasino RM. Identification of promotor region responsible for the senescence-specific expression of *sag12*. *Plant Mol. Biol.* 1999; 41:181–194.

79. Robatzek S, Somssich i.e., A new member of the *Arabidopsis* WRKY transcription factors family, *At*WRKY6, is associated with both senescence- and defence-related processes. *Plant J.* 2001; 28:123–133.

80. Robatzek S, Somssich IE. Targets of *At*WRKY6 regulation during plant senescence and pathogen defence. *Genes Dev.* 2002; 16:1139–1149.

81. Hinderhofer K, Zentgraf U. Identification of a transcription factor specifically expressed at the onset of leaf senescence. *Planta* 2001; 213:469–473.

82. Ghosh S, Mahoney SR, Penterman JN, Peirson D, Dumbroff EB. Ultrastructural and biochemical changes in chloroplasts during *Brassica napus* senescence. *Plant Physiol. Biochem.* 2001; 39:777–784.

83. Mostowska A. Response of chloroplast structure to photodynamic Herbicides and high oxygen. *Z. Naturforsch.* 1999; 54c:621–628.

84. Biswal B. Carotenoid catabolism during leaf senescence and its control by light. *J. Photochem. Photobiol.* 1995; 30:3–13.

85. Bate NJ, Rothstein SJ, Thompson JE. Expression of nuclear and chloroplast photosynthesis-specific genes during leaf senescence. *J. Exp. Bot.* 1991; 42:801–811.

86. Lu C, Lu Q, Zhang J, Kuang T. Characterization of photosynthetic pigment composition, photosystem II photochemistry and thermal energy dissipation during leaf senescence of wheat plants grown in the field. *J. Exp. Bot.* 2001; 52:1805–1810.

87. Fang Z, Bouwkamp J, Solomos T. Chlorophyllase activities and chlorophyll degradation during leaf senescence in non-yellowing mutant and wild type of *Phaseolus vulgaris* L. *J. Exp. Bot.* 1998; 49:503–510.

88. Munné-Bosch S, Penuelas J. Photo- and antioxidative protection during summer leaf senescence in *Pistacia lentiscus* L. grown under mediterranean field conditions. *Ann. Bot.* 2003; 92:385–391.

89. Charzyñska M. Programmed cell death in the ontogenesis of angiosperms. *Folia Histochem. Cytobiol.* 1996; 34 (Suppl. 2):57.

90. Havel L, Durzan DJ. Apoptosis in plants. *Bot. Acta* 1996; 109:268–277.

91. Mittler R, Lam E. Characterization of nuclease activities and DNA fragmentation induced upon hypersensitive response cell death and mechanical stress. *Plant Mol. Biol.* 1997; 34:209–221.

92. Beers EP, McDowell JM. Regulation and execution of programmed cell death in response to pathogens, stress and developmental cues. *Curr. Opin. Plant Biol.* 2001; 4:561–567.

93. Lam E, Pontier D, del Pozo O. Die and let live-programmed cell death in plants. *Curr. Opin. Plant Biol.* 1999; 2:502–507.

94. Pontier D, Balagué C, Roby D. The hypersensitive response. A programmed cell death associated with plant resistance. *C R Acad. Sci. Paris, Sciences de la vie* 1998; 321:721–734.

95. Lorrain S, Vailleau F, Balagu C, Roby D. Lesion mimic mutants: keys for deciphering cell death and defense pathways in plants? *Trends Plant Sci.* 2003; 8(6):263–271.

96. Van Breusegem F, Vranová E, Dat JF, Inzé D. The role of active oxygen species in plant signal transduction. *Plant Sci.* 2001; 161:405–414.

97. Kuran H. Changes in DNA, dry mass and protein content of leaf epidermis nuclei during aging of perennial monocotyledonous plant. *Acta Soc. Bot. Pol.* 1993; 62:149–154.

98. Barton R. Fine structure of mesophyll cells in senescing leaves of *Phaseolus*. *Planta* 1996; 71: 314–325.

99. Nii N, Kawano S, Nakamura S, Kuroiwa T. Changes in fine structure of chloroplast and chloroplast DNA of peach leaves during senescence. *J. Jpn. Soc. Hort. Sci.* 1988; 57:390–398.

100. Kołodziejek I, Kozioł J, Wałeza M, Mostowska A. Ultrastructure of mesophyll cells and pigment content in senescing leaves of maize and barley. *J. Plant Growth Regul.* 2003; 22:217–227.

101. Bortner CD, Nicklas BE, Oldenburg NBE, Cidlowski JA. The role of DNA fragmentation in apoptosis. *Trends Cell Biol.* 1995; 5:21–26.

102. Gavrieli Y, Sherman Y, Ben-Ssason SA. Identification of programmed cell death *in situ* via specific labelling of nuclear DNA fragmentation. *J. Cell Biol.* 1992; 119:493–501.

103. Wang H, Li J, Bostock RM, Gilchirst DG. Apoptosis: a functional paradigm for programmed cell death induced by a host-selective phytotoxin and invoked during development. *Plant Cell* 1996; 8:375–391.

104. Leśniewska J, Simeonova E, Sikora A, Mostowska A, Charzyñska M. Application of the "comet assay" in studies of programmed cell death (PCD) in plants. *Acta Soc. Bot. Pol.* 2000; 69:101–107.

105. Wang M, Hoekstra S, Van Bergen S, Lamers GEM, Oppedijk BJ, Van Der Heijden MW, Priester W, Schilperoort RA. Apoptosis in developing anthers and the role of ABA in this process during androgenesis in *Hordeum vulgare* L. *Plant Mol. Biol.* 1999; 39:489–501.

106. Caccia R, Delledonne M, Levine A, de Pace C, Mazzucato A. Apoptosis-like DNA fragmentation in leaves and floral organs precedes their developmental senescence. *Plant Biosyst.* 2001; 135:183–190.

107. Wood M, Power JB, Davery MR, Lowe KC, Mulligan BJ. Factors affecting single strand-preferring nuclease activity during leaf aging and dark-induced senescence in barley (*Hordeum vulgare* L.). *Plant Sci.* 1998; 131:149–159.

108. Inada N, Sakai A, Kuroiwa H, Kuroiwa T. Three-dimensional analysis of the senescence program in rice (*Oryza sativa* L.) coleoptiles. *Planta* 1998; 205:153–164.

109. Inada N, Sakai A, Kuroiwa H, Kuroiwa T. Senescence program in in rice *Oryza sativa* L. leaves: analysis of the blade of the second leaf at the tissue and cellular levels. *Protoplasma* 1999; 207:222–232.

110. Sodmergen, Kawano S, Tano S, Kuroiwa T. Preferential digestion of chloroplast nuclei (nucleoids) during senescence of the coleoptile of *Oryza sativa*. *Protoplasma* 1989; 152:565–568.

111. Sodmergen, Kawano S, Tano S, Kuroiwa T. Degradation of chloroplast DNA in second leaves of rice (*Oryza sativa*) before leaf yellowing. *Protoplasma* 1991; 160:89–98.

112. Pedroso MC, Durzan DJ. Effect of different environments on DNA fragmentation and cell death in *Kalanchoëe* leaves. *Ann. Bot.* 2000; 86:983–994.

113. Pedroso MC, Magalhaes JR, Durzan D. A nitric oxide burst precedes apoptosis in angiosperm and gymnosperm callus cells and foliar tissues. *J. Exp. Bot.* 2000; 51:1027–1036.

114. Pedroso MC, Magalhaes JR, Durzan D. Nitric oxide induces cell death in *Taxus* cells. *Plant Sci.* 2000:157:173–180.

115. Sugiyama M, Ito J, Aoyagi S, Fukuda H. Endonucleases. *Plant Mol. Biol.* 2000; 44:387–397.

116. Obara K, Kiruyama H, Fukuda H. Direct evidence of active and rapid nuclear degradation triggered by vacuole rupture during programmed cell death in *Zinnia*. *Plant Physiol.* 2001; 125:615–626.

117. Balk J, Leaver C. The *pet1-cms* mitochondrial mutation in sunflower is associated with premature programmed cell death and cytochrome c release. *Plant Cell* 2001; 13:1803–1818.

118. Jones AM. Does the plant mitochondrion integrate cellular stress and regulate programmed cell death? *Trends Plant Sci.* 2000; 5:225–230.

119. Danon A, Delorme V, Mailhac N, Gallois P. Plant programmed cell death: a common way to die. *Plant Physiol. Biochem.* 2000; 38:647–655.

120. Balk J, Leaver CJ, McCabe PF. Translocation of cytochrome *c* from the mitochondria to the cytosol occurs heat-induced programmed cell death in cucumber plants. *FEBS Lett.* 1999; 463:151–154.

121. Zhao Y, Jiang Z, Sun Y, Zhai Z. Apoptosis of mouse liver nuclei induced in the cytosol of carrot cells. *FEBS Lett.* 1999; 448:197–200.

122. Zhao Y, Wu M, Shen Y, Zhai Z. Analysis of nuclear apoptotic process in a cell-free system. *Cell Mol. Life Sci.* 2001; 58:298–306.

123. Del Pozo O, Lam E. Caspases and programmed cell death in the hypersensitive response of plants to pathogens. *Curr. Biol.* 1998; 8:1129–1132.

124. Sun Y, Zhao Y, Hong X, Zhai Z. Cytochrome c release and caspase activation during menadione-induced apoptosis in plants. *FEBS Lett.* 1999; 462:317–321.

125. Zhou J, Chen H, Jiang X, Zhu H, Dai Y. Construction of cell free system using extracts from apoptotic plant cells. *Chin. Sci. Bull.* 1999; 44:1494–1497.

126. Curtis M, Wolpert TJ. The oat mitochondrial permeability transition and its implication in victorin binding and induced cell death. *Plant J.* 2002; 29:295–312.

127. Mostowska A. Effect of 1,10-phenanthroline, a photodynamic herbicide on development and structure of chloroplasts. *Acta Physiol. Plant.* 1998; 20:419–424.

128. Nayak L, Biswal B, Ramaswamy NK, Iyer RK, Nair JS, Biswal UC. Ultraviolet-A induced changes in photosystem II of thylakoids: effects of senescence and high growth temperature. *J. Photochem. Photobiol. B* 2003; 70:59–65.

129. Binyamin L, Falah M, Portnoy V, Soudry E, Gepstein S. The early light-induced protein is also produced during leaf senescence of *Nicotiana tabacum*. *Planta* 2001; 212:591–597.

130. Miller JD, Arteca RN, Pell EJ. Senescence-associated gene expression during ozone-induced leaf senescence in *Arabidopsis*. *Plant Physiol.* 1999; 120:1015–1023.

131. Oh SA, Lee SY, Chung IK, Lee CH, Nam HG. A senescence associated gene of *Arabidopsis thaliana* is distinctively regulated during natural and artificially induced leaf senescence. *Plant Mol. Biol.* 1996; 30: 739–754.

132. Yang SH, Yamaguchi Y, Koizumi N, Kusano T, Sano H. Promoter analysis of *tbz*F, a gene encoding a bZIP-type transcription factor, reveals distinct variation in cis-regions responsible for transcriptional activation between senescing leaves and flower buds in tobacco plants. *Plant Sci.* 2002; 162(6):973–980.

133. Singh KB, Foley RC, Onate-Sanchez L. Transcription factors in plant defence and stress responses. *Curr. Opin. Plant Biol.* 2002; 5:430–436.

134. Jabs T. Reactive oxygen intermediates as mediators of programmed cell death in plants and animals. *Biochem. Pharmacol.* 1999; 57:231–245.

135. Noodén LD, Singh S, Lethan DS. Correlation of xylem sap levels with monocarpic senescesne in soybean. *Plant Physiol.* 1990: 93:33–39.

136. Gan S, Amasino RM. Cytokinins in plant senescence: from spray and pray to clone and play. *BioEssays* 1996; 18:557–565.

137. Gan S, Amasino RM. Inhibition of leaf senescence by autoregulated production of cytokinin. *Science* 1995; 270:1966–1967.

138. John I, Drake R, Farrell A, Cooper W, Lee P, Horton P, Grieson D. Delayed leaf senescence in ethylene-deficient ACC-oxidase antisense tomato plants — molecular and physiological analysis. *Plant J.* 1995; 7:483–490.

139. Grbic V, Bleecker AB. Ethylene regulates the timing of leaf senescence in Arabidopsis. *Plant J.* 1995; 8: 595–602.

140. Ma QH, Wang XM. Characterization of an ethylene receptor homologue from wheat and its expression during leaf senescence. *J. Exp. Bot.* 2003; 54:1489–1490.

141. Chen YT, Lee YR, Yang CY, Wang YT, Yang SF, Shaw JF. A novel papaya ACC oxidase gene (CP-ACO2) associated with late stage fruit ripening and leaf senescence. *Plant Sci.* 2003; 164:531–540.

142. Woo HR, Chung KM, Park JH, Oh SA, Ahn T, Hong SH, Jang SK, Nam HG. ORE9, an F-box protein that regulates leaf senescence in Arabidopsis. *Plant Cell* 2001; 13:1779–1790.

143. Ranwala AP, Miller WB. Perventive mechanisms of gibberellin$_{4+7}$ and light on low-temperature-induced leaf senescence in *Lilium* cv. Stargazer. *Postharvest Biol. Technol.* 2000; 19:85–92.

144. Rossato L, MacDuff JH, Laine P, Le Deunff E, Ourry A. Nitrogen storage and remobilization in *Brassica napus* L. during the growth cycle: effects of methyl jasmonate on nitrate uptake, senescence, growth, and VSP accumulation. *J. Exp. Bot.* 2002; 53:1131–1141.

145. Park JH, Oh SA, Kim YH, Woo HR, Nam HG. Differential expression of senescence-associated mRNAs during senescence induced by different senescence-inducing factors in *Arabidopsis. Plant Mol. Biol.* 1998; 37:445–454.

146. Weaver LM, Gan S, Quirino B, Amasino RM. A comparison of the expression patterns of several senescence-associated genes in response to stress and hormone treatment. *Plant Mol. Biol.* 1998; 37:455–469.

147. He Y, Gan S. Identical promoter elements are involved in regulation of the *opr1* gene by senescence and jasmonic acid in *Arabidopsis. Plant Mol. Biol.* 2001; 47:596–605.

148. He Y, Fukushige H, Hildebrand DF, Gan S. Evidence supporting a role of jasmonic acid in *Arabidopsis* leaf senescence. *Plant Physiol.* 2002; 128:876–884.

149. Wasternack C, Parthier B. Jasmonate-signalled plant gene expression. *Trends Plant Sci.* 1997; 2:302–307.

150. Xiao W, Sheen J, Jang JC. The role of hexokinase in plant sugar signal transduction and growth and development. Plant Mol. Biol. 2000; 44:451–461.

151. Overmyer K, Brosch M, Kangasj J. Reactive oxygen species and hormonal control of cell death. *Trends Plant Sci.* 2003;8:335–342.

152. Leshem YY, Wills RBH, Ku VVV. Evidence for function of the free radical gas — nitric oxide (NO) — as an endogenous maturation and senescence regulating factor in higher plants. *Plant Physiol. Biochem.* 1998; 36:825–833.

153. Leshem YY, Haramaty E. The characterization and contrasting effects of the nitric oxide free radical in vegetative stress and senescence of *Pisum sativum* Linn. foliage. *J. Plant Physiol.* 1996; 148:258–263.

154. Leshem YY. Nitric oxide in biological systems. *Plant Growth Regul.* 1996; 18:155–159.

155. Haramaty E, Leshem YY. Ethylene regulation by nitric oxide (NO) free radical: A possible mode of action of endogenous NO. In: Kanellis AK, Chang C, Kende H, eds. *Biology and Biotechnology of the Plant Hormone Ethylene*. Dordrecht, The Netherlands: Kluwer Academic Publishers, 1997:253–258.

156. Corpas FJ, Barroso JB, del Río LA. Peroxisomes as a source of reactive oxygen species an nitric oxide signal molecules in plant cells. *Trends Plant Sci.* 2001; 6:145–150.

157. Huang X, Kiefer E, von Rad U, Ernst D, Foissner I, Durner J. Nitric oxide burst and nitric oxide4-dependet gene induction in plants. *Plant Physiol. Biochem.* 2002; 40:625–631.

Section XIII

Photosynthesis under Environmental Stress Conditions

37 Photosynthesis in Plants under Stressful Conditions

Rama Shanker Dubey
Department of Biochemistry, Faculty of Science,
Banaras Hindu University

CONTENTS

I. INTRODUCTION

In order to increase the agricultural productivity within the limited land resource, it is essential to ensure the stability of yield against adverse environmental factors. Soil salinity, drought, flood, heat, cold, anaerobiosis, gaseous pollutants, radiations, and high levels of heavy metals in the soil are the important environmental stress factors that lead to severe crop loss every year.

Photosynthesis is essentially one of the key plant processes that directly determine crop productivity. The decline in productivity in many plant species subjected to harsh environmental conditions is often associated with a reduction in photosynthetic capacity [1]. Supreme importance has been assigned to photosynthesis because its products: (i) a high energy reduced form of organic carbon (carbohydrate) and (ii) molecular oxygen, support the life of all organisms on this planet and without which life would cease to exist. Various stressful environmental conditions reduce the photosynthetic capacity of growing plants due to their influence on any one or more of the events associated with the photosynthetic process. The influence of stresses may include decreased utilization of light energy [2], alteration in pigment composition and destruction of fine structure of chloroplast [3], impaired photophosphorylation and ATP synthesis [4], downregulation of photosystem II (PSII) [5], decreased stomatal conductance leading to closure of stomata and decreased availability of CO_2 at the site of its fixation [6], and alteration in the amount and activity of enzymes associated with CO_2 assimilation [7].

Sometimes the effects of many stresses are common and they influence the same parameter of photosynthesis whereas all different stresses may influence different events associated with photosynthesis depending on the type and extent of stress [8]. Water and salt stresses lower leaf water potential leading to decreased stomatal conductance, stomatal closure, altered chlorophyll fluorescence, photoinhibition of photosystem II, impaired ATP synthesis and RUBP regeneration, conformational changes in membrane bound ATPase enzyme complex, as well as decrease in both activity and concentration of Rubisco enzyme [2,6]. Salt stress, in addition to osmotic effects, exerts specific ion effects due to Na^+ and Cl^- penetrating in the chloroplasts, leading to ion toxicity and resulting in nutritional imbalance due to competition of salt ions and nutrients [2]. Heat stress increases membrane fluidity, leads to disorganization of chloroplast thylakoid membranes, dissociation of PS II complex, destacking of grana lamellae, and inactivation of Rubisco [9,10]. Chilling of plants leads to disorganization of thylakoids, changes in membrane fluidity, decline in Rubisco activity, disruption in circadian regulation of key photosynthetic enzymes, and inhibition in the translocation of carbohydrates [7,11]. Waterlogged conditions lead to anaerobic environment, decreased nutrient absorption, reduced stomatal conductance, and decreased level of ATP and chlorophylls [12]. Polluting gases such as SO_2, NO_2 H_2S, O_3 enter leaves and inhibit stomatal movements [12].

Heavy metal pollutants like Cd^2, Ni^{2+}, Pb^{2+}, Cu^{2+}, Hg^{2+} directly affect PSII activity, alter photosynthetic partitioning, and inhibit Rubisco activity [13–16]. The overall impact of various environmental stresses on different components of photosynthetic process may be described according to the scheme presented in Figure 37.1. This chapter presents our current status of knowledge related to the effects of different environmental stresses on the individual components associated with the process of photosynthesis in crop plants. The mechanism of stress injury and the ways in which the plants respond to the stresses have been discussed.

II. STRESSFUL CONDITIONS AND PHOTOSYNTHESIS

The common stressful conditions of the environment to which plants are exposed include excess of soluble salts in the soil, soil water deficits, heat, chilling, water logging and poor aeration of the soil, heavy metals, gaseous pollutants, etc. Under these conditions, plant growth, metabolism and more specially, photosyn-

thesis is severely affected. The extent of effect depends on the plant species, the developmental stage of the plant as well as the type, intensity, and duration of stress.

A. SALINITY

Salinity of soils is one of the most important environmental factors that limits plant growth and productivity in many parts of the world and more specially in arid and semiarid regions [12]. Accumulation of soluble salts from poor quality irrigation water, irrigation of soil with saline water, improper or restricted drainage system to flush out accumulated salts often lead to a high level of salt buildup in the soil. It is estimated that about one third of the irrigated land on earth is affected by salt.

The predominant salts in saline soils are chlorides and sulfates of Na^+, Mg^{2+}, and Ca^{2+}. NaCl contributes substantially to salinity due to its exceptionally high solubility. Plants growing in saline environments suffer injury due to osmotic stress, specific ion toxicities, and ionic imbalance [12]. Osmotic stress results due to lowering of soil water potential as salt content of soil rises. Specific ion toxicities result due to accumulation of injurious concentrations of Na^+, Cl^-, or SO_4^{2-} in the cells. Ionic imbalance or nutritional imbalance results in salt-stressed plants due to competition of salt ions with the nutrients.

Salt stress affects photosynthesis due to reduction in stomatal conductance as well as decreased intercellular partial pressure of CO_2 in leaves, reduction in chlorophyll content [17], changes in ultrastructure of chloroplasts [18,19], decreased photochemical and carboxylation reactions [5,20], and increased level of soluble sugars in the tissues [21,22].

1. Stomatal Closure and Gas Exchange Processes

Salt-stressed plants show significant decline in stomatal conductance [2,17,23]. Reduced photosynthesis in plants under salt-stressed condition is primarily attributed to decreased stomatal conductance [2,17], which results due to combined effects of osmotic stress as well as Na^+ toxicity [2]. It is believed that stomatal closure as observed under salt stress is as a result of accumulation of abscisic acid in leaves of salt-stressed plants [12]. Kaylie and coworkers [24] observed that in cheat grass (*Bromus tectorum* L.) salinity led to stunted growth through reduced leaf initiation and expansion and reduced photosynthetic rates, primarily due to stomatal limitation.

A decrease in CO_2 fixation rate per unit of leaf area is observed in plants grown under saline conditions. Though the extent of decrease in CO_2 exchange

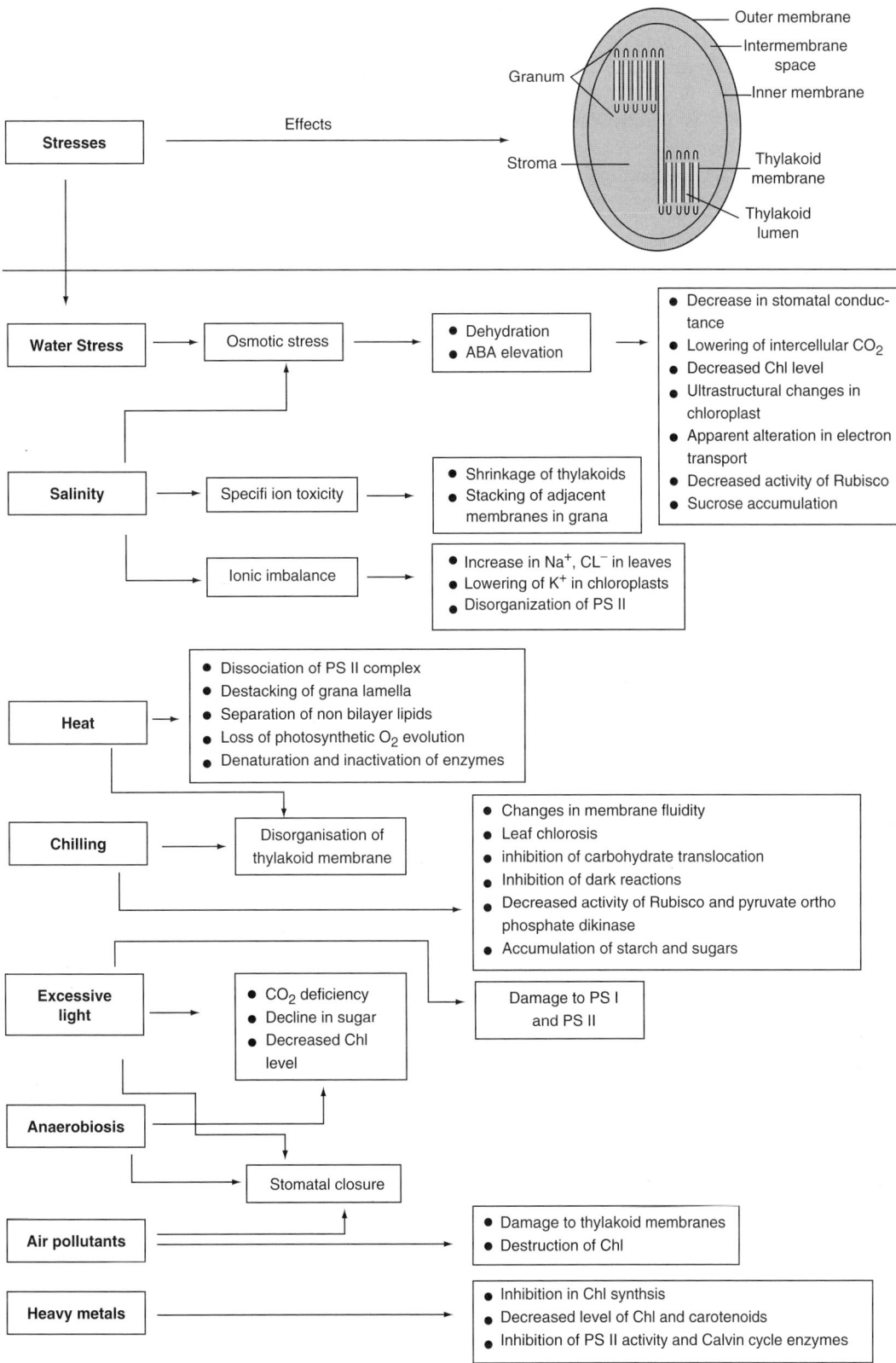

FIGURE 37.1 Effect of different stresses on photosynthetic components and related processes. Salinity and water stresses have many effects in common as both cause dehydration and increase in the level of abscisic acid. Heat and chilling cause disorganization of thylakoid membranes. The primary effect of anaerobiosis, excessive light and many air pollutants is stomatal closure.

rate (CER) under similar level of salinization varies widely in different plant species, salinity stress invariably leads to stomatal closure and decrease in intercellular CO_2 concentration C_i in leaves [2]. The immediate effect of salinity is mainly osmotic and continuous exposure further leads to specific ion toxicities. However, varied observations have been reported regarding accumulation of salinity ions in leaf tissues and alteration in the rate of photosynthesis [17,25]. According to Downton and coworkers [17], in spinach leaves, stomatal conductance and intercellular partial pressure of CO_2 decreased due to salinity but this had little effect on photosynthetic rate and on the other hand improved water use efficiency. Other workers also observed that in spinach NaCl concentrations upto 200 or even 350 mol/m^3 did not inhibit the rate of photosynthesis [17,26].

When plants of four rice (*Oryza sativa* L.) cultivars differing in salt tolerance were stressed for 1 week under 60 and 120 mmol NaCl, substantial reduction in carbon assimilation rate and stomatal conductance was observed [23]. Similarly, in pepper plants CO_2 assimilation decreased under 100 to 150 mmol NaCl but not under 50 mmol concentration [2]. In isolated mesophyll cells of cowpea leaves, the CO_2 fixation rate decreased by 30% in the medium containing 130 mol/m^3 NaCl, whereas under 173 mol/m^3 NaCl, photosynthetic rate was severely and irreversibly inhibited [25]. These observations suggest that under higher salinity level, the observed reduction in CER is primarily an osmotic effect and that concentration of salinity ions as well as duration of exposure also become important in determining gas exchange rate, CO_2 concentration in leaves, and in turn, the rate of photosynthesis.

2. Chloroplast Structure and Pigment Composition

Salinity leads to destruction of fine structure and swelling of chloroplast [19,27], instability of pigment protein complex [27,28] degradation of chlorophylls [17,29], and alteration in the content and composition of carotenoids [5,29]. Chloroplasts isolated from leaves of salt-stressed spinach plants showed about 80% of the photosynthetic capacity compared to chloroplasts from control leaves [17]. Salinity-induced swelling of thylakoid membranes appears to be due to a change in the ionic composition of the stroma [19]. Under salinity, plants accumulate higher levels of Na^+ and Cl^- ions within the chloroplasts, which leads to shrinking of thylakoid membranes [30] and stacking of adjacent membranes in grana [31]. In salinized barley, wheat, and pea plants, a marked loosening between the chlorophyll and the protein was observed in the chloroplasts [27].

Various workers have observed decreased level of chlorophyll pigments in salt-sensitive plants grown under NaCl salinity stress [1,17,19,29]. Downton and coworkers [17] observed that leaves of spinach plants grown under 200 mM NaCl contained about 73% of the chlorophyll per unit area of control plants. The decrease in the level of total chlorophylls in salt-stressed plants is mainly attributed to the destruction of chlorophyll *a*, which is supposed to be more sensitive to salinity than chlorophyll *b* [28]. Decrease in chlorophyll level in salinized plants is also partly attributed to the increased activity of chlorophyll degrading enzyme chlorophyllase [29]. Mature trees of *Prunus salicina*, acclimated to salinity under field conditions, showed reduced leaf chlorophyll content, which was apparently related to increased leaf chloride content and decreased CO_2 assimilation capacity [1]. In such trees, if leaf chloride level exceeded 0.25 mol/kg dry weight, a significant reduction in Chl content as well as visual leaf damage was apparent [1]. While performing stress studies in lentil (*Lens esculenta* Moench), Tewari and Singh [32] observed a continuous decrease in chlorophyll *a* and *b* content in the leaves of plants with increasing exchangeable sodium percentage in the soil.

Cultivars of crop species differing in salt tolerance when grown under saline conditions show different degrees of reduction in chlorophyll level. Chlorophyll in salt-tolerant cultivars are more effectively protected against the deleterious effects of Na^+ because such plants show higher accumulation of vacuolar Na^+ and osmolytes like putrescine and quaternary ammonium compounds in the chloroplasts [19].

Seedlings of rice cultivars differing in salt tolerance, when raised under increasing levels of NaCl salinity in sand culture experiments showed significant decrease in the level of chlorophylls with greater decrease in salt-sensitive cultivars than the tolerants [33]. An assessment of total chlorophyll level (Chl *a* + *b*) in rice seedlings of differing salt tolerance, raised under increasing levels of NaCl salinity over 5- to 20-day growth period, indicates that in salt-tolerant *cvs.* CSR-1 and CSR-3 with moderate salinity level of 7 dS/m NaCl, almost no change occurs in total chlorophyll level whereas with a higher salinity level of 14 dS/m NaCl, a marked decline in Chl level is observed (Figure 37.2). Whereas, in salt sensitive rice *cvs.* Ratna and Jaya, with increase in salinity, a concomitant decrease in total chlorophyll level can be seen. Salt-stressed seedlings of tolerant rice cultivars maintain higher level of total chlorophylls compared to the sensitive ones under similar level of salinization (Figure 37.2). Similar to Chl *a*, decreased level of Chl *b* is noticed in plants grown in salinized medium [19,32,33]. Salama and coworkers [19] noted an

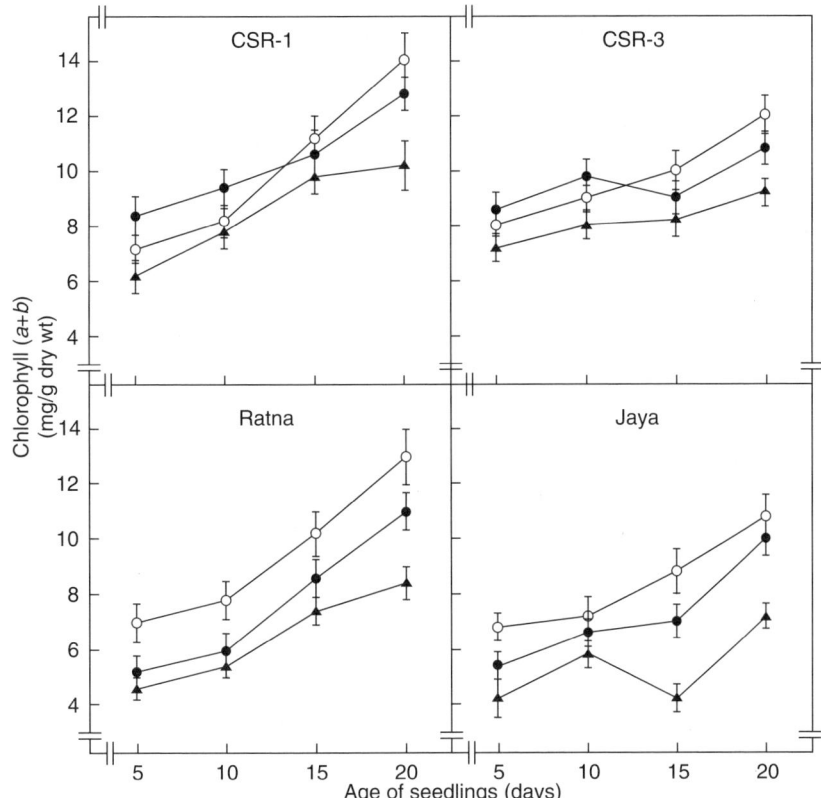

FIGURE 37.2 Chlorophyll ($a + b$) level in shoots of salt tolerant rice *cvs*. CSR-1, CSR-3, and sensitive *cvs*., Ratna and Jaya, at different days of growth under increasing conductivities of NaCl salinity (○, control; ●, 7 dS/m NaCl; ▲, 14 dS/m NaCl). Values are mean ± standard deviation based on three replicates and bars indicate standard deviations. With a higher salinity level of 14 dS/m NaCl, values of Chl level in salt-sensitive cultivars are much lower than the values in tolerant cultivars compared to respective controls.

increase in Chl *a*/Chl *b* ratio in salt-sensitive wheat plants due to salinity, although separately the levels of Chl *a* as well as that of Chl *b* decreased. Salinity induces genotype specific change in the level of carotenoids [33]. Under 14 dS/m NaCl level of salinization, seedlings of salt-sensitive rice cultivars showed more decrease in the level of carotenoids compared to the tolerants [33]. It is suggested that salt stress causes an increase in zeaxanthin content and degradation of β-carotene, which are apparently involved in protection against photoinhibition [5].

3. Photosystems and Photochemical Activities

Salinity stress enhances the susceptibility of plants to photoinhibition, a phenomenon which leads to the formation of toxic singlet oxygen in chloroplasts and degradation of the quinone-binding protein, now known as D_1 protein in the PSII complex and is caused by excess light [34]. Impairment of D_1 results in disruption of the light-dependent separation of charge between P680 and pheophytin a, and this phenomenon is associated with interruption of the transport of electrons that is medicated by PSII, ultimately leading to decreased photosynthetic activity [18]. Under natural conditions, in the field, salt stress very often occurs in combination with light stress and

it has been observed that the combination of light and salt stress is synergistic is inactivating PSII [18].

According to Kyle and coworkers [34], the 32-kDa D_1 protein, which is one of the two reaction center proteins of PSII, is the primary site of damage due to photoinhibition. The level of photoinhibition can be determined by the extent of damage and repair of D_1 protein [35]. Photodamaged PSII is repaired in a process involving the rapid turnover of D_1, with degradation of damaged D_1 and subsequent light-dependent synthesis of precursor to D_1 termed as pre-D_1. The damaged D_1 is replaced by newly synthesized pre-D_1 [18]. Evidences suggest that salt stress inhibits the transcription and translation of D_1 protein genes and in this way it inhibits the repair of photodamaged PSII [18]. In wheat plants, it was shown by Mishra and coworkers [36] that the inhibition of protein synthesis including the D_1 protein and closure of stomata by salt stress are responsible for the exacerbation of photoinhibition by salt stress.

Salinization is reported to have little effect on chlorophyll fluorescence characteristics [37] and has no significant effect on whole chain electron transport activity or on the activity of PSI [36]. The fluorescence intensity of chlorophyll in plants, algae, and cyanobacteria depends on the state of PSII reaction centers [38]. Larcher and coworkers [37], while examining the combined effects of salt and temperature stresses on chlorophyll fluorescence characteristics of cowpea (*Vigna unguiculata* L.) plants, observed that appreciable differences between controls and the various salt levels could be seen in only a few of the fluorescence characteristics. These workers observed that the fluorescence indicators such as the ratio of variable fluorescence to maximal fluorescence (F_V/F_m), steady-state levels of photochemical quenching coefficient (q_p) and nonphotochemical quenching coefficient (q_n) remained practically unaffected, whereas the peak of the induction transient, F_p (expressed as fraction of F_m) was 20% higher for salt-stressed plants than for controls. Salt-stressed cowpea plants showed an R_{fd} value (ratio between fluorescence decrease and steady-state fluorescence at saturating light) of 2.5 compared to 3.4 for controls [37].

Mishra and coworkers [36], while examining the effects of salt and light stress on wheat plants, observed that with NaCl treatments intrinsic chlorophyll fluorescence level (F_0) did not change whereas a gradual reduction in variable chlorophyll fluorescence (F_v) occurred, which was as a result of decrease in maximal fluorescence (F_m) upon salt treatment. However, no significant difference in F_v/F_m ratio could be observed between salt-treated and control plants [36]. A decrease in room temperature fluorescence of chlorophylls associated with PSII was observed in salt-stressed sorghum plants, which appeared to be due to photoinhibition of PSII activity [5].

Varying opinions exist regarding salinity effects on photosynthetic electron transport activities [2,5,33,36]. In leaves and isolated chloroplasts from barley (C_3) and sorghum (C_4) plants, electron transport activity did not decrease with increase in salt concentration [39]. In wheat and spinach, activities of whole chain electron transport, PSI and PS II in thylakoids of salt-stressed plants were similar to those from control grown plants [5,36]. This suggests that in salt-stressed field-grown plants, which are often prone to high light stress, decreased PS II activity in isolated thylakoids is mainly due to photoinhibition and not due to salt stress [39].

Chloroplasts isolated from salt-stressed seedlings of rice cultivars of differing salt tolerance, however, showed different levels of electron transport activities compared to control grown plants [33]. Results of an experiment conducted to examine electron transport reactions in chloroplasts isolated from 20-day grown seedlings of a salt-tolerant rice *cv*. CSR 1 and a sensitive *cv*. Ratna are shown in Table 37.1. As it is evident from the table, about 53% decrease in whole chain electron transport activity and 71% decrease in PSII activity can be seen in chloroplasts isolated from 14 dS/m NaCl grown seedlings of salt-sensitive *cv*. Ratna compared to electron transport reactions in chloroplasts isolated from nonsalinized seedlings of this cultivar. Extent of inhibition in electron transport activities due to salinity was less in the tolerant cultivar than in the sensitive one.

Alteration in the photochemical activity of salinity exposed plants might possibly be due to more absorption of potentially toxic ions Na^+ and Cl^- in these species, which could penetrate the chloroplasts and exert its adverse effects [19]. Another possible explanation appears to be salt stress induced photodamage to PSII; however, the mechanism by which salt stress enhances the photodamage to PSII remains unclear [18].

TABLE 37.1

Effect of Increasing Levels of NaCl Salinity *In Situ* on Electron Transport Reactions in Chloroplasts Isolated from 20-Day Rice Seedlings

Rice Cultivar	Assay	Salinity		
		Control (without NaCl)	7 ds/m NaCl	14 ds/m NaCl
CSR-1 (T)	PS I+II	430.90	370.40	295.60
	Whole chain ($H_2O \rightarrow MV$)	± 30.60	± 28.00	± 22.60
	PS II	286.19	213.09	145.18
	($H_2O \rightarrow Pd_{ox}$)	± 27.80	± 23.70	± 20.20
Ratna (S)	PS I+II	384.84	280.50	162.50
	Whole chain ($H_2O \rightarrow MV$)	± 42.50	± 32.80	± 28.50
	PS II	250.84	129.30	72.58
	($H_2O \rightarrow Pd_{ox}$)	± 36.80	± 24.60	± 18.40

Reaction rates are expressed as μmol H_2O consumed or evolved per mg chlorophyll per h. Values are mean $\pm$ S.D. based on three independent observations. T and S in parentheses indicate tolerant and sensitive rice cultivar, respectively.

4. Carboxylation

Ribulose-1,5-bisphosphate carboxylase/oxygenase (Rubisco, EC 4.1.1.39) is a key enzyme in the photosynthetic carbon reduction in all plants, and its level as well as carboxylating capacity decreases in plants subjected to salt stress [20,40]. In bean (*Phaseolus vulgaris*) plants 100 mM NaCl reduced photosyntheic efficiency, which was as a consequence of decreased Rubisco activity and decrease in pool size of RUBP [40]. In winged bean (*Phosphocarpus tetragonolobus*) plants, NaCl salinity decreased the activities of Rubisco as well as phosphoenolpyruvate carboxylase (PEPCase, EC 4.1.1.31) and also the rate of photosynthetic CO_2 fixation [41]. In leaves of 7-day-old barley seedlings, grown in the presence of 100 mM NaCl, Rubisco level was only about 20% of the control plants [20]. Miteva and coworkers [20] suggested that in barley plants NaCl salinity inhibited the synthesis of total soluble protein with a more pronounced inhibition of Rubisco synthesis. It is suggested that the reduction in the amount of Rubisco protein under salt stress might imply an effect of salt at the level of transcription, translation, or gene regulation [1].

Rubisco isolated from many plants species appears to be sensitive to NaCl [22]. It is believed that under salt stress conditions, compatible solutes like proline and it analogs accumulate in the cytoplasm/chloroplasts and provide possible protection to this enzyme against osmotic and toxic effects of salinity [42]. Under *in vitro* conditions, the proline-related analogs *N*-methyl-L-proline and *N*-methyl-*trans*-4-hydroxy-L-proline have been shown to ameliorate the inhibition of the activity of Rubisco by NaCl [42]. Though it is observed that salinity decreases the activity of Rubisco in many plants species [20], contrary to this, NaCl-adapted plants of *Tamarix jordanis* showed production of higher level of Rubisco as well as of compatible solutes [42]. Higher content of Rubisco protein in salt-adapted plants might contribute toward better adaptation of plants to salinity [43]. The activity of PEPCase has been shown to rise considerably in salt-stressed plants compared to control grown plants [44]. NaCl-stressed barley plants showed four times higher PEPCase activity than unstressed plants [44]. PEPCase isolated from many halophytes like *Suaeda monoica*, *Chloris gayana*, and *Cakile maritima* was shown to be not only a salt-tolerant enzyme but also a salt-requiring enzyme [45].

5. Level of Photosynthates

The principal end products of leaf photosynthesis are starch and sucrose. It has been observed by various groups of investigators that plants under salinity stress shown higher starch content and accumulate soluble sugars more specially sucrose [22]. The accumulation of photosynthates starch and sugars under saline conditions is mainly attributed to the impaired carbohydrate utilization as respiration rate decreases at high salinity levels [46]. The accumulation of soluble sugars under stressful condition of salinity might contribute to a favourable osmotic potential and render a protective role to biomolecules [22].

The responses of NaCl salinity on the level of starch and sugars depends on the plant organs as well as the genotypes of plants studied [47,48]. Rice genotypes of differing salt tolerance accumulate varying levels of sugars in plant parts when subjected to saline stress. An examination of the levels of total, reducing, and nonreducing sugars in shoots of rice seedlings of salt-sensitive *cvs*. Ratna, Jaya and tolerant *cvs*. CSR-1, CSR-3 grown in the presence of 14 dS/m NaCl indicated that in the both sets of cultivars, salinity caused increase in the level of sugars with more increase in the sensitive rice cultivars than in the tolerant ones (Table 37.2). It was observed that at 14 dS/m NaCl salinity level, shoots of sensitive rice cultivars maintained about 2.52 to 3.14 times total sugars level compared to nonsalinized seedlings.

B. WATER STRESS

Shortage of water or water deficit leads to water stress in growing plants. Plants are often subjected to period of soil and atmospheric water deficits during their life cycle. Water stress reduces plant growth and affects photosynthesis by reducing leaf area, enhancing stomatal closure, decreasing water status in the leaf tissues, reducing the rate of CO_2 assimilation, causing ultrastructural changes in chloroplasts, affecting electron transport and CO_2 assimilation reactions impairing ATP synthesis and RUBP generation, and altering the level of photosynthates in the tissues. Water stress causes an imbalance in the hormone level in plants. Due to alteration in hormonal balance, concentrations of many key enzymes of photosynthesis decline in water-stressed plants.

1. Leaf Area and Stomatal Conductance

As a result of decrease in water content of the leaves, cells shrink, cell volume decreases, and the solutes within the cell become more concentrated. The plasma membrane becomes thicker and compressed resulting in inhibition of cell expansion. Leaf area as well as size of individual leaves and the number of

TABLE 37.2
Levels of Nonreducing, Reducing, and Total Sugars (mg/g dry wt.) in Shoots of 20-Day-Old Nonsalinized (Control) and Salt-Stressed (14 dS/m NaCl) Seedlings of Salt-Sensitive Rice *cvs.* Ratna and Jaya, and Tolerant *cvs.* CSR-1 and CSR-3

Rice Cultivar	Nonreducing Sugars		Reducing Sugars		Total Sugars	
	Control (without NaCl)	$14\,dS/m^{-1}$ NaCl	Control (without NaCl)	$14\,dS/m^{-1}$ NaCl	Control (without NaCl)	$14\,dS/m^{-1}$ NaCl
Ratna	6.0	26.2	8.1	17.8	14.1	44.0
Jaya	8.2	19.5	11.5	30.2	19.7	49.7
CSR-1	5.8	12.2	10.2	14.1	16.0	26.3
CSR-3	7.1	15.2	9.8	18.6	16.9	33.8

Values are mean based on three independent determinations.

total leaves are reduced under water stress. Decreasing relative water content and water potential of leaves progressively decrease stomatal conductance, leading to decline in CO_2 molar fraction in chloroplasts, decreased CO_2 assimilation, and reduced rate of photosynthesis [49].

Stomatal closure is among the earliest responses of plants subjected to water stress and it is generally assumed to be the main cause of drought-induced decrease in photosynthesis, since stomatal closure leads to decrease in CO_2 intake by mesophyll cells, leading to decreased intercellular CO_2 partial pressure (C_i), decreased chloroplastic CO_2 concentration (C_c) and thereby decreased CO_2 assimilation and net photosynthesis [50]. As guard cells are exposed to the atmosphere, in air of low humidity, guard cells lose water too rapidly by evaporation causing the stomata to close by a mechanism called hydropassive closure [12]. In a different mechanism of stomatal closure, called hydroactive closure, the whole leaf gets dehydrated under water stress and increased synthesis of abscisic acid takes place in mesophyll cells and it accumulates in the chloroplast [12]. Stored abscisic acid is then released to the apoplast (cell wall space) from where it reaches to the guard cells through the transpiration stream. Redistribution of stored abscisic acid from the mesophyll chloroplasts to the apoplasts initiates the closure of stomata [51]. Under water deficit conditions in the soil, messengers from the root system like root drying or increased delivery of abscisic acid from root to leaves via transpiration stream also induce stomatal closure [12].

2. Ultrastructural Changes in Chloroplasts

Water stress leads to decreased volume of the chloroplast, permanent adhesions occur within the grana, partitions become thinner, lipid droplets increase in number and size, many thylakoid proteins are oxidatively damaged, and structural changes occur in light-harvesting chlorophyll protein complexes [52–54]. Maroti and coworkers [52], while investigating ultrastructural changes in chloroplasts of different plant species due to drought, observed that contraction of stroma, swelling, and blistering of thylakoids were characteristic features of the chloroplasts of Crassulacean acid metabolism (CAM) succulent plant *Sedum sexangulare* and mesophyll chloroplasts of C_4 sclerophyllous plant *Testuca vaginata*. These workers noted that under naturally induced drought the chloroplasts elongated and contracted along the cell wall, storma aggregated and were found along the inside surface of the envelope. Stromal lamella and stroma stuck closely in a sheet like manner. The effect of drought on *Sedum sexangulare* chloroplast was marked by shriveling of the cells and chloroplasts with a decrease in the size of electrondense granules and the electron density of the whole cytoplasm [52]. Aggregation of stroma occurred forming sheets and large plastoglobules. The number as well as the size of plastoglobules increased [52].

It was suggested by Poljakoff-Mayber [55] that swelling, distortion of stroma and grana lamellae regions, and the appearance of lipid droplets were common features of chloroplasts in conditions of water stress. Decrease in the volume of cells and chloroplasts has been noted by other workers in plant tissues undergoing dehydration due to long drought [56,57]. It is regarded that stromal aggregation under water stress is a reversible process as the normal structure is restored after drought recovery, whereas the accumulation of lipid droplets in the intrathylakoidal space may play an adaptive role during drought conditions [52]. An examination of the structural changes of bundle sheath and mesophyll chloroplasts of a C$_4$ plant *Testuca vaginata* that underwent water stress suggests that chloroplasts of bundle sheath cells are more resistant to water stress than those of mesophyll cells [52].

It has been shown that in chloroplasts, chiral macroaggregate formation of the light-harvesting chlorophyll *a/b* pigment protein complexes (CHC IIs) occurs, which is involved in the lateral separation of the two photosystems, protects the photosynthetic apparatus against photoinhibitory damage and plays an important role in the structure and function of the chloroplast [54]. Imposition of drought stress leads to disruption of the chiral macroaggregates and severe dehydration conditions cause full disruption of such aggregates [54].

In plants subjected to water stress, breakdown of photosynthetic apparatus may result due to oxidative damage of chloroplast lipids, pigments and proteins [53]. Exposure to water stress leads to increased production of reactive oxygen species (ROS) that cause damage to membranes and build up elevated level of lipid peroxides that ultimately affect photosynthesis. Chloroplasts are the major source of production of ROS in plants [58]. The application of moderate water deficit (water potential of -1.3 MPa) to pea leaves led to a 75% inhibition of photosynthesis and to increases in zeaxanthin, malondialdehyde, oxidized proteins and mitochondrial, cytosolic and chloroplastic superoxide dismutase activities [58], whereas severe water deficit (-1.9 MPa) almost completely inhibited photosynthesis and decreased the levels of chlorophylls, β-carotene, neoxanthin, etc. ROS include superoxide radical ($O_2^{\cdot-}$), hydroxyl radical ($^{\cdot}OH$), singlet oxygen, H_2O_2, etc. In chloroplasts $O_2^{\cdot-}$ is produced by photoreduction of O_2 at PSI and PSII, and singlet oxygen is formed by energy transfer to O_2 from triplet excited state chlorophyll [58], followed by spontaneous generation of H_2O_2. Fortunately, chloroplasts are the organells that have the highest antioxidative protection due to presence of carotenoids, tocopherols, and antioxidative enzymes that scavenge ROS and minimize oxidative damage [59] but dehydrative conditions greatly enhance the production of ROS thereby leading to oxidative damages within chloroplasts. In droughted wheat, sunflower, and pea plants increased production of ROS was observed primarily due to increased photoreduction of O_2 by the photosynthetic electron transport system [53]. Imposition of -2.0 MPa water stress on 4-week-old wheat plants led to oxidative damage of 68, 54, 41, and 24 kDa thylakoid polypeptides and accumulation of many crosslinked high molecular weight proteins with the substantial decrease in photosynthetic electron transport activity [53].

3. Chlorophyll Fluorescence and Photochemical Reactions

Water stress leads to characteristic changes in chlorophyll fluorescence curves [60]. However, PSII photochemistry is only marginally affected even under condition of severe water stress [60]. *In vivo* chlorophyll fluorescence of dark-adapted leaf, which can be elicited by very dim light beam modulated at high frequency, represents the minimum chlorophyll fluorescence (F_0) and is not significantly modified by water stress [61]. After illumination with nonmodulated white light of higher intensity, the fluorescence increases rapidly to a peak point. The fluorescence between F_0 and peak point is termed as variable fluorescence (F_V). Maximum fluorescence (F_m) can be induced by a short pulse of saturating white light. The value of F_V reflects the reduction of the primary electron acceptor Q_A of PS II. In the oxidized state Q_A quenches fluorescence. Quenching of F_V reflects the working of entire photosynthetic process, more specially primary photochemical events and it depends on reduction–oxidation of Q_A, light-induced proton gradient across the thylakoid membrane, or the light energy distribution between the two photosystems. Depending on the degree of oxidation or reduction of the electron transport chain, F_V is quenched or enhanced [62]. On the acceptor side of PSII, the quinone and plastoquinone pools are possibly responsible for fluorescence quenching [62]. Knowing the values of F_0, F_V, and F_m, the value of the photochemical component of fluorescence quenching (q_Q) can be calculated as suggested by Schreiber et al. [63] as $q_Q = (F_m - F_V)/(F_m - F_0)$.

Water stress causes drastic changes in the different modulated fluoresence levels, resulting in severe reduction in q_Q value [60]. Dehydration of leaves or extreme water losses caused alterations in chlorophyll fluorescence in many plant species [64]. At extreme water deficit, fluorescence changes are more pronounced. In oak leaves change in chlorophyll fluorescence was detected only when water deficit values exceeded 30% [2]. At more severe dehydration, increase in nonphotochemical quenching was observed whereas photochemical quenching remained unaffected [2]. Seven potato genotypes grown under water stress showed decline in variable fluorescence (F_V) of leaves with a concomitant decrease in net photosynthetic rate (P_N) [62]. In potatoes, total dry matter production in water-stressed plants could be correlated with F_V, which is a measure of the capacity of primary photochemical event [62]. It was suggested by Zrust and coworkers [62] that decreasing values of F_V under water stress indicate diminishing photosynthetic activity in potato leaves and that F_V values provide a method for the study of changes in the photosynthetic capacity of the potatoes in response to water stress.

It has been shown by certain workers that thylakoid membrane-related photochemical activities decline under water stress, with PSII activity being more drought sensitive than PSI [61,65]. In two native Mediterranean plants, rosemary (*Rosmarinus officinalis L.*) and lavender (*Lavandula stoechas L.*) water stress led to decrease in the relative quantum efficiency of PSII photochemistry and decreased the efficiency of energy capture by open PSII reaction centers. These events were associated with downregulation of electron transport [65]. Similarly, in water stress-exposed wheat leaves, kinetics of the Hill reaction activity declined significantly [3]. In a study on metabolic consumption of photosynthetic electron transport in tomato plants, Haupt-Herting and Fock [66] observed downregulation of PSII under water stress. Such observations, however, appear to be relevant with only certain drought-sensitive species. The observed inhibition of PSII activity under water stress might not be due to the direct effect of stress on photochemical activity but due to photoinhibition [64]. When leaves of *Lycopersicon esculentum, Solanum tuberosum,* and *Solanum nigrum* plants were illuminated with intense white light at 25°C, photoinhibition damage of PSII was more pronounced in water-stressed leaves compared to undesiccated controls [64]. In tomato and potato, water stress created by treatment of intact leaves did not significantly alter the PSII functioning in dark- and light-adapted leaf samples [64]. In these plants, PSII was shown to be highly drought resistant; rather water stress conditions provided protection to PSII against heat injury. Cornic and Fresneau [67] similarly observed that in many C_3 plants, PSII functioning and its regulation are not qualitatively changed during desiccation and that variations in PSII photochemistry could simply be understood by changes in substrate availability under these conditions.

Impaired photophosphorylation, decreased ATP synthesis by the enzyme ATP synthase and loss of ATP content have been observed in plants subjected to water stress [2,6,49]. Among the different events during photophosphorylation, electron transport activity and uncoupling or thylakoid energization are not affected due to water stress whereas the possible effect of water stress appears to be decreased ATP synthesis by the chloroplastic enzyme ATP synthase (coupling factor, cf) [2,49]. Water stress conditions retard chloroplastic ATP synthase activity. At low relative water content (RWC) of leaves inhibition in ATP synthesis occurs due to progressive inactivation or loss of ATP synthase resulting from increasing Mg^{2+} concentration in chloroplasts [49]. During water stress, Mg^{2+} concentration increases in the chloroplasts. Sunflower plants grown at high Mg^{2+} levels in nutrient medium maintained lower photosynthetic rate than plants grown at lower Mg^{2+} level [68]. Decreased ATP content and imbalance with reductant status affect cell metabolism substantially, limit RUBP biosynthesis and decrease photosynthetic potential of plants under water stress [49]. Evidences indicate that even under mild drought impaired ATP synthesis is the main factor limiting photosynthesis [6].

4. Carboxylation under Water Stress

Drought stress leads to stomatal closure, restricts CO_2 entry into leaves and thereby decreases CO_2 assimilation [69]. Several studies have suggested that decreased photosynthetic capacity under drought results from impaired regeneration of ribulose-1,5-bisphosphate as well as decreased availability and activity of CO_2 assimilating enzyme ribulose-1,5-bisphosphate carboxylase/oxygenase (Rubisco) [67,69]. The amount of Rubisco in leaves is controlled by the rate of its synthesis and degradation. Though the Rubisco holoenzyme is relatively stable even under drought stress with a half life of several days, in many plant species such as tomato, arabidopsis, and rice, a rapid decrease in the abundance of steady-state level of Rubisco small subumit (*rbc* S) transcripts has been observed under drought which indicates decreased synthesis of this enzyme under water stress [69,70].

It is suggested that the activity of Rubisco in leaves is independent of stomatal conductance, however dehydration has a direct effect on Rubisco activity [69]. In fact, Rubisco activity is modulated *in vivo* either by reaction with CO_2 and Mg^{2+}, leading to carbamylation of a lysine residue at the catalytic site that is essential for activity or by the binding of inhibitors within the catalytic site, leading to inhibition of enzyme activity [69]. In tobacco plants, it has been shown that drought leads to decrease in Rubisco activity and this decrease is due to the presence of greater amounts of tight-binding Rubisco-inhibitors in droughted leaves [69].

Plants with C_4 metabolism can use water more efficiently than C_3 plants and need less Rubisco to achieve a given rate of photosynthesis [12]. The carboxylating enzyme of C_4 plants, PEPCase, which is a cytosolic enzyme, also gets inhibited under water stress [2]. It is suggested that under water-stressed conditions reduction in chloroplast volume may lead to desiccation within the chloroplast. This may ultimately lead to conformational changes in Rubisco and inhibition of its activity [71]. Anions like sulfate and phosphate increase in the stroma of dehydrated chloroplasts and may become inhibitory to Rubisco [72]. According to Berkowitz and Gibbs [73], water stress conditions led to acidification of chloroplast stroma which also might contribute for inhibited Rubisco activity as observed in water stressed plants. Depending on the extent of accumulation of osmolytes as well as transport of ions and low molecular weight osmolytes, changes in chloroplast/protoplast volume occur, leading to alteration in the behaviour of Rubisco [72]. It has been specifically shown that drought leads to limited ribulose-1,5-bisphosphate (RuBP) regeneration and RuBP concentration decreases in droughted plants [6]. This decreased level of RuBP might be due to progressive downregulation of metabolic processes in mesophyll cells under drought and might be one of the factors for decreased Rubisco activity and thereby decreased photosynthetic efficiency under water stress [6].

5. Levels of Carbohydrates and Related Enzymes

Water stress alters the ratio of the two end products of photosynthesis starch and sucrose. Due to low carbon supply under water-stressed conditions, chloroplastic starch may be remobilized to provide carbon in favor of more sucrose synthesis [74]. The rate of sucrose synthesis is regulated by the two enzymes cytosolic fructose-1,6-bisphosphatase (FBPase) and sucrose phosphate synthase (SPS) which are subject to various types of metabolic regulations. Activities of these two enzymes decline in water-stressed leaves. In drought-stressed leaves of sugar beet (*Beta vulgaris* L.), when water potential decreased from -0.8 to -4.3 MPa, activities of FBPase and SPS declined and also starch content declined by 10% whereas about threefold increase in sucrose level was observed [74]. In bean leaves, water stress caused a decline in the partitioning ratio of starch/sucrose with no change in FBPase activity whereas SPS activity was reduced by 60% [75]. In bean leaves feedback-limited photosynthesis after water stress was highly correlated with a loss of extractable SPS activity [75]. Vassey and coworkers [76], while estimating SPS activity in leaves of bean plants subjected to water stress, observed that a mild water stress of -0.9 MPa reduced SPS activity by 50% and this effect was a consequence of the inhibition of photosynthesis caused by stomatal closure. Water stress decreases photosynthesis and the consumption of assimilates in expanding leaves as a result the amount of photosynthate exported from leaves decreases, however, translocation of the assimilates is relatively unaffected during water stress [12].

C. Heat Stress

When temperatures exceed the normal growing range of plants heat injury takes place. Most of the crop plants generally grow in the 15°C to 45°C temperature range. An increase in temperature of 10°C to 15°C above normal growth temperature leads to disorganization of chloroplast thylakoid membranes, dissociation of PSII light-harvesting complex, destacking of grana lamellae, separation of nonbilayer lipids of thylakoid membranes, loss of photosynthetic O_2 evolution activity, denaturation and inactivation of many enzymes and thereby ultimately limiting photosynthesis. Even a moderate degree of heat stress slows the growth of the whole plant and causes decrease in photosynthetic rate much faster than respiratory rate [12]. When photosynthetic rate is plotted against temperature, a characteristic bell shaped curve is obtained (Figure 37.3). Ascending side of the curve A to B shows stimulation of photosynthesis with increase in temperature. The temperature range B to C represents optimal temperature where the highest photosynthetic rates are seen. When the temperature exceeds optimum temperature, decline in photosynthetic rate is observed. This decline region C to D is associated with adverse effects of temperature on photosynthetic capacity. During optimum temperature range, the capacities of various components of photosynthetic machinery are optimally balanced [77]. Plant species growing in different habitats have different optimal temperatures for photosynthesis.

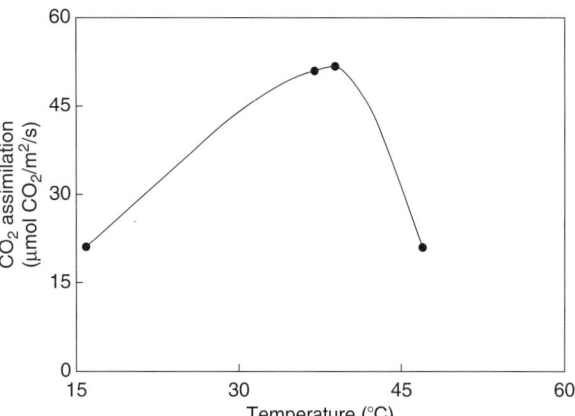

FIGURE 37.3 A typical curve showing temperature dependence of photosynthesis at saturating CO_2 concentrations. Photosynthesis changes with the change in temperature at the concentrations that saturate photosynthetic CO_2 assimilation. (Redrawn from Berry J, Bjorkmann O. *Annu. Rev. Plant Physiol.* 1980; 31:491–543. With permission.)

High temperatures modify membrane composition and structure and cause leakage of ions. Disorganization of chloroplast thylakoid membrane and inactivation of PSII takes place [9]. With rise in temperature, increasing fluidity of the membrane lipids takes place, the strength of hydrogen bonds, and electrostatic interactions between polar groups of proteins within the aqueous phase of the memrbane decrease [12]. As a result, integral membrane proteins tend to associate more strongly with lipid phase and nonbilayer lipids of the thylakoid membrane form aggregates of cylindrical inverted micelles [78].

PSII is more sensitive to elevated temperatures than PSI [77]. In intact pea leaves with high-temperature treatment, PS II activity was inhibited or down-regulated whereas PSI activity was stimulated [79]. In dark-adapted pea leaves, heat treatment caused inhibition of photosynthetic oxygen evolution and decrease in photochemical energy storage, which were correlated with a marked loss of variable PSII chlorophyll fluorescence emission whereas the capacity of cyclic electron flow around PS I increased [79]. In cold-adapted C_4 *Atriplex sabulosa* plants, electron transport activity of PSII was more sensitive to high temperature than in heat-adapted C_4 *Tidestromia oblongifolia* plants [80]. In both of these species, decline in CO_2 fixation under heat stress paralleled with decline in PSII activity. Experiments have shown that thermal inactivation of PSII is due to extraction of divalent ions Ca^{2+} and Mn^{2+} from the oxygen evolving complex of PSII as well as due to dissociation of the 32 kDa extrinsic polypeptide that is involved in the stabilization of the Mn-cluster [9].

High temperatures induce the sysnthesis of heat shock proteins (HSPs) in plants and it has been observed that in soluble portion of the chloroplast almost 19 small molecular weight HSPs (sHSPs) are synthesized at elevated temperatures and these sHSPs play an important role is photosynthetic and whole plant thermotolerance [9,81]. Chloroplast sHSPs are regarded as the most abundant and heat responsive of the plastid HSPs [81]. *In vivo* and *in vitro* experiments from *Agrostis stolonifera* genotypes indicate that chloroplast sHSPs could associate with thylakoid and protect PS II during heat stress, possibly by stabilizing the O_2 evolving complex [81].

Heat injury leads to decreased quantum yield of photosynthesis, lowered electron transport chain activity, changes in membrane properties, uncoupling of energy transfer mechanisms in chloroplasts, denaturation of proteins, as well as loss of enzyme activities. High temperature raises membrane fluidity, causes peroxidation and lateral diffusion of membrane lipids, increased membrane permeability leading to decreasing proton gradient formation across the thy-

lakoid membrane [77]. The level and activation state of the carboxylating enzyme Rubisco decreases with rise in temperature. In maize plants, temperatures exceeding 32.5°C caused decline in activation state of Rubisco and the enzyme was nearly completely inactivated at 45°C [10]. In maize leaves, the inactivation of Rubisco appears to be the primary constraint on the rate of net photosynthesis at temperature above 30°C [10]. The carboxylating enzymes Rubisco as well as PEPCase isolated from cold-adapted C_4 *A. sabulosa* plants were less stable to high temperature than the enzymes isolated from heat-adapted C_4 *T. oblongifolia* plants. In both of these plants photosynthetic rate declined at a temperature lower than that caused denaturation of carboxylating enzymes [80]. These observations indicate that due to heat stress, reduction in photosynthetic capacity is more associated with disorganisation of chloroplast membrane and uncoupling of energy transfer mechanism in chloroplasts than the inactivation of enzymes.

D. Chilling

Growth and development of most plants growing in tropical and subtropical regions are greatly inhibited by chilling temperatures [82]. In a variety of plant species, exposure to temperatures between 0°C and 15°C causes chilling injury, leading to inhibition in photosynthetic processes, decreased enzymatic activities, changes in membrane fluidily, decrease in protoplasmic streaming, swelling of chloroplasts, inhibition in the activities of photosystems, and increased susceptibility to photoinhibition of photosynthesis [11,82]. The extent of injury due to chilling depends on the duration of chilling, irradiance level, relative humidity, and the plant species [11]. Crop species like maize, bean, rice, tomato, cucumber, sweet potato, and cotton are chilling-sensitive. Even in the same crop species certain cultivars are chilling sensitive and others are tolerants [11]. Chilling injury occurs in sensitive plant species at temperatures that are too low for normal growth but not so low that ice formation could take place [12].

Photosynthetic processes are often the first to be inhibited at chilling temperaturtes and for a majority of crop species examined, photosynthesis is significant lower at temperatures around 10°C relative to that at 20°C to 25°C [83]. Low-temperature treatment alters the properties of chloroplast membrane. In chill-sensitive plants, the lipids in the bilayer have a high percentage of saturated fatty acid chains and such membranes tend to solidify into a semicrystalline state at a temperature well above 0°C [12]. In high-yielding indica rice varieties, chilling led to decline in photosynthetic rate, swelling of chloroplasts, and accumulation of starch grains within the chloroplasts [84].

In maize, chilling enhanced the distribution of excitation energy to PSI, which in part, accounted for observed decrease in the quantum yield of photosynthetic oxygen evolution [85]. Low temperature also induced alterations in the amount of excitation energy transferred from Chl *b* to Chl *a* in maize [86]. When black alder (*Alnus glutinosa*) seedlings fertilized with different doses of nitrate were acclimated in a growth chamber for 2 weeks and were exposed to 2.5 h of nighttime chilling temperatures of $-1°C$ to $4°C$, net photosynthesis declined by 17% for plants receiving low nitrate fertilizer ($0.36 \, mM$) and 19% for plants receiving high nitrate fertilizer ($7.14 \, mM$). It was suggested that in black alder chilling stimulated stomatal closure only at high nitrate level and that the major impact of chilling on photosynthesis involved interference with biochemical functions [87].

The impact of low temperature on photosynthesis is dependent on the concurrent light intensity. When high light intensities are experienced simultaneously with chilling, PS II is uniquely damaged and photoinhibition of photosysnthesis takes place [82]. Photoinhibition can occur at low temepratures even at low light intensities, leading to injury to PS II [29]. Photoinhibition of photosynthetic CO_2 assimilation occurs in many plants following chilling treatment. In *Zea mays* when either lamina of the second leaf or the whole plant was subjected to chilling treatment, significant photoinhibiton of PSII occurred [88]. Second leaves of *Zea mays* grown at $25°C$ when exposed to a photon flux of $800 \, \mu mol/m^2/s$ at $6.5°C$ for 6 h showed marked photoinhibition with 50% decrease in the quantum yield of CO_2 assimilation [88]. Plants exposed to low temperatures for a longer period show sustained downregulation of PSII complexes with low intrinsic efficiency of PSII electron transport (F_V/F_M) [89]. In intact leaves of an Australian mistletoe *Amyema miquelli*, efficiency of excitation energy transfer from light-harvesting pigments to Chl *a* molecules in PSII core complexes was markedly reduced in winter [89]. Chilling leads to degradation of photosynthetic pigments which is more pronounced in chilling-sensitive species like *Cucumis sativa* and maize compared to chilling-tolerant species like *Pisum sativum* [29]. Sensitization of photosynthesis to photoinhibition at low temeprature appears to be due to decreased activity of oxygen-scavenging enzymes, slowdown of physiological processes and the inhibition of PSII repair cycle [90]. It has been shown that presence of a large proportion of *cis*-unsaturated fatty acids in phosphatidyl glycerol (PG) of chloroplast membranes is correlated to chilling resistance in plants [91]. Transgenic rice seedlings showing 29.4% and 32%

cis-unsaturated fatty acids compared to wild-type seedlings with 19.3% fatty acids had improved chilling tolerance [91]. Cold treatment of plants leads to uperegulation of certain genes within chloroplasts, the products of which help in adaptation of plants to extreme enviornmental conditions [90]. In alfalfa (*Medicago sativa* L.) leaves, one such cold-induced mRNA for a specific chloroplast protease has been identified, the synthesis of which is induced only under low temperature and not under other stresses [90]. The enzyme pyruvate orthophosphate dikinase plays a crucial role in the declined photosynthetic capacity observed due to chilling [92]. Activity of this enzyme declines under cold treatment. In maize, at $11°C$ or below, this enzyme reversibly dissociates to less active dimeric and monomeric forms [93].

In chilling-sensitive plant species, low temperature exposure causes significant loss of the activity of certain carbon reduction cycle enzymes like Rubisco, sedoheptulose-1,7-bisphosphatase (SBPase), and chloroplastic fructose-1,6-bisphosphatase (FBPase). In *Zea mays* genotypes, growth at $14°C$ resulted in a 75% decrease in Rubisco activity and a 50% decrease in the activity of C_4 enzyme, NADP-malate dehydrogenase, compared to plants grown at $24°C$, whereas no change was observed in the activity of PEPCase [94]. An overnight chilling between $5°C$ and $7°C$ of the subtropical fruit tree mango (*Mangifera indica L*) led to substantial decline in CO_2 assimilation, which was associated with increase in stomatal limitation and lower Rubisco activity [7]. Similarly, in herbaceous chilling-sensitive crop tomato, overnight chilling caused severe disruption in the circadian regulation of key photosynthetic enzymes, leading to dysfunction of photosynthesis [7]. Contrary to these observations, in spinach leaves, cold treatment at $10°C$ for 10 days caused increase in the activity of many enzymes of carbon metabolism including Rubisco, stromal F-1, 6-BPase, sedoheptulose-1,7-bisphosphatase, phosphoglucoisomerase, malate dehydrogenase, pyruvate kinase, etc. [95]. It is suggested that in spinach leaves increased activity of carbon metabolising enzymes under low temperature exposure conditions is compensatory in nature, in an effort to increase the capacity of carbon metabolism to function under adverse kinetic constraints [95].

Among the photosynthates, accumulation of both starch and sucrose is observed in plants exposed to low temperature [96]. The most abundant and most commonly accumulated sugar is sucrose, which may accumulate upto tenfold in certain plants [96]. In spinach leaves, when plants were transferred from $25°C$ to $5°C$ conditions, a sudden increase in sucrose level was observed with a concomitant increase in the activity of its biosynthetic enzyme sucrose phosphate synthase; however, the activities of the enzymes sucrose synthase and invertase remained unaffected [96]. As a cryoprotectant, sucrose accumulation has adaptive significance for cold-exposed plants [96]. Accumulation of starch under low-temperature treatment is mainly due to production of this photosynthate in excess of its needs [96].

E. ANAEROBIOSIS

Due to poor drainage, excessive irrigation or rain, soil becomes water logged and oxygen gets depleted from the bulk of soil water, leading to anaerobiosis. Plants growing under such conditions show depressed growth and reduced photosynthesis with severe losses in yield [12]. In some plant species like pea and tomato, flooding leads to stomatal closure without significant change in leaf water potential [12]. It is believed that oxygen shortage in roots stimulates abscisic acid (ABA) production and movement of ABA to leaves can account for the stomatal closure [97].

The factors associated with low photosynthetic rate under submerged conditions include CO_2 deficiency in water, low irradiances in muddy water, settling of silt on the leaves, as well as factors related to slow diffusion of gases in solution [98]. Slow diffusion results in restriction of CO_2 influx during photosynthesis. Complete submergence is a common feature associated with low-land rice crop in Southeast Asian flood plains where deep water and floating rice cultivars are grown [99]. Setter and coworkers [98], while examining the effect of submergence on photosynthetic capacity of rice cultivars, observed that due to stagnation of water, supply of CO_2 to the chloroplasts was restricted and this was the prime reason for decreased photosynthesis of plants. CO_2 enrichment of water increased the rate of photosynthesis. Long period of submergence makes the leaves chlorotic and chloroplasts lose the capacity to fix CO_2 [99]. Concentration of soluble sugars decreases in plants after submergence, a more decrease is observed in submergence-sensitive cultivars than the tolerant ones [98]. O_2 deficiency tends to accelerate breakdown of carbohydrates and therefore a high rate of photosynthesis is required in submerged plant parts in order to compensate for the carbon loss [98]. These observations suggest that anaerobic conditions reduce photosynthetic rate in submerged plant parts with a marked decline in sugar level, which appears to be primarily due to low CO_2 level in the water environment.

F. AIR POLLUTANTS

The combustion of coal, oil, gasoline, as well as industrial activities release many gases such as CO_2, CO, SO_2, NO, NO_2, H_2S, HF, and ethylene as well as a variety of many hydrocarbons in atmosphere, which in excess concentrations are inhibitory to plant growth and have deleterious effect on photosynthesis. Ozone, produced as a result of reaction between oxygen, nitrogen oxide (NO, NO_2) hydrocarbons, and sunlight in a chain of atmospheric events is considered as one of the most potent phytotoxic air pollutants. Due to higher concentration of CO_2 and other "greenhouse gases" in the atmosphere, increased absorption of infrared radiation takes place [12], which is posing a serious threat of global warming. This may ultimately have serious impact on plant health. Elevated CO_2 level causes stomatal closure and reduces uptake of other pollutants [12]. SO_2 enters leaves through stomata and causes stomatal closure. It gets dissolved in the cell and produces bisulfite and sulfite ions, the later is toxic for the cell [12]. NO or NO_2 also reach the cells through stomata and when present in air in concentration greater than 0.1 μl/l inhibit photosynthesis. The concentration of the polluting gases varies depending on location, direction of wind, rainfall, sunlight humidity, temperature, etc. [12]. Table 37.3 shows common air pollutants, visible morphological changes that occur in plants due to these pollutants, and the associated metabolic implications.

Ozone may be present in high concentrations in urban and nearby areas. It binds to plasma membrane. Regulation of stomatal aperture by guard cells is disturbed. Both SO_2 and ozone inhibit the translocation of photosynthetic products via a disturbed pholem loading due to inactivation of the plasmalemma-bound ATPase, which ultimately leads to increased starch accumulation and finally bleaching of the photosynthetic pigments [100]. Due to its highly reactive nature, ozone damages chloroplast envelope and thylakoid membranes and thereby disrupts chemiosmotic balance [15]. Ozone leads to decrease in the level and inhibition in the activity of the carboxylating enzyme Rubisco [15,101]. The impairment of carboxylation efficiency is regarded as the initial effect of ozone on photosynthesis [101]. Presence of ozone promotes photoinhibition even when the light intensity is moderate [101]. Ascorbate, DNA, and lipids are very sensitive to ozone. Destruction of chlorophyll due to ozone has also been reported [102]. Decomposition of ozone spontaneously in aqueous medium within the cell or its reaction with a number of compounds such as phenolics and other organic molecules produces reactive oxygen species including superoxide anion (O_2^-) singlet oxygen ($^1O_2^*$), hydroxyl radical ($\cdot$OH) and peroxides that denature proteins, damage nucleic acids and cause peroxidation of membrane lipids [12]. Free $-SH$ groups pre-

TABLE 37.3
Common Air Pollutants and Their Effects on Photosynthesis

Pollutants	Morphological Changes	Metabolic Alterations
1. SO_2 and derivatives	1. Chlorophyll bleaching 2. Leaf discoloration 3. Epinasty 4. Growth Retardation	1. Alteration in $FAD/FADH_2$ and NAD^+ functions 2. Decrease in ATP pool 3. Peroxidation of thylakoid membranes 4. Inhibition in translocation of photosynthetic products
2. NO and NO_2	1. Change in leaf color 2. Growth Retardation	1. Reaction with olefins 2. Peroxidation of membrane lipids
3. Elevated CO_2	1. Stomatal closure 2. Growth Retardation 3. Abscision	1. Reduced uptake of nutrients 2. Decreased root permeability
4. Ozone	1. Decreased stomatal conductance 2. Increased starch accumulation and bleaching of photosynthetic pigments 3. Damage to chloroplast envelope and disruption of thylakoid membrane 4. Abscision	1. Spilitting of olefinic bonds and reaction with thiols 2. Oxidation of glutathione and proteinic -SH gps 3. Inhibition of lipid synthesis in mitochondria and microsomes 4. Inactivation of several key enzymes 5. Inactivation of α-1-proteinase inhibitor 6. Inactivation of plasmalemma bound ATPase 7. Uncoupling of photophosphorylation
5. Peroxides and PAN (peroxyacetyl nitrate)	1. Epinasty 2. Necrosis of leaves 3. Browning 4. Early ripening 5. Abscision	1. Ozone formation 2. Reaction with NADPH 3. Lipid peroxidation 4. Acetylation of amines 5. Reaction with thiols of enzymes

sent on enzymes are highly susceptible to oxidation by ROS.

The penetration of increasing amounts of ultra-violet-B (UV-B) radiation to the earth surface due to depletion of stratospheric ozone is a matter of greater concern to plant health. UV-B is injurious to photosynthetic apparatus and inhibits photosynthesis in both C_3 and C_4 plants [103]. Due to UV-B radiation damage to PSII occurs, marked by increase in variable chlorophyll fluorescence [104]. In rice and pea leaves, the quantum yield of photosynthetic oxygen evolution decreased with a concomitant decrease in the ratios of variable to maximum chlorophyll fluorescence yield due to UV-B radiation [103,105]. Elevated UV-B irradiance levels also cause stomatal closure, reduction in efficiency of electron transport, photophosphorylation, and carbon fixation, and thereby limit photosynthesis [105,106]. Destruction of chlorophyll and corotenoids occurs due to UV-B radiation in sensitive plant species [105]. It has been observed that UV-B radiation suppresses the expression and synthesis of photosynthetic proteins Rubisco large (*rbc* L) and small (*rbc* S) subunits and chlorophyll *a/b*-binding proteins [106]. The extent of down-regulation is dependent on the severity of UV-B exposure.

G. HEAVY METALS

Heavy metal ions such as Cd^{2+}, Ni^{2+}, Hg^{2+}, Cu^{2+}, Zn^{2+}, Pb^{2+}, Al^{3+} have been increasing in the environment and spread to the soil as a result of industrial waste, sewage sludge, agricultural runoff, mining activities, or via airborne pollution. Many of these elements have serious adverse effects on growth and metabolic processes in plants including reduction in chlorophyll content, degeneration of chloroplasts, disorganization of chloroplast thylakoids, reduction in photosynthesis, and inhibition in the activities of many enzymes.

Cadmium, which is a long-range transported heavy metal pollutant, inhibits the synthesis of chlorophylls and carotenoids and affects the ultrastructure of developing chloroplasts in many plant species [107,108]. Exposure of 7-day-old etiolated *Vigna sinensis* L. (savi) leaf segments to heavy metals Cu^{2+} and Cd^{2+} for 24 h caused inhibition in the synthesis of chlorophylls, with more inhibition of Chl *a* than of

Chl *b* [107]. The extent of inhibition was more with Cd^{2+} compared to Cu^{2+}. Wheat seedlings grown in Cd^{2+} containing medium showed a decline in total chlorophyll content as well as Chl *a/b* ratio [108]. Similarly, Pb^{2+} alters photosynthetic pigment compositon and disturbs the granal structure of chloroplasts [13]. Pb^{2+} reduces the concentrations of total chlorophylls (Chl *a* + *b*) in rice (*Oryza sativa* L.) plants. Figure 37.4 shows the observations related to the level of total chlorophyll (Chl *a* + *b*) in leaves of two rice cultivars, Ratna and Jaya, during 5- to 20-day growth period when the seedlings were raised in sand cultures containing nutrient solutions supplanted with 500 or $1000\,\mu M$ $Pb(NO_3)_2$. As evident from the figure, seedlings grown under $1000\,\mu M$ Pb^{2+} in the medium shown 57% to 67% reduced chlorophyll level compared to control-grown seedlings. Ni^{2+} reduces pigment content in various photosynthetic organisms and affects both photosystems [13].

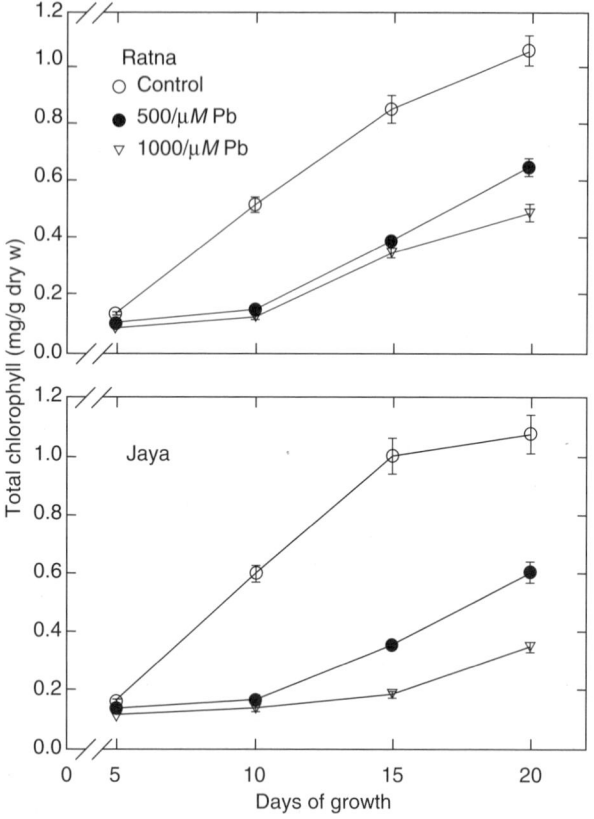

FIGURE 37.4 Level of Chl *a* + *b* in the shoots of two rice *cvs.*, Ratna and Jaya, during 5- to 20-day growth period when seedlings were raised either in nutrient solution (control) or nutrient solution containing $500\,\mu M$ or $1000\,\mu M$ $Pb(NO_3)_2$. Values are mean $\pm$ standard deviation based on three replicates and bars indicate standard deviations. A marked decline in chlorophyll level is observed in Pb^{2+}-treated seedlings compared to controls.

In general, most of the heavy metals preferentially inhibit PSII activity. In chloroplasts isolated from Cd^{2+}-treated *Triticum aestivum* seedlings, 70% decline in oxygen evolution and inhibition in PSII-mediated electron transport activity was observed [108]. It is suggested that Cd^{2+} affects electron transport on the oxidizing site of PSII [108]. Cadmium is also shown to reduce the turnover rate of the D1 protein of the reaction center of PSII [109]. Cu^{2+}, which is an integral part of plastocyanin, inhibits the electron transport at a site connecting both PSII and PSI. Excess copper induces changes in the lipid composition and fluidity of PS(II)-enriched membranes in wheat [15]. Under *in vitro* conditions, high Cu(II) levels significantly modify the oxygen evolving complex of PSII by dissociating the Mn cluster and associated cofactors in PSII-enriched oxygenic and nonoxygenic thylakoid membranes [110]. Hg^{2+} inhibits both photosystems, the inhibition in PSI is reported at the donor side beyond the cytochrome *b/f* complex, whereas PSII is affected on both donor and acceptor sides [13]. Tripathy and coworkers [111] demonstrated that Ni^{2+} affected both photosystems and toxicity was more on PSII than on PSI. Hg^{2+} binds to thylakoid membrane proteins, reacts directly with plastocyanine, replaces copper and alters the enzyme ferredoxin:NADP-reductase by reacting with –SH group [13]. Mn^{2+} toxicity reduces photosynthesis in rice bean seedlings due to peroxidative impairment of thylakoid membrane function [112]. Al^{3+}, together with kinetin, delayed the loss of pigment and protein contents and the activities of PSII and PSI in detached wheat primary leaves [113].

Cadmium and Ni^{2+} lead to decline in CO_2 fixation rates and have pronounced effects on the Calvin cycle enzymes [13,14]. A reduced CO_2 assimilation rate in *Helianthus annus* plants subjected to Cd(II) treatment, in addition to reduced Rubisco activity, photochemical quenching, and quantum efficiency of PS II was observed [114]. In pigeon pea (*Cajanus cajan L*) plants, due to Cd^{2+} and Ni^{2+} treatments, *in vivo* CER decreased and marked inhibition in the activities of the Calvin cycle enzymes occurred [115]. Rice plants grown over a 30-day period in nutrient solution containing increasing copper levels ranging from 0.002 to 6.25 mg/l showed a progressive decrease in Rubisco activity [116]. It is concluded that in rice plants, Cu-led inhibition in photosynthetic activity is primarily due to decreased Rubisco activity [116].

There is increasing evidence that many heavy metals like Cu, Cd, Pb, and Al induce formation of free radicals in cells, which cause severe oxidative damage to different cell organelles and biomolecules including thylakoid membranes and associated proteins [13,117,118]. In mung bean (*Phaseolus vulgaris*) seedlings, Cd toxicity elevates level of lipid peroxides and the decreasesd chlorophyll level observed in such seedlings appears to be due to peroxide-mediated degradation [119]. Heavy metals affect photosynthate partitioning within the different organs of plants. Cd toxicity in rice limits the availability of the photoassimilate sucrose in the cells by favoring its enhanced degradation due to invertase and sucrose synthase activities [16]. Due to heavy metals, the Calvin cycle reactions are slowed down and limitation of ATP and NADPH consumption occurs, which leads to inhibition of photosynthetic electron transport [14]. These observations suggest that inhibition in photosynthetic capacity of plants exposed to heavy metals is both due to inhibition of electron transport activities as well as of the Calvin cycle enzymes.

III. CONCLUDING REMARKS

Abiotic environmental factors like salinity, drought, heat, chilling, water logging, polluting gases, radiations, and heavy metals present in the soil strongly limit photosynthetic efficiency and crop productivity. Photosynthesis is essentially the only mechanism of energy input into the living world and represents a dominant physiological process in plants that is highly sensitive to environment. Under natural field conditions, many stresses interact and this interaction becomes so complicated that it becomes difficult to analyse the effect of a particular stress in isolation. Despite the extensive studies conducted on the effects of various stressful conditions on different photosynthetic parameters in growing plants, our knowledge is still incomplete regarding detection and quantification of the very precise changes that occur at different sites during the photosynthetic process under the influence of a particular stress. Various steps involved in the overall process of photosynthesis associated with conversion of transient energy of a photon into stable chemical energy like sucrose and other photosynthates within the photosynthetic apparatus are so tightly linked that any impairment at a particular step would influence the complete series of events ultimately limiting photosynthesis. Our scientific knowhow and devices are to be advanced to exactly identify and monitor the slightest change occurring due to a stressful condition on the different photosynthetic parameters.

As lack of water (drought) and salinity are major problems because they affect the otherwise most productive agricultural areas, increased drought and salt tolerance with better photosynthetic efficiency have been major objectives in plant breeding programs where irrigation water in limiting, water quality is poor or salinity is high. Water availability is the single greatest constraint on crop productivity as stomata frequently close to conserve water and in turn limit photosynthesis. Therefore, to accelerate crop improvement programs, it is essential to understand how plants cope with stressful environments. A major effort is needed to identify the specific molecular mechanisms that endow the plants with the capacity to adapt to a stressful condition with better photosynthetic efficiency.

With the advancement in molecular techniques several classes of genes have been identified which have been used to engineer plants tolerant to salinity, drought and cold stresses with better yield and photosynthetic efficiency. Overexpression of genes-encoding enzymes that synthesize osmoprotectants and genes-encoding transcription factors that regulate metabolic pathways leading to drought-adaptation has helped in producing transgenic drought-tolerant plants. Transgenic plants overexpressing mitochondrial superoxide dismutase (Mn-SOD) show improved tolerance to drought, freezing, and many herbicides. Tolerance to oxidative stress is being realized as an important factor in providing tolerance to a wide range of environmental stresses. As stress tolerance is a multigenic phenomenon and only a few traits have been understood at the molecular level in plants that can be associated with stress tolerance; identification, characterization, and assessment of many more complex mechanisms involving interplay of many gene products which govern many complex traits like water use efficiency stomatal conductance, ability to exclude salt, and maintenance of optimal photochemical and carboxylation reactions are essential. Once detailed information regarding the metabolic and physiological changes that the place on exposure to stress, the complexity of genes involved in stress tolerance, the signaling pathways leading to the activation of specific transcription factors are available, using powerful biotechnological tools it may be possible to transfer specific stress-tolerant genes to produce transgenic crop species with improved tolerance to stressful environments showing optimum capacity for photosynthesis.

REFERENCES

1. Ziska LH, Seemann JR, DeJong TM. 1990. Salinity induced limitations on photosynthesis in *Prunus salicina*, a deciduous tree species. *Plant Physiol.* 1990; 93:864–870.
2. Plaut Z. Photosynthesis in plant/crops under water and salt stress. In: Pessarakli M, ed. *Handbook of Plant and Crop Physiology.* New York: Marcel Dekker, 1995:587–603.

3. Behera RK, Mishra PC, Choudhury NK. High irradiance and water stress induce alterations in pigment composition and chloroplast activities of primary wheat leaves. *J. Plant Physiol.* 2002; 159:967–973.

4. Medrano H, Escalona JM, Bota J, Gulias J, Flexas J. Regulation of photosynthesis of C₃ plants in response to progressive drought: stomatal conductance as a reference parameter. *Ann. Bot.* 2002; 89:895–905.

5. Sharma PK, Hall DO. Changes in carotenoid composition and photosynthesis in sorghum under high light and salt stresses. *J. Plant Physiol.* 1992; 140:661–666.

6. Flexas J, Medrano M. Drought inhibition of photosynthesis in C₃ plants: stomatal and non-stomatal limitations revisited. *Ann. Bot.* 2002; 89:183–189.

7. Allen DJ, Ratner K, Giller YE, Gussakovsky EE, Shahak Y, Ort DR. An overnight chill induces a delayed inhibition of photosynthesis at midday in mango (*Mangifera indica* L.). *J. Exp. Bot.* 2000; 51:1893–1902.

8. Shah K, Dubey RS. Environmental stresses and their impact on nitrogen assimilation in higher plants. In: Hemantranjan A, ed. *Advances in Plant Physiology*, Vol. 5. Jodhpur, India: Scientific Publishers, 2003:397–431.

9. Carpentier R. Effect of high temperature stress on the photosynthetic apparatus. In: Pessarakli M, ed. *Handbook of Plant and Crop Stress*, 2d ed. New York: Marcel Dekker, 1999:337–348.

10. Crafts-Brandner SJ, Salvucci ME. Sensitivity of photosynthesis in a C₄ plant, maize, to heat stress. *Plant Physiol.* 2002; 129:1773–1780.

11. Ting CS, Owens TG, Wolfe DW. Seedling growth and chilling stress effects on photosynthesis in chilling sensitive and chilling tolerant cultivars of *Zea mays*. *J. Plant Physiol.* 1991; 137:559–564.

12. Taiz L, Zeiger E. *Plant Physiology*, 2d ed. Redwood City, CA: Benjanmin Cummings, 1998:725–757.

13. Carpentier R. The negative action of toxic divalent cations on the photosynthetic apparatus. In: Pessarakali M, ed. *Handbook of Plant and Crop Physiology*. New York: Marcel Dekker, 2002:763–772.

14. Krupa Z, Siedlecka A, Maksymiec W, Baszynski T. *In vivo* response of photosynthetic apparatus of *Phaseolus vulgaris* to nickel toxicity. *J. Plant Physiol.* 1993; 142:664–668.

15. Quartacci MF, Pinzino C, Sgherri CLM, Vecchia, FD, Navari-Izzo F. Growth in excess copper induces changes in the lipid composition and fluidity to PSII-enriched membranes in wheat. *Physiol. Plant.* 2000; 108:87–93.

16. Verma S, Dubey RS. Effect of cadmium on soluble sugars and enzymes of their metabolism in rice. *Biol. Plant.* 2001; 44:117–123.

17. Downton WJS, Grant WJR, Robinson SP. Photosynthetic and stomatal responses of spinach leaves to salt stress. *Plant Physiol.* 1985; 78:85–88.

18. Suleyman IA, Nishiyama Y, Miyairi S, Yamamoto H, Inagaki N, Kaneasaki Yu, Murata M. Salt stress inhibits the repair of photodamaged photosystem II by suppressing the transcription and translation of psb A

19. Salama S, Trivedi S, Busheva M, Arafa AA, Garab G, Erdei L. Effects of NaCl salinity on growth, cation accumulation, chloroplast structure and function in wheat cultivars differing in salt tolerance. *J. Plant Physiol.* 1994; 144:241–247.

20. Miteva TS, Zhelev NZ, Popova LP. Effect of salinity on the synthesis of ribulose-1,5-bisphosphate carboxylase/oxygenease in barley leaves. *J. Plant Physiol.* 1992; 140:46–51.

21. Dubey RS, Singh AK. Salinity induces accumulation of soluble sugars and alters activity of sugar metabolizing enzymes in rice plants. *Biol. Plant.* 1999; 42: 233–239.

22. Flowers TJ, Troke PF, Yeo AR. The mechanism of salt tolerance in halophytes. *Annu. Rev. Plant Physiol.* 1977; 28:89–121.

23. Dionisio-Sese ML, Tobita S. Effect of salinity on sodium content and photosynthetic responses of rice seedlings differing in salt tolerance. *J. Plant Physiol.* 2002; 157:54–58.

24. Kaylie E, Rasmuson, Anderson JE. Salinity affects development, growth and photosynthesis in cheat grass. *J. Range Manag.* 2002; 55:80–87.

25. Plaut Z, Grieve CM, Federman E. Salinity effects on photosynthesis in isolated mesophyll cells of cowpea leaves. *Plant Physiol.* 1989; 91:493–499.

26. Kaiser WM, Webb H, Sauer M. Photosynthetic capacity, osmotic response and solute content of leaves and chloroplasts from *Spinacia oleracea* under salt stress. *Z. Pflanzenphysiol.* 113(5):15–17.

27. Lapina IP, Popov BA. Effect of sodium chloride on the photosynthetic apparatus of tomatoes. *Fiziol. Rast.* 1970; 17:580–585.

28. Reddy MP, Vor AB. Changes in pigment composition, hill reaction activity and saccharide metabolism in Bajra (*Pennisetum typhoids H*) leaves under NaCl salinity. *Photosynthetica* 1986; 20:50–55.

29. Bertrand M, Schoefs B. Photosynthetic pigment metabolism in plants durin g stress. In: Pessarakli M, ed. *Handbook of Plant and Crop Stress*, 2d ed. New York: Marcel Dekker, 1999:527–543.

30. Rottenberg H. Proton and ion transport across the thylakoid membranes In : Trebst A, Avron M, eds. *Photosynthesis I Encyclopedia of Plant Physiology* N.S. Vol. 5. Berlin: Springer-Verlag, 1977:338–349 (ISBN 3-540-07962-9).

31. Barber J. Influence of surface changes on thylakoid structure and function. *Annu. Rev. Plant Physiol.* 1982; 33:261–295.

32. Tewari TN, Singh BB. Stress studies in lentil (*Lens esculenta* Moench) II. Sodicity induced changes in chlorophyll, nitrate and nitrite reductase, nucleic acids, proline, yield and yield component in lentil. *Plant Soil* 1991; 136:225–230.

33. Singh AK, Dubey RS. Changes in chlorophyll a and b contents and activity of photosystem I and II in rice seedlings induced by NaCl. *Photosynthetica* 1995; 31:489–499.

genes in Synechocystis. *Plant Physiol.* 2002 ; 130:1443–1453.

34. Kyle DJ, Ohad I, Arntzen CJ. Membrane protein damage and repair: selective loss of a quinone-protein function in chloroplast membranes. *Proc. Natl. Acad. Sci. USA* 1984; 81:4070–4074.

35. Ohad I, Kyle DJ, Arntzen CJ. Membrane protein damage and repair: removal and replacement of inactivated 32-kilodalton polypeptide in chloroplast membranes. J. *Cell Biol.* 1984; 99:481–485.

36. Mishra SK, Subahmanyam D, Singhal GS. Interrelationship between salt and light stress on the primary process of photosynthesis. *J. Plant Physiol.* 1991; 138:92–96.

37. Larcha W, Wagner J, Thammathaworn A. Effect of superimposed temperature stress on *in vivo* chlorphyll fluorescence of *Vigna unguiculata* under saline stress. *J. Plant Physiol.* 1990; 136:92–102.

38. Govindjee. Sixty three years since Kautsky: chlorophyll *a* fluorescence. *Aust. J Plant Physiol.* 1995; 22:131–160.

39. Sharma PK, Hall DO. Interaction of salt stress and photoinhibition on photosynthesis in barley and sorghum. 1991; 138:614–619.

40. Seemann JR, Sharkey TD. Salinity and nitrogen effects on photosynthesis, ribulose-1,5-bisphosphate carboxylase and metabolic pool size in *Phaseolus vulgaris* L. *Plant Physiol.* 1986; 82:555–560.

41. Rajmane NA, Karadge BA. Photosynthesis and photorespiration in winged bean (*Posphocarpus tetragonolobus* L.) grown under saline conditions. *Photosynthetica* 1986; 20:139–145.

42. Solomon A, Beer S, Waisel Y, Jones GP, Paleg LG. Effect of NaCl on the carboxylating activity of Rubisco from *Tamerix Jordonis* in the presence and absence of proline-related compatible solutes. *Physiol. Plant.* 1994; 90:198–204.

43. Takabe T, Incharoensakdi A, Arakawa K, Yokota S. CO_2 fixation rate and Rubisco content increase in the halotolerant cyanobacterium *Aphanotheca halophytica* grown in high salinites. *Plant Physiol.* 1988; 88:1120–124.

44. Miteva T S, Vaklinova S G. Salt stress and activity of some photosynthetic enzymes. *C. R. Acad. Bulg. Sci.* 1989; 42:87–89.

45. Shomer-Ilan A, Moualem-Beno D, Waisel Y. Effect of NaCl on the properties of phosphoenol pyruvate carboxylase from *Suaeda monoica* and *Chloris gayana*. *Physiol. Plant.* 1985; 65:72–78.

46. Das N, Misra M, Mishra AN. Sodium chloride salt stress induced metabolic changes in callus cultures of pearl millet (*Pennisetum americanum* L. Leeke): free solute accumulation. *J. Plant Physiol.* 1990; 137:244–246.

47. Dubey RS. Biochemical changes in germinating rice seeds under saline stress. *Biochem. Physiol. Pflanzen* 1982; 177:523–535.

48. Dubey RS. Effects of sodium chloride salinity on enzyme activity and biochemical constituents in germinating salt tolerant rice seed. *Oryza* 1984; 21:213–217.

49. Lawlor DW. Limitation to photosynthesis in water stressed leaves: stomata vs. metabolism and the role of ATP. *Ann. Bot.* 2002; 89:871–885.

50. Flexas J, Bota J, Escalona JM, Sampol B, Medrano H. Effects of drought on photosynthesis in grapevines under field conditions: an evaluation of stomatal and mesophyll limitations. *Funct. Plant Biol.* 2002; 29:461–471.

51. Cornish K, Zeevaart JAD. Movement of abscisic acid into the apoplast in response to water stress in *Xanthium strumarium* L. *Plant Physiol.* 1985; 78:623–626.

52. Maroti I, Tuba Z, Csik M, Changes of chloroplast ultrastructure and carbohydrate level in Festuca, Achillea and Sedum during drought and after recovery. *J. Plant Physiol.* 1984; 116:1–10.

53. Tambussi EA, Bartoli CG, Beltrano J, Guiamet JJ, Araus JC. Oxidative damage to thylakoid proteins in water stress leaves of wheat (*Triticum aestivum*). *Physiol. Plant.* 2000; 108:398–404.

54. Gussakovsky EE, Salakhutdinov BA, Shahak Y. Chiral macroaggregates of LHC II detected by circularly polarized luminescence in intact pea leaves are sensitive to drought stress. *Funct. Plant Biol.* 2002; 29:955–963.

55. Poljakoff-Mayber A. Ultrastructural consequences of drought. In: Paleg LG, Aspinall D eds: *Physiology and Biochemistry of Drought Resistance in Plants*. Sydney: Academic Press, 1981:389–403.

56. Kaiser WM, Heber U. Photosynthesis under osmotic stress: effect of high solute concentrations on the permeability properties of the chloroplast envelope and on activity of stroma enzymes. *Planta* 1981; 153:423–429.

57. Sharkey TD, Murray RB. Effect of water stress on photosynthetic electron transport, photophosphorylation and metabolite levels of *Xanthium stumarium* meshophyll cells. *Planta* 1982; 156:199–206.

58. Ormaetxe II, Escuredo PR, Igor CA, Becana M. Oxidative damage in pea plants exposed to water deficit or paraquat. *Plant Physiol.* 1998; 116:173–181.

59. Battle LA, Bosch M. Regulation of plant responses to drought: function of plant hormones and antioxidants. In: Hemantrajan A, ed. *Advances in Plant Physiology*, Vol 5. Jodhpur, India: Scientific Publishers, 2003:267–285.

60. Havaux M, Ernez M, Lannoye R. Correlation between heat tolerance and drought tolerance in cereals demonstrated by rapid chlorophyll fluorescence tests. J. *Plant Physiol.* 1988; 133:555–560.

61. Havaux M, Canaani O, Malkin S. Photosynthetic responses of leaves to water stress expressed by photoacoustic and related methods. *Plant Physiol.* 1986; 82:827–839.

62. Zrust J, Vacek K, Hala J, Janackova I, Adamec F, Ambroz M, Dian J, Vacha M. Influence of water stress on photosynthesis and variable chlorophyll fluorescence of potato leaves. *Biol. Plant.* 1988; 36:209–214.

63. Schreiber V, Schliwa U, Bilger W. Continuous recording of photochemical and non photochemical quenching with a new type of modulation fluorometer. *Photosynth. Res.* 1986; 10:51–62.

64. Havaux M. Stress tolerance of photosystem II *in vivo*: antagonistic effects of water, heat and photoinhibition stresses. *Plant Physiol.* 1992; 100:424–432.

65. Nogues S, Alegre L. An increase in water deficit has no impact on the photosynthetic capacity of field grown Mediterranean plants. *Funct. Plant Biol.* 2002; 29:621–630.

66. Haupt-Herting S, Fock HP. Oxygen exchange in relation to carbon assimilation in water stressed leaves during photosynthesis. *Ann. Bot.* 2002; 89:851–859.

67. Cornic G, Fresneau C. Photosynthetic carbon reduction and carbon oxidation cycles are the main electron sinks for photosystem II activity during a mild drought. *Ann. Bot.* 2002; 89:887–894.

68. Rao IM, Sharp RE, Boyer JS. Leaf magnesium alters photosynthetic response to low water potentials in sunflower. *Plant Physiol.* 1987; 84:1214–1219.

69. Parry MAJ, Andraloj PJ, Khan S, Lea PJ, Keys AJ. Rubisco activity: effects of drought stress. *Ann. Bot.* 2002; 89:833–839.

70. Srivastava LM. Abscisic acid and stress tolerance in plants In: Srivastava LM, ed. *Plant Growth and Development: Hormones and Environment.* New York: Academic Press, 2002:381–412.

71. Sengupta A, Berkowitz GA. Chloroplast osmotic adjustment and water stress effects on photosynthesis. *Plant Physiol.* 1988; 88:200–206.

72. Kaiser WM. Effects of water deficit on photosynthetic capacity. *Physiol. Plant.* 1987; 71:142–149.

73. Berkowitz GA, Gibbs M. Reduced osmotic potential inhibition of photosynthesis: site specific effects of osmotically induced stromal acidification. *Plant Physiol.* 1983; 72:1100–1109.

74. Harn C, Daie J. Regulation of the cytosolic fructose-1,6-bisphosphatase by post-translational modification and protein level in drought-stressed leaves of sugar beet. *Plant Cell Physiol.* 1992; 33:763–770.

75. Vassey TL, Shartey TD. Mild water stress of *Phaseolus vulganis* plants leads to reduced starch synthesis and extractable sucrose phosphate synthase activity. *Plant Physiol.* 1989; 89:1066–1070.

76. Vassey TL, Quick WP, Sharkey TD, Stitt M. Water stress, carbon dioxide and light effects on sucrose phosphate synthase activity in *Phaseolus vulgaris. Physiol. Plant.* 1991; 81:37–44.

77. Berry J, Bjorkmann O. Photosynthetic response and adaptation to temperature in higher plant. *Annu. Rev. Plant Physiol.* 1980; 31:491–543.

78. Gounaris K, Brain APR, Quinn PJ, Williams WP. Structural and functional changes associated with heat induced phase separations of non bilayer lipids in chloroplast thylakoid membranes. *FEBS Lett.* 1983; 153:47–52.

79. Havaux M, Greppin H, Strasser RJ. Functioning of photosystem I and II in pea leaves exposed to heat stress in the presence or absence of light. *Planta* 1991; 186:88–98.

80. Bjorkman O, Badger MR, Armodn PA. Response and adoptation of photosynthesis to high temperature. In: Turner NC, Kramer PJ, eds. *Adaptation of Plants to Water and High Temperature Stress.* New York: Wiley, 1980:233–249.

81. Heckathorn SA, Ryan SL, Baylis JA, Wang D, Hamilton EW, Cundiff L, Luthe DS. *In vivo* evidence from an *Agrostis stolonifera* selection genotype that chloroplast small heat shock proteins can protect photosystem II during heat stress. *Funct. Plant Biol.* 2002; 29:933–944.

82. Yu J, Zhaou Y, Houang L, Allen D. Chill induced inhibition of photosynthesis: genotypic variation within *Cucumis sativus. Plant Cell Physiol.* 2002; 43:1182–1188.

83. Laing WA., Greer DH, Campbell BD. Story responses of growth and photosynthesis of five C_3 pasture species to elevated CO_2 at low temperatures. *Funct. Plant Biol.* 2002; 29:1089–1096.

84. Park IK, Tsunoda S. Effect of low temperature on chloroplast structure in cultivars of rice. *Plant Cell Physiol.* 1979; 20:1449–1453.

85. Baker NR, East TM, Lon SP. Chilling damage to photosynthesis in young *Zea mays* : Photochemical function of thylakoids *in vivo. J. Exp. Bot.* 1983; 34:189–197.

86. Hayden DB, Baker NR, Percival MP, Beckwithz PB. Modification of the photosystem II light-harvesting Chl a/b protein complex in maize during chill induced photoinhibition. *Biochim. Biophys. Acta* 1986; 85:86–92.

87. Vogel CS, Dawson JO. Nitrate reductase activity, nitrogenase activity and photosynthesis of black alder exposed to chilling temperatures. *Physiol. Plant.* 1991; 82:551–558.

88. Nie GY, Long SP, Baker NR. The effects of development and suboptimal growth temperatures on photosynthetic capacity and susceptibility to chilling dependent photoinhibition in *Zea mays. Physiol. Plant.* 1992; 85:554–560.

89. Matsubara S, Gilmore A, Ball MC, Anderson JM, Osmond CB. Sustained down regulation of photosystem II in mistletoes during winter depression of photosynthesis. *Funct. Plant Biol.* 2002; 29:1157–1169.

90. Ivashuta S, Imai R, Uchiyama K, Gau M, Shimamoto Y. Changes in chloroplast FtsH-like gene during cold acclimation in alfalfa (*Medicago sativa*). *J. Plant Physiol.* 2002; 159:85–90.

91. Ariizumi S, Kishitani S, Inatsugi R, Nishida I, Murata N, Toriyama K. An increase in unsaturation of fatty acids in phosphatidyl glycerol from leaves improves the rates of photosynthesis and growth at low temperatures in transgenic rice seedlings. *Plant Cell Physiol.* 2002; 43:751–758.

92. Long SP. C_4 Photosynthesis at low temperatures. *Plant Cell Environ.* 1983; 6:345–363.

93. Sugiyama T, Bocu K. Differing sensitivity of pyruvate orthophosphate dikinase to low temperature in maize cultivars. *Plant Cell Physiol.* 1976; 17:851–854.

94. Stamp P. Photosynthetic traits of maize genotypes at constant and at fluctuating temperatures. *Plant Physiol. Biochem.* 1987; 25:729–733.

95. Holaday AS, Martindale W, Alred R, Brooks A, Leegood RC. Changes in activities of enzymes of carbon

metabolism in leaves during exposure of plants to low temperature. *Plant Physiol.* 1992; 98: 1105–1114.

96. Guy CL, Huber JLA, Huber SC. Sucrose phosphate synthase and sucrose accumulation at low temperature. *Plant Physiol.* 1992; 100:502–508.

97. Zhang, J, Zhang X. Can early wilting of old leaves account for much of the ABA accumulation in flooded pea plants. *J. Exp. Bot.* 1994; 45:1335–1342.

98. Setter TL, Waters I, Wallace I, Bhekasut P, Greenway H. Submergence of rice. I. Growth and photosynthetic response to CO_2 enrichment of flood water. *Aust. J. Plant Physiol.* 1989; 16:251–263.

99. Lambers DHR, Seshu DV. Some ideas on breeding procedures and equirements for deepwater rice improvement. *Proceedings of the 1981 International Deepwater Rice Workshop*, IRRI, Los Banos, 1982:29–44.

100. Dominy PJ, Heath RL. Inhibition of the K^+-stimulated ATPase of the plasmalemma of pinto bean leaves by ozone. *Plant Physiol.* 1985; 77:43–45.

101. Guidi L, Degl'Innocenti E, Soldatini GF. Assimilation of CO_2, enzyme activation and photosynthetic electron transport in bean leaves, as affected by high light and ozone. *New Phytol.* 2002; 156:377–388.

102. Ormond DP, Hale BA. Physiological responses of plant and crops to ozone stress. In: Pessarakli M, ed. *Handbook of Plant and Crop Physiology*. New York: Marcel Dekker, 1995:753–760.

103. Ormond DP, Hale BA. Physiological responses of plants and crops to ultraviolet-B radiation stress. In: Pessarakli M, ed. *Handbook of Plant and Crop Physiology*. New York: Marcel Dekker, 1995:761–770.

104. Stapleton AE, Thornber CS, Walbot V. UV-B component of sunlight causes measurable damage in field grown maize: developmental and cellular heterogeneity of damage and repair. *Plant Cell Environ.* 1997; 20:279–290.

105. He J, Huang LK, Witecross MI. Chloroplast ultrastructure changes in *Pisum sativum* associated with supplementary ultraviolet (UV-B) radiation. *Plant Cell Environ.* 1994; 17:771–775.

106. Mackerness SAH, Jordan BR. Changes, in gene expression in response to ultraviolet B-induced stress. In: Pessarakli M, ed. *Handbook of Plant and Crop Stress*, 2nd ed. New York: Marcel Dekker, 1999:749–768.

107. Muthuchelian K, Maria S, Rani V, Paliwal K. Differential action of Cu^{2+} and Cd^{2+} on chlorophyll biosynthesis and nitrate reductase activity in *Vigna sinensis* L. *Indian J. Plant Physiol.* 1988; 31:169–173.

108. Bhardwaj R, Moscarehas C. Cadmium induced inhibition of photosynthesis *in vivo* during development of chloroplast in *Triticum aestivum* L. *Plant Physiol. Biochem. (India)* 1989; 16:40–48.

109. Geiken B, Masojidek J, Rizzuto M, Pompili ML, Giardi MT. Incorporation of [^{35}S] methionine in higher plants reveals that stimulation of the D_1 reaction centre II protein turn over accompanies tolerance to heavy metal stress. *Plant Cell Envion.* 1998; 21:1265–1273.

110. Yruela I, Alfonso M, Baron M, Picorel R. Copper effect on the protein composition of photosystem II. *Physiol. Plant.* 2000; 110:551–557.

111. Tripathy BC, Bhatia B, Mohanty P. Inactivation of chloroplast photosynthetic electron transport activity by Ni^{2+}. *Biochim. Biophys. Acta* 1981; 638:217–224.

112. Subramanyam D, Rathore VS. Influence of manganese toxicity on photosynthesis in rice bean (*Vigna umbellata*) seedlings. *Photosynthetica* 2000; 38: 449–453.

113. Subhan D, Murthy SDS. Synergistic effect of $AlCl_3$ and kinetin on chlorophyll and protein contents and photochemical activities in detached wheat primary leaves during dark incubation. *Photosynthetica* 2000; 38:211–214.

114. Cagno RD, Guidi L, Gara LD, Soldatini GF. Combined cadmium and ozone treatments affect photosynthesis and ascorbate dependent defenses in sunflower. *New Phytol.* 2001; 151:627–636.

115. Sheoran IS, Singal HR, Singh R. Effect of cadmium and nickel on photosynthesis and the enzymes of photosynthetic carbon reduction cycle in pigeon pea (*Cajanus cajan* L). *Photosynth. Res.* 1990; 23:345–351.

116. Lidon FC, Henriques FS. Limiting step on photosynthesis of rice plants treated with varying copper levels. *J. Plant Physiol.* 1991; 138:115–118.

117. Shah K, Kumar RG, Verma S, Dubey RS. Effect of cadmium on lipid peroxidation, superoxide anion generation and activities of antioxidant enzymes in growing rice seedlings. *Plant Sci.* 2001; 61:1135–1144.

118. Verma S, Dubey RS. Lead toxicity induces lipid peroxidation and alters the activities of antioxidant enzymes in growing rice plants. *Plant Sci.* 2003; 164:645–655.

119. Somashekaraiah BV, Padmaja K, Prasad ARK. Phytotoxicity of cadmium ions on germinating seedlings of mung bean (*Phaseolus vulgaris*): involvement of lipid peroxides in chlorophyll degradation. *Physiol. Plant.* 1992; 85:85–89.

38 Photosynthetic Response of Green Plants to Environmental Stress: Inhibition of Photosynthesis and Adaptational Mechanisms

Basanti Biswal
Laboratory of Biochemistry and Molecular Biology,
School of Life Sciences, Sambalpur University

CONTENTS

I. INTRODUCTION

In nature, plants frequently experience a wide range of stresses, both biotic and abiotic, that adversely affect their growth and development. Although the literature on plant responses to biotic stress is not rich, the abiotic stress factors are extensively studied and are shown to result in several responses, both detrimental and adaptive, at different levels of plant organization.

Reviews are available on the effects of stress factors like high light [1,2], UV radiation [2–4], water deficit [2], and extreme temperatures [5,6] on photosynthetic activities. The literature in these areas is extensive and the factors could have been included in detail, with description of possible mechanisms responsible for reduction in photosynthetic efficiency. The present review, however, attempts to briefly but critically discuss the recent findings and thoughts in these areas with particular emphasis on photoinhibition and molecular biology of environmental stress response in green plants, the areas that have drawn our serious attention in recent days.

II. CHLOROPLAST, THE TARGET OF STRESS IN GREEN PLANTS

Chloroplast in green leaves of higher plants is considered as the major target of environmental stress. The organelle has potential to absorb light, split water molecules to liberate oxygen, and initiate electron transfer reactions, resulting in production of ATP and NADPH, which are used for fixation of carbon dioxide. Under normal conditions, there is a perfect coordination between light-induced electron transfer reactions associated with the thylakoid membrane and carbon dioxide fixation by the Calvin cycle in stroma. But during stress, a loss of coordination may cause leakage of electron to oxygen, which may produce toxic oxygen-free radicals. Therefore, light absorbed by the pigments and oxygen liberated by chloroplast in this condition become harmful to plants. Second, the excited reaction center of photosystem II (PS II) results in the production of a strong oxidant (P_{680}^+) required for the liberation of oxygen. But under stress condition, the strong oxidant becomes long-lived and can oxidize lipids, pigments, and proteins of the membrane. Thus, the potential of chloroplast for photoexcitation leading to liberation of oxygen, primarily responsible for survival of higher organisms in this planet, makes the leaves a major target of stress in green plants.

III. STRESS SIGNALS

The precise mechanism of stress signal perception and signal processing leading to the stress response in green plants is not known. There could be several sites for perception of stress signals. Both PS I and PS II in thylakoid membranes can be considered as the major stress perception systems. The signals received by the photosystems may subsequently be transduced through the changes in the status of plastoquinone, NADPH, ΔpH, $\Delta\psi$ and the efficiency of the Calvin cycle for carbon assimilation. In fact, photosynthetic plastoquinone pool through the changes in its redox status has been shown as the major sensor in green plants for regulation of several defense genes associated with stress. The changes may further be transmitted to various short-term stress adaptations or to the expression level of specific genes that control long-term adaptation to effectively counter the stress effect [7]. The signal transduction for the stress response in chloroplasts at the donor and acceptor sides of both the photosystems has critically been discussed recently by Biswal et al. [2].

IV. MAJOR ENVIRONMENTAL STRESSES AND PHOTOSYNTHETIC RESPONSE

A. PHOTOINHIBITION

Sunlight, which is converted to chemical energy in the process of photosynthesis by green plants, is the ultimate source of energy to sustain the biosphere on Earth. But the light, which is the substrate for photosynthesis, dramatically reduces the efficiency of the process when absorbed in excess. This process, leading to downregulation of photosynthesis by excess of unutilized light, is known as photoinhibition, a highlight stress syndrome.

1. Photoinhibition of PS II

It is well established that PS II is a major site of photoinhibition and widely studied as the most susceptible component of thylakoid membranes (for review, see Refs. [1,2]).

a. Turnover of D1 protein

Among the components studied so far, D1 protein has been extensively examined during photoinhibition of chloroplasts [5]. In fact, D1 turnover in PS II is established to be the central event of photoinhibitory changes of the chloroplasts. In high-light conditions, D1 is rapidly degraded and synthesized and its level reflects the net balance between photodamage and the repair process. The nature of damage of electron transport components responsible for induction of D1 protein degradation is still unclear. The free radical mediated induction of the degradation has been suggested by many authors (for review, see Refs. [2,8,9]). The possibility of two different mechanisms, one operating at the donor and the other at the acceptor side of PS II for degradation of the protein, has been critically discussed [5,10]. The mechanism of acceptor side inactivation operates at high irradiance when carbon dioxide fixation becomes limiting, which results in saturation of the reduced plastoquinone pool, a condition leading to overexcitation of PS II. In this situation, oxygen may receive electrons resulting in the formation of oxygen radicals that could possibly damage D1 protein in secondary reactions [5]. Alternatively, the reduced plastoquinone pool may favor a back reaction with PS II reaction center resulting in the formation of triplet reaction center. The triplet reaction center subsequently is likely to form singlet oxygen, which can bring changes in reaction center proteins including D1 protein and other nonproteinaceous components, leading to inhibition of PS II photochemistry [2,5]. A change in D1 protein conformation may result in its degradation. The donor side photoinhibition leading to D1 loss is suggested to be a consequence of stress induced destabilization of the water splitting system, which causes a slowdown of electron transfer from water to P_{680}. Under this condition, there is a possibility of the formation of strong oxidizing radicals including P_{680}^+. These radicals with high oxidizing potential accumulate at the donor side and may oxidize the proteins, including D1 and its subsequent proteolytic degradation [5].

The turnover of D1 protein in PS II during photoinhibitory stress is considered as an adaptive mechanism of chloroplast against stress. High light induced degradation of the protein leads to disassembly of PS II resulting in protection of the rest of its components against photodamage. Immediate replacement of a new copy of D1 leads to the reassembly of a fully functional photosystem.

b. Zeaxanthin, an effective xanthophyll for thermal dissipation of excess quanta during photoinhibition

Thermal dissipation of light energy, although a protective response, essentially results in a significant reduction in photosynthetic efficiency. The molecular mechanism of the process of dissipation is not clearly understood in spite of extensive literature available in the field in the last few years. Nevertheless, operation of the xanthophyll cycle as the possible mechanism for the harmless dissipation is suggested. The cycle is proposed to be operating at the light harvesting system [2].

The operation of the xanthophyll cycle in light harvesting system involves the interconversion of three xanthophylls like violaxanthin, antheraxanthin, and zeaxanthin under a specified physiological condition of the photosynthetic apparatus induced by high-light irradiance [5,11]. High light induced electron transport and consequently low lumen pH induces enhanced activity of de-epoxidase enzyme converting violaxanthin to zeaxanthin. In dark or limiting light conditions, an enhancement in the activity of epoxidase enzyme reverses the reaction, resulting in conversion of zeaxanthin to violaxanthin. Formation of zeaxanthin from violaxanthin is established to be a major event in the energy dissipation process. The role of zeaxanthin as an effective quencher of excited singlet chlorophyll and consequently a quencher of chlorophyll *a* fluorescence is well known [1,2]. The quenching of fluorescence is correlated with the thermal dissipation that brings down the level of excess quanta absorbed by photosynthetic pigments.

Although the role of zeaxanthin in the dissipation of excess excitation energy in the light harvesting system of PS II of chloroplast is established, the mode of its action still remains confusing. The xanthophyll action may be direct, with zeaxanthin directly quenching singlet chlorophyll, or indirect, with zeaxanthin behaving as an allosteric modulator of light harvesting complex (LHC) II b aggregation, which favors quenching of excess quanta through chlorophyll–chlorophyll interaction. In either case, zeaxanthin is required to have specific proximity and orientation in the LHC of PS II. The studies with mutants reveal that the thermal dissipation in photosynthetic organisms requires the synthesis of specific polypeptides in addition to a pH gradient across the membrane and formation of zeaxanthin from violaxanthin by de-epoxidase activity. Recently, Elrad et al. [12] have shown a gene, namely, *Lhcbm1*, that codes for a light harvesting polypeptide participating in the process of thermal dissipation. But the coordination between zeaxanthin, pH gradient, and the polypeptide of LHC involved in the dissipation largely remains unclear. We have proposed a model for a possible coordination among the three [13,14]. In the model, a molecular quenching complex involving chlorophyll, zeaxanthin, and glutamic acid side chain of a light-harvesting antenna polypeptide is proposed (Figure 38.1). The model explains thermal dissipation by zeaxanthin in the light-harvesting antenna only when there is formation of a proton gradient. The glutamic acid of light harvesting protein at the lumenal side is proposed to be the site for binding of zeaxanthin. The negatively charged carboxylate ion of the amino acid may form a ligand to Mg^{2+} of chlorophyll at pH >5. In this case, the negative charge of the carboxylate group is delocalized on both the oxygen atoms, and this does not permit its binding to the xanthophyll. On the other hand, the carboxyl group becomes protonated and neutral when pH <5 under quenching condition. The change in pH leads to a change in conformation of the glutamic acid residue. The carbonyl group of carboxylic acid of the side chain forms a ligand to chlorophyll *a*, and its hydroxy group forms a hydrogen bond with the hydroxy group of zeaxanthin. The complex in this model has been shown to behave as a sink for thermal dissipation of excess light quanta absorbed at the level of the light-harvesting system of photosynthetic tissues. In addition to light-harvesting antenna systems, the quenching complex also acts as a sink for dissipation of harmful quanta from reaction center core protein complex. We have suggested the location of a series of β-carotenes (both *cis* and *trans* forms) present in the reaction center and proximal antenna proteins. These β-carotenes are likely to provide a safe channel to drain out the harmful energy from the reaction center and send it to zeaxanthin in the quenching complex for safe dissipation (Figure 38.2).

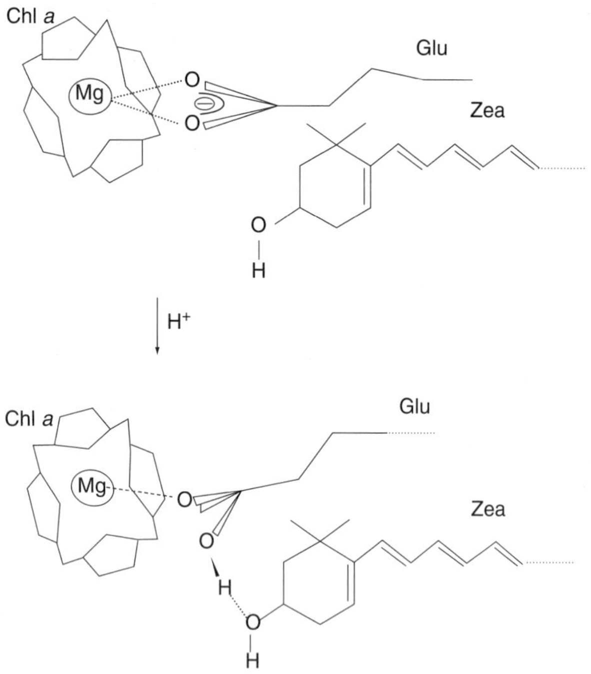

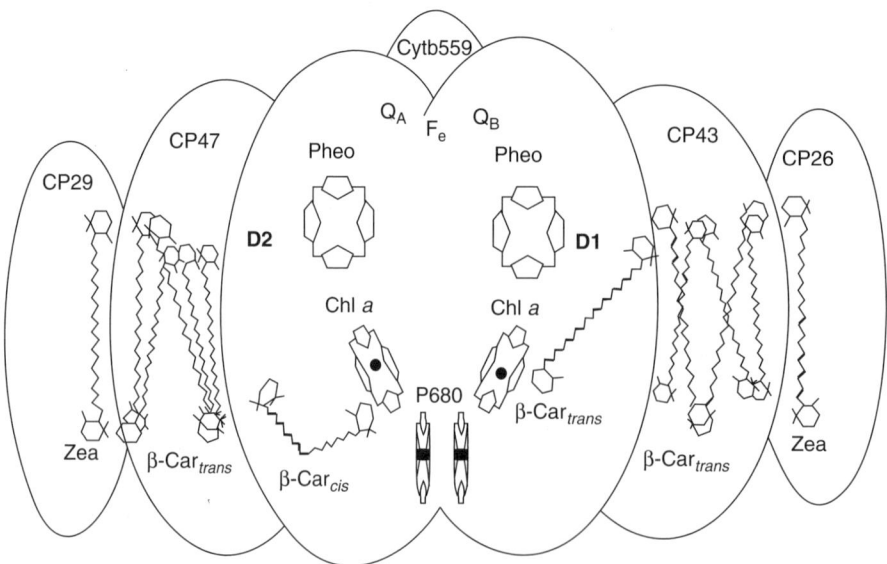

FIGURE 38.1 The proposed mechanism for low lumen pH induced formation of a quenching complex involving chlorophyll *a* (Chl *a*), zeaxanthin (Zea), and glutamic acid (Glu) side chain of light harvesting protein. The quenching complex nonradiatively dissipates the excess quanta of light energy that otherwise would have been harmful for PS II of chloroplasts. (From Nayak L, Raval MK, Biswal B, Biswal UC. *Photochem. Photobiol. Sci.* 2002; 1: 629–631. With permission.)

Other possible modes of zeaxanthin action in the protection of chloroplasts against excess light have been reviewed earlier by Biswal and Biswal [5]. Recently, Baroli et al. [15] have examined several mutants of *Chlamydomonas* and have discussed the photoprotective role of constitutively accumulated zeaxanthin in the background of its antioxidant role in quenching oxygen-free radicals in chloroplasts.

2.   PS I Photoinhibition

Photoinhibition is normally referred to as inhibition of PS II photochemistry of thylakoids. Reports are, however, available on photoinhibition of PS I (for review, see Ref. [16]). The inhibition is triggered by the inactivation of some components at the acceptor side of the photosystem. The nature of photoinhibition of PS I in many ways is different from that of PS II. In the case of PS I, photoinhibition occurs at chilling stress and the inhibition takes place in the presence of oxygen at relatively low-light conditions. It is proposed that the temperature dependent loss of protective mechanism at the acceptor side of the photosystem results in the production of oxygen-free radicals, which attack the iron–sulfur centers of the complex and ultimately degrade the product of the *psaB* gene, one of the two major subunits of the PS I reaction center.

FIGURE 38.2 A schematic representation of the organization of carotenoids in PS II protein complex. P680: primary donor; Pheo: pheophytin; Q_A, Q_B: quinone in D2 and D1 proteins, respectively; Chl *a*: accessory chlorophyll *a*; β-Car: β-carotene (the subscripts indicate the *cis* or all-*trans* isomers); Zea: zeaxanthin; Cyt*b*559: cytochrome *b*559. (From Nayak L, Raval MK, Biswal B, Biswal UC. *Photochem. Photobiol. Sci.* 2002; 1: 629–631. With permission.)

B. Ultraviolet Irradiation and Modification in Thylakoid Complexes

In addition to visible light, UV radiation is known to significantly modify the structure and function of the photosynthetic apparatus [3,4]. Increased influx of ultraviolet radiation to the surface of the Earth has, therefore, drawn a great deal of our attention in recent days. Molecular events associated with the response of green plants to the radiation have been recently reviewed by Biswal et al. [2] and Brosche and Strid [4]. While UV-B and UV-C are demonstrated to be highly detrimental, UV-A brings about mild, but significant, damaging and nondamaging effects on plant growth and photosynthesis. For example, we have recently demonstrated that UV-A does not have much effect on photosynthetic pigments but brings about a significant reduction in the efficiency of PS II photochemistry of chloroplasts during leaf senescence of wheat plants [17]. The radiation induced modification in thylakoid structure [18–20] and the primary photochemical reactions of chloroplasts [17,19,21] have been well examined. The PS II complex of thylakoids, however, has been demonstrated to be the main target of UV radiation [18–22]. The irradiation of plants exhibits a significant loss of oxygen evolution [17] and a decline in the F_v/F_m ratio [17,21]. Although the specific target molecules or components of PS II complex as the primary site of radiation-induced damage are not identified, the damage of reaction-center II including degradation [22,23] and structural modification [17,21,24] of D1 protein has been suggested.

The plants in the natural environment, however, exhibit certain adaptive responses and develop various protective mechanisms against hazardous UV radiation. Flavonoids are considered to be the potential UV protective components because of their capacity to screen UV radiation (for review, see Refs. [25,26]). The other photoreceptors that absorb light in the visible region of the spectrum are also known to show UV protective action. In this context, our work on the involvement of phytochrome in retarding UV-induced damage of chloroplasts of wheat leaves is worth mentioning [20]. We have suggested the action of phytochrome primarily through the suppression of UV-induced lipid peroxidation. Altered gene expression as an adaptive mechanism in response to UV irradiation has also been reported [4].

C. Water Stress Induced Loss of Photosynthetic Activity

Plants may experience stress both during drought (water deficit) and flooding (too much water). Both the extreme conditions are known to bring about damage of cellular organelles. Flooding stress that causes oxygen deficit is known to significantly affect cellular respiration [27]. But not much is known about the photosynthetic response to this stress although reports on the response of the photosynthetic organelle to water deficit are extensive [2,5]. Water stress causes a significant reduction in the content of photosynthetic pigments, proteins, and lipids [28,29]. Severe water stress resulting in desiccation of leaves causes not only a dramatic loss of the pigments but also disorganization of the thylakoid membranes [28]. Photochemical reactions associated with PS II have been shown to be more susceptible to stress than those associated with PS I [30]. Although the precise mechanism of stress-induced PS II inactivation is not clear, Sundari et al. [31] have suggested the possibility of depletion of manganese with the consequent effect of the loss of oxygen evolution. On the other hand, quantitative loss of D1 protein as observed by Western blot analysis in cotyledons of cluster bean under stress may lead to inactivation of PS II [28]. These changes associated with the reaction center II complex may significantly contribute to the loss of PS II photochemistry. The data on the stress induced decline in carbon dioxide fixation by the leaves of higher plants are rather controversial. The loss of chloroplast capacity to fix carbon dioxide in green leaves under stress is attributed at least partly to stress-induced closure of leaf stomata, a response exhibited by abscisic acid treatment of plants [32]. Graan and Boyer [32] have proposed that one component of photosynthetic inhibition during dehydration is nonstomatal. This component is suggested to be a metabolic one, but its precise nature still remains to be worked out. Kanechi et al. [33], however, have shown high sensitivity of Rubisco activity to dehydration condition.

The protective mechanisms developed by plants against stress have been examined in different laboratories. At the molecular level, changes in the gene expression as a response of plants to water stress have been reported, and the possible functions of gene products induced by the stress have been critically reviewed [27,34].

D. Temperature Extremes: Changes in the Structure and Function of Chloroplasts

Temperature at extremes behaves as one of the major physical stress factors that brings changes in the quantity of pigments, organization of thylakoids, primary photochemistry, and carbon dioxide fixing ability of chloroplasts [5,6]. Reports are available on the effect of chilling temperature on the loss of thylakoid photochemical reactions, and the reactions associated with PS II inactivation are suggested to be the main target of the stress [35]. The cold-induced inhibition of D1 protein synthesis during the repair cycle of the protein may be a major reason for such inactivation [35]. In addition to light reactions, low temperature also results in loss of efficiency of the Calvin cycle [36]. High-temperature stress is demonstrated to cause a decline in electron transport efficiency of PS I, PS II, and the whole chain between PS I and PS II [6]. The stress-induced denaturation of membrane proteins and uncoupling of light harvesting antenna complex from the reaction center complexes may be the possible factors contributing to the loss [37].

Acclimation of plants to high or low temperature is a complex process, but data are available on the molecular biology of the adaptation of green plants to heat [38] and cold [39,40] stress.

E. Photosynthetic Response to Interacting Stress Factors: A Simulation of the Natural Environment

1. Significance of Multistress Factors

The plants in the natural environment experience several stress factors, which may operate simultaneously or in different combinations. This is unlike the laboratory conditions where these factors are examined mostly in an isolated manner. The interacting stress factors also exhibit significant seasonal or diurnal variations with increase or decrease in the intensity of individual stress components. Only during the last few years, attempts have been made to simulate the natural stress conditions in the laboratory and examine the multistress effects on photosynthesis.

We have examined the photosynthetic response of laboratory-grown plants to multistress factors in different combinations like drought plus weak light [28], osmotic stress plus high light [29], and UV plus high temperature [17]. In the case of multistress factors, the stress effects are shown to be additive [17]. Plants also exhibit a kind of cross-adaptation establishing the links between various stresses [29,41,42]. When plants experience different stresses in a sequence, their exposure to one stress develops provision of tolerance to another stress [29].

2. Multistress Factors with Light as a Common Factor: Sensitized Photoinhibition

Since green plants grow in light, other stress factors are investigated mostly in combination with light. Photooxidative damage of chloroplasts seems to be the major and common response of all stress factors that operate in the field in the presence of light [43]. Most of the environmental stress factors are known to cause a decline of the water splitting system and in the efficiency of carbon dioxide fixation. These changes in the presence of light may lead to the formation of several oxygen radicals that can damage the photosynthetic apparatus. Here, the environmental factors other than light are normally referred to as the primary stress factors, which can sensitize chloroplast for further damage by light through photoinhibitory processes. Some of the important stress factors and their sensitizing effects for photoinhibition are described below.

Chilling stress: Extensive reports are available on how the chilling of green leaves brings about inactivation of PS II, a major syndrome of photoinhibitory damage of chloroplasts in the presence of light [35,44]. The chilling-induced decline of oxygen evolution with consequent accumulation of P_{680}^+, reduced utilization of excitation energy because of low temperature induced inactivation of carbon dioxide fixing enzymes, reduction in the efficiency of the xanthophyll cycle because of a check in the conversion of violaxanthin to zeaxanthin, and decline of D1 synthesis during protein turnover could lead to photoinhibitory damage of PS II.

Water stress, UV radiation and other abiotic stresses: These stress factors have been demonstrated to either alter oxygen evolution capacity or reduce the capacity of chloroplasts to effectively utilize the absorbed quanta for carbon dioxide fixation, leading ultimately to the formation of free radicals in the presence of light, and thus destroy the photosynthetic organelle. Recently, Deo and Biswal [28] have demonstrated water stress sensitized photoinhibitory damage of D1 protein of PS II of chloroplast isolated from cotyledons of cluster bean seedlings grown under moderate light conditions.

However, it appears that the nature of interacting stress factors in the natural environment is still more complex than their simulation under laboratory conditions. The data available so far do not permit us to have a very close look at what really happens during the operation of these stress factors in the field conditions.

V. MOLECULAR BIOLOGY OF STRESS RESPONSE

For quite a long time, the studies of plant responses to environmental stress were limited to physiological and biochemical parameters. However, owing to the dramatic upsurge in the area of plant molecular genetics in the last few years, the stress response is now being examined from quite a different angle. Attempts have been made to identify and characterize the stress responsive genes, their expression and regulation during stress treatments. Some of the recent findings in these areas are discussed below.

A. Limited or Overexpression of Genes, a Molecular Stress Response for Readjustment of Chloroplast Components

In this case, the genes code for certain structural proteins and enzymes that are required for readjustment of cellular components in order to minimize the damaging effect of stress. The stress itself induces either overexpression or limited expression of the genes of these proteins in a highly regulated manner for the readjustment. The gene products are not therefore stress specific. High light induced enhancement of the production of transcripts for the synthesis of D1 protein replacing the photodamaged protein is one of the examples of these proteins [45]. Similarly, the level of transcripts for the synthesis of LHCs remains at a low level during high-light condition [46], which results in lowering of the antenna size for minimum absorption of light quanta. Similar studies are made on stress induced regulation of gene expression in cyanobacteria. The genes responsible for the synthesis of phycobiliproteins are switched off [47] with the subsequent loss of biliproteins [48] under nitrogen stress. On the other hand, the genes that produce the proteins for the formation of gas vacuoles for buoyancy regulation are differentially activated depending on the availability of light in aqueous environment [47]. These findings clearly suggest that a stress can induce either reduction or enhancement of the accumulation of transcripts in order to produce the desired amount of certain specific proteins for adaptation under adverse environmental conditions.

B. Stress-Specific Proteins

The second category of proteins synthesized in response to stress is stress-specific and the proteins are not necessarily structural components of chloroplasts. These are believed to play a role in modulating the stress effect. Among these proteins, heat shock protein (HSP) and early light inducible protein (ELIP) are well studied with particular reference to chloroplast. In addition, expression of osmotins in response to water stress has also been examined. The plants including green algae are reported to produce HSP in response to heat stress [38]. There is a range of HSPs varying in their molecular weight. HSPs are known to

be the members of multigene subfamilies. Although heat is a major stress inducing the synthesis of these proteins, other stress factors also may cause their synthesis. The stress proteins, homologous to major classes of HSP are located in chloroplasts. Although the nature of function of these proteins is still not clear, their role in the transport of chloroplast proteins, the stability of essential proteins against thermal stress, mediating in the reassembly of various protein complexes and the repair processes, has been reported [6,38].

ELIP is a class of stress responsive proteins, so named because they were initially thought to be produced during the early stage of chloroplast development. But recently, detailed studies of these proteins reveal their association with high-light stress conditions. The formation of transcripts of ELIP has been reported to be significantly enhanced with increase in the intensity of light [46]. These proteins are nuclear-encoded, synthesized in cytoplasm, imported to chloroplast, and finally inserted into thylakoids. The proteins remain stable as long as the leaves are exposed to high irradiation but start degrading when the intensity of light is lowered [46]. The precise photoprotective nature of ELIP is still to be worked out. During light stress, the breakdown of pigment protein complexes may release free chlorophylls, which on absorption of light may produce oxygen-free radicals, toxic to chloroplast. The ELIPs may absorb these free pigments, providing the proper surface and thus may protoprotect the organelle against stress.

Literature on the protective role of late embryogenesis abundant (LEA) proteins against water stress in green plants is extensive [27]. Overexpression of LEA proteins in transgenic plants has been shown to develop resistance of the plants to drought [27]. The LEA proteins are classified into five groups, but their precise function during stress remains unclear.

In addition to water stress related proteins, the proteins synthesized in response to cold stress have been well characterized, but their precise function in modulating the stress effect on photosynthesis is not known.

C. Molecular Biology of Oxidative Stress

Oxidative stress develops in plants in response to several environmental and biotic stress factors. Environmental stress factors including high or low light, drought, extreme temperatures, and UV radiations experienced by plants are known to promote the formation of oxygen-free radicals that damage or kill cellular components including chloroplasts. The possible signaling systems associated with oxidative stress are shown in Figure 38.3. As discussed earlier, chloroplast is the major cellular organelle for production of these free radicals when green leaves experience stress. The formation of oxygen-free radicals is known to induce expression of genes for the synthesis of antioxidant enzymes and many other nonenzymatic antioxidants that constitute the antioxidant defense system against oxidative stress. The precise molecular mechanism of oxygen-free radical induced signal transduction that results in defense gene expression has not been clarified. Hydrogen peroxide is suggested to act as a stress signaling molecule in the signal transduction pathway [49]. The participation of protein kinases in the signal transduction pathway is possible [50]. The data on possible signal transduction from oxygen-free radicals down to gene expression through the action of a family of protein disulfide oxidoreductases have recently been reviewed [50]. The effects of oxygen-free radicals at the biochemical and transcriptional levels are likely to be regulated by thioredoxin, a small protein with disulphide reducing ability [50].

D.   GENE MANIPULATION AND CHLOROPLAST RESISTANCE TO STRESS

We have information available on a wide range of photosynthetic and other physiological responses to different kinds of environmental stresses. Most of the adaptive responses are well characterized. Due to gene manipulation, it is now possible to have plants with the capacity of stress tolerance in terms of these parameters. Many of the stress adaptive genes have been identified [2]. Attempts have been made to transfer these stress genes by effective transfer techniques and the subsequent regulated transcription in stress sensitive plants to make them stress resistant. Some of the specific findings relating to stress resistance of chloroplasts by genetic manipulation are worth mentioning.

The transgenic plants with the bacterial coding sequence of Mn-superoxide dismutase (Mn-SOD) have been demonstrated to exhibit high resistance against oxidative stress and other stresses like drought and metal stress [43]. Transformed plants with overexpression of chloroplast Cu–Zn-superoxide dismutase (Cu–Zn-SOD) are also found to be resistant to photoinhibition [43]. The most interesting piece of work conducted by Murata et al. [51] is the production of plants by genetic manipulation, which can effectively resist chilling stress. Since the level of unsaturated fatty acids in membranes determines their sensitivity to chilling stress, they have been successful in manipulating the level of unsaturated fatty acids in chloroplast membranes by the transformation of plants with cDNA to plants with glycerol-3-phos-

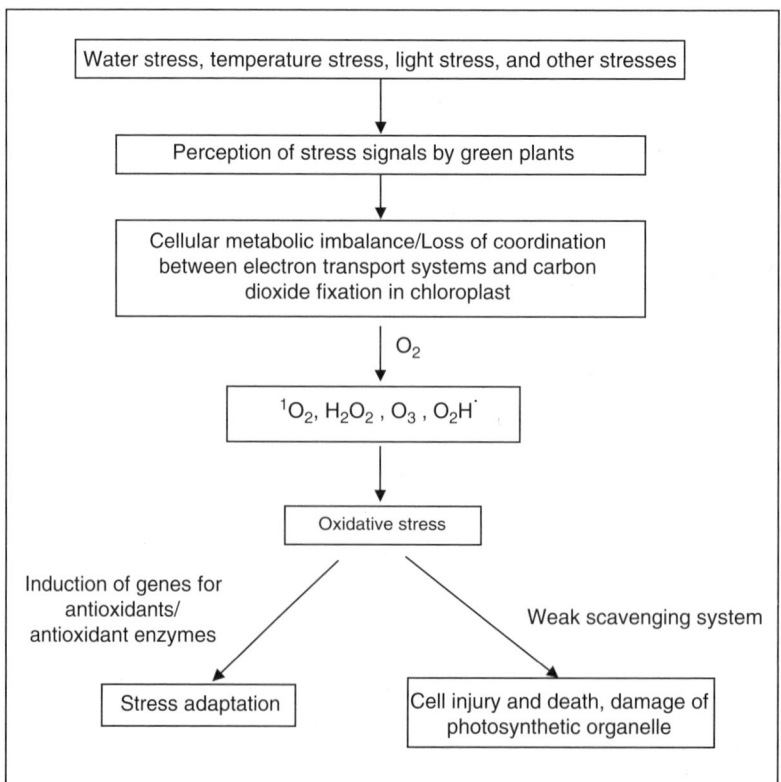

FIGURE 38.3 Perception of signals and their transmission during oxidative stress.

phate acetyl transferases from stress resistant species. Similarly, Hayashi et al. [52] have demonstrated enhanced photosynthetic tolerance of *Arabidopsis* plants to salt and cold stress with transformation of plants with *CodA* gene for choline oxidase. From recent literature it appears that genetic manipulation for developing photosynthetic resistance to stress is going to be a major area of research in plant stress physiology.

VI. CONCLUSIONS AND PERSPECTIVES

The stress biology of plants has gained impetus recently in the background of advances made in the field of molecular biology, mutational studies, and the development of new analytical techniques. This review, therefore, while describing the recent developments in the field, has raised several unanswered questions and has posed some challenging problems in the area, particularly the molecular aspects of stress response including the molecular mechanism of stress adaptation. Some of these problems are briefly described below:

1. We have vast literature available in the area of photoinhibition. The high turnover of D1, a reaction center II protein, has been suggested to be the major response of photoinhibition. Different mechanisms have been proposed to explain the rapid degradation of this protein during the process. However, the question still remains, how is the degradation of the protein initiated? The mechanism of tight coupling between D1 degradation and immediate synthesis of its copy still remains unclear. Similarly, the role of individual components of thylakoids for photoinhibition of photosynthesis has yet to be worked out.

2. Photoinhibition has recently been shown to cause inhibition of PS I photochemistry although the nature of photoinhibition of the photosystem appears to be different from that of PS II of thylakoids. The mechanism of PS I photoinhibition has not been worked out in detail. The proposed redox signaling system associated with cleavage of one of the reaction center proteins during the inhibition process requires more experimentation and clarification.

3. Formation of zeaxanthin as a response to high light stress and its photoprotective role are well established. However, there is a great deal of controversy about the mechanism of its protection. Different views on the role of the pigment

in thermal dissipation of excess energy, its action as a quencher of oxygen free radicals and as an agent in modifying the fluidity of thylakoids membranes, are expressed by different authors. It appears that more critical and extensive investigations are needed to evolve a generalized mechanism of the role of zeaxanthin that could explain the existing literature in this area.

Recently the requirement of a polypeptide in LHC of PS II has been suggested for thermal dissipation of excess light by zeaxanthin. But the coordinate action of the polypeptide and zeaxanthin in quenching singlet chlorophyll during the dissipation process has not been clarified.

4. Molecular biology of stress response in plants is still in its infancy. Nevertheless, significant progress has been made in the area. The stress responsive genes, both stress resistant and stress sensitive, have been identified in some cases. However, the details of molecular and biochemical mechanisms of natural stress resistance have to be precisely understood before proper manipulation of gene transfer technology can effectively be applied in the production of stress resistant plants.

5. Another molecular approach to examine stress response is to characterize the precise function of the proteins, namely, HSP, ELIP, osmotin, and many other stress proteins. Some of these proteins have been well investigated. Their genes have been cloned and sequenced. Unfortunately, we have failed to unravel the mechanism or the way they modulate the stress effect or protect the photosynthetic systems against stress.

6. The stress signaling system has become the focus of stress studies in recent years. If we successfully investigate the molecular and genetic aspects of stress response with its subsequent manipulation, and hopefully if we do it in the next few years, the next field of research would be the study of stress signals and their transduction in plants. We have a couple of questions to ask in this field. What are stress signals? How is the unfavorable environment perceived and recognized by plants? These questions do not have any definite answers. Although the stress response is explained at physiological, biochemical, and even molecular levels, the initial events that receive the stress signal, process it, and finally result in the response still remain unclear. The search for

signal molecules that mediate the stress response is going to be a very fascinating area of study in the future.

REFERENCES

1. Demmig-Adams B, Adams III WW. Harvesting sunlight safely. *Nature* 2000; 403: 371–374.
2. Biswal UC, Biswal B, Raval MK. *Chloroplast Biogenesis: From Proplastid to Gerontoplast*. Dordrecht, The Netherlands: Kluwer Academic Publishers, 2003.
3. Vass I. Adverse effects of UV-B light on the structure and function of the photosynthetic apparatus. In: Pessarakli M, ed. *Handbook of Photosynthesis*. New York: Marcel Dekker, Inc., 1997: 931–949.
4. Broschi M, Strid Å. Molecular events following perception of ultraviolet-B radiation by plants. *Physiol. Plant.* 2003; 117: 1–10.
5. Biswal B, Biswal UC. Photosynthesis under stress: stress signals and adaptive response of chloroplasts. In: Pessarakli M, ed. *Handbook of Plant and Crop Stress*. New York: Marcel Dekker, Inc., 1999: 315–336.
6. Carpentier R. Effect of high-temperature stress on the photosynthetic apparatus. In: Pessarakli M, ed. *Handbook of Plant and Crop Stress*. New York: Marcel Dekker, Inc., 1999: 337–348.
7. Anderson JM, Chow WS, Park Y. The grand design of photosynthesis: acclimation of the photosynthetic apparatus to environmental cues. *Photosynth. Res.* 1995; 46: 129–139.
8. Barber J. Molecular basis of vulnerability of photosystem II to damage by light. *Aust. J. Plant Physiol.* 1995; 22: 201–208.
9. Krause GH. The role of oxygen in photoinhibiton of photosynthesis. In: Foyer CH, Mullineaux PM, eds. *Causes of Photooxidatative Stress and Amelioration of Defense Systems in Plants*. Boca Raton, Fl: CRC Press, 1994: 43–76.
10. De Las Rivas J, Andersson B, Barber J. Two sites of primary degradation of the D1 protein induced by acceptor or donor side photoinhibition in PSII core complexes. *FEBS Lett.* 1992; 301: 246–252.
11. Demmig-Adams B, Adams III WW, Ebbert V, Logan BA. Ecophysiology of the xanthophyll cycle. In: Frank HA, Young AJ, Britton G, Cogdell RJ, eds. *The Photochemistry of Carotenoids*. Dordrecht, The Netherlands: Kluwer Academic Publishers, 1999: 245–269.
12. Elrad D, Niyogi KK, Grossman AR. A major light harvesting polypeptide of photosystem II functions in thermal dissipation. *Plant Cell* 2002; 14: 1801–1816.
13. Nayak L, Raval MK, Biswal B, Biswal UC. Photoprotection of green leaves by zeaxanthin, a two channel process. *Curr. Sci.* 2001; 81: 1165–1166.
14. Nayak L, Raval MK, Biswal B, Biswal UC. Topology and photoprotective role of carotenoids in photosystem II of chloroplast: a hypothesis. *Photochem. Photobiol. Sci.* 2002; 1: 629–631.
15. Baroli I, Do AD, Yamane T, Niyogi KK. Zeaxanthin accumulation in the absence of a functional xantho-

16. Sonoike K. Photoinhibition of photosystem I: its physiological significance in the chilling sensitivity of plants. *Plant Cell Physiol.* 1996; 37: 239–247.
17. Nayak L, Biswal B, Ramaswamy NK, Iyer RK, Nair JS, Biswal UC. Ultraviolet-A induced changes in photosystem II of thylakoids: effects of senescence and high growth temperature. *J. Photochem. Photobiol. B* 2003; 70: 59–65.
18. Kulandaivelu G, Nedunchezhian N. Effect of UV-B enhanced radiation on growth and photosynthetic activities on higher plants and their defense mechanisms. In: Biswal UC, Britton G, eds. *Trends in Photosynthesis Research*. Bikaner, India: Agro Botanical Publishers, 1989: 215–229.
19. Biswal B, Kulandaivelu G. Responses of aging chloroplasts to UV radiation. In: Baltsheffsky M, ed. *Current Research in Photosynthesis*. Dordrecht, The Netherlands: Kluwer Academic Publishers, 1989: 813–816.
20. Joshi PN, Biswal B, Biswal UC. Effect of UV-A on aging of wheat leaves and role of phytochrome. *Environ. Exp. Bot.* 1991; 31: 267–276.
21. Joshi PN, Biswal B, Kulandaivelu G, Biswal UC. Response of senescing wheat leaves to ultraviolet A light: changes in energy transfer efficiency and PS II photochemistry. *Radiat. Environ. Biophys.* 1994; 33: 167–176.
22. Wilson MI, Greenberg BM. Protection of the D1 photosystem II reaction center protein from degradation in ultraviolet radiation following adaptation of *Brassica napus* L. to growth in ultraviolet-B. *Photochem. Photobiol.* 1993; 57: 556–563.
23. Friso G, Spetea C, Giocometti GM, Vass I, Barbato R. Degradation of photosystem II reaction center D1 protein induced by UV B radiation in isolated thylakoids. Identification and characterization of C- and N-terminal breakdown products. *Biochim. Biophys. Acta* 1994; 1184: 78–84.
24. Joshi PN, Ramaswamy NK, Raval MK, Desai TS, Nair PM, Biswal UC. Response of senescing leaves of wheat seedlings to UV A radiation; inhibition of PSII activity in light and darkness. *Environ. Exp. Bot.* 1997; 38: 237–242.
25. Beggs CJ, Wellmann E. Photocontrol of flavonoid biosynthesis. In: Kendrick RE, Kronenberg GHM, eds. *Photomorphogenesis in Plants*. Dordrecht, The Netherlands: Kluwer Academic Publishers, 1994: 733–751.
26. Teramura AH, Ziska LH. Ultraviolet-B radiation and photosynthesis. In: Baker NR, ed. *Photosynthesis and the Environment*. Dordrecht, The Netherlands: Kluwer Academic Publishers, 1996: 435–450.
27. Bray EA, Bailey-Serres J, Weretilynk E. Responses to abiotic stresses. In: Buchanan B, Gruessem W, Jones R, eds. *Biochemistry and Molecular Biology of Plants*. Rockville, MD: American Society of Plant Physiologists, 2000: 1158–1203.
28. Deo PM, Biswal B. Response of senescing cotyledons of clusterbean to water stress in moderate and low light: possible photoprotective role of β-carotene. *Physiol. Plant.* 2001; 112: 47–54.

29. Behera SK, Nayak L, Biswal B. Senescing leaves possess potential for stress adaptation: the developing leaves acclimated to high light exhibit increased tolerance to osmotic stress during senescence. J. *Plant Physiol.* 2003; 160: 125–131.

30. Sundari DS, Raghavendra AS. Sensitivity of photosynthesis by spinach chloroplast membranes to osmotic stress *in vitro* : Rapid inhibition of O_2 evolution in presence of magnesium. *Photosynth. Res.* 1990; 23: 325–330.

31. Sundari DS, Saradadevi K, Raghavendra AS. Suppression of oxygen evolving system in spinach chloroplast membranes due to release of manganese on exposure to osmotic stress *in vitro* in presence of magnesium. J. *Plant Biochem. Biotech.* 1994; 3: 137–140.

32. Graan T, Boyer JS. Very high CO_2 partially restores photosynthesis in sunflower at low water potentials. *Planta* 1990; 181: 378–384.

33. Kanechi M, Uchida N, Yasuda T, Yamaguchi T. Non-stomatal inhibition associated with inactivation of Rubisco in dehydrated coffee leaves under unshaded and shaded conditions. *Plant Cell Physiol.* 1996; 37: 455–460.

34. Bray EA. Plant responses to water deficit. *Trends Plant Sci.* 1997; 2: 48–54.

35. Krause GH. Photoinhibition induced by low temperatures. In: Baker NR, Bowyer JR, eds. *Photoinhibition of Photosynthesis. From Molecular Mechanism to the Field.* Oxford: Bios Scientific Publishers, 1994: 331–348.

36. Grafflage S, Krause GH. Alterations of properties of ribulose bisphosphate carboxylase related to cold acclimation. In: Lee PL, Christersson L, eds. *Advances in Cold Hardiness.* Boca Raton, FL: CRC Press, 1993: 113–124.

37. Oquist G. Environmental stress and photosynthesis. In: Biggins G, ed. *Progress in Photosynthesis Research.* Vol. IV. Dordrecht, The Netherlands: Martinus Nijhoff Publishers, 1987: 1–10.

38. Vierling E. The role of heat shock proteins in plants. *Annu. Rev. Plant Physiol. Plant Mol. Biol.* 1991; 42: 579–620.

39. Nishida I, Murata N. Chilling sensitivity in plants and cyanobacteria: the crucial contribution of membrane lipids. *Annu. Rev. Plant Physiol. Plant Mol. Biol.* 1996; 47: 541–568.

40. Jung SH, Lee JY, Lee DH. Use of SAGE technology to reveal changes in gene expression in *Arabidopsis* leaves undergoing cold stress. *Plant Mol. Biol.* 2003; 52: 553–567.

41. Arora R, Pitchay DS, Bearce BC. Water stress induced heat tolerance in geranium leaf tissues; a possible linkage through stress proteins? *Physiol. Plant.* 1998; 103: 24–34.

42. Ladjal M, Epron D, Ducrey M. Effects of drought preconditioning on thermotolerance of photosystem II and susceptibility of photosynthesis to heat stress in cedar seedlings. *Tree Physiol.* 2000; 20: 1235–1241.

43. Foyer CH, Lelandais M, Kunert KJ. Photooxidative stress in plants. *Physiol. Plant.* 1994; 92: 696–717.

44. Eggert A, Van Hasselt PR, Breeman AM. Chilling-induced photoinhibition in nine isolates of *Valonia utricularis* (Chlorophyta) from different climate regions. J. *Plant Physiol.* 2003; 160: 881–891.

45. Shapira M, Lers A, Heifetz PB, Irihimovitz V, Osmond CB, Gillham NW, Boynton JE. Differential regulation of chloroplast gene expression in *Chlamydomonas reinhardtii* during photoacclimation: light stress transiently suppresses synthesis of Rubisco LSU protein while enhancing synthesis of the PS II D1 protein. *Plant Mol. Biol.* 1997; 33: 1001–1011.

46. Potter E, Kloppstech K. Effects of light stress on the expression of early light — inducible proteins in barley. *Eur. J. Biochem.* 1993; 214: 779–786.

47. Marsac NT, Houmard J. Adaptation of cyanobacteria to environmental stimuli: new steps towards molecular mechanisms. *FEMS Microbiol. Rev.* 1993; 104: 119–190.

48. Biswal B, Smith AJ, Rogers LJ. Changes in carotenoids but not in D1 protein in response to nitrogen depletion and recovery in a cyanobacterium. *FEMS Microbiol. Lett.* 1994; 116: 341–348.

49. Foyer CH, Lopez-Delgado H, Dat JF, Scott IM. Hydrogen peroxide and glutathione associated mechanisms of acclimatory stress tolerance and signaling. *Physiol. Plant.* 1997; 100: 241–254.

50. Mahalingam R, Fedoroff N. Stress response, cell death and signaling: the many faces of reactive oxygen species. *Physiol. Plant.* 2003; 119: 56–68.

51. Murata N, Ishizaki-Nishizawa O, Higashi S, Hayashi H, Tasaka Y, Nishida I. Genetically engineered alteration in the chilling sensitivity of plants. *Nature* 1992; 356: 710–713.

52. Hayashi H, Alia, Mustardy L, Deshnium P, Ida M, Murata N. Transformation of *Arabidopsis thaliana* with the *Cod A* gene for choline oxidase; accumulation of glycinebetaine and enhanced tolerance to salt and cold stress. *Plant J.* 1997; 12: 133–142.

39 Salt and Drought Stress Effects on Photosynthesis

Enzyme Cohesion and High Turnover Metabolite Shuttling, Essential for Functioning of Pathways, Is Impaired by Changes in Cytosolic Water Potential

B. Huchzermeyer
Plant Developmental Biology and Bioenergetics Laboratory,
Department of Plant Sciences, University of Hannover

H. W. Koyro
Institute of Plant Ecology, Justus-Liebig University

CONTENTS

I. INTRODUCTION

Salinity is one of the most serious factors that limits agricultural productivity [1]. Salinity occurs not only in natural habitats, at the coastline for instance, but is also a result of inadequate irrigation strategies on about 20% of the farmland in arid and semiarid areas all over the world. Under natural conditions, multiple environmental stresses co-occur frequently. It has been reported that the responses of plants to several simultaneous stresses are usually not predictable by single-factor analysis and a combination of different environmental stress factors can result in intensification, overlapping or antagonistic effects [2].

There are large differences in stress responses between species. For example, apple trees will die when permanently exposed to soil salinities higher than $50\,mM$ NaCl, while a 50% yield reduction will occur for beans at $60\,mM$, and for sugar beet 50% yield reduction will not occur before a NaCl concentration of $260\,mM$ NaCl is attained [3].

Different experimental approaches are in use to analyze drought and salt stress effects on plants [4]. As pointed out very clearly in the reviews of Lawlor [5,6], there is a general way of drought and salt stress response common to all plants. Differences in stress response among plant species is due to their special capacities in metabolic pathways, rather than completely different strategies. In this review, redundance

with respect to earlier work is due to the fact that we have to summarize main observations made with different plants to point out the general strategies. The current experimental data show that any strategy helping to stabilize at cellular level water potential and keeping adverse ions out of the cytosol, will contribute to a better plant performance under stress.

II. GAS EXCHANGE

Measuring leaf gas exchange during illumination and dark incubation is among the classic physiological methods. This technique allows to predict stress effects on plant growth rate as well as on crop yield. But it does not provide any direct proof, which biochemical reaction is the primary target of the stress.

A. SHORT- AND LONG-TERM RESPONSE

When analyzing salt and drought stress effects, the situation is complicated by the fact that short- and long-term stress responses have been observed. Sugar release into the cytosol, for instance, is a short-term stress response found in many plants, while cytosolic sugar concentrations tend to normalize during stress adaptation. The extent of sugar release as well as the period of time during which enhanced sugar concentrations can be detected varies with both, the responsiveness of the individual plant and the extent of stress applied. Therefore, short- and long-term effects are not easy to be discriminated and in most experiments an overlap of effects is observed. Interactions between leaf area and growth rate are examples of such multiple effects on plant performance. Photon receipt has not only an influence on growth but also on leaf morphology and CO_2 fixation capacity [7]. For example, in radish (*Raphanus sativus*) it has been observed that about 80% of the growth reduction at high salinity could be attributed to reduction of leaf area expansion and hence to reduction of light capture [8]. The small leaf area at high salinity was related to an increase in tuber/shoot weight ratio. The latter could be attributed to tuber formation starting at a smaller plant size at high salinity. The remaining 20% of the salinity effect on growth was most likely explained by a decrease in stomatal conductance [8].

B. Genetic Variation, Environmental, and Developmental Effects on Stress Perception and Response

Another problem is due to genetic variation of salt and drought stress tolerance observed with most plants [9]. Depending on plant species, susceptibility to salt and drought stress may vary with the developmental stage of an individual plant as well as environmental factors [10]. However, to develop cultivars with drought and salt tolerance it is essential to understand tolerance mechanisms and their direct or indirect effects on development, yield and quality of crops [11,12].

In experiments with rice it has been observed that stomatal conductance is interactively affected by salinity and atmospheric drought, thus explaining some varietal differences in sodium uptake. Sodium distribution within the plant is not homogeneous. The leaf sheaths have the highest, and the youngest leaf blades the lowest sodium concentrations. These differences can be explained with differences in leaf sheath retention of sodium, and current concentrations reflect the transpiration history of the individual leaf blade [13].

C. Limitation of Assimilation Rate by Low Internal Leaf CO_2 Concentration

In many natural locations, the shortage of water is an important environmental constant limiting plant productivity [14]. The plant reacts to water deficit with a closure of stomata to avoid further loss of water through transpiration [15]. As a consequence, the diffusion of CO_2 into the leaf is restricted [14]. The decrease in net CO_2 fixation under drought stress observed in many studies is explained by a lowered

internal CO_2 concentration that results in a limitation of photosynthesis [16].

There are several reports which underline the stomatal limitation of photosynthesis under drought stress as a primal event, which is then followed by adequate changes of photosynthetic reactions [17–20]. Assimilation rate in leaves decreases with decreasing relative water content. This observation is independent of the type of stress, drought or salt stress, applied. But, in more detailed investigations it has been found that stomatal conductance is not affected by a reduction of relative water content from 100% to 90% while a reduction of assimilation rate can be observed. This result shows that there are effects other than CO_2 limitation controlling photosynthesis as well. An overview on salt and drought stress induced response reactions is given in Figure 39.1.

CO_2 concentrations have been measured in the air, inside the leaf apoplast, and inside the chloroplasts: as a rule of thumb, under conditions of maximal stomatal conductance (relative leaf water contents between 90% and 100%) the CO_2 concentrations inside the apoplast and inside the plastids were 70% and 50% of the atmospheric CO_2 concentrations, respectively [2]. In more detailed investigations it was observed under saturating light intensity that the inhibited CO_2-dependent O_2 evolution and the net CO_2 assimilation induced by drought stress can be recovered by high external CO_2 concentration [21], implying that the perturbations of the biochemical processes are not responsible for the inhibited CO_2 assimilation, and stomata instead may play a dominant role in the decreased CO_2 assimilation under water stress. The conclusion can be further supported by the fact that photosystem II (PSII) photochemistry is hardly affected by water stress [17,21]. From these observations it can be concluded that assimilation rate is limited by CO_2 supply and the term assimilation potential, i.e., maximal assimilation rate under optimal CO_2 supply, has been introduced [6,22].

D. Limitation of Assimilation Rate by Factors Other Than Low Internal Leaf CO_2 Concentration

Nonstomatal limitation of photosynthesis under drought and salt stress, in sunflower leaves, for instance, has been attributed to inhibited coupling factor activity [23], reduced carboxylation efficiency [24], reduced RuBP regeneration [25], or to reduced amount of functional Rubisco [26].

Under conditions of high light absorption, an inhibition of $NADP^+$ regeneration in the Calvin cycle can result in an over-reduction of the

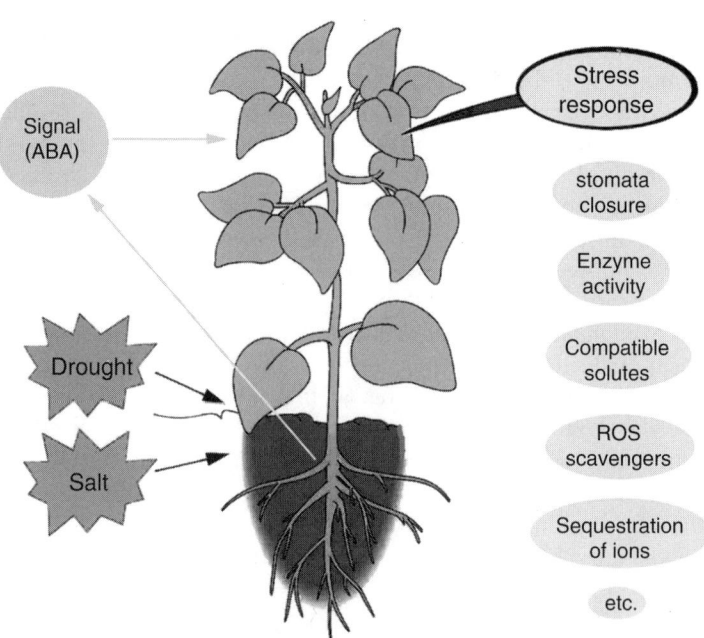

FIGURE 39.1 Salt- and drought-stress-induced response reactions. The roots are the primary plant organs of stress perception. Abscisic acid (ABA) is the most important signal molecule released from the roots in response to this type of stress. In the leaves ABA can initiate a variety of stress response reactions. The type of stress response depends on cell differentiation and metabolic status as well as ABA concentration, and the duration and extent of stress. ABA-induced stomata closure is among the macroscopic stress response reactions. This way gas exchange and respiration are regulated. But there are further stress effects affecting photosynthesis by factors other than limitation of CO_2 supply. Some of them, which are discussed in this chapter, are listed here.

photosynthetic electron transport chain and damage to the photosynthetic apparatus. There are several mechanisms known to protect plants against photodamage, including the emission of surplus radiant energy as heat in parallel to decreased PSII activity [27] and the use of electron acceptors instead of CO_2, such as oxygen in photorespiration or Mehler reaction [28] (Figure 39.2). Thermal dissipation of absorbed light energy, which can be determined by nonphotochemical fluorescence quenching in relation to CO_2 gas exchange, is well established in studies with excessive irradiance and reduced CO_2 [19].

Drought also leads to an increase in nonphotochemical quenching in leaves [29,30]. One factor contributing to nonphotochemical quenching is ΔpH-dependent high energy quenching, which is correlated with a downregulation of PSII activity [31]. When comparing drought and salt stress effects on stomatal conductance as well as on assimilation rate, it was observed that stomata are closed below a critical value of water potential, stomatal conductance adopted a minimum, but assimilation rate still was reduced with further reduction of relative water content. From such results it may be concluded that stomatal conductance and, hence, CO_2 supply is not the only factor controlling the assimilation rate. In many experiments the contribution of stomatal effects and the inhibition of other reactions to the total effect of salt and drought stress response have been investigated. One such approach is to compare plant performance under normal and enhanced CO_2 supply, respectively. Based on such data Lawlor defined two

types of plant stress response, he named them *type 1* and *type 2* [22].

In type 1 response low stomatal conductance causes apoplastic CO_2 concentration to fall and elevated apoplastic CO_2 concentration restores assimilation rate to assimilation potential. Only after substantial loss of relative water content is metabolism and thus assimilation potential impaired. Sunflowers and spinach [32] are good examples for plants showing a type 1 response. While assimilation rate and stomatal conductance of intact leaves decreased with water deficit, maximum quantum efficiency of PSII and relative quantum yield of PSII did not change significantly in stressed leaves. Under such experimental conditions photosynthesis obviously is limited by stomatal factors.

In type 2 response, apoplastic CO_2 concentration also decreases with initial loss of relative water content but assimilation potential and metabolism are progressively impaired as relative water content falls, and elevated apoplastic CO_2 concentration cannot fully restore assimilation potential.

Depending on the plant species, reduction of assimilation potential can be observed when leaf relative water content is reduced below a critical value: With some species reduction of assimilation potential is not detectable unless relative water content becomes reduced by more than 10% while others show such an effect very early, i.e., with minor reductions of leaf water potential. Moreover, under severe stress, with low relative water content assimilation rate cannot be significantly stimulated by increased CO_2 supply any more: actual assimilation rate and assimi-

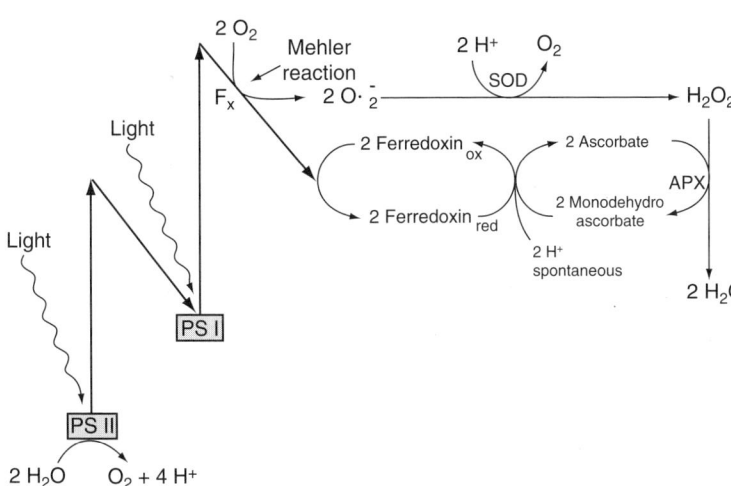

FIGURE 39.2 The Mehler reaction is an electron sink competing with ferredoxin for electrons from PSI. It is generally accepted that photosynthetic electron transport leads from the water splitting system of PSII to ferredoxin, which distributes electrons to the final sinks (CO_2 fixation, nitrate and sulfate reduction, for instance). The production of reactive oxygen species (ROS) by the Mehler reaction apparently shows a lower affinity to electrons as compared to the pathway, leading to ferredoxin. This is due to its longer half-life time (lower turnover number). But, if the ferredoxin pool becomes over-reduced because of inhibited CO_2 fixation rate or high light intensity, the percentage of electrons quenched by the Mehler reaction increases and ROS scavenging reactions have to safe chloroplasts from damage by ROS attack.

lation potential become identical. With values of relative leaf water content below 50%, no assimilation is observed any more in most plants.

From the abovementioned observations it becomes obvious that the contributions of CO_2 limitation and "additional effects" brought about by reduction of leaf water content on assimilation rate are not easy to be discriminated. Moreover, plant species apparently differ in sensitivity to reduction of relative leaf water content of these "additional effects." With respect to their stress sensitivity, light capture, photosynthetic electron transport, ATP synthesis, and CO_2 assimilation have been extensively investigated both *in vivo* and *in vitro*. But, in experimental investigations each of these reactions has been analyzed separately, and this isolated treatment of parts of the total photosynthetic reaction chain holds true for many reviews as well.

III. LIGHT CAPTURE

Light capture and photosynthetic electron transport occurs inside the chloroplast thylakoids. Pigments as well as redox systems mostly are bound to protein complexes integral to the thylakoid membrane. Light-harvesting complexes, PSII, the cytochrome $b_6 f$-complex photosystem I (PSI), and the NADP reductase have been found to directly interact [33]. Partners become lined up in direct neighborhood according to their function. The direct contact of reaction partners is essential for optimal functioning of the electron transport system, because several reaction partners (not only pigments and the water splitting system but also tyrorsine, phaeophytin and quinones, for instance) show very short half-lives of their acti-

vated states [34,35]. Subsequent reactions have to occur within even shorter periods of time in order to allow high yields of light capture. Such fast reaction sequences call for direct neighborhood of reaction partners, they cannot be achieved if diffusion of reaction partners is involved [36–38]. Moreover, a limitation of turnover as it may occur under stress, apparently results in a stimulated production of reactive oxygen species.

From *in vitro* experiments with both, isolated intact chloroplasts and thylakoid membranes, it may be concluded that light harvesting as well as photosynthetic electron transport are not affected by the enhanced cellular salt concentrations and lowering water potential to values above 50% of the control, respectively [39,40]. The functioning of the water splitting complex leading evolution of O_2 is not impaired [17,21,41]. Maintenance of high rates of O_2 evolution in stressed leaf samples in the oxygen electrode at 5% CO_2 is prime evidence that assimilation potential is unimpaired when assimilation rate (measured as net CO_2 assimilation) is inhibited and cannot be stimulated by elevated CO_2 [22].

On the other hand, several *in vivo* studies apparently demonstrated that water stress resulted in damage to the oxygen-evolving complex of PSII [42] and to the PSII reaction centers [43]. Other studies have shown that the inhibited CO_2-dependent O_2 evolution and the net CO_2 assimilation induced by water stress can be recovered by high external CO_2 concentration [21], implying that the perturbations of the biochemical processes are not responsible for the inhibited CO_2 assimilation [21,42,44], and stomata instead may play a dominant role in the decreased CO_2

assimilation under water stress [45,46]. This discrepancy calls for more detailed analysis.

A. Stress Resistance Observed under Low Light Conditions

Data from most publications agree with the expectation that primary reactions of photosynthesis are highly resistant to salt and drought stress [16], although there may be some damage at unphysiological low relative water content [47]. These latter observations may be explained in accordance with findings from experiments with cyanobacteria: In *Synechocystis* sp. strong light induces photodamage to PSII, whereas salt stress inhibits the repair of photodamaged PSII and has no directly damaging effect [48]. Because of this synergetic effect, the joint occurrence of both stresses results in a very rapid inactivation of PSII.

As it will be an important point for our discussion, we want to point out clearly that stress tolerance at low light intensities and enhanced sensitivity to drought and salt stress in the presence of high light has been reported from experiments with several plants. For instance, the maximal efficiency of tomato PSII measured in the dark is not affected by drought; however, in the light O_2 evolution decreases under water deficit [14]. Under drought stress, light capture in wheat was observed to be affected in high light, whereas no inhibition could be observed in low light [2].

In agreement with the assumption of salt resistant primary reactions, recent studies have shown that there were no changes in PSII photochemistry in salt-stressed cowpea, wheat, sorghum, and barley when they were grown under relative low light conditions [49–52], suggesting that salt stress per se has no effect on PSII photochemistry and that it is the interaction between high light and other environmental stresses which result in damage to PSII [49,51,53]. Indeed, some studies have shown that salt stress and low cytosolic water potential showed no effects on PSII photochemistry if stress treatments were carried out under relative low light but predisposed photoinhibition when salt-stressed plants were exposed to high light [5,6]. These data correlate with observations made with plant cell cultures during freezing for conservation purpose: High light intensities during freezing and thawing, when water potential is reduced in the presence of ice, reduce cell yield and dead cells show symptoms of peroxidation. To our knowledge, such results can be explained by deleterious effects of reactive oxygen species produced in the light under conditions impairing photosynthesis.

B. Induction of Photoinhibition

Another explanation is provided by authors, who discuss the observed effects as indicating downregulation of PSII to match reduced CO_2 fixation rate [42,54,55]. If the demand for products of photosynthetic electron transport is lower than its rate of production, then the imbalance can lead to overexcitation within the reaction centers and, in some circumstances, these can become inactivated. External factors leading to such a downregulation of photosynthesis and photoinhibition of electron transport rate are: low temperatures, drought stress, surplus of nutrients, or overexcitation of the reaction centers by high light [7]. Feedback inhibition by overproduction of photosynthesate has also been shown to cause photoinhibition of photosynthesis [7].

Photoinhibition of photosynthesis originally has been defined by the decline in net photosynthesis that occurs when leaves are exposed to high light for several hours [7]. The extent of photoinhibition is dependent on photon flux density, cytosolic (plastidic) water potential, and temperature [7]. Correlated with these changes in photon yield and net photosynthesis are changes in the efficiency of PSII photochemistry, as measured by the fluorescence ratio. Thermal dissipation of absorbed light energy, which can be determined by nonphotochemical fluorescence quenching in relation to CO_2 gas exchange, is well established in studies with excessive irradiance and reduced CO_2, and salt and drought stress [7,29,30].

C. Electron Consumption by Reactions Other Than CO_2 Fixation

Reduction of CO_2 to carbohydrates by the Calvin cycle consumes by far the largest proportion of electrons derived from water oxidation. In annual crops, the reduction of nitrate consumes the second biggest part. At high relative water content photorespiration will use only a small proportion of the electrons and close to none will be consumed by the Mehler ascorbate peroxidase reaction [56,57]. In the context of our discussion, it has to be stressed that nitrate uptake as well as reduction are strictly regulated by the amount of C-skeletons available. Therefore, nitrate assimilation will consume a constant proportion of electrons as compared to carboxylation. However, when assimilation rate is decreased, as at low relative water content, the electron transport chain becomes strongly reduced and electron transfer to O_2 increases, producing reactive oxygen species [58,59], which are very damaging unless alternative pathways for removal are available.

IV. CALVIN CYCLE

A. METABOLIC CAUSES OF DECREASED ASSIMILATION RATE

What metabolic reactions may be affected by reduction of relative leaf water content and impaired ion homeostasis? We will focus here on sites at which photosynthesis has been discussed in the literature to be impaired: (1) Rubisco concentration and enzyme activity; (2) regeneration of RuBP by the Calvin cycle; (3) supply of ATP and NADPH to the Calvin cycle; and (4) export and use of assimilation products as well as the involvement of ATP and assimilates in stress response.

B. LIMITATION BY RUBISCO

Rubisco is the most abundant protein contributing 50% of the total leaf protein. It has been discussed that some portion of the total Rubisco protein is not catalytical active but functions as a CO_2 buffer. Different functions of Rubisco, CO_2 binding, and CO_2 turnover, agree with the observation that for full catalytic activity in the light, Rubisco needs to be activated. Activation of Rubisco in the light requires Rubisco activase and ATP [60,61]. Inhibitors are generally analogues of the enzyme's substrate ribulose-1,5-bis phosphate (RuBP) [62]. They bind to the enzyme in the absence of RuBP. Activase releases tight-binding inhibitors from the Rubisco active sites, thus increasing specific activity. The reaction requires ATP [60], so decreased activity and activation state of Rubisco at low relative water content may be related to inadequate ATP supply to the protein complex [23,63].

The amount of Rubisco protein is generally little affected by moderate or severe salt and drought stress [64], even if experienced over a period of many days [23,44,46,65]. This means that specific Rubisco activity rather than protein concentration is decreased under drought and salt stress. Restoration of assimilation potential to values measured with control plants by rehydration also suggests that Rubisco (and, of course, other potential limitations) is not impaired irreversibly [63,65].

C. RuBP CONCENTRATIONS AND RUBISCO ACTIVITY

The rate of photosynthesis depends on synthesis of RuBP and activity of Rubisco. Therefore, the decrease in RuBP content of leaves at low relative water content [23,25,46] is significant. In stressed sunflower assimilation rate correlated with RuBP concentration [25]. This suggests that assimilation potential in these experiments was determined by RuBP content, not by CO_2.

Under salt and drought stress, there is a general decline in Calvin cycle intermediates during the phase of plant adaptation to the applied stress. Decreased RuBP concentration might be caused by the general rundown of the Calvin cycle and decrease in assimilation rate. The large ratio of 3PGA/RuBP suggests limitation in the ru-bisphosphate regeneration part of the Calvin cycle, either caused by enzyme limitations or inadequate ATP, although Tezara et al. [23] interpreted it as evidence of 3PGA production by mitochondria. In laboratory experiments with salt-tolerant plants (*Aster tripolium*, for instance), however, metabolite pattern recovered within less than 2 to 3 days under constant stress conditions.

Interaction between Rubisco activity and RuBP supply is well illustrated by studies on unstressed tobacco leaves with normal amounts of Rubisco [66,67]. The RuBP pool increased in leaves in bright light when the CO_2 concentration was decreased transiently, so that on return to normal CO_2 concentrations, assimilation rate was much greater than the steady-state rate for a short time, until the RuBP was consumed. Thus, under steady-state conditions, RuBP supply limited the rate of CO_2 assimilation.

Gunasekera and Berkowitz [46] also identified RuBP synthesis as a limitation but assumed that ATP was not limiting, so, accordingto them, the cause was inhibition of enzyme activity. Decrease in assimilation potential with stress was explained by loss of Calvin cycle enzyme activity in early studies [68]. The large 3PGA/RuBP ratio suggests that Rubisco is not limiting, otherwise the ratio should decrease. If the decrease in assimilation rate were due only to CO_2 depletion, then RuBP content should increase, as observed by von Caemmerer and Edmondson [69] at low CO_2 and saturating irradiance.

Interpretation of the available data is complicated by two aspects: (i) the duration of stress might affect results and interpretation, e.g., the decrease in ATP and RuBP observed by Tezara et al. [23] on leaves held at low relative water content for some days, may be caused by enhanced transport activities during stress adaptation on the expense of ATP. However, the decrease in ATP shown by Lawlor and Khanna-Chopra [70] was a rapid response to decreased relative water content. (ii) ATP concentrations were not clearly assigned in terms of compartmentation in these measurements.

Experiments of the above described type, but using plant species known to grow under severe salt and drought stress, currently are under investigation in several laboratories. They will help to identify strategies to cope with these adverse condition.

D. CALVIN CYCLE ENZYMES

Reduction of RuBP content at low relative water content could result from a limitation in one or more enzymes of the Calvin cycle. There is little direct evidence regarding the response of the individual enzymes of the regenerative part of the Calvin cycle to increasing cytosolic mineral content, subsequent to drought or increased salinity. In very preliminary experiments with leaf crude extracts, we did not find differences between enzyme preparations from glycophytes and halophytes, respectively [71–73]. Another point is that in most cases isolated enzymes did not show a degree of salt inhibition, sufficient to explain the extent of inhibition of the respective metabolic reaction in *in vivo* experiments. Therefore, we doubt that high resistance to salt stress of halophytes is due to changed catalytic properties of their Calvin cycle enzymes. In the literature some enzymes have been discussed to be affected by salt or drought stress, but in most cases only preparations from a single plant species have been analyzed. Here we will commend on some such examples.

The inhibition of photosynthetic electron transport and the activity of Calvin cycle enzymes under water stress have been studied in *Casuarina equisetifolia*: Total NADP-dependent malate dehydrogenase activity increased and total stromal fructose-1,6-bisphosphatase activity decreased under drought, while the activation state of these enzymes remained unchanged [74]. Water stress did not alter the activity and the activation state of Rubisco.

Thus, Sharkey and Seeman [44] concluded that low CO_2, and not enzymes, decreased the Calvin cycle activity. In contrast, Gunasekera and Berkowitz [46] concluded that the Calvin cycle activity limited assimilation potential. Similarly, decreased ATP with low relative water content observed in some studies [23,70] suggests that the Calvin cycle is not substantially impaired: if low apoplastic CO_2 concentration were the cause and no other ATP sink is activated under stress, then ATP content should rise. This conclusion may be supported by results from experiments with tobacco with antisense to phosphoribulose kinase (PRK) [75]: a 95% decrease in PRK substantially decreased assimilation rate and RuBP content but increased ATP content. Thus, there is strong evidence that the Calvin cycle per se is not the cause of decreased assimilation rate with low relative water content.

E. LIMITATION OF RuBP REGENERATION BY NADPH AND ATP AVAILABILITY

Synthesis of RuBP depends on ATP and NADPH concentration, and on the Calvin cycle activity, or more specifically on PRK activity, and concentration of the substrates ATP and ribose 5-phosphate [76]. In the Calvin cycle, NADPH is the substrate of the glyceraldehyde 3-phosphate dehydrogenase. If NADPH were limiting at low relative water content then it would decrease glyceraldehyde 3-phosphate and thus ribose 5-phosphate, the same effect as decreasing the activity of an enzyme in that portion of the cycle. In leaf mesophyll cells under salt and drought stress, it was found that NADPH content remained relatively constant [23,70] and NADH increased [70] as relative water content decreased, indicating that the electron transport capacity is sufficient to maintain and increase the reduction state of these pyridine nucleotides. Thus, with respect to high salinity tolerance of photosynthetic electron transport system in both glycophytes (pea, spinach, barley, and maize) and halophytes (*Aster tripolium* and *Beta maritima*), it is unlikely that availability of NADPH to the Calvin cycle limits its capacity to form RuBP [71,77].

The rate of ATP synthesis depends on the light reactions, generation of the *trans*-thylakoid pH gradient (ΔpH), availability of ADP and Pi, and activity of the chloroplast coupling factor (CF_1) [78–83]. Inadequate ATP concentration would decrease the Calvin cycle's ability to regenerate RuBP by PRK, so glyceraldehyde-3-phosphate would increase and RuBP decrease. This was the effect of decreased PRK activity [75], but ATP increased in the transgenics, in sharp contrast to the case of water stress in some studies [5,6,22].

Loss of photophosphorylation capacity with decreased relative water content was observed by Meyer and de Kouchkovsky and Meyer et al. [84,85]. This result may be explained by the finding that the amount of CF_1 decreased in stressed leaves [23]. Further studies concluded that limitation of assimilation potential was not caused by loss of ATP [24,45,86]. The effects of leaf water potential on photophosphorylation were measured *in vivo* with a spectrophotometer, which detects the rapid relaxation of the 518 nm electrochromic signal from carotenoids, located in thylakoids. The signal changes as the pH of the thylakoid lumen alters. Wise et al. [24] did not detect inhibition of ATP synthesis in watered and wilted field-grown sunflower. However, assimilation rate decreased substantially during the mid-afternoon when apoplastic CO_2 concentration increased, i.e., there was nonstomatal inhibition. Therefore, Wise et al.

[24] concluded that assimilation rate was downregulated as an adaptive response to drought.

F. TRIOSE EXPORT

In light, photosynthesate is exported from the chloroplasts into the cytosol as triose-phosphates as indicated in Figure 39.3. In the cytosol, sucrose, and other oligosaccharides are formed to be transported to the plant sink organs. (The oligosaccharide patterns found in the phloem sap are characteristic of each plant family.) Transport of triose-phosphate out of the chloroplast, via the phosphate translocator in counterexchange for Pi, is the principal route for Pi transport [87,88]. Therefore, triose-phosphate export and inorganic phosphate import are directly linked and affecting one of these metabolite pools will interfere with trans-membrane transport rates of the respective other species [89].

The chloroplast coupling factor has a low affinity towards inorganic phosphate with a K_m value in the range of 150 μM [79]. For maximal catalytic activity, phosphate concentration in the chloroplast stroma should exceed 2 to 5 mM in the presence of more than 1 mM Mg^{2+} [79,82,90]. Therefore, ATP synthe-

sis rate *in vivo* is highly sensitive to inadequate Pi supply [22]. In summary, as ATP synthesis is crucial for the functioning of the Calvin cycle, a strict dependence of CO_2 assimilation on chloroplast Pi supply via the phosphate translocator can be postulated. But only little data on salt and drought stress effects on phosphate translocator activity are available.

The concentrations of phosphorylated Calvin cycle intermediates in the chloroplast stroma have been reported to fall under stress, internal Pi pool is not reduced this way [44]. Analysis of Pi in chloroplasts from leaves over a range of relative water potential in relation to assimilation rate, phosphorylated intermediates, etc., is required. But there is no standard isolation protocol for intact chloroplasts applicable for different plant species. Moreover, recovery of intact organelles from one species varies with differentiation stage of leaves as well as stages of stress adaptation [91].

V. DISSIPATION OF SURPLUS ENERGY

A. PHOTORESPIRATION

Photorespiration plays a key role in consumption of reduced ferredoxin, when the O_2/CO_2 ratio inside the

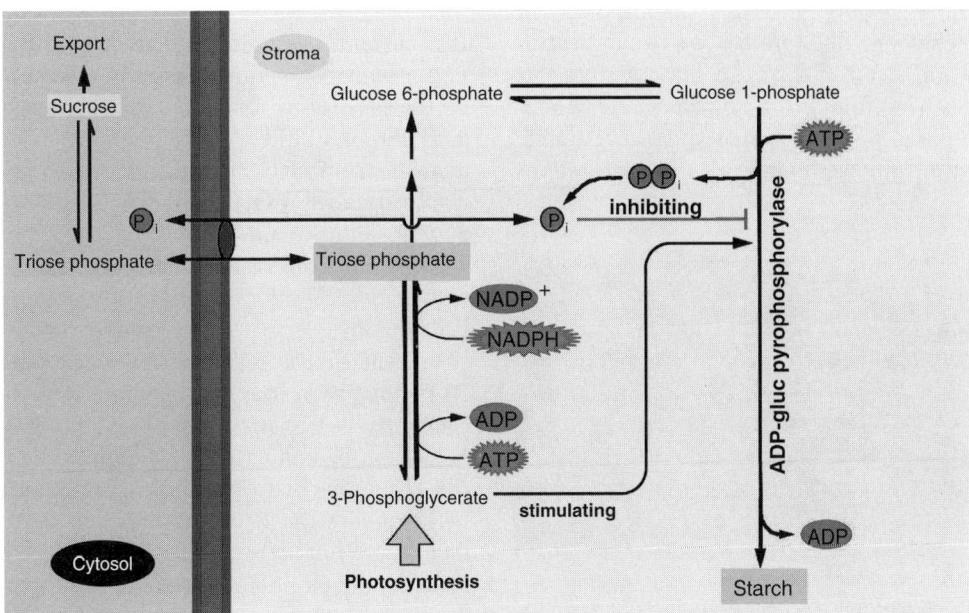

FIGURE 39.3 The phosphate–triose-phosphate counterexchange via the triose-phosphate translocator helps to keep phosphate homeostasis of the chloroplast matrix. The low affinity towards phosphate ($K_m = 150 \mu M$) of the F-type ATPase makes photophosphorylation, and this way the Calvin cycle, highly sensitive to a reduction of chloroplast phosphate content. The triose-phosphate translocator has a high affinity towards dihydroxy-acetone phosphate, but the V_{max} is low as compared to the V_{max} of CO_2 fixation. On the other hand, the K_m values of the enzymes of starch synthesis are generally low, but their V_{max} values exceed those of the Calvin cycle enzymes. This way, there is a constant export of triose-phosphates while, at high light intensities, surplus triose-phosphates are consumed by chloroplast starch production. During starch production as well as sucrose synthesis, liberated phosphate becomes available for ATP synthesis again.

leaves increases under drought [54,67,92,93]. Rubisco oxygenase activity catalyzes the reaction of RuBP with O_2, resulting in 3-PGA and phosphoglycolate synthesis. Metabolism of phosphoglycolate via the glycolate pathway, involves three cell compartments, chloroplasts, peroxisomes, and mitochondria. Functioning of this pathway obviously requires well-defined subcellular structures and aggregates of the three compartments are observed (Figure 39.4).

Decreasing relative water content has long been known to increase the ratio of photorespiration to assimilation [98,99]. Results clearly prove that the CO_2 concentration inside the chloroplasts is crucial for regulation, because in contrast to an increase in the O_2 to CO_2 ratio, a limitation in RuBP supply would reduce both, assimilation and photorespiration rates. However, it was quite a surprise when from analysis of O_2 exchange it became obvious that photorespiration is a substantial sink for electrons in leaves at high and low relative water content [14,58,59,100]. But, while a significant increase of the photorespiration/assimilation ratio was found [101], the absolute rate of photorespiration did not compensate for stress induced limitation of assimilation rate, and total consumption of reductants was inhibited as compared to control plants [58]. Therefore, electron transport rate was impaired under stress. Using chlorophyll fluorescence as an indicator, photorespiration electron fluxes in salt- or drought-stressed leaves were found to be much smaller under stress as compared to control plants. As an estimate photorespiration plus assimilation rates account for

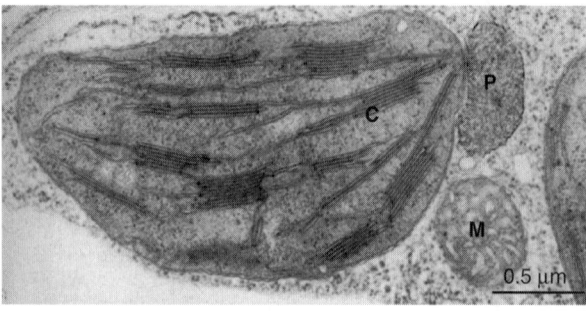

FIGURE 39.4 Electron micrograph of a chloroplast–mitochondrium–peroxisome aggregate from a mature barley leaf cell. Aggregates of cell organells are rare in young developing cells. It has been observed that compartmentation of metabolic pathway changes with organelle aggregate formation [94–96]. It was convincingly shown that these changes depend on direct metabolite shuttling among aggregated cell organelles [97]. Moreover, stability of aggregates apparently depends on cytosolic salt and osmoprotectant contents.

45% of energy consumption at large relative water content and 20% at small, and the proportion decrease under large photon flux [20].

B. MITOCHONDRIAL RESPIRATION

In leaves, mitochondrial respiration contributes to CO_2 release not only in the dark but also in the light. Mathematical simulation of the C fluxes in illuminated leaves [102] showed that with a small increase in stress, Rubisco characteristics could simulate the changes in CO_2 turnover. But at low relative water content a source of CO_2 additional to photorespiration is required. The contribution of mitochondrial respiration agrees with the increased apoplastic CO_2 concentration and equilibrium compensation points observed at low relative water content [23,98,103].

These results may be interpreted in terms of the need of additional energy supply (to drive active membrane transport, for instance) under severe salt and drought stress. Moreover, these data point out that gas exchange measurements call for very careful interpretation in drought- and salt-stress experiments, because there is no constant ratio of photorespiration and tricarbon acid cycle activities.

C. PRODUCTION OF REACTIVE OXYGEN SPECIES

It is currently assumed that the negative effect of the various environmental stresses is at least partially due to the generation of reactive oxygen species (ROS) and the inhibition of the system which defends against them. ROS include superoxide (O_2^-), hydroxyl radicals ($^{\cdot}OH^-$), hydrogen peroxide (H_2O_2), and singlet oxygen ($^1O_2$).

The functions of ROS and other radicals, NO, for instance, are under intensive investigation, since it became obvious that they are involved in signaling in plant cells alike in cells of animals and fungi [104,105]. ROS have been found to regulate growth and development as well as stress responses [106–110]. However, H_2O_2 can be converted via Fenton-type reactions to the dangerous hydroxyl radical, and failure to quench or inactivate the ROS may lead to degeneration of membrane lipids, proteins, and DNA [111,112]. Enzymes coding for defense systems are apparently under the control of promoters activated by ROS or their products, lipid peroxides, for instance.

D. ENZYMATIC DETOXIFICATION OF ROS

Apart from the xanthophyll cycle, photorespiration and other changes in metabolic activity, which may protect the chloroplast from oxidative damage [54,93,113,114], a number of enzymatic and nonenzymatic antioxidants are present in chloroplasts that control oxygen toxicity [106,112,115]. Catalase (CAT) and ascorbate peroxidase (APX) detoxify hydrogen peroxide and yield water and oxygen. The expression of both enzymes seems to be induced by oxidative products [116,117]. Glutathione reductase (GR) acts by recycling oxidized glutathione using NADPH as a cofactor. Many studies have reported changes in antioxidant enzyme activities in response to salinity, suggesting that the increases in these activities can be the basis for salt-stress tolerance [108,118].

More recently, a third type of plastidic thioredoxin has been described designated chloroplastic drought-induced stress protein of 32 kDa (CDSP32) [119,120]. The protein is composed of two typical thioredoxin modules, with only one redox active disulfide center in the C-terminal domain [121], and has been found to participate in the protection of the photosynthetic apparatus against oxidative damage [122]. CDSP32 is located in the stroma and displays a substantially increased abundance under severe osmotic and photooxidative stresses [120,121,123–125]. The *CDSP32* gene is induced at the transcript level under drought and photo-oxidative treatments [120,121]. But, the exact function of the CDSP32 protein is yet to be determined.

E. Scavengers Protecting from ROS Effects

Recently, it has been proposed that flavonoids and polyphenolic compounds act as reducing agents either as enzyme cofactors or as electron donors. Accumulation of these agents could be pronounced in tissues under stress conditions. Carotenoids, α-tocopherol, ascorbate, and glutathione (Figure 39.5) help to maintain the integrity of the photosynthetic membranes under oxidative stress [112,126–129]. Besides, some *Labiatae* plants, including rosemary (*Rosmarinus officinalis*) and sage (*Salvia officinalis*), contain the diterpene carnosic acid (CA), which displays high antioxidant properties *in vitro* [130–132], have shown that CA is present in chloroplasts, where it is oxidized to rosmanol and isorosmanol [133]. From a comparison of *in vivo* and *in vitro* effects during drought stress, Shalata and Tal [107] concluded that CA in combination with other low molecular weight antioxidants helps prevent oxidative damage under stress.

F. Experimental Approaches to Measure ROS Stress Status

Estimations of the redox state of low molecular weight antioxidants may allow us to better under-

stand the relationship between drought and oxidative stress in plants. Although levels of antioxidants indicate the potential extent of antioxidative protection and the balance between their synthesis, oxidation, and regeneration, their redox state indicates an oxidative load toward these compounds and provides us with a reliable estimation of the oxidative stress in the cell. Changes in the redox state of ascorbate and glutathione [134,135] and in that of CA [136] have been studied in drought-stressed plants. By contrast, to our knowledge, drought-induced changes in the oxidation products of α-tocopherol or carotenoids, and therefore, in their redox states, have not been reported so far in plants.

G. Physiological Importance of ROS Production as an Electron Sink

In order to generally prove the permanent production of ROS in illuminated chloroplasts, mass-spectrometric measurements of isolated chloroplasts, mesophyll cells, or whole leaves have been performed. This way, O_2 uptake has been observed in the light, which could not be accounted for by Rubisco oxygenation or mitochondrial respiration [54,137]. On the other hand, combined measurements of leaf gas exchange and chlorophyll fluorescence *in vivo* have not always found evidence of significant extra electron transport [17,138].

It is thought that O_2 photoreduction increases with increasing reduction of the ferredoxin pool, thus allowing linear electron flow to continue when NADP is scare. It has been suggested that O_2 photoreduction can assist in maintaining a high *trans*-thylakoid pH gradient, which in turn, enhances nonradiative dissipation of light energy and protects light reactions from photodamage [139].

In this context it has to be borne in mind that the slight alkalization of the chloroplast stroma in the light is an essential prerequisite for functioning of most plastidic metabolic pathways. Phosphorylation potential of ATP, like group transfer potential of many other metabolites, depends on pH (and cation concentrations, especially Mg^{2+}) and reaction equilibria within pathways would not allow metabolism to take place in other cell compartments because of their lower pH values as compared to illuminated chloroplasts. In addition, the stoichiometries of NADPH and ATP production and consumption in chloroplasts are different and the alternative sinks are considered necessary to enable ATP production without NADP reduction [127].

A great number of experiments using a variety of plant species indicate that these findings, in general, hold true for all plants. For instance, drought and salt

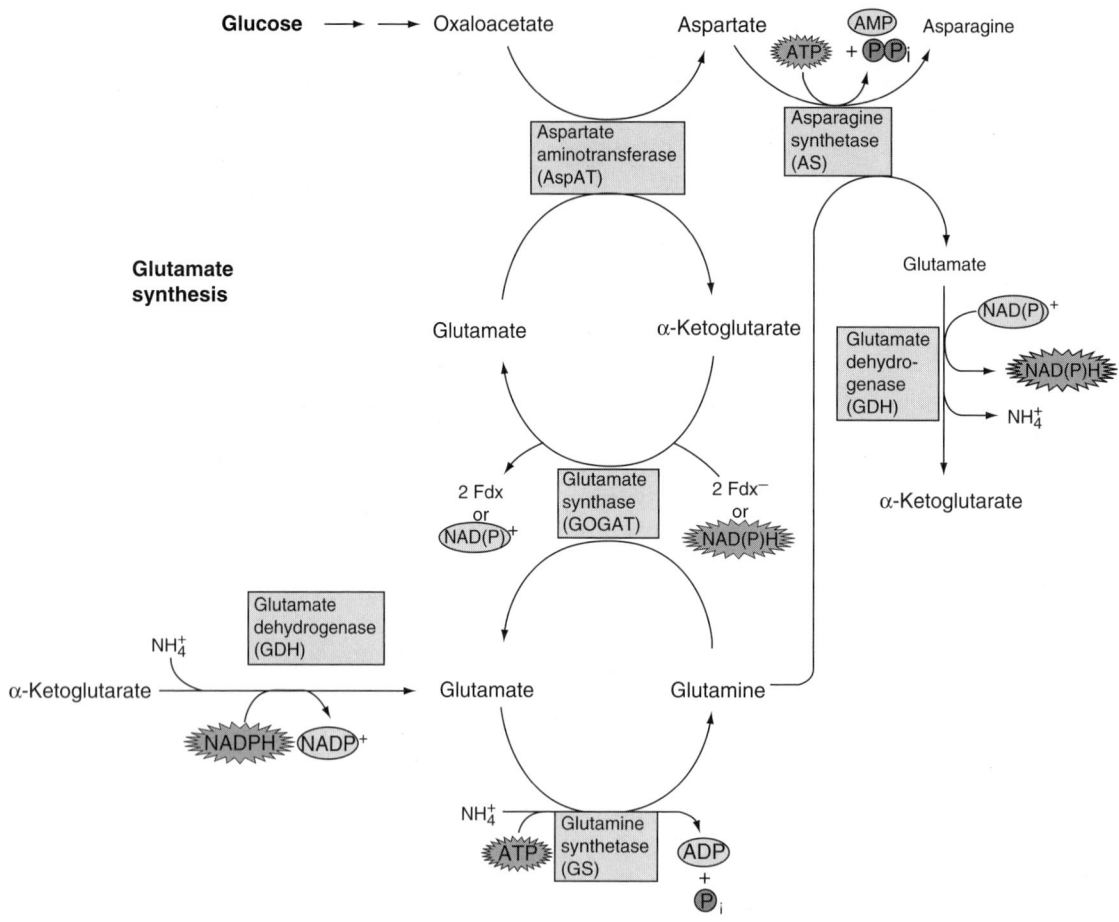

FIGURE 39.5 ROS scavenging reactions protect from damages and can act as electron sinks. ROS synthesis by the Mehler reaction, for instance, occurs when turnover of intermediate pools is inhibited. This may occur under stress, when the export or consumption of photosynthesates is blocked, and the acceptors of the electron transport become over-reduced. As shown for the example of the ascorbate–glutathione cycle, ROS scavengers have a twofold effect: (i) they protect the cell by ROS consumption; and (ii) NADH is used for their regeneration. This way, they act as electron sinks and contribute to the reduction of ROS formation.

stress alter the amounts and the activities of enzymes involved in scavenging oxygen radicals and their corresponding steady-state level of mRNA. Activities of cytosolic and chloroplastic Cu/Zn-SOD isozymes and cytosolic ascorbate peroxidase, as well as their corresponding mRNA transcripts, were increased by drought treatment of pea plants [140]. In tomato cultivars differing in salt resistance, a correlation between growth performance and antioxidant content was found [107]. Moreover, it was observed that salt-stress-mediated peroxidation in tomato leaves did not occur in the dark but during subsequent illumination of stressed plants. Additional ascorbic acid partially inhibited this response but did not significantly reduce sodium uptake or plasma membrane leakiness. Other organic solutes without known antioxidant activity

were not effective [108]. Water-stress-induced ABA accumulation triggers the increased generation of ROS and upregulates the activities of antioxidant enzymnes in maize leaves. Rates of production of ROS with decreasing relative water concentration have not been quantified, but because of their reactivity and potential for damaging thylakoids, etc., even small fluxes may be critical and rapid removal is essential [5]. It is thought that low molecular weight antioxidants cooperate to provide protection against oxidative damage to plants [132].

Data from other plant systems indicate that the stress induced accumulation of abscisic acid (ABA) promotes stomatal closure which decreases CO_2 intake. This leads to a loss in photosynthetic efficiency and oxidative stress [118].

VI. OSMOPROTECTANTS AND N METABOLISM

Biochemical studies have shown that plants under salinity stress accumulate a number of metabolites, which are termed osmoprotectants or compatible solutes because they do not interfere with biochemical reactions [141,142]. These metabolites include polyalcohols, such as mannitol, sugars like glucose, sucrose, trehalose, raffinose, and other oligosaccharides, as well as nitrogen-containing compounds, such as amino acids, betaines, and polyamines.

It has been observed that polyalcohols are accumulated under stress in some *rosaceae*, which may be salt-sensitive like apple trees, for instance. On the other hand, oligosaccharides are found to accumulate in the more salt-tolerant *Leguminosae*. From such differences no ranking in terms of protection effectiveness of osmoprotectants may be deduced. This example rather indicates that under stress plants will respond within their genetically determined spectrum and take advantage of metabolic capacity they have. As an example, the metabolic pathways leading to the synthesis of mannitol and proline are shown in Figure 39.6. It is obvious that accumulation of osmoprotectants can be interpreted as a side effect of enzyme and translocator regulation. Plants differ in metabolism as well as in the type of carbohydrates used for far distance transport. Osmoprotectants may become concentrated in the cytosol if their export is inhibited by external stress factors. Therefore, the positive action of osmoprotectants may be a side effect of photosynthesate storage in source tissues.

A. FUNCTION OF OSMOPROTECTANTS

The function of compatible solute accumulation is often associated with osmotic adjustment, by lowering the water potential to improve the uptake of water against the external gradient. But in most cases it has been observed that osmoprotectants account for less than 50% of changes of cytosolic water potential under stress [143]. Therefore, a number of other roles for these compounds have been hypothesized in the literature [115]. Possible roles include: serving as a readily available energy source or as a nitrogen source during limited growth and photosynthesis, detoxification of excess ammonia under periods of stress, and stabilization of enzymes and/or membranes [22]. We will comment on this in Section X.B in more detail.

Detailed understanding of the biochemical pathways leading to the formation of amino acids and compatible solutes will be important in engineering plants to tolerant saline environments in the future. Moreover, it has to be analysed prior to genetic manipulation that the wild type has the metabolic capacity to produce, at a sufficient rate, precursors of the desired osmoprotectant. Until now, most of the studies of compatible solute accumulation have focused on the photosynthetic tissues, or source tissues. Here, we will discuss some results observed with mannitol and proline, for example, and then focus on N metabolism.

B. MANNITOL

The osmolyte mannitol is synthesized in numerous plant species (Figure 39.6c), but not in wheat (*Triticum aestivum*). But, expression of the *mtlD* gene in transgenic wheat plants has demonstrated that cellular accumulation of mannitol can alleviate drought and salinity stress in this crop as well [144]. In celery (*Apium graveolens*), mannitol is synthesized in equal proportion to that of sucrose [145] and, therefore, celery has been chosen to study stress effects on mannitol synthesis and transport. Salt stress inhibits sucrose synthesis but does not affect the enzymes for mannitol biosynthesis. Moreover, the rate of mannitol use in sink tissues decreases during salt stress mainly because of the suppression of the NAD^+-dependent mannitol dehydrogenase, which oxidizes mannitol to mannose [146,147]. Mannitol accumulation increases when plants are exposed to low water potential irrespective of whether this is brought about by drought or salt treatment [148]. Studies using transgenic tobacco (*Nicotiana tabacum*) and *Arabidopsis* also showed improved growth of mannitol-accumulating plants under stress [149–151]. A more detailed analysis of data indicated that improved growth performance of mannitol-accumulating wheat calli and mature leaves was due to stress-protective functions of mannitol, although osmotic effects on growing regions of wheat could not been ruled out. Similar effects have been observed with *Arabidopsis* and tobacco [149–151].

C. PROLINE

Proline accumulation in response to stress is widely reported. Other nitrogenous compounds that accumulate in response to stress may have important roles in tolerance, but have received little attention compared to proline accumulation [152]. The pattern of solute accumulation during extended exposure to stress has not been widely studied. It is important that pulse chase experiments clearly showed that proline as well as other aminoacids and amides accumulating under salt and drought stress are not degraded from

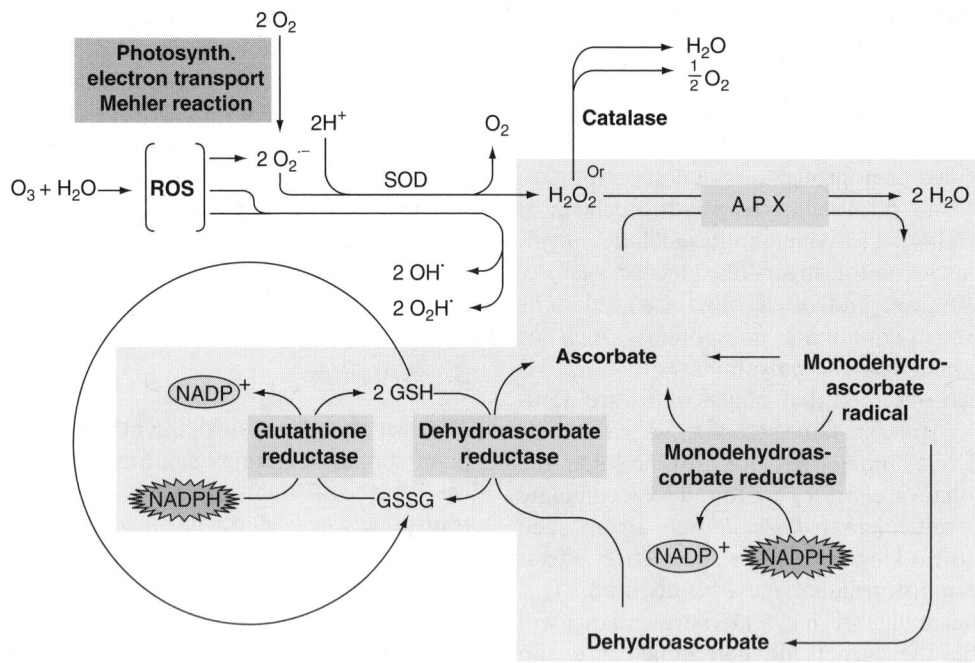

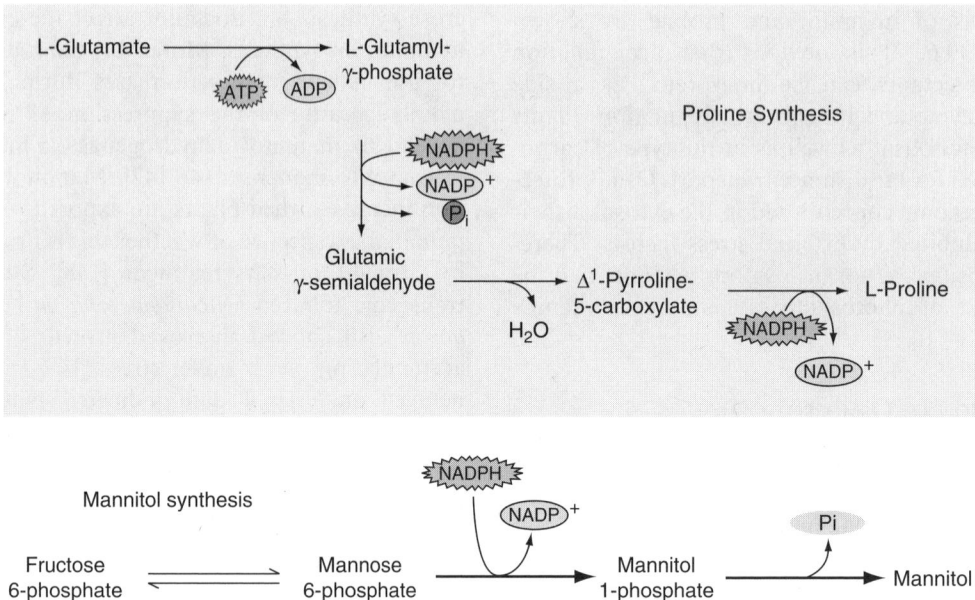

FIGURE 39.6 (a) Glutamate biosynthesis during photosynthesis. Glutamate is a substrate for proline biosynthesis. Here we show the cytosolic pathway leading to the production of glutamate in leave cells. This consumption of photosynthesates is an important sink under stress, when sugar export is inhibited. (b) Proline biosynthesis. Proline production is stimulated when cytosolic glutamate concentration is enhanced by inhibited sugar export to sink organs. This way, salt and drought stress induce the production of an important compatible solute which helps to reduce toxic effects of incoming salt. (c) Mannitol biosynthesis. Glucose is the substrate of the pathway leading to mannitol biosynthesis. Mannitol, like sorbitol and some oligosaccharides, has been found in phloem sap and therefore is understood to be a substrate for photosynthesate export to sink organs. Mannitol export is inhibited under stress, resulting in enhanced cytosolic mannitol concentrations. Therefore, the protective action of compatible solutes may be a secondary but very useful effect.

proteins but are synthesized de novo upon stress [152].

Proline concentrations increase by several orders of magnitude at low relative water content, where assimilation potential approaches zero but respiration continues. Proline is derived from glutamate (which accumulates and is not derived from recently formed carbohydrates); the reactions require ATP and NADPH [153]. Synthesis is also possible from ornithine, although it may be a minor pathway: it does not require ATP or NADPH. If redox components are very reduced at low relative water content, this may enhance or even trigger proline formation from glutamate [153,154]. This may be an important sink for redox equivalents and ATP under salt and drought stress [155]. Advantages of increased cytosolic proline concentrations are documented for bacteria and for plant membranes in vitro, for instance, and increased proline accumulation in genetically altered plants confers osmotolerance [154] although the mechanisms have been disputed.

Another aspect becomes obvious from investigations on the regulatory capacity of intermediates of proline synthesis in rice (Oryza sativa L.). It was demonstrated that intermediates of proline biosynthesis are stimulating the expression of genes characteristic for osmotic stress. Among the genes under metabolite control are: the gene for hsp70, the gene for S-adenosyl-methionine synthase, salT, Em, and dhn4. A concentration of $1 mM$ proline or $75 mM$ NaCl resulted in a minor activation of these genes, while substrates of proline synthesis, glutamine and Δ-pyrroline-5-carboxylic acid (PC5), as well as its analog 3,4-dehydroproline (3,4dPro), significantly increased the expression of these genes. Unlike NaCl, gene induction by these intermediates of proline synthesis was not paralleled by an increased level of ABA. Plants treated with PC5 or 3,4dPro accumulated osmolytes typical for osmotic stress, had increased cytosolic levels of NADH, while NADPH levels were reduced, and O_2 consumption was lower as compared to control plants [156].

D.   NITRATE REDUCTION

The effects of stress on plant nitrogen metabolism has been frequently studied, with increases in protein degradation, inhibition of protein synthesis and the accumulation or depletion of protein and nonprotein amino acids reported in a variety of monocts and dicots [152]. If plants are capable of adapting to the degree of stress, the accumulation of low molecular weight N-containing compounds is observed. This effect will peak within a few days of exposure to salinity, for instance, and then decline, but remain elevated. In general, several different compounds will occur with a few dominant ones. The interactions between CO_2 and NO_3^- assimilation and their dynamics are of key importance for crop production. An adequate supply of NO_3^- stimulates leaf growth and photosynthesis [22].

Nitrate reduction is considered the rate-limiting and regulatory step in the nitrogen assimilation pathway, and is mainly related to the activity of the cytosolic nitrate reductase (cNR, EC 1.6.6.1) [157,158] although several reports indicate that a distinct proportion of the total NR may be associated either with various organelle membranes [159,160] or with plasma membrane [161–165].

Water deficits substantially alter all aspects of nitrogen assimilation. Indeed, accumulation of amino acids is characteristic of low relative water content [166,167], particularly proline increasing greatly below a well-defined threshold due to increased rates of synthesis relative to consumption [168]. In crops, this threshold appears to be at or below 75% relative water potential, and there is no evidence of correlation between the changing response of assimilation potential and amino acid accumulation. Changes in amino acid content are loosely associated with very small assimilation rate (and assimilation potential) and with rising internal CO_2 concentration and compensation point, suggesting that there is a relationship between altered CO_2 assimilation and cell energetics.

Regulation of NR amount is complex, with interactions between nitrate, sucrose, organic acids, and amino acids determining the transcription and translation of the gene and protein activation: NR is rapidly turned over. Carbon metabolism is intimately linked with NO_3^- assimilation, and increasing sucrose and glucose concentrations stimulate NR-gene transcription and accumulation of transcripts [169] and activation.

VII.   MEMBRANES AND LIPIDS

Maintenance of membrane integrity and the selective uptake of essential minerals as well as ion compartmentation are some of the parameters that were previously related with salt tolerance acquisition [170]. Membrane lipids play a fundamental role in regulating ion movement through cell membranes since they are key determinants in the control of fluidity and the environment surrounding proteins, which influences bilayer permeability, carrier-mediated transport and the activity of membrane-bound enzymes (see Ref. [171] and references therein).

In addition to changes in membrane fluidity, lipids may modify the activity of membrane enzymes by changing their substrate affinity, activation energy, and turnover number [172]. In common with other membrane-bound enzymes, the plant plasma membrane proton-pumping *p-type* ATPase (EC 3.6.1.35), which produces an electrogenic proton gradient across this membrane, has an absolute requirement for lipids [173]. Tolerance to NaCl induces changes in plasma membrane lipid composition, fluidity, and H^+-ATPase activity of tomato calli. Plasma membrane phospholipids and sterols, lysophosphatidylcholine, phosphatidylinositol mono- and biphosphate, and free fatty acids have been shown to activate plasma membrane H^+-ATPase in several plant species [171].

VIII. CYTOSOLIC ATP

A. MITOCHONDRIAL RESPIRATION

The mitochondrial electron transport chain accepts electrons from the TCA cycle and exogenous NADH, with O_2 as the terminal acceptor. This is coupled to ATP synthesis by means of a transmembrane electric potential [90,174]. Permanent supply of redox equivalents to the active respiratory chains is essential, because intermediates, especially at the $1e^-/2e^-$ steps, are unstable and radicals might be formed and cause damages. Electron supply has to occur within the half-life time of redox intermediates. This demand of mitochondrial respiratory chains is met by a rapid exchange of reducing equivalents between the different forms and pools. The relatively large photorespiratory flux will maintain a large mitochondrial $NADH/NAD^+$ ratio and the malate/oxalacetate shuttle may maintain the large tissue NADH concentration while the NADPH content in chloroplasts is kept relatively constant at low relative water content [70]. The magnitude of mitochondrial respiration is generally small when compared to assimilation rate at moderate light intensities in unstressed leaves [98,100].

B. ATP SYNTHESIS IN MITOCHONDRIA

Regulation of ATP synthesis in the mitochondria is complex [175] and there is little understanding of what might occur at low relative water content. ATP from mitochondria contributes to the ATP/ADP ratio in the cytosol, but there is no significant transfer of ATP from mitochondria to chloroplasts. Therefore, mitochondria do not directly contribute to the energy status of chloroplasts.

Obviously there is considerable difference among plant species in maintenance of energy status under stress. ATP concentration decreased in osmotically stressed mesophyll cells of *Xanthium* [176]. Sharkey and Seeman [44] observed no differences in mildly stressed leaves of bean. The decrease in ATP concentration observed was interpreted as an indicator of an inhibition of ATP synthesis [23,70]. The possibility of increased ATP consumption by active transport, for instance, or a feedback inhibition of electron transport, were not considered by these authors. The cytosolic ATP pool will be reduced if V-type and P-type ATPase activities are stimulated under stress.

ATP content decreased in stressed wheat [70] and in sunflower [23]. But ATP was still present at the smallest relative water content. In plant species with limited control of phosphorylation potential, the decrease in ATP content of leaves occurs with relatively small loss of relative water content [47], although ATP content is not reduced to zero even at very low relative water content when assimilation rate has virtually stopped [23]. The correlation between decreasing leaf ATP content and decreasing relative water content is strong and appears to be linear [23], suggesting increased consumption of ATP during the adaptation phase to salt and drought stress as well as at reduction of relative water content below a value not tolerable by the plant.

IX. STRESS EFFECTS ON ENZYMES

The adaptation to saline stress is accompanied by alterations in the levels of numerous metabolites, proteins and mRNAs [177]. Various genes whose expression is activated in response to salt stress have been identified [178–180]. While some of these genes encode for protective proteins such as the osmotin [181], aquaporins [182], late embryogenesis abundant (LEA) proteins [183], and ion transporters [184], others code for enzymes that participate in metabolic processes specifically triggered by saline stress [185–187]. Because high salinity conditions promote plant cell dehydration [188], many genes activated by saline stress are also activated by drought. The expression of most of these genes is regulated by ABA, a hormone produced in response to both saline and drought stresses [188–190].

A microarray study with salt-stressed rice showed that the initial differences between control and stressed plants continued for hours but became less pronounced as the stressed plants adapted over time. Moreover, an analysis of salt stress response of salt-sensitive rice showed that in sensitive plants the immediate response exhibited by salt-tolerant plants was delayed and later resulted in downregulation of transcription and death [178].

A. UPREGULATED ENZYMES

Accumulation of dehydrins, for example, is considered to protect metabolic functions at low relative water content [191], as judged by enhanced plant performance under stress conditions. Aquaporins are proteins integral to cell membranes which facilitate diffusion of water (and also other small molecules) into the cell. Their functions are regulated by metabolic processes, such as reversible phosphorylation, which depend on apoplastic water potential [192]. Heat-shock proteins are synthesized and accumulate in very dehydrated tissues; they are molecular chaperones [193,194]. In wheat, salt stress results in an upregulation of the gene locus Kna1, which plays a key role in the control of K^+/Na^+ discrimination [195]. Some proteins induced by low relative water content are also induced by ABA [196]. This applies, for instance, to RAB17 which may be part of nuclear protein transport, and ASR(s714), which may function in the maintenance of nucleic acid structure. At moderate drought and salt stress, enzymes involved in primary metabolism are induced: triose-phosphate isomerase [197], enolase, and NAD-malate dehydrogenase, which are involved in glycolysis, TCA cycle, and the oxidative pentose phosphate pathway. Riccardi et al. [198] point out the importance of maintaining cellular homeostasis, emphasizing the role of such proteins in energy production.

B. DOWNREGULATED ENZYMES

Protein synthesis is generally substantially inhibited and a decrease in polyribosome content can be observed to correlate with the degree of this inhibition [182,199], but the relative water content at which inhibition occurs varies with plant species as well as its developmental stage.

Loss of sucrose phosphate synthase activity at low relative water content was among the observed effects. Native sucrose phosphate synthase is a relatively unstable dimeric protein; perhaps the changed cellular environment, e.g., increased ionic concentration, causes instability or increased amino acids inhibit activity [200]. Also, altered protein turnover might be responsible for loss of activity in the long term [201]. Similar considerations apply to loss of nitrate reductase activity in stressed leaves [202], and the loss of PSII proteins. Salt stress inhibits de novo synthesis of PSII proteins, especially the transcription and translation of the *psbA* genes which code for the D1 protein [48].

C. PROTEIN STRUCTURE

Production of correctly structured proteins and their maintenance is clearly of great importance for efficient cellular function. The concept that low relative water content impairs protein structure explains the importance of molecular chaperones, which accumulate under a range of stresses, including drought. Chaperones have an important role in the folding and assembly of proteins during synthesis, and in the removal and disposal of nonfunctional and damaged proteins.

D. V-ATPASE

Under conditions allowing crops to survive and finish their life cycle, the cytosolic Na^+ concentration in both halophytes and nonhalophytes does not exceed about $150\,mM$, because otherwise a variety of metabolic reactions would be inhibited [203–205]. Membrane-bound transport systems regulating cytosolic ion homeostasis (Na^+, K^+, Ca^{2+}) and ion accumulation in the vacuole can be considered of crucial importance for adaptation to saline conditions [206,207]. The capacity of Na^+ transport from the cytoplasm into the vacuole via the tonoplast Na^+/H^+ antiport [208] is dependent on the activity of V-ATPase (EC 3.6.1.35) and V-PPase (EC 3.6.1.1), which establish an electrochemical H^+-gradient across the tonoplast that energizes the transport of Na^+ against the concentration gradient [209–212]. The V-ATPase is the dominant H^+-pump, both in terms of protein amount and activity [213].

Under stress conditions, survival depends strongly on maintaining and adjusting of the V-ATPase activity. Regulation of gene expression and activity are involved in adapting the V-ATPase on long- and short-term bases. In general, in halophytes salinity leads to an increase in V-ATPase activity, while it does not significantly respond or decreases in many nonhalophytes. This general outline can be specified by some examples from plants differing in their specific stress responses.

The effects of salinity on tonoplast proteins have been intensively studied in the halophyte and salt-includer *Mesembryanthemum crystallinum*. Salt-tolerant cell lines from this plant showed a salt-induced increase in Na^+ transport and V-ATPase activities, while osmotic (drought) stress did not induce any changes in ATPase activity [214]. V-ATPase amounts did not change in response to salinity stress in juvenile plants that are not salt tolerant. In halotolerant mature plants the transcript levels increase in leaves of *Mesembryanthemum* [215]. Similar observations have been made in tomato plants, where the mRNA coding for the catalytic 70 kDa V-ATPase subunit was found to accumulate in leaf mesophyll cells after 24 h of tolerable salt stress [216]. This was a transient stress response that leveled off with adaptation after about 1 week of incubation. In *Mesembryanthemum* as well as in tomato salt stress induced accumulation of v-ATPase was found to be tissue specific: accumulation occurred in leaves but not in the roots [215–217].

These results suggest that increased expression of the tonoplast ATPase is an early response to salinity stress and may be associated with survival mechanisms, rather than with long-term adaptive processes. From experiments with other plant species it became obvious that V-ATPase and V-PPase genes, though the enzymes have comparable function, are not under the control of the same promoter: according to our information, the increase in V-ATPase activity is not obtained by structural changes of the enzyme, but by an increase in V-ATPase protein amount.

Previous studies demonstrate a positive correlation between increased plant size, Na^+ accumulation, and increased activity of the vacuolar H^+-ATPase as a function of Na^+ concentration during growth [218]. Increased vacuolar Na^+/H^+ exchange activity was identified. It is sodium concentration dependent, specific for Na^+ and Li^+, sensitive to methyl-isobutyl amiloride, and independent of an electric potential. The affinity of the transporter for Na^+ is almost three times higher in plants grown in high levels of salt than for control plants ($K_m = 3.8$ and $11.5\,mM$).

E. P-ATPASE

The control of Na and Cl accumulation involves not only intracellular processes, such as vacuolar compartmentation, but also integration of uptake and distribution between specific cells and tissues throughout the plant [216]. The importance of plasma membranes in plant salt tolerance has been suggested from studies on both glycophytes and halophytes [219]. The regulation of ions across the plasma membrane is thought to be achieved by an electrochemical gradient generated by the plasma membrane p-type H^+-ATPase [209].

The plasma membrane H^+ pump in plant cells plays central roles in a wide spectrum of physiological processes through the generation of a H^+ electrochemical potential gradient across the plasma membrane by the active transport of H^+ from the cytoplasm to the outside [220]. The activity of the plasma membrane H^+ pump is regulated by various external and internal factors such as phytohormones, fungal toxins, elicitors, and light [220].

In contrast to the mRNA coding for the catalytic subunit of the V-ATPase, which accumulates in leave cells exclusively, NaCl-induced accumulation of the plasma membrane P-ATPase message occurred in both roots and expanded leaves [216]. In many cases, an initial accumulation of the plasma membrane ATPase message is greater in root tissues than in expanded leaves, but increased to higher levels in the expanded leaves after several days [216].

Although Na^+ exclusion is typically correlated with increased salt tolerance in glycophytes [216], even the most resistant individuals exhibit increases in net transport rates and internal salt concentrations [216] and thus are highly dependent upon processes integral to ion regulation. Consequently, transport processes associated with plant response to salinity, such as Na^+/H^+ exchange [221], may increase the demand for the generation and maintenance of H^+-electrochemical potential gradient.

In plants, Na^+/H^+ exchangers in the plasma membrane are critical for growth in high levels of salt, removing toxic Na^+ from the cytoplasm by transport out of the cell. The molecular identity of a plasma membrane Na^+/H^+ exchanger in *Arabidopsis* (SOS1) has been determined. SOS1 is localized to the plasma membrane of leaves and roots [222]. The apparent K_m of the transporter for Na^+ is $22.8\,mM$. One Na^+ is exchanged for one H^+. The Na^+/H^+ exchange is unaffected by the presence of a membrane potential.

In extreme halophytes, many different strategies, each of them contributing to salt tolerance, are active in a concerted and well controlled way. For example, the salinity tolerance mechanisms of *Spartina* species include cellular, organizational, and whole plant adaptations, such as ion compartmentation, presence of salt glands on the leaves, ion exclusion at the root, and ion partitioning in different organs [219]. Control of ion movement across the tonoplast and plasma membranes in order to maintain a low Na^+ concentration in the cytoplasm is the key cellular factor in salinity tolerance.

X. A GENERAL MECHANISM UNDERLYING STRESS EFFECTS

It was recognized quite early that plants do not have salt-tolerant metabolism even if they, halophytes for instance, thrive in saline soil or even sea water [200,205,223]. Amino acid patterns of enzymes from glycophytes and halophytes did not differ significantly [224]. Moreover, enzymes from plant primary metabolism of halophytes, apparently salt tolerant *in vivo*, were inhibited by salt treatment of the isolated enzyme [204,223,224]. From these observations it was concluded that salt and drought tolerance means: tolerant species are capable of keeping a defined microenvironment inside the cytoplasm, which meets the requirements of their enzymes. We have outlined above that primary events of photosynthesis, the Calvin cycle, and export of photosynthesate have to be tightly linked to allow maximal turnover rates, and the rates of energy dissipation by photosynthesate consumption have to exactly match those of excitation of the photosynthetic apparatus. Absorption of light at levels in excess to the requirements for photosynthesis can be potentially damaging due to the formation of ROS. Inhibition of photosynthesis by drought and salt stress in bright sunlight is among such situations. Therefore, mild salt and drought stress can be overcome *in vitro* by stimulating photosynthesis by increased CO_2 concentrations [22] as outlined in the first part of this chapter.

When comparing data from glycophytic crops to those of laboratory experiments with halophytes, it becomes obvious that the control of the enzyme's microenvironments brought about by compatible solutes is most important for a plant's salt and drought tolerance. The results presented in the second part of this chapter indicate that plant performance under stress is improved by any effect supporting substrate shuttling among metabolic pathways. The occurrence of substrate leakage, overproduction of intermediates, and synthesis of products, like ROS, which are eventually toxic to the cells, will be limited, if there is a close neighborhood or a direct contact between enzyme and cell compartments, as shown in Figure 37.4, for example.

A. PROTEIN HYDRATION

Sap isolated from plant tissues is not an ideal Van't Hoff solute. Proteins and other organic molecules have a pronounced water binding capacity, therefore their osmotic pressure Π significantly differs from the value calculated by the Van't Hoff equation $\Pi = c_m RT M_w^{-1}$ (c_m is the concentration of mol-equivalents, the molal concentration, R is the gas constant, T is the temperature given in Kelvin, and M_w is the molecular mass). Generally, proteins will bind under physiological conditions 0.3 mg of water per mg [225]. As extensively described during the analysis of F-type ATPase function, water may bind in an exchangeable way to the protein surface, can be found in "pockets" or binding sites accessible only to small (substrate) molecules, or binds in inaccessible way inside the protein structure [90,226,227].

Cell structure and enzyme activity depend on aqueous environments. Generally, substrates and products of metabolism are dissolved in water and water is a product of many reactions. Moreover, hydratation of proteins, forming bonds to ligands and filling space are among the most important functions of water [228]. But, the functions of water are not easy to be analysed, because water activity cannot

be increased but only decreased, and any change will bring about changes in viscosity, dielectric constant, pH value, etc. [229].

In many cases, it has been observed that membrane spanning transporters are inhibited by high as well as low water potential. This has been explained by the assumption that there are at least two rate-limiting steps in a complete reaction cycle of such transporters. At least one of these involves water release and binding, respectively [229]. If water binding and release is part of the reaction cycle, altered water activity will affect steady-state distribution of intermediates, and there should be a value of water activity optimal for turnover. This assumption is supported by the detection of water pockets in membrane integral protein complexes known to depend on optimal ionic strength [78,90].

B. FUNCTION OF COMPATIBLE SOLUTES

Both adverse conditions, drought and salt stress, will reduce relative water content (water activity) inside leaf cells. As outlined below, this will affect the stability of enzyme–enzyme, enzyme–membrane, and organelle–organelle aggregates [200,230]. In several publications, stimulation of chaperone expression is described as an early stress response [5]. The function of chaperones is to keep enzymes and membranes in a functionally intact structure. The problem obviously is that protein hydratation is impaired under stress and compensation for reduced availability of water molecules is needed.

Plants have to adapt to their environment and it is observed that some specialists can survive under extreme conditions. With respect to water activity plants can be found in the range of 1.00 (pure water) to 0.750 (salt lakes). Most plants live at relative constant water activity, while plants living in tidal zones have to be capable of accepting extreme variations from 0.980 (sea water) to 1.00 (after rain fall) to 0.75 in full sunshine at low tide. Such highly tolerant plants have developed different strategies to handle surplus salt. For instance, they produce substances helping to overcome adverse salt effects. Such compounds are called compatible solutes or osmoprotectants. They are organic compounds (sugars, polyalcohols, amino acids, polyamines, betaines, etc.). Different compatible solutes are found in different plants and some plant families have specialised on their typical set of such compounds. Modulation of the synthesis and degradation of these compatible solutes is a primary goal of stress adaptation.

All compatible solutes can mimic water molecules in terms of partial charge of the H and HO residues presented at the molecule's surface. Therefore, it was postulated that compatible solutes may interact with enzymes, replacing hydration, and by this stabilize the conformation active in catalysis. Such a model does not answer the question concerning hydration of substrate metabolites. In most cases, the charge density of enzymes is lower than the one of the substrates but diffusion between dissolved, isolated enzymes would be significantly hampered.

C. METABOLITE CHANNELING

Tight coupling of reductant production by photosynthetic electron transport, reductant consumption in CO_2 and nitrate reduction, and photosynthesate consumption by sink tissues is well known. For instance, remote sensing techniques using chlorophyll fluorescence as an indicator of general plant performance are based on this knowledge. The general understanding is that there are limited intermediate pools in the compartments involved in photosynthesis and these pools function as a buffer to allow a constant rate of photosynthesate transfer to the sink organs. For the chloroplast compartment, the internal nucleotide pool and phosphate supply to the coupling factor are among the substrates discussed as limiting factors in most publications [5]. But, in terms of limitations of substrate flow both overworking of a pool and impairing of substrate channeling would produce similar results as long as turnover is measured at the cell- or tissue-level. The two alternatives can be distinguished only by measurements at single enzyme level. But, for biothermokinetic calculations of metabolite fluxes, we still lack adequate *in vivo* data. Here we will summarize arguments concerning the idea of substrate channeling.

As outlined above, electron sinks have to accept reductants at rates of their generation by the photosystems. Otherwise there is the risk of radical formation and spill off of radicals. Therefore, intermediates of photosynthesis ideally should be directly channeled between enzymes, and long distance diffusion has to be excluded. Short distance diffusion, substrate channeling, or metabolite hopping as it is called by some authors, is fast enough to meet all requirements. This situation ideally can be achieved, if enzymes belonging to a certain pathway are localized adjacent to each other. Interestingly, aggregates of Calvin cycle enzymes have been observed already by many authors. Calvin cycle enzymes as well as enzymes from other metabolic pathways have been found to form aggregates [231]. By means of the immunogold technique such complexes have been identified, and other enzymes of the chloroplast stroma were found to be membrane associated or formed complexes with membrane proteins [231,232]. Moreover, these aggre-

gates have been observed to be highly sensitive to ionic strength [231].

It has been observed that both enzyme aggregation and formation of multiorganell complexes is favored within a certain range of cytosolic ion concentration. But, aggregates will tend to dissociate at lower as well as higher ionic strength. This can be explained on the assumption that hydrate water from the surfaces of the aggregating compounds has to be extruded to form the aggregates. In diluted solutions, hydrostatic pressure will favor the dissociation of complexes and the partners will become separated by hydration. As described by Kornblatt and Kornblatt [229], water bound to surfaces of membranes or proteins occupies less volume than bulk water. Therefore, at ion concentrations stabilizing bound water at the aggregate interface in a sequestered form, hydrostatic pressure will favor aggregate formation. Increasing ion concentrations will interfere with aggregation as inorganic ions, due to their higher charge density, successfully compete for hydratation with proteins.

REFERENCES

1. Komori, T., Yamada, S., Myers, P. & Imaseki, H. Biphasic response to elevated levels of NaCl in *Nicotiana occidentalis* subspecies *obliqua* Burbidge. *Plant Sci.* 165, 159–165 (2003).
2. Lu, C. & Zhang, F. Effects of water stress on photosystem II photochemistry and its thermostability in wheat plants. *J. Exp. Bot.* 50, 1199–1206 (1999).
3. Maas, E. V. & Hoffman, G. J. Crop salt tolerance — current assessment. *ASCE J. Irrig. Drain Div.* 103, 115–134 (1977).
4. Parks, G. E., Dietrich, M. A. & Schumaker, K. S. Increased vacuolar Na^+/H^+ exchange activity in *Salicornia bigelovii* Torr. in response to NaCl. *J. Exp. Bot.* 2002, 1055–1065 (2002).
5. Lawlor, D. W. Limitation to photosynthesis in water-stressed leaves: stomata vs. metabolism and the role of ATP. *Ann. Bot.* 89, 871–885 (2002).
6. Lawlor, D. W. Carbon and nitrogen assimilation in relation to yield: mechanisms are the key to understanding production systems. *J. Exp. Bot.* 53, 773–787 (2002).
7. Greer, D. H. Photoinhibition of photosynthesis in dwarf bean (*Phaseolus vulgaris* L.) leaves: effect of sink-limitations induced by changes in daily photon receipt. *Planta* 205, 189–196 (1998).
8. Marcelis, L. F. M. & van Hooijdonk, J. Effect of salinity on growth, water use and nutrient use in radish (*Raphanus sativus* L.). *Plant Soil* 215, 57–64 (1999).
9. Handley, L. et al. Chromosome 4 controls potential water use efficiency ($\delta^{13}C$) in barley. *J. Exp. Bot.* 45, 1661–1663 (1994).
10. Pakniyat, H. et al. AFLP variation in wild barley (*Hordeum spontaneum* C. Koch) with reference to salt

tolerance and associated ecogeography. *Genome* 40, 332–341 (1997).

11. Munns, R. et al. Avenues for increasing salt tolerance of crops, and the role of physiologically based selection traits. *Plant Soil* 247, 93–105 (2002).

12. Munns, R. Comparative physiology of salt and water stress. *Plant Cell Environ.* 25, 239–250 (2002).

13. Asch, F., Dingkuhn, M., Wittstock, C. & Doerffling, K. Sodium and potassium uptake of rice panicles as affected by salinity and season in relation to yield and yield components. *Plant Soil* 207, 133–145 (1999).

14. Haupt-Herting, S. & Fock, H. Exchange of oxygen and its role in energy dissipation during drought stress in tomato plants. *Physiol. Plant.* 110, 489–495 (2000).

15. Lawlor, D. W. The effect of water deficit on photosynthesis (ed. Smirnoff, N.) *Environment and Plant Metabolism.* pp. 129–160 (BIOS Scientific Publishers, Oxford, 1995).

16. Cornic, G., Ghashghaie, J., Genty, B. & Briantais, J.-M. Leaf photosynthesis is resistant to a mild drought stress. *Photosynthetica* 27, 295–309 (1992).

17. Cornic, G. & Briantais, J.-M. Partitioning of photosynthetic electron flow between CO_2 and O_2 reduction in a C_3 leaf (*Phaseolus vulgaris* L.) at different CO_2 concentrations and during water stress. *Planta* 183, 178–184 (1991).

18. Cornic, G., Le Gouallec, J.-L., Briantais, J.-M. & Hodges, M. Effects of dehydration and high light on photosynthesis of two C_3 plants (*Phaseolus vulgaris* L. and *Elastostema repens* (Lour.) Hall f.). *Planta* 177, 84–90 (1989).

19. Dietz, K. J. & Heber, U. Carbon dioxide gas exchange and the energy status of leaves of *Primula paliunuri* under water stress. *Planta* 158, 349–356 (1983).

20. Brestic, M., Cornic, G., Fryer, M. J. & Baker, N. R. Does photorespiration protect the photosynthetic apparatus in French bean leaves from photoinhibition during drought stress? *Planta* 196, 450–457 (1995).

21. Cornic, G. Drought stress and high light effects on leaf photosynthesis (eds. Baker, N. R. & Bowyer, J. R.) *Photoinhibition of Photosynthesis.* pp. 297–313 (BIOS Scientific Publishers, Oxford, 1994).

22. Lawlor, D. W. & Cornic, G. Photosynthetic carbon assimilation and associated metabolism in relation to water deficits in higher plants. *Plant Cell Environ.* 25, 275–294 (2002).

23. Tezara, W., Mitchell, V. J., Driscoll, S. P. & Lawlor, D. W. Water stress inhibits plant photosynthesis by decreasing coupling factor and ATP. *Nature* 401, 914–917 (1999).

24. Wise, R. R. et al. Investigation of the limitations to photosynthesis induced by leaf water deficit in field grown sunflower (*Helianthus annuus* L.). *Plant Cell Environ.* 13, 923–931 (1990).

25. Giménez, C., Mitchell, V. J. & Lawlor, D. W. Regulation of photosynthetic rate of two sunflower hybrids under water stress. *Plant Physiol.* 98, 516–524 (1992).

26. Kanechi, M., Kunitomo, E., Inagaki, N. & Maekawa, S. *Water Stress Effects on Ribulose-1,5-bisphosphate Carboxylase and its Relationship to Photosynthesis in*

Sunflower Leaves (ed. Mathis, M.) (Kluwer Academic Publishers, Dodrecht, 1995).

27. Horton, P., Ruban, A. & Walters, R. Regulation of light harvesting in green plants. *Annu. Rev. Plant Physiol. Plant Mol. Biol.* 47, 655–684 (1996).

28. Tjus, S., Scheller, H., Anderson, B. & Møller, B. Active oxygen produced during selective excitation of photosystem I is damaging not only to photosystem I, but also to photosystem II. *Plant Physiol.* 125, 2007–2015 (2001).

29. Scheuermann, R., Biehler, K., Stuhlfauth, T. & Fock, H. Simultaneous gas exchange and fluorescence measurements indicate differences in the response of sunflower, bean and maize to water stress. *Photosynth. Res.* 27, 189–197 (1991).

30. Biehler, K., Haupt, S., Beckmann, J., Fock, H. & Becker, T. W. Simultaneous CO_2 and $^{16}O_2/^{18}O_2$-gas exchange measurements indicate differences in light energy dissipation between wild type and phytochrome-deficient aurea mutant of tomato during water stress. *J. Exp. Bot.* 48, 1439–1449 (1997).

31. Havaux, M., Tardy, F. & Lemoine, Y. Photosynthetic light-harvesting functions of carotenois in higher plant leaves exposed to high light irradiances. *Planta* 205 (1998).

32. Delfine, S., Alvino, A., Villani, M. C. & Loreto, F. Restrictions to carbon dioxide conductance and photosynthesis in spinach leaves recovering from salt stress. *Plant Physiol.* 119, 1101–1106 (1999).

33. Allen, J. Protein phosphorylation in regulation of photosynthesis. *Biochim. Biophys. Acta* 1098, 275–335 (1992).

34. Holzwarth, A. R., Wendler, J. & Haegnel, W. Time-resolved picosecond fluorescence spectra of the antenna chlorophylls in *Chlorella vulgaris*. Resolution of photosystem I fluorescence. *Biochimica et Biophysica Acta* 807, 155–167 (1985).

35. Junge, W. & Witt, H. T. On the ion transport system of photosynthesis. Investigations on a molecular level. *Z. Naturforsch.* 23b, 244–254 (1968).

36. Barber, J. Biophysics of photosynthesis. *Reports on Progress in Physics* 41, 1157–1199 (1978).

37. Gräber, P. *Primary Charge Separation and Energy Transduction in Photosynthesis* (eds. Milazzo, G. & Blanks, M.) (Plenum Publishing Corp., New York, 1987).

38. Allen, J. F. How does protein phosphorylation regulate photosynthesis? *Trends Biochem. Sci.* 17, 12–17 (1992).

39. Allakhverdiev, S. I., Sakamoto, A., Nishiyama, Y. & Murata, N. Inactivation of photosystems I and II in response to osmotic stress in *Synechococcus*. Contribution of water channels. *Plant Physiol.* 122, 1201–1208 (2000).

40. Gräber, P. & Witt, H. T. Relations between the electrical potential, pH gradient, proton flux and phosphorylation in the photosynthetic membrane. *Biochim. Biophys. Acta* 423, 141–163 (1976).

41. Badger, M. R. Photosynthetic oxygen exchange. *Annu. Rev. Plant Physiol. Plant Mol. Biol.* 36, 27–53 (1985).

42. Giardi, M. T. et al. Long-term drought stress induces structural and functional reorganization of photosystem II. *Planta* 199, 118–125 (1996).

43. Havaux, M. & Davaud, A. Photoinhibition of photosynthesis in chilled potato leaves is not correlated with a loss of photosystem II activity: preferential inactivation of photosystem I. *Photosynth. Res.* 40, 75–92 (1994).

44. Sharkey, T. D. & Seemann, J. R. Mild water stress effects on carbon-reduction-cycle intermediates, ribulose bisphosphate carboxylase activity, and spatial homogeneity of photosynthesis in intact leaves. *Plant Physiol.* 89, 1060–1065 (1989).

45. Ortiz-Lopez, A., Ort, D. R. & Boyer, J. S. Photophosphorylation in attached leaves of *Helianthis annuus* at low water potentials. *Plant Physiol.* 96, 1018–1025 (1991).

46. Gunasekera, D. & Berkowitz, G. A. Use of transgenic plants with ribulose 1,5-bisphosphate carboxylase/oxygenase antisense DNA to evaluate the rate limitation of photosynthesis under water stress. *Plant Physiol.* 103, 629–635 (1993).

47. Flexas, J. & Medrano, H. Drought inhibition of photosynthesis in C_3 plants: stomatal and non-stomatal limitations revisited. *Ann. Bot.* 89, 183–189 (2002).

48. Allakhverdiev, S. I. et al. Salt stress inhibits the repair of photodamaged photosystem II by suppressing the transcription and translation of psbA genes in Synechocystis. *Plant Physiol.* 130, 1443–1453 (2002).

49. Mishra, S. K., Subrahmanian, D. & Singhal, G. S. Interactionship between salt and light stress on the primary process of photosynthesis. *J. Plant Physiol.* 138, 92–96 (1991).

50. Morales, F., Abadia, A., Gomez-Aparis, J. & Abadia, J. Effects of combined NaCl and $CaCl_2$ salinity on photosynthesis parameters of barley grown in nutrient solution. *Physiol. Plant.* 86, 419–426 (1992).

51. Larcher, W., Wagner, J. & Thamathaworn, A. Effects of super-imposed temperature stress on *in vivo* chlorophyll fluorescence of *Vigna unguiculata* under saline stress. *J. Plant Physiol.* 136, 92–102 (1990).

52. Lu, C.-M. & Zhang, J. Thermostability of photosystem II is increased in salt stressed sorghum. *Aust. J. Plant Physiol.* 25, 317–324 (1998).

53. Masojídek, J. & Hall, D. O. Salinity and drought stress are amplified by high irradiance in sorghum. *Photosynthetica* 27, 159–171 (1992).

54. Osmond, C. B., Badger, M., Maxwell, K., Björkman, O. & Leegood, R. Too many photons: photorespiration, photoinhibition and photooxidation. *Trends Plant Sci.* 2, 119–120 (1997).

55. Lu, C., Zhang, J., Zhang, Q., Li, L. & Kuang, T. Modification of photosystem II photochemistry in nitrogen deficient maize and wheat plants. *J. Plant Physiol.* 158, 1423–1430 (2001).

56. Ruuska, S. A., Andrews, T. J. & Badger, M. R. Interplay between limiting processes in C3 photosynthesis studied by rapid-response gas exchange using transgenic tobacco impaired in photosynthesis. *Aust. J. Plant Physiol.* 25, 859–870 (1998).

57. Ruuska, S. A., Badger, M. R., Andrews, T. J. & von Caemmerer, S. Photosynthetic electron sinks in transgenic tobacco with reduced amounts of Rubisco: little evidence for significant Mehler reaction. *J. Exp. Bot.* 51, 357–368 (2000).

58. Biehler, K. & Fock, H. Evidence for the contribution of the Mehler-peroxidase reaction in dissipating excess electrons in drought-stressed wheat. *Plant Physiol.* 112, 265–272 (1996).

59. Haupt-Herting, S. & Fock, H. Oxygen exchange in relation to carbon assimilation in drought stressed leaves during photosynthesis. *Ann. Bot.*, 851–859 (2002).

60. Robinson, S. P. & Portis, A. R. Involvement of stromal ATP in the light activation of ribulose-1,5-bisphosphate carboxylase/oxygenase in intact chloroplasts. *Plant Physiol.* 86, 293–298 (1988).

61. Salvucci, M. E. & Ogren, W. L. The mechanism of rubisco activase: insights from studies of the properties and structure of the enzyme. *Photosynth. Res.* 47, 1–11 (1996).

62. Edmonson, D. L., Kahne, H. J. & Andrews, T. J. Substrate isomerization inhibits ribulosebisphosphate carboxylase-oxygenase during catalysis. *FEBS Lett.* 260, 62–66 (1990).

63. Parry, M. A. J., Andolojc, J. P., Khan, S., Lea, P. J. & Keys, A. J. Rubisco activity: effects of drought stress. *Ann. Bot.* 89, 833–839 (2002).

64. Flexas, J. et al. Steady-state chlorophyll fluorescence (Fs) measurements as a tool to follow variations of net CO_2 assimilation and stomatal conductance during water-stress in C_3 plants. *Physiol. Plant.* 114, 231–240 (2002).

65. Medrano, H., Parry, M. A. J., Socias, X. & Lawlor, D. W. Long term water stress inactivates Rubisco in subterranean clover. *Ann. Appl. Biol.* 131, 491–501 (1997).

66. Laisk, A. & Oja, V. Leaf photosynthesis in short pulses of CO_2. The carboxylation reaction *in vivo*. *Sov. Plant Physiol.* 21, 1123–1131 (1974).

67. von Caemmerer, S. *Biochemical Models of Leaf Photosynthesis* (CSIRO Publishing, Collingwood, 2000).

68. Chaves, M. M. Effects of water deficits on carbon assimilation. *J. Exp. Bot.* 42, 1–46 (1991).

69. von Caemmerer, S. & Edmundson, D. L. The relationship between steady state gas exchange *in vivo* RuP_2 carboxylase activity and some carbon cycle intermediates in *Raphanus sativus*. *Aust. J. Plant Physiol.* 13, 669–688 (1986).

70. Lawlor, D. W. & Khanna-Chopra, R. Regulation of photosynthesis during water stress (ed. Sybesma, C.) *Advances in Photosynthetic Research*, Vol. IV, pp. 379–382 (Martinius-Nijhoff/Dr. W. Junk Publishers, The Hague, 1984).

71. Huchzermeyer, B. & Heins, T. *Energy Metabolism and Salt Stress* (eds. Lieth, H. & Moschenko, M.) (INCO Reports, Osnabrück, 2000).

72. Huchzermeyer, B. *Some Biochemical and Physiological Aspects of Salt Stress Response in Halophytes* (ed. Lieth, H.) (University of Osnabrück, Osnabrück, 1999).

73. Huchzermeyer, B. *Biochemical Principles of Salt Tolerance* (ed. Lieth, H.) (University of Osnabrück, Osnabrück, 2000).

74. Sánchez-Rodriguez, J., Martinez-Carrasco, R. & Pérez, P. Photosynthetic electron transport and carbon-reduction-cycle enzyme activities under long-term drought stress in *Casuaria equisetifolia* Forst. & Forst. *Photosynth. Res.* 52, 255–262 (1997).

75. Paul, M. J., Knight, J. S., Habash, D., Parry, M. A. J., Lawlor, D. W., Barnes, S. A., Loynes, A. & Gray, J. C. Reduction in phosphoribulokinase activity by antisense RNA in transgenic tobacco: effect on CO_2 assimilation and growth in low irridiance. *Plant J.* 7, 535–542 (1995).

76. Lawlor, D. W. *Photosynthesis* (BIOS Scientific Publishers, Oxford, 2001).

77. Koyro, H.-W. & Huchzermeyer, B. Salt and drought stress effects on metabolic regulation in maize (ed. Pessarakli, M.) *Handbook of Plant and Crop Stress*. pp. 843–878 (Marcel Dekker, New York, 1999).

78. Richter, M. L., Hein, R. & Huchzermeyer, B. Important subunit interactions in the chloroplast ATP synthase. *Biochim. Biophys. Acta* 1458, 326–342 (2000).

79. Huchzermeyer, B. Phosphate binding to isolated chloroplast coupling factor (CF_1). *Z. Naturforsch.* 43c, 213–216 (1988).

80. Huchzermeyer, B. Nucleotide binding and ATPase activity of membrane bound chloroplast coupling factor (CF_1). *Z. Naturforsch.* 43c, 133–139 (1988).

81. Löhr, A. & Huchzermeyer, B. The role of ADP in regulation of chloroplast coupling factor (ed. Sybesma, C.) *Advances in Photosynthesis Research*. pp. 535–538 (Martinus Nijhoff/Dr. W. Junk Publishers, The Hague, 1984).

82. Löhr, A. & Huchzermeyer, B. *Regulation of Photophosphorylation Efficiency by Some Component of Chloroplast Electron Transport Chain* (Cambridge University Press, Cambridge, UK, 1985).

83. Löhr, A., Willms, I. & Huchzermeyer, B. A regulatory effect of the electron transport chain on the ATP synthase. *Arch. Biochem. Biophys.* 236, 832–840 (1985).

84. Meyer, S. & de Kouchkovsky, Y. ATPase state and activity in thylakoids from normal and water-stressed lupin. *FEBS Lett.* 303, 233–236 (1992).

85. Meyer, S., Hung, S. P. N., Trémolières, A. & de Kouchkovsky, Y. Energy coupling, membrane lipids and structure of thylakoids of lupin plants submitted to water stress. *Photosynth. Res.* 32, 95–107 (1992).

86. Ort, D. R., Oxborough, L. & Wise, R. R. Depressions of photosynthesis in crops with water deficits (eds. Baker, N. R. & Bowyer, J. R.) *Photoinhibition of Photosynthesis*. pp. 315–329 (BIOS Scientific Publishers, Oxford, 1994).

87. Stitt, M. *Metabolic Regulation of Photosynthesis* (ed. Baker, N.) (Academic Press, New York, 1997).

88. Heldt, H. & Flügge, U. Intrazellulärer Transport in grünen Pflanzenzellen. *Naturwissenschaften* 73, 1–7 (1986).

89. Jacob, J. & Lawlor, D. W. Dependence of photosynthesis of sunflower and maize leaves on phosphate supply, ribulose-1,5-bisphosphate carboxylase/oxygenase activity, and ribulose-1,5-bisphosphate pool size. *Plant Physiol.* 98, 801–807 (1992).

90. Boyer, P. D. The ATP synthase — a splendid molecular machine. *Annu. Rev. Biochem.* 66, 717–749 (1997).

91. Hausmann, N., Werhahan, W., Huchzermeyer, B., Braun, H.-P. & Papenbrock, J. How to document the purity of mitochondria prepared from green tissue of pea, tobacco and *Arabidopsis thaliana*. *Phyton* 43, 215–229 (2003).

92. Farquhar, G. D. Models describing the kinetics of ribulose bisphosphate carboxylase-oxygenase. *Arch. Biochem. Biophys.* 193, 456–468 (1979).

93. Kozaki, A. & Takeba, G. Photorespiration protects C_3 plants from photooxidation. *Nature* 384, 557–560 (1996).

94. Heintze, A., Riedel, A., Aydogdu, S. & Schultz, G. Formation of chloroplast isoprenoids from pyruvate and acetate by chloroplasts from young spinach plants. Evidence for a mevalonate pathway in immature chloroplasts. *Plant Physiol. Biochem.* 32, 791–797 (1994).

95. Schulze-Siebert, D. & Schultz, G. Full autonomy in isoprenoid synthesis in spinach chloroplasts. *Plant Physiol. Biochem.* 25, 145–153 (1987).

96. Preiss, M., Rosidi, B., Hoppe, P. & Schultz, G. Competition of CO_2 and acetate as substrates for fatty acid synthesis in immature chloroplasts of barley seedlings. *J. Plant Physiol.* 142, 525–530 (1993).

97. Schultz, G. *Assimilation of Non-Carbohydrate Compounds* (ed. Raghavendra, A. S.) (Cambridge University Press, Cambridge, UK, 1998).

98. Lawlor, D. W. Water stress induced changes in photosynthesis, photorespiration, respiration and CO_2 compensation concentration of wheat. *Photosynthetica* 10, 378–387 (1976).

99. Lawlor, D. W. & Fock, H. Photosynthesis and photorespiratory CO_2 evolution of water-stressed sunflower leaves. *Planta* 126, 247–258 (1975).

100. Tourneux, C. & TPeltier, G. Effect of water deficit on photosynthetic oxygen exchange measured using $^{18}O_2$ and mass spectroscopy in Solanum tuberosum leaf discs. *Planta* 1995, 570–577 (1995).

101. Stuhlfauth, T., Scheuermann, R. & Fock, H. P. Light energy dissipation under water stress conditions: contribution of reassimilation and evidence for additional processes. *Plant Physiol.* 92, 1053–1061 (1990).

102. Lawlor, D. W. & Pearlman, J. G. Compartmental modelling of photorespiration and carbon metabolism of water-stressed leaves. *Plant Cell Environ.* 4, 37–52 (1981).

103. Brodribb, T. Dynamics of changing intercellular CO_2 concentrations (c_i) during drought abd determination of minimal functional c_i. *Plant Physiol.* 111, 179–185 (1996).

104. Wilken, M. & Huchzermeyer, B. Suppression of mycelia formation by NO produced endogenously in *Candida tropicalis*. *Eur. J. Cell. Biol.* 78, 209–213 (1999).

105. Ninnemann, H. & Maier, J. Indications for the occurrence of nitric oxide synthases in fungi and plants and the involvement in photoconidiation of *Neurospora crassa*. *Photochem. Photobiol.* 64, 393–398 (1996).

106. Foyer, C. H., Lelandais, M. & Kunert, K. J. Photooxidative stress in plants. *Physiol. Plant.* 92, 696–717 (1994).

107. Shalata, A. & Tal, M. The effect of salt stress on lipid peroxidation and antioxidants in the leaf of the cultivated tomato and its wild salt-tolerant relative *Lysopersicon pennellii*. *Physiol. Plant.* 104, 169–174 (1998).

108. Shalata, A. & Neumann, P. M. Exogenous ascorbic acid (vitamin C) increases resistance to salt stress and reduces lipid peroxidation. *J. Exp. Bot.* 52, 2207–2211 (2001).

109. Doke, N. *The Oxidative Burst: Role in Signal Transduction and Plant Stress* (eds. Pell, E. J. & Steffen, K. L.) (Cold Spring Harbor Press, Cold Spring Harbor, NY, 1997).

110. Foyer, C. H. & Noctor, G. Leaves in the dark see the light. *Science* 284, 599–601 (1999).

111. Halliwell, B. & Gutteridge, J. M. C. *Free Radicals in Biology and Medicine*. (Oxford University Press, Oxford, 1989).

112. Asada, K. The water–water cycle in chloroplasts: scavenging of active oxygen and dissipation of excess photons. *Annu. Rev. Plant Physiol. Plant Mol. Biol.* 50, 601–639 (1999).

113. Demming-Adams, B. & Adams, W. W. I. The role of xanthophyll cycle carotenoids in the protection of photosynthesis. *Trends Plant Sci.* 1, 21–26 (1996).

114. Eskling, M., Arvidsson, P. O. & Akerlund, H. E. The xanthophyll cycle, its regulation and components. *Physiol. Plant.* 100, 806–816 (1997).

115. Smirnoff, N. The role of active oxygen in the response of plants to water deficit and desiccation. *New Phytol.* 125, 27–58 (1993).

116. Lopez, F., Vansuyt, G., Derancourt, J., Fourcroy, P. & Casse-Delbart, F. Identification by 2D-page analysis of salt-stress induced proteins in radish (*Raphanus sativus*). *Cell. Mol. Biol. (Noisy-le-grand)* 40, 85–90 (1994).

117. Jiang, M. & Zhang, J. Water stress-induced abscisic acid accumulation ttriggers the increased generation of reactive oxygen species and up-regulates the activities of antioxidant enzymes in maize leaves. *J. Exp. Bot.* 53, 2401–2410 (2002).

118. Hernandez, J. A. & Almansa, M. S. Short-term effects of salt stress on antioxidant systems and leaf water relations of pea leaves. *Physiol. Plant.* 115, 251–257 (2002).

119. Broin, M., I, B. & Rey, P. Evidence for post-translational control in the expression of a gene encoding a plastidic thioredoxin during leaf development in *Solanum tuberosum* plants. *Plant Physiol. Biochem.* 41, 303–308 (2003).

120. Broin, M., Cuiné, S., Peltier, G. & Rey, P. Involvement of CDSP32, a drought-induced thioredoxin, in the response to oxidative stress in potato plants. *FEBS Lett.* 467, 245–248 (2000).

121. Rey, P. et al. A novel thioredoxin-like protein located in the chloroplast is induced by water deficit in *Solanum tuberosum* L. plants. *Plant J.* 13, 97–107 (1998).

122. Broin, M., Cuiné, S., Eymery, F. & Rey, P. The plastidic 2-cysteine peroxiredoxin is a target for a thioredoxin involved in the protection of the photosynthetic apparatus against oxidative damage. *Plant Cell* 14, 1417–1432 (2002).

123. Eymery, F. & Rey, P. Immunocytolocalization of two chloroplastic drought-induced stress proteins in well-watered or wilted *Solanum tuberosum* L. plants. *Plant Physiol. Biochem.* 37, 305–312 (1999).

124. Pruvot, G., Massimino, J., Peltier, G. & Rey, P. Effects of low temperature, high salinity and exogenous ABA on the synthesis of two chloroplastic drought-induced proteins in *Solanum tuberosum*. *Physiol. Plant.* 97, 123–131 (1996).

125. Rorat, T. et al. PSII-S gene expression, photosynthetic activity and abundance of plastid thioredoxin-related and lipid-associated proteins during chikking stress in Solanum species differing in freezing resistance. *Physiol. Plant.* 113, 72–78 (2001).

126. Havaux, M. Carotenoids as membrane stabilizers in chloroplasts. *Trends Plant Sci.* 3, 147–151 (1998).

127. Noctor, G. & Foyer, C. H. Ascorbate and glutathione: keeping active oxygen under control. *Annu. Rev. Plant Physiol. Plant Mol. Biol.* 49, 249–279 (1998).

128. Smirnoff, N. & Wheeler, G. L. Ascorbic acid in plants: biosynthesis and function. *Crit. Rev. Plant Sci.* 19, 267–290 (2000).

129. Munné-Bosch, S. & Alegre, L. The function of tocopherols and tocotrienols in plants. *Crit. Rev. Plant Sci.* 21, 31–57 (2002).

130. Schwarz, K., Ternes, W. & Schmauderer, E. Antioxidative constituents of *Rosmarinus officinalis* and *Salvia officinalis*: III. Stability of phenolic diterpenes of rosemary extracts under thermal stress as required for technological processes. *Z. Lebensm-Unters-Forsch.* 195, 104–107 (1992).

131. Arouma, O. I., Halliwell, B., Aeschbach, R. & Löliger, J. Antioxidant and peroxidant properties of active rosemary constituents: carnosol and carnosic acid. *Xenobiotica* 22, 257–268 (1992).

132. Munné-Bosch, S. & Alegre, L. Drought-induced changes in the redox state of α-Tocopherol, ascorbate, and the diterpene carnosic acid in chloroplasts of labiatae species differing in carnosic acid contents. *Plant Physiol.* 131, 1–10 (2003).

133. Munné-Bosch, S. & Alegre, L. Subcellular compartmentation of the diterpene carnosic acid and its derivatives in the leaves of rosemary,. *Plant Physiol.* 125, 1094–1102 (2001).

134. Boo, Y. C., Lee, K. P. & Jung, J. Rice plants with a high protochlorophyllide accumulation show oxidative stress in low light that mimics water stress. *J. Plant Physiol.* 157, 405–411 (2000).

135. Robinson, J. M. & Bunce, J. a. Influence of drought-induced water stress on soybean and spinach leaf ascorbate-dehydroascorbate level and redox status. *Int. J. Plant Sci.* 161, 271–279 (2000).

136. Munné-Bosch, S. & Alegre, L. Changes in carotenooids, tocopherols and diterpenes during drought and recovery, and the biological significance of chloro-

phyll loss in *Rosmarinus officinalis* plants. *Planta* 210, 925–931 (2000).

137. Gerbaud, A. & André, M. Effect of CO_2, O_2 and light on photosynthesis and photorespiration in wheat. *Plant Physiol.* 66, 1032–1036 (1980).

138. Genty, B., Briantais, J.-M. & Baker, N. R. The relationship between the quantum yield of photosynthetic electron transport and quenching of chlorophyll fluorescence. *Biochim. Biophys. Acta* 990, 87–92 (19989).

139. Lu, Q., Wen, X., Lu, C., Zhang, Q. & Kuang, T. Photoinhibition and photoprotection in senescent leaves of field-grown wheat plants. *Plant Physiol. Biochem.* 41, 749–754 (2003).

140. Gueta-Dahan, Y., Yaniv, Z., Zilinskas, B. & Ben-Hayyim, G. Salt and oxidative stress: similar and specific responses and their relation to salt in *Citrus*. *Planta* 203, 460–469 (1997).

141. Bohnert, H. J., Jensen, R., Tarczynski, M. C. & National Research Initiative Competitive Grants Program (United States. Cooperative State Research Service). *Plant Tolerance to Salt and Drought Stress* (National Research Initiative Competitive Grants Program U.S. Department of Agriculture Cooperative State Research Service, Washington, D.C., 1994).

142. Bohnert, H. J. & Jensen, R. G. Strategies for engineering water-stress tolerance in plants. *Trends Biotechnol.* 14, 89–97 (1996).

143. Karakas, B., Ozias-Akins, P., Stushnoff, C., Suefferheld, M. & Rieger, M. Salinity and drought tolerance of mannitol-accumulating transgenic tobacco. *Plant Cell Environ.* 20, 609–616 (1997).

144. Abebe, T., Guenzi, A., Martin, B. & Cushman, J. Tolerance of mannitol-accumulating transgenic wheat to water stress and salinity. *Plant Physiol.* 131, 1748–1755 (2003).

145. Loester, W., Tyson, R., Everard, J., Redgewell, R. & Bieleski, R. Mannitol synthesis in higher plants: evidence for the role and characterization of a NADPH-dependent mannose-6-phosphate reduchtase. *Plant Physiol.* 98, 1396–1402 (1992).

146. Pharr, D. et al. The dual role of mannitol as osmoprotectant and photoassimilate in celery. *Hort. Sci.* 30, 1182–1188 (1995).

147. Stoop, J. & Pharr, D. Effect of different carbon sources on relative growth rate, internal carbohydrates, and mannitol-1-oxidoreductase activity in celery suspension cultures. *Plant Physiol.* 103, 1001–1008 (1996).

148. Patonnier, M., Peltier, J. & Marigo, G. Drought-induced increase in xylem malate and mannitol concentration and closure of *Fraxinus excelsior* L. stomata. *J. Exp. Bot.* 50, 1223–1229 (1999).

149. Tarczynski, M., Jensen, R. & Bohnert, H. Stress protection of transgenic tobacco by production of the osmolyte mannitol. *Science* 259, 508–510 (1993).

150. Tarczynski, M. C., Jensen, R. G. & Bohnert, H. J. Stress protection of transgenic tobacco by production of the osmolyte mannitol. *Science* 259, 508–510 (1993).

151. Thomas, J., Sepahi, M., Arendall, B. & Bohnert, H. Enhancement of seed germination in high salinity by engineering mannitol expression in *Arabidopsis thaliana*. *Plant Cell Environ.* 18, 801–806 (1995).

152. Gilbert, G. A., Gadush, M. V., Wilson, C. & Madore, M. Amino acid accumulation in sink and source tissues of *Coleus blumei* Benth. during salinity stress. *J. Exp. Bot.* 49, 107–114 (1998).

153. Morot-Gaudry, J.-F., Job, D. & Lea, P. *J. Amino Acid Metabolism* (eds. Lea, P. J. & Morot-Gaudry, J.-F.) (Springer Verlag, Berlin, 2001).

154. Kishor, P. B. K., Hong, Z., Miao, G., Hu, C. A. & Verma, D. P. S. Overexpression of Δ^1-pyrroline-5-carboxylate synthase increases proline production and confers osmotolerance in transgenic plants. *Plant Physiol.* 108, 1387–1394 (1995).

155. Ellis, R. P. et al. Phenotype/genotype associations for yield and salt tolerance in a barley mapping population segregating for two dwarfing genes. *J. Exp. Bot.* 53, 1163–1176 (2002).

156. Iyer, S., Caplan, A. Products of proline catabolism can induce osmotically regulated genes in Rice. *Plant Physiol.* 116, 203–211 (1998).

157. Solomonson, L. & Barber, M. Assimilatory nitrate reductase: functional properties and regulation. *Annu. Rev. Plant Physiol. Mol. Biol.* 41, 225–253 (1990).

158. Campbell, W. Nitrate reductase structure, function and regulation: bridging the gap between biochemistry and physiology. *Annu. Rev. Plant Physiol. Mol. Biol.* 50, 277–303 (1999).

159. Miflin, B. Distribution of nitrate and nitrite reductase in barley. *Nature* 214, 1133–1134 (1967).

160. Kamachi, K., Anemiya, Y., Ogura, N. & Nakagawa, H. Immunogold localization of nitrate reductase in spinach (*Spinacea oleracea*) leaves. *Plant Cell Physiol.* 28, 333–338 (1987).

161. Meyerhoff, P., Fox, T., Travis, R. & Huffaker, R. Characterization of the association of nitrate reductase with barley (*Hordeum vulgare* L.) root membrane. *Plant Physiol.* 104, 925–936 (1994).

162. de Marco, A., Jia, C., Fischer-Schliebs, E., Varanini, Z. & Lüttge, U. Evidence for two different nitrate reducing activities at the plasma-membrane in roots of *Zea mays* L. *Planta* 194, 557–564 (1994).

163. Kunze, M., Riedel, J., Lange, U., Hurwits, R. & Tischner, R. Evidences for the presence of GPI-anchored PM-NR in leaves of *Beta vulgaris* and for PM-NR in barley leaves. *Plant Physiol. Biochem.* 35, 507–512 (1997).

164. Stöhr, C. & Ullrich, W. A succinate-oxidising nitrate reductasc is located at the plasma membrane of plant roots. *Planta* 203, 129–132 (1997).

165. Ward, M., Grimes, H. & Huffaker, R. Latent nitrate reductase activity is associated with the plasma membrane of corn roots. *Planta* 177, 470–475 (1989).

166. Yamaya, T., Oaks, A., Rhodes, D. & Matsumoto, H. Synthesis of [^{15}N]H^+_4 and [^{15}N]glycine by mitochondria isolated from pea and corn shoots. *Plant Physiol.* 81, 754–757 (1985).

167. Delauney, A. J. & Verma, D. P. S. Proline biosynthesis and osmoregulation in plants. *Plant J.* 4, 215–223 (1993).

168. Lawlor, D. W. & Fock, H. Water stress induced changes in the amounts of some photosynthetic assimilatio products and respiratory metabolites of sunflower leaves. *J. Exp. Bot.* 28, 329–337 (1977).

169. Foyer, C. H., Valadier, M.-H., Migge, A. & Becker, T. W. Drought-induced effects on nitrate reductase activity and mRNA and coordination of nitrogen and carbon metabolism in maize leaves. *Plant Physiol.* 117, 283–292 (1998).

170. Santos, C. L. V., Campos, A., Azevedo, H. & Caldeira, G. *In situ* and *in vitro* senescence induced by KCl stress: nutritional imbalance, lipid peroxidation and antioxidant metabolism. *J. Exp. Bot.* 52, 351–360 (2001).

171. Kerkeb, L., Donaire, J. P., Venema, K. & Rodríguez-Rosales, M. P. Tolerance to NaCl induces exchanges in plasma membrane lipid composition, fluidity and H⁺-ATPase activity of tomato calli. *Physiol. Plant.* 113, 217–224 (2001).

172. Opekarova, M. & Tanner, W. Specific lipid requirements of membrane proteins - a putative bottleneck in heterologous expression. *Biochim. Biophys. Acta.* 1610, 11–22 (2003).

173. Briskin, D. The plasma membrane H+-ATPase of higher plant cells: biochemistry and transport function. *Biochim. Biophys. Acta* 1019, 95–109 (1990).

174. Siedow, J. N. & Umbach, A. L. Plant mitochondrial electron transfer and molecular biology. *Plant Cell* 7, 821–831 (1995).

175. Vedel, F., Lalanne, È., Dabar, M., Chétrit, P. & De Pape, R. The mitochondrial respiratory chain and ATP synthase complex: composition, structure and mutational studies. *Plant Physiol. Biochem.* 37, 629–643 (1999).

176. Sharkey, T. D. & Badger, M. Effects of water stress on photosynthetic electron transport, photophosphorylation and metabolite levels of *Xanthium strumarium* cells. *Planta* 156, 199–206 (1982).

177. Serrano, R. Salt tolerance in plants and microorganisms: toxicity targets and defense responses. *Int. Rev. Cytol.* 165, 1–52 (1996).

178. Kawasaki, S. et al. Gene expression profiles during the initial phase of salt stress in rice. *Plant Cell* 13, 889–905 (2001).

179. Shinozaki, K. Plant response to drought and salt stress: overview. *Tanpakushitsu Kakusan Koso* 44, 2186–2187 (1999).

180. Skriver, K. & Mundy, J. Gene expression in response to abscisic acid and osmotic stress. *Plant Cell* 2, 503–512 (1990).

181. Zhu, B., Chen, T. & Li, P. Expression of three osmotin-like protein genes in response to osmotic stress and fungal infection in potato. *Plant Mol. Biol.* 28, 17–26 (1995).

182. Deleu, C., Coustaut, M., Niogret, M.-F. & Larher, F. Three new osmotic stress-regulated cDNAs identified by differential display polymerase chain reaction in rapeseed leaf discs. *Plant Cell Environ.* 22, 979–998 (1999).

183. Espelund, M. et al. Late embryogenesis-abundant genes encoding proteins with different numbers of hydrophilic repeats are regulated differentially by abscisic acid and osmotic stress. *Plant J.* 2, 241–252 (1992).

184. Blumwald, E. Sodium transport and salt tolerance in plants. *Curr. Opin. Cell Biol.* 12, 431–434 (2000).

185. Gong, Z. et al. Genes that are uniquely stress regulated in salt overly sensitive (sos) mutants. *Plant Physiol.* 126, 363–375 (2001).

186. Cushman, J. C., Meyer, G., Michalowski, C. B., Schmitt, J. M. & Bohnert, H. J. Salt stress leads to differential expression of two isogenes of phosphoenolpyruvate carboxylase during Crassulacean acid metabolism induction in the common ice plant. *Plant Cell* 1, 715–725 (1989).

187. Cushman, J. C. & Bohnert, H. J. Salt stress alters A/T-rich DNA-binding factor interactions within the phosphoenolpyruvate carboxylase promoter from *Mesembryanthemum crystallinum*. *Plant Mol. Biol.* 20, 411–424 (1992).

188. Yamaguchi-Shinozaki, K. & Shinozaki, K. Improving plant drought, salt and freezing tolerance by gene transfer of a single stress-inducible transcription factor. *Novartis Found. Symp.* 236, 176–186; discussion 186–189 (2001).

189. Wilkinson, S. & Davies, W. ABA-based chemical signaling: the co-ordination of responses to stress in plants. *Plant Cell Environ.* 25, 195–210 (2002).

190. Gupta, S., Chattopadhyay, M., Chatterjee, P., Ghosh, B. & SenGupta, D. Expression of abscisic acid-responsive element-binding protein in salt-tolerant indica rice (*Oryza sativa* L. cv. Pokkali). *Plant Mol. Biol.* 37, 629–637 (1998).

191. Close, T. J. et al. *Dehydrin: The Protein* (eds. Close, T. J. & Bray, E. A.) (American Society of Plant Physiologists, Rockville, MD, 1993).

192. Maurel, C. & Chrispeels, M. J. Aquaporins. A molecular entry into plant water relations. *Plant Physiol.* 125, 135–138 (2001).

193. Kabakov, A. E. & Gabai, V. L. *Heat Shock Proteins and Cryoprotection: ATP-Deprived Mammalian Cells* (Springer Verlag, Berlin, 1997).

194. Schöffl, F., Prändl, R. & Reindl, A. Regulation of the heat shock response. *Plant Physiol.* 117, 1135–1141 (1998).

195. Gao, M.-J., Dvorak, J. & Travis, R. L. Expression of the extrinsic 23-kDa protein of photosystem II in response to salt stress is associated with the K⁺/Na⁺ discrimination locus Kna1 in wheat. *Plant Cell Rep.* 20, 774–778 (2001).

196. Bray, E. A. Classification of genes differentially expressed during water-deficit stress in *Arabidopsis thaliana*: an analysis using microarray and differential expression data. *Ann. Bot.* 89, 803–811 (2002).

197. Umeda, M. et al. Expressed sequence tags from cultured cells of rice (*Oryza sativa* L.) under stress conditions: analysis of genes in ATP-generating pathways. *Plant Mol. Biol.* 25, 469–478 (1994).

198. Riccardi, F., Gazeau, P., de Vienne, D. & Zvy, M. Protein changes in response to progressive water deficit in mayze. *Plant Physiol.* 117, 1253–1263 (1998).

199. Kramer, P. J. & Boyer, J. S. *Water Relation of Plants and Soils* (Academic Press, London, 1995).

200. Huber, S. C. & Huber, J. L. Role and regulation of sucrose-phosphate synthase in higher plants. *Annu. Rev. Plant Physiol. Plant Mol. Biol.* 47, 431–444 (1996).

201. Vassey, T. J. & Sharkey, T. D. Mild water stress of *Phaseolus vulgaris* plants leads to reduced starch synthesis and extractable sucrose phosphate synthase activity. *Plant Physiol.* 89, 1066–1070 (1989).

202. Kaiser, W. M. & Foster, J. Low CO_2 prevents nitrate reduction in leaves. *Plant Physiol.* 91, 970–974 (1989).

203. Flowers, T. J. Salt tolerance in *Sueda maritima* L. Dum. The effect of sodium chloride on growth, respiration and soluble enzymes in a comparative study with *Pisum sativum. J. Exp. Bot.* 23, 310–321 (1972).

204. Greenway, H. & Munns, R. Mechanism of salt tolerance in nonhalophytes. *Annu. Rev. Plant Physiol.* 31, 149–190 (1980).

205. Greenway, H. & Osmond, C. B. Salt responses of enzymes from species differing in salt tolerance. *Plant Physiol.* 49, 256–259 (1972).

206. Hasegawa, P. M., Bressan, R. A. & Pardo, J. M. The dawn of plant salt tolerance genetics. *Trends Plant Sci.* 5, 317–319 (2000).

207. Serrano, R. & Rodriguez-Navarro, A. Ion homeostasis during salt stress in plants. *Curr. Opin. Cell Biol.* 13, 399–404 (2001).

208. Barkla, B. J., Zingarelli, L., Blumwald, E. & Smith, J. A. C. Tonoplast Na^+/H^+ antiport activity and its energization by the vacuolar H^+-ATPase in the halophytic plant *Mesembryanthemum crystallinum* L. *Plant Physiol.* 109, 549–556 (1995).

209. Sze, H. H^+-translocating ATPases: advances using membrane vesicles. *Annu. Rev. Plant Physiol.* 36, 175–208 (1985).

210. Sze, H., Li, X. & Palmgren, M. G. Energization of plant cell membranes by H^+-pumping ATPases: regulation and biosynthesis. *Plant Cell* 11, 677–689 (1999).

211. Rea, P. A. & Poole, R. J. Chromatographic resolution of H^+-translocating pyrophosphatase from H^+-translocating ATPase of higher plant tonoplast. *Plant Physiol.* 81, 126–129 (1986).

212. Rea, P. A. & Sanders, D. Tonoplast energization: two pumps, one membrane. *Physiol. Plant.* 71, 131–141 (1987).

213. Dietz, K. J. et al. Significance of the V-type ATPase for the adaptation to stressful growth conditions and its regulation on the molecular and biochemical level. *J. Exp. Bot.* 52, 1969–1980 (2001).

214. Vera-Estrella, R., Barkla, B. J., Bohnert, H. J. & Pantoja, O. Salt stress in *Mesembryanthemum crystallinum* L. cell suspensions activates adaptive mechanisms similar to those observed in the whole plant. *Planta* 207, 426–435 (1999).

215. Golldack, D. & Dietz, K. J. Salt-induced expression of the vacuolar H^+-ATPase in the common ice plant is developmentally controlled and tissue specific. *Plant Physiol.* 125, 1643–1654 (2001).

216. Binzel, M. L. NaCl-induced accumulation to tonoplast and plasma membrane H^+-ATPase message in tomato. *Physiol. Plant.* 94, 722–728 (1995).

217. Golldack, D., Quigley, F., Michalowski, C. B., Kamasani, U. R. & Bohnert, H. J. Salinity stress-tolerant and -sensitive rice (*Oryza sativa*) regulate AKT1-type potassium channel transcripts differently. *Plant Mol. Biol.* 51, 71–81 (2003).

218. Ayala, F., O'Leary, J. W. & Schumaker, K. S. Increased vacuolar and plasma membrane H^+-ATPase activities in *Salicornia bigelovii* Torr. in response to NaCl. *J. Exp. Bot.* 47, 25–32 (1996).

219. Wu, J. & Seliskar, D. M. Salinity adaptation of plasma membrane H^+-ATPase in the salt marsh plant Spartina patens: ATP hydrolysis and enzyme kinetics. *J. Exp. Bot.* 49, 1005–1013 (1998).

220. Harada, A., Okazaki, Y. & Takagi, S. Photosynthetic control of the plasma membrane H^+-ATPase in *Vallisneria* leaves. I. Regulation of activity during light-induced membrane hyperpolarization. *Planta* 214, 863–869 (2002).

221. Blumwald, E. & Poole, R. J. Na^+/H^+ antiport in isolated tonoplast vesicles from storage tissues of *Beta vulgaris. Plant Physiol.* 78, 163–167 (1985).

222. Qiu, Q.-S., Barkla, B. J., Vera-Estrella, R., Zhu, J.-K. & Schumaker, K. S. Na^+/H^+ exchange activity in the plasma membrane of Arabidopsis. *Plant Physiol.* DOI 10.1104, 102–104 (2003).

223. Flowers, T. J. The effect of NaCl on enzyme activities from four halophyte species of Chenopodiaceae. *Phytochemistry* 11, 1881–1886 (1972).

224. Huchzermeyer, B. *Some Biochemical and Physiological Aspects of Salt Stress Response in Halophytes* (eds. Lieth, H. & Moschenko, M.) (INCO Reports, Osnabrück, 1999).

225. Nicholls, P. Water, water, everywhere. *Biochemist* 19, 7 (1997).

226. Pedersen, P. L. & Carafoli, E. Ion motive ATPases. I. Ubiquity, properties, and significance to cell function. *Trends Biochem. Sci.* 12, 146–150 (1987).

227. Pedersen, P. L. & Carafoli, E. Ion motive ATPases. II. Energy coupling and work output. *Trends Biochem. Sci.* 12, 186–189 (1987).

228. Gronenborn, A. & Clore, M. Water in and around proteins. *Biochemist* 19, 18–21 (1997).

229. Kornblatt, J. & Kornblatt, J. The role of water in recognition and catalysis by enzymes. *Biochemist* 19, 14–17 (1997).

230. Vassey, T. J., Quick, W. P., Sharkey, T. D. & Stitt, M. Water stress, carbon dioxide, and light effects on sucrose phosphate synthase activity in *Phaseolus vulgaris. Physiol. Plant.* 81, 37–44 (1991).

231. Süss, K.-H., Arkona, C., Manteuffel, R. & Adler, K. Calvin cycle multienzyme complexes are bound to chloroplast thylakoid membranes of higher plants *in situ. Proc. Natl. Acad. Sci. USA* 90, 5514–5518 (1993).

232. Adler, K., Arkona, C., Manteuffel, R. & Süss, K.-H. Electron-microscopical localization of chloroplast proteins by immunogold labelling on cryo-embedded spinach leaves. *Cell Biol. Int.* 17, 213–220 (1993).

40 Photosynthetic Carbon Metabolism of Crops under Salt Stress

Bruria Heuer

Institute of Soils, Water and Environmental Sciences, Agricultural Research Organization, Volcani Center

CONTENTS

I. INTRODUCTION

During the onset and development of salt stress within a plant, all the major processes such as photosynthesis, protein synthesis, energy, and lipid metabolism are affected. The earliest response is a reduction in the rate of leaf surface expansion, followed by a cessation of expansion as the stress intensifies. Growth resumes when the stress is relieved. Photosynthesis is the source of organic carbon and energy required by plants for their growth, biomass production, and yield. Photosynthetic rates are usually lower in plants exposed to salinity and especially to NaCl. It is still not clear whether these low photosynthesis rates are responsible for the reduced growth observed in salinized plants or if stunted plants control assimilation through a negative feedback of a reduced sink activity.

Salt effects on photosynthetic processes fall into two major categories: (1) stomatal closure, the usual response of stomata to salinization of salt-sensitive plants, and (2) effects on the capacity for CO_2 fixation apart from the altered diffusion limitations.

II. DIRECT AND INDIRECT EFFECTS OF SALT ON PHOTOSYNTHESIS

The reduction in the photosynthesis rate of plants exposed to salinity [1–3] usually depends on two aspects of salinization: the total concentration of salts (osmotic effect) and their ionic composition (specific ion effect).

A. LONG VS. SHORT-TERM EFFECTS

In arid and semiarid regions, insufficient precipitation necessitates extensive irrigation, resulting in salinization problems. High salt concentrations accumulating in the soil solution create high osmotic potentials,

779

which reduce the availability of water to the plants. Although photosynthesis has long been known to be partially or completely suppressed by sufficiently severe water stress [4], studies on the effect of salinity on photosynthesis have produced contradictory conclusions. Some investigators have shown that photosynthesis was hardly slowed down by salinity and was sometimes even stimulated by low salt concentrations [5–8]. Others have shown a significant decrease in photosynthesis in plants exposed to salinity [9–11], and we also obtained similar results with tomato plants exposed to several different salinity levels in nutrient solutions (Figure 40.1). Genotypic differences in the response of plant photosynthesis to salinity have previously been reported [12,13]. Only a few reports are available on the short-term effects of salt stress on photosynthesis, mainly after a few hours or within 1 to 2 days of the onset of exposure [14–19]. However, plant exposure to an osmotic shock can be encountered in field crops grown in arid zones and irrigated with brackish water; therefore, the short-term response is very important. For example, although sugarbeet plants are considered to be semi-tolerant to salinity, visual observations followed by measurements of CO_2 fixation have revealed an almost complete cessation of photosynthesis within 60 min [15]. Photosynthesis then started to recover and returned to the initial values equal to those in the control treatment within 24 h. The experiment was repeated with tomato plants, the results of which are shown in Figure 40.2. One hour after exposure of the plants to 135 mM NaCl, CO_2 assimilation was reduced by 64%; it had only partially recovered after 24 h.

Long-term salt effects on plant growth have often been related to direct ion toxicity due to the accumu-lation of high ion concentrations in plant tissue. Bean plants exposed to salinity showed an increase in Na$^+$ and abscisic acid (ABA) leaf concentrations with an accompanying decrease in growth and photosynthesis as salt exposure progressed, suggesting the participation of leaf ABA in the regulation of leaf growth [20]. In rice, an initial reduction in CO_2 fixation has been associated with water supply restriction, whereas long-term reduction resulted from NaCl accumulation in developing leaves [21]. Salinity reduced carbon isotope discrimination in cheatgrass, indicating long-term effects on conductance and carbon gain [22]. In C$_4$ plants, long-term exposure to salinity correlates photosynthesis potential with plant succulence [23].

B. IONIC EFFECTS AND TOXICITY AND ION COMPARTMENTATION

Plants that survive under saline conditions accumu-late high concentrations of ions, mainly sodium and chloride. The uptake of NaCl competes with that of other nutrient ions, especially K$^+$, leading to potassium deficiency [24,25]. Although leaves of gray mangrove have been reported to accumulate high NaCl concentrations, changes in photosynthesis were associated with changes in leaf K$^+$ concentrations [26]. A strong positive correlation has been found between the photosynthetic capacity of leaves and their nitrogen content, most of which is used for synthesis of components of the photosynthetic apparatus [27,28]. A specific ion effect of chloride on photosynthesis has been found in tomato plants [29], and Cl$^-$ has also been found to closely associate with the inhibition of photosynthesis in bell pepper plants [30]. A direct effect of NaCl on the photosynthesis process has

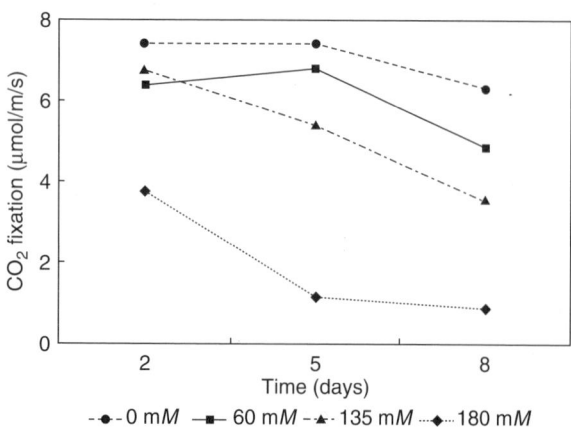

- ● -0 mM - ■ - 60 mM - ▲ -135 mM ⋯ ◆ ⋯180 mM

FIGURE 40.1 CO_2 fixation in tomato plants exposed to different levels of salinity in their nutrient media.

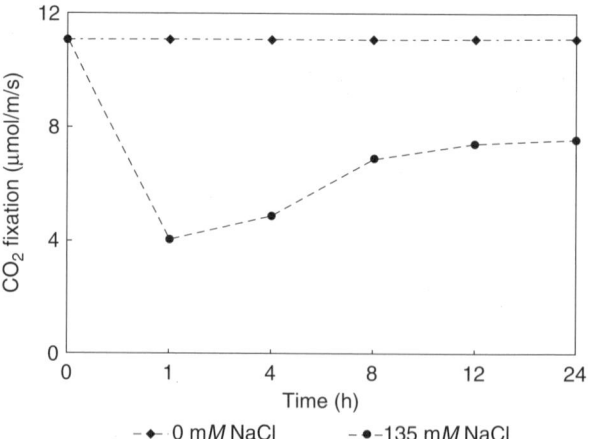

- ◆ - 0 mM NaCl   - ● -135 mM NaCl

FIGURE 40.2 Short-term response of CO_2 fixation in to-mato plants exposed to an osmotic shock.

also been found in pea plants [31]. In chickpea, photosynthetic rates were reduced more by chloride than by sulfate salinity [32]. Reports on the indirect effects of salinity on photosynthesis are available [33]. A substantial decrease in the photosynthetic capacity of spinach leaves has been attributed to the reduction in K^+ supply to the roots under high-salinity conditions [34], and under such conditions of high salinity and low K^+ supply, a reduction in the quantum yield of oxygen evolution, due to malfunction of photosystem II, has also been reported [34]. In salt stressed barley plants, reduced Mn concentrations have been correlated with a reduced CO_2 assimilation rate [35]. Photosynthetic activity in rice has been significantly increased by potassium application [36] and the net photosynthetic rate of barley was remarkably increased by nitrogen nutrition [37,38].

In the case of glycophytes, it has generally been concluded that the sensitivity of many of these species to salt may be a consequence of the failure to keep Na^+ and Cl^- out of the cytoplasm [39]. As neither halophytes nor glycophytes tolerate large amounts of salts in the cytoplasm, the current belief is that the ions involved in osmotic adjustment (Na^+, K^+, Cl^-) are largely restricted to the vacuoles [40]. Robinson et al. [41] reported that the ionic composition of isolated chloroplasts from spinach plants subjected to salt stress was different from that of the whole leaf; this also suggests compartmentation of these ions within the cell. The level of Cl^- in chloroplasts from the salt-grown plants was only slightly increased despite the large increase in leaf Cl^-, suggesting that the additional Cl^- in the leaf was restricted to the vacuoles. Isolated vacuoles of *Atriplex* have been found to contain nearly the same concentrations of Na^+ and Cl^- as the protoplasts, again suggesting that in the leaves of halophytes, NaCl is sequestered into the vacuoles [42]. Ion compartmentation among tissues and cell organelles was observed in mature *Prunus salicina*, enabling survival at low salinity levels for several years [43]. In *Triticum* genotypes, salt tolerance was achieved trough two independent mechanisms: a low rate of Na^+ accumulation and ion compartmentation within leaves [44]. Enhanced Na^+ uptake and ion compartmentation and redistribution resulted in increased tolerance to salinity in cotton [45]. On the other hand, poor ion compartmentation in plants was also reported. In bean, Cl^- concentrations were high in both cell vacuoles and chloroplast cytoplasm, indicating a lack of effective intracellular ion compartmentation in this species [46]. It seems that the longevity of the leaves and photosynthetic activity of bean depend on the balance between supply and uptake into the vacuoles, since they constitute the only compartment within the protoplast with significant storage capacity. Similar results were shown in lemon trees and annona [47,48].

C. CHLOROPLAST STRUCTURE AND CHLOROPHYLL CONTENT

Salinity causes chloroplasts to aggregate and leads to ultrastructural changes of the assimilating organs [49]. These include dilatation of thylakoid membranes, almost no sign of grana, and enlarged mesophyll cells [50–53]. In some species starch content in chloroplasts decreased [54,55], while in others it increased [54,56] following salinity. Salt stress significantly reduced chlorophyll content in pea and wheat [55,57], while it increased it in tomato [58], as well as the ratio of chlorophyll *a* to chlorophyll *b* [59,60]. In faba beans, ultrastructural damage of chloroplasts was observed in sensitive cultivars even at low salinity levels [61].

D. STOMATAL CLOSURE AND PLANT WATER STATUS

The effect of salinity on CO_2 assimilation has been the topic of many papers, but there is still uncertainty concerning the relative importance of stomatal closure and changes in mesophyll capacity in causing the observed reduction in photosynthesis, and the results are still controversial. In sunflower grown under saline conditions, parallel changes in stomatal and mesophyll conductance to CO_2 occurred [62]. Stomatal conductance was greatly reduced in lemon, sugarcane, cucumber, sorghum, wheat, rice, lucerne, guava, and celery by salinity [63–71]. The extent to which stomatal closure affects photosynthetic capacity is indicated by the magnitude of the reduction in the partial pressure of CO_2 inside the leaf (C_i) [72,73]. The reduction in C_i in response to salt stress in glycophytes contrasts with the response of halophytes, in which salinity has a smaller effect on the extent to which stomata limit photosynthesis [74]. Nonstomatal inhibition of photosynthesis under salt stress has also been reported in citrus, pepper, bean, sugarbeet, kiwifruit, soybean, and sunflower [1,30, 46,75–79].

E. PHOTOSYNTHATE PARTITIONING AND OSMOTIC ADJUSTMENT

Turgor in salt-stressed plants is maintained mainly by means of the accumulation of organic and inorganic solutes in plant organs, usually leaves, through osmoregulation [75]. It has already been mentioned that the ions involved in osmotic adjustment are largely confined to the vacuoles. The resulting decreased osmotic potential in the vacuole has to be balanced by the synthesis in the cytoplasm of noninhibitory organic solutes such as sugars, glycinebetaine, proline, and organic acids. Under salinization, the incorporation of CO_2 into organic compounds and amino acids in C_3 plants increases their resemblance to C_4 plants with respect to photosynthetic metabolism.

The synthesis of these organic solutes, which are required for osmoregulation, requires sources of carbon and energy, derived mainly from photosynthesis. In sugarbeet, NaCl has been found to decrease the proportion of carbohydrates and organic acids and to increase the proportion of amino acids [76]. In pomegranate, salinity has been reported to increase the amount of reducing sugars in the leaves and to decrease it in the roots, and to have the opposite effect on nonreducing carbohydrates [80]. Similarly, in bean leaves under saline conditions, C assimilation in sucrose was reduced and that in amino acids and sugar phosphate fractions was increased [81]. Sodium chloride has been found to increase CO_2 fixation into malate and to decrease it into aspartate in maize seedlings and chickpea plants [82,83]. Salt-treated pistachio plants accumulated high sucrose and starch concentrations in the stem and high concentrations of sucrose, reducing sugars, and starch in the main roots [84]. In celery, mannitol and sucrose are the primary photosynthetic products. Increasing salinities increased mannitol accumulation in celery and decreased sucrose and starch pools in leaf tissues, suggesting that mannitol accumulation plays a role in adaptation and tolerance to salt stress [85]. This is true for celery petioles [86] and olive plants [87]. Salinity significantly reduced starch content in soybean, while sucrose was slightly increased [88]. Salinity reduced sugar synthesis and translocation from leaves to grains in sorghum [66]. Changes in biomass partitioning have also been shown in mangroves [89]. C_4 halophytes from the Ararat valley showed a strong negative correlation between the rate of assimilate export and Cl^- content in leaves [90]. Starch biosynthesis was inhibited and more C was incorporated into amino and organic acids, particularly alanine and malate. The difference in the distribution and use of photoassimilates might help to explain the difference in salt tolerance in cultivars of species [91].

Osmoregulatory mechanisms enable plants to maintain plant water status and positive carbon balance. Sugarbeet cultivars exposed to salinity showed full osmotic adjustment [13,92]. High salinity counterbalanced with high nitrogen nutrition stimulated accumulation of sugar but not of proline in tomato leaves [29]. When tomato plants were irrigated with saline water (EC_i = 8 dS/m), proline accumulation was significantly increased in all plant organs (Table 40.1). Several reports indicate that the reduction in

TABLE 40.1
Accumulation of Proline in Tomato Plants Irrigated with Saline Water

Plant Organ	Proline Content (umol/g FW)	
	Nonsaline	Saline
Leaves, young	0.57 ± 0.06	6.27 ± 0.05
Leaves, old	1.13 ± 0.08	16.02 ± 0.12
Stems	0.52 ± 0.04	6.70 ± 0.44
Roots	1.80 ± 0.04	27.02 ± 0.18

FW: fresh weight

the photosynthetic capacity reflects a decreased allocation of photosynthate enzymes of the carbon fixation pathways [50].

F. RIBULOSE BISPHOSPHATE CARBOXYLASE/ OXYGENASE

Ribulose bisphosphate carboxylase/oxygenase (Rubisco) is the main enzyme in plants, responsible for the initial step in the C_3 photosynthetic carbon reduction cycle. Contradictory information is available concerning the effect of salt on enzyme activities. In *Phaseolus vulgaris*, salinity caused a reduction in the efficiency of the carboxylation reaction, mainly because of large reductions in RuBP pool sizes by causing regeneration of RuBP [46]. In rice, NaCl increased the activity of Rubisco-degrading endoproteinases, resulting in a significant decrease in Rubisco [93]. Following salinization, the Rubisco level decreased also in barley, spinach, and *Amaranthus tricolor* [94–96], while it increased in wheat [97]. Rubisco activity was similarly reduced by salinity [89,98–103]. Salinity did not affect Rubisco activity in *Atriplex* and olives [104,105] and even increased it in tomato [106]. Increased levels of substrate or compatible solutes have been found to mitigate enzyme inhibition by NaCl [39,107]. Although the specific activities of Rubisco in stressed and nonstressed plants are equal, salinity might decrease the activation state of the enzyme *in vivo*, which could account for the decline in the assimilation vs. internal CO_2 (A/C_i curve). In tomato leaves, salinity did not affect Rubisco activity, but reduced the potential of enzyme activation (Figure 40.3). It might also be that salt can directly inhibit catalysis *in vivo* by acting as a competitive inhibitor of RuBP. The biochemical basis for the reduction in RuBP regeneration capacity resulting from salt stress and the mechanism by which Rubisco activity is reduced at the threshold level of limiting supply of RuBP are not known.

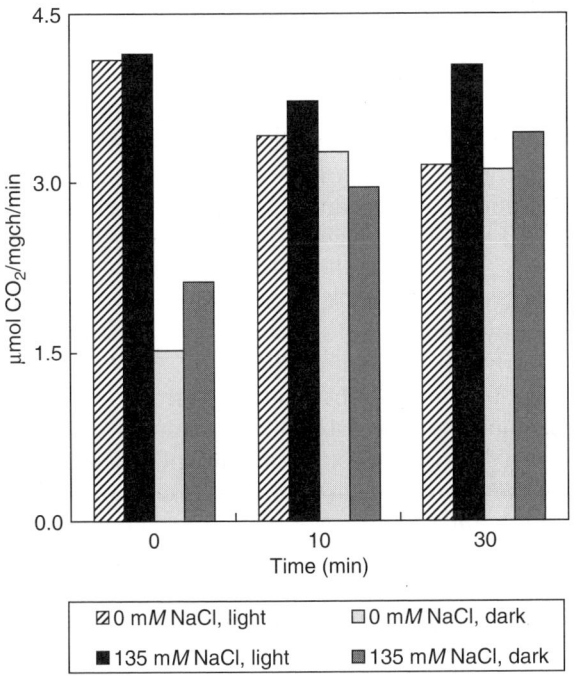

FIGURE 40.3 Rubisco activation in salt-treated tomato plants.

G. BIOCHEMICAL AND ENZYMATIC ACTIVITY AND GENE EXPRESSION

Contradictory information concerning the effect of salt on activities of other enzymes of the photosynthetic process is also available. In some cases, *in vitro* studies have shown that soluble enzymes from halophytes and glycophytes have similar sensitivities to electrolytes [39,108]. ATP production, which must be higher in salt-treated than in nontreated plants, may become limiting even when excess substrate is available [109]. Another possibility is that the spectrum of synthesized enzymes may change under stress. Differences in three proteins have been shown in salt-adapted tobacco cells; two of these differences are unique to the adapted cells [110]. Salinity increased the activity of mannose-6-phosphate reductase in celery [85] but decreased the activity of fructose-1,6-bisphosphatase (FBPase) in rice [111]. Mild salt concentrations have been found to induce conformational changes in sugarbeet Rubisco [112].

The control of photosynthetic gene expression in plants is often complex, with regulation occurring at many levels. Winicov and Seemann [113] showed that the response of alfalfa cells to salt was associated with large increases in two photosynthesis-related messenger RNAs, and a substantial increase in the activity of Rubisco. The mRNA levels from other chloroplast genes necessary for photosynthesis as well as from several nuclear genes encoding polypeptides participating in photosynthesis also increased [114]. In response to salinity or drought stress, the facultative halophyte ice plant, *Mesembryanthemum crystallinum,* switches from C3 photosynthesis to crassulacean acid metabolism (CAM). During this switch, the transcription rate of many genes encoding glycolytic, gluconeogenic, and malate metabolism enzymes is increased. In particular, transcription of the *Ppc1* and *Gap1* genes encoding a CAM-specific isozyme of phosphoenolpyruvate carboxylase and NAD-dependent glyceraldehyde-3-phosphate dehydrogenase, respectively, is increased by salinity stress [115]. The leaf-specific expression of Pgm1 contributes to the maintenance of efficient carbon flux through glycolysis/gluconeogenesis in conjuction with the stress-induced shift to CAM photosynthesis [116]. NaCl-treated *M. crystallinum* plants have been reported to accumulate high levels of PEP carboxylase mRNA, indicating that gene expression of this enzyme was affected by the salt [117]. Salt stress also leads to differential expression of two isogenes of PEP carboxylase in ice plant [118]: salinity induced *Gpd1* genes encoding the photosynthesis-related enzyme glyceraldehydes-3-phosphate dehydrogenase and *Imt1* encoding methyl transferase [119]. The gene expression of pyruvate–phosphate dikinase, malic enzyme, and tonoplast ATPase is also changed by salinity [120]. Four out of six genes of the Rubisco small subunit multigene family (RbcS) were differentially expressed in leaves of ice plant during the transition from C3 photosynthesis, but were regulated in a coordinate fashion [121].

H. REDUCTION OF PHOTOSYNTHETIC CAPACITY BY FEEDBACK INHIBITION

Feedback inhibition of photosynthesis as a result of decreased sink demand and the accumulation of sucrose or sugar-phosphate intermediates in source leaves is a long-known phenomenon [122–124]. Assimilates function as a link between source and sink tissues. Plants both produce and utilize carbohydrates and have developed mechanisms to regulate their sugar status and coordinate carbohydrate partitioning. High sugar levels result in a feedback inhibition of photosynthesis and an induction of storage processes. Thus, a negative correlation exists between photosynthesis and accumulation of soluble photosynthates in source leaves. However, some reports that could not show this relationship are also available [125].

Stress alters the type of carbohydrates that are synthesized and exported by the source tissues. A reduction in photosynthetic capacity resulting from salinity may also be a consequence of the inhibition of certain carbon metabolism processes by feedback from other salt-inhibited reactions [14,108]. There is much evidence that carbohydrates accumulate after exposure of a plant to salinity [109]. When the amount of reserve carbohydrates was measured, it was found that a salt treatment that reduces growth causes either no change or an increase in the total concentration of reserve carbohydrates (sugars and starch). These osmoregulating compounds are the principal end products of photosynthetic carbon metabolism. Carbon flow to sucrose is the main metabolic pathway in photosynthesis; for instance, sucrose synthesis has been shown to account for 60% to 70% of the total photosynthetic fixation products formed by isolated barley mesophyll protoplasts [126]. It is generally agreed that sucrose biosynthesis in the cytoplasm and photosynthetic CO_2 fixation in the chloroplast are metabolically coordinated [127]. The equilibrium between these two compartments is maintained by cytosolic phosphorus (P_i). The rate of triose-P export from the chloroplast to the cytosol and its conversion to sucrose must be regulated with respect to CO_2 assimilation. Without such regulation, stromal metabolites could be rapidly depleted and photosynthesis strongly inhibited under suboptimal conditions. In other words, the amount of photosynthetic substrate available for distribution is determined, via a feedback control of photosynthesis, by the sink requirement. Thus, the inhibition of photosynthesis under salt stress could be a consequence of the inhibition of certain photosynthetic reactions by an altered sink–source relationship. FBPase is a ubiquitous enzyme that, in nonphotosynthetic tissues, regulates the rate of gluconoegenesis, while in photosynthetic tissues, two FBPase isoenzymes (chloroplastic and cytoplastic) play a key role in carbon assimilation and metabolism. The cytosolic FBPase is one of the regulatory enzymes in the sucrose biosynthetic pathway and its activity is regulated by both fine and coarse control mechanisms [128,129]. Micromolar concentrations of F2,6BP markedly inhibit the activity of FBP and cause a switch from hyperbolic to sigmoidal saturation kinetics [130,131]. Since six other enzymes are also regulated by F2,6BP [132], this enzyme seems to play a central role in the regulation of carbon metabolism in plants by activating glycolysis and inhibiting gluconeogenesis. Metabolite studies have indicated that under reduced water potential, stromal levels of the substrate FBP accumulate and the level of the FBPase product, fructose-6-phosphate (F6P), is reduced, so that FBPase becomes rate limiting to photosynthesis [133].

III. HORMONAL INTERACTION WITH PHOTOSYNTHESIS

Change in phytohormonal balance in plants, such as lower cytokinin or giberellin and higher ABA, results in decreased photosynthetic activity. High salt concentrations trigger an increase in the endogenous ABA level, which may be responsible for the activation of salt-stress-induced genes [20,134–137]. These high concentrations also stimulate osmotic adjustment by regulating the accumulation of solutes, particularly sugars and proline [138]. ABA has been found to eliminate the inhibitory effects of NaCl on pea photosynthesis [31], and exogenous GA_3 has been shown to promote cotton growth under saline conditions and to counteract the salt-induced inhibition of growth, photosynthesis, and translocation of assimilates in bean and wheat [139,140]. Gibberellin and kinetin counteracted the adverse effect of irrigation of *Vicia faba* with seawater [141]. Similarly, kinetin and zeatin also increased the rates of photosynthesis and assimilate export from bean leaves and jute species [142]. When the water plant *Trianea bogotensis* was exposed to salinity, kinetin was found to participate in the regulation of the activity of the photosynthetic apparatus [143]. Application of auxin to NaCl-stressed maize and safflower plants resulted in significant reductions in transpiration, stomatal frequency, and dry matter production [144] and leaf growth of *V. faba* [141].

It is known that certain species of succulent plants can shift from the C_3 mode of photosynthesis to the CAM mode in response to salt stress, and it may be that this expression of CAM is promoted by ABA [145,146]. Phosphoenolpyruvate carboxylase (PEPcase), the key enzyme of CAM, is induced by salinity in leaves of *M. crystallinum* [147], and cytokinin applied exogenously to these leaves suppressed PEPcase mRNA accumulation, enzyme activity, and CAM induction. Phosphoglyceromutase mRNA was induced in response to treatment with either ABA or cytokinin [116].

IV. PHOTOSYNTHETIC EFFICIENCY AND PRODUCTIVITY UNDER STRESS CONDITIONS

Long-term photosynthetic production refers to the partitioning of carbohydrates into leaves and other plant organs such as roots or fruits. Water deficit may affect photosynthate utilization by altering either the efficiency with which photosynthates are converted to new growth or the rate at which they are used in maintaining the existing dry matter. The production of the photosynthetically active leaf area by plants is the most important factor affecting crop productivity. It is also the component of growth that is most sensitive to water stress and salinity. This topic was reviewed in 1986 [148], and many reports on leaf expansion under salt stress are available [65,66,77, 149–152]. The same is true for crop productivity under saline conditions [153–157]. For this reason, the issue of photosynthetic efficiency and productivity will not be further discussed in this chapter.

V. ROLE OF PHOTOSYNTHESIS IN COMBATING STRESS

A. ADAPTATION OR ACCLIMATION

So far, many adaptation mechanisms of plants to salt stress are known, among them osmoregulation and compartmentation of inorganic ions. Proline and other osmolytes may act as a signaling/regulatory molecule able to activate multiple responses that are part of the adaptation process. The accumulation of compatible solutes (sucrose, proline, and glycinebetaine) contributes to decreases in leaf osmotic potential under NaCl stress, allowing plants to keep a positive cell turgor by continuing water uptake so that seedlings can grow under salt stress.

Short-term adaptation to stress is primarily linked to stomatal regulation by reducing water loss by transpiration and maximizing CO_2 uptake [16]; this tends to lead to a constant ratio between transpiration and photosynthesis. Long-term adaptation includes changes in biomass allocation, specific anatomical modifications, or sophisticated physiological mechanisms. Acclimation is considered to be a midterm response; it involves osmotic adjustment, changes in cell wall elasticity, and morphological changes. Halophytes can grow under saline conditions because of their ability to maintain a high salt concentration within their cells [39]. In mangroves, the chlorophyll *a/b* ratio was increased to enable the high-energy demands for adaptation to salinity [158].

The most obvious mechanism of adaptation to salinity is a morphological one. Such mechanisms include the development of fewer and smaller leaves, increased succulence, thickening of leaf cuticle, etc. [23,41,159]. Allocation and partitioning of assimilated carbon provide resources for adaptation to stress [75,160]. Plants can be induced to adapt to salinity by gradual exposure to salt [5,161] or by reducing the rate of stress development [162]. Adaptation of cotton to salinity was adjusted by increasing mesophyll surface area to ensure normal exchange of gases and photosynthetic activities [163]. Metabolic acclimation via the accumulation of compatible solutes is regarded as a basic strategy for the protection and survival of plants [97,164]. Adaptation of wheat, maize, and duckweed to salinity was achieved by rearrangements in the oxidative metabolism through oscillations in the activity of the enzymes involved in oxidative metabolism [165].

Suspensions of plant cells have also been adapted to NaCl by successive transfers to solutions with increasing salt concentrations [166–168]. The carbon use efficiency of NaCl-adapted cells has been found to be higher than that of unadapted cells [169,170]. Amzallag et al. [171] postulated that this adaptation is achieved during a specific treatment and involves changes in the plant behavior, including the expression of properties that were not evident before the treatment. Consequently, growth under saline conditions is restored more or less to the pre-salt-exposure rate, similar to that of nontreated plants, and the full life cycle is completed. According to Singh et al. [172], the process of cellular adaptation to osmotic stress in a saline environment involves specific alteration in the gene expression of salt-adapted cells, leading to synthesis of several novel proteins.

B. CARBON DIOXIDE ENRICHMENT

Plants can respond to changes in environmental conditions by increasing dry mass allocation to assimilating organs. Enhanced demand for energy and photosynthate is observed under saline conditions. Thus, treatments that increase photosynthate supplies, such as CO_2 enrichment, were thought to be beneficial to salt-stressed plants. In general, C_3 species are more responsive to CO_2 enrichment than C_4 plants. Increased relative growth at elevated CO_2 primarily relates to increased net assimilation rate, enhancement of photosynthesis, and reduced photorespiration [173]. It stimulates plant growth and yield under high salt concentrations but can only partially relieve the deleterious effects of salinity [174–178]. In roses, salinity tolerance has been found to increase at high CO_2 levels so that some combinations of CO_2 and salt gave higher yields than those obtained with CO_2 alone, due to a decrease in blindness [179]. Elevated CO_2 improved dry matter production, leaf area, and tillering of wheat grown under saline conditions [180], whereas the photosynthesis of *Plantago maritima* leaves has been found to be insensitive to elevated CO_2 or seriously decreased by it, depending on the salt level [181].

C. INDUCTION OF CAM

Some plants respond to osmotic stress by switching from the C_3 photosynthesis pathway to CAM [182].

Most of the information about this phenomenon is derived from *M. crystallinum,* the common ice plant. During this switch, the transcription rate of many genes encoding glycolytic, gluconeogenic, and malate metabolism enzymes is increased [116,117,121,183]. The development of CAM-type photosynthesis is one of the adaptation mechanisms for severe water deficit. It provides plants with carbon dioxide and permits efficient water spending under extreme environments. Transformation of photosynthesis into a CAM pathway was also observed in *Euphorbia paralias* [184] and cacti [185]. CAM induction is dependent on organized leaf tissue and cannot be elicited in suspension culture cells [186]. The switch involves stress-initiated upregulation of the mRNAs, which encode CAM enzymes [120], specifically (1) gene expression of glycolysis and malic acid synthesis, with phosphoenolpyruvate carboxylase as the key enzyme; (2) gluconeogenesis with pyruvate phosphate dikinase as the key enzyme; and (3) transport across the membrane by the tonoplast ATPase [185,187]. A similar response has been found in *Chlorophytum* cultivars [188].

D. Selection, Breeding, and Genetic Engineering

In the last 20 years, efforts have been invested in the development of new salt-tolerant genotypes via selection and breeding. Exploitation of the genetic diversity available in crops has proved its effectiveness for this purpose [189]. Epstein et al. [190] showed that the ability of conventional crops to tolerate saline conditions could be improved by imposing appropriate selections based on the osmotic adjustment ability of plants. Selection and breeding approaches to increase tolerance are more efficient if selection is based on physiological and biochemical characters. Noble and Rogers [191] claimed that the selection and breeding approaches to increased tolerance might be more successful if selection was based directly on the physiological mechanisms or characters that confer tolerance. Use of physiological traits in breeding programs was also recommended by Flowers et al. [192] and Munns et al. [193]. However, this approach might encounter difficulties in identifying important traits that may contribute to improved yield and in having these traits expressed within the various genetic contexts that result in increased yield. Nevertheless, the traditional engineering approach of manipulating the environment for the benefit of plants is no longer adequate, and genetic adaptation of plants to saline conditions is more likely to succeed in improving agricultural productivity [157]. Attempts to improve photosynthesis by altering stomatal conductance through selecting for stomatal number have failed;

the stomata have become more frequent but smaller, with no effect on stomatal conductance, and the rate of photosynthesis has declined with the increase in productivity and yield. Significant improvement in salt stress tolerance was obtained by genetic engineering [194,195]. Transgenic tomato plants overexpressing a vacuolar Na^+/H^+ antiport were able to grow and yield in the presence of high salinity [196].

Molecular biology contributes to improving stress resistance. The development in recent years of maps of molecular markers (RFLP, RAPD, isozymes, etc.) for many crops is revolutionizing the efficiency with which breeders can introduce specific genes and so develop new crop varieties by crossing and selection.

VI. SUMMARY

Rates of photosynthesis are usually lower in salt-treated plants, but the photosynthetic potential is not greatly affected when the rates are expressed on a chlorophyll or leaf area basis. The decrease in photosynthesis may be related to several factors: (1) reduction of the CO_2 supply because of hydroactive closure of the stomata, (2) salt toxicity, (3) dehydration of cell membranes, which reduce their permeability to CO_2, (4) enhanced senescence induced by salinity, (5) changes in enzyme activity induced by changes in cytoplasmic structure, and (6) negative feedback by reduced sink activity.

The understanding of the mechanisms by which salinity affects photosynthesis would aid the improvement of growth conditions and crop yields and would provide useful tools for future genetic engineering.

REFERENCES

1. Walker RR, Torokfalvy E, Downton WJS. Photosynthetic responses of the citrus varieties Rangpur lime and Etrog citrus to salt treatment. *Aust. J. Plant Physiol.* 1982; 9:783–790.
2. Kalaji H, Nalborczyk E. Gas-exchange of barley seedlings growing under salinity stress. *Photosynthetica* 1991; 25:197–202.
3. Drew MC, Hole PS, Picchioni GA. Inhibition by NaCl of net CO_2 fixation and yield of cucumber. *Am. J. Hort. Sci. Soc.* 1990; 115:472–477.
4. Boyer JS. In: Kozlowski TT, ed. *Water Deficits and Plant Growth.* Academic Press, London, 1976:153–190.
5. Heuer B, Plaut Z. Carbon dioxide fixation and RuBP carboxylase activity in intact leaves of sugar beet plants exposed to salinity and water stress. *Ann. Bot.* 1981; 48:261–268.
6. Downton WJS, Millhouse J. Turgor maintenance during salt stress prevents loss of variable fluorescence in grapevine lea. *Plant Sci. Lett.* 1983; 31:1–7.

7. Rawat JS, Banerjee SP. The influence of salinity on growth, biomass production and photosynthesis of Eucalyptus camaldulensis Dehnh. and Dalbergia sissoo Roxb. seedlings. *Plant Soil* 1998; 205:153–169.

8. Yeo AR, Capron SJM, Flowers TJ. The effect of salinity upon photosynthesis in rice (*Oryza sativa* L.): gas exchange by individual leaves in relation to their salt content. *J. Exp. Bot.* 1985; 36:1240–1248.

9. Kapulnik Y, Heuer B. Forage production of 4 alfalfa (*Medicago sativa*) cultivars under salinity. *Arid Soil Res. Rehab.* 1991; 5:127–135.

10. Leidi EO, Silberbush M, Lips SH. Wheat growth as affected by nitrogen type, pH and salinity. 2. Photosynthesis and transpiration. *J. Plant Nutr.* 1991; 14:247–256.

11. Francois LE. Yield and quality response of salt-stressed garlic. *HortScience* 1994; 29:1314–1317.

12. Heuer B, Plaut Z. Photosynthetic activity of sugarbeet cultivars of different tolerance. *Adv. Photosynth. Res.* 1984; 4:423–425.

13. Heuer B, Plaut Z. Photosynthetic activity of sugarbeet cultivars of different tolerance. *J. Exp. Bot.* 1989; 40:437–440.

14. Rawson HM, Munns R. Leaf expansion in sunflower as influenced by salinity and short-term changes in carbon fixation (*Helianthus annuus*). *Plant Cell Environ.* 1984; 7:207–213.

15. Heuer B. Response of photosynthesis, translocation and ^{14}C partitioning in sugarbeet plants exposed to an osmotic shock. *Isr. J. Bot.* 1991; 40:275–282.

16. Golombek SD, Luedders P. Effects of short-term salinity on leaf gas exchange of the fig (*Ficus carica* L.). *Plant Soil* 1993; 148:21–27.

17. Kartens GS, Ebert G, Ludders P. Long-term and short-term effects of salinity on root respiration, photosynthesis and transpiration of citrus rootstocks. *Angew. Bot.* 1993; 67:3–8.

18. Hamilton EW III, McNaughton SJ, Coleman JS. Molecular, physiological and growth responses to sodium stress in C_4 grasses from a soil salinity gradient in the Serengeti ecosystem. *Am. J. Bot.* 2001; 88:1258–1265.

19. Hernandez JA, Almansa MS. Short-term effects of salt stress on antioxidant systems and leaf water relations of pea leaves. *Physiol. Plant.* 2002; 115:251–257.

20. Sibole JV, Montero E, Cabot C, Poschenrieder C, Barcelo J. Role of sodium in the ABA-mediated long-term growth response of bean to salt stress. *Physiol. Plant.* 1998; 104:299–305.

21. Lee KS. Effects of salinity on leaf growth and photosynthesis in rice. *Korean J. Crop Sci.* 1991; 36:22–33.

22. Rasmunson KE, Anderson JE. Salinity affects development, growth and photosynthesis in cheatgrass. *J. Range Manag.* 2002; 55:80–87.

23. Voronin PY, Manzhulin AV, Myasoedov NA, Balkonin YV, Terenteva EI. Morphological types and photosynthesis of C4 plant leaves under long-term soil salinity. *Russ. J. Plant Physiol.* 1995; 42:310–320.

24. Mozafar A, Goodin JR, Oertli JJ. Sodium and potassium interactions in increasing the salt tolerance of *Atriplex halimus* L.: II. Na$^+$ and K$^+$ uptake characteristics. *Agron. J.* 1970; 62:481–484.

25. Ball MC, Farquhar GD. Photosynthetic and stomatal responses of the grey mangrove, *Avicenia marina*, to transient salinity conditions. *Plant Physiol.* 1984; 74:7–11.

26. Ball MC, Chow WS, Anderson JM. Salinity induced potassium deficiency causes loss of functional photosystem II in leaves of the grey mangrove, *Avicennia marina*, through depletion of the atrazine binding polypeptide. *Aust. J. Plant Physiol.* 1987; 14:351–361.

27. Evans JR, Terashima I. Effects of nitrogen nutrition on electron transport components and photosynthesis in spinach. *Aust. J. Plant Physiol.* 1987; 14:59–68.

28. Sugiharto B, Miyata K, Nakamoto H, Sasakawa, H, Sugiyama T. Regulation of expression of carbon-assimilating enzymes by nitrogen in maize leaf. *Plant Physiol.* 1990; 92:963–969.

29. Heuer B, Feigin A. Interactive effects of chloride and nitrate on photosynthesis and related growth parameters in tomatoes. *Photosynthetica* 1993; 28:549–554.

30. Bethke PC, Drew MC. Stomatal and nonstomatal components to inhibition of photosynthesis in leaves of *Capsicum annum* during progressive exposure to NaCl salinity. *Plant Physiol.* 1992; 99:219–226.

31. Fedina JS, Tsonev Td, Guleva EI. ABA as a modulator of the response of *Pisum sativum* to salt stress. *J. Plant Physiol.* 1994; 143:245–249.

32. Datta KS, Sharma KD. Effect of chloride and sulphate types of salinity on characteristics of chlorophyll content, photosynthesis and respiration of chickpea (*Cicer arietinum*). *Biol. Plant.* 1990; 32:391–395.

33. Rawson HM, Long MJ, Munns R. Growth and development in NaCl-treated plants. I. Leaf Na$^+$ and Cl$^-$ concentrations do not determine gas exchange on leaf blades in barley. *Aust. J. Plant Physiol.* 1988; 15:519–527.

34. Chow WS, Ball MC, Anderson JM. Growth and photosynthetic responses of spinach to salinity: Implication of K$^+$ nutrition for salt tolerance. *Aust. J. Plant Physiol.* 1990; 17:563–578.

35. Cramer GR, Nowak RS. Supplemental manganese improves the relative growth, net assimilation and photosynthetic rates of salt-treated barley. *Physiol. Plant.* 1992; 84:600–605.

36. Bohra JS, Doerffling K. 1993; Potassium nutrition of rice *Oryza sativa* L. varieties under sodium chloride salinity. *Plant Soil* 152:299–303.

37. Shen Z, Shen Q, Liang Y, Liu, YL. Effect of nitrogen on the growth and photosynthetic activity of salt-stressed barley. *J. Plant Nutr.* 1994; 17:787–799.

38. Shen ZG, Shen Q, Guan HY, Wang ZY, Shen K. The relationship between nutrition and growth and ion balance under NaCl stress in barley. *J. Nanjing Agric. Univ.* 1994; 17:22–26.

39. Flowers TJ, Troke PF, Yeo AR. The mechanism of salt tolerance in halophytes. *Annu. Rev. Plant Physiol.* 1977; 28:89–121.

40. Munns R, Greenway H, Kirst GO. Halotolerant eukaryotes. In: Lange OL, Nobel PS, Osmond CB, Zeig-

ler HH, eds. *Physiological Plant Ecology. III. Responses to the Chemical and Biological Environment.* Encyclopedia of Plant Physiology, New Series. Vol. 12C. Springer-Verlag, Berlin, 1983:59–135.

41. Robinson SP, Downton WJS, Millhouse J. Photosynthesis and ion content of leaves and isolated chloroplasts of salt-stressed spinach. *Plant Physiol.* 1983; 73:238–242.

42. Matoh T, Watanabe J, Takahashi E. Sodium, potassium, chloride and betaine concentrations in isolated vacuoles from salt-grown *Atriplex gomelini* leaves. *Plant Physiol.* 1987; 84:173–177.

43. Ziska LH, Dejong TM, Hoffman GF, Mead RM. Sodium and chloride distribution in salt-stressed *Prunus salicina*, a deciduous tree species. *Tree Physiol.* 1991; 8:47–57.

44. Schachtman DP, Munns R. Sodium accumulation in leaves of triticum species that differ in salt tolerance. *Aust. J. Plant Physiol.* 1992; 19:331–340.

45. Leidi EO, Saiz JF. Is salinity tolerance related to Na accumulation in Upland cotton (Gossipum hirsutum) seedlings? *Plant Soil* 1997; 190:67–75.

46. Seemann JR, Critchley C. Effects of salt stress on the growth, ion content, stomatal behaviour and photosynthetic capacity of a salt-sensitive species, *Phaseolus vulgaris* L. *Planta* 1985; 164:151–162.

47. Walker RR, Blackmore DH, Sun Q. Carbon dioxide assimilation and foliar concentrations in leaves of lemon (*Citrus limon* L.) trees irrigated with NaCl or Na_2SO_4. *Aust. J. Plant Physiol.* 1993; 20:173–185.

48. Marler TE, Zozor Y. Salinity influences photosynthetic characteristics, water relations, and foliar composition of *Annona squamosa* L. *J. Am. Soc. Hortic. Sci.* 1996; 121:243–248.

49. Glagoleva TA, Chulanovskaya MV, Pakhomova MV, Voznesenskaya EV, Gamalei YV. Effect of salinity on the structure of assimilating organs and 14C labelling patterns in C3 and C4 plants of Ararat plain. *Photosynthetica* 1992; 26:363–369.

50. Brugnoli E, Bjorkman O. Growth of cotton under continuous salinity stress: influence on allocation pattern, stomatal and non-stomatal components of photosynthesis and dissipation of excess light energy. *Planta* 1992; 187:335–347.

51. Keiper FJ, Chen DM, De Filippis LF. Respiratory, photosynthetic and ultrastructural changes accompanying salt adaptation in culture of *Eucalyptus microcorys*. *J. Plant Physiol.* 1998; 152:564–573.

52. Khavari-Nejad RA, Mostofi Y. Effects of NaCl on photosynthetic pigments, saccharides, and chloroplast ultrastructure in leaves of tomato cultivars. *Photosynthetica* 1998; 35:151–154.

53. Mitsuya S, Takeoka Y, Miyake H. Effects of sodium chloride on foliar ultrastructure of sweet potato (*Ipomoea batatas* Lam.) plantlets grown under light and dark conditions *in vitro*. *J. Plant Physiol.* 2000; 157:661–667.

54. Locy RD, Chang CC, Nielsen BL, Singh NK. Photosynthesis in salt-adapted heterotrophic tobacco cells and regenerated plants. *Plant Physiol.* 1996; 110:321–328.

55. Hernandez JA, Olmos E, Corpas FJ, Sevilla F, Del Rio LA. Salt-induced oxidative stress in chloroplasts of pea plants. *Plant Sci.* 1995; 105:151–167.

56. Maiti RK, Wesche Ebeling P, Nunez Gonzalez A, Moreno Limon SL, Hernandez Pinero JL, Cadenas Avila ML. Some biotic and abiotic factors affecting chloroplast structure and chlorophyll content in bean (*Phaseolus vulgaris* L.) — review. *Agric. Rev.* 2000; 21:255–260.

57. Zhu XG, Wang Q, Zhang QD, Lu CM, Kuang TY. Response of photosynthetic functions of winter wheat to salt stress. *Plant Nutr. Fertilizer Sci.* 2002; 8:180–185.

58. Romero Aranda R, Soria T, Cuartero J. Tomato plant-water uptake and plant-water relationships under saline growth conditions. *Plant Sci.* 2001; 160:265–272.

59. Salama S, Trivedi S, Busheva M, Arafa AA, Garab G, Erdei L. Effects of NaCl salinity on growth, cation accumulation, chloroplast structure and function in wheat cultivars differing in salt tolerance. *J. Plant Physiol.* 1994; 144:241–247.

60. El-Shintinawy F. Photosynthesis in two wheat cultivars differing in salt susceptibility. *Photosynthetica* 2000; 38:615–620.

61. Melesse T, Caesar K. Stomatal and non-stomatal effects of salinity on photosynthesis in faba beans (*Vicia faba* L.). *J. Agron. Crop Sci.* 1992; 168:345–353.

62. Soldatini GF, Monetti C, Waggan MR. Photosynthesis and related parameters in sunflower seedlings as affected by sodium chloride and polyethylene glycol. *Biochem. Physiol. Pflanz.* 1989; 184:49–53.

63. Garcia Legaz MF, Ortiz JM, Garcia Lidon AG, Cerda A. Effect of salinity on growth, ion content and CO_2 assimilation rate in lemon varieties on different rootstocks. *Physiol. Plant.* 1993; 89:427–432.

64. Meinzer FC, Plaut Z, Saliendra NZ. Carbon isotope discrimination, gas exchange, and growth of sugarcane cultivars under salinity. *Plant Physiol.* 1994; 104:521–526.

65. Chartzoulakis KS. Photosynthesis, water relations and leaf growth of cucumber exposed to salt stress. *Sci. Hortic.* 1994; 59:27–35.

66. Garg BK, Gupta IC. Physiology of salt tolerance of arid zone crops. VIII. Sorghum. *Curr. Agric.* 2000; 24:9–22.

67. Ouerghi Z, Cornic G, Roudani M, Ayadi A, Brulfert J. Effect of NaCl on photosynthesis of two species (*Triticum durum* and *T. aestivum*) differing in their sensitivity to salt stress. *J. Plant Physiol.* 2000; 156:335–340.

68. Dionisio Sese ML, Tobita S. Effects of salinity on sodium content and photosynthetic responses of rice seedlings differing in salt tolerance. *J. Plant Physiol.* 2000; 157:54–58.

69. Anjali A, Baig MJ, Anuradha M, Mandal PK, Anand A. Growth and photosynthetic characteristics of Lucerne (*Medicago sativa* L.) genotypes as influenced by salinity of irrigation water. *Indian J. Plant Physiol.* 2001; 6:158–161.

70. Ali-Dinar HM, Ebert G, Ludders P. Growth, chlorophyll content, photosynthesis and water relations in guava (*Psidium guajava* L.) under salinity and different nitrogen supply. *Gartenbauwissenschaft* 1999; 64:54–59.

71. Pardossi A, Malorgio F, Oriolo D, Gucci R, Serra G, Tognoni F. Water relations and osmotic adjustment in *Apium graveolens* during long-term NaCl stress and subsequent relief. *Physiol. Plant.* 1998; 102:369–376.

72. Slatyer RO. Plant response to climatic factors. In: *Proceedings of Uppsala Symposium.* UNESCO, 1973:271–276.

73. Farquhar GD, Sharkey TD. Stomatal conductance and photosynthesis. *Annu. Rev. Plant Physiol.* 1982; 33:317–345.

74. Osmond CB, Bjorkman O, Anderson DJ. *Physiological Processes in Plant Ecology: Toward a Synthesis with Atriplex.* Springer-Verlag, Berlin, 1980.

75. McCree KJ. Whole-plant carbon balance during osmotic adjustment to drought and salinity stress. *Aust. J. Plant Physiol.* 1986; 13:33–43.

76. Shumilova AA, Magomedov IM. Change in photosynthetic metabolism of sugarbeet with increased concentration of sodium chloride in nutrient solution. *Fiziol. Biokim. Kul't. Rast.* 1989; 21:13–17.

77. Chatzoulakis KS, Therios IN, Misopolinos ND, Noitsakis BI. Growth, ion content and photosynthetic performance of salt-stressed kiwifruit plants. *Irrig. Sci.* 1995; 16:23–28.

78. Umezawa T, Shimizu K, Kato M, Ueda T. Effects of non-stomatal components on photosynthesis in soybean under salt stress. *Jpn. J. Trop. Agric.* 2001; 45:57–63.

79. Steduto P, Albrizio R, Giorio P, Sorrentino G. Gas-exchange response and stomatal and non-stomatal limitations to carbon assimilation of sunflower under salinity. *Environ. Exp. Bot.* 2000; 44:243–255.

80. Doring J, Ludders P. Influence of different salts on chlorophyll content, photosynthesis and carbohydrate metabolism pf *Punica granatum* L. *Gartenbauwissenchaft* 1986; 51:21–26.

81. Bhivare VN, Nimbalkar JD, Chavan PD. Photosynthetic carbon metabolism in French bean leaves under saline conditions. *Environ. Exp. Bot.* 1988; 28:117–121.

82. Soldatini GF, Bonicoli A. Variations in $^{14}CO_2$ fixation and gas exchange in leaves induced by moderate stress using polyethylene glycol and sodium chloride in maize seedlings. *Agrochimica* 1988; 32:482–490.

83. Murumkar CV, Chavan PD. Alternations in photosynthetic carbon metabolism of chickpea (*Cicer arietinum* L.) due to imposed NaCl salinity. *Agrochimica* 1993; 37:26–32.

84. Walker RR, Torokfalvy E, Behboudian MM. Photosynthetic rates and solute partitioning in relation to growth of salt-treated pistachio plants. *Aust. J. Plant Physiol.* 1988; 15:787–798.

85. Everard JD, Gucci R, Kann SC, Flore JA, Loescher WH. Gas exchange and carbon partitioning in the leaves of celery (*Apium graveolens* L.) at various levels of root zone salinity. *Plant Physiol.* 1994; 106:281–292.

86. Stoop JMH, Pharr DM. Growth substrate and nutrient salt environment alter mannitol-to-hexose partitioning in celery petioles. *J. Am. Soc. Hortic. Sci.* 1994; 119:237–242.

87. Gucci R, Moing A, Gravano E, Gaudillere JP. Partitioning of photosynthetic carbohydrates in leaves of salt-stressed olive plants. *Aust. J. Plant Physiol.* 1998; 25:571–579.

88. Wieneke J, Fritz R. Influence of salinity on assimilate partitioning in soybeans. *Acta Univ. Agric. Brno* 1985; 33:653–657.

89. Pezeshki SR, DeLaune RD, Patrick Jr WH. Differential response of selected mangrove to soil flooding and salinity: gas exchange and biomass partitioning. *Can. J. For. Res.* 1990; 20:869–874.

90. Glagoleva TA, Chulanovskaya MV. Photosynthetic metabolism and assimilate translocation in C_4 halophytes inhabiting the Ararat valley. *Russ. J. Plant Physiol.* 1996; 43:349–357.

91. Balibrea ME, Dell'Amico J, Bolarin MC, Perez-Alfocea F. Carbon partitioning and sucrose metabolism in tomato plants growing under salinity. *Physiol. Plant.* 2000; 110:503–511.

92. Katerji N, VanHoorn JW, Hamdy A, Mastrorilli M, Karzel EM. Osmotic adjustment of sugarbeets in response to soil salinity and its influence on stomatal conductance, growth and yield. *Agric. Water Manag.* 1997; 34:57–69.

93. Kang SM, Titus JS. Increased proteolysis of senescing rice leaves in the presence of NaCl and KCl. *Plant Physiol.* 1989; 91:1232–1237.

94. Miteva TS, Zhelev ZH, Popova LP. Effect of salinity on the synthesis of ribulose bisphosphate carboxylase/oxygenase in barley leaves. *J. Plant Physiol.* 1992; 140:46–51.

95. Delfine S, Alvino A, Zacchini M, Loreto F. Consequences of salt stress on conductance to CO_2 diffusion, Rubisco characteristics and anatomy of spinach leaves. *Aust. J. Plant Physiol.* 1998; 25:395–402.

96. Wang Y, Nii N. Changes in chlorophyll, ribulose bisphosphate carboxylase-oxygenase, glycine betaine content, photosynthesis and transpiration in *Amaranthus tricolor* leaves during salt stress. *J. Hortic. Sci. Biotechnol.* 2000; 75:623–627.

97. Kasai K, Fukayama H, Uchida N, Mori N, Yasuda T, Oji Y, Nakamura C. Salinity tolerance in *Triticum aestivum-Lophppyrum elongatum* amphiploid and SE disomic addition line evaluated by NaCl effects on photosynthesis and respiration. *Cereal Res. Commun.* 1998; 26:281–287.

98. Martino C-Di, Delfine S, Alvino A, Loreto F. Photorespiration rate in spinach leaves under moderate NaCl stress. *Photosynthetica* 1999; 36:233–242.

99. Rivelli AR, Lovelli S, Perniola M. Effects of salinity on gas exchange, water relations and growth of sunflower (*Helianthus annuus*). *Funct. Plant Biol.* 2002; 29:1405–1415.

100. Taleisnik EL. Salinity effects on growth and carbon balance in *Lycopersicon esculentum* and *L. pennellii*. *Physiol. Plant.* 1987; 71:213–218.

101. Bowman WD. Effect of salinity on leaf gas exchange in two populations of a C_4 nonhalophyte. *Plant Physiol.* 1987; 85:1055–1058.

102. Thind SK, Malik CP. Carboxylation and related reactions in wheat seedlings under osmotic stress. *Plant Physiol. Biochem.* 1988; 15:58–63.

103. Ziska LH, Seemann JR, DeJong TM. Salinity induced limitations on photosynthesis in *Prunus salicina*, a deciduous tree species. *Plant Physiol.* 1990; 93:864–870.

104. Zhu J, Meinzer FC. Efficiency of C_4 photosynthesis in *Atriplex lentiformis* under salinity stress. *Aust. J. Plant Physiol.* 1999; 26:79–86.

105. Loreto F, Centritto M, Chartzoulakis K. Photosynthetic limitations in olive cultivars with different sensitivity to salt stress. *Plant Cell Environ.* 2003; 26:595–601.

106. Makela P, Karkkainen J, Somersalo S. Effect of glycinebetaine on chloroplast ultrastructure, chlorophyll and protein content, and Rubisco activities in tomato grown under drought or salinity. *Biol. Plant.* 2000; 43:471–475.

107. Solomon A, Beer S, Waisel Y, Jones GP, Paleg LG. Effects of NaCl on the carboxylating activity of Rubisco from *Tamarix jordanis* in the presence and absence of prolinerelated compatible solutes. *Physiol. Plant.* 1994; 90:198–204.

108. Greenway H, Munns R. Mechanism of salt tolerance in nonhalophytes. *Annu. Rev. Plant Physiol.* 1980; 31:149–190.

109. Munns R, Termaat A. Whole plant responses to salinity. *Aust. J. Plant Physiol.* 1986; 13:143–160.

110. Ericson MC, Alfinito SH. Proteins produced during salt stress in tobacco cell culture (*Nicotiana tabacum*), quantitative and qualitative changes during stress resistance. *Plant Physiol.* 1984; 74:506–509.

111. Ghosh S, Bagchi S, Majumder AL. Chloroplast fructose-1,6-bisphosphatase from *Oryza* differs in salt tolerance property from the *Porteresia* enzyme and is protected by osmolytes. *Plant Sci.* 2001; 160:1171–1181.

112. Heuer B, Plaut Z. Activity and properties of RuBP carboxylase of sugarbeet plants grown under saline conditions. *Physiol. Plant.* 1982; 54:505–510.

113. Winicov I, Seeman JR. Expression of genes for photosynthesis and the relationship to salt tolerance of alfalfa (*Medicago sativa*) cells. *Plant Cell Physiol.* 1990; 31:1155–1162.

114. Winicov I, Button JD. Accumulation of photosynthesis gene transcript in response to sodium chloride by salt-tolerant alfalfa cells. *Planta* 1991; 183:478–483.

115. Schaffer HJ, Forsthoefel NR, Cushman JC. Identification of enhancer and silencer regions involved in salt-responsive expression of CAM genes in the facultative halophyte *Mesembryanthemum crysttalinum*. *Plant Mol. Biol.* 1995; 28:205–218.

116. Forsthoefel NR, Vernon DM, Cushman JC. A salinity-induced gene from the halophyte *Mesembryanthemum crystallinum* encodes a glycolic enzyme, cofactor-independent phosphoglyceromutase. *Plant Mol. Biol.* 1995; 29:213–226.

117. McElwain EF, Bohnert HJ, Thomas JC. Light moderates the induction of phosphoenolpyruvate carboxylase by NaCl and abscisic acid in *Mesembryanthemum crysttalinum*. *Plant Physiol.* 1992; 99:1261–1264.

118. Cushman JC, Meyer G, Michalowski CB, Schmitt JM. Salt stress leads to differential expression of two isogenes of PEP carboxylase during CAM induction in the common ice. *Plant Cell* 1989; 1:715–725.

119. Vernon DM, Bohnert HJ. Increased expression of a myoinositol methyl transferase in *Mesembryanthemum crystallinum* is part of a stress response distinct from Crassulacean acid metabolism induction. *Plant Physiol.* 1992; 99:1695–1698.

120. Slesak I, Miszalski Z. Stress reactions in *Mesembryanthemum crystallinum* L. *Wiadomosci. Bot.* 1999; 43:47–58.

121. DeRocher EJ, Bohnert HJ. Development and environmental stress employ different mechanisms in the expression of a plant gene family. *Plant Cell* 1993; 5:1611–1625.

122. Morcuende R, Perez P, MartinezCarrasco R, DelMolino IM, DelaPuente L. Long- and short-term effects of decreased sink demand on carbohydrate levels and photosynthesis in wheat leaves. *Plant Cell Environ.* 1996; 19:1203–1209.

123. Foyer C. Feedback inhibition of photosynthesis through source-sink regulation in leaves. *Plant Physiol. Biochem.* 1998; 26:483–492.

124. Iglesias DJ, Lliso I, Tadeo FR, Talon M. Regulation of photosynthesis through source: sink imbalance in citrus is mediated by carbohydrate content in leaves. *Physiol. Plant.* 2002; 116:563–572.

125. Geiger DR. Effect of translocation and assimilate demand on photosynthesis. *Can. J. Bot.* 1976; 54:2337–2345.

126. Kaiser G, Martinoia E, Wiemken A. Rapid appearance of photosynthetic products in the vacuoles isolated from barley mesophyll protoplasts by a new fast method. *Z. Pflanzenphysiol.* 1982; 107:103–113.

127. Sicher RC. Sucrose biosynthesis in photosynthetic tissue: rate-controlling factors and metabolic pathway. *Physiol. Plant.* 1986; 67:118–121.

128. Stitt M, Kurzel B, Heldt HW. Control of photosynthetic sucrose synthesis by fructose 2,6-bisphosphate. *Plant Physiol.* 1984; 75:554–560.

129. Trevanion SJ. Regulation of sucrose and starch synthesis in wheat (*Triticum aestivum* L.) leaves: role of fructose-2,6-bisphosphate. *Planta* 2002; 215:653–665.

130. Kruger NJ, Beevers H. Effect of fructose 2,6-bisphosphate on the kinetic properties of cytoplasmic fructose 1,6-bisphosphatase from germinating castor bean endosperm. *Plant Physiol.* 1984; 76:49–54.

131. Stitt M, Heldt HW. Control of photosynthetic sucrose synthesis by fructose 2,6-bisphosphate. *Plant Physiol.* 1985; 79:599–608.

132. Huber SC. Fructose 2,6-bisphosphate as a regulatory metabolite in plants. *Annu. Rev. Plant Physiol.* 1986; 37:233–246.

133. Boag S, Portis AR. Metabolite levels in the stroma of spinach chloroplasts exposed to osmotic stress: effects of the pH of the medium and exogenous dihydroxyacetone phosphate. *Planta* 1985; 165:416–423.

134. Skriver K, Mundy J. Gene expression in response to abscisic acid and osmotic stress. *Plant Cell* 1990; 2:503–512.

135. Godoy JA, Pardo JM, Pintor-Toro JA. A tomato cDNA inducible by salt stress and abscisic acid: nucleotide sequence and expression pattern. *Plant Mol. Biol.* 1990; 15:695–705.

136. Ke YQ, Pan TG. Effects of NaCl stress on water metabolism, photosynthetic rate and endogenous ABA in sweet potato leaves. *Plant Nutr. Fertilizer Sci.* 2001; 7:337–342.

137. He T, Cramer GR. Abscisic acid concentrations are correlated with leaf area reductions in two salt stressed rapid-cycling *Brassica* species. *Plant Soil* 1996; 179: 25–33.

138. LaRosa PC, Hasegava PM, Rhodes D, Clithero JM, Watad A-EA, Bressan RA. Abscisic acid stimulated osmotic adjustment and its involvement in adaptation of tobacco cells to NaCl. *Plant Physiol.* 1987; 85:174–181.

139. Starck Z, Karwowska R. Effect of salt-stresses on the hormonal regulation of growth, photosynthesis and distribution of C_{14} assimilates in bean plants. *Acta Soc. Bot. Pol.* 1978; 47:245–267.

140. Ashraf M, Fakhra K, Rasul E. Interactive effects of gibberellic acid (GA_3) and salt stress on growth, ion accumulation and photosynthetic capacity of two spring wheat (*Triticum aestivum* L.) cultivars differing in salt tolerance. *Plant Growth Regul.* 2002; 36:49–59.

141. Aldesuquy HS, Gaber AM. Effect of growth regulators on *Vicia faba* plants irrigated by seawater. Leaf area, pigment content and photosynthetic activity. *Biol. Plant.* 1993; 35:519–527.

142. Chaudhuri K, Choudhuri MA. Effects of short-term NaCl stress on water relations and gas exchange of two jute species. *Biol. Plant.* 1998; 40:373–380.

143. Allakhverdiev SR, Ganieva R, Stoyanov I. Effect of kinetin on the kinetics of delayed and variable fluorescence of *Triana bogotensis* K. leaves in salt stress. *Fiziol. Rast. (Sofia)* 1992; 18:73–78.

144. Radi F, Heikal MM, Abdel-Rahman AM, El-Deep BAA. Interactive effects of salinity and phytohormones on growth and plant water relationship parameters in maize and safflower plants. *Rev. Roum. Biol. Ser. Biol. Veg.* 1988; 27–37.

145. Chu C, Dai Z, Ku MSB, Edwards GE. Induction of crassulacean acid metabolism in the facultative halophyte *Mesembryanthemum crystallinum* by abscisic acid. *Plant Physiol.* 1990; 93:1253–1260.

146. Cushman JC, Vernon DM, Bohnert HJ. Abscisic acid and the transcriptional control of CAM induction during salt stress in the common ice plant. In: Verma DPS, ed. *Control of Plant Gene Expression.* CRC Press, Boca Raton, FL, 1993:287–300.

147. Schmitt JR, Piepenbrock M. Regulation of phosphoenolpyruvate carboxylase and crassulacean acid metabolism induction in *Mesembryanthemum crystallinum* L. by cytokinin. *Plant Physiol.* 1992; 99:1664–1669.

148. Kriedemann PE. Stomatal and photosynthetic limitations to leaf growth. *Aust. J. Plant Physiol.* 1986; 13:15–31.

149. Waldron L, Terry N, Nemson J. Diurnal cycles of leaf extension in unsalinized and salinized *Beta vulgaris.* *Plant Cell Environ.* 1985; 8:207–211.

150. Thiel G, Lynch J, Lauchli A. Short-term effects of salinity stress on the turgor and elongation of growing barley leaves. *J. Plant Physiol.* 1988; 132:38–44.

151. Neumann PM, van Volkenburgh E, Cleland RE. Salinity stress inhibits bean leaf expansion by reducing turgor not wall extensibility. *Plant Physiol.* 1988; 88:233–237.

152. Hirota O, Villavicencio E, Chikushi J, Takeuchi S, Nakano Y. Effect of water irrigation at fruit maturity stage on transpiration rate and growth in sweet pepper (*Capsicum annuum*). *J. Fac. Agric.* 1999; 44:39–47.

153. Boyer JS. Plant productivity and environment. *Science* 1982; 218:443–448.

154. Heuer B, Meiri A, Shalhevet Y. Salt tolerance of eggplant. *Plant Soil* 1986; 95:9–13.

155. Pessarakli M, Tucker TC, Nakabayashi K. Growth response of barley and wheat to salt stress. *J. Plant Nutr.* 1991; 14:331–340.

156. Cordovilla MP, Ligero F, Lluch C. The effect of salinity on N fixation and assimilation in *Vicia faba.* *J. Exp. Bot.* 1994; 45:1483–1488.

157. White JW, Izquierdo J. Physiology of yield potential and stress tolerance. In: van Schoonhoven A, Voysest O. eds. *Common Beans: Research for Crop Improvement.* Wallingford, U.K., C.A.B. International, 1991: 287–382.

158. Wegmann L. The Influence of Salt Stress on the Morphology, Physiology and Economic Use of the White Mangrove *Laguncularia racemosa.* Thesis, Tierarztliche Hochschule, Hannover, Germany, 1998:153.

159. Robinson SP, Downton JW, John D, Millhouse JA. Photosynthesis and ion content of leaves isolated chloroplasts of salt-stressed spinach. *Plant Physiol.* 1983; 73:238–242.

160. Fox TC, Geiger DR. Osmotic response of sugar beet source leaves at CO_2 compensation point. *Plant Physiol.* 1986; 80:239–241.

161. Hassanein AA. Physiological responses induced by shock and gradual salinization in rice (*Oryza sativa* L.) seedlings and the possible roles played by glutathione treatment. *Acta Bot. Hung.* 1999–2000; 42:139–159.

162. Plaut Z, Federman E. Acclimation of CO_2 assimilation in cotton leaves to water stress and salinity. *Plant Physiol.* 1991; 97:515–522.

163. Jafri AZ, Ahmad R. Effect of soil salinity on leaf development, stomatal size and its distribution in cotton (*Gossypium hirsutum* L). *Pak. J Bot.* 1995; 27:297–303.

164. Sakamoto A, Murata N, Forde B. Genetic engineering of glycinebetaine synthesis in plants: current status and implications for enhancement of stress tolerance. *J. Exp. Bot.* 2000; 51:81–88.

165. Mittova VO, Igamberdiev AU. Influence of salt stress on respiration metabolism in higher plants. *Izv. Akad. Nauk Ser. Biol.* 2000; 3:322–328.

166. Dix PJ, Street HE. Sodium chloride-resistant cultured cell lines from *Nicotiana sylvestris* and *Capsicum annum. Plant Sci. Lett.* 1975; 5:231–237.

167. Hasegawa PM, Bressan RA, Handa AK. Growth characteristics of NaCl-selected and non-selected cells of *Nicotiana tabacum* L. *Plant Cell Physiol.* 1980; 21:1347–1355.

168. Lerner HR. Adaptation to salinity at the plant cell level. *Plant Soil* 1985; 89:3–14.

169. Schnapp SR, Bressan RA, Hasegawa PM. Carbon use efficiency and cell expansion of NaCl-adapted tobacco. *Plant Physiol.* 1990; 93:384–388.

170. Plaut Z, Bachmann E, Oertli JJ. The effect of salinity on light and dark CO_2 fixation of salt-adapted and unadapted cell culture of *Atriplex* and tomato. *J. Exp. Bot.* 1991; 42:531–535.

171. Amzallag GN, Lerner HR, Poljakoff-Myber A. Exogenous ABA as a modulator of the response of sorghum to high salinity. *J. Exp. Bot.* 1990; 41:1529–1534.

172. Singh NK, Handa AK, Hasegawa PM, Bressan RA. Proteins associated with adaptation of cultured tobacco cells to NaCl. *Plant Physiol.* 1985; 79:126–137.

173. Rozema J. Plant responses to atmospheric carbon dioxide enrichment interactions with some soil and atmospheric conditions. *Vegetatio* 1993; 104:173–190.

174. Schwartz M, Gale J. Growth response to salinity at high levels of carbon dioxide (carbon-3 metabolic pathway plants, *Phaseolus vulgaris*, kidney beans, and *Xanthium strumarium*, carbon-4 pathway salt-sensitive *Zea mays*, maize, and the carbon-4 pathway halophyte *Atriplex halimus*). *J. Exp. Bot.* 1984; 35:193–196.

175. Mavrogianopoulus GN, Spanakis J, Tsikalas P. Effect of carbon dioxide enrichment and salinity on photosynthesis and yield in melon. *Sci. Hortic.* 1999; 79:51–63.

176. Chen K, Hu GQ, Keutgen N, Janssens MJJ, Lenz F. Effects of NaCl salinity on pepino (*Solanum muricatum Ait.*). I. Growth and yield. *Sci. Hortic.* 1999; 81:25–41.

177. Li JH, Sagi M, Gale J, Volokita M, Novoplansky A. Response of tomato plants to saline water as affected by carbon dioxide supplementation. I. Growth, yield and fruit quality. *J. Hortic. Sci. Biotechnol.* 1999; 74:232–237.

178. Bray S, Reid DM. The effect of salinity and CO_2 enrichment on the growth and anatomy of the second trifoliate leaf of *Phaseolus vulgaris. Can. J. Bot.* 2002; 80:349–359.

179. Zeroni M, Gale J. Response of 'Sonia' roses to salinity at three levels of ambient CO_2. *J. Hortic. Sci.* 1989; 64:503–511.

180. Nicolas ME, Munns R, Samarakoon AB, Gifford RM. Elevated CO_2 improves the growth of wheat under salinity. *Aust. J. Plant Physiol.* 1993; 20:349–360.

181. Flanagan LB, Jefferies RL. Photosynthetic and stomatal response of the halophyte, *Plantago maritime* L.

to fluctuations in salinity. *Plant Cell Environ.* 1989; 12:559–568.

182. Cushman JC, Michalowski CB, Bohnert HJ. Developmental control of crassulacean acid metabolism inducibility by salt stress in the common ice plant. *Plant Physiol.* 1990; 94:1137–1142.

183. Kholodova VP, Neto DS, Meshcheryakov AB, Borisova NN, Aleksandrova SN, Kuznetsov V. Can stress-induced CAM provide for performing the developmental program in *Mesembryanthemum crystallinum* plants under long-term salinity? *Russ. J. Plant Physiol.* 2002; 49:336–343.

184. Elhaak MA, Migahid MM, Wegmann K. Ecophysiological studies on *Euphorbia paralias* under soil salinity and sea water spray treatments. *J. Arid Environ.* 1997; 35:459–471.

185. Luttge U. The role of crassulacean acid metabolism (CAM) in the adaptation of plants to salinity. *New Phytol.* 1993; 125:59–71.

186. Adams P, Thomas JC, Vernon DM, Bohnert HJ, Jensen RG. Distinct cellular and organismic responses to salt stress. *Plant Cell Physiol.* 1992; 33:1215–1223.

187. Saitou K, Nakamura Y, Kawamitsu Y, Matsuoka M, Samejima M, Agata W. Changes in activities and levels of pyruvate, orthophosphate dikinase with induction of crassulacean acid metabolism in *Mesembryanthemum crystallinum* L. *Jpn. Italic NOT ALLOWEDJ. Crop Sci. 1991; 60:146–152.*

188. Nandan PR, Karmarkar SM. Study of photosynthetic enzymes from control and sodium chloride treated *Chlorophytum* cultivars. *Geobios* 1992; 19:256–259.

189. Sprague GF, Alexander DE, Dudle JW. Plant breeding and genetic engineering perspective. *BioScience* 1980; 30:17–21.

190. Epstein E, Norlyn JD, Rush DW, Kingsbury RW, Kelley DB, Cunningham GA, Wrona AF. Saline culture of crops — a genetic approach. *Science* 1980; 210:399–404.

191. Noble CL, Rogers ME. Arguments for the use of physiological criteria for improving the salt tolerance in crops. *Plant Soil* 1992; 146:99–107.

192. Flowers TJ, Garcia A, Koyama M, Yeo AR. Breeding for salt tolerance in crop plants — the role of molecular biology. *Acta Physiol. Plant.* 1997; 19:427–433.

193. Munns R, Husain S, Rivelli AR, James RA, Condon AG, Lindsay MP, Lagudah ES, Schachtman DP, Hare RA. Avenues for increasing salt tolerance of crops, and the role of physiologically based selection traits. *Plant Soil* 2002; 247:93–105.

194. Yoshida K. Plant biotechnology — genetic engineering to enhance plant salt tolerance. *J. Biosci. Bioeng.* 2002; 94:585–590.

195. Rathinasabapathi B. Metabolic engineering for stress tolerance: Installing osmoprotectant synthesis pathways. *Ann. Bot.* 2000; 86:709–716.

196. Zhang HX, Hodson JN, Williams JP, Blumwald E. Engineering salt-tolerant Brassica plants: characterization of yield and seed oil quality in transgenic plants with increased vacuolar sodium accumulation. *Nat. Biotechnol.* 2001; 19:765–768.

41 Photosynthesis under Drought Stress

Habib-ur-Rehman Athar
Department of Botany, Institute of Pure and Applied Biology,
Bahauddin Zakariya University

Muhammad Ashraf
Department of Botany, University of Agriculture

CONTENTS

I. INTRODUCTION

In arid and semiarid regions, drought stress usually occurs together with light and high-temperature stress. It has been widely reviewed in literature that water deficit is a major constraint for plant survival. Considerable progress has so far been made to determine the various physiological and biochemical processes that are essential to understand the competitive ability of plants to survive under different stress environments [1]. In nature, plants are either subjected to slowly developing water shortage (days to weeks) or face short-term water deficits (hours to days). To cope with these water stress conditions, plants use different types of strategies. However, generally, plant resistance to drought has been divided into escape, avoidance and tolerance strategies [2,3]. Escape strategy is more important in plants with short life cycle, high growth rate and gas exchange, using maximum available resources [4,5]. Plants can also resist drought conditions by avoiding tissue dehydra-

tion. It involves minimization of water loss by closing stomata and reduced light capture, and maximization of water uptake [1]. Water deficit reduces the growth of plants by reducing the photosynthetically active leaf area, the most important factor affecting crop productivity, by altering either the efficiency with which photosynthates are converted to aid new growth or the rate at which they are used in maintaining the existing dry matter [6].

Drought stress lowers the water potential of the growing plants, leading to dehydration, decreased stomatal conductance, altered chlorophyll fluorescence, photoinhibition of photosystem II (PSII), conformational changes in membrane-bound ATPase enzyme complex, as well as decrease in both activity and concentration of Rubisco enzyme [6]. The reduction in photosynthesis can be attributed to the reduced intercellular CO_2 (C_i) due primarily to stomatal limitations and partly to metabolic factors, which may be indicated by increase in C_i [7]. The reduced utilization of electrons due to reduced PCR cycle may result in reduced electron transport. In this case, thermal dissipation of excitation energy (non-photochemical quenching or NPQ) may play an important role in protecting the photosynthetic apparatus from oxidative stress [8]. Photochemical reactions associated with PSII are more susceptible to water stress than those of PSI [6]. Furthermore, it causes a significant reduction in the content of photosynthetic pigments, proteins, and lipids while under severe water stress, thylakoid membranes are disorganized and disappear as well.

In this chapter, we will focus on how drought stress affects stomatal and nonstomatal factors responsible for the regulation of photosynthesis in plants. The changes occurring in gaseous exchange properties and carbon metabolism due to water deficit are discussed. Comparison between plants differing in mode of CO_2 fixation is also made so as to assess how these different types of plants respond to drought stress.

II. PHOTOSYNTHESIS

A. CONTRIBUTION TO PLANT GROWTH AND YIELD

Photosynthesis is an important process that is carried out in all living land plants. It involves the conversion of light energy into chemical energy [9,10]. It is a primary process in plant productivity. The site of photosynthesis in plants is predominantly in the green leaf and the productivity of plants directly depends upon the chlorophyll bearing surface area, irradiance, and their potential to utilize CO_2 [11]. Plant biomass production depends upon the amount of water use for growth as well as on water use efficiency (WUE). Productivity in crop plants may be increased by WUE, and one of the major factors for enhanced WUE is net CO_2 assimilation rate [12]. Thus, final biological or economical yield can be increased by increasing the net CO_2 assimilation rate [9].

III. DROUGHT EFFECTS ON PHOTOSYNTHESIS

A. STOMATAL LIMITATIONS

1. Stomatal Closure — A Major Response to Drought

Stomatal closure is one of the earliest responses to drought protecting the plant from extensive water loss, which might otherwise result in cell dehydration, xylem cavitation, and death [1]. Moreover, stomata often close in response to drought before any change in leaf water potential or leaf water content is detect-able [13,14]. It is now well established that there is a drought-induced root-to-leaf signaling, promoted by soil drying and reaching the leaf through the transpiration stream, which induces the closure of stomata. This chemical signal has been shown to be abscisic acid (ABA), which is synthesized in roots in response to soil drying [15]. However, its role is not simple. The complex regulation of stomatal conductance is related to interspecific differences in response of stomatal conductance to leaf water potential, relative water content, ABA and other factors [9,16].

Stomatal closure in response to drought is generally assumed to be the main cause of drought-induced decrease in photosynthesis, since stomatal closure decreases CO_2 availability in the mesophyll. Moreover, the CO_2/O_2 ratio drops [9,16–21] and photorespiration increases under drought [22], whereas under severe drought, where complete stomatal closure occurs, photosynthesis and photorespiration both decline, and thermal dissipation increases to account for up to 90% of the total dissipation [23] and photoinhibition or photodamage of PSII. However, there is a strong evidence that drought-induced stomatal closure also affects mesophyll metabolism thereby reducing the photosynthetic capacity. Stomatal limitations that cause inhibition in photosynthesis have been frequently described and reviewed by various scientists [1,16,20,24–26].

Several researchers have proposed the use of stomatal conductance (g_s) as an indicator to assess the inflexion point between stomatal and nonstomatal limitations to photosynthesis under drought [16,22, 26–28]. For example, Flexas et al. [22] reported that the most common parameters, reflecting photosynthetic activity, show a more comparable response when g_s is taken as an indicator of water stress. These parameters include net CO_2 assimilation rate (A_n), internal CO_2 concentration (C_i), the estimated gross photosynthesis (A_g), electron transport rate (ETR), the ratios of ETR/A_n and ETR/A_g, dark leaf respiration (R_D), predawn ratio of variable to maximal chlorophyll fluorescence (F_v/F_m), NPQ of chlorophyll fluorescence at midday, and parameters derived from analysis of A_n/C_i curves, such as apparent corboxylation efficiency (ε), leaf respiration (L_R), CO_2 compensation point (Γ), and the CO_2 saturation rate of photosynthesis (A_{Sat}). All these parameters were found to be strongly correlated to g_s in both field grown and potted grapevine plants [22]. Similarly, other authors also found similar relationships in different crops [20,21]. Therefore, downregulation of photosynthesis depends more on CO_2 availability in the chloroplast (i.e., stomatal closure and mesophyll resistance) than on leaf water potential or leaf water content [22,26,29]. However, relationship between different photosynthetic parameters and g_s was not observed with relative water content (RWC) or leaf water potential, i.e., decreased photosynthesis caused by drought occurred at different leaf water status in different species, albeit at similar stomatal conductance.

2. Net CO₂ Assimilation Rate (A) and Photosynthetic Potential (A_{pot}) as Affected by Drought

The maximum rate of A under saturating CO_2 (C_a (ambient CO_2), C_i (intercellular CO_2), and C_c (chloroplastic CO_2)) and light in fully hydrated leaves is called photosynthetic potential (A_{pot}). Despite the evidence that stomata play a major role in the limitation of carbon assimilation under mild stress, conflicting results are also reported [7]. These discrepancies, which are mainly related to the onset of metabolic changes induced by water deficit, may be explained by differences in the rate of imposition and severity of stress, plant developmental stage and leaf age, and type of species studied. Also, the approaches used to assess limitation may contribute to the variability in responses obtained, e.g., A/C_i curves, which are generally due to high C_i [1].

The photosynthetic rate (A) of leaves of both C_3 and C_4 plants decreases as their RWC and water potential decrease (Table 41.1) [9,17–19,30]. Stomatal limitation is considered to reduce both A and CO_2 concentrations in the intercellular spaces of the leaf (C_i), which inhibit metabolism [31–33]. The use of large C_a to overcome small g_s so as to increase C_i and C_c, and restore A to A_{pot} has been a key point in water relation studies. Restoration of A to A_{pot} occurs over a wide range of RWC from 100% to 80%, but at more severe stress, A_{pot} decreases and cannot be restored by elevated C_a. According to Lawlor and Cornic [7] there are two general types of relations of A_{pot} to RWC, which are called as type 1 [34] and type 2. Type 1 has two phases. In the first phase, as RWC decreases from 100% to 75%, A_{pot} is unaffected, but decreased stomatal conductance results in smaller A, C_i, and C_c. Chloroplastic CO_2 (C_c) may reach to compensation point [35]. Downregulation of electron transport occurs by energy quenching mechanisms, and changes in carbohydrate and nitrogen metabolism caused by low C_i are considered acclamatory. This phase is reversible since it can restore A to A_{pot} at elevated level of CO_2 [7,19,36]. In the second phase, below 70% RWC, there is a metabolic inhibition which results in smaller A_{pot}. In contrast, in type 2 response, A_{pot} and stimulation of A by elevated CO_2 decreases progressively at RWC 100% to 80%. In this process, g_s leads to lower

C_i and C_c, but g_s becomes less important, and metabolic limitations are more important. This is not regarded as a consequence of the effect of low C_i or C_c on nitrate reductase or sucrose phosphate synthase activity [7,36].

Decrease in A_{pot} is considered to be caused by limited RuBP synthesis resulting from decreased ATP synthesis. Recently, Medrano et al. [41] demonstrated that RuBP content was reduced (Table 41.1) due to water shortage and a very strong correlation was observed between RuBP content and leaf water content. Moreover, RuBP response to water deficit was quite similar to that of A_{sat}, which strongly suggests that A_{sat} was downregulated by decreased capacity for RuBP regeneration, and not by decreased electron transport [41]. This is consistent with other reports that RuBP regeneration is strongly affected by drought [58,59]. Working with sunflower, Tezara et al. [37] reported that decreased capacity for RuBP regeneration might be related to drought-induced impairment of chloroplastic ATPase. Furthermore, impaired ATP synthesis even under mild water deficit has been well documented [37]. Decrease in ATP synthesis either due to ATP synthase activity [37,70] or due to increased ion (specifically Mg^{2+}) concentrations occurs in chloroplasts as RWC falls [71]. Therefore, CO_2 depletion is not a primary effect [72]. Thus, in this type of response (type 2), limitations to A_{pot} is caused by inadequate ATP supply, but not due to CO_2.

By summarizing, it is suggested that either water deficit has no effect on photosynthetic metabolism until a threshold is reached, below which it is impaired (type 1), or caused a progressive inhibition of metabolism [7].

3. Internal CO₂ Concentration (C_i) under Drought Stress

Flexas and Medrano [26] showed that decreased C_i confirms the predominance of stomatal limitations in restricting photosynthetic rate in early phase of water loss. However, metabolic changes are responsible for loss of photosynthetic potential during this phase [7]. A number of studies showing inhibition of enzyme activities under water stress and restoration of activities of these enzymes by placing water-stressed plants in high CO_2 for a number of hours strongly suggest that CO_2 availability in the chloroplast is mainly regulated by g_s [19,29,73]. Similarly, with 90% to 75% RWC, increasing ambient CO_2 to 5% restores A fully to A_{pot} of the control leaves [19,31,35]. In other studies, restoration of A to A_{pot} required 15% ambient CO_2, whereas this high ambient CO_2 inhibits metabolism in C_4 plants [74], and reduces A in sunflower at high as well as low RWC [37].

TABLE 41.1
Effect of Drought Stress on Photosynthesis as Reflected from Various Photosynthetic Parameters and Their Source

Serial No.	Photosynthetic Parameter	Effects	Crop	Ref.
1	Net CO_2 assimilation rate (A)	Decreased	*Helianthus annuus*	[37,38]
			Vitis vinifera	[21,22,39–41]
			Triticum aestivum	[42–44]
			Abutilon theophrasti	[45]
			Amaranthus retroflexus	[45]
			Betula ermanii	[46]
			Lycopersicon esculentum	[47,48]
2	Internal CO_2 (C_i)	Decreased	*Vitis vinifera*	[22]
			Triticum aestivum	[44]
			Betula ermanii	[46]
			Pachyrhizus ahipa	[49]
		Increased	*Helianthus annuus*	[38]
			Lycopersicon esculentum	[48]
3	Stomatal conductance (g_s)	Decreased	*Helianthus annuus*	[37,38]
			Vitis vinifera	[21,22,26,39–41]
			Triticum aestivum	[42–44]
			Abutilon theophrasti	[45]
			Amaranthus retroflexus	[45]
			Betula ermanii	[46]
			Lycopersicon esculentum	[47,48]
4	Water use efficiency (WUE)	Decreased	*Helianthus annuus*	[38]
			Vitis vinifera	[41]
			Triticum aestivum	[43]
			Hordeum vulgare	[50]
5	Transpiration (E)	Decreased	*Triticum aestivum*	[43]
			Lycopersicon esculentum	[47]
			Brassica napus	[47]
6	Photochemical quenching (qP)	Decreased	*Sorghum bicolor*	[51]
			Eragrostis curvula	[52]
		Unchanged	*Helianthus annuus*	[37,42]
			Hordeum vuulgare	[50]
		Increased	*Betula ermanii*	[46]
7	Nonphotochemical quenching (qNP)	Increased	*Vitis vinifera*	[22,26]
			Triticum aestivum	[42,44]
			Sorghum bicolor	[51]
			Talinum triangulare	[53]
		Decreased under severe stress	*Vitis vinifera*	[23]
			Eragrostis curvula	[52]
			Betula ermanii	[46]
8	F_v/F_m	Decreased	*Eragrostis curvula*	[52]
			Pachyrhizus ahipa (Wedd.) Parodi	[49]
		Unchanged	*Vitis vinifera*	[20,22]
			Triticum aestivum	[42]
			Hordeum vulgare	[50]
			Sorghum bicolor	[51]
9	F'_v/F'_m	Decreased	*Vitis vinifera*	[40]
			Pachyrhizus ahipa (Wedd.) Parodi	[49]
			Triticum aestivum	[42]
			Eragrostis curvula	[52]
		Increased	*Betula ermanii*	[46]
10	Quantum yield of PSII electron transport (Φ_{PSII})	Decreased	*Sorghum bicolor*	[51]

			Talinum triangulare	[53]
			Vitis vinifera	[40]
			Triticum aestivum	[42]
			Eragrostis curvula	[52]
		Unchanged	*Hordeum vulgare*	[50]
		Increased	*Betula ermanii*	[46]
11	Electron transport rate (ETR)	Decreased	*Phaseolus vulgaris*	[54]
			Helianthus annuus	[38]
			Vitis vinifera	[22,39]
			Triticum aestivum	[55]
		Unchanged	*Helianthus annuus*	[56]
			Evergreen bush	[57]
		Increased	*Betula ermanii*	[46]
			Gossypium hirsutum	Lei and Kitao (2003, personal communication)
12	RuBP content	Decreased	*Helianthus annuus*	[37,38,58]
			Nicotiana tabaccum	[59]
			Gossypium hirsutum	[60]
		Unaffected	*Helianthus annuus*	[61]
13	Rubisco protein	Decreased	*Helianthus annuus*	[38]
			Glycine max	[62]
		Unaffected	*Helianthus annuus*	[37]
			Trifolium spp.	[63]
			Vitis vinifera	[26]
14	Rubisco activity	Decreased	*Helianthus annuus*	[38]
			Vitis vinifera	[40]
			Trifolium spp.	[63]
			Triticum aestivum	[64]
			Gossypium hirsutum	[60]
			Lycopersicon esculentum	[65]
		Little affected	*Helianthus annuus*	[58]
		Unaffected	–	[66]
			French bean	[67]
15	ATP	Decreased	*Helianthus annuus*	[37,38]
			–	[68]
			Lupin	[69]
		Unaffected	*Helianthus annuus*	[61]
16	NADPH	Unaffected	–	[68]
17	Xanthophyll cycle	Increased	*Vitis vinifera*	[23]

As the stomata close, the CO_2 inside the leaf (C_i) initially declines with increasing stress and then increases as drought becomes more severe [9]. If C_i is high, this reflects the inaccuracies in the C_i calculations under drought, i.e., heterogenous stomatal closure and cuticular conductance [36], which tend to overestimate C_i. The decrease in C_i indicates that stomatal limitations dominate with moderate drought, irrespective of any metabolic impairment. However, at a certain stage of water stress, shown by a threshold value of g_s, C_i frequently increases, indicating the predominance of nonstomatal limitations to photosynthesis. In most cases, the point at which C_i starts to increase, which we call the C_i inflexion point, occurs at g_s around 50 mmol $H_2O/$ m²/sec. However, in *Medicago sativa* and *Abutilon theoprasti* the C_i inflexion point was observed at higher g_s [27,75].

B. Nonstomatal Limitations

1. Effects of Water Shortage on Photochemical Events

a. Light reaction

Under well-watered conditions, C_3 plants use a large fraction of absorbed light through photosynthesis and photorespiration [76]. As the light intensity increases, photorespiration in C_3 plants also increases. Photorespiration also increases in droughted plants

[77]. Under water stress, requirement of light decreases and low light can saturate the photosynthetic apparatus [9]. This excess light can cause severe damages to photosynthetic apparatus. To get rid of excess light, plants prevent its absorption, lose chlorophyll, or can divert the absorbed light as thermal dissipation or photorespiration. If plants are unable to cope with excess energy, the production of highly reactive molecules is exacerbated. These molecules, generated within the chloroplast, can cause oxidative damage to photosynthetic apparatus [1,78–80]. Furthermore, it has been suggested that tolerance to drought in plants depends on the ability of plants to process active oxygen species [81].

As described earlier, drought stress decreases the requirement of light energy, and excess light absorbed by the photosynthetic apparatus may cause photoinhibition. Although drought stress causes a wide range of structural and functional modifications of the photosynthetic apparatus, certain specific sites of the photosynthetic apparatus are identified as the main targets of damage [82]. It is generally accepted that the primary target of damage is PSII. Photoinhibition results in the impairment of electron transport followed by the selective degradation of D1 reaction center protein [83]. The term photoinhibition (PI) is defined as decrease in photosynthetic activity that occurs upon excess illumination. Photoinhibition in PSII and PSI was reported long ago [84–87]. If the absorbed light energy that ultimately reaches the reaction centers exceeds its consumption, the photosynthetic apparatus can be injured [88,89]. There is strong evidence which indicates that PSII is more responsible for it, because D1 protein has long been considered the primary target to PI [90–92]. It is also because PSI is more stable than PSII under strong light [93].

As described earlier, excessive illumination particularly under water deficit causes photo-oxidative damages. There are two types of photo-oxidative damage:

1. Reversible photoinhibition
2. Irreversible photoinhibition

Reversible photoinhibition. PSII can reversibly downregulate the quantum efficiency at higher light intensities [94], because energy gradient between excited chlorophylls in the antennae and reaction center chlorophyll is small. In addition, charge separation is itself a reversible process [95], and excitation energy can be transferred back to antennae. Excitation energy can be nonradiatively transformed into heat [88,94,96,97]. It involves the operation of xanthophylls cycle [83]. Zeaxanthin is produced by de-epoxidation of violaxanthin [88,97,98] and is assumed to

be a reversible process. Furthermore, back reaction is retarded under stress conditions, which completely inhibits photosynthesis; zeaxanthin levels stay at an elevated level, thereby allowing harmless energy dissipation [18,67].

Irreversible photoinhibition. According to different reviews, light can damage or inactivate PSII by two different mechanisms, which are called acceptor-side- and donor-side-induced inhibition [83,91,92,99–101]:

Acceptor-side-induced photoinhibition of PSII. Excess light keeps the plastoquinone pool fully reduced and Q_B binding site remains unoccupied and electron flow between Q_A and Q_B is inhibited. As a consequence, highly stable doubly reduced and protonated Q_A is produced and thereafter released from its binding site [102,103]. This leads to increased reaction center chlorophyll$_{P680}$ triplet formation, which strongly interacts with oxygen [104] to form the highly reactive singlet oxygen. This singlet oxygen can damage pigments and the protein structure [103]. Nonheme iron located between Q_A and Q_B acceptors at the interface of D1 and D2 proteins, is modified or released [105,106]. Photoinhibition of PSII electron transport is followed by degradation of D1 reaction center protein. D1 protein degradation is a two-step process. In the first step, the protein is damaged by photochemical event and in the second step, the damaged protein is degraded by serine-type proteolytic activity [107].

Donor-side-induced photoinhibition of PSII. If the water-oxidizing complex of PSII becomes inactive before photoinhibitory illuminance, donor-side-induced photoinhibition occurs. In this case, rate of electron donation from the water-oxidizing system is unable to keep up with the rate at which electrons are transferred from P_{680} towards acceptor side components. As a consequence, long-lived and strong oxidants are produced, which have potential to induce rapid inactivation of PSII electron transport and protein damage [100,101,108].

b. Chlorophyll fluorescence

Using chlorophyll fluorescence we can estimate the photochemical activity of PSII. Chlorophyll fluorescence is elicited by a very dim light beam modulated at high frequency called minimum chlorophyll fluorescence (F_0). After illumination with nonmodulated white light of higher intensity, the fluorescence increases rapidly to a peak point (F_m). The fluorescence between (F_0) and peak point (F_m) is called variable fluorescence (F_v). Maximum fluorescence can be induced by short pulse saturating white light. The value of F_v reflects the reduction of primary electron acceptor Q_A of PSII. In the oxidized state, Q_A quenches fluorescence. Quenching F_v reflects the working of entire photosynthetic process, especially the primary photochemical events. This process depends on the reduction and oxidation of Q_A. Depending on the degree of oxidation reduction of electron transport chain, F_v is quenched or enhanced [109].

Severe water-deficit conditions cause changes in chlorophyll fluorescence in many plant species. Change in chlorophyll fluorescence was detected in oak leaves only when water deficit values of the growth medium exceeded 30% [6]. Similarly, RWC from 100% to 75% caused changes in fluorescence parameters at equivalent values of A that were the same as in unstressed leaves, but with altered CO_2 [54]. The basal rate of electron transport (F_0) decreases in some experiments [110], but increased in some other experiments [74]. It suggests that transfer of excitation energy to the reaction centers may be altered. However, in most experiments it is unaffected [7]. Maximum quantum yield of PSII (F_v/F_m) is about 0.8 in healthy leaves and may decrease under conditions of photoinactivation [111]. There is almost no evidence indicating that shortage of water had effects on the primary photochemistry of PSII in wheat and other plants [7,42]. There is evidence that drought stress decreases the photosystem activity and alters the structure of PSII [112]. For example, field grown, high light acclimated grapevines showed a slight decline in F_v/F_m (down to 0.7) after high photon exposure at midday, regardless of water availability [113–116]. Even the low light acclimated, severely stressed plants grown in greenhouses showed F_v/F_m > 0.5 [117,118]. At predawn (i.e., after a whole night of relaxation of photoinactivation), F_v/F_m in field-grown grapevines was usually close to 0.8, and declined to 0.74 only under conditions of extremely severe drought where diurnal photosynthesis was almost zero [20,22]. In low light acclimated, drought-stressed grapevines, predawn F_v/F_m declined to 0.6 [118]. Thus, under normal conditions, F_v/F_m in grapevines remained higher than 0.5 to 0.6. However, Flexas et al. [119] found a curvilinear relationship between F_v/F_m and actual number of functional PSII units, so that the loss of 40% to 50% of functional PSII resulted in slightly reduced values of F_v/F_m (0.6 to 0.7). Comparing these results with F_v/F_m values measured in field- and greenhouse-grown grapevines, it can be inferred that up to 40% to 50% of PSII centers may be photoinactivated during day. Reduction of 40% of total functional PSII would not affect the maximum photosynthetic rate of leaves, since photosynthetic capacity is not limited by PSII concentration until about 50% of PSII centers are lost [120]. However, in the light-adapted leaves, drought stress reduced the efficiency of excitation energy-capture by open PSII reaction centers (F_v'/F_m') and the quantum yield of PSII electron transport (Φ_{PSII}), increased the NPQ (qN), and had no effects on photochemical quenching (qP) [42]. This suggests that water deficit caused modification in the PSII photochemistry in the light-adapted leaves and such modification could be a mechanism to downregulate the photosynthetic electron transport so as to decrease CO_2 assimilation.

The captured photon energy is used to excite the reaction center, and initiate electron transport and chemical reaction, or dissipated as heat via the xanthophyll cycle. Electrons are ultimately transported to CO_2 for carbohydrate production, excess electrons are transferred to O_2 and photorespiration occurs. Therefore, we can measure the rate of electron transport by the measurement of oxygen evolution [7]. Badger [121] demonstrated that high rates of O_2 evolution are maintained in stressed leaves while net CO_2 assimilation rate (A) is inhibited and cannot be stimulated by elevated CO_2. It is suggested that electrons must be transferred to O_2 instead of CO_2. In another experiment where F_v/F_m and qP remained unaffected, suggested that relative concentrations of oxidized PSII are unaffected due to shortage of water [19,37]. Generally, the quantum efficiency of electron transport in PSII and qP, decreases only at RWC below ca. 75%. In addition, water stress also modified the responses of PSII to heat stress when temperature was 35°C; thermostability of PSII was strongly enhanced in water-stressed leaves, which was reflected in less decrease in F_v/F_m, qP, (F_v'/F_m'), and Φ_{PSII} in water-stressed leaves than in well-watered leaves. There was no significant variation in the abovementioned fluorescence parameters between moderately and severely water stressed plants, indicating that the moderate water stress treatment caused the same effects on thermostability of PSII as the severe treatment. It was found that increased thermostability of PSII might be associated with an improvement of resistance of the O_2-evolving complex and the reaction centers in water-stressed plants to high temperature [42].

2. Effects of Drought on Metabolic Factors

The places where photosynthetic metabolism may be impaired include Rubisco enzyme activity, regeneration of RuBP by PCR cycle, supply of ATP and NADPH to PCR cycle, electron transport, light capture, and use of assimilation products [36].

a. Effects of drought stress on RUBP content, Rubisco activity, and PCR cycle

The amount of Rubisco protein is generally little affected by moderate or severe stress [26], even if plants experience drought over a period of many days [37,63]. Restoration of photosynthetic potential (A_{pot}) to maximum photosynthesis (A_{max}) by rehydration also suggests that Rubisco is not impaired irreversibly. However, more prolonged, severe stress often decreases Rubisco activity. Loss of Rubisco activity is probably more related to inhibition or to nonactivation of enzyme active site [63,122]. Decrease in Rubisco protein by antisense genetic modification resulted in lowering total enzyme activity and A_{pot} [123]. However, with 75% decrease in enzyme activity, A_{pot} fell only by 50%, suggesting that large change in A_{pot} under stress would require substantial reduction of Rubisco protein and activity. Photosynthetic rate also depends on synthesis of RuBP and activity of Rubisco. Therefore, decrease in RuBP content of leaves at low RWC [37,58,59] is significant. A strong sigmoidal relation between A and RuBP was demonstrated by Gimenez et al. [58] in stressed sunflower leaves where A_{pot} was progressively inhibited by falling RWC, suggesting that A depends on RuBP supply, and not on CO_2. Similarly, von Caemmerer [124] suggested that under steady-state conditions, RuBP supply limited the CO_2 assimilation. According to Lawlor [36], limited RuBP may result from inadequate supply of ATP or NADPH to the PCR cycle turnover caused by low enzyme activity.

Rubisco activase is an abundant protein [125] that regulates the conformational structure of active site of Rubisco and releases tight binding inhibitors from the Rubisco active site, thereby increasing the activity. This reaction requires ATP [126]. Thus, decreased activity and active state of Rubisco at low RWC could be due to inadequate ATP concentrations [36]. There is an evidence that Rubisco activase activity decreases at low RWC consistently with decrease in ATP concentration [122].

If water-stress-induced decrease in Rubisco protein and activity causes photosynthetic inhibition, an increase in the level of RuBP content would be associated with water-stress-induced photosynthetic inhibition. For example, when Rubisco activity decreased in tobacco plants, an increase in the steady-state level of RuBP was observed [59,127]. However, some scientists suggest that reduced Rubisco activity may not necessarily be evidenced by a build up of steady-state RuBP levels, because RuBP regeneration capacity can be downregulated in response to Rubisco activity [128]. This could lead to the maintenance of constant steady-state RuBP at varying rates of carbon flow through the system. In contrast, photosynthetic rate declined with decline in RuBP content in transgenic and nontransgenic tobacco plants showing different levels of Rubisco protein and activity, suggesting that RuBP regeneration rather than Rubisco activity rate limits photosynthesis in water stressed plants [59]. In other experiments, wild plants with a mean Rubisco activity of 70.4 μmol/m²/sec was compared with transformed plants with a mean Rubisco activity of 23.1 μmol/m²/sec. Photosynthetic sensitivity to water stress was again found to be identical in wild type and transformed plants, and RuBP level decreased with increased water stress [59]. However, it is concluded that stress effects on an enzymic step involved in RuBP regeneration cause impaired chloroplast metabolism and photosynthetic inhibition in plants exposed to water deficit.

Reduction of RuBP content at low RWC could result from a limitation in one or more enzymes of PCR cycle. The large ratio of 3PGA/RuBP suggests limitation in RuBP regeneration part of PCR cycle either caused by enzyme limitation or inadequate ATP supply [58]. But there is little direct evidence regarding the response of individual enzymes of regenerative part of PCR cycle to decreasing RWC. However, Sharkey and Seemann [66] concluded that low CO_2, not enzymes, decrease PCR cycle activity. Moreover, unstresssed transgenic plants (plants have antisense Rubisco gene to reduce Rubisco protein amount and activity, and hence reduce PCR cycle activity) show that reduced PCR cycle activity reduces A, but increases RuBP content [123]. It provides evidence that PCR cycle activity is not a cause of low A_{pot} at low RWC.

b. Drought effects on ATP synthesis

Inhibition of photophosphorylation, reduction in chloroplast ATPase activity, and low ATP content have been observed in plants subjected to water deficit [6]. Water deficit reduces the ATPase activity in various crop plants and hence the level of available ATP [6]. Similarly, many researchers have observed a progressive decrease in ATP as RWC fell [37,68,69]. Thus, the observed decrease in ATP at low RWC is due to inhibition of ATP synthesis [36].

The rate of ATP synthesis depends on various factors including light reactions, generation of transthylakoid pH gradient (ΔpH), ADP and Pi concentrations, and activity of ATP synthase or coupling factor [36]. Photophosphorylation by ATP synthase was inhibited at low RWC in isolated chloroplasts from water-stressed leaves of sunflower [70,129] due to high Mg^{2+} concentration in chloroplast [71]. Activity of CF_1 (part of ATP synthase) inhibited by high

Mg^{2+} is likely to increase in chloroplast stroma as RWC falls [71]. Additional evidence by Meyer and de Kouchkovsky [130] and Meyer et al. [69] confirms the sensitivity of CF_1 to stress conditions and loss of photphosphorylation capacity at low RWC. However, loss of ATP synthase protein was not considered as the sole cause [37]. Photophosphorylation requires ΔpH, CF_0, and active CF_1 [131]. In most studies, ΔpH is large and sufficient for ATP synthesis even at low RWC [69,130,132], and hence ATP synthesis should occur and ATP content should not decrease.

The role of changes in ATP content and ATP synthesis in reducing A_{pot} is controversial [33,37]. Boyer and his coworkers [70,129] concluded that inhibition of ATP synthesis and photophosphorylation was considered to be the main cause of metabolic limitation of A in water stressed leaves of sunflower. However, others [61,133,134] concluded that limitation of A_{pot} caused a decrease in ATP content, and loss of ATP synthase protein.

c. Drought effects on NAD(P)H (reductant)

Dynamics of NADH, NADPH, and other pyridine nucleotides in photosynthesizing leaves with different conditions are complex [135]. However, there is little information in the literature on the effect of drought stress on the dynamics of NAD(P)H. Pyridine nucleotides are in reduced state under stress; NADPH content remained relatively constant [37,68], which indicates that electron transport capacity is sufficient to maintain and increase the reduction state of the pyridine nucleotides. Increased NADH may be explained by increased mitochondrial activity and respiration (both photorespiration and dark respiration) as A is decreased and a relative increase in photorespiration would increase the peroxisomal NADH pool. NADPH is consumed only by the triosephosphate dehydrogenase reaction in the PCR cycle, for the reduction of 1,3-bisphosphate to 3-phosphoglycerate (3PGA), which is essential for RuBP synthesis [136]. Thus, inadequate reductant supply is not likely to reduce A at low RWC [7].

d. Carbohydrate metabolism as affected by water deficit

The principal end products of leaf photosynthesis are starch and sucrose. Due to low C_i under water stress, chloroplastic starch may be remobilized to provide carbon in favor of more sucrose synthesis [6]. However, sucrose content in leaves fell in rapidly stressed leaves at RWC <80%, due to low net CO_2 assimilation rate (A) and continued respiration, plus synthesis of amino acids [137]. The rate of sucrose synthesis is regulated by two enzymes, cytosolic fructose 1,6-bisphosphatase (FBPase) and sucrose phosphate synthase (SPS), which are subjected to various types of regulations [6]. Activity of SPS is greatly decreased by even small loss of RWC [7]. SPS is deactivated by protein phosphorylation [138]. As a result, sucrose synthesis decreases and phosphorylated metabolites increase in the cytoplasm, whereas phosphate (Pi) concentration decreases. Export of triosephosphate is reduced by phosphate limitation [7]. Thus, reduced flux of triosephosphate and SPS activity seems to be the cause of sucrose synthesis and low content.

Under drought stress, massive changes in gene expression [139,140] lead to an accumulation of sucrose, polyols, and fructans in source leaves [139,141]. Such changes may be adaptive as the low molecular weight carbohydrates together with amino acids and their derivatives are effective osmoprotectants [83].

IV. EFFECTS OF DROUGHT ON C4 PLANTS

One of the most intriguing plant metabolic adaptations to drought occurs in plants possessing C_4 or Crassulacean acid metabolism (CAM) photosynthesis [1]. In C_4 plants, a metabolic pump has evolved that concentrates CO_2 in bundle sheath cells where Rubisco is located [142]. The CO_2 fixation process is separated in mesophyll cells and bundle sheath cells, while reduced CO_2 concentrations may have been driving force for the evolution of C_4 plants [143]. This specialized photosynthesis led to greater WUE and ecological success in arid environments. However, the sensitivity of the photosynthetic metabolism to water deficit in C_4 plants is similar to that in C_3 plants [19].

Increasing drought stress severity caused a decrease in photosynthetic rate, and increase in PSII photochemical efficiency (F_v/F_m) of two C_4 grass cultivars. Drought tolerant (*Eragrostis curvula* cv Consol) showed small variation in these parameters. Generally, decrease in PSII quantum yield can result from photoprotective increase in thermal energy dissipation [52] induced by excess absorbed light [88,97]. However, in drought-sensitive *E. curvula* cv. Ermelo, PSII thermal energy dissipation (NPQ) was strongly inhibited due to damage to PSII structure and functionally as reflected from reduction of PSII energy-capture efficiency (F_v'/F_m'). It was found that under severe water shortage NPQ of *E. curvula* was reduced which was ascribed to downregulation of PSII activity [144].

As described earlier, C_4 plants are capable of concentrating CO_2 in bundle sheath cells to levels that have been estimated to exceed 3 to 20 times the atmospheric CO_2 concentration [145–148]. Therefore, the ratio of CO_2 to O_2 increases in bundle sheath cells, and photorespiration is considered insignificant because of the suppression of the oxygenase reaction of Rubisco [11,146,147,140,150]. However, measurable rate of photorespiration has been observed in C_4 plants [151–154]. Due to the high resistance of bundle sheath cells to gas diffusion [145,148,150,155], it is generally accepted that CO_2 released during photorespiration will be partially refixed by Rubisco. However, estimates of leakage rates of CO_2 from bundle sheath cells vary from 10% to 50% of the C_4 cycle flux [147,148].

Simultaneous gas exchange and chlorophyll fluorescence measurements under different CO_2 partial pressures suggested that above the optimal O_2 partial pressure, inhibition of net photosynthesis is associated with photorespiration. Below the optimum O_2 partial pressure, inhibition of net photosynthesis is associated with reduced PSII activity and electron transport, and open PSII centers (oxidized PSII) [45]. It might be due to decrease in ATP supply to the C_4 cycle [156]. Data from C_4-cycle limited mutant of *Amaranthus edulis* and C_3-cycle limited trasformants of *Flaveria bidentis* at varying concentration of oxygen showed that when the C_4 cycle is deficient, photorespiration is increased, and when the C_3 cycle is deficient overcycling of C_4 pathway with increased CO_2 leakage was observed. It was suggested that C_4 photosynthesis requires coordinated function of the C_3 and C_4 cycles for maximum efficiency [45].

Although, O_2 inhibits C_4 photosynthesis, especially at low CO_2 concentrations, Γ remains low [146], which reflects an efficient refixation of photorespiratory CO_2. The degree of inhibition of photosynthesis by O_2 depends on C_i at the site of Rubisco. C_i around Rubisco in bundle sheath cells of maize is 3.2-fold higher than C_i around Rubisco in mesophyll cells of wheat. However, leaf stomatal conductance was lower (391 mmol/m^2/sec) in maize than in wheat (681 mmol/m^2/sec). These differences allow maize (C_4) to have higher WUE than wheat (C_3). Similarly, some investigators observed that C_4 drought-tolerant grass cultivar (*E. curvula* cv. Consol) in comparison with drought-sensitive cultivar (*E. curvula* cv. Ermelo) had greater ability to save water during drought stress [52], and consequently reduced limitations to CO_2 uptake and photosynthetic biochemical processes were ascribed to reduced photoinhibition and photodamage to PSII systems.

In conclusion, high uptake to CO_2 at reduced stomatal conductance, concentrating CO_2 at site of Rubisco, photoprotective increase in thermal energy dissipation and high WUE are key adaptations in C_4 plants to hot and arid conditions. It has strongly been suggested, based on geological evidence, that major selective force for the evolution of C_4 photosynthesis

was decline in atmospheric level of CO_2 [157], whereas according to Dai et al. [146] low level of CO_2 combined with water stress or higher temperature likely accounts for the adaptation of C_4 plants to hot and arid environment.

V. EFFECTS OF DROUGHT ON CAM PLANTS

Crassulacean acid metabolism (CAM), a key adaptation of photosynthetic carbon fixation to limited water availability, is characterized by nocturnal CO_2 fixation and daytime CO_2 reassimilation, which generally results in improved WUE. However, CAM plants display a remarkable degree of photosynthetic plasticity within a continuum of diel gas exchange patterns. Genotypic, ontogenic, and environmental factors combine to govern the extent to which CAM is exposed [158]. Some species can switch freely and reversibly between C_3 and CAM cycle regardless of plant ontogeny. Diversity of CAM inducibilty is marked within the genus *Clusia* [159–161]. In *Clusia minor*, opposite leaves on the same node are capable of expressing either C_3 or full CAM characteristics, depending on the leaf–air vapor pressure difference [162], whereas in *Clusia uvitana*, rapid and reversible switching between C_3 photosynthesis and CAM can occur within 24 h in response to environmental changes [163]. Spatial separation of CAM inducibility is found in *Cissus quadrangularis*, a species with succulent CAM stem bearing small, short lived leaves that can switch from CAM cycling to CAM under conditions of moderate stress [164].

Ting and Sipes [165] reported two modifications of CAM. One modification, termed CAM-idling, occurs when CAM plants experience severe water deficit so that stomata close both day and night, and a low rate of cycling of organic acids through CAM pathway occurs. With this downregulation of metabolism, biochemical activities of the CAM plants are maintained until water becomes available and the plants recover again [166]. In the second modification, termed CAM-cycling, gas exchange occurs mainly during the day as in C_3 plants, yet a diurnal cycling of organic acids similar to that of CAM is observed [165].

Inducible CAM greatly decreases water loss during drought due to stomatal closure in the light when atmospheric humidity and leaf temperature favor water loss [167]. Once CAM is induced, decarboxylation of organic acids in the day increases internal CO_2 and reduces stomatal conductance [168]. The carboxylation of RuBP when stomata are closed [167] maintains linear electron flow through PSII reaction centers. Induction of CAM is accompanied by important changes in PSII photochemistry [169–171].

The xanthophyll cycle is of great importance in this regard. Pieters et al. [53] suggest that increased xanthophyll cycle activity and induction of CAM by drought are linked, in an unknown manner, and may be triggered by similar cellular conditions elicited by water deficit. In *Talinum triangulare*, NPQ of chlorophyll a fluorescence increased with water deficit, but decreased with more severe drought, when CAM activity is low [53]. Quantum yield of PSII photochemistry Φ_{PSII}, and intrinsic quantum yield of PSII (F_v/F_m) were lower in severe water deficit. Under high light and moderate drought, the D1 content in leaves was identical to control, whereas under severe stress, D1 content decreased. From this, it is concluded that under water deficit, CAM activity in plants like *T. triangulare* plays a central role in protection of photosynthetic machinery from photoinhibition. At maximum CAM activity, a relatively high intercellular CO_2 concentration and the capacity for energy dissipation by xanthophylls cycle are sufficient to prevent damage to and net degradation of D1. When CAM activity is limited and the capacity of energy dissipation by xanthophyll cycle is exhausted after prolonged drought, inactive reaction centers accumulate with the subsequent degradation of D1 [53].

VI. CONCLUSIONS AND FUTURE PROSPECTS

In view of the many reports cited earlier in the text, it is evident that water deficit adversely affects photosynthesis and plant growth. Despite a lot of research devoted during the last decade to examine the effects of water deficit on photosynthetic parameters, further work is still needed to elucidate the specific changes occurring in photochemical, gas exchange, and metabolic phenomena during photosynthesis under water-deficit conditions. Different photochemical, gas exchange, and metabolic processes of photosynthesis are so tightly linked with one another that slight change in any process may change the series of events that ultimately inhibits the overall rate of photosynthesis [1,6,7,16,23,26,36,41].

There is a large body of literature available reflecting stomatal closure as a major response to shortage of water, which leads to reduction in the net CO_2 assimilation rate. Stomatal closure decreases the photochemical efficiency and it seems to be the basic strategy of plants to cope with water deficit. The view that implicates a primary role for nonstomatal effects of water deficit on photosynthesis is in conflict with the evidence suggesting that ATP synthesis and RuBP regeneration impair the photosynthesis at mild water stress [1,7,16,23,26,36,37,41]. Therefore, mechanisms of stomatal and nonstomatal limitations are not yet clear.

When absorbed light exceeds that can be used by the photosynthetic apparatus, specific processes such as xanthophylls cycle, Mehler-peroxidase reaction, photorespiration, etc. seems to effectively retard photodegradation of photosynthetic apparatus. However, these photoprotection mechanisms are not sufficient, because in C_3, C_4, and CAM plants photoinhibition and photo-oxidation do occur under severe water deficit as reported earlier in the text [52,53,83,100,101]. PSII is more sensitive in comparison to PS I toward photoinhibition. The research for how photoinhibition starts, why PSII is more inhibited than PS I, and how violaxanthin and zeaxanthin dissipate thermal energy is a fascinating area of research.

It is now becoming evident that the relationship between photosynthesis and carbohydrate metabolism is not a simple one. Therefore, it is essential to elucidate that plants have some means by which photosynthetic activity can be coordinated with the actual needs of nonphotosynthetic plant parts under drought stress. Enhanced demand for energy and photosynthates was observed under water deficit, and strategies that increase photosynthate supply were thought to be beneficial.

Global circulation models have predicted that aridity of some regions will increase in the coming years. Furthermore, due to global warming, CO_2 concentrations will also increase. Interactive effects of CO_2 and water availability may alter the relative performance of C_3, C_4, and CAM plants. Therefore, it is possible that net CO_2 uptake and productivity will be altered in the future [45,158]. Although the patterns of effects of drought stress in C_3, C_4, and CAM plants are almost the same, water use efficiency and rate of photosynthesis are different. However, C_3 species give better response to elevated CO_2 under water deficit as compared to C_4 ones [45]. Finer details of these differences are still very unclear. Therefore, attempts to improve photosynthesis by enhancing CO_2 under drought stress will be helpful in elucidating the effect of global warming and aridity on different crop plants in the near future.

ACKNOWLEDGMENTS

The authors thank all those who provided us their valuable reprints, especially Drs. J. Flexas, H. Medrano, and David Lawlor.

REFERENCES

1. Chaves MM, Maroco JP, Pereira JS. Understanding plant responses to drought — from genes to the whole plant. *Funct. Plant Biol.* 2003; 30:239–264.
2. Levitt J. *Responses of Plants to Environmental Stresses.* New York: Academic Press, 1972.
3. Turner NC. Crop water deficits: a decade of progress. *Adv. Agron.* 1986; 39:1–51.
4. Mooney HA, Pearcy RW, Ehleringer J. Plant physiological ecology today. *BioScience* 1987; 37:18–20.
5. Maroco JP Pereira JS, Chaves MM. Growth, photosynthesis and water-use efficiency of two C_4 Sahelian grasses subjected to water deficits. *J. Arid Environ.* 2000; 45:119–137.
6. Dubey RS. Photosynthesis in plants under stressful conditions. In: Pessarakli M, ed. *Handbook of Photosynthesis.* New York: Marcel Decker, 1997: 859–875.
7. Lawlor DW, Cornic G. Photosynthetic carbon assimilation and associated metabolism in relation to water deficits in higher plants. *Plant Cell Environ.* 2002; 25:275–295.
8. Demmig B, Winter K, Kruger A, Czygan FC. Zeaxanthin and heat dissipation of excess light energy in *Nerium oleander* exposed to a combination of high light and water stress. *Plant Physiol.* 1988; 87:17–24.
9. Lawlor DW. The effects of water deficit on photosynthesis. In: Smirnoff N, ed. *Environment and Plant Metabolism.* Oxford: BIOS Scientific Publishers, 1995:129–160.
10. Taize L, Zeiger E. *Plant Physiology,* 3rd ed. Sunderland, MA: Sinauer Associates Inc. Publishers, 2002.
11. Edwards GE, Walker GA. C_3, C_4: *Mechanisms, and Cellular and Environmental Regulation of Photosynthesis.* Oxford: Blackwell Scientific, 1983.
12. Ehleringer JR, Monson RK. Evolutionary and ecological aspects of photosynthetic pathway variation. *Annu. Rev. Ecol. Syst.* 1993; 24:411–439.
13. Gollan T, Turner NC, Schulze ED. The responses of stomata and leaf gas exchange to vapour pressure deficits and soil water content III. In the sclerophyllous woody species *Nerium oleander. Oecologia* 1985; 65:356–362.
14. Socias X, Correia MJ, Medrano H. The role of abscisic acid and water relations in drought responses of subterranean clover. *J. Exp. Bot.* 1997; 48(311): 1281–1288.
15. Davies WJ, Zhang J. Root signals and the regulation of growth and development of plants in drying soils. *Annu. Rev. Plant Physiol. Plant Mol. Biol.* 1991; 42:55–76.
16. Medrano H, Escalona JM, Bota J, Gulías J, Flexas J. Regulation of photosynthesis of C_3 plants in response to progressive drought: the stomatal conductance as a reference parameter. *Ann. Bot.* 2002; 89: 895–905.
17. Chaves MM. Effects of water deficits on carbon assimilation. *J. Exp. Bot.* 1991; 42:1–46.
18. Cornic G. Drought stress and high light effects on leaf photosynthesis. In: Baker NR, Bowyer JR, eds. *Photoinhibition of Photosynthesis.* Oxford: BIOS Scientific Publishers, 1994:297–313.
19. Cornic G, Massacci A. Leaf photosynthesis under drought stress. In: Baker NR, ed. *Photosynthesis and the Environment.* Dordrecht, The Netherlands: Kluwer Academic Publishers, 1996:347–366.
20. Flexas J, Escalona JM Medrano H. Down-regulation of photosynthesis by drought under field conditions in grapevine leaves. *Aust. J. Plant Physiol.* 1998; 25: 893–900.
21. Escalona JM, Flexas J, Medrano H. Stomatal and non-stomatal limitations of photosynthesis under water stress in field-grown grapevines. *Aust. J. Plant Physiol.* 1999; 26:421–433.
22. Flexas J, Bota J, Escalona JM, Sampol B, Medrano H. Effects of drought on photosynthesis in grapevines under field condition: an evaluation of stomatal and mesophyll limitations. *Funct. Plant Biol.* 2002; 29: 461–471.
23. Medrano H, Bota J, Abadia A, Sampol B, Escalona JM, Flexas J. Effects of drought on light-energy dissipation mechanisms in high-light-acclimated, field grown grapevines. *Funct. Plant. Biol.* 2002; 29:1197–1207.
24. Kriedemnn PE, Smart RE. Effects of irradiance, temperature, and leaf water potential on photosynthesis of vine leaves. *Photosynthetica* 1969; 5:15–19.
25. Correia MJ, Pereira JS, Chaves MM, Rodrigues ML, Pacheco CA. ABA xylem concentrations determine maximum daily leaf conductance of field-grown *Vitis vinifera* L. plants. *Plant Cell Environ.* 1995; 18: 511–521.
26. Flexas J, Medrano H. Drought-inhibition of photosynthesis in C_3 plants: stomatal and non-stomatal limitations revisited. *Ann. Bot.* 2002; 89:183–189.
27. Luo Y. Changes of C_i/C_a in association with non-stomatal limitation to photosynthesis in water stressed *Abutilon theophrasti. Photosynthetica* 1991; 25: 273–279.
28. Brodribb T. Dynamics of changing intercellular CO_2 concentration C_i during drought and determination of minimum functional C_i. *Plant Physiol.* 1996; 111: 179–185.
29. Sharkey TD. Water stress effects on photosynthesis. *Photosynthetica* 1990; 24:651.
30. Kramer PJ, Boyer JS. *Water Relation of Plants and Soils.* London: Academic Press, 1995.
31. Kaiser WM. Effect of water deficit on photosynthetic capacity. *Physiol. Plant.* 1987; 71:142–149.
32. Downton WJS, Loveys BR, Grant WJR. Non-uniform stomatal closure induced by water stress causes putative non-stomatal inhibition of photosynthesis. *New Phytol.* 1988; 110:503–509.
33. Cornic G. Drought stress inhibits photosynthesis by decreasing stomatal aperture — not by affecting ATP synthesis. *Trends Plant Sci.* 2000; 5:187–188.
34. Graan T, Boyer JS. Very high CO_2 partially restores photosynthesis in sunflower at low water potentials. *Planta* 1990; 181:378–384.

35. Tourneux C, Peltier G. Effect of water deficit on photosynthetic oxygen exchange measured using $^{18}O_2$ and mass spectroscopy in *Solanum tuberosum* leaf discs. *Planta* 1995; 195:570–577.

36. Lawlor DW. Limitation to photosynthesis in water stressed leaves: stomatal versus metabolism and the role of ATP. *Ann. Bot.* 2002; 89:1–15.

37. Tezara W, Mitchell VJ, Driscoll SP, Lawlor DW. Water stress inhibits plant photosynthesis by decreasing coupling factor and ATP. *Nature* 1999; 401:914–917.

38. Tezara W, Mitchell V, Driscoll SP, Lawlor DW. Effects of water deficit and its interaction with CO_2 supply on the biochemistry and physiology of photosynthesis in sunflower. *J. Exp. Bot.* 2002; 53(375):1781–1791.

39. Flexas J, Badger M, Chow WS, Medrano H, Osmond CB. Analysis of the relative increase in photosynthetic O_2 uptake when photosynthesis in grapevine leaves is inhibited following low night temperaures and/or water stress. *Plant Physiol.* 1999; 121:675–684.

40. Maroco JP, Rodrigues ML, Lopes C, Chaves MM. Limitations to leaf photosynthesis in grapevine under drought — metabolic and modelling approaches. *Funct. Plant Biol.* 2002; 29:1–9.

41. Medrano H, Escalona JM, Cifre J, Bota J, Flexas J. A ten-year study on the physiology of two Spanish grapevine cultivars under filed conditions: effects of water availability from leaf photosynthesis to grape yield and quality. *Funct. Plant Biol.* 2003; 30:607–619.

42. Lu C, Zhang J. Effects of water stress on photosystem II photochemistry and its thermostability in wheat plants. *J. Exp. Bot.* 1999; 50(336):1199–1206.

43. Yordanov I, Tsonev T, Velikova V, Georgieva K, Ivanov P, Tsenov N, Petrova T. Changes in CO_2 assimilation, transpiration and stomatal resistance of different wheat cultivars experiencing drought under field conditions. *Bulg. J. Plant Physiol.* 2001; 27 (3–4):20–33.

44. Molnár I, Gáspár L, Stéhli L, Dulai S, Sárvári E, Király I, Galiba G, Molnár-Láng M. The effects of drought stress on the photosynthetic processes of wheat and of *Aegilops biuncialis* genotyps originating from various habitats. *Acta Biol. Szegediensis* 2002; 46(3–4):115–116.

45. Ward JK, Tissue DT, Thomas RB, Strain BR. Comparative responses of model C_3 and C_4 plants to drought in low and elevated CO_2. *Global Change Biol.* 1999; 5:857–867.

46. Kiato M, Lei TT, Koike T, Tobita H, Maruyama Y. Higher electron transport rate observed at low intercellular CO_2 concentration in long-term drought acclimated leaves of Japanese mountain birch (*Betula ermanii*). *Physiol. Plant.* 2003; 118:406–413.

47. Mäkelä P, Kontturi M, Pehu E, Somersalo S. Photosynthetic response of drought- and salt stressed tomato and turnip rape plants to foliar-applied glycinebetaine. *Physiol. Plant.* 1999; 105:45–50

48. Srinivasa Rao NK, Bhatt RM, Sadashiva AT. Tolerance to water stress in tomato cultivars. *Photosynthetica* 2000; 38(3):465–467.

49. Matos MC, Campos PS, Ramalho JC, Medeira MC, Maia MI, Semedo JM, Marques NM, Matos A. Photosynthetic activity and cellular integrity of the Andean legume *Pachyrhizus ahipa* (Wedd.) Parodi under heat and water stress. *Photosynthetica* 2002; 40(4):493–501.

50. Pshibytko NL, Kalitukho LN, Kabashinokova LF. Effects of high temperature and water deficit on photosystem II in *Hordeum vulgare* leaves of various ages. *Russ. J. Plant Physiol.* 2003; 50(1):44–51.

51. Cousins AB, Adam NR, Wall GW, Kimball BA, Pinter PJ Jr, Ottman MJ, Leavitt SW, Webber AN. Photosystem II energy use, non-photochemical quenching and xanthophylls cycle in *Sorghum bicolour* grown under drought and free-air CO_2 enrichment (FACE) conditions. *Plant Cell Environ.* 2002; 25:1551–1559.

52. Colom MR, Vazzana C. Photosynthesis and PSII functionality of drought-resistant and drought-sensitive weeping lovegrass plants. *Environ. Exp. Bot.* 2003; 49:135–144.

53. Pieters AJ, Tezara W, Herrera A. Operation of xanthophyll cycle and degradation of D1 protein in the inducible CAM plant, *Talinum triangulare*, under water deficit. *Ann. Bot.* 2003; 92:1–7.

54. Cornic G, Briantais JM. Partitioning of photosynthetic electron flow between CO_2 and O_2 reduction in a C_3 leaf (*Phaseolus vulgaris* L.) at different CO_2 concentrations and during water stress. *Planta* 1991; 183:178–184.

55. Haupt-Herting S, Fock HP. Exchange of oxygen and its role in energy dissipation during drought stress in tomato plants. *Physiol. Plant.* 2000; 110:489–495.

56. Panković D, Sakač Z, Kevrešan S, Plesničar M. Acclimation to long-term water deficit in the leaves of two sunflower hybrids: photosynthesis, electron transport and carbon metabolism. *J. Exp. Bot.* 1999; 50:127–138.

57. Munné-Bosch S, Nogués S, Alegre L. Diurnal variations of photosynthesis and dew absorption by leaves in two evergreen shrubs growing in Mediterranean field conditions. *New Phytol.* 1999; 144:109–119.

58. Giménez C, Mitchell V.J, Lawlor DW. Regulation of photosynthetic rate of two sunflower hybrids under water stress. *Plant Physiol.* 1992; 98:516–524.

59. Gunasekera D, Berkowitz GA. Use of transgenic plants with ribulose 1,5-bisphosphate carboxylase/oxygenase antisense DNA to evaluate the rate limitation of photosynthesis under water stress. *Plant Physiol.* 1993; 103:629–635.

60. Faver KL, Gerik TJ, Thaxton PM, El-Zik KM. Late season water stress in cotton: II. Leaf gas exchange and assimilation capacity. *Crop Sci.* 1996; 36:922–928.

61. Ortiz-Lopez A, Ort DR, Boyer JS. Photophosphorylation in attached leaves of *Helianthus annuus* at low water potentials. *Plant Physiol.* 1991; 96:1018–1025.

62. Majumdar S, Ghosh S, Glick BR, Dumbroff EB. Activities of chlorophyllase, phosphoenolpyruvate carboxylase and ribulose 1,5-bisphosphate carboxylase in the primary leaves of soybean during senescence and drought. *Physiol. Plant.* 1991; 81:473–480.

63. Medrano H, Parry MAJ, Socias X, Lawlor DW. Long-term water stress inactivates Rubisco in subterranean clover. *Ann. Appl. Biol.* 1997; 131:491–501.

64. Martin B, Ruiz-Torres NA. Effects of water-deficit stress on photosynthesis, its components and component limitations, and on water use efficiency in wheat (*Triticum aestivum* L). *Plant Physiol.* 1992; 100: 733–739.

65. Mäkelä P, Kärkkäinen J, Somersalo S. Effects of glycinebetaine on chloroplast ultrastructure, chlorophyll and protein content, RuBPCO activities in tomato grown under drought and salinity. *Biol. Plant.* 2000; 43(3):471–475.

66. Sharkey TD, Seeman JR. Mild water stress effects on carbon-reduction-cycle intermediates, ribulose bisphosphate carboxylase activity, and spatial homogeneity of photosynthesis in intact leaves. *Plant Physiol.* 1989; 89:1060–1065.

67. Brestic M, Cornic G, Fryer MJ, Baker NR. Does photorespiration protect the photosynthetic apparatus in French bean leaves from photoinhibition during drought stress? *Planta* 1995; 196:450–457.

68. Lawlor DW, Khanna-Chopra R. Regulation of Photosynthesis during water stress. In: Sybesma C, ed. *Advances in Photosynthetic Research.* Vol. IV. The Hague: Martinus Nijhoff/Dr W. Junk Publishers, 1984:379–382.

69. Meyer S, Hung SPN, Trémolières A, de Kouchkovsky Y. Energy coupling, membrane lipids and structure of thylakoids of lupin plants submitted to water stress. *Photosynth. Res.* 1992; 32:95–107.

70. Keck RW Boyer JS. Chloroplast response to low leaf water potentials. III. Differing inhibition of electron transport and photophosphorylation. *Plant Physiol.* 1974; 53:474–479.

71. Younis HM, Boyer JS, Govindjee. Conformation and activity of chloroplast coupling factor exposed to low chemical potential of water in cells. *Biochim. Biophys. Acta* 1979; 548:328–340.

72. Tang AC, Kawamitsa Y, Kanechi M, Boyer JS. Photosynthesis at low water potentials in leaf discs lacking epidermis. *Ann. Bot.* 2002; 89:861–871.

73. Vassey TL, Sharkey TD. Mild water stress of in *Phaseolus vulgaris* plants leads to reduced starch synthesis and extractable sucrose phosphate synthase activity. *Plant Physiol.* 1989; 89:1066–1070.

74. Saccardy K, Cornic G, Brulfert J, Reyss A. Effect of drought stress on net CO_2 uptake by *Zea* leaves. *Planta* 1996; 199:589–595.

75. Nicolodi C, Massacci A, Di Marco G. Water status effects on net photosynthesis in field grown alfalfa. *Crop Sci.* 1988; 28:944–948.

76. Maroco JP, Ku MSB, Furbank RT, Lea PJ, Leegood RC, Edwards GE. CO_2 and O_2 dependence of PSII activity in C_4 plants having genetically produced deficiencies in the C_3 and C_4 cycle. *Photosynth. Res.* 1998; 58:91–101.

77. Wingler A, Quick WP, Bungard RA, Bailey KJ, Lea PJ, Leegood RC. The role of photorespiration during drought stress: ananalysis utilising barley mutants

78. with reduced activities of photorespiratory enzymes. *Plant Cell Environ.* 1999; 22:361–373.

78. Foyer CH, Mullineaux PM. *Causes of Photo-oxidative Stress and Amelioration of Defence Systems in Plants.* Boca Raton, FL: CRC Press, 1994.

79. Smirnoff N. Plant resistance to environmental stress. *Curr. Opin. Biotechnol.* 1998; 9:214–219.

80. Niyogi KK. Photoprotection revisited: genetic and molecular approaches. *Annu. Rev. Plant Physiol. Plant Mol. Biol.* 1999; 50:333–359.

81. Seel WE, Hendry GAF, Lee JE. The combined effects of desiccation and irradiance on mosses from xeric and hydric habitats. *J. Exp. Bot.* 1992; 43:1031–1037.

82. Biswal B. Chloroplast, pigments and molecular responses of photosynthesis under stress. In: Pessarakli M, ed. *Handbook of Photosynthesis.* New York: Marcel Dekker, 1997:877–896.

83. Godde D. Adaptations of photosynthetic apparatus to stress conditions. In: Lerner HR. ed. *Plant Responses to Environmental Stresses: From Phytohormones to Genome Reorganization.* New York: Marcel Dekker, 1999:449–472.

84. Satoh K. Mechanism of photoinactivation in photosynthetic systems. I. The dark reaction in photoinactivation. *Plant Cell Physiol.* 1970; 11:15–27.

85. Satoh K. Mechanism of photoinactivation in photosynthetic systems. II. The occurrence and properties of two different types of photoinactivation. *Plant Cell Physiol.* 1970; 11:29–38.

86. Satoh K. Mechanism of photoinactivation in photosynthetic systems. III. Site and mode of photoinactivation in photosystem I. *Plant Cell Physiol.* 1970; 11:187–197.

87. Satoh K, Fork DC. Photoinhibition of reaction centres of photosystem I and II in intact *Bryopsis* chloroplasts under anaerobic conditions. *Plant Physiol.* 1982; 70:1004–1008.

88. Demmig-Adams B, Adams WW III Photoprotection and other responses of plants to high light stress. *Annu. Rev. Plant Physiol. Plant Mol. Biol.* 1992; 43:599–626.

89. Horton PA, Ruban V, Walters RG. Regulation of light harvesting in green plants. *Plant Physiol.* 1994; 106:415–420.

90. Andersson B, Styring S. 1991. Photosystem II: molecular organisation, function, and acclimation. *Curr. Top. Bioenerg.* 1991; 16:1–81.

91. Barber J, Andersson B. Too much of a good thing can be bad for photosynthesis. *Trends Biochem. Sci.* 1992; 17:61–66.

92. Aro EM, Virgin I, Andersson B. Photoinhibition of photosystem II. Inactivation, protein damage and turnover. *Biochim. Biophys. Acta* 1993; 1143:113–134.

93. Powles, SB. Photoinhibition of photosynthesis induced by visible light. *Annu. Rev. Plant Physiol.* 1984; 35:15–44.

94. Horton PA, Ruban AV, Walters RG. Regulation of light harvesting in green plants. *Annu. Rev. Plant Physiol. Plant Mol. Biol.* 1996; 47:655–684.

95. Schatz GH, Brock H, Holzwarth AR. Kinetic and energetic model for the primary processes of photosystem II. *Biophys. J.* 1988; 54:397–405.

96. Ruban AV, Horton P. Spectroscopy of non-photochemical and photochemical quenching of chlorophyll fluorescence in leaves; evidence for a role of the light harvesting complex of photosystem II in the regulation of energy-dissipation. *Photosynth. Res.* 1994; 40:181–190.

97. Demmig-Adams B, Adams WW III, Barker DH, Logan BA, Bowling DR, Verhoeven AS. Using chlorophyll fluorescence to assess the fraction of absorbed light allocated to thermal dissipation of excess excitation. *Physiol. Plant.* 1996; 98:253–264.

98. Pfündel E, Bilger W. Regulation and possible function of the violaxanthin cycle. *Photosynth. Res.* 1994; 42:89–109.

99. Prasil O, Adir N, Ohad I. Dynamics of photosystem II. In: Barber J, ed. *Topics in Photosynthesis.* Vol. 11. The Photosystems: Structure, Function, and Molecular Biology. Amsterdam, The Netherlands: Elsevier, 1992:293–348.

100. Vass I. Adverse effects of UV-B light on the structure and function of the photosynthetic apparatus. In: Pessarakli M, ed. *Handbook of Photosynthesis.* New York: Marcel Decker, 1997:931–949.

101. Hideg E. Free radical production in photosynthesis under stress conditions. In: Pessarakli M, ed. *Handbook of Photosynthesis.* New York: Marcel Decker, 1997:911–929.

102. Vass I, Styring S, Hundal T, Koivuniemi A, Aro E, Andersson B. Reversible and Irreversible Intermediates during photoinhibition of photosystem II: stable reduced Q_A species promote chlorophyll triplet formation. *Proc. Natl. Acad. Sci. USA* 1992; 89(4):1408–1412.

103. Telfer A, Barber J. Elucidating the molecular mechanisms of photoinhibition by studying isolated photosystem II reaction centers. In: Baker NR, Bowyer JR, eds. *Photoinhibition of Photosynthesis — From Molecular Mechanisms to the Field.* Oxford: BIOS Scientific Publishers, 1994:25–50.

104. Durrant JR, Giorgi LB, Barber J, Klug DR, Porter G. Characterization of triplet states in isolated photosystem II reaction centres: oxygen quenching as a mechanism of photodamage. *Biochim. Biophys. Acta* 1990; 1017:167–175.

105. Gleiter HM, Haag E, Shen JR, Eaton-Rye JJ, Inoue Y, Vermaas W, Renger G. Functional characterization of mutant strains of the *Cyanobacterium synechocystis* sp. PCC 6803 lacking short domains within the large, lumen-exposed loop of the chlorophyll protein CP47 in photosystem II. *Biochemistry* 1994; 33:12063–12071.

106. Vass I, Sanakis Y, Spetea C, Petrouleas V. Effects of photoinhibition on the QA-Fe^{2+} complex of photosystem II studied by EPR and Mossbauer spectroscopy. *Biochemistry* 1995; 34(13):4434–4440.

107. Virgin I, Salter AH, Ghanotakis DF, Andersson B. Light-induced D1 protein degradation is catalyzed by a serine-type protease. *FEBS Lett.* 1991; 287 (1–2):125–128.

108. Blubaugh DJ, Cheniae GM. Kinetics of photoinhibition in hydroxylamine-extracted photosystem II membranes: relevance to photoactivation and sites of electron donation. *Biochemistry* 1990; 29:5109–5118.

109. Zrůst J, Vacek K, Hála J, Janáčková I, Adamec F, Ambrož M, Dian J, Vácha M. Influence of water stress on photosynthesis and variable chlorophyll fluorescence of potato leaves. *Biol. Plant.* 1994; 36(2):209–217.

110. Haupt-Herting S, Fock HP. Oxygen exchange in relation to carbon assimilation in water stressed leaves during photosynthesis. *Ann. Bot.* 2002; 89:851–859.

111. Björkman O, Demmig-Adams B. Regulation of photosynthetic light energy capture, conversion, and dissipation in leaves of higher plants. In: Schulze ED, Caldwell MM eds. *Ecophysiology of Photosynthesis.* Berlin: Springer-Verlag, 1994:17–47.

112. Giardi MT, Cona A, Geiken B, Kuoera, Masjidek J, Mattoo AK. Long-term drought stress induces structural and functional reorganizationof photosystem II. *Planta* 1996; 199:118–125.

113. Gamon JA, Pearcy RW. Leaf movements, stress avoidance and photosynthesis in *Vitis californica. Oecologia* 1989; 79:475–481.

114. Iacono F, Sommer KJ. Photoinhibition of photosynthesis in *Vitis vinifera* under field conditions: effects of light climate and leaf position. *Aust. J Grape Wine Res.* 1996, 2:10–20.

115. Schultz HR. Water relations and photosynthetic responses of two grapevine cultivars of different geographical origin during water stress. *Acta Hort.* 1996; 427:251–266.

116. Chaumont M, Oso'rio ML, Chaves MM, Vanacker H, Morot-Gaudry JF, Foyer C. The absence of photoinhibition during the mid-morning depression of photosynthesis in *Vitis vinifera* grown in semi-arid and temperate climates. *J. Plant Physiol.* 1997; 150:743–751.

117. Quick WP, Chaves MM, Wendler R, David M, Rodrigues ML, Passaharinho JA, Pereira JS, Adcock MD, Leegood RC, Stitt M. The effect of water stress on photosynthetic carbon metabolism in four species grown under field conditions. *Plant Cell Environ.* 1992; 15:25–35.

118. Flexas J, Briantais JM, Cerovic Z, Medrano H, Moya I. Steady-state and maximum chlorophyll fluorescence responses to water stress in grapevine leaves: a new remote sensing system. *Remote Sens. Environ.* 2000; 73:283–297.

119. Flexas J, Gulias J, Jonasson S, Medrano H, Mus M. Seasonal patterns and control of gas exchange in local populations of the Mediterranean evergreen shrub *Pistacia lentiscus* L. *Acta Oecol.* 2001; 22:33–43.

120. Lee HY, Chow WS, Hong YN. Photoinactivation of photosystem II in leaves of *Capsicum annuum. Physiol. Plant.* 1999; 105:377–384.

121. Badger MR Photosynthetic oxygen exchange. *Annu. Rev. Plant Physiol.* 1985; 36:27–53.

122. Parry MAJ, Androlojc JP, Khan S, Lea PJ, Keys A.J. Rubisco activity: effects of water stress. *Ann. Bot.* 2002; 89:833–839.

123. Hudson GS, Evans JR, von Caemmerer S, Arvidsson YBC, Andrews TJ. Reduction of ribulose-1,5-bisphosphate carboxylase/oxygenase content by antisense RNA reduces photosynthesis transgenic tobacco plants. *Plant Physiol.* 1992; 98:294–302.

124. Von Caemmerer S. *Biochemical Models of Leaf Photosynthesis.* Collingwood: CSIRO Publishing, 2000.

125. Salvucci ME, Ogren WL. The mechanism of Rubisco activase: insights from studies of the properties and structure of the enzyme. *Photosynth. Res.* 1996; 47: 1–11.

126. Robinson SP, Portis AR. Involvement of stromal ATP in the light activation of ribulose-1,5-bisphosphate carboxylase/oxygenase in intact isolated chloroplasts. *Plant Physiol.* 1988; 86:293–298.

127. Quick WP, Schurr U, Scheibe R, Schulze ED, Rodermel SR, Bogorad L, Stitt M. Decreased ribulose-1, 5-bisphosphate carboxylase-oxygenase in transgenic tobacco transformed with "antisense" rbcS I. Impact on photosynthesis in ambient growth conditions. *Planta* 1991; 183:542–554.

128. Servaites JC, Shieh WJ, Geiger DR. Regulation of photosynthetic carbon reduction cycle by ribulose bisphosphate and phosphoglyceric acid. *Plant Physiol.* 1991; 97:1115–1121.

129. Boyer JS, Ort DR, Ortiz-Lopez A. Photophosphorylation at low water potentials. *Curr. Top. Plant Biochem. Physiol.* 1987;69–73.

130. Meyer S, de Kouchkovsky Y. ATPase state and activity in thylakoids from normal and water-stressed lupin. *FEBS Lett.* 1992; 303:233–236.

131. Haraux F, de Kouchkovsky Y. Energy coupling and ATP synthase. *Photosynth. Res.* 1998; 57:231–251.

132. De Kouchkovsky Y, Meyer S. Inactivation of chloroplast ATPase by *in vivo* decrease of water potential. In: Murata N, ed. *Research in Photosynthesis.* Vol. 2. Dordrecht: Kluwer Academic Publishers, 1992:709–712.

133. Wise RR, Sparrow DH, Ortiz-Lopez A, Ort DR. Spatial distribution of photosynthesis during drought in field-grown and acclimated and non-acclimated growth chamber-grown cotton. *Plant Physiol.* 1991; 100:26–32.

134. Ort DR, Oxborough K, Wise RR. Depressions of photosynthesis in crops with water deficits. In: Baker NR, Bowyer JR, eds. *Photoinhibition of Photosynthesis.* Oxford: BIOS Scientific Publishers, 1994:315–329.

135. Siedow JN, Umbach AL. Plant mitochondrial electron transfer and molecular biology. *Plant Cell* 1995; 7:821–831.

136. Lawlor DW. *Photosynthesis*, 3rd ed. Oxford: BIOS Scientific Publishers, 2001.

137. Lawlor DW, Fock H. Photosynthesis and photorespiratory CO_2 evolution of water-stressed sunflower leaves. *Planta* 1975; 126:247–258.

138. Huber SC, Huber JLA. Role of sucrose-phosphate synthase in sucrose metabolism in leaves. *Plant Physiol.* 1992; 99:1275–1278.

139. Bohnert HJ, Nelson DE, Jensen RG. Adaptation to environmental stress. *Plant Cell* 1995; 7:1099–1111.

140. Ingram J, Bartels D. Molecular basis of dehydration tolerance in plants. *Annu. Rev. Plant Physiol. Plant Mol. Biol.* 1996; 47:377–403.

141. Bianchi G, Gamba A, Murelli C, Salamini F, Bartels D. Novel carbohydrate metabolism in the resurrection plant *Craterostigma plantagineum. Plant J.* 1991; 1:355–359.

142. Edwards GE, Furbank RT, Hatch MD, Osmond CB. What does it take to be C_4? Lessons from the evolution of C_4 photosynthesis. *Plant Physiol.* 2001; 125: 46–49.

143. Maroco JP, Ku M, Edwards G. Utilisation of O_2 in the metabolic optimisation of C_4 photosynthesis. *Plant Cell Environ.* 2000; 23:115–121.

144. DiBlasi S, Puliga S, Losi L and Vazzana C. *S. stapfianus* and *E. curvula* cv. Consol *in vivo* photosynthesis, PSII activity and ABA content during dehydration. *Plant Growth Regul.* 1998; 25:97–104.

145. Jenkins CLD, Furbank RT, Hatch MD. Mechanism of C_4 photosynthesis. A model describing the inorganic carbon pool in bundle sheath cells. *Plant Physiol.* 1989; 91:1372–1381.

146. Dai Z, Ku MSB, Edwards GE. C_4 photosynthesis. The CO_2-concentrating mechanism and photorespiration. *Plant Physiol.* 1993; 103:83–90.

147. Hatch MD, Agostino A, Jenkins CLD. Measurement of the leakage of CI2 from bundle-sheath cells of leaves during C_4 photosynthesis. *Plant Physiol.* 1995; 108:173–181.

148. He D, Edwards GE. Estimation of diffusive resistance of bundle sheath cells to CO_2 from modeling of C_4 photosynthesis. *Photosynth. Res.* 1996; 49:195–208.

149. Hatch MD. C_4 photosynthesis: a unique blend of modified biochemistry, anatomy and ultrastructure. *Biochim. Biophys. Acta* 1987; 895:81–106.

150. Byrd GT, Brown H, Bouton JH, Basset CL, Black CC. Degree of C_4 photosynthesis in C_4 and C_3–C_4 *Flaveria* species and their hybrids. I. CO_2 assimilation and metabolism and activities of phosphoenolpyruvate carboxylase and NADP-malic enzyme. *Plant Physiol.* 1992; 100:939–946.

151. Farineau J, Lelandais M, Morot-Gaundry JD. Operation of the glycolate pathway in isolated bundle sheath strands of maize and *Panicum maximum. Physiol. Plant.* 1984; 60:208–214.

152. De Veau EJ, Burris JE. Photorespiratory rates in wheat and maize as determined by ^{18}O-labelling. *Plant Physiol.* 1989; 90:500–511.

153. Dever LV, Blackwell RD, Fullwood NJ, Lacuesta M, Leegood RC, Onek LA, Pearson M, Lea PJ. The isolation and characterization of mutants of the C_4 photosynthetic pathway. *J. Exp. Bot.* 1995; 46:1363–1376.

154. Lacuesta M, Dever LV, Muñoz-Rueda A, Lea PJ. A study of photorespiratory ammonia in the C_4

plant *Amaranthus edulis*, using mutants with altered photosynthetic capacities. *Physiol. Plant.* 1997; 99:447–455.

155. Furbank RT, Jenkins CL, Hatch MD. CO_2 concentrating mechanism of C_4 photosynthesis. Permeability of isolated bundle sheath cells to inorganic carbon. *Plant Physiol.* 1989; 91:1364–1371.

156. Maroco JP, Ku MSB, Edwards GE. Oxygen sensitivity of C_4 photosynthesis: evidence from gas exchange and fluorescence analysis with different C_4 sub-types. *Plant Cell Environ.* 1997; 20:1525–1533.

157. Ehleringer JR, Sage RF, Flanagan LB, Pearcy RW. Climate change and the evolution of C_4 photosynthesis. *Trends Ecol. Evol.* 1991; 6:95–99.

158. Cushman JC, Borland AM. Induction of crassulacean acid metabolism by water limitation. *Plant Cell Environ.* 2002; 25:295–310.

159. Lüttge U. Day–night changes of citric acid levels in crassulacean acid metabolism (CAM) in the adaptation of plants to salinity. *New Phytol.* 1993; 125: 59–71.

160. Lüttge U. *Clusia*: Plasticity and diversity in a genus of C_3/CAM intermediate tropical trees. In: Winter K, Smith JAC, eds. *Crassulacean Acid Metabolism: Biochemistry, Ecophysiology and Evolution.* Vol. 114. Berlin: Springer-Verlag, 1996:296–311.

161. Borland AM, Tecsi LI, Leegood RC, Walker RP. Inducibility of crassulacean acid metabolism (CAM) in *Clusia* species: physiological/biochemical characterisation and intercellular localization and decarboxylation process in three species, which exhibit different degrees of CAM. *Planta* 1998; 205:342–351.

162. Schmitt AK, Lee HSJ, Lüttge U. The response of C_3–CAM tress, *Clusia rosea*, to light and water stress: I.

Gas exchange characteristics. *J. Exp. Bot.* 1988; 39:1581–1590.

163. Zotz G, Winter K. Short-term regulation of crassulacean acid metabolism activity in a tropical hemiepiphyte, *Clusia uvitana. Plant Physiol.* 1993; 102:835–841.

164. Virzo de Santo AV, Bartoli G. Crassulacean acid metabolism in leaves and stems of *Cissus quadrangularis* In: Winter K, Smith JAC eds. *Crassulacean Acid Metabolism. Biochemistry, Ecophysiology and Evolution.* Berlin: Springer-Verlag, 1996.

165. Ting IP, Sipes D. Metabolic modifications of crassulacean acid metabolism in CAM-idling and CAM-cycling. In: Luden PW, Burris JE, eds. *Night Fixation and CO_2 Metabolism.* Amsterdam: Elsevier, 1985.

166. Rayder L, Ting IP. CAM-idling in *Hoya carnosa* (*Asclepiadaceae*). Photosynth. Res. 1983; 4:203–211.

167. Lüttge U. *Physiological Ecology of Tropical Plants.* Berlin: Springer-Verlag, 1997.

168. Osmond CB. Crassulacean acid metabolism: a curiosity in context. *Annu. Rev. Plant Physiol.* 1978; 29:379–414.

169. Adams WW III, Osmond B, Sharkey T. Responses of two CAM species to different irradiances during growth and susceptibility to photoinhibition by high light. *Plant Physiol.* 1987; 83:213–218.

170. Adams WW III, Osmond B. Internal CO_2 supply during photosynthesis of sun and shade grown plants in relation to photoinhibition. *Plant Physiol.* 1988; 86:117–123.

171. Mattos EA, Lüttge U. Chlorophyll fluorescence and organic acid oscillations during transitions from CAM to C_3 photosynthesis in *Clusia minor* L. (Clusiaceae). *Ann. Bot.* 2001; 88:457–463.

42 Role of Plant Growth Regulators in Stomatal Limitation to Photosynthesis during Water Stress

Jana Pospíšilová
Institute of Experimental Botany,
Academy of Sciences of the Czech Republic

Ian C. Dodd
Lancaster Environment Centre, Department of Biological Sciences,
University of Lancaster

CONTENTS

I. INTRODUCTION

Although water is the most abundant molecule on the Earth's surface, the availability of water greatly restricts terrestrial plant production. Thus, if we want to increase productivity of agriculture and forestry we need to understand how plants regulate their water status and the physiological consequences of water stress.

During water stress, most plants avoid shoot desiccation by closing their stomata to decrease transpiration (E). However, this action can also limit CO_2 influx into the leaf and consequently decrease photosynthesis (P_N). To understand how water stress decreases photosynthesis, it is important to understand the regulation of stomatal behavior. Stomata respond to many environmental signals in their aerial (external) environment such as light, CO_2 concentration, temperature, and vapor pressure. Stomata also respond to changes in their internal environment (the leaf apoplast) such as ionic composition and the concentrations of natural plant growth regulators (PGRs), known as plant hormones. Since water stress influences both hormone synthesis and response, it is likely that plant hormones are involved in the limitation of photosynthesis during water stress. Although most of the work on plant hormones and stomatal responses emphasizes the hormone abscisic acid (ABA), stomata have also been shown to respond to all hormone classes (Figure 42.1).

This chapter aims to illustrate

Changes in endogenous phytohormone content induced by water stress.

Stomatal responses to endogenous phytohormones and application of natural or synthetic PGRs.

The occurrence of photosynthetic limitations and water deficits in plants.

Possibilities of ameliorating the negative effects of water stress by application of natural or synthetic PGRs.

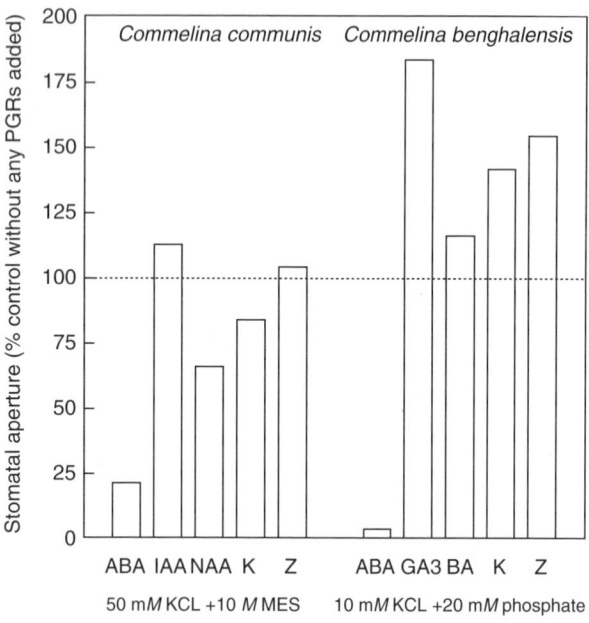

FIGURE 42.1 Responses of *Commelina* stomata to incubation of epidermal strips on 10-μM solutions of various PGRs, made up in two different incubation solutions. PGRs applied were ABA, the auxins IAA and NAA, the CKs benzyladenine (BA), kinetin (K), and Z, and GA$_3$. Data for hormone solutions made up in a buffer containing 50 mM KCl and 10 mM MES at pH 6.15 were taken from Ref. [116] (ABA), Ref. [65] (IAA), Ref. [62] (NAA), and Ref. [1] (K, Z). Data for hormone solutions made up in 10 mM KCl and 20 mM phosphate buffer at pH 6.7 were taken from Ref. [90].

II. ABSCISIC ACID

A. BIOSYNTHESIS AND COMPARTMENTATION

Although well-watered plants contain some ABA, water stress stimulates ABA biosynthesis in both roots and leaves. Many schemes representing the pathway of ABA biosynthesis start with the carotenoids of the xanthophyll cycle. The first committed step of ABA biosynthesis is the oxidative cleavage of the carotenoids 9'-*cis*-violaxanthin or 9'-*cis*-neoxanthin to xanthoxin by plastid enzymes 9-*cis*-epoxycarotenoid dioxgenases. In the second step, xanthoxin is converted to abscisic aldehyde by xanthoxin oxidase. In the last step, abscisic aldehyde oxidase catalyzes conversion of abscisic aldehyde to ABA (for recent reviews see Refs. [2,3]). During water stress, activities of the above-mentioned enzymes as well as their mRNA transcript abundance increase in both leaves and roots. In the roots, xanthophylls are in low abundance and zeaxanthin epoxidation to violaxanthin might be a further regulatory step of water stress-induced ABA biosynthesis [2,3].

There is still little information about the signaling pathway from water stress perception to activation of genes encoding the key enzymes of ABA biosynthesis, but the gene *ATHK1* and an MAP kinase cascade are thought to participate in it [4]. Interactions between the plasmalemma and cell wall seem essential to trigger water stress-induced ABA accumulation [5]. The changes in cellular volume in response to dehydration, rather than cellular water relations parameters such as water potential or pressure potential, stimulate ABA biosynthesis [6]. The ability of leaf tissue to synthesize additional ABA varies between different genotypes [7], and can be enhanced by nutrient stress and decreased by high (40°C) temperature [8].

The main catabolic pathway of ABA, degradation to phaseic acid and dihydrophaseic acid, is probably cytosolic since the enzymes are located in the endoplasmic reticulum [8]. The rate of catabolism of xylem-delivered ABA is decreased in water-stressed plants and when the xylem sap is alkalized [10,11].

ABA can move rapidly through the plant in both the xylem and the phloem in the active free form or as inactive conjugated forms (predominantly ABA glucose ester) [12,13]. In roots and leaves ABA can be transported in apoplast or symplast. Since ABA is a weak acid (pK_a = 4.5), its distribution in plant tissues will be governed by the Henderson–Hasselbach equation. At a cytosolic pH commonly found in many well-watered plants (pH 6.5), ABA present in the protonated form moves into the alkaline chloroplast stroma (pH 7.5) where it dissociates to form an anion that is not as readily permeable. For this reason, most of the ABA in unstressed leaves is assumed to be in the chloroplasts (which act as an anion trap for ABA). However, under osmotic stress the intracellular pH gradients are much smaller and a greater proportion of ABA is found in the cytosol [14]. Leaf dehydration alkalizes the apoplast, increasing apoplastic ABA concentrations [15]. Redistribution of ABA among different compartments of the leaf provides an attractive possibility for stomatal regulation in response to drought. The role of xylem ABA concentration in stomatal regulation is considered in Section IV.

B. SIGNAL TRANSDUCTION

Isolation of ABA receptors has proved elusive, as initial reports [16] are yet to be substantiated. Recently, a 42-kDa ABA-specific binding protein has been purified from the epidermis of broad bean [17]. Interest has focused on where in the guard cells an ABA receptor might reside. ABA-induced stomatal closure in isolated epidermes incubated at pH 8, when ABA is not readily able to cross the cell membrane, has been taken as evidence of extracellular ABA receptors [18]. At more acidic pHs, guard cells appear to have a significant carrier-mediated uptake of ABA [19]. The presence of intracellular ABA receptors has also been suggested, as injection of ABA into individual guard cells causes stomatal closure, indicating that stomata can perceive symplastic hormone concentrations [20,21].

ABA can regulate stomatal aperture by promoting stomatal closure or inhibiting stomatal opening, induced by changing the osmotic potential of guard cells, the mechanical properties of guard cells, or gene expression [22]. ABA-induced decreases in stomatal opening involve both inhibition of channels facilitating K^+ entry and activation of channels controlling efflux of K^+ and anions. ABA also inhibits blue light dependent H^+ efflux in *Arabidopsis thaliana* [23,24]. Ca^{2+} is a second messenger in some, but not all, ABA-induced changes in guard cell ion channels. ABA-induced inactivation of the plasmalemma inward K^+ channel is usually Ca^{2+} mediated, whereas ABA-induced activation of the plasmalemma outward K^+ channel is Ca^{2+} independent [19,25–28]. In the latter case, intracellular pH changes are probably important [19,26,29]. ABA enhances cytosolic Ca^{2+} calcium concentration by stimulating both Ca^{2+} entry across the plasma membrane and Ca^{2+} release from intracellular stores by Ca^{2+} channels sensitive to inositol-1,4,5-triphosphate or cyclic ADP-ribose [19,24,29]. Protein kinases and phosphatases may also participate in ABA signal transduction (e.g., [30,31]). H_2O_2-mediated ABA-induced inhibition of inward K^+ currents has also been suggested [32,33]. Nitric oxide can be another component of ABA-mediated stomatal closure [34].

Cytoskeleton reorganization may also be involved in the ABA-dependent or ABA-independent regulation of stomatal opening under water stress [31] by changing the mechanical properties of guard cells (modulus of elasticity). ABA treatment caused reorganization of the actin structure of guard cells from a radial pattern to a randomly oriented and short-fragmented pattern [35]. The small guanosine triphosphatase protein *AtRac1* was identified in *Arabidopsis* as a central component in the ABA-mediated disruption of the guard cell actin cytoskeleton [36]. ABA also disrupted cortical microtubules of *Vicia faba* guard cells, but not epidermal cells [37]. This effect was reversible and arrays of microtubules reappeared within 1 h of removal of ABA.

The ABA signal can also be relayed to the guard cell nucleus to alter the pattern of gene expression, leading to changes in the content of proteins involved in water transport, ion transport, or carbon metabolism [38].

C. ENVIRONMENTAL MODULATION OF STOMATAL SENSITIVITY TO ABA

The sensitivity of stomata to ABA varies widely in different species and cultivars, with leaf age, time of day, temperature, irradiance, air humidity, ambient CO_2 concentrations, plant nutritional status, ionic composition of the xylem sap, and leaf water status (reviewed in Ref. [39]). Varying stomatal responses to ABA may be important in the minute-by-minute control of stomatal aperture in a fluctuating environment. In the whole plant, differences in stomatal response to xylem ABA concentration may simply reflect differences in the amount of ABA reaching the active sites at the guard cell. However, this explanation cannot hold where isolated epidermal strips are floated on hormonal solutions.

Water stress commonly co-occurs with high temperatures and vapor pressure deficits, which will promote water loss. In maize epidermal strips, application of ABA stimulated stomatal opening below a threshold temperature, yet caused stomatal closure as the temperature increased [40]. Similarly, stomata of several species (*Bellis perennis*, *Cardamine pratensis*, *Commelina communis*) were relatively insensitive to ABA when incubated at 10°C and in some cases showed stomatal opening, but showed normal ABA-induced stomatal closure at 20°C or 30°C [41]. The temperature dependence of ABA action allows drought-stressed plants to open their stomata to maximize photosynthesis under conditions (lower temperature and vapor pressure deficit) where transpirational losses can be minimized.

In field-grown maize crops, the slope of the relationship between xylem ABA concentration and stomatal conductance (g_s) varied diurnally, with the most sensitive stomatal closure occurring at lower leaf water potentials (Ψ_{leaf}) [42]. Since an increased Ψ_{leaf} increases the rate of catabolism of xylem-supplied ABA [10], such a result might be explained in terms of differences in the amounts of ABA reaching the guard cells. However, this does not explain the increased stomatal sensitivity to ABA seen when *Commelina* epidermes were incubated on ABA solutions of decreasing osmotic potential. This interaction may be thought of as a sensitive dynamic feedback control mechanism to ensure homeostasis of leaf water status. Any decrease in leaf water status (e.g., the sun appearing from behind a cloud) will enhance stomatal response to ABA, thus decreasing E and returning leaf water status to its original value.

Previous water stress can influence stomatal response to ABA independently of current plant water status, although the effects can be variable. In *V. faba* the stomata of previously water-stressed plants were more sensitive to ABA applied through the petiole or sprayed onto leaf surfaces than the stomata of well-watered plants [43]. In contrast, stomata of previously water-stressed plants (in which almost all leaves have wilted) were less sensitive to ABA applied through the petiole than stomata of well-watered plants [44]. To try to reconcile these conflicting observations, *Commelina* plants were subjected to a slow soil drying treatment (over 15 days), and every day epidermal strips were removed to determine the sensitivity of stomatal closure to ABA using a bioassay [45]. Initially, water stress sensitized stomata to ABA (at the time that stomatal closure occurred in intact plants), but a later desensitization of stomata to ABA occurred when leaf relative water content began to decline and stomata had effectively closed completely. Stomatal sensitivity to ABA was thus greatest when stomatal closure was trying to ensure homeostasis of leaf water status, and then declined when hydraulic influences would ensure continued stomatal closure.

III.   OTHER HORMONES AND STOMATAL BEHAVIOR

A.   AUXINS

The most important natural auxins seem to be indole-3-acetic acid (IAA), indole-3-butyric acid (IBA), and phenoxyacetic acid (PAA). Many man-made auxin analogs have been synthesized and some of them (e.g., napthyl-acetic acid [NAA] or 2,4-dichloro-

phenoxyacetic acid [2,4-D]) are practically important. Auxin concentrations are highest in regions of active cell division such as the apical meristems, the cambium, the developing fruit, the embryo, and the endosperm, and in young leaves. The transport of IAA is strictly polar from the apex to the organ base. IAA can be synthesized via tryptophan-dependent and tryptophan-independent pathways [46]. Plants store most of their IAA in conjugated forms (which are probably inactive) such as ester conjugates (predominantly in monocotyledonous plants) or amide conjugates (predominantly in dicotyledonous plants) (for review, see Ref. [47]). IAA can be quickly broken down by oxidative decarboxylation.

Relatively little information is available on the changes in auxin content induced by water stress. Osmotic stress (150–300 mM NaCl) decreased IAA content in tomato roots, but the leaf IAA content remained relatively unchanged [48]. In *Fatsia japonica* leaves, IAA content increased as Ψ_{leaf} decreased during the day but only slightly increased during drought [49,50]. Although root drying can decrease root auxin concentration by up to 70% [51], it has not been investigated whether this changes xylem auxin concentration. Dehydration of detached leaves did not alter xylem auxin concentration [52].

The auxin-binding protein (ABP1) is an IAA receptor located at the plasma membrane. Binding of auxin causes a conformational change affecting the C terminus of ABP1 and this change probably activates the signal transduction pathway, which may involve activation of phospholipase and plasma membrane H$^+$-ATPase (for review, see Ref. [53]). Activation of guard cell H$^+$-ATPase by IAA may stimulate H$^+$ extrusion and stomatal opening [24]. Stomatal opening induced by IAA in epidermal strips of *Paphiopedilum tonsum* was preceded by a reduction of cytosolic pH [54]. Exogenous auxins (IAA and NAA) can also affect inwardly and outwardly rectifying K$^+$ channels in a dose- and pH-dependent manner [55–57]. Low auxin concentrations promote inward movement of K$^+$, while higher concentrations inhibit it [58]. These effects may be mediated by second messengers such as changes in cytosolic Ca^{2+} concentration [54,56]. Cyclic guanosine monophosphate (GMP) was suggested as a mediator within the Ca^{2+} signaling cascade for IBA signal transduction in *C. communis* [59,60].

Stomatal responses to exogenous auxins are dependent not only on the auxin used and the concentration, but also on plant species, age, environmental conditions, and source of the epidermis (adaxial or abaxial). High concentrations of auxins such as PAA [61] and NAA [62] can suppress stomatal opening. Stomata in epidermal strips from the adaxial leaf surface are often more responsive to auxin [63]. The effects of auxins can depend on atmospheric CO_2 concentration. In *Pisum sativum* and *Phaseolus vulgaris*, IAA increased g_s in the presence of CO_2 but not in absence of CO_2 [64]. IAA also inhibited the closing effect of high CO_2 concentration in *C. communis* [65] and *V. faba* [66].

Several reports suggest that auxins can antagonize ABA-induced stomatal closure. IAA alleviated the closing effect of ABA in epidermal peels of *C. communis* [65] and *V. faba* [66,67]. Similarly, PAA reduced the closing effect of ABA in *C. communis* [61], and vice versa (ABA reduced PAA-induced abaxial stomata closure). However, it is not known to what extent variation in endogenous auxin concentration influences stomatal sensitivity to ABA in planta.

B. CYTOKININS

Most naturally occurring cytokinins (CKs) are N^6-substituted adenine molecules with a branched five-carbon side chain, such as *trans*-zeatin (Z) and isopentenyladenine. Riboside and ribotide derivatives are less active than the free bases, and N- and O-linked glucosides are mostly inactive (e.g., [67,68]). The pathways of CK biosynthesis have not yet been completely solved. The important step is probably the formation of N^6-(Δ^2-isopentenyl) adenosine-5'-mono-phosphate from Δ^2-isopentenyl pyrophosphate and adenosine-5'-mono-phosphate catalyzed by isopentenyltransferase [70]. Another possibility is the degradation of tRNA and the isomerization of *cis*-zeatin to Z by *cis–trans* isomerase [70]. CKs are produced in plant meristematic regions including the roots [71] and transported in both the xylem and the phloem. CK metabolism is very complex and reflects the existence of many of the above-mentioned compounds with different activities (for a recent review, see Ref. [72]). Irreversible degradation of CKs by N^6-side chain cleavage is catalyzed by CK oxidase (which may also be considered to be CK dehydrogenase) [73]. Endogenous CK contents are also regulated by other plant hormones, in particular by auxins [e.g., 67,73].

Decreased leaf CK concentration in response to drought stress has been observed (for reviews, see Refs. [75,76]), although it is difficult to predict the actual change of any given CK species. For example, dehydration of wheat seedlings by 15–30 min of air drying decreased shoot concentrations of zeatin nucleotide and zeatin 9-*N*-glucoside, but the total content of Z derivatives as well as the content of free base of Z remained almost constant [77]. Mild water deficit (Ψ_{leaf} = −0.32 MPa) had no effect on sunflower xylem zeatin riboside (ZR) concentration, yet the decrease in E (caused by stomatal closure) decreased

ZR flux to the shoot. More severe water deficit (Ψ_{leaf} = −0.97 MPa) decreased both concentration and flux of ZR [78].

The cellular site and molecular mechanism of CK action are poorly understood. They probably act at the plasma membrane in concert with other signals [69,79]. The mechanism of CK action on guard cells might involve direct induction of membrane hyperpolarization by stimulation of electrogenic H^+-pump; stimulation of adenylate cyclase activity which could lead to an increase in intracellular adenosine 3',5'-cyclic monophosphate content, stimulation of guanylate cyclase activity, or interaction with a calcium–calmodulin system [80–82].

Stomatal responses to naturally occurring or synthetic CKs are variable [e.g., 82,83] although CKs can increase stomatal aperture. The apparent insensitivity of stomata to CK application may be because CK concentration is already optimal for stomatal opening [83]. In the context of stomatal limitation under water stress, CKs are often considered as antagonists of ABA action. Alleviation of ABA-induced stomatal closure by CKs has been reported in maize epidermal strips [85], detached flax leaves [86], and leaves detached from N-deprived cotton [8]. In isolated systems, such antagonism may result from interactions in the signal transduction pathways of both compounds, perhaps involving cytosolic calcium concentration [87]. *In planta*, metabolic interactions may be involved as CKs partially share a common biosynthetic origin with ABA [88].

C. GIBBERELLINS

Gibberellins (GAs) are diterpenes constituted of four isoprene units. They derive from *ent*-kaurene formed by cyclization of geranylgeranyl pyrophosphate. Many plants contain a mixture of different GAs, and at least 70 GAs have been isolated from natural sources. Cleavage of the ring system results in loss of activity. They are easily transported in both xylem and phloem.

The little that is known about changes in endogenous GA content under water stress has been previously reviewed [76], with either no change or decreases in GA content reported. The effects of foliar application of gibberellic acid (GA_3) are variable [76] although retardation of stomatal closure in water-stressed lettuce leaves following GA_3 treatment has been observed [89]. This is consistent with a report that GA_3 could reverse triazole-induced stomatal closure in isolated epidermal strips of *Commelina benghalensis* [90]. Some nonstomatal effects of GA_3 application (increased ribulose-1,5-bisphosphate carboxylase activity [91]) have been reported, thus any report of GA_3 effects on photosynthesis should carefully analyze whether stomatal or nonstomatal effects are important.

D. JASMONATES

Both jasmonic acid (JA; 3-oxo-2-(2-cis-pentenylcyclopentane-1-acetic acid)) and its methyl ester (MeJA) occur in plants. JA is formed from linoleic acid, and the first step is catalyzed by lipoxygenase (e.g., Ref. [91]). Little is known about changes in jasmonate content during water stress. Despite this, JA and MeJA have been applied to intact plants of many species, and stomatal closure is a common response [76]. The possible mechanism of JA or MeJA action on stomatal opening is probably similar to that of ABA with a suppression of H^+ efflux and K^+ influx occurring [93]. In Paphiopedilum, JA and MeJA caused intracellular alkalization, which preceded stomatal closure [94].

E. ETHYLENE

Ethylene is a single, gaseous compound synthesized by the conversion of methionine to S-adenosylmethionine, then to 1-aminocyclopropane-1-carboxylic acid (ACC) by ACC synthase, and further to ethylene by ACC oxidase [95]. Ethylene biosynthesis is enhanced under extreme temperatures, wounding, and mechanical stresses, but conspicuously, not when intact plants were droughted [96]. Exogenous application of gaseous ethylene [97] or liquid Ethrel [98] (a phosphonic acid that liberates ethylene in planta) can inhibit leaf gas exchange. Manipulation of plant ethylene production in isolated systems such as epidermal strips or detached leaves, using precursors or inhibitors of ethylene biosynthesis or action, has yielded variable results [99], and evidence that ethylene affects stomatal behavior during water stress is lacking.

F. BRASSINOSTEROIDS

Brassinosteroids (BRs) are a group of steroid-like compounds isolated from various plants. Brassinolide and castasterone are the most abundant biologically active compounds and synthetic homobrassinolide has a similar biological activity. Exogenous application of brassinolides induces a broad spectrum of responses, including proton pump activation and reorientation of cellulose microtubules (for review, see Ref. [100]). Little is known about changes in BR content during water stress. Foliar applications of brassinolide decreased stomatal opening and E in sorghum leaves and enhanced the effect of simultaneously applied ABA [101]. Pretreatment of jack pine

seedlings with homobrassinolide delayed stomatal closure induced by water stress [102].

IV. WATER STRESS AND STOMATAL LIMITATIONS TO PHOTOSYNTHESIS

A. DEFINING STOMATAL LIMITATIONS

By regulating water loss, stomata play a dominant role in the control of plant water status. Over the course of a day, plant water status fluctuates as stomata respond to various environmental signals. Light-induced stomatal opening decreases plant water status at the start of the day. As solar noon approaches, water status decreases as E increases to keep the leaves cool, and then increases as temperatures (and E) decrease toward the end of the day. Stomatal conductance and P_N also exhibit considerable diurnal fluctuation. Given such a variation in plant water status, how should water stress be defined?

Traditionally, water stress has been characterized by decreases in Ψ_{leaf} or relative water content. However, in some species, considerable stomatal closure can occur without decreases in daytime Ψ_{leaf} [103]. Since plant water status equilibrates with soil water status overnight, measurement of predawn Ψ_{leaf} can be useful to define the degree of water stress experienced since soil water status affects the magnitude of diurnal changes in Ψ_{leaf}, g_s, and P_N.

Under mild water deficits, stomatal closure to reduce water efflux simultaneously decreases the CO_2 influx, which limits photosynthesis. Decreased g_s is accompanied by a reduction in internal CO_2 concentration (c_i) and decreased diffusion of CO_2 via mesophyll cell walls, membranes, cytoplasm, and chloroplast envelope, leading to decreased chloroplastic CO_2 concentration [104,105]. When stomatal limitation occurs, there is often a linear dependence of P_N on the internal to ambient CO_2 concentration ratio (c_i/c_a; Figure 42.2). To confirm that photosynthetic limitation is exclusively stomatal in nature, it is necessary to raise the CO_2 concentration (to 1–5% CO_2) around the leaves of droughted plants, and show that this overcomes any limitation of photosynthesis.

More severe water deficit directly affects the photosynthetic capacity of mesophyll causing decreases in carboxylation as well as in electron transport chain activities, and induces ultrastructural changes in chloroplasts (for review, see e.g., Ref. [106]). Depression of P_N under high CO_2 concentrations is indicative of nonstomatal limitations.

Several mathematical models have been developed for calculation of the stomatal and nonstomatal limi-

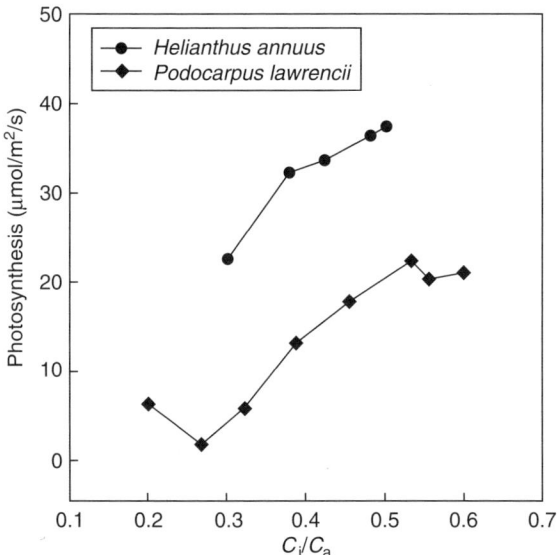

FIGURE 42.2 Dependence of P_N on the c_i/c_a ratio for droughted *Helianthus annuus* (•) and *Podocarpus lawrencii* (♦) plants. (Redrawn from Sharp RE, Boyer JS. *Plant Physiol.* 1986; 82:90–95 (•) and Brodribb T. *Plant Physiol.* 1996; 111:179–185 (♦).)

tations of P_N (e.g., Ref. [107]). However, they are usually based on the implicit assumption of uniform g_s and uniform mesophyll photosynthetic capacity over the leaf surface, which in some cases may over-estimate nonstomatal limitations (e.g., Refs. [108, 109]). Differences among species and in the rates of imposition of water deficits, as well as the interactions with other environmental stresses, play a role in the relative importance of stomatal and nonstomatal limitations of photosynthesis under drought (e.g., Ref. [110]).

B. ABA AND STOMATAL LIMITATION

Although stomata can respond to all classical hormone classes (Sections II and III), most interest has centered on the ability of ABA to induce stomatal closure. At the time that the effects of ABA on stomata were identified, it was widely assumed that stomatal closure occurred in response to decreased leaf water status. An attractive hypothesis was that drought-induced changes in Ψ_{leaf} liberated ABA from the mesophyll chloroplasts where it is normally sequestered in unstressed leaves, and that this ABA would move to the guard cells to initiate stomatal closure [111]. This hypothesis emphasized the importance of leaf ABA concentrations in determining g_s. However, drought-induced stomatal closure is not

always well correlated with bulk leaf ABA concentration, and ABA accumulation often occurs only after g_s has declined [112].

In many circumstances, xylem ABA concentration increases earlier and to a greater magnitude than changes in bulk leaf ABA concentration (Figure 42.3). The origin of this ABA is subject to debate. During some drying cycles, root ABA concentration and xylem sap concentration increase in parallel [113], suggesting that xylem ABA is root-derived. However, considerable recirculation of ABA between xylem and phloem can occur [114], thus not all ABA in a xylem sap sample is likely to be root-derived. In some species, even ABA found in the rhizosphere can be efficiently transferred across the root tissues into the xylem [115].

Irrespective of whether ABA is root- or leaf-sourced, the apoplastic ABA concentration in the vicinity of guard cells seems important in regulating stomatal opening. Apoplastic ABA concentrations can be increased by increased xylem delivery of ABA to the leaf, decreased metabolism, or sequestration by mesophyll cells [116], or redistribution of existing leaf ABA to the apoplast. Water evaporation from guard cell walls also increases the ABA concentration in the guard cell apoplast [117–119]. The concentration of free ABA in the vicinity of the guard cells may also depend on apoplastic β-glucosidase activity, which releases ABA from the physiologically inactive ABA–glucose conjugate pool [13].

While apoplastic ABA concentration may regulate stomatal opening, it is difficult to measure directly and several comprehensive studies indicate an excellent correlation between xylem ABA concentration and g_s (when xylem sap was collected from the same leaves in which g_s was measured) in species such as maize [42], sunflower [120], and tobacco [121]. Since ABA-induced stomatal closure limits leaf photosynthesis, there is also a negative relationship between P_N and xylem ABA concentration (Figure 42.4). Importantly, P_N is often less negatively affected than g_s, increasing water use efficiency (WUE).

Although xylem ABA concentration can account for stomatal closure in many experiments, sometimes stomata close prior to any increase in xylem ABA concentration [122]. Detached leaf transpiration studies have suggested the presence of other antitranspirant compounds in wheat and barley xylem sap [123], since the antitranspirant activity of xylem sap could not be explained in terms of its ABA concentration. Alkalization of xylem sap is a common response to various edaphic stresses including soil drying [124], and supplying detached *Commelina* and tomato leaves with neutral or

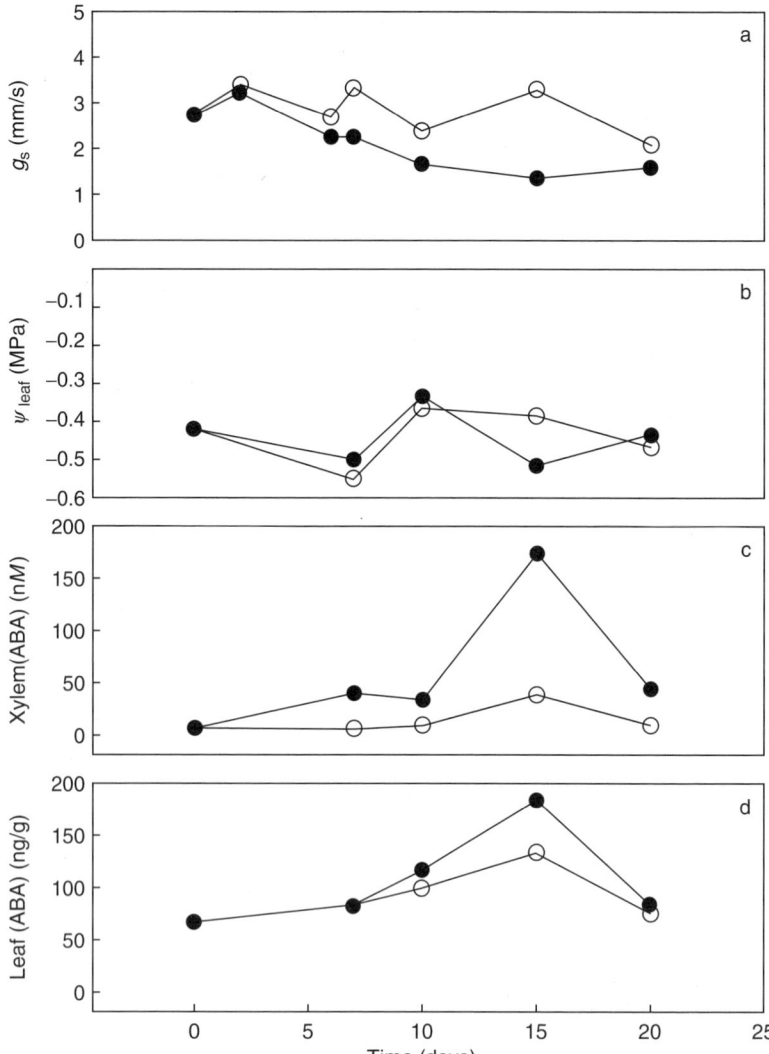

FIGURE 42.3 Stomatal conductance (a), leaf water potential (b), xylem ABA concentration (c), and leaf ABA concentration (d) of plants that were well watered (○) or remained unwatered from Day 0 (●). (Redrawn from Zhang J, Davies WJ. *Plant Cell Environ.* 1990; 13:277–285.)

alkaline buffers (pH ≥ 7) via the transpiration stream can restrict *E*. These alkaline buffers increased apoplastic pH, thus decreasing sequestration of ABA by mesophyll cells, causing increased apoplastic ABA concentrations, which closed the stomata [125]. Stomatal closure in response to xylem-supplied alkaline buffers was ABA dependent, as leaves detached from an ABA-deficient mutant (*flacca*) did not show stomatal closure when fed pH 7 buffers, and in some cases *E* actually increased. Stomatal closure in response to sap alkalization may explain observations where stomatal closure could not be readily explained in terms of ABA concentration. Thus, plants do not necessarily need to increase their ABA concentration to initiate stomatal closure (and limit photosynthesis), as the ABA concentration present in well-watered wild-type plants can be redistributed to the guard cells following an increase in apoplastic pH.

Over the course of a soil drying cycle, different mechanisms may operate to maintain increased apoplastic ABA concentrations at the guard cells. An early response to soil drying might be sap alkalization, which might initiate stomatal closure by redistributing ABA already present in the leaf to the apoplast. As root tips start to wilt, additional ABA is synthesized in the roots and augments xylem ABA concentrations. If soil drying is prolonged or severe enough, leaves may wilt, stimulating leaf ABA synthesis. It is under these conditions that nonstomatal effects of water stress may occur, and the possible involvement of ABA in this response is considered below. Following relief of stress, stomata may remain partially closed even when Ψ_{leaf} has returned to prestress values. Rewetting previously dried roots may release a pulse of ABA into the transpiration stream, and xylem ABA concentrations can remain elevated up to 48 h after rewatering [126].

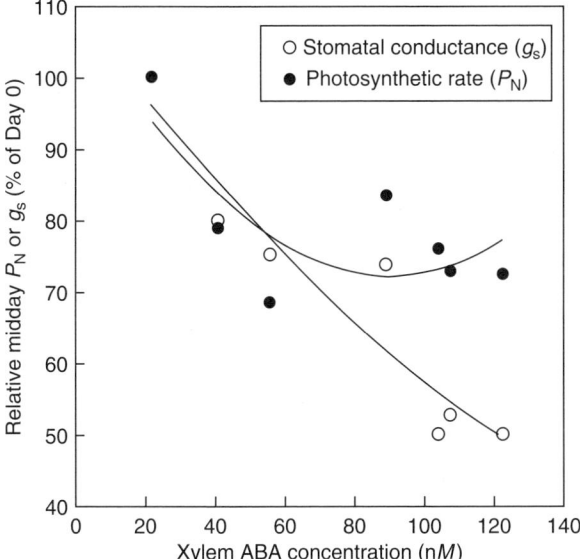

FIGURE 42.4 Responses of stomatal conductance, g_s (○), and net photosynthetic rate, P_N (•), to changes in xylem ABA concentration in *Acacia confusa* plants from which water was withheld on Day 0. (Redrawn from Liang J, Zhang J, Wong MH. *Plant Cell Environ*. 1996; 19:93–100.)

V. NONSTOMATAL EFFECTS OF PLANT GROWTH REGULATORS ON PHOTOSYNTHESIS

Although many studies indicate that the principal effect of ABA on photosynthesis is due to stomatal closure decreasing intercellular CO_2 concentration (e.g., Ref. [127]), ABA-induced depression of P_N at constant internal CO_2 concentration is a recurrent theme in the literature [128,129], suggesting a possible effect of ABA on carboxylation capacity. The possible mechanisms might be decreased activity of ribulose-1,5-bisphosphate carboxylase, decreased regeneration of ribulose-1,5-bisphosphate, or inhibition of ATP-synthase.

Nonstomatal effects of CKs have been reported including altered chlorophyll and photosynthetic protein synthesis and degradation, chloroplast composition and ultrastructure, electron transport, and enzyme activities (for review see, [130]). Exogenously applied CKs alleviated the negative effects of water stress on chlorophyll and carotenoid contents, photochemical activities of photosystems 1 and 2, and content and activity of ribulose-1,5-bisphosphate carboxylase or phosphoenolpyruvate carboxylase [131–134]. However, it is not known whether improved photosynthetic performance is due to direct, specific effects of CKs on enzyme activity, or due to delayed leaf senescence caused by CK treatment. Analysis

of senescence-induced genes that are unrelated to photosynthetic performance may resolve this distinction.

VI. AMELIORATING EFFECTS OF WATER STRESS USING PLANT GROWTH REGULATORS

Antitranspirant compounds may be useful at critical stages of the crop life cycle when it is desirable to decrease plant water use, such as after transplanting of greenhouse-grown seedlings to the field or hardening off of tissue-culture-grown plants. The potent antitranspirant effect of ABA suggests that it would be ideal for such use, and seedlings whose roots were dipped in ABA prior to transplanting showed greater survival than those dipped in water [135]. However, exogenous ABA applications often give only short-term effects due to metabolism of ABA in the plant and light-induced breakdown on plant and soil surfaces. For this reason, ABA analogs that are more resistant to inactivation have been synthesized. As we have seen, application of antitranspirants will also decrease P_N. Fortunately, the inhibitory effect of ABA on E is much greater than its inhibitory effect on P_N (e.g., Ref. [136]), thus increasing plant WUE. Similarly, application of these ABA analogs has also increased WUE [137].

In other circumstances, it might be advantageous to override photosynthetic limitation caused by ABA-induced stomatal closure. As noted above, in isolated systems (detached epidermes and leaves) IAA and some CKs can antagonize ABA-induced stomatal closure. There are also cases where foliar applications of these PGRs have increased g_s. In intact cotton plants, foliar sprays of 50 μM IAA, GA$_3$, or benzylaminopurine partially counteracted the effect of water deficit on g_s, P_N, and E [138]. However, foliar CK application generally has little consistent effect on g_s, P_N, and E of water-stressed plants [139,140], although in some cases P_N can be increased due to a delay in leaf senescence [141]. One of the inherent difficulties is knowing to what extent the substance of interest enters the leaves, and its fate in the leaf. While foliar CK application can prevent ABA-induced photosynthetic limitation, the effects can be transient and of little consequence in the long term (Figure 42.5). Consequently, transgenic approaches to alter CK status in planta may be more reliable.

VII. SUMMARY

Although prolonged water stress can decrease Ψ_{leaf} and close stomata via hydraulic influences, in many

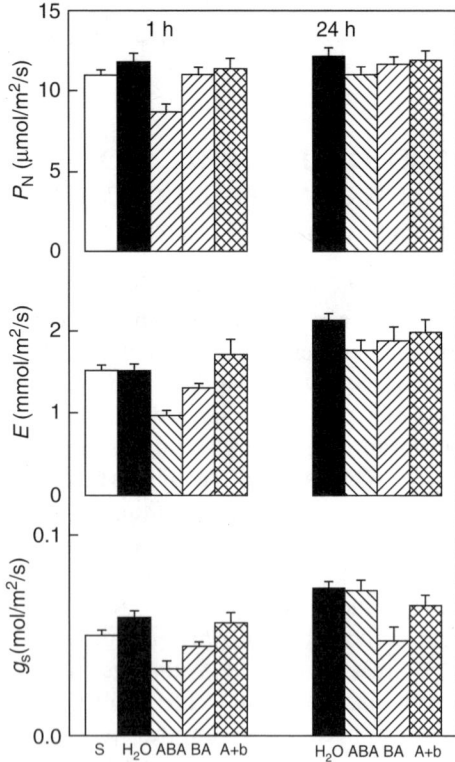

FIGURE 42.5 Net photosynthetic rate (P_N), transpiration rate (E), and stomatal conductance (g_s) in primary bean leaves 1 and 24 h after spraying with H_2O (control), $100 \mu M$ ABA, $10 \mu M$ BA, or a combination of $100 \mu M$ ABA and $10 \mu M$ BA (A + B). Means $\pm$ SE, $n = 18$

photosynthesis have been demonstrated, in the majority of cases changes in endogenous hormone concentrations during water stress affect g_s, thus modifying intercellular CO_2 concentration and then photosynthesis.

Application of synthetic or natural plant growth regulators may modify stomatal response *in vivo*, but the effects of a given application can vary according to uptake and degradation of the compound of interest, and the effect of the compound on endogenous phytohormone contents. Transgenic technologies give considerable scope for manipulating endogenous phytohormone contents, and may provide a way of reproducibly modifying stomatal responses to ABA. Under dryland agriculture where soil moisture is depleted as the crop nears maturity, enhancement of ABA-induced stomatal closure may allow the crop to survive for a sufficient period to produce some yield from stored photosynthate. Alternatively, under irrigated environments where water supply is assured, suppression of ABA-induced stomatal closure may minimize photosynthetic limitations and maximize crop yield. Given that much of the photosynthetic limitation that occurs under managed agriculture is mostly stomatal in nature, understanding how variation in plant hormone status affects photosynthesis seems important.

ACKNOWLEDGMENTS

The first author acknowledges the financial support of the Grant Agency of the Czech Republic (grant No. 522/02/1099).

cases, plants use chemical signals traveling in the xylem to initiate stomatal closure thus preventing any decrease in Ψ_{leaf}. Such signals can operate prior to any increase in bulk leaf hormone concentrations. Most interest has centred on ABA as the most probable root-to-shoot chemical signal regulating stomatal aperture. Apoplastic ABA concentration around the guard cells is crucial in determining stomatal responses, and this will depend on many factors including xylem ABA delivery to the leaf, leaf mesophyll ABA catabolism, leaf ABA synthesis, and apoplastic pH. Even when apoplastic ABA concentration is constant, environmental and physiological variables such as CO_2 concentration, temperature, and current leaf water status can alter stomatal response to ABA. Other plant hormones are also important in modifying stomatal response to a given ABA concentration. The most probable candidates for alleviating ABA effects seem to be CKs and auxins, and for stimulating ABA effects, JA and MeJA. Although some nonstomatal effects of hormone application on

REFERENCES

1. Blackman PG, Davies WJ. The effects of cytokinins and ABA on stomatal behaviour of maize and *Commelina*. *J. Exp. Bot.* 1983; 34:1619–1626.
2. Taylor IB, Burbidge A, Thompson AJ. Control of abscisic acid synthesis. *J. Exp. Bot.* 2000; 51:1563–1574.
3. Schwartz SH, Qin X, Zeevaart JAD. Elucidation of the indirect pathway of abscisic acid biosynthesis by mutants, genes and enzymes. *Plant Physiol.* 2003; 131:1591–1601.
4. Bray EA. Abscisic acid regulation of gene expression during water-deficit stress in the era of the *Arabidopsis* genome. *Plant Cell Environ.* 2002; 25:153–161.
5. Zeevaart JAD. Abscisic acid metabolism and its regulation. In: Hooykaas PJJ, Hall MA, Libbenga KR, eds. *Biochemistry and Molecular Biology of Plant Hormones.* Amsterdam: Elsevier, 1999:189–207.
6. Jia W, Zhang J, Liang J. Initiation and regulation of water deficit-induced abscisic acid accumulation in maize leaves and roots: cellular volume and water relations. *J. Exp. Bot.* 2001; 52:295–300.

7. Quarrie SA. Genetic variability and heritability of drought-induced abscisic acid accumulation in spring wheat. *Plant Cell Environ.* 1981; 4:147–151.

8. Radin JW, Parker LL, Guinn G. Water relation of cotton plants under nitrogen deficiency. V. Environmental control of abscisic acid accumulation and stomatal sensitivity to abscisic acid. *Plant Physiol.* 1982; 70:1066–1070.

9. Daeter W, Hartung W. Stress-dependent redistribution of abscisic acid (ABA) in *Hordeum vulgare* L. leaves: the role of epidermal ABA metabolism, tonoplast transport and the cuticle. *Plant Cell Environ.* 1995; 18:1367–1376.

10. Jia W, Zhang J. Comparison of exportation and metabolism of xylem-delivered ABA in maize leaves at different water status and xylem sap pH. *Plant Growth Regul.* 1997; 21:43–49.

11. Zhang J, Jia W, Zhang D-P. Effect of leaf water status and xylem pH on metabolism of xylem-transported abscisic acid. *Plant Growth Regul.* 1997; 21:51–58.

12. Sauter A, Hartung W. Radial transport of abscisic acid conjugates in maize roots: its implication for long distance stress signals. *J. Exp. Bot.* 2000; 51:925–935.

13. Sauter A, Dietz K-J, Hartung W. A possible stress physiological role of abscisic acid conjugates in root-to-shoot signalling. *Plant Cell Environ.* 2002; 25:223–228.

14. Heilmann B, Hartung W, Gimmler H. The distribution of abscisic acid between chloroplasts and cytoplasm of leaf cells and the permeability of the chloroplast envelope for abscisic acid. *Z. Pflanzenphysiol.* 1980; 97:67–78.

15. Hartung W, Radin JW, Hendrix DL. Abscisic acid movement into the apoplastic solution of water-stressed cotton leaves. Role of apoplastic pH. *Plant Physiol.* 1988; 86:908–913.

16. Hornberg C, Weiler EW. High-affinity binding sites for abscisic acid on the plasmalemma of *V. faba* guard cells. *Nature* 1984; 310:321–324.

17. Zhang D-P, Wu Z-Y, Li X-Y, Zhao Z-X. Purification and identification of a 42-kilodalton abscisic acid-specific-binding protein from epidermis of broad bean leaves. *Plant Physiol.* 2002; 128:714–725.

18. Hartung W. The site of action of abscisic acid at the guard cell plasmalemma of *Valerianella locusta*. *Plant Cell Environ.* 1983; 6:427–428.

19. Leung J, Giraudat J. Abscisic acid signal transduction. *Annu. Rev. Plant Physiol. Plant Mol. Biol.* 1998; 49:199–222.

20. Allan AC, Fricker MD, Ward JL, Beale MH, Trewavas AJ. Two transduction pathways mediate rapid effects of abscisic acid in *Commelina* guard cells. *Plant Cell* 1994; 6:1319–1328.

21. Schwartz A, Wu WH, Tucker EB, Assmann SM. Inhibition of inward K+ channels and stomatal response by abscisic-acid — an intracellular locus of phytohormone action. *Proc. Natl. Acad. Sci. USA* 1994; 91:4019–4023.

22. Hetherington AM. Guard cell signalling. *Cell* 2001; 107:711–714.

23. Roelfsema MRG, Staal M, Prins HBA. Blue-light-induced apoplastic acidification of *Arabidopsis thaliana* guard cells: inhibition by ABA is mediated through protein phosphatases. *Physiol. Plant.* 1998; 103:466–474.

24. Schroeder JI, Allen GJ, Hugouvieux V, Kwak JM, Waner D. Guard cell signal transduction. *Annu. Rev. Plant Physiol. Plant Mol. Biol.* 2001; 52:627–658.

25. MacRobbie EAC. Signalling in guard cells and regulation of ion channel activity. *J. Exp. Bot.* 1997; 48:515–528.

26. Allen GJ, Amtmann A, Sanders D. Calcium-dependent and calcium-independent K^+ mobilization channels in *Vicia faba* guard cell vacuoles. *J. Exp. Bot.* 1998; 49:305–318.

27. Assmann SM, Shimazaki K-I. The multisensory guard cell. Stomatal responses to blue light and abscisic acid. *Plant Physiol.* 1999; 119:337–361.

28. Lemtiri-Chlieh F, MacRobbie EAC, Brearley CA. Inositol hexakisphosphate is a physiological signal regulating the K^+-inward rectifying conductance in guard cells. *Proc. Natl. Acad. Sci. USA* 2000; 97:8687–8692.

29. Blatt M. Cellular signaling and volume control in stomatal movements in plants. *Annu. Rev. Cell Dev. Biol.* 2000; 16:221–241.

30. Burnett EC, Desikan R, Moser RC, Neill SJ. ABA activation of an MBP kinase in *Pisum sativum* epidermal peels correlates with stomatal responses to ABA. *J. Exp. Bot.* 2000; 51:197–205.

31. Luan S. Signalling drought in guard cells. *Plant Cell Environ.* 2002; 5:229–237.

32. Zhang X, Yu CM, An GY, Zhou Y., Shangguan ZP, Gao JF, Song CP. K^+ channels inhibited by hydrogen peroxide mediate abscisic acid signaling in *Vicia* guard cells. *Cell Res.* 2001; 11:195–202.

33. Zhang X, Zhang L, Dong F, Gao J, Galbraith DW, Song C-P: Hydrogen peroxide is involved in abscisic acid-induced stomatal closure in *Vicia faba*. *Plant Physiol.* 2001; 126:1438–1448.

34. Neill SJ, Desikan R, Clarke A, Hancock JT. Nitric oxide is a novel component of abscisic acid signalling in stomatal guard cells. *Plant Physiol.* 2002; 128:13–16.

35. Eun SO, Lee Y. Actin filaments of guard cells are reorganized in response to light and abscisic acid. *Plant Physiol.* 1997; 115:1491–1498.

36. Lemichez E, Wu Y, Sanchez J-P, Mettouchi A, Mathur J, Chua N-H. Inactivation of AtRac1 by abscisic aid is essential for stomatal closure. *Genes Dev.* 2001; 15:1808–1816.

37. Jiang C-J, Nakajima N, Kondo N. Disruption of microtubules by abscisic acid in guard cells of *Vicia faba* L. *Plant Cell Physiol.* 1996; 37:697–701.

38. Webb AAR, Larman MG, Montgomery LT, Taylor JE, Hetherington AM. The role of calcium in ABA-induced gene expression and stomatal movements. *Plant J.* 2001; 26:351–362.

39. Dodd IC, Stikic R, Davies WJ. Chemical regulation of gas exchange and growth of plants in drying soil in the field. *J. Exp. Bot.* 1996; 47:1475–1490.

40. Rodriguez JL, Davies WJ. The effects of temperature and ABA on stomata of *Zea mays* L. *J. Exp. Bot.* 1982; 33:977–987.

41. Honour SJ, Webb AAR, Mansfield TA. The responses of stomata to abscisic acid and temperature are interrelated. *Proc. R. Soc. Lond. B* 1995; 259:301–306.

42. Tardieu F, Davies WJ. Stomatal response to abscisic acid is a function of current plant water status. *Plant Physiol.* 1992; 98:540–545.

43. Davies WJ. Some effects of abscisic acid and water stress on stomata of *Vicia faba* L. *J. Exp. Bot.* 1978; 29:175–182.

44. Eamus D, Narayan AD. The influence of prior water stress and abscisic acid foliar spraying on stomatal responses to CO_2, IAA, ABA, and calcium in leaves of *Solanum melongena. J. Exp. Bot.* 1989; 40:573–579.

45. Peng ZY, Weyers JDB. Stomatal sensitivity to abscisic acid following water deficit stress. *J. Exp. Bot.* 1994; 45:835–845.

46. Cohen JD, Slovin JP, Hendrickson AM. Two genetically discrete pathways convert tryptophan to auxin: more redundancy in auxin biosynthesis. *Trends Plant Sci.* 2003; 8:197–199.

47. Slovin JP, Bandurski RS, Cohen JD. Auxin. In: Hooykaas PJJ, Hall MA, Libbenga KR, eds. *Biochemistry and Molecular Biology of Plant Hormones.* Amsterdam: Elsevier, 1999:115–140.

48. Dunlap JR, Binzel ML. NaCl reduces indole-3-acetic acid levels in the roots of tomato plants independent of stress-induced abscisic acid. *Plant Physiol.* 1996; 112:379–384.

49. Lopez-Carbonell M, Alegre L, Van Onckelen H. Effects of water stress on cellular ultrastructure and on concentrations of endogenous abscisic acid and indole-3-acetic acid in *Fatsia japonica* leaves. *Plant Growth Regul.* 1994; 14:29–35.

50. Lopez-Carbonell M, Alegre L, Van Onckelen H. Changes in cell ultrastructure and endogenous abscisic acid and indole-3-acetic acid concentrations in *Fatsia japonica* leaves under polyethylene glycol-induced water stress. *Plant Growth Regul.* 1994; 15: 29–35.

51. Masia A, Pitacco A, Braggio L, Giulivo C. Hormonal responses to partial drying of the root system of *Helianthus annuus. J. Exp. Bot.* 1994; 45:69–76.

52. Hartung W, Weiler EW, Radin JW. Auxin and cytokinins in the apoplastic solution of dehydrated cotton leaves. *J. Plant Physiol.* 1992; 140:324–327.

52. MacDonald H. Auxin perception and signal transduction. *Physiol. Plant.* 1997; 100:423–430.

54. Gehring CA, McConchie RM, Venis MA, Parish RW. Auxin-binding-protein antibodies and peptides influence stomatal opening and alter cytoplasmic pH. *Planta* 1998; 205:581–586.

55. Blatt MR, Thiel G. K^+ channels of stomatal guard cells: bimodal control of the K^+ inward-rectifier evoked by auxin. *Plant J.* 1994; 5:55–68.

56. Grabov A, Blatt MR. Co-ordination of signalling elements in guard cell ion channel control. *J. Exp. Bot.* 1998; 49:351–360.

57. Bauly JM, Sealy IM, MacDonald H, Brearley J, Dröge S, Hillmer S, Robinson DG, Venis MA, Blatt MR, Lazarus CM, Napier RM. Overexpression of auxin-binding protein enhances the sensitivity of guard cells to auxin. *Plant Physiol.* 2000; 124:1229–1238.

58. Estelle M. Auxin perception and signal transduction. In: Hooykaas PJJ, Hall MA, Libbenga KR, eds. *Biochemistry and Molecular Biology of Plant Hormones.* Amsterdam: Elsevier, 1999:411–421.

59. Cousson A, Vavasseur A. Putative involvement of cytosolic Ca^{2+} and GTP-binding proteins in cyclic GMP-mediated induction of stomatal opening by auxin in *Commelina communis* L. *Planta* 1998; 206:308–314.

60. Cousson A. Pharmacological evidence for the implication of both cyclic GMP-dependent and -independent transduction pathways within auxin-induced stomatal opening in *Commelina communis* (L.). *Plant Sci.* 2001; 161:249–258.

61. Pemadasa MA. Effects of phenylacetic acid on abaxial and adaxial stomatal movements and its interaction with abscisic acid. *New Phytol.* 1982; 92:21–30.

62. Snaith PJ, Mansfield TA. Studies of the inhibition of stomatal opening by naphth-1-ylacetic acid and abscisic acid. *J. Exp. Bot.* 1984; 35:1410–1418.

63. Pemadasa MA. Differential abaxial and adaxial stomatal responses to indole-3-acetic acid in *Commelina communis* L. *New Phytol.* 1982; 90:209–219.

64. Eamus D. Further evidence in support of an interactive model in stomatal control. *J. Exp. Bot.* 1986; 37:657–665.

65. Snaith PJ, Mansfield TA. Control of the CO_2 responses of stomata by indol-3ylacetic acid and abscisic acid. *J. Exp. Bot.* 1982; 33:360–365

66. Rícánek M, Vicherková M. Stomatal responses to ABA and IAA in isolated epidermal strips of *Vicia faba* L. *Biol. Plant.* 1992; 34:259–265.

67. Dunleavy PJ, Ladley PD. Stomatal responses of *Vicia faba* L. to indole acetic acid and abscisic acid. *J. Exp. Bot.* 1995; 46:95–100.

68. Zažímalová E, Kamínek M, Brezinová A, Motyka V. Control of cytokinin biosynthesis and metabolism. In: Hooykaas PJJ, Hall MA, Libbenga KR, eds. *Biochemistry and Molecular Biology of Plant Hormones.* Amsterdam: Elsevier, 1999:141–160.

69. Hooley R. Progress towards the identification of cytokinin receptors. In: Sopory SK, Oelmüller R, Maheshwari SC, eds. *Signal Transduction in Plants. Current Advances.* New York: Kluwer Academic Publishers, 2001:193–199.

70. Mok DWS, Mok MC. Cytokinin metabolism and action. *Annu. Rev. Plant Physiol. Plant Mol. Biol.* 2001; 52:89–118.

71. Chen CC, Ertl JR, Lesiner SM, Chang CC. Localisation of cytokinin biosynthetic sites in pea plants and carrot roots. *Plant Physiol.* 1985; 78:510–513.

72. Emery RJN, Atkins CA. Roots and cytokinins. In: Waisel Y, Eshel A, Kafkafi U, eds. *Plant Roots. The Hidden Half.* 3rd ed. New York: Marcel Dekker, 2002:417–434.

73. Galuszka P, Frébort I, Šebela M, Sauer P, Jacobsen S, Pec P. Cytokinin oxidase or dehydrogenase? Mechanism of cytokinin degradation in cereals. *Eur. J. Biochem.* 2001; 268:450–461.

74. Kamínek M, Motyka V, Vanková R. Regulation of cytokinin content in plant cells. *Physiol. Plant.* 1997; 101:689–700.

75. Naqvi SSM. Plant hormones and stress phenomena. In: Pessarakli M, eds. *Handbook of Plant and Crop Stress.* New York: Marcel Dekker, 1994:383–400.

76. Pospíšilová J. Participation of phytohormones in the stomatal regulation of gas exchange during water stress. *Biol. Plant.* 2003; 46:491–506.

77. Mustafina A, Veselov S, Valcke R, Kudoyarova G. Contents of abscisic acid and cytokinins in shoots during dehydration of wheat seedlings. *Biol. Plant.* 1997/98; 40:291–293.

78. Shashidhar VR, Prasad TG, Sudharshan L. Hormone signals from roots to shoots of sunflower (*Helianthus annuus* L.). Moderate soil drying increases delivery of abscisic acid and depresses delivery of cytokinins in xylem sap. *Ann. Bot.* 1996; 78:151–155.

79. Brault M, Maldiney R. Mechanisms of cytokinin action. *Plant Physiol. Biochem.* 1999; 37:403–412.

80. Incoll LD, Ray JP, Jewer PC. Do cytokinins act as root to shoot signals? In: Davies WJ, Jeffcoat B, eds. *Importance of Root to Shoot Communication in the Responses to Environmental Stress.* Bristol: British Society for Plant Growth Regulation, 1990:185–197.

81. Morsucci R, Curvetto N, Delmastro S. Involvement of cytokinins and adenosine 3′,5′-cyclic monophosphate in stomatal movement in *Vicia faba*. *Plant Physiol. Biochem.* 1991; 29:537–547.

82. Pharmawati M, Billington T, Gehring CA. Stomatal guard cell responses to kinetin and natriuretic peptides are cGMP-dependent. *Cell. Mol. Life Sci.* 1998; 54:272–276.

83. Incoll LD, Jewer PC. Cytokinins and stomata. In: Zeiger E, Farquhar GD, Cowan IR, eds. *Stomatal Function.* Stanford: Stanford University Press, 1987:281–292.

84. Pospíšilová J, Synková H, Rulcová J. Cytokinins and water stress. *Biol. Plant.* 2000; 43:321–328.

85. Blackman PG, Davies WJ. Age-related changes in stomatal response to cytokinins and abscisic acid. *Ann. Bot.* 1984; 54:121–125.

86. Drüge U, Schönbeck F. Effect of vesicular-arbuscular mycorrhizal infection on transpiration, photosynthesis and growth of flax (*Linum usitatissimum* L.) in relation to cytokinin levels. *J. Plant Physiol.* 1992; 141:40–48.

87. Hare PD, Cress WA, Van Staden J. The involvement of cytokinins in plant responses to environmental stress. *Plant Growth Regul.* 1997; 23:79–103.

88. Cowan AK, Cairns ALP, Bartels-Rahm B. Regulation of abscisic acid metabolism: towards a metabolic basis for abscisic acid-cytokinin antagonism. *J. Exp. Bot.* 1999; 50:595–603.

89. Aharoni N, Blumenfeld A, Richmond AE. Hormonal activity in detached lettuce leaves as affected by leaf water content. *Plant Physiol.* 1977; 59:1169–1173.

90. Santakumari M, Fletcher RA. Reversal of triazole-induced stomatal closure by gibberellic acid and cytokinins in *Commelina benghalensis*. *Physiol. Plant.* 1987; 71:95–99.

91. Yuan L, Xu DQ. Stimulation effect of gibberellic acid short-term treatment on leaf photosynthesis related to the increase in Rubisco content in broad bean and soybean. *Photosynth. Res.* 2001; 68:39–47.

92. Cleland RE. Introduction: nature, occurrence and functioning of plant hormones. In: Hooykaas PJJ, Hall MA, Libbenga KR, eds. *Biochemistry and Molecular Biology of Plant Hormones.* Amsterdam: Elsevier, 1999:2–22.

93. Raghavendra AS, Bhaskar Reddy K. Action of proline on stomata differs from that of abscisic acid, G-substances, or methyl jasmonate. *Plant Physiol.* 1987; 83:732–734.

94. Gehring CA, Irving HR, McConchie RM, Parish RW. Jasmonates induce intracellular alkalinization and closure of *Paphiopedilum* guard cells. *Ann. Bot.* 1997; 80:485–489.

95. Ievinsh G, Dreibante G, Kruzmane D. Changes of 1-aminocyclopropane-1-carboxylic acid oxidase activity in stressed *Pinus sylvestris* needles. *Biol. Plant.* 2001; 44:233–237.

96. Morgan PW, Drew MC. Ethylene and plant responses to stress. *Physiol. Plant.* 1997; 100:620–630.

97. Taylor GE, Gunderson CA. The response of foliar gas exchange to exogenously applied ethylene. *Plant Physiol.* 1986; 82:653–657.

98. Singh P, Srivastava NK, Mishra A, Sharma S. Influence of etherel and gibberellic acid on carbon metabolism, growth, and essential oil accumulation in spearmint (*Mentha spicata*). *Photosynthetica* 1999; 36:509–517.

99. Dodd IC. Hormonal interactions and stomatal responses. *J. Plant Growth Regul.* 2003; 22:32–46.

100. Yokota T. Brassinosteroids. In: Hooykaas PJJ, Hall MA, Libbenga KR, eds. *Biochemistry and Molecular Biology of Plant Hormones.* Amsterdam: Elsevier, 1999:277–293.

101. Xu H-L, Shida A, Futatsuya F, Kumura A. Effects of epibrassinolide and abscisic acid on sorghum plants growing under soil-water deficit. II. Physiological basis for drought resistance induced by exogenous epibrassinolide and abscisic acid. *Jpn. J. Crop Sci.* 1994; 63:676–681.

102. Rajasekaran LR, Blake TJ. New plant growth regulators protect photosynthesis and enhance growth under drought of jack pine seedlings. *J. Plant Growth Regul.* 1999; 18:175–181.

103. Tardieu F, Simonneau T. Variability among species of stomatal control under fluctuating soil water status and evaporative demand: modelling isohydric and anisohydric behaviours. *J. Exp. Bot.* 1998; 49:419–432.

104. Flexas J, Bota J, Escalona JM, Sampol B, Medrano H. Effects of drought on photosynthesis in grapevines under field conditions: an evaluation of stomatal and mesophyll limitations. *Funct. Plant Biol.* 2002; 29:461–471.

105. Terashima I, Ono K. Effects of HgCl$_2$ on CO$_2$ dependence of leaf photosynthesis: evidence indicating involvement of aquaporins in CO$_2$ diffusion across the plasma membrane. *Plant Cell Physiol.* 2002; 43:70–78.

106. Cornic G, Massacci A. Leaf photosynthesis under drought stress. In: Baker NR, ed. *Photosynthesis and the Environment.* Dordrecht: Kluwer Academic Publishers, 1996:347–366.

107. Wilson KB, Baldocchi DD, Hanson PJ. Quantifying stomatal and non-stomatal limitations to carbon assimilation resulting from leaf aging and drought in mature deciduous tree species. *Tree Physiol.* 2000; 20:787–797.

108. Pospíšilová J, Šantruček J. Stomatal patchiness: effects on photosynthesis. In: Pessarakli M, ed. *Handbook of Photosynthesis.* New York: Marcel Dekker, 1997:427–441.

109. Buckley TN, Farquhar GD, Mott KA. Carbon-water balance and patchy stomatal conductance. *Oecologia* 1999; 118:132–143.

110. Maroco JP, Rodrigues ML, Lopes C, Chaves MM. Limitations to leaf photosynthesis in field-grown grapevine under drought-metabolic and modelling approaches. *Funct. Plant Biol.* 2002; 29:451–459.

111. Mansfield TA, Davies WJ. Stomata and stomatal mechanisms. In: Paleg LG, Aspinall D, eds. *The Physiology and Biochemistry of Drought Resistance.* Sydney: Academic Press, 1981:315–346.

112. Davies WJ, Zhang J. Root signals and the regulation of growth and development of plants in drying soil. *Annu. Rev. Plant Physiol. Plant Mol. Biol.* 1991; 42:55–76.

113. Zhang J, Davies WJ. Abscisic acid produced in dehydrating roots may enable the plant to measure the water status of the soil. *Plant Cell Environ.* 1989; 12:73–81.

114. Wolf O, Jeschke WD, Hartung W. Long distance transport of abscisic acid in NaCl-treated intact plants of *Lupinus albus. J. Exp. Bot.* 1990; 41:593–600.

115. Freundl E, Steudle E, Hartung W. Apoplastic transport of abscisic acid through roots in maize: effect of exodermis. *Planta* 2000; 210:222–231.

116. Trejo CL, Davies WJ, Ruiz LMP. Sensitivity of stomata to abscisic acid. An effect of the mesophyll. *Plant Physiol.* 1993; 102:497–502.

117. Zhang SQ, Outlaw WH Jr, Aghoram K. Relationship between changes in the guard cell abscisic-acid content and other stress-related physiological parameters in intact plants. *J. Exp. Bot.* 2001; 52:301–308.

118. Zhang SQ, Outlaw WH Jr. The guard-cell apoplast as a site of abscisic acid accumulation in *Vicia faba* L. *Plant Cell Environ.* 2001; 24:347–355.

119. Zhang SQ, Outlaw WH Jr. Abscisic acid introduced into the transpiration stream accumulates in the guard-cell apoplast and causes stomatal closure. *Plant Cell Environ.* 2001; 24:1045–1054.

120. Tardieu F, Lafarge T, Simonneau T. Stomatal control by fed or endogenous xylem ABA in sunflower: interpretation of correlations between leaf water potential and stomatal conductance in anisohydric species. *Plant Cell Environ.* 1996; 19:75–84.

121. Borel C, Frey A, Marion-Poll A, Simonneau T, Tardieu F. Does engineering abscisic acid biosynthesis in *Nicotiana plumbaginifolia* modify stomatal response to drought? *Plant Cell Environ.* 2001; 24:477–489.

122. Trejo CL, Davies WJ. Drought-induced closure of *Phaseolus vulgaris* L. stomata precedes leaf water deficit and any increase in xylem ABA concentration. *J. Exp. Bot.* 1991; 42:1507–1515.

123. Munns R, King RW. Abscisic acid is not the only stomatal inhibitor in the transpiration stream of wheat plants. *Plant Physiol.* 1988; 88:703–708.

124. Wilkinson S, Corlett JE, Oger L, Davies WJ. Effects of xylem pH on transpiration from wild-type and *flacca* tomato leaves: a vital role for abscisic acid in preventing excessive water loss even from well-watered plants. *Plant Physiol.* 1998; 117:703–709.

125. Wilkinson S, Davies WJ. Xylem sap pH increase: a drought signal received at the apoplastic face of the guard cell that involves the suppression of saturable abscisic acid uptake by the epidermal symplast. *Plant Physiol.* 1997; 113:559–573.

126. Correia MJ, Pereira JS. Abscisic acid in apoplastic sap can account for the restriction in leaf conductance of white lupins during moderate soil drying and after rewatering. *Plant Cell Environ.* 1994; 17:845–852.

127. Meyer S, Genty B. Mapping intercellular CO$_2$ mole fraction (C_i) in *Rosa rubiginosa* leaves fed with abscisic acid by using chlorophyll fluorescence imaging. *Plant Physiol.* 1998; 116:947–957.

128. Raschke K, Hedrich R. Simultaneous and independent effects of abscisic acid on stomata and the photosynthetic apparatus in whole leaves. *Planta* 1985; 163:105–118.

129. Šantruček J, Hronková M, Kveton J, Sage RF. Photosynthesis inhibition during gas exchange oscillations in ABA-treated *Helianthus annuus*: relative role of stomatal patchiness and leaf carboxylation capacity. *Photosynthetica* 2003; 41:241–252.

130. Synková H, Wilhelmová N, Šesták Z, Pospíšilová J. Photosynthesis in transgenic plants with elevated cytokinin contents. In: Pessarakli M, ed. *Handbook of Photosynthesis.* New York: Marcel Dekker, 1997:541–552.

131. Metwally A, Tsonev T, Zeinalov Y. Effect of cytokinins on the photosynthetic apparatus in water-stressed and rehydrated bean plants. *Photosynthetica* 1997; 34:563–567.

132. Chernyad'ev II, Monakhova OF. The activity and content of ribulose-1,5-bisphosphate carboxylase/oxygenase in wheat plants as affected by water stress and kartolin-4. *Photosynthetica* 1998; 35:603–610.

133. Pandey DM, Goswami CL, Kumar B, Jain S. Hormonal regulation of photosynthetic enzymes in cotton under water stress. *Photosynthetica* 2000; 38:403–407.

134. Singh DV, Srivastava GC, Abdin MZ. Amelioration of negative effect of water stress in *Cassia angustifolia* by benzyladenine and/or ascorbic acid. *Biol. Plant.* 2001; 44:141–143.

135. Berkowitz GA, Rabin J. Antitranspirant associated abscisic acid effects on the water relations and yield of transplanted bell peppers. *Plant Physiol.* 1988; 86:329–331.

136. Loveys BR. Diurnal changes in water relations and abscisic acid in field-grown *Vitis vinifera* cultivars. III. The influence of xylem-derived abscisic acid on leaf gas exchange. *New Phytol.* 1984; 98:563–573.

137. Fuchs EE, Livingston NJ, Rose PA. Structure-activity relationships of ABA analogs based on their effects on the gas exchange of clonal white spruce (*Picea glauca*) emblings. *Physiol. Plant.* 1999; 105:246–256.

138. Kumar B, Pandey DM, Goswami CL, Jain S. Effect of growth regulators on photosynthesis, transpiration and related parameters in water stressed cotton. *Biol. Plant.* 2001; 44:475–478.

139. Rulcová J, Pospíšilová J. Effect of benzylaminopurine on rehydration of bean plants after water stress. *Biol. Plant.* 2001; 44:75–81.

140. Vomáčka L, Pospíšilová J. Rehydration of sugar beet plants after water stress: effects of cytokinins. *Biol. Plant.* 2003; 46:57–62.

141. Čatsky J, Pospíšilová J, Kamínek M, Gaudinová A, Pulkrábek J, Zahradníček J. Seasonal changes in sugar beet photosynthesis as affected by exogenous cytokinin N^6-(*m*-hydroxybenzyl)adenosine. *Biol. Plant.* 1996; 38:511–518.

43 Adverse Effects of UV-B Light on the Structure and Function of the Photosynthetic Apparatus

Imre Vass, András Szilárd, and Cosmin Sicora
Institute of Plant Biology, Biological Research Center,
Hungarian Academy of Sciences

CONTENTS

I. INTRODUCTION

Photosynthetically relevant solar radiation that reaches the surface of Earth is divided into three main spectral regions: ultraviolet-B (UV-B) (290 to 315 nm), UV-A (315 to 400 nm), and photosynthetically active radiation (PAR) (400 to 700 nm). Among those, the UV-B region is selectively attenuated by the stratospheric ozone layer [1,2]. In contrast, the UV-A and PAR radiations have no selective absorber and are affected mainly by light scattering. The biologically most damaging wavelengths below 290 nm, such as the UV-C (200 to 290 nm) region, are absorbed almost completely by the atmosphere and are therefore unimportant for biological processes under natural conditions. Thus, depletion of stratospheric ozone, which occurs as a consequence of human activities, specifically enhances the UV-B radiation reaching the Earth [3–7].

UV radiation, terrestrial life, and ozone depletion have a notable relationship with oxygenic photosynthesis. The present-day ozone shield that protects terrestrial life from damaging UV-B radiation is formed in the stratosphere from oxygen by short-wavelength ($\lambda < 242$ nm) UV radiation. In the prebiotic phase of the Earth's evolution, the atmosphere contained only a low amount of oxygen and ozone, thus damaging short-wavelength UV radiation could reach the surface without significant attenuation [8]. It seems highly probable that the primary life forms have developed in the oceans protected by the water layer against UV radiation. After invention of the oxygen-evolving capacity by photosynthesizing bacteria, the biotically produced oxygen started to accumulate in the atmosphere between 2200 and 2400 million years ago and was partly converted into ozone in the upper layers by UV radiation [9]. The gradually formed ozone layer served as a protective shield against the biologically damaging UV-B radiation and made it possible for the marine life forms to conquer terrestrial habitats. As a result of this process land plants appeared about 500 million years ago and their photosynthetically produced oxygen contributed further to the increased oxygen content of the atmosphere. In a paradoxical way, it appears that one highly UV-sensitive site in plants is the very same water-oxidizing machinery that facilitated the formation of the present-day oxygen atmosphere and ozone shield, with the latter in potential danger due to recent human activities leading to stratospheric ozone breakdown [3–6].

II. THE MAIN TARGETS OF UV-B RADIATION IN PLANTS

A. NUCLEIC ACIDS

DNA is one of the most notable targets of UV radiation in cells of living organisms. Irradiation, both in the UV-C and UV-B region, results in a multitude of DNA photoproducts [10], which may cause mutations during replications [11]. The most common DNA photoproducts are cyclobutane-type pyrimidine dimers and the pyrimidine(6,4)pyrimidone dimer [12]. In addition, DNA strand breaks, DNA–protein crosslinks, and insertion or deletion of base pairs can also be induced by UV exposure [13]. These effects are studied in detail in humans, other mammals, fungi, yeast, and bacteria. In case of plants and plant cell cultures, mainly cyclobutane dimer formation has been measured directly [14–20]. However, the

UV-induced formation and blue-light-dependent elimination of pyrimidine(6,4)pyrimidone photoproducts have also been observed in plant cells [21,22]. The reader is advised to consult Chapter 8 of this book, as well as recent detailed reviews regarding this topic [23–26].

B. AMINO ACIDS AND PROTEINS

Proteins have strong absorption at about 280 nm, and also at higher wavelengths of the UV-B region due to absorption by the aromatic amino acids tyrosine, phenylalanine, and tryptophane as well as cysteine, and thus can be direct targets of UV-B radiation. UV-induced destruction of tyrosine and tryptophane have been observed both in the free amino acid form and in proteins [27]. UV-B can also induce photooxidation of tyrosine to 4,4-dihydroxyphenylalanine (DOPA) [27] and the formation of dityrosine [27–29]. Photobiological changes initiated by tryptophan are often attributed to the formation of N-formyl kynurenine through photooxidation [27,28,30]. Cysteine is a relatively poor absorber in the UV-B region but undergoes UV-induced photolysis at a high quantum efficiency [27,31]. Disulfide bridges between cysteine residues, which are important for the tertiary structure of many proteins, can also be split by UV-B radiation [31,32] and can strongly influence protein structure and function.

UV irradiation causes not only the modification or destruction of amino acid residues, but also leads to inactivation of whole proteins and enzymes. Characteristic examples for this effect include trypsin, pepsin, lyzozyme, insulin, myosin [27], Rubisco [33–35], ATP synthase [36,37], violaxanthin deepoxidase [38], as well as the protein subunits of the photosystem II and I complexes, as discussed in detail in Sections IV and V. Inactivation of proteins and enzymes can be caused directly by UV photolysis of aromatic amino acids or the S–S groups if the affected residues are included in the active site. Alternatively, the formation of dityrosine, or the breakup of disulfide bridges may lead to significant changes in the conformation of the affected protein and thereby induce inactivation. It is also important to note that UV absorption within the protein matrix can sensitize damage far from the actual absorption site via energy migration to functionally important amino acids of the active center, as suggested for the sensitization of cysteine destruction by aromatic residues [27].

C. LIPIDS

Lipids with isolated or conjugated double bonds can also be photochemically modified by UV absorption.

Phospho- and glycolipids, which are the main components in plant cell membranes, contain unsaturated fatty acids which are destroyed by UV-B radiation in the presence of oxygen [39,40]. The consequent lipid peroxidation may have a direct effect on membrane structure, and lipid peroxy radicals may induce further damage by participating in free radical cascades [41] (see also Chapter 8 for further details). Association of lipids with proteins have also been reported to enhance the UV-B sensitivity of the plasma membrane ATP synthase [36].

D. QUINONES

Quinones are important components of various redox complexes of plant membranes, and have special role in photosynthesis as electron carriers within and between the photosystems, Photosystem II (PSII) and photosystem I (PSI) (see Chapters 8 and 9 for further details). The main absorption of plastoquinone is at about 250, 280, and 320 nm for the quinone (PQ), quinol (PQH_2), and semiquinone (PQ^-) forms, respectively [42]. Direct UV-induced destruction has been reported for plastoquinones by UV-C radiation [43–45]. UV-B irradiation has also been shown to decrease the amount plastoquinones in irradiated thylakoids [46]. In addition, the redox function of quinones in the PSII complex is impaired [47,48] (see detailed discussion below in Section IV.B.1).

E. PIGMENTS

Pigments of the photosynthetic apparatus can be destroyed by UV radiation, with concomitant loss of the photosynthetic capacity [35,49–53]. Decrease in the amount of pigment content in UV-irradiated plants may also be the consequence of reduced synthesis of the main chlorophyll pigment complexes encoded by the *cab* gene family [53,54]. In cyanobacteria, harvesting of photosynthetically active light is performed by the so-called phycobiliproteins, which contain open chain tetrapyrrole pigments and can be destroyed by UV radiation [55–59].

III. UV-B EFFECTS ON NET PHOTOSYNTHESIS

From over 300 species studied so far, about 50% have been considered sensitive, 20% to 30% moderately sensitive or tolerant, and the rest insensitive to UV-B radiation [60–62]. Typical sensitive species include pea, bean, sunflower, soybean, cucumber, squash, maize, and barley [61].

Despite the variety of UV-B targets in plants it appears that the photosynthetic apparatus is among the prime action sites of UV-B, and its damage contributes significantly to the overall UV-B effect. The most common consequences of exposure to enhanced UV-B radiation on the photosynthetic functions are as follows: (i) decreased CO_2 fixation and oxygen evolution [63–73]; (ii) impairment of PSII, and to a lesser extent, of PSI (as discused below); (iii) reduction in dry weight, secondary sugars, starch, and total chlorophyll [33,74]; (iv) decrease in Rubisco activity [33–35,64,65]; and (v) inactivation of ATP synthase [37].

In addition to direct effects of UV-B radiation on the photosynthetic apparatus, photosynthesis may also be indirectly affected. Induction of stomatal closure occurs in UV-B exposed plants, as demonstrated in cucumber seedlings, in bean and oilseed rape leaves [65,75–78]. This phenomenon may reduce photosynthetic activity by decreasing the efficiency of gas exchange. Changes in leaf thickness or anatomy may alter the penetration of photosynthetically active light into the leaf and thus indirectly impair photosynthesis [79]. UV-B irradiation may also indirectly alter whole plant photosynthesis by changes in canopy morphology [80].

The extent of UV-induced damage on plant productivity is somewhat controversial. Experiments performed under laboratory/greenhouse conditions using relatively high UV levels and high UV/visible ratios in the applied illumination tend to show larger decrease of productivity as compared to field studies simulating the effects of predicted levels of ozone depletion [64,74]. However, even under conditions that are close to the natural situation significant effects of UV were observed in agriculturally important species like maize and sunflower [81–83]. UV-induced loss of productivity can be clearly significant in plants, which have decreased repair capacity of damaged DNA as was shown for the commercially important rice cultivar Norin 1 [19]. The extent of gross damage is also strongly influenced by the co-occurrence of other environmental factors like low temperatures, which increase the damage [81] most likely by inhibiting the repair of damaged DNA or the photosythetic apparatus. Drought, on the other hand, can ameliorate the damaging effect probably due to preconditioning of antioxidant enzyme systems [84], which are important for the defence against the oxidative damaging agents common in the water- and UV stress. Interestingly, UV-B radiation can also reduce the severity of drought stress through reductions in water loss rates [85].

Further details regarding the effects of UV-B radiation on net photosynthesis of terrestrial plants are provided by extensive reviews [25,60–62,64,74,86–91]. The effects of UV-B radiation on marine algae and phytoplankton are given in Refs. [92–97].

IV. UV-B EFFECTS ON PSII

Since the mechanistic aspects of UV-induced damage on the photosynthetic apparatus are best characterized in case of the PSII complex, a detailed overview of these effects is provided below. Although under natural conditions, the relatively low intensity UV-B light is accompanied with moderate or high intensities in the visible spectral range, studies using higher than physiological UV-B intensities without visible components are inevitable to clarify the mechanistic details and molecular background of UV action in the photosynthetic apparatus. However, extrapolation of the knowledge thus gained to explain effects under physiological conditions needs special caution and thorough verification. In this section, mainly data from *in vitro* measurements will be presented in comparison to *in vivo* results whenever possible.

A. The Structure and Function of PSII

In order to provide the nonspecialist reader with the basic knowledge regarding the structure and function of PSII, a brief summary is given below. More detailed information on this topic can be found in Chapters 8 and 9 of this book and in extensive reviews [98–102].

PSII is a multifunctional pigment–protein complex embedded in the thylakoid membrane of oxygenic photosynthetic organisms (Figure 43.1). The PSII complex contains over 20 protein subunits and the redox components that mediate light-induced electron transport. The main function of PSII is the light-induced splitting of water to molecular oxygen and protons, which is unique to PSII in nature. The electrons that are liberated from water are transferred by PSII to mobile plastoquinone electron acceptors that form a pool in the hydrophobic phase of the membrane.

The reaction center of PSII is composed of a heterodimer of two hydrophobic proteins, called D1 and D2, in close association with cytochrome *b*-559 [103]. The reaction center heterodimer binds or contains all the redox electron carriers of PSII electron transport. Two chlorophyll-binding proteins, called CP43 and CP47, form the inner light-harvesting antenna of PSII, which are complemented with outer antenna systems, the LHCII in higher plants and algae, and the phycobilisomes in cyanobacteria; for reviews see Refs. [98,104]. The three-dimensional structure of the PSII complex is known to a considerable detail through computer-assisted modeling [105,106] and X-ray structure determinations [107–109].

Light absorption in PSII results in the excitation of a special reaction center chlorophyll, P680, which

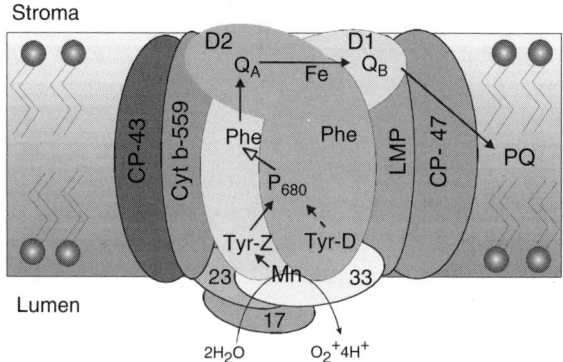

PSII

FIGURE 43.1 The structure and function of the PSII complex. The reaction center of PSII consists of the D1 and D2 protein subunits, which bind the redox cofactors of light-induced electron transport: the Mn cluster of water oxidation, the redox-active tyrosine electron donors (Tyr-Z and Tyr-D), the reaction center chlorophyll (P_{680}), the primary electron acceptor phyophytin (PheO), and the first and second quinone electron acceptors Q_A and Q_B, respectively. The reaction center heterodimer is closely associated with cytochrome *b*-559, and surrounded by chlorophyll-binding antenna (CP43 and CP47). The PSII complex also contains various low molecular mass polipeptides (LMPs).

is followed by a very fast transfer of an electron from P680* to the first electron acceptor, a pheophytin molecule (Pheo). The primary charge separation is stabilized by a series of electron transport reactions both at the reducing and oxidizing sides of PSII. The electron from Pheo is transferred to the first, Q_A, and then to the second, Q_B, quinone electron acceptor. Q_A is bound by the D2 protein and cannot be easily exchanged with plastoquinones from the lipid phase of the membrane. In contrast, Q_B is located in a binding niche formed by the D1 protein. Q_B is bound strongly in the semireduced state (Q_B^-), but can be easily exchanged with plastoquinones from the pool in both the oxidized (Q_B) and fully reduced (Q_BH_2) state. On the oxidizing side of PSII, $P680^+$ is re-reduced by a redox-active tyrosine, called Tyr-Z. The D2 protein also contains a redox-active tyrosine, called Tyr-D, but in contrast to Tyr-Z this residue does not participate in steady-state electron transfer. The final electron donor of PSII is water, whose oxidation is catalyzed by four Mn ions (for reviews, see Refs. [98,99]) bound by the D1 protein [107,108]. The proper conformation of the catalytic Mn cluster is expected to be maintained by a 33 kDa hydrophylic protein attached to the lumenal side of the D1/D2 heterodimer [110,111].

B. Inhibition of PSII Electron Transport by UV-B

There is a general consensus that the PSII complex is a highly sensitive target of UV radiation and many crucial components of PSII electron transport: the Q_A, Q_B, and PQ quinone electron acceptors, the Tyr-Z and Tyr-D redox-active tyrosine residues, as well as the Mn cluster of water oxidation have been suggested as actual target sites (Figure 43.2). Critical comparison of the literature data is often complicated by the largely different experimental conditions: different light intensities, the presence or absence of visible light, and the spectral composition of the applied UV source (contributions from UV-C and UV-A besides UV-B). However, it seems to be well established that the redox functioning of the above components are all affected by UV-B to a smaller or larger degree and the idea of multiple UV target sites in PSII is generally accepted [112].

1. The Quinone Electron Acceptors

The original suggestion for the UV effect on the quinone acceptors comes from observations showing that the action spectrum of PSII damage peaks at 250 to 260 nm [63,70], where oxidized PQ absorbs [42,113],

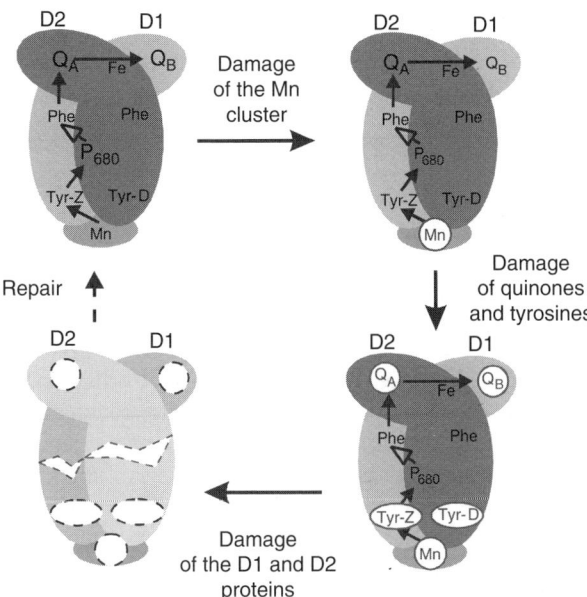

FIGURE 43.2 The sequence of UV radiation induced damaging events in PSII. The primary effect of UV radiation is the inactivation of the Mn cluster of the water-oxidizing complex. This is followed by the damage of quinone electron acceptors and tyrosine donors. Finally, both the D1 and D2 reaction center subunits are degraded. In intact cells, the damage can be repaired via resynthesis of the damaged subunits.

and also that plastoquinones are destroyed by UV-C radiation [43–45]. This idea was adapted for the UV-B-induced damage of PSII on the basis of selective absorption of plastosemiquinones in the UV-B range [42,114]. Damage of PSII quinone electron acceptors seems to be supported by a number of observations: decreased extent of flash-induced absorption change at 320 nm reflecting Q_A reduction [69,71,115]; decreased yield of absorption change at 263 nm reflecting plastoquinone reduction in the PQ pool [115]; loss of flash-induced chlorophyll-a fluorescence rise reflecting Q_A reduction [116,117]. However, these measurements all monitor the ability of PSII to transfer an electron from the donor side to the quinone acceptors. Thus, decreased yield of Q_A and PQ reduction does not necessarily reflect direct destruction of the quinone acceptors, but may also arise from an effect on the Mn_4-Tyr-P680-Pheo section of the electron transport chain. Damage of the Q_A redox function, independent of possible limitations on donor-side electron transport, has been confirmed by showing the loss of the electron paramagnetic resonance (EPR) signal arising from the $Q_A^- Fe^{2+}$ complex when Q_A reduction was induced chemically [47,118]. Similar effect was observed after UV-A-induced inhibition of PSII function [48].

UV-B and UV-A radiation retards electron transfer from Q_A to Q_B as indicated by slowed down decay of flash-induced chlorophyll fluorescence [48,116, 119]. This effect most likely indicates an UV-induced modification of the Q_B binding protein niche as also revealed by the lowered affinity of atrazine [72] and DCMU to occupy the Q_B site in UV-irradiated thylakoids and cyanobacterial cells [71,119,120].

2. The Redox-Active Tyrosines

UV-B sensitivity of tyrosine residues is based on their absorption in the UV region, which peaks at around 280 nm in the neutral form. In addition, tyrosines also absorb at around 250 and 300 nm in the oxidized radical form, as demonstrated for the Tyr-Z component of PSII [121–123]. The absorption of the oxidized Tyr-D•radical has not been measured directly, but expected to be identical with that of Tyr-Z•. Deleterious effects of UV-B radiation on the redox function of Tyr-Z and Tyr-D are revealed by the loss of the EPR signals arising from Tyr-Z• and Tyr-D• [47,117,118]. Similar effect was also observed for UV-A radiation [48].

3. The Water-Oxidizing Complex

Inhibition of the water-oxidizing complex by UV-B radiation has been suggested by a number of observations. (i) Retarded rise of variable fluorescence, typical for donor-side limited electron transport was reported in Refs. [68–70,124]. (ii) Conversion of the re-reduction kinetics of P680$^+$ from the nanosecond to the microsecond range [72,125,126], indicating retarded electron donation from the Mn cluster to P680. (iii) Faster inhibition of charge recombination of the S$_2$ state of the water-oxidizing complex with Q$_A^-$ as compared to that of Tyr-D$^{\bullet}$ with Q$_A^-$ [120]. (iv) Restoration of PSII activity by artificial electron donors, which can maintain electron transport in PSII centers that are deprived of their oxygen-evolving capacity by UV-B radiation was found by some authors [70,72], but not by others [115]. (v) The loss of the multiline EPR signal arising from the S$_2$ redox state of the Mn cluster, which was observed for both UV-B [47,118] and UV-A radiation [48]. (vi) A compelling evidence for the impaired function of the water-oxidizing complex comes from time-resolved EPR measurements showing that the stability of Tyr-Z$^{\bullet}$ is increased in UV-B [47] or UV-A [48] irradiated PSII membranes from the microsecond to the millisecond time range, which demonstrates the block of electron transfer between the catalytic Mn cluster and Tyr-Z$^{\bullet}$. This effect was corroborated by the observation of a fast decaying phase of flash-induced chlorophyll fluorescence in the presence of DCMU, which shows that Tyr-Z$^{\bullet}$ becomes the recombination partner of Q$_A^-$ after UV-induced impairment of the Mn cluster of water oxidation [119,127].

It is also of note that the reaction center of purple bacteria whose structure and function shows large homology with PSII, as far as the quinone acceptors are concerned, but does not possess water-oxidizing complex [128], is highly resistant against UV-B radiation [129], providing a further proof for the primary UV-B damage at the water-oxidizing site in PSII. Since characteristic consequences of inhibited water-oxidizing activity, such as slow rise of variable chlorophyll fluorescence, are observed not only in isolated thylakoids but also in intact leaves [79], the water-oxidizing complex appears to be the primary action site of UV-B radiation under both *in vitro* and *in vivo* conditions.

C. Damage by UV-A Radiation

In contrast to the wealth of information available on the damaging mechanism of UV-B radiation our knowledge is more limited on the effects of UV-A (315 to 400 nm) radiation. The intensity of this spectral range in the natural sunlight is at least 10 times higher than that of UV-B, and its penetration to the Earth is not attenuated significantly by the ozone layer or other components of the atmosphere [97].

Thus, the damaging effects of UV-A radiation could be highly significant [130]. UV-A radiation has been shown to damage PSII to a considerably larger extent than PSI [127]. Within PSII, the slow rise of variable chlorophyll fluorescence together with the modified oscillatory pattern of flash-induced oxygen evolution indicates a damage of PSII donor-side components [127]. Further support for the primary effect of UV-A on the water-oxidizing complex is provided by low-temperature EPR measurements [48]. The so-called multiline signal, which arises from the S$_2$ state of the water-oxidizing complex is lost much faster than the EPR signal that arises from the interaction of Q$_A^-$ with the nonheme Fe^{2+} or from Tyr-D$^{\bullet}$. Thus, the immediate cause for the loss of oxygen evolution is the inactivation of electron transport between the catalytic Mn cluster and the tyrosine electron donors. However, the loss of Q$_A$ function points to additional UV-A-induced alteration of PSII acceptor-side components. This observation is in agreement with flash-induced thermoluminescence and chlorophyll fluorescence measurements, which showed that the Q$_B$-binding pocket on the acceptor side is also modified [127]. Comparison of the characteristics of PSII damage induced by UV-A and UV-B radiation shows that the two spectral ranges inhibit PSII by very similar or identical mechanisms, which target primarily the water-oxidizing complex. Although the damaging efficiency of UV-A is much smaller than that of UV-B, due to its higher intensity the UV-A component of sunlight appears to have the same overall potential to inactivate the light reactions of photosynthesis as UV-B.

D. Possible Mechanisms of UV Action on PSII Redox Components

The mechanisms by which the PSII redox components are impaired by UV radiation are not completely clear. In case of the quinone acceptors and tyrosine donors, a direct destruction of the molecules could occur. However, the possibility that the redox function of these components is impaired due to damage of their protein environment cannot be excluded.

Damage or alteration of the protein binding site of the catalytic Mn cluster is also a likely scenario for the inactivation of the water-oxidizing function, since the Mn ions themselves are not expected to be modified by UV light. The sensitivity of PSII in different S states of the water-oxidizing complex was studied by synchronizing PSII into specific S states by short pulses of visible light, which were then illuminated with monochromatic UV-B laser flashes of 308 nm. The damage induced by the UV-B flashes showed a clear S-state dependence indicating that the water-oxidizing complex is most prone to UV damage in the S_2 and S_3 oxidation states [131]. During the S-state transitions the catalytic Mn cluster of water oxidation is sequentially oxidized, see Refs. [99,101,102]. Mn ions bound to organic ligands (such as amino acids) have pronounced absorption in the UV-B and UV-A regions in the Mn(III) and Mn(IV) oxidation states, which dominate the higher S-states, but not in the Mn(II) oxidation state, which occur in the lower S-states [132]. As a consequence the $S_1 \rightarrow S_2$ and $S_2 \rightarrow S_3$ redox transitions of the Mn cluster are accompanied by absorption changes in the UV region [122,133]. Thus, the high UV sensitivity of PSII in the S_2 and S_3 states indicates that UV absorption by the Mn(III) and Mn(IV) ions could be the primary sensitizer of UV-induced damage of the water-oxidizing machinery.

A further possibility for the inactivation of the Mn site is related to the specific structure of the 33 kDa water-soluble protein subunit of the water-oxidizing complex, that is expected to maintain the functional conformation of the Mn cluster [109–111]. The 33 kDa protein is unique among the PSII subunits in the sense that its proper conformation is likely to be stabilized by a disulfide bridge [134] whose breakup by dithiothreitol (DTT) inhibits the water-oxidizing complex [135]. Since disulfide bridges can also be split by UV-B [31], this effect may lead to the inactivation of the catalytic Mn cluster.

The action spectrum of PSII inhibition peaks at around 250 to 260 nm for both the Hill reaction [63], which measures electron transport through PSII and for the slow-down of variable fluorescence rise [70], which directly reflects the inhibition of water-oxidizing activity. The match between the action spectrum and the absorbance of single potential targets is satisfactory only for limited spectral ranges (Figure 43.3). The absorption by Mn(III) and Mn(IV) ions follows well the action spectrum of UV-induced inactivation of PSII in the whole UV-A and UV-B ranges, in contrast to PQ^- and Tyr-Z•, which give only limited match. However, in the short-wavelength (UV-C) region there is a close similarity between the action spectrum by the absorbance of the oxidized tyrosine radicals and, to a lesser extent, by oxidized plastoquinone, which indicate the involvement of these species in sensitizing UV damage in the UV-C region. Thus, while it is obvious that no single target could be responsible for the whole UV action in the UV-B plus UV-C range, it is also evident that absorption related to redox transitions of Mn ions in the water-oxidizing complex could explain most of the UV-B and UV-A effects.

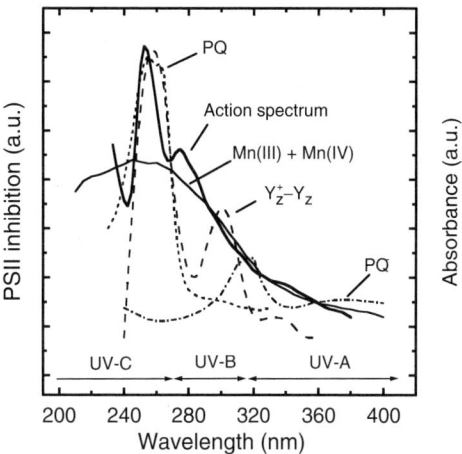

FIGURE 43.3 The action spectrum of UV damage in comparison with the absorption spectra of the main UV targets in PSII. Action spectrum of UV damage (thick solid line [63]). Absorption spectra: PQ (dotted line [42]) PQ^- (dashed-dotted line [42]), oxidized minus reduced Tyr-Z (dashed line [122]), Mn(III) + Mn(IV) bound to an organic ligand (thin solid line [132]).

E. DAMAGE OF PSII PROTEIN STRUCTURE BY UV RADIATION

An important consequence of UV irradiation is the damage of the protein backbone of the PSII reaction center. This effect is characteristic mainly for the D1 and D2 subunits that form the heart of the reaction center of PSII, and was observed both *in vivo* [46,136–139] and in isolated thylakoid preparations [115,140–145]. Other protein components of PSII seem to be damaged much later than D1 and D2, and this may be an indirect consequence of the breakup of the D1/D2 reaction center heterodimer [145].

Based on the assumption that the Q_A and Q_B acceptors are potential targets of UV-B radiation, these quinones have frequently been suggested as sensitizers of D1 and D2 protein damage [136–140]. The main argument in favor of this idea is the similarity of the action spectrum of the D1 protein degradation and of the absorption spectrum of plastosemiquinones [136]. However, the absorption of plastosemiquinones and oxidized tyrosine radicals are very similar in the UV-B range, which makes rather ambiguous the distinction between the two species. In addition, the primary UV-B-induced cleavage site of D1 appears to be located at the middle, or close to the lumenal end of the second transmembrane helix [141], which is closer to the putative binding site of the catalytic cluster of water oxidation [105,107,108], than to the Q_B site. The specific UV-B-induced D1 cleavage in the second helix can also be observed in PSII preparations from which Q_B is selectively depleted by heptane–isobutanol extraction, but retain active water-oxidizing complex [145]. In contrast to this, the 20-kDa C-terminal fragment of the D1 protein is not detected in Tris-washed PSII membranes that lack Mn ions but retain Q_B and Q_A [46]. The latter series of data point to the importance of the donor side components of PSII, primarily of the Mn cluster of water oxidation, in sensitizing D1 protein cleavage.

The loss and fragmentation of the D2 protein has not been characterized to the same extent as that of the D1 protein. The available data indicate that degradation of the D2 protein is considerably accelerated when visible light, that enhances the reduction level of Q_A, is applied together with the UV-B irradiation [146]. Furthermore, the D2 protein is not fragmented in isolated PSII reaction center complexes that lack Q_A [142]. However, in the presence of the quinone analog DBMIB a 22-kDa N-terminal breakdown product is formed, indicating a cleavage at around the Q_A site [142]. These data are considered as evidence for the role of Q_A^- in sensitizing D2 damage by UV-B. On the other hand, D2 protein degradation is retarded in the presence of the electron transport inhibitor DCMU [146], which blocks the reoxidation of Q_A^- by Q_B, thereby enhancing the accumulation of Q_A^-. Thus, although the semiquinone form of Q_A is a possible sensitizer of UV-B induced D2 protein degradation, other factors are likely to be involved in the overall UV-B damage of D2.

UV-B-induced loss and cleavage of the reaction center subunits seems to be specific to PSII, since the protein structure of the reaction center from the purple bacterium *Rhodobacter sphaeroides* R-26 is not affected by UV-B [129]. The purple bacterial reaction center lacks Mn ions and water oxidation, but binds Q_B and Q_A by its L and M subunits, respectively, in very similar protein environments as in D1 and D2 [128]. The insensitivity of the bacterial reaction center proteins to UV-B supports the above considerations that quinones are unlikely to act as primary sensitizers of D1 (and also D2) damage.

As regards the actual mechanism of UV-B-induced protein degradation, in isolated preparations D1 and D2 protein damage is not retarded by low temperature [115,141,144] and not affected by adding a cocktail of protease inhibitors [142,143], thus pointing to a mechanism that does not involve proteases in the UV-B-induced polypeptide cleavage.

It is noteworthy, however, that a *clp* type protease is required for the UV-induced exchange of two different D1 protein forms in *Synechococcus* 7942 [147]. In *Synechocystis* 6803, UV-B radiation strongly induces the gene that encodes an *ftsH* homologue protease that is involved in the repair of D1 following photodamage by visible light [148]. These results indicate that the turnover of the D1 protein, which involves the removal of the damaged D1 copies, probably requires proteases in intact cells.

V. UV EFFECTS ON OTHER COMPONENTS OF THE PHOTOSYNTHETIC APPARATUS

A. PHOTOSYSTEM I

The effects of UV-B and UV-A radiation do not seem to be evenly distributed between the two photosystems. Most studies found minor or no effect on PSI as compared to PSII [68,69,127,149]. Impairment of PSI is usually observed as a decrease in the amplitude of absorption change at 700 nm that reflects the amount of oxidized reaction center chloropyll (P700) of PSI, in cases where high intensity UV-B radiation was applied [150]. Loss of PSI activity was also observed by UV-C irradiation, but even in that case, its inhibition was much less pronounced as that of PSII [151].

The possible targets within PSI, and damage to its protein structure are not studied in detail. Since both PSI and PSII contains quinone electron acceptors, but only PSII possess the water-oxidizing complex and redox-active tyrosines see Ref. [99], the immensely different sensitivities of the two photosystems against UV-B can most likely be explained by the above discussed vulnerability of the water-oxidizing complex in PSII or by the presence of the redox-active tyrosines.

It is of note that DNA microarray experiments indicated a significant downregulation of many genes that encode PSI protein subunits in UV-B-exposed cells of the cyanobacterium *Synechocystis* 6803 [152]. Although the corresponding decrease in PSI activity has not been reported yet, this effect may indicate an acclimation response, which could readjust the PSI/PSII ratio upset by the UV damage of PSII centers.

B. THE CYTOCHROME B_6/F COMPLEX

Electron transport between the two photosystems is mediated by the cytochrome b_6/f complex: it oxidizes plastoquinol produced by PSII, and reduces plastocyanin, which serves as electron donor to PSI; for a review, see Ref. [153]. From the studies of Strid et al. [35,37], it appears that the cytochrome b_6/f complex, together with PSI, is the least affected thylakoid component by UV-B. This resistance to UV-B is notable since the cytochrome b_6/f complex contains two quinone binding sites; one where quinol oxidation occurs, and the other where quinone reduction occurs [153]. Consequently, the observed low UV-B sensitivity of the cytochrome b_6/f complex can be considered as another piece of evidence against the importance of quinones in mediating UV-B-induced damage in the photosynthetic apparatus.

C. ATP SYNTHASE AND RUBISCO

ATP synthase and ribulose 1,5-biphosphate carboxylase (Rubisco) are among the thylakoid membrane components, which are adversely affected by UV-B radiation. During supplemental UV-B irradiation of pea plants, both the amount and the activity of CF1-ATP synthase of thylakoids decreased [37]. ATP synthase from nonphotosynthetic cells can also be inactivated by UV-B [36], but no mechanistic details of this effect have been clarified.

Rubisco is the main CO_2-fixing enzyme in C3 plants, which consist of two subunits (14 and 55 kDa). Rubisco activity declines with enhanced levels of UV-B radiation [33–35]. The activity decline is correlated with the decreased amounts of both subunits and the corresponding mRNA levels [154]. Some studies actually propose Rubisco as the major potential candidate for the primary action site of inhibition by UV-B radiation of the photosynthetic apparatus in intact plant systems [64,65,155].

D. THE LIGHT-HARVESTING SYSTEMS

The light-harvesting complex of PSII (LCHII) plays an important role in light absorption and energy transfer to the reaction center as well as in thylakoid organization (for further details, see Chapter 9 of this book). UV-B radiation appears to have adverse affects on LCHII: it may lead to the functional disconnection of LHCII from PSII in isolated thylakoids [71] and decreases the RNA transcript level of the *cab* genes responsible for the synthesis of the chlorophyll *a/b* binding proteins of LHCII [53,54]. In cyanobacteria, light harvesting is performed by phycobilisomes, which are profoundly affected by UV radiation. The phycobiliproteins can be destroyed by UV-B, or the energy transfer towards the photosynthetic reaction centers can be impaired [55–59]. In the absence of protein repair, UV-B-induced damage of phycobilisomes occurs much slower than that of PSII. However, in cells, which are capable of de novo protein synthesis, PSII is efficiently repaired when the UV-B radiation is removed from the illumination protocol. In contrast, restoration of phycobilisomes is a very slow process and it is quite likely that recovery of phycobilisome function requires the development of new cells, via cell division [131]. Comparison

of PSII and phycobilisomes provides an interesting example for a highly UV-sensitive, but well-repaired component in contrast to a less sensitive, but inefficiently repaired component.

E. THE THYLAKOID MEMBRANE

UV-B radiation seems to exert adverse effects not only on various protein or pigment–protein complexes of the photosynthetic apparatus, but also on the structure of the thylakoid membrane that contains these complexes. An early consequence of UV-B irradiation is the leakage of the thylakoid membrane, i.e., an increase in ion permeability [69,156]. UV-induced membrane leakage has also been observed with plasma membranes [157] and cultured cells of higher plants [158], and explained with effects on specific ion channels [157]. UV-B-induced loss of K^+ from guard cells may be responsible for the observed loss of stomatal conductance in irradiated plants [75–77].

VI. PROTECTION, ADAPTATION, AND REPAIR

Plants possess various defense mechanisms which greatly modulate the sensitivity of the photosynthetic apparatus to UV-B radiation. These protective mechanisms include morphological changes such as increased length of epidermal cells [159], production of a vaxy cuticle [160], accumulation of UV-B absorbing compounds, particularly phenylpropanoids in the epidermal layer [79,161–167], and activation of different scavenging systems of various active oxygen species [84,168–171]. These protective defense mechanisms are discussed in detail in Chapters 8 and 9 of this book, and by extensive reviews [25,60–62,86,89].

In addition to the protective defense mechanisms by which plants can lessen the impact of UV-B radiation on the photosynthetic system by attenuating the intensity of UV-B before it reaches crucial targets, plants have also developed active defense systems by which the cells can repair the damage that has occurred. The overall UV-B sensitivity of the photosynthetic apparatus and that of the whole cell is eventually determined by the balance of damage occurred and of the efficiency of repair processes that can restore the impaired functions.

As regards repair of UV-B-induced DNA damage, a blue light requiring repair enzyme, photolyase, can directly split pyrimidine dimers, whereas, other types of DNA damage can be repaired by excision repair in the dark [10,24,172,173]. Although DNA repair in plant cells has not been studied to the same extent as in mammalian, yeast, or bacterial systems, dark repair and light-reactivation of damaged DNA has been demonstrated in *Arabidopsis thaliana* [14,18,19,21,22].

Repair is also important at the level of the photosynthetic apparatus [71,93,156]. Experiments with the cyanobacterium *Syenechocystis* 6803 and with the higher plant *Arabidopsis thaliana* have demonstrated that the inhibited PSII activity can be restored via de novo synthesis of the damaged D1 and D2 protein subunits [174]. These proteins are usually encoded by small multigene families in cyanobacteria [175], whose members respond differentially to UV-B light. In *Synechocystis* 6803, there are three *psbA* genes called *psbA1*, *psbA2*, and *psbA3* [176]. Among these, *psbA2* and *psbA3* encode identical D1 proteins, whereas, *psbA1* is not expressed. Under normal light conditions, the majority (>90%) of the *psbA* transcript is produced from *psbA2* [177]. However, in the presence of UV-B light, the expression of *psbA3* is preferentially enhanced [178] and the protein made from this gene is incorporated into the PSII complex [179]. A similar differential UV-B response of the *psbA* genes is observed in *Synechoccus* 7942. In this species there are two different D1 forms: D1:1 is encoded by *psbAI* and D1:2 is encoded by *psbAII* and *psbAIII*. This cyanobacterium exchanges D1:1 for D1:2 upon UV-B irradiation, which provides protection against the detrimental UV effects [180]. It appears that the protective effect arises not only from the different UV sensitivity of PSII containing the D1:1 and D1:2 protein forms, but also from the decreased rate of repair of D1:1 [181]. The D2 subunit is also encoded by two genes, called *psbD1* and *psbD2*, which result in identical polypeptide sequences under the influence of different promoters [182]. As a result of UV-B radiation, the expression level of *psbD2*, which produces only a small fraction of the *psbD* transcripts under normal light conditions, is significantly enhanced in *Synechocystis* 6803 [183]. Thus, it appears that an important physiological role of multiple *psbA* and *psbD* gene copies in cyanobacteria is to ensure rapid increase of the *psbA* and *psbD* transcript levels, respectively, under conditions of UV exposure when there is an increased demand for rapid D1 and D2 protein synthesis.

VII. INTERACTIONS OF VISIBLE AND UV-B LIGHT

Detrimental effects on the photosynthetic apparatus are induced not only by UV-B irradiation, but also by the visible spectral range of solar radiation (for reviews, see Refs. [98,184–191]). In their natural habi-

tat, plants are exposed simultaneously to visible and UV-B irradiances, and the interaction of the two different light regimes can greatly modulate the light sensitivity of the photosynthetic apparatus.

Since both UV-B and visible light can influence the function of PSII electron transport, the interaction of the two light regimes can lead to a wide range of effects. Earlier observations indicated examples for both a synergistic enhancement of photodamage to the function and protein structure of PSII [67,90,192] and amelioration of damaging effects of UV-B radiation by visible light [193]. More recent results demonstrated that in isolated systems or in the absence of protein repair capacity, UV-B and visible light damages PSII by independent mechanisms without synergistic interaction [194]. However, the situation is quite different in intact cells, which are capable of de novo protein synthesis. In *Synechocystis* 6803 cells the presence of low intensity visible light was shown to prevent the UV-induced loss of PSII activity by enhancing the efficiency of the protein repair process. However, at high light intensities, the UV-induced damage is not prevented, or even got enhanced due to the additional photodamage induced by visible light [194].

VIII. CONCLUDING REMARKS

Intensive research during the last two decades has yielded significant improvement in our understanding of the molecular background and physiological significance of ultraviolet radiation plant photosynthesis, which is highly important for terrestrial and aquatic ecosystems. Further research will be needed to clarify the rather complex interactions of UV radiation and other stress factors, like elevated and low temperatures, drought, visible light, which influence the protective and repair systems under conditions of present-day and predicted UV-B levels. Another important topic of interest will be the elucidation of the significance of UV damage exerted on the photosynthetic apparatus in relation to the damage caused at the level of nucleic acids. An emerging field of UV research concerns the role of UV light in signal transduction events in cells of photosynthetic organisms. It will be highly important to explore the connection of these signaling events with the adaptation and acclimation processes occurring under ultraviolet exposure.

ACKNOWLEDGMENTS

This work was supported by the Hungarian Scientific Research Fund (OTKA T-034321).

REFERENCES

1. Green AES, Sawada T, Shettle EP. The middle ultra-violet reaching the ground. *Photochem. Photobiol.* 1974; 19:251–259.
2. Madronich S. The atmosphere and UV-B radiation at ground level. In: Young AR, Moan J, Björn LO, Nultsch W, eds. *Environmental UV Photobiology*. New York: Plenum Press, 1993:1–40.
3. Molina MJ, Rowland FS. *Nature*. 1974; 249:810.
4. Stolarski R, Bojkov R, Bishop L, Zeferos C, Staehelin J, Zawodny J. Measured trends in stratospheric ozone. *Science* 1992; 256:342–349.
5. Frederick JE. Ultraviolet sunlight reaching the earth's surface: a review of recent research. *Photochem. Photobiol.* 1993; 57:175–178.
6. Kerr JB, McElroy CT. Evidence for upward trends of ultraviolet-B radiation linked to ozone depletion. *Science* 1993; 262:1032–1034.
7. McKenzie RL, Bjorn LO, Bais A, Ilyasd M. Changes in biologically active ultraviolet radiation reaching the Earth's surface. *Photochem. Photobiol. Sci.* 2003; 2:5–15.
8. Kasting JF. Earth's early atmosphere. *Science* 1993; 259:920–926.
9. Catling DC, Zahnle KJ, McKay CP. Biogenic methane, hydrogen escape, and the irreversible oxidation of early Earth. *Science* 2001; 293:839–843.
10. Sancar A, Sancar GB. DNA repair enzymes. *Annu. Rev. Biochem.* 1988; 57:29–67.
11. Jiang N, Taylor J-S. *In vivo* evidence that UV-induced C → T mutations at dipyrimdine sites could result from the replicative bypass of *cis-syn* cyclobutane dimers or their deamination products. *Biochemistry* 1993; 32:472–481.
12. Hutchinson F. A review of some topics concerning mutagenesis by ultraviolet light. *Photochem. Photobiol.* 1987; 45:897–903.
13. Smith KC. Ultraviolet radiation effects on molecules and cells. In: Smith KC, ed. *The Science of Photobiology*. New York: Plenum Press, 1977:113–142.
14. Pang Q, Hays JB. UV-B-inducible and temperature-sensitive photoreactivation of cyclobutane pyrimidine dimers in *Arabidopsis thaliana*. *Plant Physiol.* 1991; 95:536–543.
15. Quaite FE, Sutherland BM, Sutherland JC. Action spectrum for DNA damage in alflafa lowers predicted impact of ozone depletion. *Nature* 1992; 358:576–578.
16. McLennan AG. DNA damage, repair, and mutagenesis. In: Bryant JA, Dunham VL, eds. *DNA Replication in Plants*. Boca Raton, FL: CRC Press, 1987:135–186.
17. Britt AB. Repair of DNA damage induced by ultraviolet radiation. *Plant Physiol.* 1995; 108:891–896.
18. Sutherland BM, Takayanagi S, Sullivan JH, Sutherland JC. Plant responses to changing environmental stress: cyclobutyl pyrimidine dimer repair in soybean leaves. *Photochem. Photobiol.* 1996; 64:464–468.
19. Hidema J, Kumagai T, Sutherland JC, Sutherland BM. Ultraviolet B-sensitive rice cultivar deficient in cyclobutyl pyrimidine dimer repair. *Plant Physiol.* 1997; 113:39–44.
20. Britt AB. An unbearable beating by light? *Nature* 2000; 406:30–31.
21. Chen J-J, Mitchell DL, Britt AB. A light-dependent pathway for the elimination of UV-induced pyrimidine (6-4) pyrimidinone photoproducts in *Arabidopsis*. *Plant Cell* 1994; 6:1311–1317.
22. Jiang C-Z, Yee J, Mitchell DL, Britt AB. Photorepair mutants of *Arabidopsis*. *Proc. Natl. Acad. Sci. USA* 1997; 94:7441–7445.
23. Stapleton AE. Ultraviolet radiation and plants: burning questions. *Plant Cell* 1992; 4:1353–1358.
24. Sinha RP, Hader D-P. UV-induced DNA damage and repair: a review. *Photochem. Photobiol. Sci.* 2002; 1:225–236.
25. Strid A, Chow WS, Anderson JM. UV-B damage and protection at the molecular level in plants. *Photosynth. Res.* 1994; 39:475–489.
26. Britt AB. Molecular genetics of DNA repair in higher plants. *Trends Plant Sci.* 1999; 4:20–25.
27. Vladimirov YA, Roshchupkin DI, Fesenko EE. Photochemical reactions in amino acids residues and inactivation of enzymes during U.V.-irradiation. A review. *Photochem. Photobiol.* 1970; 11:227–246.
28. Creed D. The photophysics and photochemistry of the near-UV absorbing amino acids — II. Tyrosine and its simple derivates. *Photochem. Photobiol.* 1984; 39:563–575.
29. Malencik DA, Anderson SR. Dityrosine formation in calmodulin. *Biochemistry* 1987; 26:695–704.
30. Kochevar IE. UV-induced protein alterations and lipid oxidation in erythrocyte membranes. *Photochem. Photobiol.* 1990; 52:795–800.
31. Creed D. The photophysics and photochemistry of the near-UV absorbing amino acids — III. Cystein and its simple derivates. *Photochem. Photobiol.* 1984; 39:577–583.
32. Tevini M. Molecular biological effects of ultraviolet radiation. In: Tevini M, ed. *UV-B Radiation and Ozone Depletion. Effects on Humans, Animals, Plants, Microorganisms, and Materials*. Boca Raton, FL: Lewis Publishers, 1993:1–15.
33. Basiouny FM, Van TK, Biggs RH. Some morphological and biochemical characteristics of C3 and C4 plants irradiated with UV-B. *Physiol. Plant.* 1978; 42:29–32.
34. Vu CV, Allen LH, Garrard LA. Effects of UV-B radiation (280–320 nm) on ribulose-1,5-bisphosphate carboxylase in pea and soybean. *Environ. Exp. Bot.* 1984; 24:131–143.
35. Strid A, Chow WS, Anderson JM. Effects of supplementary ultraviolet-B radiation on photosynthesis in *Pisum sativum*. *Biochim. Biophys. Acta* 1990; 1020:260–268.
36. Murphy TM. Membranes as targets of ultraviolet radiation. *Physiol Plant.* 1983; 58:381–388.
37. Zhang J, Hu X, Henkow L, Jordan BR, Strid A. The effects of ultraviolet-B radiation on the CF_0F_1-ATPase. *Biochim. Biophys. Acta* 1994; 1185:295–302.

38. Pfündel EE, Pan R-S, Dilley RA. Inhibition of violaxantin deepoxidation by ultraviolet-B radiation in isolated chloroplasts and intact leaves. *Plant Physiol.* 1992; 98:1372–1380.

39. Panagopoulos I, Bornman JF, Björn LO. Effetcs of ultraviolet radiation and visible light on growth, fluorescence induction, ultraweak luminescence and peroxidase activity in sugar beet plants. *J. Photochem. Photobiol.* 1990; 8:73–87.

40. Kramer GF, Norman HA, Krizek DT, Mirecki RM. Influence of UV-B radiation polyamines, lipid peroxidation and membrane lipids in cucumber. *Phytochemistry* 1991; 30:2101–2108.

41. Halliwell B, Gutteridge JMC. *Free Radicals in Biology and Medicine.* New York: Oxford University Press, 1989.

42. Amesz J. Plastoquinone. In: Trebst A, Avron M, eds. *Encyclopedia of Plant Physiology.* Vol. 5. Berlin: Springer-Verlag, 1977:238–246.

43. Bishop NI. The possible role of plastoquinone (Q-254) in the electron transport system of photosynthesis. *CIBA Symp.* 1961; 385–404.

44. Trebst A, Pistorius E. Photosynthetische reaktionene in UV-bestrahlten Chloroplasten. *Z. Naturforsch.* 1965; 20:885–889.

45. Shavit N, Avron M. The effect of UV light on phosphorylation and the Hill reaction. *Biochim. Biophys. Acta* 1963; 66:187–195.

46. Barbato R, Frizzo A, Friso G, Rigoni F, Giacometti GM. Degradation of the D1 protein of photosystem-II reaction centre by ultraviolet-B radiation requires the presence of functional manganese on the donor side. *Eur. J. Biochem.* 1995; 227:723–729.

47. Vass I, Sass L, Spetea C, Bakou A, Ghanotakis D, Petrouleas V. UV-B induced inhibition of photosystem II electron transport studied by EPR and chlorophyll fluorescence. Impairment of donor and acceptor side components. *Biochemistry* 1996; 35:8964–8973.

48. Vass I, Turcsányi E, Touloupakis E, Ghanotakis D, Petroluleas V. The mechanism of UV-A radiation-induced inhibition of photosystem II electron transport studied by EPR and chlorophyll fluorescence. *Biochemistry* 2002; 41:10200–10208.

49. Döhler G. Effect of UV-B radiation (290–320 nm) on the nitrogen metabolism of several marine diatoms. *J. Plant Physiol.* 1985; 118:391–400.

50. Nultsch W, Agel G. Fluence rate and wavelength dependence of photobleaching in the cyanobacterium *Anabena variabilis. Arch. Microbiol.* 1986; 144:268–271.

51. Hader D-P, Hader MA. Effects of solar and artificial radiation on motility and pigmentation in *Cyanophora paradoxa. Arch. Microbiol.* 1989; 152:453–457.

52. El-Sayed SZ, Stephens FC, Bidigare RR, Ondrusek ME. Effect of ultraviolet radiation on antactic marine phytoplankton. In: Kerry KR, Hempel G, eds. *Antarctic Ecosystems. Ecological Change and Conservation.* Heidelberg: Springer, 1990:379–385.

53. Jordan BR, JAmes PE, Strid A, Anthony RG. The effect of ultraviolet-B radiation on gene expression and pigment composition in etiolated and green pea leaf tissue: UV-B induced changes are gene-specific and dependent upon the developmental stage. *Plant Cell Environ.* 1994; 17:45–54.

54. Jordan BR, Chow WS, Strid A, Anderson JM. Reduction in cab and psbA RNA transcripts in response to supplementary ultraviolet-B radiation. *FEBS Lett.* 1991; 284:5–8.

55. Sinha RP, Lebert M, Kumar A, Kumar HD, Hader D-P. Disintegration of phycobilisomes in a rice field cyanobacterium *Nostoc* sp. following UV irradiation. *Biochem. Mol. Biol. Int.* 1995; 37:697–706.

56. Lao K, Glazer AN. Ultraviolet-B photodestruction of a light-harvesting complex. *Proc. Natl. Acad. Sci. USA* 1996; 93:5258–5263.

57. Banerjee M, Hader D-P. Effect of UV radiation on the rice field cyanobacterium, *Aulosira fertilissima. Environ. Exp. Bot.* 1996; 36:281–291.

58. Nedunchezhian N, Ravindran KC, Abadia A, Abadia J, Kulandaivelu G. Damage of photosynthetic apparatus in *Anacystis nidulans* by ultraviolet-B radiation. *Biol. Plant.* 1996; 38:53–59.

59. Pandey R, Chauhan S, Singhal GS. UVB-induced photodamage to phycobilisomes of *Synechococcus* sp. PCC 7942. *J. Photochem. Photobiol.* 1997; 40:228–232.

60. Tevini M. UV-B effects on terrestrial plants and aquatic organisms. *Prog. Bot.* 1994; 55:174–190.

61. Tevini M, Teramura AH. UV-B effects on terrestrial plants. *Photochem. Photobiol.* 1989; 50:479–487.

62. Teramura AH, Sullivan JH. Effects of UV-B radiation on photosynthesis and growth of terrestrial plants. *Photosynth. Res.* 1994; 39:463–473.

63. Jones LW, Kok B. Photoinhibition of chloroplast reactions. I. Kinetics and action spectra. *Plant Physiol.* 1966; 41:1037–1043.

64. Allen DJ, Nogués S, Baker NR. Ozone depletion and increased UV-B radiation: is there a real threat to photosynthesis? *J. Exp. Bot.* 1998; 49:1775–1788.

65. Allen DJ, McKee IF, Farage PK, Baker NR. Analysis of limitations to CO_2 assimilation on exposure of leaves of two *Brassica napus* cultivars to UV-B. *Plant Cell Environ.* 1997; 20:633–640.

66. Jones LW, Kok B. Photoinhibition of chloroplast reactions. II. Multiple effects. *Plant Physiol.* 1966; 41:1044–1049.

67. Teramura AH, Biggs RH, Kossuth S. Effects of ultraviolet-B irradiances on soybean. II. Interaction between ultraviolet-B and photosynthetically active radiation on net photosynthesis, dark respiration, and transpiration. *Plant Physiol.* 1980; 65:483–488.

68. Kulandaivelu G, Noorudeen AM. Comparative study of the action of ultraviolet-C and utraviolet-B radiation on photosynthetic electron transport. *Physiol. Plant.* 1983; 58:389–394.

69. Iwanzik W, Tevini M, Dohnt G, Voss M, Weiss W, Graber P, et al. Action of UV-B radiation on photosynthetic primary reaction in spinach chloroplasts. *Physiol. Plant.* 1983; 58:401–407.

70. Bornman JF, Björn LO, Akerlund H-E. Action spectrum for inhibition by ultraviolet radiation of photo-

system II activity in spinach thylakoids. *Photobiochem. Photobiophys.* 1984; 8:305–313.

71. Renger G, Voss M, Graber P, Schulze A. Effect of UV radiation on different partial reactions of the primary processes of photosynthesis. In: Worrest C, Caldwell MM, eds. *Stratospheric Ozone Reduction, Solar Ultraviolet Radiation and Plant Life.* Berlin: Springer, 1986:171–184.

72. Renger G, Völker M, Eckert HJ, Fromme R, Hohm-Veit S, Graber P. On the mechanism of photosystem II deterioration by UV-B irradiation. *Photochem. Photobiol.* 1989; 49:97–105.

73. Desai TS. Studies on thermoluminescence, delayed light emission and oxygen evolution from photosynthetic materials: UV effects. *Photosynth. Res.* 1990; 25:17–24.

74. Fiscus EL, Booker FL. Is increased UV-B a threat to crop photosynthesis and productivity? *Photosynth. Res.* 1995; 43:81–92.

75. Teramura AH, Tevini M, Iwanzik W. Effects of ultraviolet-B irradiance on plants during mild water stress. I. Effects on diurnal stomatal resistance. *Physiol. Plant.* 1983; 57:175–180.

76. Negash L, Björn LO. Stomatal closure by ultraviolet radiation. *Physiol. Plant.* 1986; 66:360–364.

77. Negash L, Jensen P, Björn LO. Effect of ultraviolet radiation on accumulation and leakage of 86Rb$^+$ in guard cells of *Vicia faba*. *Physiol. Plant.* 1987; 69:200–204.

78. Nogués S, Allen DJ, Morison JIL, Baker NR. Characterization of stomatal closure caused by ultraviolet-B radiation. *Plant Physiol.* 1999; 121:489–496.

79. Bornman JF, Vogelman TC. The effect of UV-B radiation on leaf optical properties measured with fiber optics. *J. Exp. Bot.* 1991; 42:547–554.

80. Ryel RJ, Barnes PW, Beyschlag W, Caldwell MM, Flint SD. Plant competition for light analyzed with a multispecies canopy model. I. Model development and influenced of enhanced UV-B conditions on photosynthesis in mixed wheat and wild oat canopies. *Oecologia* 1990; 82:304–310.

81. Mark U, Tevini M. Combination effects of UV-B radiation and temperature on sunflower (*Helianthus annuus* L, cv Polstar) and maize (*Zea mays* L, cv Zenit 2000) seedlings. *J. Plant Physiol.* 1996; 148:49–56.

82. Mark U, Saile-Mark M, Tevini M. Effects of solar UVB radiation on growth, flowering and yield of central and southern European maize cultivars (*Zea mays* L). *Photochem. Photobiol.* 1996; 64:457–463.

83. Stapleton AE, Thornber CS, Walbot V. UV-B component of sunlight causes measurable damage in field-grown maize (*Zea mays* L.): developmental and cellular heterogeneity of damage and repair. *Plant Cell Environ.* 1997; 20:279–290.

84. Hideg E, Nagy A, Oberschall A, Dudits D, Vass I. Detoxification function of aldose/aldehyde reductase during drought and ultraviolet-B (280–320 nm) stresses. *Plant Cell Environ.* 2003; 26:513–522.

85. Nogués S, Allen DJ, Morison JIL, Baker NR. Ultraviolet-B radiation effects on water relations, leaf development, and photosynthesis in droughted pea plants. *Plant Physiol.* 1998; 117:173–181.

86. Bornman JF. Target sites of UV-B radiation in photosynthesis of higher plants. *J. Photochem. Photobiol.* 1989; B4:145–158.

87. Caldwell MM, Teramura AH, Tevini M. The changing solar ultraviolet climate and the ecological consequences for higher plants. *TREE* 1989; 4:363–367.

88. Middleton EM, Teramura AH. Understanding photosynthesis, pigment and growth responses induced by UV-B and UV-A irradiances. *Photochem. Photobiol.* 1994; 60:38–45.

89. Tevini M. Effects of enhanced UV-B radiation on terrestrial plants. In: Tevini M, ed. *UV-B Radiation and Ozone Depletion. Effects on Humans, Animals, Plants, Microorganisms, and Materials.* Boca Raton, FL: Lewis Publishers, 1993:125–153.

90. Bornman JF, Sundby-Emanuelson C. Response of plants to UV-B radiation: some biochemical and physiological effects. In: Smirnoff N, ed. *Environment and Plant Metabolism. Flexibility and Acclimation.* Oxford: BIOS Scientific Publishers,1995:245–262.

91. Bornman JF, Teramura AH. Effects of ultraviolet-B radiation on terrestrial plants. In: Young AR, Björn LO, Moan J, Nultsch W, eds. *Environmental UV Photobiology.* New York: Plenum Press, 1993:427–479.

92. Smith RC, Prézelin BB, Baker KS, Bidigare RR, Boucher NP, Coley T, et al. Ozone depletion: ultraviolet radiation and phytoplankton biology in Antarctic waters. *Science* 1995; 255:952–959.

93. Larkum AWD, Wood WF. The effect of UV-B radiation on photosynthesis and respiration of phytoplankton, benthic microalgae and seagrass. *Photosynth. Res.* 1993; 36:17–23.

94. Neale P, Lesser MP, Cullen JJ. Effects of ultraviolet radiation on the photosynthesis of phytoplankton in the vicinity of McMurdo station, Antarctica. *Antarct. Res. Ser.* 1993; 62:125–142.

95. Cullen JJ, Neale PJ. Ultraviolet radiation, ozone depletion, and marine photosynthesis. *Photosynth. Res.* 1994; 39:303–320.

96. Vassiliev IR, Prasil O, Wyman KD, Kolber Z, Hanson AK, Prentice JE, et al. Inhibition of PSII photochemistry by PAR and UV radiation in natural phytoplankton communities. *Photosynth. Res.* 1994; 42:51–64.

97. Holm-Hansen O, Lubin D, Helbling EW. Ultraviolet radiation and its effects on organisms in aquatic environments. In: Young AR, Björn LO, Moan J, Nultsch W, eds. *Environmental UV Photobiology.* New York: Plenum Press, 1993:379–426.

98. Andersson B, Styring S. Photosystem II: molecular organization, function, and acclimation. *Curr. Top. Bioenerg.* 1991; 16:1–81.

99. Hansson Ö, Wydrzynski T. Current perceptions of photosystem II. *Photosynth. Res.* 1990; 23:131–162.

100. Debus RJ. The manganese and calcium ions of photosynthetic oxygen evolution. *Biochim. Biophys. Acta* 1992; 1102:269–352.

101. Debus RJ. Amino acid residues that modulate the properties of tyrosine Y_Z and the manganese cluster in the water oxidizing complex of photosystem II. *Biochim. Biophys. Acta* 2001; 1503:164–186.

102. Diner BA. Amino acid residues involved in the coordination and assembly of the manganese cluster of photosystem II. Proton-coupled electron transport of the redox-active tyrosines and its relationship to water oxidation. *Biochim. Biophys. Acta* 2001; 1503:147–163.

103. Nanba O, Satoh K. Isolation of a photosystem II reaction center consisting of D1 and D2 polypeptides and cytochrome b-559. *Proc. Natl. Acad. Sci. USA* 1987; 84:109–112.

104. Andersson B, Franzén L-G. The two photosystems of oxygenic photosynthesis. In: Ernster L, ed. *Molecular Mechanisms in Bioenergetics.* Amsterdam: Elsevier Science Publishers, 1995:121–143.

105. Svensson B, Vass I, Cedergren E, Styring S. Structure of donor-side components in photosystem II predicted by computer modelling. *EMBO J.* 1990; 9:2051–2059.

106. Svensson B, Etchebest C, Tuffery P, Van Kan P, Smith J, Styring S. A model for the photosystem II reaction center core including the structure of the primary donor P_{680}. *Biochemistry* 1996; 35:14486–14502.

107. Zouni A, Witt HT, Kern J, Fromme P, Kraus N, Saenger W, et al. Crystal structure of photosystem II from *Synechococcus elongatus* at 3.8 Å resolution. *Nature.* 2001; 409:739–743.

108. Kamiya N, Shen J-R. Crystal structure of oxygen-evolving photosystem II from *Thermosynechococcus vulcanus* at 3.7 Å resolution. *Proc. Natl. Acad. Sci. USA* 2003; 100:98–103.

109. Ferreira KN, Iverson TM, Maghlaoui K, Barber F, Iwata S. Architecture of the photosynthetic oxygen-evolving center. *Science* 2004; 303:1831–1838.

110. Ono TA, Inoue Y. Ca^{2+}-dependent restoration of O_2-evolving activity in $CaCl_2$-washed PSII particles depleted of 33, 26 and 16 kDa polypeptides. *FEBS Lett.* 1984; 168:281–286.

111. Vass I, Cook KM, Deák Zs, Mayes SR, Barber J. Thermoluminescence and flash-oxygen characterization of the IC2 deletion mutant of *Synechocystis* sp. PCC 6803 lacking the photosystem II 33 kDa protein. *Biochim. Biophys. Acta* 1992; 1102:195–201.

112. Jansen MAK, Gaba V, Greenberg BM. Higher plants and UV-B radiation: balancing damage, repair and acclimation. *Trends Plant Sci.* 1998; 3:131–135.

113. Crane LF. *Plant Physiol.* 1959; 34:546–551.

114. Bensasson R, Land EJ. Optical and kinetic properties of semireduced plastoquinone and ubiquinone: electron acceptors in photosynthesis. *Biochim. Biophys. Acta* 1973; 325:175–181.

115. Melis A, Nemson JA, Harrison MA. Damage to functional components and partial degradation of photosystem II reaction center proteins upon chloroplast exposure to ultraviolet-B radiation. *Biochim. Biophys. Acta* 1992; 1100:312–320.

116. Tevini M, Grusemann P, Fieser G. Assessment of UV-B stress by chlorophyll fluorescence analysis. In: Lichtenthaler HK, ed. *Applications of Chlorophyll Fluorescence.* Dordrecht, the Netherlands: Kluwer Academic Publishers, 1988:229–238.

117. Yerkes CT, Kramer DM, Fenton JM, Crofts AR. UV-photoinhibition: studies *in vitro* and in intact plants. In: Baltscheffsky M, ed. *Current Research in Photosynthesis.* Vol. II. Dordrecht, The Netherlands: Kluwer Academic Publishers, 1990:II.6.381–II.6.384.

118. Vass I, Sass L, Spetea C, Hideg É, Petrouleas V. Ultraviolet-B radiation induced damage to the function and structure of photosystem II. In: Mathis P, ed. *Photosynthesis: From Light to Biosphere.* Vol. IV. Dordrecht, The Netherlands: Kluwer Academic Publishers, 1995:553–556.

119. Vass I, Kirilovsky D, Etienne A-L. UV-B radiation-induced donor- and acceptor-side modifications of Photosystem II in the cyanobacterium *Synechocystis* sp. PCC 6803. *Biochemistry* 1999; 38:12786–12794.

120. Hideg É, Sass L, Barbato R, Vass I. Inactivation of oxygen evolution by UV-B irradiation. A thermoluminescence study. *Photosynth. Res.* 1993; 38:455–462.

121. Diner BA, de Vitry C. Optical spectrum and kinetics of the secondary electron donor, Z, of photosystem II. In: Sybesma C, ed. *Advances in Photosynthesis Research.* Vol.I. The Hague: Martinus Nijhoff/Dr.W. Junk, 1995:407–411.

122. Dekker JP, van Gorkom HJ, Brok M, Ouwehand L. Optical characterization of photosystem II electron donors. *Biochim. Biophys. Acta* 1984; 764:301–309.

123. Gerken S, Brettel K, Schlodder E, Witt HT. Optical characterization oft he immediate electron donor to chlorophyll a^+_{II} in O_2-evolving photosystem II complexes. Tyrosine as possible electron carrier between chlorophyll a_{II} and the water-oxidizing complex. *FEBS Lett.* 1988; 237:69–75.

124. Tevini M, Pfister K. Inhibition of photosystem II by UV-B-radiation. *Z. Naturforsch.* 1984; 40c:129–133.

125. Post A, Lukins PB, Walker PJ, Larkum AWD. The effects of ultraviolet irradiation on $P680^+$ reduction in PS II core complexes measured for individual S-states and during repetitive cycling of the oxygen-evolving complex. *Photosynth. Res.* 1996; 49:21–27.

126. Larkum AWD, Karge M, Reifarth F, Eckert H-J, Post A, Renger G. Effect of monochromatic UV-B radiation on electron transfer reactions of photosystem II. *Photosynth. Res.* 2001; 68:49–60.

127. Turcsányi E, Vass I. Inhibition of photosynthetic electron transport by UV-A radiation targets the photosystem II complex. *Photochem. Photobiol.* 2000; 72:513–520.

128. Michel H, Deisenhofer J. Relevance of the photosynthetic reaction center from purple bacteria to the structure of photosystem II. *Biochemistry* 1988; 27:1–7.

129. Tandori J, Máté Z, Vass I, Maróti P. The reaction centre of the purple bacterium *Rhodopseudomonas sphaeroides* R-26 is highly resistant against UV-B radiation. *Photosynth. Res.* 1996; 50:171–179.

130. Cullen JJ, Neale P, Lesser MP. Biological weighting function for the inhibition of phytoplankton photosynthesis by ultraviolet radiation. *Science* 1992; 258:646–650.

131. Vass I, Máté Z, Turcsányi E, Sass L, Nagy F, Sicora C. Damage and repair of photosystem II under exposure to UV radiation. In: PS2001 Proceedings. 12th International Congress on Photosynthesis. Collingwood, Australia: CSIRO Publishing, 2001: S8-001.

132. Bodini ME, Willis LA, Riechel TL, Sawyer DT. Electrochemical and spectroscopic studies of manganese(II), -(III), and -(IV) Gluconate complexes. 1. Formulas and oxidation–reduction stoichiometry. *Inorg. Chem.* 1976; 15:1538–1543.

133. Lavergne J. *Biochim. Biophys. Acta* 1991; 1060:175–188.

134. Tanada S, Wada K. The status of cysteine residues in the 33 kDa protein of spinach photosystem II complexes. *Photosynth. Res.* 1988; 17:255–266.

135. Irrgang KD, Geiken B, Lange B, Renger G. Disulfide bridge modifiers and sulfhydryl group blockers are inactivating the oxygen evolving enzyme of PSII from spinach. In: Murata N, ed. *Research in Photosynthesis*. Vol.II. Dordrecht, The Netherlands: Kluwer Academic Publishers, 1992:417–420.

136. Greenberg BM, Gaba V, Canaani O, Malkin S, Mattoo AK, Edelman M. Separate photosensitizers mediate degradation of the 32-kDa photosystem II reaction centre protein in visible and UV spectral regions. *Proc. Natl. Acad. Sci. USA* 1989; 86:6617–6620.

137. Greenberg BM, Gaba V, Mattoo AK, Edelman M. Degradation of the 32 kDa photosystem II reaction center protein in UV, visible and far red light occurs through a common 23.5 intermediate. *Z. Naturforsch.* 1989; 44c:450–452.

138. Jansen MAK, Depka B, Trebst A, Edelman M. Engagement of specific sites in the plastoquinone niche regulates degradation of the D1 protein in photosystem II. *J. Biol. Chem.* 1993; 268:21246–21252.

139. Jansen MAK, Gaba V, Greenberg BM, Mattoo AK, Edelman M. UV-B driven degradation of the D1 reaction center protein of photosystem II proceeds via plastosemiquinone. In: Yamamoto HY, Smith CM, eds. *Photosynthetic Responses to the Environment*. Washington, DC: American Society of Plant Physiology, 1993:142–149.

140. Trebst A, Depka B. Degradation of the D-1 protein subunit of photosystem II in isolated thylakoids by UV light. *Z. Naturforsch.* 1990; 45c:765–771.

141. Friso G, Spetea C, Giacometti GM, Vass I, Barbato R. Degradation of photosystem II reaction center D1-protein induced by UVB irradiation in isolated thylakoids. Identification and characterization of C- and N-terminal breakdown products. *Biochim. Biophys. Acta* 1993; 1184:78–84.

142. Friso G, Barbato R, Giacometti GM, Barber J. Degradation of D2 protein due to UV-B irradiation in the reaction centre of photosystem II. *FEBS Lett.* 1994; 339:217–221.

143. Friso G, Vass I, Spetea C, Barber J, Barbato R. UB-B-induced degradation of the D1 protein in isolated reaction centres of photosystem II. *Biochim. Biophys. Acta* 1995; 1231:41–46.

144. Spetea C, Hideg É, Vass I. The quinone electron acceptors are not the main senzitizers of UV-B induced protein damage in isolated photosystem II reaction centre- and core complexes. *Plant Sci.* 1996; 115:207–215.

145. Spetea C, Hideg É, Vass I. Q_B-indpendent degradation of the reaction centre II D1 protein in UV-B irradiated thylakoid membranes. In: Mathis P, ed. *Photosynthesis: From Light to Biosphere*. Vol. IV. Dordrecht, The Netherlands: Kluwer Academic Publishers, 1995:219–222.

146. Jansen MAK, Greenberg BM, Edelman M, Mattoo AK, Gaba V. Accelerated degradation of the D2 protein of photosystem II under ultraviolet radiation. *Photochem. Photobiol.* 1996; 63:814–817.

147. Clarke AK, Schelin J, Porankiewicz J. Inactivation pf the *clp*P1 gene for the proteolytic subunit of the ATP-dependent Clp protease in the cyanobacterium *Synechococcus* limits growth and light acclimation. *Plant Mol. Biol.* 1998; 37:791–801.

148. Silva P, Thompson E, Bailey S, Kruse O, Mullineaux CW, Robinson C, et al. FtsH is involved in the early stages of repair of photosystem II in *Synechocystis* sp. PCC 6803. *Plant Cell* 2003; 15:2152–2164.

149. Brandle JR, Campbell WF, Sisson WB, Caldwell MM. Net photosynthesis, electron transport capacity, and ultrastructure of *Pisum sativum* L. exposed to ultraviolet-B radiation. *Plant Physiol.* 1977; 60:165–169.

150. Renger G, Graber P, Dohnt G, Hagemann R, Weiss W, Voss R. The effect of UV irradiation on primary processes of photosynthesis. In: Bauer H, Caldwell MM, Tevini M, Worrest RC, eds. *Biological Effects of UV-B Radiation*. München: Gesellschaft für Strahlen- und Umweltforschung mbH, 1982:110–116.

151. Okada M, Kitajima M, Butler WL. Inhibition of photosystem I and photosystem II in chloroplasts by UV radiation. *Plant Cell Physiol.* 1976; 17:35–43.

152. Huang L, McCluskey MP, Ni H, Larossa RA. Global gene expression profiles of the cyanobacterium *Synechocystis* sp. Strain PCC 6803 in response to irradiation with UV-B and white light. *J. Bacteriol.* 2002; 184:6845–6858.

153. Hope AB. The chloroplast cytochrome bf complex: a critical focus on function. *Biochim. Biophys. Acta* 1993; 1143:1–22.

154. Jordan BR, He J, Chow WS, Anderson JM. Changes in mRNA levles and polypeptide subunits of ribulose 1,5-bisphosphate carboxylase in response to supplementary ultraviolet-B radiation. *Plant Cell Environ.* 1992; 15:91–98.

155. Nogués S, Baker NR. Evaluation of the role of damage to photosystem II in the inhibition of CO_2 assimilation in pea leaves on exposure to UV-B radiation. *Plant Cell Environ.* 1995; 18:781–787.

156. Chow WS, Strid A, Anderson JM. Short-term treatment of pea plants with supplementary ultraviolet-B

radiation: recovery time-courses of some photosynthetic functions and components. In: Murata N, ed. *Research in Photosynthesis*. Vol. IV. Dordrecht, The Netherlands: Kluwer Academic Publishers, 1993:361–364.

157. Doughty JC, Hope AB. Effects of ultraviolet radiation on the plasma membranes of *Chara corallina*. III. Action spectra. *Aust. J. Plant Physiol.* 1976; 3:693–699.

158. Murphy TM, Wilson C. UV-stimulated K^+ efflux from cells: counterion and inhibitor studies. *Plant Physiol.* 1982; 70:709–713.

159. Haupt W, Scheuerlein R. Chloroplast movement. *Plant Cell Environ.* 1990; 13:595–614.

160. Mulroy TW. Spectral properties of heavily glaucous and non-glaucous leaves of a succulent rosette-plant. *Oecologia* 1979; 38:349–357.

161. Hrazdina G, Marx GA, Hoch HC. Distribution of secondary plant metabolites and their biosynthetic enzymes in pea (*Pisum sativum* L.) leaves. Anthocyans and flavonol glycosides. *Plant Physiol.* 1982; 70:745–748.

162. Schmelzer E, Jahnen W, Hahbrock K. *In situ* localization of light-induced chalcone synthase mRNA, chalcone synthase, and flavonoid end products in epidermal cells of parsley leaves. *Proc. Natl. Acad. Sci. USA* 1988; 85:2989–2993.

163. Tevini M, Iwanzik W, Teramura AH. Effects of UV-B radiation on plants during mild water stress. II. Effects on growth, protein and flavonoid content. *Z. Pflanzenphysiol.* 1983; 110:459–467.

164. Beggs CJ, Schneider-Ziebert U, Wellmann E. UV-B radiation and adaptive mechanisms in plants. In: Worrest RC, Caldwell MM, eds. *Stratospheric Ozone Reduction, Solar Ultraviolet Radiation and Plant Life*. NATO ASI Series G: Ecological Sciences, Vol. 8. Berlin: Springer-Verlag, 1986:235–250.

165. Robberecht R, Caldwell MM. Leaf optical properties of *Rumex patentia* L. and *Rumex obtusifolia* L. in regard to a protective mechanism against solar UV-B radiation injury. In: Worrest RC, Caldwell MM, eds. *Stratospheric Ozone Reduction, Solar Ultraviolet Radiation and Plant Life*. NATO ASI Series G: Ecological Sciences. Berlin: Springer-Verlag, 1986:251.

166. Tevini M, Braun J, Fieser G. The protective function of the epidermal layer of rye seedlings against ultraviolet-B radiation. *Photochem. Photobiol.* 1991; 53:329–333.

167. Strid A, Porra J. Alterations in pigment content in leaves of *Pisum sativum* after exposure to supplementary UV-B. *Plant Cell Physiol.* 1992; 33:1015–1023.

168. Asada K, Takahashi M. Production and scavenging of active oxygen in photosynthesis. In: Kyle DJ, Osmond CB, Arntzen CJ, eds. *Topics in Photosynthesis*. Vol.9. Photoinhibition. Amsterdam: Elsevier, 1987:227–288.

169. Asada K. Production and scavenging of active oxygen in chloroplasts. In: Scandalios JG, ed. *Molecular Biology of Free Radical Scavenging Systems*. New York: Cold Spring Harbor Laboratory Press, 1995:173–192.

170. Foyer CH. Ascorbic acid. In: Alscher RG, Hess JL, eds. *Antioxidants in Higher Plants*. Boca Raton, FL: CRC Press, 1993:31–58.

171. Hideg E, Mano J, Ohno C, Asada K. Increased levels of monodehydroascorbate radical in UV-B irradiated broad bean leaves. *Plant Cell Physiol.* 1997; 38:684–690.

172. Freiberg EC. *DNA Repair*. New York: W.H. Freeman, 1985.

173. Soyfer VN, Ceminis KG. Excision of thymine dimers from the DNA of UV irradiated plant seedlings. *Environ. Exp. Bot.* 1977; 17:135–143.

174. Sass L, Spetea C, Máté Z, Nagy F, Vass I. Repair of UV-B induced damage of photosystem II via *de novo* synthesis of the D1 and D2 reaction centre subunits in *Synechocystis* sp. PCC 6803. *Photosynth. Res.* 1997; 54:55–62.

175. Golden SS. Light-responsive gene expression in cyanobacteria. *J. Bacteriol.* 1995; 177:1651–1654.

176. Jansson C, Debus RJ, Osiewacz HD, Gurevitz M, McIntosh L. Construction of an obligate photoheterotrophic mutant of the cyanobacterium *Synechocystis* 6803. *Plant Physiol.* 1987; 85:1021–1025.

177. Mohamed A, Jansson C. Influence of light on accumulation of photosynthesis-specific transcripts in the cyanobacterium *Synechocystis* 6803. *Plant Mol. Biol.* 1989; 13:693–700.

178. Máté Z, Sass L, Szekeres M, Vass I, Nagy F. UV-B induced differential transcription of *psbA* genes encoding the D1 protein of photosystem II in the cyanobacterium *Synechocystis* 6803. *J. Biol. Chem.* 1998; 273:17439–17444.

179. Vass I, Kirilovsky D, Perewoska I, Máté Z, Nagy F, Etienne A-L. UV-B radiation induced exchange of the D1 reaction centre subunits produced from the *psbA2* and *psbA3* genes in the cyanobacterium *Synechocystis* sp. PCC 6803. *Eur. J. Biochem.* 2000; 267:2640–2648.

180. Campbell D, Erikson M-J, Öquist G, Gustafsson P, Clarke AK. The cyanobacterium *Synechochoccus* resists UV-B by exchanging photosystem II reaction-center D1 proteins. *Proc. Natl. Acad. Sci. USA* 1998; 95:364–369.

181. Tichy M, Lupínková L, Sicora C, Vass I, Kuvikova S, Prasil O, et al. *Synechocystis* 6803 mutants expressing distinct forms of the Photosystem II D1 protein from *Synechococcus* 7942: relationship between the *psbA* coding region and sensitivity to visible and UV-B radiation. *Biochim. Biophys. Acta* 2003; 1605:55–66.

182. Chisholm D, Williams JGK. Nucleotide sequence of *psbC*, the gene encoding the CP43 chlorophyll a-binding protein of photosystem II, in the cyanobacterium *Synechocystis* 6803. *Plant Mol. Biol.* 1988; 10:293–301.

183. Viczián A, Máté Z, Nagy F, Vass I. UV-B induced differential transcription of *psbD* genes encoding the D2 protein of photosystem II in the cyanobacterium *Synechocystis* 6803. *Photosynth. Res.* 2000; 64:257–266.

184. Powles SB. Photoinhibition of photosynthesis induced by visible light. *Annu. Rev. Plant Physiol.* 1984; 35:15–44.

185. Critchley C. The chloroplast thylakoid membrane system is a molecular conveyor belt. *Photosynth. Res.* 1988; 19:265–276.

186. Andersson B, Salter H, Virgin I, Vass I, Styring S. Photodamage to photosystem II — primary and secondary events. *J. Photochem. Photobiol.* 1992; 15B:15–31.

187. Aro E-M, Virgin I, Andersson B. Photoinhibition of photosystem II. Inactivation, protein damage and turnover. *Biochim. Biophys. Acta* 1993; 1143:113–134.

188. Critchley C, Russel AW. Photoinhibition of photosynthesis *in vivo*: the role of protein turnover in photosystem II. *Physiol. Plant.* 1994; 92:188–196.

189. Long SP, Humpries S. Photoinhibition of photosynthesis in nature. *Annu. Rev. Plant Physiol.* 1994; 45:633–662.

190. Barber J, Andersson B. Too much of a good thing: light can be bad for photosynthesis. *Trends Biochem. Sci.* 1992; 17:61–66.

191. Adir N, Zer H, Shochat S, Ohad I. Photoinhibition — a historical perspective. *Photosynth. Res.* 2003; 76:343–370.

192. Jensen MAK, Mattoo AK, Edelman M. The D1–D2 heterdimer of PSII is a major target for UV-B irradiation. In: The rapidly-metabolized herbicide binding protein of the thylakoids. Relationship to photosynthesis and crop protection. Proceedings of the SEB Symposium, Wye College, 1993.

193. Warner CW, Caldwell MM. Influence of photon flux density in the 400–700 nm waveband on inhibition of photosynthesis by UV-B (280–320 nm) irradiation in soybean leaves: separation of indirect and immediate effects. *Photochem. Photobiol.* 1983; 38:341–346.

194. Sicora C, Máté Z, Vass I. The interaction of visible and UV-B light during photodamage and repair of photosystem II. *Photosynth. Res.* 2003; 75:127–137.

44 Heavy Metal Toxicity Induced Alterations in Photosynthetic Metabolism in Plants

Shruti Mishra and *R. S. Dubey*
Department of Biochemistry, Faculty of Science, Banaras Hindu University

CONTENTS

I. INTRODUCTION

Agricultural plants are exposed to various environmental stresses from planting to marketing. Growing plants require a balanced soil environment in which all components should be present in a definite ratio. The stressful conditions of the environment such as water stress, soil salinity, heat, chilling, anaerobiosis, pathogenesis, wounding, gaseous pollutants, heavy metals, etc. drastically affect plant growth and metabolism and in turn limit crop productivity. In the present-day situation, the stress factors have multiplied in an exponential manner with the advent of modern agricultural and industrial practices.

Heavy metal contamination of agricultural land is a widely recognized problem and studies on the harmful effects caused by heavy metals on crop plants are receiving increasing attentions [1,2]. Frequently, heavy metals causing toxicity in plants are biologically nonessential. Such metals include cadmium (Cd), mercury (Hg), lead (Pb), aluminum (Al), silver (Ag), tin (Sn), arsenic (As), etc., and are important environmental pollutants. Toxic levels of some heavy metals occur naturally in some soils, however increasing human activities have modified global cycle of heavy metals, leading to widespread contamination of our environment with the toxic nonessential elements like Cd, Pb, Hg, and Al [3,4].

A fundamental factor that heightens the concern over the presence of potentially toxic heavy metals in the environment, is their nonbiodegradability and persistence in the food chain [1,5]. Heavy metals are difficult to remove from the environment and unlike many other pollutants, cannot be chemically or biologically degraded and are ultimately indestructible. Due to high affinity of heavy metals for organic matters, even low inputs lead to high levels in soils especially in humus layer [6]. In the soil they function as stress factors for growing plants and after absorption by the root system, they cause various physiological constraints inside the plant [7,8].

Heavy metals causing toxicity in plants fall into two groups — the first group includes essential metals for plants, which function as micronutrients such as Fe, Zn, Cu, and are involved in numerous physiological processes, but at high concentrations they are strongly toxic and impair plant growth. The key essential heavy metals and their toxic effects on various photosynthetic parameters are given in Table 44.1. The heavy metals of the second group include nonessential metals, which are major pollutants of the environment such as Cd, Pb, Hg, As and are very toxic even at low concentrations and for them no biological functions are known [9].

Heavy metals generally inhibit normal physiological processes. This could be due to their interference with activities of a number of enzymes essential for normal metabolic and developmental processes as well as due to their direct interactions with proteins, pigments, etc. [10,11]. The concentration causing toxicity varies with the type of ion, plant, and conditions of growth [12].

Photosynthesis, an important process for plant growth and biomass production is negatively affected due to increasing levels of heavy metals in soil environment or air emissions [2]. Heavy metals reduce photosynthesis due to their effects at various levels. The major effects include changes in chloroplast ultrastructure and pigment composition, inhibition in net photosynthetic rate, decreased carboxylation efficiency of RUBISCO, inhibition in photosystem II (PSII) activity, and electron transport.

Plants exposed to high levels of heavy metals show altered functional state of chloroplast thylakoid membranes as well as altered shape of chloroplast, size of plastoglobuli, and starch grains [13–15]. Heavy metals generally reduce chlorophyll content and decrease Chl a/b ratio and enhance chlorophyllase (Chlase) activity [14,16,17]. Chlorophyllase causes degradation of chlorophylls. Under *in vitro* conditions, chlorophyllase catalyses the removal of the phytol chain from the porphyrin head [18]. Although both Chl and Chlase are components of thylakoid membranes, due to effective compartmentation within membranes, the role of chlorophyllase in normal chlorophyll turnover appears to be limited [19,20]. Inhibition in net photosynthetic rate (P_N) has been observed in maize, soybean, tomato, pea, pigeon pea, sugar beet, barley, and maize plants grown under elevated levels of heavy metals [21–26].

RUBISCO is a bifunctional enzyme that catalyzes both carboxylation of ribulose-1,5-bisphosphate, the initial step in photosynthetic carbon reduction in C_3 plants and its oxygenation, the first reaction of the photorespiratory metabolism [27]. Inhibition of RUBISCO activity and thereby decreased carboxylation *in vivo* has been frequently observed due to heavy metals and this inhibition could be explained by either substitution for Mg^{2+} in the ternary enzyme–CO_2–metal^{2+} complex or by reaction with enzyme –SH group [28,29].

PSII is a multisubunit pigment–protein complex with the enzymatic activity of light-dependent water-oxidizing plastoquinone reductase, leading to the release of electrons, protons, and molecular oxygen [30]. Most of the heavy metals inhibit PSII activity [31]. They affect the oxygen-evolving complex (OEC) with the loss of all or part of the Mn^{2+} cluster together with some of the extrinsic polypeptides associated with the water oxidation mechanism [32]. A larger number of studies have been conducted during recent years to examine the effects of individual essential and nonessential metal ions on various processes associated with photosynthesis in plants. Figure 44.1 describes a generalized view of various parameters of

---

TABLE 44.1
Essential Metals that, in Excess, Cause Damage to Various Components of Photosynthesis

Metal	Toxic Effects
Cu	1. Disturbs architecture of thylakoid membranes and alters overall chloroplast ultrastructure
	2. Inhibits photosynthetic electron transport of both PSI and PSII
	3. Inhibits RUBP carboxylase activity
Mn	1. Inhibits chorophyll biosynthesis
	2. Decreases Chl a and b levels
	3. Reduces net photosynthetic rate (P_N)
	4. Inhibits RUBP carboxylase
Zn	1. Decreases total chlorophyll content and Chl a/b ratio
	2. Inhibits CO_2 assimilation
	3. Hampers activity of oxygen evolving compelex (OEC)
Fe	1. Impairs photosynthetic electron transport
	2. Induces oxidative stress

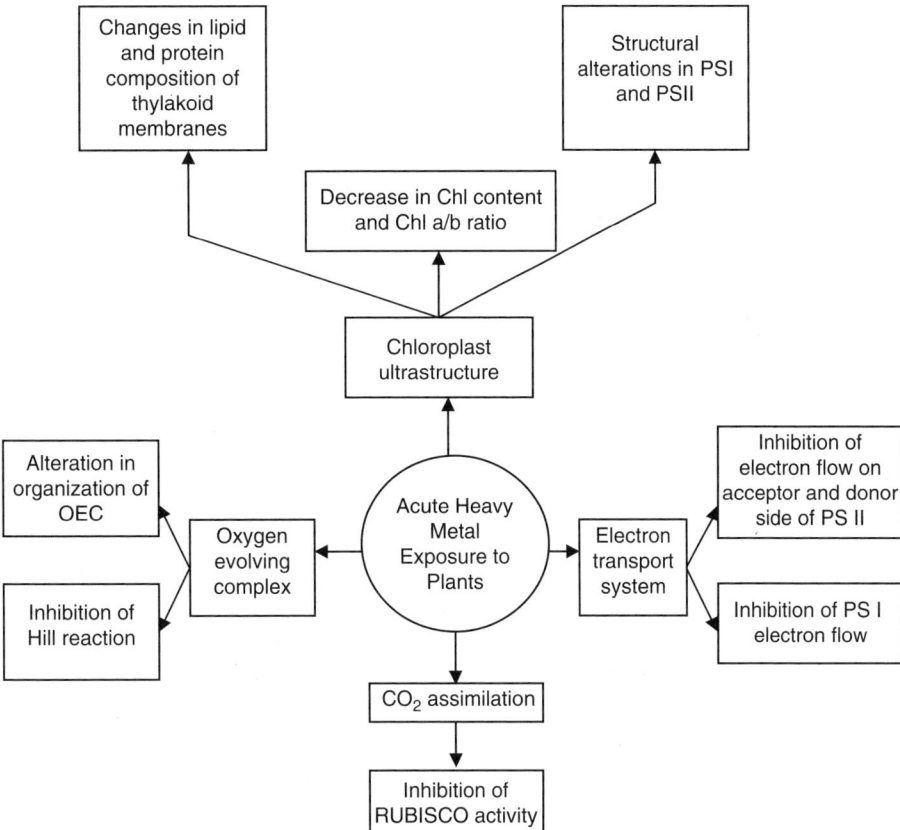

FIGURE 44.1 An overview of the effects of acute heavy metal exposure to plants on different parameters associated with photosynthesis.

photosynthesis that are affected when plants are exposed to excess levels of heavy metals. In the following sections, the effects of different metal ions on various parameters and metabolic processes associated with photosynthesis are reviewed.

II. HEAVY METALS AND PHOTOSYNTHETIC ALTERATIONS

A. COPPER

Copper is spread to natural ecosystems by agriculture, industry, and mining. Cu has long been known as an essential micronutrient for higher plants but its role in plant metabolism has been studied in detail during the recent years [33–35]. As a component of enzymes involved in several important metabolic and physiological processes, its function as a plant nutrient is based mainly on the participation of enzymatically bound copper in redox reactions [36].

Cu is a redox-active metal and is a constituent of water-soluble, blue-colored 10.5 kDa protein plastocyanin that transfers electrons between the cyto-

chrome b_f complex and P700 and serves as putative electron carrier between PSII and PSI [37]. Cu also acts as prosthetic group of chloroplastic antioxidant enzyme Cu/Zn superoxide dismutase. The role of Cu in the regulation of PSII-mediated electron transport as either a part of polypeptide involved in electron transport, or as a stabilizer of the lipid environment close to electron carriers of PSII complex has been suggested by various investigators [38,39].

The role of cupric ions in photosynthetic organisms mainly depends on its concentration within the tissues. At endogenous concentrations slightly above optimum, Cu can induce a number of deleterious effects at the physiological, biochemical, and structural levels [40].

In general, about 10% of the excess heavy metals absorbed by plants is accumulated in the leaves and only 1% enters chloroplasts [41]. In chloroplasts, excess copper may exert its direct effects by inducing structural changes in proteins and lipids and indirectly by acting as efficient generator of reactive oxygen species [42,43]. PSII contains binding sites for excess Cu both on the oxidizing and reducing side

TABLE 44.2
Essential Metals Causing Phytoxicity

Metal	Form Absorbed	Soil Conditions Causing Contamination	Minimum Foliar Level (ppm)	Maximum Foliar Level (ppm)
Cu	Cu^{2+}	Not readily lost from soil, too much can cause toxicity	5	20–100
Mn	Mn^{2+}	Acid soils may increase toxicity	30	500
Zn	Zn^{2+}	Occurs on eroded soils, least available at pH 5.5 to 7.0. Lower pH can cause more availability to the point of toxicity	20	100–200
Fe	Fe^{2+} and Fe^{3+}	Leached by water and held in lower parts of soil, low pH in soil creates toxicity	50	Rarely accumulates in excess

[44,45]. Cu is not readily lost from the soil and within the leaves of the plants, its level may reach up to 20 to 100 ppm. Table 44.2 describes the soil conditions that lead to excess levels of essential metals and foliar levels of these metals.

Macdowall [46] was the first to demonstrate the sensitivity of the photosynthetic apparatus to excess Cu and since then, the inhibitory effect of Cu on both photosystems has been confirmed in a number of publications. The toxicity symptoms of Cu are dependent on the plant growth stage at which the element is added to the nutrient solution [47]. After binding of the divalent cation Cu^{2+}, a chain of events is introduced that finally ends up with functional degradation of the photosynthetic apparatus.

1. Ultrastructural Changes in Chloroplasts

Cu commonly exerts its toxic effects on the photosynthetic apparatus by decreasing photochemical activities (mainly PSII), thereby causing damage to the structure and composition of the thylakoid membrane [48,49]. Cu disturbs the architecture of thylakoid membranes and causes changes in lipid and pigment composition [50]. In Cu-treated plants degraded intergranal thylakoid membranes, fine starch grains and numerous plastoglobuli are seen in place of normal intergranal thylakoids [50].

Valcke and Voc Poucke [51] observed swollen thylakoids and the occurence of pseudocrystalline structure in the thylakoids when plants were grown in the presence of Cu in the medium. Similar effects were seen in expanded leaves of spinach and wheat after Cu treatment [52]. Alteration in the structure and composition of the thylakoid membranes caused by Cu(II) influences the conformation and function of photosystems [53]. The thylakoid intrinsic pigment–protein complexes of higher plants are distributed within the lipid bilayer, so that any change in the

lipid composition and fluidity alter the conformation, orientation, and function of proteins involved in the photosynthetic electron flow [54]. Prolonged treatment with copper causes gradual collapse of the thylakoid structure by increasing degradation of thylakoid proteins [55].

Cu-treated plants showed a lower level of acyl lipids as structural constituents of the thylakoid membranes. In runner bean plants (*Phaseolus coccinieus* L.), Cu increased acyl lipids during the initial stages of growth. However, significant decrease in acyl lipid content with dramatic reduction in the levels of monogalactosyl diacylglycerol (MGDG) to about 55%, followed by sulfloquinovasyl diacylglycerol (SODG) to 85%, phosphatidyl glycerol (PG) to 71%, and phosphatidyl choline (PC) to 85% was observed by the end of the intensive growth period [47]. Decrease in the acyl lipid content and changes in the lipid and fatty acid composition was also noticed in the chloroplast membranes of Cu-treated spinach plants [38,56]. Regardless of the time of its application, Cu caused a relative increase in linolenic acid (18:3) and corresponding decrease in palmitic acid (16:0) in thylakoids [47]. Low levels of PG and PC in Cu treated plants impair the photosynthetic activity [47]. These lipids are shown to be responsible for grana stacking of the thylakoid membranes [57,58]. Cu-induced alteration of chloroplast ultrastructure was associated with a decrease of MGDG/DGDG ratio [50]. Degradation of the polar lipid leading to accumulation of free fatty acid (FFA) accompanied by decreased MGDG/DGDG ratio has been observed in many plants grown in presence of excess Cu [42]. MGDG is indispensable for PSII complex and its lower level disturbs the organization of PSII complex and as a consequence, decreases PSII activity [59]. Most likely Cu causes enhancement of MGDG degradation [60]. In yellowing leaves of wheat plants it was shown that the lipid matrix degradation in the thylakoid membranes was accompanied by a high activity of galactolipase, which preferentially degraded MGDG [60]. Increased fatty acid unsaturation level of membrane lipids with decreased MGDG content in PSII complex results in local modification of reaction centers, which leads to inhibition of PSII [47]. Smith and coworkers [61] suggested that Cu might interfere with the unsaturation and elongation processes of lipids, both in brown and red algae. Alterations in membrane structural components might also be due to enhanced peroxidative processes under Cu toxicity [62], which would induce disturbances in lipid–protein–pigment complexes associated with the photosystems [47,63]. It was shown by Gora and Clijsters [64] that peroxidation of lipids in primary leaves of Cu-treated *Phaseolus vulgaris* plants resulted from increased lipoxygenase activity.

A decline in the level of chlorophylls and carotenoids has been observed in plants grown under high Cu concentrations. Though, there is no certainty whether toxic copper levels affect chlorophyll synthesis via aminolevulinic acid dehydratase or via modification of chlorophyllase activity, or by stimulating chlorophyll peroxidation by inducing production of hydroxyl radicals [14,29]. It has been shown that Cu toxicity leads to formation of hydrogen peroxide, superoxide radicals and hydroxyl radicals, which could cause chlorophyll destruction in Cu-treated plants [29,62].

Carotenoids normally shield chlorophylls from peroxidation [65]. As carotenoid contents tend to decrease after Cu treatment, chlorophyll destruction can not be prevented and in turn energy is preferentially transformed to oxygen, giving rise to an oxygen singlet, which could cause further chlorophyll degradation [65]. Some authors believe that Cu inhibits the synthesis of 5-aminolevulinic acid as well as protochlorophyllide reducatase activity [66]. Excess Cu triggered Fe and Mn deficiency, which blocked synthesis of protochlorophyllide and phytoene, thus decreasing the contents of chlorophyll and carotenoids [67]. According to some investigators, Cu(II) reduced photosynthetic pigments by interfering with terpenoid biosynthesis prior to the formation of C_{20} geranyl–geranyl pyrophosphate [68,69].

2. Photosystem II

There are some experimental evidences suggesting the functional involvement of Cu in PSII-mediated electron transport. Anderson and coworkers [70] were the first to find Cu in the PSII-enriched fraction obtained from digitonin-fractionated spinach chloroplasts. Sibbald and Green [71] reported that about 75% of Cu in PSII preperations from barley and spinach was

bound to the major antenna complex of PSII (LHC II). The involvement of Cu in the water-splitting system was experimentally determined by several workers [72,73].

One of the pronounced features of PSII is its susceptibility to damage by high concentrations of toxic metal ions and by excessive light. It was shown that isolated PSII particles bind two to four copper atoms per 300 chlorophyll molecules, suggesting that most Cu^{2+} ions entering the chloroplast in a healthy plant are bound to PSII [39]. It is suggested that light is required for the expression of the toxic effect of Cu [55,74].

Extensive *in vitro* studies have shown that PSII is more susceptible to Cu toxicity than PSI [38,75]. So far most of the work has been done under *in vitro* conditions. The *in vivo* physiology of Cu toxicity is much more complex due to compartmentalization, inactivation, translocation, etc. However, the precise location of the Cu(II) binding site on PSII and the underlying mechanisms of copper inhibition are still the subject of debate. Both the acceptor and donor side have been proposed as copper-inhibitory sites.

Maksymiec and coworkers [47] attributed decrease in PSII activity to the inhibitory effect of Cu on the acceptor side of PSII, which was due to induced inhibition of the Calvin cycle and downregulation of electron transport. On the PSII reducing side, the Q_B binding site and the Pheo-Fe-Q_A domain have been reported as the most sensitive sites for Cu(II) toxicity [45,76]. Mohanty and coworkers [76] also considered the possibility that Cu may interact with the nonheme iron located in the vicinity of the Q_A and Q_B acceptors. Modification of Q_A to Q_B electron transfer was reported as a consequence of Cu treatment [74,77]. Arellano and coworkers [74] showed that this modification was due to the direct effect of Cu on the PSII donor side, however, *in vivo* experiments showed that it was due to indirect effect of Cu ions [77]. Cu diminished reoxidation of Q_A. The reduction state of Q_A is a result of the imbalance between the rate of Q_A reduction by PSII activity and the rate of Q_A reoxidation by PSI activity [78]. In addition, Cu was shown to impair the function of the oxidizing/donor side of PSII [79]. Cu inhibits photosynthetic electron transport mainly at PSII and this is at least in part, due to a decline in chlorophyll biosynthesis or to an increase in its degradation [29].

3. Electron Transport and RUBISCO Activity

Cu toxicity has a direct negative influence on the photosythetic electron transport [29]. Figure 44.2 shows different Cu-inhibitory sites associated with

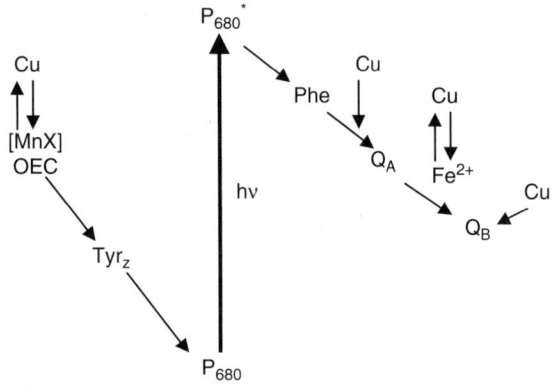

FIGURE 44.2 Cu inhibitory sites in PS II mediated electron transport.

PSII-mediated electron transport. Kinetics of P_{680} reduction in isolated PSII particles reaction centers and Tris-washed PSII particles was markedly slower in presence of Cu and it was confirmed that Cu specifically inhibited the electron donation from Tyr_Z to $P_{680}*$ on the donor side of PSII, either by a modification of this amino acid in D1 protein or by a modification of its microenvironment [80]. Using electron paramagnetic resonance (EPR) spectroscopy, it was confirmed that electron flow from tyrosine (Tyr_Z) to $P_{680}*$ is blocked at toxic Cu(II) concentrations [74]. It was shown that Q_A-Fe^{2+} EPR signal was not changed by Cu, indicating that the charge separation remained functional [80]. However, it has been observed that Cu(II) interacts not only with Tyr_Z, but also with Tyr_D on D2 protein [81].

The presence of high Cu(II) concentration can significantly modify the oxygen evolving complex of PSII, dissociating the Mn cluster and associated cofactors [82]. Fluorescence and EPR spectroscopy studies have revealed that Cu(II) ions bound with tridentate Schiff base ligand affect Mn cluster in OEC by chelation of Cu(II) ions with tryptophan and tyrosine constituents of proteins situated in photosynthetic centers [81]. High Cu(II) caused the loss of extrinsic proteins of 32, 24, and 17 kDa of the OEC of PSII and 43 as well as 47 kDa antenna proteins from the PSII core complex [55,82]. As Cu(II) has a high affinity for amine triazole or imidazole nitrogen atoms, it is suggested that Cu(II) could interact with the amino acids destabilizing 47- and 43-kDa proteins [82,83].

A possible direct interaction between Cu and Ca at the oxidizing side of PSII has also been shown both under *in vitro* and *in vivo* conditions [84,85]. Calcium appears to be indispensable for the normal functioning of the photosynthetic apparatus. *In vitro* supplementation of Cu can substitute for Ca^{2+} in OEC and

in CF_0CF_1, leading to a decrease of phosphorylation processes and PSII activity [81,86].

Inhibitory effect of Cu has also been observed on chlorophyll fluorescence parameters. In oat plants, altered fluorescence signal with increasing Cu was noted, which was due to decrease of both maximal (F_m) and variable ($F_v = F_m - F_0$) chlorophyll fluorescence of dark-adapted leaves [87]. Inhibitory effect of Cu is exhibited by strongly reduced F_v/F_m ratio [47,53]. Decrease of F_v/F_0 that was noticed in Cu-treated older leaves was possibly due to injury of the thylakoid membranes that affected photosynthetic electron transport [88].

A noncompetitive action of Cu and DCMU on Q_B binding site has been observed. DCMU blocks the reduction of Q_B and in consequence causes an increase in initial fluorescence (F_0) [89,90]. In young plants, excess Cu(II) causes stronger inhibition of the quantum efficiency of O_2 evolution than the quantum yield of electron transport [75]. Through the displacement of Ca^{2+} from its functional sites, excess Cu(II) can trigger processes of "high energy quenching," leading to acidification of the thylakoid lumen and thereby limiting photochemical processes [86].

Plants grown at elevated levels of Cu show a decline in RUBISCO activity [91]. Cu toxicity inhibits both carboxylase and oxygenase activities of RUBISCO [29]. This effect appears to be due to a metal-induced interaction with essential cysteine residues of the enzyme [92].

B. MANGANESE

Mn is one of the most abundant metals in the Earth's crust and it is also an essential micronutrient for most living organisms. It is a constitutive element of the water-splitting system that provides electrons to PSII, as well as a cofactor for different enzymes involved in redox reactions such as Mn-containing isozyme of Mn-SOD, an essential enzyme involved in protection against oxidative stress in plants [93,94].

Mn toxicity is one of the limiting factors for crop yield in acid and volcanic soils where soil conditions often lead to Mn toxicity in growing plants [95]. Solubility of Mn is strongly affected by soil pH [96]. In the process of soil acidification, increased rate of leaching of cations such as Mg and Ca enhances the solubility of the metals Mn and Al in the soil [97].

Mn toxicity involves a broad array of physiological responses. Morphological symptoms of Mn toxicity include chlorosis of leaves, brown speckles, foliar necrotic spots, etc. [36,98]. However, the decline in photosynthesis with excess leaf Mn was proposed as one of the mechanisms that constitute Mn toxicity

[99]. Various parameters of photosynthesis have been studied to explain Mn-toxicity-induced decline of photosynthetic activities under Mn toxicity conditions.

1. Chlorophyll Level

Using NMR spectroscopy, it has been observed that after uptake by plants Mn gets localized in different organelles including chloroplast and also in vacuole [100]. In the leaves of Mn-sensitive cvs of *Vigna unguiculata*, Mn gets localized as deposits of Mn oxides, whereas, in the Mn-tolerant cultivar it is uniformly distributed in an easily extractable form [101]. A decrease in chlorophyll level has been observed in plants growing under Mn toxicity conditions. In *Nicotiana tabacum* plants Mn had a direct effect on either chlorophyll synthesis or degradation which resulted in interveinal chlorosis [102].

Mn directly influences the biosynthesis of chlorophylls. Several enzymes which are involved in chlorophyll synthesis including the enzymes of isoprenoid biosynthetic pathway (which produces plant pigments) are sensitive to both Mn deficiency and toxicity [103]. A 50% decline in the level of Chl a and 35% to 55% decline in the level of Chl b was observed in the leaves of wheat plants grown under Mn toxicity conditions [104]. Similar decline in total chlorophyll was observed after exposure to excess Mn in *Phaseolus vulgaris*, *Zea mays*, and *Glycine max* plants [98]. Despite a decrease in concentration of both Chl a and b with increasing Mn in the solution, the ratio of Chl a/b increased with Mn concentration in certain plants [105].

A significant inhibition in chlorophyll biosynthesis was observed with $10 \, mM$ Mn in *Nicotiana tabacum* [106]. Inhibition in chlorophyll biosynthesis was also reported in blue green alga *Anacystis nidulans*, resulting from interrupted insertion of Mg into protoporphyrin due to Mn toxicity and thereby leading to reduced synthesis of chlorophyll [107].

Under *in vitro* conditions, enhanced degradation of chlorophyll was observed due to Mn [14]. Mn plays a role in protecting chlorophyll from photooxidation, however oxidized Mn in the leaf is believed to cause either oxidation of chlorophyll or of other chloroplast components [108]. Other workers have also proposed that Mn-induced chlorosis is not caused by inhibition of chlorophyll synthesis but due to photooxidation of chlorophyll [109].

2. Photosynthetic Rate

High concentrations of Mn inhibit photosynthesis at a variety of physiological levels [110]. In leaves of *Nicotiana tabacum*, Mn inhibited photosynthesis when data were recorded on a dry matter and per unit chlorophyll basis, without inhibiting activity of the Hill reaction, PSI and PSII [102]. With increasing Mn concentration in the nutrient solution. The net photosynthetic rate (P_N) was reduced in *Vigna umbellata*, *Phaseolus vulgaris*, *Betula ermanii*, *Alnus hirsuta*, *Ulmus davidiana* and *Acer mono*, *Triticum aestivum*, *Nicotiana tabacum*, and *Glycine max* plants [98,102, 104,105,111,112]. It is suggested that Mn-induced reductions in P_N is a direct result of reduced pigment level; however, reduced RUBISCO activity also appears to be responsible for reduction in P_N at high Mn concentration [112,114].

3. CO$_2$ Assimilation

Decreased rate of carboxylation is observed in plants exposed to high concentration of Mn in the growth medium. In tobacco plants decreased CO_2 assimilation was observed before any chlorosis and other damages were perceived in leaves due to excess Mn [102]. Carboxylation efficiency decreases concomitantly with increased level of leaf Mn [114,115]. In tobacco plant, it was shown that reduction in CO_2 assimilation during Mn toxicity was due to reduced carboxylase activity [116]. High level of Mn in tobacco leaves affects the activity rather than the amount of RUBISCO [116]. McDaniel and Toman [99], however, observed that despite a rapid accumulation of Mn in leaf tissues of tobacco, RUBISCO activity declined only after 48 h of Mn treatment. In contrast, Chatterjee and coworkers [117] reported no change in RUBISCO activity in their *in vitro* studies on wheat plants treated with excess Mn, although tissue Mn levels were considerably lower than those reported by Houtz et al. [116]. Other researchers also observed that leaf Mn accumulation reduced RUBISCO carboxylation activity and also physical presence of Mn in the leaf chloroplasts caused a reduction of RUBP regeneration capacity [114].

RUBISCO shows enhanced oxygenase activity in the presence of excess Mn in leaves [118]. Different hypotheses have been proposed for the mechanism of Mn-induced decline in RUBISCO activity. One possible hypothesis is that under conditions of Mn toxicity, Mn replaces Mg from the active center of RUBISCO, i.e., replacement of RUBISCO–Mg^{2+} with RUBISCO–Mn^{2+} occurs and this results in higher ratio of oxygenase to carboxylase activity [119].

The decline in photosynthesis with excess leaf Mn is also attributed to peroxidative impairment of thylakoid membrane function [120]. In wheat chloroplasts, Mn induced lipid peroxidation, which, in turn, inhibited electron transport and decreased the activities of photosynthetic enzymes due to polyphenol oxidation products [121,122]. Increased activity of polyphenol oxidase is regarded as the most sensitive indicator of Mn toxicity preceding chlorophyll loss and the occurrence of visible symptoms [102].

4. Fluorescence Parameters

The photochemical and nonphotochemical processes that bring about the relaxation of the excited chlorophyll molecules to ground state are measured as coefficients of photochemical (q_P) and nonphotochemical (q_N) quenching of variable fluorescence, respectively. Plants of *Vigna umbellata* treated with higher Mn in the medium showed a significant reduction in q_P with a concomitant increase in q_N values [111]. The decreasing trend of q_P with increasing leaf Mn concentration suggests that Mn in leaves causes an increase in the reduction state of PSII primary electron acceptor, Q_A, indicating a decrease in the fraction of open PSII [123]. Photoinhibition is closely associated with the decrease in q_P [124]. The decrease in q_P observed in the leaves that had accumulated Mn is also suggestive of possible photoinhibition in excess of leaf Mn [114]. The potential maximum efficiency of PSII photochemistry as represented by F_v/F_m was little affected by Mn accumulation in white birch leaves, *Vigna umbellata* and *Betula platyphylla* [111,114,125].

C. ZINC

Zinc is a major industrial pollutant of the terrestrial and aquatic environment [126]. It is an essential micronutrient involved in numerous physiological processes and has wider roles in plants, but at high concentration, it becomes strongly toxic and impairs plant growth and metabolism. Zn in an essential component of the enzymes oxidoreductases, transferases, hydrolases, lyases, isomerases, and ligases [127]. Since, Zn(II) does not undergo reduction under any conditions compatible with life, its role as metalloenzyme is inherently different from that of other metals like Cu and Fe, which are capable of redox reactions [126]. Due to similarities of ion radius of bivalant cations (Mn, Fe, Cu, Mg, and Zn), excess Zn can shift certain physiological equilibria by local competition at various sites [128].

1. Chlorophyll Content

Decreased total chlorophyll content and decline in Chl *a*/*b* ratio were observed when *Chlorella* and *Euglena gracilis* were grown in presence of Zn [129,130].

Reduction in the level of Chl, particularly Chl *b*, was observed in *Oryza sativa* grown under Zn toxicity [16]. The reduction of chlorophyll under Zn toxicity appears to be due to the sensitivity of enzymes of chlorophyll biosynthesis towards heavy metal ions [131]. Stimulation in chlorophyll degradation due to enhanced activity of chlorophyllase was observed even when Zn was supplied in millimolar concentrations [132].

2. Hill Reaction and RUBISCO Activity

Zn affects water oxidizing complex due to the local competition between Zn^{2+} and Mn^{2+} on the water splitting of PSII and substitution of Mn^{2+} by Zn^{2+} [133]. In membrane preparations from *Anacystis nidulans*, inhibition in Hill activity and oxygen evolution was observed due to Zn [134]. *In vitro* experiments related to Zn(II) toxicity showed dissociation of the OEC proteins and displacement of the native cofactors Ca^{2+}, Cl^-, Mn^{2+} due to Zn^{2+} [135]. In submembrane fractions treated with Zn, the extrinsic polypeptides with molecular weights 16 and 24 kDa associated with OEC of PSII get dissociated [32]. In rice plants a significant inhibition in Hill reaction activity was noticed under Zn toxicity [16].

A decline in photochemical activities associated with PSII observed under Zn toxicity is related to an alteration of the inner structure and composition of the thylakoid membranes [136]. A direct correlation exists between Zn in the leaves and capacity of PSII to capture and use light energy; however, the relationship is not linear [137]. Zn affects the quantum yield (ϕ) of electron flow through PSII; however, in the presence of endophytic *Neotyphodium lolii* in *Lolium perenne*, increased values of F_v/F_m and ϕ PSII were observed.

Zn inhibits CO_2 assimilation at relatively low concentrations. In *Phaseolus vulgaris* plants, Zn inhibits RUBISCO activity and a decrease in net photosynthetic rate is observed which is linked with the increase in Zn concentration in the leaves [138]. As bivalent cations are involved in both formation and catalytic function of the ternary RUBISCO–CO_2–metal^{2+} complex, Zn excess significantly diminishes RUBP carboxylase capacity by substitution of Zn^{2+} for Mg^{2+} [133].

D. IRON

Excess accumulation of Fe in plant tissues is a rare phenomenon. However, increase in leaf Fe content many cause severe cellular damage. Soil features that create Fe toxicity, apart from low pH, are low cation exchange capacity, low base status, low levels of K, PO_4^{3-}, Zn, and a lower supply of easily reducible Mn. Fe toxicity is often associated with a deficiency of Zn and Mn. It is often associated with a marked imbalance of nutrients or due to the presence of H_2S.

Elevated levels of Fe in leaf lead to an increased uptake of Fe in chloroplasts and thus, a dramatic impairment of total photosynthetic electron transport capacity. Fe uptake in dicot roots requires a reduction step and the subsequent translocation of Fe^{2+} across the cytoplasmic membrane via a presumably unknown transport protein [139].

Bronze spots on leaves are generally associated with iron toxicity [140]. In some structural studies, it was observed that chloroplasts of healthy tissues surrounded by the necrotic zone were most sensitive to metal excess [141]. In studies with *Nicotiana plumbaginifolia*, it was found that Fe excess decreased photosynthetic rate by 40% and there was increased reduction of PSII and higher thylakoid energization [139]. Iron, due to its participation in oxidation–reduction reactions within the cells, is believed to generate oxidative stress in plants when taken in excess, thereby leading to increased activity of antioxidative enzymes [142]. Fe toxicity may also cause stimulation of photorespiration [139].

E. CADMIUM

Cadmium is a nonessential potentially toxic element and is an important environmental contaminant [143]. The presence of Cd in the environment has increased with time in some areas to levels which threaten the health of aquatic and terrestrial organisms because its addition becomes greater than its removal through leakage and plant harvesting. The toxic levels of Cd are caused by natural soil characteristics or by agriculture, manufacturing, mining, and other waste disposal practices or by use of metal containing pesticides and fertilizers in agricultural soils [143]. Table 44.3 describes common nonessential heavy metal pollutants of the environment and their sources.

The higher concentration of Cd in soil environment results in enhanced Cd uptake by plant roots. Cd is compartmentalized into chloroplasts in a process that may involve the transport of free Cd and the participation of thiol-peptides [144]. The most common effect of Cd toxicity in plants is stunted growth, leaf chlorosis accompanying retardation of plastid development, and degradation of ultrastructure of chloroplasts [145,146]. Various sites of Cd-inhibitory effects on chlorophyll (Chl) content and biosynthesis have been suggested [17]. Cd exposure was shown to result in a reduction in chlorophyll content with de-

TABLE 44.3
Nonessential Metal Pollutants and their Sources

Metal	Sources
Cd	Metal working industries, mining, as a by product of mineral fertilizers, coal-fired power plants
Pb	Mining and smelting activities, Pb-containing paints, gasoline, explosives, disposal of municipal sewage and sludges enriched with Pb
Ni	Combustion of coal and oil, incineration of waste and sewage sludge, mining and electroplating, cement manufacturing, coinage etc.
Hg	Dental amalgams industrial applications, pharmaceuticals and medicines including vaccinations and laxatives, fabric softener, inks, antiseptic creams, and lotions, etc.
Al	Al-related industries, in acid soils availability of Al increases

crease in Chl *a/b* ratio [147]. Cd altered the aggregation state of phycobilisomes in blue green alga *Anacystis nidulans* [148]. An overview of the effects of nonessential heavy metal pollutants on various photosynthetic parameters is presented in Table 44.4.

Cd decreases chlorophyll formation by interacting with –SH group of enzymes δ-ALA dehydratase and porphobilinogen deaminase, leading to the accumulation of intermediates of chlorophyll synthesis like ALA and prophyrins [149]. Reports also suggest that Cd inhibits chlorophyll biosynthesis by reacting with protochlorophyllide reductase, which causes photoreduction of protochlorophillide into chlorophyllide [150]. Cd causes transformation of the long wavelength protochlorophyllide form into short wavelength ones and in this way inhibits the formation of chlorophyllide [17,150]. Fluorescence spectroscopy analysis at 77 K, which was used to study Cd-induced changes in molecular organization of protochlorophyllides in the etioplast inner membrane, revealed that irradiance of Cd-treated wheat leaves and membranes resulted in the appearance of a small amount of cholorophyllide with a characteristic band at 678 nm and appearance of a high-intensity band at 633 nm, suggesting that considerable amount of protochlorophyllide was in the inactive form [17]. The tetrapyrrole biosynthetic pathway in plants is common for chlorophyll (in chloroplast) and heme (in mitochondria). ALA synthesis is the rate-limiting and regulatory step in both organelles [151]. Cd inhibited ALA synthesis at the site of availability of glutamate for ALA synthesis [152]. Cd also induced iron deficiency [153], which later caused inhibition of chlorophyll biosynthesis and several other reactions

TABLE 44.4
Nonessential Heavy Metal Pollutants and Their Effects on Photosynthesis

Pollutants	Effects
Cd	1. Reduces both chlorophyll content and Chl *a*/*b* ratio
	2. Inhibits chlorophyll formation
	3. Decreases RUBISCO activity
	4. Inhibits both PSI and PSII
	5. Enhances lipoxygenase activity
Pb	1. Reduces both chlorophyll content and Chl *a*/*b* ratio
	2. Changes lipid composition of thylakoid membranes
	3. Influences both PSI and PSII
	4. Inhibits RUBP carboxylase
Ni	1. Reduces chlorophyll concentration
	2. Affects both PSI and PSII
	3. Causes complete inactivation of electron transport system
Al	1. Lowers chlorophyll content
	2. Decreases net photosynthetic rate (P_N)
Hg	1. Modifies chloroplast protein
	2. Affects both PSI and PSII
	3. Alters organization of OEC

associated with photosynthesis [153,154]. Interaction of Cd with functional –SH groups of enzymes has been proposed as the mechanism of inhibition of several physiological reactions due to this heavy metal [62]. By interacting with –SH groups of sulfhydryl requiring enzymes such as ALA synthase, ALA dehydratase, PBG deaminase, and protochlorophyllide reductase, Cd interferes with heme biosynthesis and chlorophyll formation [155].

Elevated Cd levels in the nutrient solution decreased RUBISCO activity in *Erythrina variegata* seedlings [156]. Cd appears to form mercaptides with thiol groups of RUBISCO, thereby decreasing its activity [92]. Cd inhibits PSI and PSII activity [157]. PSII is more sensitive to Cd than PSI and it is the primary site of action of Cd in photosynthetic electron flow in isolated spinach chloroplasts and *Nostoc linckia* [157,158]. Cd acts on the donor side of PSII [156,159]. Muthuchelian et al. [156] in their study on seedlings of *Erythrina variegata* found that higher Cd levels in nutrient solution decreased $^{14}CO_2$ fixation. In agreement with this Husaini and Rai [157] reported Cd-induced inhibition of carbon fixation in *Nostoc linckia* and suggested that such inhibition was due to decrease in ATP content by Cd.

Since ATP and NADPH are the primary requirements for CO_2-fixation, it was shown that diminished

PSII activity, at both low and high Cd concentrations whereas diminished PSI activity at high Cd concentration were responsible for a decrease in ATP and reductant pool [157]. Cd accumulation in *Euglena* chloroplasts led to inhibition of photophosphorylation [144]. Cd ions decrease the proton source for various reduction reactions and also inhibit the enzymes of several different metabolic processes where NADPH is used as H donor [17,160].

Cd inhibits photosynthesis but stimulates respiration [161]. It induces TCA cycle activities and also activities of other pathways of carbohydrate utilization. This is related to increased demand for ATP production by oxidative phosphorylation to compensate for deficits in photophosphorylation [162]. High galactolipase activity with diminished level of the thylakoid membrane lipid content has been observed under Cd toxicity [163]. Cd enhances activity of lipoxygenase which in turn might damage chloroplast membrane and cellular constituents such as proteins, DNA, and chlorophylls [164,165]. Lipoxygenase causes breakdown of biological membranes in plants [166]. It mediates oxidation of polyunsaturated fatty acids and produces free radicals which in turn destroy chloroplast membrane and this has been proposed as a general mechanism for inhibition of photosynthesis under Cd toxicity [167].

Free radicals can also be produced in chloroplasts due to blockage of electron flow in PSI by Cd. This leads to the formation of excited chlorophylls and generation of free radicals, which, in turn, initiate peroxidation reactions. Either excited chlorophyll or oxygen species derived from superoxide anion radical can initiate peroxidation reactions [143,167]. Enhanced peroxidation activity contributes significantly to the decreased level of chlorophyll and decreased photosynthetic rate observed under Cd treatment [164].

F. LEAD

Lead, a nonredox active metal, is a nonessential element for plants and animals and is considered as one of the hazardous heavy metal pollutant of the environment [168]. Pb-contaminated soils adversely affect various plant processes and lead to sharp decrease in crop productivity. Pb affects photosynthetic apparatus in multiple ways. An alteration in the photosynthetic pigment composition, reduction in total chlorophyll content, and a decrease in Chl *a*/*b* ratio have been observed in plants growing in the presence of Pb [87,169]. Pb is reported to disturb the granal structure of the chloroplasts [170]. Retardation in Chl content may be due to Pb-induced inhibition of δ-ALA dehydratase, an enzyme catalyzing the conversion of δ-ALA into porphobilinogen in the synthesis of chlorophyll [171]. Pb is known to inhibit enzyme activities due to its interaction with –SH groups or due to Pb-induced deficiency of elements essential to enzymes, e.g., Zn [172]. An enhancement in *in vitro* degradation of chlorophyll was observed by Pb^{2+} [14].

Pb toxicity causes changes in lipid composition especially in monogalactosyl diacyl glycerol (MGDG), which is concerned with membrane permeability in chloroplasts. Pb ions stimulate dehydrogenation of fatty acids. Incubation of chloroplasts with Pb ions results in the decrease of saturated fatty acids while an increase in unsaturated fatty acid linolenic acid ($C_{18:3}$) is observed [173].

Pb is considered to influence both PSI and II although PSII is more sensitive [174]. In detached pea leaves, a 2-h exposure to lead reduced PSII efficiency by about 10% [175]. Under *in vitro* conditions when assayed with isolated photosynthetic membranes, Pb produces a decline of variable chlorophyll fluorescence, indicating an inhibition on the donor side of PSII [176]. Such inhibition was partly restored by using specific electron donors such as hydroxylamine and $MnCl_2$. Donor side inhibition of PSII and possible recurrence of cyclic electron transport around PSII under Pb-toxicity conditions have also been observed [43]. It was shown by Rashid and Popovic [176] that Pb competes for binding near the calcium and chloride binding sites in the water-oxidizing complex and that Ca^{2+} and Cl^-, which are essential cofactors for oxygen evolution could protect against Pb-induced inhibition. Further experiments have confirmed that in lead-treated PSII submembrane fractions there was loss of the extrinisic polypeptides of 17 and 24 kDa [177].

Other researchers believe that Pb is less effective in damaging the photosynthetic apparatus [87,178]. It is agreed that Pb is not very well translocated in plants and its deleterious effects on photosynthesis are seen only after prolonged exposure [32].

Pb does not seem to destroy photosynthetic apparatus but results in decreased coordination among the components associated with light reaction [87]. Pb inhibits photosynthesis by inhibiting the carboxylase activity of RUBISCO. Irreversible binding of Pb with the enzyme RUBISCO dissociates it into its subunits and thus activity of the enzyme is lost [92].

Pb contamination in the soil decreases the quantum yield of photosynthesis in plants as observed by F_v/F_m ratio. Initial chlorophyll fluorescesnce (F_0) showed little decrease in oat plants grown in Pb-contaminated site, suggesting a decrease of energy transfer from the light-harvesting chl *a/b* protein complex (LHC) to PSII [87,179]. The increased half-rise time ($t_{1/2}$) from the initial (F_0) to maximal (F_m) chlorophyll fluorescence observed under Pb toxicity suggests that the amount of active pigments associated with photochemical apparatus decreased and that the functional chlorophyll antennae size of photosynthetic apparatus was smaller compared to the control grown plants [87].

G. NICKEL

Nickel is discharged in the water or soil environment with waste disposals, municipal and industrial sewage in the form of mobile organic chelates [180,181]. Ni is strongly phytotoxic at high concentrations and has a destructive effect on plant growth and physiology although it is considered as an essential micronutrient for plants [182,183]. It is readily absorbed by plant roots and then translocated to various plant parts and gets accumulated in the vacuoles, cell walls, and epidermal trichomes [184,185]. In plants, Ni is complexed with organic compounds such as amino acids and organic acids [186,187]. Ni inhibits photosynthesis and damages the photosynthetic apparatus on almost every level of its organization. Ni(II)–Glu and Ni(II) citrate treatment caused reduction in leaf chlorophyll content in cabbage plants [188]. Ni reduced the chlorophyll concentration in the leaves of plants that were grown in the presence of its inorganic forms [189,190]. Reduced chlorophyll content due to Ni is possibly attributed to inhibition of chlorophyll biosynthesis or by induction of its degradation catalyzed by increased chlorophyllase activity [14,189]. Ni affects electron transport and may cause its complete inactivation [191]. Ni-treated cabbage plants showed a significant reduction of grana structures. In such plants, Ni was shown to be localized in the chloroplasts [192]. Inside the chloroplasts, ^{63}Ni was largely associated with the lamellar fraction and to a lesser extent with the stroma and envelope membrane [193]. On treatment with organic Ni complexes, electron density of chloroplast stroma and the number of grana were reduced and appearance of thylakoid also got changed in cabbage plants [188]. The degree of advancement of these changes increased with the exposure level. Control plants had typical lens-shaped normal chloroplast but Ni-treated plants had elliptical/oblong shaped chloroplast with reduced volume, very little or no starch and with few small plastoglobuli [188].

A reduction in photosynthesis and alteration in the activities of many enzymes associated with photosynthetic carbon reduction cycle are observed in pigeon pea (*Cajanus cajan* L.) plants due to Ni [24]. Ni affects both photosystems PSI and PSII. The inhibitory site for Ni appears to be on the donor side of

PSII, as Ni caused inhibition of the reduction of PSII artificial electron acceptor [194]. Certain workers have suggested that Ni-induced phytotoxicity in plants is mediated by peroxidation of membrane lipids due to the induction of free radical reactions by Ni [190,195].

H. ALUMINUM

Indiscriminate use of acid forming nitrogenous fertilizers lead to acidity in the soil. Al toxicity is considered to be the most common cause of limited plant growth and production in acid soils [196,197]. Al shows a number of adverse effects on physiological and biochemical processes in plants [198]. Al affects photosynthesis by lowering the chlorophyll content and reducing electron flow [199]. Al-stress-induced loss in chlorophyll has been reported in many plant species like lemna, sorghum, wheat, and tobacco [200,201]. Decline in Chl a/b ratio was observed in *Oryza sativa* grown in the presence of excess Al [202]. Al taken up by plants accumulates mostly in roots, which are regarded as primary target sites of Al toxicity and the retardation in shoot growth or decrease in chlorophyll content appears to be only a secondary event of Al toxicity [203,204]. Al toxicity caused a decline in photosynthetic rate in *Oryza sativa* and *Sorghum bicolor* and led to ATP depletion [200,202,205]. At low concentration Al has been shown to stimulate PSII-mediated oxygen evolution in cyanobacteria and in isolated chloroplasts [206].

I. MERCURY

Mercury is an important environmental contaminant and is highly toxic to photosynthetic organisms. Both photosystems are affected due to Hg [32]. Mercury binds to –SH groups present in proteins [207]. Nahar and Tajamir-Riahi [208] observed a strong interaction between PSII submembrane fractions and mercury due to the formation of metal protein binding through peptide SH, C = O and C–N groups. Mercury has been shown to react directly with plastocyanine, replacing copper [209]. Electron paramagnetic resonance studies also indicate that the reaction centre of PSI is oxidized by mercury in the dark [210]. Using simultaneous fluorescence and photoacoustic measurement studies with isolated thylakoid membranes it has been observed that PSII is also affected on both donor and acceptor sides by mercury [43]. On the acceptor side, the inhibition was proposed between quinone and acceptors Q_A and Q_B [211]. On the donor side, using PSII submembrane fractions, it was shown that the inhibition could be reversed by chloride ions that act as a cofactor for the OEC and that mercury selectively removed the 33-kDa extrinsic polypeptide associated with OEC [212]. Mercury is also known to form organometallic complexes with amino acids present in chloroplast protein [210].

III. CONCLUSIONS

Proliferation of industrial activities and metallurgical operations release huge quantities of heavy metals into the environment. These heavy metals are readily absorbed from the soil by the growing plants. Among the essential micronutrients, which exert strong phytotoxicity at high concentrations include Cu, Mn, and Zn, whereas common nonessential heavy metals that are major pollutants of the environment are Cd, Pb, Ni, Al, and Hg. The heavy metals interfere with photosynthesis due to their effects at various levels. The central metal atom of chlorophyll, Mg, can be substituted with Cu, Zn, Cd, Pb, Hg, or Ni. This substitution prevents light harvesting by the affected chlorophyll molecules. Most of these metals enhance the activity of chlorophyll degrading enzyme chlorophyllase. When these metal cations reach the photosynthetic apparatus they cause ultrastructural changes in thylakoid membranes, affect the activities of PSI and PSII and inhibit the carboxylation activity of RUBP carboxylase. Most of these metals inhibit photosynthetic electron transport due to their direct inhibitory action at the level of PSII. Oxygen-evolving complex is clearly affected with the loss of Mn cluster and some extrinsic polypeptides associated with the water oxidation mechanism. Some of these metals interact strongly with the functional –SH groups present on enzymes of chlorophyll biosynthesis and thylakoid membrane proteins. A common response to heavy metal toxicity in plants involves generation of reactive oxygen species, which, in turn, leads to peroxidation of lipids of thylakoid membrane. As a result, certain specific isoforms of antoxidant enzymes, more specially of ascorbate peroxidase appear in chloroplasts. Induced synthesis of the antioxidant enzymes ascorbate peroxidase, superoxide dismutase under metal toxicity seemingly serves as protective mechanism for chloroplasts and its internal constituents against metal toxicity induced oxidative damage. However, more investigations are needed to unveil the exact sites and mode of actions of the different heavy metals on individual components of the photosynthetic process, the extent of damage caused on the photosynthetic parameters by the heavy metals and the possible components of the plant system which would confer metal stress tolerance.

REFERENCES

1. Salt DE, Rauser WE. Mg-ATP dependent transport of phytochelatins across the tonoplast of oat roots. *Plant Physiol.* 1995; 107:1293–1301.
2. Masarovicova E, Cicak A, Stefanick I. Plant responses to air pollution and heavy metal stresses. In: Pessarakli M, ed. *Handbook of Plant and Crop Stress.* 2d ed. New York: Marcel Dekker, 1999:569–598.
3. Alia, Saradhi PP. Proline accumulation under heavy metal stress. *J. Plant Physiol.* 1991; 138:554–558.
4. Rout GR, Samantaray S, Das P. The role of nickel on somatic embryogenesis in *Setaria italica L. Italic NOT ALLOWEDin vitro.* Euphytica *1998; 101:319–324.*
5. Nellesson H, Fletcher JS. Assessment of published literature on the uptake, accumulation, and translocation of heavy metals by vascular plants. *Chemosphere* 1993; 9:1669–1680.
6. Woolhouse HW. Toxicity and tolerance in the response of plants to metals. In : Lange OL, Nobel PS, Osmond CB, Ziegler H, eds. *Encyclopedia of Plant Physiology.* Vol. 12C: Physiological Plant Ecology. Berlin: Springer Verlag, 1983:245–300.
7. Van Steveninck RFM, Van Steveninck ME, Wells AJ, Fernando DR. Zinc tolerance and the binding of zinc as zinc phytate in *Lemna minor.* X-ray microanalytical evidence. *J. Plant Physiol.* 1990; 137:140–146.
8. Ouzounidou G. Changes in variable chlorophyll fluorescence as a result of Cu-treatment: Dose–response relations in *Silene* and *Thlaspi. Photosynthetica* 1993; 29:445–462.
9. Marschner H. *Mineral Nutrition in Higher Plants.* London: Academic Press/Harcourt B & Company Publishers, 1986.
10. Stoyanova DP, Tschakalova ES. The effect of lead and copper in the photosynthetic apparatus in *Elodea canadensis* Rich. *Photosynthetica* 1993; 28:63–74.
11. Bertrand M, Guary JC. How plants adopt their physiology to an excess of metals. In: Pessarakli M, ed. *Handbook of Plant and Crop Physiology.* 2d ed. New York: Marcel Dekker, 2002:751–761.
12. Hirschi KD, Korenkov VD, Wilganowski NL, Wagner GJ. Expression of *Arabidopsis CAX2* in tobacco. Altered metal accumulation and increased manganese tolerance. *Plant Physiol.* 2000; 124:125–133.
13. Krause GH, Weis E. Chlorophyll fluorescence and photosynthesis: the basics. *Annu. Rev. Plant Physiol. Plant Mol. Biol.* 1991; 42:313–349.
14. Abdel-Basset R, Issa AA, Adam MS. Chlorophyllase activity: effects of heavy metals and calcium. *Photosynthetica* 1995; 31:421–425.
15. Molas J. Changes of chloroplast ultrastructure and total chlorophyll concentration in cabbage leaves caused by excess of organic Ni(II) complexes. *Environ. Exp. Bot.* 2002; 47:115–126.
16. Ajay, Rathore VS. Effect of Zn^{2+} stress in rice (*Oryza sativa* cv. Manhar) on growth and photosynthetic processes. *Photosynthetica* 1995; 31(4):571–584.
17. Böddi B, Oravecz AR, Lehoczki E. Effect of cadmium on organization and photoreduction of protochlorophyllide in dark-grown leaves and etioplast inner membrane preparations of wheat. *Photosynthetica* 1995; 31:411–420.
18. Majumdar S, Ghosh S, Glick BR, Dumbroff EB. Activities of chlorophyllase, phophoenolpyruvate carboxylase and ribulose-1,5-biophosphate carboxylase in the primary leaves of soyabean during senescence and drought. *Physiol. Plant.* 1991; 81:473–480.
19. Hirschfeld KR, Goldschmidt EE. Chlorophyllase activity in chlorophyll-free citrus chromoplasts. *Plant Cell. Rep.* 1983; 2:117–118.
20. Drazkiewicz M. Chlorophyllase: Occurrence, functions, mechanism of action, effects of external and internal factors. *Photosynthetica* 1994; 30:321–331.
21. Bazzaz FA, Rohlfe GL, Carlson RW. Effect of Cd on photosynthesis and transpiration of excised leaves of corn and sunflower. *Physiol. Plant.* 1974; 32:373–377.
22. Baszynski T, Wajda L, Król M, Wolinska D, Krupa Z, Tukendorf A. Photosynthetic activities of cadmium-treated tomato plants. *Physiol. Plant.* 1980; 48:365–370.
23. Angelov M, Tsonev T, Uzunova A, Gaidardjieva K. Cu^{2+} effect upon photosynthesis, chloroplast structure, RNA and protein synthesis of pea plants. *Photosynthetica* 1993; 28:341–350
24. Sheoran IS, Singal HR, Singh R. Effect of cadmium and nickel on photosynthesis and the enzymes of the photosynthetic carbon reduction cycle in pigeonpea (*Cajanus cajan* L.). *Photosynth. Res.* 1990; 23:345–351.
25. Greger M, Ögren E. Direct and indirect effects of Cd^{2+} on photosynthesis in sugar beet (*Beta vulgaris*). *Physiol. Plant.* 1991; 83:129–135.
26. Stiborova M, Ditrichova M, Březinova A. Effect of heavy metal ions on growth and biochemical characteristics of photosynthesis of barley and maize seedlings. *Biol. Plant.* 1987; 29:453–467.
27. Andrews TJ, Lorimer GH. Rubisco: Structure, mechanisms, and prospects for improvement. In : Sttumpf PK, Conn EE, eds. *The Biochemistry of Plants.* Vol 10. San Diego, CA: Academic Press, 1987:131–218.
28. Konečna B, Frič F, Masarovičova E. Ribulose-1,5-biphosphate carboxylase activity and protein content in pollution damaged leaves of three oak species. *Photosynthetica* 1989; 23:566–574.
29. Lidon FC, Henriques FS. Limiting step on photosynthesis of rice plants treated with varying copper levels. *J. Plant Physiol.* 1991; 138:115–118.
30. Katoh H, Ikeuchi M. Targeted disruption of *psbX* and biochemical characterisation of photosystem II complex in the thermophilic cyanobacterium *Synechococcus elongatus. Plant Cell Physiol.* 2001; 42(2):179–188.
31. Purohit S, Singh VP. Uniconazole (S-3307) induced protection of *Abelmoshus esculentus* L. against cadmium stress. *Photosynthetica* 1999; 36:597–599.
32. Carpentier R. The negative action of toxic divalent cations on the photosynthetic apparatus. In : Pessarakali M, ed. *Handbook of Plant and Crop Physiology.* 2d ed. New York: Marcel Dekker, 2002:763–772.
33. Sommer A. Copper as an essential for plant growth. *Plant Physiol.* 1931; 6:339–345.

34. Lipman C, Mackiney E. Proof of the essential nature of copper for higher plants. *Plant Physiol.* 1931; 6:593–599.

35. Arnon D, Stout P. The essentiality for certain elements in minute quantity for plants with special reference to copper. *Plant Physiol.* 1939; 9:371–375.

36. Marschner H. *Mineral Nutrition of Higher Plants.* London: Academic Press, 1995, ISBN 0-12-473543-6.

37. Kaim W, Schwederski B. *Bioorganic Chemistry: Inorganic Elements in the Chemistry of Life. An Introduction and Guide.* Chichester, UK: John Willey & Sons, 1995.

38. Baron M, Arellano JB, Gorgé JL. Copper and photosystem II: a controversial relationship. *Physiol. Plant.* 1995; 94:174–180.

39. Maksymiec W. Effect of copper on cellular processes in higher plants. *Photosynthetica* 1997; 34:321–342.

40. Luna CM, Gonzalez C, Trippi VS. Oxidative damage caused by an excess of copper in oat leaves. *Plant Cell Physiol.* 1994; 35:11–15.

41. Ernst WHO, Verkleij JAC, Schat H. Metal tolerance in plants. *Acta Bot. Neerl.* 1992; 41:229–248.

42. Quartacci MF, Pinzino C, Sgherri CLM, Dalla Vecchia F, Navari-Izzo F. Growth in excess copper induces changes in the lipid composition and fluidity of PSII-enriched membranes in wheat. *Physiol. Plant.* 2000; 108:87–93.

43. Boucher N, Carpentier R. Hg^{2+}, Cu^{2+}, and Pb^{2+}-induced changes in photosystem II photochemical yield and energy storage in isolated thylakoid membranes: a study using simultaneous fluorescence and photoacoustic measurements. *Photosynth. Res.* 1999; 59:167–174.

44. Samuelsson G, Öquist G. Effects of copper chloride on photosynthetic electron transport and chlorophyll protein complexes of S*pinacea oleracea. Plant Cell Physiol.* 1980; 21:445–454.

45. Yruela I, Gatzen G, Picorel R, Holzwarth AR. Cu(II)-inhibitory effect on photosystem II from higher plants. A picosecond time-resolved fluorescence study. *Biochemistry* 1996; 35:9469–9474.

46. Macdowall FD. The effect of some inhibitors of photosynthesis upon the chemical reduction of a dye by isolated chloroplasts. *Plant Physiol.* 1949; 24:464–480.

47. Maksymiec W, Russa R, Urbanik-Sypniewska T, Baszynski T. Effect of excess Cu on the photosynthetic apparatus of runner bean leaves at two different growth stages. *Physiol. Plant.* 1994; 91:715–721.

48. Clijsters H, Van Assche F, Gora L. Physiological responses of higher plants to soil contamination with metals. In : Rozema J, Verkleij JAC, eds. *Ecological Responses to Environmental Stresses.* Dordrecht: Kluwer Academic Publishers, 1991, ISBN 0-7923-0762-3.

49. Ouzounidou G. The use of photoacoustic spectroscopy in assessing leaf photosynthesis under copper stress: correlation of energy storage to photosystem II fluorescence parameters and redox change of P700. *Plant Sci.* 1996; 113:229–237.

50. Maksymiec W, Bednara J, Baszynski R. Responses of runner bean plants to excess copper as a function of plant growth stages: effects on morphology and structure of primary leaves and their chloroplast ultrastructure. *Photosynthetica* 1995; 31(3):427–435.

51. Valcke R, Van Poucke M. The effect of water stress on greening of primary barley (*Hordeum Vulgare* L. cvs. Menuet) leaves. In: Marcelle R, Clijsters H, Van Poucke M, eds. *Effects of Stress on Photosynthesis.* The Hague: Martinus Mijhoff/Dr. W. Junk Publishers, 1983, ISBN 90-247-279945.

52. Eleftheriou EP, Karataglis S. Ultrastructural and morphological characteristics of cultivated wheat growing on copper-polluted fields. *Bot. Acta* 1989; 102:134–140.

53. Lidon FC, Ramalho JC, Henriques FS, Copper inhibition of rice photosynthesis. *J. Plant Physiol.* 1993; 142:12–17.

54. Quartacci MF, Pinzino C, Sgherri CLM, Navari-Izzo F. Lipid composition and protein dynamics in thylakoids of two wheat cultivars differently sensitive to drought. *Plant Physiol.* 1995; 108:191–197.

55. Pätsikkä E, Aro EM, Tyystjärvi E. Mechanism of copper-enhanced photoinhibition of thylakoid membranes. *Physiol. Plant.* 2001; 113:142–150.

56. Maksymiec W, Russa R, Urbanik-Sypniewska T, Baszynski T. Changes in acyl lipid and fatty acid composition in thylakoids of copper non-tolerant spinach exposed to excess copper. *J. Plant Physiol.* 1992; 140:52–55.

57. Siegenthaler PA, Rawyler A. Acyl lipids in thylakoid membranes: distribution and involvement in photosynthetic functions. In: Staehelin LA, Arntzen C, eds. *Encyclopedia of Plant Physiology*, New Series. Vol. 19. Photosynthesis III. Photosynthetic Membranes and Light Harvesting Systems. Berlin: Springer-Verlag, 1986: 693–703, ISBN 3-540-16140-6.

58. Krupa Z. Acyl lipids in the supramolecular chlorophyll–protein complexes of photosynthesis-isolation artifacts or integral components regulating their structure and functions? *Acta Soc. Bot. Pol.* 1988; 57:401–418.

59. Murata N, Higashi S-I, Fujimura Y. Glycerolipids in various preparations of photosystem II from spinach chloroplasts. *Biochim. Biophys. Acta* 1990; 1019:261–268.

60. O'Sullivan JN, Dalling MJ. The effect of a thylakoid associated galactolipase on the morphology and photochemical activity of isolated wheat leaf chloroplasts. *J. Plant Physiol.* 1989; 134:504–509.

61. Smith KL, Bryan GW, Harwood JL. Changes in endogenous fatty acids and lipid synthesis associated with copper pollution in *Fucus* spp. *J. Exp. Bot.* 1985; 36:663–669.

62. Sandmann G, Bogar P. Copper-mediated lipid peroxidation processes in photosynthetic membranes. *Plant Physiol.* 1980; 66:797–800.

63. Lidon FC, Henriques S. Changes in the thylakoid membrane polypeptide patterns triggered by excess Cu in rice. *Photosynthetica* 1993; 28:109–117.

64. Gora L, Clijsters H. Effect of copper and zinc on the ethylene metabolism in *Phaseolus vulgaris* L. In: Clijsters H, De Proft M, Marcelle R, Van Poucke M, eds. *Biochemical and Physiological Aspects of Ethylene Production in Lower and Higher plants.* Dordrecht: Kluwer Academic Publishers, 1989:219–228, ISBN 0-7923-0201-x.

65. Mishra AN, Biswal UC. Changes in chlorophylls and carotenoids during aging of attached and detached leaves and of isolated chloroplasts of wheat seedlings. *Photosyhthetica* 1981; 15:75–79.

66. van Assche F, Clijsters H. Effects of metals on enzyme activity in plants. *Plant Cell Environ.* 1990; 13:195–206.

67. Lidon cf., Henriques SF. Effects of excess copper on the photosynthetic pigments in rice plants. *Bot. Bull. Acad. Sin.* 1992; 33:141–149.

68. Henriques F. Effects of copper deficiency on the photosynthetic apparatus of sugar beet (*Beta vulgaris* L.). *J. Plant Physiol.* 1989; 135:453–458.

69. Droppa M, Horvath G. The role of copper in photosynthesis. *Crit. Plant Sci.* 1990; 9:111–123.

70. Anderson JM, Boardman NK, David DJ. Trace metal composition of fractions obtained by digitonin fragmentation of spinach chloroplasts. *Biochem. Biophys. Res. Commun.* 1964; 17:685–690.

71. Sibbald PR, Green BR. Copper in photosystem II: association with LHC II. *Photosynth. Res.* 1987; 14:201–209.

72. Holdsworth ES, Arshad JH. A Mn–Cu–pigment complex isolated from the PSII of *phaeodactylum tricornutum. Arch. Biochem. Biophys.* 1977; 183:361–377.

73. Ono T, Nakatani HY, Johnson E, Arntzen CJ, Inove Y. Comparative biochemical properties of oxygen evolving photosystem II particles and of chloroplasts isolated from intermittently flashed wheat leaves. In: Sybesma C, ed. *Advances in Photosynthesis Research.* Vol. 1. The Hague: Dr. W. Jumk Publishers, 1984:383–386, ISBN 90-247-2946-7.

74. Arellano JB, Lázaro JJ, López Gorge J, Barón M. The donor side of photosystem II as the copper-inhibitory binding site. *Photosynth. Res.* 1995; 45:127–134.

75. Ouzounidou G, Moustakas M, Strasser RJ. Sites of action of copper in the photosynthetic apparatus of maize leaves: kinetic analysis of chlorophyll fluorescence, oxygen evolution, absorption changes and thermal dissipation as monitored by photoacoustic signals. *Aust. J. Palnt Physiol.* 1997; 24:81–90.

76. Mohanty N, Vass I, Demeter S. Copper toxicity affects photosystem II electron transport at the secondary quinone acceptor (Q_B). *Plant Physiol.* 1989; 90:175–179.

77. Ciscato M, Valcke R, Van Loven K, Clijsters H, Navari-Izzo F. Effects of *in vivo* copper treatment on the photosynthetic apparatus of two *Triticum durum* cultivars with different stress sensitivity. *Physiol. Plant.* 1997; 100:901–908.

78. Maksymiec W, Baszynski T. The role of Ca ions in changes induced by excess Cu^{2+} in bean plants. Growth parameters. *Acta Physiol. Plant.* 1998; 20:411–417.

79. Samson G, Morissette JC, Popovic R. Copper quenching of variable fluorescence in *Dunaliella tertiolecta*. New evidence for a copper inhibition effect on PS II photochemistry. *Photochem. Photobiol.* 1988; 48:329–332.

80. Schröder WP, Arellano JB, Bittner T, Eckert H-J, Barón M, Renger G. Flash induced absorption spectroscopy studies of copper interaction with photosystem II in higher plants. *J. Biol. Chem.* 1995; 269:32865–32870.

81. Seršen F, Kráľová K, Bumbálová A, Švajlenová O. The effect of Cu(II) ions bound with tridentate Schiff base ligands upon the photosynthetic apparatus. *J. Plant Physiol.* 1997; 151:299–305.

82. Yruela I, Alfonso M, Barón M, Picorel R. Copper effect on the protein composition of photosystem II. *Physiol. Plant.* 2000; 110 :551–557.

83. Vacha F, Joseph DM, Durrent JR, Tefler A, Klug DR, Porter G, Barber J. Photochemistry and spectroscopy of a five-chlorophyll reaction center of photosystem II isolated by using a Cu affinity column. *Proc. Natl. Acad. Sci. USA* 1995; 92:2929–2933.

84. Sabat SC. Copper ion inhibition of electron transport activity in sodium chloride washed photosystem II particle is partially prevented by calcium ion. *Z. Naturforsch.* 1996; 51c:179–184.

85. Maksymiec W, Baszynski T. The role of Ca^{2+} ions in modulating changes induced in bean plants by an excess of Cu^{2+} ions. Chlorophyll fluorescence measurements. *Physiol. Plant.* 1999; 105:562–568.

86. Krieger A, Weis E. The role of calcium in the pH-dependent control of photosystem II. *Photosynth. Res.* 1993; 37:117–130.

87. Moustakas M, Lanaras T, Symeonidis L, Karataglis S. Growth and some photosynthetic characteristics of field grown *Avena sativa* under copper and lead stress. *Photosynthetica* 1994; 30(3):389–396.

88. Moustakas M, Ouzounidou G, Lannoye R. Rapid screening for aluminium tolerance in cereals by use of the chlorophyll fluorescence test. *Plant Breed.* 1993; III:343–346.

89. Trebst A. The three-dimensional structure of the herbicide binding niche on the reaction center polypeptides of photosystem II. *Z. Naturforsch.* 1987; 42c:742–750.

90. Öquist G, Wass R. A portable microprocessor operated instrument for measuring chlorophyll fluorescence kinetics in stress physiology. *Physiol. Plant.* 1988; 73:211–217.

91. Schäfer C, Simper H, Hofmann B. Glucose feeding results in coordinated changes of chlorophyll content, ribulose-1,5-biphosphate carboxylase-oxygenase activity and photosynthetic potential in photoautrophic suspension cultured cells of *Chenopodium rubrum. Plant Cell Environ.* 1992; 15:343–350.

92. Siborova M. Cd^{2+} ions affect the quaternary structure of ribulose-1,5-bisphosphate carboxylase from barley leaves. *Biochem. Physiol. Pflanzen* 1988; 183:371–378.

93. Burnell JN. The biochemistry of manganese in plants. In: Graham RD, Hannam J, Uren NC, eds.

Manganese in Soils and Plants. Dordrecht, The Netherlands: Kluwer Academic Publishers, 1988:125–133.

94. Bowler C, VanCamp W, VanMontagu M, Inze D. Superoxide dismutase in plants. *Crit. Rev. Plant Sci.* 1994; 13:199–218.

95. Carver BF, Ownby JD. Acid soil tolerance in wheat. *Adv. Agron.* 1995; 54:117–173.

96. Gilkes RJ, McKenzie RM. Geochemistry and mineralogy of manganese in soils. In: Graham RD, Hennam RJ, Uren NC, eds. *Manganese in Soils and Plants.* Dordrecht, The Netherlands: Kluwer Academic Publishers, 1988:23–25.

97. Fernandez IJ. Effects of acidic precipitation on soil productivity. In: Adriano DC, Johnson AH, eds. *Acidic Precipitation.* Vol. 2. Biological and Ecological Effects. New York: Springer-Verlag, 1989:61–83, ISBN 0-387-97000-2.

98. González A, Lynch P. Effects of manganese toxicity on leaf CO_2 assimilation of contrasting common bean genotypes. *Phsyiol. Plant.* 1997; 101:872–880.

99. McDaniel KL, Toman FR. Short term effects of manganese toxicity on ribulose-1,5-bisphosphate carboxylase in tobacco chloroplasts. *J. Plant Nutr.* 1994; 17:523–536.

100. McCain DC, Markley JL. More manganese accumulates in maple sun leaves than in shade leaves. *Plant Physiol.* 1989; 90:1417–1421.

101. Horst WJ. Factors responsible for genotypic manganese tolerance in cowpea (*Vigna unguiculata*) *Plant Soil* 1983; 72:213–218.

102. Nable RO, Houtz RL, Cheniae GM. Early inhibition of photosynthesis during development of Mn toxicity in tobacco. *Plant Physiol.* 1988; 86:1136–1142.

103. Wilkinson RE, Ohki K. Influence of manganese deficiency and toxicity on isoprenoid synthesis. *Plant Physiol.* 1988; 87:841–846.

104. Ohki K. Manganese deficiency and toxicity effects on photosynthesis, chlorophyll and transpiration in wheat. *Crop Sci.* 1985; 25:187–191.

105. Macfie SM, Taylor GJ. The effects of excess manganese on photosynthetic rate and concentration of chlorophyll in *Triticum aestivum* grown in solution culture. *Physiol. Plant.* 1992; 85:467–475.

106. Clairmont KB, Hagar WG, Davis EA. Manganese toxicity to chlorophyll synthesis in tobacco callus. *Plant Physiol.* 1986; 80:291–293.

107. Csatorday K, Gombos Z, Szalontai B. Mn^{2+} and Co^{2+} toxicity in chlorophyll biosynthesis. *Proc. Natl. Acad. Sci. USA* 1984; 81:476–478.

108. Horiguchi T. Mechanisms of manganese toxicity and tolerance of plants. VII. Effect of light intensity on manganese-induced chlorosis. *J. Plant Nutr.* 1988; 11:235–246.

109. Gerresten FC. Manganese in relation to photosynthesis. III. Uptake of oxygen by illuminated crude chloroplasts suspensions. *Plant Soil* 1950; 2:323–342.

110. Clijsters H, Van Assche F. Inhibition of photosynthesis by heavy metals. *Photosynth. Res.* 1985; 7:31–40.

111. Subrahmanyam D, Rathore VS. Influence of manganese toxicity on photosynthesis in ricebean (*Vigna umbellata*) seedlings. *Photosynthetica* 2000; 38(3):449–453.

112. Kitao M, Lei TT, Koike T. Comparison of photosynthetic responses to manganese toxicity of deciduous broad leaved trees in northern Japan. *Environ. Pollut.* 1997; 97:113–118.

113. Ohki K. Manganese critical levels for soyabean growth and physiological processes. *J. Plant Nutr.* 1981; 3:271–284.

114. Kitao M, Thomas T, Lei TT, Koike T. Effects of manganese toxicity on photosynthesis of white birch (*Betula platyphylla* var. *japonica*) seedlings. *Physiol. Plant.* 1997; 101:249–256.

115. Ohki K. Manganese deficiency and toxicity effects on growth, development and nutrient composition in wheat. *Agron. J.* 1984; 76:213–218.

116. Houtz RL, Nable RO, Cheniae GM. Evidence for effects on the *in vivo* activity of ribulose-bisphosphate carboxylase/oxygenase during development of Mn toxicity in tobacco. *Plant Physiol.* 1988; 86:1143–1149.

117. Chatterjee C, Nautiyal N, Agarwala SC. Influence of changes in manganese and magnesium supply on some aspects of wheat physiology. *Soil Sci. Plant Nutr.* 1994; 40:191–197.

118. Jordan DB, Ogren WL. Species variation in kinetic properties of Rubisco. *Arch. Biochem. Biophys.* 1983; 227:425–433.

119. Jordan DB, Ogren WL. A sensitive assay procedure for simultaneous determination of ribulose-1,5-bisphosphate carboxylase and oxygenase activities. *Plant Physiol.* 1981; 67:237–245.

120. Panda S, Mishra AK, Biswal UC. Manganese induced peroxidation of thylakoid lipids and changes in chlorophyll-*a* fluorescence during aging of cell free chloroplasts in light. *Phytochemistry* 1987; 26:3217–3219.

121. Panda S, Raval MK, Biswal UC. Manganese-induced modification of membrane lipid peroxidation during aging of isolated wheat chloroplasts. *Photobiochem. Photobiophys.* 1986; 13:53–61.

122. Vaughn KC, Duke SO. Function of polyphenol oxidase in higher plants. *Physiol. Plant.* 1984; 60:106–112.

123. Krause GH, Weis E. Chlorophyll fluorescence and photosynthesis: the basics. *Annu. Rev. Plant Physiol. Plant Mol. Biol.* 1991; 42:313–349.

124. Gray GR, Boese SR, Huner NPA. A comparison of low temperature growth vs low temperature shifts to induce resistance to photoinhibition in spinach (*Spinacea oleracea*). *Physiol. Plant.* 1994; 90:560–566.

125. Butler WL. Energy distribution in the photochemical apparatus of photosynthesis. *Annu. Rev. Plant Physiol.* 1978; 29:345–378.

126. Chaney RL. Zn toxicity. In: Robson AD, ed. *Zn is Soils and Plants. Developments in Plants and Soil Sciences.* Vol. 55. Dordrecht, The Netherlands: Kluwer Academic Publishers, 1993:45–57.

127. Barak P, Helmke PA. The chemistry of Zn. In: Robson AD, ed. *Zn in Soils and Plants. Developments in Plants and Soils Sciences.* Vol. 55. Dordrecht, The Netherlands: Kluwer Academic Publishers, 1993:1–13.

128. De Fillipis LF, Ziegler H. Effect of sublethal concentration of zinc, cadmium and mercury on the photosynthetic carbon reduction cycle of *Euglena*. *J. Plant Physiol.* 1993; 142:167–172.

129. De Fillipis LF, Pallaghy CK. The effect of sub-lethal concentrations of mercury and zinc on *Chlorella*. II. Photosynthesis and pigment composition. *Z. Pflanzenphysiol.* 1976; 78:314–322.

130. De Fillipis LF, Hampp R, Ziegler H. The effects of sublethal concentrations of zinc, cadmium and mercury on *Euglena*. Growth and pigments. *Z. Pflanzenphysiol.* 1981; 101:37–47.

131. Sharma SD, Chopra RN. Effect of lead nitrate and lead acetate on growth of the moss *Semibarbula orientalis* (Web.) Wijk. et Mary — growth *in vitro*. *J. Plant Physiol.* 1987; 129:242–249.

132. Nag P, Paul AK, Mukherji S. Heavy metal effects in plant tissues involving chlorophyll, chlorophyllase, Hill reaction activity and gel electrophoretic patterns of soluble proteins. *Indian J. Exp. Biol.* 1981; 19:702–706.

133. Van Assche F, Clijsters H. Inhibition of photosynthesis in *Phaseolus vulagris* by treatment with toxic concentration of zinc:effect on ribulose-1,5-bisphosphate carboxylase/oxygenase. *J. Plant Physiol.* 1986; 125:355–360.

134. Singh DP, Singh SP. Action of heavy metals on Hill activity and O_2 evolution in *Anacystis nidulans*. *Plant Physiol.* 1987; 83:12–14.

135. Rashid A, Camm EL, Ekramoddoullah KM. Molecular mechanism of action of Pb^{2+} and Zn^{2+} on water oxidising complex of photosystem II. *FEBS Lett.* 1994; 350:296–298.

136. Baszynski T, Tukendorf A, Ruszkowska M, Skorzynska E, Maksymiec W. Characteristic of the photosynthetic apparatus of copper non-tolerant spinach exposed to excess copper. *J. Plant Physiol.* 1988; 132:708–713.

137. Monnet F, Vaillant N, Hitmi A, Coudret A, Sallanon H. Endophytic *Neotyphodium lolii* induced tolerance to Zn stress on *Lolium perenne*. *Physiol. Plant.* 2001; 113:557–563.

138. Monnet F, Vaillant N, Vernay P, Coudret A, Sallanon H, Hitmi A. Relationship between PSII activity, CO_2 fixation, and Zn, Mn and Mg contents of *Lolium perenne* under zinc stress. *J. Plant Physiol.* 2001; 158:1137–1144.

139. Kampfenkel K, Montagu MV, Inze D. Effects of iron excess in *Nicotiana plumbaginifolia* plants. *Plant Physiol.* 1995; 107:725–735.

140. Poonamperuma FN, Bradfield R, Peech M. Physiological disease of rice attributable to iron toxicity. *Nature* 1955; 175:265.

141. Morghan JT, Freeman TJ. Influence of Fe EDDHA on growth and manganese accumulation influx. *Soil Sci. Soc. Am. J.* 1978; 42:455–460.

142. Halliwell B, Gutteridge JMC. Oxygen toxicity, oxygen radicals, transition metals and disease. *Biochem. J.* 1984; 219:1–14.

143. Shah K, Kumar RG, Verma S, Dubey RS. Effect of cadmium on lipid peroxidation, superoxide anion generation and activities of antioxidant enzymes in growing rice seedlings. *Plant Sci.* 2001; 161:1135–1144.

144. Mendoza-Cozatl D, Devars S, Loza-Tavera H, Moreno-Sanchez R. Cadmium accumulation in the chloroplast of *Euglena gracilis*. *Physiol. Plant.* 2002; 115:276–283.

145. Ghoshroy S, Nadakavukaren MJ. Influence of cadmium on the ultrastructure of developing chloroplasts in soyabean and corn. *Environ. Exp. Bot.* 1990; 30(2):187–192.

146. Stoyanova DP, Tchakalova ES. Cadmium induced ultrastructural changes in chloroplast of the leaves and stems parenchyma in Myriophyllum spicatum L. *Photosynthetica* 1997; 34(2):241–248.

147. Ouzounidou G, Moustakas M, Eleftherou EP. Physiological and ultrastructural effects of cadmium on wheat (*Triticum aestivum* L.) leaves. *Arch. Environ. Contam. Toxicol.* 1997; 32:154–160.

148. Gupta A, Singhal GS. Effects of heavy metals on phycobili proteins of *Anacystis nidulans*. *Photosynthetica* 1996; 32(4):545–548.

149. Padmaja K, Prasad DDK, Prasad ARK. Inhibition of chlorophyll synthesis in *Phaseolus vulgaris* L. seedlings by cadmium acetate. *Photosynthetica* 1990; 24:399–405.

150. Stobart AK, Griffiths WT, Bukhari IA, Sherwood RP. The effect of Cd^{2+} on the biosynthesis of chlorophyll in leaves of barley. *Physiol. Plant.* 1985; 63:293–298.

151. Porra RJ, Meisch H. The biosynthesis of chlorophyll. *Trends Biochem. Sci.* 1984; 9:99–104.

152. Parekh D, Puranik RM, Srivastava HS. Inhibition of chlorophyll biosynthesis by cadmium in greening maize leaf segments. *Biochem. Physiol. Pflanzen* 1990; 186:239–242.

153. Marschner H. General introduction to the mineral nutrition of plants. In: Läuchli A, Beileski RL, eds. *Inorganic Plant Nutrition*. Berlin: Springer Verlag, 1983:5–60.

154. Siedlecka A, Baszynski T. Inhibition of electron flow around photosystem I in chloroplasts of Cd treated maize plants is due to Cd-induced iron deficiency. *Physiol. Plant.* 1993; 87:199–202.

155. Prasad DDK, Prasad ARK. Effect of lead and mercury on chlorophyll synthesis in mung bean seedlings. *Phytochemistry* 1987; 26: 881–883.

156. Muthuchelian K, Bertamini M, Nedunchezhian N. Triacontanol can protect *Erythrina variegata* from cadmium toxicity. *J. Plant Physiol.* 2001; 158:1487–1490.

157. Husaini Y, Rai LC. Studies on nitrogen and phosphorus metabolism and the photosynthetic electron transport system of *Nostoc linckia* under cadmium stress. *J. Plant Phsyiol.* 1991; 138:429–435.

158. Yang DH, Xu CH, Zhao FH, Dai YL. The effect of cadmium on photosystem II in spinach chloroplasts. *Acta Bot. Sin.* 1989; 31:702–707.

159. Voigt J, Nagel K. The donor side of photosystem II is impaired in a Cd^{2+}-tolerant mutant strain of the unicellular green alga *Chlamydomonas reinhardtii*. *J. Plant Physiol.* 2002; 159: 941–950.

160. Ferretti M, Ghisi R, Merlo L, Dalla Vecchia F, Passera C. Effect of cadmium on photosynthesis and enzymes of photosynthetic sulfate and nitrate assimilation pathways in maize (*Zea mays* L.). *Photosynthetica* 1993; 29:49–54.

161. Mendelssohn IA, McKee KL, Kong T. A comparison of physiological indicators of sub-lethal Cd stress in wetland plants. *Environ. Exp. Bot.* 2001; 46:263–275.

162. Ernst WHO. Biochemical aspects of cadmium in plants. In: Nriagu JO, ed. *Cadmium in the Environment.* New York: John Wiley & Sons, 1980:639–653.

163. Skórzyñska E, Urbanik-Sypniewska T, Russa R, Baszynski T. Galactolipase activity in Cd-treated Runner bean plants. *J. Plant Physiol.* 1991; 138:454–459.

164. Somashekaraiah BV, Padmaja K, Prasad ARK. Phytotoxicity of cadmium ions on germinating seedlings of mung bean (*Phaseolus vulgaris*): Involvement of lipid peroxides in chlorophyll degradation. *Physiol. Plant.* 1992; 85:85–89.

165. Summerfield FW, Tappel AI. Effects of dietary polyunsaturated fats and vitamin E on aging and peroxidative damage to DNA. *Arch. Biochem. Biophys.* 1984; 282:408–416.

166. Grossmann S, Leshem Y. Lowering of endogenous lipoxygenase activity in *Pisum sativum* foliage by cytokinin as related to senescence. *Physiol. Plant.* 1978; 43:359–362.

167. Kato M, Simizu S. Chlorophyll metabolism in higher plants. VI. Involvement of peroxidase in chlorophyll degradation. *Plant Cell Physiol.* 1985; 26:1291–1301.

168. Salt DE, Smith RD, Raskin I. Photoremediation. *Annu. Rev. Plant Physiol. Plant Mol. Biol.* 1998; 49:643–668.

169. Geebelen W, Vangronsveld J, Adriano DC, Van Poucke LC, Clijsters H. Effects of Pb-EDTA and EDTA on oxidative stress reactions and mineral uptake in *Phaseolus vulgaris*. *Physiol. Plant.* 2002; 115:377–384.

170. Rebechini HM, Hanzley L. Lead induced ultrastructural changes in the chloroplasts of the hydrophyte *Ceratophyllum demersum*. *Z. Pflanzenphysiol.* 1974; 73:377–386.

171. Prassad DDK, Prassad ARK. Altered δ-aminolaevulinic acid metabolism by lead and mercury in germinating seedlings of Bajra (*Pennisetum typhoideum*). *J. Plant Physiol.* 1987; 127:241–249.

172. Seregin IV, Ivanov VB. Physiological aspects of cadmium and lead toxic effects on higher plants. *Russ. J. Plant Physiol.* 2001; 48:523–544.

173. Stefanov K, Seizova K, Popova I, Petkovv, Georgi K, Popov S. Effect of lead ions on the phospholipid composition in leaves of *Zea mays* and *Phaseolus vulgaris*. *J. Plant Physiol.* 1995; 147:243–246.

174. Becerril JM, Munoz-Rueda A, Aparicio-Tejo P, Gonzalez-Murua C. The effects of cadmium and lead on photosynthetic electron transport in clover and lucerne. *Plant Physiol. Biochem.* 1988; 26:357–363.

175. Parys E, Romanovska E, Siedlecka M, Poskuta JW. The effect of lead on photosynthesis and respiration in detached leaves and in mesophyll protoplasts of *Pisum sativum*. *Acta Physiol. Plant.* 1998; 20:313–322.

176. Rashid A, Popovic R. Protective role of $CaCl_2$ against Pb^{2+} inhibition in photosystem II. *FEBS Lett.* 1990; 271:181–184.

177. Rashid A, Camin EL, Ekramoddoulah AKM. Molecular mechanism of action of Pb^{2+} and Zn^{2+} on water oxidising complex of photosystem II. *FEBS Lett.* 1994; 350:296–298.

178. Ahmed A, Tajmir-Riahi HA. Interaction of toxic metal ions Cd^{2+}, Hg^{2+}, and Pb^{2+} with light harvesting proteins of chloroplast thylakoid membranes. An FTTR spectroscopic study. *J. Inorg. Biochem.* 1993; 50:235–243.

179. Bilger W, Schreiber U. Energy-dependent quenching of dark level chlorophyll fluorescence in intact leaves. *Photsynth. Res.* 1986; 10:303–308.

180. Barcan V, Kovnatsky E. Soil surface geochemical anomaly around the copper–nickel metallurgical smelter. *Water Air Soil Pollut.* 1998; 103:197–218.

181. Karam NS, Ereifej KI, Shibli RA, AbuKundais H, Alkofahi A, Malkawi Y. Metal concentrations, growth and yield of potato produced from *in vitro* plantlets or microtubers and grown in municipal solid-waste-amended substrates. *J. Plant Nutr.* 1998; 21:725–739.

182. Ankel-Fuchu D, Thauer RK. Nickel in biology: Nickel as an essential trace element. In: Lancaster JR, ed. *The Bioinorganic Chemistry of Nickel.* Weinheim, Germany: VCH, 1988:93–110.

183. Madhav Rao KV, Sresty TVS. Antioxidative parameters in the seedlings of pigeonpea (*Cajanus cajan* L. Millspaugh) in response to Zn and Ni stresses. *Plant Sci.* 2000; 157:113–128.

184. Krämer U, Pickering IJ, Prince RC, Raskin I, Salt DE. Subcellular localization and speciation of nickel in hyperaccumulator and non-hyperaccumulator *Thlaspi* species. *Plant Physiol.* 2000; 122:1343–1353.

185. Küpper H, Lombi E, Zhao F-J, Weishammer G, McGrath SP. Cellular compartmentation of nickel in the hyperaccumulators *Alyssum lesbiacum*, *Alyssum bertolonii* and *Thlaspi goesingense*. *J. Exp. Bot.* 2001; 52:2291–2300.

186. Krämer U, Cotter-Howells JD, Charnock JM, Baker AJM, Smith JA. Free histidine as a metal chelator in plants that accumulate nickel. *Nature* 1996; 379:635–638.

187. Tartar E, Mihucz VG, Varga A, Zaray G, Cseh E. Effect of lead, nickel and vanadium contamination on organic acid transport in xylem sap of cucumber. *Inorg. Biochem.* 1999; 75:219–223.

188. Molas J. Changes of chloroplast ultrastructure and total chlorophyll concentration in cabbage leaves caused by excess of organic Ni(II) complexes. *Environ. Exp. Bot.* 2002; 47:115–126.

189. Krupa Z, Siedlecka A, Maksymiec W, Baszynski T. *In vivo* response of photosynthetic apparatus of *Phaseolus vulgaris* L. to nickel toxicity. *J. Plant Physiol.* 1993; 142:664–668.

190. Molas J. Changes in morphological and anatomical structure of cabbage (*Brassica oleracia* L.) outer leaves and in ultrastructure of their chloroplasts caused by an

in vitro excess of nickel. *Photosynthetica* 1997; 34:513–522.

191. Tripathy BC, Bhatia B, Mohanty P. inactivation of chloroplast photosynthetic electron transport activity by Ni^{2+}. *Biochim. Biophys. Acta* 1981; 638:217–224.

192. L' Huillier L, d' Auzac J, Durand M, Michaud-Ferriere N. Nickel effects on two maize (*Zea mays*) cultivars: growth, structure, Ni concentration, and localization. *Can. J. Bot.* 1996; 74:1547–1554.

193. Veeranjaneyulu K, Das VSR. Intrachloroplast localisation of ^{65}Zn and ^{63}Ni in a Zn-tolerant plant: *Ocimum basilicum* Benth. *J. Exp. Bot.* 1982; 137:1161–1165.

194. Singh DP, Khare P, Bisen PS. Effect of Ni^{2+}, Hg^{2+} and Cu^{2+} on growth, oxygen, evolution and photosynthetic electron transport in *Cylindrospermum* IU942. *J. Plant Physiol.* 1989; 134:406–412.

195. Boominathan R, Doran PM. Ni-induced oxidative stress in roots of the Ni hyperaccumulators, *Alyssum bertolonii*. *New Phytol.* 2002; 156:205–215.

196. Foy CD. Plant adaptation to acid, aluminium-toxic soils. Commun. *Soil Sci. Plant Anal.* 1988; 19:959–987.

197. Ritchie GS. Role of dissolution and precipitation of minerals in controlling soluble aluminum in acidic soils. *Adv. Agron.* 1994; 53:47–83.

198. Delhaize E, Craig S, Beaton CD, Bennet RJ, Jagdish VC, Randall PJ, Aluminum tolerance in wheat (*Triticum aestivum* L.) I. Uptake and distribution of aluminum in root apices. *Plant Physiol.* 1993; 103:685–693.

199. Barnabás B, Kovács G, Hegedüs A, Erdei, S, Horváth G. Regeneration of doubled haploid plants from *in vitro* selected microspores to improve aluminum tolerance in wheat. *J. Plant Physiol.* 2000; 156:217–222.

200. Ohki K. Photosynthesis, chlorophyll and transpiration responses in aluminum stressed wheat and sorghum. *Crop Sci.* 1986; 26:572–575.

201. Severi A. Aluminum toxicity in *Lemna minor* L. Effects of citrate and kinetin. *Environ. Exp. Bot.* 1997;37:53–61.

202. Sarkunan V, Biddappa CC, Nayak SK. Physiology of aluminum toxicity in rice. *Curr. Sci.* 1984; 53:822–824.

203. Kochian LV, Cellular mechanisms of aluminum toxicity and resistance in plants. *Annu. Rev. Plant Mol. Biol.* 1995; 46:237–260.

204. Moustakas M, Ouzounidou G, Lannoze R. Aluminum effects on photosynthesis and elemental uptake in an aluminum tolerant and non-tolerant wheat cultivar. *J. Plant Nutr.* 1995; 18:669–683.

205. Yamamoto Y, Kobayashi Y, Rama-Devi, Rikiishi S, Matsumoto H. Al toxicity is associated with mitochondrial dysfunction and the production of reactive oxygen species in plant cells. *Plant Physiol.* 2002; 128:63–72.

206. Wavare RA, Mohanty P. Aluminum stimulation of photoelectron transport in spheroplasts of cyano bacterium *Synechococcus cedrorum*. *Photobiochem. Photobiophys.* 1982; 3:327–335.

207. Bernier M, Popovic R, Carpentier R. Mercury inhibition at the donor side of photosystem II is reversed by chloride. *FEBS Lett.* 1993; 321:19–23.

208. Nahar S, Tajmir-Riahi HA. Complexation of heavy metal cations Hg, Cd, and Pb with proteins of PSII: evidence for metal-sulfur binding and protein conformational transition by FTIR spectroscopy. *J. Cell Interface Sci.* 1996; 178:648–656.

209. Kimimura M, Kotoh S. Studies on electron transport associated with photosystem II functional site of plastocyanin: inhibitory effects of $HgCl_2$ on electron transport and plastocyanin in chloroplasts. *Biochim. Biophys. Acta* 1972; 283:279–292.

210. Sersen F, Kral'ova K, Bumbalova A. Action of mercury on the photosynthetic apparatus of spinach chloroplasts. *Photosynthetica* 1998; 35:551–559.

211. Prokowski Z. Effects of $HgCl_2$ on long-lived delayed luminescence in *Scendesmus quadricauda*. *Photosynthetica* 1993; 28:563–566.

212. Bernier M, Carpentier R. The action of mercury on the binding of the extrinsic polypeptides associated with the water oxidising complex of photosystem II. *FEBS Lett.* 1995; 360:251–254.

45 Effects of Heavy Metals on Chlorophyll–Protein Complexes in Higher Plants: Causes and Consequences

Éva Sárvári

Department of Plant Physiology, Eötvös Loránd University

CONTENTS

I. INTRODUCTION

Photosynthesis is a complex process that transforms light into chemical energy. The primary processes of photosynthesis are driven by pigment-binding proteins called simply chlorophyll (Chl)-proteins (CP). They are organized into supercomplexes/particles, namely photosystem I (PSI) and PSII embedded in the thylakoid membranes in chloroplasts of higher plants. Built from antenna and core parts, they serve as light-harvesting and trapping units; the final traps being the primary reductants in the photosynthetic electron transport pathway. PSII and PSI operate mostly in series, and PSI also operated in a complementary cyclic way, to produce reducing power and energy in the form of highly reducing electrons or

NADPH and ATP in a suitable ratio for the biochemical processes carried out in the chloroplast/cell. Environmental changes in light quality and intensity could affect the amount of these end products, while changes in temperature and water availability influence their use by the chloroplast metabolism. Overloading by light produces reactive radicals, which are harmful for the system. Therefore, excitation must be properly balanced by modulation of the organization and stoichiometry of PS cores and their antennae or diverted into other channels by protective mechanisms in order to optimize the efficiency of photosynthesis and avoid the damage of the photosynthetic apparatus. Stressors, including heavy metals (HMs), can be considered as special environmental factors that, by causing serious imbalance or damage in different metabolic steps, switch on multiple protective mechanisms and regulatory responses in plants. Revealing physiological responses and regulatory processes under environmental extremes has both scientific and practical importance. The direct and indirect effects of HM stressors on Chl–protein complexes of higher (mainly vascular) plants will be reviewed here with special emphasis on the reasons of their changed accumulation and organization.

II. CHLOROPHYLL–PROTEIN COMPLEXES IN HIGHER PLANTS

Chl–proteins are multicofactor proteins that bind pigments such as Chl *a*, Chl *b*, and carotenoids in green plants [1]. Some of them are also associated with other cofactors participating in the electron transport from water to $NADP^+$. Membrane-intrinsic proteins with α-helical structure serve as dynamic scaffolds to bind and arrange all pigments and other cofactors in a suitable distance and orientation. They influence cofactor properties (spectroscopic characteristics or redox potentials) by providing metal ligands, H-bonds, π-interactions, or hydrophobic pockets. The Mg atoms of Chls are coordinated to side-chain nitrogens (His, Gln, Asn) or side-chain (Glu, Asp, Tyr) or backbone oxygens directly or through water molecules. Carotenoids are bound to proteins by weaker interactions. Though from the point of view of photosynthetic process, the main function of Chl–proteins is light gathering or primary photochemistry, some complexes are switched into energy-dissipating mode to prevent the damage of the system if light is present in relative excess [2].

Chl–proteins of both PSs can be classified as Chl *a*-proteins and Chl *a/b*-proteins (Table 45.1). Chl *a*-proteins binding Chl *a* and β-carotene constitute the inner part of complexes and function as reaction centers (RCs) and inner/core light-harvesting antennae [3–6]. In PSI, the RC and most of inner antenna pigments are bound to the same proteins (PSI-A,B). The pigment cofactors of PSI-A/B include the six Chl *a*, which take part in the electron transport: P700, the primary acceptors (A_0), and accessory Chl *a*-s. Other electron transport cofactors, two quinones and one Fe-S cluster, are also bound to the complex. Light energy captured by the antenna is collected by the red Chls with long wavelength absorption/emission maxima localized near to the RCs. By analogy with cyanobacteria, some small core proteins (PSI-F,G,H,K,L) may also bind Chl or β-carotene as evidenced from biochemical and mutant studies on higher plants [7,8]. In contrast, the RC (D1/D2) and inner antennae (CP43 and CP47) are separate

proteins in PSII, and no Chl bound to small subunits has been shown [9]. In addition to the pigments, P680, two Chl *a*-s and the two pheophytins, and the quinones (Q_A and Q_B) participating in the electron transport, the RC binds two β-carotenes and two Chl *a*, called Chl_Z, which may direct the excitation energy from the antenna to the RC or dissipate it in an oxidized form [10]. The inner antenna of PSII binds much less pigments than that of PSI [11,12].

Chl *a/b* (Cab)-proteins binding Chl *a*, Chl *b*, and different xanthophyll pigments are parts of the peripheral light-harvesting antennae or serve as connecting antenna between the core and peripheral antennae [13–15]. They are all three-helix proteins with similar sequence and pigment complement [16]. Four Lhca proteins forming homo-(LHCI-680A,B) or heterodimers (LHCI-730) build up the antenna of PSI. Interestingly, Chls of the longest wavelength maxima are found in the peripheral PSI antenna, and their exact role is debated [17–19]. The antenna of PSII contains more components: Lhcb1, Lhcb2, and Lhcb3 form the peripheral LHCII complexes. Trimers of different composition were isolated: Lhcb1 homotrimers and heterotrimers composed of Lhcb1/Lhcb2 (2/1), Lhcb1/Lhcb3 (2/1) [20] or Lhcb1/Lhcb2/Lhcb3 [21]. CP29 (Lhcb4), CP26 (Lhcb5), and CP24 (Lhcb6), however, are mostly monomeric, and bind less Chl than LHCII [22–25].

Chl–proteins are organized into more complicated Chl–protein complexes/photosystems (PSI + LHCI and PSII + LHCII) for efficient light harvesting and protected photochemistry. Supramolecular organization of PSII in higher plants was revealed by electron microscopy and image analysis [26]. CP43 is located closer to D1, while CP47 is closer to D2 constituting the core of PSII together with other nonpigmented proteins. CP29 connects trimeric LHCII complexes of different rotational orientation (S/M — strongly/moderately bound) to the core. In addition, L — loosely bound LHCII trimers — as well as CP26 and CP24 are also bound to the core. Highly active PSII is dimeric, while PSI is a monomer in higher plants. Crystal structure studies on mutants, and single-particle analysis indicated that PSI binds LHCI dimers connected by PSI-K, PSI-G, and PSI-F [27–29] on one side of the complex [15,30]. LHCII can also be attached to PSI through PSI-H [31]. The photosystems are located in different domains of the thylakoid membrane [32]. PSII with its LHCII antenna were mainly found in the grana core, while PSI + LHCI particles were shown both in grana margins and stroma thylakoids. Heterogeneity of both PSI and PSII according to antenna size, localization, and function has been discovered [33–35].

TABLE 45.1

Characteristics of Chl-Proteins in Higher Plants

	Proteins			Pigments						Fluor.em. (77 K) (nm)	Ref.
	Mass (kDa)	TMH	Gene	Chl	Chl a/b	β-car	Lutein	Neox	Violax		
PSII core											
D1	38	5	$psbA^C$	3	–	1	–	–	–		[3,6]
D2	39	5	$psbD^C$	3	–	1	–	–	–		[3,6]
CP47	56	6	$psbB^C$	16[a]	–	3	–	–	–	695	[11,12]
CP43	50	6	$psbC^C$	13[a]	–	5	–	–	–	683	[11,12]
PSII CA											
CP29	29	3	$lhcb4^N$	8	3.0	–	1.0	0.35	0.65	680	[23]
CP26	26	3	$lhcb5^N$	8	2.2	–	1.0	0.61	0.38	680	[24]
CP24	24	3	$lhcb6^N$	10	1.0	–	1.0	1.00	–	680	[22]
LHCII											
Lhcb1	25	3	$lhcb1^N$	12	1.3	–	1.9	1.0	0.2	680	[16,25]
Lhcb2	25	3	$lhcb2^N$			–					
Lhcb3	24	3	$lhcb3^N$			–					
PSII-S	22	4	$psbS^N$	5	6.0	Tr	0.6	0.3	0.25	675	[39]
PSI core				96[a]	–	22[a]				720	[5,17]
PSI-A	83	11	$psaA^C$	43[a]	–		–	–	–		[4,5]
PSI-B	82	11	$psaB^C$	42[a]	–		–	–	–		[4,5]
LHCI											
Lhca1	22	3	$lhca1^N$	10	4.0	–	2.0	–	1	686	[18]
Lhca2	23	3	$lhca2^N$	10	1.9	–	1.5	–	0.5	701	[19]
Lhca3	25	3	$lhca3^N$	10	6.0	0.6	1.4	–	0.7	725	[19]
Lhca4	22	3	$lhca4^N$	10	2.3	–	1.5	–	0.5	732	[17,18]
Elip-like	7–17	1–3		+					+	674	[38,44]

[a] Determined in cyanobacterial complexes.

Notes: TMH, transmembrane helix; car, carotene; Neox, neoxanthine; Violax, violaxanthine; Tr, traces; N/C, encoded in the nucleus/chloroplast.

Protein data were obtained from Refs. [9,15].

The changes in the function and dynamic interactions between Chl–protein complexes can be followed by the fluorescence emission characteristics of the system studied. The intensity of room temperature fluorescence emission is influenced by the fate of absorbed light energy: both functionally important photochemistry and protective heat dissipation decrease the fluorescence yield, thus the latter can be used to estimate the former processes [36]. Variations in the amount or physical environment/spatial interactions of complexes are represented by wavelength shifts or changed ratio of the 77 K fluorescence emission maxima (Table 45.1).

In addition to the abovementioned ones, other Chl-binding proteins were also discovered in higher plants the exact function of which is debated [37,38]. They all show sequence similarity to Cab proteins, but have 1-4 transmembrane α-helices (Table 45.1). They are stable in the absence of pigments, but may

bind Chl and carotenoids. However, the Chls are only weakly attached to the protein, and there is only poor excitonic coupling between the chromophores, which excludes a light-harvesting function. The four-helix PSII-S (CP22) is constitutively expressed like the other antenna proteins, is present in stoichiometric amount in PSII (2/PSII), and shows extreme lateral heterogeneity as it is found almost exclusively in granal PSII [39]. Its location is not exactly known, biochemical evidences place it close to the PSII core [40]. New arguments seem to underline its regulatory, photoprotective role [41]. Elip-like proteins include one-helix (Hlips — high light induced proteins, Scps — small Cab-like proteins, Ohps — one-helix proteins), two-helix (Seps — stress-enhanced proteins, Lils — light-harvesting-like), and three-helix proteins (Elips — early light-induced proteins, Cbr — carotenoid biosynthesis related), which all have an Elip consensus motif in Helix 1 [38,42,43]. They are transiently expressed during greening and under light and dehydration stresses, and are present in substoichiometric amount in stroma thylakoids. They probably fulfill photoprotective function by binding newly synthesized Chls or those released during the turnover of pigment-proteins or, alternatively, they may stabilize the proper assembly of complexes or act as sinks for excess excitation energy [38,44,45].

The biogenesis of Chl–proteins takes place in different compartments of the cell. The genes of Chl a-proteins are localized, transcribed, and translated in the chloroplast, and they are cotranslationally inserted into the thylakoids, while Chl a/b-proteins and other Cab-like proteins are encoded in the nucleus by gene families, synthesized in the cytoplasm, and posttranslationally inserted into the thylakoids [37,38,46]. Pigment binding is probably cotranslational (Chl a-proteins) or coinsertional (Cab-proteins). Transcription and translation of different Chl–proteins are redox regulated according to the actual environmental requirements [47,48]. Moreover, gene expression differences and posttranslational modifications may influence the properties of the antennae.

III. EFFECTS OF HEAVY METALS ON CHLOROPHYLL–PROTEIN COMPLEXES

Heavy metals (HMs) are defined as metals with density higher than $5 \, g/cm^3$. Among those HMs that are available for plants, Fe, Mn, and Mo are important micronutrients, Zn, Ni, Cu, Co, and Cr are toxic but have some importance as trace elements, and As, Hg, Cd, and Pb have no known importance and are mostly toxic. They are naturally occurring components in soils. Toxicity problems come into prominence due to human activity. Mining, coal-firing, intensive road traffic, different industrial activities, and agronomical practice such as the use of phosphate fertilizers, sewage sludge deposited in lands, pesticides, and seed coat dressing lead to the emission of HMs and their accumulation in the environment [49,50].

HMs can affect plant growth and production in a multiple way by inhibiting a number of physiological processes in plants [49,51–54]. They were shown to cause disturbance in plant ion-[55–57] and water-balance [58], to interfere with protein metabolism through influencing nitrate and sulfate reduction [59–61]. Though only a small part of toxic HMs (around 1% of leaf content) reaches the chloroplasts, photosynthetic light reactions and enzymatic processes are the main targets of HMs [62–64].

A. MODIFICATIONS OF THE COMPOSITION, ORGANIZATION, AND FUNCTION OF CHLOROPHYLL–PROTEIN COMPLEXES BY TOXIC HEAVY METALS

1. PSII Core

Numerous studies have demonstrated that PSII is the main target of HM stress. However, the exact mechanism has not been unambiguously elucidated yet despite a wealth of information accumulated. Though both Cu at equimolar concentration to PSII RC [65] and Cd [66] were shown to stimulate O_2 evolution, the most frequently reported effect of Cd, Cr, Cu, Hg, Ni, Pb, and Zn was its inhibition in both *in vitro* and *in vivo* experiments (for reviews, see Refs. [63,64,67–71]). On the basis of the results of experiments carried out with isolated chloroplasts or PSII particles treated with Cu, Pb, Zn, or Hg, the inhibition was attributed to the dissociation of the oxygen-evolving complex (OEC) proteins and to the displacement or substitution of the cofactors (Ca^{2+}, Cl^-, and Mn) necessary for water splitting [72–75]. In accordance, Ca excess partly eliminated the symptoms of Cu stress by stabilizing the PSII complex and increasing its electron transport activity [74]. Yruela et al. [76] showed that <230 Cu(II)/PSII RC, which inhibited O_2 evolution and variable Chl a fluorescence around 50%, did not affect the polypeptide composition of PSII, and only higher Cu concentration ($300 \, \mu M$, Cu(II)/PSII RC = 1400) caused the release of OEC polypeptides. Flash-induced absorption difference spectroscopy of the partial reactions of PSII electron transport revealed that Cu (25 to $100 \, \mu M$) specifically modified Tyr_Z and its microenvironment so that the electron transfer to $P680^+$ was extremely slowed down [77]. Electron paramagnetic resonance (EPR) studies on isolated chloroplasts also indicated the interaction of Cu and Hg with Tyr_Z^+/Tyr_D^+, and the appearance of free Mn^{2+} due to copper treatment [72,73].

Inhibition sites of HMs were also shown at the acceptor side of PSII. Electron transport between Pheo and Q_A was impaired in PSII core and RC particles by Cu ($80 \, \mu M$) treatment [78]. Thermoluminescence measurements identified the electron transport between Q_A and Q_B as the site of action of Ni, Zn, and Co in isolated thylakoids [79]. Displacement of nonheme Fe by Cu in PSII RC (1000 Cu/PSII) and Zn (10,000 Zn/PSII) was detected by EPR spectroscopy [80]. Conformational modification of the Q_B pocket by Cd was supposed on the basis of herbicide binding studies [66]. Coordination of Cd, Zn, Ni, Co, and Cu to the residues on the protonation pathway was shown by x-ray crystallography in the purple bacterial RC [81].

The decreased rate of electron transfer was also reflected in the fluorescence induction parameters of HM-treated plants. Reduced fluorescence yield due to degradation or inactivation of PSII centers and increased fluorescence quenching have been reported in many cases [82–84]. Though the above results suggest that different HMs act similarly, Cd ($5 \mu M$), Zn ($50 \mu M$), and Cu ($50 \mu M$) were found to have different mechanism of action when fast fluorescence transients of developing bean plants were investigated [85]. While Cu influenced only the re-oxidation, and Zn only the reduction of Q_A, Cd influenced both processes. However, Cd slowed down the re-oxidation of Q_A in contrast to the effect of Cu. These changes may be indirect consequences of the inhibition of the Calvin cycle, but the direct effect of Cd on Q_A–Q_B electron transport due to its possible interaction with nonheme Fe cannot be excluded. In addition, the response of Cu and Zn seemed to be elastic, the plants were able to adjust to the altered supply of Cu and Zn, while that of Cd was plastic, i.e., a new steady-state level was reached.

Less information is available in connection with the effects of HMs on the composition and organization of PSII core, except the changes caused in the phosphorylation state and turnover/degradation of D1 protein [86,87], the details of which will be discussed later. In addition to the release of the OEC proteins (see above), high concentration Cu treatment of PSII (BBY) particle also destabilized and liberated antenna proteins CP47 and CP43 *in vitro* [76]. Cd caused similar effect on the polypeptide content of chloroplasts *in vivo* [88,89]. Newly synthesized proteins in the region of 20 to 29 kDa (putative stress proteins) were found in isolated PSII particles after longer stress [90].

2. LHCII

Coordination of Hg, Cd, and Pb to protein nitrogen and carbonyl oxygen atoms, and in the case of Hg also to sulfur donor sites was shown with Fourier transform infrared (FTIR) spectroscopy in isolated LHCII [91]. At higher concentrations (10 to 20 mM), Hg and Cd, but not Pb caused strong conformational changes of the protein decreasing the α-helix content heavily.

Cd (0.2 mM) diminished the ratio of oligomeric to monomeric LHCII in thylakoids of greening radish seedlings [92,93]. Isolated LHCII of Cd-treated radish cotyledons contained somewhat lower amount of Chls, xantophylls, and acyl lipids, particularly MGDG, DGDG, and PG [93]. The most dramatic change, strictly parallel to the changes in the reduction of the oligomeric forms or LHCII, was observed in *trans*-Δ^3-hexadecenoic acid content specifically bound to PG, which plays an important role in the association of the protein subunits in the trimer [16,94]. It may be connected with the inhibition by Cd of the palmitate desaturase activity [93]. However, the LHCII was exclusively present as a trimer in our Cd-treated cucumber [95] and also in poplar plants after mild solubilization with glucosidic detergents (Figure 45.1). Therefore, the LHCII may be present in the form of less stable trimers in Cd-treated plants or the stabilization is only retarded during greening and takes place under longer development.

Concerning the organization of LHCII antenna, the shape of the 77 K fluorescence emission spectrum did not change in bean plants developed with 0.3 to $15 \mu M$ Cu (10 to 110 Cu/PSII) [96] or in cucumber treated with $10 \mu M$ Pb, therefore the characteristics of the PSs remained like those in controls. The lower Chl

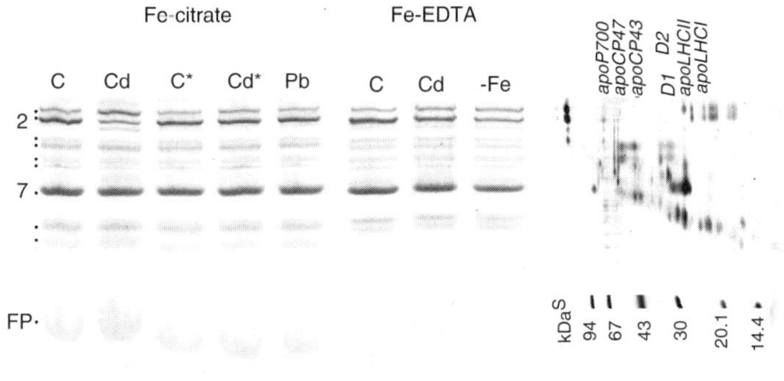

FIGURE 45.1 Chl-protein patterns of iron-deficient and HM-treated plants. Left: Green bands (marked with points and numbers) separated by Deriphat PAGE after mild solubilization of thylakoids by glucosidic detergents [95]. FP — free pigment. Poplar plants grown hydroponically in the presence of $10 \mu M$ Fe-EDTA or Fe-citrate as iron source up to four-leaf stage were treated with Cd or Pb ($10 \mu M$) or by withdrawal of iron (-Fe) for another 14 days. * — iron supply was increased to $50 \mu M$ during the treatment. Right: Method of identification of bands by 2D gel electrophoresis: native gel electrophoresis followed by denaturing gel electrophoresis showing the polypeptide pattern of green bands. Protein standards (in kDa): phosphorylase b (94), bovine serum albumin (67), ovalbumin (43), carbonic anhydrase (30), soybean tripsin inhibitor (20.1), α-lactalbumin (14.4).

a/b ratio, and the accelerated PSII closure referred to larger PSII antenna at higher Cu concentrations (0.2 to 1 mM) in barley plants [97]. Cd-treated (30 to 120 μM) wheat seedlings exhibited reduced PSII emissions [98], and a maximum at 680 nm replaced the one at 685 nm together with some increase around 700 nm in severely inhibited cucumber plants [95], i.e., LHCII antenna decreased and perhaps detached from the core under strong Cd stress. Therefore, antenna changes were different depending on the circumstances.

3. PSI Holocomplex

PSI electron transport activity *in vitro* was more tolerant to HM treatment than that of PSII (see Ref. [63] and references therein, [99,100]). Pb (45 Pb/Chl) decreased the active P700 content by 28%, and changed the kinetics of the dark re-reduction of photo-oxidized P700 [101]. In isolated thylakoids Cu did not affect the function of P700 [72]. However, in the presence of ascorbate rapid destruction of Chl and P700 Chl–protein was detected probably due to the formation of H_2O_2 in a Cu(II)-catalyzed auto-oxidation of ascorbate [102]. Oxidation of P700 in the dark with Hg (Chl:$HgCl_2$ = 1:1), was indicated by EPR spectroscopy [73].

Lower sensitivity of PSI than PSII activity was reported even *in vivo*. Moderately reduced PSI electron transport activity was measured in Cd-treated (30 to 120 μM) greening wheat seedlings [98] and in Ni-treated (100 to 500 μM) more or less mature bean plants [103]. PSI activity measured by photochemical energy storage in far red light and P700 photo-oxidation in developing maize plants was less affected with Cu (8, 80 μM) treatment than PSII activity [83]. However, 10-day chromate (20 to 40 μM) treatment reduced considerably the P700 oxidation in Spirodela [104]. Ferredoxin-dependent $NADP^+$ reduction was much more inhibited than methylviologen reduction with dichlorophenol-indophenol as electron donor in chloroplasts isolated from Cd-treated (10, 20, 30 μM) maize seedlings, and was correlated with the decreased ferredoxin and Fe content [99].

The amount of long wavelength fluorescence (77 K) decreased in developing, Cd-treated (10 to 120 μM) wheat and cucumber plants [95,98] referring to disturbed PSI antenna organization. The long wavelength emission was blue shifted in severely inhibited plants [95]. In addition, a population of PSI binding both LHCI and LHCII came into prominence (Figure 45.1: band 1 on gels) compared with PSI containing only LHCI (band 2). Comparable change in PSI composition was observed under light stress [105]. Moreover, the amount of the LHCI antenna in PSI particles isolated from Cd-treated plants was a little reduced relative to that of the control (Sárvári et al., unpublished) similar to the phenomenon observed in iron- and copper-deficient *Chlamydomonas* [106,107].

B. CHANGES IN THE ACCUMULATION OF CHLOROPHYLL–PROTEINS UNDER HEAVY METAL STRESS

HMs influenced not only the composition and function of complexes but also their relative accumulation and consequently the ultrastructure of thylakoids.

In Cu-treated (315 μM) spinach plants, the amount of CPa (CP47, CP43) decreased significantly [108]. In 6/12/18-day-old bean plant in intensive/intermediate/stationary growth phase Cu (315 μM) treatment for another 10 days increased (but the highest inhibition of growth was observed)/did not change/decreased the Chl content, decreased the Chl a/b, and Chl/carotenoid ratios in the primary leaves [109]. Higher tolerance to Cu based on O_2 evolution was shown in young leaves. In agreement, fluorescence induction showed that the primary photochemical processes in PSII did not change in plants in the intensive and intermediate growth stages, but energy-dissipating processes were activated [110]. F_v/F_m decrease was only found in plants treated in the stationary phase and was related to senescence processes. Analogous changes were observed under similar circumstances, but using Cd (25 μM) stress [111]. Similarly, investigating the effect of Cd (50 to 200 μM) toxicity on the different sections of rye leaves in different developmental stages, the strongest decrease in the Chl content and F_v/F_m, and increase of nonphotochemical quenching (NPQ) occurred in the more mature leaf sections [112].

The abovementioned changes were reflected in the effects of Cd (10 μM) and Pb (10, 50 μM) on the accumulation of Chl–protein complexes in cucumber [95,113]. Cd treatment from one-leaf stage up to the control plants had seven leaves (around 4 weeks) gave an order of sensitivity of the complexes PSI > PSII core > LHCII corresponding to the sequence of degradation of complexes in senescing leaves [114]. When the Pb and Cd treatment started later, in four-leaf stage of plants, the PSII core was the most sensitive complex in the mature leaves. In the newly emergent ones, however, Pb and Cd reduced the amount of complexes in the order of LHCII > PSI > PSII core, and PSI > LHCII > PSII core, respectively [95]. In accordance with these data, a decreased effective antenna size of PSII was suggested on the basis of fluorescence induction parameters in detached pea leaves greening in the presence of Pb [115]. Chl–

protein patterns similar to those found in Cd-treated cucumber were also obtained in barley plants greening with Ni (0.4 to 1 mM), in *Salix* treated with Cd (90 μM) or Cu (45 μM) (Sárvári et al., unpublished), and with poplar plants treated with Cd (10 μM) from four-leaf stage (Figure 45.1) [116]. The maximal, actual, and intrinsic efficiency of PSII decreased moderately and NPQ rose during the long-term Cd stress (Table 45.2). Furthermore, the symptoms became stronger with increasing light intensity during the treatment [117]. However, higher sensitivity of PSII compared to the other complexes was observed in chromate-treated (100, 500 μM) Spirodela [118] in agreement with the conclusion of the measured fluorescence transients [119], and in the more tolerant *Phragmites* leaves emerged under Cd (90 μM) or Cu (45 μM) treatment (Sárvári et al., unpublished). PSII accumulation was retarded even in bean plants treated with very low concentrations (0.5 to 1 μM) of Cd, Ni, and Pb, in spite of the fact that the accumulation of Chl and other complexes was stimulated [120]. The Cd-induced changes usually lowered the Chl a/b ratio due to a relative decrease in the amount of PSII core or to the lower relative sensitivity of LHCII than PSI.

HM treatment reduced the size and sometimes the number of chloroplasts and the amount of the thylakoid system. Either the stroma lamellae were more markedly destroyed with irregularly spaced grana or on the contrary, the higher sensitivity of grana structures was observed [88,121–124]. In the last stage, swelling of thylakoid membranes, numerous plastoglobuli, and sometimes crystal-like bodies were seen. In young leaves, metal toxicity had a severe inhibitory effect on the development of thylakoids [125].

In conclusion, the Chl–protein patterns obtained during HM treatments could be classified into four stages (Figure 45.2). The first and second stages are similar in character, the only difference being that the Chl and Chl–protein accumulation is stimulated compared to the control in the first stage and inhibited in the second one. The first stage was observed in mild stress treatments given in the intensive or intermediate growth phase of leaves, and the PSII efficiency did not change. The second stage varied in strength depending on the dose of the stressor. It was characteristic to leaves treated in intermediate growth phase, as well as to some stressors such as chromate, or to more tolerant plants treated in both the intensive or inter-

TABLE 45.2
Chl content, Chl a/b ratio, CO$_2$ Fixation, Ion Content, and Fluorescence Induction Parameters of Cd-Treated and Iron-Deficient Poplar Plants. Plants Grown in Hydroponics with 10 μM Fe–EDTA (iron depleted during the treatment) Fe–Citrate (Cd treatments) Were Treated from Four-Leaf Stage, and Parameters of Leaves Emerged before (+) or during the 2-Week Treatment Were Used. Values of a Representative Experiment Are Expressed as the Percentage of the Control (except Cd contents).

Parameters	0 μM Fe	50 μM Fe	10 μM Fe 10 μM Cd$^+$	10 μM Fe 10 μM Cd	50 μM Fe 10 μM Cd
Chl $a+b$ (μg/cm^2)	62.1	98.8	118.0	58.6	104.6
Chl a/b	84.8	101.5	96.8	90.6	97.0
^{14}CO$_2$ fixation (cpm/cm^2)	44.0	117.0	77.5	21.5	106.7
F_0	164.0	94.2	114.5	165.2	95.0
F_v/F_m*	80.8	100.6	100.2	81.1	100.0
Φ_{PSII}**	79.2	101.2	100.3	81.2	104.8
F_v'/F_m'***	80.5	101.2	99.6	74.4	104.0
q_P	108.3	100.0	100.8	116.6	100.8
NPQ	356.5	106.1	101.4	314.2	82.5
Cd (nmol/cm^2)	–	–	4.43	5.78	nd
Fe (nmol/cm^2)	50.7	nd	117.1	55.4	nd
Mn (nmol cm^2)	134.0	nd	127.5	79.7	nd

Note: F_0, initial Chl fluorescence, *maximal, **actual, and ***intrinsic (excitation capture) efficiency of PSII, q_P, photochemical quenching coefficient, NPQ, nonphotochemical quenching, nd, not determined.

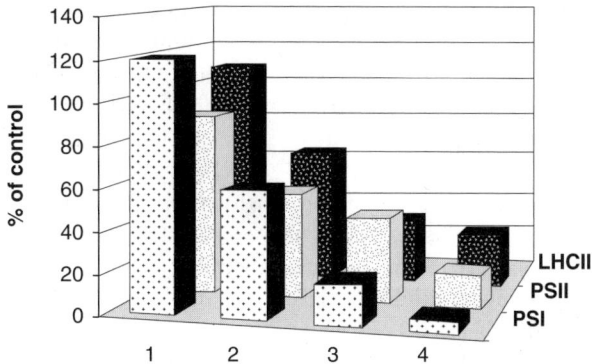

FIGURE 45.2 Patterns of Chl–protein complexes in different stages of HM stress. Amount of Chl–proteins calculated in μg Chl per cm² leaf material are expressed as the percentage of control values.

mediate growth phase. Stage 3 was observed most frequently in leaves of sensitive plants emerging under HM stress with stronger stressors, which influenced PSI heavily. The strength of the stress also varied as in stage 2. Changes in the maximal efficiency of PSII were slight to moderate in Stages 2 and 3 depending on the circumstances. Stage 4 was characteristic of plants with totally exhausted stores, showing very low maximal efficiency of PSII, and dying if the treatment continued.

Since the relative rates of both biosynthetic and degradation processes determine the accumulation of complexes, studies on the course of events during

greening or acclimation may give clues to the explanation of the final pattern. Cd (1 to 100 μM) slowed down the rate of accumulation of Chls and carotenoids, and reduced synthesis of LHCII, LHCI, and OEC polypeptides, as well as a delay in the appearance of PSII activity and grana stacking was observed in greening radish seedlings [92]. The slower rate of accumulation of LHC proteins can be explained by Cd suppression of the transcription of Lhcb1 [126] and inhibition of the activity of the protease, which cleaves the precursor of LHCII apoprotein to its mature form [92]. While Cd (2.5 to 10 mM) had a more pronounced effect on PSII activity during the initial stages of Cd treatment, PSI activity was also equally affected in pea plants after longer treatment [100]. During regreening of cucumber plants etiolated by iron deficiency in the presence of Cd (1 μM) and Fe (10 μM), the accumulation rate of each complex was differently reduced compared to the control (Figure 45.3). The recovery of complexes in control plants was similar to the iron nutrition-mediated chloroplast development in sugar beet plants [127]. A lag period of 24 h was observed before the bulk Chl was accumulated, and the accumulation rate of PSI was the highest. In plants greening in the presence of Cd, first the level of PSI grew the most rapidly relative to the control, the accumulation rate of PSII was significantly more inhibited, and the increase in the amount of LHCII lagged behind the RC complexes. Subsequently, the destruction of complexes was observed, which

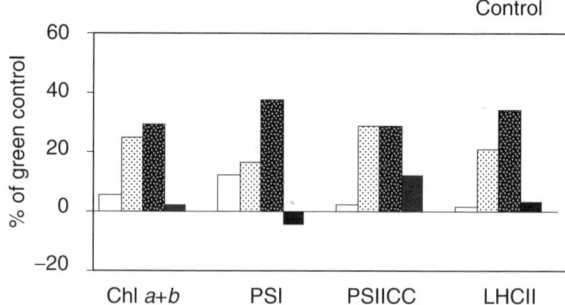

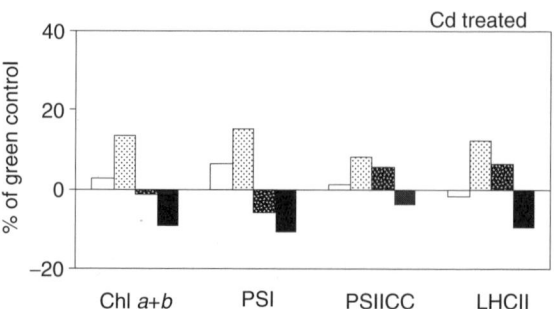

FIGURE 45.3 Accumulation of Chl and Chl–protein complexes during iron resupply induced greening of cucumber leaves in the presence and absence of 1 μM CdNO₃. Leaves were etiolated by iron deprivation. Subsequent periods of greening (white, dotted, gray, and black) represent 24 h each. All values measured as μg Chl per cm² leaf material are expressed as the percentage of values measured in the green control.

started earlier in the case of PSI, later in LHCII, and PSII core was the most stable complex hardly degrading during the investigated period (120 h). The early degradation of PSI can be related to the inactivation of the protective mechanisms [128] due to Cd inhibition of protective enzymes [129]. Therefore, Cd influenced both the synthesis and the degradation of all complexes, but degradation of PSII core lagged behind that of the other complexes in the later stages of development. Alteration of D1 turnover was also found in different stresses, and it was suggested that it might represent stress adaptation response [86].

IV. SOME REASONS AND REMEDIES OF HEAVY METAL EFFECT

Concerning the mechanism of action of HMs, direct effects as blocking functional groups in biologically important molecules by HM binding-induced activity/conformational changes or by displacing/substituting essential prosthetic groups, as well as indirect effects such as HM-induced disturbances of mineral metabolism and oxidative stress in plants have been assumed [53,56,57,130]. From the point of view of the pigment–protein complexes, HMs may also cause regulatory changes triggered by the modified balance of photosynthetic functions due to inhibition of the biogenesis of PSs or functional damage of the existing complexes.

A. INFLUENCE OF HEAVY METALS ON PHOTOSYNTHETIC PIGMENTS AND MEMBRANE LIPIDS

Though Chl accumulation have been shown to be stimulated by micromolar and submicromolar concentrations of Ni, Mn, Fe, Co, Cd, Pb, and Cr in the nutrient solution [120,131], the most frequently observed symptom of HM ($10 \mu M$ to $1 mM$) treatment or metal deficiency was the chlorosis of leaves (see Ref. [132] and references therein). The extent of chlorosis was strongly influenced by the experimental conditions (composition of the nutrient solution, age of plants, time of the treatment), and the plant species. It is the relative rates of leaf growth and Chl accumulation that have particularly great influence, and can explain the sometimes contradictory results concerning the sensitivity of leaves of different age [95,109,111]. For this reason, it should be more reliable to determine the Chl contents on a whole leaf basis or in a given number of chloroplasts, which is usually not the case.

Concerning the inhibition of Chl accumulation, different affectivity of HMs employed under the same circumstances was observed such as Cd > Cu > Pb [133], Cu > Co,Ni > Mn,Zn [134], Co > Ni > Cd > others [135], and Co > Cu > Cr [136]. The inhibition was concentration and dose dependent [103,131,133,136]. The relative sensitivity of Chl a, Chl b, and carotenoids differed considerably from experiment to experiment (see Ref. [132] and references therein). This is probably caused by the different effects of HMs on the accumulation of Chl–proteins depending on the actual circumstances.

The reason of chlorosis is not unequivocally determined. Several authors found evidence for the direct inhibition of enzymes of the Chl biosynthesis pathway by HMs. Inhibition of the accumulation of δ-amino-levulinic acid (ALA) [131,137,138] and that of the activity of ALA-dehydratase (EC 4.2.1.24) was shown by Cd, Fe, Co, Ni, Mn, Hg, Pb, and Cr [131,138–140] at similar concentrations which inhibited the Chl accumulation in vivo. It was frequently supposed that HMs made their impact on enzymes by directly complexing to –SH groups in the catalytic center. Feedback inhibition of ALA accumulation by the high Mg–protoporphyrin monomethyl ester content can also be evoked by HM induced iron deficiency [141,142]. Other enzymes of Chl biosynthesis such as porphobilinogenase [131] and protochlorophyllide oxidoreductase [137,143,144] could be inhibited only at very high metal concentrations. Neither the phototransformation of protochlorophyllide nor its dark accumulation was influenced by Cd in vivo in barley leaves [145]. Instead, the presence of highly fluorescent, not stably assembled Chl could be detected. At the same time, greening pattern of Cd-treated bean seedlings under intermittent illumination referred to some inhibitory effect of Cd on protochlorophyllide accumulation, regeneration, or phototransformation [126]. Even marked reduction, in chloroplast density, supposedly due to inhibition of chloroplast division by Cu and Cd treatment, may lead to the development of chlorosis in certain cases [146,147].

On the other hand, decreased Chl content can be attributed to increased degradation of the existing Chl. Enhancement of Chlase activity by HMs was found in vitro [148]. Cd-induced lipid peroxidation, and peroxide-mediated oxidative degradation of Chls was also suggested [149,150]. Accelerated senescence induced by Cd was proposed as the main cause of Chl breakdown in bean leaves treated in an older age [111]. In more mature, but still growing leaves, the Chl concentration hardly changed, but the Chl a/b ratio varied suggesting an acclimative response [95,121].

In conclusion, HM inhibition of Chl biosynthesis may contribute to the decreased accumulation of

Chl–proteins in young leaves. However, the presence of imperfectly bound Chl in developing plants under HM treatment points out that the importance of other processes, affecting membrane biogenesis and the proper assembly of complexes, is more significant [92,95,97,145,151]. HM-induced inhibition of the required aggregation (stabilization) of complexes was evidenced from FTIR and EPR results [152]. During stronger and longer HM treatment or in older leaves, HM-induced decrease in Chl content and changes in the Chl a/b ratio can be also connected with acclimation and degradation processes.

Another effect of HMs observed in higher plants *in vivo* is the substitution of the magnesium in the Chl molecule by toxic HMs (HM-Chls) such as mercury, copper, cadmium, nickel, zinc, and lead [153,154]. Most HM-Chls are unsuitable for photosynthesis. Their energy transfer (and fluorescence emission) efficiency is strongly decreased because of their blue-shifted absorption spectra and rather unstable first excitation state, which relaxes thermally [155]. In addition, the photochemical capacity of HM-Chls, i.e., the ability to release electrons from the singlet excited state, is also decreased relative to Mg-Chl [156]. The toxicity of metals could be ordered as Hg > Cu > Cd > Zn > Ni > Pb, and it was proportional with their complex formation rate and not with the thermodynamic stability of the complexes formed [153]. Under extreme shade circumstances (1 to $5 \, \text{W/m}^2$), great part of the Mg atoms of Chls were replaced by HMs, and either formed long-lived, stable complexes with blue shifted absorption spectra (color change) and low fluorescence emission (copper, nickel) or, in the case of unstable complexes, pronounced bleaching of Chl was observed. Under intense light (100 to $150 \, \text{W/m}^2$), only a low fraction of Chl magnesium (<2%) was replaced, and large amount of Chl broke down with the appearance of allomerization products. The possibility of HM–Chl formation, particularly in the pigments participating in the electron transport, could also explain the observed functional changes of Chl–protein complexes under the effect of HMs. Results presented in other papers such as blue-shifted spectrum or direct observation of substituted Chls strengthen the view of the *in vivo* occurrence of HM-Chls (see Ref. [153] and references therein).

The Chl–protein complexes are embedded in the lipid bilayer of the thylakoids. Any change in the lipid composition and fluidity may alter the conformation, orientation, and function of the complexes. HMs were shown to affect the lipid composition and fluidity of thylakoids [157]. Cd lowered the amount of the acyl lipids, particularly MGDG and PG, and decreased the amount of *trans*-Δ^3-hexadecenoic acid

[93]. Lower toxic Cu and Pb doses decreased the amounts of phospholipids (PG and PC), higher doses also affected that of MGDG [151,158]. MGDG is necessary to PSII activity [159], and together with PG is important for building up the oligomeric LHCII and for grana stacking [16], thereby influencing the energy transfer towards PSII. Fatty acids of thylakoid lipids showed the tendency of saturated species to increase replacing the unsaturated ones, which made the membranes more rigid [152,160]. Similar changes in the lipid content and fatty acid composition were also observed in Fe-deficient pea plants [161].

B. DISTURBANCE OF MINERAL METABOLISM BY HEAVY METALS

One of the most important effects of HMs on photosynthesis is their influence on the mineral nutrition of plants. Excess trace or toxic metals very often induce iron deficiency [55–57]. Cd and Cu decreased the amount of both Fe and Mn [96,162–164]. Cd-treated cucumber did not show the signs of element deficiency [95]. However, the total leaf iron content is not always a suitable marker of physiological iron deficiency [165].

The reduced iron accumulation has been attributed to either HM inhibition of uptake or translocation processes [166]. Excess (1 mM) Cu, Mn, Zn, and Co (but not Ni) reduced ^{59}Fe translocation to young barley leaves [134]. Cd interference with either iron uptake [167] or translocation of Fe and Mn [168,169] was also reported. In addition to causing element deficiency, HM treatment simultaneously altered cation balance in plants due to competition of metals in uptake or translocation processes. Effectiveness of Cd, Pb, and Zn depended on the Fe and Mn concentration in the nutrient solution [57,171,172].

Studies on the alterations of the Chl–protein pattern under selected ion deficiencies may shed light on their importance in the HM effect. Manganese diminished the amount of CPa (CP47 and CP43), i.e., that of the PSII core, and it was accompanied by a decrease in the Chl a/b ratio, the short wavelength fluorescence emission measured at 77 K, and PSII electron transport activity [172–174]. Lamellar structure of Mn-deficient chloroplasts did not differ from that of the control [174]. Deficient plants were recovered totally within two days by re-addition of manganese [174]. Restoration of PSII activity and some recovery of grana stacking were achieved in Cd-treated (20 μM) tomato plants by increasing the manganese concentration to 1 mM in the nutrient solution [175]. Therefore, manganese deficiency was connected mainly with PSII activity changes, and it could not be responsible for the HM effects on Chl–protein pattern, because symptoms of severe manganese deficiency were slight compared to the effects of HMs, and HM-induced changes were hardly recovered by the addition of extra amount of manganese.

In contrast, severe Fe-deficiency syndrome was highly similar to a strong HM effect. Iron deficiency induced different biochemical and activity changes in the photosynthetic apparatus [176,177]. It diminished the pigment content, with carotenoids less sensitive than Chl a [161,178,179]. The changes in the Chl a/b ratio depended on the extent of the iron deficiency. Both PSI and PSII donor- and acceptor-side-related activity changes were reported, and Q_A was reduced even in the dark [179–181]. The decreased actual efficiency of PSII under iron deficiency could not be simply explained with a lower number of photosynthetic units, but it was related to the closure of PSII centers at moderate iron deficiency, whereas it was associated with a decreased intrinsic PSII efficiency at severe iron deficiency [182]. The stronger inhibition of O_2 evolution than the actual PSII efficiency suggests the existence of changed electron pathways.

Concerning the compositional changes in Chl–protein complexes, iron deficiency ($<1\,\mu M$ Fe), which was followed by the expression of a marker gene (*Fox1*) encoding an iron assimilation component, diminished the amount of Lhca3 in the isolated PSI particles in *Chlamydomonas* [106]. It was present in the free LHCI pool as it was evidenced from the blueshifted 77 K fluorescence spectra and the high yield of fluorescence. Loosening of LHCI connection was due to marked loss of PSI-K. Severe iron limitation resulted in loss of PSI, LHCI (loss of long wavelength fluorescence), PSII (lack of F_v), and residual LHCII with 77 K fluorescence emission at 680 nm was shown. The iron-deficiency-induced changes in Chl–protein complexes were not the consequence of inhibition of Chl synthesis, rather they were suggested to be specific regulatory events occurring at posttranscriptional level, which started well before the decline in the Chl content. The first step was the expression of Fox1, which accumulated about 80-fold within 24 h, excluding the possibility of general protein synthesis inhibition and referring to a very sensitive Fe-sensing mechanism. The decrease in the ferredoxin content (within 1 day) was followed by a decrease in PSI-K (within 2 days), then, with a distinct lag period, the amount of Lhca3 that is connected to the core through PSI-K [27,28] was reduced, and later on a decrease in PSI core polypeptides was observed. LHCII, CF1, and D1 were found to be stable within the 5-day experiment. Furthermore, disappearance and formation of new LHCI and LHCII polypeptide spots were detected by 2D PAGE during this period.

The observed processes proved to be mostly posttranscriptional events (different gene procession, specific proteolysis) though translational control of the synthesis of some components could not be ruled out. As iron-deficient plants were not light-sensitive and light-sensitive mutants (e.g., PSI-F) grew better in iron-limited conditions, it was suggested that these antenna changes made the cells able to optimize photosynthesis and minimize the damaging effect of excess light. Differential processing may influence antenna associations by shifting the antenna function from efficient light-harvesting to dissipation of light, as it was supposed in the case of the iron deficiency induced antenna rearrangement in cyanobacteria [183].

Antenna disconnection under iron deficiency was also shown in PSII. The use of phase fluorometry and time-correlated single photon counting indicated the presence of a long lifetime component (3.3 ns) contributing to the total fluorescence by approximately 15%, and persisting during the whole induction period, which was interpreted as a disconnected internal PSII antenna component [184].

Iron deficiency induced more or less the same changes in the Chl–protein pattern as HM treatment. Mild iron deficiency in young expanding leaves of previously Fe–citrate-grown poplar plants slightly decreased the amount of PSII core [116]. Strong iron deficiency in developing leaves of previously Fe-EDTA grown poplar plants, however, caused similar change in the Chl–protein pattern to Cd, decreasing the accumulation of complexes in the order of PSI > LHCII > PSII core (Figure 45.1). In accordance, the long wavelength emission was blueshifted and decreased as in cucumber either Fe-deficient or Cd-treated [95,185]. Changes of the same character were observed in the Chl–protein pattern of pea [161]. Reductions in the amount of PSI core, LHCI-680, D1, and CP43 were also reported in iron deficient tomato, but LHCII remained stable [179]. The extreme sensitivity of PSI is not surprising in the light of its high iron content (12/PSI). Furthermore, the assembly of the F_x cluster is a critical requirement for the stability of the PSI core [186].

In addition, Chl content and the amount of all the complexes decreased in parallel with the reduction of the iron content of leaves both in iron-deficient and Cd-treated poplar plants [163]. While the slopes of curves in the case of Chl, PSI, and LHCII were more or less the same in Fe-deficient and Cd-treated plants, the decline of PSII core with the decreasing iron content of leaves was much stronger in the latter treatment showing that other factors than iron deficiency also influenced its accumulation. The maximal, actual, and intrinsic efficiency decreased and NPQ rose under both treatments referring to both photo-inhibition and high nonphotochemical energy dissipation (Table 45.2).

The chlorosis of iron-deficient leaves was easily reversible upon Fe resupply [127,187], as well as the deleterious effects of Cu and Cd on photosynthesis and Chl–protein pattern were reduced considerably by simultaneous higher Fe supply (Figure 45.1, Table 45.2) [164,188]. Reconnection of unconnected LHCI to the newly synthesized PSI centers was observed in red algae during the readdition of iron [189].

Therefore, it is evident that many aspects of iron deficiency are very similar to the HM syndrome concerning the Chl–protein complexes. The most important ones are: (i) the similar changes in the Chl–protein pattern, (ii) similarity of the iron deficiency and Cd-induced rearrangement of antennae and alterations of 77 K fluorescence spectra, and (iii) the same slopes of the curves showing the dependence between the Fe content of leaves and the amounts of Chl and PSI and LHCII in both types of treatment. It means that iron deficiency plays an essential role in eliciting the HM symptoms. The most important effects of iron deficiency on the accumulation of Chl–protein complexes may be the inhibition of Chl synthesis, having its greatest impact on the accumulation of LHCII, and slowing down the assembly of PSI or decreasing its stability. The reduction in the amount of PSII core, however, is influenced not only by Fe deficiency, but also by other factors, e.g., Mn deficiency and degradation rate in Cd-treated plants.

C. Effects of Light Excess Due to Inhibition of Photosynthesis by Heavy Metals

Stressors, including HMs, result in an imbalance between the light absorption and light energy utilization even under moderate irradiance [190]. Such excess light relative to the capacity of photosynthesis is sensed through alterations in the PSII excitation pressure or the more reduced/energized state of the system, and protective mechanisms are switched on to avoid damage of the photosynthetic apparatus. If the state is a long-lasting one, signal transduction pathways are initiated to coordinate photosynthesis-related gene expression. Depending on the severity of the stress, however, more or less damage of the system is unavoidable.

1. Effects of Reactive Oxygen Species on Chl–Protein Complexes

Plants are more exposed to oxidative stress because one of the major sources of reactive oxygen species (ROS) is the photosynthetic electron transport [191].

The water-splitting enzyme produces H_2O_2 *in vitro* by deactivation of the S_3 state of the enzyme. Some electron leakage to O_2 at the acceptor side of PSII results in H_2O_2 liberation. The main site of ROS production is, however, the reducing side of PSI, where superoxide is formed by electron transport to O_2 (Mehler reaction). Superoxide is transformed into H_2O_2 by superoxide dismutase (SOD). Inhibition of the electron flow may also lead to the formation of singlet oxygen, because triplet Chl is formed due to back reactions in the PSII RC, which reacts with O_2 [192]. Singlet oxygen production in the cytochrome b_6/f complex was also detected [193]. ROS cause nonenzymatic breakdown of lipids (lipid peroxidation) and may oxidize aromatic and sulfur-containing amino acids. Both nonenzymic antioxidants, glutathione and ascorbate localized in the chloroplast stroma, and α-tocopherol and carotenoids in the membranes, as well as enzymes such as SOD and ascorbate peroxidase (Apx) take part in the elimination of ROS [191]. The enzymic and nonenzymic antioxidants are organized into an antioxidant network called the water–water cycle [194]. The superoxide formed by photosynthetic electron transport from H_2O to O_2 is disproportioned to H_2O_2 and O_2 by SOD, H_2O_2 is then reduced to water by Apx and ascorbate, which is regenerated by dehydroascorbate reductase, glutathione, and glutathione reductase using photosynthetically produced NADPH.

While plants are able to cope with the normal rate of ROS production, stress factors may seriously disturb the balance of the detoxification mechanisms. HMs may stimulate the formation of ROS as a consequence of inhibition of metabolic reactions, and thereby increasing the electron transport to oxygen, which leads to higher superoxide production [130]. In addition, transition metal catalysts (Fe, Cu) promote the production of biologically far the most reactive ROS, the hydroxyl radical from superoxide or H_2O_2 (Haber–Weiss or Fenton-type reaction). Though the defense mechanisms are frequently stimulated by HM treatment at the beginning of application, after longer treatment or at higher concentration HMs may also damage the detoxification system [129,130,195].

Excessive level of ROS formation results in damage to the photosynthetic apparatus causing photoinhibition of both PSII and PSI. PSII photoinhibition can be caused either by acceptor- or donor-side mechanisms [196]. In the acceptor-side mechanism, singlet oxygen is produced due to the overreduction of Q_A, which is able to oxidize nearby pigments, redox components and amino acids. The donor-side, oxygen-independent inactivation results in prolonged lifetime of $P680^+$ and Tyr_Z^+ causing irreversible oxidation in the surroundings. Both processes damage D1 protein, which is usually rapidly (within hours) and efficiently repaired by replacing the damaged protein with a new one [196,197]. The degradation and resynthesis take place in the stroma lamellae. The damaged, phosphorylated D1 protein is triggered to degradation by conformational change in the granum. The inactive PSII becomes monomerized, LHCII, and OEC is displaced. The core migrates into the stroma lamellae, where D1 is degraded by specific proteases after dephosphorylation. Concomitantly with its degradation, a new polypeptide is synthesized and cotranslationally inserted into the thylakoid membrane and ligated by cofactors. The repaired PSII moves back to the granum where the native complex reassembles. However, at excessive inactivation of PSII, the complexes cannot be all repaired, and most of the damaged phosphorylated centers stay in the grana for longer time. Phosphorylation probably stabilizes the complex, and protects all the components from degradation. Active oxygen species were reported to play a role even in the degradation of LHCII in high light [198].

Photoinhibition of PSI is not so frequently observed than that of PSII, but happens *in vivo*. The process needs oxygen, and the primary targets are the iron–sulfur centers [128]. Hydroxyl radicals, produced by Fenton-type reaction between the photoreduced Fe of the degraded centers and superoxide, are the cause of the damage of stromal extrinsic proteins and that of PSI-B. The recovery of PSI from photoinhibition is very slow process, lasts for several days, so it must be prevented. PSII electron transport inhibitors suppress PSI inactivation, thus the photoinhibition of PSII can be regarded as a protective mechanism. Other protective processes may be the Mehler reaction in the presence of protective enzymes and the cyclic electron transport, which also downregulates PSII, by creating a pH gradient [195,199]. Furthermore, the $P700^+$ species, accumulating at high light and being quenchers, may also dissipate energy as heat.

PSI activity does not seem to be particularly sensitive to stress. However, a slight loss of PSI was observed under high light [200,201] or mineral deficiency stress [202]. In contrast, PSII photoinhibition, measured as inhibition of oxygen evolution or decrease in F_v/F_m, was often found in HM stress [83,84,96,164]. It may be enhanced by weakly coupled Chl–proteins [203,204], which were reported to be present in HM-stressed plants [97,145]. Photoinhibition due to iron excess (50 to 900 μM) was accompanied by light-induced oxidative degradation of D1 protein, i.e., that of PSII [205,206]. Changes in D1 amounts during short and strong Cd (5 mM) stress in pea and bean plants detected by pulse-chase experiment with radiolabelled ^{35}[S]methionin showed that D1 turnover was stimulated in the first hours of treatment, and later it was inhibited [66,90].

The stimulation or inhibition of D1 turnover was found in different stresses, and may represent a stress adaptation response [86,87]. Our knowledge about the regulation of D1 protein turnover is rather incomplete. It was inhibited if the PQ pool was mostly reduced [207]. Under chronic mineral deficiency photoinactivated PSII did not accumulate in the first phase, i.e., there was a rapid degradation of PSII, which also induced the degradation of LHCII (chlorosis), but after losing considerable amount of PSII D1 degradation became limiting, the level of phosphorylated D1 protein even in the dark was enhanced, and inactivated PSII centers could accumulate [208,209]. HM induced PSII core "stabilization" after a considerable loss of Chl may be connected with such type of processes.

2. Regulatory Processes under Excess Light

Because of the importance of photosynthesis as energy source, plants developed a number of strategies to avoid damage of the photosynthetic apparatus and maintain photosynthetic efficiency as high as possible even under adverse conditions. These involve activation of different protecting mechanisms, and changed gene expression both to optimize photosynthesis or to further enhance protective mechanisms.

Plants use a wide range of regulatory and protective mechanism to get rid of excess light and to avoid the production of ROS. Acclimation of that type occurs on a time scale of minutes. They include state transitions, photochemical and nonphotochemical quenching processes. State transitions balance the light supply of PSII and PSI [211]. Photochemical mechanisms consume the excess electrons or reduced substances in a more or less dissipative way, such as cyclic electron transport, photorespiration, chlororespiration, and indirectly also the respiration [2,210]. Among nonphotochemical mechanisms, which dissipate excess light in the form of heat thereby decreasing not only the quantum yield of photosynthesis but also quenching the fluorescence emission of the system, there exist antenna and RC-related processes [190]. Suggested mechanisms of action of qE-type antenna quenching induced by the highly energized state of thylakoids were the deepoxidation of violaxanthin to zeaxanthin [212], and zeaxanthin binding and conformational/aggregation state changes of Chl–proteins such as LHCII [190,213,214], minor antenna of PSII (CP29, CP26) [215, but see 216], or PsbS [41,217], which generate antenna traps able to deactivate by heat production. PSII RC inactivation may induce futile cyclic electron transport around PSII [218]. In addition, photoinhibited PSII centers also seem to dissipate energy as heat under sustained excess light [219–221].

From the examples studied so far it is clear that any one combination of these mechanisms can be used in plants to lower the excitation pressure. Elevated photochemistry protected winter wheat or rye, and LHCII aggregates containing zeaxanthin were shown to dissipate excess energy in *Pinus sylvestris* during winter [222]. Excess Mn (0.18 to 1.8 mM), Al (1 mM), and Cd (50 μM) treatment all decreased the yield of PSII photochemistry with a parallel increase in q_N (Table 45.2) [66,82,84,123]. Using Cu-treated PSII (BBY) particles, the F_m quenching could be connected with the Cu-induced oxidation of both forms of cyt b_{559} (LP and HP) and that of Chl$_Z$ [223], which is an efficient quencher of antenna fluorescence in its oxidized state [10]. Increased cyclic electron transport around PSI and less inhibited respiration than photosynthesis has also been reported under HM stress [83,115,224]. The fact that a residual part of the maximal quantum yield of PSII photochemistry and a quite large portion of energy storage measured by phototacoustic spectroscopy was unaffected in isolated thylakoids (not able to perform PSI cyclic electron transport) by Hg, Cu, and Pb at full inhibition of the O_2 evolution can be explained by the occurrence of cyclic electron transport activity around PSII [225]. However, HMs can also interfere with some of these regulatory processes: Cu treatment (8, 80 μM) was shown to highly downregulate the change from state 1 to state 2 [83].

Alternatively, plants can acclimate to the new environmental conditions by means of changed gene expression, leading to biogenesis or degradation of Chl–protein components, the result of which is their changed ratio better suited to the given circumstances [47,48,222,226,227]. According to a recent hypothesis, the signal is the altered redox state of some components (PQ, cyt b_6/f complex) of the photosynthetic machinery (perceptional control) or that of small molecules (thioredoxin, glutathione) in the chloroplasts, or the appearance of ROS (transductional control) [48]. Regulator components are active in the range of light when their redox state is variable. At low light intensity this is the PQ pool, the function of which is the fine-tuning of the operation of the photosynthetic apparatus. At higher light, when the PQ pool is totally reduced, the thioredoxin, the redox state of which depends on both the linear and cyclic electron transport, become the most important redox regulator. At high light intensity, where thioredoxin is also reduced, the concentrations of oxidized glutathione and H_2O_2 is increased, and act as activation

signal for induction of light stress defense mechanisms. Regulatory mechanisms control gene expression at all levels, transcription, posttranscriptional processes and translation can be affected. PQ-regulated transcription of *psaAB* (gene of P700 apoprotein), and *psbA* (gene of D1) [228], and the nuclear *Lhcb* (genes of PSII antenna components) [229,230] were described. Binding of a translation-activating protein complex to the *psbA* mRNA 5′ untranslated region was enhanced by means of activation by reduced thioredoxin and a disulfide isomerase-like enzyme [231,232]. Redox state of glutathione is thought to play an important role in enhancing the expression of the *psbA* gene under high light stress, i.e., when the amount of oxidized glutathione increased [233]. High light induces the repression of *Lhcb* transcription [234] and the proteolysis of the existing polypeptides [235,236].

Alterations in the abundance and organization of pigment-proteins were also reported under photoinhibitory light. Disconnection of LHCII from PSII was shown independently of phosphorylation under photoinhibitory conditions [237]. LHCII content of PSI increased [105]. A detailed investigation of changes in Chl–proteins under different light regimes revealed that during rising the irradiance level from normal to high light the amount of PSI core was constant, that of PSII core doubled, while that of Lhcb1,2 dramatically decreased, and that of the Lhcb4,5,6 did not change, but a new Lhcb4 polypeptide of higher molecular weight (posttranslational modification) appeared [238]. Reduction in transcript abundance of *Lhcb1*, *Lhcb4*, and *Lhca* under high light was also shown in *Chlamydomonas* [239]. In Tris-washed thylakoids irradiated with photoinhibitory light and solubilized by digitonin, the dissociation of PSI core and LHCI was observed [240]. Decreased PSI antenna was also observed in *Porphyridium cruentum* with increasing irradiance [241]. In green algae reduction in the amount of Lhcb polypeptides and increased level of a carotenoid-binding protein, Cbr was detected, which was attributed to the concomitant repression and de-repression of their nuclear genes, respectively [222]. Elips can protect plants from high light induced photo-oxidative stress [45,242]. These changes may all contribute to the protection against photoinhibition.

In conclusion, it can be said that the changes of the photosynthetic apparatus during HM treatment are mostly of acclimatory character, in which the excess light generated by the stress-induced damage of photosynthetic function plays an important role. In stage 1 (Figure 45.2), when the system still has reserves, the elevation in the amount of PSI is probably due to an increased requirement of ATP for synthesis/regeneration of proteins, the alteration of which was induced by the HM stress. The increased LHCII/PSII may help the functioning of the undamaged PSII centers. The situation remains the same as long as the relative damage of PSI is smaller than that of PSII even if the synthesis/accumulation of complexes is inhibited (Stage 2). However, when the amount of PSI becomes limiting, e.g., due to HM-induced iron deficiency or increased photoinhibitory damage caused by the inhibition of protective enzymes, other control mechanisms are switched on, which probably also works under strongly photoinhibitory conditions (stage 3). This involves decreased accumulation/proteolytic degradation of antenna complexes, and relative stabilization of PSII centers, a part of which is photoinhibited. These centers may participate in energy dissipation [220,243]. In stage 4, senescence becomes prevalent. The advanced stage of senescence is characterized by massive Chl breakdown accompanied by a strong decrease in PSII efficiency, and the LHCII is the most stable component [114,244].

V. CONCLUDING REMARKS

For the interpretation of the effects of HMs on Chl–proteins, it is important to bear in mind that (i) their toxicity and mechanism of action are different in mature and developing plants, and (ii) the toxicity of the essential metals needed as micronutrient (Fe, Mn) or that of the trace element (Cu, Zn, Ni) and nonessential metals (Hg, Cd, Pb) may be expressed in quite different concentration range, because plants can tolerate the essential micronutrients in wide concentration range. Furthermore, though the effects of HMs are basically dose dependent, the time factor is of primary importance, i.e., the effects obtained during a very short time treatment cannot be compared to those of long treatments. In addition, different plant species can significantly modulate the response *in vivo* due to the different, and not only photosynthesis related, protective mechanisms.

In the case of mature leaves, HMs' influence on the element uptake or Chl synthesis does not seem to be the main reason of the changes observed in the organization and function of Chl–protein complexes. The direct HM binding including substitution of important ions (Ca, Mg, Mn), oxidative damage of proteins and lipids can be the primary effects, which induce compositional (e.g., HM-Chls) and consequently functional changes, and variations in the amount of complexes. Usually, the PSII is the most sensitive complex. Under strong stress, increased degradation of D1 leading to decreased amount of PSII is the result. Induction/strengthening of protective mechanisms is usually observed during the stress-resisting period, which involve the reorganization of the antenna for increased nonphotochemical quenching, induction of cyclic electron transport around PSI and PSII, and increased expression of stress-related proteins.

HM effect on the photosynthetic apparatus of developing leaves is even more complicated. The direct inhibition of Chl synthesis, depression of nitrate reduction (influencing protein synthesis), and most importantly the inhibition of uptake and translocation of ions (Fe, Mn, Ca) can be involved in the HM effects retarding the synthesis and assembly of complexes. After the photosynthetic apparatus begins functioning, regulatory mechanisms are activated. Under mild stress, upregulated synthesis of the complex having functional defects can occur. Strong stress evokes responses in the Chl–protein pattern characteristic to excess light. The higher the light intensity, the stronger the stress response. Processes mentioned in mature plants can also contribute.

Therefore, long HM treatment mainly affects the synthesis or stability of apoproteins during the development of the photosynthetic apparatus. The resulted changes under not too severe stress, when PSII efficiency was only moderately affected, seems to be of acclimative character, which may help to optimize photosynthesis under adverse conditions. However, the details and the progress of processes, namely the exact causes and consequences in the subsequent steps in a developing or a more mature system, are to be discovered. The most promising experimental setup seems to be studying the progress of events from gene expression to compositional, organizational, and functional changes step by step during the *in vivo* HM treatment. The dynamism in pigment binding and interactions among the Chl–proteins seems to be of prime importance. This kind of work has just started in the recent years. We practically do not have any information concerning "helper" pigment-proteins (PsbS, Elips) under HM stress. In addition, more detailed study of the isolated complexes is necessary to find out the direct effects such as cofactor substitution and compositional or conformational changes of functional importance.

ACKNOWLEDGMENTS

I am grateful to Zsuzsanna Ostorics for technical assistance. Research in the lab was financially supported by grants of IC15-CT98-0126 and OTKA T-043646.

REFERENCES

1. Green BR, Durnford DG. The chlorophyll-carotenoid proteins of oxygenic photosynthesis. *Annu. Rev. Plant Physiol. Plant Mol. Biol.* 1996; 47:685–714.

2. Niyogi KK. Safety valves for photosynthesis. *Curr. Opin. Plant Biol.* 2000; 3:455–460.

3. Zouni A, Witt HT, Kern J, Fromme P, Krauss N, Saenger W, Orth P. Crystal structure of photosystem II from *Synechococcus elongatus* at 3.8 angstrom resolution. *Nature* 2001; 409:739–743.

4. Jordan P, Fromme P, Witt HT, Klukas O, Saenger W, Krauss N. Three-dimensional structure of cyanobacterial photosystem I at 2.5 angstrom resolution. *Nature* 2001; 411:909–917.

5. Fromme P, Jordan P, Krauss N. Structure of photosystem I. *Biochim. Biophys. Acta* 2001; 1507:5–31.

6. Fromme P, Kern J, Loll B, Biesiadka J, Saenger W, Witt HT, Krauss N, Zouni A. Functional implications on the mechanism of the function of photosystem II including water oxidation based on the structure of photosystem II. *Philos. Trans. R. Soc. Lond. B* 2002; 357:1337–1345.

7. Preiss S, Peter GF, Anandan S, Thornber JP. The multiple pigment-proteins of the photosystem I antenna. *Photochem. Photobiol.* 1993; 57:152–157.

8. Ihalainen JA, Jensen PE, Haldrup A, van Stokkum IHM, van Grondelle R, Scheller HV, Dekker JP. Pigment organization and energy transfer dynamics in isolated photosystem I (PSI) complexes from *Arabidopsis thaliana* depleted of the PSI-G, PSI-K, PSI-L, or PSI-N subunit. *Biophys. J.* 2002; 83:2190–2201.

9. Hankamer B, Barber J, Boekema EJ. Structure and membrane organization of photosystem II in green plants. *Annu. Rev. Plant Physiol. Plant Mol. Biol.* 1997; 48:641–671.

10. Schweitzer RH, Brudvig GW. Fluorescence quenching by chlorophyll cations in photosystem II. *Biochemistry* 1997; 36:11351–11359.

11. Alfonso M, Montoya G, Cases R, Rodríguez R, Picorel R. Core antenna complexes, CP43 and CP47, of higher plant photosystem II. Spectral properties, pigment stoichiometry, and amino acid composition. *Biochemistry* 1994; 33:10494–10500.

12. Vasilev S, Orth P, Zouni A, Owens TG, Bruce D. Excited-state dynamics in photosystem II: insights from the x-ray crystal structure. *Proc. Natl. Acad. Sci. USA* 2001; 98:8602–8607.

13. Jansson S. The light-harvesting chlorophyll *a/b*-binding proteins. *Biochim. Biophys. Acta* 1994; 1184:1–19.

14. Sandona D, Croce R, Pagano A, Crimi M, Bassi R. Higher plants light harvesting proteins. Structure and function as revealed by mutation analysis of either protein or chromophore moieties. *Biochim. Biophys. Acta* 1998; 1365:207–214.

15. Scheller HV, Jensen PE, Haldrup A, Lunde C, Knoetzel J. Role of subunits in eukaryotic photosystem I. *Biochim. Biophys. Acta* 2001; 1507:41–60.

16. Kühlbrandt W. Structure and function of the plant light-harvesting complex, LHCII. *Curr. Opin. Struct. Biol.* 1994; 4:519–528.

17. Knoetzel J, Bossmann B, Grimme LH Chlorina and viridis mutants of barley (*Hordeum vulgare* L.) allow assignment of long-wavelength chlorophyll forms to individual Lhca proteins of photosystem I *in vivo*. *FEBS Lett.* 1998; 436:339–342.

18. Croce R, Morosinotto T, Castelletti S, Breton J, Bassi R. The Lhca antenna complexes of higher plants photosystem I. *Biochim. Biophys. Acta* 2002; 1556:29–40.

19. Castelletti S, Morosinotto T, Robert B, Caffarri S, Bassi R, Croce R. Recombinant Lhca2 and Lhca3 subunits of the photosystem I antenna system. *Biochemistry* 2003; 42:4226–4234.

20. Jackowski G, Jansson S. Characterization of photosystem II antenna complexes separated by non-denaturing isoelectric focusing. *Z. Naturforsch.* 1998; 53c:841–848.

21. Jackowski G, Kacprzak K, Jansson S. Identification of Lhcb1/Lhcb2/Lhcb3 heterotrimers of the main light-harvesting chlorophyll a/b-protein complex of photosystem II (LHC II). *Biochim. Biophys. Acta* 2001; 1504:340–345.

22. Bassi R, Pineau B, Dainese P, Marquardt J. Carotenoid-binding proteins of photosystem II. *Eur. J. Biochem.* 1993; 212:297–303.

23. Crimi M, Dorra D, Bosinger CS, Giuffra E, Holzwarth AR, Bassi R. Time-resolved fluorescence analysis of the recombinant photosystem II antenna complex CP29 — effects of zeaxanthin, pH and phosphorylation. *Eur. J. Biochem.* 2001; 268:260–267.

24. Croce R, Canino G, Ros F, Bassi R. Chromophore organization in the higher-plant photosystem II antenna protein CP26. *Biochemistry* 2002; 41:7334–7343.

25. Ruban AV, Lee PJ, Wentworth M, Young AJ, Horton P. Determination of the stoichiometry and strength of binding of xanthophylls to the photosystem II light harvesting complexes. *J. Biol. Chem.* 1999; 274:10458–10465.

26. Boekema EJ, van Roon H, van Breemen JFL, Dekker JP. Supramolecular organization of photosystem II and its light-harvesting antenna in partially solubilized photosystem II membranes. *Eur. J. Biochem.* 1999; 266:444–452.

27. Ben-Shem A, Frolow F, Nelson N. Crystal structure of photosystem I. Nature 2003; 426:630–635.

28. Jensen PE, Gilpin M, Knoetzel J, Scheller HV. The PSI-K subunit of photosystem I is involved in the interaction between light-harvesting complex I and the photosystem I reaction center core. *J. Biol. Chem.* 2000; 275:24701–24708.

29. Jensen PE, Rosgaard L, Knoetzel J, Scheller HV. Photosystem I activity is increased in the absence of the PSI-G subunit. *J. Biol. Chem.* 2002; 277:2798–2803.

30. Boekema EJ, Jensen PE, Schlodder E, van Breemen JFL, van Roon H, Scheller HV, Dekker JP. Green

plant photosystem I binds light-harvesting complex I on one side of the complex. *Biochemistry* 2001; 40:1029–1036.

31. Lunde C, Jensen PE, Haldrup A, Knoetzel J, Scheller HV. The PSI-H subunit of photosystem I is essential for state transitions in plant photosynthesis. *Nature* 2000; 408:613–615.

32. Albertsson P-A. The structure and function of the chloroplast photosynthetic membrane — a model for domain organization. *Photosynth. Res.* 1995; 46:141–149.

33. Melis A, Guenther GE, Morissey PJ, Ghirardi ML. Photosystem II heterogeneity in chloroplasts. In: Lichtenthaler HK, ed. *Applications of Chlorophyll Fluorescence.* Dordrecht: Kluwer Academic Publishers, 1988:33–43.

34. Wollenberger L, Weibull C, Albertsson P-A. Further characterization of the chloroplast grana margins: the non-detergent preparation of granal photosystem I cannot reduce ferredoxin in the absence of $NADP^+$ reduction. *Biochim. Biophys. Acta* 1995; 1230:10–22.

35. Jansson S, Stefánsson H, Nyström U, Gustafsson P, Albertsson P-A. Antenna protein composition of PS I and PS II in thylakoid sub-domains. *Biochim. Biophys. Acta* 1997; 1320:297–309.

36. Krause GH, Weis E. Chlorophyll fluorescence and photosynthesis: the basics. *Annu. Rev. Plant Physiol. Plant Mol. Biol.* 1991; 42:313–349.

37. Funk C. The PsbS protein: a Cab-protein with a function of its own. In: Aro EM, Andersson B, eds. *Regulation of Photosynthesis.* Dordrecht: Kluwer Academic Publishers, 2001:453–467.

38. Adamska I. The Elip family of stress proteins in the thylakoid membranes of pro- and eukaryota. In: Aro EM, Andersson B, eds. *Regulation of Photosynthesis.* Dordrecht: Kluwer Academic Publishers, 2001:487–505.

39. Funk C, Schröder WP, Napiwotzki A, Tjus SE, Renger G, Andersson B. The PSII-S protein of higher plants: a new type of pigment-binding protein. *Biochemistry* 1995; 34:11133–11141.

40. Nield J, Funk C, Barber J. Supermolecular structure of photosystem II and location of the PsbS protein. *Philos. Trans. R. Soc. Lond. B* 2000; 355:1337–1343.

41. Li XP, Bjorkman O, Shih C, Grossman AR, Rosenquist M, Jansson S, Niyogi KK. A pigment-binding protein essential for regulation of photosynthetic light harvesting. *Nature* 2000; 403:391–395.

42. Jansson S, Andersson J, Kim SJ, Jackowski G. An *Arabidopsis thaliana* protein homologous to cyanobacterial high-light-inducible proteins. *Plant Mol. Biol.* 2000; 42:345–351.

43. Heddad M, Adamska I. Light stress-regulated two-helix proteins in *Arabidopsis thaliana* related to the chlorophyll *a/b*-binding gene family. *Proc. Natl. Acad. Sci. USA* 2000; 97:3741–3746.

44. Adamska I, RoobolBoza M, Lindahl M, Andersson B. Isolation of pigment-binding early light-inducible proteins from pea. *Eur. J. Biochem.* 1999; 260:453–460.

45. Hutin C, Nussaume L, Moise N, Moya I, Kloppstech K, Havaux M. Early light-induced proteins protect *Arabidopsis* from photooxidative stress. *Proc. Natl. Acad. Sci. USA* 2003; 100:4921–4926.

46. Paulsen H. Pigment assembly — transport and ligation. In: Aro EM, Andersson B, eds. *Regulation of Photosynthesis.* Dordrecht: Kluwer Academic Publishers, 2001:219–233.

47. Carlberg I, Rintamaki E, Aro EM, Andersson B. Thylakoid protein phosphorylation and the thiol redox state. *Biochemistry* 1999; 38:3197–3204.

48. Pfannschmidt T, Allen JF, Oelmüller R. Principles of redox control in photosynthesis gene expression. *Physiol. Plant.* 2001; 112:1–9.

49. Lepp NW. *Effect of Heavy Metals on Plant Function.* Vols. 1, 2. London: Applied Science Publishers Ltd, 1981.

50. Nriagu JO, Pacyna JM. Quantitative assessment of worldwide contamination of air, water and soils with trace metals. *Nature* 1988; 333:134–139.

51. Prasad MNV, Hagemeyer *J. Heavy Metal Stress in Plants. From Molecules to Ecosystems.* Berlin: Springer-Verlag, 1999.

52. Prasad MNV, Strzalka K. *Physiology and Biochemistry of Metal Toxicity and Tolerance in Plants.* Dordrecht: Kluwer Academic Publishers, 2002.

53. Van Assche F, Clijsters H. Effects of metals on enzyme activity in plants. *Plant Cell Environ.* 1990; 13:195–206.

54. Fodor F. The physiology of heavy metal toxicity in higher plants. In:. Prasad MNV, Strzalka K, eds. *Physiology and Biochemistry of Metal Toxicity and Tolerance in Plants.* Dordrecht: Kluwer Academic Publishers, 2002:149–177.

55. Wallace A, Wallace GA, Cha JW. Some modifications in trace metal toxicities and deficiencies in plants resulting from interactions with other elements and chelating agents — the special case of iron. *J. Plant Nutr.* 1992; 15:1589–1598.

56. Siedlecka A. Some aspects of interactions between heavy metals and plant mineral nutrients. *Acta Soc. Bot. Pol.* 1995; 3:265–272.

57. Siedlecka, A, Krupa Z. Cd/Fe interaction in higher plants — its consequences for the photosynthetic apparatus. *Photosynthetica* 1999; 36:321–331.

58. Barceló J, Poschenrieder C. Plant water relations as affected by heavy metal stress: a review. *J. Plant Nutr.* 1990; 13:1–37.

59. Nussbaum S, Schmutz D, Brunold C. Regulation of assimilatory sulfate reduction by cadmium in *Zea mays* L. *Plant Physiol.* 1988; 88:1407–1410.

60. Hernández LE, Gárate A, Carpena-Ruiz R. Effects of cadmium on the uptake, distribution and assimilation of nitrate in *Pisum sativum. Plant Soil* 1997; 189:97–106.

61. Gouia H, Ghorbal MH, Meyer C. Effects of cadmium on activity of nitrate reductase and on other enzymes of the nitrate assimilation pathway in bean. *Plant Physiol. Biochem.* 2000; 38:629–638.

62. Clijsters H, van Assche F. Inhibition of photosynthesis by heavy metals. *Photosynth. Res.* 1985; 7:31–40.

63. Krupa Z, Baszynski T. Some aspects of heavy metals toxicity towards photosynthetic apparatus — direct and indirect effects on light and dark reactions. *Acta Physiol. Plant.* 1995; 17:177–190.

64. Mishliwa-Kurdziel B, Prasad MNV, Strzalka K. Heavy metal influence on the light phase of photosynthesis. In: Prasad MNV, Strzalka K, eds. *Physiology and Biochemistry of Metal Toxicity and Tolerance in Plants.* Dordrecht: Kluwer Academic Publishers, 2002:229–255.

65. Burda K, Kruk J, Strzalka K, Schmid GH. Stimulation of oxygen evolution in photosystem II by copper(II) ions. *Z. Naturforsch.* 2002; 57c:853–857.

66. Geiken B, Masojidek J, Rizzuto M, Pompili ML, Giardi MT. Incorporation of [S-35]methionine in higher plants reveals that stimulation of the D1 reaction centre II protein turnover accompanies tolerance to heavy metal stress. *Plant Cell Environ.* 1998; 21:1265–1273.

67. Droppa M, Horváth G. The role of copper in photosynthesis. *Crit. Rev. Plant Sci.* 1990; 9:111–123.

68. Baron M, Arellano JB, Gorgé JL. Copper and photosystem II: a controversial relationship. *Physiol. Plant.* 1995; 94:174–180.

69. Maksymiec W. Effect of copper on cellular processes in higher plants. *Photosynthetica* 1997; 34:321–342.

70. Samantaray S, Rout GR, Das P. Role of chromium on plant growth and metabolism. *Acta Physiol. Plant.* 1998; 20:201–212.

71. Patra M, Sharma A. Mercury toxicity in plants. *Bot. Rev.* 2000; 66:379–422.

72. Sersen F, Král'ová K, Bumbálová A, Svajlenova O. The effect of Cu(II) ions bound with tridentate Schiff base ligands upon the photosynthetic apparatus. *J. Plant Physiol.* 1997; 151:299–305.

73. Sersen F, Král'ová K, Bumbálová A. Action of mercury on the photosynthetic apparatus of spinach chloroplasts. *Photosynthetica* 1998; 35:551–559.

74. Maksymiec W, Baszynski T. The role of Ca^{2+} ions in modulating changes induced in bean plants by an excess of Cu^{2+} ions. Chlorophyll fluorescence measurements *Physiol. Plant.* 1999; 105:562–568.

75. Pätsikkä E, Aro EM, Tyystjärvi E. Mechanism of copper-enhanced photoinhibition in thylakoid membranes. *Physiol. Plant.* 2001; 113:142–150.

76. Yruela I, Alfonso M, Baron M, Picorel R. Copper effect on the protein composition of photosystem II. *Physiol. Plant.* 2000; 110:551–557.

77. Schröder WP, Arellano JB, Bittner T, Baron M, Eckert H-J, Renger G. Flash-induced absorption spectroscopy studies of copper interaction with photosystem II in higher plants. *J. Biol. Chem.* 1994; 30:32865–32870.

78. Yruela I, Alfonso M, de Zarate IO, Montoya G, Picorel R. Precise location of the Cu(II)-inhibitory binding site in higher plant and bacterial photosynthetic reaction centers as probed by light-induced absorption changes. *J. Biol. Chem.* 1993; 268:1684–1689.

79. Mohanty N, Vass I, Demeter S. Impairment of photosystem 2 activity at the level of secondary electron acceptor in chloroplasts treated with cobalt, nickel and zinc ions. *Physiol. Plant.* 1989; 76:386–390.

80. Jegerschold C, MacMillan F, Lubitz W, Rutherford AW. Effects of copper and zinc ions on photosystem II studied by EPR spectroscopy. *Biochemistry* 1999; 38:12439–12445.

81. Axelrod HL, Abresch EC, Paddock ML, Okamura MY, Feher G. Determination of the binding sites of the proton transfer inhibitors Cd^{2+} and Zn^{2+} in bacterial reaction centers. *Proc. Natl. Acad. Sci. USA* 2000; 97:1542–1547.

82. Krupa Z, Öquist G, Huner NPA. The effects of cadmium on photosynthesis of *Phaseolus vulgaris* — a fluorescence analysis. *Physiol. Plant.* 1993; 88:626–630.

83. Ouzounidou G, Moustakas M, Strasser RJ. Sites of action of copper in the photosynthetic apparatus of maize leaves: kinetic analysis of chlorophyll fluorescence, oxygen evolution, absorption changes and thermal dissipation as monitored by photoacoustic signals. *Aust. J. Plant Physiol.* 1997; 24:81–90.

84. Kitao M, Lei TT, Koike T. Effects of manganese toxicity on photosynthesis of white birch (*Betula platyphylla* var. japonica) seedlings. *Physiol. Plant.* 1997; 101:249–256.

85. Ciscato M, Vangronsveld J, Valcke R. Effects of heavy metals on the fast chlorophyll fluorescence induction kinetics of photosystem II: a comparative study. *Z. Naturforsch.* 1999; 54c:735–739.

86. Giardi MT, Masojídek J, Godde D. Effects of abiotic stresses on the turnover of the D1 reaction centre II protein. *Physiol. Plant.* 1997; 101:635–642.

87. Godde D. Adaptations of the photosynthetic apparatus to stress conditions. In: Lerner HR, ed. *Plant Responses to Environmental Stresses. From Phytohormones to Genom Organization.* New York: Marcel Dekker, 1999:449–474.

88. Skórzynska E, Baszynski T. The changes in PSII complex polypeptides under cadmium treatment — are they of direct or indirect nature? *Acta Physiol. Plant.* 1993; 15:263–269.

89. Nedunchezhian N, Kulandaivelu G. Effect of Cd and UV-B radiation on polypeptide composition and photosystem activities of *Vigna unguiculata* chloroplasts. *Biol. Plant.* 1995; 37:437–441.

90. Franco E, Alessandrelli S, Masojídek J, Margonelli A, Giardi MT. Modulation of D1 protein turnover under cadmium and heat stresses monitored by (^{35}S(methionine incorporation. *Plant Sci.* 1999; 144:53–61.

91. Ahmed A, Tajmir-Riahi HA. Interaction of toxic metal ions Cd^{2+}, Hg^{2+}, and Pb^{2+} with light-harvesting proteins of chloroplast thylakoid membranes. An FTIR spectroscopic study. *J. Inorg. Biochem.* 1993; 50:235–243.

92. Krupa Z, Skórzynska E, Maksymiec W, Baszynski T. Effect of cadmium treatment on the photosynthetic apparatus and its photochemical activities in greening radish seedlings. *Photosynthetica* 1987; 21:156–164.

93. Krupa Z. Cadmium-induced changes in the composition and structure of the light-harvesting complex

II in radish cotyledons. *Physiol. Plant.* 1988; 73:518–524.

94. Trémolières A, Dubacq JP, Duval JC, Lemoine Y, Rémy R. Role of phosphatidylglycerol containing trans-hexadecenoic acid in oligomeric organisation of the light-harvesting chlorophyll protein (LHCP). In: Wintermans JFGM, Kuiper PJC, eds. *Biochemistry and Metabolism of Plant Lipids.* Amsterdam: Elsevier Biomedical Press, 1982:369–372.

95. Sárvári É, Fodor F, Cseh E, Varga A, Záray G, Zolla L. Relationship between changes in ion content of leaves and chlorophyll-protein composition in cucumber under Cd and Pb stress. *Z. Naturforsch.* 1999; 54c:746–753.

96. Pätsikkä E, Aro EM, Tyystjärvi E. Increase in the quantum yield of photoinhibition contributes to copper toxicity *in vivo. Plant Physiol.* 1998; 117:619–627.

97. Caspi V, Droppa M, Horváth G, Malkin S, Marder JB, Raskin VI. The effect of copper on chlorophyll organization during greening of barley leaves. *Photosynth. Res.* 1999; 62:165–174.

98. Atal N, Saradhi PP, Mohanty P. Inhibition of the chloroplast photochemical reactions by treatment of wheat seedlings with low concentrations of cadmium: Analysis of electron transport activities and changes in fluorescence yield. *Plant Cell Physiol.* 1991; 32:943–951.

99. Siedlecka A, Baszynski T. Inhibition of electron flow around photosystem I in chloroplasts of Cd treated maize plants is due to Cd-induced iron deficiency. *Physiol Plant.* 1993; 87:199–202.

100. Chugh LK, Sawhney SK. Photosynthetic activities of *Pisum sativum* seedlings grown in presence of cadmium. *Plant Physiol. Biochem.* 1999; 37:297–303.

101. Wong D, Govindjee. Effects of lead ions on photosystem I in isolated chloroplasts: studies on the reaction center P700. *Photosynthetica* 1976; 10:241–254.

102. Samuelsson G, Öquist G. Effects of copper chloride on photosynthetic electron transport and chlorophyll-protein complexes of *Spinacia oleracea. Plant Cell Physiol.* 1980; 21:445–454.

103. Krupa Z, Siedlecka A, Maksymiec W, Baszynski T. *In vivo* response of photosynthetic apparatus of *Phaseolus vulgaris* L. to nickel toxicity. *J. Plant Physiol.* 1993; 142:664–668.

104. Susplugas S, Srivastava A, Strasser RJ. Changes in the photosynthetic activities during several stages of vegetative growth of *Spirodela polyrhiza*: effect of chromate. *J. Plant Physiol.* 2000; 157:503–512.

105. Herrmann B, Kilian R, Peter S, Schäfer C. Light-stress-related changes in the properties of photosystem I. *Planta* 1997; 201:456–462.

106. Moseley JL, Allinger T, Herzog S, Hoerth P, Wehinger E, Merchant S, Hippler M. Adaptation to Fe-deficiency requires remodeling of the photosynthetic apparatus. *EMBO J.* 2002; 21:6709–6720.

107. Moseley JL, Page MD, Alder NP, Eriksson M, Quinn J, Soto F, Theg SM, Hippler M, Merchant S. Reciprocal expression of two candidate di-iron enzymes

affecting photosystem I and light-harvesting complex accumulation. *Plant Cell* 2002; 14:673–688.

108. Baszynski T, Tukendorf A, Ruszkowska M, Skórzynska E, Maksymiec W. Characteristics of the photosynthetic apparatus of copper non-tolerant spinach exposed to excess copper. *J. Plant Physiol.* 1988; 132:708–713.

109. Maksymiec W, Baszynski T. Different susceptibility of runner bean plants to excess copper as a function of the growth stages of primary leaves. *J. Plant Physiol.* 1996; 149:217–221.

110. Maksymiec W, Baszynski T. Chlorophyll fluorescence in primary leaves of excess Cu-treated runner bean plants depends on their growth stages and the duration of Cu-action. *J. Plant Physiol.* 1996; 149:196–200.

111. Skórzynska-Polit E, Baszynski T. Differences in sensitivity of the photosynthetic apparatus in Cd-stressed runner bean plants in relation to their age. *Plant Sci.* 1997; 128:11–21.

112. Krupa Z, Moniak M. The stage of leaf maturity implicates the response of the photosynthetic apparatus to cadmium toxicity. *Plant Sci.* 1998; 138:149–156.

113. Sárvári É, Gáspár L, Fodor F, Cseh E, Kröpfl K, Varga A, Baron M. Comparison of the effects of Pb treatment on thylakoid development in poplar and cucumber plants. *Acta Biol. Szeged.* 2002; 46:163–165.

114. Humbeck K, Quast S, Krupinska K. Functional and molecular changes in the photosynthetic apparatus during senescence of flag leaves from field-grown barley plants. *Plant Cell Environ.* 1996; 19:337–344.

115. Lukaszek M, Poskuta JW. Development of photosynthetic apparatus and respiration in pea seedlings during greening as influenced by toxic concentration of lead. *Acta Physiol. Plant.* 1998; 20:35–40.

116. Sárvári É, Fodor F, Cseh E, Szigeti Z, Láng F. Comparison of the effects of Cd stress and Fe-deficiency on the thylakoid development in poplar offsprings. *Plant Physiol. Biochem.* 2000; 38S:180.

117. Láng F, Sárvári É, Gáspár L, Fodor F, Cseh E. Influence of light intensity on thylakoid development under Cd stress in poplar. 12th Congress on Photosynthesis, CSIRO Publ., Brisbane, 2001:S3–23.

118. Appenroth K-J, Keresztes Á, Sárvári É, Jaglarz A, Fischer W. Multiple effects of chromate on *Spirodela polyrhiza*: Electron microscopy and biochemical investigations. *Plant Biol.* 2003; 5:315–323.

119. Appenroth KJ, Stöckel J, Srivastava A, Strasser RJ. Multiple effects of chromate on the photosynthetic apparatus of *Spirodela polyrhiza* as probed by OJIP chlorophyll a fluorescence measurements. *Environ. Pollut.* 2001; 115:49–64.

120. Nyitrai P, Bóka K, Gáspár L, Sárvári É, Lenti K, Keresztes Á. Characterization of the stimulating effect of low-dose stressors in maize and bean seedlings. *J. Plant Physiol.* 2003; 160:1175–1183.

121. Barceló J, Vázquez MD, Poschenrieder C. Structural and ultrastuctural disorders in cadmium-treated bush bean plants (*Phaseolus vulgaris* L.). *New Phytol.* 1988; 108:37–49.

122. Maksymiec W, Bednara J, Baszynski T. Responses of runner bean plants to excess copper as a function of plant growth stages: effects on morphology and structure of primary leaves and their chloroplast ultrastructure. *Photosynthetica* 1995; 31:427–435.

123. Moustakas M, Ouzounidou G, Eleftheriou EP, Lannoye R. Indirect effects of aluminium stress on the function of the photosynthetic apparatus. *Plant Physiol. Biochem.* 1996; 34:553–560.

124. Stoyanova DP, Tchakalova ES. Cadmium-induced ultrastructural changes in chloroplasts of the leaves and stems parenchyma in *Myriophyllum spicatum* L. *Photosynthetica* 1997; 34:241–248.

125. Barceló J, Poschenrieder C. Structural and ultrastructural changes in heavy metal exposed plants. In: *Heavy Metal Stress in Plants. From Molecules to Ecosystems.* Prasad MNV, Hagemaeyer J, eds. Berlin: Springer-Verlag, 1999:183–205.

126. Tziveleka L, Kaldis A, Hegedüs A, Kissimon J, Prombona A, Horváth G, Argyroudi-Akoyunoglou J. The effect of Cd on chlorophyll and light-harvesting complex II biosynthesis in greening plants. *Z. Naturforsch.* 1999; 54c:740–745.

127. Nishio JN, Abadía J, Terry N. Chlorophyll-proteins and electron transport during iron nutrition-mediated chloroplast development. *Plant Physiol.* 1985; 78:296–299.

128. Hihara Y, Sonoike K. Regulation, inhibition and protection of photosystem I. In: Aro EM, Andersson B, eds. *Regulation of Photosynthesis.* Dordrecht: Kluwer Academic Publishers, 2001:507–531.

129. Gallego SM, Benavídes M-P, Tomaro ML. Effect of heavy metal ion excess on sunflower leaves: evidence for involvement of oxidative stress. *Plant Sci.* 1996; 121:151–159.

130. Schützendübel A, Polle A. Plant responses to abiotic stresses: heavy metal-induced oxidative stress and protection by mycorrhization. *J. Exp. Bot.* 2002; 53:1351–1365.

131. Shalygo NV, Kolesnikova NV, Voronetskaya VV, Averina NG. Effects of Mn^{2+}, Fe^{2+}, Co^{2+} and Ni^{2+} on chlorophyll accumulation and early stages of chlorophyll formation in greening barley seedlings. *Russ. J. Plant Physiol.* 1999; 46:496–501.

132. Mishliwa-Kurdziel B, Strzalka K. Influence of metals on biosynthesis of photosynthetic pigments. In: Prasad MNV, Strzalka K, eds. *Physiology and Biochemistry of Metal Toxicity and Tolerance in Plants.* Dordrecht: Kluwer Academic Publishers, 2002:201–227.

133. Stiborova M, Doubravova M, Brezinova A, Friedrich A. Effect of heavy metal ions on growth and biochemical characteristics of photosynthesis of barley. *Photosynthetica* 1986; 20:418–425.

134. Agarwala SC, Bisht SS, Sharma CP. Relative effectiveness of certain heavy metals in producing toxicity and symptoms of iron deficiency in barley. *Can. J. Bot.* 1977; 55:1299–1307.

135. Imai I, Siegel SM. A specific response to toxic cadmium levels in red kidney bean embryo. *Physiol. Plant.* 1973; 29:118–120.

136. Chatterjee J, Chatterjee C. Phytotoxicity of cobalt, chromium and copper in cauliflower. *Environ. Pollut.* 2000; 109:69–74.

136. Milivojevic DB, Stojanovic DD, Drinic SD. Effects of aluminium on pigments and pigment-protein complexes of soybean. *Biol. Plant.* 2000; 43:595–597.

137. Stobart AK, Griffiths WT, Ameen-Bukhari I, Sherwood RP. The effect of Cd on the biosynthesis of chlorophyll in leaves of barley. *Physiol. Plant.* 1985; 63:293–298.

138. Padmaja K, Prasad DDK, Prasad ARK. Inhibition of chlorophyll synthesis in *Phaseolus vulgaris* L. seedlings by cadmium acetate. *Photosynthetica* 1990; 24:399–405.

139. Prasad DDK, Prasad ARK. Effect of lead and mercury on chlorophyll synthesis in mung bean seedlings. *Phytochemistry* 1987; 26:881–883.

140. Vajpayee P, Tripathi RD, Rai UN, Ali MB, Singh SN. Chromium accumulation reduces chlorophyll biosynthesis, nitrate reductase activity and protein content of *Nymphea alba. Chemosphere* 2000; 41:1075–1082.

141. Spiller SC, Castelfranco AM, Castelfranco PA. Effects of iron and oxygen on chlorophyll biosynthesis. I. *In vivo* observations on iron and oxygen-deficient plants. *Plant Physiol.* 1982; 69:107–111.

142. Beale SI. Enzymes of chlorophyll biosynthesis. *Photosynth. Res.* 1999; 60:43–73.

143. Böddi B, Oravecz AR, Lehoczki E. Effect of cadmium on organization and photoreduction of protochlorophyllide in dark-grown leaves and etioplast inner membrane preparations of wheat. *Photosynthetica* 1995; 31:411–420.

144. Lenti K, Fodor F, Böddi B. Mercury inhibits the activity of the NADPH:protochlorophyllide oxidoreductase (POR). *Photosynthetica* 2002; 40:145–151.

145. Horváth G, Droppa M, Oravecz Á, Raskin VI, Marder JB. Formation of the photosynthetic apparatus during greening. *Planta* 1996; 199:238–243.

146. Ouzounidou G, Eleftheriou EP, Karataglis S. Ecophysiological and ultrastructural effects of copper in *Thlaspi ochroleucum* (Cruciferae). *Can. J. Bot.* 1992; 70:947–957.

147. Baryla A, Carrier P, Franck F, Coulomb C, Sahut C, Havaux M. Leaf chlorosis in oilseed rape plants (*Brassica napus*) grown on cadmium-polluted soil: causes and consequences for photosynthesis and growth. *Planta* 2001; 212:696–709.

148. Abdel-Basset R, Issa AA, Adam MS. Chlorophyllase activity: effects of heavy metals and calcium. *Photosynthetica* 1995; 31:421–425.

149. Sandmann G, Böger P. Copper mediated lipid peroxidation process in photosynthetic membranes. *Plant Physiol.* 1980; 66:779–800.

150. Somashekaraiah BV, Padmaja K, Prasad ARK. Phytotoxicity of cadmium ions on germinating seedlings of mung bean (*Phaseolus vulgaris*): involvement of lipid peroxides in chlorophyll degradation. *Physiol. Plant.* 1992; 85:85–89.

151. Maksymiec W, Russa R, Urbanik-Sypniewska T, Baszynski T. Changes in acyl lipid and fatty acid compos-

ition in thylakoids of copper non-tolerant spinach exposed to excess copper. *J. Plant Physiol.* 1992; 140:52–55.

152. Szalontai B, Horváth LI, Debreczeny M, Droppa M, Horváth G. Molecular rearrangements of thylakoids after heavy metal poisoning, as seen by Fourier transform infrared (FTIR) and electron spin resonance (ESR) spectroscopy. *Photosynth. Res.* 1999; 61:241–252.

153. Küpper H, Küpper F, Spiller M. Environmental relevance of heavy metal-substituted chlorophylls using the example of water plants. *J. Exp. Bot.* 1996; 295:259–266.

154. Küpper H, Küpper F, Spiller M. *In situ* detection of heavy metal substituted chlorophylls in water plants. *Photosynth. Res.* 1998; 58:123–133.

155. Watanabe T, Kobayashi M. Chlorophylls as functional molecules in photosynthesis. Molecular composition *in vivo* and physical chemistry *in vitro*. Special Articles on Coordination Chemistry of Biologically Important Substances 1988; 4:383–395.

156. Watanabe T, Machida K, Suzuki H, Kobayashi M, Honda K. Photoelectrochemistry of metallochlorophylls. *Coord. Chem. Rev.* 1985; 64:207–224.

157. Devi SR, Prasad MNV. Membrane lipid alterations in heavy metal exposed plants. In: *Heavy Metal Stress in Plants. From Molecules to Ecosystems.* Prasad MNV, Hagemaeyer J, eds. Berlin: Springer-Verlag, 1999:99–116.

158. Stefanov KL, Pandev SD, Seizova KA, Tyankova LA, Popov SS. Effect of lead on the lipid metabolism in spinach leaves and thylakoid membranes. *Biol. Plant.* 1995; 37:251–256.

159. Murata N, Higashi SI, Fujimura Y. Glycerolipids in various preparations of photosystem II from spinach chloroplasts. *Biochim. Biophys. Acta* 1990; 1019:261–268.

160. Quartacci MF, Pinzino C, Sgherri CLM, Dalla Vecchia F, Navari-Izzo F. Growth in excess copper induces changes in the lipid composition and fluidity of PSII-enriched membranes in wheat. *Physiol. Plant.* 2000; 108:87–93.

161. Abadía A, Lemoine Y, Trémolières A, Ambard-Bretteville F, Rémy R. Iron deficiency in pea: effects on pigment, lipid and pigment-protein complex composition of thylakoids. *Plant Physiol. Biochem.* 1989; 27:659–687.

162. Hernández LE, Lozano-Rodrígez E, Gárate A, Carpena-Ruiz R. Influence of cadmium on the uptake, tissue accumulation and subcellular distribution of manganese in pea seedlings. *Plant Sci.* 1998; 132:139–151.

163. Sárvári É, Szigeti Z, Fodor F, Cseh E, Tussor K, Záray Gy, Veres Sz, Mészáros I. Relationship of iron deficiency and the altered thylakoid development in Cd treated poplar plants. 12th Congress on Photosynthesis, CSIRO Publ., Brisbane, 2001:S3–25.

164. Pätsikkä E, Kairavou M, Sersen F, Aro EM, Tyystjärvi E. Excess copper predisposes photosystem II to photoinhibition *in vivo* by outcompeting iron and

causing decrease in leaf chlorophyll. *Plant Physiol.* 2002; 129:1359–1367.

165. Römheld W. The chlorosis paradox: Fe inactivation as a secondary event in chlorotic leaves of grapevine. *J. Plant Nutr.* 2000; 23:1629–1643.

166. Cseh E. Metal permeability, transport and efflux in plants. In: Prasad MNV, Strzalka K, eds. *Physiology and Biochemistry of Metal Toxicity and Tolerance in Plants.* Dordrecht: Kluwer Academic Publishers, 2002:1–36.

167. Alcántara E, Romera FJ, Canete M, de la Guardia MD. Effects of heavy metals on both induction and function of root Fe(III) reductase in Fe-deficient cucumber (*Cucumis sativus* L.) plants. *J. Exp. Bot.* 1994; 45:1893–1898.

168. Fodor F, Sárvári É, Láng F, Szigeti Z, Cseh E. Effects of Pb and Cd on cucumber depending on the Fe-complex in the culture solution. *J. Plant Physiol.* 1996; 148:434–439.

169. Varga A, Martinez RMG, Záray G, Fodor F. Investigation of effects of cadmium, lead, nickel and vanadium contamination on the uptake and transport processes in cucumber plants by TXRF spectrometry. *Spectrochim. Acta B* 1999; 54:1455–1462.

170. Thys C, Vanthomme P, Schrevens E, de Proft M. Interactions of Cd with Zn, Cu, Mn and Fe for lettuce (*Lactuca sativa* L.) in hydroponic culture. *Plant Cell Environ.* 1991; 14:713–717.

171. Symeonidis L, Karataglis S. Interactive effects of cadmium, lead and zinc on root growth of two metal tolerant genotypes of *Holcus lanatus* L. *Biometals* 1992; 5:173–178.

172. Anderson JM, Thorne SW. The fluorescence properties of manganese-deficient spinach chloroplasts. *Biochim. Biophys. Acta* 1968; 162:122–134.

173. Abadía J, Nishio JN, Terry N. Chlorophyll-protein and polypeptide composition of Mn-deficient sugar beet thylakoids. *Photosynth. Res.* 1986; 7:237–245.

174. Simpson DJ, Robinson SP. Freeze-fracture ultrastructure of thylakoid membranes in chloroplasts from manganese-deficient plants. *Plant Physiol.* 1984; 74:735–741.

175. Baszynski T, Wajda L, Król M, Wolinska D, Krupa Z, Tukendorf A. Photosynthetic activities of cadmium-treated tomato plants. *Physiol. Plant.* 1980; 48:365–370.

176. Abadía J. Leaf responses to Fe deficiency: A review. *J. Plant Nutr.* 1992; 15:1699–1713.

177. Terry N, Zayed AM. Physiology and biochemistry of leaves under iron deficiency. In: Abadía J, ed. *Iron Nutrition in Soils and Plants.* Dordrecht: Kluwer Academic Publishers, 1995:283–294.

178. Morales F, Abadía A, Abadía J. Characterization of the xanthophylls cycle and other photosynthetic pigment changes induced by iron deficiency in sugar beet (*Beta vulgaris* L.). *Plant Physiol.* 1990; 94:607–613.

179. Ferraro F, Castagna A, Soldatini GF, Ranieri A. Tomato (*Lycopersicon esculentum* M.) T3238FER and T3238fer genotypes. Influence of different iron

concentrations on thylakoid pigment and protein composition. *Plant Sci.* 2003; 164:783–792.

180. Bertamini M, Muthuchelian K, Nedunchezhian N. Iron deficiency induced changes on the donor side of PS II in field grown grapevine (*Vitis vinifera* L. cv. Pinot noir) leaves. *Plant Sci.* 2002; 162:599–605.

181. Belkhodja R, Morales F, Quilez R, LopezMillan AF, Abadía A, Abadía J. Iron deficiency causes changes in chlorophyll fluorescence due to the reduction in the dark of the photosystem II acceptor side. *Photosynth. Res.* 1998; 56:265–276.

182. Morales F, Abadía A, Abadía J. Photosynthesis, quenching of chlorophyll fluorescence and thermal energy dissipation in iron-deficient sugar beet leaves. *Aust. J. Plant Physiol.* 1998; 25:403–412.

183. Sandström S, Park Y-I, Öquist G, Gustafsson P. CP43', the *isiA* gene product, functions as an excitation energy dissipator in the cyanobacterium *Synechococcus* sp. PCC 7942. *Photochem. Photobiol.* 2001; 74:431–437.

184. Morales F, Moise N, Quilez R, Abadía A, Abadía J, Moya I. Iron deficiency interrupts energy transfer from a disconnected part of the antenna to the rest of photosystem II. *Photosynth. Res.* 2001; 70:207–220.

185. Fodor F, Böddi B, Sárvári É, Záray G, Cseh E, Láng F. Correlation of iron content, spectral forms of chlorophyll and chlorophyll-proteins in iron deficient cucumber (*Cucumis sativus*). *Physiol. Plant.* 1995; 93:750–756.

186. Vassiliev IR, Yu JP, Jung YS, Schulz R, Ganago AO, McIntosh L, Golbeck JH. The cysteine-proximal aspartates in the F-X-binding niche of photosystem I — effect of alanine and lysine replacements on photoautotrophic growth, electron transfer rates, single-turnover flash efficiency, and EPR spectral properties. *J. Biol. Chem.* 1999; 274:9993–10001.

187. Platt-Aloia KA, Thomson WW, Terry N. Changes in plastid ultrastructure during iron nutrition-mediated chloroplast development. *Protoplasma* 1983; 114:85–92.

188. Siedlecka A, Krupa Z. Interaction between cadmium and iron. Accumulation and distribution of metals and changes in growth parameters of *Phaseolus vulgaris* L. seedlings. *Acta Soc. Bot. Pol.* 1996; 65:277–282.

189. Doan JM, Schoefs B, Ruban AV, Etienne AL. Changes in the LHCI aggregation state during iron repletion in the unicellular red alga *Rhodella violacea*. *FEBS Lett.* 2003; 533:59–62.

190. Horton P, Ruban AV, Walters RG. Regulation of light harvesting in green plants. *Annu. Rev. Plant Physiol. Plant Mol. Biol.* 1996; 47:655–684.

191. Baier M, Dietz KJ. The costs and benefits of oxygen for photosynthesizing plant cells. *Prog. Bot.* 1999; 60:282–314.

192. Hideg É, Kálai T, Hideg K, Vass I. Photoinhibition of photosynthesis *in vivo* results in singlet oxygen production detection *via* nitroxide-induced fluorescence quenching in broad bean leaves. *Biochemistry* 1998; 37:11405–11411.

193. Suh HJ, Kim CS, Jung J. Cytochrome b(6)/f complex as an indigenous photodynamic generator of singlet oxygen in thylakoid membranes. *Photochem. Photobiol.* 2000; 71:103–109.

194. Asada K. The water-water cycle in chloroplasts: scavenging of active oxygens and dissipation of excess photons. *Annu. Rev. Plant Physiol. Plant Mol. Biol.* 1999; 50:601–639.

195. Clijsters H, Cuypers A, Vangronsveld J. Physiological responses to heavy metals in higher plants; defence against oxidative stress. *Z. Naturforsch.* 1999; 54c:730–734.

196. Andersson B, Aro EM. Photodamage and D1 protein turnover in photosystem II. In: Aro EM, Andersson B, eds. *Regulation of Photosynthesis.* Dordrecht: Kluwer Academic Publishers, 2001:377–393.

197. Baena-Gonzales E, Aro EM. Biogenesis, assembly and turnover of photosystem II units. *Philos. Trans. R. Soc. Lond. B* 2002; 357:1451–1460.

198. Zolla L, Rinalducci S. Involvement of active oxygen species in degradation of light-harvesting proteins under light stresses. *Biochemistry* 2002; 41:14391–14402.

199. Makino A, Miyake C, Yokota A. Physiological functions of the water-water cycle (Mehler reaction) and the cyclic electron flow around PSI in rice leaves. *Plant Cell Physiol.* 2002; 43:1017–1026.

200. Godde D, Buchhold J, Ebbert V, Oettmeier W. Photoinhibition in intact spinach plants: effect of high light intensities on the function of the two photosystems and on the content of the D1 protein under nitrogen. *Biochim. Biophys. Acta* 1992; 1140:69–77.

201. Schmid V, Peter S, Schäfer C. Prolonged high light treatment of plant cells results in changes of the amount, the localization and the electrophoretic behavior of several thylakoid membrane proteins. *Photosynth. Res.* 1995; 44:287–295.

202. Godde D, Dannehl H. Stress induced chlorosis and increase in D1 protein turnover precedes photoinhibition in spinach suffering under combined magnesium and sulfur deficiency. *Planta* 1994; 195:291–300.

203. Santabarbara S, Cazzalini I, Rivadossi A, Garlaschi FM, Zucchelli G, Jennings RC. Photoinhibition *in vivo* and *in vitro* involves weakly coupled chlorophyll-protein complexes. *Photochem. Photobiol.* 2002; 75:613–618.

204. Santabarbara S, Neverov KV, Garlaschi FM, Zucchelli G, Jennings RC. Involvement of uncoupled antenna chlorophylls in photoinhibition in thylakoids. *FEBS Lett.* 2001; 491:109–113.

205. Kim CS, Jung J. The susceptibility of mung bean chloroplasts to photoinhibition is increased by an excess supply of iron to plants: a photobiological aspect of iron toxicity in plant leaves. *Photochem. Photobiol.* 1993; 58:120–126.

206. Suh HJ, Kim CS, Lee JY, Jung J. Photodynamic effect of iron excess on photosystem II function in pea plants. *Photochem. Photobiol.* 2002; 75:513–518.

207. Hollinderbäumer R, Ebbert V, Godde D. Inhibition of CO_2-fixation and its effect on the activity of photosys-

tem II, on D1-protein synthesis and phosphorylation. *Photosynth. Res.* 1997; 52:105–116.

208. Dannehl H, Herbik A, Godde D. Stress-induced degradation of the photosynthetic apparatus is accompanied by changes in thylakoid protein turnover and phosphorylation. *Physiol. Plant.* 1995; 93:179–186.

209. Giardi MT, Cona A, Geiken B, Kucera T, Masojídek J, Mattoo AK. Long-term drought stress induces structural and functional reorganization of photosystem II. *Planta* 1996; 199:118–125.

210. Niyogi KK. Photoprotection revisited: genetic and molecular approaches. *Annu. Rev. Plant Physiol. Plant Mol. Biol.* 1999; 50:333–359.

211. Allen JF, Forsberg J. Molecular recognition in thylakoid structure and function. *Trends Plant Sci.* 2001; 6:317–326.

212. Demmig-Adams B, Adams WW III. The role of xanthophylls cycle carotenoids in the protection of photosynthesis. *Trends Plant Sci.* 1996; 1:21–26.

213. Horton P. Are grana necessary for regulation of light harvesting? *Aust. J. Plant Physiol.* 1999; 26:659–669.

214. Garab G, Cseh Z, Kovács L, Rajagopal S, Várkonyi Z, Wentworth M, Mustárdy L, Der A, Ruban AV, Papp E, Holzenburg A, Horton P. Light-induced trimer to monomer transition in the main light-harvesting antenna complex of plants: thermo-optic mechanism. *Biochemistry* 2002; 41:15121–15129.

215. Bassi R, Caffarri S. Lhc proteins and the regulation of photosynthetic light harvesting function by xanthophylls. *Photosynth. Res.* 2000; 64:243–256.

216. Andersson J, Walters RG, Horton P, Jansson S. Antisense inhibition of the photosynthetic antenna proteins CP29 and CP26: implications for the mechanism of protective energy dissipation. *Plant Cell* 2001; 13:1193–1204.

217. Aspinall-O'Dea M, Wentworth M, Pascal A, Robert B, Ruban A, Horton P. *In vitro* reconstitution of the activated zeaxanthin state associated with energy dissipation in plants. *Proc. Natl. Acad. Sci. USA* 2002; 99:16331–16335.

218. Stewart DH, Brudvig GW. Cytochrome b(559) of photosystem II. *Biochim. Biophys. Acta* 1998; 1367:63–87.

219. Öquist G, Chow WS, Anderson JM. Photoinhibition of photosynthesis represents a mechanism for the long-term regulation of photosystem II. *Planta* 1992; 186:450–460.

220. Chow WS, Lee HY, Park YI, Park YM, Hong YN, Anderson JM. The role of inactive photosystem-II-mediated quenching in a last-ditch community defence against high light stress *in vivo*. *Philos. Trans. R. Soc. Lond. B* 2002; 357:1441–1449.

221. Peterson RB, Havir EA. Contrasting modes of regulation of PSII light utilization with changing irradiance in normal and *psbS* mutant leaves of *Arabidopsis thaliana*. *Photosynth. Res.* 2003; 75:57–70.

222. Huner NPA, Öquist G, Sarhan F. Energy balance and acclimation to light and cold. *Trends Plant Sci.* 1998; 3:224–230.

223. Burda K, Kruk J, Schmid GH, Strzalka K. Inhibition of oxygen evolution in photosystem II by Cu(II) ions is associated with oxidation of cytochrome b(559). *Biochem. J.* 2003; 371:597–601.

224. Romanowska E, Igamberdiev AU, Parys E, Gardeström P. Stimulation of respiration by Pb^{2+} in detached leaves and mitochondria of C_3 and C_4 plants. *Physiol. Plant.* 2002; 116:148–154.

225. Boucher N, Carpentier R. Hg^{2+}, Cu^{2+}, and Pb^{2+}-induced changes in photosystem II photochemical yield and energy storage in isolated thylakoid membranes: a study using simultaneous fluorescence and photoacoustic measurements. *Photosynth. Res.* 1999; 59:167–174.

226. Anderson JM, Chow WS, Park Y-I. The grand design of photosynthesis: acclimation of the photosynthetic apparatus to environmental cues. *Photosynth. Res.* 1995; 46:129–139.

227. Walters RG, Rogers JM, Shephard F, Horton P. Acclimation of *Arabidopsis thaliana* to the light environment: the role of photoreceptors. *Planta* 1999; 209:517–527.

228. Pfannschmidt T, Nilsson A, Allen JF. Photosynthetic control of chloroplast gene expression. *Nature* 1999; 397:625–628.

229. Escoubas J-M, Lomas M, LaRoche J, Falkowski PG. Light intensity regulation of *cab* gene transcription is signaled by the redox state of the plastoquinone pool. *Proc. Natl. Acad. Sci. USA* 1995; 92:10237–10241.

230. Durnford DG, Falkowski PG. Chloroplast redox regulation of nuclear gene transcription during photoacclimation. *Photosynth. Res.* 1997; 53:229–241.

231. Danon A, Mayfield SP. Light-regulated translation of chloroplast messenger RNAs through redox potential. *Science* 1994; 266:1717–1719.

232. Trebitsh T, Levitan A, Sofer A, Danon A. Translation of chloroplast *psbA* mRNA is modulated in the light by counteracting oxidizing and reducing activities. *Mol. Cell. Biol.* 2000; 20:1116–1123.

233. Karpinski S, Escobar C, Karpinska B, Creissen G, Mullineaux PM. Photosynthetic electron transport regulates the expression of cytosolic ascorbate peroxidase genes in *Arabidopsis* during excess light stress. *Plant Cell* 1997; 9:627–640.

234. Teramoto H, Nakamori A, Minagawa J, Ono T. Light-intensity-dependent expression of *Lhc* gene family encoding light-harvesting chlorophyll-*a/b* proteins of photosystem II in *Chlamydomonas reinhardtii*. *Plant Physiol.* 2002; 130:325–333.

235. Lindahl M, Yang DH, Andersson B. Regulatory proteolysis of the major light-harvesting chlorophyll *a/b* protein of photosystem II by a light-induced membrane-associated enzymic system. *Eur. J. Biochem.* 1995; 231:503–509.

236. Yang, DH, Webster J, Adam Z, Lindahl M, Andersson B. Induction of acclimative proteolysis of the light-harvesting chlorophyll *a/b* protein of photosystem II in response to elevated light intensities. *Plant Physiol.* 1998; 118:827–834.

237. Schuster G, Dewit M, Staehelin LA, Ohad I. Transient inactivation of the thylakoid photosystem II light-harvesting protein kinase system and concomitant changes in intramembrane particle size during photoinhibition of *Chlamydomonas reinhardtii. J. Cell Biol.* 1986; 103:71–80.

238. Bailey S, Walters RG, Jansson S, Horton P. Acclimation of *Arabidopsis thaliana* to the light environment: the existence of separate low light and high light responses. *Planta* 2001; 213:794–801.

239. Durnford DG, Price JA, McKim SM, Sarchfield ML. Light-harvesting complex gene expression is controlled by both transcriptional and post-transcriptional mechanisms during photoacclimation in *Chlamydomonas reinhardtii. Physiol. Plant.* 2003; 118:193–205.

240. Barbato R, Friso G, Rigoni F, Dalla Vecchia F, Giacometti, GM. Structural changes and lateral distribution of photosystem II during donor side photoinhibition of thylakoids. *J. Cell Biol.* 1992; 119:325–335.

241. Tan S, Wolfe GR, Cunningham X Jr, Gantt E. Decrease of polypeptides in the PSI antenna complex with increasing growth irradiance in the red alga *Porphyridium cruentum. Photosynth. Res.* 1995; 45:1–10.

242. Montané M-H, Dreyer S, Triantaphylides C, Kloppstech K. Early light-inducible proteins during long-term acclimation of barley to photooxidative stress caused by light and cold: high level of accumulation by posttranscriptional regulation. *Planta* 1997; 202:293–302.

243. Anderson JM, Aro EM. Grana stacking and protection of photosystem II in thylakoid membranes of higher plant leaves under sustained high light irradiance: an hypothesis. *Photosynth. Res.* 1994; 41:315–326.

244. Biswal B. Chloroplast metabolism during leaf greening and degreening. In: Pessarakli M, ed. *Handbook of Photosynthesis*. New York: Marcel Dekker, 1997: 71–81.

Section XIV

Photosynthesis in the Past, Present, and Future

46 Origin and Evolution of C₄ Photosynthesis

Bruce N. Smith
Brigham Young University

CONTENTS

I. EVOLUTION OF PHOTOSYNTHESIS AND THE ATMOSPHERE

A. ATMOSPHERE OF THE EARLY EARTH

All the elements, including those common in living things, were synthesized from primordial hydrogen in the interior of stars [1]. As a result of supernovas and other stellar instabilities, many elements were spewn into space. Since hydrogen and the noble gases are greatly depleted on Earth as compared with their cosmic abundances [2], it is likely that the chunks of matter giving rise to the protoplanet did not carry with them gaseous shells of their own. As a result of contraction and redistribution of materials in the developing plant, an atmosphere of water and CO_2 was released with lesser amounts of CO, N_2, H_2, CH_4, H_2S, NH_3, HF, HCl, and others [3,4]. In time, this highly reduced atmosphere has become our present gaseous environment of nitrogen (78%), oxygen (20.9%), argon (0.9%), and a small amount of carbon dioxide (0.03%), and other gases [3]. Some have suggested [5,6] that gradual oxidation of the atmosphere has been due entirely to physical dissociation of water vapor. Most evidence, however, points to a biological origin for the gradually increasing oxygen content of the atmosphere [3,7]. As outlined in Figure 46.1, photosynthetic oxygen was used initially to oxidize other components of the atmosphere, only later to become an ever-increasing portion of the atmosphere.

Evidence [8] now indicates that several times in the history of the Earth there have been major outflows

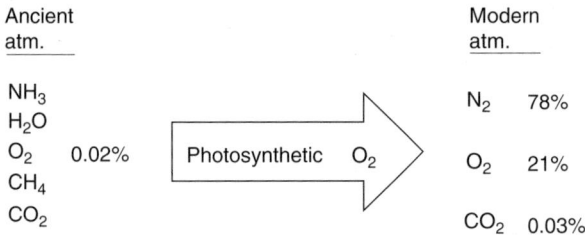

FIGURE 46.1 Scheme for transforming the ancient atmosphere of the Earth into the present atmosphere by photosynthethic oxygen. (Adapted from *Cosmos, Earth, and Man,* Yale University Press, New Haven, CT, 1978.)

of magma from the mantle accompanied by massive exhalations of CO_2, giving an atmospheric CO_2 concentration tenfold greater than at present. This led to an increase in O_2 (from photosynthesis) and warming of the Earth's surface by a greenhouse effect as well as massive deposits of carbonates or burial of organic matter under anoxic conditions, which became coal and oil [9].

B. PROKARYOTIC PHOTOSYNTHESIS

Given water, a reducing atmosphere, and energy sources, organic molecules can be synthesized abiotically [4]. Organic molecules have even been reported in deep space. The first organisms were probably anaerobic heterotrophs similar to modern archaeobacteria. Bacterial photosynthesis [10] probably developed early with light energy used to produce organic matter with H_2S, NH_3, or organic substrates serving as hydrogen donors. Since water was abundant, it soon became the major source of hydrogen, with oxygen released as a by-product.

Several lines of evidence point to increasing levels of oxygen with time. Fossil microalgae over 3 billion years old have been found [3,11]. In morphology, ancient fossils are very similar to recent cyanobacteria. Some of these forms are unicellular, others are filamentous, and still others are colonial. Presumably the first photosynthetic organisms were anoxic autotrophs, and oxygen may well have been an objectionable by-product. Cloud [12] has postulated that dissolved ferrous iron could have been a convenient oxygen acceptor and that deposition of oxidized iron in sediments must have taken place long before oxygen could have entered the atmosphere in significant quantities. Indeed, extensive deposition of banded iron sediments occurred 2 to 3 billion years ago [12], and it is only in the last 2 billion years that atmospheric oxygen has been present in significant amounts. This analysis is supported by studies of sulfur in Precam-

brian rocks [13], which indicate that the oxygen pressure in the Earth's atmosphere must have been very low at the time of sulfur deposition. Boychenko [14] has noted that change from fermentation to more recent aerobic respiration involved developments by organisms of various metal-containing enzymes. The evolution in organisms of oxidation functions catalyzed by these enzymes paralleled the increase in redox potentials of reactions occurring in the biosphere during successive geological eras. Thus, the pattern of respiration is that expected if the most primitive organisms evolved in a reduced environment and more recent forms in a more oxidized environment.

Urey [15] proposed that in the Earth's early atmosphere, oxygen was kept below 0.02% of the total atmosphere by the freezing of water vapor in the so-called cold trap at around 10-km altitude and the circumstance that the same wavelengths of ultraviolet sun rays, which dissociate water and form free oxygen, are also absorbed by the same oxygen to form ozone. Hence, there is competition for the use of this part of the spectrum, and the more free oxygen there is in the atmosphere, the less light of the proper wavelength is available for further dissociation of water. Thus, 0.02% is an important level that cannot be broken by any inorganic process but could be broken by photosynthesis [15].

Direct evidence that most of our present atmospheric oxygen came from photosynthesis is seen in the Dole effect, illustrated in Figure 46.2. The two most common stable isotopes of oxygen are ^{18}O and ^{16}O,

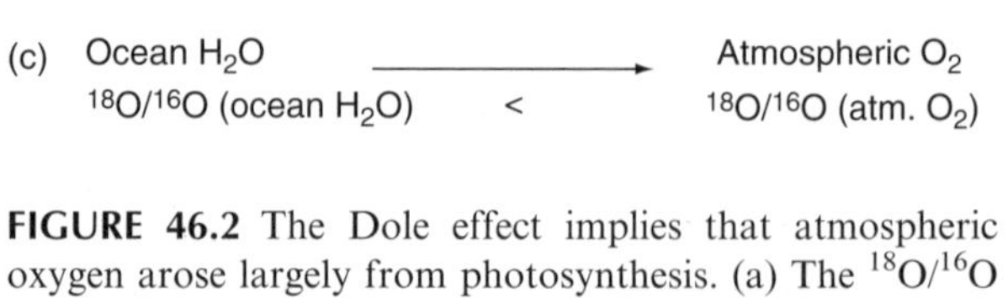

(a) $2 H_2O$ $\xrightarrow[\text{or U V}]{\text{Electrolysis}}$ $2 H_2 + O_2$

$^{18}O/^{16}O (H_2O)$ $>$ $^{18}O/^{16}O (O_2)$

(b) $CO_2 + 2 H_2O$ $\xrightarrow{\text{Photosynthesis}}$ $(CH_2O) + O_2 + H_2O$

$^{18}O/^{16}O (H_2O)$ $<$ $^{18}O/^{16}O (O_2)$

(c) Ocean H_2O $\xrightarrow{\hspace{2cm}}$ Atmospheric O_2
$^{18}O/^{16}O$ (ocean H_2O) $<$ $^{18}O/^{16}O$ (atm. O_2)

FIGURE 46.2 The Dole effect implies that atmospheric oxygen arose largely from photosynthesis. (a) The $^{18}O/^{16}O$ ratio of oxygen released from water by electrolysis or ultraviolet radiation is less than the $^{18}O/^{16}O$ ratio of the water itself. (b) The $^{18}O/^{16}O$ ratio of oxygen released during photosynthesis is greater than the $^{18}O/^{16}O$ ratio of the source water. (c) Modern atmospheric oxygen has a larger $^{18}O/^{16}O$ ratio than ocean water. (Adapted from M. Dole and G. Jenks, *Science, 100*:409 [1944] and G. A. Lane and M. Dole, *Science, 123*:574 [1956]. With permission.)

which normally occur on earth in a ratio of one atom of ^{18}O for every 250 atoms of ^{16}O. This proportion can be altered slightly by various biological and physical processes. For instance, water may be dissociated into oxygen and hydrogen by physical means such as ultraviolet radiation or electrolysis. The O_2 released is depleted in the naturally occurring stable isotope ^{18}O with respect to the water reservoir from which it came. On the other hand, oxygen liberated from water during photosynthesis is slightly enriched in ^{18}O relative to the source water [16]. This effect is enhanced by preferential uptake of ^{16}O in respiration [17]. The $^{18}O/^{16}O$ ratio of ocean water is quite constant everywhere, as is the ratio for atmospheric oxygen. However, oxygen from the atmosphere is considerably enriched in the heavy isotope with respect to ocean water.

Thus, for the first half of the history of the Earth, the atmosphere was reduced with an oxygen content less than 0.02% and a carbon content (CH_4, CO, CO_2, etc.) as much as 16-fold higher than our present atmosphere [8]. The oxygen content of the atmosphere then began to increase with a concomitant decrease in carbon [3]. Most of the carbon moved from the atmosphere into the ocean as dissolved CO_2 and bicarbonate. There the carbon was deposited as calcium carbonate by marine organisms [18]. This resulted (Table 46.1) in more than 78% of the carbon on Earth deposited as carbonate sediments, with another 21% in sedimentary shales and sandstones. With the relatively small amounts of coal and petroleum, the total amount of carbon that is bound in sedimentary rocks and thus largely unavailable to organisms is

99.7% of the total carbon on Earth. Only 0.3% by contrast is labile or available for the carbon cycle in the biosphere. Since there is so little available, the turnover time for the labile carbon on Earth is a very rapid 100 years. Today, by contrast with the early Earth, we have an atmosphere with a great deal of oxygen (21%) and very little CO_2 (0.03%). This change may have been gradual [3] or punctuated by fluctuations in carbon from outgassing of the mantle and the consequent buildup of oxygen with fluctuations in O_2 ranging from 14% to 35% of the atmosphere [8]. These changes have been mediated by organisms with important consequences for them.

II. C$_3$ PHOTOSYNTHESIS AND PHOTORESPIRATION

A. CHLOROPLASTS

Visible light is transformed into chemical potential energy in the thylakoid membranes, which make up the grana of the chloroplast. This chemical energy, such as ATP and NADPH, is then used to reduce CO_2 to the level of carbohydrates, fat, and protein. The source of carbon for all terrestrial plants is atmospheric CO_2. Algae and submerged spermatophytes utilize either dissolved CO_2 or bicarbonate. The overall reaction for photosynthesis can be written as follows:

$$CO_2 + 2H_2O \xrightarrow{\text{light}} (CH_2O)n + O_2 + H_2$$

Calvin [19] and coworkers have elucidated very elegantly the mechanism by which inorganic carbon is synthesized into complex organic molecules. The initial fixation step involves the enzyme ribulose bisphosphate carboxylase (Rubisco) in the stroma of the chloroplast. This enzyme can make up as much as half of the soluble protein in some leaves. The reaction requires CO_2 and a five-carbon sugar (ribulose-1,5-bisphosphate), and the product is two molecules of a three-carbon acid (3-phosphoglyceric acid [PGA]). This mechanism for carbon fixation is found in virtually all green tissues. Rubisco, however, has another property — it also reacts with oxygen [20]. As shown in Figure 46.3, the products of this reaction with oxygen are PGA and phosphoglycolate. Oxygen and carbon dioxide thus compete for the same active site on the enzyme.

B. MICROBODIES AND MITOCHONDRIA

After removal of the phosphate in the chloroplast (Figure 46.3), the glycolate moves into the microbody (peroxisome or glyoxysome) where it is oxidized with hydrogen peroxide to glyoxylate [21] and then trans-

TABLE 46.1
Distribution of Carbon on Earth Expressed as Percentage of Total Global Carbon

	% Total Carbon
Nonlabile	99.7
Carbonates and shales	78.46
Sandstones	21.22
Coals	0.037
Petroleum	0.0012
Labile	0.3
Oceans (CO_3 and HCO_3)	2.43
Oceans (dissolved organic matter)	0.020
Biosphere	0.011
Atmosphere	0.004

Note: Labile carbon is that available for the carbon cycle.

Source: Adapted from E. I. Hamilton, *Applied Geochronology*, Academic Press, New York, 1965. With permission

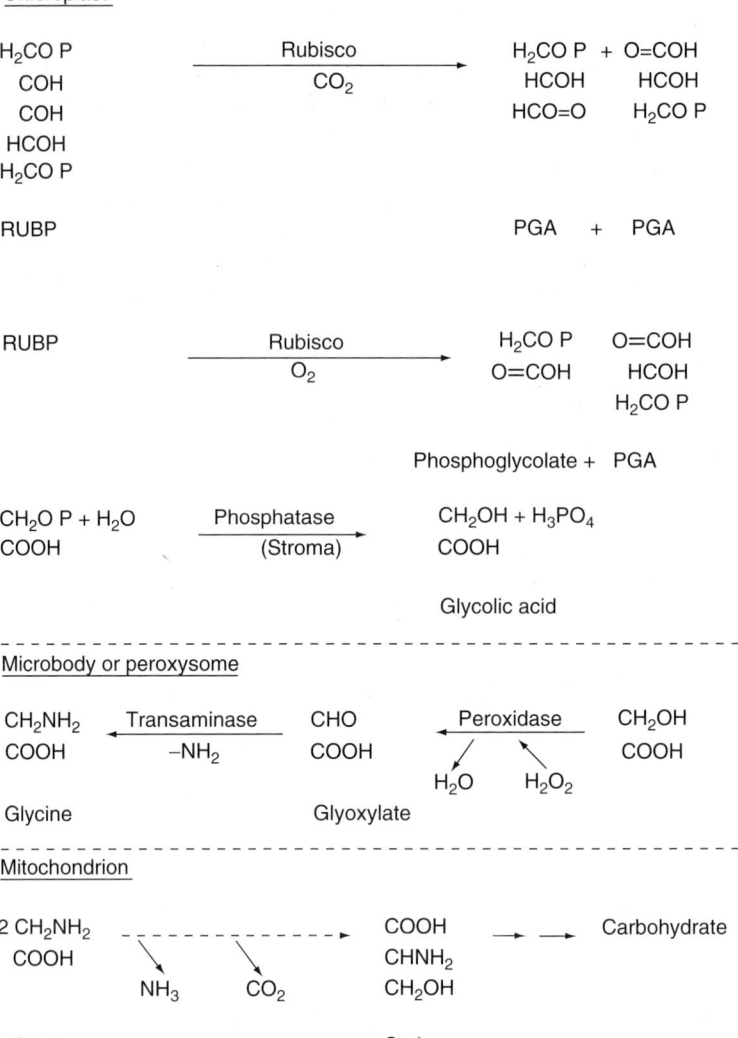

FIGURE 46.3 Photorespiration begins in the stroma of the chloroplast with O_2 rather than CO_2 binding with Rubisco and ribulose bisphosphate to give phosphoglycolate and PGA. The phosphoglycolate is subsequently dephosphorylated to glycolic acid, which is subsequently transported into the microbody or peroxisome. There the glycolic acid is oxidized by hydrogen peroxide to glyoxylate, which is then transaminated to form glycine. The amino acid moves into the mitochondrion, where two molecules of glycine form one molecular of serine with loss of ammonia and CO_2. The serine can be further metabolized. Photorespiration is thus the light-dependent and oxygen-dependent evolution of carbon dioxide and ammonia. (Adapted from N. E. Tolbert and J. Preiss (eds.), *Regulation of Atmospheric CO_2 and O_2 by Photosynthetic Carbon Metabolism,* Oxford University Press, Oxford, 1994. With permission.)

aminated to produce glycine. Glycine is transported into the mitochondrion, where two molecules of glycine produce a molecule of serine with the evolution of ammonia and CO_2 [22]. Much of the ammonia and CO_2 are refixed [23]. The serine can then enter any of several metabolic pathways [21].

Photorespiration is light-dependent, oxygen-dependent CO_2 evolution. The amount of carbon lost in the light can be as much as 50-fold greater than the carbon lost in the dark by the same tissues [24]. Photosynthesis exceeds photorespiration by a factor of 3 or more, so special techniques are necessary to measure CO_2 evolution in the light. Since Rubisco has a greater affinity for CO_2 than for O_2, both photosynthesis and growth are stimulated in O_2-depleted or CO_2-enriched atmospheres [25].

Rubisco, regardless of source, has oxygenase activity [26]. Indeed, even Rubisco from anaerobic photosynthetic bacteria can show oxygen inhibition *in vitro*

[27]. Rubisco activity has been conserved during evolution possibly because of the key and ubiquitous role played by the enzyme in photosynthetic CO_2 fixation. Plant species have characteristic and different rates of photorespiration, sometimes expressed as compensation points [28]. These differences probably do not represent differences in Rubisco itself but in stomatal function and leaf anatomy [29].

Early in the history of the Earth, the environment was rich in carbon but poor in oxygen. Under these circumstances, photosynthesis evolved with a key step catalyzed by Rubisco. In the absence of oxygen this was a very efficient process. However, as oxygen concentration in the atmosphere increased due to photosynthetic splitting of water and release of oxygen, O_2 competition for the active site of Rubisco became important enough to decrease the efficiency of carbon fixation in photosynthesis. Because the challenge presented by rising atmospheric oxygen concentration could not be met by modification of the enzyme to prevent oxygenation, adequate means for the disposal of the phosphoglycolate waste product would have been required. In aquatic organisms dephosphorylation and excretion of glycolate may have been employed, and many modern algal species still do this [30].

III. C₄ PHOTOSYNTHESIS

A. DISCOVERY AND IDENTIFICATION

A small number of plant species have relatively recently evolved a mechanism to protect Rubisco from oxygen and thus improve the efficiency of photosynthesis. The mechanism includes a syndrome of distinguishing anatomical and physiological characters associated with light harvesting and fixation of carbon [31]. Kortschak et al. [32] described malate and aspartate as the first products of photosynthesis in sugarcane. Since the first products have four carbons, the syndrome has been called C₄ photosynthesis. A number of sugarcane researchers and weed scientists soon expanded our knowledge of C₄ species [33]. C₄ plants differ from C₃ plants in anatomy, physiology, biochemistry, carbon isotopic ratios, and ecology [34].

B. KRANZ ANATOMY

Anatomical features include a chlorenchymatous sheath of large, thick-walled cells (Kranz cells) surrounding vascular bundles of leaves [35]. In turn, mesophyll cells form a cylinder around the bundle sheath cells. The function of bundle sheath cells and mesophyll cells is to channel CO_2 and keep Rubisco in a high-CO_2, low-O_2 environment and thus minimize photorespiratory carbon loss [36].

C. C₃–C₄ INTERMEDIATES AND GENETICS

A number of species have been identified that lack fully developed Kranz anatomy and have C₃-like carbon isotopic ratios but reduced rates of photorespiration. *Flaveria* in the Asteraceae and *Panicum* in the Poaceae are genera that contain C₃, C₄, and C₃–C₄ intermediate species. Breeding experiments indicate that many genes are probably involved, as is a degree of maternal inheritance [37].

D. EVOLUTION OF C₄ PHOTOSYNTHESIS

A fossil C₄ bunchgrass from the early Pliocene has been described [38]. Based on carbon isotopic ratios of terrestrial plant carbon in Bengal fan sediments, it was concluded that in the late Miocene there was a rapid increase in C₄ plants, particularly in the lowlands [39,40]. Since the subfamily Eragrostoideae of the family Poaceae is entirely C₄ and is found on all continents [41], C₄ plants may have made an appearance before the breakup of Pangaea. C₃ photosynthesis is the only mode found to occur in algae, bryophytes, pteridophytes, gymnosperms, and the great majority of angiosperms [42]. C₄ photosynthesis has been described in at least 18 families and occurred more than once in several families (Table 46.2). These families, for the most part, are not closely related to one another, and it is most likely that C₄ photosynthesis is an adaptation to warm, arid, semitropical to tropical conditions.

IV. C₄ BIOCHEMISTRY

A. LIGHT-HARVESTING EFFICIENCY

C₄ plants at optimum temperatures are often not saturated even under full sunlight [43]. Under different conditions other factors may be limiting. While the light-harvesting unit in C₃ plants often contains 250 to 350 chlorophyll molecules, in C₄ plants the

---

TABLE 46.2
Families in Which the C₄ Photosynthetic Pathways Is Known To Occur

Monocotyledonae
  Cyperaceae
  Liliaceae
  Poaceae

Dicotyledonae
  Acanthaceae
  Aizoaceae
  Amaranthaceae
  Asteraceae
  Boraginaceae
  Capparadaceae
  Caryophyllaceae
  Chenopodiaceae
  Clemaceae
  Euphorbiaceae
  Molluginaceae
  Nyctaginaceae
  Polygonaceae
  Portulacaceae
  Scrophulariaceae
  Zygophyllaceae

Source: From B. N. Smith, *BioSystems*, 8:24 (1976), J. R. Ehleringer, *HortScience*, 14:217 (1979), and J. Lloyd and G. D. Farquhar, *Oecologia*, 99:201 (1994). With permission.

units may be smaller (75 chlorophyll molecules) but more numerous [44].

B. CARBON FIXATION

In the cytoplasm of mesophyll cells, CO_2 is combined with phosphoenolpyruvate (PEP) by means of PEP carboxylase to form oxaloacetic acid (OAA). What happens next depends on the species of plant (see Table 46.3). For NADP — malic enzyme (NADP–ME) plants, the OAA moves to malic acid, which then moves through plasmodesmata to the bundle sheath cells, where in the chloroplast malate is decarboxylated with CO_2, combining with Rubisco and entering the Calvin cycle. The remaining pyruvate is transported back into the mesophyll cell, where pyruvate Pi dikinase removes the terminal

two phosphates from ATP and transforms pyruvate to PEP. In NADP–ME species, OAA undergoes transamination to form aspartate, which is transported to the bundle sheath cell and deaminated back to malate, which in a bundle sheath mitochondrion is decarboxylated. The CO_2 moves to the chloroplast and enters the Calvin cycle, while the pyruvate is transaminated to alanine for transport back to the mesophyll cell. In the mesophyll cell, the sequence is alanine to pyruvate to PEP via pyruvate Pi dikinase. In the third group the OAA is transaminated to aspartate. The aspartate moves to the bundle sheath cell where an aminotransferase again produces OAA, which in the cytoplasm is decarboxylated with ATP yielding PEP [45].

C. NITRATE AND SULFATE REDUCTION

In C_3 plants, nitrate reductase is found largely in the cytoplasm of mesophyll and palisade parenchyma cells of the leaf. Nitrite reductase is found in the chloroplasts of the same cells. Sulfate reduction also takes place in chloroplasts of leaf cells.

In C_4 plants Rubisco is found in bundle sheath cells, while PEP carboxylase is found in the mesophyll cells. A similar division of labor is noted for nitrate and sulfate reduction. Nitrate and nitrite reduction occurs in mesophyll cells and chloroplasts [46], while sulfate reduction takes place in bundle sheath cells [47].

V. CRASSULACEAN ACID METABOLISM

Succulent desert plants often exhibit a mode of carbon fixation called crassulacean acid metabolism (CAM). CAM plants under conditions of drought, short hot days, and long cool nights, will fix CO_2 at night via PEP carboxylase and store it as malic acid in the vacuole [48]. During the day they are obligate and can only fix carbon in this way. Others are facultative and only use the CAM mode under conditions of water stress. Under favorable conditions they can utilize the C_3 mode of photosynthesis. Although certain superficial similarities do exist between C_4 and CAM, the differences are fundamental enough so that they probably had a separate origin.

The CAM pathway has now been identified in 27 families [34,42] including some xeric ferns and the gymnosperm *Welwitschia* (see Table 46.4). Some families contain all three photosynthetic types. CAM plants appear to be more united by adaptations to xeric environments than by phylogeny. Several separate evolutionary origins are evident. For instance, in the large and diverse genus *Euphorbia*, both CAM and C_4 species arose from C_3 ancestors but from very different sections of the genus [49].

TABLE 46.3
Differences in C_4 Plants

NADP–ME type

  Malate + $NADP^+$?CO_2 + NADPH + H^+ +
  pyruvate (chloroplast)[a]
  Chloroplasts centrifugal in BSC, grana greatly reduced[b]
  Transport forms: malate into BSC, pyruvate into MC[c]
  Energy requirement: 5ATP + 2NADPH/CO_2 fixed[d]
  Corn, sugarcane, sorghum, crabgrass (*Digitaria*)[e]

NAD–ME type

  Malate + NAD+ ?CO_2 + NADH + H^+ +
  pyruvate (mitochondrion)
  Chloroplasts centripetal in BSC, grana abundant
  Transport forms: aspartate into BSC, alanine into MC
  Energy requirement: 5ATP + 2NADPH/CO_2 fixed
  Pigweed (*Amaranthus*), purslane (*Portulaca*), millet
  (*Panicum miliacium*)

PEP–CK type

  OAA + ATP ?CO_2 + PEP + ADP (cytoplasm)
  Chloroplasts centrifugal in BSC, grana abundant
  Transport forms: OAA or malate into BSC, PEP into MC
  Energy requirement: 4ATP + 2NADPH/CO_2 fixed
  Guinea grass (*Panicum maximum*), Rhoades grass
  (*Chloris guyana*)

C_3 type

  Energy requirement: 3ATP + 2NADPH/CO_2 fixed

[a]Differences in biochemistry.
[b]Differences in anatomy.
[c]Differences in transport.
[d]Differences in energy requirement.
[e]Examples.
BSC = Bundle sheath cell, MC = mesophyll cell.

Source: From G. E. Edwards and S. C. Huber, in *The Biochemistry of Plants*, Vol. 8 (M. D. Hatch and N. K. Boardman, eds.), Academic Press, New York, 1981, p. 237. With permission.

TABLE 46.4
Families in Which CAM Is Known To Occur

Filicinae
  Polypodiaceae

Gymnospermae
  Welwitschiaceae

Angiospermae
  Monocotyledonae

Agavaceae	Liliaceae
Bromeliaceae	Orchidaceae

  Dicotyledonae

Aizoaceae	Euphorbiaceae
Asclepiadaceae	Geraniaceae
Asteraceae	Labiatae
Bataceae	Oxalidaceae
Cactaceae	Passifloraceae
Capparaceae	Piperaceae
Caryophyllaceae	Plantaginaceae
Chenopodiaceae	Portuloacaceae
Crassulaceae	
Didiereaceae	

Source: From B. N. Smith, *BioSystems, 8*:24 (1976), J. R. Ehleringer, *HortScience, 14*:217 (1979), and J. Lloyd and G. D. Farquhar, *Oecologia, 99*:201 (1994). With permission.

CAM plants protect Rubisco from oxygen by keeping the stomates closed in the daytime. The partial pressure of oxygen must be low in the vicinity of Rubisco when CAM is operative. Water loss is, of course, greatly restricted with the stomates closed, but so is gas exchange. The only source of CO_2 for photosynthesis in the absence of communication with the atmosphere is from malate decarboxylation. Under these conditions, rates of photosynthesis are very low, but the strategy of CAM plants seems to be survival rather than rapid growth.

VI. C$_4$ ECOLOGY

A. GEOGRAPHIC DISTRIBUTION

An estimated 21% of total global photosynthesis is by C$_4$ plants [50]. Most C$_4$ plants are found between 45°N and 35°S latitudes. They are more abundant in warm, dry climates and seem to be rather recent invaders of the colder temperate zones. C$_4$ species in both grasses and dicots in North America are more abundant in the warmer latitudes and progressively less so with movement north [51,52].

B. COMPETITION AND HABITAT SELECTION

In warm deserts where plants can grow most of the year, C$_3$ plants tend to be winter and spring active, while C$_4$ plants grow in the late spring and summer [42]. In a study on short-grass prairies of Wyoming, C$_3$ grasses were predominant early in the growing season and at high altitudes, while C$_4$ grasses grew later in the year and showed a preference for lower altitudes [53].

VII. FUTURE OF C$_4$ PHOTOSYNTHESIS

A. GREENHOUSE EFFECT

Carbon dioxide concentrations in the atmosphere undergo seasonal and even daily fluctuations but have shown an increase of at least 70 ppm in the past 60 years due to human activity [25]. The atmospheric CO_2 concentration has fluctuated throughout time, with recent levels being very low [54]. Any increase in CO_2 obviously will benefit C$_3$ photosynthesis. However, since the greenhouse effect will also raise the temperature and may change weather patterns, C$_4$ plants may still have some advantage.

B. GENETICS AND SELECTION

As mentioned above, we are still far from understanding the genetic basis for C$_4$ photosynthesis. However, given the rapid progress in so many areas of genetics, we may soon know enough to attempt hybridization or genetic manipulation. Care must be taken as no pathway has the advantage under all situations. Ecotypes always grow best where they are well adapted, whether they are C$_3$ or C$_4$ plants [55].

VIII. SUMMARY

The carbon content of the atmosphere has decreased and the oxygen content has increased, perhaps more than once, during the history of life on Earth. C$_3$ photosynthesis became established first. As the oxygen content of the atmosphere increased, O_2 competition with CO_2 for Rubisco became large enough to reduce the efficiency of photosynthesis in C$_3$ plants. Recently, the C$_4$ syndrome has evolved through convergent evolution not just once but many times. C$_4$ photosynthesis is efficient because the Rubisco is sequestered in the bundle sheath cells and is thus not exposed to atmospheric oxygen. One would expect to see this adaptation become more widespread in the plant kingdom with continued high O_2/CO_2 ratios. CAM plants, however, seem to show primarily an adaptation for water conservation in extremely arid environments.

REFERENCES

1. W. A. Fowler, *Nuclear Astrophysics,* American Philosophical Society, Philadelphia, PA, 1967.
2. H. E. Suess, *J. Geol., 57*:600 (1949).
3. P. Cloud, *Cosmos, Earth, and Man,* Yale University Press, New Haven, CT, 1978.
4. S. L. Miller and L. E. Orgel, *The Origins of Life on the Earth,* Prentice Hall, Englewood Cliffs, NJ, 1974.
5. R. T. Brinkman, *J. Geophys. Res., 74*:5355 (1969).
6. K. M. Towe, *Precambrian Res., 16*:1 (1981).
7. A. P. Woodford, *J. Geol. Educ., 20*:276 (1972).
8. T. J. Algeo, R. A. Berner, J. B. Maynard, and S. E. Scheckler, *GSA Today, 5*:45 (1995).
9. T. E. Cerling, Y. Wang, and J. Quade, *Nature, 361*:344 (1993).
10. W. F. Loomis, *Four Billion Years,* Sinauer, Sunderland, MA, 1988.
11. J. W. Schopf, *Biol. Rev., 45*:319 (1970).
12. P. E. Cloud, Jr., *Science, 148*:27 (1965).
13. E. C. Perry, Jr., J. Monster, and T. Reimer, *Science, 171*:1015 (1971).
14. E. A. Boychenko, *Geokhimiya, 8*:971 (1967).
15. H. C. Urey, *The Planets, Their Origin and Development,* Yale University Press, New Haven, CT, 1952.
16. M. Dole and G. Jenks, *Science, 100*:409 (1944).
17. G. A. Lane and M. Dole, *Science, 123*:574 (1956).
18. E. I. Hamilton, *Applied Geochronology,* Academic Press, New York, 1965.
19. M. Calvin, *Science, 135*:879 (1962).
20. G. Bowes and W. L. Ogren, *J. Biol. Chem., 247*:2171 (1972).
21. N. E. Tolbert and J. Preiss (eds.), *Regulation of Atmospheric CO₂ and O₂ by Photosynthetic Carbon Metabolism,* Oxford University Press, Oxford, 1994.
22. D. J. Oliver, *Annu. Rev. Plant Physiol., 13*:689 (1972).
23. J. C. Hsu and B. N. Smith, *Plant Cell Physiol., 13*:689 (1972).
24. I. Zelitch, *Photosynthesis, Photorespiration, and Plant Productivity,* Academic Press, New York, 1971.
25. G. Bowes, *Annu. Rev. Plant Physiol., 44*:309 (1993).
26. G. H. Lorimer and T. J. Andrews, *Nature, 243*:359 (1973).
27. F. J. Ryan, S. O. Jolley, and N. E. Tolbert, *Biochim. Biophys. Res. Commun., 59*:1233 (1974).
28. D. M. Moss, *Nature, 193*:587 (1962).
29. G. D. Farquhar, M. H. O'Leary, and J. A. Berry, *Aust. J. Plant Physiol., 9*:121 (1982).
30. G. E. Fogg, C. Nalewajko, and W. D. Watt, *Proc. R. Soc. (Lond.) Ser. B, 162*:517 (1965).
31. J. R. Ehleringer, R. F. Sage, L. B. Flanagan, and R. W. Pearcy, *Trends Ecol. Evol., 6*:95 (1991).
32. H. P. Kortschak, C. E. Hartt, and G. O. Burr, *Plant Physiol., 40*:209 (1965).
33. M. D. Hatch and C. R. Slack, *Annu. Rev. Plant Physiol., 21*:141 (1970).
34. B. N. Smith, *BioSystems, 8*:24 (1976).
35. W. M. Laetsch, *Annu. Rev. Plant Physiol., 25*:27 (1974).
36. Z. Dai, M. S. B. Ku, and G. E. Edwards, *Plant Physiol., 107*:815 (1995).
37. M. S. B. Ku, J. Wu, Z. Dai, R. A. Scott, C. Chu, and G. E. Edwards, *Plant Physiol., 96*:518 (1991).
38. E. M. V. Nambudiri, W. D. Tidwell, B. N. Smith, and N. P. Hebbert, *Nature, 276*:816 (1978).
39. C. France-Lanord and L. A. Derry, *Geochim. Cosmochim. Acta, 58*:4809 (1994).
40. M. E. Morgan, J. D. Kingston, and B. D. Marino, *Nature, 367*:162 (1994).
41. W. V. Brown, *Memoirs Torrey Bot. Club, 23*:1 (1977).
42. J. R. Ehleringer, *HortScience, 14*:217 (1979).
43. T. M. Chen, R. H. Brown, and C. C. Black, Jr., *Plant Physiol., 44*:649 (1969).
44. C. R. Benedict, K. J. McCree, and R. J. Kohel, *Plant Physiol., 49*:968 (1972).
45. G. E. Edwards and S. C. Huber, in *The Biochemistry of Plants,* Vol. 8 (M. D. Hatch and N. K. Boardman, eds.), Academic Press, New York, 1981, p. 237.
46. R. Moore and C. C. Black, Jr., *Plant Physiol., 64*:309 (1979).
47. B. C. Gerwick and C. C. Black, Jr., *Plant Physiol., 64*:590 (1979).
48. M. Kluge and I. P. Ting, *Crassulacean Acid Metabolism,* Springer-Verlag, Berlin, 1978.
49. G. L. Webster, W. V. Brown, and B. N. Smith, *Taxon, 24*:27 (1975).
50. J. Lloyd and G. D. Farquhar, *Oecologia, 99*:201 (1994).
51. J. A. Teeri and L. G. Stowe, *Oecologia, 23*:1 (1976).
52. L. G. Stowe and J. A. Terri, *Am. Nat., 112*:609 (1978).
53. T. W. Boutton, A. T. Harrison, and B. N. Smith, *Oecologia, 45*:287 (1980).
54. J. W. C. White, P. Cials, R. A. Figge, R. Kenny, and V. Markgraf, *Nature, 367*:153 (1994).
55. C. McMillan, *BioScience, 19*:131 (1969).

Index

A